Practical Handbook of
ENVIRONMENTAL SITE CHARACTERIZATION
AND
GROUND-WATER MONITORING

SECOND EDITION

Edited by
DAVID M. NIELSEN

Taylor & Francis
Taylor & Francis Group

Boca Raton   London   New York

A CRC title, part of the Taylor & Francis imprint, a member of the
Taylor & Francis Group, the academic division of T&F Informa plc.

Published in 2006 by
CRC Press
Taylor & Francis Group
6000 Broken Sound Parkway NW, Suite 300
Boca Raton, FL 33487-2742

© 2006 by Taylor & Francis Group, LLC
CRC Press is an imprint of Taylor & Francis Group

No claim to original U.S. Government works
Printed in the United States of America on acid-free paper
10 9 8 7 6 5 4 3 2 1

International Standard Book Number-10: 1-56670-589-4 (Hardcover)
International Standard Book Number-13: 978-1-56670-589-9 (Hardcover)
Library of Congress Card Number 2005043909

This book contains information obtained from authentic and highly regarded sources. Reprinted material is quoted with permission, and sources are indicated. A wide variety of references are listed. Reasonable efforts have been made to publish reliable data and information, but the author and the publisher cannot assume responsibility for the validity of all materials or for the consequences of their use.

No part of this book may be reprinted, reproduced, transmitted, or utilized in any form by any electronic, mechanical, or other means, now known or hereafter invented, including photocopying, microfilming, and recording, or in any information storage or retrieval system, without written permission from the publishers.

For permission to photocopy or use material electronically from this work, please access www.copyright.com (http://www.copyright.com/) or contact the Copyright Clearance Center, Inc. (CCC) 222 Rosewood Drive, Danvers, MA 01923, 978-750-8400. CCC is a not-for-profit organization that provides licenses and registration for a variety of users. For organizations that have been granted a photocopy license by the CCC, a separate system of payment has been arranged.

Trademark Notice: Product or corporate names may be trademarks or registered trademarks, and are used only for identification and explanation without intent to infringe.

Library of Congress Cataloging-in-Publication Data

Practical handbook of environmental site characterization and ground-water monitoring, second edition / edited by David M. Nielsen.--2nd ed.
      p. cm.
   Includes bibliographical references and index.
   ISBN 1-56670-589-4 (alk. paper)
   1. Groundwater--Quality--Measurement. I. Nielsen, David.

TD426.P73 2005
628.1'61--dc22   2005043909

T&F informa

Taylor & Francis Group
is the Academic Division of T&F Informa plc.

Visit the Taylor & Francis Web site at
http://www.taylorandfrancis.com
and the CRC Press Web site at
http://www.crcpress.com

Preface

An enormous amount of progress has been made in the fields of environmental site characterization and ground-water monitoring since the first edition of this book was published in 1991. Tremendous advances in technology and in methodologies used to define site environmental conditions, in particular ground-water quality, have put our knowledge of subsurface processes light years ahead of what it was barely a decade ago. Clearly these advances made the publication of this significantly updated and expanded second edition necessary and worthwhile.

It is certainly not out of line to say that the fields of environmental site characterization and ground-water monitoring have evolved from a state of relative infancy to one of maturity in the course of the last 15 years. Thousands of sites have been characterized and have been (and are being) monitored, with increasing levels of confidence in the data produced because of advances in technology and advances in our understanding of environmental processes and our role in altering those processes. One significant indication of the degree of maturity of these areas of scientific study is the rate of publication of papers on these and related subjects. The major scientific journals have gotten significantly thinner, and the number of periodicals devoted to publishing non-refereed articles on these subjects has steadily declined.

The progress we have made in these areas has come with very few changes in the regulatory arena. All the regulatory programs that had been enacted and implemented by 1991 are still in place, but without many changes in scope or coverage. In addition, few new regulatory programs (except perhaps at the local level) have been created. Thus, regulatory compliance has become less of a driver — economics seem to have taken over as the major force providing impetus for improving the way we conduct environmental investigations and ground-water monitoring programs.

"Cheaper, faster, better" was the mantra of the 1990s, resulting in more streamlined approaches to both environmental site characterization and ground-water monitoring, but also creating a struggle between the application of good science vs. getting a project done as quickly and inexpensively as possible. In the decade of the 2000s, the challenge is for those in a regulatory role and field practitioners to strike a balance between good science and economics. This text provides unbiased technical discussions of the tremendously powerful tools that have been developed since the first edition of this book was published, to help environmental professionals meet that challenge.

We have spent the last 20 years developing standards (through ASTM International) to bring order to fields that were formerly quite disorganized. Where very few standards that could be applied to environmental projects existed prior to 1991, more than 300 new standards have since been written for a wide variety of tasks that are routinely conducted in environmental investigations. Where a dearth of companies qualified to conduct environmental investigations existed prior to 1991, in the period since then the number of companies grew tremendously, then declined as mergers and acquisitions ensued. With this change, there is now fierce competition for what new work is available.

Whereas in 1991 we had few technologies and methodologies that were truly applied exclusively to environmental problem solving (i.e., not pilfered from some other field), we now have many. Where we had few colleges and universities producing graduates qualified to work in the environmental field, in the period since 1991 we have seen a sharp increase in the number of graduates, followed by a gradual decline as other more lucrative fields have siphoned off scientific talent. Where regulatory agency staff were once swamped with work, they now seem to have stabilized.

I described the situation in the environmental field in 1991 as "catching up," but I think we can all agree that we have caught up and we now have a handle on things. But, as in 1991, much more remains to be learned. The decade and a half that has passed produced some extraordinary technology, yet even more exciting and robust technology is on the horizon. It will be very interesting to see what the future holds for those of us involved in environmental site characterization and ground-water monitoring. If it is anything like what has happened in the field since the first edition of this book was published I, for one, cannot wait for it to arrive.

David M. Nielsen
Nielsen Ground-Water Science, Inc.
The Nielsen Environmental Field School
Galena, Ohio

Contributors

James W. Ashley
Green Mountain Ground Water
PO Box 222
West Danville, Vermont 05873

Thomas Ballestero
Department of Civil Engineering
University of New Hampshire
238 Gregg Hall
Durham, New Hampshire 03824
tom.ballestero@unh.edu

Michael J. Barden
Geoscience Resources, Inc.
7827 Republic Drive NE
Albuquerque, New Mexico 87109
mjbarden@geoscienceresources.com

Richard C. Benson
Technos, Inc.
10430 NW 31st Terrace
Miami, Florida 33172
info@technos-inc.com

Ken Bradbury
Wisconsin Geological and Natural
  History Survey
University of Wisconsin — Extension
3817 Mineral Point Road
Madison, Wisconsin 53705
krbradbu@facstaff.wisc.edu

Olin C. Braids
O.C. Braids and Associates, LLC
9119 Corporate Lake Drive, Suite 150
Tampa, Florida 33634
obraids@gfnet.com

Thomas M. Christy
Geoprobe Systems
601 N. Broadway
Salina, Kansas 67401
christyt@geoprobe.com

Matthew G. Dalton
Dalton, Olmsted and Fuglevand, Inc.
10827 NW 68th Street, Suite B
Kirkland, Washington 98033
mdalton@dofnw.com

W. Zachary Dickson
Todd H. Wiedemeier & Associates, LLC
6608 S. Kiem Road
Evergreen, Colorado 80439
zach@thwa.com

Murray Einarson
Geomatrix Consultants, Inc.
2101 Webster Street
Oakland, California 94612
meinarson@geomatrix.com

O. D. Evans
Praxair Services, Inc.
3755 N. Business Center Drive
Tucson, Arizona 85705
odevans@tracerresearch.com

Jeffrey Farrar
U.S. Bureau of Reclamation
Earth Sciences and Research Laboratory
PO Box 25007, D-8340
Denver, Colorado 80225
jfarrar@do.usbr.gov

Stephen P. Farrington
Applied Research Associates, Inc.
415 Waterman Road
South Royalton, Vermont 05068
sfarrington@ara.com

Robert Gibbons
Center for Health Statistics
University of Illinois at Chicago
1601 W. Taylor
Chicago, Illinois 60612
rdgib@uic.edu

Patrick E. Haas
Mitretek Systems, Inc.
San Antonio, Texas
pehaas@mitretek.com

Beverly Herzog
Environmental Geoscience Center
Illinois State Geological Survey
615 E. Peabody Drive
Champaign, Illinois 61820
Herzog@isgs.uiuc.edu

Brent E. Huntsman
Terran Corp.
4080 Executive Drive
Beavercreek, Ohio 45430
behuntsman@terrancorp.com

Curtis A. Kraemer
Shaw Environmental, Inc.
111 Van Cedarfield Road
Colchester, Connecticut 06415
ckraemer@adelphia.net

Mark L. Kram
Groundswell Technologies, Inc.
3940 Maricopa Drive
Santa Barbara, California 93110
mark.kram@groundswelltech.com

Charles T. Kufs
Terrabyte, Inc.
137 Summit Avenue
Willow Grove, Pennsylvania 19090
terabyte@speakeasy.net

Kathryn S. Makeig
Waste Science Inc.
1411 Fallswood Drive
Rockville, Maryland 20854
katy@makeig.com

Carol J. Maslansky
Maslansky Geoenvironmental, Inc.
8500 Buchanan Drive
Prescott, Arizona 86305
maslanskygeo@earthlink.net

Steven P. Maslansky
Maslansky Geoenvironmental, Inc.
8500 Buchanan Drive
Prescott, Arizona 86305
maslanskygeo@earthlink.net

Wesley McCall
Geoprobe Systems
601 N. Broadway
Salina, Kansas 67401
mccallw@geoprobe.com

David M. Nielsen
Nielsen Ground-Water Science, Inc.
The Nielsen Environmental
    Field School
4686 State Route 605 South
Galena, Ohio 43021
BMW02Racer@aol.com or
    info@envirofieldschool.com

Gillian L. Nielsen
Nielsen Ground-Water Science, Inc.
The Nielsen Environmental
    Field School
4686 State Route 605 South
Galena, Ohio 43021
racinrabit@earthlink.net or
    info@envirofieldschool.com

Lynne M. Preslo
GeoEco
Reno, Nevada
lpreslo@aol.com

Tom Ruda
Diedrich Drill, Inc.
5 Fisher Street
LaPorte, Indiana 46350
dditr@diedrichdrill.com

Martin N. Sara
ERM, Inc.
704 N. Deerpath Drive
Vernon Hills, Illinois 60061
marty.sara@erm.com

Ronald Schalla
Battelle Memorial Institute
Pacific Northwest National Lab Division
902 Battelle Boulevard
Richland, Washington 99354
ron.schalla@pnl.gov

John Sevee
Sevee & Maher Engineers, Inc.
4 Blanchard Road
Cumberland Center, Maine 04021
info@smemaine.com

James A. Shultz
EA Engineering Science and Technology
333 Turnpike Road
Southborough, Massachusetts 01772
js7@eaest.com

Glenn Thompson
Praxair Services, Inc.
3755 N. Business Center Drive
Tucson, Arizona 85705
gthompson@tracerresearch.com

Rock J. Vitale
Environmental Standards, Inc.
1140 Valley Forge Road
Valley Forge, Pennsylvania 19482
RVitale@envstd.com

Todd H. Wiedemeier
Todd H. Wiedemeier & Associates, LLC
6608 S. Kiem Road
Evergreen, Colorado 80439
todd@thwa.com

Contents

1. **Regulatory Mandates for Ground-Water Monitoring** 1
   Kathryn S. Makeig and David M. Nielsen

2. **Environmental Site Characterization** 35
   David M. Nielsen, Gillian L. Nielsen, and Lynne M. Preslo

3. **Monitoring and Sampling the Vadose Zone** 207
   Thomas Ballestero, Beverly Herzog, and Glenn Thompson

4. **Remote Sensing and Geophysical Methods for Evaluation
   of Subsurface Conditions** 249
   Richard C. Benson

5. **Environmental Drilling for Soil Sampling, Rock Coring, Borehole
   Logging, and Monitoring Well Installation** 297
   Tom Ruda and Jeffrey Farrar

6. **Use of Direct-Push Technologies in Environmental Site
   Characterization and Ground-Water Monitoring** 345
   Wesley McCall, David M. Nielsen, Stephen P. Farrington, and Thomas M. Christy

7. **DNAPL Characterization Methods and Approaches:
   Performance and Cost Comparisons** 473
   Mark L. Kram

8. **Ground-Water Monitoring System Design** 517
   Martin N. Sara

9. **Designing Monitoring Programs to Effectively Evaluate the
   Performance of Natural Attenuation** 573
   Todd H. Wiedemeier, Michael J. Barden, Patrick E. Haas, and W. Zachary Dickson

10. **Design and Installation of Ground-Water Monitoring Wells** 639
    David M. Nielsen and Ronald Schalla

11. **Multilevel Ground-Water Monitoring** 807
    Murray Einarson

12. **Monitoring Well Post-Installation Considerations** 849
    Curtis A. Kraemer, James A. Shultz, and James W. Ashley

13. Acquisition and Interpretation of Water-Level Data 883
    Matthew G. Dalton, Brent E. Huntsman, and Ken Bradbury

14. Methods and Procedures for Defining Aquifer Parameters 913
    John Sevee

15. Ground-Water Sampling 959
    David M. Nielsen and Gillian L. Nielsen

16. Ground-Water Sample Analysis 1113
    Rock J. Vitale and Olin C. Braids

17. Organization and Analysis of Ground-Water Quality Data 1135
    Martin N. Sara and Robert Gibbons

18. Diagnosis of Ground-Water Monitoring Problems 1193
    Charles T. Kufs

19. Health and Safety Considerations in
    Ground-Water Monitoring Investigations 1219
    Steven P. Maslansky and Carol J. Maslansky

20. Decontamination of Field Equipment Used in Environmental
    Site Characterization and Ground-Water Monitoring Projects 1263
    Gillian L. Nielsen

Index 1281

1

Regulatory Mandates for Ground-Water Monitoring

Kathryn S. Makeig and David M. Nielsen

CONTENTS

The Need for Ground-Water Monitoring: Protection of a Resource at Risk 1
Federal Regulatory Mandates for Ground-Water Monitoring 2
   Resource Conservation and Recovery Act 3
      RCRA Subtitle D 4
      RCRA Subtitle C 4
      RCRA Subtitle I 7
   Comprehensive Environmental Response, Compensation and Liability Act 8
   Toxic Substances Control Act 10
   Clean Water Act 10
   Safe Drinking Water Act 11
      Drinking Water Quality Standards 11
      Sole-Source Aquifer Program 12
      Wellhead Protection Program 12
      Underground Injection Control Program 26
   Surface Mining Control and Reclamation Act 26
   Brownfields 28
   Federal Ground-Water Protection Strategy 29
      Ground-Water Classification 30
Discussion of Ground-Water Quality Standards 30
Mechanisms for a Workable Federal Ground-Water Program 32
Comprehensive State Ground-Water Protection Programs 32
Ground Water and Terrorism 33
Ground Water and Development 33
References 33

The Need for Ground-Water Monitoring: Protection of a Resource at Risk

Ground water has been described as one of the world's most valuable natural resources. People around the world have long depended on ground water for many uses, but primarily for drinking water. In the United States alone, more than 125 million people — nearly half the population — depend on ground water for their drinking water supply. Approximately 80% of public water supply systems providing drinking water in the United States, including those of one third of the nation's largest cities, depend at least partly on ground water (U.S. EPA, 2004a). Additionally, 95% of all domestic water needs in rural areas is served by ground-water resources. Ground water is also used extensively in the western U.S. for irrigation, in the northern U.S. for residential and commercial heating, in the southern U.S. for cooling, and across the nation for various industrial purposes. National reliance on ground water has increased dramatically over the past few decades and will continue to increase as consumption and use of water increases in the future. This reliance will be underscored if surface-water shortages, caused by prolonged droughts, continue to occur and development of arid land continues at its current pace.

The need for the regulation of activities that pose a threat to the quality of ground water has become an overriding concern when communities face commercial and private development. In many areas of the world, this once-pristine, widely available resource is in a delicate balance between supply and demand. The quantity of useable ground water in any given area is closely linked to the quality of the water available for various uses. The apparent ignorance of humans about the finite nature of this resource has led to its exploitation, its abuse as a dumping ground for unwanted waste materials, and its excessive mining, particularly in the western United States. Since the mid-1970s, there have been increased efforts to protect this resource from further degradation and there are now regulatory mandates in place both to protect useable ground water and to clean up ground water that has suffered from the effects of short-sighted waste-management practices. However, efforts to protect the quantity of ground water continue to lag behind development. This promises to be a challenge to 21st century planners in North America and abroad.

In spite of numerous uses of ground water, there are limitations and constraints placed on the appropriation and quality of this resource. Both Federal and state legislators have attempted to address the evolving requirements for ground water that, in some cases, must be clean enough to drink and, in other cases, must be only relatively free of chemicals that could affect the performance of an industrial process. Legislation has addressed the problem of potential contamination from the use of ground water both as a resource and as a means of disposal. Additional constraints are being placed on this resource as Tribal Nations make demands for ground-water and surface-water quality that surpass requirements dictated by risk. Finally, legislation associated with Brownfields and Superfund call for the clean up of ground water that has been already contaminated.

This chapter discusses the role of ground-water monitoring within the framework of existing environmental and resource regulations, focusing on protection of the resource from over-development and contamination. It places the discussion in the context of the levels of protection that will keep this resource abundant and free of unhealthy contaminants.

Federal Regulatory Mandates for Ground-Water Monitoring

There exist a variety of federal agencies whose missions include the protection of ground water. Among them are the Nuclear Regulatory Commission, the Office of Surface Mining (OSM), the Department of Energy, the Department of Defense, and the U.S. Environmental Protection Agency (U.S. EPA). By far, the largest body of environmental regulations involving ground-water protection and requiring ground-water monitoring has been promulgated by the U.S. EPA. Copies of these regulations are readily available from a number of sources, but the primary sources are the U.S. EPA (www.epa.gov) and the Government Printing Office. The primary emphasis of the following discussion is on the Federal mandates for ground-water monitoring included in documents issued by the aforementioned agencies.

The major Federal regulatory programs that involve the implementation of ground-water monitoring include the following:

- The Resource Conservation and Recovery Act (RCRA) including the Hazardous and Solid Waste Amendments (HSWA) which, in turn, include the Underground Storage Tank (UST) Program
- The Comprehensive Environmental Response, Compensation, and Liability Act (CERCLA or "Superfund") including the Superfund Amendments and Reauthorization Act (SARA)
- The Toxic Substances Control Act (TSCA)
- The Clean Water Act (CWA) and CWA Amendments
- The Safe Drinking Water Act (SDWA) and SDWA Amendments, including the Underground Injection Control (UIC) Program and the Wellhead Protection Program (WHPP)
- The Surface Mining Control and Reclamation Act (SMCRA)

Each of these pieces of legislation is described briefly, and the ground-water monitoring provisions of each program are summarized in the following sections.

Resource Conservation and Recovery Act

The RCRA (Public Law 94-580) was passed by the Congress in October 1976, as an amendment to the 1965 Solid Waste Disposal Act. Its purpose was to address the problem of how to safely dispose of the huge volumes of solid and hazardous waste generated nationwide each year (U.S. EPA, 1986). RCRA has evolved from a relatively limited program dealing with nonhazardous solid waste to a far-reaching program that focuses primarily on hazardous waste. Solid and hazardous waste generators, transporters, and owners or operators of treatment, storage, and disposal facilities (TSDFs) comprise the RCRA-regulated community. On November 8, 1984, the Congress passed the HSWA to RCRA, thereby greatly expanding the nature and complexity of activities covered under RCRA.

The objectives of RCRA, as set forth by the Congress, are:

- The improvement of solid-waste disposal practices to protect human health and the environment
- The regulation of hazardous wastes, from initial generation to ultimate disposal
- The establishment of resource conservation as the preferred approach to solid and hazardous wastes management

Section 1003 of RCRA, which outlines these objectives, clearly indicates the applicability of the Act to ground-water protection, as does Section 1004, which defines the terms used in the Act.

To achieve RCRA's goals, three programs were established by U.S. EPA. The first program, termed Subtitle D, encourages states to formulate comprehensive solid-waste management plans, primarily for nonhazardous waste. The second program, Subtitle C, establishes a program to control hazardous waste from the time it is generated until its ultimate disposal — the so-called "cradle-to-grave" concept. The third program, Subtitle I, regulates certain underground storage systems.

RCRA Subtitle D

Subtitle D establishes a voluntary program under which participating states receive Federal financial and technical support to develop and implement solid-waste management plans. This program is primarily a planning tool used to clarify state, local, and regional roles in the management of solid waste. One of the objectives of this portion of the act is to identify those facilities that are "open dumps." Although originally there were no specific regulations within this program requiring the monitoring of ground water, the HSWA now contains rules governing land-disposal units. The current version has specific ground-water monitoring requirements.

RCRA Subtitle C

Subtitle C is the backbone of RCRA. It calls for the management of hazardous waste from the time it is generated until its ultimate disposal, through a complex system of standards applicable to generators and transporters of hazardous waste and to owners and operators of hazardous waste TSDFs. Subtitle C clearly defines what is considered a hazardous waste and what is not and defines the types of facilities that fall under these regulations. The purpose of these regulations is to protect human health and the environment, with an emphasis on the protection of ground water. EPA has set performance criteria that apply to most forms of land disposal including landfills, surface impoundments, waste piles, and land-treatment units. The siting, design, and operating specifications developed for hazardous waste facilities require that the owner and operator employ natural geologic or engineering design features and waste management practices that minimize adverse effects on ground water and surface water. The basic purposes of these requirements are to minimize the production of leachate and to avoid situations that could compromise the integrity of the facility's liner and final cover (landfills) or its natural ability to ameliorate waste migration (land-treatment facilities).

Subtitle C also has set forth requirements for the installation and operation of ground-water monitoring systems as a means of evaluating the performance of TSDFs. These ground-water monitoring requirements outline procedures for (1) installing ground-water monitoring systems, (2) developing a ground-water sampling and analysis program, and (3) preparing a ground-water quality assessment plan. Exempt from these requirements are those TSDFs which can demonstrate that there is a low potential for migration of hazardous waste from the facility via the uppermost aquifer to water-supply wells or surface water. Such a facility may apply for a waiver from ground-water monitoring requirements.

There are a number of parts to Subtitle C. Sections containing requirements for ground-water monitoring include:

Part 264: Regulations for Owners and Operators of Permitted Hazardous Waste Facilities

Part 265: Interim Status Standards for Owners and Operators of Hazardous Waste Facilities

Part 267: Interim Standards for Owners and Operators of New Hazardous Waste Management Facilities*

---

*This part was only temporary until Part 264 was finalized, but has never been removed from Subtitle C. Specific requirements under Part 267 are no longer applicable.

Part 270: Regulations for Federally Administered Hazardous Waste Permit Programs (Part B permits)

Part 271: Requirements for Authorization of State Hazardous Waste Programs

The regulatory scheme established under RCRA is to grant permits to all TSDFs that are in compliance with RCRA requirements. The standards set forth in Part 264 apply to these permitted facilities. Because there were thousands of facilities that applied for and were awaiting permits early in the administration of the program, RCRA provided the means to regulate nonpermitted facilities prior to their final permitting. New facilities waiting to be built or in the process of being built fall under Part 264. Established facilities operating without a final permit, but under the regulatory framework, fall under Part 265. The information needed to submit an application for status as a permitted facility is detailed under Part 270.

Most ground-water monitoring requirements included in Subtitle C apply to the water quality in the "uppermost aquifer," although, at some sites with known contamination, monitoring of other connected hydrogeologic units may be required to characterize the extent of the contaminant plume.

Part 264: For facilities operating under a RCRA permit, there are generally three types of ground-water monitoring that may be required. The monitoring scheme is based on a phased approach, so that facilities that have not released contaminants into the ground water have different requirements than those that have released contaminants. The most rudimentary monitoring scheme is the Detection Monitoring Program (40 CFR 264.98). This program must consist of

> A sufficient number of wells, installed at appropriate locations and depths, to yield ground-water samples from the uppermost aquifer that:
>
> - Represent the quality of ground water that has not been affected by leakage from the regulated unit, and
> - Represent the quality of ground water passing the point of compliance (roughly the boundary of the waste management unit or units, such as individual or adjacent groups of impoundments or landfills).
>
> (40 CFR 264.97)

The Part 264 regulations essentially require that each TSDF must have installed detection-monitoring wells both hydraulically upgradient and hydraulically downgradient of the limit of the waste management area. The number, location, depth, and construction details of the upgradient wells must be sufficient to yield ground-water samples, which are representative of background water quality in the uppermost aquifer beneath the facility. The number, location, depth, and construction details of downgradient wells must ensure that these wells can detect any wastes that migrate from the waste management area to the uppermost aquifer. Both upgradient and downgradient wells must be cased in a manner that maintains the integrity of the well, screened and packed with sand to enable the collection of ground-water samples, and the annular space above the sampling depth sealed to prevent contamination of samples and ground water. Regulations require a minimum of one upgradient well and three downgradient wells. Monitoring is required during the active life of the facility, during its closure period, and during any postclosure period that is applicable.

The ground-water sampling and analysis plan developed for compliance with Part 264 regulations must include procedures and techniques for sample collection, sample preservation, analytical procedures, and chain of custody control. The owner and operator must monitor ground water in all wells for a period of 1 yr, on a quarterly basis. Samples must be analyzed for three separate sets of indicator parameters including:

- Parameters characterizing the suitability of the ground water as a drinking water supply, including all water quality parameters mandated for analysis under the SDWA (Table 1.3)
- Parameters establishing ground-water quality, including chloride, iron, manganese, phenols, sodium, and sulfate
- Parameters used as indicators of ground-water contamination, including pH, specific conductance, total organic carbon, and total organic halogens

After the first year, all monitoring wells must be sampled and the samples analyzed, such that all parameters used to establish ground-water quality are sampled and analyzed annually and parameters used as indicators of ground-water contamination are sampled and analyzed semiannually.

Part 264 also requires owners and operators of TSDFs to determine the extent to which wastes may have entered the uppermost aquifer in the event that a statistically significant change in the concentrations of the monitored chemical parameters indicates a release from the regulated unit during the Detection Monitoring Program. This second phase of monitoring is called the Compliance Monitoring Program (40 CFR 264.99). The Compliance Monitoring Program applies to units in which there is a reason to believe that concentrations of certain chemicals in the ground water exceed the established ground-water protection standards (40 CFR 264.92). The U.S. EPA Regional Administrator has a certain amount of discretion in identifying the parameters to be monitored, as set forth in the permit.

If the Compliance Monitoring Program establishes that there is a release of a type and magnitude to be of concern at the compliance point of the facility, then a Corrective Action Program must be implemented (40 CFR 264.100). The Corrective Action Program requires that the owner or operator remove or treat the wastes that are causing the release, so that the ground-water quality complies with the ground-water protection standards. In this program, the primary purpose of the ground-water monitoring network is to monitor the effectiveness of the corrective action. Ground-water cleanup criteria are usually determined either by the individual states or within a state on a case-by-case basis. In all cases, the cleanup criteria must be as stringent as, or more stringent than, various standards set by the Federal government.

After the TSDF ceases operation, the ground-water monitoring network may still be required to monitor the facility during the closure and postclosure periods. The closure period usually runs from the time the facility receives the final volume of waste until all activities at the facility cease (40 CFR 264.112 and 264.113). Postclosure monitoring, usually a period of 30 yr after closure, is required at facilities in which all of the waste or waste constituents are not removed from the facility at closure. This applies primarily to landfills and land-treatment facilities, but can also apply to surface impoundments that are closed with waste constituents remaining in the ground (40 CFR 264.117). Certain demonstrations can be made to reduce the duration of the postclosure monitoring period. Table 1.1 lists the citations associated with Part 264 ground-water monitoring requirements.

Part 265: Part 265 of RCRA addresses facilities that are under interim status. This applies to existing TSDFs that are waiting to obtain a final permit. The ground-water monitoring requirements under Part 265 are much narrower in scope than those under Part 264 and are explained under 40 CFR 265.91 through 265.93. For interim status, a facility needs

TABLE 1.1

Ground-Water Monitoring Citations for RCRA Part 264

Citation	Description
40 CFR 264.97	General ground-water monitoring requirements
40 CFR 264.98	Detection Monitoring Program
40 CFR 264.99	Compliance Monitoring Program
40 CFR 264.100	Corrective Action Program
40 CFR 264.112	TSDF closure
40 CFR 264.117	TSDF postclosure
40 CFR 264.221	Design and operation of surface impoundments
40 CFR 264.228	Closure and postclosure of surface impoundments
40 CFR 264.310	Closure and postclosure of landfills

only to perform one type of ground-water monitoring, similar in some respects to the detection monitoring of Part 264. However, unlike Part 264 requirements, there is no phased approach, and if a release from the facility is detected by the monitoring system, a Ground-Water Quality Assessment Program is implemented (40 CFR 265.93). There are no provisions that clearly spell out the procedures, once in the Ground-Water Quality Assessment Program, to determine whether ground-water remediation is required. Table 1.2 lists the citations associated with Part 265 ground-water monitoring requirements.

Part 270: Owners and operators of hazardous waste management facilities are required to file a Part A and Part B permit application to receive their facility permit to operate. A Part A notification serves to notify the U.S. EPA Regional Administrator of the existence of the facility and the wastes that are associated with it. A Part B permit application requires the generation of a substantial amount of information about the facility and the activities that take place at the facility. As part of a Part B application for owners of landfills, surface impoundments, waste piles, and land treatments units, information regarding the protection of the ground water is necessary (40 CFR 270.14 and 270.97).

Part 271: Part 271 of RCRA deals with the authorization of state programs. The regulations require that states seeking authority for their programs have regulations similar to those promulgated under RCRA for TSDFs (40 CFR 271.12 and 271.128).

RCRA Subtitle I

In 1985, the U.S. EPA estimated that as many as 100,000 to 300,000 USTs could be leaking their contents to the environment and polluting ground water (U.S. EPA, 1985). To address this problem, the Congress created a program under HSWA, entitled Subtitle I, to prevent the leakage of stored products from USTs. These amendments to RCRA were

TABLE 1.2

Ground-Water Monitoring Requirements for RCRA Part 265

Citation	Description
40 CFR 264.90 through 264.94	Ground-water monitoring program
40 CFR 264.112	Closure of Interim Status TSDF
40 CFR 264.117 and 264.118	Postclosure of Interim Status TSDF
40 CFR 264.221	Interim Status surface impoundments
40 CFR 264.301	Interim Status landfill design
40 CFR 264.310	Interim Status landfill closure and postclosure

significant in that they marked the first time that RCRA regulations were applied to raw product as well as to waste. Subtitle I is limited to regulating the underground storage of petroleum or hazardous chemicals, while Subtitle C regulates the underground storage of hazardous wastes.

Although there is no specific language in Subtitle I that requires the monitoring of ground water, there are references to a tank owner having the ability to detect releases. Subtitle I also authorizes the Federal and state personnel to monitor the surrounding soils, air, surface water, and ground water (U.S. EPA, 1985). There is also specific language in a number of state UST programs that ground-water monitoring wells shall be installed adjacent to each new and existing tank or tank field.

Final rules covering technical standards and requirements for new and existing USTs containing petroleum and hazardous chemicals took effect in December 1988. The purposes of these rules are to regulate the vast numbers of underground tanks and to minimize the environmental impact of leakage from these tanks by implementing early detection techniques, ground-water monitoring, and physical protection of the tanks themselves. The use of ground-water monitoring wells is one of the specified methods that can achieve the required monthly monitoring for releases from these tanks.

The schedule for technically upgrading and monitoring requirements for existing tanks is dependent on the tank age. However, after 1993, all existing tanks were required to perform monthly leak-detection monitoring, by means of in-tank gaging, vapor monitoring, interstitial monitoring, or ground-water monitoring.

Tanks that are confirmed to be leaking must initiate corrective action. The rules do not specify the types of measurements or site assessment techniques that must be employed. However, it is implied that soil and ground-water samples should be obtained. If there has been a confirmed release that requires corrective action, then a corrective action plan must be submitted, which will address the remediation of soil and ground water, as required, and the means to verify the success of these actions. It is important to note that many states and local municipalities have additional requirements that may regulate the monitoring or remediation of a petroleum hydrocarbon release.

Comprehensive Environmental Response, Compensation and Liability Act

The CERCLA, more popularly known as Superfund, was passed by the Congress in December 1980 to deal with threats posed to the public by abandoned waste sites. With the SARA of 1986, CERCLA assumed a greater role in the cleanup of hazardous waste sites. The main objectives of CERCLA, as established by the Congress, are:

- To develop a comprehensive program to set priorities for cleaning up the worst existing hazardous waste sites
- To make responsible parties pay for those cleanups wherever possible
- To set up a Hazardous Waste Trust Fund for the twofold purpose of performing remedial cleanups in cases where responsible parties could not be held accountable and responding to emergency situations involving hazardous substances
- To advance scientific and technical capabilities in all aspects of hazardous waste management, treatment, and disposal (U.S. EPA, 1987a)

There are several steps involved in completing a Superfund cleanup. The initial report of the existence of a site may come from a private individual or a facility manager, either to EPA's National Response Center or to a local or state official. After EPA learns of the site, it conducts a site assessment, during which it collects all available background information to determine the potential hazards posed by the site. In the preliminary assessment step, EPA not only tries to identify the size of the problem and the types of wastes at the site, but also attempts to identify any and all PRPs associated with the wastes. If the preliminary assessment reveals evidence that the site may pose a significant threat to human health or the environment, then a site inspection is performed to define more precisely which media have been impacted, which contaminants are present at the site (and at what levels), contaminant migration potential, and threats posed by the site to drinking water, soil, and air quality. The site-inspection step may involve the installation of short-term ground-water monitoring points. The site is scored and then ranked using the EPA Hazard Ranking System. If the ranking is high enough to place the site on the National Priorities List (NPL), then the next phase of the site investigation, site characterization, is warranted. As of April 2004, 1,238 waste sites had been listed on the NPL by EPA, with another 65 sites proposed for the list (U.S. EPA, 2004b). This is increased from 1,010 sites as of January 1990. However, EPA estimated that in 1980, there were 9,000 "problem" hazardous waste sites. In 1989, more than 30,000 sites had been entered into EPA's computerized database (CERCLIS).

The ultimate objective of placing sites on the NPL is their permanent cleanup. As of April 2004, 583 sites on the NPL were listed as "construction completed," with the remainder listed as "deleted" (267) or "construction needed or ongoing" (388). To identify a cleanup strategy that best suits a particular situation, each of the sites on the NPL undergoes a Remedial Investigation/Feasibility Study (RI/FS). The RI/FS is a process of site characterization and remedy evaluation, which facilitates the selection of remedial measures that will most effectively eliminate, reduce, or control risks to human health and the environment. Ground-water monitoring is a critical element of the RI, as it is necessary to establish the nature and extent of ground-water contamination at the site and whether or not ground water serves as a pathway for waste constituents to migrate away from the site. The FS is often heavily dependent on the data gathered during the RI, so that the optimal remedial technology (or combination of technologies) may be implemented at the site. Ground-water monitoring is also a critical factor in evaluating whether the remedial activities implemented at the site are successful in abating ground-water contamination.

Guidance documents available from U.S. EPA set forth the procedures that should be followed to conduct a RI in support of a FS (U.S. EPA, 1988). The focus of the RI effort depends on the quality of the existing data, key site problems, the need to provide sufficient technical data to support the FS, and enforcement needs. These factors dictate the study parameters and the types and amount of sampling that will be sufficient to meet the needs of the study. Therefore, unlike RCRA, CERCLA does not set up any specific ground-water monitoring program requirements — the investigator must address each site individually. Although the purpose of the RI is to characterize the hydrogeologic setting and any contamination present at the site, there are several other important aspects to conducting a ground-water monitoring program that are required for the FS.

The collection of data that will help in the evaluation of remedial technology alternatives is essential during the RI. These data may not directly aid in the definition of the problem, but could predict interactions between water quality and certain alternatives. For example, although the level of iron present in the ground water is not an essential piece of information to establish the presence or extent of ground-water contamination, it may be useful in the FS portion of the project. If air stripping is proposed in the FS as a candidate remedial technology, then the concentrations of iron must be known to devise methods of preventing scale buildup on the air-stripping unit, which would reduce its effectiveness.

Ground-water monitoring is also essential during the cleanup of a contaminated site. After a remedy has been implemented at a site, ground-water conditions must be monitored to assess the effectiveness of remediation efforts and, ultimately, to determine when the remediation effort can be discontinued and the site can be declared clean.

Toxic Substances Control Act

The TSCA (Public Law 94-469), enacted by the Congress in 1976, brought about significant changes in the day-to-day operation of the U.S. chemical industry. With TSCA, the U.S. EPA was given the power to prohibit or regulate the manufacture, processing, distribution, use, or disposal of chemical products that pose an unreasonable risk to human health or the environment. TSCA also provides the U.S. EPA with the authority to require premarket testing of a wide range of chemicals to evaluate the health effects that they may cause. To enable EPA to monitor the marketing of new chemicals, TSCA requires manufacturers to submit premanufacture notices on new chemical substances and to keep records identifying the new uses of existing chemicals. To be included in these records are data such as the amounts of chemicals produced, how and where the chemicals are stored and transported, any known or projected occupational exposures, and the methods used to dispose of the chemicals.

The U.S. EPA is authorized to take a variety of steps to protect against threats to human health or the environment by the introduction or unrestricted use of new chemicals. Such steps include publication of the chemical inventory, information gathering authority, and permitting access to manufacturing data, which could assist in the development of source inventories for ground-water protection planning and investigation. For example, any RCRA facility that handles hazardous wastes, which contain more than 50 ppm of polychlorinated biphenyls (PCBs) is regulated under both the RCRA and TSCA; initial ground-water monitoring for background data at PCB disposal sites is also required.

Clean Water Act

The CWA of 1972 (Public Law 92-500) and the CWA Amendments of 1977 (Public Law 95-217) established a major milestone in water pollution control law. At that time, the CWA was one of the most far-reaching Federal laws ever enacted. The objective of the CWA was to "restore and maintain the chemical, physical, and biological integrity of the Nation's waters." However, the language of the Act stresses the need to protect "navigable waters," thus the major emphasis and the thrust of enforcement have been toward surface water. To the extent that surface water and ground water are hydraulically connected, protection of surface-water quality will also protect ground-water quality and vice versa. Several specific provisions of the CWA have served to enhance ground-water protection.

The potentially most effective means for controlling ground-water contamination under the CWA is found in Section 208, which provides for statewide and areawide planning for pollution control, including funding to set up and implement water-quality management planning programs. The water-quality management program required by Section 208 has served as a catalyst for the development of several state ground-water management programs. The most powerful means for controlling ground-water contamination under Section 208 requires water-quality management plans to include a process to control the disposal of pollutants on land or in subsurface excavations to protect ground- and surface-water quality. For example, where CWA funds are used to construct municipal sewage treatment plants that use land-application techniques, the municipality is required to design the plant to ensure protection of ground water (40 CFR 35, Appendix A). The primary responsibility for preparing plans and implementing programs is in the hands of state, regional, and local agencies. It is within U.S. EPA's power to withhold approval of a plan or program that does not adequately provide for ground-water protection, but not within its power to act if the ground-water provisions of the plan are not implemented.

Section 304 of the CWA requires EPA to develop and issue guidelines for identifying and evaluating the nature and extent of nonpoint sources of pollutants. Guidelines have also been developed for processes, procedures, and methods to control pollution resulting from, among others, "the disposal of pollutants in wells or in subsurface excavations, saltwater intrusion resulting from reduction of freshwater flow from any cause, including extraction of ground water, and changes in movement, flow and circulation of any navigable waters or ground waters."

Section 402, which describes the National Pollutant Discharge Elimination System (NPDES), empowers the U.S. EPA to issue permits for the discharge of any point-source pollutant or combination of pollutants to navigable waters. Individual states may issue NPDES permits if they develop programs and are authorized by the EPA to do so. A trust fund that was the precursor to Superfund was set up to deal with problems stemming from NPDES discharges. However, no provision was made to deal with damages to land resources resulting from contamination by hazardous wastes. One specific requirement for approval of a state NPDES program is that the state must provide for control of the disposal of pollutants into wells.

Finally, Sections 104 and 106 provide for the establishment and funding of national and state programs to equip and maintain both surface-water and ground-water surveillance systems. This is the strongest provision relating specifically to ground-water monitoring, but the systems that are authorized under this program would be primarily large-scale in nature. In addition, while the authority exists under Section 106 for the use of funds to establish regional or statewide ground-water monitoring networks, most money has been channeled to surface-water programs at the state level.

The formation of the National Contingency Plan for dealing with emergencies from hazardous waste was an important offshoot of the Clean Water Act. This plan remains the guiding principle behind the implementation of Superfund.

Safe Drinking Water Act

The SDWA (Public Act 93-523) was passed by the Congress in 1974 in response to accumulating evidence that unsafe levels of contaminants in public drinking water supplies, including ground water, were posing a threat to the public health. The amendments to the SDWA, which were passed in June 1986, established the first nationwide program to protect ground-water resources used for public water supplies from a wide range of potential threats. The goal of the SDWA, as its name implies, is to ensure the provision of a safe supply of drinking water to all Americans served by public water supply systems. Several major provisions to the SDWA relate specifically to ground-water quality. The SDWA provides protection to ground water through:

- The establishment of drinking water standards (40 CFR 141; Fed. Reg. Vol. 43[243])
- Sole-source aquifer designation (42 U.S.C. 300f, Sec. 1427)
- The establishment of the WHPP (42 U.S.C. 300f, Sec. 1428)
- The UIC Program (42 U.S.C. 300f, Sec. 1424; 40 CFR 144)

Drinking Water Quality Standards

Promulgation of drinking-water quality standards to apply to public water supply systems (those which supply water to 25 or more people or have more than 15 service connections) is required by Section 1412 of the SDWA. Standards known as National Primary Drinking Water Standards (NPDWSs) and National Secondary Drinking Water Standards (NSDWSs) were developed by the U.S. EPA to meet this requirement. The NPDWSs are legally enforceable health-related standards that set maximum contaminant levels (MCLs) for bacteria, turbidity, and a variety of inorganic and organic chemicals and radionuclides in public water supplies (Table 1.3) (U.S. EPA, 2003). This list, current as of June 2003, has expanded significantly since the standards were first issued as interim standards in June 1977. The NSDWSs are non-enforceable guidelines regulating contaminants that may cause cosmetic effects (such as skin or tooth discoloration) or aesthetic effects (such as taste, odor, or color) in drinking water (Table 1.4) (U.S. EPA, 2003). The MCLs set under SDWA may also be used for enforcement purposes in ground-water monitoring programs conducted both at RCRA interim status and at RCRA permitted facilities and for establishing ground-water cleanup levels at RCRA or CERCLA site.

Sole-Source Aquifer Program

Another provision of the SDWA related to protection of ground water is the Sole-Source Aquifer Program, also known as the Gonzales Amendment. This program provides local, regional, or state agency with a legal mechanism to protect the recharge zones of specially designated aquifers. It establishes a procedure whereby the U.S. EPA, either on its own initiative or upon petition, may designate an aquifer as a sole or principal source of drinking water for an area. After such a designation, no Federal financial assistance may be granted to a project that EPA determines could contaminate the aquifer through its recharge zone so as to create a "significant hazard to public health." This is defined as

> Any level of contaminant which causes or may cause the aquifer to exceed any MCL set forth in any national drinking water standard at any point where the water may be used for drinking water purposes or which may otherwise adversely affect the health of persons or which may require a public water system to install additional treatment to prevent such adverse effects.

As of April 2004, the U.S. EPA had made 73 sole-source aquifer designations across the USA (Table 1.5). A limiting factor in the sole-source aquifer provision is that it protects aquifer recharge zones only from federally funded projects that might contaminate an aquifer — nonfederally funded projects are not regulated. Although there are no specific provisions for ground-water monitoring in the Sole-Source Aquifer Program, data from ground-water monitoring wells and systems are used extensively to support petitions for sole-source aquifer designation and would be used to detect contamination from existing contaminant sources located in recharge zones of these important aquifers.

Wellhead Protection Program

Part of U.S. EPA's goal of providing protection for ground-water resources was accomplished by the establishment of state WHPPs, which protect wellhead areas within their jurisdiction from contaminants that may have any adverse effect on the health of persons. One of the major elements of a WHPP is the determination of zones within which contaminant source assessment and management are addressed. These zones, designated as Wellhead Protection Areas (WHPAs), are defined in the SDWA as

> The surface and subsurface area surrounding a water well or wellfield, supplying a public water system, through which contaminants are reasonably likely to move toward and reach such water well or wellfield.

TABLE 1.3
National Primary Drinking Water Standards

	Contaminant	MCL or TT[a] (mg/l)[b]	Potential Health Effects from Exposure above the MCL	Common Sources of Contaminant in Drinking Water	Public Health Goal
OC	Acrylamide	TT[c]	Nervous system or blood problems; increased risk of cancer	Added to water during sewage or wastewater treatment	0
OC	Alachlor	0.002	Eye, liver, kidney, or spleen problems; anemia; increased risk of cancer	Runoff from herbicide used on row crops	0
R	Alpha particles	15 pCi/l	Increased risk of cancer	Erosion of natural deposits of certain minerals that are radioactive and may emit a form of radiation known as alpha radiation	0
IOC	Antimony	0.006	Increase in blood cholesterol; decrease in blood sugar	Discharge from petroleum refineries; fire retardants; ceramics; electronics; solder	0.006
IOC	Arsenic	0.010 as of 01/23/06	Skin damage or problems with circulatory systems and may have increased risk of cancer	Erosion of natural deposits; runoff from orchards, glass and electronics production wastes	0
IOC	Asbestos (fibers >10 μm)	7 MFL	Increased risk of developing benign intestinal polyps	Decay of asbestos cement in water mains; erosion of natural deposits	7 MFL
OC	Atrazine	0.003	Cardiovascular system or reproductive problems	Runoff from herbicide used on row crops	0.003
IOC	Barium	2	Increased blood pressure	Discharge of drilling wastes; discharge from metal refineries; erosion of natural deposits	2
OC	Benzene	0.005	Anemia; decrease in blood platelets; increased risk of cancer	Discharge from factories; leaching from gas storage tanks and landfills	0
OC	Benzo(a)pyrene polycyclic aromatic hydrocarbons (PAHs)	0.0002	Reproductive difficulties; increased risk of cancer	Leaching from linings of water storage tanks and distribution lines	0

(Table continued)

TABLE 1.3 Continued

	Contaminant	MCL or TT[a] (mg/l)[b]	Potential Health Effects from Exposure above the MCL	Common Sources of Contaminant in Drinking Water	Public Health Goal
IOC	Beryllium	0.004	Intestinal lesions	Discharge from metal refineries and coal-burning factories; discharge from electrical, aerospace, and defense industries	0.004
R	Beta particles and photon emitters	4 mrem/yr	Increased risk of cancer	Decay of natural and man-made deposits of certain minerals that are radioactive and may emit forms of radiation known as photons and beta radiation	0
DBP	Bromate	0.010	Increased risk of cancer	Byproduct of drinking water disinfection	0
IOC	Cadmium	0.005	Kidney damage	Corrosion of galvanized pipes; erosion of natural deposits; discharge from metal refineries; runoff from waste batteries and paints	0.005
OC	Carbofuran	0.04	Problems with blood, nervous system, or reproductive system	Leaching of soil fumigant used on rice and alfalfa	0.04
OC	Carbon tetrachloride	0.005	Liver problems; increased risk of cancer	Discharge from chemical plants and other industrial activities	0
D	Chloramines (as Cl$_2$)	MRDL = 4.0[a]	Eye or nose irritation; stomach discomfort; anemia	Water additive used to control microbes	MRDLG = 4[a]
OC	Chlordane	0.002	Liver or nervous system problems; increased risk of cancer	Residue of banned termiticide	0
D	Chlorine (as Cl$_2$)	MRDL = 4.0[a]	Eye or nose irritation; stomach discomfort	Water additive used to control microbes	MRDLG = 4[a]
D	Chlorine dioxide (as ClO$_2$)	MRDL = 0.8[a]	Anemia; nervous system effects in infants and young children	Water additive used to control microbes	MRDLG = 0.8[a]
DBP	Chlorite	1.0	Anemia; nervous system effects in infants and young children	Byproduct of drinking water disinfection	0.8
OC	Chlorobenzene	0.1	Liver or kidney problems	Discharge from chemical and agricultural chemical factories	0.1

Regulatory Mandates for Ground-Water Monitoring

IOC	Chromium (total)	0.1	Allergic dermatitis	Discharge from steel and pulp mills; erosion of natural deposits	0.1
IOC	Copper	TT[d] action level = 1.3	Short-term exposure: gastrointestinal distress; long-term exposure: liver or kidney damage. People with Wilson's disease should consult their personal doctor if the amount of copper in water exceeds the action level	Corrosion of household plumbing systems; erosion of natural deposits	1.3
M	*Cryptosporidium*	TT[e]	Gastrointestinal illness (e.g., diarrhea, vomiting, cramps)	Human and animal fecal waste	0
IOC	Cyanide (as free cyanide)	0.2	Nerve damage or thyroid problems	Discharge from steel and metal factories; discharge from plastic and fertilizer factories	0.2
OC	2,4-D	0.07	Kidney, liver, or adrenal gland problems	Runoff from herbicide used on row crops	0.07
OC	Dalapon	0.2	Minor kidney changes	Runoff from herbicide used on rights of way	0.2
OC	1,2-Dibromo-3-chloropropane	0.0002	Reproductive difficulties; increased risk of cancer	Runoff and leaching from soil fumigant used on soybeans, cotton, pineapples, and orchards	0
OC	*o*-Dichlorobenzene	0.6	Liver, kidney, or circulatory system problems	Discharge from industrial chemical factories	0.6
OC	*p*-Dichlorobenzene	0.075	Anemia; liver kidney or spleen damage; changes in blood	Discharge from industrial chemical factories	0.075
OC	1,2-Dichloroethane	0.005	Increased risk of cancer	Discharge from industrial chemical factories	0
OC	1,1-Dichloroethylene	0.007	Liver problems	Discharge from industrial chemical factories	0.007
OC	*cis*-1,2-Dichloroethylene	0.07	Liver problems	Discharge from industrial chemical factories	0.07
OC	*trans*-1,2-Dichloroethylene	0.1	Liver problems	Discharge from industrial chemical factories	0.1

(Table continued)

TABLE 1.3 Continued

	Contaminant	MCL or TT[a] (mg/l)[b]	Potential Health Effects from Exposure above the MCL	Common Sources of Contaminant in Drinking Water	Public Health Goal
OC	Dichloromethane	0.005	Liver problems; increased risk of cancer	Discharge from drug and chemical factories	0
OC	1,2-Dichloropropane	0.005	Increased risk of cancer	Discharge from industrial chemical factories	0
OC	Di(2-ethylhexyl)-adipate	0.4	Weight loss; liver problems; possible reproductive difficulties	Discharge from chemical factories	0.4
OC	Di(2-ethylhexyl)-phthalate	0.006	Reproductive difficulties; liver problems; increased risk of cancer	Discharge from rubber and chemical factories	0
OC	Dinoseb	0.007	Reproductive difficulties	Runoff from herbicide used on soybeans and vegetables	0.007
OC	Dioxin (2,3,7,8-TCDD)	0.00000003	Reproductive difficulties; increased risk of cancer	Emissions from waste incineration and other combustion; discharge from chemical factories	0
OC	Diquat	0.02	Cataracts	Runoff from herbicide use	0.02
OC	Endothall	0.1	Stomach and intestinal problems	Runoff from herbicide use	0.1
OC	Endrin	0.002	Liver problems	Residue of banned insecticide	0.002
OC	Epichlorohydrin	TT[c]	Increased cancer risk; over a period of time, stomach problems	Discharge from industrial chemical factories; an impurity of some water treatment chemicals	0
OC	Ethylbenzene	0.7	Liver or kidney problems	Discharge from petroleum refineries	0.7
OC	Ethylene dibromide	0.00005	Problems with liver, stomach, reproductive system, or kidneys; increased risk of cancer	Discharge from petroleum refineries	0
IOC	Fluoride	4.0	Bone disease (pain and tenderness of the bones); children may get mottled teeth	Water additive that promotes strong teeth; erosion of natural deposits; discharge from fertilizer and aluminum factories	4.0

Regulatory Mandates for Ground-Water Monitoring 17

M	*Giardia lamblia*	TT[e]	Gastrointestinal illness (e.g., diarrhea, vomiting, cramps)	Human and animal fecal wastes	0
OC	Glyphosate	0.7	Kidney problems; reproductive difficulties	Runoff from herbicide use	0.7
DBP	Haloacetic acids (HAA5)	0.060	Increased risk of cancer	Byproduct of drinking water disinfection	N/A[f]
OC	Heptachlor	0.0004	Liver damage; increased risk of cancer	Residue of banned termiticide	0
OC	Heptachlor epoxide	0.0002	Liver damage; increased risk of cancer	Breakdown of heptachlor	0
M	Heterotrophic plate count (HPC)	TT[e]	HPC has no health effects; it is an analytic method used to measure the variety of bacteria that are common in water. The lower the concentration of bacteria in drinking water, the better maintained the water system is	HPC measures a range of bacteria that are naturally present in the environment	N/A
OC	Hexachlorobenzene	0.001	Liver or kidney problems; reproductive difficulties; increased risk of cancer	Discharge from metal refineries and agricultural chemical factories	0
OC	Hexachlorocyclo-pentadiene	0.05	Kidney or stomach problems	Discharge from chemical factories	0.05
IOC	Lead	TT[d] action level = 0.015	Infants and children: delays in physical or mental development; children could show slight defects in attention span and learning abilities; adults: kidney problems; high blood pressure	Corrosion of household plumbing systems; erosion of natural deposits	0
M	*Legionella*	TT[e]	Legionnaire's disease, a type of pneumonia	Found naturally in water; multiplies in heating systems	0
OC	Lindane	0.0002	Liver or kidney problems	Runoff and leaching from insecticide used on cattle, lumber and gardens	0.0002

(Table continued)

TABLE 1.3 Continued

	Contaminant	MCL or TT[a] (mg/l)[b]	Potential Health Effects from Exposure above the MCL	Common Sources of Contaminant in Drinking Water	Public Health Goal
IOC	Mercury (inorganic)	0.002	Kidney damage	Erosion of natural deposits; discharge from refineries and factories; runoff from landfills and croplands	0.002
OC	Methoxychlor	0.04	Reproductive difficulties	Runoff and leaching from insecticide used on fruits, vegetables, alfalfa, and livestock	0.04
IOC	Nitrate (as nitrogen)	10.0	Infants less than the age of 6 months who drink water containing nitrate in excess of the MCL could become seriously ill and, if untreated, may die. Symptoms include shortness of breath and blue-baby syndrome	Runoff from fertilizer use; leaching from septic tanks and sewage; erosion of natural deposits	10.0
IOC	Nitrite (as nitrogen)	1.0	Infants less than the age of 6 months who drink water containing nitrite in excess of the MCL could become seriously ill and, if untreated, may die. Symptoms include shortness of breath and blue-baby syndrome	Runoff from fertilizer use; leaching from septic tanks and sewage; erosion of natural deposits	1.0
OC	Oxamyl (Vydate)	0.2	Slight nervous system effects	Runoff and leaching from insecticide used on apples, potatoes, and tomatoes	0.2
OC	Pentachlorophenol	0.001	Liver or kidney problems; increased cancer risk	Discharge from wood preserving factories	0
OC	Picloram	0.5	Liver problems	Herbicide runoff	0.5
OC	Polychlorinated biphenyls (PCBs)	0.0005	Skin changes; thymus gland problems; immune deficiencies; reproductive or nervous system difficulties; increased risk of cancer	Runoff from landfills; discharge of waste chemicals	0

Regulatory Mandates for Ground-Water Monitoring

R	Radium 226 and 228 (combined)	5 pCi/l	Increased risk of cancer	Erosion of natural deposits	0
IOC	Selenium	0.05	Hair or fingernail loss; numbness in fingers or toes; circulatory problems	Discharge from petroleum refineries; erosion of natural deposits; discharge from mines	0.05
OC	Simazine	0.004	Problems with blood	Herbicide runoff	0.004
OC	Styrene	0.1	Liver, kidney, or circulatory system problems	Discharge from rubber and plastic factories; leaching from landfills	0.1
OC	Tetrachloroethylene	0.005	Liver problems; increased risk of cancer	Discharge from factories and dry cleaners	0
IOC	Thallium	0.002	Hair loss; changes in blood; kidney, intestine, or liver problems	Leaching from ore-processing sites; discharge from electronics, glass, and drug factories	0.0005
OC	Toluene	1.0	Nervous system; kidney or liver problems	Discharge from petroleum factories	1.0
M	Total coliforms (including fecal coliform and *Escherichia coli*)	5.0%[g]	Not a health threat in itself; it is used to indicate whether other potentially harmful bacteria may be present[h]	Coliforms are naturally present in the environment as well as feces; fecal coliforms and *E. coli* only come from human and animal fecal waste	0
DBP	Total trihalomethanes	0.08	Liver, kidney, or central nervous system problems; increased risk of cancer	Byproduct of drinking water disinfection	0
OC	Toxaphene	0.003	Kidney, liver, or thyroid problems; increased risk of cancer	Runoff and leaching from insecticide used on cotton and cattle	0
OC	2,4,5-TP (Silvex)	0.05	Liver problems	Residue of banned herbicide	0.05
OC	1,2,4-Trichlorobenzene	0.07	Changes in adrenal glands	Discharge from textile finishing factories	0.07
OC	1,1,1-Trichloroethane	0.2	Liver, nervous system, or circulatory problems	Discharge from metal degreasing sites and other factories	0.2
OC	1,1,2-Trichloroethane	0.005	Liver, kidney, or immune system problems	Discharge from industrial chemical factories	0.003
OC	Trichloroethylene	0.005	Liver problems; increased risk of cancer	Discharge from metal degreasing sites and other factories	0

(Table continued)

TABLE 1.3 Continued

	Contaminant	MCL or TT[a] (mg/l)[b]	Potential Health Effects from Exposure above the MCL	Common Sources of Contaminant in Drinking Water	Public Health Goal
M	Turbidity	TT[e]	Turbidity is a measure of the cloudiness of water. It is used to indicate water quality and filtration effectiveness (e.g., whether disease-carrying organisms are present). Higher turbidity levels are often associated with higher levels of disease-causing micro-organisms such as viruses, parasites, and some bacteria. These organisms can cause symptoms such as nausea, cramps, diarrhea, and associated headaches	Soil runoff	N/A
R	Uranium	30 µg/l	Increased risk of cancer; kidney toxicity	Erosion of natural deposits	0
OC	Vinyl chloride	0.002	Increased risk of cancer	Leaching from PVC pipes; discharge from plastic factories	0
M	Viruses (enteric)	TT[e]	Gastrointestinal illness (e.g., diarrhea, vomiting, cramps)	Human and animal fecal wastes	0
OC	Xylenes (total)	10	Nervous system damage	Discharge from petroleum factories; discharge from chemical factories	10

Note: D, disinfectant; DBP, disinfection byproduct; IOC, inorganic chemical; M, microorganism; OC, organic chemical; R, radionuclide; MFL, million fibers per liter.

[a] MCLG: the level of a contaminant in drinking water below which there is no known or expected risk to health. MCLGs allow for a margin of safety and are non-enforceable public health goals; MCL: the highest level of a contaminant allowed in drinking water. MCLs are set as close to MCLGs as feasible using the best available treatment technology and taking cost into consideration. MCLs are enforceable standards; MRDLG: the level of a drinking water disinfectant below which there is no known or expected risk to health. MRDLGs do not reflect the benefits of the use of disinfectants to control microbial contaminants; MRDL: the highest level of a disinfectant allowed in drinking water. There is convincing evidence that addition of a disinfectant is necessary for control of microbial contaminants; TT: a required process intended to reduce the level of a contaminant in drinking water.

[b] Units are in milligrams per liter (mg/l) unless otherwise noted. Milligrams per liter are equivalent to parts per million (ppm).

[c] Each water system must certify, in writing, to the state (using a third party or manufacturers' certification) that when it uses acrylamide or epichlorohydrin to treat water, the combination (or product) of dose and monomer level does not exceed the levels specified as follows: acrylamide = 0.05% dosed at 1 mg/l (or equivalent); epichlorohydrin = 0.01% dosed at 20 mg/l (or equivalent).

[d] Lead and copper are regulated by a TT that requires systems to control the corrosiveness of their water. If more than 10% of tap water samples exceed the action level, water systems must take additional steps. For copper, the action level is 1.3 mg/l; for lead, the action level is 0.015 mg/l.

[e] EPA's surface water treatment rules require systems using surface water or ground water under the direct influence of surface water to (a) disinfect their water and (b) filter their water or meet criteria for avoiding filtration, so that the following contaminants are controlled at the following levels:

- *Cryptosporidium* (as of 01/01/02 for systems serving more than 10,000 and 01/04/05 for systems serving less than 10,000) 99% removal.
- *Giardia lambia*: 99.9% removal or inactivation.
- Viruses: 99.99% removal or inactivation.
- *Legionella*: No limit, but EPA believes that if *Giardia* and viruses are removed or inactivated, *Legionella* will also be controlled.
- Turbidity: At no time can turbidity (cloudiness of water) go above 5 nephelometric turbidity units (NTU); systems that filter must ensure that the turbidity go no higher than 1 NTU (0.5 NTU for conventional or direct filtration) in at least 95% of the daily samples in any month. As of 01/01/02 for systems serving more than 10,000 and 01/14/05 for systems serving less than 10,000, turbidity may never exceed 1 NTU, and must not exceed 0.3 NTU in 95% of daily samples in any month.
- HPC: No more than 500 bacterial colonies per milliliter.
- Long Term 1 Enhanced Surface Water Treatment (effective date: 01/14/05): Surface water systems or GWUDI systems serving fewer than 10,000 people must comply with the applicable Long Term 1 Enhanced Surface Water Treatment Rule provisions (e.g., turbidity standards, individual filter monitoring, *Cryptosporidium* removal requirements, updated watershed control requirements for unfiltered systems).
- Filter backwash recycling: The Filter Backwash Recycling Rule requires systems that recycle to return specific recycle flows through all processes of the system's existing conventional or direct filtration system or at an alternate location approved by the state.

[f] Although there is no collective MCLG for this contaminant group, there are individual MCLGs for some of the individual contaminants: (a) Haloacetic acids: dichloroacetic acid (0), trichloroacetic acid (0.3 mg/l); (b) Trihalomethanes: bromodichloromethane (0); bromoform (0); dibromochloromethane (0.06 mg/l).

[g] No more than 5.0% samples can be total coliform-positive per month. (For water systems that collect fewer than 40 routine water samples per month, no more than one sample can be total coliform-positive per month.) [Every sample that has total coliform must be analyzed for either fecal coliform or *E. coli*. If two consecutive samples are TC-positive and one is also positive for *E. coli* fecal coliforms, this is defined as an acute MCL violation.]

[h] Fecal coliform and *E. coli* are bacteria whose presence indicates that the water may be contaminated with human or animal wastes. Disease-causing microbes (pathogens) in these wastes can cause diarrhea, cramps, nausea, headaches, or other symptoms. These pathogens may pose a special health risk for infants, young children, and people with severely compromised immune systems.

Source: U.S. EPA Office of Water (4606M), EPA 816-F-03-016, June 2003 (www.epa.gov/safewater).

TABLE 1.4

National Secondary Drinking Water Standards[a]

Contaminant	Secondary Standard
Aluminum	0.05 to 0.2 mg/l
Chloride	250 mg/l
Color	15 (color units)
Copper	1.0 mg/l
Corrosivity	Noncorrosive
Fluoride	2.0 mg/l
Foaming agents	0.5 mg/l
Iron	0.3 mg/l
Manganese	0.05 mg/l
Odor	3 threshold odor number
pH	6.5 to 8.5
Silver	0.10 mg/l
Sulfate	250 mg/l
Total dissolved solids	500 mg/l
Zinc	5 mg/l

[a] NSDWSs are non-enforceable guidelines regulating contaminants that may cause cosmetic effects (such as skin or tooth discoloration) or aesthetic effects (such as taste, odor, or color) in drinking water. EPA recommends secondary standards to water systems but does not require systems to comply. However, states may choose to adopt them as enforceable standards.
Source: U.S. EPA Office of Water (4606M), EPA 816-F-03-016, June 2003 (www.epa.gov/safewater).

Hence, the law establishes the concept of protecting some of the recharge areas for these points of public drinking water withdrawal. Ground-water monitoring systems are not specifically required under the WHP, but there are few other reliable methods that can be used to generate the data required to support delineation of WHPAs. The states have been given flexibility in determining appropriate operational approaches to WHPA delineation.

The U.S. EPA published guidelines for delineation of WHPAs to assist the states in developing their programs (U.S. EPA, 1987b). The delineation guidelines assume that WHPA delineation and protection are targeted to three general threats. The first threat is the direct introduction of contaminants to the area immediately contiguous to the well through improper casing, road runoff, spills, and accidents. The second basic threat is from microbial contaminants such as bacteria and viruses. The third major threat is from the broad range of chemical contaminants including inorganic and naturally occurring or synthetically derived organic chemicals.

U.S. EPA's WHPA delineation policy is generally based on the analysis of criteria, criteria thresholds, and delineation methods. The criteria, or conceptual standards on which WHPA delineation may be based, include distance, drawdown, travel time, flow system boundaries, and the capacity of the aquifer to assimilate contaminants. Choice of the criteria to be applied in any particular program is based on both technical and nontechnical considerations. Criteria and criteria thresholds define the general technical basis of the WHPA. Selecting appropriate criteria thresholds is a key decision point, which must be done in conjunction with establishing the management elements of the WHPP. The WHPA delineation methods are used to translate or apply these criteria, and to develop on-the-ground or on-the-map WHPA boundaries. The specific methods to be used in delineating a WHPA range from simple radius-of-influence estimation techniques to

TABLE 1.5
Sole-Source Aquifer Designations made by the U.S. EPA as of April 2004

State	Sole-Source Aquifer Name	Federal Register Citation	Publication Date	GIS Map
U.S. EPA Region I				
CT	Pootatuck Aquifer	55 FR 11056	03/26/90	Yes
MA	Cape Cod Aquifer	47 FR 30282	07/13/82	Yes
MA	Nantucket Island Aquifer	49 FR 2952	01/24/84	Yes
MA	Martha's Vineyard Aquifer	53 FR 3451	02/05/88	Yes
MA	Head of Neponset Aquifer System	53 FR 49920	12/12/88	Yes
MA	Plymouth-Carver Aquifer	55 FR 32137	08/07/90	Yes
MA	Canoe River Aquifer	58 FR 28402	05/13/93	Yes
MA	Broad Brook Basin of the Barnes	60 FR 20989	04/28/95	Yes
ME	Monhegan Island	53 FR 24496	06/29/88	Yes
ME	Vinalhaven Island Aquifer System	54 FR 29779	07/14/89	Yes
ME	North Haven Island Aquifer System	54 FR 29934	07/17/89	Yes
ME	Isleboro Island Aquifer System	64 FR 186	09/27/99	No
RI	Block Island Aquifer	49 FR 2952	01/24/84	Yes
RI/CT	Pawcatuck Basin Aquifer System	53 FR 17108	05/13/88	Yes
RI	Hunt-Annaquatucket Pettaquamscutt	53 FR 19026	05/26/88	Yes
U.S. EPA Region II				
NJ	Buried Valley Aquifers, Central Basin, Essex, and Morris Counties	45 FR 30537	05/08/80	Yes
NJ	Upper Rockaway River Basin	49 FR 2946	01/24/84	Yes
NJ	Ridgewood Area Aquifers	49 FR 2943	01/24/84	Yes
NJ/NY	Highlands Aquifer System, Passaic, Morris, and Essex Counties — NJ Orange County — NY	52 FR 37213	10/05/87	Yes
NJ[a]/DE/PA	NJ Coastal Plain Aquifer System	53 FR 23791	06/24/88	Yes
NJ/NY	NJ Fifteen Basin Aquifers	53 FR 23685	06/23/88	Yes
NJ/NY	Ramapo River Basin Aquifer Systems	57 FR 39201	08/28/92	Yes
NY	Nassau/Suffolk Counties, Long Island	43 FR 26611	06/21/78	Yes
NY	Kings/Queens Counties	49 FR 2950	01/24/84	Yes
NY	Schenectady/Niskayuna	50 FR 2022	01/14/85	Yes
NY	Clinton Street-Ballpark Valley Aquifer System, Broome, and Tioga Counties	50 FR 2025	01/14/85	Yes
NY	Cattaraugus Creek Basin Aquifer, WY, and Allegany Counties	52 FR 36100	09/25/87	Yes
NY	Cortland-Homer-Preble Aquifer System	53 FR 22045	06/13/88	Yes

(Table continued)

TABLE 1.5 Continued

State	Sole-Source Aquifer Name	Federal Register Citation	Publication Date	GIS Map
U.S. EPA Region III				
DE/PA/NJ[b]	New Jersey Coastal Plain Aquifer	53 FR 23791	06/24/88	Yes
MD	Maryland Piedmont Aquifer, Montgomery, Howard, and Carroll Counties	45 FR 57165	08/27/80	Yes
MD	Poolesville Area Aquifer Extension of the Maryland Piedmont Aquifer	63 FR 3042	02/06/98	Yes
PA	Seven Valleys Aquifer, York County	50 FR 9126	03/06/85	Yes
VA	Prospect Hill Aquifer, Clark County	52 FR 21733	06/09/87	Yes
VA	Columbia and Yorktown, Eastover Multi-Aquifer System, Accomack, and North Hampton Counties	62 FR 17187	04/09/97	Yes
U.S. EPA Region IV				
FL	Biscayne Aquifer, Broward, Dade, Monroe, and Palm Beach Counties	44 FR 58797	10/11/79	No
FL	Volusia-Floridan Aquifer, Flagler, and Putnam Counties	52 FR 44221	11/18/87	No
LA[c]/MS	Southern Hills Regional Aquifer System	53 FR 25538	07/07/88	No
U.S. EPA Region V				
IN	St. Joseph Aquifer System	53 FR 23682	06/23/88	No
MN	Mille Lacs Aquifer	55 FR 43407	10/29/90	No
OH	Pleasant City Aquifer	52 FR 32342	08/27/87	Yes
OH	Bass Island Aquifer, Catawba Island Aquifer	52 FR 37009	10/02/87	Yes
OH	Miami Valley Buried Aquifer	53 FR 15876	05/04/88	Yes
OH	OKI Extension of the Miami, Buried Valley Aquifer	53 FR 25670	07/08/88	Yes
OH	Allan County Area Combined Aquifer System	57 FR 53111	11/06/92	Yes
U.S. EPA Region VI				
LA	Chicot Aquifer System	53 FR 20893	06/07/88	Yes
LA[d]/MS	Southern Hills Aquifer System	53 FR 25538	07/07/88	Yes
OK	Arbuckle-Simpson Aquifer, South Central Oklahoma	54 FR 39230	09/25/89	Yes
TX	Edwards Aquifer, San Antonio Area	40 FR 58344	12/16/75	Yes
TX	Edwards Aquifer, Austin Area	53 FR 20897	06/07/88	Yes
U.S. EPA Region VII				
There are no designated sole-source aquifers in Region VII as of April 2004				
U.S. EPA Region VIII				
MT	Missoula Valley Aquifer	53 FR 20895	06/07/88	No
UT	Castle Valley Aquifer System	66 FR 41027	08/06/01	No

UT	Western Uinta Arch Paleozoic Aquifer System at Oakley, UT	65 FR 232	12/01/00	No
UT	Glen Canyon Aquifer System	67 FR 736	01/07/02	No
WY[e]	Eastern Snake River Plain Aquifer Stream Flow Source Area	56 FR 50638	10/07/91	Yes
WY	Elk Mountain Aquifer	63 FR 38167	07/15/98	No
U.S. EPA Region IX				
AZ	Upper Santa Cruz and Avra Basin Aquifer	49 FR 2948	01/24/84	Yes
AZ	Bisbee-Naco Aquifer	53 FR 38337	09/30/88	Yes
CA	Fresno County Aquifer	44 FR 52751	09/10/79	Yes
CA	Santa Margarita Aquifer, Scotts Valley	50 FR 2023	01/14/85	Yes
CA	Campo/Cottonwood Creek Aquifer	58 FR 31024	05/28/93	Yes
CA	Ocotillo-Coyote Wells Aquifer	61 FR 47752	09/10/96	Yes
GU	Northern Guam Aquifer System	43 FR 17867	04/26/78	Yes
HI	Southern Oahu Basal Aquifer	52 FR 45496	11/30/87	Yes
HI	Molokai Aquifer	58 FR 23063	04/20/93	Yes
U.S. EPA Region X				
ID/WY[f]	Eastern Snake River Plain Aquifer	56 FR 50638	10/07/91	Yes
OR	North Florence-Dunal Aquifer	52 FR 37519	10/07/87	Yes
WA/ID	Spokane Valley Rathdrum Prairie Aquifer	42 FR 5566	02/09/78	Yes
WA	Camano Island Aquifer	47 FR 14779	04/06/82	Yes
WA	Whidbey Island Aquifer	47 FR 14779	04/06/82	Yes
WA	Cross Valley Aquifer	52 FR 18606	05/18/87	Yes
WA	Newberg Area Aquifer	52 FR 37215	10/05/87	Yes
WA	Cedar Valley (Renton Aquifer)	53 FR 38779	10/03/88	Yes
WA/ID	Lewiston Basin Aquifer	53 FR 49920	12/12/88	Yes
WA	Central Pierce County Aquifer System	59 FR 224	01/03/94	Yes
WA	Marrowstone Island Aquifer System	59 FR 28752	06/02/94	Yes
WA	Vashon-Maury Island Aquifer System	59 FR 34468	07/05/94	Yes
WA	Guemes Island Aquifer System	62 FR 5928-3	12/01/97	Yes

[a] The New Jersey Coastal Plains Aquifer is jointly managed with Region III. While listed in both regions, it is counted only once in the national total of 73.
[b] The New Jersey Coastal Plains Aquifer is jointly managed with Region II. While listed in both regions, it is counted only once in the national total of 73.
[c] The Southern Hills Regional Aquifer system is jointly managed with Region VI. While listed in both regions, it is counted only once in the national total of 73.
[d] The Southern Hills Regional Aquifer System is jointly managed with Region IV. While listed in both regions, it is counted only once in the national total of 73.
[e] The Eastern Snake River Plan Aquifer is jointly managed with Region X. While listed in both regions, it is counted only once in the national total of 73.
[f] The Eastern Snake River Plan Aquifer is jointly managed with Region VIII. While listed in both regions, it is counted only once in the national total of 73.
Source: U.S. EPA Office of Ground Water and Drinking Water, April 2004 (www.epa.gov/safewater/ssanp.html).

highly complex and comprehensive numerical modeling techniques. Regardless of the method used, data from existing or yet to be installed ground-water monitoring systems are critical to proper delineation of WHPAs.

Underground Injection Control Program

The UIC Program was developed under SDWA to protect current and future sources of drinking water (defined as all ground water with a total dissolved solids content of less than 10,000 mg/l), from endangerment by underground injection of fluids. The basic concept of the UIC Program is to prevent contamination of fresh-water aquifers by ensuring that injected fluids are confined within the injection wells and the intended injection zone. The need for such a program was compelling when it was first conceived, as the EPA estimated that in 1980, there were more than 600,000 wells injecting more than 850 billion gallons of fluid per year beneath the surface (U.S. EPA, 2002). Considering the types of fluids injected (ranging from storm water runoff to hazardous wastes), the number of facilities in operation, and the complexity and diversity of geology in areas where underground injection is practiced, the task of regulating this industry is quite complex.

To ensure effective regulation of injection wells, standards have been set for each of five types or classes of injection wells, which are described in Table 1.6. The UIC regulations establish minimum standards for injection well design, construction, operation, monitoring, and decommissioning procedures, and state program requirements. Wells in Classes I to III come under rigid permitting requirements; Class IV wells are forbidden; and Class V wells are permitted or forbidden on a case-by-case basis. The U.S. EPA Regional Administrator may require ground-water monitoring at an underground injection point to evaluate whether an underground source of drinking water may be endangered by injection of fluids into Class II enhanced recovery wells, Class IV wells, and some Class V wells. In addition, the owner or operator of a Class I, II, or III well can be required

> To install and use monitoring wells within the area of review if required by the Director (of the U.S. EPA), to monitor any migration of fluids into and pressure in the underground sources of drinking water.
>
> (40 CFR 144.28)

The type, number, and location of the wells; the parameters to be measured; and the frequency of monitoring must be approved by the EPA.

Under the UIC Program, aquifers that do not currently serve as a source of drinking water or could not in the future are exempted from protection, because they are (1) mineral, hydrocarbon, or geothermal-energy producing; (2) situated at a depth or location that makes recovery of the water for drinking economically or technologically impractical; or (3) so contaminated that it would be infeasible to render the water fit for human consumption. Moreover, in keeping with SDWA policy, only ground water that supplies or could supply in the future any public water supply system is protected. Consequently, the UIC Program does not apply to either ground water used for purposes other than drinking or ground-water supplying nonpublic water systems.

Surface Mining Control and Reclamation Act

The SMCRA of 1977, under the administration of the U.S. Department of the Interior (specifically the OSM), provides authority for various levels of government to control

Regulatory Mandates for Ground-Water Monitoring 27

environmental impacts resulting from all mining activities, even though the title of the Act refers only to surface mining. Among the purposes of the Act are to:

- Establish a nationwide program to protect society and the environment from the adverse effects of mining operations
- Ensure that mining operations are conducted so as to protect the environment

TABLE 1.6

U.S. EPA Injection Well Classifications Under the UIC Program of the SDWA

Class of well	Description
Class I	Wells used by generators of hazardous wastes or owners or operators of hazardous waste management facilities to inject hazardous waste. In addition, industrial and municipal disposal wells used to inject fluids beneath the lowermost formation containing, within 0.25 mi of the well, an underground source of drinking water
Class II	Wells that inject fluids, which are brought to the surface in connection with conventional oil and natural gas production, those used for enhanced recovery of oil or natural gas, and those used for storage of liquid hydrocarbons
Class III	Wells that inject for the purpose of extracting minerals or energy, including solution mining wells, wells used for *in situ* combustion of fossil fuel, wells used for recovery of geothermal energy, and wells used in the mining of sulfur by the Frasch process
Class IV	Wells used by generators of hazardous wastes or radioactive wastes, by owners and operators of hazardous waste management facilities, or by owners and operators of radioactive waste disposal sites to dispose of hazardous wastes or radioactive wastes into or above a formation that, within 0.25 miles of the well, contains an underground source of drinking water
Class V	Any injection well not included in the Classes I to IV, including: 1. Air-conditioning return-flow wells used to return to the supply aquifer, the water used for heating or cooling in a heat pump 2. Cesspools or other devices that receive wastes, which have an open bottom and sometimes have perforated sides. The UIC requirements do not apply to single family residential cesspools 3. Cooling-water return-flow wells used to inject water previously used for cooling 4. Drainage wells used to drain surface fluid, primarily storm water runoff into a subsurface formation 5. Dry wells used for the injection of wastes into a subsurface formation 6. Recharge wells used to replenish the water in an aquifer 7. Saltwater intrusion barrier wells used to inject water into a freshwater aquifer to prevent the intrusion of saltwater into the freshwater 8. Sand backfill wells used to inject a mixture of water and sand, mill tailings, or other solids into mined out portions of subsurface mines 9. Septic system wells used: • To inject the waste or effluent from a multiple dwelling, business establishment, community, or regional business establishment septic tank • For a multiple dwelling, community, or regional cesspool. The UIC requirements do not apply to single family residential waste disposal systems 10. Subsidence control wells (not used for the purpose of oil or natural gas production) used to inject fluids into a nonoil or gas-producing zone to reduce or eliminate subsidence associated with the overdraft of freshwater 11. Wells used for the storage of hydrocarbons which are gases at standard temperature and pressure 12. Geothermal wells used in heating and aquaculture 13. Nuclear disposal wells

Source: U.S. EPA Underground Injection Control Program regulations as outlined in the Federal Register, Vol. 45(123), June 24, 1980.

- Ensure that adequate procedures are followed to reclaim surface areas as contemporaneously as possible with surface mining operations
- Promote the reclamation of mined areas left without adequate reclamation prior to enactment of this Act and which continue, in their un-reclaimed condition, to substantially degrade the quality of the environment, prevent or damage the beneficial use of land or water resources, or endanger the health or safety of the public

Several sections of SMCRA deal specifically with ground water. For new mines, permit applications described under Section 507 must include the determination of the probable hydrologic consequences of the mining and the reclamation proposed. Of particular concern is the determination of the impact of mining and reclamation on the quantity and quality of water in both surface-water and ground-water systems. All permit applications must be accompanied by geologic maps and cross-sections of the land to be affected showing, among others, the locations of aquifers and estimated water levels (Lehr et al., 1984).

Reclamation plan requirements outlined in Section 508 compel mine operators to provide a detailed description of the measures to be taken during the mining and reclamation process to ensure the protection of the quality of surface water and ground water from the adverse effects of the mining and reclamation process. In addition, the operator must recognize the rights of the present users to this water and must ensure the protection of the quantity of surface water and ground water from the mining and reclamation operation or provide alternate sources of water where such protection cannot be assured.

Section 515 of SMCRA outlines general environmental protection performance standards applicable to mining and reclamation operations that require the operation to

> Minimize the disturbances to the prevailing hydrologic balance at the mine site and to the quality and quantity of water in surface-water and ground-water systems both during and after mining operations and during reclamation.

Water quality is to be preserved by avoiding acid or other toxic mine drainage through such means as preventing or removing water from contact with toxic deposits, treating drainage to reduce its toxic content, or burying or otherwise disposing of acid-forming or toxic materials in a manner to prevent contamination of both surface water and ground water. Mine operators are required to maintain the hydrologic balance of the area by restoring the recharge capacity of the mined area to approximate premining conditions.

Under Section 517, the OSM may require mine operators to install, use, and maintain ground-water monitoring systems. The preparation of a ground-water monitoring plan is required under 30 CFR 780, which deals with the application for a surface mining permit. For those mining and reclamation operations that disturb aquifers, the OSM has the power to specify monitoring sites to record the level, amount, and quality of ground water in aquifers affected or potentially affected by the mining operation. The OSM has set forth standards and procedures for the collection and analysis of data generated by ground-water monitoring programs required under SMCRA. As part of the minimum requirements for the required Reclamation and Operations Plan, hydrogeologic information must be supplied concerning the quality of the surface water and ground water in the permitted area and adjacent areas.

Brownfields

In 1995, the U.S. EPA addressed the problem associated with former industrial and urban sites with minor contamination by creating the Brownfields Initiative. This was a new approach to management of contaminated property which, unlike the RCRA and CERCLA regulations, was based on a partnership model. Prior to the initiative, marginally contaminated sites were largely ignored by developers because the magnitude of contamination often was unknown and the liability for this contamination was not something developers wanted to assume. Under this initiative, the U.S. EPA has established the Brownfields National Partnership and provides local communities with seed money to encourage local governments to develop these properties and manage any contamination associated with them. The local governments, in turn, create 2 year pilot programs that are used to build local capabilities, with technical guidance provided at the Federal level.

Brownfields sites fall into one of the following several categories:

- Brownfields Assessment Pilots provide funding for environmental assessments and community outreach.
- Brownfields Cleanup Revolving Loan Fund Pilots provide funding to capitalize loans that are used to clean up brownfields.
- Brownfields Job Training and Development Demonstration Pilots provide environmental training for residents of brownfields communities.
- RCRA and Brownfields Prevention Pilots utilize the inherent flexibility in RCRA regulations to prevent brownfields from being developed on RCRA properties.
- Clean Air/Brownfields Partnership Pilots help to determine the potential air quality and other environmental and economic benefits of redeveloping urban brownfields.
- Brownfields Showcase Communities serve as national models for successful brownfields assessment, cleanup, and redevelopment.
- Targeted Brownfields Assessments provide funding or technical assistance for environmental assessments at selected brownfields sites not targeted by EPA Assessment Pilots.

While ground-water monitoring is not specifically required under the Brownfields Initiative, some level of monitoring (typically short-term monitoring) is generally necessary to determine the presence or absence and types and levels of contaminants in ground water at each site.

Federal Ground-Water Protection Strategy

When the U.S. EPA established a Ground-Water Protection Strategy in August 1984 (U.S. EPA, 1984), it concluded that state governments have the primary responsibility for ground-water protection policies and implementation, yet it set national goals and management strategies for implementing existing federal laws.

The strategy sets a policy framework to guide U.S. EPA's programs affecting ground water. This framework involves developing a system for classification of the nation's ground water. The agency uses this classification system to evaluate the siting of RCRA facilities and will continue to use the immediacy of a threat to ground water as a factor in selecting sites for Superfund cleanup.

Specifically the policy calls for EPA to:

- Provide financial support to states for ground-water protection program development and institution building

- Assess the problems that may exist from sources of ground-water contamination, which were previously not addressed
- Study the need for further regulation of land disposal facilities including surface impoundments and landfills
- Issue guidelines for agency decisions affecting ground-water protection and cleanup
- Establish an Office of Ground-Water Protection within EPA to coordinate agency policies (Bird, 1985)

The classification of ground water is the backbone of the policy, which helps to provide consistency in agency decisions.

Ground-Water Classification

The Environmental Protection Agency released a draft document of guidelines for ground-water classification as part of its Ground-Water Protection Strategy. The document established three classifications for ground water. Class I, special ground water, is ecologically vital or irreplaceable as a source of drinking water. Ecologically vital ground water is defined as that "which supports habitats for species listed or proposed for listing under the Endangered Species Act or which provides the base flow for a particularly sensitive ecological system that, if polluted, would destroy a unique habitat." Class I water is considered to be highly vulnerable to contamination because of the hydrologic characteristics of the area in which the ground water occurs. With its authority under RCRA, EPA will ban the siting of new disposal facilities and require very stringent cleanup levels (involving cleanup to background or drinking water levels) to be applied to existing facilities above Class I ground water. EPA also has considered developing special permit conditions under the UIC Program to protect these waters.

Class II ground water includes current or potential sources of drinking water and water having other beneficial uses. Class III ground water includes water not considered to be a potential source of drinking water and water that may be contaminated naturally (e.g., highly saline ground water, with total dissolved solids levels over 10,000 mg/l) or by human activity, beyond levels that allow cleanup using methods reasonably employed in public water system treatment.

Essentially, Class I ground water would receive the highest level of protection, Class II ground water would receive less protection, and Class III ground water would receive the least protection under this ground-water classification system. There is a provision for variances to lower the protection levels.

Discussion of Ground-Water Quality Standards

Ground-water monitoring only becomes meaningful when the results of the analyses for water quality are compared to some useful reference point. In many cases, ground-water quality standards are applied to water used for consumptive purposes as it leaves the tap. In other cases, standards are applied to ground water after it has been cleaned up or as it discharges to a surface water body or in terms of the risk posed by a specified exposure. Further complicating the issue is the fact that many states have different or more restrictive standards than the Federal government.

To achieve a better understanding of the standards that can be applied, a definition of some of the basic terms is appropriate. The SDWA states:

> The term "Primary Drinking Water Regulation" means a regulation which (1) applies to public water systems, (2) specifies contaminants which may have an adverse effect on the health of persons, (3) specifies for each contaminant either a maximum contaminant level or a reduced level based on treatment, and (4) contains criteria and procedures to assure a supply of drinking water that will comply with the maximum contaminant levels (MCLs) and the requirements for the minimum quality of water that can be taken into the system.

The Secondary Drinking Water Regulations, also defined in the SDWA, are described as follows:

> The term "Secondary Drinking Water Regulation" means a regulation which applies to public water systems and which specifies the maximum contaminant levels which are requisite to protect the public welfare. This applies to any contaminant in drinking water which may adversely affect the odor or appearance of the water so that a significant number of users discontinue its use.

The term MCL refers to the maximum permissible level of a contaminant in water which is delivered to any user of a public water system. These are enforceable standards that are set as close to maximum contaminant level goals (MCLGs) as feasible. These standards are often applied to ground water that is used for drinking water purposes, regardless of whether it is supplied by a public system or a private well. These standards also consider the best technology that is available, treatment technologies that can be applied, and associated costs.

The MCLG, previously called a recommended maximum contaminant level, is the maximum level of a contaminant in drinking water at which no known or anticipated adverse effect on the health of persons would occur, which allows for an adequate margin of safety. These are non-enforceable health goals (40 CFR 141.2, July 1987).

The CWA also has established water-quality criteria that are not limited to ground water, known as the 304(a)(1) criteria. They are not rules and are not enforceable. Rather, these criteria present specific data and guidance on the environmental effects of pollutants, which can be useful in deriving regulatory requirements based on considerations of water-quality impacts. They are, therefore, comparable to the MCLGs, as they are not based on technology or cost, but are on health goals. These standards can be used when protection of a drinking water source is not the sole objective, and they can be applied to water-quality-based effluent limitations and toxic pollutant effluent standards (Federal Register, 1980). Although these standards were derived for surface water, they have application to ground water, particularly where other standards for certain chemicals have not yet been set.

Lists of various national and state standards and criteria are available from the U.S. EPA and various state regulatory agencies. Extreme caution should be exercised in applying these criteria and standards to specific site conditions. These criteria, for the most part, do not take into account some other important factors that should be considered when applying standards to ground water used for consumptive purposes. These considerations include the population that will be using the water, the exposure from other sources that could contribute to the risk, and some of the other risks of exposure other than carcinogenic effects. Clearly, the application of water-quality standards and criteria is neither simple nor straightforward and requires expertise in other fields, particularly toxicology.

Mechanisms for a Workable Federal Ground-Water Program

Even the best-laid plans have elements that make implementation more difficult than it first appears — environmental regulations are no exception. Now that major Federal waste-management regulations have been in place for more than two decades, there are functional goals that must be kept in mind if the mandated protection of ground water is going to succeed.

The first element is communication. The regulations will do no good if the regulated community will not follow them, even under the threat of civil or criminal penalty. Enforcement bodies have limited resources for informing the regulated community of their obligations under the law. As a result, most cases of noncompliance are the result of ignorance rather than malice or the profit motive. It is essential for the regulations to be communicated to those parties that they directly affect. Industrial facilities must be made aware of the limitations placed on their practices, such as management of waste, discharge of process wastewater, standards that treatment works must meet, and the permits that must be obtained. Even individual homeowners must be made aware that they are responsible for protecting their small portion of the ground-water resource. Federal, state, and local regulations generally can be easily accessed on the Internet or by calling the appropriate regulatory agency. Every effort must be made to disseminate this information to the people who need it.

The second element is the establishment of standardized evaluation and protection practices and the application of the same basic standards to similar situations. It is well known that ground water is a dynamic resource and the hydrogeologic settings in which it occurs very widely. However, there are sound scientific and engineering practices that can be applied to the evaluation of ground water to ensure that suitable and appropriate conclusions and recommendations concerning its potential use or abuse can be drawn. Similarly, the standards that are applied to the protection and cleanup of an aquifer should be made clear and not left to the whimsy of an individual regulator. There is a broad spectrum of standards and policies, ranging from nondegradation to inaction, routinely being applied by regulators who lack direction. Standards that are health- or risk-based, technology-sensitive, and use-directed are in the process of being developed and will do much to bring this element into focus. However, a standard baseline for protection and cleanup would do much to minimize the uncertainty currently associated with site evaluation.

Finally, there should be a mechanism at all levels for changing and amending the regulations as conditions change. This is found to some degree at Federal and state levels with the owner's ability to request a waiver to a portion of a regulation or to apply alternate standards in particular cases. This ability should be expanded and streamlined as much as possible to address the changing nature of the resource. If new practices pose new health risks, or if the chemical or physical nature of an aquifer changes substantially over time, the mechanisms must be in place to revise regulations and applicable standards.

Comprehensive State Ground-Water Protection Programs

Comprehensive State Ground-Water Protection Programs (CSGWPPs) are partnerships between states and the U.S. EPA and Tribal Nations to implement EPA's ground-water protection goals and principles. The purpose is to achieve a more efficient, coherent, and comprehensive approach to protecting and managing the nation's ground-water resources. This program will be used to prevent contamination and to consider use, value, and vulnerability in setting priorities for both protection and remediation.

This program is used to establish goals and set priorities based on local needs and to clarify the roles and responsibilities for ground-water management across federal, state, and local jurisdictions. CSGWPPs can be used as a template to plan and implement ground-water protection and remediation strategies.

Ground Water and Terrorism

Owing to the terrorist attacks of September and October 2001, which included the use of commercial airliners as weapons and using the U.S. Postal Service to deliver pathogens through the mail, federal, regional, state, and local governments all are taking a closer look at the safety of our water supply. Surface-water sources would seem to be the most vulnerable target, due to easier access and faster dispersion of contaminants in that medium. Ground water, by its very nature, is protected, both by its slow flow rate and by the natural cleansing action of the porous media through which it passes. In fact, ground water could prove to be a "safe haven" for the storage and retrieval of uncontaminated drinking water. Comprehensive emergency preparedness plans should include guidelines and procedures for monitoring and maintaining the quality of the nation's ground-water supply. Additional security at the well head should be undertaken for water systems supplying large populations.

Ground Water and Development

One of the most stressful impacts on natural resources in the late 20th and early 21st centuries has been the exploitation of ground water. Anxiety about this limited resource is especially pronounced in the Western U.S., where water can be scarce, heavily laden with salts, or deep and expensive to retrieve. Land-use decisions are beginning to take water resource availability into account. A law that took effect in the state of California in January 2002 may have far-reaching impacts on development. In essence, this law states that all developers are required to provide detailed proof that an ample water supply exists and can be tapped for at least 20 years for every housing development involving over 500 homes. If this cannot be demonstrated, the developers are not allowed to break ground (Anonymous, 2001). This law may serve as a model for other states seeking to protect potable ground-water resources. It may also be used to avoid the battles that sometimes ensue over water used for domestic supply, agricultural use, ranching and that needed to sustain and protect endangered species.

References

Anon., *New California Law Seeks to Curb Runaway Sprawl*, The Washington Post, Dec. 23, 2001.
Bird, J.C., *Ground Water Protection: Emerging Issues and Policy Challenges*, Environmental and Energy Study Institute, Washington, DC, 1985.
CERCLA. Comprehensive Environmental Response, Compensation, and Liability Act 1980 & 1986 (Public Law 96-510, as amended by Public Law 99-499).

CWA. Clean Water Act 1970 & 1977 (Public Law 92-500, as amended by Public Law 92-217 and Public Law 95-576; 33 U.S.C. 1251 et seq.).

Federal Register, Vol. 45(231), Nov. 28, 1980.

Lehr, J.H., Nielsen, D.M. and Montgomery, J.J., U.S. Federal legislation pertaining to ground water protection, in *Ground-Water Pollution and Microbiology*, Bitton G. and Gerba, C.P., Eds., John Wiley and Sons, New York, NY, 1984, chap. 14, pp. 353–371.

SDWA. Safe Drinking Water Act 1980 (Public Law 95-523, as amended by Public Law 95-1900; 42 U.S.C. 3001 et seq.).

SWDA Solid Waste Disposal Act, as amended by the Resource Conservation and Recovery Act (Public Law 94-580, as amended by Public Law 96-609 and Public Law 96-482; 42 U.S.C. 6901 et seq.).

U.S. Environmental Protection Agency, *Ground Water Protection Strategy*, U.S. Environmental Protection Agency, Office of Ground-Water Protection, Washington, DC, 1984, 75 pp.

U.S. Environmental Protection Agency, *Underground Storage Tanks (UST), the New Federal Law*, U.S. Environmental Protection Agency, Office of Solid Waste and Emergency Response, Washington, DC, 1985.

U.S. Environmental Protection Agency, *RCRA Orientation Manual*, U.S. Environmental Protection Agency, Office of Solid Waste and Emergency Response, Washington, DC, 1986.

U.S. Environmental Protection Agency, *The New Superfund — Protecting the People and Their Environment*, Washington, DC, Vol. 13(1), U.S. Environmental Protection Agency, Office of Public Affairs, 1987a.

U.S. Environmental Protection Agency, *Guidelines for Delineation of Wellhead Protection Areas*, U.S. Environmental Protection Agency, Office of Ground-Water Protection, Washington, DC, 1987b.

U.S. Environmental Protection Agency, Guidance for Conducting Remedial Investigations and Feasibility Studies under CERCLA, October, 1988, U.S. Environmental Protection Agency, Office of Solid Waste and Emergency Response, Washington, DC, EPA/540/G-89/004, Directive 9355.3-01, 1988.

U.S. Environmental Protection Agency, *Summary of the History of the UIC Program*, U.S. Environmental Protection Agency, Office of Water, Washington, DC, 2002.

U.S. Environmental Protection Agency, *List of Drinking Water Contaminants and Their MCLs*, U.S. Environmental Protection Agency, Office of Ground Water and Drinking Water, Washington, DC, EPA/816/F-03-016, Jun. 2003, (www.epa.gov/safewater/mcl.html).

U.S. Environmental Protection Agency, *National Priority List Sites*, U.S. Environmental Protection Agency, Office of Superfund, Washington, DC, Apr. 2004a (www.epa.gov/superfund/sites/query/queryhtm/npltotal.htm).

U.S. Environmental Protection Agency, *Public Drinking Water Systems in the United States*, U.S. Environmental Protection Agency, Office of Ground Water and Drinking Water, Washington, DC, Apr. 2004b (www.epa.gov/safewater/pws/pwss.html).

2

Environmental Site Characterization

David M. Nielsen, Gillian L. Nielsen, and Lynne M. Preslo

CONTENTS

Introduction	36
Background	36
The Importance of Environmental Site Characterization	40
Historical Problems with Environmental Site Characterization	41
The Problem of Heterogeneity	42
The Problem of Sample and Data Representativeness	48
Objectives of Environmental Site Characterization	51
Data Required from an Environmental Site Characterization Program	54
Approaches to Environmental Site Characterization	56
The Conventional or Traditional (Phased) Approach	56
Phase I	57
Phase II and Beyond	65
Improved Approaches to Environmental Site Characterization	66
Background	66
Descriptions of Improved Approaches to Environmental Site Characterization	68
Accelerated Site Characterization	69
Expedited Site Characterization	72
The Triad Approach	73
Dynamic Field Activities	75
Other Improved Approaches to Environmental Site Characterization	78
Elements of an Environmental Site Characterization Program	79
Systematic Project Planning	79
Selecting Key Project Personnel	82
Identifying Project Goals and Objectives	84
Defining Decisions to be Made	85
Establishing DQOs	86
Reviewing Existing Site Information	89
Types of Information	89
Sources of Information	96
Methods of Collecting Existing Information	97
Traditional Literature Searches	97
Computer Literature Searches	99
Review of Aerial Photography and Imagery	100
Fracture Trace Analysis	109
Review of State and Local Regulatory Agency Files	111
Review of Site Owner and Operator Files	113
Review of Other Available Records	113

Interviews	114
Conducting Site Reconnaissance	114
Developing an Initial CSM	121
Designing a Data Collection Program	135
Evaluating and Selecting Site Characterization Field Methods	137
Sample Collection Methods	138
DP Technologies	138
Sonic Drilling	143
Conventional Drilling Methods	143
Sample Analysis Methods	148
Field-Based Methods	148
Other Investigation Methods	151
Surface Geophysics	151
Soil-Gas Surveys	153
Dynamic Work Plans	154
Adaptive Sampling and Analysis Plans	161
Supporting Work Plans	166
Sampling and Analysis Plan	166
Quality Assurance Project Plan	168
Health and Safety Plan	169
Data Management Plan	170
Data Analysis, Evaluation, and Interpretation: Revising the CSM	172
Preparing a Final Report	175
Field-Based Analytical Technologies	176
The Role of Regulatory Agencies in Adopting the Use of Field Analytical Technologies	177
Advantages of Generating Real-Time Data through Field Sample Analysis	178
Limitations to Developing Strategies that Rely on Real-Time Data	179
Development of an Effective Field Analytical Program	179
Field Analysis QA/QC Programs	180
Calibration Standards	181
Method Blanks	181
Instrument Blanks	182
Duplicate Samples	182
Matrix Spikes and Surrogate Spikes	182
Control Samples	183
Confirmatory Samples	183
Selection of Field Analytical Technologies	183
Summary of Available Field Analytical Technologies	185
Reagent Test Kits	185
Field-Portable X-Ray Fluorescence	190
Headspace Screening, GC, and GC/MS	192
References	197

Introduction

Background

The use, storage, handling, and disposal of organic chemical compounds such as halogenated solvents, mixtures such as petroleum products, additives such as methyl tertiary

FIGURE 2.1
Leaking UST systems, in which mostly petroleum products are stored, are a major source of environmental contamination. These systems have been regulated by the U.S. EPA under authority of the RCRA (Subtitle I) since 1986.

butyl ether (MTBE), and inorganic materials such as heavy metals and perchlorate have resulted in thousands of cases of surface and subsurface contamination during the last few decades. Causes of this contamination have included leaking underground storage tanks (USTs) and associated piping (Figure 2.1); accidental spills during storage, handling or transportation of chemicals (Figure 2.2), or during filling of USTs; discharges from industrial and municipal sewer systems (Figure 2.3); and past storage and waste disposal practices that were either considered acceptable industry standards at one time or simply unregulated (Figure 2.4).

The U.S. Environmental Protection Agency (U.S. EPA) has estimated that there are more than 30,000 hazardous waste sites nationwide (1,238 of which are on U.S. EPA's National Priorities list [U.S. EPA, 2004a]) and more than 360,000 UST sites at which releases are either suspected or have been confirmed but not cleaned up (U.S. EPA, 1997a). The Government Accounting Office projects that the U.S. EPA will spend in excess of $150 billion and the U.S. Departments of Energy and Defense will spend nearly $1 trillion on these sites over the next 20 to 30 years. The cost of dealing with these sites threatens to become one of the largest domestic expenditures in the nation's history.

Dealing with the subsurface contamination caused by these and other sites becomes increasingly important when one considers that the use of ground water is increasing nationwide and, in fact, doubled from 1975 to 1985. This increase is partly due to declines in surface-water quality and availability and partly due to population growth, but the use of ground water in the United States is growing at a rate even faster than that of population growth. Along with this increasing usage, environmental regulations have evolved to protect this highly vulnerable resource. It is due to these regulations that the need has surfaced in recent years for establishment of ground-water monitoring programs, preparation of environmental and engineering feasibility studies, and design and implementation of selected remedial management alternatives at contaminated sites. However, before the monitoring or cleanup of any environmentally contaminated site can begin, the site must be thoroughly and accurately characterized to build an understanding of the environmental conditions and the nature and extent of contamination that exists at the site so that monitoring and cleanup efforts can be successful.

FIGURE 2.2
Transportation accidents are less important sources of environmental contamination, but they can still have a significant effect on public water supplies. Some local government agencies have established well-head or watershed protection areas that require spills from such accidents to be reported immediately to the appropriate local response authority to prevent movement of contamination into the water supply.

Historically, a large percentage of the expenditures for dealing with contaminated sites has been associated with environmental site characterization activities, primarily because of the approaches that have been followed and the methods that have been used, which have resulted in a laborious, time consuming, and expensive process. The U.S. Department of Energy (U.S. DOE) estimated that in 1992, of the $6 billion spent on environmental consulting fees for hazardous waste site cleanup, about 25%, or $1.7 billion, was spent on site characterization (U.S. DOE, 1998). In 1997, the U.S. EPA estimated that site characterization costs accounted for between 10% and 50% of the total remediation costs at petroleum-contaminated sites and that site characterization costs made up an even higher percentage of remediation costs at sites where remediation by natural attenuation was an appropriate remedy (U.S. EPA, 1997a).

Environmental site characterization has evolved significantly since environmental investigations began uncovering the litany of problems that existed at sites storing, managing, and disposing of wastes, chemicals, and petroleum products in the late 1970s. Nearly 30 years of advances in site characterization technology and field instrumentation have led to significant changes in the approach that many investigators take to environmental site characterization. Much of the technology and instrumentation used in environmental site characterization is described in other chapters in this book and in many other sources in the scientific and engineering literature; some, including field-based analytical methods, will be covered in this chapter. The primary focus of this chapter is on describing the

FIGURE 2.3
Industrial discharges to surface-water systems have been regulated by the Clean Water Act since 1970. Despite this, some unpermitted discharges still occur. These discharges can contaminate surface water and ground water, through leakage in the sewer line between the point of origin and the discharge point.

FIGURE 2.4
Most uncontrolled hazardous waste sites that once stored thousands of drums of toxic materials have been identified and the worst have been placed on U.S. EPA's National Priority list, a program initiated under CERCLA (better known as Superfund). However, at least a few dozen new sites are discovered every year, requiring site characterization in preparation for cleanup.

approaches that are currently followed by forward-thinking environmental professionals to use these improved technologies to conduct efficient, cost effective, and accurate environmental site characterization programs.

The Importance of Environmental Site Characterization

Site characterization is the foundation of most environmental projects, whether they are focused on long-term monitoring, risk assessment, remediation, or some other goal. It is during site characterization that the most basic information regarding site conditions, and the very important data that guide decisions made during these projects, must be generated. Therefore, it is critical that site characterization be conducted only after a great deal of forethought and planning and carried out with an eye toward achieving specific objectives, usually related to advancing to the next step of the project. Inadequate preparation and planning often lead to incomplete site characterization, resulting in the need to return to the site to gather additional data, which leads to inefficiency and cost overruns. Generating insufficient data, or data of substandard quality, often results in developing inaccurate or misleading conclusions regarding site environmental conditions, which can delay appropriate responses to the problem and result in an increased risk to human health and the environment. This can also result in poorly conceived monitoring or remediation program designs and, ultimately, increased costs or even complete failure of the monitoring or remediation program. By nature, there are always gaps in the information provided in site characterization programs, because it is not feasible to sample and analyze every grain of soil or every drop of ground water or surface water at a site. It is, therefore, not always obvious when a site characterization program is complete or when the information has been accurately interpreted. To reduce uncertainty regarding site conditions and to increase confidence in decisions based on the data collected during an environmental site characterization program, it is important to collect a large quantity of high-quality data focused on meeting the objectives of the project. It is also important to have experienced environmental professionals to interpret these data to construct a three-dimensional conceptual site model (CSM), which accurately depicts site environmental conditions.

As the foundation for monitoring, risk assessment, and remediation projects, environmental site characterization must provide critical sets of data to allow efficient and cost-effective design and implementation of these projects. The data collected during site characterization are generally keyed to establishing environmental conditions at a site, both ambient and man-impacted, in either a specific medium (i.e., soil, ground water, surface water, or air) or multiple media, over space at a single point in time. This "snapshot" view of site conditions serves as a baseline and is the basis for further project work at the site. If further work is required at a site, as is often the case, the next step is normally monitoring. The goal of monitoring is to provide information on changes in environmental conditions (i.e., ground-water levels or variations in concentrations of specific analytes) at the site or in a specific medium, over time, usually at fixed locations (i.e., monitoring wells for ground water, NPDES discharge points for surface water). This information, in turn, is often used in making decisions regarding the need to do additional work at the site, usually based on potential risks posed by site conditions that may require remedial action (e.g., a plume of ground-water contamination moving toward a water-supply well). This reflects a maturity of the site investigation from characterization and risk assessment to long-term monitoring and remedial action to address risk.

Environmental Site Characterization

Historical Problems with Environmental Site Characterization

The primary reasons for failure of long-term ground-water monitoring and environmental remediation programs include the following:

- Inexact or incomplete definition of site geology and hydrogeology, which results in improper positioning of monitoring wells, or selection of inefficient remediation methods
- Poor definition of contaminant distribution, which results in placement of too few (or too many) monitoring wells to accomplish project objectives, or incomplete site cleanup
- Inadequate collection of chemical data (i.e., incorrect analytes or wrong detection limits), resulting in monitoring for too few chemical parameters, selection of inappropriate analytical methods, or selection of an inappropriate remedial approach

Historically, the root cause of all of these problems has been the high cost of collecting and analyzing samples, which often limits the amount of data that can be collected to describe site conditions. For a long-term ground-water monitoring program to be successful, the environmental site characterization program must generate very specific and detailed information on ground-water conditions to support decisions on where to position monitoring wells (Figure 2.5) and well screens, how to design wells to collect representative samples (Figure 2.6), and when to collect samples to accurately depict temporal changes in ground-water chemistry. For an environmental remediation program to be successful, the environmental site characterization program must generate very specific and detailed information on contaminant-related conditions to support decisions on which remedial options are best suited to deal with the site-specific problems (Figure 2.7 to Figure 2.9) and where and when they should be applied to get the "best bang for the buck."

FIGURE 2.5
One of the important uses of data from an environmental site characterization program is positioning ground-water monitoring wells in preparation for conducting a long-term monitoring program. These wells have screens at two different depths to monitor two different types of contamination — LNAPLs and dissolved-phase hydrocarbons. Properly positioning wells and well screens under these conditions requires a substantial amount of data.

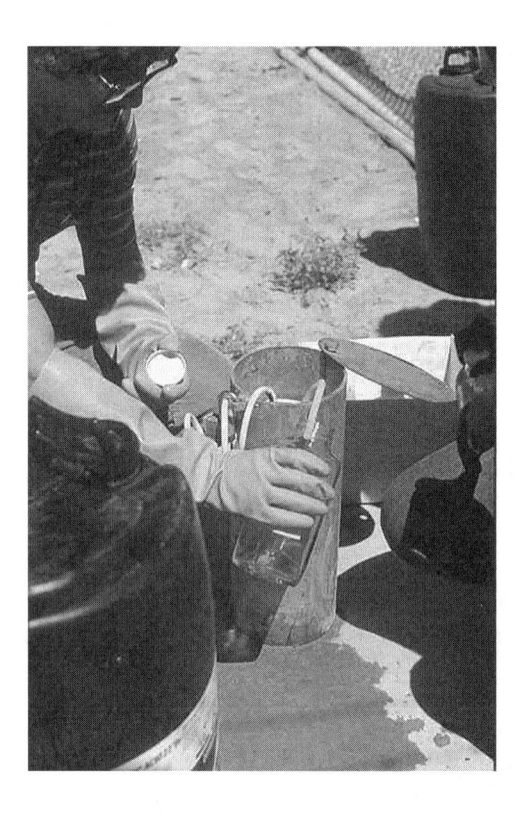

FIGURE 2.6
Extensive site characterization data are also required to define the parameters for ground-water sampling programs, including frequency of sampling, chemical constituents for which to sample and analyze, and types of equipment and procedures to use to collect representative samples.

The specific types of data required for both applications will be discussed in detail in later sections of this chapter.

The Problem of Heterogeneity

For the vast majority of site investigations, contaminant data have historically been generated by taking a few small volume samples from an environmental medium (Figure 2.10) and analyzing them for trace-level contaminants. The per-sample costs for trace-level analyses are high because satisfactory analytical performance requires sophisticated instrumentation (Figure 2.11), along with experienced and properly trained operators. Therefore, there is a strong financial incentive to minimize the number of samples to be analyzed, resulting in a data set that is, in many cases, nonrepresentative of actual site conditions (Crumbling, 2002; Crumbling et al., 2003). Compounding the potential for a nonrepresentative data set is the fact that, especially for soil samples, an even smaller volume of the sample (a subsample; Figure 2.12) is analyzed to generate the result. Consequently, the volume of matrix actually analyzed is very small when compared with the volume of the parent matrix to which the analytical results are typically extrapolated, increasing the risk of obtaining highly variable results and skewed data sets (Gilbert and Doctor, 1985).

If contaminants occurred at nearly constant concentrations throughout the parent matrix (i.e., if both the matrix and the contaminant distribution within the matrix were homogeneous), then drawing conclusions about the parent matrix based on just a few

FIGURE 2.7
Site characterization data must be sufficient to enable remedial design specialists to determine whether soil excavation and off-site treatment or disposal is an appropriate approach to deal with site contamination. This alternative is best suited to sites where contamination is confined to soils and where the volume and depth of contamination is limited.

samples would be straightforward and valid. However, environmental media are not homogeneous — they range from moderately to highly heterogeneous — and, under these conditions, investigators extrapolate beyond the available evidence at great risk.

Field studies have shown that matrix heterogeneity (the combination of environmental and contaminant heterogeneity) severely limits the confidence with which analytical results can be justifiably extrapolated beyond the very small samples that are analyzed in most investigations (Crumbling et al., 2003). Environmental heterogeneity is inherent in soils (Figure 2.13) and geological materials (Figure 2.14), as well as soil gas, soil pore water, ground water, surface water, and even atmospheric air. It is well established that environmental heterogeneity strongly affects contaminant heterogeneity, resulting in contaminant distributions for many chemicals of concern (COCs) (particularly non-aqueous phase liquids [NAPLs]) that may vary by several orders of magnitude over vertical and horizontal distances of only a few feet (Ronen et al., 1987; Cherry, 1992; Puls and McCarthy, 1995). Contaminant heterogeneity is also a consequence of the release mechanisms, the partitioning behavior of the analytes, and the transport and transformation mechanisms produced by interactions with environmental media, all of which are site-specific.

FIGURE 2.8
Where contamination is more widespread or deep, where contaminants are volatile, and where contaminants may have reached ground water, *in situ* treatment methods such as soil vapor extraction (note the manifold of PVC pipe connecting several vapor extraction wells) may be appropriate. Again, extensive site characterization data are required to make these decisions.

FIGURE 2.9
If the site characterization program uncovers the presence of LNAPLs at the site, a separate-phase recovery system, such as this skimmer floating on the water surface in a recovery well, is usually called for. Determining the extent of LNAPL contamination to enable proper recovery well positioning requires a comprehensive sampling effort.

FIGURE 2.10
Traditional environmental site characterization programs rely heavily on collection of small-volume samples (such as this split-spoon sample) collected at variable depth intervals (usually one sample every 5 ft) from widely spaced boreholes. Given the heterogeneity of soils and geologic materials, such a sampling program is highly unlikely to be successful in producing samples representative of the complex nature of the medium or contaminant distribution in the medium.

The impact of heterogeneity on data uncertainty is well known. Prior to the 1980s, environmental investigators recognized that matrix heterogeneity compromised their ability to draw reliable conclusions from analytical data. In 1991, the U.S. EPA published the conclusions of an expert panel convened to explore the ramifications of environmental variability (Homsher et al., 1991). The panel noted that studies showed that 70% to 90% of data variability at contaminated sites was caused by natural, in-place variability, with only 10% to 30% of variability being contributed by the data generation process (such as sample collection procedures, field sample handling, laboratory sample handling and cleanup,

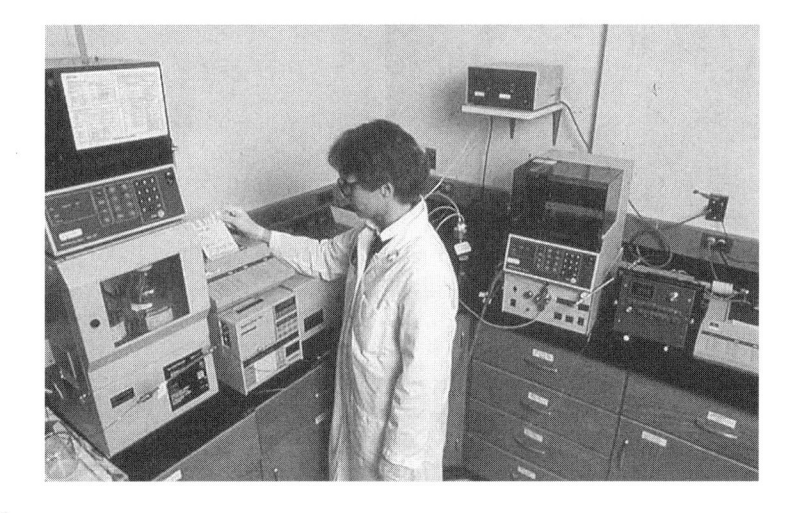

FIGURE 2.11
Many state regulatory agencies still require that all sample analyses be conducted by a fixed-based laboratory using sophisticated analytical equipment, strictly following well-documented analytical methods and U.S. EPA Contract Laboratory Program (CLP) protocols. Such analyses are expensive and the protocols are much more stringent than those required to produce useful high-quality data.

FIGURE 2.12
A subsample of the split-spoon sample extracted from the formation is collected by the field technician, placed in a sample jar, and sent to the laboratory for analysis. When this sample is analyzed in the lab, a lab technician takes a subsample of it to run through the analytical process. The actual volume of sample analyzed is less than 5 g for most parameters. It is very difficult to imagine that this small sample could be representative of the large volume of heterogeneous formation materials to which the analytical results are often extrapolated.

laboratory analyses, data handling, data reporting, and data interpretation). However, most of the efforts spent in improving data quality have focused on improving the data generation process, rather than on increasing the number and density of samples used to describe environmental conditions.

FIGURE 2.13
Heterogeneous soils are the rule rather than the exception. This figure shows a soil derived from glacial outwash, with highly variable grain sizes, which leads to tremendous variability in water transmission and contaminant transport characteristics.

FIGURE 2.14
The geological materials, which are the parent materials for soils, are themselves moderately to highly heterogeneous. This thick sequence of glacial and alluvial materials is composed of alternating layers of silty fine to medium sands and clean sands and gravels, which have very different contaminant transport potential.

Decision errors about risk and remediation are an unavoidable consequence of the fact that traditional or conventional site characterization programs rely on static, limited-scope sampling programs, and expensive fixed laboratory analyses. In conventional environmental site characterization programs, budgeting constraints often mean that relatively few samples can be analyzed when compared with the number of samples needed to accurately characterize the heterogeneous media and heterogeneous contaminant distributions that exist at most sites. Very high analytical quality is seldom required to provide the data needed to refine the CSM, which is typically developed for the site as a tool to understand site conditions (Crumbling et al., 2003). However, without a reliable CSM to support the representativeness of expensive, high analytical quality data points provided by traditional site characterization programs, those data may be misleading and result in incorrect decisions.

When the sampling point density (the number of samples per unit volume of an environmental medium) is insufficient to accurately represent the degree of heterogeneity of the medium, incomplete or inaccurate CSMs are produced and decision errors are the result. Estimates regarding the nature and extent of contamination may be strongly biased, and interpretations of the importance of exposure pathways (and the risk they represent) may be wrong. Decisions regarding the three-dimensional positioning of long-term monitoring wells, particularly with regard to well screen length, may result in an inaccurate picture of contaminant extent, concentrations, and movement patterns. Remedial designs may fail to achieve cleanup to a required level within a required time frame, requiring another round of characterization to establish the reason for the failure and another round of cleanup when unexpected contamination is discovered.

Generating representative data is not a simple matter when evaluating heterogeneous environmental media, such as geologic materials. Although the data collected from the medium may be correct in the sense that the analytical results are accurate for the very small samples analyzed, extrapolating the results from those very small samples to a much larger volume of the medium represented by the CSM often creates a misleading picture. This is termed "sampling error." Sampling error occurs when the analysis is accurate but the sample that is analyzed is not representative of what the data user believes it represents. Because environmental media are nearly always heterogeneous, sampling errors can contribute to highly inaccurate and misleading CSMs which, in turn, can lead to erroneous decisions. The factors that contribute to sampling errors are termed "sampling uncertainties."

Spatial heterogeneity, at the scale of most traditional grid-based sampling strategies, is a large contributor to sampling uncertainty. The problem is that collecting very few high-quality samples causes investigations to miss important areas of contamination and thus to fail in defining the true extent of contamination, particularly discrete contamination "hot spots." When only a very few small samples are collected, data interpreters have little choice but to attempt to extrapolate the results of those few samples analyzed in a fixed laboratory (often as small as a few grams) to volumes of the medium the samples are required to represent, which may be six orders of magnitude (or more) larger. Statistical calculations (such as the calculation of a mean) include the assumption that the result from a very small sample within a grid block represents the contamination concentration for the entire grid block. The degree to which this is a valid assumption depends on how the CSM was constructed (i.e., how the data interpreter thinks the contamination got there and whether the release mechanism is likely to produce uniform contaminant concentrations) (Crumbling, 2002). Ill-conceived site characterization, which makes it appear that contamination is more widespread than it actually is, needlessly increases the cost of cleanup when clean environmental media are lumped together with contaminated material, unnecessarily increasing the volume of media to be remediated, while artificially decreasing the efficiency of the remedial approach (ITRC, 2003).

Overall uncertainty in the data set used to develop and revise the CSM is best managed using less-expensive analyses (such as field-based analyses) that allow an increase in the number of samples collected for the same budget (Crumbling et al., 2001). In a site characterization program employing one of the improved approaches described later in this chapter, high numbers of less-expensive, field-based analyses are used to develop the CSM and greatly reduce sampling-related uncertainties. Fewer carefully selected, more expensive fixed-laboratory analyses are then used to manage analytical uncertainty. These analyses are reserved for samples of known representativeness to answer questions that the less-expensive field analyses cannot address. In this way, the improved approaches to site characterization use a second-generation data quality model that departs from the traditional practice of using analytical uncertainty as a surrogate for overall data uncertainty, which is flawed. By carefully and expressly managing sampling uncertainty, these improved approaches to site characterization keep the project team focused on all sources of data uncertainty and guide the selection of sampling point locations and investigation techniques to minimize decision errors (Crumbling et al., 2003).

The Problem of Sample and Data Representativeness

In evaluations of environmental data quality by application of the PARCC parameters (Precision, Accuracy, Representativeness, Completeness, and Comparability), the criterion of representativeness is often overlooked or misunderstood. Representativeness is of paramount importance to data quality and is defined as the degree to which sample data accurately and precisely represent a characteristic of a population, parameter variations at a sampling point, or an environmental condition (U.S. EPA, 1987). Representativeness is a qualitative parameter that depends primarily on proper design of the sampling program (Jenkins, 1996) — the sampling design must be structured so that data can be confidently extended from a sampling point to a larger volume of material. The planners of environmental investigations have historically understood representativeness as narrowly relating only to parameter variation at a sampling point, thus placing more emphasis on analytical accuracy and precision and on completeness and comparability of chemical data.

Environmental data may be accurate, precise, complete, and comparable, but if they are not representative of site conditions, they become meaningless. If such data are used with the belief that they are representative, they could be relied upon to design a remedial approach that turns out to be ineffective, which could be financially disastrous. The principal reasons for overestimating or underestimating the extent of contamination at a site usually stem from improper design of the sampling and analysis program, including:

- Nonrepresentative samples analyzed for the correct contaminants
- Representative samples analyzed for the incorrect contaminants
- Nonrepresentative samples analyzed for the incorrect contaminants

All three of these situations present a distorted view of the conditions at the site under investigation and are equally useless for planning monitoring or remediation activities (Popek and Kassakhian, 1998).

The concept of representativeness demands that the scale (spatial, temporal, chemical speciation, etc.) of the supporting data be the same (within tolerable uncertainty bounds) as the scale needed to make the intended decisions (i.e., does acceptable risk exist or not; how much contamination must be removed or treated; what treatment alternative is appropriate; what environmental matrix requires monitoring; what analytes to monitor for; and where, when, and how to sample) (Crumbling, 2002). Because of the effects of heterogeneity previously described, collecting samples or data that are truly representative at the scale of decisions about risk, monitoring, or remediation demands thinking on different scales, which is not commonly done in traditional site characterization programs. For example, discrete contamination patterns (such as hot spots) may only be discovered if the investigation is conducted using a sample spacing of only a few meters in the horizontal plane and less than a meter (or continuous sampling) in the vertical plane. Hot spots are rarely detected if the sample spacing is 50 to 100 m horizontally and more than a meter vertically, which is often the case using the traditional approach (Figure 2.15). However, it is not resource effective to characterize all relevant properties at a site in a representative way at all possible scales, so there must be a rationale applied to decide which scales are important. The purposes of project planning are to develop an understanding of the scale over which decision making will occur, to identify what uncertainties need to be resolved for defensible decision making to occur, and to design a data collection scheme that provides the information to manage those uncertainties (Crumbling, 2002).

In spite of the fact that environmental investigators have accumulated significant experience in environmental site characterization, at least one U.S. EPA report indicates that many investigators historically had a poor understanding of, or ignored, the data quality objectives (DQO) process (U.S. EPA, 1994). According to this EPA document, in their work plans, environmental investigators adhered to strict analytical protocols and data validation procedures (i.e., focusing on analytical accuracy and precision) to achieve data quality goals, rather than focusing on the overall project objectives and the means to fulfill them. In addition, because of financial and project scheduling pressures, investigators often reduced the number of samples and sampling locations (thereby reducing data representativeness), while substituting the correct analytical methods with irrelevant tests. Placing undue emphasis on expensive analytical protocols and data validation invariably leads to misleading conclusions about site conditions and

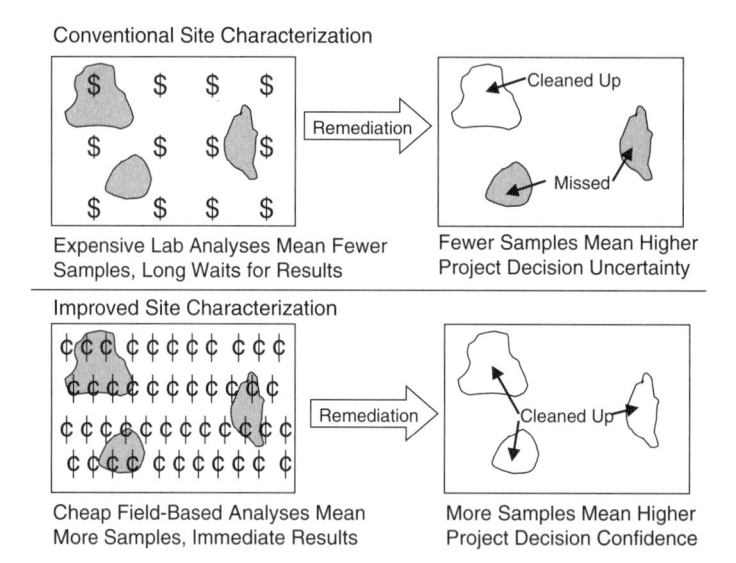

FIGURE 2.15
Collecting a few widely spaced samples (with regard to both horizontal and vertical separation) for expensive fixed-based lab analysis is rarely sufficient to accurately depict complex site conditions or even to achieve a goal as simple as locating contaminant hot spots. A much more representative data set and, therefore, a better understanding of site conditions, can be generated by collecting a larger number of more closely spaced samples for less-expensive analyses, without sacrificing data quality.

ill-conceived plans for monitoring or remediation. Instead, investigators should focus on understanding the objectives of the investigation, identifying the intended use of the data, developing representative sampling program designs, and using field analytical techniques capable of generating high-quality results. Greater emphasis on geostatistical analysis, which is better suited to modeling spatial patterning, will produce more cost-effective sampling designs than classical statistical models.

Careful management of sample representativeness was inconceivable from a cost standpoint, when data from standard fixed-laboratory analyses were the only data that regulatory agencies would accept. It was much easier to oversee data quality if that concept was defined in terms of analytical method and laboratory performance. The problem with defining data quality in those terms is that analytical data are generated from environmental samples, which are collected from environmental media that are inherently heterogeneous. Even perfect analytical quality is no guarantee that sample collection will produce data that are representative of site conditions. The more heterogeneous the matrix, the more likely it is that a data set will be skewed by sampling program design and collection of a small number of samples relative to the volume of the matrix the samples are required to represent (Crumbling et al., 2003).

To be successful in terms of providing representative data, a site characterization program must be designed to provide spatially dense three-dimensional coverage of critical data over portions of the site that are of particular interest to investigators (i.e., contamination source zones, preferential flow pathways, and exposure points). The three-dimensional approach, employing a variety of investigative methods and tools (including many types of field analytical methods that produce high-quality data), allows very accurate delineation of subsurface contamination, critical physical characteristics, important features controlling contaminant movement, and accurate estimations of contaminant mass in all phases. All of these factors must be well defined for investigators to evaluate the applicability of remedial alternatives and the effectiveness of the remedial approach selected (Barcelona, 1994).

Another complicating factor is that decisions about risk are usually based on an estimate of average conditions within a matrix to which receptors are exposed. In contrast, decisions about risk reduction (remediation) strategies are usually based on discriminating between zones with higher levels of contamination requiring remedial attention and zones with lower levels of concentration that may not require treatment or removal. If data are not representative, in terms of the decision being made (averages in one case and extremes in the other case), using the data will lead to flawed decisions (Crumbling et al., 2003). The older data-quality models, which consider data quality only in terms of analytical method performance, ignore sampling uncertainties and the importance of matrix heterogeneity. Given what we know today, this cannot be considered sound science. It is imperative that environmental professionals update their data-quality model to reflect current scientific thinking. The technology, tools, field methods, and sampling strategies now exist to cost-effectively implement a sounder data quality model.

An even more compelling reason to update old data-quality models is to reduce the financial and liability risks created when nonrepresentative data lead to erroneous decisions. Attempts to save resources in the short run by skimping on the site characterization program ultimately wastes far more resources than could possibly be saved when erroneous decisions result in constant revisions to remedial plans. Popek (1997) observed that reductions in the comprehensiveness of the field investigation, based on budgetary considerations, schedule-driven approval of incomplete plans, and superficial or protocol-oriented reviews by technically unqualified regulatory agency personnel, all come back to haunt stakeholders at remediation time. Remedial action case histories have, in fact, proved that the perception of site conditions based on traditional site investigation approaches does not reflect reality. Use of incomplete site investigation data often leads to underestimating or overestimating the extent of contamination, sometimes on an alarming scale. In either case, ramifications may be substantial with respect to remediation budgets and public perception of the environmental industry.

Objectives of Environmental Site Characterization

The major objectives of most environmental site characterization programs are to provide an understanding of site physical conditions (soils, geography, geology, hydrology, and biology), to assess the type, distribution, and extent of surface and subsurface contamination, and to define contaminant transport pathways, locations of potential receptors, and routes and points of exposure. This allows investigators to evaluate regulatory compliance, determine the risks posed by the site to human health and the environment, assess the appropriateness of long-term monitoring, evaluate the need and responsibility for remediation, and determine the appropriate cleanup levels. The types of sites requiring characterization typically include controlled and uncontrolled hazardous waste sites (Figure 2.16); controlled and uncontrolled industrial and municipal solid waste and other nonhazardous waste disposal sites (Figure 2.17); petroleum product refining, transmission, and storage sites (Figure 2.18); and sites involved in real-estate transactions (Figure 2.19). Each site investigated will have a unique set of circumstances surrounding the problems that must be uncovered, and investigators will have to establish and meet project-specific objectives to define those circumstances.

Some of the more common (and some not so common) project-specific objectives of environmental site characterization programs include:

- Determining ambient environmental conditions at a small site in preparation for a property transfer or for preparing a landfill permit application

FIGURE 2.16
U.S. EPA or state-designated Superfund sites are among the types of sites at which environmental site characterization programs are routinely conducted to determine the optimum remedial approach for the site.

- Determining ground-water conditions (depth to ground water, presence of water-table or confined conditions, flow direction and rate, gradient, etc.) in preparation for establishment of an ambient monitoring program
- Determining both ground-water conditions and the presence or absence of contaminants in ground water, for the purpose of designing a compliance ground-water monitoring program under Resource Conservation Recovery Act (RCRA)
- Determining the extent of contamination in a ground-water system, for the purpose of establishing an assessment monitoring program under RCRA

FIGURE 2.17
Municipal and industrial solid-waste landfills, regulated under RCRA Subtitle D, require site characterization in preparation for installation of detection ground-water monitoring wells. If routine monitoring detects the presence of contamination, then additional characterization is required to establish an assessment monitoring program to determine the extent, rate, and direction of movement of the contamination and then to clean it up if necessary.

FIGURE 2.18
The many potential source areas for contamination at petroleum product storage and distribution terminals include the pipeline supplying the product, the above-ground tanks storing the products, the above and below-ground tanks storing the additives, the piping that distributes products and additives to the loading rack, and the loading rack where the additives are mixed with products and tanker trucks are filled. The products and additives that are typically stored at these facilities have significantly different characteristics and behave very differently when released to the environment.

- Determining the nature and three-dimensional extent of several phases of environmental contamination for complex mixtures of chemical contaminants in several environmental media at a site, in preparation for a Superfund site remediation program
- Assessment of site suitability for disposal or land treatment of industrial or domestic liquid or solid wastes

FIGURE 2.19
Environmental site characterization programs are widely used at sites where property transfers are conducted with the hope that the site is not contaminated or that it can be cleaned up with a minimum of effort and sold for a profit. The need to conduct a cost-effective, yet thorough site investigation is emphasized in these cases.

- Assessment of site suitability for some future land use, which may be compromised by site characteristics such as flooding, seismic activity, or landslides
- Delineation of ground-water or well-head protection areas
- Evaluating soil suitability for agricultural practices to minimize soil erosion and contamination from agricultural chemicals

Data Required from an Environmental Site Characterization Program

Conducting an environmental site characterization program with the project objectives and the eventual endpoint in mind helps investigators to focus on collecting the types, quantity, and quality of data required, which results in a more cost-effective investigation. Successful monitoring or remediation of a contaminated site depends in large part on the ability of the site characterization program to collect sufficient data to accurately define a few important factors including:

- The location and extent of the primary and secondary source areas for the contaminants released at the site, the likely volume released, and the manner in which it was released (rate and duration for continuous releases and cycle for intermittent releases)
- The nature (physical and chemical properties) of the contaminants present in the various environmental media at the site and the major chemical, physical, and biological processes that may affect contaminant distribution (i.e., dissolution, advection, dispersion, diffusion, sorption [adsorption, absorption, and desorption], precipitation, volatilization, biotransformation, and biodegradation) and how the contaminants are likely to behave (in terms of fate and transport) in a subsurface environment
- The three-dimensional distribution and concentrations of the contaminants in all environmental media at the site in all phases (residual phase [adsorbed onto and trapped between soil particles], dissolved phase [dissolved in soil pore water, ground water, and surface water], vapor phase [in the pore space of unsaturated soils], and non-aqueous phase [either LNAPL above the water table or DNAPL resting on the top of the first confining layer]), to allow quantification of the mass of contaminants present
- The soil, geological, and hydrogeological conditions at the site, particularly the degree of heterogeneity that exists (focusing on the presence of preferential pathways and barriers to movement) and how that may influence contaminant behavior
- The presence of potential exposure pathways and receptors including water supply wells (municipal, domestic, agricultural, and industrial), surface-water intakes, buildings with basements, utility corridors, and sensitive ecological areas

A comprehensive list of the specific types of data that are typically required to be generated by an environmental site characterization program to address these factors is included in Table 2.1.

The site characterization program must be structured to collect a sufficient quantity of these types of data that are of a quality sufficient to meet the program objectives — this is the key to establishing useful long-term monitoring programs, to establishing realistic cleanup goals, and to selecting appropriate remediation technologies.

Environmental Site Characterization

TABLE 2.1

Types of Data Typically Required by an Environmental Site Characterization Program

Soil and unconsolidated geological material parameters
  Type or nature
    Texture, grain size distribution, degree of sorting (gradation), bulk density, degree of heterogeneity, sedimentary structures (lamination, cross-bedding, erosional features, etc.), degree of weathering, degree of induration, nature of origin (alluvial, glacial, marine, lacustrine, aeolian, etc.)
  Distribution
    Thickness, areal extent, topographic location
  Physical properties
    Air permeability, capillarity, temperature, color
  Hydraulic properties
    Hydraulic conductivity (saturated and unsaturated), permeability, porosity (type and amount; total and effective), matric potential, wettability, moisture content, specific retention, transmissivity, storativity, specific yield, infiltration, or percolation rate
  Chemistry
    Mineralogy, cation exchange capacity, organic carbon content, pH, nutrient content, redox potential, major ions
  Microbiology
    Microbial population (type and numbers)

Geological parameters (bedrock)
  Type
    Lithology (rock type — granular vs. fractured vs. solution channeled), stratigraphy, grain size distribution (in sedimentary rocks)
  Distribution
    Thickness, areal extent, boundaries, outcrop areas
  Physical properties
    Structure (fractures, faults, folds, discontinuities)
  Hydraulic properties
    Hydraulic conductivity, transmissivity, porosity (type and amount; total and effective)
  Chemistry
    Mineralogy

Ground-water parameters
  Conditions of occurrence
    Confined/semiconfined/unconfined/perched, depth to water table/capillary fringe, water-level fluctuations, relationships with surface water, recharge and discharge areas and amounts, thickness and areal extent of each aquifer and each confining bed, interconnections between aquifers
  Conditions of movement
    Flow direction, gradients (horizontal and vertical), flow velocity, natural variations (i.e., seasonal and tidal)
  Physical properties
    Temperature and turbidity
  Chemistry
    pH, major ions (nitrate, sulfate, iron, manganese, etc.); dissolved oxygen, methane, carbon dioxide and hydrogen sulfide content; organic carbon content; redox potential; specific conductance; total dissolved solids; salinity; background (upgradient) levels of contaminants of concern; seasonal fluctuations in chemistry
  Microbiology
    Microbial population (type and number)
  Patterns of use
    Type (municipal, residential, commercial, industrial, agricultural), amount, locations of points of use

Surface-water parameters
  Conditions of occurrence
    Static vs. dynamic, drainage pattern and area, width and depth, elevation, presence of obstructions to flow, stratification, relationships with ground water
  Conditions of movement
    Flow direction, gradient, flow velocity, inflow and outflow volumes, sediment transport or deposition regime, flood frequency and duration
  Physical properties
    Temperature, turbidity and suspended solids

(Table continued)

TABLE 2.1 *Continued*

Chemistry
  pH, major ions, dissolved oxygen content, specific conductance, total dissolved solids, BOD, COD, background (upstream) levels of contaminants of concern
Microbiology
  Microbial population (type and number)
Patterns of use
  Type, amount, locations of intakes

Contaminant parameters
  Type
    Inorganic vs. organic vs. biological
  Physical properties
    Solubility or miscibility, density or specific gravity, viscosity, surface tension, volatility (vapor pressure and Henry's law constant), adsorption coefficient (K_d for inorganic materials; K_{oc} for organic compounds), dielectric constant, mobility, toxicity, reactivity, ignitability, corrosivity, biodegradability, persistence
  Chemistry
    Chemical composition, concentration (in all media, in all phases — vapor, dissolved, NAPL, residual), speciation (metals), degradation products or pathways
  Distribution
    Media impacted, areal extent, vertical extent, phases present
  Details of release
    Type of release (catastrophic/periodic/long-term), location, volume, time since release, source type (point/diffuse)

Facility parameters
  Type
    UST, AST, landfill, surface impoundment, etc.
  Location
    Above grade, below grade, location relative to property boundary, accessibility
  Design and construction features
    Liners, leachate collection systems, overfill protection, berms, dispensers
  Operational details
    Waste and product types handled, throughput (volume), treatment and discharge points
  History and period of use

Other important parameters
  Area involved
  Geomorphology and topography
  Climatic conditions
    Water balance (precipitation vs. evapotransportation), temperature, prevailing wind direction and speed, frequency of climatic extremes
  Vegetative cover
    Types, area covered, diversity, seasonal changes
  Surrounding land uses
    Types (residential, commercial, industrial, agricultural, etc.), history of use, activities (present and past)
  Presence and proximity of receptors
    Man (buildings with basements, public and private water supply wells, utility corridors), wildlife (surface water, wetlands, sensitive ecological areas)
  Presence and proximity of anthropogenic influences
    Pumping wells, injection wells, recharge basins, dewatering operations (quarries, sand and gravel operations, mines, excavations)

Approaches to Environmental Site Characterization

The Conventional or Traditional (Phased) Approach

Until the mid-1990s, environmental site characterization programs were almost exclusively conducted using a phased or staged approach, in which the field work was carried out and data were collected in a piecemeal fashion over several mobilizations to the site (Nielsen, 1995). In the early 1980s, when this approach was first applied to sites with environmental contamination, there were good reasons to adopt a carefully staged approach to site characterization. First, there was a need to build a baseline of knowledge in this relatively new field, then there was a need to deal with the tremendous difficulty involved when attempting to predict contaminant behavior in complex and highly heterogeneous hydrogeologic settings. In addition, the analytical methods of the time — established in U.S. EPA's SW-846 (U.S. EPA, 1996a) — required the use of carefully documented analytical procedures and the controlled environment of fixed laboratories for proper implementation of quality control oversight. When these factors were combined with the periodic budgeting cycles for most government-funded and private-sector work, it is not difficult to understand how multiple phases of work became the accepted approach, even though it proved to be very expensive and time consuming (ITRC, 2003).

The objective of the phased approach is to gradually learn enough about site conditions by progressing from an initial phase of the investigation, where the understanding of site conditions is very limited, through a second phase, where the understanding is better but not optimum, to a third phase, and so on, until a sufficient level of understanding of site conditions has been achieved. Some site characterization programs using this approach have lasted for several years or more, many without generating the type, quantity, or quality of data required to satisfy project objectives. While many traditional environmental site characterization programs have eventually succeeded, they have generally done so at an unnecessarily high cost to the site owner or operator. The typical sequence of events followed using the phased approach is described in detail in the following two sections.

Phase I

In Phase I, a review of available site background information (Figure 2.20) is conducted by the project staff to provide a basis for developing a sampling and analysis plan. Most investigators using the phased approach start by using a grid-based sampling strategy in an attempt to maximize coverage of the entire site, while keeping analytical costs in line. For most media, the sampling strategy is focused on the plan view. The number and locations of all sampling points and the analytical methods to be used are predetermined by the project manager (who manages the project from the office), and the work plan is rigid. The work plan usually requires regulatory approval prior to implementation and is static in its application, often containing no provisions for changes in direction based on what is learned in the field investigation (e.g., the locations of buried utilities or other obstacles to drilling or the discovery of contamination hot spots). After approval, which may take weeks or months, the work plan is set into motion and samples of relevant environmental media (soil, soil gas, ground water, surface water, or sediment) that may be affected by site operations, and other relevant field data are then collected by junior field staff. In investigations conducted for the purpose of preparing for long-term ground-water monitoring or soil and ground-water remediation, the sampling program focuses on sampling soil, geologic formation material, and ground water.

Depending on the size of the site, anywhere from a few to a few dozen soil borings may be drilled, normally using a hollow-stem auger (Figure 2.21), typically on 100 to 200 ft centers to define shallow soil conditions and site geology. Borings are generally sampled using a standard split-tube sampler (Figure 2.22) either every 5 ft in depth or at every change in formation material detected by the drilling contractor. Samples are examined in the field to determine physical characteristics (i.e., grain size, color, degree

FIGURE 2.20
In both the traditional environmental site characterization approach and the more modern characterization approaches, a review of existing site information is conducted to provide investigators with information upon which decisions on how to conduct the field investigation are based. This important step is used to gain insight into general site conditions and identify important data gaps that must be filled by the field investigation.

FIGURE 2.21
Hollow-stem auger drilling is the most widely used drilling method for shallow soil sampling and well installation in traditional site characterization programs. This 4.25 in. I.D. auger makes an 8 in. diameter borehole and produces one 55 gallon drum of drill cuttings for every 17 ft of drilling.

FIGURE 2.22
Standard 18 to 24 in. long, 2 in. O.D. split-spoon samplers are the most widely used soil sampling method in traditional site characterization programs. Depending on the materials sampled, recovery can range from 100% (shown here in a sand and fine gravel) to less than 15 or 20% (in dense, stiff clays or saturated fine sands).

of sorting, and moisture content) (Figure 2.23) and to detect the presence of visually evident soil contamination (Figure 2.24). Samples are usually screened in the field using a device appropriate for the contaminants of concern (i.e., a flame ionization detector [FID] or a photoionization detector [PID] for volatile organic compounds [VOCs]) (Figure 2.25). A selected number of soil samples (usually one per soil boring) are then packaged (usually crammed into large sample jars with screw lids) and shipped to an off-site (fixed) laboratory for chemical analysis, with a waiting period of 6 to 10 weeks for analytical results (depending on the analytical protocols used). The analytical methods used would be specified based either on the review of existing information and best professional judgment or on specific regulatory requirements. Drums of

FIGURE 2.23
Samples or portions of a sample not destined for lab analysis are examined in the field to determine physical characteristics, using one of several available soil classification systems and a standard set of sample descriptors including color, moisture content, organic matter content, and other descriptors as noted in Table 2.1.

FIGURE 2.24
If a sample exhibits obvious signs of contamination (such as the dark zone approximately 4 in. below the top of this sample), then that portion of the sample is typically containerized and set aside for lab analysis.

potentially contaminated drill cuttings (approximately one drum of cuttings [Figure 2.26] for every 17 ft of drilling in a hole drilled using a 4.25 in. I.D. hollow-stem auger) are stored and sampled. These samples are also packaged and shipped to a fixed laboratory and analyzed to determine the proper disposal method for the cuttings.

A minimum number of ground-water monitoring wells (Figure 2.27) are usually installed in a few (normally four) preselected locations, with only existing information (usually very limited general information) to use as guidance and with the hope that most of the wells will be downgradient of the area of interest. Wells are usually installed in some of the same soil borings from which the soil samples were collected, partly because of the convenience of having the hole already drilled. Well design usually consists

FIGURE 2.25
Depending on the types of contaminants expected at the site, one of several types of field screening tools, such as this PID (just right of center in the photo), is typically used to determine the presence or absence of contamination in the samples collected at the site.

Environmental Site Characterization 61

FIGURE 2.26
Drums of investigation-derived waste (IDW) produced by drilling soil borings at a small service station site with a hollow-stem auger. Sampling, analyzing, storing, and disposing of these materials can consume a significant portion of the budget in a traditional site characterization program.

of 2 in. PVC casing and screen, with slotted casing of the same slot size and length (normally 10 ft) and filter pack materials of the same grain size for all wells at the site, regardless of thickness or grain size of the zone of interest. Ground-water levels are collected from the wells (Figure 2.28) to determine ground-water flow direction and gradient only in the horizontal plane, without regard to the potential for vertical flow or vertical gradients. Ground-water samples are usually collected with a bailer (Figure 2.29) after purging three to five well volumes of water and thoroughly agitating and mixing the water column in the well and incorporating formation solids into the

FIGURE 2.27
Ground-water monitoring wells are often used to characterize ground-water conditions during traditional site characterization programs. For small sites (and even some large sites), the convention has been to install one well in an upgradient position (if gradient is known) and three wells in downgradient positions. This minimalistic approach is a carry over from early requirements for monitoring at RCRA Subtitle D sites and is often inadequate to depict true ground-water conditions, even at small sites.

FIGURE 2.28
Water-level measurement is required to determine the hydraulic gradient and ground-water flow direction across a site. Such measurements in individual, short-screened wells (with less than 10 ft long screens) are adequate for making such determinations where ground-water flow is strictly horizontal. However, if there is a vertical component to flow, wells screened at different depths (or multilevel monitoring systems) are required to discern vertical gradients. (Photo courtesy of Jim Quince.)

sample. Drums of potentially contaminated purge water (sometimes more than one drum per well) (Figure 2.30) are stored and later sampled and the samples analyzed to determine an appropriate disposition for the purge water. Ground-water samples may or may not be filtered in the field (Figure 2.31) to remove formation solids. Samples are packaged and sent to an off-site laboratory (Figure 2.32) for analysis for a range of parameters determined by examining existing site information or by consulting regulatory requirements. In some cases, the methods specified do not match the matrix to be analyzed,

FIGURE 2.29
The bailer is the ground-water sampling device most often used in traditional ground-water sampling programs, even though it is very difficult to collect a representative sample with this device (see Chapter 15).

Environmental Site Characterization 63

FIGURE 2.30
Using traditional well-volume purging, as suggested by many regulatory agencies, results in the generation of large volumes of purge water. Managing this IDW, like that generated during drilling, adds unnecessary cost to the investigation.

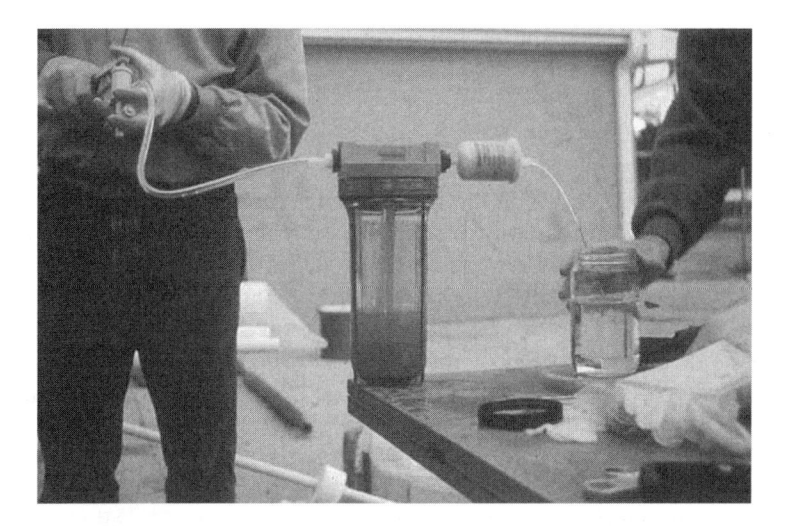

FIGURE 2.31
Bailers typically produce samples loaded with suspended sediment that do not represent the condition of *in situ* ground water. In an attempt to "fix" the sample, many samplers use filtration to remove the sediment. For the reasons discussed in Chapter 15, this is not the way to collect a representative sample.

FIGURE 2.32
Following ground-water sample collection, sample containers are labeled, packed into an appropriate container (usually a cooler), and shipped with a chain-of-custody form to a fixed-base analytical laboratory. A 6 to 10 week waiting period for analytical results then begins.

and little attention is paid to whether the analytical method is appropriate to produce the data required to meet project objectives. Sampling personnel are demobilized and return to the office before any analytical results are available. After another 6 to 10 week waiting period for analytical results, all of the analytical data (soil and ground water) are assembled and evaluated in the office in conjunction with other data collected during the review of existing information and the field investigation. Data interpretation is conducted months after the field investigation, anomalous results are noted, data gaps are identified, and a report is prepared to document the results of the investigation.

The results of the Phase I investigation are usually focused on mapping the boundaries of the contamination, rather than on defining the source area (where contamination levels are highest) or locating the most significant contaminant mass. Because of the generally small number of data collection points specified in the work plan for Phase I, results usually indicate a need to expand the scope of work including the collection of more samples and the installation of more wells in different locations. The next phase of the investigation is then planned to resolve anomalies and uncertainties raised during Phase I, to fill in the data gaps, and to ensure that all aspects of the soil and ground-water system (background, upgradient, downgradient, and other areas) are adequately characterized. In some cases, the lack of sampling points in areas where contamination actually is present may falsely indicate that no contamination exists, thereby signaling a premature end to the investigation. This can lead to incorrect decisions regarding the need for further work at the site and a false sense of security on the part of the site owner or operator. This may, in turn, lead to legal consequences for the investigator or the site owner or operator when contamination that should have been detected by the investigation is uncovered in the future.

Phase II and Beyond

Phase II of the site characterization program, which requires the preparation and regulatory approval of a second work plan, would build on the data collected in the Phase I field investigation and the review of existing information. Phase II would focus on those areas at the site where contamination was detected, particularly if those areas represented potential preferential contaminant migration pathways or exposure pathways for receptors. The same sampling strategy, the same field investigation methods, and the same analytical methods would generally be used in this and subsequent phases. This process continues, through Phase III, Phase IV, and beyond, until the extent of contamination is adequately defined (as determined by the project manager, who may not have visited the site at all) and a sufficient number of wells are installed to constitute a long-term ground-water monitoring system.

The number of mobilizations required to implement the phased approach and the need to analyze samples in an off-site, fixed laboratory significantly increase the cost of the investigation. Those who favor the phased approach usually do so because it is what they and their predecessors have always used. They contend that the expense of investigating the site in a number of phases can be recovered through progressively more selective sampling, based on data gathered in previous phases. However, employing the phased approach requires a great deal of time (often 6 months to more than a year, even at a small site) and, during this time, contamination can spread, making it more expensive and difficult to monitor or clean up. Because the project manager, who is often the only one authorized to make changes in the work plan (if changes are allowed), is in the office during the investigation, the junior field staff have to contact project manager for approval of any changes in the work plan or the scope of work made necessary by unexpected site conditions. This makes the process very inefficient. Additionally, many investigators using this approach tend to think in only two dimensions and focus on defining the horizontal limits of a contaminant plume, rather than looking at it as a three-dimensional problem or locating and quantifying contaminant mass, which is much more important from a remedial design standpoint. Because of the increased cost and difficulty of continuously sampling geological materials (and sampling at greater depths) and installing nested or clustered wells or multilevel sampling devices to define the vertical extent of soil and ground-water contamination, many phased investigations ignore the three-dimensional nature of subsurface contamination and only partially define the problem. The site conditions reported are thus often incomplete or incorrect, resulting in ineffective designs for long-term monitoring or remediation programs.

With increasing emphasis on making environmental site characterization projects (and the risk assessment, monitoring, or remediation programs that follow) "cheaper, faster, and better" beginning in the early 1990s, and with concurrent advances in the technology applied to site characterization projects', the phased approach began to fall out of favor (Nielsen, 1995). In particular, technologies developed to allow rapid acquisition of soil, soil-gas, and ground-water samples (often continuously) and to collect subsurface profile information on a variety of important parameters for describing soil and ground-water conditions *in situ* and in real time, without generating large quantities of investigation-derived wastes (IDW) (i.e., direct-push [DP] technologies [see Chapter 6] or sonic drilling [see Chapter 5]). At the same time, technologies developed to analyze samples for a broad range of environmental contaminants in the field and in real time (i.e., field-portable GCs for organic chemicals and x-ray fluorescence [XRF] equipment for metals, see later sections in this chapter). Combining these capabilities allowed compressing the phased approach into one or two mobilizations to the site. This paved the way for a new generation of site characterization approaches that promised significant reductions in the time and cost required for site characterization programs, along with improvements in the data collected and overall project efficiency.

Improved Approaches to Environmental Site Characterization

Background

Most of the new-generation approaches to environmental site characterization take their cues from one of two historical approaches to site investigation used outside the environmental arena, the method of multiple working hypotheses or the observational method. The method of multiple working hypotheses, first described by Chamberlain (1890), was applied to the study of geology. This method involves an iterative process that progresses as follows:

- Begin by observing some aspect of geology for which an explanation is sought
- Develop multiple theories or hypotheses to provide possible explanations for what is observed (i.e., create a conceptual model based on sparse information to describe a geologic feature for which more than one explanation is possible)
- Use the hypotheses to make predictions
- Take measurements and make further observations in the field to test these predictions and to confirm or refute one or more of the hypotheses
- Either draw the conclusion that one of the hypotheses is true or refine the conceptual model based on the field measurements and observations
- Repeat the process until only one plausible explanation remains to account for the geologic feature being studied

This method is well suited to use in the initial stages of environmental site characterization because the relative difficulty in making direct observations means that conceptualization of potential migration pathways for contaminants is based on a relatively limited number of observations for which more than one explanation is possible.

The observational method, a systematic approach to engineering under conditions of uncertainty, was used by Dr. Karl Terzaghi for applied soil mechanics investigations from the 1920s to 1950s and documented by Bjerrum (1960) and Peck (1969; 1975). It is an investigative process for geotechnical characterization of soils and geotechnical engineering design in which characterization, design, and construction proceed hand-in-hand. The observational method recognizes that while considerable time, expense, and effort can be devoted to attempting to characterize complex subsurface conditions, residual site uncertainties can be significant and modifications to design and construction are to be expected. Observations regarding the change and response of the soil system as construction proceeds are used to modify the engineering design. A critical element of the method is an early assessment of the most probable conditions and the most unfavorable conceivable deviations from these conditions. The observational method employs the following key elements (Peck, 1969):

- Exploration sufficient to establish at least the general nature, pattern, and properties of subsurface deposits, but not necessarily in detail

- Assessment of the most probable conditions and the most conceivable deviations from these conditions
- Establishment of the design based on a working hypothesis of behavior anticipated under the most probable conditions
- Selection of quantities to be observed as construction proceeds and calculation of the anticipated values on the basis of the working hypotheses
- Calculation of the same quantities under the most unfavorable conditions compatible with the available data concerning the subsurface conditions
- Selection in advance of a course of action or modification of design for every foreseeable significant deviation of the observational findings from those predicted on the basis of the working hypothesis
- Measurement of quantities to be observed and evaluation of actual conditions
- Modification of the design to suit actual conditions

Applications of the observational approach to the U.S. EPA Superfund Remedial Investigation/Feasibility Study (RI/FS) process have been described by Mark et al. (1989), Brown et al. (1989), and Brown (1990). The principal feature of the observational method that makes it applicable to hazardous waste site investigations is its explicit recognition of uncertainty. From a technical perspective, the subsurface environment presents substantial uncertainty that plagues accurate source characterization, assessment of contaminant distribution and chemical fate and transport, and assessment of exposure risks. The observational method fundamentally recognizes that uncertainty is present and uses a structured approach to determine the appropriateness of a design as it is being implemented. It requires planning for potential unfavorable conditions and potential design modifications. In this application, the emphasis in the RI stage is on gathering information to establish general site conditions, constructing and confirming a conceptual model, and identifying most probable conditions and reasonable deviations from those conditions as the basis for a flexible approach to remedial design. The use of the observational method in this application involves the following (Mark et al., 1989):

- Evaluate existing data and conduct an investigation sufficient to establish the general nature, pattern, and properties of the physical setting and the contaminants present. The level of site characterization depends on the site and the expected general response actions.
- Assess the most probable site conditions and maximum credible deviations from these conditions. The most probable site conditions are working hypotheses based on interpretation of available data and are not necessarily based on a statistical evaluation. The maximum credible deviations from the most probable conditions do not represent worst-case scenarios or maximum conceivable contaminant concentrations, but credible conditions based on interpretation of existing data. If a reasonable working hypothesis of the most probable site conditions cannot be developed, additional site investigation is required.
- Evaluate alternatives and establish a remedial design based on the hypothesis of the most probable site conditions.
- Calculate or estimate the physical and chemical conditions expected to be observed during implementation and operation of the remedial action, given the most probable site conditions.

- Calculate or estimate the same parameters for the remedial action given maximum credible deviations from the most probable site conditions.
- Select a course of action based on the most probable conditions and prepare contingency design modifications for foreseeable maximum credible deviations.
- Construct and operate the selected remedial action, monitor the selected parameters, and evaluate the observed conditions with respect to the working hypothesis of the most probable conditions and credible deviations.
- Modify the remedial action through the predetermined course of action to suit actual conditions, as required.

The observational method does not place any limits on the types of information to gather or provide specific guidance on sampling program design or on investigative methods that are appropriate for use for subsurface investigations. The key is to gather the information required to develop a conceptual model or working hypothesis as efficiently as possible. Information gathering not only supports conceptual model development but also confirms its underlying assumptions. Assumptions that cannot be confirmed establish the basis for identifying conceivable or reasonable deviations. The key to knowing when to stop the iterative process of site investigation and refinement of the conceptual model is finding that the remaining uncertainties can be handled as reasonable deviations (Brown, 1990).

Descriptions of Improved Approaches to Environmental Site Characterization

Several different improved approaches to environmental site characterization have evolved from the methods described earlier. These approaches have all been developed since the mid-1990s, under slightly different names and applying to different types of sites, but following the same basic principles. They include:

- The Accelerated Site Characterization (ASC) approach described in the ASTM Standard Guide for Accelerated Site Characterization for Confirmed or Suspected Petroleum Releases (ASTM Standard E 1912 [ASTM, 2004a]) and in Taylor and Erikson (1996), and the Expedited Site Assessment (ESA) approach described in U.S. EPA's "Expedited Site Assessment Tools for Underground Storage Tank Sites" (U.S. EPA, 1997a). Both of these apply primarily to small sites mainly involved with the storage and handling of petroleum products (i.e., service stations or petroleum distribution terminals).
- The Expedited Site Characterization (ESC) approach described in the ASTM Standard Practice for Expedited Site Characterization of Vadose Zone and Ground-Water Contamination at Hazardous Waste Contaminated Sites (ASTM Standard D-6235 [ASTM, 2004b]) and several U.S. Department of Energy Documents (Burton, 1993; Burton et al., 1995; U.S. DOE, 1998, 2001). This approach applies mainly to large sites known or suspected to be contaminated with hazardous wastes.
- The Triad approach described in "Technical and Regulatory Guidance for the Triad Approach: A New Paradigm for Environmental Project Management" (ITRC, 2003) and in several U.S. EPA documents (Crumbling, 2001a; Crumbling et al., 2003; U.S. EPA, 2004b). This approach is designed to apply to all types of contaminated sites.
- The Dynamic Field Activities (DFA) approach described in several U.S. EPA Superfund Program documents (U.S. EPA, 2001a, 2003). This approach applies primarily to Superfund (CERCLA) sites at which uncontrolled disposal of hazardous wastes is known to have occurred.

Although the basic principles for all of these approaches are the same and the philosophies followed in each approach are similar, there are some subtle (and not so subtle) differences in implementing these approaches which require that each approach be discussed separately. These discussions are followed subsequently.

Like the traditional approach, all of the improved approaches to environmental site characterization involve conducting a review of existing information on the site and stressing the importance of developing a comprehensive understanding of the probable site conditions so that accurate predictions regarding contaminant source areas, contaminant distribution, and presence of preferential migration and exposure pathways can be made. However, from this point on, the improved approaches differ substantially from the traditional approach. After the following discussions of each approach, the elements that are common to all approaches (and the differences in application of these elements inherent in each approach) are described in detail.

Accelerated Site Characterization

ASC is a process for rapid and accurate characterization of a site at which a confirmed or suspected petroleum product release has occurred, in one mobilization of equipment and personnel. The process requires a significant amount of up-front planning and flexibility in its application, a review of existing information, and development of a CSM to use as a basis for planning the field investigation. It makes use of rapid sampling tools and techniques, field analytical methods, and on-site interpretation of field data to refine the conceptual model as the investigation proceeds. The ASC process requires a senior on-site manager to evaluate and interpret data as they are generated and to make decisions to guide the investigation in the field. Evaluation of field data concurrent with the investigation allows the on-site manager to select subsequent sampling points and adjust the overall sampling and analysis plan or the scope of the investigation based on actual site conditions, resulting in a more comprehensive and cost-effective snapshot view of subsurface conditions. A level of communication must be established between the on-site manager, the site owner or operator, the regulatory agency, and other interested parties prior to the beginning of the site characterization program. The ASC process applies specifically to collecting and evaluating information on site soils, geology, and hydrogeology; nature and distribution of COCs; contaminant source areas; and potential exposure pathways and points of exposure.

The most important feature of the ASC process is the on-site iterative process — a logical, scientific approach to site investigation that senior personnel in the field follow to meet site characterization objectives. A flow chart of the ASC process is presented in Figure 2.33. While many of the steps in an ASC program are similar to those used in a traditional site characterization program, the iterative approach to field activities requires the use of a flexible or dynamic work plan and includes the following activities:

- Analysis, evaluation, and interpretation of field-generated geologic, hydrogeologic, and chemical data as they are collected
- Continuous refinement of the CSM and development of an improved understanding of site conditions using field-generated data
- Modification of the sampling and analysis program to address any adjustments in the scope of work made necessary by site conditions
- Collection of additional data necessary to complete the site characterization and meet the objectives of the investigation

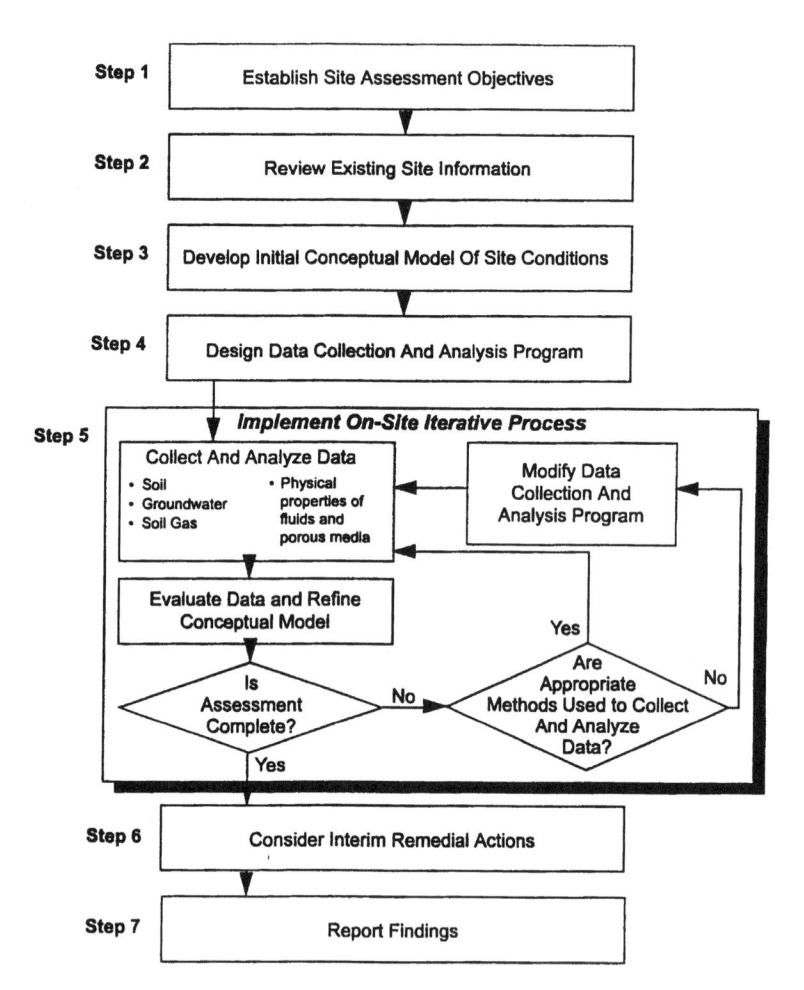

FIGURE 2.33
Flow chart for the ASC process. (From ASTM, 2004a. With permission.)

A major factor in the application of the ASC approach is selecting the right tools for the field investigation and building in the flexibility to change tools as site conditions dictate. The emphasis is on using rapid sampling tools (Figure 2.34) and advanced field analytical methods (Figure 2.35) that provide high-quality data upon which important decisions may be based. Rather than focusing on producing laboratory-quality data for a relatively small number of samples, as a traditional site characterization program does, the emphasis in an ASC program is on producing analytical data of sufficient quality for decision making for a much larger number of samples. The sheer number of samples greatly reduces uncertainty, provides greater confidence in the results of the investigation, and allows for comprehensive three-dimensional quantification of impacts on soil and ground water at a fraction of the cost of a conventional investigation.

The advantages of using the ASC process include (ASTM, 2004a):

- Immediate identification of potential risks to human or environmental receptors or potential liabilities or both
- Rapid determination of the need for interim remedial actions, site classification, and prioritization

Environmental Site Characterization

FIGURE 2.34
The emphasis of the ASC approach (and all other improved site characterization approaches) is on using rapid sampling tools that produce minimal amounts of IDW, like this DP rig. These rigs are highly versatile, mobile, and low-profile and can be easily operated by a two-person crew to produce 300 ft or more of continuously sampled borehole in a single 8 to 10 h work day. (Photo courtesy of Geoprobe Systems.)

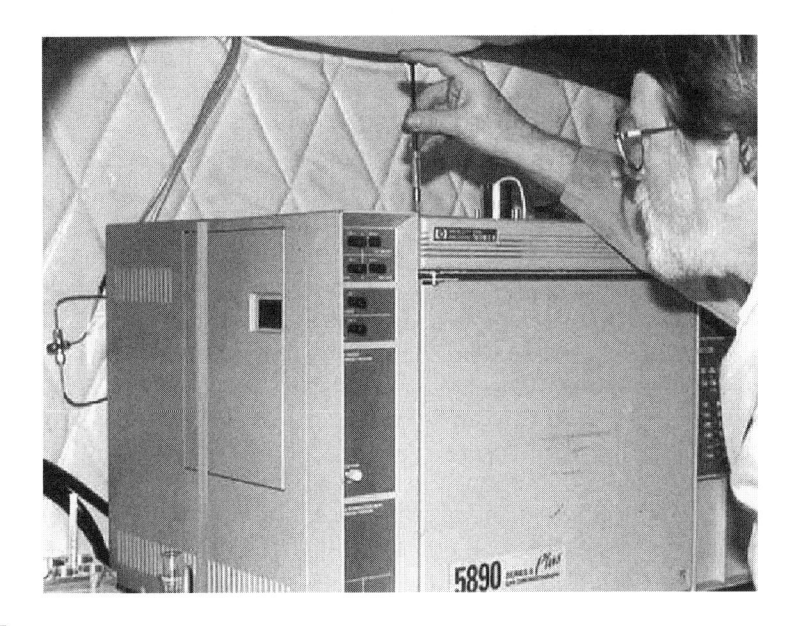

FIGURE 2.35
Highly sophisticated field-based analytical equipment can be used to produce data of the same quality as those produced by lab-based equipment, at a lower cost and higher level of convenience. Because samples are analyzed in the field immediately following collection, sample errors due to shipping, handling, and holding time exceedances are eliminated. Because analysis costs less, more samples can be analyzed for the same budget.

- Rapid sample collection and analysis, real-time or near-real-time analytical results, and maximum data comparability
- Optimization of sampling point locations and analytical methods
- Greater number of data points for resources expended
- Nearly immediate data availability for accelerating corrective action decisions
- Collection of vertical and horizontal data, allowing for three-dimensional delineation of COCs in soil, soil vapor, and ground water

Expedited Site Characterization

The ESC process originated in 1989 out of work conducted at the U.S. Department of Energy's Argonne National Laboratory in Argonne, Illinois (Burton, 1993; ASTM, 2004b). The process was first successfully applied in 1992 at several landfills in New Mexico operated by the U.S. Bureau of Land Management. It has since been used at sites operated by the U.S. Department of Agriculture (in Nebraska and Kansas), the U.S. Department of Energy (Savannah River Site in South Carolina; Pantex Plant in Texas; St. Louis Airport in Missouri), and the U.S. Department of Defense (many locations), at a former manufactured gas plant in Iowa, and at an oil refinery in Katowice, Poland (U.S. DOE, 1998; ASTM, 2004b), among other sites.

ESC is a process for identifying vadose zone, ground water, and other relevant contaminant sources and migration pathways and for defining the distribution, concentration, and fate of contaminants for the purpose of providing the necessary information to choose a course of action (i.e., long-term monitoring, risk assessment, active remediation, or no further action) that addresses the risks posed by the site to human health and the environment. Generally, the process is applicable to larger-scale projects such as Superfund remedial investigations, RCRA facility investigations, and petroleum releases at large facilities such as refineries, although it can also be applied to other contaminated sites. The process is not as useful for small petroleum release sites (e.g., service stations), real-estate property transfers, or sites where contamination is limited and does not threaten ground water, or at which the cost of remedial action is likely to be less than the cost of site characterization. At sites where it can be applied effectively, the ESC process should provide a greater quantity of higher-quality information for decision making at a lower cost and in a shorter period of time than traditional site characterization.

The ESC process focuses on collecting only the information required to meet site characterization objectives and on ensuring that characterization ends as soon as objectives are met. Central to the ESC process is the use of judgment-based sampling and measurement to characterize site contamination in a limited number of field mobilizations (usually two). An ESC program is led by senior technical personnel operating within a framework of a dynamic work plan that allows the flexibility of selecting the type and location of samples and measurements needed to optimize data collection activities and adjusting the work plan in the field to respond to site conditions. The on-site analysis, validation, and interpretation of field data and the continual integration of those data into a CSM are important features of an ESC program. ESC employs an iterative process for developing and testing multiple working hypotheses aimed at reducing uncertainty through the use of a CSM that is continually revised as data are obtained in the field.

The ESC process is based on good scientific practice and is flexible enough to accommodate a variety of different approaches to collecting environmental data, but it is not tied to any particular regulatory program, site investigation method or technique, analytical method, or data evaluation methodology. Appropriate investigation techniques used in an ESC program are highly site specific and are selected based on the professional judgment of a core technical team. Whenever feasible, noninvasive or minimally invasive methods are used. Appropriate chemical analytical methods are equally site-specific. Analyses may be conducted in the field (preferred where appropriate) or in the laboratory, depending on data quality requirements, required turnaround time, and costs.

The Triad Approach

To support the scientific and legal defensibility of decisions involving contaminated sites, the U.S. EPA has advocated the use of the Triad approach, a conceptual and strategic framework that explicitly recognizes the scientific and technical complexities of environmental site characterization, risk estimation, and remedial design (Crumbling et al., 2003; U.S. EPA, 2004b). In particular, the Triad approach acknowledges the fact that environmental media are fundamentally heterogeneous on a variety of scales, which adds complexity to sampling program design, analytical method performance, and spatial interpretation of environmental data. The approach integrates systematic planning, dynamic work plans, and real-time measurement technologies (the three elements of the "Triad") to reduce decision uncertainty and achieve more efficient and cost-effective site characterization. Most of the ideas expressed in the Triad approach are not new — what is new is the effort to comprehensively incorporate these ideas into the next-generation model for site characterization and cleanup practices supported by the U.S. EPA. Table 2.2 lists the major components of the Triad approach.

The Triad approach specifically focuses on identifying, understanding, and managing sources of decision uncertainty that could lead to decision errors. When scientific data are used to provide input into the decision-making process, the uncertainty in those data needs to be managed to a degree commensurate with the desired level of decision confidence. Because most data uncertainty stems from sampling variability and a lack of data representativeness, the Triad approach maximizes the use of new sampling, analytical, and measurement technologies to cost effectively increase sampling density so contaminant distribution and spatial heterogeneity can be characterized at the scale of project decisions. Better site characterization (leading to decreased uncertainty) is also possible because plumes can be chased and three-dimensional spatial patterns of contamination can be delineated in real time by revising the sampling program design on a daily basis as new information is acquired.

The Triad approach takes advantage of scientific and process improvements in three basic areas: systematic project planning, dynamic work strategies, and real-time measurement technologies (Figure 2.36). Systematic project planning is the most important element in the Triad approach. This element encompasses all tasks that produce clear project objectives and decisions, which describe unknowns (uncertainties) that could cause erroneous decisions, and that foster clear communication, documentation, and coordination of all project activities. Having clear project goals spelled out upfront allows project planners (a multidisciplinary team of experienced technical staff) to develop effective data-collection strategies to achieve these goals. Project planners must identify the type, quantity, and quality of data needed to satisfy project objectives. This improves the quality of the investigation activities because data collection can be done more efficiently and the uncertainties that stand in the way of achieving project goals can be more easily addressed. A key product of the systematic planning process is the initial CSM, a planning tool that organizes what is known about a site and identifies what more must be known about the site to make the decisions that will achieve the project goals. The CSM thus also serves as the basis for developing a dynamic work plan.

The word "dynamic" describes work strategies designed around consensus-derived decision logic. This element of the Triad approach is based on real-time decision making that can quickly direct and refine field work as new information becomes available.

TABLE 2.2

Major Components of the Triad Approach

Component		Answers
Systematic project planning	Project initiation	
	Assemble project team	Who
	Define project objectives	What
	Identify key decision makers	Why
	Define decisions to be made	
	Develop initial CSM	
Dynamic work strategy	Project start-up	
	Ongoing revision of the CSM	What
	Draft adaptive work plan and sampling strategy or decision logic	Why
	Develop detailed analytical strategy: field-based or fixed lab	How
	Develop data management plan	When
	Develop quality assurance project plan	Where
	Develop HSP	Who
Adaptive work plan implementation	Plan approval	
	Client/regulator/stakeholder review and approval	Who
	Refine project decision logic and finalize plans	What
		Why
		How
Real-time measurement technologies	Field program	
	Sampling and analysis to fill data gaps	When
	Data validation, verification, and assessment	Where
		Who
		What
		How
Decision making	Are project objectives met?	
	Evolve or refine CSM	Why
	Modify adaptive work plan	What
	Client/stakeholder/regulatory review and approval	How
		Who

Source: ITRC, (2003). With permission.

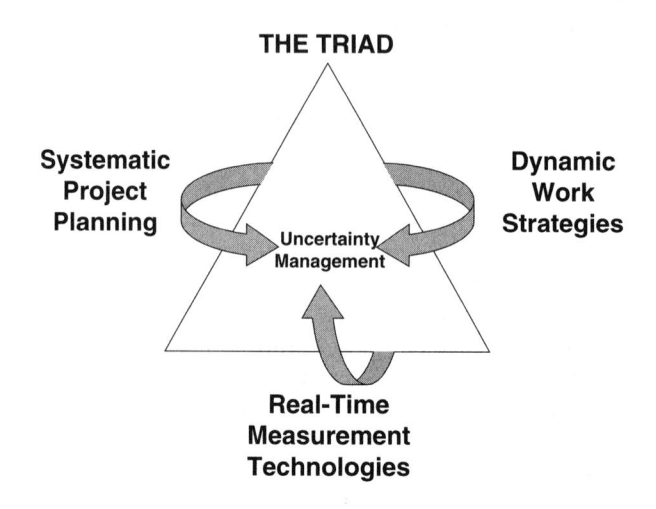

FIGURE 2.36
The three elements of the Triad approach (systematic project planning, dynamic work strategies, and real-time measurement technologies) are integrated to produce an accurate CSM upon which decisions regarding risk and remediation are based.

FIGURE 2.37
Real-time measurement technologies, including surface geophysics (in this case, electromagnetic conductivity), are a cornerstone of the Triad approach.

Real-time decision making, which requires the presence of experienced staff in the field, greatly improves the efficiency of the investigation and significantly reduces project lifetime costs and duration by making multiple mobilizations unnecessary. Overall project quality and decision confidence are also improved because more data can be collected for the same budget, and these data can be used in a rapid feedback loop to fill important gaps in the CSM. It is critical to use the CSM as a tool to avoid sampling errors and to interpret results from data sets derived from various data collection activities.

Real-time measurement technologies, the third element of the Triad approach, make real-time decision making possible. These technologies include surface geophysics (Figure 2.37) and other imaging technologies, rapid sampling and *in situ* measurement platforms (Figure 2.38), field-based analytical methods (Figure 2.35), and rapid turn-around from mobile and fixed laboratories. Another important aspect of the Triad approach is the data management program, which requires software packages for processing, displaying, and sharing data, so that the CSM can evolve while the field investigation team is at the site. Together, real-time measurement technologies and dynamic work plans work hand-in-hand so that data collection is focused and informative (ITRC, 2003).

The Triad approach is an outgrowth of the natural evolution of the environmental industry in response to imperatives that include evolving economic considerations and improved science and technology for site characterization. It is not narrowly focused on a single U.S. EPA program — it is applicable to all types of environmental programs. The concepts behind the Triad approach apply to any type of site, no matter what stage of investigation and no matter what size. The approach applies to sites in any setting, no matter how complex, and any type of contaminant, whether LNAPL, DNAPL, dissolved phase, vapor phase, or residual phase.

Dynamic Field Activities

DFAs are hazardous waste site assessment, characterization, monitoring, and remediation activities that combine on-site data generation with on-site decision making. They are called "dynamic" because activities are designed to incorporate change as new

FIGURE 2.38
Rapid sampling and *in situ* measurement platforms, such as the DP rig depicted in Figure 2.34 and this cone penetration testing (CPT) rig, can be used to collect real-time *in situ* profiling data or samples for *ex situ* analysis (in conjunction with field-based analytical methods) to speed up the site-characterization process.

information is obtained, thus accommodating the iterative nature of environmental investigations in the field and minimizing the number of mobilizations necessary to reach a site decision. Because of its flexible approach to data collection, the DFA approach is applicable to all stages of the Superfund response process (U.S. EPA, 2001a, 2003).

DFAs provide an iterative, flexible framework for collecting data and making decisions at hazardous waste sites throughout the cleanup process. Figure 2.39 illustrates the four major steps included in the DFA process, which include:

- Using a systematic planning process
- Preparing a dynamic work plan
- Conducting iterative sampling, sample analysis, and data evaluation
- Writing a final report

The key feature of this process is that it uses a dynamic work plan that is flexible enough to allow changes in sampling and analytical activities to occur in the field so that project objectives can be attained in a minimum number of mobilizations. These work plans provide the blueprint for how adjustments are to be made. It is important that senior technical personnel, who have both the experience and the authority to make important decisions in the field, conduct the field investigation so the process is seamless. In addition, important to the success of this approach is the use of field-based analytical methods as the primary source of data used in decision making. Through effective use of these resources, DFAs have the ability to significantly reduce the time and cost of the field investigation, while improving the quality of the data collected and the quality of site decisions.

As used in the DFA process, systematic planning is a common-sense approach to ensure that the level of detail in project planning is commensurate with the intended use of the data and the available resources. It requires that all interested parties (investigators, site owner or operator, regulators, and others) collaborate to establish clear project objectives and communicate during the project to ensure that objectives are still reasonable and that

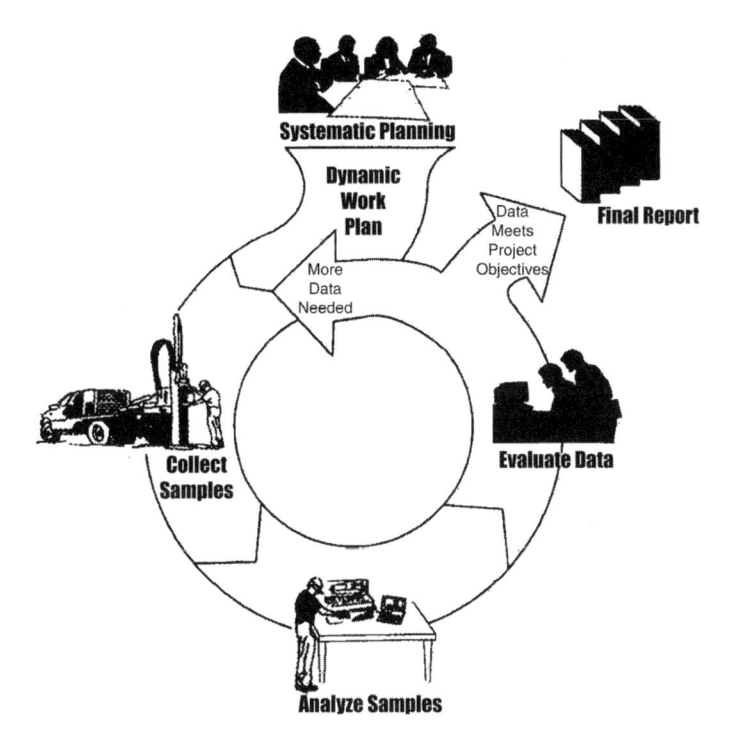

FIGURE 2.39
The four major steps in DFA approach: systematic planning; dynamic work plans; iterative sampling, analysis, and data evaluation; and the final report. (From U.S. EPA, 2003.)

they are being met. Systematic planning is an iterative process that begins at the outset of the project and continues throughout project implementation. To facilitate the use of the systematic planning process, U.S. EPA guidance recommends the use of data quality objectives (U.S. EPA, 2000a). Regardless of the formal process used, systematic planning will entail:

- Reviewing existing site information
- Selecting key personnel
- Identifying project objectives
- Developing an initial CSM and modifying it as the investigation produces additional data
- Preparing sampling and measurement strategies
- Selecting appropriate analytical methods, equipment, and contractors

The development of an initial CSM is an essential activity because it pulls together all of the existing site information in an easily understood format such as a series of maps, cross sections, and diagrams that depict soil and geological conditions, surface-water and ground-water conditions, contaminant concentrations, potential migration pathways, locations of human and environmental receptors, and other information important to understanding site conditions. The initial CSM is a valuable tool used in the selection of appropriate sampling, analytical and *in situ* measurement tools, and the creation of sampling and analysis plans. As more data are collected, both the CSM and project objectives are revised as needed so that subsequent site decisions can be based on them.

For example, if the initial CSM is based on an assumption of random contaminant releases, investigators may choose a random grid to begin sampling at the site. If the investigators discover through initial sampling that there is a pattern to the contamination, they would alter the sampling strategy and project planners may need to fine-tune project objectives to ensure that the type, quantity, and quality of data collected is appropriate. Systematic planning must also establish processes for quickly integrating collected data into the CSM and transmitting it to off-site interested parties.

The dynamic work plan is written after the initial phase of systematic planning has been completed. This document provides the project team with the lines of communication and agreed-upon criteria required to facilitate decision making in the field. It is simply an outline of a sequence of activities that accommodate decision making and involvement of interested parties to keep the project moving forward. To do this effectively, a dynamic work plan employs an adaptive sampling and analysis strategy, which consists of an initial sampling and analysis plan that is modified in the field as additional data are collected and analyzed. Dynamic work plans tend to make use of innovative technologies that produce data in real time, which is necessary for on-site decision making. In particular, dynamic work plans incorporate rapid sampling methods and field-based analytical methods because they provide a cost-effective means of reducing uncertainties by allowing more data points for the same budget. The dynamic work plan should include contingencies so that unexpected findings or unsuccessful methods that make changes in the plan necessary can be dealt with without causing delays in the field work. The dynamic work plan must be accompanied by other documents that address specific elements of the work to be conducted at the site and that support the on-site decision-making process. These documents include a Field Sampling Plan (FSP), a Quality Assurance Project Plan (QAPP), a Site Health and Safety Plan (HSP), and a Data Management Plan (DMP). These plans provide a higher level of detail than the dynamic work plan and include the standard operating procedures required to conduct daily activities at the site.

Finally, DFAs use an iterative sampling, analysis, and data evaluation strategy that allows the project team to continually revise the CSM in the field until they are satisfied that they have reached project objectives, thus minimizing the number of mobilizations. Because data are available within minutes or hours of sampling, decisions can be made in real (or nearly real) time. Experienced personnel are an essential component of this process because their knowledge is needed to evaluate and interpret results and to guide the progress of the project. Typically, a very experienced and multidisciplined field manager will supervise field activities and ensure that appropriate personnel have the information they need to generate and evaluate field data. At the conclusion of field activities, a report is written to document results and provide guidance on a subsequent course of action.

Other Improved Approaches to Environmental Site Characterization

A few other improved approaches to environmental site characterization have been developed by a variety of groups, including Federal and state government agencies and contractors, environmental consulting firms, and research labs. Because all of these approaches include most of the same elements as the approaches described earlier, but in different combinations, they will not be described in detail here. The interested reader is directed to the references cited. These approaches include the following:

- The Rapid Site Assessment approach, used by the Florida Department of Environmental Protection's Drycleaning Solvent Cleanup Program (Applegate and Fitton, 1997)

- The Field Assessment Screening Team (FAST) approach, used by Martin Marietta Energy Systems at U.S. DOE sites (Nickelson and Long, 1995)
- The Rapid Adaptive Site Characterization approach, used by Stone Environmental, Inc. (Pitkin, 2001) and The Johnson Co. (Moore, 2000)
- The QuickSite approach, used by the U.S. Department of Energy (U.S. DOE, 2004)
- The Source Area Mapping approach, used by McLaren and Hart (Gelb et al., 1998)
- The Dynamic Site Assessment approach, used by Weiss Associates (Thiesen and Weiss, 1990)
- The Technical Project Planning approach, used by the U.S. Army Corps of Engineers (U.S. Army Corps of Engineers, 1998)
- The Streamlined Approach for Environmental Restoration (SAFER), used by the U.S. Department of Energy (U.S. DOE, 1993)
- The Superfund Accelerated Cleanup Model (SACM) approach, used by the U.S. EPA (U.S. EPA, 1988, 1989a, 1992)

Elements of an Environmental Site Characterization Program

Whether an investigator chooses to use the conventional or traditional approach or one of the improved approaches to environmental site characterization, he or she will have to employ a few common elements to provide a structure for the investigation. Some investigations will use all of the elements described subsequently and others will use only a few. Some investigations will use the elements exactly as described subsequently and others may employ only portions of each element. For the sake of brevity, each element is described in the context of how it would be used as part of a typical ASC/ESC/Triad/DFA approach, with significant departures from the norm noted where they apply for each approach.

Systematic Project Planning

The ASC/ESC/Triad/DFA approaches are all applied on the principle that the quality of an investigation depends on achieving a level of decision confidence that meets the expectations of the interested parties for a successful project outcome. To reach the desired outcome, the project team makes specific technical, regulatory, economic, and engineering decisions, each with inherent uncertainty. Detailed and systematic project planning provides project staff with effective ways to ensure confidence in the project outcome, despite the presence of uncertainties that affect project decisions.

The ASC/ESC/Triad/DFA approaches all depend on some level of systematic project planning to support an effective environmental site characterization program, although it is more heavily emphasized in the Triad and DFA approaches. In the planning phase of a project, investigators specify the intended use of the environmental data to be collected and plan the management and technical activities (i.e., sampling and analysis) required to generate the required data. Systematic planning is based on the scientific method and includes such concepts as objectivity of approach and acceptability of results (U.S. EPA, 2000b). It applies equally well to small, simple sites and large, complex sites. It is a common-sense approach designed to ensure that the level of detail in planning is commensurate with the importance and intended use of the data, the decisions to be made, and the available resources. Through a systematic planning process, the project team can develop acceptance or performance criteria for the quality of the data collected and for the quality of the decision. Systematic planning is not only the first step in implementing an effective environmental site characterization program, but also an iterative process that takes place throughout project implementation.

Systematic planning for any environmental site characterization program will entail the following steps:

- Select key project personnel
- Identify project goals and objectives
- Define decisions to be made
- Establish DQOs
- Review existing site data
- Conduct site reconnaissance
- Develop an initial CSM
- Identify data gaps in the CSM and constraints to data collection
- Establish the type, quantity, and quality of data needed from the field investigation
- Design a data collection and analysis program to address how, when, and where data will be collected
- Evaluate and select site characterization field methods for data collection
- Prepare a dynamic work plan to implement the field investigation
- Prepare supporting work plans to complement the dynamic work plan, including:
  a. A Sampling and Analysis Plan (SAP)
  b. A Quality Assurance Project Plan (QAPP)
  c. A Site Health and Safety Plan (HSP)
  d. A Data Management Plan (DMP)
- Describe methods for data analysis, evaluation, and interpretation, which address the intended use of the data

Establishing a project team with a cross-section of necessary technical and project management skills and experience is of fundamental importance to successful project planning. However, technical skills alone are not enough to carry a project and the team must include interested parties with a variety of backgrounds and perspectives including regulators, representatives of the site owner or operator, and other stakeholders. All of these individuals should be involved from the outset to ensure that their input is included in the development of project goals and objectives.

Systematic project planning always involves establishing clear goals and objectives to guide the investigation — the importance of this cannot be overemphasized. Very often during the course of performing environmental investigations, insufficient attention is paid to establishing clear goals and objectives for the work required, sometimes leading to unproductive investigations that fail to efficiently gather the information necessary for scientifically and legally defensible decisions. With objectives clearly defined, available resources can be used more effectively.

Another very important component of systematic planning is developing an initial CSM and revising the CSM as data are collected during the field investigation. Planning with the desired project outcome in mind reveals which data gaps in the CSM are truly critical and require attention. This applies regardless of whether the ultimate goal of the project is to design an effective long-term monitoring system, conduct a comprehensive risk assessment, or prepare a remediation plan.

Optimization of data collection is a central theme of systematic project planning. The data collection program designed to fill data gaps in the CSM should be tailored specifically to address the decisions to be made, to avoid straying from the stated project objectives and wasting resources. Systematic planning, as applied in the Triad approach, focuses data collection efforts by starting with the desired project outcome and working backwards through the project decisions that influence project outcome, the CSM that is the basis for project decisions, the data needed to produce the initial CSM and revise it in the field, and the details of sampling and analysis needed to produce the required data (ITRC, 2003; Figure 2.40). As the site characterization program is implemented, the samples collected are used to generate the data required to revise the CSM, and the CSM is then used to make decisions about whether the desired project outcome can be achieved.

Another theme that is emphasized in systematic project planning is the need for quality control (QC). The project QC program must be comprehensive enough to detect deviations from expected performance and to allow for estimation of sampling and analytical uncertainties, as well as their impact on decision making. QC procedures will vary by sample collection and analytical technology and in accordance with the type of decision to be made. Varying the levels of analytical quality through the mixing and matching of methods offers potential cost and time savings, but the added complexity that this produces with respect to the QC program must be carefully managed.

In summary, systematic project planning combines several familiar project planning activities with a few important new tasks, such as early focus on project outcome and identification of key decision points. To successfully apply one of the improved approaches to environmental site characterization, these new tasks must be fully integrated into the planning process. Failure to include all facets of systematic project planning can result in failure to achieve the desired project outcome. With a focus on managing and reducing decision uncertainty, systematic project planning allows site characterization programs to be done right the first time.

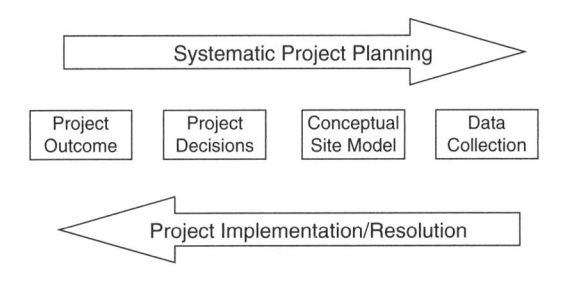

FIGURE 2.40
The relationships between systematic project planning and project implementation. Systematic planning tailors data collection by starting at the highest level (the desired outcome) and working downward into the details of sampling and analysis (arrow pointing right). As the work strategy is implemented, the data produced are used to revise the CSM which is, in turn, used to make decisions about whether the outcome can be satisfactorily achieved (arrow pointing left). (From ITRC, 2003. With permission.)

Selecting Key Project Personnel

The selection of qualified and experienced personnel is a very important element of conducting a successful environmental site characterization program and is highly project- and approach-specific. For example, guidance available for the ASC approach simply requires that an on-site manager be present during the course of the field investigation to make decisions to guide the characterization (ASTM, 2004a). No levels of experience and no fields of expertise are specified for the on-site manager. The authors' experience suggests that a senior staff person (with 10 years or more of field experience) with expertise in the field of hydrogeology (and experience examining and interpreting data from different disciplines) is the best choice for this position. The on-site manager must have sufficient experience to be able to interpret highly complex sets of geologic, hydrogeologic, geophysical, chemical, and other data as they are generated. They must also have the experience and authority to adjust sample locations or the scope of the investigation as needed. Otherwise, an ASC has little chance of successfully achieving the objective of fully characterizing the site in one mobilization.

Guidance available for the Triad approach suggests that experienced technical staff (geologists, hydrogeologists, chemists, engineers, etc.) must be involved in all aspects of the site characterization program from the outset. They should have the expertise to identify the most resource-effective characterization tools for collecting data at the site, they must be familiar with both the established and more innovative tools of their discipline, and they must be able to work together to construct a workable sampling strategy for the site. For example, the hydrogeologist must be familiar with the cost and performance issues associated with conventional drilling methods, as well as more innovative and less costly DP technologies. The geologist must understand how uncertainties due to sampling considerations (where, when, and how samples are collected) affect the representativeness of data generated from those samples and thus the ability of those samples to provide accurate site information (Crumbling, 2001b). The chemist must know the relative merits of traditional sample preservation, preparation, and analysis methods, as well as the strengths and limitations of innovative techniques, especially field-based analytical methods. The chemist must also have the experience to identify potential sources of interference that may occur during sample analysis and know how to adjust the method to correct for this. When risk assessment is part of a project, the involvement of a risk assessor at the beginning of the project is vital to ensure that meaningful data will be collected for risk assessment purposes. When project planners wish to express desired decision confidence objectively and rigorously in terms of statistical certainty levels, statistical expertise is required to translate the overall decision goal into data collection and sampling strategies. Because sampling design and analytical strategy interact to influence statistical confidence in final decisions, collaboration between the project chemist, geologist, hydrogeologist, and statistician is key to selecting a strategy that can achieve project objectives accurately and cost effectively (Crumbling, 2001a). During the investigation, these staff must be either in the field or available via telecommunications to guide the unfolding investigation in real time as directed by the project work plan (ITRC, 2003).

The ESC and DFA approaches are more specific in terms of personnel requirements and provide a better indication of what levels of experience are required for the staff selected for most environmental site characterization programs. Both the ESC approach and the DFA approach advocate formation of a core technical team, headed by a team leader and typically consisting of three or four professionals with expertise in geology, hydrogeology, chemistry, and geophysics — other areas of expertise may be included as appropriate. The technical team leader manages the investigation from the field and has overall responsibility for development of work plans; execution of field activities; data management, evaluation, and interpretation; communication with the site owner or operator, regulatory agency staff, and stakeholders; keeping the investigation moving forward at a reasonable pace; and generation of final project deliverables. This role generally requires a multidisciplined individual who exhibits a high level of professional judgment and has at least 10 years of field experience in managing and conducting site investigations. This individual has final decision-making responsibility within the framework provided by the dynamic work plan and must have experience in integrating information from multiple disciplines into a CSM to guide field investigation activities.

The core technical team members generally include a project geologist, a project hydrogeologist, a project chemist, and a project geophysicist, each with 5 to 10 years of field experience in their respective disciplines and specific knowledge regarding the operation of specialty equipment to be used on the project. The relative importance of required areas of expertise will vary somewhat from project to project, but other expertise that may need to be represented on the core technical team includes soil science, hydrology, geomorphology, stratigraphy, geochemistry, sedimentology, climatology, ecology, biology, microbiology, risk assessment, and statistics. The team members are involved, as needed, in all steps of the ESC/DFA process beginning with the establishment of project goals and objectives, and need to be integrators as well as specialists. They assist in assembling the initial conceptual model for the site and in developing sampling strategies and work plans. They either supervise or are present in the field during data collection involving their areas of expertise, they participate in data collection, processing, and interpretation and they help revise the CSM based on data collected in the field. The optimization of field investigation activities and the quality of the final CSM are strongly influenced by the interaction of the different perspectives of the core technical team members.

The core technical team operates with the support of a larger project team that includes technical personnel and equipment operators involved in data collection and sampling, as well as other support functions. Some of these team members may have a special role in the planning phase in determining the types, quantity, and quality of data required from the field investigation. The project team members should provide the following support functions (not every function will be required for every project):

- Logistics
- Geoscience technical support
- Geophysics
- Drilling and DP operations
- Surveying
- Specialty sampling support (e.g., surface water, biological, air)
- Health and safety
- Chemical analytical support (field and/or laboratory)
- Data management
- Quality assurance
- Contract management
- Statistics and geostatistics
- Fate and transport analysis and modeling
- Risk assessment
- Remediation engineering

- Community relations
- Waste management
- Information technology

For large projects, each of these roles may be filled by separate individuals. However, for small projects, one person may be able to fill several roles to decrease costs and increase integration of the team. For example, the project chemist might also fill the roles of risk assessor, statistician, and quality assurance specialist or health and safety specialist, depending upon their specific training and experience. In addition, because the total amount of time spent on each project for each function varies considerably, some of these functions can be performed on an as-needed basis, while others may require a full-time commitment.

The ASC/ESC/Triad/DFA approaches all differ from the conventional or traditional approach in that emphasis is placed on having experienced personnel in the field rather than sending junior staff to do all of the field work. The higher cost of placing highly experienced personnel in the field can generally be expected to be offset by the expert judgment that typically reduces the time and total cost to generate an accurate CSM and by the increased project efficiency, related to real-time in-field decision making, which experience affords. This expert judgment is required because the heterogeneity of subsurface materials results in highly complex contaminant distribution patterns that requires extensive experience in data interpretation to resolve. The multidisciplinary perspective of the core technical team functions as an on-site peer review of the evolving CSM and should help identify inconsistencies that might be missed by a single experienced individual.

Identifying Project Goals and Objectives

Establishing the site-specific goals and objectives of an environmental site characterization program is important to a successful, efficient, and cost-effective project, as the goals and objectives will dictate the time and resource requirements of the investigation. Defining objectives helps the investigator focus squarely on collecting the data that are required to accomplish project goals and helps the investigator avoid wasting precious time and resources gathering unnecessary information. Project objectives help determine the appropriate site characterization approach, the quantity and quality of samples that must be collected and analyzed, and the investigative techniques and analytical methods to apply and the order in which they will be used. The exact project objectives must be clearly defined and agreed upon by all interested parties (i.e., the investigator, the site owner or operator, the regulatory agency, stakeholders, and others) early in the site characterization planning process. They should be clearly expressed in writing and referred to often during the life of the study. Otherwise, as work progresses, there may be a tendency for the investigation to drift from the stated goals, resulting in the collection of non-essential information, perhaps at the expense of required information.

While this sounds straightforward, very often investigators or their clients lose sight of project objectives in an attempt to cut investigation costs. In these cases, investigators are left with sparse and inadequate data that provide more questions than answers regarding the sources, types, and extent of contaminants present at a given site. While controlling costs is important, it should not override the need to meet project objectives.

The objectives of the site characterization program should be established with the eventual endpoint of the project in mind (usually long-term monitoring, risk assessment, or remediation). In general, the primary objective of site characterization will be to provide sufficient high-quality data to either design an effective long-term monitoring program or to permit evaluation of remedial alternatives and then design the remediation program. Occasionally, the initial objectives may need to be changed as the investigation uncovers additional information. For example, the initial objective may be to define the presence and distribution of subsurface contamination for the purpose of establishing a long-term monitoring system. However, if the investigation discovers serious and widespread contamination that threatens a neighboring property or a water-supply well, the objectives may shift to evaluating the need for interim remediation and possible remedial alternatives.

After project objectives are articulated, the uncertainties that stand in the way of meeting these objectives can be addressed and the environmental data that must be collected to reduce uncertainty can be easily defined. In generating data required to make decisions, the sampling and analytical uncertainties inherent to environmental data generation must be managed to a level commensurate with project decision needs. Having clear project objectives spelled out up front improves the quality of investigation activities because data collection becomes more efficient and focused on reducing uncertainty.

Defining Decisions to be Made

To achieve the desired project outcome, a number of regulatory and technical decisions must be made during the project. In practice, project decisions are made using a combination of scientific data and other inputs including political, social, and economic considerations that may be of local, regional, state, or national significance. Different projects will require that different decisions, using different sets of information, be made. A partial list of decisions that might be made, related to using the Triad approach (ITRC, 2003) at a site, for example, includes deciding whether:

- Contamination at the site is present at levels greater than background
- Contamination at the site is present at levels greater than regulatory action levels
- Contamination has been adequately characterized in all phases (vapor, dissolved, residual, non-aqueous phases)
- The matrix heterogeneity and variability in contaminant distribution has been adequately addressed
- There is a threat to ground-water or surface-water resources
- People have been or are in danger of being exposed to the contamination and, if so, by what pathways
- Environmental (ecological) receptors have been or are in danger of being exposed to the contamination and, if so, by what pathways
- The site can be closed with no further action
- Long-term monitoring will be required at the site
- Institutional controls, such as land-use restrictions, are appropriate for the site
- A risk-based remedial strategy is appropriate to apply at the site
- Natural attenutation is occurring and, if so, at what rate
- There are cost-effective remedial options available to address the contamination
- It is possible to apply new and innovative remedial options

Making these and other decisions requires extensive knowledge of the site physical setting and site contamination issues, collectively referred to as the CSM, which is described later in this chapter. Thus, the project team's confidence in making correct decisions depends on its ability to construct an accurate CSM. Furthermore, because different decisions may require emphasis on different aspects of the data comprising the CSM, the CSM may have to be constructed in different configurations to meet the needs of the project. A complex site may require several different depictions of the CSM, each of which addresses a different subset of the decisions to be made to move the project forward.

Doubts about whether decisions are made correctly create doubt (uncertainty) about the success of the project outcome. Thus, management of decision uncertainty is a very important goal of every site characterization project. Managing decision uncertainty generally requires development of a well-conceived sampling plan that supports the generation of large quantities of data relevant to the decisions to be made. Prior to beginning a site characterization project, investigators must establish what levels of decision uncertainty are tolerable (i.e., how much contamination can be missed by the sampling program without causing undue risk) and establish a site-specific decision strategy that is appropriate for the site.

The site-specific decision strategy is determined during systematic project planning, with input from regulators, the site owner or operator, and other stakeholders. If very little information is available early in the project to know what decision strategy would be best, the systematic planning process focuses on the information needed to decide what decision strategy makes the most sense. Factors driving the selection of one strategy over another (e.g., selecting a cleanup strategy vs. a containment option) can be arrayed into a matrix or decision tree, which is refined as the needed information is gathered during the project. Selection of a decision strategy may be summarized as a series of "if ... then" statements, which capture the relationships between drivers such as cost, risk, cleanup vs. containment options, and stakeholder concerns. As long as all stakeholders agree on the decision logic, final selection of the decision strategy can be a seamless part of field implementation (ITRC, 2003).

Establishing DQOs

To facilitate the use of the systematic planning process, U.S. EPA has developed guidance that recommends the use of DQOs. DQOs are qualitative and quantitative statements that clarify project goals and objectives, define the appropriate type of data, and specify tolerable levels of potential decision errors, which will be used as the basis for establishing the type, quality, and quantity of environmental data needed to support well-informed, valid, and defensible decisions (U.S. EPA, 2000a). DQO statements do not directly set the criteria for the quality of data that will be gathered during the project — the process for determining the quality of data that will be needed to meet project goals must be done after the DQOs are established (Crumbling, 2001c). Because several different levels of data quality may be appropriate to answer the site-specific scientific and engineering questions that must be addressed in any given environmental site investigation project, the terms "sufficient" or "acceptable" data quality are meaningful only when the intended uses of the data are known. Therefore, it cannot be overemphasized that cost-effective site investigations are highly dependent on anticipating data usage during the life of the site characterization, monitoring or cleanup program.

DQOs are universally applicable, where the results of a sampling program will be used to select between alternative actions (i.e., no further action vs. long-term monitoring vs. active remediation). By helping the project team collect only those data that are needed to answer specific questions to resolve a site-specific problem, DQOs put limited project resources to best use and reduce project costs (Makeig, 1995). Because each site has different facilities, history, physical characteristics (soil, geology, hydrology, hydrogeology, geochemistry, biology), human-related influences (utilities and pumping wells), potential receptors, background conditions, neighboring land uses, and contaminants with unique sets of chemical and physical characteristics and source areas, DQOs must be determined on a site-specific basis. This means that appropriate sampling protocols and sample analysis criteria can vary considerably from project to project (Thurnblad, 1995).

The DQO process is a systematic, iterative, and flexible planning process to develop sampling designs for data collection activities that support decision making. The process is focused on generating appropriate project data and is based on the scientific method. It acknowledges that investigators do not have infinite resources to address site-specific problems and that it is impossible to have a 100% guarantee that the right conclusion regarding those problems has been reached. The DQO process attempts to weigh these issues and provide a balance that is satisfactory to all interested parties between the resources that must be committed and the uncertainty that is acceptable. It thus allows project managers to specify acceptable data quality goals by establishing acceptable limits on decision errors. By definition, decision errors occur when variability or bias in the data misleads the decision maker into choosing an incorrect course of action. By using the DQO process, the project manager provides the criteria for determining when the data are sufficient for site decisions.

The procedure for establishing DQOs is comprised three basic steps: (1) identifying data uses; (2) identifying data types; and (3) identifying data quality needs (U.S. EPA, 1993a). These steps are intended to provide data of rigorous quality sufficient to meet legal challenges (i.e., defensible data). The DQO process relates the data needs to the specific decisions that must be made at a given site and involves the following (U.S. EPA, 2000a):

State the problem(s): The investigator, in consultation with the site owner or operator, regulators, and other stakeholders, must concisely describe the problem(s) that require resolution. The investigator must define overall project objectives (outlining the scientific and engineering issues to be addressed) and review prior field studies and existing information (fusing soft information with hard data) to gain an understanding of the problem(s) so that the focus of the investigation will be clear and unambiguous.

Identify the decision: The investigator must identify the decision that will be required to solve the problem(s). If more than one problem needs to be addressed, the process can be repeated until a list of concise decision statements is formulated. This will eliminate many of the extraneous and distracting questions that are not critical to solving the problem(s).

Identify the inputs to the decision: The investigator must identify the information that needs to be collected to support the decisions, the sources of that information, and the type of data quality needed to make the required decisions.

Define the study boundaries: The investigator must specify the population of interest (all potential sampling points from which a subset will be collected), the spatial and temporal circumstances (time period and geographic area) to which decisions will apply and within which data will be collected, and the scale of decision making (defining the smallest subset of the population for which decisions will be made).

Develop decision rules: The investigator must integrate the decision outputs from previous steps into a single statement that describes the logical basis for choosing among alternative actions to solve the problem(s). The decision rule can take the form of an "if ... then" statement, which will make choosing between different alternatives straightforward.

Specify acceptable limits on decision errors: The investigator, in conjunction with all stakeholders, must set acceptable limits on decision errors, which are used to establish appropriate performance goals for limiting uncertainty in the data. This allows identification of the tolerance that stakeholders have for errors made during the sampling and analysis program.

Optimize the design: The investigator must identify the most resource-effective sampling and analysis design for generating data that are expected to satisfy the DQOs. This allows statistical (or other) optimization of the sampling and analysis program.

Steps 1 through 5 are primarily focused on identifying qualitative criteria and provide the structure to help a project team articulate project goals and decisions and the project's constraints (time, geographic area, budget, etc.). The sixth step defines quantitative criteria, expressed as limits on the probability or chance (risk) of making a decision error that the decision maker can tolerate or a statistical expression of how much uncertainty can be tolerated in the final decision. These items must all be thoroughly understood before the task of developing the data gathering plans that can meet those goals within the given constraints is begun. Step 7 consists of developing project-specific sampling and analysis plans based on the criteria developed in the first six steps. This involves determining the type and number of samples, their locations, and their volume and defining the QA/QC activities that will ensure that sampling design and measurement errors are managed sufficiently to meet the tolerable decision error rates specified in the DQOs. The DQO process is thus used to define the quantitative and qualitative criteria or determining when and where to sample, how many samples or measurements to collect, and at what desired confidence level. DQOs should express what decisions the data produced by the sampling and analysis plan will support, but they should not specify how those data will be generated (i.e., which sampling or analytical methods will be used) (Crumbling, 2001c).

An important part of the DQO process is developing an understanding of how uncertainties can impact the decision-making process. When defining DQOs, the investigator must determine the answer to the question "What is the acceptable probability of not detecting a contaminated zone at the site?" Stated another way, it is necessary to identify what degree of certainty is acceptable in determining, for example, if a certain chemical is present at a threshold concentration at a site. Specifically, it is necessary to decide how small a concentration is to be detected with what certainty and how small a volume of contaminated soil or water needs to be found. If the answer is a very small concentration and a very small volume, then it may not be possible to meet DQOs. For example, there may be no access to certain parts of the site, there may simply be no analytical method capable of sufficient resolution or reliability, or the budget may not allow the collection and analysis of enough samples (U.S. EPA, 2000b).

When technically feasible, an expression of statistical uncertainty may be desirable, because it can be more objective if it is done in a technically valid manner. However, in the environmental field, statistical treatment of uncertainty may not always be technically feasible or even necessary. Qualitative expressions of decision confidence through the exercise of professional judgment (such as a "weight of evidence" approach) may well be sufficient, and in some cases, the only option available. An important part of systematic planning is identifying the information gaps that could cause a decision to be made in error. If the existence of information gaps increases the likelihood of decision error beyond what is acceptable, then it is desirable to fill those data gaps, if it is feasible to do so. Planning how to gather environmental data that can acceptably fill data gaps is the real purpose of the DQO process (Crumbling, 2001c).

Reviewing Existing Site Information

In an environmental site characterization program, one of the first and most crucial steps after establishment of project objectives involves locating, collecting, assembling, and organizing all available background information on the site into a manageable database to initiate development of an accurate initial CSM. This information helps to determine the steps that need to be carried out to characterize the site, to define what gaps exist in the database for the site, and to determine the scope of the dynamic work plan and all other work plans for the site. This information will also be used to help determine the field methods required to generate the data needed to meet project objectives.

The review of existing information will help investigators save a significant amount of time and money by ensuring that the field investigation does not duplicate efforts that have already generated some of the data required to evaluate site conditions. It will also make it less likely that any important factors will be overlooked during the field investigation and subsequent phases of work (i.e., monitoring or remediation) at the site.

Types of Information

Many different types and formats of information relevant to site environmental conditions must be collected and reviewed, including technical and nontechnical documents (published and unpublished), diagrams, maps, tables of "hard" data, regulatory agency files, anecdotal information, and other "soft" information from a wide variety of sources. Any hard data must be critically evaluated with respect to the methods used to produce the data, and the data must be validated so that only data of adequate quality (as defined by project objectives) are used to create the initial CSM. It is particularly important to evaluate existing sampling and analytical data to determine how samples were collected, handled and analyzed, and what the laboratory QA/QC performance was like on analytical work, to ensure that the data meet stated DQOs and QA/QC controls.

The existing information that should be reviewed includes:

- Results of any previously conducted investigations at the site or on adjacent sites (as available), including environmental, geotechnical, property transfer, and other reports, to establish as much about general site conditions (especially subsurface conditions) as possible
- Regional and local geologic and hydrogeologic reports, maps (Figure 2.41), and cross-sections, to establish the nature of local geologic materials and stratigraphy, approximate ground-water levels and flow directions, elevation of the top of bedrock, thickness of unconsolidated materials, and major structural features
- Current and historical topographic maps (Figure 2.42 and Figure 2.43), to determine site topography (and how it may have changed over time), to locate the site within the regional framework, to identify possible sites to access geologic outcrops (road cuts, stream cuts, mines, quarries, sand and gravel pits, etc.), to identify major cultural features (power lines, pipelines, roads, rail lines, etc.), and to locate potentially impacted surface water bodies, wetlands, and other ecologically sensitive areas

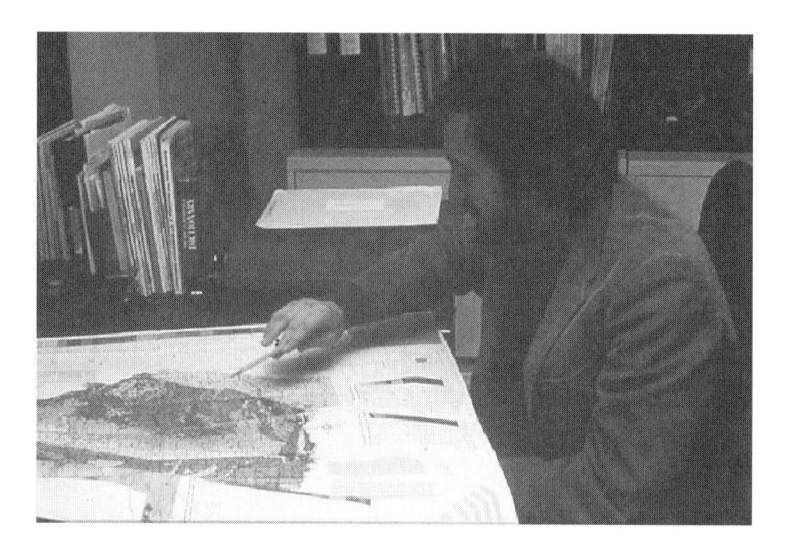

FIGURE 2.41
Regional geologic maps are useful tools to use to begin deciphering the general geology of the area in which the site is located. Investigators must be careful not to rely too heavily on this information, because the scale of these maps makes it difficult to apply the mapped information directly to the site. Site-specific geology may not be accurately represented by information on such maps.

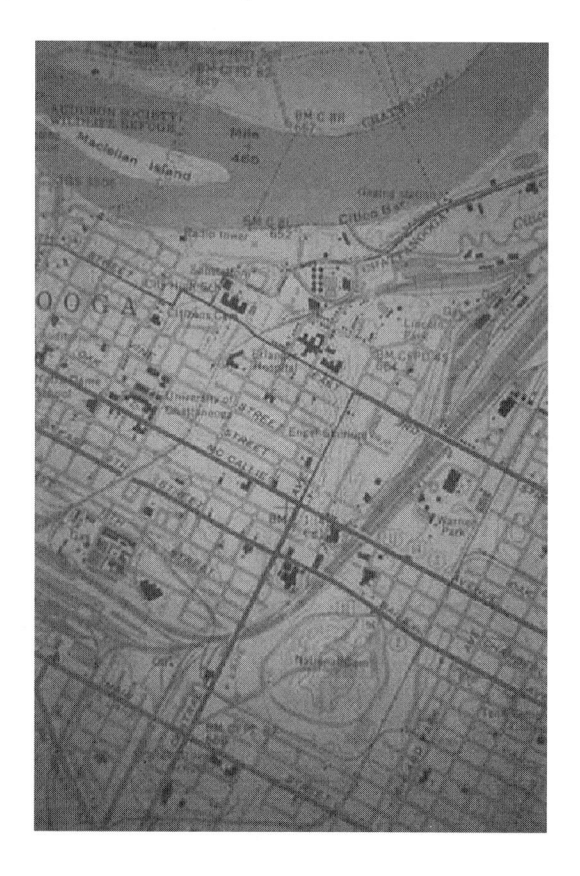

FIGURE 2.42
Current topographic maps are valuable sources of many types of regional information, not just topography. All types of cultural features, many notable geomorphic, geographic and geologic features, and locations of surface-water bodies and areas of ecological significance can be discerned by consulting these maps.

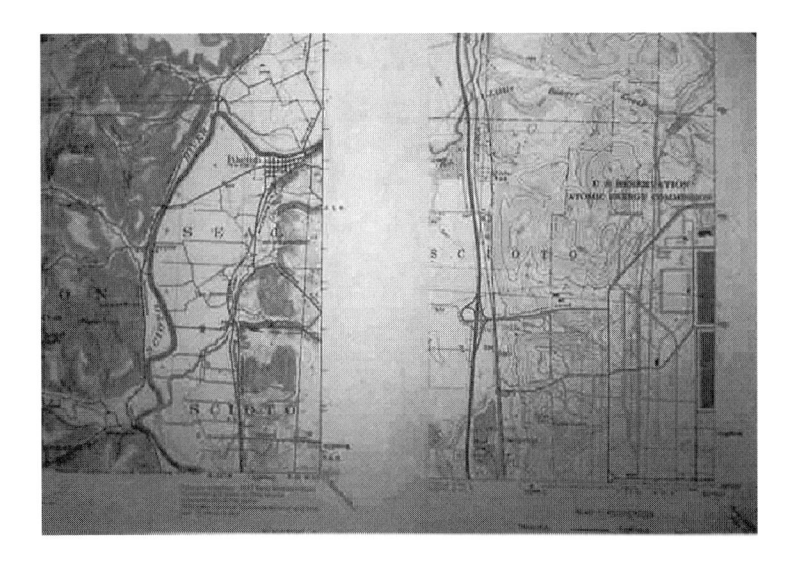

FIGURE 2.43
For areas where both current (right) and historical (left) topographic maps are available, these maps should be compared to detect any changes in cultural and natural features. These changes, such as former drainageways filled in during site construction, are often of significance in an environmental investigation. Note that these maps are at different scales, complicating the interpreter's job. The areas on these maps are depicted in the aerial photographs in Figure 2.62 and Figure 2.63.

- Soil surveys (Figure 2.44), to establish shallow soil conditions (generally to a depth of 5 or 6 ft)
- Boring and well construction logs (Figure 2.45), well completion reports or permits issued for wells (domestic, municipal, agricultural, irrigation, industrial, geothermal, monitoring, or others) either on the site or on adjacent properties, to establish geological and hydrogeological conditions, to identify possible receptors, and to identify possible influences on local ground-water flow

FIGURE 2.44
Soil surveys, available from the U.S. Department of Agriculture's Soil Conservation Service local offices (or state equivalent offices), are excellent sources of information on shallow subsurface conditions (to a depth of 6 ft).

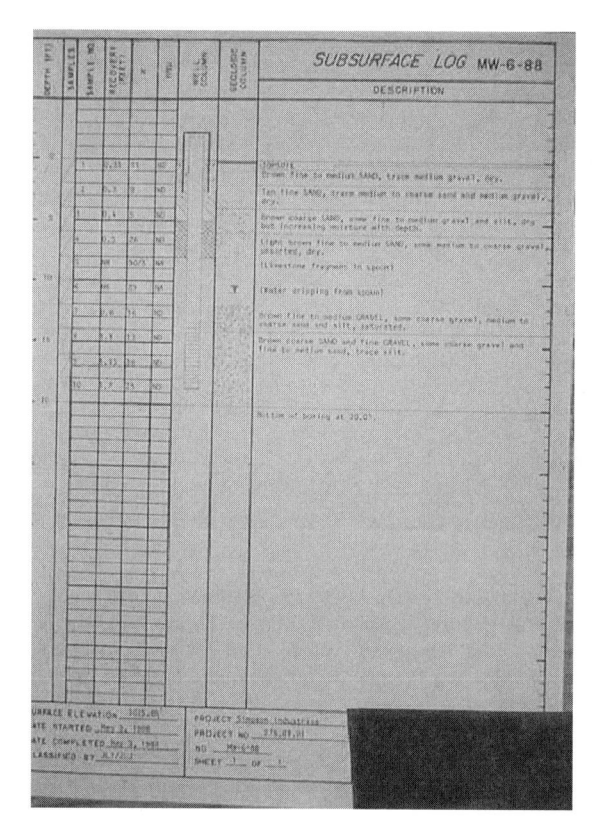

FIGURE 2.45
Soil boring and well-construction logs can help establish geologic conditions at the site and can provide valuable information on ground-water conditions.

- Site plan maps (Figure 2.46) and engineering drawings (blueprints or as-builts) from current and former site owners or operators, for on-site waste management units, product storage and transmission facilities, subsurface utilities, and product and waste monitoring units
- Historical and current land-based photographs of the site, to establish past site operational practices and to locate important site features that may no longer be evident (old drainage ditches, old waste disposal operations, etc.)
- Historical and current property ownership and land-use records, for the site and adjacent properties, to identify other possible sources of contamination and contacts for possible access agreements for off-site investigation
- For product storage facilities, records of any tank tightness or leak detection tests, inventory control records, records of pump or dispenser malfunctions, tank or piping monitoring records, and maintenance and repair records
- Records of products, chemicals or wastes manufactured, generated, stored, handled, or disposed on site, and methods and facilities used for storage, handling, and disposal, to establish COCs and potential source areas and to select appropriate analytical methods
- Reports of any spills, leaks, overflows, discharges, or releases at the site or on adjacent sites to establish potential source areas

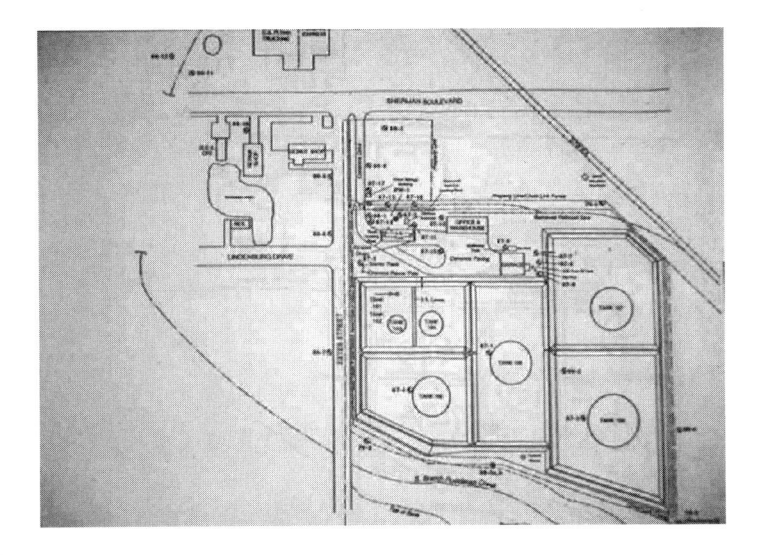

FIGURE 2.46
Site plan maps can orient investigators to the site and assist them in locating potential contaminant source areas, utility corridors, and other features important to interpret other information collected during the investigation.

- Records of any regulatory agency administrative actions (consent orders, etc.) taken against the site or sites on adjacent properties
- Local fire department records for permits issued for fuel or flammable chemical storage, which provide information on type of product or chemical stored; type volume and location of storage units; and results of periodic inspections
- Property insurance records that may identify the types of products, chemicals, or materials manufactured, stored, or used at the site and the means by which they are stored and handled
- Fire insurance maps (i.e., Sanborn maps, available from Sanborn Mapping and Geographic Information Service, Pelham, NY, USA) that were prepared for industrial and manufacturing facilities from the mid-1800s to the mid-1900s, to identify past property uses
- Utility company and municipal utility service records, to establish locations of all on-site underground and above-ground utilities (Figure 2.47), including sanitary and storm sewers, water lines, gas lines, pipelines, power lines, cable TV lines, telephone lines, septic systems, dry wells, and other conduits that may serve as either potential obstacles to subsurface investigations or potential man-made preferential flow pathways (Figure 2.48)
- Local and regional land-use maps (Figure 2.49), to establish surrounding land uses, including the presence of schools, hospitals, wetlands, and other potentially sensitive receptors within 0.25 miles of the site
- Climatic data, including data on precipitation events and patterns, evapotranspiration rates, prevailing wind direction, and temperature, which can be used to estimate infiltration rates and rates and periods of ground-water recharge
- Satellite imagery and aerial photographs (Figure 2.50) of the site and the surrounding area, to identify historical and current structures and engineered facilities at the site and on adjacent properties, to identify areas of vegetative stress, to

FIGURE 2.47
Locating underground utilities is an important step preceding the field investigation to help investigators avoid the liabilities and hazards created by drilling through these features. In most areas, one-call services are available to locate all underground utilities on public property. On-site utilities must generally be located by other means, such as a review of site owner and operator files, site plan maps, or engineering as-builts for the facility.

FIGURE 2.48
Underground utility corridors, such as this gas line trench excavated in native clay soil and backfilled with pea gravel, can serve as preferential flow pathways for contaminants and ground water. In this case, petroleum products leaked from an upgradient UST, entered the trench, and moved several hundred yards, whereas in the native clay soil, no movement was apparent.

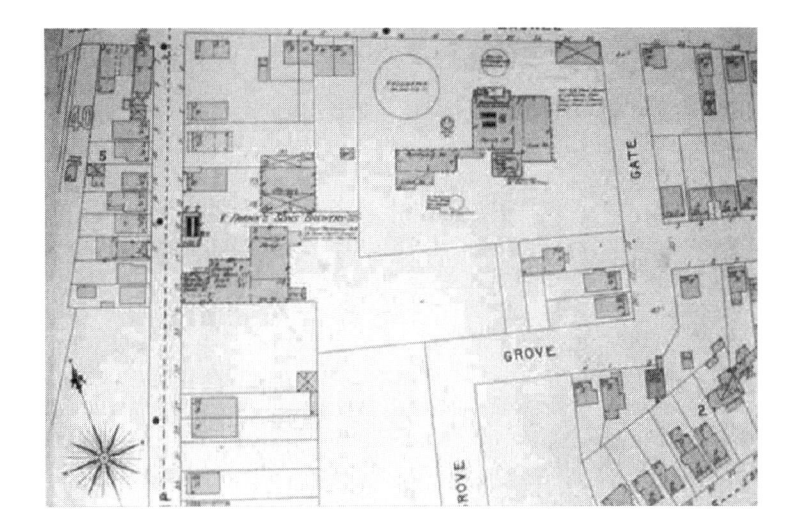

FIGURE 2.49
Local land-use and property ownership maps can alert investigators to uses of neighboring properties that may have a bearing on the investigation.

determine changes in site conditions and land use over time, and to identify lineaments that may indicate the presence of faults and major fracture or joint systems

- Production, shipping, receipt, inventory, and billing records to identify the types of products, chemicals, or materials produced, used, or handled on site.

FIGURE 2.50
Among the most valuable and easily accessible pieces of existing information are simple black and white aerial photos, which can yield a wealth of important information on surface and near-surface conditions and how these conditions have changed over time.

Discrepancies between received and inventoried and used and sold products, chemicals, or materials may indicate potential losses

- Records documenting local influences on ground-water flow, including pumping (or injection) records for on-site and off-site water supply or irrigation wells; foundation, mine, gravel pit, or quarry dewatering operations; and stream discharge records and river stage variations
- Material Safety Data Sheets (MSDSs) for all products, chemicals, and materials handled on site

It is very useful to collect existing information not just on the site under investigation, but also on surrounding properties, because they frequently affect site conditions. This is important because other means of investigating neighboring properties (on-site inspection, sampling, etc.) are often not available without obtaining access agreements, which may be a point of contention (Figure 2.51).

Sources of Information

Sources of existing site information that should be contacted to provide relevant material are many and varied and include:

- Federal government agencies, including the U.S. Department of the Interior (Geological Survey, Bureau of Land Management, Bureau of Reclamation), the U.S. Environmental Protection Agency, the U.S. Department of Agriculture (Soil/Natural Resources Conservation Service, Agricultural Stabilization and Conservation Service, Forest Service), the U.S. Fish and Wildlife Service, the U.S. Department of Commerce (National Oceanographic and Atmospheric Administration, National Weather Service), and the U.S. Army Corps of Engineers
- State government agencies, including environmental regulatory agencies, natural resource management agencies, geological surveys, water resource and water quality management agencies, health departments, and transportation departments

FIGURE 2.51
During the information-gathering stage, it is important to collect as much publicly available information on adjoining sites as possible, as access to these sites is often a point of contention, as is the case here.

- Regional, county, and local and municipal government agencies, including planning agencies, water districts, health departments, utility departments (sewer, water, electric, gas), public works and sanitation departments (landfills and roads), departments of weights and measures, engineering departments, title agencies, property tax assessors, building inspectors, fire marshals, and clerks offices
- Local Universities, including departments of geology, engineering, biology, hydrology, geography, and the university library
- Local environmental and engineering consulting firms, aerial survey firms, and drilling contractors
- Commercial information and environmental database search firms
- The site owner and operator

Methods of Collecting Existing Information

Methods of collecting existing information are straightforward (although sometimes time consuming) and can take various forms. Among these are traditional literature searches; computer database searches; review of available aerial photographs; review of regulatory and nonregulatory government agency files; searches of records of the site owner and operator; review of other local or regional information; and interviews with past and present employees and neighbors. Brief descriptions of these are discussed subsequently.

Traditional Literature Searches

One of the most frequently overlooked but most useful tools in conducting site investigation studies is the use of existing literature as a basis for constructing the initial CSM for the site. A traditional literature review involves obtaining documents through traditional channels (i.e., local experts, universities, state and Federal government environmental agency offices, and branch offices of the U.S. Geological Survey [USGS]) or through computerized searches such as those conducted over the Internet.

The easiest and best way to begin a literature search is by locating the most comprehensive and recent references that pertain to the subject being addressed. The lists of other references contained in these references usually serve as a springboard for further investigation of the literature. The challenge is to find such references without a lengthy search. The least time-consuming method to discover good and timely references on a subject is to contact an expert in that particular field who has published on the subject, who is very likely to be familiar with the recent literature, or even with the site itself. A good Internet search should also yield valuable information, but it is often difficult to sort through the volume of material that most Internet searches uncover to pick the references that are most relevant. Local experts are normally the best and most direct sources of information.

The next level of effort involves the use of a few select sources of information on the subject matter. In the area of geology and ground water, the USGS Index, state geological and water surveys, state departments of natural resources, the U.S. Soil Conservation Service (SCS), and selected water resources abstracts are all very useful sources of information on soil, geologic, and ground-water conditions in a particular geographic area. The USGS has an index of publications that is available over the Internet (www.usgs.gov) and in most libraries. Additionally, the USGS provides periodic updates on more recent publications. Many state geological surveys have similar indices, most of which are also updated periodically. For example, the New York and Illinois State Geological Surveys have indices that are updated on an annual basis.

When the subject matter being researched is not purely related to soil, ground water, or geology, other sources of information can be tapped. For example, one may be researching characteristics of a particular contaminant and its behavior in the subsurface environment. Pollution Abstracts and Environmental Abstracts Indices are good sources of information about the related and recent literature on this subject. With the Pollution Abstracts Index, the subject of interest is indexed alphabetically with references to abstracts contained in larger annually updated volumes. For example, if researchers were searching for information on chlorobenzene, they would look up chlorobenzene in the index and find references to abstract numbers. They would then refer to the particular abstracts to evaluate whether those articles pertain to the subject of the study. Environment Abstracts Index has an advantage in that the index allows the researcher to see the article, title, and subject without referring to the full abstract. Additionally, Pollution Abstracts Index contains mainstream sources of literature that are pragmatically oriented. In contrast, the Environment Abstracts Index tends to contain more esoteric information and contains more pure research than Pollution Abstracts.

Potentially very useful sources of information on geology and ground water are the unpublished university masters theses or doctoral dissertations. Most libraries have an index called "Dissertation Abstracts," which is updated periodically and lists the theses and dissertations completed at accredited universities across the country. Often if one is researching in a library in the same geographic area being investigated, pertinent theses are likely to be in either that library or one in close proximity. If a thesis is located at a university that is far away, it often can be photocopied and sent to the researcher, provided by University Microfilms International (Ann Arbor, MI, USA), or accessed over the Internet.

Sometimes the challenge is not in finding the literature, but finding a library that is open to use by persons not affiliated with a particular institution. Often universities in the same geographic area will allow persons not affiliated with the university to use library privileges either gratis or for a nominal fee. When such arrangements cannot be worked out, then a local city or county library can be used, and the information, if not available at the municipal library, can be obtained through interlibrary loan. Most library catalogs can also be accessed over the Internet, although copies of the actual documents may or may not be available via this method.

For information on practically oriented government research, the National Technical Information Service (NTIS) is a valuable source of information. NTIS has become the reproduction service for all U.S. EPA and many other government agency documents. In the past, the U.S. EPA published many documents that were provided free to the public. U.S. EPA's printing is now done only in limited numbers, but most of these documents are available either as reproductions from NTIS or on the U.S. EPA website (www.epa.gov), although locating specific documents without knowing the issuing office within the agency is often a challenging task. NTIS also carries many other publications from both private noninstitutional sources and other governmental and academic sources. An index entitled "Government Reports, Announcements, and Index" is available from NTIS on their web site (www.ntis.gov).

Several other good sources of information are available. Among them are Georef Index, Chemical Abstracts Index, and Index to Priority Pollutants. The Georef Index is a very comprehensive and long-running source of geological information. Georef has both a thesaurus to aid in finding the right indexing word and a guide to indexing. Another source of information is Chemical Abstracts Index, which is updated monthly and recompiled on an annual basis. The monthly index is much more timely, but the annual index takes less work to research a particular topic area. Finally, the Index to Priority Pollutants is a source of literature on particular contaminants.

Computer Literature Searches

The principal advantage of a computer literature search is that if done properly, it can scan the appropriate literature and a large number of different databases very quickly. The main disadvantage is that the researcher usually loses some control over the search, because there is usually at least one middleman involved in the process. Ironically, even though the computer literature search technique uses computer technology, the success or failure of such a search depends on the human element. The ability of the researcher to communicate with the research librarian is of paramount importance. Normally, most computer literature searches are conducted at a library by a research librarian, although it is possible for a researcher to tie into various databases and indexes by modem from a computer terminal located at the researcher's office. There are advantages to using a library system or an established computer literature search system, because such systems usually tie into a large number of databases and may tie into a computer search middleman operator whose business is to broker large numbers of source indices. An example of the large number of computerized databases available is shown in Table 2.3.

When a research librarian conducts a computer literature search on behalf of the researcher, it is of critical importance that the librarian understands the topic of interest to the researcher. Very often the research librarian conducts an interview with the researcher to obtain information about the topic of interest. On the basis of this interview, the research librarian will choose several key words that can be used in combination to scan the indices. The computer literature indices are indexed by key words. When the information is entered by data entry personnel, they are responsible for selecting the most important key words that pertain to a particular article. As a result, the researcher

TABLE 2.3

Representative Sampling of Databases for Computer Literature Searches

DIALOG BRS ORBIT	DIALOG, BRS, and ORBIT are large commercial systems containing hundreds of computerized databases dealing with a broad scope of disciplines including technical and chemical literature and state and federal regulations
CELDS	Computer-Aided Environmental Legislative Data System is a collection of abstracted federal and state environmental regulations and standards. CELDS provides quick access to current controls on activities that may affect the environment, as well as data for environmental impact analysis and environmental quality management
DTIC	Defense Technical Information Center system is the resource for information on Department of Defense Research Development, Test, and Evaluation activities. It provides data on all stages of Defense Research and Development planned work, work in progress, and work completed or terminated
NLM	The National Library of Medicine system contains a number of computerized databases containing toxicological and chemical information
HAZARDLINE	HAZARDLINE is a comprehensive databank providing information on over 500 hazardous workplace substances, as defined by OSHA. Also included are OSHA regulations, NIOSH criteria documents, and information necessary for protection of the worker and employer
CIS	The Chemical Information System is an integrated online system covering a wide variety of subjects related to chemistry
LEXIS/NEXIS	U.S. federal and state case law, U.S. federal statutes and regulations, tax information, daily news to annual reports, etc

depends on two levels beyond their own interpretation of a particular article. The person who enters the information into the index interprets the publication and enters key words accordingly, and the research librarian interprets the needs of the researcher and enters key words accordingly. Then, the computer matches the research librarian's key words with the key words found in the various indices.

One of the most significant conflicts confronting the researcher in conducting any computer literature search is the need to assess whether it is important for a particular search to be more comprehensive or more focused. A comprehensive search will include a wide range of articles, some of which are not pertinent to the subject addressed by the search. The focused or "relevant" search may exclude some articles that may be somewhat tangential to the subject, but also some that may be of interest. For example, if a researcher were investigating the literature related to the biodegradation of chlorobenzene in a saturated flow system, the comprehensive literature review might include only one key word "chlorobenzene" and the result might be any article that related to the characteristics of chlorobenzene. A very focused search might use the key words chlorobenzene, biodegradation, ground water, and southeastern USA. Unfortunately, when many such key words are used with a subject, the result may be that no relevant research is found. A compromise might be the use of specific subsets of keywords. For example, one might try to scan using first, chlorobenzene and ground water, and then chlorobenzene and biodegradation in combination.

For someone to do their own computer literature searches, it is necessary for that individual to connect a computer terminal via modem to either an intermediate database company or the database. A company called Dialog (a subsidiary of Lockheed) is a large intermediate computer database source. The National Ground Water Association (NGWA) also maintains a database and computer literature search system that can be accessed either by a research librarian at NGWA or directly by modem from a computer terminal. Any database system will allow the individual to set up an account number, usually with a modest annual charge and with a time charge for actual computer connect time.

For persons desiring to learn more about the art and science of computer literature searches, there are at least two good periodicals published on the subject. One is entitled "Data Base" and the other one is entitled "On Line." "Data Base" tends to dwell more on the usefulness of various databases, while "On Line" tends to focus more on the techniques used to conduct successful literature or data searches. For example, "On Line" may review searching techniques or discuss the difference between the Environmental Index and Pollution Abstracts.

Review of Aerial Photography and Imagery

Several types of aerial photographs and imagery are useful for identifying surface and near-surface features in environmental investigations. Black and white panchromatic photos (Figure 2.52), color photos (Figure 2.53), color infrared photos (Figure 2.54), and various types of satellite and low-level imagery (ERTS imagery, LANDSAT imagery, and side-looking airborne radar [SLAR] imagery, to name a few) (Figure 2.55) can be used for a variety of purposes. The major uses include:

- Identifying soil and geological material types
- Identifying geomorphological features (floodplains, stream cuts, bluffs, fault scarps, and other features) and geological structures
- Identifying the presence and extent of joints and fractures in surface exposures of bedrock

Environmental Site Characterization 101

FIGURE 2.52
Black and white panchromatic photos are the most widely available form of aerial photography. As this photo shows, many common surface features, such as the above-ground storage tanks (small white and dark circular features; note the berms around the tanks), rail lines (dark curvilinear features), and surface water (dark area in lower left), are easily identifiable.

- Identifying surface-water bodies, springs and seeps, and sensitive ecological areas (e.g., wetlands)
- Identifying cultural features (roads, buildings, pipelines, power transmission lines, railroad tracks, canals, etc.)
- Identifying changes in land use over time
- Identifying possible source areas for contaminants of concern (landfills, lagoons and surface impoundments, above-ground storage tanks, waste burial pits, dumps, etc.) and their sizes

FIGURE 2.53
This color aerial photo (reproduced here in black and white) also shows an above-ground storage tank farm adjacent to a surface-water body (the Delaware River) (bottom) and a parcel of undeveloped property (right) with a tributary stream flowing toward the river.

FIGURE 2.54
Color infrared photography (reproduced here in black and white) would show healthy vegetation as bright red. The turf in the stadium (lower right) is black, indicating that it is artificial turf. Distressed vegetation would be depicted in brown.

- Identifying past waste management or chemical handling practices
- Identifying obstacles to site characterization field work
- Identifying drainage and other surface topography alterations caused by site development or construction
- Identifying possible preferential ground-water flow pathways
- Identifying the location of shallow soil or ground-water contamination (through vegetative stress)
- Monitoring and assessing the progress of site cleanup activities

FIGURE 2.55
Satellite imagery of the Salton Sea in southern California (upper left), depicting the intensively irrigated agricultural development south of the Sea (to the right) and desert transitioning to mountainous terrain on either side.

Satellite and other imagery is covered in Chapter 4; the uses and sources of black and white, color, and color infrared photography are described below.

Black and white aerial photo coverage is available for about 90% of the United States (ASTM, 2004c) and is relatively inexpensive. The historical photographic record for most of the USA dates back to the mid-1930s. Simple black and white aerial photographs are very useful, particularly if a historical series of these photographs is available (Figure 2.56 to Figure 2.63). These photos are even more useful if they are available in stereoscopic pairs (along a flight line or on adjacent overlapping flight lines) (Figure 2.64), to allow identification of surface features in three dimensions. In almost every case, such photos are available, but finding all of the desired photos will normally require the investigator to access multiple sources. Among the multiple sources, the scales, flight altitudes, area of coverage, resolution, clarity, and other key photographic features are highly variable, so the challenge is to find a way to minimize the differences to make interpretation easier. Ordering photos that are either reduced or enlarged (depending on the scales of the original photos) to make the scales compatible is the best way to produce useful historical sequences.

FIGURE 2.56
A historical sequence of aerial photographs available for a site near Toledo, Ohio, dating from June 16, 1940 (top) to May 3, 1986. Such a historical sequence is very valuable in detecting changes in surface and near-surface conditions (particularly land use) over time.

FIGURE 2.57
This black and white aerial photo, along with those in Figure 2.58 to Figure 2.61, comprise a historical sequence for a site in the center of this photo. This photo, from July 1937, shows a major river at the top (north) (with a barge dock visible in the upper center) and a flood wall protecting the city (angular feature following the bank of the river). The site is a one-block by one-block parcel with a large building in the northwest corner, a large and tall above-ground storage tank in the north center, two small buildings in the northeast corner, a large, low circular above-ground structure on the east side with a small above-ground storage tank and a small building next to it, two more small above-ground tanks in the southeast corner, and a building in the south center. On the basis of land ownership records and interviews with former employees of the former site owner and contractors who did later construction of the site, it was determined that the site was a storage area for waste liquid from a coal gasification plant. The large, low circular structure was a covered dipping pit for wooden utility poles, containing thousands of gallons of creosote. The parcels to the east of the site are predominantly occupied by warehouses and light manufacturing; those to the south are predominantly residential.

FIGURE 2.58
This photo of the same site (center), taken in June 1946, shows the building in the northwest corner of the property and also shows that all but one of the above-ground storage tanks have been removed (the largest one leaving a dark round footprint), and the large circular structure is now visible as an uncovered pit with a dark bottom.

Environmental Site Characterization

FIGURE 2.59
This photo, taken in July 1954, shows that the pit has been filled in and is no longer visible (except for a faint circular outline) and that a long, thin building has been constructed to the east of the large building. Note that a baseball diamond has been built on the parcel just west of the site, within the floodwall.

Often these photographs will show the progression of site activity over time, such as a landfill filling individual trenches, construction eliminating some preexisting natural feature, structures or waste-management units being built at a site, or the location and operating periods of "burn pit" activities that are no longer apparent based on a site reconnaissance. Aerial photos are also excellent tools for determining historic site conditions on surrounding properties and may, in fact, be the best means of uncovering critical information on past practices on adjacent sites. When combined with historical topographic maps, aerial photos can be very useful in defining possible shallow

FIGURE 2.60
This photo, taken in August 1972 (at a different scale), shows that the large building on the site has been replaced with a larger building (now a large rectangular structure with a white roof), and all signs of the former storage tanks and dipping pit are gone. Several buildings on the parcel to the east have been removed and replaced with what appears to be an auto salvage yard.

FIGURE 2.61
The final photo in this historical sequence (taken in July 1986) shows an addition to the large structure at the site that covers the areas where the large above-ground tank and dipping pit were once located. The auto salvage yard on the site to the east is now gone (as are all but two buildings), and the site to the south has been developed. By examining this series of photos, it can be determined that the probability of residual contamination from the former above-ground storage tanks and creosote dipping pit, which would now be buried beneath the latest addition to the building, is high.

FIGURE 2.62
This photo, taken on May 13, 1951, shows predominantly agricultural land with a few woodlots in south-central Ohio. The river on the left is the Scioto River. Note the tributary drainageways, particularly those in the lower right of the photo.

Environmental Site Characterization   107

FIGURE 2.63
This photo of the area pictured in the lower right corner of the photo in Figure 2.62, from June 24, 1983, shows the U.S. DOE Portsmouth Gaseous Diffusion Plant that was constructed in the mid-1950s. The area within the ring road, some 2000 acres, was leveled during construction, and all former natural drainageways were filled in with rip rap and coarse fill. These features were later found to serve as man-made ground-water and contaminant migration pathways.

ground-water flow pathways caused by infilling of former drainageways with coarse fill material. Additionally, aerial photographs are often used for fracture trace analysis (described subsequently) to discern ground-water flow pathways in terrain where the flow is dominated by fracture flow or solution channels.

Both color and color infrared photography are useful for documenting historical developments, but because the history of color and color infrared aerial photography is relatively short, there are fewer such photos available and a historical series may be

FIGURE 2.64
A stereo pair of aerial photos, in which overlapping photos along the same flight line allow viewing features in three dimensions with a stereoscope, can be very useful in identifying features with relief and even estimating differences in elevation and heights of structures or natural features.

difficult to locate. Color infrared photography shows vegetation and changes in water quality very well, so it can be an excellent indicator of the vegetative stress that may occur when plant or tree roots encounter contaminated soil or ground water and a good indicator of surface-water quality degradation over time. Color infrared photos are also useful in locating point sources of contamination entering a surface water body, such as outfalls from power plants or industries, NPDES-permitted (or unpermitted) discharge points, or leachate seeps.

Aerial photos of all types are also very useful tools for assessing current site conditions. Even if current photos are not available from the normal government sources (listed below), they can be supplied on a contract basis by commercial services (located in all major metropolitan areas and even some smaller towns), often within 24 h of a request (provided weather conditions are favorable), to provide near-real-time data.

When compared at equivalent scales, aerial photos from historical overflights can often provide initial information to answer the following questions about sites where waste disposal practices are of interest:

- *What was the appearance of the site before it was developed or prior to deposition of waste?* Such information is critical to assess predevelopment drainage, topographic changes, and natural soils and geologic data.
- *What were the modes and times of waste deposition?* Initial information on the site size and volume can be provided so that the scope of the site investigation can be conceptualized before field work begins.

When evaluated by the site investigation team, aerial photos can often provide a great deal of qualitative information to answer many other site-specific questions and can help direct the investigation to examine areas of interest that appear on the photos. Aerial photos can also be used in combination with ground-based information on subsurface features to discern other features of interest. For example, aerial photographs can be used to identify possible source areas and, when used in conjunction with local geologic and hydrogeologic data, can help locate potential contaminant movement and exposure pathways.

Aerial photography is available through a number of different sources including the following:

- The National Archives and Records Service in Washington, DC
- The National Cartographic Information Center (NCIC) in Washington, DC (NCIC also maintains an affiliated office in each state)
- The USGS (including the Regional Mapping Centers in Reston, VA, Rolla, MO, Stennis Space Center, MS, Denver, CO, Menlo Park, CA, and Anchorage, AK, and the EROS Data Center in Sioux Falls, SD [the main source of LANDSAT and ERTS imagery])
- The U.S. Department of Agriculture (including the Agricultural Stabilization and Conservation Service's Aerial Photography Field Office in Salt Lake City, UT, the Soil Conservation Service's Aerial Photography Field Office in Dallas, TX, and the SCS offices in each state capitol)
- The U.S. Bureau of Reclamation office in Denver, CO
- The U.S. Bureau of Land Management office in Denver, CO
- The U.S. Forest Service offices in Washington, DC and Denver, CO
- The U.S. Army Corps of Engineers District Offices across the U.S. (a main source of SLAR imagery)

- The Tennessee Valley Authority office in Chattanooga, TN
- The U.S. Environmental Protection Agency's Remote Sensing Laboratory in Las Vegas, NV
- The Defense Intelligence Agency in Washington, DC
- The National Ocean Survey offices in Rockville, MD and Detroit, MI (Great Lakes Division)
- Commercial aerial-photo services (such as National Aerial Resources of Troy, NY, and local aerial photo specialists)
- State planning and natural resource management agencies
- State geological surveys, water resource agencies, and departments of transportation

Additional information on the use of aerial photography and satellite imagery in environmental investigations is available in the ASTM Standard D 5518, Standard Guide for Acquisition of File Aerial Photography and Imagery for Establishing Historic Site Use and Surficial Conditions (ASTM, 2004c), Zellmer (1995), and Finkbeiner and O'Toole (1985). Several excellent references detail the procedures used in interpretation of aerial photographs and satellite imagery for a variety of environmental purposes. These include Avery (1968), Ray (1972), Lillesand and Kiefer (1972), Deutsch et al. (1981), and ASTM (1988).

Fracture Trace Analysis

Black and white aerial photos and imagery can also be very useful in detecting the presence and location of fractures in surface and near-surface bedrock through a technique known as fracture trace analysis. Fracture traces are surface expressions of joint and fracture patterns or faults that can be located by evaluating linear features on aerial photos or satellite imagery (Figure 2.65). On aerial photos, natural linear features appear as:

- Tonal variations in surface soils (caused by differences in soil moisture and organic matter content)
- Alignment of vegetative patterns (caused by differences in availability of water)
- Straight stream segments or valleys (caused by alignment of streams along weaker zones in rock)
- Lines of springs or seeps (caused by movement of ground water along fracture zones)
- Alignments of surface depressions or sinkholes in Karst or other carbonate rock terrain (caused by weaknesses or chemical differences in the rock)
- Other features showing a linear orientation (such as swales, gullies, or sags formed due to soil settling into fractures or fault zones)

Many of the linear features detected on aerial photos or imagery are surface expressions of fractures in bedrock that may be more than 100 ft deep.

Fracture trace analysis is important at sites where bedrock (and the ground water included in it) is close enough to the surface to have been impacted by site operations, because fracture zones are often preferential ground-water flow pathways and, therefore, can also serve as pathways for contaminant transport. These fracture zones are also the

FIGURE 2.65
This photo, taken on March 11, 1970, shows an area of predominantly agricultural land in glaciated terrain underlain at a depth of about 40 ft by sedimentary bedrock (a sequence of shales and limestones) that is fractured. Fracture traces, marked in the upper right-hand corner of the photo, trend northwest and southeast (major axis of fracturing) and southwest and northeast (minor axis of fracturing). Most of the surface expressions of fracturing are soil tonal differences (dark linear features) or shallow drainage alignments (dark semi-linear features).

best locations for wells installed to characterize or monitor ground water at sites where bedrock is shallow.

Large-scale linear features identified on satellite or low-altitude imagery or high-altitude photography are called lineaments. These features, which represent zones of greater fracturing, may be on the order of tens to hundreds of miles long. They can provide the basis for more closely examining small-scale features on low-altitude black and white photos, especially with respect to fracture orientation (direction). Because of the way in which rock responds to stresses applied to it (i.e., resulting from tectonic activity), distinct fracture patterns develop in the rock. Normally, there will be one major axis of fracturing (usually nearly vertical) and one minor axis of fracturing that is approximately perpendicular to the major axis. This type of pattern is not always evident on photography because the strike and dip of the rock layers and of the fractures themselves affect the angle exposed at the bedrock surface and, therefore, the apparent orientation of the fracture traces on the photos. For this reason, all lineaments and fracture traces identified on aerial photos or imagery should be field-checked and confirmed during site reconnaissance. Investigators should examine available rock outcrops (Figure 2.66 and Figure 2.67); stream alignments; vegetation alignments (discounting those along property or fence lines); lines of springs and seeps (Figure 2.68); swales, gullies, and sags (Figure 2.69); and other linear features, eliminating all obvious nongeologic linear features (i.e., fence lines, pipelines, power lines, rail lines, jeep trails, and other anthropogenic features).

Fracture trace analysis, and the significance of fracture traces in an environmental context, is described in more detail in Ross and Frohlich (1993), Sweet and Mitchell (1990), U.S. Department of the Interior (1982), Lattman (1958), Lattman and Matzke (1961), Lattman and Parizek (1964), Parizek (1976), Setzer (1966), Wobber (1967), and Smith et al. (1982).

Environmental Site Characterization 111

FIGURE 2.66
To confirm that the fracture traces noted on aerial photographs are natural features, it is important to field-check and map those features and correlate them with the fracture traces. Fracture zones such as the one depicted in this road cut should be noted and mapped, with the strike and dip of the fractures measured with a Brunton compass.

Review of State and Local Regulatory Agency Files

In some instances, state and local government agencies may have gathered information on past uses of the site under investigation and surrounding properties, which may have included the disposal of solid or hazardous wastes or the storage of regulated products. For example, many states have kept files on landfills for at least the period of time during which the state environmental agency has been in existence. These files may be of use for indicating whether industrial or hazardous wastes have historically

FIGURE 2.67
In this photo, fractures that are serving as shallow preferential ground-water movement pathways are evident as the water comes to the surface of the road cut and freezes. Note the fairly regular pattern of fracturing, which is the rule rather than the exception in most types of rock (in this case, metamorphic rock in western Massachusetts).

FIGURE 2.68
Lines of springs and seeps such as this are a good indicator of the presence of fractures that should be noted when field-checking fracture traces.

been disposed at the site or in local municipal landfills. Additionally, with the implementation of state programs to regulate above-ground storage tanks and USTs and drycleaning solvent sites, agency files often contain records of sites that store, manage, or use petroleum products and solvents. In addition to these records, public

FIGURE 2.69
This swale, although unremarkable by itself, is a good indicator of the presence of fractures in bedrock buried beneath overburden. The absence of surface drainage in this feature indicates that subsurface drainage is probably occurring, another indication of the presence of fracturing.

agencies also maintain well records (such as detailed well logs) and ambient groundwater quality information. In most cases, all of these records are in the public domain, and therefore, accessible through a simple request for the information or through Freedom of Information Act requests. However, in some cases, the records are confidential and the agency may require the written permission of the site owner or well owner before the agency can release the files to the requesting party. Many states maintain computer databases that are regularly updated and easily accessed through the Internet search, while other states keep only hard copies requiring more time-consuming manual searches through archival files. Use of government agency records in environmental investigations is covered in more detail in Miller (1992) and Mauch (1991).

Review of Site Owner and Operator Files

The site owner or operator is often the only source of information that is critical to an environmental site investigation, including the results of any previous investigations conducted at the site, and site plan maps and engineering drawings of waste management units, product or material storage facilities, and underground utilities. Other valuable information that only the site owner and operator can provide includes records of products, chemicals, or wastes that are manufactured, generated, stored, handled, or disposed on site; reports of any spills, leaks, releases, or discharges from site facilities; and shipping, receipt, inventory, and billing records, purchase requisitions, hazardous waste manifests, and other communications that relate to deliveries or sales of materials that may become the focus of the investigation.

For very good reason, site owners and operators are often reluctant to open their files and records to anyone, including those who have been retained to help them. Legal counsel often is concerned that information about past practices that is provided to consultants may be forced from them during litigation. Although this is a difficult issue, it is one that usually can be resolved if the environmental management and legal representatives of a company are made aware of the need for their consultant to construct a complete picture of site conditions. If the issue is one of confidentiality, it usually can be handled by having the site investigators work directly with the corporate attorney under attorney-client privilege.

Review of Other Available Records

State, regional, and local planning agencies, local tax assessors' offices, county clerks' offices, utilities boards, and many other local government agencies and state and local historical societies can provide a wealth of information about past land uses, land owners, water quality, and significant occurrences that may provide clues about subsurface conditions. For example, the county clerk's office maintains records of property ownership through time, which can be uncovered by conducting a title or deed search for a particular parcel at the county courthouse or by contracting a company that specializes in these searches. Water purveyors (i.e., public utilities) in various jurisdictions maintain records of well production, well completion information, and water quality. Local historical society records may contain maps showing that a city-owned park was the site of a coal gasification plant 30 years ago. Coal gasification plants were prolific producers of various types of contamination, much of which was disposed at or adjacent to the plant site. The many sources of the many types of information available to investigators are listed in an earlier section.

Interviews

Interviews of current and former site personnel (particularly site operations and environmental staff) can help uncover otherwise difficult-to-obtain information on the existence, timing, magnitude, and duration of releases or spills; types of products, chemicals, or wastes historically handled at the site and practices used for handling them; former waste management practices used on site; construction details of product or chemical handling or waste management facilities; details on product or chemical transport and delivery or waste removal from the site; site construction or engineering activities that may have altered site conditions; locations of old wells; and other important site-specific information. Such interviews should be conducted in person by experienced interviewers in a nonadversarial manner and location to avoid intimidating the subject. Information from interviews is often inaccurate and contradictory and it should be used with caution, preferably after confirmation from some other source. Interviews with individuals who have lived or worked near the site for an extended period of time may also yield valuable anecdotal information about general operational practices and waste disposal activities that may not be formally documented. Interviews with local and regional experts (i.e., university staff, drilling contractors, consultants, environmental agency and natural resource agency personnel, and construction and excavation contractors) are often a very useful source of excellent information on local soils, geology, hydrogeology, hydrology, climate, and biology. To conserve time and effort, it is often a good idea to schedule any planned interviews to coincide with site reconnaissance activities.

Discussions with the site owner and operator may also yield valuable information. Although there may be occasions when the site owner and operator may be reluctant to divulge details about on-site activities, this information is often critical to developing an accurate understanding of site conditions and to creating an effective approach to site characterization. The past practices employed at the site may be an embarrassment to the site owner and operator, but it is often important information for guiding the site characterization program in the right direction. The best approach to obtaining this information is to inform the owner and operator why the information is needed and how it will be used and ensure them that it will be treated with confidentiality.

Conducting Site Reconnaissance

Site reconnaissance should be conducted as soon after the review of existing information as possible to ensure that the information is fresh and that gaps in the information that could be filled during a site visit can be readily identified. The main purposes of site reconnaissance are to:

- Provide an opportunity to verify the accuracy of information gathered during the review of existing data
- Allow collection of information on site-specific local and regional features not described in existing information
- Verify site location with respect to local features and neighboring sites
- Provide a first-hand inspection of the general conditions present at the site (and on adjacent properties)
- Identify site characteristics needing further investigation
- Familiarize investigators with the site and allow them to observe site operations (i.e., manufacturing operations, product and chemical storage and dispensing practices, waste-disposal practices, and other relevant operations) first-hand

- Note modifications to the site since the last site plans or maps were produced
- Check for obvious signs of contamination on-site and off-site
- Assess the potential for health and safety hazards and the condition of site security
- Identify obstacles to conducting the field investigation, including steep topography (Figure 2.70), thick vegetation, presence of bedrock outcrops at the surface, canopies (Figure 2.71), and other structures
- Locate above-ground utilities and possible locations of below-ground utilities based on the presence and location of utility vaults

To conserve time and effort, it is often a good idea to schedule utility company or one-call utility locating service visits to the site to coincide with the site reconnaissance visit.

The site should be walked over to note the condition of surface soils, vegetation, and surface water bodies. Stained, discolored, disturbed, or malodorous soil (Figure 2.72 and Figure 2.73) may indicate the presence of spill areas. Yellowed foliage, stunted growth, malformation, and dead plants and trees (Figure 2.74) are all signs of vegetative stress, which can also indicate areas of spills or releases. Seeps or discharges of colored, viscous, or malodorous fluid (Figure 2.75 and Figure 2.76) or fluid that creates a sheen on a water surface (Figure 2.77) may indicate the presence of leachate or septage outbreaks or petroleum products. It should be noted if the soil staining, vegetative stress or seeps are historical or ongoing problems. Indications that the site has been used as a dumping area (Figure 2.78 and Figure 2.79) should be studied to determine possible sources of the waste

FIGURE 2.70
One of the objectives of site reconnaissance is to locate obstacles to the field investigation, such as the steep slopes, thick vegetation, and overhead utilities depicted here.

FIGURE 2.71
Other obstacles to the field investigation include canopies, pump islands, underground product distribution lines, and rebar-reinforced concrete depicted here.

(both on-site and off-site) — low areas often attract dirty fill material. Signs of excavation at the site should be studied to determine the reason for the excavation, such as possible burial of wastes. Regular patterns of depressions or bermed areas may indicate the former presence of trenches, drainage ditches, lagoons, surface impoundments, or above-ground storage tanks. When inspecting paved surfaces, investigators should note patched or repaved areas (Figure 2.80) and utility vaults or valve boxes, as they can indicate areas where USTs and associated piping are located.

Features relevant to the objectives of the investigation, including possible contaminant sources, exposure pathways, and potential on-site and off-site receptors, should be observed and described with respect to location, condition, and dimensions. Possible

FIGURE 2.72
Discolored soil sometimes has to be uncovered to remove the weathered surficial material. In this case, the discoloration was due to dumping of textile dye, which turned the soil bright blue (a color not often found in native soils).

Environmental Site Characterization 117

FIGURE 2.73
This petroleum-stained soil is an indication that the product handling practices at this small heating oil distribution center need to be improved. Such indications can direct investigators to focus on specific areas during the field investigation.

sources are many and varied and include sumps, floor drains, septic tanks, leach fields, land treatment areas (Figure 2.81), dry wells, catch basins, lagoons (Figure 2.82), surface impoundments (Figure 2.83), outfalls or discharges (Figure 2.84), drums (Figure 2.85), transformers (Figure 2.86), dumps, landfills (Figure 2.87), waste piles, chemical storage areas (Figure 2.88), underground and above-ground storage tanks and piping (Figure 2.89), and stained soils (Figure 2.90). Possible exposure pathways and

FIGURE 2.74
Stressed or dead vegetation is often an indicator of shallow soil or ground-water contamination. In this case, the shallow plume of wood preservative contamination was very evident, as it was outlined by dead vegetation within the plume and by healthy vegetation outside the plume. (Photo courtesy of David Miller.)

FIGURE 2.75
During site reconnaissance, investigators should be on the look-out for surface indications of subsurface contamination, such as this leachate seep and the impressions of methane bubbles evident in the soft sediment in the foreground.

receptors include public water supply wells (Figure 2.91), private wells, surface-water intakes, buildings with basements, utility corridors, and sensitive ecological areas (Figure 2.92).

Geomorphic features, such as bedrock outcrops (Figure 2.93), stream cuts, flood plains, drainage divides, stream terraces, fault scarps and other natural features, should be noted. Surface topography should be examined and slopes assessed to determine potential access problems for drilling or DP equipment. Surface drainageways (creeks, streams, and rivers), wetlands, springs, reservoirs, ponds, lakes, and other features that may serve as ground-water recharge or discharge areas should be noted (Figure 2.94 and Figure 2.95), especially with respect to their topographic position in relation to potential contaminant sources.

Man-made features on and near the site, such as road cuts (Figure 2.96), open surface mines, sand and gravel operations (Figure 2.97), quarries (Figure 2.98), foundation excavations (Figure 2.99), and other exposures of soil or rock, should be described and their locations noted on a base map to permit correlation with existing information. Other man-made features, such as roads, paved areas, old foundations, rail lines, pipelines, power transmission lines, and structures of all kinds should be described and their locations noted. These may serve as constraints to the use of some of the methods proposed for the field investigation.

All of the information gathered during the site reconnaissance step should be combined with the existing site information to produce an accurate CSM.

FIGURE 2.76
Leachate seeps often take the same preferential flow pathways as ground water, such as this fracture, a fact that investigators should consider when conducting site reconnaissance.

FIGURE 2.77
A sheen of iridescent fluid on the surface of a water body, such as that shown here, can indicate a discharge of petroleum products or other immiscible fluids.

FIGURE 2.78
During the site walkover, investigators should check for both obvious and subtle signs that the site has been used for a dumping area. This low area, which has been partially filled with 5 gal pails of an unknown thick, dark liquid, was one of several areas at this site targeted for further investigation.

FIGURE 2.79
Low areas at this site have been backfilled with what appears at the surface to be innocuous materials (including a 1964 Ford Falcon), but what lurks beneath the surface may be more insidious and should attract the interest of field investigators.

Environmental Site Characterization 121

FIGURE 2.80
Patches of asphalt or concrete at the surface may be indications that an UST, septic tank, or other potential subsurface source of contamination has been removed.

Developing an Initial CSM

On the basis of existing site information and information collected during site reconnaissance, the field manager, in concert with the senior technical staff, develops an initial CSM. The CSM is the primary tool used to predict the degree of heterogeneity and the nature of spatial patterning of data and contaminant migration pathways. As the CSM evolves, it is used to verify whether the initial predictions were accurate and to assess whether the degree of heterogeneity present will affect the performance of statistical sampling plans. When it is complete, the CSM is used to integrate knowledge of heterogeneity, spatial patterning and contamination migration pathways into decisions about exposure

FIGURE 2.81
Possible sources of contaminants should always be the focus of site reconnaissance activities. In the foreground of this photo are several potential sources that appear innocuous — two fields where land treatment activities for petroleum wastes, related to the petroleum refining and storage facility in the background, are being conducted.

FIGURE 2.82
The keen observer will note that the liner in this hazardous waste lagoon is no longer functional and the contents are emptying into the ground-water system through the sandy soils from which the berms of the lagoon are constructed. (Photo courtesy of David Miller.)

pathways, and their associated human health risks, long-term monitoring strategies, and the selection and design of remedial systems (ITRC, 2003).

The CSM begins as simple abstractions in the investigator's mind, developed after examining existing data — it generally focuses on features at the site that exert controls on contaminant distribution and movement. The CSM provides a standard means of summarizing and displaying what is known about the site and identifying what must be known about the site to develop technically sound DQOs. It must be structured to address all of the essential features of the site and to incorporate all of the data elements required to meet project objectives (usually to prepare for monitoring, risk assessment, or remedial action).

FIGURE 2.83
This unlined surface impoundment, which holds a thick, viscous sludge produced by a manufacturing process at this site, was a possible (later confirmed as an actual) source of subsurface contamination at this site.

FIGURE 2.84
Unpermitted discharges to surface-water bodies, such as this one, should be investigated thoroughly to determine downstream impacts to aquatic life, wildlife, and man.

ASTM Standard E 1689 (Standard Guide for Developing Conceptual Site Models for Contaminated Sites [ASTM, 2004d]) lists the six basic activities associated with developing a CSM for a contaminated site (not necessarily listed in the order in which they should be addressed) as follows:

- Identification of potential contaminants in all environmental media
- Identification and characterization of the sources of contaminants
- Delineation of potential migration pathways through environmental media such as ground water, surface water, soils, sediment, biota, and air

FIGURE 2.85
If site reconnaissance uncovers evidence of storage of drums of potentially hazardous materials, the field investigation should be configured to determine the contents of the drums and whether the contents have leaked into soil, surface water, or ground water.

FIGURE 2.86
The presence of transformers should alert investigators to the possible presence of PCBs, which were commonly used in these devices because of their excellent insulating properties.

FIGURE 2.87
If site reconnaissance should uncover the presence of apparent solid-waste landfill areas, the next visit should be to the site owner and operator or the plant operating engineer to determine the contents of the landfill.

FIGURE 2.88
Chemical storage areas, such as this lake full of TCE (see sign), may warrant the attention of field investigators. This photo was, of course, taken on April 1.

FIGURE 2.89
Above-ground storage tanks and the associated transmission pipelines and distribution piping were the focus of this investigation, during which a combination monitoring and recovery well (foreground) was installed.

FIGURE 2.90
This stained soil beneath the distribution piping at a petroleum distribution terminal is an obvious sign of leakage, which attracted the attention of the site reconnaissance team. The field investigation was configured to sample soil and ground water beneath this facility to determine its impact on these media.

- Establishment of background concentrations of contaminants for each contaminated medium
- Identification and characterization of potential environmental receptors (human and ecological)
- Determination of the limits of the study area or system boundaries

The CSM should be an easily understood, basic narrative, and graphic compilation of the field manager's understanding and interpretation of site conditions related to the

FIGURE 2.91
The site reconnaissance team should confirm the existence and location of critical exposure pathways and receptors including public water supply wells such as this one. To determine this susceptibility of wells to contamination, information on background ground-water chemistry, pumping times and rates, and ambient ground-water flow rate and direction must also be considered.

FIGURE 2.92
Identification of nearby or on-site ecological receptors, such as wetlands and wildlife habitat areas, is another key objective of site reconnaissance.

objectives of the investigation. The narrative portion of the CSM should include the following essential elements:

- A brief site summary of the information available for the site and how this information relates to the objectives of the investigation. A brief description of current conditions at the site should be included
- Historical information concerning the site, including anything of a historical nature that may have a bearing on the present environmental condition of the site
- A description of the site physical setting including topography, soils, geology, ground-water and surface-water conditions, and biological features of note
- A description of each of the possible sources of contamination including their nature (type of source, types of contaminants associated with each source, chemical and physical properties of potential contaminants [such as solubility,

FIGURE 2.93
Noting the presence of rock outcrops on site or in adjacent areas can assist investigators in creating an accurate picture of site geology for inclusion in the CSM.

FIGURE 2.94
While they do not exhibit obvious outward signs that they are important areas, upland areas with flat or rolling topography and permeable soils are often critical ground-water recharge zones. A hydrogeologist should be part of the site reconnaissance team to identify these areas and potential threats to ground-water quality within these areas.

volatility, sorption coefficient, density, and viscosity]), condition, location, and dimensions

- Descriptions of each of the identified and potential migration and exposure pathways for each of the affected environmental media
- Descriptions of each of the potential human and ecological receptors and how they interrelate with the exposure pathways

The graphic portion of the CSM should include maps, cross-sections, tables of data, figures, and other representations of site conditions. Generally, the field manager develops a site base map with locations of roads and buildings, accurate locations and configurations of product or chemical storage, handling and disposal facilities (and other potential contaminant sources), to be used for depicting site geology, ground-water

FIGURE 2.95
Even less obvious are most ground-water discharge zones, although in humid climates, the base flow of most streams is provided by ground water. The discharge areas in the base of this stream are apparent as sand boils (white areas) that are displacing detritus (dark material) on the streambed. Again, the keen eye of a hydrogeologist should be able to discern such features.

FIGURE 2.96
Road cuts, such as natural exposures of rock, should be noted and examined to provide key information on regional geology for inclusion in the CSM.

conditions, and contamination contours (if such data are available). Depending on the size of the site, a USGS topographic map may serve as a good base map. The base map serves as the basis for planning the field investigation and will be used for developing the initial sampling and analysis plan. An example of an initial CSM represented on plan-view maps and a cross-section is provided in Figure 2.100. The graphics can be drawn by hand or generated using computer graphics programs before field work begins; the graphics are updated and revised on site as the site characterization program progresses and additional data become available. Maps and cross-sections should include a scale (and degree of exaggeration, if any, in the case of cross-sections) and direction indicator and indicate the locations of possible contaminant sources relative to the property boundaries.

FIGURE 2.97
Man-made excavations, such as this sand and gravel operation, provide excellent opportunities to gain insight into shallow regional and local geology. The character of geologic materials in this gravel pit (adjacent to the site being investigated) confirmed information on regional geologic maps and allowed investigators to get a first-hand view of the depth to ground water.

FIGURE 2.98
Quarries offer the same opportunities for observing the character of subsurface materials in bedrock terrains as sand and gravel operations do for unconsolidated materials.

On the maps, cross-sections, and other depictions of site conditions that comprise the CSM, the following information should be included:

- Topography and location of surface drainage routes and water bodies
- Known or anticipated soil and geologic conditions including the nature, degree of heterogeneity, locations, and depths of distinct subsurface geologic units
- Known or anticipated hydrogeologic conditions including ground-water depth, flow direction and velocity, possible interconnections between aquifers, and possible interactions with surface-water bodies
- Locations of man-caused alterations to geologic and hydrogeologic conditions (i.e., utility corridors and pumping wells)
- Known or suspected contaminant source areas

FIGURE 2.99
This foundation excavation, at a property adjacent to the one being investigated, provided a good look at shallow geology in an area where no regional geological information could be located in a document search. The geology at this site accurately reflected the shallow geology at the site under investigation and helped investigators to determine an investigative strategy for the site.

Environmental Site Characterization 131

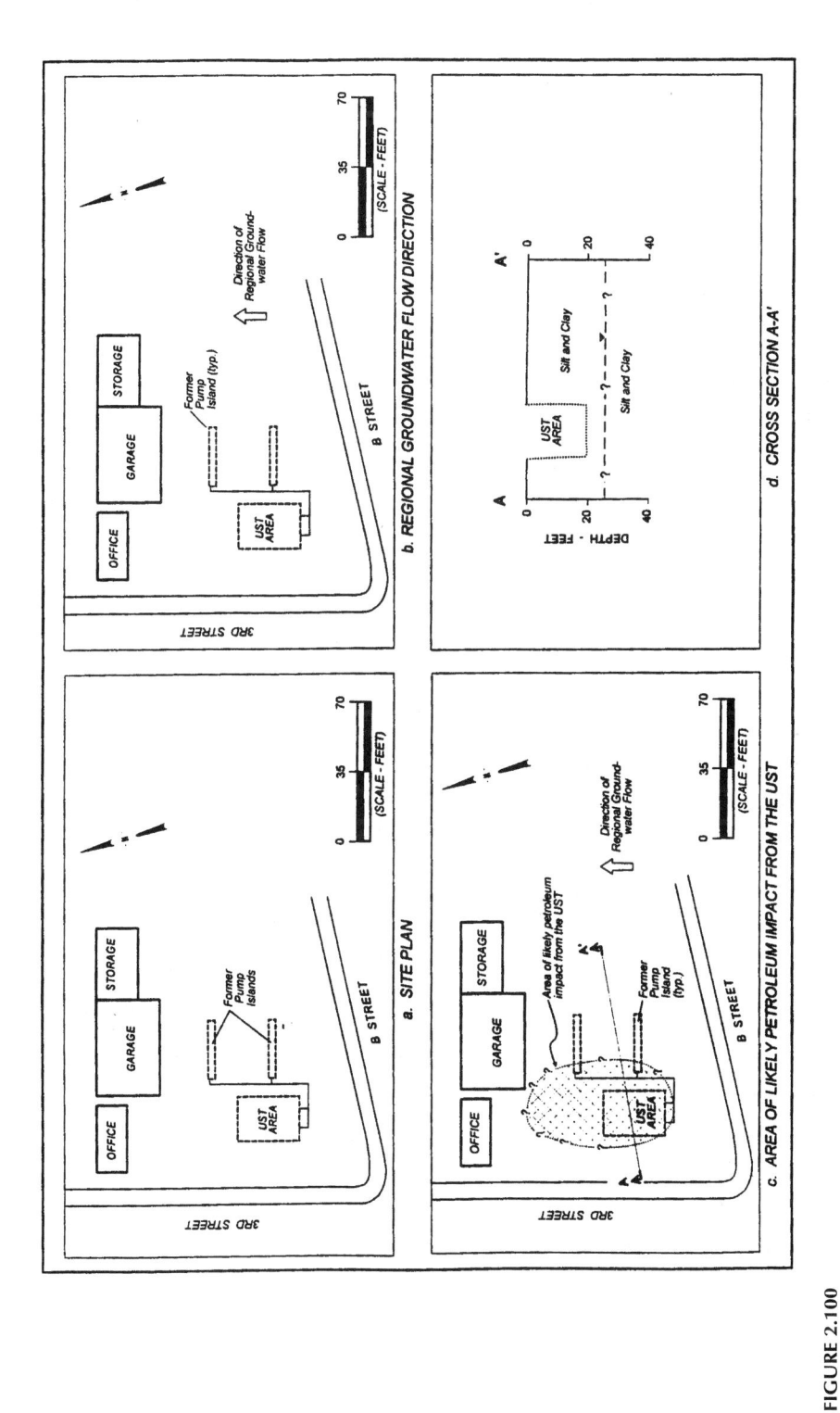

FIGURE 2.100
An example of an initial CSM constructed for a site based on a review of existing information and site reconnaissance. (From ASTM, 2004a. With permission.)

- Any existing soil, soil gas, ground water, surface water, or sediment analytical data
- Locations of any documented spills, releases, or discharges at or in the vicinity of the site (especially on surrounding properties hydraulically upgradient of the site)
- Potential three-dimensional distribution of COCs (based on behavioral characteristics and environmental conditions)
- Background geochemical conditions
- Locations of potential migration pathways, points and routes of exposure, and locations of receptors
- Constraints to the field investigation

Documenting all or most of these features and how they interrelate allows investigators to develop a targeted sampling and analysis plan to allow three-dimensional mapping of important subsurface features and contaminant distribution.

Development of the CSM is critical for determining potential migration pathways and exposure routes and for suggesting possible effects of the contaminants on human health and the environment. Uncertainties associated with the CSM need to be clearly identified so that efforts can be taken to reduce these uncertainties to acceptable levels. Early versions of the CSM, which are usually based on limited or incomplete information, will identify and emphasize the uncertainties that should be addressed.

Potential migration pathways through all environmental media should be identified for each potential source area. Complete exposure pathways should be identified and distinguished from incomplete pathways. An exposure pathway is incomplete if any of the following elements are missing: (1) a mechanism of contaminant release from a primary or secondary source; (2) a transport medium if potential receptors are not located at the source; and (3) a point of potential contact of environmental receptors with the contaminated medium (ASTM, 2004d). The potential for both current and future releases and contaminant migration along the complete pathways to the potential receptors should be determined. A diagram of exposure pathways for all sources at the site should be prepared, as tracking contaminant migration from sources to potential receptors is one of the most important uses of the CSM. Detailed guidance on identifying migration pathways, exposure pathways, and environmental (both human and ecological) receptors is included in ASTM Standard E 1689 (ASTM, 2004d) and U.S. Army Corps of Engineers (2003).

The U.S. Army Corps of Engineers (U.S. Army Corps of Engineers, 2003) suggests the development of various facility profiles as an effective means of organizing and presenting information about sources and potential receptors and the interactions between them. They describe five profile types that address specific, yet overlapping types of information. These profile types include:

- Facility profile-describes man-made features and potential sources at or near the site
- Physical profile-describes factors that may affect the release, fate and transport, and site access
- Release profile-describes the mechanism for the release and the movement and extent of contaminants in the environment
- Land use and exposure profile-provides information used to identify and evaluate the applicable exposure scenarios, receptors, and receptor locations
- Ecological profile-describes the natural habitats of the site and ecological receptors in those areas

Typical information associated with each profile type is presented in Table 2.4. These information needs are not comprehensive, and each site may require different or additional information as determined by the project team. This approach should work well for a variety of contaminated sites.

The initial CSM, in particular those aspects that address the heterogeneity of the affected environmental medium, the presence of potential contaminant movement and exposure pathways, and the projected three-dimensional contaminant distribution, is based on multiple working hypotheses and is dynamic in nature. When new data are collected during the field investigation, the hypotheses are tested and confirmed, modified, or rejected. For

TABLE 2.4

Typical Information Needs Associated with Profile Types

Profile Type	Typical Information Needs
Facility profile	All structures, sewer systems, process lines, underground utilities Physical boundaries (past and current), fencing, administrative controls, etc. Current and historical process and manufacturing areas Manufacturing activity areas Storage and waste disposal areas Historical features that indicate potential source areas (landfills or lagoons, ground scars)
Physical profile	Topographic and vegetative features or other natural barriers Surface water features and drainage pathways Surface and subsurface geology, including soil type and properties Meteorological data Geophysical data Hydrogeological data for depth to ground water and aquifer characteristics Other physical site factors that affect site activities Soil boring or monitoring well logs and locations
Release profile	Determination of contaminant movement from source areas Contaminants and media of potential concern Impact of chemical mixtures and co-located waste on transport mechanisms Locations and delineation of confirmed releases with sampling locations Migration routes and mechanisms Modeling results
Land use and exposure profile	Receptors associated with current and reasonable future land use on and near the facility (residential, recreational, commercial, agricultural, industrial, public forest, etc.) Zoning Types of current or future activities at the facility, including frequency and nature of activity (intrusive or nonintrusive) Beneficial resource determination (aquifer classification, natural resources, wetlands, cultural resources, etc.) Resource use locations (water supply wells, recreational swimming, boating or fishing areas, hiking trails, grazing lands, historical burial grounds, etc.) Demographics, including subpopulation types and locations (schools, hospitals, day care centers, site workers, etc.)
Ecological profile	Description of the property at the facility, including habitat type (wetland, forest, desert, pond, etc.) Primary use of the property and degree of disturbance, if any Identification of any ecological receptors in relation to habitat type (endangered or threatened species, migratory animals, fish, etc.) Relationship of any releases to potential habitat areas (locations, contaminants or hazards of concern, sampling data, migration pathways, etc.)

Source: U.S. Army Corps of Engineers (2003).

example, analysis of a ground-water exposure pathway will usually entail developing some hypotheses about ground-water flow velocity and direction relative to potential receptors. If these parameters are not known, they can be measured through collected data and interpreted through computer modeling or professional judgment. If the results from data collection confirm the predictions, the CSM is updated to show that the hypothesis is correct. However, if the results do not support the predicted outcome, it may indicate that the hypothesis was incorrect and should be restated. This will require revision to the existing CSM. This process is depicted in Figure 2.101.

New data are used to revise the CSM, to build an increasingly accurate understanding of site physical conditions, what contamination is present and where, whether the contamination poses current or future risks to potential receptors and, if so, how that risk can be mitigated. The CSM and the sampling and analysis plan are tightly coupled in a rapid feedback loop — the CSM guides the collection of new samples, but the CSM is also refined as the results of sampling are integrated into it. The updated CSM then guides the collection of more data, which further refines the CSM, and the process continues until the collection of additional data no longer changes the CSM. Overall project goals and objectives are also revisited throughout the field investigation to ensure that they are still compatible with the evolving CSM. For example, if the initial CSM for a site characterization program is based on the assumption that contaminant releases at the site were random (spatially and temporally), the field manager would likely choose a random systematic grid as the basis to begin sampling. However, if the field manager discovers through initial sampling that there is a pattern to the contamination, then the field manager would need to alter the sampling strategy to ensure that the project objectives will still be met. The CSM becomes sufficiently accurate when the field manager and senior staff are confident that the CSM represents actual site heterogeneity so that decisions about monitoring, exposure risk, and remediation can be correct and cost effective.

The components of the initial CSM that are most heavily emphasized and the complexity and degree of detail incorporated into the CSM depend on the objectives of the site characterization program, the complexity of the site, and the decisions that must be made. For example, decisions about ground-water contamination movement or cleanup

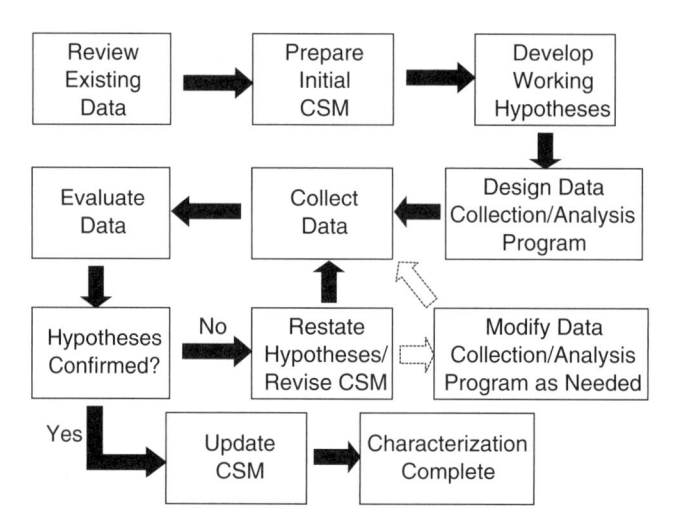

FIGURE 2.101
The process of development and revision of the CSM.

need a CSM that emphasizes hydrogeology and contaminant transport and fate information, whereas decisions about contaminant exposure require a CSM that focuses on identifying all possible receptors and exposure pathways. A complex site where multiple objectives must be met may have several different but related depictions of the CSM, each of which either addresses a different medium or subset of the decisions to be made or represents one of the multiple hypotheses that needs to be clarified by collecting more data. Because the effectiveness and efficiency of the sampling program will be directly related to the accuracy of the CSM, it typically incorporates as much detail as the evaluation of existing site information will allow. Serious data gaps should be left as blank spaces on the maps and cross-sections, and uncertain boundaries should be identified with question marks or dashed lines.

Data gaps always exist in initial CSMs. Gaps are identified by comparing what is already known about the site with what needs to be known to support appropriate regulatory and engineering decisions. Data gaps in the initial CSM will then be filled through collection and interpretation of field data. The CSM is the most valuable tool that investigators can use in making decisions on where and how to collect and analyze samples and what additional methods (e.g., geophysics and cone penetration testing [CPT]) may be used to generate essential data. As the CSM evolves, decision uncertainty decreases. Evolution of the CSM ceases when the model does not change with the incorporation of new field data. The final CSM, following completion of all field activities and data evaluation and interpretation, should be detailed and accurate enough to meet site characterization objectives and provide enough information to base important site decisions on. Ultimately, the final CSM is used to design long-term monitoring programs, to assess risks posed by the site or to select and design the best options for remediation.

Additional specific guidance on developing CSMs that apply exclusively to ground-water systems can be found in ASTM Standard D 5979, Standard Guide for Conceptualization and Characterization of Ground-Water Systems (ASTM, 2004e). Another excellent reference on the subject of CSM development that relates specifically to hazardous waste projects is available from the U.S. Army Corps of Engineers (U.S. Army Corps of Engineers, 2003).

Designing a Data Collection Program

On the basis of data gaps that are identified in the CSM and the data needs that must be satisfied for the site characterization program to meet its objectives, a data collection program is formulated. The data collection program is a general work plan that specifies the types, quantity, and quality of data that must be collected. It also specifies the investigative methods and equipment that will be used to define the site physical characteristics, the potential exposure pathways, the risk of exposure, the contaminant source areas, and the extent of any contamination that exists in the various environmental media at the site. The data collection program supplies the data required to refine the CSM and to resolve any uncertainties and observations that are inconsistent with the initial CSM (outliers). The number and location of first-round data collection points and the sample collection and analysis criteria (depth intervals, sampling protocol, contaminants of concern, data quality levels, analytical methods, and data validation techniques) are specified in the plan. Subsequent data collection points are determined in the field based on site conditions and the results of previous data collection, using the decision process described in Dynamic Work Plans. The data collection program may be documented in an informal manner or may simply consist of a discussion among the field manager, the senior project personnel, and the appropriate interested parties (site owner and operator and regulators). The field manager may need to make adjustments in the plan and the scope of work as the understanding of site conditions evolves, particularly if the preferred field methods or equipment choices prove to be inappropriate or inadequate to deal with expected site conditions.

Proper implementation of the data collection program requires that the field manager be familiar with the capabilities and limitations of a wide variety of sampling tools, field analytical methods, and *in situ* measurement methods, and that the field manager be capable of rapidly interpreting field-generated data as they become available. The field manager must also be able to implement contingencies based on reasonably anticipated deviations from expected site conditions such as shallow bedrock, presence of boulders, depth to ground water, and presence of previously unidentified contaminants. Such contingencies may include changes in field methods or equipment requirements, alterations in plans for dealing with IDW or the need to gain off-site access.

The selection of appropriate data collection methods and equipment should be based on the following criteria:

- Objectives and scope of the site characterization program
- Capabilities, limitations, speed and relative cost of each method and piece of equipment (rental vs. purchase vs. subcontracting) and of combinations of methods and equipment
- Anticipated site physical conditions (soil, geologic, and hydrogeologic conditions)
- Anticipated COCs and concentrations
- Site features and layout
- Potential for disturbance to site operations
- Constraints to use of the methods or equipment
- Potential obstacles to deployment of equipment in the field

Some of the more common field methods and equipment that can be used in an environmental site characterization program are briefly described later in this chapter; others are described in great detail in other chapters in this book and in the scientific literature. A comprehensive inventory of these methods and tools can be found in ASTM Standard D 5730 (ASTM, 2004f) and in U.S. EPA (1993b, 1993c).

One of the important objectives of the data collection program that is sometimes not given enough attention is establishment of background conditions. Background samples serve three important functions:

- Establishment of the range of concentrations of an analyte attributable to natural occurrence at a site
- Establishment of the range of concentrations of an analyte attributable to sources other than the sources that have been identified
- Determining the extent to which contamination exceeds background levels

The number and location of samples needed to establish background concentrations in each environmental medium will vary with site-specific conditions. The number and location of samples must be sufficient to distinguish contamination attributable to the sources under consideration from naturally occurring or nearby anthropogenic contamination.

Evaluating and Selecting Site Characterization Field Methods

None of the improved environmental site characterization approaches are technology specific, but they all emphasize selection and use of the most appropriate technologies for a particular site. These approaches all have a bias toward noninvasive and minimally invasive methods, but some site conditions require the use of technology that can overcome difficult conditions (e.g., shallow bedrock, boulders, and dense soils). The principle of using multiple, complementary measurements allows for better, more accurate three-dimensional physical and chemical characterization of subsurface conditions. The various noninvasive and minimally invasive technologies available today allow much more cost-effective investigations than were possible using conventional drilling, sampling, and well installation methods alone. Improvements in and miniaturization of chemical analytical methods, including automation of sample analyses and improvements in software used for processing analytical instrument signals, together with development of a wide variety of field test kits, makes real-time analytical results available for almost any chemical parameter. The use of low-cost mobile laboratories means that there are essentially no limits on the quantity or quality of analytical data that can be produced and used for on-site decision making in the ASC/ESC/Triad/DFA process.

The focus of the field investigation is on collecting accurate, reliable real-time data using minimally intrusive methods. To keep the investigation moving forward, it is important to use rapid sampling methods in conjunction with field analytical methods to collect real-time data to guide further sampling efforts. It is also important to use methods that minimize the generation of IDW (including drill cuttings and development water and purge water from wells) and that reduce the exposure of field personnel to potentially hazardous materials.

Many field investigation methods are potentially applicable to environmental site characterization programs, including the conventional methods that most investigators are familiar with and use on a regular basis. There are also a number of newly developed and innovative rapid sampling, field analytical, and *in situ* measurement technologies that can cost-effectively provide a high density of data points for the refinement of the CSM that should be considered in designing an effective data collection program. Factors to consider when selecting equipment and methods for a site investigation include:

- Objectives, data quality requirements, and anticipated scope of the investigation
- Site physical characteristics — soil types and geological material types (unconsolidated vs. bedrock)
- Depth requirements for subsurface sampling methods
- Anticipated contaminants of concern and concentrations
- Ability to produce real-time or near-real-time data (speed of sampling and analysis)
- Equipment characteristics — durability, reliability, and limitations to use
- Cost of the equipment to use — purchase vs. rental vs. contracting
- Flexibility in application of the equipment and method to a variety of environmental media and site conditions and to combined use with other complementary methods
- Potential for production of IDW
- Potential for disturbance to site operations and neighboring properties (because of noise, space requirements, etc.)

Equipment or methods selected and standard operating procedures (SOPs) for use of the equipment or methods should be detailed in the FSP.

Although many of the rapid sampling and field analytical tools have been available for use for more than a decade, their use has been relatively limited. Many government agency personnel and some consultants still consider the use of such tools as "screening" tools rather than tools capable of producing valid and defensible data. With the recent publication of several key U.S. EPA- and ITRC-generated documents (i.e., U.S. EPA 2001a, 2003; ITRC, 2003) and several relevant ASTM Standards (ASTM, 2004a, 2004b) that encourage the use of these technologies, they have started to gain acceptance on a more widespread basis as methods capable of producing data acceptable for a wide variety of applications.

The technologies and methods that are most universally applicable to use with the ASC/ESC/Triad/DFA approaches include the following:

- Sample collection methods
  a. DP technologies, including cone penetration testing (CPT) rigs
  b. Sonic drilling
- Sample analytical methods
  a. Field-based analytical methods appropriate for the COCs (many methods are available; these are discussed in detail later in this chapter)
  b. Methods used in a portable lab setting
- Other investigation methods
  a. Surface geophysics (many methods available)
  b. Soil-gas surveys

These technologies and methods are described briefly subsequently; additional detailed descriptions are provided in other chapters in this book and in references cited in the following sections of this chapter.

Sample Collection Methods

For all environmental site characterization projects, at least one environmental medium (whether soil, soil gas, ground water, surface water, or sediment) will have to be sampled. Because most programs focus on contamination of soil, soil gas, or ground water, this section focuses on methods for collecting these types of samples. Selection of appropriate sampling tools or methods for these media depends on those factors discussed earlier. With specific reference to site physical characteristics, methods appropriate for penetrating the materials present in the subsurface at any given site are highly dependent upon the character of geological formation materials. While DP methods (described briefly subsequently) are often preferred for conducting ASC/ESC/Triad/DFA sampling programs, they are limited in their application by certain geological conditions including the presence of bedrock, boulders, cobbles, dense sands or clays, and thick gravel zones. In these cases, either sonic drilling or conventional drilling methods (also discussed briefly subsequently) will be required to collect soil samples and to install piezometers or wells.

DP Technologies

DP technologies (Figure 2.102) include equipment that is used to push, drive, or vibrate profiling tools or devices into the ground to enable the rapid collection of *in situ* measurements or samples of soil, soil gas, soil pore water, or ground water. To collect samples, a string of small-diameter hollow rod (generally 1 to 2.25 in. O.D.) with one of several

Environmental Site Characterization

FIGURE 2.102
DP systems like the one pictured here allow rapid deployment of a variety of *in situ* measurement tools, as well as tools for collecting representative samples of soil, soil gas, soil pore water, and ground water. These systems use vibrational energy, static weight, or hydraulic hammers to advance tools into unconsolidated surface materials to depths in excess of 200 ft under favorable geologic conditions.

types of sampling tools (Figure 2.103 and Figure 2.104) on the bottom is either pushed into the subsurface using the static weight of the rig, driven using a pneumatic or hydraulic percussion hammer, or vibrated using a high-frequency drive head. Samples can be collected on either a discrete or continuous basis from depths ranging from less than 50 ft to more than 200 ft, depending on the size and capability of the rig.

Many DP systems, including CPT rigs (Figure 2.105) also have the capability of collecting continuous real-time *in situ* measurements of a variety of parameters. Parameters that can be

FIGURE 2.103
This soil sampling tool allows collection of 4-ft long, 1.8-in. diameter samples on either a discrete or continuous basis, normally achieving much better recovery than traditional split-spoon samplers because of the way in which the sampler is advanced.

FIGURE 2.104
Discrete ground-water sampling tools like this one (with a screen in the foreground and the sheath that slides over it behind) remain closed until they are advanced to the depth at which samples are desired. The sampler is opened by retracting the drive rod a distance that reflects the thickness of the interval from which investigators want to collect a sample (from several inches to as much as 30 in.). When the drive rod is retracted, the screen drops out of the sheath and is exposed to the formation.

FIGURE 2.105
This CPT rig (a U.S. Navy Site Characterization and Penetrometer System [SCAPS]) uses the static weight of the rig (up to 60 t) applied to a hydraulic ram to rapidly advance any of a variety of tools into unconsolidated formation materials. The standard CPT setup collects data on the resistance of the soil to penetration (at the tip of the probe) and the soil friction generated on a sleeve behind the tip. The ratio of these values allows identification of soil type. Many other tools (discussed in Chapter 6) are available to collect *in situ* measurements and samples of all subsurface media.

measured include soil electrical conductivity; resistance of the soil to penetration (which can be correlated to soil type); presence and concentrations of VOCs in soil or ground water; presence and concentrations of NAPLs in soil or ground water (Figure 2.106); and distribution of pore pressures (which can be used to determine the position of the water table and formation hydraulic conductivity). One tool available for use with CPT rigs can also provide a very detailed video image of soil penetrated by the probe and can allow identification of NAPLs present in soil pores. As samples and data are collected using DP technology, no IDW is generated and workers are not exposed to potentially hazardous materials, except when handling samples. The capability of collecting continuous samples and continuous *in situ* data (i.e., through vertical profiling) is particularly valuable because this helps develop an accurate and detailed three-dimensional CSM.

DP rigs also have the capability of installing wells (Figure 2.107) and multilevel monitoring systems (Figure 2.108) in locations that can be identified by sampling or *in situ* measurement methods as the optimum positions for either short-term or long-term monitoring, which saves a substantial amount of both time and money versus conventional approaches to well installation and positioning. Most DP rigs are small, compact, versatile, and inconspicuous, usually mounted on a pickup truck, cargo van, all-terrain vehicle, or tracked vehicle platform (Figure 2.109), although CPT rigs can be as large as a conventional drilling rig and quite heavy (10 to 60 t). Detailed descriptions of all DP technologies available for use in environmental site characterization are included in Chapter 6 and in U.S. EPA (1997a).

FIGURE 2.106
This laser-beam generator is part of a system that can be deployed on a CPT rig to detect the presence and measure the concentrations of petroleum hydrocarbons *in situ* using laser-induced fluorescence (LIF). This system has been used at hundreds of DOD, DOE, and privately owned sites with great success.

FIGURE 2.107
DP rigs are capable of installing long-term monitoring wells in diameters ranging from 1/2-inch I.D. to 2 in. I.D. This well has a prepacked well screen that allows collection of sediment-free samples from most unconsolidated formation materials.

FIGURE 2.108
Multilevel monitoring systems, like this continuous multichannel tubing (CMT) system, can be installed in the large (3.125 in.) O.D. drive casing advanced by some of the larger DP rigs. Such systems (described in Chapter 11) allow collection of samples from as many as 7 different zones from the surface to as deep as 300 ft.

Environmental Site Characterization 143

FIGURE 2.109
DP rigs are available on a variety of different platforms including those shown here and tracked vehicles. (Photo courtesy of Geoprobe Systems.)

Sonic Drilling

Sonic drilling, which is a method of drilling unlike most conventional drilling technologies, is capable of rapidly collecting continuous samples of geologic materials while generating very little IDW. Sonic drilling (Figure 2.110) uses high-frequency vibrations transmitted from the drill head through the first of two strings of drill pipe (the core barrel; Figure 2.111) to penetrate formation materials without rotation of the drill string. Through the use of an open bit, the core barrel continuously cores the formation materials penetrated in 10 ft (or longer) increments. The second string of drill pipe (the temporary casing) is then vibrated around the first string and displaces formation materials to the outside of the drill string. This string of pipe remains in place to hold the hole open as the core barrel is extracted and the continuous sample is removed (Figure 2.112 and Figure 2.113). Drilling continues in this manner until the desired depth is reached. A well or multilevel monitoring system (Figure 2.114) can be installed in the borehole and completed as the outer drill string is removed from the hole.

Sonic drilling rigs can penetrate unconsolidated materials of any type (including gravels, cobbles, boulders, dense stiff clays, and other difficult-to-drill materials) and most types of bedrock at a high production rate (often between 0.5 and 1 ft/min) to depths in excess of 800 ft. The main limitations of the method include the size of the rig and support truck and the per-foot cost of drilling, which can be at least partially offset by the substantial time savings versus conventional drilling methods. A detailed description of sonic drilling is included in Chapter 5.

Conventional Drilling Methods

Conventional drilling methods can be used in the ASC/ESC/Triad/DFA approaches as well as the traditional or conventional approach, but they are usually not as cost effective as DP because of issues related to management of IDW, speed, mobilization costs, and the size of the rigs and necessary support equipment.

FIGURE 2.110
Sonic drilling uses a vibratory drill head (top of photo) to impart a standing harmonic wave to concentric strings of drill pipe to advance them into the ground without rotation of the drill string or disaggregation of formation materials. The vibratory action displaces formation materials around the outside of the outer string of pipe and allows collection of a continuous sample in the inner string of pipe. Overall drilling rates of more (in some materials, much more) than a foot per minute are possible with sonic drilling (instantaneous rates are much higher).

Hollow-Stem Auger: The most common drilling method applied to environmental site characterization is hollow-stem auger drilling (Figure 2.115), which uses a drill string with helical flights wound around a hollow center stem of a diameter ranging from 2.25 in. I.D. to 12.25 in. I.D. and with a cutting head (bit) at the bottom. The borehole is advanced and drill cuttings are conveyed to the surface by rotation of the auger, as formation materials are disaggregated by the cutting head. Soil samples are collected and wells are installed through the hollow center stem, which provides access to the subsurface without the need for a temporary casing. The most effective and efficient soil sampling method used with hollow-stem auger is the continuous tube sampler (Figure 2.116), which provides a continuous 5 ft long sample of the material penetrated by the lead auger flight (Figure 2.117). Drilling fluids are generally not used with hollow-stem augers unless difficult drilling conditions (e.g., heaving sands) are encountered, in which case the augers may be filled with either water or a bentonite-based fluid. One of the main problems with this method is the amount of drill cuttings produced, which averages about one 55 gallon drum for every 17 ft of drilling (with a 4.25 in. I.D. auger).

Direct Mud Rotary: Direct mud rotary methods (Figure 2.118) use a drilling fluid that consists of water and bentonite, with the appropriate amount of various additives if difficult drilling conditions (i.e., loose gravels or fractured rock) are encountered.

FIGURE 2.111
The inner string of pipe advanced by a sonic drilling rig is a 10 ft long core barrel that collects a relatively undisturbed continuous sample of formation materials as the borehole is advanced.

The hole is advanced by rotational action of a drill string that consists of thick-walled hollow steel drill pipe with a drill bit on the bottom and a water swivel on the top. As the drill string is rotated, the drilling fluid is pumped down the drill pipe to exit at the bit, where it serves several important functions. It cools the bit, it brings drill cuttings to the surface as it circulates back up the hole, and the hydrostatic pressure it creates down hole holds the hole open. The circulation of the fluid can be reversed on some rigs so the fluid goes down the annular space between the borehole and the drill string and cuttings are brought up through the center of the drill string. Soil sampling is accomplished by removing the entire drill string, assembling a string of sampling rod with a sampling device (i.e., split-spoon or thin-wall tube sampler) on the end, lowering the sampling string to the bottom of the hole, and advancing it by driving or pushing it into the soil. This is a very time-consuming process that delays drilling.

Several problems occur with this type of drilling when contaminated materials are encountered. First, some of the contaminated materials are incorporated into the drilling fluid, which is circulated down the hole as drilling continues, and cross contaminates all of the formation materials penetrated by the borehole. Secondly, the drilling fluid (which may total several hundred gallons) has to be managed as a contaminated material and must be properly containerized and disposed. Because these issues add significant cost and potential liability to the drilling operation, it is generally recommended that some other method be used if the possibility of drilling through contaminated materials exists.

FIGURE 2.112
After the core barrel is retrieved from the borehole, the continuous sample is extracted either by removing a liner from the core barrel or by vibrating the sample out of the core barrel into a sample sleeve (shown here) 5 ft at a time.

Air Rotary: Air rotary methods are similar to mud rotary, but they use air or air mixed with foaming agents as the drilling fluid. Like mud rotary rigs, air rotary rigs may use either direct or reverse circulation of the drilling fluid. The most commonly used air rotary rig for environmental work is a dual-tube reverse circulation rig (Figure 2.119),

FIGURE 2.113
The size of the continuous samples from the core barrel allows collection of a complete suite of samples for physical analysis, as well as chemical analysis. For site characterization purposes, continuous sampling like this is much preferred over discrete samples such as those provided by split-spoon sampling.

FIGURE 2.114
Sonic drilling rigs can rapidly advance boreholes through most types of geological materials and can provide for installation of long-term monitoring wells and multilevel monitoring systems. Installation of a seven-channel CMT system is shown here. (Photo courtesy of Murray Einarson.)

FIGURE 2.115
The hollow-stem auger is the most commonly used conventional drilling method for conducting environmental field work. However, augers are unable to penetrate bedrock, boulders, or gravel and cobble zones, and they produce significant amounts of IDW.

FIGURE 2.116
The continuous tube sampler is preferred for collecting samples from hollow-stem augers. It is installed in the lead 5 ft auger section (with the cutting shoe ahead of the auger cutting teeth) and advanced as the auger is advanced. It is held stationary and does not rotate with the auger.

which can be used to efficiently overcome a variety of difficult drilling conditions and to sample formation materials and install wells without removing the entire drill string from the borehole. With direct air rotary methods, temporary casing is often advanced to hold the borehole open during drilling in unconsolidated materials (Figure 2.120), because air, unlike the water-based fluid used in mud rotary drilling, does not have the ability to hold the hole open during drilling. Air rotary with a down-hole hammer (Figure 2.121) is one of the best available methods for drilling through very hard (i.e., igneous or metamorphic) bedrock. Dealing with the air returned to the surface while drilling through contaminated materials can be problematic and expensive. In addition, it is important that the air compressor on the rig incorporates a filtration system to avoid contamination of the borehole by compressor oil entrained in the air used for drilling.

Other Drilling Methods: Other drilling methods are also available, including cable tool (Figure 2.122), Odex or Tubex, solid-stem augers, and bucket augers, but they are infrequently used in environmental projects. All of the conventional drilling methods are described in detail in Chapter 5.

Sample Analysis Methods
Field-Based Methods

Field-based analytical methods are those that can be applied at the same location as samples are collected (Figure 2.123). The equipment is often included within DP rigs

FIGURE 2.117
The sample collected with the continuous tube sampler is usually undisturbed and provides an excellent representation of subsurface formation materials. This sample, collected in an angle-drilled borehole, shows formation materials that are under reducing conditions, except for the thin zone adjacent to the ruler. At the center of this zone, which shows discoloration caused by oxidation, is a fracture along which water movement is occurring in the *in situ* materials. This preferential pathway, and others like it, served as movement pathways for contamination at a hazardous waste site in Illinois. (Photo courtesy of the Illinois State Geological Survey.)

(Figure 2.124). The methods available include methods that can be applied with field test kits (i.e., colorimetric and immunoassay kits) (Figure 2.125 and Figure 2.126) and easy-to-operate hand-held equipment (PIDs, FIDs, [Figure 2.127], GCs [Figure 2.128], and XRF detectors [Figure 2.129]), as well as more rigorous methods that require the controlled environment of a mobile laboratory (GC/mass spectrometers [GC/MS], inductively coupled plasma, directly coupled plasma, etc.) (Figure 2.130). The analytical methods selected for any given project will depend on the following:

- The COCs or other analytes to be measured
- The targeted environmental medium (or media)
- The method's ability to measure contaminants in the targeted medium
- The data quality level achievable with the method
- Ability to use the method in combination with other complementary methods
- The relative speed and cost of the method

Of all the tools applied in an ASC/ESC/Triad/DFA program, field-based analytical methods are perhaps the most important because they allow the generation of real-time or near-real-time analytical data, which supports the on-site decision making process that moves the project forward. Used properly, some field-based analytical methods are capable of generating the same high-quality data as fixed laboratory-based methods. Other methods are capable of generating varying levels of data quality and can be applied to situations where the highest analytical quality data are not necessary to

FIGURE 2.118
Direct mud rotary drilling is another commonly used conventional drilling method used in environmental investigations.

accomplish project objectives. All of these methods allow rapid generation of analytical results for large quantities of samples, enabling the collection of high-density data that helps reduce uncertainty in the CSM. To be assured of the highest quality data possible with each method, an experienced operator must be available to operate the equipment and an experienced project chemist must be available for selection of the appropriate methods, QA/QC, review of analytical results, and data interpretation.

An ASC/ESC/Triad/DFA project often makes use of a variety of field-based analytical methods of different types, to generate data of differing quality that matches project requirements. For example, an inexpensive yet reliable qualitative or semiquantitative method, such as a PID, FID, or immunoassay kit, may be used to delineate a contaminant hot spot and a quantitative method, such as a field-portable GC, can be used to identify a specific contaminant at a specific concentration in that hot spot. A limited number of samples (generally 5% to 10%) may be submitted to a fixed lab for confirmation. The analytical approach can be selected based on the criteria outlined earlier. Contingencies should also be identified in the event that the methods selected do not produce the needed quality or quantity of data.

Perhaps the most important application of field-based analytical methods is in the production of large quantities of data of appropriate quality for the investigation. The sheer number of data points provides the investigator with greater discriminatory ability and provides a level of data quality and representativeness that extends beyond individual sample

FIGURE 2.119
The dual-tube reverse-circulation air rotary rig is an excellent choice for rapid drilling through difficult conditions including bedrock, boulders, cobble zones, heaving sands, and cavernous formations.

quality. This allows for cost-effective reduction in uncertainty by providing more data points, to effectively guide the investigation to completion. Field-based analytical methods are described in detail later in this chapter and in a variety of other references including NJ DEP (2003), U.S. EPA (1997a, 1997b, 1998a, 1998b, 1993c), ASTM (2004g), and Robbat (1997).

Other Investigation Methods

Surface Geophysics

Surface geophysical methods include a number of remote-sensing technologies for cost effectively generating a large amount of data on subsurface conditions in real time, without producing IDW and without penetrating the surface. Advances in instrument capabilities and signal processing methods have reduced the cost and turnaround time for data interpretation for many geophysical methods. These methods can be used efficiently to collect the first data at the site and to help focus later sampling efforts in the right places. They are also very useful in correlating geologic data between widely spaced boreholes and in identifying disturbed zones in the subsurface that may indicate the presence of waste disposal areas. They utilize indirect measurements of one or more subsurface material properties (i.e., electrical conductivity or resistivity and soil or rock density) to define geologic, hydrogeologic, or other physical or contaminant features that cannot be directly observed. Geophysical methods are well suited to determining the locations of subsurface objects, which may be indicators of contaminant sources, or may pose obstacles to DP or drilling efforts. Geophysical methods are capable of defining interfaces between unconsolidated materials and bedrock (providing the depth to bedrock), between unsaturated and saturated materials

FIGURE 2.120
In situations where air rotary drilling is used for drilling through overburden above bedrock, a casing driver (in this case, a hydraulic hammer) may be used to advance casing to hold the hole open during drilling.

(providing the depth to the water table), and between loose, noncohesive materials (sands and gravels) and more dense, cohesive materials (silts and clays). They can also provide very useful detailed information on hydrogeological conditions, which may indicate the presence of contaminant migration or exposure pathways such as sand and gravel lenses in a clay matrix or fracture zones or solution channels in bedrock. Finally, some geophysical methods are capable of determining the presence of some types of contaminants (typically electrically conductive dissolved-phase inorganic compounds, but also LNAPLs). Often more than one method is used because, at a given site, one method may not be as useful or successful as another, and the information gathered from each method is slightly different and, therefore, may be complementary or corroborative.

The information that can be supplied by geophysical methods can be used to delineate important subsurface features, to develop an efficient sampling plan, and to select appropriate sampling and analytical tools. Thus, it is important that these methods be used in conjunction with a dynamic work plan that will provide flexibility to allow for changes in the SAP as geophysical data are collected. Because the information supplied by geophysical methods is highly interpretive and because there are no unique solutions to geophysical problems, a project geophysicist must be available on site to select the appropriate methods, to direct the investigation and to process and interpret the data collected. Information provided by a geophysical survey should always be confirmed by direct observation. Most methods are susceptible to interference from cultural features including presence of buildings, fences, rail lines, and power lines.

The surface geophysical methods available for use in environmental site characterization programs include ground-penetrating radar (Figure 2.131), electromagnetic

FIGURE 2.121
For drilling through very hard bedrock, such as igneous or metamorphic rock, one of the most efficient methods to use is air rotary with a down-hole hammer.

conductivity (Figure 2.132), seismic refraction and reflection (Figure 2.133), electrical resistivity (Figure 2.134), magnetometry, gravimetry, and metal detection. The applications, advantages, and limitations of each of these methods as used in environmental site characterization programs are described in detail in Chapter 4 and in other references including Benson et al. (1984), California EPA (1995), U.S. EPA (1993b, 1993d, 1997a, 1997b, 2000c), and CCME (1994).

Soil-Gas Surveys

Soil-gas surveys provide a means of determining the concentration or flux of VOCs or other vapor-phase constituents (usually related to petroleum hydrocarbons or chlorinated organic compound sources) present in soil pore spaces in the vadose zone. This technology can only be used for detection of vapor-phase contaminants (i.e., those with a vapor pressure of at least 1.0 mm Hg or a Henry's law constant of at least 5×10^{-4} atm m^3/mol at 20°C [CCME, 1994]), but results may be extrapolated to infer the presence of either residual-phase materials in soil or dissolved-phase constituents in ground water. Like geophysical methods, soil-gas surveys can be used to collect a large amount of data quickly and to focus later sampling efforts in places where contamination is evident. They can also identify situations where health risks are present due to the migration of vapor-phase contaminants in shallow soils, which may pose a risk to receptors in buildings.

Soil-gas techniques are divided into two general categories — active and passive. Passive methods employ a sorbent sampling device that is buried in the ground for a specified period of time, then retrieved and submitted to a fixed-base laboratory for sample extraction

FIGURE 2.122
Cable-tool drilling is not used often in environmental drilling work because it is much slower than other methods and it produces large amounts of IDW.

and analysis. Because the time intervals for use of these methods may be several days to several weeks, these methods are less useful in a ASC/ESC/Triad/DFA approach. Active methods consist of installing probes or monitoring points into the soil (typically using DP technology) and withdrawing samples of soil gas that are usually analyzed on site to generate real-time data (Figure 2.135 to Figure 2.137). Soil-gas survey results can delineate areas of soil and ground-water contamination caused by VOCs and provide useful information to guide soil and ground-water sampling efforts. Some active methods can provide semi-quantitative data that can be used for estimating contaminant mass in vadose zone soils. Active soil-gas methods can also be used in conjunction with DP soil and ground-water sampling methods to define multiphase contaminant problems such as commonly occur with petroleum products and chlorinated solvents.

Limitations to the use of soil-gas surveys include shallow ground-water conditions (i.e., less than 5 ft deep), presence of low-permeability materials (i.e., clays or silts); presence of surface sources not related to contamination (i.e., automotive emissions and asphalt); high soil-moisture content; and presence of organic-rich soils (e.g., peat). Soil-gas surveys are described in detail in Chapter 3 and Chapter 6.

Dynamic Work Plans

A dynamic work plan provides the project team with the lines of communication and agreed-upon criteria required to facilitate decision making in the field. It outlines a sequence of activities that accommodate the decision-making process and the involvement of interested parties to keep the project moving forward. Dynamic work plans rely partly on an adaptive sampling and analysis strategy. Dynamic work plans do not dictate the

FIGURE 2.123
Field-based analytical methods are those that can be applied at or near the point of sample collection in the field. Use of these methods avoids the potential sources of sample error associated with sample handling, packaging, shipping, and holding time.

FIGURE 2.124
Some DP rigs can be equipped with sophisticated field-based analytical equipment like this GC, to allow analysis of samples as soon as they are collected.

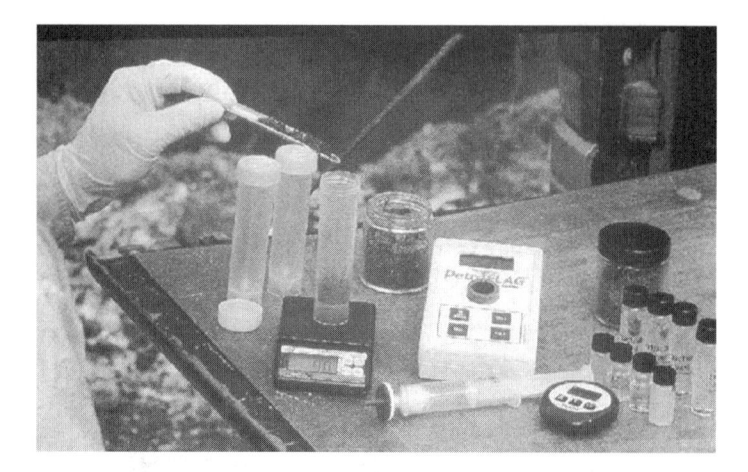

FIGURE 2.125
Field test kits, such as this turbidimetric kit, can be used to provide semi-quantitative to quantitative results for many types of petroleum hydrocarbons. (Photo courtesy of Dexsil Corp.)

exact location and number of samples to be collected or measurements to be made or the details of the sample analysis to be performed. Instead, they identify the suite of field investigation methods and measurements that may be necessary to characterize the site and specify the decision-making logic that will be used in the field to determine which chemical compounds require analysis, where to collect the samples and measurements, and when to stop sampling. The dynamic work plan may identify the maximum potential number of samples, provided there is a clear understanding that the actual number and location of samples will be determined by on-site technical decision making. The field manager adjusts the location and type of field data collection efforts in response to previous observations and data interpretation to optimize site characterization efforts.

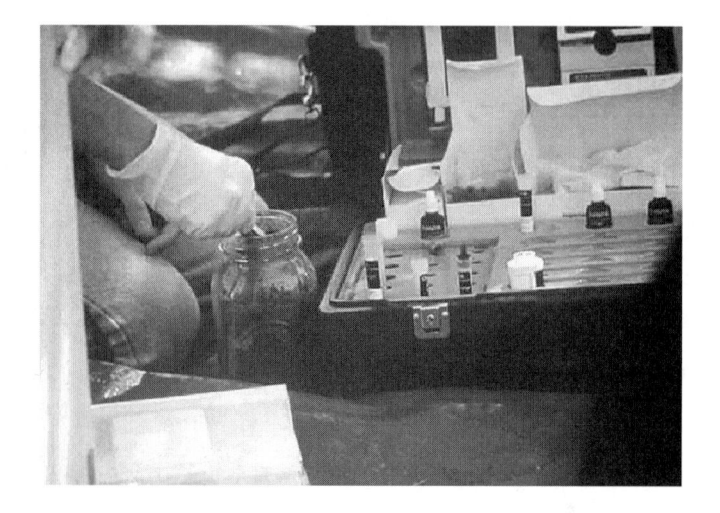

FIGURE 2.126
Immunoassay test kits are available for on-site analysis of a wide range of parameters, including pentachlorophenol, 2,4-D, PCBs, dioxin, total petroleum hydrocarbons, PAHs, BTEX, toxaphene, chlordane, DDT, TCE, and mercury. Detection limits for immunoassay are comparable to or lower than those for conventional analytical methods and are often less than MCLs. Many of the immunoassay methods are included in the current version of SW 846.

FIGURE 2.127
FIDs and PIDs (pictured) are relatively inexpensive devices that can provide qualitative information on the presence of VOCs and SVOCs, and are widely used for sample headspace analysis.

Dynamic work plans should include contingencies so that unexpected findings can be dealt with appropriately and unsuccessful field methods can be quickly modified or replaced without having to cease field work. For example, a dynamic work plan might include a contingency for an alternative sampling method or an alternative analytical method to be used if the preferred method either fails to perform as expected or is inappropriate for the contaminants discovered at the site. Although every effort should be made to ensure that the selected equipment and methods are appropriate for the expected field conditions and contaminants at the outset of the project, even thorough planning cannot account for unexpected circumstances. Thus, dynamic work plans should address procedures that would be used to replace unsuccessful methods with alternate methods if the need should arise. This discussion, most often presented in an "if . . . then" format, should be included in the FSP and the QAPP.

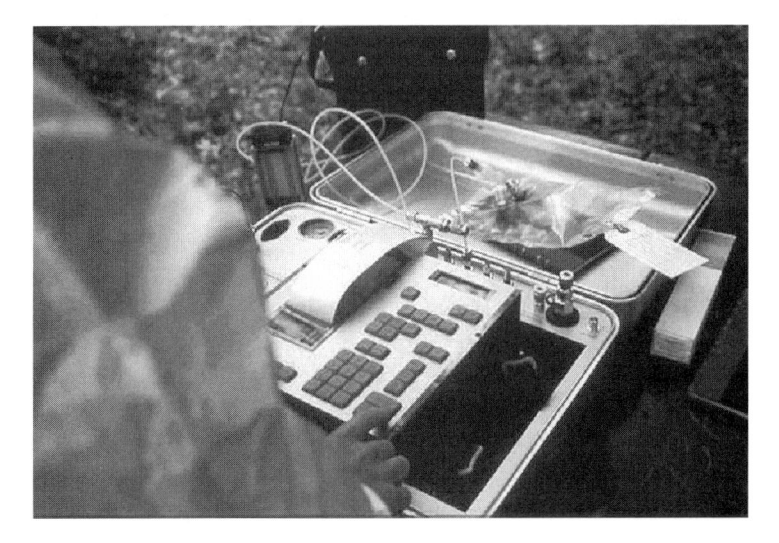

FIGURE 2.128
Field-portable GCs are capable of analyzing samples for a wide variety of organic compounds and providing quantitative data at detection limits comparable to those produced by lab-based instruments.

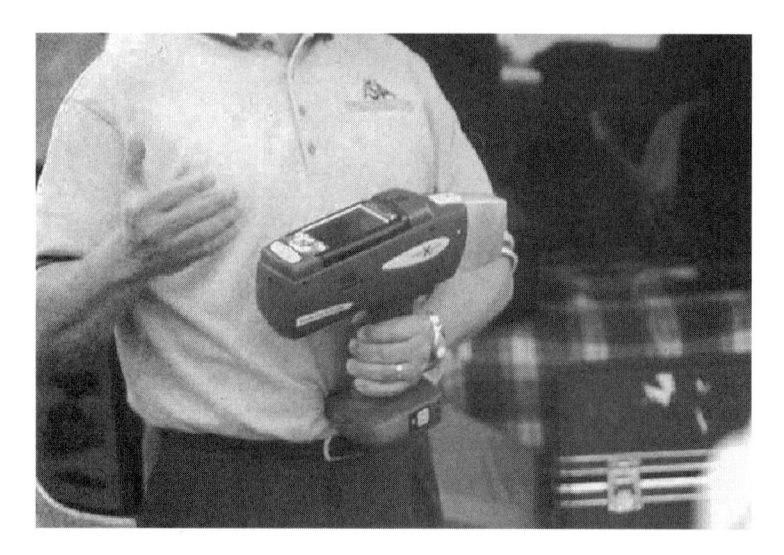

FIGURE 2.129
Field-portable XRF devices are a relatively recent development in field analysis for metals and other elements, ranging from potassium through uranium on the periodic chart. Detectors like this one use an x-ray tube to produce an excitation energy that changes the electron positions in individual elements, which causes them to fluoresce at different energy levels or wavelengths that are characteristic of those elements. Data produced are quantitative or qualitative, depending on how the unit is configured.

Dynamic work plans also tend to make use of innovative field methods and equipment because these generally produce the real-time data necessary for on-site decision making. As a result, geophysical methods, rapid sampling and *in situ* measurement technologies, and field-based analytical methods are typically integrated into the dynamic work plan in a way that makes full use of their ability to increase data density on a real-time

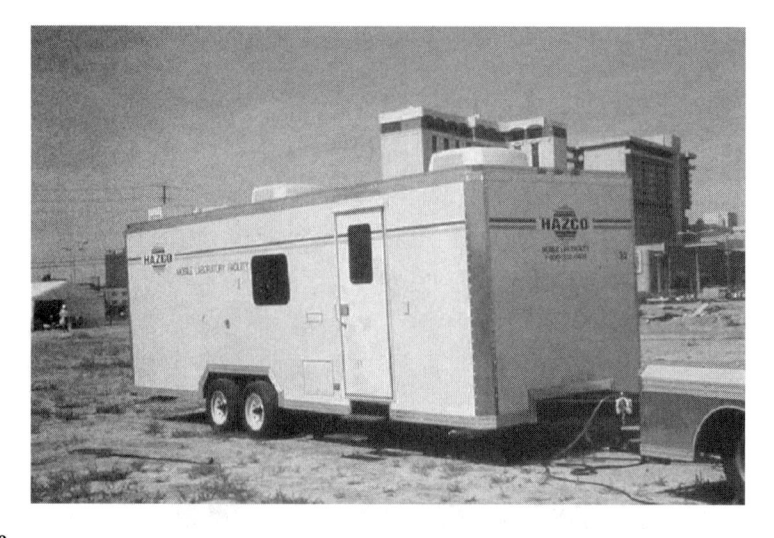

FIGURE 2.130
Mobile laboratories offer a climate-controlled environment with a power generator that can support the use of any of a number of instruments that are typically found in fixed-based laboratories. Site characterization programs that produce large quantities of samples on a daily basis are often best served by this option for field sample analysis.

FIGURE 2.131
Ground-penetrating radar can be used to detect disturbances in subsurface materials (such as excavation and backfill, utility trenches, UST pits, and buried drums) and the location of interfaces that reflect electromagnetic energy (such as the bedrock surface, confining beds, or buried foundations) in real time.

basis. Because many field-based analytical methods provide real-time to near-real-time quantitative measurements, greater confidence should be obtained in the sampling program. If semiquantitative or screening level data are produced, then a percentage of the results should be verified by quantitative methods, either in the field or in a fixed laboratory, as a QA/QC check for the real-time data. However, these analyses are not typically used as the primary data source for decision making.

A dynamic work plan contains the same kind of QC measures associated with a conventional sampling approach; however, the application may be more complex. Multiple field analytical technologies are typically used in conjunction with fixed-laboratory

FIGURE 2.132
Electromagnetic conductivity can be used to rapidly detect the presence of conductive materials (such as ferrous metals) and interfaces that have contrasting electrical conductivity (such as a clay-rich confining bed below a sandy aquifer).

FIGURE 2.133
Seismic refraction requires the installation of geophones to record the arrival of shock waves refracted off interfaces between formation materials of different types, produced by a release of energy at the surface (usually the striking of a sledge hammer on a steel plate or a small, controlled explosion). Interpretation of data on arrival times allows interpretation of the geology at the site.

analysis methods, with each managing different components of data uncertainty. It is often advisable to evaluate some QC data very early during the investigation. For example, it may be desirable to confirm that an on-site method is performing as expected soon after it is used, because real-time decisions depend on its performance. "Adaptive quality control" describes QC procedures that support higher frequencies of QC samples when the uncertainty is high and lower frequencies when there is greater confidence in the analytical performance (ITRC, 2003).

Dynamic work strategies allow a sample-by-sample evaluation of results, if desired. Results can be assessed in real time for their value to CSM development and to project decisions. If there is a conflict between a result and the current CSM, there are two possibilities: either the result or the CSM is wrong. Within a dynamic work plan, it is a simple matter to quickly double-check and resolve an incompatible data result. Something may have gone wrong with the analysis or the sampling. Perhaps an equipment problem has developed that needs to be rectified. If the result is confirmed to be correct, then the CSM needs to be modified. Incompatible results are valuable clues to detect sampling errors or false assumptions in the CSM. Quality control within the context of a dynamic work strategy is much more effective at catching mistakes than traditional work strategies relying on rigid, inflexible work plans, and fixed-laboratory analyses, because results are immediately compared to the current CSM (ITRC, 2003).

FIGURE 2.134
Electrical resistivity is a commonly used geophysical method that depends on contrasts in the ability of subsurface materials (natural or man-made) to conduct electrical current. Clay-rich materials are better conductors than sands; ground water with a high TDS content conducts current better than water with a low TDS content. (Photo courtesy of Dick Benson.)

Adaptive Sampling and Analysis Plans

Adaptive sampling and analysis plans (ASAPs) change as the CSM is refined based on the sampling and analytical results produced in the field. A successful ASAP requires analytical methods and instrumentation that are field-practical and can produce data fast enough to support the dynamic work plan process. Some large ASAPs can produce hundreds of samples a day, so sample throughput capacity has to be gauged to closely match sample production rates. If the sample production rate is significantly greater than the rate at which samples can be analyzed, either the sampling program must be delayed to allow the analysis to catch up or there will be pressure to continue sampling without the benefit of results from the previous round of sampling and the value of adaptive sampling will be lost. If the analysts can handle significantly more samples than can be produced, then per-sample analytical costs will be driven up as the analytical equipment sits idle. Managing, integrating, and displaying the information associated with sampling may pose a logistical challenge, which can interfere with the progress of the site characterization program if this is not adequately addressed. The coordination of data, including sample location, chain-of-custody records, sample results, and sample analyses, can become an issue if the logistics of the ASAP data management program have not been laid out and tested beforehand, resulting in the inability of the field manager to make timely decisions. Adaptive sampling requires a high degree of coordination and control of field-level decision making because sampling points are not predetermined. The ability to make decisions in the field in response to sampling results is what makes

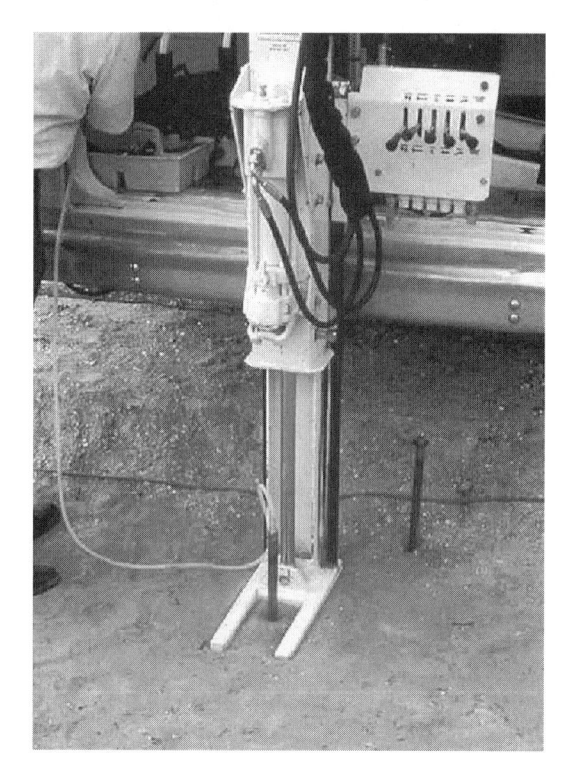

FIGURE 2.135
Soil-gas surveys often employ DP rigs to install probes in the shallow subsurface to collect samples of soil vapor.

adaptive sampling efficient. If timely decisions cannot be made, the value of adaptive sampling is diminished. A typical ASAP program includes some type of in-field database management system along with some form of GIS for data display. Good qualitative support, which includes using on-site technical staff with an accurate understanding of sampling program progress, is a prerequisite for quantitative decision making (Robbat, 1997).

In an ASAP, once the initial sample analytical data are obtained, the CSM is evaluated for accuracy. Typically, several rounds of sampling data are required before confidence in the CSM is obtained. The number of sampling rounds made during the same mobilization is dependent on the DQO specifications for confirming the absence of contaminants in areas thought to be clean (conditions for no further action) and for determining the extent, direction, concentration, and rate of contaminant movement, and the volume of contaminated soil and its risk to human health and ground water.

Quantitative decision support for ASAPs requires the ability to estimate contaminant extent based on sampling results, determines the uncertainty associated with those estimates, measures the utility expected from additional sampling (i.e., reductions in uncertainty), and finds new sampling locations that provide the most value. Quantitative decision support for ASAPs must take into account two general characteristics of contamination at hazardous waste sites. The first characteristic is that while there may be initially few, if any, hard data (i.e., results from the analysis of collected samples) available, upon which a sampling program may be based, there is typically a wealth of other pertinent "soft" information. Soft information refers to all other types of data that might be available for a site, including site maps, aerial photographs, results of nonintrusive geophysical

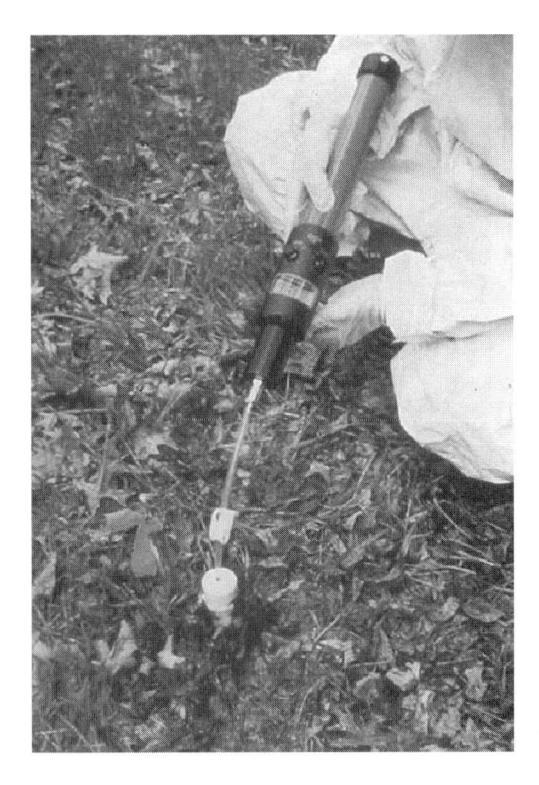

FIGURE 2.136
Soil-gas probes are usually screened with a PID or FID to determine the presence or absence of VOCs prior to collecting samples.

surveys, historical information concerning the nature and source of contamination, and other similar information. This information, while not absolutely conclusive regarding the presence or absence of contamination above action levels at any particular location, contributes significantly to the understanding about the probable location of contamination. The second characteristic is that spatial autocorrelation is usually present at hazardous waste sites and must be accounted for when drawing conclusions from discrete sample results. When sample results are correlated and the level of correlation is a function of the distance separating the samples, spatial autocorrelation exists.

U.S. DOE (2001) suggests using a combination of Bayesian analysis with geostatistics to guide ASAP design and implementation. Bayesian analysis can be used to merge soft information about the probable location of contamination with hard data that might be available for the site. Geostatistics can be used to interpolate sampling results from locations where hard data exist to other locations that lack hard data. Geostatistics is grounded in the presence of spatial autocorrelation — the fact that two samples collected very close to each other will have results that are similar, but samples separated by a large distance may have results that are totally unrelated. For the purposes of contaminant extent delineation, the primary issue is not the absolute value of a contaminant observed but whether that value exceeds some action level or cleanup goal. In this context, sample results can be reduced to a value of either 0 or 1. A value of 0 is assigned if contamination above action levels is not detected, and a value of 1 is assigned if it is detected. A specialized form of geostatistics called indicator kriging can be used to interpolate these values and determine the spatial distribution of contamination above and below action levels.

FIGURE 2.137
Soil-gas samples are collected for on-site analysis in a field-portable GC using an air sampling pump connected to a Tedlar sample bag. The sample will be analyzed immediately.

With most ASAPs, the common approach is to sample contaminated areas (i.e., hot spots) more heavily than other areas, which is rarely the case in traditional site characterization programs. Therefore, if semiquantitative or quantitative field analyses are performed, no additional confirmatory samples are necessary other than those that would typically be necessary to verify data from one fixed-base lab versus another. Off-site lab analysis would be performed only when on-site quantitative analysis is not possible or cost effective.

U.S. DOE (2001) recommends an ASAP design and implementation process for contamination delineation that follows these steps:

- A set of decision points forming a regular grid is laid across the site. Decision points are so named because at each point a decision will have to be made — based on the available information — will this point be considered clean (i.e., the probability of contamination above the prescribed action level at this point is acceptably low) or contaminated (i.e., the probability of contamination being present at this point is unacceptably high). The acceptable level of uncertainty serves as the criterion for differentiating between decision points that can be considered clean and points that must be treated as contaminated. For example, the acceptable level of uncertainty may be set at 0.2 — a decision point with probability of contamination greater than 0.2 will be considered contaminated, while decision points with probability of contamination less than 0.2 will be considered clean. This value must be selected before the program begins with mutual agreement from the stakeholders and regulators involved.

This treatment of uncertainty is consistent with the Type I and Type II error analysis advocated by the U.S. EPA DQOs approach to environmental restoration decision making (U.S. EPA, 2000a, 2000b).

- On the basis of soft information available for the site, a probability is assigned to each decision point that captures the investigator's initial beliefs about the probability of contamination above action levels at that location. In some cases, the investigator may be absolutely sure that soil contamination would be found. In other cases, the investigator may be absolutely sure that soil contamination could not exist. Yet in other areas, investigator may not be able to draw any conclusion at all concerning the likely presence or absence of contamination (i.e., there is a 50–50 chance that contamination is present).

- If sample results are initially available, the probabilities at each of the decision points are updated with these hard data. Johnson (1996) provides a detailed description of how Bayesian analysis can be combined with indicator geostatistics to accomplish the required updating. The site is then broken into three regions: (1) regions where the probability of contamination associated with decision points is below the predefined acceptable level of uncertainty — these regions are accepted as clean with perhaps only minimal confirmatory sampling; (2) regions where the probability of contamination is so high that there is no need for sampling to confirm the presence of contamination; and (3) regions where the probability of contamination above action levels is neither very low nor very high — regions that represent areas of uncertainty in the context of the presence or absence of contamination above prescribed action levels.

- The final step is actual sampling. There are several alternative decision rules that can be used to drive data collection. The U.S. DOE (2001) approach is to focus on maximizing the areas classified as clean, that is, areas that have an acceptably low probability of contamination above action levels being present.

This decision rule tends to produce an ASAP that starts at the fringe of known contamination and works its way in. As data are collected, the underlying probability model is updated, the value of collecting additional information is evaluated, and additional sampling locations are selected that maximize the area classified as clean. Sampling stops where the additional value of sampling no longer warrants the investment. This becomes a simple cost calculation that weighs sampling and analysis costs with the expected volume of soil that might be reclaimed as "clean" and hence, remediation costs that are avoided if sampling moves forward.

Regardless of the decision rule used, the process is the same. Sampling locations are selected, which provide the most benefit in the context of the selected decision rule. These would be sampled, their results analyzed, the probabilities of contamination associated with the decision point grid updated with the sample results, the extent of contamination determined again along with the number of "uncertain" decision points remaining, and a decision made regarding whether additional sampling locations are justified. If so, the next best set of locations would be selected and the process carried through another iteration. When the expected gain in information from additional sampling no longer warrants the costs of collecting and analyzing additional samples, the program stops.

Field-based analytical results will differ from off-site laboratory results for VOC-contaminated soil samples, with off-site lab results generally producing lower measurement concentrations because of analyte loss during sample transport and storage. Care must be taken when comparing these two types of data. Because site investigation and cleanup decisions using an ASAP are made based on field data, off-site laboratory analysis should be performed on no more than 10% of the samples analyzed quantitatively in the field (Robbat, 1997). Field techniques that produce different data quality with the same instrumentation offer cost advantages over analytical techniques that produce either screening level or quantitative data. Time and total project cost savings result when the sample load best matches the sample throughput rate of the instrumentation, maximizing the effectiveness of field personnel and equipment.

Finally, field work begins based on the initial conceptual model. As new data are generated, scientists and engineers may disagree over the direction taken in the field investigation. Experience has shown that this will most likely occur based first on field discipline and second on stakeholder bias. One or more changes in direction should be proposed, with start and stop decisions delineated in the dynamic work plan. New results should refine the conceptual model and dictate future directions. Clearly articulated parameters with respect to sample number and DQO specifications obtained as a function of time should be identified in the work plan to set constraints on how long a particular pathway is followed before altering the investigation direction. One member of the investigation team and one member of the regulatory agency involved in project oversight must have final site decision-making authority. Site work stops when answers to the questions posed in the work plan meet site-specific confidence levels established as part of the DQO process. To ensure that site-specific goals have been met, the project team should statistically evaluate the results of its findings (U.S. EPA, 1996b). An adaptive sampling and analysis program focuses staff, equipment, and financial resources in areas where contamination exists, while providing a cursory inspection in areas that pose no or little risk to human health and the environment.

The dynamic work plan is generally accompanied by a series of work documents that are written to follow an adaptive or dynamic decision-making strategy and target specific subjects. These include an FSP, a QAPP, a DMP, and a Site HSP. These plans provide a much higher level of detail than the dynamic work plan and often include very specific SOPs required to conduct daily activities. All of these documents must support an on-site dynamic decision-making process and discuss how overall decision uncertainty will be managed.

Supporting Work Plans

Sampling and Analysis Plan

An SAP must be site-specific and must bring the sampling and analytical procedures and protocols, the DQOs, and other project requirements together in one clear plan. The SAP should provide a record of how site access, security, contingency procedures, and management responsibilities are to be handled, document the equipment and procedures used during all sampling events conducted at the site, and describe the project requirements for all field and laboratory activities, data assessment activities, and deliverables. The procedures and protocols specified in the SAP should be consistently followed throughout the life of the project. If the SAP is modified during the life of a project, the modifications must be considered when evaluating the data generated from the project. Any deviations from the SAP, including reasons for the deviations, should always be clearly documented (Thurnblad, 1995).

The length and level of detail included in the SAP will depend on the project's complexity and any specific regulatory requirements. For a small, simple project, the SAP may be fairly short and simple and written as a single document. For a large and complex site, the SAP may include a number of other separate and distinct planning documents, including a QAPP, a FSP, a HSP, and a DMP.

In general, the SAP should include a number of important elements, grouped into sections as follows:

Project background: This section should include a brief summary of the project including the site name and location; site size; site ownership history including the name and address of the current site owner and operator; authority under which the work is to be performed; and the purpose and scope of the SAP. The inclusion of a map noting the location of the project is advisable.

Site description and history: This section should include a description of the topography, soils, geology, hydrology, hydrogeology, and climatic setting of the site. Other relevant information includes the locations of buildings and roads, product storage and transmission or waste management facilities, descriptions of site operations related to potential contaminant sources, discharge point locations, and chemical use history. This section should also include a brief history of the site in terms of present and former land use, site activities, reported spills, and product storage or waste disposal activities that may have contributed to potential contamination over the years.

Previous investigations: This section includes a discussion of any previous investigation activities and other response activities that may have been conducted at the site, including those conducted for environmental, geotechnical, engineering or other purposes.

Project objectives: This section explains the purpose of the project, the regulatory framework under which the work is being conducted, what goals and objectives are to be met, what questions are to be answered, and what decisions are to be made. This section sets the stage for the preparation of the remainder of the SAP.

CSM: This section summarizes all of the available information on the site (particularly those aspects related to potential migration and exposure pathways) in a few simple maps, diagrams, cross-sections, and perhaps a narrative format and identifies all of the data gaps that need to be filled during implementation of the FSP to make project decisions. This defines the scope of the field investigation.

Addressing data gaps: This section provides the plan for collecting data to fill the data gaps identified in the previous section.

DQOs: This section describes how data will be used to make project decisions and serves as a general scoping guide for data acquisition activities defined in the FSP.

FSP: This section provides details on the specific data and sample collection and analysis activities that are designed to support the objectives of the project. These activities may be divided into four broad categories: (1) source characterization, (2) geologic characterization, (3) hydrogeologic characterization, and (4) chemical characterization. Included in this section should be the following:

  a. A schedule for conducting the field investigation and reporting the results
  b. Information on site access and security arrangements
  c. Assigned sampling team personnel and their duties
  d. Procedures for completing sample chain-of-custody forms and other data acquisition forms
  e. A description of QA/QC procedures to be implemented during the project

f. A description of the types of IDW that will be generated during the project and how these wastes will be collected, stored, transported, and treated or disposed

g. The general sampling and analysis strategy to be used, including the following (all in the context of the dynamic work plan):

   i. The environmental media to be addressed

   ii. The types, concentrations, and forms (i.e., phases) of chemical parameters (contaminants) to be measured and sampled

   iii. The analytical methods (and their detection limits), equipment, and procedures for conducting field-based and fixed lab analyses (and contingency plans for selecting alternate methods)

   iv. The methods, equipment, procedures, and protocols for collecting data and collecting samples (and contingency plans for selecting alternate methods)

   v. The types of samples to be collected

   vi. The locations and numbers of each of the types of samples to be collected in the initial round of sampling

   vii. A description of the procedure used for determining the locations of subsequent sampling points in the field

Additional detail on preparation of SAPs is available in Chapter 15, and in a variety of other sources, including U.S. Army Corps of Engineers (2001), U.S. Air Force (1997), U.S. EPA (1985), Thurnblad (1995), and Lesnik and Crumbling (2001).

Quality Assurance Project Plan

The QAPP is a tool for project managers and investigation planners to document the type and quality of data needed for environmental decisions and to describe the methods for collecting and assessing those data. The QAPP integrates all technical and quality aspects of a project, including planning, implementation, and assessment. The purpose of the QAPP is to document planning results for the site characterization project and to provide a project-specific "blueprint" for obtaining the type and quality of environmental data needed for a specific decision or use. The QAPP documents how QA and QC are applied to the project operations to assure that the results obtained are of the type and quality needed and expected. The QAPP also describes the policy, organization, and functional activities necessary to achieve project DQOs.

The QAPP is a formal document describing in comprehensive detail the necessary QA, QC, and other technical activities that must be implemented to ensure that the results of the work performed during a project will satisfy the stated performance criteria. The QAPP must provide sufficient detail to demonstrate that:

- The project technical and quality objectives are identified and agreed upon by all interested parties
- The intended measurements, data generation, or data acquisition methods are appropriate for achieving project objectives
- Assessment procedures are sufficient for confirming that data of the type and quality needed and expected are obtained
- Any limitations on the use of the data can be identified and documented

The QAPP must integrate the contributions and requirements of everyone involved in the project into a clear, concise statement of what is to be accomplished, how it will be done, and by whom. It must provide understandable instructions to those who must implement the QAPP, such as the field sampling team, analytical chemists, modelers, and data reviewers. To be effective, the QAPP must specify the level or degree of QA and QC activities needed for the project. Because this will vary according to the purpose and type of work being done, a graded approach should be used in planning the work. This means that the QA and QC activities applied to a project will be commensurate with the purpose of the project, the type of work to be done, and the intended use of the results.

The QAPP is generally composed of standardized, recognizable elements covering the entire project, from planning, through implementation, to assessment. These elements are presented in that order in the QAPP and are arranged into four general groups, which include (U.S. EPA, 1999a):

- *Project management:* The elements in this group address the basic area of project management, including the project history and objectives and roles and responsibilities of the project staff. These elements ensure that the project has well-defined objectives, that the project staff understand the objectives and the approach to be used, and that the planning outputs have been documented.
- *Data generation and acquisition:* The elements in this group address all aspects of project design and implementation. Implementation of these elements ensures that appropriate methods for sampling, measurement and analysis, data collection or generation, data handling, and QC activities are employed and properly documented.
- *Assessment and oversight:* The elements in this group address the activities for assessing the effectiveness of the implementation of the project and associated QA and QC activities. The purpose of assessment is to ensure that the QAPP is implemented as prescribed.
- *Data validation and usability:* The elements in this group address the QA activities that occur after the data collection or generation phase of the project is completed. Implementation of these elements ensures that the data conform to the specified criteria, thus achieving the project objectives.

Additional detail on U.S. EPA requirements and guidance for QAPPs can be found in U.S. EPA (1989b, 1998c, 1999a, 2000d, 2001b). Examples of generic QAPPs assembled by other Federal agencies include U.S. Air Force (1998) and U.S. Geological Survey (1997).

Health and Safety Plan

Any environmental site characterization project conducted at a site known or suspected to be contaminated must have a site-specific HSP written for the benefit of those conducting the investigation. The very nature of the work conducted at these sites demands the creation of a plan to deal with the hazards and dangers that may be encountered, including hazardous substances, the possibility of drilling through buried utilities, the dangers of working with heavy equipment, and the possibility of creating a spark that initiates a fire, among others. A HSP serves to satisfy regulatory requirements of the Occupational Safety and Health Administration (OSHA) (29 CFR 1910) and contractual requests for work conducted at many industrial facilities but, more importantly, it serves as a tool to protect workers from potentially hazardous situations. A good HSP anticipates unexpected situations, provides a plan for dealing with those situations, and provides workers with a means of coping with those situations.

A good HSP must address a wide range of topics, including some that may seem to be unrelated to health and safety. The areas that should be addressed in the plan include:

- Safety staff organization and responsibility of key personnel
- Safety and health hazard assessment for site operations
- Personnel protective equipment requirements
- Methods to assess personal and environmental exposure
- Standard operating safety procedures, work practices, and engineering controls
- Site control measures
- Required hygiene and decontamination procedures
- Emergency equipment and medical emergency procedures
- Emergency response plan and contingency procedures
- Logs, reports, and record keeping

Chapter 19 covers these and other aspects of health and safety planning in great detail. A model HSP for environmental investigations can be found in Maslansky and Maslansky (1997); procedures for evaluating HSPs can be found in U.S. EPA (1989c); and health and safety planning for remedial investigations is covered in U.S. EPA (1985).

Data Management Plan

The ability to manage and easily use all of the data generated by the field investigation is critical to the success of an ASC/ESC/Triad/DFA approach to environmental site characterization. The DMP, which describes how the data collected during the field investigation will be managed throughout the field investigation, is the key to conducting an effective dynamic site characterization project. The on-site technical decision making that guides the dynamic field characterization and sampling activities used in an ASC/ESC/Triad/DFA approach requires a much higher level of data management than is typically found in traditional site characterization programs.

To serve its purpose, the DMP must specify the following:

- The staffing requirements for overseeing all aspects of the data management program, including the responsibilities of all staff
- The hardware and the software packages used for storing, organizing, reducing, analyzing, and presenting the data
- The means of incorporating, organizing, coordinating, and integrating a variety of types of data from a variety of sources and formats into a comprehensive site database
- The means and formats (electronic and other) used for entering data into the site database
- The procedures used for verifying and validating chemical and non-chemical (geologic, hydrologic, and other) data
- Procedures used for conducting a QA review of data collected during the field investigation

- The means for providing access to the data to both on-site field personnel and off-site interested parties (i.e., the site owner and operator, regulators, legal counsel, and other stakeholders)
- The means of analyzing and interpreting data and presenting it in easily understood formats
- The means (and frequency) of producing data summaries for use by on-site personnel to assist them in planning subsequent activities at the site
- The means of summarizing and presenting data for inclusion in the final report

The development of a DMP that can provide the necessary level of support to the field manager and the remainder of the field team is a process that requires a great deal of thought and expertise. Development of the plan begins with incorporating certain elements of the site-specific SAP, including the objectives of the investigation and information on the locations of data collection points, the types of analyses performed on samples, and the types of samples and data that must be generated and recorded to support project objectives. The DMP determines the data flow paths that must function for the site characterization program to work, including the interrelationships between the data management system and the field team and the intrarelationships within the system (Olson, 1991). The computer equipment and software must be specified and should be field tested with increasing amounts and complexity of data to ensure that the hardware and software perform as expected. After work at the site has started, the DMP is important in documenting the procedures for handling data from various sources, including the field technicians and the analytical staff.

For most dynamic environmental site characterization programs, the majority of data will be generated electronically. The level of effort required to produce rapid turnaround times will be increased if the field-generated (analytical, geologic, hydrologic, and other) data are not produced electronically and have to be entered into the database by hand. Because of the nature of much of the nonanalytical data collected, it is usually in written form. Written field records must be mated with sample chain-of-custody records to match field data collected from sampling points with analytical results from those same points. After the analytical results are received from either the on-site lab or field analysts, they can be entered into the database and linked with the sampling points. For most projects, all field-generated data will eventually need to be reduced to an electronic format for use in modeling, data visualization programs, and other data manipulation systems.

Data verification and validation is a very important element of a good DMP. Analytical data should be verified and validated by chemists to ensure that the data are verified as complete, with known ranges, and have fulfilled the requested analyses. Nonanalytical data must also be verified and validated (generally by a geologist or hydrogeologist) in the same time frame as analytical data to ensure that field decisions are based on accurate information from all sources and an accurate depiction of subsurface conditions. Data verification and validation QA/QC procedures are discussed in detail in U.S. EPA (2000e, 2001c).

Data summaries are also important, as they are used to show the status of the investigation and also as the primary tool to plan future investigative work at the site. These summaries may be issued in a variety of formats, but the most useful are graphic (i.e., maps and cross-sections depicting results of work done to date) and tabular information on all sample locations, sampling points, and analytical results. Software packages that allow the depiction of data in three-dimensional images (i.e., data visualization software) are very useful in summarizing chemical, geologic, and hydrogeologic data.

Because ASC/ESC/Triad/DFA approaches to environmental site characterization generally rely on a daily evolution of the data, data summaries may need to be prepared on a daily basis.

The importance of a good DMP cannot be overemphasized. All aspects of the DMP must be in harmony to accommodate the fact that data from different sources are interrelated and must be effectively linked to be useful to project planners. Properly developed and implemented, a good DMP allows a tremendous amount of data to be handled, interpreted, and reported effectively and allows generation of timely and informative data summaries. This allows the field manager to keep the project moving forward to completion as rapidly as possible.

Data Analysis, Evaluation, and Interpretation: Revising the CSM

The project team member in charge of data management is responsible for coordination of site activities to ensure that all data collected in a given time frame are incorporated into the site database and made available to the project team for data evaluation, analysis, and interpretation as rapidly as possible. This is important to keep the project moving forward in a timely manner. The hydrogeologic, geologic, soil, chemical, and other data collected during the field investigation must be periodically assembled by the field manager (usually on a daily basis) and evaluated on-site by the appropriate project staff. Data evaluation must be completed before data analysis and interpretation to determine whether data quality requirements are met and whether the data can be used for their intended purpose. This will lead to validation of properly collected data and exclusion of improperly collected data. Considerations for data validation include the following:

- QA/QC results (instrumentation calibration checks, duplicate analyses, field blanks, equipment blanks, etc.)
- Comparison of higher quality level data with lower quality level data
- Consistency of results among analytical methods and sampling techniques
- Comparison with results from other media
- Comparison with other COCs or indicator parameters
- Comparison against previous data (if available)
- An evaluation of whether the data fit or make sense in the context of site conditions

After the validity of the data has been assessed, the valid data can be used in data analysis and interpretation.

The primary focus of the ASC/ESC/Triad/DFA process is the use of multidisciplinary integration and interpretation of field measurements and sampling results to provide information used to refine the CSM and construct a more accurate picture of site conditions. These data (and the revised CSM) then serve as the basis for selection of the type and location of subsequent field measurements and sampling points. This process continues on an iterative basis throughout the life of the project. Data analysis and interpretation and revision of the CSM are generally done on a daily basis to provide a foundation for planning the next day's field work, although it may be done more or less frequently, depending on the size of the site and the complexity of the investigation.

While the collection of geologic, hydrogeologic, and chemical data in the field is a relatively straightforward mechanical process, the analysis and interpretation of those data is much more difficult and complex and requires special expertise and substantial experience. Interpretation of environmental data must be done by experienced professionals with specialized knowledge and extensive training in the examination of specific aspects of the data (Regan et al., 1991). For example, an experienced field geologist should be in charge of interpretation of complex geological data, an experienced analytical chemist (or geochemist) should interpret complex sets of chemical data, and an experienced field hydrogeologist should take the lead in interpreting complex hydraulic head and other hydrogeologic data. The interpretation of environmental data is an art and, by definition, requires the conscious use of skill and imagination. Because these qualities in most scientists are highly personalized, the interpretations of data among individuals are often unique and imprecise. Data must often be extrapolated between widely spaced boreholes, under situations in which correlation may seem impossible because the borehole conditions seem so different. Under these conditions, drawing lines on a map or a cross-section to indicate correlations or boundaries where few pieces of data are available requires great skill, imagination, and knowledge of environmental processes. Studying available surface exposures of subsurface materials (i.e., nearby road cuts, streamcuts, fault scarps, foundation excavations, or cuts in quarries or mines) can reduce the guesswork associated with extrapolation of subsurface geologic data. However, the most valuable asset to anyone attempting to interpret subsurface data is the ability to visualize a problem in three dimensions.

Graphical presentations of data, such as maps, cross-sections, flow nets, and graphs, are extremely helpful in data analysis and interpretation and in presenting data to interested parties. Graphical presentations of data facilitate interpretation of spatial relationships between data points and, when time-series data are available, can indicate the presence or absence of trends. Plan (map) views of data, cross-sections or fence diagrams, and data contouring methods are all very useful methods for data presentation. The degree of detail of the graphical presentations of site conditions varies according to the objectives of site characterization, the complexity of site geology, hydrogeology and geochemistry, the nature and number of source areas, and the nature and distribution of the contaminants of concern.

A map of the area being characterized provides essential information about the land surface including natural and anthropogenic topographic features, land uses, surficial geology, and the locations of sampling points and other data acquisition points (i.e., geophysical or CPT measurements). Maps also serve as the basis for data contouring efforts and can help tie in contaminant source areas with contaminant plumes.

Cross-sections should identify actual surface and subsurface observation points according to elevation and location. Cross-sections showing correlation of stratigraphic or lithologic units and interpretations of other conditions (i.e., contaminant locations) between direct subsurface observations (i.e., sampling results) and indirect observations (i.e., geophysical survey results) should be indicated as interpretations based on standard geologic procedures. Solid lines should be used where information demonstrates definite correlations, and dashed lines should be used where information is sketchy or less complete. Cross-sections should be accompanied by a narrative presentation describing anomalies or otherwise significant variations in the site conditions that might affect any data interpretations. Additional exploration should be considered if sufficient information is not available on critical parts of the subsurface (i.e., boundaries between aquifers and confining beds) to develop accurate cross-sections, with realistic descriptions of anticipated variations in subsurface conditions, to meet project requirements.

Contouring methods, such as constructing structure contours of the top of buried bedrock, potentiometric surface contours of water levels in a confined aquifer, flow nets to describe the hydraulic relationships between aquifers beneath a site, or isoconcentration contours of a particular chemical species in soil, soil gas, or ground water, are very valuable tools for data interpretation. Contours of horizontally planar (or nearly planar) surfaces may be presented on maps, and contours of chemical or hydraulic head data can be presented on both maps and cross-sections, to depict, for example, the three-dimensional distribution of hydraulic head or contaminants. Three-dimensional contouring can be done with more sophisticated computer imaging or data visualization programs, often with very impressive results. All contouring should be done using appropriate interpolation techniques and the method of interpolation documented in an accompanying narrative. It is often a good practice to contour the same data using several different interpolation methods and to compare the results.

Methods for analyzing time-series data include the use of bar charts, graphs, and piper or stiff diagrams. These are described in detail in Chapter 17.

Statistical methods used to analyze data should be appropriate for the data type. Most conventional statistical methods assume a normal distribution around the mean. Typically, environmental data do not exhibit normality because of spatial autocorrelation, the presence of outliers, or other effects. Therefore, geostatistical methods are considered best for analyzing spatially related data. Chapter 17 provides additional detail on the use of statistical methods in analysis of environmental data.

As shown in the flowchart in Figure 2.101, the CSM is refined in an iterative process of data collection, evaluation, and interpretation. Compilation of the data onto simple 2D graphics is sufficient for on-site data interpretation at small, simple sites. Using an interactive data processing program, to combine sample location information with depth information, in combination with a sophisticated 3D computer imaging or data visualization program, may be feasible for larger, more complex sites. In either case, the focus is on continually updating the graphics prepared during generation of the initial conceptual model. As the investigation proceeds, the maps, diagrams, cross-sections, and flow nets are continually revised by incorporating new data and new interpretations are made to reflect the incorporation of updated information. For example, geologic contacts are repositioned using new geologic data, cross-sections are redrawn using new borehole lithologic data, hydraulic relationships within and between aquifers are reconsidered using new hydraulic head data, and new isoconcentration contours are drawn using new geochemical data. The revised CSM can then be used to make specific predictions regarding the conditions anticipated at subsequent sampling or data collection points. Using field-generated graphics, the field manager can direct the investigation to test the predictions, fill in the data gaps, resolve anomalies or explain outliers, and resolve differences between anticipated and actual results from prior sampling rounds. As new data are collected and the investigation proceeds, differences between the initial CSM and the data obtained during site characterization are used to adjust data collection activities and the sampling and analysis program in an iterative manner until all relevant site conditions are accurately defined. The daily cycle of data collection, evaluation, and interpretation continues until the field manager, in consultation with the other project team members, the site owner and operator, and the regulatory authority determines that the objectives of the site characterization program have been satisfied or that constraints to the investigation are such that they prevent additional characterization. Typically, the site characterization is complete when the CSM no longer changes with the incorporation of additional data, when no major unexplained anomalous observations remain, and when sufficient detail in depiction of site physical and contaminant-related conditions has been achieved to fulfill the requirements of the investigation.

Preparing a Final Report

All types of site characterization projects, whether completed using the conventional or traditional approach, or an ASC/ESC/Triad/DFA approach, will require the production of a comprehensive final report to document the results of the project. Reports should be well organized and concisely written and should contain a variety of graphical and tabular displays of important information. The report for an environmental site characterization project should include the following information at a minimum:

- A description of the purposes and objectives of the site characterization program
- The location of the site investigated, in terms pertinent to the project. This may include a large-scale regional map on which the site location is identified (a USGS 1:24,000 topographic map usually works well) or a current aerial photo on which the site location is identified, or both
- A description of the regional- and site-specific physical setting (topography, geology, soils, hydrology, hydrogeology, geochemistry, climatic setting, background soil and ground-water chemistry, biology, etc.)
- A description of the history of the site, the former and current activities at the site, the site facilities related to the objectives of the investigation (source areas, including product storage and transmission facilities, waste storage and disposal facilities, discharge points, etc.), the types of chemicals, products, or wastes handled at the site (and their chemical and physical characteristics, fate and transport characteristics, maximum contaminant levels, or other regulatory action levels, as available), and the history of any releases, spills, or discharges known to have occurred at the site
- A small-scale map of the site and surrounding area, with all features pertinent to the site-characterization project (site boundaries, source areas, land uses on site and adjacent properties, buildings, roads, surface-water bodies and wetlands, man-made drainage features, underground utilities, locations of public and private wells and springs, and locations of potential receptors [human and ecological]) identified and with surface elevation contours and significant geomorphic features (sinkholes, bluffs or cliffs, fault scarps, streamcuts, etc.) marked
- A description of all of the investigation methods and procedures used during the data collection and analysis portion of the site-characterization program
- A small-scale map of the site and cross-sections in the direction of and orthogonal to the direction of ground-water flow, on which the locations of all sampling and data collection points (soil borings, DP and CPT holes for soil, soil-gas and ground-water sampling and *in situ* measurements, wells, vertical profile locations, surface geophysical survey lines, etc.) are identified
- A narrative summary of the results of the investigation, including all field measurements and observations, results of all sampling and analytical activities conducted at the site, descriptions of any spatial or temporal variations or trends in the data, descriptions of contaminant distribution in all phases, predictions on contaminant fate and transport, and any limitations to the use of the site data
- A description of the QA/QC measures implemented during the investigation

- A graphical presentation of all borehole logs, well construction logs, vertical profiling (e.g., DP or CPT) logs, and geophysical measurements
- A tabular presentation of all field measurements and observations, all sampling and analytical results, and all other data collection results
- A graphical presentation of the final CSM, including all sampling and analytical results and other data collection results for all environmental media, showing all data interpretation (geologic maps and cross-sections showing units of geological significance; hydrogeologic maps and cross-sections showing perched zones, the water table, confining zones, piezometric surfaces in confined zones, and directions of flow and gradients in and relationships between all hydrogeologic units; contaminant isoconcentration contour maps and cross-sections for soil, soil gas, and ground water; maps and cross-sections showing contaminant movement pathways and human exposure pathways; graphs of time-series hydraulic head or chemical data, etc.)
- Conclusions of the investigation, including a summary of the risks posed to identified receptors, recommendations for further work at the site, including no further action, short- or long-term monitoring, natural attenuation, risk-based corrective action, active remediation (and identification of potentially useful remedial methods) or some other course of action
- References cited in the final report, including all relevant reports of previous investigations at the site and at adjacent sites, any regional geology and hydrogeology reports, and any other references pertinent to the subject of the investigation
- Appendices, including all raw data (analytical and other data), QA/QC evaluations, and other detailed information available from the investigation

Because all of the improved approaches to site characterization provide for data evaluation, analysis, and interpretation in the field, report writing for projects using one of these approaches can often be significantly streamlined in comparison to projects using the conventional or traditional approach. This is countered somewhat by the need to include substantially more data in a report for a project using and ASC/ESC/Triad/DFA approach than is typically found in a report produced for a project using the conventional or traditional approach. However, because experienced staff are more involved with the field work in a project using an ASC/ESC/Triad/DFA approach, they generally require less time to review and become familiar with the documentation in preparation for writing the report.

Additional guidance on preparing final reports for environmental site characterization projects can be found in U.S. EPA (2003) and ITRC (2003).

Field-Based Analytical Technologies

A critical component of any successful dynamic environmental site characterization program is a well thought out field analytical program. A properly designed field analytical program will facilitate timely revision of the CSM in the field and allow real-time decision making during site characterization. When properly designed and implemented, the field analytical program will provide accurate chemical information so investigators in the field can use this information to select the next sampling location. This will minimize the number of mobilizations necessary to characterize a site which, in turn, will reduce overall project costs.

The Role of Regulatory Agencies in Adopting the Use of Field Analytical Technologies

Recognizing the importance of obtaining accurate analytical data in the field, the U.S. EPA issued a Policy Directive (9380.0-25) in April 1996 to "openly encourage the evaluation and use of new field measurement and monitoring methods" (U.S. EPA, 1996c). This Policy is embraced by a number of different branches within the U.S. EPA (Table 2.5) that support the use of innovative cleanup and field measurement technologies.

Historically, the biggest barrier to the use of established and emerging field analytical technologies has been the reluctance of end users to incorporate them into a work plan because the answer to the question "Will the regulatory agency accept the data generated by this method?" has typically been "no." There are several reasons for this response: (1) a lack of understanding of the capabilities and limitations of the methodologies on the part of regulatory agency personnel and the end user; (2) historical applications of field analytical methods were as qualitative screening tools, and compound identification was usually not possible; (3) early attempts at quantitative field analyses did not incorporate sufficient QA/QC measures to validate the data being generated; and (4) a lack of availability of many options for field sample analysis, with regard to the types of samples that could be analyzed or the parameters that could be detected. The U.S. EPA Directive is intended to increase awareness at both the end user and state and Federal regulatory agency levels and to facilitate increased approval and use of field analytical tools during site characterization. By encouraging the use of field analytical tools, the Directive seeks to improve environmental decision making, while reducing the cost and time required to remediate a site. The Directive also provides guidance on how to use these technologies effectively.

The Directive also indicates that a re-examination of DQOs is appropriate on the regulatory agency level so there is not a continued insistence on unnecessarily using more costly and overly stringent laboratory-based SW846 analytical methods for all data generated during site characterization, to the exclusion of more cost-effective methods. This is a significant change in philosophy for many environmental professionals that is taking time to come into practice. To assist in this transition, OSWER and the U.S. EPA Office of Research and Development (ORD) have created many outreach and support groups (many of which are Internet based) to encourage the use and development of new and innovative measurement and monitoring technologies including:

- Vendor Field Analytical Characterization Technology System (Vendor FACTS) (U.S. EPA, 1998a)
- Superfund Innovative Technology Evaluation (SITE) Programs

TABLE 2.5

U.S. EPA Branches Supporting Policy Directive 9380.0-25

Office of Solid Waste and Emergency Response (OSWER)
Office of Emergency and Remedial Response (OERR)
Office of Underground Storage Tanks (OUST)
Office of Chemical Emergency Preparedness and Prevention (CEPPO)
Technology Innovation Office (TIO)
Office of Enforcement and Compliance Assurance (OECA)
Federal Facility Leadership Council
EPA Brownfields Coordinators

- Environmental Technology Verification (ETV) Program (U.S. EPA, 1997c)
    a. ETV's Consortium for Site Characterization Technology (CSCT)
- REACH IT, sponsored by U.S. EPA's Office of Superfund Remediation and Technology Innovation (OSRTI) (U.S. EPA, 2004c)
- Federal Remediation Technologies Roundtable (FRTR)

By increasing collaborative efforts with vendors developing new field analytical technologies through these outreach and support groups, U.S. EPA has shortened the time frame from 30 months to less than 18 months for updates of SW846 field-based analytical technologies (U.S. EPA 1998a). This reflects the desired shift in emphasis from "prescriptive" testing procedures to a focus on "performance-based" field measurements.

CSCT was initially established as a pilot program under the ETV Program as a consensus-based group charged with the responsibility of evaluating and validating the performance of site characterization and monitoring technologies. CSCT collaborates with the Interstate Technology and Regulatory Council (ITRC). The ultimate goal of CSCT and ITRC is to fast track the acceptance of new, proven field technologies. A number of technology evaluation reports (e.g., U.S. EPA, 1997b, 1997c, 1998b, 1999b) and verification statements for field analytical technologies have been completed, including field-portable x-ray fluorescence (XRF) and field-portable GC/MS. Reports on the successful uses of field analytical technologies are also published by OERR, ORD, and TIO. Virtually all of these reports are available on-line to end users and regulatory agency personnel. On basis of this information, many state regulatory and nonregulatory agencies are now accepting the use of field sample analysis methods in specific programs, including UST and dry-cleaning facility programs, and some have incorporated field test methods into state reimbursement fund programs.

Advantages of Generating Real-Time Data through Field Sample Analysis

As already discussed in this chapter, field sample analyses provide site investigators with real-time or near-real-time data upon which they can base decisions on how to proceed with site activities during sample collection. This can assist in delineating the presence of hot spots across the site during site characterization or during source removal activities such as soil excavation, to determine when sufficient material has been excavated from an area.

It is important to define the terms "real-time" and "near-real-time" when discussing advantages of field sample analysis. "Real-time" data are generated when the results of an analysis are available either instantaneously, as in the case of PIDs when used in survey mode to determine total VOCs, or within a few minutes without data reduction, as in the case of field-portable XRF analysis of *in situ* soil samples for metals. "Near-real-time" data are data generated through field sample analysis that may require more time to process samples for analysis, to actually perform the analysis or to interpret the data. The results of analysis of samples using field-portable GC/MS instrumentation would be considered near-real-time data. In either case, the time to obtain accurate chemical data using field analytical methods is always less than using conventional off-site analysis. Typically, the turnaround for receiving results from off-site laboratories can run from several weeks to several months, depending on the analyses being performed, the level of QA/QC required during the analyses, and the work load of the laboratory.

Rapid data turnaround allows real-time identification and characterization of contaminant sources, which is the critical first step of many site characterization programs conducted at sites known to be or suspected of being contaminated. Real-time data also permit three-dimensional delineation of the extent and magnitude of the contaminant plume during a single mobilization. This allows more effective fine tuning of the CSM and optimal design of follow-up monitoring or remediation systems, if required. For example, three-dimensional soil sample collection can determine the location of potential preferential pathways for contaminant migration based on grain size and depth to ground water and, when combined with field sample analysis, can also permit identification of real migration pathways based on contaminant chemistry. This information is key to the effective positioning of ground-water monitoring wells (with respect to well-screen placement) and to targeting specific zones in the subsurface for remediation.

Limitations to Developing Strategies that Rely on Real-Time Data

Perhaps the largest hurdle to overcome when proposing to implement a field analytical program is obtaining regulatory approval for the use of the methods and acceptance of the results. With the previously discussed U.S. EPA programs in place, this should become less of a problem. However, for now, there still exists a widespread misconception that only SW846 methods are acceptable for generating valid analytical data. Closely related to this is the tendency of some investigators and regulators to judge the performance of field methods against data generated through analysis of confirmatory samples in a remote fixed laboratory. This results in delays using field data and defeats the purpose of using a dynamic work plan for the investigation. This difficulty can be overcome by implementing an effective field sample analysis QA/QC program, which will permit validation of field data as they are being generated.

An effective field analysis plan should ideally be written by an environmental professional who has both laboratory analytical experience and field sampling experience. This unique combination of experience brings together the knowledge necessary to effectively select the most appropriate analytical tools for the parameters of interest and the understanding of field sampling error and the difficulties associated with sample matrix interferences. During implementation of a field analysis program, it is recommended that trained analytical chemists be available to lead field analysis teams to ensure accuracy and precision during analysis of samples and to provide expertise for data interpretation and QA/QC troubleshooting.

From a project management perspective, the switch to an ASC/ESC/Triad/DFA approach, which, by definition, incorporates field analyses, will require that an increased amount of project resources be available for the initial part of the investigation. Rather than spread site characterization activities over a period of years, as is commonly the case when traditional approaches are implemented, use of an ASC/ESC/Triad/DFA approach could require the equivalent of half (or more) of the project resources to be available in the first few weeks or months of the project — a choice that some facility owners or operators may not find favorable from a cash-flow perspective.

Development of an Effective Field Analytical Program

When developing a field analytical program for any ASC/ESC/Triad/DFA project, a number of important criteria need to be addressed, as summarized in Table 2.6.

As with any field activity, it is important to clearly define objectives. For field analytical programs, typical objectives include determining the locations of hot spots of contamination, defining contaminant levels at property boundaries, identifying contaminants present to determine if there is a health risk to nearby receptors, and locating and

TABLE 2.6

Criteria to be Considered in the Development of a Field Analytical Program

---
Objectives of the program
Type of data required (qualitative vs. semi-quantitative vs. quantitative)
End use of the data
Required level of field QA/QC on sample analyses
Parameters to be analyzed
Matrix to be analyzed (soil, gas, or water)
Anticipated concentrations of expected contaminants (i.e., ppm vs. ppb)
Potential for matrix interference
Field operational conditions (light, relative humidity, temperature, and precipitation)
Operator skill
Reliability of method and associated instrumentation
Time required to perform analyses
Per-sample costs (disposables, equipment, time, waste disposal, etc.)

---

characterizing a source or sources of a contaminant. It is imperative that the project-specific objectives be kept in mind when collecting and interpreting data to ensure that the objectives are addressed.

Tied closely to objectives is the determination of what kind of data are required to be generated by field analyses. The qualitative data generated by some methods simply provide a "yes/no" indication of contamination. This type of data can be generated, for example, when using PIDs or FIDs to analyze samples of soil gas or to analyze the headspace of soil or water samples. When trying to locate zones of apparent VOC contamination to direct subsequent sampling efforts, this type of information may be sufficient. This approach is often referred to as sample "screening" rather than sample analysis. Qualitative screening may be appropriate as a means of focusing successive quantitative analyses by screening out large classes of compounds when results produce negative findings. Positive results or "hits" should be followed up with more quantitative analyses (Parris et al., 1993).

When more detailed information is required, methods that generate quantitative data must be used. Quantitative methods not only identify individual compounds but also determine compound-specific concentrations. This type of information is very powerful when developing an initial understanding of site conditions or when refining the CSM. To be accurate, however, more analytical skill is required, as is a higher level of QA/QC to validate the results generated.

There is also a category of methods in between — semi-quantitative. These methods may be able to provide some information on concentrations, but the concentrations may not be compound specific — they are related to a "family" of compounds. An example of this group of analytical methods is the turbidimetric-based field analysis kit called PetroFlag™. This method uses a chemical extraction process to create an extract fluid for turbidimetric analysis of total petroleum hydrocarbons. Results are compared against a calibration standard (which is generally not compound specific) using a field turbidimeter.

Field Analysis QA/QC Programs

When generating semi-quantitative and quantitative data, it is essential to implement a field sample analysis QA/QC program. There are two general components to the field analysis QA/QC program: those procedures implemented in the field during sample analysis

TABLE 2.7

QC Samples to Include in a Field Analytical Program

Calibration standards
Method blanks
Instrument blanks
Duplicate samples
Matrix spikes and surrogate spikes
Control samples
Confirmatory samples

(e.g., use of calibration blanks) and off-site confirmatory procedures (i.e., where samples are sent to an off-site fixed laboratory). The level of QA/QC required will vary with the objectives of the field analysis program, but it is critical to ensure that there is sufficient QA/QC to permit validation of any results generated, especially if results are being submitted to outside groups (such as regulatory agencies) or where results are close to action levels. It is always better to have a higher degree of QA/QC than may be warranted at the time, than to have an inadequate program, and run the risk of data not being accepted.

A comprehensive field analytical QA/QC program will include the submission of confirmatory samples to a fixed laboratory for analysis. In general, between 5% and 10% of the total number of samples collected and analyzed are submitted for confirmatory analysis.

Table 2.7 provides a list of some of the analytical QC samples that should be included in the field sampling program. There may be instances where a field technology designed to provide semi-quantitative data may not facilitate preparation of one or more of these QC samples. In those cases, QA/QC programs will be somewhat more limited.

Calibration Standards

To ensure accurate quantification of sample analyses, it is imperative that the analytical instrumentation be calibrated in accordance with instrumentation and analytical method specifications. Calibration standards are prepared per the analytical method to create a multipoint calibration curve. The more standards used to create this curve, the greater the accuracy of quantification. Ideally, three calibration standards and a "zero" calibration standard are used to create this calibration curve, and the standards selected frame the anticipated concentrations for the samples to be analyzed. It is important that this instrument calibration be performed in the field under the same operating conditions under which the samples will be analyzed. Single-point calibration curves should be avoided because of the poor degree of accuracy in calibration. It is recognized, however, that instrumentation software used in some hand-held field analytical equipment does not facilitate multipoint calibration. This factor should be considered when selecting instrumentation and methodologies during the development of the field analytical program. As samples are analyzed, it is a good practice to reanalyze the suite of calibration standards with each batch of samples. This confirms the accuracy of quantification of the samples and also ensures that there is no instrumentation drift during analysis, which can occur under field operating conditions over time.

Method Blanks

Method blanks are samples, typically prepared with deionized water from a known source, that are carried through all phases of sample preparation and analysis. The purpose of the method blank is to determine whether contaminants are potentially introduced into samples during sample preparation (e.g., cross-contamination due to poor pipetting techniques), handling (e.g., storage of samples in the vicinity of contaminants prior to analysis), and analysis (e.g., contaminants in extract solvents). It is recommended that one method blank be analyzed for every 20 field samples (U.S. EPA, 1999b). The method blank is also sometimes known as a preparation blank.

Instrument Blanks

Closely related to the method blank is the instrument blank. The purpose of an instrument blank is to determine whether there is any contamination of samples resulting from contamination within the analytical instrumentation. An example of where this QC blank should be run is when a field-portable GC is used for sample analysis. A phenomenon called "column saturation" can occur when a highly contaminated sample is analyzed and leaves behind a residual level of contaminant on the analytical column, which will be detected if another sample is analyzed shortly thereafter. This causes a false positive in the subsequent sample (or samples). An instrument blank is a sample of air, water, or solvent (which one is used depends upon the method being used) that has not undergone any sample processing (i.e., extraction or digestion). This sample is analyzed to determine whether cross-contamination of samples occurs due to contaminants present in the analytical instrument. Instrument blanks should always be analyzed immediately after samples that have high concentrations of a parameter of interest. If a contaminant is detected in an instrument blank, the analyst must implement corrective measures until nothing is detected in the blank.

Duplicate Samples

Duplicate samples are used as part of a field analytical QC program to verify the precision of results generated. To prepare a field analytical duplicate sample (not to be confused with a field sampling duplicate sample), the field analyst generates two separate aliquots of one sample submitted for analysis. Each aliquot is independently processed and analyzed for the same parameter to determine the precision of the analytical system. Duplicate samples should be run at a minimum frequency of one duplicate per 20 field samples. Duplicate results are typically compared as relative percent difference.

Matrix Spikes and Surrogate Spikes

Another check on analytical method precision is to prepare and analyze matrix spikes and surrogate spike QC samples. To prepare a spiked sample, a known concentration of a compound of interest is added to a sample to determine the accuracy of the analytical system. The spiking solution is typically purchased as a certified spiking solution which is accompanied by a certificate of analysis identifying the compound and its concentration. Matrix spike samples are used for both organic and inorganic compounds. To prepare a matrix spike, one aliquot of a sample is analyzed before being spiked to determine the baseline for the sample. A second aliquot is spiked with the parameters of interest at a known concentration. The result of analysis of the spike, reported as percent recovery, is a direct measurement of analytical accuracy.

For inorganic compounds, a third aliquot of sample is spiked in the same manner. A comparison of respective recoveries between the two spikes is a measure of analytical precision. Surrogate recovery spikes are used for organic compounds only. Surrogate recovery spikes measure how well the analytical method works on an individual sample basis. A surrogate compound is a special compound synthetically prepared, which is not found naturally but is similar to several compounds of interest. Surrogate spiking compounds are added to a sample that is treated as an unknown during analysis. Percent recovery is calculated concurrently with concentrations for analytes of interest. Matrix spikes and surrogate spikes should be analyzed at a frequency of one per 20 field samples.

Control Samples

Control samples are used to assess the accuracy of the field analyst and the field test method. Control samples are samples of known concentrations that are analyzed with each set of calibration standards before analysis of the regular samples. The control sample may be commercially prepared or may be prepared in the laboratory and taken into the field. The concentration of the control sample must fall within a specified range for the method to be considered accurate. Ideally, the control sample will have concentrations close to those found in the field samples.

Confirmatory Samples

Confirmatory samples are collected with the objective of supporting proper interpretation of the results of field test kit data and to judge the accuracy of method's results from the standpoint of making correct project decisions. A confirmatory sample is prepared in the field as a duplicate of the same sample that is analyzed on-site with the chosen field test method. The duplicate portion is sent to an off-site laboratory for formal analysis. The results of the on-site analyses are compared to the results of the analyses by the off-site laboratory. The number of confirmatory samples should be sufficient to allow for management of analytical uncertainty so that the use of the field method can be defended as scientifically valid. The number of confirmatory samples will vary between projects depending on the method, the complexity of samples being analyzed, how the end data are being used, and the likelihood of method interferences. If the field analytical QA/QC program is effective, there should be sufficient real-time checks and balances on the quality of data being generated, thus field teams should not have to rely on the results of confirmatory samples to validate information being provided. To do so would remove the benefit of field analysis as an aid in field decision making.

When the field analytical portion of the sampling and analysis plan is written, it is important to establish the appropriate ratio of field samples to QC samples. The ratio of one QC sample for every 20 field samples may be suitable for many programs. However, in cases where a higher degree of scrutiny of data will occur (e.g., during litigation), the ratio may need to be decreased to one QC sample for every 10 field samples. The lower the ratio, the higher the degree of confidence the analyst can have in the accuracy and precision of data being generated. The tradeoff, however, is a reduction in the number of samples that can be run per day and the increased cost associated with performing additional analyses on QC samples.

Selection of Field Analytical Technologies

Field analytical methods are divided into two broad categories: *in situ* methods and *ex situ* methods. *In situ* methods include those technologies that are capable of taking measurements in place without requiring collection of a sample for analysis. For example, there are many *in situ* analytical tools used in conjunction with DP equipment to deploy a sensor, such as a laser-induced fluorescence (LIF) probe, into the subsurface. Because these methods are discussed at length in Chapter 6, *in situ* measurement tools are not discussed here.

Ex situ field measurement techniques all require that a sample of the matrix of interest be collected for analysis. As a consequence, the accuracy and precision of data generated by *ex situ* methods can potentially be affected by sampling error and bias. These issues should be addressed separately in the site-specific sampling and analysis plan to minimize the impact of this source of error and variability.

When selecting the most appropriate field analytical methods for a project, it is important to identify which parameters are anticipated to occur at the site, what media will be analyzed, and what concentrations of contaminants are expected. A list of anticipated contaminants should be developed at the onset of any project to develop an effective HSP (Chapter 19), so this information can also be useful for this application. A number of field analytical technologies have been developed for single compounds or for families of related compounds, so the individual preparing the field analytical program must understand which parameters are detectable with any given technology being considered. Many methods are designed to work with only one type of sample, either water (e.g., ground water and surface water) or solids (e.g., soil, sediment, and sludge). Some technologies can be used for both soil and water samples, but the sample preparation and analytical method may vary as with the type of data generated and the detection limits. For example, if a soil-gas sample is analyzed for volatile constituents using a field-portable GC, data generated will be quantitative. If that same instrumentation is used for soil sample analysis, a headspace method must be used and the data, while still quantitative, are only an indirect indication of what may be in the soil sample, because the soil sample was not digested or extracted for analysis. There may, in fact, be other contaminants still present in the sample that did not volatilize sufficiently to be detected in the headspace sample. The third factor to consider is contaminant concentration. Many reagent kits or test kits are designed to operate within a concentration range, so it is important to estimate whether contaminants are expected to be in the parts-per-million range or parts-per-billion range, for example. In some cases, dilutions can be prepared in the field if unexpectedly high sample concentrations are encountered, but dilutions can potentially introduce a source of error and should be avoided if possible.

The experience of a trained analytical chemist will be important in identifying the potential for matrix interference to occur during sample analysis. Matrix interference can include things such as the presence of one constituent in such high concentrations that it masks one or more other contaminants of interest; the presence of a compound (that may be unrelated to the contaminants of interest) that can change the basic chemistry of the analytical method (e.g., the presence of chlorine can interfere with a number of chemical reactions); or the color of a sample may make it impossible to identify color changes indicating contaminant concentrations (when a colorimetric method is used). Some field test kits indicate known sources of interference for the test method and, in some cases, offer recommendations on how to change field sample preparation or analytical procedures to correct for this interference (e.g., sample filtration). Unfortunately, not all test kits provide helpful information regarding the potential for sample interference during analysis and, therefore, a source of error could be introduced with that particular method.

An effective field analytical program involves discussions between the project analytical chemist and the field personnel who will be performing the analyses. This communication needs to incorporate discussions related to field team experience in sample collection and analysis. In addition, field personnel need to inform project managers about anticipated field operational conditions such as temperature, relative humidity, dust, wind, sunlight, and noise. These factors can have a direct influence on sample analytical method selection.

From a project management perspective, time to perform analyses and the per-sample cost must be evaluated to determine whether the cost for field analysis is less than, equal to, or more than remote laboratory analyses. If field analytical costs are equal to or greater than laboratory analytical costs, then the value of real-time data must be included in the equation. Down time associated with methods that incorporate less-than-reliable equipment and test methods should also be considered.

Summary of Available Field Analytical Technologies

A number of field analytical technologies have been developed. Selection of one method over another depends on careful examination of a number of variables as discussed earlier. In particular, field test methods must be carefully selected with the type of sample (i.e., solid vs. liquid) and the nature of the contaminant (volatility and anticipated concentration) to be analyzed in mind. Many field test kits are differentiated based on their capability of analyzing samples that are (1) volatile (e.g., field-portable GC); (2) semi-volatile (e.g., PetroFlag); or (3) nonvolatile (e.g., XRF devices). It is critical that some background research be done during the planning phase to determine the nature of the contaminant of interest. To assist in making these difficult decisions, the reader is directed to U.S. EPA's Field Sampling and Analysis Technologies Matrix (U.S. EPA, 1998b) and U.S. EPA's on-line Field Methods Encyclopedia (U.S. EPA, 2004c). Both of these resources provide comparisons of applications for various field analytical technologies.

Reagent Test Kits

Perhaps the widest assortment of field analytical methods available for a variety of parameters is the group of methods collectively referred to as reagent test kits. Reagent test kits are self-contained kits that use a chemical reaction which produces color or turbidity to identify contaminants either qualitatively or quantitatively. Test kits provide a number of advantages to field analytical teams. They are generally highly portable, easy to use, quick and inexpensive to implement, and available for a wide range of analytes. Some test kits are designed to be operated by an individual with no analytical experience, while others are more successful when run by an individual with some analytical experience.

ASTM Standard D 5463 is a guide for the use of test kits to measure inorganic constituents (e.g., metals) in water (ASTM, 2004g). An equal number of test kits are available for organic parameters, but these are not addressed in D 5463. As indicated in this ASTM Standard, many test kits have been developed to exactly replicate an official test method of a standard-setting organization such as the U.S. EPA. In other cases, minor modifications of official test methods are made to improve performance, operator convenience, or ease of use in the field. In still other cases, the test kit may be based on an analytical method that is completely unique and not approved by any official organization. The U.S. EPA has approved a number of field colorimetric test methods and has included them in SW846. Examples are presented in Table 2.8.

Technologies included in reagent test kit options include colorimetric test kits where no sample preparation is required (e.g., Hach AccuVac-Vial test method 8171 for nitrate); colorimetric test kits where some sample preparation, such as sample extraction, is required (e.g., immunoassay test kits); and test kits in which a turbidimetric-based test is conducted following solvent extraction of contaminants from a sample (e.g., PetroFlag). The degree of quantification of data generated by colorimetric test kits is highly dependent on the method selected for determining the color intensity generated by the test method. In some kits, color intensity is determined visually by comparing final color against a color chart or photograph, as is the case with some versions of the Hanby Field Test Kit. This

TABLE 2.8

Examples of U.S. EPA-Approved Colorimetric Test Methods

EPA SW846 Test Method Number	Method Name
8510	Field method for RDX in soil
8515	Trinitrotoluene in soil by colorimetric screening
9078	Screening test method for PCB in soil
9079	Screening test method for PCB in transformer oil

results in the generation of semiquantitative data at best, because the method is highly subject to personal bias regarding interpretation of color. In other kits, including a number of immunoassay test kits, color change is compared against a reference standard that has been calibrated. Differences in color between the sample and the reference standard are determined electronically using devices such as photometer or spectrophotometer. Spectrophotometers are commonly used for determination of color intensity vs. method-specific programable reference wavelengths of light. With the use of these devices, data generated by the method are quantitative.

The Hanby Field Test Kit is a field method widely documented in U.S. EPA guidance (U.S. EPA, 1990, 1995a, 1998b, 1999b), for qualitative or semi-quantitative analysis of solid or water samples for aromatic compounds found in petroleum fuels (e.g., gasoline, diesel fuel, jet fuel, crude oil, motor oil, benzene, toluene, ethylbenzene, and xylenes [BTEX], and polynuclear aromatic hydrocarbons [PAHs]), as well as PCBs (Hewitt, 2000a, 2000b). These test kits are based on the Friedel–Crafts alkylation reaction for color formation in the presence of aromatic compounds. Typical detection limits are 1.0 mg/kg for soil and 0.10 mg/l for water, with typical analytical ranges of 1.0 to 1000 mg/kg for soil and 0.1 to 20 mg/l for water. Test methods for soil and water samples vary in the sample preparation procedures, but both require a solvent extraction procedure to liberate the hydrocarbon or PCB contaminant from the matrix and produce an extract fluid that is analyzed in the field. Through the addition of catalysts, a final extract fluid that exhibits a color is produced. The color is interpreted either visually, by comparing color intensity against a series of manufacturer-provided color photographs (qualitative data) or by using a reflective photometer to provide semi-quantitative determination of color intensity. It is necessary to have prior knowledge about which type of hydrocarbon contamination is present to avoid potential error when selecting the appropriate photo for color comparison. Color comparison is also problematic when more than one type of aromatic hydrocarbon is present (Francis et al., 1992). Evaluations of data generated by the photometer have not been supportive of manufacturer claims of increased accuracy of sample quantification (Hewitt, 2000b).

Dexsil Corporation manufactures a number of test kits that have been evaluated by the U.S. EPA for use in the field. Chlor-N-Soil and Chlor-N-Oil (U.S. EPA, 1995b, 1997e) are two kits that are designed to detect PCBs in soil, oil, and wipe samples. These test kits operate on the principle of total organic chlorine detection. With these test kits, PCB compounds are extracted from the sample using an organic solvent. The sample extract is treated with metallic sodium to strip chlorine from the biphenyl compound to form chloride ions. The chloride content in the extract fluid is measured with an indicating solution of mercuric nitrate and diphenyl carbazone, which combine to create a vivid purple color. The development of color is inversely proportional to chloride content of the extract solution (i.e., strong purple color indicates no chloride is present [and therefore no PCBs]; a yellow or clear color indicates the presence of PCBs). Colorimetric determinations are relative to a standard and are semi-quantitative in nature. Quantitative data are available from a similar method that uses a chloride-specific electrode, referred to as the L2000 chloride analyzer (Mahon et al., 2002). Using this electrode, it is possible to achieve a detection limit of 5 ppm. With any of these methods, the presence in the sample of chlorinated solvents, chlorinated pesticides, or inorganic chlorides from sources such as road salt or seawater, can cause false positives during analysis. These test kits are designed to be conservative, so false positives are more likely than false negatives. These three methods were added in Updates II or III to the U.S. EPA SW846 Methods as indicated in Table 2.8.

The Envirol Quick Test™ test kit provides quantitative results for pentachloraphenol (PCP) in soil and water, trinitrotoluene in soil, and carcinogenic PAHs in soil (U.S. EPA, 1999b, 2004c). The kits use a photochemical reaction that produces a color proportional to the concentration of the analyte of interest. In this method, soil and water samples are solvent extracted then filtered to generate an extract fluid that undergoes further preparation in the field to reduce interference. After the extract preparation is complete, color change, as the degree of absorbance of the sample, is quantified when the sample is placed in a small portable photometer. The photometer is calibrated to three standard solutions to create a calibration curve that equates the degree of absorbance to concentrations of PCP, TNT, or carcinogenic PAHs. The operating range of the method is 1.5 to 90 ppm for PCP in soil and water, 3.0 to 100 ppm for TNT in soil, and 1 to 3000 ppm for carcinogenic PAHs in soil. This is a more involved field test method that takes a significantly longer time to perform than other methods. The presence of tri- and tetrachlorophenols can result in positive interference for PCP, while creosote has been shown to interfere with the carcinogenic PAH test. Mono- and dichlorophenols must be present in relatively high concentrations to be detected by the PCP test.

The PetroFlag test kit is designed to provide semi-quantitative to quantitative results for gasoline, diesel fuel, jet fuel, fuel oil, motor oil, transformer oil, hydraulic fluid, greases, and many other types of hydrocarbons in soil (Lynn et al., 1994; Seyfried and Wright, 1995; Lynn and Lynn, 2002; U.S. EPA, 2004c). In this test method, soil samples are prepared for analysis by conducting a solvent extraction to create an extract that is filtered into a vial containing a patented development solution. Following a reaction time, the extract is placed into a hand-held turbidimeter that measures the turbidity or optical density of the final extract against a known standard. The concentration of hydrocarbons present is proportional to the turbidity of the sample. PetroFlag will detect hydrocarbons over a range of 20 to 2000 ppm, but the detection limit will vary with the type of hydrocarbon being analyzed. Detection limits for heavier hydrocarbons, such as jet fuel or motor oil, provide for a more turbid final solution, thus increasing the sensitivity of the analyzer. False positives can occur with this method if naturally occurring waxes and oils (e.g., plant resins or waxes) are present in the sample. For accurate quantification, the analyte being tested must be known so the field analyst can select the most appropriate calibration curve for the turbidimeter. When the contaminant is unknown, only semi-quantitative data can be generated. This method is proposed to be included in Update 4A in SW846 as Method 9074, "Turbidimetric Screening Method for Total Recoverable Hydrocarbons in Soil" (U.S. EPA, 2004c).

The AccuSensor field test kit was developed to permit analysis of trichloroethylene (TCE); total trihalomethanes (THMs) in chloroform equivalent; BTEX; and tetrachloroethylene, in water only (U.S. EPA 1999b, 2004c). The test kit is based on the Fujiwara reaction where geminal species react with pyridine in the presence of water, and hydroxide ions form a visible light-absorbing product. This test method has the advantage that no solvent extraction steps are required in the field. Instead, a water sample is poured into a standard 40 ml VOA vial (leaving a headspace) and the vial is sealed with an AccuSensor cap. The sample is shaken to induce volatilization of constituents into the headspace of the vial, then the cap is inserted into the AccuSensor meter and a lever on the side of the cap is turned to expose a porous Teflon membrane that will permit volatiles from the headspace to permeate through the membrane to react with the Fujiwara reagent. The degree of absorbance is measured by the meter over a 5 min period, after which the concentration (in ppb) is displayed on the meter. Because this method is based on the analysis of headspace, it is affected by analyte Henry's law constants, diffusion rates, and reaction kinetics, all of which are temperature controlled. This is addressed by a thermistor in the meter that provides temperature compensation. There are also known interferences that must be considered when this method is used. For example, if chloroform is present when analyzing for TCE, measurements may be affected by an error rate of 40% when the concentration of chloroform is equal to that of TCE. There is also a potential for cross interference between readings for TCE and THMs. The minimum detection limit of this method is 10 ppb for THM and 5 ppb for TCE.

Immunoassay methods have been used widely in the food and health-care industries for years (Dohrman, 1991) and have been applied in the environmental industry as a way to provide semi-quantitative and quantitative data for a wide range of organic and inorganic compounds in soil and water samples. The most common applications are for gasoline, diesel fuel, jet fuel, BTEX, PAH, PCP, various pesticides and herbicides, explosives and propellants, and individual Arochlors and mixtures of PCBs in soil and water (Thorne and Myers, 1997; U.S. EPA, 1995c, 1995d). Immunoassay has also been used to detect cadmium (Khosraviani et al., 1997) and mercury (Bruce et al., 1999) in soil. Immunoassay kits primarily measure lighter aromatic petroleum fractions, because straight-chain hydrocarbons do not lend themselves to eliciting immune system responses. The technology is not effective when analyzing heavy petroleum products with few aromatic compounds (e.g., heating oil) or highly degraded petroleum fuels, from which the lighter aromatic fractions have been lost.

There are four types of immunoassay: enzyme immunoassay, radioimmunoassay, fluorescent immunoassay, and enzyme-linked immunosorbent assay (ELISA) (Gee and van Emon, 2004; U.S. EPA, 2004c). Of the four, ELISA is most often used in environmental applications because it can be optimized for speed, sensitivity, and selectivity; it has a longer shelf life; it is simpler to use; and it does not require the use of radioactive materials. During the analytical procedure, a known amount of sample and a known amount of enzyme conjugate are introduced into a test tube that contains the antibodies and the target analyte present in the sample competes with the labeled antigen in the enzyme conjugate for a limited number of antibody binding sites. Then, a chromogen is added to the test tube to react with the enzymes on the labeled antigen to cause the formation of a color. The more analyte present in the sample, the more enzyme conjugate it will displace from the binding sites. The amount of bound conjugate is inversely proportional to the amount of analyte in the sample. The original concentration of the analytes can be determined by measuring the amount of enzyme conjugate bound to the antibody. Because the amount of bound enzyme conjugate determines the intensity of the color, the intensity of the color is inversely proportional to the amount of analyte present in the sample. Concentrations of analytes are identified through the use of a sensitive colorimetric reaction and are quantified by comparing the color developed by a sample of unknown concentration with the color formed by a known standard. The concentration of the analyte is determined by the intensity of color, which can be estimated visually by comparing the sample with a color chart (qualitative data) or quantitatively by using a photometer or spectrophotometer (quantitative data).

There are a number of advantages to using immunoassay in the field when compared with remote laboratory analyses, including speed of analysis, portability, ease of use,

TABLE 2.9

U.S. EPA SW846 Immunoassay-Based Field Test Methods

Method Number	Method Name
4010	Screening for PCP by immunoassay
4015	2,4-D in water and soil by immunoassay
4020	PCBs in soil by immunoassay
4025	Dioxin in water and soil by immunoassay
4030	TPH in soil by immunoassay
4035	Soil screening for PAHs by immunoassay
4040	Toxaphene in soil by immunoassay
4041	Chlordane in soil by immunoassay
4042	DDT in soil by immunoassay
4051	RDX explosives in water and soil by immunoassay
4060	TCE in soil by immunoassay
4670	Triazine herbicides as atrazine by immunoassay
4500	Mercury in soil by immunoassay

relatively low cost per sample, and availability of methods for a wide range of contaminants. U.S. EPA has approved immunoassay methods for thirteen 4000 series test methods in SW846 (U.S. EPA, 1996a) as provided in Table 2.9. Detection limits for immunoassay are comparable to or even lower than those for conventional analytical methods and are often less than maximum contaminant limits or MCLs (pesticide detection limits in water are an order of magnitude lower than MCLs). The actual detection limit will vary by test method, analyte of interest, sample matrix (soil vs. water), concentration, and interference sources. It is possible to achieve parts-per-billion detection limits in water, while soil samples typically have parts-per-million detection limits, due to the necessity of a solvent (e.g., methanol) extraction step prior to analysis. For very highly contaminated samples, it may be necessary to dilute the original sample to perform the analysis.

There are several disadvantages to applying immunoassay technology to environmental samples. The trick, in the beginning of the project, is to correctly identify the analyte or family of analytes of interest at the site because the kits are very contaminant-specific. If multiple or similar compounds are found at a site, it may be difficult to accurately quantify the individual compounds due to a phenomenon referred to as "cross-reactivity," which can occur when an antibody reacts to a substance that is similar in structure but is not its target compound. For example, a PCP test kit may also respond to tetra-, tri-, and dichlorophenol (Gerlach et al., 1997a, U.S. EPA, 1995e, 2004c). Tetrachlorophenol is most likely to cause a response, because its chemical structure is closer to that of PCP. A test kit designed for TNT may also respond to 1,3,5-trinitrobenzene and DNT (Thorne and Myers, 1997). Manufacturers of immunoassay test kits typically provide information on cross-reactivity for compounds similar to the target. This is particularly important when immunoassay kits are used to analyze BTEX compounds (Gerlach et al., 1997b). The BTEX test kit will respond to all six BTEX components (including xylene isomers) to different degrees but the test does not provide compound-specific information. This interference can result in generation of false positives. With this source of error in mind, if the objective of an environmental site characterization project is to determine the concentration of benzene in ground water or soil that is contaminated with gasoline, immunoassay may not be the best technology to use. This is especially true if compound-specific information is required to assess exposure risk. On the other hand, this cross-reactivity can be desirable if the user is looking for a number of similar constituents and is not particularly concerned with individual compound identification, but whether or not contamination is present at a site.

Many of the sample reagents, including the antibodies and chromogens, are highly sensitive to direct sunlight and temperature. Sunlight can break down the reagents or cause a change in the colorimetric reaction. Therefore, care must be taken to keep test kits out of direct sunlight. It is also important to store the kits in a cool environment (e.g., a field refrigerator or cooler) and to monitor expiration dates for each test kit to ensure that the antibodies do not expire prior to use.

Field-Portable X-Ray Fluorescence

A second major category of field analytical technologies is XRF. XRF technology has been used in industry for many years for applications such as determining metallic content of alloys. It is being increasingly applied in environmental site investigations where trace metal contamination is of primary concern. Initially, the emphasis in using XRF was on determining lead concentrations in residential paints and house dust, but it has expanded to environmental site characterization investigations and site remediation projects. In response to this growing market, field-portable XRF units have seen increased development to make the instrumentation smaller, more portable, battery operated, rugged, capable of operating with more than one type of radioisotope source, capable of analyzing the complete RCRA list of metals, and capable of analyzing soil samples, water samples, and materials such as plant tissues (Hewitt, 1995; U.S. EPA, 1996d, 2004c; Walsh, 2004).

There are two general categories of instrumentation used for XRF analysis of environmental samples: a device that requires the use of a radioisotope source and a detector and a device that utilizes a miniature x-ray tube source. In XRF analysis, a process called the photoelectric effect is the fundamental reaction that occurs during analysis. Fluorescent x-rays are produced by exposing a sample to an x-ray source that has an excitation energy similar to, but greater than, the binding energy of the inner-shell electrons of the elements in the sample. Some of the source x-rays will be scattered, but a portion will be absorbed by the elements in the sample. Because of their higher energy level, the excited outer-shell electrons will eject the inner-shell electrons. The electron vacancies that result will be filled by electrons cascading in from outer electron shells that have higher energy states than the inner-shell electrons they are replacing. This causes the outer shell electrons to give off energy in the form of fluorescent x-rays as they cascade down. It is this generation of x-rays that is referred to as XRF. Because every element has a different electron shell configuration, each element emits a unique x-ray at a set energy level or wavelength, which is characteristic of that element. The elements present in a sample can be identified by observing the energy level of the characteristic x-rays, while the intensity of the x-rays is proportional to the concentration of the element. Data generated are both qualitative (observing the energy of the characteristic x-ray) and quantitative in nature (when the intensity of the x-ray is determined relative to known reference standards).

The XRF has two basic components — the radioisotope source and the detector. The source irradiates the sample to produce the characteristic x-rays, while the detector measures both the energy of the characteristic x-rays that are emitted and their intensity to quantify concentrations. Table 2.10 presents examples of common field-portable radioisotope sources and corresponding elements that can be detected. Instrumentation with more than one radioisotope permits greater flexibility in terms of the elements that can be detected.

TABLE 2.10

Examples of Common Field-Portable Radioisotopes

Radioisotope	Elements Detected
Fe-55	Sulfur, potassium, calcium, titanium, and chromium
Cd-109	Vanadium, chromium, manganese, iron, cobalt, nickel, copper, zinc, arsenic, selenium, strontium, zirconium, molybdenum, mercury, lead, rubidium, and uranium
Am-241	Cadmium, tin, antimony, barium, and silver

Miniature x-ray tube sources are employed by a number of manufacturers. The major advantage of x-ray tube technology is that it does not require licensing or special shipping as XRF units using radioactive sources do. X-ray tube units do, however, require registration with some states — something that is important to keep in mind when traveling to different field sites in different states. Tube-based instruments commonly utilize a low-power hot filament cathode x-ray tube. The transmission anode operates at a high enough energy range (~35 keV) to simultaneously excite a large range of elements (K through U). Interferences and sensitivity problems associated with high energy sources are corrected using sophisticated software built into the XRF unit.

An XRF detector can be operated in two modes to analyze environmental samples — *in situ* or with a collected sample. Not all XRF instrumentation has capabilities to perform both types of sample analysis. DQOs that specify required detection limits or sample precision in addition to the objectives of the field analysis program will dictate which method is most appropriate for any investigation. For units that use multiple sources, after the sample has been exposed to one source, the turret is rotated to expose it to the next source. The length of time the sample is exposed to each source (measured in seconds) is referred to as the count time, which can range from 30 sec to as long as 200 sec per source depending upon the data quality needs of the investigation (longer count times equate to lower detection limits). Fluorescent and backscattered x-rays from the sample reenter the analyzer through the window and are counted by the instrument's detector. X-rays emitted by the sample at each energy level are called "counts," which are recorded by the detector. The detector also measures the energy of each x-ray and builds a spectrum of analyte peaks on a multichannel analyzer. Instrumentation software integrates the peaks to produce a readout of spectra and concentrations of analytes that can be stored for later viewing and downloading.

In situ analysis refers to the rapid screening of soils in place. In this application, the window of the XRF probe is placed in direct contact with the ground surface and a trigger is pulled to expose the sample to the radiation source. Count times for *in situ* analysis are typically very short (30 to 60 counts per source), consequently detection limits are somewhat higher than for intrusive methods. When sample heterogeneity is of concern, it is recommended that three or four measurements be taken within a small area, with an average value being reported. *In situ* measurements permit very rapid determination of metal concentrations in very shallow soils (typically less than 1 cm below ground surface) over a large area. When taking *in situ* measurements, it is important to remove any unrepresentative debris, such as rocks, pebbles, leaves, vegetation, or roots, from the surface to be analyzed. The surface should also be smooth to ensure that good contact is made between the entire window surface area and the soil surface. Used properly, *in situ* XRF can be a valuable tool on sites undergoing soil excavation activities during remediation.

Intrusive analysis is used when greater precision and lower detection limits are required to satisfy the objectives of the field analytical program. This is achieved through more extensive sample preparation and longer analysis times to reduce heterogeneity among samples and to increase the sensitivity of the instrument. For intrusive operation, a sample is collected and undergoes a series of steps to prepare it for analysis. When samples are collected, all unrepresentative debris, as described earlier for *in situ* sampling, must be removed from the sample. Field sample preparation can be time- and equipment-intensive, requiring sample homogenizing, drying if moisture content is greater than 20%, grinding, and sieving. As a consequence, sample preparation is sometimes performed in a remote laboratory, resulting in a delay in obtaining analytical results. Time is saved if sample preparation can be performed in an on-site mobile laboratory facility. Once the sample is prepared, it is transferred into a polyethylene sample cup that has a transparent Mylar window. The sample cup is then placed over the probe window for analysis.

With either method of analysis, it is important to understand the detection limits that can be obtained with XRF technology. Instrument detection limits (the absolute threshold concentration that the equipment can resolve) can range from 10 to 100 ppm in soil samples, while method detection limits, which are dependent on the analytical method (and sample preparation method) selected, can range from 40 to 200 ppm or higher, as is the case for chromium, which has a detection limit that may be as high as 900 ppm (U.S. EPA, 2004c). Therefore, if detection limit requirements for an investigation are in the parts-per-billion range, XRF technology should not be considered. Other limitations include the expense and difficulty in obtaining reference calibration standards, the expense associated with the radiation sources that must be replaced every 2 years, and the costs associated with licensing requirements for instrumentation that relies on a radioisotope source. Instrumentation using an Si (Li) detector requires liquid nitrogen and a special aluminum container (dewar) to hold the liquid nitrogen, which adds disposable supplies expenses and time to the per-sample analysis cost.

A number of factors can affect the detection and quantification of elements in a sample using XRF. Some of these interferences can be inherent in the analytical method, while others, such as calibration procedures, are instrument related. Sources of error include interference from the sample matrix, moisture content of the sample, sampling error, and detector resolution limitations. Ideally, samples will be homogenized to remove variations in the physical structure of soil samples (e.g., particle size, uniformity, homogeneity, and condition of the surface) and will be dried to a moisture content of 20% or less. As discussed earlier, detector applications and limitations must also be considered.

Headspace Screening, GC, and GC/MS

VOCs and SVOCs are the most common groups of contaminants of interest at sites undergoing environmental site characterization and remediation. Consequently, a number of field analytical methods have been developed to permit qualitative and quantitative analyses of samples for these compounds. These methods fall into three general categories: headspace screening methods for total VOC determination, GC for identification and quantification of specific compounds, and GC/MS for identification and quantification of specific compounds with a greater level of accuracy and precision than that possible through GC alone. Each technology is unique in its instrumentation requirements, level of experience required by field personnel, type of data generated, detection limits obtained, and accuracy and precision during sample analysis.

Qualitative Headspace Screening: Qualitative headspace screening of samples is the most common form of field sample analysis in the environmental industry (Fitzgerald, 1993; Hewitt and Myers, 1999). To conduct headspace screening, a sample of material (solid or liquid) is collected and placed into a gas-tight container. The container is approximately half-filled, leaving open space above the sample (i.e., headspace). A period of time (typically 30 min) is allowed to pass, during which time concentrations of volatile constituents in the sample will come into equilibrium with the headspace in the container. This technique is referred to as static headspace screening.

Qualitative headspace screening involves using instrumentation such as a PID or a FID to remove a small sample of the vapors from the headspace to determine the presence of total VOCs present (U.S. EPA, 1997d). Using instrumentation such as PID or FID in survey mode, investigators are able to detect the presence of a wide variety of organic and some inorganic compounds, but the instrumentation cannot identify what specific compounds are present and in what concentrations. Therefore, data generated are considered to be qualitative. An additional limitation to headspace screening is that not all compounds will volatilize from the sample at ambient temperatures within the same time frame. To induce volatilization, dynamic headspace screening methods that can incorporate a heating process, such as putting the sample container into a water bath for a prescribed period of time, can result in higher readings for VOCs.

Chemical analysis based on ionization has been used since the 1960s, with the first portable analyzers coming into use in the early 1970s (Driscoll and Spaziani, 1975). PIDs are compact, hand-held devices that draw gases (ambient air or headspace) into the device, where the gas is passed into a chamber housing a special ultraviolet lamp with specific eV ratings. In this chamber, contaminants are ionized as a result of being bombarded by high-energy ultraviolet light. The compounds absorb the energy of the light, causing excitation and temporary loss of an electron and forming a positively charged ion (ionization). The lamp emits energy that is sufficient to ionize any compounds contained in the gases that have an ionization potential (IP) less than the ionization energy of the lamp. The closer the IP is to the energy of the lamp (without exceeding it), the more sensitive the PID is to that compound. If a compound has a higher IP than the ionization energy of the lamp, the compound will not be detected. Examples of compounds detected by a PID and their corresponding IPs are presented in Table 2.11.

When the ions are formed in the ionization chamber, an ion current is produced. This current is amplified and then displayed on either an analog or digital display.

TABLE 2.11

Examples of Compounds Detected by PIDs

Compound	Ionization Potential (eV)
Hydrogen sulfide	10.46
Hexane	10.17
Octane	9.82
Trichloroethylene	9.45
Tetrachloroethylene	9.32
Benzene	9.25
Xylene	8.45
Toluene	8.82
Acetone	9.69
Methyl ethyl ketone	9.30
Carbon disulfide	10.08
Ammonia	10.15
Methyl mercaptan	9.44

Source: Based on Maslansky and Maslansky, 1993. With permission.

The typical operating range of PIDs is 0.1 to 2000 ppm (relative to the calibration gas, commonly isobutylene). Sensitivities vary with the lamp in use (i.e., eV rating, condition), environmental conditions at the time of use (e.g., relative humidity, dust, and presence of corrosive gases), and equipment manufacturer.

FIDs are instruments used primarily for nonhalogenated aromatic and straight-chain petroleum hydrocarbon compounds. FIDs will not detect halogenated VOCs unless they are present in very high concentrations. FIDs have a detection range of 0.2 to 1000 ppm or 1.0 to 10,000 ppm relative to the factory calibration gas, which is commonly methane. These meters are used in a fashion similar to PIDs, for determining levels of total VOCs in the headspace of samples. Only organic compounds are detected; inorganic compounds such as hydrogen sulfide, nitrogen dioxide, carbon dioxide, and carbon monoxide are not detected (Table 2.12).

The major difference between an FID and a PID is the way in which ionization occurs. In an FID, an air sample is drawn into the instrument, where it is carried to a combustion chamber. In the chamber, organic materials are burned using a hydrogen-fueled flame, creating charged ionic particles. These ions are attracted to a collecting electrode that produces a small ion current, which is amplified and translated into a meter display. The current produced is directly proportional to the number of ions formed and collected. The flame has sufficient energy to ionize any organic materials with an IP of 15.4 or less, therefore, the FID can detect some compounds with high IPs that cannot be detected by PIDs. Organic compounds burn with different efficiencies, so while FIDs are not affected by IP as PIDs are, compound detection is based on the burning efficiency of a compound. Aromatic compounds burn more readily and release a larger number of ions than oxygenated compounds such as methyl alcohol. Some compounds, such as formaldehyde, do not have the requisite bond structure (multiple carbon–hydrogen bonds) to release a sufficient number of ions to be detected. Table 2.12 presents examples of the compounds that can be detected using an FID and their relative burning efficiency. Equipment manufacturers should be consulted to confirm that a particular compound of interest can be detected with their instrumentation.

TABLE 2.12

Examples of Compounds Detected by an FID

Compound	Burning Efficiency
Methane	High
n-Butane	Moderate
Octane	High
Acetylene	Very high
Ethylene	High
Methylene chloride	High
Chloroform	Moderate
Carbon tetrachloride	Low
Benzene	Very high
Toluene	High
Methyl alcohol	Low
Ethyl alcohol	Low
Acetic acid	High
Carbon disulfide	Not detected

Source: Based on Maslansky and Maslansky, 1993. With permission.

There are a number of disadvantages with FIDs. FID readings are affected by ambient oxygen concentrations. Oxygen deficiency reduces the ability of the hydrogen fuel to burn and can cause the flame to be extinguished. The required oxygen concentration will vary by manufacturer. FIDs require a fuel to operate. Most use pure hydrogen, which is expensive, limits the ability to ship equipment from field site to field site and can be a hazard in some environments. However, intrinsically safe FIDs are available. Dust and particulate matter can cause erratic readings and decreased instrument response.

Headspace screening data generated by both PIDs and FIDs are extremely useful for a variety of field applications including locating hotspots of VOCs, determining which samples may be selected for quantitative analysis either on site or off site, providing information for health and safety of field personnel (see Chapter 19), and determination of "clean" vs. contaminated materials during excavation.

Gas Chromatography: More detailed information on identification and quantification of organic contaminants is often required in the field to facilitate decision making. When this is the case, gas chromatography may be the field analytical technology of choice. Gas chromatography is widely used for analysis of organic compounds in air, soil gas, soil, and water samples in the low parts per billion to the low part per million range. Two categories of GCs can be used in the field: field-portable GCs and lab-grade portable GCs. Field-portable GCs are small, battery-operated, and fully self-contained instruments. Being fully contained, they are very portable and have many of the same analytical capabilities as laboratory-based instrumentation. The small size and battery power limits the number and type of detectors the GC can use and also limits oven temperature control, which is necessary for some compound separation. Field-portable instruments also tend to be relatively expensive. The other alternative is to move a laboratory-grade instrument into the field. Laboratory units are able to perform all EPA methods for sample analysis by GC; however, they are less portable, require external sources of power and carrier gases, require more support equipment such as computers, and they are often more sensitive to ambient operating conditions such as temperature, wind, dust, precipitation, and sunlight. Therefore, they generally require a climate-controlled environment in which to operate.

The principles of chromatography and sample analysis are similar with both types of equipment. GC analysis of samples involves the introduction of a gas sample (collected as a headspace sample from solid or liquid samples, an actual sample of gas, or an extract gas resulting from sample preparation methods such as purge and trap) into a heated injection port, typically using a gas-tight syringe or pumped from specially designed tedlar gas bags. After the sample is introduced into the GC, it travels through a heated chromatographic column. GC columns are small-diameter, coiled tubular columns that contain a packing material specified by the analytical method for a particular compound or family of compounds. Some columns are open-tubular in design. Once in the column, the sample travels the length of the column assisted by an inert carrier gas such as nitrogen. During their travel through the packed column, individual compounds contained in the sample separate based on their affinity for the packing material. In open-tubular columns, the packing material is a liquid organic compound, which is coated on the internal surface of the fused silica column. Components in the sample that have a high affinity with the packing material are strongly retained, while other components with low affinity continue to travel through the column. It is this difference in mobility, due to differing affinities for the packing material, that separates individual compounds in the sample. The time to travel through the column is referred to as the retention time, which is compound-specific for any given column and method. After a compound has traveled the length of the column

TABLE 2.13

Examples of Method-Specific GC Detectors

EPA Method Number	Compounds Detected	Detector Required
8041	Phenols	FID or ECD
8061A	Phthalates	ECD
8070A	Amines	NPD
8081A	Chlorinated pesticides	ECD
8082	PCBs	ECD
8100	PAHs	FID
8240	VOCs	PID

Note: FID, flame ionization detector; NPD, nitrogen–phosphorous detector; ECD, electron capture detector; PID, photoionization detector.
Source: Based on U.S. EPA, 2004c.

(or "elutes" from the column), it is swept by the carrier gas into a detector that generates a measurable electrical signal referred to as a peak. Detector response is plotted on a chromatogram as a function of the time required for the analyte to elute from the column (relative to the time of sample injection) and the signal strength generated by the detector, which equates to the concentration of the compound. The position of the peaks on the time axis serves to identify the compound, while the area under the peak represents the concentration of the compound.

A number of different detectors are available for use in GC instrumentation. The most commonly used detector is the PID. Different analytical methods require the use of specific detectors as illustrated in Table 2.13.

Analysis of samples using a GC requires an operator with some analytical chemistry experience to ensure accuracy and precision in data generated.

Gas Chromatography/Mass Spectroscopy: Mass spectrometry is an established laboratory analytical technique that identifies compounds by the mass-to-charge ratio of the analyte molecule. Mass spectrometry is especially powerful as an analytical tool because the signals produced are the direct result of chemical reactions such as ionization and fragmentation, rather than energy state changes that are fundamental to most other spectroscopic techniques. Because of this distinction, mass spectrometry is the best tool to use when definitive compound identification is required (U.S. EPA, 1996e, 1999b, 2001d, 2004c). Coupling a mass spectrometer with a GC allows not only compound separation, but also definitive identification of complex compounds that is not possible through GC alone. In GC/MS analysis of samples, a sample is initially run through a GC to separate compounds on the GC column. As compounds elute from the column, they are directed into the ion source of the mass spectrometer through a heated interface. This ionization process causes compounds to lose electrons and form a charged molecular ion that has the same molecular weight as the compound molecule. An electron beam of 20 eV is used to extract an electron from the molecule. Excess energy from the beam further fragments the molecular ion into fragment (daughter) ions with lower mass-to-charge ratio. The positive ions produced by electron impact are attracted through the slits of the ion source and mass analyzer where they are analyzed for differentiation according to their mass-to-charge ratios. The mass-sorted ions are detected by an electron multiplier and the resulting signal is sent to a data system for processing. A display of the electron multiplier signal generated by the sorted molecular ions is displayed as the mass spectrum. The mass spectrum is in the form of a bar graph that relates the relative intensity of signals generated to their mass-to-charge ratios. The largest peak in each spectrum is termed the base peak. The heights of the remaining peaks are computed as a percentage of the base peak height. The spectrum is then compared to a preprogrammed spectral library for compound identification based on the fragmentation pattern and peak ratios. Detection limits have been reported in the parts-per-billion to even the parts-per-trillion and quadrillion ranges, depending upon the type of mass analyzer used. Quadrapole analyzers are most common in field portable GC/MS instrumentation.

GC/MS analysis of samples has traditionally been performed exclusively in the laboratory, however, manufacturers have been able to develop small, durable field GC/MS instruments that are capable of the same analyses. Some EPA methods that incorporate field GC/MS technologies include:

Method 8270C	SVOCs
Method 8280	Dioxins
Method 8260	MTBE
Method 8240	VOCs

Of the field analytical technologies discussed, GC/MS is potentially the most powerful tool available, but it is also one of the most expensive, and requires the highest level of analytical chemistry expertise to be used with accuracy and precision. Unlike some instrumentation that is designed to be operated by the nonchemist, GC/MS instrumentation requires the expertise of an experienced chemist. GC/MS should not be selected as the tool of choice unless adequately trained and experienced personnel are available for the duration of the field analytical program to run the tests and interpret the field data.

Spending time initially to design an effective and realistic field analytical program for a project will be rewarded. Data will be generated with confidence in its accuracy, precision and meaningfulness to the project. By selecting the best analytical technologies for the specific application and implementing them under the guidance of a strong field QA/QC program and a trained analytical chemist, decision making in the field will be effective.

References

Applegate, J.L. and D.M. Fitton, Rapid site assessment applied to the Florida Department of Environmental Protection's dry cleaning solvent cleanup program, *Proceedings of the Superfund XVIII Conference*, E.J. Krause and Associates, Washington, DC, 1997, pp. 695–703.

ASTM, Geotechnical Applications of Remote Sensing and Remote Data Transmission, *ASTM Special Technical Publication 976*, A.I. Johnson and C.B. Pettersson, Eds., American Society for Testing and Materials, Philadelphia, PA, 1988.

ASTM, Standard Guide for Accelerated Site Characterization for Confirmed or Suspected Petroleum Releases, ASTM Standard E 1912, ASTM International, West Conshohocken, PA, 2004a, 20 pp.

ASTM, Standard Practice for Expedited Site Characterization of Vadose Zone and Ground-Water Contamination at Hazardous Waste Contaminated Sites, ASTM Standard D 6235, ASTM International, West Conshohocken, PA, 2004b, 50 pp.

ASTM, Standard Guide for Acquisition of File Aerial Photography and Imagery for Establishing Historic Site Use and Surficial Conditions, ASTM Standard D 5518, ASTM International, West Conshohocken, PA, 2004c, 8 pp.

ASTM, Standard Guide for Developing Conceptual Models for Contaminated Sites, ASTM Standard E 1689, ASTM International, West Conshohocken, PA, 2004d, 8 pp.

ASTM, Standard Guide for Conceptualization and Characterization of Ground-Water Systems, ASTM Standard D 5979, ASTM International, West Conshohocken, PA, 2004e, 7 pp.

ASTM, Standard Guide for Site Characterization for Environmental Purposes with Emphasis on Soil, Rock, the Vadose Zone and Ground Water, ASTM Standard D 5730, ASTM International, West Conshohocken, PA, 2004f, 26 pp.

ASTM, Standard Guide for the Use of Test Kits to Measure Inorganic Constituents in Water; ASTM Standard D 5463, ASTM International, West Conshohocken, PA, 2004g, 5 pp.

Avery, T.E., *Interpretation of Aerial Photographs, 2nd ed.*, Burgess Publishing Co., Minneapolis, MN, 1968, 324 pp.

Barcelona, M.J., Site characterization: what should we measure, where (when) and why? EPA/600/R-94/162, *Proceedings of the Symposium on Natural Attenuation of Ground Water*, U.S. Environmental Protection Agency, Office of Research and Development, Washington, DC, 1994, pp. 20–25.

Benson, R.C., R.A. Glaccum, and M.R. Noel, Geophysical Techniques for Sensing Buried Wastes and Waste Migration, EPA/600/7-84/064, U.S. Environmental Protection Agency, Office of Research and Development, Environmental Monitoring Systems Laboratory, Las Vegas, NV, 1984.

Bjerrum, L., Some notes on Terzaghi's method of working: from theory to practice in soil mechanics, in *Selections from the Writings of Karl Terzaghi*, John Wiley and Sons, Inc., New York, NY, 1960, pp. 22–25.

Brown, S.M., Technology improves aquifer remediation, *Environmental Protection*, October, 29–70, 1990.

Brown, S.M., D.R. Lincoln, and W.A. Wallace, Application of the observational method to hazardous waste engineering, *Journal of Management in Engineering*, 6(4), 479–500, 1989.

Bruce, M.L., K.L. Richards, and L.M. Miller, Mercury in soil screening by immunoassay, *American Environmental Laboratory*, February, 1999, 30–31.

Burton, J.C., Expedited site characterization: a rapid, cost-effective process for preremedial site characterization, *Proceedings of the Superfund XIV Conference, Vol. II*, Hazardous Materials Control Research Institute, Greenbelt, MD, 1993, pp. 809–826.

Burton, J.C., J.L. Walker, P.K. Aggarwal, and W.T. Meyer, Argonne's expedited site characterization: an integrated approach to time- and cost-effective remedial investigations, *Proceedings of the 88th Annual Meeting of the AWMA, Air and Waste Management Association*, Pittsburgh, PA, 1995, 27 pp.

California EPA, *Guidelines for Hydrogeologic Characterization of Hazardous Substance Release Sites, Vol. 1: Field Investigation Manual*, California Environmental Protection Agency, Sacramento, CA, 1995.

CCME, Subsurface Assessment Handbook for Contaminated Sites, Canadian Council of Ministers of the Environment, Report # CCME-EPC-NCSRP-48E, Prepared by the Waterloo Centre for Ground Water Research, 1994, 293 pp.

Chamberlain, T.C., 1890, The method of multiple working hypotheses, *Science*, 15 (Reprinted in *Science*, 148, 754–759, 1965).

Cherry, J.A., Ground-water monitoring: some current deficiencies and alternative approaches, in *Hazardous Waste Site Investigations: Toward Better Decisions*, R.B. Gammage and B.A. Berven, Eds., Lewis Publishers, Inc., Chelsea, MI, chap. 13, 1992, pp. 119–133.

Crumbling, D.M., Current Perspectives in Site Remediation and Monitoring: Using the Triad Approach to Improve the Cost Effectiveness of Hazardous Waste Site Cleanups, EPA/542-RA-01-016, U.S. Environmental Protection Agency, Office of Solid Waste and Emergency Response, Washington, DC, 2001a.

Crumbling, D.M., Applying the Concept of Effective Data to Environmental Analyses for Contaminated Sites, EPA/542-R-01-013, U.S. Environmental Protection Agency, Office of Solid Waste and Emergency Response, Washington, DC, 2001b.

Crumbling, D.M., Clarifying DQO Terminology Usage to Support Modernization of Site Cleanup Practice, EPA/542/R-01/014, U.S. Environmental Protection Agency, Office of Solid Waste and Emergency Response, Washington, DC, 2001c.

Crumbling, D.M., In search of representativeness: evolving the environmental data quality model, *Quality Assurance*, 9, 179–190, 2002.

Crumbling, D.M., C. Groenjes, B. Lesnick, K. Lynch, J. Shockley, J. van Ee, R. Howe, L. Keith, and J. McKenna, Managing uncertainty in environmental decisions: applying the concept of effective data at contaminated sites could reduce costs and improve cleanups, *Environmental Science and Technology*, 35(9), 404A–409A, 2001.

Crumbling, D.M., J. Griffith, and D.M. Powell, Improving decision quality: making the case for adopting next-generation site characterization practices, *Remediation*, Spring, 2003, pp. 91–111.

Deutsch, M., D.R. Wiesnet, and A. Rango, Eds., Satellite Hydrology, *Proceedings of the Fifth Annual Symposium on Remote Sensing*, American Water Resources Association, Minneapolis, MN, 1981, 730 pp.

Dohrman, L., Detection of environmental contaminants by enzyme immunoassay, *American Environmental Laboratory*, October, 31–33, 1991.

Driscoll, J.N. and F.F. Spaziani, Trace gas analysis by photoionization, *Analytical Instrumentation*, 13, 111–114, 1975.

Finkbeiner, M.A., and M.M. O'Toole, Application of aerial photography in assessing environmental hazards and monitoring cleanup operations at hazardous waste sites, *Proceedings of the Sixth National Conference on Management of Uncontrolled Hazardous Waste Sites*, Hazardous Materials Control Research Institute, Silver Spring, MD, 1985, pp. 116–124.

Fitzgerald, J., On-site analytical screening of gasoline-contaminated media using a jar headspace procedure, in *Principles and Practices for Petroleum Contaminated Soils*, Lewis Publishers, Ann Arbor, MI, Chap. 4, 1993, pp. 49–66.

Francis, R.A., B.A. Martin, J.A. Cerutti, and L.J. Marler, Field assessment of hydrocarbon contamination using extraction/colorimetric techniques, in *Hydrocarbon Contaminated Soils*, Lewis Publishers, Chelsea, MI, Chap. 13, 1992, pp. 197–213.

Gee, S.J. and J.M. Van Emon, A User's Guide to Environmental Immunochemical Analysis, EPA Research Grant 891047 Report, U.S. Environmental Protection Agency, Environmental Monitoring Systems Laboratory, Las Vegas, NV, 2004.

Gelb, S.B., J.P. Mack, P.R. Baker, and N. Findley, SAM to the rescue: source area mapping streamlines site assessment and remediation, *Soil and Ground Water Cleanup*, December/January, 6–11, 1998.

Gerlach, R.W., R.J. White, M.E. Silverstein, and J.M. Van Emon, SITE evaluation of field portable pentachlorophenol immunoassays, *Chemosphere*, 35(11), 2727–2749, 1997a.

Gerlach, R.W., J.R. White, D.F. N. O'Leary, and J.M. Van Emon, Field evaluation of an immunoassay test for benzene, toluene and xylene (BTX), *Water Research*, 31(4), 941–945, 1997b.

Gilbert, R.O., and P.G. Doctor, Determining the number and size of soil aliquots for assessing particulate contaminant concentrations, *Journal of Environmental Quality*, 14(2), 286–292, 1985.

Hewitt, A.D., Rapid Screening of Metals Using Portable High-Resolution X-Ray Fluorescence Spectrometers, ERDC/CRREL Special Report 95-14, U.S. Army Corps of Engineers, Cold Regions Research and Engineering Laboratory, Hanover, NH, 1995.

Hewitt, A.D., Evaluating the Hanby Test Kits for Screening Soil and Ground Water for Total Petroleum Hydrocarbons: Field Demonstration, ERDC/CRREL technical report TR-00-7, U.S. Army Corps of Engineers, Cold Regions Research and Engineering Laboratory, Hanover, NH, 2000a, 19 pp.

Hewitt, A.D., Evaluation of H.E.L.P. Mate 2000 for the Identification and Quantification of Petroleum Hydrocarbon Products, ERDC/CRREL Technical Report TR-00-20, U.S. Army Corps of Engineers, Cold Regions Research and Engineering Laboratory, 2000b, 13 pp.

Hewitt, A.D. and K.F. Myers, Sampling and On-Site Analytical Methods for Volatiles in Soil and Ground Water: Field Guidance Manual, ERDC/CRREL Special Report 99-16, U.S. Army Corps of Engineers, Cold Regions Research and Engineering Laboratory, Hanover, NH, 1999.

Homsher, M.I., F. Haeberer, P.J. Marsden, R.K. Mitchum, D. Neptune, and J. Warren, Performance-based criteria: a panel discussion, *Environmental Laboratory*, October/November, 1991.

ITRC, *Technical and Regulatory Guidance for the Triad Approach: A New Paradigm for Environmental Project Management*, Interstate Technology and Regulatory Council, Washington, DC, 2003, 64 pp.

Jenkins, T.F., Sample representativeness: a necessary element in explosives site characterization, *Proceedings of the 12th Annual Waste Testing and Quality Assurance Symposium*, U.S. Environmental Protection Agency, Washington, DC, 1996, pp. 30–35.

Johnson, R.L., A Bayesian/geostatistical approach to the design of adaptive sampling programs, in *Geostatistics for Environmental and Geotechnical Applications*, Special Technical Publication 1283, R.M. Srivastava, S. Rouhani, M.V. Cromer, and A.I. Johnson, Eds., ASTM International, West Conshohocken, PA, 1996.

Khosraviani, M., A.R. Pavlov, G.C. Flowers, and D.A. Blake, Detection of heavy metals by immunoassay: optimization and validation of a rapid, portable assay for ionic cadmium, *Environmental Science and Technology*, 32(1), 137–142, 1997.

Lattman, L.H., Technique of mapping geologic fracture traces and lineaments on aerial photographs, *Photogrammetric Engineering*, 24, 568–576, 1958.

Lattman, L.H. and R.H. Matzke, Geological significance of fracture traces, *Photogrammetric Engineering*, 27, 435–438, 1961.

Lattman, L.H. and Parizek, R.R., Relationship between fracture traces and the occurrence of ground water in carbonate rocks, *Journal of Hydrology*, 2, 73–91, 1964.

Lesnik, B. and D.M. Crumbling, Some guidelines for preparing SAPs using systematic planning and the PBMS approach, *Environmental Testing and Analysis*, January/February, 26–40, 2001.

Lillesand, T.M. and R.W. Kiefer, *Remote Sensing and Image Interpretation*, John Wiley and Sons, Inc., New York, NY, 1979, 612 pp.

Lynn, T.B. and A.C. Lynn, Analysis of SITE Program TPH Field Trial Data for SW-846 Method 9074—The PetroFlag Hydrocarbon Analyzer, *Proceedings of the 10th International On-Site Conference*, San Diego, CA, 2002.

Lynn, T.B., S. Finch, and L. Sacramone, Evaluation of a New Non-Immunoassay Field Test Kit for Total Petroleum Hydrocarbons in Soil, *Proceedings of the 10th Annual Waste Testing and Quality Assurance Symposium*, Arlington, VA, 1994.

Mahon, J.D., D. Balog, A.C. Lynn, and T.B. Lynn, In-Field Screening Techniques for PCBs in Transformer Oil: U.S. EPA Field Trial Results for the L2000DX Analyzer, *Proceedings of MY TRANSFO*, Torino, Italy, 2002.

Makeig, K.S., Plan to get the data you need: use DQOs, *The Professional Geologist*, August, 12–14, 1995.

Mark, D.L., L.A. Holm, and N.L. Ziemba, Application of the observational method to an operable unit feasibility study — a case study, *Proceedings of Superfund '89*, Hazardous Materials Control Research Institute, Silver Spring, MD, 1989, pp. 436–442.

Maslansky, C.J. and S.P. Maslansky, *Air Monitoring Instrumentation, A Manual for Emergency, Investigatory, and Remedial Responders*, Van Nostrand Reinhold, New York, NY, 1993, 304 pp.

Maslansky, S.P. and C.J. Maslansky, *Health and Safety at Hazardous Waste Sites — An Investigators and Remediators Guide to HAZWOPER*, John Wiley and Sons, Inc., New York, NY, 1997.

Mauch, J.C., Public records in the due diligence process: what's available and how to obtain them; *Proceedings of the 34th Annual Meeting of AEG*, Association of Engineering Geologists, Denver, CO, 1991, pp. 503–518.

Miller, R.D., Use of government environmental records to identify and analyze hydrocarbon-contaminated sites, in *Hydrocarbon Contaminated Soils and Ground Water*, Lewis Publishers, Inc., Chelsea, MI, 1992, chap. 29, pp. 501–515.

Moore, M.B., Innovations in Field Analysis and Site Characterization Technologies for Contaminated Sites — Emphasizing Best Available Technologies and Data Presentation Methods, Presented at the American Bar Association Eighth Section Fall Meeting, September 20–24, New Orleans, LA, 2000.

Nickelson, M.D. and D.D. Long, Field Assessment Screening Team (FAST) Technology Process and Economics, *Proceedings of the International Symposium on Field Screening Methods for Hazardous Wastes and Toxic Chemicals, Vol. I*, Air and Waste Management Association, Pittsburgh, PA, 1995, pp. 79–82.

Nielsen, D.M., Think remediation during site assessment: the expedited approach, *International Ground Water Technology*, June, 15–21, 1995.

NJ DEP, Field sample analysis; in *Field Sampling Procedures Manual*, New Jersey Department of Environmental Protection, Trenton, NJ, 2003, chap. 7, 18 pp.

Olson, C.M., The Development and Implementation of a Remedial Investigation Work Plan and Data Management System; *Ground Water Monitoring Review*, 11(2), 145–152, 1991.

Parizek, R.R., On the nature and significance of fracture traces and lineaments in carbonate and other terranes, in *Karst Hydrology and Water Resources*, V.V. Yevjevich, ed., Water Resources Publications, Fort Collins, CO, 1976, pp. 3.1–3.62.

Parris, G.E., J.M. Schreiber, and A. Gladwell, Misuse of analytical screening tests in environmental investigations; in *Effective and Safe Waste Management*, R.L. Jolley and R.G.M. Wang, Eds., Lewis Publishers, Ann Arbor, MI, 1993, chap. 16, pp. 175–181.

Peck, R.B., Advantages and limitations of the observational method in applied soil mechanics, *Geotechnique*, 19, 171–187, 1969.

Peck, R.B., Advantages and limitations of the observational method in applied soil mechanics; in *Milestones in Soil Mechanics, The First Ten Rankin Lectures*, Thomas Telford Ltd. Publishers, Edinburgh, UK, 1975, pp. 263–279.

Pitkin, S., *Rapid adaptive site characterization: better science and technology saves time and money*; Stone Environmental Co., Montpelier, VT. Available at http://www.stone-env.com/services/rasc.html, 2001.

Popek, E.P., Investigation versus remediation: perception and reality, *Proceedings of the 13th Annual Waste Testing and Quality Assurance Symposium*, U.S. Environmental Protection Agency, Washington, DC, 1997, pp. 183–188.

Popek, E.P., and G.H. Kassakhian, Investigation versus remediation: perception and reality, *Environmental Testing and Analysis*, May/June, 24–38, 1998.

Puls, R.W. and J.F. McCarthy, Well purging and sampling (Workshop Group Summary), in *Ground-Water Sampling — A Workshop Summary*, EPA/600/R-94/205, U.S. Environmental Protection Agency, Office of Research and Development, Washington, DC, 1995, pp. 82–87.

Ray, R.G., Aerial Photographs in Geologic Interpretation and Mapping, Professional Paper 373, U.S. Geological Survey, Reston, VA, 1972, 230 pp.

Regan, D.T., M.M. Mitchell, and J.A. Conte, The improved use of geologic data in ground-water investigations, in *Current Practices in Ground Water and Vadose Zone Investigations*, D.M. Nielsen and M.N. Sara, Eds., Special Technical Publication 1118, ASTM International, West Conshohocken, PA, 1991, pp. 24–38.

Robbat, A., Jr., *A Guideline for Dynamic Work Plans and Field Analytics: The Keys to Cost-Effective Site Characterization and Cleanup*; U.S. Environmental Protection Agency, Washington, DC, 1997.

Ronen, D., M. Magaritz, H. Gvirtzman, and W. Garner, Microscale chemical heterogeneity in ground water, *Journal of Hydrology*, 92(1–2), 173–178, 1987.

Ross, A.L. and R.K. Frohlich, Fracture trace analysis with a geographic information system (GIS), *Bulletin of the Association of Engineering Geologists*, 30(1), 87–98, 1993.

Setzer, J., Hydrologic significance of tectonic fractures detectable on air photos, *Ground Water*, 4(4), 23–29, 1966.

Seyfried, J.S. and K.A. Wright, Performance Evaluation of a New Low-Cost Field Test Kit for Analysis of Hydrocarbon-Contaminated Soil at a Diesel Fuel Release Site, in *Proceedings of the 11th Annual Waste Testing and Quality Assurance Symposium*, U.S. Environmental Protection Agency, Washington, DC, 1995.

Smith, S., D. Nielsen, D. Armitage, J. Poehlman, J. Ritchie, and H. Heiss, The use of aerial and satellite imagery with field study in ground-water prospecting, in *Ground-Water Hydrology for Water Well Contractors*, National Water Well Association, Worthington, OH, 1982, chap. 24, 28 pp.

Sweet, F.R., and C.R. Mitchell, Use of fractures and lineament analyses in contaminated bedrock aquifer remedial investigations, *Proceedings of the 1990 Focus Conference on Eastern Regional Ground-Water Issues*, National Ground-Water Association, Dublin, OH, 1990, pp. 299–311.

Taylor, M.B., and J.S. Erikson, Accelerated site characterization: a little ahead of its time, *ASTM Standardization News*, June, 34–39, 1996.

Thiesen, J.P., and R.B. Weiss, Dynamic Site Assessment: A Cost-Saving Strategy for Closure; *Proceedings of the Conference on Petroleum Hydrocarbons and Organic Chemicals in Ground Water*, National Ground Water Association, Dublin, OH, 1990, pp. 627–639.

Thorne, P.G., and K.F. Myers, Evaluation of Commercial Enzyme Immunoassays for the Field Screening of TNT and RDX in Water, Special Report 97-32, U.S. Army Corps of Engineers, Cold Regions Research and Engineering Laboratory, Hanover, NH, 1997, 20 pp.

Thurnblad, T., *Development of Sampling Plans, Protocols and Reports: Ground-Water Sampling Guidance*, Ground Water and Solid Waste Division, Minnesota Pollution Control Agency, Saint Paul, MN, 1995.

U.S. Air Force, Model Field Sampling Plan, U.S. Air Force Center for Environmental Excellence (AFCEE) Model FSP Version 1.1, March 1997.

U.S. Air Force, Quality Assurance Project Plan, U.S. Air Force Center for Environmental Excellence (AFCEE) QAPP Version 3.0, March 1998.

U.S. Army Corps of Engineers, Technical Project Planning (TPP) Process, Engineer Manual #200-1-2, U.S. Army Corps of Engineers, Washington, DC, 1998.

U.S. Army Corps of Engineers, Requirements for the Preparation of Sampling and Analysis Plans; Engineer Manual #200-1-3, U.S. Army Corps of Engineers, Washington, DC, 2001.

U.S. Army Corps of Engineers, Conceptual Site Models for Ordinance and Explosives (OE) and Hazardous, Toxic, and Radioactive Waste (HTRW) Projects, Engineer Manual #1110-1-1200, U.S. Army Corps of Engineers, Washington, DC, 2003.

U.S. Department of the Interior, Use of Fracture Traces in Water Well Location: A Handbook, OWRT Publication TT/82-1, By Meiser and Earl Hydrogeologists for the U.S. Department of the Interior, Office of Water Research and Technology, Washington, DC, 1982, 55 pp.

U.S. DOE, Remedial Investigation/Feasibility Studies (RI/FS): Process, Elements and Techniques, U.S. Department of Energy, Office of Environmental Guidance, RCRA/CERCLA Division (EH 231), Washington, DC, 1993.

U.S. DOE, Expedited Site Characterization: Innovative Technology Summary Report, DOE/EM-0420, U.S. Department of Energy, Office of Environmental Management, 1998.

U.S. DOE, Adaptive Sampling and Analysis Programs (ASAPs), DOE/EM-0592, U.S. Department of Energy, Office of Environmental Management, Washington, DC, 2001.

U.S. DOE, The QuickSite Process, U.S. Department of Energy, Argonne National Laboratory, Argonne, IL, Available at http://www.quicksite.anl.gov., 2004.

U.S. EPA, Guidance on Remedial Investigations Under CERCLA, EPA/540/G-85/002, U.S. Environmental Protection Agency, Office of Solid Waste and Emergency Response, Washington, DC, 1985.

U.S. EPA, Data Quality Objectives for Remediation Response Activities, EPA/540/G-87/003, U.S. Environmental Protection Agency, Office of Emergency and Remedial Response, Washington, DC, 1987.

U.S. EPA, Guidance for Conducting Remedial Investigations and Feasibility Studies Under CERCLA, Interim Final, OSWER Directive 9355.3-01, EPA/540/G-89/004, U.S. Environmental Protection Agency, Office of Solid Waste and Emergency Response, Washington, DC, 1988.

U.S. EPA, RI/FS Streamlining, OSWER Directive 9344.3-06, U.S. Environmental Protection Agency, Office of Solid Waste and Emergency Response, Washington, DC, 1989a.

U.S. EPA, Preparing Perfect Project Plans: A Pocket Guide for the Preparation of Quality Assurance Project Plans, EPA/600/9-89/087, U.S. Environmental Protection Agency, Office of Research and Development, Risk Reduction Engineering Laboratory, Cincinnati, OH, 1989b.

U.S. EPA, Health and Safety Audit Guidelines, EPA/540/G-89/010, U.S. Environmental Protection Agency, Office of Solid Waste and Emergency Response, Washington, DC, 1989c.

U.S. EPA, Field Measurements: Dependable Data When You Need It, EPA/530/UST-90-003, U.S. Environmental Protection Agency, Office of Solid Waste and Emergency Response (OS-420), Washington, DC, 1990.

U.S. EPA, Assessing Sites Under SACM — Interim Guide, OSWER Directive 9203.1-05I, U.S. Environmental Protection Agency, Office of Solid Waste and Emergency Response, Washington, DC, 1992.

U.S. EPA, Data Quality Objectives Process for Superfund, EPA/540/R-93/071, U.S. Environmental Protection Agency, Office of Emergency and Remedial Response, Washington, DC, 1993a.

U.S. EPA, Subsurface Characterization and Monitoring Techniques, A Desk Reference Guide — Volume I: Solids and Ground Water, EPA/625/R-93/003a, U.S. Environmental Protection Agency, Office of Research and Development, Washington, DC, 1993b.

U.S. EPA, Subsurface Characterization and Monitoring Techniques, A Desk Reference Guide — Volume II: The Vadose Zone, Field Screening and Analytical Methods, EPA/625/R-93/003b, U.S. Environmental Protection Agency, Office of Research and Development, Washington, DC, 1993c.

U.S. EPA, Use of Airborne, Surface, and Borehole Geophysical Techniques at Contaminated Sites, EPA/625/R-93/007, U.S. Environmental Protection Agency, Office of Research and Development, Washington, DC, 1993d.

U.S. EPA, Guidance for Planning for Data Collection in Support of Environmental Decision Making Using the Data Quality Objectives Process, EPA QA/G4, U.S. Environmental Protection Agency, Office of Environmental Information, Washington, DC, 1994.

U.S. EPA, HNU-Hanby PCP Immunoassay Test Kit, EPA/540/R-95/515, Innovative Technology Evaluation Report, U.S. Environmental Protection Agency, Office of Research and Development, Washington, DC, 1995a.

U.S. EPA, Chlor-N-Soil PCB Test Kit L2000 PCB/Chloride Analyzer, EPA/540/R-95/518, Innovative Technology Evaluation Report, U.S. Environmental Protection Agency, Office of Research and Development, Washington, DC, 1995b.

U.S. EPA, EnviroGard PCB Test Kit, EPA/540/R-95/517, Innovative Technology Evaluation Report, U.S. Environmental Protection Agency, Office of Research and Development, Washington, DC, 1995c.

U.S. EPA, Field Analytical Screening Program: PCB Method, EPA/540/R-95/528, Innovative Technology Evaluation Report, U.S. Environmental Protection Agency, Office of Research and Development, Washington, DC, 1995d.

U.S. EPA, PCP Immunoassay Technologies, EPA/540/R-95/514, Innovative Technology Evaluation Report, U.S. Environmental Protection Agency, Office of Research and Development, Washington, DC, 1995e.

U.S. EPA, Test Methods for Evaluation of Solid Waste — Physical/Chemical Methods, SW 846, U.S. Environmental Protection Agency, Office of Solid Waste and Emergency Response, Washington, DC, 1996a.

U.S. EPA, Guidance for Data Quality: Practical Methods for Data Analysis (QA/G-9), EPA/600/R-96/084, U.S. Environmental Protection Agency, Office of Environmental Information, Washington, DC, 1996b.

U.S. EPA, Policy Directive 9380.0-25, U.S. Environmental Protection Agency, Office of Solid Waste and Emergency Response, Washington, DC, 1996c.

U.S. EPA, Field-Portable X-Ray Fluorescence (FPXRF), EPA/542/F-96/009a, U.S. Environmental Protection Agency, Office of Research and Development, Washington, DC, 1996d.

U.S. EPA, Portable Gas Chromatograph/Mass Spectrometer, EPA/542/F-96/009c, U.S. Environmental Protection Agency, Office of Research and Development, Washington, DC, 1996e.

U.S. EPA, Expedited Site Assessment Tools for Underground Storage Tank Sites, EPA/510/B-97/001, U.S. Environmental Protection Agency, Office of Solid Waste and Emergency Response, Washington, DC, 1997a.

U.S. EPA, Field Analytical and Site Characterization Technologies, Summary of Applications, EPA/542/R-97/011, U.S. Environmental Protection Agency, Office of Solid Waste and Emergency Response, Technology Innovation Office, Washington, DC, 1997b.

U.S. EPA, Environmental Technology Verification Program, EPA/600/F-97/005, Environmental Protection Agency, Washington, DC, 1997c.

U.S. EPA, Wellhead Monitoring for Volatile Organic Compounds Fact Sheet, EPA/542/F-97/023, Environmental Protection Agency, Office of Research and Development, Washington, DC, 1997d.

U.S. EPA, PCB Analysis Technologies Fact Sheet, EPA/542/F-97/021, U.S. Environmental Protection Agency, Office of Research and Development, Washington, DC, 1997e.

U.S. EPA, Vendor Field Analytical Characterization Technology System (Vendor FACTS) 3.0 Bulletin, EPA/542/N-98/002, U.S. Environmental Protection Agency, Office of Solid Waste and Emergency Response, Washington, DC, 1998a.

U.S. EPA, Field Sampling and Analysis Technologies Matrix and Reference Guide, EPA/542/B-98/002, U.S. Environmental Protection Agency, Office of Solid Waste and Emergency Response, Washington, DC, 1998b.

U.S. EPA, EPA Guidance for Quality Assurance Project Plans (QA/G-5), EPA/600/R-98/018, U.S. Environmental Protection Agency, Office of Research and Development, Washington, DC, 1998c.

U.S. EPA, EPA Requirements for Quality Assurance Project Plans (QA/R-5), U.S. Environmental Protection Agency, Office of Environmental Information, Washington, DC, 1999a.

U.S. EPA, Field-Based Site Characterization Technologies, U.S. Environmental Protection Agency, CERCLA Education Center, Technology Innovation Office, Washington, DC, 1999b.

U.S. EPA, Guidance for the Data Quality Objectives Process (QA/G-4), EPA/600/R-96/055, U.S. Environmental Protection Agency, Office of Environmental Information, Washington, DC, 2000a.

U.S. EPA, Data Quality Objectives Process for Hazardous Waste Site Investigations, EPA/600/12-00/007, U.S. Environmental Protection Agency, Office of Environmental Information, Washington, DC, 2000b.

U.S. EPA, Innovations in Site Characterization: Geophysical Investigations at Hazardous Waste Sites, EPA/542/R-00/003, U.S. Environmental Protection Agency, Office of Solid Waste and Emergency Response, Washington, DC, 2000c.

U.S. EPA, Guidance for Choosing a Sampling Design for Environmental Data Collection (QA/G-5S), U.S. Environmental Protection Agency, Office of Environmental Information, Washington, DC, 2000d.

U.S. EPA, Guidance for Data Quality Assessment: Practical Methods for Data Analysis, EPA/600/R-96/084, U.S. Environmental Protection Agency, Office of Environmental Information, Washington, DC, 2000e.

U.S. EPA, Integrating Dynamic Field Activities Into the Superfund Process, U.S. Environmental Protection Agency, Office of Emergency and Remedial Response, Washington, DC, 2001a, Available at http://www.epa.gov/superfund.

U.S. EPA, EPA Requirements for Quality Assurance Project Plans (QA/R-5), EPA/240/B-01/003, U.S. Environmental Protection Agency, Office of Environmental Information, Washington, DC, 2001b.

U.S. EPA, Guidance on Environmental Data Verification and Data Validation (QA/G-8), U.S. Environmental Protection Agency, Office of Environmental Information, Washington, DC, 2001c.

U.S. EPA, Innovations in Site Characterization Technology Evaluation: Real-Time VOC Analysis Using a Field Portable GC/MS, EPA/542/R-01/011, U.S. Environmental Protection Agency, Office of Research and Development, Washington, DC, 2001d.

U.S. EPA, Using Dynamic Field Activities for On-Site Decision-Making: A Guide for Project Managers, EPA/540/R-03/002, U.S. Environmental Protection Agency, Office of Solid Waste and Emergency Response, Washington, DC, 2003.

U.S. EPA, National Priority List Sites, U.S. Environmental Protection Agency, Office of Superfund, Washington, DC, 2004a, Available at http://www.epa.gov/superfund.

U.S. EPA, Improving Sampling, Analysis and Data Management for Site Investigations and Cleanup, EPA/542/F-04/001A, U.S. Environmental Protection Agency, Office of Solid Waste and Emergency Response, Washington, DC, 2004b.

U.S. EPA, Field Methods Encyclopedia, U.S. Environmental Protection Agency, Technology Innovation Office, Washington, DC, 2004c, Available at http://www.clu-in.org.

U.S. Geological Survey, A Quality Assurance Plan for District Ground-Water Activities of the U.S. Geological Survey, Open File Report 97-11, U.S. Geological Survey, Reston, VA, 1997.

Walsh, M.E., Field-Portable X-Ray Fluorescence Determinations of Metals in Post-Blast Ordinance Residues, ERDC/CRREL Technical Report TR-04-5, U.S. Army Corps of Engineers, Cold Regions Research and Engineering Laboratory, Hanover, NH, 2004.

Wobber, F.J., Fracture traces in Illinois, *Photogrammetric Engineering*, 33, 499–506, 1967.

Zellmer, J.T., Uses of aerial photographs in environmental site assessment investigations, *The Professional Geologist*, September, 10–11, 1995.

3

Monitoring and Sampling the Vadose Zone

Thomas Ballestero, Beverly Herzog, and Glenn Thompson

CONTENTS

Introduction	208
Characteristics of the Vadose Zone	208
Definitions and Terminology	208
Multiple-Phase Components of the Vadose Zone	209
Solid Phase	209
Sedimentary Deposits	209
Fractured Rock	211
Vadose Zone Water	212
Gas/Vapor Phase in the Vadose Zone	213
Immiscible Fluids	213
Vadose Zone Moisture and Energy	214
Hydrostatics	214
Capillarity	215
Vadose Zone Moisture	216
Vadose Zone Suction	216
Hysteresis	217
Energy Potential in the Vadose Zone	218
Vadose Zone Flow	220
Water	220
Vapors (Gases)	221
Relative Permeability	221
Vadose Zone Monitoring Methods	222
Monitoring Storage Properties	222
Tensiometers	222
Electrical Resistance Blocks	224
Thermocouple Psychrometers	225
Gamma-Ray Attenuation	225
Nuclear Moisture Logging	225
Other Methods	226
Monitoring Vadose Zone Transmission Properties	226
Field Measurements of Infiltration Rates	226
Determination of Water Flux Characteristics	229
Theoretical Perspective	229
Darcy's Law	229
Green–Ampt Wetting Front Model	229
Internal Drainage Method	230
Borehole Permeameters	231

        Measurement of Tracer Movement 231
    Monitoring Water Quality in the Vadose Zone 232
        Electrical Properties Measurements 232
        Soil Sampling and Water Sampling 232
            Pore Water Extraction 232
            Suction Lysimeters 233
            Pan Lysimeters 236
    Soil-Gas Monitoring Technology 236
        Introduction 236
        Background on Methodology 237
        Sampling and Analytical Procedures 237
        Quality Assurance and Quality Control Procedures 238
        Applications 239
            Case Study 239
            Halocarbon Solvents versus Petroleum Hydrocarbons 240
        Problems 240
            Geologic Barriers 240
            Suitability of Compounds to Soil-Gas Technology 242
            Interpretation of Soil-Gas Data 243
            Summary 244
References 244

Introduction

This chapter focuses on an aspect of hydrology that ultimately determines the quantity and quality of ground water available in any given area: the vadose (or unsaturated) zone. The vadose zone is a very important link between human-related influences (i.e., artificial recharge, septic systems, landfills, etc.), climate, and the ground-water system. The vadose zone generally exists between the land surface and the water table, whether the formations are composed of unconsolidated material or bedrock. Simply put, the vadose zone is a porous medium of incomplete saturation, and although water flux in this zone may still subscribe to Darcy's law, the fluid physics are very different than that for the saturated zone. In this chapter, vadose zone definitions and hydraulic theory are presented briefly before a discussion of vadose zone monitoring techniques. The purposes of the chapter are to enlighten the reader about vadose zone hydrology and to present the possibilities for incorporating vadose zone monitoring as an integral part of ground-water investigations.

Characteristics of the Vadose Zone

Definitions and Terminology

The word "vadose" is derived from the Latin word "vadosus," meaning "shallow." According to *Webster's Third New International Dictionary*, vadose is defined as "... of, relating to, or resulting from water or solutions in the part of the earth's crust that is above the permanent ground-water level." Thus, the vadose zone is the region of the shallow subsurface bounded on top by the earth's surface and on bottom by the water table.

The dominant terminology for the porous media and the interstitial fluids that exist within these media above the water table is either the vadose zone or the unsaturated zone. These terms are used synonymously. Other descriptors that may be seen in the literature include the tension-dominated zone (saturated and unsaturated), the soil-moisture zone, or the zone of aeration. The following sections identify the basic terms and definitions related to the vadose zone, the physics of fluid movement and, most importantly, the monitoring technology available for delineation of the vadose zone.

Multiple-Phase Components of the Vadose Zone

The two basic components of the vadose zone are the solid and nonsolid (or fluid) phases. Owing to the hydrologic and agricultural importance of the fluid phase, more descriptive categories exist for this phase than for the solid phase, including water (soil water, soil moisture), vapor (air, soil gas), and immiscible liquids (hydrophobic fluids). Each phase is described in the following subsections.

Solid Phase

The solid phase of the vadose zone is characterized predominantly by the skeletal structure, through which the fluid phases may pass or be retained. This solid skeletal structure is composed of inert to reactive particles of fractured rock, cobbles, gravel, sand, silt, and clay as well as organic matter such as roots, leaves, and waste products of micro- and macro-organisms. Very small (<10 μm) solid particles (colloids) can be transported by fluids through the interstices (pore spaces) between the solid materials. Owing to the relatively low fluid velocities found in the vadose zone, the portion of the solid phase that is mobile is extremely small and, for all practical purposes, may be ignored.

Fixed and mobile micro-organisms comprise a second element that can be included with the solid phase. A classic example of fixed micro-organisms is the mat of micro-organism buildup that occurs beneath the leach fields of subsurface wastewater disposal systems (i.e., septic system leach fields). An example of mobile micro-organisms is evident in situations where ground-water samples produce coliform bacteria cultures. In the vadose zone, fixed soil bacteria are commonly found at concentrations of 10^3 to 10^7 cells per gram of soil (Atlas and Bartha, 1998).

Terms used to describe the inert particles in the solid phase include grain-size distribution, porosity, roundness, angularity, uniformity, and specific surface. The organic matter is typically separated from the inert solids and just identified as a percentage of the total volume. Many of the physical descriptors of the inert solid phase, which are inter-related, are closely related to the characteristics, content, and mobility of the fluid phases. For example, porosity directly affects the amount of water that can exist in the voids—the lower the porosity, the lower the capacity for water content in the vadose zone. In addition, for a given grain-size distribution, lower porosity means fewer pores and lower capability for transmission of fluids through the vadose zone.

Sedimentary Deposits

The grain-size distribution of sedimentary deposits represents the cumulative probability of occurrence (percent passing a certain standard sieve size) of various grain sizes in these deposits. Grain-size distribution has a significant effect on fluid phase content and fluid transmission characteristics. Each individual grain can be physically described by three linear measurements: length, width, and breadth (in descending magnitudes). In trying to fit a grain through a square sieve opening, the smallest sized opening through which the grain may pass is roughly equivalent to the width dimension. By taking a bulk sample of formation material and sieving the sample with a sequence of sieves (sieve with largest openings first, sieve with smallest openings last), each sieve would pass or retain a certain portion of the sample. If the first sieve retains 10% of the sample, then 90% passes this sieve. If the next sieve retains 30% of the sample, then 60% of the sample passes the second sieve, and a total of 40% is retained by the first two sieves (10% was retained on the first sieve that had larger sieve openings; if particles were retained on the first sieve, they would also be retained on the second.) This standard sieve analysis procedure is well described in elementary soil mechanics textbooks and ASTM Standards (ASTM D 422 [ASTM, 2004a]).

Sieve analysis information is usually plotted on a graph as particle (sieve) diameter on the x-axis (usually a log scale) and percent passing (by weight) on the y-axis. The cumulative percent retained (100% minus the percent passing) may also be plotted on a secondary y-axis. For grain size increasing to the right and percent passing increasing upward (see Figure 3.1), the grain-size distribution is characteristically S-shaped. Selecting a particular value from the percent passing scale, say 85%, the grain-size diameter from the grain-size curve is defined as D_{85}.

Two important elements of the grain-size distribution are the median grain size (D_{50}) and the uniformity coefficient (C_u), which is the ratio of the 60% passing size (D_{60}) to the 10% passing size (D_{10}). Typically, grain-size distributions are determined by sieve analysis for particles of silt size or larger and by hydrometer analysis for clays (fines). As the grain-size distribution curve becomes more vertical, the uniformity coefficient tends toward unity. For widely varying particle sizes, the curve is flatter and the C_u is larger.

The importance of D_{50} and C_u with respect to the fluid phases is straightforward. For a constant C_u, average pore size increases with increased D_{50}. For a constant D_{50}, increases in C_u decrease the porosity and average pore size. This latter relationship results from the fact that poorly sorted or well-graded (high C_u) deposits will have smaller grains filling in the interstices between the larger grains.

Particle shape (angularity) can also have a bearing on porosity and fluid transmission. The shape of individual particles can range from spherical to very angular to flat. Depending on the packing and mixing of these particles, a wide range of porosities is possible.

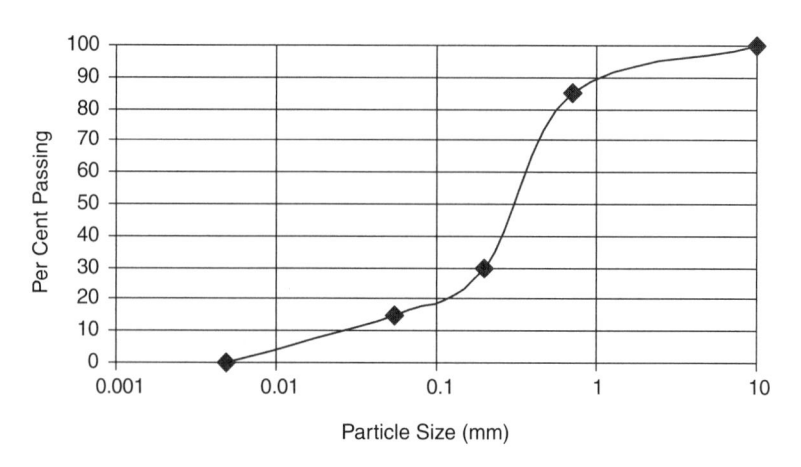

FIGURE 3.1
A typical grain-size distribution curve.

Monitoring and Sampling the Vadose Zone

The last important descriptor of the solid phase is its specific surface — the ratio of a grain's surface area to its volume (or in some cases, to its mass). Grains with the largest specific surface are flat plates (clays) and those with the smallest specific surface are spheres (weathered silicate grains). Owing to the importance of surface chemistry in contaminant transport studies, clay content of the solid portion of the medium is important, partially because of the large specific surface of clays.

Fractured Rock

For fractured rock, a massive structure is decimated by many/few large/small cracks. The cracks or fractures possess varying lengths and orientations. In general, the fractures can be statistically described by probability distributions for length, width (aperture), orientation, and density. On a very large scale, the mechanics of such a vadose zone may be no different than for a sedimentary deposit vadose zone. The perverse nature of the fractured rock medium is that the effect of scale confounds traditional thinking, monitoring, and analysis.

In revisiting what Bear (1979), Corey (1977), and McWhorter and Sunada (1981) describe as the "representative elemental volume" (REV), the primary difficulty of fractured rock vadose zone monitoring may be recognized. A REV is the volume of material that must be used in order to obtain a valid estimation of a particular parameter. This concept is illustrated for porosity in Figure 3.2 for sand. If a very small sample size volume is used for the sand, for example a sample size on the order of the size of the pores or smaller, various samples from the sand using such a small volume will yield very different estimates of porosity. Some samples will be entirely solid (porosity of zero) and others will be entirely void (porosity of unity). As the sample size increases to be larger than the sand grains, the estimates of porosity from different samples become more consistent. If samples get too large, then nonhomogeneities in the sand deposit create more inconsistency in the porosity estimates.

The range of variability of the porosity estimates as a function of the size of the REV, for the sand of Figure 3.2, is plotted in Figure 3.3. In comparison, the range of variability of porosity for a fractured rock vadose zone is also plotted in Figure 3.3. As evident in Figure 3.3, a competent fractured rock normally exhibits very low porosity and a very large sample volume may be necessary to get a reliable estimate of porosity.

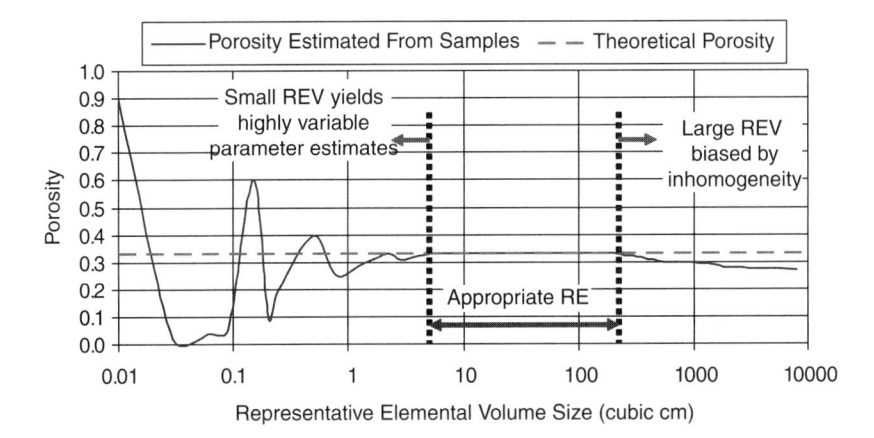

FIGURE 3.2
Example of the variability of sand porosity estimates as a function of the size of the representative elementary volume.

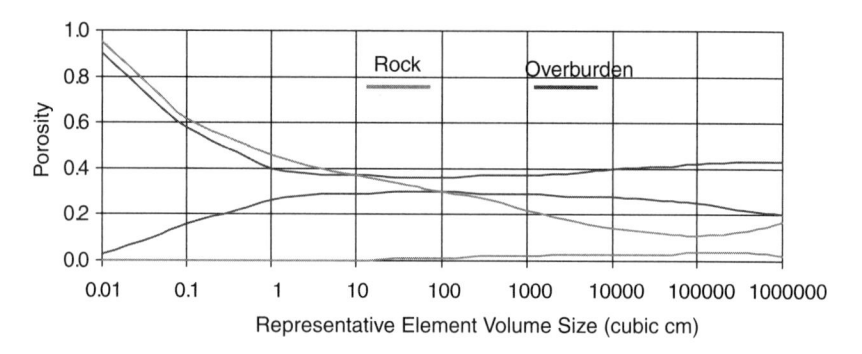

FIGURE 3.3
Scatter of porosity data for sand and fractured rock.

Extending the REV concept to a fractured rock vadose zone, a sample size on the order of thousands of cubic meters may be required for an accurate estimate of porosity (see Figure 3.3). When considering that most of our field instruments, at best, investigate between 0.01 and 5 m^3, it is easy to see why it is difficult to work with the fractured rock vadose zone. This fact alone has led practitioners toward two basic avenues of investigating fractured rock vadose zones: (a) disaggregation of the problem into that of a solid mass where little to no fluid phase occurs and there exists a continuum of interconnected pore spaces or (b) re-evaluating the REV by either taking many small samples or a few large samples. Identification of the REV is paramount in dealing with fractured rock systems. The REV has been presented here and exemplified with porosity, yet the concept is valid for any descriptive parameter, including hydraulic conductivity.

Vadose Zone Water

A given molecule of water may reside in the vadose zone from minutes to centuries, depending on the size of the particular vadose zone and its transport characteristics. Mechanisms by which water may enter the vadose zone from above include precipitation and recharge (i.e., rainfall infiltration, spreading basin, septic system, etc.). From below, water may flow from the saturated zone into the vadose zone. Lastly, and least importantly, water may enter the vadose zone from within due to any of the numerous biological or chemical reactions that have water as an end product. For example, the biodegradation of petroleum hydrocarbons leads to the production of carbon dioxide and water. Also, water vapor that enters the vadose zone may be condensed into the liquid phase via a temperature change.

Just as water may enter the vadose zone, it may also exit. At or just below the ground surface, water may exit due to evapotranspiration processes. From below, water may drain into the saturated zone. Finally, vadose zone water may be consumed by certain biological or chemical reactions within the vadose zone.

Fluid properties that are important in describing vadose zone water (as well as other vadose zone fluids) include density, specific weight, kinematic viscosity, bulk modulus of elasticity, vapor pressure, surface tension, dynamic viscosity, and wettability in the presence of air. Detailed descriptions of these fluid properties may be found in most texts on either fluid mechanics or chemistry.

An important characteristic relating any of the liquid phases (in this case soil water) to the solid phase is wettability. Wettability is the property that is characterized by the relative interfacial forces of two fluids at a solid boundary. The two fluids, practically speaking, may be liquid–liquid (i.e., water–oil) or liquid–gas (i.e., water–soil gas). Thus, wettability describes the relative affinity for one fluid over another to a solid surface. With the two fluids in the presence of the solid, one fluid will preferentially coat (wet) the solid surface. For example, with water and air on glass, water wets and thus will tend to coat the glass. This explains the shape of the meniscus in a glass capillary tube. Extending this example to the vadose zone, soil water will tend to coat the soil particles and move in very tortuous paths in a porous medium, while air will be left in the larger pore spaces. For petroleum products and soil water, water preferentially wets over the petroleum products, even if the petroleum was there first.

Gas/Vapor Phase in the Vadose Zone

The important practical characteristic of the vadose zone is that the pores contain more than one fluid. Other than soil water, the fluid of most interest is soil gas, and there is always a trade-off between the two fluids. During precipitation or recharge events, water volume in the pore space increases and the soil gas is displaced. When the vadose zone is draining water to the ground water or when it is drying out due to evapotranspiration, the volume of soil gas in the pore space increases.

Soil gas has descriptive properties similar to those of soil water. In addition, the perfect gas constant is of utility. It is important to recognize that the liquid (soil water) vapor pressure will require that the soil gas and liquid ultimately come into equilibrium. Thus, there will be a certain amount of water vapor found in the soil gas. More importantly, volatile chemicals will also be found in the vapor phase if the solid or liquid phase of the chemical is present in or near the vadose zone. Therefore, the vapor phase can be sampled and analyzed in order to make statements about liquid fluid phases in the vadose zone or below, for example, lying on top of the water table or contained within the saturated zone.

Immiscible Fluids

Fluids other than water or vapor may be found in the vadose zone. Fluids that can easily mix with water (i.e., septic system effluent or landfill leachate) are known as miscible fluids. Miscible fluids in the vadose zone typically have dissolved solids concentrations, temperature, and density similar to the existing vadose zone water; if not, the fluid may temporarily be considered immiscible. For example, a septic system effluent may act as a fluid into itself and not readily mix with vadose zone water until the effluent temperature moderates with its surroundings.

In other cases, there are fluids in which the primary composition is that of hydrophobic molecules. This type of fluid may never mix with water and is considered immiscible. Most immiscible fluids exhibit some solubility in water. For practical purposes, it is assumed that the immiscible fluid retains its original volume integrity. For example, gasoline and water do not readily mix, but after keeping the two in a closed container for a few months, traces of some gasoline constituents (i.e., benzene, toluene, ethylbenzene, xylenes) will be found dissolved in the water phase in the container.

When an immiscible fluid such as gasoline enters the vadose zone, there is increased competition for the void spaces: pore spaces will be filled with a mixture of gasoline, water, and soil gas. In addition, equilibrium thermodynamics will result in volatilization of some gasoline and vaporization of some water, both into the soil gas. A portion of the gasoline will also dissolve into the water.

Vadose Zone Moisture and Energy

Hydrostatics

For a constant density, static-fluid continuum where gravity is the only acting acceleration field, the law of hydrostatics can be derived as:

$$\frac{dp}{dz} = -\rho g \tag{3.1}$$

$$\frac{dp}{dx} = 0 \tag{3.2}$$

where p is the fluid pressure, z the vertical coordinate axis (positive upward), ρ the fluid density, and g the acceleration due to gravity. For ground water and vadose zone considerations, by setting $z = 0$ (called the vertical datum) at a location where $p = 0$ (i.e., at the water table), Equation 3.1 can be integrated to its more common form:

$$p = \rho g h \tag{3.3}$$

where p is the pressure at any distance h vertically from the zero pressure datum. Below the vertical datum, z is negative but h is positive and pressure is positive. Above the vertical datum, z is positive, h is negative, and pressure is negative. This relationship results in a linear pressure distribution above and below the vertical datum (water table) as depicted in Figure 3.4 for a glass of water. In this figure, it can be seen that where $z = 0$, $p = 0$. Below this level, z is negative and p is positive, with p increasing inearly as z decreases linearly. When above the level of $z = 0$, z is positive and p is negative. A practical analogy is utilizing a straw in a can of beverage. With no capillarity effect in the straw, the level of the fluid in the straw is at the level of fluid in the can. If you draw the beverage up into the straw and put your thumb over it, the pressure distribution in the fluid in the straw would be as described by Equation 3.3 and depicted in Figure 3.4. The zero datum here is the liquid surface of the beverage in the can. In Figure 3.4, the pressure distribution above the water surface is sketched as a dashed line. Obviously, there is no water above that in the glass; if there were and it was connected (thereby

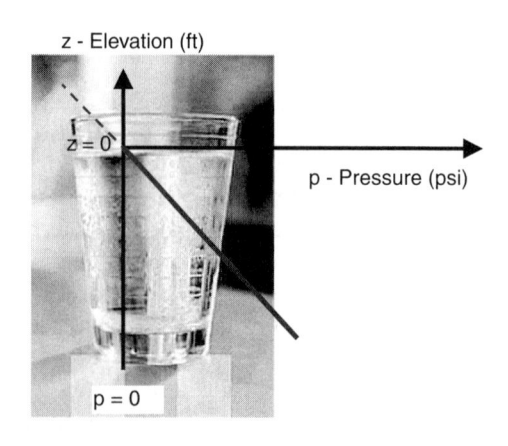

FIGURE 3.4
Pressure variation in a static fluid.

creating a continuum) to the water in the glass, it would follow the linear pressure distribution (as in the straw analogy). The important point here is that the pressure of water at the water table is atmospheric (zero gage pressure), and therefore the water above it (in the vadose zone) exists at pressures less than atmospheric.

Thus, in the vadose zone, water exists above a zero pressure datum (the water table), the fluid is at negative pressures, and Equation 3.3 accurately predicts the pressure in the vadose zone water as long as saturation is maintained. However, due to the breakdown of the fluid continuum (saturation) in the vadose zone, Equation 3.3 does not accurately describe the pressure situation above the capillary fringe (where saturation does exist).

Capillarity

When two immiscible fluids exist at an interface, there is a tendency for the molecules of each fluid to move away from the interface and be nearer to their like molecules. In order to keep molecules at the interface, energy must be expended on each and every molecule at the interface. This free surface energy is measured by the fluid property known as surface tension. The combination of immiscibility plus surface tension results in the interface acting as a membrane. Given finite fluid masses, the molecular forces that exist at the fluid interface tend to deform the interface of the finite mass fluid into a curved surface (e.g., a raindrop — water is the finite fluid mass and the air is infinite). Because the surface is curved for the finite fluid mass, with possibly more than one radius of curvature, there is an imbalance of forces at the surface. This imbalance is offset by a pressure difference across the interface between the fluids. The difference in pressure across the interface is known as the capillary pressure and can be computed in the vadose zone as:

$$p_c = p_a - p_w \tag{3.4}$$

where p_c is the capillary pressure, p_a the soil gas pressure, and p_w the soil water pressure.

Combining capillarity considerations with wettability (the affinity for a fluid to a solid surface), relationships for static conditions in the vadose zone may be developed. In a very basic analogy, for a single small-diameter glass tube (capillary tube) standing vertically and partially submerged in a tank of water, an equilibrium analysis of the forces existing on an element of the curved fluid surface in the tube yields:

$$p_c = \frac{2\sigma}{r} \tag{3.5}$$

where p_c is again the capillary pressure across the interface, σ the surface tension of the water, and r the radius of the capillary tube. If the upper end of the tube is open to the atmosphere, the pressure on top of the water surface in the tube is atmospheric. From hydrostatics, Equation 3.3 yields a pressure in the water, right at this interface, of

$$p_w = -\rho g h \tag{3.6}$$

where h is the height of capillary rise above the surface level of water in the tank (which is open to the atmosphere). Substituting Equation 3.6 into Equation 3.5 yields:

$$h = \frac{2\sigma}{\rho g r} \tag{3.7}$$

where it can be seen that the height of capillary rise is inversely proportional to the radius of the capillary tube. Typical values for water yield $h \approx 0.15/r$ (h and r in centimeter). The difficulty in using this in soils is that r is the radius of the pore spaces, which are assumed to act as a bundle of straws. For soils, r must be replaced by a measure of the representative pore size, d_n. In most practical applications, d_n is a function of the median particle size (d_{50}), where $0.155\, d_{50} < d_n < 0.414 d_{50}$ (Iwata et al., 1988).

Vadose Zone Moisture

The vadose zone pore spaces can be filled with any fluid; the most common fluids are air and water. If the total volume of pores in a sample of the vadose zone is V_p and the volume of the pore space occupied by water is V_w, the saturation (S) is calculated as:

$$S = \frac{V_w}{V_p} \tag{3.8}$$

S can range from 0 to 1.0 and is sometimes reported as a percentage. When evaluating the quantity of water in a sample compared to that of the total sample (of volume V_T), the volumetric water (moisture) content Θ is calculated as:

$$\Theta = \frac{V_w}{V_T} \tag{3.9}$$

and obviously ranges from 0 to the porosity (φ). Thus to relate the two:

$$\Theta = S\varphi \tag{3.10}$$

Under field conditions, gravity alone cannot drain the unsaturated zone because surface tension, osmotic, and molecular forces can act against it. An example of this is a sponge taken from a tub of water: initially, the sponge will drain by gravity, however, after this gravity drainage ceases, the sponge is still moist. The lower limit for S in the vadose zone is S_r — the residual saturation. At the soil surface, when evaporation dominates, S_r may approach zero.

From the early part of the last century, vadose zone investigators recognized the relationships between moisture content (or saturation) and the distance above the water table. A typical relationship, for static conditions, is shown in Figure 3.5. At distances above the water table, S approaches S_r. Moving closer to the water table, saturation increases to the field saturation level (S_s). Some residual air will remain at and just below the water table because water-table fluctuations entrap air in this region.

Vadose Zone Suction

As described in Capillarity, above the water table, negative pressures (or suction) exist in the liquid phases primarily due to the curvature of the surface of the finite-sized liquids in this region. When multiplied by -1, the negative pressures become a positive number: the soil suction (or tension). A plot of a typical vadose zone (soil) suction relationship is shown in Figure 3.6. At the water table, water pressure and soil suction are atmospheric or zero gage pressure. Above the water table, water pressure decreases or soil suction increases. Combining the soil suction information with saturation or moisture content, as both moisture content and vadose zone suction are functions of the distance from the water table, vadose zone suction can be plotted against soil moisture content for a given soil. The plot of moisture content versus soil suction is called the soil moisture characteristic curve, and is generally considered a property of the soil (although many variables affect it). Thus, by monitoring vadose zone suction (soil suction or matric potential), the moisture content can be estimated. Figure 3.7 is an example of a soil moisture characteristic curve.

Monitoring and Sampling the Vadose Zone

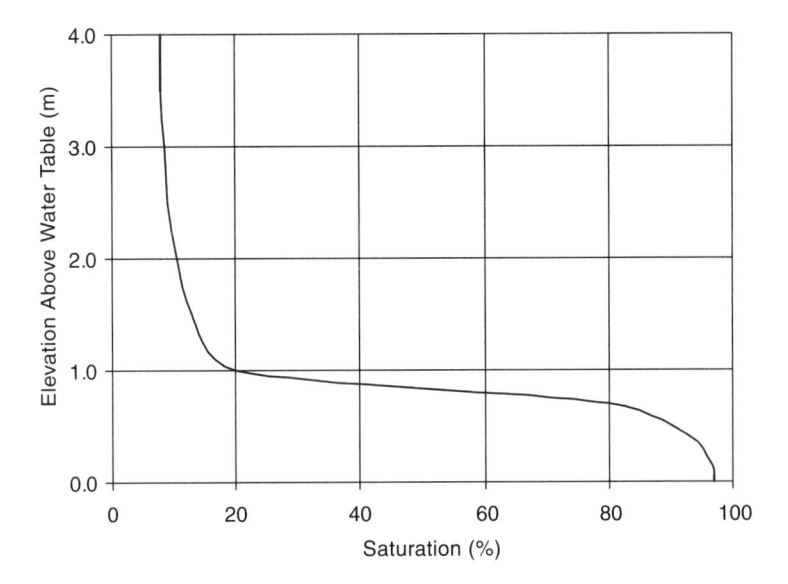

FIGURE 3.5
Example saturation condition in the vadose zone.

Hysteresis

The moisture content–suction relationship (and therefore the moisture content–elevation relationship) is not unique for a given vadose zone. Depending on whether or not the vadose zone is undergoing drainage (de-sorption) or wetting (sorption), the moisture content at a given elevation above the water table can be represented by more than one pressure. Figure 3.8 depicts this phenomenon, which is known as hysteresis. This process can be explained by analogy of the variation of the pore radii in the vadose zone to that of an ink bottle, depicted in Figure 3.9. In wetting of an initially dry soil, wettability and capillarity will allow water to move vertically into pore spaces. Capillary rise will cease when there is a balance between surface tension and gravitational forces

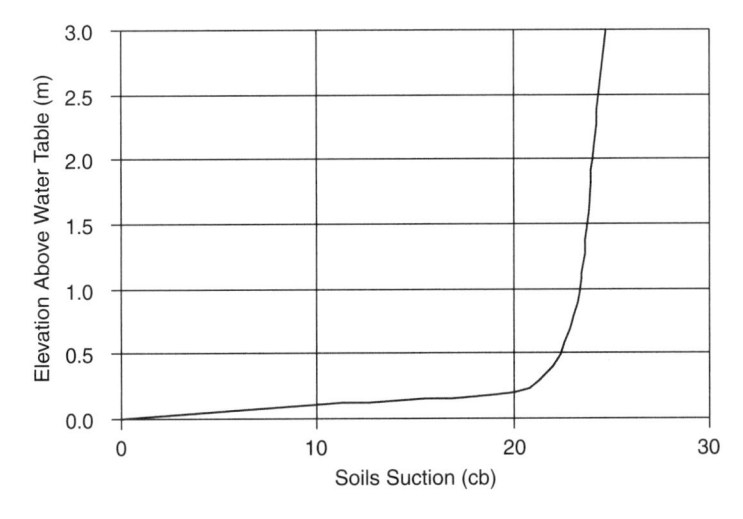

FIGURE 3.6
Vadose zone suction profile.

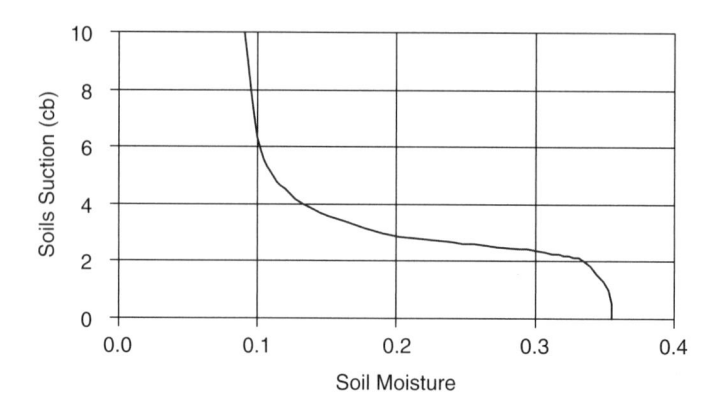

FIGURE 3.7
Vadose zone suction–moisture content relationship for a coarse soil (wetting).

for a given pore size. When the pore size changes from r to a larger value R, although the capillary rise for r could allow water to move vertically higher (h_r), the capillary rise for R may be smaller (h_R) than the amount of rise which has already occurred (h_w).

Thus, when the pore size increases and water is moving upward in the capillary space during wetting, capillary rise ceases in order to maintain the force equilibrium of the interface. This effect gives rise to the wetting curve in Figure 3.8.

If the vadose zone was initially saturated and then allowed to drain, drainage would occur in more passive heights of capillary rise, while leaving some large pores below saturated. In addition, some pores will have their connection to surrounding vadose zone water ruptured by drainage, becoming islands of water in the vadose zone. These factors result in higher moisture contents during drainage, for a given pressure, than for wetting.

Energy Potential in the Vadose Zone

The total status of energy for soil moisture is described by the total moisture potential ψ_T (L^2/T^2 in units of cm²/sec²). ψ_T is comprised of the sum of three primary potentials: gravitational (ψ_g), pressure (ψ_p), and osmotic (ψ_o) potential. These potentials are analogous

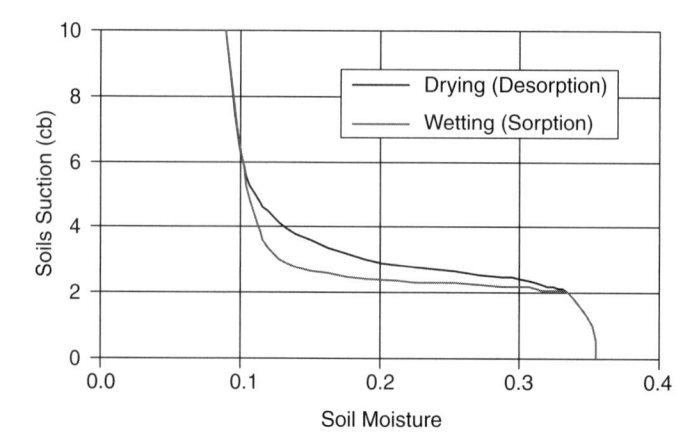

FIGURE 3.8
Hysteresis in suction–moisture content relationship.

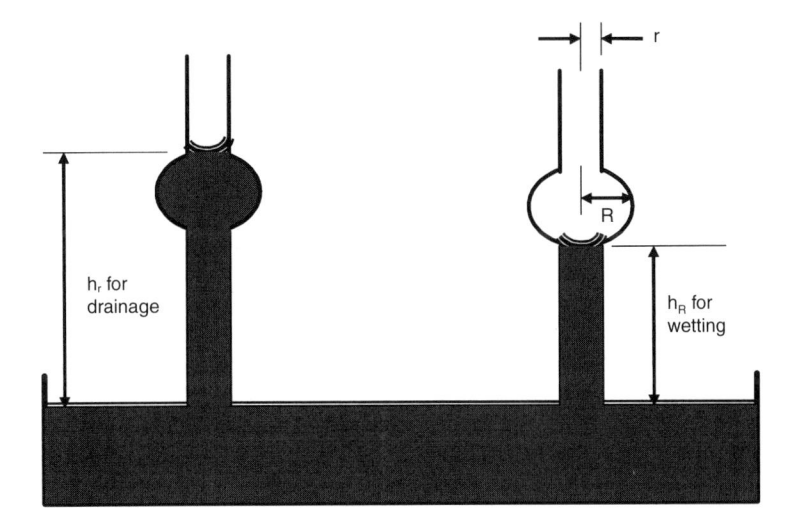

FIGURE 3.9
The ink bottle effect, illustrating hysteresis.

to the Bernoulli sums of gravitational head, pressure head, and velocity head. For vadose zone moisture flow, the velocity potential is negligible. For surface water or piping considerations, osmotic potential is negligible.

ψ_g represents a potential energy due to the vertical location of the vadose zone moisture of interest:

$$\psi_g = gz \qquad (3.11)$$

where g is the acceleration due to gravity and z is the distance above (+) or below (−) some vertical datum.

ψ_p is the hydrostatic pressure existing in the vadose zone water.

$$\psi_p = p/\rho \qquad (3.12)$$

where p is the hydrostatic pressure and ρ is the density of the liquid (water).

ψ_o represents the potential resulting from maintenance of concentration gradients of solutes in vadose moisture systems. Normally, solute molecules in a zone of high concentration diffuse to zones of lower concentration. If some barrier exists in the vadose zone, which prevents the movement of the solute to the zones of lower concentration, yet allows movement of the solvent (water) in any direction, a pressure must exist across this barrier when solvent movement through the membrane equilibrates (solvent flow in one direction is balanced by solvent flow the opposite way). The barrier is commonly referred to as a semipermeable membrane. Figure 3.10 depicts the osmotic pressure at equilibrium. In the figure, h_o is the height of solution yielding a pressure (p_o) on the membrane. In this case:

$$\psi_o = \frac{\psi_o}{\rho} gh_o = \frac{MRT}{\rho} \qquad (3.13)$$

where M is the total molar concentration of the solute, T the temperature (°K), and R the gas constant. This equation is only good for dilute solutions.

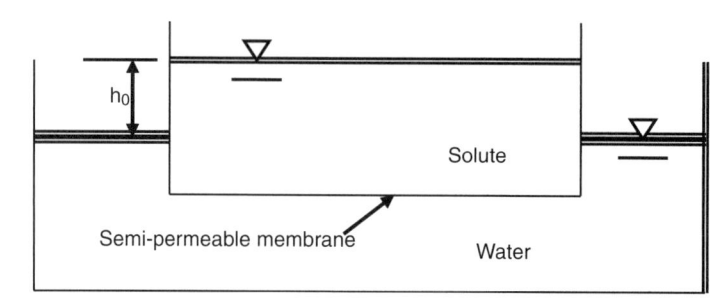

FIGURE 3.10
Definition sketch for osmotic potential.

Vadose Zone Flow

Water

Although many of the definitions and characteristics of the vadose zone presented herein may appear different than for aquifer (saturated) systems, the basic equations defining flow for the vadose zone are the same as those for aquifers: Darcy's law (energy) and continuity. In this case, it must be recognized that hydraulic conductivity (or permeability) is a function of moisture content (θ) in the vadose zone. The total potential (ψ_T) in the vadose zone can be converted to head (as is typically used in Darcy's law) by dividing by the acceleration due to gravity:

$$h_T = \frac{\psi_T}{g} \tag{3.14}$$

Thus, Darcy's law is now written as:

$$q = -K(\psi)\nabla h_T \tag{3.15}$$

where q is the flux, $K(\psi)$ the hydraulic conductivity, and ∇ the del operator for spatial vector partial differentiation.

Compared to saturated ground-water flow, where the hydraulic conductivity (K) is that at saturation, vadose zone mechanics are such that K is a function of saturation or moisture content. Figure 3.11 depicts such a relationship. Quite obviously, water flow becomes more difficult as the degree of saturation decreases. In fact, there can be dramatic decreases in hydraulic conductivity, over three to six orders of magnitude, as the formation moves from saturation to residual moisture content. This results from the fact that air takes the most advantageous pore spaces, leaving the most tortuous paths for water.

Combining continuity with Darcy's law yields the transient vadose zone flow equation:

$$\frac{\partial \theta}{\partial t} = -\nabla[K(\psi)\nabla h_T] \tag{3.16}$$

Solutions to this equation have been analytically derived for horizontal and vertical flow cases. One common extension is (θ) for total potential head (h_T) (for example, see Hillel, 1980b, pp. 204–207). Simple saturated ground-water flow scenarios cast against various boundary conditions (for example, a well pumping in an infinite-size aquifer) result in analytical equations to predict aquifer responses (drawdown) in response to

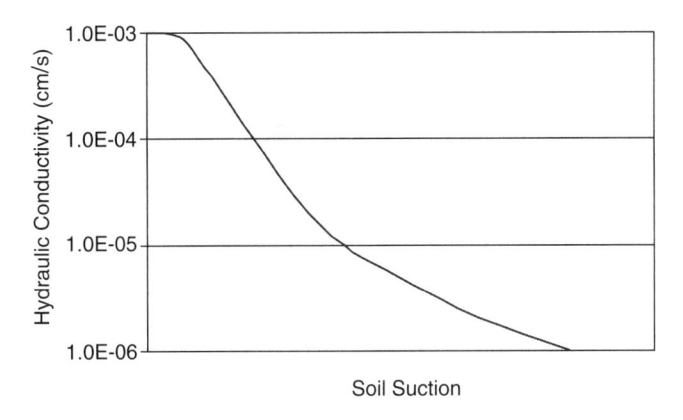

FIGURE 3.11
Example of moisture content–hydraulic conductivity relationship.

known signals (pumping). These same equations can also be employed to identify the aquifer parameters. Similarly, solutions to the vadose zone soil moisture flux theory yield equations that can be used in predictive (soil moisture) or descriptive (unsaturated hydraulic conductivity) modes.

Vapors (Gases)

Natural vapor flux in the vadose zone is dominated by diffusive (Fickian) transport. As such, the vapor flux (q_v) is computed from a diffusion equation:

$$q_v = -D_v \frac{\partial C}{\partial x} \qquad (3.17)$$

where D_v is the diffusion coefficient for water vapor in the porous media, C the concentration of water vapor, and x the spatial coordinate.

In general, if vapor flux is induced, i.e., by soil venting, then it is dominated by a pressure gradient. That is, instead of diffusion processes accounting for flux, an excessive pressure condition (suction or positive pressure) will drive vapors from regions of high pressure to low pressure. In these instances, velocity and elevation potentials are considered insignificant when compared with pressure potential. For these types of field conditions, Fick's law and Darcy's law are not valid and equations dealing with turbulent fluid transport are necessary, such as a Darcy–Weisbach formulation.

When considering multiple-phase flow, it must be recognized that there is typically a threshold value of fluid content that must be achieved before a particular fluid can move. In the case of three-phase flow (gasoline–water–vapor) there is, for practical purposes, only a small window of saturation at which all three fluids can move. With this in mind, an obvious strategy for immobilization of the gasoline would be to increase water content or, more preferably, vapor (air) content. By such an action (pumping air into the vadose zone), vapor content increases so as to push the system out of the three-phase flow "window." Once immobilization of the gasoline occurs, *in situ* or other cleanup methodologies can be addressed.

Relative Permeability

When investigating the transport of multiple fluid phases (air, water, and nonaqueous phase liquids) in the vadose zone, it is common to develop the hydraulic conductivity or permeability for each phase, for various fluid contents, and then to relate these values to the hydraulic conductivity or permeability at saturation. Such plots (relative permeability versus moisture content), for each fluid phase, assist in identifying the relative mobility of each phase for any particular moisture (or fluid) content. Permeability (k) is solely a property of the porous medium (pore size), whereas hydraulic conductivity is a property of both the porous medium and the fluid (density and viscosity). The relationship between hydraulic conductivity (K) and permeability (k) is: $k = K\mu/\rho g$, where K is the saturated hydraulic conductivity, μ the fluid viscosity, ρ the fluid density, and g the acceleration due to gravity. A good discussion of relative permeability and typical plots for air and water appears in Corey, 1977.

Vadose Zone Monitoring Methods

In monitoring the vadose zone, common objectives include determining fluid saturation (water, air, and nonaqueous phase), assessing fluid transmission capability (infiltration, unsaturated hydraulic conductivity, or relative permeability), and sampling the fluids present. Because of the inter-relationship of some variables (moisture content, pressure, hydraulic conductivity), often one of these variables is measured and used as a surrogate or estimator of the others. However, this should only be performed when there is appropriate calibration information. General sources for methods and applications relating to soil water monitoring include Wilson (1980), Wilson (1981), Everett (1980), Everett et al. (1976), and Fenn et al. (1977). Soil-gas monitoring is discussed later in this chapter.

Monitoring Storage Properties

The physical properties of the vadose zone associated with water storage include bulk density, total thickness, porosity, water content, and soil moisture versus tension relationship. Total potential water storage can be estimated from the first two properties, which are easily measured. Total porosity can be used in place of bulk density for estimating total potential storage, while pore-size distribution affects fluid transmission. Water content can be measured directly or estimated from the soil moisture characteristic curve. This section discusses measurement of tension and water content, which can be measured using tensiometers, electrical resistance blocks, thermocouple psychrometers, gamma-ray attenuation, or nuclear magnetic resonance.

Tensiometers

Tensiometers are used to measure soil matric potential (pressure). These devices create a water continuum to the vadose zone and therefore pressure can be measured anywhere in this continuum. By using Equation 3.3, pressure in the vadose zone is calculated from the point of the pressure measurement. A tensiometer consists of a porous ceramic cup (or other porous surface) attached to a pressure sensor via a tube filled with water. The porous cup is located in the vadose zone where information on the pressure (matric potential, or suction) is desired. The pressure sensor is commonly a Bourdon-type pressure gage (Figure 3.12), a pressure transducer, or in the past, a mercury-filled manometer. The principle of operation is that water can freely flow into or out of the porous cup at soil tensions that do not exceed the air-entry tension of the porous cup (usually in the range of 0.5 to 1 bar). As water moves out of the porous cup into the

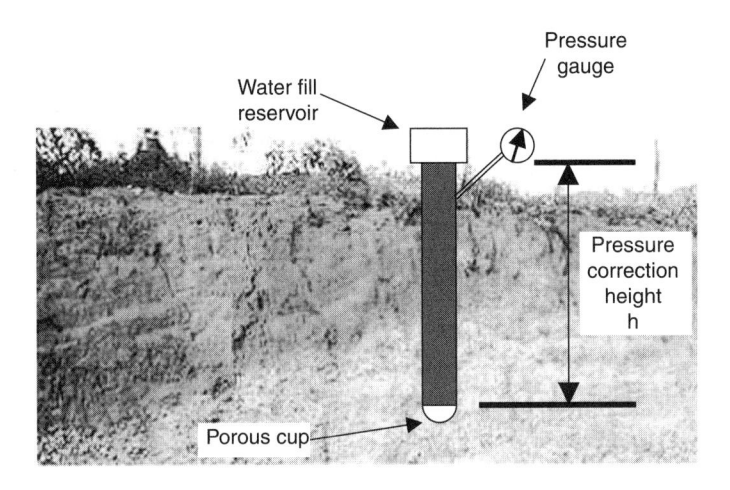

FIGURE 3.12
Schematic diagram of typical commercially available gage tensiometer.

unsaturated soil, a vacuum is formed in the water tube, which exerts a corresponding pressure at the gage, manometer, or transducer diaphragm. Because the pressure gage is not located at the porous cup, pressure readings must be corrected by adding the height (h) of the water column between the soil surface and the gage or transducer to the tensiometer reading (Figure 3.12). For example, if the pressure gage (p_g) in Figure 3.12 read 70 centibars (70 cb) of suction and the distance from the bottom of the gage to the center of the ceramic cup was 1 m, the soil moisture pressure at the ceramic cup (p) would be:

$$p = p_g + \gamma h = -70\,\text{cb} + 9800\,\text{N/m}^3(1\,\text{m})(0.001\,\text{cb/N/m}^2) = -60.2\,\text{cb}$$

This follows directly from the previous section on hydrostatics (linear increase in pressure moving downward in a liquid) and Equation 3.1 and Equation 3.3.

To assure proper operation, the tensiometer should be installed with as little disturbance to the soil as possible, and the porous cup should be hydraulically connected with the surrounding soil. The latter can be accomplished by forcing the ceramic cup into a snug hole or by placing the cup in a slurry of material removed from the hole (per ASTM Standard D 3404 [ASTM, 2004b]).

Tensiometers are inexpensive and can be purchased from any of the several manufacturers or custom-designed for special applications. Ethylene glycol solution, or a similar liquid, can be used to make them operational during periods of freezing and thawing. Stannard (1990) presents a number of designs, along with their advantages and disadvantages.

Vacuum gage tensiometers are durable and easily operated and maintained. However, they are less accurate and precise than manometer and transducer tensiometers. Response time with these instruments varies from poor to excellent. Calibration is required before installation and occasionally after installation. Data are collected manually. Gage tensiometers are not well suited to measuring hydraulic gradients, but can be used for gross measurements of moisture movement. They also cannot measure positive pressure, so they lose their usefulness as the water table moves higher than the elevation of the ceramic cup.

Manometer tensiometers are also durable and easily operated. Maintenance is dependent on tensiometer design; those with small water lines require frequent purging. Since these devices can be custom made, they are more versatile than gage tensiometers. Wilson (1990) rates the accuracy of manometer tensiometers as "excellent," precision as "good," and response time as "fair." Data are collected manually, and both positive and negative pressures can be measured. A major advantage of these instruments over the other types of tensiometers is that calibration is never required.

Tensiometers connected to transducers can measure both positive and negative pressures and provide nearly instantaneous *in situ* readings of soil-water pressure, which can be recorded electronically. As with manometer tensiometers, maintenance and versatility are dependent upon the design of the water conduits. Response time is the most rapid of the three types of tensiometers, making these the best choice for tracking wetting fronts. However, these are the most expensive of the three alternatives and periodic recalibration is required due to instrument drift.

Disadvantages for all types of tensiometers include a bottom tension limit of about 1 bar, with decreased accuracy in readings at tensions greater than about 0.8 bars. The bottom tension limit is related to the fact that the larger the suction (tension), the more likely that air will enter the porous cup and invalidate the readings. Readings are sensitive to temperature changes, atmospheric pressure changes, and air bubbles in the water lines. Tensiometers with small-diameter water conduits are especially susceptible to air bubbles and require frequent purging to assure accurate readings. In addition, because the soil-moisture characteristic curve is required for determining soil moisture from tension measurement and the curve is subject to hysteresis, it is necessary to know whether the soil is wetting or drying when the measurement is taken. Other sources of error include operator error in reading the manometer/gage and poor pressure transducer calibration.

Electrical Resistance Blocks

Electrical resistance blocks are inexpensive and can be used to measure either moisture content or soil-water pressure. Essentially, electrical resistance is used as a surrogate variable for soil moisture. The blocks consist of two metal plates imbedded in a porous material, usually gypsum, nylon, or fiberglass. Wires are attached to the plates so that changes in the electrical resistance between the two plates can be measured. As the moisture content (or tension) of the electrical resistance block changes, coincident and in equilibrium with the surrounding soil, the electrical resistance properties of the block are altered. Before use, electrical resistance blocks must be calibrated in the laboratory using soil from the installation site. Calibration produces curves of electrical resistance versus soil moisture or soil-water pressure. Because each block should produce the same curve, calibration allows the user to find faulty blocks before they are installed.

The chief advantages of electrical resistance blocks are (1) they are suited for general use in the study of soil-water relations; (2) they are inexpensive; (3) they can be used to determine either suction or moisture content; and (4) they require little maintenance.

While electrical resistance blocks present an attractive monitoring alternative, they have problems that may render them unusable in certain situations. Problems include temperature sensitivity, time-consuming calibration, slow response times, the independent effects of salinity on electrical resistance, and inaccuracy of measurements of high water contents (or low soil-water pressure). They are generally used only for suction in excess of 0.8 atm, which is the upper practical limit on suction for tensiometers. In addition, resistance blocks made of gypsum will eventually dissolve, making them unsuitable for long-term use.

Thermocouple Psychrometers

Thermocouple psychrometers are used to measure *in situ* soil-water pressure under very dry conditions, where tensiometers cannot be used because of air entry problems. They provide measurements of total water potential. Soil-water pressure is determined based on the relationship between soil-water pressure (potential) and relative humidity in the soil (Brown and van Haveren, 1972). Psychrometers are composed of a porous bulb to sample the relative humidity of the soil, a thermocouple, a heat sink, a reference electrode, and related circuitry. Calibration is required for each psychrometer unit before field installation.

The major disadvantage with this technology is that psychrometers are very sensitive to temperature fluctuations, so that it is necessary to record and correct for even diurnal temperature changes. However, where very dry conditions prevail, psychrometers may be the best monitoring choice. Psychrometers have successfully measured *in situ* suction values as high as 30 atm (Watson, 1974). Other disadvantages include the expense and complexity of these instruments (Bruce and Luxmore, 1988).

Gamma-Ray Attenuation

Gamma-ray attenuation can be used to indirectly measure moisture content by non-destructively determining soil density. Attentuation of gamma rays (commonly from a cesium source) passing through a soil column depends on the density of the soil column. If the soil density remains constant (i.e., the soil is nonswelling), changes in attenuation reflect changes in moisture content. This technique requires parallel access holes, one each for the source and the detector. Measurements can be taken as close as 2 cm, either vertically or horizontally, allowing an accurate determination of the location of the wetting front.

Major disadvantages of this technology are that gamma-ray attenuation units are expensive and difficult to use, they require special care in the handling of the radioactive source, and instrument calibration is affected by changes in bulk density (due to swelling, frost heave, etc.). In addition, this technology is unsuitable for applications in which vertical boreholes cannot be installed.

Nuclear Moisture Logging

A second nuclear method for nondestructively measuring moisture content is nuclear moisture logging (ASTM D 3017 [ASTM, 2004c] and ASTM D 6031 [ASTM, 2004d]). In this method, a probe containing a neutron source (e.g., usually americium or beryllium) and a detector is lowered down an access hole using a cable. The access hole is usually constructed of steel or aluminum. Neutrons emitted from the radioactive source interact with the hydrogen in the water of the surrounding soil. From counts of radioactivity taken at discrete intervals, moisture content can be calculated.

This method had several positive qualities (Schmugge et al., 1980). Readings are directly related to soil moisture and moisture content can be measured regardless of its physical state. Average moisture contents can be determined with depth. Repeated measurements can be taken at the same site, allowing measurement of rapid changes in moisture content as well as long-term changes. Like several of the other methods, the system can be interfaced with electrical recording equipment.

Nuclear moisture logging equipment also has several disadvantages. Equipment is very expensive. Only moisture content can be measured using this method; no information is provided on soil-water pressure or changes in density. Because of the sphere of influence, accurate measurements cannot be taken near the soil surface. In addition, the accuracy of the method is not high for detecting small changes in water content, especially for dry soils. Like gamma-ray logging, this method requires care in the handling of the radioactive source.

Other Methods

There are a variety of other methods for obtaining information on soil moisture content. These methods range from destructive to nondestructive and noninvasive. Soil samples can be removed from the field and moisture content measured destructively. Although accurate, the major disadvantages include repeatability, the need for drilling equipment to sample at significant depths, and the long time it takes to get the actual data.

A variety of nondestructive techniques operate on the electrical or magnetic properties of the soil-water system. Each method develops moisture content from the sphere of measurement of the instrument. These methods include time domain reflectometry (propagating an electromagnetic wave between electrodes to measure the dielectric properties — see ASTM D 6565 [ASTM, 2004e]), nuclear magnetic resonance (generating magnetic fields to measure induction decays — discriminates between bound and free water in the soil), soil capacitance (measuring the capacitance between two buried electrodes), and fiber optics (measuring light attenuation from a known source).

Remote sensing techniques have the ability to rapidly cover large areas, but possess much less sensitivity. These techniques include thermal infrared imagery and radar.

Monitoring Vadose Zone Transmission Properties

Vadose zone transmission properties are generally of greater interest in ground-water monitoring studies than are storage properties. Field methods for measuring the unsaturated hydraulic conductivity, which are described in detail in ASTM Standard D 5126 (ASTM, 2004f), are complicated due to its dependence on moisture content. However, one can measure flux and use this to calculate fluid transmission rates at known levels of saturation or matric potential. In a manner similar to ground-water investigations, boundary and initial conditions are prescribed (flow rate), variables are recorded (tension, moisture content), and then, from hydraulic theory (Equation 3.16), hydraulic conductivity is estimated. The rate of moisture movement is determined indirectly from infiltration rates or measurements of unsaturated flow. When using field data to estimate transmission properties in the vadose zone, it is important to remember that large variations in these parameters can easily result from soil heterogeneities.

Field Measurements of Infiltration Rates

Field measurements of infiltration rates are appropriate for estimating downward fluid transmission during the wetting cycle. Infiltration rates are affected by soil texture and structure (including soil layers), initial moisture content, entrapped air, and water salinity. Waste disposal options for which the principal component of flux is downward include surface spreading or ponding of wastes and installation of landfill liners composed of earthen materials.

Infiltration is determined using infiltrometers; infiltrometers do not directly measure hydraulic conductivity. Infiltration is the process by which water enters a permeable material. When infiltration begins, the infiltration rate is relatively high and is dominated by matric potential gradient. As the matric potential gradient decreases, the infiltration rate asymptotically decreases with time until the gravity-induced infiltration rate, called the steady-state infiltrability, is approached (Hillel, 1980a). This relationship is shown in Figure 3.13. Steady-state infiltrability is directly proportional to saturated hydraulic conductivity and hydraulic gradient. Therefore, in order to calculate saturated hydraulic conductivity from infiltration data, the hydraulic gradient and the extent of lateral flow must be known. Gradient data can be obtained using many of the instruments described in the previous section. Saturated hydraulic conductivity is of interest even in the vadose zone because it is the upper boundary for unsaturated hydraulic conductivity; use of this value provides a conservative estimate of fluid transmission time. With very long times, the gradient approaches unity and hence, from Darcy's law, the infiltration rate approaches the value of the saturated hydraulic conductivity.

While infiltrometers can be designed with either a single ring or a double ring, the double-ring method is preferred because its design minimizes lateral flow, simplifying the calculation of saturated hydraulic conductivity. The method is described in ASTM Standard D 3385 (ASTM, 2004g). The principle of operation is based on maintaining a constant head in the inner and outer rings of the infiltrometer. Both rings are sealed in the soil to prevent leakage under the rings. Water is added to the rings to maintain the constant head; if the inner ring is covered to prevent evaporation, the volume of water added to the inner ring is equal to the water infiltrating into the soil. In the design of Daniel and Trautwein (1986) water is added to the inner ring through an intravenous (IV) bag (Figure 3.14). As water from both rings enters the soil, water exits the IV bag and moves into the inner ring to maintain a constant head. The IV bag design is well suited to soils with low infiltration rates because the small amount of added water can be measured accurately by weight. For more permeable soils, the water level can be maintained by adding measured volumes of water to the inner ring. Measurements of infiltration are taken until the system reaches steady-state infiltrability. If the test is

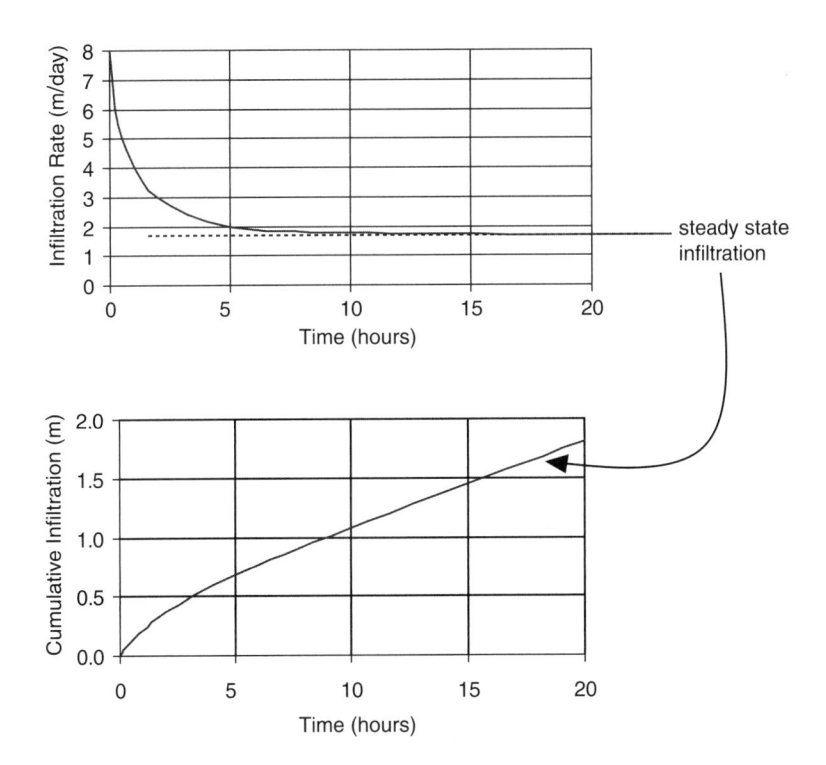

FIGURE 3.13
The relation between infiltration rate and cumulative infiltration.

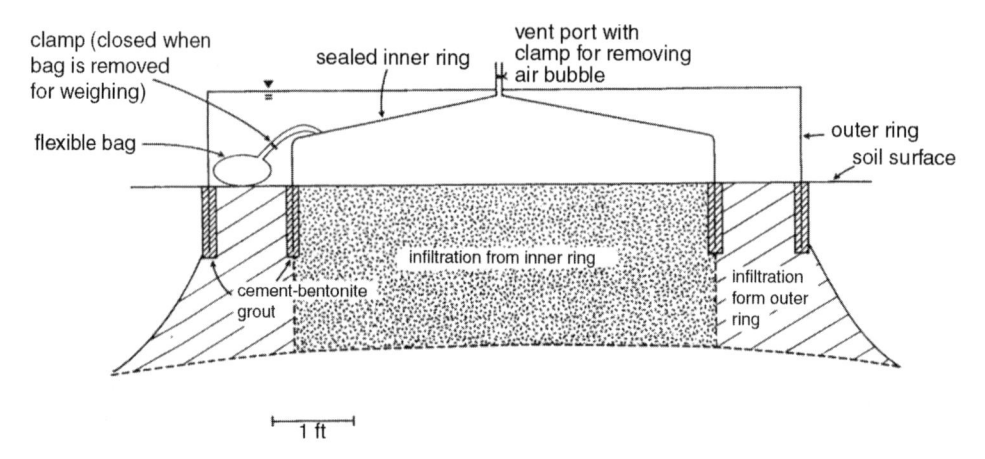

FIGURE 3.14
Schematic diagram of a sealed double-ring infiltrometer.

performed to prove that a soil meets some regulatory requirement, such as the requirement that earthen liners have a saturated hydraulic conductivity of 1×10^{-7} cm/sec or less (U.S. EPA, 1988), the test may end when this infiltration rate is achieved because the infiltration rate decreases with time and saturated hydraulic conductivity will be no more than the infiltration rate.

Infiltrometers generally range in size from less than one square foot to about 25 square feet. Large infiltrometers with IV bags were designed for soils with low infiltration rates, generally in the range of 1×10^{-5} to 1×10^{-8} cm/sec (Daniel and Trautwein, 1986). The large size is necessary to include macrostructures and to obtain measurable amounts of water loss.

Hydraulic conductivity can be calculated from infiltration rate using either Darcy's law or the Green–Ampt (1911) approximation. If Darcy's law is used, the hydraulic conductivity (K) is equal to the discharge of water flowing out of the infiltrometer (Q) divided by the product of the infiltrometer area (A) times the vertical head gradient (I): $K = Q/(AI)$. Here, Q/A is the measured, steady-state infiltration rate (infiltrability) per unit area. The head gradient, I, is measured with tensiometers at various depths. This method assumes that the flow is occurring under saturated conditions.

The Green–Ampt approximation assumes that the wetting front is sharp, the matric potential at the front is constant, and the wetted zone is uniformly wetted and of constant hydraulic conductivity. The assumption of a sharp wetting front may be reasonable for fine-grained soils, as shown by dye studies in an experimental earthen liner (Albrecht et al., 1989). The Green–Ampt approximation differs from the Darcy's law calculation in that knowledge of the depth of the wetting front is required instead of a measured hydraulic gradient. Under these assumptions, the analytical solution to vertical infiltration produces an equation that resembles the Darcy's equation:

$$K = i \left[1 + \frac{h + \psi_T}{L_f} \right]^{-1} \tag{3.18}$$

where i ($=Q/A$) is the steady-state infiltration rate and the bracketed term is the hydraulic gradient. Within the bracketed term, h is the height of the water in the infiltrometer, ψ_T is the total potential at the wetting front (also known as the wetting front suction), and L_f is the depth of the wetting front below the bottom of the infiltrometer. In many instances (especially after long times), ψ_T is assumed to be 0. This assumption results in a lower estimate of K, which may or may not be conservative, depending on the nature of the objective (irrigation, waste disposal, etc.).

In a similar fashion, a disc permeameter or tension infiltrometer also measures *in situ* sorptivity (S) and hydraulic conductivity for a prescribed potential. These devices allow potential and water content to be controlled accurately over a range of negative and positive heads and therefore have the ability to conveniently measure sorptivity (S) at selected tensions. Sorptivity is a combination of hydraulic conductivity, potential, and moisture content (Tindall and Kunkel, 1999) and basically represents the proportionality constant between infiltration rate and the inverse square root of time ($I = S/2t^{1/2}$). The disc permeameter (Perroux and White, 1988) is uncomplicated and does not greatly disturb the soil surface being measured. Methods and calculations for the disc devices may be found in references such as Ankeney et al. (1988).

Determination of Water Flux Characteristics

Hydraulic conductivity, flow velocity, and flux are important transmission parameters for the vadose zone. Measurement of these parameters has become more commonplace due to the need to understand how fluids move from the land surface to the groundwater system. This information is particularly necessary for waste disposal applications. A comparison of methods available for quantifying soil-water flux is presented in ASTM Standard D6642 (ASTM, 2004h).

Theoretical Perspective

Steady-state infiltration, discussed in the last section, is an appropriate base for determining flux during the wetting cycle. During the drying cycle, three major approaches to evaluating flux are possible (Everett, 1980). These include (1) calculating flux from mathematical formulae and empirical relationships between soil suction, soil-water content, and hydraulic conductivity; (2) measuring changes in the water content of the soil profile over time; and (3) direct measurements using flow meters.

Darcy's Law

The easiest method available for calculating saturated flow from infiltrometer data is the use of Darcy's law. This method is conservative because it assumes the soil is saturated; it is appropriate for the wetting cycle when steady-state flux is determined from an infiltration test. In simple terms, solving for average linear velocity (V_x), Darcy's law can be written as $V_x = Q/(n_e A)$, where Q is the discharge, n_e is the effective porosity, and A is the cross-sectional area of flow. Q/A is the measured steady-state infiltration rate per unit area; for use in Darcy's law, Q/A is negative because flow is downward.

Green–Ampt Wetting Front Model

The Green–Ampt wetting front model is used with infiltration data and assumes unsaturated conditions below a wetting front. Travel time (velocity times distance) is predicted from:

$$t = \left\{\frac{\theta_s - \theta_i}{K_{sat}}\right\}\left[L_f - (h + \psi_f) \ln\left\{1 + \frac{L_f}{h + \psi_T}\right\}\right] \qquad (3.19)$$

where θ_s and θ_i are initial and saturated moisture contents, L_f is the depth to the wetting front, h is the pond depth, and ψ_T is the total moisture potential just below the wetting front.

Internal Drainage Method

The internal drainage method can be used to determine the unsaturated hydraulic conductivity in the field by monitoring the transient internal drainage of a near-surface soil profile. The method is described in detail in Hillel (1980b) and was extended to layered profiles by Hillel et al. (1972). It requires simultaneous measurement of moisture content and suction under conditions of internal drainage alone; evapotranspiration must be prevented. The method also assumes that flow is vertical and that the water table is deep enough so that it does not affect the drainage process.

Tensiometers and neutron access tubes or gypsum blocks are installed near the center of the test area. Depth intervals for the instruments should not exceed 30 cm, with a desirable total depth of up to 2 m. The test area (at least 5 m by 5 m in plan view to avoid lateral disturbances on the monitoring devices) is then ponded or irrigated until the soil profile is as wet as practical (at or near saturation), then covered with plastic to avoid future fluxes across this boundary. Simultaneous measurements of soil suction and moisture content are collected until soil suction exceeds 0.5 bar; at greater suction, the drainage process may be so slow that changes become imperceptible. The measurement period for this test can be several weeks for slowly draining soils. Data are graphed as moisture content (Figure 3.15a) and suction versus time (Figure 3.15b) for each measured depth within the soil profile. Also, the snapshots of matric suction (Figure 3.15c) and total hydraulic head (suction plus depth, Figure 3.15d) assemble the data for subsequent analysis. The plots help to visually determine possible effects of nonhomogeneities. From these measurements, instantaneous values of potential gradient and flux can be obtained, allowing the calculation of hydraulic conductivity and, hence, flow velocity.

Soil moisture flux is calculated at each time and depth from:

$$q_z = \Delta z \frac{\partial \theta}{\partial t} \qquad (3.20)$$

where $\partial \theta / \partial t$ is the slope of the wetness curve at the time of interest (calculated from the data that generated Figure 3.15a) and Δz is the depth increment over which the measurements are made (each curve in Figure 3.15a). Equation 3.20 is the flux through the bottom of the uppermost increment and is due to the loss in moisture in the first zone. Flux through the bottom of each succeeding layer is obtained by summing these incremental fluxes for all layers overlying the depth of interest. The flux out of the bottom of the subsequent zone is due to the dewatering in that zone, plus the flux from above, hence the summation. Flow velocity can be calculated from Darcy's law, as discussed earlier.

Hydraulic head profiles (total potentials) are obtained using the suction (matric potential) versus time data, adding depth (gravitational potential) to suction to obtain the total hydraulic head for each time (Figure 3.15d). Hydraulic conductivity, K_z, is calculated from:

$$K_z = \frac{q_z}{(\partial H / \partial z)} \qquad (3.21)$$

where $\partial H / \partial z$ is the slope of the hydraulic head versus depth curve for the time of interest. K_z is calculated for several depths and times, each of which has a corresponding moisture content. As the final step, moisture content or soil suction is plotted against hydraulic conductivity so that flux and velocity can be calculated using field data at actual monitoring points. This yields a plot similar to Figure 3.11.

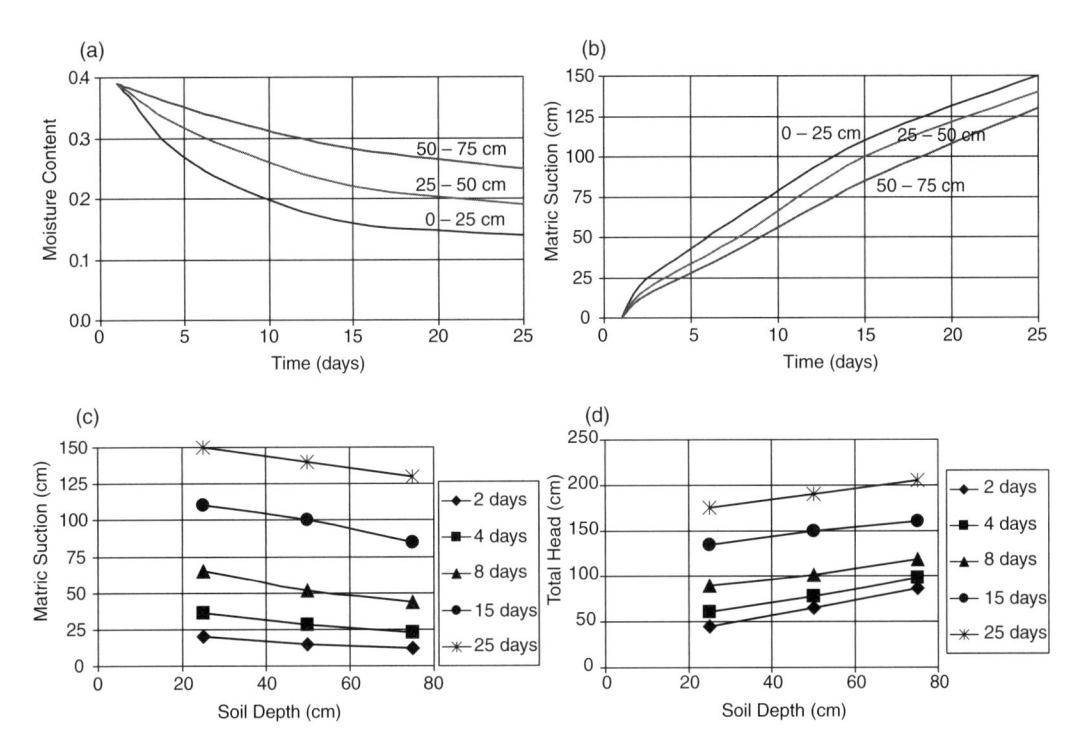

Monitoring and Sampling the Vadose Zone

FIGURE 3.15
(a) Volumetric wetness as a function of time for different depth layers in a draining profile. (b) Matric suction variation with time for different depth layers in a draining profile. (c) Matric suction variation with time and depth during drainage. (d) Total hydraulic head variation with time and depth during drainage. Head values are suction: the higher the suction, the lower the energy.

Borehole Permeameters

Some methods for determining saturated hydraulic conductivity and sorptivity evolved from borehole methods. Here, a borehole of constant radius (r) has a constant depth of water (H) maintained in it. The result is a bulb-shaped wetting front that moves away from the borehole. Theoretical treatments (Nasberg, 1951; Glover, 1953; Reynolds et al., 1983) describe the relations between the borehole water head (H) and the saturated hydraulic conductivity for a given borehole radius. In general, for these theories to be valid, the depth to the water table must be greater than three times the water depth in the borehole. In addition, the steady-state solutions are most valid for $H/r > 10$. A good comparison of the methods appears in Stephens (1996). Of note, this method has been commercialized (i.e., the Guelph Permeameter), and methods developed to also identify the hydraulic conductivity versus pressure (suction) relationship (Reynolds and Elrick, 1986).

Measurement of Tracer Movement

Tracers are matter or energy carried by ground water, which can provide information on the rate and direction of ground-water movement. Tracers can be natural, such as heat carried by hot springs; intentionally added, such as dyes; or accidentally introduced, such as oil from an underground storage tank (Davis et al., 1985). Use of tracers in the saturated zone for determining aquifer parameters is discussed in Chapter 14, so only a brief overview is presented here. The main difference between use of tracers under unsaturated versus saturated conditions is the practical problem of sampling a tracer at increasing depths under unsaturated conditions (Everett, 1980).

Davis et al. (1985) presents a thorough discussion of tracers. The most important property of any selected tracer is that its behavior in the subsurface should be well understood. Ideally, it should move at the same rate as the ground water, should not interact with the soil matrix, and should not modify the hydraulic conductivity or other properties of the medium being monitored. Concentration of the tracer should be much greater than the background concentration of the same constituent in the natural system. The tracer should be relatively inexpensive and easily detectable with widely available technology. For most applications, the tracer should also be nontoxic.

A variety of tracers have been successfully used to monitor moisture movement in the unsaturated zone. Fluorescein and rhodamine WT dyes have been successfully used to track the wetting front beneath double-ring infiltrometers and to indicate preferential flow paths in an experimental earthen liner (Albrecht et al., 1989). Tritium from a low-level radioactive waste disposal site was successfully used to determine the rate of water (and tritium) movement in the unsaturated zone at the waste disposal site (Healy et al., 1986).

Monitoring Water Quality in the Vadose Zone

The goal of most vadose-zone monitoring programs is to measure the spatial and temporal changes in water quality. Monitoring the vadose zone can provide an early warning system to detect contaminant movement so that corrective action can begin before an underlying aquifer is contaminated. Wilson (1980) presents a thorough discussion of the chemical reactions affecting contaminant migration in the vadose zone.

Three types of methods are available for monitoring water quality in the vadose zone. These include (1) indirect methods, including measurements of electrical and thermal properties; (2) direct measurement of pore water from soil cores; and (3) direct soil-water sampling.

Electrical Properties Measurements

Electrical conductivity (EC) or its inverse, resistivity, is used extensively to characterize soil salinity and to map shallow contaminant plumes. For shallow soils, electrical conductivity is primarily a function of soil solution (Wilson, 1980). The success of using electrical properties to delineate plumes is dependent upon the contrast between the conductivity of the plume and the natural water, the depth and thickness of the plume, and lateral variations in geology.

Electrical resistivity can be measured using surface geophysical techniques, as discussed in Chapter 4, or by direct-push-deployed sensors, as described in Chapter 6. It can also be measured by using electrical resistance blocks (salinity sensors) to evaluate soil salinity. Electrical resistance blocks were discussed earlier as a means of measuring *in situ* moisture content. They can be installed beneath a waste disposal site before the site becomes operational and monitored remotely. Salinity sensors must be calibrated to provide a curve of soil salinity versus electrical conductance. Electrical conductance is highly temperature dependent, so accurate measurement of soil solution temperature is a necessary companion to this device.

Soil Sampling and Water Sampling

Pore Water Extraction

Collection of soil cores is discussed in Chapter 5 and Chapter 6. Soil cores can provide pore water for water-quality analysis. Because of the difficulty and expense of obtaining soil

Monitoring and Sampling the Vadose Zone

cores, this is not the most desirable method for obtaining pore-water quality samples. However, when cores are obtained during borehole and monitoring well drilling, the added expense of collecting pore-water samples from cores is significantly reduced.

Certain parameters, including pH, Eh, and EC, are unstable and must be measured in the field. This requires either extracting the pore water or, more simply, making a saturated paste of the material and taking measurements on the paste. Pore water can be extracted by placing a soil sample in a commercially available filter press, hydraulic ram, or centrifuge and forcing the interstitial water out of the soil sample. These and other methods for pore-water extraction are presented by Fenn et al. (1977). After extraction, standard analytical techniques can be used on the water sample.

Suction Lysimeters

Suction lysimeters allow the collection of *in situ* soil water. They have a significant advantage over pore-water extraction in that repeated samples can be taken at a given location. A typical design, as shown in Figure 3.16, consists of a porous cup attached to a PVC sample accumulation chamber and two access tubes that lead to the land surface. Porous cups are commonly made of ceramic, alundum (an aluminum oxide), or PTFE; the first two are hydrophilic while the latter is hydrophobic. The sampling radius

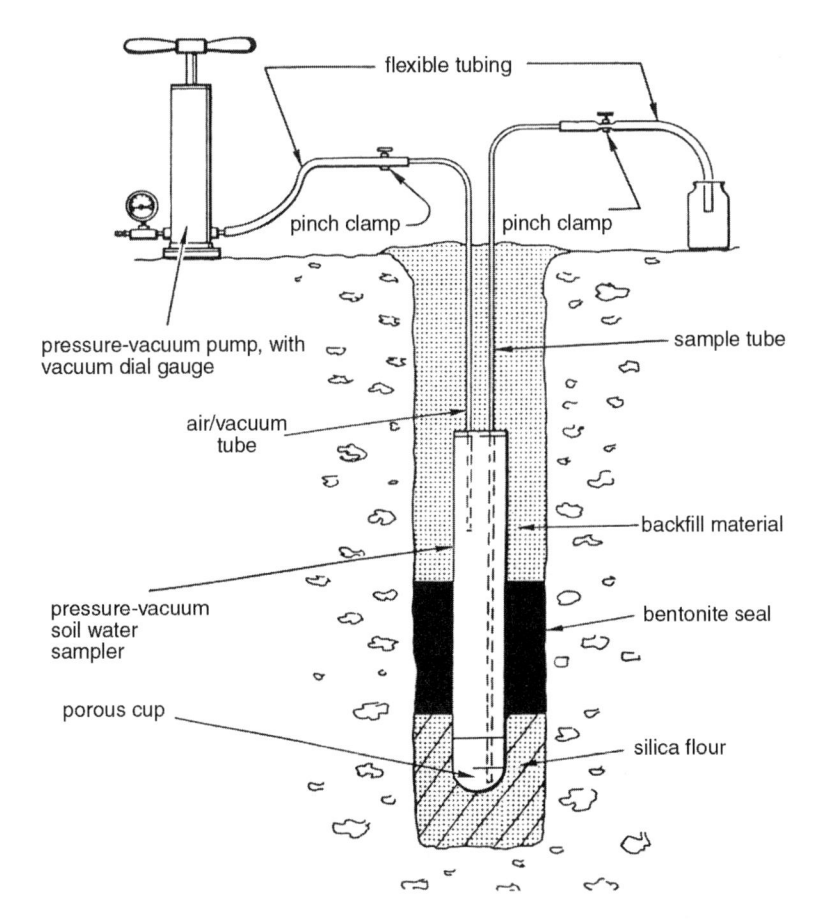

FIGURE 3.16
Schematic diagram of an installed pressure-vacuum (suction) lysimeter. [Adapted from Soil Moisture Equipment Corp., undated. With permission.]

of a lysimeter is on the order of centimeters, so that many are needed if they are to function as an effective early warning system (Morrison and Lowery, 1990). A variation to standard lysimeter design is the "well-type lysimeter" described by Ball and Coley (1986), which produces a larger sample volume than a standard lysimeter. Thorough discussions of soil-water sampling techniques are presented by Litaor (1988) and Wilson (1990). These techniques are also covered in detail by ASTM Standard D 4696 (ASTM, 2004i).

Lysimeters are installed in a borehole with silica flour packed around the porous cups. Silica flour is necessary to prevent plugging of the cup. Also, without silica flour, lysimeters with PTFE cups will not hold 10 centibars of vacuum (Everett et al., 1988). The sample tube ends at the bottom of the lysimeter, while the air tube ends near the top of the sample accumulation chamber. To collect a sample, the sample tube is clamped and a suction is applied to the lysimeter through the air tube, which is then clamped. This causes an inward gradient gradually drawing water into the sampler. Hours to days may be necessary to collect a sufficient sample volume, and the vacuum may need to be periodically re-established during this time. In order to collect the sample, first the suction is released by opening the air tube. Next, the sample tube is connected to the sample collection vessel and then opened. Air pressure is applied to the air tube thereby forcing the sample to the surface through the sample collection tube and into the collection vessel. This design has been used to depths of at least 55 ft (Apgar and Langmuir, 1971).

For deep lysimeters in which higher pressures are required to force the sample to the surface, pressure in excess of 1 atm in the sample will send the sample back through the porous cup into the soil instead of to the surface in a standard lysimeter. Wood (1973) modified the standard lysimeter design to allow sampling from any depth within the vadose zone. This design prevents the pressurization problem by including a check-valve to prevent pressurization of the porous cup. Wood (1978) was successful in collecting samples from depths in excess of 100 ft.

Questions have been raised as to the validity of samples collected from suction lysimeters. Some studies have indicated that the ceramic cups can alter the chemical composition of samples, making the samples not representative of actual water quality. Wolff (1967) found that new ceramic cups yield several milligrams per liter of Ca, Mg, Na, HCO_3, and SiO_2 even after cleaning with dilute HCl. Grover and Lamborn (1970) and Hansen and Harris (1975) found substantial bias and variability in soil-water samples of NO_3-N, PO_4-P, Na, K, and Ca. Up to a 60% change was noted in sample concentrations caused by sample intake rate, plugging of the ceramic cup and sorption and screening of some ions. Ceramic cups were the source of excessive Ca, Na, and K in samples with low solute concentrations and served to absorb P. Rinsing the cups with dilute HCl before installation reportedly reduced the problems with Na, K, and P to acceptable levels. Levin and Jackson (1977) found that Ca, Mg, and PO_4 were not altered by lysimeters that were used to collect soil-water samples from intact soil cores.

Lysimeters have also been found to screen certain contaminants. Parizek and Lane (1970) concluded that pressure-vacuum lysimeters are not useful for analysis of soil bacteria, BOD, or suspended solids because of screening. Dazzo and Rothwell (1974) found that screening and adsorption of bacteria by ceramic cups with a pore size of 3 to 8 μm rendered them unusable for fecal coliform. Because the effective pore size of the porous ceramic cup used in most lysimeters is about 1 μm, colloidal particles may pass through. Everett et al. (1988) reported that volatile organics were lost from suction lysimeters, but that the amount of loss was difficult to estimate.

The pre-1980 studies all used solutions with relatively low solute concentrations, which results in high sampling errors and sample variability. Little work had been performed prior to 1980 to determine the effects of these samplers on highly contaminated soil solutions. Despite these problems, soil-water samplers were commonly used to monitor highly contaminated soils.

Silkworth and Grigal (1981) studied the effect of porous cup size on sample chemistry. They found less alteration with large ceramic cups (4.6 cm diameter) than with small ceramic cups (2.2 cm diameter). Large ceramic samplers compared well with those collected from fritted glass samplers. Large ceramic samplers also produced more sample and had a lower failure rate than either glass or small ceramic samplers.

Crease and Dreiss (1985) studied the effects of three types of lysimeter cups on trace element and major cation concentrations. Sampler cup materials included ceramics, alundum, and PTFE, all of which are used for commercially available lysimeters. Their study indicated a low potential for significant sample bias by contaminants released from the cleaned and uncleaned samplers when the sample had been buffered to a pH of 6 to 7. They postulated that differences between their study and earlier studies of ceramic cups might be due to differences in composition of the ceramic cups tested, because many different ceramic formulations are available. They concluded that the bias introduced by porous lysimeter cups should only be significant for soil waters with low contaminant concentrations, especially given other sources of error in the collection and analysis of soil pore-water samples.

These conclusions were supported by Peters and Healy (1988), who found that major ion concentrations collected from pressure-vacuum lysimeters were representative of *in situ* chemical concentrations where total dissolved solids concentrations were greater than 500 mg/l. However, they found that trace-metal concentrations were significantly affected by sampling with lysimeters.

Based on an extensive program of suction lysimeter testing, Everett et al. (1988) made the following conclusions and recommendations:

- Prior to field installation, pressure tests should be used to check all lysimeters for leaks.
- The approximate bubbling pressure of ceramic low-flow cups is 2.38 atm (35 psi); for ceramic high-flow cups, 1.224 atm (18 psi); and for PTFE cups, 0.068 atm (1 psi).
- Low-flow ceramic cups are capable of holding their vacuum for several months.
- PTFE lysimeters must be used with silica flour slurry.
- The dead space in suction lysimeters must be determined prior to field installation or laboratory tests.
- Suction lysimeters placed in most types of soil will experience a rapid drop-off of intake rate but will stabilize after about 15 l has been pulled through the porous segments.
- Use of silica flour around the porous segments negates most plugging associated with finer particles in soils.
- The effective operating range of ceramic lysimeters is between 0 and 60 cb of suction regardless of the use of silica flour.
- The operating range of PTFE lysimeters without silica flour is extremely narrow, but with the use of silica flour is extended to about 7 cb of suction.
- Volatile organics can be obtained using a suction lysimeter where equilibrium is established and maintained.
- Volatile organics are lost from suction lysimeters if the vacuum needs to be intermittently re-established to draw sufficient sample.

Pan Lysimeters

Pan lysimeters, also called free-drainage samplers, are used to collect water samples by gravity drainage. They are used at waste disposal sites below earthen liners to provide early detection of moisture or solute movement through the liner. The typical pan lysimeter consists of a shallow cone with a drain in the center. The cone is filled with sand or gravel and the lysimeter is placed on top of the drainage layer, below a geofabric and the liner material. Less dead space for sample collection exists in the lysimeter than in the drainage layer, so that the breakthrough curve is sharper and earlier than the breakthrough curve from the bottom of the drainage layer. The principle of operation is that under unsaturated conditions, sand will have a lower hydraulic conductivity than the surrounding gravel drainage layer. Thus, as the bottom of the earthen liner approaches saturation, water will move preferentially toward the pan lysimeter, instead of into the drainage layer. At saturation, this preferential movement disappears.

Pan lysimeters are relatively inexpensive and can be homemade. For liner monitoring, they should be installed before the soil liner is emplaced. However, they have been installed in tunnels extending from trenches and in buried culverts (Wilson, 1990).

Many pan lysimeters currently in operation have been built with a misunderstanding of the theory of unsaturated zone flow, causing the lysimeters not to provide early information on breakthrough. The error is the belief that under both saturated and unsaturated conditions, water moves faster in coarser material. This has led to lysimeters filled with gravel surrounded by a sand drainage layer. In this case, the preferential flow under unsaturated conditions is away from the lysimeter — the lysimeter will not collect water before the liner becomes nearly saturated.

Soil-Gas Monitoring Technology

Introduction

Sampling and analysis of soil gas for the delineation of subsurface volatile organic compound (VOC) contamination became very popular in the 1980s. The technology has proven to be effective in a wide range of geologic settings and for many different VOCs. Several methods are used for the collection and analysis of soil-gas samples. These methods, which are described in detail in ASTM Standard D 5314 (ASTM, 2004j), are generally divided into two types — active and passive techniques. Active sampling is the term applied to those methods that physically remove the gas sample from the vadose zone, usually by pumping (Marrin and Thompson, 1987; Thompson and Marrin, 1987). Passive sampling refers to a technique of burying an absorbent material within the vadose zone and capturing the VOCs present by chemical sorption (Kerfoot and Mayer, 1986; Bisque, 1984). Active sampling techniques have become more popular because the samples can easily be analyzed in the field and actual concentrations are measured, whereas passive methods measure only relative concentrations across the site. The real-time results made possible by the rapid field analysis using active techniques are very helpful for directing the soil-gas investigation. Because the results can be used to direct the investigation, fewer unnecessary samples are collected when compared with aboratory-based investigations. This results in both time and cost savings. Owing to the greater popularity of active techniques over passive techniques, this discussion concentrates on active techniques.

Presented here are applications and limitations of soil-gas monitoring and sampling technology. Special attention is given to the variables that can impact the effectiveness of a soil-gas investigation. These variables include presence of geologic barriers, suitability of the compound to soil-gas monitoring applications, and interpretation of soil-gas data.

Monitoring and Sampling the Vadose Zone 237

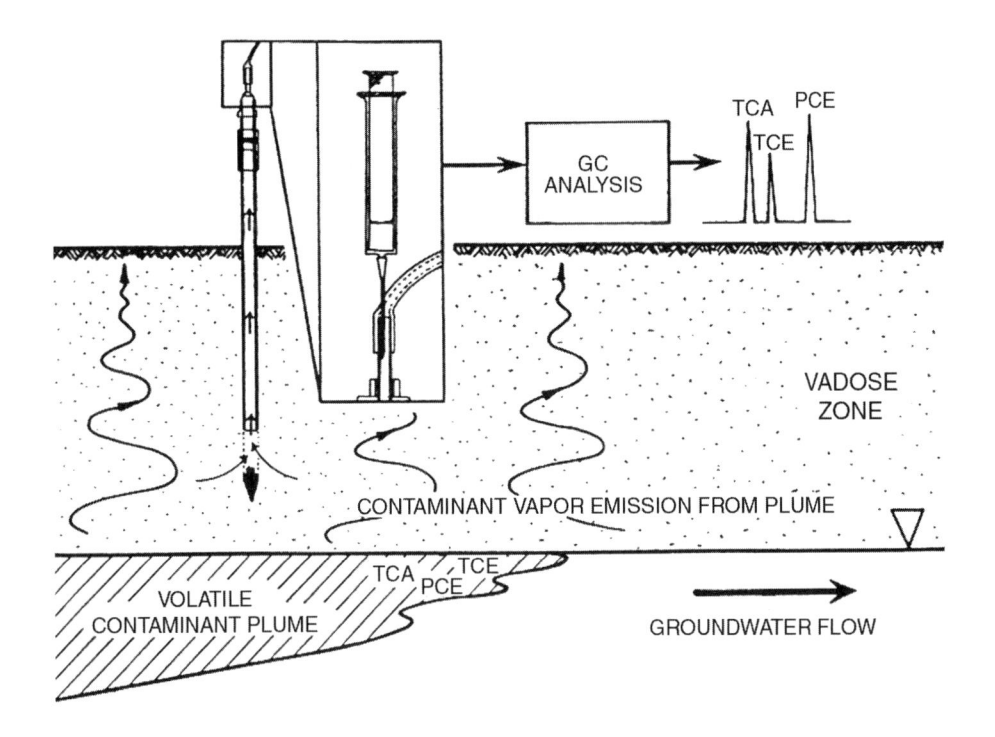

FIGURE 3.17
Schematic diagram of the soil gas contaminant investigation technology.

Background on Methodology

Figure 3.17 shows a schematic representation of the driving principles behind soil-gas technology. The presence of VOCs in shallow soil gas indicates that the observed compounds may be present either in the vadose zone or in the saturated zone below. Soil-gas technology is most effective in mapping low-molecular-weight halogenated chemical solvents and petroleum hydrocarbons possessing high vapor pressures and low aqueous solubilities. These compounds readily evaporate out of ground water and into the soil gas as a result of the favorable gas and liquid partitioning coefficients. Once in the soil gas, VOCs diffuse vertically and horizontally through the soil to the ground surface, where they dissipate in the atmosphere. The contamination acts as a source and the above-ground atmosphere acts as a sink, with a concentration gradient typically developing in between. The concentration gradient in soil gas between the source and ground surface may be locally distorted by hydrologic and geologic anomalies (e.g., clays, perched water); however, soil-gas mapping generally remains effective because distribution of the contamination is usually broader in areal extent than the local geologic barriers and is defined using a large data base. The presence of small-scale geologic obstructions tends to create anomalies in the soil-gas and ground-water correlations, but generally does not obscure the broader areal picture of contaminant distribution.

Sampling and Analytical Procedures

Soil-gas samples can be collected by mechanically advancing a hollow steel or stainless steel probe, fitted with a porous or screened tip, to a depth generally less than 5 m into the ground. The actual depth of the probe is a function of the depth to the water table or the source of the contamination. For soil-gas methods to be most effective, the probe tip should be close to the contaminant (within 2 to 10 m). When the contaminant is located at the water table, the farther the probe tip from the water table, the more likely it is that processes such as dispersion, sorption, and biodegradation will reduce the vapor concentrations of the contaminants. In addition, changes in geology may direct vapors away from a direct line between the contaminant and the probe tip. The above-ground end of the sampling probe is attached to a vacuum pump. The sampling train is purged by drawing air out of the soil through the probe. After purging, an aliquot of the evacuation stream is collected for analysis.

Several methods are currently being used for the analysis of soil-gas samples. Field-portable gas chromatographs (GCs), a variety of detectors, and laboratory-type bench-top GCs are in common usage. The most commonly used detectors are the electron capture detector (ECD), the flame ionization detector (FID), the photo-ionization detector (PID), and the Hall electrolytic conductivity detector. The ECD works well for detecting halocarbon compounds and the FID works well with hydrocarbon compounds. These detectors are highly selective to their respective categories of compounds and thus significantly reduce the problem of misidentification of unknowns. A PID may be used for detecting vinyl chloride, a compound that is not sufficiently detectable using either the ECD or FID. The Hall detector offers reasonable sensitivity to all of the halogenated compounds including vinyl chloride, but is much less sensitive than the ECD to the primary solvents such as trichloroethene (TCE), tetrachloroethane (PCE), and 1,1,1-trichloroethane (TCA).

Quality Assurance and Quality Control Procedures

A very important, yet often overlooked aspect of soil-gas investigation is quality assurance/quality control. The following are recommended procedures that have been successful in several applications of soil-gas technology.

- Steel probes and sampling train parts are used only once. Before being used, they are washed with a high-pressure soap and hot water spray or steam-cleaned to eliminate the possibility of cross-contamination.
- Prior to sampling each day, system blanks are run to check the sampling train for contamination by drawing ambient air from above ground through the system and comparing the soil-gas analysis with the concurrently sampled ambient air analysis.
- Sample containers and subsampling equipment are used for only one sample before being washed and baked to remove any residual VOC contamination.
- Sample containers and subsampling equipment are checked for contamination by running carrier gas blanks.
- Septa through which soil-gas samples are injected into the chromatograph are replaced on a daily basis to prevent possible gas leaks from the chromatographic column.
- Analytical instruments are calibrated each day. Calibration checks are also run after approximately every five soil-gas-sampling locations or a minimum of three times per day.
- Soil-gas pumping is monitored by a vacuum gage to ensure that an adequate gas flow from the vadose zone is maintained. A negative pressure (vacuum) usually indicates that a reliable gas sample cannot be obtained because the soil has a very low air permeability.

Applications

Case Study

Soil-gas investigations are most often applied either for defining the areal extent of contamination migrating from a known source or for identifying potential sources of ground-water contamination problems. Soil-gas data are typically used as a basis for more efficiently locating soil borings and monitoring wells, which are required to confirm the presence and distribution of subsurface contamination. The following case study gives a typical example of the plume mapping and source identification applications of soil-gas technology.

Figure 3.18 shows an example of the use of soil-gas technology to locate a contamination source. The depth to water was 120 ft, and the geologic materials were silty clays. Soil-gas samples were collected from a depth of 5 ft. Well I-1, in the southeast corner of the study area, was contaminated with TCA. A large industrial complex existed on the west side of the road, extending more than a mile north and south of the well. Soil-gas sampling was initiated along a transect extending several feet along a north–south road between the well and the complex. One soil-gas sample on this transect detected TCA slightly above background (Point 633, Figure 3.18). A second east–west transect was initiated along a convenient road into the complex a short distance north of Point 633. The samples along the second transect detected increasingly higher TCA concentrations. Because the soil-gas analyses were performed in the field, the sampling plan could be easily directed to "zero in" on the source area. In this case, the source was a business with a TCA tank. The long axis of the detectable TCA soil gas plume extended more than 3000 ft from the source toward the contaminated well, which was about 1 mile away. The investigation left very little doubt about the source of TCA contamination in the I-1 well.

This investigation represents an optimum usage of the soil-gas technology. The general distribution of the contaminant can be defined relatively quickly using a probe spacing between 100 and 300 ft. After the soil-gas investigation, verification drilling and soil sampling can proceed very efficiently.

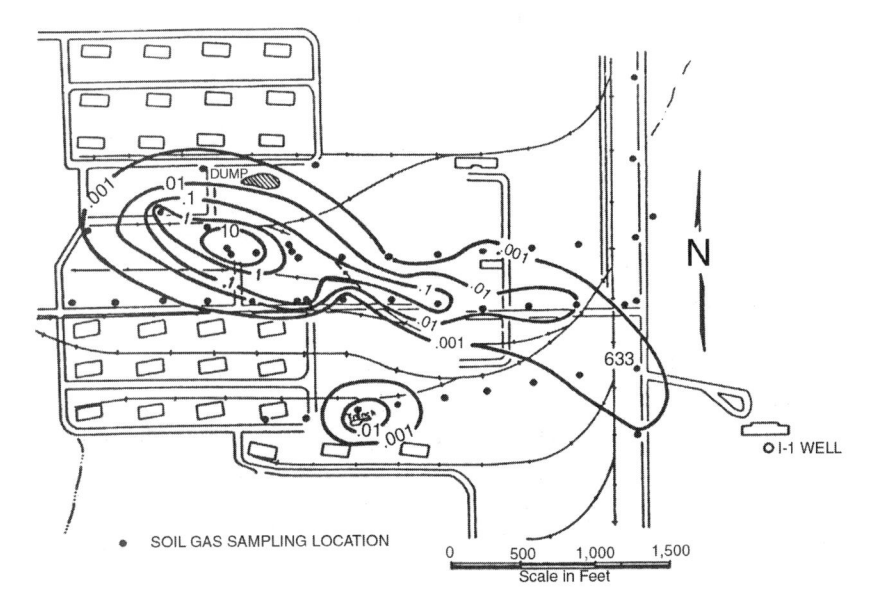

FIGURE 3.18
Representative application of the soil gas contaminant investigation technology.

Halocarbon Solvents versus Petroleum Hydrocarbons

The compounds most suited to detection by current soil-gas technology are the primary halocarbon solvents. The most common compounds in this group are TCA, TCE, PCE, and 1,1,2-trichlorotrifluorethane or Freon (F-113). These compounds readily volatilize out of ground water and into soil vapor as a result of their high gas and liquid partitioning coefficients. Good detection of these solvent vapors can be expected in most geologic settings. The exceptions are situations where there are geologic barriers to the migration of contaminant vapors. These barriers are discussed further in the following section.

There is no specific depth limitation for remote detection of the primary halocarbon solvents. These vapors tend to resist degradation and, in the absence of the geologic barriers, will migrate through a thick unsaturated zone to escape into the atmosphere. Remote detection of certain halocarbons from depths greater than 300 ft has been performed.

The application of soil-gas technology to hydrocarbons is more limited than to halocarbons. Good detection of hydrocarbon vapors is common in settings with shallow ground water (<10 m) and fairly permeable soils. A principal limitation to the application of soil-gas technology to hydrocarbon contamination is the relatively rapid degradation of hydrocarbons in well-oxygenated shallow soil. Owing to degradation, significant concentrations of the hydrocarbon vapors tend to appear and disappear abruptly in the soil-gas profile (Evans and Thompson, 1986). Table 3.1 shows the abrupt vertical change in hydrocarbon (benzene, toluene, and total hydrocarbon) concentrations compared with the smooth concentration observed for PCE, a common halocarbon solvent.

The most common problems associated with soil-gas investigations are geologic barriers, unsuitable target compounds, and the tendency to over-interpret soil-gas data. An awareness of the limitations of the technology is very important when planning and directing a soil-gas investigation.

Problems

Geologic Barriers

The most common geologic barrier to the migration of VOC vapors is water saturation of sediments in the vadose zone. A soil-gas investigation can be successful in low-permeability clay soils, but if the sediments are completely water saturated, soil-gas technology is not effective. Saturated sediments form a nearly impermeable barrier to the migration of contaminant vapors by molecular diffusion, thus preventing remote detection via shallow soil-gas samples.

Recharge of significant amounts of clean water over contaminated water commonly limits the area of effective soil-gas sampling. Clean recharge acts as a complete barrier only at sites where the recharge is significantly greater than the seasonal fluctuations in the water table. Fluctuations in the water table will allow the contaminated water to be dispersed through

TABLE 3.1

Comparison of PCE and Hydrocarbon Concentrations in Soil Gas above a Contaminated Aquifer (Values are Given in µg/l)

Depth (ft) below Ground Surface	PCE	Benzene	Toulene	Total Petroleum Hydrocarbons
5	0.006	<0.1	<0.1	<0.1
10	0.01	<0.1	<0.1	<0.1
15	0.03	220	31	600

Monitoring and Sampling the Vadose Zone 241

the clean water and capillary zone, thus maintaining vapor transport through the capillary fringe. This mechanism also explains the observation that contaminant vapors are commonly easiest to detect near water-supply wells, where pumping causes variations in the water level, enhancing transport through the capillary fringe.

Figure 3.19 shows the effect of increased soil moisture, or recharge, on the vapors emanating from a TCE contaminated aquifer. The horseshoe indentations correspond to

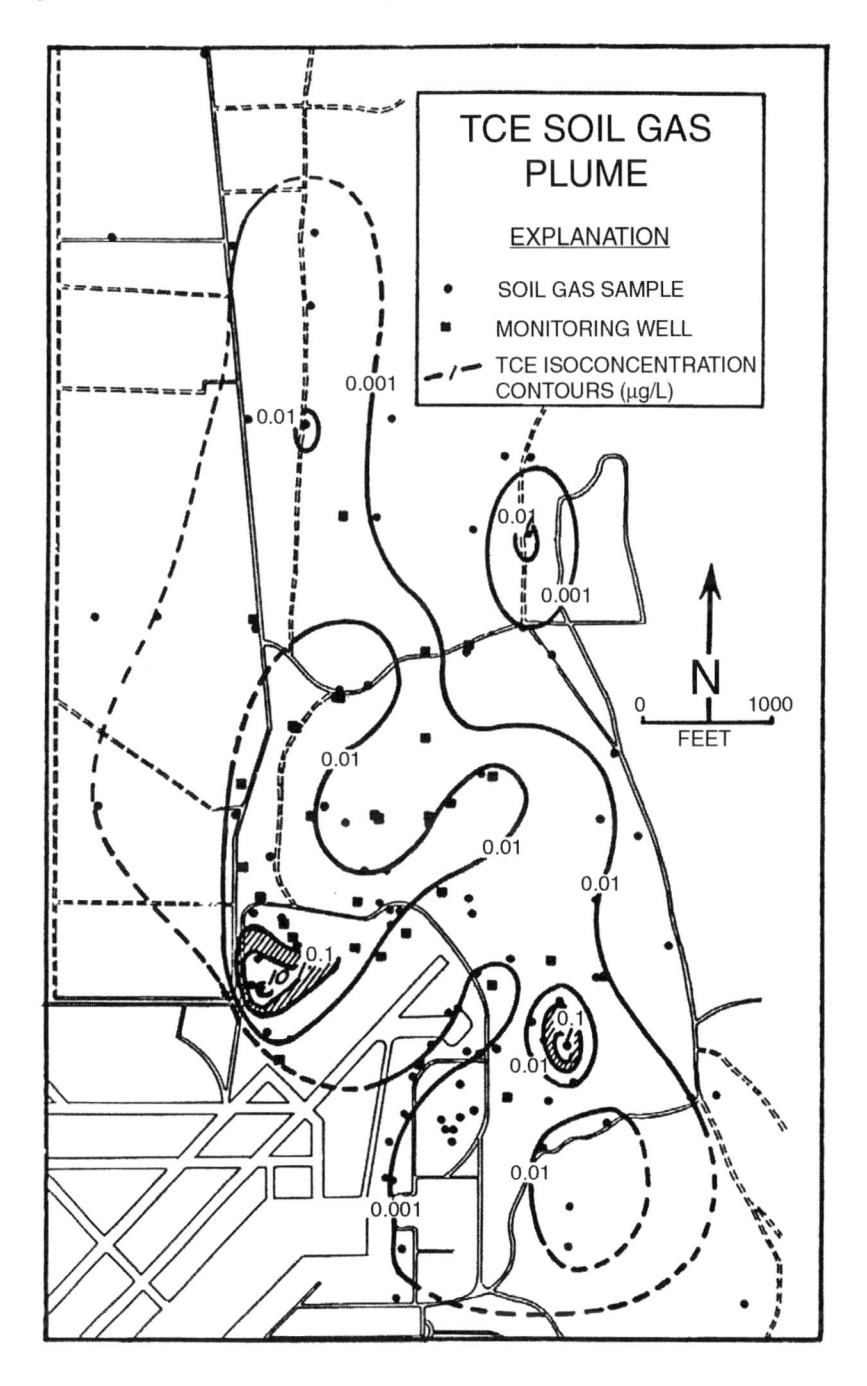

FIGURE 3.19
Example of the effect of recharge on the distribution of a TCE plume.

drainage or topographically low areas where a large amount of surface water and runoff is collected or channeled. These areas selectively received greater amounts of recharge, which reduced the concentration of the contaminant vapors detected in the shallow soil gas. Contrary to the appearance of the soil-gas map, the ground-water contamination does not necessarily diverge in a corresponding manner.

Suitability of Compounds to Soil-Gas Technology

Unsatisfactory results are often obtained from a soil-gas investigation when poor or unsuitable target compounds are chosen. The limitations are related to the compound's volatility, stability, and aqueous solubility. Suitable compounds are those that have a boiling point less than 150°C, low aqueous solubility, and relatively good resistance to degradation.

The suitability of a compound to soil-gas detection relies on the compound being present in the subsurface in the vapor phase. Compounds with boiling points greater than 150°C and vapor pressures less than 10 mmHg at 20°C probably will not be present in the vapor phase in sufficient quantities to be adequately detected in the soil gas in most applications. Compounds with boiling points greater than 150°C can commonly only be detected in the soil gas where they are present as significant residue in the soil.

Compounds that are miscible with water are poorly suited for soil-gas investigations. The high solubility of these compounds greatly reduces their vapor pressure in the presence of water. Thus, highly soluble VOCs such as alcohols and ketones will not favorably partition into the vapor phase sufficiently to be detectable in the soil gas.

The stability of a compound can also be a limiting factor to the utility of a soil-gas investigation. Nonhalogenated chemicals, particularly C5 and higher hydrocarbons, tend to degrade readily in oxygenated soil if they are present in low concentrations. This tendency to degrade limits the effectiveness of a soil-gas investigation in geologic settings where the depth to ground water is greater than 10 m or less than 2 m. In the case of ground-water depth being greater than 10 m, the limitation is being able to advance the sampling probes to an adequate depth to detect significant amounts of hydrocarbons. As shown in Table 3.1, hydrocarbons tend to appear abruptly in the soil profile. In most geologic settings, a soil-gas probe must be advanced to within about 2 m of the water-table surface to get a reliable soil-gas signal. The time required to advance soil-gas-sampling probes to depths greater than 6 m tends to reduce the cost-effective nature of a soil-gas investigation. At sites with deep hydrocarbon contamination, soil-gas technology may only be able to delineate the distribution of soil contamination in the source area. Degradation of most volatile compounds appears to be inhibited whenever vapors are present in high concentrations. Typically, vapor concentrations in the vicinity of leaking underground storage tanks are high enough to destroy soil bacteria and persist for long periods of time in shallow oxygenated soil.

Research has been conducted on techniques that may improve the means for remote detection of hydrocarbons in the situations described earlier. The occurrences of elevated levels of carbon dioxide above a dissolved plume where the primary hydrocarbons are not detectable may prove to be useful in delineating the areal extent of contamination. Preliminary work by Kerfoot et al. (1988) at the Pittman Lateral site in Nevada indicated that this approach may be successful.

Stability of halogenated chemicals is generally related to the number and type of halogens on the molecule — stability of the molecule increases with the number of halogens. Fluorine produces greater stability than chlorine, and chlorine produces greater stability than bromine. Fluorocarbons tend to persist even at low concentrations in the environment. As a result, they have accumulated in the atmosphere to the extent that they now pose a threat to the ozone layer (Zurer, 1987). Solvents having three or four chlorines on the molecule (i.e., C_2Cl_4, C_2Cl_3H, CCl_3CH_3, and CCl_4) commonly degrade to some degree in the subsurface environment, but degradation is slow enough to have little impact on their detectability in soil gas. Dichloro compounds (i.e., dichloroethene [DCE] and dichloroethane [DCA] isomers) are produced in the subsurface as the first degradation products of the primary chlorinated solvents. These products appear to degrade in the soil-gas environment slightly faster than the primary solvents (Vogel et al., 1987).

As a result, soil-gas data for the dichloro compounds are apt to be less representative of their ground-water distribution than the same data for the primary solvents. Monochlorinated vinyl chloride (C_2ClH_3), a second-stage degradation product, may be the least stable chloro compound in the soil-gas environment. Vinyl chloride has been detected in soil gas associated with landfills, but seldom detected in soil gas over contaminated ground water (Table 3.2). This indicates that it is probably an unreliable indicator of ground-water contamination.

Interpretation of Soil-Gas Data

Soil-gas data are normally regarded as remote or indirect indications of ground-water or soil contamination from volatile chemicals. As with other remote detection methods, the data are subject to limitations that may cause them to be misrepresentative or inaccurate at any particular location.

Most problems result from attempts to over-interpret the data. Usually this is evident when too much importance is placed on a single point or data anomaly in a very small area. Commonly, investigations begin by collection of soil-gas samples adjacent to wells or areas of known contamination to establish a basis for interpreting the soil-gas data. The findings are sometimes disappointing because the high, medium, and low concentrations in soil gas may not be measured at the same locations as high, medium, and low concentrations in ground water. Small-scale geologic and soil-moisture variability typically accounts for these problems and may make the data at any given point or in a small area highly misrepresentative of all subsurface conditions. In spite of an initially poor correlation, the investigation is probably worth continuing if the contamination was detectable in at least 50% of the locations where it was known to exist. Soil-gas detection of contamination is generally more successful when evaluated over a broad area and used to determine only the presence or absence of contamination in that area.

A second problem relates to the tendency of some users to include the possible effects of short-term climate changes on the soil-gas data. Typically, barometric pressure changes, recent rainfall events, and air temperature are parameters that are considered unnecessarily. Barometric pressure changes have long been known to be responsible for only a small amount of air transport into and out of the soil. Air exchange due to

TABLE 3.2

Vinyl Chloride Concentrations in Soil Gas and Ground Water (15–20 ft to Ground Water)

Water ($\mu g/l$)	Soil Gas ($\mu g/l$)
520	<0.005
110	<0.005
510	<0.01
1200	<0.005

barometric fluctuations is believed to be limited to the upper 1% of the thickness of the unsaturated zone (Buckingham, 1904). However, soil ventilation due to barometric pressure changes may be important in the immediate vicinity of a borehole, where an air conduit exists into the soil.

A single rainfall event rarely has any appreciable effect on soil-gas measurements. If the soils are normally unsaturated, even a heavy rain will not produce saturated conditions, except for a brief period of time (probably less than an hour) at the ground surface. However, soils consisting of fine marine sediments where the depth to water is 2 m or less are typically problematic. These soils remain nearly saturated due to capillary forces drawing water upward from the water table as well as high residual moisture content. As a result, soil-gas investigations are often not useful in these bay mud type environments.

Summary

In summary, soil-gas technology is an effective tool for the delineation of subsurface VOC contamination. A well-planned investigation which takes into account the effects of geologic barriers, the suitability of the compounds to the application of the soil-gas technology, and reasonable interpretation of the data will yield results that can be used to more efficiently direct a conventional soil boring and ground-water monitoring well installation program.

References

Albrecht, K.A., B.L. Herzog, L.R. Follmer, I.G. Krapac, R.A. Griffin, and K. Cartwright, Excavation of an instrumented earthen liner: inspection of dyed flow paths and soil morphology, *Hazardous Waste and Hazardous Materials*, 6(3), 269–279, 1989.

Ankeney, M.D., T.C. Kaspar, and R. Horton, Simple Field Methods for Determining the Unsaturated Hydraulic Conductivity, *Soil Science Society of America Journal*, 55, 467–470, 1988.

Apgar, M.A. and D. Langmuir, Ground-water pollution of a landfill above the water table, *Ground Water*, 9(6), 76–96, 1971.

ASTM, Standard Test Method for Particle-Size Analysis of Soils, ASTM Standard D 422, ASTM International, West Conshohocken, PA, 2004a, 7 pp.

ASTM, Standard Guide for Measuring Matric Potential in the Vadose Zone Using Tensiometers, ASTM Standard D 3404, ASTM International, West Conshohocken, PA, 2004b, 10 pp.

ASTM, Standard Test Method for Water Content of Soil and Rock in Place by Nuclear Methods (Shallow Depth), ASTM Standard D 3017, ASTM International, West Conshohocken, PA, 2004c, 6 pp.

ASTM, Standard Test Method for Logging In-Situ Moisture Content and Density of Soil and Rock by the Nuclear Method in Horizontal, Slanted, and Vertical Access Tubes, ASTM Standard D 6031, ASTM International, West Conshohocken, PA, 2004d, 5 pp.

ASTM, Standard Test Method for Determination of Water (Moisture) Content of Soil by the Time-Domain Reflectrometry (TDR) Method, ASTM Standard D 6565, ASTM International, West Conshohocken, PA, 2004e, 5 pp.

ASTM, Standard Guide for Comparison of Field Methods for Determining Hydraulic Conductivity in the Vadose Zone, ASTM Standard D 5126, ASTM International, West Conshohocken, PA, 2004f, 10 pp.

ASTM, Standard Test Method for Infiltration Rate of Soils in the Field Using a Double-RingInfiltrometer, ASTM Standard D 3385, ASTM International, West Conshohocken, PA, 2004g, 7 pp.

ASTM, Standard Guide for Comparison of Techniques to Quantify the Soil–Water (Moisture) Flux, ASTM Standard D 6642, ASTM International, West Conshohocken, PA, 2004h, 11 pp.

ASTM, Standard Guide for Pore-Liquid Sampling From the Vadose Zone, ASTM Standard D 4696, ASTM International, West Conshohocken, PA, 2004i, 31 pp.

ASTM, Standard Guide for Soil–Gas Monitoring in the Vadose Zone, ASTM Standard D 5314, ASTM International, West Conshohocken, PA, 2004j, 31 pp.

Atlas, R.M. and R. Bartha, *Microbial Ecology Fundamentals and Applications*, 4th ed., Addison-Wesley/Longman, Inc., Menlo Park, CA, 1998.

Ball, J. and D.M. Coley, A Comparison of Vadose Monitoring Procedures, *Proceedings of the Sixth National Symposium and Exposition on Aquifer Restoration and Ground Water Monitoring*, National Water Well Association, Dublin, OH, 1986, pp. 52–61.

Bear, Jacob, *Hydraulics of Ground Water*, McGraw-Hill, Inc., New York, NY, 1979.

Bisque, R.E., Migration Rates of Volatiles from Buried Hydrocarbon Sources Through Soil Media, *Proceedings of the NWWA/API Petroleum Hydrocarbons and Organic Chemicals in Ground Water: Prevention, Detection and Restoration Conference*, National Water Well Association, Dublin, OH, 1984, pp. 267–271.

Brown, R.W. and B.P. van Haveren, *Proceedings of the Symposium on Thermocouple Psychrometers*, Utah Agricultural Experiment Station, Utah State University, Logan, UT, 1972, 342 pp.

Bruce, R.R. and R.J. Luxmore, Water Retention: Field Methods, *Methods of Soil Analysis, Part 1. Physical and Mineralogical Methods*, Agronomy Monograph No. 9, American Society of Agronomy/Soil Science Society of America, Madison, WI, 1988, pp. 663–686.

Buckingham, E., Contributions to Our Knowledge of the Aeration of Soils, U.S. Department of Agriculture, Soils Bureau Bulletin 26, 1904, 52 pp.

Corey, A.T., *Mechanics of Heterogeneous Fluids in Porous Media*, Water Resources Publications, Fort Collins, CO, 1977.

Crease, C.L. and S.J. Dreiss, Soil Water Samplers: Do They Significantly Bias Concentrations in Water Samples?, *Proceedings of the NWWA Conference on Characterization and Monitoring of the Vadose (Unsaturated) Zone*, National Water Well Association, Dublin, OH, 1985, pp. 173–181.

Daniel, D.E. and S.J. Trautwein, Field Permeability Test for Earthen Liners, in: *Use of In-Situ Tests in Geotechnical Engineering*, S.P. Clemence, Ed., American Society of Civil Engineers, New York, NY, 1986.

Davis, S.N., D.J. Campbell, H.W. Bentley, and T.J. Flynn, *Ground Water Tracers*, National Water Well Association, Worthington, OH, 1985.

Dazzo, F.B. and D.F. Rothwell, Evaluation of porcelain samplers for bacteriological sampling, *Applied Microbiology*, 27(6), 1172–1175, 1974.

Evans, O.D. and G.M. Thompson, Field and Interpretation Techniques for Delineating Subsurface Petroleum Hydrocarbons Spills Using Soil Gas Analysis, *Proceedings of the NWWA/API Conference on Petroleum Hydrocarbons and Organic Chemicals in Ground Water: Prevention, Detection and Restoration*, National Water Well Association, Dublin, OH, 1986.

Everett, L.G., *Groundwater Monitoring*, General Electric Company, Technology Marketing Operation, Schenectady, NY, 1980.

Everett, L.G., K.D. Schmidt, R.M. Tinlin, and D.K. Todd, Monitoring Ground Water Quality: Methods and Costs, Solid Waste Management Series SW-616, U.S. Environmental Protection Agency, 1976.

Everett, L.G., L.G. McMillion, and L.A. Eccles, Suction Lysimeter Operation at Hazardous Waste Sites, in: *Ground-Water Contamination: Field Methods*, A.G. Collins and A.I. Johnson, Eds., ASTM Special Technical Publication 963, American Society for Testing and Materials, Philadelphia, 1988, pp. 304–327.

Fenn, D., E. Cocozza, J. Isbiter, O. Braids, B. Yare, and P. Roux, Procedures Manual for Ground Water Monitoring at Solid Waste Disposal Facilities, EPA/530/SW-611, U.S. Environmental Protection Agency, Office of Water and Waste Management, Washington, DC, 1977, 260 pp.

Glover, R.E., Flow for a Test Hole Located Above Ground-Water Level, in: *Theory and Problems of Water Percolation*, C.N. Zangar, Ed., Engineering Monograph 8, US Bureau of Reclamation, Denver, CO, 1953.

Green, W.H. and G.A. Ampt, Studies on Soil Physics: 1. Flow of Air and Water Through Soils, *J. Agric. Sci.*, 4(1), 1–24, 1911.

Grover, B.L. and R.E. Lamborn, Preparation of porous ceramic cups to be used for extraction of soil water having low solute concentrations, *Soil Science Society of America Proceedings*, 34(4), 706–708, 1970.

Hansen, E.A. and A.R. Harris, Validity of Soil-Water Samples Collected with Porous Ceramic Cups, *Soil Science Society of America Proceedings*, 39(3), 528–536, 1975.

Healy, R.W., M.P. deVries, and R.G. Striegl, Concepts and Data-Collection Techniques Used in a Study of the Unsaturated Zone at a Low-Level Radioactive-Waste Disposal Site Near Sheffield, Illinois, U.S. Geological Survey Water-Resources Investigations Report 85-4228, U.S. Geological Survey, Urbana, Illinois, 1986, 35 pp.

Hillel, D., *Applications of Soil Physics*, Academic Press, New York, NY, 1980a.

Hillel, D., *Fundamentals of Soil Physics*, Academic Press, New York, NY, 1980b.

Hillel, D., V.D. Krentos, and Y. Strylianou, Procedure and Test of an Internal Drainage Method for Measuring Soil Hydraulic Characteristics In Situ, *Soil Science*, 114, 395–400, 1972.

Iwata, S., T. Tabuchi, and B. Warkentin, *Soil–Water Interactions: Mechanisms and Applications*, Marcel Dekker, Inc., New York, NY, 1988.

Kerfoot, H.B. and C.L. Mayer, The Use of Industrial Hygiene Samplers for Soil Gas Surveying, *Ground-Water Monitoring Review*, 6(4), 74–78, 1986.

Kerfoot, H.B., C.L. Mayer, P.B. Durgin, and J.J. D'Lugosz, Measurement of Carbon Dioxide in Soil Gases for Indication of Subsurface Hydrocarbon Contamination, *Ground-Water Monitoring Review*, 8(2), 67–71, 1988.

Levin, M.J. and D.R. Jackson, A Comparison of In Situ Extractors for Sampling Soil Water, *Soil Science Society of America Proceedings*, 41(3), 535–536, 1977.

Litaor, M.I., Review of Soil Solution Samplers, *Water Resources Research*, 24(5), 727–733, 1988.

Marrin, D.L. and G.M. Thompson, Gaseous Behavior of TCE Overlying a Contaminated Aquifer, *Ground Water*, 25(1), 21–27, 1987.

McWhorter, D. and D. Sunada, *Ground Water Hydrology and Hydraulics*, Water Resources Publications, Fort Collins, CO, 1981.

Morrison, R.D. and B. Lowery, Sampling Radius of a Porous Cup Sampler: Experimental Results, *Ground Water*, 28(2), 262–267, 1990.

Nasberg, V.M., *The Problem of Flow in an Unsaturated Soil, or Injection Under Pressure*, M. Reliant transl., 1973, Izvestja Akademia Nauk. SSSR odt tekh Nauk No. 9, 1951.

Parizek, R.R. and B.E. Lane, Soil–water sampling using pan and deep pressure-vacuum lysimeters, *Journal of Hydrology*, 11(1), 1–21, 1970.

Perroux, K.M. and I. White, Designs for Disc Permeameters, *Soil Science Society of America Journal*, 52, 1205–1215, 1988.

Peters, C.A. and R.W. Healy, The Representativeness of Pore-Water Samples Collected from the Unsaturated Zone Using Pressure-Vacuum Lysimeters, *Ground-Water Monitoring Review*, 8(2), 96–101, 1988.

Reynolds, W.D. and D.E. Elrick, A Method for Simultaneous In Situ Measurement in the Vadose Zone of Field-Saturated Hydraulic Conductivity, Sorptivity, and the Conductivity–Pressure Head Relationship, *Ground-Water Monitoring Review*, 6(1), 84–95, 1986.

Reynolds, W.D., D.E. Elrick, and G.C. Topp, A Reexamination of the Constant Head Well Permeameter Method for Measuring Saturated Hydraulic Conductivity Above the Water Table, *Soil Science*, 136, 250, 1983.

Schmugge, T.J., T.J. Jackson, and H.L. McKim, Survey of Methods for Soil Moisture Determination, *Water Resources Research*, 16(6), 961–979, 1980.

Silkworth, D.R. and D.R. Grigal, Field Comparison of Soil Solution Samplers, *Soil Science Society of America Journal*, 45(2), 440–442, 1981.

Soilmoisture Equipment Corporation, Operating Instructions for the Model 1920 Pressure Vacuum Soil Water Sampler, Soilmoisture Equipment Corporation, Santa Barbara, CA, 6 pp.

Stannard, D.I., Tensiometers — Theory, Construction and Use, in: *Ground Water and Vadose Zone Monitoring*, D.M. Nielsen and A.I. Johnson, Eds., ASTM Special Technical Publication 1053, American Society for Testing and Materials, Philadelphia, 1990, pp. 34–51.

Stephens, D.B., *Vadose Zone Hydrology*, CRC Press, Boca Raton, FL, 1996.

Thompson, G.M. and D.L. Marrin, Soil–Gas Contaminant Investigations: A Dynamic Approach, *Ground-Water Monitoring Review*, 7(3), 88–93, 1987.

Tindall, J.A. and J.R. Kunkel, *Unsaturated Zone Hydrology for Scientists and Engineers*, Prentice-Hall, Inc., Englewood Cliffs, NJ, 1999.

U.S. EPA, *Design, Construction and Evaluation of Clay Liners for Waste Management Facilities*, EPA/530-SW-86-007F, U.S. Environmental Protection Agency, Risk Reduction Engineering Laboratory, Cincinnati, OH, 1988.

Vogel, T.M., C.S. Criddle, and P.L. McCarty, Transformations of halogenated aliphatic compounds, *Environmental Science and Technology*, 21(8), 722–734, 1987.

Watson, K.K., Some Applications of Unsaturated Flow Theory, in: *Drainage for Agriculture*, J. van Schilfgaard, Ed., *Agronomy*, No. 17, American Society of Agronomy, Madison, WI, 1974.

Wilson, L.G., Monitoring in the Vadose Zone: A Review of Technical Elements and Methods, EPA-600/7-80-134, U.S. Environmental Protection Agency, Environmental Monitoring Systems Laboratory, Office of Research and Development, Las Vegas, NV, 1980.

Wilson, L.G., Monitoring in the Vadose Zone, Part 1: Storage Changes, *Ground-Water Monitoring Review*, 1(3), 32–41, 1981.

Wilson, L.G., Methods for Sampling Fluids in the Vadose Zone, in: *Ground Water and Vadose Zone Monitoring*, D.M. Nielsen and A.I. Johnson, Eds., ASTM Special Technical Publication 1053, American Society for Testing and Materials, Philadelphia, 1990, pp. 7–24.

Wolff, R.G., Weathering of Woodstock Granite Near Baltimore, Maryland, *American Journal of Science*, 265(2), 106–117, 1967.

Wood, W.W., A Technique Using Porous Cups for Water Sampling at Any Depth in the Unsaturated Zone, *Water Resources Research*, 9(2), 486–488, 1973.

Wood, W.W., Use of Laboratory Data to Predict Sulfate Sorption During Artificial Ground-Water Recharge, *Ground Water*, 16(1), 22–31, 1978.

Zurer, P.S., Antarctic Ozone Hole: Complex Picture Emerges, *Chemical and Engineering News*, 65(44), 22–26, 1987.

4

Remote Sensing and Geophysical Methods for Evaluation of Subsurface Conditions

Richard C. Benson

CONTENTS

Introduction	250
Background	251
Sample Density	252
How Geophysical Methods Are Used	253
Continuous and Station Measurements	254
Site Investigation Methods Are Scale Dependent	255
Applications of Geophysical Measurements	256
Assessing Hydrogeologic Conditions	256
Detecting and Mapping Contaminant Plumes	256
Locating and Mapping Buried Wastes and Utilities	256
Airborne, Surface, and Downhole Geophysics	256
Remote Sensing and Airborne Geophysical Methods	258
Imaging Methods	258
Nonimaging Methods	260
Surface Geophysical Methods	261
Ground-Penetrating Radar	261
EM and Resistivity Methods	263
Electromagnetics	265
Resistivity	267
Direct Current Resistivity Measurements	267
2D Resistivity Imaging	267
Capacitively Coupled Resistivity	268
Comparison of EM and Resistivity Measurements	268
Seismic Methods	269
Seismic Refraction	270
Seismic Reflection	270
Surface Wave Analysis	272
Microgravity	273
Metal Detection	273
Magnetometry	274
Measurements over Water	276
Downhole Geophysical Measurements	276
Nuclear Logs	277
Natural Gamma Log	277
Gamma–Gamma (Density) Log	277
Neutron–Neutron (Porosity) Log	277

    Nonnuclear Logs 279
        Induction Log 279
        Resistivity Log 281
        Resistance Log 282
        Spontaneous-Potential Log 282
        Temperature Log 282
        Fluid Flow 282
        Fluid Conductivity 283
        Mechanical Caliper Log 283
        Imaging of Borehole Conditions 283
Applications of Geophysical Methods 283
    Assessing Hydrogeologic Conditions 283
    Detecting and Mapping Contaminant Plumes 287
    Locating and Mapping Buried Wastes and Utilities 290
Summary 291
    Selection of Geophysical Methods 292
References 292

Introduction

Remote sensing and geophysical methods encompass a wide range of airborne, surface, and downhole tools that provide a means of investigating hydrogeologic conditions and locating buried waste materials. Under certain conditions, some of the geophysical methods provide a means of detecting contaminant plumes.

Geophysical measurements can be made relatively quickly, thereby increasing sample density. Continuous data acquisition along a traverse line can be employed with certain techniques at speeds up to several miles per hour. Because of the greater sample density, anomalous conditions are more likely to be detected, resulting in a more accurate characterization of subsurface conditions.

Geophysical methods, such as any other means of measurement, have advantages and limitations. There is no single, universally applicable geophysical method, and some methods are quite site specific in their performance. Thus, the user must carefully select the method or methods and determine how they are applied to specific site conditions and project requirements.

Unlike direct sampling and analysis, such as obtaining a soil or water sample and sending it to a laboratory, the geophysical methods provide nondestructive, *in situ* measurements of physical, electrical, or geochemical properties of the natural or contaminated soil, rock, and contained fluids. The success of a geophysical method depends on the size of the target and the existence of a sufficient contrast between the measured properties of the target and background conditions. If there is no measurable contrast, the target will not be recognized. Similarly, if a layer is sufficiently thin or if the size of the target is sufficiently small, it will not be detected.

Geophysical techniques are not new. They have been used for decades in oil and gas exploration, mineral exploration, geotechnical applications, and regional water–resources development (Griffith and King, 1969; Zohdy et al., 1974; Telford et al., 1982). Geophysical methods, as applied to hazardous waste site investigations, are somewhat different in their application because they are usually required to produce higher resolution shallow data (typically less than 100 ft or so in depth). In less than one decade in the latter part of the

20th century (1975 to 1985), extensive development in geophysical field instrumentation, field methods, analytical techniques, and related computer processing resulted in a striking improvement in our capability to provide a high-resolution assessment of shallow subsurface conditions.

However, many environmental professionals still view geophysics as a "black box" technology. This is unfortunate because the methods are based upon sound principles of physics, geochemistry, and electronics. The "black box" image simply reflects a lack of understanding of the science behind the technology.

This chapter provides an overview of the various geophysical methods. The first section provides background material and identifies the three basic areas in which geophysical methods can be applied. The second section deals with airborne, surface, and downhole geophysical methods, discussing specific techniques for each. Major emphasis is placed upon the surface and downhole geophysical methods that are most commonly used for the type of applications discussed in this chapter. The chapter ends with application tables and a discussion to aid in selecting methodologies for specific field problems.

The examples of data shown within this chapter are considered to be of excellent quality. These high-quality data are presented to aid the reader in understanding the geophysical methods discussed. In practice, data will often be less clear than these examples, which requires that the skill of an experienced interpreter be employed in data evaluation.

Background

Traditional approaches to subsurface field investigations at potentially contaminated sites have often been inadequate. Site investigations have traditionally relied upon conventional direct sampling methods such as:

- Soil borings and monitoring wells for gathering hydrogeologic data and soil and ground-water samples
- Laboratory analysis of soil and ground-water samples to provide a quantitative assessment of site conditions
- Extensive interpolation and extrapolation from these points of data

This approach has evolved over many years and is commonly considered the standard approach to use when conducting environmental field investigations. However, there are numerous pitfalls associated with this approach, which can result in an incomplete or even erroneous understanding of site conditions. These pitfalls have been the subject of numerous papers and conferences over the past few decades (Lysyj, 1983; Hileman, 1984; Perazzo et al., 1984; Walker and Allen, 1984; Dunbar et al., 1985). They have also precipitated the tightening of ground-water monitoring regulations (U.S. EPA, 1986).

The single most critical factor faced in site evaluation work is accurately characterizing site hydrogeology (Benson and Pasley, 1984). If an investigator has an accurate understanding of site hydrogeology, predicting the movements of contaminants or designing a cleanup operation is reasonably straightforward. If all sites had simple, horizontally stratified geology with uniform properties, site characterization would be easy. Data from just one boring would be sufficient to characterize a site. However, in most geologic settings, this will not be the case and the investigator must be alert to variations that can cause significant errors in site characterization.

In the design of many soil and rock sampling programs and monitoring well networks, the placement of borings and wells has been done mainly by educated guesswork. The accuracy and effectiveness of such an approach is heavily dependent upon the assumption that subsurface conditions are uniform. This approach usually assumes that information on regional hydrogeology and ground-water flow conditions (as obtained from literature) is valid for the site-specific setting and that data from a few site-specific borings or monitoring wells can be used to characterize the site. These assumptions are frequently invalid, resulting in nonrepresentative locations for borings and monitoring wells and erroneous generalizations from this limited information. To improve the accuracy of the site investigation, a large number of borings would be required.

Sample Density

A soil or core sample obtained from drilling may be representative of only a limited area surrounding the hole. Fractures, bedding planes, solution cavities, bedrock channels, sand lenses, and local permeable zones can easily be missed by borehole programs.

Insight into the number of discrete samples or borings required for accurate sampling can be obtained by considering detection probability (Benson et al., 1982). Figure 4.1a shows a target area that is one tenth of the total site area. This target area (the size and location of which are usually unknown) could be a waste burial site, a plume from a chemical spill, an old sinkhole, or a buried channel. On the basis of probability

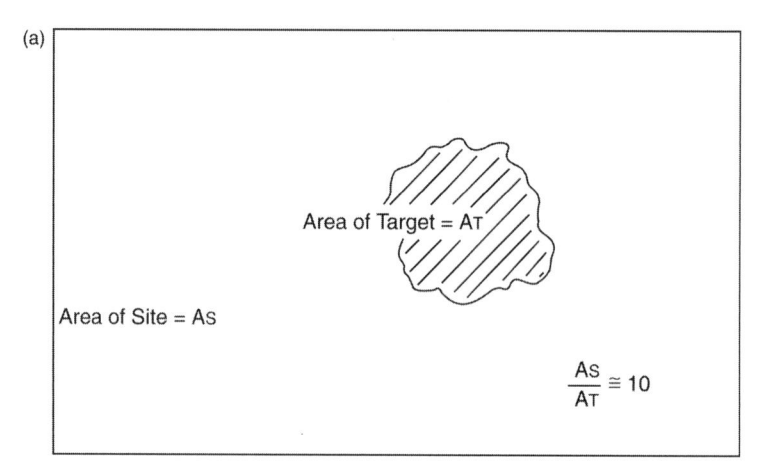

(a)

Area of Target = A_T

Area of Site = A_S

$\dfrac{A_S}{A_T} \cong 10$

Illustrates Site to Target Area Ratio
As/At = 10 in this case

(b)

Probability of Detection	As/At = 10	As/At = 100	As/At = 1000
100	16	160	1600
98	13	130	1300
90	10	100	1000
75	8	80	800
50	5	50	500
40	4	40	400
30	3	30	300

Number of Sample points required for various As/At ratios and Probability of Detection. This table assumes uniform sample grid if a random placement is used, the number of samples must be increased by a factor of 1.6

FIGURE 4.1
(a and b) Spatial sampling requirements. (From Benson and LaFountain, 1984. With permission.)

calculations, the number of borings required to achieve various detection probabilities is shown in Figure 4.1b (Benson and LaFountain, 1984). For example, a site-to-target-area ratio of 1/10 and a detection probability of 90% would require at least 16 borings spaced over a regular grid. For a smaller target, such as a narrow sand lens or fracture, the site-to-target-area ratio increases significantly. Thus, 100 to 1000 borings may be required to give a 90% confidence level in characterizing many sites, making the subsurface investigation like "looking for a needle in a haystack."

Achieving a good statistical evaluation of complex site conditions requires borings to be placed in a close-order grid, which would reduce the site to "Swiss cheese." In many cases, direct sampling alone is not sufficient to accurately characterize site conditions from a technical or cost point of view. This is the primary reason for the application of geophysical methods.

How Geophysical Methods Are Used

Data obtained from borings or monitoring wells generally represent conditions present in only a very localized area. In contrast, geophysical methods usually measure a much larger volume of the subsurface (Figure 4.2). These measurements provide an average response over a large volume of subsurface conditions; providing a means of detecting subsurface conditions such as buried channel or sand lens that a limited number of borings may miss. When geophysical methods are used in this manner, they are essentially anomaly detectors. Once an overall characterization of a site has been made using geophysical methods and anomalous zones have been identified, a better drilling and sampling plan may be designed by:

- Locating soil borings and monitoring wells to provide samples that are representative of site conditions
- Minimizing the number of samples, borings, or monitoring wells required to accurately characterize a site
- Reducing field investigation time and cost
- Significantly improving the accuracy of the overall investigation

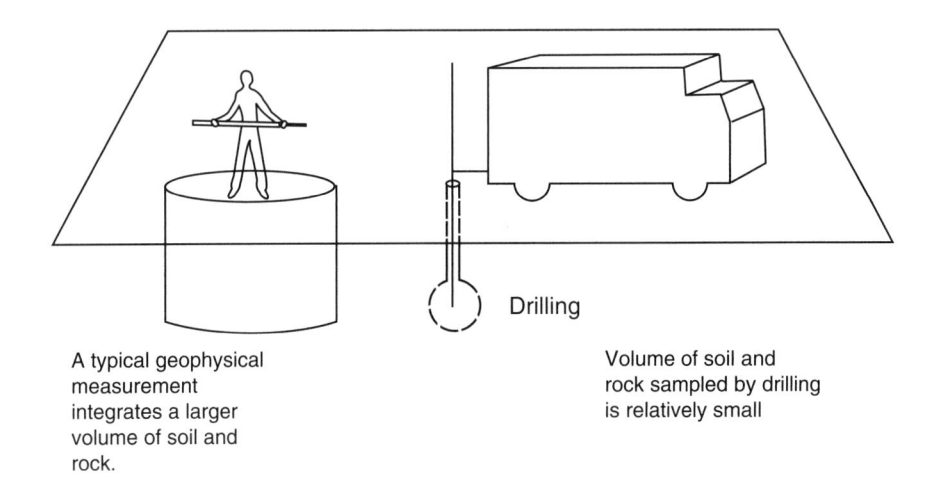

FIGURE 4.2
Comparison of volumes sampled by geophysical methods and a borehole. (From Benson et al., 1982. With permission.)

This approach yields a much greater confidence in the final results, with fewer borings or wells and an overall cost savings. The estimated cost of long-term sampling from a single monitoring well can range from $125,000 upward over a 30 year period (2004 dollars). This is nearly two orders of magnitude greater than the cost of installing the monitoring well. With this in mind, it makes good sense to minimize the number of monitoring wells at a site and optimize the locations of those installed. Using this approach, drilling is no longer used for hit-or-miss reconnaissance, but is dedicated to the specific quantitative assessment of subsurface conditions. Boreholes or wells located with this approach are called "smart holes" because they are scientifically placed, for a specific purpose, in a specific location, based on knowledge of site conditions; eliminating much of the guess work (Benson and Pasley, 1984). While smart holes might sometimes be placed without the benefit of geophysical methods, they often can be placed more reliably if geophysical methods are incorporated into the subsurface investigation program.

If borings have already been drilled or monitoring wells installed, geophysical surveys can still provide significant benefits. The location of existing borings and monitoring wells relative to anomalous site conditions can be assessed, thus providing a means of evaluating the representativeness of any existing data. Then, if additional borings or wells are needed to fill gaps in the overall site coverage, they can be accurately placed as smart holes. Assessment of site conditions will often require that an area larger than the site itself be considered. Contaminant transport by ground water and the geohydrologic factors that control flow do not stop at arbitrary site boundaries or property lines. Insight into the character of the local setting is often derived from the knowledge of the broader picture. An analogy can be drawn to the use of a camera's telescopic zoom from an overall wide angle view to a close-up telescopic view of the finer details. Omitting the broad overview will often result in a number of critical gaps in information about the setting. Geophysical methods provide a means of rapid reconnaissance over larger areas and can often be employed to obtain the big picture.

Continuous and Station Measurements

Geophysical surveys often involve making measurements of subsurface properties at discrete points over a site. That is, the instrumentation is located at a station along a survey line or a grid, and measurements are made at one point at a time. However, some techniques can provide measurement of subsurface parameters continuously as the instrumentation is moved along the survey line.

By estimating the size of geologic features or anomalies before the survey is carried out, suitable station spacing can be selected. However, if the size estimate of the geologic feature is in error, the data will not be representative and can lead to errors in the assessment of site conditions. Continuous methods should be employed whenever possible to minimize the possibility of making such errors, to achieve maximum resolution and to minimize project costs. This is particularly true when site conditions are suspected of being highly variable, and a small sample interval is required.

Although the continuous surface geophysical methods referred to in this chapter are typically limited to a depth of 50 ft or less, they are applicable to many site investigations. They can provide continuity of subsurface information that is not practically obtainable from station measurements. Continuous surface geophysical methods can be applied at speeds of 1 to 5 mph or more, resulting in a cost-effective approach for relatively shallow survey work. To illustrate the benefits of continuous measurements, a comparison between station measurements and continuous measurements is discussed subsequently.

The lower set of data in Figure 4.3 reveals the highly variable nature of a site indicated by a continuous measurement technique. The upper set of data in Figure 4.3 shows the loss

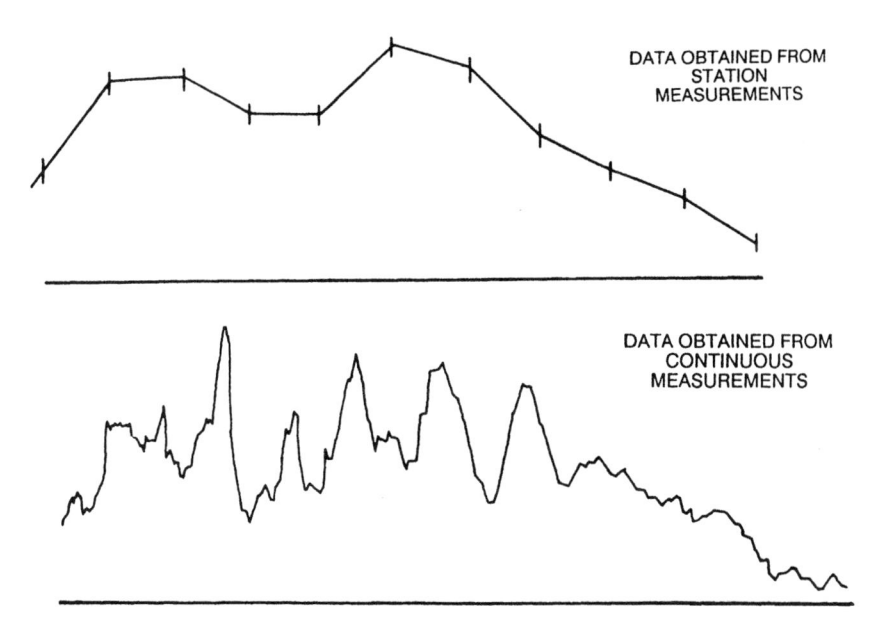

The data was obtained with an EM-34 with a 10 meter coil spacing. The higher values of electrical conductivity are caused by fractures in the underlaying gypsum rock.

FIGURE 4.3
Comparison of station and continuous measurements from the same site.

of information that can result from a limited number of station measurements and interpolating between sample points. This limited number of measurements results in a distorted set of data and leads to errors in interpretation of site conditions when target size is significantly smaller than station spacing. By running closely spaced parallel survey lines with continuous methods, subtle changes in subsurface parameters can often be mapped. Total site coverage can even be achieved if necessary.

The data in Figure 4.3 were obtained by surface electromagnetic (EM) measurements of subsurface electrical conductivity. The higher conductivity values indicate fractures within underlying gypsum rock. These fractures show up because they are more electrically conductive due to water content and weathering of the gypsum rock.

Site Investigation Methods Are Scale Dependent

All site investigation methods, including geophysics, are scale dependent. For example, aerial photography is an effective tool to be used in regional studies and for obtaining the big picture at a local site investigation. However, it will not provide any information about site-specific soil conditions at a depth of 10 ft. Conversely, a boring will provide information on soil conditions versus depth, but information from the boring is only valid for a very limited extent immediately around the borehole. Geophysical measurements made of the subsurface can be used to determine detailed soil conditions over a few hundred square feet or over many square miles. In contrast, geophysical logging measurements made down a borehole will extend the measurements from the hole itself radially to a distance of 6 in. to a few feet, depending upon the log used.

Therefore, the site investigation method must be selected to suit data and project requirements (Benson and Scaife, 1987). Typically, a subsurface investigation will include measurements of the big picture (aerial photography), intermediate picture (surface geophysical measurements), and the very local details (boring and sampling data).

Applications of Geophysical Measurements

There are three major areas for the application of geophysical methods at potentially contaminated sites. They are:

- Assessing hydrogeologic conditions
- Detecting and mapping contaminant plumes
- Locating and mapping buried wastes and utilities

Assessing Hydrogeologic Conditions

Probably, the most important task of a site investigation will be characterizing hydrogeologic conditions. A variety of geophysical methods can be used to assess natural hydrogeologic conditions, such as depth to bedrock, degree of weathering, and presence of sand and clay lenses, fracture zones, and buried relic stream channels (Keys and MacCary, 1976; Benson and Glaccum, 1979; Benson et al., 1982; Benson and Scaife, 1987). Accurately understanding the hydrogeologic conditions and anomalies can make the difference between success and failure in site characterization, because these features control ground-water flow and contaminant transport.

Detecting and Mapping Contaminant Plumes

A major objective of many site investigations is the detection and mapping of contaminant plumes. Geophysical methods can be employed in two ways to solve this problem. Some methods can be used for the direct detection of contaminants. In cases in which the contaminant cannot be detected directly, geophysical methods can be used to assess the detailed hydrogeologic conditions that control ground-water flow. Then, the location of the contaminants can be estimated and the ground-water and contaminant flow pathways can be identified (Cartwright and McComas, 1968; Benson et al., 1982, 1985; McNeill, 1982; Greenhouse and Monier-Williams, 1985).

Locating and Mapping Buried Wastes and Utilities

Geophysical methods can also be used to locate and map the areal extent, and sometimes the depth, of buried wastes in trenches and landfills (Benson et al., 1982). There are methods that can also be employed to detect buried drums, tanks, and utility lines. At many sites, the trenches associated with buried pipes and utilities will be of interest because they serve as permeable pathways for ground-water and contaminant movement.

Airborne, Surface, and Downhole Geophysics

There are three different modes in which geophysical measurements can be applied — from the air, from land, and from over water (Figure 4.4). Airborne methods are usually

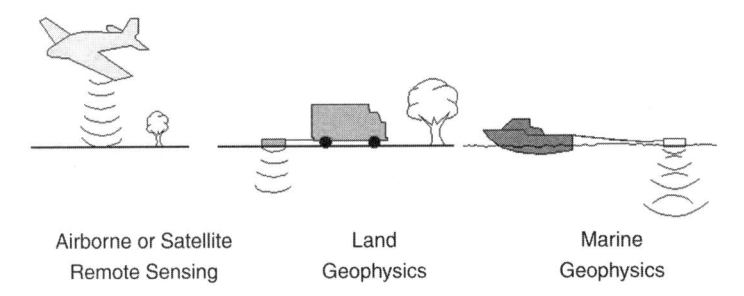

FIGURE 4.4
Three modes of surface geophysical measurements.

employed to obtain a regional overview, or the big picture, of site conditions. Land-based methods provide a means of rapid reconnaissance over a large area or a means of obtaining site-specific details. Many land-based methods are adapted for use on water. There are a variety of ways of making downhole geophysical measurements that can be used to provide very localized details down a borehole or well or between boreholes (Figure 4.5).

Airborne and satellite remote sensing clearly have merits in terms of spatial coverage per unit time and cost. Imaging methods (photographic, infrared, and others) provide a "picture" of the site and the surrounding area. They give us an excellent overview of regional conditions that let us see the pieces of the puzzle totally assembled and in perspective. However, they provide little, if any, subsurface data other than those data derived by skilled interpretation. Other airborne (nonimaging) methods, such as magnetics and radiometrics, can provide a measure of certain subsurface conditions.

While surface geophysical methods yield much less spatial coverage per unit time than airborne methods, they significantly improve resolution (the ability to detect a small feature) while providing subsurface information. Sometimes, continuous data acquisition can be obtained at speeds up to several miles per hour. In certain situations, total site coverage is technically and economically feasible. However, an inherent limitation of all surface geophysical methods is that their resolution decreases with depth.

The major benefit of downhole geophysical methods is that they provide detailed high-resolution data at depth around a borehole or well in which they are deployed. Unlike

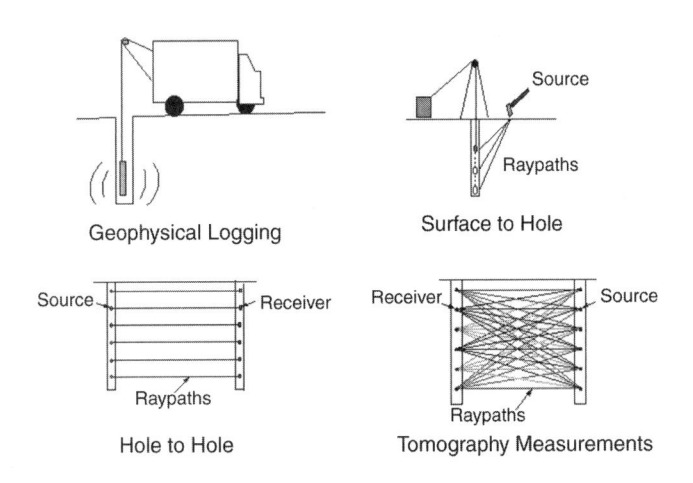

FIGURE 4.5
Four types of borehole geophysical measurements.

surface geophysical methods, where resolution decreases with depth and the resolution of downhole logging methods is independent of depth. In addition, most downhole methods provide continuous data along the depth of the hole. The volume sampled by downhole methods is usually limited to the area immediately around the boring (a cubic foot to a cubic yard). The cost per unit area of coverage for the downhole methods is obviously much higher than for surface methods, because all downhole techniques require a borehole or monitoring well. However, if holes are already in place or if they are to be drilled for other purposes, the overall cost of downhole logging is relatively low.

All of these approaches — airborne remote sensing and surface and downhole geophysical survey methods — have a place in subsurface investigations. Through the use of appropriate combinations of geophysical measurements and borehole data, an accurate 3D picture of subsurface conditions can be generated. The resulting understanding of subsurface conditions can then be used to develop an accurate conceptual site model, which incorporates the big picture through the local details.

Remote Sensing and Airborne Geophysical Methods

Airborne remote sensing and geophysical methods cover a wide range of the EM spectrum, from the lower frequency airborne EM method to the very short wavelength gamma rays measured by the radiometric method. Figure 4.6 shows the range of wavelengths employed for specific measurements. The terms airborne and remote sensing, as used in this section, include measurements made from aircraft, as well as from satellites.

Imaging Methods

Imaging methods are those that result in a "picture" of the surface. A wide range of aerial photos can be obtained, from those taken by hand-held 35 mm cameras to those obtained by complex satellite sensors. Aerial photographs are a source of geological information and provide an overview of site conditions. Aerial photos, U.S. Geological Survey topographic maps, and U.S. Department of Agriculture soil survey maps will often be the first data reviewed for a project because they provide a rapid, low-cost means of obtaining the necessary big picture overview of site conditions.

Large-scale aerial photos with a 19 × 19 in. format and a scale of 1:3600 are commonly available from a country surveyor's or tax assessor's office or through commercial firms. These photos provide a local overview of the site for project planning and a means of accurately locating survey grids, as well as buildings, roads, and other surface features.

Standard small scale (9 × 9 in.; 1:24,000) black-and-white photos are available in stereo pairs from a number of sources, including the Soil Conservation Service (U.S. Department of Agriculture), the U.S. Geological Survey, and state agencies (e.g., Departments of Tranportation). These aerial photos are often available for the past 70 years, thereby providing a historic record of site conditions. This type of photography is used to provide a very broad overview of the site and to allow photogeologic interpretation. Aerial photo interpretation can provide information on bedrock type, landforms, presence of lineaments or fractures (through a technique called fracture trace analysis), soil texture, site drainage conditions, susceptibility to flooding, and slope of the land surface.

Apart from these two relatively standard photographic formats, there are a number of other options available including high-altitude photography, color photography, false-color infrared, and a wide range of satellite imagery. Each of these can be very

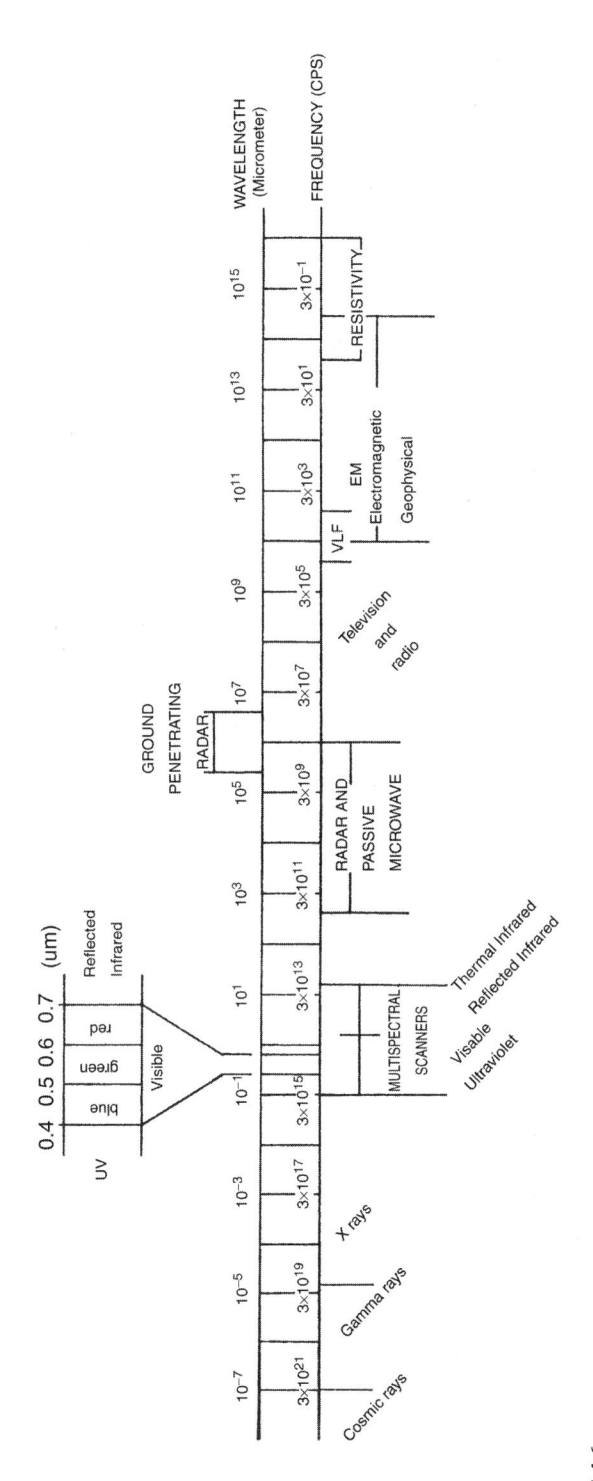

FIGURE 4.6
Portions of the EM spectrum used for geophysical measurements. (From Technos Inc. With permission.)

useful in specific applications. However, the black-and-white formats are the first and often the primary types used in most site-characterization programs, because they are low in cost and readily available.

There are two other forms of imagery that are occasionally very useful for specific problems: thermal infrared and side-looking airborne radar (SLAR). Thermal infrared is different from false-color infrared in that it is a measure of the thermal response of an area measured in the infrared spectrum. The earth's surface emits radiation in the thermal infrared wavelengths. These emissions are recorded by electronic detectors and compared to the reflected energy recorded by infrared film. The image (or thermogram) is presented on video or on film, or it can be stored on magnetic tape. This method can provide a means of locating springs, identifying seeps from a landfill, locating moist or dry areas, characterizing surface soil and rock, and identifying vegetation stress. It is also useful in a number of other conditions in which a difference in temperature is a characteristic feature.

SLAR is an electronic image-producing system that uses a radar beam transmitted off to the side of the aircraft. The result is an obliquely illuminated view of the terrain. This oblique view enhances subtle surface features and facilitates geologic interpretation. Another important property of SLAR is that it is an active system which provides its own source of illumination in the form of microwave energy. Thus, imagery can be obtained either day or night, regardless of cloud cover. The SLAR products commonly used for analysis are image strips and mosaics. SLAR imagery is available from the U.S. Geological Survey for selected areas of the United States.

Nonimaging Methods

Nonimaging methods do not result in picture, but provide a measurement of some parameter along the flight line of the aircraft. These methods include EM measurement (using frequencies up to a few kilohertz [kHz]), magnetic measurements, radiometric measurements of gamma radiation, and ground-penetrating radar using frequencies of around 100 megahertz (MHz). These are referred to as nonimaging methods because as the aircraft moves along a survey line, a series of measurements are obtained, rather than an image. However, by running parallel survey lines, a contour map of the measured parameters can be developed for the site. While imaging methods provide only a measure of surface conditions, the nonimaging methods measure subsurface conditions. These methods are normally applied to large areas or areas that are not easily accessible by land.

The EM method (which is described further in the section on surface geophysical methods) measures electrical conductivity of subsurface materials. It provides a measure of gross changes in geologic, hydrologic, and environmental conditions based upon electrical conductivity. In certain conditions, this method could be used to map soil cover, locate coastal saltwater intrusion, or even map a large leachate plume.

Magnetic measurements provide a means of determining the magnetic susceptibility of soil and rock and, therefore, can provide a geologic map. The resulting maps provide an overview of the gross geologic conditions (based upon magnetic properties) and can identify larger anomalous conditions.

Radiometric measurements provide a means of measuring the natural radioactivity (potassium-40 and daughter products of the uranium and thorium decay series) that is emitted from many rocks. Total measurements or spectral measurements can be obtained to characterize the count from specific elements. While this method has been applied to mineral exploration, a radiometric map, such as a magnetic map, provides insight into the overall geologic structure and can identify larger anomalous conditions.

Ground-penetrating radar, which is commonly used on the surface, has been used for limited airborne applications (this method is described further in the section on surface geophysical methods). Helicopter surveys have been successfully applied to obtain some soil, ice, snow, and permafrost thickness measurements. In general, where radar penetration is good and the site is clear of vegetation and cultural features, the method may be usable to obtain shallow profiles in these materials.

Surface Geophysical Methods

The surface methods discussed in this section include:

- Ground-penetrating radar
- Electromagnetics
- Resistivity
- Seismic refraction
- Seismic reflection
- Surface wave analysis
- Microgravity
- Metal detection
- Magnetics

These techniques are included, because they are used regularly and have proved effective for assessments of potentially contaminated sites. A brief description of each of these surface geophysical techniques is presented in this section.

Ground-Penetrating Radar

Ground-penetrating radar uses high-frequency EM waves from less than 100 to 1000 MHz to acquire subsurface information. Energy is radiated downward into the ground from a transmitter and is reflected back to a receiving antenna. The reflected signals are recorded and produce a continuous cross-sectional picture or profile of shallow subsurface conditions. ASTM Standard D 6432 (ASTM, 2004a) provides guidance on the use of ground-penetrating radar in environmental site characterization.

Reflections of the radar wave occur whenever there is a change in dielectric constant or electrical conductivity between two materials. Changes in conductivity and in dielectric properties are associated with natural hydrogeologic conditions such as bedding, cementation, moisture, clay content, voids, and fractures. Therefore, an interface between two soil or rock layers that have a sufficient contrast in electric properties will show up in the radar profile (Benson and Glaccum, 1979; Benson et al., 1982; Benson and Scaife, 1987). Figure 4.7 shows a radar record of a sand–clay interface. The water table can be detected in coarse-grained materials but not in fine-grained sediments with a large capillary boundary. Both metallic and nonmetallic buried pipes and drums can also be detected.

The vertical scale of the radar profile is in units of time (nanoseconds or 10^{-9} sec). The time it takes for an EM wave to move down to a reflector and back to the surface is relatively short because the waves are traveling at almost the speed of light. The time scale is

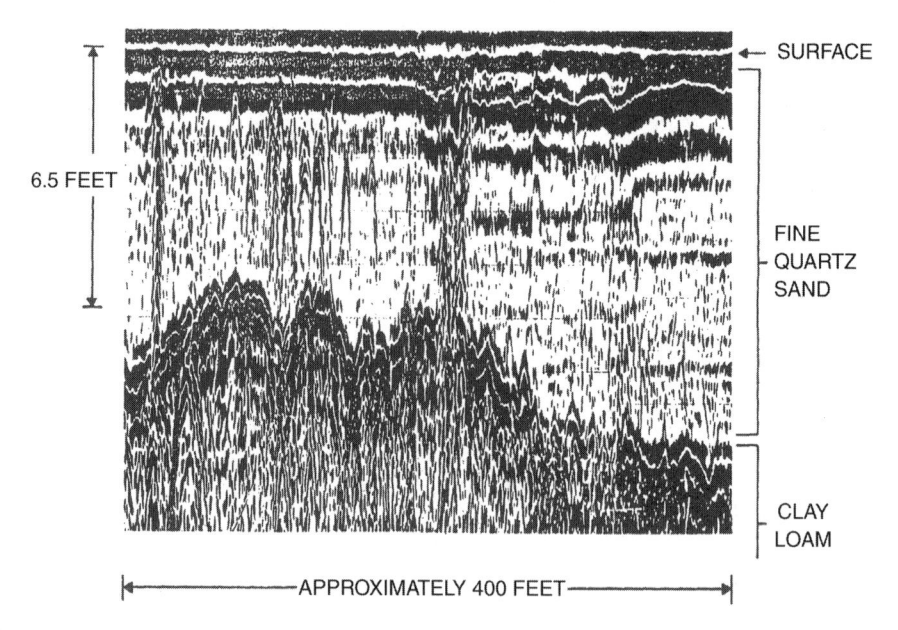

FIGURE 4.7
Radar profile of quartz sand over clay. (Note the level of detail that can be obtained.)

converted to depth by making some assumptions about the velocity of the waves in the subsurface materials.

Depth of penetration of the radar wave is highly site specific. The method is limited in depth by attenuation due to the higher electrical conductivity of subsurface materials or scattering. Generally, radar penetration is better in coarse, dry, sandy, or massive rock; poorer results are obtained in wet, fine-grained, clayey (conductive) soils. Data can be obtained in saturated materials if the specific conductance of the pore fluid is sufficiently low. Radar has been applied to map the sediments in fresh-water lakes and rivers. While radar penetration in soil and rock of more than 100 ft has been reported, penetration of 15 to 30 ft is more typical. In silts and clays, penetration may be limited to a few feet or less.

The continuous data produced by the radar method offers a number of advantages over some of the other geophysical methods. Continuous profiling permits data to be gathered much more rapidly, thereby providing a large amount of data. In some cases, total site coverage of an area can be obtained. Continuous radar data may be obtained at speeds of 5 to 10 mph or more. Very high lateral resolution data can be obtained by towing the antenna by hand at much slower speeds (less than 1 mph).

Radar has the highest resolution of all of the surface geophysical methods. Vertical resolution of radar data can range from less than an inch to several feet, depending upon the depth and the frequency used. A variety of antennas can be selected to cover frequencies from less than 100 to 1000 MHz. Lower frequencies provide greater depths of penetration with lower resolution and higher frequencies provide less penetration with higher resolution.

Preliminary field analysis of radar data is possible using the picture-like record. However, despite its simple graphic format, there are many pitfalls in the interpretation of radar data. Often, there are multiple bands within the data due to ringing — these may obscure layers and cause confusion in interpretation. Overhead reflections may appear on the record when not using shielded antennas (generally a problem with lower frequency unshielded antennas), and system noise can sometimes clutter up the record.

EM and Resistivity Methods

The EM and resistivity methods are similar in the sense that they both measure the same parameter, but in different ways. Electrical conductivity values (mhos/meter) are the reciprocal of resistivity values (ohm/meter). Electrical conductivity (or resistivity) is a function of the type of soil and rock, its porosity, and the conductivity of the fluids that fill the pore spaces. The conductivity of the pore fluids often dominates the measurement. Both methods are applicable to the assessment of natural hydrogeologic conditions and to mapping of contaminant plumes (Griffith and King, 1969; Benson et al., 1982; McNeill, 1982; Telford et al., 1982).

Natural variations in subsurface conductivity (or resistivity) may be caused by changes in basic soil or rock types, thickness of soil and rock layers, moisture content, and depth to water table. Localized deposits of natural organics, clay, sand, gravel, or salt-rich zones will also affect subsurface conductivity (or resistivity) values. Structural features such as fractures or voids can also produce changes in conductivity (or resistivity).

The absolute values of conductivity (or resistivity) for geologic materials are not necessarily diagnostic in themselves, but their spatial variations, both laterally and with depth, can be significant. It is the identification of these spatial variations or anomalies that enable the electrical methods to rapidly find potential problem areas (Figure 4.8).

Because the conductivity of the fluids in the pore spaces can dominate the measurements, detection and mapping of contaminant plumes can often be accomplished using electrical methods. Because inorganic species, in sufficient concentrations, are often more electrically conductive than clean ground water (because, in dissociated form, they are charged), both the lateral and vertical extent of an inorganic contaminant plume can often be mapped using electrical methods. Correlation between ground-water chemistry data and results using electrical methods to map inorganics from landfills has been as good as 0.96 at the 95% confidence level (Benson et al., 1985). Electrical methods provide a means of directly mapping the extent of the inorganic contaminants *in situ*, obtaining direction of flow and estimating concentration gradients (Figure 4.9). These

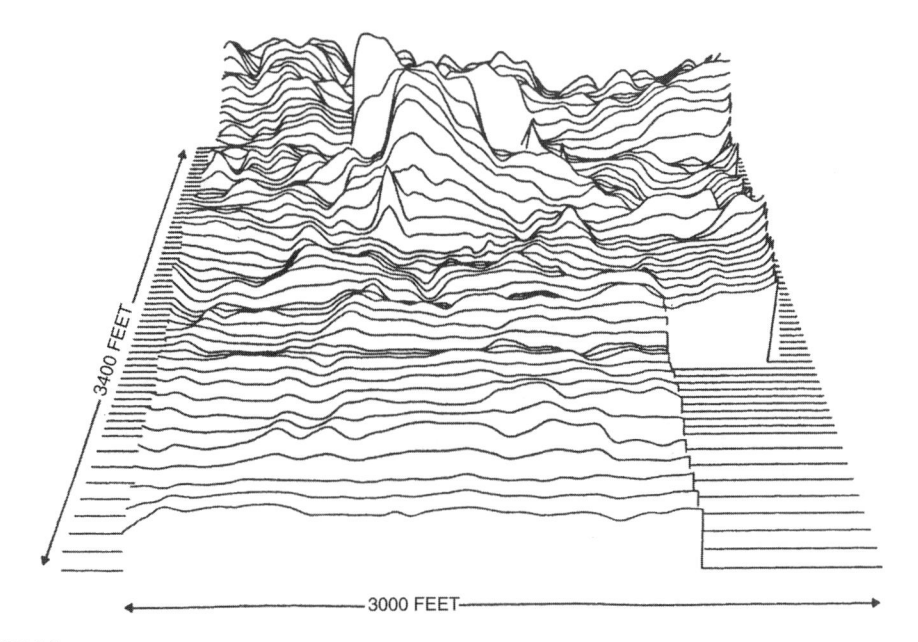

FIGURE 4.8
Continuous EM profile measurements show a large inorganic plume (center rear) and considerable natural geologic variation.

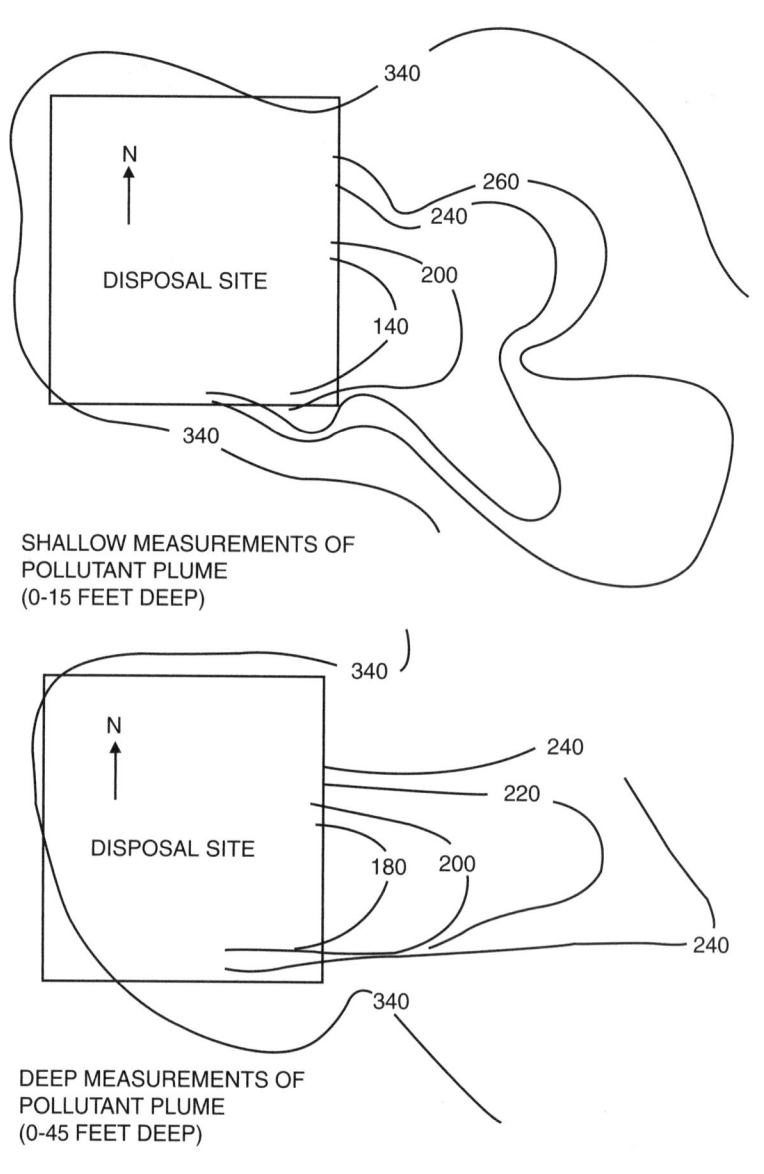

FIGURE 4.9
Resistivity map of leachate plume from a landfill (values are in ohm-feet; landfill is approximately 1 square mile).

methods can also be used for time-series measurements to obtain data on plume dynamics and thus provide vital information for modeling ground-water flow (Benson et al., 1988).

If the contaminant plume consists of a mix of organic and inorganic species, such as leachate from a landfill, a first approximation to the distribution of the organics can often be made by using electrical methods to map the more electrically conductive inorganics (Figure 4.9). Correlation between ground-water chemistry data for total organic carbon in a landfill leachate and results using electrical methods has been as good as 0.85 at the 95% confidence level (Benson et al., 1985).

In cases in which pure (nonaqueous phase) organic compounds, such as trichloroethylene exist, electrical as well as other geophysical methods can often be used to define permeable pathways or buried channels through which the contaminants may move. Since the mid-1980s, there have been significant advances in direct detection of organic compounds using radar and electrical methods (Olhoeft, 1986). For example, where hydrocarbons have been in place for a long period of time and biodegradation is taking place, a low resistivity (high conductivity) can often be detected due to the sulfides produced and an increase in total dissolved solids (Cassidy et al., 2001).

Both EM and resistivity methods may be used to obtain data by "profiling" or "sounding." Profiling provides a means of mapping lateral changes in subsurface electrical conductivity (or resistivity) to a given depth. Profiling measurements are made by obtaining data at a number of stations along a survey line. The spacings between the measurements will depend upon the variability of the setting and upon the lateral resolution desired. At each station along the profile line, data may be obtained for one depth or a number of depths depending upon project requirements. It is useful to take at least two measurements, a shallow one and a deeper one, so that the influence of highly variable shallow soils and cultural influences can be assessed. Profiling is well suited to delineation of hydrogeologic anomalies, mapping of contaminant plumes, and location of buried waste material.

The sounding method provides a means of determining the vertical changes in electrical conductivity (or resistivity) correlating with soil and rock layers. In this case, the instrument is located at one location and measurements are made at increasing depths. Interpretation of sounding data provides the depth, thickness, and conductivity (or resistivity) of subsurface layers with different electrical conductivities (or resistivities) (Figure 4.10).

Electromagnetics

Two types of EM instrumentation are in use. The most common is the frequency-domain system in which the transmitter is radiating energy at all times. This system measures changes in magnitude of the currents induced within the ground (McNeill, 1982). The time-domain system, in which the transmitter is cycled on and off, measures changes in the induced currents within the ground as a function of time. The frequency-domain and time-domain systems both induce currents into the ground by EM induction. ASTM Standards D 6820 (ASTM, 2004b) and D 6639 (ASTM, 2004c) provide guidance on the use of time-domain and frequency-domain EM conductivity, respectively, for environmental site characterization.

Because EM instruments do not require electrical contact with the ground, measurements may be made quite rapidly. Lateral variations in conductivity can be detected and mapped by profiling. Using commonly available frequency-domain EM instruments, profiling station measurements may be made to depths ranging from 2.5 to 200 ft.

Continuous EM profiling data can be obtained from 2.5 ft to a depth of 50 ft (Benson et al., 1982). These continuous measurements significantly improve lateral resolution for mapping small hydrogeologic features (Figure 4.3). Data can be recorded on an analog strip chart recorder or a digital data acquisition system. The excellent lateral resolution obtained from continuous EM profiling has been used to outline closely spaced burial pits, to reveal the migration of contaminants into the surrounding soils (Figure 4.8) or to delineate complex fracture patterns (Figure 4.3) (Benson et al., 1982).

In addition to evaluation of natural hydrogeologic conditions and mapping of contaminant plumes, some EM instrumentation can be used to locate trench boundaries, buried wastes and drums, and metallic utility lines. Frequency-domain EM instruments provide two outputs consisting of an in-phase component and an out-of-phase component. The out-of-phase component is used to measure electrical conductivity and can also be used to locate pipes. The in-phase component is a measure of the magnetic susceptibility and can be used to detect both ferrous and nonferrous metal. For example, using the in-phase component, a single 55 gal steel drum can be detected at a depth of about 6 to 8 ft.

Resistivity Sounding

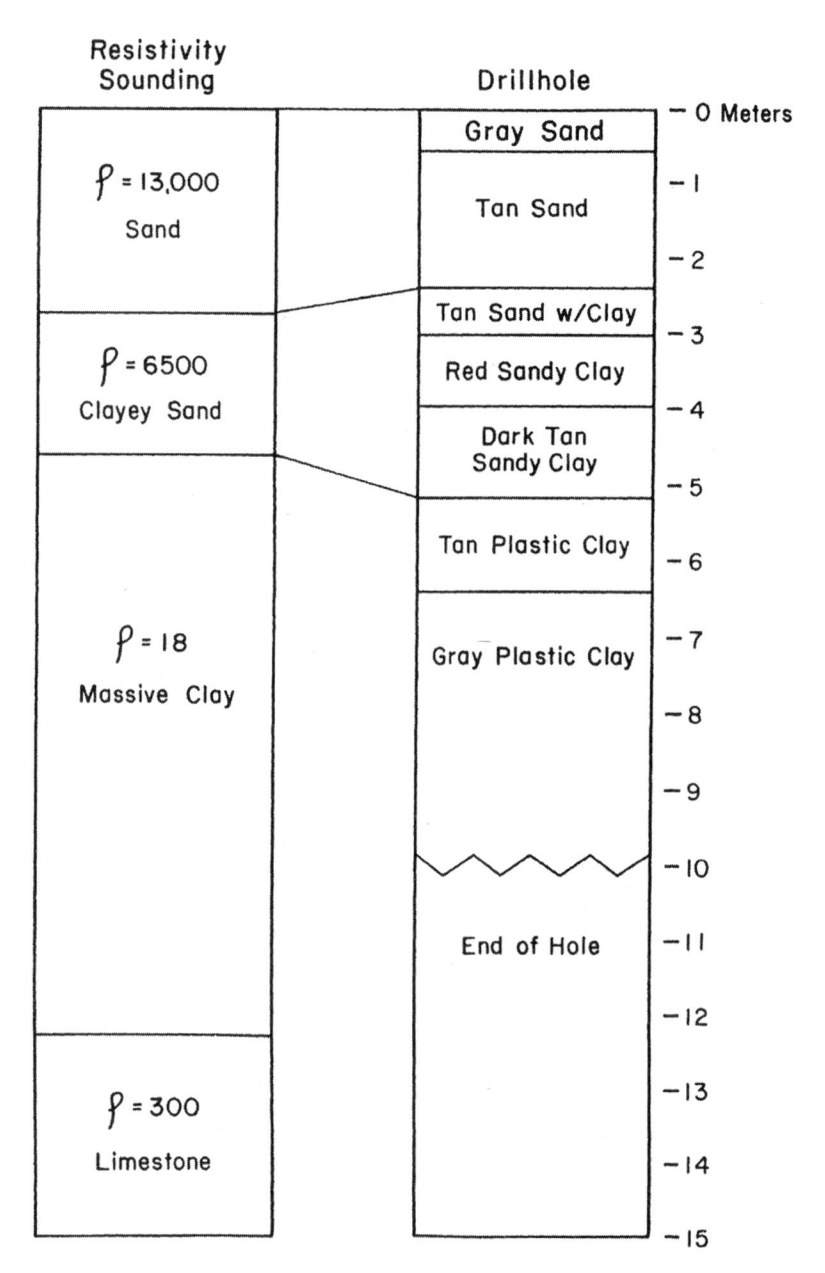

ρ = 13,000 Sand	
ρ = 6500 Clayey Sand	
ρ = 18 Massive Clay	
ρ = 300 Limestone	

Drillhole

0 Meters	Gray Sand
-1	
-2	Tan Sand
-3	Tan Sand w/Clay
-4	Red Sandy Clay
-5	Dark Tan Sandy Clay
-6	Tan Plastic Clay
-7	
-8	Gray Plastic Clay
-9	
-10	
-11	End of Hole
-12	
-13	
-14	
-15	

FIGURE 4.10
Resistivity geoelectric section showing correlation with a driller's log (resistivity values are in ohm-feet).

Vertical variations in conductivity can be determined by sounding. The instrumentation is placed at one location and measurements are made at increasing depths by a changing coil orientation or coil spacing. Data can be acquired at depths ranging from 2.5 to 200 ft by combining data from a variety of commonly available frequency-domain EM instruments. The vertical resolution of frequency-domain EM soundings is relatively poor because measurements are made at only a few depths. However, they do provide a quick means of obtaining limited vertical information. In contrast, time-domain transient EM systems are capable of providing detailed sounding data to depths of 150 ft to more than 1000 ft.

The depth of investigation of frequency-domain EM instruments is governed by coil configuration and operating frequency. EM profiling measurements are typically made with a system having a fixed coil configuration and a constant frequency, thereby providing a constant depth of investigation over a uniform subsurface. The GEM-2 (Won et al., 1996) is a multifrequency EM instrument operating in a frequency range of 90 Hz to 22 kHz. The instrument consists of a transmitter and receiver coil separated by about $5\frac{1}{2}$ ft. The user can select up to 16 different frequencies to be sampled while moving down a survey line. By recording data at different frequencies, it is theoretically possible to characterize subsurface electrical properties at multiple depths of investigation.

Resistivity

Direct Current Resistivity Measurements

As with EM measurements, electrical resistivity measurements are a function of the type of soil or rock, its porosity, and the conductivity of the fluids that fill the pore spaces. The method may be used in many of the same applications as the EM method (Cartwright and McComas, 1968; Griffith and King, 1969; Zohdy et al., 1974; Mooney, 1980; Benson et al., 1982; Telford et al., 1982). ASTM Standard D 6431 (ASTM, 2004d) provides guidance on the use of the DC resistivity method for environmental site characterization.

The resistivity method requires that an electrical current be passed through the ground using a pair of surface electrodes. The resulting voltage is measured at the surface between a second pair of electrodes. This requires that metal stakes be driven into the ground. A greater spacing between electrodes results in a greater depth of measurement. Usually, the depth of investigation is less than the spacing between electrodes. There are a number of electrode geometries that can be used, including the Wenner, Schlumberger, dipole–dipole, and many more. The simplest, in terms of geometry, is the Wenner array, which consists of four electrodes, spaced equally, all in a line. The resistivity of the soil and rock is calculated based on the electrode separation, the geometry of the electrode array, the applied current, and the measured voltage.

The resistivity technique may be used for profiling or sounding, similar to EM measurements. Profiling provides a means of mapping lateral changes in subsurface electrical properties to a given depth and is well suited to the delineation of hydrogeologic anomalies and mapping inorganic contaminant plumes (Figure 4.9).

Sounding measurements provide a means of determining the vertical changes in subsurface electrical properties. Interpretation of sounding data provides the depth, thickness, and resistivity of subsurface layers. Data can be interpreted using master curves for two to three layers (Orellana and Mooney, 1966), or computer models may be used to handle more than two or three layers (Mooney, 1980). Sounding data are used to create a geoelectric section that illustrates changes in the vertical and lateral resistivity conditions at a site. Figure 4.10 shows a geoelectric section developed from a resistivity sounding, along with drillers log showing the correlation.

One drawback to resistivity sounding is that the array requires considerable space. For example, a Wenner array sounding (with four electrodes equally spaced) may require that the spacing between the electrodes be as much as three to four times the depth of interest. Therefore, a sounding to a depth of 100 ft could require an overall array length (from current electrode to current electrode) of 900 to 1200 ft. At many sites, this much space may not be available.

2D Resistivity Imaging

In recent years, 2D resistivity imaging has been used to detect vertical as well as lateral variations in subsurface resistivity to produce a 2D geoelectric cross-section. Linear electrode arrays (usually consisting of 28 or more electrodes) are used to collect data at different electrode separations and positions along a profile line. A computer-controlled system switches which electrodes serve as the current electrodes and which electrodes serve as the voltage electrodes at any given time. This allows for automated recording of data. After the data are acquired, an inversion program can be used to fit a 2D model to the observed data to produce a geoelectric cross-section. Figure 4.11 shows a resistivity cross-section from an automated survey that used 28 electrodes in a roll along mode with a dipole–dipole array. Loke (1996) provides a complete overview of the resistivity imaging method. This technique can also be used to produce 3D geoelectric models of the subsurface by using a gridded array of electrodes. However, this method is very time consuming and is not often used.

Capacitively Coupled Resistivity

Traditional resistivity measurements use electrodes that are in direct contact with the ground to inject a DC current and to measure the resulting voltage difference. However, in areas where the surface resistivity is extremely high and in areas where driving electrode stakes into the ground is not feasible (e.g., concrete, exposed rock, etc.), traditional resistivity measurements are not easily obtained. In these conditions, it is now possible to use capacitively coupled alternating current to measure subsurface resistivity with a dipole–dipole array. The conductors in the cables act as one plate of the capacitor and the earth acts as the other plate, with the insulating sheath as the capacitor's insulator. Because an AC signal can pass between the plates of a capacitor, an AC equivalent to traditional DC resistivity measurements can be made (Geometrics, 1999). Multiple passes with different dipole spacing provide data to different depths. With sufficient data from different depths, an inversion program (Loke, 1996) can be used to model to produce a geoelectric section much like Figure 4.11. Capacitively coupled resistivity measurements are only useful in relatively resistive environments because the signal will be attenuated in conductive environments. The maximum depth of investigation is typically limited to 35 to 70 ft and will decrease with decreasing resistivity.

Comparison of EM and Resistivity Measurements

The frequency-domain EM method is often preferred for making profiling measurements because it requires less space for a measurement to a given depth. In addition, because the EM method does not require that electrodes be driven into the ground, it can be run more

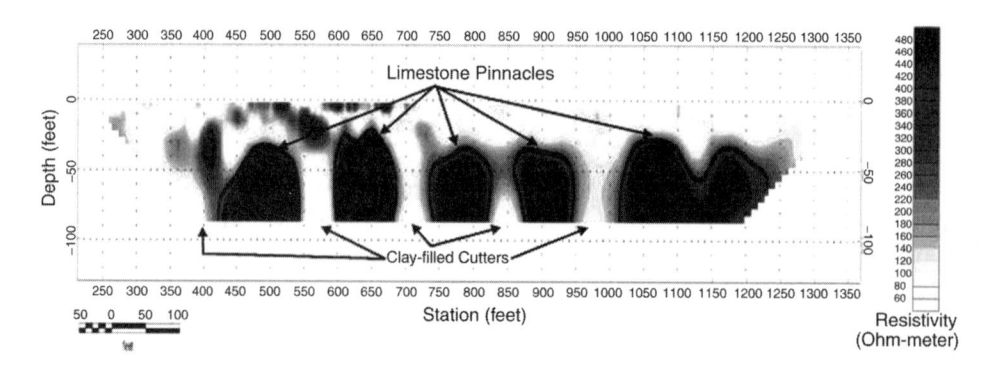

FIGURE 4.11
2D resistivity imaging showing rock cutters and pinnacles.

rapidly and is not influenced by shallow geologic noise associated with the electrodes resistivity used in measurements. In contrast, because resistivity methods provide better vertical resolution than the frequency-domain EM method, the resistivity method is commonly employed for sounding or imaging measurements. When space is limited and deep measurements are needed, there are advantages to using the time-domain EM system for soundings because it requires less space than long resistivity arrays.

EM measurements can be affected by buried metal pipes, metal fences, nearby vehicles, buildings, and power lines, as are resistivity measurements. But resistivity measurements are often less sensitive to many of these problems, permitting resistivity measurements to be made near such cultural sources of interference, where EM measurements often cannot be made.

EM conductivity and resistivity values from the same location may not agree, due to the difference in the volume of material being sampled and the differences in current distribution inherent to the two methods. Measurements will only be the same if they are made over a uniform medium.

Seismic Methods

Seismic techniques are often used to determine the top of bedrock, to determine depth to the water table, to assess the continuity of geologic strata, and to locate fractures, faults, and buried bedrock channels. These methods may also be used to characterize the type of rock, degree of weathering, and rippability based upon the seismic velocity of the rock. The seismic velocity in rock is related to thes rock's material properties such as density and hardness. By measuring both compressive (P) waves and shear (S) waves and knowing the density of a soil or rock, one can calculate the modulus properties of the materials through which the waves travel.

Seismic waves are transmitted into the subsurface by a source, which can sometimes be as simple as a sledgehammer. These waves are refracted and reflected when they pass from a soil or rock type with one seismic velocity into another with a different seismic velocity. An array of geophones placed on the surface measures the travel time of the seismic waves from the source to the geophones. The refraction and reflection techniques use the travel times of the waves and the geometry of the source-to-geophone wave paths to model subsurface conditions. The unit of time is milliseconds (10^{-3} sec). For most refraction work, the first refracted compressional wave arrivals (P-waves) are used. For reflection work, the later arriving reflected compressional waves are used. It is also possible to measure shear wave arrivals (S-waves), which can be useful in determining properties such as the elastic moduli of shallow subsurface materials (Mooney, 1977), which are important in engineering applications. Measurements of both compressional (P) and shear (S) wave velocity and density values provide the data from which we can calculate modulus of materials for engineering purposes. Crice (2001) provides an excellent discussion of shear waves.

A seismic source, geophones, and a seismograph are required to make the measurements. The seismic source may be a simple sledgehammer or other mechanical source with which to strike the ground. Explosives may be utilized for deeper applications that require greater energy. Geophones implanted into the ground surface translate the ground vibrations of seismic energy into an electrical signal. The electrical signal is displayed on the seismograph, permitting measurement of the arrival time of the seismic wave and displaying the waveforms from a number of geophones. Geophone spacing can be varied from a few feet to a few hundred feet depending upon the depth of interest and the resolution needed.

Because the seismic refraction and reflection methods measure small ground vibrations, they are inherently susceptible to vibration noise from a variety of natural (e.g., wind and waves) and cultural sources (e.g., walking, vehicles, and machinery).

Seismic Refraction

The refraction method is commonly applied to shallow investigations up to a few hundred feet deep (Griffith and King, 1969; Benson et al., 1982; Telford et al., 1982; Haeni, 1986). However, with the application of sufficient energy, surveys to a few thousand feet and more are possible. Up to three and sometimes four layers of soil and rock can normally be determined, if a sufficient velocity difference or contrast exists between adjacent layers. A typical refraction line for a shallow investigation might consist of 12 or 24 geophones set at equal spacings as close as 5 to 10 ft. Two seismic impulses at each end of the geophone array are created and their refracted waves recorded separately. The refraction survey may require a maximum source-to-geophone distance four to five times the depth of investigation. ASTM Standard D 5777 (ASTM, 2004e) provides guidance on the use of seismic refraction for environmental site characterization.

Significantly greater source energy will be required as the depth of investigation increases. Two inherent limits to the refraction method are its inability to detect a lower velocity layer beneath a higher velocity layer and its inability to detect thin layers.

Seismic refraction work can be carried out in a number of ways. The simplest approach, in terms of field and interpretation procedures, can be carried out by creating two separate seismic impulses, one at each end of the geophone array. The results of this simple measurement provide two depths and thus the dip of rock under the array of geophones. This method is described in detail by Mooney (1973) and Haeni (1986).

A more detailed refraction survey can be carried out so that depths are obtained under every geophone (Figure 4.12). This survey will produce a detailed profile of the top of rock. Lateral resolution will depend upon the geophone spacing, which might range from 5 to 50 ft. This method is described in detail by Redpath (1973). The general reciprocal method described by Palmer (1980) will accommodate varying velocities within each layer, while calculating the depth beneath each geophone.

Seismic Reflection

In comparison, the seismic reflection survey is capable of much deeper investigations with less energy than the refraction method. While reflections have been obtained from depths as shallow as 10 ft, the shallow reflection method is more commonly applied to depths of 50 to 100 ft or more. The reflection technique can be used effectively to depths of a few thousand feet and can provide relatively detailed geologic sections (Figure 4.13). As with radar reflections, the vertical scale is measured in two-way travel time — that is, the time it takes for a wave to travel down to an interface and back up to the surface again. The time scale must then be converted to depth making some assumptions regarding seismic velocity within the strata.

There are two approaches currently used to obtain shallow seismic reflection data — the common offset method, developed by Hunter et al. (1982), and the common depth point (CDP) method adapted from the oil industry by Lankston and Lankston (1983) and Steeples (1984). The common offset method uses low-cost equipment and software but has some site-specific limitations that are not inherent in the CDP method. The CDP method has fewer site-specific limitations, but is more dependent upon sophisticated hardware and software capabilities. Hardware and software for the shallow CDP method are readily available.

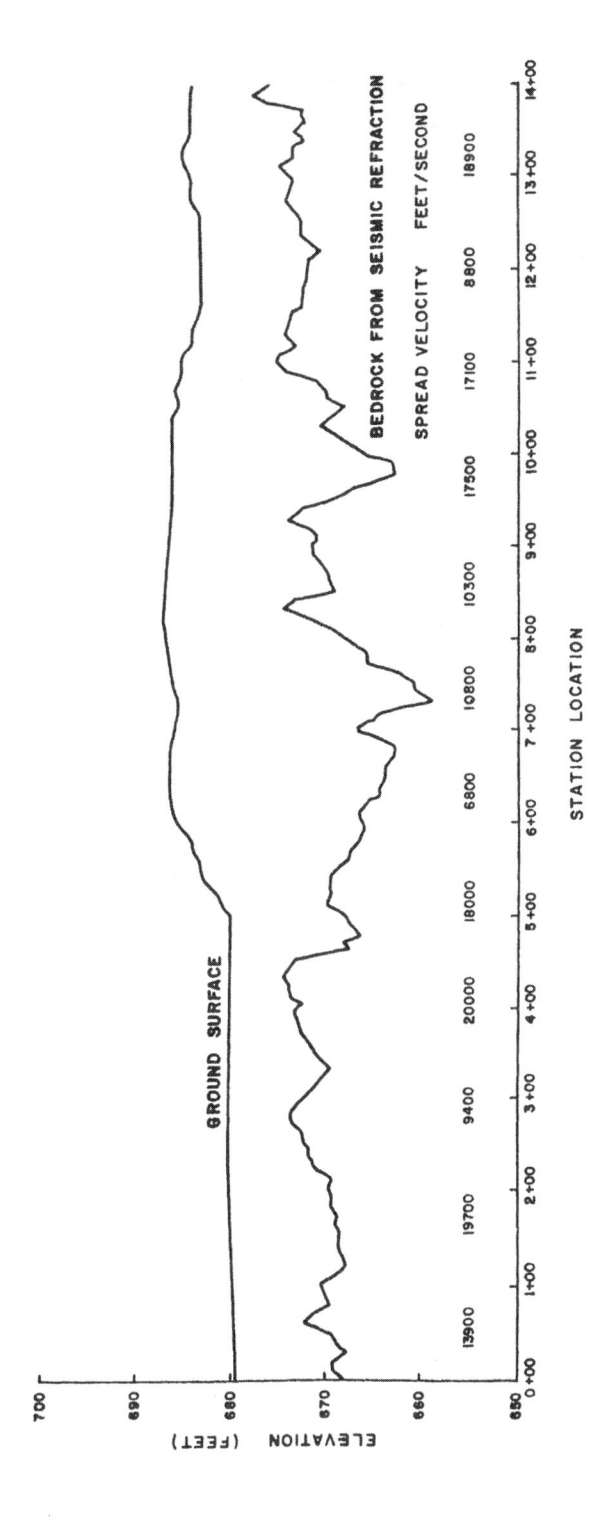

FIGURE 4.12
Profile of top of bedrock from seismic refraction survey (depth to rock determined under each geophone; geophone spacing is 10 ft).

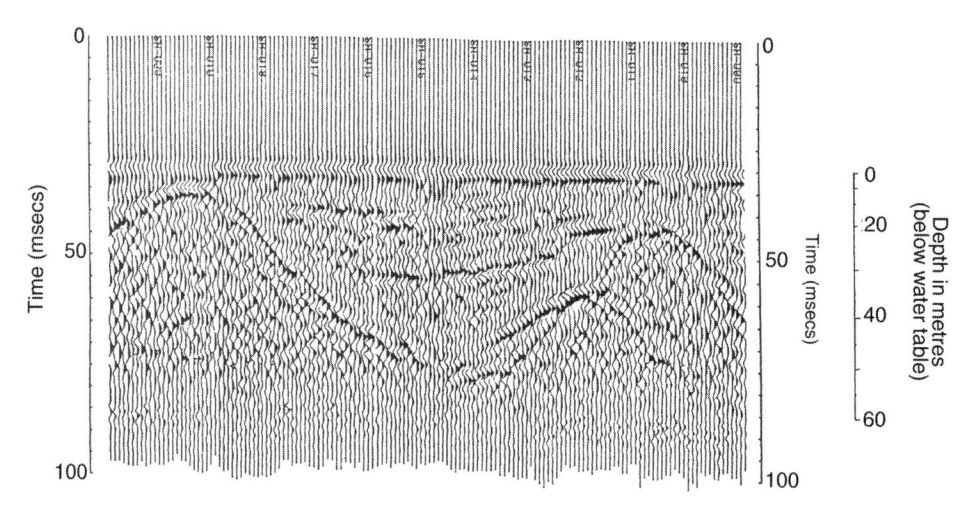

FIGURE 4.13
Common offset seismic reflection data showing channel in bedrock. (From Dr. Jim Hunter, Geological Survey of Canada. With permission.)

The shallow high-resolution reflection methods discussed here attempt to utilize the highest frequencies possible (150 to 600 Hz) to improve vertical resolution and relatively closely spaced geophones (1 to 20 ft apart) to provide good lateral resolution. Because of the need for higher frequencies, attention must be given to selection of a seismic source and its optimum coupling to soil or rock as well as to geophone placement.

The reflection method is limited by its ability to transmit energy, particularly high frequency energy, into the soil and rock. Loose soil near the surface limits the ability of the soil system to transmit high frequency energy into and out of rock, limiting the resolution that can be obtained. The most common limitation, however, will be that of acoustic noise caused by natural or cultural sources.

Surface Wave Analysis

For seismic reflection surveys, surface waves are generally considered unwanted noise. However, it is possible to exploit the sensitivity of the surface wave to changes in material velocities that are present in the subsurface. Surface wave propagation depends on frequency (depth of penetration), phase velocity (compressional and shear wave velocities), and density. Each of these properties will affect the surface wave dispersion curve (phase velocity vs. frequency) in a predictable fashion. Shear wave velocity has the greatest impact on the properties of a surface wave, so the dispersion curve can be inverted in such a way to obtain the shear wave velocity as a function of depth.

Nazarian et al. (1983) developed the concept of spectral analysis of surface waves (SASW) to estimate 1D shear wave velocities for engineering applications. It was later discovered that using SASW concepts, together with multitrace seismic acquisition methods, can be effective in detecting anomalous conditions in subsurface materials. This led to multichannel analysis of surface waves (MASW) (Miller et al., 2001), which permits the generation of 2D shear wave velocity field cross-section.

The MASW method has been shown to be effective in providing information about the horizontal and vertical continuity of shallow materials in the upper few feet to depths of more than 100 ft (Miller et al., 2001).

Microgravity

Gravity instruments respond to changes in the earth's gravitational field caused by changes in the density of the soil and rock. By measuring the spatial changes in the gravitational field, variations in subsurface geologic conditions can be determined (Griffith and King, 1969; Telford et al., 1982). There are two basic types of gravity surveys: a regional gravity survey and a local microgravity survey. A regional gravity survey employs widely spaced (a few thousand feet to a few miles) stations and is carried out with a standard gravity meter. These surveys are used to assess major geologic conditions over many hundreds of square miles. In contrast, microgravity surveys have station spacings of 5 to 20 ft (typically) and are carried out with a very sensitive microgravimeter. These surveys are used to detect and map shallow, localized geologic anomalies such as bedrock channels, fractures, and cavities. ASTM Standard D 6430 (ASTM, 2004f) provides guidance on the use of the gravity method for environmental site characterization.

The unit of acceleration used in gravity measurement is the gallon. The earth's normal gravity is 980 gal. Microgravity measurements are sensitive to within a few microgals (10^{-6} gal).

The microgravity survey results in a Bouguer anomaly, which is the difference between the observed gravity values and theoretical gravity values. The Bouguer anomaly is made up of deep-seated effects (the regional Bouguer anomaly) and shallow effects (the local Bouguer anomaly). It is the local Bouguer anomaly that is of interest in microgravity work (Figure 4.14).

A gravimeter is designed to measure extremely small differences in the gravitational field and is a very delicate instrument. The instrument is thermostatically controlled to minimize drift caused by temperature variations. Considerable care must be taken in shipment and general field use to avoid shock to the instrument. Gravity measurements may be affected by ground noise (Seismic Methods), winds, and temperature. To compensate for minor instrument drift throughout the day, measurements must be made at a base station every hour or so, so drift corrections can be applied to the data. Corrections must also be made for the constantly changing earth tides, changes in elevation (to the nearest 0.01 ft), and topography. Gravity data may be presented as a profile or as a contour map, depending upon project needs.

Metal Detection

Metal detectors are commonly used by utility and survey crews for locating buried pipes, cables, and property stakes. They can also be used for detecting buried drums and for

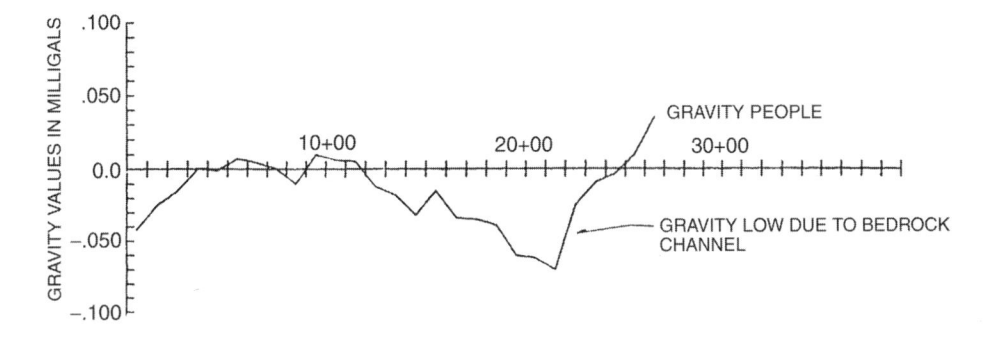

FIGURE 4.14
Microgravity profile showing bedrock channel. (From Technos Inc. With permission.)

delineating the boundaries of trenches containing metallic drums or trash (Figure 4.15) (Benson et al., 1982). Metal detectors can detect both ferrous metals such as iron and steel and nonferrous metals such as aluminum and copper.

Metal detectors have a relatively short detection range, because the detector's response is proportional to the cross-section of the target and inversely proportional to the sixth power of the distance to the target. Small metal objects, such as quart-sized containers, can be detected at a distance of approximately 2 to 3 ft. Specialized metal detectors will detect larger objects, such as 55 gal drums, at depths of 3 to 10 ft, and massive piles of 55 gal drums may be detected at depths of up to about 15 ft. The metal detector is a continuously sensing instrument used with a sweeping motion while moving forward along a survey line. It may also be held steady while a traverse line is walked and the results are recorded. The area of detection of a metal detector is approximately equal to its coil size or coil spacing (typically 1 to 3 ft). Metal detectors can be affected by nearby metallic pipes, fences, cars, buildings and, in some cases, changes in soil conditions.

Magnetometry

A magnetometer measures the intensity of the earth's magnetic field. As with gravity surveys, a magnetic survey can be used to map geologic conditions over large areas. This type of survey is useful for mapping regional geologic conditions. In certain geologic environments, magnetics can also be used to map depth to bedrock, channels, and fractures (Griffith and King, 1969; Breiner, 1973; Telford et al., 1982). The primary application of magnetic measurements at potentially contaminated sites is in detecting buried drums, tanks, and pipes (Breiner, 1973; Benson et al., 1982). A magnetometer will only respond to ferrous metals (iron and steel) and will not detect nonferrous metals. The presence of buried ferrous metals creates a local variation in the strength of the earth's magnetic field, permitting the detection and mapping of buried ferrous metal (Figure 4.16).

Two types of magnetic measurements are commonly made: total field measurements and gradient measurements. A total field measurement responds to the total magnetic field of the earth, any changes caused by a target, natural magnetic variations, and cultural magnetic noise (ferrous pipe, fences, buildings, and vehicles).

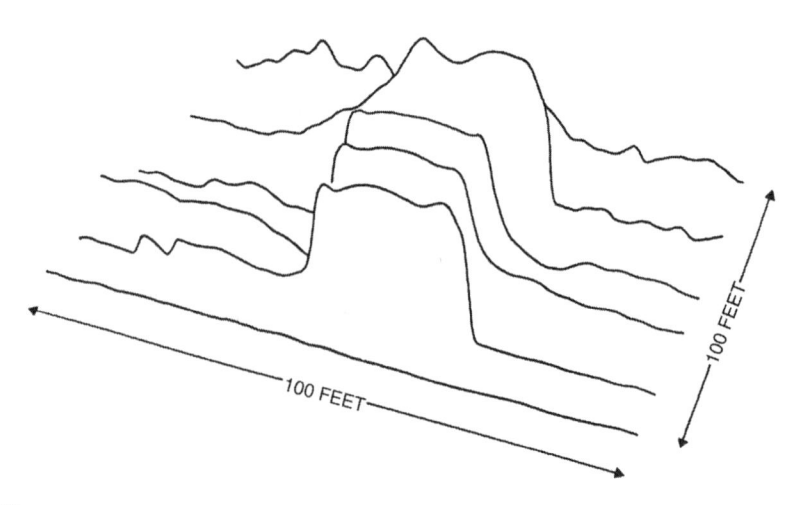

FIGURE 4.15
Results of a metal detector survey to locate a burial trench.

The effectiveness of total field magnetometers can be reduced or totally inhibited by noise or interference caused by time-variable changes in the earth's magnetic field or spatial variations due to magnetic minerals in the soil, steel debris, pipes, fences, buildings, and passing vehicles.

A base station magnetometer can be used to reduce the effects of natural noise by subtracting the base station values from those of the search magnetometer. This can minimize any errors due to natural long-period changes of the earth's field. Cultural noise, however, will remain a problem with total field measurements. Many of these problems can be avoided by use of gradient measurements and proper field techniques.

Gradient measurements are made by a gradiometer, which is simply two magnetic sensors separated vertically (or horizontally), usually by a few feet. Gradient measurements have some distinct advantages over total field measurements. They are insensitive to natural spatial and temporal changes in the earth's magnetic field and minimize most cultural effects; because the response of a gradiometer is the difference of two total field measurements, it responds only to the local gradient. As a result, a gradiometer is better able to locate a relatively small target, such as a buried drum. The disadvantage of a gradiometer is that it provides a slightly less sensitive measurement than a total field instrument.

A total field magnetometer's response is proportional to the mass of the ferrous target and inversely proportional to the cube of the distance to the target. A gradiometer's response is inversely proportional to the fourth power of the distance to the target, making it less sensitive than the total field measurement. While gradiometers are inherently less sensitive than total field instruments, they are also much less sensitive to many sources of noise. Typically, a single 55 gal drum can be detected at depths up to about 20 ft with a total field magnetometer or about 10 ft with a gradient magnetometer. Massive piles of drums can be detected at depths up to 50 ft or more with a total field magnetometer or about 25 ft with a gradient magnetometer.

A total field or gradient proton procession magnetometer normally requires the operator to stop and take a measurement, while a fluxgate gradiometer permits the continuous acquisition of data as the magnetometer is moved across the site. Continuous coverage is much more suitable for detailed (high resolution) surveys to identify local targets, such as drums, and the mapping of areas in which complex anomalies are expected.

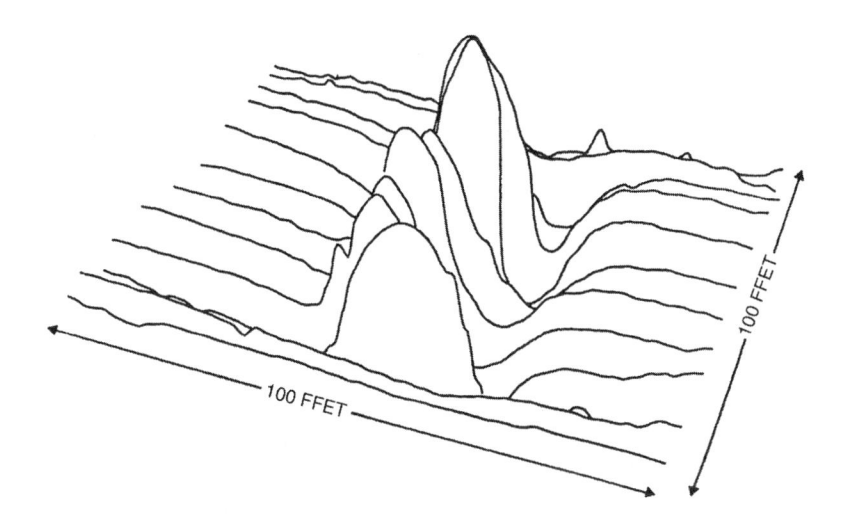

FIGURE 4.16
Magnetic gradient over a trench with buried drums (the trench is approximately 20/100 ft).

Measurements over Water

Many of the surface methods are adapted to make bottom and subbottom measurements over rivers, lakes, reservoirs, estuaries, and off-shore. For example, radar can be used to map the bottom and subbottom conditions in fresh water. EM conductivity (EM31 and EM34) measurements have been made from fiberglass or rubber boats in fresh water. Resistivity profiling, sounding, and 2D resistivity imaging measurements can be made in both fresh water and salt water. Seismic reflection methods are used to obtain bathymetry (depth to bottom) and at lower frequencies subbottom penetration of hundreds of feet to 1000 ft or more. Side-scan sonar can be used to develop an acoustic image of bottom conditions, mud, sands, reefs and to locate sunken ships and aircraft.

Downhole Geophysical Measurements

One of the most common subsurface investigation techniques is that of sampling soil and rock at discrete intervals (typically every 5 ft) as a boring is advanced. This method provides gross information on subsurface lithology but sand lenses, fractures, or other subtle changes in geology, which can affect hydraulic conductivity, can easily go undetected. Although continuous sampling or coring can improve the description of geologic conditions, it is very costly and time consuming and material description is somewhat subjective. Furthermore, 100% sample recovery is rarely achieved.

A number of downhole logging techniques are available for determining the characteristics of soil, rock, or fluid along the length of a borehole or a monitoring well (Keys and MacCary, 1976). These methods provide continuous, high-resolution *in situ* measurements that are often more representative of hydrogeologic conditions than samples obtained from borings. A number of logging techniques are available, and an adequate assessment of subsurface conditions will often require that multiple logs be used because each log responds to a different property of the soil, rock, or fluid. Some of these techniques will provide measurements from inside plastic or steel casing and some will allow measurements to be made in the unsaturated zone, as well as the saturated zone.

Downhole logging measurements can be correlated to the known geologic strata (through direct comparison with soil samples) in one hole and then can be used to identify and correlate geologic strata in other holes without sampling. Thin layers and subtleties, not readily detected in soil or core samples, can often be resolved by logging. Logging can significantly improve the ability to accurately characterize and correlate strata between borings by providing high-resolution data independent of subjective interpretations of soil and rock types.

A number of soil and rock properties can be measured *in situ*. Values for soil and rock porosity, density, seismic velocity, and elastic moduli can be obtained to facilitate engineering design. Even more important is the ability to identify the uniformity or lack of uniformity of subsurface conditions. Downhole measurements can be used to identify permeable zones such as sand lenses in glacial tills, weathered zones, and fractures or solution cavities in rock. The same measurements are also effective for identifying impermeable zones, such as aquitards, and assessing their continuity and integrity.

Monitoring wells that have been in place for years provide the basis for long-term chemical monitoring. For many of these wells, neither geologic logs nor installation records is available. Using downhole techniques, it is possible to obtain geologic information and well construction details. In addition, logging may be used to determine whether a problem exists with well construction and what type of remedial work, if any, is necessary to correct it.

By running nuclear logs in existing holes with steel or PVC casing, geologic strata outside the casing can be characterized. Under some conditions in an open borehole or PVC-cased well, contaminants outside PVC casing can be detected by running EM induction logs. A downhole television camera can be used within cased wells to assess monitoring well conditions or it can be used within an uncased borehole to assess the existence of fractures.

While each log is susceptible to both natural and cultural noise, borehole diameter will probably be of most concern. Most logs provide measurements within a radius of 6 to 12 in. from the hole. Therefore, as the borehole diameter becomes larger, the measured results become more dominated by drilling and well-construction aspects.

A description of the most commonly used logs is given subsequently. Table 4.1 lists the conditions in which these logs can be used and some of the limitations inherent in the use of each log.

Nuclear Logs

Natural Gamma Log

A natural gamma log records the amount of natural gamma radiation that is emitted by rocks and unconsolidated materials. The chief use of natural gamma logs is the identification of lithology and stratigraphic correlation in open or cased holes above and below the water table. ASTM Standard D 6274 (ASTM, 2004g) provides guidance on the use of the natural gamma method in environmental site characterization.

The gamma-emitting radioisotopes normally found in all rocks and unconsolidated materials are potassium-40 and daughter products of the uranium and thorium decay series. Because clays and shales concentrate these heavy radioactive elements through the processes of ion exchange and adsorption, the natural gamma activity of shale and clay-bearing sediments is much higher than that of quartz sands and carbonates. Therefore, the gamma log, which indicates an increase in clay or shale content by an increase in counts per second (Figure 4.17), is useful for evaluating the presence, variability, and integrity of clays and shales. The radius of investigation for the natural gamma log is from about 6 to 12 in. (Keys and MacCary, 1976).

Gamma–Gamma (Density) Log

A gamma–gamma log is used to determine the relative bulk density of the soil or rock and to identify lithology. The log can be used in open or cased holes above and below the water table (Figure 4.18).

The gamma–gamma log is an active probe containing both a radiation source and a detector. This log provides a response, in counts per second, that is averaged over the distance between the source and the detector. The radius of investigation for the gamma–gamma log is relatively small (only about 6 in.). Therefore, borehole diameter variations and well construction factors can affect this log more than other logs (Keys and MacCary, 1976).

Neutron–Neutron (Porosity) Log

A neutron–neutron log provides a measure of the relative moisture content above the water table and porosity below the water table (Figure 4.18). It can be run in open or cased holes above and below the water table. The neutron–neutron log is an active probe with both a radiation source and a detector. It provides a response, in counts per second, that is averaged over the distance between the source and the detector.

TABLE 4.1
General Characteristics and Uses of Downhole Geophysical Logs

Downhole Log	Parameter Measured (or Calculated)	Casing Uncased	Casing PVC	Casing Steel	Saturated	Unsaturated	Radius of Measurement	Effect of Hole Diameter, and Mud
Natural gamma	Natural gamma radiation	Yes	Yes	Yes	Yes	Yes	6–12 in.	Moderate
Gamma–gamma	Density	Yes	Yes	Yes	Yes	Yes	6 in.	Significant
Neutron	Porosity below water table: moisture content above water table	Yes	Yes	Yes	Yes	Yes	6–12 in.	Moderate
Induction	Electrical conductivity	Yes	Yes	No	Yes	Yes	30 in.	Negligible
Resistivity	Electrical resistivity	Yes	No	No	Yes	No	12–60 in.	Significant to minimal depending upon probe used
Single point resistance	Electrical resistance	Yes	No	No	Yes	No	Near borehole surface	Significant
Spontaneous potential	Voltage: responds to dissimilar minerals and flow	Yes	No	No	Yes	No	Near borehole surface	Significant
Temperature	Temperature	Yes	No	No	Yes	No	Within borehole	N/A
Fluid conductivity	Electrical conductivity	Yes	No	No	Yes	No	Within borehole	N/A
Flow	Fluid flow	Yes	No	No	Yes	No	Within borehole	N/A
Caliper	Hole diameter	Yes	Yes	Yes	Yes	Yes	To limit of senor typically 2–3 ft	N/A

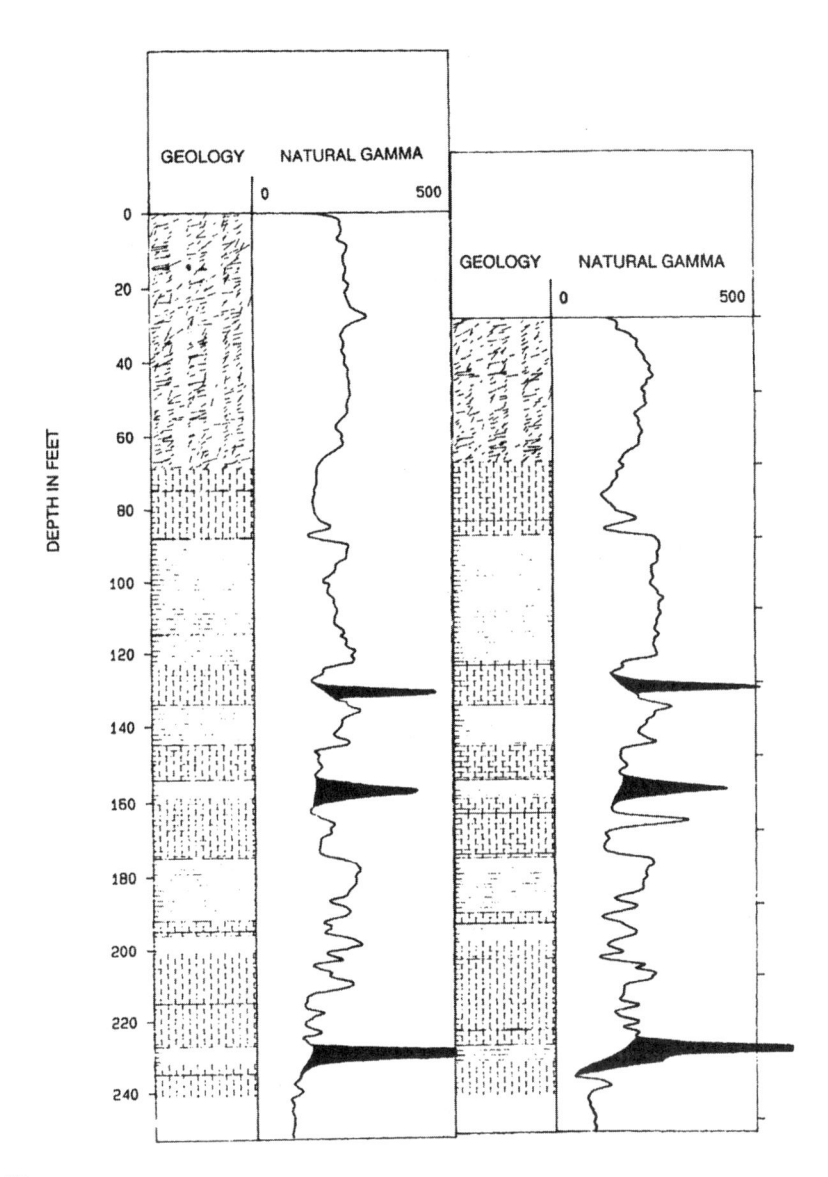

FIGURE 4.17
Natural gamma logs from two nearby boreholes, 100 ft apart (note the characterization and correlation of the shale and limestone units). (From Technos Inc. With permission.)

The radius of investigation for the neutron–neutron probe is approximately 6 in. (up to 12 in. in very porous formations). Borehole diameter variations and well construction factors can affect this log, but not as severely as the density log (Keys and MacCary, 1976). ASTM Standard D 6727 (ASTM, 2004h) provides guidance on the use of the neutron–neutron method for environmental site characterization.

Nonnuclear Logs

Induction Log

The induction log is an EM induction method for measuring the electrical conductivity of soil or rock in open or PVC-cased boreholes above or below the water table (similar to EM

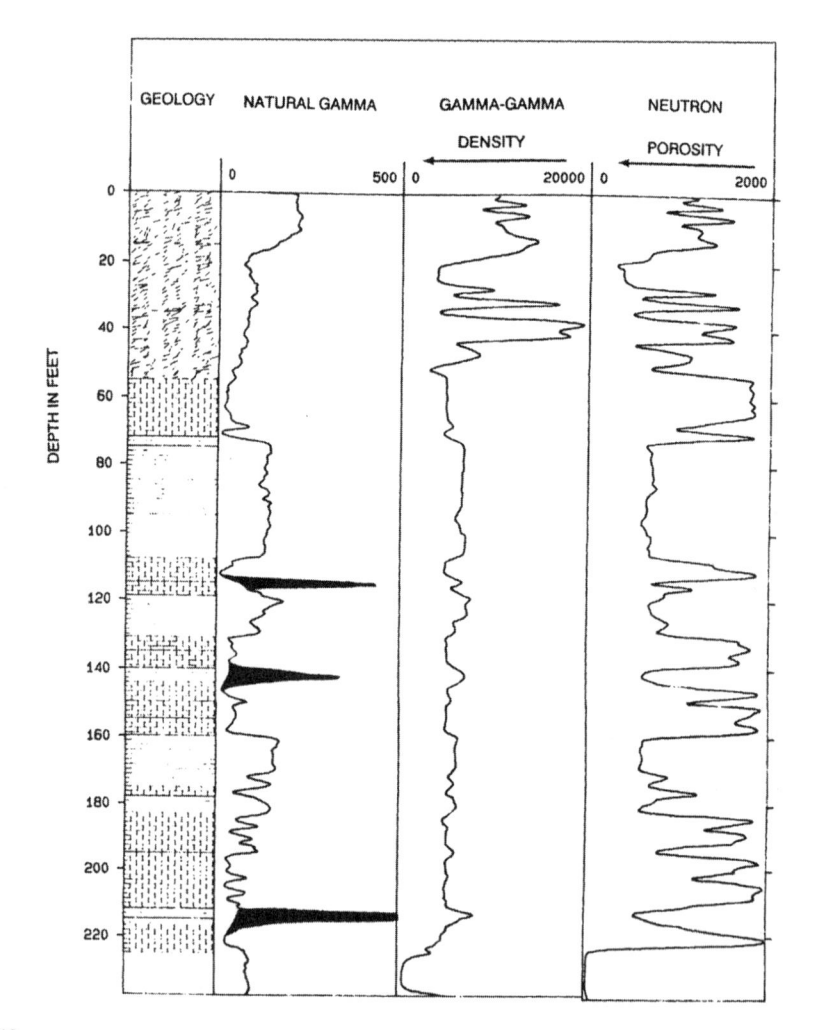

FIGURE 4.18
A suite of logs from within the same borehole (the natural gamma log provides a means of characterizing the shale, the gamma–gamma log provides a measure of density, and the neutron log provides a measure of porosity within the shale and limestone units). (From Technos Inc. With permission.)

measurements made on the surface). The induction log can be used for identification of lithology and stratigraphic correlation. Electrical conductivity is a function of soil and rock type, porosity, permeability, and the fluids filling the pore spaces. Because the response of the log (millimhos/meter) will be a function of the specific conductance of the pore fluids, it is an excellent indicator of the presence of inorganic contamination (Figure 4.19) and, in some cases (when organics are mixed with inorganics or when a thick layer of hydrocarbons is present), organic contamination. Variations in conductivity with depth may also indicate changes in clay content, permeability of a formation, or fractures. An induction log provides data similar to that provided by a resistivity log (because conductivity is the reciprocal of resistivity). However, the induction log can be run without electrical contact with the formation. Therefore, the induction log can be used in both the vadose zone and the saturated zone and it can be used to log through PVC casing.

The radius of investigation for the induction log is approximately 2.5 ft from the center of the well. Because this log has a much larger radius of investigation than other logs, it is almost totally insensitive to borehole and construction effects and as such is a good

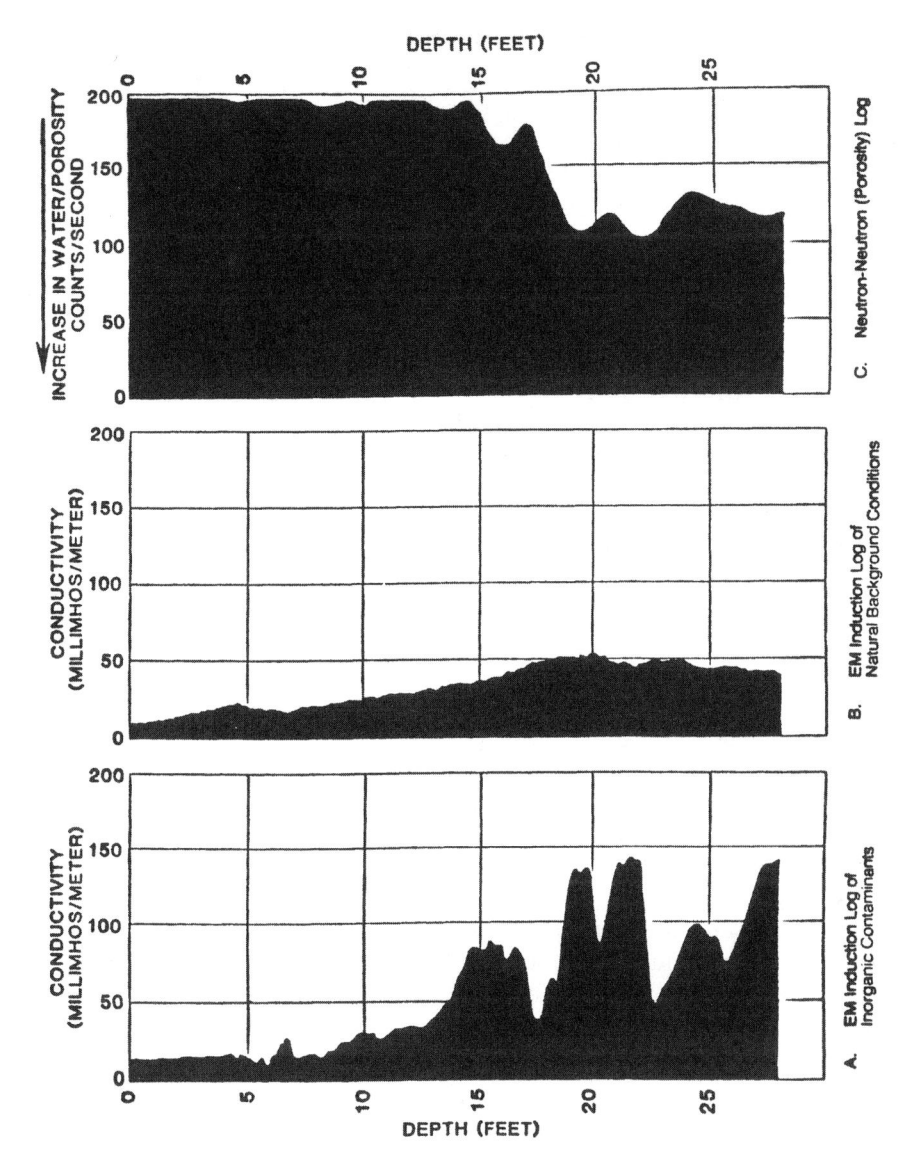

FIGURE 4.19
Induction and porosity logs are used to identify contaminants and permeable zones. (From Technos Inc. With permission.)

indicator of the overall soil and rock conditions surrounding the borehole. ASTM Standard D 6726 (ASTM, 2004i) provides guidance on the use of the EM induction method in environmental site characterization.

Resistivity Log

The resistivity log measures the apparent resistivity (measured in ohm-feet or ohm-meters) of rock and soil within a borehole. Because resistivity is the reciprocal of conductivity, which is the property measured by an induction log, the resistivity log responds to and measures the same properties and features as the induction log. However, because of the need for electrical contact with the borehole wall, the resistivity log can only be run in an uncased hole filled with water or drilling fluid.

There are a number of electrode spacings or geometries that may be used for resistivity logs. The most common is the "normal" log. Short normal probes (typically an electrode spacing of 16 or 18 in.) give good vertical resolution and measure the apparent resistivity of the formation immediately around the borehole. Long normal probes (typically an electrode spacing of 64 in.) have less vertical resolution but measure the apparent resistivity of undisturbed rock within a larger radius from the borehole, similar to the induction log (Keys and MacCary, 1976).

Resistance Log

A resistance log (sometimes referred to as single-point resistance) measures the resistance (in ohms) of the earth materials lying between a downhole electrode and a surface electrode. It can only be run in uncased holes in the saturated zone. The primary uses of resistance logs are geologic correlation and the identification of fractures or washout zones in resistive rocks. The resistance log should not be confused with the resistivity log, which provides a quantitative measure of the material resistivity.

The radius of investigation of the resistance log is quite small. It is in many cases as strongly affected by conductivity of the borehole fluid as it is affected by the resistance of the surrounding volume of rock (Keys and MacCary, 1976).

Spontaneous-Potential Log

The spontaneous-potential (SP) log measures the natural potential (in millivolts) developed between the borehole fluid and the surrounding rock materials. It can only be run in uncased holes within the saturated zone. The SP voltage consists of two components. The first component results from electrochemical potential caused by dissimilar minerals. The second component is the streaming potential caused by water moving through a permeable medium.

SP measurements are subject to considerable noise from the electrodes, hydrogeologic conditions, and borehole fluids. Even though these measurements do not provide quantitative results, they have a number of applications including:

- Characterizing lithology
- Providing information on the geochemical oxidation–reduction conditions
- Providing an indication of fluid flow

The radius of investigation of the SP log is highly variable (Keys and MacCary, 1976).

Temperature Log

A temperature log is a continuous record of the temperature of the borehole fluid immediately surrounding the sensor as it is lowered within an open borehole. The temperature log will often indicate a zone of ground-water flow within the uncased portion of a borehole. Flow is indicated when an increase or decrease in water temperature occurs. Changes in temperature can also be used to monitor leaks in casing where damage or corrosion has occurred. A temperature log may have a sensitivity of 0.5°C or better.

Fluid Flow

There are many ways of measuring fluid flow within a borehole (Keys and MacCary, 1976). The most commonly used method is the use of an impeller-type flow meter that provides counts per second. The count rate can usually be calibrated to provide results in feet per minute or gallons per minute.

Fluid Conductivity

A fluid conductivity log provides a measurement of the specific conductance of the borehole fluids (micromhos/centimeter). If accurate values are needed (as opposed to anomaly detection), a temperature log must also be run so that corrections can be made.

Mechanical Caliper Log

A mechanical caliper log provides a record of the diameter of an open borehole or of the inside diameter of a well casing. The caliper probe consists of spring-loaded arms that extend from the logging tool so that they follow the sides of the borehole or casing.

Caliper logs are utilized to measure borehole diameter, to locate fractures and cavities in an open borehole. The caliper log can be used to determine well construction details and casing diameter. It can also be used to reveal casing deterioration due to extreme corrosion or accumulation of minerals on the interior of the well casing. ASTM Standard D 6167 (ASTM, 2004j) provides guidance on the use of the mechanical caliper log in environmental site characterization.

Imaging of Borehole Conditions

Imaging of a borehole wall to characterize rock type, fractures, and voids is commonly done using a downhole video camera. Imagery to provide a detailed core-like view of the borehole from which dimensions and angles of fractures can be determined are made using an acoustic log (acoustic televiewer). This log is especially useful in holes with turbid water, where optical viewing of the borehole wall is limited. An optical televiewer can also be used to provide a high degree of detail of the borehole wall, if visibility is good. In those boreholes intersecting large cavernous zones below the water table or open water-filled mines, scanning sonar can be used to determine the size and shape of the cavity or mine.

Applications of Geophysical Methods

There is no simple, exact way to select the geophysical methods required to solve a particular problem. Tables 4.2 to 4.4 are provided to illustrate how geophysical methods may be used to carry out assessment of hydrogeologic conditions, detecting and mapping contaminants, and locating and mapping buried wastes and utilities. However, simple tables and rules of thumb often fail when considering specific project needs and site-specific conditions and, therefore, the tables presented here should only be used as an initial guide.

Assessing Hydrogeologic Conditions

The first and often the most important task of most environmental site investigations is the evaluation of natural hydrogeologic conditions. A description of overall hydrogeologic conditions and identification of any hydrogeologic anomalies is usually required. Knowledge of the natural anomalies, in relation to the overall setting, can ensure that drilling and

TABLE 4.2

Surface Geophysical Methods for Evaluation of Natural Hydrogeologic Conditions[a]

Method	General Application	Continuous Measurements	Depth of Penetration	Major Limitations
Radar	Profiling and mapping; highest resolution of any method	Yes	To 100 ft (typically less than 30 ft)	Penetration limited by soil conditions
EM (frequency domain)	Profiling and mapping; very rapid measurements	Yes (50 ft)	To 200 ft	Affected by cultural features (metal fences, pipes, buildings, vehicles)
EM (time domain)	Soundings	No	To few 1000 ft	Does not provide measurements shallower than about 150 ft
Resistivity	Soundings or profiling and mapping	No	No limit (commonly used to a few 100 ft)	Requires good ground contact and long electrode arrays. Integrates a large volume of subsurface. Affected by cultural features (metal fences, pipes, buildings, vehicles)
Seismic refraction	Profiling and mapping soil and rock	No	No limit (commonly used to a few 100 ft)	Requires considerable energy for deeper surveys. Sensitive to ground vibrations
Seismic reflection	Profiling and mapping soil and rock	No	To few 1000 ft	Shallow surveys less than 50 ft are most critical. Sensitive to ground vibrations
Microgravity	Profiling and mapping soil and rock	No	No limit (commonly used to a few 100 ft)	Slow, requires extensive data reduction. Sensitive to ground vibrations
Magnetics	Profiling and mapping soil and rock	Yes	No limit (commonly used to a few 100 ft)	Only applicable in certain rock environments. Limited by cultural ferrous metal features

[a] Applications and comments should only be used as guidelines. In some applications, an alternate method may provide better results.

sampling is done in the locations that will most likely yield information on the location or movement of a contaminant plume.

The first step in any environmental site investigation (Chapter 2) is to obtain appropriate background literature, maps, and aerial photos so that geophysical surveys and other site work can be planned. Photo imagery is almost a necessity in any serious site investigation, to assist in planning a geophysical survey and to locate the survey grid.

Table 4.2 lists possible applications of surface geophysical methods and some of their advantages and limitations in evaluating hydrogeologic conditions. Variations in the shallow natural setting are best evaluated with ground-penetrating radar, which provides

TABLE 4.3

Surface Geophysical Methods for Mapping of Contaminant Plumes[a]

Mapping permeable pathways, bedrock channels, etc.
The fundamental approach to evaluating the direction of ground-water flow and the possible extent of a contaminant plume is by determining the hydrogeologic characteristics of the site (see Table 4.2 for evaluation of natural hydrogeologic conditions)

Mapping of inorganics or mixed inorganics and organics
When inorganics are present in sufficient concentrations above background or organics are part of such an inorganic plume, they can be mapped by the electrical methods and sometimes by radar. The higher specific conductance of the pore fluids acts as a tracer by which the plume can be mapped

Mapping of hydrocarbons
When sufficient hydrocarbons have been present in the soil or floating on a shallow water table, for a sufficient period of time they may sometimes be mapped by the electrical methods or by radar. Owing to changes in dielectric constant or suppression of the capillary zone, they may sometimes be mapped by radar (in some situations where degradation of hydrocarbons is occurring, conductivity may increase). (Also see Table 4.2 for evaluation of natural hydrogeologic conditions and Table 4.4 for mapping of cultural pathways.)

Radar: Limited applications — may sometimes be used to detect shallow floaters (0 to 20 ft) to map hydrocarbons in soil. May detect thickness in some cases
EM: May be applicable to detect low conductivity at some sites or higher conductivity where biodegradation is occurring
Resistivity: May be applicable to detect high resistivity at some sites or higher conductivity where biodegradation is occurring

---

[a] Applications and comments should only be used as guidelines. In some applications, an alternate method may provide better results.

---

the greatest resolution. However, depth of penetration of the radar signal is highly site specific and is typically less than 30 ft. When silts and clays are present at the surface, penetration may be limited to only a few feet.

Even with these limitations, ground-penetrating radar can often help solve problems at a depth greater than its sensing range. For example, by looking for anomalies in shallow marker beds or by observing shallow soil piping, shallow radar data can be used to predict the presence of cavities and fractures far beyond its range. Investigation of such near surface indicators with radar and other methods to evaluate deeper conditions is a powerful technique (Benson and Yuhr, 1987).

High-resolution seismic reflection can be used in combination with radar to provide a more complete depth profile. While this method has less resolution than radar, information can be acquired to depths of hundreds of feet or more. The reflection method is often found to be ineffective at depths shallower than 25 to 50 ft, where radar is most effective. Therefore, these two methods are quite complementary for developing detailed geologic profiles. It should be noted, however, that the cost of seismic work is considerably greater than the cost for a radar survey.

Seismic refraction and resistivity soundings provide good vertical information, although they are not capable of achieving the lateral and vertical resolution of radar, or in some cases, the vertical resolution of seismic reflection. The frequency-domain EM techniques have very good lateral resolution in the continuous mode to depths of about 50 ft, but are somewhat lacking in their capability to produce vertical detail (sounding data). Yet, the EM methods can provide some relative sounding information (i.e., thick vs. thin or shallow vs. deep) very quickly and more cost effectively than resistivity or seismic refraction.

TABLE 4.4

Surface Geophysical Methods for Location and Mapping of Buried Wastes and Utilities[a]

Method	Bulk Wastes without Metals	Bulk Wastes with Metals	55 gal Drums	Pipes and Tanks
Radar	Very good if soil conditions are appropriate; sometimes effective to obtain shallow boundaries in poor soil conditions	Very good if soil conditions are appropriate; sometimes effective to obtain shallow boundaries in poor soil conditions	Good if soil conditions are appropriate (may provide depth)	Very good for metal and nonmetal if soil conditions are appropriate (may provide depth)
EM	Excellent to depths less than 20 ft	Excellent to depths less than 20 ft	Very good (single drum to 6–8 ft)	Very good for metal tanks
Resistivity	Good	Good	N/A	N/A
Seismic refraction	Fair (may provide depth)	Fair (may provide depth)	N/A	N/A
Microgravity	Fair (may provide depth)	Fair (may provide depth)	N/A	N/A
Metal detector	N/A	Very good (shallow)	Very good (shallow)	Very good (shallow)
Magnetometer	N/A	Very good (ferrous only; deeper than metal detector)	Very good (ferrous only; deeper than metal detector)	Very good (ferrous only; deeper than metal detector)

[a] Applications and comments should only be used as guidelines. In some applications, an alternate method may provide better results.

Probably, the two best techniques to map lateral variations in soil and rock, from a speed and resolution point of view, are radar and continuous EM measurements. While radar performance is highly site specific, the EM technique can be applied in almost any environment and can often provide deeper information, but with much less vertical resolution than radar. Continuous EM profiling measurements provide high lateral resolution and can be run at speeds from 1 to 5 mph, depending on the detail required. The rapid speed at which EM measurements can be obtained and the option of continuous profile measurements at depths of up to 50 ft makes EM the best choice for profile work under most situations.

The resistivity method can also be used for profile measurements by moving the electrode array in small increments to provide data at closely spaced intervals. This is a slow process relative to an EM survey, and resistivity data can be affected by near-surface geologic noise at the electrodes.

Sometimes, one method may work and another will fail under a given set of site conditions. For example, in many cases, resistivity measurements can be made adjacent to a chain link fence or a buried pipeline where EM measurements cannot.

In order for any geophysical method to work, there has to be a contrast in the parameter being measured. The best method is the one in which the parameters being measured have the greatest contrast and will be least influenced by site-specific conditions and noise. The final decision must be made on a site-by-site basis.

Once the surface methods have defined the 2D or 3D conditions reasonably well, boring locations can be selected. These locations should be selected to be representative of the normal background conditions at the site and to investigate any anomalies. Generally, if there are anomalous site conditions present, including sand lenses, fractures subtle changes in formation permeability, or geochemical anomalies, downhole logs should be run.

The drilling program should be designed to provide a means of accurately characterizing soil and rock conditions to the greatest extent possible within the budget available. If an adequate downhole logging program is used, most of the holes can be drilled without sampling. However, it is always good practice to continuously sample or core at least one or two boreholes and then log them along with any other holes. This procedure provides a reference for the logging data to compare to site-specific soil samples or rock cores. The logs can then be used to extrapolate soil and rock type and other conditions to nearby boreholes.

When the appropriate logs are combined, continuous *in situ* logging measurements can be obtained in both the vadose zone and the saturated zone to characterize hydrogeologic conditions. Geologic formations can be identified and easily correlated from hole to hole. Relative estimates of clay content, density, and porosity can be given. Permeable sand lenses and fractures can be identified, as can impermeable clay and shale zones. In addition, the continuity of impermeable zones can be assessed. The maximum amount of data should always be obtained from each borehole because borings are often few and costly.

Natural gamma logs can be used for geologic characterization and stratigraphic correlation. For example, the natural gamma logs shown in Figure 4.17 clearly show the contrast between the limestone units (low counts) and the shale units (high counts). In this case, correlation of the stratigraphy from natural gamma logs in adjacent boreholes is easily made.

Figure 4.18 shows a suite of natural gamma, density, and porosity logs from the same borehole. The density log shows variable conditions in the overlying soil, but fairly uniform density within the shale and limestone units. In contrast, the porosity log shows considerable variation throughout both the shale and limestone units. Without calibration, these logs can be used to indicate relative changes in density and porosity. By calibrating these logs, quantitative results for density and porosity may be obtained in some situations.

Detecting and Mapping Contaminant Plumes

Table 4.3 illustrates how surface geophysical methods can be applied to mapping contaminant plumes. The fundamental approach to evaluating the direction of ground-water flow and the possible extent of a contaminant plume is by determining the hydrogeologic characteristics of the site (i.e., determining the presence of preferential pathways such as buried channels, fractures, and permeable zones).

"Direct" detection of inorganics (or organic compounds mixed with inorganics) can be accomplished by electrical methods, including ground-penetrating radar and complete resistivity measurements (Olhoeft, 1986, 1992; Olhoeft and King, 1991), as shown in Table 4.3. When inorganic species are present in sufficient amounts, they can be detected by electrical methods and radar. The higher specific conductance of the pore fluids acts as a tracer by which the plume can be mapped. In cases in which inorganic plumes have a very low specific conductance, or dispersed organic compounds are encountered, they will not be detectable by electrical methods.

Where suitable penetration is possible, ground-penetrating radar can provide a means for mapping the depth to the top of and lateral extent of shallow inorganic plumes. However, because of the site-specific behavior of radar, the EM or resistivity methods are most often used. Of the two methods, EM measurements are preferred for profile work, particularly where continuous sampling can be employed. Resistivity is preferred for sounding work.

Both resistivity and EM conductivity can miss a contaminant plume if the measurements are in the wrong location. However, rapid EM profiling by either continuous or station measurements allows coverage of a site with closely spaced data. It is not unreasonable from a cost perspective to have overlapping measurements, therefore, providing total site coverage using the EM profiling method.

Both resistivity and EM are capable of vertical soundings. The frequency-domain EM method provides a depth of penetration that is limited to about 200 ft and provides less resolution than the resistivity method because measurements are made at only a few depths. The depth to which resistivity sounding data can be obtained is virtually unlimited. Depths of a few hundred feet to thousands of feet are obtainable. However, the long resistivity arrays necessary for deep measurement may not be practical in some areas due to space restrictions and cultural factors. Here, the time-domain EM transient systems, which have a smaller coil size, would be the choice for measurements to depths from 150 ft to a few thousand feet or more.

In some cases, organic compounds can be mapped because they are mixed with inorganic species. Figure 4.8 shows the inorganic plume from a chemical and drum recycling center, which also contains organic compounds. Figure 4.9 shows the inorganic plume from a landfill that contains low levels of organic compounds. The results in Figure 4.20 show an excellent comparison of EM, resistivity, and organic vapor analysis responses obtained from a mixed plume of organics and inorganics confined in a buried channel. Clearly, if inorganics are present, they can be used as an easily detectable tracer that will provide a first approximation of where the organics may be.

The delineation of hydrocarbons and light, nonaqueous phase liquids in the subsurface poses a problem for the EM and resistivity methods. EM surveys over known areas of separate-phase hydrocarbon product have shown low conductivity, high conductivity, or no detectable anomaly associated with the hydrocarbon (Monier-Williams, 1995).

Some investigators have suggested that direct detection of major hydrocarbon spills can be accomplished by looking for low EM conductivities (or high resistivities) associated with the organics. Recent spills of petroleum products do not seem to yield a high resistivity or a low EM conductivity, because the product does not displace the grain-to-grain surface tension of water. Therefore, the electrical conductivity of the water remains dominant. However, if the organics have been in the ground for some time, and there is a substantial amount of separate-phase product, the surface tension of the water may be overcome. Then, where the conductivity of the natural soil conditions is high enough (or the resistivity low enough), a reasonable electrical contrast between the hydrocarbons and the natural condition may produce an anomaly.

Recent research has shown that the biodegradation of hydrocarbons will produce acids and biosurfactants (Atekwana et al., 2000; Cassidy et al., 2001). These products will increase the total dissolved solids content of ground water over time and thereby increase the electrical conductivity. For example, laboratory experiments have shown that diesel fuel will produce an increase in total dissolved solids of over 1700 mg/l in 120 days when biodegradation is taking place (Cassidy et al., 2001). If significant biodegradation occurs within a subsurface hydrocarbon plume, it may produce a detectable high-conductivity anomaly in EM measurements.

Both electrical measurements (EM and resistivity) along with radar can be used to monitor changes with time and during remediation (Olhoeft, 1986, 1992; Olhoeft and King, 1991). Once the spatial extent of a contaminant has been mapped by surface geophysics and after boreholes have been installed, continuous downhole logging can be used to evaluate changes in the vertical hydraulic conductivity of soil and rock, as well as the distribution of contaminants. The vertical distribution and concentration of contaminants at a site can vary significantly as a result of small local changes in hydraulic conductivity. Because hydraulic conductivity can change by more than an order of magnitude in less than a foot, it can have a significant impact on test results obtained from a

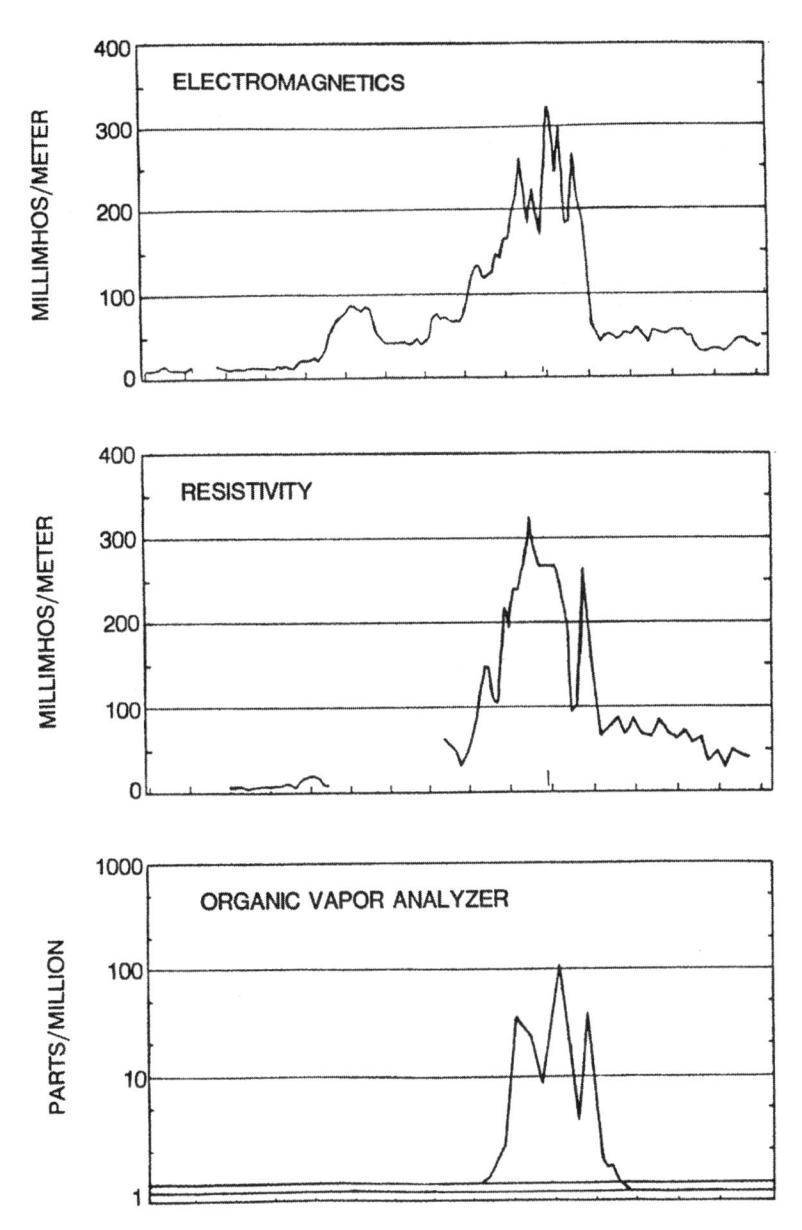

FIGURE 4.20
Organic vapor profile over a buried channel. Note correlation with resistivity and EM measurements (this is an excellent example of a buried channel controlling flow and the level of correlation between organic and inorganic contaminants).

monitoring well. The chemical concentration in a well may be low, average, or high, depending upon screen length and location. Two downhole logging techniques, particularly well suited for hydraulic conductivity evaluations are the EM induction log (or resistivity logs — only in an open borehole) and neutron (porosity) log. Both of these logs can be run in an open borehole or within an existing PVC-cased well, either above or below the water table (Figure 4.19).

The EM induction log, shown in Figure 4.19a, indicates the presence of inorganic contaminants that have preferentially migrated within five discrete zones of increased hydraulic conductivity in the limestone. These zones, which are indicated by higher electrical conductivity values, ranging from 2 to 4 ft in thickness. The presence of high hydraulic conductivity zones detected by the EM induction method was confirmed by using a downhole television camera that visually located small cavities and fractures in each zone. Figure 4.19b shows an EM induction log of natural conditions taken in a background well. No permeable zones are indicated by the log because there are no inorganic species present. Figure 4.19c shows a neutron log taken in the same background well. In this log, zones of variable porosity are revealed whether contaminants are present. Conditions shown by this log are also representative of conditions at the contaminated well.

An adequate assessment of conditions in a borehole often requires that more than one log be run. At this site, an EM induction log was used to identify the contaminated zones and a neutron log was used to identify zones of increased porosity. Once conditions at a site are understood, a reliable and representative monitoring well system can be designed and data from existing monitoring wells can be more accurately evaluated.

Locating and Mapping Buried Wastes and Utilities

Locating and mapping of buried wastes, utilities, drums, and tanks are a common application of geophysical methods. Table 4.4 lists the surface geophysical methods applicable to this problem. Locating buried bulk wastes where no metal is present can often be accomplished by ground-penetrating radar, if soil conditions are suitable. Often the shallow edges of trenches can be detected even in soil conditions that provide poor radar penetration. Shallow EM tools are also effective for most location problems. When metals are present, EM conductivity, metal detectors, and magnetometers are the primary choices. Metal detectors and magnetometers are unaffected by most soil types or by the presence of contaminants. However, EM measurements are influenced by both variations in soil and the presence of contaminants.

To locate buried 55 gal steel drums, the use of metal detectors, magnetometers, or the in-phase component of EM measurements are recommended. All three methods can be used to locate single 55 gal drums, as well as large piles of drums within their depth limitations.

Both the metal detector and the EM will respond to ferrous and nonferrous metals, while a magnetometer will respond only to ferrous metals. Therefore, it is necessary to assess what metals may be present to select the appropriate method.

While radar can be used to find drums, it will often be unable to detect a single drum if it is not oriented so that energy is reflected back to the antenna. Furthermore, many natural and man-made objects may have a radar response similar to that of a drum.

For small, discrete, critical targets such as a single 55 gal drum, continuous measurements (on closely spaced lines of about 5 ft) are required to assure detection. Radar, EM equipment, metal detectors, and certain magnetometers can provide these continuous measurements. However, there may be cases in which the proximity of other metal structures may limit the use of EM to locate drums or trenches, making radar a clear choice.

Metal detectors and radar both provide reasonably good spatial resolution to pinpoint the location of a target. However, EM equipment and magnetometers do not provide the same target resolution because the shape of their response curve is broader and often more complex.

Metal detectors, EM units, and total field magnetometers are highly susceptible to interference from nearby metallic cultural features. Any of these features can produce an erroneous response, which may be incorrectly interpreted as a subsurface target. Because metal detectors are relatively short-range devices, they can be operated closer to such sources of interference than can most magnetometers. Measurements made with a total field proton procession magnetometer are susceptible to interference from high magnetic gradients, natural changes in the earth's magnetic field, and nearby power lines, whereas fluxgate gradiometers do not suffer from these shortcomings.

Seismic, resistivity, magnetic, and gravity techniques may also be used to locate boundaries of larger trenches and landfills. These techniques are much slower and will provide less resolution than the previously described methods. However, they are often the only techniques that can be used to estimate the thickness of a landfill or trench. It should be noted that interpretation of such data should be done with caution by experienced personnel.

Summary

All of the geophysical methods discussed in this chapter are scientifically sound, and all have been proven in the field. Like other technologies, however, they may fail to provide the desired results when applied to the wrong problem or when improperly used. The techniques must be matched to site-specific conditions by a person who thoroughly understands the uses and limitations of the methods.

To improve the accuracy of environmental site characterization, adopting a broader, integrated systems approach is recommended. Geophysics is just one of many technologies that can be readily incorporated into an environmental site investigation program. An integrated systems approach provides the benefits of both direct sampling and remote sensing techniques. Airborne or surface geophysical methods are generally used as initial reconnaissance tools to cover an area in a quick search for anomalous conditions. Surface geophysical methods can then be employed for a detailed assessment of site conditions. After potential problem areas have been identified, the drilling locations for borings and monitoring wells can be selected with a higher degree of confidence to provide representative samples. Analyses of soil and water samples from properly located borings or monitoring wells will then provide the necessary quantitative measurements of subsurface parameters. Downhole geophysical methods can be applied to define details of conditions with depth. This approach delivers greater confidence in the final data interpretation with fewer borings and wells and an overall cost savings. Furthermore, the drilling operations are no longer being used for hit-or-miss reconnaissance, but rather as specific quantitative tools (smart holes).

Before selecting a method or methods, the project objectives must be clearly defined and as much as possible should be learned about site conditions. Information such as accessibility and site topography should be available. In addition, general soil and rock types and conditions, the approximate depth to water table, depth to rock, and background specific conductance of ground water should be known or estimated. If appropriate, the type of contaminant should also be defined. Finally, one should consider whether it is likely that sufficient contrast may exist in the parameters being measured. If, in fact, there is no contrast in the parameter being measured between one layer and another, the geophysical method will fail to provide a response. Similarly, if a layer is sufficiently thin or the size of the target is sufficiently small, the layer or target may not be detected.

The question of whether drilling or geophysics should be done first often arises. Because the results of geophysical work usually result in identification of anomalous conditions, geophysics should generally be done first so that anomalous areas can be identified for drilling and sampling.

However, if borings or monitoring wells have already been installed, geophysical surveys can still deliver increased accuracy. The location and data from existing boreholes and monitoring wells can be assessed using geophysical methods, thus providing a means of evaluating the validity of data already acquired. If additional boreholes or wells are needed to fill gaps in the data, they can be located with a high degree of confidence.

Because each geophysical technique measures a different parameter, the information from one method is often complemented by the information from another method. The synergistic use of multiple geophysical techniques often serves to enhance environmental site characterization. Those familiar with traditional well logging will recognize this concept, as multiple logs are commonly used to aid in interpretation of subsurface conditions.

It should be noted that the use of any geophysical technique depends on its specific application and on site conditions. Therefore, no single method should be expected to solve all site evaluation problems. Furthermore, geophysical technology is not in itself a panacea. Its successful application is dependent upon integrating the geophysical data with other sources of information. This must be done by persons with training and experience in geophysical methodology, as well as in the broader aspects of the earth sciences. Geophysical methods do not offer a substitute for borings and wells, but provide a means to minimize or optimize the number of boreholes and wells, to ensure that they are in reasonably representative locations, and to fill in the gaps between boreholes.

Selection of Geophysical Methods

ASTM has prepared a Standard Guide for Selecting Surface Geophysical Methods (ASTM D 6429; ASTM, 2004k) and a Standard Guide for Planning and Conducting Borehole Geophysical Logging (ASTM D 5753; ASTM, 2004l). Many specific Standard Guides for both surface and borehole geophysical methods are complete and others are in the process of being completed.

The expedited site characterization (ESC) process developed by DOE (Chapter 2) is summarized in the Standard Practice for Expedited Site Characterization of Vadose Zone and Ground-Water Contamination at Hazardous Waste Contaminated Sites (ASTM D 6235; ASTM, 2004m), which provides a generalized strategy for effective environmental site characterization. The process emphasizes the use of remote sensing and geophysical methods prior to the installation and sampling of boreholes. This strategy lowers the number of randomly placed boreholes and wells and provides information to position boreholes and wells in representative locations to significantly improve the accuracy of the site characterization process. Additional strategies for site characterization are summarized in Chapter 2 and in Benson (2001).

Most critical to the success of an environmental site characterization program are the senior experienced hands-on professionals who are sensitive to the issues of geologic uncertainty and who posses the skill, wisdom, and persistence to pursue them.

References

ASTM, Standard Guide for Using the Surface Ground-Penetrating Radar Method for Subsurface Investigations, ASTM Standard D 6432, ASTM International, West Conshohocken, PA, 2004a.
ASTM, Standard Guide for the Use of the Time-Domain Electromagnetic Method for Subsurface Investigations, ASTM Standard D 6820, ASTM International, West Conshohocken, PA, 2004b.
ASTM, Standard Guide for the Use of the Frequency-Domain Electromagnetic Method for Subsurface Investigations, ASTM Standard D 6639, ASTM International, West Conshohocken, PA, 2004c.

ASTM, Standard Guide for Using the Direct Current Resistivity Method for Subsurface Investigations, ASTM Standard D 6431, ASTM International, West Conshohocken, PA, 2004d.
ASTM, Standard Guide for the Using the Seismic Refraction Method for Subsurface Investigations, ASTM Standard D 5777, ASTM International, West Conshohocken, PA, 2004e.
ASTM, Standard Guide for Using the Gravity Method for Subsurface Investigations, ASTM Standard D 6430, ASTM International, West Conshohocken, PA, 2004f.
ASTM, Standard Guide for Conducting Borehole Geophysical Logging — Gamma, ASTM Standard D 6274, ASTM International, West Conshohocken, PA, 2004g.
ASTM, Standard Guide for Conducting Borehole Geophysical Logging — Neutron; ASTM Standard D 6727, ASTM International, West Conshohocken, PA, 2004h.
ASTM, Standard Guide for Conducting Borehole Geophysical Logging — Electromagnetic Induction, ASTM Standard D 6726, ASTM International, West Conshohocken, PA, 2004i.
ASTM, Standard Guide for Conducting Borehole Geophysical Logging — Mechanical Caliper, ASTM Standard D 6167, ASTM International, West Conshohocken, PA, 2004j.
ASTM, Standard Guide for Selecting Surface Geophysical Methods, ASTM Standard D 6429, ASTM International, West Conshohocken, PA, 2004k.
ASTM, Standard Guide for Planning and Conducting Borehole Geophysical Logging, ASTM Standard D 5753, ASTM International, West Conshohocken, PA, 2004l.
ASTM, Standard Practice for Expedited Site Characterization of Vadose Zone and Ground Water Contamination at Hazardous Waste Contaminated Sites, ASTM Standard D 6235, ASTM International, West Conshohocken, PA, 2004m.
Atekwanna, E.A., W.A. Sauck, and D.D. Werkema, Investigations of geoelectric signatures at a hydrocarbon contaminated site, *Journal of Applied Geophysics*, 44, 167–180, 2000.
Benson, R.C., Strategies of Site Characterization and Risk Management, Invited Presentation to the National Research Council, Committee on Coal Waste Impoundment, Failures Subcommittee on Impoundment Site Characterization, Washington DC, June 11, 2001.
Benson, R.C. and R.A. Glaccum, Radar surveys for geotechnical site assessment; in *Geophysical Methods in Geotechnical Engineering*, Specialty Session, American Society of Civil Engineers, Atlanta, GA, 1979, pp. 161–178.
Benson, R.C. and L.J. LaFountain, Evaluation of Subsidence or Collapse Potential Due to Subsurface Cavities, in *Sinkholes, Their Geology, Engineering and Environmental Impact, Proceedings of the First Multidisciplinary Conference on Sinkholes*, Orlando, FL, 1984, pp. 201–215.
Benson, R.C. and D.C. Pasley, Ground water monitoring: a practical approach for a major utility company, *Proceedings of the Fourth National Symposium and Exposition on Aquifer Restoration and Ground Water Monitoring*, National Water Well Association, Worthington, OH, pp. 270–278, 1984.
Benson, R.C. and J. Scaife, Assessment of Flow in Fractured Rock and Karst Environments, *Proceedings of the Second Multidisciplinary Conference on Sinkholes and the Environmental Impacts of Karst*, Orlando, FL, 1987.
Benson, R.C. and L. Yuhr, Assessment and Long-Term Monitoring of Localized Subsidence Using Ground Penetrating Radar, *Proceedings of the Second Multidisciplinary Conference on Sinkholes and the Environmental Impacts of Karst*, Orlando, FL, 1987.
Benson, R.C., R.A. Glaccum and M.R. Noel, Geophysical Techniques for Sensing Buried Wastes and Waste Migration, EPA/600/7-84/064, U.S. Environmental Protection Agency, Environmental Monitoring Systems Laboratory, Las Vegas, NV, 236 pp, 1982.
Benson, R.C., M.S. Turner, W.D. Vogelsong and P.P. Turner, Correlation between field geophysical measurements and laboratory water sample analysis, *Proceedings of the Conference on Surface and Borehole Geophysical Methods in Ground Water Investigations*, National Water Well Association, Worthington, OH, pp. 178–197, 1985.
Benson, R.C., M.S. Turner, P.P. Turner and W.D. Vogelsong, In situ, time-series measurements for long-term ground-water monitoring, in *Ground Water Contamination: Field Methods*, ASTM STP 963, Collins, A.G. and Johnson, A.I. Eds., American Society for Testing and Materials, Philadelphia, PA, 1988, pp. 58–72.
Breiner, S., *Applications Manual for Portable Magnetometers*, Geometrics Inc., San Jose, CA, 58 pp, 1973.
Cartwright, K. and M. McComas, Geophysical surveys in the vicinity of sanitary landfills in northeastern illinois, *Ground Water*, 6(5), 23–30, 1968.

Cassidy, D. P., D. D. Werkema, W. Sauck, E. Atekwana, S. Rossbach and J. Duris, The effects of LNAPL biodegredation products on electrical conductivity measurements, *Journal of Environmental and Engineering Geophysics*, 6(1), 47–53, 2001.

Crice, D., 2001, Borehole Shear-Wave Surveys for Engineering Site Investigations, Geostuff, Saratoga, CA. Available at www.geostuff.com

Dunbar, D., H. Tuchfeld, R. Siegel and R. Sterbentz, Ground-water quality anomalies encountered during well construction, sampling and analysis in the environs of a hazardous waste management facility, *Ground-Water Monitoring Review*, 5(3), 70–74, 1985.

Geometrics, Inc., *Ohm-Mapper TR1 Operations Manual*, Geometrics, Inc., San Jose, CA, 116 pp, 1999.

Greenhouse, J. P. and M. Monier-Williams, Geophysical monitoring of ground-water contamination around waste disposal sites, *Ground-Water Monitoring Review*, 5(4), 63–69, 1985.

Griffith, D. H. and R. F. King, *Applied Geophysics for Engineers and Geologists*, Pergamon Press, London, U.K, 1969.

Haeni, P., Application of Seismic-Refraction Techniques to Hydrologic Studies, Open File Report No. 84–746, U.S. Geological Survey, Hartford, CT, 144 pp, 1986.

Hileman, B., Water quality uncertainties, *Environmental Science and Technology*, 18(4), 124–126, 1984.

Hunter, J. A., R. A. Burns, R. L. Good, H. A. MacAulay and R. M. Cagne, Optimum field techniques for bedrock reflection mapping with the multichannel engineering seismograph, in *Current Research, Part B*, Geological Survey of Canada, Paper 82-1B, 1982, pp. 125–129.

Keys, W. S. and L. M. MacCary, Application of borehole geophysics to water-resources investigations, *Techniques of Water-Resources Investigations of the United States Geological Survey*, U.S. Geological Survey, Reston, VA, Chap. E1, 1976.

Lankston, R. W. and M. M. Lankston, *An Introduction to the Utilization of the Shallow or Engineering Seismic Reflection Method*, Geo-Compu-Graph, Inc., Denver, CO, 1983.

Loke, M. H., RES2DINV ver. 2.0, *Rapid 2-D Resistivity Inversion Using the Least-Squares Method*, Advanced Geosciences Inc, Austin, TX, 1996.

Lysyj, I., Indicator methods for post-closure monitoring of ground waters, *Proceedings of the National Conference on Management of Uncontrolled Hazardous Waste Sites*, Hazardous Materials Control Research Institute, Silver Spring, MD, pp. 446–448, 1983.

McNeill, J. D., Electromagnetic resistivity mapping of contaminant plumes, *Proceedings of the National Conference on Management of Uncontrolled Hazardous Waste Sites*, Hazardous Materials Control Research Institute, Silver Spring, MD, pp. 1–6, 1982.

Miller, R. D., J. Xia, C. B. Park and J. Ivanov, Shear wave velocity field to detect anomalies under asphalt, in *52nd Annual Highway Geology Symposium*, Maryland State Highway Administration and the Maryland Geological Society, Cumberland, MD, 2001, pp. 40–49.

Monier-Williams, M., Properties of light non-aqueous phase liquids and detection using commonly applied shallow sensing geophysical techniques, *Proceedings of the Symposium on the Application of Geophysics to Engineering and Environmental Problems*, Society of Engineering and Environmental Geophysics, Denver, CO, pp. 1–13, 1995.

Mooney, H. M., Engineering seismology, in *Handbook of Engineering Geophysics*, Vol. 1, Bison Instruments, Minneapolis, MN, 1973.

Mooney, H. M., Shear (S) waves in engineering seismology, in *Handbook of Engineering Geophysics*, Vol. 2, Bison Instruments, Minneapolis, MN, 1977.

Mooney, H. M., Electrical resistivity, in *Handbook of Engineering Geophysics*, Vol. 3, Bison Instruments, Minneapolis, MN, 1980.

Nazarian, S., K. H. Stokoe II and W. R. Hudson, Use of Spectral Analysis of Surface Waves Method for Determination of Moduli and Thickness of Pavement Systems, Transportation Research Record No. 930, pp. 38–45, 1983.

Olhoeft, G. R., Direct detection of hydrocarbons and organic chemicals with ground penetrating radar and complex resistivity, *Proceedings of the NWWA/API Conference on Petroleum Hydrocarbons and Organic Chemicals in Ground Water — Prevention, Detection and Restoration*, National Water Well Association, Dublin, OH, pp. 284–305, 1986.

Olhoeft, G. R., Geophysical detection of hydrocarbon and organic chemical contamination, *Proceedings of the Symposium on the Application of Geophysics to Engineering and Environmental*

Problems, Bell, R.S., Ed., Society of Engineering and Environmental Geophysics, Denver, CO, 1992, pp. 587–595.

Olhoeft, G. R. and T. V. V. King, Mapping subsurface organic compounds non-invasively by their reactions with clays, in *U.S. Geological Survey Toxic Substances Hydrology Program, Proceedings of the Technical Meeting*, Monterrey, California, March 11–15, 1991, U.S. Geological Survey Water Resources Investigations Publication 91-4034, pp. 552–557, 1991.

Orellana, E. and H. M. Mooney, *Master Tables and Curves for Vertical Electrical Sounding Over Layered Structures*, Interciencia, Madrid, Spain, 1966.

Palmer, D., *The Generalized Reciprocal Method of Seismic Refraction Interpretation*, Burke, K.B.S., Ed., Department of Geology, University of New Brunswick, Fredericton, NB, Canada, 104 pp, 1980.

Perazzo, J. A., R. C. Dorrler and J. P. Mack, Long-term confidence in ground water monitoring systems, *Ground-Water Monitoring Review*, 4(4), 119–123, 1984.

Redpath, B. B., Seismic Refraction Exploration for Engineering Site Investigations, U.S. Army Engineer Waterways Experiment Station, Explosive Excavation Research Laboratory Technical Report E-73-4, U.S. Army Corps of Engineers, Waterways Experiment Station, Livermore, CA, 52 pp, 1973.

Steeples, D. W., High-resolution seismic reflection at 200 Hz, *Oil and Gas Journal*, December 1984, pp. 86–92, 1984.

Telford, W. M., L. P. Geldart, R. E. Sheriff and D. A. Keys, *Applied Geophysics*, Cambridge University Press, Oxford, U.K., 1982.

U.S. Environmental Protection Agency, RCRA Ground-Water Monitoring Technical Enforcement Guidance Document, OSWER-9950.1, U.S. Environmental Protection Agency, Office of Waste Programs Enforcement, Washington, DC, 1986.

Walker, S. E. and D. C. Allen, Background ground water quality monitoring: temporal variations, *Proceedings, Fourth National Symposium and Exposition on Aquifer Restoration and Ground Water Monitoring*, National Water Well Association, Worthington, OH, pp. 226–231, 1984.

Won, I. J., D. A. Keiswetter, G. R. A. Fields and L. C. Sutton, GEM-2: a new multi-frequency electromagnetic sensor, *Journal of Engineering and Environmental Geophysics*, 1(2), 129–137, 1996.

Zohdy, A. A., G. P. Eaton and D. R. Mabey, Application of surface geophysics to ground-water investigations, *Techniques of Water-Resources Investigations of the United States Geological Survey*, U. S. Geological Survey, Reston, VA, 1974, Chap. D1, 116 pp.

5

Environmental Drilling for Soil Sampling, Rock Coring, Borehole Logging, and Monitoring Well Installation

Tom Ruda and Jeffrey Farrar

CONTENTS

Introduction	298
Drilling Methods: Drilling without Circulation Fluids	299
Probing	299
Direct-Push Displacement Boring	300
Auger Drilling	300
Bucket Auger	301
Continuous-Flight Solid-Stem Augers	301
Hollow-Stem Augers	302
Drilling Methods: Drilling with Circulation Fluids	305
Wash Boring	306
Rotary Drilling	306
Reverse Circulation Rotary Drilling	308
Dual-Tube Reverse Circulation Drilling	308
Percussion Drilling	309
Dual Rotary Drilling	311
Sonic Drilling	312
Selection of Drilling Methods	313
Introduction	313
Health and Safety	314
Access and Noise	315
Disposal of Fluids and Cuttings	315
Lithology and Aquifer Characteristics	316
Depth of Drilling	317
Sample Type	317
Cost	318
Soil Sampling	318
Types of Samples	318
Bulk Samples	319
Representative Samples	319
Undisturbed Samples	319
Composite Samples	319
Types of Soil Samplers	320
Solid-Barrel Samplers	320
Split-Barrel Samplers	320

      Thin-Wall Tube Samplers 321
      Continuous Tube Samplers 321
      Rotary Samplers 323
      Piston Samplers 323
   Methods of Sampling 324
      Driving 326
      Pushing 326
      Rotation while Pushing 326
Rock Coring 326
   Introduction 326
   Core Losses 329
   Rock Coring Logs 329
Handling Procedures for Soil and Rock Samples 331
   Introduction 331
   Rock Core Samples 332
   Cuttings or Disturbed Samples 333
Borehole Logging 333
   Introduction 333
   Log Heading Information 334
   Log Completion Information 334
   Sample Information 334
   Soil and Rock Descriptions 335
   Drilling Information 336
Drilling Contracts 338
   Introduction 338
   Agreement 339
   General Conditions 339
   Construction Drawings 340
   Specifications 341
   Special Conditions 341
   Other Documents 342
Conclusions 342
References 343

Introduction

Drilling and soil sampling for environmental site characterization and ground-water monitoring well installation utilizes much of the same technology used in geotechnical exploration, mineral exploration, oil and gas well drilling, and water well drilling. However, there are some very significant differences in how the technology is applied. For example, the primary purpose of most geotechnical exploration projects is to recover an intact physical specimen that can be tested for physical strength or inspected for material properties that may be indicative of the performance of the sampled material under projected conditions. For environmental site characterization, primary consideration must be given for collecting a sample that is representative of *in situ* physical conditions and valid for both chemical and physical analyses. The sample must not be

Environmental Drilling for Soil Sampling

contaminated by drilling fluid or its physical properties altered by the drilling or sampling procedures. Care must be taken to preserve the sample in its natural state for on-site analysis or for transport to the laboratory.

The method of drilling and sampling chosen for a given project depends upon many site-specific factors including site geology, the type of contaminants expected, the shape and size of the specimen desired, and the final disposition of the borehole. This chapter describes a variety of methods available for drilling and includes comments on the suitability of those methods for environmental site characterization and ground-water monitoring applications. The drilling methods discussed are grouped into two general categories: (1) methods that do not use a circulation medium to transport drill cuttings to the surface and (2) methods that do use a circulation medium.

Drilling Methods: Drilling without Circulation Fluids

Probing

Probing can be done with a tool as simple as a slender steel rod, 0.25 to 0.5 in. in diameter and 3 to 4 ft long, having a tee handle (Figure 5.1). This type of tool is often used to probe into the soil by hand to locate and outline shallow subsurface obstructions (e.g., boulders, utility conduits, piping, subsurface structures, survey markers) in advance of powered drilling. Resistance to penetration indicates the presence of an obstruction. Probes can also be used to profile bedrock surfaces and to establish various soil or formation interfaces, if the difference in density between penetrated formations, which affects penetration resistance, is recognizable.

When a probe is advanced or pushed into the ground, it forces the formation material out of its path by displacing the soil. Thus, a probe is a simple form of displacement boring.

FIGURE 5.1
Soil probe.

Direct-Push Displacement Boring

The word "boring" in this context may be a misnomer, but through common usage it has come to include a borehole made by sampling and removing soil material or displacing soil material out of the path of direct-push tooling. Any displaced material not removed by sampling is simply compacted and forced into the formation. Direct-push methods for sampling and well installation are presented in detail in Chapter 6.

Direct-push boring is a simple, efficient method of obtaining samples and installing wells. It can be accomplished without the need for heavy equipment or circulation fluids and without producing drill cuttings. The depth of boring and the sample size recovered are dependent on the resistance of the formation to penetration. Direct-push methods are of questionable value in dense clays or sands and in thick gravels and cobbles or bouldery formations and cannot be used to penetrate competent, unweathered bedrock. Direct-push methods are well suited for shallow borings in soft materials and where the boring location is not accessible to large, heavy equipment.

Direct-push machines are generally small units, either truck- or ATV-mounted. Direct-push tooling is forced into the soil by the application of the direct weight of the machine, by the percussive effect of a hydraulic or mechanical hammer, or by both. Most machines are equipped with a hydraulic hammer, have a down-force and retraction system, and may employ a limited (low torque) rotation capability. Some direct-push machines use a vibratory system to advance the tooling. Direct-push units can be either direct-push specific or conventional drilling rigs, which have been fitted with hydraulic or mechanical hammers or vibratory heads.

Auger Drilling

Auger drilling utilizes a spiral tool form to convey drilled borehole material to the surface (Figure 5.2). Mechanically, an auger consists of a long inclined plane with a fixed mechanical advantage. The drilling and conveying capacity of a specific auger is directly proportional to the torque applied to rotate the auger. Auger drilling does not normally require the use of circulation fluids, although fluids can be used to cope with blowing, heaving, or running sands.

An auger is essentially a conveyer that has a cutting bit at its bottom end to disaggregate formation material. While drill cuttings are generally lifted upward by the auger, some can

(a) Disc auger.

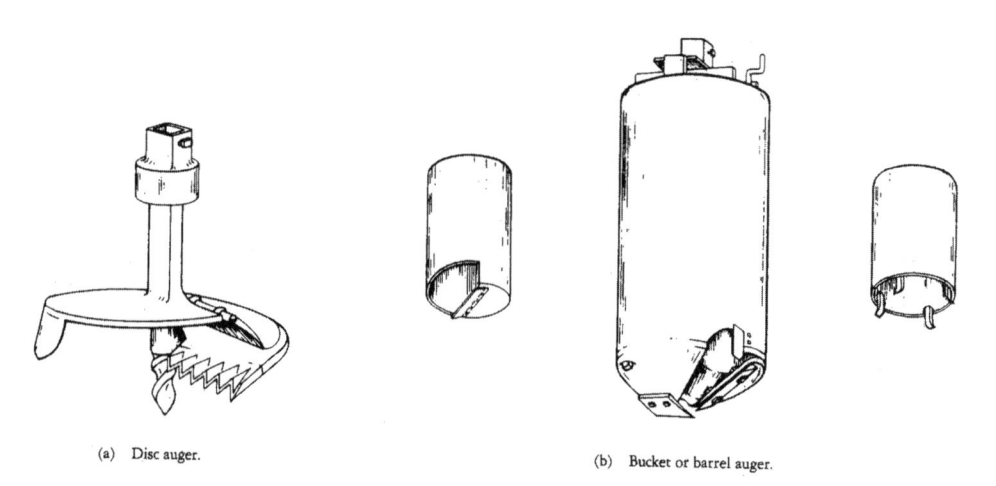

(b) Bucket or barrel auger.

FIGURE 5.2
Mechanical augers: (a) Disc auger; (b) Bucket auger.

Environmental Drilling for Soil Sampling

also be forced against or into the borehole wall during conveyance to the surface. Three basic types of auger are in use in environmental work: bucket auger, solid-stem continuous flight auger, and hollow-stem continuous flight auger.

Bucket Auger

Bucket auger machines (Figure 5.2b) utilize an auger bucket with cutting teeth attached to a square torque bar (known as a Kelly bar) that passes through a ring-type drive mechanism. Generally, the auger bucket advances into the formation by a combination of dead weight and tooth cutting angle. After the bucket is advanced 1 or 2 ft, it is withdrawn from the hole by means of a wire-line hoist cable attached to the top of the Kelly bar. When the bucket reaches the surface, it is swung to the side of the hole and the drill cuttings are dumped out through the bottom by means of a hinge-and-latch device on the bucket bottom.

The Kelly bar generally telescopes to permit digging to greater depths. The solid bar is nested within one or more square tubes. Most bucket auger machines have a depth capacity of 30 to 75 ft and most are used for large-diameter holes ranging from about 16 to 48 in. Bucket augers smaller than 16 in. in diameter are rare. Most bucket auger machines are gravity fed and are used for vertical holes. They are not normally used to drill holes for single monitoring well installations, but are sometimes used to drill holes for well nests as well as production wells and recovery wells. They are more commonly used to drill drain wells, caissons, and building footings.

Continuous-Flight Solid-Stem Augers

Continuous-flight solid-stem augers (Figure 5.3) consist of a plugged tubular steel center shaft, around which a continuous steel strip in the form of a helix is welded. An individual auger section is known as an auger "flight," and is normally 5 ft long, although other

FIGURE 5.3
Continuous-flight solid-stem auger.

lengths are available. Manufacture of flights is such that when connected to one another, the helix is continuous across the connections and throughout the depth of the borehole. Connections are normally made by means of a hex or square pin fitted to one end of a flight, which slips into a corresponding hex or square box fitted to the other end of a flight. Torque is transmitted from the drilling rig to and through the flights by the hex or square connections on each flight. Down-force is transmitted by shoulder-to-shoulder contact of the flights. Retract force is taken by a pin inserted through a hole that has been drilled through a flat face of the hex or square connections at a 90° angle to the axis of the flight. This pin is known as a "u-pin" or "drive clip."

Auger drill cutter heads are attached to the bottom auger flight in the same manner as flights are connected to each other. Most cutter heads are of the field-replaceable bit type, where a hardened or tungsten carbide steel inserted bit does the cutting. Carbide insert teeth come in different configurations including carbide-tipped fingers, conical (round) points, spade faces, or flat blades. Other types of cutter heads include the fishtail or clay bit or the one-piece carbide-tipped finger drill head.

Auger drill cutter heads are generally designed to cut a hole approximately 0.5 in. greater in diameter than the diameter of the auger to which they are attached. For example, the cutter head designed for use with 4 in. augers actually measures 4.5 in. in diameter when new. The auger that actually has a 5.5 in. diameter is known as a 6 in. auger, as the cutter head design for it is 6.0 in. in diameter. Thus, a conventional continuous-flight auger is known by the nominal diameter of the drill head.

In addition to diameter, augers are specified by the pitch of the auger and the shape and dimension of the connections. The pitch is the distance along the axis of the auger that it takes for the helix to make one complete 360° turn. The pitch of an auger used for vertical drilling will generally be 65 to 85% of the hole diameter. This gentle pitch allows easy conveyance of auger cuttings up the borehole.

Continuous solid-stem auger drilling is most successful in dry formations or cohesive materials, where the hole stays open when the auger is removed. This method is not frequently used for well installation because the borehole normally collapses below the water table, particularly in noncohesive materials.

Hollow-Stem Augers

Hollow-stem augers (Figure 5.4) are a form of continuous-flight auger in which the helix is wound around, and welded to, a hollow center tube. When flights are connected, the hollow-stem auger will present a smooth, uniform bore throughout its length and the flighting will be continuous from the top to the bottom of the hole. The hollow-stem opening is very useful because it allows for use of a sampling barrel inside and provides a protected opening for installation of monitoring wells. Hollow-stem augering can be done without drilling fluids and is the most common drilling method used for installation of shallow monitoring wells (see ASTM Standard D 5784; ASTM, 2004a). Hollow-stem auger systems can be equipped with a continuous sampler, which can take disturbed samples in a split inner barrel or the inner barrel can be equipped with acrylic liners for relatively undisturbed sampling (see ASTM Standard D 6151; ASTM, 2004b).

Connection of one auger to another is by means of a series of keys or keyways, hex-shaped box and pin, square flat-spline box and pin, or a threaded connection. Threaded connectors transmit both torque and push or retract force at the thread. In the other types of connections, torque is transmitted by the spline or key or keyway; push force is transmitted by shoulder-to-shoulder contact and retract force is carried by the connecting bolts. Hollow-stem augers are specified by the inside diameter of the hollow stem and not by the borehole diameter drilled.

Environmental Drilling for Soil Sampling

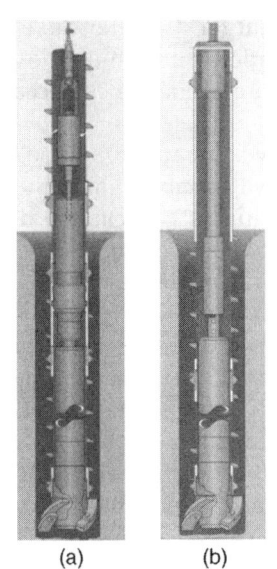

FIGURE 5.4
Hollow-stem augers: (a) wire-line type; (b) rod type. (Taken from U.S. Bureau of Reclamation Earth Manual, Part I, 3rd ed.)

The hollow-stem auger is a conveyer, but when compared with regular continuous-flight augers, the center tube is much larger. The pitch of the flighting generally follows the same formula as for solid-stem augers. The change in proportion, between axle diameter and hole size, results in a considerable change in the conveyance characteristics of the auger. Proper movement of drill cuttings to the surface must be accommodated by a change in drilling technique on the part of the driller. Essentially, this is done by more rotation, not more revolutions per minute. Some formations must be augered at very low revolutions per minute or they do not auger at all. Heavy formations, such as adobe or "fat" clays, should be auger-drilled at 30 to 50 r/min. Clean sand that will stand open during drilling can be successfully augered at 250 r/min. Care must be taken when drilling in caving noncohesive sands below the water table, because excessive rotation can remove a large amount of material from the formation and create slumped zones or voids.

Hollow-stem auger drill heads generally consist of two pieces: an outer head with cutting teeth, attached to the bottom of the lead auger, and either an inner pilot assembly with center bit that is removable through the center of the auger or an inner sampling barrel for continuous sampling. The ability to withdraw the pilot bit assembly or sampler while leaving the auger in place is the principal advantage of using hollow-stem augers. This provides an open and cased hole into which samplers, downhole hammers, instrumentation, monitoring well casing, or other items can be inserted. Replacing the pilot assembly, assuming nothing was left in the hole, allows drilling of the borehole to continue. As shown in Figure 5.4, the pilot assembly is normally held in place and retrieved from inside the auger by drill rods or hex rods. The pilot assembly can also be operated by wire-line in certain subsurface conditions.

Hollow-stem augers are available with inside diameters of 2.25, 2.5, 3.25, 3.75, 4.0, 4.25, 6.25, 6.625, 8.25, 9.25, 10.25, and 12.25 in. The most commonly used sizes for geotechnical work are 2.25 and 3.25 in. I.D. The 4.25 in. I.D. auger is the most common size for installing 2 in. nominal diameter monitoring wells. The 6.25 in. I.D. auger works well for installing 2 and 4 in. monitoring wells. Larger diameter sizes are also preferred when undisturbed samples are to be taken. The larger sizes 8.25, 9.25, 10.25, and 12.25 in. I.D. augers are generally used for installation of larger-diameter dual-purpose recovery and monitoring wells in which the use of circulation fluids must be avoided. The initial cost of these larger auger sizes and the high torque machines necessary to operate them is relatively high.

The successful use of the hollow-stem auger method depends greatly on the skill of the operator and the depth to the water table. Once the water table is encountered, the dynamics of using hollow-stem augers can change dramatically. When drilling below the water table, the pressure inside the hollow stem auger must equal or exceed the pressure of the ground water, to keep the ground water from moving into the center stem of the auger, and bringing with it formation materials. This problem is an issue more in granular, noncohesive formations than in cohesive formations. Any time formation materials flow into the center stem of the auger, the character of the formation can be compromised. The inflow of materials can also complicate the installation of monitoring wells and other instrumentation.

During monitoring well installation with hollow-stem augers, drilling fluids are generally not used. If they are necessary, they must be removed from the borehole before ambient ground-water conditions can be ascertained. Using the hollow-stem auger drilling method in clean granular soils below the water table will require equalization of borehole fluid pressures to keep formation materials from entering the auger core. There are several methods and procedures available to help alleviate this problem. Flex plugs are plastic baskets that can be fitted into the auger above the cutter head. These plugs allow the passage of samplers, and then they close by soil pressure while drilling. Dry auger systems are also available. They consist of seals for the auger bolt holes and the hollow-stem auger joints, and bottom plugs, made of wood, plastic, or stainless steel, for the hollow center stem. Advancing sealed augers prohibits the collection of formation samples, so this method is normally used as an adjunct to a separate boring in which samples are collected. When using a sealed auger, it is generally necessary to fill the center stem with clean potable water before removing the bottom plug to prevent any inflow of formation materials. Drillers develop specific techniques with which they have experienced success in the formations they commonly drill. It is prudent to take advantage of the drillers experience in these situations whenever possible.

Drilling rigs used to operate solid-stem augers and hollow-stem augers are of a top-drive design, in which all down-force is applied directly to the top of the auger. Most rigs are designed to drill with 5 ft auger lengths, although a few rigs with 10 and 15 ft strokes are available. The primary characteristic of these drills, and the main feature that distinguishes auger drills from standard rotary rigs, is relatively high torque. Most auger drills are also capable of providing rotary drilling functions, however, some may lack the high revolutions per minute capability required for diamond core drilling of rock. An auger drilling rig used for simple flight auger work might consist of a rotary and feed or retract system only. Simple exploration consists of augering a hole and collecting drill cuttings from the flights as they arrive at the surface. More often the drilling rig has a hoist, a driving device, and an off-hole and side-to-side movement mechanism. These features permit removal of augers or use of sampling tools without having to physically move the drilling rig.

Some drilling rigs have the ability to work tools through a large-bore, hollow, top-drive spindle directly into the hollow stem auger. This feature eliminates the need for disconnect and reconnect and on and off borehole manipulation. It also permits continuous access to the bottom of the borehole and permits percussion drilling or drive sampling through the center of the auger while the hollow-stem auger is being advanced. Wire-line systems are much faster, but have a higher incidence of "sanding in" below the water table. When the system sands in, the only alternative is to pull back the augers to get the system to relatch with the overshot. Pulling the augers back causes formation disturbance and should be avoided. The success of this feature is, however, contingent on the soil conditions encountered.

Drilling Methods: Drilling with Circulation Fluids

Circulation fluids are an essential element in the use of the drilling methods described in this section. A circulation medium can be a liquid, such as water or drilling mud (water with special additives) (see ASTM D 5783; ASTM, 2004c), or it can be a gas, such as air or foam (air with additives of various types) (see ASTM D 5782; 2004d). Circulation media are normally forced down through the drill rod, out through the bit, and back up the annulus between the borehole wall and drill rod. The functions of a circulation medium are to cool and lubricate the drill bit, stabilize the borehole, and remove drill cuttings. Bentonite drilling fluid can form a filter cake or seal on the borehole wall, which can prevent leakage of water from some granular formations into the borehole.

Successful removal of drill cuttings requires a minimum up-hole velocity of the circulation medium and depends on the viscosity of the fluid (if water or drilling mud is the fluid used), volume and pressure of air (if air is the fluid used), and the size and density of the cuttings. Minimum velocity can be estimated by using a modified form of Stokes law or by consulting one of many available references (e.g., Anderson, 1979). Generally, the minimum up-hole velocity needed to transport cuttings is about 150 ft/min for plain water with no additives and about 3000 ft/min for air with no additives; additives decrease the required minimum velocity. Excessive up-hole velocities can cause borehole wall erosion, which can result in premature cutting collection in created cavities and caving of the borehole.

Because air is available everywhere and water has to be hauled, it is always a good idea to at least consider the possible use of air. A good rule of thumb says that the correct pump volume (in gpm) when multiplied by 4 will be the correct volume of air (in cfm). The primary advantage of air is the quick recovery of drill cuttings at the surface due to the high velocity return. However, those cuttings are typically in a highly disturbed state.

The use of circulation fluids may involve the addition of materials or chemicals to the borehole. Additions to water to create a circulation liquid appropriate for site-specific conditions include various types of drilling muds, most of which are a form of bentonitic clay or synthetic or naturally derived polymers. Polymers, however, may contain complex chemicals. Fluids containing polymer additives are generally not allowed for environmental drilling applications, even though many polymers are approved by the National Sanitation Foundation for use in water wells. If the polymer chemistry is known and is judged not to interfere with analysis of samples from the borehole (or a monitoring well installed in the borehole), the polymers might be suitable for environmental drilling. Polymer drilling fluids can be broken down to improve well performance. Pure bentonite clay drilling products can be used, but these materials are often more difficult to clean and flush from the borehole to prepare for well installation. Compressed air usually contains a substantial amount of hydrocarbon lubricants released by the compressor. When using compressed air, it is either necessary to incorporate a coalescing HEPA filter system in the air line to remove these potential borehole contaminants or to use an oil-less air compressor. As a general rule, methods of drilling that require a circulation medium for environmental investigations are only used when absolutely necessary to complete the task. However, in some cases, proper use of fluid circulation drilling methods can prevent cross-contamination. Drilling with water, drilling mud, or air requires caution to avoid fracturing formation materials. The drill bit must not be blocked, and pumps or compressors should be equipped with pressure-relief valves to avoid fracturing.

Several types of drilling methods that use circulation media are described subsequently.

Wash Boring

Wash boring is a simple, almost obsolete method of advancing a borehole. The formation is cut by a chopping and twisting action of a bit, and disaggregated formation material is washed to the surface by a circulation fluid. Equipment can be as simple as a tripod with a sheave, a drill rod with a bit, a pump with hoses, and a hoist with rope. Circulation fluid, usually water, is pumped into the drill rod and out through the bit, which is raised and dropped using the hoist and rope, as the assembly is manually turned back and forth.

In wash boring, the chopping action cuts the formation and the turning of the bit maintains the roundness and straightness of the hole. The cut material is washed up the annulus created between the borehole and the drill rod to the surface where it is screened out or settles out of the circulation fluid in a wash tray or pit. The fluid is then recirculated back through the drill rod into the borehole. If water is used as the circulation fluid in non-cohesive sand and gravel formations, caving may occur, and casing may be required to hold the hole open.

In wash boring, samples of cuttings are generally caught in a sieve or screen held in the return stream. Samples of unconsolidated material can be obtained by driving a sampler into the bottom of the borehole using the hoist, drill rod, and a drive hammer after the bit has been removed from the drill rod. In addition, thin-wall tube samplers can be manually pushed into soft cohesive formations.

Rotary Drilling

In rotary drilling (Figure 5.5), a drill rod with an attached bit is continuously rotated against the face of the borehole to disaggregate formation material while circulation fluid is pumped through the rod and bit to flush cuttings to the surface (ASTM D 5783; ASTM, 2004c). It differs in principle from wash boring in that the drill rod is continuously rotated, while in wash boring, the drill rod is periodically turned by the driller. A rotary drill can supply continuous rotation under down-force pressure to advance the borehole more efficiently.

Rotary drilling is usually accomplished with truck-, ATV-, skid-, or trailer-mounted rig. These rigs generally carry their own pumps and operating components. A typical rotary drilling rig consists of a power unit, a rotation mechanism, a feed or retract system, drum hoists, a cathead or driving device, an on-off hole mechanism for moving the rotation mechanism away from the drilling axis, and a pump or compressor complete with pressure hose, piping, swivel, and other equipment as necessary to circulate the drilling medium.

Pumping and circulation with a rotary drill is no different in principle from that used in wash boring, but the rotary drill can drill holes that are larger in diameter and much deeper. The increased capacity is a result of (1) the rotary mechanism, which causes continuous rotation of the drill rod and the bit and (2) the feed or retract system, which allows continuous application of down-force on the bit, causing the bit to cut new formation material while the circulation fluid flushes cuttings to the surface.

For effective rotary drilling, the down-force on the bit should be great enough to cause continuous penetration of the formation. As a rule of thumb, this force should be approximately 1,500 to 2,500 lb/in. of bit diameter. If the crushing strength of the formation exceeds that achievable with the drilling rig, it is normally necessary either to use a heavier, more powerful drill or to use a more suitable method of drilling, such as diamond coring.

FIGURE 5.5
Rotary drilling rig. (Taken from U.S. Bureau of Reclamation Earth Manual, Part I, 3rd ed.)

Rotation speeds of most rotary drills are in the range of approximately 15 to 750 r/min. Rotary drills are generally used at speeds of around 250 r/min for work in most unconsolidated formations or roller-bit work in rock. Higher speeds are used for rock coring (ASTM D 2113; ASTM, 2004e). The lowest speeds (15 to 25 r/min) are used with downhole hammers, which are described later.

Specifications for drilling machines often show pull-down capabilities that are well in excess of the weight of the truck and drill combined. Although the rated pull-down may be theoretically possible, it generally cannot be achieved unless the drill is tied down to some form of anchor. As a general rule, and in most drilling applications, use of pull-down in excess of that which can be applied and safely contained by the weight on the rear axles of the truck is not advisable as it can result in the drill being misaligned with the borehole.

Rated depth capacity of a rotary drill is the length of drill rod that weighs 80% of the maximum main-hoist, single-line, bare-drum capacity. For example, a drill having a maximum main-hoist, single-line, bare-drum capacity of 10,000 lb would have a rated depth capacity that would be equivalent to 8,000 lb. This drill used with NW drill rods, which weigh 5.5 lb/ft, would have an NW-rated depth capacity of 1,455 ft (8,000/5.5).

Borehole diameter rating of a rotary drill is based on several factors, the most significant of which are the delivered volume of the circulation pump or compressor and, in deeper holes, the delivery pressure of the pump or compressor. In rating the pump or compressor, both delivered volume and delivery pressure will be specific to the type of drilling fluid in use. The drilling fluid to be used is often determined from advance knowledge of the diameter and depth of hole required. For example, if the driller knows that a 4 in. nominal (4.5 in. O.D.) well casing with a 1 in. thick filter pack must be set, a borehole of not less than 5.5 in. in diameter must be drilled. The driller will probably choose to drill at least a 7 in. diameter hole.

Three basic types of rotary drill are in common use: (1) stationary table, in which the rods are rotated by means of a square or splined Kelly bar as it passes through a fixed rotating table; (2) moving rotary box, in which the rods are rotated by means of a square or splined Kelly bar that passes through a rotating gear box, which is moved up or down by means of hydraulic cylinders; and (3) top-head drive, in which a rotating spindle travels up or down applying feed, retract, and rotation forces directly to the top of the drill string.

Reverse Circulation Rotary Drilling

Reverse circulation is a method of rotary drilling in which the circulation fluid flows from the ground surface down the annulus between the drill rod and the borehole. The fluid carries the drill cuttings back to the surface inside the drill rod. At the surface, the fluid is expelled through a swivel into the circulation pit or tank where the cuttings settle out.

Reverse circulation is especially useful in very large boreholes and in those cases where the velocity of conventional rotary circulation would erode the borehole wall. To increase the diameter of a hole drilled by conventional rotary methods, the capacity of the pump or compressor must be increased to maintain an adequate up-hole velocity to lift cuttings. With reverse circulation, the up-hole velocity is controlled by the inside diameter of the drill rod, not the borehole diameter.

Standard reverse circulation drilling has few applications in environmental work. However, one form of drilling that is often referred to as reverse circulation is widely used on environmental drilling projects. This method is described subsequently.

Dual-Tube Reverse Circulation Drilling

Dual-tube reverse circulation (Figure 5.6) (ASTM D 5781; ASTM, 2004f) is a form of rotary drilling similar to reverse circulation in which two concentric strings of drill pipe are assembled as a unit to create a controlled annulus. The circulation medium, which may consist of air, water, mist, foam, or drilling mud, is pumped through an outer swivel down through the annulus between the strings of drill pipe to the bit, where it is deflected upward into the center pipe. The bit used with dual-tube equipment is of a design that cuts

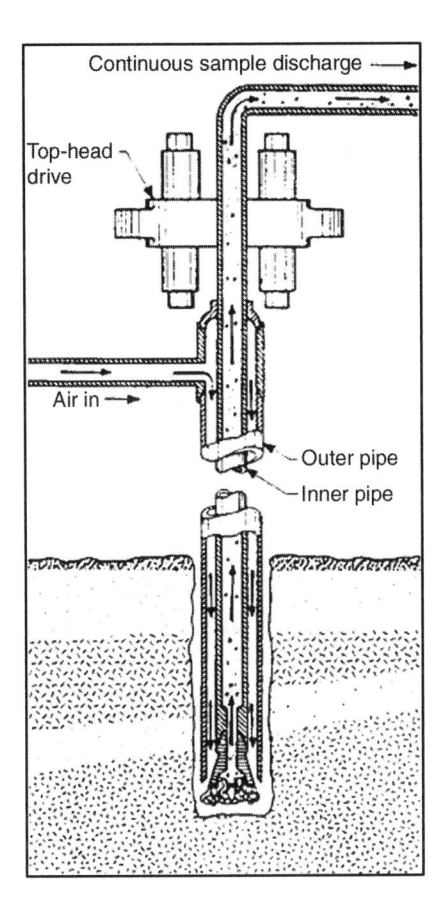

Environmental Drilling for Soil Sampling

FIGURE 5.6
Dual-tube reverse circulation drilling. (Taken from U.S. Bureau of Reclamation Earth Manual, Part I, 3rd ed.)

an annular ring or kerf, which forces all cut material to move toward the center of the hole. Cut material is returned through the inner pipe and swivel to the surface. The cuttings may be collected as a sample or otherwise collected for proper disposal, depending on the purpose of the boring. While formation materials are quickly returned to the surface, they are highly disturbed and can generally be used only for rudimentary classification. Dual-tube reverse circulation is sometimes used when water sampling during borehole advancement is desired. The dual-tube drill string must be over-drilled with overshot casing if a monitoring well or other instrumentation is to be installed in the borehole.

Percussion Drilling

Percussion drilling is a form of drilling in which the basic method of advance is hammering, striking, or beating on the formation. Rotation may also be involved but, if so, it is used primarily to maintain roundness and straightness of the percussion-drilled borehole. Three basic types of percussion drilling equipment are in use: (1) cable tool, or "churn" drilling, in which a bit, hammer, or other heavy tool is alternately raised and dropped; (2) air percussion, in which an air-actuated device with an attached bit breaks the formation; and (3) air-operated casing hammer. All three methods use impact energy to break or cut the formation.

Cable-tool drilling (ASTM D 5875; ASTM 2004g) is one of the oldest methods of drilling and is still widely used for drilling water-supply wells. Its application in environmental work is limited, mainly because the method is very slow. Drilling rates of only 10 to 20 ft per day are common. Furthermore, holes smaller than 6 in. in diameter are impractical because of the need to use a relatively large, heavy bit. Nevertheless, the method does not use large volumes of drilling fluid and allows sampling of ground water as the hole is advanced in high-yielding formations. For these reasons, the method will continue to have use in environmental drilling work.

In cable-tool drilling, the bit breaks up and pulverizes the soil or rock. Cuttings are recovered by adding water or water with additives to the borehole to form a slurry with the drill cuttings. The slurry is then periodically bailed from the borehole. In unconsolidated formations, casing is advanced behind the bit. The diameter of the casing is slightly larger than the diameter of the bit, and it is equipped with a drive shoe on the lower end. Casing is driven by retrieving the bit from the hole and equipping it with drive clamps. The weight of the bit is then applied as a hammering force to the top of the casing, driving it into the borehole.

In air percussion drilling, air is used to actuate a down-hole hammer that is connected to the end of the drill rod string. Air exhausted from the hammer is used to carry cuttings to the surface continuously as the borehole is advanced. Connection to the drill is with a rod string to provide slow rotation and sufficient feed or retract force for proper operation of the hammer. Down-hole hammers are excellent tools for drilling in rock formations that will stand open without caving and, coupled with casing drivers, are effective in cobble and boulder formations. Down-hole hammers are operated with air compressors at the surface and are often lubricated with petroleum compounds. Both hammers and compressors require lubrication that may contaminate the formation surrounding the borehole. In extreme cases, hammers can be lubricated with other materials, but the use of oil-less compressors is rare. Air compressors should be located away from drill rig exhaust and should be equipped with HEPA filters.

A third method of percussion drilling uses an air-operated, drill-through casing hammer (Figure 5.7) (ASTM D 5872; ASTM, 2004h). This device is similar to a pile-driving hammer except for a hole through its axis by which a drill rod string can be inserted. This arrangement allows drilling to proceed while the casing is being driven. The casing hammer or the drill string can also be operated independently. The casing is generally cleaned out with a rotary rock bit or a down-hole hammer. This drilling method is especially useful in cobble and boulder formations.

The air-operated casing hammer requires internal lubrication that is provided by hydrocarbon lubricants added by means of in-line oilers. This need for lubrication must be evaluated when considering use of the drill-through casing hammer for environmental work. The down-hole hammer can be operated with clean water or environmentally safe lubricants. However, as with any compressed air used in drilling, the compressor-derived contaminants must be filtered out of the air stream.

ODEX (also known as TUBEX) is an adaptation of the air-operated down-hole hammer. It uses a swing-out eccentric bit as a casing underreamer. The percussion bit is a two-piece bit consisting of a concentric pilot bit behind, which is an eccentric second bit that swings out to enlarge the borehole diameter. The driller controls the swing-out by forward or reverse rotation of the drill string. Immediately above the eccentric bit is a "drive sub," which engages a special internal shouldered drive shoe on the bottom of the ODEX casing. Thus, ODEX casing is actually pulled down by the drill string as the hole is advanced. Cuttings blow up through the drive sub and stem or casing annulus to a swivel, which conducts them to a sample collector. Casing advancers with down-hole hammer systems are rapid methods for advancing boreholes in cobble and boulder

Environmental Drilling for Soil Sampling

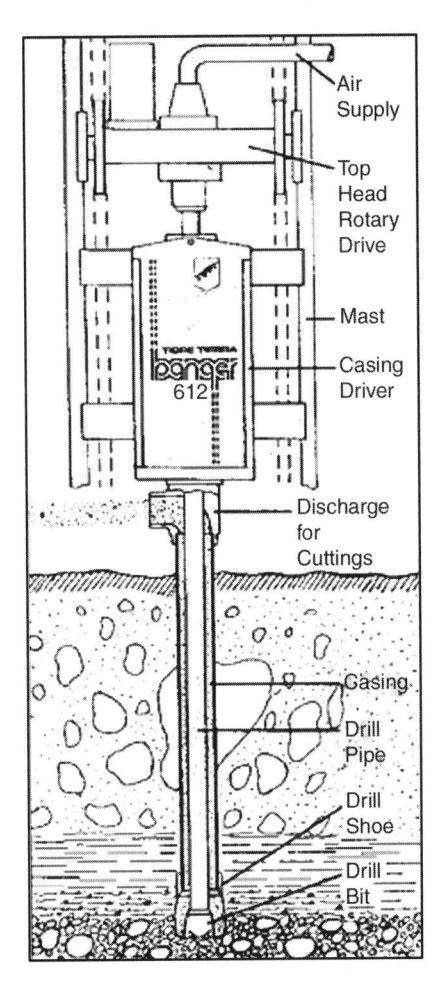

FIGURE 5.7
Drill-through casing hammer. (Taken from U.S. Bureau of Reclamation Earth Manual, Part I, 3rd ed.)

formations, in which other methods such as fluid rotary or hollow-stem augers may be less effective.

SIM-CAS is very similar to ODEX except that the casing is pushed down from a pushing head on the drill at the surface as opposed to being pulled from the bottom. The eccentric bit, drill stem sizes, hole sizes, and air requirements are essentially the same as those for ODEX. Both SIM-CAS and ODEX use compressed air or foam for operation. Therefore, proper filtration of the air stream is required for environmental work.

Dual Rotary Drilling

Dual rotary drilling is similar to the drill-through casing hammer except that the casing is rotated into the formation while the inner rod string drills out the inside of the casing. Dual rotary drills are large, heavy units generally mounted on tandem or triple-axle vehicles, and they generally require more site access preparation than other drilling methods.

Dual rotary drills have two rotary tables. The lower table rotates and applies down-force and retract force to casing in sizes from 6 to 40 in. I.D. The upper table rotates and applies down-force and retract force to the drill rod string. The rod string can accommodate conventional rotary tools or down-hole hammers in appropriate sizes for cleaning the casing. Drill rod string activity can remain inside the casing or work ahead as necessary to advance the borehole. The rotational torque of these rigs ranges from 41,672 to 262,771 ft/lb. Dual rotary drills are most effective in bouldery hard formations where large diameter wells are required.

Sonic Drilling

Sonic drilling (Figure 5.8) utilizes high-frequency vibration, aided by down pressure and rotation, to advance drilling tools through various subsurface formations. Power for tool advancement in sonic drilling is created by a sine generator, positioned at the top of the drill mast, with rapidly rotating eccentric, counter-balanced weights that are timed to direct 100% of the vibration at 0° and 180° (e.g., along the length of the drive casing). The sine generator generally operates between 0 and 185 Hz. To generate effective vibration, it requires the sine generator weights to be rotated at speeds of between 3,000 and 10,800 r/min. The vibratory effect causes the soils adjacent to the drive casing or sampling barrel to liquefy, allowing the sampler or casing to pass through.

Sonic drilling technology was developed in the mid-1970s for use in mineral exploration, but never proved very effective for this application. It was adapted to environmental drilling in the early 1990s and has since proved very effective for this application.

The sonic drilling procedure begins with advancing a sampling barrel by vibration and down-force, 10 ft into the formation. An outer casing is then vibrated over the sampling barrel to the same depth. The sampling barrel is then removed from the hole, swung out to the side, and the sample is vibrated out or slid out of the sampling barrel in liners. The sampling barrel is then reinserted into the borehole and advanced through the next 10 ft sampling interval. The casing is again advanced over the sample barrel

FIGURE 5.8
Sonic drilling rig.

and the barrel removed. This process is repeated until the boring is completed to its total depth. The sampling barrel is always advanced without drilling fluids. A drilling fluid (generally water based) may be used when advancing the casing over the sampling barrel to prevent entry of sediment into the annular space and any subsequent locking up of the sampling barrel and casing.

Sonic drilling offers some unique features that make it well suited to environmental work. Drilling in most unconsolidated formations is generally very rapid. Continuous sampling, generally in 10 ft increments (but sometimes longer) is part of the drilling process and, therefore, is not an added cost as it is with all other drilling methods. Drill cuttings are very minimal, as the only materials brought to the surface are the samples. This significantly reduces cutting disposal and drilling area cleanup costs. On completion of the boring, the hole is cased to the bottom, making monitoring well or instrumentation installation very efficient. Because the borehole is installed without or with very little drilling fluid, water sampling, well development, and pumping tests can be accomplished in less time and with less accumulation of waste fluid for disposal than with most other drilling methods. The currently available sonic equipment has the capability of driving 10 in. casing to depths of 700 ft. Bits used in sonic drilling can penetrate boulders, construction debris, and bedrock (to a limited depth). Sonic drills have the unique feature of being convertible to using down-hole hammers, conventional mud or air rotary, or diamond rock-coring tools.

Currently available sonic drills are generally large units mounted on tandem or tri-axle vehicles. Thus, they may require site access preparation. The samples recovered are, to some extent, disturbed — the vibratory action during penetration of the sampling barrel into the soil and during removal of the soil sample from the sampling barrel can stratify some formations and consolidate or loosen others. Samples can be tested with confidence for chemical compounds but, unless gathered by conventional tools (e.g., a split-barrel sampler), cannot generally be used for determining engineering properties. Sonic drilling is more efficient in unconsolidated formations than in most bedrock.

Sonic drilling is evolving with more widespread use on environmental jobs. Future developments will most likely include methods to generate engineering property values as well as better samples for physical analysis and physically smaller drilling rigs. Sonic drilling holds a great deal of promise for future improvements and refinements to the drilling practice.

Selection of Drilling Methods

Introduction

Selecting drilling methods for environmental investigations is typically a process of evaluating trade-offs. Drilling methods that allow for quick, efficient well construction may not be well suited for soil or rock sampling. Methods that accommodate geophysical logging are not necessarily well adapted to drilling in highly contaminated areas. When compared with drilling for the purpose of installing production wells or for mineral exploration, the environmental investigator usually has more options to consider when selecting drilling methods.

A higher level of field supervision and personal safety protection is normally required when drilling on environmental sites. Collection of samples for environmental site characterization and construction of monitoring wells is a relatively complicated business. It requires a high level of cooperation and communication between the driller and environmental investigator. Use of outdated techniques and drilling shortcuts may introduce contaminants into soil samples or into formations adjacent to the borehole or result in spreading of contamination from shallow zones to deeper zones or from zones of higher hydraulic head to zones of lower hydraulic head. A high degree of accuracy is required in every measurement, from tallying lengths of casing and screen to measure the depth and volume of boreholes, filter-packed intervals, and annular seals.

In selecting a drilling method for monitoring well construction, the single most important consideration is that the well be built to allow the collection of representative ground-water samples from a specified depth or interval. However, time, cost, and many other factors must also be considered. ASTM Standard D 6286, Standard Guide For Selection of Drilling Methods for Environmental Site Characterization (ASTM, 2004i), provides very helpful information on the myriad of factors influencing the selection of a drilling method appropriate for a site-specific application.

Health and Safety

At potentially contaminated sites, contamination levels in the subsurface are often unknown prior to drilling and may range from very high to non-existent. Work at any potentially contaminated site, regardless of the drilling method, used should be governed by a site-specific health and safety plan that protects both personnel and equipment (see Chapter 19 for more details). In addition to generalized safety programs to govern conduct of the drilling crew, each drilling method or rig will have some specific safety requirements that apply. The level of effort needed to provide adequate job-site protection is directed more by the requirements for working with the chemicals anticipated than by the drilling method chosen to perform the work. The drilling method chosen generally has more effect on the cost of dealing with drilling waste and project progress than on the cost of health and safety protection. For example, auger drilling, while very efficient for shallow drilling projects, generates a significant volume of drill cuttings that must be properly disposed. As drilling personnel must handle the augers and containerize and clean up the drill cuttings, the risk of dermal contact is high unless adequate safety measures are taken. For the purpose of waste minimization, methods that do not produce cuttings, such as direct-push methods or sonic drilling, have a distinct advantage. Cable-tool drilling does not require the use of large amounts of circulation fluids to remove cuttings from the borehole, but requires contact with heavy tools, cables, and casing, and it produces large volumes of drill cuttings.

Drilling methods that use air as the circulation fluid present different risks. Air drilling to install monitoring wells in contaminated, high-yield formations produces large volumes of potentially contaminated air and water, even when casing is advanced as the hole is drilled. The discharged water (sometimes, the air) must be collected and processed to avoid spreading contamination or causing a hazardous condition for the drilling crew or passers-by. The drillers must take precautions to avoid direct exposure to the fluids produced and to airborne contaminants, as well as be prepared to handle heavy containers of investigation-derived wastes (IDW).

When drilling with air through decomposing refuse that produces methane (a flammable gas), an underground fire could occur due to introduction of high volumes of air into the borehole. To minimize this possibility, a foam additive may be necessary. Augers and sonic drilling can also generate considerable heat during drilling. Therefore, where fire, heat, or explosion is a possibility, such as at a landfill, drilling with a water-based drilling fluid may be preferred.

Use of drilling fluids will require the collection and proper disposal of that drilling fluid as well as the drill cuttings generated. The drilling crew needs to be protected from splash or spray that may contain harmful contaminants. In drilling, there is no generalized protection plan that can be followed beyond a few basic common sense items. Each site offers its own unique set of circumstances that need to be addressed in a site-specific health and safety plan.

Access and Noise

Drilling and sampling in connection with environmental investigations is frequently carried out in urban areas. Access and equipment noise are important considerations in selecting drilling methods for work at urban sites. Most drilling equipment is mounted on trucks that have limited maneuverability. In congested areas near factories, power plants, refineries, and manufacturing facilities, side-to-side and overhead clearances are critical to the selection of an appropriate drilling method. Drills mounted on rubber-tired all-terrain carriers or tracked vehicles may be as effective in urban settings as they are on off-road sites. Combination of auger and rotary drilling equipment is generally smaller, lighter, and more maneuverable than large rotary drills. The mast height of these units is usually less than other types of drilling rigs. Many auger and rotary rigs have detachable masts to give added overhead clearance. For extremely tight locations, small skid-mounted, trailer-mounted, or highly maneuverable track-mounted equipment is available from some drilling contractors. Electric-powered or LP gas-powered rigs can be used for indoor work where adequate ventilation for regular gas or diesel engine exhaust is not available.

Rotary, cable tool, sonic, dual rotary, and percussion drilling equipment is usually mounted on larger trucks than auger equipment and requires more room to maneuver and operate. Drill pipe for rotary rigs is generally 20 ft long, which requires more space to handle than the 5 ft augers that are customarily used with auger drilling rigs. Circulation fluids, such as mud or water, also require the use of a portable mud pit and holding tanks that extend from the rear and side of the rig.

Cable-tool rigs may require guylines for lateral mast support, which limit their use in restricted locations. These guylines usually extend at least 15 to 20 ft to the front, rear, and sides of the mast.

Noise can be a major obstacle to environmental drilling in populated areas. Some cities have noise ordinances that restrict the operation of heavy machinery, such as drilling equipment, to specific hours. The allowable noise levels are also restricted. In municipalities that do not have noise ordinances, citizen complaints can still be filed under the provisions of local nuisance ordinances.

Some drilling equipment can be modified to control noise, but for most types of equipment, noise control is impractical or impossible. Noise from impact or percussion equipment, such as casing hammers and casing drivers, is most difficult to control. Therefore, other drilling methods may be better suited for use in residential or urban areas where noise is an issue.

Disposal of Fluids and Cuttings

Disposal of drilling fluids and cuttings (IDW) is a very important consideration in the selection of an appropriate drilling method. If soil and ground water are heavily contaminated, IDW may have to be handled as a hazardous waste. Disposal of this waste at a licensed hazardous waste landfill can add significant cost to the drilling project. Direct-push technologies and sonic drilling, which do not produce drill cuttings, are very useful alternatives to minimize IDW.

Drilling methods that use drilling mud or water as a circulation medium can produce large volumes of potentially contaminated fluids. Using air as a circulation medium below the water table can also produce large volumes of ground water. If this ground water is contaminated, the cost for transport and disposal may exceed drilling costs. Where disposal of contaminated drilling fluids is a major concern, direct-push and sonic drilling have an advantage over most other drilling methods.

Lithology and Aquifer Characteristics

Lithology and anticipated aquifer characteristics are primary factors to consider when selecting a drilling method. Certain drilling methods are better suited to drilling certain types of soils and rock formations than others. Some characteristics of subsurface formations may preclude the use of certain drilling methods altogether. For example, auger drilling is most effective in unconsolidated and semiconsolidated materials and cannot be used to drill competent, unweathered bedrock. Augers are also unable to penetrate boulders, cobbly zones, and very dense, compacted clays. Augers are also somewhat limited in their ability to drill below the water table, particularly in loose granular soils. During drilling with hollow-stem auger equipment, ground water may move inside the auger to equalize the pressure and may carry sandy formation material inside the auger (referred to as heaving, running, or blowing sands). This makes it very difficult to collect a representative formation sample or to install a monitoring well. In extreme cases, this condition can result in the augers getting stuck in the borehole.

Air- and mud-rotary drilling equipment can be used to advance a hole through most types of unconsolidated or consolidated materials. However, lost circulation zones are particularly troublesome unless casing is advanced as the hole is drilled. The use of dual-wall reverse circulation drilling can significantly reduce the severity of lost circulation problems.

Lost circulation zones are usually caused by the fractures or solution channels in bedrock or by the presence of very coarse-grained unconsolidated material (e.g., gravel or cobbles). When such a zone is encountered, drilling fluids and cuttings may not be returned to the surface, creating several problems. When drilling with drilling mud or water, the lost fluids and additives have to be replaced to allow drilling to continue. Drilling progress can be slowed or stopped if extra water has to be hauled in by truck. Without circulation return, the hole is advanced blindly — the driller and hydrogeologist will not be able to examine cuttings to assess changes in lithology.

In addition, when a borehole is advanced without circulation return for more than a few feet, the potential for stuck drill rod exists. Cuttings that are not transported to the surface can be carried into the lost circulation zone. When circulation is stopped to add a length of drill rod, the cuttings may fall back into the hole on top of the bit, causing the entire string of drill rod to become stuck in the hole. With rotary drilling, caving and lost circulation in unconsolidated materials can be effectively overcome by advancing casing closely behind the bit.

Cobbles and boulders also represent a problem for drilling with conventional rotary methods. Boulders can roll beneath the bit instead of being crushed and deflect the borehole. It is thus important to use drill rod stabilization collars when working in boulder zones. The stabilization collar is a length of rod that is just slightly smaller in diameter than the bit. The collar adds weight to the drill string and helps keep the borehole aligned using the drilled borehole to keep the bit running straight. Without using stabilizers, a 6 in. bit connected to NWJ drill rods drilling through a boulder formation can create a borehole that is in essence only 2.75 in. in diameter as the bit weaves around the boulders. This makes correctly installing a monitoring well very difficult.

When using air as a circulation medium, the anticipated formation yield must be considered. In high-yield formations, drilling with air rotary more than a few feet below the water table will produce large volumes of water. If these large volumes of water contain suspected contaminants, the water produced from the hole may need to be contained for proper disposal. Fluid containment may require special sheeting under and around the drilling rig, transfer pumps, holding vessels, and transport vehicles. In unconsolidated formations, the water moving up the borehole may erode the borehole walls unless casing is installed. Surface casing should be set to the rock surface when using air as a circulation medium.

Using air with the dual-tube reverse circulation drilling method below the water table will generally produce less fluid than open hole air rotary in high-yield aquifers. The dual-tube drill pipe seals off borehole wall inflow, as water can only enter the drill string through the bit. However, in productive formations, the rate of inflow through the bit can be substantial.

Depth of Drilling

For most monitoring well installations, depth is a major consideration in selecting a drilling method. Both rotary and cable-tool methods have been used to drill to depths of several thousand feet. Auger drilling is generally effective to depths up to about 150 ft, although hollow-stem augers have been used to drill to depths of more than 400 ft under favorable conditions. Bucket augers are generally limited to depths of less than 100 ft because of the length of the Kelly bar and caving potential. The primary limitations for drilling deep holes with hollow-stem auger equipment are the torque requirements and the time required for retrieval of the drill string.

Other drilling methods also have depth limitations. Dual-tube reverse circulation drilling has been successfully used to depths in excess of 500 ft, although depths of 300 to 400 ft are more common. In dry unconsolidated materials, casing hammers are most effective at depths of less than about 200 ft. At greater depths, the penetration rate decreases because of the increased friction between the casing and the borehole wall. Telescoping casing strings can be used to overcome this limitation to a certain extent. Sonic drilling is less cost effective for shallow borings because of setup time, but it is very effective from 50 to 300 ft and can be used to depths as great as 700 ft under favorable conditions.

Sample Type

Ability to collect soil or rock samples is one of the most important considerations in selecting a drilling method. For collecting relatively undisturbed samples of unconsolidated materials, hollow-stem auger drilling (ASTM D 6151; ASTM, 2004b) is the preferred method. No fluids are introduced into the hole, and samples remain uncontaminated by the drilling process. ASTM Standard D 6169 (ASTM, 2004j) provides very useful information about selection of soil and rock sampling devices used with drilling rigs for environmental investigations.

When drilling with conventional rotary methods in unconsolidated materials, collecting representative samples for chemical analysis is more difficult, as drilling fluids may chemically alter the sampled materials. If the hole is filled with drilling mud or water, some invasion of the sampled interval in the bottom of the hole can occur.

Undisturbed and unaltered samples of unconsolidated materials are nearly impossible to obtain with a cable-tool drilling rig unless the rig is fitted with special sampling tooling. Materials several feet below the bottom of the hole can be affected by the impact of the bit.

In rock, sampling by conventional or wire-line coring methods (ASTM D 2113; ASTM, 2004e) is the preferred method for obtaining relatively undisturbed samples. Most conventional or wire-line sampling equipment can be used with rotary drilling equipment. Obtaining undisturbed samples of rock with cable-tool drilling equipment is impossible.

Sonic drilling can provide rock samples, however, the vibratory action can cause fractures in some formations. These fractures can affect hydraulic conductivity evaluations.

Auger and rotary drilling rigs can be used for rock coring as long as they are capable of sufficient rotary speed. Hollow-stem augers can be used to sample the unconsolidated portion of the borehole, advanced to bedrock, and left in place to serve as temporary casing. The borehole can then be advanced using conventional or wire-line coring methods. However, using hollow-stem augers as casing for rock coring can have some serious drawbacks for monitoring well construction. The auger cutter head seldom provides an effective seal at the rock interface and the auger joints are not watertight. Fluids can also migrate relatively freely along the outer flights while the rock is being cored. Cross-contamination by drilling fluids or contaminated ground water can thus take place, and unconsolidated material can slough back into the borehole.

Cost

Cost is always a factor in selecting methods to carry out a drilling program. In addition to the footage or hourly rates associated with drilling, there are many other costs that should be considered. For example, the cost of collecting and analyzing water samples from a monitoring well as part of a regularly scheduled sampling program can exceed the cost of well installation after only a few rounds of samples. If excessive time is required to obtain a sample because improper drilling or development methods were used or if samples are unrepresentative because contaminants were introduced into the formation during the drilling process, then the use of the lowest cost drilling method may not provide the lowest overall project cost.

The total project cost for a drilling, soil and rock sampling, and monitoring well construction job usually includes the labor for supervisory hydrogeologist or engineer. Faster drilling methods may therefore have a cost advantage that is not included in the drilling footage rate. Cable-tool drilling, which generally has a lower per-foot rate and is slower than the most other methods, can result in excessive overall costs for monitoring well construction if supervision is provided. Sonic drilling has a high per-foot rate, but is much faster and more productive than the most other methods, which can reduce overall project costs through reductions in labor and per-diem costs.

An accurate comparison of costs between different drilling methods is difficult to make because of the number of site-specific variables that affect drilling costs. One method may have distinct cost advantages over others, but not for all conditions. For example, under certain conditions, air rotary can be an extremely fast and inexpensive method for drilling. In dry sandstones and siltstones, the cost of air-rotary drilling can be relatively low (several dollars per foot) for a 5 in. diameter hole and the penetration rate can be as high as 100 ft/h. However, in a coarse-grained alluvium in the same area, an air-rotary rig would have difficulty in achieving a penetration rate of 5 ft/h. The best advice on drilling method selection can often be obtained from local drilling companies familiar with the conditions in the area.

Soil Sampling

Types of Samples

Four basic sample types are collected in environmental site characterization work — bulk samples, representative samples, undisturbed samples, and composite samples. These are described in the following sections.

Bulk Samples

Bulk samples are simply a shovelful or handful of the drill cuttings taken from the borehole. This type of sample provides a very generalized picture of the formations penetrated by drilling, as the various formations tend to be mixed together. These samples are usually placed in containers and transported to the laboratory for physical or chemical analysis. This type of sample is considered the least accurate of the four basic types of sample and is not widely used in environmental site characterization programs.

Representative Samples

Representative samples are generally taken in some form of drive or push tube. They represent material from a specific, discrete depth interval in the borehole. For example, if a hollow-stem auger were advanced to a depth of 10 ft, the center bit removed and a 2 ft long sampler lowered down-hole and driven for 24 in., the recovered material in the sampler would be representative of the 10 to 12 ft depth interval. For the purposes of this discussion, "representative" also describes a sample in which all the constituents are present, but not necessarily in a completely undisturbed state. Thus, a representative sample is a sample taken from a specific depth interval that contains all of the constituents present in the formation at that depth interval. For chemical analyses of soils, representative samples are often collected and sub-sampled (per ASTM D 4547 [ASTM, 2004k] or ASTM D 6418 [ASTM, 2004l]) immediately.

Undisturbed Samples

Undisturbed samples are high-quality samples taken under strictly controlled circumstances to minimize the physical or chemical disturbance to the sample. The goal of undisturbed sampling is to sample all constituents of the formation without altering the presampling relationship between constituents in the sample. Undisturbed samples are generally required for hydraulic conductivity testing. In the discussion of soil samplers provided subsequently, it is important to note that even the best soil samplers generally disturb clean sand samples. If the hydraulic conductivity value for a sand sample is desired, representative (disturbed) samples can be collected and the hydraulic conductivity can be evaluated by grain-size analysis or by remolding the sand in preparation for a constant head permeability test (ASTM D 2434; ASTM, 2004m).

"*In situ*" is a Latin term which, roughly translated, means "in place." For many years, the term "*in situ*" was used to describe the highest possible quality of sample, because it was thought impossible to collect a truly undisturbed sample. Recently, and with the advent of more tests actually made in place, the term undisturbed sample is preferred and is more accurate for those samples that are removed from a borehole in a relatively undisturbed fashion. "*In situ*" generally refers to tests performed in the borehole, including those made with cone penetration testing rigs (Chapter 6).

Composite Samples

Composite samples are a blend or mix of several discrete samples or material from different sources (i.e., different boreholes). Such a sample might be a combination of a discrete sample from one depth interval in one borehole and another discrete sample from the same depth interval in a different borehole, mixed to represent similar formation materials across a site. It might also be a combination of discrete samples from different intervals in the same borehole, mixed in such a way to represent the entire borehole.

Types of Soil Samplers

Solid-Barrel Samplers

Solid-barrel samplers consist of a steel tube attached to a connector head at the top and a drive shoe at the bottom. The length is generally between 12 and 60 in. and the diameter ranges from 1 to 6 in. The sampler is normally made out of steel or stainless steel and is generally used with liners, in which the sample is collected for ease of removal. Liners may be made of brass, stainless steel, or various plastics. The type of material investigated and the tests and analyses conducted on the sample generally determine which liner material is appropriate for the specific application.

Split-Barrel Samplers

Split-barrel samplers (ASTM D 3550 [ASTM, 2004n] and ASTM D 1586 [ASTM, 2004o]) (Figure 5.9) are similar to solid-barrel samplers except that the tubular section is split longitudinally into two equal semi-cylindrical halves. Normally manufactured of high-strength steel, split-barrel samplers are also available in stainless steel for special sampling circumstances. The split-barrel sampler may be used either lined or unlined. The split-barrel sampler is also referred to as a "split-spoon" sampler and is used in the standard penetration test (SPT). The SPT (ASTM D 1586; ASTM, 2004o) is used for geotechnical engineering studies and is often not required for environmental work. This test is often specified for soil sampling based on precedent or familiarity, but it is often not the most economical way to collect environmental samples for chemical or physical analysis. For example, hollow-stem augers equipped with continuous-tube samplers can collect 5 ft long representative samples much faster than split barrel SPT samplers, which are only driven in 2 ft increments.

The split-barrel sampler is generally available in 1, 1.5, 2.0, 2.5, 3.0, and 3.5 in. O.D. The split-barrel sampler is the most commonly used sampler in both environmental and geotechnical work. For geotechnical work, the sampler can be fitted with 12 1 in. rings on the bottom and a 6 in. liner on the top to facilitate laboratory testing. For environmental work, 6 in. liners are commonly used to collect the samples. Split-barrel samples are often physically disturbed but are still suitable for soil chemistry evaluation. Storage of samples in liners may result in loss of volatile compounds with time. The split-barrel sample is often subsampled immediately after the sample is brought to the surface, using the techniques described in either ASTM D 4547 (ASTM, 2004k) or ASTM D 6418 (ASTM, 2004l).

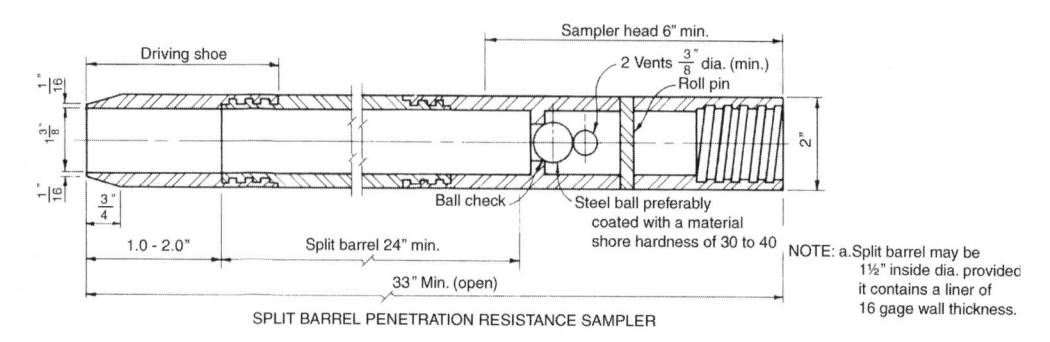

FIGURE 5.9
Thick-wall split-barrel drive sampler (SPT). (Taken from U.S. Bureau of Reclamation Earth Manual, Part II, 3rd ed.)

Thin-Wall Tube Samplers

Thin-wall tube samplers (Figure 5.10) consist of a connector head and a length of thin-wall steel, brass, or stainless steel tube (ASTM D 1587; ASTM, 2004p). The most common sampler length is 30 in., to collect a 24 in. sample. The thin-wall tube is sharpened at the cutting end and crimped to 1.0% of its inside diameter to allow free movement of the sample into the tube. The original seamless tubing was made by U.S. Steel's Shelby process; thus, this type of sampler is also known as a Shelby tube. Tubes today are generally made of electric-welded flash-controlled steel and are much stronger, straighter, and cheaper than the older type.

The thin-wall tube sampler is used primarily in cohesive, soft, or clayey formations where relatively undisturbed samples are desired. The most commonly used thin-wall tube sampler is 3 in. O.D. Other available sizes are 2, 2.5, and 5 in. O.D.

Common to all of the samplers discussed earlier is the "connector head." The connector head is the link between the sampler barrel or tube and the drill rod, which, in turn, connects the sampler to the drilling rig. For all three samplers, the connector head contains a ball check valve that allows air, water, or mud to escape through the connector head into the drill rod as the sampler is advanced to fill with soil. When sampler advance is stopped, the valve closes, preventing the expelled fluid from reentering the tube and forcing the sample out. If the check valve becomes plugged or ceases to function properly, it is difficult or impossible to collect a sample from a water- or mud-filled borehole, as these fluids are incompressible, and advancing the sampler would displace the sample to the sides of the borehole.

Continuous Tube Samplers

Continuous tube samplers, which are used exclusively with hollow-stem augers, consist of a 5 ft long steel split-barrel, with a cutting shoe at the bottom and a head at the top. The head may serve as a connector to mate with drill rod extending to the surface or it may serve as a connector onto which a wire-line latch is lowered to retrieve the sampler after the sample has been collected. In the former case, drill rod is held in place in a chuck on the rig to prevent rotation of the sampler as the auger is drilling. In the latter case, the head is equipped with a bearing system to allow the augers to rotate while the sampler is held stationary by soil friction.

Sample recovery is dependent on the positioning of the sampling barrel cutting shoe in relation to the auger cutter head and the clearance ratio of the cutting shoe inside diameter to the inside diameter of the inner liner and barrel. In the normal operating position, the sampler shoe should lead the auger cutter head teeth by at least 1 in. The shoe position can be adjusted as needed for good recovery — extended out for soft material and retracted for hard formations (Figure 5.11). After the hole has been advanced a distance equal to the length of the sampling barrel, auger rotation is stopped and the sampler is brought to the surface and emptied. A clean sampling barrel is then reinserted into the auger string, additional auger added, and the procedure repeated.

The continuous tube sampling system works well in clays and other cohesive soils and in granular materials above the water table. However, when the water table is encountered in granular formations, the problem of material inflow (heaving sand) into the auger can occur after the sampling tube is removed. This inflow can prevent the sampling barrel from being returned to the proper position at the bottom of the auger. In coarse granular materials, there tends to be some settling of materials in the sampling barrel due to the volume recovered, while in expanding clays, the sample barrel may be difficult or impossible to recover because the sampler may become wedged in place. The sampling barrels can be fitted with liners if needed. Because the volume of material recovered with the continuous tube sampler is substantial, sample handling and storage requires extra

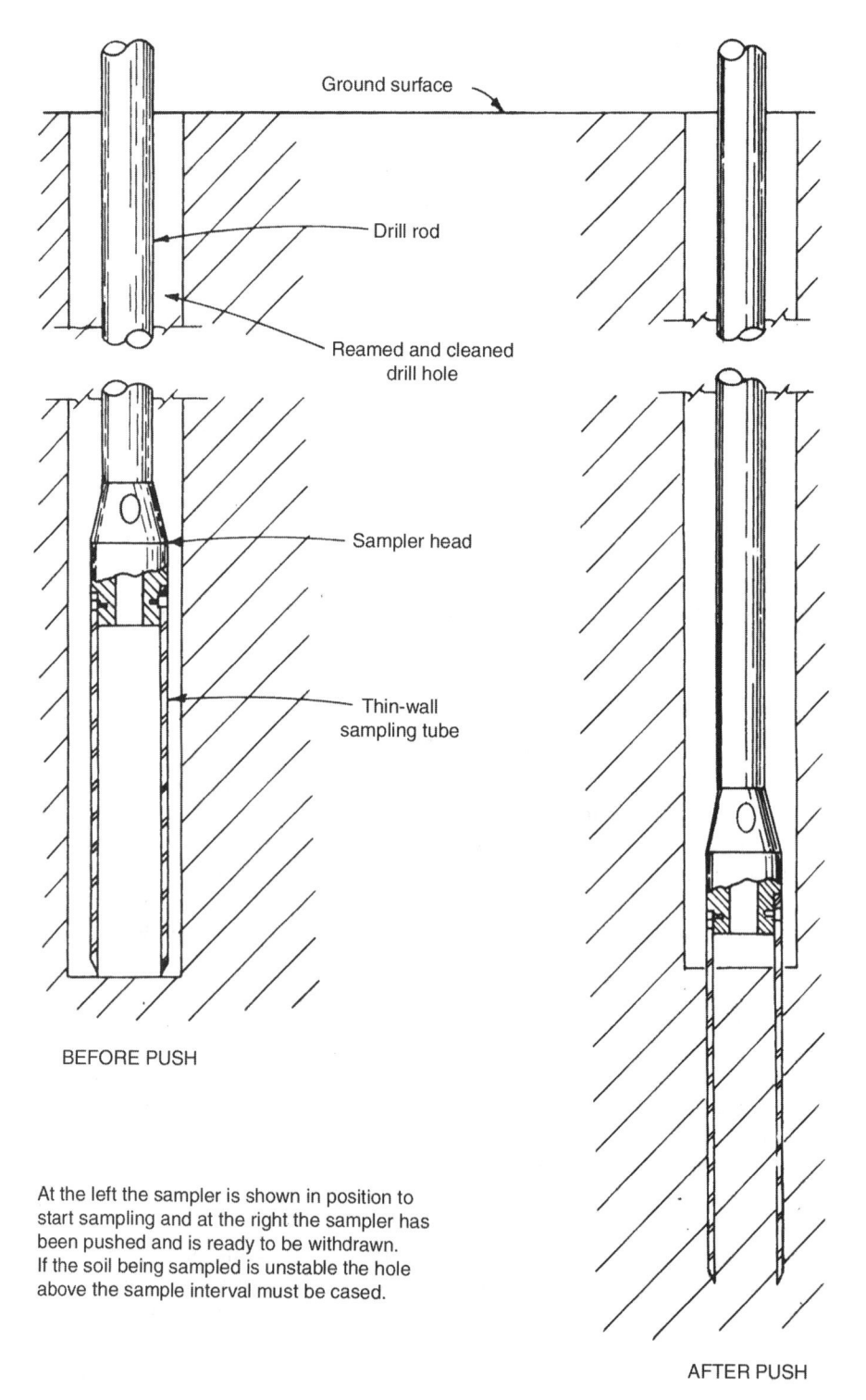

FIGURE 5.10
Thin-wall tube sampler. (a) Before push. (b) After push. (Taken from U.S. Bureau of Reclamation Earth Manual, Part II, 3rd ed.)

FIGURE 5.11
Continuous tube sampler (used with hollow-stem auger).

consideration. Continuous tube sampling systems are available for 3.25, 4.25, and 6.25 in. I.D. hollow-stem augers.

Rotary Samplers

Rotary samplers, such as the Pitcher or Denison samplers (Figure 5.12) and the Geo-barrel, are very useful for geotechnical work, but for environmental work, their usefulness is limited, as they require the use of circulation fluids. Both disturbed and undisturbed samples can be retrieved with this type of sampler.

Piston Samplers

Piston samplers typically consist of a thin-wall tube, a piston, and mechanisms for regulating movement between the tube and the piston. They are specialized tools, made specifically for use in collecting samples from soft formations. They are not widely used outside of geotechnical work, although they have been used occasionally to sample noncohesive, difficult-to-recover sediments for environmental analysis (Munch and Killey, 1985; Zapico et al., 1987). A hydraulically activated stationary piston sampler design is shown in Figure 5.13. A similar piston sampler design is often used in direct-push applications (Chapter 6), however, samples collected with this tool may undergo some disturbance because of the advancement method used (i.e., percussion hammering or vibration).

To obtain a sample with the hydraulically activated piston sampler, the sampler is lowered to the bottom of the borehole on drill rods, with the fixed piston flush with the lower end of the sampler tube. Pressurized fluid actuates the driving piston, forcing the tube into the soil while the fixed piston remains stationary. Pulling back on the sampler creates a strong suction effect between the sample and the piston that assists in holding the sample in the tube.

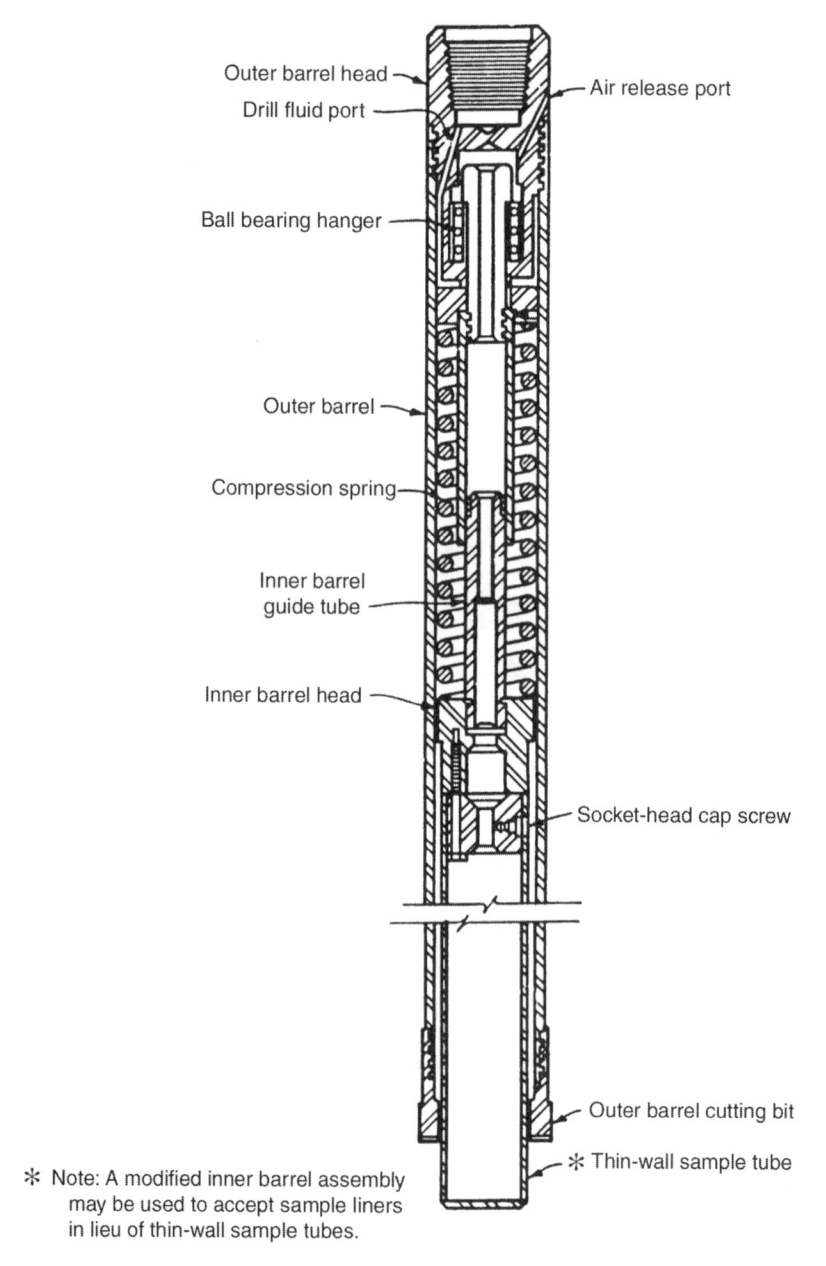

FIGURE 5.12
Pitcher rotary soil core sampler. (Taken from U.S. Bureau of Reclamation Earth Manual, Part II, 3rd ed.)

The mechanical fixed-piston sampler operates much the same as the hydraulic device except that this piston sampler relies on a series of actuating rods to hold the fixed piston in place as the thin-wall tube is pushed over the piston.

Methods of Sampling

Samples in environmental work are generally taken from the bottom of the borehole, although bulk samples are generally taken from the drill cuttings above ground. For

Environmental Drilling for Soil Sampling 325

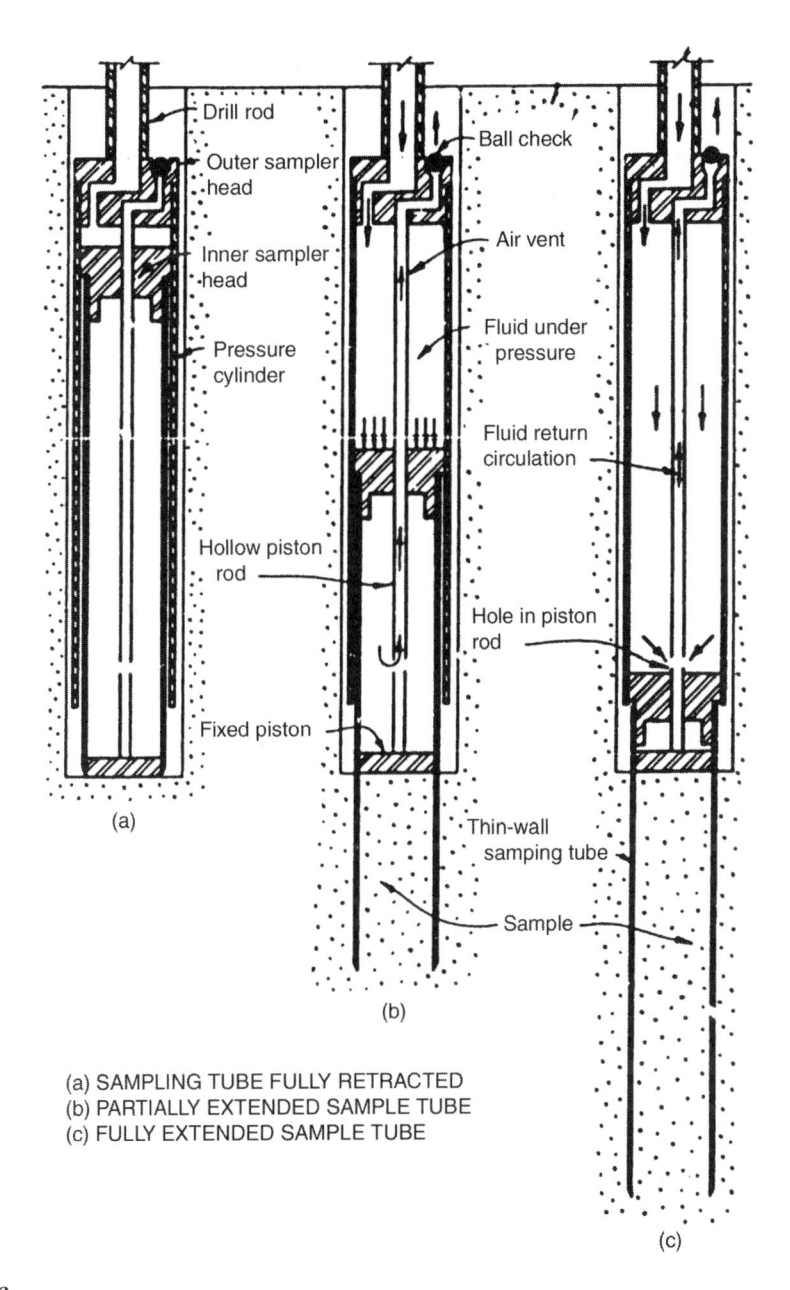

FIGURE 5.13
Piston sampler (Osterberg type). (Taken from U.S. Bureau of Reclamation Earth Manual, Part II, 3rd ed.)

example, if a sample is required from 12 to 14 ft interval, the hole is drilled to 12.0 ft, the sampler installed in the borehole, advanced 2 ft, and retrieved from the borehole. Driving or pushing is the usual method employed for advancing conventional samplers. After that, sample is taken, the borehole is advanced to the next sampling depth, and the sampling process is repeated. Under some circumstances, continuous samples are required and the sampling objective is to incrementally obtain a continuous column of formation materials penetrated by drilling from ground surface to the total depth of the hole. Incremental drilling and sampling with a conventional split-barrel or thin-wall tube sampler often misses material, so other methods are more commonly used.

Continuous samples can be more easily obtained using either a continuous tube sampler with a hollow-stem auger or the core barrel employed in sonic drilling.

Driving

Driving is the most common method of obtaining split-barrel samples. For most sampling, a 140 lb safety hammer with internal impact surfaces is used. The 140 lb safety hammer is connected to the sampler by drill rods and the hammer remains at the ground surface. Hydraulically operated automatic sampling hammers are available on many drilling rigs and are highly recommended from a safety standpoint. When the sampler has been driven to the desired depth, it is retracted by hydraulic cylinder pullback or by back-pounding with the hammer. However, back-pounding the sampler often causes the sample to fall out.

With either type of hammer, a record of the number of hammer blows needed to advance the sampler can provide useful information on the type of material that is being penetrated. The blows are usually counted for each 6 in. increment of the total drive. For a 24 in. drive, four numbers are recorded. If the sampler cannot be advanced 6 in. with a reasonable number of blows (usually 50), sampler refusal has occurred and the sampling effort at the particular depth is terminated. However, if a sample must be recovered, additional driving can be done. However, caution must be exercised, as excessive driving can cause the sampler or the drill rod to fail.

Pushing

Pushing of samplers is generally accomplished by using the drilling rig hydraulic feed system to press the sampler into the soil at a controlled rate. For safety, the drill rod string should be directly connected to the drill. With the rod string already connected, the drill's hydraulic retract system can be used to withdraw the sampler from the hole. Pushing is the normal mode of advance for the thin-wall tube sampler and is occasionally used for split-barrel or solid-barrel samplers in soft soils.

Rotation while Pushing

Rotation while pushing is a sampling technique used for Denison or Pitcher sampler or continuous-tube samplers advanced with augers. Figure 5.14 shows an example of Denison rotating soil core sampler. The sampling barrel is suspended from a bearing assembly inside the Denison barrel. The bearing assembly allows the barrel to remain stationary while the auger is drilled into the formation. Forcing the sampling barrel to remain stationary allows the formation material to enter the barrel with a minimum of disturbance. The sampler barrel is connected to the drill rods, hex rods, or a wire-line system for retrieval from the bottom of the borehole.

Rock Coring

Introduction

The purpose of rock coring is to obtain undisturbed samples of solid, fractured, or weathered rock formations by use of diamond or carbide-bit drilling methodology (ASTM D 2113; ASTM, 2004e). Coring is used in subsurface exploration for structures such as dams, tunnels, bridges, buildings, and power plants. It is also used in exploration for mineral deposits in rock, for rock quarry materials, and in scientific studies of the earth's

Environmental Drilling for Soil Sampling

DRILLING, CORING, AND SAMPLING TECHNIQUES

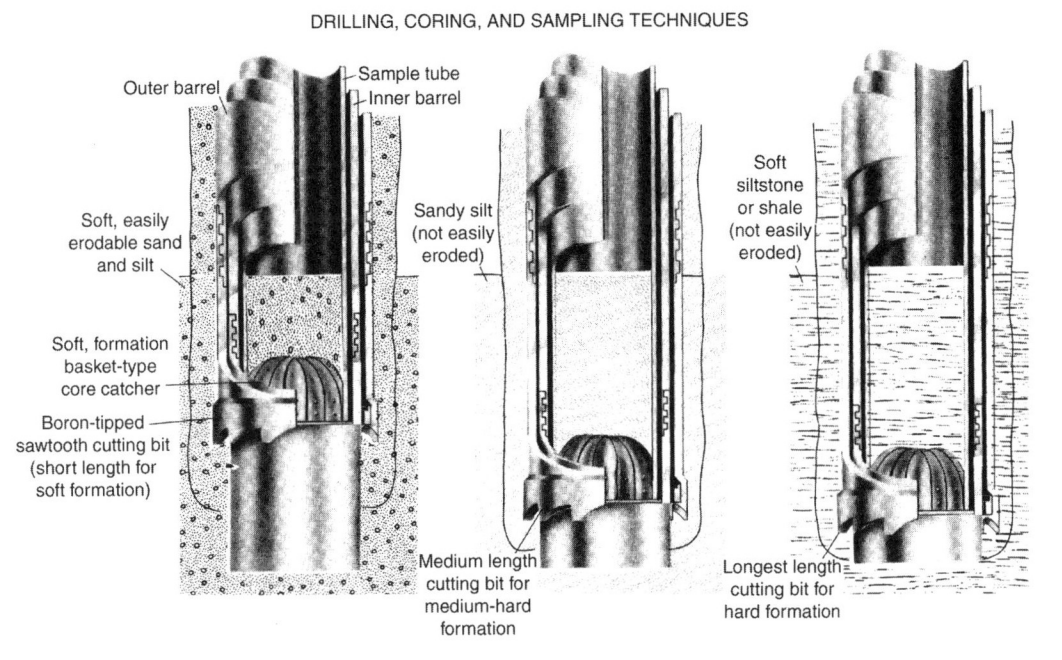

FIGURE 5.14
Relationship of the inner barrel protrusion, pressing with rotation, using Denison double-tube soil core barrel. (Taken from U.S. Geological Survey. With permission.)

structure. Coring equipment is available to drill holes at any angle desired. Core drilling can be used to retrieve a continuous sample, showing the characteristics of the entire interval that was drilled, or it can be used to obtain discrete rock samples from selected intervals.

Rock coring requires the use of drilling fluids to cool the bit and circulate cuttings. The drilling fluid may affect the chemistry of ground water in the fractures and interstices. It is very difficult to avoid drilling fluid effects if direct sampling of cores is required. Most wire-line systems used in rock work well with polymer drilling fluids. These fluids can often be successfully broken down in preparation for well installation, and chemical evaluations can be obtained from the well after well development. Most polymer fluids are approved by the National Sanitation Foundation for installation of water-supply wells in drinking water aquifers.

Samples obtained by core drilling may be tested for load bearing capacity, hydraulic conductivity, porosity, and mineral or chemical content. The types of drilling rigs and coring equipment used in rock core drilling are diverse. Figure 5.15 illustrates a typical rock core barrel utilized in the rotary drilling process. Types of core barrels vary. Most common is the dual-tube barrel. It has an outer barrel to which the bit is attached and an inner tube in which the core is collected.

There are two types of coring systems — conventional and wire-line. Conventional coring systems utilize a core barrel attached to the end of drill rod string. The entire rod string and core barrel assembly are removed from the borehole following each core cutting run. The core barrel is then emptied and returned to the borehole for the next run. Wire-line coring systems use an outer barrel and coring casing of an inner diameter that allows the inner barrel assembly to be brought to the surface through the drill string. A special overshot device is lowered into the coring casing on a wire-line to catch and retrieve the inner core barrel. Wire-line coring systems are quite efficient for deep coring projects and are more economical because they do not require multiple trips in the drill hole. Some wire-line systems, such as the Geo-barrel, can be converted from soil sampling to rock sampling.

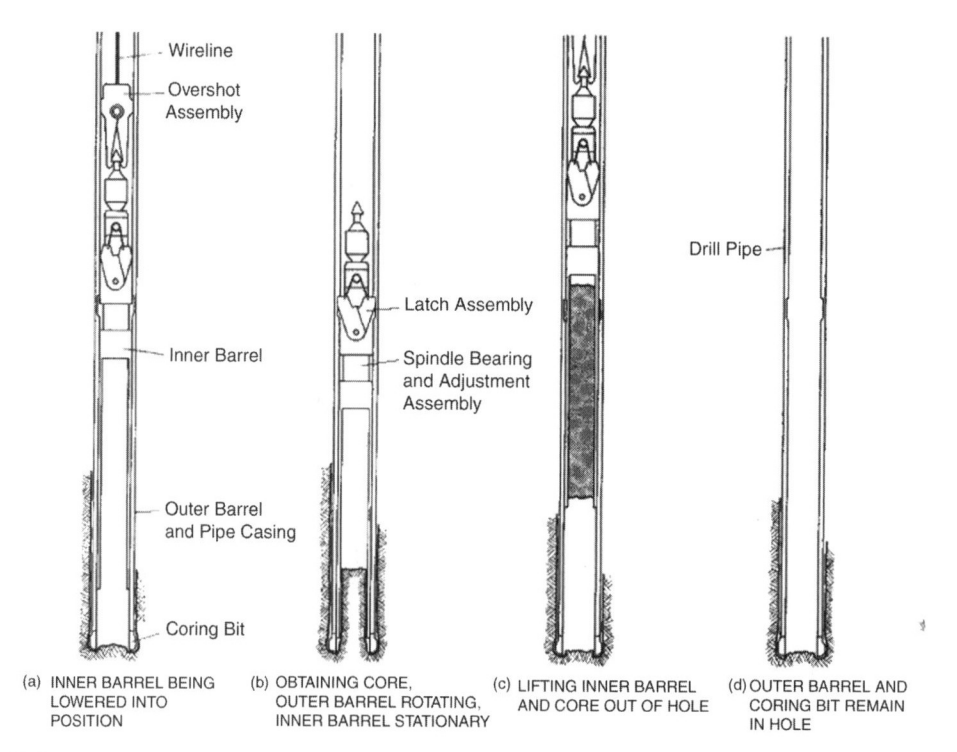

FIGURE 5.15
Wire-line rock core barrel. (Taken from U.S. Bureau of Reclamation Earth Manual, Part I, 3rd ed.)

Rock core drilling with diamond or carbide bits is done with a circulation medium. Because many rock formations are naturally fractured, loss of fluid circulation is a common occurrence. The lack of return water may indicate that rock cuttings are accumulating either in openings in the rock mass or in the annular space between the drill rods and the inside of the core hole. If cuttings are accumulating in voids, the results of any hydraulic conductivity tests, geophysical logging, or other down-hole measurements may be affected. If cuttings are accumulating in the annulus, the core barrel can become stuck or blocked in the hole, resulting in either loss of or damage to the tools or loss of the borehole. Loss of circulation must be avoided or drilling fluid can penetrate the surrounding formation. Clear water easily leaks into the surrounding formation. In some cases, either bentonite drilling mud or polymer drilling fluid can be used to avoid this problem; other additives should also be considered.

Other coring problems commonly arise when the rock mass surrounding the drill hole is not self-supporting. Fragments of rock that protrude into the hole above the bit during drilling may retard or prevent removal of the core barrel. The protruding material may also cause rapid abrasion of the sidewalls of the bit, the core barrel, or the drill rods. Sidewall caving that occurs after the rods have been withdrawn may also prevent the core barrel from reaching the bottom of the hole for the start of the next coring run.

In general, core holes that are small in diameter and close to vertical are less susceptible to wall failure than large diameter holes or holes that are drilled at lower angles. Some types of rock formations (e.g., friable sandstone or solution-channeled limestone or dolomite) are also more susceptible to wall failure than others. The main techniques for overcoming wall failures are (1) use of high-density drilling fluid additive to plug the fractures and temporarily seal the borehole wall, (2) installing casing in the unstable interval and reducing hole size below the casing, and (3) cementing the hole and redrilling after

Environmental Drilling for Soil Sampling

the cement has set. The choice of technique depends on the nature of the investigation and the severity of the problem.

Another problem frequently encountered in rock coring is rapid bit wear during the drilling of highly abrasive rock formations. However, excessive bit wear can also occur in formations that are not particularly abrasive. Impregnated diamond bits have largely replaced surface set diamond bits in coring practice. The diamond particles in an impregnated bit are cast in the matrix. A surface-set bit has the individual diamonds set on the matrix. As an impregnated bit wears, more of the diamond is exposed to do the cutting. As a surface-set bit wears, the stones loosen in the matrix and finally drop out. Impregnated bits are cheaper, more robust, resist impact better, and handle broken formations more effectively than surface-set bits. They are made for most standard sizes and types of core barrels. They are not standardized between manufacturers and thus the user is advised not to attempt interchanges without consulting the maker.

While impregnated bits can be used almost wherever surface-set bits can be used, certain formations may still react better to a surface-set bit. For example, soft shales and siltstones react better to coarse surface-set bits.

Core Losses

During diamond core drilling, losses of core can occur. Core losses in relatively unconsolidated materials are frequent and may be caused by erosion of the core during circulation of the fluid down the core barrel. These losses can be minimized if (1) the bit discharge is directed away from the core (use of bottom or face discharge), (2) the volume of drilling fluid used is kept to the minimum necessary to remove the drill cuttings from the core hole, (3) polymer compounds are added to the drilling fluid, (4) vibration and chatter of the drill rods is minimized, (5) the speed, rotation, and rate of advancement of the drill rod are controlled, and (6) appropriate core-catching devices are properly positioned at the junction between the bit and the core barrel.

Core losses in relatively consolidated materials can occur when the rock being cored is highly fractured and broken or when a fragment of the rock becomes wedged in a portion of the core bit or barrel. If a fragment becomes wedged, the only practical solution is to retrieve the core barrel and remove rock fragment at the surface. Blockages are apparent when a sudden increase in pressure of the drilling fluid is observed or when downward advance of the drill string is obstructed.

Occasionally a dropped core occurs when a segment of core remains attached to the formation, dragging the core out of the barrel. The portion of the core between the bit face and the core retainer may also fall back into the hole. At times, this core fragment can be recovered on the next core run. If the loss is written off on an initial run and then the dropped core is picked up on a subsequent run without being noticed, an error will be introduced into the record of the depth of core. Therefore, care should be taken on every core run to determine (1) percent recovery, (2) amount and location of core loss, and (3) actual depth of the beginning and end of the core run in each case. On coring runs conducted after a portion of the core is dropped, the risk of overfilling the core barrel is present.

Rock Coring Logs

The purpose of the core log is to record all relevant information obtained during drilling and coring and to record a field description of the core. Core log forms are more specialized than standard drilling log sheets and generally will contain columns for recording percent core recovery, rock quality designation (RQD), and number and orientation of fractures (Figure 5.16).

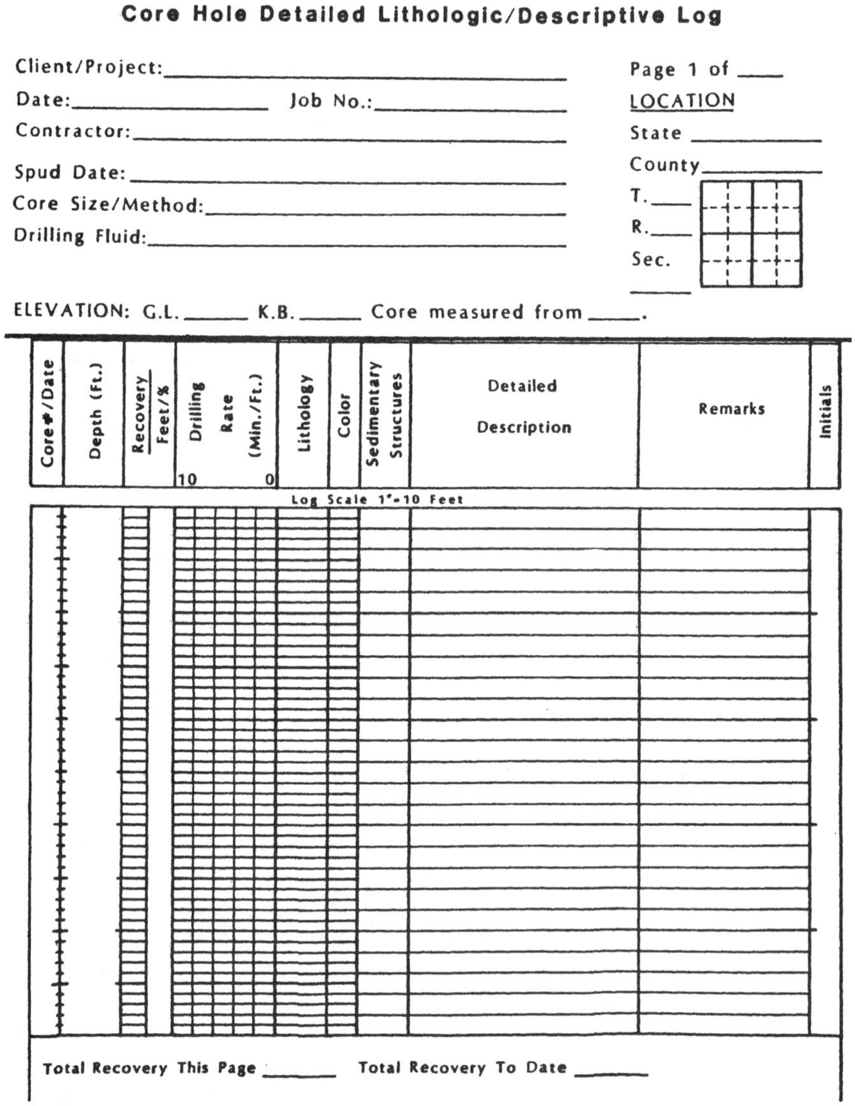
Core Hole Detailed Lithologic/Descriptive Log

Client/Project: ____ Page 1 of ____
Date: ____ Job No.: ____ LOCATION
Contractor: ____ State ____
Spud Date: ____ County ____
Core Size/Method: ____ T. ____
Drilling Fluid: ____ R. ____
Sec. ____

ELEVATION: G.L. ____ K.B. ____ Core measured from ____.

Core#/Date	Depth (Ft.)	Recovery Feet/%	Drilling Rate (Min./Ft.) 10 … 0	Lithology	Color	Sedimentary Structures	Detailed Description	Remarks	Initials

Log Scale 1"=10 Feet

Total Recovery This Page ____ Total Recovery To Date ____

FIGURE 5.16
Rock core log form.

The percent of recovery should be calculated and recorded for each core run (ASTM D 5878; ASTM, 2004q). Dividing the length of recovered core by the length of the core run and multiplying the result by 100 provides the percent recovery. Unrecovered core from one core run that is recovered on the subsequent run should be included in the run in which it was originally cored for the purpose of recovery and other measurements. Such occurrences should be noted on the coring log.

RQD should also be calculated and recorded. Although this measurement was originally developed for geotechnical work, RQD is also a useful measurement for environmental work. First, measure the total length of all sound core pieces 4 in. or longer in length. To calculate RQD, divide this total length by the length of the core run and multiply the result by 100. Both the total length of all sound core pieces 4 in. or longer and the

TABLE 5.1

Rock Quality Terms

% RQD	Descriptive Rock Quality
0 to 25	Very poor
25 to 50	Poor
50 to 75	Fair
75 to 90	Good
90 to 100	Excellent

% RQD are usually recorded on the log. Table 5.1 lists % RQD and a qualitative description of rock. In general, the higher the % RQD, the higher the quality of the rock.

Handling Procedures for Soil and Rock Samples

Introduction

The purpose of soil and rock sampling is to collect representative samples of materials in a zone of interest. A representative sample is assured by controlling the quality of the procedures associated with each phase of the sampling process. This includes collecting the sample, transferring the sample from the sampler to a container, preserving the sample (if necessary), and transporting the sample to the laboratory or storage area. The need for complete, accurate, written documentation of each step of this process cannot be overstated. ASTM Standards D 4220 (ASTM, 2004r) and D 5079 (ASTM, 2004s) provide very useful information on shipping and storage of soil and rock samples.

Each sample should be properly labeled and identified, including a sample number, project number (if applicable), and sample recovery depth interval. The sample should be properly sealed and packaged according to the type of sample, and the container should be labeled with all appropriate information. Examples of soil sample packaging include thin-wall tubes, sampling device liners, glass jars, plastic bags, and other containers such as core boxes, core tubes, or specialized wax containers. Samples are occasionally waxed in the field to preserve moisture content.

In cases where chemical analysis is required, subsamples are often obtained immediately, sealed, and preserved separately (ASTM D 4547 [ASTM, 2004k] or ASTM D 6418 [ASTM, 2004l]). Care should be taken so that the samples remain as undisturbed as possible during the shipping process. For example, thin-wall tubes are shipped vertically if they contain sensitive soils, to prevent any stratification or reshaping of the sample. Prior to packaging and shipping, samples can be photographed with information such as depth, sample number, borehole, and project number visible within the photograph. If the samples get mixed up at any point during the shipping or handling process, identification may be possible through the photographs.

Chain-of-custody forms (Figure 5.17) are frequently utilized to provide a method for identifying every individual who has custody of samples, especially when chemical analyses are to be performed. The sample collector signs the custody of the sample over to the person who is responsible for shipping, who then signs it over to the individual responsible for storage or laboratory analysis. The chain-of-custody form indicates the samples by number, time of pickup, time of delivery, the person from whom the samples were received, the person delivering the samples, the person receiving the samples, and the condition of the samples at pickup and at delivery.

Chain of Custody Record

Project Manager	Project name	Batch identification
Sampled by and title (signatures)		
Laboratory		

Field identification	Date collected	Time	Lab identification	Sample type	Number of containers									Remarks	Initials and date sample destroyed

Relinquished by (signature)	Date	Time	Relinquished by (signature)	Date	Time	Relinquished by (signature)	Date	Time
Received by (signature)	Date	Time	Received by (signature)	Date	Time	Received by (signature)	Date	Time

Point of contact: __________
Remarks: __________
DISTRIBUTION: Original: Acoompanies shipment. Pink copy: Field records. Yellow copy: Associae laboratory

Sample shipped via
☐ Priority Mail ☐ Bus ☐ Other
☐ Express Mail ☐ UPS

FIGURE 5.17
Chain-of-custody form. (Taken from U.S. Bureau of Reclamation Engineering Geology Manual, vol. 1, 2nd ed., 1998, p. 413, 3rd ed.)

Rock Core Samples

After a core has been removed from the core barrel, it should be inspected, logged, and placed into the shipping or storage containers. These may be one of several types of rigid containers with lids. Rock core is relatively fragile and represents a significant investment of time and money. It deserves special labeling, handling, and storage procedures.

Suitable containers for storing and shipping rock cores are core boxes or rigid PVC tubes. Core boxes are generally made of wood, cardboard, or reinforced plastic. Cardboard is generally suitable for temporary storage. Wooden boxes are best for long-term storage. Rock core is placed into the core box left to right starting in the upper left-hand corner so that it reads like a book from top to bottom of the core hole. It should be labeled by writing directly on the core if possible. Information should include core top and bottom, depth, position of core loss zones, and identification of fractures. Fractures that were made after the core was removed from the core barrel or by the coring process are termed mechanical breaks and should be distinguished from naturally occurring fractures.

Core boxes should also be labeled on the tops and ends. Information should include the core run numbers, depth interval of core, hole number, project name and number, and the total length of core in the box. Wooden blocks are usually used as spacers inside core boxes to mark ends of core runs and positions of core loss zones. These blocks should be labeled with the depth. The inside of the core box lid can also be labeled to show core orientation, depths, and other significant features. Figure 5.18 shows an example of core stored in a core box.

Environmental Drilling for Soil Sampling 333

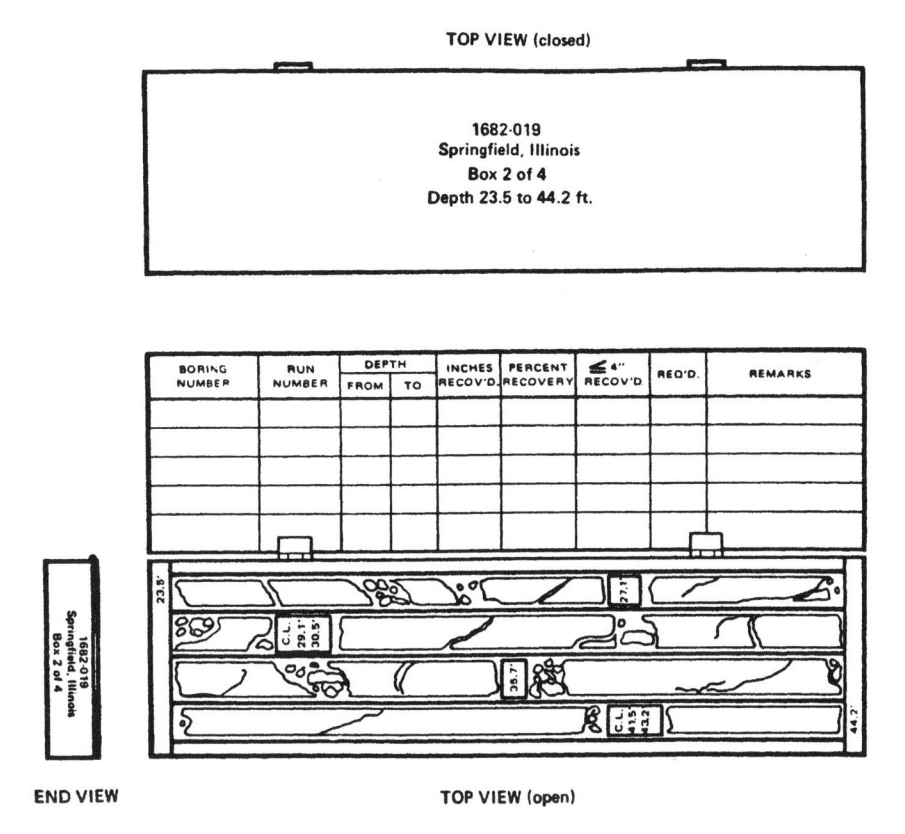

FIGURE 5.18
Arrangement of rock core and labeling of rock core boxes.

Cuttings or Disturbed Samples

Procedures for handling, labeling, and packaging samples of drill cuttings or bulk samples are similar to those used for cores. After the lithologic characterization is completed, the samples are generally packaged and labeled according to depth, project name and number, boring number, location, and date. Cuttings or disturbed samples can usually be labeled on the sample container only, and the container becomes part of the label. If the sample is separated from the container, identification is usually impossible.

Cloth bags, plastic bags, plastic pails with lids, and glass jars are the containers that are most commonly used for storing and shipping cuttings or disturbed samples. Sometimes, cuttings or disturbed samples are split, the original sample is maintained in the original container, and the split sample is shipped to a laboratory for analysis. Once a sample is split, the original sample container should be identified as holding a split sample.

Borehole Logging

Introduction

A borehole log is the written record of drilling, sampling, and well-construction activities for a given borehole prepared by the field hydrogeologist or engineer or by the driller. These reports are prepared on site as the borehole is advanced. They are often the sole record of the significant events that occurred during field work.

The field log is generally converted into a final report log. The final log generally incorporates the results of laboratory tests, geologic soil classifications, and other field tests that may have been performed. Final logs are generally typewritten or computer generated.

Every drilling, sampling, and instrumentation installation project is unique. Many different style forms for logging are used and trying to describe them all is impossible. However, certain types of information are common to most projects and most borehole logs. These are covered in the following sections. More detailed information on borehole logging procedures and desired information is available in ASTM D 5434 (ASTM, 2004t).

Log Heading Information

The log heading is the top part of the borehole log form. It usually includes, at a minimum, the project name and number, property location, surface elevation, surface conditions, type of drilling rig or pumping equipment, bit size and type, driller or logger's name, borehole physical location, and borehole coordinates. Figure 5.19 shows an example of a borehole log. Each sheet of the borehole log should contain the same header information. While it is imperative that the headers of each of these sheets be particular to the project, forms used for different purposes on the same project may be different. As an example, a standard drilling log can be different from a rock core log, which will in turn be different from a pumping test or water-quality testing form. It is important to record information such as water level, reference datum used, casing diameter, depth of hole, casing stickup, casing material, drilling methodology, and sampling methodology.

Log Completion Information

Log completion information should include the time and date that drilling started, the time and date of completion, total depth drilled, total depth cased, abandonment procedures used, if any, and final water-level measurement, among other important data.

Sample Information

Sample information is recorded as the hole is advanced and samples are collected. The types of information that should be recorded on the borehole log are:

- Sampling method (e.g., split-barrel, thin-wall tube, continuous tube sampler, or core barrel).
- Number of blows required to advance a split-barrel sampler to conduct an SPT. Blow counts are usually recorded for each 6 in. interval.
- Size and type of sampler. This information may be recorded only once on the log header if the same sampler is used for every sample, but if different types of samplers are used in the same hole, the type of sampler used for each sample should be recorded.
- Sample number and depth. Sample depth and all other depths recorded on the borehole log should always be measured from the ground surface unless a very compelling reason to do otherwise is presented. If a reference datum other than ground surface is used, it should be recorded.
- Length of drives, pushes, or core runs. In soils, sampling attempts are usually in the range of a few inches to about 10 ft, depending on the hardness of the soil and the drilling method used. Core runs in rock are usually 5 to 20 ft long.

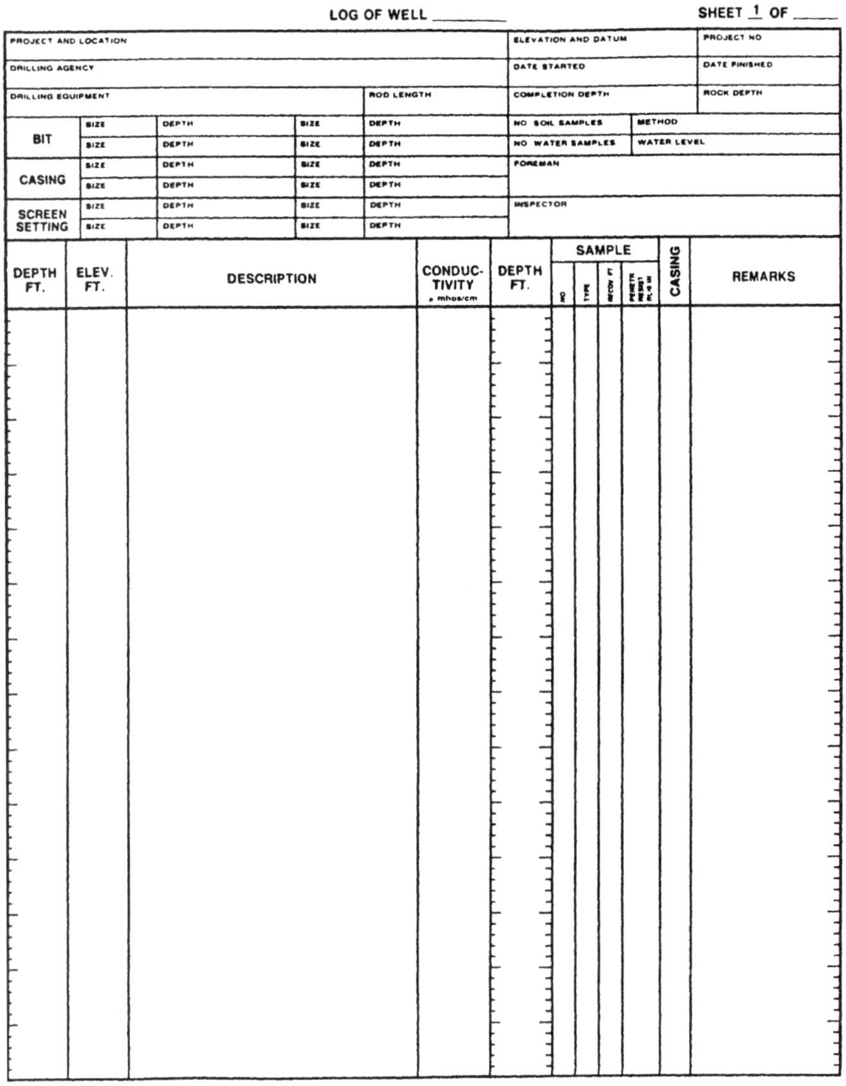

Environmental Drilling for Soil Sampling 335

FIGURE 5.19
First page of a borehole log.

- Length of the recovered sample and the calculated percent of recovery. Percent recovery is an important measure of the degree to which the sample is representative of actual subsurface conditions. Borehole logs generally have a separate column for recording recovery either as percent of the total or for listing the actual length.
- Rock quality designation.
- Portion of the sample saved or submitted to a laboratory for analysis. All of a rock core is usually saved because it stores well and is relatively expensive to obtain.

Soil and Rock Descriptions

Describing soil and rock samples in the field is as much an art as it is a science. The time available to examine each sample is usually limited, lighting conditions are variable, and it may be raining, snowing, or extremely hot. In contaminated areas, the field personnel may be working in protective clothing, including gloves and respirators. Nevertheless, no one is better prepared to provide sample descriptions than the field person who has witnessed the entire process of drilling and collecting the soil sample. Field personnel must therefore be well trained and experienced in sample classification and description.

Several methods for classifying and describing soils are in relatively widespread use. The Unified Soil Classification System (USCS) (ASTM D 2487 [ASTM, 2004u] and ASTM D 2488 [ASTM, 2004v]) is the most popular, and the guidelines for its use are reproduced in Figure 5.20. With the USCS, a soil is first classified according to whether it is predominantly coarse-grained or fine-grained. Coarse-grained soils are then further subdivided according to the dominance of sand and gravel. Fine-grained soils are subdivided on the basis of the liquid limit and the degree of plasticity. The accurate identification of silts and clays can be aided by the use of some simple field tests. Clay is sticky, will smear readily, and can be rolled into a thin thread even when the moisture content is low. When it is dry, clay forms hard lumps. On the other hand, silt has a low dry strength and can be rolled into threads only at a high moisture content. A wet silt sample will puddle and become shiny when it is tapped (dilatancy test).

In some respects, rock classification can be more difficult than soil classification. A nearly infinite number of rock lithologies have been recognized and named, and textbooks and college courses are devoted to the identification of rocks. In many instances, lithology is not the most important factor in evaluating the hydrogeologic aspects of rock, although lithology provides important clues. Features such as weathering, fractures, bedding planes, and porosity are usually more significant, and these can be easily recognized with a moderate degree of experience and training. Two excellent field guides for use in the identification and naming of rocks are the *AGI Data Sheets*, published by the American Geological Institute (American Geological Institute, 2002), and Compton's *Manual of Field Geology* (Compton, 1962) (Table 5.2 and Table 5.3).

Local knowledge and experience can be valuable in assisting in the accurate classification, identification, and logging of soil and rock. However, localized terms, such as slang, tend to take on different meanings to different field personnel. When possible, field personnel, keeping the concept of local terms in mind, should try to review existing logs prior to drilling in a new area. On drilling projects in areas where existing information is limited, local geological agency offices or a local college or university geology department may be a good source of information.

Drilling Information

A record of drilling operations provides valuable documentation for contract records and procedures for future site work. The drilling record information also helps prepare cost estimates and schedules for later drilling and sampling.

Drilling information should be recorded separately from soil or rock descriptions as the hole is advanced. A separate column should be used on the borehole logging form, and entries should be made at a position that indicates the depth. For emphasis, the depth of the entry should also be written with the entry. Types of information that are most useful include the following:

- Changes in penetration rate, rig noise, or drilling action.
- Interruptions in drilling, including breakdowns for repairs, rod trips for changing bits, interruptions to install casing, or interruptions to mix drilling fluids. Bit and casing diameter should be recorded.

FIGURE 5.20
The Unified Soil Classification System. (*Source:* ASTM, 2004u. With permission.)

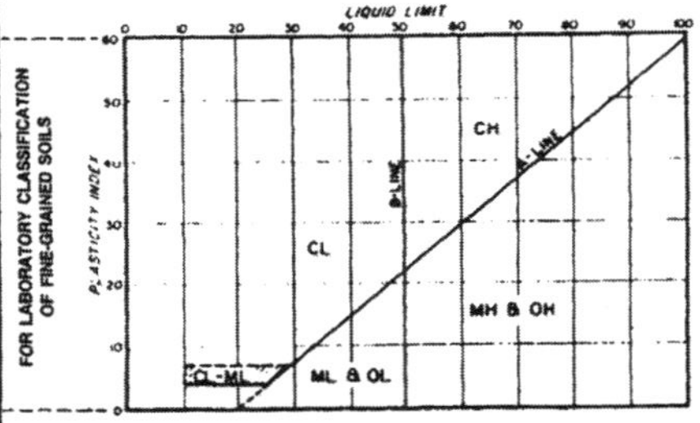

TABLE 5.2

Degree of Weathering

Term	Description
Fresh	The rock shows no discoloration, loss of strength, or any other effect of weathering
Slightly weathered	The rock is slightly discolored, but not noticeably lower in strength than the fresh rock
Moderately weathered	The rock is discolored and noticeably weakened, but 2 in. diameter drill cores cannot usually be broken up by hand, across the rock fabric
Highly weathered	The rock is usually discolored and weakened to such an extent that 2 in. diameter cores can be broken up readily by hand, across the rock fabric. Wet strength usually much lower than dry strength
Extremely weathered	The rock is discolored and is entirely changed to a soil, but the original fabric of the rock is mostly preserved. The properties of the soil depend on the composition and structure of the parent rock

TABLE 5.3

Bedding Characteristics

Average Bed Thickness (ft)	Term
Less than 0.001	Thinly laminated
0.001 to 0.01	Laminated
0.01 to 0.1	Thin bedded
0.1 to 1.0	Medium bedded
Greater than 1.0	Thick bedded

- Conditions of the circulation fluids and cuttings, including loss of circulation, appearance of water in cuttings from auger drilling or rotary drilling with air, and change in color and presence of odors. If water is uncounted in a hole drilled with air, the rate of water production should be estimated and recorded.
- Penetration rate and bit pressure. Drilling rigs are generally equipped with gages that measure hydraulic pressure. Field personnel can estimate the rate in feet per minute or, in difficult drilling conditions, minutes per foot.
- Changes in drilling personnel. For deep boring projects, drilling rigs are sometimes operated 24 hours a day, and two or three separate crews may be involved in drilling a single hole

If a borehole is finished with construction of a monitoring well, a separate form, called a well construction log, is usually used to record the well construction information (Chapter 10).

Drilling Contracts

Introduction

A contract is a binding agreement between two or more parties. A drilling contract is an agreement between a drilling contractor, who agrees to drill one or more borings, and an owner or their representative, who agrees to pay the driller for the work completed. With environmental projects, the customer may be a consultant, the owner of the facility potentially responsible for ground-water contamination, or a government agency.

Thousands of water supply wells have been drilled with nothing more than a verbal agreement and a handshake between the well driller and the customer, and many wells will continue to be drilled this way. However, because environmental drilling projects frequently involve the expenditure of large sums of money and require close adherence to technical specifications and work plans, some form of written agreement between the driller and the customer should be used. The written agreement may be as simple as a purchase order or it may be a detailed construction contract several hundred pages long.

Purchase orders issued by a company or a government agency typically have the terms of the agreement printed on the back in "fine print." Purchase orders are typically heavily loaded in favor of the issuing company or agency and, for this reason, the drilling contractor should review them carefully. The front of the purchase order is typically arranged in columns that are well suited to ordering specific quantities of materials or supplies, but the format is difficult to adapt to the purchase of drilling services without reference to additional technical specifications, proposals, or work plans and drawings. Furthermore, the terms and conditions are usually not adequate to cover the contingencies that can arise during the completion of drilling projects. However, in spite of their shortcomings, purchase orders are widely used because purchasing departments are comfortable with them. Preparing and executing a more formal contract can involve legal review and lawyer's fees.

The remainder of this section describes the important components of a comprehensive drilling contract such as one that might be prepared for a drilling program involving the installation of several monitoring wells. The purpose of this section is twofold: (1) to eliminate some of the mystery involving the language in formal written contracts and (2) to educate the driller, client, facility owner, and consultant on the important components of a standard drilling contract. The purpose of this section is not to demonstrate how to write contracts, but some of the information contained in this section should be useful to the hydrogeologist who is requested to assist in preparing a drilling contract.

Agreement

The agreement is typically the first component that makes up the drilling contract documents. It identifies the parties, it describes the work in general terms, and it specifies the documents that are part of the contract. The first few lines of the agreement usually contain the date that the agreement is signed, and the last lines are the signatures of the parties who enter into agreement.

The agreement may set the start date of work and define the time of completion. It also usually states the contract time. The agreement may include provisions for changes, termination, and method of payment. However, some of these items may be described in other documents that comprise a set of contract documents.

General Conditions

The general conditions contain the principal contract provisions that govern the parties involved. These provisions define the legal relationships among the various parties (i.e., owner, drilling contractor, subcontractors, and consultant).

The general conditions are the "fine print" of the contract. Many large companies that enter into numerous construction agreements typically have a carefully honed set of general conditions to accompany every construction contract. These conditions are usually heavily loaded in favor of the company that prepared them. A drilling contractor who accepts the standard general conditions issued by such a company should be prepared to consult an attorney for legal advice prior to signing the contract. Important clauses or articles in standard general conditions are described subsequently:

Definitions define the owner, the contractor, and subcontractors. These clauses describe the functions, rights, and obligations of each of these parties.

Indemnification requires the contractor to indemnify and hold harmless the owner, his agents, and employees from and against claims, damages, losses, and expenses arising out of the negligence or omissions of the contractor, or the contractor's agents and employees.

Surety bonds are sometimes required to protect the owner against the contractor's default. The bond provides funds to complete the project in the event of default.

Disputes are usually required to be resolved by an arbitration process.

Time for completing the project is usually stated in the formal agreement, but the method of measuring the time is usually stated in the general conditions.

Safety requirements are described. The responsibility for initiating, maintaining, and supervising safety programs may be shared or may be borne by the owner or by the contractor.

Changes may consist of additions, deletions, or other revisions. The general conditions usually contain a clause describing the mechanism for changing the scope of work.

Corrections of defective work are required under the terms of most general conditions. Defective work done by the contractor is corrected at the expense of the contractor.

Termination provisions describe the situation under which either the owner or the contractor may terminate the contract

Other clauses that may be included in the general conditions of the comprehensive drilling contract include royalties and patents; subcontracts; permits, laws, taxes, and regulations; project representatives; warranty and guarantee; insurance; and cleanup.

Construction Drawings

Drawings are generally prepared for drilling projects that are competitively bid. Along with the specifications, they describe and locate the various elements of the drilling project. For most drilling projects, the drawings will consist of one or more maps showing the locations of the borings or wells, access, and possibly utilities. It should be noted that even if utilities are shown on plans, the responsibility for field location generally lies with the drilling contractor. Construction or completion diagrams for all of the types of monitoring wells to be set on the project are generally included. These drawings will show the anticipated depth of the well, the length of the well screen, the length of the blank casing, requirements for surface casing, and position of grouted and sand-packed intervals.

The terms of some general conditions create parity among drawings and specifications. Other terms of general conditions establish priorities, so that one document prevails if the information between the two does not agree. For example, the drawings may show that the slot size of a well screen should be 0.020 in., whereas the specifications state the slot size should be 0.040 in. If the contract documents specify that the terms of the drawings prevail over the specifications, then the driller should base his bid on screen with the

0.020 in. slot size. If no priority is specifically stated, the drawings usually prevail in the event of disputes.

Specifications

Specifications provide technical information concerning materials, components, and equipment with respect to quality, performance, and results. Two types of specifications can be written — procedure style and performance style. Procedure specifications specify in detail the quality, properties, and composition of materials and the method of doing the work. Procedure specifications may specifically state wall thickness for well casing, the size of sand in the filter pack, or the brand of well screen to be installed. Performance specifications specify performance characteristics without stipulating the methods by which the characteristics are to be obtained. Performance specifications are not widely used for environmental drilling contracts.

The clause or "approved equal" is common in many procedure specifications. This clause allows the widest possible competition and permits drilling contractors to shop around for the best bargain when preparing their bid. However, use of the phrase may cause controversy. It raises the question of how to measure and judge quality. If the clause is used, the specifications should include a statement explaining who will be responsible for approving substitute items.

Important components of the technical specifications in a drilling contract may include some or all of the following items:

- Drilling methods
- Drillers logs, geophysical logs, and driller's reports
- Sampling methods and depth at which samples will be obtained
- Casing installation
- Grouting
- Well screens
- Filter pack
- Plumbness and alignment
- Well development
- Aquifer testing
- Abandonment requirements

Other clauses may also be included. The consultant or hydrogeologist usually writes the technical specification section. Along with the drawings, the technical specifications play an important role in determining the final quality of monitoring well construction. A complete set of drawings and technical specifications reduces the need for on-the-spot decision making by the owner's representative. However, because of the unknown nature of the subsurface conditions on many environmental projects, the need for regular supervision to provide answers to construction questions cannot be completely eliminated.

Special Conditions

Some drilling contracts may contain special conditions that supplement and modify the general conditions. They are typically used where a set of standard general conditions for construction contracts is used for a drilling contract. Special conditions are written individually for each project and usually govern in the event of conflict with the terms in the general conditions.

The purpose of the special conditions overlaps somewhat with the purpose of the technical specifications. Important components of the special conditions document may include:

- Scope and general description of work including well size, depth, and location
- Subsurface information including lithology and depth to the water table
- Work schedule
- Liquidated damages or a fixed sum, which is paid by the contractor for each day's delay in completion beyond an agreed-upon date
- Special permits and taxes
- Property boundaries
- Location of existing utilities

Other Documents

Noncontractual documents such as instructions for bidders, advertisements for bids, proposals prepared by bidders, statutes, local rules and regulations, and well construction codes may or may not be specifically referenced in the contract documents. Nevertheless, they are used as guides in resolving contract disputes and in determining the legal obligations of the parties. The subsurface conditions and site inspection clauses in instructions to bidders are a frequent source of contract disputes for drilling contracts. In the instructions to bidders, the clauses typically require the bidder to represent that "he has visited the site and has familiarized himself with the local conditions." Disputes arise when the plans and specifications contain information concerning site conditions that differ substantially from those encountered during drilling. The information in the plans and specifications may have been obtained from earlier drilling or it may represent an educated guess. Frequently, a clause is inserted in the general conditions, which states that data regarding subsurface conditions should not be relied upon for estimating drilling costs. The driller is expected to have sufficient experience in the area to judge drilling conditions and develop a realistic bid.

Conclusions

The successful completion of an environmental drilling project with the collection of representative soil or rock samples and the construction of functional monitoring wells is the goal of most environmental drilling programs and also should be the goal of all of the participants in the program. The greatest obstacle to the successful completion of an environmental drilling and well installation program is the "as it is written, so let it be done" mentality that exists in the minds of regulators and specification writers and the "that is the way we always do it" attitude of the contactors. Drilling is an inexact science. Mother nature, in her quest to establish mystery, did not lay everything down in an orderly fashion. Thus, none of the drilling method or technique works best everywhere. Only through an understanding of the vagaries of nature, the cooperative reasoning of all the parties and the utilization of the accumulated experiences of those doing the work can the completion of environmental drilling projects be an enjoyable and rewarding experience for all involved.

References

American Geological Institute, 2002, AGI Data Sheets, Third Edition; Compiled by J.T. Dutro, R.V. Dietrich, and R.M. Foose, American Geological Institute, Alexandria, VA, 294 pp.

Anderson, Keith E., *Water Well Handbook*; Missouri Water Well and Pump Contractors Assn., Inc., Belle, MO, pp. 281, 1979.

ASTM, 2004a, Standard Guide for Use of Hollow-Stem Augers for Geoenvironmental Exploration and Installation of Subsurface Water-Quality Monitoring Devices; ASTM Standard D 5784, ASTM International, West Conshohocken, PA.

ASTM, 2004b, Standard Practice for Using Hollow-Stem Augers for Geotechnical Exploration and Soil Sampling; ASTM Standard D 6151, ASTM International, West Conshohocken, PA.

ASTM, 2004c, Standard Guide for Use of Direct Rotary Drilling With Water-Based Drilling Fluid for Geoenvironmental Exploration and Installation of Subsurface Water-Quality Monitoring Devices; ASTM Standard D 5783, ASTM International, West Conshohocken, PA.

ASTM, 2004d, Standard Guide for Use of Direct Air Rotary Drilling for Geoenvironmental Exploration and Installation of Subsurface Water-Quality Monitoring Devices; ASTM Standard D 5782, ASTM International, West Conshohocken, PA.

ASTM, 2004e, Standard Practice for Rock Core Drilling and Sampling of Rock for Site Investigation; ASTM Standard D 2113, ASTM International, West Conshohocken, PA.

ASTM, 2004f, Standard Guide for Use of Dual-Wall Reverse Circulation Drilling for Geoenvironmental Exploration and Installation of Subsurface Water-Quality Monitoring Devices; ASTM Standard D 5781, ASTM International, West Conshohocken, PA.

ASTM, 2004g, Standard Guide for Use of Cable-Tool Drilling and Sampling Methods for Geoenvironmental Exploration and Installation of Subsurface Water-Quality Monitoring Devices; ASTM Standard D 5875, ASTM International, West Conshohocken, PA.

ASTM, 2004h, Standard Guide for Use of Casing Advancement Drilling Methods for Geoenvironmental Exploration and Installation of Subsurface Water-Quality Monitoring Devices; ASTM Standard D 5872, ASTM International, West Conshohocken, PA.

ASTM, 2004i, Standard Guide for Selection of Drilling Methods for Environmental Site Characterization; ASTM Standard D 6286, ASTM International, West Conshohocken, PA.

ASTM, 2004j, Standard Guide for Selection of Soil and Rock Sampling Devices Used With Drilling Rigs for Environmental Investigations; ASTM Standard D 6169, ASTM International, West Conshohocken, PA.

ASTM, 2004k, Standard Guide for Sampling Waste and Soils for Volatile Organic Compounds; ASTM Standard D 4547, ASTM International, West Conshohocken, PA, 10 pp.

ASTM, 2004l, Standard Practice for Using the EnCore Sampler for Sampling and Storing Soil for Volatile Organic Analysis; ASTM Standard D 6418, ASTM International, West Conshohocken, PA, 9 pp.

ASTM, 2004m, Standard Test Method for Permeability of Granular Soils (Constant Head); ASTM Standard D 2434, ASTM International, West Conshohocken, PA.

ASTM, 2004n, Standard Practice for Thick-Wall, Ring-Lined, Split-Barrel Drive Sampling of Soils; ASTM Standard D 3550, ASTM International, West Conshohocken, PA.

ASTM, 2004o, Standard Test Method for Penetration Test and Split-Barrel Sampling of Soils; ASTM Standard D 1586, ASTM International, West Conshohocken, PA.

ASTM, 2004p, Standard Practice for Thin-Walled Tube Sampling of Soils for Geotechnical Purposes; ASTM Standard D 1587, ASTM International, West Conshohocken, PA.

ASTM, 2004q, Standard Guide for Using Rock-Mass Classification Systems for Engineering Purposes; ASTM Standard D 5878, ASTM International, West Conshohocken, PA.

ASTM, 2004r, Standard Practice for Preserving and Transporting Soil Samples; ASTM Standard D 4220, ASTM International, West Conshohocken, PA.

ASTM, 2004s, Standard Practice for Preserving and Transporting Rock Core Samples; ASTM Standard D 5079, ASTM International, West Conshohocken, PA.

ASTM, 2004t, Standard Guide for Field Logging of Subsurface Explorations of Soil and Rock; ASTM Standard D 5434, ASTM International, West Conshohocken, PA.

ASTM, 2004u, Standard Practice for Classification of Soils for Engineering Purposes (Unified Soil Classification System); ASTM Standard D 2487, ASTM International, West Conshohocken, PA.

ASTM, 2004v, Standard Practice for Description and Identification of Soils (Visual-Manual Procedure); ASTM Standard D 2488, ASTM International, West Conshohocken, PA.

Compton, R.R. *Manual of Field Geology*, John Wiley and Sons, Inc., New York, NY, 1962.

Munch, J.H. and R.W.D. Killey, Equipment and methodology for sampling and testing cohesionless sediments; *Ground-Water Monitoring Review*, 5(1), 38–42, 1985.

Zapico, M.M., S. Vales and J.A. Cherry, 1987, A Wire-Line Piston Core Barrel for Sampling Cohesionless Sand and Gravel Below the Water Table; *Ground-Water Monitoring Review*, 7(3), 74–82, 1987.

6

Use of Direct-Push Technologies in Environmental Site Characterization and Ground-Water Monitoring

Wesley McCall, David M. Nielsen, Stephen P. Farrington, and Thomas M. Christy

CONTENTS

Introduction 347
What Is DP Technology? 348
Advantages and Limitations of DP Technology 350
    Advantages of DP Technology 351
        Improved Site Access 351
        Minimal Generation of Investigation-Derived Wastes 351
        Minimal Subsurface Disturbance 351
        Rapid Sampling and Logging Capability and Faster Overall Site Investigations 351
        Lower Cost 352
        Rapid, Cost-Effective Vertical Profiling Capability 352
        Acquisition of More Depth-Discrete Samples 352
    Limitations of DP Technologies 353
        Geological Constraints 353
        Depth Constraints 353
        Soil Compaction and Smearing 354
How DP Technology Is Used 354
What Are DP Systems? 355
Equipment for Advancing DP Rods and Tools 356
    Manual and Mechanically Assisted Methods 356
    Conventional Drilling Rigs 356
    DP CPT Rigs 357
    Percussion DP Machines 360
    Vibratory Heads 362
DP Rod Systems 363
    Single-Rod Systems 364
    Dual-Tube Systems 365
DP Sampling Methods 367
    Presampling Considerations 368
    DP Soil Sampling Methods 368
        Single-Rod Soil-Sampling Methods 369
            Nonsealed Soil Samplers 369
            Sealed Soil Samplers 371
        Dual-Tube Soil-Sampling Methods 373
            Split-Barrel Samplers 374

      Solid-Barrel Samplers 374
      Dual-Tube Precore System 376
      Wireline CPT™ 377
   Applications and Limitations of Soil Sampling Methods and Tools 378
      Open vs. Sealed Samplers 378
      Single-Rod vs. Dual-Tube Method 378
   Continuous vs. Discrete Sampling and Lithologic Logging 379
      Sample Integrity and Chemical Analysis 379
DP Soil-Gas Sampling 381
   ASTM Guidance for Soil-Gas Sampling 381
   Soil Matrix and Probe Insertion 381
   Common Probe Configurations 382
      Expendable Drive Point Samplers 382
      Retractable Drive Point Samplers 384
      Retraction Distance 385
      Multiple-Depth Sampling 385
      Sampling with Cased Systems 386
      Soil-Gas Implants 387
   Soil-Gas Sampling Train 391
   Quality Assurance and Quality Control Procedures for Soil-Gas Sampling ... 392
Ground-Water Sampling Methods 393
   Single-Rod Methods 394
      Exposed-Screen Sampling Tools and Profilers 395
      Sealed-Screen Samplers 398
      Combined Sampling Tools 403
   Dual-Tube Methods 404
      ConeSipper® 407
   Applications and Limitations of DP Ground-Water Sampling Tools 407
   Sample Recovery from Small-Diameter DP Rods and Tools 411
      Bailers 412
      Inertial-Lift (Tubing Check-Valve) Pumps 412
      Peristaltic and Other Suction-Lift Pumps 413
      Gas-Drive Pumps 414
      Bladder Pumps 414
In Situ Measurements Using Specialized DP Probes 416
   CPT Systems 416
     CPT Platforms 416
     CPT Probe Systems 417
     CPT Data Acquisition 418
     CPT Data Interpretation 419
     Tools for Ascertaining Soil Properties 420
        Soil Electrical Resistivity and Conductivity Probes 420
        Soil Moisture Probes 423
        Nuclear Logging Tools 424
     Tools for Delineating Subsurface Contamination 425
        Continuous Vapor Profiling 426
        LIF Spectroscopy 427
        Raman Spectroscopy 430
        Fuel Fluorescence Detector 430
     Other CPT-Deployed Probes and Sensors 432
     Membrane Interface Probe 433

MIP System Components	433
Field Operation	434
Example Logs and Interpretation	435
Other DP Tools	437
Video Imaging Systems	437
Applications of DP Technology to Monitor Well Installation	438
Requirements for Monitoring Well Construction and Use	439
Annular Seal and Grouting Requirements	440
DP Well Installation Methods: Advantages and Limitations	441
Driven Well Points	442
Open Hole Procedure	442
Mandrel-Pushed Screen and Casing	443
Cased Method	445
Naturally Developed Wells	445
Filter Packed Wells	447
Prepacked Well Screens	448
Comparisons of DP Wells to Conventional Wells	450
Results of Recent Research	450
Applications and Limitations of DP for Well Installation	454
Driven Well Point and Mandrel-Driven Screens	457
Open Hole	457
Cased Hole	457
Summary	457
Methods for Sealing DP Probe Holes	458
Gravity Pouring	459
Re-Entry Grouting	459
Retraction Grouting	460
Advancement Grouting	461
Tremie Grouting	461
Summary	462
References	463

Introduction

Passage of the Resource Conservation and Recovery Act (RCRA) by Congress in 1976 addressed management of hazardous and nonhazardous wastes generated at active industrial facilities and landfills. In 1980, passage of the Comprehensive Environmental Response Compensation and Liability Act (CERCLA, commonly called "Superfund") addressed hazardous wastes at abandoned facilities and spills of hazardous materials (U.S. EPA, 1990). Together, RCRA and CERCLA established the earliest national requirements for characterization and monitoring of sites at which contaminated soil and ground water were suspected or found. The U.S. Environmental Protection Agency (U.S. EPA) was charged with administering the rules and regulations associated with these critical pieces of environmental legislation and with overseeing the environmental investigations conducted by those seeking to comply with the regulations.

Initially, existing methods used for geotechnical investigations and installation of water-supply wells were adopted and modified to fit the environmental investigation requirements expressed in RCRA and CERCLA. However, the limitations of these methods for conducting accurate, cost-efficient environmental site investigations quickly became apparent. The sampling strategies often used in traditional geotechnical drilling (e.g., hollow-stem auger [HSA] drilling on a grid pattern with 100 ft spacing and split-barrel sampling at every 5 ft depth interval) rarely provide the detail necessary to resolve the heterogeneities in geological materials that control contaminant transport in the vadose zone and in ground water. These heterogeneities can have a profound effect on the movement and distribution of non-aqueous phase liquids (NAPLs) (Pitkin et al., 1994) and on the concentrations of aqueous-phase contaminants, which can vary by several orders of magnitude over vertical distances of only a few centimeters (a few inches) (Cherry, 1992). Furthermore, the large volume of contaminated drill cuttings and decontamination fluids generated from the use of these methods during environmental investigations is problematic and very costly. Handling and disposing of the contaminated waste materials generated from these drilling and sampling methods often consumes a large fraction of a project budget. Generating, handling, and managing these wastes also create additional and significant exposure hazards for workers, local residents, and the environment.

Equipment and methods largely developed in the late 1980s and 1990s overcame many of the limitations of conventional drilling methods for environmental investigations. Smaller, less expensive machines that advanced small-diameter probes into the subsurface by displacing rather than excavating soils were developed for soil, soil-gas, and ground-water sampling and logging changes in subsurface conditions. These new machines and the tools and methods used with them became known as "direct-push" (DP) technology.

What Is DP Technology?

DP technology includes a variety of methods for collecting samples of environmental media or information detected *in situ* from the subsurface and for installing monitoring wells. These tasks can be accomplished using DP machines that advance tools by pushing (using hydraulic rams and the static weight of the machine), driving (using hydraulic or mechanical percussion hammers), or vibrating (using a vibratory drive head) a tool string into the ground. The DP machines typically used range from small, lightweight all-terrain or tracked vehicles to cargo vans and pickup trucks to massive 10 to 40 ton (9,000 to 36,000 kg) cone-penetrometer trucks. The tools are typically advanced using either one string or two concentric strings of small-diameter rods or drive casing. Rotary drilling action is generally not used to advance the rods or tools and essentially no drill cuttings are generated from the DP sampling or logging process. Some small-diameter tools can even be advanced manually, using "slam bars" or hand-held rotary percussion hammers. In general, the smaller, lighter equipment is used for sample retrieval or logging a single subsurface variable (e.g., soil electrical conductivity) in cases where the objective is near the ground surface (generally within 100 ft [30 m]). The larger equipment can be outfitted with more complex tool strings to collect more sophisticated data on several subsurface variables simultaneously and used to access much greater depths (in excess of 250 ft [76 m]). The manual methods are usually limited to depths of 10 to 15 ft (3 to 4.5 m) and have been used mostly for shallow soil-gas sampling.

The first applications of DP technology actually date back to the early 1930s, when an *in situ* soil-testing instrument called a cone penetrometer or "Dutch cone" was developed in Holland for use in geotechnical testing (mainly determining the bearing capacity) of soils ranging from clays to coarse sands and fine gravels. The original cone penetrometers involved simple mechanical measurements of the penetration resistance to pushing a tool with a conical tip into the soil. Electronic measurements were begun in 1948 and further improved in the early 1970s (de Reister, 1971); a friction sleeve was added in the 1960s (Begemann, 1965) to measure soil cohesive strength. The present-day cone penetration testing (CPT) procedure makes use of the reaction weight of a heavy (10 to 40 t [9,000 to 36,000 kg]) truck and hydraulic rams to advance mechanical or electronic cones placed at the tip of the tool string. CPT is widely accepted for determining the engineering properties of soil and delineating soil stratigraphy, at a resolution of centimeters (<1 in.). Standardization of the geotechnical applications of the cone penetration test was established in 1986 by ASTM Standard D 3441 (ASTM, 2004a). Later ASTM Standards have standardized the use of CPT and other DP technologies for various environmental site characterization and ground-water monitoring activities. In addition to the mechanical and electronic cones, a variety of other CPT-deployed tools have been developed over the years to provide additional valuable subsurface information. These tools include:

- A piezometric cone, to directly measure formation hydraulic head and to conduct pore pressure dissipation tests for characterizing formation hydraulic conductivity (Torstensson, 1975; Wissa et al., 1975)
- A soil resistivity probe, to define soil electrical properties (which can be used to distinguish soil type and, in some cases, degree of saturation and water chemistry)
- A soil dielectric probe (with electrical resistivity measurement), to infer volumetric soil moisture content from the dielectric contrast between water, soil solids, and air
- Probes utilizing laser-induced (UV) fluorescence (LIF) analysis to detect the presence and relative concentration of petroleum hydrocarbons (Lieberman et al., 1991)
- A down-hole video system, to display, in real time, and record video images of soil and other conditions encountered during tool advancement (Lieberman and Knowles, 1998)

CPT rigs have also been used to install piezometers and monitoring wells as large as 2 in. in nominal diameter. CPT systems are discussed in more detail in later sections.

The more recent environmental applications of DP technology were developed in the mid- to late 1980s, originally to enable investigators to collect soil-gas samples from shallow soils. The first generation of this equipment was mounted on lightweight carrier vehicles (cargo vans or pickup trucks) and depended strictly on hydraulic rams, using the static weight of the vehicle to advance the tool string. Thus, depth capability was limited to less than 20 ft (6 m) for most equipment, even with the very small (less than 1 in. [25 mm]) diameter rods and sampling tools that were used. DP equipment has evolved to the point at which rigs using hydraulic or mechanical percussion hammers or vibratory systems can routinely collect samples, soil conductivity information, and *in situ* data on the presence of certain chemical compounds to depths in excess of 100 ft (30 m) in favorable soil conditions. Newer rigs can also advance drive casing as large as 3.25 to 4.5 in. (82.5 to 114 mm) outside diameter to depths greater than 60 ft (18.3 m) in favorable soil conditions. This capability enables the installation of piezometers, monitoring wells, multilevel monitoring systems, or other devices from 0.5 in. (13 mm) to as large as 2 in. (51 mm) nominal diameter.

DP methods are now used widely for soil sampling at targeted or discrete depth intervals and for continuous soil sampling, to construct detailed lithologic logs and geologic cross-sections. A variety of DP ground-water sampling tools have been developed for temporary installation (often less than 1 h) to obtain discrete samples for environmental analysis, either in a fixed lab or a mobile lab in the field. Several tools have also been designed to allow for collection of ground-water samples at multiple depths during one advancement of the tool string (vertical ground-water profiling). Profiling methods can provide the detailed three-dimensional information on contaminant distribution and variations in ground-water chemistry that is required for optimal design of ground-water monitoring systems or selection and implementation of remediation methods. Many of the DP ground-water sampling tools can be used to determine formation hydraulic conductivity using conventional slug testing field methods and data analysis methods (Chapter 14).

The logging devices advanced with DP machines can provide information on formation lithology or contaminant distribution, often on a scale of inches or centimeters. CPT piezocone measurements provide continuous profiling of soil mechanical properties that vary with lithology. Electrical logging devices provide information on the electrical conductivity or resistivity of the formation that correlates with the lithology and, in some cases, with water chemistry (e.g., where salt-water intrusion is suspected or where a highly ionized plume, such as landfill leachate, exists). A semipermeable membrane used on the membrane interface probe (MIP) allows for detection and quantification of total volatile organic chemical compounds (VOCS) in the low parts-per-million (milligrams-per-liter) range. Optical devices, such as the fuel fluorescence detector (FFD) and LIF probe can distinguish the presence or absence of light non-aqueous phase liquids (LNAPLs) directly through a sapphire window in the probe. Advances in these and other systems are under development, which may allow for detection and identification of specific compounds in the parts-per-billion (micrograms-per-liter) range.

Regulatory agency recognition of the applications and advantages of DP technology has led to increased use of the technology. U.S. EPA (1997) included DP as one of the primary means of providing thorough and rapid site assessments for underground storage tank (UST) sites. Thornton et al. (1997) concluded that DP technologies should be more widely used in CERCLA site assessments because they allow rapid, high-quality data acquisition in the field and enable investigators to use flexible work plans, which can significantly reduce costs. Applegate and Fitton (1997) also concluded that DP technologies, as an integral part of a rapid site assessment approach, could be applied to Superfund sites to allow more of the allocated resources to be directed to actual site remediation rather than to more lengthy and costly conventional site assessment approaches. The U.S. EPA Technology Innovation Office also advocates the use of innovative tools, including DP technology, to improve the cost effectiveness and quality of hazardous waste site characterization and monitoring (Crumbling, 2000).

Advantages and Limitations of DP Technology

DP technology offers a number of significant advantages over conventional investigative technologies but, as with any technology, there are also some limitations to its use. The technology is evolving to address most of the current limitations. In most situations, investigators making decisions on which methodology to use in the characterization or monitoring of a given site will find that the advantages of using DP technologies far outweigh the limitations. There are, however, some limitations that will preclude the use of DP technologies at some sites. The general advantages and limitations of DP technologies are discussed in the following sections. Additional tool- or method-specific advantages and limitations are covered later in this chapter.

Advantages of DP Technology

The advantages of DP technologies, compared with conventional investigative technologies, are discussed in the following sections.

Improved Site Access

Most DP machines (with the exception of CPT rigs) are much smaller, lighter weight, more mobile, more compact, more self-contained, and less intrusive than conventional drilling equipment, so DP sampling can be done in many areas where drilling rigs do not normally have access. DP sampling is possible in narrow alleyways, near buildings, beneath overhead obstructions (e.g., canopies or power lines), on soft soils, in residential areas, and in other difficult access areas. Manual DP sampling can be done in remote locations, heavily vegetated areas, and areas otherwise inaccessible to rigs including within buildings and in basements.

Minimal Generation of Investigation-Derived Wastes

When DP tools are advanced, they displace subsurface materials rather than excavating them, so no drill cuttings and almost no other investigation-derived wastes (IDW) are produced during the course of a site-characterization program. This produces additional benefits including significantly reduced exposure of on-site personnel and off-site receptors to potentially hazardous materials, reduced disturbance of formation materials, and reduced overall investigation costs, particularly for operations in contaminated soils (Schroeder et al., 1991; Ohio EPA, 2005). In contrast, conventional drilling methods often produce substantial quantities of IDW, which may require special and costly management and disposal practices. For example, a typical 4.25 in. (10.8 cm) I.D. by 8 in. (20.3 cm) O.D. hollow-stem auger (HSA) produces approximately one 55 gal (208 l) drum full of drill cuttings for every 16 ft (5 m) of drilling. A mud-rotary rig may use several hundred gallons of drilling fluid per hole, which requires appropriate handling and disposal in addition to the drill cuttings produced. Disposal costs for these wastes may utilize a large part of the project budget, which could otherwise be used for additional sample collection.

Minimal Subsurface Disturbance

DP sampling tools and wells can be installed with significantly less disturbance to formation materials than conventional drilling methods typically cause. In DP installations, formation materials are simply displaced rather than subjected to significant disaggregation or exposure to drilling fluids. Additionally, DP boreholes are much smaller and generally not subjected to exposure to atmospheric conditions during tool installation or sample collection, providing less opportunity for potential sample alteration (e.g., loss of volatile contaminants).

Rapid Sampling and Logging Capability and Faster Overall Site Investigations

DP machine setup and installation of sampling and logging tools is much simpler than sampling with conventional drilling technologies and allows for more rapid sample recovery and data acquisition, as well as faster installation of monitoring wells for later sample acquisition, which increases overall project speed. Production rates achievable using DP technology are greater than with conventional drilling in most unconsolidated formations (Schroeder et al., 1991; Booth et al., 1993; Ohio EPA, 2005). It is not uncommon for a competent DP contractor to be able to achieve 300 to 500 ft (91 to 152 m) of investigation in an 8 to 10 h day under favorable geological conditions (depending on whether discrete or continuous samples are collected). In contrast, conventional drilling may achieve a production rate of 150 to 200 ft per day under similar conditions, because of the increased time requirements for decontaminating equipment, managing drill cuttings, and collecting samples. Investigations conducted using DP sampling tools can generally be completed in one fourth to half of the time required for conventional investigations.

Lower Cost

The overall cost for a DP sampling program is typically between one fourth and half of the cost for a sampling program using conventional technologies because the equipment is less expensive to operate, fewer personnel are generally required, sample collection and data acquisition are faster, production rates are higher, fewer capital expenditures are required, decontamination operations take less time and require fewer materials, and much less IDW is generated. Reduced overall project cost means that more samples could be collected with the same project budget which, in turn, allows for better definition of subsurface conditions (U.S. EPA, 1997, 1998a; Ohio EPA, 2005).

Rapid, Cost-Effective Vertical Profiling Capability

Perhaps the most significant advantage of DP technologies is realized when using DP logging and sampling tools for vertical profiling. DP logging tools can be used to collect highly detailed, high-resolution data on a variety of subsurface conditions (soil stratigraphy, soil electrical resistivity or conductivity, soil moisture content, and presence and concentration of several classes of contaminants). Most DP data acquisition systems provide a minimum spatial resolution of 2 to 5 cm (1 to 2 in.) in the vertical, which greatly improves geological, hydrological, and geochemical interpretation, and these data are available in real time. This capability allows field staff to direct the course of an investigation during field operations. In contrast, the typical conventional sampling approach (split-spoon sampling at 5 ft intervals) offers very poor resolution, and analytical results (from a fixed laboratory) are typically not available until 6 or 8 weeks after the field investigation, so the sampling approach must depend on predetermined locations. The ability of DP technologies to provide real-time data has become an important component of the accelerated or expedited approach to site characterization (Chapter 2). Using this approach often ensures that zones of contamination can be located and defined in a single field mobilization, during which media samples can also be collected if field or laboratory analytical verification is desired.

DP sampling tools can also be used to collect either multiple depth-discrete samples of soil, soil gas, or ground water in a single borehole or to collect several samples with some tools as the hole is advanced, to enable easy and fast vertical profiling. Multiple profiles can be used to construct detailed cross sections that allow excellent definition of three-dimensional variations in soil, soil gas, or ground-water chemistry. This is a significant advantage at hydrogeologically complex sites and at NAPL release sites, where stratification may be significant and contaminant distribution highly complex. Comparable detailed definition of contaminant distribution using conventional technology would be exceedingly expensive if not cost prohibitive.

Acquisition of More Depth-Discrete Samples

DP sampling tools typically collect samples from intervals ranging from a few inches to 4 ft, so the data are truly depth discrete, as well as time discrete, and represent a smaller spatial scale than conventional technologies (e.g., monitoring wells). Samples collected from conventional monitoring wells are representative of averaged conditions over the length of the well screen, which may range in length from 5 to 20 ft (1.5 to 6 m) or more. Whether one sample or the other will provide the data required for any particular site depends on the objectives of the site-characterization or monitoring program. Because of the larger sampling interval, conventional monitoring wells are more likely than discrete DP samples from a short interval to detect the presence of contamination, albeit at potentially diluted levels. However, in cases where only a small portion of the screened zone is contaminated or where the contaminant is originally present at a very low level, dilution may be significant enough to prevent contaminant detection. In contrast, because the sampling interval accessed by DP sampling tools is smaller, the possibility of missing a thin contaminant zone is increased. These zones are, however, easily identified by vertical profiling, which is routinely done with DP methods. The use of DP sampling points also improves the detection of both aqueous-phase and non-aqueous-phase contaminants at representative levels when the tools are properly placed.

Limitations of DP Technologies

The limitations of DP technology are discussed in the following subsections:

Geological Constraints

Unfavorable conditions for any DP technology include gravel layers, cobbles, boulders, dense, stiff soils, highly cemented soils such as caliche, or bedrock (with the exception of highly weathered bedrock such as saprolite), all of which impede or prevent the advance of the tool string. These limitations are the same as the limitations of HSA drilling, the most common drilling method used in environmental site characterization and in the installation of conventional monitoring wells. However, because many cities and associated industrial and waste disposal facilities are situated along rivers, a large proportion of sites requiring environmental investigation are amenable to DP methods. In addition, a large proportion of the upper midwestern and northeastern USA are covered with glacial tills and glacio-fluvial deposits and much of the eastern seaboard and gulf coast are mantled by coastal plain sediments, all of which are generally amenable to investigation with DP methods.

Depth Constraints

The actual penetration depth for any DP system depends on many variables including:

- The type of driving system used (e.g., the power and mass of the equipment used)
- The diameter of the down-hole tools and drive rod or casing being advanced
- The resistance that subsurface materials offer to tool advance (which is controlled by formation density, degree of consolidation, and grain size)
- The soil friction acting on the down-hole tools and drive rod or casing

The greater the diameter (or surface area) of the rod or drive casing and the higher the plastic clay content of subsurface soils, the greater the soil friction and the more limited the penetration depth. While methods are available to decrease the friction between soil and rods (e.g., using a lead tool or a drive point that has a larger diameter than the rod string), these techniques increase the possibility of cross-contamination in some soils.

Recent advances in DP technology have significantly increased the capabilities of DP machines in terms of penetration depth and hole diameter. Most DP systems are capable of achieving depths of 50 to 100 ft (15 to 30 m), although newer, larger, and more powerful DP rigs can drive tools to depths approaching 200 ft (61 m) under favorable subsurface conditions (e.g., loose, unconsolidated materials finer than gravel). This penetration depth is comparable to the practical depth limitation for HSAs in favorable conditions. CPT rigs are capable of penetration depths in excess of 250 ft (76 m) under favorable geologic conditions, using the static weight of the rig (10 to 40 t [9,000 to 36,000 kg]) alone to provide the reaction force to advance the rods.

Soil Compaction and Smearing

The displacement of subsurface geological materials by the advance of DP tools causes some soil compaction and some minor disaggregation of formation materials that may result in localized reductions in formation hydraulic conductivity, particularly in predominantly fine-grained materials. The shape and exterior surfaces of the sampling tools and drive rod minimize the downward transport of soil (referred to as "dragdown") and water (referred to as "cross-contamination"). However, there is some evidence that advancing DP probes in predominantly fine-grained materials can cause some smearing of clays across thin zones of coarser materials. This can lead to difficulty in obtaining representative water samples or accurate water levels or hydraulic test results from either tools or wells installed in these formations. For DP well installations and some DP sampling tools, this borehole damage must be removed through development to establish a good hydraulic connection between the tool or well and the formation.

How DP Technology Is Used

Selecting a DP technology appropriate for a specific site or application requires a clear understanding of data collection goals, because many tools are designed for only one specific purpose (e.g., collection of ground-water samples). However, it is important to consider that DP tools are often used to their best advantage when used in combinations. For example, potential positions for screens of monitoring wells installed for the purpose of tracking a dissolved-phase contaminant plume can be selected most effectively by first examining continuous soil samples, stratigraphic information generated from a CPT, or a soil electrical conductivity log. This allows investigators to identify the coarser formation materials that may serve as preferential migration pathways. The zones with highest hydraulic conductivity can be identified using either a piezocone to conduct pore-pressure dissipation tests or slug test results from open boreholes or piezometers installed using DP methods. This can help investigators to identify preferential contaminant transport pathways in subsurface materials, whether or not they are contaminated. Identification of actual contaminated zones to be targeted for monitoring can be accomplished by using one of several available probes to remotely detect specific types of contaminants. For example, the CPT-deployed FFD or LIF probe may be used to determine the presence and relative concentrations of petroleum hydrocarbons in real time and to determine, through their association with these hydrocarbons, the presence of organic solvents that may exist as DNAPLs (Kram et al., 2001a). In cases where VOCs are of interest, the MIP can detect them in the vadose zone and the saturated zone. Once the zones with the contaminants of interest have been identified, they can be sampled using one or more of a variety of discrete ground-water sampling tools and the samples analyzed in the field to confirm the types and levels of contamination present. Alternatively, ground-water sampling tools can be used to conduct vertical profiling of ground-water chemistry to locate the most highly contaminated zones and produce a three-dimensional depiction of the contaminant plume. All of this work can assist greatly in optimizing the positions of monitoring wells and well screens (or multilevel monitoring systems), and, ultimately, the number of wells (or multilevel sampling ports) required to monitor the contaminant plume. The wells or multilevel monitoring systems themselves can be installed using DP machines and used for either short-term or long-term monitoring of site ground-water conditions.

DP technologies are very powerful when used in conjunction with quantitative and semiquantitative field analytical methods, ranging from relatively simple colorimetric methods to immunoassay methods, spectrophotometric methods, x-ray fluorescence (XRF), LIF, field-portable gas chromatography, and other methods. Some of these methods are capable of providing data quality equivalent to that available from fixed laboratory equipment. The availability of high-quality analytical information in the field enables the investigator to modify the sampling plan in response to real-time results. The combination of innovative sampling and analytical technologies allows much more rapid, efficient and cost-effective three-dimensional characterization of subsurface conditions than is possible using traditional drilling methods. Rapid sampling capabilities make DP technology ideally suited for use in accelerated or expedited site-characterization programs (Chapter 2), during which a comprehensive site characterization effort can often be completed in one mobilization to the site (Connecticut DEP, 2000). Investigators using DP methods are able to collect many more samples in a shorter period of time and at a lower cost than is possible using conventional methods (U.S. EPA, 1998a). Because DP investigations produce high volumes of samples at rapid rates, the use of mobile laboratories for chemical and physical analyses of samples is economically feasible for many site-characterization programs.

What Are DP Systems?

A DP system generally consists of a sampling or logging tool (composed of a drive head; tool body; tip; for some tools, a sample liner or chamber; and, for some ground-water sampling tools, a screen), drive rod or casing extending from the sampling zone to the surface, and the equipment used to advance the tool or logging device and rod or casing. The keys to using DP technology to collect the highest quality samples and subsurface data possible are:

- Understanding the capabilities and limitations of each of the tools and systems available and how these tools and systems are deployed, individually and in combinations, to their best advantage
- Following appropriate procedures for using the tools and systems and collecting samples and data from them
- Understanding the appropriate uses and limitations of the samples and data collected from the tools and systems

The following sections describe in detail the DP tools and systems used in environmental investigations, their capabilities and limitations, and the procedures that should be followed to collect the best possible samples and data. The equipment used in DP systems is also described in detail in U.S. EPA (1997) and ASTM (2004b, 2004c, 2004d, 2004e).

Equipment for Advancing DP Rods and Tools

Equipment available to advance DP sampling tools and logging devices into the subsurface ranges from simple, manual slide hammers to 40 t (36,000 kg) CPT trucks. Some of the more popular DP units are operated by hydraulics and are mounted in conventional pickup trucks or on other small truck chassis. Local soil and geological conditions and any access limitations should be evaluated before selecting methods and equipment to use for any site investigation. In general, low-density, fine-grained materials that are poorly consolidated are easy to penetrate with smaller DP equipment, while densely compacted coarse sands with gravel may require use of a larger DP machine or a unit with drilling capability. DP equipment is not designed for penetration of consolidated rock.

Manual and Mechanically Assisted Methods

In the mid-1980s, manual methods were often used to drive a hollow pipe several feet into the soil to collect soil-gas samples. A slam bar or slide hammer commonly used for driving small metal fence posts was adapted and improved upon for DP applications (Figure 6.1). These manual rod drivers are now used for advancing small-diameter soil and ground-water sampling tools, as well as soil-gas sampling devices. Several models of hand-held electric or pneumatic rotary hammers can also be used to advance the smaller DP tools for sampling (Figure 6.2). These hammers can make advancing the tools quicker and easier but do not assist with the retrieval operation.

Rods and tools that are driven into the ground must eventually be extracted. Simple mechanical jack assemblies (Figure 6.3) have been adapted or designed specifically for this operation. Some manufacturers have developed a more elaborate system for manual sampling activities that integrates the driving and retrieval systems into a compact unit (Figure 6.4).

Manual and mechanically assisted DP sampling methods work best in fine-grained soils as these soils are usually easier to penetrate. These methods can be useful for sampling in remote or otherwise inaccessible locations such as heavily forested terrain or in the basement of a building. However, the capabilities of these methods are typically limited in both penetration depth and sample size. Manual sampling for soil gas, soil, and ground water with smaller diameter tools (1 to 1.5 in. [25 to 38 mm] O.D.) has been conducted at depths ranging from just a few feet (1 m) below grade to more than 30 ft (10 m) in favorable soil conditions. Manual methods generally use single-rod systems and small-bore sampling tools that limit the volume of sample that can be recovered.

Conventional Drilling Rigs

Conventional drilling rigs (Figure 6.5a) have been used to conduct a hybrid type of DP sampling for many years. After a drilling rig excavates a borehole to the intended sampling depth, the cathead and a 140 lb (63.5 kg) hammer (or a hydraulic safety hammer) is used to drive either a split-spoon soil sampler or other tools into the undisturbed soil below the drilled borehole. Some drillers have added a percussion hammer to their rigs to facilitate DP sampling. Alternatively, the weight and the pull-down of the drilling rig have been used to press samplers, such as thin-wall (or Shelby) tube samplers, into the formation beneath the drilled borehole. These methods combine conventional drilling with DP methods to conduct sampling and testing activities.

FIGURE 6.1
Using a 30 lb (13.6 kg) manual driver to advance probe rods with small-diameter tools may be successfully conducted to depths of more than 20 ft under amenable conditions. Samples of soil, soil gas, and ground water may be collected with manual methods.

Having rotary drilling and DP capabilities combined in one machine can be advantageous under difficult drilling and sampling conditions. The drill can be used to penetrate difficult formation materials, such as thick caliche layers in arid areas, and DP sampling methods can then be used below this zone. However, many of the advantages of DP (e.g., elimination of drill cuttings, smaller and more mobile equipment) are sacrificed. These combination machines have been used to conduct DP sampling for soil gas, soil, and ground water and to install monitoring wells at various sites and soil conditions at depths approaching 200 ft (61 m).

DP CPT Rigs

CPT systems use the static weight of a large CPT truck (Figure 6.5b) combined with hydraulic rams to advance the CPT cone or other sampling or logging devices into the

FIGURE 6.2
Several varieties of electrical or pneumatic hammers have been adapted for use in advancing DP tools. These hammers do not assist with extraction of the tools.

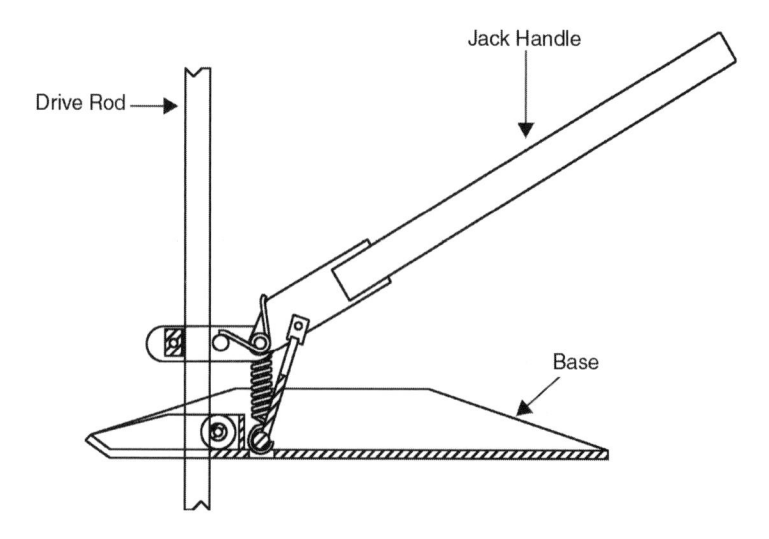

FIGURE 6.3
A simple mechanical jack may be used to extract manually driven tools used for soil, ground-water, or soil-gas sampling.

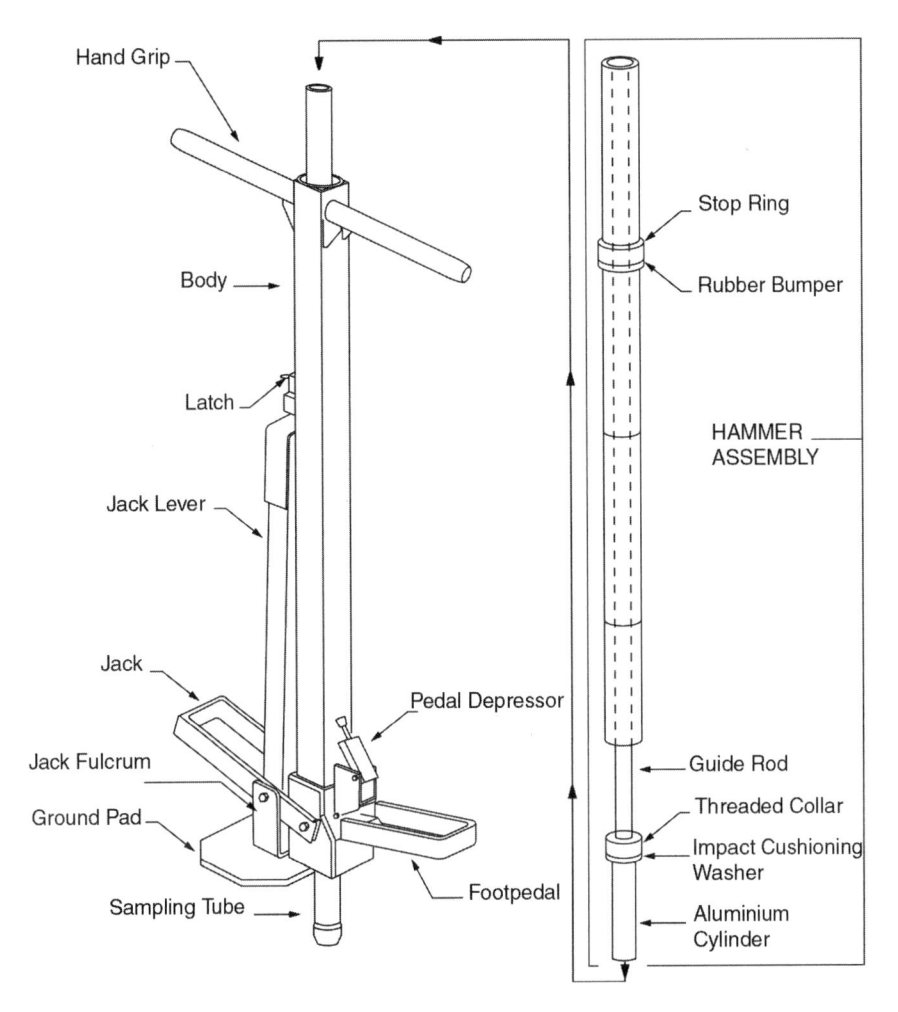

FIGURE 6.4
At least one manufacturer has designed an integral system for advancing and retrieving soil sampling tools. This manual system may be used for continuous or discrete-depth sampling. (Modified from U.S. EPA, 1998c.)

subsurface. Additional weight may be added to large 10 to 40 t (9,000 to 36,000 kg) CPT trucks via tanks of water or lead blocks to increase their penetration capability. CPT systems have been used for DP soil sampling, ground-water sampling, and for installing piezometers and monitoring wells. The primary advantages of CPT units are their ability to reach great depths to conduct standard cone penetration testing to define the physical or geotechnical parameters and stratigraphy *in situ*, in addition to sampling, chemical sensing, video inspection, and sample collection. Later sections describe these and other capabilities of CPT systems.

CPT trucks are used to conduct sensing, sampling, and logging operations to depths of over 250 ft (76 m) in favorable soil conditions. Medium-duty (up to 20 t) CPT systems are similar in size and maneuverability to a conventional drilling rig, but heavyweight (30 t and larger) systems can have difficulty in accessing very rugged locations, sites with soft surface soils, or sites where working space is limited. CPT systems are frequently mounted on low-ground-pressure tracked vehicles for use in rough terrain and areas of weak bearing capacity, such as bogs and marshes.

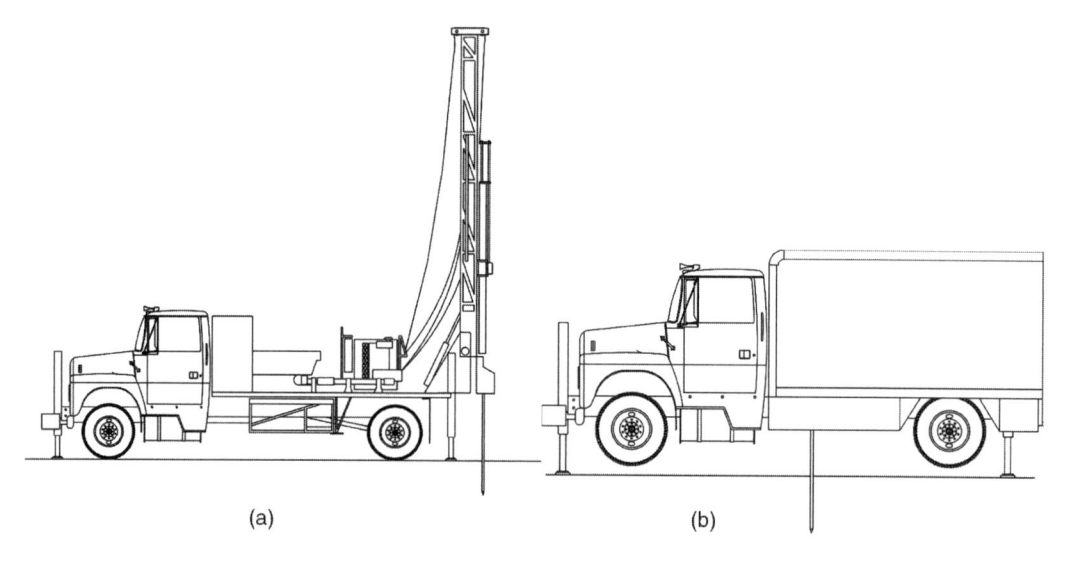

360 Handbook of Environmental Site Characterization and Ground-Water Monitoring

FIGURE 6.5
Both conventional drilling rigs (a) and CPT units (b) may be used to advance many DP tools and logging devices. Access with these larger vehicles may be problematic at some locations. Overhead electrical utilities are an additional consideration with the mast on conventional drilling equipment. Drilling rigs may weigh 5 to 10 t and CPT units may weigh as much as 20 to 40 t.

Percussion DP Machines

DP sampling using hydraulic (Figure 6.6) or mechanical percussion hammers has been used widely for soil and ground-water sampling, electrical logging and contaminant detection, and installation of small-diameter monitoring wells. The impact of the

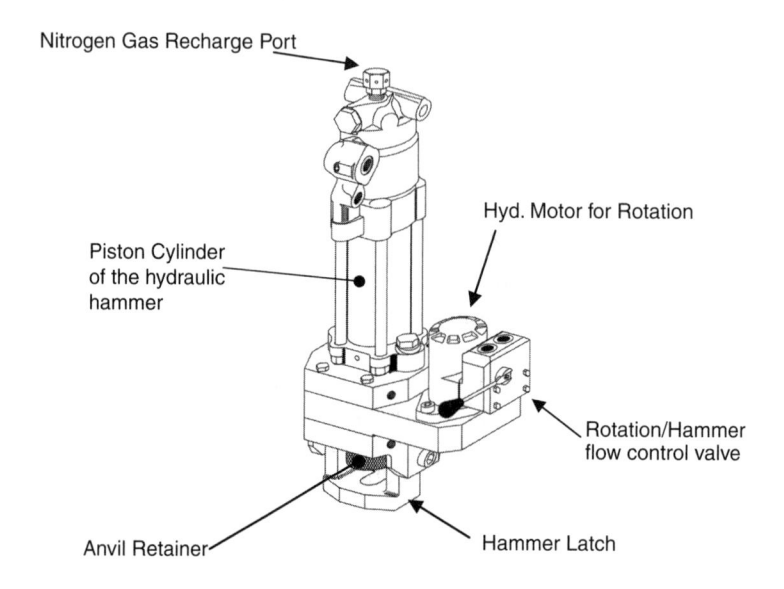

FIGURE 6.6
Hydraulic-powered percussion hammers have been modified to meet the needs of DP sampling and logging. These hammers operate at blow frequencies of 30 Hz or more and impact force ranging from about 5 to 50 t. A hydraulic motor to provide rotary action in the head facilitates penetration of surface pavement such as concrete or asphalt.

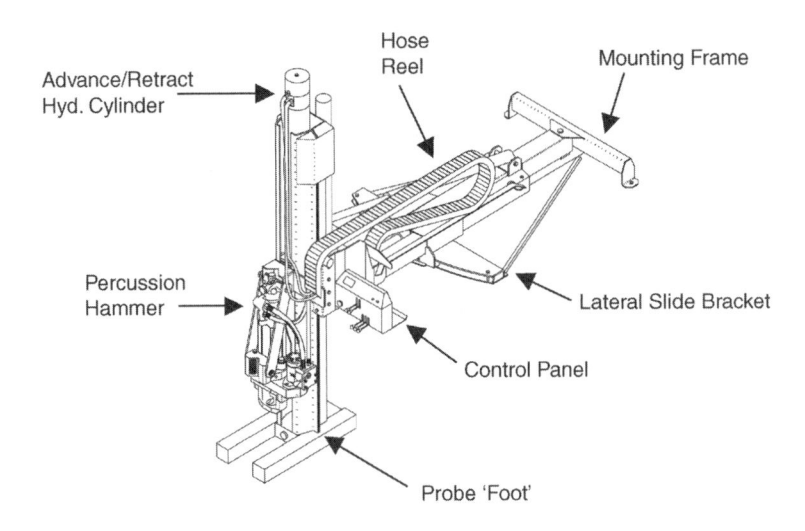

FIGURE 6.7
A typical percussion-type DP machine that is often mounted in a heavy-duty pickup truck. Hydraulic power for these smaller machines is provided by a hydraulic pump powered by the vehicle engine. A hydraulic fluid reservoir with optional fluid cooler is mounted in the vehicle. These units have advanced sampling and logging tools to depths of over 100 ft (about 30 m) under amenable soil and geologic conditions.

percussion hammer reduces the static weight needed for successful penetration and sampling under many soil conditions. The reduced weight and size requirements for the percussion-type DP machines means that smaller, lighter, and less-expensive equipment (Figure 6.7) is required, improving mobility and site access and lowering overall investigation costs. Many percussion-operated DP units have been mounted in conventional pickup trucks or cargo vans (Figure 6.8). Percussion-type DP units also have been mounted on track-driven machines (Figure 6.9) that allow operation in rugged areas not accessible to conventional drilling equipment. Some of the larger, more powerful percussion-type DP units may also be equipped with rotary drill heads, which provide the capability to use auger drilling when needed.

Depending on the formation conditions and the size of the tools being driven, the smaller percussion-operated DP units have been able to penetrate to depths ranging

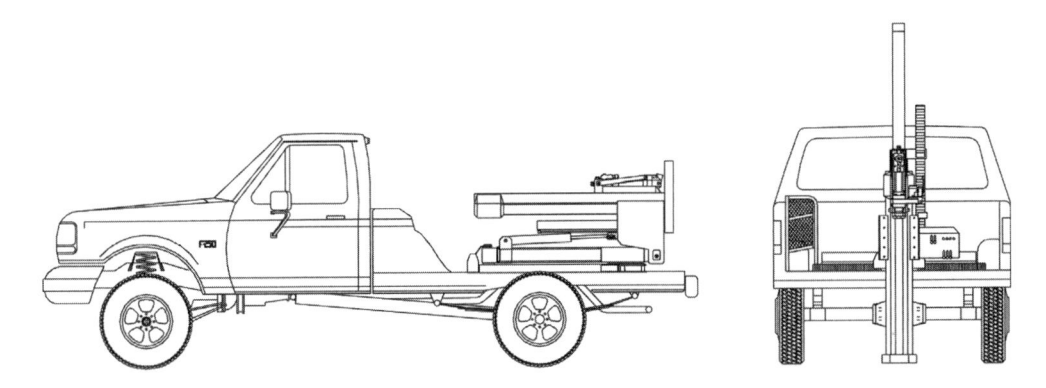

FIGURE 6.8
Diagrams of a pickup truck-mounted DP unit. Side view cutaway shows the unit folded in the vehicle bed for transport. Rear view shows the DP unit extended in upright position prepared for tool advancement.

FIGURE 6.9
Using a track-mounted DP machine to drive 3.25 in. (83 mm) probe rod for the purpose of installing a monitoring well. Track-mounted DP machines are popular as their compact size allows them to access sampling locations with limited space. They also offer low ground pressure and can maneuver in rugged terrain. The DP unit may be equipped with an optional auger head. This allows for rotary drilling when field conditions or project specifications require this option.

from 50 ft (15 m) to more than 100 ft (30 m). Under favorable formation conditions, the larger, more powerful percussion-type DP units have penetrated to depths exceeding 200 ft (61 m). These machines can be used to operate single-rod and dual-tube soil- and ground-water sampling devices, to collect soil-gas samples, and to conduct logging operations. Using larger diameter drive casing, the more powerful percussion-type DP units have been used to install small-diameter monitoring wells (0.5 to 1.5 in. [13 to 38 mm] I.D.) to depths exceeding 100 ft (30 m). The percussion of the hydraulic or mechanical hammers enables these units to easily penetrate clays, silts, sands, and fine gravels at many locations. However, penetration of thick layers of caliche, dense sands or sediments with coarse gravel, cobbles or boulders is often very difficult to impossible. Under some conditions, the larger DP units with an optional drill head can be used to penetrate difficult soils to facilitate DP sampling below these zones.

Vibratory Heads

Some percussion DP machines use a vibratory head that clamps onto the tool string and vibrates it at relatively high frequencies to advance the tool string and sampling devices or drive casing into the subsurface. The vibratory action, which is essentially a standing harmonic wave transmitted through the tool string, greatly reduces the side-wall friction on the tool string and results in an increased rate of penetration and greater sampling depths than would be possible using only the static weight of the rig. Many of these systems are mounted on small, highly mobile tracked rigs or Bobcat-like vehicles. These systems are capable of penetrating to depths of about 100 ft (30 m) in favorable soil conditions (no cobbles, boulders, or bedrock).

Some manufacturers have combined a more powerful vibratory head with rotary drilling capability to develop the sonic or rotasonic drilling technique (Figure 6.10) (Chapter 5). These units are typically large, powerful machines capable of penetrating very dense soils and difficult formations that contain abundant cobbles or boulders. This technology has excellent capability for recovering continuous soil samples from formations that were considered difficult or impossible to sample using other techniques. Depending on formation conditions, the rotasonic equipment may be able to penetrate to depths in excess of 700 ft (215 m).

DP Rod Systems

In early applications of DP methods to environmental investigations, simple tools were advanced into the subsurface with small-diameter rods for sampling activities. These simple systems used a single string of rods for collecting soil, soil-gas, and ground-water samples. If continuous soil sampling was conducted, the tool string was tripped in and out of the open borehole multiple times to collect the samples. While this is a simple time- and cost-effective method, there are some sampling and data-quality concerns that arise from working through an open borehole. To overcome some of the limitations associated with this single-rod method, dual-tube systems were developed, which allowed for DP sampling through a cased borehole. This method uses a larger diameter outer rod string (or drive casing) to case the hole and a smaller diameter inner rod string to recover samples. Sampling through a cased borehole this way eliminates concerns with sloughing of soils that may carry contaminants down hole, but has its own limitations.

FIGURE 6.10
The rotasonic drilling method is usually considered a DP method even though rotary action is combined with the push and sonic vibration. Currently, rotasonic drilling is conducted with two trucks, one being the rotasonic rig and the other a support truck with drilling steel and water supply. While these systems are expensive to mobilize and operate, they have successfully recovered samples from difficult formations at depths approaching 800 ft (242 m).

Single-Rod Systems

Single-rod systems use a single string of hollow steel rods to connect the driving equipment at the surface to the sampling tools or sensing equipment being driven into the subsurface. The drive rods used in single-rod systems range in size from 1 in. (25 mm) O.D. × 0.5 in. (13 mm) I.D. to as large as 2.125 in. (54 mm) O.D. × 1.5 in. (38 mm) I.D. In general, smaller diameter tool and rod strings have less tip resistance to penetration and less side-wall friction, so they will penetrate the same formation materials more easily and to greater depths than larger diameter tool and rod strings. Thread designs on the rods vary from simple single-lead threads to double- or even triple-lead threads on larger diameter tooling. Designs with multiple-lead threads are more robust and usually require fewer rotations on the rod joint to complete the connection, so they are more efficient in the field. Basic components of a single-rod sampling system include the drive rods, drive cap, pull cap, and down-hole sampling or logging device (Figure 6.11).

Single-rod sampling systems have been used widely for soil, soil-gas, and ground-water sampling. An example of single-rod soil sampling with an open-tube sampler (Figure 6.12) shows the basic steps for continuous soil sampling. As sampling is conducted, the open borehole may provide a conduit for material at shallower depths to slough down the

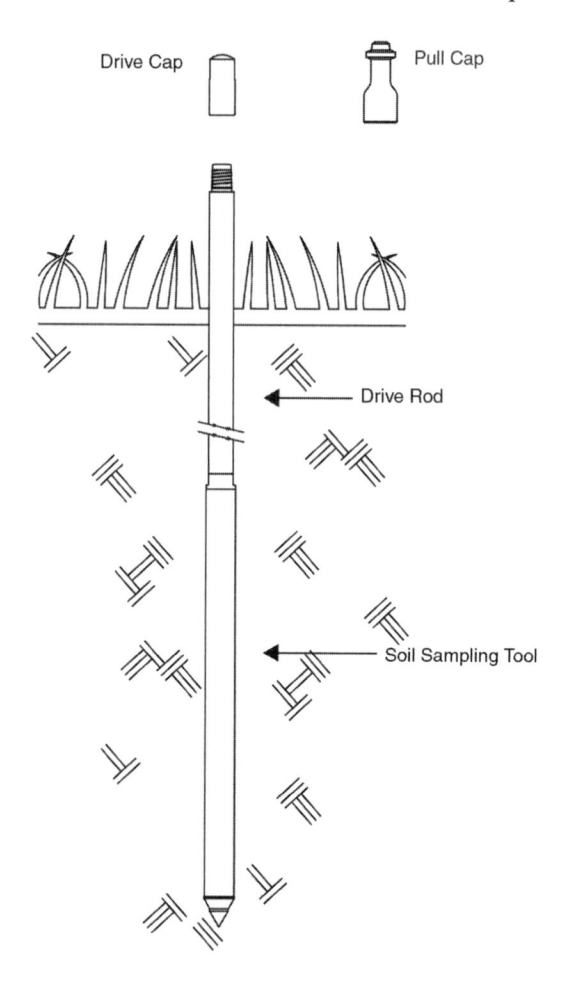

FIGURE 6.11
Components of a single-rod sampling system include the drive rods, drive cap, pull cap, and the down-hole sampling or logging device. A piston-operated soil sampler is shown in this diagram.

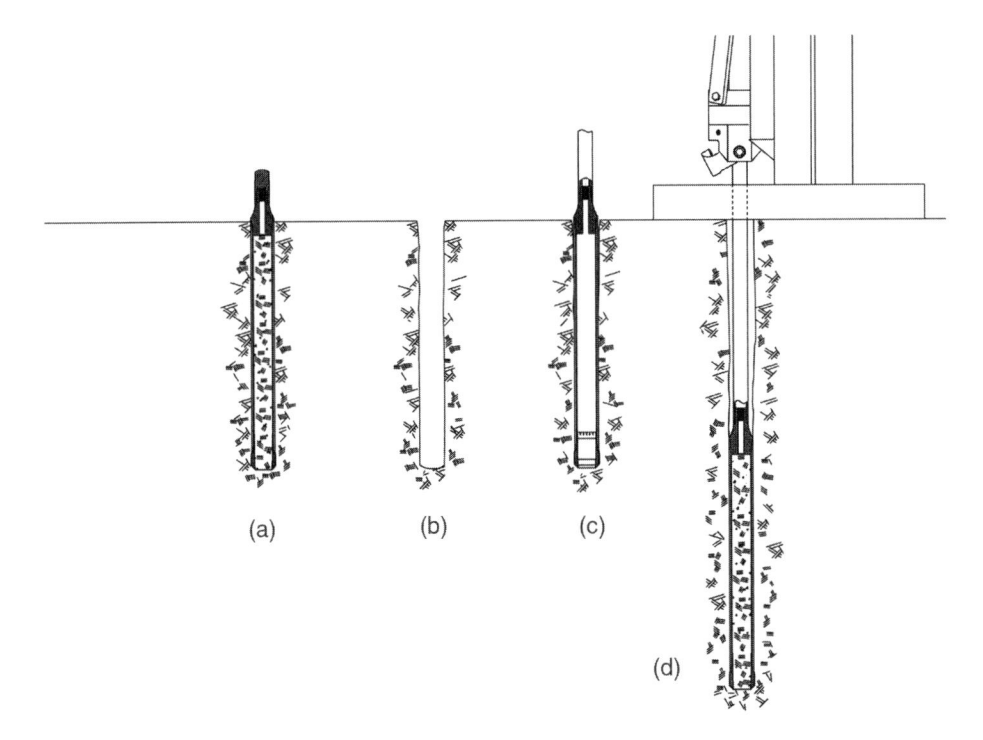

FIGURE 6.12
The primary steps for operation of an open-barrel, single-tube soil sampler by DP methods. Continuous coring must be conducted when open-barrel samplers are used. (a) The assembled sampler is advanced from the surface to a depth equal to the length of the core barrel. (b) The sampler is retracted, leaving an open borehole that can cave or collapse in noncohesive soils. (c) A decontaminated sampler is advanced to the bottom of the open borehole, and drive rods are added as necessary. (d) Once the sampler is at the base of the borehole, it is advanced through the next sample interval and retrieved to collect the sample. Under most soil conditions, use of the hydraulic hammer during sample collection improves sample recovery.

hole, resulting in the potential for cross-contamination. Another potential concern, also resulting from sloughing of material down hole, is the possible misidentification of lithology and incorrect geologic logs and cross-sections. When sampling fine-grained, cohesive soils, sloughing may not be a significant problem at shallow depths. However, in sandy noncohesive soils, particularly below the water table, sloughing and collapse may restrict the use of this method. Single-rod methods have also been used to collect ground-water samples from discrete intervals and to conduct ground-water profiling. The single-rod method can provide high-integrity ground-water samples when a protected screen device is used. In addition, essentially all down-hole logging conducted with DP equipment is accomplished with single-rod methods. This includes electrical conductivity or resistivity logging, CPT, and contaminant detection with systems such as the MIP or LIF devices.

Dual-Tube Systems

The basic components of a dual-tube system include the inner and outer drive rods and drive caps and down-hole sampling devices (Figure 6.13). The inner rod string is generally of the same dimensions and construction as the single-rod systems described earlier. The outer rod string (or drive casing) ranges in diameter from 2.125 in. (54 mm) O.D. × 1.5 in. (38 mm) I.D. to 4.5 in. (114 mm) O.D. × 3.875 in. (98.5 mm) I.D. The dual-tube system is most often used for continuous coring of soils and sediments (Figure 6.14). Use of the

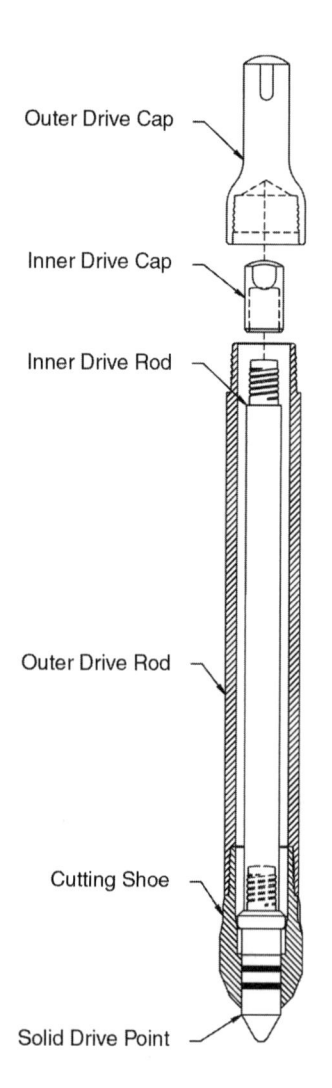

FIGURE 6.13
Basic components of a DP dual-tube system. The inner rods can be used to hold a solid drive point in position to prevent soils from entering the annulus of the system when the tool string is advanced without sampling. At the desired sampling depth, the inner rods are retracted and the solid drive point is replaced with the sampling device. Soil samples, ground-water samples, and soil-gas samples may be collected with this system. Slug tests to measure formation hydraulic conductivity may also be performed.

outer rods to stabilize the borehole prevents sloughing of material down the borehole and provides a controlled method for insertion and recovery of sampling tools. The dual-tube system is also used for ground-water sampling and slug testing, as well as for installation of piezometers, monitoring wells, or multilevel monitoring systems, and for other sampling or monitoring activities best conducted through a cased borehole.

While the dual-tube method does prevent sloughing of material down hole, it suffers from at least one major limitation. When soil sampling is done below the water table in noncohesive materials, the bore of the outer rods is subject to heaving or blow-in of formation material. Heaving or blow-in most commonly occurs in saturated loose sands when samples are collected from below the water table. As the sample (or drive point) is removed from the cutting shoe under these conditions (Figure 6.14e), the higher hydraulic head outside of the rods, and the suction created by removing the sampler, work to

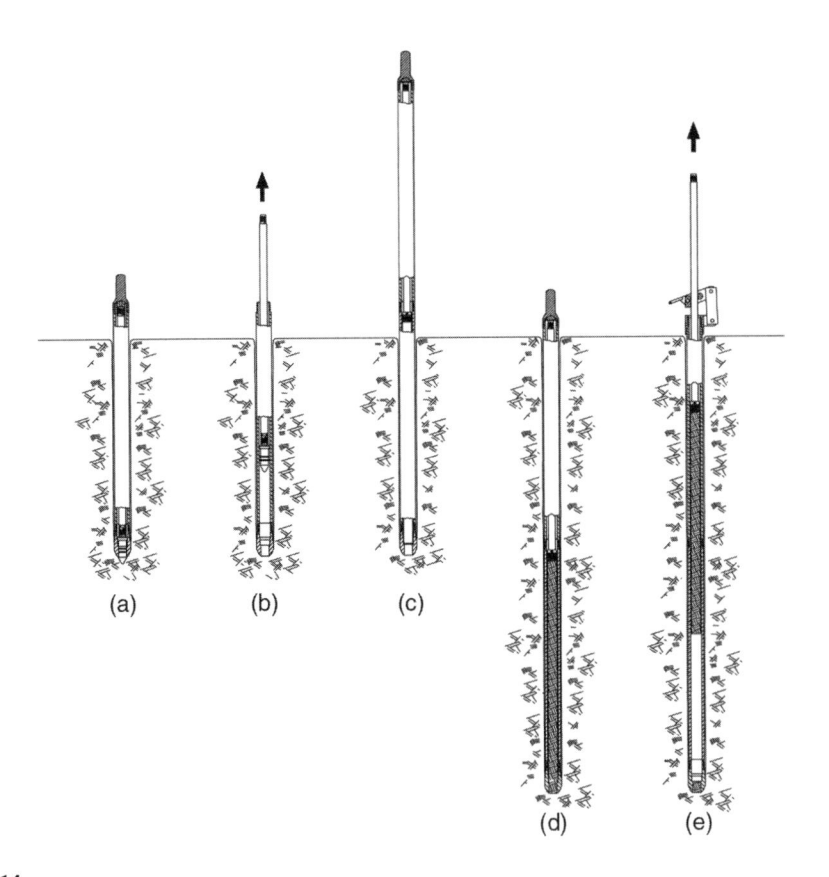

Use of Direct-Push Technologies in Environmental Site 367

FIGURE 6.14
The dual-rod system is used to control the borehole wall, especially in noncohesive formations, and minimize the potential for cross-contamination. (a) The dual-rod system may be advanced with a solid drive tip to the desired depth if shallow samples are not desired. (b) The inner rods are removed and the sample tube is inserted. (c) A length of outer rod is added to begin sample collection. (d) With the inner rods holding the sample tube in place, the dual-tube system is advanced through the sample interval. Use of the hydraulic hammer during sample collection usually increases recovery. (e) Once the sample is collected, the inner rods are raised to recover the sample and the outer rods stay in place. The sequence is repeated with drive rods added incrementally to conduct continuous coring.

simultaneously push and pull the noncohesive sands into the bore of the outer rods. When heaving or blow-in does occur, it may be difficult to clear the material (often several feet [1 m]) from the rod bore so that sampling may continue. In the worst case, the boring may have to be abandoned. Adding water to the annulus of the dual-tube system to maintain positive hydraulic head may prevent blow-in, but it is usually a slow and time-consuming process. In addition, adding water to the boring may dilute or otherwise alter the chemistry of the sample being collected. Under these conditions, a sealed single-rod soil sampler may provide higher integrity samples.

DP Sampling Methods

Some of the earliest applications of DP equipment were for soil-gas sampling, followed by soil and ground-water sampling (Christy and Spradlin, 1992). Most of the early soil-sampling tools were small-diameter (<2 in. [5 cm]) single-rod systems often used for depth-discrete or targeted sampling. As the use of DP equipment for soil sampling increased, dual-tube soil-sampling systems were developed to improve the integrity of samples for environmental investigations. Use of DP methods and tools for collection of ground-water samples for environmental site characterization also developed quickly. Initial ground-water sampling tools included the Hydropunch®, a device with a short screen and a long tool body that houses a sample chamber, the BAT sampler, a porous-tipped tool behind which is an evacuated sample container, and a simple mill-slotted rod used to collect water samples from sandy formations. Protected-screen systems were developed to allow collection of high-integrity depth-discrete samples from what are essentially temporary well points. More recently, single-rod (Pitkin et al., 1999) and dual-tube (McCall et al., 2002) ground-water profiling systems have been developed to aid in the vertical definition of ground-water chemistry (Schulmeister et al., 2001), contaminant distribution, and vertical variations in hydraulic conductivity (Butler et al., 2002; Geoprobe Systems, 2002). The following sections describe in more detail the DP sampling tools and methods used for environmental investigations.

Presampling Considerations

As with any sampling tool or sampling system, if the tools are not used correctly and applied under appropriate field conditions, the sample quality and resulting analytical data may be suspect. Selection of the appropriate DP tools to obtain high-integrity samples must be guided by the site-specific conditions and objectives of the sampling program.

Equipment decontamination is also a vital step in ensuring that representative samples are obtained. If appropriate equipment decontamination (Chapter 20) is not practiced, cross-contamination can become a serious problem. Appropriate decontamination may consist of simply scraping off gross soil material followed by a soap and water wash and tap water or deionized water rinse. Parker and Ranney (1997) found that heating of ground-water sampling equipment significantly improves the effectiveness of decontamination procedures. The site-specific project work plan should provide guidance on decontamination procedures for any given project. Additional information on appropriate decontamination procedures is provided in ASTM Standards D 5088 and D 5608 (ASTM, 2004f, 2004g).

Health and safety of the field personnel at any potentially contaminated facility is also of primary importance (Chapter 19), so development and use of a site-specific health and safety plan is necessary to ensure that worker injury or exposure to hazardous substances does not occur. While the machines and tools used in DP technology generally pose fewer health-and-safety issues than conventional drilling and sampling methods, it is important to understand those hazards that do exist, so they may be avoided.

DP Soil Sampling Methods

The two primary DP methods used for soil sampling are the single-rod and dual-tube (or cased) methods. In the single-rod method, a soil sampler is attached to the lead drive rod and advanced into the soil to collect the sample (Figure 6.15). The sampler (with sample) is retrieved and the sample is removed, then the sampler is decontaminated, reassembled, and advanced back down the open borehole to sample the next interval. With the dual-tube (or cased) method, a larger diameter outer tube and smaller diameter inner tube with attached sampler are simultaneously advanced into the subsurface (Figure 6.16). Once the sample interval is penetrated, the inner rod and sampler (with sample) are

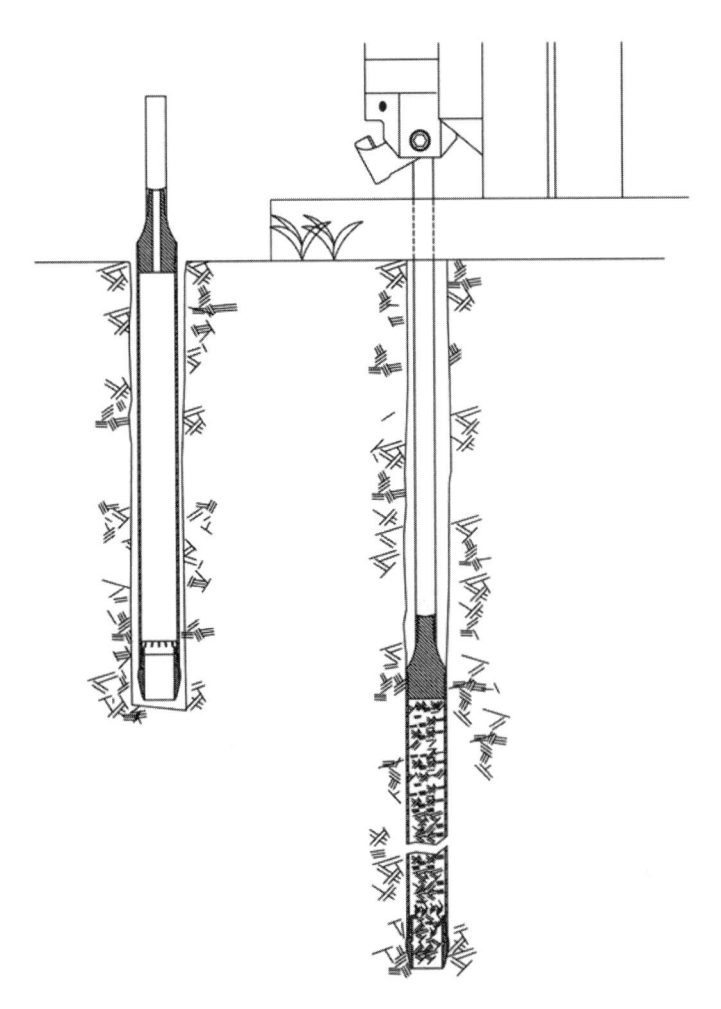

FIGURE 6.15
Single-tube soil sampling through an open borehole can be a cost- and time-efficient method to obtain samples for lithologic identification and chemical analysis under appropriate conditions. In noncohesive soils, formation collapse may be problematic and the potential for migration of contaminants down the open borehole must be evaluated to assure sample integrity and to ensure that DQOs are met.

retracted to the surface, while the outer rod (or casing) remains in the borehole to hold the hole open and prevent sloughing of formation material. The dual-tube method generally improves sample integrity for environmental investigations.

Single-Rod Soil-Sampling Methods

Two types of soil sampler designs are commonly used in the single-rod method — the nonsealed or open-barrel sampler and the sealed or closed-piston sampler. Each of these is described in detail subsequently.

Nonsealed Soil Samplers

Open-barrel samplers have been widely used to conduct continuous coring from ground surface to the desired depth with DP machines. These relatively simple tools consist of a drive head, sample tube, and cutting shoe and are usually equipped with a plastic sample

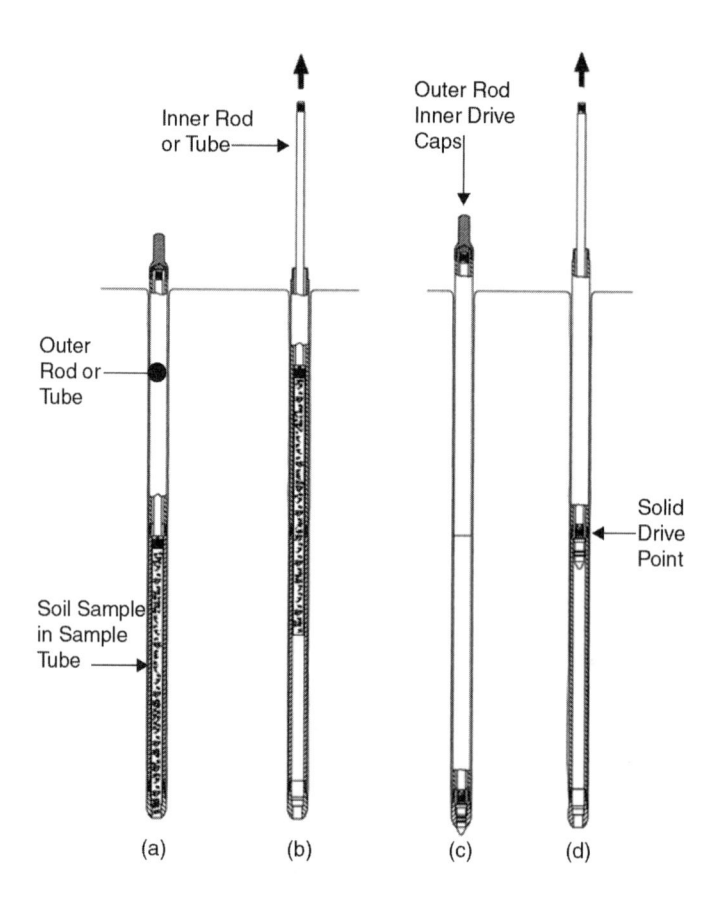

FIGURE 6.16
Dual-tube sampling provides a cased borehole to eliminate concerns about borehole collapse and minimize concerns about cross-contamination of samples. (a) Continuous soil sampling is most often conducted with the dual-tube system. (b) The sample and sample tube are "tripped out" with the inner rods, while the outer rods remain in place. (c) If sampling through a specific interval is not desired, a solid drive point may be installed with the inner rods to allow for advancement to the desired sampling interval. (d) The inner rods and drive point are tripped out and a fresh sample tube is installed to begin sampling. Flow or "heave" of saturated sands into the bore of the outer rods may occur when the sample or drive point is extracted. Under these conditions, single-tube closed-piston sampling may be more effective.

liner (Figure 6.17). Sample catchers may also be used to improve sample recovery in non-cohesive soils. Open-barrel samplers usually provide a sample core ranging from 1.5 to 2 in. (38 to 51 mm) in diameter and from 2 to 5 ft (0.6 to 1.5 m) in length. These tools are sometimes used for depth-discrete sampling. This is done by first driving a preprobe, of a diameter the same as or larger than the sampler, to the top of the desired sample interval. The preprobe is removed from the borehole, and the sampler is lowered through the open hole and then advanced through the sample interval (Figure 6.15). Additional information on application and operation of the open-barrel samplers is available in ASTM Standard D 6282 (ASTM, 2004c) and other sources (U.S. EPA, 1997; Geoprobe Systems, 1998a).

Split-barrel (or split-spoon) samplers have been widely used with conventional drilling equipment and methods for many years to collect "representative" soil samples, especially for geotechnical purposes (ASTM, 2004h). Some of these tools have been modified or specially adapted for use with DP methods. A typical split-barrel sampler (Figure 6.18) has a drive head, longitudinally split sample barrel, and cutting shoe. These tools normally have a check valve in the drive head to allow air to escape from the sampler

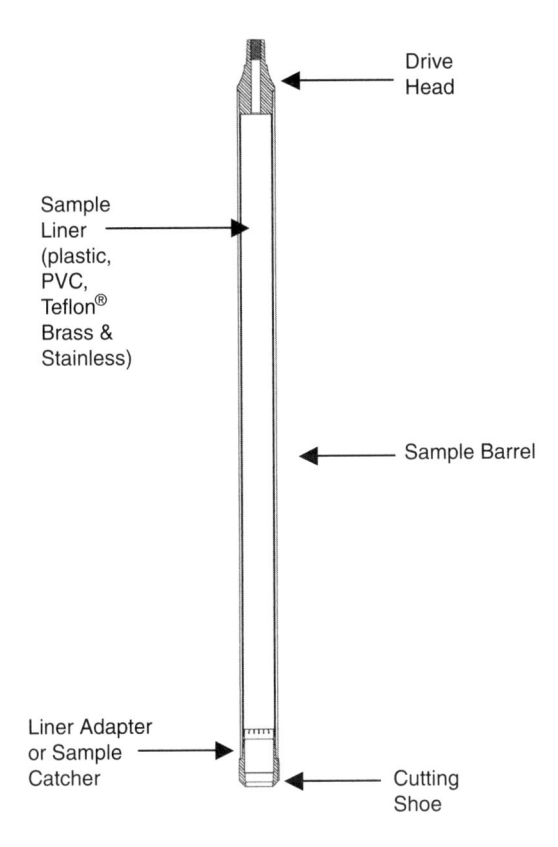

Use of Direct-Push Technologies in Environmental Site 371

FIGURE 6.17
Basic components of an open barrel, single-tube soil sampler. These tools are most commonly used for continuous soil coring in cohesive formations.

as the sample barrel is filled and to help with sample retention (to prevent wash-out). They may also be equipped with a sample liner and sample catcher as needed. The sampling operation for these tools is conducted in a manner similar to that for open-barrel samplers when operated with DP equipment.

Thin-wall (or Shelby-tube) samplers have been used extensively with conventional drilling methods to collect physically "undisturbed" samples for geotechnical and hydrologic testing purposes (U.S. EPA, 1991; ASTM, 2004i, 2004j). A typical thin-wall sampler (Figure 6.19) should be designed with an inside clearance ratio of at least 1%, while larger clearance ratios may be required as the sample plasticity increases (ASTM, 2004i). Thin-wall samplers have been adapted for use with DP equipment for environmental and geotechnical sampling. Thin-wall samplers are operated in a manner similar to that for open-barrel samplers (Figure 6.17) for environmental sampling operations.

Sealed Soil Samplers

Closed-piston samplers are the only sealed soil samplers commonly used with single-rod DP methods. In general, closed-piston samplers consist of a drive head, sample tube, piston rod, and point (Figure 6.20). These samplers are usually equipped with a plastic or metal sample liner and may be equipped with a sample catcher to enhance sample recovery in noncohesive soils. The piston rod may be only the length of the sample tube and held in position with a threaded locking pin or the piston rod may be extended the entire length

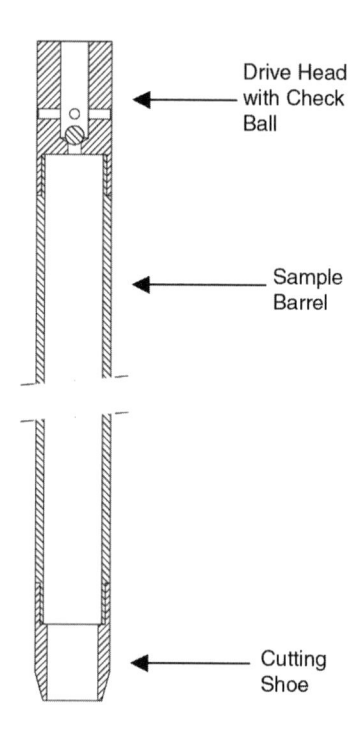

FIGURE 6.18
Components of a typical split-spoon sampler. These simple tools have been widely used for geotechnical testing and some have been modified for use with DP machines to conduct environmental sampling. The longitudinal split in the sample barrel provides for easy access to the sample in the field.

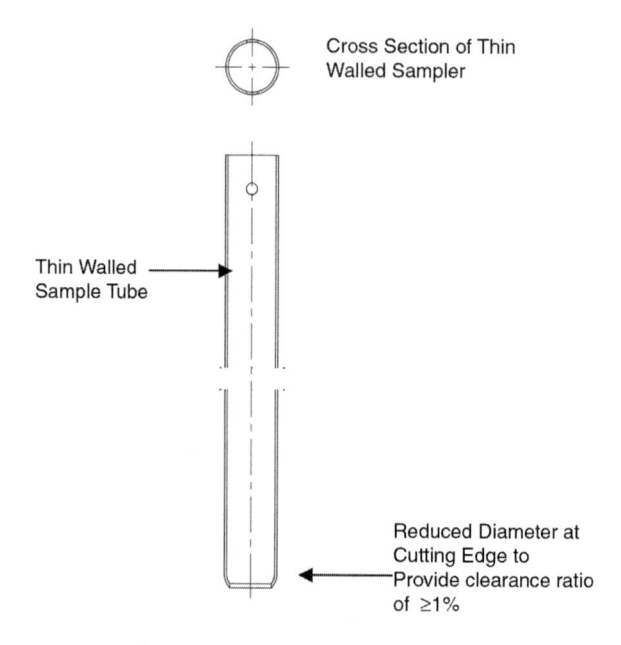

FIGURE 6.19
Thin-walled or Shelby-tube samplers were designed for collection of geotechnical samples where minimal physical disturbance (compression, etc.) of the soil is required for obtaining representative data on physical parameters such as bulk density. Thin-walled samplers may be used with DP machines when needed. The wall thickness-to-diameter ratio must meet specific criteria for a sampler to be considered a thin-walled device. These simple tools have been used occasionally for environmental sampling.

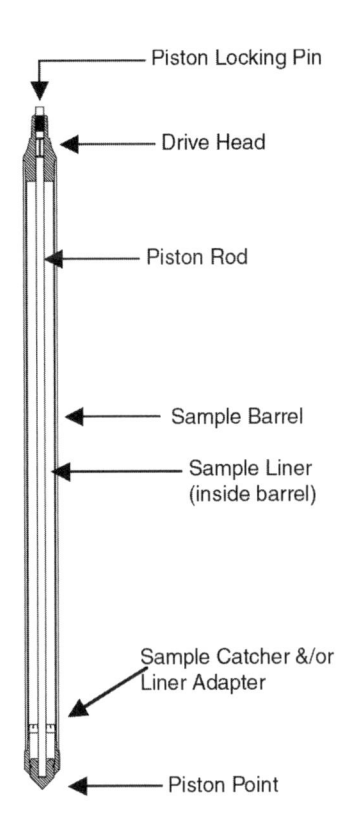

FIGURE 6.20
Components of a sealed, single-tube soil sampling tool. When operated correctly, a closed-piston sampler can significantly increase sample integrity and eliminate collection of collapsed material during open-hole sampling. Closed-piston samplers may provide an option for sampling in saturated, noncohesive sands when dual-tube methods are ineffective.

of the sampler and drive rod tool string in some designs. In either case, the piston rods should be removed to minimize compression of the sample and optimize sample recovery. As with most DP soil-sampling methods, use of a percussion hammer or vibratory head during sample collection will generally improve sample recovery. Conversely, use of a sample catcher in soft materials (e.g., saturated clays) may reduce sample recovery.

The smaller, more rugged closed-piston samplers are commonly used for depth-discrete sampling and are routinely driven through 20 to 50 ft (6 to 15 m) or more of soil and sediment to collect a sample from a targeted zone at a specific depth. The smaller closed-piston tools usually provide a sample 1 to 2 ft (0.3 to 0.6 m) in length and 1 in. (25 mm) or less in diameter. The larger diameter closed-piston tools are generally recommended for continuous coring operations, but depth-discrete sampling may be conducted under favorable soil conditions. General information on application and operation of closed-piston samplers is provided in ASTM Standard D 6282 (ASTM 2004c); additional information is available from other sources (U.S. EPA, 1997; Geoprobe Systems, 1998a, 2000).

Dual-Tube Soil-Sampling Methods

Two types of sampler designs, the split barrel and solid barrel, are used in dual-tube soil-sampling activities. A third type of sampler, a small thin-wall sampler, is used to core ahead of the outer rods or drive casing so that a minimally disturbed sample may be recovered.

Split-Barrel Samplers

Split-barrel samplers used in DP dual-tube soil sampling generally are smaller diameter adaptations of split-barrel samplers used with conventional drilling techniques. The split-barrel samplers consist of a drive head, longitudinally split sample tube and cutting shoe; some may be equipped with a sample liner (Figure 6.21). Sample catchers may be used with some models to enhance recovery when sampling in noncohesive soils such as saturated sands. Once the dual-tube system is at the desired sampling depth, the split-barrel sampler is assembled and lowered through the outer rods (casing) with the inner rods. The tool is then advanced through the interval to be sampled and the inner rods with sampler and sample are then retrieved.

Solid-Barrel Samplers

Solid-barrel samplers are the most widely used dual-tube soil-sampling tools for environmental investigations. In this type of system, the drive casing may be 2 to 3 in. (51 to 76 mm) O.D. with an inner rod typically not more than 1 to 1.5 in. (25 to 38 mm) O.D. (Figure 6.22). The solid-barrel sampler, with an outer diameter slightly smaller than the drive casing inner diameter, may consist of plastic, plastic with metal sheath, or thin-walled steel. The samples recovered typically range from 1 to 2 in. (25 to 51 mm) O.D. with length varying from 2 to 5 ft (0.6 to 1.5 m). The proportion of sample recovery with longer samplers, especially those with smaller diameters, will generally decrease; using a shorter sampler with larger diameter usually increases the percentage of sample recovery. Some solid-barrel samplers may be equipped with sample catchers or sample liners designed with integral catchers to improve recovery in noncohesive materials.

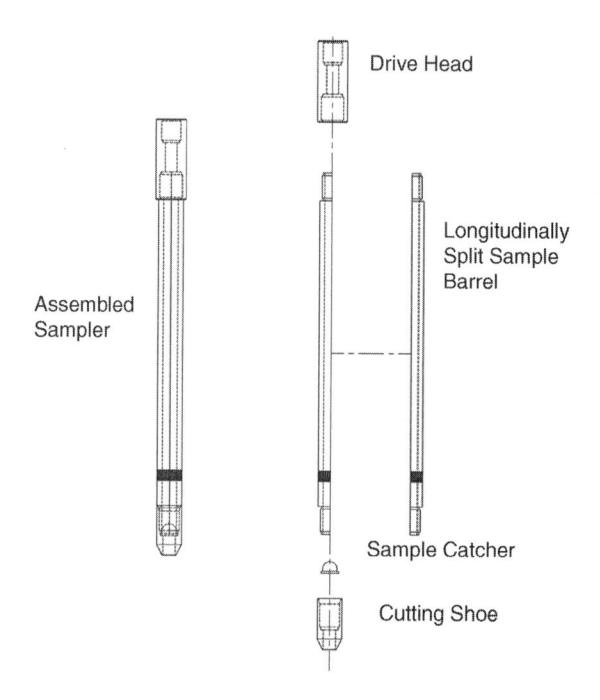

FIGURE 6.21
Split-barrel samplers were designed for use in the standard penetration test conducted for geotechnical purposes. Some samplers have been designed specifically for use in some DP dual-tube systems. This option provides the convenience of a split-barrel sampler for use in environmental applications.

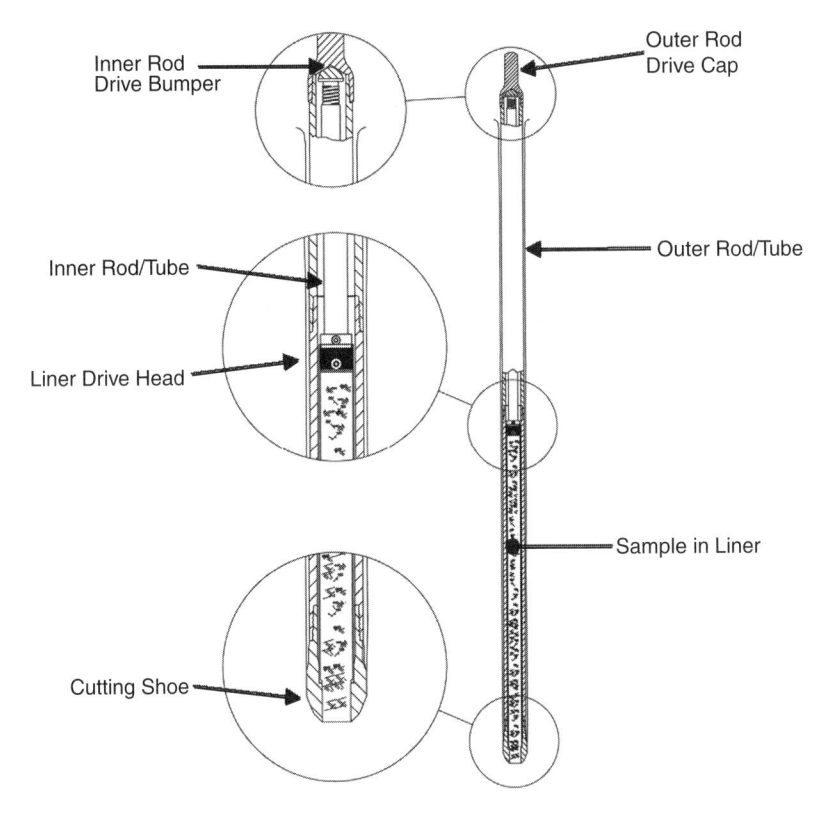

Use of Direct-Push Technologies in Environmental Site 375

FIGURE 6.22
Dual-tube systems are most often operated with plastic or clear PVC sample tubes held in position with a drive head and the inner rod string. The sample tubes are cut open to access the sample for testing and subsampling. Using a shorter sample interval will generally improve the percentage of sample recovery, but increases the number of trips in and out of the borehole, and therefore, the time required to sample the same interval.

The sequence of steps involved in continuous soil sampling with the dual-tube solid-barrel sampler system is shown in Figure 6.23. The outer drive casing, with a hardened steel drive shoe, and inner sample barrel, attached to an internal set of rods, are simultaneously driven into the ground. After advancing the assembly the length of the sample barrel, the inner sample barrel is retrieved to the surface, while the drive casing is left in place to prevent the borehole from collapsing. The presence of the drive casing ensures that the subsequent soil sample is collected from the next sampling interval, rather than potentially contaminated slough from higher up in the borehole. A clean sample barrel is lowered to the bottom of the borehole, an additional section of drive casing and inner rod is added, and the drive casing and sample barrel are driven the length of the sampler. This process is repeated until the desired depth is reached, allowing the collection of continuous soil cores. Alternately, with the inner drive rods, the sample barrel, and a drive point in place, the dual-tube system may be advanced to a target sampling depth before sampling is conducted. The drive point can then be released, and the assembly advanced the length of the sample barrel so that a sample from a discrete depth can be collected. This process can be repeated at multiple depths as needed. Additional information on the application and operation of solid-barrel samplers may be found in ASTM Standard D 6282 (ASTM, 2004c), Einarson (1995) and Geoprobe Systems (1998b).

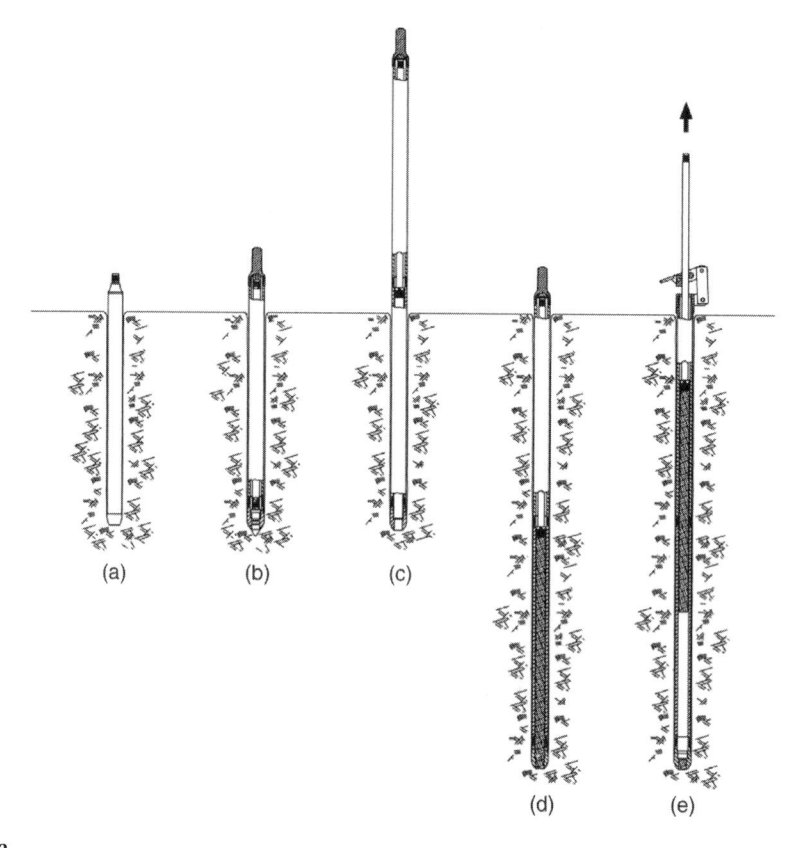

FIGURE 6.23
Operational steps of dual-tube DP soil sampling. (a) Often the first sample at ground surface is collected with an open-barrel single-tube sampler to start a guide hole for the dual-tube system. (b) The dual-tube system, with a solid drive point in place, is advanced to the depth where dual-tube sampling is to begin. (c) The solid drive point is removed and replaced with a sample liner and additional inner and outer drive rods. (d) The dual-tube system is advanced through the sample interval — hammering during advancement usually enhances sample recovery. (e) The inner rods, liner, and sample are tripped out, leaving the outer rods in place. The outer rods prevent borehole collapse, helping to preserve sample integrity. Steps (c) through (e) are repeated incrementally to conduct continuous sampling.

Dual-Tube Precore System

Thin-wall samplers are generally used to obtain soil and sediment samples whose physical properties are relatively undisturbed (U.S. EPA, 1991; ASTM, 2004i). Hvorslev (1949) determined that to obtain a physically undisturbed sample, the thickness of the sample tube wall must be less than 2.5% of the outside diameter of the sample tube. One small thin-wall sampler (1 in. [25 mm] O.D. × 12 in. [305 mm] length) has been developed for use with the dual-tube system. The sample tube is assembled onto a drive head and lowered on the inner rods through the outer rods to the base of the cutting shoe (Figure 6.24). The DP machine's hydraulic system is then used to push the thin-wall tube into the soil to collect the sample. This tool was designed to allow a soil core to be removed ahead of the dual-tube rods so that minimal disturbance occurs in fine-grained formations. This also allows for the insertion of a small screen through the dual-tube system that can be used for water sampling and slug testing (McCall et al., 2002) from a minimally disturbed formation, which is especially important in fine-grained materials. While this tool does not strictly meet the stringent wall thickness ratio outlined in ASTM Standard D 1587 (ASTM, 2004i), it does minimize compaction and disturbance of the formation.

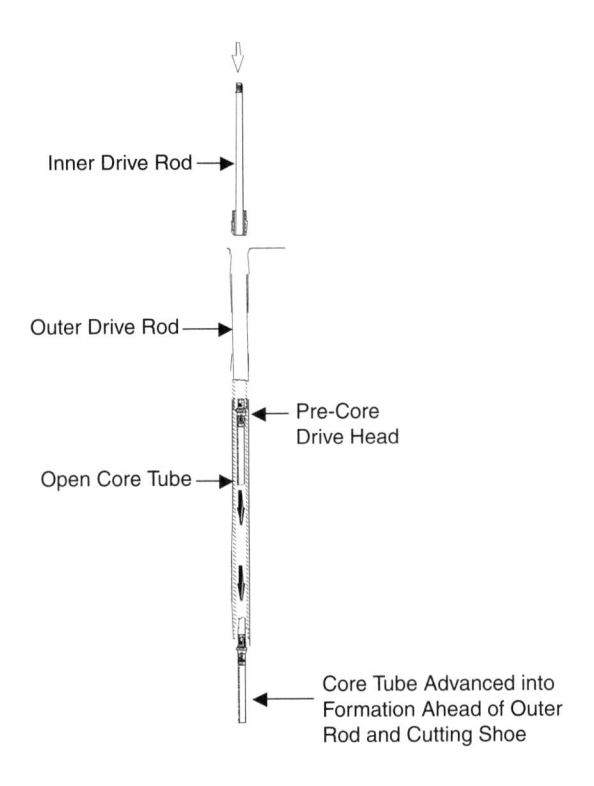

FIGURE 6.24
A simple open-tube sampler may be advanced beyond the outer casing of the dual-tube system to collect relatively undisturbed soil cores. This sampling procedure is most often used in fine-grained formations to permit placement of a screen in minimally disturbed formation materials, especially for slug testing.

Wireline CPT™

The Wireline CPT system consists of an assortment of tools that can be swapped at, or retrieved from, any depth without retracting the rods from the ground (Farrington et al., 2000). In contrast to conventional CPT, which can require multiple penetrations at a single location to collect various types of information or samples, the wireline system allows nearly all characterization work to be accomplished in a single penetration. A complete wireline rod string can be advanced into the subsurface using a standard, medium-duty CPT rig. This results in a significant reduction in the time required to accomplish various site characterization tasks, such as multiple depth sample collection, and an increase in the versatility of CPT for accomplishing a variety of tasks in combination, such as piezometric or optical measurements, sampling, and sealing the hole upon retraction. Thus, the Wireline CPT system offers increased utility and cost savings over conventional CPT.

The most commonly used wireline tool is the wireline soil sampler, depicted in Figure 6.25. This sampler allows the collection and retrieval of core samples from multiple depths during a penetration without requiring retraction of the CPT rods from the ground. The tool's relatively low cost, replaceable sample barrel produces a 1 in. (25 mm) diameter, 12 in. (305 mm) long core of soil. It accommodates the use of a plastic sample retainer for use in loose soils and is easily separable from the locking mechanism and cutting ring. The inexpensive cutting ring receives the wear of operation and is easily replaceable. It also holds the disposable basket in place. Both ends of the barrel are threaded identically to connect to either the lock and cutting ring or to end plugs used for sealing the sample. The U.S. Department of Energy (U.S. DOE, 2001) reported that

FIGURE 6.25
The wireline CPT soil sampler represents a dramatic cost savings by not requiring tripping up and down the borehole with heavy inner rods.

the wireline soil sampler offers dramatic cost savings over other CPT-based soil-sampling systems for continuous core sampling.

Applications and Limitations of Soil Sampling Methods and Tools

As with many soil-sampling methods and tools, DP methods and tools are suitable for use under a variety of geologic and site conditions and knowing their applications and limitations is important in selecting the appropriate tool for any given situation.

Open vs. Sealed Samplers

In general, sealed soil sampling tools will provide higher integrity samples for chemical analysis, but chemical analysis is not always the primary objective of the sampling program. When used correctly under favorable conditions, open-tube samplers can also provide high-quality samples for chemical analyses. When an open-barrel sampler is used, either continuous coring must be conducted or preprobing to the desired sampling depth is required. In cohesive soils that do not slough into the open borehole, open-barrel samplers have been used to successfully sample to depths exceeding 30 ft (9 m). Close inspection of samples is required to determine whether sloughing has occurred. Disturbed or loose material at the top of the sample indicates that sloughing has occurred and material other than that from the desired interval is present. Cross-contamination and movement of contaminants down hole must be considered when open-tube sampling is conducted and sample analyses are to be performed. Sealed or closed-piston samplers can eliminate many of the concerns with spurious material from borehole sloughing being incorporated into the sample. The field operator must be sure that the sampler is advanced through any slough before releasing the piston to initiate sample collection.

Single-Rod vs. Dual-Tube Method

The dual-tube method is generally considered to provide samples of higher integrity for lithologic definition and chemical analyses, particularly in cases where continuous sampling is desired. However, for depth-discrete sampling, the closed-piston single tube system can provide samples of equal quality at a significantly lower cost. The objectives of the sampling program and any budget limitations must be considered to select the best sampling system and tools for the project. Some of the benefits and limitations of the single-rod and dual-tube systems are discussed subsequently.

Single-Rod Soil-Sampling System

- Simpler to operate, usually with fewer parts.
- Generally lower cost than dual-tube systems.

- Many designs of open-barrel and closed-piston samplers are available for use with single-rod systems in a variety of sizes.
- Closed-piston samplers, the most efficient method for targeted or depth-discrete sampling with single-rod systems, may be used above or below the water table for sample collection.
- Because smaller drive rods are generally used with single-rod systems, depth of penetration may exceed that of dual-tube systems under the same soil conditions.
- Samples are collected through an open borehole, which may allow sloughing formation material to move down hole, resulting in increased potential for contaminant migration down hole and cross-contamination of samples.

Dual-Tube Soil-Sampling System

- Generally provides higher integrity samples than the single-rod system, especially samples used for chemical analyses.
- Usually more complicated to operate than the single-rod system.
- Initial cost is higher than the single-rod system because it requires two rod strings.
- Long-term operating costs may be lower because the system is generally more robust.
- Depth of penetration is usually less than the single-rod system under the same conditions because larger diameter rods are required, resulting in more side-wall friction.
- May be difficult to use below the water table, especially when heaving sands are present. Heaving sands can compromise sample integrity or prevent sample collection.

Continuous vs. Discrete Sampling and Lithologic Logging

Both the single-rod and dual-tube DP soil-sampling methods can be used for continuous or discrete-interval sampling. Continuous sampling is used to obtain a complete vertical record of the site-specific lithology and to accurately define zones of contamination using field sample analytical methods. Discrete-interval sampling is an efficient way to target specific depths for tracking contaminant plumes or important geologic strata (e.g., aquitards or thin sand layers). Discrete-interval sampling is most efficiently conducted with single-rod closed-piston samplers and is often conducted after several continuously sampled boreholes have been installed and site stratigraphy is well defined. Discrete-interval sampling may be used in combination with CPT logging, electrical logging, or contaminant logging methods and is an efficient way to verify or further quantify DP logging results when required.

Sample Integrity and Chemical Analysis

The primary purpose of most environmental investigations is to obtain samples to determine whether contaminants of some type are present. An accurate determination of the concentration of contaminants in samples collected from the site under investigation is vital to correct data interpretation and determination of the appropriate response or remedial methods to be applied (U.S. EPA, 1997, 1998a). To this end, it is important to select and correctly use the best sampling method to meet the sampling program objectives. The cost of sampling and the labor required to complete the project must also be considered. Under many conditions, the dual-tube soil-sampling method will provide higher integrity samples for chemical analyses. However, if soil samples for chemical analyses are to be collected below the water table (especially in noncohesive materials), the single-tube closed-piston sampler may be the preferred method. When the dual-tube system is used to sample through a free-product layer (LNAPL) at the water table, there is a potential for free product or high concentrations of contaminants to be left in the fluid in the bore of the outer rods. This condition can lead to contamination of the dual-tube sampler and potential for cross-contamination of any samples collected after this zone is sampled.

The U.S. EPA conducted a Superfund Innovative Technology Evaluation (SITE) demonstration to compare various DP soil samplers to conventional drilling and sampling methods (U.S. EPA, 1998b, 1998c, 1998d, 1998e). Hollow-stem augering coupled with split-spoon sampling were the conventional drilling and sampling methods used. The DP soil sampling tools tested in the demonstration included one manually advanced open-tube sampler (U.S. EPA, 1998c) and three DP machine-advanced systems including two single-rod closed samplers (U.S. EPA, 1998d, 1998e) and one dual-tube system (U.S. EPA, 1998b). One test site, in Iowa, was characterized by over-compacted, generally fine-grained glacial till, while the second test site, in Colorado, was characterized by dense, dry sands at all but the deepest sampling interval. One of the sampling locations at the Iowa site was grossly contaminated and free product was present. At this location in Iowa, and at one highly contaminated location at the Colorado site, sampling tools were advanced through the highly contaminated zones and recovered for analysis for determination of chemical integrity. The chemical integrity samples consisted of potting soil (very high in organic matter) packed within the sample chamber of the tool as it is normally operated. Results of the integrity tests (Table 6.1) reveal that the DP dual-tube method provides higher integrity samples than the single tube methods. Note that for the split-spoon samples the HSAs were drilled through the contaminated zone and then the prepared split-spoon integrity sample was lowered through the open bore of the augers and removed. Conversely, for the DP single-rod samplers, the prepared integrity sample tube was driven through the soil in intimate contact with the grossly contaminated medium. It is also of interest that the only DP

TABLE 6.1

Results of DP Soil Sampler Integrity Tests

Sampler Type	Sampler Name	Number of Contaminated Integrity Samples — Iowa Site	Number of Contaminated Integrity Samples — Colorado Site	Contaminant Type and Range of Concentrations
Open tube	Subsoil probe	2	0	1,2-DCE at 114 to 5700 µg/kg; TCE at 3.17 to 4070 µg/kg; PCE at 212 µg/kg (one sample)
Closed barrel	Core barrel	6	2	1,2-DCE at 2.1 to 4410 µg/kg; TCE at 5.3 to 1960 µg/kg; PCE at <1 to 602 µg/kg; 1,1,1-TCA at 108 to 218 µg/kg
Closed piston	Large bore	5	0	1,2-DCE at 3.4 to 295 µg/kg; TCE at 14.4 to 46.3 µg/kg
Dual tube	Dual tube liner sampler	0	1	1,2 DCE at 6.1 µg/l
HSA	Split spoon	0	0	None

Note: DCE, dichloroethylene; TCE, trichloroethylene; PCE, perchloroethylene; TCA, trichloroethane.

system able to recover samples from the deepest test interval (45 ft) was the single-tube locking piston sampler (U.S. EPA, 1998e). In each case, the U.S. EPA reports concluded that there was no significant or consistent analytical bias between soil samples collected with DP methods and those collected using HSAs and split-spoon samplers.

DP Soil-Gas Sampling

DP sampling tools are used extensively for soil-gas sampling. In its simplest form, DP soil-gas sampling entails inserting a hollow probe into unsaturated soils, applying a vacuum, and sampling the soil gas, which enters the probe in response to the applied vacuum. However, many different and much more complex systems have been developed for soil-gas sampling. It is not possible to cover every one of the myriad techniques that have been developed, so this section focuses on typical methods of probe insertion and soil-gas sample extraction using DP tools. The emphasis of this section is on features and practices that are common to the industry and on quality control procedures that should be employed regardless of the probe configuration involved. Additional information on the theory behind soil-gas sampling and monitoring is included in Chapter 3.

ASTM Guidance for Soil-Gas Sampling

ASTM Standard D 5314, Standard Guide for Soil-Gas Monitoring in the Vadose Zone (ASTM, 2004k), covers the basic principles of soil-gas occurrence and distribution in soil, sampling methods, quality control, and sample analysis. None of the sampling methods or procedures and none of the specific types of equipment are recommended by this guide, as consideration of site-specific factors preclude such recommendations. The guide recognizes that soil-gas collection methods must be adapted to meet site conditions and discusses the inherent difficulty in obtaining repeatable soil-gas results.

The majority of DP systems for soil-gas sampling are classified in ASTM Standard D 5314 as "whole air-active" systems. This means that soil gas is forced to move into a probe by the application of reduced pressure and that the sample is then contained for subsequent analysis. This sampling method is explained in the following sections.

Soil Matrix and Probe Insertion

The vadose zone soil matrix is a complex system, the basic components of which are illustrated in Figure 6.26. Soil-gas sampling for environmental purposes is normally concerned with determination of the concentration of a specific chemical compound. This compound may be distributed in several phases within the soil matrix, including the vapor phase in the soil air, the dissolved phase in the soil pore water, the sorbed phase primarily on the soil organic fraction, and, in some cases, the free-product phase. The distribution of the compound among these various phases is primarily dependent upon the physical properties of the compound.

DP soil-gas sampling first involves insertion of a probe into the soil matrix, which is accomplished by displacing and compacting the soil immediately adjacent to the probe hole (none of the soil is removed from the hole). The compaction process results in the rapid movement or shifting of soil particles into surrounding void spaces. Void space in the soil matrix is reduced for some distance around the inserted probe as shown in Figure 6.27. The net result of this process is that gases are rapidly displaced from the void spaces adjacent to the probe hole. Pore water in the soil, being less mobile than soil gas, remains in the void spaces adjacent to the probe hole. In fact, relative pore water content in the soil next to the probe can increase beyond saturation and serve as

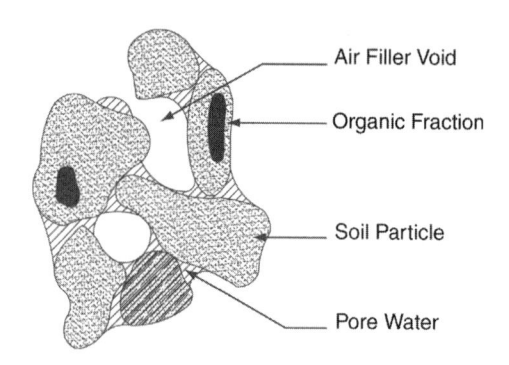

FIGURE 6.26
An illustration of the soil mass in the vadose zone. Successful soil-gas sampling requires the presence of continuous air-filled voids in the soil.

a limiting factor in compaction of the soil. In certain cases, insertion of the probe results in a thin, liquefied soil mass next to the probe that actually aids in probe insertion; the near-probe soil mass reacts as a liquid rather than a solid matrix when compacted.

Not all soils will exhibit localized saturation as a result of probe insertion. This phenomenon occurs primarily in clay and silt soils with high initial moisture content; sandy soils will not exhibit this behavior. This localized soil liquefaction does not preclude collection of soil-gas samples. However, all field practitioners should recognize that these conditions exist and are, in fact, common. To prevent entry of flowing, liquified soil into the sampling tool as the probe is advanced, probe tips may have to be sealed with o-rings. One result of soil compaction is that soil gas may not enter the probe until a sufficient vacuum is applied to overcome pore water tension in the saturated layer around the probe. In addition, the mechanical disturbance will result in a localized movement or redistribution of the contaminant between the various soil phases.

Another aspect of probe insertion is that the pressure at the interior of the probe (atmospheric) may be different than the pressure in the soil pores. Depending on the direction of the pressure differential, removal of the tip from the probe may result in either the flow of atmospheric gases into the soil or flow of the venting of soil gas into the probe. ASTM Standard D 5314 recognizes this potential for loss or dilution of soil gas and discourages use of methods that leave the soil open to free exchange with the atmosphere. Regardless of the method or tools used for soil-gas sampling, the sampling procedure should be designed to minimize the time during which the probe is allowed to remain in an unsealed condition, exposed to free exchange of atmospheric gas with soil gas.

Common Probe Configurations

Expendable Drive Point Samplers

The most common type of soil-gas sampling tool is the expendable-point configuration shown in Figure 6.28. This configuration has been used in a variety of sizes and materials. Rod materials range from 0.5 in. (13 mm) diameter steel tubing to 0.75 in. (19 mm) water pipe to 1.5 in. (38 mm) probe rod with hardened, threaded ends. All have been used successfully and all have their advantages in soil-gas sampling. Figure 6.29 shows a typical expendable point system prior to insertion in the soil.

Once driven into the soil to the desired sampling depth, the probe rod is retracted. Retraction leaves behind the expendable point and exposes the soil profile for soil-gas sampling. Sampling is performed by applying a vacuum to the rod, which ideally induces a flow of soil gas into the sampling train.

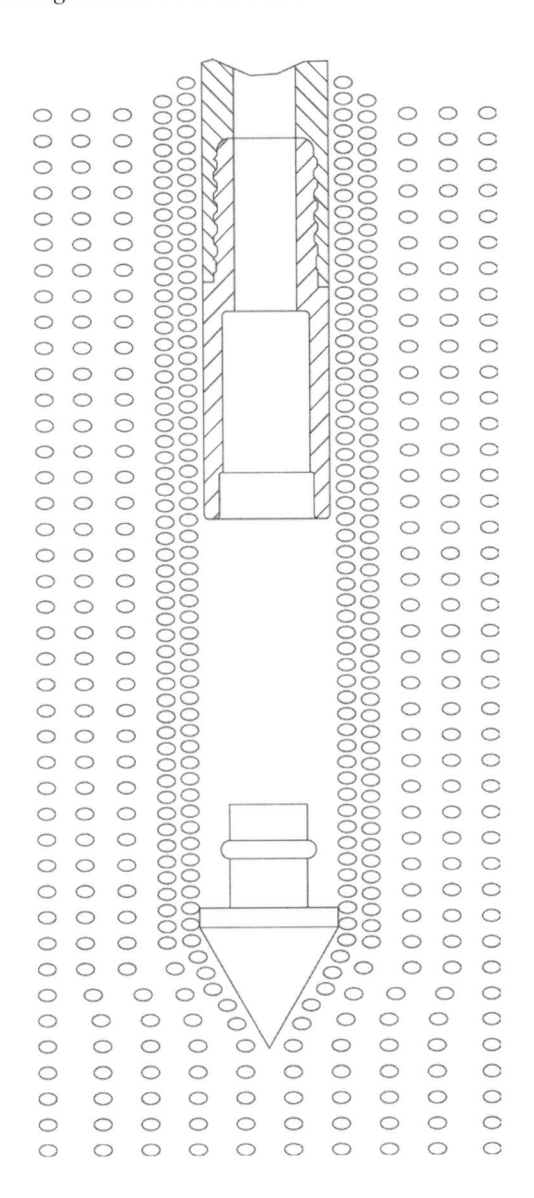

Use of Direct-Push Technologies in Environmental Site 383

FIGURE 6.27
This figure depicts soil compaction around a soil-gas sampling point and rod. Soils are compacted as a result of probe insertion. Soils immediately adjacent to the probe undergo the most compaction. Compaction occurs by rearrangement of the soil and, in most cases, results in a reduction in void space in the disturbed soil. Soil compaction in this manner can reduce or preclude soil-gas flow from fine-grained soils.

One common modification to sampling through the rod is to equip the probe rod with a fitting at the bottom (point) end that will allow tubing to be coupled directly to the point holder as shown in Figure 6.30. The advantages of this arrangement are that the purge volume is reduced and the problem of leakage through the rod joints is eliminated. Tubing materials used for this purpose include polyethylene, various types of Teflon®, and stainless steel. It is important to choose tubing materials that will not sorb or interfere with the soil-gas analytes of interest. ASTM Standard D 5314 encourages the use of sampling systems with low internal volume to reduce the volume of soil gas that must be purged from the sampling train prior to sampling.

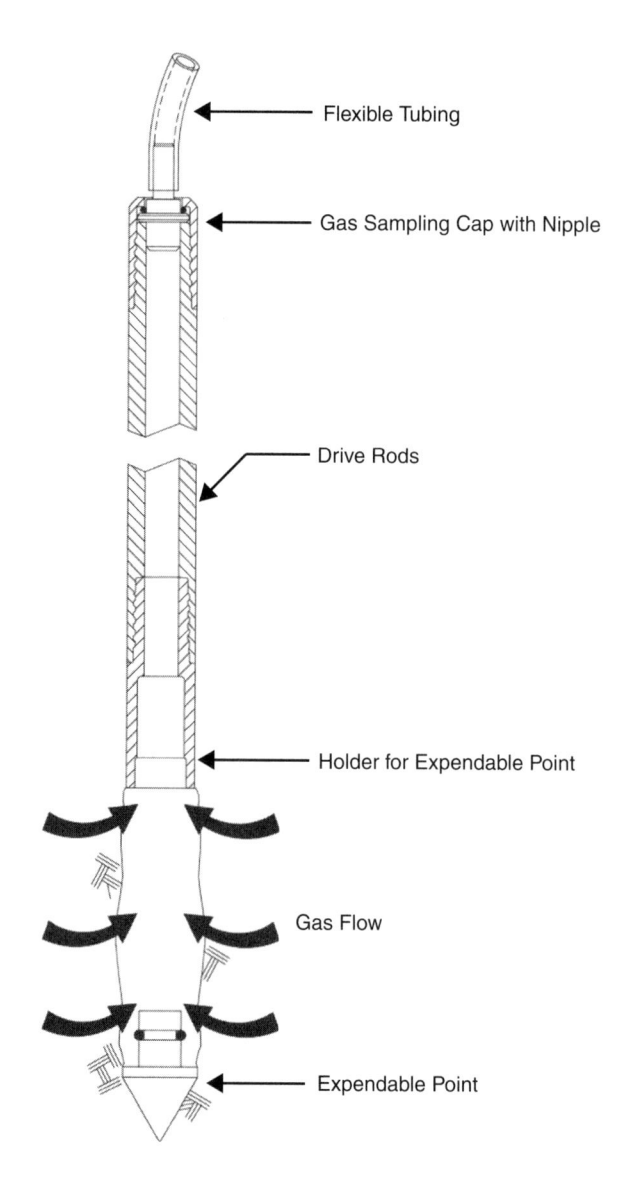

FIGURE 6.28
A typical expendable or "lost" point probing tool configuration for soil-gas sampling. This configuration is popular for soil-gas sampling, owing to its simplicity and ease of use. The drive point remains in the ground after sampling. Vacuum is applied to the flexible tubing connected to the top of the rod system.

Retractable Drive Point Samplers

Many field practitioners prefer to use retractable drive points (Figures 6.29 and 6.31) for soil-gas sampling. Various configurations are available from manufacturers. Retractable drive points are attached to lead end of the drive rod and remain with the tool upon retraction. The advantage of this type of point is that it saves the cost of an expendable point at each location sampled. The disadvantages of retractable points are that they have a higher initial cost and they require more time to decontaminate between sampling points. It is important that the outside diameter of the retractable drive point be smaller than the outside diameter of the probe rod. This will allow the flow of soil gas around the point if the assembly is retracted an extended distance (Figure 6.31).

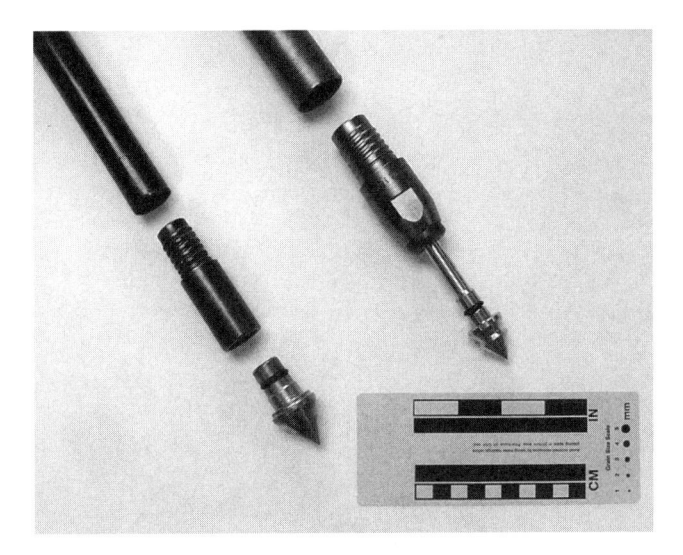

FIGURE 6.29
Typical DP soil-gas sampling point configurations. On the left is an expendable drive point and point holder which threads into a 1 in. (25 mm) probe rod. On the right is a retractable soil-gas sampling point adapted to 1.25 in. (32 mm) probe rod. (Photo courtesy of Geoprobe Systems.)

Retraction Distance

The distance to retract the sampling assembly from the probe hole depends on the purpose of the survey and the soil types present in the subsurface. In sandy soils, a point could be retracted 1 in. (25 mm) and yield sufficient soil gas for all sampling purposes. In clay and silt soils, the point could be retracted 12 in. (305 mm) and still not yield soil gas after application of vacuum.

No specific distance for probe retraction (and, thus, length of sampling interval) can be recommended here, nor is one specified in ASTM Standard D 5314. However, Standard D 5314 does specify "strict adherence to a standard operating procedure" for soil-gas sampling at any site. The project planner is therefore well advised to investigate the composition of site soils and devise a sampling protocol appropriate for those conditions. A retraction distance of 6 in. (152 mm) is common in a dry sandy soil. A silt or clay soil may require a retraction distance of 48 in. (1220 mm) or more.

One method of determining the appropriate retraction distance in fine-grained soils is to pull the drive point back a short distance and then apply a vacuum to the sampling train and check the results. As an example, the rod could be retracted a distance of 6 in. (152 mm) and an initial vacuum of 50 kPa (7.25 psi) then applied to the sampling train. With the vacuum pump shut off, record the decay (if any) in the vacuum level. It is common in fine-grained formations for the vacuum to remain at its initial level with no decay, which indicates that no soil gas is flowing into the probe. There is a benefit to this condition as it indicates to the operator that the sampling train is free from leakage of atmospheric air into the probe. If this condition persists, the probe rods can be retracted a further distance until a soil zone that yields soil gas is encountered, a condition that will be indicated by rapid decay of the vacuum in the shut-in sampling train. Figure 6.32 shows probe rods being retracted while the static vacuum in the sampling train is monitored.

Multiple-Depth Sampling

For some soil-gas sampling purposes, it may be desirable to sample soil gas at multiple depths. Various sampling tips and porous filter sampling devices have been designed to

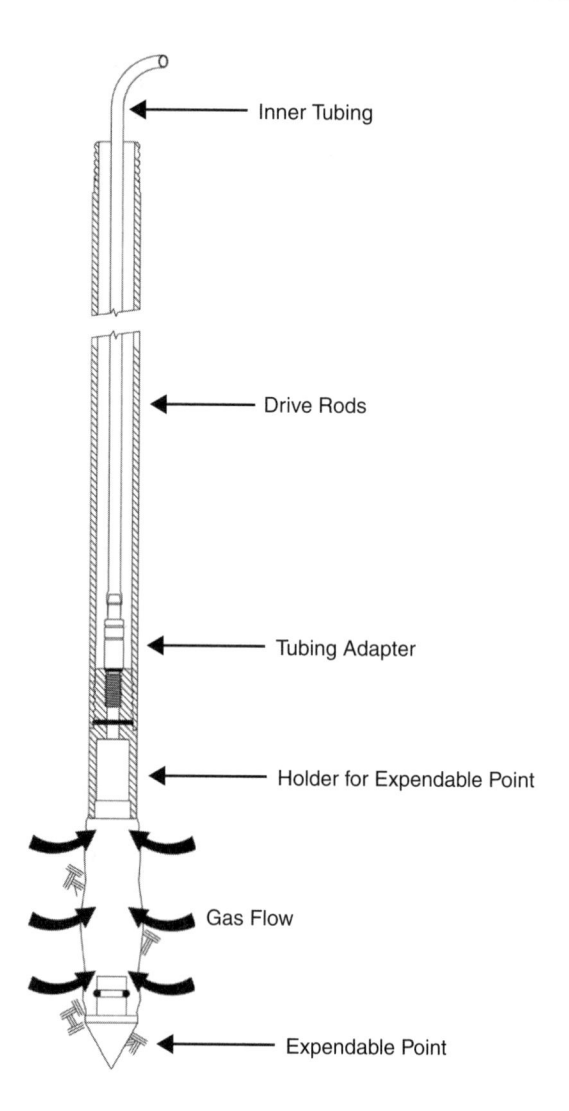

FIGURE 6.30
Cross-section of a soil-gas sampling system using a sampling tube through the bore of the probe rods. Such systems are advantageous in that they reduce the volume that must be purged from the rods and they eliminate concerns with rod joint leakage. Inner tubing materials of Teflon, polyethylene, and stainless steel have been used for this application.

perform multiple-depth soil-gas sampling and have been successfully employed on many projects. The use of retractable drive points for this purpose is generally not recommended due to the frequently encountered problem of clogging the sampling tip with soil when pushing down on a previously opened sampler. Because porous and screen-type devices are subject to clogging in fine-grained soils, their use is limited to sandy materials. The most common method for performing multiple-depth sampling is simply to retract the entire sampling assembly, reload it with clean components, and drive a new probe hole.

Sampling with Cased Systems

A variety of cased systems have been used for soil-gas sampling. The primary use of these systems is for deep applications where multiple intervals are to be sampled in the same probe hole. These systems are especially useful in sandy soils. Soil samples from these

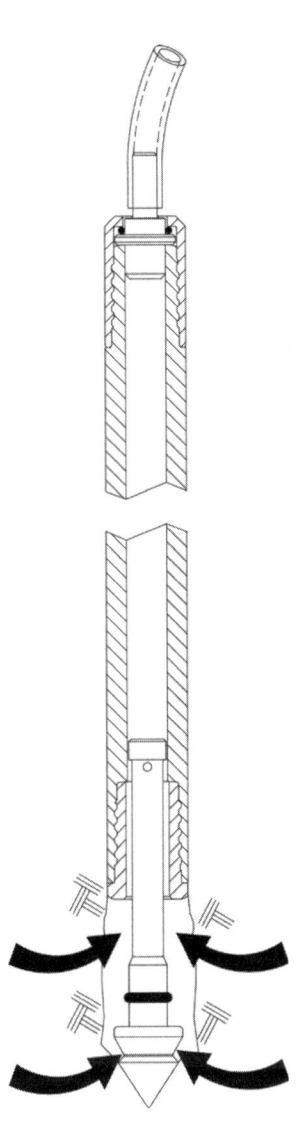

FIGURE 6.31
Cross-section of a retractable drive point soil-gas sampler, shown in the extended or pull-back position. Once the point has been extended, it should be removed from the hole, cleaned, and reset before probing at the next location.

soils will often show diminished or no detectable levels of volatile contaminants due to losses that occur during transfer of the sample from the sampling tool to the container. Soil-gas profiles in these same soils can, however, be successful in identifying the same contaminants *in situ*. Figure 6.33 shows a sequence of soil and soil-gas sampling through a driven casing. Cased systems have been used for soil-gas sampling to depths exceeding 100 ft (30 m).

Soil-Gas Implants

All of the systems discussed earlier are used for one-time or "grab" sampling of soil gas. However, it is often necessary to perform long-term monitoring of the soil atmosphere. DP tooling can be used to install soil-gas sampling implants to meet this objective. Typical applications of soil-gas sampling implants include landfills where methane is often

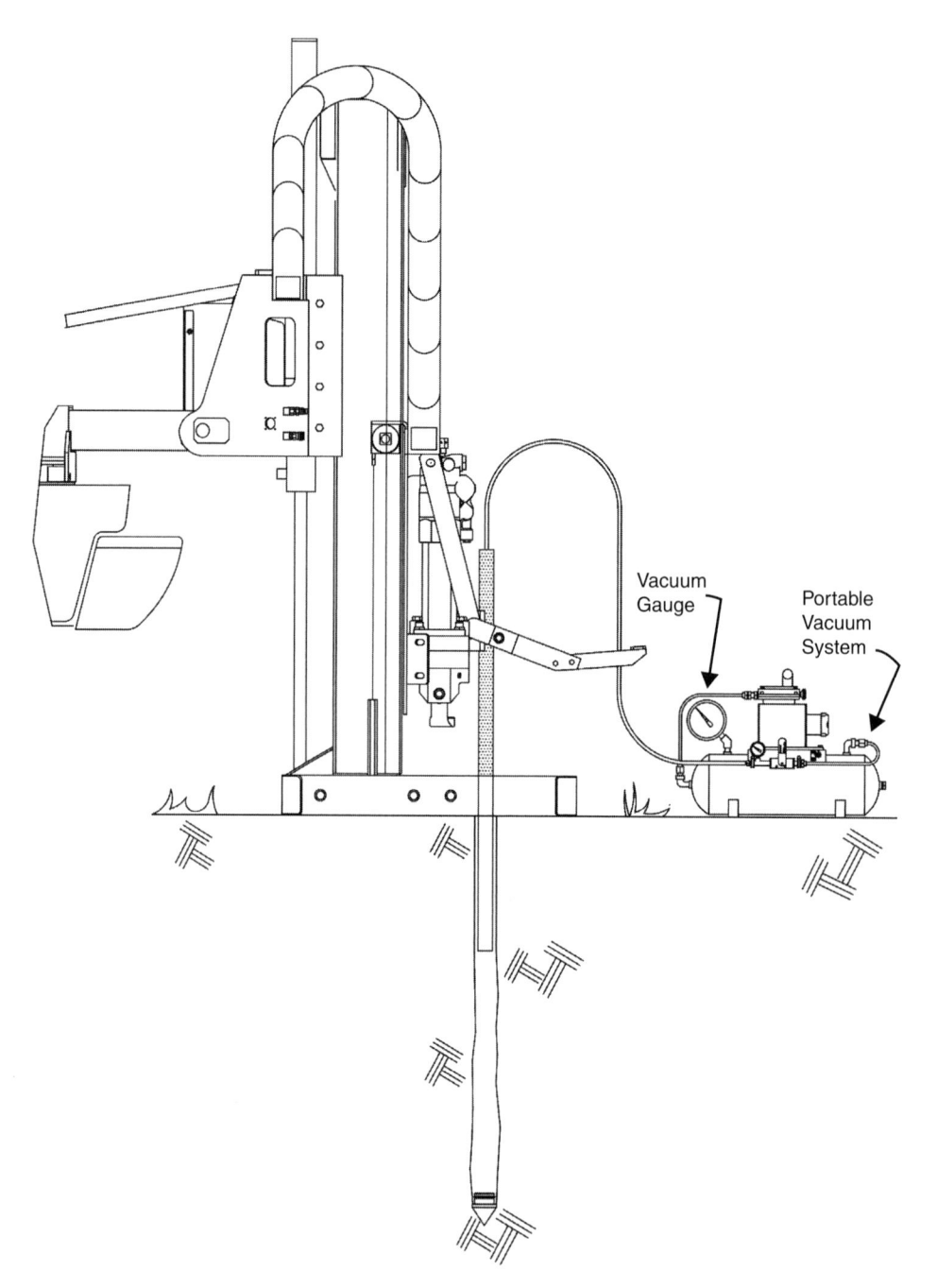

FIGURE 6.32
Pulling back the sampling rod to find a zone that will yield soil gas. In this case, a vacuum has been applied to the rods and the valve closed between the rod and vacuum source. Yield of soil gas from the sampling rod will be indicated by rapid decay of vacuum on the line gage.

monitored, remediation sites where cleanup is being monitored or residences are being protected, or UST sites where long-term soil-gas monitoring may be used for leak detection. An example of data collected from a sampling implant installed at a hydrocarbon spill site is shown in Figure 6.34.

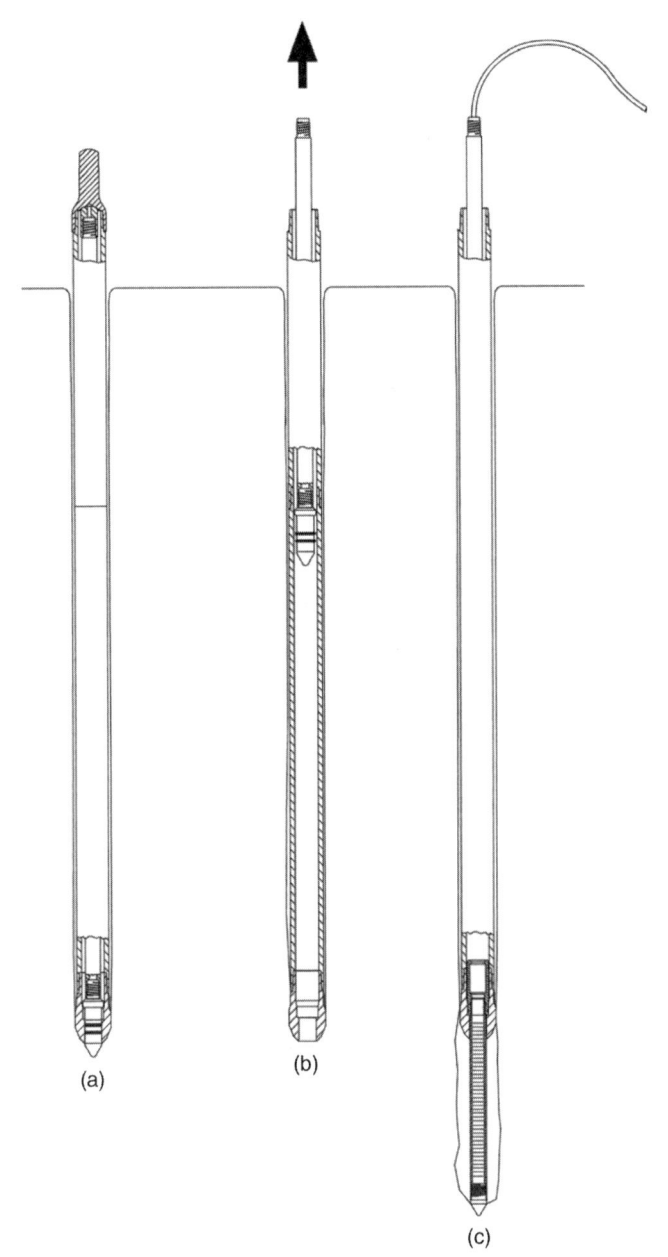

Use of Direct-Push Technologies in Environmental Site 389

FIGURE 6.33
Soil-gas sampling using a cased system. (a) The outer casing with solid drive point is driven to the top of the desired sampling interval. (b) The drive point and inner rod string are removed from the casing. (c) A screen section attached to an inner rod string is driven beyond the casing shoe into fresh soil. A mechanical seal must be used between the casing shoe and the sampling drive point.

A variety of soil-gas sampling implants have been used for long-term soil-gas monitoring. This includes the use of polyvinyl chloride (PVC) points constructed in the same manner as a DP-installed ground-water monitoring well, but completed in the vadose zone. Standard carbon-steel pipe also has been driven into the ground and used for this purpose. However, the amount of soil gas required for the sampling event is normally small (typically only 1 or 2 l), and this demand can normally be met with small-diameter

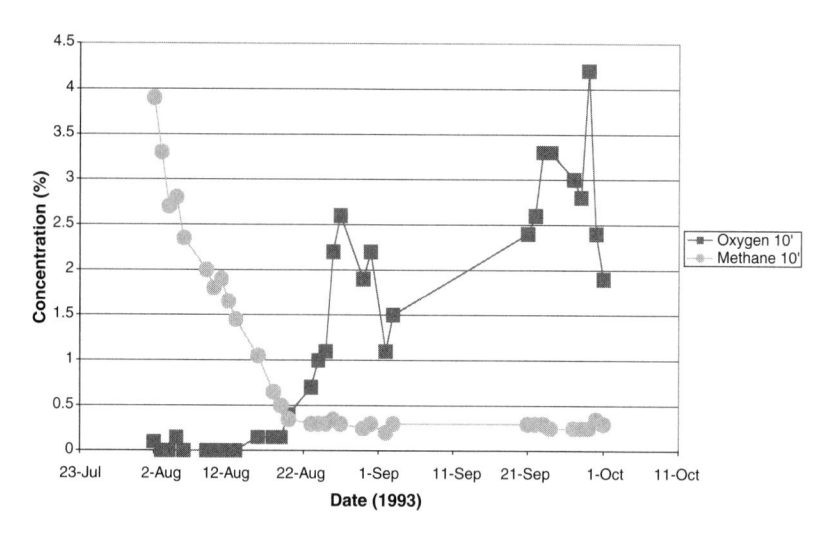

FIGURE 6.34
Oxygen and methane concentrations measured in samples pumped from a screen implant placed 10 ft (3 m) below the ground surface at a fuel spill site. Note that methane concentrations are decreasing at the same time that oxygen concentrations are increasing.

implants placed in the ground through the center of the DP rod. A typical stainless-steel implant is shown in Figure 6.35. A variety of soil-gas sampling implants are available commercially. Care should be taken to install implants in soil zones that will yield soil gas, and the installation should be tested to ensure recovery of soil gas at the time of

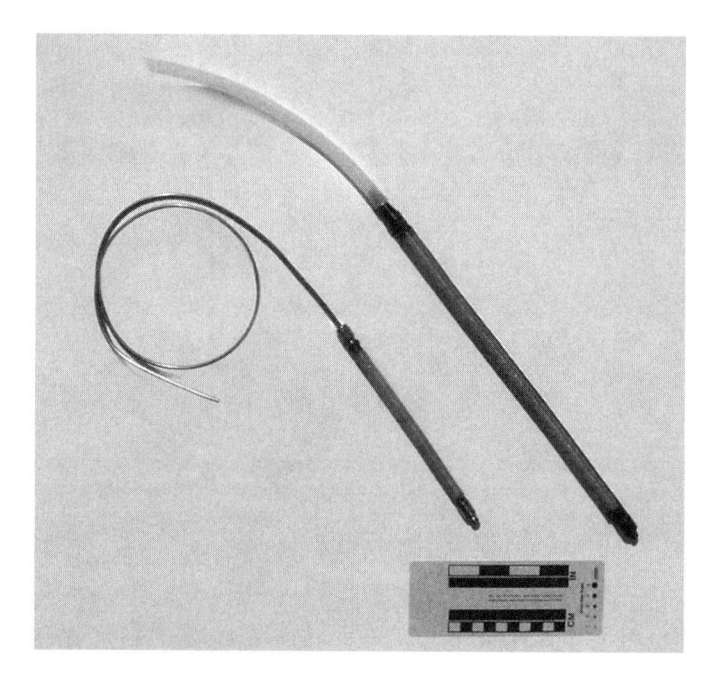

FIGURE 6.35
Soil-gas sampling implants. These implants are constructed of fine-mesh (0.15 mm pore size) stainless-steel screen. The implant on the left is attached to 0.125 in. (3 mm) stainless-steel tubing. The implant on the right is attached to 0.25 in. (6 mm) I.D. polyethylene tubing. These implants are designed for placement through the bore of DP rods. (Photo courtesy of Geoprobe Systems.)

Use of Direct-Push Technologies in Environmental Site 391

completion of the implant. It is a good practice to first use a DP probe to grab a sample from the zone to be monitored to ensure that flow conditions are adequate prior to installing an implant. Any open volume around the implant should be filled with sand or glass beads prior to installing seal materials in the probe hole. The annular space between the sampling tube and the probe hole can be sealed with granular bentonite, which should be hydrated at 2 in. intervals to promote proper sealing. A typical implant installation is shown in Figure 6.36.

Soil-Gas Sampling Train

The sampling train consists of all components required to transport the soil gas from the sampling zone in the subsurface to the sample container at the surface. No specific

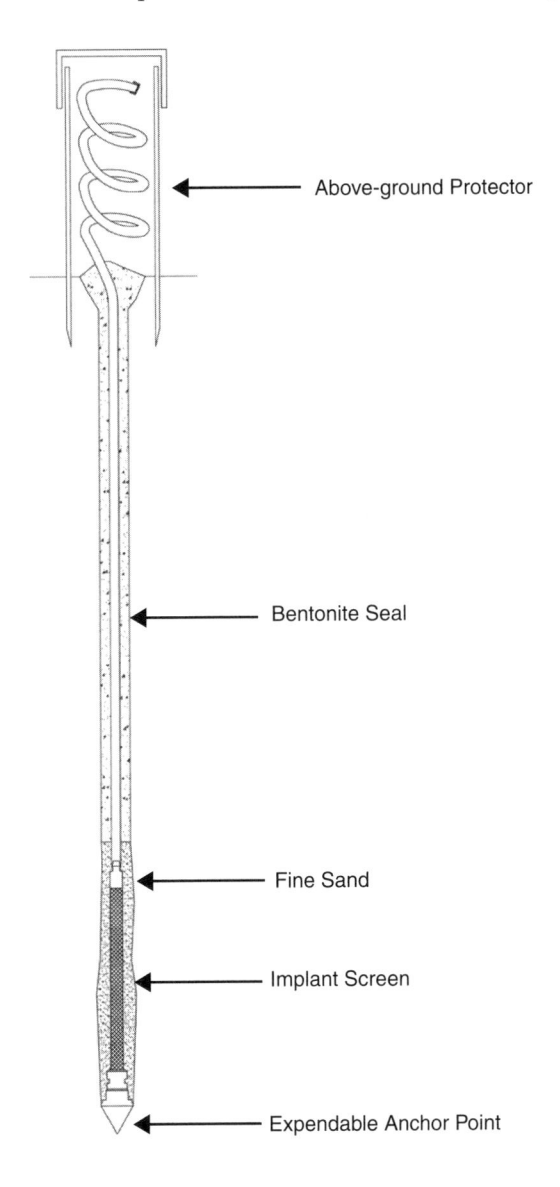

FIGURE 6.36
A typical soil-gas implant installation for long-term monitoring. Note that the upper end of the sample tube must be kept sealed with a gas-tight cap between sampling events.

vacuum source, vacuum application procedure, or sample container system is specified by ASTM Standard D 5314. It is recognized that many different systems are in use and that selection of the best equipment, materials, and procedure is the responsibility of the person conducting the soil-gas sampling program. Typical elements of the soil-gas sampling train are shown in Figure 6.37.

Every sampling train normally includes a vacuum source (a pump or evacuated tank) and a container or in-line analyzer (common in landfill applications). It is important that the investigator be able to confirm that the sampling train is air tight (that soil gas rather than atmospheric air is being sampled) and that soil gas is flowing from the soil-gas sampling point at depth.

A pressure gage and line valve should always be included in the sampling train. These devices allow rapid testing for movement of soil gas and aid in testing the tightness of the system. Before each use, the sampling train (including tubing, container, and vacuum source) should be assembled and checked for leaks. Leak checks are normally performed by applying a vacuum to the system and then closing the valve to the sampling train. There should be no decay in vacuum pressure for at least 1 min. A line pressure gage is also used when a syringe is used for sampling. To capture a full aliquot with a sampling syringe, the sampling train must be allowed to return to atmospheric pressure.

The sampling train should always include a means for confirming that flow is occurring from the subsurface. This is done with a bag sampler by simply observing the inflation of the bag. Measuring the change in pressure on an evacuated canister will also confirm flow. Flow from in-line vacuum pumps can be confirmed by attaching a flow meter to the pump discharge.

Quality Assurance and Quality Control Procedures for Soil-Gas Sampling

The topic of quality assurance and quality control (QA/QC) for soil-gas sampling is addressed at great length in ASTM Standard D 5314. Field practitioners are encouraged

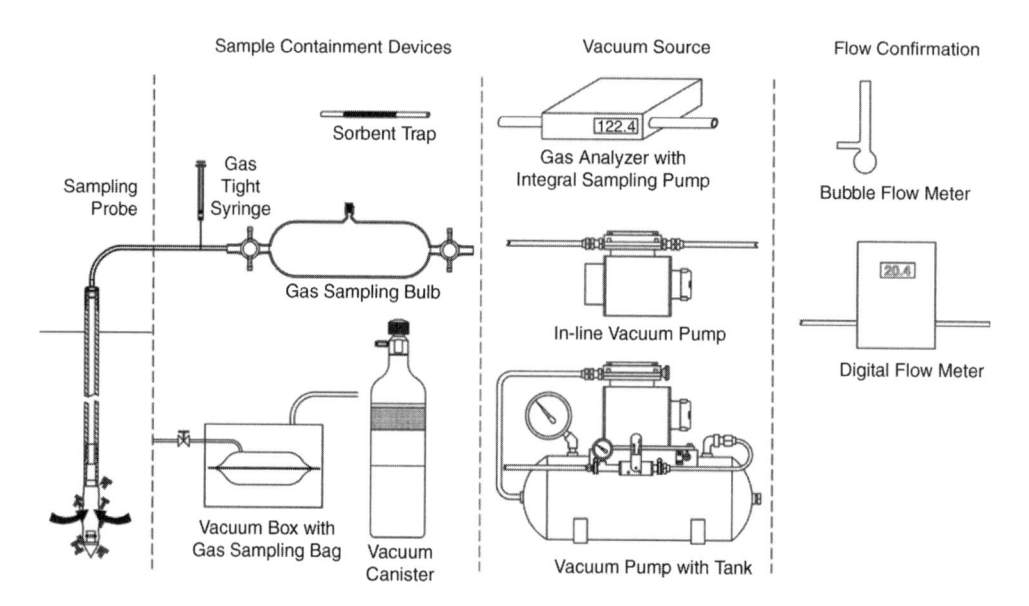

FIGURE 6.37
General elements of the soil-gas sampling train. Many types of sample containers and vacuum sources are used by field practitioners. Regardless of the components selected, care must be taken to assure that the sampling train is leak-free and that soil gas is flowing from the sample point.

to consult that document to ensure that their sampling programs conform to the standard. Practical QA/QC procedures for soil-gas sampling are briefly addressed in the following paragraphs, which reflect the authors' field experience in performing soil-gas sampling. The points and procedures discussed here are only for the sampling portion of a soil-gas sampling event. The analytical portion of a soil-gas sampling event should have separate QA/QC procedures.

The following general topics should be addressed in the field QA/QC plan for any soil-gas sampling event:

- Sampling procedures should be standardized in writing. This should include what tools will be used for sampling, intended depths, level of vacuum to be applied, volume to be removed or purged, and sample containers to be used. ASTM Standard 5314 specifically states that the container type used in the soil-gas survey should not vary during any given sampling event.
- Include a means for testing the sampling train for leaks. A leak check of the sampling train should be used between sampling points. This test is very quick and will prevent the analysis of a series of atmospheric air samples.
- If the purpose of the survey is to detect low levels of VOCs, a standard of the expected concentration should be run through the sampling train, using the standard sampling procedure, and the resulting sample analyzed. This can be easily done by preparing a standard concentration of the analyte in a gas-sampling bag, attaching the gas-sampling bag to the sampling train, and running the standard through the sampling train. Recovery should be compared to sample aliquots taken directly from the sample bag. This check can be used to eliminate sorptive or leaking elements from the sample train. The gas standard should contact every element of the sampling train in the same manner as a soil-gas sample.
- The sampling system should include a means of measuring the volume purged from the probe hole.
- Periodic sample blanks of carrier gas or atmospheric air should be taken to ensure the efficacy of decontamination procedures.
- A form should be created and used by the field staff to record sample conditions at each point, leak checks, decontamination procedures, and any QA/QC checks made.

Ground-Water Sampling Methods

The availability of DP methods for ground-water sampling has significantly affected the way in which investigations of contaminated ground water are conducted (Christy and Spradlin, 1992; Thornton et al., 1997; U.S. EPA, 1997, 1998a). DP methods allow installation of temporary ground-water sampling points in unconsolidated materials to collect representative samples at a fraction of the cost required for the installation of permanent wells by conventional drilling methods (Ehrenzeller et al., 1991; Cordry, 1995; Applegate and Fitton, 1997). Samples can be collected from DP ground-water sampling tools immediately after they are installed at the target depth, unlike conventional monitoring wells, which often must be sampled several weeks after well installation and development to avoid the effects of well installation trauma (Walker, 1983). DP sampling tools are typically used during initial site-characterization studies to collect single samples at discrete depths or to collect multiple samples from different vertical intervals in the same borehole (vertical profiling). These tools are often used in conjunction with field analytical methods, which makes it possible to define not only the presence of contamination but also its vertical and lateral extent (plume definition) in a time- and cost-efficient manner. Thus, these ground-water sampling tools are widely used to conduct accelerated (ASTM, 2004l) and expedited (Connecticut DEP, 1997; U.S. EPA, 1997, 1998a; ASTM, 2004m) site-characterization projects. DP sampling tools are also used to assist in selecting optimum locations for short-term and long-term monitoring wells and positions and appropriate lengths for well screens, and they can be used to conduct formation hydraulic conductivity tests (McCall et al., 2002).

DP ground-water sampling tools typically collect depth-discrete samples from intervals ranging from a few inches (a few centimeters) to 3 or 4 ft (1 to 1.3 m), allowing excellent vertical resolution when compared with conventional monitoring wells. The ability to collect multiple depth-discrete samples or continuous samples with some tools distinguishes DP technologies from conventional wells and is a significant advantage at hydrogeologically complex sites, where contaminant stratification may be significant. Relatively minor textural variations in geological materials, which can occur over vertical distances of only a few inches (a few centimeters), often have profound effects on the distribution and movement of NAPLs (Pitkin et al., 1994). At sites with heterogeneous geology, the concentrations of aqueous-phase contaminants can vary by several orders of magnitude over vertical distances of only a few inches (a few centimeters) (Cherry, 1992). Conventional monitoring wells typically average contaminant concentrations over the length of the well screen (Robbins and Martin-Hayden, 1991; Church and Granato, 1996; Hutchins and Acree, 2000) and can dilute contaminant concentrations significantly. In highly heterogeneous materials, DP ground-water sampling tools can be used to provide multiple samples or continuous samples in the same borehole to allow detailed spatial resolution of complex contaminant distribution, which is not possible with conventional wells. Investigators are cautioned, however, that it is possible to miss thin zones of contamination if a DP ground-water sampling tool is installed to collect only a single sample in a complex hydrogeologic setting and the tool is not positioned properly.

Because these tools are designed for multiple reuse, thorough decontamination (Chapter 20) (Parker and Ranney, 1997; ASTM, 2004f) must be conducted to prevent cross-contamination. Some DP ground-water sampling tools are designed with disposable screens that may be discarded after use to reduce decontamination activities. Disposable screens may be particularly useful in zones in which free product (LNAPL or DNAPL) is present. The probe holes in which many of these tools are installed can be pressure grouted from the bottom up to meet state requirements and U.S. EPA guidelines to protect ground-water resources and prevent cross-contamination.

Because there is a large variety of equipment currently available and because new tools are continually being developed, it is not possible to provide a detailed description of all available DP ground-water sampling tools. Instead, descriptions, deployment options, applications and limitations of general categories of samplers are presented and unique features of selected tools are described subsequently. For purposes of discussion, DP ground-water sampling methods are divided into single-rod and dual-tube methods and further categorized as exposed-screen or protected-screen devices. ASTM Standard D 6001 (ASTM, 2004b) explains the use and application of many DP ground-water sampling tools in additional detail.

Single-Rod Methods

The single-rod methods include both exposed-screen tools and protected-screen (or sealed) tools. In an exposed-screen tool, the screen of the sampler is exposed to the

Use of Direct-Push Technologies in Environmental Site 395

formation as it is advanced, while in a protected-screen tool, the screen is covered by a sheath as it is advanced into the formation.

Exposed-Screen Sampling Tools and Profilers

Exposed-screen tools have a short (e.g., 6 in. to 3 ft [15.2 cm to 1 m]) interval of exposed fine-mesh screen, narrow slots, or small holes at the terminal end of the tool or in the tool body that remains open to the formations penetrated as the tool is advanced. This allows samples to be collected either continuously or periodically to profile ground-water chemistry. This type of profiling can be done in a single DP probe hole without withdrawing the tool or DP rods from the hole. Exposed-screen tools can also be used to collect water-level measurements from discrete depths to assist in defining vertical hydraulic head distribution (and, therefore, vertical gradients) and in determining the potential for vertical flow within the same formation or between different formations. Additionally, some of these tools can be used to conduct hydraulic tests (e.g., slug or bail tests) at specific intervals to define the distribution of hydraulic conductivity in formation materials and to locate preferential flow pathways and barriers to flow (Butler et al., 2002).

Mill-slotted well points were some of the first DP ground-water sampling tools used for environmental investigations (Christy and Spradlin, 1992). These simple tools consist of a vertically slotted rod with an attached or expendable point; they range in diameter from 1 in. (25 mm) O.D. to over 3 in. (76 mm) O.D. (Figure 6.38). Some of these tools are the same diameter as the drive rods and attach directly to them (Figure 6.38a). Other designs are larger diameter than the drive rods and require a drive head for attachment to the tool string (Figure 6.38b). The vertical slots on the rods are typically 0.020 in. (0.5 mm) wide. These tools perform best when used in sandy formations. Under some conditions, they can be advanced to multiple depths in one borehole to conduct ground-water profiling.

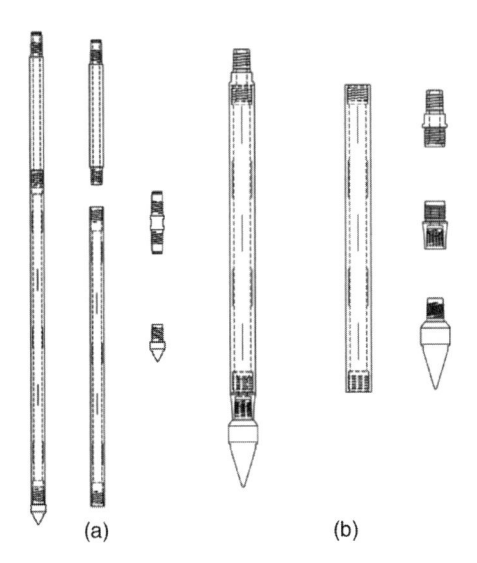

FIGURE 6.38
Simple mill-slotted well points (a, b) may be used for single-depth sampling or profile sampling at multiple depths in sandy formations. These tools can become clogged when advanced through a thin layer of clay, making sample collection difficult. Because the screens are exposed as they are advanced, there is also some concern for cross-contamination. When used as screening tools under good formation conditions, these tools can provide a time- and cost-efficient sampling option.

Wire-screen well points may be advanced directly into the formation in a manner similar to mill-slotted well points. These devices are constructed with stainless-steel wire-wound screens (Figure 6.39) 1 in. (25 mm) or greater in diameter and are typically 1 to 5 ft (0.3 to 1.5 m) in length. The screen slot size may be as fine as 0.004 in. (0.1 mm) in some models, but is more typically 0.020 or 0.010 in. (0.5 or 0.25 mm). Under favorable conditions, these tools may be used to conduct vertical profiling of ground-water chemistry (U.S. EPA 1998a; Schulmeister et al., 2001).

Another version of the wire-screen well point is designed with an inner ported rod to give the tool greater strength (Figure 6.40a). When sands are present from the surface to the desired sampling depths this tool may be advanced directly into the formation from the surface. Where fine-grained soils overlie a sandy zone to be sampled, this profiling tool may be telescoped through larger diameter casing into the sandy formation (Figure 6.40b). This prevents clogging of the screen with fines from the shallow soils.

One exposed-screen tool (the Waterloo Profiler®) is composed of a 6 in. (15 cm) long, uniform diameter, stainless steel drive tip in which several 0.25 in. (6 mm) sampling ports have been drilled and covered with fine-mesh stainless-steel screen (Figure 6.41). The sampling ports are accessed from the surface via small-diameter stainless-steel, Teflon, or polyethylene tubing that runs through the drive rods (Figure 6.41). As the tool is advanced, the operator slowly pumps clean (e.g., distilled or deionized) water down the tubing and out the ports using a peristaltic pump. This keeps formation water from entering the tool as it is advanced. The uniform tool diameter ensures that there is a good seal with the formation above the sampling ports and thus no cross-contamination or drag-down occurs. As the target sampling zone is reached, the pump flow is reversed and samples are collected following purging of the sampling lines. Potentially long purging times (more than 10 min) may be required to eliminate the possibility

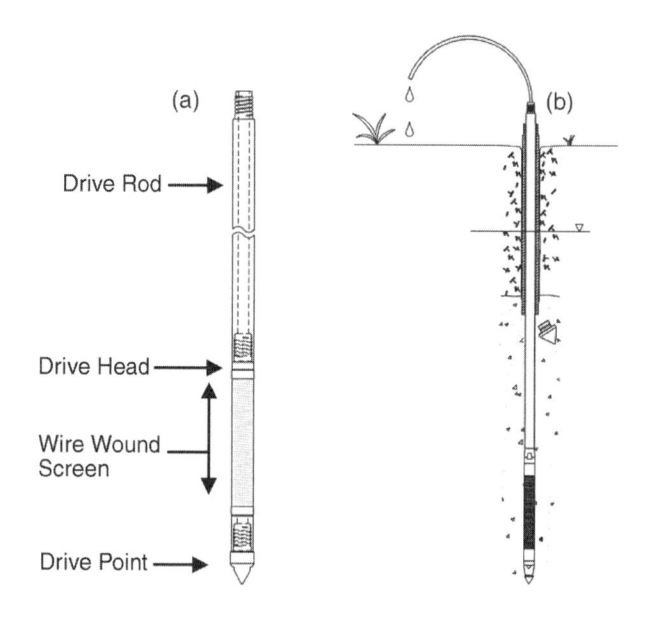

FIGURE 6.39
(a) To improve the percentage of open area on simple exposed-screen samplers, continuous-slot wire-wound screens have been incorporated. In sandy formations, these screens improve yield when compared with most mill-slotted designs. These screens are still subject to clogging in fine-grained materials and the potential for cross-contamination must be evaluated. (b) Telescoping exposed-screen profilers through larger diameter rods is one method to prevent clogging of the screen in shallow fine-grained soils.

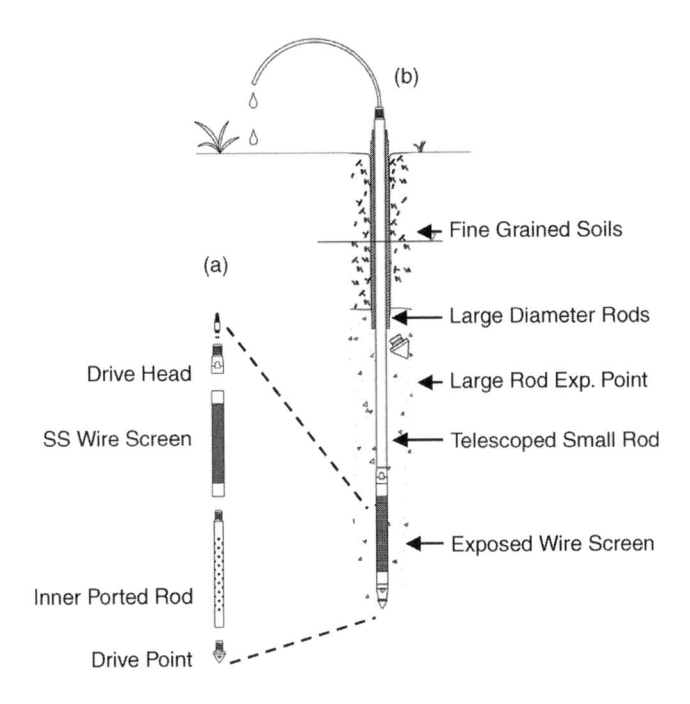

Use of Direct-Push Technologies in Environmental Site 397

FIGURE 6.40
(a) Some wire-wound screens are supported by an inner rod to transmit driving forces and provide additional strength to the assembly. (b) When fine-grained soils are present above a sandy zone, these tools may be telescoped through larger diameter rods to prevent clogging of the exposed screen and minimize the potential for cross-contamination.

that the clean water may have diluted formation water samples. After samples are collected, the pump is reversed again and clean water is pumped through the sampling ports to flush out the prior sample and to keep the tool from clogging, and the tool is then advanced to the next target sampling zone, where the process is repeated (Figure 6.42). Several field studies (Cherry et al., 1992; Pitkin et al., 1994, 1999) have demonstrated that this tool is capable of providing a very detailed view of the anatomy of contaminant plumes (Figure 6.43), particularly in complex stratified geological

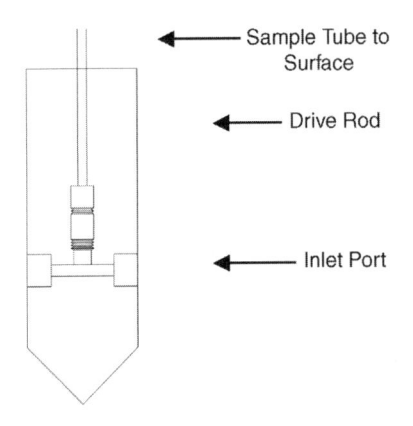

FIGURE 6.41
One of the earliest unprotected-screen profiling tools was designed with screened inlet ports attached to the surface by stainless-steel tubing. A peristaltic pump could be used for sampling with flow reversed during advancement to minimize clogging of the ports. (From Pitkin et al., 1994. With permission.)

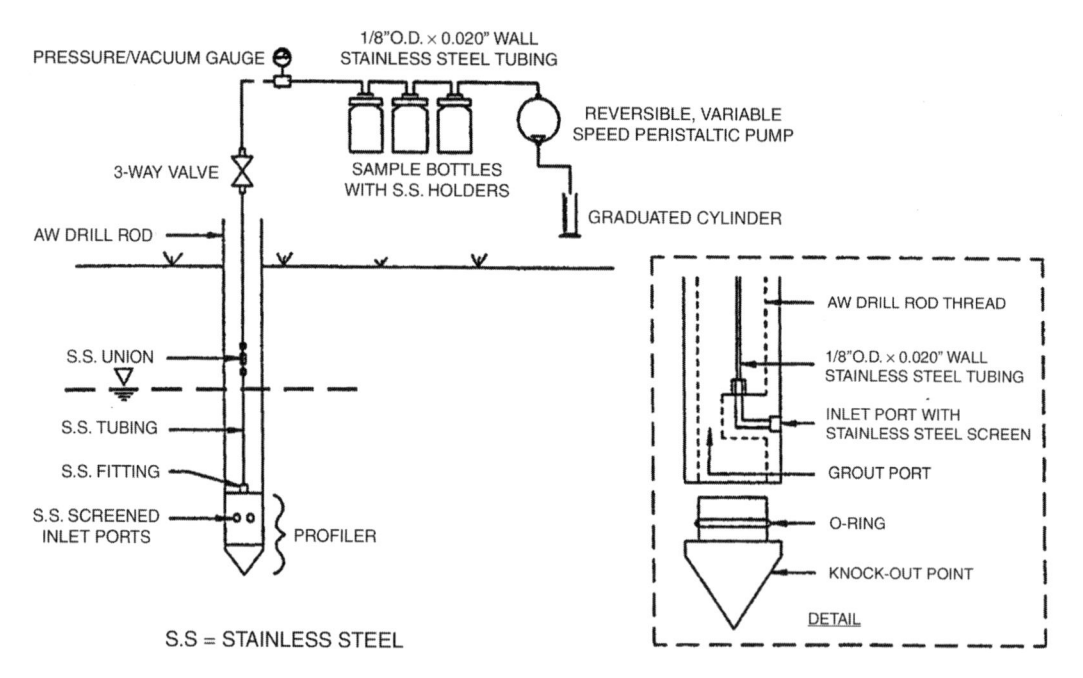

FIGURE 6.42
The Waterloo Profiler is flushed as it is advanced. When it reaches a zone to be sampled, the flow of the pump is reversed, and formation water enters the tool and is brought to the surface to be collected as a sample. The process is repeated for multiple zones, providing a chemical profile of a contaminant plume. (From Pitkin et al, 1999. With permission.)

materials, without the effects of drag-down or cross-contamination. However, because a peristaltic (suction-lift) pump is typically used to collect samples, this device is limited to sampling from lifts of less than 30 ft, and there may be a negative bias in samples collected for analysis of VOCs, dissolved gases, or other pressure-sensitive parameters. As Pitkin et al. (1994) point out, the advantage of being able to delineate plumes in detail and, in particular, to locate zones of high concentration may override the disadvantage posed by this source of sample bias at many sites.

Sealed-Screen Samplers

Sealed or protected-screen samplers have an outer sheath or housing that protects the screen from contact with the formation as the tool string is advanced to the selected sampling depth. When the tool is within the desired sampling zone, the screen is exposed to the formation by retracting the probe rods and water flows into the tool (and, in some cases, up the rod string) under the hydraulic head conditions that exist at that depth. O-ring seals are generally used between the drive tip and the tool body to ensure that the sampler is water tight as it is advanced. If the drive rod is a smaller diameter than the tool body (which is generally the case), the only seal to prevent cross-contamination from zones penetrated by the tool in stiff, cohesive soils will be between the formation and the part of the tool body above the screen. Thus, care should be taken not to retract the tool more than the length of the tool body itself in these conditions.

This type of sampling tool can be used for collecting accurate, depth-discrete samples under *in situ* hydrostatic pressure, with minimal aeration and agitation (Edge and Cordry, 1989; Smolley and Kappmeyer, 1991; Zemo et al., 1995). However, most samplers of this type deployed using single-rod DP systems are limited to collecting one sample per

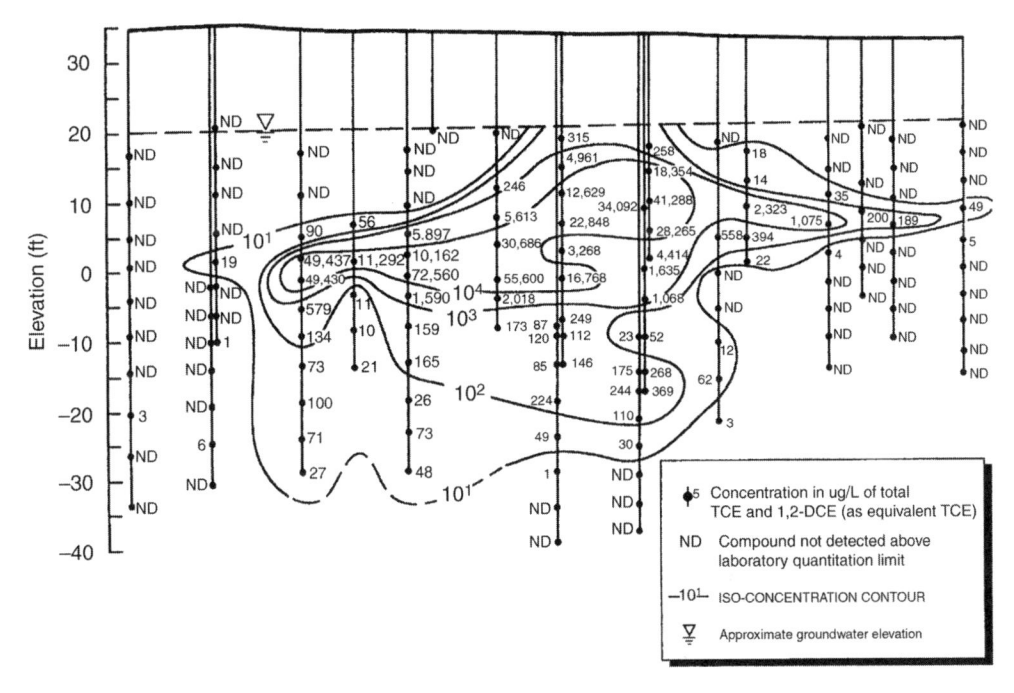

FIGURE 6.43
A transect across a plume of dissolved-phase chlorinated solvents, displaying the results of 14 profiles conducted with the Waterloo Profiler. (From Pitkin et al., 1999. With permission.)

advance of the tool. Multidepth sampling can only be accomplished with multiple tool advances, but the tool must be brought to the surface after each sample has been collected, decontaminated, and reassembled to permit advancement of the tool to the next depth. In addition, because part of the original borehole may stand open after removal of the tool and rod string, the possibility of cross-contamination within the borehole exists between tool advances (Zemo et al., 1994).

One protected-screen ground-water sampling tool (Figure 6.44) is designed with a wire-wound stainless-steel screen that is enlarged at the top. In this design, when the tool body is retracted and the screen is fully extended, the enlarged screen top is retained by a reduced diameter in the lower end of the tool body. This allows for recovery of the screen for decontamination and reuse at multiple depths or multiple locations. This design also provides the field operator with a tool with a variable-length screen. The screen sheath may be retracted enough to expose only a few inches (a few centimeters) of the screen, 1 or 2 ft (0.3 to 0.6 m) of the screen, or the entire screen length (40 in. [102 cm]). This can be very useful when sampling from a small interval, such as a thin sand layer, is required to define a contaminant migration pathway. This feature can also be useful for conducting slug tests over discrete intervals for defining vertical variations in hydraulic conductivity (Butler et al., 2002). The steps for installation and operation of a simple protected-screen tool are illustrated in Figure 6.45. This protected-screen tool is designed with a grout plug, so it is possible to conduct bottom-up grouting with a tremie tube to meet state requirements and U.S. EPA and ASTM guidance for borehole grouting when sampling in a given probe hole is concluded.

Another protected-screen sampling tool (the Hydropunch) is designed for operation in two sampling modes (Cordry, 1991). In one mode, the tool is advanced to the target sampling depth, a short (6 in. [15 cm]) stainless-steel mesh screen is exposed to the formation, and the

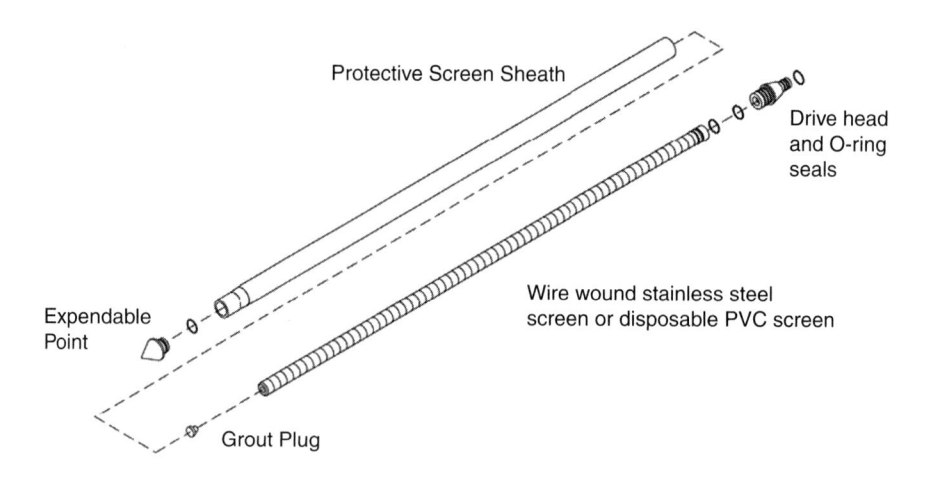

FIGURE 6.44
Single-tube, protected-screen ground-water sampling tools provide an efficient means to collect high-integrity ground-water samples at single depth intervals. Ability to vary screen length from a few centimeters to a meter allows for targeting a specific interval for sampling. Bottom-up grouting seals the borehole, thus preventing contaminant migration.

tool and rods above the tool fill with water under hydrostatic pressure. A sample is collected from the tool using a device inserted down the rod string from the surface (e.g., a bailer, inertial-lift pump, peristaltic pump, or other appropriate device). A second design of this tool (Figure 6.46) incorporates a longer (30 to 42 in. [76 to 107 cm] long) expendable

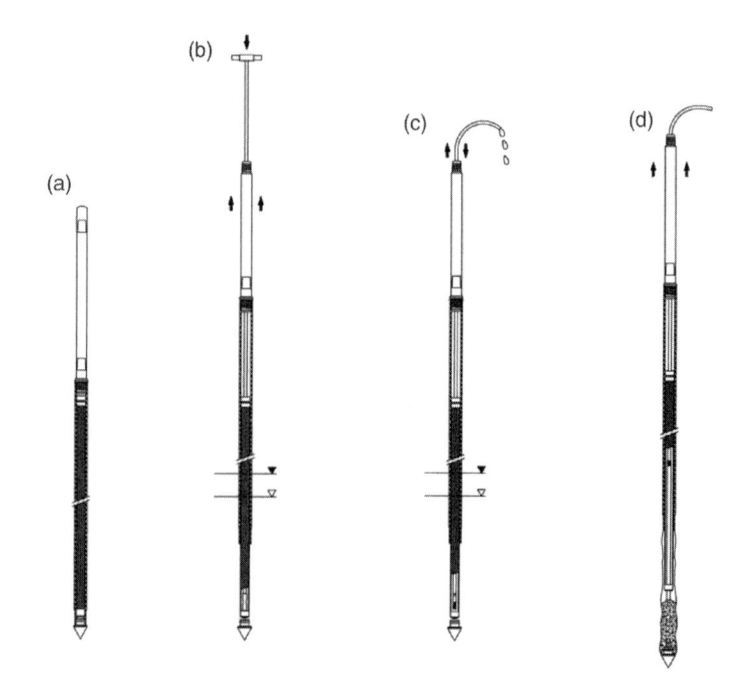

FIGURE 6.45
(a) The protected-screen ground-water sampler is advanced to the desired sampling depth. (b) The screen is exposed to the formation over the sample interval. (c) Development and sampling are conducted. (d) After sampling is completed, bottom-up grouting may be conducted to seal the probe hole.

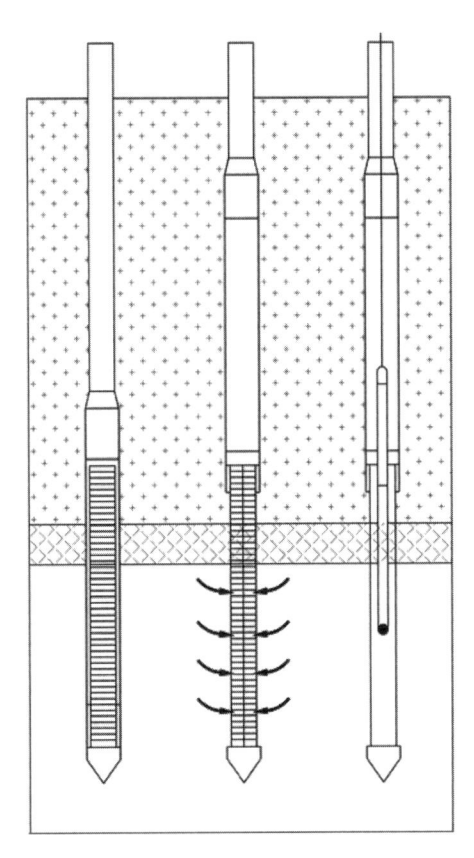

FIGURE 6.46
Some DP ground-water sampling tools, such as the Hydropunch (pictured), recover the sample within the sampler body. The sampler is retrieved and the sample transferred from the sampler to appropriate sample vials.

polyethylene or slotted PVC screen intended to straddle the water-table surface to enable detection of LNAPLs and collection of samples of ground water or LNAPL. After sampling is concluded, the tool body is retrieved to the surface, leaving the expendable screen and drive point in the probe hole. Additional samples can be collected by reloading the tool with a new drive point and screen, advancing to the next zone of interest and repeating the process. In either of these designs, an unlimited volume of sample may be collected; formation hydraulic tests may be conducted in the version of this tool with the longer screen. When the expendable screen is used, pressure grouting with a high-pressure injection pull cap may be conducted (Figure 6.47) after sampling is concluded in the probe hole. In the second mode of operation, after the tool (with a 6 in. [15 cm] screen) is advanced to the target sampling depth and the screen is exposed to the formation, integral check valves are used to trap the sample in the tool body as the tool fills with water. The tool must then be brought back to the surface and the sample decanted into an appropriate sample container. In this mode, the volume of sample collected is limited to the volume of the inside of the tool (either 500 ml or 1.2 l, depending on tool design). To collect samples at multiple depths, the tool must be decontaminated and driven to the next depth, and the sampling process repeated. To decommission the hole following sampling, re-entry grouting is generally necessary.

In another sealed-screen sampling tool (the BAT Enviroprobe) (Figure 6.48), a septum behind the screen of the tool keeps the tool from filling with water as the screen is

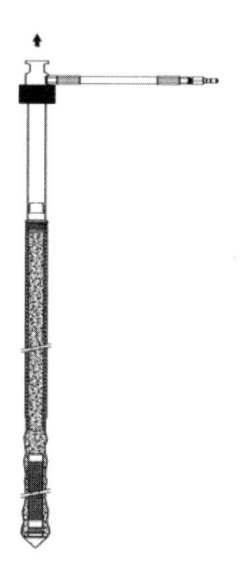

FIGURE 6.47
When an expendable PVC screen is used with the protected-screen sampler, pressure grouting may be conducted directly through the drive rods to seal the probe hole from the bottom up.

exposed. To operate this system, the tool is driven to the target sampling depth and the rods pulled back to expose the very short (2 to 3 in. [5 to 7.6 cm]) screen. An evacuated (under vacuum) glass sample vial is then lowered down the drive rods and coupled, via a hypodermic needle that penetrates the septum, to the chamber behind the screen. The sample is drawn under vacuum into the sample vial, which may be either 50 or 100 ml, and maintained at *in situ* pressure conditions until the point at which the

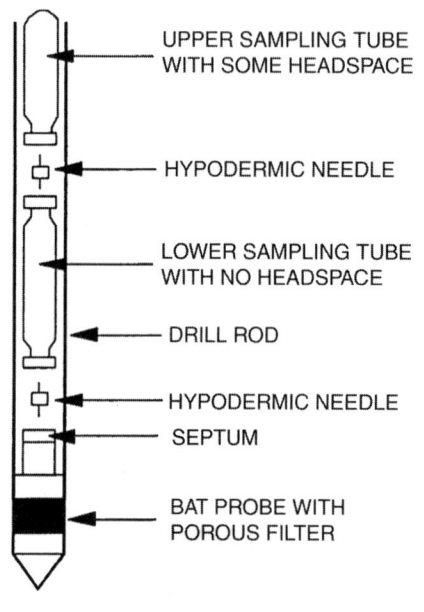

FIGURE 6.48
The BAT Enviroprobe, a sealed-screen tool used to profile ground-water systems.

sample is analyzed (Berzins, 1992). The sample is not exposed to atmospheric conditions (pressure, temperature, and atmospheric air) and thus should be more representative of *in situ* conditions from that perspective. However, because the sample is collected under vacuum, any headspace left in the container will lead to partitioning of volatiles or gases out of the sample and potential for significant sample bias. Additional sample volume may be collected by stringing multiple vials together or by lowering additional sample vials down the rods. Probe holes made with this tool must be decommissioned by re-entry grouting.

Combined Sampling Tools

A combined sampling tool (the Simulprobe®) (U.S. EPA, 1998e) provides the field operator with the ability to sample soil, ground water, and soil gas with the same tool (Figure 6.49). The tool is made of hardened steel, is 3.38 in. (8.6 cm) O.D. and 26.5 in. (67.3 cm) long, and contains a split-barrel soil core sampler (similar to a split-spoon sampler) that is 2.5 in. (6.4 cm) I.D. and 18 in. (45.7 cm) long. Ports on the side of the tool lead to a sample canister behind the soil core sampler, from which soil-gas or ground-water samples may be collected from the surface. A drive tip seals the sample barrel until the target sampling depth is reached. At the target sampling depth, a latch holding the tip in place is released and the sampler is advanced to collect the soil core. At the same interval, either a soil-gas sample (in the vadose zone) or a ground-water sample (below the water table) can be

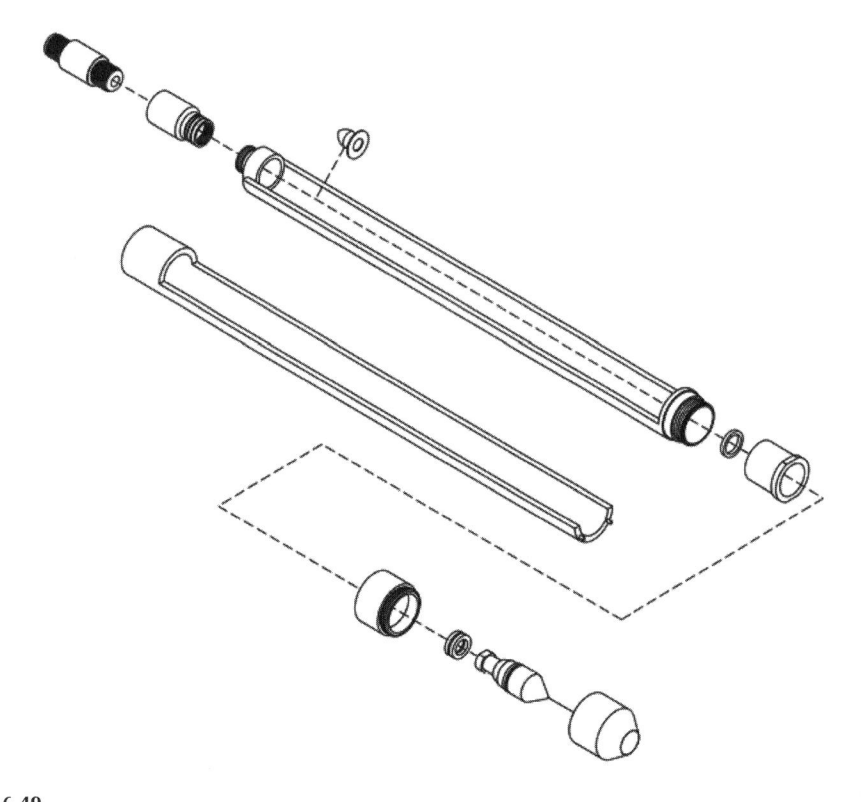

FIGURE 6.49
Some specialized DP tools have been designed to collect samples of soil, ground water, or soil gas, depending on the location-specific requirements. While providing an array of sampling options, these tools are more complicated to operate and may be subject to a higher rate of failure in the field.

collected from the geologic materials surrounding the core barrel by opening the ports in the side of the tool. This allows spatial and temporal correlation of any gas or liquid phase contamination present in the formation with the geologic material in the soil core and better estimates of total contaminant mass. Soil-gas samples are collected via a vacuum pump, which draws a continuous stream of gas to the surface through the probe rods. Ground-water samples can be collected in a 19 in. (48.3 cm) long, 2 l (0.5 gal) stainless-steel canister, located behind the soil core barrel, which fills under hydrostatic pressure and is brought to the surface with the soil core. Alternately, ground-water samples can be collected after the water passes through a check valve above the sample canister and into the probe rods, with a sampling device (e.g., bailer, inertial-lift pump, peristaltic pump, or bladder pump) lowered through the probe rods. This tool can be used on either a single-rod system or a dual-tube system or with conventional drilling methods. When sampling more than one environmental medium at a target depth is required, this tool provides a cost-efficient means of accomplishing project goals.

Dual-Tube Methods

The simplest type of dual-tube system employs an outer casing advanced to the desired sampling depth and an inner drive rod with a screen at the end that is lowered to the base of the borehole (Figure 6.50) (ASTM, 2004b). The screen is driven into the formation ahead of the outer casing with the inner drive rod. Depending on the specific design of the system, it may be operated as a depth-discrete sampler or as a profiling tool for sampling at multiple depths in one borehole.

One dual-tube system has been designed specifically for vertical profiling of ground-water contamination under various formation conditions. This system consists of a nominal 0.75 in. (19 mm) PVC with 0.010 in. (0.25 mm) slotted screen with screen head and tapered point (Figure 6.51). The dual-tube rods with drive point are advanced to the desired sampling depth and the inner rods and drive point are removed

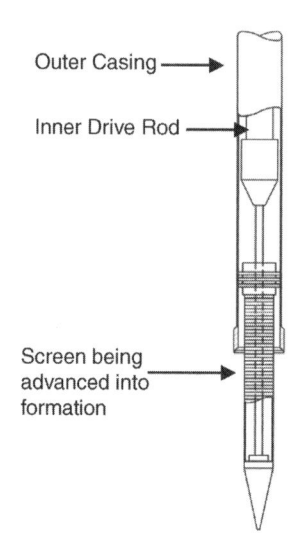

FIGURE 6.50
A simple dual-tube ground-water sampling device uses the inner drive rods to advance a screen beyond the outer casing. The inner drive rods are removed for screen development and sample collection.

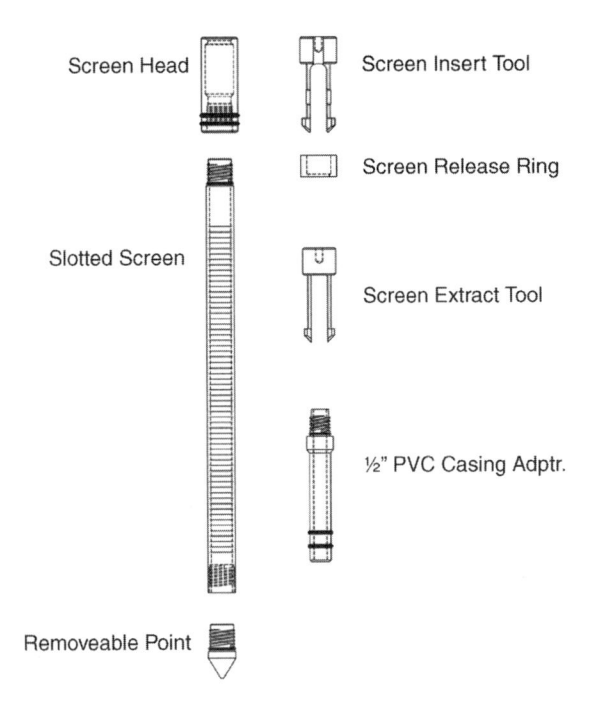

FIGURE 6.51
Components of a dual-tube ground-water profiling tool. The screen, screen head, and point are assembled for installation through the outer drive rods of the dual-tube system. The insert tool and release ring are used to lower and hold the screen in position for deployment. After development, sampling and slug testing are completed; the extract tool is used to recover the screen for later use. The PVC adapter may be used to attach PVC casing to the screen and isolate the water being sampled from the bore of the drive rods if sample parameters may be impacted by exposure to the steel drive rods. This also provides a smaller casing radius for slug testing in lower hydraulic conductivity formations.

(Figure 6.52). The screen assembly is lowered through the drive rods and held in position at the bottom of the borehole with small extension rods. The outer rods are retracted to expose the screen to the formation and the extension rods and insert tool are then removed. The screen may then be developed and ground-water samples collected. Additionally, slug tests may be performed at each depth interval to determine the formation hydraulic conductivity (McCall et al., 2002). After sampling and testing are completed, the extension rods and a retrieval tool are used to extract the PVC screen. The inner rods and drive point are reinstalled and the dual-tube rods are advanced to the next sampling interval. The cycle (Figure 6.52) is repeated for sampling and slug testing at multiple depths.

Another dual-tube sampling system (the Enviro-Core®) consists of a 2.4 in. (6 cm) O.D. drive casing with hardened steel drive shoe and a 1.8 in. (4.6 cm) O.D. inner soil core barrel that can be used first to core soil either continuously or at discrete depths. At the point in the subsurface at which a ground-water sample or hydraulic test is desired, the soil core barrel is removed and replaced with a 1.5 in. (38 mm) nominal diameter temporary well screen (Figure 6.53). While the screen is held in place, the drive casing is retracted to expose the screen to formation materials, which collapse around the screen (if the materials are noncohesive). The temporary screen is constructed of a short length (1 to 4 ft [0.3 to 1.2 m]) of either stainless-steel wire-wound screen or slotted PVC attached to

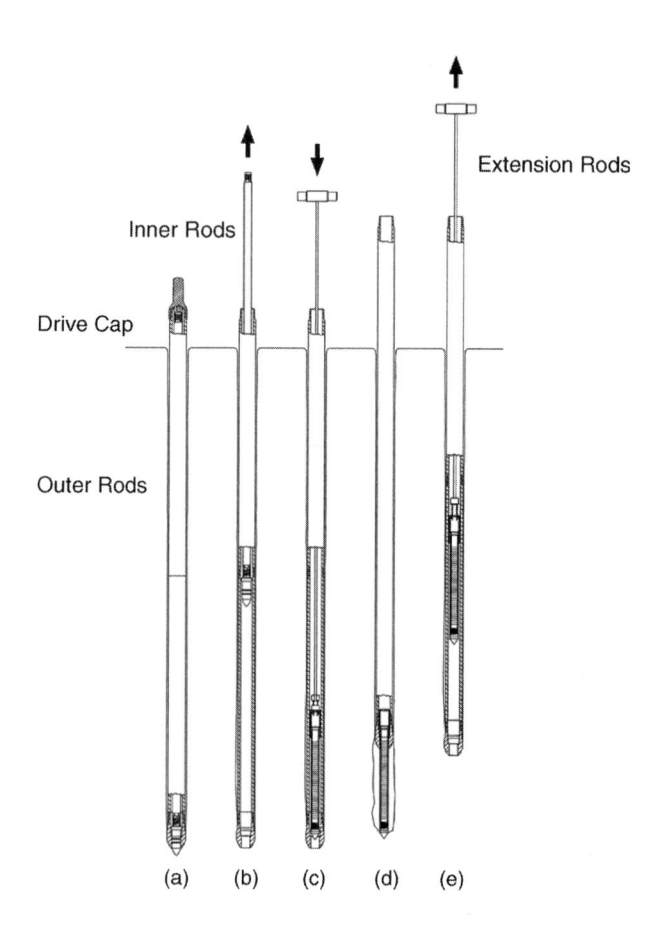

FIGURE 6.52
The basic steps for installation and operation of the dual-tube ground-water profiler system. (a) The dual-tube system is advanced to depth with inner drive rods and solid point. Alternatively, soil sampling may be conducted as the tool string is advanced. (b) The inner rods and drive point or soil sample are tripped out of the boring and the outer rods remain in place. In saturated sands, formation heave will have to be controlled by adding water to increase water pressure inside the rods. (c) The assembled screen is inserted to the bottom of the borehole and held in position with extension rods. (d) Once the outer drive rods are retracted, exposing the screen to the formation, the extension rods are removed. Screen development, sampling, and slug testing may now be conducted. (e) After sampling and testing are completed, the screen is removed with the extraction tool. The inner rods are re-inserted and the tool string is advanced to the next desired depth, where steps (a) through (e) are repeated. In this manner, ground-water sampling (profiling) and slug testing at multiple depths may be accomplished.

an inflatable packer above the screen. The length of the well screen is selected to match the length of the target sampling interval. After the drive casing is retracted the length of the screen, the packer is inflated to seal the annular space between the screen and the drive casing. The screen may then be developed using either surging and pumping or bailing or overpumping and backwashing. Water-level information can be collected to determine formation hydraulic head after the water level has stabilized in the drive casing. Slug or bail tests can be conducted by adding or removing water; ground-water samples can be collected using an appropriate sampling device lowered from the surface. The advantage of this method is that there is generally less soil compaction surrounding the screen, because soil is collected as a sample from the target zone rather than displaced (Einarson, 1995). Formation hydraulic conductivity should therefore be less affected than with other DP methods.

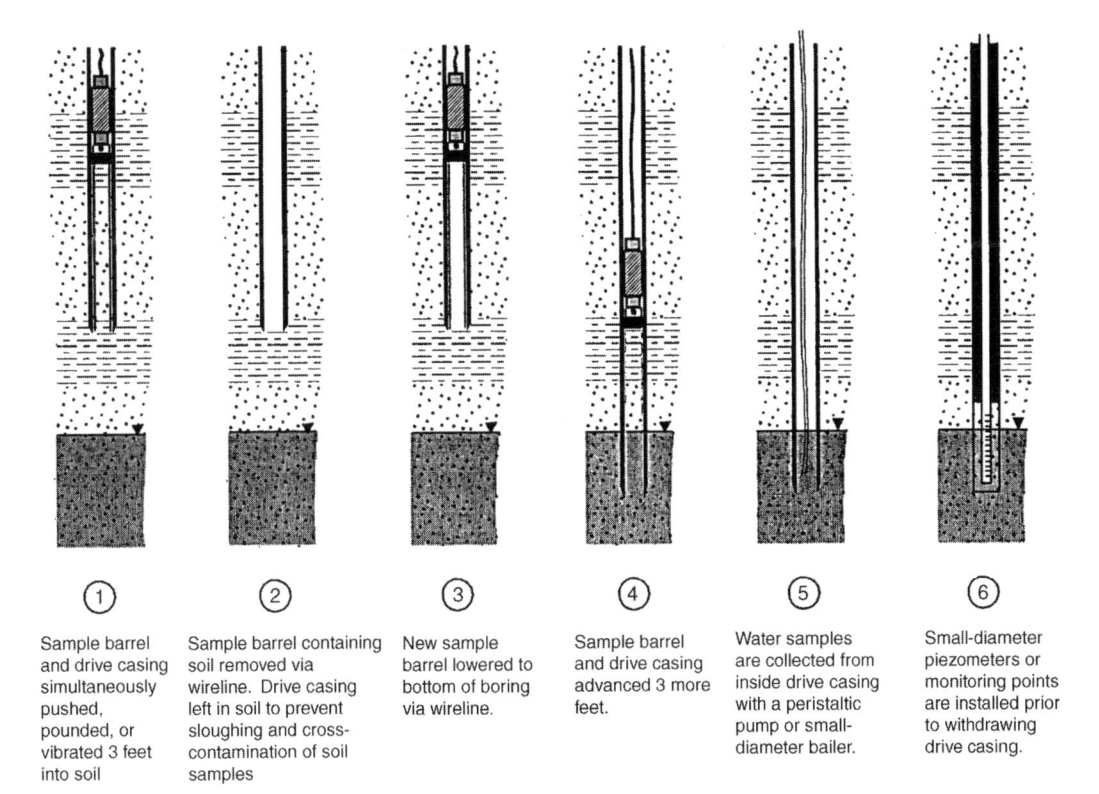

Use of Direct-Push Technologies in Environmental Site 407

①	②	③	④	⑤	⑥
Sample barrel and drive casing simultaneously pushed, pounded, or vibrated 3 feet into soil	Sample barrel containing soil removed via wireline. Drive casing left in soil to prevent sloughing and cross-contamination of soil samples	New sample barrel lowered to bottom of boring via wireline.	Sample barrel and drive casing advanced 3 more feet.	Water samples are collected from inside drive casing with a peristaltic pump or small-diameter bailer.	Small-diameter piezometers or monitoring points are installed prior to withdrawing drive casing.

FIGURE 6.53
Enviro-Core with screen.

ConeSipper®

The ConeSipper (Figure 6.54) is a tool for collecting discrete ground-water and vapor samples as well as performing continuous vapor profiling during a CPT penetration. It can be used in conjunction with a piezocone or other sensor modules. An important feature of the this tool is its ability to draw discrete ground-water samples into an internal chamber where they can either be lifted to the ground surface for laboratory analysis or left in contact with down-hole sensors for *in situ* analysis. Lifting to the surface is achieved by charging the chamber with inert gas (e.g., nitrogen) from the surface. The unique configuration of valves, tubing, inlets, and outlets enables the pressurized gas to lift the water sample from the chamber through tubing to the surface. The chamber and sample inlet port can be purged by increasing the gas pressure above the release threshold of a down-hole purge valve. With modification, this tool has also been used to perform down-hole gas purging of water samples for VOC analysis, suction lysimetry for vadose zone sample collection, and *in situ* pH, oxidation–reduction potential (ORP), and temperature sensing.

Applications and Limitations of DP Ground-Water Sampling Tools

Single-rod exposed-screen sampling tools are generally the least expensive and simplest to operate in the field. However, these tools are susceptible to clogging of the screen, particularly if it passes through fine-grained materials (e.g., silts or clays), but sometimes even in relatively clean sandy formations (Butler et al., 2002). Development and unclogging of the

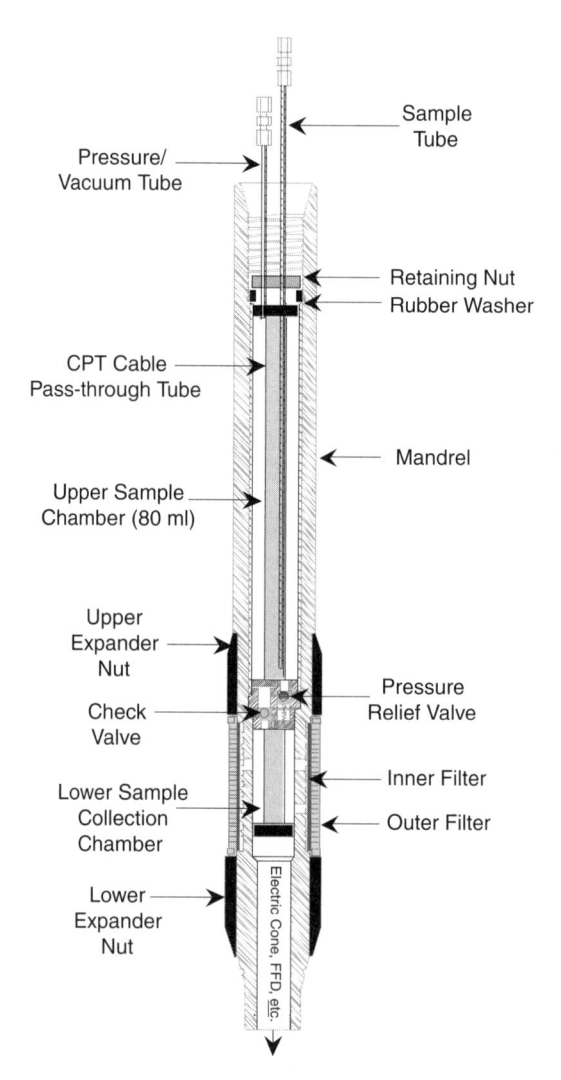

FIGURE 6.54
The ConeSipper delivers multiple discrete ground-water or vapor samples to the surface during a single push and is purgeable *in situ*.

exposed screen may be difficult or impossible if the tool is advanced through a clay layer. If the material that clogs the screen is contaminated, it may not be possible to obtain representative ground-water samples at succeeding depths if profiling is conducted.

Protected-screen single-rod samplers are usually more expensive than exposed-screen tools and often are more complicated to operate, but they do eliminate the clogging problems encountered with the exposed-screen tools. When o-rings or other appropriate methods are used to seal protected-screen samplers and drive rods, high-integrity, depth-discrete samples may be obtained with these tools. One limitation of protected-screen samplers is that they can not be advanced further after the screen is opened for sampling. However, some field operators have advanced these tools to the bottom of the proposed profiling interval and then incrementally retracted the tool and sampled at each desired interval as the tool is withdrawn from the probe hole. This approach is most applicable in sandy formations that collapse as the tool is retracted.

Dual-tube ground-water sampling systems generally have a higher initial cost because of the need for two sets of drive rods. These systems are more rugged than single-rod systems, and some tooling may allow the field operator to conduct soil sampling, ground-water profiling, and formation hydraulic testing at multiple depths in one borehole without having to retract the drive casing. Because larger diameter rods and drive casing are required to operate the dual-tube systems, they may not be able to penetrate to as great a depth as smaller diameter single-rod samplers under the same formation conditions.

The use of any DP ground-water sampling tool may be affected by smearing of fine-grained materials across the sampling zone or possible drag-down of NAPLs, contaminated soil, or ground water from zones above the desired sampling interval. Although the shape and exterior surfaces of most sampling tools and drive rod minimize the downward transport of soil and water, there is some evidence that advancing DP probes in predominantly fine-grained materials can cause some smearing of fines across thin zones of coarser materials (Henebry and Robbins, 2000). Minor drag-down of contaminated soil or water has also been noted in some cases, particularly when sampling from an exposed-screen tool after it has been advanced through highly concentrated contaminants (Pitkin et al., 1999). An effective solution to this problem is to develop the tool in place to remedy any smearing that may be present and to purge the tool of water for a short period of time to remove any effects of drag-down prior to collecting any samples for chemical analysis.

One of the consequences of using DP ground-water sampling tools is that the advance of the tool causes some minor disturbance of formation materials, including compaction and disaggregation of formation materials and associated breakage of grain coatings and cementing agents such as iron oxy-hydroxides and carbonates. The effect is that when an exposed-screen tool is sampled or a sealed-screen tool is opened, the initial water removed from the tool is usually quite turbid, containing substantial particulate matter that is not mobile under ambient ground-water flow conditions. Many of these particles are highly surface-reactive and, because of their high surface area per unit mass and volume, have very high sorptive capacities and strong binding capabilities for selected groups of analytes. These include metal ions and hydrophobic organic compounds (polyaromatic hydrocarbons [PAHs], PCBs, pesticides, and dioxin) that adsorb onto organic particulate matter; metal ions that may form organometallic complexes with organic colloidal matter; and radionuclides and metal ions that may adsorb or otherwise be bound onto inorganic matter (clays, metal oxides, and inorganic precipitates). Some of the particulate matter (the largest fraction) will settle out quickly, but much of it will remain suspended in the water column and would be collected as part of the sample if the sampling tool were not purged first. Developing the sampling tool and using low-flow sampling methods (ASTM, 2004n) generally results in significantly lower turbidity levels and improved sample quality. However, the impact that turbidity may have on samples collected from DP sampling points depends on the objectives of sampling, the data-quality objectives (DQOs) (U.S. EPA, 2000), and the analytes of interest. Table 6.2 (modified from U.S. EPA, 1997) provides some guidelines on sampling protocols that should be followed to meet specific DQOs when using DP ground-water sampling tools.

If the objective of sampling is simply to determine the presence or absence (and approximate concentrations) of highly soluble, weakly sorbed VOCs, as in a screening-level investigation, the presence of turbidity will probably have little or no effect on analytical results (Paul and Puls, 1997). If the objective is to quantify the soluble fraction of analytes, which are also major constituents of the formation materials (such as Fe, Al, Ca, Na, Mg, Mn, and Si), inclusion of particulate matter in samples poses a serious obstacle to proper interpretation of sampling results (Powell and Puls, 1997). If the objective is to determine or predict

TABLE 6.2

Data Quality Levels and Ground-Water Sampling Protocol

Data Quality Level	General Field Application	Analytical Protocol	Sampling Protocol for Ground Water[a]
1. Qualitative screening	Presence of contamination (e.g., yes or no: low, medium, and high concentration)	Hand held PID/FID, Hach kit, XRF for metals in soils	Tubing bottom check valve, or bailer, ground-water sampler or profiler with no development required
2. Semiquantitative screening	Approximation of contaminant zone or level (e.g., 10s, 100s, 1000s, free product?)	Immunoassay kits, Hach kits, hand held GC, XRF for metals in soils (single point standardization, duplicates?)	Tubing bottom check valve, bailer, or peristaltic pump, minimal development prior to sampling
3. Quantitative delineation	Define specific contaminants and accurate concentrations (e.g., TCE = 7.3 µg/l; Cr = 87 µg/l)	Field or lab GC or AA/ICP methods with defined quality control (standards, spikes, duplicates, etc.)	Develop SP15/16 sampler or DP well to lower turbidity, sample with peristaltic pump or possibly bladder pump
4. Quantitative clean zone	Regulatory monitoring, determining clean samples, closure sampling (e.g., <5.0 µg/l benzene)	Approved laboratory methods with stringent quality control for legally defensible data	Develop GW sampler or well and quantitatively measure turbidity to confirm quality, sample with bladder pump

Note: PID, photoionization detector; FID, flame ionization detector; GC, gas chromatography; XRF, x-ray fluorescence analysis; AA, atomic absorption spectroscopy; ICP, inductively coupled plasma emission spectroscopy.
Source: Modified from U.S. EPA, 1997.
[a] These are generalized sampling protocols for the respective data quality level and will need to be optimized for the specific site conditions and regulatory requirements. Sampling protocols were not specified in the UST Guidance Document (U.S. EPA, 1997).

transport behavior of trace metals or other analytes that may be bound to particulate matter or organic species that may be sorbed to particulate matter, sample preservation may liberate these species from the particle surfaces, producing a strong positive bias in the final analytical results. In contrast, if the sample is filtered, filtration may produce an underestimation of the concentrations of other analytes (Chapter 15). For example, analytes in the aqueous phase that have a strong tendency toward ion exchange or adsorption onto solid surfaces may partition onto the particles and be removed through sample filtration, producing a negative bias. Large concentration differences in filtered and unfiltered samples at a variety of sites have been observed for metals (Puls and Barcelona, 1989; Puls et al., 1992; Pohlmann et al., 1994), PAHs (Backhus et al., 1993), and radionuclides (Buddemeier and Hunt, 1988; Penrose et al., 1990). Thus, it is important to follow appropriate development and sampling protocol to ensure collection of representative samples from DP sampling points.

DP ground-water sampling tools of any type should be equipped with drive tips that are less than or equal to the diameter of the tool body (or drive casing). This avoids creating an annulus between the tool body (or drive casing) and the soil that may stay open during tool advance in some stiff, cohesive soils. The presence of such an annulus may result in cross-contamination within the time frame of sampling (usually immediately following tool advance to the target sampling depth). An exception to this rule is where tools are advanced through plastic, cohesive soils where it is desirable to reduce friction on the tool and drive rod or casing during tool advance. Because plastic soils will conform to the shape and diameter of the tool body soon after the tip has made the initial opening in the soil, cross-contamination is not usually an issue in this situation. The investigator should have site-specific samples to verify soil types and conditions.

Most tips are made of either steel or aluminum to withstand driving forces. Some sampler tips are expendable and are left in the ground after sampling is concluded and the tool is withdrawn from the hole, while others are part of the sampling tool and are retrieved with the sampler. In either case, o-ring seals should be used between the drive tip and the tool body to ensure that the sampler is water-tight as it is driven to the target sampling depth so cross-contamination is avoided.

Owing to the rigorous mode of installation of DP sampling tools, the tool body is typically made of some type of steel (hardened steel, stainless steel, or some other alloy). Screens and other internal parts may be made of one or more of a variety of materials including stainless steel and many types of plastics (PVC, polyethylene, polypropylene, Teflon, or other fluoropolymers). The types of materials selected for DP sampling tools used for a particular sampling program should be based on possible chemical interactions with the water to be sampled. Drive rod and drive casing is always made of hardened steel so that it can withstand driving forces and pull-back on the rod string as the tool is withdrawn from the ground. For drive rod and drive casing used above a DP sampler, the joints should be water tight to avoid possible cross-contamination of samples from zones above the sampling tool. O-ring seals or Teflon tape may be used on rod and casing joints to provide a seal to prevent leakage; some specially designed rods, with precision-machined, tapered, water-tight threads, may also be used.

Ground-water investigation and remediation programs have begun to focus on details of ground-water geochemistry and their relationships to and influence on natural degradation of contaminants (monitored natural attenuation) and engineered remedial actions (Wiedemeier et al., 1995, 1996; U.S. EPA, 1998a; ASTM, 2004o). Some of the water-quality parameters measured for these projects are very sensitive and may be easily altered during sampling if adequate care is not taken. DP ground-water sampling tools are very useful in conducting vertical profiling of the ground-water quality parameters, which are indicators of natural attenuation. However, preliminary field research (Schulmeister et al., 2001) has found that the steel tools commonly used in DP ground-water sampling devices can significantly alter some of these sensitive parameters. The observed DO levels, ORP, and dissolved iron concentration in ground-water samples collected with steel DP tools may be significantly different when compared with samples collected from monitoring devices constructed without steel materials (e.g., PVC and polyethylene). Field tests have shown that isolating the sample from contact with the steel rods during sample collection significantly reduces the effect. The use of stainless-steel and gaseous nitrided steel parts where sample contact does occur, together with isolation from the probe rods, was found to provide samples with DO and ORP values equivalent to samples collected from the plastic devices (Schulmeister et al., 2001).

Sample Recovery from Small-Diameter DP Rods and Tools

Sampling devices that are available for use in small-diameter (e.g., <1 in. [25 mm] I.D.) ground-water sampling tools include bailers, inertial-lift (or tubing check-valve) pumps, peristaltic and other suction-lift pumps, gas-drive pumps, and bladder pumps. These devices and their appropriate applications and limitations are discussed in detail in Chapter 15 and in ASTM Standard D 6634 (ASTM, 2004p); their specific applications to

DP technology are briefly described subsequently. Selecting the appropriate sampling device for any given project will depend on DQOs, field conditions, and budgetary constraints. Some general guidelines for ground-water sampling to meet particular DQOs are provided in Table 6.2. Modification of these guidelines for site-specific conditions and project-specific sampling and analytical objectives will be required.

Bailers

Bailers capable of sampling through DP rod as small as 0.5 in. (13 mm) I.D., available in plastics and stainless steel (Figure 6.55), are the simplest devices for sampling from small-diameter DP tools. Volumes recovered from these "microbailers" range from about 20 to 75 ml (0.7 to 2.5 fl oz). When only small sample volumes are required, bailers can be an acceptable sampling option. Gently lowering the bailer into and removing the bailer from the water column is necessary to avoid surging the tool, increasing sample turbidity and potentially causing loss of any volatile contaminants (Baerg et al., 1992). Bailers cannot be used for low-flow purging and sampling (Puls and Barcelona, 1996; ASTM, 2004n).

Inertial-Lift (Tubing Check-Valve) Pumps

Inertial-lift pumps are very inexpensive to use and can provide significant volumes of water at flow rates that range from a few milliliters (less than 1 fl oz) to more than 1 gal (3.8 l) a minute. The tubing used in these devices can be made of virtually any rigid or semirigid material, although Teflon and polyethylene are the most commonly used. Check valves can be made of stainless steel or plastics (Figure 6.56). Check valve sizes

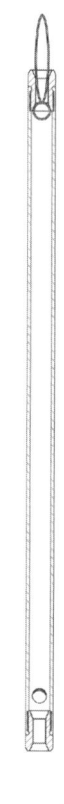

FIGURE 6.55
Small-diameter bailers constructed of stainless steel or various polymers may be used to sample small-diameter DP wells.

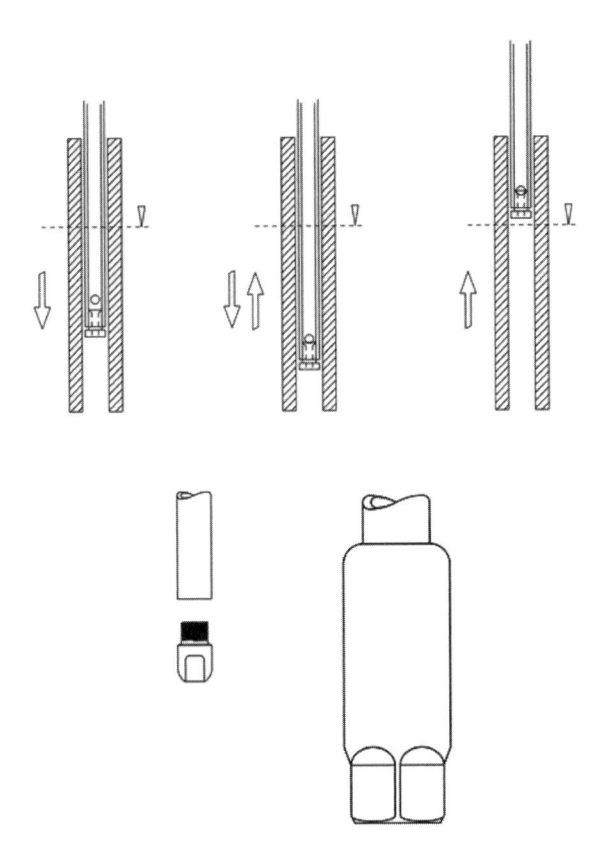

FIGURE 6.56
Several companies provide small inertial-lift pumps that are useful for development and purging of small-diameter ground-water sampling tools and wells. These devices use a simple check ball and check valve at the end of a length of tubing. This simple pump is oscillated up and down to bring water to the surface. Agitation of the water column from use of these pumps may cause loss of volatiles and increased sample turbidity.

range from less than 0.5 in. (13 mm) to more than 1 in. (25 mm) diameter and are attached to tubing ranging from 0.375 in. (9 mm) to 1 in. (25 mm) diameter. To operate an inertial-lift pump, the tubing with check valve is lowered to the depth at which purging and sampling is to be conducted. The tubing is oscillated up and down manually or mechanically. In fine-grained formations (e.g., silty sands, sandy silts), the formation yield will control the achievable flow rate. Practical research (Baerg et al., 1992; Puls and Barcelona, 1996) indicates that oscillation of inertial-lift pumps increases sample turbidity and loss of VOCs. Optionally, the tubing with check valve can be operated like a bailer, when it is lowered through the desired sample interval and retrieved to the surface without oscillation. This should minimize loss of volatiles and potential increases in turbidity. However, any water column overlying the desired sample interval will also be recovered in the tube.

Peristaltic and Other Suction-Lift Pumps

When the static water level in the sampling tool is less than about 25 ft (7.6 m) below the top of the DP rod, a peristaltic pump (Figure 6.57) may be used for sample collection. The pump head on the peristaltic pump contains rollers that squeeze tubing strung through the pump head, creating a suction on the end of the tubing submerged in the well.

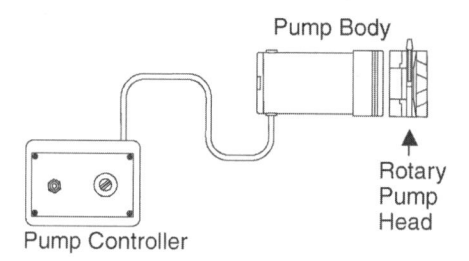

FIGURE 6.57
Peristaltic pumps may be used to sample from wells when the static water level is less than about 25 ft below grade. Flexible tubing is placed under the rollers on the rotary head and attached to a tube inserted below the water level. Negative pressure is applied to lift the water, so loss of volatile components and dissolved gases may occur. Easily adjustable flow rate is one of the advantages of most peristaltic pumps.

The suction slowly pulls the ground water to the surface for sampling. Peristaltic pumps are available in various sizes, with smaller units providing flow rates adjustable from a few milliliters per minute (less than 1 fl oz) to several hundred milliliters (tens of fluid ounce) per minute. Some large peristaltic pumps can provide flow rates exceeding a gallon (3.8 L) per minute. At lifts exceeding 6 or 8 ft (1.8 or 2.5 m), the suction used by peristaltic pumps to raise the sample to the surface can cause loss of dissolved gases (e.g., carbon dioxide and oxygen) and VOCs from the sample (Ho, 1983; Devlin, 1987; Barker and Dickhout, 1988). Other effects on samples include pH shifts that can affect concentrations of trace metals (Houghton and Berger, 1984).

Suction-lift pumps equipped with a trap to capture the sample have been used to sample from small-diameter DP ground-water sampling tools. As with peristaltic pumps, however, the negative pressure placed on the water as it is drawn from the tool can cause significant degassing of the sample and loss of volatile constituents and other analytes. Suction-lift pumps may be acceptable when sampling for relatively inert, non-volatile components (e.g., major ions or salts).

Gas-Drive Pumps

Gas-drive pumps (Figure 6.58) are available in diameters as small as 0.5 in. (13 mm) for use in DP rods and tools. A gas-drive pump forces a discrete column of water to the surface via pressure-induced lift without mixing of the drive gas and water. Hydrostatic pressure opens the inlet check valve and fills the pump chamber in the fill cycle. An inert pressurized gas (e.g., ultra-pure nitrogen) is applied to the chamber, closing the inlet check valve, opening the outlet check valve, and displacing water up the discharge line in the discharge cycle. By releasing the gas pressure back up the gas pressure supply tube, the cycles can be repeated. Within gas-drive pumps, there is a limited interface between the drive gas and the water. There is, however, a potential for loss of dissolved gases and VOCs across this interface (Gillham et al., 1983; Barcelona et al., 1985). This potential greatly increases if the pump is allowed to discharge more than the pump volume, which would cause drive gas to be blown up the discharge line with water. Contamination of the sample may also result from using a noninert drive gas (e.g., compressed air) or from impurities in the drive gas. Typical lifts for gas-drive pumps are between 100 and 250 ft (30 and 76 m).

Bladder Pumps

Bladder pumps are now manufactured in diameters small enough to fit inside DP rods as small as 0.5 in. (13 mm) I.D. (Figure 6.59). These pumps are capable of consistently

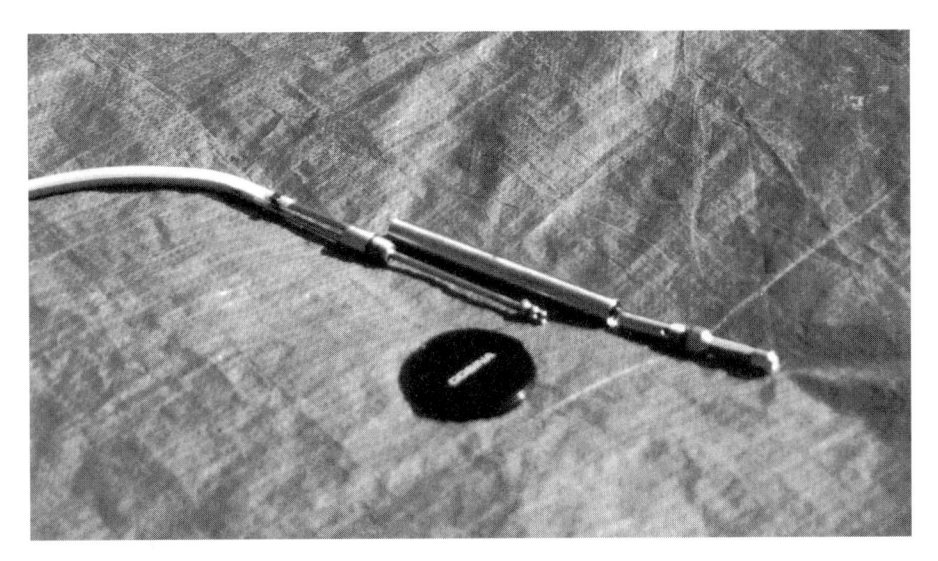

FIGURE 6.58
A small-diameter gas-drive pump, disassembled for cleaning.

providing the highest integrity ground-water samples (Barcelona et al., 1984; Pohlmann et al., 1990; Parker, 1994; Nielsen and Yeates, 1985). They operate at flow rates from as low as a few milliliters (less than 1 fl oz) per minute to more than 1 gal (3.8 l) per minute in models that can be used in larger diameter DP rods. Low flow rates (typically

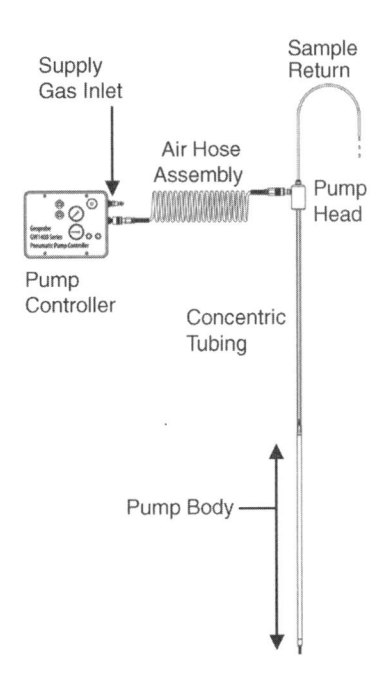

FIGURE 6.59
Several pump manufacturers have developed small-diameter bladder pumps for use in small-diameter DP tools and wells. The smallest bladder pump available will fit inside nominal 0.5 in. diameter PVC casing. These pumps enable the collection of water quality samples following the low-flow sampling protocol. Flow rates range from less than 100 ml/min to over 500 ml/min, depending on the pump design and depth to water.

between 100 and 500 ml/min [3.4 and 17 fl oz/min]) are preferred during purging and sampling to minimize disturbance of the sample, drawdown in the well, and loss of volatile constituents (Puls and Barcelona, 1996; ASTM, 2004n). Bladder pumps are the preferred devices for use in conducting low-flow ground-water purging and sampling programs (Puls and Barcelona, 1996).

In Situ Measurements Using Specialized DP Probes

DP methods, particularly CPT systems, have been used for decades to advance probes into the ground to collect data *in situ* about subsurface properties. Over the years, a number of technologies have been linked to CPT platforms to provide specialized information, many focused on obtaining real-time data on one or more subsurface variables. CPT is capable of delivering sensing probes as well as down-hole sampling and analytical devices to depths exceeding 200 ft in favorable geological conditions. The level of resolution and real-time data acquisition capabilities, joined with the capabilities of computer data analysis and interpretation, have taken site characterization to a new level. Probes are now available to provide high-resolution data on site stratigraphy, pore-pressure distribution, moisture content, porosity, hydraulic conductivity, and presence and concentrations of virtually every class of chemical contaminants in soil, soil gas, and ground water (Kram et al., 2001c). Video probes are also available to add the visual dimension to subsurface remote sensing. Experience at many sites has proven that these tools can be deployed at substantial cost savings over conventional drilling and sampling technologies while providing much more detailed, high-resolution data than are available to the investigator immediately, to help guide the course of the investigation in the field. This section describes a number of tools that can be used to take *in situ* measurements to accomplish a variety of site-characterization objectives.

CPT Systems

Over the years, CPT systems have evolved from simple geotechnical tools limited to use in low-strength soils to sophisticated platforms that can be used to obtain a wide variety of environmental measurements in soils of widely varying strength and composition, including lightly cemented soils. Since the early 1990s, CPT has become an important alternative to and, in many cases, a replacement for conventional drilling in the performance of environmental site-investigation and characterization programs (U.S. DOE, 1996). In the initial phase of a site-characterization program, CPT can make important contributions to the development of a site conceptual model by providing detailed mapping of physical features such as stratigraphy, water-table elevation, and hydraulic conductivity, as well as delineating the distribution of contaminants.

CPT Platforms

Major components of a CPT system include (a) the instrumented probe, (b) the instrumentation conditioning and recording system (these are generally self-contained and computerized), (c) the hydraulic push system, and (d) the load frame on which the push system is mounted. The load frame can be incorporated in a variety of embodiments including heavy-weight trucks (from 10 to 40 t [9,000 to 36,000 kg] push capacity), tracked vehicles (generally with push capacities of 10 to 15 t [9,000 to 13,500 kg]), trailers, and man-portable and other specialty devices. A few of these embodiments are depicted in Figure 6.60.

FIGURE 6.60
CPT vehicles can take many embodiments: (a) a 20 t truck; (b) a lightweight track rig; (c) a specialty platform.

For trucks and heavy tracked vehicles, the reaction weight of the vehicle provides the force necessary to overcome resistance to penetration. For lighter platforms, earth anchors or anchoring to a structure are used to provide the necessary reaction force. Generally with a truck-mounted CPT, the up-hole equipment is mounted inside a van body attached to a ten-wheel truck chassis with a diesel engine. Ballast in the form of metal weights or water is carried separately and added to the push rig at the investigation site to achieve an overall push capability of up to 80,000 lb (36,000 kg). Penetration force is supplied by the hydraulic push system — a pair of large hydraulic cylinders bolted to a reinforced frame attached to the vehicle chassis.

CPT Probe Systems

ASTM Standard D 3441 (ASTM, 2004a) specifies two standard probe diameters, 1.44 in. (3.658 cm) and 1.75 in. (4.445 cm). The area on the tip of these probes is 10 and 15 cm^2 (1.55 and 2.33 in.^{2}), respectively, and each cone has a 60° point angle. The friction sleeve behind the conical tip of the probe is a standard 150 cm^2 on the smaller diameter tool and 200 cm^2 on the larger diameter tool. The larger diameter probe is most common in environmental site investigation because it is better able to accommodate the variety of sensors and attachments that have been developed for environmental purposes.

A typical CPT probe is shown in Figure 6.61 with the major components highlighted. Two load cells are incorporated into the probe to measure the resistance to penetration encountered vertically against the conical tip and the friction exerted along the surface of the friction sleeve. Each load cell comprises a cross-sectionally uniform cylinder inside the probe, instrumented with four strain gages in a full-bridge circuit. Output from the load cells is transmitted from the probe assembly via a cable running through the push rod. The analog signals generated by the probe may be digitized down-hole or at the surface and are generally recorded and plotted by computer located in the truck body. ASTM Standard D 3441 specifies a maximum penetration rate of 2 cm/sec

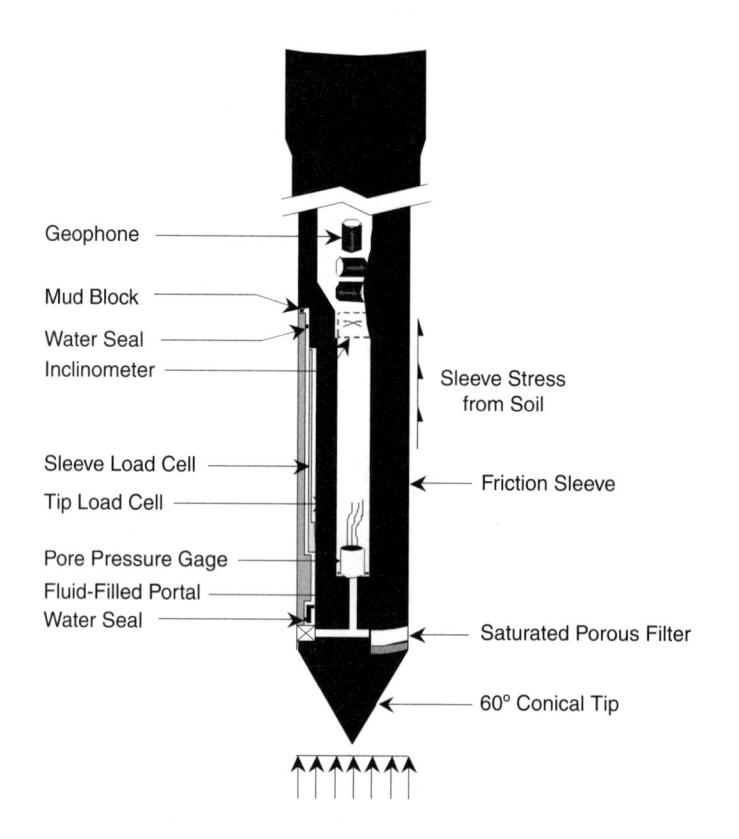

FIGURE 6.61
A typical CPT probe provides the ability to obtain detailed stratigraphic information. This piezocone also provides data on *in situ* water pressures and formation hydraulic conductivity.

(0.8 in./sec) and data are most commonly recorded at a rate of once per second, thus providing a minimum spatial resolution of 2 cm (0.8 in.) in the vertical. The depth of penetration is measured using a linear displacement transducer.

A common measurement added to the standard tip and sleeve stress is pore pressure, which is very useful in soil classification, determining water-table depth, locating perched water zones, and estimating hydraulic conductivity. A probe incorporating the three more-or-less standard elements of tip stress, sleeve friction, and pore fluid pressure is generally referred to as a piezocone. Other sensors, such as seismic geophones, inclinometers, and temperature sensors have also been incorporated into various models of piezocone. Many additional commercially available modules can be mounted behind the cone. These modules, which are described subsequently, house sensors for properties including electrical resistivity, volumetric moisture, and chemical constituent concentrations. Other cones and modules are available for collecting soil, ground-water, and soil-gas samples.

CPT Data Acquisition

Measurements from sensors in the CPT probe are transmitted from the probe, through cabling, to up-hole electronic data acquisition equipment, which is usually integrated with a field-portable computer and graphics monitor. Two types of data acquisition systems — analog and digital — are in general use. As the name implies, the first type of system transmits analog signals, such as voltage or current, from the CPT probe to the surface via multiconductor cabling. Up-hole signal conditioners amplify and filter these signals and feed them to analog-to-digital converters whose output is recorded on a portable computer, typically at a rate of once per second.

Digital cone systems employ down-hole circuitry and a microcontroller embedded in the probe to perform signal conditioning and analog-to-digital conversion "in the hole." Digital data are then transmitted to the surface for recording using a standard asynchronous serial communication protocol such as RS-232. Digital cone systems offer several advantages over analog cone systems in that they (1) limit the number of conductors required to transmit data from multiple down-hole sensors, (2) reduce or eliminate susceptibility to induced electrical noise, and (3) simplify operation for the field technicians, as these probes typically store all calibration factors on-board the microcontroller embedded in the probe.

CPT Data Interpretation

The data acquired by the CPT probe are evaluated and reported in real time using a geotechnical soil behavior classification chart (Robertson and Campanella, 1986) and numerical algorithms that were developed to automate the interpretation of geologic layering indicated by CPT data (Davis, 1986). Profiles obtained during a CPT push can also be readily interpreted by experienced personnel in the field. An example of data obtained from a CPT profile is given in Figure 6.62. This penetration was conducted at Cape Canaveral, Florida, USA as part of an environmental investigation into DNAPL contamination. Instruments deployed on the push string included a standard piezocone and a soil moisture and resistivity module. The tip, sleeve, and pore pressure data provided by the piezocone appear in the first, second, and fourth columns of Figure 6.62, respectively. These data are used in a soil classification scheme to yield the interpretation of the geologic layering, shown in the fifth column. The interpreted profile indicates that the geological material consists of clean sand to a depth of 35 ft (10.7 m), at which point the fine-grained content of the soil begins to increase. Below 35 ft (10.7 m), layers of sand mixed with fine-grained material and clay layers are observed. As this example illustrates, a significant advantage of the CPT is that thin layers of varying soil type are readily distinguished.

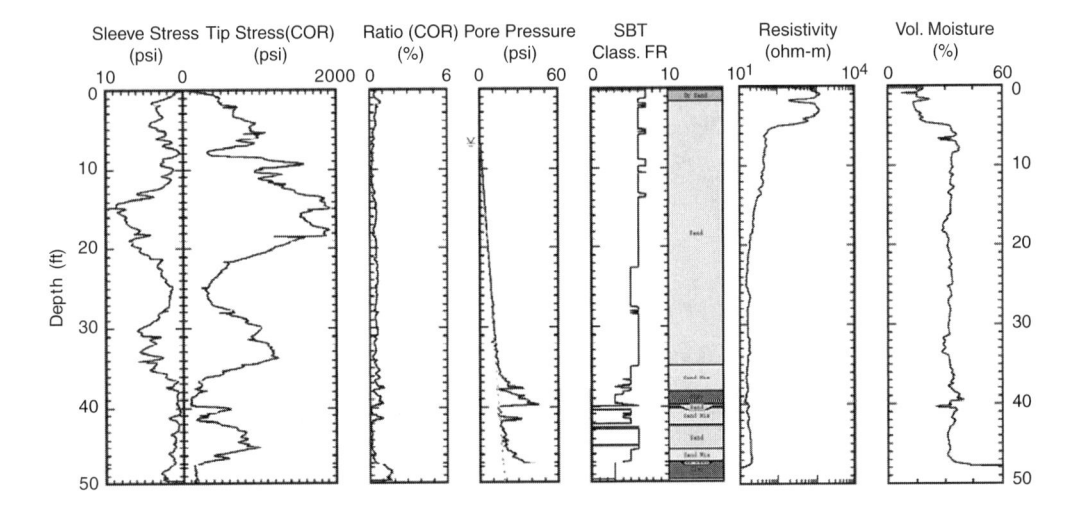

FIGURE 6.62
CPT provides a plethora of data for understanding subsurface conditions. This is an example of environmental data from an investigation at Cape Canaveral. Data shown were obtained from a piezocone and a soil-moisture and resistivity module.

The upper layers show variations in the tip and sleeve resistance related to the bearing strength of the soils present. The friction ratio (column 3 in Figure 6.62), which is the ratio of the sleeve stress to the tip stress (both in units of pressure), is a key determinant of the soil type. Sands, being noncohesive but with strong bearing capacity, will typically exhibit a low friction ratio, whereas clays, being more cohesive but weaker in bearing capacity, will exhibit a higher friction ratio. As Figure 6.62 shows, the friction ratio is essentially less than 1 to a depth of 35 ft (10.7 m), indicating that the material is a sand. This is further implied by the pore pressure profile, which shows an essentially hydrostatic pore pressure response with depth, as would be expected in a well-drained material such as clean sand. At a depth of 35 ft (10.7 m), the friction ratio and pore pressure response increase, indicating an increase in the fine-grained content of the soil. Geotechnical properties and hydraulic conductivity can also be estimated from CPT data. In low-permeability materials, hydraulic conductivity can be estimated *in situ* using the CPT pore pressure dissipation test (Baligh and Levadoux, 1980; Robertson and Campanella, 1988). In higher permeability materials, CPT-installed wells can be used to perform traditional formation hydraulic conductivity testing (e.g., slug or bail tests and pumping tests). A more complete description of the uses of CPT to determine soil types and properties is provided in Lunne et al. (1997). Several ASTM standards (Table 6.3) have been developed for conducting CPT investigations.

Tools for Ascertaining Soil Properties

Soil Electrical Resistivity and Conductivity Probes

Electrical resistivity (or the inverse measurement, conductivity) surveying of soils and rocks has been used extensively to locate and define the distribution of ground water, ore bodies, and geologic structure using surface and borehole techniques. The property of electrical resistivity or conductivity in soils is a function of the soil type (mineralogy and grain size), fine-grained material content, soil porosity, and conductivity of the pore fluid (Gardner et al., 1991). While the advent of commercial electrical resistivity and conductivity probes is relatively recent, electrical resistivity profiling has been used in the geotechnical community to identify formation materials, especially to distinguish fine-grained materials (clays and silts) from coarse-grained materials (sands and gravels). It is also useful in estimating the potential corrosiveness of site geochemistry to metallic well casing and underground pipes. In addition, resistivity (or conductivity) is very useful for investigating water quality, particularly in identifying leachates from landfills, and zones of salt-water intrusion, which typically have distinctively high concentrations of dissolved ionic species.

TABLE 6.3

ASTM Standards Applicable to CPT

ASTM Designation	Title
D 3441	Test method for deep, quasi-static, cone and friction-cone penetration tests of soil (ASTM, 2004a)
D 5778	Test method for performing electronic friction cone and piezocone penetration testing of soils (ASTM, 2004q)
D 6067	Practice for electronic CPT for environmental site characterization (ASTM, 2004r)
D 6187	Practice for cone penetrometer technology characterization of petroleum contaminated sites with nitrogen LIF (ASTM, 2004s)

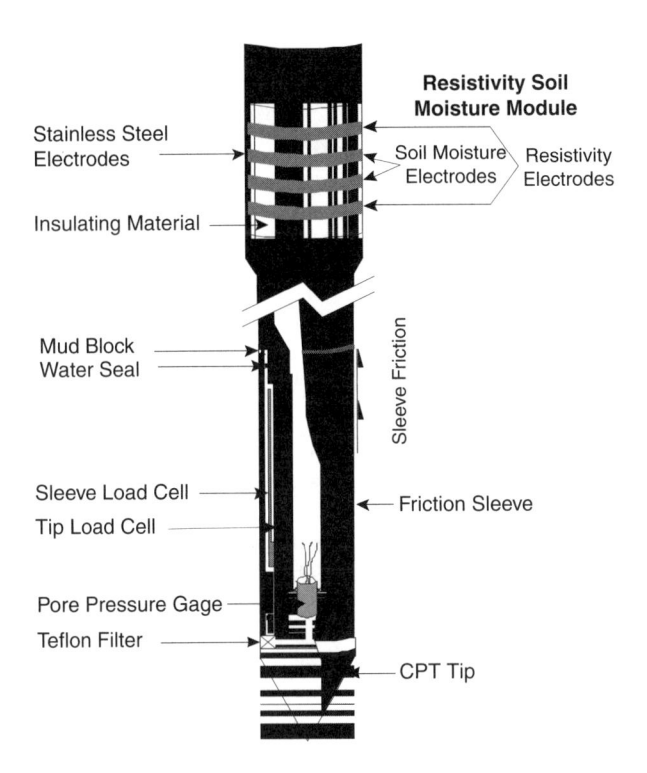

FIGURE 6.63
CPT resistivity module.

An example of a resistivity module is shown in Figure 6.63. On a CPT system, the module is usually installed directly behind the CPT probe, but it can be used without the CPT probe by installing a dummy tip in the end of the resistivity module. There are several styles of resistivity probes — some use a four-ring or two-ring configuration and others use a four-pole or two-pole configuration. In both of these configurations, independent measurements of the current and voltage potential are required. While the resistivity in profiles is usually referred to as a DC resistivity profile, in fact, the resistivity is measured using a low-frequency (10 to 1000 Hz) AC system. The alternating current is required to avoid potential polarization of the soil. An electrical resistivity profile from the Cape Canaveral site is shown in the sixth column of Figure 6.62. The electrical resistivity is fairly high (1000 Ωm) in the dry soils near the ground surface, but decreases rapidly in the capillary fringe and below the water table. Above the water table, a thin band of lower resistivity is evident, possibly indicating a thin layer of fine-grained material. Below a depth of 20 ft, the resistivity is relatively constant at about 20 Ωm.

At least one electrical conductivity (or e-log) system is designed for advancement with percussion hammer DP machines (McCall, 1999; U.S. EPA, 1997; McCall and Zimmerman, 2002) and has been used successfully for high-resolution hydrostratigraphic characterization (Schulmeister et al., 2003). The components for operation of this system include a string pot for tracking depth, the electrical conductivity probe, a cord set, and data acquisition system (Figure 6.64). Logs are viewed on-screen as the probe is advanced and files are saved for printing and easy download to spreadsheet and data presentation software systems. An example log from an alluvial aquifer (Figure 6.65) reveals that interpretation of the logs is relatively simple, with higher electrical conductivity correlating with an increasing proportion of clay. After obtaining the initial log, depth-targeted soil samples

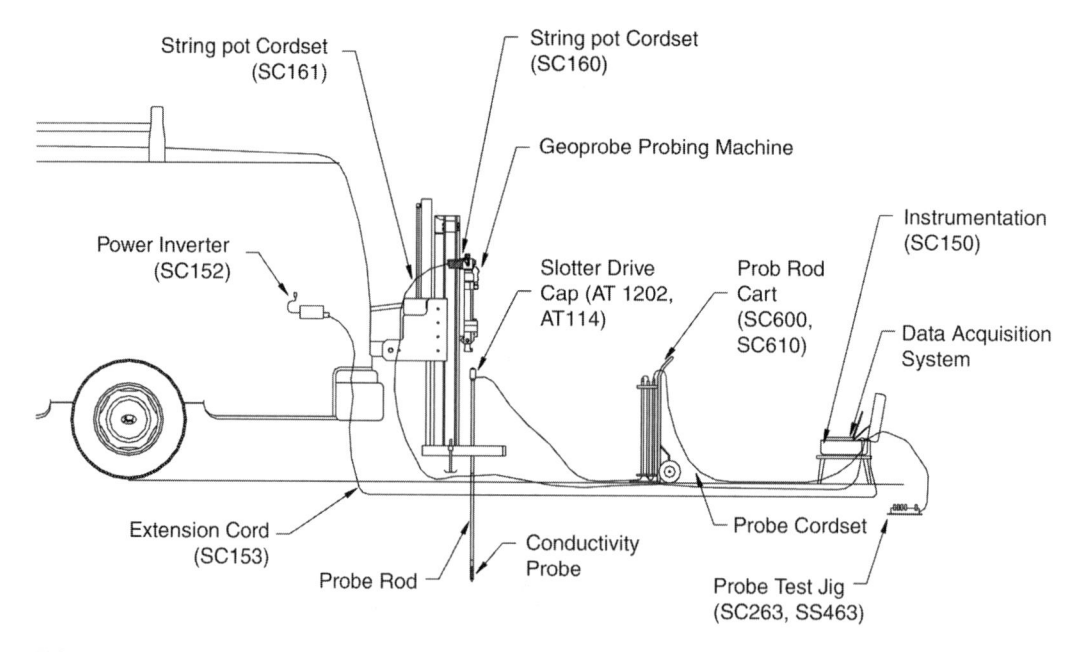

FIGURE 6.64
Major components of the DP electrical conductivity logging system. The DP machine used to advance the electrical conductivity probe into the subsurface may be mounted in a van, pickup truck, or track unit. The rod cart is used to hold probe rods that have the signal cable prestrung through them to make the logging operation more efficient. The instrumentation case receives raw data signals from the probe and string pot for processing and sends results to the laptop computer for real-time viewing of the e-log.

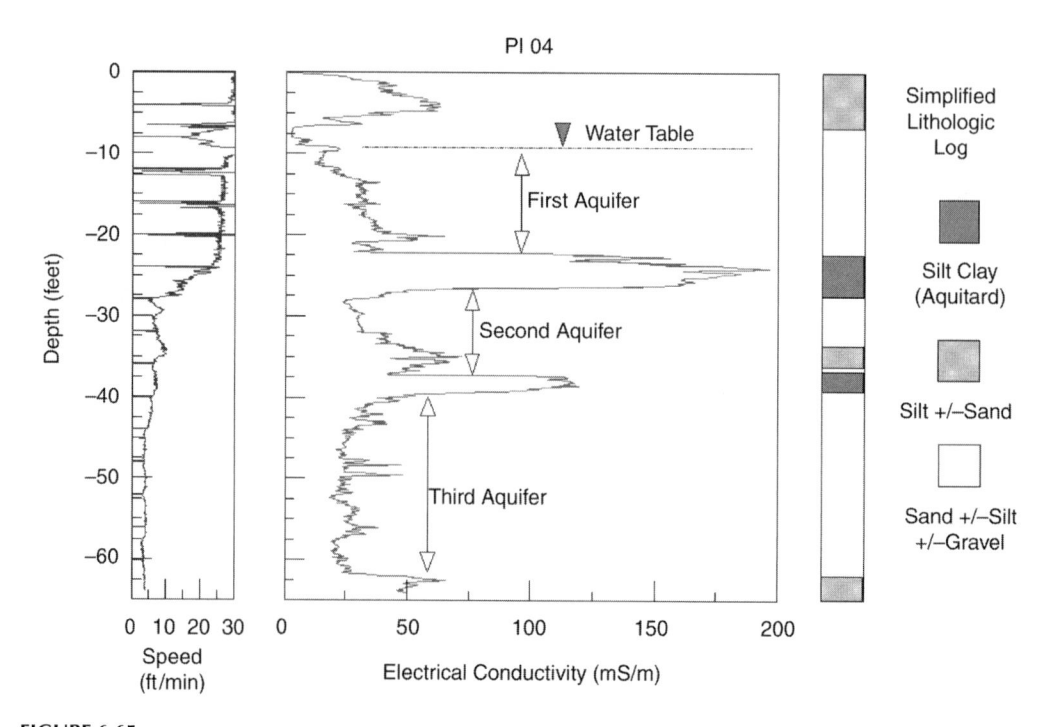

FIGURE 6.65
Example DP electrical conductivity log (PI04) from an alluvial aquifer system. The depth/speed graph is on the left, and electrical conductivity results are on the right. A simplified lithologic log is provided for ease of interpretation. In general, higher electrical conductivity indicates an increase in clay content. Additionally, a decrease in the speed of advancement at about 28 ft indicates an increase in formation density below that depth.

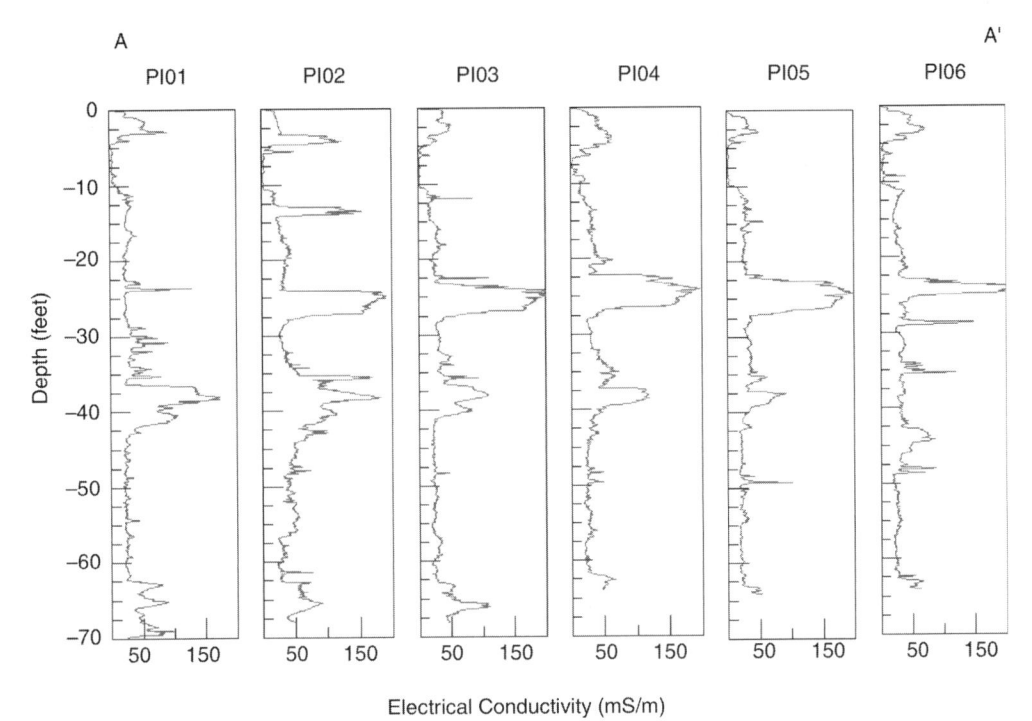

FIGURE 6.66
Six DP e-logs that were obtained in a transect across an alluvial aquifer are shown here. The high electrical conductivity observed from about 24 to 27 ft on each log corresponds to a silt/clay aquitard. Notice this aquitard thins and appears to be pinching out at the PI01 location. A second silt/clay aquitard is indicated by an increase in electrical conductivity on five of the e-logs at a depth near 40 ft. Notice that this layer is absent in PI06.

may be collected to confirm the log results and site-specific interpretation. Additional targeted sampling is necessary only when significant changes are observed at other logging locations. Several logs obtained across this site (Figure 6.66) on a spacing of 100 ft intervals can be correlated to construct a detailed geologic cross-section. Software programs for data contouring (e.g., Surfer) can be used to construct cross-sections based on electrical conductivity (Figure 6.67), providing the investigator with accurate details of the subsurface stratigraphy in these unconsolidated aquifer systems.

When accurate elevations are obtained at each log location, the data may be used to define the elevation of formation contacts. When multiple e-logs are obtained over a study area, elevations on the surface of a clay layer or sand layer may be determined and a contour map of the surface plotted. A contour map on top of a clay layer based on e-logs (Figure 6.68) can be used to determine migration and collection points for DNAPLs. Using the e-logs and depth or elevation data, it is also possible to determine the thickness of sand units and contour these data (Figure 6.69) to derive estimates of the unit volume, saturated volume, and potential contaminant mass for remedial actions.

Soil Moisture Probes

A relatively recent development in CPT instrumentation is the combined soil moisture/resistivity probe. This probe uses sharp contrasts in the dielectric permittivity of the three components of the soil matrix (e.g., air, water, and soil) to infer moisture content from soil dielectric. Topp et al. (1980; see also Topp and Davis, 1985) first explored the relationship between soil dielectric and water content using time domain reflectometry.

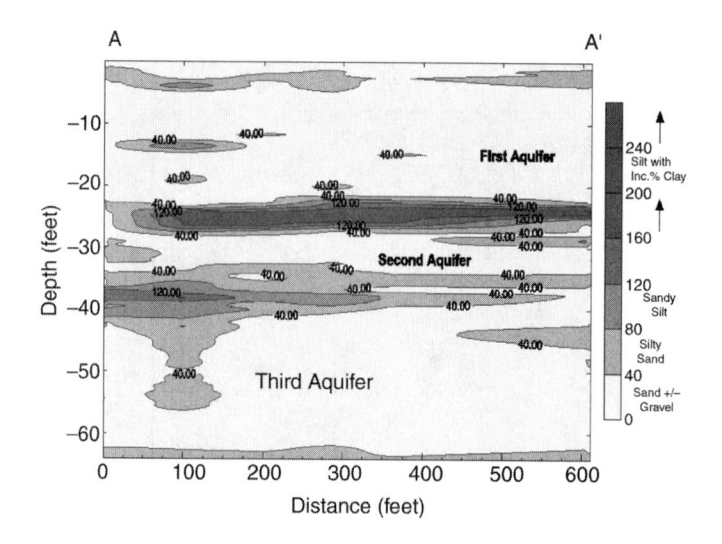

FIGURE 6.67
A geologic cross-section of the alluvial aquifer site based on contouring of the electrical conductivity logs in Figure 6.66. Darker colors correspond to higher electrical conductivity and thus finer grained materials. Lighter colors correspond to lower electrical conductivity and thus coarser grained materials. The darker zones at about 25 and 40 ft correspond with the aquitards, while the lighter zones form local aquifers at the site. These features control migration of contaminants at this site.

Later developments employed a simpler measurement apparatus to determine the soil dielectric based on capacitance.

Soil moisture content is critically important in many environmental site investigations, particularly in the evaluation of contaminant transport in the vadose zone. As a geotechnical parameter, soil moisture content is used to evaluate the performance of geotechnical structures such as pavements, foundations, earth dams, and retaining walls. Previously, soil moisture measurements were made using laboratory techniques on samples retrieved from a borehole or test pit. An advantage of a CPT soil moisture probe is that a continuous soil moisture and stratigraphy profile can be obtained simultaneously, at a dramatic cost and time savings over sampling combined with laboratory testing.

An example soil moisture profile is shown in Figure 6.62 for the Cape Canaveral site. The moisture content above the water table is roughly 10% and begins to increase in the capillary fringe as the water table is approached. Below the water table, the moisture content is about 35% in the sands and approaches 60% going into the clay aquitard at the base of the profile. Minor variations in the water content are observed throughout, most likely due to small changes in the porosity of the soil. Below the water table, volumetric soil moisture is equivalent to porosity. Therefore, this probe can be used in the saturated zone to estimate porosity.

Nuclear Logging Tools

Nuclear logging tools are geophysical instruments that detect natural formation radiation or that emit radiation and measure the response of the formation. These tools can be advanced with DP probes, including CPT systems, single-rod DP systems, and inside the drive casing of dual-tube DP systems, to define site stratigraphy and to locate the water table and certain types of contaminants. Unlike other sensing tools, nuclear logging tools can be used to record data on formation characteristics through metal rod or casing. The three primary methods of nuclear logging include natural gamma, gamma–gamma, and neutron logging.

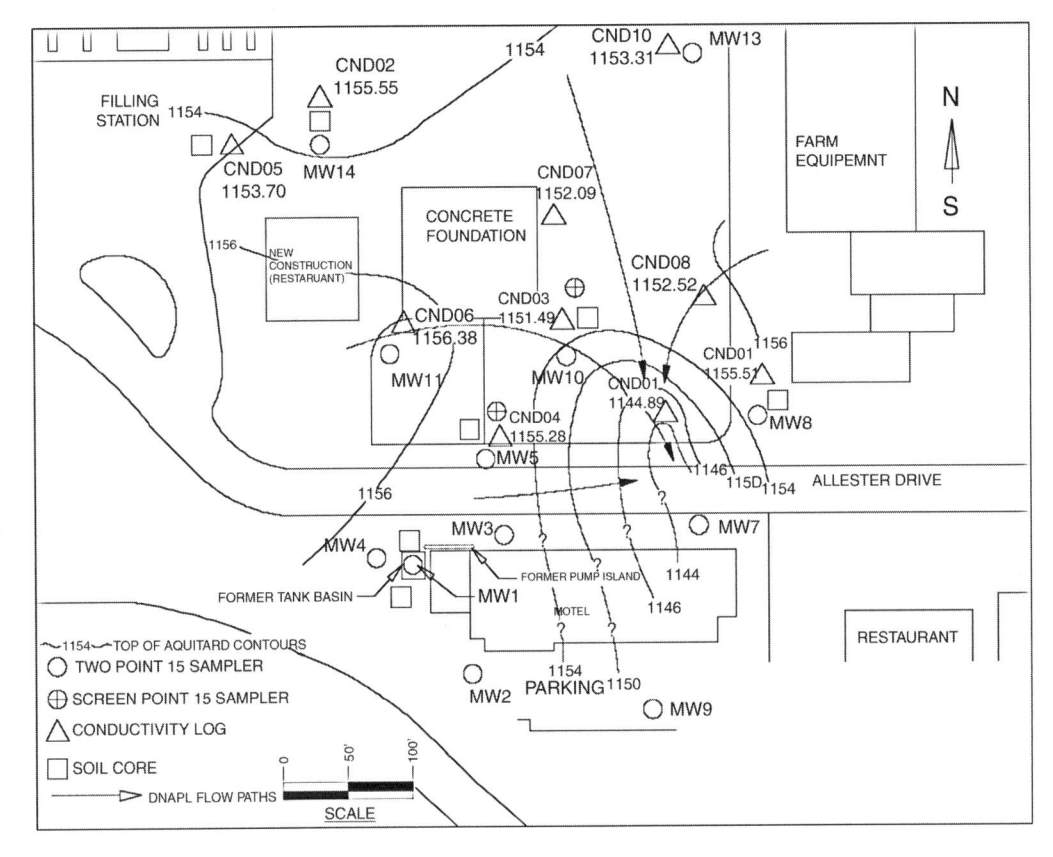

FIGURE 6.68
Surface elevations measured at points where electrical logs are obtained allow determination of the elevation of a formation contact at depth by simple subtraction of the depth measurement. Plotting and contouring these elevations can provide a map on the upper surface of a clay layer as shown here. This map shows a valley-like feature in the upper surface of this clay layer. This clearly indicates how DNAPLs would flow over this surface and migrate toward the low point in the valley.

Natural gamma tools log the amount of natural gamma particles emitted by natural formation materials. Because clays tend to contain more radioactive isotopes than sand, they are generally easily distinguished on a natural gamma log. Thus, by logging changes in natural gamma radiation, it is possible to determine stratigraphy, particularly with respect to clay content. Gamma–gamma tools emit gamma radiation and measure the response of the formation, which is closely related to formation density and porosity. Thus, this method can provide stratigraphic information, as well as information on soil porosity. Neutron methods emit neutrons into formation materials and measure the response, which is dependent primarily on moisture content. This makes neutron logging useful in defining the position of the water table. If other methods can be used to define stratigraphy and degree of saturation of formation materials, neutron logging may be used to detect the presence and actual thickness of separate-phase petroleum hydrocarbons or other LNAPLs (U.S. EPA, 1997).

Tools for Delineating Subsurface Contamination

Over the past decade, a number of research programs sponsored by the U.S. Department of Defense, the U.S. Department of Energy, and the private sector have led

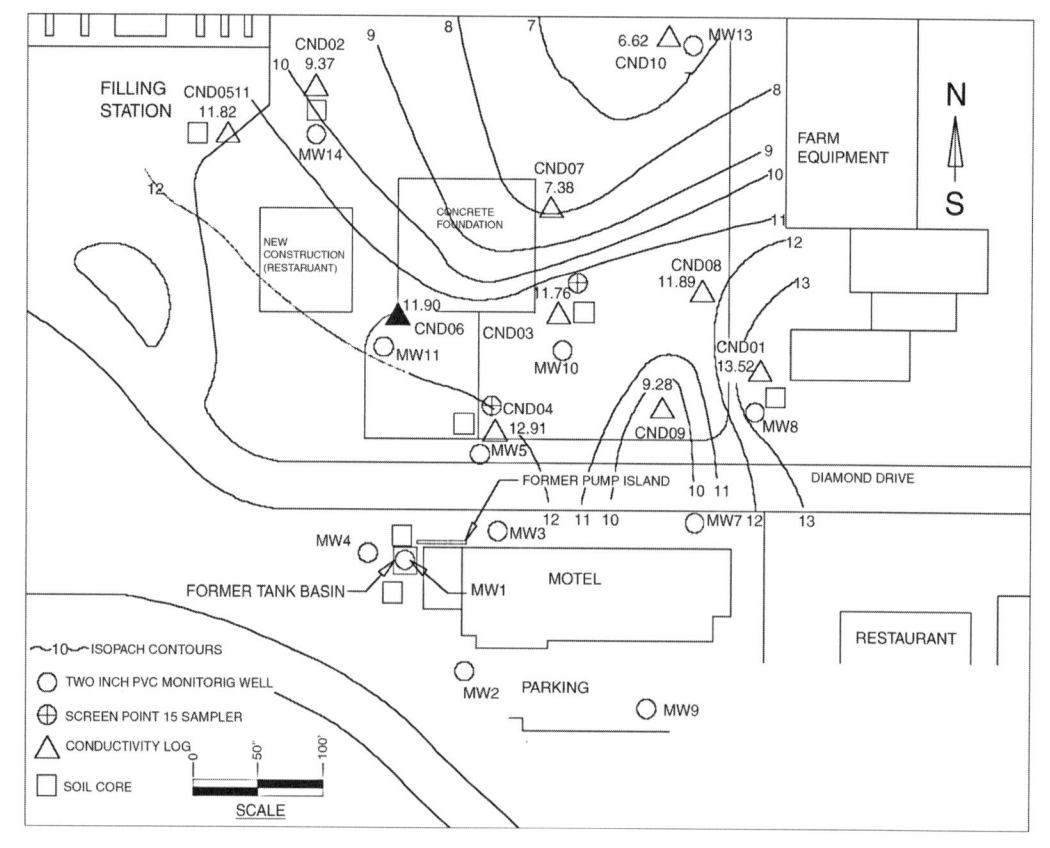

FIGURE 6.69
Depth measurements from an electrical log to the top and bottom of a sand layer can be used to determine thickness of the unit. When several logs are obtained across the site, the thickness of the unit at each location can be plotted on a site map and contoured to show how thickness of the unit varies, creating an isopach map. This information can be used to determine the volume of contaminated media that requires remediation.

to the development of CPT-based and other DP-deployed sensors, which can detect and quantify subsurface contamination *in situ* and in real time. As with laboratory analytical instruments, each probe is selective for a specific class or species of contamination. In addition, most of these probes do not achieve the sensitivity *in situ* that can be attained using laboratory instrumentation. Therefore, the most appropriate use of chemical constituent data obtained *in situ* via DP methods is for delineation of contaminant distribution with screening-level quantification. A brief description of some of the more popular DP probes used to locate contamination *in situ* follows.

Continuous Vapor Profiling

The use of vapor detection technology is a logical choice for real-time, *in situ* detection of VOCs (e.g., carbon tetrachloride, trichloroethylene, or aromatic hydrocarbons such as benzene) in the vadose zone. This is the approach most widely used with CPT to develop continuous, high spatial resolution VOC contaminant distribution profiles vs. depth (Buttner et al., 1995; Frye et al., 1995; Palusky et al., 1995; Rossabi et al., 1995; Farrington and Bratton, 1997). Among the vapor detection devices that have been interfaced to the cone penetrometer are conductivity detectors, surface acoustic wave (SAW) detectors, IR absorption spectrometers, ECD, photoionization detectors (PID), direct sampling ion trap mass spectrometers (DSITMS), and fast gas chromatographs (GCs) equipped with ECD, PID, SAW, or MS detectors.

To perform continuous soil-gas measurements, vapors must be drawn into the cone penetrometer through a port in the probe. Several devices have been developed for this purpose. The simplest of these is similar in construction to the piezocone, except that there are no sensors in the probe. As with the piezocone, a cylindrical porous filter is positioned behind the replaceable conical tip. Rather than transmitting saturated zone excess pore water pressure to a sealed pressure transducer, the filter instead allows vadose zone soil vapor to be drawn into the hollow center of the tool. Under a slight vacuum provided by an up-hole pump, a continuous stream of soil vapor is drawn to a gas analyzer at the surface via a gas sampling tube. Soil gas can be continuously sampled and analyzed while advancing the cone, providing a continuous record of contaminant concentration with depth.

Another such tool combines a piezocone and soil-gas sampling tool into a single device. The probe measures tip stress and sleeve stress, while permitting the continuous collection of soil-gas samples through a filter element that would otherwise be used to transmit pore fluid pressure. In this way, continuous vertical profiles of vapor analysis can be generated simultaneously with geotechnical data used to determine soil stratigraphy. One of the limitations of this tool is that its use is confined to soils that have moderate to high air permeability and that will transmit vapor readily.

Any of the analyzers described earlier could be employed with a gas-sampling tool. However, none of the GC-based devices is rapid enough to operate in a continuous sampling mode. While responding faster, the stand-alone detectors generally provide detection limits in the low parts-per-million (volume) to high parts-per-billion (volume) range, with the exception of the DSITMS (U.S. DOE, 1998) and photoacoustic IR analyzer (Bruel and Kjaer, 1994), which provide quantitation of concentrations in the tens of parts-per-billion. Considerably lower detection limits can be achieved using a preconcentration technique (e.g., trap and desorb) in front of the sensor, but this approach compromises speed and spatial resolution.

LIF Spectroscopy

The CPT-deployed LIF tool uses either a wavelength-tunable dye laser (Shinn et al., 1994; St. Germain and Gillespie, 1995; U.S. EPA, 1995a; Taer et al., 1996) or a fixed-wavelength nitrogen laser (Lieberman et al., 1991; U.S. EPA, 1995b; ASTM, 2004s) for detection of petroleum hydrocarbons in soil and ground water. The nitrogen laser, which has a 0.8 nsec pulse width and a pulse energy of 1.4 mJ, fires pulses of 337 nm wavelength ultraviolet light into a 100 m long, 500 μm diameter silica-clad optical fiber that runs down an umbilical cord through the probe rods to the tool (Kram et al., 1997) (Figure 6.70). The tool, which has a 6.4 mm diameter and 2 mm thick sapphire window mounted flush with the outside of the probe rod, is generally located 0.6 m (2 ft) behind the standard CPT cone or piezocone to allow correlation of contaminant detection with stratigraphy. As the laser pulses pass into the soil, petroleum hydrocarbons in the soil respond by giving off a characteristic fluorescence. This emitted light is carried back to the surface over a second optical fiber and measured with a spectrograph which disperses the signal. The resulting energy is distributed as a function of wavelength and measured using a photodiode array. This information is then computer recorded and compared to a standard curve to provide a measurement of the fluorescent response (wavelength and intensity). The wavelengths detected indicate the types of hydrocarbons present, while the fluorescent spectral intensity is directly related to the concentration of petroleum products in the soil.

The LIF system thus provides a continuous profile of petroleum hydrocarbon distribution in subsurface materials in real time as the probe is advanced. To ensure consistent performance, the system is calibrated using a laboratory standard at the beginning and end of each push. The 337 nm nitrogen laser is best suited to detecting petroleum hydrocarbons which contain three-ring (polynuclear) aromatic compounds such as diesel fuel, kerosene, some jet fuels, creosote, and heating oils (Lieberman et al., 1995). Lasers with higher excitation energies (or lower wavelengths), such as the neodymium: yttrium–Aluminum–Garnet (Nd:YAG) laser, can be used to detect lighter (single- and double-ring) aromatic hydrocarbons in automotive and aviation gasoline. While it is not currently possible to directly detect DNAPL constituents using LIF, there is evidence

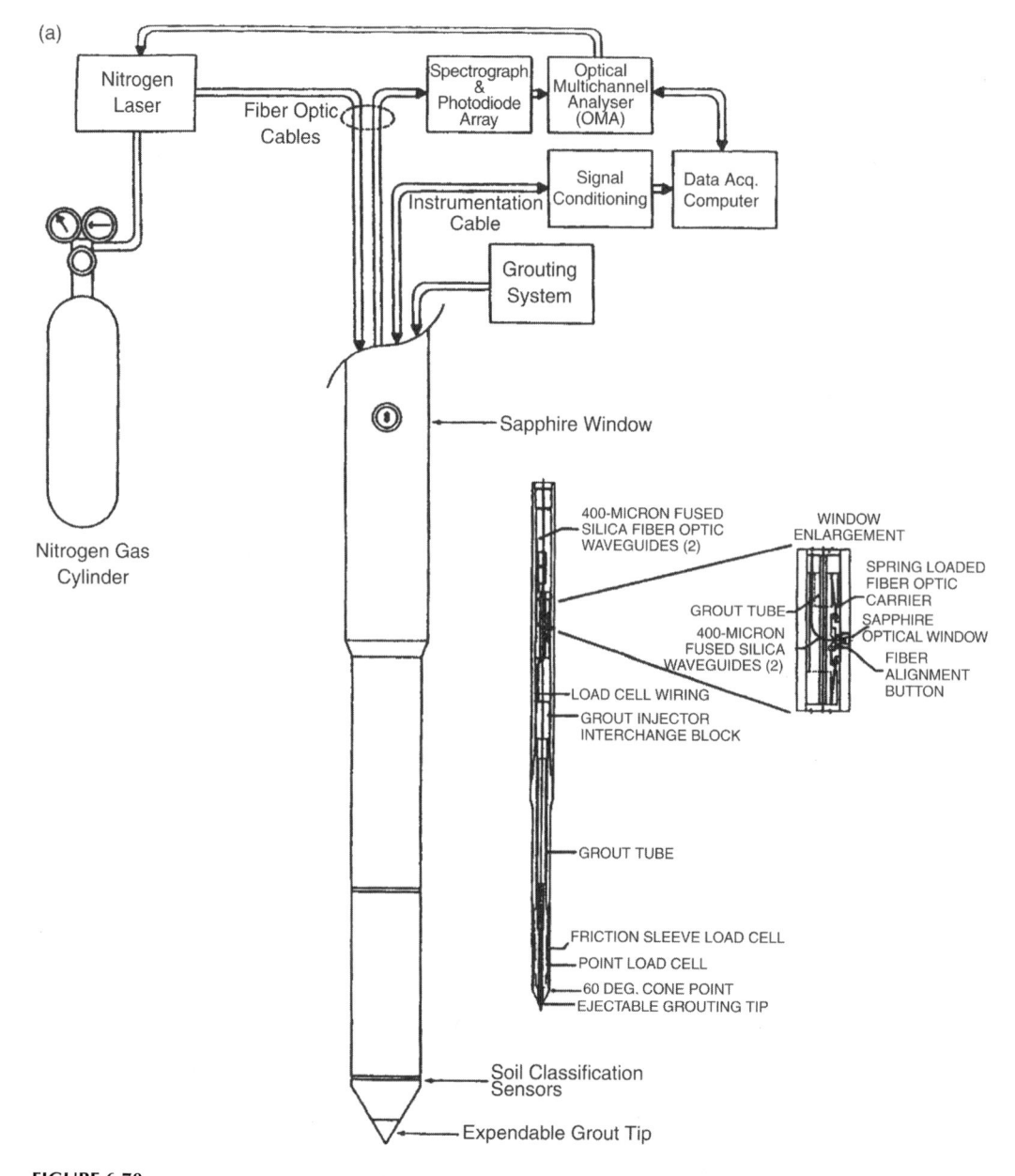

FIGURE 6.70
(a) Schematic of an LIF tool. (b) The down-hole LIF probe, showing the sapphire window (center). (c) The nitrogen laser.

Use of Direct-Push Technologies in Environmental Site

(b)

(c)

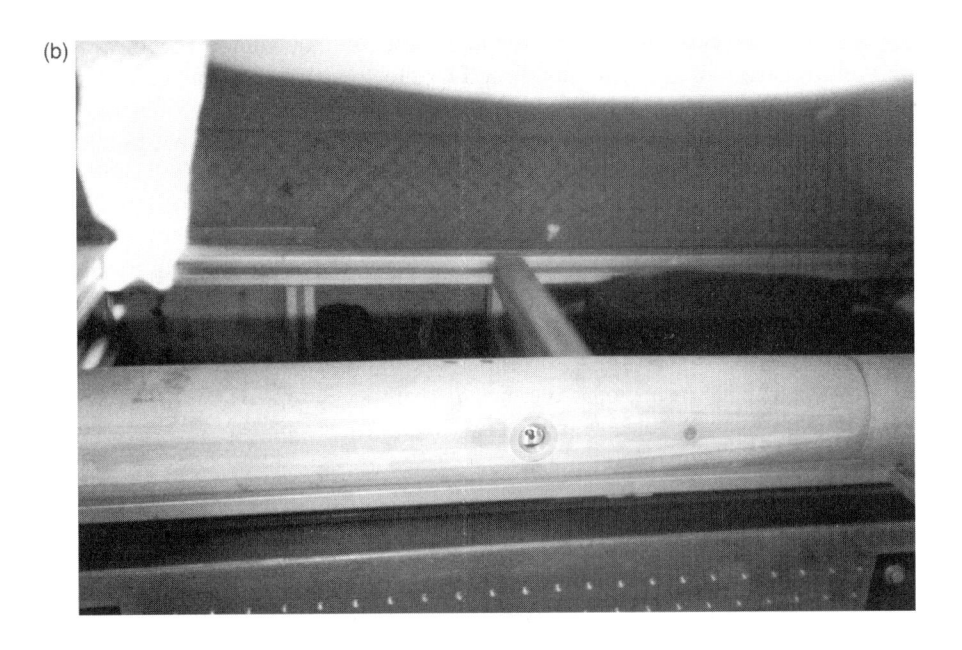

FIGURE 6.70
Continued.

that their presence can be detected indirectly at some sites if they are commingled with petroleum hydrocarbons that can be directly detected (Kram et al., 2001a). The LIF response can be affected by the presence of naturally occurring fluorescent minerals, humic and fulvic acids, organic debris and calcium carbonate (Kram et al., 1997). The borehole created by probe advance can be grouted as the probe is extracted, with a special mixture of grout pumped at a high pressure through a tube running down the inside of the penetrometer rods. Decontamination involves cleaning the rods as they are removed from the hole, which is done in a self-contained compartment below the truck body. All waste water from the decontamination process is collected by a vacuum system and stored in drums for later disposal. A cost-benefit analysis conducted by the U.S. Department of Energy (Schroeder et al., 1991) indicates that a 25 to 35% cost savings can be realized with the LIF system vs. conventional drilling, sampling, and sample analysis methods.

Raman Spectroscopy

CPT-based Raman spectroscopy is very similar in concept and deployment to LIF spectroscopy — the down-hole tool is constructed similarly and many of the other components of the system are the same. The difference is in the laser used to raise the excitation state of the contaminant molecules (an infrared laser with a longer [785 nm] wavelength) and the analytical method used to examine the compounds of interest. Raman spectroscopy has been used for many years to detect a variety of organic and inorganic chemicals. It does not involve a radiative process, like LIF, but rather it measures the light inelastically scattered from the incident light radiation (Tobin, 1971). The energy shifts in the scattered light are correlated to the vibrational modes of the particular compound and constitute the Raman spectrum for the compound. The vibrational modes of the compound depend on the elemental constituents and energy state of the molecule. The number of modes and associated energies of these modes are unique to each molecule and therefore a unique Raman spectrum is produced for each compound (Colthup et al., 1990). As the material outside the sapphire window of the probe is exposed to laser light, the molecules in the compounds present scatter light, vibrate in a distinctive way, and create a vibrational "fingerprint." This fingerprint is captured and transmitted via fiber optic cables to the analyzer, where it is compared to known vibrational signals stored in a computer database. The Raman system has been used to detect metals, metal complexes, organic compounds, oxidizers and radionuclides in complex mixtures of waste (U.S. DOE, 1999), DNAPLs such as trichloroethylene (TCE) and perchloroethylene (PCE) (Rossabi et al., 2000), and a variety of other compounds.

Fuel Fluorescence Detector

The FFD is another valuable tool for *in situ* detection of fluorescent compounds. Although the FFD will detect any chemical that fluoresces when excited at a wavelength of 254 nm, the primary application for the instrument is sensing petroleum hydrocarbons. A commercially available FFD system (Figure 6.71) consists of a down-hole CPT module coupled through electrical cables to an up-hole controller module. The down-hole module contains a mercury lamp excitation source, fluorescence emission collection optics, and two to three down-hole photomultiplier tubes (PMTs). The fluorescence emission is delivered through a short fiber-optic cable to each PMT. Interchangeable optical filters configured in front of the PMTs allow the user to select specific wavelength ranges (i.e., spectral "bands") to control the selectivity of contaminant detection. If the fluorescence emission profile of a contaminant is known, then a filter with band-pass closely matched to the wavelength of maximum emission intensity can be used to enhance sensitivity and discriminate against background fluorescence

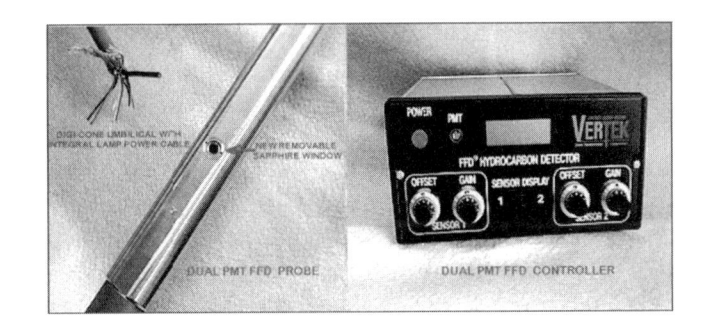

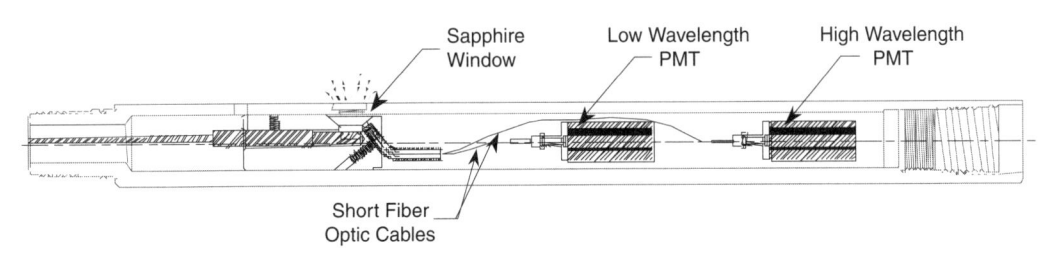

FIGURE 6.71
A commercial FFD provides an inexpensive and very effective means for delineating NAPL petroleum products.

outside the wavelength region of interest. This approach is comparable to that used in LIF systems in which a single, narrow band of fluorescence emission is isolated with a spectrometer and detected with a PMT during the CPT push.

With its 254 nm excitation wavelength, the FFD is highly versatile because it induces a fluorescent response in virtually all aromatic hydrocarbon components of fuels. Differential transport, biodegradation, and chemical "weathering" can all produce a change in a fuel's composition and its corresponding fluorescence spectrum. Light aromatic compounds, such as benzene and its alkyl derivatives, which fluoresce at lower wavelengths (about 300 nm), migrate faster and are more easily degraded than larger polycyclic aromatic compounds which fluoresce at longer wavelengths (up to 500 nm). Differentiating contaminant fluorescence from soil background fluorescence can be difficult with the standard FFD. It is also difficult to determine changes in fuel composition using single channel (band-pass filter) detection. During CPT deployment, a decrease in conventional FFD response with a band-pass filter designed for light aromatic hydrocarbon detection could indicate that fuel concentration is decreasing, but could also be caused by preferential degradation of the light aromatics. Obtaining a full fluorescence emission spectrum through the use of an up-hole spectrometer can be used to discriminate between these effects. This is available as an optional accessory for use with the FFD. In this configuration, an optical fiber transmits the fluorescence emission spectrum to the up-hole spectrometer.

There are many other situations in which a spectrometer can augment the information provided by the basic FFD system and enhance its performance. The spectrometer provides the user with a full fluorescence emission spectrum from the sample, rather than just a measure of fluorescence intensity over a limited spectral band pass. A spectrometer affords the user a number of useful capabilities such as the ability to identify unknown fuels and to select an optimal band-pass filter for a particular contaminant and site. Rather than conducting many time-consuming standard FFD experiments with different filters in front of the PMT, a single fluorescence spectrum collected quickly with a spectrometer can identify a fuel and guide the choice of a filter.

Other CPT-Deployed Probes and Sensors

A variety of specialized CPT-based probes and sensors have been developed primarily by the U.S. Department of Defense and the U.S. Department of Energy to improve the cost effectiveness of site-characterization programs at large sites with complex waste release histories. These probes and sensors include the following (Adams and Robitaille, 2000):

- An explosives sensor that details the presence and extent of a variety of explosive materials (nitroaromatic compounds) in soil. This probe incorporates an external heating element or pyrolizer that transforms explosive compounds in the soil into electroactive nitrogen-containing vapors. A pneumatic system draws the vapors from the soil through an internal vapor gas sweep stream into the probe. Electrochemical sensors in the probe examine the vapor for presence and identification of explosive nitrogen-containing compounds. Increase in current output of the sensor electrode are directly related to the concentration of explosives in soil. Other materials in soil (e.g., fertilizers) may contain some of the same compounds characteristic of explosives. Therefore, the explosives sensor is equipped with detectors that differentiate between compounds containing organic nitrogen (explosives) and inorganic nitrogen (such as fertilizer).

- A metals sensor that uses laser-induced breakdown spectroscopy (LIBS) to detect, identify, and delineate heavy metals in soil. The LIBS sensor uses light from a high-power pulsed laser (typically a Nd:YAG laser operating at 1.06 μm wavelength) to generate a diagnostic plasma in the soil. When the output of the laser is focused on the surface of the soil, the soil (and contained contaminant) heats up rapidly to become an electronically excited plasma. When the excitation energy from the laser is removed, the excited electrons drop to lower energy levels with the emission of characteristic photons for a brief period of time. The plasma emission spectrum from the soil is transmitted up a fiber optic cable to a spectrometer, which breaks this light into its constituent colors. Elemental analysis is conducted by observing the wavelengths and intensities of the spectral emission lines. The wavelengths of light in the plasma correspond to specific metal elements and the intensity of the light indicates concentrations present. LIBS is able to detect metals in the single parts-per-million (milligrams-per-kilogram) range (U.S. DOE, 2000).

- A metals sensor that uses XRF to detect metal elements at levels below 100 ppm (100 mg/kg) in both the vadose and saturated zones. This probe can detect elements with atomic numbers higher than 20 (e.g., higher than Ca), which includes most metals and radionuclides. XRF technology is a well-established, nondestructive laboratory and field-screening method for determining elemental concentrations in complex samples. The CPT-deployed XRF unit operates by detecting the characteristic x-ray emissions produced by metal atoms in soil exposed to an x-ray source. The sensor is advanced to a selected depth, at which point an x-ray source in the probe tip bombards the surrounding soil with incident x-rays. Metal atoms in the soil are excited and emit fluorescent x-rays with an energy characteristic of specific elements. The emitted x-rays are detected at the probe tip and provide an individual peak for each metal species present in the soil. The data are sent up-hole to an analyzer, where the elemental signatures are recorded in real time, providing data on metals present in the soil and their concentrations.

Membrane Interface Probe

The MIP is a DP tool used for the detection of VOCs in the subsurface. In this tool, a heated, semipermeable membrane acts as an interface between the formation materials and the detector system (Figure 6.72). This membrane is housed in a 1.5 in. (38 mm) diameter probe that is advanced incrementally by DP methods into unconsolidated formations. A regulated carrier gas, such as ultra-pure nitrogen (N_2), is transmitted through an inert supply tube from the surface to the probe at depth. The clean carrier gas sweeps across the interior surface of the heated membrane, and VOCs are transferred to the clean gas under a concentration gradient (Figure 6.73). The regulated gas flow carries the VOCs to the surface through the return tube to the detectors. The MIP system can be operated with a PID, FID, ECD, or a dry electrolytic conductivity detector. Two or more detectors may be operated in sequence. The MIP system provides semiquantitative data on the concentration of total volatiles in the formation (Christy, 1996). GC and GC/MS systems have also been coupled with the MIP system to allow for identification and quantification of specific analytes.

MIP System Components

The components of the MIP system (Figure 6.74) include a stringpot that provides for accurate tracking of probe depth and speed. The control module and field computer are used to operate the system and acquire and store the MIP data. The trunk line, which contains the gas supply and return tubes and electrical lines, is prestrung through drive rods that are stored on a mobile rack. These components, along with the detectors or GC, comprise the MIP system. The MIP probe includes an electrical conductivity dipole (Figure 6.72), which provides an electrical log of the formations penetrated by the tool string. Electrical logs may be used to define lithology.

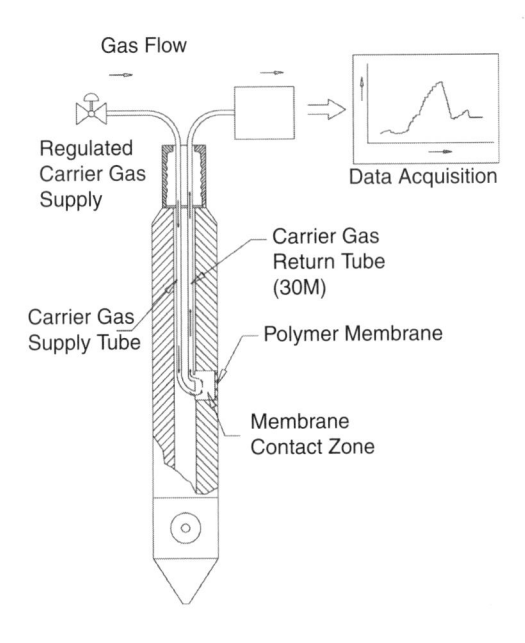

FIGURE 6.72
Cutaway view providing a schematic of the MIP. A regulated carrier gas is supplied down hole through a trunk line. The carrier gas sweeps across the interior surface of the polymer membrane and returns to the surface detector system via the trunk line. The semipermeable membrane is heated to 100 to 125°C to enhance the rate of permeation through the membrane. A field computer with data acquisition software provides real-time display and saves the detector logs, electrical log, and temperature log vs. depth.

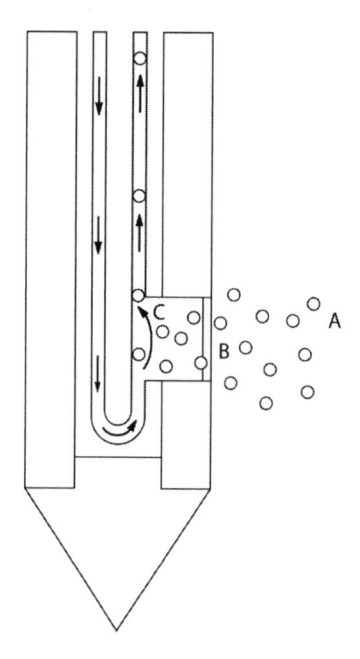

FIGURE 6.73
This schematic of the MIP provides a representation of the movement of volatile contaminants from elevated concentration in the formation (A) into and through the semipermeable membrane (B) and then uptake by the carrier gas (C) for transport to the detector at the surface. The semipermeable membrane is a rugged composite of steel and polymer that prevents movement of liquids across the interface but does allow gases to permeate, including some water vapor.

Field Operation

For successful and efficient application of the MIP system in the field, the operators must be familiar with running both a DP machine and a GC with associated computer hardware and software. The MIP probe and detector system is field calibrated with a standard solution containing one of the analytes of interest. When petroleum contaminants are being investigated, distilled water spiked with a known concentration of benzene or toluene may be used as a field standard. Field standardization serves several functions. First, this standard run verifies the travel time of the carrier gas and contaminants from the membrane through the return tube to the detectors (Figure 6.72 and Figure 6.74). Secondly, the standard run will also verify that the system is operating properly and no leaks or obstructions to carrier gas flow exist. Finally, the standardization provides a quantitative measure of the detector response (in microvolts) to a known contaminant concentration in solution. This may be used to provide a semiquantitative estimate of contaminant mass in the formation being tested. Once MIP logs are obtained in a suspect area, targeted samples of soil and ground water may be collected for laboratory or field analysis. These samples may be used to verify that nondetects by the MIP are indeed clean zones and that elevated responses by the MIP system correlate with highly contaminated areas.

After field calibration is completed, a slotted drive cap is placed on the MIP probe and the probe is advanced incrementally into the subsurface (Figure 6.75). Depending on the level of vertical detail required in the contaminant profile, the increment of probe advancement may be adjusted. Typically, an increment of 1 ft (30 cm) is used, with the probe being advanced an additional foot every minute. The probe is advanced to depth by adding the pre-strung rods until the log is completed. The re-entry grouting method (discussed in later section of this chapter) is used to properly decommission the MIP borings. The drive rods and MIP probe may be decontaminated with an alconox-and-water wash and water rinse.

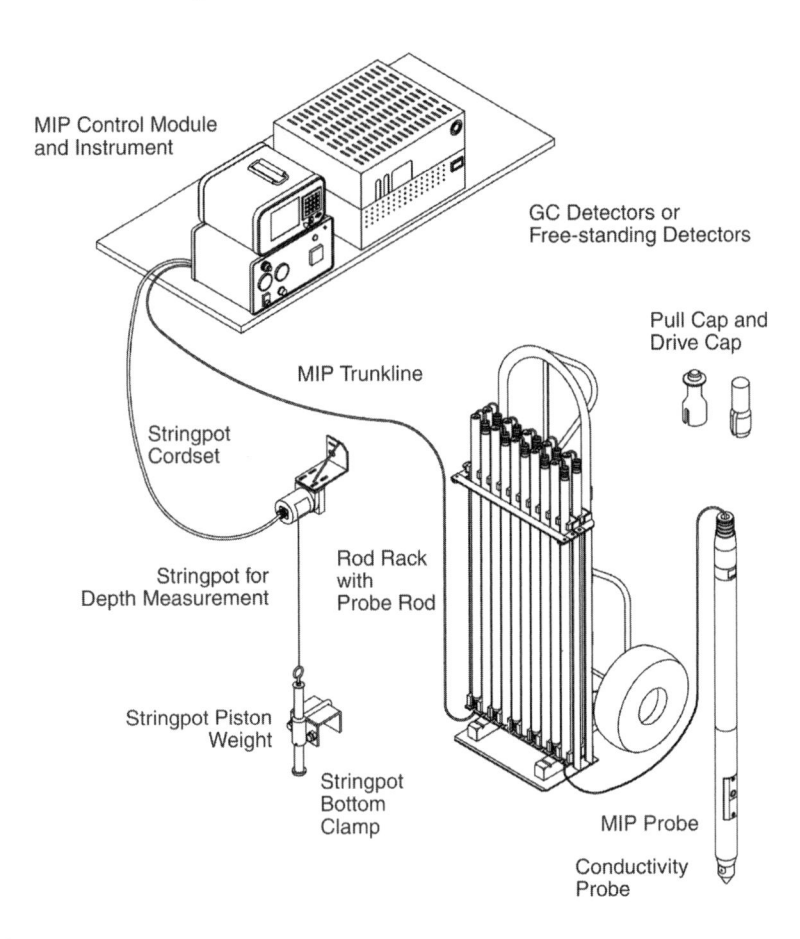

Use of Direct-Push Technologies in Environmental Site 435

FIGURE 6.74
The primary components of the MIP system include the items specified in this diagram along with an appropriate carrier gas supply. The carrier gas used may vary depending on the detectors and analytes of interest. Nitrogen is the most commonly used carrier gas and, when an FID is operated, hydrogen and air must be supplied for the detector. The stringpot and piston weight are mounted on the DP unit to track depth and speed of advancement.

Example Logs and Interpretation

Advancing the MIP provides the operator with logs of contaminant detection and lithology, which are stored in ASCII file format and which may be downloaded to many software systems for presentation and display. Plotting of the contaminant log next to the electrical conductivity or lithology log can provide information about contaminant migration pathways or lithologic boundaries to contaminant migration. An MIP log obtained using an FID detector where petroleum contamination was suspected found elevated levels of hydrocarbons between approximately 25 ft (7.6 m) and 32 ft (9.8 m) below grade (Figure 6.76). In an adjacent borehole, continuous soil samples were collected and samples were analyzed for total gasoline range organics (TGRO) from each 2 ft interval beginning at one foot below grade. When the soil sample analytical results are plotted adjacent to the MIP–FID log (Figure 6.76), a clear correlation is observed between the MIP log and the soil sample data. Nondetect MIP results correlate with nondetect soil sample results and contaminated soil sample results correspond to elevated detector readings from the MIP system.

When the electrical log is plotted adjacent to the MIP–FID log, it is easily seen that the electrical conductivity of the formation is near or slightly below 100 mS/m over the zone where contamination was observed (Figure 6.77). Soil samples collected from the 25 to

FIGURE 6.75
MIP logging operations in the field. For this project, the MIP system, including the probe rods, trunk line, and detector system, is housed in the bed of the truck shown in the background. The MIP probe is being pushed into the ground using the track-mounted DP machine in the foreground. A trunk line containing gas and electrical lines runs from the truck to the probing machine.

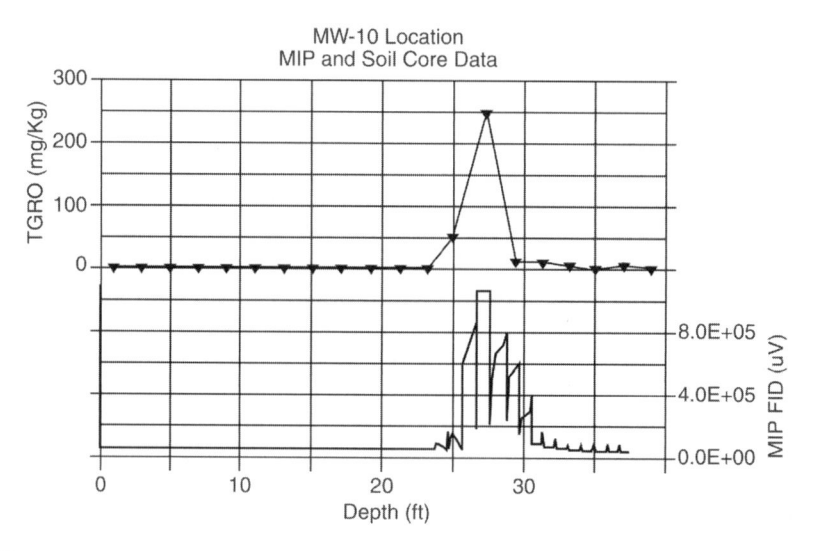

FIGURE 6.76
An MIP–FID log obtained at a petroleum contaminated site is displayed in the lower half of this diagram. Probing depth is on the bottom (x) axis and detector response in microvolts is on the right side (y) axis. One-foot increments were used in probe advancement with no contamination detected until a depth of about 25 ft. Significant detector response was observed at the 25 to 32 ft depth interval, below which the detector response returned to baseline. The DP unit was moved about 3 ft (1 m) from the log location, and continuous soil sampling was conducted. Subsamples from each odd-foot interval were analyzed for GRO and the results are plotted (triangles) on the upper half of this graph with concentration in milligrams per kilograms on the left (y) axis. Soil sample results consistently verified the results of the MIP–FID log.

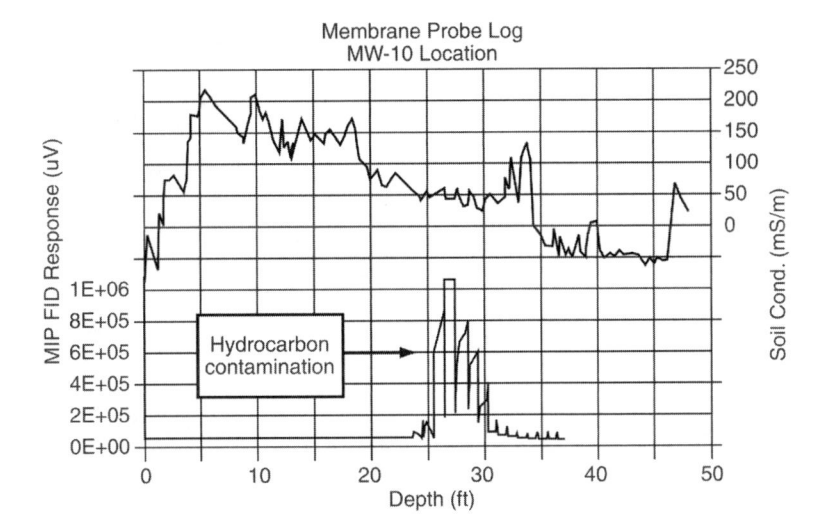

Use of Direct-Push Technologies in Environmental Site 437

FIGURE 6.77
The electrical log obtained during advancement of the MIP probe is shown plotted above the MIP–FID detector results presented in Figure 6.76. Where the electrical log response is below 100 mS/m (between 25 and 30 ft deep), samples were found to contain some sand and have increased permeability compared with the formation materials above and immediately below this interval. Multiple logs across the site confirmed this interval as the preferential migration pathway. Elevated electrical conductivity at the 32 to 35 ft depth in each log was found to be a clay or silt layer with low permeability. This layer behaves as a barrier to movement of contaminants into the underlying sand aquifer.

30 ft (7.6 to 9 m) interval revealed that the clayey silt contained a small proportion of sand in this zone. The higher electrical conductivity of the formation observed between about 32 and 35 ft (9.8 and 10.7 m) (just below the contaminated zone) consisted of a clayey silt with no sand. Several MIP logs obtained across this site showed very similar results, indicating that the clayey silt layer between 32 and 35 ft (9.8 and 10.7 m) was preventing movement of the contaminants into the sandy aquifer below 35 ft (10.7 m) (low electrical conductivity). In nearby monitoring wells (screened from 10 to 30 ft [3 to 9 m]), free product was occasionally observed on the water table and the static water level was consistently measured at approximately 18 ft (5.5 m) below grade. The MIP logs clearly defined the vertical extent of the petroleum as being confined to the sandy zone (migration pathway) between the 25 and 30 ft (7.6 and 9 m) depth interval. Typically, the MIP system can detect volatile compounds such as benzene, tetrachloroethylene, and trichloroethylene at concentrations of approximately 500 ppb (500 mg/l) in ground water.

Other DP Tools

Video Imaging Systems

A relatively new CPT-deployed probe that is finding increased use at environmental sites is an *in situ* video microscope, known variously as the Videocone™ or the GeoVis system. This technology has been developed at several research organizations including the Geo-Delft Laboratory in the Netherlands, the University of Michigan, the U.S. Navy Research Laboratory, and Applied Research Associates, Inc. The *in situ* video imaging system provides the capability to collect real-time video images of subsurface materials from which determinations of grain size, moisture content (degree of saturation), and contaminant distribution can be made. Research is ongoing to expand the capabilities of this system for determining soil types and color and for identifying contaminants *in situ*.

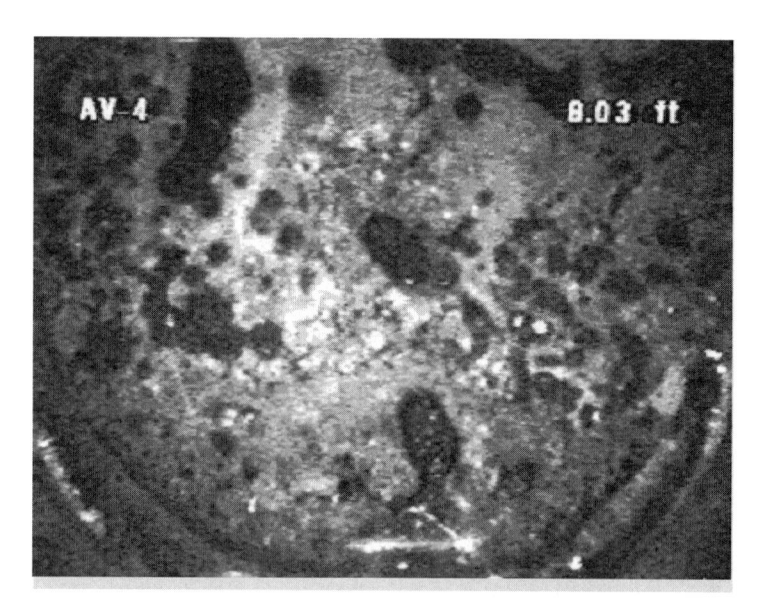

FIGURE 6.78
In situ image of coal tar DNAPL from a Videocone, obtained at Sidney, NY.

The *in situ* video imaging system consists of a miniature color CCD camera coupled with a magnification and focusing lens system that provides approximately a 100× magnification factor when viewed on a standard 13 in. monitor. White light-emitting diodes or high-intensity lamps are integrated into a standard cone penetrometer rod housing to provide soil illumination. Illumination and visualization are conducted through a non-scratching sapphire window on the side of the CPT rod. The video signal is usually transmitted to the surface via a coaxial cable and then fed into a video recorder and monitor or digitized. Images are displayed in real time as the CPT probe is pushed into the ground. Owing to the quantity of data being sent, video CPT soundings are usually conducted at a push rate slower than the standard 2 cm/sec (0.8 in./sec). Figure 6.78 is an image from the Videocone at a site in Sidney, New York, USA.

The video signal is delivered through a video cable to a text inserter module at the surface. Standard CPT cabling is used to deliver power from an up-hole camera control module to the down-hole camera and light source. A text inserter is interfaced with both the CPT DAS and standard video recorder (VCR). The text inserter automatically superimposes the borehole inner diameter and depth of the video cone on the image being recorded (Figure 6.78). A cable pass-through channel in the down-hole module allows a standard CPT cone or other probes to be configured ahead of the video unit. In the standard configuration, video images have an area approximately 2.5 × 2.0 mm. Objects as small as about 10 μm can be resolved (Lieberman et al., 1997, 1998).

Applications of DP Technology to Monitor Well Installation

The advantages of DP methods, especially the significant reduction in potentially contaminated IDW, the reduced disturbance to the formation, and the generally smaller, more mobile equipment that provides improved site accessibility relative to conventional drilling, make it an attractive alternative for installing monitoring wells. In the early 1990s, DP field operators began installing temporary PVC piezometers through open DP borings to obtain water levels and ground-water samples for field screening purposes, usually in formations that were slow to recharge. Some field personnel made attempts to construct small-scale monitoring wells with gravity installation of sand and bentonite chips around small-diameter (e.g., less than 1 in. [25 mm] nominal diameter) PVC screens and casings. However, because of the small diameters and small annular space between the DP rods and the well casing and screen, these attempts were difficult and time consuming at best. Methods to effectively install monitoring wells and conduct bottom-up grouting with DP equipment that met the essential monitoring well construction requirements of the day (U.S. EPA, 1991, 1986; ASTM 2004t) were first accomplished in the mid-1990s (Geoprobe Systems, 1996a; McCall et al., 1997). To speed up the DP well-installation process and ensure proper placement of filter-pack sand and annular seals, prepacked screens and modular well construction components were developed. Other field procedures to permit DP installation of long-term monitoring wells have also been developed. The typical requirements for ground-water quality monitoring well construction and different DP well installation methods are reviewed in the following sections.

Requirements for Monitoring Well Construction and Use

A ground-water quality monitoring well must serve only a few basic functions to meet most regulatory requirements. The functions are to:

- Provide for collection of representative water-quality samples
- Allow measurement of total well depth
- Allow measurement of static water levels
- Provide for detection of NAPLs (dense or light)
- Provide for collection of samples of NAPL as needed (U.S. EPA, 1986)

General construction requirements for a ground-water quality monitoring well have also been established (Chapter 10; U.S. EPA, 1986, 1991, 1992; ASTM, 2004t) (Figure 6.79). The essential requirements include:

- Well construction should result in only minimal disturbance to the formation.
- Well construction materials must be compatible with the chemical conditions in the formation.
- Wells must have an adequate annular seal composed of materials that will not interfere with formation water-quality sampling.
- Wells must be developed so that natural ground-water flow conditions are re-established after the well is installed (development must also remove any drilling fluids used in borehole installation).
- The well screen interval must be installed in the target monitoring zone.
- Surface protection must be adequate to prevent tampering and damage to the well.
- Well diameter must be sufficient to allow well development, aquifer testing, and water-quality sampling to be successfully conducted.

Individual state regulations and guidance for ground-water quality monitoring well construction vary somewhat but, in purpose and intent, are consistent with these requirements. ASTM Standard D 5092 (ASTM, 2004t) provides guidance on construction of conventionally drilled and installed monitoring wells. The well construction requirements

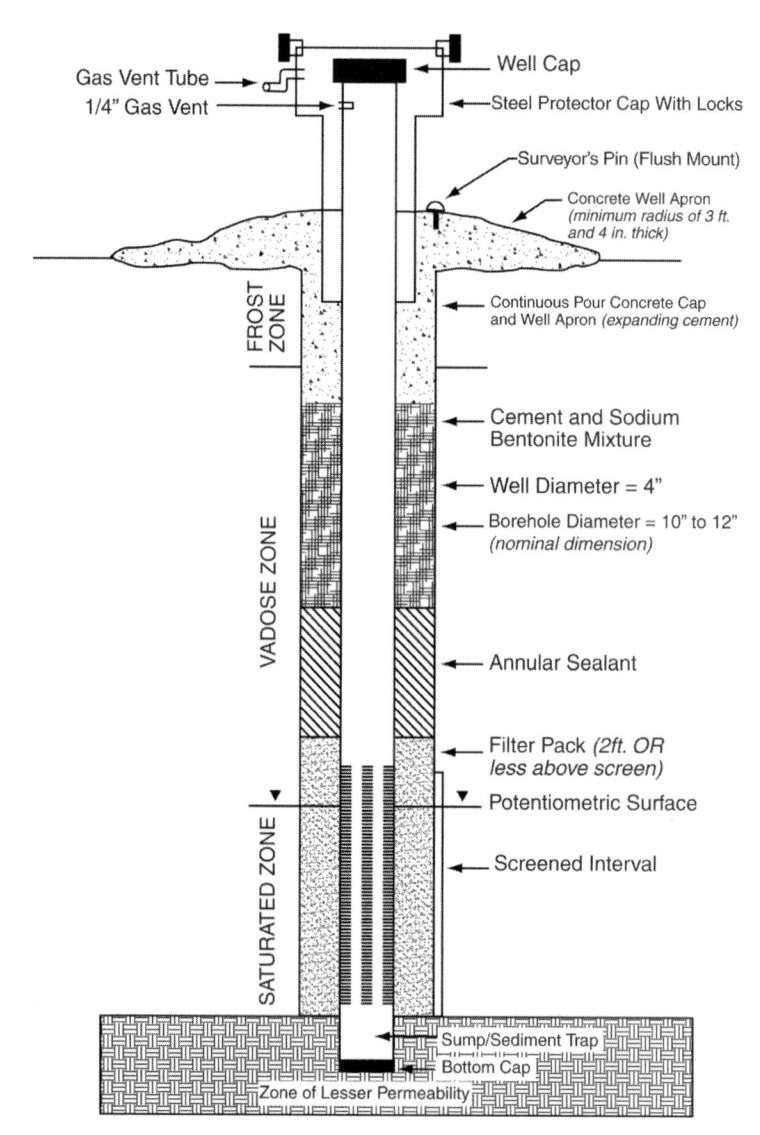

FIGURE 6.79
Generalized cross-section of a typical water-quality monitoring well. (After U.S. EPA, 1986).

in Standard D 5092 are similar to those outlined earlier but are more specific and detailed. Two additional ASTM Standards, D 6724 and D 6725 (ASTM, 2004d, 2004e), provide detailed guidance related specifically to construction of DP-installed monitoring wells that meet the earlier requirements. The following discussion summarizes the information included in these two standards and demonstrates how typical DP well installation methods differ from conventional methods.

Annular Seal and Grouting Requirements

The monitoring well construction requirement that is most vigorously enforced by state regulatory agencies is the annular seal and grouting requirement. Protection of ground-water resources is a very important issue in all states. Without an adequate annular seal, either surface contaminants or contaminants in a shallow zone in the subsurface may infiltrate down a DP probe hole or conventionally installed borehole and contaminate a formerly clean zone. Most existing state regulations and guidance for monitoring well construction, annular seals, and grouting are based on conventional drilling procedures and grouting equipment available 20 or more years ago. A 1 in. (25 mm) diameter (or larger) tremie tube is usually required to successfully pump grout from the bottom up in a conventional well completion. For this reason, the majority of state regulations require an annular space of 2 to 3 in. (51 to 76 mm) between the well casing outer diameter and the drill pipe or borehole inner diameter. Because most DP monitoring wells are installed through drive rods less than 2.5 in. (64 mm) O.D., a 2 or 3 in. (51 or 76 mm) annular space is not available for grouting. However, grouting equipment capable of pumping viscous (20 to 30% solids) bentonite slurries through a 0.25 in. (6.4 mm) I.D. tremie tube has been developed (Geoprobe Systems, 1996b). This system (Figure 6.80) has successfully conducted bottom-up tremie grouting in wells exceeding 100 ft (30 m) in depth (McCall, 2000a). Because of the cost-efficient installation, waste minimization, and improved grouting equipment available for DP well completions, many state agencies regulating well construction now permit installation of DP monitoring wells. Some state agencies require a regulatory variance for installation of DP wells because of the smaller annulus. State and local regulatory requirements should be verified before proceeding with any DP monitoring well installation.

DP Well Installation Methods: Advantages and Limitations

Several different methods for DP installation of monitoring wells have been developed and employed. These range from simple driven well points that have been used for many years to newer DP methods that use a cased borehole for high-integrity construction. The basic installation procedures and important limitations and advantages of each are discussed subsequently.

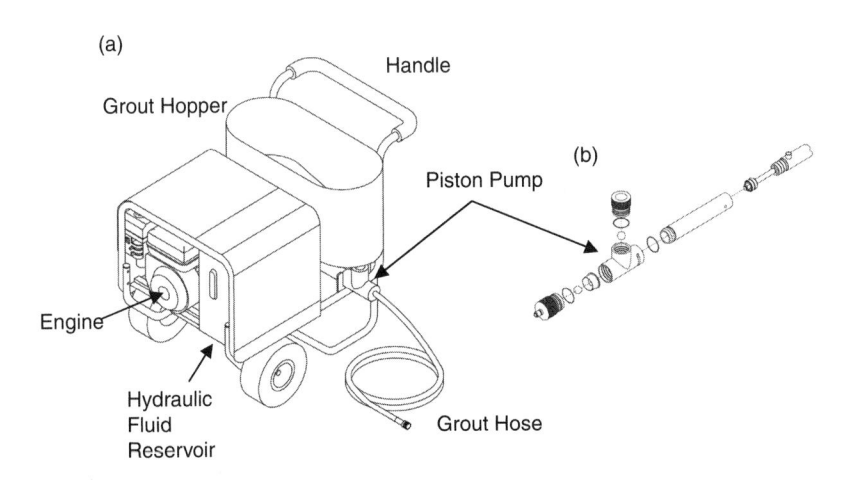

FIGURE 6.80
(a) High-pressure (1000 psi or more) grout pumps have been developed to enable placement of 25% or more solids bentonite slurries through a 0.25 in. I.D. tremie tube. Nylon tubing with a bursting pressure of 1500 psi is used to pump in bentonite or neat cement grouts by the bottom-up tremie method to meet the most stringent regulatory requirements for well construction. (b) A simple piston pump is used to pump the viscous grout slurries. Bottom-up tremie grouting has been completed with this system to depths exceeding 120 ft.

Driven Well Points

Simple driven well points (Figure 6.81) have been installed using manual methods, traditional drilling equipment with hammers, and DP techniques (Driscoll, 1986; U.S. EPA, 1991; Christy and Spradlin, 1992). In some types of driven well points, the screen is exposed to the formation as it is advanced. These devices are best used in locations where sandy materials extend to the surface so that clogging of the screen does not occur as it is advanced to depth. If contamination is present above the zone to be monitored, cross-contamination is a significant concern. Because of these limitations, exposed-screen driven well points are not often used for long-term ground-water quality monitoring.

Protected-screen well points (Figure 6.82) are usually not as susceptible to clogging of the screen and cross-contamination as exposed-screen well points. However, with either well-point design, it is not possible to install an annular seal, conduct grouting, or install a filter pack. Without the ability to install a filter pack, the well points can be installed only in formations that can be naturally developed (discussed subsequently), which limits their use under many formation conditions. Additionally, without an annular seal or grout, the potential for cross-contamination by movement of contaminants down the space between the well casing and formation is increased. These factors significantly limit the use of driven well points for ground-water quality monitoring at potentially contaminated sites.

Open Hole Procedure

Another procedure that has been used for installing monitoring wells by DP methods is the simple open-hole method (Figure 6.83). Soil sampling is conducted to depth and, once sampling is completed, well screen and casing are assembled and lowered through the open borehole. Sand and granular bentonite are often poured down the open hole to build a filter pack and well seal. This is effective only in cohesive formations where the borehole remains open. Even under good formation conditions, sloughing may occur and, if contamination is present above the screened interval, cross-contamination can be a problem. If fine-grained materials collapse around the screen during installation

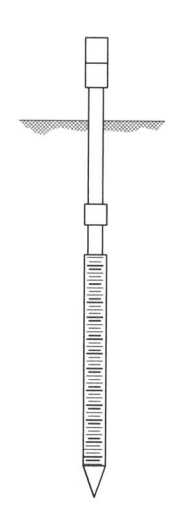

FIGURE 6.81
Well point driven from the surface for ground-water sampling. This is one of the least expensive means of installing a well screen but is limited to use in sandy formations.

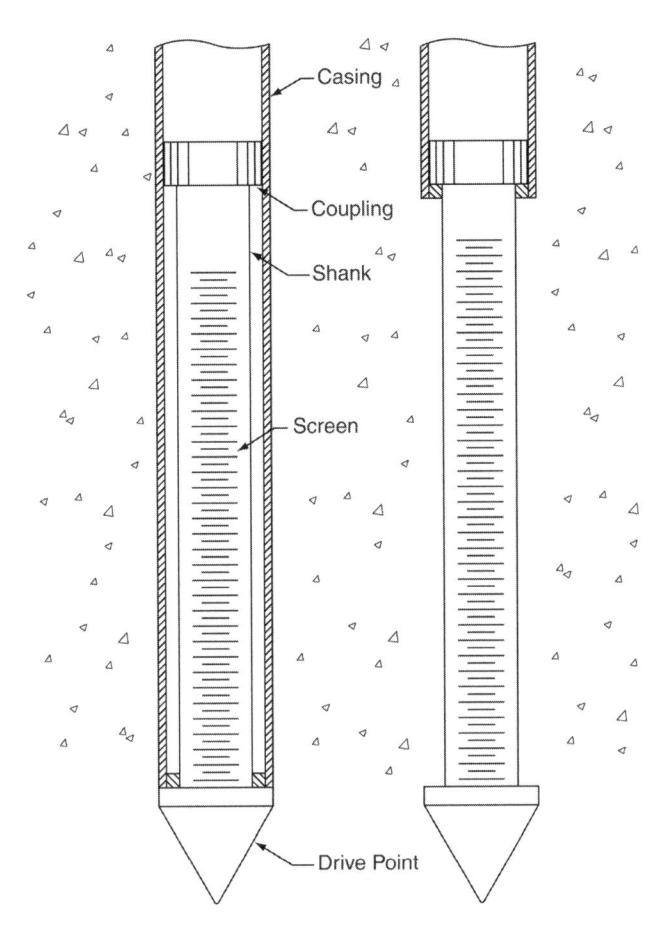

FIGURE 6.82
Installation of a protected-screen well point. The well screen is driven to depth inside a metal casing. Retraction of the casing exposes the screen to the formation.

of the sand pack, well development, to ensure low turbidity samples, may also be difficult. Because of the formation collapse and cross-contamination potential, monitoring wells installed through open DP boreholes are generally not used for long-term ground-water monitoring applications. They are often used as piezometers to obtain water-level measurements and to obtain samples to meet screening-level DQOs. Appropriate grouting procedures must be used when these wells are decommissioned to eliminate the potential for migration of contaminants down hole.

Mandrel-Pushed Screen and Casing

Another procedure developed for DP well installation uses an internal steel rod or mandrel with PVC screen and casing over the outside of the mandrel to advance the screen to depth (Figure 6.84). This method was developed specifically for use with CPT trucks in sandy formations (Kram et al., 2001b; ASTM 2004d). A slightly oversized drive point threads into the bottom end of the slotted PVC screen and the drive rod (or mandrel) slides inside the screen and PVC casing. The drive rod is advanced to depth with a CPT unit or other DP machine with the screen exposed to the formation. Once the base of the screen is at the desired depth, the inner drive rod is retracted and the

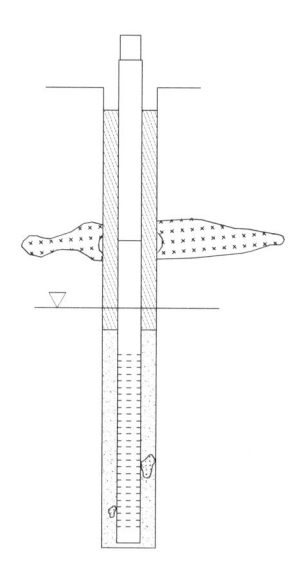

FIGURE 6.83
Open-hole DP well construction. Well casing is inserted down the open hole created by soil sampling. This method often results in low integrity well construction and is not recommended for long-term monitoring wells.

casing remains in place. This is a relatively fast and inexpensive method for DP installation of an exposed-screen monitoring well.

Limitations with this method are the same as those for other exposed-screen tools. If fine-grained materials overlie the sandy zone to be monitored, clogging of the screen may occur and development may be difficult. In addition, if contamination is present above the monitoring zone, the screen may be contaminated during installation, resulting in cross-contamination of any water-quality samples collected from the well. Because there is little or no annular space, it is not possible to install a filter pack or conduct grouting operations. Thus, these wells must be naturally developed (discussed subsequently).

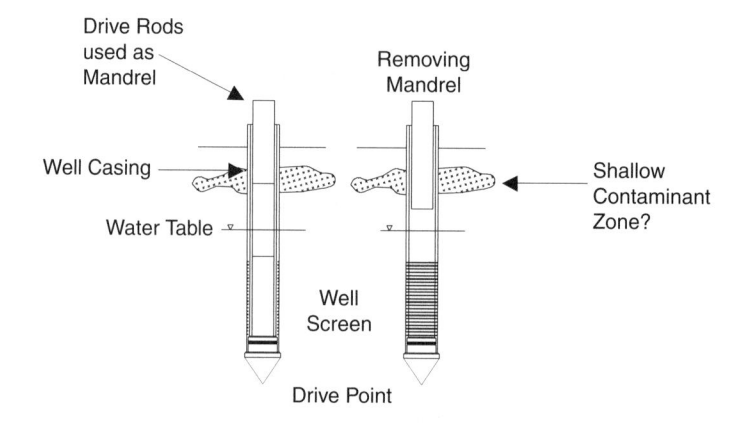

FIGURE 6.84
Installation of well screen and casing using the drive rods as a mandrel with the casing and screen exposed to the formation as the tool string is advanced. This method works well where the sandy formation extends to the surface so that fine-grained material will not clog the well screen during installation. The possibility of cross-contamination of the screen from shallow depths must be considered. Annular grouting is not possible using this method.

Proponents of this well installation method note that the formation is compacted as the riser and casing are pushed to depth and relief of this compaction will result in a seal between the formation and riser, eliminating the need for an annular seal or grout. No research is available showing that rebound of the compacted formation will be sufficient to provide an adequate well seal, especially in sandy formations in which this method is best applied.

Cased Method

With the development of more powerful DP machines, it became possible to install larger diameter drive casing that permitted DP installation of monitoring wells through a cased borehole. Wells installed in this manner can either be naturally developed, filter-packed, or installed with prepacked well screens. Many formations that are typically monitored are ill-suited for natural development because of their grain-size distributions (U.S. EPA, 1991; ASTM, 2004t). For this reason, filter-packed wells, in which the filter media is gravity installed through the drive casing to build a filter pack, or wells with prepacked well screens are more often installed. These installations are very similar to wells constructed using conventional drilling methods (Chapter 10).

Naturally Developed Wells

The installation of screen and well casing for naturally developed wells is relatively simple. Continuous soil coring may be conducted to depth (Figure 6.85) or drive casing is advanced to depth with an expendable point. The screen and casing are lowered through the drive casing to the desired depth and then the drive casing is retracted. Time is allowed for the natural formation to collapse around and at least 1 to 2 ft above the top of the well screen. With the drive casing retracted above the screen, but still in place, the annular seal may be gravity installed and grout tremied into the annulus. Even in the larger drive casing, enough annular space between the riser and drive casing inner diameter must be available to allow for installation of the annular seal and grout. For example, if a drive casing with a 1.5 in. (38 mm) I.D. is used, nominal 0.5 in. (13 mm) screen and casing will be the largest materials that can be used in well construction. Nominal 0.5 in. (13 mm) PVC has an outer diameter of approximately 0.8 in. (20 mm), leaving an annular space of approximately 0.375 in. (9 mm) to install annular seal and grout materials. In larger drive casing (e.g., 2 or 2.5 in. [51 or 64 mm] I.D.), nominal 0.75 in. (19 mm), 1 in.(25 mm), and possibly 1.5 in. (38 mm) well screen and casing may be installed.

Probably the greatest limitation to the use of naturally developed wells is the relatively rigorous requirement established for the grain-size distribution of the natural formation (U.S. EPA, 1991). Determining whether a formation is acceptable for natural development is commonly based on two parameters:

> *Effective grain size*: the sieve size that retains 90% of formation material (sieve opening in thousands of an inch [tenths of a millimeter]).
>
> *Uniformity coefficient*: the ratio of the sieve size that retains 40% of formation material to the effective grain size.

For example, if the sieve size opening that retains 90% of the formation material is 0.015 in. (0.4 mm) and the sieve size opening that retains 40% of the material is 0.045 in. (1.1 mm), the uniformity coefficient is 3.0.

It is recommended that a formation have an effective grain size ≥ 0.010 in. (U.S. EPA, 1991) to be acceptable for natural development. A uniformity coefficient of <5.0 is

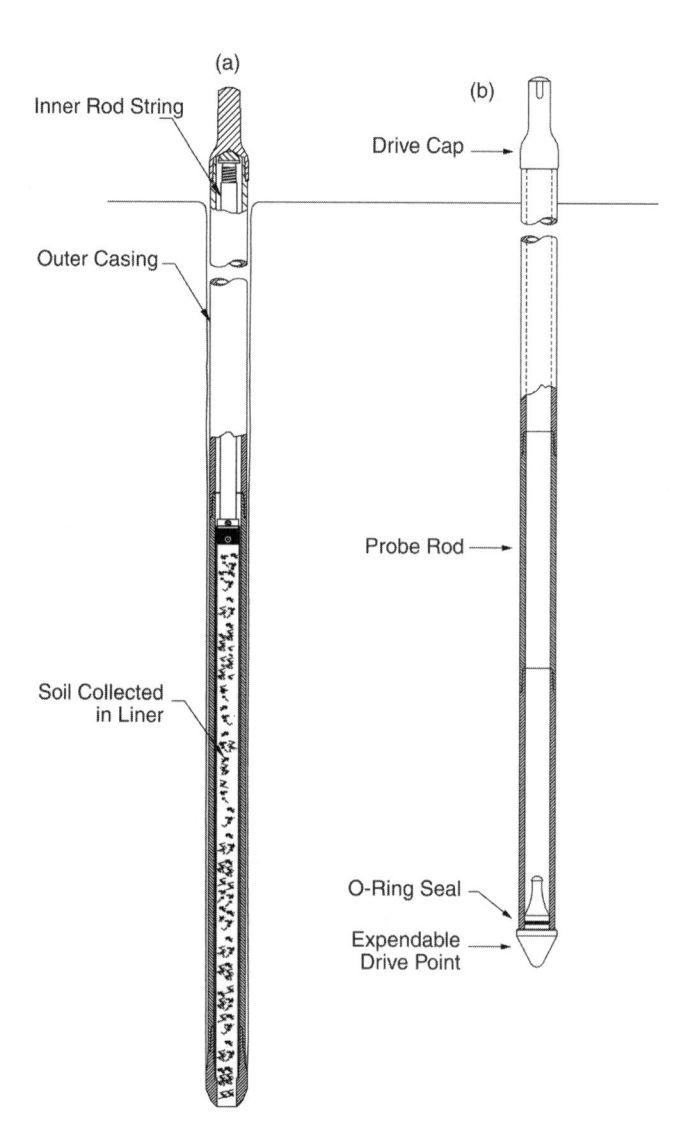

FIGURE 6.85
Two methods of driving casing (probe rods) through which a monitoring well screen and riser will be set. (a) Shows continuous cores being collected as the probe rods are advanced. (b) Shows the probe rods being driven with an expendable point closing the lower end.

considered to represent a well sorted formation (ASTM, 2004t). While the example above would meet the requirements for natural development, many formations, especially fine-grained materials, do not meet these criteria. Field research (Kram et al., 2001b) indicates that naturally developed wells in formations that would not meet these strict criteria can still provide low-turbidity samples when low-flow purging and sampling protocol (Puls and Barcelona, 1996; ASTM, 2004n) is followed for DP-installed wells. The criteria for naturally developed wells were originally established for large-diameter production wells that are often pumped at flow rates of tens to hundreds of gallons (liters) per minute. It may be that these criteria are not as well suited to smaller diameter wells that are sampled using low-flow protocol, where flow rates less than 500 ml/min (17 fl oz/min) are typically used.

Use of Direct-Push Technologies in Environmental Site 447

When the natural formation is used as the filter medium, well development, to establish flow into the well and to allow collection of low-turbidity samples, is essential. Appropriate well development methods (U.S. EPA, 1991; ASTM, 2004u) must be applied to optimize well performance and to obtain the sample quality required. Appropriate well development methods may include surging and pumping or bailing or overpumping and backwashing. Care must always be taken during well development to prevent damage to the well and possible reduction in well yield.

Filter Packed Wells

The most widely used monitoring well construction method for conventional drilled wells involves gravity installation (pouring) of filter-pack sand down the annulus between the well casing and the drill casing or open borehole. When large-diameter drilling equipment is used, the annular space is large enough so that bridging of gravity-installed filter-pack sand can be avoided or easily corrected. Pouring filter-pack sand can also be done in larger DP drive casing (Figure 6.86), but care must be exercised to prevent bridging of the filter-pack sand in the annulus and incorrect well installation or loss of the well altogether from complications. When sufficient annular space is available, a tremie tube may be used to install the filter pack in the screened interval. The first steps for installing a filter-packed monitoring well are the same as the first two steps for a naturally developed well. Gravity installation of the filter pack sand follows until the sand pack extends about

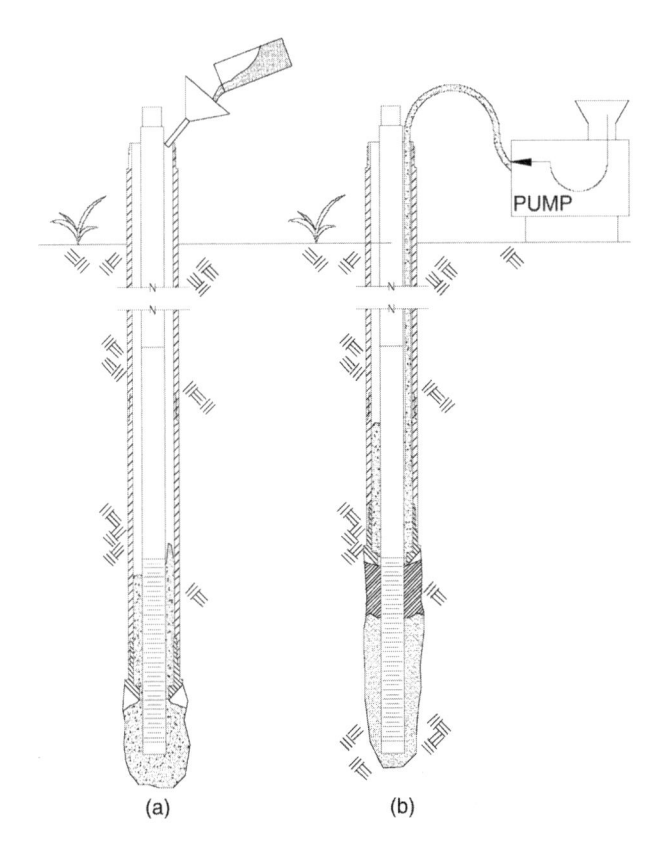

FIGURE 6.86
(a) When large drive rods are used for DP well installation, the filter pack may be poured through the annulus between the rods and well casing. Sand bridging may occur in the annular space, especially when the screen is well below the water table. (b) The annular seal and grout may be pumped with a tremie tube placed in the annular space.

2 ft (0.6 m) above the screened interval (Figure 6.86). It is important to monitor the depth to sand with a weighted tape to ensure correct placement and installation of the filter pack. The annular seal may be placed as bentonite pellets or chips or installed as a grout via a tremie tube and grout pump (Figure 6.86) as the drive casing is slowly retracted. The grout must be kept inside the drive casing as it is retracted to prevent formation collapse against the casing and possible compromise of the annular seal.

While gravity or tremie installation of filter-pack sand may work relatively well in larger DP drive casing (>2 in. [51 mm] I.D.), in smaller DP rods or drive casing, the available annular space is more limited. This limited annulus hinders and slows the gravity or tremie installation of filter-pack sand and increases the possibility of sand bridging. To overcome these limitations, small-diameter prepacked well screens are used widely to facilitate the installation of DP wells.

Prepacked Well Screens

Prepacked well screens have been in use for some time in larger diameter conventional wells, but availability of these screens has been limited (U.S. EPA, 1991). Some early prepacked screens used a polymer material to bind the filter sand together and to a slotted inner screen. However, the polymer binders were found to cause significant analytical interference with many environmental parameters, so they could not be used for water-quality sampling under most conditions. Some manufacturers designed dual-wall prepacked screens having both an inner and outer wall of slotted PVC to hold the filter-pack sand in place. Other designs for small-diameter screens have used a stainless-steel mesh as the outer component of the dual wall screen (Figure 6.87). Prepacked screens simplify and speed up the process of well installation and provide assurance of placement and thickness of the filter-pack sand, which is often not achieved with standard means of filter-pack installation. The concentration and types of contaminants at a site, as well as the ambient ground-water chemistry, should be evaluated to select the appropriate

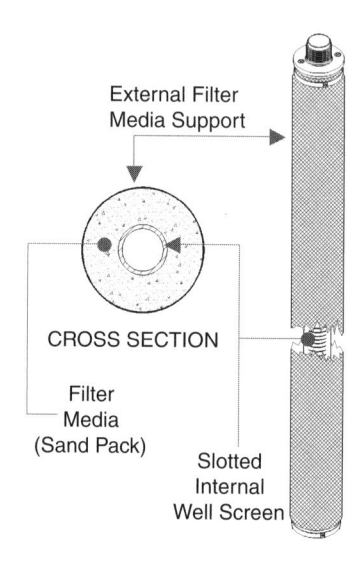

FIGURE 6.87
A typical prepacked well screen. The internal well screen is usually constructed of Schedule 40 or 80 PVC with factory-cut 0.010 in. slots — some screens are available with 0.25 in. slots. The external filter media support is usually constructed of stainless-steel wire cloth with a pore size of approximately 0.011 in. Graded silica sand or equivalent material is used for the filter media. Some prepacked screens are available as sleeves or jackets that slide over factory-available slotted PVC.

materials for prepacked screens (Hewitt, 1989a, b, 1991; Parker et al., 1990, 1991; Parker and Ranney, 1994, 1995, 1996).

Prepacked screen wells may be installed through a cased borehole following continuous sampling or by advancing drive casing with an expendable point to the desired installation depth (Figure 6.88). Most small-diameter prepacked screens are available in lengths of 2, 3, or 5 ft (0.6, 0.9, or 1.5 m) and two or more screens may be threaded together to achieve longer screened intervals when needed. After the drive casing is at the desired depth, the prepacked screens and casing are assembled and lowered into position. The drive casing is slowly retracted to expose the screen to the formation (Figure 6.89). When the screens are installed in fine sands or silty formations below the water table, some of the fine sand or silt may flow in between the screen and drive casing, causing some binding. Where this occurs, small-diameter steel extension rods may be lowered through the well casing and screen used to hold the assembly in place. Sometimes gentle tapping on the base of the screens with the extension rods will loosen the screen, and drive-casing retraction may be completed. To prevent flow or heave of material between the screens and drive casing, water (of known quality) may be added to the annulus between the well casing and the drive casing. This creates a positive hydraulic head, which will prevent the rapid influx of water and fine materials from the formation.

As the drive casing is retracted above the top of the screen, the annular seal and grout barrier are installed. If the existing formation collapses around the well casing, this may

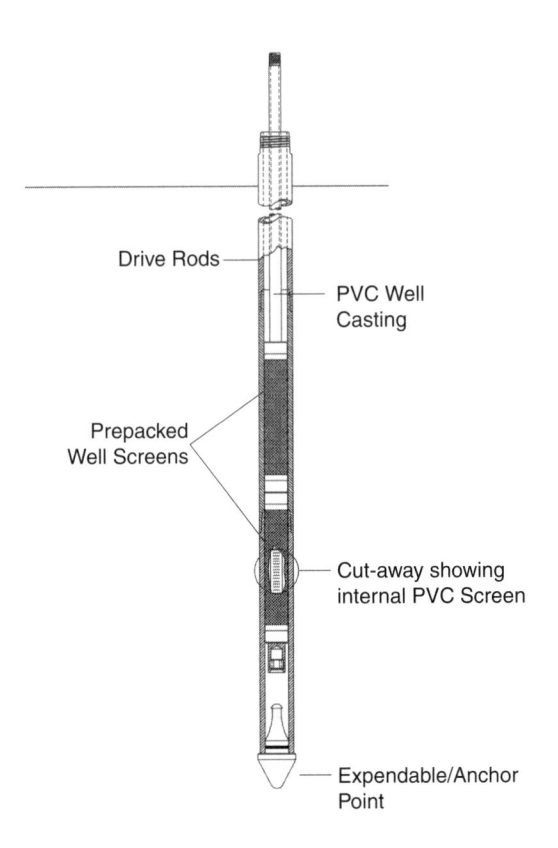

FIGURE 6.88
After the drive rods and expendable point are advanced to the desired depth, the prepacked screens and casing are assembled and lowered into position. Some expendable points latch onto a special adapter at the base of the screen to help anchor the prepacked screens in position.

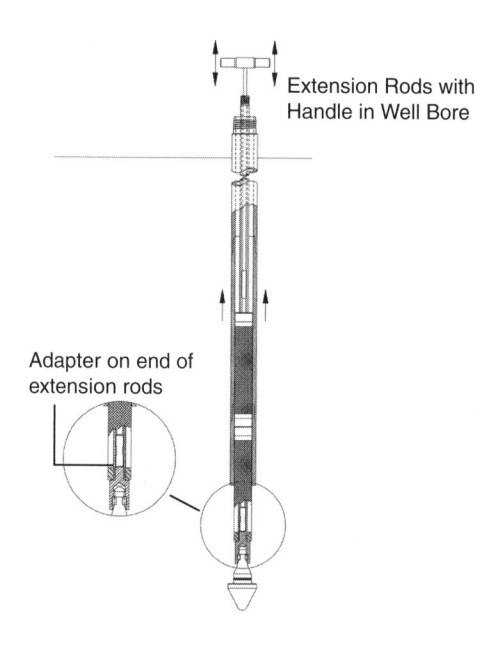

FIGURE 6.89
Once the prepacked screens are in position, the drive rods are slowly retracted to expose the screen to the formation. Extension rods with an adapter may be used to help hold the screens in position as the rods are retracted if any binding occurs.

form an acceptable grout barrier. However, some formations do not readily collapse. Project or regulatory guidelines may require installation of a specific grout barrier material such as fine sand. In those situations, fine sand for the grout barrier may be installed by pouring from the surface (Figure 6.90). Some manufacturers have developed a modular grout barrier design (Figure 6.91), which may eliminate the need for gravity installation of the barrier material. Compatibility of the modular components with the site-specific contaminants and water chemistry should be verified to prevent the potential for biased sample results.

Some of the modular well systems also include a modular well seal or bentonite sleeve (Figure 6.91). The modular systems can be effective and efficient to use under the appropriate conditions. When the top of the well screen is close to or above the water table, the annular seal should be constructed of granular bentonite tremied down the annular space. Tremie installation of the annular seal and grout (Figure 6.92) will work under almost all conditions encountered in the field and provides high-integrity well construction. DP-installed monitoring wells are typically completed at the surface with the same type of flush-to-grade enclosures (Figure 6.93) or above-ground protective casing used for conventional monitoring wells.

Comparisons of DP Wells to Conventional Wells

Results of Recent Research

Several studies have been published comparing the performance of DP monitoring wells to conventional drilled monitoring wells, with respect to their ability to provide representative samples. However, the representativeness of ground-water samples collected from any device is difficult to demonstrate because of the complexity and heterogeneity of the system being monitored and sensitivity of many analytes to the sample

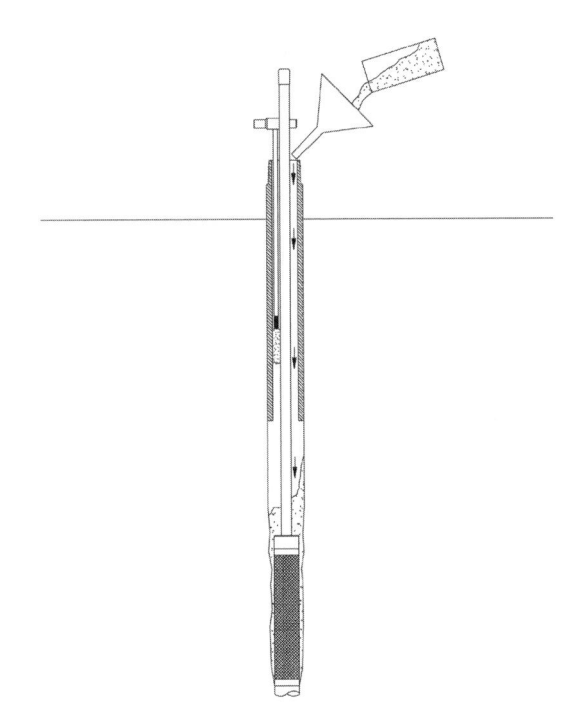

FIGURE 6.90
If formation collapse does not occur above the screens, fine sand may be installed to build a barrier above the screened interval. This barrier prevents grout slurries from entering and altering the water quality in the screened zone.

collection process itself (Puls and Barcelona, 1996). Thus, there is no absolute standard against which to measure the performance of DP-installed wells. By default, and really the only feasible option at present, results from DP wells are compared to results from conventionally installed monitoring wells. To be valid, comparisons between these two technologies must carefully consider a number of complex site-specific issues. Until relatively recently, no studies have been designed to account for the number of variables that make comparisons between the two technologies difficult or to provide sufficiently rigorous statistical analysis of sample analytical results to overcome the spatial and temporal uncertainties in the data. Particularly difficult in earlier studies was accounting for the fact that most DP wells, by design, had shorter well screens than conventionally installed wells and thus the samples from these wells were more depth-discrete, whereas samples collected from conventional monitoring wells, with longer screens, were more representative of conditions averaged over larger intervals. However, several field comparisons have overcome this problem through superior experimental design and have demonstrated conclusively that DP wells and conventionally installed wells of similar construction produce samples that are essentially equivalent.

Kram et al. (2001b) conducted a number of detailed statistical analyses of samples collected from several sets of co-located DP-installed wells and conventionally (HSA) installed wells, with screens of equivalent lengths installed at the same depths in an alluvial formation contaminated by a large MTBE plume. Statistical comparisons of analytical results for MTBE and suites of major ions (cations including Ca, Na, K, Mn, Mg, Fe, and Ba; anions including SO_4, NO_3, Cl, and Fl) from the two types of wells showed no significant performance differences and no strong systematic variations based on well type. In this study, spatial and temporal variations in chemical concentrations were considerably larger than any variations associated with well type.

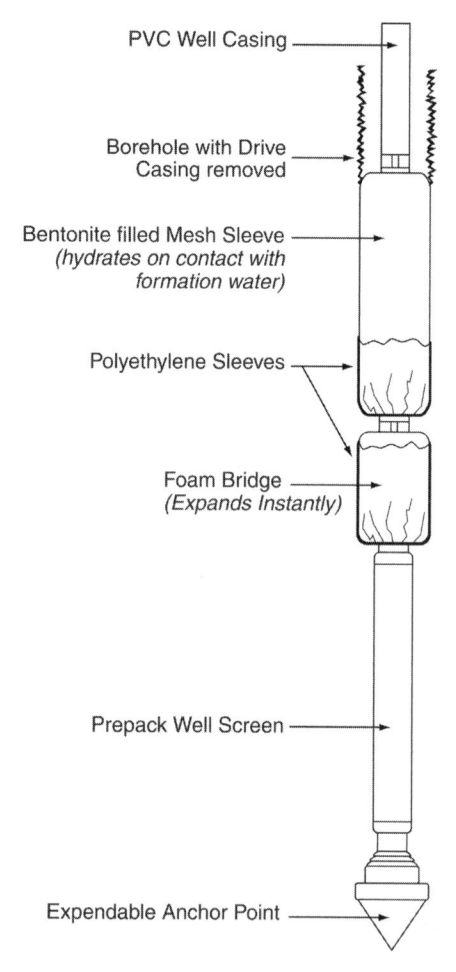

FIGURE 6.91
Foam bridge type modular well seal expands to fill probe hole following retraction of drive casing. (Courtesy of GeoInsight Inc.)

McCall et al. (1997) compared analytical results for several chlorinated VOCs (CCl$_3$, CCl$_4$, TCE, PCE, 1,1-DCE, 1,2-DCA, and 1,1,1-TCA) for samples collected from five paired sets of wells. Co-located, DP-installed 0.5 in. (13 mm) diameter PVC prepacked screen wells and HSA-installed 2 in. (51 mm) diameter PVC wells with nearly equivalent screen lengths were installed in an alluvial aquifer. Analytical results for the paired wells (Figure 6.94) showed a very good correlation ($r^2 = 0.82$, and slope $= 0.98$) for the VOCs between the DP-installed and HSA-installed wells, indicating that DP wells provide water-quality samples that are at least the equivalent of those from conventional wells.

McCall (2000b) conducted a study to evaluate the performance of 0.5 in. (13 mm) prepacked screen wells relative to 2 in. (51 mm) diameter HSA-installed PVC wells for metal analytes. The primary analyte of concern was hexavalent chromium, but data for total and dissolved iron, barium and chromium and several major cations (Ca, Na, and Mg), and sulfate were also evaluated. Results of the total chromium, dissolved chromium, and hexavalent chromium analyses (Figure 6.95) revealed a very good correlation between the paired wells for each of these parameters. The correlation coefficients (r^2) were >0.99 for each chromium species and regression line slope values were between 0.90 and 0.95 for

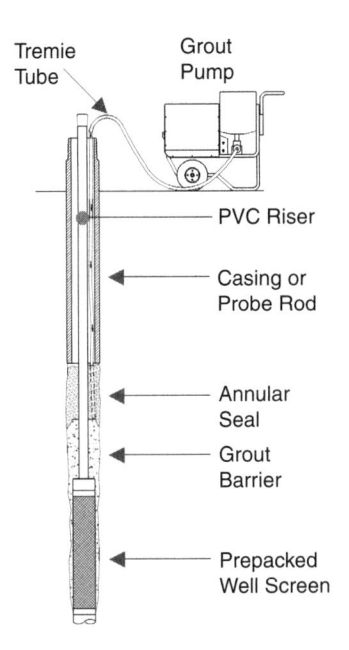

FIGURE 6.92
Use of a high-pressure grout pump and nylon tremie tube makes it possible to perform bottom-up grouting in the small annular spaces of DP equipment. Grout slurries of 20 to 30% bentonite or neat cement are most commonly used to meet state regulatory requirements.

the various chromium species. Results for the major cations (Figure 6.96) also indicated very good correlation between the DP and HSA well designs. For both the dissolved and total metals analyses of these cations, the correlation coefficients were >0.98 and the regression line slope values were >0.96, indicating excellent agreement for results from the paired wells. The variance between DP wells was consistently lower than between the HSA wells, indicating that DP wells can yield ground-water samples of at least equivalent quality and probably higher quality than conventional wells when they are properly designed and installed. Turbidity measurements (Figure 6.97) also indicated that turbidity was lower in the DP wells than in the HSA wells. This may have been partly due to the use of coarser filter pack sand in construction of the HSA wells.

Farrington et al. (2003) conducted detailed statistical analyses to compare analytical results for 13 VOCs (PCE, TCE, *cis*-1,2-DCE, *trans*-DCE, vinyl chloride, benzene, toluene, ethylbenzene, *o*-xylene, *m*-xylene, *p*-xylene, 1,4-DCB, TCA, and MTBE) and 18 inorganic water-quality parameters in samples obtained from DP and conventional wells. The study included a range of DP well sizes (0.5 to 2 in. diameter) and packing arrangements at five field sites representing a variety of geologic settings. DP wells were installed adjacent to existing HSA drilled wells and the relative well locations and sampling sequences were carefully designed to eliminate potential systematic effects. With one exception (*cis*-1,2-DCE) (in 0.5 in. diameter wells), differences in concentrations between DP wells and HSA wells were found to be either not statistically significant or the DP wells were found to be more conservative than HSA wells (e.g., they produced higher VOC results). No significant differences were detected for any of the 18 inorganic parameters.

Farrington et al. (2003) also conducted a series of seven slug tests on each of a cluster of five DP and HSA wells at one site to identify the influence of different well geometries and installation techniques on the measured hydraulic conductivity of the formation. The data indicated that the DP wells produced more consistent results than the HSA wells and

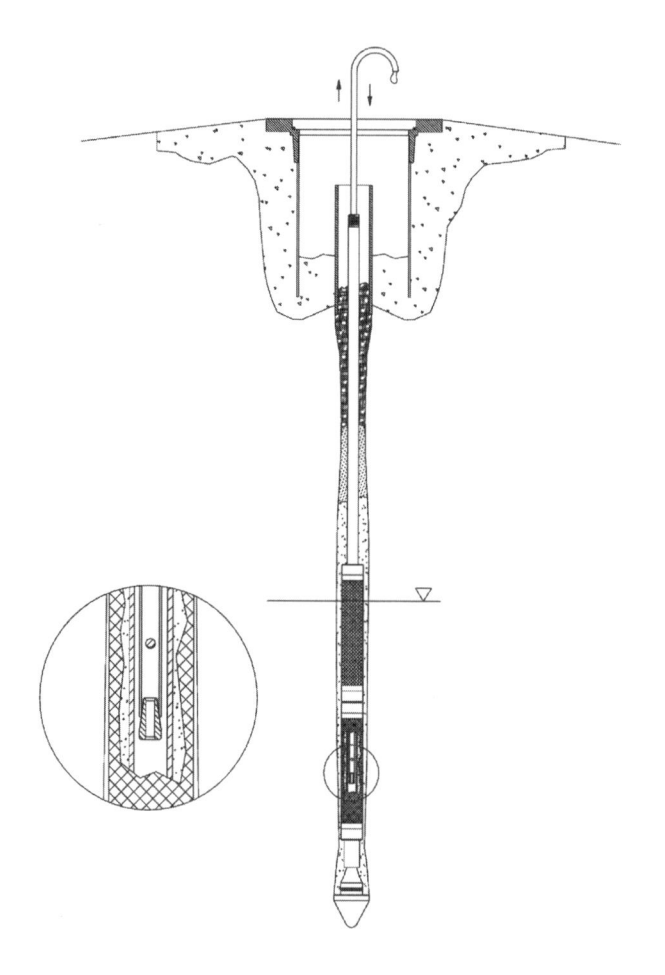

FIGURE 6.93
Conventional types of flush-mount or above-ground well protection are used to protect DP installed wells from physical damage or tampering. Small locking well caps are also available for even 0.5 in. nominal PVC casing. The insert shows the use of an inertial-lift pump (check valve and check ball) to purge and develop small-diameter wells. Well development is an essential part of the construction process and assures that representative, low-turbidity samples are obtained.

found agreement with other literature (Butler et al., 2002) concluding that small-diameter wells appear to constrict flow, reducing the apparent hydraulic conductivity of formation materials with hydraulic conductivities exceeding 70 m/day. The data from this study added that this phenomenon may occur at hydraulic conductivities as low as 22 m/day.

Other research projects have evaluated the use of DP-installed monitoring wells for ground-water quality sampling, but the experimental designs for some of these studies are not as well thought out, nor are the data analyses as rigorous as those in the studies cited earlier. Despite this, each of these studies (e.g., Foster et al., 1995; Thornton et al., 1997; BP Corporation North America, 2002) indicates that the differences in data quality between DP-installed wells and conventionally installed wells are statistically insignificant.

Applications and Limitations of DP for Well Installation

DP well-installation methods are designed for use in unconsolidated formation materials ranging from clays and silts to sands and gravels, and they work well in these settings.

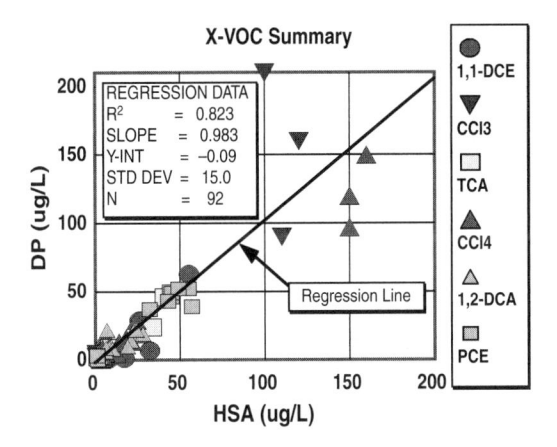

Use of Direct-Push Technologies in Environmental Site 455

FIGURE 6.94
Results from five sets of paired DP and HSA installed wells. DP wells have 1.4 in. O.D. prepacked screens with nominal 0.5 in. PVC casing and HSA wells are 2 in. PVC. The wells have similar screened intervals and are completed at similar depths (60 ft) in a sandy alluvial aquifer (Smoky Hill River). The slope of the regression line and correlation coefficient (R^2) indicate that the wells are providing equivalent quality samples. (After McCall et al., 1997. With permission.)

They are not designed for penetration of consolidated bedrock, sediments that contain large cobbles or boulders, or cemented soils such as caliche, which may be difficult or impossible to penetrate with DP tools. A little background research on local geology will clarify whether DP well installation methods are applicable for a particular site.

DP-installed monitoring wells generally range in size from 0.5 to 1 in. (13 to 25 mm) diameter, although some 1.5 and 2 in. (38 and 51 mm) wells can be installed with larger tooling. The smaller size of these wells usually precludes the use of electric submersible

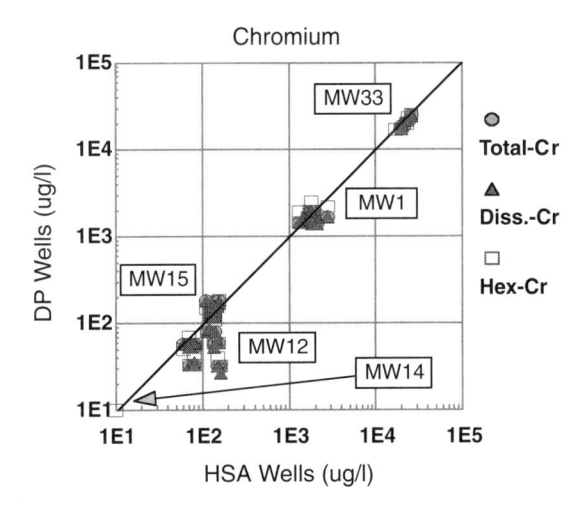

FIGURE 6.95
Plot of chromium results from five sets of paired DP and HSA wells from a site in the Arkansas River alluvial aquifer. DP wells are constructed with prepacked screens with nominal 0.5 in. PVC casing and HSA wells are constructed with nominal 2 in. PVC via 8.25 in. O.D. augers. Results for total and dissolved metals analyses and hexavalent chromium are plotted. The correlation coefficients are better than 0.99 and regression line slopes are between 0.90 and 0.95 for chromium analyses. These results indicate that the DP installed wells are providing equivalent quality samples as compared to conventionally installed monitoring wells for the sensitive chromium species. (After McCall, 2000. With permission.)

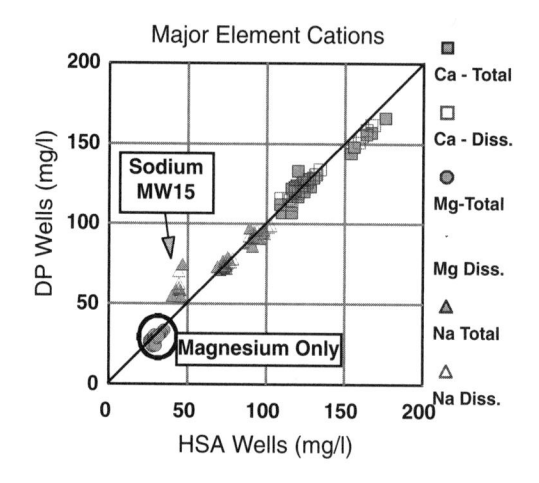

FIGURE 6.96
Plot of major cation results from five sets of paired DP and HSA wells from the Arkansas River alluvial aquifer site. DP wells are constructed with prepacked screens with nominal 0.5 in. PVC casing and HSA wells are constructed with 2 in. nominal PVC via 8.25 in. O.D. augers. Results for both total and dissolved metals analyses are plotted. The correlation coefficients are better than 0.98 and regression line slopes above 0.96 for both the total and dissolved metals analyses. These results again indicate that the DP installed wells are providing equivalent quality samples as compared to conventionally installed monitoring wells. (After McCall, 2000. With permission.)

pumps, but other equally or more appropriate devices are available in small diameters for ground-water sampling. The small diameter of most DP wells may also restrict the use of some down-hole measurement devices such as multiparameter water-quality sondes and conventional geophysical logging tools.

Each of the DP well installation methods has advantages and limitations for its use. Some of these are discussed in the following sections.

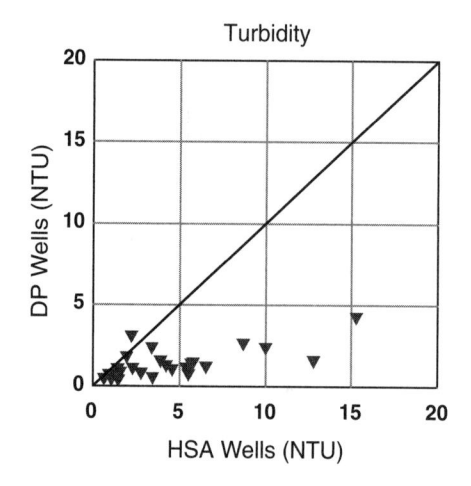

FIGURE 6.97
Turbidity measurements were made on samples collected from the paired DP and HSA wells installed at the Arkansas River alluvial site. Turbidity is reported in nephelometric turbidity units (NTU). U.S. EPA RCRA guidance specifies that turbidity be less than 5 NTU in environmental wells. Coarser filter media was used in constructing the HSA wells and resulted in higher turbidity levels in these wells. When properly designed and developed, the DP prepacked screen wells can yield low turbidity samples. (After McCall, 2000. With permission.)

Driven Well Point and Mandrel-Driven Screens

- Quick and inexpensive installation method.
- Grouting is not possible, so the well seal may not be effective.
- No filter pack is used in construction, so well development may be a prolonged process and sample turbidity in some silty formations may be problematic.
- Exposed screens are subject to damage, clogging, and cross-contamination during installation.
- Because of the inability to grout, these wells are generally not acceptable under state regulations as long-term monitoring wells, but they may be acceptable as temporary piezometers.
- Few or no drill cuttings are produced and minimal quantities of liquid waste result from decontamination, well development, and sampling.

Open Hole

- Quick, easy, low-cost installation method.
- Potential for sloughing of contaminated materials down the open hole during installation is high; therefore, well and sample integrity may be suspect and potential for cross-contamination is increased.
- Limited control during installation of filter pack and potential for formation collapse on bare screen can result in turbid samples.
- Limited control during installation of annular seal and grout and potential for formation collapse can compromise seal integrity and increase potential for migration of contaminants down hole.
- Few or no drill cuttings are produced and minimal quantities of liquid waste result from decontamination, well development, and sampling.

Cased Hole

- Somewhat more time consuming and expensive than simpler DP methods, but still offers significant savings over conventional drilling methods.
- Provides highest integrity well construction and sample quality of all the DP methods.
- Smaller wells require more sophisticated high-pressure grouting equipment to conduct bottom-up tremie grouting.
- Tremie grouting provides a high-quality seal, high-integrity construction, and higher confidence in ground-water sample quality.
- Allows for deeper well construction than other DP methods because of the cased hole installation procedure.
- Few or no drill cuttings are produced and minimal quantities of liquid waste result from decontamination, well development, and sampling.

Summary

Of the various DP well construction methods discussed earlier, only the cased-hole methods meet the general and specific grouting requirements of the state regulations, U.S. EPA guidance, and ASTM standards. This includes naturally developed wells (when formation conditions are acceptable), filter-packed wells, and prepacked screen wells. Drive-point wells, open-borehole wells, and mandrel-installed wells do not meet current state, EPA, or ASTM recommendations for annular seal or grouting requirements.

Field research indicates that properly installed DP monitoring wells provide representative water-level measurements and ground-water quality samples for organic and inorganic analytes, as well as water-quality indicator parameters such as DO, ORP, pH, specific conductance, and turbidity. Additional field studies have demonstrated that DP monitoring wells may be used to conduct slug tests to provide accurate estimates of formation hydraulic parameters (Henebry and Robbins, 2000; Butler et al., 2002). Properly constructed DP monitoring wells can be used to accomplish the same objectives as conventionally installed wells. Their uses include:

- Water-quality sampling and monitoring for RCRA, CERCLA, UST, Monitored Natural Attenuation, Brownfields, and other environmental investigations
- Detection and assessment monitoring programs at RCRA facilities
- Water-level and hydraulic head measurements for construction of water-table and potentiometric surface maps
- Determination of formation hydraulic conductivity for use in ground-water flow velocity calculations
- Monitoring progress of remedial actions

Additionally, because DP methods generate significantly less solid and liquid waste than conventional drilling methods during well installation, well development, and sampling, significant cost savings and reduction in exposure hazards are usually gained.

Methods for Sealing DP Probe Holes

Regulations that require sealing or decommissioning of boreholes vary significantly from state to state but every state has some minimum requirement. Additionally, the U.S. EPA has established guidelines for grouting of boreholes and wells under the RCRA program (U.S. EPA, 1986). ASTM has also published a Standard (ASTM, 2004v) that provides information on appropriate grouting procedures and materials for decommissioning wells and boreholes. In some states, the regulations require any borehole that penetrates 10 ft (3 m) or more below grade to be grouted. Other states may require grouting only if the water table is penetrated by the boring. Regardless, if probe holes are not properly sealed, the site investigator may be held liable for migration of contaminants from the surface down the open hole or from one horizon to another within the open hole. Fines can be as high as several thousand dollars per day per improperly sealed borehole. This provides a strong incentive to take the time and effort required to properly decommission or seal probe holes when work is completed.

State regulations also vary regarding the type and density of grout required and the method of placement that must be used. Most often, a grout slurry of bentonite or neat cement is recommended; some states suggest adding as much as 5% bentonite to neat cement for grouting purposes. Specifications for grout slurries are usually given in percent weight of dry material to water. To make up a 25% solids bentonite slurry, it is necessary to mix about 2.8 lb (1.3 kg) of bentonite per gallon (3.8 l) of water. Neat cement grouts are usually prepared by mixing 6 to 7 gal (22.7 to 26.5 l) of water with each 94 lb (42.6 kg) bag of ASTM C 150 Portland cement (ASTM, 2004w). Methods for placement of grout range from simply pouring the slurry down the borehole to placement

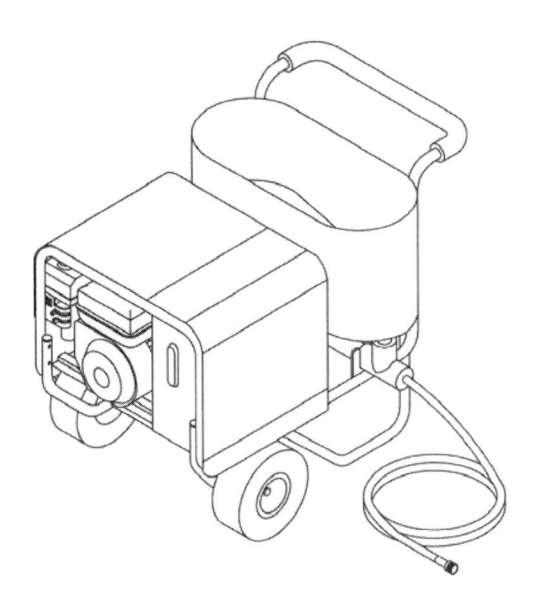

FIGURE 6.98
High-pressure grouting equipment is used to pump slurries of bentonite or cement through small-diameter tremie tubes to complete bottom-up grouting. Grout slurries may also be injected directly through drive rods after sampling activities are completed.

with a grout pump and side-port tremie tube through the drive casing or drill string (Lutenegger and Degroot, 1995). Some states permit pouring bentonite chips, pellets, or granules through drilling rods or augers to construct a seal above a monitoring well screen when the top of the screen is near or above the water table. However, this practice may be difficult or impractical in many of the smaller diameter DP tools.

For many applications in small-diameter DP tools, a high-pressure piston pump (Figure 6.98) is required to successfully place the viscous slurries required by some regulations. During installation of small-diameter DP monitoring wells, tremie tubes as small as 0.25 in. I.D. × 0.375 in. O.D. (0.6 cm I.D. × 0.9 cm O.D.) are required. To withstand the pressures generated during tremie grouting (500 to 1000 psi), nylon tremie tubes are often used.

Gravity Pouring

When manual methods were used to drive soil-gas sampling tools to depths of 4 or 5 ft (1.2 or 1.5 m) and later extract the tools, it was common practice to pour granular or powdered bentonite down the open probe hole to create a barrier to prevent downward migration of contaminants. Usually a gallon (3.8 l) of water was added as each foot or two (0.3 or 0.6 m) of bentonite was placed to ensure that the material hydrated and sealed the probe hole. This method may still be useful in shallow probe holes in which cohesive soils are present. As DP methods improved and larger tools were advanced to greater depths, gravity pouring often did not provide an acceptable seal and higher integrity methods for grout placement were developed.

Re-Entry Grouting

When single-rod soil sampling is conducted, the borehole is open to depth unless sloughing of noncohesive materials occurs. One simple way to seal this type of probe hole from the bottom up is to re-enter the boring with a string of drive rods equipped with an

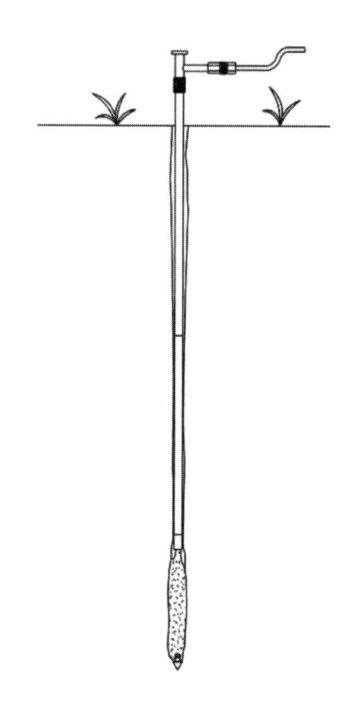

FIGURE 6.99
Re-entry grouting is most often conducted after single-tube soil sampling is completed. A secondary tool string with an expendable point is advanced back down the probe hole and grout is injected directly through the drive rods to conduct bottom-up decommissioning.

expendable point on the lead rod (Figure 6.99). A grout slurry is prepared and pumped directly through the rod bore using a grout pull cap to fill the probe hole from the bottom to the surface. Pressure from the grout pump pushes the expendable point out as the tool string is retracted. As each rod is removed from the tool string, the grout cap is placed on the succeeding rod and the retraction process continues as grout is pumped through the rods. This is continued until the last rod is removed from the probe hole. This grouting method provides an excellent procedure for placement of viscous grout slurries to depths of 100 ft (30 m) or more because of the high pressures that can be supplied through the drive rods. In cohesive formations in which the borehole remains open to depth, this method can provide a high-integrity seal. The primary limitation of re-entry grouting is that the borehole may partially or completely collapse after the tool string is removed if the formation is poorly cohesive. If collapse occurs, the secondary tool string may not follow the original borehole, so the integrity of the seal may not be satisfactory.

Retraction Grouting

This method is often used for grouting when soil-gas or ground-water samples are collected with DP methods or dual tube soil or ground-water sampling is conducted. After the sample is collected, grout is pumped directly through the primary rod string (Figure 6.100). This eliminates the need to remove the primary rod string and insert a secondary rod string to conduct grouting, as is done in re-entry grouting. Not only is this more time efficient, but it also eliminates any concerns with regard to achieving a high-integrity seal in the original probe hole. Retraction grouting can be effective at depths exceeding 100 ft (30 m), but the field operator must take steps to insure that heaving sands do not clog the probe rods before injection of the grout begins. The sampling operation must leave an open tool string in place for this method to be used

Use of Direct-Push Technologies in Environmental Site 461

FIGURE 6.100
Retraction grouting may be conducted through single-tube or dual-tube sampling systems. As shown, bottom-up grouting with a tremie tube may be conducted through single-tube ground-water samplers equipped with a grout plug. Alternatively, grout is pumped directly through the primary drive rods. This method provides a high-integrity seal using either a bentonite or cement slurry.

for grouting. Single-rod soil sampling is an example of a sampling method that could not be followed by retraction grouting through the primary rod string.

Advancement Grouting

Some specialized DP sampling or logging methods use advancement grouting to assist with penetration through the formation and to seal the bore hole as tooling is advanced. The best example of advancement grouting is when grout is injected behind a CPT cone to lubricate the borehole and maximize the potential depth of penetration (Figure 6.101). This is possible with CPT systems that do not use cables to transmit the data signal to the surface, but instead use an acoustic signal through the drive rods. In this situation, grout may be pumped down the open bore of the drive rods as the CPT cone is advanced for testing. A ported adapter behind (up hole of) the cone allows the grout to be injected in the probe hole just behind the cone for lubrication of the borehole walls. This also minimizes or prevents the potential for any down-hole migration of contaminants during the logging process. Grouting as the CPT rods are retracted is recommended to be sure a high integrity seal is achieved. Advancement grouting is a relatively specialized method used for limited applications but it can provide the capability to grout to depths exceeding 100 ft (30 m) when the need arises.

Tremie Grouting

To ensure the highest integrity placement of grout during well construction, many states require the use of the tremie grouting method. This method can also be used to grout DP

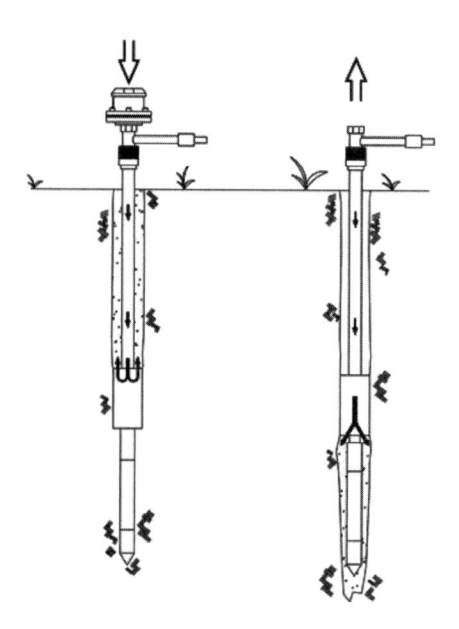

FIGURE 6.101
Advancement grouting is a specialized method most often used during CPT. A specially designed grouting tool is placed above the CPT probe so a grout slurry may be injected as the tool string is advanced. This minimizes friction between the drive rods and formation and helps prevent downward movement of any contaminants during logging operations. Grout is usually injected during the retraction operation to assure that a high-integrity seal is obtained for decommissioning.

probe holes after the rods and tools are removed from the hole. A 0.5 to 1 in. (13 to 25 mm) diameter tremie tube (usually PVC pipe) is lowered to the bottom of the open probe hole, and a grout slurry is prepared and pumped under pressure down the tremie tube (Figure 6.102). This method assures that undiluted grout is placed from the base of the hole to the top of the hole. Grouting can be done either with the tremie down hole during grout placement or as the tremie is removed from the hole. The disadvantage of using this method is that in noncohesive formation materials, the borehole will often collapse (particularly below the water table) and not allow insertion of the tremie to the total depth of the hole.

Tremie grouting may also be done through the center of DP tool strings as they are removed from the probe hole. Much smaller diameter tremie tubes are required for this application, as the tube must fit into the opening in the probe rods, which may be as small as 0.5 in. (13 mm). Tremie tubes as small as 0.375 in. O.D. by 0.25 in. I.D. (0.9 cm O.D. × 0.6 cm I.D.) are available for use in grouting through DP tools. These tremie tubes are usually made of nylon or other materials that can withstand pressures of 1000 psi or greater, which occur when thick grout slurries are pumped through more than 50 ft (15 m) of the small-diameter tube. High-pressure grout pumps (Figure 6.98) and nylon tremie tubes make it possible to conduct bottom-up tremie grouting in small-diameter DP rods and in monitoring wells installed with DP equipment (Figure 6.102).

Summary

The techniques that may be used for properly decommissioning DP probe holes range from simply pouring bentonite chips, pellets, or granules down shallow open holes to high-integrity grouting of deep DP holes using a tremie tube to conduct bottom-up

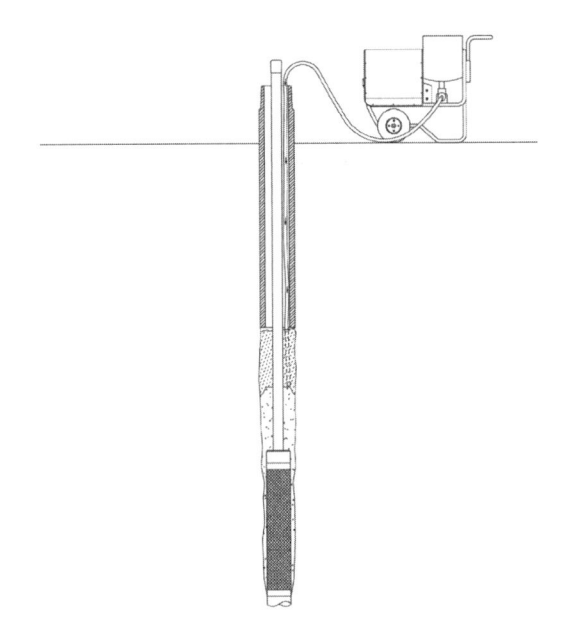

FIGURE 6.102
The use of small-diameter nylon tremie tubes and high-pressure grouting equipment enables bottom-up grouting during installation of small-diameter wells with DP methods. High-integrity seals are obtained with 20 to 30% solids bentonite slurries or neat cement grouts.

grouting. Pumping grout slurries directly through small-diameter DP drive rods as they are retracted also provides an effective method for decommissioning small-diameter probe holes. For some DP sampling techniques (e.g., soil-gas and ground-water sampling), grout may be pumped through the primary rod string. In other situations, especially single-rod soil sampling, a secondary string of drive rods must be driven back down the original probe hole to conduct bottom-up pressure grouting. State and local regulations should be consulted for clarification of grouting requirements and appropriate grout mixes. Bentonite slurries can dry and crack if used above the water table in arid climates. Some research has also demonstrated that bentonite materials can dessicate and crack in the presence of non-aqueous phase petroleum hydrocarbons and solvents. Neat cement grout may be more appropriate in these situations.

References

Adams, J.W. and G. Robitaille, The Tri-Services Site Characterization and Analysis Penetrometer System (SCAPs), U.S. Army Environmental Center Report SFIM-AEC-ET-TR-99073, U.S. Army Environmental Center, Aberdeen Proving Ground, MD, 2000, 40 pp.

Applegate, J.L. and D.M. Fitton, Rapid Site Assessment Applied to the Florida Department of Environmental Protection's Drycleaning Solvent Cleanup Program, *Proceedings of the Superfund XVIII Conference*, Vol. II, E.J. Krause and Associates, Washington, DC, 1997, pp. 695–703.

ASTM, Standard Method of Deep Quasi-Static Cone and Friction-Cone Penetration Tests of Soil, ASTM Standard D 3441, ASTM International, West Conshohocken, PA, 2004a, 7 pp.

ASTM, Standard Guide for Direct-Push Water Sampling for Geoenvironmental Investigations, ASTM Standard D 6001, ASTM International, West Conshohocken, PA, 2004b, 13 pp.

ASTM, Standard Guide for Direct-Push Soil Sampling for Environmental Site Characterization, ASTM Standard D 6282, ASTM International, West Conshohocken, PA, 2004c, 19 pp.

ASTM, Standard Guide for Selection and Installation of Direct-Push Ground Water Monitoring Wells, ASTM Standard D 6724, ASTM International, West Conshohocken, PA, 2004d, 9 pp.

ASTM, Standard Practice for Direct-Push Installation of Pre-Packed Screen Monitoring Wells in Unconsolidated Aquifers, ASTM Standard D 6725, ASTM International, West Conshohocken, PA, 2004e, 15 pp.

ASTM, Standard Practice for Decontamination of Field Equipment Used at Non-Radioactive Waste Sites, ASTM Standard D 5088, ASTM International, West Conshohocken, PA, 2004f, 3 pp.

ASTM, Standard Practice for Decontamination of Field Equipment Used at Low-Level Radioactive Waste Sites, ASTM Standard D 5608, ASTM International, West Conshohocken, PA, 2004g, 8 pp.

ASTM, Standard Test Method for Penetration Test and Split-Barrel Sampling of Soils, ASTM Standard D 1586, ASTM International, West Conshohocken, PA, 2004h, 5 pp.

ASTM, Standard Practice for Thin-Walled Tube Sampling of Soils, ASTM Standard D 1587, ASTM International, West Conshohocken, PA, 2004i, 3 pp.

ASTM, Standard Test Method for Measurement of Hydraulic Conductivity of Saturated Porous Materials Using a Flexible Wall Permeameter, ASTM Standard D 5084, ASTM International, West Conshohocken, PA, 2004j, 8 pp.

ASTM, Standard Guide for Soil Gas Monitoring in the Vadose Zone, ASTM Standard D 5314, ASTM International, West Conshohocken, PA, 2004k, 31 pp.

ASTM, Standard Guide for Accelerated Site Characterization for Confirmed or Suspected Petroleum Releases, ASTM Standard E 1912, ASTM International, West Conshohocken, PA, 2004l, 20 pp.

ASTM, Standard Practice for Expedited Site Characterization of Vadose Zone and Ground Water Contamination at Hazardous Waste Contaminated Sites, ASTM Standard D 6235, ASTM International, West Conshohocken, PA, 2004m, 50 pp.

ASTM, Standard Practice for Low-Flow Purging and Sampling for Wells and Devices Used for Ground-Water Quality Investigations, ASTM Standard D 6771, ASTM International, West Conshohocken, PA, 2004n, 7 pp.

ASTM, Standard Guide for Remediation of Ground Water by Natural Attenuation at Petroleum Release Sites, ASTM Standard E 1943, ASTM International, West Conshohocken, PA, 2004o, 43 pp.

ASTM, Standard Guide for the Selection of Purging and Sampling Devices for Ground-Water Monitoring Wells, ASTM Standard D 6634, ASTM International, West Conshohocken, PA, 2004p, 14 pp.

ASTM, Standard Test Method for Performing Electronic Friction Cone and Piezocone Penetration Testing of Soils, ASTM Standard D 5778, ASTM International, West Conshohocken, PA, 2004q, 18 pp.

ASTM, Standard Guide for Using the Electronic Cone Penetrometer for Environmental Site Characterization, ASTM Standard D 6067, ASTM International, West Conshohocken, PA, 2004r, 7 pp.

ASTM, Standard Practice for Cone Penetrometer Technology Characterization of Petroleum Contaminated Sites with Nitrogen Laser-Induced Fluorescence, ASTM D 6167, ASTM International, West Conshohocken, PA, 2004s, 7pp.

ASTM, Standard Practice for Design and Installation of Ground Water Monitoring Wells, ASTM Standard D 5092, ASTM International, West Conshohocken, PA, 2004t, 14 pp.

ASTM, Standard Guide for Development of Ground-Water Monitoring Wells in Granular Aquifers, ASTM Standard D 5521, ASTM International, West Conshohocken, PA, 2004u, 15 pp.

ASTM, Standard Guide for Decommissioning of Ground Water Wells, Vadose Zone Monitoring Devices, Boreholes and Other Devices for Environmental Activities, ASTM Standard D 5299, ASTM International, West Conshohocken, PA, 2004v, 16 pp.

ASTM, Standard Specification for Portland Cement, ASTM Standard C 150, ASTM International, West Conshohocken, PA, 2004w, 8 pp.

Backhus, D.A., J.N. Ryan, D.M. Groher, J.K. MacFarlane, and P.M. Gschwend, Sampling colloids and colloid-sssociated contaminants in ground water, *Ground Water*, 31(3), 466–479, 1993.

Baerg, D.L., R.C. Storr, J.A. Cherry, and D.J.A. Smyth, Performance Testing of Conventional and Innovative Down-Hole Samplers and Pumps for VOCs in a Laboratory Monitoring Well, *Proceedings of the National Ground Water Sampling Symposium*, Washington DC, Grundfos Pumps Corp., Clovis, CA, 1992, pp. 71–76.

Baligh, M.M, and J.-N. Levadoux, Pore Pressure Dissipation After Cone Penetration, Sea Grant Program, Massachusetts Institute of Technology, Report MITSG 80-13, April, 1980.

Barcelona, M.J., J.A. Helfrich, E.E. Garske, and J.P. Gibb, A laboratory evaluation of ground-water sampling mechanisms, *Ground-Water Monitoring Review*, 4(2), 32–41, 1984.

Barcelona, M.J., J.P. Gibb, J.A. Helfrich, and E.E. Garske, Practical Guide for Ground-Water Sampling, Illinois State Water Survey Contract Report 374, Illinois State Water Survey, Champaign, IL, 1985, 94 pp.

Barker, J.F. and R. Dickhout, An evaluation of some systems for sampling gas-charged ground water for volatile organic analysis, *Ground-Water Monitoring Review*, 8(4), 112–120, 1988.

Begemann, H.K.S., The friction jacket cone as an aid in determining the soil profile, *Proceedings of the Sixth ICSMFE*, Montreal, Quebec, Canada, Vol. I, 1965, pp. 17–20.

Berzins, N.A., Use of the cone penetrometer test and bat ground-water monitoring system to assess deficiencies in monitoring well data, *Proceedings of the Sixth National Outdoor Action Conference*, National Ground Water Association, Westerville, OH, 1992, pp. 327–339.

Booth, S.R., C.J. Durepo, and D.L. Temer, Cost-Effectiveness of Cone Penetrometer Technology (CPT), Los Alamos National Laboratory Report LA-UR-93-3383, August 1993.

BP Corporation North America, Inc., Monitoring Well Comparison Study: An Evaluation of Direct-Push Versus Conventional Monitoring Wells, Study Conducted By BP Corporation, North America, Inc. and the UST Programs of U.S. EPA Region 4, Atlanta, GA and Region 5, Chicago, IL, 2002, 80 pp.

Bruel, D. and P. Kjaer, *Gas Detection Limits*, ASTM Special Technical Publication, 1994.

Buddemeier, R.W. and J.R. Hunt, Transport of colloidal contaminants in ground water: radionuclide migration at the nevada test site, *Applied Geochemistry*, 3, 535–548, 1988.

Butler, J.J., Jr., J.M. Healey, G.W. McCall, E.J. Garnett, and S.P. Loheide II, Hydraulic tests with direct-push equipment, *Ground Water*, 40(1), 25–36, 2002.

Buttner, W., W. Penrose, J. Stetter, C. Christy, and C. Nakaishi, A hand-portable instrument system for the real-time analysis of chlorinated organic compound contamination, *Proceedings of the Fourth International Symposium on Field Screening Methods for Hazardous Wastes and Toxic Chemicals*, U.S. Environmental Protection Agency, Las Vegas, NV, 1995, pp. 702–712.

Cherry, J.A., Ground-water monitoring: some current deficiencies and alternative approaches, in *Hazardous Waste Site Investigations: Toward Better Decisions*, Gammage, R.B. and Berven, B.A. Eds., Lewis Publishers, Boca Raton, FL, 1992, chap. 13, pp. 119–134.

Cherry, J.A., R.A. Ingleton, D.K. Solomon, and N.D. Farrow, Low-technology approaches for drive-point profiling of contaminant distributions, In *Proceedings of the National Ground Water Sampling Symposium*, Grundfos Pumps Corp., Clovis, CA, 1992, pp. 109–111.

Christy, T.M., A driveable permeable membrane sensor for the detection of volatile compounds in soil, *Proceedings of the Tenth National Outdoor Action Conference and Exposition*, National Ground-Water Association, Westerville, OH, 1996, pp. 169–177.

Christy, T.M. and S.C. Spradlin, The use of small-diameter probing equipment for contaminated site investigation, *Proceedings of the Sixth National Outdoor Action Conference*, National Ground Water Association, Dublin, Ohio, 1992, pp. 87–101.

Church, P.E. and G.E. Granato, Bias in ground-water data caused by well-bore flow in long-screen wells, *Ground Water*, 34, 262–273, 1996.

Colthup, N.B., L.H. Daly, and S.E. Wimberley, *Introduction to Infrared Raman Spectroscopy*, 3rd ed., Academic Press, San Diego, CA, 1990.

Connecticut DEP, Recommended Guidelines for Multilevel Sampling of Soil and Ground Water in Conducting Expedited Site Investigations at Underground Storage Tank Sites in Connecticut, LUST Trust Fund Program, Connecticut Department of Environmental Protection. Hartford, CT, 1997.

Connecticut DEP, Expedited Site Assessment: The CD, Connecticut Department of Environmental Protection, Hartford, CT, 2000.

Cordry, K.E., HydroPunch II — the second generation: a new *in situ* ground water sampling tool, *Proceedings of the Fifth National Outdoor Action Conference*, National Ground-Water Association, Dublin, Ohio, 1991, pp. 715–723.

Cordry, K.E., Powerpunch: a self completing direct push well, *Proceedings of the Ninth National Outdoor Action Conference and Exposition*, National Ground-Water Association, Dublin, Ohio, 1995, pp. 363–370.

Crumbling, D., Improving the Cost-Effectiveness of Hazardous Waste Site Characterization and Monitoring, Special Report 6, Failsafe, January, 2000, 10 pp.

Davis, J.C., *Statistics and Data Analysis in Geology, 2nd ed.*, John Wiley & Sons, Inc., Toronto, Ontario, Canada, 1986, 656 pp.

Devlin, J.F., Recommendations concerning materials and pumping systems used in the sampling of ground water contaminated with volatile organics, *Water Pollution Research Journal of Canada*, 22(1), 65–72, 1987.

Driscoll, F.G., *Ground Water and Wells*, Johnson Screens Division, St. Paul Minnesota, 1986, 1089 pp.

Edge, R.W. and K.E. Cordry, The hydropunch: an *in situ* sampling tool for collecting ground water from unconsolidated sediments, *Ground-Water Monitoring Review*, 9(3), 177–183, 1989.

Ehrenzeller, J.L., F.G. Baker, and V.E. Keys, Using cone penetrometer and hydropunch methods to collect reconnaissance ground water samples, *Proceedings of the Fifth National Outdoor Action Conference*, National Ground-Water Association, Dublin, OH, 1991, pp. 733–746.

Einarson, M.D., Enviro-Core — a new direct-push technology for collecting continuous soil cores, *Proceedings of the Ninth National Outdoor Action Conference and Exposition*, National Ground-Water Association, Westerville, OH, 1995, pp. 419–433.

Farrington, S.P. and W.L. Bratton, A thermal desorbing CPT probe for sampling volatile organic compounds, *Proceedings of the Fifth International Symposium on Field Screening Methods for Hazardous Wastes and Toxic Chemicals*, U.S. Environmental Protection Agency, Las Vegas, NV, 1997, pp. 530–545.

Farrington, S.P., M.L. Gildea, J.C. Bianchi, W.L. Bratton, and W.C. Dickerson, Development of a Wireline CPT System for Multiple Tool Usage, Final Technical Report Under DOE Award DE-AR26-95FT40366, October, 2000.

Farrington, S.P., M.L. Gildea, and J. Shinn, Demonstration/Validation of Long-Term Monitoring Using Wells Installed by Direct-Push Technologies, Final Report for the Environmental Security Technology Certification Program, April, 2003.

Foster, M., J. Stefanov, T. Bauder, J. Shinn, and R. Wilson, Piezometer installation using a cone penetrometer, *Ground-Water Monitoring Review*, 15(4), 70–73, 1995.

Frye, G., D. Gilbert, C. Colburn, R. Cernosek, and T. Steinfort, Above-ground and *in situ* field screening of VOCs using portable acoustic wave sensor (PAWS) systems, *Proceedings of the Fourth International Symposium on Field Screening Methods for Hazardous Wastes and Toxic Chemicals*, Air and Waste Management Association, Pittsburgh, PA, 1995, pp. 715–727.

Gardner, C.H.K., T.J. Dean et al., *Soil Analysis — Physical Methods*, Marcel Dekker, New York, NY, 1991, pp. 1–73.

Geoprobe Systems, Geoprobe Pre-Packed Screen Monitoring Well: Standard Operating Procedure, Geoprobe Technical Bulletin 96-2000, Kejr, Inc. Salina, KS, 1996a.

Geoprobe Systems, GS-1000 Series Grout System: Technical Sales Bulletin, Kejr, Inc., Salina, KS, 1996b.

Geoprobe Systems, Geoprobe Macro-Core® Soil Sampler: Standard Operating Procedure, Geoprobe Technical Bulletin 95-8500, Revised September 1998, Kejr, Inc., Salina, KS, 1998a.

Geoprobe Systems, Geoprobe DT21 Dual Tube Soil Sampling System: Standard Operating Procedure, Geoprobe Technical Bulletin 98-2100, September, 1998, Kejr, Inc., Salina, KS, 1998b.

Geoprobe Systems, Geoprobe Large-Bore Soil Sampler: Standard Operating Procedure, Geoprobe Technical Bulletin 93-660, Revised March, 2000, Kejr, Inc., Salina, KS, 2000.

Geoprobe Systems, Pneumatic Slug Test Kit (GW 1600) Installation and Operation Instructions, Geoprobe Technical Bulletin 19344, Kejr, Inc., Salina, KS, 2002.

Gillham, R.W., M.J.L. Robin, J.F. Barker, and J.A. Cherry, Ground-Water Monitoring and Sample Bias, American Petroleum Institute Publication 4367, 1983, 206 pp.

Henebry, B.J. and G.A. Robbins, Reducing the influence of skin effects on hydraulic conductivity determinations in multi-level samplers installed with direct-push methods, *Ground Water*, 38(6), 882–886, 2000.

Hewitt, A.D., Influence of Well Casing Composition on Trace Metals in Ground Water, Special Report 89-9, U.S. Army Corps of Engineers Cold Regions Research and Engineering Laboratory (CRREL), Hanover, NH, 1989a.

Hewitt, A.D., Leaching of Metal Pollutants from Four Well Casings Used for Ground-Water Monitoring, Special Report 89-32, U.S. Army Corps of Engineers Cold Regions Research and Engineering Laboratory (CRREL), Hanover, NH, 1989b.

Hewitt, A.D., Potential Influences of Common Well Casings on Metal Concentrations in Well Water with Low Dissolved Oxygen, Special Report 91-13, U.S. Army Corps of Engineers Cold Regions Research and Engineering Laboratory (CRREL), Hanover, NH, 1991.

Ho, James, S.Y., Effect of sampling variables on recovery of volatile organics in water, *Journal of the American Water Works Association*, 75(11), 583–586, 1983.

Houghton, R.L. and M.E. Berger, Effects of well-casing composition and sampling method on apparent quality of ground water, *Proceedings of the Fourth National Symposium on Aquifer Restoration and Ground Water Monitoring*, National Water Well Association, Worthington, OH, 1984, pp. 203–213.

Hutchins, S.R. and S.D. Acree, Ground water sampling bias observed in shallow conventional wells, *Ground-Water Monitoring and Remediation*, 20(1), 86–93, 2000.

Hvorslev, M.J., Subsurface Exploration and Sampling of Soils for Civil Engineering Purposes, U.S. Army Corps of Engineers, Waterways Experiment Station (WES), Vicksburg, MS, 1949, 465 pp.

Kram, M.L., Use of SCAPS petroleum hydrocarbon sensor technology for real-time indirect DNAPL detection, *Journal of Soil Contamination*, 17(1), 73–86, 1998.

Kram, M.L., M. Dean, and R. Soule, The ABCs of SCAPS, *Soil and Ground Water Cleanup*, May, pp. 20–22, 1997.

Kram, M.L., S.H. Lieberman, J. Fee, and A.A. Keller, Use of LIF for real-time *in situ* mixed NAPL source zone detection, *Ground-Water Monitoring and Remediation*, 21(1), 67–76, 2001a.

Kram, M.L., D. Lorenzana, J. Michaelsen, and E. Lory, Performance Comparison: Direct-Push Wells Versus Drilled Wells, U.S. Navy, Naval Facilities Engineering Service Center (NFESC) Technical Report TR-2120-ENV, NFESC, Port Hueneme, CA, 2001b, 55 pp.

Kram, M.L., A.A. Keller, J. Rossabi, and L.G. Everett, DNAPL characterization methods and approaches, Part 1: performance comparisons, *Ground-Water Monitoring and Remediation*, 21(4), 109–123, 2001c.

Lieberman, S.H. and D.S. Knowles, Cone penetrometer deployable *in situ* video microscope for characterizing subsurface soil properties, *Field Analytical Chemistry and Technology*, 2(2), 127–132, 1998.

Lieberman, S.H., G.A. Theriault, S.S. Cooper, P.G. Malone, R.S. Olsen, and P.W. Lurk, Rapid subsurface *in situ* field screening of petroleum hydrocarbon contamination using laser induced fluorescence over optical fibers, *Proceedings of the Symposium on Field Screening Methods For Hazardous Wastes and Toxic Chemicals*, U.S. Environmental Protection Agency, Washington, DC, 1991, pp. 57–63.

Lieberman, S.H., D.S. Knowles, W.C. McGinnis, M. Davey, P.M. Stang, and D. McHugh, Intercomparison of *in situ* measurements of petroleum hydrocarbons using a cone penetrometer deployed laser induced fluorescence (LIF) sensor with conventional laboratory-based measurements, *Proceedings of the Symposium on Field Screening Methods For Hazardous Wastes and Toxic Chemicals*, Vol. 2, Air and Waste Management Association, Pittsburgh, PA, 1995, pp. 1223–1235.

Lieberman, S.H., D.S. Knowles, J. Kertesz, P.M. Stang, and D. Mendez, Cone penetrometer deployed *in situ* video microscope for characterizing subsurface soil properties, *Proceedings of the Symposium on Field Screening Methods For Hazardous Wastes and Toxic Chemicals*, Air and Waste Management Association, Pittsburgh, PA, 1997, pp. 579–587.

Lieberman, S.H., G.W. Anderson, V. Games, J. Costanza, and A. Taer, Use of a cone penetrometer deployed video-imaging system for *in situ* detection of NAPLs in subsurface soil environments, *Proceedings of the Conference on Petroleum Hydrocarbons and Organic Chemicals in Ground Water*, National Ground-Water Association, Westerville, OH, 1998, pp. 384–389.

Lunne, T., P.K. Robertson, and J.J.M. Powell, *Cone Penetration Testing in Geotechnical Practice*, Spon Press, New York, NY, 1997.

Lutenegger, A.J. and D.J. DeGroot, Techniques for sealing cone penetrometer holes, *Canadian Geotechnical Journal*, 32(5), pp. 880–891, 1995.

McCall, W., Field comparison of paired direct-push and HSA monitoring wells, *Proceedings of the Battelle Conference on Natural Attenuation of Chlorinated Solvents, Petroleum Hydrocarbons, and Other Organic Compounds*, Alleman and Leeson, Eds., Battelle Press, Columbus, OH, 1999.

McCall, W., TRIP REPORT: Installation of Geoprobe Systems Pre-Packed Screen Monitoring Wells at the Cold Regions Research and Engineering Lab, Hanover NH, Geoprobe Systems Inc., Salina, KS, October, 2000a.

McCall, W., Innovative direct-push ground-water monitoring compared to conventional methods, *Proceedings of the Second International Conference on Remediation of Chlorinated and Recalcitrant Compounds*, Vol. 2, Battelle Press, Columbus, OH, 2000b, pp. 71–78.

McCall, W., S. Stover, C. Enos, and G. Fuhrmann, Field comparison of direct-push pre-packed screen monitoring wells to paired HSA 2″ PVC Wells, *Proceedings of the Superfund XVIII Conference*, Vol. 2, E.J. Krause and Associates, Washington, DC, 1997, pp. 647–655.

McCall, W., J.J. Butler, Jr., J. Healey, A. Lanier, S. Sellwood, and E. Garnett, A dual-tube direct-push method for vertical profiling of hydraulic conductivity in unconsolidated formations, *Environmental and Engineering Geoscience*, VIII(2), 75–84, 2002.

McCall, W. and P. Zimmerman, Direct-push electrical and CPT logging; An introduction, *Proceedings of the Seventh International Symposium on Borehole Geophysics for Minerals, Geotechnical and Ground-Water Applications*, The Society of Professional Well Log Analysts, Houston, TX, 2000, pp. 103–114.

Nielsen, D.M. and G.L. Yeates, A comparison of sampling mechanisms available for small-diameter ground-water monitoring wells, *Ground-Water Monitoring Review*, 5(2), 83–99, 1985.

Ohio EPA, Use of Direct-Push Technologies for Soil and Ground-Water Sampling; Chapter 15 in Technical Guidance for Ground-Water Investigations, Ohio Environmental Protection Agency Division of Drinking and Ground Water, Columbus, OH, 2005, 25 pp.

Palusky, M., L. Waters, R. Counts, and R. Jenkins, Adapting the Bruel and Kjaer multi-gas monitor Type 1302 to measure selected volatile organic compounds (VOCs) in soil, *Proceedings of the Fourth International Symposium on Field Screening Methods for Hazardous Wastes and Toxic Chemicals*, Air and Waste Management Association, Pittsburgh, PA, 1995, pp. 357–369.

Parker, L., Suggested guidelines for the use of PTFE, PVC and stainless steel in samplers and well casings, in *Current Practices in Ground Water and Vadose Zone Investigations*, D.M. Nielsen and M.N. Sara, Eds., ASTM Special Technical Publication 1118, American Society for Testing and Materials, Philadelphia, PA, 1991, pp. 217–229.

Parker, L., The effects of ground-water sampling devices on water quality: a literature review, *Ground-Water Monitoring and Remediation*, 14(2), 130–141, 1994.

Parker, L.V. and T.A. Ranney, Softening of Rigid PVC by Aqueous Solutions of Organic Solvents, Special Report 94-27, U.S. Army Corps of Engineers Cold Regions Research and Engineering Laboratory (CRREL), Hanover, NH, 1994.

Parker, L.V. and T.A. Ranney, Additional Studies on the Softening of Rigid PVC by Aqueous Solutions of Organic Solvents, Special Report 95-8, U.S. Army Corps of Engineers Cold Regions Research and Engineering Laboratory (CRREL), Hanover, NH, 1995.

Parker, L.V. and T.A. Ranney, Further Studies on the Softening of Rigid PVC by Aqueous Solutions of Organic Solvents, Special Report 95-26, U.S. Army Corps of Engineers Cold Regions Research and Engineering Laboratory (CRREL), Hanover, NH, 1996.

Parker, L.V. and T.A Ranney, Decontaminating Ground-Water Sampling Devices, Special Report 97-23, U.S. Army Corps of Engineers Cold Regions Research and Engineering Laboratory (CRREL), Hanover, NH, 1997, 20 pp.

Parker, L.V., A.D. Hewitt, and T.F. Jenkins, Influence of casing materials on trace-level chemicals in well water, *Ground-Water Monitoring Review*, 10(2), 146–156, 1990.

Paul, C.J. and R.W. Puls, Impact of turbidity on TCE and degradation products in ground water, *Ground-Water Monitoring and Remediation*, 17(1), 128–133, 1997.

Penrose, W.R., W.L. Polzer, E.H. Essington, D.M. Nelson, and K.A. Orlandin, Mobility of plutonium and americium through a shallow aquifer in a semi-arid region, *Environmental Science and Technology*, 24, 228–234, 1990.

Pitkin, S., R.A. Ingleton, and J.A. Cherry, Use of a drive-point sampling device for detailed characterization of a PCE plume in a sand aquifer at a dry cleaning facility, *Proceedings of the Eighth National Outdoor Action Conference and Exposition*, National Ground Water Association, Dublin, Ohio, 1994, pp. 395–412.

Pitkin, S.E., J.A. Cherry, R.A. Ingleton, and M. Broholm, Field demonstrations using the Waterloo ground-water profiler, *Ground-Water Monitoring and Remediation*, 19(2), 122–131, 1999.

Pohlmann, K.F., R.P. Blegen, and J.W. Hess, Field Comparison of Ground-Water Sampling Devices for Hazardous Waste Sites: An Evaluation Using Volatile Organic Compounds, EPA/600/4-90/028, U.S. Environmental Protection Agency, Office of Research and Development, Las Vegas, NV, 1990, 102 pp.

Pohlmann, K.F., G.A. Icopini, R.D. McArthur, and C.G. Rosal, Evaluation of Sampling and Field-Filtration Methods for Analysis of Trace Metals in Ground Water, EPA/600/R-94/119, U.S. Environmental Protection Agency, Office of Research and Development, Las Vegas, NV, 1994.

Powell, R.M. and R.W. Puls, Hitting the bull's-eye in ground-water sampling, *Pollution Engineering*, July, 1997, pp. 51–54.

Puls, R.W. and M.J. Barcelona, Ground-Water Sampling for Metals Analysis, EPA/540/4-89/001, Ground Water Issue, U.S. Environmental Protection Agency, Office of Solid Waste and Emergency Response, Washington, DC, 1989, 6 pp.

Puls, R.W. and M.J. Barcelona, Low-Flow (Minimal Drawdown) Ground-Water Sampling Procedures, EPA/540/S-95/504, EPA Ground-Water Issue, U.S. Environmental Protection Agency, Office of Solid Waste and Emergency Response, Washington, DC, 1996, 12 pp.

Puls, R.W., D.A. Clark, B. Bledsoe, R.M. Powell, and C.J. Paul, Metals in ground water: sampling artifacts and reproducibility, *Hazardous Waste and Hazardous Materials*, 9(9), 149–162, 1992.

de Reister, J., Electric penetrometer for site investigations, *Journal of SMFE Division, SCE*, 97(SM-2), 457–472, 1971.

Robbins, G.A. and J.M. Martin-Hayden, Mass balance evaluation of monitoring well purging, Part 1: theoretical models and implications for representative sampling, *Journal of Contaminant Hydrology*, 8, 203–224, 1991.

Robertson, P.K. and R.G. Campanella, Guidelines for the Use, Interpretation, and Application of the CPT and CPTU, Soil Mechanics Series 105, University of British Columbia, Vancouver, British Columbia, 1986, Canada.

Robertson, P.K. and R.G. Campanella, 1988, Guidelines for Using CPT, CPTU, and Marchetti DMT for Geotechnical Design, Volume II — Using CPT and CPTU Data, U.S. Department of Transportation, Federal Highway Administration, Report FHWA-PA-87-023-84-24, February 1988.

Rossabi, J., B.D. Riha, C. May, B. Pemberton, T. Jarosch, C.A. Eddy-Dilek, B. Looney, and R. Raymond, The Savannah River environmental technology field test platform: phase II, *Proceedings of the Fourth International Symposium on Field Screening Methods for Hazardous Wastes and Toxic Chemicals*, Air and Waste Management Association, Pittsburgh, PA, 1995, pp. 83–97.

Rossabi, J., B.D. Riha, J.W. Haas III, C.A. Eddy-Dilek, A.G. Lustig-Kreeger, M. Carrabba, W.K. Hyde, and J. Bello, Field tests of a DNAPL characterization system using cone penetrometer-based Raman spectroscopy, *Ground-Water Monitoring and Remediation*, 20(4), 72–81, 2000.

Schroeder, J.D., S.R. Booth, and L.K. Trocki, Cost effectiveness of the site characterization and analysis penetrometer system (SCAPS), Los Alamos National Laboratory Report LA-UR-91-4016, December, 1991.

Schulmeister, M.K., J.J. Butler Jr., D.O. Whittmore, S. Birk, J.M. Healey, G.W. McCall, S.M. Sellwood, and M.A. Townsend, A New Direct-Push Based Approach for the Chemical Investigation of Stream-Aquifer Interactions, Geological Society of America Annual Meeting, 2001 (Abstract 26024).

Schulmeister, M.K., J.J. Butler Jr., J.M. Healey, L. Zheng, D.A. Wysochi and G.W. McCall, Direct-push electrical conductivity logging for high-resolution hydrostratigraphic characterization, *Ground-Water Monitoring and Remediation*, 23(3) 52–62, 2003.

Shinn, J.D., W.L. Bratton, G. Gillispie, and R. St. Germain, Air Force Site Characterization and Analysis Penetrometer System (AFSCAPS): Laser-Induced Fluorescence Cone Penetrometer System Development and Evaluation, Air Force Civil Engineering Support Agency, Publication AL/EQ-TR-1993-0009, December, 1994.

Smolley, M. and J.C. Kappmeyer, Cone Penetrometer tests and HydroPunch sampling: a screening technique for plume definition, *Ground-Water Monitoring Review*, 11(2), 101–106, 1991.

St. Germain, R.W. and G.D. Gillespie, Real-time continuous measurement of subsurface petroleum contamination with the rapid optical screening tool (ROST), *Proceedings of the Symposium on Field Screening Methods For Hazardous Wastes and Toxic Chemicals*, Air and Waste Management Association, Pittsburgh, PA, 1995, pp. 467–477.

Taer, A.D., R.F. Farrell, and B. Ford, Use of the cone penetrometer and laser-induced fluorescence in an oil-field site characterization, *Proceedings of the Conference on Petroleum Hydrocarbons and Organic Chemicals in Ground Water*, National Ground Water Association, Westerville, OH, 1996, pp. 257–271.

Thornton, D., S. Ita, and K. Larsen, Broader use of innovative ground water access technologies, *Proceedings, Superfund XVIII Conference, Vol. II*, E.J. Krause and Associates, Washington, DC, 1997, pp. 639–646.

Tobin, M.C., *Laser Raman Spectroscopy*, John Wiley and Sons, New York, NY, 1971.

Topp, G.C. and J.L. Davis, Measurement of soil water content using time domain reflectometry (TDR): a field application, *Journal of the Soil Science Society of America*, 49, 19–24, 1985.

Topp, G.C., J.L. Davis, and A.P. Annan, Electromagnetic determination of soil water content: measurements in coaxial transmission lines, *Water Resources Research*, 16(3), 574–582, 1980.

Torstensson, B.A., Pore pressure sounding instrument, *Proceedings of the ASCE Specialty Conference on In Situ Measurement of Soil Properties, Vol. II*, ASCE, New York, NY, 1975, pp. 48–54.

U.S. DOE, Innovative Technology Summary Report: Cone Penetrometer, U.S. Department of Energy, Office of Environmental Management, Publication DOE/EM-0309, April, 1996.

U.S. DOE, Innovative Technology Summary Report: Direct Sampling Ion Trap Mass Spectrometry, U.S. Department of Energy, Office of Environmental Management, OST Reference 69, December, 1998.

U.S. DOE, Raman Probe, U.S. Department of Energy, Office of Environmental Management Report DOE/EM-0442, OST Reference 1544, 1999, 20 pp.

U.S. DOE, Fiber Optic/Cone Penetrometer System for Subsurface Heavy Metals Detection, U.S. Department of Energy Office of Science and Technology, Report OST/TMS ID 319, 2000, 21 pp.

U.S. DOE, Innovative Technology Summary Report: Wireline Cone Penetrometer for Multiple Tool Usage, U.S. Department of Energy, Office of Environmental Management, Publication DOE/EM-0617, September, 2001.

U.S. EPA, RCRA Technical Enforcement Guidance Document, Publication 9950.1, U.S. Environmental Protection Agency, Office of Solid Waste and Emergency Response, Washington, DC, 1986, 208 pp.

U.S. EPA, RCRA Orientation Manual: 1990 Edition, EPA/530-SW-90-036, U.S. Environmental Protection Agency, Office of Solid Waste, Washington, DC, 1990.

U.S. EPA, Handbook of Suggested Practices for the Design and Installation of Ground-Water Monitoring Wells, EPA/600/4-89/034, U.S. Environmental Protection Agency, Office of Research and Development, Washington, DC, 1991, 221 pp.

U.S. EPA, RCRA Ground-Water Monitoring: Draft Technical Guidance, EPA/530-R-93-001/PB93-139-350, U.S. Environmental Protection Agency, Office of Solid Waste, Washington DC, 1992.

U.S. EPA, Rapid Optical Screening Tool, EPA/540/MR-95/519, U.S. Environmental Protection Agency, Superfund Innovative Technology Evaluation Program, Washington, DC, 1995a.

U.S. EPA, Site Characterization Analysis Penetrometer System (SCAPS) LIF Sensor, EPA/540/MR-95/520, U.S. Environmental Protection Agency, Superfund Innovative Technology Evaluation Program, Washington, DC, 1995b.

U.S. EPA, Expedited Site Assessment Tools for Underground Storage Tank Sites: A Guide for Regulators, EPA 510-B-97-001, U.S. Environmental Protection Agency, Office of Underground Storage Tanks/OSWER, Washington, DC, 1997.

U.S. EPA, Seminars: Monitored Natural Attenuation for Ground Water, EPA/625/K-98/001, U.S. Environmental Protection Agency, Office of Research and Development, Washington, DC, 1998a.

U.S. EPA, Environmental Technology Verification Report — Soil Sampling Technology: Art's Manufacturing and Supply AMS Dual Tube Liner Sampler, EPA/600/R-98/093, U.S. Environmental Protection Agency, Office of Research and Development, Washington, DC, 1998b.

U.S. EPA, Environmental Technology Verification Report — Soil Sampling Technology: Clements Associates, Inc. JMC Environmentalist's Subsoil Probe, EPA/600/R-98/091, U.S. Environmental Protection Agency, Office of Research and Development, Washington, DC, 1998c.

U.S. EPA, Environmental Technology Verification Report — Soil Sampling Technology: Geoprobe Systems, Inc. Large-Bore Soil Sampler, EPA/600/R-98/092, U.S. Environmental Protection Agency, Office of Research and Development, Washington, DC, 1998d.

U.S. EPA, Environmental Technology Verification Report — Soil Sampling Technology: Simulprobe Technologies, Inc. Core Barrel Sampler, EPA/600/R-98/094, U.S. Environmental Protection Agency, Office of Research and Development, Washington, DC, 1998e.

U.S. EPA, Data Quality Objectives Process for Hazardous Waste Site Investigations: EPA QA/G-HHW Final, EPA/600/R-00/007, U.S. Environmental Protection Agency, Office of Environmental Information, Washington, DC, 2000.

Walker, S.E., Background ground-water quality monitoring: well installation trauma, *Proceedings of the Third National Symposium on Aquifer Restoration and Ground-Water Monitoring*, National Water Well Association, Worthington, Ohio, 1983, pp. 235–246.

Wiedemeier, T.H., J.T. Wilson, D.H. Kampbell, R.N. Miller, and J.E. Hansen, Technical Protocol for Implementing Intrinsic Remediation With Long-Term Monitoring for Natural Attenuation of Fuel Contamination Dissolved in Ground Water, U.S. Air Force Center for Environmental Excellence, Technology Transfer Division, Brooks Air Force Base, San Antonio, TX, 1995.

Wiedemeier, T.H., J.T. Wilson, D. Kampbell, J.E. Hansen, and P. Haas, Technical Protocol for Evaluating the Natural Attenuation of Chlorinated Ethenes in Ground Water, *Proceedings of the Conference on Petroleum Hydrocarbons and Organic Chemicals in Ground Water*, National Ground-Water Association, Westerville, OH, 1996.

Wissa, A.E.Z., R.T. Martin, and J.E. Garlanger, The piezometer probe, *Proceedings of the ASCE Specialty Conference on in situ Measurement of Soil Properties, Vol. I*, American Society of Civil Engineers, New York, NY, 1996, pp. 536–545.

Zemo, D.A., Y.G. Pierce, and J.D. Gallinatti, Cone penetrometer testing and discrete-depth ground water sampling techniques: a cost-effective method of site characterization in a multiple-aquifer setting, *Ground-Water Monitoring and Remediation*, 14(4), 176–182, 1994.

Zemo, D.A., T.A. Delfino, J.D. Gallinatti, V.A. Baker, and L.R. Hilpert, Field comparison of analytical results from discrete-depth ground water samplers, *Ground-Water Monitoring and Remediation*, 15(1), 133–141, 1995.

7

DNAPL Characterization Methods and Approaches: Performance and Cost Comparisons

Mark L. Kram

CONTENTS

Introduction	474
Descriptions of DNAPL Site-Characterization Techniques	475
Baseline Methods	475
Soil-Gas Surveys	483
Partitioning Interwell Tracer Tests	483
Radon Flux Rates	484
Back-Tracking Dissolved VOC Concentrations in Wells	484
Geophysical Surveys	485
Cone Penetrometer Testing Methods: General	485
CPT Methods: Permeable Membrane	486
CPT Methods: Hydrosparge	486
CPT Methods: Fluorescence Techniques	487
CPT Methods: GeoVIS	488
CPT Methods: LIF/GeoVIS	488
CPT Methods: Raman Spectroscopy	489
CPT Methods: LIF/Raman	489
CPT Methods: Electrochemical Sensor	489
CPT Methods: Waterloo (Ingleton) Profiler	490
CPT Methods: Co-Solvent Injection/Extraction (PIX)	490
FLUTe Membrane	491
Cost Analysis	491
Baseline Approach	501
Soil-Gas Surveys	501
Partitioning Interwell Tracer Tests	501
Radon Flux Rates	503
Back-Tracking Using Dissolved Concentrations in Wells	503
Geophysics	504
CPT Approaches: General	504
CPT Approaches: MIP	505
CPT Approaches: Hydrosparge	505
CPT Approaches: Fluorescence Techniques	505
CPT Approaches: GeoVIS	505
CPT Approaches: LIF/GeoVIS	506
CPT Approaches: Raman Spectroscopy	506
CPT Approaches: LIF/Raman	506

CPT Approaches: Electrochemical Sensor 507
CPT Approaches: Waterloo (Ingleton) Profiler 507
CPT Approaches: PIX 508
Ribbon NAPL Sampler FLUTe 508
Discussion and Conclusions 508
References 512

Introduction

Contamination of soil and ground water by the release of dense nonaqueous phase liquids (DNAPLs), including halogenated solvents, has posed serious environmental problems for many years. To be able to remediate a site contaminated with DNAPLs, it is necessary to remove or isolate undissolved (nonaqueous phase) product remaining in the subsurface. Failure to remove residual (held in the vadose zone under capillary forces) or free-phase (mobile) product may result in continued, long-term contamination of the surrounding ground water. The marginally soluble organic contaminants can partition into the aqueous phase at rates slow enough to continue to exist as a residual phase, yet rapid enough to render water supplies a threat to public health.

DNAPLs can migrate to depths well below the water table. As they migrate, they leave behind trails of microglobules in the pore spaces of the soil matrix, which effectively serve as long-term sources of ground-water pollution. Current conceptual DNAPL transport models suggest that when vertically migrating free-phase DNAPL encounters a confining layer (e.g., a competent clay or bedrock zone), it can accumulate, or "pool," and spread laterally until it encounters a fracture or an alternative path of relatively low flow resistance toward deeper zones. In addition, globules can enter pores and be held as a residual phase in capillary suspension. This complex mode of subsurface transport results in unpredictable heterogeneous distribution of nonaqueous product that is difficult to delineate. The current lack of appropriate methods for detecting and delineating widely dispersed microglobules of DNAPL product has been identified as one of the most significant challenges limiting effective cleanup of sites contaminated with these pollutants (Feenstra et al., 1996).

This chapter describes and compares many of the best approaches and methods currently used to detect and delineate DNAPL contaminant source zones and presents a cost comparison using several contrived conceptual sites, each exhibiting a particular set of physical characteristics. The objective is to allow readers to determine which site-characterization approaches are best to pursue based on site characteristics, method performance capabilities, and method costs. A distinction between specific methods and site-management approaches will be necessary when comparing options. An approach signified by a method descriptor (such as "soil gas survey" or "surface geophysics") implies that the approach includes the method as part of the overall characterization effort, as well as the necessary pre- and post-characterization activities, which might include additional methods. Selected candidate methods are grouped into sets of approaches that represent site-management options for achieving cost-effective DNAPL source zone characterization. Inherent in these characterization approaches is the goal of identifying and quantifying physical- and chemical-site characteristics that lead toward effective remediation alternatives. Approach comparisons based on the level of chemical and hydrogeologic resolution, associated costs, and the need for additional data requirements are generated to assist with selection of appropriate site remediation management options.

Although cost comparison studies have been conducted in the past by federal agencies (i.e., Federal Remediation Technologies Roundtable [Field Analysis Technologies Matrix], U.S. EPA [Hazardous Waste Cleanup Information], etc.), a method comparative analysis that includes unit costs for several model scenarios has not yet been performed. In general, costs associated with characterization activities are generated for a specific site and competitive methods are not usually directly compared under identical conditions. At sites where several methods have been compared side-by-side, bias becomes an important issue due to the heterogeneity of the soils and distribution of DNAPL, leading to inconsistent comparison conditions.

Environmental characterization efforts for contaminated sites typically evolve through a series of stages. Initially no information is available. This stage will be referred to as t_0. At t_1, some preliminary (generally nonintrusive) information becomes available, which indicates the potential for risks associated with contaminant exposure. This information would include items typically contained in a Preliminary Site Assessment. At t_2, data collection activities related to subsurface characterization are sufficient to initiate design of a remediation system. At t_3, the site is considered remediated and monitoring is established to determine whether there is further risk. At t_4, monitoring ceases and regulatory closure is achieved, thereby requiring no further action. The approaches discussed in this chapter are composed of multiple methods applied in a logical sequence with the goal of reaching stage t_2.

Descriptions of DNAPL Site-Characterization Techniques

The techniques described in this chapter were originally compared in Kram et al. (2001a, 2002). These particular techniques were selected because they have been used at several sites to identify DNAPL source zones. In addition, each of the methods described has demonstrated potential for successful DNAPL plume delineation, either directly or indirectly. Some of the methods have been extensively tested (e.g., sample collection and analysis, soil gas surveys, seismic surveys, and other geophysical surveys), while others are considered relatively new techniques (e.g., flexible liner underground technologies everting [FLUTe], ultraviolet [UV] fluorescence using a cone penetrometer, and precision injection/extraction [PIX]). Brief descriptions of the methods investigated in this effort are presented subsequently. Table 7.1 describes some of the positive and negative attributes associated with each of these site-characterization options and identifies pertinent references for obtaining additional information.

Baseline Methods

A baseline characterization method typically consists of sample collection from the surface and at frequent intervals. For simplicity, 1.5 m (5 ft) depth intervals are considered, although this is not a recommendation. Samples are typically collected using conventional drilling and split-spoon sampling equipment and are analyzed using U.S. EPA-approved methods for identifying volatile organic constituents (VOCs). Because VOCs can be easily and rapidly liberated during sample handling and transport, this can lead to an underestimation of the concentration. Several actions can be taken to improve the baseline method. For example, samples can be immediately immersed in methanol to inhibit the amount of volatilization due to handling and transport. In addition, samples can be subjected to field

TABLE 7.1
DNAPL Site-Characterization Methods

Methods	Advantages	Disadvantages	Ref.
Baseline methods			
Disposal witness	Direct evidence via observation of disposal incident	Best-guess approach for sampling location and depth; volume not easily quantifiable; generally small source quantities	Cohen and Mercer, 1993; Pankow and Cherry, 1996
Chemical analysis of soil, rock, and water samples (including fault planes in consolidated regimes)	Direct evidence; vertically continuous soil samples can lead to reliable identification; UV fluorescence, soil and water shake tests, shake tests with hydrophobic dyes, sponge coring, and swab tests can be used	Lack of reliable sampling methodology; best-guess approach for sampling location and depth and spacing; subsample selection for analysis can be biased; potential for loss of volatiles; improper collection methods can lead to vertical migration of contaminants; drilling fluids (including air) can sometimes result in loss of DNAPLs before samples are recovered; logistics for handling and transferring consolidated rock or cohesive clay samples can be complex	Cohen et al., 1992; Cohen and Mercer, 1993; Pankow and Cherry, 1996; MSE, 2002
Visual field evidence	Direct evidence via soil and fluid centrifuge, dye enhancement or field analytical results	Best-guess approach for sampling location and depth; volume not quantifiable; small source quantities require careful centimeter-by-centimeter examination	Cohen and Mercer, 1993; Pankow and Cherry, 1996
Enhanced visual identification — shake-tests	Direct evidence	Best-guess approach for sampling location and depth; volume not easily quantifiable	Cohen et al., 1992; Cohen and Mercer, 1993; Pankow and Cherry, 1996
Enhanced visual I.D. — UV fluorescence with portable light	Indirect evidence (co-mingled NAPL source)	Best-guess approach for sampling location and depth; volume not easily quantifiable; can have false positives	Cohen et al., 1992; Pankow and Cherry, 1996
Enhanced visual I.D. — dye addition with Sudan IV or Oil Red O	Direct evidence; excellent screening tool	Best-guess approach for sampling location and depth; volume not easily quantifiable; Sudan IV listed as a possible mutagen; soil type and moisture condition may influence accuracy; qualitative	Cohen et al., 1992; Cohen and Mercer, 1993; Pankow and Cherry, 1996
Vapor analysis while sampling sediments or drilling	Indirect evidence (while drilling or via head-space analysis of samples) if readings of 1000 to 2000 ppm vapor (may infer DNAPL)	Questionable vertical control; water can skew or inhibit volatile detection; false positives possible due to equipment exhaust; could liberate volatile constituents if sample integrity is disrupted; semiquantitative; drilling can lead to vertical migration of contaminants	Cohen and Mercer, 1993; Pankow and Cherry, 1996
Drilling water analysis	Indirect evidence; can help to avoid drilling through vertical lithologic barriers	Questionable vertical control; concentrations can be diluted; not quantifiably representative of subsurface conditions; some drilling methods not capable of yielding water samples that reflect composition of ground water; drilling can lead to vertical migration of contaminants	Cohen and Mercer, 1993; Pankow and Cherry, 1996; Taylor and Serafini, 1988

DNAPL Characterization Methods and Approaches: Performance and Cost Comparisons 477

Observation wells	Direct evidence if product recovered; indirect evidence if concentrated dissolved phase constituents are detected (see Back-Tracking Dissolved VOC Concentrations in Wells)	Difficult to determine DNAPL volume and vertical distribution; DNAPL may not easily flow into well, especially if present at residual saturation or if hydraulic potential of DNAPL is insufficient to overcome capillary pressure in the filter pack; relatively large DNAPL volumes must enter the boring to be detected in wells; DNAPLs that enter annulus may exit boring below end cap if formation is permeable, leading to vertical transmission of contaminants without detection in the well; sampling from bottom of the well can be logistically challenging	Cohen and Mercer, 1993; Pankow and Cherry, 1996
Test Pits	Direct evidence based on observation of materials while excavating; can be good for obtaining detailed lithologic information; can observe relationship between DNAPL distribution and lithologic characteristics	Difficult to determine DNAPL volume and vertical distribution; DNAPL may not easily flow into pit; depth limited (to approximately 5 to 8 m below ground surface); can be difficult to keep pit open in saturated conditions; potentially hazardous working conditions	Pankow and Cherry, 1996
Soil-gas surveys	Indirect evidence based on VOC detection in vadose zone; very high concentrations (approaching saturated vapor concentrations) may be indicative of DNAPLs present in vadose zone adjacent to the sampling point	Subaqueous DNAPL may not easily volatilize; not generally depth specific due to migration characteristics of materials; preferential pathways can lead to misinterpretation; poor correlations between soil-gas concentration and soil concentrations; false negatives possible since vapor concentrations can rapidly decline due to transport by diffusion	Cohen and Mercer, 1993; Marrin, 1988; Marrin and Kerfoot, 1988
Partitioning interwell tracer tests	Indirect evidence; can be used for volume estimates and evaluation of remediation method efficiency	Tracer migration may follow different pathway other than DNAPL; split flow paths and meandering can lead to inaccurate measurements; in organic rich soils, may have partitioning into organics other than DNAPL; inadequate tracer detection limits may lead to underestimation of NAPL saturations, especially in low permeability layers; tracers may not partition out of solution in low permeability soils that inhibit ground-water flow; porous-media heterogeneity and variable DNAPL saturation can decrease accuracy; an inferential volume integrating estimate	Jin et al., 1995; Nelson and Brusseau, 1996; Annable et al., 1998; Burt et al., 1998; Knox et al., 1998; Meinardus et al., 1998; Payne et al., 1998; Dwarakanath et al., 1999; Nelson et al., 1999; Wise, 1999; Yoon et al., 1999
Radon flux rates	Indirect evidence based on aqueous Rn concentration deficits due to partitioning into organic phase; rapid equilibration of Rn; passive sampling (as opposed to injection); can assist with evaluation of remedial effectiveness	Logistically difficult; lack of reliable sampling methodology; specialized sampling and analytical procedures required; site-specific NAPL to water Rn-222 partition coefficients difficult to obtain; best-guess approach for sampling location and depth; areas displaying highly variable background Rn concentrations may prove challenging; geologic factors may lead to low correlation between Rn concentration and NAPL presence	Semprini et al., 1998

(Table continued)

TABLE 7.1 Continued

Methods	Advantages	Disadvantages	Ref.
Back-tracking using dissolved concentrations in wells	Indirect evidence provided conditions are ideal (significant source volume; conditions conducive to impede dissolved contaminant degradation); spatial distribution of dissolved materials can sometimes provide information about spatial distribution of DNAPL; 10 or 1% of saturated aqueous concentration "rules of thumb" have been suggested for inferring the presence of a DNAPL phase; if ground-water samples are collected in close proximity to DNAPL zones and monitoring well intake zone is comparable to the size of the DNAPL residual zone, rule of thumb dissolved concentrations can be expected; saturated concentrations in ground water found only immediately above the source and in a thin layer at the elevation of the source in the near-downgradient area; drive-point devices used to collect detailed vertical profiles of dissolved concentrations provide the highest probability for detecting peak concentrations; extreme temporal variations in dissolved concentrations observed in a monitoring well may indicate that the well is located along the margin of dissolved plume	Concentrations may not be indicative of how close to source sample was collected; lower than 1% of effective aqueous solubility concentrations do not preclude the presence of NAPL; active sampling, spacing of monitoring wells, and well screen length may dilute concentrations; the 1% rule of thumb must be cautiously applied, since the dissolved plume emanating from large source zones may exhibit dissolved concentrations above 1% for a substantial distance downgradient of source; best-guess approach for sampling location and depth; conventional monitoring wells not likely to allow for detection of peak dissolved concentrations at DNAPL sites since well screens are generally too long, not placed in proper locations, and in insufficient quantity; highly conductive zones can demonstrate lower concentrations in course-grained materials that are well flushed	Feenstra et al., 1991; Newell and Ross, 1991; Anderson et al., 1992; Johnson and Pankow, 1992; Cohen and Mercer, 1993; Pankow and Cherry, 1996; Feenstra and Cherry, 1988
Surface geophysics	Indirect evidence based on potential migration pathways; may provide direct evidence based on acoustic or electromagnetic contrasts provided that conditions are ideal (significant product volumes; sufficient contrasts between source area and background)	Anomalies may not be indicative of contrasts between source area and background; depths to suspected source zones not known without intrusive "truth-shots"; resolution not adequate to detect ganglia on a centimeter scale or less; cultural interference possible	Cohen and Mercer, 1993; Pankow and Cherry, 1996; Adams et al., 1998; Sinclair and Kram, 1998

DNAPL Characterization Methods and Approaches: Performance and Cost Comparisons 479

Subsurface geophysics	Indirect evidence based on potential migration pathways; may provide direct evidence based on acoustic, electromagnetic, gamma, or neutron contrasts provided that conditions are ideal (significant product volumes; sufficient contrasts between source area and background and porosity and moisture content)	Anomalies may not be indicative of contrasts between source area and background; resolution not adequate to detect "ganglia" on a centimeter scale or less; cultural interference possible; porosity or moisture content can interfere with some methods (e.g., neutron logging)	Brewster et al., 1992; Cohen and Mercer, 1993; Pankow and Cherry, 1996
CPT methods			
Permeable membrane sensor; MIP	Indirect evidence based on VOC partitioning into metal–polymer membrane; can be coupled with lithologic sensors for correlation; can use different types of detectors (FID, PID, XSD, ITMS, etc.); excellent screening method with good resolution; can be deployed on smaller direct push units	When operating with a noncontinuous configuration, user required to determine appropriate depths while "on the fly," which can be difficult in zones of "ganglia"; bulk fluids can not travel across membrane; semiquantitative; clogging can occur; VOC carryover can occur in transfer line and membrane; Teflon membrane coating can be damaged by sand and gravel; ITMS cannot distinguish between analytes with identical mass fragments; limited by lithology; heat front or pressure front may inhibit membrane contact with contaminant	Christy, 1998; Myers et al., 2002 a, b
Hydrosparge	Indirect evidence based on VOC partitioning into carrier gas; can be coupled with lithologic sensors for correlation; can use different types of detectors (FID, PID, ITMS, etc.)	User required to determine appropriate depths while "on the fly," which can be difficult in zones of "ganglia"; system purge not always rapid; clogging can occur; limited by lithology	Davis et al., 1997, 1998
Florescence (e.g., LIF) techniques	Indirect evidence based on fluorescence of co-mingled materials (naturally occurring organics, multi-ring fuel compounds, etc.); rapid measurement in real-time; depth discreet signals; can be coupled with lithologic sensors for correlation; good screening method with high resolution; can use several off-the-shelf energy sources	Limited by lithology; false negatives and positives possible; co-mingled fluorophores required; semiquantitative so requires confirmation samples; not yet fully matured; detection level depends on co-mingled NAPL composition, soil type, and excitation source wavelength; pressure or heat front may force droplets away from window	Kram, 1996, 1997, 1998, 2002; Kram et al., 1997; 2001b, 2002, 2003; Kram and Keller, 2003a, b; Lieberman et al., 2000; MSE, 2002

(Table continued)

TABLE 7.1 Continued

Methods	Advantages	Disadvantages	Ref.
GeoVIS	Direct evidence based on video image processing; can be coupled with lithologic sensors for correlation; data easy to interpret in light colored soil matrix	Limited by lithology; rate of data collection limited by ability to visibly process information; transparent NAPL droplets not detectable; pressure or heat front may force droplets away from window	Lieberman and Knowles, 1998; Lieberman et al., 2000
LIF/GeoVIS	Direct evidence based on video image processing; can be coupled with lithologic sensors for correlation; data easy to interpret in light colored soil matrix; when droplets are transparent, LIF can often indirectly locate source zones	Limited by lithology; rate of GeoVIS data collection limited by ability to visibly process information; co-mingled fluorophores may be required; transparent NAPL droplets not detectable by GeoVIS; pressure or heat front may force droplets away from window	Lieberman and Knowles, 1998; Lieberman et al., 1998, 2000
Raman spectroscopy	Direct evidence based on Raman scatter; fluorescence may be due to co-mingled materials (indirect evidence for DNAPL); sensitivity may be enhanced through surface coating (requires sample in contact with substrate for this configuration)	Noncontinuous stream of data; fluorescence due to organic materials can interfere; detection threshold dependent upon probability of droplets appearing on sapphire window, amount of contaminants in sediment, type of sediment, soil moisture content, and degree of heterogeneity; pressure or heat front may force droplets away from window; detection enhancement can require longer analytical times	Mosier-Boss et al., 1997; Rossabi et al., 2000
LIF/Raman	Indirect evidence based on fluorescence of co-mingled materials (naturally occurring organics, multi-ring fuel compounds, etc.); rapid measurement; depth discreet signals; can be coupled with lithologic sensors for correlation; good screening method with high resolution; several off-the-shelf energy sources available; direct evidence based on Raman scatter	Limited by lithology; false negatives and positives possible; semiquantitative so requires confirmation samples; not yet fully matured; pressure or heat front may force droplets away from window	Kenny et al., 1999

Electrochemical sensor probe	Indirect evidence based on soil vapor; 10 ppm detection levels; sensor is small, has simple circuit requirements, low power needs, and high selectivity	Subaqueous DNAPL may not easily volatilize; not generally depth specific due to migration characteristics of materials; preferential pathways can lead to misinterpretation	Adams et al., 1997
Waterloo (Ingleton) Profiler	Indirect evidence based on use of direct-push tool to collect aqueous samples from small-spaced ports, determine hydraulic head distribution and hydraulic conductivity distribution; inverse model dissolved concentrations to generate concentration profiles, hydraulic conductivity versus concentration comparisons, and map distribution of percent ratio of dominant solvent species to total chlorinated compounds	Dissolved concentrations may not be indicative of proximity to NAPL source; concentrations lower than 1% of effective aqueous solubility do not preclude the presence of NAPL; active sampling may affect concentrations; best-guess approach for sampling location and depth; highly conductive zones can demonstrate lower concentrations in well-flushed course-grained materials; clogging can occur; limited by lithology	Sudicky, 1986; Pitkin, 1998
Co-solvent injection/extraction; PIX probe	Potential direct evidence of presence of DNAPL; can be coupled with lithologic sensors	Difficult to insure direct contact between co-solvent and DNAPL; density differences between co-solvent and DNAPL could pose challenges; best-guess approach for sampling location and depth; requires relatively long sampling times (approximately 2 h or more per sample)	Looney et al., 1998; MSE, 2002
FLUTe membrane	Direct evidence; can be deployed using CPT; good screening method with good resolution	Qualitative; requires confirmation sampling; may be difficult to apply in consolidated materials	MSE, 2002

"shake-tests" where density differences between the relatively heavier DNAPL and water are qualitatively identified. They can also be exposed to UV fluorescence with a portable meter to try to qualitatively identify potential fluorophores in an oil phase. Another useful method includes the addition of a dye such as Sudan IV or Oil Red O to help qualitatively identify separate phases. Sudan IV is a solid, hydrophobic dye which turns orange-red in the presence of NAPLs. It is also common to collect and analyze soil vapors and cutting fluids generated during drilling or to excavate a test pit and analyze soils, fluids, and vapors within the pit or along a trench wall. Some investigators place a small amount of soil or water sample in a closed container, wait for equilibration, and collect a sample of the vapors that have partitioned into the nonliquid portion in the container. This "head-space" sample is analyzed for VOC concentrations. Headspace analyses can be useful for concentration measurements and for identifying DNAPL sources. The baseline method and subsequent variations can be useful for identifying DNAPL source areas as long as the samples are collected from appropriate locations. Since globules and pathways can be extremely small, identifying DNAPL source zones with a 1.5 m (5 ft) sampling frequency can prove to be very difficult. Although commonly used, the probability for misidentifying DNAPL ganglia and microglobules using this sampling frequency is considerable. As the sampling frequency is increased, the probability of detection increases, but the cost also increases significantly. In addition, using this method as part of a site-characterization approach may not be sufficient to reach the t_2 characterization stage. Permeability tests, well installations, and evaluation of residual-phase versus free-phase product may be necessary depending on site conditions and initial findings.

Pore-water concentration can be used to evaluate whether constituent concentrations exhibited by the pore water and soil are indicative of DNAPL presence. If it is assumed that the sample is in equilibrium with respect to partitioning between solid, fluid, and vapor phases, then the pore-water concentration (C_w) can be expressed relative to the soil concentration (C_t) of a particular constituent by the following relationship (Pankow and Cherry, 1996):

$$C_w = \frac{C_t \rho_b}{K_d \rho_b + \theta_w + H_c \theta_a} \quad (7.1)$$

where ρ_b, dry bulk density of the soil sample (g/cm^3); θ_w, water-filled porosity (volume fraction); θ_a, air-filled porosity (volume fraction); K_d, water–soil partition coefficient for compound of interest (cm^3/g); and H_c, Henry's law constant for compound of interest (dimensionless).

If the value C_w for a given sample is near the solubility limit of the component in water (or the estimated effective solubility of a particular component from a mixture), DNAPL is presumed to be located in the vicinity of the sample collection location. The largest uncertainty in C_w is generally caused by uncertainty in the estimate of K_d, which depends upon the fraction of organic carbon content present, the octanol–water partition coefficient for the compound of interest, and complications due to unaccountable sorption, intergranular diffusion, dissolved organic matter, and co-solvency. Using equilibrium calculations, Feenstra et al. (1991) show that at soil concentrations greater than several thousand milligrams per kilogram trichloroethylene (TCE) in a sandy soil, the majority of the component is present in the DNAPL phase. Because several components are generally present, solubility estimates are not easy to determine with a high level of certainty. In addition, depending on the rate of ground-water flow, dissolved component concentrations adjacent to NAPL can vary considerably. Therefore, there is a significant need to search for superior, more direct NAPL detection methods than those categorized as baseline.

Soil-Gas Surveys

Soil-gas surveys have been successfully used to screen DNAPL sites for more than a decade (Marrin, 1988; Marrin and Kerfoot, 1988). Soil-gas surveys consist of insertion of soil vapor collection devices into the subsurface (most commonly using a direct-push approach), application of a slight vacuum to the soil, collection of a vapor sample, and on-site measurement of VOCs in the vapor sample using a gas chromatograph. Because these methods can only be used in the vadose zone, they are typically used to try to identify DNAPL source release areas.

Using soil-gas surveys as part of a site-characterization approach may not be sufficient to reach the t_2 characterization stage. Permeability tests, well installations, and evaluation of residual-phase versus free-phase product may be necessary depending on site conditions and initial findings. Soil-gas survey data can help determine optimal locations for further intrusive efforts beneath the water table.

Partitioning Interwell Tracer Tests

The partitioning interwell tracer test (PITT) is based on transport properties of several tracers, each with different partitioning characteristics (Jin et al., 1995). At least one of the tracers is nonreactive (e.g., nonpartitioning and nonabsorbing) with respect to the DNAPL organic liquid, while the other tracers partition, to various levels, into the organic liquid. The organic liquids retain the partitioning tracers and retard their migration, thereby leading to differential recovery times corresponding to the strength of partitioning and amount of DNAPL encountered (Nelson et al., 1999). In practice, identification of a DNAPL zone is necessary prior to setting up the PITT. In addition, a forced flow-field is established to transport tracers across the zone of investigation. A suite of tracers is introduced to the subsurface within a target DNAPL zone and recovered from a different location, typically using injection and recovery wells. Breakthrough curves are generated for each tracer depicting concentration versus time for a particular recovery well. The conservative, nonabsorbing tracer is initially recovered, followed by the partitioning tracers. DNAPL saturation calculations depend upon determination of a retardation factor for each tracer, which is typically calculated using a comparative moment analysis with the nonreactive tracer (Jin et al., 1995). In the simplest case, the retardation factor equals the quotient of the travel times of the partitioning (t_p) and nonpartitioning (t_{np}) tracers. For two phases, the retardation factor (R), DNAPL water-partition coefficient (K_{dw}), and DNAPL saturation (S_n) are related by the following equation:

$$R = 1 + \left[\frac{S_n}{(1-S_n)}\right] K_{dw} = \frac{t_p}{t_{np}} \tag{7.2}$$

This relationship assumes no tracer sorption is occurring. In addition, partition coefficient variability due to differences in NAPL composition can introduce errors in the estimation of NAPL saturation (Dwarakanath et al., 1999). It is also important to recognize that thin fractures in Karst can skew the results due to the behavior of DNAPLs in fractured media (Keller et al., 2000).

An approach that includes the PITT technique requires several additional components to reach the t_2 characterization stage (e.g., preliminary identification and delineation of DNAPL zones, confirmation efforts, hydrologic control, post-PITT modeling, etc.). However, since saturation volumes can be estimated, the method affords detailed information that can be used for remedial design and evaluation of remedial effectiveness, provided the site lithology is accommodating (e.g., of medium to high permeability with low levels of organic matter) and the preliminary site-characterization efforts are sufficient.

Radon Flux Rates

Radon-222 (Rn-222) is often present as a dissolved gas in subsurface fluids. Rn-222 is a naturally occurring chemically inert radioactive gas resulting from the decay of uranium-238. As with several of the tracers used in the PITT approach, Rn has a strong affinity to partition into organic fluids. By observing a relative deficit in the aqueous Rn concentration, one can surmise that partitioning into a NAPL phase has occurred. According to Semprini et al. (1998), this Rn deficit is directly related to the residual NAPL saturation as follows:

$$\frac{C_{NAPL}}{C_{bkgrd}} = \frac{1}{1 + S_{NAPL}(K - 1)} \quad (7.3)$$

where C_{NAPL}, Rn concentration in ground water collected from a NAPL contaminated region; C_{bkgrd}, Rn concentration in ground water collected from an adjacent noncontaminated region; S_{NAPL}, residual volumetric NAPL saturation; and K, NAPL to water Rn partition coefficient (typically ranging from 20 to 70).

This equation predicts that the Rn concentration within a NAPL contaminated zone decreases when compared with a background value as the NAPL saturation increases. Due to preferential partitioning into the NAPL phase, Rn is retarded more in the presence of NAPL, which can help with evaluation of site flow characteristics. In addition, according to the model, as residual NAPL saturation increases, Rn concentration in the ground water adjacent to the NAPL will greatly decrease relative to the background Rn concentrations. This implies high sensitivity with respect to identification of suspected DNAPL locations, whereby even small quantities of residual NAPL will lead to a significant Rn deficit. Although useful as a DNAPL source zone screening and characterization method, the effect of remediation can be gaged by monitoring Rn concentrations in the treatment zone. Increases in Rn concentration can provide a semiquantitative estimate of the NAPL removed.

As with the PITT method, an approach that includes the Rn flux rate technique requires several additional components to reach the t_2 characterization stage (e.g., preliminary identification and location of DNAPL zones, possible well installations, confirmation efforts, etc.). However, since Rn-222 is already present in the subsurface, there is no need to inject materials. The technique affords detailed information that can be used for remedial design and evaluation of remedial effectiveness as long as Rn concentrations are fairly homogeneous on a local scale. Areas with radioactive waste or natural uranium deposits would not be appropriate.

Back-Tracking Dissolved VOC Concentrations in Wells

One method commonly used to help identify DNAPL source zones is to analyze dissolved VOC concentrations in monitoring wells. Johnson and Pankow (1992) and Anderson et al. (1992) describe the use of downgradient solute concentrations to locate DNAPL source zones through the application of physical and analytical models. The source zone location is estimated by observing the well pattern distribution, noting the localized ground-water flow patterns, and back-tracking upgradient against the direction of ground-water flow. Computer simulations based on measured hydrogeologic properties, such as hydraulic conductivity (or permeability), hydraulic gradient, and velocity, can be used to generate flow nets or particle tracking simulations. Flow nets and particle tracking simulations may then be used to elucidate the most probable location of DNAPL source zones. Partitioning calculations comparing pore water concentrations of components to pure phase aqueous solubility can be conducted to assess the possible presence of residual DNAPL contamination when visible evidence does not exist. While the back-tracking approach is often employed in environmental investigations, it is not considered a baseline method in this chapter, because monitoring wells are generally installed following conventional soil and water sampling and analysis approaches. An approach that includes the use of a back-tracking technique requires several additional components (e.g., permeability tests, well installations, confirmation of DNAPL presence, evaluation of residual-phase versus free-phase product, etc.) to reach the t_2 stage.

Geophysical Surveys

Several geophysical techniques have been used to try to locate DNAPL source zones (Pankow and Cherry, 1996; Adams et al., 1998; Sinclair and Kram, 1998). Geophysical surveys are not generally capable of directly detecting DNAPLs, but can assist with determination of geologic structure (Chapter 4). These types of approaches can be separated into two main categories: surface and subsurface geophysical surveys, which refer to the location of the energy source used to interrogate the subsurface and the location of receivers. Surface geophysical surveys generally consist of an energy source (transmitter or impact apparatus) and receivers located at the ground surface. Energy impulses, such as electromagnetic or acoustic impulses, are transmitted to and propagated through the subsurface, either reflected or refracted off the interfaces between layers or between materials with differing signal transmission properties, and the reflected signals are received at several locations on the ground surface. In a three-dimensional survey, a grid of geophones and energy source points are configured to generate data that are sampled from a range of different angles and distances. The signal receiving time data can be analyzed to identify anomalies which may represent possible pathways and traps for DNAPLs. The locations attributed to the anomalies are discernible by the signal return times recorded at each receiver. Confirmation samples must be collected from discrete depths. Therefore, wave propagation rates (acoustic or electromagnetic) for each of the subsurface strata must be known to convert time increments to units of length. Subsurface geophysical surveys are similar to surface geophysical surveys except that they are more intrusive in that the source or receivers may be located below the ground surface. As with most screening methods, confirmation drilling, sampling, and analyses are essential to determine whether the anomalies are representative of pathways containing DNAPLs. Therefore, all geographical methods are components requiring intrusive validation approaches to reach t_2 stage when using a geophysical technique in the overall DNAPL source zone characterization approach. Spatial resolution will depend on type of method used, spacing of receivers, soil and rock types, and several additional factors. Typical resolution is on the order of meters to tens of meters, which may preclude determination of ganglia and microglobule location for most cases.

Cone Penetrometer Testing Methods: General

Cone penetrometer testing (CPT) methods consist of techniques that utilize a direct-push apparatus to deliver the sensor probes and sampling and analytical devices to depths of interest (Chapter 6). Lithology will dictate whether CPT methods can be used at a particular site. In general, CPT methods can be used at sites consisting of relatively loosely compacted, noncemented, fine- to medium-textured sedimentary deposits (ASTM D 3441; ASTM, 2004). Recent development of robust real-time *in situ* sensor technologies has revolutionized the chemical and physical site-assessment arena. The level of resolution and rapid data acquisition capabilities, coupled with simultaneous technical advances in computer capabilities, have generated new ways to represent and interpret data streams and have reduced the level of uncertainty in fate and transport models. These methods have led to revised conceptual models regarding subsurface flow, pathway configuration, and transport dynamics associated with the release of chemicals to the environment (Lieberman et al., 1998). The current trend is to mount several sensors onto a single probe, thereby allowing for coupling of lithologic and chemical information (Lieberman et al., 1991; Lieberman and Knowles, 1998). This allows for rapid site characterization and remedial design. With respect to DNAPL detection, some CPT methods provide for direct quantitative detection, while others serve as screening techniques that require confirmation analyses. The following paragraphs describe some of the most promising DNAPL detection methods that use the CPT as a delivery platform.

CPT Methods: Permeable Membrane

The permeable membrane interface probe (MIP) was developed to allow for near real-time evaluation of subsurface volatile constituents (Christy, 1998). The MIP probe consists of a thin composite (metal and polymer) membrane mounted along the outside of a push rod which is in contact with a carrier gas line within the probe. The carrier gas line can be connected to several types of detectors, including flame ionization detectors, photo-ionization detectors, ion trap mass spectrometers (ITMSs) (Myers et al., 2002a, b), and a halogen-specific detector (XSD). The probe can be advanced with either a hammer probe or a hydraulic ram system. In practice, the MIP membrane is heated to 80 to 125°C as it is advanced through the subsurface. VOCs present in the subsurface can partition into the membrane and migrate through it by diffusive flux. Once through the membrane, the carrier gas sweeps the VOCs to the detector. Rapid site characterization can be possible with this tool. When combined with soil classification sensors, such as soil conductivity, piezocone, or electronic load cells, correlations between subsurface lithology, contaminant location, and migration potential can be generated. Confirmation samples will be required when using the MIP probe for DNAPL source zone evaluation. However, a DNAPL source zone characterization approach incorporating the MIP probe technique, when coupled with lithologic sensors, will allow investigators to rapidly reach the t_2 stage.

CPT Methods: Hydrosparge

The Tri-Services (U.S. Army, U.S. Navy, and U.S. Air Force) Site Characterization and Analysis Penetrometer System (SCAPS) Hydrosparge system integrates a customized CPT probe with a small sampling port, a sparging device, and an aboveground detector situated in the truck. The probe is advanced into the groundwater to a target depth and a liquid sample is allowed to enter the sampling port. A direct sparging device bubbles inert helium carrier gas through the sample to purge VOCs. The stripped VOCs are carried to the surface for analysis using an ITMS or gas chromatograph and mass spectrometer. The ITMS Hydrosparge system has demonstrated good correlation ($R^2 = 0.87$) with EPA Method 8260 for dissolved halogenated contaminant concentrations ranging from one to several thousand nanograms per milliliter (Davis et al., 1997). Confirmation samples will be required when using the Hydrosparge probe for DNAPL source zone evaluation. However, a DNAPL source zone characterization approach incorporating the Hydrosparge probe technique, when coupled with lithologic sensors, will allow investigators to rapidly reach the t_2 stage.

CPT Methods: Fluorescence Techniques

Molecular absorption in the UV and visible regions of the electromagnetic spectrum is dependent on the electronic structure of the organic molecule. Absorption of energy is quantized, resulting in the elevation of electrons from orbitals in the ground state to higher energy orbitals in an excited state. When the excited electrons cascade toward the ground state, light energy is released as fluorescence emission spectra, which can be readily measured and analyzed (Silverstein et al., 1991). While a detailed treatment of the theoretical basis for electronic spectra is beyond the scope of this section, it is important to note that compounds consisting of double carbon bonds with weakly attached electrons (specifically polyaromatic hydrocarbons) are susceptible to excitation and can therefore be identified using low energy bombardment techniques. Several relatively low energy lasers and lamps are capable of exciting electrons from polyaromatic fuel hydrocarbons. Source energy will dictate which specific analytes and compounds can be detected. In general, lighter aromatic compounds, such as benzene, require relatively higher energy sources (290 nm wavelength or lower) for detection, while many laser sources (e.g., 290 nm sources from dye lasers, 308 nm excimer lasers, and 337 nm nitrogen lasers) are well suited for polyaromatic compound detection. Kram (2002) demonstrated that an optimal excitation source for obtaining superior detection levels can be determined for specific analytes and mixtures, that available systems can be ranked according to detection capabilities, and that higher energy excitation sources do not necessarily yield superior detection levels. Several techniques have been utilized to analyze the absorption and emission spectra to identify specific analytes and suites of analytes (e.g., fuel types) and to determine relative fluorophore concentrations based on fluorescence intensity (Kenny et al., 1999).

Several energy sources, such as lasers and mercury lamps, have been deployed using the CPT platform. For instance, the Tri-Service SCAPS deploys a fiber optic based laser induced fluorescence (LIF) chemical detection system, which allows for real-time, *in situ* subsurface detection of polyaromatic fuel hydrocarbon contaminants (Lieberman et al., 1991). Naturally occurring organic materials, such as humic and fulvic acids, will also fluoresce when exposed to the SCAPS laser system. UV light is launched into a silica clad optical fiber connected to a sapphire window mounted flush with the outside of the penetrometer rod. A second fiber is used to collect the fluorescent emission energy from the soil in contact with the sapphire window as the probe is advanced and returns it to the detector system at the surface. While the SCAPS LIF system is not presently capable of directly detecting pure DNAPLs, many of the hydrocarbon constituents SCAPS can detect are miscible with DNAPLs and can become commingled with these materials. For instance, TCE is often used to clean oil-soaked metal parts. SCAPS is often capable of detecting many of the polyaromatic compounds in the oil. Because petroleum hydrocarbons are miscible with DNAPLs, they can be carried to depths beneath the water table. Detection of hydrocarbons located at depths beneath the water table can assist with the delineation of subaqueous free phase DNAPLs at sites where both contaminants are present (Kram, 1996). Keller and Kram (1998) and Kram (2002) have demonstrated that fluorophore (i.e., fluorescing compounds) concentrations less than 1% in chlorinated solvent are detectable with currently available instrumentation. The SCAPS LIF system has been used to indirectly locate DNAPL source zones at several sites by identifying co-mingled petroleum constituents beneath the water table (Kram, 1998; Kram et al., 2001b;

Kram, 2002). LIF techniques have also been coupled with other types of sensors (e.g., Raman, GeoVIS, load cells, etc.) for analytical and visible confirmation and for identifying potential contaminant pathways. Confirmation samples will be required when using the LIF probe for DNAPL source zone evaluation. However, a DNAPL source zone characterization approach incorporating the LIF probe technique, when coupled with lithologic and visible confirmation sensors, will allow investigators to rapidly reach the t_2 stage.

CPT Methods: GeoVIS

The GeoVIS is a real-time, *in situ*, microscopic soil video imaging system developed by the U.S. Navy (Lieberman et al., 1998). The system consists of a miniature charge-coupled device video camera coupled with magnification and focusing lens systems integrated into a CPT platform. Soil in contact with the probe is illuminated with an array of white light emitting diodes and imaged through a sapphire window mounted on the probe. The video signal from the camera is returned to the surface and displayed in real-time on a video monitor, recorded on a videocassette recorder, and can be captured digitally with a frame grabber installed in a microcomputer system. The digital image data can be incorporated into the SCAPS operation and data processing software to allow for depth-specific video clip recall. In its current configuration, the system images an area that is 2×2.5 mm^2, providing a magnification factor of approximately $\times 100$ when viewed on a 13 in. monitor. This particular system can be advanced at a rate of approximately 4 in./min. A newer system has been developed for advancing a probe capable of delivering a 5×6.5 mm^2 image at a rate of approximately 18 in./min. For DNAPL investigations, the GeoVIS probe has been pushed into soils known to yield fluorescence responses using a LIF probe. In addition, the GeoVIS has been combined with a standard LIF probe (see description subsequently). For the GeoVIS to be most successful, a recognizable color or textural contrast must exist between the DNAPL globules and the soil matrix. The GeoVIS has helped researchers better conceptualize the complex multicomponent contaminant transport regime and may soon allow for more realistic modeling scenarios. New data processing approaches are under development to estimate NAPL saturation, as well as soil porosity. Confirmation samples will be required when using the GeoVIS probe for DNAPL source zone evaluation. However, a DNAPL source zone characterization approach incorporating the GeoVIS probe technique, when coupled with chemical and lithologic sensors, will allow investigators to rapidly reach the t_2 stage. In 1999, the combined LIF/GeoVIS probe was used to map out a commingled fuel and chlorinated hydrocarbon NAPL plume, establish remediation design parameters, and served as the tool of choice for evaluating remedial effectiveness at the former Naval Air Station at Alameda, California (Lieberman et al., 2000; Udell et al., 2000).

CPT Methods: LIF/GeoVIS

The coupling of direct push sensors can provide for conclusive evidence of the presence of DNAPL. For instance, the use of LIF coupled to the GeoVIS and soil lithology sensors has been successfully demonstrated at several sites (Lieberman et al., 1998, 2000; Kram et al., 2001b). In practice, each of the sensor systems collects *in situ* data that are displayed in real-time. Correlation between indirect DNAPL identification using the LIF and direct detection GeoVIS information has been strong where visible contrasts between soil color and DNAPL color are recognizable and where co-mingled LNAPL and DNAPL materials are present. This "dual-probe" system has challenged conventional conceptual models regarding configuration of DNAPL plumes, depths of plume migration in seemingly unobstructed alluvial deposits, and liberation of breakdown products based on either biological or inorganic processes. Confirmation samples may be required when using the LIF/GeoVIS probe for chlorinated DNAPL source zone evaluation. However, a DNAPL source zone characterization approach incorporating the LIF/GeoVIS probe technique, when coupled with lithologic sensors, will allow investigators to rapidly reach the t_2 stage.

CPT Methods: Raman Spectroscopy

Raman spectroscopy is similar to LIF in that it relies on the detection of light wavelength shifts from compounds of interest. However, the Raman technique is capable of direct identification of several chlorinated DNAPL constituents, while the fluorescence techniques are not. As mentioned earlier, fluorescence spectra are the result of an electronic transition caused by the quantum absorption and subsequent release of energy (in the form of light) as an excited electron cascades towards lower energy states. Raman spectroscopy is used to detect light inelastically scattered from the incident radiation and does not involve a radiative process dependent upon electron energy transition. Energy shifts in the scattered light are correlated to the vibrational modes of particular compounds, so compound-specific spectra are generated. The number of vibrational modes and associated energies of these modes are unique to each compound. When performing Raman spectroscopy with a monochromatic light source such as a laser, both fluorescence and scattering occur. The fluorescent signal can potentially obscure the Raman spectrum. Because fluorescence emission is fixed in wavelength, the incident light source wavelength is often altered to elucidate the Raman signal. Standard signal processing (i.e., edge detection and filtering) has also been effective at extracting the Raman signal out of a fluorescent background (Mosier-Boss et al., 1997). A Raman device has been coupled to a CPT platform and successfully used to identify subsurface DNAPL constituents by their unique spectral signatures at the Savannah River Site in Aiken, South Carolina (Rossabi et al., 2000). Confirmation samples will be required when using the Raman probe for DNAPL source zone evaluation. However, a DNAPL source zone characterization approach incorporating the Raman probe technique, when coupled with lithologic sensors, can allow investigators to reach the t_2 stage.

CPT Methods: LIF/Raman

The coupling of LIF/Raman techniques into a direct-push probe has proven useful at a former drycleaner site in Jacksonville, Florida (Kenny et al., 1999). The probe can lead to successful identification of DNAPL source zones whether the DNAPL is pure, co-mingled with petroleum based compounds, or co-mingled with naturally occurring organic materials. Incorporation of soil lithology sensors into this probe can be useful for identifying potential migration pathways and for generating remediation approaches. Confirmation samples will be required when using the LIF/Raman probe for chlorinated DNAPL source zone evaluation. However, a DNAPL source zone characterization approach incorporating the LIF/Raman probe technique, when coupled with lithologic sensors, will allow investigators to rapidly reach the t_2 stage.

CPT Methods: Electrochemical Sensor

Electrochemical sensors that respond to chlorine have been used to detect chlorinated hydrocarbon organic vapors in soils (Adams et al., 1997). In practice, the probe is advanced to the maximum depth of interest (generally based on probe soil sensors). Soil vapors pass through a vapor sampling port in contact with the soil and are pneumatically transported to the sensor inside the probe. Chlorine gas levels are measured as the probe is retracted to the surface. The sensor is calibrated on a periodic basis to allow for semiquantification. The sensor signal is proportional to the chlorine concentration in the vapors. Using electrochemical sensors as part of a vadose zone characterization approach may not be sufficient to reach the t_2 characterization stage. Permeability tests, well installations, and evaluation of residual-phase versus free-phase product are necessary depending on site conditions and initial findings. Electrochemical sensor data can help determine optimal locations for further intrusive efforts beneath the water table.

CPT Methods: Waterloo (Ingleton) Profiler

The Waterloo Profiler was developed at the University of Waterloo Center for Ground Water Research (Pitkin, 1998). The system consists of a stainless-steel drive point with small-diameter (typically 0.156 in.) circular ports fitted with 25-mesh stainless-steel screen. The ports are each connected to a common reservoir in the tip of the profiler, which is connected to a delivery system composed of stainless-steel tubing within the profiler and a peristaltic pump at the surface. The system allows for ground-water sampling from several discrete depths with inch-scale vertical resolution. In addition, depth-discrete aquifer testing can be conducted to generate a vertical profile of hydraulic conductivity and hydraulic head. The device has been successfully used to map DNAPL plumes by profiling in transects normal to the axis of the hydraulic gradient (Pitkin, 1998). In addition, solute concentrations along transects and hydraulic head profiles have been used to "back-track" to identify potential DNAPL source areas upgradient of the profiling regions. Confirmation samples will be required when using the Waterloo Profiler for DNAPL source zone evaluation. However, a chlorinated DNAPL source zone characterization approach incorporating the Waterloo Profiler technique, when coupled with analysis of depth discrete solute concentrations, piezometric head values, and estimates of hydraulic conductivity, will allow investigators to reach the t_2 stage.

CPT Methods: Co-Solvent Injection/Extraction (PIX)

The co-solvent injection/extraction (or PIX) method functions by solubilizing, mobilizing, and recovering the NAPL in contact with either a single well or a specialized probe. In practice, the probe is advanced to a target depth (or a well is fitted with a packer to isolate the screened zone). A known amount of water with a tracer of fixed concentration is injected a few inches into the formation and recovered by overextraction (extracting a larger volume than was originally injected). Then a known amount of alcohol is injected and overextracted. Differences in component concentrations, alcohol concentrations, and tracer concentrations are compared to determine the potential presence of DNAPL using a mass balance approach (Looney et al., 1998). Variations include the incorporation of lithologic sensors to help identify candidate DNAPL zones based on potential migration conduits. The target depth is identified, the advancement of the probe is suspended, and the test is conducted. This technique has been successfully implemented at the Interagency DNAPL Consortium site in Cape Canaveral, Florida (MSE, 2002). Confirmation samples will be required when using the PIX technique for DNAPL source zone evaluation. However, a DNAPL source zone characterization approach incorporating the PIX technique, when coupled with relative permeability data resulting from extraction observations, can allow investigators to reach the t_2 stage.

FLUTe Membrane

The FLUTe device consists of an inflatable membrane used to deploy a hydrophobic absorbent ribbon to the subsurface. The ribbon is forced against the side of a borehole or penetrometer push hole in zones of suspected DNAPL contamination. If DNAPLs are present, they will wick into the ribbon. The membrane device is retracted using a tether connected to the deepest portion of the liner, and the ribbon is visually inspected and analyzed for DNAPLs (MSE, 2002). Analysis consists of extraction and measurement of the concentration of contaminants from the ribbon or use of Sudan IV dye. A Sudan IV-impregnated ribbon was successfully used at the Savannah River Site in Aiken, South Carolina and at the Cape Canaveral Air Station in Florida (MSE, 2002). The membrane was deployed using a cone penetrometer whereby the rods were advanced to a target depth, the membrane was strung through the rods, and the membrane was incrementally inflated as the rods were retrieved. Confirmation samples may be required when using the FLUTe technique for chlorinated DNAPL source zone evaluation. However, a DNAPL source zone characterization approach incorporating the FLUTe technique, when coupled with lithologic information and permeability analyses on soil samples, can allow investigators to reach the t_2 stage.

Cost Analysis

Comparable cost data for DNAPL site-characterization methods and approaches are limited. Rarely, several methods are compared to each other on a systematic basis at the same site. Typically, when data are available for a particular approach or method, it is usually compared to a set of confirmation data collected and analyzed using standardized field laboratory methods. The data collection locations for confirmation samples are typically dictated by previous results. For example, when one uses a field screening technique, confirmation samples are collected from locations identified as polluted or clean based on the field-screening method results. Because each method and approach varies in terms of spatial resolution and completeness with respect to requirements for remedial design, corresponding confirmation approaches will also vary. Due to the lack of comparable cost data, the lack of resources for conducting method comparisons in the field under various scenarios, and the differences associated with confirmation approaches anticipated for particular methods, selected DNAPL site-characterization methods and approaches were evaluated using synthetic site scenarios. Three "unit model scenarios" (UMSs) were used to compare the selected site-characterization techniques and approaches.

Descriptions of the three UMSs and specific parameters are presented in Table 7.2. Although the scenarios are not comprehensive, they provide a general framework for technology evaluation. The scenarios each represent sites with relatively shallow water tables. Cost estimates can be adjusted by normalizing (e.g., based on depth, area, or estimated contaminant volume) or by adjusting the assumptions presented for each approach (Table 7.3). For instance, each UMS consists of volumes of approximately 785,400 ft^3 (22,250 m^3). The "depth of resolution" values refer to the maximum depth of characterization required. For an equal volume with a 1 acre (4,047 m^2 or 43,560 ft^2) footprint at the surface, the depth of resolution would be approximately 18 ft (5.5 m). As described in Table 7.2, the following is considered:

- A depth of resolution of 100 ft (30.5 m) for each UMS.
- All releases initiated at the same time and within 10 yr of the initial investigation.

TABLE 7.2
Synthetic UMSs and Predetermined Parameters

Unit Model Scenario	Map Area	Depth to Ground water/Depth of Resolution	Soil Type	Volume of DNAPL
1	50 ft (15.2 m) radius	15 ft and 100 ft (4.6 m/ 30.5 m)	Alluvial, medium to fine grained	264 gal (1000 l)
2	50 ft (15.2 m) radius	15 ft/100 ft (4.6 m/ 30.5 m)	Gravel or bedrock deposits	264 gal (1000 l)
3	50 ft (15.2 m) radius	15 ft/100 ft (4.6 m/ 30.5 m)	Karst	264 gal (1000 l)

- NAPL penetrated the subsurface to depths beyond the water table.
- DNAPL is distributed heterogeneously within the UMS volume, with the majority located between approximately 65 to 75 ft (19.9 and 22.9 m) below ground surface (identified using the screening and confirmation efforts).
- Depth to ground water, depth of resolution, and volume of DNAPL released are identical for each UMS.

The main cost differences between the approaches were due to differences in soil type, which has an effect on the potential for data or sample accessibility and resolution due to lithologic properties (competence, penetrability, acoustic or electromagnetic signal transmission, etc.).

This portion of the section presents cost analyses for each of the methods discussed earlier. To generate a useful cost comparison, several cost and approach assumptions were required. These assumptions are presented in Table 7.3 and discussed subsequently. Each approach was compared to a common baseline approach, which consists of sample collection from the surface and from consecutive discrete 1.5 m (5 ft) depth intervals. The intent is not to imply that a 1.5 m (5 ft) level of resolution is valid for all sites; rather this was a sampling increment considered representative of typical investigations. Although commonly used, the chance for misidentifying DNAPL ganglia and microglobules using this type of approach is very high. In addition, penetration of zones containing free-phase DNAPL using the baseline approach could lead to vertical migration of contaminants to deeper zones, exacerbating the problems associated with the release. Table 7.4 presents cost estimates and an estimate of savings based on comparisons with baseline approaches. A negative savings value indicates that the approach is more costly than the baseline approach. Where possible, references to previous studies were incorporated into the cost analyses for each scenario.

It is important to recognize that each method (or approach component) presents specific advantages and disadvantages and that due to the nature of each method and the sequence with which it can be applied in the overall site management process, direct comparisons involve some uncertainty. A project manager who knows little about the location of DNAPL at a site yet is interested in the most cost-effective approach, must place each candidate method in the proper context within the characterization process. Comparison of characterization components in isolation tends to bias the cost estimate, thereby rendering the comparison fallible. In an attempt to normalize the comparison, each method is evaluated in a manner consistent with the niche fulfilled. It is assumed that little is known and that the project manager wants to obtain enough information to determine whether the site is clean or how to properly design a remediation system based on specific

TABLE 7.3
Approach and Cost Assumptions

Approaches	Approach Assumptions	Cost Assumptions
Baseline approaches	Collection and analysis of 21 soil samples per hole, for five holes; field observations based on shake-tests, UV lamp, addition of Sudan IV, drill cutting fluids soils and vapors, and other screening activities conducted simultaneously with sample collection activities; chemical and physical laboratory analyses; mobilization–demobilization; reporting	$1,500/day for the drilling equipment; $10/1 ft (0.3 m) for drilling (UMS 1); $20 per sample for collection; $3/1 ft (0.3 m) for grouting; $1,000/day for mobilization and demobilization; $300/day for per diem for a three-person crew; $170/h for standby labor; $100/h for decontamination; $40/1 ft^3 (0.03 m^3) for drilling waste disposal (approximately 35 ft^3 (1 m^3)/8 in. [20.3 cm] diameter hole); $150 per sample for laboratory chemical analyses; $200 per sample (15 total) for laboratory physical analyses; $3,000 for reporting the results (including boring logs and chemical data)
Soil-gas surveys	25 pushes to 15 ft in 5 × 5 grid; pore vapor samples collected every meter (approximately 3 ft; five samples per push); confirmation sampling and analyses from two pushes (three soil samples each) to 20 ft (6.1 m); mobilization-demobilization for push rig; two additional sample borings (using a drill rig) collecting samples at 5 ft (1.5 m) intervals from 20 to 100 ft (6.1 to 30.5 m); chemical and physical laboratory analyses; mobilization–demobilization of drill rig; reporting	$150 per push; $20 per soil-gas sample analyzed; $1/ft for grouting (1,500 ft total); $200/day for mobilization and demobilization of the push rig; $200/day for per diem for a two-person push rig crew (6 days); $170 per confirmation sample (6 total) for collection and analyses; $1,500/day for the drilling equipment; $10/1 ft (0.3 m) for drilling; $20 per sample for collection; $3/1 ft (0.3 m) for grouting; $1,000/day for mobilization and demobilization; $300/day for per diem for a three-person crew; $170/h for standby labor (3 h); $100/h for decontamination (10 h); $40/0.03 m^3 (1 ft^3) for drilling waste disposal (approximately 35 ft^3 [1 m^3]/8 in. [20.3 cm] diameter hole; two drilled holes); $150 per soil sample (48 total) for laboratory chemical analyses; $200 vper sample (6 total) for laboratory physical analyses; and $3,000 for reporting results (chemical data and a map depicting VOC plumes indicative of potential DNAPL vadose zone sources)
Partitioning interwell tracer tests	Initial field screening (via FLUTe) and confirmation efforts; eight wells (four injection, three extraction; one monitoring) to 75 ft (22.9 m); laboratory efforts to determine tracer attributes; aquifer tests; conservative tracer tests; hydraulic control; modeling; PITT; mobilization–demobilization; reporting	$28,800 to deploy the FLUTe (UMS 1) and collect and analyze six soil samples (chemical and physical analyses); $4,000 (80 h at $50/h) for laboratory tests to assess initial residual saturation, select candidate tracers and determine corresponding partition coefficients; $1,500/day for 4 days for the eight wells drill rig

(*Table continued*)

TABLE 7.3 *Continued*

Approaches	Approach Assumptions	Cost Assumptions
		expenses for well installation efforts; $100/h (2 h for each well) for well development; $300/day for field work (6 days) and assessment (3 days) for a conservative tracer test; $19,520 for well development disposal costs and $1,600 for aquifer testing waste disposal; $2,000 for GAC treatment equipment; $3,500 for mobilization, demobilization, treatment system breakthrough analyses (18 water samples total [two sample per day]), operational supplies, and discharge permits; $10,000 for tracer breakthrough analytical expenses; $4,000 for modeling expenses for PITT design and flow regime assessment; PITT labor expenses of $400/day (plus $200/day for per diem) over the period of 9 days for hydraulic control, injection of tracers for 0.5 days, injection of potable water for 6 days to flood and recover tracers, sampling, and on-site chemical analyses; reporting (summary of the preliminary characterization efforts, well logs, aquifer test results, analytical results, interpretation, and modeling results) costs of $5,000
Radon flux rates	Initial field screening (via FLUTe) and confirmation efforts; five monitoring wells to 75 ft (22.9 m); an aquifer pump test; sampling for radon; mobilization–demobilization; modeling; reporting	$28,800 to conduct a FLUTe survey (UMS 1) and collect and analyze confirmation soil samples; $1,500/day for drill rig expenses for well installation plus $10/1 ft (0.3 m) for the five wells described earlier; approximately $1,000/day for 2 days for mobilization and demobilization; $100/h for 2 h each well (10 h total) for well development; $2,800 for disposal of wastes for well installation and $1,400 for aquifer testing waste disposal; approximately $200 each for Rn analytical costs for five samples (total cost $1,000 per round); approximately $3,000 (60 h at $50/h) for modeling expenses (displaying Rn data distribution superimposed on aquifer test data); $300/day (for 3 days total) for the Rn survey labor; $4,000 for reporting (to include an overall data summary package, estimate of residual NAPL saturation distribution, presentation of modeling results, and well design descriptions [construction and development details])

(Table continued)

TABLE 7.3 *Continued*

Approaches	Approach Assumptions	Cost Assumptions
Back-tracking using dissolved concentrations in wells	Collection and analysis of 21 soil samples per hole, for five holes; installation of five monitoring wells to 100 ft (30.5 m); one round of ground-water sampling at ten depths per well; mobilization–demobilization; reporting	$1,500/day for the drilling equipment; $10/1 ft (0.3 m) for drilling (UMS 1); $20 per sample for collection; $3/1 ft (0.3 m) for grouting; $1,000/day for mobilization and demobilization; $300/day for per diem for a three-person crew; 75 ft (22.9 m) of blank PVC riser at a rate of $3/1 ft (0.3 m); 425 ft (129.5 m) of PVC screen at a rate of $4/1 ft (0.3 m); 50 sacks of graded filter pack material at $6 per sack; five traffic boxes at $75 each; 10 sacks of bentonite at $6 per sack; five sacks of concrete at $3 per sack; solid waste disposal (175 ft^3 [5 m^3] at $40/1 ft^3 [0.03 m^3]) for drilling waste disposal (approximately 35 ft^3 [1 m^3]/8 in. [20.3 cm] diameter hole), well development costs (including generation of 262.5 ft^3 [7.4 m^3] of aqueous wastes per hole [three well volumes] at $10/1 ft^3 [0.03 m^3], and sampling costs for 10 isolated depths per well (for a total of 50 additional samples) for the first round of water sampling; five of the soil samples from each boring evaluated for grain size distribution ($60 per sample) to determine filter pack grain size and corresponding screen slot sizes; $170/h for standby labor; $100/h for decontamination; $150 per sample for laboratory chemical analyses, $200 per sample (15 total) for laboratory physical analyses; $4,000 for reporting (to include sampling logs, boring logs, well construction and development logs, sampling results [soil and water], grain size distribution data, and results from modeling scenarios which depict particle-tracking flow paths in reverse direction [in time increments] based on assumed aquifer properties corresponding to soil types identified in the soil sampling efforts)
Surface geophysics	Extensive predeployment site planning; field measurements; vertical seismic profiles (requiring boreholes); data processing and interpretation; attribute analysis; confirmation sampling from five borings (eight samples each) to 100 ft (30.5 m); mobilization–demobilization; reporting	$2,000 for the field survey; $9,545 for data processing and interpretation; five additional sampling borings consisting of eight samples each, ranging from 30 to 100 ft (9.1 to 30.5 m) below grade; $6,000 (40 samples at $150 each) for chemical analytical costs; $1,000 (five samples at $200 each) for physical analytical

(Table continued)

TABLE 7.3 *Continued*

Approaches	Approach Assumptions	Cost Assumptions
		efforts; $4,000 for reporting (processed geophysical information, confirmation results, and specific predictions for NAPL location)
Subsurface geophysics	Extensive predeployment planning; field measurements; vertical seismic profiles; data processing and interpretation; attribute analysis; confirmation sampling from five borings (10 soil samples each) to 100 ft (30.5 m); installation of a well to 100 ft (30.5 m); sampling from the well (ten water samples total); mobilization–demobilization; reporting	In addition to costs articulated for the surface geophysical approach earlier, costs for an additional well (fully screened from 15 to 100 ft [4.6 to 30.5 m], omitting sampling and analyses) at a rate of $7,690 (UMS 1), $9,610 (UMS 2), and $11,530 (UMS 3); one additional day of drilling and well installation was required for UMS 1, 2 days for UMS 2, and 3 days for UMS 3; plus two additional days for mobilization and demobilization
CPT approaches Permeable membrane sensor; MIP	Field measurements for five pushes to 100 ft (30.5 m); data processing and interpretation; confirmation sampling analyses from two pushes (three soil samples each) to 75 ft (22.9 m); mobilization–demobilization; reporting	$3,500/day for a three-man crew, pushing, near-real-time analyses with an ITMS, ground-water sampling (90 samples in total), grouting, standby, and decontamination; approximately $150 each for soil confirmation analyses (6 total); $1,000/day for 2 days for mobilization and demobilization; approximately $300/day for 7 days total for per diem; approximately $3,000 for reporting (field and confirmation data and limited interpretation)
Hydrosparge	Field measurements on a 5 ft interval for five pushes to 100 ft (30.5 m); data processing and interpretation; confirmation sampling analyses from two pushes (three soil samples each) to 75 ft (22.9 m); mobilization–demobilization; reporting	$3,500/day for a three-man crew, pushing, near-real-time analyses with an ITMS or gas chromatograph, confirmation sampling, grouting, standby, and decontamination; approximately $150 each for soil confirmation analyses (6 total); $1,000/day for 2 days for mobilization and demobilization; approximately $300/day for 9 days total for per diem; approximately $3,000 for reporting (field and confirmation data and limited, interpretation)
Fluorescence (e.g., LIF) techniques	Field measurements for five pushes to 100 ft (30.5 m); data processing and interpretation; confirmation sampling analyses from two pushes (three soil samples each) to 75 ft (22.9 m); mobilization–demobilization; reporting	$3,500/day for a three-man crew, pushing, real-time analyses with a LIF system, confirmation sampling, grouting, standby, and decontamination; approximately $150 each for soil confirmation analyses (6 total); $1,000/day for 2 days for mobilization and demobilization; approximately $300/day for 6 days total for per diem; approximately

(Table continued)

DNAPL Characterization Methods and Approaches: Performance and Cost Comparisons   497

TABLE 7.3 *Continued*

Approaches	Approach Assumptions	Cost Assumptions
GeoVIS	Field measurements for five pushes to 100 ft (30.5 m); data processing and interpretation; confirmation sampling analyses from two pushes (three soil samples each) to 75 ft (22.9 m); mobilization–demobilization; reporting	$3,000 for reporting (field and confirmation data and limited interpretation) $3,500/day for a three-man crew, pushing, real-time analyses with a LIF system, confirmation sampling, grouting, standby, and decontamination; approximately $150 each for soil confirmation analyses (6 total); $1,000/day for 2 days for mobilization and demobilization; approximately $300/day for 10 days total for per diem; approximately $2,000 for reporting (field and confirmation data and limited interpretation)
LIF/GeoVIS	Field measurements for five pushes to 100 ft (30.5 m); data processing and interpretation; confirmation sampling analyses from two pushes (three soil samples each) to 75 ft (22.9 m); mobilization–demobilization; reporting	$3,500/day for a three-man crew, pushing, real-time analyses with a LIF system, confirmation sampling, grouting, standby, and decontamination; approximately $150 each for soil confirmation analyses (6 total); $1,000/day for 2 days for mobilization and demobilization; approximately $300/day for 10 days total for per diem; approximately $3,000 for reporting (field and confirmation data and limited interpretation)
Raman spectroscopy	Field measurements on a 5 ft interval for five pushes to 100 ft (30.5 m); data processing and interpretation; confirmation sampling analyses from two pushes (three soil samples each) to 75 ft (22.9 m); mobilization–demobilization; reporting	$3,500/day for a three-man crew, pushing, real-time analyses with a Raman system, confirmation sampling, grouting, standby, and decontamination; approximately $150 each for soil confirmation analyses (6 total); $1,000/day for 2 days for mobilization and demobilization; approximately $300/day for 9 days total for per diem; approximately $3,000 for reporting (field and confirmation data and limited interpretation)
LIF/Raman	Field measurements for five pushes to 100 ft (30.5 m); data processing and interpretation; confirmation sampling analyses from two pushes (three soil samples each) to 75 ft (22.9 m); mobilization–demobilization; reporting	$3,500/day for a three-man crew, pushing, real-time analyses with a Raman system, confirmation sampling, grouting, standby, and decontamination; approximately $150 each for soil confirmation analyses (6 total); $1,000/day for 2 days for mobilization and demobilization; approximately $300/day for 9 days total for per diem; approximately $3,000 for reporting (field and confirmation data and limited interpretation)

(Table continued)

TABLE 7.3 *Continued*

Approaches	Approach Assumptions	Cost Assumptions
Electrochemical sensor probe	25 pushes to 15 ft in 5×5 grid; pore vapor samples collected every meter (approximately 3 ft; five samples per push); confirmation sampling and analyses from two pushes (three soil samples each) to 20 ft (6.1 m); mobilization–demobilization for push rig; two additional sample borings (using a drill rig) collecting samples at 5 ft (1.5 m) intervals from 20 to 100 ft (6.1 to 30.5 m); chemical and physical laboratory analyses; mobilization–demobilization of drill rig; reporting	$3,500/day for a two-man CPT crew, pushing, real-time analyses with a field gas chlorine sensor, confirmation sampling, grouting (through probe tip), standby, and decontamination; approximately $150 each for soil confirmation analyses (6 total); $1,000/day for 2 days for push rig mobilization and demobilization; approximately $200/day for 6 days total for per diem (two-person crew); $1,500/day for the drilling equipment; $10/1 ft (0.3 m) for drilling; $20 per sample for collection; $3/1 ft (0.3 m) for grouting; $1,000/day for drill rig mobilization and demobilization; $300/day for per diem for a three-person crew; $170/h for standby labor (3 h); $100/h for decontamination (10 h); $40/0.03 m^3 (1 f^3) for drilling waste disposal (approximately 1 m^3 [35 f^3]/20.3 cm [8 in.] diameter hole; two drilled holes); $150 per soil sample (48 total) for laboratory chemical analyses; $200 per sample (6 total) for laboratory physical analyses; $3,000 for reporting results (chemical data and a map depicting VOC plumes indicative of potential DNAPL vadose zone and ground-water sources)
Waterloo (Ingleton) Profiler	Field measurements on a 5 ft (1.5 m) saturated interval for five pushes to 100 ft (30.5 m); data processing and interpretation; confirmation sampling analyses from two pushes (three soil samples each) to 75 ft (22.9 m); mobilization–demobilization; reporting	Approximately $2,000/day (assuming that a GeoProbe-type rig is used for this deployment) for a two-man crew, pushing, real-time aquifer analyses with the profiler components, sample collection, confirmation soil sampling, grouting, standby, and decontamination; approximately six field days to complete the five pushes, confirmation sampling, and grouting operations; approximately $150 each (96 total) for laboratory analyses; approximately $200/day for 2 days for mobilization and demobilization; approximately $200/day for 8 days total for per diem; approximately $3,000 for reporting (field and confirmation data and limited interpretation)

(Table continued)

DNAPL Characterization Methods and Approaches: Performance and Cost Comparisons

TABLE 7.3 *Continued*

Approaches	Approach Assumptions	Cost Assumptions
Co-solvent injection/ extraction; PIX probe	Field measurements on a 5 ft (1.5 m) interval for five pushes to 100 ft (30.5 m); data processing and interpretation; confirmation sampling analyses from two pushes (three soil samples each) to 75 ft (22.9 m); mobilization–demobilization; reporting	Approximately $3,500/day for a three-man crew, pushing, near-real-time analyses with a field gas chromatograph, confirmation sampling, grouting, standby, and decontamination; approximately $150 each for soil confirmation analyses (6 total); $1,000/day for 2 days for mobilization and demobilization; approximately $300/day for 19 days total for per diem; approximately $3,000 for reporting (field and confirmation data and limited interpretation)
FLUTe membrane	Deployment of five FLUTe liners to 100 ft (30.5 m); data processing and interpretation; confirmation sampling analyses (chemical and physical laboratory analyses) from two locations (three soil samples each) to 75 ft (22.9 m), mobilization–demobilization; reporting UMS 1 implemented using CPT while UMS 2 and UMS 3 implemented using conventional drilling equipment	UMS 1: $3,500/day for a two-man crew, pushing, retraction, and analyses using the FLUTe system, confirmation sampling, grouting, standby, and decontamination; approximately $150 each for soil confirmation analyses (6 total); $200 per soil sample physical analyses (grain size distribution and permeability); $1,000/day for 2 days for mobilization and demobilization; approximately $200/day for 5 days total for per diem; approximately $3,000 for reporting (field and confirmation data and limited interpretation UMS 2 and UMS 3: $1,500/day for the drilling equipment; $10/1 ft (0.3 m) for drilling ($20/1 ft [0.3 m] for UMS 3); $20 per sample for collection; $3/1 ft (0.3 m) for grouting; $1,000/day for mobilization and demobilization; $200/day for per diem for a two-person crew (6 days for UMS 2; 7 days for UMS 3); $170/h for standby labor; $100/h for decontamination; $40/1 ft^3 (0.03 m^3) for solid drilling waste disposal (approximately 35 ft^3 (1 m^3)/8 in. (20.3 cm) diameter hole to 100 ft (30.5 m); 26.25 ft^3 (0.74 m^3)/8 in. (20.3 cm) diameter hole to 75 ft); $150 per sample (6 total) for laboratory chemical analyses; $200 per sample (6 total) for laboratory physical analyses; and $2,000 for reporting

Note: Numbers in tables refers to number of samples assumed in cost estimate; Numbers in "CPT Approaches" section refers to number of CPT pushes.

site constraints. As mentioned earlier, a distinction between specific methods and site-management approaches is employed. Therefore, the approaches described subsequently consist of the specific methods of interest, as well as the required confirmation methods. For several cases, preliminary-characterization efforts are also considered in the cost

TABLE 7.4

Cost Comparisons for Each Approach

Approaches	UMS 1 ($) (Savings)	UMS 2 ($) (Savings)	UMS 3 ($) (Savings)
Baseline approaches	46,160 (0)	50,300 (0)	59,440 (0)
Soil-gas surveys	38,360 (7,800)	N/A (N/A)	N/A (N/A)
Partitioning interwell tracer tests	113,580 (−67,420)	126,130 (−75,830)	144,740 (−85,300)
Radon flux rates	70,870 (−24,710)	89,745 (−39,445)	104,425 (−44,985)
Back-tracking using dissolved concentrations in wells	62,290 (−16,130)	66,430 (−16,130)	75,570 (−16,130)
Surface geophysics	54,773 (−8,613)	59,163 (−8,863)	70,444 (−11,004)
Subsurface geophysics	62,613 (−16,453)	68,973 (−18,673)	82,224 (−22,784)
CPT approaches			
Permeable membrane sensor; MIP	25,500 (20,660)	N/A (N/A)	N/A (N/A)
Hydrosparge	33,100 (13,060)	N/A (N/A)	N/A (N/A)
Florescence (e.g., LIF) techniques	21,700 (24,460)	N/A (N/A)	N/A (N/A)
GeoVIS	35,900 (10,260)	N/A (N/A)	N/A (N/A)
LIF/GeoVIS	36,900 (9,260)	N/A (N/A)	N/A (N/A)
Raman spectroscopy	33,100 (13,060)	N/A (N/A)	N/A (N/A)
LIF/Raman	33,100 (13,060)	N/A (N/A)	N/A (N/A)
Electrochemical sensor probe	47,070 (−910)	N/A (N/A)	N/A (N/A)
Waterloo (Ingleton) Profiler	31,400 (14,760)	N/A (N/A)	N/A (N/A)
Co-solvent injection/extraction; PIX probe	76,100 (−29,940)	N/A (N/A)	N/A (N/A)
FLUTe membrane	28,600 (17,560)	39,550 (10,750)	48,290 (11,150)

Note: N/A, not applicable.

estimates. Including all preliminary and confirmation efforts in the cost estimates allows one to more adequately evaluate and compare the site-management options.

Because several approaches consist of similar activities, it is important to use consistent cost estimates for common line items. Table 7.5 lists cost estimates for generic line item approach components. To be consistent for confirmation and hydraulic control costs for applicable methods (e.g., PITT, radon flux, and FLUTe), it will be assumed that a zone

TABLE 7.5

Generic Cost Estimates for Approach Line Item Components

Item	Cost ($)	Per Unit
Drill rig	10 (UMS 1 and UMS 2) 20 (UMS 3)	Foot
Push rig	3500	Day
Sampling	20	Sample
Grouting	3	Foot
Mobilization–demobilization	1000	Day
Per diem ($100pp/day)	300	Day
Standby labor	170	Hour
Decontamination labor	100	Hour
Drilling waste disposal	40	Cubic foot
Laboratory chemical analyses	150	Sample
Laboratory physical analyses	200	Sample
Drilling waste disposal (sed)	40	Cubic foot
Drilling waste disposal (water)	10	Cubic foot
Per diem	100	Person-day
Reporting	2000 to 5000	Report

of DNAPL has been identified at a depth ranging from 19.9 to 22.9 m (65 to 75 ft) using the screening and confirmation efforts. Since these costs are highly dependent on the actual depths required, caution should be exercised when applying these values to dependent cost components.

Baseline Approach

Samples are typically collected from consecutive depth intervals using conventional drilling equipment and are analyzed using U.S. EPA-approved methods for identifying VOCs. Rapid field evaluations, such as shake-tests, use of a UV lamp, addition of Sudan IV, and observations of drill cutting fluids, soils, and vapors (e.g., head-space analyses), are also incorporated into this baseline. We assume that soil sampling from five locations will be conducted to depths of 100 ft (30.5 m) below grade, at 5 ft (1.5 m) intervals. Therefore, 21 samples per hole, for a total of 105 samples, would be collected for each UMS. Samples exhibiting high concentrations would be further analyzed for grain size distribution and permeability.

Cost differences between each scenario are attributed to time requirements based on drilling difficulties. For UMS 1, it is assumed that the project requires 3 days to complete plus 1 day each to mobilize and demobilize. It is also assumed that there will be 1 h of standby each day, 1 h to decontaminate the equipment used each day, and each workday consists of 10 h. The total anticipated cost for UMS 1 is $46,160 for this effort.

For UMS 2, the same UMS 1 assumptions are used, with the exception that it is assumed that the project requires 5 days to complete (given that additional time will be required to drill through resistant materials) plus 1 day each to mobilize and demobilize. The total anticipated cost for UMS 2 is $ 50,300 for this effort.

For UMS 3, the same UMS 1 assumptions are used, with the exception that it is assumed that the costs for drilling will be $20/ft (0.3 m) and that the project will require 7 days to complete (given that additional time is required to drill through competent materials) plus 1 day each to mobilize and demobilize. The total anticipated cost for UMS 3 is $59,440 for this effort.

Soil-Gas Surveys

It is assumed that a 5 × 5 grid of pushes (20 ft [6.1 m]) apart in north and south directions) to depths of 15 ft (4.6 m) will be required to characterize the potential DNAPL source zone based on vadose zone soil pore vapor chemistry. Soil-gas samples are to be collected with a direct-push system every three vertical feet. For the 25 pushes, 125 soil-gas samples will be analyzed over four field days. Two additional confirmation sampling pushes, collecting three soil samples each to 20 ft (6.1 m), will be included in the investigation. Assuming that a "hot spot" is identified in the vadose zone, an additional confirmation sampling effort consisting of two soil borings will be conducted to depths of 100 ft (30.5 m) below grade, collecting samples at 5 ft (1.5 m) intervals with a conventional drilling rig over the course of 3 days. Therefore, 21 samples per hole, for a total of 42 samples, would be collected and analyzed. Samples exhibiting high concentrations would be further analyzed for grain size distribution and permeability. It is assumed that only UMS 1 is feasible using this approach, as penetration through gravels and consolidated units is prohibitive. Well installation efforts will require 4 days. The total anticipated cost is presented in Table 7.4.

Partitioning Interwell Tracer Tests

While the PITT method affords useful data related to DNAPL volume present, it serves as perhaps the second or third characterization phase in an approach aimed at getting to the

t_2 design level. A PITT requires several preliminary steps that include:

- Location and delineation of the NAPL source
- Soil sampling
- Conventional laboratory analyses
- Laboratory tests to evaluate initial residual saturation levels in soil samples
- Laboratory tests to select candidate tracers and determine corresponding partition coefficients via column studies (often but not always a requirement)
- Aquifer testing to determine hydraulic data specific to the aquifer volume to be tested (e.g., sustainable injection and extraction rates and calibration data for a design model)
- A conservative interwell tracer test using bromide or chloride
- Flow and design modeling of the site

In addition, several injection, extraction, and monitoring wells must be installed prior to running the PITT. For sites composed of large source zones, several PITTs may be conducted.

For this assessment, it is assumed that preliminary field screening, confirmation, and well installation efforts are conducted using methods and associated costs described in this chapter. Details are provided subsequently:

- Field screening will include use of the FLUTe membrane to 100 ft (30.5 m) depth at five locations.
- Confirmation will include collection and analysis of six samples from two locations to a total depth of 75 ft (22.9 m).
- Wells will be emplaced in a configuration similar to that described in Meinardus et al. (1998) and screened at depths approximately 65 to 75 ft (19.8 to 22.9 m) beneath the water table.
- Four injection wells (three for tracer introduction and one for hydraulic control), three extraction wells (for tracer recovery), and one interwell monitoring point will be installed.
- Water extracted for hydraulic control and sample collection will be treated with granular activated carbon (GAC).

Owing to the amount of data processing required, reporting requirements will be more extensive, and therefore more expensive, than for the baseline approach. Further details and cost summaries are provided in Table 7.3 and Table 7.4, respectively.

A significant portion ($9000) of the total cost is due to treatment and disposal of liquid wastes generated during aquifer control. In addition, use of conventional laboratory methods to analyze tracer concentrations during the PITT can increase anticipated expenses depending on tracers used and frequency of sampling. Use of a field analytical system could significantly reduce analytical costs. Several PITTs have been conducted using only one injection well and one extraction well. This approach would cost less to conduct than the example provided. However, the savings may only represent a small percentage of the total, because costs are dominated by the preliminary site-characterization efforts, which would probably not differ greatly for the single versus multiple extraction options. Information derived from the PITT approach provides additional remediation design information, as residual NAPL volume can be estimated.

Radon Flux Rates

As with the PITT approach, several assumptions are required to adequately assess the radon flux rate approach. In practice, samples for Rn-222 measurements can be obtained using conventional water sampling approaches from installed wells (Semprini et al., 1998), use of direct-push discrete ground-water sampling equipment, and from multiple depth-discrete sampling equipment such as the Waterloo Profiler. For this exercise, it is assumed that several wells will be required for evaluating the distribution of Rn-222 levels at a DNAPL site. Several preliminary steps are required, including a field screening technique (such as the soil-gas survey method), confirmation soil sampling and analyses, installation of wells in appropriate locations, and aquifer testing to determine hydraulic data specific to the aquifer being tested. For UMS 1, UMS 2, and UMS 3, the following assumptions are made:

- Field screening approach includes use of the FLUTe membrane to 100 ft (30.5 m) depth at five locations.
- Confirmation will include collection and analysis of six samples from two locations to a total depth of 75 ft (22.9 m).
- Five wells will be installed to 75 ft (22.9 m) with screens installed from 65 to 75 ft (19.8 to 22.9 m) beneath the water table.

Owing to the amount of data processing involved, reporting requirements will be more extensive, and therefore more expensive, than for the baseline approach. However, because hydraulic control (and corresponding level of data processing detail) will generally not be required, the report will be less expensive than the PITT report. Rn-222 flux information may assist with remediation design, because residual NAPL volume estimates can be derived. Further details and cost summaries are provided in Table 7.3 and Table 7.4, respectively.

Back-Tracking Using Dissolved Concentrations in Wells

It is assumed that soil samples are collected during installation at the same frequency specified in the baseline approach. In addition, well installation costs are incurred at rates presented subsequently. Because well screens are to be installed over the entire saturated thickness, packers will be necessary for isolating specific sampling depths. A potentially cost-effective alternative is to use clusters or nests of direct-push wells, screened at selected discrete depth ranges for UMS 1. In addition, the Waterloo Profiler or FLUTe multilevel sampler can also be a cost-effective alternative for UMS 1. For this section, it is assumed that the wells are emplaced using conventional drilling techniques.

For UMS 1, the same expenses presented in the baseline soil sampling and analysis approach are incurred, with the exception that grouting requirements will be replaced by well installation costs (five 10 cm [4 in.] diameter wells) and 7 days will be required (plus 1 day each for mobilization and demobilization) for the soil sampling and well installation efforts. An aquifer test (not included here) is generally conducted to identify hydraulic conductivity, transmissivity, and aquifer storage properties. However, because the screened zone is very long (approximately 85 ft [26 m]) for each well, it may not be very practical to attribute one averaged value to each of these parameters.

UMS 2 will require approximately 7 days for the well installation efforts. UMS 3 will require approximately 9 days for the well installation efforts. These costs only include one round of water sampling. Subsequent sampling rounds will run approximately

$8250 per round for the analytical costs ($7500) and labor ($750). Information gained from this investigation may be useful for site remediation. To obtain useful information, wells must be placed in the appropriate locations adjacent to NAPL sources. Although not considered here, a more appropriate (and costlier in the short term) approach would include use of a screening technique (such as soil-gas survey or CPT sensor method for UMS 1 and FLUTe for UMS 2 and 3) prior to selection of well installation locations.

Geophysics

Three-dimensional seismic surveying technology was evaluated to delineate DNAPL source zones at three specific military sites over the past 4 years (Sinclair and Kram, 1998). The main differences between the sites were lithologic characteristics and contaminant areal extent. Total costs included expenses for conducting the field measurements, generating vertical seismic profiles, data processing and interpretation, attribute analyses, confirmation drilling and sampling, laboratory analyses, and generation of plans and reports. Two of the sites consisted of alluvial deposits (similar to UMS 1 and UMS 2), while the other was consisted of dense fractured limestone and dolomite (similar to UMS 3). The average total cost incurred for the study was approximately $230,000 per site for each of the three sites investigated (Trotsky, 1999). Costs and assumptions presented in Table 7.3 are normalized to account for the smaller study footprint for each UMS. As with all surface geophysical methods, it is extremely important to consider costs associated with "ground-truthing," which is required for converting time units to depth units based on acoustic wave transmissive properties of the site-specific stratigraphy. Costs for obtaining these data can be substantial, because intrusive activities (e.g., soil borings) are required for every application of this approach. Without this critical step, the surface geophysical survey data are useless for characterizing DNAPL. For this reason, surface geophysical surveys should not be classified as "nonintrusive" approaches.

For subsurface geophysical approaches, it is assumed that a well will be necessary to lower the transmitting device and generate a more accurate subsurface lithologic characterization. Costs for an additional well (fully screened from 15 to 100 ft [4.6 to 30.5 m]), omitting sampling and analyses) were added to each of the corresponding costs for the surface geophysical approach presented earlier. It was assumed that one additional day of drilling and well installation was required for UMS 1, 2 days for UMS 2, and 3 days for UMS 3, plus 2 days for mobilization and demobilization. Additional assumptions and cost estimates are presented in Table 7.3.

CPT Approaches: General

Because penetration using CPT through gravels and consolidated units is not feasible with current platforms, it is assumed that only UMS 1 can be characterized using the CPT approaches. Innovative developments, such as sonic head CPT and laser drilling, may soon allow for CPT applications in more consolidated materials. For this cost analysis, both the conventional CPT push rigs, which consist of reaction forces of 13,620 kg (15 t) or greater, and the lighter truck- and van-mounted direct-push rigs are considered. Although some smaller direct-push rigs are capable of advancing sensor probes with a hydraulic ram system, most of these lighter weight systems operate via a hammer technique and, therefore, cannot advance many of the sensor systems available. The smaller rigs can be less expensive to operate than the larger CPT systems and services are generally charged on a per-foot or per-push rate. The larger CPT rig services are typically charged at a per-day rate, which sometimes includes reporting. For this comparison, it is assumed that the soil-gas survey and the Waterloo (Ingleton) Profiler survey are conducted with a smaller rig, while all the other CPT approaches are conducted with the larger rig. Assumption details and cost summaries are provided in Table 7.3 and Table 7.4, respectively.

CPT Approaches: MIP

It is assumed that five MIP pushes to 100 ft (30.5 m) will be required to screen the site, plus two additional pushes for confirmation sampling. Grouting will require additional pushes (7 total) with a grout probe. Because soil lithologic data will be collected along with the chemical screening information, this level of effort may be enough to identify potential remediation options. At a minimum, determination of required data gaps is feasible with this level of effort, because data profiles are relatively continuous with a resolution of a few centimeters. Use of a smaller truck- or van-mounted CPT system could save approximately $7500.

CPT Approaches: Hydrosparge

Hydrosparge field sampling and analytical operations will require that probe advancement be stopped every 5 ft (1.5 m), resulting in 20 events for each push. It is assumed that approximately 15 Hydrosparge sampling events can be accomplished per day. As with the MIP approach, grouting will require additional pushes (7 total) with a grout probe. Therefore, it will require approximately seven field days to complete the 100 Hydrosparge sampling events. Soil lithologic data will be collected using soil sensors, along with the chemical screening information. Therefore, this level of effort may be enough to identify potential remediation options. Chemical data profiles are spaced at 5 ft (1.5 m) intervals, while soil type profiles are relatively continuous with resolution of a few centimeters. In practice, lithologic observations can be used to optimize the chemical data collection depths.

CPT Approaches: Fluorescence Techniques

This method assumes that the DNAPL contains fluorescing co-constituents, which is often, but not always, the case. Grouting for the fluorescence pushes will not require additional pushes, because the probe is equipped with grouting capabilities through the tip as the device is retracted. However, additional pushes will be required to grout the two sampling holes.

This level of effort may be enough to identify potential remediation options, because soil lithologic data will be collected along with the chemical screening information. At a minimum, identification of data gaps is feasible with this level of effort, since data profiles are relatively continuous with resolution of a few centimeters.

CPT Approaches: GeoVIS

GeoVIS operations will require that probe advancement be run relatively slower than conventional CPT operations to be able to observe images in real-time. It is assumed that approximately one run of the GeoVIS to 100 ft can be accomplished per day. As with the LIF approach, two additional pushes will be required to collect confirmation samples (3 per push) to depths of approximately 75 ft (22.9 m). In addition, grouting will require additional pushes (7 total) with a grout probe. Therefore, it will require approximately eight field days to complete the five pushes, confirmation sampling, and grouting operations. Reporting costs are less than for the baseline approach, because the level of effort is relatively less.

This level of effort may be enough to identify potential remediation options. Soil images are continuous with resolution greater than a fraction of a centimeter, while soil type profiles are relatively continuous with resolution greater than one third of a centimeter.

CPT Approaches: LIF/GeoVIS

The considerations are the same as for the individual LIF or GeoVIS CPT approaches. Grouting will require additional pushes (7 total) with a grout probe, because the current configuration does not allow for grouting through the tip. In addition, LIF/GeoVIS operations will require that probe advancement be run relatively slower than conventional CPT operations to be able to observe images in real-time. It is assumed that approximately one run of the LIF/GeoVIS to 100 ft can be accomplished per day. Therefore, it will require approximately eight field days to complete the five pushes, confirmation sampling, and grouting operations.

Soil images and fluorescence data are continuous with resolution greater than a centimeter for the video and the fluorescence data, while soil type profiles are relatively continuous with resolution about one third of a centimeter. Costs are comparable to the GeoVIS approach, but the data set is more complete (requiring additional reporting time) and the potential for false negatives is reduced.

CPT Approaches: Raman Spectroscopy

It is assumed that probe advance is stopped every 5 ft (1.5 m), resulting in 20 events for each push. In practice, operators often couple Raman data with real-time lithologic sensor data and stop only for Raman data collection activities when a potential vertical barrier is encountered. Raman pushes will be grouted through the probe tip upon retraction. Sampling efforts will require additional pushes (2 total) with a grout probe. We assume that approximately seven field days will be required to complete the Raman pushes and confirmation sampling.

Because soil lithologic data will be collected along with the chemical screening information, this level of effort may be enough to identify potential remediation options. Chemical data profiles are spaced at 5 ft (1.5 m) intervals, while soil type profiles are relatively continuous with resolution greater than one third of a meter. For very detailed investigations, Raman spectra are sometimes acquired every 0.5 to 3 ft (0.2 to 0.9 m) as the penetrometer is advanced (Rossabi et al., 2000). For sediments likely to contain DNAPL based on knowledge of disposal, previous work, or lithologic characteristics indicative of potential contaminant migration pathways, the 0.5 ft (0.2 m) frequency is used. Therefore, the time requirements outlined in this hypothetical case will need to be adapted to site specific observations while in the field.

CPT Approaches: LIF/Raman

As with the Raman approach, it is assumed that the LIF/Raman probe advancement is stopped every 5 ft (1.5 m), resulting in 20 events for each push, and that grouting can be completed through the probe tip. Sampling efforts will require additional pushes (2 total) with a grout probe. It is assumed that approximately seven field days will be required to complete the LIF/Raman pushes and confirmation sampling.

Raman data profiles are spaced at 5 ft (1.5 m) intervals, and LIF and soil-type data profiles are generated with relatively continuous resolution greater than one third of a meter. Costs are comparable to the Raman approach, since the Raman measurement is the rate-limiting step. The data set generated by coupled LIF/Raman is more complete (potentially requiring additional reporting time) and the potential for false positives and false negatives is reduced. As mentioned earlier, Raman spectra are sometimes acquired every 0.5 to 3 ft (0.2 to 0.9 m) as the penetrometer is advanced (Rossabi et al., 2000). For sediments likely to contain DNAPL based on knowledge of disposal, previous work, or lithologic or LIF characteristics indicative of potential contaminant migration pathways, the 0.5 ft (0.2 m) frequency is often used. Therefore, as with several other approaches described, the time requirements outlined in this hypothetical case will need to be adapted to site-specific observations while in the field.

CPT Approaches: Electrochemical Sensor

It is assumed that a 5 × 5 grid of pushes 20 ft (6.1 m) apart in north and south directions) to depths of 15 ft (4.6 m) will be required to characterize the potential DNAPL source zone based on vadose zone soil pore vapor chlorine concentrations. Soil-gas samples are to be collected with a 15 t or greater CPT rig every meter (approximately 3 ft). Therefore, for the 25 pushes required, 125 soil-gas samples will be analyzed over four field days. Two additional confirmation sampling pushes, collecting three samples each to 20 ft (6.1 m), will be included in the investigation. As with the soil-gas survey example, it is assumed that a hot spot is identified in the vadose zone. Therefore, an additional sampling effort consisting of two soil collection borings will be conducted to depths of 100 ft (30.5 m) below grade, collecting samples at 5 ft (1.5 m) intervals with a drilling rig over the course of 3 days. Therefore, 21 samples per hole, for a total of 42 samples, would be collected and analyzed. Samples exhibiting high concentrations would be further analyzed for grain size distribution and permeability.

Using the CPT for sampling may reduce costs, since an additional mobilization–demobilization charge will not be incurred and less solid waste will be generated. In addition, some smaller direct-push rigs may be used (at a reduced cost) for both the vadose zone screening and sampling activities beneath the water table.

CPT Approaches: Waterloo (Ingleton) Profiler

To be consistent with the other approaches evaluated, it is assumed that five pushes to advance the Waterloo Profiler will be used to screen the site. In addition, it is assumed that field sampling and analytical operations are conducted at a 5 ft (1.5 m) frequency in the saturated zone, resulting in 18 samples (or sampling events) for each push. Two additional pushes will be required to collect confirmation soil samples (3 per push) to depths of approximately 75 ft (22.9 m). Waterloo Profiler pushes will be grouted through the probe tip upon retraction. Sampling efforts will require additional pushes (2 total) with a grout probe. Hydraulic conductivity via constant head analysis requires only a few minutes for each test. Ground-water sampling requires variable amounts of time, depending upon the formation. We assume that approximately one run of the Waterloo Profiler to 100 ft can be accomplished per day.

Because soil hydrogeologic data are collected along with the chemical information, this level of effort may be enough to identify potential remediation options. Concentration versus hydraulic conductivity, concentration versus depth, and piezometric surface can be useful for this purpose. Chemical and hydrogeologic data profiles are spaced at 5 ft (1.5 m) intervals for this scenario. However, the probe is capable of resolution down to a fraction of a meter. For very detailed investigations, profiler data are acquired every 5 to 7.5 cm (2 to 3 in.) as the probe is advanced (Pitkin, 1998). This will require more time, and therefore more costs, than the scenario described earlier. For sediments likely to contain DNAPL based on knowledge of disposal details, previous work, or hydrogeologic characteristics indicative of potential contaminant migration pathways, the 5 to 7.5 cm (2 to 3 in.) frequency may be used. Therefore, the time requirements outlined in this hypothetical case will need to be adapted to site specific observations while in the field.

CPT Approaches: PIX

It is assumed that PIX probe analytical operations will require that probe advance is stopped every five saturated feet (1.5 m), resulting in 18 events for each push. It is further assumed that approximately six PIX events can be accomplished per day, due to the solvent–solute equilibrium requirements. Grouting will require additional pushes (7 total) with a grout probe. Therefore, it will require approximately 15 field days to complete the 90 sampling events and two additional days for confirmation sampling.

In practice, use of the PIX approach for UMS 1 may require a less extensive effort than that described earlier, because operators generally try to identify potential barriers to vertical NAPL migration prior to running the PIX, thereby focusing on candidate source zones. If vertical barriers are readily apparent using soil classification sensors, costs for the PIX method could be significantly less expensive (by as much as 50%) than the estimate provided. Because soil lithologic data will be collected along with the chemical screening information, this level of effort may be enough to identify potential remediation options. Chemical data profiles are spaced at 5 ft (1.5 m) intervals (for this scenario), while soil-type profiles are relatively continuous with greater than one third of a meter resolution.

Ribbon NAPL Sampler FLUTe

The Ribbon NAPL Sampler FLUTe method can be implemented using either a direct-push rig or a conventional drilling rig. For this assessment, it is assumed that a direct-push rig (13,620 kg [15 t] or greater capacity) is used for UMS 1 and a conventional drilling rig is used for UMS 2 and UMS 3. Two additional pushes or borings will be required to collect confirmation samples (3 per push or installation) to depths of approximately 75 ft (22.9 m). In addition, grouting requirements will be carried out by advancing the CPT grout probe (UMS 1) or auger flights (UMS 2 and UMS 3) to total depths attained for the seven holes. Reporting requirements will be relatively minimal when compared with approaches requiring more intensive data processing and presentation.

For UMS 1, it is assumed that the project requires 3 days to complete plus 1 day each to mobilize and demobilize. It can also be assumed that there will be 1 h of standby each day, 1 h to decontaminate the equipment used each day, and each workday consists of 10 h. For UMS 2, it is assumed that the FLUTe is advanced using a conventional drilling rig and that the project requires 4 days to complete plus 1 day each to mobilize and demobilize. For UMS 3, the same UMS 2 assumptions are used, with the exception that it is assumed that the costs for drilling will be $20/ft and that the project requires 5 days to complete plus 1 day each to mobilize and demobilize.

Discussion and Conclusions

This chapter describes and compares many of the methods and approaches currently used to detect and delineate DNAPL contaminant source zones. General performance comparisons were generated to identify potential site management considerations required to reach a level of site understanding adequate to initiate remediation design efforts. Specific advantages and disadvantages for several methods were presented in Table 7.1. In addition, characterization approach cost comparisons for conceptual sites exhibiting particular sets of physical characteristics were generated. Perhaps, the most important issue raised deals with the recognition that each candidate method must be placed in its proper context within the characterization process. The process is therefore considered an approach which consists of several methods, each serving to complement individual method components. It is through this recognition that a true assessment of the anticipated site management costs and project duration can be derived.

Methods described as baseline in this chapter are clearly not valid for most cases. The level of resolution and detail required for assessment and initiation of remedial design are not generally achievable using these techniques. However, these types of approaches can serve as confirmation efforts provided a specific DNAPL source location is suspected based on more rigorous alternatives such as those described in this chapter.

Because each method has specific advantages and disadvantages, several methods can be complementary in an overall site-management plan, each serving a particular niche. This can be considered a "hybrid" approach, whereby the strengths of individual characterization components are exploited at the most appropriate and logical times in the site-management process. An example characterization approach at an unconsolidated alluvium site begins with the generation of a lithologic profile followed by deployment of the direct-push FLUTe or LIF/GeoVIS method, then analysis of confirmation samples. After determining the location of the DNAPL source zone, discretely screened or multilevel wells can be installed and a Radon flux rate survey or PITT survey can be used to estimate the amount of NAPL present. Likewise, for sites composed of fractured crystalline rock or karst, one can initially screen the site with a geophysical survey (including vertical profiling to convert units of time to units of length or depth). Following the geophysical survey, a FLUTe method (deployed via drill rig) and confirmation effort can be conducted to determine the location of the DNAPL source area. Discretely screened or multilevel wells can then be installed and a Radon flux rate survey or PITT can be used to estimate the amount of NAPL present. The number of available method combinations and potential options are extensive. Approach cost estimates presented subsequently can be used to estimate anticipated costs for these hybrid management strategies.

Table 7.4 presents the cost and savings estimates for each approach included in the analysis. Figure 7.1 to Figure 7.3 display the cost values for each UMS graphically. The savings were derived by subtracting the cost estimate for each approach from the cost estimate for the corresponding baseline approach. A negative savings value indicates that the approach incorporating the particular DNAPL characterization method is more costly than the baseline approach. Approaches that cannot be implemented in gravelly or consolidated geologic materials, such as CPT approaches, are not included in the UMS 2 and UMS 3 comparisons.

The least expensive approaches for UMS 1 include several CPT sensor approaches, such as fluorescence and MIP, and the FLUTe approach. Note that the FLUTe approach was installed with a CPT device for UMS 1. The fluorescence and MIP approaches must always include confirmation efforts, either by use of conventional analyses or by coupling to additional sensors such as the GeoVIS. However, MSE (2002) believe that the FLUTe approach may not require chemical confirmation once a larger database has been generated. If supported by regulators, this will substantially reduce the costs (by close to $6000 for this scenario) associated with the FLUTe approach. However, it is believed that regulators will require confirmation efforts for at least the next few years. The FLUTe approach may be more definitive with respect to identifying DNAPL source zones. While the fluorescence and MIP approaches generate soil classification data,

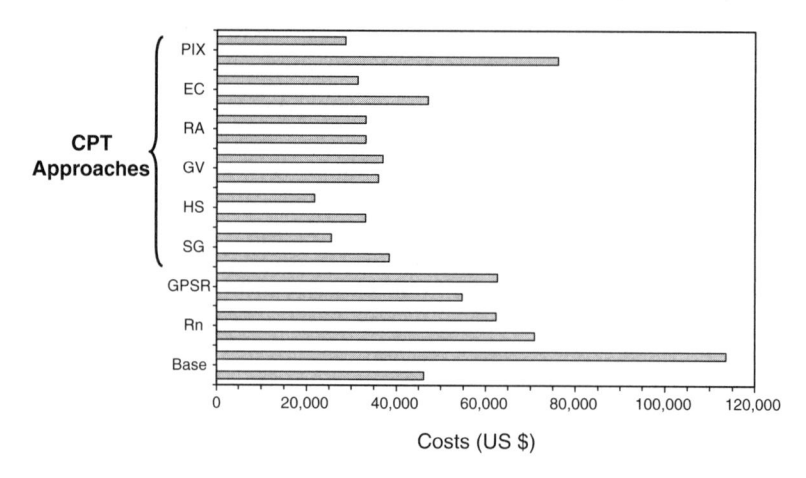

FIGURE 7.1
Costs for each DNAPL characterization approach. (Base, baseline; PITT, partition interwell tracer test; Rn, radon flux survey; BT, solute back-track; GPSR, surface geophysical; GPSB, subsurface geophysical; SG, soil-gas survey; MIP, membrane interface probe; HS, hydrosparge; FL, fluorescence probe; GV, GeoVIS probe; FL/GV, fluorescence probe with GeoVIS; RA, raman probe; FL/RA, fluorescence with Raman probe; EC, electrochemical sensor probe; WP, Waterloo Profiler; PIX, precision injection extraction; FLUTe, flexible liner underground everting membrane; CPT, cone penetrometer testing.)

the FLUTe approach will either require that lithology sensors are operated during the preliminary pushes or that additional laboratory tests are conducted on soil samples to determine soil type and hydraulic properties. Several additional approaches, including soil gas, Hydrosparge, GeoVIS, fluorescence-GeoVIS, Raman, fluorescence-Raman, and the Waterloo Profiler are very competitive (ranging from $20,000 to $40,000) for UMS 1. The baseline approach was estimated to be approximately $46,000 for UMS 1.

The most expensive approach for UMS 1 is the PITT survey. While this approach yields detailed hydrologic information and DNAPL volume estimates, water treatment costs associated with hydraulic control, and costs associated with preliminary site characterization and setup (e.g., aquifer testing, well installation, etc.) can be very high. Once a site has been adequately characterized and wells are properly installed and screened in optimal locations, the PITT approach can be a useful endeavor. PITT

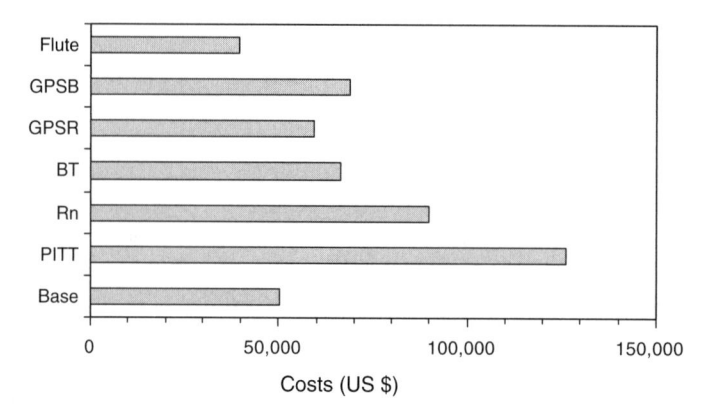

FIGURE 7.2
Costs for each applicable approach using Unit Model Scenario 2.

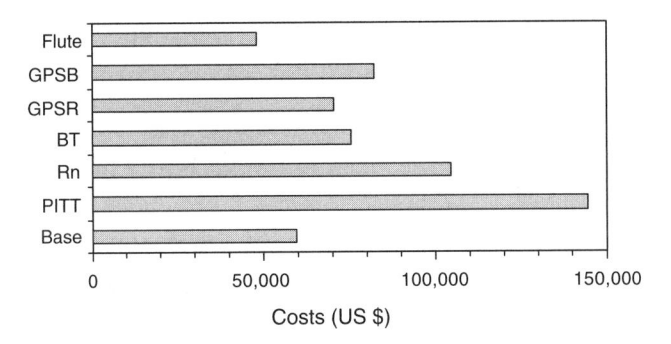

FIGURE 7.3
Costs for each applicable approach using Unit Model Scenario 3.

approaches for evaluation of remediation effectiveness have been successfully demonstrated with remarkably accurate mass removal estimates (Meinardus et al., 1998). During one particular test conducted at Hill Air Force Base, Utah, a PITT was used to estimate that approximately 1310 l (346 gal) of residual DNAPL remained in a test area prior to removal with use of a surfactant. A postremediation PITT indicated that 1291 l (341 gal) had been recovered, with approximately 19 l (5 gal) remaining in the swept volume. The effluent treatment system recorded 1374 l (363 gal) recovered.

The PIX approach was very expensive under the assumptions used for UMS 1. In practice, the PIX method would not generally be used to screen at frequent depth intervals. Provided that potential traps or vertical migration barriers can be adequately recognized and injection–extraction tests can be performed at fewer depth locations, thereby leading to lower costs than those presented in this chapter. Although not considered in the cost analyses, a back-tracking approach could be coupled with radon analyses, potentially resulting in better indirect DNAPL source area resolution and estimates of NAPL saturation. The PIX and back-tracking approaches each include confirmation steps, unless NAPL is recovered in the wells or during extraction.

The geophysical approaches cost more than the baseline approach for UMS 1, because they require confirmation steps roughly equal in cost to baseline efforts. Although not generally capable of identifying DNAPL source areas, geophysical approaches have been used to assist with locating appropriate sample collection zones based on interpretation of lithology to predict potential flow pathways. Direct DNAPL detection is often not possible under conditions presented in UMS 1, which consists of unconsolidated soils. This is because DNAPL commonly occurs as discrete blobs or pools (often adjacent to vertically confining layers which can act as reflectors) that are generally smaller than the spatial resolution of the geophysical technique or not capable of yielding a detectable contrast.

The FLUTe approach (with confirmation efforts) is the least expensive of the approaches evaluated for UMS 2. Only the FLUTe approach resulted in costs lower than the baseline approach for this scenario. The Radon flux rate, back-tracking, and geophysical approaches range in costs from approximately $50,000 to approximately $70,000. The FLUTe approach will generally provide more NAPL location detail and depth resolution than the other approaches under conditions presented in UMS 2. The most expensive approach for UMS 2 is the PITT survey. As mentioned earlier, the PITT approach yields detailed hydrologic information and volume estimates. However, water treatment costs associated with hydraulic control and costs associated with preliminary site characterization and setup (e.g., aquifer testing, well installation, etc.) can be prohibitive. If a site has been adequately characterized and wells are properly installed and screened in optimal locations, the PITT approach can be used to determine target removal volumes. Although current enhancement efforts are underway, CPT approaches cannot currently penetrate soils characteristic of UMS 2.

For UMS 3, the FLUTe approach (with confirmation efforts) is the least expensive of the candidate approaches and is the only approach costing less than the baseline for this scenario. As with UMS 1 and UMS 2, the most expensive approach for UMS 3 is the PITT survey.

This chapter compares many of the methods and approaches currently used to detect and delineate DNAPL contaminant source zones. General performance comparisons were generated to identify potential site-management considerations required to reach a level of site understanding adequate to initiate remediation design efforts. Specific advantages and disadvantages for several methods were presented. For this effort, characterization approach cost comparisons for conceptual sites exhibiting particular sets of physical characteristics were generated. While this chapter describes and compares the specific DNAPL characterization approaches, it will be up to the reader to determine which approach is most appropriate for the specific-site conditions and concerns. In general, cost will most likely be the determining factor for approach selection (Kram et al., 2002). However, several approach limitations should weigh heavily in the ultimate selection of the most appropriate site-management strategy. For instance, CPT methods cannot be used in gravel or highly consolidated soils. Similarly, approaches such as soil-gas surveys and surface geophysical surveys generally require relatively more confirmation sampling due to the limited depth resolution provided by the field data. These factors, as well as others presented in this chapter, should be carefully considered prior to making the financial commitment to a DNAPL characterization site-management strategy.

Caveat: The opinions stated in this chapter are those of the author and do not constitute endorsements of particular approaches or methods, nor are they representative of the opinions of the author's employer. The cost data used in this analysis were synthesized from a variety of sources including the author's experience, several commercial vendors, consultants and government employees within the environmental industry.

References

Adams, J.W., W.M. Davis, E.R. Cespedes, W.J. Buttner, and M.W. Findlay, Development of cone penetrometer electro-chemical sensor probes for chlorinated solvents and explosives, *Proceedings of a Specialty Conference on Field Analytical Methods for Hazardous Wastes and Toxic Chemicals*, Air and Waste Management Association, Pittsburgh, PA, 1997, pp. 667–670.

Adams, M.L., B. Herridge, N. Sinclair, T. Fox, and C. Perry, 3-D seismic reflection surveys for direct detection of DNAPL, in *Non-Aqueous Phase Liquids, Remediation of Chlorinated and Recalcitrant Compounds*, G.B. Wickramanayake and R.E. Hinchee, Eds., Battelle Press, Columbus, OH, 1998, pp. 155–160.

Anderson, M.R., R.L. Johnson, and J.M. Pankow, Dissolution of dense chlorinated solvents into ground water. 3. Modeling contaminant plumes from fingers and pools of solvent, *Environmental Science and Technology*, 26, 901–908, 1992.

Annable, M.D., J.W. Jawitz, P.S.C. Rao, D.P. Dai, H. Kim, and A.L. Wood, Field evaluation of interfacial and partitioning tracers for characterization of effective NAPL-water contact areas, *Ground Water*, 36(3), 495–502, 1998.

ASTM, Standard Method for Deep Quasi-Static, Cone and Friction-Cone Penetration Tests of Soil, ASTM Standard D 3441, ASTM International, West Conshohocken, PA, 2004.

Brewster, M.L., A.P. Annan, J.P. Greenhouse, G.W. Schneider, and J.D. Redman, Geophysical detection of DNAPLs: field experiments, *Proceedings of the International Association of Hydrogeologists Conference*, Hamilton, Ontario, May, 1992, pp. 176–194.

Burt, R.A., R.D. Norris, and D.J. Wilson, Modeling mass transport effects in partitioning inter-well tracer tests, in *Non-Aqueous Phase Liquids, Remediation of Chlorinated and Recalcitrant Compounds*, G.B. Wickramanayake and R.E. Hinchee, Eds., Battelle Press, Columbus, OH, 1998, pp. 119–124.

Christy, T.M., A permeable membrane sensor for the detection of volatile compounds in soil, *Proceedings of the Symposium on the Application of Geophysics to Environmental and Engineering Problems*, Society for Environmental and Engineering Geophysics, Denver, CO, 1998, pp. 65–72.

Cohen, R.M. and J.W. Mercer, *DNAPL Site Characterization*, C.K. Smoley, Ed., New York, NY, 1993.

Cohen, R.M, A.P. Bryda, S.T. Shaw, and C.P. Spalding, Evaluation of visual methods to detect NAPL in soil and water, *Ground-Water Monitoring Review*, 12(4), 132–141, 1992.

Davis, W.M., J.F. Powell, K. Konecny, J. Furey, C.V. Thompson, M. Wise, and G. Robitaille, Rapid *in-situ* determination of volatile organic contaminants in ground water using the site characterization and analysis penetrometer system, *Proceedings of the Conference on Field Analytical Methods for Hazardous Wastes and Toxic Chemicals*, Air and Waste Management Association, Pittsburgh, PA, 1997, pp. 464–469.

Davis, W.M., M.B. Wise, J.S. Furey, and C.V. Thompson, Rapid detection of volatile organic compounds in ground water by *in-situ* purge and direct-sampling ion-trap mass spectrometry, *Field Analytical Chemistry and Technology*, 2(2), 89–96, 1998.

Dwarakanath, V., N. Deeds, and G.A. Pope, Analysis of partitioning inter-well tracer tests, *Environmental Science and Technology*, 33, 3829–3836, 1999.

Federal Remediation Technologies Roundtable Field Analysis Technologies Matrix, 2005. Available at http://www.frtr.gov/site/.

Feenstra, S., D.M. Mackay, and J.A. Cherry, Presence of residual NAPL based on organic chemical concentrations in soil samples, *Ground-Water Monitoring Review*, 11(2), 128–136, 1991.

Feenstra, S., J.A. Cherry, and B.L. Parker, Conceptual models of the behavior of dense non-aqueous phase liquids (DNAPLS) in the subsurface, in *Dense Chlorinated Solvents and other DNAPLs in Ground Water: History, Behavior, and Remediation*, Waterloo Press, Waterloo, Ontario, Canada, 1996, pp. 53–88.

Feenstra, S. and J.A. Cherry, Subsurface contamination by dense non-aqueous phase liquid (DNAPL) chemicals, *Proceedings, International Ground Water Symposium*, International Association of Hydrogeologists, Halifax, Nova Scotia, May 1–4, 1988, 62–69.

Jin, M., M. Delshad, V. Dwarakanath, D.C. McKinney, G.A. Pope, K. Sepehrnoori, C.E. Tilburg, and R.E. Jackson, Partitioning tracer test for detection, estimation and remediation performance assessment of subsurface non-aqueous phase liquids, *Water Resources Research*, 31(5), 1201–1211, 1995.

Johnson, R.L., and J.M. Pankow, Dissolution of dense chlorinated solvents into ground water. 2. Source functions for pools of solvents, *Environmental Science and Technology*, 26, 896–901, 1992.

Keller, A.A. and M.L. Kram, Use of fluorophore/DNAPL mixtures to detect DNAPLs *in-situ*, in *Non-Aqueous Phase Liquids, Remediation of Chlorinated and Recalcitrant Compounds*, G.B. Wickramanayake and R.E. Hinchee, Eds., Battelle Press, Columbus, OH, 1998, pp. 131–136.

Keller, A.A., M.J. Blunt, and P.V. Roberts, Behavior of dense non-aqueous phase liquids in fractured porous media under two-phase flow conditions, *Transport in Porous Media*, 38, 189–203, 2000.

Kenny, J.E., J.W. Pepper, A.O. Wright, Y.M. Chen, S.L. Schwartz, and C.G. Skelton, *Subsurface Contamination Monitoring Using Laser Fluorescence*, K. Balshaw-Biddle, C.L. Oubre, and C.H. Ward, Eds., Lewis Publishers, Boca Raton, FL, 1999, 160 pp.

Knox, R.C., D.A. Sabatini, M. Goodspeed, M. Hasegawa, and L. Chen, Hydraulic Considerations for Advanced Subsurface Characterization and Remediation Technologies, IAHS Publication No. 250, pp. 391–399, 1998.

Kram, M.L., Framework for Successful SCAPS Deployment, Proceedings of the Sixth Annual West Coast Conference on Contaminated Soils and Ground Water: Analysis, Fate, Environmental and Public Health Effects, and Remediation, AEHS, Amherst, MA, 1996.

Kram, M.L., Use of SCAPS Petroleum Hydrocarbon Sensor Technology for Real-Time Indirect DNAPL Detection, Proceedings of the Seventh Annual West Coast Conference on Contaminated Soils and Ground Water: Analysis, Fate, Environmental and Public Health Effects, and Remediation, AEHS, Amherst, MA, 1997.

Kram, M.L., Use of SCAPS petroleum hydrocarbon sensor technology for real-time indirect DNAPL detection, *Journal of Soil Contamination*, 7(1), 73–86, 1998.

Kram, M.L., DNAPL Detection Using Optimized Fluorescence Methods; Ph.D. thesis, University of California at Santa Barbara, 2002, 255 pp.

Kram, M.L. and A.A. Keller, Complex NAPL site characterization using fluorescence, Part 2: Analysis of soil matrix effects on the excitation/emission matrix, *Soil and Sediment Contamination: an International Journal*, March/April, 13(2), 119–134, 2004.

Kram, M.L. and A.A. Keller, Complex NAPL site characterization using fluorescence, Part 3: Detection capabilities for specific excitation sources, *Soil and Sediment Contamination: an International Journal*, March/April, 13(2), 135–148, 2004.

Kram, M.L., M. Dean, and R. Soule, The ABCs of SCAPS, *Soil and Groundwater Cleanup*, May, 1997, pp. 20–22.

Kram, M.L., A.A. Keller, J. Rossabi, and L.G. Everett, DNAPL characterization methods and approaches, Part 1: performance comparisons, *Ground Water Monitoring and Remediation*, 21(4), 109–123, 2001a.

Kram, M.L., S.H. Lieberman, J. Fee, and A.A. Keller, Use of LIF for real-time *in-situ* mixed NAPL source zone detection, *Ground-Water Monitoring and Remediation*, 21(1), 67–76, 2001b.

Kram, M.L., A.A. Keller, J. Rossabi, and L.G. Everett, DNAPL characterization methods and approaches, Part 2: Cost comparisons, *Ground-Water Monitoring and Remediation*, 22(1), 46–61, 2002.

Kram, M.L., A.A. Keller, S.M. Massick, and L.E. Laverman, Complex NAPL site characterization using fluorescence, Part 1: Selection of excitation wavelength based on NAPL composition, *Soil and Sediment Contamination: an International Journal*, March/April, 13(2), 103–118, 2004.

Lieberman, S.H. and D.S. Knowles, Cone penetrometer deployed *in-situ* video microscope for characterizing subsurface soil properties, *Field Analytical Chemistry and Technology*, 2(2), 127–132, 1998.

Lieberman, S.H., G.A. Theriault, S.S. Cooper, P.G. Malone, R.S. Olsen, and P.W. Lurk, Rapid subsurface, *in-situ* field screening of petroleum hydrocarbon contamination using laser induced fluorescence over optical fibers, *Proceedings of the Second International Symposium on Field Screening Methods for Hazardous Wastes and Toxic Chemicals*, U.S. Environmental Protection Agency, Las Vegas, NV, 1991, pp. 57–63.

Lieberman, S.H., G.W. Anderson, and A. Taer, Use of a cone penetrometer deployed video-imaging systems for *in-situ* detection of NAPLs in subsurface soil environments, *Proceedings of the Petroleum Hydrocarbon and Organic Chemicals in Ground Water: Prevention, Detection, and Remediation*, National Ground Water Association, Westerville, OH, 1998, pp. 384–390.

Lieberman, S.H., P. Boss, G.W. Anderson, G. Heron, and K.S. Udell, Characterization of NAPL Distributions Using *in-situ* Imaging and LIF, Proceedings of the Second International Conference on Remediation of Chlorinated and Recalcitrant Compounds, Battelle Press, Columbus, OH, 2000.

Looney, B.B., K.M. Jerome, and C. Davey, Single-well dnapl characterization using alcohol injection/extraction, in *Non-Aqueous Phase Liquids, Remediation of Chlorinated and Recalcitrant Compounds*, G.B. Wickramanayake and R.E. Hinchee, Eds., Battelle Press, Columbus, OH, 1998, pp. 113–118.

Marrin, D.L., Soil-gas sampling and misinterpretation, *Ground-Water Monitoring Review*, 8(2), 51–54, 1988.

Marrin, D.L. and H.B. Kerfoot, Soil-gas surveying techniques, *Environmental Science and Technology*, 22, 740–745, 1988.

Meinardus, H.W., R.E. Jackson, M. Jin, J.T. Londergan, S. Taffinder, and J.S. Ginn, Characterization of a DNAPL zone with partitioning interwell tracer tests, in *Non-Aqueous Phase Liquids, Remediation of Chlorinated and Recalcitrant Compounds*, G.B. Wickramanayake and R.E. Hinchee, Eds., Battelle Press, Columbus, OH, 1998, pp. 143–148.

Mosier-Boss, P.A., R. Newbery, and S.H. Lieberman, Development of a cone penetrometer deployed solvent sensor using a SERS fiber optic probe, *Proceedings of a Specialty Conference on Field Analytical Methods for Hazardous Wastes and Toxic Chemicals*, U.S. Environmental Protection Agency, Las Vegas, NV, 1997, pp. 588–599.

MSE Technology Applications, Inc., Cost Analysis of Dense Non Aqueous Phase Liquid Characterization Tools, prepared for U.S. Department of Energy, Contract No. DE-AC22-96EW96405, 2002.

Myers, K.F., J. Costanza, and W.M. Davis, Cost and Performance Report for Tri-Services Site Characterization and Analysis Penetrometer System (SCAPS) Membrane Interface Probe, USACE ERDC/EL TR-02-01, January, 2002a, 40 pp.

Myers, K.F., W.M. Davis, and J. Costanza, Tri-Services Site Characterization and Analysis Penetrometer System Validation of the Membrane Interface Probe, U.S. Army Corps of Engineers, ERDC/EL TR-02-16, July, 2002b, 57 pp.

Nelson, N.T. and M.L. Brusseau, Field study of the partitioning tracer method for detection of dense nonaqueous phase liquid in a trichloroethene-contaminated aquifer, *Environmental Science and Technology*, 30(9), 2859–2863, 1996.

Nelson, N.T., M. Oostrom, T.W. Wietsma, and M.L. Brusseau, Partitioning tracer method for the *in-situ* measurement of DNAPL saturation: influence of heterogeneity and sampling method, *Environmental Science and Technology*, 33, 4046–4053, 1999.

Newell, C. and R.R. Ross, Estimating Potential for Occurrence of DNAPL at Superfund Sites, Quick Reference Guide Sheet, U.S. Environmental Protection Agency Publication Number 9355.4-07FS, U.S. Environmental Protection Agency, Washington, DC, 1991.

Pankow, J.F., and J.A. Cherry, *Dense Chlorinated Solvents and other DNAPLs in Groundwater: History, Behavior, and Remediation*, Waterloo Press, Portland, OR, 1996, 522 pp.

Payne, T., J. Brannon, R. Falta, and J. Rossabi, Detection limit effects on interpretation of NAPL partitioning tracer tests, in *Non-Aqueous Phase Liquids, Remediation of Chlorinated and Recalcitrant Compounds*, G.B. Wickramanayake and R.E. Hinchee, Eds., Battelle Press, Columbus, OH, 1998, pp. 125–130.

Pitkin, S.E., Detailed subsurface characterization using the Waterloo Profiler, *Proceedings of the Symposium on the Application of Geophysics to Environmental and Engineering Problems*, Society of Environmental and Engineering Geophysics, Denver, CO, 1998, pp. 53–64.

Rossabi, J., B.D. Riha, C.A. Eddy-Dilek, A. Lustig, M. Carrabba, W.K. Hyde, and J. Bello, Field tests of a DNAPL characterization system using cone penetrometer-based Raman spectroscopy, *Ground-Water Monitoring and Remediation*, 20(4), 72–81, 2000.

Semprini, L., M. Cantaloub, S. Gottipati, O. Hopkins, and J. Istok, Radon-222 as a Tracer for quantifying and monitoring NAPL remediation, in *Non-Aqueous Phase Liquids, Remediation of Chlorinated and Recalcitrant Compounds*, G.B. Wickramanayake and R.E. Hinchee, Eds., Battelle Press, Columbus, OH, 1998, pp. 137–142.

Silverstein, R.M., G.C. Bassler, and T.C. Morrill, *Spectrometric Identification of Organic Compounds*, 5th ed. John Wiley and Sons, Inc., New York, NY, 1991, 419 pp.

Sinclair, N. and M. Kram, High-resolution 3-D seismic reflection surveys for characterization of hazardous waste sites, Proceedings of the Third Tri-Services ESTCP Workshop, August 18, 1998.

Sudicky, E.A., A natural gradient experiment on solute transport in a sand aquifer: spatial variability of hydraulic conductivity and its role in the dispersion process, *Water Resources Research*, 22(13), 2069–2082, 1986.

Taylor, T.W. and M.C. Serafini, Screened auger sampling: the technique and two case studies, *Ground-Water Monitoring Review*, 8(3), 145–152, 1988.

Trotsky, J., Personal communication, December 8th, 1999.

Udell, K.S., G. Heron, S. Collins, M. Begas-Landeau, S. Kaslusky, H. Liang, M.L. McDonald, W. Mabey, and T. Heron, Field Demonstration of Steam-Enhanced Extraction at Alameda Point, California, Proceedings of the Second International Conference on Remediation of Chlorinated and Recalcitrant Compounds, Battelle Press, Columbus, OH, 2000.

U.S. Environmental Protection Agency, Hazardous Waste Cleanup Information, 2005. Available at http://www.clu-in.org/char1.cfm.

Wise, W.R., NAPL characterization via partitioning tracer tests: quantifying effects of partitioning nonlinearities, *Journal of Contaminant Hydrology*, 36(1–2), 167–183, 1999.

Yoon, S., I. Barman, A. Datta-Gupta, and G.A. Pope, *In-situ* characterization of residual NAPL distribution using streamline-based inversion of partitioning tracer tests, Proceedings of the 1999 Exploration and Production Environmental Conference, SPE/EPA, Austin, TX, March 1–3, 1999, pp. 391–400.

8

Ground-Water Monitoring System Design

Martin N. Sara

CONTENTS

Introduction	517
Regulatory Concepts in Facility Monitoring	519
Data Analysis Required for Monitoring System Design	525
Selecting the Target Monitoring Zones	526
Locating Background and Downgradient Wells	528
Gradients	528
Steep or Flat Gradients	531
Procedures for Gradient Controlled Sites	534
Geologic Controls	543
Single Homogeneous Aquifer	545
Single Aquifer of Variable Hydraulic Conductivity	546
Multiple Aquifers	549
Low-Hydraulic Conductivity Environments	551
Geologic Structural Control	554
Perched Ground Water	556
Secondary Hydraulic Conductivity	559
Density Control	563
Separation of Adjacent Monitoring Programs	566
Simple Gradient Control	566
Gradient and Lithology Control	567
Complex Ground-Water Flow Conditions	569
Monitoring System Design Criteria	571
References	571

Introduction

At least three different types of ground-water monitoring programs are commonly applied to environmental site assessment and remediation projects. These include broad-based programs, such as reconnaissance monitoring of wide areas for the purpose of resource evaluation, and detection and assessment monitoring of ground-water quality beneath a specific facility, such as a solid- or hazardous-waste landfill or surface impoundment. Detection and assessment monitoring include program elements from the Resource Conservation and Recovery Act (RCRA) Subtitles C and D, and many aspects of ground-water monitoring under the Superfund program. This chapter begins with a detailed discussion of detection monitoring programs, then addresses design aspects of assessment monitoring programs. Area-wide reconnaissance programs, which generally have very different design goals compared with the more regulation-driven detection and assessment programs, are not discussed in detail.

To assist in the design of an optimum or ideal ground-water monitoring system for a detection monitoring program, a list of desired system attributes can be formulated. These attributes include:

- Installation of a three-dimensional array of monitoring points for discrete sampling, water-level measurement, and hydraulic testing
- Ability to provide continuous real-time measurements of chemical parameters and hydraulic head at each monitoring point
- Installation of as few boreholes as possible penetrating the facility area, to minimize the potential for cross-contamination
- Installation of a sufficient number of monitoring points so that complex hydrogeologic conditions will not confound interpretation of data or prevent detection of potential releases from the facility
- Ability to immediately detect significant releases by sufficiently frequent measurements of indicator parameters
- Installation of monitoring points that are sufficiently reliable to maintain reproducibility and representativeness of ground-water sampling data
- Ability to conduct convenient maintenance and quality control auditing

These seven attributes of an ideal ground-water monitoring system are not obtainable using currently available technology. However, many of these attributes can be achieved by a monitoring system that is thoughtfully designed through use of a detailed conceptual understanding of the hydrogeologic conditions present at the target site.

Because ground water moves slowly and in predictable pathways, nearly continuous monitoring at many points can be effectively achieved if the target monitoring zone is carefully selected and a reasonable sampling schedule is established. Techniques provided in other chapters of this book, combined with those described in ASTM Standard D 5092 (ASTM, 2002), effectively address the remaining monitoring system attributes listed earlier.

Selection of the proper locations for monitoring wells should be based on a holistic approach to the evaluation of a specific site. Decisions regarding placement of wells in this process must weigh and balance data collected and analyzed in the field, laboratory, and office.

The question "How much detection monitoring is enough?", when answered in the context of the number of monitoring wells required at a site, will be entirely site-specific. In general, the monitoring system designer should ensure that convincing evidence is established to validate each assumption and to demonstrate the basic capability of the system to produce groundwater samples representative of both upgradient (background) and downgradient conditions. General rules of thumb are provided in this chapter, but the reader should bear in mind that "enough" is a subjective determination to address the questions of how much monitoring is necessary to provide a monitoring system capable of detecting ground-water contamination and how much demonstration is required to convince a regulatory agency of that capability.

The key to complying with most regulatory programs that require ground-water monitoring is demonstrating that the system is capable of addressing a few important items. The owner or operator of a facility required to monitor ground water must install and implement a monitoring system that is capable of determining the facility's impact on ground water. The monitoring system must be capable of yielding representative ground-water samples for chemical analyses. The number, locations, and depths of detection monitoring wells must be such that the system is capable of prompt detection of any statistically significant differences in indicator parameters.

The monitoring system designer must base decisions on numbers and locations of monitoring wells on performance criteria that describe what comprises a sufficient monitoring system. Some very simple geologic environments may be effectively monitored with U.S. EPA's suggested minimum system of one upgradient well and three downgradient wells (U.S. EPA, 1982). However, this level of monitoring is adequate for very few sites. It is not uncommon for monitoring systems to utilize dozens or even hundreds of sampling points to achieve the objectives of detection monitoring. This is especially true for sites located in state in which regulations are aggressively enforced. This is also true for facilities that have been in operation over long periods of time or that consist of multiple regulated units or expansions.

Regulatory Concepts in Facility Monitoring

Hazardous and solid-waste management facilities must comply with U.S. EPA ground-water-monitoring requirements set forth in RCRA Subtitles C and D (40 CFR 264.97). The owner or operator of a hazardous-waste management facility must comply with the following requirements for any ground-water monitoring program developed to satisfy §264.98, §264.99, or §264.100:

1. The ground-water monitoring system must consist of a sufficient number of wells, installed at appropriate locations and depths, to yield ground-water samples from the uppermost aquifer that:
   a. Represent the quality of background (upgradient) ground water that has not been affected by possible leakage from a facility
   b. Represent the quality of ground water downgradient of the facility
2. If a facility contains more than one regulated unit, separate ground-water-monitoring systems are not required for each regulated unit provided that provisions for sampling the ground water in the uppermost aquifer will enable detection and measurement at the compliance point of hazardous constituents from the regulated units that have entered the ground water in the uppermost aquifer.

Many millions of dollars and thousands of words in reports and meetings have been spent on defining exactly what these relatively few lines of text really mean in the context of monitoring of hazardous waste sites. Both RCRA Subtitle C (hazardous waste) and Subtitle D (solid municipal waste) facilities are required to meet these basic points of detection monitoring programs. This Federal rule can be depicted in a single figure that illustrates the concept of detection monitoring. Figure 8.1 shows a conceptual presentation of the §264.97 guidance on placement of detection monitoring wells.

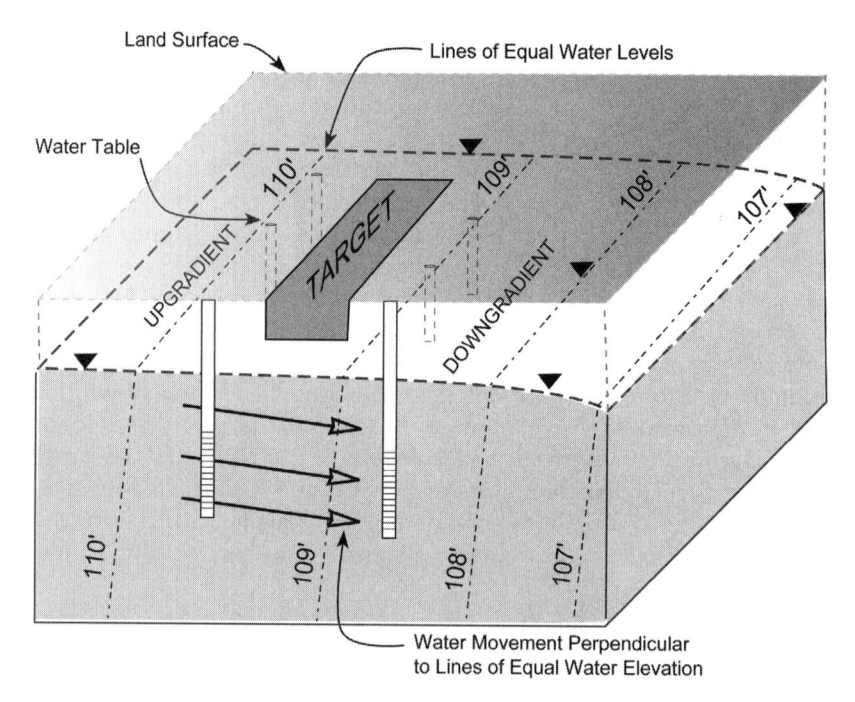

FIGURE 8.1
Regulatory context of detection monitoring.

The RCRA Ground-Water Monitoring Technical Enforcement Guidance Document (TEGD) (U.S. EPA, 1986) provides additional guidance on placement and number of upgradient or background wells by recommending that these wells are:

- Located beyond the upgradient extent of possible contamination from the hazardous waste management unit so that they reflect background water quality
- Screened at the same stratigraphic horizons as downgradient wells to ensure comparability of data
- Of sufficient number to account for natural spatial variability in background groundwater quality

If the conceptual homogeneous unconfined uppermost aquifer (Figure 8.1) were present at every site, it would be relatively easy to meet these three TEGD requirements. However, this conceptual hydrogeologic condition is seldom observed in the field, as such a simple unconfined aquifer flow system is rare. RCRA Subtitle D includes many components specific to ground-water monitoring. Section 258.51 (ground-water monitoring systems) requires that:

> A ground-water monitoring system must be installed that consists of a sufficient number of wells, installed at appropriate locations and depths, to yield groundwater samples from the uppermost aquifer (as defined in Section 258.2) that:
>
> 1. Represent the quality of background ground water that has not been affected by leakage from a unit. A determination of background quality may include sampling of wells that are not hydraulically upgradient of the waste management area where:

a. Hydrogeologic conditions do not allow the owner or operator to determine what wells are hydraulically upgradient
b. Sampling at other wells will provide an indication of background ground-water quality that is as representative or more representative than that provided by the upgradient wells

2. Represent the quality of ground water passing the relevant points of compliance specified by the Director of an approved State under Section 258.40(d) or at the waste management unit boundary in unapproved States. The downgradient monitoring system must be installed at the relevant point of compliance specified by the Director of an approved State under Section 258.40(d) or at the waste management unit boundary in unapproved States that ensures detection of ground-water contamination in the uppermost aquifer. When physical obstacles preclude installation of groundwater-monitoring wells at the relevant point of compliance at existing units, the downgradient monitoring system may be installed at the closest practicable distance hydraulically downgradient from the relevant point of compliance specified by the Director of an approved State under Section 258.40 that ensures detection of ground-water contamination in the uppermost aquifer.

The Director of an approved State, under Section 258.51(d), may approve a multi-unit ground-water monitoring system instead of separate ground-water monitoring systems that meet the requirement of Section 258.5(a) and will be as protective of human health and the environment as individual monitoring systems for each municipal solid waste landfill (MSWLF) unit, based on the following factors:

- Number, spacing, and orientation of the MSWLF units
- Hydrogeologic setting
- Site history
- Engineering design of the MSWLF units
- Type of waste accepted at the MSWLF units

RCRA Subtitle D, Section 258.51(d)(2), goes on to require that the number, spacing, and depths of monitoring wells shall be:

Determined based upon site-specific technical information that must include thorough characterization of:
1. Aquifer thickness, ground-water flow rates, ground-water flow direction, including seasonal and temporal fluctuations in ground-water flow
2. Saturated and unsaturated geologic units and fill materials overlying the uppermost aquifer, materials comprising the uppermost aquifer, and materials comprising the confining units defining the lower boundary of the uppermost aquifer, including, but not limited to: thicknesses, stratigraphy, lithology, hydraulic conductivities, porosities, and effective porosities.

The earlier-mentioned Subtitle D technical requirements may need a full definition of site geologic and hydrogeologic conditions. Many waste disposal facilities are located in complex geologic environments in which very extensive site investigations are required to properly locate the wells required for detection ground-water monitoring systems as described earlier. Layering of geologic units of significantly different hydraulic conductivity complicates the simple conceptual model described by the Federal rules. Figures 8.2a and b show a two-layer system with the uppermost aquifer consisting of homogeneous isotropic sand below a near-surface silt of clay unit of lower hydraulic conductivity.

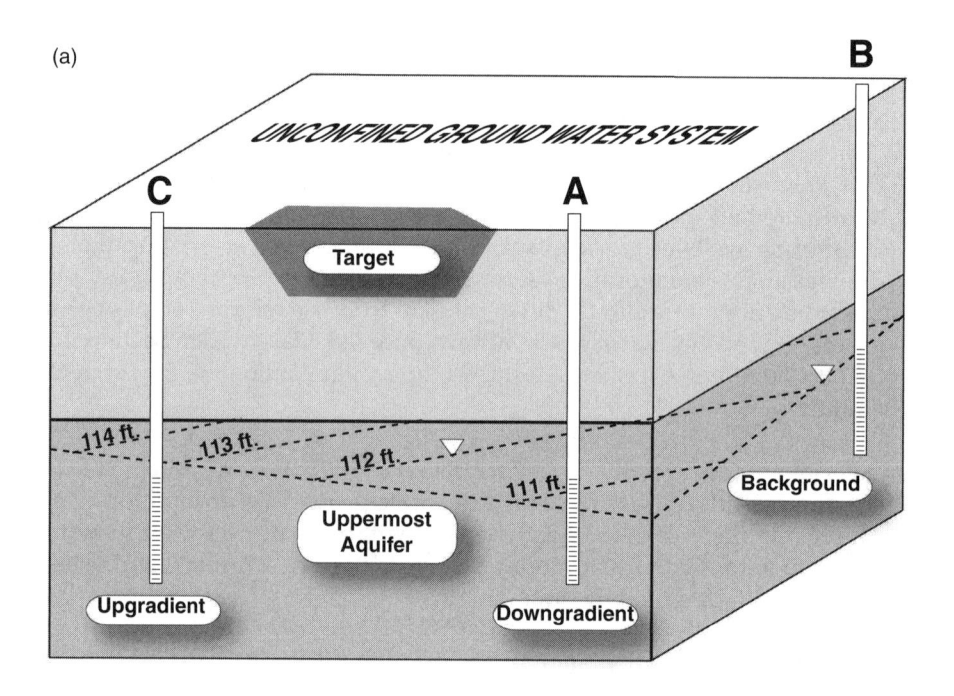

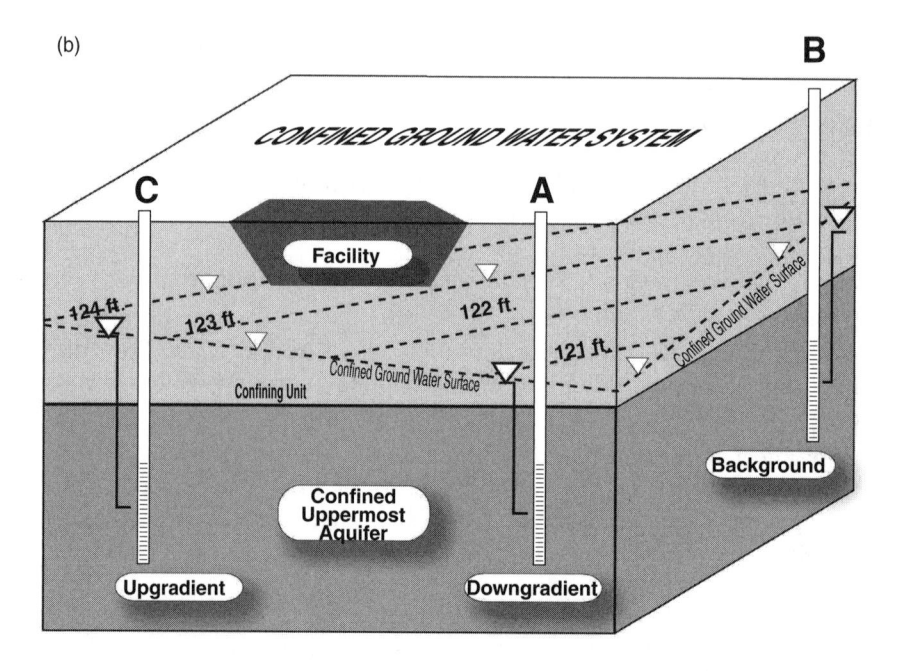

FIGURE 8.2
(a) Unconfined, (b) confined, and (c) unconfined and confined, ground-water systems.

Ground-Water Monitoring System Design   523

(c)

Note: Ground Water Level
Above Bottom of Confining Unit

CONFINED/UNCONFINED WATER TABLE

C — Confining Unit — *Confined Ground Water Surface*

123 ft.
122 ft.
121 ft.
120 ft.
119 ft.
118 ft.
Unconfined Ground Water Surface

Target

A

B — Background

Uppermost Aquifer

Upgradient   Downgradient

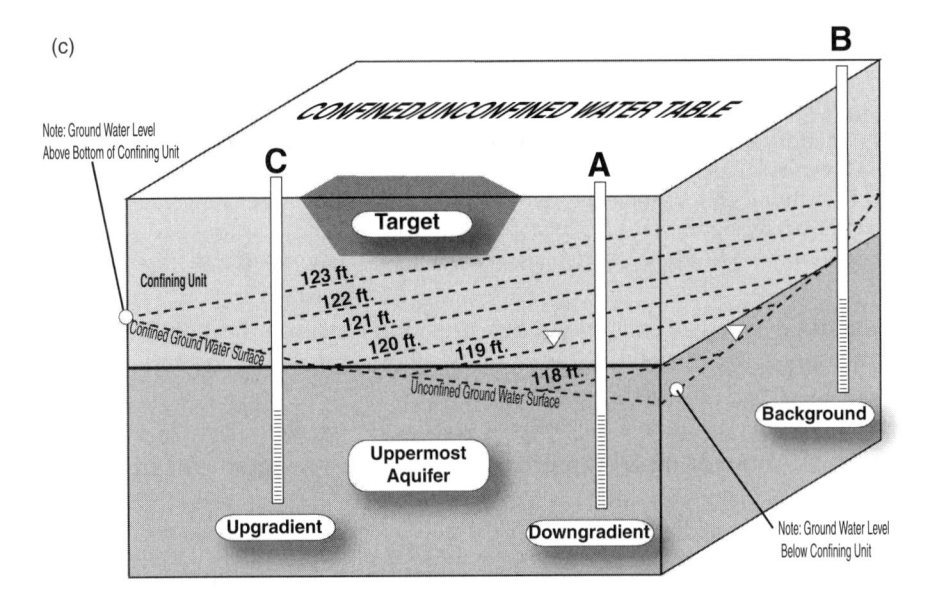

Note: Ground Water Level Below Confining Unit

FIGURE 8.2
Continued.

In Figure 8.2a, the uppermost aquifer is unconfined, in that the upper surface of the saturated zone (the water table) is free to rise and fall in response to changes in recharge conditions. Where an aquifer is overlain by a low hydraulic conductivity unit (a confining bed), as shown in Figure 8.2b, the aquifer is said to be confined by that unit. Downgradient well positions are shown as point A in both figures. Both upgradient and background wells are also shown in these figures. The concept of background representing not hydraulically upgradient locations but reflecting general water quality of the uppermost aquifer is represented by point B. In each case, the sand unit should be considered as the uppermost aquifer for the following reasons:

- The sand unit has regional areal extent and is saturated.
- The sand unit has sufficient hydraulic conductivity to yield usable quantities of water to springs or wells.
- The sand unit would be the zone in which leachate from the facility could migrate horizontally away from the site to potentially affect human health and the environment.

Much of the early concern of regulatory agencies with respect to Subtitle C detection monitoring programs is with meeting Federal regulations in 40 CFR 265.91, which describes ground-water monitoring system requirements for interim-status hazardous-waste disposal facilities. These regulations state:

> A ground-water monitoring system must be capable of yielding ground-water samples for analysis and must consist of:
>
> (1) Monitoring wells (at least one) installed hydraulically upgradient (i.e., in the direction of increasing static head) from the limit of the waste management area. Their number, locations and depths must be sufficient to yield ground-water samples that are:

(a) representative of background ground-water quality in the uppermost aquifer near the facility
(b) not affected by the facility
(2) Monitoring wells (at least three) installed hydraulically downgradient (i.e., in the direction of decreasing static head) at the limit of the waste management area. Their number, locations and depths must ensure that they immediately detect any statistically significant amounts of hazardous waste or hazardous waste constituents that migrate from the waste management area to the uppermost aquifer.

This interim-status rule has several key features different from Section 264.97 rules that have been widely used in defining what a detection monitoring system should consist of, specifically:

- At least one monitoring well upgradient and three downgradients from a facility
- Immediate detection capabilities

While immediate detection is open to widely variable interpretation, especially considering the slow movement of ground water, the TEGD (U.S. EPA, 1986) provides some additional guidance on how to meet the "immediate" criteria by placing detection monitoring wells immediately adjacent to the waste management unit. The Federal Subtitle D regulations proposed for nonhazardous solid-waste sites set 150-m buffer zones (or property boundary, whichever is less) for placement of monitoring wells. This buffer zone was also included in U.S. EPA's final rules (U.S. EPA, October 1991) and might include a compliance boundary set at the edge of the waste management unit. Reducing these Federal regulations to a series of criteria based on concepts presented in this text results in a series of technical points. The detection monitoring system should have:

- Sufficient wells, both upgradient (background) and downgradient, to detect discharges from the regulated facility
- Wells located within a flow path from the regulated facility in the uppermost aquifer

Furthermore, the uppermost aquifer should have sufficient hydraulic conductivity and extent so that sampling could be conducted within the waste unit boundary for both Subtitle C hazardous-waste facilities and Subtitle D solid-waste sites. An adequate detection monitoring program can be designed for any geologic or hydrogeologic environment using the earlier-mentioned criteria. The following sections present conceptual models for detection monitoring programs for a wide variety of hydrogeologic environments.

Prior to selecting the locations and depths for the screened intervals for ground-water monitoring wells, the ground-water monitoring system designer must have, at a minimum:

- Performed a complete site characterization program
- Established a conceptual hydrogeologic model for the site
- Constructed ground-water flow nets in three dimensions
- Located the facility boundaries and waste disposal areas

Each of these tasks provides data that will be used to select the target monitoring zones for the monitoring system. The remaining sections describe the monitoring system design

Ground-Water Monitoring System Design

BASE MAP

LOCATION OF SITE FEATURES
- Topographic map format
- Plot all pertinent site features on map
- Surficial soil units should be sampled for analysis
- Define recharge and discharge areas on map

CONCEPTUAL MODEL

CONCEPTUAL MODELS & CROSS-SECTIONS
- Prepare cross-section with lithology
- Define facility base grades
- Compare base grades with permeable units
- Establish most likely uppermost aquifer

FLOWNET

GROUND-WATER FLOW DIRECTIONS
- Plot piezometric and potentiometric heads
- Define relative head differences between aquifers
- Check for interconnection between aquifers
- Calculate rates of ground-water movement
- Plot flow directions using flow lines and equipotentials
- Establish if vertical gradients would predict target zone

TARGET MONITORING ZONE

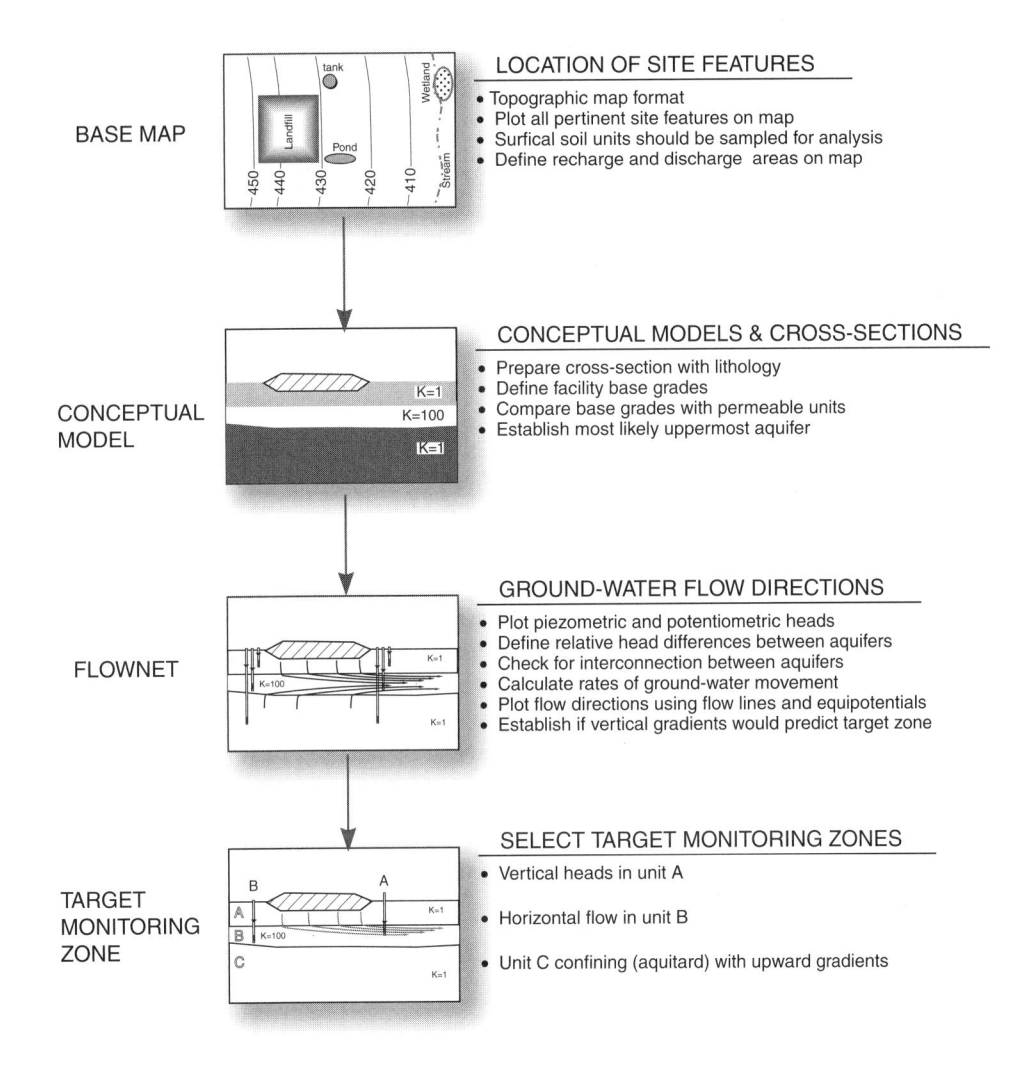

SELECT TARGET MONITORING ZONES
- Vertical heads in unit A
- Horizontal flow in unit B
- Unit C confining (aquitard) with upward gradients

FIGURE 8.3
Monitoring system design summary.

process summarized in Figure 8.3. Examples of the process are included to assist in the design conceptualization process.

Data Analysis Required for Monitoring System Design

Geologic factors (related chiefly to geologic formations and their water-bearing properties) and hydrologic factors (related to the movement of water in these formations) must be known in some detail to properly design a ground-water monitoring system. These data are normally developed in a field investigation, conducted as described in Chapter 2.

The geologic framework of a site includes the lithology, texture, structure, mineralogy, and distribution of the unconsolidated and consolidated earth materials through which ground water flows. The hydraulic properties of these earth materials depend upon the geologic framework. Thus, the geologic framework of the facility heavily influences the design of the ground-water monitoring system. Elements of the hydrogeologic framework and the site hydrogeology that should be considered in ground-water monitoring system design include:

- The spatial location and configuration of the uppermost aquifer and its hydraulic properties (e.g., horizontal and vertical hydraulic conductivity, depth and location of the ground-water surface, seasonal fluctuations of groundwater surface elevation)
- Hydraulic gradient (vertical and horizontal) within the geologic materials underlying the facility
- Discharge and recharge areas of the site
- Facility operational considerations

These data are used to establish the locations of both upgradient and downgradient wells in the uppermost aquifer. Both upgradient and downgradient wells should be located in the direction of ground-water flow along flow pathways most likely to transport ground-water and any potential contaminants contained in ground water. These pathways should be identified from data gained from existing information and the field investigation. The objective of the field investigation and subsequent data analysis and interpretation is to provide some or all of the following information:

- Lithologic characteristics of the subsurface, including:
    a. Stratigraphic and hydrogeologic units
    b. Classification of hydrogeologic units
    c. Extent of hydrogeologic units
- Key hydrogeologic characteristics used to describe the site, including:
    a. Hydraulic conductivity (vertical and horizontal)
    b. Porosity
    c. Gradient (vertical and horizontal)
    d. Specific yield
- Aquifer characteristics including:
    a. Boundaries
    b. Type of aquifer (unconfined or confined)
    c. Saturated and unsaturated conditions

Each piece of data is an important building block in establishing the conceptual hydrogeologic model and in targeting the zones to be monitored. These data are used in combination to define the uppermost aquifer and hydraulic gradients and to allow the construction of a flow net that will provide identification of aquifer flow pathways so that the target monitoring zones can be selected.

Selecting the Target Monitoring Zones

The first task in the design of a detection ground-water monitoring system is the selection of the target monitoring zones (see ASTM, 2002). The logic used in selection of a target monitoring zone is illustrated in Figure 8.4. A review of features of the facility to be

Ground-Water Monitoring System Design

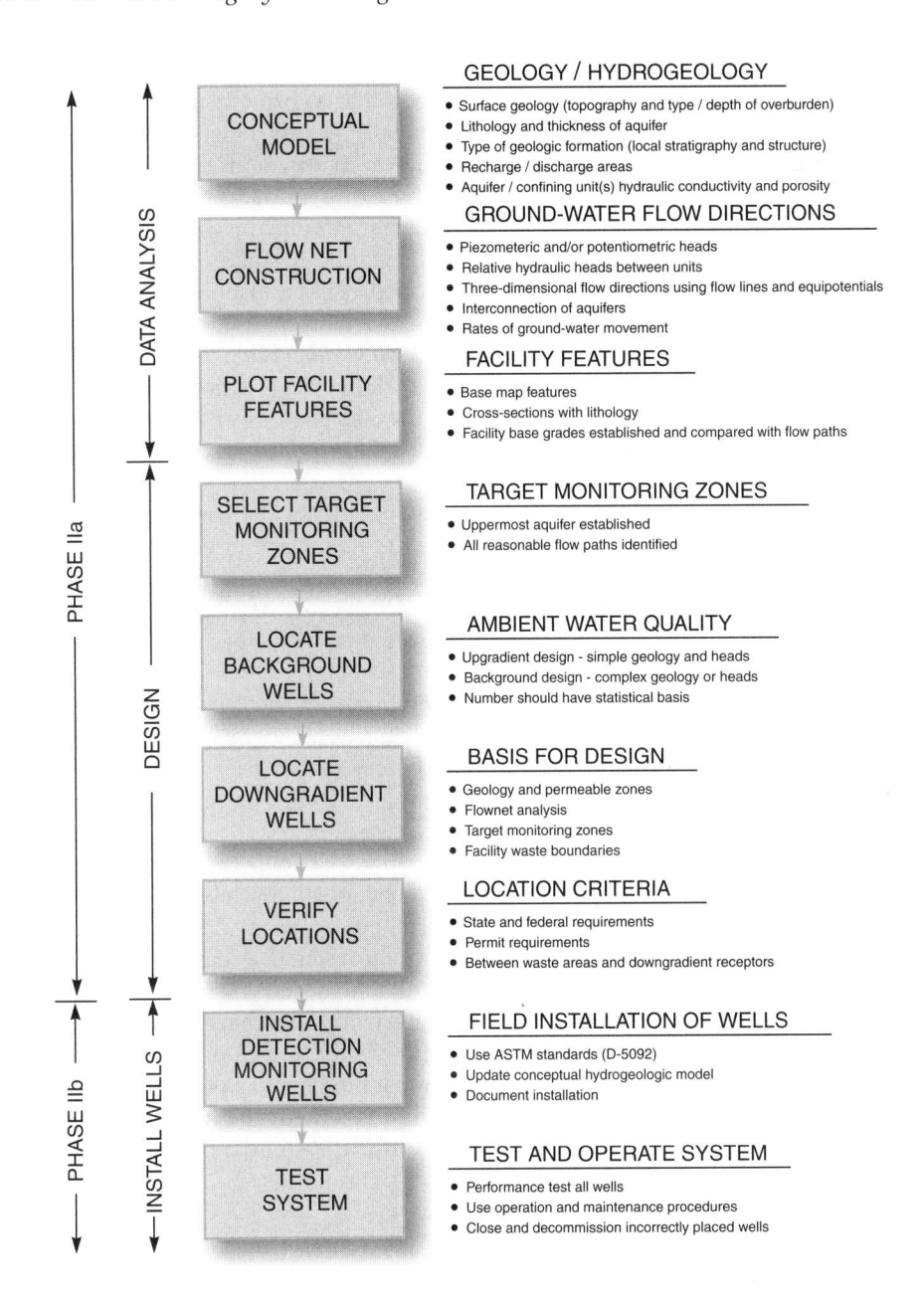

FIGURE 8.4
Flow diagram of monitoring system design.

monitored, used in combination with conceptual models and flow nets, provides the system designer with the information to select those zones that will provide a high level of certainty that releases from the facility will be immediately detected. The concept of the target monitoring zone was developed as a means of directing the ground-water-monitoring system designer toward placement of well screens in the uppermost aquifer at locations and depths that would have the highest likelihood of detecting leakage from a facility. "Target monitoring zone" is defined in ASTM Standard D-5092 (ASTM, 2002) as the ground-water flow path from a particular area or facility in which monitoring wells will be installed. The target monitoring zones should be a stratum (or strata) in which there is a reasonable expectation that a correctly placed well will intercept migrating contaminants. This target zone usually lies in the saturated geologic unit in which ground-water flow rates are the highest because it possesses the highest hydraulic conductivity of the material underlying or adjacent to the facility of interest. Figure 8.4 illustrates the process of selection of a target monitoring zone using information on facility features, geologic characteristics, and hydraulic characteristics gathered during the preliminary field investigation. This selection process can be described as a series of steps.

Step 1. Locate Site Features on a Topographic Base Map Format: Site features should be compared with information on geologic and soil maps to define the location of important facility components in relation to the distribution of surficial geological materials. Any likely recharge or discharge areas (streams, wetlands, or other surface-water) should be located.

Step 2. Cross-Section Construction and Conceptual Model Development: Cross-sections should be constructed, based on boring logs or geophysical traverses. These sections should be compared with the location of site features and facility components. The base grade of the facility should be plotted on cross-sections of sensitive geologic units or ground-water flow pathways. A conceptual model should be constructed to establish the site geological framework and to illustrate distribution of geologic materials of differing hydraulic conductivity.

Step 3. Use Flow Nets to Define Likely Direction of Ground-Water Flow: Construction of flow nets assists in defining the gradient and direction of ground-water flow in the uppermost aquifer. The rates of flow along flow paths can be calculated from the information provided by the flow net using equations from Chapter 14. Vertical gradients can be used to predict target zones by comparison of relative heads between units. Interconnections between aquifers can be predicted from relationships between hydraulic conductivities and hydraulic heads for the units defined in the conceptual models.

Step 4. Select Target Monitoring Zones: The unit meeting the regulatory definition of the uppermost aquifer, which also shows primarily horizontal ground-water movement under or adjacent to the facility, would represent the primary target monitoring zone. This unit would probably consist of permeable material that discharges to other permeable units or to local discharge areas. The system designer should locate the flow paths within the uppermost aquifer that would represent the most likely zones of ground-water movement away from the facility. These zones, typically those with the highest hydraulic conductivity, would be the focus of the detection monitoring system. If the uppermost aquifer is interconnected with other aquifers, these aquifers should also be monitored to provide safeguards for downgradient ground-water users.

This four-step procedure for selecting the target monitoring zones must be flexible enough to accommodate environmental effects due to seasonal changes in gradient or due to future plans to expand or alter the configuration of the facility. The target monitoring zone might include only a portion of a very thick aquifer (e.g., the top 30 ft) or it might span several geologic units (as in the case of a thin, permeable, unconsolidated unit overlying weathered, or fractured bedrock). These target zones represent the proper locations for placement of monitoring wells.

Locating Background and Downgradient Wells

Gradients

The basis for detection monitoring programs is knowledge of the hydraulically upgradient and downgradient direction from the site to be monitored. Figure 8.1 illustrates a simple relationship of ground-water movement from higher potentiometric surface elevations (upgradient) to lower potentiometric surface elevations (downgradient). This simple conceptual model of a homogeneous aquifer is the basis for much of the regulatory thought on ground-water monitoring. Unfortunately, it is very rare to find such a simple flow configuration in the real world.

After selection of the target monitoring zones, the next step in design of a ground-water-monitoring system is locating upgradient or background monitoring wells within those zones. The conceptual geologic model and flow net construction will have defined the uppermost aquifer and the relative direction of ground-water flow, both vertically and horizontally. Location of upgradient wells should be based not only on this information, but also on other factors mainly relating to the physical presence of the facility. The numbers of upgradient or background wells installed at a site must be based on the size of the facility, the geologic or hydrogeologic environment, and the ability to satisfy statistical criteria for analysis of water-quality data (see Chapter 17). As a general guidance, it is very difficult to conduct any type of statistical test (applicable under RCRA) unless several upgradient wells are used in the monitoring system. This is due to the natural spatial variability observed in geologic environments and ground-water chemistry. This spatial variability must be anticipated and evaluated during the design process so that sufficient background water-quality data are available for background-to-downgradient water-quality statistical comparisons.

The TEGD (U.S. EPA, 1986) defines upgradient wells as "one or more wells that are placed hydraulically upgradient of the site and are capable of yielding ground-water samples that are representative of regional conditions and not affected by the regulated facility." This usage of the term upgradient is consistent with 40 CFR 265.91, which links background and upgradient for interim RCRA sites. Background wells would meet the 40 CFR 264.97 test to "represent the quality of background water that has not been affected by leakage from a regulated unit and represent the quality of ground water passing the point of compliance." The term upgradient can be a difficult concept to demonstrate in ground-water monitoring system design, because field conditions may not match the simple regulatory models. As a closing statement on the relationship between upgradient and downgradient wells, a correctly located detection monitoring well will only be placed within the flow paths from the base grades of the facility to be monitored. This is due to the placement of the downgradient well screen within the flow path from the facility. The designer must carefully consider site-specific hydraulic conditions to accurately locate upgradient monitoring wells because ground water does not always flow as expected in a simple regulatory model (horizontally, from upgradient to downgradient areas).

Simple single-aquifer flow systems can only be established through a clear understanding of the directional movement of ground water and through evaluation of the ground-water gradients across a site. Figure 8.5 and Figure 8.6 illustrate, in plan view and cross-section, the flow in the proximity of a gaining stream, where discharging ground water provides the stream's base flow. Figures 8.7a and b illustrate flow in the proximity of a losing stream, where surface water supports adjacent ground-water levels.

In each case, this simple system provides directional components to allow the positioning of ground-water monitoring wells. Figure 8.5 and Figure 8.6 illustrate a facility (B) located in a recharge area that discharges to streams on either side of the facility. Ground-water flow lines are shown in plan and cross-section. Because the facility is located directly on top of the recharge area, the downgradient flow zone is composed of a wide arc around the facility. This provides perhaps the simplest example of a gradient-controlled system.

Potential target zones for a detection monitoring system are shown in Figure 8.5. Background water-quality target zones should be sufficiently upgradient of the facility,

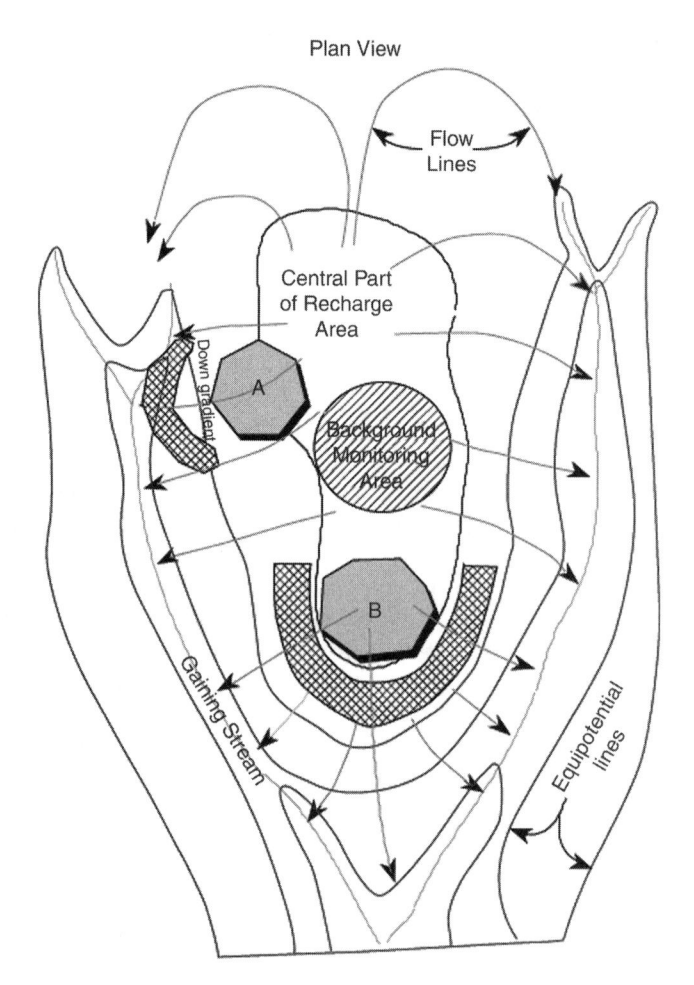

FIGURE 8.5
Potential target monitoring areas.

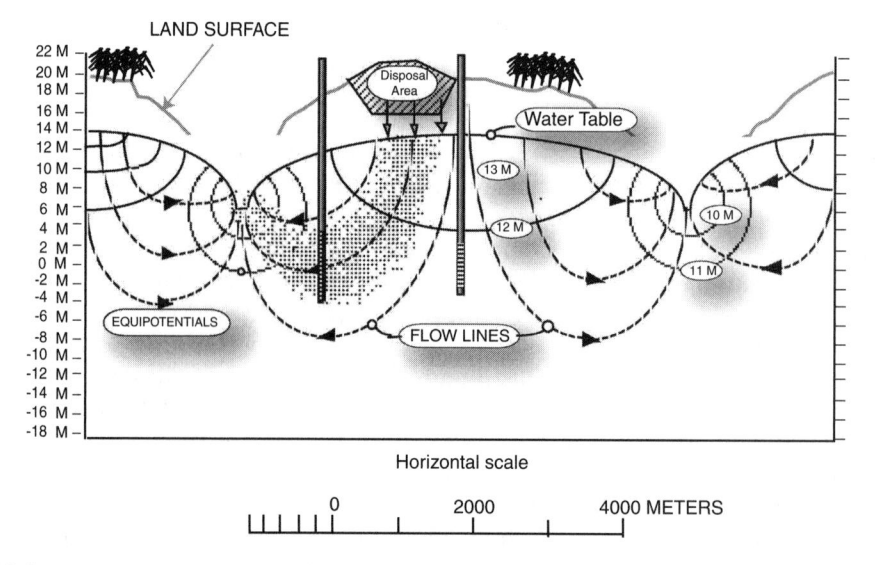

FIGURE 8.6
Cross-section of target monitoring zones.

so they are not affected by the facility. Several conclusions can be drawn from Figure 8.5 and Figure 8.6:

- Facility A would have its downgradient monitoring wells located within the ground-water flow lines shown. This facility location would have background monitoring wells located in the central recharge area.
- Facility B would have an upgradient or background well in the area indicated. Because the facility is located directly within the local recharge area, this would not be considered an upgradient well, but rather a background well that represents water quality similar to that for a well that would be upgradient from the facility.

Actual flow conditions would result in a water table significantly flatter than that shown in Figure 8.6. Vertical exaggeration (approximately 125 to 1) makes the flow lines appear to travel deeper than would be represented in the real world. The vertical scale indicates that the monitoring wells installed at the site should be screened from 19 to 24 m below ground surface to intercept the ground-water flow (and any contained contaminants) emanating from beneath the site.

Figures 8.7a and b illustrate a losing stream condition and the resultant monitoring target zones for Facility A. Because the stream in this illustration is recharging ground-water, and thus represents the highest point of upgradient ground water, target monitoring zones are located along the flow lines shown in Figure 8.7. Depths of screen placement must be based on the projected vertical gradients in the area. One can observe from the example provided that the location of wells in a detection monitoring system is particularly sensitive to whether the stream is gaining or losing. This relatively simple complication can lead to incorrect location of downgradient detection monitoring wells. Piezometers located perpendicular to the stream and careful evaluation of stream flow can provide the basic data to define the recharge and discharge relationship of the surface and ground-water system.

Steep or Flat Gradients

Even simple single-aquifer systems require consideration of local gradients adjacent to the facility of interest. In an area with a relatively flat gradient it is necessary to consider possible ground-water flow in what would normally be an upgradient direction. In an area with a steep gradient on the water-table surface, as shown in Figure 8.8 (typical of low-hydraulic conductivity materials), there is little potential for reversal of flow directions. The target monitoring zone in an area with such a steep gradient would normally be much narrower than in a flat gradient environment. The relationship between horizontal and vertical gradients is still required to establish the depth of the detection monitoring well screens. Figure 8.8 shows placement of two upgradient piezometers (C and D) and two downgradient piezometers (A and B). A monitoring well screened at point B would meet the regulatory criteria of being downgradient from the facility (and from upgradient piezometer C). However, the flow path screened by a well at point B would be too deep to intercept flow from the unlined facility. A detection monitoring well located at the depth of piezometer A would be correctly placed to monitor conditions downgradient of the facility.

Figure 8.9 and Figure 8.10 show an unlined landfill within an area where the hydraulic gradient is low. The target monitoring zone is characteristically thicker than it would be in an area with a high hydraulic gradient (as shown in Figure 8.8). The discharge directions shown in these figures represent a common hydraulic condition for unlined facilities. The cross-sectional view (Figure 8.9) shows both intermediate and shallow flow cells. The

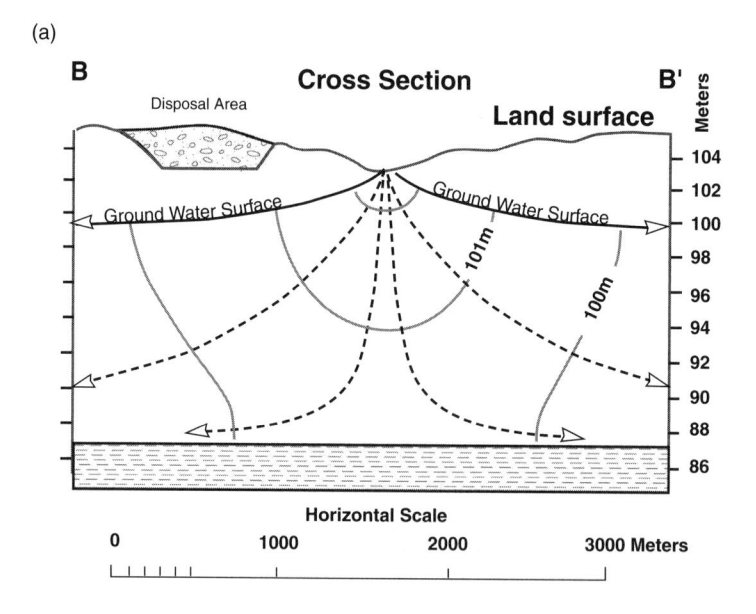

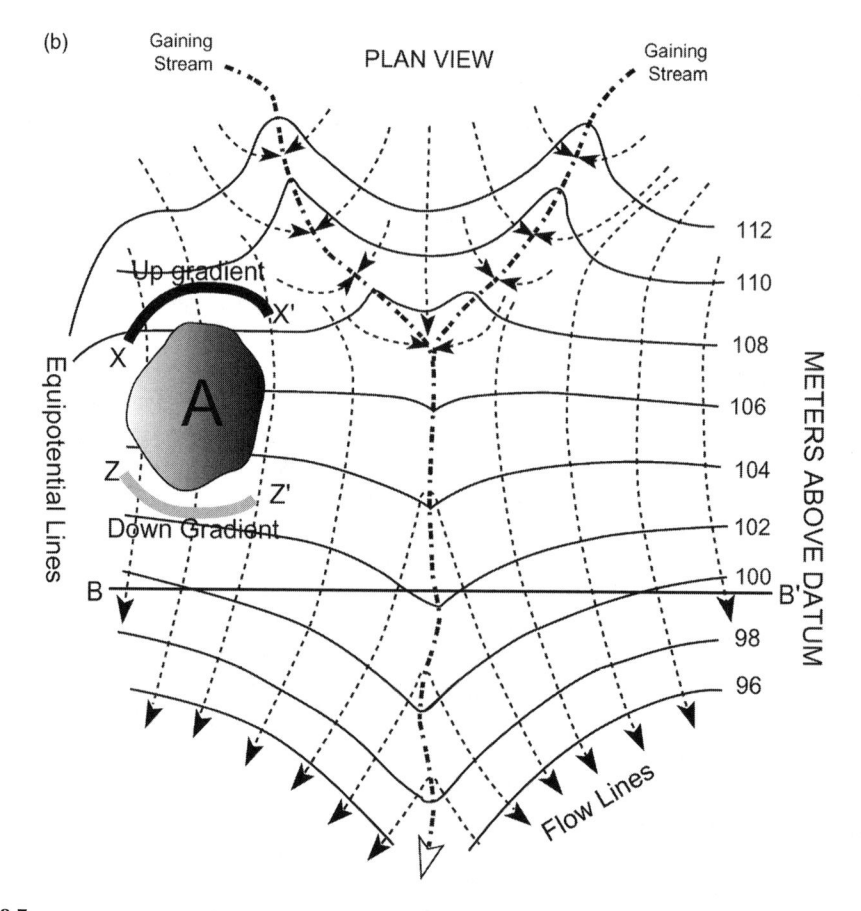

FIGURE 8.7
Losing stream target monitoring zones: (a) cross-sectional view and (b) plan view.

Ground-Water Monitoring System Design

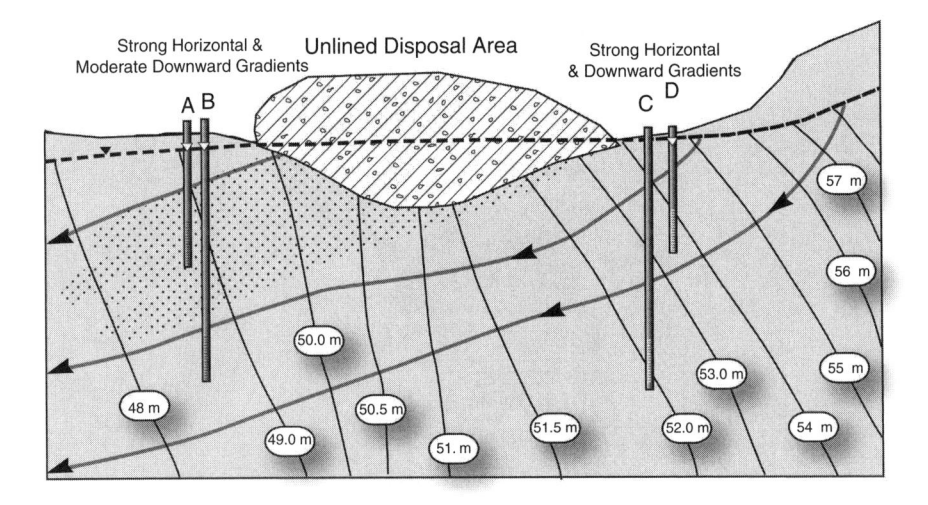

FIGURE 8.8
Steep gradient facilities example.

intermediate flow system, at least in this case, is not affected by the facility. The shallow flow cell is discharging in what could be viewed as both a downgradient and a perceived upgradient direction. The upgradient component is due to the higher heads observed at piezometer D (52.0 m) and lower heads in the other three piezometers (A, B, and C). Establishing these discharge–recharge cell relationships requires sophisticated interpretative skills by the monitoring system designer, as well as sufficient field piezometric data. A plan view of this type of system is shown in Figure 8.10. The shallow local ground-water flow system discharging from the facility causes a disturbance in the regional ground-water system. Without very detailed potentiometric data, the effect of the

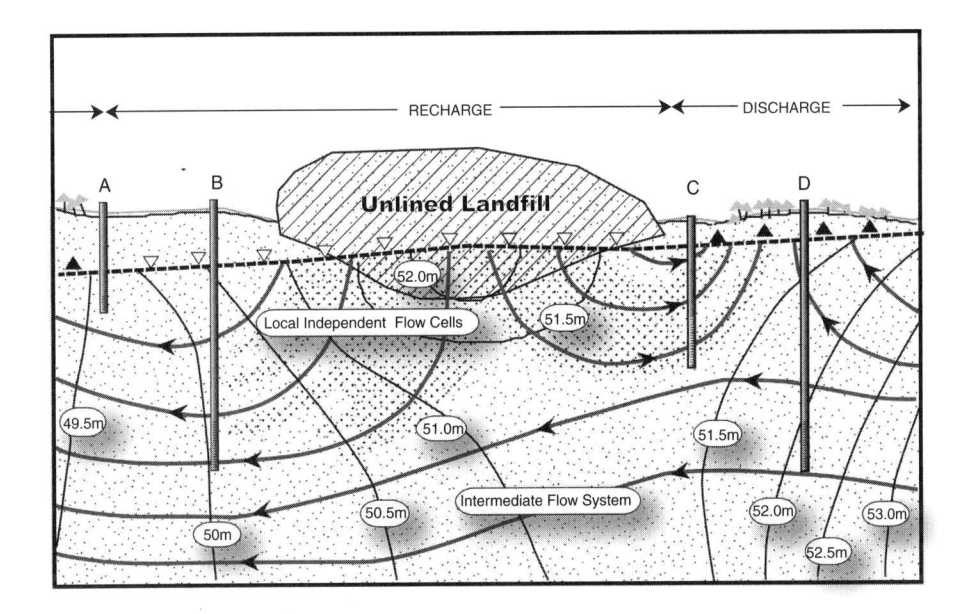

FIGURE 8.9
Conceptual model of local flow cells.

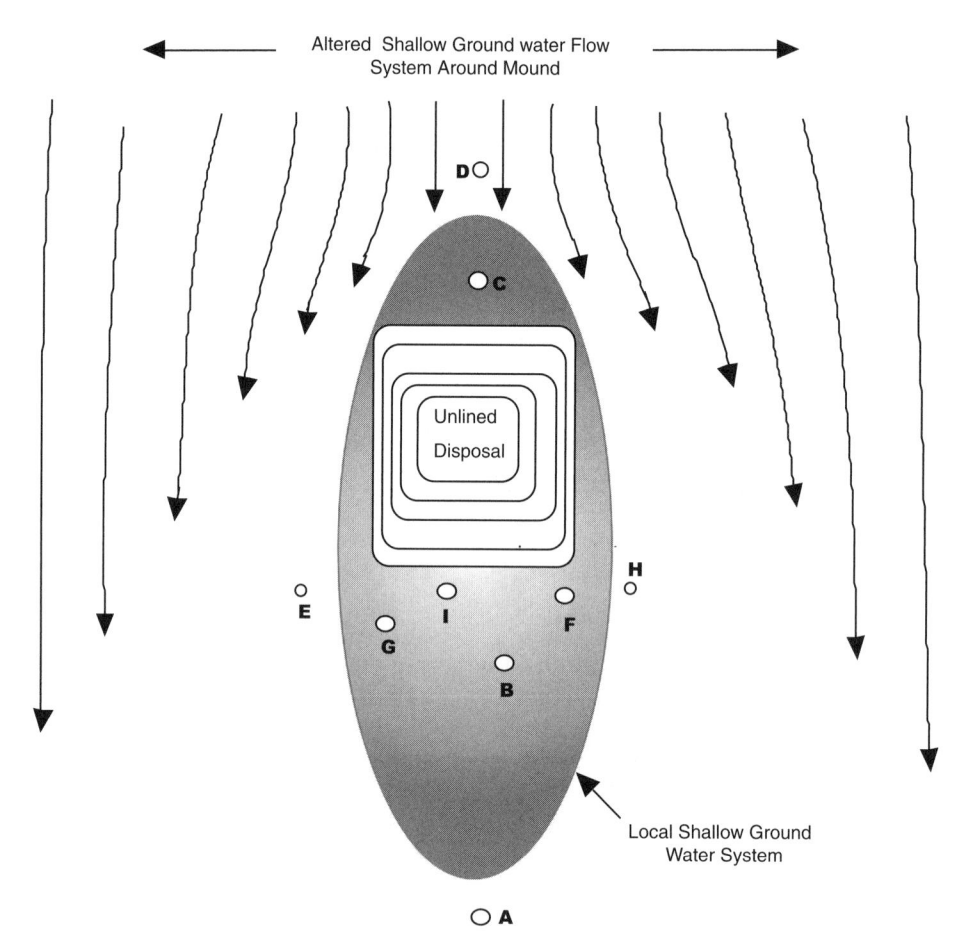

FIGURE 8.10
Map view of local flow cells.

disturbance may be difficult to establish in the field, so sufficient care should be exercised to locate background monitoring wells out of the area of influence of the local cell. The flow cells depicted in Figure 8.9 and Figure 8.10 represent a typical flow net for a low-gradient site. The local flow cells discharging around the topographically higher site would result in downgradient monitoring wells located in what would typically be thought of as an upgradient location. Positioning of detection monitoring well screens at locations B or C would place the wells in the intermediate flow cell. As such, they would not represent conditions truly downgradient of the facility.

Procedures for Gradient Controlled Sites

Even with simple homogeneous (single hydraulic conductivity) environments, care must be taken to fully understand the three-dimensional nature of ground-water flow. As a general guidance, the monitoring system designer should:

- Establish lithology and gradients as with single-aquifer systems
- Compare natural (baseline) gradients across the site and determine the hydraulic conductivity of the aquifer

- Select positions for upgradient monitoring wells, as in position D of Figure 8.9 and Figure 8.10.

Gradient Control or Flow Nets: Unfortunately, most real-world geologic systems are not composed of simple single layers. When observed field conditions include layers of material of variable hydraulic conductivity, more complex evaluations of how ground-water movement is affected by the variable geologic materials must be conducted.

Figure 8.11 shows an unconfined aquifer separated from a confined aquifer by a low-hydraulic conductivity confining bed. Ground-water movement through this system involves flow not only through the aquifers but also across the confining bed. The hydraulic conductivities of aquifers are tens to thousands of times greater than those of confining beds. For a given rate of flow, the head loss per unit of distance along a flow line is tens to thousands of times less in aquifers than it is in confining beds. Consequently, lateral flow in confining beds usually is negligible and flow in aquifers tends to be parallel to aquifer boundaries, as shown in Figure 8.11.

Differences in the hydraulic conductivity of aquifers and confining beds cause refraction or bending of flow lines at their boundaries. As flow lines move from aquifers into confining beds, they are refracted toward the direction perpendicular to the boundary. In other words, they are refracted in the direction that produces the shortest flow path in the confining bed. As the flow lines emerge from the confining bed, they are refracted toward the direction parallel to the boundary (Figure 8.11). Hence, ground water tends to move horizontally in aquifers and vertically in confining beds or low-hydraulic conductivity materials. This observation is important in determining the locations and depths of the wells comprising the facility's detection monitoring system.

Lateral flow components in aquifers have direct relevance to ground-water monitoring system design, because the physical location and depth of the wells must correspond to the overall three-dimensional components of flow typically at the edge of the facility. Most detection monitoring programs concentrate on establishing target monitoring zones in the uppermost aquifers beneath a site. These target monitoring zones are directly correlated with the hydrostratigraphic zone that has the highest rate of flow away from the facility so that immediate detection of leakage from the facility could be accomplished. Some assessment monitoring programs may involve monitoring the uppermost aquifer, deeper aquifers, and zones between the uppermost and deeper aquifers. As a general statement, the three-dimensional ground-water flow established by hydraulic head relationships is necessary for either detection or assessment monitoring system design.

Movement of water through aquifer or confining unit systems is controlled by the vertical and horizontal hydraulic conductivity, the thicknesses of the aquifers and confining

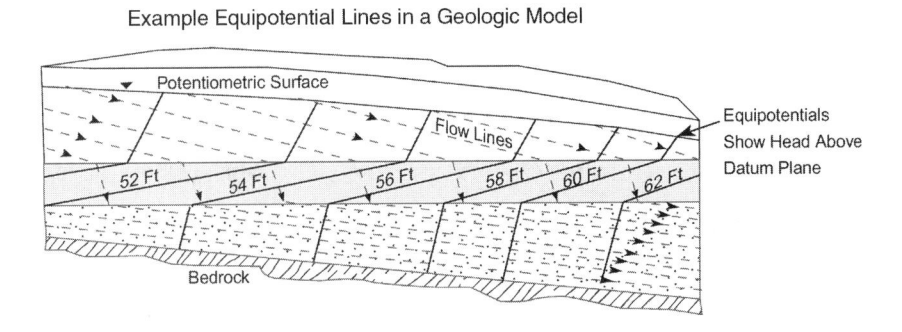

FIGURE 8.11
Unconfined and confined flow nets.

beds, the recharge and discharge (boundary) conditions and the hydraulic gradients. Because of the relatively large head loss that occurs as water moves across confining beds, the most vigorous circulation of ground water normally occurs through the shallowest aquifers. Movement generally becomes slower as depth increases (Heath, 1984). The uppermost aquifers will usually show contamination first (unless a direct conduit for downward movement exists into deeper aquifers), and thus must be served by monitoring efforts. The concentration of flow lines in aquifers is illustrated further by Figures 8.12a and b (from Freeze and Witherspoon, 1967). Aquifers may be bounded by a sloping confining layer and a flat-lying confining unit, a common situation in glaciated regions where low-hydraulic conductivity clay-rich tills overlie higher hydraulic conductivity, outwash sand, and gravel aquifers. Nearly vertical flow occurs through the generally thick, low-hydraulic conductivity materials, while nearly horizontal flow occurs within the underlying aquifer. The aquifer represents the only zone in which ground water moving away from a facility could be properly intercepted and monitored and thus should be considered the target monitoring zone.

This concept is further illustrated in Figure 8.13, in which piezometers installed at increasing depths in the confined aquifer and in the confining zone indicate that a strong downward gradient exists in the fine-grained overburden material. Monitoring wells located in Figure 8.13 at A3 and B3 would represent background and downgradient locations, respectively. The target zone should be screened at both these locations. Figure 8.14 illustrates an unconfined flow system in a recharge area. Recharge areas with strong downward gradients may require special consideration of local shallow flow cells. Depth–location relationships are especially important in such situations. For example, downgradient monitoring wells in the unconfined aquifer, shown in Figure 8.14, should be located in a target zone screened at or below the interval screened by piezometer B2.

Figure 8.15 illustrates the potential ground-water flow paths to a discharging stream. Both upgradient (A and B) and downgradient wells (C) are shown in this simple conceptual illustration. However, even this relatively simple conceptual model can demonstrate how a shallow downgradient well (C) would not intercept potential leachate flow from the unlined waste disposal area. The downgradient ground-water monitoring point for facilities located in discharge areas must be designed on the basis of shallow, near-surface discharge

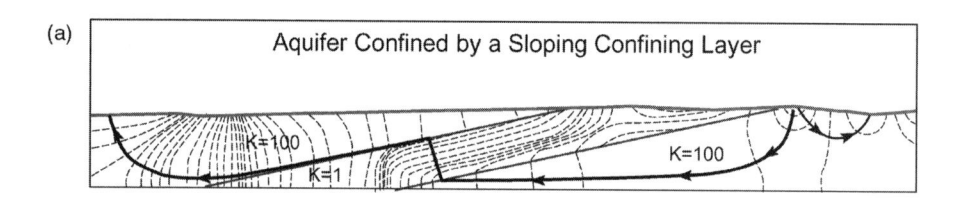

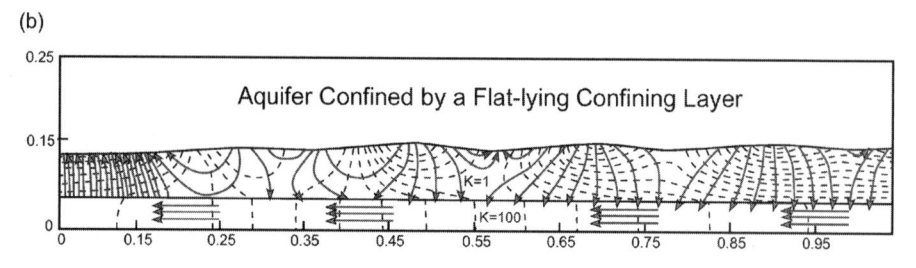

FIGURE 8.12
Regional ground-water flow in a confined aquifer.

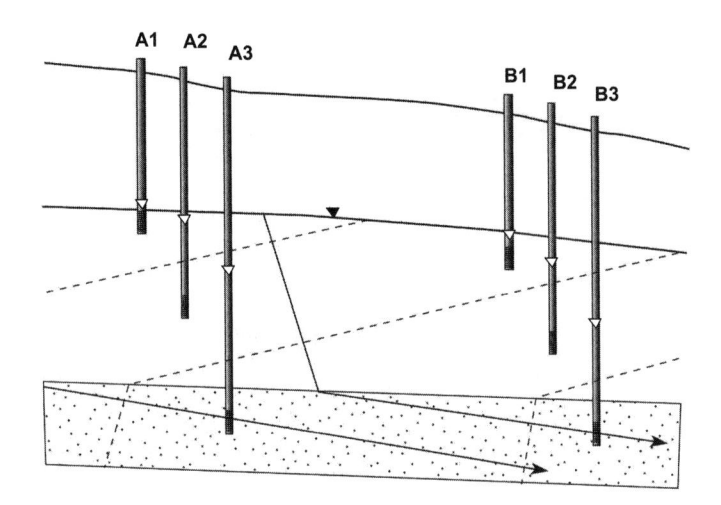

FIGURE 8.13
Confined aquifer piezometer nest.

to wetlands or streams. Upgradient wells should be screened in shallow flow paths, as illustrated by well B. Deeper upgradient wells (as illustrated by well A) would probably suffice, but may not represent ground water flowing in the target monitoring zone.

Ground-water monitoring in complex alluvial deposits often presents difficult problems with respect to identification of target monitoring zones. These deposits often have sandy zones of limited areal extent encapsulated within a matrix of low-hydraulic conductivity sediments. Sand tank experiments have shown that these discontinuous sandy deposits do not affect the downward movement of ground water when strong downward gradients exist. Figure 8.16 shows such a conceptual situation. Shallow permeable zones contained within the low-hydraulic conductivity materials do not have significant horizontal gradients; vertical gradients usually dominate in such environments. Monitoring points

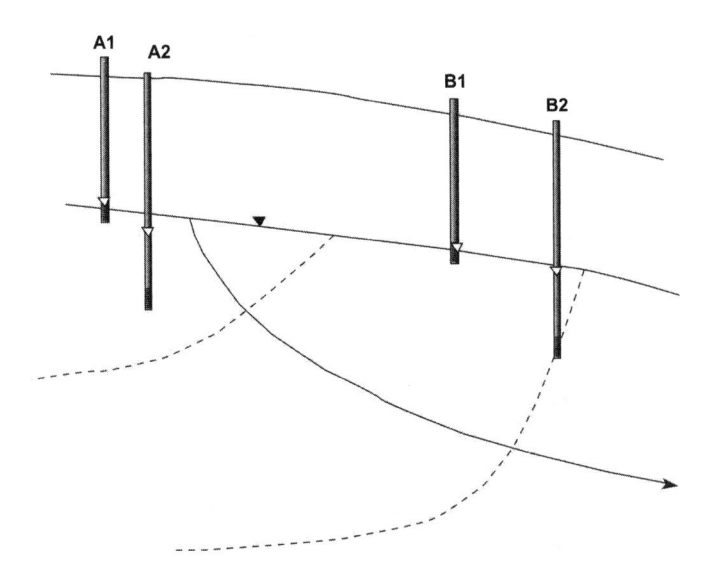

FIGURE 8.14
Unconfined aquifer piezometer nest.

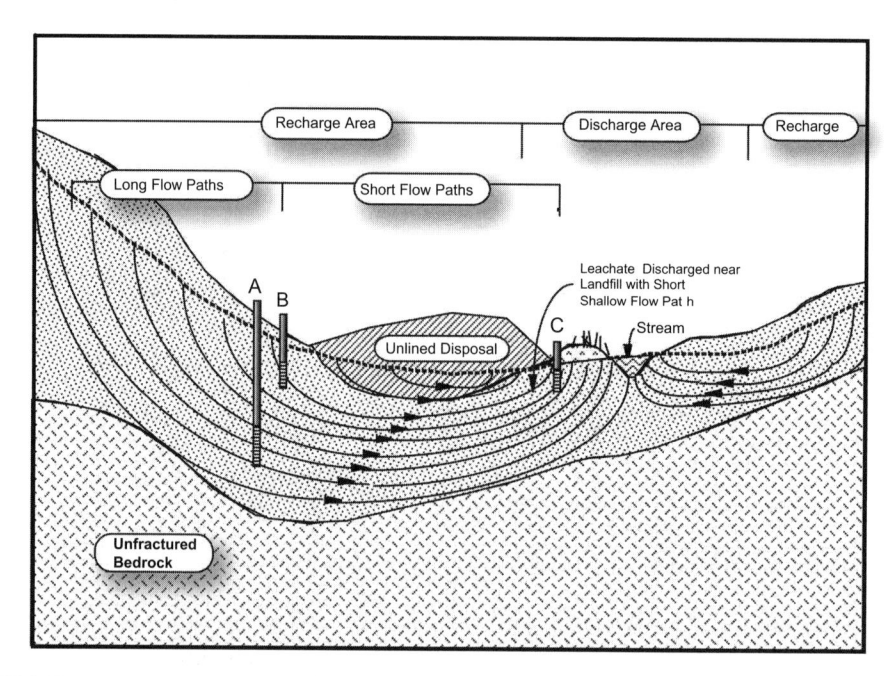

FIGURE 8.15
Shallow discharging ground-water system.

located adjacent to a facility located in these deposits (such as well A) may not represent a target monitoring zone. Only where significant horizontal flow exists, as in the regional (uppermost) aquifer, would a horizontally downgradient target flow path be found. Well B represents a correct downgradient monitoring point for this situation. However, upper permeable units may represent uppermost aquifers if they have sufficient hydraulic conductivity and are of sufficient areal extent to serve as a water source for off-site groundwater users. These more permeable sandy lenses, channels, and tabular deposits have been observed in many types of geologic environments. These materials can range from recent glacial deposits, such as tills with interlayered outwash sands, to unconsolidated overbank deposits associated with alluvial materials, to consolidated claystone deposits with interbedded channel sandstone deposits. The five important criteria for establishing the need to monitor saturated sand units located within lower hydraulic conductivity materials are:

- Differential hydraulic conductivity
- Directional hydraulic heads
- Unit prevalence
- Unit thickness
- Use of water from the unit

Differential hydraulic conductivity refers to the variation in hydraulic conductivity observed between geologic units. Directional hydraulic head refers to the potential flow directions observed from piezometers located within individual units. Unit prevalence is a qualitative judgment based on the overall site stratigraphic characterization. Unit thickness is defined from the field investigation program and is based on observed

Ground-Water Monitoring System Design 539

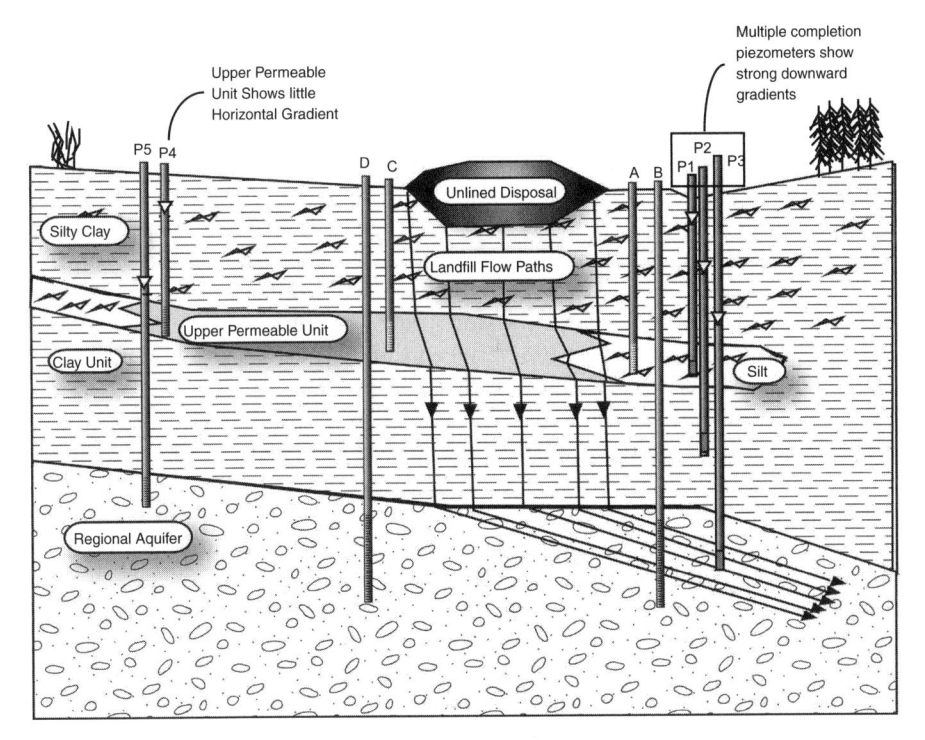

FIGURE 8.16
Low-hydraulic conductivity environments with non-discharging sand lenses.

thickness of the sandy units in soil borings conducted at the site. Water use refers to the presence of human receptors who tap the unit with water-supply wells and use the water from the wells for some purpose (drinking, irrigation, etc.).

Each of these criteria must be considered in order for a monitoring system designer to decide if a particular permeable unit would require monitoring as a target monitoring zone. Evaluation of differential hydraulic conductivity involves an order-of-magnitude comparison of the sandy units to the adjacent matrix materials. Freeze and Cherry (1979, p. 173) state that, "In aquifer–aquitard systems with permeability contrasts of two orders of magnitude or more, flow lines tend to become almost horizontal in the aquifers and almost vertical in the aquitards." This flow pattern requires that the aquifer either discharges into other permeable units, discharges to surface water, or is pumped from the system.

Directional hydraulic heads provide an indication as to the discharge potential of the sandy units. If vertical directional heads are discharging upward (from below the unit) and downward (from above the unit) into the sandy layers, it is likely that the unit discharges into adjacent lower head areas.

The unit prevalence criterion provides an indication of how continuous the layer is in the subsurface. Data required to support this determination are gathered during borehole drilling activity to demonstrate the continuity of the unit in the site area. As a general guidance, if 100% of the boreholes drilled at a site indicated the presence of the definable unit at roughly equivalent elevations, it is likely that the geologic stratum is continuous, and it should be monitored. If the percentage of boreholes in which the unit is present falls to 50% or shows an elevation variability, the unit is much less likely to represent a continuous feature that should be monitored. An understanding of the depositional history of the geologic unit represents the best method for evaluating the continuity of more permeable deposits that could discharge ground water to downgradient, off site areas. Alluvial channel deposits may have been cut off by aggrading streams during the geologic past. However, sufficient stratigraphic data should be established to confirm such assumptions before ruling out discharge through such linear features.

The drilling program also establishes the unit thickness. If the saturated permeable units are very thick (e.g., 100 ft or more), it is likely that the unit would require monitoring. As the units become thinner, the other factors become more important in the overall decision to monitor or not to monitor the unit as the uppermost aquifer.

The last criterion, use of water from the units, can outweigh all the other factors, assuming that there is a hydraulic connection between the facility and the downgradient water users. Each factor must be weighed in the decision process.

A monitoring system designer is often required by a regulatory agency (or multiple agencies) to monitor all potential pathways for ground water and contaminant movement. Rather than blindly installing monitoring wells in every permeable unit, the author recommends using technical reasoning for flow path interception. Monitor only those geologic units that have a reasonable chance to provide flow toward downgradient receptors (either human or ecological). Stick to detection monitoring in the classical uppermost aquifer that is discharging off-site. In the majority of cases this approach will both meet the letter of the law and limit long-term liability issues.

If a thin (e.g., 1-ft thick), discontinuous sand unit with a differential hydraulic conductivity of one order of magnitude shows potentiometric heads passing through the unit (i.e., heads continued downward through the sand unit) and few borings contact the unit, it would not be considered as a target monitoring zone. If that unit is somewhat thicker (e.g., 10-ft thick) and is contacted by only a few borings, it may or may not be necessary to monitor the unit as the uppermost aquifer. However, if the saturated unit is 20-ft thick, it is penetrated by most borings and shows potentiometric heads discharging into the unit from above and below, the unit would probably have to be monitored as the uppermost aquifer.

Figures 8.17a–d illustrate the use of this concept with a series of conceptual models with various levels of discharge from sandy units. The levels of discharge range from almost none in Figure 8.17a to significant discharge between the unconsolidated and bedrock systems in Figure 8.17d. The interpretation of site hydrogeologic conditions, and thus the design of the monitoring system in each case, would be based on the following key points:

- The lateral extent and thickness of the various geologic materials present
- The hydraulic conductivity of each of the individual lithologic units
- The gradients obtained from piezometers placed in each of the permeable units
- The discharge and recharge potentials of the geologic units present on site.

The conceptualization and flow net construction should be based on illustrations with 1-to-1 scales. Figures 8.17a–d provide some additional keys to determining appropriate monitoring locations. The conditions depicted in Figures 8.17a and b would indicate that detection monitoring only need to be conducted in the regional (uppermost) aquifer. Ground water discharges down through the two sandy layers (one discontinuous unit in Figure 8.17b), into the regional bedrock aquifer. The conditions depicted in Figure 8.17d indicate that sandy unit 2 would represent the better detection monitoring target as the uppermost aquifer. The first sandy unit in Figure 8.17d would not represent an effective monitoring location because of its thin, limited-discharge nature.

Groundwater samples obtained from this unit would only be representative of conditions along the edge of the facility within the flow path shown in Figure 8.17d. While a case could be made that a monitoring well located at point D may be necessary to evaluate the area along one side of the facility, approximately 95% of the area would be monitored if wells were placed at downgradient locations in sandy units.

Figure 8.17c represents a situation in which both the sandy units and the regional system should be monitored. The decision to monitor both sandy units should be weighed on the basis of additional site characterization work to determine the regional extent and current or future use of water within the sandy units. If the second sandy unit represents a likely

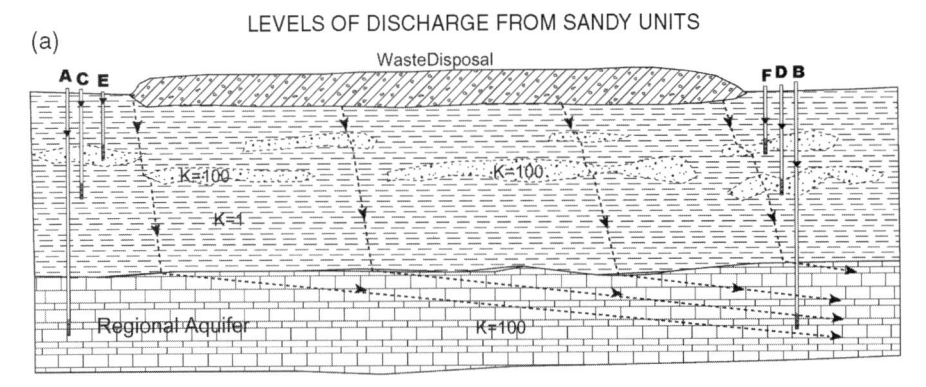

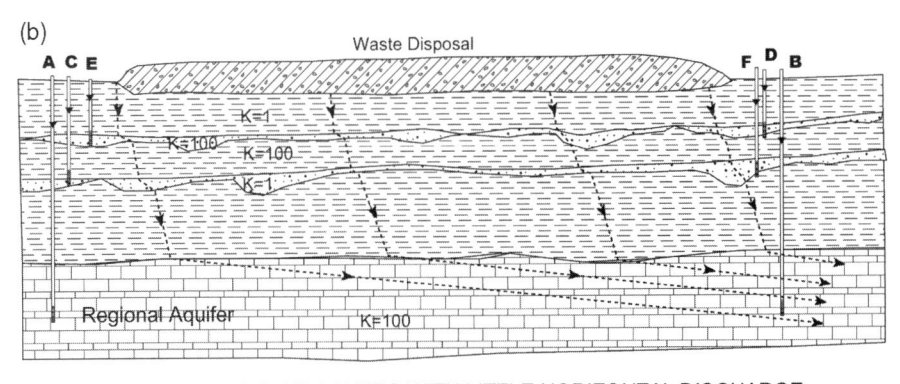

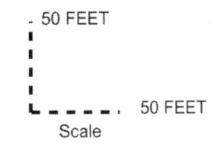

FIGURE 8.17
Conceptual model with (a) non-discharging sand lenses, (b) non-discharging sand layers, (c) discharging sand lenses, and (d) discharging sand layers.

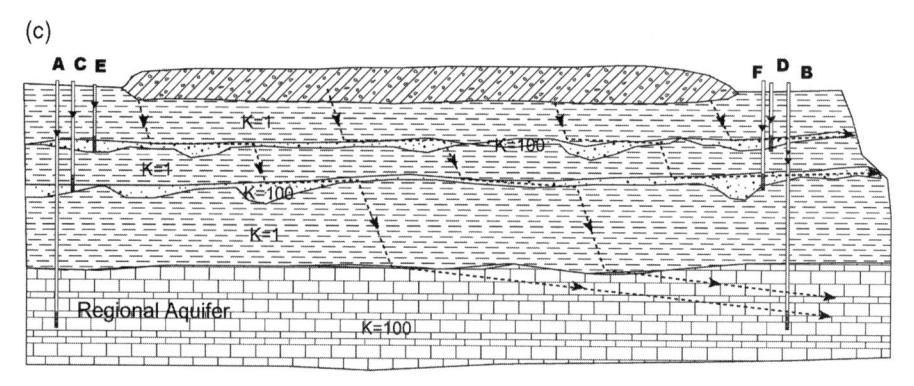

(c) SAND LAYERS WITH SIGNIFICANT HORIZONTAL DISCHARGE

KEYS IN UNDERSTANDING CONCEPTUAL MODEL
- Confirmed sand layers in drilling program
- Verticle gradients in only fine grained units
- Regional aquifer known as productive unit
- Sand layers used locally or discharge to streams

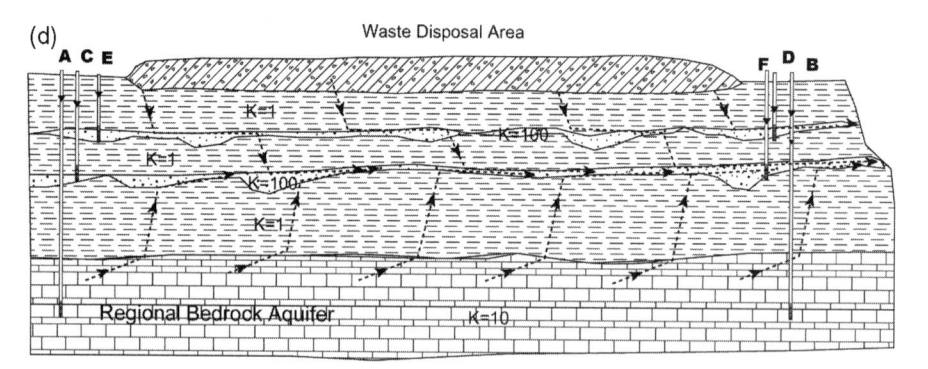

(d) SAND LAYERS DISCHARGING BOTH BEDROCK AND OVERBURDEN

KEYS TO UNDERSTANDING CONCEPTUAL MODEL
- Well defined sand units present in all boreholes
- Sand layers have regional use as water supply
- Gradients discharge into sand layers
- Regional aquifer discharges into sand layer

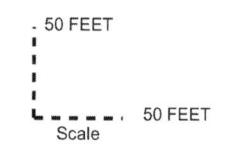

FIGURE 8.17
Continued.

flow path, and hence a target monitoring zone from the facility, it should be included in the monitoring program.

Detailed evaluations of layered geologic units can be used to define the specific discharging, more permeable strata next to a waste disposal area. Figure 8.18 shows an evaluation of a cross-sectional area 40-ft deep and 200-ft wide. The waste disposal area is just to the left of piezometer C. The flow net superimposed on the cross-section is based on information from both piezometers and wells screened along the cross-section line. Recognizing that long-screened wells provide a hydraulic head value that is averaged over the length of the well screen, more validity should be placed on hydraulic head data gathered

Ground-Water Monitoring System Design   543

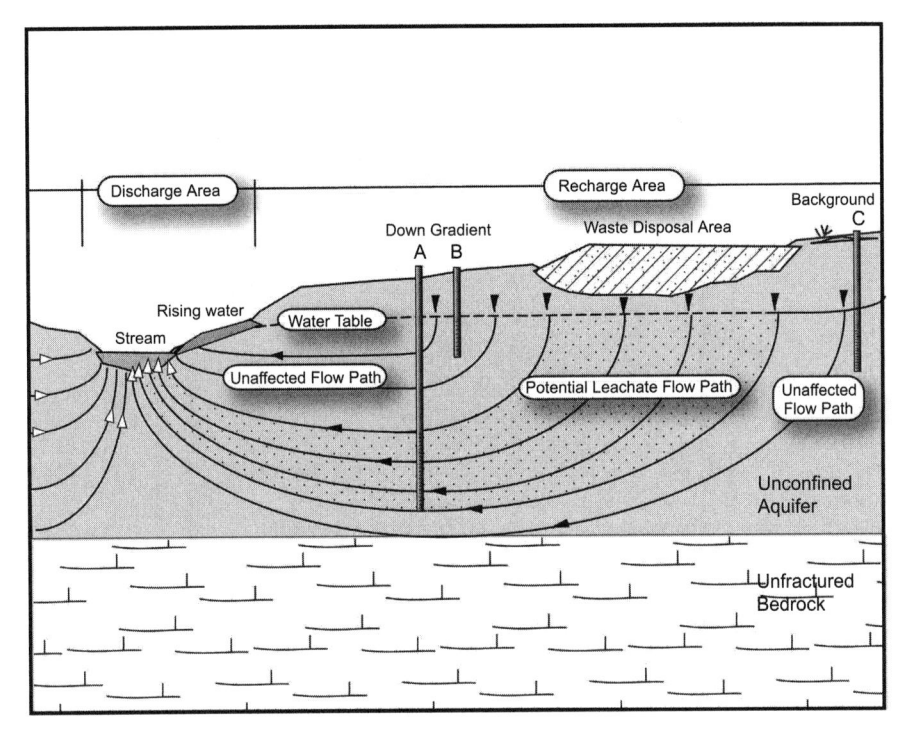

FIGURE 8.18
Conceptual model of discharging sand units.

from short-screened piezometers. The results of this linked cross-section and flow net shows the discharging nature of the shallow continuous sandy layer above the unfractured bedrock. A decision was made to monitor at a location within the relatively thin discharging sandy zone. Although there may be some upward movement of ground water from deeper, less permeable units, the flow lines that bound the base grades of the waste disposal areas would probably represent the optimum location and depth for detection monitoring.

Multiple Piezometers to Establish Flow Relationships: Hydraulic heads established by multiple piezometers can identify the potential flow paths from a facility in homogeneous materials. Figure 8.19 illustrates an upgradient area of recharge and downgradient discharge point as defined by water levels measured in piezometers. The downgradient piezometers show an upward vertical gradient, while upgradient piezometers show a downward gradient. Figure 8.20 illustrates a recharge condition both in background and downgradient piezometers. The heads shown in monitoring wells A and B represent the average of the hydraulic heads spanned by the well screens.

Geologic Controls

Geologic controls over ground-water movement represent the most critical factors that should be considered in ground-water monitoring network design (see Figure 8.20). The goal, in most cases, is to define the most likely zone in which ground water moves beneath a facility and, therefore, the most likely zone for any possible contaminant movement to occur and be detected. The following discussion first addresses simple geologic systems where design of the monitoring system is relatively straightforward, based on

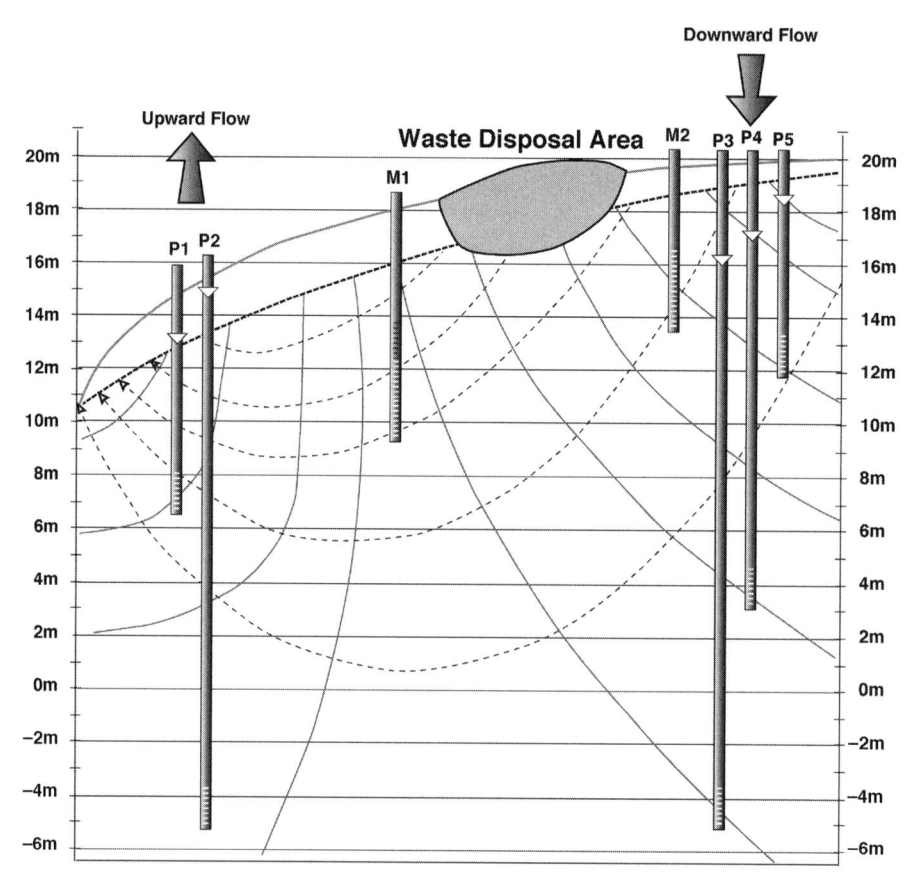

FIGURE 8.19
Gradient comparisons for recharging and discharging areas.

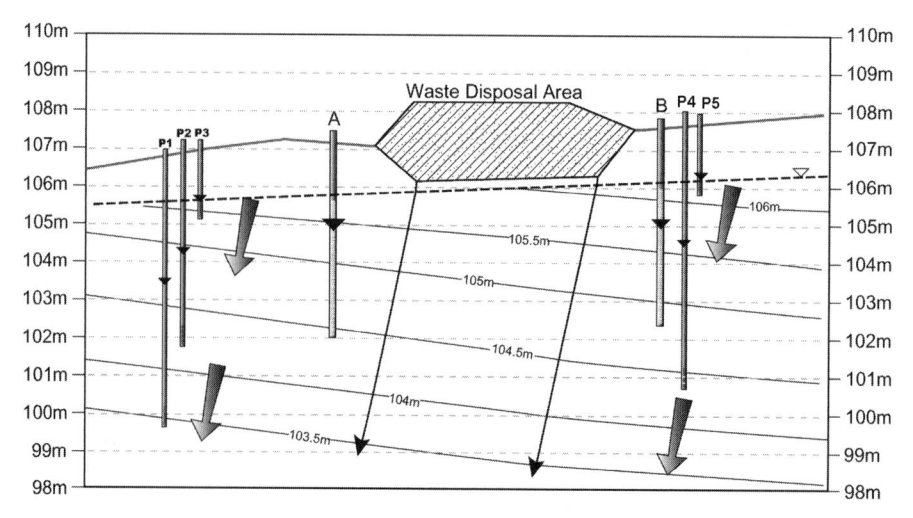

FIGURE 8.20
Conceptual recharging conditions.

Ground-Water Monitoring System Design

the geology and ground-water flow directions. The discussion then moves to more complex geologic systems that require significant site assessment and conceptualization to design an appropriate monitoring system. The discussion also includes a design for perched water conditions. Some of the following examples include unlined waste disposal sites where leachate movement is shown to dramatize the potential flow paths and target monitoring zones.

Single Homogeneous Aquifer

The single homogeneous aquifer represents the simplest geologic environment in which to design a detection monitoring system. The single homogeneous aquifer system requires only the following steps to define the target monitoring zone:

- Evaluate aquifer geometry, thickness, and vertical and horizontal hydraulic conductivity variability by way of continuously sampled stratigraphic borings logged to confirm homogeneous and isotropic conditions within each layer.
- Prepare a conceptual geologic or hydrogeologic model and plot potential target monitoring zones.
- Construct flow nets using water level or hydraulic head information from piezometers or observations wells to confirm target monitoring zones.
- Install wells to monitor potential contaminant flow paths.

Figure 8.20 illustrates the subsurface movement of leachate from an unlined solid-waste facility in a humid environment. Selection of appropriate screen positions for downgradient wells is relatively simple using the procedure earlier. Figure 8.21 (from Freeze and Cherry, 1979) represents isoconcentrations of chloride next to an unlined solid-waste landfill. The contours are based on water quality obtained from numerous, closely spaced sampling points screened at various depths. The location of the target monitoring zone here would be the centerline of the chloride plume. The centerline, with the highest chloride concentrations, represents the most direct flow path away from the landfill. Monitoring wells located in this zone (along the highest chloride contour) would provide the earliest detection of leachate excursion away from the facility. Figure 8.21 (field-determined flow fields) and Figure 8.20 (constructed from a flow net) provide

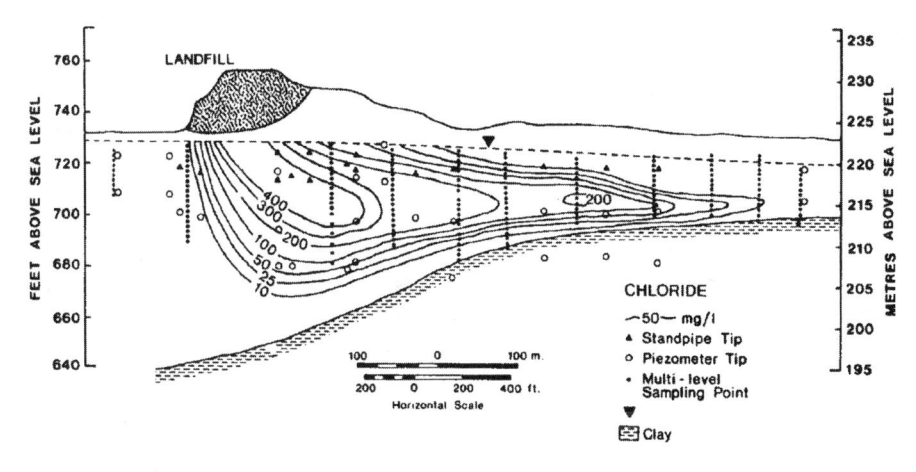

FIGURE 8.21
Example leachate water quality plume, field-determined.

essentially the same solution to ground-water flow for this particular hydrogeologic environment.

Single Aquifer of Variable Hydraulic Conductivity

Differences in hydraulic conductivity due to changes in stratigraphy with depth (anisotropy) can influence the design of an effective monitoring system. The procedure for monitoring system design in this type of hydrogeologic setting would include the following steps:

- Determine the horizontal extent and thickness of individual geologic units by evaluating geologic logs of continuously sampled stratigraphic borings to a depth of least 25 ft below the base grade of the facility. This suggested depth is used as a rule of thumb and actual depths may vary based on site conditions.
- Establish hydraulic conductivity for each unit from results of field and laboratory tests confirming anisotropic conditions.
- Construct a flow net based on observed hydraulic heads (from piezometers) and prepare a conceptual geologic or hydrogeologic model to select target monitoring zones.
- Install monitoring wells based on defined target monitoring zones that represent primarily horizontal movement of ground water.

If anisotropic conditions are observed in analysis of laboratory and field hydraulic conductivity test results, it will first be necessary to establish the true degree of anisotropy to properly draw the flow net. The differences between hydraulic conductivity measurements observed in the field and those observed in the laboratory are not always the result of anisotropic formation conditions. Because laboratory measurements are made from relatively small volume samples and field hydraulic conductivity measurements are based on very large volumes of the formation, some variability in test results should be expected. It is common for such results to vary by more than an order of magnitude. The values obtained from field and laboratory tests should be compared with both the descriptions of the geologic samples collected and the lithologic logs of the boreholes. If the comparison shows little reason for a wide variation in hydraulic conductivity, an inspection of the samples provided to the laboratory should be conducted. Special care should be taken to ensure that the samples collected in the field represent the typical lithology, rather than exceptions to the typical formation conditions. In any case, it is rare to obtain exactly the same hydraulic conductivity values from an individual field test and a laboratory test. The smaller the laboratory sample tested, the more likely the sample is to provide a different value from an average value obtained from field tests conducted in a screened well. For these reasons it is recommended that sufficient numbers of laboratory samples be collected from each hydrostratigraphic unit to provide representative test values. As a rule of thumb, for small or geologically simple sites, three laboratory samples for each hydrostratigraphic unit should provide sufficient information to allow a comparison with field-obtained hydraulic conductivity results. For large or geologically complex sites, many additional laboratory determinations will generally be necessary.

Figure 8.22 depicts a time sequence for a leachate plume from an unlined solid-waste facility. Leachate movement in the system is represented by primarily vertical flow in the lower hydraulic conductivity units and horizontal flow in more permeable silty sands and gravels. The monitoring system for this facility would consist of wells screened in the silty sand unit directly next to the facility, in the gravel or in both units. The extent of the geologic units, the potential for off-site leachate movement and the current or potential

Ground-Water Monitoring System Design

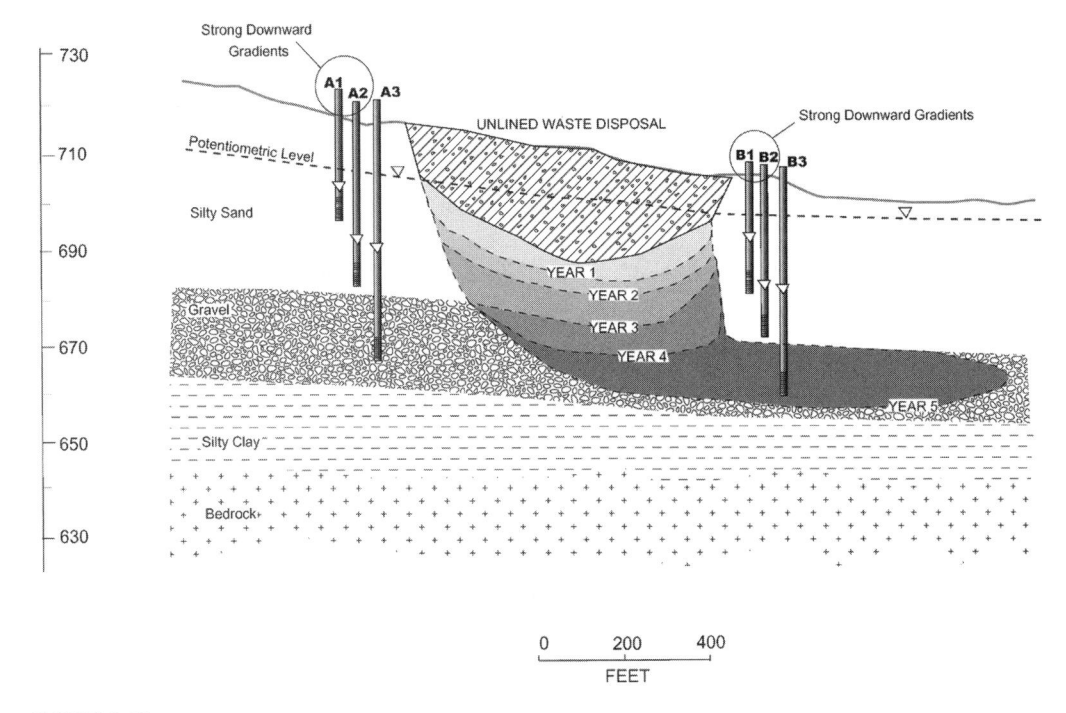

FIGURE 8.22
Time sequence for a leachate plume.

use of the water contained in the units are some of the deciding factors in the actual system design. If the silty sand were discontinuous, the gravel would be the primary monitoring target zone. However, if the silty sand extends beyond the site boundaries and sufficient horizontal flow exists to allow this zone to be monitored effectively at the edge of the facility, both the silty sand and the gravel would be targets for ground-water monitoring. The silty sand represents the probable first affected unit and the gravel would most likely represent a water supply for off-site downgradient water users. An important key is the potential for horizontal leachate movement in the silty sand unit. If flow nets show discharge from the silty sand unit to downgradient receptors, this unit would likely represent the uppermost aquifer. Therefore, detection monitoring would be required within this unit. Conversely, if the silty sand shows strong downward gradients, and it does not discharge to downgradient receptors (as shown in Figure 8.22), then little would be gained from monitoring this unit.

Figure 8.23 shows a sand and gravel unit as the uppermost aquifer beneath two clay-rich tills. Typical of near-surface, low-hydraulic conductivity units, ground-water flow is nearly vertical in the tills. Ground water then flows horizontally in the much higher hydraulic conductivity sand and gravel aquifer. This sand and gravel unit is the only potential target monitoring zone for a facility located in this type of environment. The dominance of vertical flow in low-hydraulic conductivity deposits and horizontal flow in continuous, permeable zones is very typical. In glaciated regions, deeper sand and gravel valley fill or outwash deposits are often in direct contact with underlying weathered or highly fractured bedrock. Such systems would represent a composite target monitoring zone. Small lenses of sand within a mass of low-hydraulic conductivity material, however, do not represent adequate targets for monitoring. Thin or discontinuous sand lenses do not provide the hydraulic heads necessary for horizontal movement of ground water away from a facility. Figure 8.24 and Figure 8.25 represent the idealized

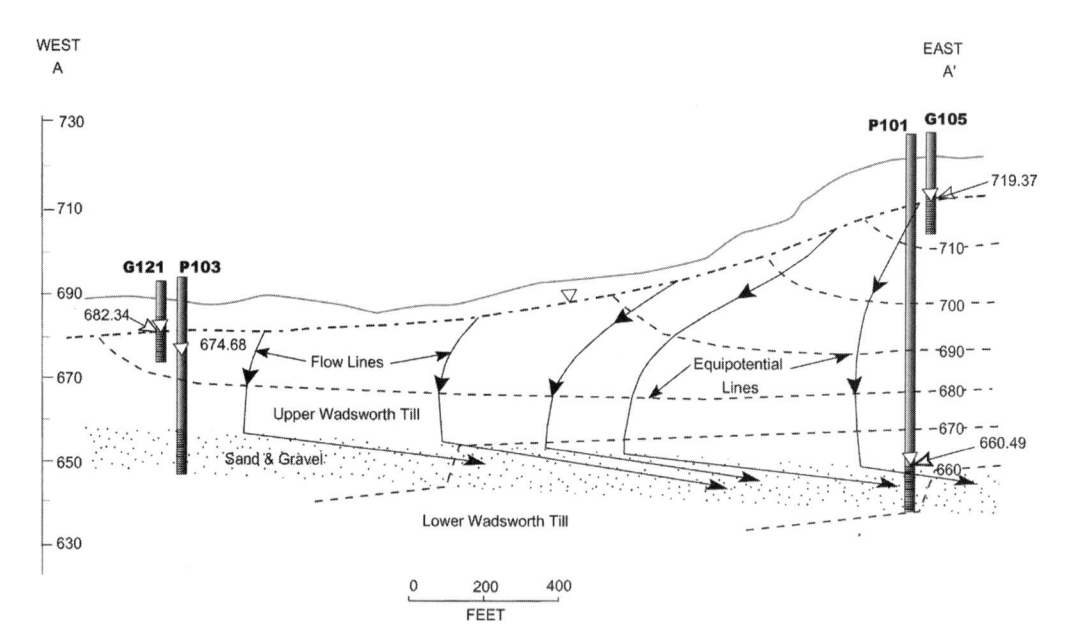

FIGURE 8.23
Conceptual flow in layered deposits.

cross-section of a facility located in a clay-rich till above a bedrock aquifer. A series of discontinuous sand seams is present within the clay till. Numerical modeling of the system provided the velocity vector and concentration contour plots shown in Figure 8.25. A point source of contamination was simulated in the modeling project. The point source produced a plume that moved horizontally in near-surface material (the jointed till), vertically downward through the clay till and sand lenses, and finally horizontally in the underlying

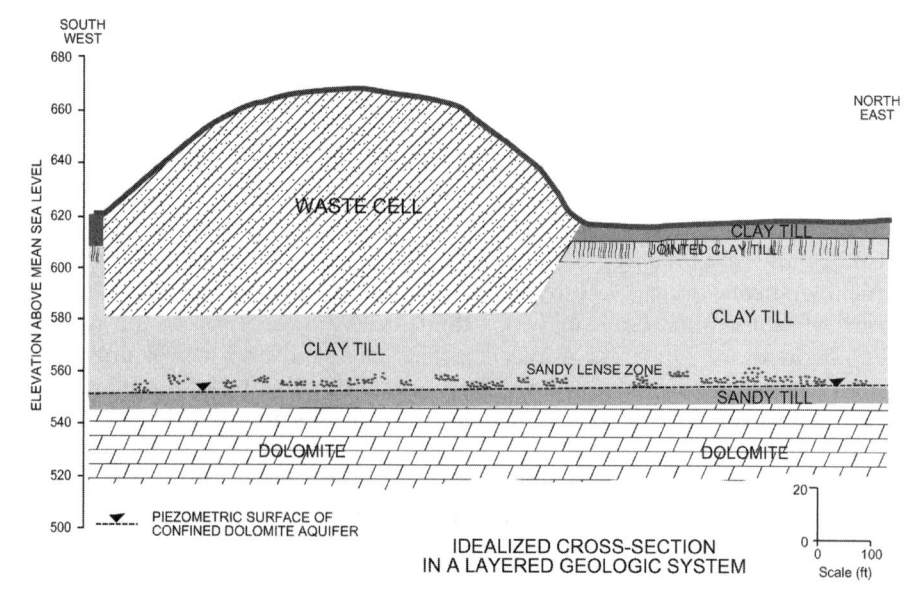

FIGURE 8.24
Conceptual model in layered deposits.

Ground-Water Monitoring System Design

SW   POINT OF SOURCE OF CONTAMINATION   **NE**

[Cross-section diagram showing: JOINTED CLAY TILL, CLAY TILL, SANDY LENSE ZONE, SANDY TILL, DOLOMITE, with normalized concentration contours (1E-6, 1E-3, 0.5) at distances 100-500 ft]

VELOCITY VECTOR AND
NORMALIZED CONCENTRATION
CONTOUR PLOT

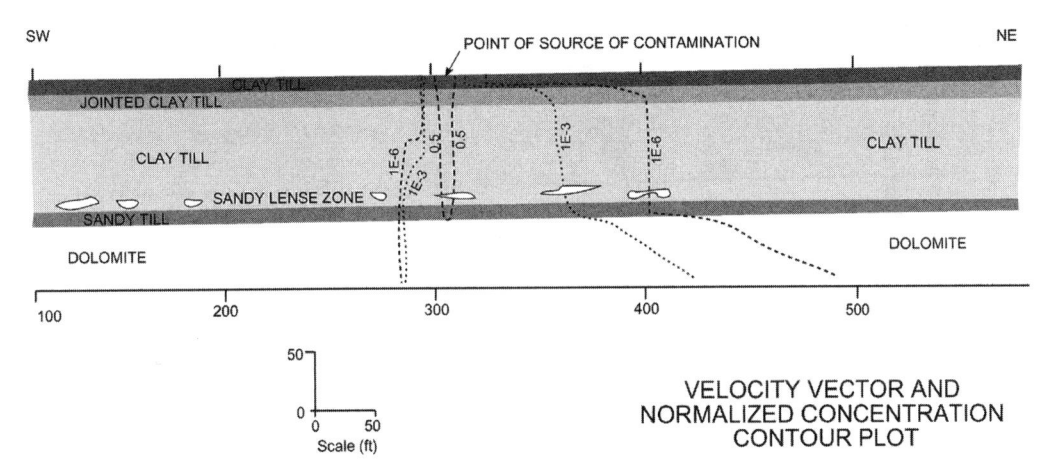

FIGURE 8.25
Computer model of flow in layered deposits.

dolomite bedrock. The dolomite represents the target monitoring zone in this situation, due to the following factors:

- The near-surface, jointed till is shallow and does not represent a flow path away from the base of the facility.
- The near-surface tills can be influenced by vertical recharge events that are not associated with ground-water passing beneath the facility (i.e., not in the flow path).
- The thick clay tills and the minor sand lenses do not represent aquifers.
- The thick clay till and enclosed sand lenses, when considered as composite units, have primarily vertical ground water flow components.
- The dolomites can yield sufficient water to wells to be considered aquifers, and do represent a horizontal flow path away from the facility.

Therefore, the dolomites would represent the target monitoring zone for the facility.

Multiple Aquifers

Multiple aquifers represent a challenge to the ground-water monitoring system designer. Ground-water in layered aquifers often moves in different directions. Thus, monitoring multiple aquifers requires a more detailed knowledge of the three-dimensional hydrogeologic system to accurately establish a capable monitoring system.

Figure 8.26 shows a two-aquifer system with ground-water flow in opposite directions. Such a geologic environment would require sufficiently detailed geologic and hydrogeologic characterization to establish target flow paths from the facility. The following procedure is recommended to establish a ground-water monitoring system for a two-aquifer system as shown in Figure 8.26:

- Install stratigraphic borings using continuous sampling techniques from the surface through all overburden units down to competent bedrock.

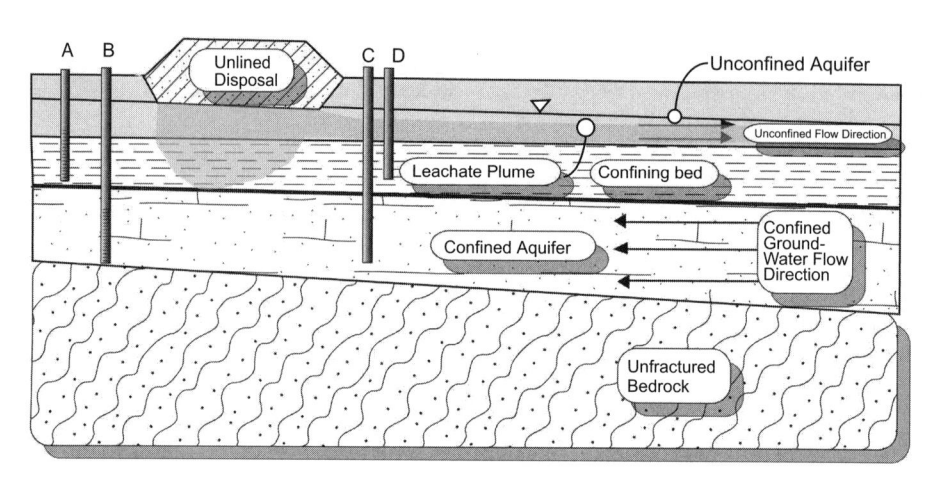

FIGURE 8.26
Two-layer flow model.

- Install piezometers in each discrete unit so that vertical and horizontal gradients can be established for each unit and between units.
- Establish hydraulic conductivity for each unit by conducting field *in situ* and laboratory hydraulic conductivity tests.
- Construct a flow net for the entire system.
- Develop a geologic or hydrogeologic conceptual site model and establish the target monitoring zones.
- Install monitoring wells.

If the goal of the monitoring system were to provide immediate detection of any contamination released from a facility (i.e., as in a detection monitoring program), the target monitoring zone should be the unconfined uppermost aquifer. If the goal of the system is to assess the extent of contamination emanating from a site (i.e., as in an assessment monitoring program), defining the rate and extent of contaminant movement would require monitoring in both the upper and lower aquifers. If a nearby surface stream serves as a base-flow discharge point for one of the aquifers, the stream would probably also require water quality monitoring. The monitoring program should also define if there is ground-water flow beneath the stream.

Figure 8.27 shows a three-aquifer system including a deep, interconnected, fractured bedrock aquifer. As with the two-aquifer system, the assessment technique should be as follows:

- Install borings to take soil samples sufficient to characterize the unconsolidated materials down to competent bedrock. Determine if continuous sampling and logging techniques are necessary for the geologic environment. Rock core drilling would be required to evaluate bedrock fractures and hydraulic conductivity.
- Install a series of piezometers in each geologic unit to establish hydraulic gradients.
- Establish hydraulic conductivity (horizontal and vertical) for each geologic unit, including confining units.
- Construct a flow net and piezometric contour map for each aquifer.

Ground-Water Monitoring System Design   551

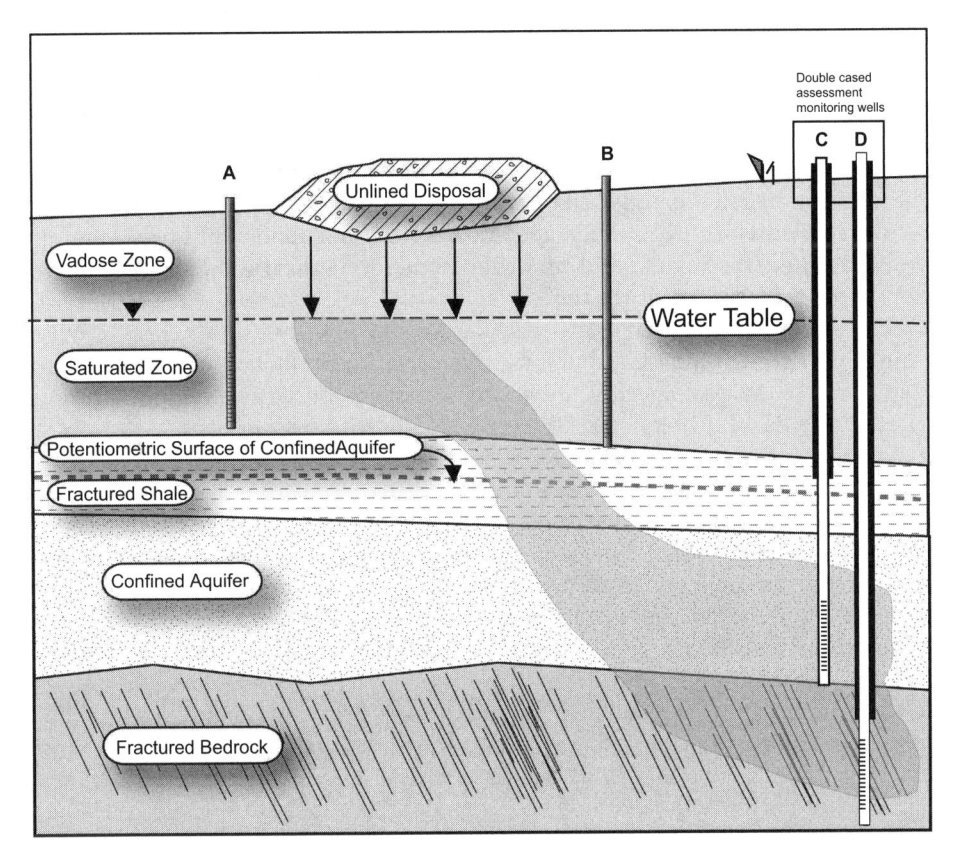

FIGURE 8.27
Three-layer flow model.

- Develop a geologic or hydrologic conceptual site model and establish target monitoring zones.
- Install monitoring wells. Assessment monitoring wells in deeper units should be double-cased through the overlying units as necessary to prevent cross-communication between units.

As with the two-aquifer system, a monitoring system installed for the purpose of detecting contamination would focus on the uppermost aquifer to provide immediate detection of leachate from the facility.

The shaded area in Figure 8.27 represents widespread contamination that provides many challenges in assessment monitoring programs. Typically, if assessment programs require full project planning at the project start (such as in the Superfund program), these deeper contaminated zones are often not included in sampling programs. A phased program that includes full geologic conceptualization and flow net construction should be completed before development of ground-water quality sampling plans. Chapter 2 provides additional guidance for assessment monitoring evaluations.

Low-Hydraulic Conductivity Environments

Probably the most difficult geologic environment in which to design a ground-water-monitoring system is thick, low-hydraulic conductivity materials overlying an aquifer at depth. Much of the uncertainty surrounding ground-water monitoring of potentially contaminated sites is a result of the difficulty in interpreting ground-water movement in low-hydraulic conductivity environments. Figure 8.28 illustrates a facility located in a thick low-hydraulic conductivity clay overlying a high-hydraulic conductivity sand. The sand is confined and the clay contains minor sand lenses, acting as a subdrain to the adjacent low-hydraulic conductivity clay, thus masking the directional components of shallow ground-water flow. Piezometers must be installed within the clay and the uppermost aquifer (the lower sand) in order to define vertical gradients and to assist in selection of the target monitoring zones.

Geologic environments that consist of primarily low-hydraulic conductivity units containing higher hydraulic conductivity materials of significant lateral extent require comprehensive hydrogeologic investigations to define the target monitoring zones. An example of the kind of conceptual geologic descriptions necessary for evaluating low-hydraulic conductivity environments is provided later.

The shallow, unconfined ground-water surface is affected by a facility leachate collection system that acts as an underdrain. Figure 8.29 illustrates a conceptual model of a buried channel located in much less permeable claystone. One example of this type of lithologic system is the Cretaceous Dawson Formation in the Denver, Colorado, which was deposited in a fluvial, deltaic environment. The Dawson stratigraphic sequence consists of depth-uncorrelatable, vertically stacked sandstone channel deposits, which are isolated within a fine-grained claystone that originated as a backswamp deposit in the Cretaceous delta. Thin, isolated sandstone lenses (as viewed in cross-sections) are present in the sequences that are characteristic of levee splay deposits and minor overbank deposits. The majority of geologic materials in such a sequence are matrix-supported diamicts that have very low-hydraulic conductivity. The channel deposits represent clast-supported units. These channel deposits can provide discharge pathways both to recent alluvial materials present in ephemeral stream channels and to adjacent claystone units.

On the basis of an evaluation of the depositional environment, through detailed core analysis using facies codes, the Dawson Formation deposits were determined to have been laid down in a delta that was gradually uplifted by ancestral Rocky Mountain tectonics in early Tertiary time. Different depositional characteristics of each sand sequence observed in cored boreholes emphasize that the sands were deposited by separate and

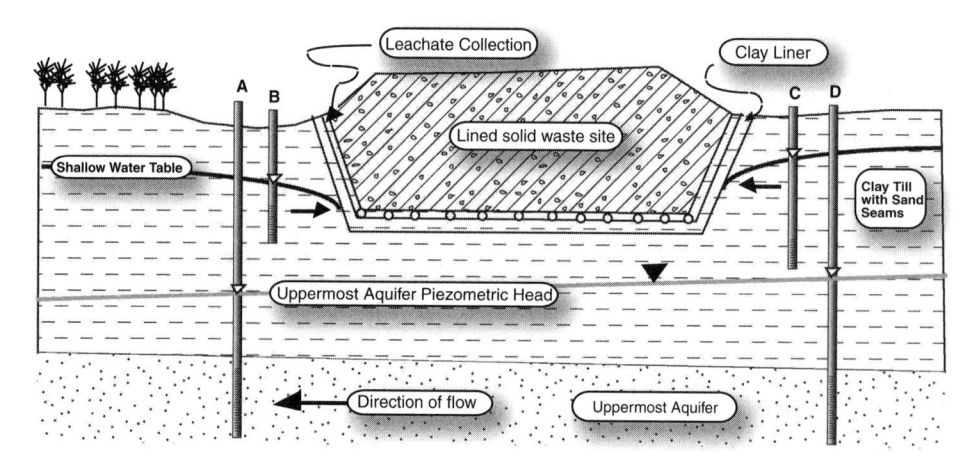

FIGURE 8.28
Effect of drains on low-hydraulic conductivity materials.

Ground-Water Monitoring System Design 553

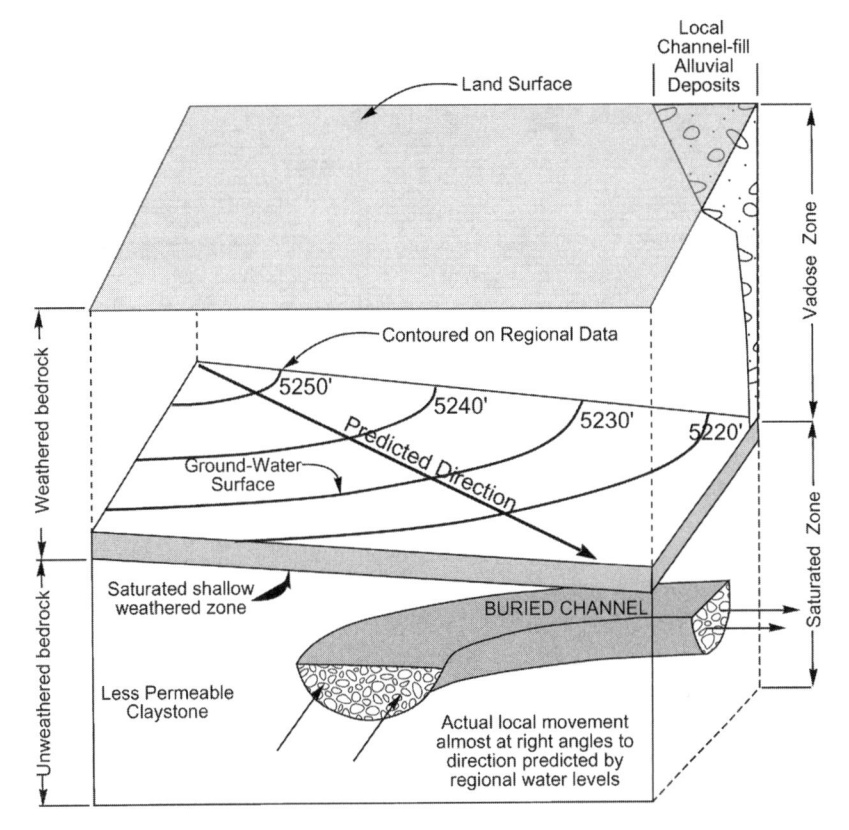

FIGURE 8.29
Channel deposits in low-hydraulic conductivity materials.

different stream systems and, therefore, were not vertically interconnected. Minor sand lenses, such as the levee splay deposits or overbank matrix-supported sands that were deposited in backswamps, also have limited area extent. They are connected horizontally over short distances and are vertically separated from other sandstones in the system by the intervening claystone.

Near-surface Dawson claystones are typically weathered and can become seasonally saturated as a perched ground-water system with sufficient hydraulic conductivity to comprise a target monitoring zone. This weathered zone can be easily defined by shallow (<30 ft) borings and piezometers. The deeper sand channel deposits, however, present a more difficult directional flow analysis problem. In this environment, each sand channel deposit has its primary component of flow in the stratigraphically down-dip direction. Structural warping of the geologic units tilted the channel deposits back toward the original source direction. These deposits are thus very difficult to evaluate in terms of discharge direction. Monitoring such a heterogeneous geologic environment requires very detailed drilling and rock coring, and boring log information must be located through geophysical methods to define the depth and location of these channels. These channels may be secondary target monitoring zones, as they likely serve as subdrain systems for the shallow, unweathered claystones.

Low-hydraulic conductivity environments may also have a more permeable upper unit discharging locally to a stream or river. Figure 8.30 illustrates a thick clay unit confining a regional uppermost aquifer. The near-surface weathered silty clay unit has relatively low hydraulic conductivity with minor sandy units and the unlined landfill discharges

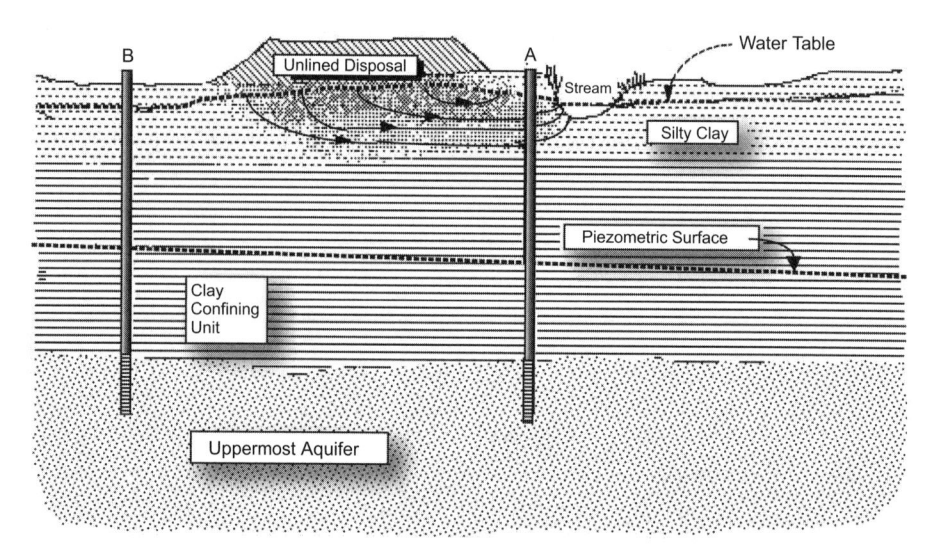

FIGURE 8.30
Low-hydraulic conductivity materials with weathered upper units.

leachate as seeps or springs near the landfill base and into the stream. These sites typically look bad due to the surface discharge of leachates; however, these surface discharges represent leachate that did not move into subsurface pathways. Monitoring of such an environment would probably include alternative sampling of the stream. Visual inspection of local streams can provide insight into springs and small discharge areas.

Because the deep aquifer is confined by the thick clay unit and the unlined landfill discharges to the stream, monitoring wells for a detection monitoring program would probably be located between the stream and the landfill. However, the relatively low-hydraulic conductivity of the near-surface materials will make monitoring difficult in practice due to long (days to weeks) recovery times for the wells. Additional piezometers should be located across the stream to verify that ground-water discharge occurs along both sides of the stream.

Geologic Structural Control

Geologic structures such as dipping beds, faults, cross-bedding, and facies changes can greatly affect the rate and direction of ground-water movement. The monitoring system designer must consider geologic structural controls throughout the entire site investigation to ensure that the site is adequately characterized. The development of a conceptual model is the key to successful ground-water monitoring system design in structurally controlled environments. Geologic structures affect ground-water movement in several ways:

- Acting as more permeable flow paths, because of higher primary porosity (i.e., cross-bedded sands) or through secondary hydraulic conductivity enhancement (natural fractures)
- Acting as either barriers to ground-water flow or as conduits for ground-water flow, as do many fault zones, depending upon the nature of the material in the fault zone

If the fault zone fill material consists of finely ground rock and clay (gouge), the material may have a very low (e.g., $<10^{-6}$ cm/sec) hydraulic conductivity. Significant differences in ground-water levels can occur across such faults. The hydrogeologist should be alert to large (e.g., >20%) unexplained differences in water levels across a site in faulted environments. These differences may be due to fault gouge retarding ground-water flow across the fault. Impounding faults can occur in unconsolidated clays as well as sedimentary and even igneous rocks. For example, faulted sequences of interbedded shales, which normally would not hinder lateral ground-water flow, may have weathered clay materials smeared along the fault, which can act as barriers to flow. Fault-zone flow barriers are relatively common in the ground-water basins of southern California. However, it is important to note that faults can also act as conduits for ground-water flow. These systems should be evaluated by careful observation of water levels next to the faulted units. If the potentiometric surface flattens over the faulted area, then it is likely that this area represents a higher hydraulic conductivity zone that is discharging to other aquifers or surface discharge points.

Definition of geologic structures as considerations in ground-water monitoring system design should include the following points:

- Identification of major geologic structures, regionally and site-specifically, early in the site investigation
- Identification of potential fault areas through literature surveys and aerial photo review
- Identification of springs, vegetation changes, and surface geology through site reconnaissance prior to drilling
- Development of an initial conceptual hydrogeologic model
- Installation of borings placed to define geologic structure, variable water levels, and gradients in each geologic unit
- Reconciliation of logs of stratigraphic borings and piezometer water levels with the conceptual geologic model
- Interpretation of structural contour, water-table elevation, and piezometric level contour maps to develop a linked hydrogeologic conceptual model with flow nets to identify target monitoring zones

The monitoring system should be designed only after consideration of all of the information gathered during the review of existing information and the evaluation of data collected during the field investigation.

The effect of geologic structures on leachate movement, such as the simple dipping orientation of the bedrock, is illustrated in Figure 8.31. Steeply dipping, alternating beds of sandstone and shale can have significantly different (e.g., 1×10^{-4} to 1×10^{-7} cm/sec) hydraulic conductivity. The three-dimensional view (Figure 8.31) shows preferential movement of ground water along strike of the sandstone. The view illustrates the down-dip movement of contaminants. A detection monitoring system located in the shales would not be capable of early detection of leakage from the unlined site. One indication of the highly variable hydraulic conductivity of the rock mass is the overland flow of leachate. Because the leachate cannot move rapidly into the sandstone (and less so in the shales), leachate is rejected to the surface over the shale outcrop. A leachate seep occurs at the contact between the sandstone and shale. Surface seeps (and springs) are excellent indicators of changes in formation hydraulic conductivity and should always be carefully considered in the development of conceptual hydrogeologic models.

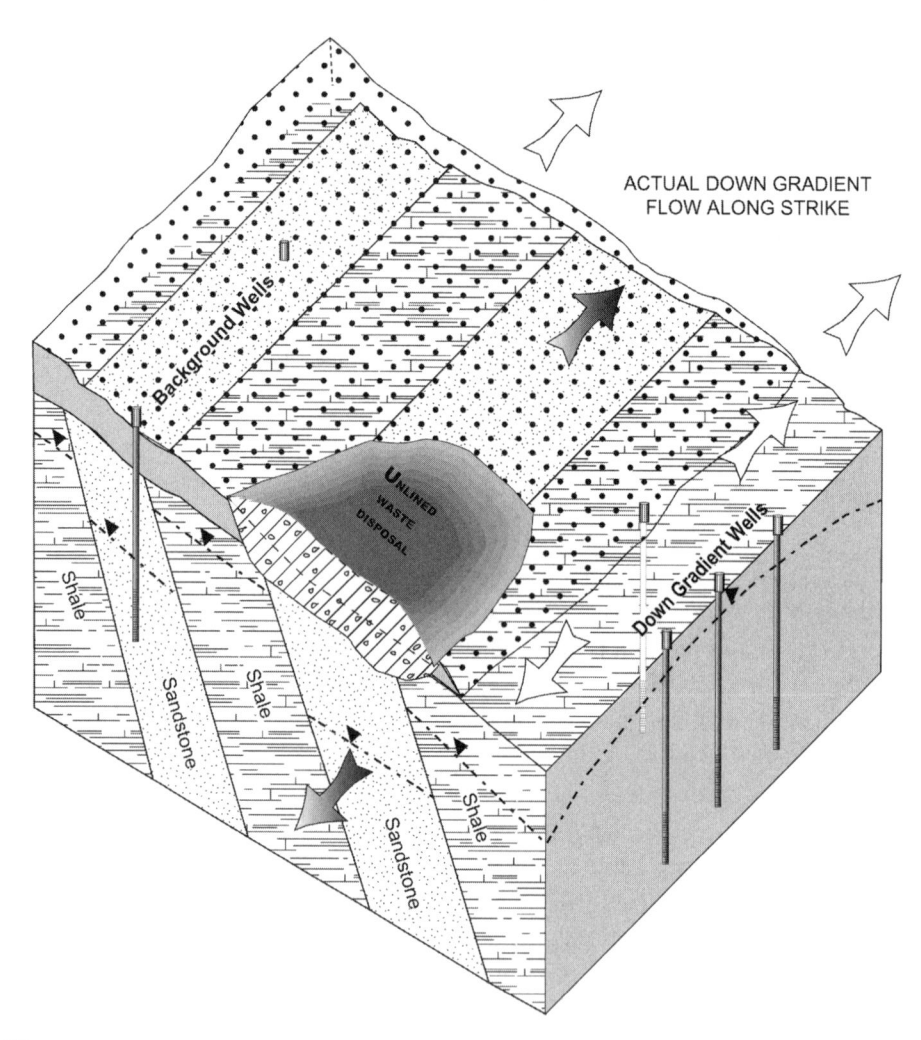

FIGURE 8.31
Structural control of ground-water flow.

Perched Ground Water

Monitoring programs in perched ground-water environments present a number of complications to the monitoring system designer. Figure 8.32 shows potential leachate pathways from an unlined waste disposal site. Perched ground water does not follow regional ground-water gradients, but rather will flow along a hydraulic conductivity interface in response to gravity, as shown in Figure 8.33. The approach used to design a monitoring system for a perched water condition would include the following:

- Evaluate the lateral extent and thickness of various geologic units down to at least 25 ft below the base grade of the facility through continuously sampled soil borings. Particular attention must be paid to the presence of saturated zones above fine-grained low-hydraulic conductivity layers. A rule of thumb is that potential perched zones may occur at a hydraulic conductivity contrast of two orders of magnitude. A three-order-of-magnitude hydraulic conductivity difference between two adjacent units will almost always result in perched ground water. Contrasts between sand and silt or sand and clay will likely show such

Ground-Water Monitoring System Design 557

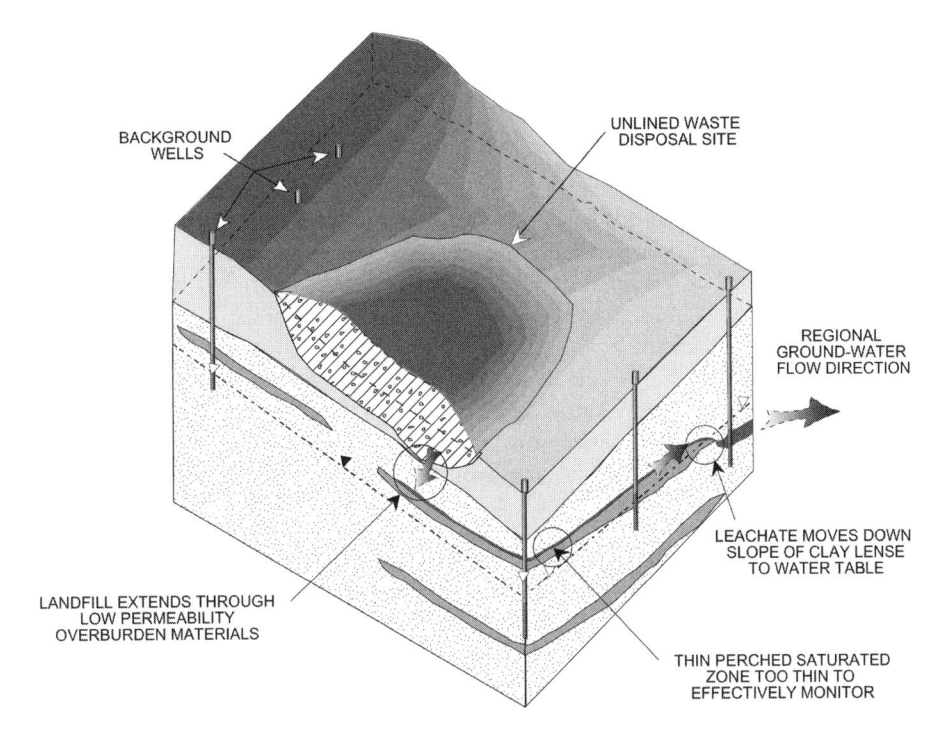

FIGURE 8.32
Perched structural control of ground-water flow.

a three-order-of-magnitude variation. Such contrasts in hydraulic conductivity will result in a perched water zone that will make interpretation of flow direction very difficult if not properly recognized. Geologic cross-sections can help identify potential locations of perched water.

- Carefully evaluate road cuts and stream cuts in the vicinity of the site for the presence of low-hydraulic conductivity geologic units that could cause perched conditions. The units in the cuts may already have been drained; however, the

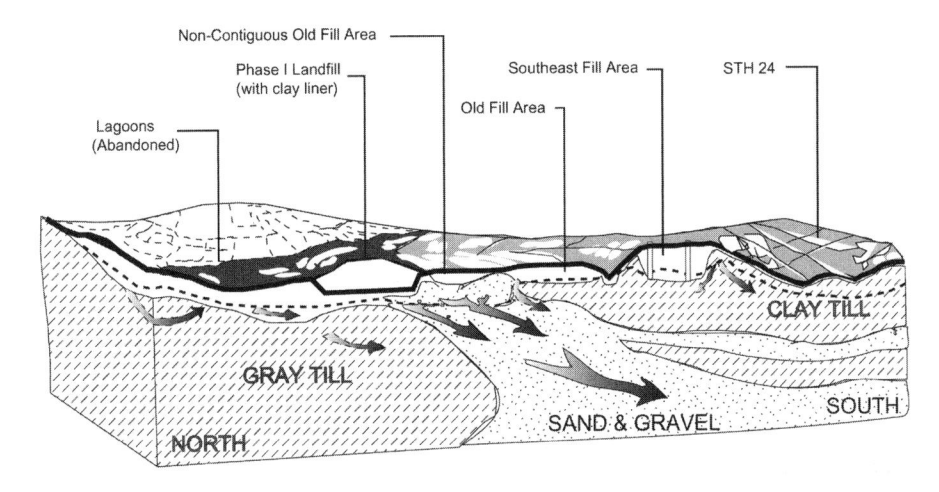

FIGURE 8.33
Ground-water discharge through more permeable units.

perched water zones may be located by vegetation concentration above the perching unit or, during winter, frozen ground-water discharges.

- Sufficient continuously sampled borings should be drilled to define the horizontal extent and elevations of potential low-hydraulic conductivity zones above the regional ground-water surface.
- Piezometers (at least three) should be completed in each geologic unit, including permeable units above potential perching units to establish the presence of perched water zones. Care must be taken to complete the piezometers so that the bottom of the screened zone is situated at the interface between the perching unit and the overlying more permeable unit. If the perching unit is very thin (e.g., less than 1 ft) or if it has a very thin saturated area (2–4 in.), the need to monitor the zone should be reevaluated. Very thin perching units are often discontinuous and it may be more important to understand the geologic history of the system to predict the orientation of the perching systems rather than to focus on one individual unit.
- A contour map of the top surface of the low-hydraulic conductivity unit should be constructed to define potential perched water flow directions. The contour map should be combined with cross-sections showing water levels to establish perched water flow paths.
- If the perched saturated zone is below the base of the facility and sufficiently thick to be characterized as an uppermost aquifer, or if it is sufficiently thick to allow collection of adequate samples to serve as an early detection location, it should be considered a target monitoring zone.
- If the perched saturated zone is too thin (e.g., <2 ft) to be saturated year-round, the monitoring system should be installed in the first permanently saturated zone (the uppermost aquifer) beneath the perched zone, as shown in Figure 8.32.

A detection monitoring system could be installed in either the perched water body, if a sufficiently thick (e.g., >10 ft) saturated zone exists, or within the deeper uppermost aquifer, based on hydraulic gradients. A key point in determining if there is a potential for perched water bodies is the stratigraphy present at the site. If clay or other fine-grained materials are present near the surface, the potential for occurrence of perched water bodies is greatly reduced, due to limited recharge. If highly permeable (e.g., $>1 \times 10^{-3}$) material exists near the surface, with less permeable material below, perched water bodies are more likely to occur. The lower the amount of recharge, the less likely that the hydraulic conductivity contrast will act as a significant perching mechanism. Monitoring for rate and extent determinations would concentrate on the first aquifer beneath the perched zone rather than on the perched zone itself.

Although the potential may be present at a site for discontinuous individual clay units to perch ground water, the limited extent or lack of continuity of the perching units may not require definition of the individual units. Figure 8.33 illustrates a conceptual cross-section and flow net of unconsolidated deposits overlying a regional confining unit. In this example, the individual clay units are limited horizontally, so ground water flows through windows between the clay deposits. A single unconfined water surface was established with few perching conditions recognized. Flow path A discharges from around piezometers D-15 toward the southeast, working through and around the various flat-lying clay units. Alternative flow paths could be constructed from the D-15 area to the northwest, as this location represents an upland recharge area. Discharge is toward streams (gaining streams) cutting into the Coopers formation to the northwest and southeast of the facility area. The projected flow path A would be best monitored in the area of D-13, as the majority of flow for the system passes below the clay unit at this location. To the northwest, monitoring of the zone between 50 and 60 ft above National Geodetic Vertical Datum (NGVD) would provide a secondary flow path target from the site area.

Secondary Hydraulic Conductivity

The three basic types of ground-water occurrence and movement are shown in Figures 8.34a–c. Figure 8.34a shows primary porosity where ground water moves through the interstices (voids) between sand-sized grains. Figure 8.34b shows ground-water movement through fractures that represent secondary porosity. Figure 8.34c shows ground-water movement through solution channels developed in a carbonate rock, another type of secondary porosity. A site-specific geologic environment could consist of any or all of these media. The field investigation should determine the dominant flow mechanism beneath the facility to be monitored so that the appropriate locations for monitoring wells can be selected.

Fractured or solution-channeled carbonate rocks provide a special problem in ground-water monitoring system design. Often there will be highly directional ground-water movement along discontinuities or dissolution-widened joints. The success of any monitoring system in a fractured or solution-channeled environment requires detailed knowledge of the joint, fracture, or solution channel patterns in the rock. In some instances, remote image or aerial photo interpretation (e.g., fracture-trace analysis) and special field techniques (e.g., tracer tests) can identify the target monitoring zone in a secondary porosity environment.

Most consolidated rocks (with the exception of some sandstones and conglomerates) have few well-connected primary intergranular openings available for ground-water flow and usually have much lower hydraulic conductivity values than their unconsolidated equivalents. Groundwater flow in bedrock aquifers often takes place through secondary openings such as fractures (joints, bedding planes) and solution channels. The investigator designing a monitoring system should fully identify areas in which this factor is important and should do so at an early stage in the site investigation. Although regional flow patterns should be well established in the site investigation, it is often very difficult to predict ground-water flow through a set of fractures or solution channels on a site-specific scale (e.g., in the vicinity of a monitoring well). Thus, facilities located over bedrock aquifers should employ additional investigative techniques (e.g., fracture-trace analysis [see Chapter 2], geophysical investigations [see Chapter 4], detailed coring

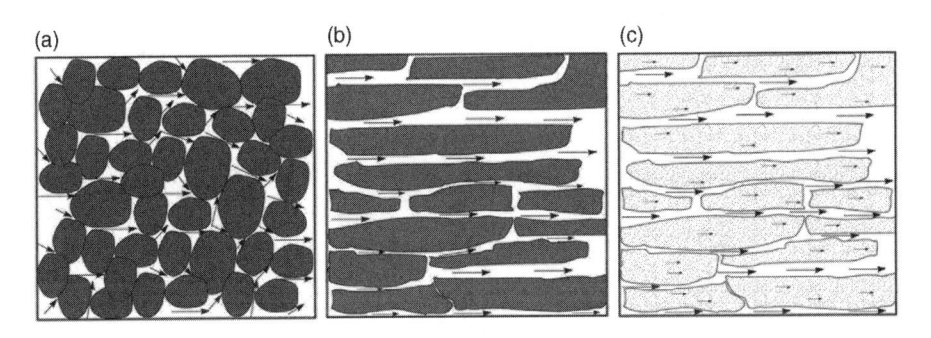

FIGURE 8.34
Various forms of ground-water flow.

[see Chapter 5] and geologic mapping, and pumping tests specifically designed to evaluate anisotropy [see Chapter 14]) to adequately determine likely ground-water flow pathways.

Fractured rock environments require consideration of specific flow paths to define the target zone monitoring system design. Chapter 2 discusses the field investigation tasks appropriate for a fractured-rock environment. Figure 8.35 shows the individual groundwater flow paths in a fractured rock environment of a single rock type. Leachate is shown moving down from an unlined landfill toward a series of fracture sets that control local groundwater flow. Detection monitoring system design would involve placing well screens both upgradient and downgradient of the facility. Individual screen depths must be based on the results of the rock coring program and the observed fractures or weathered zones rather than on only observed hydraulic heads. In an assessment monitoring situation, long (e.g., >15 ft) well screens should be avoided to reduce the potential for cross-contamination caused by leachate entering the well in an upper zone and moving downward through the screen into formerly uncontaminated zones. Fracture patterns can be highly localized and unpredictable, as shown in Figure 8.36, or more evenly distributed and predictable, as illustrated in Figure 8.37. Often, both primary and secondary porosities are present in bedrock units, as illustrated by Figure 8.38, and so the site investigation must include measurement of the hydraulic and geologic parameters for each medium present at the site. The approach for ground-water monitoring system design in a fractured geologic environment should follow these procedures:

- Evaluate fracture patterns using background information, aerial photographs (fracture-trace analysis), and measurement of fractures at surface exposures.
- Establish a core drilling program at the site that may include use of rock quality designation (RQD), fracture orientation of cores, borehole video logging, and detailed visual logging. Packer hydraulic conductivity tests should be considered for use in the investigation, in which packer intervals should be selected to test specific, observed discontinuities. Consideration should also be given to including angle core drilling in those areas where vertical fractures may be present.

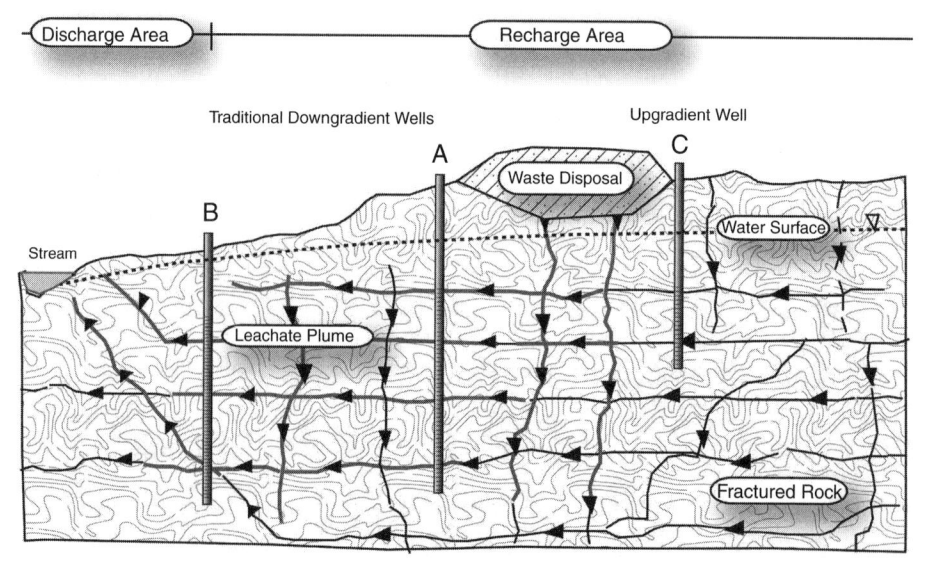

FIGURE 8.35
Control of ground-water flow in fractured rock.

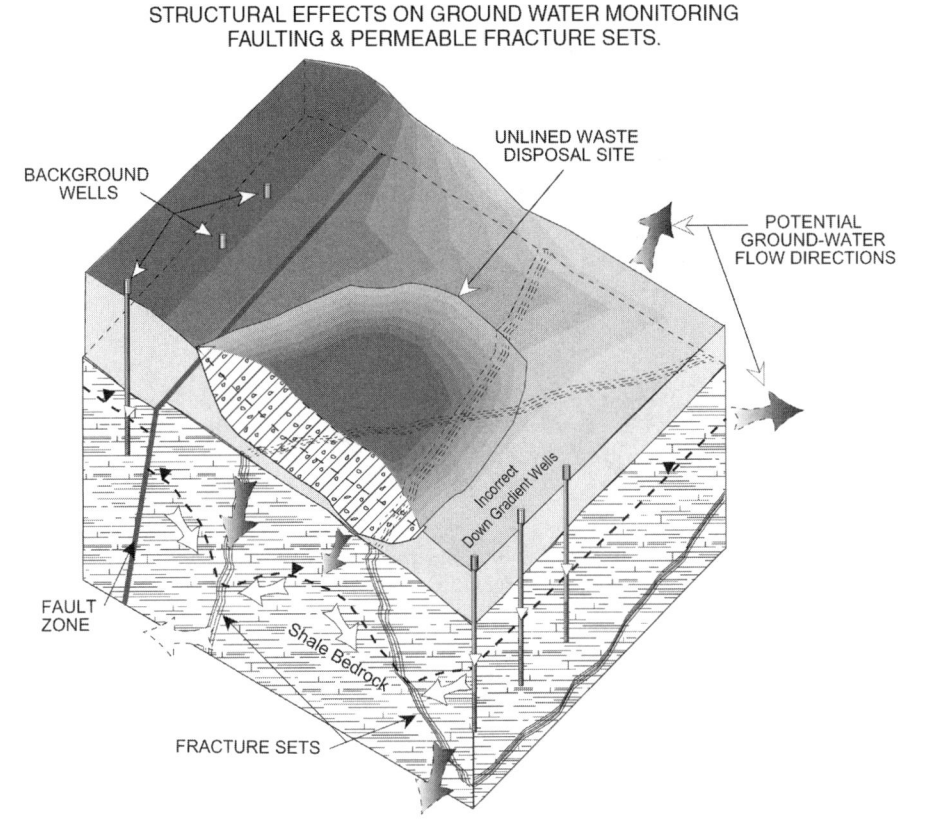

STRUCTURAL EFFECTS ON GROUND WATER MONITORING
FAULTING & PERMEABLE FRACTURE SETS.

FIGURE 8.36
Local flow pathways in fractured rock.

- Implement borehole geophysical surveys, such as caliper logs, flow logs, acoustic televiewer logs, and temperature surveys.
- Install multiple piezometers or multi-level monitoring systems for assessment of hydraulic conditions in individual fracture zones detected in the coring and geologic logging process.
- Measure piezometric heads and gradients in relationship to joint patterns or fracture sets.
- Establish a conceptual model defining the target flow zones in plan view and in cross-section.

A detection monitoring system can be effective in a fractured-rock environment if the wells are screened in highly permeable fractures (those flowing into the borehole) down-gradient from the facility. Wells in these locations can very quickly (e.g., within days to weeks) and detect leachate releases from the facility.

Solution-channeled bedrock (Karst) terrain presents additional challenges to the ground-water monitoring system designer because monitoring wells can easily miss permeable solution channels and may even end up as dry holes. Quinlan (1990) provides a full description of the requirements for ground-water monitoring in Karst terrain. Karst environments require careful consideration of where to locate both background and

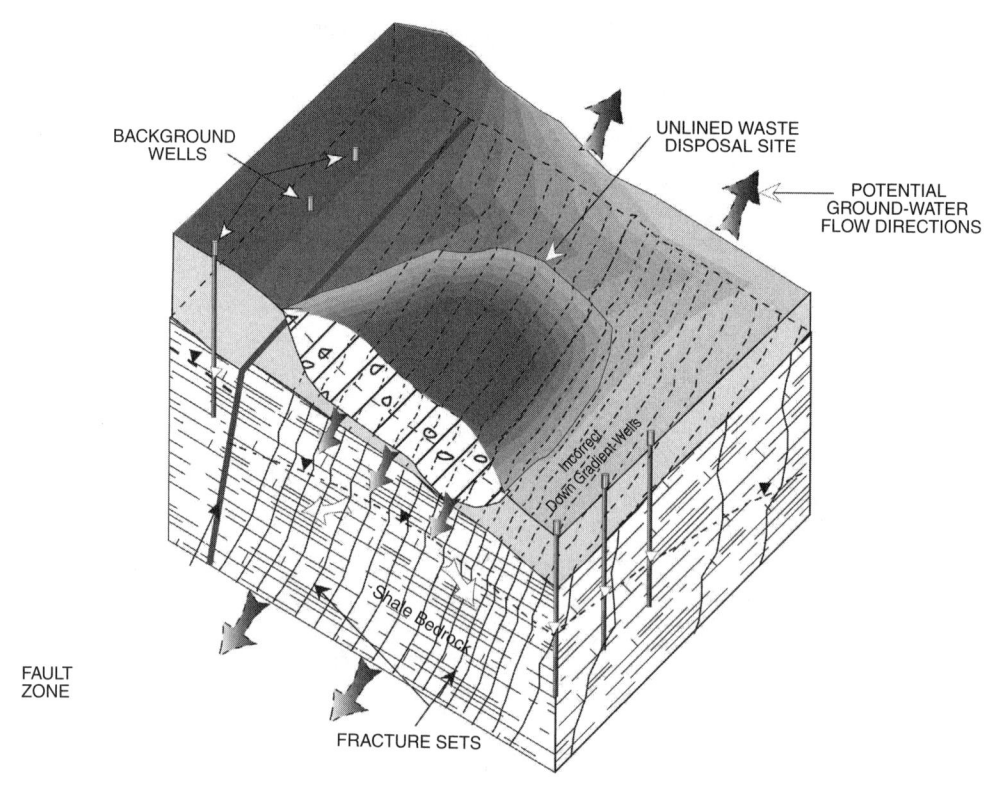

FIGURE 8.37
Evenly distributed flow pathways in fractured rock.

downgradient wells and springs as well as when to monitor the extremely fast reacting system. Quinlan (1990) recommends the following procedures for design of monitoring systems in Karst terrain:

- Review the regional and local geologic and hydrogeologic literature for the area in question.
- Evaluate topographic and geologic maps.
- Conduct a survey of local springs.

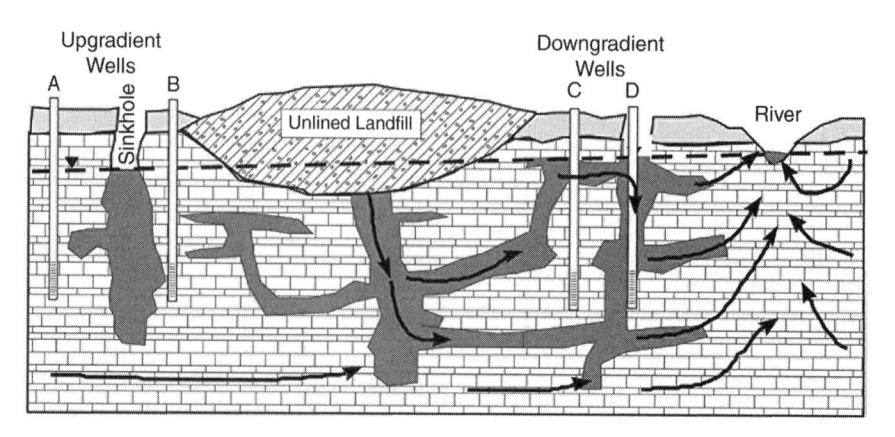

FIGURE 8.38
Local flow pathways in Karst terrain.

- Map the regional and local potentiometric surface.
- Conduct a dye-tracing study based on data collected above.
- Perform the first dye-tracing study, preferably during moderate flow conditions.
- Evaluate the results of the first dye-tracing study and modify the design of the tracing study if necessary.
- Collect samples from springs and analyze indicator parameters (turbidity, specific conductance, pH, temperature, dissolved oxygen, oxidation–reduction potential) to determine whether local springs at or near the facility are characterized by conduit or diffuse flow.
- Perform additional dye-tracing studies during moderate flow conditions, always modifying the tracing plan, as necessary, in light of the results of the previous trace results. For most facilities, it is necessary to perform at least two dye-tracing studies during moderate flow conditions.
- Repeat selected dye traces during base-flow and flood-flow conditions.
- Integrate dye-tracing results, available hydraulic head data, and indicator parameter data used to discriminate between conduit and diffuse flow into a monitoring plan.
- Review the entire project area to select appropriate locations for long-term monitoring wells.

General guidance for detection monitoring in Karst terrain must include careful consideration of background well locations. In general terms, background well locations must be based on:

- Negative results from dye tracing tests
- Locations in similar geology and geochemistry as downgradient sites
- Locations in similar cultural environments

Sampling for water quality in Karst terrain also does not meet the typical regulatory model for biannual or quarterly sampling periods. Quinlan (1990) recommends sampling based on storm and meltwater events. Figure 8.38 illustrates monitoring of an unlined facility in a Karst area, where sinkholes are present beneath and next to the facility. Normally, such a setting is not easy to monitor, but a monitoring system can be developed to determine the facility's impact on the environment.

The assessment monitoring procedure would be similar to that used above with the possible addition of other surface geophysical surveys. Ground-penetrating radar, electromagnetic conductivity, and seismic refraction surveys (see Chapter 4) can help identify some zones of solution channeling and deep weathering within the rock mass.

Identification of target monitoring zones in the plan view and in cross-section should be done before any monitoring wells are installed at the site. In most Karst systems, gradients are typically very low, and ground-water flow direction is difficult to determine. This requires that very accurate surveys be conducted to allow for definition of ground-water flow direction.

Density Control

Several monitoring system designs are reviewed here to illustrate some of the major implications of contaminant density and immiscibility in the design of ground-water-monitoring systems.

In the preceding discussions of monitoring system design, it was assumed that the potential contaminants were soluble in water and that the density of the water was not altered by the presence of the solutes. Although typical leachates from codisposal, hazardous-waste, and solid-waste disposal facilities fit this profile, there are a number of monitoring situations for which these neutral density assumptions are not appropriate. In particular, wastes such as brines or other water-based industrial effluents can be soluble in water but may have a density significantly greater than that of water. Petroleum hydrocarbons (e.g., gasoline, kerosene, diesel fuel, and fuel oil) and organic solvents (e.g., trichloroethylene, perchloroethylene, methyl ethyl ketone) are only poorly soluble in water, with a large percentage of these fluids remaining in a nonaqueous phase (NAPLs) after they are released to the environment. Varying solubilities and densities of stored products and waste materials can produce transport characteristics quite unlike those normally associated with neutral-density leachates.

The effect of density in contaminant fate and transport is considered in Bear (1972), while several authors, including Schwille (1988), Corey (1977), and Collins (1961) consider the mechanics of subsurface flow of immiscible fluids, which are represented schematically in Figure 8.39.

Figure 8.39a shows a contaminant plume developing as a result of seepage of a dense immiscible fluid (DNAPL) into the ground water system. The fluid tends to move vertically downward to the bottom of the uppermost aquifer (or the top of the first confining unit). Once on the bottom of the aquifer, the movement of the DNAPL is governed by the topography of the top of the confining unit, thus the direction of flow will not necessarily be the same as the direction of regional ground-water flow. Owing to dispersion, contaminants that solubilize from the DNAPL will be contributed in aqueous phase to the local ground-water flow system. Thus, several areas of contamination may be present — the major plume of DNAPL and the adjacent ground-water zone contaminated by soluble levels of the dense fluid as a result of dispersion. Establishing the area of contaminated ground water would require use of the typical methods of investigation employed in any assessment monitoring situation. Locating the major pool of DNAPL would require knowledge of the position and inclination of the surface of the first confining unit and the installation of sampling points at the top of this surface (or at the bottom of the aquifer). Over a period of time, the density of the fluid mass would decrease to the point at which the plume would migrate according to local ground-water flow conditions (Cherry and Feenstra, 1991). Soluble constituents would continue to be contributed to the local flow system; however, the DNAPL would likely remain at the bottom of the aquifer for a prolonged period of time, moving according to gravity.

Figure 8.39b illustrates the movement of an immiscible fluid having a density less than that of water (LNAPL; see API, 1989). In this case, the major contamination occurs near the top of the saturated zone, and movement is controlled by the directional slope of the water table. As a result of dispersion, soluble contaminants would be contributed to the regional flow system and a dissolved-phase plume would gradually be distributed downgradient. In the case of LNAPL product spills, monitoring points should be concentrated in the upper part of the aquifer if vertical flow components are not significant.

Figure 8.39c depicts the infiltration and movement of a fluid having a density less than or similar to that of water. In this case, the main area of contamination occurs near the top of the saturated zone and the plume is dispersed in the direction of ground-water flow. As the immiscible phase moves through the porous medium, a residual amount of fluid is retained in the pore spaces and on the grains of soil of the medium in a relatively immobile state, to slowly leak into ground water as it flows past the residual fluid. Immiscible fluids have soluble constituents that are leached from the fluid and move along with the regional ground-water flow system. In addition, volatile constituents could be contributed to the gas phase in the zone above the water table.

Ground-Water Monitoring System Design

VARIABLE TYPES OF PLUME GENERATION

(a) DENSE - MISCIBLE FLUID

(b) LOW-DENSITY MISCIBLE FLUID

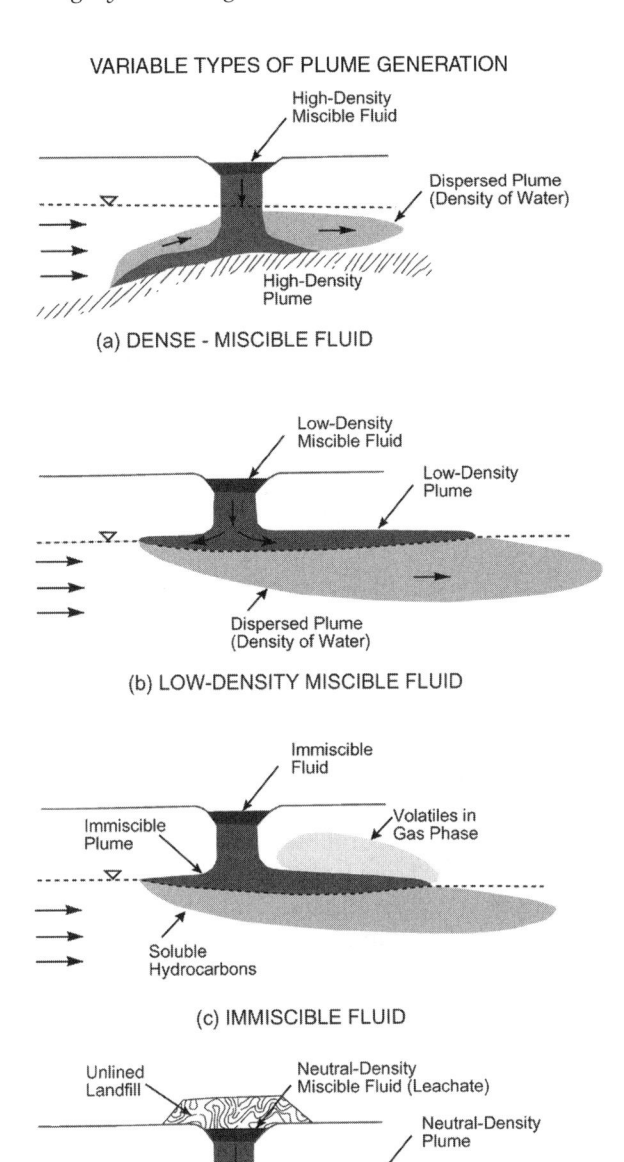

(c) IMMISCIBLE FLUID

(d) NEUTRAL-DENSITY MISCIBLE FLUID

FIGURE 8.39
Density considerations in facility monitoring.

Figure 8.39d presents a neutral-density soluble fluid moving from an unlined landfill into a local aquifer. Because the leachate is soluble, the resultant plume moves along with ground-water flow in the aquifer, and monitoring would be based on target monitoring zones and three-dimensional ground-water flow components. Ground-water monitoring system design for this type of contaminant should use both conceptual models and flow-net construction along with consideration of the leachate density.

Separation of Adjacent Monitoring Programs

New landfills built to state-of-the-art design criteria are commonly being constructed next to traditional disposal areas, many of which do not have liners and leachate collection systems. Even with extensive double composite liners, the new facilities must demonstrate the long-term engineering performance of the new cells through ground-water detection monitoring programs. Selection of the proper locations for detection monitoring wells should be based on a holistic approach to the evaluation of a specific site. The placement of the ground-water monitoring wells in this process must weigh and balance data collected in the field, laboratory, and office. This is especially true for multiple facilities located on adjacent properties.

The target monitoring zones are further useful for the separation of detection ground-water-monitoring systems from adjacent facilities that may have been unlined or are currently impacting water quality. The following case histories illustrate how ground-water flow concepts can be effectively used to evaluate the optimum locations of wells and positions of well screens for a detection monitoring system.

Simple Gradient Control

Facility A represents a relatively simple condition where an existing 20-acre landfill is located in a downgradient position from a 25-acre expansion as shown in Figure 8.40. The expansion has a 60-mil High Density Polyethylene (HDPE) liner and a leachate collection system. Ground-water surface contours, shown in Figure 8.40, when combined with the hydrogeological cross-section (Figure 8.41), show the target monitoring zone to be the Pleistocene terrace deposits of interlayered sands and clayey sands. The underlying Choctawhatchee Formation was not judged to be significantly more permeable than the terrace deposits and both formations are underlain by the locally nonwater-bearing Hawthorn formation. Ground water would not have a significant downward movement component

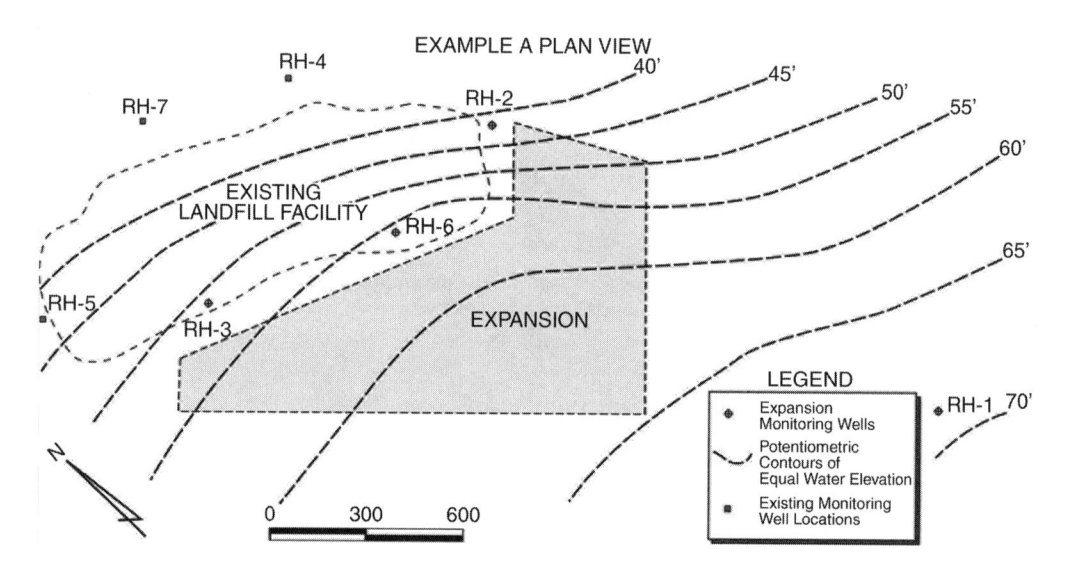

FIGURE 8.40
Gradient-controlled ground-water contours for the example A facility.

Ground-Water Monitoring System Design

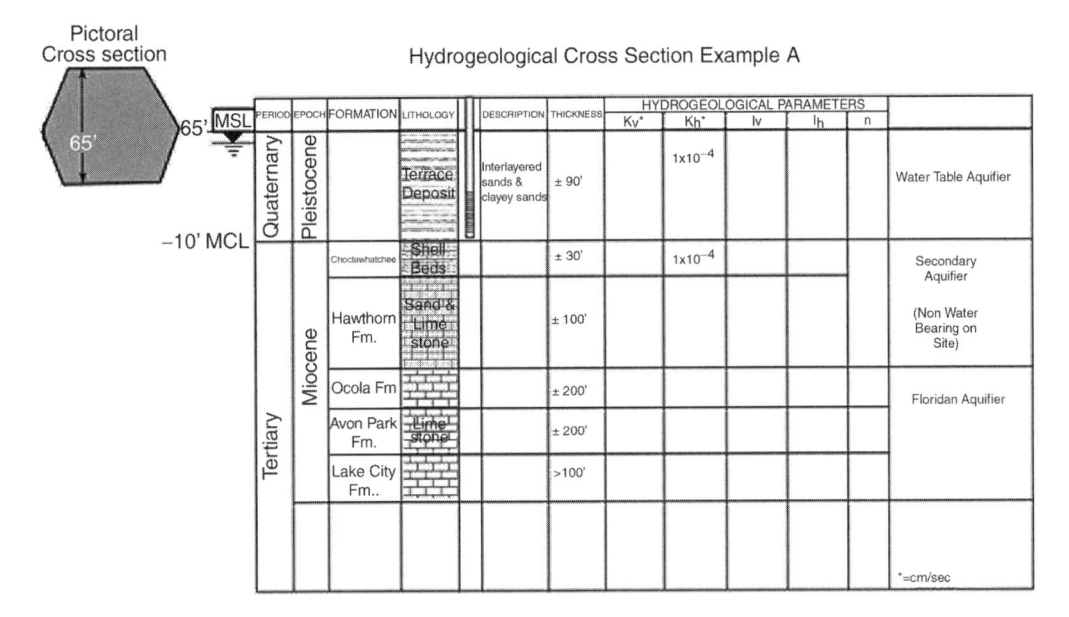

FIGURE 8.41
Hydrogeological cross-section of the example A facility.

adjacent to the site, and detection monitoring wells should be located between the two facilities. The flatness of the ground-water contours as they pass beneath the site shows that there is no significant mounding in the existing facility. This would not cause the area between the two facilities to be a discharge point for the existing site. Wells located between the facilities, with the observed ground-water flow conditions, would be adequate to monitor the expansion. At least two background wells are located between 65′ and 70′ ground-water contour line. The well depths should be completed between the base grade of the expansion and the base of the terrace deposits.

Gradient and Lithology Control

Site B represents a slightly more complex example where a ground-water divide and lithology complicate ground-water monitoring conditions. Figure 8.42 shows a 16-acre lined cell with leachate collection next to a 5-acre closed commercial and large closed county landfill. The hydrogeologic cross-section (Figure 8.43) illustrates the lithology and base grade configurations for the site. The site hydrogeology is dominated by the thick 100-ft Cooper formation, which acts as a regional confining unit. The shallow Pleistocene aquifer has a generally unsaturated upper sand unit with an intervening clay unit overlying a saturated lower sand unit.

The hydrogeologic cross-section, as presented, is insufficient to fully evaluate localized flow conditions for the purpose of selecting locations for detection monitoring wells. Cross-section G–G′ (Figure 8.44) must be evaluated in combination with water-table contours shown in Figure 8.42. The water-table contours show a ground-water divide occurring along a natural ridge area. The new lined cell is located along the nose of the ridge and on top of the ground-water divide. Monitoring of sites located on ground-water divides requires a three-sided approach. If the cell was located directly on top of a hill, downgradient could be in all four directions and monitoring background ground-water quality would be more complicated. Background water quality for the example B site

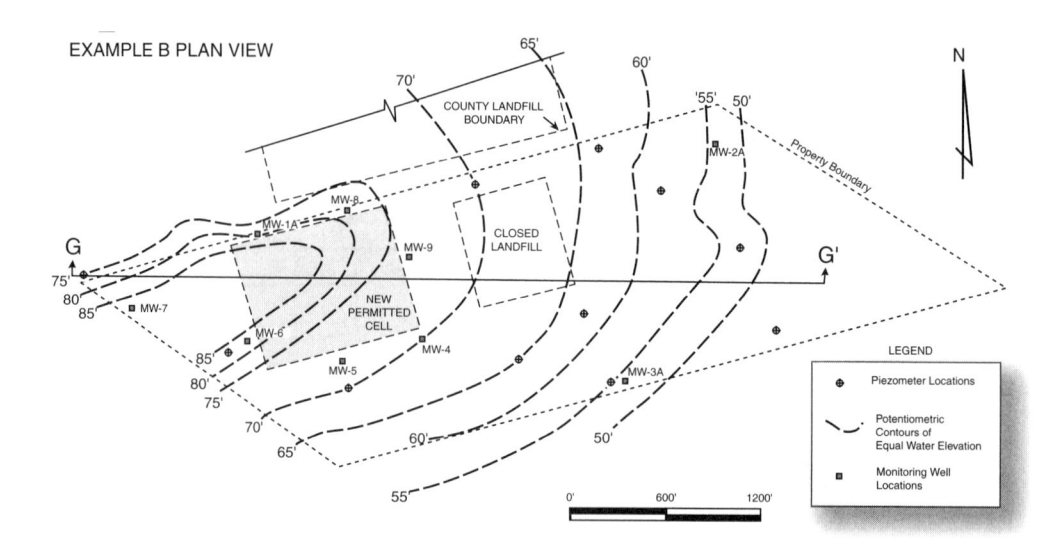

FIGURE 8.42
Plan view of the example B facility.

would be along the ridge line in the locations shown between MW-7 and MW-6. Downgradient locations would be located from MW-1A clock-wise around through MW-5. Because the site has sufficient ground-water gradients from the new cell toward the old closed county and commercial sites, local discharge between the old and new sites would not be considered a problem. The cross-section of G–G' illustrates a combined conceptual model of the geologic site cross-section and the observed ground-water flow paths. Flow path A represents a potential flow line from below the lined facility in one direction toward a local creek discharge point. A flow path could also be drawn in the opposite direction (as represented by the ground-water divide shown in Figure 8.44).

The well screen depths that would most effectively cover the facility in a detection monitoring system should be located in the sandy unit above the Coopers Formation, directly

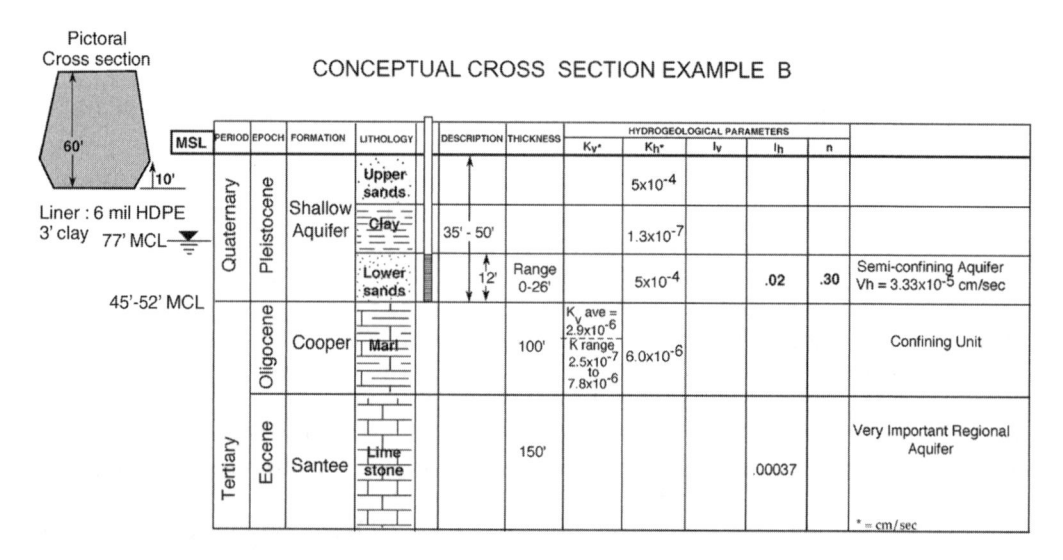

FIGURE 8.43
Conceptual view of the example B facility.

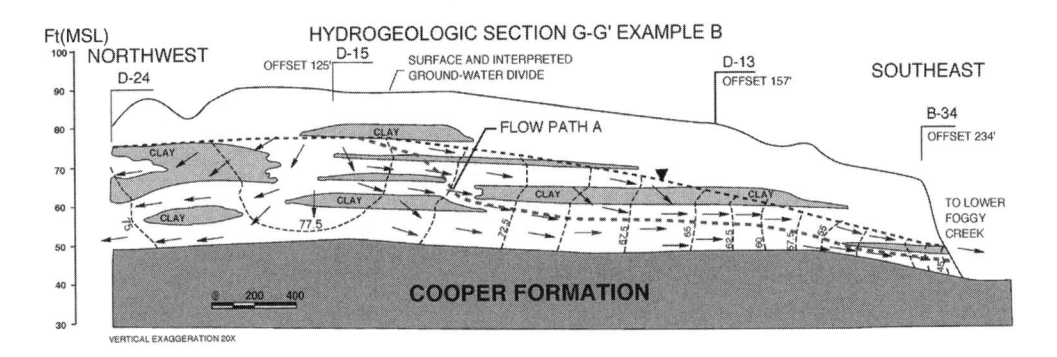

FIGURE 8.44
Cross-section of the example B facility.

below the first saturated Pleistocene clay unit. Downgradient detection monitoring wells should be located in both cross-section directions, as indicated by the ground-water flow arrows. Information on local lithology and construction of ground-water flow nets makes selection of potential depths for well screens a relatively straightforward task.

The previous two examples represent primarily horizontal ground-water flow conditions. The following example, however, represents a more complex three-dimensional problem requiring complete linkage of flow and lithology.

Complex Ground-Water Flow Conditions

Site C represents a new 70-acre lined site with leachate collection next to a closed 80-acre MSW landfill (Figure 8.45). The closed site is unlined without leachate collection facilities. The uppermost aquifer, as shown in Figure 8.46, is Pleistocene alluvium (35-ft thick). The underlying Garber-Wellington Formation (300–400-ft thick) is the main regional aquifer. The Garber-Wellington Formation shows hydraulic conductivity similar to the overlying alluvium, and so simple horizontal flow conditions cannot be assumed for detection monitoring purposes. The ground-water contour plan (Figure 8.45) shows that complex localized flow conditions exist at the site due to the effects of the inactive sand and gravel operation (now full of water) and the river to the west of the site. Ground-water movement is generally from the southwest to the west of the new facility. However, local discharges also occur to the east of the closed facility. The physical condition of having an unlined site generally upgradient of a new facility can make interpretation and comparison of background and downgradient water quality very difficult. The ground-water flow conditions are further complicated by the absence of a confining unit to separate the uppermost aquifer (alluvium) from deeper regional aquifers.

This site must be evaluated through use of a linked conceptual model and flow-net construction cutting an east–west cross-section. Figure 8.46 illustrates site ground-water flow through the uppermost aquifer into the Garber-Wellington and discharging in two directions (west and east). The lined site's potential flow path discharges to the river, and can be monitored with wells completed between 1080 and 1130 ft above msl. This 50-foot thick zone represents the most likely flow path for potential discharges from the lined facility. Completion of deeper monitoring wells would intercept flow paths coming from the unlined facility and would not represent conditions truly downgradient from the active disposal site.

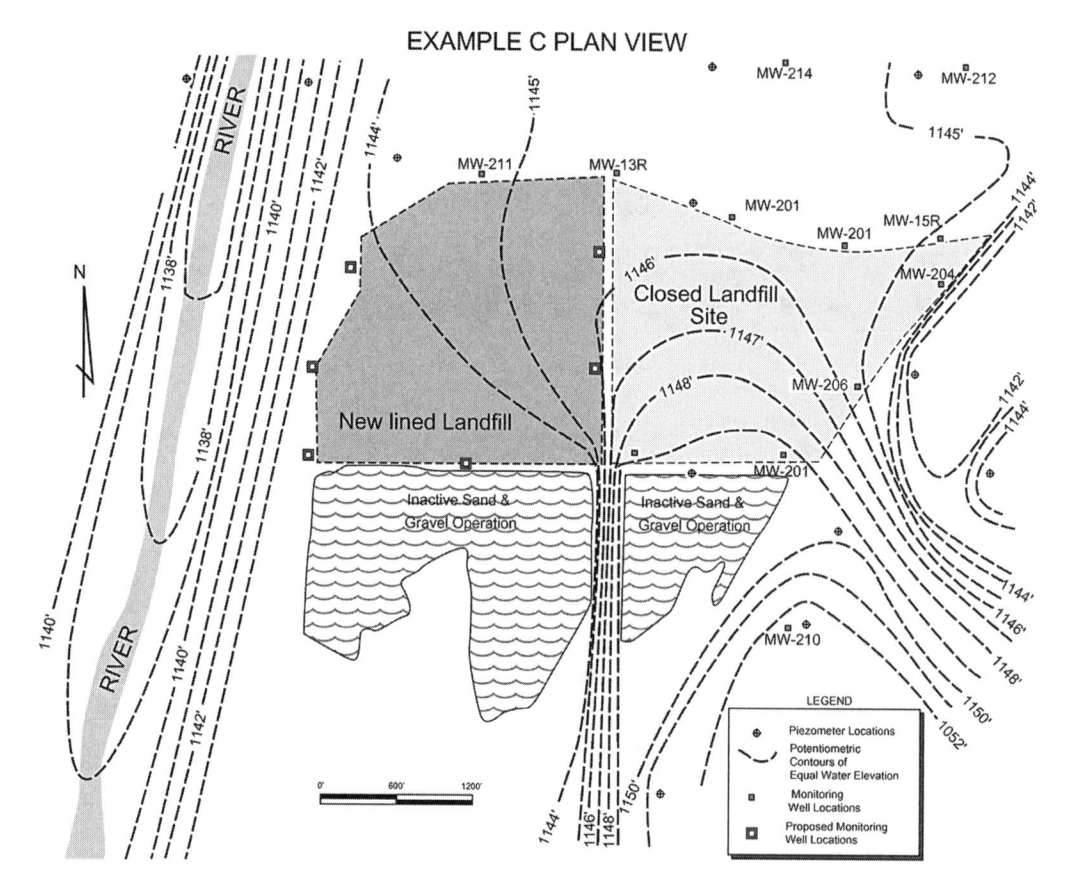

FIGURE 8.45
Ground-water contour map for the example C facility.

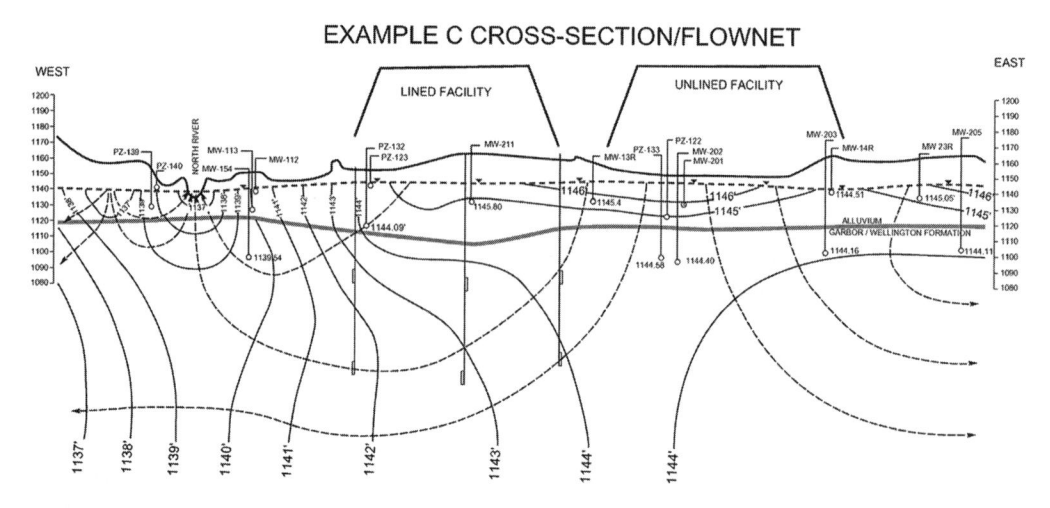

FIGURE 8.46
Flow-net construction for the example C facility.

Locations for background wells must also be carefully chosen relative to adjacent facilities and the overall geology and ground-water flow conditions. The example C monitoring wells are located between the new facility and the river in an arc down to the sand and gravel operation south of the lined cell.

Background wells are located between the closed site and the new facility. As with the downgradient detection monitoring wells, the depths of the well screens for background water-quality monitoring points must be carefully selected. The screens should probably be no deeper than between elevation 1140 and 1130 ft above msl. As with any detection ground-water monitoring system located next to an unlined facility, the wells must be entirely screened below the seasonal low water table. This is to ensure that landfill gas potentially moving through the vadose zone from the unlined site would not enter a monitoring well and cross-contaminate ground-water samples taken from the well. This type of cross-contamination can confound detection monitoring analytical results. Gas movement in the vadose zone should be monitored through a separate gas-monitoring network designed especially for this purpose.

Monitoring System Design Criteria

The successful implementation of an effective environmental monitoring system is dependent upon a number of technical factors. For example, monitoring wells must be properly located, constructed of appropriate materials, and properly installed. Similarly, samples must be carefully obtained, properly analyzed, and the data clearly reported and interpreted.

These factors collectively represent the criteria which must be considered in the design of a comprehensive and effective monitoring system. The design criteria identified and utilized in the development of a ground-water monitoring plan are outlined as follows:

- Monitoring Network Design Criteria — Recommended locations and specifications for monitoring wells and surface water sampling stations based upon an analysis of potential migration pathways.
- Sampling Protocol — Sampling instrumentation and approved methodologies, sampling frequency, sample storage and preservation requirements, and chain-of-custody procedures.
- Analytical Protocol — Parameter selection, approved field and laboratory methods, and quality assurance or quality control procedures.
- Data Management Criteria — Recommended procedures for data statistical analyses and data reporting.

References

American Petroleum Institute, *A Guide to the Assessment and Remediation of Underground Petroleum Releases*, 2nd ed., API Publication #1628, American Petroleum Institute, Washington, DC, 1989.

American Society for Testing & Materials (ASTM), 2002, Standard Practice for Design and Installation of Ground Water Monitoring Wells in Aquifers, D5092-02, Annual Book of ASTM Standards, Vol. 04.08, Philadelphia, PA, 14 pp.

Bear, J., Dynamics of Fluids in Porous Media, American Elsevier Publishing Co., New York, NY, 1972.

Cherry, J.A. and S. Feenstra, Identification of DNAPL sites: An eleven point approach, draft document in Dense Immiscible Phase Liquid Contaminants in Porous and Fractured Media, Short Course notes, Waterloo Center for Ground Water Research, Kitchener, Ontario, 1991.

Collins, R.E., *Flow of Fluids in Porous Materials*, Van Nostrand/Reinhold Publishing Co., New York, NY, 1961.

Corey, A.T., Mechanics of Heterogeneous Fluids in Porous Media, Water Resources Publications, Fort Collins, CO, 1977.

Freeze, R.A. and J.A. Cherry, Groundwater, Prentice-Hall, Englewood Cliffs, NJ, 1979, p. 604.

Freeze, R.A. and P.A. Witherspoon, Theoretical analysis of regional ground water flow, *Water Resources Research*, 3, 623–634, 1967.

Heath, R.C., 1984, Ground water regions of the United States: U.S. Geological Survey Water Supply Paper 2242, 78 pp.

Quinlan, J.F., Special problems of ground-water monitoring in Karst terrains, in *Ground Water and Vadose Zone Monitoring*, D.M. Nielsen and A.I. Johnson, Eds., ASTM STP 1053, ASTM International, West Conshohocken, PA, 1990, pp. 275–307.

Schwille, F., Dense Chlorinated Solvents in Porous and Fractured Media, J.F. Pankow Transl., Lewis Publishers, Chelsea, MI, 1988.

U.S. EPA, 1982, Code of Federal Register 40 CFR part 264, 26 July, 1982.

U.S. EPA, RCRA Ground-Water Monitoring Technical Enforcement Guidance Document, OSWER-9950.1, U.S. Environmental Protection Agency, Office of Waste Programs Enforcement and Office of Solid Waste and Emergency Response, 1986.

9

Designing Monitoring Programs to Effectively Evaluate the Performance of Natural Attenuation

Todd H. Wiedemeier, Michael J. Barden, Patrick E. Haas, and W. Zachary Dickson

CONTENTS

Introduction	574
Purpose of Monitoring for Natural Attenuation	576
Types of Monitoring for Natural Attenuation	577
Site Characterization Monitoring	577
Validation Monitoring	578
Long-Term Monitoring	578
Performance Monitoring	578
Compliance or Contingency Monitoring	578
Essential Design Elements of a Monitoring Plan	579
Location and Placement of Monitoring Points	580
Plumes that do not Discharge to Surface-Water Bodies	582
Plumes that Discharge to Surface Water Bodies	586
Analytical Protocols — What to Analyze for and When	586
Typical Ground-Water Analytes for Evaluating the Long-Term Performance of Natural Attenuation	588
Sampling in the NAPL Source Area	588
Contaminants and Transformation Products	594
Naturally Occurring Electron Acceptors and Metabolic Byproducts	594
General Water-Quality Parameters	594
Supplemental Monitoring Parameters	594
Ground-Water Sampling Techniques	595
Sampling with Peristaltic Pumps	595
Sampling with Submersible Pumps	595
Sampling with Bailers	596
Diffusion Samplers	597
General Ground-Water Sampling Considerations	597
Sampling Frequency	598
Evaluation and Interpretation of Monitoring Data	599
Evaluating Contaminant Data	600
Graphical Methods for Evaluating Plume Behavior	600
Statistical Methods for Evaluating Plume Behavior	604
Nature of Ground-Water Concentration Data and Appropriate Statistical Methods	607
Tests for Trend	607
Tests for Differences between Groups of Data	612

Using Statistical Results	615
Evaluation of Ground-Water Geochemical and Supplemental Data	616
Evaluation of Daughter Product Data	617
Evaluation of Electron Acceptor Data	617
Dissolved Oxygen	618
Nitrate	619
Sulfate and Sulfide	619
Evaluation of Metabolic Byproduct Data	620
Fe(II)	620
Methane	620
Ethene and Ethane	620
Evaluation of General Ground-Water Monitoring Parameters	620
Oxidation–Reduction Potential (ORP)	620
pH	621
Temperature	621
Conductivity	621
Evaluation of Supplemental Data	621
Supplemental Daughter Product Data	622
Supplemental Geochemical Data	622
Microcosm Studies	624
Volatile Fatty Acids (VFAs)	625
Phospholipid Fatty Acids (PLFAs)	625
Denaturing Gradient Gel Electrophoresis (DGGE)	626
Practical Considerations for Microbial Characterization Techniques	627
Stable Isotopes	627
Contingency Plans	627
Monitoring Duration and Exit Strategies	631
References	634

Introduction

Natural attenuation processes affect the migration and fate of organic compounds in all hydrologic systems. Over the past several years, regulatory agencies and environmental professionals have come to recognize the importance of these natural processes in affecting contaminant attenuation. When they are shown to be protective of human health and the environment, and when a well-designed monitoring program is in place to document the efficacy of these processes, they can be a valuable component of site remediation strategies. In April 1999, the Office of Solid Waste and Emergency Response (OSWER) of the U.S. Environmental Protection Agency (U.S. EPA) published a directive on the use of natural attenuation, entitled *Use of Monitored Natural Attenuation at Superfund, RCRA Corrective Action, and Underground Storage Tank Sites* (U.S. EPA, 1999). As implied by the title of this policy document, monitoring will be required to evaluate the long-term effectiveness of natural attenuation and to assure protection of human health and the environment. According to U.S. EPA (1999), the monitoring program designed for each site should specify the location, frequency, and types of samples and measurements necessary to evaluate if the remedy is performing as expected and if it is capable of attaining remediation objectives.

Designing an effective monitoring program involves locating ground-water monitoring wells and developing a site-specific ground-water sampling and analysis strategy and contingency plan. The monitoring program should be designed to monitor contaminant plume behavior over time and to verify that natural attenuation is occurring at rates sufficient to protect potential downgradient receptors. All available site-specific data and information developed during site characterization, conceptual model development, and ground-water modeling (as appropriate) should be used when preparing a monitoring program. The design of the monitoring program should include consideration of existing receptor exposure pathways, as well as exposure pathways arising from potential future use of the ground water and land. The results of a natural attenuation evaluation as described by U.S. EPA (1998) and Wiedemeier et al. (1995, 1999) are critical to the design of a monitoring program. For those sites where the ground-water flow field cannot be determined with certainty (e.g., fractured bedrock), the evaluation of natural attenuation, and the design of a monitoring program, can be problematic.

The monitoring strategy for a given site will depend upon several primary and secondary factors and will likely be modified over time as new information is obtained. Primary technical factors to consider include (at a minimum) distance to potential receptor exposure points, ground-water seepage velocity and direction, types of contaminants, aquifer heterogeneity, the three-dimensional distribution of constituents of concern; areas of unique geochemical conditions; surface-water impacts, and the effects of engineered remediation systems. In addition, primary factors can include the level of understanding of historical plume behavior and site complexity. In other words, if one has 10 yr of defensible data demonstrating a stable or shrinking plume and site conditions that are unlikely to change, the monitoring strategy can be optimized to focus on monitoring critical areas. Primary regulatory factors may include points of compliance, alternate concentration limits, or requirements identified under the Resource Conservation and Recovery Act (RCRA) or site-specific records of decision, remedial action plans, or decision documents. Secondary factors to consider include (at a minimum) access issues, property lines, and contaminant contributions from off-site sources. Each of these factors will influence the final design of the monitoring program. Perhaps the most critical factors to consider when developing a monitoring program are the distance to potential receptor exposure points and the seepage velocity of ground water. The combination of these two factors will influence well spacing and sampling frequency. Typically, the greater the ground-water seepage velocity and the shorter the distance to potential receptors, the greater the sampling frequency. The use of seepage velocity usually (if not always) overestimates the rate of solute movement because some sorption, dispersion, and biodegradation of dissolved contaminants likely are occurring which will retard the downgradient movement of the contaminants. The analytical protocol developed for a site should be influenced mainly by the type of contamination and the geochemical conditions that affect the fate of the chemicals of concern. Sites with chlorinated solvent contamination likely will require a more diverse suite of analytical parameters (e.g., chloride, ethene, ethane, known solvent breakdown products, etc.) than sites contaminated with fuel hydrocarbons. This is because of the differences in the patterns of biodegradation between different contaminants. For example, it is now well known that fuel hydrocarbons almost invariably biodegrade in the shallow subsurface. This is in contrast to chlorinated solvents, which exhibit varying degrees of biodegradation potential under unique geochemical conditions. The degree of aquifer heterogeneity also will influence the placement of the monitoring wells, with more heterogeneous sites typically requiring a more elaborate sampling network. If surface water is impacted, several factors must be considered, including the amount of contaminant flux into the body of water, the regulatory status (e.g., impaired), and the physical characteristics of the water body. Placement of sample collection points, the analytical protocols to be used for monitoring, and the determination of sampling frequency, are described later in this chapter.

One of the most important purposes of long-term monitoring is to confirm that the contaminant plume is behaving as predicted with no unacceptable impacts to human health or the environment. Graphical and statistical methods can be used to evaluate plume stability and behavior. When evaluating the stability of a contaminant plume, it is important that the historical data demonstrate a clear and meaningful trend at appropriate monitoring points. Graphical and statistical techniques that can be used to evaluate plume stability are described later in this chapter.

Changing site conditions can result in variable plume behavior over time. To circumvent potential problems, a contingency plan should be an integral part of the monitoring program. Contingency plans are used to help ensure protection of human health and the environment should a contaminant plume begin to migrate farther or faster than predicted, and typically involve some kind of engineered remediation. It is prudent to update the contingency plan on a periodic basis as the plume attenuates or as new remediation technologies are developed. Although some engineered remediation systems may be effective in achieving plume containment, it should be kept in mind when developing the contingency plan that some remediation systems may have an adverse impact on contaminant degradation. The development of contingency plans is discussed subsequently.

As with any remedial option for sites contaminated with organic compounds, remediation goals and an exit strategy should be established early in the regulatory negotiation process. This will help establish clear objectives for long-term monitoring, and should help define the length of time that monitoring will be required. Exit strategies are discussed later in this chapter.

Decisions regarding remedy effectiveness and the adequacy of the monitoring program will generally result in either continuation of the program, program modification, implementation of a contingency or alternative remedy, or termination of the performance monitoring program (U.S. EPA, 2004). Such decisions are appropriately based on site-specific, quantifiable performance criteria defined in the monitoring plan. Continuation of the program without modification should not be considered a default, but would be best supported by contaminant concentrations behaving according to remedial expectations while ground-water flow and geochemical parameters remain within acceptable ranges. Monitoring programs should be subjected to periodic review and optimization to ensure that goals are being met. Modification of the program, including increases or decreases in monitoring parameters, frequency, or locations, may be warranted to reflect changing conditions or improved understanding of natural attenuation processes at the site. Situations that may trigger implementation of a contingency or alternative remedy are discussed later in this chapter.

The material presented in this chapter is intended for use in conjunction with the Air Force Center for Environmental Excellence (AFCEE), U.S. EPA, and U.S. Department of Energy (DOE) technical protocols for evaluating and monitoring natural attenuation (Wiedemeier et al., 1995, 1999; U.S. EPA, 1998, 1999, 2004). The approach specified herein can lower monitoring costs by reducing the number of monitoring wells, the frequency of sampling, and the number of analytes required to demonstrate the continuing efficacy of natural attenuation.

Purpose of Monitoring for Natural Attenuation

Although the purpose of natural attenuation monitoring and, thus, the monitoring program will be site-specific, all monitoring programs should be designed to accomplish the following minimum goals (U.S. EPA, 1999):

- Demonstrate that natural attenuation is occurring according to expectations
- Detect changes in environmental conditions (e.g., hydrogeologic, geochemical, microbiological, or other changes) that may reduce (or enhance) the efficacy of the natural attenuation processes
- Identify any potentially toxic and mobile transformation products
- Verify that the dissolved contaminant plume is not expanding
- Verify that there has been no unacceptable impact to downgradient receptors
- Detect new releases of contaminants to the environment that could create an unacceptable risk to receptors or impact the effectiveness of the natural attenuation remedy
- Demonstrate the efficacy of institutional controls that were put in place to protect potential receptors
- Verify progress toward attainment of cleanup objectives

In addition to meeting all of these requirements, a site-specific contingency plan must be specified as a backup remedy in the event that natural attenuation fails to perform as anticipated.

Types of Monitoring for Natural Attenuation

In order to meet the objectives required by the U.S. EPA described earlier, three types of environmental monitoring are described, including:

- Site characterization monitoring (i.e., baseline monitoring), to describe the disposition of contamination and forecast its future behavior
- Validation monitoring, to determine if predictions based on site characterization are accurate
- Long-term monitoring, to ensure that the behavior of the contaminant plume does not change over time

Each type of monitoring has specific objectives that are defined and discussed briefly in the following subsections.

Site Characterization Monitoring

Site characterization monitoring includes monitoring activities conducted during the initial site characterization of a remedial investigation or a natural attenuation evaluation (feasibility study) that provide data on the contaminant distribution and the hydrogeologic and geochemical conditions at a site. This information is used to identify and quantify the natural attenuation processes involved, to evaluate the geochemical conditions that may govern contaminant transformation or degradation processes, and to determine if monitored natural attenuation is viable as a remediation approach at a site. The collection and interpretation of characterization monitoring data for petroleum hydrocarbons is described by Wiedemeier et al. (1995, 1999). The collection and interpretation of characterization monitoring data for chlorinated compounds is described by U.S. EPA (1998) and Wiedemeier et al. (1999, 2005).

Validation Monitoring

Validation monitoring is used to ensure that the analytical results obtained from the baseline (i.e., site characterization) sampling events are accurate. Validation monitoring consists of collecting the complete analytical suites specified by Wiedemeier et al. (1995, 1999) and U.S. EPA (1998) for one or two sampling rounds after completion of site characterization. In addition, Wiedemeier et al. (2005) lists site-specific supplemental analytes such as acetylene, isotopes, microbial analyses, and mineralogical analyses for iron minerals which may be useful for validation monitoring at more complex sites.

Long-Term Monitoring

Long-term monitoring involves collecting a subset of the parameters specified by Wiedemeier et al. (1995, 1999), U.S. EPA (1998), and Wiedemeier et al. (2005). Ultimately the subset of parameters selected for analysis on an ongoing basis will be site-specific. This chapter describes how to effectively and efficiently specify the location, frequency, and types of samples and analyses required to meet the objectives of long-term monitoring. In addition, guidance is provided on developing contingency remedies that mitigate unacceptable conditions without adversely impacting the natural biodegradation reactions occurring at a site, should engineered remediation or additional land-use control be required. Two types of monitoring (and monitoring wells) are utilized for long-term monitoring: performance monitoring and compliance, or contingency, monitoring.

Performance Monitoring

Performance monitoring is intended to ensure that the behavior of the contaminant plume does not change over time and that the remedial action is progressing appropriately. It involves collecting a subset of the parameters used in site characterization monitoring that focus on the most significant parameters appropriate to the site. This information is used to evaluate and explain solute plume behavior and any changes in conditions that may affect the efficacy of the natural attenuation remedy.

Performance monitoring wells (PMWs) should be located upgradient from, within, transverse to, and just downgradient from the solute plume. These wells are used to verify that the concentrations of individual constituents of concern, plume boundaries, and overall progression toward remedial goals are acceptable over time and space.

Compliance or Contingency Monitoring

Compliance, or contingency, monitoring is intended to ensure compliance with regulatory requirements associated with a monitored natural attenuation remedy. These include ensuring that the plume does not expand past preestablished boundaries and identifying situations that will "trigger" a change in the monitoring plan or implementation of a contingency plan. It involves collecting data from appropriate locations that focuses on detecting and recognizing "unacceptable" solute plume behavior that indicates potential or real failure of a monitored natural attenuation remedy and allows sufficient time to reevaluate the remedy and implement contingency measures. Statistically significant detection of unacceptable concentrations of contaminants at the contingency monitoring wells may trigger implementation of the contingency remedy.

Essential Design Elements of a Monitoring Plan

The ability to design an appropriate and adequate monitoring plan for natural attenuation is entirely dependent upon the quality of site characterization. The information developed during site characterization defines the spatial distribution of constituents of interest and provides an understanding of the hydrogeological setting and underlying natural attenuation processes. If these aspects are not understood, a monitoring program cannot be designed effectively — if you do not understand the problem, you cannot monitor it.

Adequate site characterization and a sound conceptual model of the site are essential to the design of a long-term monitoring plan and it is important to remember that solute plumes are dynamic, three-dimensional entities. Effective monitoring of natural attenuation processes involves a three-dimensional approach to monitoring network design and clearly defined performance criteria based on site-specific remedial action objectives. A well-designed long-term monitoring program should provide all data necessary to document and evaluate the effectiveness and protectiveness of the current remedy. Periodic evaluations are often required under various regulatory programs (e.g., CERCLA 5-yr reviews) or under site-specific agreements.

The degree of aquifer heterogeneity also will influence the placement of the monitoring wells, with more heterogeneous sites possibly requiring a more elaborate sampling network. If surface water is impacted, several factors must be considered, including the amount of contaminant flux into the body of water. For those sites where the ground-water flow field cannot be determined with certainty (e.g., fractured bedrock), the evaluation of natural attenuation, or any remedial action, and the design of a monitoring program can be problematic.

Designing an effective monitoring program requires the proper placement of ground-water monitoring wells and developing a site-specific ground-water sampling and analysis strategy. The monitoring program should be designed to monitor solute plume behavior over time and to verify that natural attenuation is occurring at rates sufficient to protect potential downgradient receptors. All available site-specific data and information developed during site characterization, conceptual model development, ground-water modeling (as appropriate), and regulatory negotiations should be used when preparing a monitoring program. The monitoring program designed for each site must specify the purpose, location, sampling frequency, and types of samples and measurements necessary to evaluate if the remedy is performing as expected (U.S. EPA, 1999). The data collected during long-term monitoring are used to evaluate changes in three-dimensional solute plume boundaries, contaminant mass and concentrations in the solute plume, and hydrological and geochemical changes that may indicate changes in remedy performance. The design of the monitoring program also must include consideration of existing receptor exposure pathways, as well as those that may arise from potential future land use and ground-water use (U.S. EPA, 1999).

The monitoring strategy for a given site will depend upon a variety of factors and will likely be modified over time as new information is acquired. If adequate data to define seasonal variation in contaminant concentrations, geochemical parameters and water levels (ground-water flow patterns) are not available from the site characterization or natural attenuation evaluation (feasibility study) for the site, monitoring of these parameters should be continued to determine the short-term variation and to verify that data collected from any new monitoring points are consistent with the site conceptual model (validation monitoring). Quarterly ground-water level and contaminant monitoring are often used to determine if a seasonal variation exists. However, one should consider the use of newly available tools like dedicated water-level loggers to first determine if naturally occurring conditions like water-level fluctuations occur throughout the year or in response to specific climatic changes. This allows the environmental professional to not only identify significant changes in hydrologic conditions, but also to schedule contaminant or geochemical monitoring during these significant events (e.g., snow melt, river ice breakup, seasonal ground-water flow reversals, etc.). Arbitrarily scheduled quarterly monitoring may lead to improperly timed sampling or looking for effects that do not exist. Once this information is available, the sampling parameters and frequency can be adjusted to optimize data collection.

A variety of technical, institutional, and regulatory factors affect the design of a long-term monitoring program for natural attenuation. The technical factors include distance to potential receptor exposure points; ground-water seepage velocity and direction; types of contaminants; aquifer heterogeneity; the three-dimensional distribution of constituents of concern; areas of unique geochemical conditions; surface-water impacts; and the effects of engineered remediation systems. In addition to the technical issues involved, institutional and regulatory factors must also be considered. These include issues of access to the necessary locations, property boundaries, regulatory framework (e.g., RCRA) or site-specific requirements, and contaminant contributions from offsite sources.

In addition, primary factors can include the level of understanding of historical plume behavior and site complexity. In other words, if one has 10 yr of defensible data demonstrating a stable or receding plume and site conditions that are unlikely to change, the monitoring strategy can be optimized to focus on monitoring critical areas.

The use of existing monitoring wells from the site characterization as part of the long-term monitoring program must be considered in light of their location, current condition, and construction. The location of existing monitoring points should be carefully evaluated to determine whether the data obtained will be useful as part of the long-term monitoring program. Monitoring points that are not located along flow paths can provide information on spatial relationships in the solute plume, but the resulting data may be difficult to interpret without the use of detailed spatial analysis. Also, the length of the screened interval of existing monitoring wells may not provide the necessary resolution to provide unequivocal data.

Location and Placement of Monitoring Points

Effective monitoring of natural attenuation processes involves a three-dimensional approach to monitoring network design (as required) and clearly defined performance criteria based on site-specific remedial action objectives. Ideally, long-term monitoring points should be located along ground-water flow paths so that the data generated from upgradient monitoring points can be related to the data obtained from downgradient monitoring points. The post-characterization monitoring strategy for a given site will depend upon several factors. Primary factors to be considered when locating monitoring points include (at a minimum) distance to potential receptor exposure points, ground-water seepage velocity and direction, types of contaminants, aquifer heterogeneity, the three-dimensional distribution of constituents of concern; areas of unique geochemical conditions; surface-water impacts, and the effects of engineered remediation systems. In addition, primary factors can include the level of understanding of historical plume behavior and site complexity. In other words, if available information on plume behavior and land use support the position that a significant or unacceptable change in trends or conditions is not plausible, then monitoring frequency can be reduced. Monitoring programs should not be used solely to confirm the obvious. They are more appropriately designed to provide critical updates and to provide for contingency action in the event that it is required. Secondary factors to consider include (at a minimum) access issues, property lines, and contributing off-site contaminant sources. Each of these factors will influence the final design of the monitoring program. Perhaps the most critical factors to consider when developing a monitoring program are the distance to potential receptors and the seepage velocity of ground water. These two factors will strongly influence monitoring well spacing and sampling frequency. Typically, the faster the ground-water seepage velocity and the shorter the distance to potential receptor exposure points, the greater the sampling frequency. The use of a range of site-specific seepage velocity estimates is conservative because some sorption and biodegradation are likely retarding contaminant migration relative to ground-water flow.

The placement of monitoring wells and the frequency of sampling must yield useful data and allow detection of significant changes in plume configuration and definition of trends in contaminant concentrations over time. In many cases it may be possible to utilize some of the existing monitoring wells at a site, thereby reducing the cost of implementing the long-term monitoring plan. However, it is important that these wells are located in appropriate locations. Not all wells installed during site characterization may be appropriate or necessary for long-term monitoring. Because monitoring wells installed for site characterization purposes will not necessarily provide meaningful long-term monitoring data, it is important to be selective in determining which of the existing wells to sample. The locations and screened intervals of long-term monitoring wells should be based on site stratigraphy and plume behavior as revealed during site characterization. This requires a detailed understanding of the three-dimensional relationship between contaminants and stratigraphy to ensure that monitoring wells are screened in the same hydrogeologic unit as the contaminant plume, and that they are in the path of contaminated ground-water flow. The geologic complexity of the site and ground-water seepage velocity ultimately will dictate the density of the sampling network.

Two types of wells, PMWs and contingency wells, are used for validation monitoring and long-term monitoring after the initial site characterization and baseline evaluation of natural attenuation. The PMWs, located upgradient from, within, and just downgradient from the plume (Figure 9.1), are used to verify the predictions made during the evaluation of natural attenuation (Wiedemeier et al., 1995; U.S. EPA, 1998). Contingency monitoring wells are placed beyond the maximum predicted lateral and downgradient boundaries of the plume, and typically upgradient from known or potential receptor

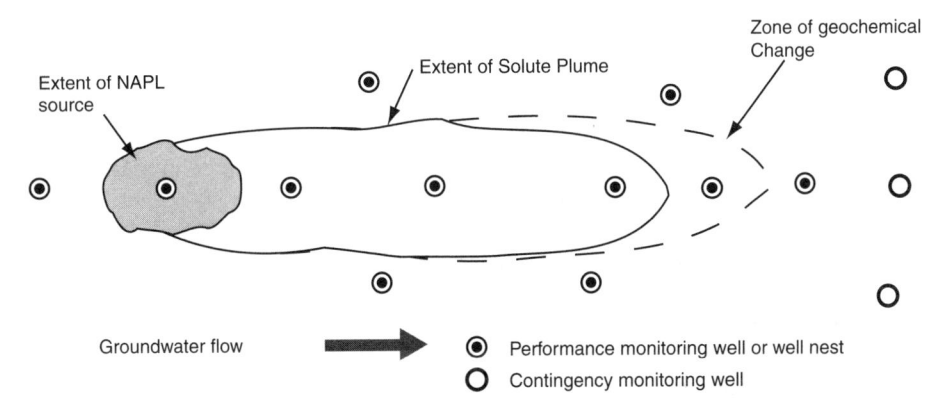

FIGURE 9.1
Conceptual diagram of monitoring point locations for a monitored natural attenuation remedy.

exposure points, to ensure that the plume does not threaten human health or the environment (Figure 9.1). If preestablished trigger levels are exceeded at the contingency monitoring wells, they should be verified prior to the implementation of the contingency plan.

Where possible, contaminant, geochemical and hydrogeological data should be used to locate and design monitoring wells, especially those wells downgradient from the plume. For example, geochemical parameters such as dissolved oxygen, nitrate, Fe(II), sulfate, and methane can be used in conjunction with contaminant data to ensure the proper placement of downgradient contingency monitoring wells in locations with "treated" ground water. "Treated" ground water exhibits a predicable change in geochemistry even though it may lack detectable concentration of site contaminants. This approach ensures that the downgradient monitoring network is in the flow path of the contaminant plume. The frequency of sampling will depend on the location of potential receptor exposure points and the seepage velocity of ground water. To evaluate the behavior of the dissolved contaminant plume over time and to estimate cleanup time frames, statistical methods should be employed.

Plumes that do not Discharge to Surface-Water Bodies

For plumes that do not discharge to a surface-water body, the monitoring program includes PMWs and contingency monitoring wells. Geochemical data should be used when possible to confirm that downgradient wells are sampling ground water that was once contaminated with organic compounds. Wells downgradient from a contaminant plume, and completed in the same stratigraphic horizon, that do not contain organic compounds but have depleted electron acceptor (e.g., dissolved oxygen, nitrate, sulfate) and elevated metabolic byproduct (e.g., iron(II), methane, chloride, alkalinity) concentrations relative to background levels provide good evidence that the ground water being sampled flowed through the contaminant plume and has been treated. Such wells have been termed "smoking guns" because they provide fairly conclusive evidence that the ground water was contaminated at one time and has since been treated (Wiedemeier et al., 1995, 1999). Because concentrations of electron acceptors and metabolic byproducts typically will return to background concentrations at some distance downgradient from the contaminant plume, it is important to locate at least one PMW close to the downgradient edge of the contaminant plume. This also will allow better resolution of the behavior of the leading edge of the plume to determine if the plume is at steady-state equilibrium, is receding, or is expanding. Figures 9.2 and 9.3 illustrate how geochemical data can be used to place monitoring wells for solute plumes emanating from non-aqueous phase liquid (NAPL) sources. Figure 9.2 illustrates a hypothetical monitoring network for a solute plume from a light non-aqueous phase liquid (LNAPL) source. Figure 9.3 illustrates a hypothetical monitoring network for a solute plume from a dense non-aqueous phase liquid (DNAPL) where the NAPL materials have migrated to a lower confining layer. In contrast, a contaminant source and dissolved phase plume comprised solely of dissolved phase DNAPL materials is more likely to behave like the plume depicted in Figure 9.2b. These figures depict: (1) upgradient (PMW-1A) and crossgradient (PMW-1B and PMW-1C) wells in unimpacted ground water; (2) wells in the NAPL source area (PMW-2); (3) wells downgradient from the NAPL source area in the plume (PMW-3 and PMW-4); (4) a well located downgradient from the plume where contaminants are not detectable, soluble electron acceptors are depleted, and metabolic byproducts are elevated with respect to unimpacted ground water (PMW-5); (5) a well (PMW-6) in treated ground water; and (6) contingency wells. Note that these figures are only examples of monitoring well placement. The actual location and number of monitoring wells must be determined on a site-specific basis.

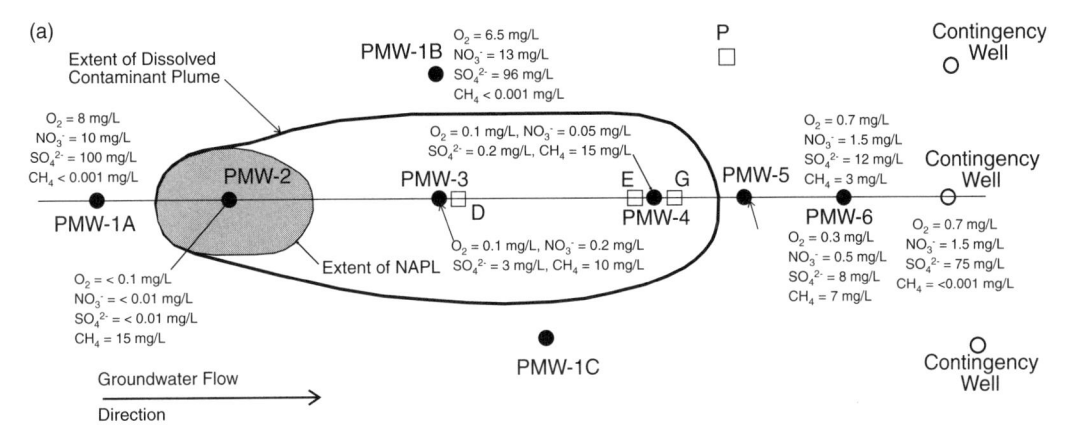

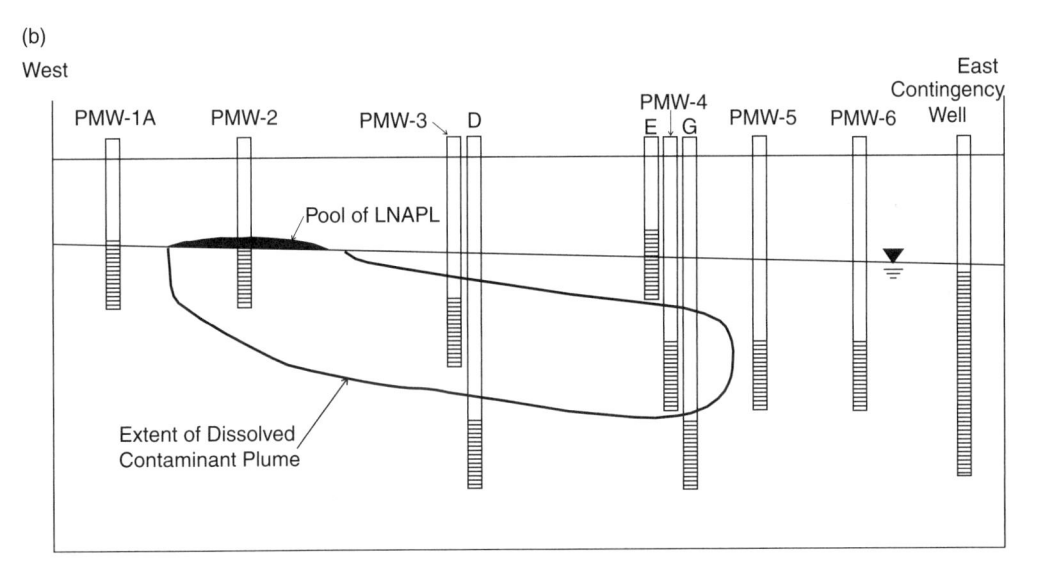

Designing Monitoring Programs to Evaluate the Performance of Natural Attenuation 583

FIGURE 9.2
Locating monitoring wells using contaminant and geochemical data (a) Plan view of LNAPL. (b) Cross-sectional view of LNAPL. (Modified from Wiedemeier et al., 1999. With permission.)

Table 9.1 summarizes sampling locations. The upgradient and crossgradient PMWs are intended to monitor for changes in background water quality that can provide an indication of changing conditions that could affect natural attenuation. The need for, and placement of, crossgradient PMWs is related to whether ground-water flow directions change due to site-specific seasonal or other hydrological conditions. In contrast, if the ground-water flow direction, plume configuration, and behavior are well-established and unlikely to change, then the number and sampling frequency of upgradient and crossgradient monitoring points can be reduced. The PMWs in the NAPL source area are intended to monitor changing apparent NAPL thickness, distribution, or composition

FIGURE 9.3
Locating monitoring wells using contaminant and geochemical data (a) Plan view of DNAPL. (b) Cross-sectional view of DNAPL. (Modified from Wiedemeier et al., 1999. With permission.)

over time and to give an indication of the changing solute concentration in ground water. PMWs downgradient from the NAPL source area are intended to monitor plume behavior and changing contaminant concentrations over time. Ideally, these wells will be aligned parallel to the direction of ground-water flow and the center line of the plume. It should be kept in mind that this requires good definition of the plume and fairly uniform (unchanging) hydraulic gradients. The PMWs located downgradient from the dissolved contaminant plume are intended to provide early detection of contaminant plume migration toward a contingency well. These wells should be located in the flow path of the contaminant plume. The placement and spacing of the PMWs located in the

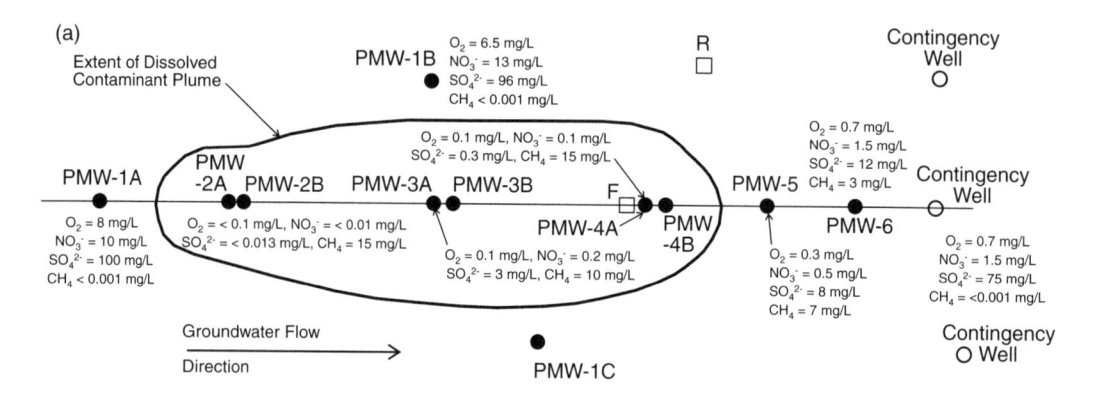

TABLE 9.1
Sampling Locations, Purpose, and Analytical Parameters for Validation Monitoring and Long-Term Monitoring of Ground Water

Type of Well	Location	Purpose	Validation Sampling	Long-Term Monitoring
PMW-1 (A, B, C)	Upgradient/crossgradient	Monitor background water quality	Contaminants, daughter products, and full suite of geochemical parameters[a]	Limited suite of geochemical parameters[b,c]
PMW-2	NAPL source area	Monitor changing NAPL composition/source strength and plume behavior over time	Contaminants, daughter products, and full suite of geochemical parameters[a,d]	Contaminants and daughter products in ground water beneath NAPL and limited suite of geochemical parameters[b–d] collection of LNAPL samples on a periodic basis may be warranted
PMW-3 and PMW-4	Downgradient from NAPL source area along plume centerline	Monitor plume behavior over time	Contaminants, daughter products, and full suite of geochemical parameters[a]	Contaminants, daughter products, and limited suite of geochemical parameters[b,c]
PMW-5	Immediately downgradient from plume	Early detection of plume migration	Contaminants, daughter products, and full suite of geochemical parameters[a]	Contaminants, daughter products, and limited suite of geochemical parameters[b,c]
PMW-6	Between contingency wells and the other PMWs	Early detection of plume migration	Contaminants, daughter products, and full suite of geochemical parameters[a]	Contaminants, daughter products, and limited suite of geochemical parameters[b,c]
Contingency wells	Downgradient from most downgradient PMW well (PMW-6 in this case) and upgradient from receptor exposure point	Monitor for plume migration toward a potential receptor and trigger contingency plan	Contaminants, daughter products, and full suite of geochemical parameters[a]	Contaminants, daughter products, and limited suite of geochemical parameters[b,c]
Surface water	At and upgradient and downgradient from discharge point	Determine surface-water impacts	Contaminants and daughter products	Contaminants and daughter products

[a] For fuel hydrocarbon plumes, the full suite of geochemical parameters should include dissolved oxygen, nitrate, Fe (II), sulfate, methane, temperature, pH, conductivity, and ORP. For chlorinated solvent plumes ethane or ethene and sulfide should be added to the full suite of geochemical parameters recommended for fuel hydrocarbon plumes. In addition, certain supplemental data may be useful for evaluating complex plumes, most notably solvent plumes. See Section 4.2 for more information on analytes.
[b] At a minimum, the limited suite of geochemical parameters should include dissolved oxygen, ORP, temperature, and pH.
[c] If plume behavior changes or is suspected of changing, then analyze for contaminants and the full suite of geochemical parameters plus any supplemental data that may be appropriate.
[d] LNAPL thickness measurements should be obtained.

downgradient portion of the plume (PMW-4 in this example) and the well located downgradient from the contaminant plume (PMW-5 in this example) are particularly important. This is because the closer the downgradient well (i.e., PM-5) is to the contaminant plume, the less time required to confirm that the plume is at steady-state equilibrium, or is receding. For example, if wells PMW-4 and PMW-5 in Figure 9.2 are 500 ft apart and ground water is flowing at 50 ft per yr, it will take at least 10 yr of monitoring data to show that the contaminant plume is not migrating at the seepage velocity of the ground water. It will take even longer to show that the contaminant plume is not migrating downgradient at some retarded solute transport velocity. If, on the other hand, wells PMW-4 and PMW-5 in Figure 9.2 are 100 ft apart, then it will take about 2 yr of monitoring data to show that the contaminant plume is not migrating at the seepage velocity of the ground water, and is thus being retarded by some mechanism of natural attenuation.

Contingency wells are intended to monitor unexpected plume migration and to trigger implementation of the contingency plan. All of the contingency wells should be located in the flow path or potential flow path of the contaminant plume. The distance between downgradient PMWs and contingency wells and the density of the monitoring network should be based on the ground-water seepage velocity, solute transport velocity, and the distance to potential receptor exposure points. Contingency wells should be placed a sufficient distance upgradient from potential exposure points in the flow path of the solute plume to ensure that a contingency plan can be implemented before potential receptors are impacted. To be conservative, these distance calculations should be made based on a representative seepage velocity of the ground water rather than on the solute transport velocity.

Plumes that Discharge to Surface Water Bodies

For sites where contaminated ground water discharges to surface water, the monitoring strategy must be highly customized to factor in all the physical, chemical, and biological processes that occur at and beyond the ground-water and surface-water interface. Figure 9.4 is a hypothetical monitoring strategy for a contaminant plume discharging to a body of surface water. This figure depicts (1) an upgradient (PMW-1A) well and cross-gradient wells (PMW-1B and PMW-1C) in unimpacted ground water; (2) a well in the NAPL source area (PMW-2); (3) wells downgradient from the NAPL source area in the zone of anaerobic treatment (PMW-3 and PMW-4); and (4) surface-water collection points. The purpose of the first three sampling locations is the same as that discussed earlier for contaminant plumes that do not discharge to a surface water body. The fourth type of sampling location is intended to provide information on the impact of the contaminant plume on the surface water body. Mass flux calculations can be completed to estimate the amount of contamination entering the surface water body and the resultant contaminant concentrations in the surface water. In many cases, the relationship between mass flux into the surface water and dilution (and volatilization) will be such that the contamination is not detectable or is quickly diluted or volatilized to nondetectable concentrations a short distance from the point of discharge.

Analytical Protocols — What to Analyze for and When

The analytical protocol for a long-term monitoring program defines the specific parameters that will be analyzed, as well as the locations where samples for these parameters will be collected and when the samples will be collected. The specific analytical parameters that should be collected in a long-term monitoring program for natural attenuation

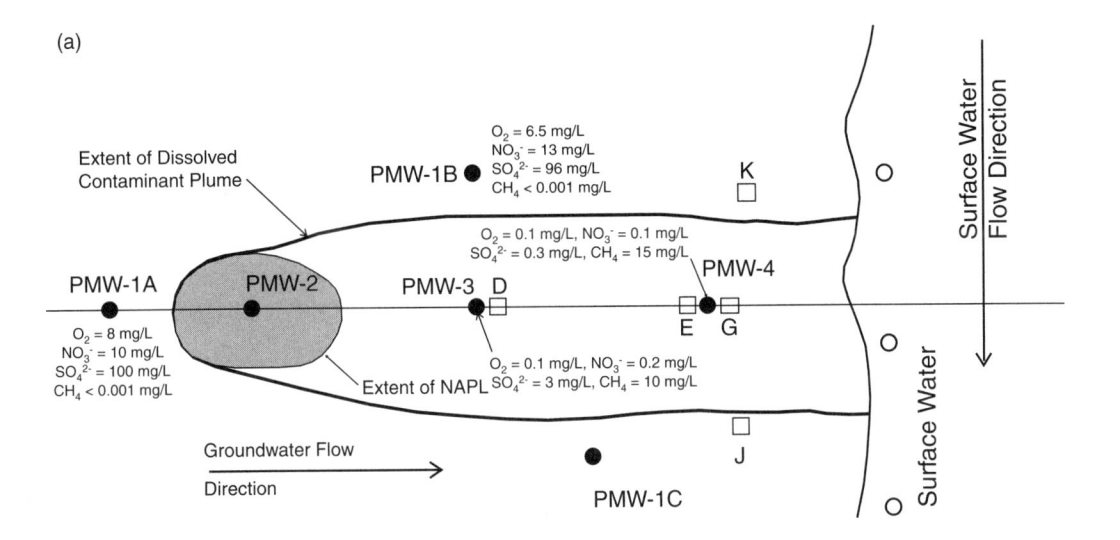

Designing Monitoring Programs to Evaluate the Performance of Natural Attenuation 587

(a)

PMW-1B: $O_2 = 6.5$ mg/L, $NO_3^- = 13$ mg/L, $SO_4^{2-} = 96$ mg/L, $CH_4 < 0.001$ mg/L

Extent of Dissolved Contaminant Plume

PMW-3 area: $O_2 = 0.1$ mg/L, $NO_3^- = 0.1$ mg/L, $SO_4^{2-} = 0.3$ mg/L, $CH_4 = 15$ mg/L

PMW-1A: $O_2 = 8$ mg/L, $NO_3^- = 10$ mg/L, $SO_4^{2-} = 100$ mg/L, $CH_4 < 0.001$ mg/L

Extent of NAPL: $O_2 = 0.1$ mg/L, $NO_3^- = 0.2$ mg/L, $SO_4^{2-} = 3$ mg/L, $CH_4 = 10$ mg/L

Groundwater Flow Direction →

Surface Water Flow Direction ↓

Wells D, E, G, J, and K have geochemistry similar to wells PMW-1 (A, B, and C) (i.e., background) so they probably are not screened across the flowpath of the contaminant plume and therefore are not being used for this hypothetical monitoring program.

LEGEND
● Performance Monitoring Well
○ Surface-Water Sampling Location
□ Site Characterization Well

(b)

West ——— East

PMW-1A, PMW-2, Pool of LNAPL, PMW-3 D, PMW-4 E/G, Surface Water Body

Extent of Dissolved Contaminant Plume

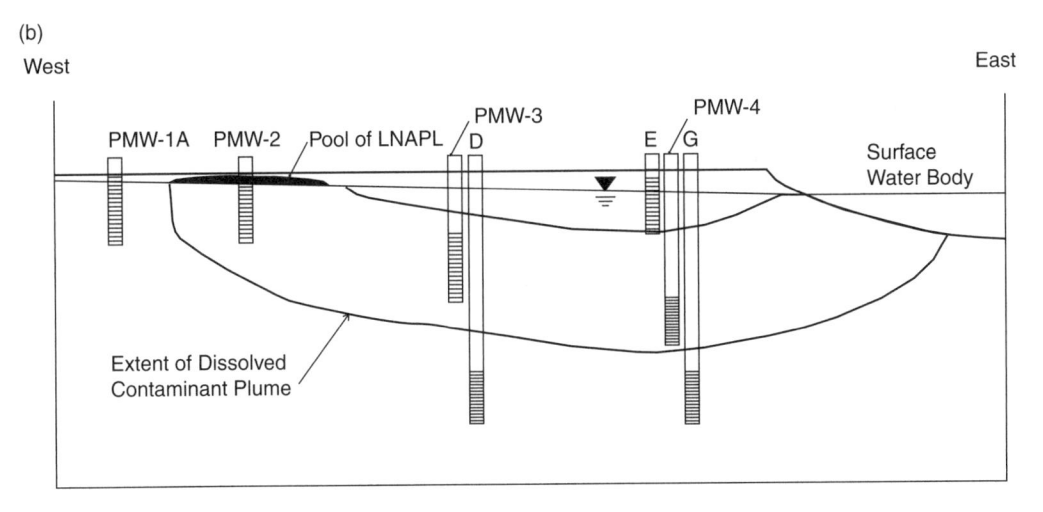

FIGURE 9.4
Locating monitoring wells and surface water sampling locations for a discharging plume. (a) Plan view. (b) Cross-sectional view. (Modified from Wiedemeier et al., 1999. With permission.)

will depend upon both the contaminants of interest and the particular transformation or degradation mechanisms that are involved. For example, sites with chlorinated solvent contamination will likely require a different suite of analytical parameters than petroleum hydrocarbons. This is because of the differences in the patterns of biodegradation between these different types of contaminants. It is now widely accepted that petroleum hydrocarbon compounds are almost invariably mineralized through oxidation by bacteria in the shallow subsurface. This is in contrast to chlorinated solvents, which exhibit varying degrees of biodegradation potential involving oxidation (mineralization) and reduction (reductive dechlorination). In addition to the biological reactions, chlorinated solvents can be degraded by abiotic reductive dechlorination reactions. Both biological and abiotic reactions depend upon the specific compound and the site-specific geochemistry. This section describes both typical and supplemental ground-water analytes that are useful for monitoring natural attenuation.

Typical Ground-Water Analytes for Evaluating the Long-Term Performance of Natural Attenuation

Typical ground-water analytical parameters for monitoring natural attenuation are summarized in Table 9.2. The suggested list of analytes presented in Table 9.2 includes contaminants and geochemical parameters. As summarized in Table 9.1, some of the analytical parameters are for validation monitoring, some are for long-term monitoring, and some are for both. There also are different geochemical analyses suggested for plumes of chlorinated solvents. This is because these plumes are particularly sensitive to changes in ground-water geochemistry, such as depletion of organic carbon or increasing dissolved oxygen concentrations. Such changes may inhibit reductive dechlorination. Any Federal- or State-specific analytical requirements not listed in Tables 9.1 and 9.2 also should be addressed in the sampling and analysis plan to ensure that all data required for regulatory decision-making are collected. In addition, water-level and, if present, light nonaqueous-phase liquid (LNAPL) measurements, should be made during each sampling event to ensure that the ground-water flow direction has not changed.

The analytes listed in Table 9.2 fall into several broad categories, including source-term parameters, contaminants and daughter products, electron acceptors, metabolic byproducts, and general water-quality parameters. These analytes are useful for: (1) estimating the composition and strength of a NAPL source, (2) demonstrating that natural attenuation is occurring, and (3) evaluating the relative importance of the various natural attenuation mechanisms. It should be kept in mind that it may be necessary to modify Table 9.2 on a site-specific basis. In addition to the parameters listed in Table 9.2, the supplemental parameters summarized in Table 9.3 may be useful for monitoring natural attenuation at sites where degradation mechanisms are not apparent.

Sampling in the NAPL Source Area

NAPL in the subsurface, whether present at a residual saturation or in quantities sufficient to cause formation of a mobile or immobile pool of NAPL, acts as a continuing source of ground-water contamination. Thus, as long as NAPL remains in the subsurface at concentrations sufficient to impact ground water, the solute plume will persist. This has several implications for natural attenuation and the length of time that monitoring must be conducted. The degree and rate of weathering of the NAPL, and hence its composition and strength, dictate the amount of aqueous-phase contamination at a site. Significant reductions in soluble and toxic constituents in NAPLs can occur due to natural or enhanced destructive processes (AFCEE, 2003).

Collection and analysis of NAPL samples allows the investigator to determine the composition and physical properties of the NAPL. In some cases, it may be possible to complete NAPL-to-water partitioning calculations to show that the effective solubility of a compound is no longer high enough to impact ground water at concentrations above regulatory guidelines. Additionally, NAPL samples collected over time can help define the compositional changes and allow estimates of source decay (weathering) rates to be made.

TABLE 9.2
Typical Ground-Water Analytes for Evaluating the Long-Term Performance of Natural Attenuation[a]

Analysis	Method/Reference	Comments	Data Use	Sample Volume, Sample Container, Sample Preservation	Field or Fixed-Base Laboratory
Chemicals of concern	SW8260B	Handbook method	Used to determine presence of parent and daughter compounds and rates of attenuation	Collect 3 × 40 ml VOA vials, preserve with HCL and cool to 4°C	Fixed-base laboratory
Dissolved oxygen	E360.1 — Dissolved oxygen membrane electrode.	Avoid exposure to atmospheric oxygen	Concentrations less than 1 mg/l generally indicate an anaerobic pathway	Measure dissolved oxygen onsite using a flow-through cell	Field
Nitrate	IC[b] method E300	Method E300 is a Handbook method	Substrate for microbial respiration if oxygen is depleted. Absence is required for Fe(III) reduction to occur	Collect 1 l poly container and cool to 4°C	Fixed-base laboratory
Iron(II) (Fe^{2+})	Colorimetric hach method	Filter with 0.45 μ inline filter	Indicates an anaerobic degradation process due to depletion of oxygen, nitrate, and manganese. Required for abiotic reductive dechlorination	Collect 100 ml of water in a headspace-free container to eliminate introduction of oxygen and analyze as soon as possible	Field
Sulfate (SO$_4^{2-}$)	IC method E300	Method E300 is a Handbook method	Substrate for anaerobic microbial respiration	Collect 1 l poly container and cool to 4°C	Fixed-base laboratory
Sulfide	E376.1		Required for abiotic reductive dechlorination	Collect 500 ml in plastic or glass container, preserve with NaOH to pH < 9, cool to 4°C, no headspace	Fixed-base laboratory
ORP	Direct-reading probe	Avoid introduction of oxygen during sampling	The ORP of ground water influences and is influenced by the nature of biologically mediated reactions	Measure ORP onsite using a flow-through cell	Field

(Table continued)

TABLE 9.2 Continued

Analysis	Method/Reference	Comments	Data Use	Sample Volume, Sample Container, Sample Preservation	Field or Fixed-Base Laboratory
Methane, ethane, and ethene	RSK-175 (Kampbell and Vandegrift, 1998)	Method published by researchers at the U.S. EPA	The presence of methane suggests biodegradation via methanogenesis. Ethane and ethene are daughter products of complete dechlorination	Collect 6 × 40 ml VOA vials, preserve with HCL, and cool to 4°C	Fixed-base laboratory
pH	E150.1 — Field probe with direct reading meter	Field	Fundamental measurement which is critical for interpretation of carbonate data. Used as a well stabilization criterion	Measure in flow-through cell during well purging	Field
Temperature	170.1 — Field probe with direct reading meter	Field only	Fundamental measurement required in all thermodynamic calculations	Measure in flow-through cell during well purging	Field
Conductivity	E120.1/SW9050, direct reading meter	Protocols/Handbook methods	General water quality parameter that is proportional to the dissolved ions present in solution	Measure in flow-through cell during well purging	Field

[a] Not all analytes will be required for every site or every sampling event.
[b] Ion chromatography.

TABLE 9.3
Supplemental Ground-Water Analytes for Evaluating the Long-Term Performance of Natural Attenuation

Analysis	Method/Reference	Comments	Data Use	Sample Volume, Sample Container, Sample Preservation	Field or Fixed-Base Laboratory
Manganese	Colorimetric Hach method	Filter with 0.45 μ inline filter	May indicate an anaerobic degradation process due to depletion of oxygen, and nitrate. Interferences can occur if hydrogen sulfide and high concentrations of calcium are present	Collect 100 ml of water in a headspace-free container to eliminate introduction of oxygen and analyze as soon as possible	Field
Bicarbonate and Carbonate Alkalinity	Hach digital titrate	Field filter with 0.45 μ inline filter. Carbonate alkalinity only significant at pH >8.5	General water quality parameter used (1) to measure the buffering capacity of ground water and (2) as a marker to verify that all site samples are obtained from the same ground-water system	Collect 100 ml of water in glass container. Analyze as soon as possible	Field
DOC	E415.1	Field filter with 0.45 μ inline filter. Minimize aeration and fill sample container completely	Used to classify plume and to evaluate the potential for biologic and biologically predicated abiotic degradation	Collect 250 ml glass amber container, preserve with H_2SO_4 and cool to 4°C	Fixed-base laboratory
DIC	E415.1	Filter in the field with 0.45 μ inline filter. Minimize aeration of sample and fill sample container completely to avoid loss of CO_2	An increase of DIC above background concentrations provides a footprint in ground water that has been remediated by biological processes. Carbon dioxide is the most universal end product of chlorinated hydrocarbon biodegradation. DIC is the sum of dissolved carbon dioxide, carbonic acid, bicarbonate and carbonate	Collect 250 ml glass amber container, preserve with H_2SO_4 and cool to 4°C. (same sample bottle as DOC)	Fixed-base laboratory

(Table continued)

TABLE 9.3 Continued

Analysis	Method/Reference	Comments	Data Use	Sample Volume, Sample Container, Sample Preservation	Field or Fixed-Base Laboratory
Anions — Cl, Fl, and SO_4, NO_3, HCO_3, CO_3, Br	Method E300 (Cl, Fl, SO_4, NO_3, and Br). Method E310.1 (HCO_3 and CO_3)	Filter in the field with 0.45 μ inline filter	Can be used graphically (e.g., Piper and Stiff diagrams) with cations to identify different hydrogeologic units and identify areas impacted by contamination	Collect 1 l poly container and cool to 4°C	Fixed-base laboratory
Cations — Ca, Mg, K, and Na, Mn, Fe	SW6010	Filter in the field with 0.45 μ inline filter	Can be used graphically with anions to identify different hydrogeologic units and identify areas impacted by contamination	Collect 500 ml poly container, preserve with HNO_3 and cool to 4°C	Fixed-base laboratory
Chloride[a]	IC method E300[a]	Method SW9050 may also be used	Final product of chlorinated solvent reduction. Can be used as a tracer	Collect 1 l poly container and cool to 4°C	Fixed-base laboratory
Hydrogen	Equilibration with gas in the field analyzed with a reducing gas detector in the lab	Supplemental specialized analysis to be completed on select wells	Determine current terminal electron accepting process and if sufficient hydrogen is available for reductive dechlorination	Sampled at well head. Requires the production of 100 ml per min of water for 30 min	Fixed-base laboratory

Acetylene			Product of abiotic reductive dechlorination by iron sulfides	Fixed-base laboratory
VFAs	Light hydrocarbon analysis	GC-FID method[b]	Biomarkers of anaerobic metabolism. Anaerobic bacteria produce these compounds by fermentation	Collect 2 × 44 ml VOA vials, no headspace and cool to 4°C	Fixed-base laboratory
PLFAs	IC[c]	Can be a useful indicator of microbial metabolism of added substrate	Provides microbial biomass, community structure and physiological status data	Collect 1–2 l of ground water in a sterile widemouth poly bottle and cool to 4°C	Fixed-base laboratory
DGGE	White et al. (1997)	PLFA data can be readily correlated with contaminant and geochemical trends	Identifies most dominant microorganisms in the ground water	Collect 500–1000 ml of ground water in a sterile widemouth poly bottle and cool to 4°C	Fixed-base laboratory
Isotopes	Muyzer et al. (1993)	Mainly for use in forensic or failure analyses	Helps elucidate biotic versus abiotic dechlorination pathways	Collect 44 ml VOA vial no headspace and cool to 4°C	Fixed-base laboratory
	Sherwood-Lollar et al. (1999)	May need more than one quarter of data and interpretation may become complicated if there is more than one source			

[a]Included in major anion analysis.
[b]Gas chromatography — flame ionization detector.
[c]Ion chromatography.

Contaminants and Transformation Products

Clearly, the chemicals of concern identified in the site characterization must be part of the analytical protocol. The appropriate analytical methods will depend upon the specific contaminants involved. Analytical methods are often specified by the governing regulatory agency. In addition, the analytical methods used must identify any potentially toxic and mobile transformation products (U.S. EPA, 1999).

Naturally Occurring Electron Acceptors and Metabolic Byproducts

The purpose of sampling geochemical parameters as part of a long-term monitoring program for natural attenuation is to provide salient information regarding changes in conditions that may affect the behavior of the solute plume and the efficacy of natural attenuation. The measurement of geochemical parameters associated with naturally occurring oxidation–reduction processes is useful for evaluating the occurrence and relative importance of the various terminal electron-accepting processes. The monitoring of geochemical parameters during the long-term monitoring program should focus on the specific parameters that are of significant importance at the site as identified during site characterization. For example, if nitrate is present at only very low concentrations in the plume and in background locations, then ground water should not be analyzed for nitrate on a routine basis.

Naturally occurring electron acceptors that are typically monitored for natural attenuation include dissolved oxygen, nitrate, and sulfate. Table 9.2 summarizes analytical methods and data uses for these compounds. The interpretation of electron acceptor data is also discussed later.

Metabolic byproduct data that can be collected during natural attenuation monitoring include Fe(II), sulfide, and methane. Table 9.2 summarizes analytical methods and data uses for these compounds. The interpretation of metabolic byproduct data is discussed later.

General Water-Quality Parameters

General water-quality parameters, including pH, temperature, conductivity, oxidation-reduction potential (ORP), and conductivity, should be a part of every sampling event. Table 9.2 summarizes analytical methods and data uses for these compounds. The interpretation of general water-quality data is discussed later. Because the pH, temperature, and conductivity of a ground-water sample can change significantly within a short-time following sample acquisition, these parameters, along with dissolved oxygen and ORP, must be measured in the field in unfiltered, unpreserved, "fresh" water. The measurements are best made either downhole, or directly from a flow-through cell, and the measured values should be recorded in the ground-water sampling record.

Supplemental Monitoring Parameters

In addition to the analytes described earlier, additional lines of evidence can be collected in the form of supplemental or confirmatory parameters discussed subsequently. These data can be particularly useful if negative indicators of contaminant attenuation or conflicting results are present and the plume is no longer behaving as expected. Some of the potential supplemental analytes include manganese, alkalinity, dissolved inorganic carbon (DIC), anions, cations, hydrogen, acetylene, volatile fatty acids (VFAs), phospholipid fatty acids (PLFAs), denaturing gradient gel electrophoresis (DGGE), and isotope analyses. Table 9.3 summarizes analytical methods and data uses for these compounds. The interpretation of supplemental data is discussed later.

Ground-Water Sampling Techniques

The ground-water sampling procedures presented in this section are important, because the quality of several of the biogeochemical indicators used to evaluate degradation can be significantly affected by poor sampling technique. Poor data quality can result in erroneous conclusions regarding the efficacy, or even the occurrence, of degradation, so care must be taken during ground-water sample collection. Because of the accuracy required for many of the analytical procedures required to evaluate natural attenuation, care must be exercised when extracting ground water from the sampling device. Varied equipment and methods are available for the extraction of ground water (see Chapter 15). The approach used should be determined on the basis of application (purging or sampling), hydrogeologic conditions, monitoring location dimensions, and regulatory requirements.

Portable ground-water extraction devices from four generic classifications are commonly used to collect ground-water samples: grab samplers, suction lift samplers, submersible samplers, and passive samplers. Sampling devices discussed in this chapter include peristaltic pumps, electric submersible pumps, positive-displacement pumps, bailers, and diffusion samplers.

Sampling with Peristaltic Pumps

Suction-lift sampling technology is best represented in environmental investigations by the peristaltic pump. A peristaltic pump extracts water using a vacuum created by cyclically advancing a sealed compression along flexible tubing. This pumping technique means that extracted water contacts nothing other than tubing that can be easily replaced between sampling locations. This reduces the possibility of cross-contamination. Furthermore, peristaltic pumps can be used to extract minimally disturbed ground water from any diameter monitoring point at variable low-flow rates; however, because of the limited flow rate, peristaltic pumps are impractical for purging and sampling wells that are larger than 2 in. in diameter. Because of the features of the peristaltic pump, representative samples are simple to collect, and reliable flow-through cells are simple to establish. The biggest drawback of sampling with a peristaltic pump is the maximum achievable pumping depth which is equivalent to the height of water column that can be supported by an imperfect vacuum. This effectively limits the use of a peristaltic pump to monitoring locations with ground water depths of less than approximately 25 ft, depending on the altitude of the site. Also, sample degassing can occur in the tubing as a result of the vacuum applied to the sample and the high-rate of cyclical loading. If bubbles are observed in the tubing during purging or sampling, the flow rate of the peristaltic pump should be slowed. If bubbles are still apparent, the lift is probably too great to maintain the dissolved gas content in the sample, and sampling should not be attempted (see Chapter 15 for more detail). The final potential disadvantage with a peristaltic pump is the low flow rate. Although advantageous for sampling, this can be inappropriate during purging at locations requiring large extraction volumes.

To prevent downhole aeration of the sample in wells screened across the water table, well drawdown should not exceed about 5–10% of the height of the standing column of water in the well. The pump tubing should be immersed alongside the dissolved oxygen probe beneath the water level in the sampling container (Figure 9.5). This will minimize aeration and keep water flowing past the dissolved oxygen probe's sampling membrane.

Sampling with Submersible Pumps

Submersible pumps, some of which are positive-displacement pumps, include bladder pumps, progressing cavity pumps (i.e., the Keck® pump), centrifugal electric submersible

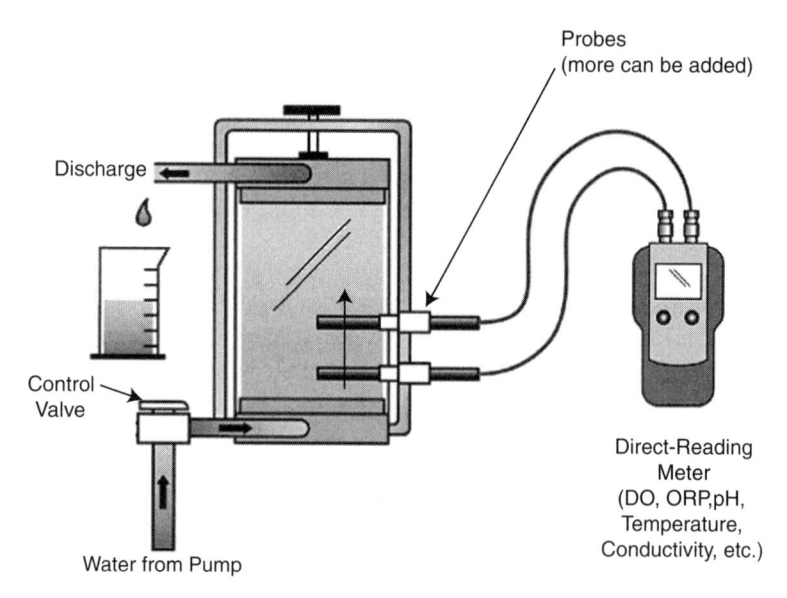

FIGURE 9.5
Schematic diagram of a flow-through cell.

pumps (i.e., the Grundfos Redi-Flo II® pump), electric submersible gear pumps (i.e., the Fultz® pump), double-acting piston pumps (i.e., the Bennett® pump) and pumps of other designs (i.e., the Enviro-Tech Purger ES® pump). Most of these pumps operate downhole at lifts of up to a few hundred feet (with the exception of the Bennett pump, which will operate at lifts of more than 1000 ft) and at pumping rates of between $\frac{1}{2}$ gallon and several gallons per minute. Some submersible pumps are particularly useful for applications requiring the extraction of large volumes of water, and most can be used for the extraction of ground water from depths in excess of 100 ft. Because the pumps operate downhole, they require appropriately sized wells. A well diameter of at least 2 in. is typically required; however, larger well diameters can be required depending on the selected pump type, extraction depth, and extraction rate. It is important that cavitation is not introduced while using a submersible pump. Because the typical submersible pump design results in contact between the ground water and internal as well as external surfaces of the pump, rigorous decontamination procedures must be implemented to avoid cross-contamination if a pump that is not dedicated to the well is used for sampling.

Sampling with Bailers

Bailers are the most common sampling devices in use in most monitoring programs. Bailers can be used at any depth in wells with an inside diameter of at least 0.5 in. However, ground-water sample collection becomes less efficient as the well diameter (and hence the bailer diameter) decreases. Disposable bailers can be used to avoid decontamination expenses and potential cross-contamination problems. Drawbacks of bailers include significant agitation and aeration of the water column in the well, mixing of the water column above the screen with the water column in the screen, and the inability to maintain steady, nonturbulent flow in the well. Agitation and aeration can be minimized, but not eliminated, through careful immersion into and extraction from the standing column of water in the well or sampling point. Aeration also can be an issue during transfer of the sample from the bailer to the sample container. Once again, this aeration can be minimized, but not eliminated. Because of aeration, accurate dissolved oxygen and ORP measurements can be difficult or impossible to obtain when using a bailer. Bailers cannot be used with flow-through cells.

When using a bailer, the bailer should be slowly immersed in the standing column of water in the well to minimize aeration. After sample collection, the water should be drained from the bottom of the bailer (using bottom-emptying tubing) into the sampling container. The tubing used for this operation should be immersed alongside the dissolved oxygen probe beneath the water level in the sampling container (Figure 9.5). This will minimize aeration and keep water flowing past the dissolved oxygen probe's sampling membrane.

Diffusion Samplers

Diffusion samplers can be useful for sampling low-solubility, low vapor pressure volatile organic compounds (VOCs) (such as benzene and tetrachloroethylene), but not high-solubility, high vapor pressure VOCs (such as methyl tertiary butyl ether [MTBE] and acetone), during long-term monitoring. The diffusion sampler technology utilizes a deionized water-filled, low-density polyethylene bag to collect water samples from ground-water monitoring wells for VOC laboratory analyses. The bag allows selected VOCs in ground water to diffuse into the deionized water. Chemical equilibrium between the selected VOCs in ground water and the deionized water in the sampler will occur over time, resulting in a water sample (from the diffusion sampler) that is representative of the concentrations of those selected VOCs in the ground water. Diffusion samplers can be used to rapidly and inexpensively obtain ground-water samples for selected VOCs in monitoring wells in which horizontal flow dominates (Vroblesky et al., 1996; Vroblesky and Hyde, 1997). They should not be used in wells in which vertical flow occurs within the well screen. When used appropriately, representative samples can be obtained without well purging to identify temporal changes in ground-water chemistry for selected VOCs (Vroblesky, 2001). Potentially large cost savings in long-term ground-water monitoring efforts may be realized due to the simplicity of the diffusion samplers compared with traditional purge-and-sample techniques. One drawback of the diffusion sampler is that insufficient water is collected to allow measurement of many of the parameters listed in Table 9.2.

General Ground-Water Sampling Considerations

Purging consists of the evacuation of water from the monitoring location prior to sampling, so that "fresh" formation water will enter the monitoring location and be available for sampling. Because sampling can occur immediately upon completion of purging, it is best to limit ground-water agitation, and consequently, aeration of the ground water and volatilization of contaminants. Two sources for agitation include the purging device and the cascading of water down the screen as drawdown occurs in the well. To avoid agitation, a low-disturbance device such as a peristaltic pump (at low lifts) or a positive-displacement pump is recommended for purging, while equipment such as bailers should be avoided. To avoid aeration, wells or sampling points screened below the water table should be pumped at a rate that prevents lowering of the water table to below the top of the screen. If practical, wells or sampling points screened across the water table should be pumped at a rate that lowers the total height of the water column no more than 5–10%.

A flow-through cell, such as the simple one pictured in Figure 9.5, should be used for the measurement of ground-water quality indicator parameters such as pH, temperature, specific conductance, dissolved oxygen, and ORP. Measurements of these parameters should be taken during well purging and immediately before sample acquisition using a multi-parameter sonde or a direct-reading meter (see Chapter 15). Because most well purging techniques can allow aeration of collected ground-water samples, it is important to minimize potential aeration by taking the following precautions:

(1) Use a submersible pump of some type to purge the well when possible. To prevent downhole aeration of the sample in wells screened across the water table, drawdown should not exceed about 5–10% of the height of the standing column of water in the well. The discharge end of the pump tubing should be attached to a flow-through cell or immersed alongside the dissolved oxygen and ORP probes beneath the water level in a sampling container (Figure 9.5). This will minimize aeration and keep water flowing past the dissolved oxygen probe's sampling membrane.

(2) If bubbles are observed in the tubing during purging using a peristaltic pump, the flow rate of the peristaltic pump must be slowed. If bubbles are still apparent, sampling should be discontinued.

(3) When using a bailer, the bailer should be slowly immersed in the standing column of water in the well to minimize aeration. After sample collection, the water should be drained from the bottom of the bailer through tubing into the sampling container. The tubing used for this operation should be immersed alongside the dissolved oxygen and ORP probes beneath the water level in the sampling container (Figure 9.5). This will minimize aeration and keep water flowing past the dissolved oxygen probe's sampling membrane.

(4) Downhole dissolved oxygen probes are preferred for dissolved oxygen analyses, but such probes must be thoroughly and carefully decontaminated between wells. Some decontamination solutions can be harmful to the dissolved oxygen probe (see Chapter 15).

Samples should be collected directly from the pump discharge tubing or bailer into a sample container of appropriate size, style, and preservation for the desired analysis. Water should be directed down the inner walls of the sample bottle to minimize aeration of the sample. All samples to be analyzed for volatile constituents (e.g., SW8010, SW8020, SW8240, SW8260, and TPH-g) or dissolved gases (e.g., methane, ethane, and ethene) must be filled and sealed so that no headspace remains in the container.

Sampling Frequency

The determination of appropriate sampling frequency requires a balancing of several factors including, among others, the chemical characteristics of the contaminants of concern, distance to potential receptors, ground-water seepage velocity, solute transport velocity, the amount of historical data, and how well the contaminant plume is understood.

In the past, the monitoring of dissolved contaminant plumes typically was needlessly time- and location-intensive and, in many cases, involved the quarterly sampling of every monitoring well at a site. On the basis of our current understanding of the behavior of dissolved contaminant plumes, this may not be necessary in many cases. However, quarterly sampling of long-term monitoring wells during the first year of sampling may be useful to help confirm the direction of plume migration and to better establish baseline conditions and seasonal variability. If variability due to seasonal and climatic events is suspected, the installation of dedicated water-level loggers in selected wells may be the best technique to determine the presence and timing of significant events. The probability the prescheduled quarterly monitoring will occur during a unique temporal event may be low at many sites. Quarterly monitoring may represent more of a misplaced tradition than a technically sound approach. Thus, information should be compiled to identify the nature, probability, and timing of significant seasonal or climatic events. If significant variability is encountered during the first year, then more frequent and precisely timed sampling may be required. On the basis of the results of the first year's sampling, the sampling frequency may be reduced to annual (or less frequent) sampling during the period showing the highest contaminant concentrations or the greatest extent of the plume.

At a minimum, the frequency of long-term monitoring should be related to:

(1) The natural variability in contaminant concentrations.
(2) The distance and travel time from the source to the location where acceptance criteria are applied.
(3) The reduction in contaminant concentrations required to meet the acceptance criteria.
(4) The occurrence of a significant seasonal or climatic event (e.g., snow melt, rainy season, river ice breakup, spring thaw, etc.).

Ideally, the number of wells to be sampled and the frequency of sampling will be based on plume behavior and the variability in contaminant concentrations, the distance and estimated time of contaminant travel between long-term monitoring wells, and the distance and estimated time of contaminant travel between PMWs and contingency wells. Sampling frequency should be determined by the final placement of the PMWs and contingency monitoring wells and the ground-water seepage and contaminant transport velocity.

One method of estimating sampling frequency is to divide the distance between a point just downgradient from the leading edge of the contaminant plume and a downgradient contingency well located in the plume's flow path by the seepage velocity of ground water. For example, consider the contaminant plume depicted in Figure 9.2. If the distance between well PMW-5 and the center contingency well is 500 ft, and the seepage velocity of ground water is 250 ft per yr, then a sampling frequency of 2 yr (500 or 250 ft per yr) may be appropriate for this site. Because the exact location of the leading edge of a dissolved contaminant plume generally is not known, some professional judgment may be required when making these calculations.

According to U.S. EPA (1999), flexibility for adjusting the monitoring frequency over the life of the remedy should be included in the monitoring plan. For example, it may be appropriate to decrease the monitoring frequency at some point in time, once it has been determined that natural attenuation is progressing as expected and very little change is observed from one sampling round to the next. Conversely, the monitoring frequency may need to be increased if unexpected conditions (e.g., plume migration) are observed. Remedial process optimization (RPO) is a value-added process aimed at determining that monitoring programs and remedial alternatives are effective, protective, and cost-effective. This approach should be applied to long-term monitoring programs (AFCEE, 2001).

Evaluation and Interpretation of Monitoring Data

One of the essential components of a long-term monitoring plan is the evaluation and interpretation of the resulting data. Too often, monitored natural attenuation remedies are proposed and implemented with little or no consideration of why the monitoring data are being collected. The evaluation and interpretation of long-term monitoring data focuses on detection of spatial and temporal changes, the relation of these changes to natural attenuation processes and plume behavior, and assessment of their impacts on the achievement of site-specific goals.

Long-term monitoring data should be examined in the context of relevant natural attenuation processes and the hydrogeologic setting. This includes evaluation and interpretation of contaminant concentration data, geochemical data, and other parameters to provide empirical evidence for how the ground-water solute plume is behaving over time, to provide insight into how the natural attenuation processes are affecting this behavior, and to allow the identification and explanation of changes that may alter solute plume behavior and affect the performance of the natural attenuation remedy. The purpose is to explain what is happening in the solute plume, not just observe it.

Of particular interest are changes in conditions that may affect the efficacy of the natural attenuation remedy or signal a change in solute plume behavior. These can include indications of additional contaminant releases, changes in geochemical conditions (e.g., redox conditions) that may alter contaminant transformation processes and rates, detections of contaminants at the horizontal and vertical plume boundaries that may indicate plume expansion, and changes in ground-water flow velocities or directions that may move contaminants into previously unaffected areas.

Evaluating Contaminant Data

The fundamental reason for monitoring of natural attenuation is to establish the behavior of the ground-water solute plume so it can be evaluated in relation to remediation objectives for the site. This evaluation typically relies on ground-water monitoring data for locations within the solute plume that indicate how concentrations of constituents of interest are changing over time.

On a conceptual level, the behavior of ground-water solute plumes is a continuum with three major phases: plume expansion, plume stabilization, and plume recession (Figure 9.6). These phases represent the interaction between mass loading to ground water from the source and the action of various attenuation mechanisms in the aquifer. Different portions of a solute plume, as well as different individual constituents, may exhibit different behaviors that can change over time in response to changes in environmental conditions (Figure 9.6).

Evaluation of solute plume behavior can be qualitative or quantitative, or both, depending upon the availability of the necessary data. The particular methods used depend on the availability and quality of monitoring data and will change over time as more monitoring data are generated. The approaches for evaluating solute plume behavior can be separated into two classes: (1) graphical methods that rely on visual interpretation of monitoring data and (2) statistical methods that rely on quantitative analysis of the monitoring data.

Graphical Methods for Evaluating Plume Behavior

Graphical methods are essential tools for evaluating plume behavior. There are several ways to present data to illustrate changes in contaminant concentrations and plume configuration over time. The most common graphical techniques include: (1) preparing isopleth maps of contaminant concentration over time; (2) plotting contaminant concentrations versus time for individual monitoring wells; and (3) plotting contaminant

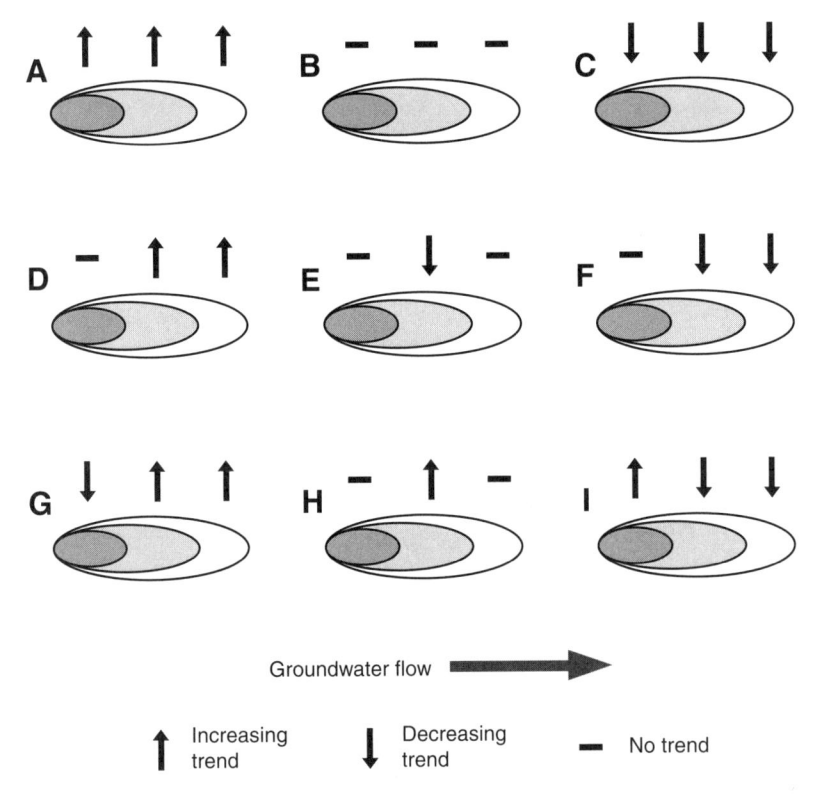

FIGURE 9.6
Solute plume behavior illustrated by concentration trends over time for monitoring points in the vicinity of the source, mid-plume, and the distal part of the plume.

concentrations versus distance for several wells along the ground-water flow path over several sampling events.

Isopleth maps of contaminant concentrations prepared for successive monitoring rounds are useful for depicting spatial distribution of the solute plume over time. The example in Figure 9.7 shows isopleth maps for total VOC concentrations in ground water at the depth of greatest contaminant concentration. Note that the plotted contaminant data were collected during the same season. This is important because seasonal variations in recharge can cause significant changes in contaminant concentrations and ground-water geochemistry, and an apparent change in plume size and contaminant concentrations could simply be the result of seasonal dilution.

Another method that can be used to present data showing changes in contaminant concentrations and plume configuration over time is to plot contaminant concentrations versus time for individual monitoring wells, or to plot contaminant concentrations versus distance downgradient for several wells along the ground-water flow path over several sampling events. It is important when plotting data in this manner that a least one data point be located a short distance downgradient from the contamination in the ground-water flow path. This ensures that contaminant concentrations in the aquifer as a whole are decreasing and that a pulse of contaminant is not simply migrating downgradient from the observation wells. To ensure that contaminants are not moving downgradient, it is important that downgradient wells are located in the path of contaminated ground-water flow. Geochemical data can be used to confirm that downgradient wells are sampling ground water that was once contaminated with organic compounds.

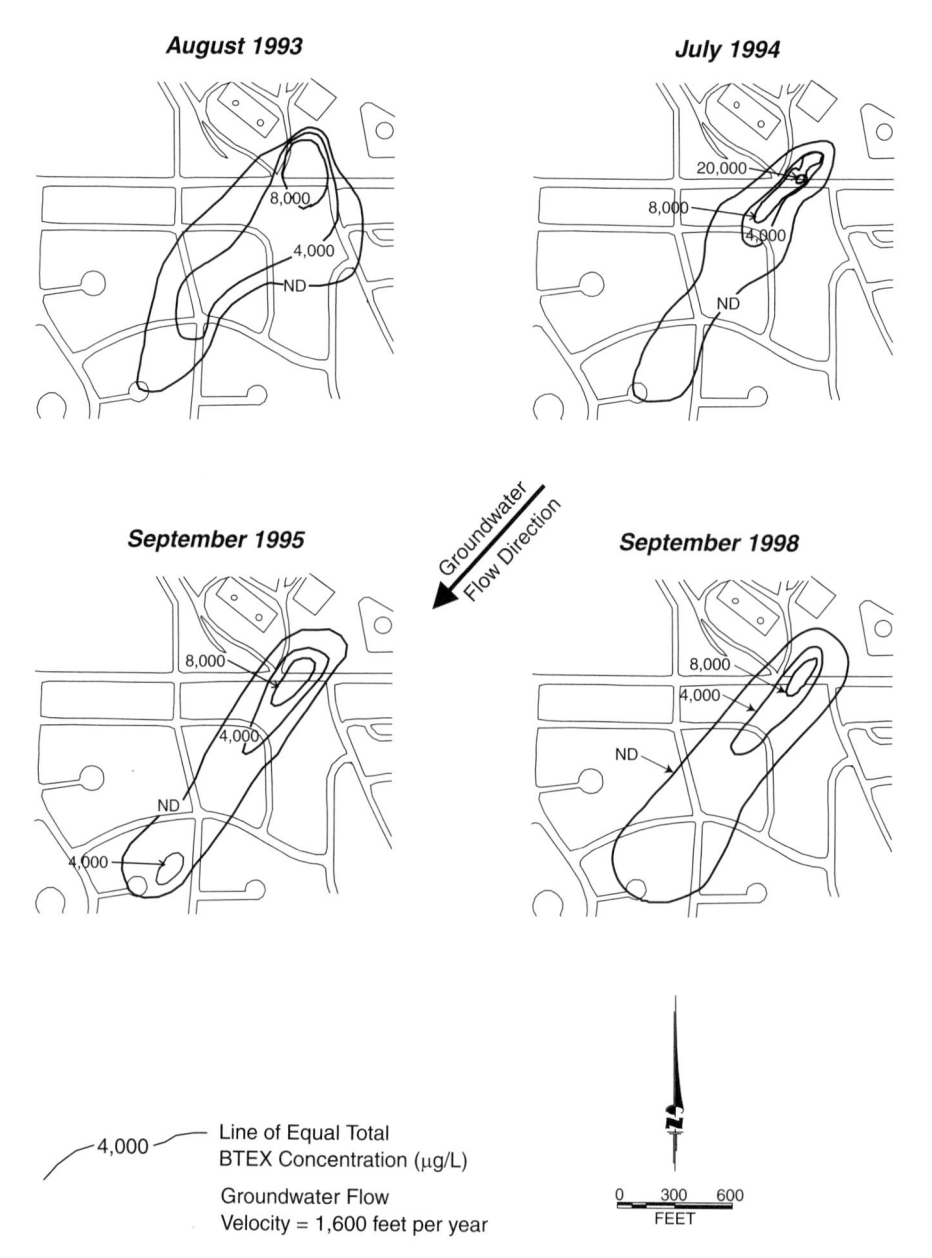

FIGURE 9.7
Isopleth maps showing contaminant distribution over time.

Contaminant concentration versus time plots should only be completed using data from events that are considered comparable. Data from different events may not always be comparable due to abnormal conditions like high water levels due to 100-yr flood events or other unique events.

Figure 9.8 presents a plot of contaminant concentration versus time in one well, and contaminant concentrations versus distance downgradient along the flow path for several sampling events. On the basis of the geochemical data presented in this figure, it is reasonably certain that well H is in the plume's flow path. Therefore, if the plume were migrating downgradient, this migration should be detected. Wells F and H are

Designing Monitoring Programs to Evaluate the Performance of Natural Attenuation 603

spaced 100 ft apart, and the ground-water seepage velocity is 50 ft per yr; with 8 yr of sampling data from the same season, we can conclude with reasonable certainty that the plume is not migrating downgradient. The combination of decreasing contaminant concentrations shown by the plots in Figure 9.8b, and the lack of contaminant migration provide converging lines of evidence for natural attenuation and contaminant mass destruction. The chemical and geochemical data discussed by Wiedemeier et al. (1995, 1999) and U.S. EPA (1998) can be used to show that this loss of contaminant mass is the result of degradation.

While plotting concentration data versus time is recommended for any plume stability analysis, discerning trends in the plotted data can be a subjective process, particularly if the data do not display a uniform trend, but show some variability over time (Figure 9.9).

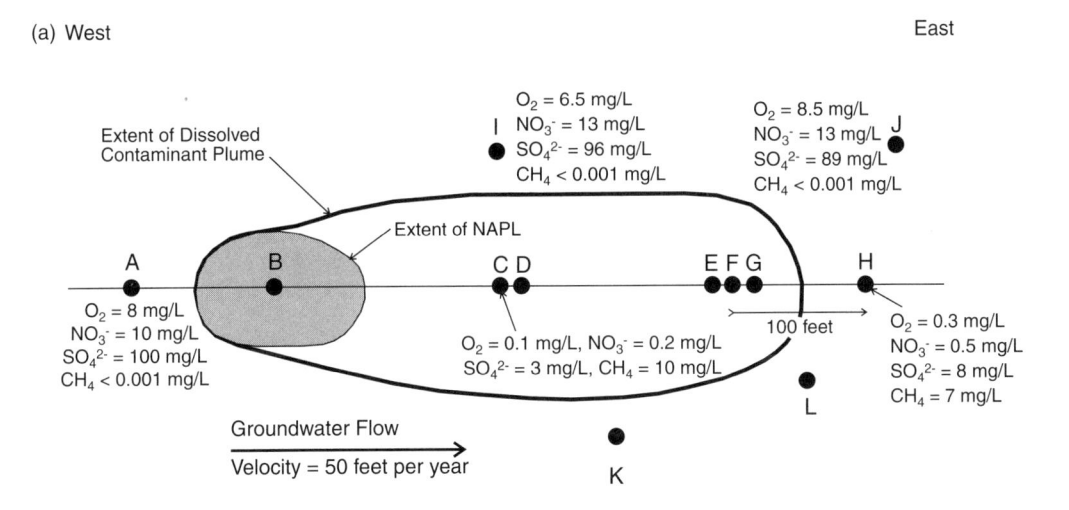

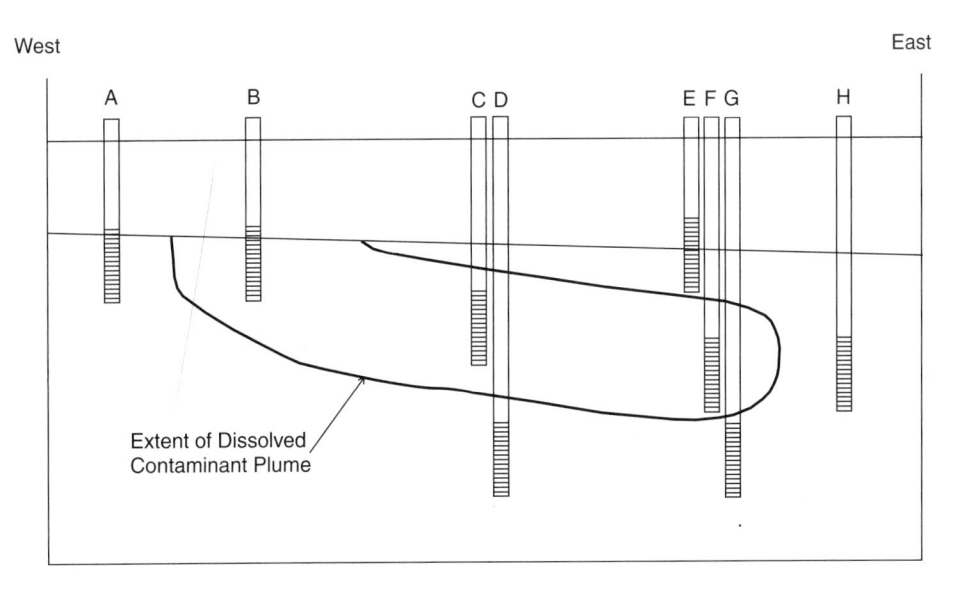

FIGURE 9.8
(a) Sampling locations for the (b) plots of contaminant concentration versus time and distance downgradient.

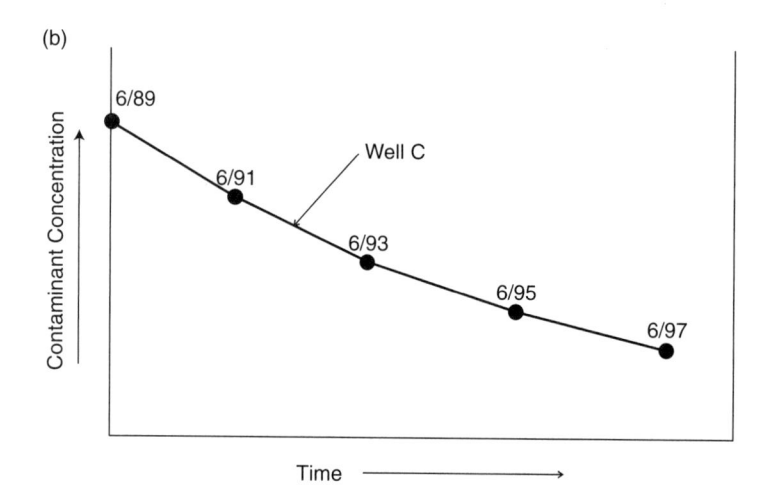

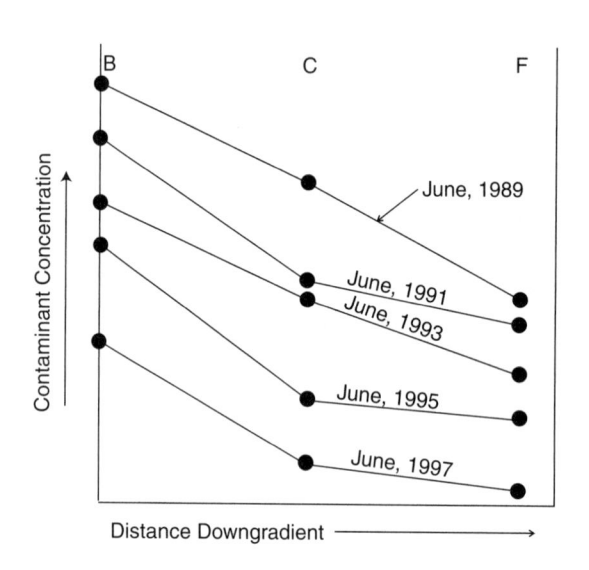

FIGURE 9.8
Continued.

Statistical Methods for Evaluating Plume Behavior

Statistical methods are powerful tools for identifying significant changes and trends in ground-water concentration data. They provide for an objective evaluation of the data and allow statements to be made about the confidence in results. This provides a quantitative indication of the likelihood that conclusions drawn from the data are correct. In evaluating natural attenuation, statistical methods are used to assess ground-water monitoring data for the presence of significant trends or changes in concentrations over time that can provide insight into solute plume behavior. Once again, it is paramount to verify that the monitoring events and data subject to statistical analyses are comparable. If high water levels correlate with higher contaminant concentrations, then data from high and low water table events may not be comparable. If sampling was conducted during an extreme weather event (e.g., a 100-yr flood), then it may not be comparable to previous events. A more detailed discussion of the concept of comparability is found in Gilbert (1987).

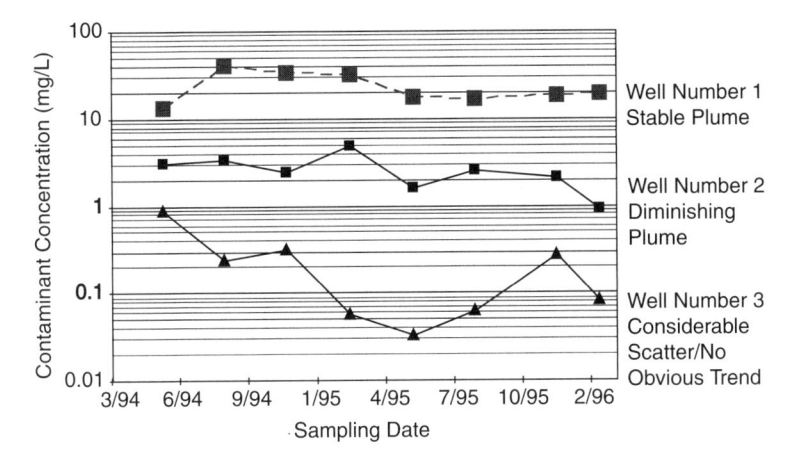

FIGURE 9.9
Example plume stability plots.

The application of statistics requires an understanding of the underlying assumptions of the tests and nature of the data because these determine the selection of appropriate methods and interpretation of the results. While a detailed review of the statistical analysis of concentration data and its application are beyond the scope of this chapter, a brief discussion of the significant factors and some methods that are applicable in the majority of situations is provided. More detailed discussion is available in several statistics texts (e.g., Gilbert, 1987; Gibbons, 1994; U.S. EPA, 2000; Helsel and Hirsch, 2002). This is not a theoretical discussion; rather, it provides practical considerations where statistical results are used in decision-making.

Statistical tests are a form of hypothesis testing and their basis is the comparison of what statisticians call the "null hypothesis" (H_0) to an alternative hypothesis (H_1). The null hypothesis is the statistical hypothesis being tested; generally that the test results are merely a product of chance factors. For example, to test for a trend in a concentration time series, H_0 would be that there is no change in concentration over time, and H_1 would be that the concentration is either increasing or decreasing with time. The two hypotheses are compared using a test statistic that is calculated from the data series being tested.

Most statistical tests are intended to detect a significant difference between a group of samples or from a predefined condition. This is determined by comparing the value for the test statistic calculated from the data set to the probability of obtaining that value purely due to chance. The probability values are determined from the "null" distribution for the test statistic that is the distribution of values for the test statistic under the null hypothesis (H_0). The significance level is a means of determining whether the test statistic is "significantly" different from values that would typically occur under H_0. If the probability for the test statistic value calculated from the data set is less than the level of significance, the null hypothesis (H_0) can be rejected in favor of the alternative hypothesis (H_1).

There are two possible types of decision errors associated with statistical hypothesis testing. A Type I error is when H_0 is incorrectly rejected. A Type II error is when H_0 is accepted when H_1 is true. Both types of decision errors have implications for the conclusions drawn from results of statistical tests.

A Type I error is rejecting the null hypothesis (H_0) when it is in fact true. This is essentially equivalent to a "false positive" result, such as concluding that there is an increasing or decreasing trend in concentration over time when no trend is actually present. The probability of incorrectly rejecting H_0 is the "significance level" (α) of the test. Type I errors are controlled by selecting an appropriate α-value to reduce the likelihood of drawing an incorrect conclusion from the test.

The inability to reject the null hypothesis (failure to accept the alternative hypothesis) at some level of significance does not imply that the null hypothesis is true. A Type II error is failing to reject (accepting) the null hypothesis (H_0) when it is false and the alternative hypothesis (H_1) is true. This is essentially equivalent to a "false negative" result, such as concluding that there is no trend in concentration over time when an increasing or decreasing trend is actually present. The probability of this occurring is β and the power of a statistical test to detect a significant difference is $1 - \beta$. The statistical power of a test is related to both the α-value selected and the sample size (n).

Ideally, we would like to minimize both Type I and Type II errors in using statistical tests, but this is difficult in practice. The importance of either type of decision error should be evaluated in terms of the ultimate use of the results of the statistical test. A pragmatic approach is to specify an acceptable value for α and concurrently reduce β by (1) increasing the sample size and (2) using a statistical test with the greatest power for the type of data being evaluated (Helsel and Hirsch, 2002).

Statistical tests are described as one- or two-sided depending upon the specific alternative hypothesis involved. A two-sided test is used when a difference in either direction from H_0 would cause H_0 to be rejected, such as a test for detecting the *presence* of a trend or change in concentration. For example, if there is no reason to assume that concentrations are not stable or that departures from H_0 in only one direction are of interest, a two-sided test is appropriate. A one-sided test is used when a change in only one direction from H_0 would cause H_0 to be rejected, such as a test for detecting an *increase* (or *decrease*) in concentration over time. For example, if only evidence that concentration is increasing (or decreasing) over time is considered important, H_0 would be stated as "the change in concentration over time is less (or greater) than or equal to zero (0)" and H_1 would be "the change in concentration over time is greater (less) than zero (0)."

The null distributions for most test statistics are symmetrical and the probability values for only one "tail" of the distribution are given. For detecting an increase (or decrease), only the difference in one direction is important and the critical test statistic value at α is used (one-sided tail). For detecting the presence of a trend or change in concentration, both a positive or negative difference is important and the critical test statistic value at $\alpha/2$ is used (two-sided tail).

The issue of confidence levels, or significance levels, and their meaning is a source of considerable confusion on the part of users. The practical implication of the confidence level is that there is error associated with the decision to reject the null hypothesis. If the calculated value of the test statistic leads you to reject the null hypothesis, it does not mean that the value for the test statistic you obtained could not have occurred by chance. It means that the probability of obtaining that value by chance alone is sufficiently small that it is reasonable to conclude that the result is not due to chance and that the decision to reject the null hypothesis is correct. The confidence level simply quantifies the likelihood that rejecting the null hypothesis is appropriate.

The confidence level for a statistical test is related to the significance level (α) and is simply described by the value $1 - \alpha$, typically expressed as a percentage. The significance level (α) is specified in advance of the test and defines the "acceptable" level of Type 1 error that the user is willing to tolerate in deciding to reject the null hypothesis. For example, if the desired confidence level for a statistical test is 95% (0.95), the significance level would be specified as 0.05 and the null hypothesis would be rejected if the calculated test statistic value has a probability ≤ 0.05. This means that the likelihood of making an incorrect decision to reject the null hypothesis is 5 in 100 (1 in 20) and, conversely, the likelihood that the decision to reject the null hypothesis is correct is 95 in 100 (19 in 20).

The confidence level simply quantifies the "confidence" associated with obtaining a "significant" result for a statistical test, such as concluding that there is a trend in concentration over time or a difference in concentrations. There is no magic to defining the appropriate confidence level and adjusting the confidence level simply changes the tolerance for Type I error in decision-making. In most scientific applications, a 95% confidence level is used as there is general concurrence that the associated error (5%) is sufficiently small. Decreasing the confidence level for a statistical test will increase the likelihood of obtaining a "significant" result, but will also increase the chances that the null hypothesis will be incorrectly rejected. The specified confidence level is simply a reflection of the user's willingness to accept a mistaken conclusion for a statistical test.

Nature of Ground-Water Concentration Data and Appropriate Statistical Methods

Issues involved with the statistical analysis of ground-water concentration data are myriad, but most commonly involve missing values, nondetect (censored) values, small number of data points, and the lack of certain knowledge of the underlying distribution. All of these complicate the application of statistical methods and either require significant data manipulation or the use of methods that are little affected by these data characteristics. Trend analysis, in particular, is sensitive to these issues, as well as to changes in sampling and analytical procedures, seasonal or other cyclic variation in the data, and correlated data (Gilbert, 1987).

Statistical approaches can be separated into parametric and nonparametric methods. The familiar parametric statistics, such as regression analysis, rely on data conforming to an underlying distribution, such as normal (Gaussian) or log-normal. Parametric statistics are sensitive to missing data points and outliers, how nondetect values are handled, and departures from the assumed distribution. Nonparametric statistical methods do not depend on assumptions regarding the underlying data distribution and are also known as "distribution-free" methods. They can accommodate missing data points and nondetect values that are common in ground-water concentration data sets. These methods rely on the ranks or relative magnitudes of the data rather than the actual values and are fairly straightforward to use. In many situations, particularly those involving small data sets, nonparametric methods perform as well or better than parametric ones (Helsel and Hirsch, 2002).

The selection of statistical methods is frequently limited by the availability of sufficient data. Aside from the issues mentioned earlier, parametric methods are sensitive to sample size and their power is reduced for small data sets, such as are common in ground-water concentration data. Nonparametric methods typically are equally or more powerful for discerning trends and changes for small data sets.

Due to the issues associated with most ground-water concentration data, the use of nonparametric techniques are generally preferred (Gilbert, 1987; Gibbons, 1994) and some commonly used methods are described briefly below. Additional information on these nonparametric methods is provided in Hollander and Wolfe (1999), Conover (1999), and Helsel and Hirsch (2002).

Tests for Trend

The Mann–Kendall test for trend (Mann, 1945; Kendall, 1975) is used to determine the presence or absence of a trend in concentration over time for individual monitoring points. It is a test for zero slope of time-ordered data that is based on a nonparametric analog of linear regression. The basic methodology and its variants (such as the Seasonal

TABLE 9.4

Example Calculation of the Mann–Kendall Statistic for TCE Concentrations in a Monitoring Well with Ten Sampling Events

Event	1	2	3	4	5	6	7	8	9	10	
Concentration (µg/l)	56	78	63	43	45	36	38	40	46	42	Row sums
		1	1	−1	−1	−1	−1	−1	−1	−1	−5
			−1	−1	−1	−1	−1	−1	−1	−1	−8
				−1	−1	−1	−1	−1	−1	−1	−7
					1	−1	−1	−1	1	−1	−2
						−1	−1	−1	1	−1	−3
							1	1	1	1	4
								1	1	1	3
									1	1	2
										−1	−1

Mann–Kendall Statistic $S = $ −17

Mann–Kendall test) are described in Gilbert (1987) and Helsel and Hirsch (2002) and four or more independent sampling events are required. The results of the Mann–Kendall test indicate the presence or absence of a statistically significant increasing or decreasing trend in concentrations over time at a monitoring point. These results can be used to help evaluate whether the solute plume is receding, expanding, or stable.

The Mann–Kendall test for between 4 and 40 data points is very straightforward to apply and an example calculation is provided in Table 9.4. Concentration data are ordered sequentially over time and a matrix is constructed comparing each data value to subsequent values. Starting with the earliest data point, each subsequent data point is compared and a value entered into the matrix: +1 if the later value is greater, −1 if the later value is less, and 0 if the later value is equal to the earliest data point. The process is repeated for the next data point in the sequence, comparing its value to subsequent ones, until all data points in the sequence have been compared and appropriate values entered into the matrix. The values in each row in the matrix are then summed and the row sums are then summed to generate the Mann–Kendall statistic (S).

Once the S-statistic has been calculated, it is compared with the table of null probability values of S for the number of data points (n) in the series (Table 9.5). If the probability value for the calculated S-statistic and the number of data points (n) is less than the specified significance level for the test (α for one-sided; $\alpha/2$ for two-sided), the result is significant at the $1 - \alpha$ confidence level and a trend is present. The calculated S-statistic (−17) and n (10) for the example calculation in Table 9.4 correspond to a probability of 0.078 in Table 9.5. For a one-sided test, this result is less than the α for the 90% confidence level ($\alpha = 0.1$), indicating a significant result, but is greater than the α for the 95% confidence level ($\alpha = 0.05$), indicating that the result is not significant at this level of confidence.

Designing Monitoring Programs to Evaluate the Performance of Natural Attenuation

TABLE 9.5

Null Probabilities for the Mann–Kendall Statistic, $n = 4\text{--}20$

S	4	5	6	7	8	9	10	11	12	13	14	15	16	17	18	19	20
0	0.625	0.592			0.548	0.540			0.527	0.524			0.518	0.516			0.513
±1			0.500	0.500			0.500	0.500			0.500	0.500			0.500	0.500	
±2	0.375	0.408			0.452	0.460			0.473	0.476			0.482	0.484			0.487
±3			0.360	0.386			0.431	0.440			0.457	0.461			0.470	0.473	
±4	0.167	0.242			0.360	0.381			0.420	0.429			0.447	0.452			0.462
±5			0.235	0.281			0.364	0.381			0.415	0.423			0.441	0.445	
±6	0.042	0.117			0.274	0.306			0.369	0.383			0.412	0.420			0.436
±7			0.136	0.191			0.300	0.324			0.374	0.385			0.411	0.418	
±8		0.042			0.199	0.238			0.319	0.338			0.378	0.388			0.411
±9			0.068	0.119			0.242	0.271			0.334	0.349			0.383	0.391	
±10		0.008			0.138	0.179			0.273	0.295			0.345	0.358			0.387
±11			0.028	0.068			0.190	0.223			0.295	0.313			0.354	0.365	
±12					0.089	0.130			0.230	0.255			0.313	0.328			0.362
±13			0.008	0.035			0.146	0.179			0.259	0.279			0.327	0.339	
±14					0.054	0.090			0.190	0.218			0.282	0.299			0.339
±15			0.001	0.015			0.108	0.141			0.225	0.248			0.300	0.314	
±16					0.031	0.060			0.155	0.184			0.253	0.271			0.315
±17				0.005			0.078	0.109			0.194	0.218			0.275	0.290	
±18					0.016	0.038			0.125	0.153			0.225	0.245			0.293
±19				0.001			0.054	0.082			0.165	0.190			0.250	0.267	
±20					0.007	0.022			0.098	0.126			0.199	0.220			0.271
±21				0.000			0.036	0.060			0.140	0.164			0.227	0.245	
±22					0.002	0.012			0.076	0.102			0.175	0.196			0.250
±23							0.023	0.043			0.117	0.141			0.205	0.223	
±24					0.001	0.006			0.058	0.082			0.153	0.174			0.230
±25							0.014	0.030			0.096	0.120			0.184	0.203	
±26					0.000	0.003			0.043	0.064			0.133	0.154			0.211
±27							0.008	0.020			0.079	0.101			0.165	0.184	
±28						0.001			0.031	0.050			0.114	0.135			0.193
±29							0.005	0.013			0.063	0.084			0.147	0.166	
±30						0.000			0.022	0.038			0.097	0.118			0.176
±31							0.002	0.008			0.050	0.070			0.130	0.149	
±32									0.016	0.029			0.083	0.102			0.159
±33							0.001	0.005			0.040	0.057			0.115	0.133	
±34									0.010	0.021			0.070	0.088			0.144
±35							0.000	0.003			0.031	0.046			0.100	0.119	
±36									0.007	0.015			0.058	0.076			0.130
±37								0.002			0.024	0.037			0.088	0.105	
±38									0.004	0.011			0.048	0.064			0.117
±39								0.001			0.018	0.029			0.076	0.093	
±40									0.003	0.007			0.039	0.054			0.104
±41								0.000			0.013	0.023			0.066	0.082	
±42									0.002	0.005			0.032	0.046			0.093
±43											0.010	0.018			0.056	0.072	
±44									0.001	0.003			0.026	0.038			0.082
±45											0.007	0.014			0.048	0.062	
±46									0.000	0.002			0.021	0.032			0.073
±47											0.005	0.010			0.041	0.054	
±48										0.001			0.016	0.026			0.064
±49											0.003	0.008			0.034	0.047	
±50										0.001			0.013	0.021			0.056
±51											0.002	0.006			0.029	0.040	
±52										0.000			0.010	0.017			0.049
±53											0.002	0.004			0.024	0.034	
±54													0.008	0.014			0.043
±55											0.001	0.003			0.020	0.029	
±56													0.006	0.011			0.037
±57											0.001	0.002			0.016	0.025	

(Table continued)

TABLE 9.5 Continued

S	4	5	6	7	8	9	10	11	12	13	14	15	16	17	18	19	20
±58													0.004	0.009			0.032
±59											0.000	0.001			0.013	0.021	
±60													0.003	0.007			0.027
±61											0.001				0.011	0.017	
±62													0.002	0.005			0.023
±63											0.001				0.009	0.014	
±64													0.002	0.004			0.020
±65											0.000				0.007	0.012	
±66													0.001	0.003			0.017
±67															0.005	0.010	
±68													0.001	0.002			0.014
±69															0.004	0.008	
±70													0.001	0.002			0.012
±71															0.003	0.006	
±72													0.000	0.001			0.010
±73															0.003	0.005	
±74														0.001			0.008
±75															0.002	0.004	
±76														0.001			0.007
±77															0.001	0.003	
±78														0.000			0.006
±79															0.001	0.003	
±80																	0.005
±81															0.001	0.002	
±82																	0.004
±83															0.001	0.002	
±84																	0.003
±85															0.000	0.001	
±86																	0.002
±87																0.001	
±88																	0.002
±89																0.001	
±90																	0.002
±91																0.001	
±92																	0.001
±93																0.000	
±94																	0.001
±95																	
±96																	0.001
±97																	
±98																	0.001
±99																	
±100																	0.000

Source: Adapted from Hollander and Wolfe (1999). Used with permission.

Because the S value is negative, we can conclude that a decreasing trend in concentration over time is present at the 90% confidence level. Whether this result is "significant" would depend upon the significance level (α) specified for the test.

The Mann–Kendall test is robust to missing data points and nondetect values. Missing data points are simply ignored because they do not influence the test result. Nondetect values are replaced with a common value less than the smallest concentration value in the data series. If multiple detection limits are involved, the data must be further censored at the highest detection limit (Helsel and Hirsch, 2002). This decreases the power of the test to detect trends due to the increased number of tied values, but the impact in most situations involving small data sets is not significant. If the number of tied values is a significant proportion of the data series, the tie correction for the large-sample approximation described subsequently can be used.

In the unusual circumstance that more than 40 data points are available, a modification of the Mann–Kendall test based on the normal approximation can be used. This version of the Mann–Kendall test uses "Z" as the test statistic. The test is performed by calculating the S-statistic for the data set as described earlier. The variance of the S-statistic is then calculated as:

$$\text{VAR}(S) = \frac{1}{18}\left[n(n-1)(2n+5) - \sum_{p=1}^{q} t_p(t_p-1)(2t_p+5)\right]$$

where n is the number of data points in the data set, q is the number of groups of tied values, and t_p is the number of data points in pth group of tied values. If the calculated S is 0, the Z-statistic is also 0. Otherwise, the Z-statistic is calculated as follows:

$$Z = \begin{cases} \dfrac{S-1}{\sqrt{\text{VAR}(S)}} & \text{if } S > 0 \\ \dfrac{S+1}{\sqrt{\text{VAR}(S)}} & \text{if } S < 0 \end{cases}$$

The sign of the calculated Z indicates whether a trend is increasing (positive) or decreasing (negative). Once the Z-statistic has been calculated, it is compared with the table of null probability values for Z that can be found in most statistics texts. Critical values for the Z-statistic at probabilities for the commonly used significance levels for one-sided ($p = \alpha$) and two-sided ($p = \alpha/2$) tests are 1.29 ($p = 0.1$), 1.64 ($p = 0.05$), and 1.96 ($p = 0.025$).

A general consideration for using the Mann–Kendall test is that a nonsignificant result does not demonstrate stability because the result could be due to concentrations at the monitoring point actually being at steady-state (stable) or to the data set being inadequate to provide a statistically significant result (Barden, 2003). Failing to reject H_0 does not mean that it was "proven" that there is no trend. Rather, it is a statement that the evidence available is not sufficient to conclude that there is a trend at the specified confidence level.

A suggested approach to dealing with the issue of a nonsignificant result for the Mann–Kendall test is to use the coefficient of variation as an indication, or "test," of stability (GSI, 1998; Wiedemeier et al., 1999; Ling et al., 2003). The coefficient of variation (CV) measures the spread of a set of data as a proportion of its mean and the suggested approach concludes that a Mann–Kendall test that is not significant at the 90% confidence level where CV < 1 indicates stability. However, the coefficient of variation is a relative measure of variation described by the ratio of the sample standard deviation to the sample mean. Thus, it depends upon both values and has no implicit meaning. If the mean value is large, even a small CV can include significant variation. Data series with "low" values for CV certainly show less scatter in the data, but there is no objective basis for using a particular value of CV to determine "stability."

A useful variation on the Mann–Kendall test is a test for "homogeneity of stations" (Gilbert, 1987; Helsel and Hirsch, 2002). This test essentially pools the results for Mann–Kendall tests at individual monitoring points and allows statements to be made about consistency of trends throughout the plume or portions of the plume (e.g., whether the trends at all monitoring points are in the same direction — all increasing or all decreasing). Such a general statement about the presence or absence of monotonic trends is useful for making interpretations of the overall behavior of the entire plume or specific portions of the plume. For chlorinated solvent solute plumes, these results can be used in combination with geochemical data to discern different types of environments.

The presence of seasonal variability in ground-water concentration time series data can make discerning trends difficult because it contributes short-term variation, caused by water-level fluctuations and other seasonal effects, that appear as background noise in a Mann–Kendall test for the whole time series. If the source of the seasonal effect can be identified, one way to "remove" the effect is to normalize the concentration data to the source variable. For example, if ground-water concentrations are shown to be correlated with water levels in monitoring wells, they could be "normalized" by dividing concentrations by water levels. This is a simplistic approach and more sophisticated data normalization techniques can be used (Helsel and Hirsch, 2002).

The "Seasonal Kendall test" (Hirsch et al., 1982; Hirsch and Slack, 1984) is a modification of the Mann–Kendall test that addresses this short-term variability due to seasonality and allows evaluation of overall trends in the time series. In a seasonal Kendall test, the Mann–Kendall test is applied to each season (e.g., quarter) separately and then the results are combined for an overall test (Hirsch et al., 1982). Each season by itself may show a positive trend, none of which is significant, but the overall seasonal Kendall statistic can be quite significant. The test has all the advantages of the Mann–Kendall test, but is more robust because it removes short-term variability caused by seasonality. When successive seasons are correlated, a correction must be used based on the covariance among seasons (Hirsch and Slack, 1984).

The seasonal Kendall test consists of calculating the Mann–Kendall statistic, S, and its variance, $VAR(S)$, for the data from each season collected over a period of years. These "seasonal" statistics are then summed and the test statistic Z is calculated as described earlier using the summed values. As with the normal approximation described earlier, the sign of the calculated Z indicates whether a trend is increasing (positive) or decreasing (negative). The calculated Z-statistic then is compared with the table of null probability values for Z that can be found in most statistics texts. There is some question regarding the direct application of the standard Z table values for a small number of "seasons" and few years of sampling data (Gilbert, 1987). However, the exact distribution for the test statistic can be determined using the technique described in Hirsch et al. (1982).

A practical limitation on the use of the seasonal Kendall test for evaluating ground-water data in long-term monitoring of natural attenuation is that seasonal (e.g., quarterly) data must be available. If the monitoring frequency is changed to annual or semi-annual basis, these seasonal data may be lost. If seasonal effects are identified during site characterization, or in the early stages of the long-term monitoring program, continued quarterly monitoring may be warranted to adequately define the impact of seasonal effects on trend results and to determine the appropriate frequency for later monitoring. Additionally, the number of data points for each season and the number of seasons considered can impact the results of the seasonal Kendall test. Generally, at least 3 yr of monitoring data should be included in the analysis.

Tests for Differences between Groups of Data

Another type of statistical test that is commonly suggested for evaluating ground-water concentration data for natural attenuation is a test for significant differences between groups of data. Several nonparametric methods are available for performing such comparisons and the appropriate method depends upon the number of groups to be compared and whether the data are paired (Gilbert, 1987; Helsel and Hirsch, 2002). All of these methods are nonparametric analogs of the Student's *t*-test. These methods test whether measurements from one data set are consistently larger or smaller than those from another data set, either using relative ranks of the data or the differences.

Two-sample tests are typically used for comparing earlier data sets to those from later time periods. These can include comparing concentrations for several monitoring points at two time points or comparing concentrations from an individual monitoring point for one time period to those for another time period (e.g., quarterly monitoring results for 1 yr to those for another year). Such a comparison can essentially identify the presence or absence of a step trend in concentrations over time. Two-sample procedures should only be used when the data sets being analyzed can be naturally broken into two distinct time periods or when a known event has occurred that is likely to have resulted in a significant change in concentrations (Helsel and Hirsch, 2002). In general, the monotonic trend methods discussed previously are more appropriate.

The Mann–Whitney U-test (Mann and Whitney, 1947), also called the Wilcoxon rank sum test, is commonly suggested for the purpose of identifying step trends and has been specified in some States' regulations (e.g., New Jersey, Wisconsin). The typical application of this test is to compare concentrations from individual monitoring points for one time period to those for another time period (e.g., quarterly monitoring results for 1 yr to those for another year). The Mann–Whitney U-test is based on the assumption that the two data sets are independent, meaning that there is no natural way to pair the data. However, in the typical use of this test for evaluating natural attenuation, the data for the two groups can be considered paired by "seasons" and are not really independent. Use of the Mann–Whitney U-test should be limited to the situations noted above and where data set independence can be assured.

Data are considered paired when there is a natural way to spatially or temporally associate data values in each group. In many cases, the data involved in evaluating natural attenuation will be paired by location or by season (e.g., quarterly data). In such situations, a paired-sample test, such as the "sign test" or the "Wilcoxon signed rank test" (not to be confused with the Wilcoxon rank sum test), is more appropriate (Gilbert, 1987).

The sign test is more versatile than the Wilcoxon signed rank test because it has no distributional assumptions and can accommodate a few nondetect values. However, it has less ability to detect differences between populations. The test statistic is the number of data pairs where $x_{1i} > x_{2i}$ (the number of positive differences). However, at small sample sizes the sign test has limited utility. The Wilcoxon signed rank test is a more powerful alternative to the sign test that is more likely to detect significant differences between data sets. However, it does require that the underlying distribution is symmetrical. In some cases where the differences are not symmetric in the original units, but a logarithmic transformation of the two data sets produces symmetric differences, the Wilcoxon signed rank test is also appropriate (Helsel and Hirsch, 2002).

The Wilcoxon signed rank test involves calculating and ranking the differences (D_i) of the data pairs. The H_0 for the test is the median of the differences is zero (0). Example calculations are shown in Table 9.6 for quarterly concentration data in a monitoring well from 2 yr, and in Table 9.7 for concentration data from multiple monitoring wells

TABLE 9.6

Example Calculations for the Wilcoxon Signed-Rank Test Comparing Groups of Paired Data for Quarterly Concentration Data in a Single Monitoring Well for 2 yr (μg/l)

Quarter	Year 1 (x)	Year 2 (y)	Difference	Rank
1st	32	27	5	4
2nd	46	42	4	2.5
3rd	28	30	−2	−1
4th	30	26	4	2.5
				$W^+ = 9$

TABLE 9.7

Example Calculations for the Wilcoxon Signed-Rank Test Comparing Groups of Paired Data for Concentrations in Several Monitoring Wells for 2 yr (µg/l)

			Raw Values		Log of Values	
Well	Year 1 (x)	Year 2 (y)	Difference	Rank	Difference	Rank
MW-1	1045	890	155	8	0.070	5
MW-2	352	241	111	7	0.165	8
MW-3	256	287	−31	−6	−0.050	−3
MW-4	132	128	4	2.5	0.013	1
MW-5	46	40	6	5	0.061	4
MW-6	28	30	−2	−1	−0.030	−2
MW-7	30	25	5	4	0.079	6
MW-8	10	14	−4	−2.5	−0.146	−7
				$W^+ = 26.5$		$W^+ = 24$

for 2 yr. The difference between each pair of values ($x_i - y_i$) in the two data sets is calculated and the absolute value of the differences ($|D_i|$) is then ranked from smallest to largest. The test uses only nonzero differences, so tied values ($x_i - y_i = 0$) are deleted and the sample size is reduced by the number of tied values. When two nonzero differences are tied, the average of the ranks involved is assigned to the tied values.

The signed rank (R_i) for each pair is determined by the sign of the difference for each pair ($x_i - y_i$); "+" for a positive difference and "−" for a negative difference. The test statistic W^+ is then calculated as the sum of the positive ranks. The W^+-statistic is compared with a table of critical values for W^+ quantiles (Table 9.8). For the appropriate sample size

TABLE 9.8

Critical Test Statistic Values for the Signed-Rank Statistic W^+, $n = 4$–20

	[Reject H_0: at One-Sided α When $W^+ \leq w$ (Table Entry) (Small W)]			[Reject H_0: at One-Sided α When $W^+ \geq w'$ (Table Entry) (Large W)]		
	α-Level			α-Level		
n	0.025	0.05	0.1	0.025	0.05	0.1
4			0			10
5		0	2		15	13
6	0	2	3	21	19	18
7	2	3	5	26	25	23
8	3	5	8	33	31	28
9	5	8	10	40	37	35
10	8	10	14	47	45	41
11	10	13	17	56	53	49
12	13	17	21	65	61	57
13	17	21	26	74	70	65
14	21	25	31	84	80	74
15	25	30	36	95	90	84
16	29	35	42	107	101	94
17	34	41	48	119	112	105
18	40	47	55	131	124	116
19	46	53	62	144	137	128
20	52	60	69	158	150	141

Source: Adapted from McCornack (1965). Used with permission.

in Table 9.8 the critical values (w and w') are obtained for the significance level of the test. For a two-sided test ($p = \alpha/2$), the null hypothesis is rejected if $W^+ \leq w$ or $W^+ \geq w'$ (x tends to be larger or smaller than y). For a one-sided test ($p = \alpha$), the null hypothesis is rejected if either $W^+ \leq w$ (x tends to be smaller than y; concentrations increase) or $W^+ \geq w'$ (x tends to be larger than y; concentrations decrease).

The calculated W^+-statistic (9) for the example shown in Table 9.6 is greater than the critical value for a significant increase ($w = 0$) or less than the critical value for a significant decrease ($w' = 10$) for a one-sided test at the 90% confidence level ($\alpha = 0.1$) for the sample size, n (4). The null hypothesis of no increase, or decrease, of concentration in this monitoring well cannot be rejected and no significant change in overall concentration is indicated at this confidence level. For the sample size in this example, the 95% confidence level for a one-sided test cannot be resolved and neither the 90 or 95% confidence levels can be resolved for a two-sided test. This illustrates the limitation of small sample sizes for such tests.

In the example shown in Table 9.7, the symmetry of the differences for the data pairs is questionable. Recalculating the differences using the logarithms of the data values, $\log(x_i) - \log(y_i)$, gives a distribution of differences that is more symmetrical. These differences are then ranked as described earlier and the W^+-statistic is calculated. The calculated W^+-statistic (24) for the example shown in Table 9.7 is greater than the critical values for w and less than the critical values for w' at the 90% ($\alpha = 0.1$) and 95% ($\alpha = 0.05$) confidence levels for the sample size, n (8). This indicates a non-significant result for either a one- or two-sided test at these confidence levels so the null hypotheses would be accepted and no significant change in overall concentrations for these monitoring wells is indicated.

Using Statistical Results

The use of results from statistical tests in evaluating the performance of a natural attenuation remedy allows quantifiable patterns in contaminant concentrations over time to be determined. These can provide insight into solute plume behavior and changes over time in different parts of the solute plume that reflect the performance of natural attenuation. An important note is that none of the statistical tests described earlier are tests for solute plume stability; none presently exist. In evaluating solute plume stability, it is important to combine statistical results with observations of the solute plume boundaries. The presence or absence of statistically significant trends in concentration over time at monitoring points do not necessarily translate into spatial changes in solute plume configuration. The lack of statistically significant trends in concentration over time can generally be taken to represent a steady-state condition at a given monitoring point, but this implies nothing about solute plume behavior. Consideration of results at all the monitoring points is necessary.

In evaluating statistical results for concentration data, it is necessary to consider all of the performance monitoring points. Depending on the dynamics of mass transfer from the source and the specific natural attenuation processes involved, different portions of a solute plume may exhibit different types of behavior (Figure 9.6). No single monitoring point can provide statistical results that are definitive because different monitoring points will be located in different geochemical environments that impact the ambient degradation and transformation processes.

A general consideration for the use of statistical methods in identifying trends and evaluating solute plume behavior is that statistical significance does not necessarily imply real-world significance and statistical test results can provide a false sense of assurance regarding conclusions (Barden, 2003). It is important to always relate statistical results and evaluation back to the physical problem in the field to ensure that the results are meaningful. Changes in concentration and trends in concentration time series should be evaluated in the context of the scientific understanding of the relevant natural attenuation processes. The point is to be able to explain why the observed patterns

(Figure 9.6) indicated by the statistical results are occurring. The reason for a "statistically significant" change in concentrations is not provided by the statistics themselves.

As an example, consider the results from tests for step trends. Comparison of concentration data for two successive years does not imply that the result is meaningful. The fact that concentrations in the second year are lower (or higher) than those from the previous year only demonstrates a "statistically significant" difference. This does not imply that data from subsequent years would produce the same result. A fundamental flaw in this sort of analysis is that 2 yr of data in most hydrogeologic settings is not a very large amount and the resulting evaluation may not be substantive in the real world.

A consideration with the seasonal Mann–Kendall test is that trends of opposite sign in different seasons may offset each other, giving the impression that no trends are present. This is typically not a substantive concern because the point of the test is to determine overall trends in the data series that may help to describe solute plume behavior. However, the individual seasonal trends may be of importance for helping to unravel relationships between parameters, in which case they could be examined individually in more detail.

Similarly, it is common for the test for "homogeneity of stations" to show no significant overall trend, even though trends are significant within contiguous portions of the solute plume. Careful consideration of how monitoring points should be grouped is necessary to evaluate portions of the solute plume. Graphical evaluation of the data combined with a scientific understanding of the problem should be a good guide on how to group contiguous monitoring points for statistical analysis.

Evaluation of Ground-Water Geochemical and Supplemental Data

The ground-water geochemical data collected during validation monitoring and subsequent long-term monitoring should be evaluated to:

(1) Demonstrate that natural attenuation, and specifically degradation, is occurring according to expectations.
(2) Detect changes in environmental conditions (e.g., hydrogeologic, geochemical, microbiological, or other changes) that may reduce the efficacy of the natural attenuation process.

The interpretation of geochemical data as they apply to degradation of fuel hydrocarbons is discussed in detail by Wiedemeier et al. (1995, 1999). The interpretation of geochemical data as they apply to degradation of chlorinated solvents is discussed in detail by U.S. EPA (1998) and Wiedemeier et al. (1999, 2005).

The evaluation of ground-water geochemical data during long-term monitoring of natural attenuation is similar to that in site characterization monitoring. The same basis for interpretation is used that is described in the various protocols for evaluating natural attenuation (e.g., Wiedemeier et al., 1995, 1999, 2005; U.S. EPA, 1998). However, the focus during long-term monitoring is on using the geochemical parameters to help explain observed changes in contaminant concentrations and solute plume behavior.

Ground-water geochemical data (Table 9.2) and supplemental data (Table 9.3) can be useful for providing ongoing information on conditions in and around the solute plume. This information is used to provide a mechanistic interpretation of plume behavior (i.e., why the observed changes in contaminant concentration are occurring).

Geochemical data collected during long-term monitoring should be evaluated to determine changes in parameters associated with significant site-specific transformation or degradation processes and changes in background conditions that might impact natural attenuation processes. The interpretation of geochemical data relies on the availability of an established baseline that includes the range of variation in the parameter values. The significance of changes in geochemical parameter values depends on the expected variation based on existing observations. If such a baseline is not available from the site characterization or natural attenuation evaluation (feasibility study), ground-water geochemical data collected during validation monitoring should be evaluated to determine consistency with the data collected during the site characterization and expected variability in the parameters.

Sampling of geochemical parameters in upgradient, background locations can provide an "early warning" of changes that might adversely impact solute plume stability. In most cases, sampling of geochemical parameters alone is adequate because they are more sensitive to potential problems than sampling for contaminants. For example, if a hydrocarbon solute plume undergoing MNA is stable or receding and a new hydrocarbon release occurs upgradient, the development of the new solute plume could deplete available dissolved oxygen, nitrate, and sulfate in ground water. The observation of a sustained reduction in concentrations of dissolved oxygen, nitrate, or sulfate or the sustained increase in Mn(II), Fe(II), alkalinity or methane concentrations in upgradient, background monitoring locations would indicate that the geochemical "shadow" of the new solute plume is encroaching on the original plume. Such a situation would be expected to affect the efficacy of natural attenuation in the original hydrocarbon solute plume by affecting the dynamics of biodegradation. Whether this would substantively affect the MNA remedy would depend on the site-specific circumstances, but some readjustment of the solute plume would likely occur in response to the changed conditions and reevaluation would be warranted.

Similarly, the geochemical indicators of a sustained reduction in concentrations of dissolved oxygen, nitrate, or sulfate or the sustained increase in Mn(II), Fe(II), methane, or alkalinity concentrations in sentry or point-of-action monitoring locations can provide an "early warning" that a solute plume may be moving downgradient. A sustained increase in chloride concentrations could provide a similar indicator for chlorinated solvent solute plumes.

Evaluation of Daughter Product Data

Concentrations of chlorinated solvents and their transformation products give a direct indication of the presence or absence of transformation processes. In many cases the production of *cis*-1,2-dichloroethene (*cis*-1,2-DCE), vinyl chloride (VC), and chloride ions along ground-water flow paths is direct evidence of biodegradation. For example, if trichloroethane (TCE) was the only contaminant released at a site, then any *cis*-1,2-DCE or VC present at the site must have come from the degradation of the parent TCE. In some cases, the presence of *cis*-1,2-DCE and lack of VC may be indicative of abiotic degradation.

Evaluation of Electron Acceptor Data

Naturally occurring electron acceptors affect the degradation of petroleum hydrocarbons and chlorinated solvents in different ways. In general, the more electron acceptors present in ground water contaminated with petroleum hydrocarbons, the better. This is because microbes consume these compounds while degrading the hydrocarbons. In contrast, naturally occurring electron acceptors can compete with reductive dechlorination of chlorinated solvents, thus reducing the efficiency of the reaction.

The stabilization of hydrocarbon solute plumes typically is controlled by naturally occurring biodegradation processes that use naturally occurring inorganic electron acceptors. The scenario of a hydrocarbon solute plume "running out" of electron acceptors to support biodegradation is a common concern on the part of regulators.

For all intents and purposes, electron acceptors dissolved in ground water, such as dissolved oxygen, nitrate and sulfate, will be readily available unless there is a change in background ground-water geochemistry that depletes these constituents. Similarly, methanogenesis is effectively self-perpetuating because it is driven by fermentation reactions that only require reduced organic carbon, so methane would be expected to be present as long as fermentable organic matter is available. However, if the major biodegradation reaction is iron reduction or manganese reduction that relies on bioavailable solid-phase Fe(III) or Mn(IV), there is a limited *in situ* supply available. Depending upon the mass of hydrocarbon present, it is possible to use up the bioavailable electron acceptor, resulting in a cessation of these reactions. This can cause a solute plume that was stable to shift position downgradient (Cozzarelli et al., 2001). A similar, though much less likely, situation could be envisioned for a solute plume with sulfate reduction as the predominant terminal electron-accepting process where the sulfate is produced by *in situ* dissolution of sulfate minerals.

Such a situation can cause consternation if the underlying natural attenuation processes are not understood and the significant biodegradation processes monitored using geochemical parameters. The depletion of bioavailable solid-phase Fe(III) or Mn(II) would be accompanied by a steady decrease in measured Fe(II) or Mn(II) concentrations in parts of the plume where they were previously elevated as the soluble, reduced ions are transported downgradient and precipitated as mineral phases.

An important consideration for evaluating geochemical parameters is that Fe(II), Mn(II), and methane are mobile in ground water. Therefore, the detection of these constituents at a given monitoring location is not necessarily indicative of iron-reducing, manganese-reducing, or methanogenic conditions at that location; instead, detection of these constituents could indicate that such conditions are present upgradient from the monitoring location.

In the case of solute plumes derived from chlorinated solvents, the redox conditions in the aquifer are of paramount importance for controlling what transformation reactions will occur. Nitrate, Fe(III) and sulfate are electron acceptors that compete with dehalorespiration. The presence of nitrate, Fe(II) and sulfate could indicate conditions where biological reductive dechlorination may not occur or be inefficient. However, less-chlorinated compounds, such as DCE and VC, can be mineralized through direct oxidation by bacteria under iron-reducing conditions (Bradley and Chapelle, 1996, 1997). The occurrence of elevated Fe(II) together with depleted sulfate concentrations is indicative that the ground water is sulfate reducing. In this case, biological and abiotic reductive dechlorination reactions may be important.

Dissolved Oxygen

Dissolved oxygen is the favored electron acceptor used by microbes for the biodegradation of many forms of organic carbon. Strictly anaerobic bacteria generally cannot function at dissolved oxygen concentrations greater than about 0.5 mg/l and hence Fe(III) reduction, sulfate reduction, methanogenesis, and reductive dechlorination (biological or abiotic) cannot occur. This is why it is important to have a source of carbon in the aquifer that can be used by aerobic microorganisms as a primary substrate. During aerobic respiration, dissolved oxygen concentrations decrease and the aquifer quickly becomes anaerobic. The concentration of dissolved oxygen in an aquifer is a very important parameter for determining if the system is capable of supporting the degradation of chlorinated solvents.

Dissolved oxygen measurements should be taken during well purging and immediately before sample acquisition using a direct-reading meter, preferably in a flow-through cell. Each of these measurements should be recorded. Because many well purging techniques can allow aeration of collected ground-water samples, it is important to minimize the potential for aeration. Because of the difficulty in obtaining accurate dissolved oxygen measurements, especially when the concentration falls below about 1 mg/l, these measurements should be used in a qualitative manner. One use of dissolved oxygen measurements is during well purging. Stabilization of dissolved oxygen concentrations, in conjunction with pH, temperature, and conductivity, can be useful during well purging to determine when the well has been purged sufficiently to provide representative samples.

Measurements of dissolved oxygen should always be interpreted with an eye toward possible sampling errors and should never be relied upon alone or interpreted without consideration of other geochemical parameters, particularly Fe(II), sulfate, methane, and ORP. Inconsistencies between these parameters and dissolved oxygen measurements almost invariably indicate aeration of the sample. In the authors' experience, more time has been wasted in dealing with misinterpretation of spurious dissolved oxygen measurements than any other single parameter. Due to its reactivity with dissolved oxygen, the presence of Fe(II) is a strong indicator of anaerobic conditions in the aquifer. With these observations in mind, if Fe(II) concentrations are elevated, sulfate concentrations are depleted, and methane concentrations are elevated within the solute plume and dissolved oxygen concentrations greater than between about 0.5 and 1 mg/l were measured within the plume, then the dissolved oxygen measurements should be viewed with a high degree of skepticism and in many cases should be discarded.

If dissolved oxygen is present in the aquifer, the measurement of reduced dissolved gasses such as sulfide and methane should not be undertaken. The reason for this is that the presence of dissolved oxygen precludes the formation of these gasses.

Nitrate

After dissolved oxygen has been depleted in the microbiological treatment zone, nitrate is used as an electron acceptor for anaerobic biodegradation of organic carbon via denitrification. During denitrification, nitrate concentrations measured in ground water decrease. Thus, nitrate concentrations below background in areas with dissolved contamination provide evidence for denitrification. Denitrification is a reaction that competes with reductive dechlorination. The absence of nitrate is a prerequisite for iron and sulfate reduction, so it is important that this compound is absent in ground water for biological and abiotic reactions to proceed.

Sulfate and Sulfide

Sulfate is used as an electron acceptor for anaerobic biodegradation during sulfate reduction wherein sulfate (SO_4^{2-}) is reduced to sulfide (HS^- or H_2S). During this process, sulfate concentrations measured in ground water decrease and sulfide is produced. The sulfide produced during sulfate reduction is very reactive and in most cases is quickly complexed with Fe(II) and solid-phase iron minerals. From the standpoint of chlorinated solvent degradation, sulfate reduction is important for two reasons: (1) reductive dechlorination caused by biological processes does not become efficient until the dominant terminal electron accepting process is sulfate reduction or methanogenesis and (2) sulfate reduction is important for abiotic mechanisms of reductive dechlorination because it results in the production of sulfide. High sulfate concentrations will likely have the following ramifications:

(1) They will reduce the efficiency of biological reductive dechlorination because sulfate is a competing electron acceptor.
(2) They will increase the efficiency of abiotic reductive dechlorination, especially if appreciable amounts of Fe(II) are present.

Evaluation of Metabolic Byproduct Data

Fe(II)

When Fe(III) is used as an electron acceptor during anaerobic biodegradation of organic carbon, it is reduced to Fe(II), which is somewhat soluble in water. Elevated Fe(II) concentrations are an indication that anaerobic degradation of organic carbon has occurred via Fe(III) reduction. The presence of Fe(II) (and sulfide) is required in order for many of the abiotic reactions described elsewhere in this document to occur. In addition, Bradley and Chapelle (1996, 1997) have shown that VC and DCE can be biologically oxidized under iron-reducing conditions. Fe(III) reduction is a reaction that competes with dehalorespiration.

Methane

As implied by the name, methanogenesis results in the production of methane during the biodegradation of organic carbon. The presence of methane in ground water is indicative of strongly reducing conditions and biologically mediated reductive dechlorination is typically very efficient under these conditions. Analysis of methane concentrations in ground water should be conducted by a qualified laboratory. It is important that the detection limit for methane be on the order of 1 µg/l, especially when evaluating the degradation of chlorinated solvents.

The presence of methane generally is indicative of a strongly reducing environment where reductive dechlorination of chlorinated ethenes to *cis*-1,2-DCE and VC and then to ethene or ethane is likely. If no VC is present then abiotic reactions should be evaluated. Methane can also be transported by advective ground-water flow. Because of this, its presence in ground water does not ensure that the immediate environment is methanogenic; only that methanogenic conditions exist in the vicinity. Evaluating the presence of methane in concert with the other geochemical indicators (e.g., Fe[II] and SO_4^{2-}) is essential.

Ethene and Ethane

Ethene and ethane are the end products of reductive dechlorination. Because these compounds are extremely transitory, their concentrations typically remain low with concentrations at sites with active reductive dechlorination in the order of hundreds of micrograms per liter.

Evaluation of General Ground-Water Monitoring Parameters

Oxidation–Reduction Potential (ORP)

The ORP of ground water is a measure of electron activity and is an indicator of the relative tendency of a solution to accept or transfer electrons. Oxidation–reduction reactions in ground water containing organic compounds (natural or anthropogenic) are usually biologically mediated, and therefore, the ORP of a ground-water system depends on and influences rates of degradation (both biological and abiotic). The ORP of ground water generally ranges from -400 to $+600$ mV. ORP readings should only be used on a qualitative basis. In general, the lower the ORP of ground water, the more reducing the system is, and the more likely that reductive dechlorination will be efficient.

ORP measurements can be used to provide real-time data on the location of the contaminant plume, especially in areas undergoing anaerobic biodegradation. Mapping the ORP of the ground water in the field helps the field scientist determine the approximate location of the contaminant plume. To map the ORP of the ground water in the field, it is important to have at least one ORP measurement (preferably more) from a well located upgradient from, or peripheral to, the plume. ORP measurements should be taken during well purging and immediately before and after sample acquisition using a direct-reading meter. Because most well purging techniques can allow aeration of collected ground-water samples that can affect ORP measurements, it is important to minimize potential aeration by using a flow-through cell.

pH

Bacteria generally prefer environments with a neutral or slightly alkaline pH. The optimal pH range for most microorganisms is between 6 and 8 standard units; however, many microorganisms can tolerate pHs well outside of this range. For example, pH values may be as low as 4 or 5 in aquifers with active oxidation of sulfides, and pH values as high as 9 may be found in carbonate-buffered systems (Chapelle, 1993). In addition, pH values as low as 3 have been measured for ground water contaminated with municipal waste leachates, which often contain elevated concentrations of organic acids (Baedecker and Back, 1979). In ground water contaminated with sludges from cement manufacturing, pH values as high as 11 have been measured (Chapelle, 1993).

Temperature

Ground-water temperature directly affects the solubility of oxygen and other geochemical species. For example, dissolved oxygen is more soluble in cold water than in warm water. Ground-water temperature also affects the metabolic activity of bacteria. Rates of hydrocarbon biodegradation roughly double for every 10°C increase in temperature (the "Q_{10}" rule) over the temperature range between 5 and 25°C. However, in the authors' experience, the temperature of ground water rarely is a limiting factor for degradation of organic compounds. For example, degradation of these compounds has been observed at ground-water temperatures as low as 34°F and as high as 85°F.

Conductivity

Conductivity is a measure of the ability of a solution to conduct electricity. The conductivity of ground water is directly related to the concentration of ions in solution; conductivity increases as ion concentration increases. The conductivity of ground water emanating from a landfill or other waste unit may be significantly different from that of native ground water or surface water. Thus, the conductivity of ground water in the plume may be a useful indicator of the ground-water flow path and may indicate that a plume-resident tracer is present.

Evaluation of Supplemental Data

Supplemental data should only be collected for sites where the operant mechanisms of natural attenuation are not obvious. For sites contaminated with petroleum hydrocarbons the collection of supplemental data will only very rarely, if ever, be required. For sites contaminated with chlorinated solvents, the collection of supplemental data will be required only on rare occasions. One example of where supplemental data may be useful is for a site where the degradation of PCE appears to "stall" at *cis*-1,2-DCE (i.e., no VC or ethene and ethane are being produced). This could be caused by at least two scenarios: (1) the system does not contain the microbial consortium required to completely degrade the PCE to ethene; or (2) the *cis*-1,2-DCE that is being produced may be degraded by abiotic mechanisms that bypass the production of VC and convert the chlorinated compounds to acetylene and ethene. Even without supplemental data, one may be able to deduce the operant mechanisms by evaluating plume stability if an adequate historical database is available. For sites without significant historical data, supplemental data may be valuable.

Supplemental Daughter Product Data

Acetylene: Acetylene is a product of the abiotic dechlorination of chlorinated aliphatic hydrocarbons (e.g., PCE and TCE) by iron sulfides. Although the exact pathway has not been fully determined, it is thought that the pathway for TCE oxidation is via the *cis*-dichlorovinyl radical directly to acetylene (Butler and Hayes, 1999). Therefore, its presence suggests that abiotic dechlorination is occurring. Practical field experience has shown that the volatile and labile nature of acetylene often precludes its detection. Therefore, the absence of detectable concentrations of acetylene does not indicate that abiotic reactions are not occurring. Research is underway to provide a means of preserving samples to be analyzed for acetylene so that laboratory analysis is possible.

Supplemental Geochemical Data

In some cases additional geochemical data can be useful for evaluating the predominant geochemical environment in ground water. Table 9.3 summarizes some of the supplemental data that may be useful for evaluating natural attenuation.

Mn(II): When Mn(IV) is used as an electron acceptor during anaerobic biodegradation of organic carbon, it is reduced to Mn(II). Mn(II) concentrations can be used as an indicator that anaerobic degradation of organic carbon has occurred via Mn(IV) reduction. Changes in Mn(II) concentrations inside the contaminant plume versus background concentrations can be used to estimate the mass of contaminant that has been biodegraded by Mn(IV) reduction. Mn(IV) reduction is a reaction that competes with reductive dechlorination. In addition, manganese can react with the hydrogen sulfide created during sulfate reduction, which could result in the formation of abiotically reactive manganese sulfide minerals.

Carbon Dioxide: Metabolic processes operating during biodegradation of organic compounds leads to the production of carbon dioxide (CO_2). However, CO_2 released into ground water rapidly reacts to form carbonic acid (H_2CO_3) and its dissociated ions. Accurate measurement of the amount of carbon dioxide produced during biodegradation is difficult because carbonate in ground water (measured as alkalinity) serves as both a source and sink for free carbon dioxide. If the carbon dioxide produced during metabolism is not completely removed by the natural carbonate buffering system of the aquifer, carbon dioxide concentrations higher than background may be observed. However CO_2 measurements alone typically are uninformative.

Alkalinity: Biologically active portions of a dissolved contaminant plume typically can be identified by an increase in alkalinity. This increase in alkalinity is brought about by the production of carbon dioxide during the biodegradation of organic carbon. Alkalinity results from the presence of hydroxides, carbonates, and bicarbonates of cations such as calcium, magnesium, sodium, and potassium. These species result from the dissolution of rock (especially carbonate rocks), the transfer of carbon dioxide from the atmosphere, and respiration of microorganisms. Alkalinity is important in the maintenance of ground-water pH because it buffers the ground-water system against acids generated during both aerobic and anaerobic biodegradation. In general, areas with reduced organic carbon exhibit a total alkalinity that is higher than that seen in those areas with low organic carbon concentrations. This is expected because the microbially mediated reactions involved in biodegradation of organic carbon cause an increase in the total alkalinity in the system. Changes in alkalinity are most pronounced during aerobic respiration, denitrification, Fe(III) reduction, and sulfate reduction, and less pronounced during methanogenesis (Morel and Hering, 1993).

Dissolved Organic Carbon: Dissolved organic carbon of anthropogenic or natural origin represent an important parameter at sites impacted with chlorinated solvents because it is a necessary ingredient in chlorinated solvent degradation. Thus, its presence and relative concentration is an important parameter for periodic monitoring. A statistically significant decline over time or space may indicate that conditions less conducive to biotic or abiotic reductive dechlorination may be forthcoming.

Dissolved Hydrogen: Concentrations of dissolved hydrogen can be used to evaluate terminal electron-accepting processes in ground-water systems (Lovley and Goodwin, 1988; Lovley et al., 1994; Chapelle et al., 1995). Because each terminal electron-accepting process has a characteristic hydrogen concentration associated with it, hydrogen concentrations can be an indicator of predominant terminal electron-accepting processes. These characteristic ranges are as follows:

Aerobic respiration	0 nM
Denitrification	0.03–0.1 nM
Iron reduction	0.2–1 nM
Sulfate reduction	1–5 nM
Methanogenesis	>5 nM

ORP measurements are based on the concept of thermodynamic equilibrium and, within the constraints of that assumption, can be used to evaluate terminal electron-accepting processes in ground-water systems. The use of dissolved hydrogen to classify the system is based on the ecological concept of interspecies hydrogen transfer by microorganisms and, within the constraints of that assumption, can also be used to evaluate terminal electron-accepting processes. These methods, therefore, are fundamentally different. A direct comparison of these methods (Chapelle et al., 1997) has shown that while ORP measurements were effective in delineating oxic from anoxic ground water, they could not reliably distinguish between nitrate-reducing, Fe(III)-reducing, sulfate-reducing, or methanogenic zones in an aquifer. In contrast, the measurement of dissolved hydrogen could readily distinguish between different anaerobic zones. At those sites where distinguishing between different anaerobic processes is important (such as at sites contaminated with chlorinated solvents), hydrogen measurements can be useful for delineating the distribution of terminal electron-accepting processes.

In practice, it is preferable to interpret hydrogen concentrations in the context of electron acceptor [dissolved oxygen, nitrate, Mn(IV), Fe(III), sulfate] availability and the presence of the final products [Mn(II), Fe(II), hydrogen sulfide, methane] of microbial metabolism (Chapelle et al., 1995). For example, if sulfate concentrations in ground water are less than 0.5 mg/l, methane concentrations are greater than 0.5 mg/l, and hydrogen concentrations are in the 5–20 nM range, it can be concluded with a high degree of certainty that methanogenesis is the predominant terminal electron-accepting process in the aquifer. Similar logic can be applied to identifying denitrification (presence of nitrate, hydrogen <0.1 nM), Fe(III) reduction [production of Fe(II), hydrogen 0.2–0.8 nM], and sulfate reduction (presence of sulfate, production of sulfide, hydrogen 1–4 nM).

Chapelle et al. (1997) compare three methods for measuring hydrogen concentrations in ground water; a downhole sampler, a gas-stripping method, and a diffusion sampler. The downhole sampler and gas-stripping methods gave similar results. The diffusion sampler appeared to overestimate hydrogen concentrations. Of these methods, the gas-stripping method is better suited to field conditions because it is faster (approximately 30 min for a single sample collection as opposed to 2 h for the downhole sampler and 8 h for the diffusion sampler), the analysis is easier (less sample manipulation is required), and the data computations are more straightforward (hydrogen concentrations need not be corrected for water sample volume) (Chapelle et al., 1997). At least one commercial laboratory uses the gas-stripping method (called the "bubble-strip" method) for hydrogen sampling and analysis.

Chloride: During biodegradation of chlorinated hydrocarbons dissolved in ground water, chloride is released into the ground water, resulting in the accumulation of biogenic chloride. This results in chloride concentrations in ground water in the contaminant plume that are elevated relative to background concentrations. In aquifers with low background concentrations of chloride, the concentration of this material in the solute plume can be seen to increase as chlorinated solvents are degraded. Although site-specific, chlorinated solvent concentrations must be above about 10 mg/l and display significant reductions to raise dissolved chloride concentrations above background levels at sites with "low" background concentrations of chloride. Other anthropogenic sources of elevated chloride (e.g., road salt, landfill, evaporation ponds, brine disposal, etc.) can still be useful in assessing ground-water flow paths and dispersion effects.

Elemental chlorine is the most abundant of the halogens. Although chlorine can occur in oxidation states ranging from Cl^- to Cl^{7+}, the chloride ion (Cl^-) is the only form of major significance in natural waters (Hem, 1985). Chloride forms ion pairs or complex ions with some of the cations present in natural waters, but these complexes are not strong enough to be of significance in the chemistry of fresh water (Hem, 1985). The chemical behavior of chloride is neutral. Chloride ions generally do not enter into oxidation–reduction reactions, form no important soluble complexes with other ions (unless the chloride concentration is extremely high), do not form salts of low solubility, are not significantly adsorbed on mineral surfaces, and play few vital biochemical roles (Hem, 1985). Thus, physical processes control the migration of chloride ions in the subsurface. Kaufman and Orlob (1956) conducted tracer experiments in ground water, and found that chloride moved through most of the soils tested more conservatively (i.e., with less retardation and loss) than any of the other tracers tested. Because of the neutral chemical behavior of chloride, it can be used as a conservative tracer to estimate biodegradation rates.

Microcosm Studies

Although several types of microbiological data may be used, the most common type of data collected for evaluating the degradation of organic contaminants in aquifer material is the laboratory microcosm study. If properly designed, implemented, and interpreted, microcosm studies can provide very convincing documentation of the potential for biodegradation and abiotic reductive dechlorination. Microcosm studies are the only "line of evidence" that allows an unequivocal mass balance on the biodegradation of environmental contaminants. If the microcosm study is properly designed, it will be easy for decision makers with nontechnical backgrounds to understand. The results of a microcosm study are strongly influenced by the nature of the geological material submitted for study, the physical properties of the microcosm, the sampling strategy, and the duration of the study. Therefore, relating laboratory microcosm results back to *in situ* field conditions can be difficult. Additionally, microcosm studies are time consuming and expensive to conduct. For these reasons, microcosm studies should be used very selectively in assessing the efficiency of natural attenuation and enhanced remediation.

There are some circumstances, however, when laboratory studies are useful. When specific questions are raised concerning *conditions* under which degradation processes occur or do not occur, controlled laboratory studies are often required. For example, if concentrations of a particular compound are observed to decrease in the field, it is often not clear whether this decrease is due to sorption, dilution, or biological or abiotic degradation. Laboratory studies in which the effects of each process can be isolated and controlled (they usually cannot be controlled in the field) are the only available method of answering these questions.

Volatile Fatty Acids (VFAs)

The VFAs pyruvate, lactate, formate, acetate, propionate, and butyrate are used as biomarkers of anaerobic metabolism. Anaerobic bacteria produce these compounds by fermentation, while under aerobic conditions these compounds are rapidly oxidized for carbon and energy by aerobic bacteria. The VFAs are analyzed by ion chromatography and represent a specialized method. The presence of these compounds is an indication that fermentation is occurring and that the environment may be conducive for reductive dechlorination.

Phospholipid Fatty Acids (PLFAs)

Examining the PLFAs in environmental samples provides an indication of the different types of bacteria that may be present at a site. Distinct classes of microbes have different cell membrane compositions. PLFAs are essential components of the membranes of all cells (except for the Archaea), so their sum includes most of the important actors in microbial communities. Methanogens are members of the Archaea and are not included in this analysis. There are four different types of information in PLFA profiles — biomass, community structure, diversity, and physiological status. Thus, PLFA analyses may be useful indicator of the presence of certain classes of microbes. This information may be used qualitatively to correlate that the observed phospholipid profile observed at the site is consistent with a particular class of microorganism with a unique and interesting metabolic capability (e.g., sulfate reduction).

Biomass: PLFA analysis is purported to be a reliable and accurate method available for the determination of viable microbial biomass. Because phospholipids break down rapidly upon cell death (White et al., 1979; White and Ringelberg, 1995), the PLFA biomass does not contain "fossil" lipids of dead cells. The sum of the PLFAs, expressed as picomoles (1 pmol = 1×10^{-12} mol), is proportional to the number of cells. The proportions used typically are taken from cells grown in laboratory media, and vary somewhat with the type of organism and environmental conditions. Starving bacterial cells have the lowest cells/pmol, and healthy eukaryotic cells have the highest. Biomass can be useful for evaluating the possibility for reductive dechlorination. If biomass appears low, but evidence of active reductive declination is high and credible, then PLFA data is uninformative. If evidence of active reductive dechlorination is low and PLFA data suggests low biomass, then there may be microbial limitations and unfavorable geochemical conditions present.

Community Structure: The PLFAs in an environmental sample is the sum of the microbial community's PLFAs, and reflects the proportions of different organisms in the sample. PLFA profiles are routinely used to classify bacteria and fungi (Tighe et al., 2000) and are one of the characteristics used to describe new bacterial species (Vandamme et al., 1996). Broad phylogenic groups of microbes have different fatty acid profiles, making it possible to distinguish among them (Edlund et al., 1985; Dowling et al., 1986; White et al., 1996, 1997). Because reductive chlorination results from the work of a microbial consortium, community structure can be useful for evaluating the possibility for reductive dechlorination.

Diversity: The diversity of a microbial community is a measure of the number of different organisms and the evenness of their distribution. Natural communities in an undisturbed environment tend to have high diversity. Contamination with toxic compounds will reduce the diversity by killing all but the resistant organisms. The addition of a large amount of a food source will initially reduce the diversity as the opportunists (usually Proteobacteria) over-grow organisms less able to reproduce rapidly. The formulas used to calculate microbial community diversity from PLFA profiles have been adapted from those applied to communities of macro-organisms (Hedrick et al., 2000). Because reductive dechlorination results from the work of a microbial consortium, an analysis of microbial diversity can be useful for evaluating the possibility for reductive dechlorination.

Physiological Status: The membrane of a microbe must adapt to the changing conditions of its environment, and these changes are reflected in the PLFA. Toxic compounds or environmental conditions that disrupt the membrane cause some bacteria to make *trans* fatty acids from the usual *cis* fatty acids (Guckert et al., 1986). Many Proteobacteria and others respond to starvation or highly toxic conditions by making cyclopropyl (Guckert et al., 1986) or mid-chain branched fatty acids (Tsitko et al., 1999). The physiological status biomarkers for toxic stress and starvation or toxicity are formed by dividing the amount of the stress-induced fatty acid by the amount of its biosynthetic precursor.

Denaturing Gradient Gel Electrophoresis (DGGE)

The recovery of DNA and RNA and its subsequent analysis after amplification by polymerase chain reaction (PCR) provides a powerful tool for characterizing microbial community structure that complements the PLFA analysis. As with PLFA analysis, numerous studies have used PCR amplification of ribosomal RNA genes (rDNA) to characterize microbial populations in a number of different environments and have demonstrated that the dominant microorganisms isolated by culture frequently do not match those identified by molecular techniques (Amann et al., 1995). Given that often only 0.1–10% of visually countable bacteria in samples are cultured and previous studies have demonstrated that organisms obtained from culturing are not necessarily the numerically dominant organisms *in situ*, it is apparent that the results from culture-based community structure assessments can be noticeably incomplete.

DGGE analysis can be used to detect and identify organisms from a whole community of organisms and thus can be used to determine if the requisite microbes for reductive dechlorination are present. The DGGE approach directly determines the species composition of complex microbial assemblages based on the amplification of conserved gene sequences (16S rDNA fragments for prokaryotes, 18S or 28S rDNA for eukaryotes). In DGGE analysis, differences in gene sequences among organisms allow DNA from various organisms to be physically separated in a denaturing gradient gel, thereby allowing one to generate profiles of numerically dominant bacterial community members for a sample. The profiles are visible as bands (or lines) in a gel. The banding patterns and relative intensities of the bands provide a measure of difference among the communities. Gel bands from dominant species, which constitute at least 1% of the total bacterial community, can be excised and sequenced. Sequence analysis of individual bands is used to infer the identity of the source organism based on database searches and phylogenetic methods. Phylogenetic affiliations are determined by comparing the rDNA sequences retrieved from samples to rDNA sequences of known bacterial sequences in national databases, such as the Ribosomal Database Project (RDP) or GenBank.

Practical Considerations for Microbial Characterization Techniques

Microbial characterization techniques like those mentioned earlier are only indicated when quantitative evidence in the form of contaminant mass loss over time and space and confirmatory geochemical data are limited, conflicting, or indicate that a site-specific deficiency exists. Microbes tend to exist and thrive on the solid matix. Interestingly, ground-water sampling techniques have evolved to collect a clean, clear, sediment-free sample. Ground water from monitoring wells is the most common sample material for the microbial characterization techniques discussed earlier. In summary, a "dirty," turbid, sediment-rich sample represents a better source of microbial biomass. Thus, a deliberate effort should be made to collect some sediment to increase the chances that a sufficient and representative amount of biomass is collected. Practitioners should verify with the laboratory staff conducting these specialized analyses that sufficient biomass and microbial DNA were obtained from site samples to complete an acceptable analysis. In other words, a sample with insufficient microbial biomass will give an inconclusive or negative result.

Stable Isotopes

Analysis of stable isotope ratios between parent and daughter compounds can be useful for identifying the biodegradation of chlorinated compounds because isotopic fractionation commonly occurs during biodegradation. This fractionation results in a characteristic pattern of isotope ratios between parent compounds and daughter products. For the chlorinated ethenes, non-destructive subsurface processes such as dissolution, sorption, and volatilization do not involve isotopic fractionation greater than 0.5‰ (Slater et al., 2001). This is the typical accuracy and reproducibility of continuous flow isotope analysis techniques (Slater et al., 2001).

Hunkeler et al. (1999) studied the occurrence of stable carbon isotope ($^{13}C/^{12}C$) fractionation during the reductive dechlorination of PCE to ethene in the field and in the laboratory using aquifer material from the same site located in Toronto, Ontario, Canada. According to these researchers, all dechlorination steps in the microcosm were accompanied by stable carbon isotope fractionation with similar results for the field study. In the microcosm study the largest fractionation occurred during dechlorination of *cis*-1,2-DCE and VC, resulting in a large enrichment of ^{13}C in the remaining *cis*-1,2-DCE and VC. Stable carbon isotope ratios ($\delta^{13}C$) of *cis*-1,2-DCE and VC increased from −25.7 to 1.5‰ and −37 to −2.5‰, respectively. The $\delta^{13}C$ of ethene was initially −60.2‰ and approached the $\delta^{13}C$ of the added PCE (−27.3‰) as dechlorination came to completion. On the basis of their work, they conclude that strong enrichment of ^{13}C in *cis*-1,2-DCE and VC during microbial dechlorination may serve as a powerful tool to monitor the last two steps of dechlorination. These steps frequently determine the rate of dechlorination of chlorinated ethenes at field sites where degradation is occurring.

Contingency Plans

A contingency plan is an integral part of a monitored natural attenuation remedy as per the U.S. EPA OSWER Directive 9200-4.17 (U.S. EPA, 1999). Interestingly, contingency remedies are specifically requested in guidance for other remedial alternatives. However, it makes good technical sense to actively evaluate remedial performance and to have a well-formulated contingency plan when a remedy fails to achieve the desired level of effectiveness or protectiveness. The purpose of a contingency plan is to define the appropriate actions to be taken in the event that natural attenuation proves inadequate to achieve remedial goals. Changing site conditions can result in variable plume behavior over time. To circumvent potential problems, a contingency plan that specifies a contingency remedy should be an integral part of the monitoring program. A contingency remedy is a cleanup technology or approach specified in the site remedy decision document that functions as a backup remedy in the event that the selected remedy fails to perform as anticipated. A contingency remedy may specify a technology (or technologies) that is (are) different from the selected remedy, or it may simply call for modification and enhancement of the selected remedy, if needed. Contingency remedies generally should be flexible to allow for the incorporation of new information about site risks and technologies. Contingency remedies should be developed where the selected technology is not proven for the specific site application, where there is significant uncertainty regarding the nature and extent of contamination at the time the remedy is selected, or where there is uncertainty regarding whether or not a proven technology will perform as anticipated under the particular circumstances of the site. The U.S. EPA (1999) recommends that remedies employing monitored natural attenuation be evaluated to determine the need for including one or more contingency measures that would be capable of achieving remediation objectives. The U.S. EPA believes that a contingency measure may be particularly appropriate for a monitored natural attenuation remedy that has been selected based primarily on predictive analysis rather than on historical trends from actual monitoring data.

One or more criteria ("triggers") that will signal unacceptable performance of the selected remedy and indicate when to implement contingency measures should be established. Such criteria might include the following (U.S. EPA, 1999, 2004):

- Increasing contaminant concentrations or trends not predicted during remedy selection or indicative of new releases
- Contaminant migration beyond established plume or compliance boundaries
- Contaminants not decreasing at a rate sufficient to meet remediation objectives
- Changes in land or ground-water use that have the potential to reduce the protectiveness of the remedy
- Contaminants observed at locations posing or having the potential to pose unacceptable risks to receptors

Care is needed when establishing triggers for contingency remedies to ensure that sampling variability or seasonal fluctuations do not unnecessarily trigger implementation of a contingency remedy. For example, an anomalous spike in dissolved concentrations at wells, that may set off a trigger, might not be a true indication of a change in trend. Trends in contaminant concentrations can be analyzed using statistical techniques.

The most common remedial systems for complementing natural attenuation are source reduction technologies. Source reduction can be an important element of site remediation if site closure or shortened monitoring time frames are desired.

It is prudent to update the contingency plan on a periodic basis as the plume attenuates or as new remediation technologies are developed. Although some engineered remediation systems may be effective in achieving plume containment, other remediation systems may have an adverse impact on degradation. Table 9.9 summarizes some of the potential interactions between remediation systems and natural attenuation. For example, the introduction of oxygen via air sparging into an aquifer contaminated with chlorinated solvents may alter the geochemistry of the ground water to the point that reductive dechlorination can no longer occur and the natural treatment system is destroyed.

TABLE 9.9
Interactions between Active Remediation Technologies and Natural Attenuation (Wiedemeier and Chapelle, 2000)

	Possible Benefits		Possible Detriments	
Technology	Petroleum Hydrocarbons	Chlorinated Solvents	Petroleum Hydrocarbons	Chlorinated Solvents
Bioslurping	Source removal, volatilization, enhanced oxygen delivery or aerobic biodegradation	Source removal, volatilization, enhanced oxidation of DCE and VC, possible enhanced aerobic cometabolism	None	Enhanced oxygen delivery or decreased reductive dechlorination
Pump and treat	Plume containment, enhanced oxygen delivery or aerobic biodegradation	Plume containment, enhanced oxidation of DCE and VC, possible enhanced aerobic cometabolism	None	Enhanced oxygen delivery or decreased reductive dechlorination
Air sparging	Volatilization, enhanced oxygen delivery or aerobic biodegradation	Volatilization, enhanced oxidation of DCE and VC, possible enhanced aerobic cometabolism	None	Enhanced oxygen delivery or decreased reductive dechlorination
Soil vapor extraction (SVE) or Bioventing	Source reduction, particularly BTEX	SVE reduces source in unsaturated zone	None	Air injection can spread recalcitrant volatiles
In-well circulation or stripping	Volatilization, enhanced oxygen delivery or aerobic biodegradation	Volatilization, enhanced oxidation of DCE and VC, possible enhanced aerobic cometabolism	None	Enhanced oxygen delivery or decreased reductive dechlorination
Landfill caps	Source containment or isolation	Source containment or isolation, reduced oxygen delivery through elimination of recharge or stimulation of reductive dechlorination	Reduced oxygen delivery or aerobic biodegradation	Decreased oxidation of DCE and VC, decreased aerobic cometabolism

(Table continued)

TABLE 9.9 Continued

Technology	Possible Benefits — Petroleum Hydrocarbons	Possible Benefits — Chlorinated Solvents	Possible Detriments — Petroleum Hydrocarbons	Possible Detriments — Chlorinated Solvents
Phytoremediation	Plant-specific transpiration or enzymatically mediated degradation, enhanced biodegradation in the rhizosphere, and plume containment	Plant-specific transpiration or enzymatically mediated degradation, enhanced biodegradation in the rhizosphere, and plume containment	None	Unknown
Excavation or backfilling	Source removal, enhanced oxygen delivery or aerobic biodegradation	Source removal, enhanced oxidation of DCE and VC, possible enhanced aerobic cometabolism	None	Enhanced oxygen delivery or decreased reductive dechlorination
Chemical oxidation (e.g., Fenton's reagent, potassium permanganate, etc.)	Enhanced oxidation	Enhanced oxidation	None	Enhanced oxygen delivery or decreased reductive dechlorination through oxidation and removal of fermentable carbon substrates. Lowered pH possibly inhibits microbial activity
Chemical reduction (e.g., sodium dithionate)	Unknown	Scavenges inorganic electron acceptors or enhanced reductive dechlorination	Scavenges inorganic electron acceptors or decreased oxidation	Decreased oxidation of DCE and VC, decreased aerobic cometabolism
Oxygen-releasing materials	Enhanced oxygen delivery or aerobic biodegradation	Enhanced oxidation of DCE and VC	None	Decreased reductive dechlorination through oxidation and removal of fermentable carbon substrates
Carbon substrate addition	None	Stimulation of reductive dechlorination	Competing carbon source	Decreased oxidation of DCE and VC, decreased aerobic cometabolism at injection point
Zero-valent iron barrier walls	Unknown	Enhanced reductive dechlorination	Unknown	None
Biological barrier walls	Unknown	Enhanced reductive dechlorination	Unknown	None

A ground-water pump-and-treat system can have the same effect by drawing oxygen-rich ground water through the contaminant plume. Because of these potential adverse affects, the impacts of any proposed remediation system on naturally occurring processes should be evaluated when developing a contingency plan.

Monitoring Duration and Exit Strategies

The duration of monitoring and the exit strategy for a long-term monitoring program are interrelated issues. Because the long-term monitoring of natural attenuation is effectively the implementation of the remedial action, the exit strategy consists of the decision criteria that will allow the long-term monitoring program to end. Defining the decision points and criteria will depend upon the specific remedial action objectives for a given site and, thus, the regulatory framework. This discussion does not purport to address all the considerations or situations that might arise at a particular site; rather it presents a practical approach to the issues that provides a framework for developing an exit strategy for a site.

In general, the objectives of performance monitoring for natural attenuation are derived from the site-specific remedial action objectives (e.g., what the remedy is intended to accomplish) and applicable target concentrations (U.S. EPA, 2004). As with any remediation option for sites with ground-water contamination, remedial goals should be established early in the process. This will help define the specific purposes for the long-term monitoring program and should help define the length of time that monitoring will be required. Long-term monitoring should continue until remediation objectives have been achieved, and longer if necessary to verify that the site no longer poses a threat to human health or the environment (U.S. EPA, 1999, 2004). Typically, *verification* monitoring is continued for a specified period (e.g., 3–5 yr) after remediation objectives have been achieved to ensure that concentrations are stable and remain below target levels (U.S. EPA, 1999, 2004). While this sounds relatively straightforward, it means different things in different regulatory settings and presumes that the decision criteria are concentration-based.

The duration of a long-term monitoring program for natural attenuation is perhaps the most perplexing and uncertain aspect from a design standpoint. For sites with a NAPL source (the typical case), the major control on persistence of a solute plume is the source mass available to dissolve into ground water. Unfortunately, this is commonly one of the more uncertain parameters. Additional complications arise from mass-transfer limitations on source decay (Chapelle et al., 2003). Projections regarding solute plume duration, regardless of how they are developed, are only as good as the quality of the available data and, in the vast majority of cases, uncertainties of an order of magnitude are the norm. These estimates can be better refined over the course of long-term monitoring as data are developed that characterize the source decay rate (AFCEE, 2003). Even small differences in estimates of mass-transfer from the source can have significant impacts on monitoring duration in practical terms. For example, a factor of two difference in the ground-water flow velocity can change the estimated plume duration from 30 to 60 yr (Chapelle et al., 2003). This presents a quandary for practitioners who are faced with regulatory requests, as well as those from site owners, to define the remediation time frame, or how long the cleanup will take.

Typically, the remediation time frame for a site is based on achieving compliance with some concentration-based target level. While methodologies for making such estimates are available (e.g., Chapelle et al., 2003), the reality is that in most cases the results are uncertain and the time frame will be several decades, or longer, to approach concentration-based target levels. However, the ultimate remedial goals for a site (e.g., attaining MCLs throughout the solute plume) can be different than the decision criteria for continuing the long-term monitoring program. From the regulatory standpoint, the question is one of whether the site continues to warrant regulatory concern and oversight. Site "closure" is not the hard and fast determination that many perceive; rather it is a decision on the part of the regulatory agency that, based on the available information, the site warrants "no further action," with various qualifications and stipulations. Viewed in this context, the necessary duration of the long-term monitoring program is the time needed to unequivocally support a decision regarding management of the site. Simply stated, long-term monitoring should continue until the data gathered adequately support a decision for closure of the site (no further action) or the need to implement another remediation option. The monitoring program should focus on providing the data needed to support decision-making and address outstanding questions or concerns. Continuing a monitoring program past the point where the data collected are useful in supporting decisions becomes an exercise in collecting data for the sake of collecting data rather than providing necessary information. In essence, if the additional data will not change decisions regarding management of the site, there is little point in collecting it; if it will, it should be collected.

The single difference between a natural attenuation remedy and other remediation approaches is that, *if natural attenuation is effective and the conditions do not change*, it will continue to operate whether the solute plume is monitored or not. The underlying question then becomes why is the solute plume being monitored? The answer to this question will depend upon both the technical basis for the monitoring (e.g., what information is needed) and the site-specific remediation objectives and regulatory requirements. However, an important point is that the fundamental cleanup objective for most sites, the reduction of contaminant concentrations in ground water to specified levels, is not changed. The question is only whether continued monitoring is needed.

The duration of long-term monitoring and the criteria for ending the monitoring program are directly related to the purpose for monitoring natural attenuation and specific remedial goals (remedial action objectives) for a given site. These involve both technical and institutional considerations. The technical considerations involve the specific objectives for the performance monitoring; what information is needed and what changes in conditions are of interest and importance for success of the remedy. The institutional considerations involve issues related to land use, ground-water use, preventing exposure to the contaminants, and management of the site.

The technical basis for long-term monitoring will depend on the specific chemicals of concern and the natural attenuation processes that act upon them. These factors determine the conditions necessary for the various degradation and transformation processes to occur, the controls on those processes, and the changes in conditions that might impact the efficacy of the natural attenuation remedy. If the factors are well understood and solute plume behavior is adequately defined at a site, a case may be made for ending the monitoring program. For example, a solute plume from a petroleum hydrocarbon release, where all of the constituents of concern are readily mineralized by direct microbial oxidation, presents one situation. If a case can be made that environmental conditions are consistent and the available long-term monitoring data unequivocally demonstrate that the solute plume is stable or receding, the decision to end the long-term monitoring program can be supported on a technical basis. Biodegradation of the constituents of concern [e.g., benzene, toluene, ethylbenzene, and xylenes (BTEX)] to innocuous products would continue under both anaerobic and aerobic conditions and, over time, ultimately eliminate the solute plume as the source is depleted. Barring a change in conditions that directly alters the mass balance of the solute plume, such as a new release, or causes an increase in the potential for exposure to the contaminants, a cogent technical argument can be made that continued monitoring of remedy performance is not needed.

A practical way to address the issue of monitoring duration is to consider the specific concerns that are important for the efficacy of natural attenuation in a given situation. For example, consider the case of chlorinated solvent solute plumes undergoing reductive dechlorination. In the case of a solute plume where natural attenuation appears to be effective due to a strongly reducing environment resulting from a source of anthropogenic organic carbon (e.g., the Type 1 environment of Wiedemeier et al., 1999), one of the major considerations is whether there is adequate reduced organic carbon available to maintain the strongly reducing environment until all of the chlorinated solvent constituents are gone. In such a situation, long-term monitoring must continue until this can be unequivocally established; most likely until contaminant concentrations reach MCLs. The situation would be different for a similar solute plume where the reduced organic carbon is naturally occurring in the aquifer (e.g., the Type 2 environment of Wiedemeier et al., 1999) and the availability of an adequate supply of reduced organic carbon to maintain the strongly reducing conditions can be reasonably assured. In this setting, a case might be made that environmental conditions are consistent and, similar to the petroleum hydrocarbon example above with the same qualifications, a cogent technical argument might be made that continued monitoring of remedy performance is not needed. Clearly, the weight of evidence needed to support such an argument must be available from the specific site and is not trivial. These examples illustrate on a conceptual level the type of considerations that should be incorporated into the development of an exit strategy.

In many, if not most cases, the decision to stop the long-term monitoring program and "close" the site depends more on the availability and efficacy of land use and other institutional controls for site management than on the technical aspects. A sound technical case may be made for ceasing long-term monitoring of the solute plume, but long-term management questions may necessitate its continuing in some form.

In some cases, the concern may have management and technical connotations. For example, a solute plume from a petroleum hydrocarbon product release could be demonstrably stable and contained within the property boundaries of the site. If the point of compliance is the property boundary, there is no potential for another release at the site, and adequate institutional controls are in place, such a site could meet the criteria for "closure" in some situations. However, the proximity of another facility immediately upgradient of the site and the potential for a release of petroleum hydrocarbons there could raise concerns over the long-term efficacy of the natural attenuation remedy. A new release at the upgradient facility could place the site in the geochemical "shadow" of depleted electron acceptors. This may not have a substantive impact on the overall efficacy of natural attenuation at the site, but could cause a shift in location of the plume boundaries that might put it out of compliance. Such a situation could warrant continued monitoring of the site in some form, particularly at upgradient locations.

The reliability of land use and other institutional controls are an essential consideration for an exit strategy that provides for an end to long-term monitoring before the target concentrations for a site (e.g., MCLs) are reached. Because the solute plume will still be present at the site, measures to ensure that exposure to the contaminants is prevented and the site is managed appropriately are necessary. Due to the lack of uniformity in the way institutional controls are handled in different legal and regulatory jurisdictions, the specific approach to this issue at a given site will likely vary. However, effective controls on land use and ground-water use must be implemented. In the case where verification of the attainment of target ground-water concentrations is necessary, either for regulatory compliance or to justify the removal of institutional controls, an "event-driven" approach to future monitoring should be developed. Such an approach would schedule additional monitoring events when estimated plume behavior suggests that cleanup goals will be achieved.

To summarize, the focus of developing an exit strategy should be on identifying what are the specific questions and concerns at a given site and tailoring the long-term monitoring program to address those questions and concerns. The monitoring data collected should have a specific purpose in terms of elucidating plume behavior and supporting decisions regarding management of the site. As a final note, long-term monitoring should continue until it is certain that protection of human health and the environment is ensured.

References

Amann, R.I., W. Ludwig, and K.H. Schleifer, Phylogenetic identification and in-situ detection of individual microbial cells without cultivation, *Microbiological Reviews*, 59, 143–169, 1995.

AFCEE, *Remedial Process Optimization Handbook*, U.S. Air Force Center for Environmental Excellence, Brooks Air Force Base, San Antonio, TX, 2001. Available at: http://www.afcee.brooks.af.mil/products/rpo/docs/rpohandbook.pdf.

AFCEE, *Light Non-Aqueous-Phase Liquid Weathering at Various Fuel Release Sites*, U.S. Air Force Center for Environmental Excellence, Brooks Air Force Base, San Antonio, TX, 2003. Available at: http://www.afcee.brooks.af.mil/products/techtrans/download/fuelweatheringreport.pdf.

ASTM, Standard Guide for Remediation by Natural Attenuation in Ground Water at Petroleum Release Sites, ASTM Standard E 1943, ASTM International, West Conshohocken, PA, 2004, 43 pp.

Baedecker, M.J. and W. Back, Hydrogeological processes and chemical reactions at a landfill, *Ground Water*, 17(5), 429–437, 1979.

Barden, M.J., Practical Use of Statistics for Natural Attenuation Trend Analysis, Environmental Institute for Continuing Education Online Seminar #EST-0101, June, 2003.

Bradley, P.M. and F.H. Chapelle, Anaerobic mineralization of vinyl chloride in Fe(III)-reducing aquifer sediments, *Environmental Science and Technology*, 30, 2084–2086, 1996.

Bradley, P.M. and F.H. Chapelle, Kinetics of DCE and VC mineralization under methanogenic and Fe(III)-reducing conditions, *Environmental Science and Technology*, 31, 2692–2696, 1997.

Butler, E.C. and K.F. Hayes, Kinetics of the transformation of trichloroethylene and tetrachloroethylene by iron sulfide, *Environmental Science and Technology*, 33, 2021–2027, 1999.

Butler, E.C. and K.F. Hayes, Factors influencing rates and products in the transformation of trichloroethylene by iron sulfide and iron metal, *Environmental Science and Technology*, 35, 3884–3891, 2001.

Chapelle, F.H., *Ground-Water Microbiology and Geochemistry*, John Wiley & Sons, Inc., New York, NY, 1993, 424 pp.

Chapelle, F.H., P.B. McMahon, N.M. Dubrovsky, R.F. Fujii, E.T. Oaksford, and D.A. Vroblesky, Deducing the distribution of terminal electron-accepting processes in hydrologically diverse groundwater systems, *Water Resources Research*, 31, 359–371, 1995.

Chapelle, F.H., D.A. Vroblesky, J.C. Woodward, and D.R. Lovley, Practical considerations for measuring hydrogen concentrations in ground water, *Environmental Science and Technology*, 31, 2873–2877, 1997.

Chapelle, F.H., M.A. Widdowson, J.S. Brauner, E. Mendez III, and C.C. Casey, Methodology for Estimating Times of Remediation Associated With Monitored Natural Attenuation, Water Resources Investigations Report 03-4057, U.S. Geological Survey, Reston, VA, 2003, 51 pp.

Conant, B., Jr., J.A. Cherry, and R.W. Gillham, A PCE ground-water plume discharging to a river: influence of the streambed and near-river zone on contaminant distributions, *Journal of Contaminant Hydrology*, 73, 249–279, 2004.

Conover, W.J., *Practical Nonparametric Statistics*, 3rd ed., John Wiley & Sons, Inc., New York, NY, 1999, 584 pp.

Cozzarelli, I.M., B.A. Bekins, M.J. Baedecker, G.R. Aiken, R.P. Eganhouse, and M.E. Tuccillo, Progression of natural attenuation processes at a crude-oil spill site–I. Geochemical evolution of the plume, *Journal of Contaminant Hydrology*, 53(3–4), 369–385, 2001.

Dowling, N.J.E., F. Widdel, and D.C. White, Phospholipid ester-linked fatty acid biomarkers of acetate-oxidizing sulfate reducers and other sulfide-forming bacteria, *Journal of General Microbiology*, 132, 1815–1825, 1986.

Edlund, A., P.D. Nichols, R. Roffey, and D.C. White, Extractable and lipopolysaccharide fatty acid and hydroxy acid profiles from desulfovibrio species, *Journal of Lipid Research*, 26, 982–988, 1985.

Gibbons, R.D., *Statistical Methods for Ground-Water Monitoring*, John Wiley & Sons, Inc., New York, NY, 1994, 286 pp.

Gilbert, R.O., *Statistical Methods for Environmental Pollution Monitoring*, Van Nostrand Reinhold, New York, NY, 1987, 320 pp.

Guckert, J.B., M.A. Hood, and D.C. White, Phospholipid ester-linked fatty acid profile changes during nutrient deprivation of vibrio cholerae — increases in the *trans/cis* ratio and proportions of cyclopropyl fatty acids, *Applied Environmental Microbiology*, 52, 794–801, 1986.

GSI, Remediation by Natural Attenuation (RNA) ToolKit Users Manual, Groundwater Services, Inc., Houston, TX, 1998.

Helsel, D.R. and R.M. Hirsch, Statistical Methods in Water Resources, U.S. Geological Survey: Techniques of Water-Resources Investigations, Book 4, Chap. A3, 2002, 510 pp. Available at: http://water.usgs.gov/twri/twri4a3/.

Hedrick, D.B., A. Peacock, J.R. Stephen, S.J. Macnaughton, J. Brüggemann, and D.C. White, Measuring soil microbial community diversity using polar lipid fatty acid and denatured gradient gel electrophoresis data, *Journal of Microbiological Methods*, 41, 235–248, 2000.

Hem, J.D., Study and Interpretation of the Chemical Characteristics of Natural Water, Water Supply Paper 2254, U. S. Geological Survey, Reston, VA, 1985, 264 pp.

Hirsch, R.M. and J.R. Slack, A nonparametric trend test for seasonal data with serial dependence, *Water Resources Research*, 20, 727–732, 1984.

Hirsch, R.M., J.R. Slack, and R.A. Smith, Techniques of trend analysis for monthly water quality data, *Water Resources Research*, 18, 107–121, 1982.

Hollander, M. and D.A. Wolfe, *Nonparametric Statistical Methods*, 2nd Ed., John Wiley & Sons, Inc., New York, NY, 1999, 787 pp.

Hunkeler, D., R. Aravena, and B.J. Butler, Monitoring microbial dechlorination of tetrachloroethene (PCE) in ground water using compound-specific stable carbon isotope ratios — microcosm and field studies, *Environmental Science and Technology*, 33, 2733–2738, 1999.

Kampbell, D.H., and S.A. Vandegrift, Analysis of dissolved methane, ethane, and ethylene in ground water by a standard gas chromatographic technique. *Journal of Chromatographic Science*, 36, 253–256, 1998.

Kaufman, W.J. and G.T. Orlob, Measuring ground-water movement with radioactive and chemical tracers, *American Water Works Association Journal*, 48, 559–572, 1956.

Kearl, P., N. Korte, and T. Cronk, Suggested modifications to ground-water sampling procedures based on observations from the collodial borescope, *Ground Water Monitoring Review*, 12(2), 155–166, 1992.

Kendall, M.G., *Rank Correlation Methods*, 4th ed., Charles Griffith, London, U.K., 1975.

Kennedy, L., J. Everett, and J. Gonzales, Aqueous and Mineral Intrinsic Bioremediation Assessment (AMIBA) Protocol, U.S. Air Force Center for Environmental Excellence, Brooks Air Force Base, San Antonio, TX, 2000.

Ling, M., H.S. Rifai, C.J. Newell, J.J. Aziz, and J.R. Gonzales, Ground-water monitoring plans at small-scale sites: an innovative spatial and temporal methodology, *Journal of Environmental Monitoring*, 5, 126–134, 2003.

Lovley, D.R. and S. Goodwin, Hydrogen concentrations as an indicator of the predominant terminal electron-accepting reaction in aquatic sediments, *Geochimica et Cosmochimica Acta*, 52, 2993–3003, 1988.

Lovley, D.R., F.H. Chapelle, and J.C. Woodward, Use of dissolved H_2 concentrations to determine distribution of microbially catalyzed redox reactions in anoxic ground water, *Environmental Science and Technology*, 28(7), 1205–1210, 1994.

Mann, H.B., Nonparametric tests against trend, *Econometrica*, 13, 245–259, 1945.

Mann, H.B. and D.R. Whitney, On a test of whether one or more random variables is statistically larger than the other, *Annual of Mathematical Sciences*, 18, 52–54, 1947.

McCornack, R.L., Extended tables of the Wilcoxon matched pair signed rank statistic. *Journal of the American Statistical Association*, 60(311), 864–871, 1965.

Morel, F.M.M. and J.G. Hering, *Principles and Applications of Aquatic Chemistry*, John Wiley & Sons, Inc., New York, NY, 1993.

Muyzer, G., E.C. De Waal, and A.G. Uitterlinden. Profiling of complex microbial populations by denaturing gradient gel electrophoresis analysis of polymerase chain reaction-amplified genes coding for 16S rRNA. *Applied and Environmental Microbiology*, 59, 695–700, 1993.

Sherwood-Lollar, B., G.F. Slater, J. Ahad, B. Sleep, J. Spivack, M. Brennan, and P. MacKenzie, Contrasting carbon isotope fractionation during biodegradation of TCE and toluene — implications for intrinsic bioremediation, *Organic Geochemistry*, 30, 813–818, 1999.

Slater, G.F., B.S. Lollar, B.E. Sleep, and E.A. Edwards, Variability in carbon isotopic fractionation during biodegradation of chlorinated ethenes — Implications for field applications, *Environmental Science and Technology*, 35, 901–907, 2001.

Tighe, S.W., P. de Lajudie, K. Dipietro, K. Lindström, G. Nick, and B.D.W. Jarvis, Analysis of cellular fatty acids and phenotypic relationships of Agrobacterium, Bradyrhizobium, Mesorhizobium, Rhizobium and Sinorhizobium species using the Sherlock microbial identification system, *International Journal of Systems Evolution Microbiology*, 50, 787–801, 2000.

Tsitko, I.V., G.M. Zaitsev, A.G. Lobanok, and M.S. Salkinoja-Salonen, Effect of aromatic compounds on cellular fatty acid composition of *Rhodococcus opacus*; *Applied and Environmental Microbiology*, 65(2), 853–855, 1999.

U.S. EPA, Technical Protocol for Evaluating the Natural Attenuation of Chlorinated Solvents, EPA/600/R-98/128, U.S. Environmental Protection Agency, Office of Solid Waste and Emergency Response, Washington, DC, 1998.

U.S. EPA, Use of Monitored Natural Attenuation at Superfund, RCRA Corrective Action, and Underground Storage Tank Sites, OSWER Directive 9200.4-17P, U.S. Environmental Protection Agency, Office of Solid Waste and Emergency Response, Washington, DC, 1999. Available at: http://www.epa.gov/ada/download/reports/600R04027/600R04027.pdf.

U.S. EPA, Guidance for Data Quality Assessment, Practical Methods for Data Analysis, EPA QA/G-9 (QA00 Update), U.S. Environmental Protection Agency, Office of Environmental Information, Washington, DC, 2000.

U.S. EPA, Performance Monitoring of MNA Remedies for VOCs in Ground Water, EPA/600/R-04/027, U.S. Environmental Protection Agency, Office of Research and Development, National Risk Management Research Laboratory, Cincinnati, OH, 2004, 73 pp. Available at: http://www.epa.gov/ada/download/reports/600R04027/600R04027.pdf.

Vandamme, P., B. Pot, M. Gillis, P. de Vos, K. Kersters, and J. Swings, Polyphasic taxonomy, a consensus approach to bacterial systematics, *Microbiolical Review*, 60(2), 407–438, 1996.

Vroblesky, D.A., User's Guide for Polyethylene-Based Passive Diffusion Bag Samplers to Obtain Volatile Organic Compound Concentrations in Wells. Part 1: Deployment, Recovery, Data Interpretation, and Quality Control and Assurance. U.S. Geological Survey, Water-Resources Investigations Report 01-4060, 2001, 18 pp.

Vroblesky, D.A. and W.T. Hyde, Diffusion samplers as an inexpensive approach to monitoring VOCs in ground water, *Ground Water Monitoring and Remediation*, 17(3), 177–184, 1997.

Vroblesky, D.A., M.M. Loray, and P. Trimble, Mapping zones of contaminated ground water discharge using creek-bottom sediment vapor samples, Aberdeen Proving Grounds, Maryland, *Ground Water*, 25(1), 7–12, 1991.

Vroblesky, D.A., L.C. Rhodes, and J.F. Robertson, Locating VOC contamination in a fractured rock aquifer at the ground-water/surface-water interface using passive vapor collectors, *Ground Water*, 34(2), 223–230, 1996.

White, D.C. and D.B. Ringelberg, Utility of signature lipid biomarker analysis in determining in-situ viable biomass, community structure, and nutritional/physiological status of the deep subsurface microbiota, in *The Microbiology of the Terrestrial Subsurface*, P.S. Amy and D.L. Halderman, Eds., CRC Press, Boca Raton, FL, 1995.

White, D.C., W.M. Davis, J.S. Nickels, J.D. King, and R.J. Bobbie, Determination of the sedimentary microbial biomass by extractable lipid phosphate, *Oecologia*, 40, 51–62, 1979.

White, D.C., J.O. Stair, and D.B. Ringelberg, Quantitative comparisons of in-situ microbial biodiversity by signature biomarker analysis; *Journal of Industrial Microbiology*, 17, 185–196, 1996.

White, D.C., H.C. Pinkart, and D.B. Ringelberg, Biomass measurements: biochemical approaches, In *Manual of Environmental Microbiology*, 1st ed., C.H. Hurst, G. Knudsen, M. McInerney, L.D. Stetzenbach and M. Walter, Eds., American Society for Microbiology Press, Washington DC, 1997, pp. 91–101.

Wiedemeier, T.H. and F.H. Chapelle, Technical Guidelines for Evaluating Monitored Natural Attenuation of Petroleum Hydrocarbons and Chlorinated Solvents in Ground Water at Naval and Marine Corps Facilities, U.S. Department of the Navy, 2000.

Wiedemeier, T.H., J.T. Wilson, D.H. Kampbell, R.N. Miller, and J.E. Hansen, Technical Protocol for Implementing Intrinsic Remediation with Long-Term Monitoring for Natural Attenuation of Fuel Contamination Dissolved in Ground Water, US Air Force Center for Environmental Excellence, Brooks Air Force Base, San Antonio, TX, 1995.

Wiedemeier, T.H., H.S. Rifai, C.J. Newell, and J.T. Wilson, *Natural Attenuation of Fuels and Chlorinated Solvents in the Subsurface*, John Wiley & Sons, Inc., New York, NY, 1999, 617 pp.

Wiedemeier, T.H., M.J. Barden, W.Z. Dickson, and D. Major, Multiple Lines of Evidence Supporting Natural Attenuation: Lines of Inquiry Supporting Monitored Natural Attenuation of Chlorinated Solvents. U.S. Department of Energy, WSRC-TR-2003-00331, 2005.

10

Design and Installation of Ground-Water Monitoring Wells

David M. Nielsen and Ronald Schalla

CONTENTS

Introduction	640
Purposes and Objectives of Monitoring Wells	648
Site Characterization	650
Types of Monitoring Well Completions	657
Single-Casing, Single-Screen Wells	657
Multiple-Casing, Single-Screen Wells	659
Bedrock Completions	659
Monitoring Multiple Vertically Separated Zones	661
Well Clusters	661
Single-Casing, Multiple-Screen Wells	662
Nested Wells	662
Single-Casing, Long-Screen Wells	664
Multilevel Monitoring Systems	664
Borehole and Drilling Impacts on Monitoring Well Installations	667
Design Components of Monitoring Wells	677
Monitoring Well Casing and Screen Materials	677
Requirements of Casing and Screen Materials	678
Physical Strength of Well Casing and Screen	678
Chemical Resistance of Casing and Screen Materials	682
Chemical Interference from Casing and Screen Materials	683
Types of Casing and Screen Materials	683
Thermoplastic Materials	683
Metallic Materials	692
Coupling Procedures for Joining Casing	697
Joining PVC Casing	698
Joining Steel Casing	703
Factors Affecting Selection of Well Casing Diameter	705
Down-Hole Equipment	706
Well-Installation Method	708
Anticipated Well Depth and Casing Strength	708
Ease of Well Development	708
Purge Volume	708
Rate of Recovery	709
Unit Cost of Casing Materials and Drilling	709
Centering Casing in the Borehole	709

Sediment Sumps	712
Cleaning Requirements for Casing and Screen Materials	713
Costs of Casing Materials	714
Hybrid Wells	715
Monitoring Well Intakes — Well Screen and Filter Pack	716
Naturally Developed Wells	717
Well Screen Design in a Naturally Developed Well	718
Filter-Packed Wells	723
Filter-Pack and Well Screen Design for Filter-Packed Wells	725
Well Screens	741
Types of Well Screens Suitable for Monitoring Wells	742
Prepacked and Sleeved Well Screens	748
Monitoring Well Screen Open Area	751
Well Screen Length	751
Annular Seal Design and Installation	754
Annular Seal Materials	755
Bentonite	756
Neat Cement	763
Methods of Installation of Annular Seal Materials	769
Bentonite	769
Neat Cement	775
Surface Completion for Monitoring Wells: Protective Measures	777
Surface Seals	777
Above-Grade Completions	779
Flush-to-Grade Completions	784
Well Identification, Surveying, and Alignment Testing	792
Documentation of Well Construction Details	797
References	797

Introduction

Installing ground-water monitoring wells to detect trace (i.e., micrograms per liter [µg/l] or parts per billion [ppb]) levels of organic and inorganic contaminants in ground-water systems is a common practice at a variety of sites, including landfills, industrial facilities, service stations, Superfund sites, waste-water treatment facilities, and petrochemical plants. Hundreds of thousands of monitoring wells have been installed since the late 1970s and tens of thousands more are installed each year. Unfortunately, many of these wells were and are designed and installed by consultants and contractors who are not aware of correct monitoring well design and construction practices. As a result, many existing monitoring wells and some wells currently being installed have critical design flaws, and were or are being installed using methods and materials that may adversely affect the quality of samples collected from those wells.

The objective of most ground-water monitoring programs is to obtain "representative" ground-water data, including water-level data, hydraulic conductivity test data, and ground-water sample analytical data. To obtain the latter, it is necessary to be able to collect ground-water samples that retain both the physical and chemical properties of the ground water, and that are minimally affected by the sample acquisition process. As part of the sample acquisition process, proper ground-water monitoring well design and installation techniques are necessary to ensure that potential chemical alteration of samples is minimized and that representative samples can be collected.

Most ground-water monitoring well design and installation problems can be traced back to a mistaken belief in a "cookbook" approach or a "one-size-fits-all" philosophy that ignores site-specific geologic, hydrologic, geochemical, biological, and contaminant-related conditions. The fact is that each site at which monitoring wells are installed is geologically, hydrologically, geochemically, and microbiologically unique, and these factors dictate a unique design for each well. The well designer must develop well design and installation specifications that are flexible, and that take into account site-specific conditions and accommodate changes made necessary by unanticipated geologic conditions at any given well location.

Lack of a sufficient number of professionals adequately trained and experienced in proper monitoring well design and construction practices and procedures contributes to other ground-water monitoring well installation problems. Additionally, modern analytical laboratory capability is now reaching the parts per trillion detection level for many classes of analytes, while our means of gaining subsurface access to obtain ground-water samples is comparatively crude, although it is improving with time. Most potential sources of sample chemical alteration inherent in monitoring well installation are known or can be anticipated and thus can be avoided or controlled.

The basic requirement for proper ground-water monitoring well design and installation is a set of workable, flexible guidelines adaptable to a wide variety of hydrogeologic settings and geochemical environments and usable by both the consultants who design the wells and the contractors who install them. ASTM Standard D5092 — Standard Practice for Design and Installation of Ground-Water Monitoring Wells (ASTM, 2004a) — provides such guidelines and is an excellent example of how specifications for monitoring wells should be developed. One step toward developing these guidelines is identifying the most common problem areas in well design and construction. Among the most common monitoring well design flaws and installation problems are the following:

- Use of well casing or well screen materials that are not compatible with the hydrogeologic environment, the known or anticipated contaminants, or the specific requirements of the ground-water sampling program, resulting in chemical alteration of samples or failure of the well (Figure 10.1).
- Use of well screen that is not commercially produced (i.e., field-slotted, drilled, or perforated casing) or incorrect well screen slot-sizing practices, resulting in well sedimentation and the acquisition of turbid samples throughout the life of the monitoring program (Figure 10.2).
- Use of a single well-screen and filter-pack combination (e.g., a 0.010 in. well-screen slot size with a 20–40 sand) for all wells installed at a site (or multiple sites), regardless of formation grain-size distribution. This often results in siltation of the well, damage to pumps, and significant turbidity in samples when applied to formations finer than this design is appropriate for (Figure 10.3), and loss of filter pack to the formation, invasion of overlying well construction materials (e.g., annular seals) and lower-than-expected well yields when applied to formations coarser than this design is appropriate for.
- Improper length and placement of the well screen so that acquisition of water-level, hydraulic conductivity or water-quality data from discrete zones is impossible (Figure 10.4)

FIGURE 10.1
Using well construction materials inappropriate for site conditions can compromise sample integrity. In this case, corrosion of the galvanized steel casing and screen contributed metals (iron, zinc) to the samples, which resulted in anomalously high concentrations in analytical results.

- Improper selection and placement of filter-pack materials, resulting in well sedimentation, well-screen plugging, ground-water sample chemical alteration, or potential well failure (Figure 10.5).
- Improper selection and placement of annular seal materials, resulting in alteration of sample chemical quality, plugging of the filter pack and well screen, or cross-contamination from geologic units that have been improperly sealed off (Figure 10.6).
- Inadequate surface protection measures, resulting in surface water entering the well, alteration of sample chemical quality, and damage to or destruction of the well (Figure 10.7).

Any one or a combination of these monitoring well design and installation problems could cause a well or series of wells to be unsuitable for collecting representative ground-water data. On the basis of an examination of sampling results from thousands of wells, the authors estimate that more than 65% of ground-water monitoring wells installed in North America since the late 1970s suffer from more than one of the aforementioned problems, and thus are improperly designed for their intended purpose. As a result, many of these wells are producing water-level data, hydraulic conductivity test data, and ground-water samples that are not representative, in terms of the data expected from them. The consequences are: (1) inaccurate and misleading water-table or potentiometric surface maps and depictions of ground-water flow directions and hydraulic gradients; (2) inaccurate and misleading ground-water flow rate calculations; and (3) inaccurate and misleading depictions of ground-water chemistry and maps of contaminant plume

Design and Installation of Ground-Water Monitoring Wells 643

FIGURE 10.2
Using nonstandard materials for well screens can compromise sample integrity. In this case, use of septic tank drain pipe as a well resulted in inclusion of surface runoff and high levels of sediment (and high turbidity) in samples, and anomalously high levels of creosote (a hydrophobic constituent sorbed to sediment particles) in analytical results.

concentrations and extent. Improperly designed or installed wells often must be decommissioned and replaced, which is costly and time consuming. Proper monitoring well design and installation practices are thus essential to ensure cost-effective acquisition of representative ground-water samples and other ground-water data.

FIGURE 10.3
Using a "one-size-fits-all" approach to well screen and filter-pack design (i.e., a 0.010 in. screen slot and a 20–40 sand) usually results in very high turbidity in samples, which often requires that the samples be filtered. However, filtration cannot reverse the damage done to the samples, and may even exacerbate the problem.

FIGURE 10.4
Long well screens can make it impossible to collect discrete information on water levels and water chemistry from target-monitoring zones. This well screen, which extends 2 ft above ground surface, is clearly inappropriate for a monitoring well.

FIGURE 10.5
Using inappropriate filter-pack material can compromise sample integrity. This material is composed of a variety of mineral matter that could alter sample chemistry through dissolution, by contributing constituents to the sample that are not present in the ground water, or through sorption of some constituents that are present in the ground water.

Design and Installation of Ground-Water Monitoring Wells 645

FIGURE 10.6
Using inappropriate annular seal materials can result in cross-contamination between zones in the subsurface, or infiltration of surface runoff down the borehole. Drill cuttings (shown here) should never be returned to the borehole as part of an annular seal.

Proper design and installation of ground-water monitoring wells requires a detailed knowledge of site-specific geologic, hydrologic, geochemical, and microbiological conditions, which can only be obtained from a thorough site-characterization program (see Chapter 2). An up-to-date knowledge of well design and installation practices and procedures is also important. Site-specific design considerations include the following:

- Purpose or objective of the ground-water monitoring program (i.e., water-quality monitoring versus water-level monitoring)
- Surficial conditions, including topography, drainage, seasonal variations in climate, and site access
- Hydrogeologic setting, including type of geology (unconsolidated or consolidated, grain sizes, mineralogy), aquifer physical characteristics (preferential flow pathways, degree of heterogeneity, type of porosity, hydraulic conductivity), type of aquifer (confined or unconfined), recharge or discharge conditions, and ground-water and surface-water interrelationships
- Ambient ground-water chemistry and microbiology (e.g., presence of iron bacteria)
- Characteristics of site-specific contaminants, including chemistry, density, viscosity, reactivity, and concentration
- Anthropogenic influences (e.g., man-induced changes in hydraulic conditions)
- Any applicable regulatory requirements

FIGURE 10.7
Inadequate surface seals can compromise sample integrity. In this case, the thin layer of cement that was used as a surface seal for this well was easily damaged by yard maintenance equipment, resulting in surface runoff pouring directly down the borehole during precipitation events.

A unique set of site-specific design considerations exists for each site and for each individual well installation, and this requires that each well be designed as a unique structure.

To develop a knowledge of proper monitoring well design practices, it is first necessary to understand the individual design components of monitoring wells and how they combine to produce the final structure — the well itself. While it is not practical to describe a "typical" monitoring well in which the design components are fixed, it is possible to describe each of the individual design components, which include the following:

- Well casing
- Well screen
- Filter pack
- Annular seal
- Surface protection

These individual design components can be tailored and assembled during well construction to suit the site-specific considerations described earlier. Figure 10.8 illustrates the design components typical of most monitoring wells, and how they are assembled to produce a well. ASTM Standard D5092 (ASTM, 2004a) demonstrates the flexibility in design criteria that is necessary to accommodate site-specific conditions, provides insight into why wells must be designed as unique structures, and outlines proper monitoring well design practices.

Design and Installation of Ground-Water Monitoring Wells 647

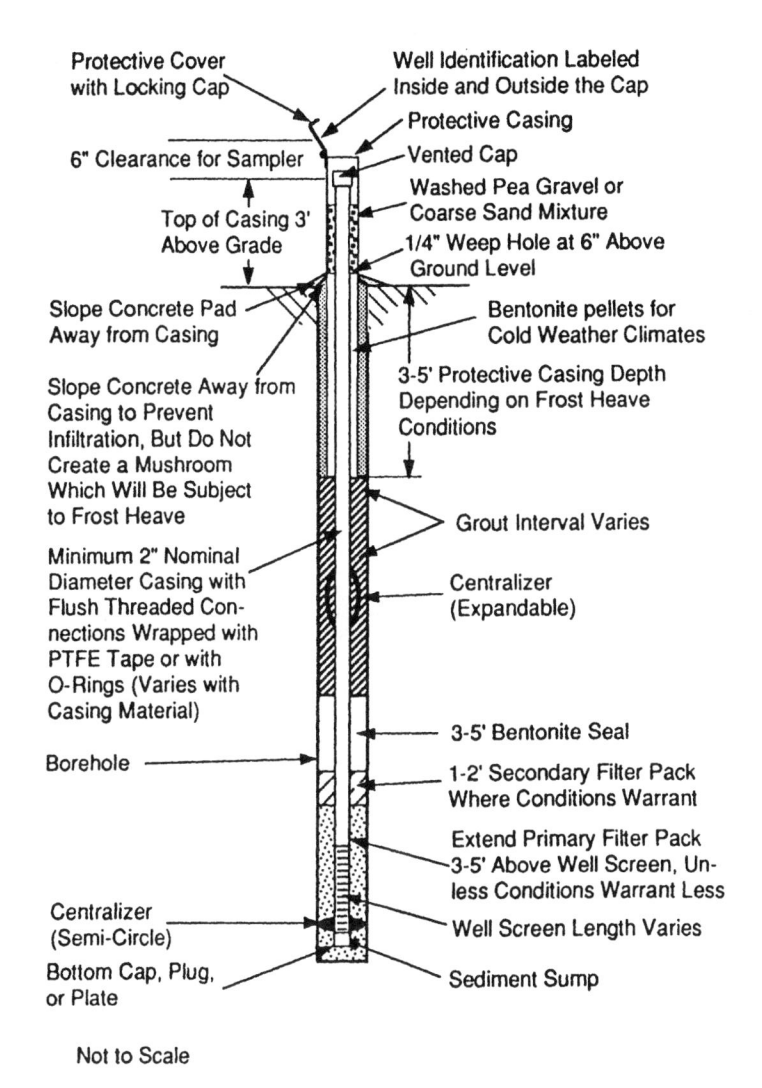

FIGURE 10.8
Design components of a typical monitoring well.

Proper installation of ground-water monitoring wells requires knowledge of state-of-the-art practices for environmental drilling (see Chapter 5) to ensure borehole integrity, to minimize or eliminate damage to the borehole wall, and to avoid potential contamination of the borehole or well caused by the drilling process. Proper selection of well construction materials, and use of proper coupling and placement techniques for well casing and screen, slot-sizing procedures for screens, placement and sizing techniques for filter packs, placement procedures for annular seals, and installation of surface protective measures must all be applied to ensure that a monitoring well will perform as intended.

There are limitations to the current technology for monitoring well design and installation. For example, most of the technology developed to date pertaining to well design and installation has been intended for application to geologic materials that are considered aquifers (i.e., water-bearing geologic units that yield significant quantities of water to wells). Therefore, most of the techniques described herein are effectively applied to monitoring wells constructed in geologic units that have less than 50% by weight fine-grained materials (i.e., materials passing the #200 U.S. sieve size, which includes silt and clay), but not to materials that are predominantly finer. Technology is just developing that will allow the installation of sediment-free monitoring wells in predominantly fine-grained geologic materials, and some of these designs are described.

Purposes and Objectives of Monitoring Wells

Before installing a monitoring well or system, it is very important to establish the purposes and objectives of the well and system. The purposes and objectives will, in many cases, dictate the design parameters for the well (including well diameter, well casing and screen materials, well screen length and placement, and well screen slot size and open area), and the drilling or direct-push method used to install the borehole and well. For example, for wells installed for the purpose of collecting representative ground-water samples, it is important that the materials of construction do not interact with formation water chemistry or with any contaminants present. The materials should neither leach constituents into the samples collected from the well (creating false positives), nor remove constituents from samples (creating false negatives). For wells installed for the purpose of defining and monitoring the three-dimensional extent of a contaminant plume and the detailed chemistry of the plume, it is important that the well screen lengths be appropriate to conduct sampling of very discrete intervals (typically between 2 and 5 ft). For wells installed for the purpose of conducting hydraulic conductivity tests, the well screen open area must be sufficient to avoid interfering with the test results (typically at least 8 to 10% open area); the diameter of the well may need to be large enough to accommodate a pump of sufficient capacity to stress the formation (typically at least 2 to 4 in. inside diameter). Similar constraints are placed on well design by other monitoring purposes and objectives.

Although ground-water monitoring wells are used to accomplish many different purposes and objectives, the most common purposes of these wells are to collect:

- Representative ground-water quality samples from a target-monitoring zone for chemical analysis, accurate to the limits of detection for many parameters (especially volatile organic compounds and trace metals), to allow detection and monitoring of contaminant plumes (Figure 10.9).

FIGURE 10.9
An important objective of nearly all monitoring wells is to collect representative samples of ground water. (Photo courtesy of Severn Trent/QED Environmental Systems.)

Design and Installation of Ground-Water Monitoring Wells 649

FIGURE 10.10
Collecting accurate and precise water-level data is another important objective of most monitoring wells. (Photo courtesy of Jim Quince.)

- Accurate ground-water level (or hydraulic head) data at a specific location in the ground-water flow system, to permit construction of water-table or potentiometric surface contour maps or flow nets, and to allow definition of ground-water flow direction in the horizontal plane (and, for some situations, in the vertical plane) (Figure 10.10).
- Accurate and representative hydraulic parameter data, especially hydraulic conductivity test data from pumping tests, slug and bail tests, and pressure tests, to allow definition of preferential flow pathways and calculation of ground-water flow velocity (Figure 10.11).

Although, ideally, the method used to install the wells, the materials used to construct the wells, and the other design features of the wells are selected so they do not affect the data quality required from the wells, in cases in which the boring or the well is used to satisfy other purposes (e.g., to collect borehole geophysical data), compromises in installation method, well design, and construction may be necessary. For example, although water

FIGURE 10.11
Many monitoring wells are also used to collect data from formation hydraulic conductivity tests, including slug tests, bail tests, and pumping tests (shown here). (Photo courtesy of In-Situ Inc.)

quality monitoring wells are generally installed using drilling methods that do not require the use of drilling fluids (to avoid compromising water quality in the well and the adjacent formation), some borehole geophysical logs must be run in uncased, fluid-filled boreholes to produce valid results. Thus, such wells may have to be drilled with mud rotary methods to accommodate the use of borehole geophysical tools. In such a case, the driller must take great care to control the drilling fluid as the borehole is installed, and to remove the drilling fluid from the borehole and the formation screened by the well before the well is installed or during the well development process.

Site Characterization

A thorough knowledge of site-specific geologic, hydrologic, geochemical, and microbiological conditions is necessary to properly locate monitoring wells and position well screens, to select appropriate well construction materials, and to confidently apply conventional monitoring well design and installation practices and procedures. A monitoring well that is improperly located, designed or installed is of little or no value, and may produce data that completely misrepresent the hydraulic head and water-quality conditions of the monitored interval, leading to inaccurate conclusions about site conditions. However, a properly positioned, designed and installed well that detects contamination at the earliest possible time could help save significant amounts of time and money in remediation, and prevent extensive contamination of ground water. The key to proper monitoring well location, design, and installation is developing a thorough understanding of site-specific conditions through environmental site characterization (see Chapter 2).

Prior to the installation of any monitoring well or system, a conceptual site model, that identifies potential ground-water flow pathways and the target-monitoring zones (the zones most likely to convey contaminants or to be impacted by a release from the facility to be monitored), should be developed. It should be noted that, while some regulatory programs focus only on the "uppermost aquifer," there may be several target-monitoring zones beneath any given site (either within the uppermost aquifer or in that formation and formations below), and each of these zones should be monitored with equal care.

Development of the conceptual site model is normally accomplished in two phases — a literature search and initial site reconnaissance, after which a preliminary conceptual site model is created, and a detailed field investigation, after which a revised conceptual site model is formulated. When the hydrogeology of a project area is relatively uncomplicated and well documented (a rare situation indeed), the initial site reconnaissance may provide sufficient information to identify preferential flow pathways and the target-monitoring zone. Where little or no background data are available or where the geology is complex (the more common situation), a field investigation will be required to develop the necessary conceptual site model.

Every effort should be made to collect and review all available literature pertaining to the project area, including all field and laboratory data from previous investigations (Figure 10.12). Information such as (but not limited to) topographic maps, aerial photographs, satellite imagery, historical ownership and land use records for the site and adjacent properties, soil surveys, geologic and hydrogeologic reports and maps, water well and soil boring logs, information from local well drillers, geotechnical investigation reports, and other reports and maps related to the project area, should be reviewed to locate information relevant to well design and installation. The data needs that must be satisfied to allow proper monitoring well design and installation are outlined in Table 10.1.

Design and Installation of Ground-Water Monitoring Wells 651

FIGURE 10.12
Evaluating data from previous site investigations and from existing reports and maps is an important element of site characterization that can provide relevant information to begin the well design process.

The relevant data should be verified during an initial site reconnaissance, during which surface conditions at the site are noted, roadcuts, streamcuts, and other geologic outcrops are examined, topography and local drainage routes are noted, above-ground and below-ground utilities are located, and potential locations for monitoring wells are selected. The distribution of the geologic materials likely to be found in the subsurface and the ground-water flow pathways that are likely to be located during a field investigation may be hypothesized in a preliminary conceptual site model using information obtained in the literature search and site reconnaissance.

TABLE 10.1

Data Needs and Uses for Ground-Water Monitoring Well Design and Installation

Data Needed	Data Uses
Geologic material type (unconsolidated versus bedrock)	Selecting an appropriate drilling method Determining well completion type (cased well versus open bedrock borehole)
Presence of difficult drilling conditions (heaving sands, boulders, cobble zones, large voids, or caverns)	Selecting an appropriate drilling method Determining the need for special well completion procedures
Presence of preferential ground-water flow pathways	Defining the target-monitoring zones
Position, depth, and thickness of the target-monitoring zones	Defining proper well placement, well screen placement and length, and well depth
Grain size of the formation material in the target-monitoring zones	Selecting well screen slot size and filter pack grain size
Position and degree of fluctuation of the water table	Defining well screen placement and length for wells monitoring light non-aqueous phase liquids (LNAPLs)
Depth of frost penetration	Defining depth of the surface seal
Ambient ground-water geochemistry (especially pH, O_2, CO_2, TDS, Cl^-, H_2S)	Selecting appropriate well casing and screen materials Selecting appropriate annular seal materials
Contaminant types and concentrations (especially presence of NAPLs)	Selecting appropriate well casing and screen materials Defining well screen placement and length
Site microbiology (especially iron bacteria)	Selecting appropriate well casing and screen materials Determining the need for special well maintenance

The goal of the field investigation is to refine the preliminary conceptual site model so that the target-monitoring zones can be identified prior to monitoring well installation. Characterization of the preferential flow pathways for ground water and contaminants that make up the target-monitoring zones involves defining the porosity (type and amount), hydraulic conductivity, lithology, stratigraphy, structure, hydraulic head distribution, and geochemistry of each hydrologic unit present beneath the site. These characteristics are determined by conducting an exploratory program that may include soil borings and direct-push investigations, surface and borehole geophysical investigations, piezometers to collect hydraulic head information, hydraulic conductivity tests, and ground-water sampling to detect the possible presence of contaminants.

Soil borings and direct-push probe holes (Figure 10.13) should be deep enough to provide the required geologic data (including lithology, stratigraphy, and structure) and hydraulic parameter data (including type and amount of porosity, and permeability or hydraulic conductivity). At least a few of these boreholes should be continuously sampled to provide good stratigraphic correlation from boring to boring, so no potentially important intervals are missed. In cases in which cone penetration testing (CPT)

FIGURE 10.13
Direct-push (DP) technologies, such as the CPT rig shown in (a) or the smaller percussion hammer DP machine shown in (b), are capable of collecting soil, soil–gas, and ground-water samples, and generating data on soil type, soil electrical conductivity, and presence of specific types of contaminants, all of which can assist in well positioning and well design. These rigs are also capable of installing small-diameter DP wells.

FIGURE 10.14
Soil description in the field provides important information on the character of geologic material that can be used to position and design wells.

equipment (see Chapter 6) is used to directly generate site stratigraphic information, it is necessary to collect continuous samples from probe holes immediately adjacent to several representative CPT holes to provide good geologic correlation.

For samples collected from soil borings or direct-push probe holes, geologic material properties should not be based solely on field sample description or classification (Figure 10.14), but should be confirmed by laboratory and field tests made on the samples. At least one soil boring or direct-push probe hole (preferably more) should be continuously sampled (Figure 10.15), to provide an indication of the degree of heterogeneity of subsurface materials. Sample collection should be conducted according to the appropriate ASTM method (e.g., see Table 10.2), given the characteristics of the geologic materials. An accurate boring log and soil sampling record should be compiled for each exploratory boring or direct-push probe hole. Boring logs should include the location, geotechnical data (e.g., blow counts), and sample description information for each geologic material identified in the borehole. Description and identification of geologic materials should be done consistently according to ASTM Standard D 2488 (Standard Practice for Description and Identification of Soils [ASTM, 2004b]); classification of soils should be in accordance with one of the common soil classification systems (e.g., the Unified Soil Classification System or ASTM D 2487 [ASTM, 2004c]; the USDA Classification System; or the AASHTO Classification System).

Surface and borehole geophysical surveys (Figure 10.16) may be used to supplement soil-boring data and to aid in interpretation of subsurface conditions between soil borings. For example, direct-push electrical conductivity profiles (see Chapter 6) can be correlated with continuous soil boring information at one or two locations, then used to fill the gaps in the data in between soil borings, at a much lower cost than conducting continuous soil sampling at numerous locations. Surface geophysical methods such as seismic refraction (ASTM D 5777 [ASTM, 2004d]), electrical resistivity (ASTM D 6431 [ASTM, 2004e]), ground-penetrating radar (ASTM D 6432 [ASTM, 2004f]), gravity (ASTM D 6430 [ASTM, 2004g]), and electromagnetic conductivity (ASTM D 6639 [ASTM, 2004h] and ASTM D 6820 [ASTM, 2004i]) (also, see Chapter 4) can be particularly valuable for defining geology when distinct differences in the properties of subsurface materials are noted. On the basis of the results of the combined soil boring and geophysical investigations, geologic cross-sections and fence diagrams should be constructed to identify the zones of coarsest geologic materials.

FIGURE 10.15
Continuously cored boreholes provide the best opportunities for detailed description and correlation of subsurface materials and features, which can be invaluable in positioning and designing wells.

Ground-water flow direction must be determined in three dimensions by measuring the horizontal and vertical hydraulic gradients in the geologic materials present at the site. However, because ground water will flow along the pathways of least resistance (i.e., within the highest hydraulic conductivity formation materials present at the site), actual ground-water flow direction at any position in the subsurface may be oblique to the hydraulic gradient (e.g., within buried stream channel gravel bodies in a silt or clay matrix of alluvial materials). Flow direction must be determined by installing piezometers or short-screened wells in soil borings that penetrate the zones of interest at the site, in configurations that allow the determination of both horizontal and vertical gradients (Figure 10.17). The depths and locations of the piezometers or wells will depend on the

TABLE 10.2

ASTM Standards Related to Soil Sample Collection

D 6282	Guide for direct-push soil sampling for environmental site characterization
D 4700	Guide for soil sampling from the vadose zone
D 6169	Guide for selection of soil and rock sampling devices used with drill rigs for environmental investigations
D 1452	Practice for soil investigation and sampling by auger borings
D 1586	Test method for standard penetration test and split-barrel sampling of soils
D 1587	Practice for thin-walled tube sampling of soils for geotechnical purposes
D 3550	Practice for thick-wall, ring-lined, split-barrel, drive sampling of soils
D 6519	Practice for sampling of soil using the hydraulically operated stationary piston sampler
D 4220	Practice for preserving and transporting soil samples
D 5079	Practice for preserving and transporting rock core samples

FIGURE 10.16
Surface geophysical methods, such as ground-penetrating radar (shown here) can provide detailed information on geologic conditions and subsurface features (i.e., underground tanks, buried drums, etc.) over large areas in a relatively short period of time, and in a cost-effective and noninvasive manner. (Photo courtesy of Dick Benson, Technos, Inc.)

FIGURE 10.17
Bundle piezometers can provide vital information on the 3D distribution of hydraulic heads and dissolved-phase contaminants, which is important in positioning wells and determining appropriate well screen length for long-term monitoring wells.

anticipated hydraulic connections between conceptualized flow pathways and the locations of the flow pathways with respect to the facility or activity to be monitored. Following the acquisition of water-level data from the piezometers or wells, potentiometric surface maps and cross-sectional flow nets should be prepared. At this point, it is important to compare geologic cross-sections with flow nets to identify the locations of preferential flow pathways to plot flow directions in three dimensions.

Hydraulic conductivity tests (either field-based tests, such as pumping tests, slug and bail tests, or pressure tests, or laboratory tests) should be conducted on formation materials to identify the materials with the highest hydraulic conductivity and to confirm the presence of preferential flow pathways (Figure 10.18). See Chapter 14 and ASTM D 4043 (ASTM, 2004j) for more detailed information on hydraulic conductivity testing alternatives and methods.

Where appropriate, ground-water samples should be collected during the field investigation, either from direct-push sampling tools (see Chapter 6) or from piezometers or wells installed in preferential flow pathways to detect the presence of contaminants. See Chapter 15 for additional detail on appropriate ground-water sampling practices.

The preliminary conceptual site model should be revised based on the results of the field investigation. The geologic cross-sections and fence diagrams, the potentiometric surface maps and cross-sectional flow nets, the hydraulic conductivity test results and the ground-water sample analytical results should all be considered together to draw conclusions regarding which flow pathways are the appropriate target-monitoring zones. The monitoring wells to be installed to monitor these zones can then be located, designed and installed properly to provide the required information on these zones for the long-term monitoring program.

FIGURE 10.18
Hydraulic conductivity testing, including slug testing (shown here) provides important data on formation hydraulic conductivity, which helps identify preferential ground-water and contaminant flow pathways, and aids in proper well-screen positioning for long-term monitoring wells.

Design and Installation of Ground-Water Monitoring Wells

Types of Monitoring Well Completions

The types of possible monitoring well completions range from simple single screened interval or open-borehole bedrock wells to more complex multiple-casing or multiple-screen wells. Each type of well completion has its applications, advantages and disadvantages, each can meet specific sets of objectives, and each is described briefly below. General recommendations for the application of each well completion type are included in Table 10.3.

Single-Casing, Single-Screen Wells

The simplest and most common type of well completion is the single-casing, single-screen well, which typically consists of a short well screen at the bottom of the well and a single

TABLE 10.3

Recommendations on Applications of Various Well Completion Types

1. Single-casing, single-screen wells with short well screens
   - Monitoring discrete zones (preferential flow pathways, such as sand and gravel lenses in a fine-grained matrix)
   - Collecting discrete water-level data (i.e., from a pumping test)
2. Multiple-casing single-screen (telescoping casing) wells
   - Monitoring discrete zones beneath confining beds or beneath known or suspected contaminated zones
3. Bedrock completions
   a. Single-casing, single-screen (short screen) wells
      - Monitoring discrete zones (preferential flow pathways, such as fracture zones or solution channels)
   b. Single-casing, single-screen (short screen) wells with surface casing
      - Monitoring discrete zones beneath confining beds or beneath known or suspected contaminated zones
      - Monitoring the zone immediately beneath unconsolidated overburden
   c. Open-bedrock boreholes
      - Use as a screening tool to monitor thick sequences where only horizontal flow occurs
      - Not recommended where vertical gradients are present, or where data from a discrete zone are desired
4. Well cluster (multiple single-casing, single-screen (short screen) wells completed at different depths in individual boreholes)
   - Monitoring multiple discrete zones (i.e., multiple thin formations in a sequence of alternating coarse-grained and fine-grained materials)
   - Monitoring multiple levels in a single thick formation
   - Determining vertical gradients
   - Evaluating chemical stratification
5. Multiple-screen well (with packers between screened zones)
   - Monitoring multiple discrete zones
   - Monitoring multiple levels in a single thick formation
   - Determining vertical gradients
   - Evaluating chemical stratification
   - Not recommended where zones of interest are separated by only a few feet
6. Nested wells (multiple single-casing, single-screen wells completed at different depths in a single borehole)
   - Monitoring multiple discrete zones
   - Monitoring multiple levels in a single thick formation
   - Not recommended where zones of interest are separated by only a few feet
7. Single-casing, single-screen wells with long well screens
   - Use as a screening tool to monitor thick sequences where only horizontal flow occurs
   - Not recommended where vertical gradients are present, or where data from a discrete zone are desired
8. Multilevel monitoring system
   - Monitoring multiple discrete zones
   - Monitoring multiple levels in a single thick formation
   - Determining vertical gradients
   - Evaluating chemical stratification, or measuring small-scale features of contaminant distribution

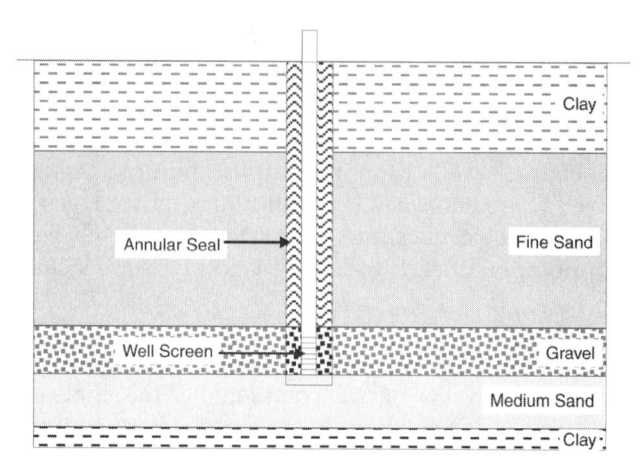

FIGURE 10.19
A single-casing, single-screen (short screen) well.

string of casing that extends to ground surface (Figure 10.19 and Figure 10.20). In the annulus between the casing or screen and the borehole, filter-pack sand is placed around and just above the screen, and annular seal material extends from the top of the filter pack to ground surface, where some type of protective structure is installed. This type of completion is most appropriate for situations in which the objective is to monitor the zone of water-table fluctuation (or LNAPLs at the water-table surface) or a single discrete interval, such as a thin sand seam within a matrix of silt and clay.

FIGURE 10.20
A single-casing, single-screen well.

Multiple-Casing, Single-Screen Wells

Multiple-casing, single-screen wells (Figure 10.21), sometimes referred to as telescoping casing wells, are often used in situations where it is necessary to drill through one or more contaminated zones to complete a well in a formation below. This type of completion may also be used where a difficult drilling condition (e.g., heaving or caving sands or solution-channeled rock) makes it difficult to use a preferred drilling method to reach the zone of interest. In this type of completion, a large-diameter (minimum 6- to 8-in. diameter) pilot borehole is drilled to just below the contaminated zone or the difficult drilling zone, using a drilling method appropriate for the subsurface conditions (see Chapter 5). In most cases, the borehole is terminated in the top of a confining layer or competent bedrock, where a large-diameter surface or conductor casing is installed in the borehole and pressure-grouted in place, typically using ASTM C 150 Type I or II Portland cement (ASTM, 2004k). After the grout has completely set (usually 48 to 72 h), a smaller diameter borehole is advanced using the preferred drilling method, from the bottom of the pilot borehole to the zone of interest. The monitoring well is then completed in this borehole, usually in the same manner as a single-casing, single-screen well. The difference is that the surface casing remains in place, to ensure that there is no hydraulic communication between the upper zone and the zone of interest.

Bedrock Completions

Bedrock completions (Figure 10.22) can generally be done in one of three ways. The first method is to drill the borehole through overburden (if present) and through bedrock to the zone of interest, and then complete the well in the same manner as a single-casing, single-screen well. A variant of this type of completion is to drill a large-diameter pilot borehole through the overburden into competent bedrock, and then to install a surface or conductor casing in the borehole and pressure-grout it in place. As in the multiple-casing well described earlier, a smaller diameter borehole is drilled through bedrock to the zone of interest, and a single-casing, single-screen well is completed in this borehole. This type of completion provides the same assurance that the zone of interest is isolated from other zones penetrated by the borehole, and interzonal flow is minimized or eliminated. The third and most common method for installing monitoring wells in bedrock is to

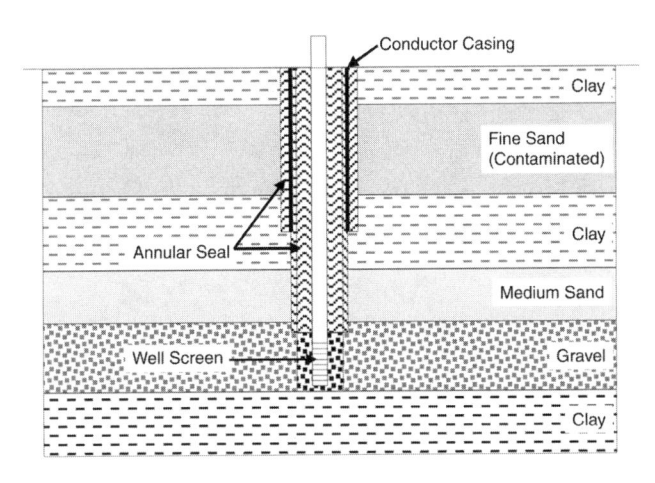

FIGURE 10.21
A multiple-casing (telescoping), single-screen well.

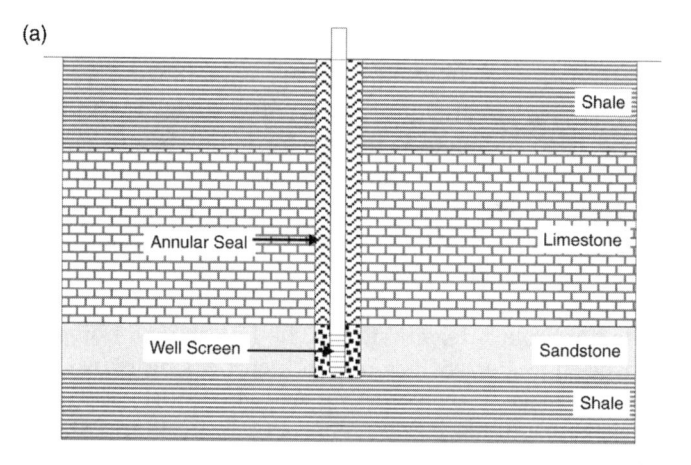

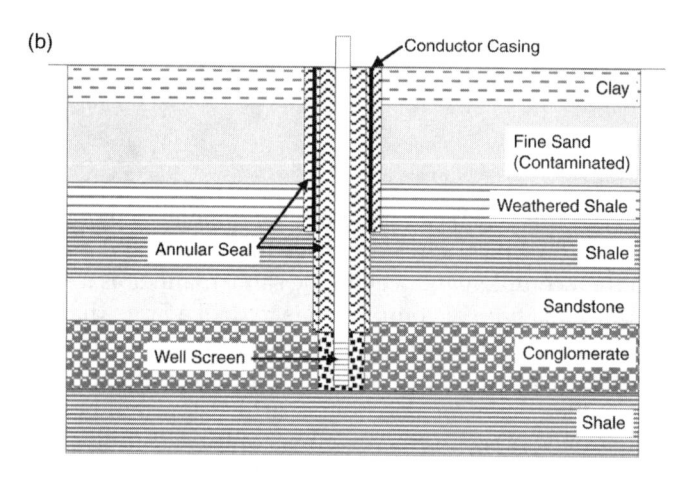

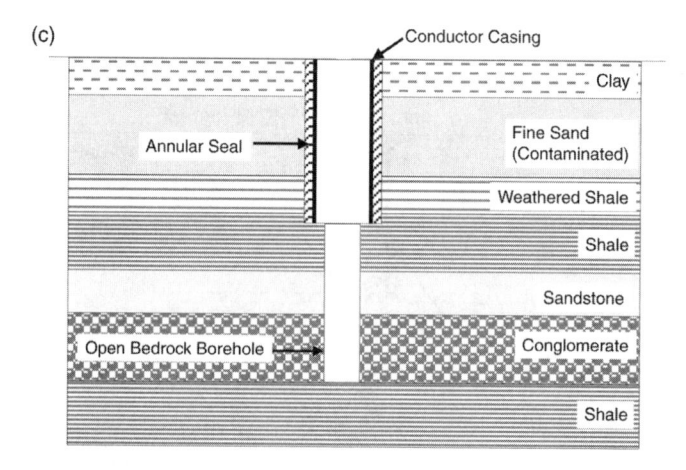

FIGURE 10.22
Well completion types used in bedrock. (a) A single-casing, single-screen well installed through overburden and into bedrock. (b) A single-casing, single-screen well installed after a hole has been drilled and a conductor casing installed to the top of bedrock and grouted in place. (c) An open bedrock borehole drilled after a conductor casing has been installed and grouted in place.

Design and Installation of Ground-Water Monitoring Wells

drill a large-diameter pilot borehole through overburden into competent bedrock, set and grout a surface or conductor casing in place, and continue drilling to the zone of interest after the grout has set. Perhaps the best method to use when drilling the open borehole through bedrock is rock coring, because it provides the best samples (upon which judgments about which zone to monitor may be based) and the most consistent borehole (a smooth, round hole without constrictions or washouts). When the drilling is complete, the finished well consists of a cased borehole from ground surface to the top of competent bedrock, and an open bedrock borehole from the bottom of the surface casing to the zone of interest. There is no well casing, no well screen, and no filter pack and the only annular seal is that between the surface casing and the overburden. One potentially significant problem with this type of completion is that the entire open borehole interval can contribute water to the well and, thus, is the interval monitored by the well. It is very difficult or impossible to monitor a specific zone, unless specialized equipment (e.g., a dual packer set-up with a pump in between) is used to isolate specific discrete sampling zones. However, even the use of packers does not guarantee isolation of a fracture, a fracture zone or a solution channel from the rest of the open borehole. A fracture or solution channel in the packed-off zone may be connected to a fracture or solution channel above or below the packed-off zone by connecting fractures or solution channels some distance from the borehole.

Monitoring Multiple Vertically Separated Zones

Well Clusters

For situations in which the objective is to monitor several different vertical intervals in the same location, either within the same formation or in different formations, or where the objective is determining the vertical distribution of hydraulic head or vertical differences in water quality, several alternatives are available. The simplest solution is to monitor each interval of interest with a single-casing, single-screen well (each in its own borehole, with a short screen), though this may end up costing more than other alternatives. This type of configuration, depicted in Figure 10.23 and Figure 10.24, is termed a "well cluster." Well clusters can be used to reliably determine vertical gradients and to monitor discrete zones or to evaluate chemical stratification within a single thick zone.

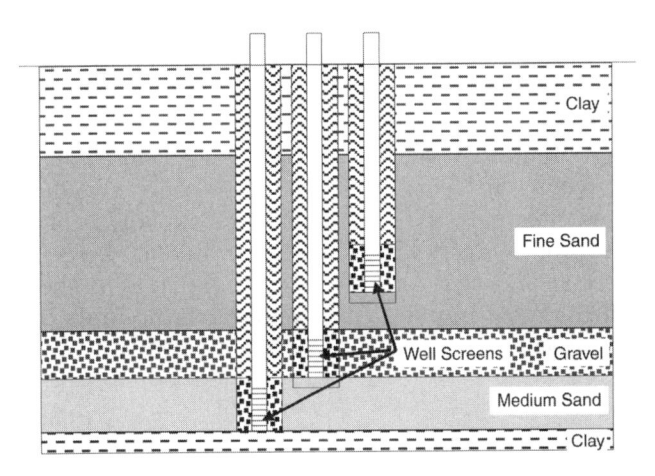

FIGURE 10.23
Multiple single-casing, single-screen wells in closely-spaced individual boreholes, screened at different depths (well cluster).

FIGURE 10.24
A well cluster consisting of two wells installed at different depths. In this case, the well in the foreground is completed with the well screen at the bottom of the uppermost aquifer (the top of the first confining bed), to monitor a DNAPL and the associated dissolved-phase plume. The well in the background is completed with the well screen straddling the water table, to monitor an LNAPL and the associated dissolved-phase plume from the same facility.

Single-Casing, Multiple-Screen Wells

Another possible completion to satisfy this objective is the single-casing, multiple-screen well, which consists of alternating sections of well screen (adjacent to the zones of interest) and well casing (between the zones of interest) in a single borehole (Figure 10.25). In this type of completion, filter-pack sand is installed around and just above and below each screened interval, and annular seal material is installed between the screened, filter-packed intervals to inhibit hydraulic communication between zones of interest. To inhibit movement of water between screened zones within the well, inflatable or mechanically actuated packers must be installed in the cased portions of the well. To allow collection of discrete hydraulic head (water-level) data and ground-water samples from the screened zones, it is common to install dedicated pressure transducers and sampling pumps in these zones. Though it may be possible with this type of completion to save money in well construction compared to installing multiple single-screened wells at different depths, these savings are often consumed by the purchase of the in-well equipment necessary to seal off discrete sampling zones and gather accurate hydraulic head data and ground-water samples.

Nested Wells

A third type of completion designed to monitor several different vertical intervals is termed a "nested well", in which several small-diameter single-casing, single-screen wells are installed in a single large-diameter borehole (Figure 10.26 and Figure 10.27).

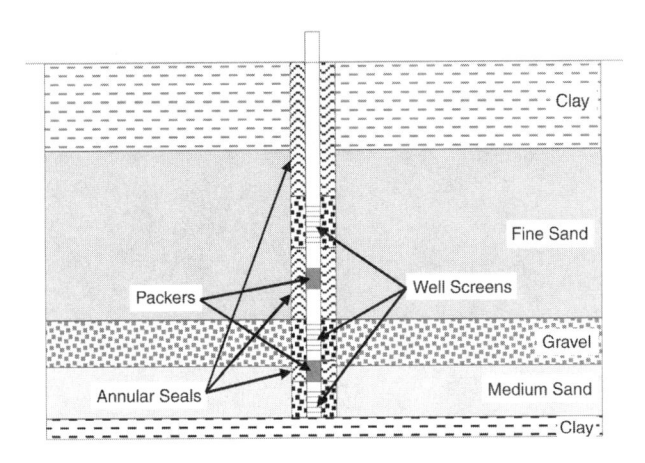

Design and Installation of Ground-Water Monitoring Wells 663

FIGURE 10.25
A single-casing, multiple-screen well. In order to prevent hydraulic communication between screened zones, the borehole behind blank casing must be filled with annular seal material, and packers must be installed in the well in each of the blank casing zones. This usually requires the installation of dedicated pumps and pressure transducers to collect samples and hydraulic head data from each of the screened zones.

These nested wells are similar to the bundle piezometers described by Cherry et al. (1983) and Barker et al. (1987) and depicted in Figure 10.17. The individual screened intervals are filter packed, and the filter-packed intervals are separated by annular seal material. It is often difficult to ensure that these well nests are completed as designed, with well screens, filter packs, and annular seals placed properly and functioning. Lapham et al. (1996) report that as many as five 2- to 3-in. diameter monitoring wells have been successfully installed in a single 10-in. borehole in situations in which the annular seals are several tens of feet thick. To ensure that the annular seals are effective in isolating the screened zones, it is recommended that the completion be tested. This can be done by pumping each individual well in sequence, while the hydraulic response is recorded in the other wells, or by introducing a tracer into one well screen and monitoring the other intervals for the presence of that tracer (Meiri, 1989; LeBlanc et al., 1991). Because the costs for this type of completion, which include increased costs for drilling the larger borehole

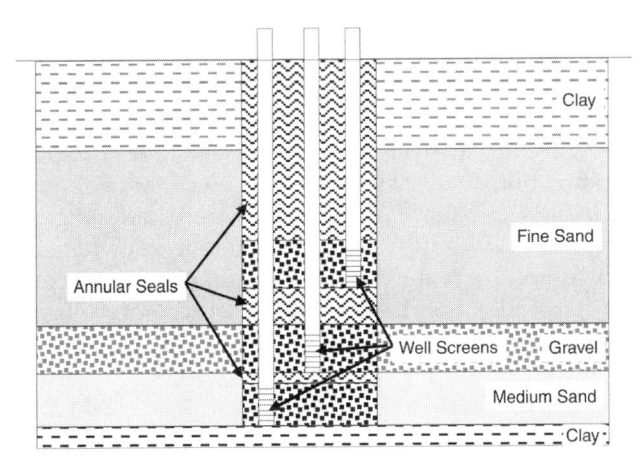

FIGURE 10.26
Multiple single-screen wells in a single borehole (well nest).

FIGURE 10.27
A well nest, with two small-diameter wells with short well screens installed to different depths in the same borehole. (Photo courtesy of Illinois State Geological Survey.)

required, are comparable to those for clustered wells, there are few advantages for using this type of completion over others.

Single-Casing, Long-Screen Wells

A much less desirable, though still commonly employed alternative is to install a single-casing, long-screen well (Figure 10.28), in which the screen spans all of the vertical intervals of interest, or the entire saturated thickness of a formation. The well designer's expectations are usually that flow through the well will be exclusively horizontal and that positioning a pump adjacent to a specific zone in the screen will allow sampling just that portion of the screen. Such expectations are usually not realized, as there are nearly always differences in hydraulic head from one part of the well screen to another, which result in water movement within the screen, from zones of high hydraulic head to zones of lower hydraulic head (Figure 10.29). In this situation, the well screen serves as a conduit for movement of ground water (and, potentially, contaminants). The hydraulic head data and ground-water samples collected from the well are not representative of any one zone (despite efforts to isolate specific zones), but a composite of the conditions throughout the screen (see Well Screen Length section and Chapter 11). For these reasons, this type of well completion is usually looked upon unfavorably by regulatory agencies.

Multilevel Monitoring Systems

Perhaps the best alternative for monitoring multiple vertical intervals in one location is to use a multilevel monitoring system. Several of the different designs that are currently

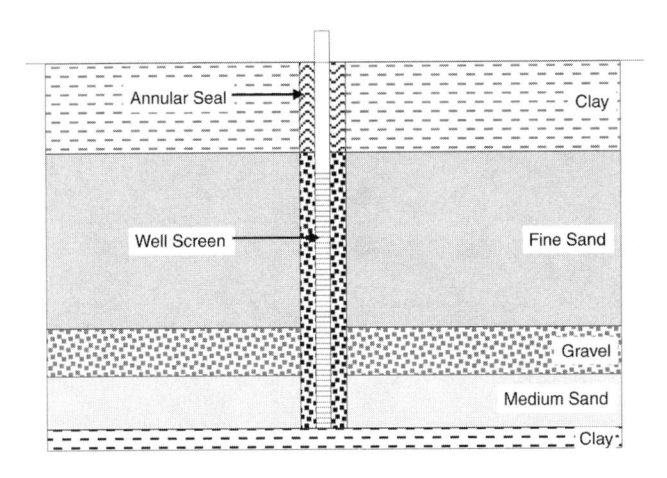

FIGURE 10.28
A single-casing, single-screen (long screen) well. Such a well is inappropriate if the objective of the well is to collect water-level, water-chemistry, or hydraulic conductivity test data from a discrete zone, such as a relatively thin preferential flow pathway. Data from this type of well completion represent average values across the entire screened zone, which may distort the conceptual model for the site.

available are depicted in Figure 10.30 and Figure 10.31; these systems are described in detail in Chapter 11.

The decision to use one of these alternatives over another to monitor different vertical intervals at a site depends on a number of factors related to project objectives. For example, the interpretation of changes in chemical concentration with depth at the centimeter scale can be problematic if well screens are not located in the same borehole (Gibs et al., 1993). Consequently, a single-borehole configuration might initially be preferred if measurement of small-scale features of a contaminant plume is a primary project objective. However, ensuring the chemical integrity of the sampled interval could be very difficult with this type of completion. Isolating sampling intervals by

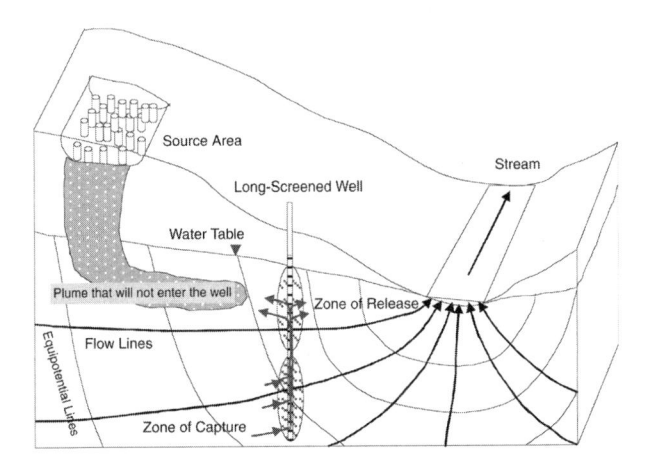

FIGURE 10.29
Flow through a long well screen in a thick unconsolidated formation. The well is situated near a ground-water discharge zone (the stream), so there is a difference in hydraulic head that forces water to move from the bottom of the screen (where hydraulic head is highest) toward the top of the screen (where hydraulic head is lowest). Movement of water in the well may actually push the shallow contaminant plume around the well and, as a consequence, it may not be detected in this well. (*Source*: McIlvride and Rector, 1988.)

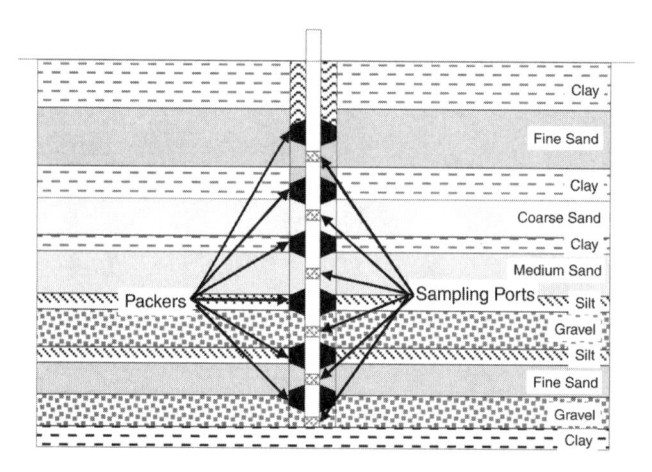

FIGURE 10.30
A multilevel monitoring system.

constructing each well in its own borehole will produce more reliable data, especially when the depth-interval scale of interest is on the scale of meters (Lapham et al., 1996).

In addition to factors related to project objectives, logistical and other factors (including cost-effectiveness, borehole size, and well design, installation and development simplicity) must also be considered. For example, well clusters can be installed and developed relatively easily in small-diameter boreholes in nearly any geologic conditions, are very simple to design, and are cost-effective at depths of less than 100 ft or so. On the other hand, well nests require a larger diameter borehole, are more complex to design, and

FIGURE 10.31
Installation of a multilevel monitoring system in a bedrock borehole. (Photo courtesy of Westbay Systems Inc.)

are more difficult to install and develop properly, particularly in situations in which thin annular seals are required between individual wells. However, well nests may be less expensive to install compared with other alternatives, particularly in cases where more than two or three zones are the targets of monitoring.

Borehole and Drilling Impacts on Monitoring Well Installations

Any borehole in which a monitoring well is intended to be installed should be straight (vertical) and plumb, and of sufficient diameter so that the well can be constructed within it without any major difficulties. Borehole alignment can be assessed through use of a borehole deviation survey, which determines the direction and distance of the bottom of the borehole relative to the top of the borehole and points in between, or a borehole dipmeter. Misalignment is usually not significant for shallow boreholes (e.g., less than 50 ft deep) in relatively homogeneous geological materials, but it can become a problem in deep boreholes or where difficult drilling conditions (boulders, cobbles, caliche, fractures or voids in bedrock, or significant contrasts in formation hardness) are encountered. In these cases, a borehole deviation survey is recommended.

Although some state regulations or guidance manuals require a minimum 2-in. annular space between the well casing or screen and the borehole wall (meaning that the borehole diameter must be at least 4 in. larger than the well diameter), there are really no valid reasons for such a requirement. It is commonly cited that the purpose of this requirement is to ensure that a minimum 2-in.-thick filter pack completely surrounds the well screen. However, without an accompanying requirement for centering devices along the screen (which is very rare), there is no assurance that well installation will produce this result. Furthermore, as Driscoll (1986) points out, a filter pack need only be a few grain diameters thick to achieve its intended purpose (mechanical filtration to remove formation fines), so 2-in. thick filter packs are not necessary to ensure a good well completion. Finally, in wells of 2 in. I.D. and smaller, well development is difficult to accomplish if the filter pack is more than 1 or 2 in. thick. It is difficult to develop sufficient energy in such wells to reach back to the borehole wall to rectify formation damage and produce a good hydraulic connection with the formation.

It is also occasionally cited in state regulations or guidance documents that a 2-in. annular space is required to use a tremie pipe to install filter-pack materials and grout mixtures in the annular space. Prepacked well screens (Figure 10.32) offer sufficient filter-pack thickness and an assurance that the entire screen is surrounded by filter-pack material, and do not require a tremie pipe for filter-pack installation. Also, both high-solids bentonite grout and neat cement grout can be successfully installed using tremie pipes as small as 0.375 in. O.D. in annular spaces as small as 0.5 in. Several demonstrations of direct-push well installations (McCall et al., 1997; Kram et al., 2001) have documented that wells as small as 0.5 in. nominal diameter and as large as 1.5 in. nominal diameter can be installed successfully with an annular space much less than 2 in. surrounding the well (Figure 10.33).

In the other extreme, if the well is a relatively small diameter (2 in. nominal diameter or less) and the annular space is too large (e.g., more than 2 in. surrounding the casing or screen), well development through the thick filter-pack material will almost certainly be ineffective. Additionally, if the well casing material is polyvinylchloride (PVC) and the annular space above the filter pack is grouted with neat cement, the higher heat of hydration and the greater weight of the larger mass of neat cement in the larger borehole may combine to cause failure of the PVC casing, particularly at the joints.

FIGURE 10.32
Prepacked well screens are viable alternatives to conventional well designs. This type of screen is routinely used in direct-push well installations.

FIGURE 10.33
Installation of a prepacked well screen using a direct-push machine.

Design and Installation of Ground-Water Monitoring Wells

In some situations, it is desirable to drill the borehole several feet deeper than the well screen will be set to provide room below the screen for a sump (a blank piece of casing below the screen, with a plug on the bottom) (Figure 10.34). The sump can be used to collect sediment that is brought into the well during development so that the sediment does not fill up and clog the bottom of the screen. Sumps can also be used as traps to collect dense non-aqueous phase liquids (DNAPLs) in cases where the bottom of the well screen is installed at the top of the first confining layer. However, boreholes that are over-drilled in error should be backfilled with bentonite and then with filter-pack sand to the desired well depth and the well completed above so the borehole does not serve as a conduit for contaminant movement.

The type of drilling equipment required to produce a stable, open, vertical borehole for installation of a monitoring well depends upon the site geology, hydrology, and the purposes and objectives of the monitoring program. Engineering and geological judgment and some knowledge of the subsurface conditions to be encountered during drilling is required for the selection of the appropriate method used for drilling both the exploratory boreholes used for site characterization and the boreholes in which monitoring wells will be installed. Chapter 5 of this book and ASTM Standard D 6286 (Standard Guide to the Selection of Drilling Methods for Environmental Site Characterization [ASTM, 2004l]) should be consulted for additional detail on drilling method selection. Appropriate drilling methods for installing monitoring wells may include any one or a combination of several of the following methods: hollow-stem auger (ASTM D 5784 [ASTM, 2004m]); direct mud rotary (ASTM D 5783 [ASTM 2004n]); direct air rotary (ASTM D 5782 [ASTM, 2004o]); direct rotary casing advancement (ASTM D 5876 [ASTM, 2004p]); dual-wall reverse-circulation rotary (ASTM D 5781 [ASTM, 2004q]); cable tool (ASTM D 5785 [ASTM, 2004r]); or various casing advancement methods (ASTM D 5872 [ASTM, 2004s]), including sonic drilling (see Figure 10.35–Figure 10.40). Wherever feasible, it is advisable to use drilling procedures that do not require the introduction of drilling fluids, and that minimize the production of drill cuttings. Where the use of a drilling fluid is necessary to cope with site-specific drilling conditions, the fluid used should have as little impact as possible on water chemistry in the vicinity of the borehole. In addition, care should be taken to remove as much drilling fluid as possible from the

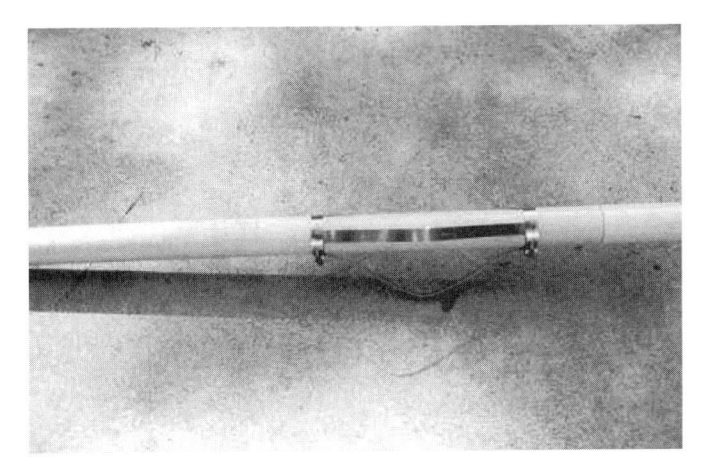

FIGURE 10.34
A sump positioned below the well screen (with a centralizer installed to help center it in the borehole) to collect sediment brought into the well during well development or to collect DNAPLs moving at the interface between an aquifer and a confining bed.

FIGURE 10.35
Hollow-stem auger drilling is the most widely used method for installing shallow monitoring wells in unconsolidated materials.

FIGURE 10.36
Direct mud rotary drilling is a commonly used method for installing both shallow and deep monitoring wells in unconsolidated materials and bedrock, but it cannot be used effectively in cavernous formations.

FIGURE 10.37
Air rotary with a casing hammer can be used to install monitoring wells in both unconsolidated materials and bedrock.

FIGURE 10.38
Dual-tube reverse-circulation rotary drilling is a useful method to overcome difficult drilling conditions, including cavernous or highly fractured bedrock and bouldery unconsolidated materials.

FIGURE 10.39
Cable-tool drilling is a slow but still reliable method for installing wells in all types of materials, especially where cased boreholes are required.

FIGURE 10.40
Sonic drilling is perhaps the most effective and cost-efficient method to use for installing monitoring wells because it is very fast, it does not require the use of drilling fluid, it produces almost no drilling waste, and continuous formation samples are collected as part of the drilling process.

well and the surrounding formation during the well development process, so that the potential impact on future collection of ground-water samples is minimized.

Some state guidelines or regulations and some Federal guidelines (e.g., Aller et al., 1991) require that the drilling method used to install monitoring wells must not introduce foreign materials (including drilling fluids) into the borehole, and that the disturbance to the formation caused by the drilling method must be minimized. Related additional requirements often specify that if a drilling fluid must be used, it must be water or air. The water must be clean water, chemically analyzed for the same suite of parameters as ground-water samples in the monitoring program for which the wells are installed. However, water alone cannot perform all of the necessary functions of a drilling fluid, so bentonite is often used, and is one of the few additives that regulatory agencies allow (Figure 10.41). Usually only pure bentonite (without polymeric additives) is specified, to stabilize the borehole or to control down-hole fluid losses or heaving sands. Polymeric drilling fluids or additives may contain organic compounds that enhance biological degradation of the drilling fluid (Figure 10.42), but the biological activity can cause long-term variations in the chemistry of ground-water samples that could be difficult to reverse (Lapham et al., 1996). If air is used, it is often required to be filtered by a high-efficiency in-line oil filter or an oil trap to reduce or remove any oil discharged into the air stream by the compressor. However, in most cases, air filters can reduce down-hole contamination (to low parts per million [ppm] levels in air) but not eliminate it. Oil-free compressors should be specified to prevent possible oil contamination of the borehole.

FIGURE 10.41
Bentonite drilling mud is a water-based drilling fluid that contains sufficient powdered bentonite to create a fluid viscous enough to entrain drill cuttings. The bentonite also creates a filter cake on the walls of the borehole that keeps formation water from invading the borehole and helps to hold the hole open during drilling. This filter cake must be removed from the borehole either prior to well installation or during well development or it may prevent formation water from entering the borehole and well.

FIGURE 10.42
Using organic polymer drilling fluids prevents the build-up of filter cake material on the borehole wall, but these fluids and their breakdown additives can alter water chemistry in the vicinity of the borehole for an extended period of time.

Some lubricants used on the drill string with some drilling methods (in particular, any of the rotary methods) can affect the chemistry of samples collected from the borehole or the well (Figure 10.43). For this reason, hydrocarbon-based lubricants are often not allowed in drilling programs overseen by regulatory agencies. Synthetic lubricants or other lubricants, such as canola oil-based lubricants with Teflon flakes (e.g., King Stuff, Hydro-Lube), can perform the same functions as petroleum-based compounds, and are environmentally acceptable (Figure 10.44). If drilling fluids or lubricants of any type are used, the type and amount of drilling fluid or lubricant used should be documented and samples retained for possible future analysis.

Although the drilling of a borehole is discussed in Chapter 5, the impact of the borehole on well installation requires additional discussion here. Borehole characteristics may dramatically affect the integrity and mechanical strength of the annular seal. The effects of annular seal failure on long-term well performance are discussed later in this chapter.

FIGURE 10.43
Petroleum hydrocarbon-based drill string lubricants should not be used in drilling monitoring wells because of the potential for alteration of water chemistry in the borehole.

Design and Installation of Ground-Water Monitoring Wells 675

FIGURE 10.44
Synthetic drill-string lubricants, such as this canola oil-based lubricant that contains Teflon flakes, are appropriate for use in drilling monitoring wells.

Examples of the relationship of the borehole to annular seal performance and other issues of concern are provided here.

Borehole characteristics are generally controlled by two primary factors: (1) the natural stratigraphic and geologic characteristics of the formation materials being drilled (e.g., rock hardness or sediment cohesiveness) and (2) the modifications of formation materials caused by the drilling technique.

The natural characteristics of the geologic media greatly influence the shape of the borehole wall. Drilling through well-lithified rock is most likely to produce the straight, smooth, symmetrical borehole wall most commonly envisioned by the layman, and most commonly portrayed in reports. Even some partially lithified or cemented sediment can conform nicely to drill-bit size to form this type of borehole. However, contrasting lithologies of irregular cementation, mechanical stability, or hardness typically causes highly irregular borehole wall shapes. Mechanically weak or comparatively softer formation materials commonly wash out or cave into the borehole, causing a localized increase in the amount of cuttings removed per foot drilled and a larger (often more than double the bit size) borehole diameter, which can be much larger under some conditions.

Furthermore, borehole irregularities are commonly not symmetrical. Diagrams in reports commonly portray boreholes as symmetrical, yet caliper logging and experience in drilling most sediment clearly indicates that irregular borehole shape is more common. Most illustrations indicate the geologic fabric of the drilled sediments or rock to be perpendicular to the axis of the borehole. In contrast, geologic fabric is often inclined to the borehole axis, whether due to depositional fabric (e.g., dipping bedding planes) or post-depositional features such as folding, faulting, joints, fractures, slumping, or clastic dikes. The interaction between geologic fabrics not perpendicular to the borehole and the drill bit commonly causes irregular cross-sectional shapes. These irregular shapes increase the difficulty of creating an effective annular seal, and confirming that a well has a good annular seal. Drilling sediments with geologic fabrics that are inclined to the borehole axis and that display contrasting mechanical properties to the bit can cause the drill bit to deviate, or "wander," resulting in a borehole that is neither straight nor plumb.

A borehole that is neither straight nor plumb may meet all specifications with regard to annular dimensions, but along its length does not maintain the same annular separation from installed casing. In such cases, it is common for casing to alternately press against and be remote from the borehole walls (unless centralizers are very fortuitously placed). Having casing in contact with the borehole wall affects the distribution and mechanical strength of annular seal materials.

The walls of the borehole are also affected by the drilling technique. Cable-tool drilling, using a "hard-tool" bit, routinely creates a concentric ring of compacted materials at the face of the borehole; other casing advancement methods create a similar effect (Figure 10.45a). Compaction lowers the hydraulic conductivity of formation materials and will affect both the entry of water into the borehole and well, and the results of formation hydraulic conductivity tests. Auger drilling often creates a "skin" on the borehole wall that reflects the characteristics of the last interval drilled, particularly if it contains clay. Smearing of clay on the borehole walls (Figure 10.45b) seals off pore spaces, fractures and other openings, greatly reducing flow into the borehole and well. Direct mud rotary drilling with bentonite-based drilling fluid creates a filter cake that coats the borehole wall and, in coarser formations, penetrates outward from the borehole wall into the formation to various extents. The filter cake is formed as the liquid portion of the drilling fluid penetrates the formation, leaving behind the bentonite solids (platelets) that stack up on top of each other. This filter cake significantly reduces formation hydraulic conductivity, even though it may only measure less than 0.125 in. thick (Figure 10.45c). Being composed of bentonite, it also has a very high cation exchange capacity, and thus has the ability to alter water chemistry for trace metals and major ions. Similar effects can be created by air rotary, depending on additives (soap-surfactants, misted mud, etc.). Reverse-circulation

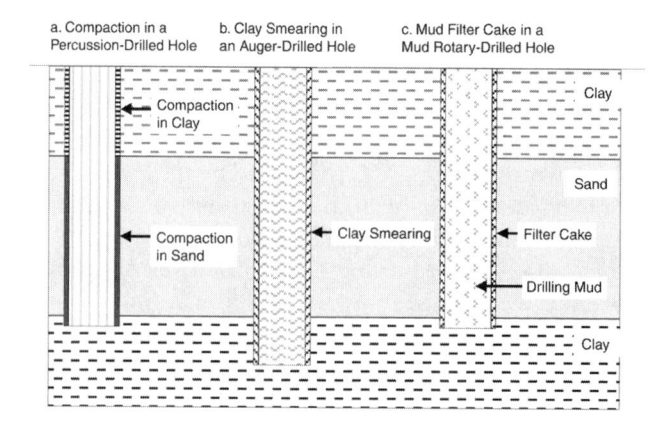

FIGURE 10.45
All drilling methods cause some form of damage to the borehole wall that must either be repaired before well installation or corrected during well development. (a) Compaction caused by percussion drilling methods, such as cable-tool or air rotary with a casing hammer. (b) Clay smearing on the borehole wall caused by auger drilling. (c) Drilling mud filter cake on the borehole wall caused during direct mud rotary drilling.

rotary methods remove drill cuttings from the borehole through the movement of fluid first down the borehole-drill string annulus, then up through the drill string. This may alter the makeup of the material and the size range of the material lining the borehole wall. All of these forms of drilling damage may reduce flow through permeable intervals penetrated by the borehole, and should be removed either at the end of the drilling process or during well development (see Chapter 12).

All drilling methods leave a short-term "signature" on the borehole. Perhaps the most significant impact on boreholes in which monitoring wells are installed is caused by direct mud rotary drilling. For large-diameter water-supply wells, if the mud weight and viscosity are controlled, and vigorous, long-duration well development is conducted, removing the mud filter cake is usually not a problem. However, in small-diameter boreholes and well screens, particularly those with artificial filter packs and low-open-area slotted casing as well screen, removing the filter cake and restoring formation hydraulic conductivity through well development alone is very difficult and requires persistence. Non-steel (PVC) casing and screen typically used in monitoring wells can also be easily damaged by vigorous development that would not harm more robust large-diameter wells made of more durable materials.

Also, drilling mud greatly inhibits (and, in many cases, may prevent) the annular seal from bonding directly with the material forming the borehole wall. If the mud breaks down over time and dissipates through the pores of the formation material, the integrity of the annular seal is reduced. Furthermore, if problems in hole stability or fluid loss occur during drilling, lost-circulation material may be introduced into the drilling fluid to increase the stability of the borehole wall, but this may have potentially negative consequences. These materials can never be removed completely from the borehole following drilling, or during well development. The short-term physical and chemical "signature" produced by these materials may affect baseline hydraulic conductivity testing results and initial ground-water sample analytical results obtained from those wells. However, in the long term, these materials tend to break down, allowing flow to occur into the borehole in zones initially not identified as flowing (or greatly flowing). Thus, over the long term, flow in some zones adjacent to the borehole may improve while, at the same time, well construction materials designed to confine the flow into the well to specific flow zones (e.g., annular seal materials) may deteriorate.

Design Components of Monitoring Wells

Monitoring Well Casing and Screen Materials

The purpose of casing in a ground-water monitoring well is to provide a means of access from the surface to some zone of interest in subsurface saturated geologic materials. Well casing prevents the collapse of geologic materials into the borehole, and allows access to the ground water in the target-monitoring zone, by means of a well screen generally attached to the terminal end of the casing, for determinations of ground-water quality and potentiometric head. Casing also prevents (in conjunction with a proper annular seal) hydraulic communication between separate water-bearing zones penetrated by the borehole.

Historically, the selection of casing material for water-supply wells and other types of wells focused on the material's structural strength, ease of handling, and durability in long-term exposure to natural subsurface conditions. As ground-water samples taken from monitoring wells began to be chemically analyzed at the ppb (or $\mu g/l$) level, however, the focus shifted to the potential impact that casing materials may have on the chemical integrity, or "representativeness" of the ground-water samples. Additionally, durability in long-term exposures to potentially hostile man-induced chemical environments is a real concern. The selection of appropriate materials for monitoring well casing and screen must consider all these factors to ensure that the well will produce representative samples, and that it will endure for the life of the monitoring program.

Unique site-specific and logistical factors should be the controlling criteria in the selection of monitoring well casing and screen materials. Site-specific factors include the geologic environment, natural geochemical environment, anticipated well depth, and types of contaminants present or anticipated. Logistical factors include well drilling method, ease of handling and cleaning, and cost (for materials and shipping). Because no single casing or screen material can be used reliably over the wide range and variety of natural and man-induced site-specific conditions that may be encountered, it is critical that these conditions be evaluated thoroughly before selecting a material for monitoring well casing and screen. The selection of monitoring well casing and screen materials must be based on the ability of three primary casing characteristics — physical strength, chemical resistance, and chemical interference potential — to meet site-specific conditions.

Requirements of Casing and Screen Materials

Physical Strength of Well Casing and Screen

Monitoring well casing and screen materials must have the structural strength to withstand the forces exerted on them by the surrounding geologic materials and the forces imposed on them during well installation and development (Figure 10.46). The material

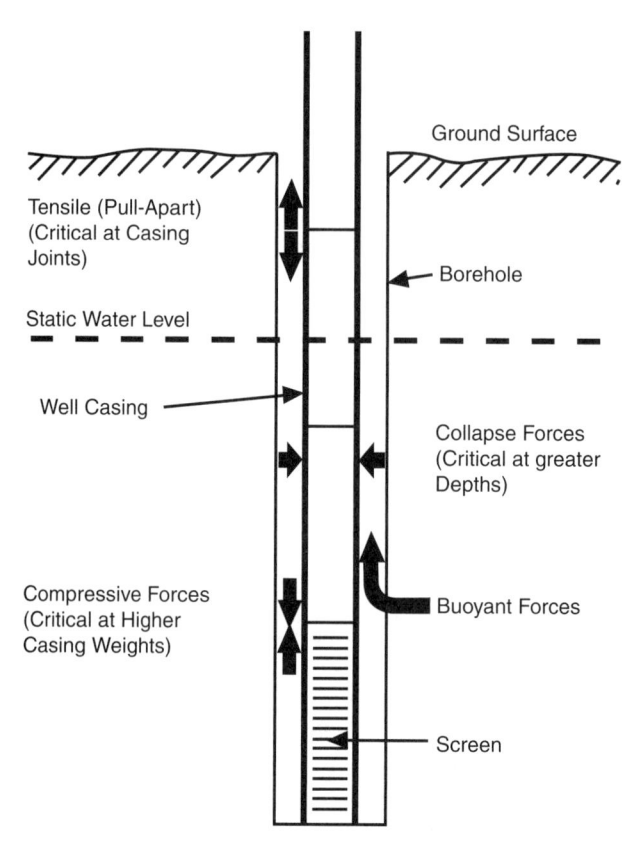

FIGURE 10.46
Forces exerted on a monitoring well casing and screen during well installation.

should be able to retain its structural integrity for the expected duration of the monitoring program, under both natural and man-induced subsurface conditions. The three components of casing and screen structural strength are tensile strength, compressive (column) strength, and collapse strength. Each property must be evaluated for a particular application. Relative strengths of stainless steel and PVC casing are presented in Table 10.4 and Table 10.5, respectively. The values in these tables are for materials available from one manufacturer, though relative strengths of materials supplied by other manufacturers should be similar, except for the tensile strength of the casing joint, which varies with the coupling design. A comparison of the relative tensile and collapse strengths of small-diameter casing, for five materials that have historically been used in monitoring well construction, is presented in Table 10.6. The weight per unit length, which is used along with tensile strength to calculate maximum permissible string length for a casing material, is presented in Table 10.7.

The tensile strength of a casing or screen material, defined as the load required to pull the casing apart, is the most significant strength-related property of casing or screen materials. Tensile strength varies according to casing composition, manufacturing technique, and the type of casing joint used; it is closely related to the strength of the parent material as well as the casing dimensions (diameter and wall thickness). For a monitoring well installation, the casing material selected should have, as a minimum, enough tensile strength to support its own weight when suspended from the surface in an air-filled

TABLE 10.4

Material Strength Data for Type 304 Stainless Steel[a]

Nominal Size	O.D. (in.)	I.D. (in.)	Wt (lb/ft)	Collapse (psi)	Tensile (lb)	Column[b] (lb)	Joint Tensile (lb)
2-in. schedule 40 casing	2.375	2.067	3.653	3526	85,900	6350	15,900
2-in. schedule 5 casing	2.375	2.245	1.604	986	37,760	3000	15,900
2-in. wire-wound screen	2.375	1.900	4.0	1665	10,880	810	15,900
4-in. schedule 40 casing	4.500	4.026	10.790	2672	254,400	69,000	81,750
4-in. schedule 5 casing	4.500	4.334	3.915	315	92,000	26,800	81,750
4-in. wire-wound screen	4.500	4.000	6.0	249	16,320	4500	81,750
5-in. schedule 40 casing	5.563	5.047	14.6	2231	343,200	145,490	91,500
5-in. schedule 5 casing	5.563	5.345	6.4	350	148,800	66,660	91,500
5-in. wire-wound screen	5.560	5.030	4.8	134	38,600	13,040	91,500
6-in. schedule 40 casing	6.625	6.065	19.0	1942	444,800	270,000	94,500
6-in. schedule 5 casing	6.625	6.407	7.6	129	178,400	113,660	94,500
6-in. wire-wound screen	6.620	6.090	5.5	176	54,000	19,170	94,500

[a] Information provided by Johnson Filtration Systems Inc.
[b] For all column calculations, the span = 20 ft, hinged at one end, and fixed at the other end.

TABLE 10.5

Material Strength Data for PVC[a]

Nominal Size	O.D. (in.)	I.D. (in.)	Wt (lb/ft)	Collapse (psi)	Tensile (lb)	Column[b] (lb)	Joint Tensile (lb)
2-in. schedule 40 casing	2.375	2.067	0.64	307	7500	90	1800
2-in. schedule 80 casing	2.375	1.939	0.88	947	9875	125	1800
2-in. wire-wound screen	2.375	1.875	0.8	99	1800	25	1800
4-in. schedule 40 casing	4.500	4.026	1.9	158	22,200	1030	6050
4-in. schedule 80 casing	4.500	3.826	2.6	494	30,850	1375	6050
4-in. wire-wound screen	4.620	4.000	1.7	79	2250	150	6050
5-in. schedule 80 casing	5.563	4.813	3.9	324	42,780	2940	6050
5-in. wire-wound screen	5.560	4.810	2.5	79	4610	307	6050
6-in. schedule 80 casing	6.625	5.761	5.4	292	58,830	5760	4000
6-in. wire-wound screen	6.620	5.680	3.7	87	5770	552	4000

[a]Information provided by Johnson Filtration Systems Inc.
[b]For all column calculations, the span = 20 ft, hinged at one end, and fixed at the other end.

TABLE 10.6

Comparative Strengths of Well Casing Materials[a]

Material	Casing Tensile Strength (lb) 2-in. Nominal	Casing Tensile Strength (lb) 4-in. Nominal	Casing Collapse Strength (lb/in.^{2}) 2-in. Nominal	Casing Collapse Strength (lb/in.^{2}) 4-in. Nominal
PVC	7500	22,000	307	158
PVC casing joint[b]	2800	6050	300	150
Stainless steel[c]	37,760	92,000	896	315
Stainless steel casing joint[b]	15,900	81,750	No data	No data
Polytetrafluoroethylene (PTFE)	3800	No data	No data	No data
PTFE casing joints[b]	540	1890	No data	No data
Epoxy fiberglass	22,600	56,500	330	250
Epoxy casing joints[d]	14,000	30,000	230	150
Acrylonitrile–butadiene–styrene (ABS)	8830	22,000	No data	No data
ABS casing joints[d]	3360	5600	No data	No data

[a]Information provided by E.I. du Pont de Nemours & Company, Wilmington, DE.
[b]All joints are flush-threaded.
[c]Stainless steel casing materials are schedule 5 with schedule 40 joints; other casing materials (PVC, PTFE, epoxy, ABS) are schedule 40.
[d]Joints are not flush-threaded, but are a special type that is thicker than schedule 40.

TABLE 10.7

Weight per Unit Length and Weight Ratios of Well Casing Materials (2-in. Nominal)

Material	Weight by Schedule Number (lb/ft) #5	#10	#40	#80	Approximate Weight Ratios[a]
PVC	—	—	0.65	0.91	1.3
Stainless steel	1.62	2.06	3.65	5.07	3.4[b]
PTFE	—	—	1.50	1.90	3.0
Epoxy fiberglass	—	—	0.50	—	1.0
ABS	—	—	0.60	—	1.2

[a] Weight ratio is obtained by multiplying each schedule 40 weight by 2.
[b] Schedule 5 casing with schedule 40 coupling for a 10-ft length of pipe.

borehole (Figure 10.47). The maximum installation depth can be calculated by dividing the tensile strength for a given casing material by the linear weight of the casing. In most cases, the casing will encounter water in the borehole during installation; the buoyant force of the water increases the length of casing that can be suspended in the borehole by a factor that depends on the specific gravity of the casing material. The tensile strength of the casing joints is more important than the tensile strength of the casing itself, because the joints are usually the weakest points in the casing string. Therefore, joint strength is more commonly used to determine the maximum axial load that can be placed on a casing string.

The compressive or column strength of a casing or screen material is defined as the load required to deform the material by compressing it. The properties of the casing or screen material, specifically the yield strength and stiffness, are more significant in determining

FIGURE 10.47
Installation of well casing and screen materials in a drilled borehole.

compressive strength than are the dimensional parameters, although casing wall thickness is also important.

Another significant strength-related property of casing and screen materials is collapse strength, or the capability of a casing to resist collapse caused by any and all external loads to which it is subjected, both during and after installation. The collapse strength of a casing material is determined principally by dimensional parameters. Most notably, the collapse strength of a piece of casing is proportional to the cube of its wall thickness. Therefore, a small increase in wall thickness provides a significant increase in collapse strength. Casing and screen are most susceptible to collapse during installation, when the casing string has not yet been confined and restrained by the placement of filter pack or annular seal materials around it. Once a casing string is properly installed and confined, its resistance to collapse is enhanced so that collapse is no longer a concern (NWWA/PPI, 1980).

Among the external loadings on casing that may contribute to casing collapse are the following:

- Net external hydrostatic pressure produced when the static water level outside the casing is higher than that on the inside
- Asymmetrical loads on the casing resulting from uneven placement of backfill (e.g., annular seal) materials
- Uneven collapse of unstable formation materials
- Weight of grout on the outside of a partially water-filled casing
- Forces associated with well development that produce large differential pressures on the inside and outside of the casing

Of these, only the first, external hydrostatic pressure, can be predicted and calculated with any accuracy. To provide sufficient margin against possible collapse by all normally anticipated external loadings, a casing material is selected so that its resistance to collapse is greater than that required to resist external hydrostatic pressure alone. Generally, a safety factor of at least two is recommended (NWWA/PPI, 1980). In well installations in difficult drilling or geologic conditions (e.g., heaving sands), a safety factor of at least three should be employed.

Except for joint strength, all of the strength characteristics of a piece of casing are reduced when the casing is slotted to produce well screen. Continuous-slot wire-wound well screen varies in strength depending on the configuration of the vertical columns and the wire-wrap screen, and the type of material. In general, however, the strength of continuous-slot wire-wound screen is greater than that of slotted casing.

Chemical Resistance of Casing and Screen Materials

Materials used for monitoring well casing and screen must be durable enough to withstand potential chemical attacks from either natural chemical constituents or contaminants in ground water. In particular, metallic casing materials should be resistant to corrosion (galvanic or electrochemical) and plastic casing materials should be resistant to chemical degradation (Parker, 1991). Because the extent to which chemical attacks occur is primarily dependent on the presence and concentration of certain chemical constituents in ground water, the casing material should be selected after considering existing or anticipated ground-water chemistry. Not only may natural or man-induced ground-water chemistry affect the structural integrity of monitoring well casing or screen, but by-products of casing deterioration also may adversely affect the chemistry of water samples taken from monitoring wells.

Chemical Interference from Casing and Screen Materials

Materials used for monitoring well casing and screen must not remove chemicals from ground water by adsorption on to the material surface or by absorption into the material matrix or pores. Loss of chemical constituents from a ground-water sample may create "false negatives," which produce the false impression that those chemical constituents are not present, or are present below their actual concentration in solution. Additionally, the well casing and screen materials must not desorb or leach chemical constituents from them into the ground water to be sampled. The addition of leached or desorbed chemicals to a ground-water sample may produce "false positives," which indicate possible ground-water contamination when, in fact, none is present. Therefore, in the selection of monitoring well materials, the potential interactions between casing or screen materials and the natural and man-induced geochemical environment must be carefully considered (Parker, 1991).

Types of Casing and Screen Materials

Casing used in monitoring wells could conceivably be made of almost any rigid tubular material, although experience dictates that the choices are limited to only a few materials. Casing materials historically or currently used in ground-water monitoring wells can be categorized into four general types:

- Thermoplastic materials, including PVC and ABS
- Fluoropolymer materials, including polytetrafluoroethylene (PTFE), tetrafluoroethylence (TFE), fluorinated ethylene propylene (FEP), perfluoroalkoxy (PFA), and polyvinylidine fluoride (PVDF)
- Metallic materials, including carbon steel, low-carbon steel, galvanized steel, and stainless steel (particularly types 304 and 316)
- Fiberglass-reinforced materials, including fiberglass-reinforced epoxy (FRE) and fiberglass-reinforced plastic (FRP)

Each of these materials has physical and chemical characteristics that influence its use in site-specific hydrogeologic and contaminant-related conditions. PVC, low-carbon steel, galvanized steel and stainless steel casing and screen materials are discussed in greater detail in the following sections. Other materials (e.g., fluoropolymer materials, ABS, FRE and FRP) are either infrequently or no longer used in monitoring wells, or not manufactured in the sizes typically used for monitoring wells. In some cases, too little practical application-related information is available on which to base decisions on selection of these materials for use in ground-water monitoring wells.

Thermoplastic Materials

Thermoplastics are man-made materials that consist of varying formulations of plastics that can be formed and reformed repeatedly; they are softened by heating and harden upon cooling. This characteristic allows thermoplastics to be molded or extruded into rigid well casings.

Nearly all thermoplastic well casing is one of two materials: PVC or ABS. The strength, rigidity, and temperature resistance characteristics of both of these materials are generally sufficient to allow well casings and screens made from them to withstand the typical stresses of handling, installation, and loading for most well installations. In addition, rigid, hardened thermoplastics offer complete resistance to galvanic and electrochemical corrosion, high resistance to abrasion, high strength-to-weight ratios, light weight, durability in most natural ground-water environments, low maintenance, partial flexibility, workability, and low cost, making them ideal for many monitoring well applications. The use of ABS casing and screen materials in monitoring wells, however, is very rare; for this reason, the remainder of the discussion in this section will focus on PVC.

Rigid PVC well casing (Figure 10.48) is produced by combining PVC resin with various types of stabilizers, lubricants, pigments, fillers, and processing aids. The amounts of these additives can be varied to produce different PVC plastics with properties tailored to specific applications. PVC used for well casing is composed of a rigid hardened (unplasticized) polymer formulation (PVC Type 1) that has high tensile, compressive and collapse strength, and good chemical resistance except to low molecular weight ketones, aldehydes, and chlorinated solvents (Barcelona et al., 1983).

Strength-Related Characteristics of PVC: Typical physical properties of PVC well-casing materials are provided in Table 10.8. Included in this table are minimum tensile strength and compressive strength values for several cell classes of PVC, as required under ASTM Standard D 1785 (ASTM, 2004t). Dimensions, hydraulic collapse pressure, and unit weight of PVC well casing are provided in Table 10.9.

In comparison to metallic materials, the tensile, compressive, and collapse strength of PVC is relatively low. With respect to tensile strength, the light weight of PVC offsets the low strength so that for most installations of PVC well casing, the axial loading is not a limiting factor. Assuming a dry borehole and a safety factor of two (equivalent to a tensile strength of half the value given in Table 10.8), the theoretical maximum permissible string length for PVC casing materials is in excess of 4000 ft. This maximum string length is for fully cured, solvent-cemented connections under conditions of short-term loading only. In monitoring well applications, the use of solvent-cemented joints is

FIGURE 10.48
PVC is the most commonly used material for monitoring well casing and screen because of its low weight per unit length, relatively high strength, excellent chemical resistance characteristics, and low cost.

TABLE 10.8

Typical Physical Properties of PVC Well Casing Materials at 73.4°F[a]

Property	ASTM Test Method	PVC Cell Class, per ASTM D-1784	
		12454-B&C	14333-C&D
Specific gravity	D-792	1.40	1.35
Tensile strength [lb/in.^{2}]	D-638	7000[b]	6000[b]
Tensile modulus of elasticity [lb/in.^{2}]	D-638	400,000[a]	320,000[b]
Compressive strength [lb/in.^{2}]	D-695	9000	8000
Impact strength, izod, ft-lb/in. notch	D-256	0.65	5.0
Deflection temperature under load (264 psi) [°F]	D-648	158[b]	140[b]
Coefficient of linear expansion [in./in. °F]	D-696	3.0×10^{-5}	5.0×10^{-5}

[a]*Source*: From NWWA/PPI, 1980. With permission.
[b]These are minimum values set by the corresponding ASTM Cell Class designation. All others represent typical values.

discouraged because of the high potential for sample chemical alteration; threaded joints are much more commonly used. For threaded connections, the maximum string length is reduced substantially and, for long-term loading, the maximum permissible string length is further reduced. Although the degree of reduction of maximum permissible string length depends on the type of joint used, it can be expected that flush-joint, threaded connections will reduce the theoretical maximum permissible string length by 30 to 70%, or to about 1200–2000 ft.

Because the specific gravity of PVC (1.4) is not much higher than that of water, PVC well casings are relatively light when immersed in water. The buoyant force of water for PVC is, therefore, very high, increasing the maximum string length by about 40% for that portion of the casing in contact with water.

Chemical Resistance and Chemical Interference Characteristics of PVC: With respect to chemical resistance, PVC is superior in some respects to metallic materials because it is a nonconductor and thus totally immune to electrochemical or galvanic corrosion. In addition, PVC is resistant to biological attack, and to chemical attack by soil, water, and other naturally occurring substances present in the subsurface. The resistance of PVC to common hazardous materials applies for most acids, oxidizing agents, salts, alkalies, oils, and fuels (NWWA/PPI, 1980). Even after long-term (6 months) immersion in common types of gasoline containing high concentrations of aromatic hydrocarbons (benzene, toluene, ethylbenzene, and xylenes), rigid PVC does not exhibit any swelling or other alteration effects (Schmidt, 1987). PVC is, however, susceptible to chemical attack by certain organic solvents. These solvents can produce an effect called solvation, which is the physical degradation of the plastic. Solvent cementing of thermoplastic well casing is based on solvation, which occurs in the presence of specific organic solvents. If these solvents (which include tetrahydrofuran [THF], methyl ethyl ketone [MEK], methyl isobutyl ketone [MIBK], cyclohexanone [CH], methylene chloride, acetone, dimethylformamide [DMF], and pyridine) are present in the subsurface in pure compound form or in very high aqueous-phase concentrations, they will chemically degrade PVC well casing to some degree. In general, the chemical attack on the PVC polymer matrix will increase as the organic content of the solution with which it is in contact increases.

TABLE 10.9
Dimensions, Hydraulic Collapse Pressure, and Unit Weight of PVC Well Casing

Outside Diameter (in.) Nominal	Actual	SCH#	Wall Thickness (in.)	SDR	Weight in Air (lb/100 ft) PVC 12454	PVC 14333	Weight in Water (lb/100 ft) PVC 12454	PVC 14333	Hydraulic Collapse Pressure (psi) PVC 12454	PVC 14333
2	2.375	SCH80	0.218	10.9	94	91	27	24	947	758
		SCH40	0.154	15.4	69	66	20	17	307	246
2½	2.875	SCH80	0.276	10.4	144	139	41	36	1110	885
		SCH40	0.203	14.2	109	105	31	27	400	320
3	3.500	SCH80	0.300	11.7	193	186	55	48	750	600
		SCH40	0.216	16.2	143	138	41	36	262	210
3½	4.000	SCH80	0.316	12.6	235	227	67	59	589	471
		SCH40	0.226	17.7	172	176	49	43	197	158
4	4.500	SCH80	0.337	13.3	282	272	80	70	494	395
		SCH40	0.237	19.0	203	196	58	51	158	126
4½	4.950	SCH80	0.248	20.0	235	226	67	58	134	107
		SCH40	0.190	26.0	182	176	52	46	59	47
5	5.563	SCH80	0.375	14.8	391	377	112	98	350	280
		SCH40	0.258	21.6	276	266	79	69	105	84
6	6.625	SCH80	0.432	15.3	538	519	154	134	314	171
		SCH40	0.280	23.7	358	345	102	89	78	62

Barcelona et al. (1983) and Berens (1985) list the groups of chemical compounds that may cause degradation of the thermoplastic polymer matrix and the release of compounding ingredients which otherwise would remain in the solid material. These chemical compounds include low molecular weight ketones, aldehydes, amines, and chlorinated alkenes and alkanes.

Ranney and Parker (1995, 1997) studied the degradation of rigid, hardened Type I PVC well casing caused by a variety of neat chemical compounds (pure products) and specific chemical conditions. They found that PVC has excellent resistance to alkaline and acidic conditions, aliphatic hydrocarbons, alcohols and gasoline, and that the classes of neat chemical compounds that degraded PVC included aromatic hydrocarbons, aromatic and aliphatic chlorinated solvents, ketones, anilines, aldehydes and nitrogen-containing organic compounds. The specific compounds studied by Ranney and Parker are listed in Table 10.10. Parker and Ranney (1994b, 1995, 1996) determined that when the relative solubility of a single organic solute or the sum of the relative solubilities of several organic solutes that are known PVC solvents is less than 0.1 (10% of the solubility limit of the solvents in water), there is no measurable degradation effect (either softening or weight gain) on PVC. At slightly higher relative solubilities of known PVC solvents (0.2 and 0.4, or 20 and 40% of the solubility limit), they found very slight changes in hardness of the PVC. For chemical compounds that are considered to be swelling agents for PVC (e.g., trichloroethylene, chlorobenzene, nitrobenzene, nitrotoluene, bromodichloromethane, 1,1,2-trichloroethane, and 1,2-dichloroethane), they found that the relative solubilities had to exceed 0.5 for degradation effects to occur. This work is in agreement with that of Berens (1985), who determined that at activities less than about 0.25, the calculated permeation of most organic compounds through the wall of PVC pipe is effectively zero for many centuries, and that rigid PVC will be softened only by strong PVC solvents or swelling agents at activities greater than 0.5. The conclusion that can be derived from these studies is that, in general, PVC should not be used in situations in which the aqueous-phase concentrations of known solvents or swelling agents for PVC exceed 25% of the solubility limit of the solvent or swelling agent.

TABLE 10.10

Compounds Studied in the PVC Degradation Studies Conducted by Ranney and Parker (1995, 1997)

Hydrocarbons (aliphatic and aromatic)	Oxygen-containing compounds
Benzene	(either a ketone, alcohol, aldehyde, or ether)
Gasoline (93 octane, unleaded)	Acetone
Hexane (85% N-hexane)	Benzaldehyde
Kerosene (K−1)	Benzyl alcohol
Toluene	Cyclohexanone
o-Xylene	Methyl alcohol
Chlorinated solvents	Methyl ethyl ketone
(aliphatic and aromatic)	Tetrahydrofuran
Bromochloromethane	Nitrogen-containing compounds
Carbon tetrachloride	N-Butylamine
Chlorobenzene	Diethylamine
Chloroform	Dimethylformamide
1,2-Dichlorobenzene	Nitrobenzene
1,2-Dichloroethane	Acids and bases
trans-1,2-Dichloroethylene	Acetic acid (glacial)
Methylene chloride	Hydrochloric acid (25% w/v)
Tetrachloroethylene	Sodium hydroxide (25% w/v)
Trichloroethylene	

Several studies have been conducted to address the leaching or desorption effects that aqueous solutions of organic compounds and trace metals may have on PVC, and the effects that PVC may have on sorption of low concentrations of organic solutes or trace metals. In one study, Ranney and Parker (1998a) determined that PVC was the least active material of six polymeric materials studied (PVC, FRE, FRP, ABS, FEP and PTFE), with respect to sorption and leaching, when exposed to ppm level concentrations of 11 organic solutes. All of the materials studied sorbed some of the solutes to some extent, but PVC sorbed the analytes studied most slowly and to the least extent compared with the other materials, and did not leach any organic constituents in the presence of the solutes tested. In a follow-up study to test for sorption and leaching of trace levels of metals (Ranney and Parker, 1998b), PVC was one of the most inert materials tested, sorbing only statistically insignificant levels of Cd and only slightly higher levels of Pb. They concluded that PVC is a good choice for most monitoring applications where both organic and inorganic analytes are of concern. Hewitt (1992, 1994) found that, of six metals examined (Cr, Ni, Cu, Cd, Fe, and Pb), only Pb leached at a level higher than that from the control, and concluded that PVC ranked with PTFE as the best materials of construction for wells intended to monitor trace levels of metals.

In other laboratory studies of leaching associated with PVC well-casing material, Curran and Tomson (1983), Ranney and Parker (1994) and Parker and Jenkins (1986) determined that little or no leaching occurred. In the former study, it was found that, in testing several different samples (brands) of rigid PVC well casing, trace organics either were not leached or were leached only at the sub-ppb level. In the latter study, which was conducted using ground water in contact with two different brands of PVC, it was concluded that no chemical constituents were leached at sufficient concentrations to interfere with reverse-phase HPLC analysis for low ppb levels of 2,4,6-trinitrotoluene (TNT), hexahydro-1,3,5-trinitro-1,3,5,7-tetrazocine (HMX), or 2,4-dinitrotoluene (DNT) in solution. Parker et al. (1990) found that PVC casings sorbed lead and leached cadmium in very small quantities over several hours; Ranney and Parker (1996) had similar results for cadmium. The study by Curran and Tomson (1983) confirmed previous field work at Rice University (Tomson et al., 1979) which suggested that PVC well casing did not leach significant amounts (i.e., at the sub-ppb level) of trace organics into ground-water samples.

Miller (1982) conducted a laboratory study to determine whether several plastics, including rigid PVC well casing, exhibited any tendency to sorb potential contaminants from solution. Under the conditions of his test, Miller found that PVC moderately adsorbed tetrachloroethylene and adsorbed lead, but did not adsorb trichlorofluoromethane, trichloroethylene, bromoform, 1,1,1-trichloroethane, 1,1,2-trichloroethane, or chromium. In this experiment, sorption was measured weekly for 6 weeks and compared with a control; maximum sorption of tetrachloroethylene occurred at 2 weeks. Although Miller (1982) attributed the losses of tetrachloroethylene and lead strictly to adsorption, the anomalous behavior of tetrachloroethylene compared with that for other organics of similar structure (i.e., trichloroethylene) is not explained. In a follow-up study to determine whether or not the tetrachloroethylene could be desorbed and recovered, only a small fraction of the tetrachloroethylene was recovered. Thus, whether strong adsorption or some other mechanism (i.e., enhanced biodegradation in the presence of PVC) accounts for the difference is not clear.

In another laboratory study, Reynolds and Gillham (1985) determined that losses of selected organics (specifically 1,1,1-trichloroethane, 1,1,2,2-tetrachloroethane, bromoform, hexachloroethane, and tetrachloroethylene) in the presence of PVC and other polymeric casing materials could be a source of bias to ground-water samples collected from water standing in the well bore. PVC was found to adsorb four of the five compounds studied (all except 1,1,1-trichloroethane), but it was concluded that the rate of adsorption was sufficiently slow that adsorption bias would not be significant for the adsorbed compounds if well purging and sampling were done on the same day. However, losses due to biodegradation in this study cannot be ruled out because nothing was done to prevent such losses (i.e., no biocide was added to the solutions). In fact, there is evidence that biotransformation occurred in the samples exposed to PVC. After 3 weeks, the authors noticed that these samples contained several additional peaks that were similar to peaks observed in degraded stock solutions of two of the organic compounds studied (bromoform and hexachloroethane).

In the laboratory study of Parker and Jenkins (1986), it was found that significant losses of TNT and HMX from solution occurred in the presence of PVC well casing. However, a follow-up study to determine the mechanism for the losses led them to attribute the losses to increased microbial degradation rather than to adsorption. These results raise questions regarding whether losses found in other laboratory or field studies which did not consider biodegradation as a loss mechanism should, in fact, be attributed to biodegradation rather than to either adsorption or absorption.

Gillham and O'Hannesin (1990) studied the sorption of six monoaromatic hydrocarbons (benzene, toluene, ethylbenzene, o-, m-, and p-xylenes) onto five well casing materials (PVC, PTFE, PVDF, FRE, and stainless steel) and two tubing materials (flexible PVC and polyethylene). They determined that, while some degree of sorption was observed for all compounds on all polymer well-casing materials, rigid PVC showed the least sorption, followed by PVDF, FRE, and PTFE (which showed significant sorption at rapid rates); no uptake of any compound onto stainless steel was noted. They concluded that, based on relative sorption characteristics of polymer well casing materials, PVC was a better material to use than PTFE.

A field study conducted to determine the potential for sorption of low concentrations (about 100 ppb) of volatile aromatic hydrocarbons onto various casing materials demonstrated that there was no significant difference in adsorption among PVC, PTFE, and stainless steel (Sykes et al., 1986).

In two separate studies of sorption effects caused by several different types of well casing materials (PVC, PTFE, and stainless steel 304 and 316), Parker et al. (1990) and Parker and Ranney (1993) determined that while PVC sorbed several organic compounds, the rates of sorption were always considerably slower than the rates observed for PTFE (stainless steel materials did not sorb any organic compounds). They concluded that the rates of sorption of organic compounds onto PVC were slow enough that they would not be of any concern in monitoring wells that are purged prior to sampling. Also, they found that while PVC sorbed Pb and leached Cd, PTFE also sorbed Pb, and the stainless steels sorbed Cr, As, Pb, and leached Cd and were subject to surface oxidation (rust). This led them to conclude that PVC was a good choice for monitoring metal species and that, for situations in which both metals and organic compounds were of concern, PVC was the best compromise of the four materials tested. In a follow-up study, Parker and Ranney (1994a) concluded that organic compound concentrations (at the ppm and ppb levels) did not affect the percentage loss relative to controls and, again, that PTFE sorbs organic compounds at rates faster than PVC. Notably, they determined that the rate and extent of loss due to sorption associated with PVC was not large enough to be of concern in ground-water monitoring, and that there is no basis for using stainless steel or PTFE casing materials over PVC for monitoring trace levels of organic compounds.

Extensive research has been conducted in the laboratory (specifically on water-supply piping) to evaluate vinyl chloride monomer leaching from new and old PVC pipe. The data generated in these studies support the conclusion that, under conditions in which PVC is in contact with water, the level of trace vinyl chloride leaching from PVC pipe is extremely low compared to residual vinyl chloride monomer (RVCM) content in PVC pipe. Since 1976, when the National Sanitation Foundation (NSF) established an RVCM monitoring and control program for PVC pipe used in potable water supplies and well casing, process control of RVCM levels in PVC pipe has improved markedly. According to Barcelona et al. (1983), the level of RVCM allowed in NSF-certified PVC products (less than or equal to 10 ppm RVCM) limits potential leached concentrations of vinyl chloride monomer to less than 1 ppb. Leachable amounts of vinyl chloride monomer have decreased as RVCM levels in piping products continue to be reduced. No documented instances of residual vinyl chloride monomer occurrence in ground-water samples are known.

Plasticizers are not added to PVC formulations used for Type I PVC well casing because casing is a rigid, hardened material. By contrast, flexible PVC tubing (also known as Tygon tubing) may contain from 30 to 50% plasticizers by weight. The presence of these high levels of plasticizers in flexible PVC tubing has been documented by several researchers (Junk et al., 1974; Barcelona et al., 1983; Barcelona et al., 1985b) to produce significant chemical interference effects. However, because PVC well casing contains no plasticizers, no plasticizer-induced chemical interference problem should exist for PVC well casing.

Rigid PVC may contain other additives, primarily stabilizers, at levels approaching 5% by weight. Some representative chemical classes of additives that have been used in the manufacture of rigid, hardened PVC well casing are listed in Table 10.11. Boettner et al. (1981) determined through a laboratory study that several of the PVC heat-stabilizing compounds, notably dimethyltin and dibutyltin species, could potentially leach out of rigid PVC at very low (sub-ppb) levels. This leaching was found to decrease dramatically over time. Factors that influenced the leaching process in this study included solution pH, temperature and ionic composition, and exposed surface area and surface porosity of the pipe material. No leaching of stabilizing compounds into ground-water samples collected from PVC wells has been documented in the scientific literature.

In addition to setting a limit on RVCM, the NSF has set specifications for certain chemical constituents in PVC formulations. The purpose of these specifications, outlined in NSF Standard 14 (National Sanitation Foundation, 2004), is to control the amount of chemical additives in both PVC well casing and pipe used for potable water supply. The maximum

TABLE 10.11

Representative Classes of Additives in Rigid PVC Materials Used for Pipe or Well Casing[a]

(Concentration in wt.%)

Heat Stabilizers (0.2–1.0%)	Fillers (1–5%)
Dibutyltin diesters of lauric and maleic acids	$CaCO_3$
Dibutyltin *bis* (laurylmercaptide)	Diatomaceous earth
Dibutyltin-β-mercaptopropionate	Clays
Di-*n*-octyltin maleate	Pigments
Di-*n*-octyltin-S,S,-*bis* isoctyl mercaptoacetate	TiO_2
Di-*n*-octyltin-β-mercaptopropionate	Carbon black
Various other alkyltin compounds	Iron and other metallic oxides
Various proprietary antimony compounds	Lubricants (1–5%)
	Stearic acid
	Calcium stearate
	Glycerol monostearate
	Montan wax
	Polyethylene wax

[a] *Source*: From Barcelona et al., 1983. With permission.

contaminant levels permitted in a standardized leach test on NSF-approved PVC products are given in Table 10.12. Most of these levels correspond to those set by the Safe Drinking Water Act for chemical constituents in water covered by the National Interim Primary Drinking Water Standards. Only PVC products that carry either the "NSF wc" (well casing) or "NSF pw" (potable water) designation have met the specifications set forth in Standard 14; only these products should be used for casing or screen in monitoring wells. Other non-NSF-listed products may include in their formulation chemical additives not addressed by the specifications, or may carry levels of the listed chemical parameters higher than those permitted by the specifications. As an example, even though neither lead nor cadmium have been permitted as compounding ingredients in U.S.-manufactured, NSF-listed PVC well casing since 1970, PVC manufactured in other countries (and imported to the U.S.) may be stabilized with lead or cadmium compounds that have been demonstrated to leach from the PVC (Barcelona et al., 1983).

Most of the work that has been done to determine chemical interference effects of PVC well casing (whether by leaching of chemical constituents from or sorption of chemical constituents to PVC) has been conducted under laboratory conditions. Furthermore, in most of the laboratory work the PVC has been exposed to a solution (usually distilled, deionized, or "organic-free" water) over prolonged periods of time (several days to several months), thus allowing the PVC an extended period of time in which to exhibit sorption or leaching effects. This may be comparable to a field situation in which ground water is exposed to the PVC well casing as it may be between quarterly or monthly sampling rounds. However, only a few studies consider the fact that, prior to sampling, the well is usually purged of stagnant water stored in the casing between sampling rounds, and that ground water flows through the well screen between sampling events. Thus, the water that would have been affected by any sorption or leaching effects

TABLE 10.12

Chemical Parameters Covered by NSF Standard 14 for Finished Products[a] and in Standard Leach Tests

Parameter	Maximum Contaminant Level (mg/l)
Antimony (Sb)	0.05[b]
Arsenic (As)	0.05
Barium (Ba)	1.0
Cadmium (Cd)	0.01
Chromium (Cr)	0.05
Lead (Pb)	0.05
Mercury (Hg)	0.002
Phenolic substances	0.05[b]
RVCM[a]	10[b]
Selenium (Se)	0.01[b]
Tin (Sn)	0.05

Note: Tabulated values are the maximum levels permissible in NSF-listed products after standardized leach testing in weakly acidic aqueous solution. [Carbonic acid solution with 100 mg/l hardness as $CaCO_3$ with 0.5 mg/l chlorine; pH 5.0 to 0.2; and surface to solution ratio of 6.5 cm^2/ml.]
[a] Total residual after complete dissolution of polymer matrix.
[b] Not covered under National Interim Primary Drinking Water Regulations.
Source: From Barcelona et al., 1983. With permission.

(if they were present at all) would ideally have been removed from the well during purging. Because samples are generally taken immediately after the purging of stagnant water in contact with the casing, the water sampled will have had a minimum of time (seconds or fractions of a second) with which to come in contact with casing or screen materials. Because of this, Barcelona et al. (1983) and Reynolds and Gillham (1985) suggest that the potential sample bias effects due to interactions with well casing materials may be disregarded.

Metallic Materials

Metallic well casing and screen materials available for use in monitoring wells include carbon steel, low carbon steel, galvanized steel, and stainless steel. Well casings made of any of these steels are stronger, more rigid, and less temperature sensitive than PVC casing materials. The strength and rigidity characteristics of steel casing materials are sufficient to meet virtually any subsurface condition encountered in a ground-water monitoring situation. However, all steels are subject to corrosion, a chemical resistance and chemical interference problem that may also affect casing strength in long-term exposures to certain subsurface geochemical environments.

Corrosion of steel well casings and screens can both limit the useful life of the monitoring well installation and result in ground-water sample analytical bias. It is therefore important to select both casing and screen materials that are resistant to the corrosion potential of the conditions to which they are exposed.

Corrosion is defined as the weakening or destruction of a metallic material by chemical action. Several well-defined forms of corrosive attack on steels have been observed and defined. In all forms, corrosion proceeds by electrochemical action, and water in contact with the metal is an essential factor. The forms of corrosion typical in environments in which steel well casing and screen materials are installed include (Johnson Division, 1966):

- General oxidation or "rusting" of the steel surface, resulting in uniform destruction of the surface with occasional perforation in some areas
- Selective corrosion or loss of one element of an alloy (i.e., dezincification), leaving a structurally weakened material
- Bi-metallic corrosion, caused by the creation of a galvanic cell at or near the juncture of two different metals
- Pitting corrosion, or highly localized corrosion by pitting or perforation, with little loss of metal outside of these areas
- Stress corrosion, or corrosion induced in areas where the metal is highly stressed

To determine the potential for corrosion of steels, it is first necessary to determine natural geochemical conditions. The following list of geochemical indicators can help recognize potentially corrosive conditions (modified from Johnson Division, 1966):

- Low pH: if ground-water pH is less than 7.0, water is acidic and corrosive conditions exist.
- High dissolved oxygen content: if dissolved oxygen content exceeds 2 ppm, corrosive water is indicated.
- Presence of hydrogen sulfide (H_2S): presence of H_2S in quantities as low as 1 ppm can cause corrosion.
- Total dissolved solids (TDS): if TDS is greater than 1000 ppm, the electrical conductivity of the water is great enough to cause electrolytic corrosion.

- Carbon dioxide (CO_2): corrosion is possible if the CO_2 content of the water exceeds 50 ppm.
- Chloride ion (Cl^-) content: if Cl^- content exceeds 500 ppm, corrosion can be expected.

Combinations of these corrosive conditions generally increase the corrosive effect. To date, however, no specific data have been generated on the expected life of steel well-casing or screen materials exposed to natural subsurface geochemical conditions, perhaps because the range of subsurface conditions is so wide and unpredictable.

Carbon steels were produced primarily to provide increased resistance (compared with iron) to atmospheric corrosion. Achieving this increased resistance requires that the material be subjected to alternately wet (saturated) and dry (open to the atmosphere) conditions. In most monitoring wells, water fluctuations, which may be extreme during sampling, are usually not sufficient in either duration or occurrence to provide these conditions, so corrosion of carbon steel casing and screen materials is a common problem. The difference between the corrosion resistance of carbon and low-carbon steels is negligible under conditions in which the materials are buried in soils or in the saturated zone, so both materials may be expected to corrode approximately equally. Corrosion products include iron, manganese, and trace-metal oxides as well as various metal sulfides (Barcelona et al., 1983). Under oxidizing conditions, the principal corrosion products are solid hydrous metal oxides, while under reducing conditions, high levels of dissolved metals (principally iron and manganese) can be expected (Barcelona et al., 1983). While the electroplating process of galvanizing (application of a zinc coating) somewhat improves the corrosion resistance of either carbon or low-carbon steel, in many environments the improvement is only slight and short-term. The products of corrosion of galvanized steel include iron, manganese, zinc, and cadmium (Barcelona et al., 1983). These constituents can be contributed to ground-water samples in wells in which galvanized steel casing or screen is used and corrosion is evident.

Clearly the presence of corrosion products represents a high potential for alteration of ground-water sample chemical quality. The surfaces on which corrosion occurs also present potential sites for a variety of chemical reactions (i.e., formation of organometallic complexes) and adsorption to occur. These surface interactions can cause significant changes in dissolved metal or organic compound concentrations in ground-water samples. According to Barcelona et al. (1983), even flushing the stored water from the well casing prior to sampling may not be sufficient to minimize this source of sample bias because the effects of the disturbance of surface coatings or accumulated corrosion products in the bottom of the well would be difficult, if not impossible, to predict. On the basis of these observations, the use of carbon steel, low-carbon steel, and galvanized steel in wells used for ground-water quality monitoring should be discouraged in most natural geochemical environments. These materials may, however, be well suited to use in piezometers or monitoring wells that are used strictly for water-level monitoring or for sampling ground water for constituents other than metals or organic compounds.

On the other hand, stainless steel performs very well in most corrosive environments, particularly under oxidizing conditions. In fact, stainless steel requires exposure to oxygen in order to attain its highest corrosion resistance. Oxygen combines with part of the stainless steel alloy to form a thin, invisible protective film on the surface of the metal (Johnson Division, 1966). As long as the film remains intact, the corrosion resistance of stainless steel is very high. However, several studies (Jones and Miller, 1988; Parker et al., 1990; Hewitt, 1994) have cited the formation of an iron oxide coating on the surface of stainless steel casing materials in long-term exposures to ground water as a detriment to accurate water-quality determinations. This coating can have unpredictable and changeable effects on the adsorption capacity of the material.

Several different types of stainless steel alloys are available for use in monitoring wells. The most common alloys are Type 304 and Type 316, which are part of the 18-8 or 300 series of stainless steels. Both types of stainless steel are available in low-carbon forms, designated by an "L" after the number (i.e., 304L), which are more easily welded than the normal carbon types. Table 10.13 describes dimensions, hydraulic collapse pressure, burst pressure, and unit weight of stainless steel casing.

Type 304 stainless steel is perhaps the most practical choice from a corrosion resistance and cost standpoint. As indicated in Table 10.14 (modified from Johnson Division, 1966), Type 304 is composed of slightly more than 18% chromium and more than 8% nickel, with about 72% iron and not more than 0.08% carbon. The chromium and nickel content give the Type 304 alloy excellent resistance to corrosion; its low carbon content improves its weldability. Table 10.14 demonstrates that Type 316 stainless steel is compositionally similar to Type 304 with two exceptions — a 2 to 3% molybdenum content and a higher nickel content (replacing the equivalent percentage of iron). This compositional difference gives Type 316 stainless steel an improved resistance to sulfur-containing species as well as sulfuric acid solutions (Barcelona et al., 1983), so it performs better under reducing conditions than Type 304. According to Barcelona et al. (1983), Type 316 stainless steel is less susceptible to pitting or pinhole corrosion caused by organic acids or halide solutions. However, they also point out that for either formulation of stainless steel, long-term exposure to corrosive conditions may result in corrosion (Figure 10.49 and Figure 10.50) and the subsequent chromium or nickel contamination of ground-water samples and that insoluble halogen and sulfur compounds may also form as a result of corrosion of stainless steel.

Parker et al. (1990) showed that both Type 304 and Type 316 stainless steel casing materials sorbed arsenic, chromium and lead while leaching cadmium. The same study demonstrated that there were no detectable sorption-related losses of any of 10 organic compounds, including chlorinated alkanes, chlorinated aromatics, nitroaromatics, and nitroamines, in exposures to either Type 304 or Type 316 stainless steel. Hewitt (1992) found that both Type 304 and Type 316 stainless steel leached lead, cadmium, chromium, copper, nickel, and iron (Type 304 significantly more iron and cadmium than the control; Type 316 significantly more iron, nickel, and cadmium than the control). The same study determined that Type 304 sorbed cadmium, chromium, lead, and copper (significantly more chromium and lead than the control), and that Type 316 sorbed significantly more lead than the control. In a later study, Hewitt (1994) found that both Type 304 and Type 316 well screens exposed to flowing ground water for periods ranging from 0.25 to 8 h can leach significant amounts of iron, nickel, and chromium, and sorb significant amounts of lead and cadmium. At several field sites, Type 304 and 304L stainless steel wells have been found to leach chromium in quantities of 10 to 30 ppb for more than a year (Smith, 1988; Schalla et al., 1988; Smith et al., 1989). Jones and Miller (1988) also cite the potential for stainless steel casing and screen materials to leach chromium, nickel, molybdenum, and iron into the aqueous environment.

Oakley and Korte (1996), in a study conducted at Williams Air Force Base in Mesa, Arizona, found that Type 304 well screen materials were the source of elevated nickel and chromium levels in ground-water samples. The presence of chloride at levels between 600 and 900 mg/l and TDS as high as 3000 mg/l apparently caused crevice corrosion of the stainless steel screens, releasing both chromium (in a precipitate) and nickel (in dissolved form). Interestingly, they found that a difference in quality between manufacturers was the reason that one set of wells at the site produced high levels of chromium and nickel, while a second set did not. Royce (1991) reported levels of nickel and

Design and Installation of Ground-Water Monitoring Wells

TABLE 10.13
Dimensions, Hydraulic Collapse and Burst Pressure, and Unit Weight of Stainless Steel Well Casing

Nominal Size (in.)	Schedule No.	Outside Diameter (in.)	Wall Thickness (in.)	Inside Diameter (in.)	Internal Cross-Section Area (sq. in.)	Internal Pressure (psi) Test	Internal Pressure (psi) Bursting	External Pressure (psi) Collapsing	Weight (lb/ft)
2	5	2.315	0.065	2.245	3.958	0.820	4.105	0.896	1.619
	10	2.375	0.109	2.157	3.654	1.375	6.884	2.196	2.063
	40	2.375	0.154	2.067	3.356	1.945	9.726	3.526	3.087
	80	2.375	0.218	1.939	2.953	2.500	13.766	5.418	5.069
2½	5	2.875	0.083	2.709	5.761	0.865	4.330	1.001	2.498
	10	2.875	0.120	2.635	5.450	1.250	6.260	1.905	3.564
	40	2.875	0.203	2.469	4.785	2.118	10.591	3.931	5.347
3	5	3.500	0.083	3.334	8.726	0.710	3.557	0.639	3.057
	10	3.500	0.120	3.260	8.343	1.030	5.142	1.375	4.372
	40	3.500	0.216	3.680	7.389	1.851	9.257	3.307	7.647
3½	5	4.000	0.083	3.834	11.540	0.620	3.112	0.431	3.505
	10	4.000	0.120	3.760	11.100	0.900	4.500	1.081	5.019
	40	4.000	0.226	3.548	9.887	1.695	8.475	2.941	9.194
4	5	4.500	0.083	4.334	14.750	0.555	2.766	0.315	3.952
	10	4.500	0.120	4.260	14.250	0.800	4.000	0.845	5.666
	40	4.500	0.237	4.026	12.720	1.580	7.900	2.672	10.891
5	5	5.563	0.109	5.345	22.430	0.587	2.949	0.350	6.409
	10	5.563	0.134	5.295	22.010	0.722	3.613	0.665	7.842
	40	5.563	0.258	5.047	20.000	1.391	6.957	2.231	14.754
6	5	6.625	0.109	6.407	32.220	0.494	2.467	0.129	7.656
	10	6.625	0.134	6.357	31.720	0.606	3.033	0.394	9.376
	40	6.625	0.280	6.065	28.890	1.288	6.340	1.942	19.152

Source: ARMCO, Inc. stainless steel pipe specifications.

TABLE 10.14

Composition of Stainless Steel Well Casing and Screen Materials

Chemical Component[a]	SS 304	SS 316
Carbon	0.08	0.08
Manganese	2.00	2.00
Phosphorous	0.04	0.045
Sulfur	0.03	0.03
Silicon	0.75	1.00
Chromium	18.0–20.0	16.0–18.0
Nickel	8.0–11.0	10.0–14.0
Molybdenum	—	2.0–3.0
Iron	Remainder	Remainder

[a] All chemical components measured in percentage.

chromium up to 25 times higher in samples collected from stainless steel wells when compared with analytical results from PVC wells at the same site, constructed in the same zone in the same formation. In this case, the high levels of nickel and chromium were attributed to microbiologically induced corrosion, which is a fairly common occurrence, particularly in natural waters containing culprit species of microorganisms (e.g., iron bacteria — see Figure 10.51). Driscoll (1986) indicates that iron bacteria have been found in wells in all of the conterminous United States, but are a significant problem in the southeastern states, the upper midwest, and southern California.

The resistance to corrosion of both types of stainless steel can be improved by treatment with nitric acid and potassium dichromate solutions. These treatment processes, usually done at steel mills or manufacturing facilities, are referred to as Mil-Spec or QQ-P-35C. The passivation of stainless steel substantially increases corrosion resistance in ground water with high concentrations of halides in solution. Passivation may also be important for reducing adsorption of certain radionuclides if a site being monitored contains mixed waste (Raber et al., 1983).

FIGURE 10.49
Corrosion can be an issue even in stainless steel wells. In this case, corrosion was present at the weld between the well screen and the threaded joint prior to installation in the borehole. Galvanic corrosion often occurs first at welds and joints, or where two dissimilar metals are joined.

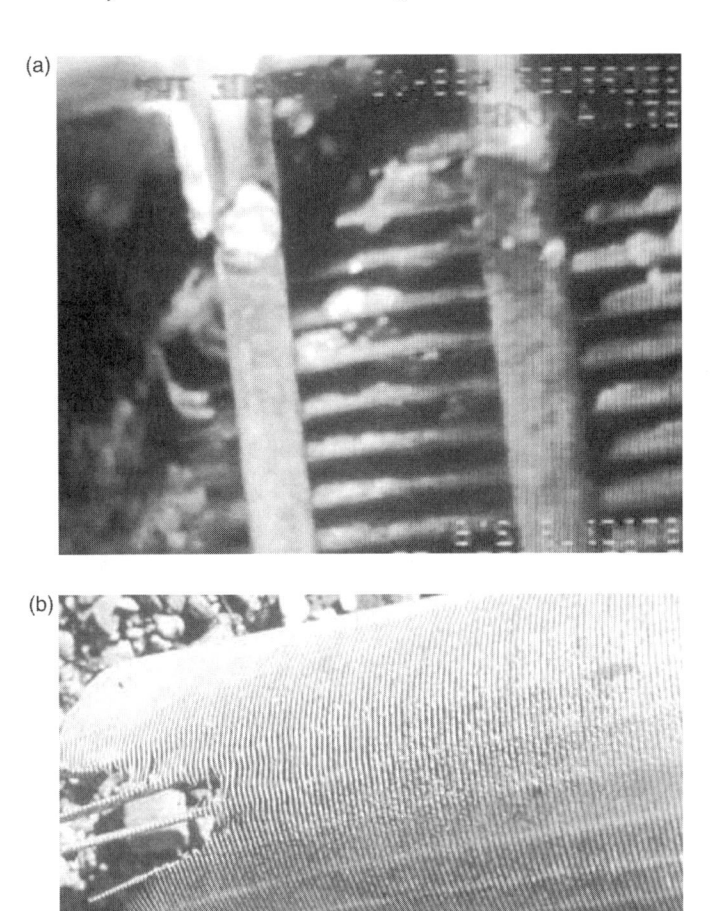

Design and Installation of Ground-Water Monitoring Wells 697

FIGURE 10.50
(a) A down-hole camera photo of corrosion in a 4-in. diameter stainless steel 316 well screen. Corrosion occurred after the well was installed in an aquifer in which the natural pH of the water was less than 5.0, and resulted in failure of the well screen. (b) The same well screen after excavation of the well, showing severe pitting corrosion throughout the well screen. (Photos courtesy of John Oneacre.)

Coupling Procedures for Joining Casing

Rigid monitoring well casing and screen is produced in various lengths (usually 5, 10, or 20 ft) that are joined by one of several types of coupling methods during installation. The coupling method used is dependent on the type of casing and casing joint type desired. Irrespective of which technique is used, a uniform inner casing diameter should be maintained in monitoring well installations; a uniform outer casing diameter is also commonly desirable. Inconsistent inner diameters result in problems in using tight-fitting down-hole equipment (i.e., development tools, borehole geophysical tools, purging or sampling devices), while an uneven outer diameter creates potential problems (i.e., bridging) with filter pack and annular seal placement. The latter problems tend to promote vertical water migration in the annular space between the borehole and the casing at the casing–annular seal interface to a greater degree than is experienced with uniform outer diameter casing (Morrison, 1984). Figure 10.52 illustrates some common types of joints used for assembling lengths of casing.

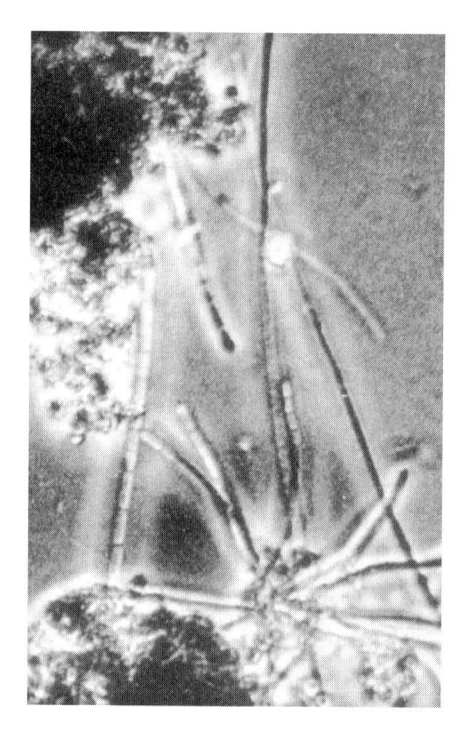

FIGURE 10.51
Iron bacteria are common in ground water across the U.S., and can cause corrosion and bacterial fouling in wells built of any type of steel material, including stainless steel. (*Source*: Driscoll, 1986. With permission.)

Joining PVC Casing

There are two basic methods of joining sections of PVC well casing: solvent cementing (using slip joints) and mechanical joining (using threaded joints). In solvent cementing, a solvent primer is generally used to clean the two pieces of casing to be joined and solvent cement is then spread over the cleaned surface areas (Figure 10.53). The two sections are assembled while the cement is wet, allowing the active solvent agents to penetrate and soften the two casing surfaces that are joined. As the cement cures, the two pieces of casing are fused together; a residue of chemicals from the solvent cement remains at the joint. There are many different formulations of PVC solvent cement, but most cements consist of two or more of the following organic chemical constituents: THF, MEK, MIBK, CH, and DMF (Sosebee et al., 1983).

Clearly, the cements used in solvent welding, which are themselves organic chemicals, could produce some impact on the integrity of ground-water samples. Boettner et al. (1981) noted the leaching of MEK, CH, and THF into water in solvent-cemented PVC pipe at levels ranging from 10 ppb to 10 ppm in a short-term (15-day) leaching test. Wang and Bricker (1979), Sosebee et al. (1983) and Martin and Lee (1989) each demonstrated that one or more of the aforementioned PVC solvents do, in fact, appear in ground-water samples collected from monitoring wells in which those solvents are used to join well casing sections, probably due to diffusion of the solvents from the glued joints into water contained in the cased portion of the well. These studies indicate that even though the well may be developed and purged prior to sampling, detectable levels of these compounds may be found in samples months to years after well installation. The presence of these solvents may mask the possible presence of these and other organic compounds in ground water from possible contaminant sources.

Design and Installation of Ground-Water Monitoring Wells 699

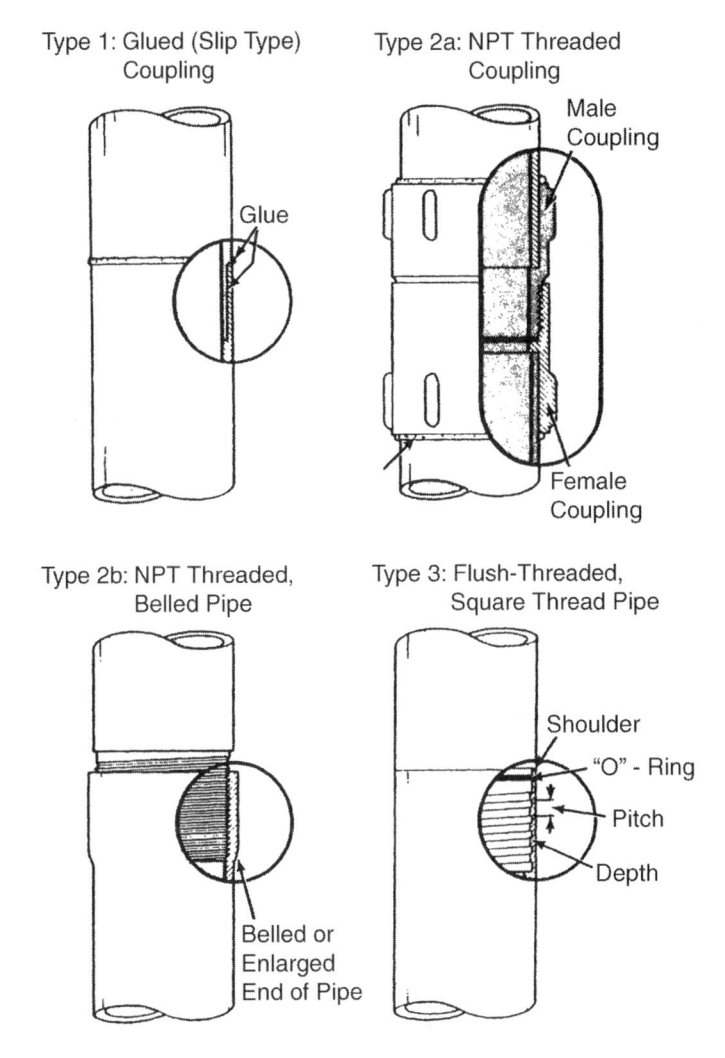

FIGURE 10.52
Common types of joints used for assembling lengths of well casing and screen.

Barcelona et al. (1983) note that even minimal solvent cement application is sufficient to result in consistent levels of primer and cement components at levels exceeding 100 ppb in ground-water samples, despite proper well development and purging prior to sampling. They indicate that these effects may persist for months after well construction, even after repeated attempts to develop the wells. Dunbar et al. (1985) cite a case in which THF was found at low levels (10 to 200 ppb) in samples taken from several PVC monitoring wells in which PVC solvent cement was used, more than 2 yr after the wells were installed. In samples from adjacent monitoring wells in which threaded PVC casing was used, no THF was found, prompting the conclusion that the THF concentrations were a relict of solvent cement use in well construction. Compounds from PVC primers and adhesives can also create problems in the analysis of ground-water samples for VOCs by coeluting with other VOCs during sample analysis, thereby masking the identification of other VOCs (Lapham et al., 1996). All of these results point to the fact that solvent cementing is not appropriate for use in joining sections of PVC casing used in ground-water monitoring wells used for determinations of ground-water quality.

FIGURE 10.53
Solvent cementing (gluing) is often used to join lengths of PVC casing and screen in water-supply wells, and may be used in piezometers used only for water-level measurement. However, solvent welding should never be done in monitoring wells used for the purpose of collecting water-quality samples.

The most common method of mechanical joining of PVC materials is by threaded connections. Molded and machined threads are available in a variety of thread configurations, including acme, buttress, standard National Pipe Thread (NPT), and flat square threads (Figure 10.54). A good threaded joint must be easy to start, have a method of alignment (to avoid cross-threading), be easy to make up or complete, and have a sealing feature (such as an O-ring). Several of the available thread types are hard-starting and easy to cross-thread, while others tend to gall, or the edges bend over (Foster, 1989). Casing buyers should be aware of the fact that most manufacturers have their own thread type and that threaded casing from one manufacturer may not be compatible with threaded casing from another manufacturer. If threads do not match and a joint is forced, it is likely that the joint will fail or leak during or after casing installation.

FIGURE 10.54
A number of different thread types are used in threaded joints. It is important not to mix thread types, as joint failure will result.

Two types of threaded connections are common — flush-joint threaded casing and pipe joined with threaded couplings. The latter type of connection is not widely used because the joint has a larger outside diameter than the casing, which can create problems with bridging during filter pack and annular seal installation, particularly in wells installed using hollow-stem augers.

Flush-joint threaded casing is by far the most widely used type of casing. Flush-joint threaded casing has male and female threaded ends that, when threaded together, produce a union internally and externally equal in diameter to the casing sections that are joined. This internal and external flush joint is designed to be free of gaps or irregularities at the connection when hand tightened. The smooth interior surface permits unobstructed insertion and removal of well development and sampling equipment. This smooth surface is significant because the clearance provided for most equipment (in, e.g., 2-in. nominal diameter Schedule 40 casing) is usually less than 0.3 in., and sharp edges could cause plastic-coated cables, chambers or tubing on pumps, and surge blocks to snag or become temporarily or permanently lodged in the casing. The flush exterior surface permits easy installation and completion using temporary steel driven casing or hollow-stem augers. Threaded joints should not be supplemented with metal fasteners (e.g., rivets or screws), as these materials can reduce the effective inside diameter of the well and damage pumps and other devices used in the well. Fasteners, usually made of steel, can also corrode, causing possible chemical anomalies or leaving behind holes in the casing through which leakage can occur.

Effective seals are needed at casing joints to prevent neat cement or bentonite grout or contaminated water (i.e., cross-contamination in multiple completion wells) from entering the well. Some flush-threaded coupling designs include O-rings to ensure an effective coupling. For casing that does not come with O-rings, O-rings can be ordered directly from O-ring manufacturers in a greater variety of materials and sizes than are available from casing suppliers. The most common elastomer bases are nitrile (Buna-N), neoprene, ethylene propylene, butyl, Viton, polyacrylate, polysulfide, silicone, and fluorosilicone. Of these, the best choices for monitoring wells usually would be either nitrile, ethylene propylene, or Viton; however, materials should be selected on the basis of chemical resistance, temperature stability, strength, and other properties relevant to site conditions. Low tensile strength and the lack of tear and abrasion resistance make silicone and fluorosilicone poor choices for most applications. Some common ground-water contaminants and the relative suitability of elastomer materials are presented in Table 10.15. The O-ring elastomers are listed in decreasing order of preference for each chemical medium. Additional information on compatibility of elastomers for many more chemical media is available from O-ring manufacturers.

In lieu of using O-rings, some threaded couplings (NPT-type tapered threaded couplings) can be sealed using Teflon tape; however, Teflon tape is difficult to apply to threaded couplings in cold, wet weather. Gaskets or O-rings, particularly those installed at the factory, can save time during installation and ensure a good seal, which is not always true when Teflon tape is wrapped on threads.

Most current designs of the flush-joint threaded coupling have flat square threads, which are easier to screw together, less likely to become cross threaded, and easier to unscrew than the NPT-type tapered, V-shaped threads.

ASTM Standard F-480 (ASTM, 2004u) specifies a particular type of joint for thermoplastic casing materials (Figure 10.55). The ASTM F-480 joint (Figure 10.56) features interlocking angled faces, an O-ring seal, a two threads per inch square-form thread and an easy-starting lead (Foster, 1989). The thread is quick and easy to field-assemble, is reusable, and the O-rings are readily available in inert materials. The joint, after it is made up, is designed to resist internal pressures of at least 88 psi (204 ft of head

TABLE 10.15

O-Ring Elastomer Materials Suitable for Common Chemical Media

Chemical Media	Elastomers[a]	Chemical Media[a]	Elastomers[a]
Arsenic (acidic)	N, E, V, C	Potassium dichromate	N, E, V, C
Benzene	V	Salt water	N, E
Carbon tetrachloride	V	Stoddard solvent	N, V, P
Chrome plating	E, V, B	Tetrahydrofuran	E
Solutions	N, V, P	Toluene	V
Creosote	V	Trichloroethane	V
Dry-cleaning fluids	N, V	Trichloroethylene	V
Gasoline	E, B	TNT	V
Hydrazine	N, V	Type I fuel (MIL-S-3136)	N, V
JP4 (MIL-J5624)	S	Type II fuel (MIL-S-3136)	V, N
Lacquer solvents	E, B	Vinyl chloride	Metal
Methylene chloride	E, B	Xylene	V
Methyl ethyl ketone	V	Zinc salt solutions	N, E, V, C
Phenol	—	—	—

[a] Elastomer codes: B, butyl; N, nitrile (Buna-N); C, neoprene (chloroprene); E, ethylene propylene; V, fluorocarbon ("Viton"); P, polyacrylate; S, polysulfide.

[water]) and external pressures of at least 28 psi (65 ft of head [water]). The tensile strength at the joint is sufficient to hold 2050 ft of schedule 40 PVC casing. Most manufacturers now produce the F-480 joint, and it is available in PVC and stainless steel. This type of joint should not be Teflon-taped, as taping can significantly reduce the tensile strength at the joint (Foster, 1989), and can lead to leakage at the joint.

Because all joints in a monitoring well casing should be watertight, the extent of tightening of joints should comply with manufacturers' recommendations. Caution should be

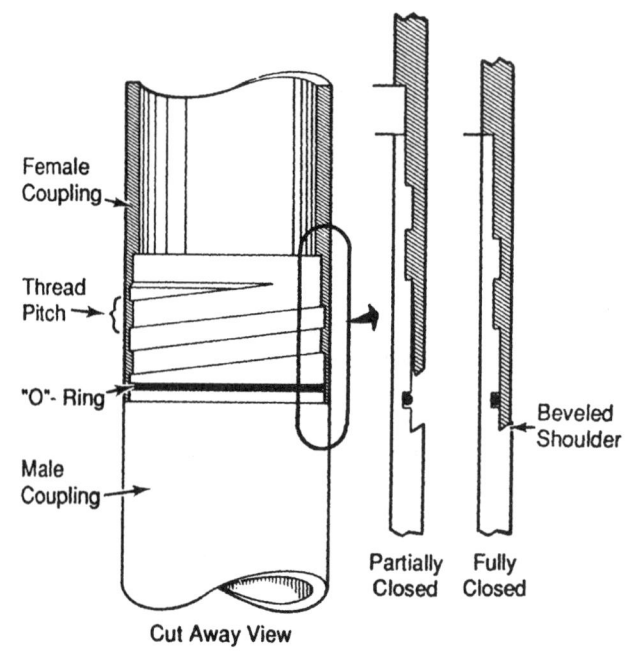

FIGURE 10.55
The ASTM F-480 joint is the preferred joint for assembling PVC well casing and screen.

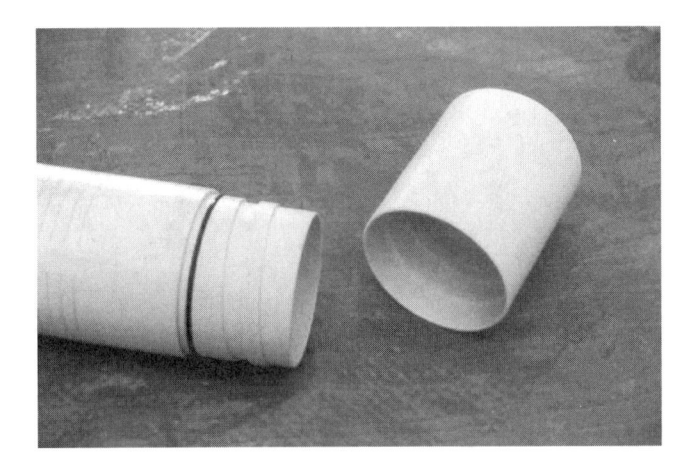

FIGURE 10.56
The ASTM F-480 joint has a flat, square thread, two threads per inch, with beveled shoulders and an O-ring seal that makes a water-tight joint upon assembly.

exercised in tightening joints in PVC casing, as over-tightening could lead to structural failure of the joint (NWWA/PPI, 1980) and a nonflush inside or outside diameter.

When using PVC well casing and screen, the ASTM F-480 threaded joint is preferred. The problems associated with the use of solvent primers and cements are thus avoided. Casing with threads machined or molded directly onto the pipe (without use of larger diameter couplings) provides a flush joint between both inner and outer diameters and is thus best suited to use in monitoring well construction. Though this type of joint slightly reduces the tensile strength of the casing string compared to a solvent-cemented joint, in shallow (i.e., less than 200 ft) monitoring well installations this is usually not a critical concern.

Some manufacturers of flush-joint threaded casing provide sections of casing in exact lengths (i.e., compensated lengths), so that each length when threaded to the next is exactly 5.0 or 10.0 ft in length. This eliminates the potential for screening the wrong interval during placement of the well screen and casing. Less exact casing lengths can vary by as much as 0.5 ft, and typically are from 0.05 to 0.2 ft short of the specified dimension. It is thus important to measure the length of each section of casing and screen before it is installed, so that an accurate well log can be produced.

Joining Steel Casing

There are generally two options available for joining steel well casings — resistance welding or threaded joints (either with or without couplings). Welding (Figure 10.57) produces a casing string with a relatively smooth inner and outer diameter, while threaded joints may or may not, depending upon whether or not couplings are used. With welding, it is generally possible to produce joints that are as strong as, or stronger than, the casing, thus enhancing the tensile strength of the casing string. The disadvantages of welding include greater assembly time, the difficulty of properly welding casing in the vertical position, enhancement of corrosion potential in the vicinity of the weld, and the danger of ignition of potentially explosive gases which may be encountered in some monitoring situations.

Because of these disadvantages, threaded joints are much more commonly employed in monitoring well installations completed with steel casing and screen. Threaded joints (Figure 10.58) provide inexpensive, fast, and convenient connections, and greatly reduce

FIGURE 10.57
Welding steel casing is rarely done at sites at which monitoring wells are installed, such as landfills or gasoline spill sites, because of the potential for ignition of potentially explosive gases.

potential problems with chemical resistance or chemical interference (due to corrosion) and ignition of explosive gases. One limitation to using threaded joints is that the tensile strength of the casing string is reduced. If threaded couplings are used, the larger outside diameter at the coupling increases the probability of bridging of materials installed in the annulus. Because strength requirements for small-diameter wells typical of monitoring installations are not as critical, and because of the high initial tensile strength of steel casings, the reduced tensile strength at the threaded joint usually does not pose a significant problem.

Most of the stainless steel casing produced for use in monitoring wells is thin-walled casing in which threads cannot be machined. The joints in thin-walled (i.e., schedule 5 or 10) stainless steel casing are usually made in a short (2- to 3-in.) length of schedule 40 pipe that is welded on to the end of the thin-walled casing at the factory (Figure 10.59). While the outer diameter of this type of threaded joint is flush, the inner diameter is not. The inside diameter of the joint in this type of casing is thus the effective inside diameter of the casing, which may be critical in the use of some down-hole

FIGURE 10.58
Threaded joints are the most commonly used joints for steel casing and screen used in monitoring wells.

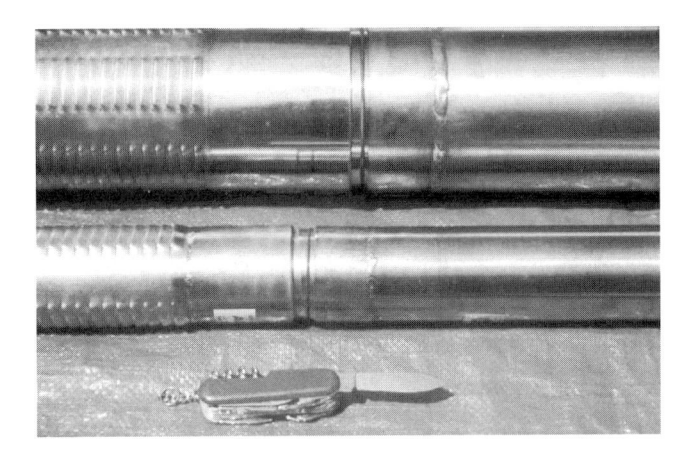

FIGURE 10.59
Threaded joints in stainless steel well casing and screen are machined in schedule 40 (thick-walled) material, which is then welded to screen or thinner walled (i.e., schedule 5 or 10) casing, producing a ledge at the weld and a smaller inside diameter at the joint, which can damage development tools and sampling devices used in the well.

equipment. Additionally, because the inner joint between the thin-walled casing and the schedule 40 joint is not smooth, the use of some development techniques (i.e., use of a surge block) may be restricted. Extra care should be exercised in these wells when using purging and sampling devices so these devices are not damaged during installation in or removal from the well.

Factors Affecting Selection of Well Casing Diameter

The nominal diameter of most existing ground-water quality monitoring wells is either 2 or 4 in., which is smaller than that of most water wells or wells used in water-resource studies. The nominal diameter of wells installed by direct-push methods (see Chapter 6) ranges from as large as 2 in. to as small as 0.5 in. The advantages and disadvantages of small-diameter versus large-diameter wells have been debated (Schmidt, 1982, 1983; Rinaldo-Lee, 1983; Schalla and Oberlander, 1983; Voytek, 1983). Important reasons for using large-diameter wells include their suitability for determining large-scale aquifer characteristics (i.e., transmissivity, storativity) and boundary conditions of high-yield aquifers via pumping tests. However, in situations in which such high-yield conditions are not important to achieving the objectives of monitoring, the reasoning has been that small-diameter wells are better (Voytek, 1983). Small-diameter wells are less expensive because: (1) smaller materials are installed; (2) costs per foot are lower because borehole diameters are smaller and less-costly drilling or direct-push methods can be used; and (3) the quantities of potentially contaminated drill cuttings, well-development-related water and sediment, and purge water that may require special handling and disposal practices are smaller. For direct-push wells, the latter costs are further reduced because drill cuttings are essentially eliminated.

Care should be exercised in the selection of casing diameter. While casing outside diameter is standardized, variations in wall thickness (i.e., casing schedule) result in variations in casing inside diameter (Figure 10.60). As illustrated by Table 10.16, 2-in. nominal casing is a standard 2.375 in. O.D.; wall thickness varies from 0.065 in. for schedule 5 to 0.218 in. for schedule 80. This means that inside diameters for 2-in. nominal casing vary from 2.245 in. for schedule 5 thin-walled casing (e.g., typical of stainless steel) to 2.067 in. for

FIGURE 10.60
Variations in well casing wall thickness due to specification of different casing schedules (i.e., schedule 80 [left] versus schedule 40 [right]) cause variations in casing inside diameter, which may make installation of some pumps and other down-hole equipment difficult.

schedule 40 to only 1.939 in. for schedule 80 thick-walled casing (typical of PVC) (Figure 10.61). This factor must be taken into consideration when determining the proper casing size for a particular monitoring program.

The diameter of the casing selected for a monitoring well is dependent on the purpose of the well, the volume of the well, and the need for the well to accommodate down-hole equipment (Figure 10.62). Additional casing diameter selection criteria include: well installation method used (drilling or direct-push), anticipated depth of the well and associated strength requirements, ease of well development, volume of water required to be purged prior to sampling, rate of recovery of the well after purging, and cost.

Down-Hole Equipment

A variety of down-hole equipment may be utilized in a monitoring well, including well development tools, borehole geophysical tools, pumps for conducting pumping tests, water-level measuring devices, and purging and sampling devices. In general, large-diameter wells (i.e., those with 4-in. or greater inside diameter) can be developed by a wider variety of methods with commercially available development tools, although most well-development tools have been adapted for use in smaller diameter wells, including those as small as 0.5 in. nominal diameter. The same applies to borehole geophysical equipment — wells 4-in. or greater in inside diameter will accommodate most borehole

TABLE 10.16

Outside Diameter, Wall Thickness, and Inside Diameter of Well Casing

Casing Size [in.] (Nominal)	Outside Diameter (Standard)	Wall Thickness				Inside Diameter			
		Sch 5	Sch 10	Sch 40	Sch 80	Sch 5	Sch 10	Sch 40	Sch 80
2	2.375	0.065	0.109	0.154	0.218	2.245	2.157	2.067	1.939
3	3.500	0.083	0.120	0.216	0.300	3.334	3.260	3.068	2.900
4	4.500	0.083	0.120	0.237	0.337	4.334	4.260	4.026	3.826
5	5.563	0.109	0.134	0.258	0.375	5.345	5.295	5.047	4.813
6	6.625	0.109	0.134	0.280	0.432	6.407	6.357	6.065	5.761

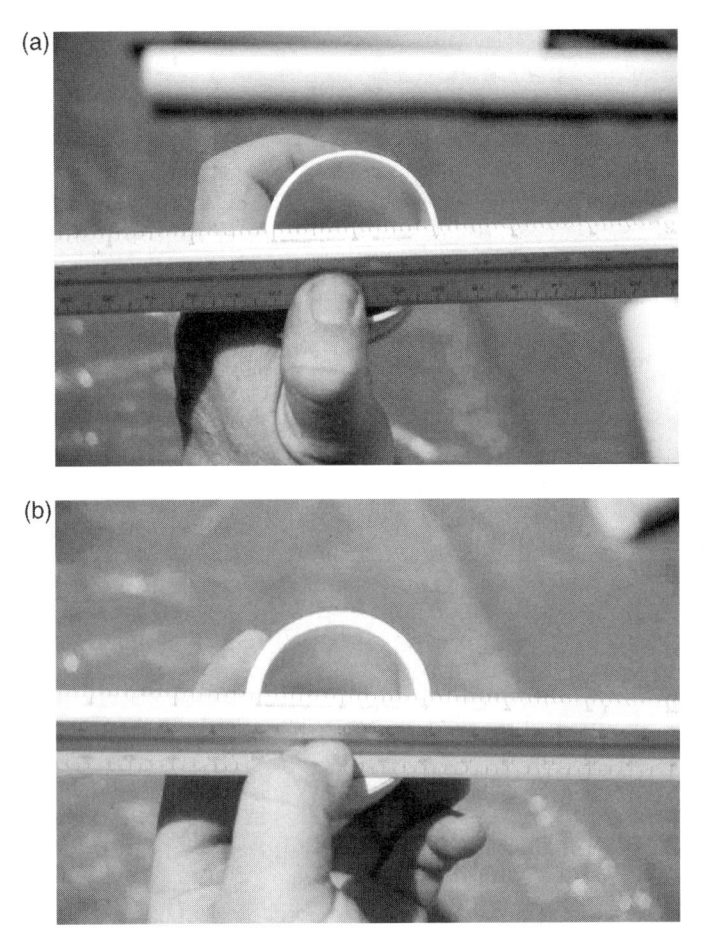

Design and Installation of Ground-Water Monitoring Wells

FIGURE 10.61
Compare the inside diameters of the schedule 40 PVC well casing (a), at 2.067 in., and the schedule 80 PVC well casing (b), at 1.939 in.

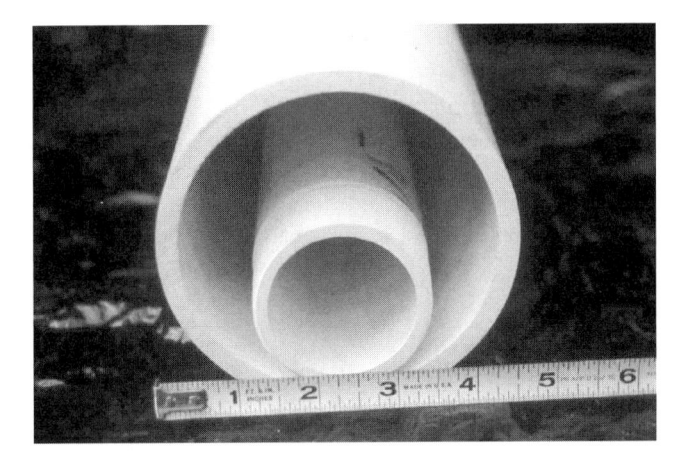

FIGURE 10.62
The diameter of a monitoring well must be selected based on a variety of factors, including volume, which may affect the amount of water that must be purged from the well. The volume of a 2-in. nominal casing (0.163 gal/ft) is $1/4$ the volume of a 4-in. nominal casing (0.65 gal/ft).

geophysical tools; only a limited number of slim-hole well logging tools are available for use in wells as small as 2 in. in diameter.

Generally, wells 4 in. in inside diameter or larger are best able to accommodate pumps of sufficient capacity to conduct a pumping test in most formations. However, pumps capable of pumping at rates of 8 gal/min at low lifts are available for 2-in. nominal diameter wells (see Chapter 15). In geologic materials of low hydraulic conductivity in which pumping tests would not be successfully applied, smaller diameter wells will meet the requirements of other hydraulic testing methods such as slug tests, bail tests, or pressure tests.

For most water-level measuring methods (chalked tapes, electric probes, etc.) small-diameter wells do not pose a problem. Some wells or piezometers used strictly for measuring water levels may, in fact, be as small as 0.5 in. inside diameter. However, only large-diameter wells (4-in. inside diameter or greater) will accommodate the installation of most float-type continuous water-level recorders.

With respect to purging and sampling devices, a greater variety of high-capacity pumping devices is available for large-diameter wells than for small-diameter wells. Recently, devices for both purging and sampling from very small-diameter wells (i.e., wells as small as 0.5 in. nominal diameter) have become available, but most of these are limited to relatively low pumping rates. This does not pose a problem at sites at which sampling methods such as low-flow purging and sampling are used.

Well-Installation Method

While almost any diameter well can be installed with drilling methods such as air or mud rotary or cable tool, a wider variety of well-installation methods is available for small-diameter wells. For example, generally only wells of 6-in. nominal diameter or less can be installed though a hollow-stem auger; only wells of 2-in. nominal diameter or less can be installed using direct-push methods. Additionally, small-diameter wells can be more easily driven or jetted into place.

Anticipated Well Depth and Casing Strength

For shallow monitoring wells, the strength characteristics of nearly all diameters of all casing materials are adequate. The greater strength requirements for deeper well installations — to prevent well casing failure, severe bends in the casing, and difficulties in well construction procedures — favor larger diameter, thicker-walled casing.

Ease of Well Development

With regard to ease of well development, smaller diameter wells usually take less time to develop because smaller volumes of water and sediment are involved. However, the number of development methods that can be used in smaller diameter wells is more limited. Also, because the energy generated by well development methods is much less in smaller diameter wells, the development process may not be as effective in these wells as it is in larger diameter wells; development effects may only extend a few inches or less beyond the well screen.

Purge Volume

Because it is sometimes necessary to remove water standing in a well prior to taking a ground-water sample, the volume of water that must be purged from the well should be considered. Because volume increases as the square of the radius, it increases significantly with well diameter. For example, the volume of water in a 4-in. nominal diameter well (0.65 gallons per foot of casing) is four times the volume in a 2-in. nominal diameter well (0.163 gallons per foot of casing). The larger storage volume in large-diameter wells can significantly increase the time required for sampling by increasing the volume of water that must be purged from the well prior to sampling. Additionally, with increased volume, purging costs increase, especially if the purged water is contaminated and thus must be treated or contained, and later disposed as a hazardous material.

Rate of Recovery

Rate of recovery of the well is also an important consideration. In formations with very low hydraulic conductivities, the greater storage volume of a large-diameter well can result in a much slower recovery of water levels after purging. Small-diameter wells are usually preferred in low hydraulic conductivity formations because the well volume is small and because the time of recovery is directly proportional to well volume, which increases with well diameter. In a formation with a given hydraulic conductivity, it takes less time for a small-diameter well to recover when a slug of water is removed than for a large-diameter well (Rinaldo-Lee, 1983).

Unit Cost of Casing Materials and Drilling

Unit (per foot) costs of both casing materials and drilling increase as well diameter increases. Depending upon the material selected, unit costs for a 4- or 6-in. nominal diameter well can be from two to ten times the unit cost for a 2-in. nominal diameter well (Richter and Collentine, 1983). Several shallow small-diameter wells can commonly be constructed for the same price as a single large diameter well of the same depth; this is not always the case with deep wells. The collection of a greater amount of data over a large area can sometimes be accomplished with small-diameter wells for the same price, particularly if more cost-effective installation methods (e.g., direct-push methods) can be used.

The highly variable geologic and contaminant conditions found at sites at which monitoring wells are constructed, and the varied purposes for which monitoring wells are installed, dictate that a flexible approach to selection of well diameter is required, and that no one well diameter is optimum.

Centering Casing in the Borehole

It is important to install the casing string (casing and screen) so that it is centered in the borehole. The primary reason for this is to ensure that, if the well is to be filter-packed, the filter-pack material fills the annular space evenly around the screen. Perhaps equally important is ensuring that annular seal materials fill the annular space evenly around the casing, to prevent movement of water in the borehole between formations.

Centering can be accomplished effectively in one of several ways. For wells that are installed in cohesive materials in which the borehole stands open during well construction, or wells in which a temporary casing is used to hold the borehole open during well construction, centralizers (also called centering guides) can be affixed to the outside of the casing string before or as the string is installed. Centralizers are available in several different designs, with several different methods of attachment to casing or screen possible. Four of the most common designs are depicted in Figure 10.63. Centralizers must be constructed of a suitable material that has the strength, rigidity, chemical resistance, and other characteristics appropriate for the specific application. Materials commonly available include low-carbon, galvanized and stainless steel; PVC; CPVC, polyethylene, FRE, and FRP. Selection of an appropriate design and material should be based upon resistance to corrosion or chemical degradation, method of borehole installation used, condition and shape of the borehole wall, type of fluid in the borehole, and

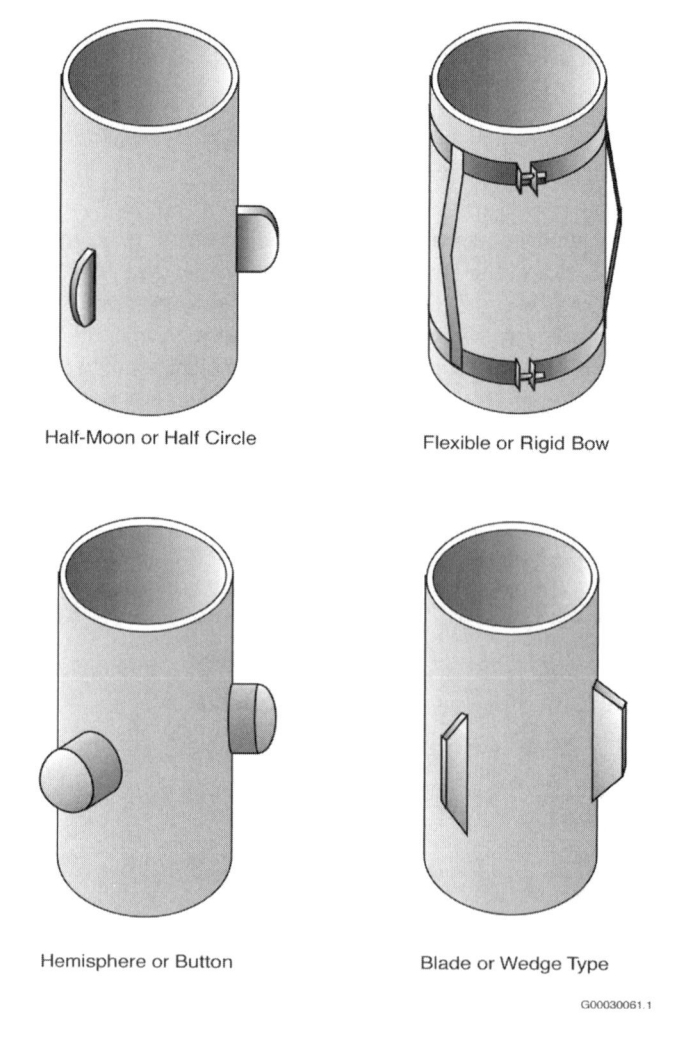

FIGURE 10.63
Four types of centering guides (centralizers) commonly used on casing or pipe.

method to be used for installation of filter pack and annular seal materials. Centralizers are generally attached immediately above the well screen (and, in some cases, at the bottom of the well screen or on the sump below the well screen) and at 10- or 20-ft intervals along the casing to the surface (depending upon the rigidity of the casing material), but not within the well screen. Centralizers should never be attached at or within two pipe diameters of casing joints (particularly in threaded, flush-joint casing), as these are the weakest points in the casing string. Centralizers should be sized to project a diameter just less than that of the borehole or temporary casing, so that they fit in the borehole and the casing string is kept an equal distance from all sides of the borehole. The casing string should remain centered in the borehole if centralizers are installed in this manner. Presuming that the borehole is straight and plumb, the well installed with centralizers within the borehole should also be straight and plumb.

Three types of centralizers (half-moon shaped, button-shaped, and wedge-shaped centralizers; see Figure 10.63) are designed to be permanently attached to the casing string, either by welding (for steel materials) or by adhesives (for plastic materials).

Design and Installation of Ground-Water Monitoring Wells 711

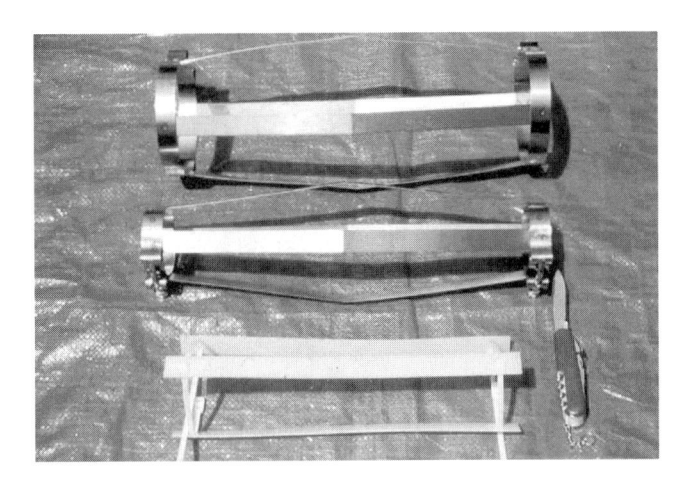

FIGURE 10.64
Flexible, bow-type centralizers are very commonly used in monitoring wells.

These designs are generally spaced every 90° or 120° around the casing, and are not likely to cause bridging of filter pack or annular seal materials as they are installed in the annular space. Half-moon shaped centralizers have the advantage that they are least likely to cause snagging or entanglement of cables or tapes used to measure depths of annular fill materials because of their smooth shape. While button-shaped centralizers are slightly more likely and blade-shaped centralizers much more likely to snag cables and tapes, the smooth surface of button-type centralizers allows easier extrication and less potential for damage to cables and tapes than the blade-shaped design. Each of these three designs allows easy installation and removal of tremie pipes, but each can be easily damaged or broken off if hit with a tremie pipe or if the tremie is forced to bend around the centralizers.

Flexible or rigid bow-type centralizers are probably the most common types of centralizers used in monitoring well installations (Figure 10.64). They are attached to the casing string either by bolts or straps that are tightened against the outer surface of the casing (Figure 10.65). Bolts are not as desirable an attachment method as straps, because they can more easily deform or penetrate the casing as the centralizer is

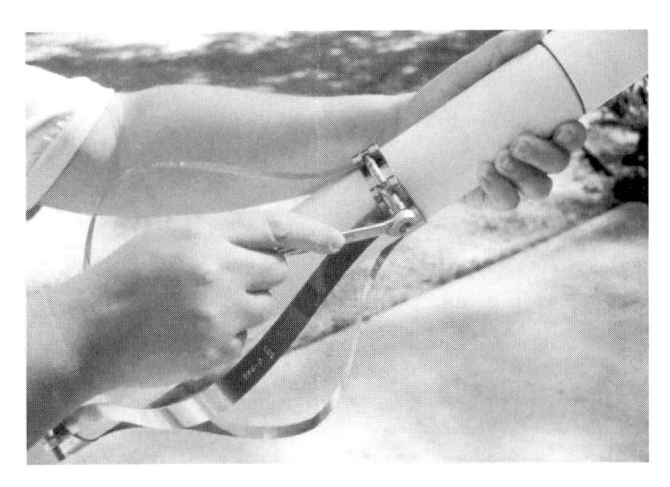

FIGURE 10.65
Attaching a bow-type centralizer to well casing using straps tightened by a bolt.

secured. These centralizers are more likely to foul or damage tapes and cables lowered into the annulus, and more likely to cause bridging than other types of centralizers. This is because of the additional surface areas and ledges that are created by the shape of the centralizer and the method of attachment to the casing. These centralizers are easy to attach in the field (no welding or adhesives are required), and allow the casing string to be easily lowered into boreholes that are less than straight.

A casing string can also be installed centered in the borehole (or nearly so) without centralizers if it is installed through the center of a hollow-stem auger or a small-diameter direct-push drive casing. The auger or drive casing acts as a centering device, maintaining a relatively consistent annular space on all sides of the well casing. Because the inside of a hollow-stem auger or drive casing is always somewhat larger than the well casing string installed within it, it is possible for the well casing to be installed slightly off-center, with the casing resting against one side of the auger or drive casing. However, in direct-push wells in which prepacked well screens are used, and the annulus is grouted with high-solids bentonite grout, this generally does not materially affect the well installation.

Sediment Sumps

It is a common practice in water-supply wells to install a sediment sump or trap (also called a "tailpipe" or "rat trap"), or a piece of blank casing installed below the well screen with a bottom plug (Figure 10.66). The purpose of a sump is to collect sediment either brought into the well during development or carried into the well by continued pumping over time. Some contractors have carried this practice over to the installation of monitoring wells. Some monitoring well designers install a sump below the screen to allow for the collection of samples of DNAPLS in situations where the bottom of the screen is set at the top of the first confining layer. The blank piece of pipe below the well screen also provides a place to install a centralizer to help keep the screen centered in the borehole.

In a properly installed, filter-packed monitoring well, relatively little sediment should be developed into the well, whereas in a naturally developed well, a great deal of sediment may be brought into the well during development. The sediment sump provides a place for developed sediment to collect so that the bottom of the well screen is not filled with and potentially clogged with sediment. Yu (1989) points out that any sediment brought into the well during development must be removed prior to ground-water sampling. This avoids the phenomenon of chemicals sorbing onto and then desorbing from the sediments that

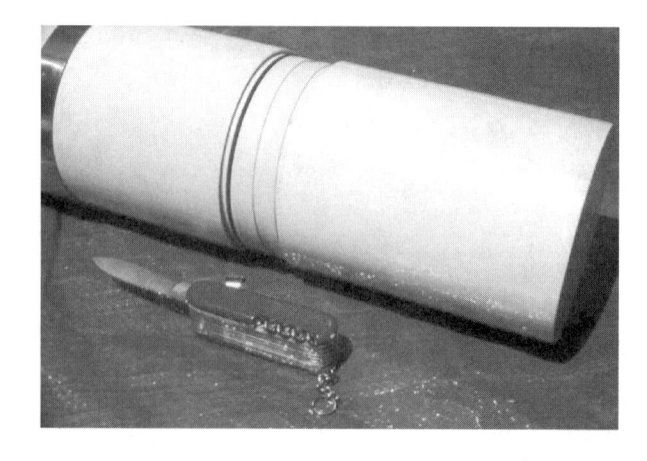

FIGURE 10.66
Sediment sumps are often installed below the well screen to capture sediment brought into the well during well development.

may have collected in the sump. Yu (1989) also contends that the two suggested uses of a sump are mutually exclusive (i.e., if the sump traps sediment, it cannot trap DNAPLS). However, if sediment is removed from the sump following development, there should be no reason why the sump cannot be used to trap DNAPLS. The use of a sediment sump is at the discretion of the well designer.

Cleaning Requirements for Casing and Screen Materials

During the production of any casing or screen material, certain chemical substances may be used to assist in the extrusion, molding, machining, or stabilization of the casing material. For example, during the production of steel casing, considerable quantities of oils and solvents are used in various manufacturing stages, as during the machining of threads. In the manufacturing of PVC well casing, a wax layer can develop on the walls of the casing; additionally, protective coatings of natural or synthetic waxes, fatty acids, or fatty acid esters may be added to enhance the durability of the casing (Barcelona et al., 1983). All of these represent potential sources of chemical interference, and these chemical substances must be removed prior to installation of the casing in a borehole. If trace amounts of these materials remain on the casing after installation, they may affect the chemical integrity of samples taken from those wells.

Careful preinstallation cleaning of casing materials (Figure 10.67) is essential to avoid potential chemical interference problems from the presence of such substances as cutting oils, cleaning solvents, lubricants, threading compounds, waxes, or other chemical residues. Chapter 20 and ASTM Standards D 5088 (ASTM, 2004v) and D 5608 (ASTM, 2004w) provide details on cleaning protocols typically used to remove these residues. The cleaning protocol to be used for any given site will depend upon: (1) data collection requirements, including the analytes targeted for sampling and analyte concentrations; (2) local, state, and Federal regulations that may be enforced at the site; and (3) the contaminants expected to be present on the casing and screen materials (Lapham et al., 1996). For PVC, Curran and Tomson (1983) suggest washing the casing with a strong, low-sudsing, laboratory-grade nonphosphate detergent solution (e.g., Liquinox; Micro; Detergent 8) and then rinsing with water before installation. Barcelona et al. (1983) suggest the same procedure for all casing materials. To accomplish the removal of some cutting oils, lubricants or solvents, it may be necessary to steam-clean casing materials or employ a high-pressure hot water wash. Care should also be taken to ensure that

FIGURE 10.67
Well casing materials should be thoroughly cleaned before installation.

casing materials are protected from contamination while they are onsite awaiting installation in the borehole. This is usually done by providing a clean storage area way from any potential contaminant sources (air, water, or soil) or using plastic sheeting spread on the ground for temporary storage adjacent to the work area.

Factory cleaning of casing or screen in a controlled environment by standard detergent washing, rinsing, and air-drying procedures is superior to most cleaning efforts attempted in the field. Factory cleaned and sealed casing and screen (Figure 10.68) can be certified by the supplier. Individually wrapped sections in a common shipping crate are easiest to keep clean and to install.

For some investigations, quality control samples may be collected to evaluate the effectiveness of a cleaning procedure. The only way to demonstrate convincingly that the well construction materials are free of contaminants of interest following cleaning is to collect and analyze quality control samples. Records of cleaning procedures used at any given site should be maintained to provide documentation for use in the event that concerns arise following well installation.

Costs of Casing Materials

As Scalf et al. (1981) point out, the dilemma for the field investigator is the relation between cost and accuracy. PVC is approximately 1/4 the cost of Type 304 stainless steel, which may be a major consideration when a ground-water monitoring project entails the installation of a large number of wells, or deep wells. On the other hand, if the particular compounds of interest in a monitoring program are also components of the lower-cost casing, or if the lower-cost casing may leach materials in the presence of

FIGURE 10.68
Factory-cleaned and packaged well casing and screen materials may be a better alternative to cleaning materials in the field.

site-specific geochemical conditions, and analytical sensitivity is in the ppb range, the data generated on contaminants detected may be suspect.

In many situations, it may be possible to use the lowest-cost casing material without compromising accuracy. For example, if the contaminants to be monitored are already defined and they do not include chemical constituents that could potentially leach from or sorb onto the lowest-cost well casing (as defined by laboratory or field studies), it should be possible to use the less expensive casing without compromising analytical accuracy. Or, as Scalf et al. (1981) suggest, wells constructed of "less than optimum" materials, or materials that strike a good compromise might be used with a reasonable level of confidence for sampling if at least one identically constructed well was available in an uncontaminated part of the monitored formation to provide ground-water samples as "blanks" for comparison.

Hybrid Wells

The preceding discussions on chemical interference are all related to situations in which the casing or screen material is in contact with the saturated portion of the subsurface. For materials that are not in contact with the saturated zone, the arguments regarding chemical attack and sorption/leaching phenomena are generally not valid. The inside of the casing in such situations is exposed only to air, and the annular seal surrounds the outside of the casing. Thus, it may be possible to utilize less chemically resistant or less chemically inert casing materials in the unsaturated zone, coupled with a more chemically resistant–inert material in the saturated zone (Figure 10.69). It should be noted, however, that seasonal or more frequent (i.e., tidal or pumping-influenced) variations in ground-water levels must be taken into account when determining the depth at which the more

FIGURE 10.69
In some situations, a hybrid well (i.e., using stainless steel screen and PVC casing) may be desirable. In this case, DNAPLs (some of which were PVC softening agents) were present in the zone to be screened, but not above, requiring the use of a screen material resistant to chemical degradation, but allowing the use of PVC casing.

resistant or inert material should be used. In addition, it should be noted that some types of casing material might not be compatible. For example, the joining of two different types of steel casing (i.e., galvanized and stainless steel) may result in the creation of a galvanic cell and subsequent corrosion at the joint because the two metals have different electromotive potentials. Thus, materials should be considered for compatibility prior to installation.

Monitoring Well Intakes — Well Screen and Filter Pack

Proper design of a hydraulically efficient monitoring well in unconsolidated geologic materials and in certain types of poorly consolidated geologic materials requires that a well screen be placed within the target-monitoring zone. The screen should be surrounded by materials that are coarser and of higher hydraulic conductivity than natural formation material (i.e., filter pack). This allows ground water to flow freely into the well from the adjacent formation while minimizing or eliminating the entrance of fine-grained formation materials (clay, silt, and fine sand) into the well between sampling rounds and during sampling. Thus, the well will not silt up between sampling rounds, and it can provide ground-water samples that are free of suspended sediment and low in turbidity, provided that appropriate sampling methods are used. Sediment-free ground-water samples allow for dramatically shortened filtration times or the elimination of sample filtration, and greatly reduce the potential for sample bias or interference.

Much of the technology applied to the design and installation of monitoring wells has been derived from the water-supply well industry. It should be noted that while production or water-supply wells and monitoring wells are similar in many ways, there are some distinct differences between the two types of wells. For example, one significant difference between monitoring wells and water-supply wells is that the intake section, or screen, of monitoring wells is often purposely situated in a zone of poor quality water and poor yield. The quality of water entering a monitoring well can range from drinking water to a hazardous waste or concentrated leachate. In contrast, water-supply wells are normally designed to obtain water from highly productive zones containing good-quality water. The screen of a monitoring well often extends only a short length to obtain water from, or to monitor conditions within, a relatively short interval of material within a formation. Water-supply wells are frequently designed with longer or multiple screened intervals to obtain as much water as possible, often from multiple water-bearing strata. Although there are usually differences between the design and function of monitoring wells and water-supply wells, water-supply wells are sometimes used as monitoring wells and vice versa.

There are two basic types of wells and well intake designs for wells installed in unconsolidated or poorly consolidated geologic materials. The first type is a well in which natural formation materials are allowed to collapse around the well screen and the filter pack is derived directly from formation materials through proper well development (i.e., "naturally developed" wells). The second type is a well in which filter-pack sand of a selected grain size and composition is installed around the screen from the surface (i.e., "filter-packed" wells). In both types of wells, the effective diameter of the well is increased by surrounding the well screen with an envelope of coarse material with higher hydraulic conductivity than the natural formation material. Additionally, the material surrounding the well screen performs a mechanical filtration function to strain out fine-grained sediment from the surrounding formation.

The major drawback of current-day well intake technology is that it is based on mechanical filtration. Mechanical filtration, using the grain sizes that are practical and most easily developed or installed, has limits with respect to formation grain sizes to which it can be applied. The limit of mechanical filtration for monitoring wells is defined by the finest filter-pack material that can be practically installed, either via conventional means (i.e., a tremie pipe) or via a prepacked or sleeved well screen. The finest filter-pack material that can be conventionally installed is a 40 × 70 sand (fine to medium sand), which can be used with a well screen slot as small as 0.008 in. This design will work for formations that have a D_{30} (70% retained) size of 0.002 in., or fine sand with some silt. The finest filter-pack material that can be installed via a prepacked or sleeved well screen is a silica flour with a grain size of 0.003 in. (200 mesh), which will retain formation material as fine as silt. Formations with a small fraction of clay (up to about 20%) (e.g., a silty fine sand with some clay or a silt with some clay) can be successfully monitored with this design as long as the well is properly developed. Formations with more than 50% passing a #200 sieve, and having more than 20% clay content, should not be monitored using conventional well designs.

Naturally Developed Wells

In a naturally developed well, formation materials are allowed to collapse around the well screen after it has been installed in the borehole. A high hydraulic conductivity envelope of coarse materials (a natural filter pack) is developed adjacent to the well screen *in situ* by removing the fine-grained fraction from the natural formation materials during the well development process.

As described by Johnson Division (1966), the envelope of coarse, graded material created around a well screen during the development process can be visualized as a series of cylindrical zones. In the zone immediately adjacent to the well screen, development removes all particles smaller than the screen openings, leaving only the coarsest materials in place. Slightly farther away, some medium-sized grains remain mixed with coarse materials. Beyond that zone, the material grades back to the original character of the water-bearing formation. By creating this succession of coarse, graded zones around the screen (Figure 10.70), development stabilizes the formation so that no further

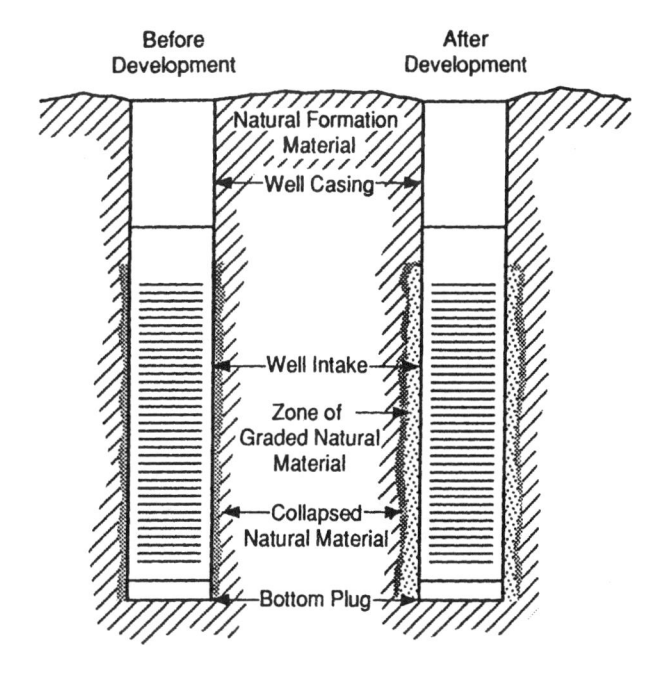

FIGURE 10.70
The gradation of coarse to fine materials away from the well screen is required for natural formation material to serve as a mechanical filter in a naturally developed well. This gradation is produced by careful well development.

movement of fine-grained materials should take place, and the well should yield sediment-free water.

The decision on whether or not a natural filter pack can be developed is generally based on geologic conditions, specifically the grain-size distribution of natural formation materials in the zone of interest. For this reason, the importance of obtaining accurate formation samples (Figure 10.71) cannot be overemphasized. Naturally developed wells are generally recommended in situations in which natural formation materials are relatively coarse-grained, permeable, and uniform in grain size. Grain-size distribution must be determined by conducting a mechanical (sieve) analysis of a sample or samples taken of the formation materials from the intended screened interval (Figure 10.72–Figure 10.76).

After samples of formation material are sieved, a plot of grain size versus cumulative percentage of sample retained on each sieve is made (Figure 10.77). On the basis of this grain-size distribution, and specifically upon the effective size and uniformity coefficient of the formation materials, well screen slot sizes are selected. The effective size is equivalent to the sieve size that retains 90% (or passes 10%) of the formation material (Figure 10.78); this is termed the D_{10} size. The uniformity coefficient is the ratio of the sieve size that retains 40% (or passes 60%, the D_{60} size) of the formation material to the effective size (Figure 10.79). A naturally developed well can normally be justified if the effective grain size of the formation material is greater than 0.010 in. and the uniformity coefficient is greater than 3.0 (Johnson Division, 1966). These criteria are directly applicable to monitoring wells.

Well Screen Design in a Naturally Developed Well

Proper sizing of monitoring well screen slot size is the most important aspect of this type of monitoring well design. There has been an unfortunate tendency among many monitoring well designers to install one screen slot size (i.e., 0.010-in. slots) in every well, regardless of formation characteristics. As Williams (1981) points out, this can lead to difficulties with well development and poor well performance — it can also result in acquisition of sediment-laden samples for the life of the monitoring well.

FIGURE 10.71
Collecting accurate formation samples from the portion of the formation to be screened by the well, in this case using a split-spoon sampler, is required for proper well design.

Design and Installation of Ground-Water Monitoring Wells 719

FIGURE 10.72
The first step in conducting a field sieve analysis of a formation sample is to dry the sample so it can be easily separated by running it through the sieves. In this case, the sample is being dried with a propane torch.

FIGURE 10.73
After the formation sample is dried, it must be weighed to obtain a total sample weight.

FIGURE 10.74
The sample is then poured into the top of the stack of sieves. A range of 6 to 10 sieve sizes appropriate for the formation materials is selected based on visual examination of the sample, with the largest sieve size on top, smallest on the bottom, and a pan below. The sieves are then shaken vigorously to separate the sample on the various sieves, from coarsest on top to finest on the bottom or in the pan.

FIGURE 10.75
The amount of the total sample retained by each of the sieves is then weighed and compared with the total weight. Starting with the largest sieve, the cumulative percent retained for each successive sieve is calculated and the results are plotted on special sieve analysis graph paper, which plots grain size in thousandths of an inch (or U.S. Standard sieve size) versus cumulative percent retained.

Design and Installation of Ground-Water Monitoring Wells 721

FIGURE 10.76
The graphed results of the sieve analysis are then analyzed to determine whether a naturally developed well design or a filter-packed well design is appropriate. For naturally developed wells, the well screen slot size is then chosen, and for filter-packed wells, the filter pack grain size and the well screen slot size are selected to match the formation grain-size distribution.

Well screen slot sizes are generally selected based upon the following criteria:

- Where the uniformity coefficient of the formation material is greater than 6, the slot size should be that which retains no less than 50% of formation materials
- Where the uniformity coefficient of the formation material is greater than 3 but less than 6, the slot size should be that which retains no less than 60% of formation materials
- Where the uniformity coefficient of the formation material is less than 3, the slot size should be that which retains no less than 70% of formation materials

The slot size determined from a sieve analysis is seldom that of commercially available screen slot sizes (Table 10.17), so the nearest smaller commercially available slot size is generally used. Because optimum yield from the well is not as critical to achieve as it is in water-supply wells, screens for monitoring wells are usually designed to have smaller slot sizes than indicated by the above design criteria. This means that much less than 50% (generally less than 30%) of the formation material adjacent to the well will be drawn into the well during development.

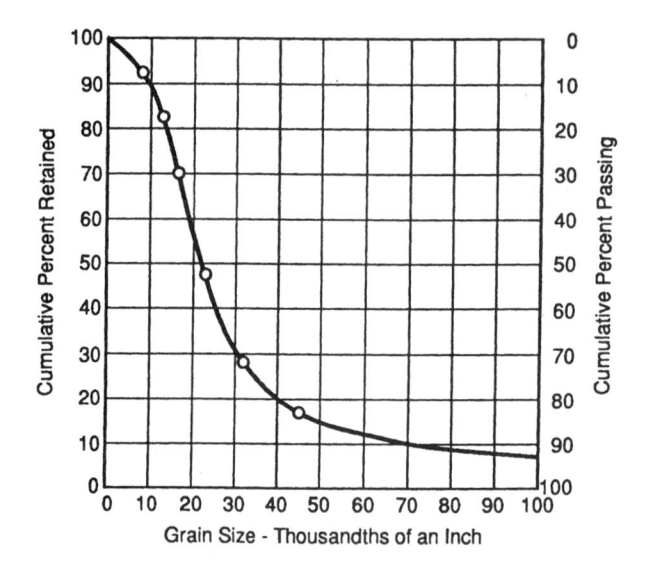

FIGURE 10.77
A typical grain-size distribution curve produced by a sieve analysis of a formation sample. The results usually plot in an "S"-shaped curve, with a steep curve indicating more uniform materials, and a flat curve indicating poorly sorted (or well graded) materials.

Installing a naturally filter-packed well is advantageous in formations comprised predominantly of coarse materials, particularly if mud rotary drilling is used. The absence of an artificial filter pack allows for maximum effectiveness for developing the formation and for removing the fine drill cuttings and drilling fluids from the borehole in the screened interval. Perhaps the biggest drawback for naturally developed wells may be the time required for well development to remove fine-grained formation material. Because the design of a monitoring well screen may allow more than 30% of the formation materials near the well screen to enter the well, development can often be a long, drawn-out process. Increased development has other disadvantages: (1) it delays installation of the

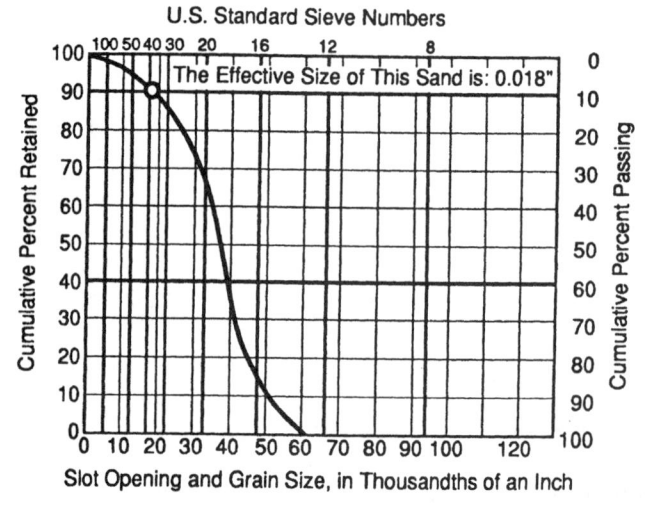

FIGURE 10.78
The effective size of a formation sample is equal to 90% retained size.

Design and Installation of Ground-Water Monitoring Wells

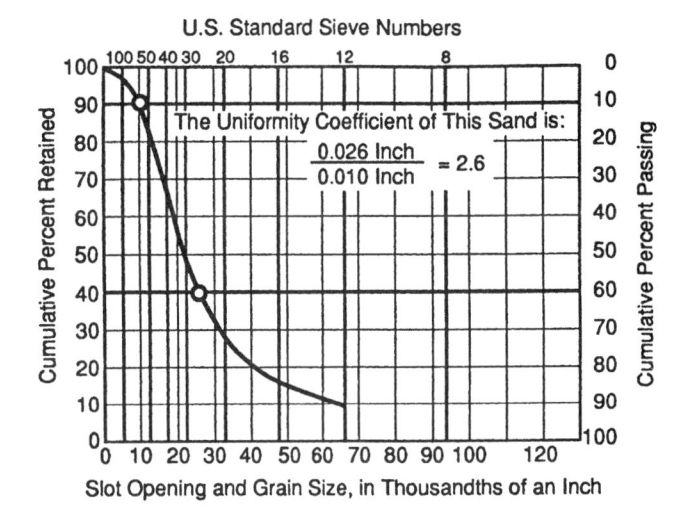

FIGURE 10.79
The uniformity coefficient of a formation sample is the ratio of the 40% retained size to the 90% retained size (the effective size).

overlying well seal; (2) it may require the handling of large quantities of contaminated sediment and water; and (3) it allows settlement adjacent to the screen, which may result in the invasion of overlying sediments. Also, unless a formation is fairly coarse grained (i.e., medium sand or coarser), developing a natural filter pack for a monitoring well is difficult, because the small diameter and the short screen length limits the rate of water withdrawal, and the removal of formation fines from the well is sometimes difficult.

Filter-Packed Wells

Filter-packed wells are wells in which the natural formation materials surrounding the well intake are deliberately replaced by coarser granular material introduced from the surface. In much of the literature describing water-supply well design (i.e., Johnson Division, 1966; Driscoll, 1986; Roscoe Moss Company, 1990), the term "gravel pack" is used to describe the material added to the borehole to act as a filter. Because the term "gravel" is classically used to describe large-diameter granular material (>0.08 in. or >2 mm), and because nearly all coarse material placed in monitoring wells is from fine to medium to coarse sand-sized material and not gravel, the use of the term "sand pack" or "filter pack" is preferred. True gravel-sized material is rarely used as filter pack in a monitoring well.

TABLE 10.17

Typical Commercially Available Slotted Casing Slot Widths

0.006	0.016	0.040
0.007	0.018	0.050
0.008	0.020	0.060
0.010	0.025	0.070
0.012	0.030	0.080
0.014	0.035	0.100

Source: From NWWA/PPI, 1980. With permission.

The introduction of filter-pack material into the annular space between a centrally positioned well screen and the borehole wall serves a variety of purposes. The primary purposes of an installed filter pack are to retain formation material while allowing ground water to enter the well, and to stabilize aquifer materials in order to avoid excessive settlement or collapse of materials above the well screen during well development. The introduction of material coarser than natural formation materials also results in an increase in the effective diameter of the well and in an accompanying increase in the amount of water that flows toward and into the well.

Filter packs have been used extensively by water well contractors to construct efficient, large-diameter wells to provide water for irrigation and for municipal and industrial uses. Monitoring wells serve a different purpose and thus have different filter-pack design requirements. Because monitoring wells are designed to serve as sampling points in a formation, they are typically smaller in diameter and screened only in the portion of the formation that is of specific interest to the investigator (the target-monitoring zone). Furthermore, the design properties for monitoring well filter packs are more stringent than those for water-supply well filter packs, because the disturbance of water chemistry must be minimized.

Until the first edition of this book was published in 1991, water-supply well design practices were historically used in the design of well screens and filter packs for monitoring wells. This is primarily because available well-design texts, which pertain almost exclusively to water well technology (Johnson Division, 1966; Anderson, 1971; Campbell and Lehr, 1973; Driscoll, 1986; Roscoe Moss Company, 1990), did not adequately address the subject of well screen and filter-pack requirements for monitoring wells. Prior to 1991, information on monitoring well design appeared in only a limited number of research papers. In the years since 1991, a great deal of experience has been gained in the field of monitoring well design and installation. An ASTM Standard on the subject is now available (ASTM D 5092 [ASTM, 2004a]), and much more data are now available on the effectiveness of various designs in site-specific applications.

The decision regarding whether a filter pack should be used in the construction of a monitoring well is based primarily on geologic considerations. There are several geologic situations in which the use of a filter-pack material is required for a monitoring well:

- Where the natural formation material is unconsolidated and consists primarily of fine-grained sand, silt, or clay-sized particles
- Where a long screened interval is required, and the well screen spans highly stratified geologic materials of widely varying grain sizes
- Where the formation in which the intake is to be placed is poorly cemented, such as a friable sandstone
- Where the formation is a fractured or solution-channeled rock in which particulate matter may be carried through large fractures or solution openings

The use of a filter pack in a fine-grained formation material allows the screen slot size to be considerably larger than if the screen were placed in the formation material without the filter pack. This is particularly true where fine-grained sands, silts, and clays predominate in the zone of interest, and small enough slot sizes in well screens to hold out formation materials are impractical or commercially unavailable. The larger screen slot size afforded by use of a filter pack thus allows for the collection of adequate volumes of sediment-free samples, and results in both decreased head loss and increased well efficiency.

Filter packs are especially well suited to use in highly stratified formations, in which thin layers of fine-grained materials alternate with layers of coarser materials or vice versa. In such geologic environments, it is often difficult to precisely determine the position and thickness of each individual stratum and to choose the correct position and slot size for a well screen. Completing the well with a filter pack, sized to suit the finest layer of a stratified sequence, resolves the latter problem and helps to ensure that the well will produce water free of suspended sediment.

Quantitative criteria exist with which decisions can be made concerning whether a natural or an installed filter pack should be used in a water-supply well (Johnson Division, 1966; Campbell and Lehr, 1973; U.S. EPA, 1975; Williams, 1981). In water-supply wells, the use of an installed filter pack is recommended in situations where the effective grain size of the natural formation materials is smaller than 0.010 in. and the uniformity coefficient is less than 3.0. For monitoring wells, ASTM Standard D 5092 (ASTM, 2004a) recommends a different approach and suggests that an installed filter pack be employed if a sieve analysis of formation materials indicates that a screen slot size of 0.020 in. or less is required to retain 50% of the natural material.

Economic considerations may also affect decisions concerning the appropriateness of an installed filter pack. Costs associated with filter-packed wells are generally higher than those associated with naturally developed wells, primarily because specially sized and washed sand must be purchased and transported to the site. Additionally, larger boreholes may be required for conventionally filter packed wells in which a tremie pipe is used (usually a 4- to 6-in. diameter borehole for a 2-in. inside diameter well, or a 6- to 8-in. borehole for a 4-in. diameter well. The increased costs for larger diameter boreholes may be particularly significant if drilling is done with a hollow-stem auger rig.

Filter-Pack and Well Screen Design for Filter-Packed Wells

To achieve the purposes of creating a high hydraulic conductivity envelope around the well screen, retaining formation materials while allowing ground water to flow into the well, and not interfering with ground-water chemistry, the filter-pack material for a monitoring well must have certain characteristics. While some of these characteristics are the same as those desirable for water-supply wells, monitoring wells also have some unique requirements. Filter-pack design factors for monitoring wells include: (1) filter-pack grain size and well-screen slot size; (2) filter-pack grain size distribution properties (i.e., uniformity coefficient, effective size, kurtosis, and skewness); (3) filter-pack grain shape properties (i.e., roundness, sphericity); (4) filter-pack dimensions (thickness and length); and (5) filter-pack material type. When a filter pack is dictated by sieve analysis or by geologic conditions, the filter-pack grain size and well screen slot size are generally designed together. After well development, a monitoring well with a correctly designed and installed filter pack and well screen combination should produce samples free of artifactual turbidity.

Filter-Pack Grain Size and Well Screen Slot Size: Filter-pack grain size and well-screen slot size are based on the grain-size distribution of the formation material, as determined by sieve analysis (ASTM D 422 [ASTM, 2004x]). The filter pack, which is the interface with the formation, is the principal hydraulic structure of the well and is designed first. The first step in designing the filter pack is to perform sieve analyses on formation samples collected from the target-monitoring zone. This is the same procedure followed for sizing well screen slot sizes in a naturally developed well. The filter-pack grain size is then selected on the basis of the finest formation materials present. Because the design theory of filter-pack gradation is based on the mechanical retention of formation materials, following the procedures below (outlined in ASTM Standard D 5092 [ASTM, 2004a]) is very important in the proper design of a well intake.

For formation materials that are relatively coarse grained (e.g., fine, medium, and coarse sands and gravels), the grain-size distribution of the primary filter pack is determined by calculating the D_{30} (30% finer) size, the D_{60} (60% finer) size, and the D_{10} (10% finer) size of the filter pack. The first point plotted on the filter-pack grain-size distribution curve is the D_{30} size. The primary filter pack is usually selected to have a D_{30} grain size that is about three to six times greater than the D_{30} grain size of the formation material being retained (see Figure 10.80). A multiplication factor of 3 is used if the formation material is relatively fine-grained and well sorted or uniform (small range in grain sizes); a multiplication factor of 6 is used if the formation is relatively coarse grained and poorly sorted or nonuniform (large range in grain sizes). Thus, 70% of the filter pack will have a grain size that is three to six times larger than the D_{30} size of the formation materials. This ensures that the filter pack is coarser (with a higher hydraulic conductivity) than the formation material, and allows for unrestricted ground-water flow from the formation into the monitoring well, while formation materials are mechanically filtered out.

The next two points on the filter-pack distribution curve are the D_{60} and D_{10} grain sizes. Because the well screen slots have uniform openings, the filter pack should be composed of particles that are as uniform in size as is practical. Ideally, the uniformity coefficient (the quotient of the D_{60} size divided by the D_{10} size [effective size]) of the filter pack should be 1.0. However, a more practical and achievable uniformity coefficient for a wide range of filter pack sizes is 2.5. This value, which ensures that the filter pack has a small range in grain sizes and is uniform, should represent a maximum value, not an ideal.

The D_{60} and D_{10} grain sizes of the filter pack are calculated by a trial and error method using grain sizes that are close to the D_{30} size of the filter pack. After the D_{30}, D_{60}, and D_{10} sizes of the filter pack are determined, a smooth curve is drawn through these points. The final step in filter-pack design is to specify the limits of the grain size envelope, which defines the permissible range in grain sizes for the filter pack. The permissible range on either size of the grain size curve is 8%. The boundaries of the grain size envelope are drawn on either side of the filter-pack grain-size distribution curve, and filter-pack design is complete. A filter medium having a grain size distribution within the boundaries

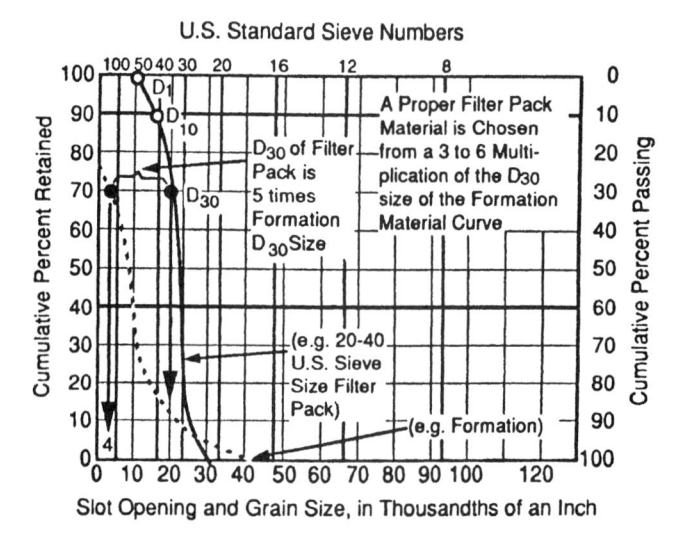

FIGURE 10.80
The filter-pack material selected to surround the well screen should have a 70% retained size that is three to six times greater than the 70% retained size of the formation materials. In this example, a multiplier of 5 was selected, based on formation material characteristics.

of this envelope (preferably as close to the ideal grain-size distribution curve as possible) is then obtained from a local sand supplier.

The size of well screen slots is selected after the filter-pack grain-size distribution is specified. In monitoring wells it is desirable for the well screen to retain 90 to 99% of filter-pack materials (Figure 10.81), because development is generally done after the well has been completed, and it is very important to avoid excessive settling of the materials surrounding the well. To retain 99% of the filter pack, the well screen slot size should be approximately equal to the D_1 size of the filter pack (Table 10.18 and Figure 10.81). To retain 90% of the filter pack, the well screen slot size should be approximately equal to the D_{10} size of the filter pack (Table 10.18).

In formation materials that are predominantly fine-grained (finer than fine to very fine sands), soil piping can occur when a hydraulic gradient exists between the formation and the well (as would be the case during well development and sampling). To prevent soil piping in these materials, the following criteria are used for designing granular filter packs:

$$\frac{D_{15} \text{ of filter}}{D_{85} \text{ of formation}} \leq 4 \text{ to } 5 \quad \text{and} \quad \frac{D_{15} \text{ of filter}}{D_{15} \text{ of formation}} \geq 4 \text{ to } 5$$

The left-hand side of this equation is the fundamental criterion for the prevention of soil piping through a granular filter, while the right-hand side of the equation is the hydraulic conductivity criterion. This latter criterion serves the same purpose as multiplying the D_{30} grain size of the formation by a factor of between 3 and 6 for coarser formation materials. Filter-pack materials suitable for retaining formation materials in formations that are predominantly fine-grained are themselves, by necessity, relatively fine-grained (e.g., fine to very fine sands), presenting several problems for well designers and installers. First, well screen slot sizes suitable for retaining such fine-grained filter-pack materials are not widely available (the smallest commercially available slotted well casing is 0.006 in. [6 slot]; the smallest commercially available continuous-slot wire-wound screen is 0.004 in. [4 slot]). Second, the finest filter-pack material practical for conventional (tremie tube) installation is a 40 × 70 (0.008 × 0.018 in.) sand, which can be used with a

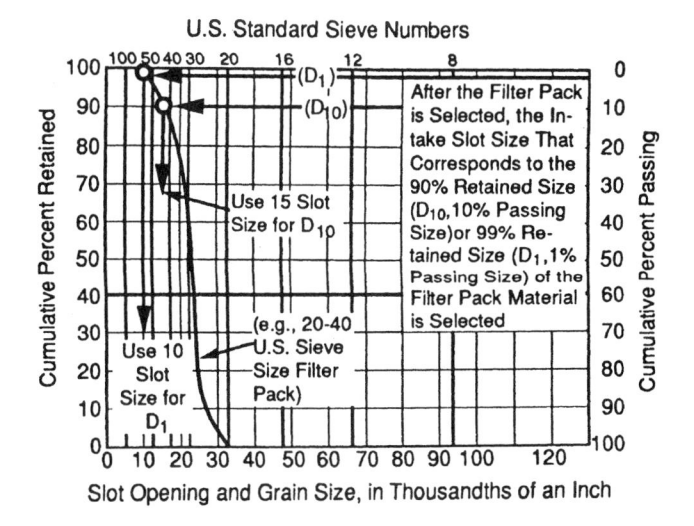

FIGURE 10.81
The well screen slot size is selected to retain between 90 and 99% of the selected filter-pack material.

TABLE 10.18

Recommended (Achievable) Filter-Pack Characteristics for Common Screen Slot Sizes

Size of Screen Opening (mm [in.])	Slot No.	Sand Pack Mesh Size	1% Passing Size (D_1) (mm)	Effective Size (D_{10}) (mm)	30% Passing Size (D_{30}) (mm)	Range of Uniformity Coefficient	Roundness (Powers Scale)	Fall Velocities[a] (cm/sec)
0.125 (0.005)	5	40–140	0.09–0.12	0.14–0.17	0.17–0.21	1.3–2.0	2–5	6–3
0.25 (0.010)	10	20–40	0.25–0.40	0.4–0.5	0.5–0.6	1.1–1.6	3–5	9–6
0.50 (0.020)	20	10–20	0.7–0.9	1.0–1.2	1.2–1.5	1.1–1.6	3–6	14–9
0.75 (0.030)	30	10–20	0.7–0.9	1.0–1.2	1.2–1.5	1.1–1.6	3–6	14–9
1.0 (0.040)	40	8–12	1.2–1.4	1.6–1.8	1.7–2.0	1.1–1.6	4–6	16–13
1.5 (0.060)	60	6–9	1.5–1.8	2.3–2.8	2.5–3.0	1.1–1.7	4–6	18–15
2.0 (0.080)	80	4–8	2.0–2.4	2.4–3.0	2.6–3.1	1.1–1.7	4–6	22–16

[a] Fall velocities in centimeters per second are approximate for the range of sand pack mesh sizes named in this table. If water in annular space is very turbid, fall velocities may be less than half the values shown here. If a viscous drilling mud is still in the annulus, fine sand particles may require hours to settle.

well screen slot as small as 0.008 in. (8 slot). Finer grained filter-pack materials cannot be placed practically by either tremie tubes or pouring down the annular space or down augers. Thus, the best method for ensuring proper installation of filter packs in predominantly fine-grained formation materials is to use prepacked or sleeved screens, which are described in detail in ASTM Standard D 6725 (ASTM, 2004y). A 50 × 100 (0.011 × 0.006 in.) filter-pack sand can be used with a 0.006-in. slot size prepacked or sleeved screen, and a 60 × 120 (0.0097 × 0.0045 in.) filter-pack sand can be used with a 0.004-in. slot size prepacked or sleeved screen. Filter packs that are finer than these (e.g., sands as fine as 100 × 120 [0.006 × 0.0045 in.], or silica flour as fine as 200 mesh [0.003 in.]) can only be installed within stainless steel or nylon mesh sleeves that can be placed over pipe-based screens. While these sleeves, or the space between internal and external screens in a prepacked well screen may be as thin as 0.5 in., the basis for mechanical retention dictates that a filter-pack thickness of only two or three grain diameters is needed to contain and control formation materials. Laboratory tests have demonstrated that a properly sized filter-pack material with a thickness of less than 0.5 in. (12.7 mm) successfully retains formation particles regardless of the velocity of water passing through the filter pack (Driscoll, 1986).

The finest filter-pack material that can be practically installed via a prepacked or sleeved screen is silica flour with a grain size of 0.003 in. (200 mesh). This fine a filter-pack material will retain formation material as fine as silt, but not clay. Formations with a small fraction of clay (up to about 20%) can be successfully monitored, as long as the wells installed in these formations are properly developed. The clay nearest the wall of the borehole, and any drilling damage to the formation, is removed during well development, and a good hydraulic connection is established between the filter pack and the formation. For mechanical filtration to be effective in formations with more than 50% fines, the filter-pack design would have to include silt-sized particles in the filter pack in order to meet the design criteria. This is impractical, as placement would be impossible and screen mesh fine enough to retain the material is not commercially available. Therefore, formations with more than 50% passing a #200 sieve, and having more than 20% clay content, should not be monitored using conventional well designs. Alternative monitoring technologies, such as pressure-vacuum lysimeters (see Chapter 3) should be used in these formations.

Because many formations have uniformity coefficients from 3 to 10 or higher, the formation's coarsest particles will probably be coarser than the coarsest particle in the filter pack if these design criteria are used. The formation's finest particles may be three to six times finer than the finest particles in the filter pack. The multiplier of three to six mentioned above is based on the assumption that, for uniform filter packs, the largest pore space will be 1/3 to 1/6 the average filter-pack particle size. Because retaining the bulk of formation particles is very important for monitoring wells, the same multipliers can be applied more conservatively for less uniform formations (i.e., uniformity coefficients of 6 to 10) using the D_{10} (10% passing) size for selecting filter packs (Table 10.18).

Filter Pack Grain Size Distribution Properties: Uniformity Coefficient, Kurtosis, and Skewness: Two types of filter packs are in common use in water-supply wells and in monitoring wells — the uniform filter pack and the graded filter pack. Uniform filter packs are generally preferred to graded packs for monitoring wells. Graded packs are more susceptible to the invasion of formation materials at the formation and filter pack interface, resulting in a partial filling of voids between grains and reduced hydraulic conductivity. With a uniform filter pack, the fine-grained formation materials can travel between the grains of the filter pack and be drawn into the well during development, thereby increasing formation permeability while retaining the highly permeable nature of the filter-pack material.

Furthermore, the filter pack is more efficient at achieving its purposes if it is composed of uniformly sized particles because the slots in a given well screen have uniform openings. Ideally, the uniformity coefficient (the 60% passing [D_{60}] size divided by the 10% passing [D_{10}] size [effective size]) of the filter pack should be as close to 1.0 as possible (i.e., the D_{60} size and the D_{10} size should be nearly identical), though this is rarely achieved. A low uniformity coefficient is very desirable, particularly if the tails of the particle-size distribution curve are also uniform (i.e., mesokurtic or platykurtic). The importance of kurtosis and skewness for filter packs is discussed by Schalla and Walters (1990).

Commercial filter-pack materials that meet uniformity coefficient requirements typically have suitable kurtosis and skewness values. These characteristics, which describe the distribution of particle sizes, are important because during installation in the borehole, particles falling at terminal velocity in the quiescent fluid surrounding the well screen will be influenced by a number of variables, including fluid and particle density, fluid viscosity, and particle diameter, shape, and surface roughness. Assuming that all variables except particle size are constant, the fall velocity for sand-sized particles composed of quartz will be approximately proportional to the square root of the particle diameter (Simons and Senturk, 1977). For example, a coarse sand grain 4 mm (0.156 in.) in diameter falls twice as fast as a medium sand grain 1 mm (0.039 in.) in diameter. Approximate fall velocities for sand-sized particles are shown in Table 10.18.

If a coarse or medium sand filter-pack material was poured into a well annulus through a few feet of water, the particles would segregate according to size, with the finest particles at the top of the screen and the coarsest particles at the bottom (Figure 10.82). During well development, fine formation particles would pass through the filter pack and well screen at the bottom and become lodged in the slots, as detailed in the bottom window in Figure 10.82. At the top of the screen, most, or possibly all, of the filter-pack material would be lost during development if the screen were designed to retain 90% of the filter pack, because the finest 10% of the filter-pack material would be concentrated at the top of the screened interval (Figure 10.82a). If sand were added in increments, a series of fine layers alternating with coarse layers would form at intervals along the length of the screen, and much of the volume of the fine layers would be removed during well development.

The problem with particle segregation would be eliminated through the use of a uniform filter pack if the design of the filter pack and well screen allowed less than 5% of the filter-pack material to pass through the screen slots. With a more uniform filter pack, the segregation problems with the coarse fraction would be diminished, because the tail of the coarse portion of the distribution curve would consist of a higher percentage of smaller, coarse-grained particles.

Uniformity of the filter-pack material is important because the screen slot size is uniform and the particles in a filter pack that are not uniform will segregate during placement through water, and will not be distributed evenly. Through the use of elaborate and time-consuming circulation processes, the particles in a nonuniform filter pack can be distributed somewhat more evenly (Campbell and Lehr, 1973). However, using uniform filter-pack materials is more cost effective and does not require the introduction of foreign fluids into the borehole. Grain-size distribution characteristics, their ideal values, and desirable ranges for filter-pack materials are shown in Table 10.18.

Filter-Pack Grain Shape Properties: Roundness and Sphericity: Roundness and sphericity are important parameters for filter-pack design because particles that are less round and less spherical tumble and oscillate as they fall through water. Tumbling and oscillating slow the rate at which the particles fall. Sand grains generally become less rounded and less

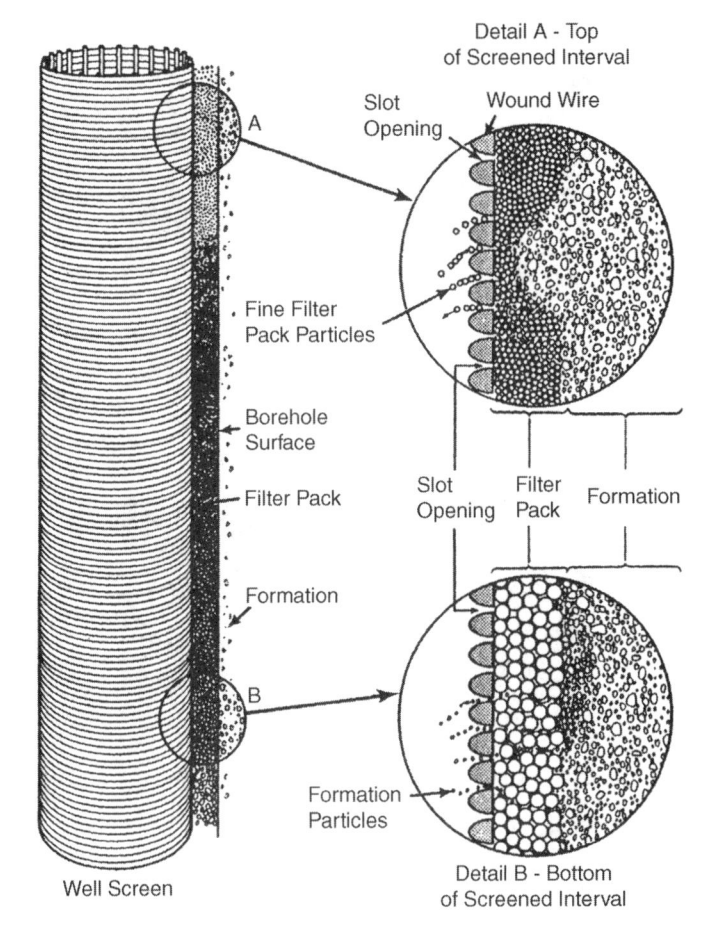

Design and Installation of Ground-Water Monitoring Wells 731

FIGURE 10.82
Pouring filter-pack material through a standing water column in a borehole results in segregation of the filter-pack material and, potentially, incorporation of sloughing formation material into the filter pack.

spherical as the particle size decreases. The greater angularity of the smaller grains tends to slow them, thus increasing the difference between the velocities of the large and small particles and increasing the likelihood of segregation of the particles.

Particle roundness (i.e., the curvature of the edges of a particle) is also important because minor changes in roundness can increase the potential for sand bridging during well development and well screen clogging. A common method used to define particle roundness is the Powers scale (based on the visual comparison of particles to photographic charts), which ranges from 1 (very angular) to 6 (well-rounded) (Figure 10.83) (Powers, 1953). The potential for bridging and well screen clogging is particularly high between very angular (Powers scale 1) and sub-angular (Powers scale 2 to 3). Roundness values of between 4 and 6 on the Powers scale are preferred for monitoring well filter-pack sand material.

Sphericity can define quantitatively how nearly equal are the three dimensions of a particle (Schalla and Walters, 1990). Sphericity, like roundness, can also reduce the potential for sand bridging in monitoring wells. However, sphericity is rarely included in well specifications that define the shape of sand particles, and numerical values are infrequently mentioned by producers of filter-pack materials.

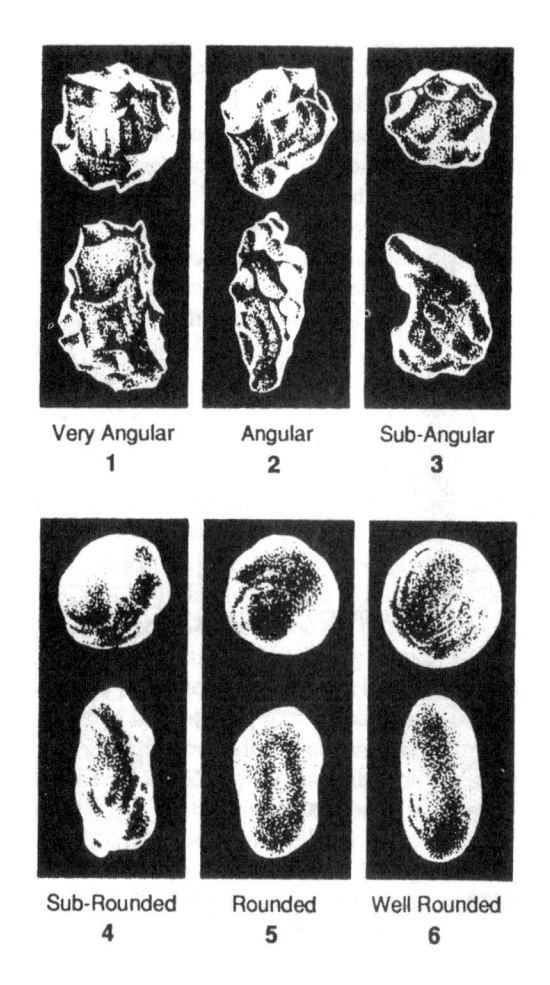

FIGURE 10.83
The Powers scale of particle roundness. (*Source*: Powers, 1953; Courtesy of the Society of Economic Paleontologists and Mineralogists.)

Filter-Pack Dimensions: With respect to filter-pack length, the pack should generally extend from just below the bottom of the well screen to at least 3 ft above the top of the well screen. This serves to account for any settling of filter-pack material that may occur during well development and allows a sufficient "buffer" between the well screen and the annular seal above.

The filter pack must be at least thick enough to surround the well screen completely, but thin enough to minimize resistance caused by the filter pack to the flow of fine-grained formation material and water into the well during development. To accommodate the filter pack, the well screen should be centered in the borehole and the annulus should be large enough and approximately symmetrical to preclude bridging and uneven placement of filter-pack material around the screen. A thicker filter pack does not materially increase the yield of the well, nor does it reduce the amount of fine material in the water flowing to the well (Ahrens, 1957). In fact, thicker filter packs prevent the effective development of the formation and the removal of residual materials (i.e., drill cuttings, drilling fluid) produced by the installation of the borehole. Most references in the literature (Johnson Division, 1966; Walker, 1974 and others; U.S. EPA, 1975; Williams, 1981) suggest that a filter-pack thickness of around 2 to 3 in. is appropriate for water-supply wells.

However, Driscoll (1986) suggests that in order for the filter pack to serve its intended purpose (mechanical filtration), it need be only 2 to 3 grain diameters in thickness. Thin filter packs (e.g., less than 0.5 in. thick) have proven successful in prepacked monitoring well screens that are installed using direct-push methods.

Filter-Pack Materials: In order to provide the minimum potential for alteration of ground-water sample chemistry, the materials comprising the filter pack in a monitoring well should be as chemically inert as possible. For example, Barcelona et al. (1985a) suggest that filter-pack materials should be composed primarily of clean quartz sand (Figure 10.84). The individual grains of the filter-pack materials should consist of less than 5% nonsiliceous material (Paul et al., 1988; ASTM, 2004a). Siliceous (quartz) material is preferred because it is nonreactive under nearly all subsurface conditions and it is readily available. In no case should filter-pack materials comprised of crushed stone be utilized because of the irregular nature of the particles and the potential for chemical alteration of ground water that would come in contact with the filter-pack material. This can occur as a result of the exposure of fresh surfaces of reactive minerals in the rock, on which chemical reactions can occur. This is particularly true regarding the use of crushed limestone. Limestone ($CaCO_3$) may significantly raise the pH of water with which it comes in contact; this, in turn, can affect the presence of other chemical constituents, including dissolved trace metals. Such material is not appropriate for use as a filter pack in a monitoring well.

Filter-pack materials should be washed, dried, and packaged at the processing facility. Most sand comes in standard 100-lb bags (Figure 10.85), which equal approximately 1 ft^3. It is recommended that bags with plastic (i.e., polyethylene) liners sandwiched between paper be requested to minimize bag breakage and sand loss and contamination during shipping or during storage in wet weather.

Methods of Filter-Pack Installation: Several methods of placing filter packs in the annular space of a monitoring well are available, including gravity (free-fall) placement, placement by tremie pipe, reverse-circulation placement, and backwashing.

Placement of filter packs by gravity or free-fall (i.e., pouring sand down the annulus) (Figure 10.86) can be accomplished only in very shallow wells (e.g., less than 20 ft total depth), with an annular space greater than 2 in. The potential occurrence of bridging or segregation of the filter-pack material is minimized in shallow wells. Bridging can

FIGURE 10.84
Filter-pack material should be composed of clean quartz sand (less than 5% nonsiliceous material).

FIGURE 10.85
A standard 100-lb bag of clean quartz filter-pack sand.

FIGURE 10.86
Placement of filter-pack sand by pouring down the annular space should only be attempted in wells less than 20-ft deep installed in relatively cohesive formations that will not collapse into the borehole during placement.

result in the occurrence of large unfilled voids in the filter pack, or in the failure of filter-pack materials to reach their intended depth. Segregation of filter-pack materials can result in a well that consistently produces sediment-laden samples. This problem is particularly likely to occur in deep wells in which a shallow static water level is present in the borehole. In this situation, as the sand falls through the column of water, the greater drag exerted on smaller particles due to their greater surface-area-to-weight ratio causes finer grains to fall at a slower rate than coarser grains. Thus, coarser materials end up comprising the lower portion of the filter pack, and finer materials make up the upper part. This is usually not a problem in the placement of truly uniform filter packs, where the uniformity coefficient is between 1.0 and 2.5 (Johnson Division, 1966). However, in most cases, placement by gravity or free-fall is not recommended. Another drawback to gravity placement is that formation materials may slough off the borehole wall and become incorporated into the filter pack, thus reducing its effectiveness and, potentially, contaminating the filter pack.

With the tremie pipe placement method (Figure 10.87), the filter-pack material is introduced through a rigid or partially flexible tube or pipe via gravity directly to the interval adjacent to the well screen (Figure 10.88), thus eliminating the potential for bridging in the annulus. Initially, the pipe is positioned so that its terminal end is at the bottom of the casing-borehole annulus. The filter-pack material is then poured down the tremie or slurried into the tremie with water and the tremie is raised periodically so that the filter-pack material can fill the annular space around the intake. The preferred minimum diameter of a tube used for a tremie pipe is 1 in.; larger diameter tremie pipes are advisable for filter-pack materials that are coarse-grained or characterized by a uniformity coefficient exceeding 2.5 (California Department of Health Services, 1985).

If the filter pack is being installed in a temporarily cased borehole, the temporary casing is pulled back progressively to expose the screen as the filter-pack material builds up

FIGURE 10.87
The tremie-pipe method of filter-pack placement is recommended in most situations.

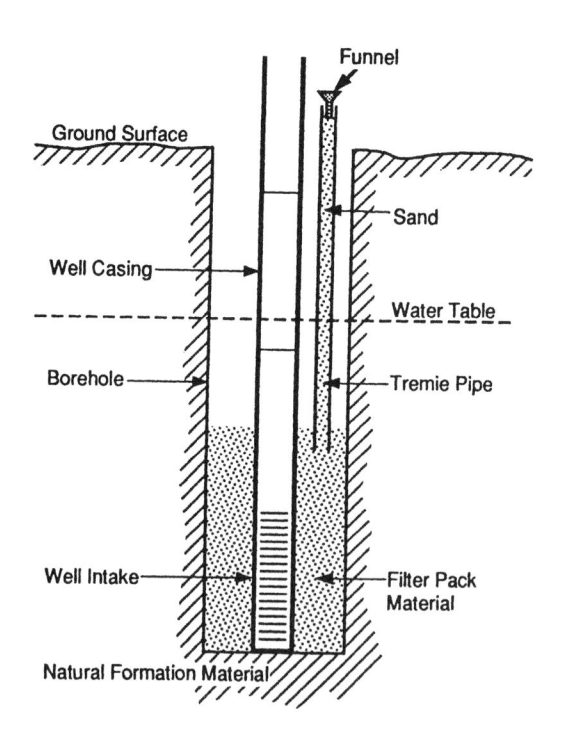

FIGURE 10.88
The tremie-pipe method of filter-pack placement uses a rigid 1- to 1½-in. diameter tube to convey the sand to the zone directly adjacent to the well screen.

around the well screen. The same approach is recommended for hollow-stem auger installations in noncohesive unconsolidated formations. Raising the temporary casing (or the augers) before beginning filter-pack placement is undesirable because formation materials may enter the borehole and cave against the well screen. This may result in a well that later produces sediment-laden samples. Pulling the casing or augers back 1 or 2 ft at a time, while adding filter-pack material (Figure 10.89), is a safer, more conservative approach

FIGURE 10.89
Leaving hollow-stem augers in place during placement of well screen and casing, and pulling back the augers during filter-pack placement is recommended in noncohesive formation materials.

in noncohesive unconsolidated formations that are prone to caving. The technique of filling the space between the temporary casing and the well screen with filter-pack material and then pulling back the temporary casing to expose the entire screen length at one time may result in the well screen, filter pack and temporary casing becoming locked together. In consolidated or well-cemented formations, or in cohesive unconsolidated formations (i.e., predominantly clay or silt), the temporary casing or hollow-stem auger can usually be raised well above the filter pack prior to filter-pack placement without the hole collapsing on the screen. The progress of filter-pack placement should be continually checked with a weighted measuring tape, accurate to the nearest 0.1 ft, to determine when the filter pack has reached the desired height in the borehole (Figure 10.90). The volumes of sand that are needed to fill an annular space of a given size can be determined using Table 10.19 and Table 10.20. Table 10.21 provides a specific set of values (e.g., density, packaging, and volume yield) for commonly used filter pack and other annular fill materials.

For deep wells (i.e., greater than 250 ft) and nonuniform filter-pack materials (uniformity coefficient greater than 2.5), a variation of the standard tremie method, employing a pump to pressure-feed the materials into the annulus, is suggested by California Department of Health Services (1985).

In the reverse-circulation method, a sand and water mixture is fed into the annulus around the well screen, and a return flow of water passes into the well screen. The water is then pumped up through the casing to the surface (Figure 10.91). The filter-pack material is generally introduced into the annulus at a maximum rate of 1.5 ft^3/min to allow for an even distribution of material around the screen (Johnson Division, 1966). Because of the potential for alteration of sample chemical quality posed by the introduction into the borehole of water from a surface source, this method is infrequently used for water-quality monitoring wells.

Backwashing filter-pack material into place is done by allowing sand to free-fall down the annulus while concurrently pumping clean fresh water down the inside of the casing, through the well screen, and back up the annulus (Figure 10.92). This allows for the placement of a more uniform filter pack, because the coarser materials settle out and remain in place while the finer materials will be washed back up the annulus. Backwashing is a

FIGURE 10.90
The progress of filter-pack placement in hollow-stem auger completions should be checked continuously with a weighted tape to ensure that filter-pack material is falling out the bottom of the augers and filling the borehole as expected.

TABLE 10.19

Calculating Minimum Filter Pack and Annular Seal Volumes

Diameter of Casing or Hole (in.)	Cubic Feet per Foot of Depth	Cubic Meters per Meter of Depth ($\times 10^{-3}$)
1	0.0055	0.509
1.5	0.0123	1.142
2	0.0218	2.024
2.5	0.0341	3.167
3	0.0491	4.558
3.5	0.0668	6.209
4	0.0873	8.110
4.5	0.1104	10.26
5	0.1364	12.67
5.5	0.1650	15.33
6	0.1963	18.24
6.5	0.2304	21.42
7	0.2673	24.84
8	0.3491	32.43
9	0.4418	41.04
10	0.5454	50.67
11	0.6600	61.31
12	0.7854	72.96
14	1.0690	99.35
16	1.389	129.09

particularly effective method of filter-pack placement in noncohesive heaving sands and silts, but it is also effective in cohesive, noncaving geologic materials. It is not commonly used in water-quality monitoring wells because of its potential for alteration of ground-water sample chemical quality.

TABLE 10.20

Minimum Annulus Volumes for 2-, 4-, and 6-in. Casings in Boreholes of 4 to 16 in.

O.D. of Casing/Inside I.D. of Borehole (in.)	Cubic Feet per Foot of Depth	U.S. Gallons per Foot of Depth	Liters per Meter of Depth	Cubic Meters per Meter of Depth
2.5 in 4	0.053	0.398	4.59	0.005
2.5 in 6	0.162	1.213	14.00	0.014
2.5 in 8	0.315	2.356	27.19	0.027
2.5 in 10	0.511	3.825	44.13	0.044
2.5 in 12	0.751	5.620	64.84	0.065
4.5 in 6	0.086	0.642	7.41	0.007
4.5 in 8	0.239	1.785	20.60	0.021
4.5 in 10	0.435	3.254	37.55	0.038
4.5 in 12	0.675	5.049	58.26	0.058
4.5 in 14	0.959	7.170	82.74	0.083
4.5 in 16	1.279	9.564	110.36	0.110
6.5 in 8	0.119	0.888	10.25	0.010
6.5 in 10	0.315	2.356	27.19	0.027
6.5 in 12	0.555	4.151	47.90	0.048
6.5 in 14	0.839	6.273	72.38	0.072
6.5 in 16	1.159	8.666	100.00	0.100

Design and Installation of Ground-Water Monitoring Wells 739

TABLE 10.21

Volume Calculation Information for Materials Commonly Used in Monitoring Well Construction

Material Description	Density	Packaging Unit	Yield per Packaging Unit (ft^3)
Colorado silica sand[a] 8–12 mesh	98.5 lb/ft^3	100-lb bag	1.01
Colorado silica sand 10–20 mesh	92.9 lb/ft^3	100-lb bag	1.07
Colorado silica sand 20–40 mesh	89.0 lb/ft^3	100-lb bag	1.12
Colorado silica sand 40–100 mesh	86.6 lb/ft^3	100-lb bag	1.15
Sodium bentonite[b] 8 mesh	70 lb/ft^3	100-lb bag	1.43
Sodium bentonite[b] 8 mesh	70 lb/ft^3	50-lb bag	0.71
Sodium bentonite − 10 + 100 mesh	69 lb/ft^3	100-lb bag	1.45
Sodium bentonite 200 mesh	58–59 lb/ft^3	100-lb bag	171
Volclay[c] grout	9.4 lb/gal	52-lb bag	3.50
Volclay 1/4-in. pellets	80 lb/ft^3	50-lb bucket	0.62
Premix concrete	136 lb/ft^3	90-lb bag	0.66

[a]Colorado Silica Sand, Inc., Colorado Springs, CO.
[b]Wyo-Ben, Inc., Billings, MT.
[c]American Colloid Co., Arlington Heights, IL.

Secondary Filter Packs: The main purpose of a secondary filter pack is to prevent neat cement or bentonite grout or other annular seal materials from infiltrating through the primary filter pack to the screened interval of the well. Such infiltration would partially or totally seal the well from the formation to be monitored, and alter the quality of ground-water samples taken from the well. The necessity for a secondary filter depends on the particle-size distribution of the formation. The secondary filter pack must be

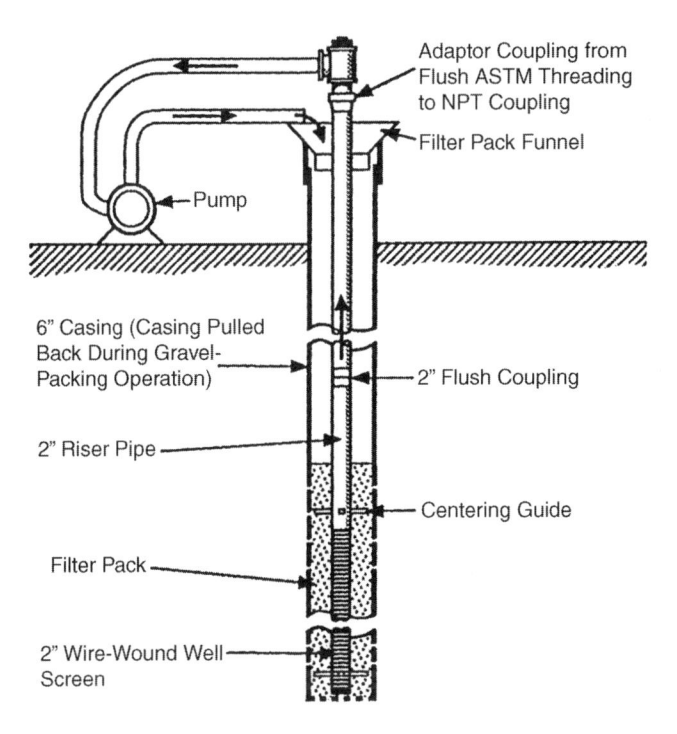

FIGURE 10.91
The reverse-circulation method of placement of filter-pack material.

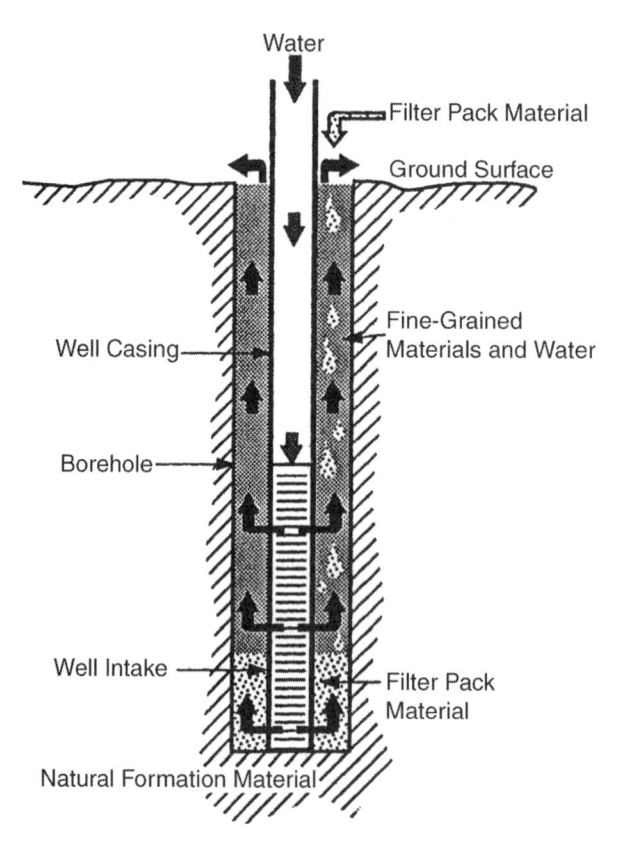

FIGURE 10.92
The backwashing method of placement of filter-pack materials.

composed of well-graded (not uniform), preferably positively skewed sand, with the coarsest fraction equal to the 90% retained size of the filter pack, and with less than 2% by weight passing the No. 200 mesh sieve. Although the particles in the secondary filter need not be as uniform and round as those in the primary filter pack, they should have mineralogical characteristics similar to, or at least compatible with, those of the primary filter pack.

Three important surfaces must be considered in designing the secondary filter pack: the bottom surface (or the interface with the primary filter pack), the top surface (or the interface with the grout or other annular seal materials), and the outer surface (or the interface with the formation).

For a properly designed primary filter pack–secondary filter pack interface, the D_{10} (10% passing) size of the secondary filter must be larger than the voids (interstices) in the primary filter pack. This prevents the fine materials of the secondary filter pack from invading the primary filter pack. Therefore, the D_{10} size of the secondary filter should be between one-third and one-fifth the D_{10} size of the primary filter pack. Referring to Table 10.18, a primary filter pack of 8- to 12-mesh sand would require a secondary filter pack consisting of a layer of 20- to 40-mesh sand with 40- to 140-mesh sand on top.

At the secondary filter pack–annular seal interface, the filter-pack material should be as fine as possible, so that neat cement or bentonite grout will not infiltrate the secondary filter pack significantly, and will not infiltrate the primary filter pack at all. Although the particles need to be fine grained, they should not be so fine grained that the time required for them to settle is significantly influenced by fluid viscosity or minor turbulence caused by the placement of the secondary filter pack. The smallest particle size of the secondary filter should be no larger than U.S. Standard Mesh No. 140 and no smaller than No. 200 to reduce grout infiltration through the filter pack. This recommendation is based not on the size of the neat cement or bentonite particles in the grout slurry but on the viscosity of the grout (which should have a Marsh funnel viscosity of at least 80 sec) and on the height of the grout column above the sand pack, which can exert a significant hydraulic head on the top of the secondary filter-pack material.

The secondary filter pack should extend at least 1 to 2 ft above the top of the primary filter pack. The upper half of the secondary filter pack should be in contact with a lithologic layer of equal or lower permeability and thickness to prevent grout from migrating around the secondary filter pack and into the coarser primary filter pack.

Well Screens

The primary purposes of a well screen are to provide access to a specific portion of subsurface materials (the target-monitoring zone) for sample collection and data gathering, and to provide designed openings for ground water to flow through the well. The well screen also provides structural support for the primary filter pack, and prevents movement of filter-pack material into the well.

The well screen design is based on either the grain-size distribution of the formation (in the case of a naturally developed well), or the grain-size distribution of the primary filter-pack material (in the case of a filter-packed well). The screen openings must be small enough to retain most (if not all) of the formation or filter-pack materials, yet large enough to maintain ground-water flow through the screen. In naturally developed wells, the slot size of the screen should retain at least 70% of formation materials; the finest 30% of formation materials will be brought into the well during well development. For this reason, naturally developed well completions are generally only recommended in predominantly coarse, granular, noncohesive strata that will fall in easily around the screen. In such formations, the objectives of filter packing, to increase hydraulic conductivity in the materials immediately surrounding the well screen and to promote easy flow of water into and through the screen, will be met by native materials. In wells in which a filter-pack material of a selected grain-size distribution is introduced from the surface, the well screen slot size selected should retain at least 90% and preferably as much as 99% of the filter-pack material. This will avoid excessive settlement of other annular fill materials installed above the filter pack, and their possible intrusion into the screened interval.

In selecting a well screen slot size, it is also important to choose a size that is large enough to maintain ground-water flow velocities during water removal from the well, from the well-screen and filter-pack interface back to the natural formation materials, of less than 0.10 ft/sec. This criterion, which is empirically derived, describes the point at which turbulent flow conditions occur (Driscoll, 1986). If well screen entrance velocities exceed 0.10 ft/sec, turbulent flow may occur during purging and sampling, resulting in the mobilization of sediment from the formation (and increased sample turbidity) and reductions in well efficiency. Generally a slot size that is less conservative rather than more conservative (in terms of the slot size selected via the process described above) should be used to satisfy this criterion.

The practice of using a single well-screen slot size and filter-pack grain size combination (e.g., a 0.010-in. slot size with a 20–40 sand) for all wells, regardless of formation grain-size distribution, is not recommended and should be avoided. This practice will result in siltation of the well between sampling rounds and significant turbidity in samples when applied to formations finer than the design dictated by formation grain-size analysis.

It will also result in the loss of filter pack, possible collapse of the well screen, and invasion of overlying annular fill materials (e.g., secondary filter pack, annular seal materials and grout) when applied to formations coarser than the design dictated by formation grain-size analysis.

The well screen should be constructed of a material compatible with the environment in which it is installed. The screen is the part of the monitoring well that is most susceptible to corrosion and chemical degradation, and provides the highest potential for sorption or leaching to occur. Screens have a larger surface area of exposed material than casing, are placed in a position designed to be in contact with potential contaminants (the saturated zone), and are placed in an environment where reactive materials are constantly being renewed by flowing water. To avoid corrosion, chemical degradation, sorption, and leaching problems, the materials from which screens are made should be selected using the same guidelines as for casing materials. In fact, the well screen should normally be made of the same material as the well casing to avoid material incompatibility problems (e.g., galvanic corrosion caused by the presence of two dissimilar metals, such as stainless steel screen and galvanized steel casing). Immediately prior to installation, the well screen should be cleaned (if it is not material that has been certified as clean from the manufacturer and maintained in a clean environment at the site) using a protocol appropriate for the site under investigation (see Chapter 20).

The well screen must have the structural strength to withstand well installation and development stresses without being damaged; it should always be plugged at the bottom (unless a sediment sump is used, in which case the sump should be plugged at the bottom). The minimum nominal diameter of the well screen should be chosen based on factors specific to the particular application (such as the outside diameter of the purging and sampling devices to be used in the well). Well screens as small as 0.50 in. and as large as 16 in. nominal diameter are available for use in monitoring well applications.

The length of the well screen is very important, and it should reflect the thickness of the target-monitoring zone, as determined by the site characterization program. Long well screens are discouraged unless they are specifically necessary to accomplish the objectives of the monitoring program. Well designers should keep in mind the fact that hydraulic head data and ground-water samples represent average values along the length of the screened interval and, in long-screened wells, vertical flow often occurs because of head differences along the length of the screen. It is impossible to isolate sections of the screen through the use of mechanical devices such as packers or sampling techniques such as low-flow purging and sampling, to enable collection of discrete data or samples from specific portions of the screened interval. And, if vertical flow does occur, as is normally the case in ground-water recharge zones and ground-water discharge zones, the well screen may act as a vehicle for redistribution or spreading of contaminants, and possible cross-contamination of screened formation materials. This should be avoided at all costs.

Types of Well Screens Suitable for Monitoring Wells

The hydraulic efficiency of a well screen depends primarily upon the amount of open area available per unit length of screen. While hydraulic efficiency is of secondary concern in monitoring wells, increased open area in monitoring well screens also permits easy flow of water from the formation into the well (and faster recovery of the well), allows for effective development of the well, and allows for more accurate hydraulic conductivity testing results, particularly from slug tests or bail tests. The amount of open area in a well screen is controlled by the type of well screen.

A number of different types of screens are available for use in water-supply wells; several of these are also suitable for use in monitoring wells. Commercially manufactured

Design and Installation of Ground-Water Monitoring Wells 743

well screens similar to those used for water-supply wells are recommended for use in monitoring wells even though hydraulic efficiency is not a primary concern. The primary reason for this recommendation is the strict quality control followed by most commercial screen manufacturers. Hand-slotted, drilled or torch-cut casings should never be used as monitoring well screens because of the poor control over screen slot size, the lack of open area, and the fact that hand-sawed, drilled or torch-cut openings provide fresh surfaces for sorption, leaching, corrosion, or other chemical problems to occur (Figure 10.93). Likewise, perforated casings, produced by either the application of a casing knife or a perforating gun to blank casing installed down-hole, are not recommended. In this type of intake, screen openings cannot be closely spaced, the percentage of open area is low, the opening sizes are highly variable, and opening sizes small enough to control fine-grained materials are impossible to produce. Additionally, perforation tends to hasten corrosion attack on steel casing, because the jagged edges and rough surfaces of the perforations are susceptible to selective corrosion.

Commercially manufactured well screens, including louvered screen, bridge-slot screen, machine-slotted well casing, and continuous-slot wire-wound screen are available for use in water-supply wells, but the latter two types of screens (Figure 10.94a and b) predominate by far in monitoring wells. This is probably because they are the only types of screens widely available in the small (i.e., 0.5- to 4-in. nominal) diameters that are more commonly used in monitoring wells and piezometers.

Slotted well casing (Figure 10.95) is the most widely used type of well screen in monitoring wells. Such a well screen is fabricated from standard well casing, into which circumferential slots of predetermined widths are cut at a regular vertical spacing (typically 0.25 or 0.125 in.) by machining tools. Slotted well casing can be manufactured from PVC and stainless steel and is available in nominal diameters ranging from 0.5 to 16 in. Table 10.22 lists the most common slot widths available for slotted casing. The slot openings are designated by numbers that correspond to the width of the opening in thousandths of an inch. A #10 slot, for example, refers to an opening of 0.010 in., while a #40 slot refers to an opening of 0.040 in.

The slots in slotted well casings are a consistent width for the entire wall thickness of the casing, which can lead to significant clogging of the screen when irregularly shaped

FIGURE 10.93
Hand-slotted casing should never be used as well screen. Poor quality control in cutting the slots and a significant amount of fresh surface area where sorption of some soluble constituents may occur make such materials unsuitable for use in monitoring wells.

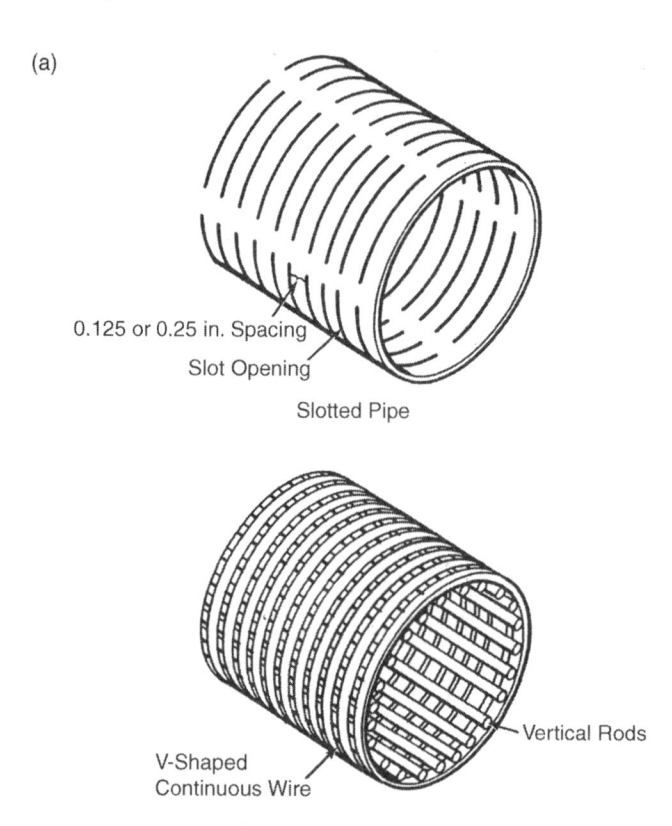

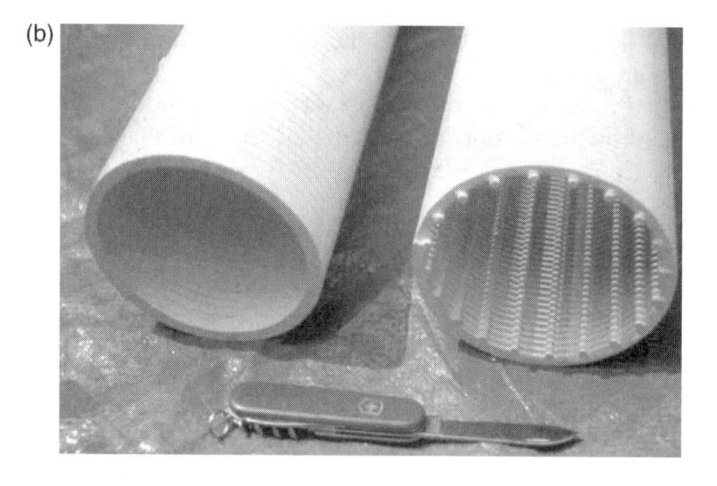

FIGURE 10.94
The two primary types of well screen used in monitoring wells — slotted casing and continuous-slot wire-wound screen.

formation particles are brought through the screen during well development and sampling. This, in turn, can result in declining performance (with respect to well efficiency, yield and recovery) of the well over time. Furthermore, in most slotted casing, the length of the slot on the inside of the casing is much less than the apparent length of the slot as viewed from the outside. The effect of this is that the effective open area of the screen is

Design and Installation of Ground-Water Monitoring Wells

FIGURE 10.95
Slotted well casing is the most widely used type of well screen used in PVC monitoring wells, but it has a very low percentage of open area, and is generally not suitable for wells in which hydraulic conductivity testing must be performed.

significantly reduced. These disadvantages are overcome through the use of continuous slot, wire-wound screen.

Continuous-slot wire-wound screen (Figure 10.96) is manufactured by winding cold-drawn wire, approximately triangular or V-shaped in cross section, spirally around a circular array of longitudinally arranged support rods (Figure 10.97). At each point where the wire crosses a rod, the two pieces are securely joined by welding, creating a one-piece rigid unit (Johnson Division, 1966). Continuous-slot screens can be fabricated of any metal that can be resistance-welded, including bronze, silicon red brass, stainless steel (304 and 316), galvanized and low-carbon steel, and any thermoplastic that can be sonic-welded, including PVC and ABS.

The slot openings of continuous-slot wire-wound screens are produced by spacing the successive turns of the wire as desired. This configuration provides significantly greater open area per unit length and diameter than is available with any other screen type. For

TABLE 10.22

Intake Area (Square Inches per Lineal Foot of Intake) for Continuous-Slot Wire-Wound Screen

Screen Size (in.)	6-Slot (0.006)	8-Slot (0.008)	10-Slot (0.010)	12-Slot (0.012)	15-Slot (0.015)	20-Slot (0.020)	25-Slot (0.025)	30-Slot (0.030)	35-Slot (0.035)	40-Slot (0.040)	50-Slot (0.050)
1¼ PS	3.0	3.4	4.8	6.0	7.0	8.9	10.8	12.5	14.1	15.6	18.4
1½ PS	3.4	4.5	5.5	6.5	8.1	10.2	12.3	14.2	16.2	17.9	20.1
2 PS	4.3	5.5	6.8	8.1	10.0	12.8	15.4	17.9	20.3	22.4	26.3
3 PS	5.4	7.1	8.8	10.4	12.8	16.5	20.0	23.2	26.5	29.3	34.7
4 PS	7.0	9.0	11.3	13.5	16.5	21.2	25.8	30.0	33.9	37.7	44.5
4 Spec	7.4	9.7	11.9	14.2	17.2	22.2	27.1	31.3	35.5	39.7	46.8
4½ PS	7.1	9.4	11.7	13.8	17.0	21.9	26.8	31.0	35.2	39.4	46.5
5 PS	8.1	10.6	13.1	15.5	19.1	24.7	30.0	34.9	39.7	44.2	52.4
6 PS	8.1	10.6	13.2	15.6	19.2	25.0	30.5	35.8	40.7	45.4	54.3
8 PS	13.4	17.6	21.7	25.7	31.5	40.6	49.7	57.4	65.0	72.3	85.6

The maximum transmitting capacity of screens can be derived from these figures. To determine GPM per ft of screen, multiply the intake areas in square inches by 0.31. This is the maximum capacity of the screen under ideal conditions with an entrance velocity of 0.1 ft/sec.

Source: Johnson Filtration Systems Inc. Environmental Equipment Catalog.

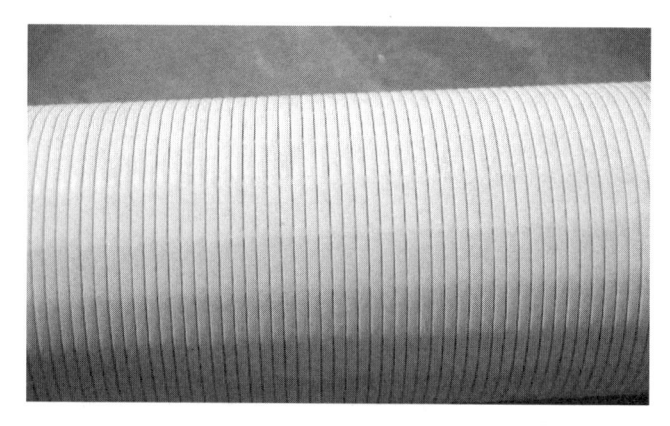

FIGURE 10.96
Continuous-slot wire-wound well screen, available in PVC, stainless steel and galvanized steel, has significantly more open area than slotted casing.

example, for 2-in. nominal diameter PVC well screen, the open area ranges from about 4 in.^{2}/ft of screen for the 0.006-in. slot size to more than 26 in.^{2}/ft of screen for the 0.050-in. slot size (Table 10.22). The percentage of open area in a continuous-slot screen is often more than twice that provided by standard slotted well casing in the smaller slot sizes (0.010 and 0.020 in.), even if the slots in the slotted casing are placed 0.125 in. apart, rather than the standard 0.25 in. (Table 10.23). Continuous-slot screens also provide a wider range of available slot sizes than any other type of screen and have slot sizes that are accurate to within ±0.002 in. The continuous-slot screen is also more effective in preventing formation materials from becoming clogged in the openings during well

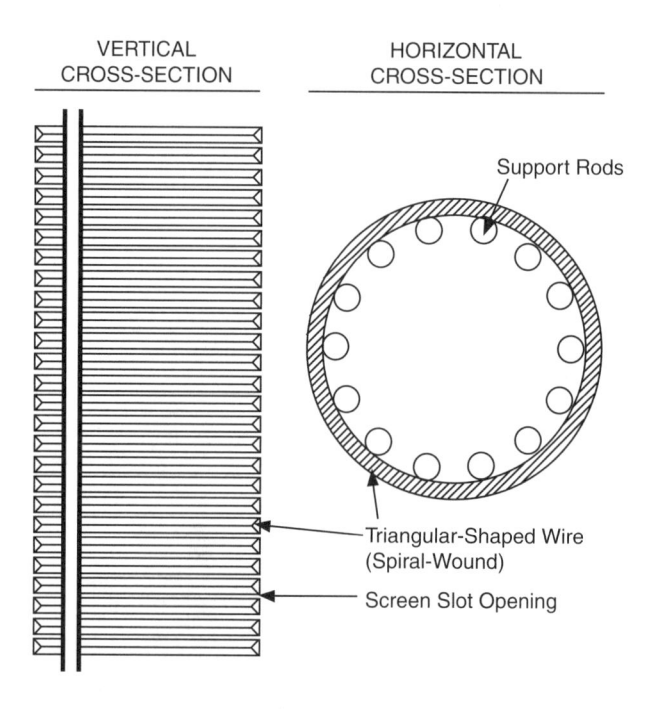

FIGURE 10.97
Continuous-slot wire-wound well screen is constructed of a V-shaped wire-wound around and welded to a series of vertical support rods.

TABLE 10.23

Comparison of Screen Open Area (%) for Continuous-Slot Screen and Slotted Pipe

	Continuous Slot[a]		Slotted Pipe[b]	
Screen Diameter (in.)	10 Slot	20 Slot	10 Slot	20 Slot
2	7.6	14.4	2.9–5.1	5.5–9.4
4	6.8	12.7	2.4–4.3	4.6–7.8
6	5.3	10.0	2.0–3.6	3.9–6.6

[a] Data are for PVC; in stainless steel, the open area will be twice as great.
[b] Because pipe slotting is performed in many different ways, a range from low to high is given.
Source: Information provided by Johnson Filtration Systems, Inc.

development. The triangular-shaped wire is wound so that the slot openings between adjacent wraps of the wire are V-shaped, with sharp outer edges; the slots are narrowest at the outer face of the screen and widen inward. This allows particles slightly smaller than the openings to pass freely into the well without wedging in the opening, making these screens nonclogging.

Louvered (shutter-type) screen (Figure 10.98) has openings in the form of louvers that are manufactured in solid-wall metal tubing by stamping with a punch outward against dies that limit the size of the openings (Helwig et al., 1984). The number and sizes of openings that can be made depends on the series of die sets used by individual manufacturers. Because a complete range of die sets is impractical, the opening sizes of commercially available screens are somewhat limited. Additionally, because of the large blank spaces that must be left between adjacent openings, the percentage of open area on louvered intakes is limited. Also, the shape of the louvered openings is such that the shutter-type intakes cannot be used successfully for naturally developed wells, so their use is confined almost exclusively to filter-packed wells (Johnson Division, 1966). This type of screen is available in steel materials in diameters ranging from 1 to 16 in.; it is rarely used in monitoring wells, though it has proven useful for recovery wells.

Bridge-slot screen (Figure 10.99) is manufactured on a press from flat sheets or plates of metallic material that are rolled into cylinders and seam-welded after being perforated. The slot is usually vertical, and produces two parallel openings longitudinally aligned

FIGURE 10.98
Louvered (shutter-type) well screen is also available for monitoring wells in stainless steel and galvanized steel.

FIGURE 10.99
Bridge-slot well screen is only available in diameters 6 in. and larger, and is used more in water-supply and recovery wells than in monitoring wells.

to the well axis. Normally, 5-ft sections of bridge-slot screen are available, and these can be welded into larger screen sections if desired. The chief advantages of this type of screen are reasonably high open area, minimal frictional head losses, and low cost. One important disadvantage is its low collapse strength caused by the presence of a large number of vertically oriented slots. Bridge-slot screen is usually installed in filter-packed water-supply wells; its use in monitoring wells is limited because it is only produced in diameters 6 in. and larger and because it is available only in metallic materials.

Prepacked and Sleeved Well Screens

An alternative to designing and installing filter-pack material and well screen separately is to use a prepacked or sleeved screen assembly (Figure 10.100). Such designs are particularly useful in formations that are predominantly fine-grained, including silts, silty

FIGURE 10.100
Prepacked well screens are used where borehole conditions make normal installation of screen and filter-pack materials difficult or impossible.

fine sands, silts with some clay, silty clayey fine sands and fine sands. A prepacked well screen consists of an internal well screen, an external well screen (or filter medium support structure, such as stainless steel or nylon mesh material), and the filter medium contained between the screens (or within the mesh material), which together comprise an integrated structure (Figure 10.101a and b). The internal and external screens (or mesh material) are constructed of materials that are compatible with the monitored environment, and are usually of a common slot size specified by the well designer to retain the formation or filter-pack material (depending on whether the well is naturally developed or filter packed. The filter pack is normally an inert granular material (e.g., silica sand or glass beads) that has a grain-size distribution chosen to retain formation materials. For example, Gillespie (1991) describes a prepacked screen fabricated by welding a 2-in. diameter 0.008-in. slot, continuous-slot screen inside a 3-in. diameter, 0.008-in. slot continuous-slot screen, and packing the space between the screens with a 40×60 mesh-size sand. This design is effective in minimizing sediment production and turbidity in wells installed in formation materials having an effective size of 0.004 in. or less (e.g., a silty fine sand). Similar designs, with well screen slot sizes as small as 0.004 in. and filter-pack materials as fine as 60×120 mesh-size sand, can be used in formation materials with an effective size smaller than 0.002 in. (e.g., silts or silts with some clay). This type of well design minimizes the velocity of materials moving into the well during sampling, but may also inhibit well development, particularly if development is attempted using only a bailer (which is not recommended — see Chapter 12).

A sleeved well screen consists of a slotted, pipe-base screen over which a sleeve of stainless steel or nylon mesh material, filled with a selected filter medium, is installed (Figure 10.102). Filter-pack materials as fine as 100×120 mesh-size sand, or silica flour as fine as 200-mesh, can be installed within stainless steel or nylon mesh sleeves. While these sleeves, or the space between the internal and external screens in a prepacked screen, may be as thin as 0.25 to 0.5 in., the basis for mechanical retention dictates that a filter-pack thickness of only 2 or 3 grain diameters is needed to contain and control formation materials. Laboratory tests have demonstrated that a properly sized filter-pack material, with a thickness of less than 0.5 in., successfully retains formation particles regardless of the velocity of water passing through the filter pack (Driscoll, 1986).

Prepacked or sleeved well screens may be used for any formation conditions, but they are most often used where heaving sands make accurate placement of conventional screens and filter packs impossible, or where predominantly fine-grained formation materials are encountered. In the latter case, using prepacked or sleeved screens is the only practical means of ensuring that filter-pack materials of the selected grain-size distribution (generally fine to very-fine sands) can be installed completely surrounding the well screen.

Prepacked and sleeved well screens are installed in the borehole in the same manner as a conventional well screen. If the borehole does not collapse on the screen following withdrawal of the temporary casing, drive casing or hollow-stem auger (which is common in cohesive formation materials like silts and clays), the annular space should be backfilled with a fine to medium sand so that the prepacked or sleeved screen is completely surrounded by sand. If the formation collapses on the prepacked or sleeved screen (which is typical in noncohesive sands and gravels), the temporary casing, drive casing or auger should be retrieved to a point several feet above the screen, and the remainder of the well completed with a bentonite seal and grout as described in the following sections of this chapter. Additional detail on installation of prepacked or sleeved well screens, specifically for direct-push well installations, can be found in Chapter 6 and ASTM Standard D 6725 (ASTM, 2004y).

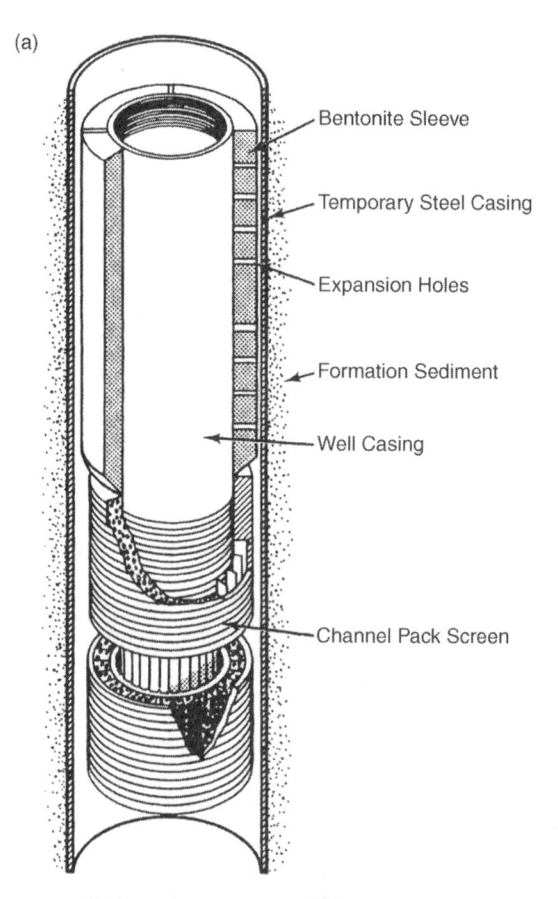

FIGURE 10.101
This prepacked well screen consists of two concentric well screens welded together, with filter-pack sand installed in between. Small-diameter versions are also widely used in direct-push well installations.

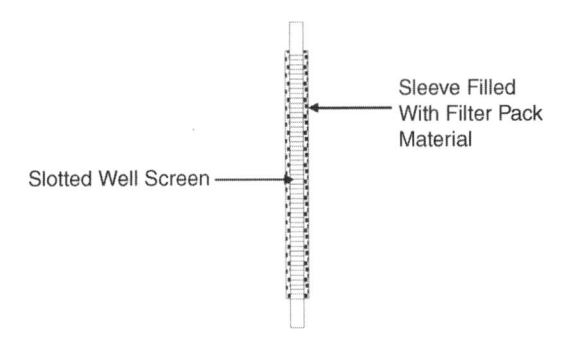

FIGURE 10.102
A sleeved well screen consists of a slotted pipe-base screen over which a sleeve of stainless steel or nylon mesh material, filled with a selected filter-pack material, is installed.

Monitoring Well Screen Open Area

The hydraulic performance of well screens has been stated as being independent of screen design (Clark and Turner, 1983), provided that the open area of the screen exceeds a threshold of about 10%. This effective limit of about 10% is related to the minimum open area of rhomboidal packing of spherical particles, which is 9.2%. However, other studies indicate that screen design is important and that the open-area thresholds can be as low as 8% (Bikis, 1979; Jackson, 1983). Others (Ahmad et al., 1983) suggest that screens with open areas of 8 to 38% do not differ significantly in regard to total drawdown, even in fluvial deposits rich in silt and clay. Well screen open area, as a design criterion for reducing head loss and entrance velocities, is important in water-supply wells, but is not as important in monitoring wells. However, Driscoll (1986) indicates that the open area in a water-supply well screen should be at least equal to the effective porosity of the formation material and the filter pack. This important guideline, which is also valid for monitoring wells, means that screens with open areas of greater than 8% would be needed in most situations and would require the exclusive use of continuous-slot wire-wound well screens. It is also important that the well screen open area approaches or exceeds the formation's effective porosity, so that the screen is not the limiting factor in formation hydraulic testing.

In choosing between types of well screen, another factor to consider is the speed, efficiency, and effectiveness of well development. Field experience with both large- and small-diameter wells indicates that screens with a higher percentage of open area greatly reduce the time and effort required for well development (Schalla, 1986). Similar findings on the importance of the percentage of open area have been reported by others (Clark and Turner, 1983; Ericson et al., 1985; Kill, 1990; Gillespie, 1991). A high percentage of open area is particularly important where smaller slot sizes and fine-grained filter packs must be used to retain the bulk of the formation sediments, as is the case with many monitoring wells.

Well Screen Length

Of all the factors that affect the ability of a monitoring well to provide representative ground-water samples, representative hydraulic head data, representative hydraulic conductivity test data, and other information, the length of the well screen is the most important and perhaps the least well understood. It is important to understand that, whatever the length of the screen, the data collected from the well, whether they are hydraulic head data, ground-water chemistry data, or hydraulic conductivity test data, will generally represent an average of the conditions that exist along the length of the screen. Therefore, before deciding on screen length for a well or a set of wells in a monitoring program, it is critical to define the objectives that the wells must satisfy.

Most monitoring wells are installed for the purpose of detecting, assessing, and monitoring the occurrence of ground-water contamination. While this may seem straightforward, it is important to specify whether the objective of the monitoring program is to simply detect contamination at any level, or to measure absolute concentrations of contaminants that may be present in a specific zone. Many investigators attempt to install wells that achieve both objectives, but this is difficult at best and, more often than not, impossible. In the case of the former objective, a long-screened well may be appropriate as a screening tool; in the case of the latter objective, a short-screened well, targeting a specific zone known or suspected to be contaminated, would be required. The differences in chemical concentrations in samples collected from these two types of wells would likely be profound, and might prompt very different decisions on the part of those reviewing the data. Additionally, in most monitoring situations, wells are installed to double as ground-water sampling points to monitor water quality (for the purpose of detecting changes that may be the result of a release), and piezometers to monitor water levels or hydraulic head in a discrete part of a formation (for the purpose of determining ground-water flow direction). To accomplish these objectives, well screens must generally be short (i.e., between 2 and 5 ft long), and always focused on the target-monitoring zone. Long-screened wells, which average ground-water chemistry and hydraulic head data over the length of the screen, would not provide the discrete data required to accurately depict contaminant distribution or flow directions. To understand the importance of screen length in monitoring wells, it is critical to first consider how contaminants behave in the ground water system.

A number of researchers have studied the distribution of soluble and immiscible contaminants in the subsurface (see, e.g., LeBlanc, 1984; Mackay et al., 1986; LeBlanc et al., 1991; Smith et al., 1991; Cherry, 1992; Pitkin et al., 1994). Many have concluded that, because of the heterogeneities in geologic material that control contaminant transport, contaminant concentrations often vary by one to three orders of magnitude over vertical distances ranging from a few inches to a few feet. Other researchers have demonstrated that dense, non-aqueous phase liquids and some plumes of dissolved-phase contaminants (e.g., some landfill leachates and oil-field brines) often plunge deep beneath the water table because of density-driven flow (see, e.g., Freeze and Cherry, 1979; Pettyjohn, 1979; Anderson et al., 1987; Murphy et al., 1988; Huling and Weaver, 1991). Still other researchers have shown that, at contaminated sites at which natural attenuation processes are important, there are often several geochemically distinct zones present in the ground-water system in which different microbiological processes are taking place (see, e.g., McAllister and Chiang, 1994; Weidemeier et al., 1995; Christensen et al., 2000; Azadpour-Keeley et al., 2001; Weidemeier and Haas, 2002). Additionally, there are situations in which contaminant plumes may be driven deep below the water table because of the vertical differences in hydraulic head that occur in ground-water recharge areas. There are also situations in which a plume may emerge from the ground-water system to discharge in surface-water bodies because of the vertical differences in hydraulic head that occur in ground-water discharge zones. To develop an understanding of the true nature of ground-water chemistry and contaminant distribution in all of these situations, it is extremely critical to define both the three-dimensional distribution of hydraulic head and the spatial differences in water chemistry. The length of well screens in wells installed to define these conditions is the most important element in the success of a contaminant detection and monitoring program.

Researchers have also studied the effects of well screen length on sample chemistry and concluded that the concentrations of chemical constituents in samples collected from wells are composited over the length of the screen, typically representing a weighted average of concentrations across the screen (see, e.g., Robbins, 1989; Martin-Hayden et al., 1991; Robbins and Martin-Hayden, 1991; Chiang et al., 1995; Church and Granato, 1996). Concentrations are normally skewed toward the zones of highest hydraulic conductivity, which will yield more water to the well when it is purged and sampled. Because the highest hydraulic conductivity zones are the most important contaminant transport pathways, it may be rationalized that such samples are acceptable in terms of accurately representing conditions in the formation. However, significant dilution of samples, caused by screens penetrating zones in which contaminants may not be present (e.g., lower hydraulic conductivity zones) and by inappropriate purging and sampling practices (e.g., purging large volumes of water prior to sampling) is bound to occur. In fact, Robbins (1989), Martin-Hayden et al. (1991) and Robbins and Martin-Hayden (1991) found that contaminant concentrations in water table wells can vary by several orders of magnitude, depending on well screen placement and length. Seasonal variations in concentrations of dissolved-phase hydrocarbons can be extreme, because the vertical profiles of contamination below the water table essentially remain constant as the water table rises (when concentrations are typically more dilute) and falls (when concentrations are typically higher). Complicating this situation is the fact that in water table wells, samples represent a smaller interval of the saturated zone when the water table is lower, and a larger interval when the water table is higher. This makes accurate interpretation of sampling results, in terms of defining contaminant plumes, very difficult at best. Martin-Hayden (2000) noted that increasing awareness of the complexities of ground-water flow and contaminant transport occurring at highly variable concentrations and at small scales makes it important that the artificial variability in contaminant distribution imposed by well screen length and the sampling process does not obscure the natural complexities of contaminant distribution.

Wells with long screens are not capable of providing data of sufficient quality to define the three-dimensional distribution of hydraulic head or ground-water chemistry because of the averaging effects that occur in well screens. If the objective of a monitoring program is to define the true nature and distribution of ground-water contamination and hydraulic heads at a site where complex geologic and hydraulic conditions and contaminant distribution patterns occur, research demonstrates that multiple wells with short screens placed at close vertical intervals, or multilevel monitoring systems, are needed (see Chapter 11). Well screens should generally be between 2 and 5 ft, rarely exceeding 10 ft in length. Short screens provide more specific information about contaminant distribution, hydraulic head, and flow in the zone of interest. On the other hand, if the objective of the well is to monitor for the gross presence of contaminants in an aquifer, as it might be in a RCRA detection monitoring program, a longer screen, perhaps designed to monitor the entire saturated thickness of a formation, might be selected. This type of well would provide both an integrated water sample and an integrated hydraulic head measurement, and would thus serve only as a screening tool.

For wells installed specifically to monitor the presence of LNAPLs, well screen length must be determined by the degree of water-table fluctuation. It is important that the screen be long enough to keep the water table within it during extreme highs as well as extreme lows, which means the well designer must consider historical water-level data for the site or surrounding area. If the water table rises above the top of the screen, or falls below the bottom of the screen, it is not possible to use the well for LNAPL detection. Additionally, if a sediment sump is used on a well in which the bottom of the screen is above the water table, the sump may remain filled with water and the well may provide a false indication of the absence of LNAPL. Therefore, the well screen must be long enough to extend above the historical high (at least 3 ft), and below the historical low (at least 2 ft) and, if a sediment sump is used, it should have a drain hole to allow water to escape in the event that the water level drops below the bottom of the screen. It should be noted that wells that are used for LNAPL detection, and in which LNAPLs are found, should not be used to collect ground-water samples for determination of dissolved-phase concentrations (see Chapter 15).

Annular Seal Design and Installation

Any annular space that is produced as the result of the installation of well casing in a borehole provides a potential channel for vertical movement of water and contaminants, unless the annular space is properly sealed. In any casing or borehole system, there are several potential pathways for water and contaminants to follow (Figure 10.103). One pathway is through the sealing material — if the material is not properly formulated and installed, or if it cracks, shrinks, or deteriorates after placement, high-permeability vertical channels could cause significant migration of formation water or contaminants from zones of high hydraulic head to zones of lower hydraulic head. Because well casing is relatively smooth, another potential pathway exists between the casing and sealing material. This pathway could occur for several reasons, including temperature changes between the well casing and the seal material (principally neat cement) during curing or setting, shrinkage of the seal material during curing or setting, or poor bonding between the seal material and the casing (Kurt and Johnson, 1982). A third pathway, resulting from bridging due to improper placement of seal materials, is also possible. All of these pathways can be anticipated and avoided with proper annular

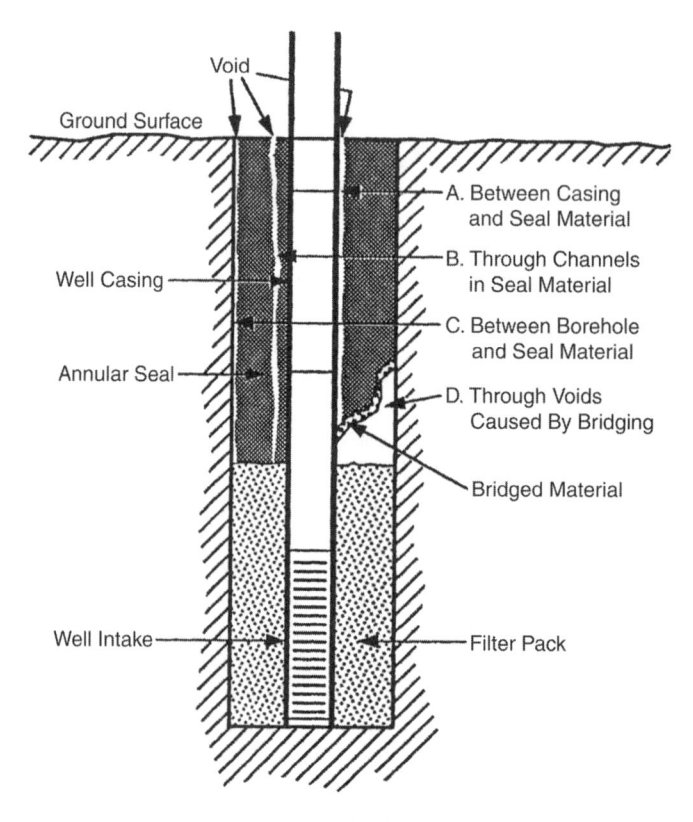

FIGURE 10.103
Potential pathways that water and contaminants may follow in improperly installed annular seal materials.

seal selection, formulation, and placement. Because monitoring wells are often located near or within areas affected by contaminants, an adequate annular seal is especially critical to both the protection of ground-water quality and to the integrity of samples and water-level and other data collected from the well.

The annular seal in a monitoring well (i.e., the seal material placed above the filter pack in the annulus between the borehole and the well casing) serves several important purposes. These include: (1) providing protection against infiltration of surface water and potential contaminants from the ground surface down the casing or borehole annulus; (2) sealing off discrete sampling zones both hydraulically and chemically; and (3) prohibiting vertical movement of water in the casing or borehole annulus from one aquifer to another or from zones of high hydraulic head to zones of lower hydraulic head. Such vertical movement can cause cross-contamination, which can greatly influence the representativeness of ground-water samples, or cause an anomalous hydraulic response of the monitored zone, resulting in incorrect conclusions regarding formation water quality, and distorted maps of potentiometric surfaces. The annular seal also provides an element of structural integrity for the well, and may increase the life of the casing by protecting it against exterior corrosion or chemical degradation. A satisfactory annular seal should have a lower hydraulic conductivity than the surrounding formations, should have strength and deformation characteristics similar to the surrounding formations, and should be compatible with both native materials and well casing materials. The seal must completely fill the annular space and envelop the entire length of the well casing above the secondary filter pack to ensure that no vertical movement of fluids occurs within the borehole.

Annular Seal Materials

The annular seal may be composed of several different types of permanent, stable, low hydraulic conductivity materials, including bentonite (pellets, chips, granules, or powdered material mixed with water to make bentonite grout) neat cement grout, and variations of both. The most effective seals are made by using expanding or nonshrinking materials that will not pull away from either the casing or the borehole during or after curing or setting. Bentonite, nonshrinking (also termed "expanding") neat cement, or neat cement with shrinkage-compensating additives are among the most effective materials for this purpose (Johnson et al., 1980; Barcelona et al., 1983, 1985a; Calhoun, 1988). If the casing or borehole annulus is backfilled with other material (i.e., recompacted drill cuttings, sand, or borrow material), a low hydraulic conductivity annular seal cannot be ensured, and the borehole may then act as a conduit for vertical movement of potentially contaminated ground water. This is especially true regarding the use of drill cuttings. First, it is not possible to recompact drill cuttings to achieve a hydraulic conductivity lower than that of the undisturbed native material. Also, as drill cuttings are shoveled into the annular space, the irregular shapes and sizes of the materials often cause bridging to occur, leaving significant gaps in the seal. Additionally, recompacted drill cuttings, which may contain contaminants encountered during drilling, may interfere with the water chemistry of the water samples collected from the well. Thus, drill cuttings should never be used as annular seal materials in monitoring wells.

The selection of an appropriate annular seal material for any given borehole depends on a number of site-specific factors, including:

- Soil and geological conditions present in the borehole
- Well drilling method used
- Depth of the borehole

- Height of the water column in the borehole
- Size of the annular space between the borehole and the well casing
- Water chemistry of formation water
- Presence of contaminants, including NAPLs, either lighter or denser than water (LNAPLs or DNAPLs)

No single annular seal material is suitable for all types of well installations across a wide range of site-specific conditions.

Bentonite

Bentonite is a hydrous aluminum silicate composed principally of the clay mineral montmorillonite. It is a member of the smectite clay mineral group that contains cations and interlamellar surfaces that can be readily hydrated (Papp, 1996). Bentonite expands significantly when hydrated, due to the incorporation of water molecules into the clay lattice. Expansion of bentonite in fresh water can be on the order of 8 to 15 times the volume of dry bentonite (Colangelo and Upadhyay, 1990). Two types of bentonite are available as annular seal materials — sodium bentonite and calcium bentonite. Sodium bentonite exhibits greater swelling capacity than calcium bentonite — it is characterized as capable of absorbing at least five times its weight in water, and expanding when fully saturated with water to a volume 12 to 15 times its original dry volume (Papp, 1996). In contrast, calcium bentonite may only expand to 8 to 10 times its original volume after it is hydrated. Either form of bentonite, when hydrated, forms a very dense clay mass, which sets up with an in-place hydraulic conductivity typically in the range of 1×10^{-7} to 1×10^{-9} cm/sec. Under favorable water-quality conditions, bentonite expands sufficiently to provide a tight seal between the well casing and the adjacent formation material.

Bentonite used for the purpose of sealing the annulus of monitoring wells is generally sodium bentonite, which is widely used in North America because of its availability and its greater swelling capacity. Calcium bentonite is also available, though less commonly used. Both sodium bentonite and calcium bentonite are available as pellets, granules, chips, or in powdered form. Bentonite pellets (Figure 10.104) are uniformly shaped and sized (generally $1/4$, $3/8$, $1/2$ and $3/4$ in. diameters), and simply consist of highly compressed montmorillonite granules or powder with a moisture content ranging from 9 to 12%. Pellets expand at variable rates, controlled mainly by water quality and temperature, when exposed to fresh water. Some pellets are coated with a wax-like food-grade coating that retards swelling for several minutes and allows the pellets to be dropped through a water column of up to 150 ft without bridging. Uncoated pellets generally cannot be dropped through a water column of more than 30 ft ($1/4$ in. pellets) to 50 ft ($1/2$ in. pellets) without bridging.

Granules are irregularly shaped (angular to subangular) small-diameter ($1/16$ to $1/4$-in., or 8–20 mesh, 12–40 mesh and 30–50 mesh) polymer-free processed particles of montmorillonite (Figure 10.105). Granular bentonite expands at a much faster rate than pellets when exposed to fresh water, because of the high surface area and low moisture content (<10%) of the granules. It generally cannot be placed in dry form through a column of standing water for this reason, as it tends to clump and bridge in the annular space. It can be placed through a tremie pipe under gas pressure, or mixed into a grout with a polymer additive (to retard the swelling time) and water and pumped through a tremie pipe to its intended depth in the borehole.

Chips are larger-sized, irregularly shaped (angular to subangular) chunks of native material that are mechanically separated into different sizes, ranging from $1/4$ to $3/4$ in.

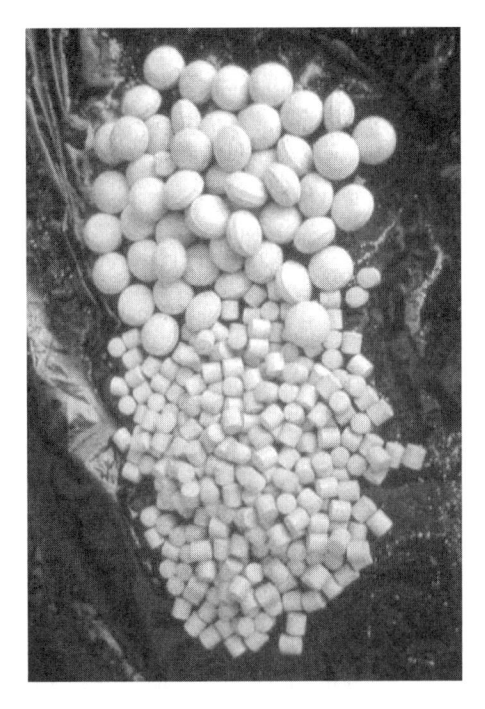

Design and Installation of Ground-Water Monitoring Wells 757

FIGURE 10.104
Bentonite pellets used in constructing the annular seal in a shallow monitoring well. Pellets must be used in the saturated zone, or they will not hydrate properly to form an annular seal.

diameter (Figure 10.106). Because of their high moisture content (ranging between 14 and 18%), chips expand at a much slower rate than either pellets or granules when exposed to fresh water, and can be dropped through a standing water column more easily than uncoated pellets or granules. Papp (1995) suggests that larger bentonite chips may be poured through a standing water column of up to 500 ft without fear of bridging.

FIGURE 10.105
Granular bentonite used in constructing the annular seal in a shallow monitoring well. Granular bentonite is used primarily where a seal has to be constructed in the vadose zone — it cannot be poured through a standing column of water in a borehole. It is installed in 2- to 3-in. thick layers in the unsaturated portion of the borehole (i.e., in a well constructed to monitor LNAPLs), and it must be hydrated by pouring water from the surface.

FIGURE 10.106
Bentonite chips used in constructing the annular seal in a deep monitoring well. Chips must be used in the saturated zone, where they are able to hydrate properly to form an annular seal.

Powdered bentonite is the pulverized material (finer than 200 mesh) produced by the processing plant after mining (Figure 10.107a and b). Powdered bentonite is generally mixed with water (or water and a polymer) and made into a grout to allow its placement as an annular seal. The key to the success of a bentonite grout lies in the percentage of solid material in the grout. The percentage of solids is calculated using the following formula:

$$\frac{\text{Weight of material}}{\text{Weight of material} + \text{Weight of water}} \times 100 = \% \text{ solids.}$$

While bentonite used for drilling fluids is generally a high-viscosity mixture with a low-solids content, just the opposite is desirable for bentonite used for an annular seal material. Bentonite drilling fluids are not made to be mixed at high solids contents, so they are unsuitable by themselves as annular seal materials. In these fluids, the bentonite often separates and settles out of the water over time, causing serious subsidence problems and leaving an inadequate seal (Edil et al., 1992).

Bentonites manufactured specifically for use as grouting materials are designed to be mixed as high-solids, low-viscosity slurries (Figure 10.108). While a solids content of 30% is optimum (yielding a weight of 10.2 lb/gal), most mixtures are in the range of 20 to 25% solids (yielding weights from 9.2 to 9.6 lb/gal). Such mixtures result in a seal with a hydraulic conductivity between 1×10^{-7} to 1×10^{-9} cm/sec — the higher the solids content, the lower the hydraulic conductivity. High-solids bentonite grouts were first used in the mid-1980s as a substitute for neat cement as an annular seal in monitoring wells. These purpose-made grouts are preferred by many contractors because they form a very-low hydraulic conductivity seal, they do not produce heat as they set, they have a more neutral pH than cement, they weigh much less than cement, they are easy to mix and pump, they cause less wear on mixing and pumping equipment, and they remain plastic after they set.

High-solids bentonite slurries can also be created through the use of an inhibitor to retard the swelling of bentonite powder while suspending it as a colloid in solution. The inhibitor is a viscosifying agent that coats the bentonite powder, allows it to partially hydrate and, after placement, disperses to allow the bentonite to absorb additional water and swell to create a positive seal. Another means of creating a high-solids grout is to

FIGURE 10.107
Powdered bentonite must be mixed with water and made into a grout to allow its placement as an annular seal material. (a) Powdered bentonite comes in standard 50-lb bags. (b) The powdered bentonite prior to mixing with water.

premix a very high-quality bentonite drilling fluid from a pure powdered bentonite and then add a granular bentonite, just prior to placing the mixture via a tremie pipe. A 20% solids (9.5 lb/gal) mixture can be achieved in this manner. Still another means of creating a high-solids grout is to blend various grades of sodium and calcium montmorillonite. Blended bentonites, such as specially processed, coarse-ground nondrilling mud grade granular bentonite used for grouting, entrained into a slurry of finely ground, naturally occurring, chemically unaltered bentonite used for drilling fluid, have been engineered to build less viscosity and produce a grout with a higher weight and solids content. Such blended bentonites, with solids contents of about 24 or 25% and weights of 9.5 to 9.6 lb/gal, have demonstrated low hydraulic conductivity, good swelling characteristics and flexibility, and have been judged to provide an excellent annular seal (Edil et al., 1992).

A 30% solids bentonite grout is generally prepared by mixing dry bentonite powder into fresh water in a ratio of approximately 50 lb of bentonite to 14 gal of water, to yield 2.2 ft^3 of grout. A 20% solids grout would be mixed in the proportion of 50 lb of bentonite to 24 gal of water, yielding 3.5 ft^3 of grout. Ideally, the mix water should be cool, with a near-neutral pH (temperature <70°F and pH between 7 and 8). If too little water is

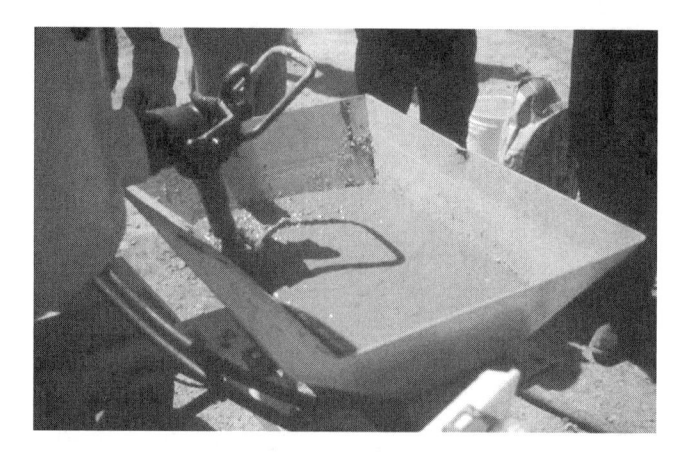

FIGURE 10.108
This high-solids bentonite grout has been mixed with a high-shear paddle mixer and dumped into the hopper in preparation for pumping down the annular space.

used (and the grout is too thick), the grout may be very difficult or impossible to mix and pump, and it may set prematurely. If too much water is used (and the grout is too thin), the grout may never set up properly, and large amounts of grout may be lost to the formation. Once the grout is mixed, it should remain workable for at least 20 to 30 min (up to 2 h if extenders are used or if the grout is continuously sheared during mixing). Some bentonite grouts will require the addition of either a premix additive to condition the makeup water (i.e., to raise or lower pH), or a polymer additive (organic or inorganic) to delay wetting of the bentonite and prevent premature set-up, allowing extended working time. If the mix water used to prepare a bentonite grout is too warm (>90°F), the set-up time of the grout may be accelerated significantly, making it very difficult to pump.

Bentonite grouts should be tested by weighing them, both during mixing and upon completion of the grouting, using a mud balance (Figure 10.109). Obtaining the same weight of the grout after mixing and as the grout exits the top of the borehole during grouting ensures that the grout has not been diluted in the hole, and that there is a complete seal from the bottom of the borehole to the top.

FIGURE 10.109
Weighing bentonite grout with a mud balance to confirm that the solids content is correct. The high-solids bentonite grout being tested is a 30% solids grout, which should weigh 10.1 lb/gal.

Because it is a clay mineral, bentonite possesses a high cation exchange capacity (CEC). This allows the bentonite to trade off cations that make up the chemical structure of the bentonite (principally Na, Al, Fe, and Mn) with cations that exist in the aqueous solution. The degree of cation exchange depends primarily on the chemistries of both the bentonite and the aqueous solution and on the pH and redox potential of the aqueous solution. In addition to having a high CEC, bentonite will generally set up with a moderate to high pH (i.e., grouts from 8.0 to 8.5; pellets and chips from 8.5 to 10). Thus, bentonite may have an impact on the quality of the ground water with which it comes in contact, particularly with regard to pH and metallic ion content. If a bentonite seal is placed too close to the top of the well screen, the potential exists for collecting water samples that have been affected by the presence of the bentonite and that are not truly representative of formation-quality water. Because of this, the recommended practice is to place any bentonite sealing material used in the well no closer than 3 to 5 ft above the top of the well screen, or to use a secondary filter pack above the primary filter pack in a filter packed well.

Other chemical considerations include the potential presence of additives (i.e., organic or inorganic polymers) in the bentonite material. The chemistry of the specific bentonite product that is used must be scrutinized closely to ensure that the bentonite does not contain an additive that could affect the representativeness of ground-water samples.

The use of dry bentonite (pellets, chips, or granules) as a sealing material depends on its efficient hydration and maintaining that hydration following placement. Hydration requires the presence of water (of sufficient quantity and quality) within the geologic materials penetrated by the borehole. Generally, efficient hydration of bentonite pellets or chips placed in dry form will occur only in the saturated zone. Efforts to hydrate chips or pellets placed in the vadose zone (e.g., in wells used to monitor for separate-phase LNAPLs), which usually consist of pouring several 5-gal buckets of water down the annular space, are nearly always unsuccessful. The water usually either passes through the seal materials and into the filter pack below, or is lost to the adjacent formation, and infiltrates downward to the water table. For these materials to hydrate sufficiently to form an effective seal (Figure 10.110), they must be exposed to water for a minimum of 1 to 2 h. In contrast, granular bentonite will hydrate almost immediately upon contact with water, and will remain hydrated in most soils in humid and semi-arid climates, where the soil moisture content is generally more than 18 to 20%. Granular bentonite can form an effective seal in the vadose zone if it is installed in lifts of 2 to 3 in., then hydrated, and the process repeated until the desired seal thickness is reached. Dry-placed bentonite of any kind is generally not appropriate for use in the vadose zone in arid climates, where sufficient soil moisture content generally is not available during all times of the year to allow for complete and continuous hydration of the bentonite.

High-solids bentonite grouts are not suitable for use at the surface in arid climates (Figure 10.111), but they may be suitable for use in the annular space below the surface seal in all types of climates, including arid climates. Work by Papp (1997) demonstrated that a 30% solids bentonite grout could be subjected to repeated hydration–dehydration cycles without failure of the seal. Visual observations made during laboratory and field tests indicated that the high-solids grout rehydrated sufficiently to maintain the integrity of the annular seal even after severe dehydration.

Under certain water quality conditions, notably in water with a high total dissolved solids (TDS) content (>5000 ppm) or a high chloride ion content, the swelling and continued hydration of bentonite is inhibited. The degree of inhibition is dependent upon the type of bentonite used and the level of TDS or chloride in the water. The result is that the bentonite may not swell to its anticipated volume, or may not swell at all, and an ineffective annular seal may be formed. At very high chloride ion contents (>8000 ppm),

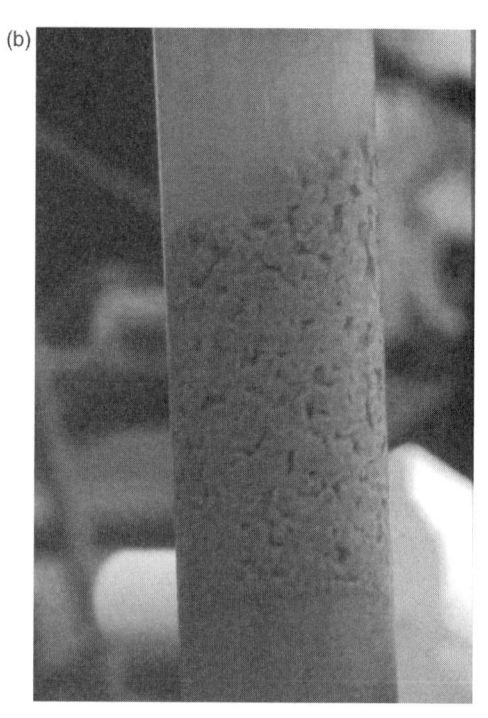

FIGURE 10.110
Bentonite pellets require sufficient time to hydrate before they can be expected to hold up the weight of a column of grout placed above the annular seal. (a) Pellets after 20 min of hydration (only slightly hydrated), (b) pellets after 40 min of hydration (still not hydrated sufficiently). Under these water chemistry and temperature conditions, these pellets should be allowed to hydrate another 20 min or more before grout is placed on top of them. Complete hydration usually requires 48 to 72 h of contact with fresh water.

bentonite will flocculate and not form a seal at all. An alternative seal material (e.g., neat cement) should be used in these situations.

Several studies have been conducted to determine the effects of organic solvents and other chemicals (i.e., xylene, acetone, acetic acid, aniline, ethylene glycol, methanol, and heptane) on hydrated clays, including bentonite (Anderson et al., 1982; Brown et al., 1983). This research has demonstrated that bentonite and other clays may lose their effectiveness as low-permeability barrier materials in the presence of highly concentrated solutions of selected chemical substances. These studies have demonstrated that the hydraulic conductivity of bentonite and other clays subjected to high concentrations of organic acids, basic and neutral polar organic compounds, and neutral nonpolar organic compounds may increase by several orders of magnitude. The increase in hydraulic conductivity is attributed to desiccation and dehydration of the clay material, which can potentially form conduits for vertical movement of fluid within boreholes in which bentonite is used as a sealing material. In contrast, a study done by McCaulou and Huling (1999) demonstrated that saturated dissolved-phase concentrations of trichloroethylene, methylene chloride, and creosote did not affect the hydration, swelling, or the final hydraulic conductivity of bentonite pellet seals. These results indicate that bentonite seals in zones that are heavily contaminated with DNAPL constituents, but outside the DNAPL zone, may resist chemical desiccation and failure.

Villaume (1985) points to possible attack on, and loss of integrity of bentonite seals due to dehydration and shrinkage of the clay by LNAPLs (petroleum hydrocarbons) in the

Design and Installation of Ground-Water Monitoring Wells 763

FIGURE 10.111
Bentonite grouts are not suitable for use at the surface in arid climates.

free-product phase. McCaulou and Huling (1999) demonstrated that unhydrated bentonite pellets submersed in separate-phase DNAPLs (trichloroethylene, methylene chloride, and creosote) retained their original shape, did not swell, and did not perform as a barrier to fluid movement. They also showed that hydrated bentonite that was exposed to separate-phase DNAPLs developed desiccation cracks up to 5 mm wide, leading to significant increases in hydraulic conductivity. Thus, where these NAPLs exist in the subsurface, bentonite will not perform as an effective seal material, and another material (e.g., cement grout) should be chosen.

In summary, factors that should be considered in evaluating the use of bentonite as an annular seal material include:

- Position of the static water level in a given borehole, taking into account seasonal and other natural and man-induced fluctuations
- Ambient water quality (especially with respect to total dissolved solids content and chloride content)
- Types and potential concentrations of contaminants (particularly NAPLs) expected to be encountered in the subsurface

Neat Cement

Neat cement is a mixture of ASTM Standard C-150 (ASTM, 2004k) Portland cement and water (without the addition of aggregate or sand), in the proportion of 5 to 6 gal of clean water per bag (94 lb or 1 ft^3) of cement (Figure 10.112). Cement mixtures with more than about 6 or 7 gal of water per bag of cement are not recommended, as they may develop

FIGURE 10.112
ASTM C-150 Type I/Type II Portland cement comes in standard 94-lb (1 ft^3) bags. Each bag of cement is mixed with 5$\frac{1}{2}$ to 6 gal of cool fresh water (no aggregate) to create neat cement grout.

voids which contain only water and may generate "bleed water" or "free water," which contains very high concentrations of soluble mineral matter from the cement. This may adversely affect water quality in the well for prolonged periods of time. Five types of ASTM Portland cement are produced: Type I, for general use; Type II, for moderate sulfate resistance or moderate heat of hydration; Type III, for high early strength; Type IV, for low heat of hydration; and Type V, for high sulfate resistance (Moehrl, 1964). ASTM Types I, II, and III correspond to API (American Petroleum Institute) Classes A, B, and C, respectively. Type I Portland cement is by far the most widely used cement in ground-water related work. In cold-weather climates, cements with air-entraining agents are generally preferred due to their water tightness and freeze-thaw resistance. Air-entrained cements are designated with an "A" after the ASTM cement Type (i.e., Type IA).

Portland cement mixed with water in the proportions above yields a grout that weighs from 14.5 to 15.2 lb/gal. A typical 15.2-lb/gal Type I neat cement slurry would have a mixed volume of about 1.5 ft^3/sack and a set volume of about 1.2 ft^3/sack, and would remain workable for up to 1 h. The volumetric shrinkage of the set cement would be about 17%, the porosity about 54%, and the hydraulic conductivity about 1×10^{-5} cm/sec (Moehrl, 1964). The time required for such a cement mixture to develop structural strength would range from about 24 to 48 h, depending primarily on water content.

Shrinkage is one of the primary problems with Portland cement used as an annular seal material. When cement shrinkage occurs, the cement may pull away from either the casing or the borehole wall. Cement shrinkage may also lead to cracking along the length of the grout column. Any of these problems destroy the integrity of the seal and results in opening channels or pathways for contaminant migration or migration of surface water down the annular space. Most of the problems associated with shrinkage (and other problems, including long set times) can be corrected through the use of additives or the use of shrinkage-compensated cements. ASTM Standard C-845 (ASTM, 2004z) describes three types of "expanding" cement — Types K, M, and S — which have characteristics similar to Type I or Type II Portland cement, but which contain different combinations of shrinkage-compensating additives. These cements have proven to be very useful in ground-water monitoring well construction. However, many contractors still prefer to use additives in an attempt to produce a cement with desirable borehole-sealing qualities.

A variety of additives may be mixed with the cement slurry to change the properties of the cement. The more common additives, their ranges of proportions (measured as percent by volume), and their effects on the cement mixture include the following:

- Bentonite (3–8%), which improves the workability of the cement slurry, reduces the slurry weight and density, and produces a lower unit cost sealing material. Bentonite also reduces the set strength of a cement seal, and lengthens the set time considerably. Normally, bentonite is prehydrated before it is mixed with neat cement, because if it is added dry, it tends to lump and it will not hydrate properly. However, Mikkelsen (2002) advocates mixing the cement and water first, then adding dry powdered bentonite, through a jet nozzle and using a recirculation mixing procedure, as a means of controlling the water-to-cement ratio and making the strength or modulus of the set grout more predictable. It is commonly believed that adding bentonite to cement will cause the mixture to expand as the bentonite hydrates and swells. However, bentonite has been shown to be chemically incompatible with cement (Colangelo and Lytwynyshyn, 1987; Calhoun, 1988; Listi, 1993), which results in bentonite not swelling in a cement mixture. During setting, cement releases Ca^{2+} ions, which can replace the Na^{2+} ions in the bentonite, converting it to a subbentonite and significantly reducing its ability to swell. Cement also releases OH^- ions during setting, which causes flocculation of the bentonite. Bentonite is thus not appropriate for use as an additive in cement if the goal is to produce an expanding mixture.
- Calcium chloride (1–3%), which provides for an accelerated setting time and a higher early strength, particularly useful features in cold climates. This additive also aids in reducing the amount of slurry that might enter into zones of coarse material, which in turn avoids bridging of the seal.
- Gypsum (3–6%), which produces a quick-setting, very hard cement that expands upon setting. The high cost of gypsum as an additive, however, limits its use to special operations.
- Aluminum power (1%), which also produces a stronger, quick-setting cement that expands upon setting, thus providing a tighter seal (Ahrens, 1970).
- Fly ash or pozzolans (10–20%), which increase sulfate resistance and early compressive strength.
- Hydroxylated carboxylic acid, which retards setting time and improves workability without compromising set strength.
- Diatomaceous earth, which reduces slurry density, increases water demand and thickening time, and reduces set strength.

Water used to mix neat cement should be cool, clean fresh water free of oil, soluble chemicals, silt, organic material, alkalies and other contaminants, and the total dissolved mineral content should be less than 500 ppm (Smith, 1987). A high sulfate content is particularly undesirable (Campbell and Lehr, 1975). Water with a high chloride content can either accelerate or retard the setting time of cement slurries, depending upon the type of cement used. Inorganic materials (sulfates, hydroxides, carbonates, and bicarbonates) can also accelerate setting time. Using high-temperature mix water (>90°F), or exposing cement to high temperatures (e.g., pumping cement through grout hose lying in the hot sun) can cause cement to flash set, which can cause significant damage to mixing and pumping equipment.

When cement slurry is prepared, the hydration process begins immediately as the mix water chemically reacts with the cement. The cement develops gel strength as a direct result of the hydration process. As gel-strength development progresses, the cement transitions from a fluid to a solid material that exhibits compressive strength (Smith, 1987). Depending upon the chemical formulation of the cement, the fineness of the grind, and the conditions under which the slurry is placed, there is an optimum range for the amount of mix water that will completely react or combine with the cement. For normal ASTM Type I Portland cement mixtures, 5 to 6 gal of water is recommended. Excess water that does not combine chemically with the cement, referred to as "bleed water," is very highly alkaline. This bleed water can separate from the slurry, percolate through or along the cement seal surrounding the casing, and infiltrate through or bypass the bentonite chip or pellet seal and secondary filter-pack sand, to contaminate water collected as a sample from the well (Evans and Ellingson, 1988). Bleed water can be minimized or eliminated by strictly controlling the amount of mix water used during cement preparation, and measuring the cement slurry density with a mud balance.

If too much water is used (i.e., more than 6 gal), excessive shrinkage will occur upon setting, which means that the annulus will not be completely filled after the grouting operation. The voids in the annulus may not be seen from the surface, but they will be present along the length of the casing (Kurt, 1983). These voids lead to the creation of channels and pockets of free water behind the casing. This, in turn, creates zones of lower density and increased permeability. The less dense or the greater the permeability of the cement, the less resistant the cement is to chemical attack. Proper water-to-cement ratios are thus very important to the success of a cement seal. Control of these ratios can be achieved by gauging the water tanks and by continuously weighing the slurry, which will help to assure the desired properties of the set cement. Although the cement volume is increased by increasing the water-to-cement ratio, so is the permeability of the cement. Because protection of ground-water quality is one of the most important reasons for cementing wells, increasing the water-to-cement ratio is not a recommended design alternative (Kurt and Johnson, 1982). Proper mix-water ratios should be adhered to as part of a documentable quality control program. Preferably, a mud balance, Marsh funnel, or some type of viscosity meter should be used to determine if proper ratios have been achieved. A slurry with too much water may create a permanent water-quality problem which may lead to the need to decommission the monitoring well (Williams and Evans, 1987).

The mixing of neat cement grout may be accomplished manually (Figure 10.113) or with a mechanical mixer (Figure 10.114). Mixing must be continuous so that the slurry can be placed without interruption. Prolonged mixing should be avoided because it disrupts the hydration process and reduces the ultimate strength and quality of the cement. Prolonged mixing will also cause unnecessary wear and tear on the equipment used to mix and pump the grout, because cement is a very abrasive material. Cement grout should be mixed to a fairly stiff consistency and immediately pumped via a tremie pipe to its intended position in the annulus. The type of mixer recommended is a paddle mixer. The types of pumps recommended for use with neat cement grout include reciprocating (piston) pumps, diaphragm pumps, centrifugal pumps or moyno pumps, all commonly used by well drilling contractors.

Neat cement, because of its chemical nature [lime (calcium carbonate), aluminum, silica, magnesium, ferric oxide, and sulfur trioxide], is a highly alkaline substance (pH from 10 to 13), and thus introduces the potential for significantly raising the pH of water with which it comes in contact. Raising the pH, in turn, can affect other chemical constituents in the water (e.g., causing dissolved metals to precipitate from solution). In addition, because the mixture is a slurry and because it is generally placed in a column which imparts a high hydraulic pressure, it may tend to infiltrate into the coarse materials that comprise

Design and Installation of Ground-Water Monitoring Wells 767

FIGURE 10.113
Neat cement can be mixed manually in large troughs, and recirculated using positive-displacement pumps to smooth out the mixture prior to pumping down the borehole.

either the secondary or primary filter pack. This is particularly true of thinner slurries (i.e., those mixed with more than 6 gal of water per sack of cement). The cement infiltration problem can be aggravated in this situation if well development is attempted before the cement has completely set.

All of these issues can result in severe and persistent effects on both the performance of the monitoring well (in terms of yield) and the quality of samples taken from the monitoring well. Placement of a thin grout directly on top of the primary filter pack, with subsequent infiltration, will result in the plugging of the filter pack (and potentially the well screen) with cementitious material upon setting. Additionally, the presence of high pH cement within or adjacent to the filter pack will cause anomalous pH readings in subsequent water samples collected from the well. Dunbar et al. (1985) reported an incident attributed to this phenomenon, in which several wells completed in this manner in low-permeability geologic materials consistently produced samples with a pH greater than 9 for 2.5 yr, despite repeated attempts at well development. Neat cement should never

FIGURE 10.114
Neat cement is mixed most efficiently with a high-shear paddle mixer.

be placed directly on top of the primary filter pack in a monitoring well. It has been suggested in this chapter and by others that a very fine-grained secondary filter pack, from 2 to 3 ft thick, be placed atop the primary filter-pack material before placement of the neat cement grout, to minimize or eliminate the grout infiltration potential (Ramsey and Maddox, 1982; Barcelona et al., 1985a). A 2- to 5-ft thick bentonite pellet or chip seal above the primary filter pack would accomplish the same purpose but would require additional time during well construction to allow the bentonite to hydrate sufficiently to hold back the column of cement grout. Either of these procedures should minimize the impairment of well performance and the potential chemical interference effects caused by the proximity of neat cement to the well screen.

Another potential problem related to the use of neat cement as an annular seal material concerns the heat generated by the cement as it sets. When water is mixed with any type of Portland cement, a series of spontaneous chemical reactions, called hydration reactions, occur. If allowed to continue to completion, these reactions transform the cement slurry into a rigid, solid material. As the hydration reactions progress and the cement cures, heat is given off as a byproduct. This heat is known as the heat of hydration (Troxell et al., 1968). The rate of generation of the heat of hydration is a function of curing temperature, time, cement chemical composition, and the presence of chemical additives (Lerch and Ford, 1948). Generally, the heat of hydration is of little concern. However, if large volumes of cement are used or if the heat is not rapidly dissipated (as it would not be in a borehole because of the insulating properties of geologic materials), relatively large temperature increases may occur (Verbeck and Foster, 1950; Molz and Kurt, 1979; Jackson, 1983). These temperature increases, coupled with the weight of the grout column (at 15.2 lb/gal) may compromise the structural integrity of some types of well casing, notably PVC. PVC characteristically loses strength and stiffness as the temperature of the casing increases. Because the collapse pressure resistance of a casing is proportional to material stiffness, a sufficient rise in casing temperature, coupled with the increased hydrostatic pressure due to the presence of the cement, could cause unanticipated casing failure (Johnson et al., 1980). However, because the boreholes for most monitoring wells are generally small diameter (8 in. or less), a heat of hydration high enough to damage PVC casing is generally not created. Moreover, because many monitoring wells are shallow, excessive hydrostatic pressures caused by the presence of grout outside the casing, are generally not created. The use of setting time accelerators such as calcium chloride, gypsum, or aluminum powder can, however, increase the heat of hydration, causing PVC casing to become very hot while the grout is curing, and resulting in increased potential for casing failure.

Several methods can minimize heat of hydration. Adding agents such as bentonite or diatomaceous earth to the grout mix to retard the setting time will reduce peak temperatures. Other methods for retarding the setting time include adding inert materials such as silica sand to the grout, circulating cool water inside the casing during grout curing, and increasing the water-to-cement ratio of the grout mix (Kurt, 1983). The latter option, however, results in increased shrinkage and decreased strength upon setting, and is not recommended.

The high weight of cement grout may contribute not only to the potential for failure of PVC casing, but also to the increased loss of grout material to the formations penetrated by the borehole, particularly where sand and gravel and fractured rock are present. The more pressure exerted on the formation, the higher the rate and amount of loss. To determine the grout pressure on the formation, use the following formula:

Grout weight [lb/gal] × Height of grout column [ft] × 0.052 = Grout pressure [psi]

Using the formula, a 100-ft column of cement grout weighing 15.2 lb/gal would have a grout pressure of 79 psi. By comparison, a 100-ft column of 30% solids bentonite grout, weighing 10.2 lb/gal, would have a grout pressure of only 53 psi. Because of the relatively lower viscosity and higher grout pressure of the cement grout, it is much more likely to be lost to coarse-grained formations or fractured rock.

Methods of Installation of Annular Seal Materials

Bentonite

Bentonite may be placed in the borehole as a dry solid material or as a grout. Bentonite pellets, chips, or granules may be placed dry; either granular or powdered bentonite may be mixed with water at the surface to form a grout and then pumped into the annular space. Before bentonite materials are placed in the borehole, it is necessary to calculate the amount of material needed to fill the annular space. Table 10.21 will help the user calculate volumes for all types of bentonite materials. If it turns out that the volume of material actually required to fill the space is less than the calculated amount, bridging of seal material or sloughing of formation materials may have occurred in the hole. If the volume of material actually required to fill the space is more than the calculated amount, there may be washouts in the borehole or losses to the formation. In any case, these conditions should be noted and appropriate action taken to correct the problem.

In relatively shallow monitoring wells, with water columns less than about 30 ft and with sufficient annular space (i.e., more than 2 in.) on all sides of the casing, uncoated bentonite pellets or bentonite chips may be placed by the gravity (free-fall) method. The pellets or chips are simply poured down the borehole from the ground surface (Figure 10.115). Uncoated pellets, particularly the smallest diameter ($1/4$ in.) pellets, become sticky almost immediately upon contact with water, and they tend to bridge most easily (Kaempfer, 2003). This is typically caused by the bentonite pellets sticking to the borehole wall, the casing, or each other before reaching their intended depth. Coated pellets may be poured through 150 ft of water, and $1/4$ and $3/8$ in. diameter chips may be placed by gravity through water columns of more than 300 ft. Larger diameter chips ($1/2$ to $3/4$ in.) can be placed by pouring through more than 500 ft of water without

FIGURE 10.115
Placement of bentonite pellets or chips by pouring down the annulus is the most practical way of installing these materials. However, this can only be done effectively in cased holes installed in noncohesive materials or in uncased boreholes installed in cohesive materials.

fear of bridging (Papp, 1995). Neither bentonite pellets nor chips are effectively installed through a tremie pipe (Figure 10.116), as the restricted diameter of the tremie promotes bridging.

In deep wells, especially where static water levels are shallow, placing dry bentonite via the gravity method introduces both a high potential for bridging and the possibility that sloughing material from the borehole wall will be included in the seal. If bridging occurs, significant gaps may occur in the seal. The bentonite should therefore be tamped with a tamping rod after placement to ensure that no bridging has occurred. If sloughing material is included in the seal, "windows" of high-permeability material may develop. This situation results in an ineffective annular seal, and there is no remedial measure that can correct this problem.

The rate of placement of bentonite pellets or chips and the presence of fines in the poured material are important factors in the ability of the bentonite to reach the intended depth without bridging. Generally, pellets or chips should be poured at a rate of no more than 2 or 3 min per 50-lb bag — pouring at a faster rate promotes clumping and bridging, particularly with small-diameter uncoated pellets. The fine dust or broken particles that are unavoidably mixed in with chips and pellets increases the possibility of bridging. Dust or broken particles may be present as a result of the packaging process or due to breakage or abrasion of the surfaces of the materials during transport. The dust increases the viscosity of the water in the borehole, slowing the rate of fall of the material; the broken particles tend to swell readily, sticking to the borehole wall or the casing. Sieving the bentonite pellets or chips with a coarse screen before placing them in the borehole, to remove the dust and broken particles, is always a good practice (Kaempfer, 2003).

FIGURE 10.116
Using a tremie pipe to place bentonite pellets or chips is impractical, because as soon as the materials hit the static water level in the pipe, they tend to bridge, unless the tremie can be kept clear of water (i.e., by placing it under air pressure).

Design and Installation of Ground-Water Monitoring Wells 771

It generally requires at least 1 to 2 h for bentonite pellets or chips to hydrate sufficiently to hold back a column of grout (Figure 10.117), and 48 to 72 h to reach a hydraulic conductivity of less than 1×10^{-7} cm/sec. Longer hydration times result in lower hydraulic conductivity (down to a low of about 1×10^{-9} cm/sec).

Usually, granular bentonite is only placed effectively as a dry material in the unsaturated portion of a borehole, and hydrated from the surface in lifts of 2 to 3 in. This process must be repeated until the desired seal thickness is achieved. However, there are situations in which granular bentonite is desired as a seal material in the saturated zone. In these cases, the bentonite can be installed through a bentonite injection system that uses compressed air to displace water from an injection line (similar to a tremie pipe) and to carry the bentonite to the intended placement zone beneath the water table (Boyle, 1992; Thompson et al., 1994). This system is comprised of 3 tanks that allow granular bentonite, and filter-pack sand, to be fed from separate pressure cylinders into the injection line (Figure 10.118). The third tank holds a supply of high-pressure air used to clear any clogs that may develop in the injection line. Continuous airflow through the injection line keeps the line dry at all times through the injection process. The injection line is typically placed 1 to 3 ft above the bottom of the intended placement zone, and is raised slowly as the granular bentonite is injected into place. Slight turbulence at the point of injection distributes the bentonite uniformly around the annular space, as well as into bedrock fractures (if present). The injection line tubing can be used to sound the top of the bentonite material as it is injected into the borehole. The injection of bentonite with this system allows for very accurate placement of bentonite seals and filter packs, both in wells and in multilevel monitoring device installations. Placement of granular bentonite with this system is unique in that expansion does not take place until the bentonite is

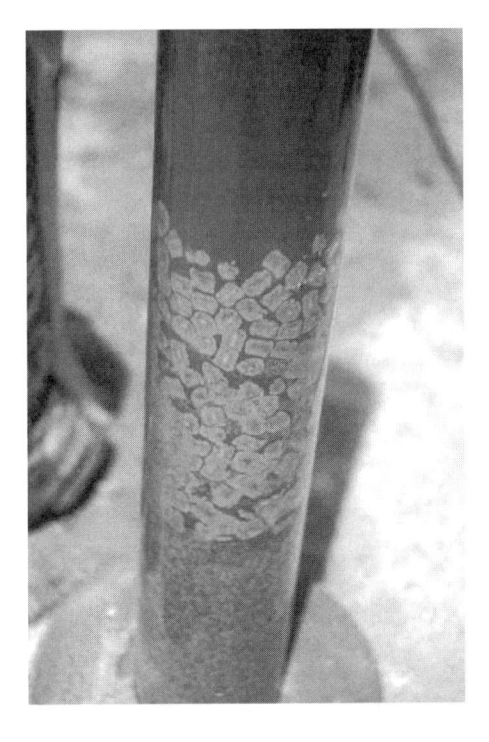

FIGURE 10.117
These bentonite pellets were not allowed to hydrate sufficiently (only 25 min) before placement of neat cement grout on top of them. Note that the grout has channeled through the bentonite pellets and into the filter-pack sand below.

FIGURE 10.118
A dry injection system for the placement of annular seal (and filter-pack) materials in monitoring wells.

injected at the specific location desired. In this way, the full swelling capacity of the granular bentonite can be realized at the zone of placement, rather than during the mixing or pumping process. This ensures a very dense, low hydraulic conductivity annular seal (Thompson et al., 1994).

Granular bentonite may also be mixed with a liquid polymer (organic or inorganic) and water and conveyed as a grout through a tremie pipe from the surface directly to its intended depth in the annulus (Figure 10.119). A pipe with an inside diameter of at

FIGURE 10.119
Granular bentonite may be mixed with water and an organic polymer (to retard hydration) into a grout that can be tremied into the borehole.

least 1.5 in. should be used with granular bentonite grout to avoid bridging of the bentonite and clogging of the tremie. Because granular bentonite hydrates very quickly, its hydration must be slowed by a polymeric wetting agent. However, the use of organic polymers may impact the chemical integrity of ground-water samples later collected from the well, and is not recommended.

Successful use of bentonite grout as an effective well seal can be achieved only by calculating the correct volume required, and then using proper mixing, pumping, and placement methods. A simple formula can be used to calculate the volume of grout required:

$$\frac{(\text{Borehole diameter [in.]})^2 - (\text{Casing diameter [in.]})^2 \times 0.0408 \times \text{Length of seal [ft]}}{7.48}$$

$$= \text{Grout volume required [ft}^3\text{]}$$

As an example, in a grouting job in which a hollow-stem auger was used to install a 2-in. nominal diameter well to a depth of 65 ft, the borehole diameter was 8 in. The well screen was 10 ft long, the filter pack extended to 3 ft above the top of the screen, and a 2-ft bentonite pellet seal was installed above the filter pack, leaving 50 ft of borehole to be grouted. The volume of grout required to fill the borehole, therefore, was:

$$\frac{(8)^2 - (2)^2 \times 0.048 \times 50}{7.48} = 16.4 \text{ ft}^3.$$

Using a 30% solids grout, 7½ bags of grout were required (16.4 ft³/2.2 ft³ per bag) to completely grout the borehole.

With regard to mixing, bentonite grouts are generally mixed in a batch mixer, and the resulting grout is pumped using a positive-displacement pump. The grout should never be poured directly down the annulus using the gravity or free-fall method because the grout would either be diluted by water in the borehole, or the bentonite would segregate out of the grout mixture. In either case, the grout would not form an effective annular seal. Furthermore, grout that is poured down the annulus cannot effectively displace water or loose formation materials up the borehole, and thus it is not possible to ensure that the seal occupies the entire borehole.

Bentonite grouts should be mixed using equipment that will thoroughly shear the bentonite, dispersing the bentonite platelets in the water and allowing the bentonite to hydrate most efficiently. High-shear paddle mixers, which mechanically agitate the grout slurry using paddles or blades that rotate in a barrel-like container, are the most effective mixers. Complete portable grouting units that include a mixing tank, a hopper, a grout pump, and a portable power unit, are available from many sources (Figure 10.120). While grout mixing with jetting equipment and recirculation with diaphragm or moyno-type pumps can be done, it is not as effective, as it does not shear the bentonite and it takes more time. After the grout has been mixed, it is ready to be pumped down the borehole.

Grout pump selection is an important factor when using a tremie pipe to pump a bentonite grout down hole. Grout must be pumped down the tremie pipe and up the annulus to the surface, displacing drill cuttings, drilling fluid, water, and loose formation materials from the annulus. The pump must therefore be capable of developing sufficient pressure to accomplish this. Positive displacement pumps are generally the best suited to pumping viscous fluids like grout. They move a specific volume for each cycle of the pump, and provide a constant flow rate at a constant speed, regardless of the discharge pressure present at the outlet. Shearing action in these pumps is minimized, which keeps the

FIGURE 10.120
A complete portable grouting unit includes a high-shear paddle mixer, a hopper, a grout pump, high-pressure hose and a portable power unit.

rate of additional grout hydration low and allows a longer working time before the grout sets up (Piasecki, 2002). Several types of positive displacement pumps are widely used in grouting, including moyno-type (progressing cavity) pumps, diaphragm pumps, piston pumps, and gear pumps. Moyno-type pumps are well suited to pumping one-step grouts (bentonite and water only), but not two-step grouts (bentonite–polymer mixtures). The grinding action in these pumps, between the rotor and the stator, wears off the polymer from the bentonite and allows the bentonite to swell within and plug the pump. Gear pumps are not as well suited to pumping one-step grouts, because as grout viscosity increases, excessive resistance to pumping is created, slowing the pumping rate significantly. These pumps are better suited to pumping two-step grouts because they do not add any shearing that would remove the polymer from the bentonite particles.

After the grout has been properly mixed, it should be pumped under positive pressure through a side-discharge tremie pipe (Figure 10.121) down the annular space so the grout is not directed under pressure directly onto the top of the filter pack (Figure 10.122). All hoses, tubes, pipes, water swivels, and other passageways through which the grout must pass should have a minimum inside diameter of 1 in.; a 1.25- or 1.5-in. tremie pipe is preferred. If a bottom-discharge tremie pipe is used, it may cause severe erosion of the filter pack or result in grout being injected directly into the filter pack. The side-discharge tremie pipe should be run to the bottom of the annular space (i.e., just above the filter pack or the level to which noncohesive material has collapsed in the borehole in a naturally-developed well) and should be left there during placement, so the grout fills the annulus from the bottom up. The tremie can be removed after the grout has been placed to its intended level in the annulus.

FIGURE 10.121
After grout has been mixed, it should be pumped under positive pressure through a tremie pipe to the zone just above the bentonite seal.

The setting of bentonite grout generally requires between 8 and 48 h, during which time the grout mass builds gel strength. Bentonite grouts do not become rigid and develop structural strength as does neat cement. A fully set bentonite grout seal will generally have the consistency of a very thick peanut butter or putty (Figure 10.123). Well development should not be attempted until the bentonite has fully set, to avoid the possibility of pulling some of the grout through the top of the primary filter pack and into the screen.

Bentonite grout is already hydrated before placement, but its ability to maintain an effective seal depends upon constant hydration after placement. Unless the geologic materials in which the grout is emplaced are either saturated or have sufficient moisture content to maintain hydration of the bentonite, the seal may desiccate and crack, affecting its integrity. Most soils in humid and semiarid climates have more than sufficient soil moisture to maintain bentonite hydration, but some soils in arid climates may be too dry and may actually extract moisture from the bentonite seal.

The moderate to high pH and high cation exchange capacity of various bentonite materials poses a potential for chemical interference with ground-water samples. Therefore, it is recommended that a bentonite seal be placed well above the top of the well screen (i.e., at least 3 to 5 ft), and that a secondary filter pack be used on top of the primary filter pack. This should result in a minimal impact by the bentonite on ground-water sample integrity.

Neat Cement

As with a bentonite grout, a neat cement grout must be properly mixed, pumped, and placed in the borehole to ensure an effective annular seal. Neat cement should not be

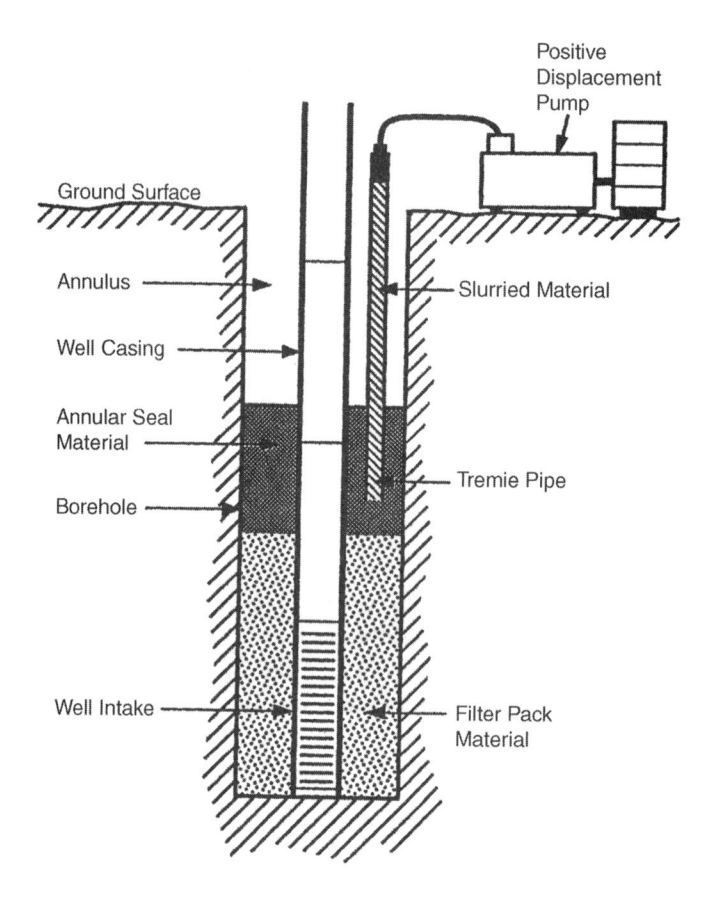

FIGURE 10.122
Tremie-pipe placement of grout (either bentonite or neat cement). The tremie pipe used for grouting should discharge to the side so grouting under pressure does not force grout into the bentonite seal or the filter pack.

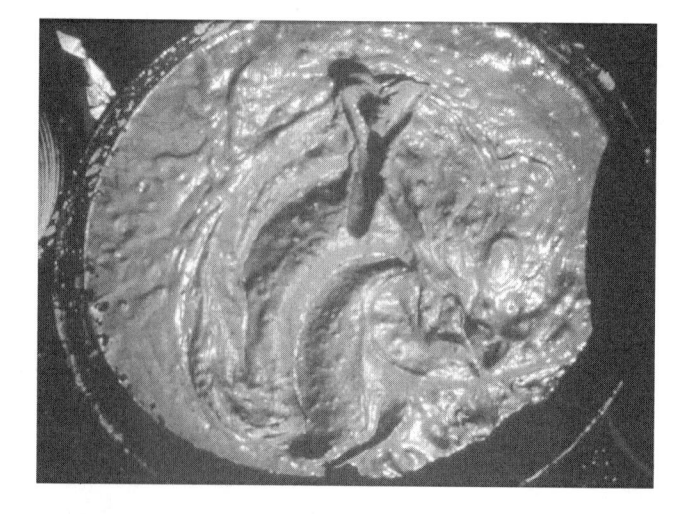

FIGURE 10.123
The consistency of fully hydrated bentonite grout, after setting for 72 h, should be similar to thick peanut butter or putty.

placed in the annulus by pouring (gravity) unless there is adequate clearance (i.e., at least 3 in.) between the casing and the borehole, the annulus is dry, and the bottom of the annular space to be filled is clearly visible from the surface and not more than 30 ft deep (U.S. EPA, 1975). Pouring a neat cement grout from the surface through standing water in the annulus introduces a high potential for the mixture to be diluted, to segregate, or to bridge after it reaches the level of standing water and before it reaches its intended depth of placement. In addition, in its free-fall down the annulus, the grout may pick up sloughing material from the walls of the borehole, causing a breach in the seal.

For neat cement grout, the mixing, pumping and placement methods of choice are the same as for bentonite grout. Assuming the annular space is large enough, a tremie pipe with a minimum inside diameter of 1.5 in. should be inserted in the annulus to within a few inches of the bottom of the space to be sealed. The tremie pipe discharge port should be located on the side of the terminal end of the pipe. Grout should be pumped through the tremie pipe, discharging at the bottom of the annular space and flowing upward around the casing until the annular space is completely filled. This procedure allows the grout to displace drill cuttings, drilling fluid, water and loose formation materials ahead of the grout, thus minimizing both contamination and dilution of the grout, which can reduce its bonding strength. This procedure also minimizes potential bridging of the grout with formation material. The tremie pipe may be moved upward as the grout is placed, or be left in place at the bottom of the annulus until grouting is completed. However, the end of the tremie pipe should always remain in the grout column to avoid formation of air pockets. After placement, the tremie pipe should be removed immediately to avoid the possibility of the grout setting around the pipe, causing difficulty in removing the pipe or creating a channel in the grout as the pipe is removed. To avoid the formation of cold joints, the grout should be placed in one continuous mass before initial setting of the cement or before the mixture loses its fluidity. The curing time required for Type I Portland cement to reach its maximum strength is a minimum of 48 to 72 h, though setting time accelerators can reduce this time significantly.

Surface Completion for Monitoring Wells: Protective Measures

Two types of surface completions are common for ground-water monitoring wells: the above-grade completion, which is generally preferred wherever practical, and the flush-to-grade completion, which may be required or desired under some site conditions. The primary purposes of either type of completion are to prevent surface runoff from entering and infiltrating down the annulus of the well, and to protect the well from accidental damage or vandalism. The level of protection for any individual well or set of wells should be based on the importance of data from the wells, the desired life of the wells, the physical location of the wells (considering the probability of and vulnerability to damage due to natural events or vandalism), and the cost of protecting the wells from damage versus the cost of replacing the wells (ASTM D 5787 [ASTM, 2004aa]).

Surface Seals

Regardless of which type of completion is selected for any given well, a surface seal of cement or concrete should surround the well casing and fill the annular space between the casing and the borehole at the surface. The surface seal may be an extension of the annular seal installed above the filter pack, or a separate seal emplaced atop the annular seal. If it is separate from the annular seal, it should generally extend from 3 to 5 ft below ground surface, unless conditions do not allow, and no more than 6 in. above ground level inside the protective casing in an above-grade completion. In some states, well installation regulations developed for water-supply wells, but also applied to monitoring wells, require that the cement surface seal extend to greater depths (i.e., 10 to 20 ft or more) to ensure sanitary protection of the well.

Because the annular space is generally larger and the surface material adjacent to the borehole is more highly disturbed from drilling at the surface than at depth, the surface seal will generally extend from 1 to 2 ft away from the well casing at the surface (Figure 10.124). The walls of the borehole at the surface should be vertical, so that the seal is cylindrical in cross section. Some well installers prefer to mound the cement surface seal slightly around the well casing or protective casing to allow for shrinkage of the cement, and to provide a slope away from the well that discourages surface runoff from entering the well. Any mound, however, should be limited in size and slope so access to the well is not impaired. Some well installers construct a cement or concrete pad, generally from 2- to 6-ft^2, around the well to provide a clean working area for field personnel to sample the well (Figure 10.125). While this practice may work well in warm-weather climates, it is not advised in cold-weather climates.

Cement used for surface seals in cold-weather climates should be resistant to cracking induced by alternating freezing and thawing conditions. The use of air-entrained cements is generally preferred in these situations, and can substantially reduce damage to or destruction of surface seals. The cement surface seal should be extended below the frost line (the depth below ground surface that reaches 32°C (0°F) for an extended period of time) — this will prevent potential well damage caused by frost heave (Figure 10.126a and b). The cement surface seal should not extend beyond the diameter of the borehole (i.e., no large pads should be built) and should not be tapered downward, to avoid increased potential for frost-heave damage. Gates (1989) recommends using bentonite pellets, chips or granules as a surface seal in areas where frost heave is a problem (Figure 10.127). Because bentonite freezes at a much lower temperature than the surrounding soil (Hoekstra, 1969) and, even below the freezing point, bentonite has a high percentage of unfrozen pore water, it shears much easier than frozen silt or sand. To avoid creating a slippery, messy surface around the well, Gates (1989) suggests applying a

FIGURE 10.124
The surface seal of a monitoring well in cold climates should not extend more than 2 ft beyond the diameter of the borehole, to avoid damage from frost heave.

Design and Installation of Ground-Water Monitoring Wells

FIGURE 10.125
In warm climates, some well installers prefer to construct a pad around the well, to provide a level working area.

layer of medium to coarse "crusher run" gravel over the bentonite. The gravel provides a clean surface area, allows water through to keep the bentonite hydrated, and provides a thermal blanket that further reduces frost-heave potential.

Above-Grade Completions

In an above-grade completion (Figure 10.128), a protective casing is set into the cement or concrete surface seal while it is still wet and uncured, and is installed around the well casing (Figure 10.129). The protective casing discourages unauthorized entry into the well, prevents damage from contact with vehicles and, in the case of wells in which PVC well casing is installed, protects the casing from degradation caused by direct exposure to sunlight (i.e., from photodegradation in the presence of ultraviolet rays). The protective casing should be made of corrosion-resistant material (i.e., stainless steel or anodized aluminum) (Figure 10.130), although it is more common to use carbon steel, which is painted to inhibit corrosion. Steel casing should be painted and the paint

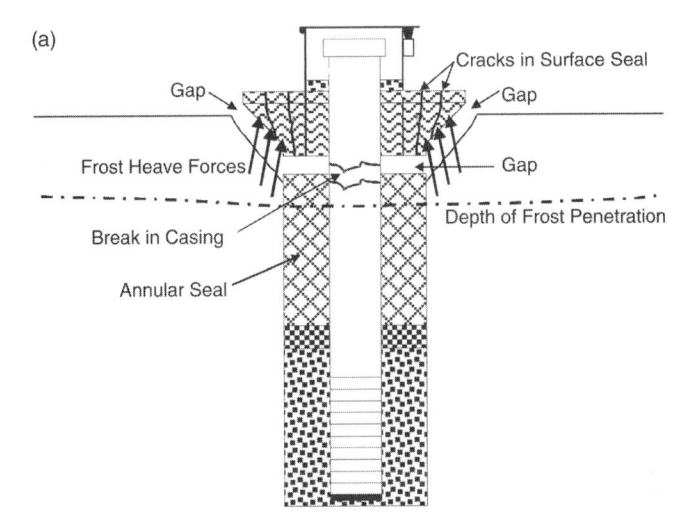

FIGURE 10.126
Damage caused by frost heave can compromise the surface seal and allow surface runoff to enter the borehole.

FIGURE 10.126
Continued.

dry before installation to avoid the possibility of getting paint on or in the well. As with the well casing, the protective casing should be thoroughly cleaned before installation to ensure removal of any chemicals or coatings (other than paint). The inside diameter should be at least 2 in. larger in diameter than the well casing, including allowances for the size of the inner casing cap, to allow easy access to the well for sampling (Figure 10.131). The protective casing may be fitted with a locking lid, in which case it should be installed to provide adequate clearance (3 to 6 in.) between the top of the

FIGURE 10.127
In cold climates, a bentonite pellet or chip surface seal may be a better choice than cement or concrete.

Design and Installation of Ground-Water Monitoring Wells 781

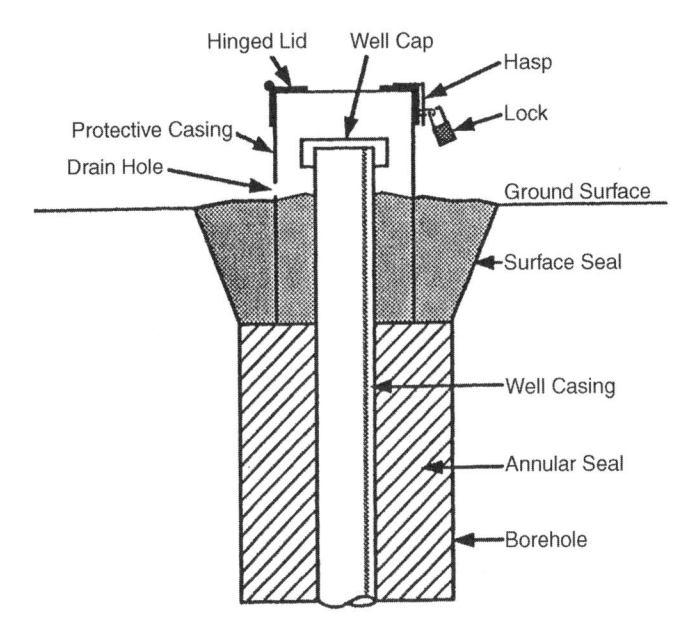

FIGURE 10.128
A typical above-grade monitoring well completion.

in-place well casing cap and the bottom of the protective casing's locking lid. This is particularly true for wells in which dedicated pumps are installed, where tubing may protrude through the well cap. During installation, the protective casing should be positioned and maintained in a plumb position. It is usually installed so that approximately half of the protective casing is anchored into the cement surface seal, and half extends above the seal to protect the well casing. It is advisable that the protective casing extend at least 2.5 to 3 ft above ground surface, so the well can be easily located.

In areas not subject to flooding, both the inner well casing and the outer protective casing should be vented to prevent the accumulation and entrapment of potentially

FIGURE 10.129
Installing a protective casing in the concrete surface seal.

FIGURE 10.130
Stainless steel protective casing provides corrosion resistance superior to that of painted mild steel materials.

FIGURE 10.131
The inside diameter of the protective casing should be at least 2 in. larger than that of the well casing, to allow easy access to the well for water-level measurement and ground-water sampling.

explosive gases, and to allow water levels in the well to respond to changes in barometric pressure and hydraulic head. Additionally, the outer protective casing should have a drain hole (from 1/4 to 1/2 in. diameter) installed just above the top of the cement level in the space between the protective casing and the well casing to allow the drainage of any water accumulating in this space. This is particularly critical in cold climates, where the freezing of water trapped between the inner well casing and the outer protective casing can cause the inner casing to buckle and fail. A 60-mesh screen or a coarse sand or pea gravel blanket, no more than 6 in. thick, should be installed between the well casing and the protective casing to prevent insects and small animals from entering the drain hole.

In areas subject to flooding, it is advisable to extend the well casing and protective casing above the anticipated flood level (Figure 10.132). If this is not possible or desirable, either the locking lid on the protective casing or the well casing cap should be water-tight to prevent water leakage into the well (Figure 10.133). In this situation, it is advisable to wait for a period of time after removing the well casing cap to allow the static water level in the well to stabilize before taking water-level measurements, as the water level may not have equilibrated with atmospheric pressure.

A case-hardened steel lock is generally installed on the protective casing lid to provide well security (Figure 10.134), but weather-induced corrosion often causes the locking mechanism to jam. Because lubricants (i.e., graphite, petroleum-based sprays, and silicon) provide a potential source of ground-water sample chemical alteration, their use in lubricating locks or for freeing corroded locking mechanisms is not recommended. Rather, the use of some type of protective measure to shield the lock from the elements (i.e., a plastic covering) should be considered. As a practical matter, all locks for a set of monitoring wells should be keyed alike to avoid having to carry and sort through multiple keys for individual locks. Access to keys should be controlled to prevent unauthorized access to the well.

Construction of the lid, hinges, and hasps or locking lugs on the protective casing should be sufficiently rugged, made of sturdy metal, and welded to prevent unauthorized access to the casing by prying or hammering. Both the protective casing and the lock should be heavy enough to resist penetration by bullets in areas where shooting may occur. In extreme cases, locked chain-link fences or small structures may need to be

FIGURE 10.132
In areas prone to flooding, the top of the protective casing should extend above the flood level. This well is in the flood plain of a major river, where flood stage often reaches 6 to 8 ft above ground surface.

FIGURE 10.133
A water-tight locking cap on the well casing should prevent entry of unauthorized personnel into the well. However, it may also prevent the water level in the well from responding to increases in hydraulic head in the formation by trapping an air column in the well casing. Wells fitted with this type of cap should be properly vented.

built around the well to protect it from damage or vandalism (Figure 10.135 and Figure 10.136).

In high-traffic areas, such as parking lots or in areas where heavy equipment may be working (i.e., a landfill or industrial facility), additional protection is often required to prevent damage to the well from vehicles (Figure 10.137). Above-grade completions are often protected by the installation of "bumper guards," or brightly painted steel posts (Figure 10.138). Normally, three or four 3-ft high (after installation), 3- to 4-in. diameter posts are installed surrounding and within 3 to 4 ft of the well, in a configuration designed to prevent vehicles from striking the protective casing. These posts are typically filled with cement or concrete and set in 2-ft deep post holes backfilled with concrete.

Flush-to-Grade Completions

The use of flush-to-grade completions has become increasingly popular in the last decade for a variety of applications. In a flush-to-grade completion (Figure 10.139), a protective structure resembling a utility vault (Figure 10.140), meter box or manhole is generally set into a cement or concrete surface seal before it has cured, and is installed around the well casing, which has been cut off below grade (Figure 10.141). The structures used may be round, square, or rectangular, with dimensions from 6 to 36 in. across, and 6 in. to more than 24 in. deep. This type of completion is generally used in high-traffic areas such as streets, parking lots, and service stations, where an above-grade completion would severely disrupt traffic patterns or facility operations. For this reason, the structure must generally be constructed steel, aluminum, or a high-strength composite plastic material,

Design and Installation of Ground-Water Monitoring Wells 785

FIGURE 10.134
A case-hardened steel lock on the lid of the protective casing provides adequate security under most conditions.

FIGURE 10.135
In areas where a high level of security is required, it may be necessary to provide a locked, fenced-in enclosure for the well.

FIGURE 10.136
A structure built to protect the well may be necessary in extreme cases. If the structure selected appears to be similar to another type of structure, it should be clearly identified as a well enclosure (note the marking on the structure on the right).

and traffic rated. The depth of such completions should extend below the gravel subgrade beneath the asphalt or concrete pavement (generally at least 1 ft), and the concrete or cement extended slightly deeper, to keep water from the subgrade material from backing up inside the enclosure. Completions in public rights-of-way may also have to meet specific construction requirements of local highway departments, and may be subject to municipal permitting or other government regulation. Flush-to-grade completions are also used in areas where such completions are required by municipal easements, where vandalism is a problem, or where aesthetics or other factors dictate the use of a low-profile well installation.

Because of the potential for surface runoff to enter any below-grade protective structure, this type of completion must be carefully designed and installed. For example, the seal between the protective structure and its removable cover should be watertight. Installing a flexible O-ring or gasket at the point where the cover fits over the protective structure

FIGURE 10.137
In high-traffic areas, particularly where heavy construction vehicles are active, protective casing alone is often not sufficient to prevent damage to the well.

Design and Installation of Ground-Water Monitoring Wells 787

FIGURE 10.138
Bumper guards — brightly painted cement-filled steel posts set in concrete — can provide protection from damage by vehicles.

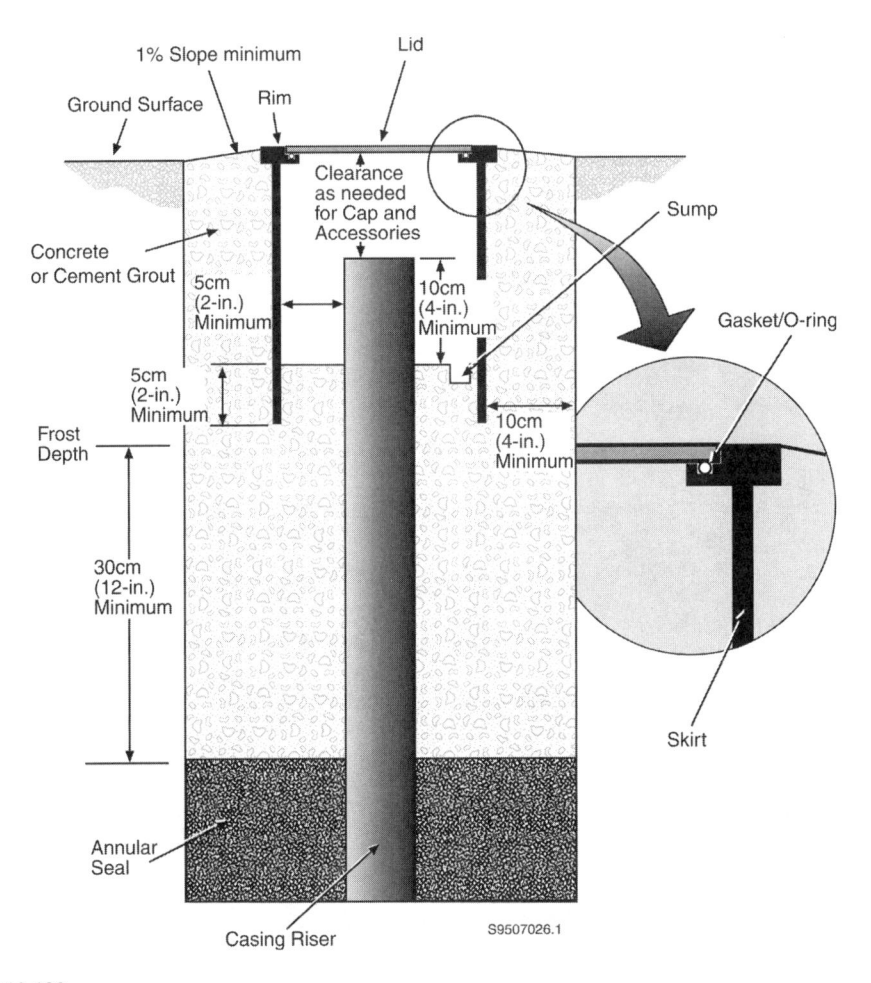

FIGURE 10.139
A typical flush-to-grade monitoring well completion.

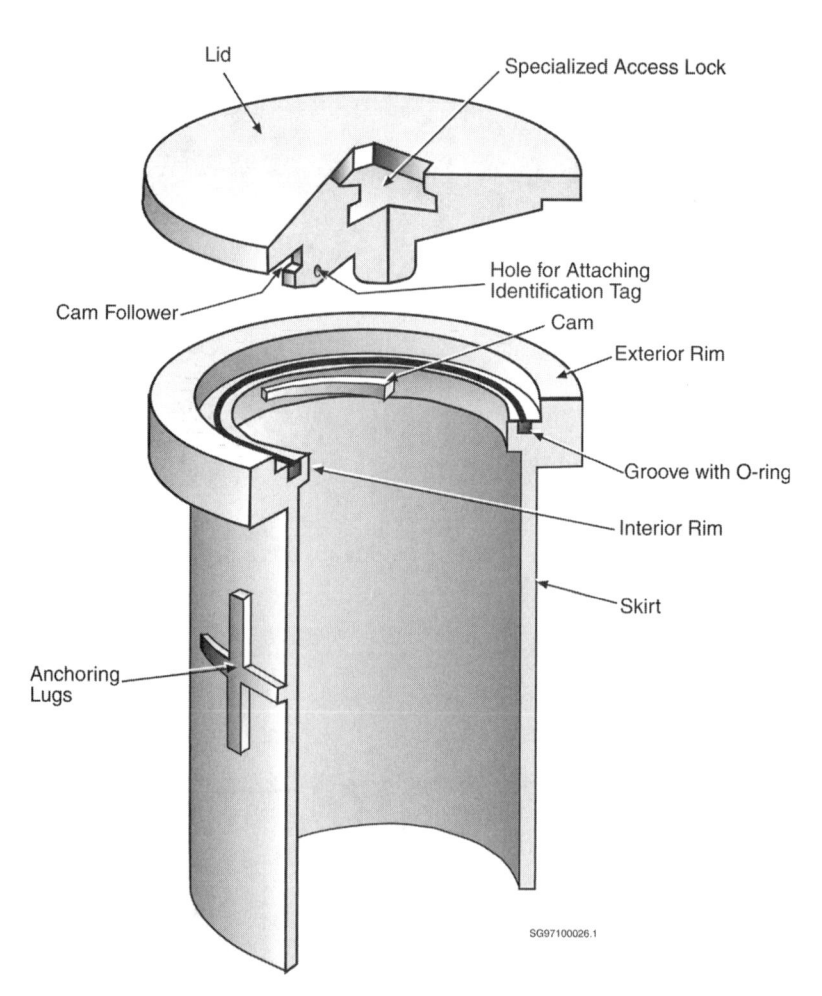

FIGURE 10.140
Construction details of a typical flush-to-grade well vault.

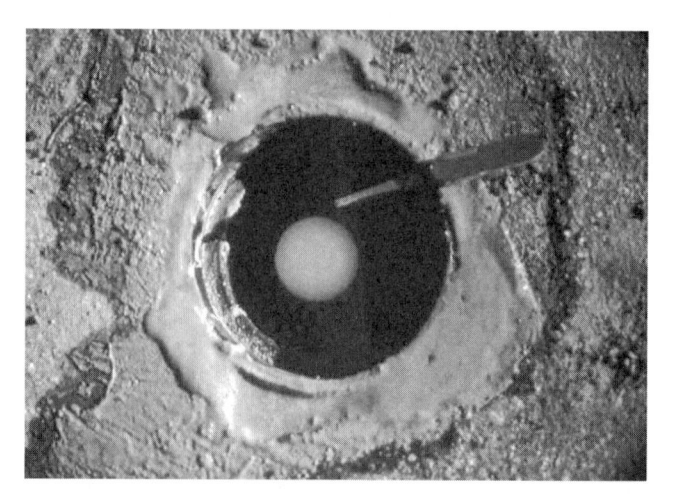

FIGURE 10.141
Well casing cut below grade so the casing cap will fit beneath the lid of the well vault.

will suffice for sealing the protective structure under most conditions, but the gasket or O-ring must be properly maintained (Figure 10.142a and b). The bond between the cement or concrete surface seal and the protective structure must also be watertight. Use of expanding cement should ensure that the cement surface seal bonds tightly to the protective structure. Concrete mixtures should contain cement and aggregate consistent with suitable load-bearing pavement design if the structures are installed in traffic areas. The cement or concrete seal should extend at least 4 to 6 in. from the edge of the protective structure (Figure 10.143). In areas of significant street runoff, additional safeguards, such as building a low, gently sloping mound of cement around the protective structure and placing the structure slightly above grade, may be necessary to prevent entry of surface runoff (Figure 10.143). In cold-weather climates, where parking lots and roads are cleared of snow using snowplows, well vaults may need to be set slightly below the surrounding cement or concrete to prevent the blade of the plow from shearing off the lid of the vault. In this type of completion, drainage of any water that may accumulate on the lid can be accomplished using curved drainage channels that will not allow snowplow blades to drop into a straight groove and damage the lid (Figure 10.144 and Figure 10.145).

Flush-to-grade completions should not be installed in low-lying areas that receive surface runoff and in which ponding or flooding occurs (Figure 10.146). Under these conditions, water can easily infiltrate the protective enclosure, and may enter the well. To prevent

FIGURE 10.142
(a) The gasket on a flush-to-grade well vault is designed to provide a water-tight seal. (b) Unless the gasket is checked, cleaned, and periodically replaced, however, it often allows water into the enclosure.

FIGURE 10.143
The concrete seal around the well vault should extend at least 4 to 6 in. from the edge of the vault and should slope away from the vault to discourage entry of surface runoff.

water leakage into the well, flush-to-grade wells are usually not vented, and an expanding well casing cap is used to seal the casing. However, an expanding well cap will prevent air pressure inside the well casing from equilibrating with atmospheric pressure, which can affect the water level in the well for a period of time after removing the cap. Thus, water-level measurements in these wells should be made only after the static water level in the well has stabilized (as determined from replicate water-level measurements).

For completions in which dedicated pumping equipment is installed, or in which downhole probes or data loggers may need to be used, it is advisable to have sufficient room in the enclosure to keep the required equipment. Vertical clearance between the well casing and the structure's lid is important in these situations. In some situations, it may be necessary to construct a sump on one side of the enclosure, to allow collection and removal of any fluids that may enter the enclosure. In some soil conditions, it may be possible to install a drain in place of the sump, to allow any water that collects in the enclosure to drain out into the soil below. However, such drains may allow soil-dwelling insects to enter and infest the enclosure, so they should be used with care.

All flush-to-grade completions should have a lid that is tamper-resistant and, optimally, lockable (Figure 10.147). A tamper-resistant lid may be opened with a specially sized or designed tool that is not available to the general public (Figure 10.148a–c). The openings for such tools should be kept free of degradation, corrosion, and sediment, which may impair authorized access to the well. Lockable lids are not common in flush-to-grade completions because of the difficulty in maintaining locks under high-traffic conditions. For this reason, it is very important to provide a means to lock the inner well casing to restrict access to the well. Lockable well caps should be designed so they cannot be removed without first unlocking the cap — only a few designs meet this important criterion. If a locking cap can be easily removed without using a key, there is a good chance that the cap will be removed or displaced; subsequent entry of fluids into the well could compromise sample integrity. In areas in which other below-grade enclosures exist (i.e., at a service station), it is very important to clearly identify the flush-to-grade completion as a monitoring well to set it apart from other subsurface structures (i.e., underground storage tank fill pipes), so that the well is not mistaken for something else (Figure 10.149).

Design and Installation of Ground-Water Monitoring Wells 791

Cross Section Detail of Well Vault & Concrete Pad

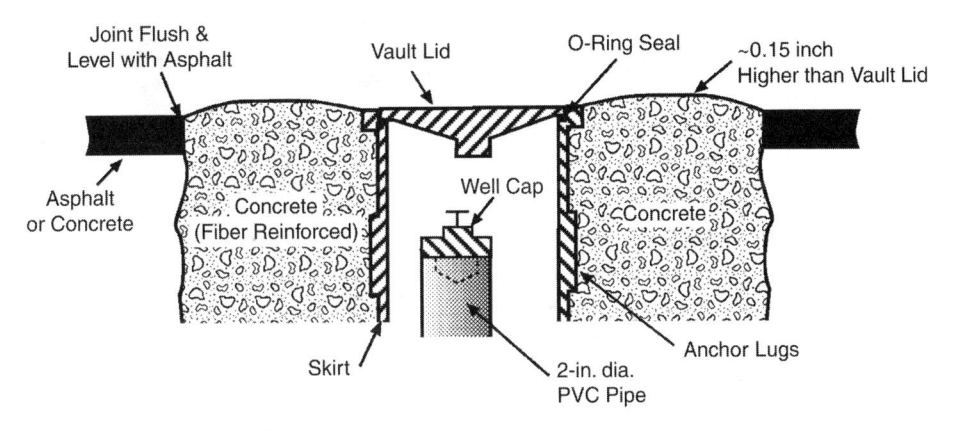

Plan View Detail of Well Vault & Concrete Pad

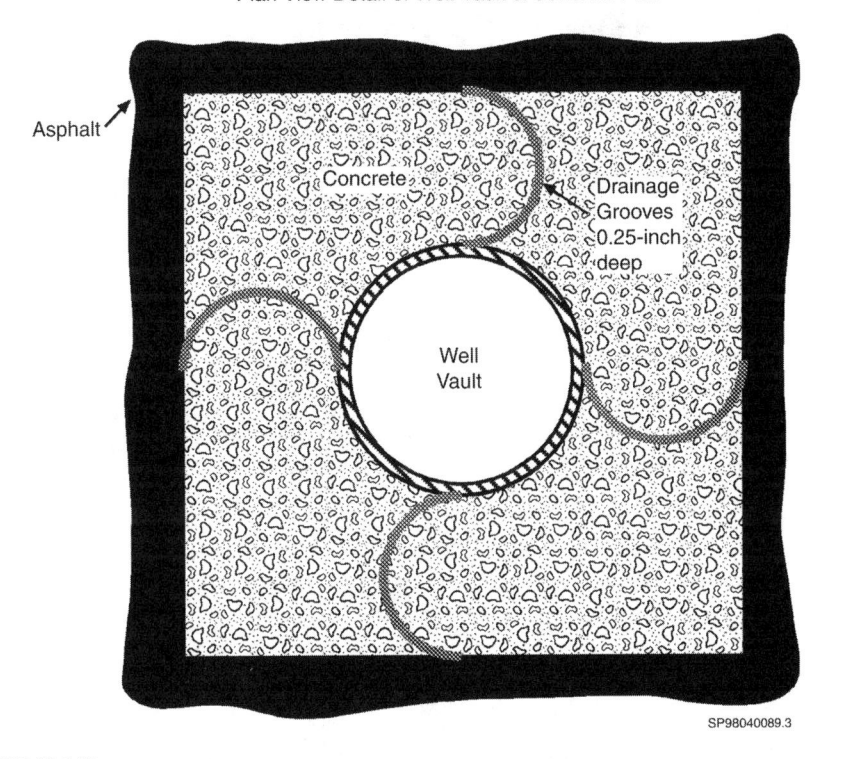

FIGURE 10.144
A flush-to-grade completion in an area where snow plows could damage the well vault. Note that the vault is slightly below the surface seal, and drainage channels are cut into the seal to allow water to drain away from the vault.

Maintenance is required on all flush-to-grade enclosures to maintain their integrity. Perhaps the most important maintenance item, and the one most frequently ignored, is the flexible gasket or O-ring installed between the lid and the enclosure. The gasket or O-ring should be inspected and replaced if necessary, and the area where it seats cleaned to remove debris each time the enclosure is opened. Neglecting to do so may cause the gasket or O-ring to fail as a watertight seal. The cement or concrete surface seal and the locking casing cap should also be inspected, particularly in cases in

FIGURE 10.145
Drainage channels cut into the concrete surface seal in an asphalt paved area.

which it is evident that water has infiltrated into the enclosure. Any presence of water or other fluids in the enclosure should be documented so that if sampling results from the well appear to be anomalous, a possible explanation is provided.

Well Identification, Surveying, and Alignment Testing

All monitoring wells should be marked with some form of clear, permanent identification on a nonremovable part of the well (Figure 10.150). Various methods of identification have been successfully used, including painting the number of the well on the protective

FIGURE 10.146
Flush-to-grade completions should not be used in areas where ponding or flooding occur.

Design and Installation of Ground-Water Monitoring Wells　　　　　793

FIGURE 10.147
A high-security lockable well vault that may be installed either flush-to-grade or above grade.

casing with the aid of a painting stencil (Figure 10.151), imprinting or engraving the number directly on the protective casing, marking the well number in the cement or concrete surface seal (Figure 10.152), and attaching a noncorroding imprinted metal tag to various parts of the well. Information that may be included on imprinted tags includes the following:

- Well identification number or permit number
- Well depth

FIGURE 10.148
A special tool is required to open this lockable well vault. This sequence of three photos (a–c) shows the steps necessary to access the well.

FIGURE 10.148
Continued.

- Elevation of the surveyed measuring point
- Surveyed geographic coordinates
- Screen elevation and length
- Date of installation
- Owner's name

Identification requirements may vary by regulatory jurisdiction, so the well installer should always check with the appropriate agency to determine how the well should be marked.

The physical location of every monitoring well should be surveyed by a licensed professional surveyor. Spatial surveying requirements vary by regulatory jurisdiction, but they generally involve surveying with reference to the state plane coordinate system to an accuracy of ± 1 ft. The elevation of a clearly marked reference measuring point on the top of the well casing (not the protective casing) should be surveyed in to accuracy of ± 0.01 ft, to allow accurate measurement of water levels. The elevation survey should be done either in reference to mean sea level (which is, in turn, established by reference to an established National Geodetic Vertical Datum), or in reference to a site-specific benchmark. Additional information on surveying requirements is found in Chapter 12.

Design and Installation of Ground-Water Monitoring Wells 795

FIGURE 10.149
All flush-to-grade completions should be clearly identified as monitoring wells so they are not mistaken for other subsurface structures that are similar in appearance.

Well alignment may need to be assessed to check for proper screen placement and to ensure smooth passage of sampling and hydraulic testing equipment into the well. The preferred method is to conduct a borehole deviation survey, which determines the direction and distance of the bottom of the borehole relative to the top of the borehole and points in between. A borehole dipmeter survey may also be conducted. Well misalignment is usually not significant for shallow boreholes (e.g., less than 50 ft deep) in relatively homogeneous geological materials, but it can become a problem in deep boreholes or where difficult drilling conditions were encountered during drilling or well construction. An alternative method that can be used in very shallow wells (i.e., <40 ft deep) is to pass a length of steel pipe, no less than 0.5 in. smaller in outside diameter than the inside diameter of the well, through the casing and screen. The pipe should be able to be lowered to the bottom of the well without binding. AWWA (1984) discusses several procedures for assessing alignment and plumbness of deep wells.

FIGURE 10.150
Monitoring well identification should never be placed on a removable part of the well, such as the well cap.

FIGURE 10.151
Monitoring well identification is usually painted or etched on the outer part of the protective well casing.

FIGURE 10.152
Occasionally, the monitoring well identification may be marked into the concrete surface seal.

TABLE 10.24

Items to Document in a Well Construction Report

Date and time of start and completion of well construction
Boring/well number
Drilling method and drilling fluid used (if applicable)
Borehole diameter and well casing and screen diameter
Latitude and longitude (GPS location)
Surveyed well location (± 0.5 ft) with sketch of location (map location)
Total borehole depth (± 0.1 ft)
Total well depth (± 0.1 ft)
Casing length and material
Screened intervals
Screen materials, length, design, and slot size
Casing and screen joint type
Depth and elevation of top and bottom of screen
Filter-pack material/grain size, volume calculations, and placement method
Depth and elevation of top and bottom of filter pack
Annular seal composition, volume (calculated and actual), and placement method
Surface seal composition, placement method, and volume (calculated and actual)
Surface seal and well apron design/construction
Depth and elevation of water level in the well
Well development procedure and ground-water turbidity
Type and design of protective casing
Well cap and lock type
Surveyed ground surface elevation (± 0.01 ft)
Surveyed reference point elevation (± 0.01 ft) on well casing
Detailed drawing of well (with dimensions)
Point where ground water was encountered in the borehole
Water level after completion of well development

Documentation of Well Construction Details

The details of well construction should be documented by those installing the wells during and after the installation of any monitoring well or set of wells. This will allow anyone who uses the wells in the future (i.e., for water-level measurement, sampling, or hydraulic testing) to judge the utility of the well for their particular use. It will also allow anyone attempting to determine the cause for anomalies in data collected from the well (i.e., anomalous water-level readings, unusual water chemistry, lower than expected yield, unexpectedly low hydraulic conductivity test results) to determine if the anomalies could be attributed to well construction. Most states and some local government agencies now require that the installation of all wells in their jurisdiction (including monitoring wells) be thoroughly documented, although the requirements for documentation are highly variable. Table 10.24 provides a list of the most important details of well construction that should be documented for monitoring wells.

References

Ahmad, M.U., E.B. Williams, and L. Hamdan, Commentaries on experiments to assess the hydraulic efficiency of well screens, *Ground Water*, 21(3), 282–286, 1983.

Ahrens, T.P., Well design criteria: Part One, *Water Well Journal*, 11(4), 13–30, 1957.

Ahrens, T.P., Basic considerations of well design: Part III, *Water Well Journal*, 24(3), 47–51, 1970.

Aller, L., T.W. Bennett, G. Hackett, R. Petty, J. Lehr, H. Sedoris, D.M. Nielsen, and J.E. Denne, Handbook of Suggested Practices for the Design and Installation of Ground-Water Monitoring Wells, EPA/600/4-89/034, U.S. Environmental Protection Agency, Office of Research and Development, Las Vegas, NV, 1991, 221 pp.

Anderson, K.E., *Water Well Handbook*, Missouri Water Well and Pump Contractors Association, Inc., Rolla, MO, 1971.

Anderson, D.C., K.W. Brown, and J.W. Green, Effect of Organic Fluids on Permeability of Clay Soil Liners, EPA-600/9-82-002, U.S. Environmental Protection Agency Report, Land Disposal of Hazardous Waste: Proceedings, 1982.

Anderson, M.R., R.L. Johnson, and J.F. Pankow, The Dissolution of Residual Non-Aqueous Phase Liquid (DNAPL) From a Saturated Porous Medium, Proceedings of the NWWA/API Conference on Petroleum Hydrocarbons and Organic Chemicals in Ground Water, National Water Well Association, Dublin, OH, 1987, pp. 409–427.

ASTM, Standard Practice for Design and Installation of Ground-Water Monitoring Wells, ASTM Standard D 5092, ASTM, West Conshohocken, PA, 2004a, 20 pp.

ASTM, Standard Practice for Description and Identification of Soils (Visual–Manual Procedure), ASTM Standard D 2488, ASTM, West Conshohocken, PA, 2004b, 11 pp.

ASTM, Standard Classification of Soils for Engineering Purposes (Unified Soil Classification System), ASTM Standard D 2487, ASTM, West Conshohocken, PA, 2004c, 11 pp.

ASTM, Standard Guide for Using the Seismic Refraction Method for Subsurface Investigation, ASTM Standard D 5777, ASTM, West Conshohocken, PA, 2004d, 14 pp.

ASTM, Standard Guide for Using the Direct Current Resistivity Method for Subsurface Investigation, ASTM Standard D 6431, ASTM, West Conshohocken, PA, 2004e, 14 pp.

ASTM, Standard Guide for Using the Surface Ground Penetrating Radar Method for Subsurface Investigation, ASTM Standard D 6432, ASTM, West Conshohocken, PA, 2004f, 18 pp.

ASTM, Standard Guide for Using the Gravity Method for Subsurface Investigation, ASTM Standard D 6430, ASTM, West Conshohocken, PA, 2004g, 10 pp.

ASTM, Standard Guide for Using the Frequency Domain Electromagnetic Method for Subsurface Investigation, ASTM Standard D 6639, ASTM, West Conshohocken, PA, 2004h, 14 pp.

ASTM, Standard Guide for Using the Time Domain Electromagnetic Method for Subsurface Investigation, ASTM Standard D 6820, ASTM, West Conshohocken, PA, 2004i, 14 pp.

ASTM, Standard Guide for Selection of Aquifer Test Method in Determining Hydraulic Properties by Well Techniques, ASTM Standard D 4043, ASTM, West Conshohocken, PA, 2004j, 5 pp.

ASTM, Standard Specification for Portland Cement, ASTM Standard C 150, ASTM, West Conshohocken, PA, 2004k, 14 pp.

ASTM, Standard Guide for Selection of Drilling Methods for Environmental Site Characterization, ASTM Standard D 6286, ASTM, West Conshohocken, PA, 2004l, 16 pp.

ASTM, Standard Guide for Use of Hollow-Stem Augers for Geoenvironmental Exploration and Installation of Subsurface Water-Quality Monitoring Devices, ASTM Standard D 5784, ASTM, West Conshohocken, PA, 2004m, 7 pp.

ASTM, Standard Guide for Use of Direct Rotary Drilling With Water-Based Drilling Fluid for Geoenvironmental Exploration and Installation of Subsurface Water-Quality Monitoring Devices, ASTM Standard D 5783, ASTM, West Conshohocken, PA, 2004n, 7 pp.

ASTM, Standard Guide for Use of Direct Air-Rotary Drilling for Geoenvironmental Exploration and Installation of Subsurface Water-Quality Monitoring Devices, ASTM Standard D 5782, ASTM, West Conshohocken, PA, 2004o, 7 pp.

ASTM, Standard Guide for Use of Direct Rotary Wireline Casing Advancement Methods for Geoenvironmental Exploration and Installation of Subsurface Water-Quality Monitoring Devices, ASTM Standard D 5876, ASTM, West Conshohocken, PA, 2004p, 11 pp.

ASTM, Standard Guide for Use of Dual-Wall Reverse Circulation Drilling for Geoenvironmental Exploration and Installation of Subsurface Water-Quality Monitoring Devices, ASTM Standard D 5781, ASTM, West Conshohocken, PA, 2004q, 8 pp.

ASTM, Standard Guide for Use of Cable-Tool Drilling and Sampling Methods for Geoenvironmental Exploration and Installation of Subsurface Water-Quality Monitoring Devices, ASTM Standard D 5875, ASTM, West Conshohocken, PA, 2004r, 9 pp.

ASTM, Standard Guide for Use of Casing Advancement Drilling Methods for Geoenvironmental Exploration and Installation of Subsurface Water-Quality Monitoring Devices, ASTM Standard D 5872, ASTM, West Conshohocken, PA, 2004s, 8 pp.

ASTM, Standard Specification for Poly (Vinyl Chloride) Plastic Pipe, Schedules 40, 80, and 120, ASTM Standard D 1785, ASTM, West Conshohocken, PA, 2004t, 11 pp.

ASTM, Standard Specification for Thermoplastic Well Casing Pipe and Couplings Made in Standard Dimension Ratios (SDR), Schedule 40 and Schedule 80, ASTM Standard F 480, ASTM, West Conshohocken, PA, 2004u, 14 pp.

ASTM, Standard Practice for Decontamination of Field Equipment Used at Non-Radioactive Waste Sites, ASTM Standard D 5088, ASTM, West Conshohocken, PA, 2004v, 3 pp.

ASTM, Standard Practice for Decontamination of Field Equipment Used at Low-Level Radioactive Waste Sites, ASTM Standard D 5608, ASTM, West Conshohocken, PA, 2004w, 8 pp.

ASTM, Standard Test Method for Particle-Size Analysis of Soils, ASTM Standard D 422, ASTM, West Conshohocken, PA, 2004x, 7 pp.

ASTM, Standard Practice for the Installation of Pre-Packed Screen Monitoring Wells in Unconsolidated Aquifers, ASTM Standard D 6725, ASTM, West Conshohocken, PA, 2004y, 15 pp.

ASTM, Standard Specification for Expansive Hydraulic Cement, ASTM Standard C 845, ASTM, West Conshohocken, PA, 2004z, 3 pp.

ASTM, Standard Practice for Monitoring Well Protection, ASTM Standard D 5787, ASTM, West Conshohocken, PA, 2004aa, 4 pp.

AWWA, Standard Specifications for Deep Wells, AWWA Standard A100, American Water Works Association, Denver, CO, 1984.

Azadpour-Keeley, A., J.W. Keeley, H.H. Russell, and G.W. Sewell, Monitored natural attenuation of contaminants in the subsurface: processes, *Ground-Water Monitoring & Remediation*, 20(2), 97–107, 2001.

Barcelona, M.J., J.P. Gibb, and R.A. Miller, A Guide to the Selection of Materials for Monitoring Well Construction and Ground-Water Sampling, ISWS Contract Report #327, Illinois Department of Energy and Natural Resources, Water Survey Division, Champaign, IL, 1983, 78 pp.

Barcelona, M.J., J.P. Gibb, J.A. Helfrich, and E.E. Garske, A Practical Guide for Ground-Water Sampling, ISWS Contract Report #374, Illinois Department of Energy and Natural Resources, Water Survey Division, Champaign, IL, 1985a, 95 pp.

Barcelona, M.J., J.A. Helfrich, and E.E. Garske, Sample tubing effects on ground water samples, *Analytical Chemistry*, 5, 460–464, 1985b.

Barker, J.F., G.C. Patrick, L. Lemon, and G.M. Travis, Some biases in sampling multilevel piezometers for volatile organics, *Ground-Water Monitoring Review*, 7(2), 48–54, 1987.

Berens, A.R., Prediction of organic chemical permeation through PVC pipe, *Journal of the American Water Works Association*, November, 57–64, 1985.

Bikis, E.A., A laboratory and field study of fiberglass and continuous-slot screens, *Ground Water*, 17(1), 111, 1979.

Boettner, E.A., G.L. Ball, Z. Hollingsworth, and R. Aquino, Organic and Organotin Compounds Leached from PVC and CPVC Pipes, EPA-600/1-81-062, U.S. Environmental Protection Agency, Office of Research and Development, Washington, DC, 1981, 102 pp.

Boyle, D.R., A dry injection system for the emplacement of filter packs and annular seals in groundwater monitoring wells, *Ground-Water Monitoring Review*, 12(1), 120–125, 1992.

Brown, K.W., J.W. Green, and J.C. Thomas, The Influence of Selected Organic Liquids on the Permeability of Clay Liners, EPA-600/9-83-018, U.S. Environmental Protection Agency Report, Land Disposal of Hazardous Waste: Proceedings, 1983, pp. 114–125.

California Department of Health Services, The California Site Mitigation Decision Tree, California Department of Health Services, Toxic Substances Control Division Draft Working Document, 1985.

Calhoun, D.E., Sealing well casings: an idea whose time has come, *Water Well Journal*, 42(2), 25–29, 1988.

Campbell, M.D. and J.H. Lehr, *Water Well Technology*, McGraw-Hill Book Company, New York, NY, 1973, 681 pp.

Campbell, M.D. and J.H. Lehr, Well cementing, *Water Well Journal*, 29(7), 39–42, 1975.

Cherry, J.A., Ground-water monitoring: some current deficiencies and alternative approaches, in *Hazardous Waste Site Investigations: Toward Better Decisions*, Lewis Publishers, Boca Raton, FL, 1992, Chap. 13, pp. 119–134.

Cherry, J.A., R.W. Gillham, E.G. Anderson, and P.E. Johnson, Migration of contaminants in ground water at a landfill: a case study, *Journal of Hydrology*, 63, 31–49, 1983.

Chiang, C., G. Raven, and C. Dawson, The relationship between monitoring well and aquifer solute concentrations, *Ground Water*, 33(5), 718–726, 1995.

Christensen, T.H., P.L. Bjerg, and P. Kjeldsen, Natural attenuation: a feasible approach to remediation of ground-water pollution at landfills, *Ground-Water Monitoring & Remediation*, 19(1), 69–77, 2000.

Church, P.E. and G.E. Granato, Bias in ground-water data caused by well-bore flow in long-screen wells, *Ground Water*, 34(2), 262–273, 1996.

Clark, L. and P.A. Turner, Experiments to assess the hydraulic efficiency of wells, *Ground Water*, 21(3), 1983.

Colangelo, R.V. and G.R. Lytwynyshyn, Cement Bentonite Grout and its Effect on Water Quality Samples: A Field Test of Volclay Grout, Proceedings of the First National Outdoor Action Conference and Exposition, National Water Well Association, Dublin, OH, 1987, pp. 345–358.

Colangelo, R.V. and Upadhyay, H.D., The Origin and Physical Properties of Bentonite and Its Usage in the Ground-Water Monitoring Industry, Proceedings, Superfund 90, the 11th National Conference, Hazardous Materials Control Research Institute, Washington, DC, 1990, pp. 308–313.

Curran, C.M. and M.B. Tomson, Leaching of trace organics into water from five common plastics, *Ground-Water Monitoring Review*, 3(3), 68–71, 1983.

Driscoll, F.G., *Ground Water and Wells*, 2nd ed., Johnson Division, UOP, Inc., St. Paul, MN, 1986.

Dunbar, D., H. Tuchfeld, R. Siegel, and R. Sterbentz, Ground water quality anomalies encountered during well construction, sampling and analysis in the environs of a hazardous waste management facility, *Ground-Water Monitoring Review*, 5(3), 70–74, 1985.

Edil, T.B., M.M.K. Chang, L.T. Lan, and T.V. Riewe, Sealing characteristics of selected grouts for water wells, *Ground Water*, 30(3), 351–361, 1992.

Ericson, W.A., J.E. Brinkman, and P.S. Darr, Types and usages of drilling fluids utilized to install monitoring wells associated with metals and radionuclide ground water studies, *Ground-Water Monitoring Review*, 5(1), 30–33, 1985.

Evans, L.G. and S.B. Ellingson, The Formation of Cement Slurry Bleed-Water and Minimizing its Effects on Water Quality Samples, Proceedings of the Ground-Water Geochemistry Conference, Denver, CO, National Water Well Association, Dublin, OH, 1988, pp. 377–389.

Foster, S., Flush-joint threads find a home, *Ground Water Monitoring Review*, 9(2), 55–58, 1989.

Freeze, R.A. and J.A. Cherry, *Groundwater*, Prentice-Hall, Inc., Englewood Cliffs, NJ, 1979, 604 pp.

Gates, W.C.B., Protection of ground-water monitoring wells against frost heave, *Bulletin of the Association of Engineering Geologists*, 26(2), 241–251, 1989.

Gibs, J., G.A. Brown, K.S. Turner, C.L. MacLeod, J.C. Jelinski, and S.A. Koehnlein, Effects of small-scale vertical variations in well-screen inflow rates and concentrations of organic compounds on the collection of representative ground-water quality samples, *Ground Water*, 31(2), 201–208, 1993.

Gillespie, G.A., An Effective Monitoring Well Screen for Fine Sand Aquifers, *Current Practices in Ground-Water and Vadose Zone Investigations*, ASTM Special Technical Publication #1118, ASTM, West Conshohocken, PA, 1991, pp. 241–255.

Gillham, R.W. and S.F. O'Hannesin, Sorption of Aromatic Hydrocarbons by Materials Used in Construction of Ground Water Sampling Wells, *Ground-Water and Vadose Zone Monitoring*, ASTM Special Technical Publication #1053, 1990, pp. 108–124.

Helwig, O.J., V.H. Scott, and J.C. Scalmanini, *Improving Well and Pump Efficiency*, American Water Works Association, Denver, CO, 1984, 46 pp.

Hewitt, A.D., Potential of common well casing materials to influence aqueous metal concentrations, *Ground-Water Monitoring & Remediation*, 12(2), 131–136, 1992.

Hewitt, A.D., Dynamic study of common well screen materials, *Ground-Water Monitoring & Remediation*, 14(1), 87–94, 1994.

Hoekstra, P., Water Movement and Freezing Pressures, Soil Science Society of America Proceedings, 33, 1969, pp. 512–518.

Huling, S.G. and J.W. Weaver, Dense Non-Aqueous Phase Liquids, EPA/540/4-91-002, U.S. Environmental Protection Agency, Office of Research and Development, Washington, DC, 1991.

Jackson, P.A., A laboratory and field study of well screen performance and design, *Ground Water*, 12(6), 771–772, 1983.

Johnson Division, U.O.P., *Ground Water and Wells*, Edward E. Johnson, Inc., St. Paul, MN, 1966, 440 pp.

Johnson, R.C., Jr., C.E. Kurt, and G.F. Dunham, Jr., Well grouting and casing temperature increases, *Ground Water*, 18(1), 7–13, 1980.

Jones, J.N. and G.D. Miller, Adsorption of Selected Organic Contaminants Onto Possible Well Casing Materials, *Ground-Water Contamination: Field Methods*, ASTM Special Technical Publication #963, ASTM, Philadelphia, PA, 1988, pp. 185–198.

Junk, G.A., H.J. Svec, R.D. Vick, and M.J. Avery, Contamination of water by synthetic polymer tubes, *Environmental Science and Technology*, 8(13), 1100–1106, 1974.

Kaempfer, T.T., Update on bentonite chips and pellets for sealing piezometers in boreholes, *Geotechnical Instrumentation News*, December, 32–37, 2003.

Kill, D.L., Monitoring Well Development — Why and How, *Ground-Water and Vadose Zone Monitoring*, ASTM Special Technical Publication #1053, ASTM, Philadelphia, PA, 1990, pp. 82–90.

Kram, M., D. Lorenzana, J. Michaelsen, and E. Lory, Performance Comparison: Direct-Push Wells Versus Drilled Wells, NFESC Technical Report TR-2120-ENV, U.S. Navy, Naval Facilities Engineering Command, Port Hueneme, CA, 2001, 55 pp.

Kurt, C.E., Cement-based seals for thermoplastic water well casings, *Water Well Journal*, 37(1), 38–40, 1983.

Kurt, C.E. and R.C. Johnson, Jr., Permeability of grout seals surrounding thermoplastic well casing, *Ground Water*, 20(4), 415–419, 1982.

Lapham, W.W., F.D. Wilde, and M.T. Koterba, Guidelines and Standard Procedures for Studies of Ground-Water Quality: Selection and Installation of Wells, and Supporting Documentation, U.S. Geological Survey, Water Resources Investigations Report 96-4233, 1996.

LeBlanc, D.R., Sewage Plume in a Sand and Gravel Aquifer, Cape Cod, Massachusetts, U.S. Geological Survey Water Supply Paper 2218, 1984.

LeBlanc, D.R., S.P. Garabedian, K.M. Hess, L.W. Gelhar, R.D. Quadri, K.G. Stollenwerk, and W.W. Wood, Large-scale natural gradient tracer tests in sand and gravel, Cape Cod, Massachusetts — experimental design and observed tracer movement, *Water Resources Research*, 27(5), 895–910, 1991.

Lerch, W. and C.L. Ford, Long-time study of cement performance in concrete, Chapter 3, Chemical and Physical Tests of the Cements, *Journal of the American Concrete Institute*, 19(8), 1948.

Listi, R., Monitoring well grout: Why I think bentonite is better, *Water Well Journal*, May, 5–6, 1993.

Mackay, D.M., J.A. Cherry, D.L. Freyberg, and P.V. Roberts, A natural gradient experiment on solute transport in a sand aquifer, Part I: approach and overview of plume movement, *Water Resources Research*, 22, 2017–2029, 1986.

Martin, W.H. and C.C. Lee, Persistent pH and Tetrahydrofuran Anomalies Attributable to Well Construction, Proceedings of the Third National Outdoor Action Conference, National Water Well Association, Dublin, OH, 1989, pp. 201–214.

Martin-Hayden, J.M., G.A. Robbins, and R.D. Bristol, Mass balance evaluation of monitoring well purging, Part II: Field tests at a gasoline contamination site, *Journal of Contaminant Hydrology*, 8, 225–241, 1991.

Martin-Hayden, J.M., Sample concentration response to laminar well bore flow: Implications to ground water data variability, *Ground Water*, 38(1), 12–19, 2000.

McAllister, P.M. and C.Y. Chiang, A practical approach to evaluating natural attenuation of contaminants in ground water, *Ground-Water Monitoring & Remediation*, 14(2), 161–173, 1994.

McCall, W., S. Stover, C. Enos, and G. Fuhrmann, Field Comparison of Direct-Push Pre-Packed Screen Wells to Paired HSA 2 in. PVC Wells, Proceedings, Superfund XVIII, E.J. Krause & Associates, Washington, DC, 1997, pp. 647–655.

McCaulou, D.R. and S.G. Huling, Compatibility of bentonite and DNAPLs, *Ground-Water Monitoring & Remediation*, 19(2), 78–86, 1999.

McIlvride, W.A. and B.M. Rector, Comparison of Short- and Long-Screened Wells in Alluvial Sediments, Proceedings of the Second Outdoor Action Conference, National Ground-Water Association, Dublin, OH, 1988, pp. 375–390.

Meiri, D., A tracer test for detecting cross contamination along a monitoring well column, *Ground-Water Monitoring Review*, 9(2), 78–81, 1989.

Mikkelsen, P.E., Cement–bentonite grout backfill for borehole instruments, *Geotechnical Instrumentation News*, December, 38–42, 2002.

Miller, G.D., Uptake and Release of Lead, Chromium and Trace-Level Volatile Organics Exposed to Synthetic Well Casings, Proceedings of the Second National Symposium on Aquifer Restoration and Ground Water Monitoring, National Water Well Association, Worthington, OH, 1982, pp. 236–245.

Moehrl, K.E., Well grouting and well protection, *Journal of the American Water Works Association*, 56(4), 423–431, 1964.

Molz, F.J. and C.E. Kurt, Grout-induced temperature rises surrounding wells, *Ground Water*, 17(3), 264–269, 1979.

Morrison, R.D., *Ground Water Monitoring Technology: Procedures, Equipment and Applications*, Timco Mfg., Inc., Prairie Du Sec., WI, 1984, 111 pp.

Murphy, E.C., A.E. Kehew, G.H. Groenewold, and W.A. Beal, Leachate Generated By an Oil-And-Gas Brine Pond Site in North Dakota, *Ground Water*, 26(1), 31–38, 1988.

National Sanitation Foundation (NSF), Plastic Piping Components and Related Materials, Standard Number 14, National Sanitation Foundation, Ann Arbor, MI, 2004, 8 pp.

National Water Well Association and Plastic Pipe Institute (NWWA/PPI), *Manual on the Selection and Installation of Thermoplastic Water Well Casing*, National Water Well Association, Worthington, OH, 1980, 64 pp.

Oakley, D. and N.E. Korte, Nickel and chromium in ground-water samples as influenced by well construction and sampling methods, *Ground-Water Monitoring & Remediation*, 16(1), 93–99, 1996.

Papp, J.E., Prevent unwanted migration with bentonite, *International Ground-Water Technology*, December, 18–20, 1995.

Papp, J.E., Sodium bentonite as a borehole sealant, *Sealing of Boreholes and Underground Excavations in Rock*, Chapman and Hall Publishers, London, U.K., 1996, Chap. 12.

Papp, J.E., Case history featuring high-solids bentonite grout, *National Drillers Buyers Guide*, February, 1997.

Parker, L.V., Suggested Guidelines for the Use of PTFE, PVC, and Stainless Steel in Samplers and Well Casings, *Current Practices in Ground-Water and Vadose Zone Investigations*, ASTM Special Technical Publication #1118, ASTM, West Conshohocken, PA, 1991, pp. 217–229.

Parker, L.V. and T.F. Jenkins, Suitability of polyvinyl chloride well casings for monitoring munitions in ground water, *Ground-Water Monitoring Review*, 6(3), 92–98, 1986.

Parker, L.V., A.D. Hewitt, and T.F. Jenkins, Influence of casing materials on trace-level chemicals in well water, *Ground-Water Monitoring Review*, 10(2), 146–156, 1990.

Parker, L.V. and T.A. Ranney, Effect of Concentration on Sorption of Dissolved by Well Casings, Special Report 93-8, U.S. Army Corps of Engineers, Cold Regions Research & Engineering Laboratory, Hanover, NH, 1993, 17 pp.

Parker, L.V. and T.A. Ranney, Effect of concentration on sorption of dissolved organics by PVC, PTFE, and stainless steel well casings, *Ground-Water Monitoring & Remediation*, 14(3), 139–149, 1994a.

Parker, L.V. and T.A. Ranney, Softening of Rigid PVC by Aqueous Solutions of Organic Solvents, Special Report 94-27, U.S. Army Corps of Engineers, Cold Regions Research & Engineering Laboratory, Hanover, NH, 1994b, 17 pp.

Parker, L.V. and T.A. Ranney, Additional Studies on the Softening of Rigid PVC by Aqueous Solutions of Organic Solvents, Special Report 95-8, U.S. Army Corps of Engineers, Cold Regions Research & Engineering Laboratory, Hanover, NH, 1995, 15 pp.

Parker, L.V. and T.A. Ranney, Further Studies on the Softening of Rigid PVC by Aqueous Solutions of Organic Solvents, Special Report 96-26, U.S. Army Corps of Engineers, Cold Regions Research & Engineering Laboratory, Hanover, NH, 1996, 22 pp.

Paul, D.G., C.D. Palmer, and D.S. Cherkauer, The effect of construction, installation, and development on the turbidity of water in monitoring wells in fine-grained glacial till, *Ground-Water Monitoring Review*, 7(1), 73–82, 1988.

Pettyjohn, W.A., Ground-Water Pollution — An Imminent Disaster, EPA-600/9-79-029, Proceedings of the Fourth National Ground-Water Quality Symposium, U.S. Environmental Protection Agency, Office of Research and Development, Ada, OK, 1979, pp. 18–24.

Piasecki, J., Pressure grouting the annular space: mixing and pumping equipment, *National Driller*, December, 2002.

Pitkin, S., R.A. Ingleton, and J.A. Cherry, Use of a Drive-Point Sampling Device for Detailed Characterization of a PCE Plume in a Sand Aquifer at a Dry-Cleaning Facility, Proceedings of the Eighth National Outdoor Action Conference and Exposition, National Ground-Water Association, Dublin, OH, 1994, pp. 395–412.

Powers, M.C., A new roundness scale for sedimentary particles, *Journal of Sedimentary Petrology*, 23, 117–119, 1953.

Raber, E., J. Garrison, and V. Oversby, The sorption of selected radionuclides on various metal and polymeric materials, *Radioactive Waste Management and the Nuclear Fuel Cycle*, 4(1), 41–52, 1983.

Ramsey, R.H. and G.E. Maddox, Monitoring Ground-Water Contamination in Spokane County, Washington, Proceedings of the Second National Symposium on Aquifer Restoration and Ground Water Monitoring, National Water Well Association, Worthington, OH, 1982, pp. 198–204.

Ranney, T.A. and L.V. Parker, Sorption of Trace-Level Organics by ABS, FEP, FRE, and FRP Well Casings, Special Report 94-15, U.S. Army Corps of Engineers, Cold Regions Research & Engineering Laboratory, Hanover, NH, 1994, 31 pp.

Ranney, T.A. and L.V. Parker, Susceptibility of ABS, FEP, FRE, FRP, PTFE, and PVC Well Casing to Degradation by Chemicals, Special Report 95-1, U.S. Army Corps of Engineers, Cold Regions Research & Engineering Laboratory, Hanover, NH, 1995, 11 pp.

Ranney, T.A. and L.V. Parker, Sorption and Leaching of Trace-Level Metals by Polymeric Well Casings, Special Report 96-8, U.S. Army Corps of Engineers, Cold Regions Research & Engineering Laboratory, Hanover, NH, 1996, 15 pp.

Ranney, T.A. and L.V. Parker, Comparison of fiberglass and other polymeric well casings, Part I: Susceptibility to degradation by chemicals, *Ground-Water Monitoring & Remediation*, 17(1), 97–103, 1997.

Ranney, T.A. and L.V. Parker, Comparison of fiberglass and other polymeric well casings, Part II: Sorption and leaching of trace-level organics, *Ground-Water Monitoring & Remediation*, 18(2), 107–112, 1998a.

Ranney, T.A. and L.V. Parker, Comparison of fiberglass and other polymeric well casings, Part III: Sorption and leaching of trace-level metals, *Ground-Water Monitoring & Remediation*, 18(3), 127–133, 1998b.

Reynolds, G.W. and R.W. Gillham, Adsorption of Halogenated Organic Compounds by Polymer Materials Commonly Used in Ground Water Monitoring, Proceedings of the Second Canadian/American Conference on Hydrogeology, National Water Well Association, Dublin, OH, 1985, pp. 125–132.

Richter, H.R. and M.G. Collentine, Will My Monitoring Wells Survive Down There? Installation Techniques for Hazardous Waste Studies, Proceedings of the Third National Symposium on Aquifer Restoration and Ground Water Monitoring National Water Well Association, Worthington, OH, 1983, pp. 223–229.

Rinaldo-Lee, M.B., Small vs. large diameter monitoring wells, *Ground-Water Monitoring Review*, 3(1), 72–75, 1983.

Robbins, G.A., Influence of using purged and partially penetrating monitoring wells on contaminant detection, mapping, and modeling, *Ground Water*, 27(2), 155–162, 1989.

Robbins, G.A. and J.M. Martin-Hayden, Mass balance evaluation of monitoring well purging, Part I: Theoretical models and implications for representative sampling, *Journal of Contaminant Hydrology*, 8, 203–224, 1991.

Roscoe Moss Company, *Handbook of Ground-Water Development*, John Wiley and Sons, New York, NY, 1990, 493 pp.

Royce, K.L., Selection of Well Construction Materials, *Water Well Journal*, August, 1991, 30–32.

Scalf, M.A., J.F. McNabb, W.J. Dunlap, R.L. Crosby, and J. Fryberger, *Manual of Ground Water Quality Sampling Procedures*, National Water Well Association, Dublin, OH, 1981.

Schalla, R., A Comparison of the Effects of Rotary Wash and Air Rotary Drilling Techniques on Pumping Test Results, Proceedings of the Sixth National Symposium and Exposition on Aquifer Restoration and Ground Water Monitoring, National Water Well Association, Dublin, OH, 1986, pp. 7–26.

Schalla, R. and P.L. Oberlander, Variation in the diameter of monitoring wells, *Water Well Journal*, 37(5), 56–57, 1983.

Schalla, R., R.W. Wallace, R.L. Aaberg, S.P. Airthart, D.J. Bates, J.V.M. Carlile, C.S. Cline, D.I. Dennison, M.D. Freshley, P.R. Heller, E.R. Jensen, K.B. Olsen, R.G. Parkhurst, J.T. Reiger, and E.J. Westergard. Interim Characterization Report for the 300 Area Process Trenches, PNL-6716, Pacific Northwest Laboratory, Richland, WA, 1988, p. 677.

Schalla, R. and W.H. Walters, Rationale for the Design of Monitoring Well Screens and Filter Packs, *Ground-Water and Vadose Zone Monitoring*, ASTM Special Technical Publication #1053, 1990, pp. 64–75.

Schmidt, G.W., The use of PVC casing and screen in the presence of gasoline on the ground water table, *Ground-Water Monitoring Review*, 7(2), 94, 1987.

Schmidt, K.D., The case for large-diameter monitor wells, *Water Well Journal*, 36(12), 28–29, 1982.

Schmidt, K.D., How Representative Are Water Samples Collected From Wells? Proceedings of the Second National Symposium on Aquifer Restoration and Ground Water Monitoring, National Water Well Association, Dublin, OH, 1983, pp. 177–128.

Simons, D.B. and F. Senturk, *Sediment Transport Technology*, Water Resources Publications, U.S. Geological Survey, Fort Collins, CO, 1977, 807 p.

Smith, D.K., *Cementing*, Society of Petroleum Engineers Monograph Series, Vol. 4, Society of Petroleum Engineers of AIME, Dallas, TX, 1987, 184 pp.

Smith, R.L., R.W. Harvey, and D.R. LeBlanc, Importance of closely spaced vertical sampling in delineating chemical and microbial gradients in ground-water studies, *Journal of Contaminant Hydrology*, 7, 285–300, 1991.

Smith, R.M., Resource Conservation and Recovery Act Ground-Water Monitoring Projects for Hanford Facilities: Progress Report for the Period April 1 to June 31, 1988, PNL-6675/UC-11, 41, Pacific Northwest Laboratory, Richland, WA, 1988.

Smith, R.M., D.J. Bates, and R.E. Lundgren, Resource Conservation and Recovery Act Ground-Water Monitoring Projects for Hanford Facilities: Progress Report for the Period January 1 to March 31, 1989, PNL-6957/UC-11, 41, Pacific Northwest Laboratory, Richland, WA, 1989.

Sykes, A.L., R.A. McAllister, and J.B. Homolya, Sorption of organics by monitoring well construction materials, *Ground-Water Monitoring Review*, 6(4), 44–47, 1986.

Sosebee, J.B., P.C. Geiszler, D.L. Winegardner, and C. Fisher, Contamination of Ground Water Samples With Poly (Vinyl Chloride) Adhesive and Poly (Vinyl Chloride) Primer From Monitor Wells, Proceedings of the ASTM Second Symposium on Hazardous and Industrial Solid Waste Testing, ASTM Special Technical Publication #805, 1983, pp. 38–49.

Thompson, I., C. Bodimeade, P. Kapteyn, and J. Pianosi, Placement of Bentonite Seals in Boreholes With Multiple Piezometers Using an Injector Device, Proceedings of the Eighth National Outdoor Action Conference and Exposition, National Ground-Water Association, Dublin, OH, 1994, pp. 663–671.

Tomson, M.B., S.R. Hutchins, J.M. King, and C.H. Ward, Trace Organic Contamination of Ground Water: Methods for Study and Preliminary Results, Third World Congress on Water Resources, Mexico City, Mexico, 8, pp. 3701–3709, 1979.

Troxell, G.E., H.E. Davis, and J.W. Kelly, *Composition and Properties of Concrete*, McGraw-Hill Book Co., New York, NY, 1968.

U.S. EPA, Manual of Water Well Construction Practices, EPA-570/9-750-001, U.S. Environmental Protection Agency, Washington, DC, 1975, 156 pp.

Verbeck, G.J. and C.W. Foster, Long-Time Study of Cement Performance in Concrete with Special Reference to Heats of Hydration, Proceedings of the American Society for Testing and Materials, Vol. 50, 1950.

Villaume, James F., Investigations at sites contaminated with dense non-aqueous phase liquids (DNAPLS), *Ground-Water Monitoring Review*, 5(2), 60–74, 1985.

Voytek, J.E., Jr., Considerations in the design and installation of monitoring wells, *Ground-Water Monitoring Review*, 3(1), 70–71, 1983.

Walker, W.H., Tube wells, open wells and optimum ground water resource development, *Ground Water*, 12(1), 10–15, 1974.

Wang, T. and J.L. Bricker, 2-Butanone and tetrahydrofuran contamination in the water supply, *Bulletin of Environmental Contamination and Toxicology*, 23, 620–623, 1979.

Wiedemeier, T.H., J.T. Wilson, D.H. Kampbell, R.N. Miller, and J.E. Hansen, Technical Protocol for Implementing Intrinsic Remediation With Long-Term Monitoring for Natural Attenuation of Fuel Contamination Dissolved in Ground Water, U.S. Air Force Center for Environmental Excellence, Technology Transfer Division, Brooks Air Force Base, San Antonio, TX, 1995.

Wiedemeier, T.H. and P.E. Haas, Designing monitoring programs to effectively evaluate the performance of natural attenuation, *Ground-Water Monitoring & Remediation*, 21(3), 124–135, 2002.

Williams, C. and L.G. Evans, Guide to the Selection of Cement, Bentonite and Other Additives for Use in Monitor Well Construction, Proceedings of the First National Outdoor Action Conference, National Water Well Association, Dublin, OH, 1987, pp. 325–343.

Williams, E.B., Fundamental concepts of well design, *Ground Water*, 19(5), 527–542, 1981.

Yu, J.K., Should we use a well foot (sediment trap) in monitoring wells? *Ground-Water Monitoring Review*, 9, 59–60, 1989.

11

Multilevel Ground-Water Monitoring

Murray Einarson

CONTENTS

Introduction	808
Why Three-Dimensional Plume Delineation is Necessary	812
Measurement of Vertical Hydraulic Heads	814
One Time Sampling versus Permanent Multilevel Monitoring Devices	814
Where You Monitor is as Important as How You Monitor	816
Options for Multilevel Ground-Water Monitoring	819
Multilevel Sampling within Single-Interval Monitoring Wells	819
Multiple Diffusion Samplers Installed inside Single-Interval Monitoring Wells	819
Diffusion Multilevel System	820
Passive Diffusion Bag Samplers	821
Active Collection of Samples from Multiple Depths within a Single-Interval Well Using Grab Samplers or Depth-Discrete Pumping	821
Grab or Thief Samplers	822
Collecting Depth-Discrete Samples by Pumping from Different Depths in Well Screens	822
Nested Wells (Multiple Tubes or Casings in a Single Borehole)	823
Bundle Wells Installed in Collapsing Sand Formations	824
Nested Wells Installed with Seals between Monitored Zones	824
Well Clusters (One Well per Borehole)	829
Dedicated Multilevel Ground-Water Monitoring Systems	830
Drilling and Installation Considerations	833
Installations in Open Boreholes	833
Installations in Unconsolidated Sedimentary Deposits	833
Minimizing Cross-Contamination	836
Development of Multilevel Wells	836
Westbay MP System	837
Solinst Waterloo System	839
Solinst CMT System	841
Water FLUTe System	843
References	845

Introduction

One of the most important discoveries made during the last four decades of ground-water research is that the distribution of dissolved contaminants in the subsurface is spatially complex, especially in the vertical dimension. This is due to a number of factors, including the labyrinthine distribution of residual contamination in most non-aqueous-phase liquid (NAPL) source zones, geologic heterogeneity, and mixing mechanisms (e.g., mechanical mixing and molecular diffusion), that are relatively weak in most ground-water flow systems (National Research Council, 1994). This discovery was made possible by the use of multilevel sampling devices that facilitated the collection of discrete ground-water samples from up to 20 different depths in a single borehole (Cherry et al., 1981; MacFarlane et al., 1983; Reinhard et al., 1984; Smith et al., 1987; Robertson et al., 1991; van der Kamp et al., 1994).

Assessment and monitoring of ground-water contamination at nonresearch sites in North America began in earnest in the late 1980s following passage of the Resource Conservation and Recovery Act (RCRA) and the Comprehensive Environmental Response, Compensation, and Liability Act (CERCLA), commonly known as "Superfund." At nonresearch sites, however, environmental consultants — following early guidance from U.S. EPA and some State regulatory agencies — installed single-interval monitoring wells with screen lengths ranging from 10 to 30 ft to collect ground-water samples. Since then, the use of such wells (referred to in this chapter as "conventional" monitoring wells) to collect ground-water samples for chemical analysis has become standard practice in North America. Analysis of samples from single-interval, conventional monitoring wells, however, has led to a common misconception by ground-water practitioners that contaminant plumes are vertically homogeneous because, lacking data to the contrary, most assume that the concentrations of solutes measured in the samples are representative of concentrations within the entire portion of the aquifers screened by the wells.

In the late 1980s, ground-water researchers began to study the biases and apparent plume distortion caused by conventional, single-interval monitoring wells (see Sidebar). The studies show that conventional monitoring wells yield composite samples that mask the true vertical distribution of dissolved contaminants in the aquifer. Further, the composite samples are strongly biased by the position and length of the well screens, the pumping rate during sampling, and ambient vertical flow in the well (see Sidebar). Continued industry reliance on conventional monitoring wells for site assessment and monitoring has prolonged the misconception that the distribution of dissolved contaminants in the subsurface is more homogeneous than it really is. This can have serious consequences for health risk assessments and the performance of *in situ* remediation systems, as discussed later in this chapter.

The bias caused by compositing in monitoring wells is shown conceptually in Figure 11.1. In Figure 11.1a, several monitoring wells are shown. The well labeled "L" is a single-interval well with a relatively long screen. Wells labeled "M" make up a cluster of three wells completed at different depths in the aquifer. Well "N" is a multilevel monitoring well that yields ground-water samples from seven discrete depths. In Figure 11.1b, the concentrations of a hypothetical dissolved contaminant in the aquifer are depicted in a heavy dashed line. Well "L" (the well with a relatively long screen) yields a sample that is a mixture of water containing high concentrations of the contaminant (entering the well from the upper part of the well screen) and water that has lower concentrations of the solute (entering the well from deeper portions of the aquifer). The sample from well

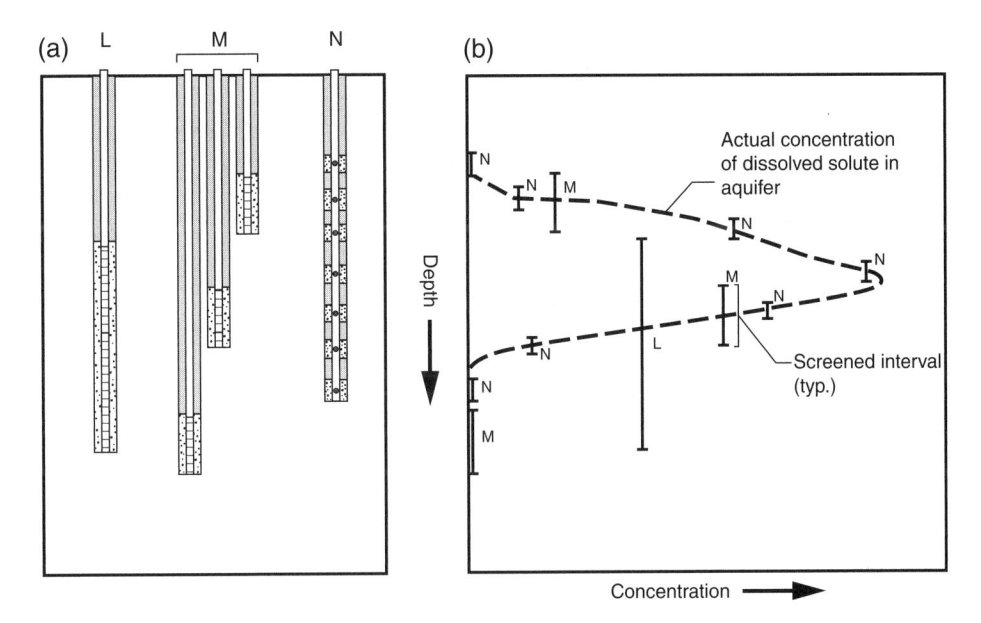

FIGURE 11.1
Effect of well screen length on sample concentrations. (a) Three types of monitoring well completions – single-zone, long-screen well (well "L"); cluster of three wells completed to different depths (wells "M"); and multilevel well (well "N"). (b) Heavy dashed line shows actual concentration of a dissolved solute in the aquifer. Single-zone, long-screen well (well "L") yields a sample that is a mixture of high concentrations of the solute entering the upper portion of the well screens and low concentrations entering the lower portion of the well. Multilevel monitoring well (well "N") yields samples that most closely represent the true distribution of the dissolved solute in the aquifer. See text for further discussion. (From John Cherry. With permission.)

"L" is therefore a composite that: (1) understates the peak concentrations in the portion of the aquifer screened by the upper part of the well and (2) overstates the presumed depth of dissolved-phase contamination in the aquifer. The cluster of three wells with shorter well screens (well cluster "M") yields samples that more closely reflect the actual distribution of the dissolved-phase contaminants in the aquifer than the sample from the single long-screened well. The multilevel well (well "N") provides samples that most closely resemble the actual distribution of the dissolved-phase contaminants in the aquifer.

A real-life example of the bias caused by sample compositing can be seen in data collected from a multilevel monitoring well that was installed in Santa Monica, CA to monitor a dissolved plume of methyl tert butyl ether (MTBE). The multilevel well was located within 20 ft of a pair of 4-in. diameter conventional monitoring wells (Wells MW-14 and MW-16) in order to compare the concentrations of MTBE in water samples collected from the multilevel well with samples collected from the conventional wells (Einarson and Cherry, 2002). A summary of the stratigraphy and construction of the CMT well and the nearby conventional monitoring wells is shown in Figure 11.3. A graph of MTBE concentrations versus depth for all three wells is shown on the right of the figure. Comparison of the MTBE concentrations measured in samples from the multi-level well with data from the conventional wells provides an example of contaminant mixing in monitoring wells described earlier. It is clear from the figure that the conventional wells yield ground-water samples that are a composite of ground water within the vertical interval of the aquifer screened by the wells. Analysis of a sample from Zone 3 of the multilevel well shows that MTBE is present in the aquifer at concentrations as high as 5300 μg/l. However, the concentration of MTBE measured in samples from the

Sample Biases and Cross-Contamination Associated with Conventional Single-Interval Monitoring Wells

Several field, laboratory, and modeling studies have been performed in the last 15 yr to evaluate whether ground-water samples collected from conventional, single-interval monitoring wells (i.e., wells having a single-screened interval ranging from 10 to 30 ft long) accurately reflect the concentration of dissolved contaminants in the portion of the aquifer screened by the wells (Robbins, 1989; Martin-Hayden et al., 1991; Robbins and Martin-Hayden, 1991; Gibs et al., 1993; Akindunni et al., 1995; Chiang et al., 1995; Conant Jr. et al., 1995; Church and Granato, 1996; Reilly and Gibs, 1996; Martin-Hayden and Robbins, 1997; Reilly and LeBlanc, 1998; Hutchins and Acree, 2000; Martin-Hayden, 2000a; 2000b; Elci et al., 2001). From these studies, it is clear that water samples collected from conventional monitoring wells are actually blended or composite samples. If the dissolved contaminants are stratified within the aquifer, which, based on detailed vertical ground-water sampling at several field research sites, appears to be the rule rather than the exception, compositing in long-screened wells during sampling results in underestimation of the maximum concentrations present in the aquifer. Robbins (1989) calculated that the negative bias caused by in-well blending could be up to an order of magnitude. Gibs et al. (1993) performed a field study and concluded that the contaminant concentration in a vertically averaged sample would be 28% of the maximum concentration in the aquifer. Moreover, if the wells partially penetrate the aquifer, an additional bias is introduced due to ground water (either clean or contaminated) flowing into the well from above and below the well screens (Akindunni et al., 1995; Conant Jr. et al., 1995; Chiang et al., 1995). Further, modeling performed by Martin-Hayden and Robbins (1997) showed that vertical concentration averaging in monitoring wells can result in significant overprediction of contaminant retardation factors and apparent decay constants.

Other researchers have focused on the biases caused by ambient vertical flow of ground water in wells when they are not being pumped (McIlvride and Rector, 1988; Reilly et al., 1989; Church and Granato, 1996; Hutchins and Acree, 2000; Elci et al., 2001; Elci et al., 2003). In areas with vertical hydraulic gradients, installation of a monitoring well may set up a local vertical flow system because of the natural vertical hydraulic gradient at the well location. The well then acts as a "short circuit" along this gradient, with the resulting flow in the wellbore often of sufficient magnitude to compromise the integrity of any samples collected from the well (Elci et al., 2001). Reilly et al. (1989) concluded that ambient vertical flow renders long-screen wells "almost useless." They also noted that borehole flow and transport of contaminants in long-screen wells may contaminate parts of the aquifer that would not otherwise become contaminated in the absence of a long-screen well. Church and Granato (1996) concluded that "long-screen wells will fail even in a relatively ideal setting, and therefore, cannot be relied upon for accurate measurements of water-table levels, collection of water-quality samples, or fluid-conductance logging." Hutchins and Acree (2000) found that ambient vertical flow of less contaminated ground water into a monitoring well with only 10 ft of well screen caused a significant negative bias that could not be negated by purging the well prior to sampling. Elci et al. (2001) used a numerical model to simulate ambient vertical flow in a fully screened well at the Savannah River Site near Aiken, SC (see Figure 11.2). The site has an upward hydraulic gradient, so flow within the well was upward. Tracer transport simulations showed how a contaminant located initially in a lower portion of the aquifer ("A" in Figure 11.2) was transported into the upper portion and diluted throughout the entire well by inflowing water. Even after full purging, samples from such a well will yield misleading and ambiguous data concerning solute concentrations, location of a contaminant source, and plume geometry (Elci et al., 2001). Not only are the samples from the well biased, but also, as shown in the figure, the well itself has created a vertical conduit that has cross-contaminated the aquifer. There are also other significant implications of the ambient flow condition depicted in the Figure 11.2. Imagine that clean water and not a tracer or contaminant plume entered the well at location "A" in the figure. Clean water would therefore be flowing up the wellbore and would be discharging in the upper portion of the aquifer. What if in this scenario the source of contamination was higher up in the aquifer near the location of "B" in the figure (e.g., a plume emanating from a fuel release site)? The plume emanating from source "B" would actually flow *around* the dome of clean water being discharged from the monitoring well and would completely escape detection. Samples collected from the well, even samples carefully collected with depth-discrete bailers or diffusion bag samplers, would be sampling clean water entering the well from the bottom of the well screen. Elci et al. (2001) point out that ambient ground-water flow in monitoring wells is not atypical. They report that significant ambient vertical flow occurred in 73% of 142 wells that had been tested using sensitive borehole flowmeters. It is for these reasons that Elci et al. (2001) conclude that the "use of long-screened monitoring wells should be phased out unless an appropriate multilevel sampling device prevents vertical flow."

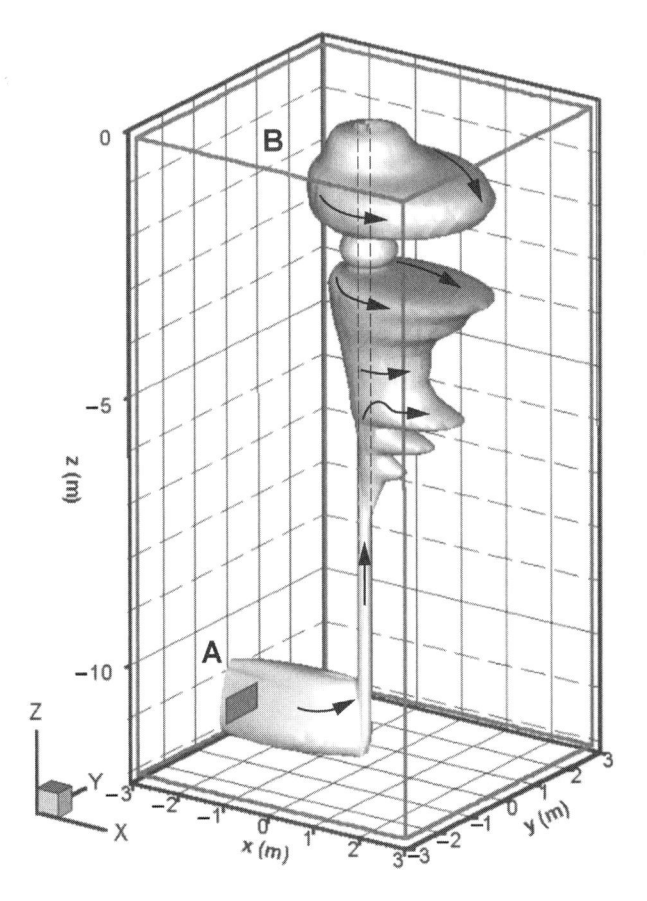

FIGURE 11.2
Simulation of the hydraulic capture of a deep contaminant plume by an unpumped, fully screened monitoring well and transport up and out of the wellbore under ambient flow conditions. See Sidebar for further discussion. (Adapted from Elci et al. [2001]. With permission.)

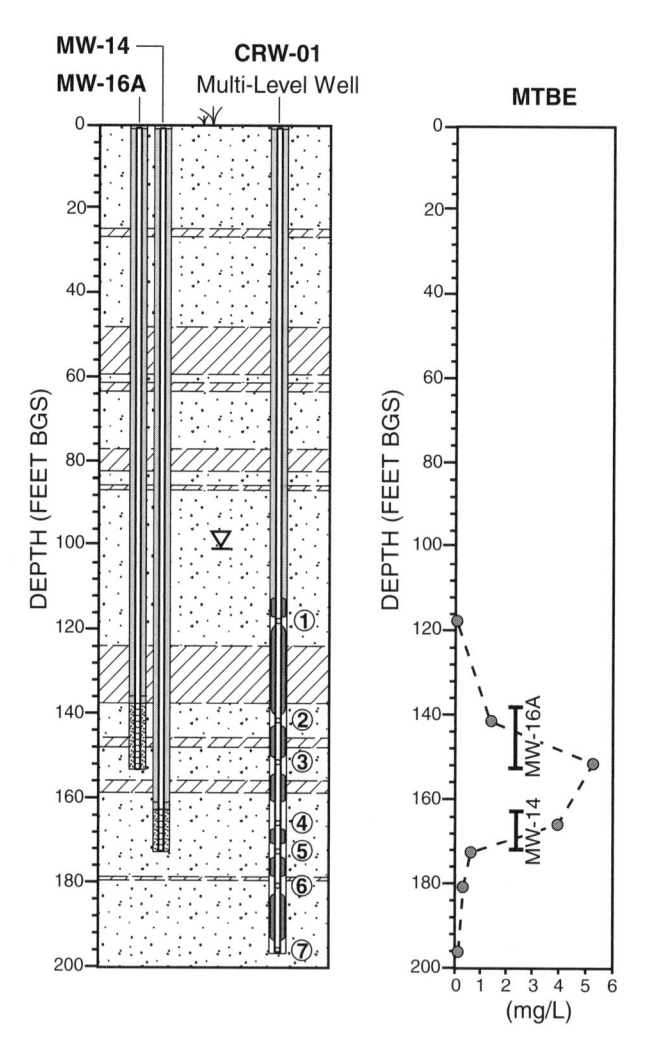

FIGURE 11.3
Construction details and MTBE concentration profile from a multilevel well plotted next to data from two nearby conventional monitoring wells, Santa Monica, CA. (Adapted from Einarson and Cherry [2002]. With permission.)

conventional wells is much lower (approximately 2300 μg/l) because relatively clean water (entering the upper portion of MW-16's well screen and the lower portion of MW-14's well screen) mixes with the water containing high concentrations of MTBE when these wells are pumped.

Why Three-Dimensional Plume Delineation is Necessary

Defining the true distribution of dissolved contaminants is arguably the most important part of an environmental site assessment. The risk to downgradient receptors is commonly estimated by calculating the future concentration at the receptor's location. The calculations are typically performed by estimating (using analytical or numerical equations) the attenuation of the contaminant from some starting concentration near the release site. If the starting concentration is underestimated (e.g., by using results obtained from composite samples from long-screened monitoring wells), the risk to the downgradient receptor (typically a water-supply well) may be underestimated. Similar arguments can be made for predictions of the risks associated with exposures to vapors emanating from residual contamination near source areas or flowing in shallow contaminant plumes. Vapor migration is dominated by molecular diffusion. Because diffusion is driven by concentration gradients, underestimating the peak contaminant concentrations in the subsurface will result in an underestimation of the risk posed to the vapor receptors. However, in other cases, data from long-screened wells can *overestimate* the risk to vapor receptors. For example, ground-water recharge at a site may create a layer of clean water atop a deeper dissolved contaminant plume. The layer of clean water may constitute an effective diffusion barrier that impedes the upward migration of volatile contaminants from the dissolved plume (Rivett, 1995). The layer of clean ground water overlying the contaminant plume could only be identified if multilevel ground-water monitoring wells or direct-push (DP) samplers were used. The same layer of clean ground water would be completely missed by collecting a composite ground-water sample from a single-zone well screened over the same depth interval.

Finally, effective remediation systems can be designed only if the concentration and distribution of the contaminants are accurately defined. This is especially true for passive *in situ* remediation technologies, such as permeable reactive barriers (PRBs). PRBs treat contaminants *in situ* by trapping or degrading the contaminants as they flow through them under natural gradient conditions. Complete removal or treatment of the contaminants requires sufficient residence time within the PRB. In all PRBs, the requisite residence time is a function of the concentration of the dissolved contaminants flowing through the PRBs. If the peak concentrations of the contaminant in the aquifer are not defined (e.g., because of sample blending in conventional wells), the PRB may be under-designed, leading to insufficient residence time and contaminant breakthrough.

It should also be noted that there are likely many instances where PRBs (or wells used for pump-and-treat remediation) have been installed deeper than they need to be. When conventional single-interval monitoring wells are used to define the maximum depth of contamination at a site, it is usually assumed that the contamination extends to the portion of the aquifer corresponding to the bottom of the well screens. Depth-discrete multilevel monitoring may show, however, that the contamination is limited to much shallower depths. Thus, the PRB may not need to extend to as great a depth as otherwise thought. Because the installation costs of PRBs rise considerably with depth, significant cost savings can be had by accurately defining the vertical extent of contamination using multilevel monitoring wells or depth-discrete DP ground-water samplers.

Site assessment technologies and practices have been changing rapidly in the last decade. As the biases associated with long-screened monitoring wells have become recognized, many practitioners have been installing monitoring wells with shorter well screens. It is not uncommon now to see monitoring wells being installed with screen intervals as short as 2 or 3 ft. While this is a favorable development as it reduces the sampling biases associated with long screens, it also increases the likelihood that high-concentration zones may be missed if only one monitoring well is installed at a particular location. In fact, depending on the depth of the monitoring wells, the contamination can sometimes be missed altogether (e.g., if the well screens are positioned too high and yield samples of clean water above a diving plume). Consequently, one short-screened monitoring well per location is not sufficient to define the vertical extent of dissolved contamination. Depth-discrete sampling devices should be installed at several depths at each location to accurately map the vertical extent of dissolved contamination. Sampling devices should also be installed to depths where they extend beneath dissolved plumes, that is, where the deepest samples no longer detect contamination, or detect it at concentrations that are below a particular threshold value.

Measurement of Vertical Hydraulic Heads

The foregoing discussion focused on the importance of accurately mapping contaminant concentrations in three dimensions. Depth-discrete measurement of hydraulic pressures (heads) is also a necessary part of environmental site assessments. Mapping the hydraulic head distribution in three dimensions allows site investigators to make accurate predictions about the movement and future location of dissolved contaminants. Vertical hydraulic gradients are present at most sites, and the magnitudes of vertical gradients often exceed horizontal hydraulic gradients. Upward hydraulic gradients occur in ground-water discharge areas; conversely, downward hydraulic gradients exist where ground-water recharge occurs, and can be exacerbated by pumping of nearby remediation and water-supply wells. Defining the vertical hydraulic head distribution at a contaminated site is an essential part of developing the site conceptual model, and is most often depicted using flow nets or three-dimensional ground-water flow models.

Hydraulic heads are determined by measuring the depth-to-water in a piezometer or short-screened well and subtracting the distance from a known datum (in North America, typically the top-of-casing elevation referenced to feet above mean sea level). Hydraulic pressures can also be monitored continuously using electronic pressure transducers. Pressure transducers as small as 0.39 in. in outside diameter now exist (e.g., Druck Model PDCR 35/D) for use in small-diameter wells and piezometers. If the focus of a particular study is solely on measuring hydraulic heads and not collecting ground-water samples, the pressure transducers can be buried directly to provide single- or multiple-depth hydraulic head data.

Definition of vertical hydraulic gradients is also necessary to judge whether or not ambient vertical flow of ground water is likely occurring in conventional single-interval monitoring wells at a particular site. As discussed in the Sidebar, ambient vertical flow of ground water may occur in monitoring wells and other long-screened wells (e.g., remediation wells or water-supply wells) whenever (1) vertical hydraulic gradients exist in the aquifer and (2) the wells are not being pumped. Ambient vertical ground-water flow in wells can redistribute dissolved solutes in the subsurface, which can result in cross-contamination of the aquifer and chemically biased samples being collected from the wells. If no vertical hydraulic gradients exist in the portion of the aquifer screened in a particular well, however, ground-water flow can be assumed to be horizontal through the well and vertical flow and redistribution of contaminants may not be a problem. If there is reason to believe that ground water flows horizontally through the well, the well can sometimes be sampled in a way that sheds light on the natural vertical distribution of dissolved contaminants in the portion of the aquifer screened by the monitoring well. A discussion of techniques that can be used to collect depth-discrete samples from single-interval monitoring wells is presented later in this chapter.

One Time Sampling versus Permanent Multilevel Monitoring Devices

There has been a growing trend in the last decade to collect one-time ground-water samples at sites underlain by unconsolidated sedimentary deposits using single-interval direct-push (DP) samplers such as the Hydropunch™, BAT sampler, and other DP ground-water sampling tools generically referred to as "sealed-screen samplers" (U.S. EPA, 1997). These tools allow site investigators to collect ground-water samples from discrete depths without having to install permanent monitoring wells. Most of the tools are, however, designed to collect samples from single depths. If samples are desired from multiple zones, the tools usually must be retrieved, emptied of their contents, cleaned, and re-advanced to the next sampling depth. Thus, obtaining a vertical profile of contaminant concentrations from many depths can be a time-consuming process with most DP ground-water sampling tools. Another tool, the Waterloo Ground-Water Profiler, allows for the collection of discrete ground-water samples from multiple depths without having to retrieve and re-deploy the sampling tool between different depths (Pitkin et al., 1999). A similar tool, the Cone-Sipper™ is typically used with cone penetrometer testing rigs. Another comparable tool, the Geoprobe Ground-Water Profiler, is also available. All these DP ground-water sampling tools are described in detail in Chapter 6.

One-time DP ground-water sampling tools have some advantages over permanent multilevel-monitoring wells. First, it is generally faster to collect depth-discrete ground-water samples using DP sampling tools than to install, develop, and sample permanent multilevel ground-water monitoring wells. Secondly, many site owners dislike having permanent or semipermanent monitoring devices installed on their properties. The wells must be protected during site demolition and reconstruction activities, tracked through all property transfers, and then decommissioned when they are no longer needed. Also, many responsible parties (RPs) fear that if they have permanent monitoring wells on their property, the regulatory agency overseeing the work will require them to monitor the wells for an indeterminate and possibly protracted period of time.

DP ground-water sampling tools, however, often do not tell the whole story. For example, they do not provide information about the vertical hydraulic head distribution at a particular site. Also, one of their main advantages — the fact that they are used to collect one-time samples — is a drawback at many sites. Monitoring a plume over time with DP sampling equipment requires remobilization of the DP contractor and re-advancement of the DP sampling tools each time another round of samples is desired. This becomes costly if long-term ground-water monitoring is needed. Also, the samples are collected with driven probes and the resulting probe holes are usually grouted after the last sample has been collected. It is therefore not possible to obtain samples from exactly the same points in the aquifer at a later date. Consequently, exclusive use of DP ground-water sampling tools is generally not cost-effective at sites where ongoing ground-water monitoring is needed.

So, when and where should permanent multilevel ground-water monitoring systems be installed? First, they should be installed whenever and wherever it is necessary to determine the vertical hydraulic head distribution. Because measuring vertical hydraulic heads is fundamental in the development of a site conceptual model, installation of multilevel monitoring wells or piezometers that allow for measurement of hydraulic heads at multiple depths is needed at virtually every contaminated site. Measuring temporal changes in hydraulic heads at a site is particularly important in understanding the ground-water flow system, mixing mechanisms, and contaminant distribution. Secondly, any time that ongoing, long-term multilevel water quality monitoring is needed, permanent multilevel ground-water monitoring devices should be installed. Considering that ongoing ground-water monitoring (of hydraulic heads and chemistry) is needed and required at most contaminated sites, permanent multilevel monitoring devices should play an important role at most sites. For example, long-term ground-water monitoring is often necessary to verify the effectiveness of active remediation. At other sites, time-series samples may need to be collected to document suspected seasonal fluctuations in the concentration or flux of contaminants emanating from a residual NAPL source zone. And, of course, long-term multilevel monitoring is necessary at sites where monitored natural attenuation is the selected remediation method (see Chapter 9). Permanent multilevel monitoring wells should therefore be utilized at most contaminated sites.

Careful planning should be undertaken to select the optimal locations and depths for the multilevel devices. In unconsolidated sedimentary deposits, it is usually good practice to first define the general location and depth of the dissolved contaminant plume using DP ground-water sampling tools. Then, multilevel monitoring devices can be installed at the locations and depths that provide the maximum information.

This chapter focuses on permanent multilevel monitoring devices, and Chapter 6 presents a discussion of DP methods for collecting one-time samples. Both are important technologies used to characterize contaminated sites in three dimensions.

Where You Monitor is as Important as How You Monitor

The locations of ground-water monitoring wells installed at contaminated sites in the United States have historically been selected in order to provide data used to construct plume maps. Conventional plume maps are two-dimensional, plan-view contour maps of contaminant concentrations obtained from laboratory analyses of ground-water samples collected from monitoring wells. Unfortunately, such maps rarely provide an accurate depiction of the true three-dimensional contaminant distribution due to several factors. These include: (1) the complexity of most dissolved plumes of contaminants; (2) the wide spacing of most monitoring well networks relative to the high-strength plume cores that are often thin and narrow; and (3) variations in concentrations in samples from the wells caused by differences in well depths, screened intervals, and pumping rates (see Sidebar for a discussion of biases associated with conventional monitoring wells).

Ground-water researchers have utilized high-resolution ground-water sampling networks to characterize dissolved plumes at both controlled and accidental release sites in unconsolidated aquifers. A particularly useful approach has utilized transects of closely spaced multilevel monitoring wells or DP sampling points oriented perpendicular to the plume axes (Semprini et al., 1995; Borden et al., 1997; Devlin et al., 2001; Einarson and Mackay, 2001; Kao and Wang, 2001; Newell et al., 2003; Guilbeault et al., 2005) (Figure 11.4). The wells or sampling points are often spaced 20 ft (or less) apart horizontally and facilitate the collection of discrete ground-water samples from multiple depths. The optimal vertical spacing of monitoring points in a sampling transect is a function of many factors (e.g., the purpose of the monitoring, the type of contamination, the nature and geometry of the source zone, subsurface geology, distance from the contaminant source, etc.) and is the subject of ongoing research (e.g., see Guilbeault et al., 2005). A minimum of one transect is installed downgradient from the source zone to define the strength and temporal variability of the contaminant source, or to assess the effectiveness of remediation efforts. Multiple sampling transects are used to evaluate the natural attenuation of contaminants (see U.S. EPA, 1998; Chapter 9 of this book). Recent advances in monitoring technologies described in this and other chapters have made these sampling technologies accessible to environmental consultants and cost-effective for use at non-research sites.

Transects of multilevel wells are superior to monitoring networks comprised of spatially distributed conventional monitoring wells for several reasons. First and foremost, the dense grid or "fence" of sampling points makes it far more likely to detect and accurately

Multi-Level Well Transect

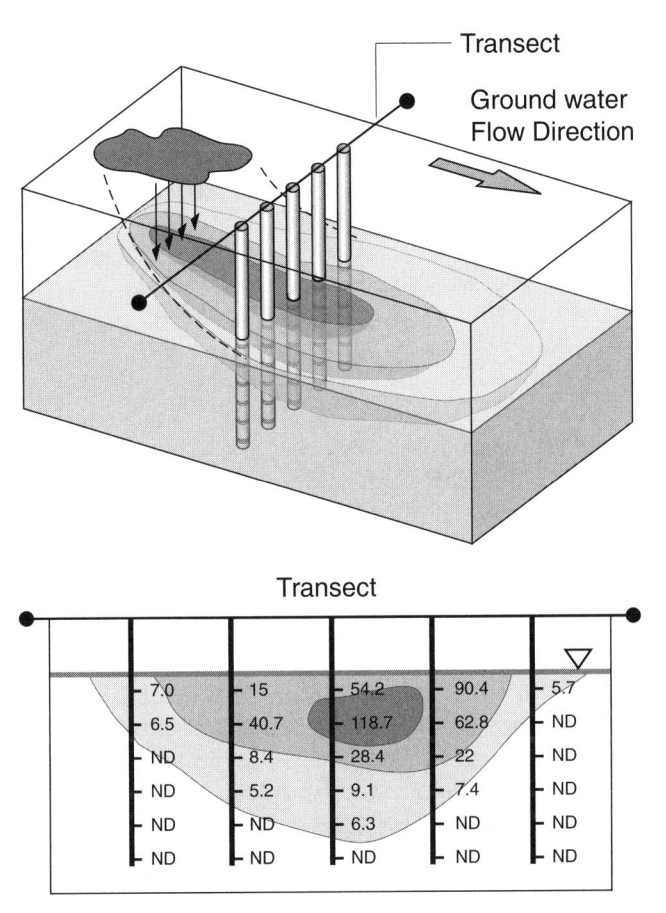

FIGURE 11.4
Transect of multilevel wells. (Illustration courtesy of LFR Levine–Fricke and the American Petroleum Institute.)

delineate dissolved-phase plumes of contaminants (especially high-strength zones or "plume cores") than if sparse networks of conventional monitoring wells were used. This is particularly advantageous when the characterization is being performed to determine the optimal width, depth, and thickness of PRBs (Figure 11.5), or the locations and screen intervals of extraction wells used in conjunction with pump-and-treat remediation. Secondly, detailed plume definition may show that plumes that were thought to be co-mingled are actually separate. This is clearly important for fair cost allocation associated with regional cleanup efforts. Thirdly, transects of closely spaced multilevel wells are much less sensitive to slight shifts in the lateral and vertical position of dissolved plumes than sparse networks of conventional wells. For example, in areas where the hydraulic flow systems change over time (e.g., seasonal changes in flow direction), dissolved plumes may shift laterally and vertically in the aquifer. Take, for instance, a well that is screened in a high-strength part of a narrow dissolved plume (or in a single plume core within a larger plume with multiple cores). Samples collected initially from the well would contain high concentrations of the target contaminant. What if the plume core then shifted slightly away from the well (either laterally or vertically) in

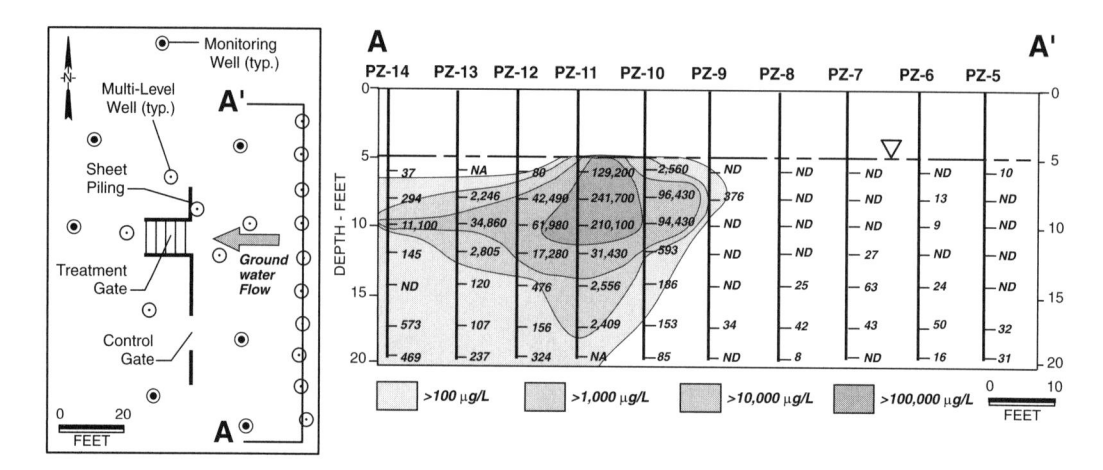

FIGURE 11.5
Contours of total chlorinated VOC concentrations along a sampling transect installed upgradient from a funnel-and-gate PRB, Alameda Naval Air Station, CA. (From Einarson and Cherry [2002]. With permission.)

response to a gradual change in lateral or vertical ground-water flow direction? Samples taken over time from the well would contain progressively lower concentrations of the target contaminant simply because the well is sampling lower concentration parts of the same dissolved plume over time. A plot of sampling results for the well would show declining concentrations over time. This trend could logically (but incorrectly) be attributed to source depletion or natural biodegradation. If, on the other hand, the same plume was monitored with a dense network of multilevel wells arranged in a transect across the plume, lateral and vertical shifts in the plume location could be easily recognized. Shifts in the position of the plume are obvious if the data are contoured in a vertical cross-section drawn across the plume (i.e., along the transect) as shown in Figure 11.5. Finally, sampling transects facilitate the calculation of the rate of contaminant migration, referred to as contaminant mass discharge or total mass flux. Feenstra et al. (1996) defined the plume mass discharge as the amount of contaminant mass migrating through cross-sections of the aquifer orthogonal to ground-water flow per unit of time. Contaminant mass discharge is a powerful site characterization parameter that, at some sites, may allow site investigators to predict the potential impact a plume may have if it were to be captured by a downgradient water supply well (Einarson and Mackay, 2001). Monitoring changes in contaminant mass discharge along the flow path has also been advocated as a way to perform more quantitative evaluations of natural attenuation (U.S. EPA, 1998). Characterizing dissolved plumes on the basis of contaminant mass discharge, therefore, allows site owners and regulators to focus cleanup efforts on the sites that pose the most significant threat to downgradient receptors (Feenstra et al., 1996; U.S. EPA, 1998; Einarson and Mackay, 2001; Newell et al., 2003).

The above discussion notwithstanding, there are times when individual multilevel wells or individual clusters of monitoring wells distributed more broadly over a site are appropriate. For example, individual multilevel wells or well clusters may be areally distributed at a site to provide information regarding the three-dimensional distribution of hydraulic head. Definition of the hydraulic head in three dimensions is needed to understand the ground-water flow system, calibrate numerical models, and estimate the probable location and trajectory of a dissolved plume prior to installing detailed sampling transects.

Options for Multilevel Ground-Water Monitoring

More options and technologies exist now than ever before for measuring hydraulic heads and collecting discrete ground-water samples from multiple depths at contaminated sites. Technologies for multilevel ground-water monitoring include nests of wells installed in single boreholes and clusters of wells completed to different depths. Several specialized multilevel monitoring systems are also commercially available. These technologies are described in the following sections. Also, it may be possible in some cases to obtain information regarding the vertical distribution of dissolved contamination by carefully collecting depth-discrete samples from within conventional single-interval monitoring wells. The next section begins with a discussion of techniques for performing depth-discrete sampling in conventional single-interval monitoring wells and explains when those techniques can and cannot be relied upon to yield data that accurately depict the concentrations and distribution of contaminants in the portion of the aquifer screened by the wells.

Multilevel Sampling within Single-Interval Monitoring Wells

In recent years there has been a growing trend toward measuring vertical contaminant "profiles" within conventional single-interval wells. In some cases, it may be possible to collect multidepth ground-water samples from single-interval monitoring wells that shed light on the vertical distribution of contaminants in an aquifer. However, as discussed, this is not necessarily a simple task and conventional sampling equipment and approaches often do not yield satisfactory results. New technologies such as passive diffusion samplers may yield better results but they can easily be misapplied, resulting in data that can be misinterpreted.

Multiple Diffusion Samplers Installed inside Single-Interval Monitoring Wells

A thorough discussion of passive diffusion samplers is presented in Chapter 15. The information in this section therefore augments the material presented in Chapter 15, specifically as it relates to the placement of multiple diffusion samplers in a single monitoring well in an attempt to gain information regarding the vertical distribution of contaminants in the subsurface. The first step in this effort consists of installing diffusion samplers at multiple depths in the screened interval of a monitoring well. The diffusion samplers are made of either dialysis cells or polyethylene bags (further discussion of each of these types of samplers is presented). The sample bags or dialysis cells contain deionized, organic-free water, which is physically isolated from ground water in the monitoring well by a thin sheet or membrane of polyethylene, or, in the case of the dialysis chamber sampler, a cellulose membrane. In theory, dissolved contaminants flowing through the well under natural flow conditions diffuse through the membrane and into the water inside the polyethylene bags or dialysis cells. The rate of diffusion is controlled by Fick's law, which incorporates both the diffusion coefficient of the contaminant through the membrane material and the concentration gradient. The samplers are left in the well for a period of up to several weeks and then removed. Samples of the water within the sample bags or dialysis cells are collected and analyzed for the contaminants of interest.

As discussed in Chapter 15, several factors affect the performance of diffusion samplers. These include:

- *The target analyte.* For example, hydrophobic organic compounds like halogenated ethenes and ethanes and aromatic hydrocarbons rapidly diffuse through polyethylene. However, hydrophilic compounds like MTBE and most charged inorganic solutes do not.
- *The exposure period.* The samplers must remain in the well until the concentrations of the target compounds in the polyethylene bags or dialysis cells have equilibrated with the concentrations in the ground water. Because molecular diffusion is a function of compound-specific diffusion coefficients and concentration gradients, the exposure period required to reach equilibration varies for different target compounds and different sites (because dissolved concentrations in ground water differ between sites and even between the depths of the different sample bags or containers in the same well).
- *Well construction.* It is assumed that ground water flows unobstructed through the well under ambient flow conditions. This may not be the case for wells that are not in good hydraulic connection with the borehole. Poor hydraulic connection may occur due to smearing of clays on the borehole wall during drilling, compaction of displaced soil (in the case of DP well installation), or inadequate well development.

There is an additional factor that must be considered when multiple diffusion samplers are placed inside single-interval monitoring wells in an effort to define the vertical distribution and extent of contamination in an aquifer. The factor is the assumption that ground water is flowing horizontally through the well. If there are vertical hydraulic gradients in the aquifer (even small ones), there may be ambient vertical flow of ground water in the monitoring well (see Sidebar). In that case, the multidepth diffusion samplers will come in contact with ground water flowing both horizontally and vertically within the well and not ground water flowing solely horizontally in the aquifer at the depth where the samplers are placed. Samples collected from the passive samplers may therefore accurately reflect the concentrations of the solute of interest *in the well* at the depths of the samplers, but they would not reflect the actual distribution of contaminants *in the aquifer* at these depths. The resulting data may therefore be ambiguous and misleading. To avoid this, the use of multiple diffusive samplers placed in a single well screen to obtain depth-discrete samples should be done only in aquifers where ground water is known to be flowing horizontally. Before diffusion sampling devices are installed in the well, site data should be reviewed to ensure that there are no vertical gradients in the formation. As discussed earlier, this can be done by examining vertical head data from multilevel wells or well clusters. Alternatively, borehole flowmeter surveys can sometimes be performed in the well prior to installing the samplers to directly measure whether or not ambient vertical flow of ground water is occurring in the well.

Diffusion Multilevel System

The diffusion multilevel system (DMLS) was the first diffusion sampler designed to collect multidepth samples from single-interval monitoring wells. Developed by researchers at the Weizmann Institute of Science in Israel in the 1980s, the DMLS utilizes multiple 20 ml dialysis chambers positioned at different depths in the well to collect samples containing dissolved solutes that flow through the monitoring well under ambient conditions (Ronen et al., 1987). Deionized water is placed in the chambers prior to insertion of the

DMLS into the well. Solutes in the ground water flowing through the well diffuse into the dialysis chambers. After a few weeks, the DMLS is removed from the well and samples from the various chambers are collected and analyzed. The DMLS can be used to collect samples containing a variety of inorganic and organic compounds, including chloride, nitrate, sulfate, dissolved oxygen, tetrachloroethylene, and 1,1,1-trichloroethane. Rubber or Viton washers are placed between the various dialysis chambers to reduce or eliminate vertical flow of ground water within the well. More detailed descriptions of the development and testing of the DMLS are presented in Ronen et al. (1987). An evaluation of multidepth ground-water sampling that included the DMLS is presented in Puls and Paul (1997).

The system became commercially available in the U.S. when the patent rights were acquired by Johnson Well Products, Inc. Johnson sold the DMLS worldwide between 1994 and 1998, but discontinued its sale of the DMLS in 1998 when Johnson was acquired by the Weatherford Company. Ownership of the DMLS reverted to the Margan Corporation, an Israeli company with offices in the U.S. Information regarding the availability of the DMLS can be obtained by contacting the Margan Corporation (www.margancorporation.com).

Passive Diffusion Bag Samplers

As discussed in Chapter 15, diffusion bags made of polyethylene have recently become available for passive sampling of dissolved volatile organic compounds (VOCs). An early application of the bags was to delineate the location of a VOC plume discharging to surface water (Vroblesky et al., 1996). Passive diffusion bag (PDB) samplers have subsequently been used to collect ground-water samples from monitoring wells (Vroblesky and Hyde, 1997). One of the claimed advantages of using PDB samplers for collecting ground-water samples from monitoring wells is that there is essentially no disruption of the flow in the well during sample collection, because no pumping occurs. There is, of course, disruption and mixing of water in the well when the PDB samplers are being inserted into the well. But, the mixed water in the well is usually flushed away by natural flow through the well during the week or two that the PDB samplers are left to equilibrate in the monitoring well.

Several PDB samplers can be tied together and suspended in a monitoring well to obtain information regarding the stratification of contaminants in the well (Vroblesky and Hyde, 1997). While this is appealing in concept, the data must be interpreted with the awareness that ambient vertical flow in the well may have created a vertical distribution of the target VOCs in the well that differs significantly from that which exists in the aquifer (see Sidebar). Consequently, the results may be misleading and can result in either underestimating or overestimating the risks to potential receptors and improper remediation system design.

Active Collection of Samples from Multiple Depths within a Single-Interval Well Using Grab Samplers or Depth-Discrete Pumping

The earlier discussion describes passive methods of collecting depth-discrete samples from monitoring wells using PDB samplers. There are also "active" methods for collecting ground-water samples from various depths in a single-interval monitoring well. These include grab or "thief" samplers (e.g., pressurized bailers, the Kabis Water Sampler™, the Hydrasleeve™) and pumping methods. Like PDB samplers, however, these active sampling methods simply yield samples from multiple depths in the well, which may or may not represent the distribution of the target solutes in the aquifer due to possible ambient vertical flow of ground water in the well as discussed earlier.

Grab or Thief Samplers

Grab or "thief" samplers (e.g., the Discrete Interval Sampler™, Kabis Water Sampler, Hydrasleeve, Pneumo-Bailer™, etc.) are nonpumping devices used to collect depth-discrete samples of ground water from a well. The devices are lowered into a well to a target depth and then actuated to collect a ground-water sample from specific depth. In the case of the Discrete Interval Sampler™, the sampler is pressurized at the ground surface, which seats a check valve in the sampler, thereby preventing water from entering it. When the sampler is at the target depth, the pressure is released. This opens the check valve and allows ground water from the target depth to flow into the sampler. The sampler is then re-pressurized, thereby preventing the introduction of ground water from other intervals into the sampler while it is being retrieved. The procedure is repeated to collect samples from other depths in the well. For more information about the samplers, the reader is referred to an evaluation of five discrete interval ground-water sampling devices performed by the U.S. Army Corps of Engineers (Parker and Clark, 2002) and to Chapter 15 of this book. Grab or thief samplers are also used to collect depth-discrete samples from wells (both monitoring wells and water-supply wells) that are being pumped as the samples are being collected. Collecting depth-discrete samples from wells as they are being pumped has been shown to be a useful technique to determine where contaminants are entering the wells (Foote et al., 1998; Jansen, 1998; Gossell et al., 1999; Sukop, 2000).

Using grab or thief samplers to collect depth-discrete samples under non-pumping conditions may sometimes yield ambiguous results. First, ambient vertical flow in the well may have redistributed contaminants in the well prior to sample collection (see Sidebar and earlier discussion). Secondly, the process of lowering the sampler to the target depth(s) may cause considerable mixing in the well. Thus, the sample collected may be a mixture of water from other zones, even if the contaminant distribution in the well closely matched that in the aquifer prior to lowering the sampler into the well. Also, lowering the sampler into the well and removing it may create a plunging action that can significantly increase the turbidity of water in the well. This can cause a significant sampling bias, especially when the target analytes include dissolved metals (Parker and Clark, 2002). If time allows, it is desirable to let sufficient time pass after lowering the sampler to the desired depth, but before collecting the sample, to restore the natural flow condition in the well. From single-well tracer-test theory, the time needed for the mixed water to be purged from the well by natural ground-water flow (assuming flow is horizontal through the well) is approximately 0.5 times the effective diameter of the well, divided by the Darcy velocity (Drost et al., 1968; Freeze and Cherry, 1979).

Collecting Depth-Discrete Samples by Pumping from Different Depths in Well Screens

There have been many instances where site investigators have attempted to gain insight into the vertical distribution of dissolved contaminants in an aquifer by sequentially pumping at low flow rates from different depths in a long well screen. Typically, "profiles" of solute concentrations have been obtained by collecting a series of samples obtained with the sampling pump placed at different depths in the well screen interval. The sampling pumps used for this purpose have included submersible pumps, bladder pumps, or simply small-diameter "drop tubes" attached to a peristaltic pump at the ground surface. Whether or not the samples collected in this manner yield insight into the vertical distribution of solutes in the adjacent aquifer is neither certain nor straightforward to evaluate. The data would, of course, be strongly biased if ambient vertical flow within the well has redistributed contaminants in the well as discussed earlier. However, even for wells where vertical gradients are absent and ground water flows horizontally through the well, pumping at low rates from different depths in the well screens may yield equivocal data depending on when the samples are collected after pumping begins. Studies by Martin-Hayden (2000a, 2000b) show that the water extracted immediately after pumping begins is derived from the region nearest the pump intake. As pumping proceeds, water pumped from the well becomes a mixture of water stored in the well and ground water entering the well screen from the formation. Therefore, the very first volume of water pumped from the well is most representative of the water quality adjacent to the pump intake. This initial volume of water is what should be sampled and analyzed if the goal is to obtain a sample that is most representative of water quality in the aquifer at the depth of the pump intake. As pumping proceeds, the extracted water becomes less and less representative of ground water near the pump because it contains water that has been transported from portions of the well screen further and further away from the pump intake. Given sufficient time and continued pumping, the well will be fully purged and the sample collected will be a flow-weighted composite of the ground water flowing into the entire well screen. Recent simulations of steady-state low-rate flow into a long-screened monitoring well support the hypothesis that under steady-state pumping conditions (i.e., when the well has been fully purged), the depth of the pump intake has no effect on the quality of water extracted during pumping (Varljen et al., 2004).

Nested Wells (Multiple Tubes or Casings in a Single Borehole)

Nested wells are multilevel monitoring wells in which multiple tubes or casings are installed at different depths within the same borehole (Figure 11.6). In order to measure depth-discrete hydraulic heads and collect depth-discrete ground-water samples, each well screen in the nested well should be no more than 2 or 3 ft in length. Types of nested wells include bundles of small-diameter tubing or PVC casing where physical separation between the intakes of the sampling tubes or pipes is provided by sand that collapses around the tubing or pipes as soon as the insertion pipe is withdrawn. In noncollapsing formations, annular seals must be installed inside the borehole to prevent hydraulic

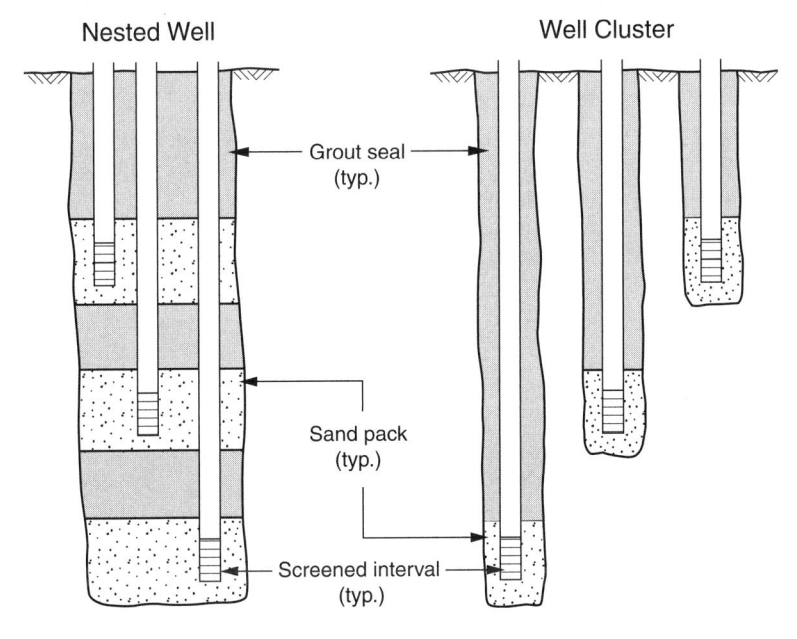

FIGURE 11.6
Nested well and well cluster. (Adapted from Johnson [1983]. With permission.)

connection between the various monitored zones. Installation of the annular seals in nested wells must be done carefully to prevent hydraulic connection between the different monitoring zones. Nested wells with annular seals between monitored zones were the most popular types of multilevel monitoring wells in the 1970s and early 1980s. However, several well-publicized failures of nested wells caused many state and Federal regulatory agencies to ban or discourage their construction. Nested wells are still being installed and, in fact, are experiencing a renaissance due to the growing awareness of the importance of multilevel ground-water monitoring. Important issues related to annular seals in nested wells, including methods for improving the quality of the seals, are discussed.

Bundle Wells Installed in Collapsing Sand Formations

Ground-water researchers studying unconsolidated sedimentary aquifers have used bundles of small-diameter flexible tubing for over 30 yr to collect depth-discrete ground-water samples from as many as 20 different depths in the same borehole (Cherry et al., 1983; Reinhard et al., 1984; Mackay et al., 1986). A typical bundle well design is provided by Cherry et al. (1983) and is depicted in Figure 11.7. Each tube in the bundle has a maximum intake length (i.e., screen length) of approximately 10 cm. A variation of this design, using multiple 0.5-in. PVC pipes, has been used successfully to collect depth-discrete ground-water samples during recent comprehensive studies of a dissolved MTBE plume in Long Island, NY (Haas and Sosik, 1998) (see Figure 11.8).

The bundles of tubing or pipe are typically installed inside a driven insertion tube or pipe that has been advanced to the maximum depth of the well. When the insertion tube is withdrawn, sand collapses around the tubing bundle. Whether or not every void space between every tube or pipe is filled with sand is not certain, but experience gained from many hundreds of such installations in collapsing sand formations at detailed field research sites shows that vertical flow of contaminants along the well bundles is not significant. Nonetheless, bundle wells should only be used when and where the site investigator is confident that the formation will fully collapse around the tubing bundle and where strong vertical hydraulic gradients are absent. Bundle wells are easily installed using DP sampling equipment.

Water samples are usually collected from these types of wells using peristaltic pumps or small-diameter tubing check-valve pumps (e.g., Waterra™ pumps). If the tubing or pipe is large enough, small-diameter water-level meters can be used to measure the depth to water inside the tubes or pipes. If the tubes are too small to measure water levels using electronic water-level meters and the static depth to water is less than 25 ft or so, a sufficient vacuum can be applied simultaneously to all of the tubes to raise the water levels to an elevation above the ground surface. Relative hydraulic heads in the various tubes can be measured using sight tubes. Absolute head values for each zone can be obtained by subtracting the applied vacuum (converted to units of feet or meters of water) from the elevation of the water levels in the sight tubes.

Nested Wells Installed with Seals between Monitored Zones

A conceptual design of a nested well is shown in Figure 11.6. In the diagram, there are bentonite or grout seals between the various screen and sand pack intervals. These seals are installed by pouring bentonite chips (or pumping cement or bentonite grout) into the borehole as the well is being built. Building the well therefore starts with pouring sand into the borehole until the sand rises to a depth above the deepest well screen. Then, the bentonite or grout seal is placed in the borehole annulus up to a depth just below the next deepest well screen. Next, sand is poured into the borehole to cover the screen for that zone. The process of adding alternating layers of sand and bentonite (or

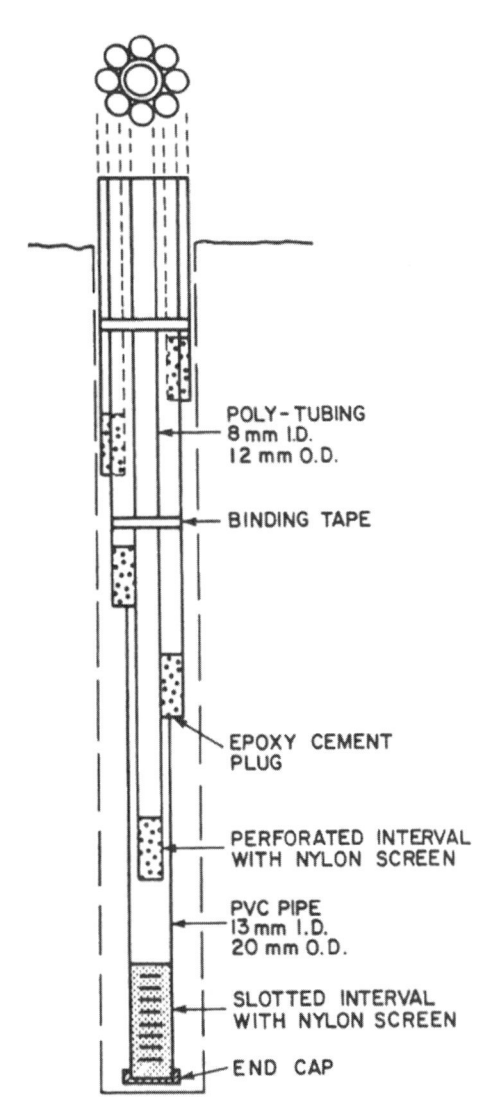

FIGURE 11.7
Bundle well. (From Cherry et al. [1983]. With permission.)

cement grout) continues until the well is fully built. Building a well like this is time consuming, and particular attention must be paid to avoid adding too much sand or bentonite. If too much sand is added, the thickness of the overlying bentonite seal may be inadequate and the seal jeopardized. If too much bentonite (or cement) is added, the screens of the next monitoring zone may be covered and rendered useless. Consequently, when building a nested well, the depth of the sand or bentonite should be measured frequently as the annular materials are being placed to avoid adding too much sand or seal material. One of the most important tools a driller has when building nested wells is a weighted measuring line or "tag line" which allows him to accurately measure the depth of the annular fill materials as the well is being built. Weighted measuring lines used for well construction are often home made or can be purchased commercially.

Even if the annular seals are placed to the exact depths specified in the well design, there are other reasons why the seals between the monitored zones may be compromised. Few

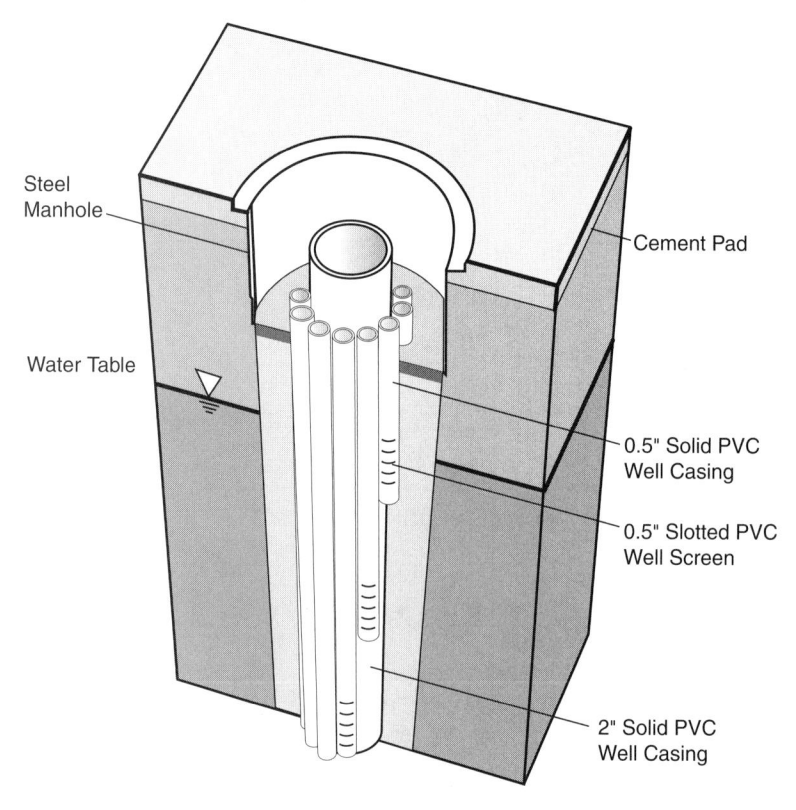

FIGURE 11.8
Bundle well made of 0.5-in. PVC pipes surrounding 2-in. PVC well casing. (From New York State Department of Environmental Conservation.)

nested wells are actually constructed like the one depicted in Figure 11.6. A more realistic construction diagram is shown in Figure 11.9a. No borehole is perfectly plumb and straight. Consequently, unless specialized centralizers are used, it is difficult to keep multiple casings centered and separate from one another in the borehole during well construction. If the casings are not centered and separate in the borehole, void spaces can exist in the seal between the various casings and borehole wall. The void spaces can then allow vertical movement of ground water within the borehole between zones. Flow (and therefore cross-contamination) can occur between zones during purging and sampling when strong vertical hydraulic gradients are induced by pumping. Ambient flow and cross-contamination can also occur between zones if vertical hydraulic gradients naturally exist in the formations being monitored.

The likelihood of vertical leakage through the annular seals of a nested well increases with the number of separate casings within the borehole. Also, the likelihood of vertical leakage is higher with shallow nested wells where only a few feet of an annular seal exists between the various monitored zones. It is for these reasons that the installation of nested wells is discouraged or prohibited by many governmental or regulatory agencies. For example, nested wells are prohibited in the State of Washington (State of Washington, 2004). The California Department of Water Resources notes that it can be difficult to install effective seals in nested wells (California Department of Water Resources, 1990). The U.S. Army Corps of Engineers prohibits the use of nested wells (U.S. Army Corps of Engineers, 1998). And, the U.S. EPA notes that "data may be erroneous and the use of nested wells is discouraged" (U.S. EPA, 1992).

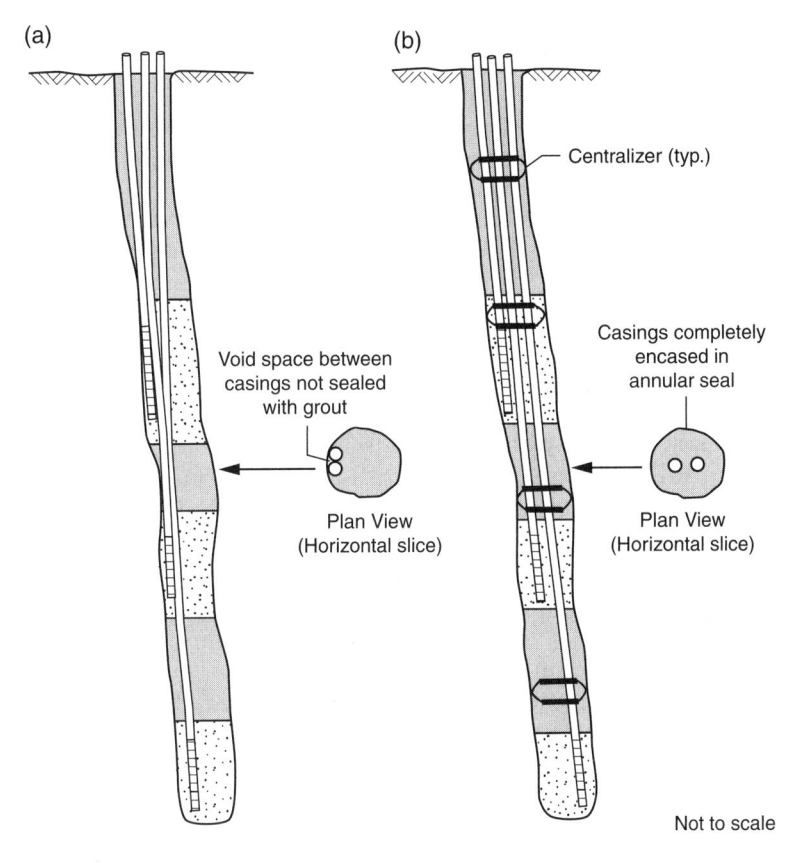

Multilevel Ground-Water Monitoring

FIGURE 11.9
Nested wells. (a) Installation without centralizers may result in imperfect seals between monitored zones. (b) Centralizers keep casings separate and centered in the borehole, resulting in superior seals between the monitored zones.

Further, Johnson (1983) notes that:

> The existence of several pipes or tubes in a single borehole and the utilization of shorter seals to accommodate the spacings between the monitoring points makes single-borehole completions more difficult to seal than the individual wells

Aller et al. (1989) state in the *Handbook of Suggested Practices for the Design and Installation of Ground Water Monitoring Wells* that:

> A substantial problem with this type of construction is leakage along the risers as well as along the borehole wall. The primary difficulty with multiple completions in a single borehole is that it is difficult to be certain that the seal placed between the screened zones does not provide a conduit that results in interconnection between previously non-connected zones within the borehole. Of particular concern is leakage along the borehole wall and along risers where overlying seals are penetrated. It is often difficult to get an effective seal between the seal and the material of the risers.

The above cautions and caveats notwithstanding, not everyone installing nested monitoring wells has experienced failed seals between the monitoring zones. The U.S. Geological

Survey (USGS) has reportedly had success installing nested wells even without the use of spacers or centralizers to keep the casings separate in the borehole (Hanson et al., 2002). The USGS installations typically use bentonite slurry to seal between zones. Other reasons why the USGS nested wells have been more successful than others may be that their wells are often very deep (several hundreds to thousands of feet deep), resulting in seals that are several tens to hundreds of feet thick. Also, the USGS drills relatively large boreholes (12 in. or larger) and rarely installs more than three casings in a single hole. A diagram of a nested well constructed by USGS is shown in Hanson et al. (2002).

There are often suggestions that spacers or centralizers be used to keep the various casings separate and centered in the borehole. Some regulations even require it (e.g., California Department of Water Resources, Santa Clara Valley Water District). As shown in Figure 11.9b, centralizers keep the casings separate and centered and can greatly enhance the integrity of the annular seals between the monitored zones. So, why are not spacers or centralizers more widely used during the installation of nested wells? The answer may be that there are no commercially available spacers or centralizers designed for installing nested wells. Conventional well centralizers are designed to center a single casing in a borehole. One type of centralizer for nested wells was used to install nested monitoring wells to depths over 200 ft in California, but those centralizers had to be welded to the various casings, necessitating the use of steel casing for the wells instead of PVC (Nakamoto et al., 1986).

Many drillers have found that using custom-made centralizers to center multiple casings in a single borehole often makes it more difficult, rather than easier, to install reliable annular seals. This is because the centralizers form obstructions to sand and bentonite that is being poured from the surface, causing bridging. Also, there is often no room to insert a tremie pipe into the borehole when such centralizers are used. And, measuring or "tag" lines can become tangled on the centralizers during well construction.

Figure 11.10 shows the design of a well centralizer designed for nested wells.[1] The centralizer assembly uses two 1.5-in.-thick PVC spacer discs that are attached to a conventional 6-in. "lantern" style steel or PVC centralizer. The centralizer assembly is designed for installing three 1-in. PVC wells within a borehole 8 in. or larger in diameter. A novel feature of this centralizer is that it has a hole in the center of each spacer disc to facilitate the use of a 2-in. tremie pipe during well construction. A three-zone centered nested well is constructed as follows. First, a 2-in. tremie pipe is inserted to the bottom of the borehole. Next, two of the PVC spacer discs are threaded over the 2-in. tremie pipe. The first (deepest) 1-in. well screen is attached to the discs by pushing it into the 1-in. cutouts in the discs. The lantern centralizer is then attached to the two discs, securing the 1-in. PVC to the disc or centralizer assembly, and the centralizer and 1-in. PVC are lowered into the borehole. At the depth corresponding to the next centralizer, the process is repeated. At the depth corresponding to the middle monitoring zone, the second well screen is attached to one of the other cutouts in the centering discs. Centering discs and centralizers are assembled and sections of 1-in. PVC casing are attached in this way until the entire three-zone nested well has been fully inserted to the bottom of the borehole. The sand and bentonite seals are then installed by pouring the materials through the 2-in. tremie pipe as it is removed from the borehole. The 2-in. tremie is sufficiently large to pour sand and bentonite pellets through it. A measuring line can also be run inside of the 2-in. tremie to measure the depths of the sand and bentonite lifts as the well is being constructed. The tremie pipe is incrementally removed from the borehole as the well is constructed.

---

[1] The centralizer assembly described here is not commercially available but can be easily fabricated by most drilling contractors.

Multilevel Ground-Water Monitoring

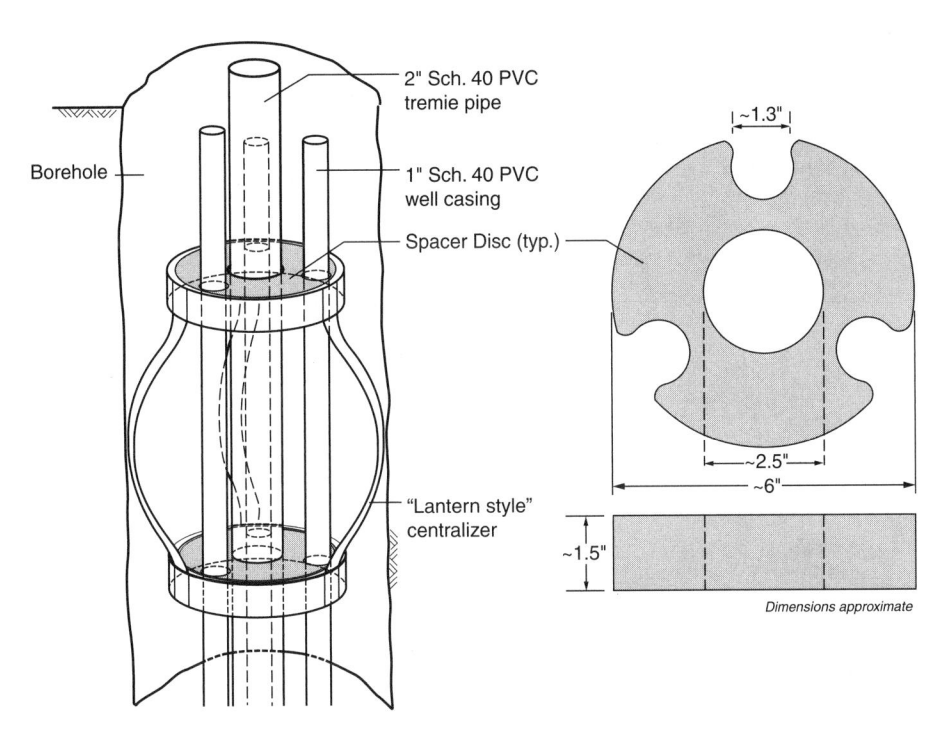

FIGURE 11.10
Design of a centralizer for a three-zone nested well. See text for further discussion.

Well Clusters (One Well per Borehole)

A cluster of monitoring wells is a grouping of individual wells, each completed to a different depth (Figure 11.6). The main advantage of well clusters over nested wells is that the seals are easier to install and more reliable because there is only one casing in each borehole. It is for this reason that well clusters are widely recommended by governmental and regulatory agencies. As with nested wells, the screened interval of each well in the cluster should be no more than 2 or 3 ft long so that the head measurements and ground-water samples from each well will be depth discrete and not composited over a larger part of the aquifer.

The main disadvantage of clusters of wells is the increased cost of drilling separate boreholes for each well. Costs for well clusters are especially high if each borehole needs to be continuously cored. In some cases it is sufficient to continuously core the deepest boring and then design the entire well cluster based on the data obtained from the single core. However, if one expects significant variations in the geology, even over short horizontal distances (e.g., in fractured bedrock or fluvial deposits), then each borehole in the cluster should be cored. This can add significant cost to the well cluster installation.

In plan view, the individual wells in the cluster should be installed close together, on the order of 10 ft apart or less, so that the head data obtained from them is a result of variations in the vertical head and not horizontal gradients. Also, care should be taken to avoid installing clusters of monitoring wells with overlapping screens. As shown in Figure 11.11, overlapping screens can allow vertical movement of contaminant plumes if vertical hydraulic gradients are present. Finally, clusters of wells should be installed with the wells oriented in a line perpendicular to the flow direction or with the deeper wells located progressively in the downgradient direction. This avoids the possibility

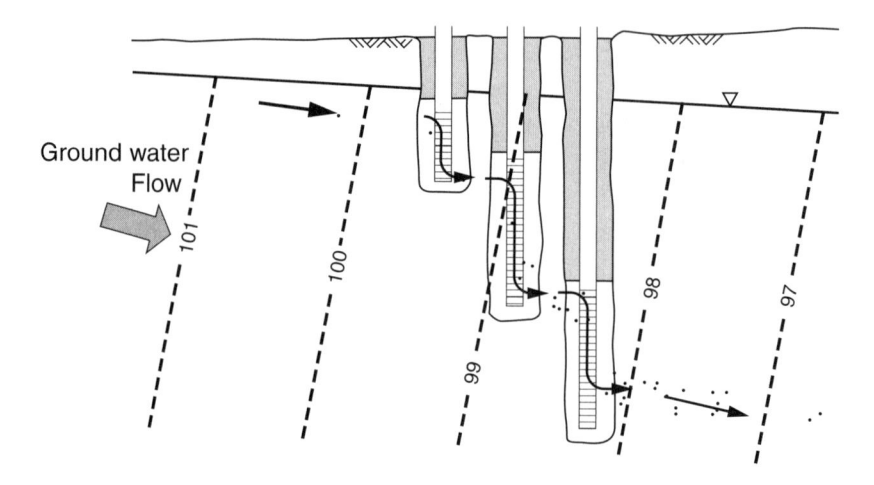

FIGURE 11.11
Cluster of monitoring wells with overlapping well screens. If vertical gradients are present, well clusters installed like this can lead to short-circuiting of the contaminant plume and cross-contamination of the aquifer.

that the wells will be sampling ground water that is affected by contact with the annular seal of an upgradient monitoring well.

At sites underlain by unconsolidated sedimentary deposits, the use of clusters of individual wells for multilevel monitoring is becoming more and more economical (and therefore more popular) due to the use of DP installation methods and small-diameter monitoring wells with prepacked well screens. At many sites, several clusters of small-diameter wells can be installed in a single day using powerful DP rigs.

Dedicated Multilevel Ground-Water Monitoring Systems

There are several dedicated multilevel ground-water monitoring systems currently on the market. Four commercially available systems that have seen relatively widespread use are: the Westbay MP® system; the Solinst Waterloo™ system; the Solinst CMT™ system; and the Water FLUTe™ system. A comparison of these systems is presented in Table 11.1; each system is also described in detail below. These dedicated multilevel systems offer the following advantages.

- They facilitate the collection of ground-water samples and measurement of hydraulic heads from many more discrete depths than is practical with nested wells or well clusters (e.g., 10 or more discrete depths can be monitored with most dedicated multilevel monitoring systems).
- Only one pipe (or tube) is placed in the borehole. This simplifies the process of installing annular seals between the monitored zones and improves the reliability of the seals (e.g., compared with nested wells).
- Total project costs can be significantly lower due to reduced drilling costs, less secondary waste, less time spent monitoring and sampling, and fewer wells for decommissioning.
- The volume of purge water produced during routine sampling is decreased or eliminated, reducing costs related to storage, testing, transport and disposal of purged fluids.

TABLE 11.1
Comparison of Four Dedicated Multilevel Ground-Water Monitoring Systems

Description	Westbay MP® System	Solinst Waterloo™ System	Solinst CMT™ System	Water FLUTe™ System	Comments
Materials	PVC, polyurethane, Viton, and stainless steel	PVC, stainless steel, Viton, rubber, and Teflon or polyethylene tubing	Polyethylene and stainless steel	Polyurethane-coated nylon, stainless steel or brass, and polyethylene, PVDF, or Teflon tubing	Materials vary depending on sealing and pumping options
Maximum depth (ft)	4000	750	300	1000	Maximum depth for routine installations
Maximum number of sampling points	20 per 100 ft of well	15	7	20+	With exception of Westbay system, depends on diameter of system and size of sampling tubes
Allows use of pressure transducers to monitor hydraulic pressure	x	x	x	x	Westbay MP system uses a specialized tool for sample collection and pressure measurement (see text) Dedicated pressure sensors can also be installed
Maximum sampling points when dedicated pressure transducers are used in each monitored zone	See comments	8	3	20+	With Westbay MP system, dedicated pressure sensors must be removed prior to collecting ground-water samples from the same zones
Sampling methods	See comments	Peristaltic pump, inertial-lift pump, double-valve pump, bladder pump	Peristaltic pump, inertial-lift pump, double-valve pump	Peristaltic pump, inertial-lift pump, double-valve pump, bladder pump	Westbay system uses specialized tool for sample collection and pressure measurement (see text)
Optimal borehole diameter (in.)	3–6	3–6	3–6	3–10	

(Table continued)

TABLE 11.1 *Continued*

Description	Westbay MP® System	Solinst Waterloo™ System	Solinst CMT™ System	Water FLUTe™ System	Comments
Built-in features for well development and hydraulic testing	×	—	—	—	
Can be installed immediately after well designed; that is, no delay due to shipping customized well components to site from factory	×	—	×	—	
Removable system	×	×	—	×	Solinst Waterloo system removable when deflatable packers used. Deflatable packers under development for Solinst CMT system will make it removable. Successful removal of any multi-level system depends on borehole conditions
Can be installed in open holes in bedrock and massive clay deposits	×	×	×	×	
Can be installed in unconsolidated deposits	×	×	×	×	
Can be installed in multiscreened wells	×	×	×	×	
Seals and sand pack can be installed by backfilling from surface	×	×	×	—	FLUTe system seals borehole; other annular seals are therefore not needed
Inflatable packers available for sealing borehole in bedrock or multiscreened wells	×	×	—	—	Inflatable packers under development for Solinst CMT system. Water FLUTe system can be thought of as one long packer
Can be installed with DP (e.g., Geoprobe) equipment	—	—	×	×	

- The small volume of water stored in each monitoring zone or tube minimizes the time required for heads in the well to equilibrate with formation pressures. This is particularly advantageous when multilevel monitoring is performed in low-yield formations and aquitards.
- A single multilevel monitoring well has a much smaller "footprint" at the ground surface than a cluster of individual wells. A single multilevel well is therefore less noticeable and obtrusive than a large cluster of wells.

Dedicated multilevel systems also have some disadvantages, including the following:

- Fewer options exist for sampling dedicated multilevel systems than for conventional monitoring wells. This is due to the design of the wells and the relatively small diameter of sampling tubes installed inside the multilevel wells. Several small-diameter pumps have been developed, however, to facilitate collection of ground-water samples from small-diameter wells and tubing (see following text).
- Owing to the specialized nature of some of the components or monitoring tools used in multilevel systems, some training or technical assistance is generally recommended, at least for first-time installers of the systems.
- It may be more difficult to decommission specialized multilevel monitoring systems than conventional single-interval PVC monitoring wells.

Drilling and Installation Considerations

Installations in Open Boreholes

Boreholes drilled into bedrock or silt and clay deposits usually stay open after the hole is drilled and the drill string has been removed. Multilevel wells can therefore be constructed directly inside of the open boreholes. Oftentimes, it is not necessary to have a drilling rig on site during the construction of the multilevel well if the multilevel well casing[2] can be lowered into the borehole by hand or using a winch. Because the boreholes stay open, however, the annular space between the well casing and the boreholes must be sealed to prevent vertical flow of ground water between the various monitored zones. With some multilevel systems (e.g., Westbay MP, Solinst Waterloo), inflatable rubber, polyurethane, or Viton packers can be used to seal the annular space between the monitored zones. The annular space can also be sealed by backfilling the annulus with alternating lifts of sand (at the depths of the intake ports) and clay or cement (in the intervals between the various intake ports). Finally, the novel design of the Water FLUTe system also seals the borehole between the sampling ports.

Installations in Unconsolidated Sedimentary Deposits

Unlike boreholes drilled into competent bedrock, most boreholes drilled in unconsolidated deposits will not stay open when drilling has been completed and the drill rods are removed. Consequently, some method of keeping the borehole open is necessary while the multilevel well casing is inserted and the well constructed. One way to accomplish this is by advancing steel drive casing as the borehole is drilled. The steel drive casing is left in the borehole while the well casing is inserted, and is then pulled back incrementally as the multilevel well is constructed. If the formation will collapse completely around the multilevel well casing, it is usually not necessary to install annular seals between the monitored zones because the collapsing sand restores the original per-

---

[2]Or "tubing" in the case of the Solinst CMT system; "liner" in the case of the FLUTe system. "Casing" is used generically in this discussion.

meability of the formation. If the formation will not collapse completely around the multi-level well casing, however, gaps can exist in the annular space, allowing vertical flow of ground water between different monitoring zones. In this case, alternating layers of sand and bentonite or cement must be emplaced by backfilling as the steel drive casing is withdrawn from the borehole. Drilling methods that employ driven casing include air-rotary casing advance and rotasonic (Barrow, 1994). Rotasonic drilling (also referred to simply as sonic drilling) is ideal for installing multilevel monitoring wells because (1) steel drive casing is advanced as drilling progresses; (2) continuous cores are routinely collected (logs of the cores can then be used to design the multilevel wells); and (3) the rate of penetration is usually high.

Two of the multilevel monitoring systems, Solinst CMT and Water FLUTe, can be installed with DP drilling equipment. Both the multilevel systems can be inserted into small-diameter (approximately 3 in. OD) steel casing that has been driven to the target depth. Use of a dual-tube DP system facilitates collection of continuous cores while advancing an outer drive casing that can then be retracted as the multilevel well is constructed (Einarson, 1995). Because of the relatively small size of most DP sampling rigs, however, the maximum depth of multilevel wells installed with this drilling method is approximately 50 ft in most sedimentary deposits.

Multilevel monitoring wells can also be installed in boreholes drilled with hollow-stem augers and mud rotary drilling methods, but these drilling methods have some significant drawbacks. Hollow-stem augers keep the borehole open while allowing the multilevel well casing to be inserted through the augers to the bottom of the borehole. Sand packs and annular seals are then emplaced as the augers are incrementally removed from the borehole. The action of the augers during drilling, however, often creates a skin of smeared fine-grained soil which can seal some thin, permeable strata or fractures in clay (D'Astous et al., 1989) and generally reduce the permeability of the formation along the entire length of the borehole. Also, if the augers penetrate soil containing high concentrations of contaminants (either residual NAPL or sorbed mass), the contaminants can be smeared against the borehole wall from the depth that they were penetrated up to the ground surface. This can impart a long-lived positive bias to ground-water samples collected from a multilevel well subsequently installed in the borehole.

Multilevel monitoring wells can be installed in boreholes drilled with mud rotary drilling equipment, but this drilling method too has undesirable effects when it comes to installing multilevel wells. With mud rotary drilling, the borehole is kept open by (1) the hydrostatic pressure of the drilling fluid (drilling mud) and (2) the creation of a tough, pliable filter cake or clay "skin" that develops from exfiltration of the drilling fluid through the borehole wall. Circulation of the drilling fluid, however, can cross-contaminate the borehole if contaminants in the drilling fluid penetrate the formation (by advection or diffusion) or sorb onto the borehole wall. This can cause a lingering chemical bias similar to the one described above for wells installed with hollow-stem auger drilling equipment. Also, it is often more difficult to place sand packs and annular seals in mud-filled boreholes than boreholes containing air or clear water. This is because the high density and viscosity of the drilling fluid makes it difficult to pour sand and bentonite pellets through the drilling fluid (many contractors will therefore thin the drilling mud with water prior to building a well). In most cases, though, the sand and bentonite or cement must be pumped through a tremie pipe. Finally, the drilling fluid and filter cake may be difficult to remove after the multilevel well has been constructed. With the exception of the Westbay MP system, none of the multilevel systems described in this section facilitate robust well development to remove the drilling mud and filter cake. Therefore, the Westbay MP system would be a good choice for a multilevel well installed in a mud-rotary drilled borehole. Other multilevel systems have been

Multilevel Ground-Water Monitoring 835

installed successfully in boreholes drilled using biodegradable drilling fluids (e.g., guar-based slurries), however. The use of a biodegradable drilling fluid reduces the need for vigorous well development to remove the drilling mud and filter cake.

Another way to install the dedicated multilevel systems described in this section is inside of multiscreened wells instead of directly in boreholes (Figure 11.12). With this type of installation, the multilevel monitoring system is installed inside a steel or PVC well that has been constructed with short screens at multiple depths. The depths of the well screens correspond to the depths of the ports in the multilevel monitoring system. This adds another step to the well installation process (i.e., first installing a multiscreened well), but has several advantages. First, installing conventional steel or PVC wells is straightforward and routine for most drilling contractors. Hence, it is not necessary that the drilling contractor have expertise in installing multilevel monitoring systems. Once the multiscreened wells have been installed and developed, the drilling contractor's job is done, and the multilevel systems can be installed by field technicians, often at a lower cost. Secondly, the various monitoring zones can be developed using standard well development equipment and procedures before the multilevel monitoring systems

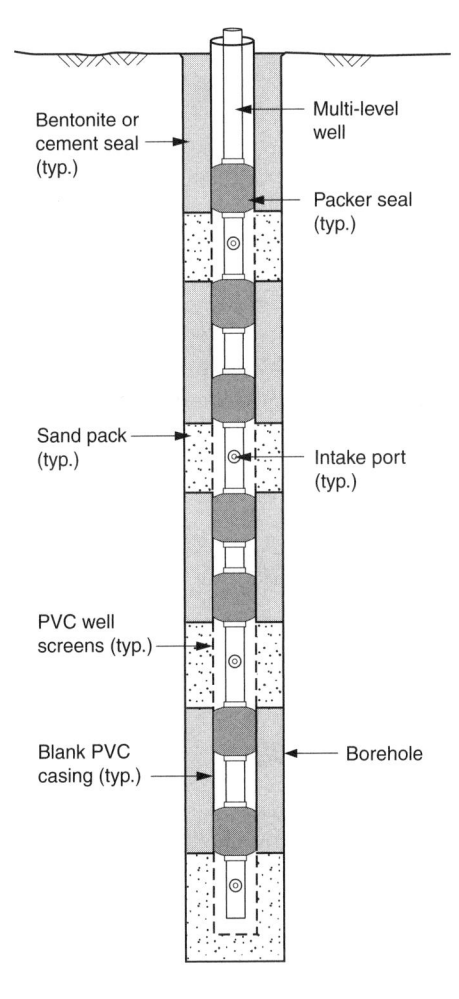

FIGURE 11.12
A dedicated multilevel monitoring system installed inside a steel or PVC well constructed with multiple well screens.

are installed in the wells. Finally, installing multilevel systems inside multiscreened wells may simplify the task of decommissioning the wells once they are no longer needed. Most of the multilevel systems can be constructed so that they can be easily removed from the wells. Then, the multi-screened wells can be pressure-grouted or drilled out using standard well decommissioning procedures (see Chapter 12).

Minimizing Cross-Contamination

A properly constructed multilevel monitoring well should clearly prevent vertical "short circuiting" of ground water between different monitored zones. As discussed earlier, however, cross-contamination can occur in the borehole before the well is constructed. Cross-contamination can occur if NAPL is penetrated and becomes incorporated in the drilling fluid or flows into and along the borehole wall. This severe form of cross-contamination (and ways to avoid it during drilling) is described elsewhere (see Pankow and Cherry, 1996) and is therefore not discussed further in this chapter.

The cross-contamination discussed in this section is related to the redistribution of dissolved solutes within the borehole both during and after drilling — but before the well is constructed. Cross-contamination of fluids in the borehole during drilling has already been discussed and recommendations made to avoid it. In short, when drilling in unconsolidated deposits (both sand and gravel aquifers and low permeability clay deposits), advancing steel casing while drilling is the best way to minimize the potential for cross-contamination of dissolved solutes in the borehole. The drive casing stays in the ground until the multilevel well is ready to be constructed, and is retracted incrementally as the multilevel well is being built. In boreholes drilled in rock, however, it is usually not possible to advance steel casing, and some degree of cross-contamination in the borehole should be expected due to the circulation of fluids (either drilling mud, water, or compressed air). Note that the potential bias caused by circulation of fluids during drilling is not restricted to boreholes drilled for multilevel wells but can occur with all types of monitoring wells.

Further, if a multilevel well is not installed in an open borehole immediately after drilling ceases, vertical flow of potentially contaminated ground water can occur in the borehole from zones of high head to low head during the time that the borehole has been drilled and the multilevel well installed. To minimize potential chemical biases caused by this intra-borehole flow, the multilevel well should be installed in the borehole as quickly as possible. If this is not possible, the borehole can be temporarily sealed to prevent ambient vertical flow. This has been done at several sites using blank FLUTe liners. (Several technologies to temporarily seal boreholes drilled in fractured rock [including FLUTe] are currently being evaluated by researchers at the University of Waterloo [Cherry, 2004].) Partial mitigation of this bias may be accomplished by pumping from the various monitoring zones immediately after the well has been constructed, but low-level contamination may linger for months or years if the contaminants have sorbed onto or diffused into the aquifer matrix (Sterling et al., 2005). The likelihood (and potential longevity) of a positive ground-water sampling bias occurring due to circulation of drilling fluids and intra-borehole ground-water flow after drilling depends on many factors, including the nature and concentration of the contaminant, the nature of the geologic material, the time of exposure, and extent of penetration into the formation, and must be evaluated on a case-by-case basis.

Development of Multilevel Wells

The purpose of multilevel monitoring wells is to provide depth-discrete samples of ground water and accurate depth-discrete measurements of hydraulic head. They are not designed to provide large volumes of water as are water supply or remediation wells. Consequently, the requirements for developing multilevel monitoring wells are different than for other types of wells. In general, as long as there is good hydraulic connection between the monitoring ports and the formation and the samples collected from the wells are sediment-free and exhibit turbidity within reasonable levels, the above requirements are met. With each of the dedicated multilevel systems described in this chapter, this level of well development can usually be achieved simply by over-pumping the various ports with the pumps used for sampling. In the case of wells installed in boreholes drilled with mud rotary methods, however, more rigorous development is necessary. This can be accomplished best with any of the four multilevel systems provided that they are installed in multiscreened wells that have already been developed using traditional development methods. Over-pumping is also often done to remove water added to the borehole during drilling and well construction. This is due to widely held concern that if this water is not removed, it could cause a negative bias in the samples subsequently collected from the well. If the volume of water that needs to be removed is small, the water can be removed by pumping the zones using the same pumps used for sampling (the Westbay MP system allows for use of higher capacity pumps for well purging). Air lift techniques have also been used successfully to pump water at relatively high rates from small-diameter sampling tubes (see Einarson and Cherry, 2002). Finally, in most flowing aquifers, it is usually sufficient to simply allow some time to pass before collecting the first samples in order to allow the added water to drift away from the intake ports of the well. In most cases the added water will have drifted away from the intake ports of the multilevel wells in several days and samples collected from the well will be ground water. Some site investigators have added an inert tracer (e.g., potassium bromide) to the water used during drilling and well construction. They then pump water from the various ports (or let sufficient time pass for the added water to drift away from the sampling ports) until the tracer is no longer detected. They can then be confident that ground-water samples collected thereafter consist entirely of ground water and not water added during drilling or well construction.

Westbay MP System

Schlumberger produces the Westbay MP system, a modular instrumentation system for multilevel ground-water monitoring. The MP system can be divided into two parts: (1) the casing system and (2) portable probes and tools that provide a compatible data acquisition system.

The Westbay casing system (Figure 11.13) is designed to allow the monitoring of multiple discrete levels in a single borehole. One single string of water-tight Westbay casing is installed in the borehole. Each level or monitoring zone has valved couplings to provide a selective, controlled connection between the ground water outside the casing and instruments inside the casing. Westbay packers or backfill are used to seal the borehole between monitoring zones to prevent the unnatural vertical flow of ground water and maintain the natural distribution of fluid pressures and chemistry. The Westbay system can be installed in either open boreholes or cased wells with multiple screens.

Westbay system packers are individually inflated with water to pressures of 100–200 psi above ambient. Westbay packers accommodate a range of borehole sizes (Table 11.1) and, according to the manufacturer, withstand significant gradients along the borehole.

Data are obtained using one or more wireline probes with sensors that are lowered inside the casing to each monitoring zone. The probes locate and open the valved ports to measure fluid pressure, collect fluid samples or test hydrogeologic parameters. Multiple probes can be connected in series to provide continuous multilevel data. Software permits

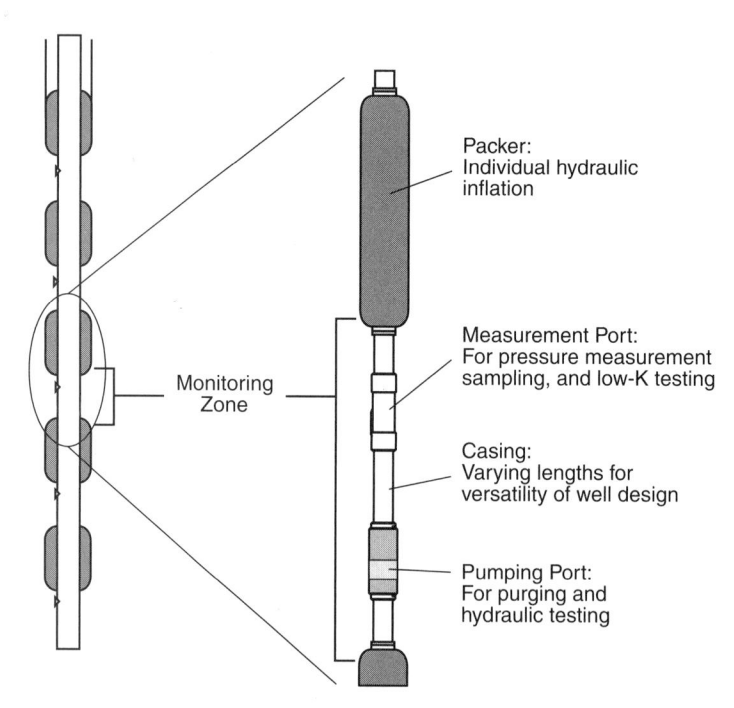

FIGURE 11.13
The Westbay MP® system.

notebook computers to interface with the probes and collect data at the surface or from a remote location.

The design of the MP system results in no restriction to the number of zones that can be completed in one borehole, apart from the physical ability to fit the length of the components in the well. The user can have materials on site ahead of time as it is not necessary to know the precise size of the borehole or the desired location of seals and monitoring zones before the equipment is shipped. Users also have access to a wide range of monitoring and testing capabilities such as manual or automated monitoring of pressure (water level), discrete sampling without repeated purging, pulse testing of low-permeability environments, rising- or falling-head (slug) or constant-head hydraulic conductivity testing, vertical interference testing, and cross-well testing (including injection and withdrawal of tracers) (Figure 11.14). Pressure measurements are made under shut-in conditions, making the system responsive to pressure changes. Ground-water samples are collected at formation pressure without repeated purging.

The Westbay system has been in use since 1978 and has been installed in a variety of geologic environments ranging from soft seabed sediments to unconsolidated alluvial deposits, to highly fractured bedrock. Examples of project applications include environmental characterization related to ground-water contamination (e.g., Raven et al., 1992; Gernand et al., 2001; Taraszki et al., 2002) to ground-water resource management (Black et al., 1988), and characterization and monitoring related to nuclear waste repositories (Delouvrier and Delay, 2004). Depths of installation varied from 100 ft (30 m) to greater than 4000 ft (1200 m).

Westbay instrumentation is sold as a complete system and Westbay technicians assist with initial installations and provide on-site training of local personnel. Field quality-control procedures permit the quality of the well installation and the operation of the testing and sampling equipment to be verified at any time.

Multilevel Ground-Water Monitoring 839

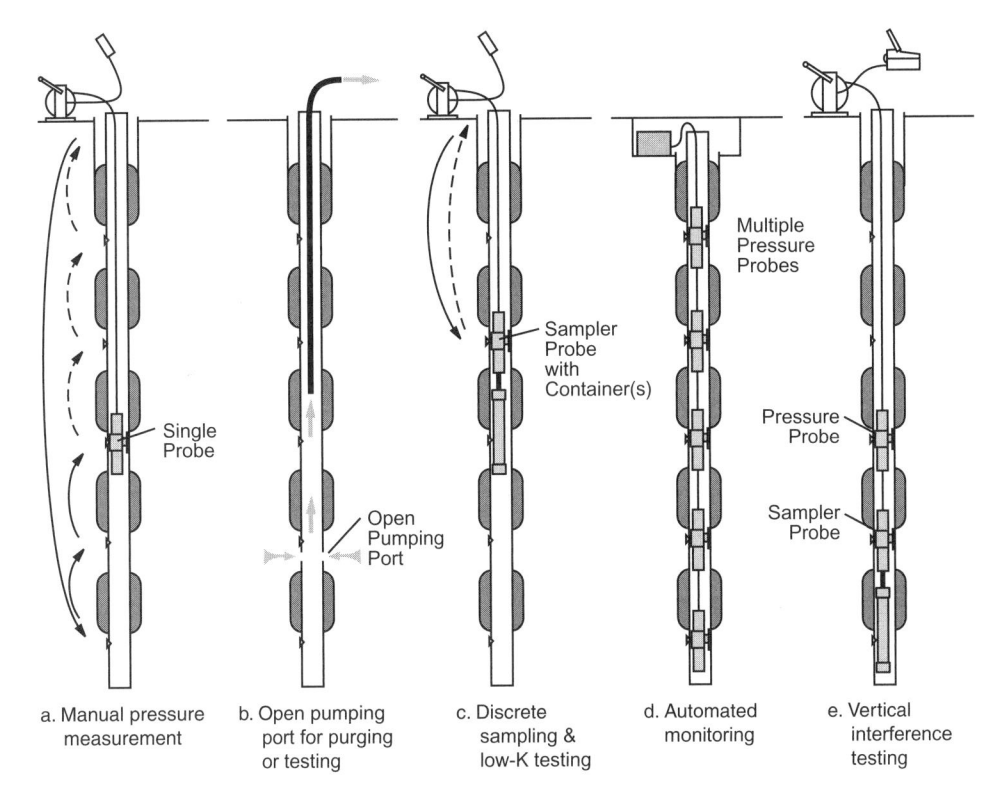

FIGURE 11.14
Options for pumping, testing, and monitoring with the Westbay MP system. See text for discussion.

A detailed technical description of the Westbay multilevel monitoring system is presented by Black et al. (1986). Further information about the Westbay multilevel system can be obtained from Westbay Instruments Inc. (www.westbay.com) and from its parent company, Schlumberger Water Services (www.slb.com/waterservices).

Solinst Waterloo System

The Solinst Waterloo Multilevel ground-water monitoring system is a modular multilevel monitoring system manufactured by Solinst Canada, Ltd. to collect ground-water data from multiple depths within a single drilled borehole. Originally developed by researchers at the University of Waterloo (Cherry and Johnson, 1982), it consists of a series of monitoring ports positioned at specific intervals along 2-in. Schedule 80 PVC casing (Figure 11.15). The ports are typically isolated in the borehole either by in-line packers (permanent or removable), or by alternating layers of sand and bentonite backfilled from the surface. The Solinst Waterloo Multilevel system can also be installed inside multi-screened wells.

The ports and packers are connected to the 2-in. Schedule 80 PVC casing with a special water-tight joint. Monitoring ports are constructed of stainless steel or PVC and have the same water-tight joint to connect with the other system components. Water is added to the inside of the 2-in. PVC casing to overcome buoyancy during installation and to inflate permanent or deflatable packers (if used). A case study in which a removable Waterloo multilevel monitoring system equipped with deflatable packers was used is presented by Sterling et al. (2005).

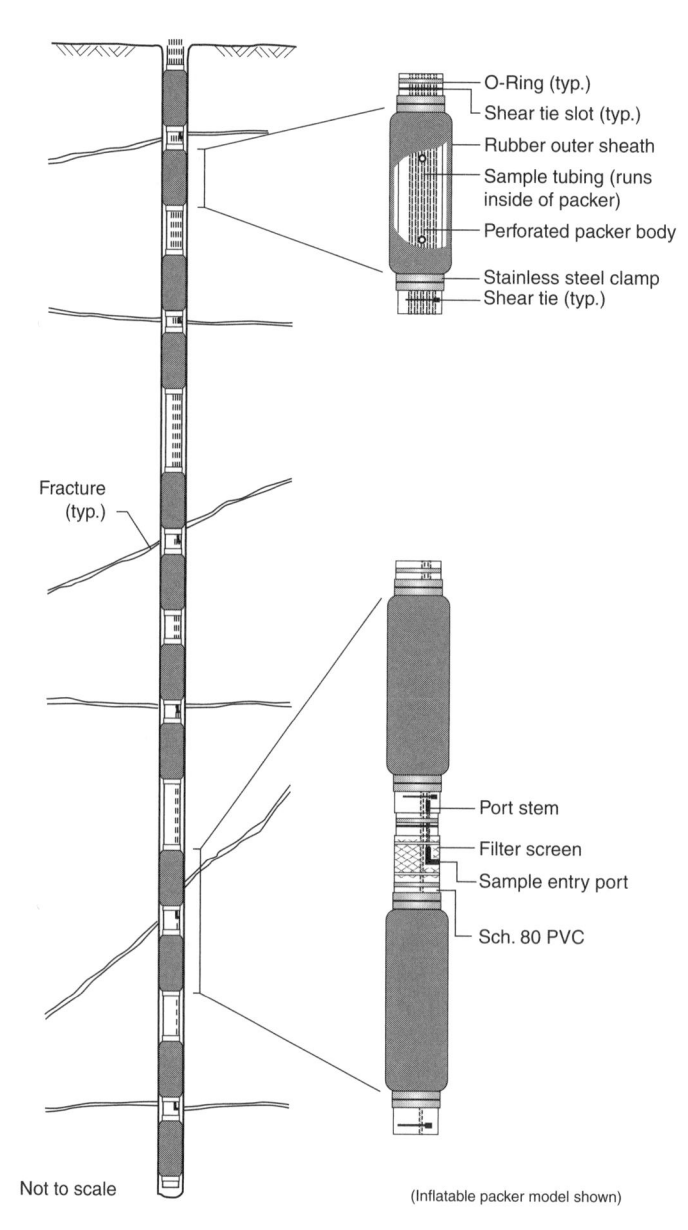

FIGURE 11.15
The Solinst Waterloo Multilevel ground-water monitoring system. (Adapted from Cherry and Johnson [1982], With permission.)

Each monitoring port has either a single or dual stem. Each stem is connected to either: (1) an open tube that runs inside the 2-in. PVC casing to the ground surface; (2) a double valve pump; (3) a bladder pump; or (4) a pressure transducer. Pressure transducers can be connected to a data logger for continuous recording of water levels. If open tubes are connected to the port stems, samples can be obtained from inside the tubes using a peristaltic pump, an inertial-lift (i.e., check-valve) pump, or a double-valve gas-drive (positive displacement) pump. Water levels can also be measured in the open tubes using small-diameter water-level meters. Because each port is plumbed to some type of monitoring device, contact between ground water entering the ports and water added to the inside of the 2-in. PVC casing is prevented. If a single stem is used, only one monitoring device can be used per monitored zone. If dual stems are used, two devices (e.g., a bladder pump and pressure transducer) can be used per zone.

Depending on the monitoring options chosen, the number of zones that can be monitored typically ranges from three to eight, although systems with as many as 15 sampling ports have been installed. Systems installed in fractured rock formations are typically installed in 3- or 4-in.-diameter core holes. A wellhead that facilitates simultaneous purging and sampling of all monitored zones is available. More information about the Solinst Waterloo System is available from Solinst Canada, Ltd. (www.solinst.com).

Solinst CMT System

The Solinst continuous multichannel tubing (CMT) system is a multilevel ground-water monitoring system that uses custom-extruded flexible 1.6-in. OD multichannel HDPE tubing to monitor as many as seven discrete zones within a single borehole in either unconsolidated sedimentary deposits or bedrock (Figure 11.16). Prior to inserting the tubing in the borehole, ports are created that allow ground water to enter six outer pie-shaped channels (nominal diameter = 0.5 in.) and a central hexagonal center channel (nominal diameter = 0.4 in.) at different depths, facilitating the measurement of depth-discrete piezometric heads and the collection of depth-discrete ground-water samples.

The multichannel tubing can be extruded in lengths up to 300 ft and is shipped in 4-ft-diameter coils. The desired length of tubing, equal to the total depth of the multilevel well, is cut from a coil, and the well is built at the job site based on the hydrogeologic data obtained from the exploratory boring or other methods (e.g., CPT or geophysical data). The tubing is stiff enough to be easily handled, yet light and flexible enough to allow site workers to insert the multilevel well hand-over-hand into the borehole.

Construction of the intake ports and screens is done before the CMT tubing is inserted into the borehole. A small continuous mark along the outside of one of the channels facilitates identification of specific channels. Depth-discrete intake ports are created by cutting ports through the exterior wall of the tubing into each of the channels at the desired depths. Channel 1 ports correspond to the shallowest monitoring interval; channel 2 ports are created further down the tubing (i.e., to monitor a deeper zone), and so forth. The central channel, channel 7, is open to the bottom of the multilevel well. In this way, the ports of the various channels are staggered both vertically and around the perimeter of the multichannel tubing (Figure 11.16). For most of the installations performed as of 2004, an intake interval of approximately 6 in. has been created. The depth interval of the intake ports can be increased by cutting more ports in the tubing.

Stagnant water in the tubing below the intake ports is hydraulically isolated by plugging the channels a few inches below each intake port. This has been done by inserting and expanding a mechanical plug into each channel. Expanding mechanical plugs are also inserted into each of the outer six channels at the very bottom of the tubing. This effectively seals the various channels from just below the intake ports to the bottom of the tubing. Small vent holes are drilled directly beneath the upper polyethylene plugs (i.e., the plugs located just below the intake ports) to allow air to vent out of the sealed channels during installation. The seventh (internal) channel is open to the bottom of the tubing.

Well screens are constructed by wrapping synthetic or stainless steel fabric mesh completely around the tubing in the interval containing the ports. The mesh is secured to the tubing using stainless steel clamps. The size of the mesh openings can be selected based on the grain-size distribution of the particular water-bearing zone being monitored. A guide-point cap containing stainless steel mesh is attached to the bottom of the tubing to enable the central channel to be used as the deepest monitoring zone.

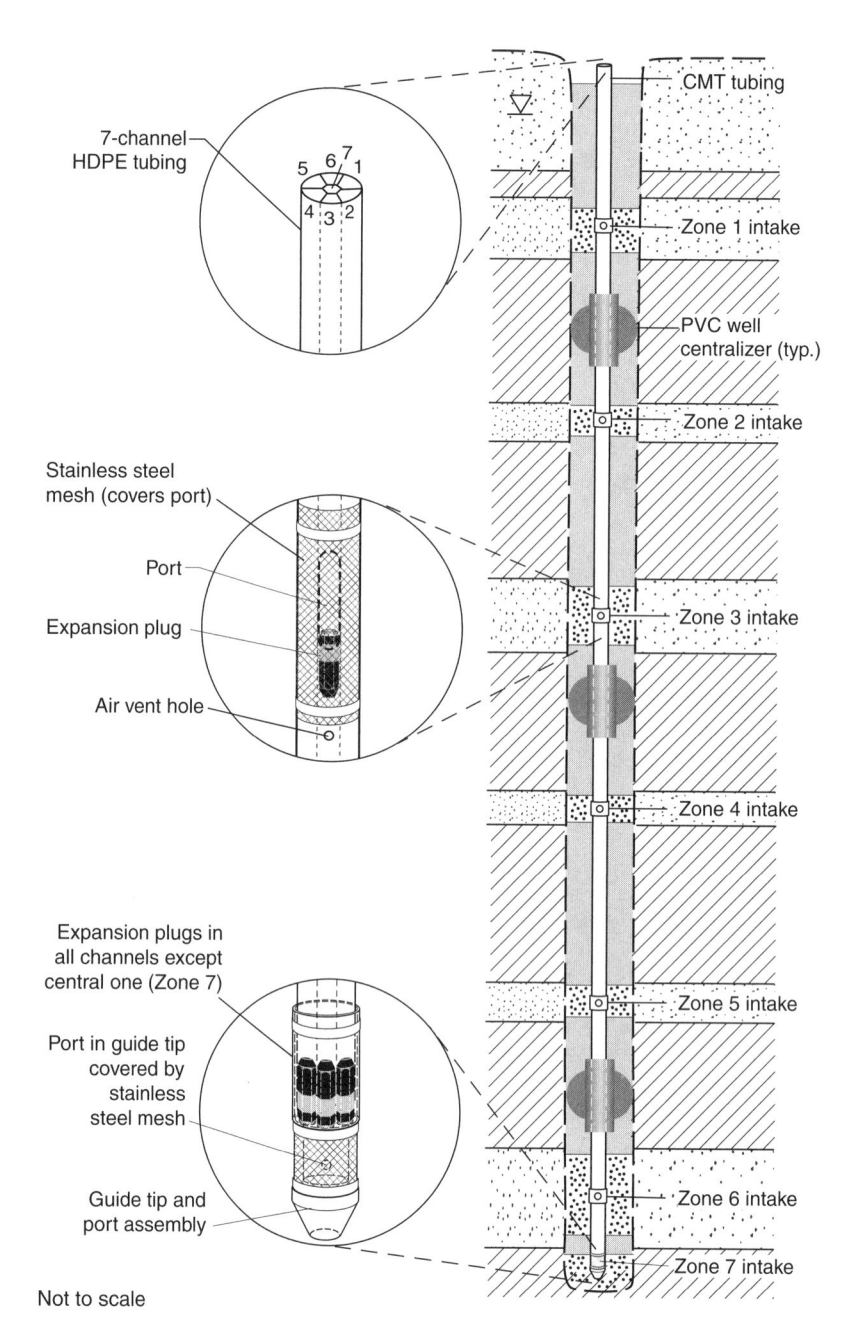

FIGURE 11.16
The Solinst CMT system.

Sand packs and annular seals between the various monitored zones can be installed by backfilling the borehole with alternating layers of sand and bentonite. Inflatable rubber packers for permanent or temporary installations in bedrock aquifers and multiscreen wells are also under development (see Johnson et al., 2002).

Hydraulic heads are measured with conventional waterlevel meters or electronic pressure transducers to generate vertical profiles of hydraulic head. Ground-water samples are collected using peristaltic pumps, small-diameter bailers, inertial lift pumps, or small-diameter double-valve pumps.

CMT multilevel wells have been installed to depths up to 300 ft below ground surface, although most systems have been installed to depths under 200 ft. These wells have been installed in boreholes created in unconsolidated deposits and bedrock using a wide range of drilling equipment including rotasonic, air rotary, diamond-bit coring, and hollow-stem auger.

A small (1.1 in.) diameter three-channel CMT system has also been developed for installation with DP sampling equipment. Sand pack and bentonite cartridges have also been developed for the three-channel CMT system and are undergoing field trials (unpublished results).

The CMT multilevel monitoring system is described in detail in Einarson and Cherry (2002). A case study in California where CMT wells were installed to depths of 200 ft using sonic drilling equipment is presented by Lewis (2001). The use of CMT wells to assess the fate and transport of MTBE in a chalk aquifer in the U.K. is described by Wealthall et al. (2001). More information about the CMT multilevel monitoring system is available from Solinst Canada, Ltd. (www.solinst.com).

Water FLUTe System

The Water FLUTe (Flexible Liner Underground Technology) is a multilevel ground-water monitoring system that uses a flexible impermeable liner of polyurethane-coated nylon fabric to isolate more than 20 discrete intervals in a single borehole. The system comes in various sizes and can monitor boreholes from 2 to 20 in. diameter (most installations are in 4- to 10-in. diameter boreholes). The system is custom-made at the factory to the customer's specifications. Sampling ports are created in the liner at the specified depths and small-diameter tubing (0.17 and 0.5 in. OD) is connected to the sampling ports. Pressure transducers and cables (if used) are also installed at the appropriate positions in the liner. The system is pressure tested to 300 psi at the factory. The system is shipped to the job site on a reel and is lowered to the bottom of the borehole by spooling the liner, sampling tubes, transducer cables, etc. off of the reel (Figure 11.17a). The system is shipped "inside out" which facilitates everting the liner and tubes into the borehole. Once the liner is everted, the sampling tubes and cables are inside the liner. The force required to evert the liner comes from hydrostatic pressure that is created by filling the liner with water at the ground surface. Ground water in the borehole is either displaced by the liner or can be pumped out during the installation. The borehole is sealed over its entire length by the pressurized liner. The system is removable by reversing the installation procedure, and may be installed in open boreholes or multiscreened wells.

Samples are collected by applying gas pressure to the sampling tubes, which forces the ground-water sample to the surface (Figure 11.17b). Two check valves are installed in each of the sampling tubes. One of the check valves prevents the water sample from being forced back out of the sampling port when the pressure is applied. The second check valve prevents the ground-water sample from falling back down the sampling line between pressure applications. The system is pumped in three strokes with two purge-pressure applications and one lower-pressure application for sampling. The two purge strokes completely remove all stagnant water from the system. All ports can be purged and sampled simultaneously because the dedicated pump system for each port is essentially the same length regardless of the port depth. Hence, each port produces the same purge and sample volume.

Depth-to-water measurements can be made inside the sampling tubing using small-diameter water-level meters. Optional dedicated pressure transducers facilitate continu-

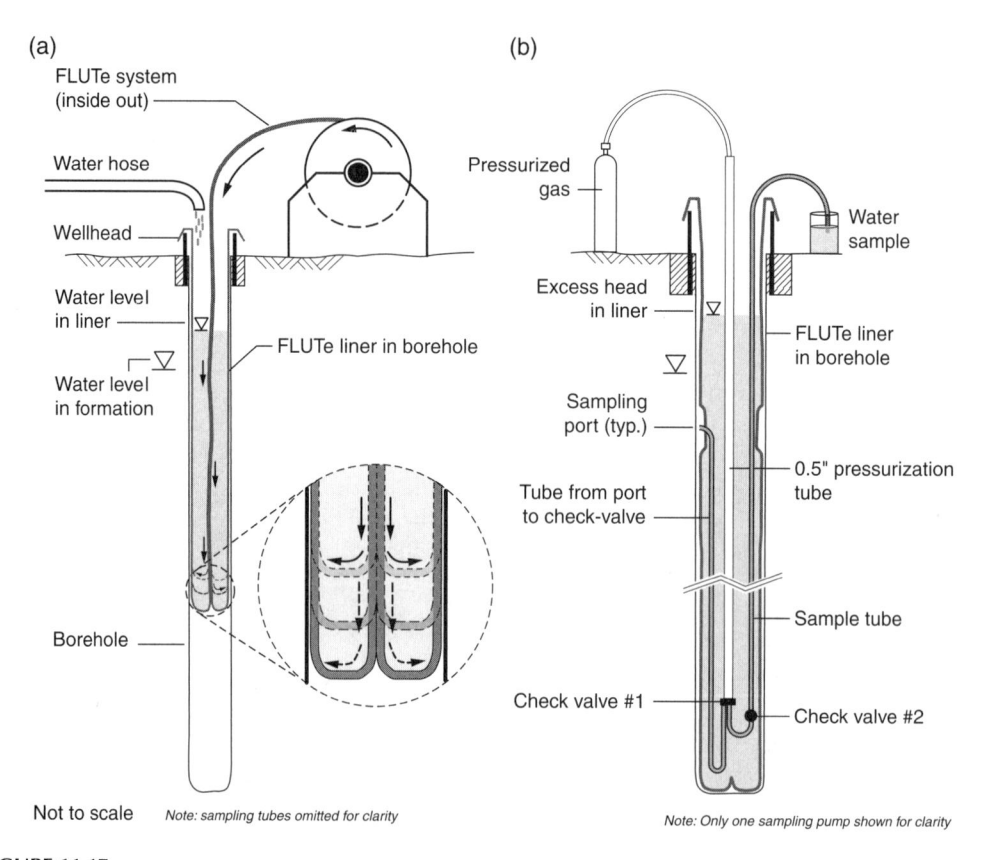

FIGURE 11.17
The Water FLUTe system. (a) Installation of a Water FLUTe system; (b) Collecting ground water samples with a Water FLUTe multilevel system.

ous, long-term pressure monitoring. The pressure transducers do not interfere with sampling or manual water-level measurement, or limit the number of ports on the system.

The eversion installation procedure allows installation into nearly horizontal angled holes. A smaller diameter Water FLUTe system has been successfully installed in DP holes with five ports to 60 ft. The seal of the hole is provided by the pressurized liner; no sealing backfill or hole collapse is typically required.

According to the manufacturer, other FLUTe flexible liner systems are used for the following hydrologic applications:

- Sealing of boreholes with blank liners
- Hydraulic conductivity profiling of a borehole while installing a sealing liner
- Multilevel sampling in the vadose zone
- Color reactive mapping of LNAPL and DNAPL in boreholes and cores
- Liner augmentation of horizontal drilling
- Towing of logging tools and cameras into boreholes

More information about the FLUTe system can be obtained from Flexible Liner Underground Technologies, Ltd. (www.flut.com).

References

Akindunni, F.F., R.W. Gillham, B. Conant, Jr., and T. Franz, Modeling of contaminant movement near pumping wells: saturated-unsaturated flow with particle tracking, *Ground Water*, 33(2), 264–274, 1995.

Aller, L., T.W. Bennett, G. Hackett, R.J. Petty, J.H. Lehr, H. Sedoris, D.M. Nielsen, and J.E. Denne, Handbook of Suggested Practices for the Design and Installation of Ground-Water Monitoring Wells, EPA/600/4-89/034, U.S. Environmental Protection Agency, Las Vegas, NV, Cooperative Agreement CR-812350-01, 1989, 245 pp.

Barrow, J.C., The resonant sonic drilling method: An innovative technology for environmental restoration programs, *Ground Water Monitoring and Remediation*, 14(2), 153–160, 1994.

Black, W.H., H.R. Smith, and F.D. Patton, Multiple-Level Ground-Water Monitoring With the MP System, Proceedings of the Conference on Surface and Borehole Geophysical Methods and Ground Water Instrumentation, National Water Well Association, Dublin, OH, 1986, pp. 41–60.

Black, W.H., J.A. Goodrich, and F.D. Patton, Ground-Water Monitoring for Resource Management, paper presented at ASCE International Symposium on Artificial Recharge of Ground Water, Anaheim, CA, August 21–28, 1988, pp. 1–7.

Borden, R.C., R.A. Daniel, L.E.I. LeBrun, and C.W. Davis, Intrinsic biodegradation of MTBE and BTEX in a gasoline-contaminated aquifer, *Water Resources Research*, 33(5), 1105–1115, 1997.

California Department of Water Resources, California Water Well Standards, Bulletin 74-90, California Department of Water Resources, Sacramento, CA, 1990.

Cherry, J.A., personal communication, 2004.

Cherry, J.A. and P.E. Johnson, A multi-level device for monitoring in fractured rock, *Ground-Water Monitoring Review*, 2(3), 41–44, 1982.

Cherry, J.A., J.F. Barker, P.M. Buszka, J.P. Hewetson, and C.I. Mayfield, Contaminant Occurrence in an Unconfined Sand Aquifer at a Municipal Landfill, Proceedings of the Fourth Annual Madison Conference on Applied Research and Practice on Municipal and Industrial Waste, University of Wisconsin, Madison, Wisconsin, 1981, pp. 1–19.

Cherry, J.A., R.W. Gillham, E.G. Anderson, and P.E. Johnson, Migration of contaminants in ground water at a landfill: a case study 2. Ground-water monitoring devices, *Journal of Hydrology*, 63, 31–49, 1983.

Chiang, C., G. Raven, and C. Dawson, The relationship between monitoring well and aquifer solute concentrations, *Ground Water*, 33(5), 718–726, 1995.

Church, P.E. and G.E. Granato, Bias in ground-water data caused by well-bore flow in long-screen wells, *Ground Water*, 34(2), 262–273, 1996.

Conant, B., Jr., F.F. Akindunni, and R.W. Gillham, Effect of well-screen placement on recovery of stratified contaminants, *Ground Water*, 33(3), 445–457, 1995.

D'Astous, A.Y., W.W. Ruland, R.G. Bruce, J.A. Cherry, and R.W. Gillham, Fracture effects in the shallow ground water zone in weathered Sarnia-Area Clay, *Canadian Geotechnical Journal*, 26, 43–56, 1989.

Delouvrier, J. and J. Delay, Multi-Level Ground-Water Pressure Monitoring at the Meuse/Haute-Marne Underground Research Laboratory, France, Proceedings of EurEnGeo 2004, Liege, Belgium, May, 2004, pp. 377–384.

Devlin, J.F., M.L. McMaster, D.J. Katic, and J.F. Barker, 2001, Evaluating Natural Attenuation in a Controlled Field Experiment by Mass Balances, Flux Fences and Snapshots: A Comparison of Results, Paper presented at Ground Water Quality 2001, Sheffield, U.K., July 18–21, 2001, pp. 282–285.

Drost, W., D. Klotz, A. Koch, H. Moser, F. Neumaier, and W. Rauert, 1968, Point dilution methods of investigating ground water flow by means of radioisotopes, *Water Resources Research*, 4(1), 125–146, 1968.

Einarson, M.D., EnviroCore: A new dual-tube direct push system for collecting continuous soil cores, *Proceedings of the Ninth National Outdoor Action Conference*, National Ground Water Association, Dublin, OH, 1995, pp. 419–433.

Einarson, M.D. and J.A. Cherry, A new multi-level ground-water monitoring system utilizing multi-channel tubing, *Ground-Water Monitoring and Remediation*, 22(4), 52–65, 2002.

Einarson, M.D. and D.M. Mackay, Predicting impacts of ground-water contamination, *Environmental Science & Technology*, 35(3), 66A–73A, 2001.

Elci, A., F. Molz, and W.R. Waldrop, Implications of observed and simulated ambient flow in monitoring wells, *Ground Water*, 39(6), 853–862, 2001.

Elci, A., G.P. Flach, and F. Molz, Detrimental effects of natural vertical head gradients on chemical and water-level measurements in observation wells: identification and control, *Journal of Contaminant Hydrology*, 28, 70–81, 2003.

Feenstra, S., J.A. Cherry, and B.L. Parker, Conceptual models for the behavior of nonaqueous phase liquids (DNAPLs) in the subsurface, in *Dense Chlorinated Solvents and Other DNAPLs in Ground Water*, James F. Pankow and John A. Cherry, Eds., Waterloo Press, Waterloo, Ontario, Canada, pp. 53–88.

Foote, G.R., N.T. Bice, L.D. Rowles, and J.D. Gallinatti, TCE and flow monitoring methods using an existing water supply well, *Journal of Environmental Engineering*, June, 564–571, 1998.

Freeze, R.A. and J.A. Cherry, *Groundwater*, Prentice-Hall, Englewood Cliffs, NJ, 1979, 604 pp.

Gernand, J., B. Rundell, and C. Yen, Practical Bedrock Aquifer Characterization Using Borehole Geophysics and Multilevel Wells, Practice Periodical of Hazardous, Toxic, and Radioactive Waste Management, 2001, pp. 111–118.

Gibs, J., G.A. Brown, K.S. Turner, C.L. MacLeod, J.C. Jelinski, and S.A. Koehnlein, Effects of small-scale vertical variations in well-screen inflow rates and concentrations of organic compounds on the collection of representative ground-water quality samples, *Ground Water*, 31(2), 201–208, 1993.

Gossell, M.A., T. Nishikawa, R.T. Hanson, J.A. Izbicki, M.A. Tabidian, and K. Bertine, Application of flowmeter and depth-dependent water quality data for improved production well construction, *Ground Water*, 37(5), 729–735, 1999.

Guilbeault, M.A., B.L. Parker, and J.A. Cherry, Mass flux distributions from DNAPL Zones in Sandy Aquifers, *Ground Water*, 43(1), 70–86, 2005.

Haas, J.E. and C. Sosik, Smart Pump-and-Treat Strategy for MTBE Impacting a Public Water Supply Well Field, Proceedings of the Fourteenth Annual Conference on Contaminated Soils, AEHS, Amherst, MA, 1998, pp. 1–31.

Hanson, R.T., M.W. Newhouse, C.M. Wentworth, C.F. Williams, T.E. Noce, and M.J. Bennett, Santa Clara Valley Water District Multi-Aquifer Monitoring Well Site, Coyote Creek Outdoor Classroom, San Jose, California, U.S. Geological Survey, Open-File Report 02-369, 2002, 4 pp.

Hutchins, S.R. and S.D. Acree, Ground-water sampling bias observed in shallow conventional wells, *Ground-Water Monitoring and Remediation*, 20(1), 86–93, 2000.

Jansen, J., Geophysical well logging and discrete water sampling methods to identify sources of elevated barium in a water well in Northern Illinois, Proceedings of the Symposium on the Application of Geophysics to Environmental and Engineering Problems, Society for Environmental and Engineering Geophysics, Denver, CO, 1998, pp. 315–322.

Johnson, T.L., A comparison of well nests vs. single-well completions; *Ground-Water Monitoring Review*, 3(1), 76–78, 1983.

Johnson, C.D., F.P. Haeni, and J.W. Lane, Importance of discrete-zone monitoring systems in fractured-bedrock wells — a case study from the University of Connecticut Landfill, Storrs, Connecticut, Proceedings of the Symposium on the Application of Geophysics to Engineering and Environmental Problems, Society for Environmental and Engineering Geophysics, Denver, CO, 2002, pp. 1–4.

Kao, C.M. and Y.S. Wang, Field investigation of natural attenuation and intrinsic biodegradation rates at an underground storage tank site, *Environmental Geology*, 40(4–5), 622–631, 2001.

Lewis, M., Installing continuous multi-chamber tubing using sonic drilling, *Water Well Journal*, July, 16–17, 2001, pp. 16–17.

MacFarlane, D.S., J.A. Cherry, R.W. Gillham, and E.A. Sudicky, Migration of contaminants in ground water at a landfill: a case study 1. ground water flow and plume delineation, *Journal of Hydrology*, 63, 1–29, 1983.

Mackay, D.M., J.A. Cherry, D.L. Freyberg, and P.V. Roberts, A natural gradient experiment on solute transport in a sand aquifer 1. Approach and overview of plume movement, *Water Resources Research*, 22(13), 2017–2029, 1986.

Martin-Hayden, J.M., Controlled laboratory investigations of wellbore concentration response to pumping, *Ground Water*, 38(1), 121–128, 2000a.

Martin-Hayden, J.M., Sample concentration response to laminar wellbore flow: implications to ground-water data variability, *Ground Water*, 38(1), 12–19, 2000b.

Martin-Hayden, J.M. and G.A. Robbins, Plume distortion and apparent attenuation due to concentration averaging in monitoring wells, *Ground Water*, 35(2), 339–346, 1997.

Martin-Hayden, J.M., G.A. Robbins, and R.D. Bristol, Mass-balance evaluation of monitoring well purging, Part II. Field tests at a gasoline contamination site, *Journal of Contaminant Hydrology*, 8, 225–241, 1991.

McIlvride, W.A. and B.M. Rector, Comparison of short and long-screened wells in alluvial sediments, Proceedings of the Second Outdoor Action Conference, National Ground-Water Association, Dublin, OH, 1988, pp. 375–390.

Nakamoto, D.B., F.R. McLaren, and P.J. Philips, Multiple completion monitor wells, *Ground-Water Monitoring Review*, 6(2), 50–55, 1986.

National Research Council, *Alternatives for Ground Water Cleanup*, National Research Council, Washington, DC, National Academy Press, 1994, 314 pp.

Newell, C.J., J.A. Conner, and D.L. Rowen, Ground Water Remediation Strategies Tool, Publication Number 4730, American Petroleum Institute, Washington, DC, 2003, 80 pp.

Pankow, J.F. and J.A. Cherry, *Dense Chlorinated Solvents and other DNAPLs in Groundwater*, Waterloo Press, Waterloo, Ontario, Canada, 1996, 522 pp.

Parker, L.V. and C.H. Clark, Study of Five Discrete Interval-Type Ground Water Sampling Devices, ERDC/CRREL Publication #TR-02–12, U.S. Army Corps of Engineers, Hanover, NH, 2002, 49 pp.

Pitkin, S.E., J.A. Cherry, R.A. Ingleton, and M. Broholm, Field demonstrations using the Waterloo ground-water profiler, *Ground-Water Monitoring and Remediation*, 19(2), 122–131, 1999.

Puls, R.W. and C.J. Paul, Multi-layer sampling in conventional monitoring wells for improved estimation of vertical contaminant distribution and mass, *Journal of Contaminant Hydrology*, 25(1), 85–111, 1997.

Raven, K.G., K.S. Novakowski, R.M. Yager, and R.J. Heystee, Supernormal fluid pressures in sedimentary rocks of southern Ontario–Western New York State, *Canadian Geotechnical Journal*, 29, 80–93, 1992.

Reilly, T.E. and J. Gibs, Effects of physical and chemical heterogeneity on water quality samples obtained from wells, *Ground Water*, 31(5), 805–813, 1993.

Reilly, T.E. and D.R. LeBlanc, Experimental evaluation of factors affecting temporal variability of water samples obtained from long-screened wells, *Ground Water*, 36(4), 566–576, 1998.

Reilly, T.E., O.L. Franke, and G.D. Bennett, Bias in ground-water samples caused by well-bore flow, *Journal of Hydrologic Engineering*, 115, 270–276, 1989.

Reinhard, M., N.L. Goodman, and J.F. Barker, Occurrence and distribution of organic chemicals in two landfill leachate plumes, *Environmental Science & Technology*, 18, 953–961, 1984.

Rivett, M.O., Soil-gas signatures from volatile chlorinated solvents: borden field experiments, *Ground Water*, 33(1), 84–98, 1995.

Robbins, G.A., Influence of purged and partially penetrating monitoring wells on contaminant detection, mapping and modeling, *Ground Water*, 27(2), 155–162, 1989.

Robbins, G.A. and J.M. Martin-Hayden, Mass-balance evaluation of monitoring well purging, Part I. Theoretical models and implications for representative sampling, *Journal of Contaminant Hydrology*, 8, 203–224, 1991.

Robertson, W.D., J.A. Cherry, and E.A. Sudicky, Ground-water contamination from two small septic systems on sand aquifers, *Ground Water*, 29(1), 82–92, 1991.

Ronen, D., M. Magaritz, and I. Levy, An in-situ multilevel sampler for preventive monitoring and study of hydrochemical profiles in aquifers, *Ground-Water Monitoring Review*, 7(4), 69–74, 1987.

Semprini, L., P.K. Kitanidis, D.H. Kampbell, and J.T. Wilson, Anaerobic transformation of chlorinated aliphatic hydrocarbons in a sand aquifer based on spatial chemical distributions, *Water Resources Research*, 31(4), 1051–1062, 1995.

Smith, R.L., R.W. Harvey, J.H. Duff, and D.R. LeBlanc, Importance of Close-Interval Vertical Sampling in Delineating Chemical and Microbiological Gradients in Ground-Water Studies, U.S. Geological Survey, Open File Report 87-109, 1987, pp. B33–B35.

State of Washington, Washington State Administrative Code (WAC), 173-160-420(3), State of Washington, 2004.

Sterling, S.N., B.L. Parker, J.A. Cherry, J.H. Williams, J.W. Lane, and F.P. Haeni, Vertical cross contamination of TCE in a borehole in fractured sandstone, *Ground Water*, in press.

Sukop, M.C., Estimation of vertical concentration profiles from existing wells. *Ground Water*, 38(6), 836–841, 2000.

Taraszki, M., C. Merey, and G. Mitchell, Ground-water quality evaluation using Westbay monitoring well systems, Former Fort Ord, California, Proceedings of the 2002 National Monitoring Conference, National Water Quality Monitoring Council, Washington, DC, 2002.

U.S. Army Corps of Engineers, Monitoring Well Design, Installation and Documentation at Hazardous, Toxic, and Radioactive Waste Sites, Engineer Manual 1110–1-4000, U.S. Army Corps of Engineers, Washington, DC, 1998, 69 pp.

U.S. EPA, RCRA Ground-Water Monitoring: Draft Technical Guidance. Office of Solid Waste, U.S. Environmental Protection Agency, November 1992. Update to Chapter 11 of SW-846 (Revision 0, September 1986) and the Technical Enforcement Guidance Document (TEGD), 1992.

U.S. EPA, Expedited Site Assessment Tools for Underground Storage Tank Sites — A Guide for Regulators, EPA/510-B-97-001, U.S. Environmental Protection Agency, Washington, DC, 1997.

U.S. EPA, Monitored Natural Attenuation for Ground Water, Seminar Notes, EPA/625/K-98/00, U.S. Government Printing Office, Washington, DC, 1998.

van der Kamp, G., L.D. Luba, J.A. Cherry, and H. Maathuis, Field study of a long and very narrow contaminant plume, *Ground Water*, 32(6), 1008–1016, 1994.

Varljen, M.D., M.J. Barcelona, J. Obereiner, and D. Kaminski, Numerical simulations to assess the monitoring zone achieved during low-flow purging and sampling, *Ground-Water Monitoring and Remediation*, in press.

Vroblesky, D.A. and W.T. Hyde, Diffusion samplers as an inexpensive approach to monitoring VOCs in ground water, *Ground-Water Monitoring and Remediation*, 17(3), 177–184, 1997.

Vroblesky, D.A., L.C. Rhodes, J.F. Robertson, and J.A. Harrigan, Locating VOC contamination in a fractured-rock aquifer at the ground-water/surface-water interface using passive vapor collectors, *Ground Water*, 34(2), 223–230, 1996.

Wealthall, G.P., S.F. Thornton, and D.N. Lerner, 2001, Assessing the Transport and Fate of MTBE-Amended Petroleum Hydrocarbons in the UK Chalk Aquifer, paper presented at Ground Water Quality 2001, *Third International Conference on Ground-Water Quality*, University of Sheffield, U.K., June 18–21, 2001, pp. 369–374.

12

Monitoring Well Post-Installation Considerations

Curtis A. Kraemer, James A. Shultz, and James W. Ashley

CONTENTS
Introduction 849
Monitoring Well Development 850
   General Considerations 850
   AirLift Surging and Pumping with Compressed Air 855
   Mechanical Surging and Pumping with a Surface Centrifugal Pump 857
   Mechanical Surging and Pumping with a Submersible Pump 857
   Mechanical Surging with a Surge Block 858
   Valved and Air-Vented Surge Plunger 859
   Mechanical Surging with a Bailer 859
   Manual Development 860
   High-Pressure Water Jetting 860
   Decontamination 861
   Surveying 861
   Well Identification 864
Reporting Well Construction Details 866
Monitoring Well Maintenance and Rehabilitation 868
Monitoring Well and Borehole Decommissioning 872
   Objectives 872
   Planning for Decommissioning 872
   Location and Inspection 875
   Decommissioning Materials 875
   Placing Abandonment Materials 877
   Procedures for Decommissioning 878
   Records and Reports 880
References 881

Introduction

Following the installation of a monitoring well or monitoring system, several important issues must be addressed to ensure the well's or system's integrity, identity, and long-term operation. This chapter covers a variety of important monitoring well post-installation considerations including: well development, surveying, identification, reporting of construction details, maintenance and rehabilitation, and abandonment. Monitoring well development is very important in terms of ensuring collection of representative data from a well because it is the activity performed in the well to correct damage done during drilling to the formation surrounding the borehole, and to remove fine materials (silt, clay, fine sand) and drilling fluids from the filter pack and the formation in the immediate vicinity of the well. This is done to provide maximum efficiency and hydraulic communication between the well and the adjacent formation to ensure that future formation hydraulic conductivity test results are of maximum value and to ensure that representative ground-water samples may be collected in the future. Surveying of monitoring wells to a common datum is necessary to obtain accurate water-level data, to construct ground-water contour maps, and to allow the correlation of stratigraphic horizons from well to well and the subsequent development of hydrogeologic cross-sections. Proper well identification and complete reporting of monitoring well construction details are necessary elements of documentation. Monitoring well maintenance and rehabilitation are necessary to keep wells operational and to maximize the life of a well system. Monitoring well decommissioning (also called "abandonment") is required to mitigate the potential for an unused, unnecessary, or malfunctioning well to become a vertical conduit for contaminant migration. Each important element of a ground-water monitoring program is discussed in the following sections.

Monitoring Well Development

General Considerations

The purpose of this section is to provide a summary of monitoring well development guidelines and methods and refer the reader to the published literature in which the details are well presented. This summary does not attempt to address all related safety concerns. It is the responsibility of the practitioner to establish appropriate health and safety practices.

All drilling and direct-push well installation methods create at least some amount of damage (clogging, coating, smearing, and compaction) to the borehole wall and the natural formation materials (unconsolidated overburden or bedrock) immediately adjacent to the borehole wall, resulting in a localized reduction in formation hydraulic conductivity. This drilling damage must be rectified to allow the well to produce water representative of that in the formation, and to allow accurate determinations of formation hydraulic conductivity via slug or bail tests or pumping tests. Where highly permeable material is encountered, there can be a significant loss of drilling fluid to the adjacent formation, altering the quality of the ground water in the vicinity of the well. This lost drilling fluid must be recovered to allow collection of representative ground-water samples in the future. Additionally, during the installation of filter-packed monitoring wells, fine materials from the adjacent formation can mix with the filter pack material as it is placed around the well screen, reducing its permeability. These fine materials must be removed so they do not become entrained in future ground-water samples or result in reduced well yield.

The goals of monitoring well development are to:

- Remove fine materials (silt, clay, fine sand) and water lost during drilling from the filter-pack materials and the formation materials in the immediate vicinity of the screened interval of a monitoring well (for wells completed in unconsolidated overburden or incompetent [weathered or very fractured] bedrock) or an open bedrock borehole (for wells completed in competent rock).
- Correct damage to the borehole wall and adjacent native geological (natural formation) material caused during drilling.

- Stabilize the filter pack and formation material in the immediate vicinity of the well screen.
- Maximize the hydraulic communication between the well and the adjacent formation material.

During well installation, extraneous materials (e.g., grout, bentonite, sand) may inadvertently be dropped into the well, or material from higher levels in the borehole may slough into the screened zone. During air-rotary drilling of bedrock, cuttings may stick to the borehole wall above the water table or, during hollow-stem auger drilling, cuttings from silt or clay units penetrated by the drill can smear the borehole wall above such units. Proper installation of a monitoring well, including pumping or bailing to remove accumulated drilling fluids and fines from the well during installation or immediately after it has been installed, will minimize the amount of well development required. Installation of a monitoring well should not be considered complete until it has been properly developed. Prior to developing a monitoring well, the interior of the well casing or open bedrock borehole above the water table should be rinsed using water from the well, if possible. The purposes of this preliminary operation are to provide a well free of extraneous materials and to mitigate the amount of debris and fine materials that may have collected in the well during installation. Well development can then be conducted.

The well development activity is typically composed of:

- The application of sufficient energy to create ground-water flow reversals (surging) into and out of the well and the filter pack and formation to release and draw fine materials into the well.
- Pumping to draw water lost to the formation during drilling out of the borehole and adjacent formation, along with the fines that have been brought into the well during surging.

The development of monitoring wells should be performed either after the well casing or screen and filter pack have been installed (but before installation of the annular seal or grout materials), or as soon as practical after the well installation has been completed and the annular seal materials have cured. Maximizing well efficiency and hydraulic communication between the well and the adjacent formation improves the value of the data from future formation hydraulic conductivity tests in the well and the representativeness of ground-water samples collected from the well. Additionally, the removal of fines during development minimizes the potential for clogging and damaging of pumping equipment used during future formation hydraulic conductivity testing (e.g., slug or bail testing) and purging prior to sampling.

Since the first edition of this book was published in 1991, a large amount of the published literature regarding well development has focused on the development of water-supply wells. One important document published since the first edition is ASTM Standard D 5521, Standard Guide for Development of Ground-Water Monitoring Wells in Granular Aquifers (ASTM, 2004a), which details several development methods specifically applied to monitoring wells. Additionally, several other documents and journal articles have been published recently regarding monitoring well development methods; these will be referenced in the eight subsections that follow.

Problems can occur in monitoring wells when trying to use some of the development methods that are designed for use in water-supply wells. Water-supply well screens are generally 6 in. or larger in diameter and have a high percentage of open area, to allow them to extract a large volume of water from an aquifer. In contrast, monitoring wells are not installed to provide a high yield, but primarily to allow collection of accurate water-level data and representative samples of ground water from specific depth intervals. Furthermore, many monitoring wells are purposely installed in poorly yielding, predominantly fine-grained geological materials that may not be considered aquifers. Many monitoring wells are constructed with 2–4-in. diameter, machine-slotted PVC well screens that have a very low percentage of open area through which ground water and fine materials can move. This adds considerably to the challenge of developing a well, as such well screens and the filter packs surrounding them are essentially at odds with what would be ideal for well development to be effective, and they hinder rather than facilitate the process. Standard water-supply well-development methods are not typically useful in these wells without some modification. Water-quality issues associated with some well development methods (i.e., hydraulic jetting using water foreign to the formation) also have to be considered. Additionally, contamination present in the discharged development water from monitoring wells can create a containment and disposal problem. Discharge of such water to the ground surface versus containment (e.g., in drums or tanks for appropriate future disposal) must be addressed on a site-by-site basis, along with the related issue of health and safety considerations for field personnel.

Other factors that affect the proper completion of monitoring well development are time and cost. Monitoring wells are often installed in response to a regulatory requirement to investigate potential or known ground-water contamination and, therefore, they do not generate revenue like many water-supply wells (e.g., $ per 1000 gal of water use). Monitoring well development is usually charged as an hourly fee and wells installed in low-yielding formations (composed of very fine sand, or with a high percentage of silt or clay) can require considerable time for water-level recovery after water is evacuated from the well casing. The well development activity can be long and tedious, and is often left until all of the monitoring wells have been installed at a site. The well development process is then often performed with man-portable equipment that can be handled by one or two people.

Ideally, all monitoring wells should be developed for a sufficient period of time (often many hours) to remove all fine material from the filter pack and adjacent formation to allow for the collection of turbidity-free, representative samples of ground water for chemical analysis. However, there is often a cost–benefit trade off during well development, and development is often discontinued after a short period of time, when much of the visible turbidity appears to have been removed, but before it is completely removed. If a monitoring well is not completely developed so that representative samples of the ground water can be collected for analysis, then all of the money spent installing the monitoring well and analyzing ground-water samples may result in analytical data that are not truly representative of subsurface conditions at the site. These analytical data often form the basis for important decisions on whether a site must be remediated, on selection of site remediation methods and on effectiveness of an implemented remedial action, and must be representative of site subsurface conditions. Therefore, the cost–benefit ratio should strongly favor proper well development. Monitoring well development should continue until:

- Visibly clear water is discharged during the active (surging) portion of the development process.
- Field-measured quality (e.g., pH, Eh, conductivity) of the discharged water stabilizes and the turbidity is reduced to less than 10 Nephelometric Turbidity Units (NTU).
- The total volume of water discharged from the well is at least equal to the estimated volume of fluid lost to the formation during drilling and well installation (ASTM, 2004a).

These criteria should be applied to all wells, and to all development methods described herein. Treadway (1992) further suggested that those developing monitoring wells should not set time limits on any part of the well development program, but use information from the well (quantity and quality) for evaluating the effectiveness of the well development program.

Monitoring well development is, in part, accomplished by actively agitating (surging) the water column in a well, forcing water back and forth through the well screen and the filter pack and formation to release fine materials (silt, clay, fine sand) from the formation and bring them into the well. This material is then removed, along with drilling and well installation fluids, by pumping or bailing. Passive well development, using only pumping (especially at low flow rates) without surging, results in the flow of water in only one direction through the well screen. This will not effectively stabilize the filter pack or the adjacent formation, nor will it effectively remove fines or drilling fluids from the formation. Although visibly clear water may eventually be discharged as a result of such pumping, the next activity that creates a surge in the well (a slug or bail test or pumping test, purging prior to sampling, or sampling with a bailer) can release considerable turbidity. Because of their intended use, monitoring wells (unlike water-supply wells) spend most of their time in a dormant (unpumped) condition. Therefore, there is no activity in a monitoring well to continue removal of small amounts of fines over an extended period of time. No matter how complete the passive development of a monitoring well appears to be at the time of development, there is a high probability (especially for wells completed in fine-grained formations) that future introduction of pumps or bailers for testing or sampling will create a surge, rendering the water produced from the well at least somewhat turbid. It is, therefore, imperative to adequately develop monitoring wells to minimize future turbidity problems and to mitigate the need for redevelopment.

It is important to note that a field investigation by Paul et al. (1988) indicates that surging during well development increased the turbidity of monitoring wells installed in fine-grained glacial till at two sites in Wisconsin. The well development study consisted of two events of surging and bailing some of the wells, and only bailing the other wells. The results of the second event were compared with the results of the first event, to determine if turbidity was reduced and formation hydraulic conductivity was improved. Although the method of well installation was found to have an impact on turbidity increase, wells installed by the same method, but whose development included surging, yielded water samples with turbidity between 3 and 100 times greater than water samples from wells that were simply bailed. The conclusion of this study was that surging should be avoided for monitoring wells completed in fine-grained till because: (1) it substantially increases the turbidity of water samples; (2) it does not significantly improve hydraulic well response; and (3) it adds unnecessary cost to the overall sampling program. This appears to indicate that for monitoring wells constructed in some very fine-grained formations, well development will not appreciably decrease turbidity or improve well communication with the natural formation. Additionally, Paul et al. (1988) indicate that there was no significant advantage in the use of continuous-slot screens or filter wrap over the use of less expensive factory-slotted screens in predominantly fine-grained formations.

Because monitoring wells are installed to obtain representative samples for water-quality analysis, no foreign material or fluids should be introduced during development that could alter existing subsurface chemical conditions including, for example, dispersing agents, acids, and disinfectants. Therefore, the use of chemicals in the well development process will not be addressed here. Also, the addition of water to aid in development is generally not recommended. The introduction of foreign water to a monitoring well has the potential to add contaminants not previously present at a site, or to dilute contamination present in the vicinity of a monitoring well. However, if it is necessary to add water, the source of the development water must be of good quality and documented by laboratory analysis for at least the analytical parameters included in the site-specific monitoring program. Under such conditions, the amount of water discharged from the well during development should at least equal the amount of water added.

The selection of an appropriate method or combination of methods for development of a monitoring well should include consideration of the following factors:

- Drilling and well installation method employed
- Condition at the bottom of the well casing (capped or plugged? How?)
- Well casing and screen diameter
- Screen length, slot size, and percent of open area
- Depth to static water level
- Height of the water column within the well
- Presence, type, and thickness of the filter pack
- Character (e.g., silt, gravel, bedrock) and hydraulic conductivity of the natural formation
- Site accessibility
- Type of personnel to perform the development
- Type of equipment available, e.g., personally portable or truck-mounted or drill rig
- Type of suspected or known contaminants present
- Need for appropriate disposal of the discharged water
- Time available and cost effectiveness of the methods
- Health and safety requirements for field staff
- Appropriate regulatory agency approval of selected methods

The following sections provide a summary of guidance for eight methods applicable to the development of 2–4-in. diameter monitoring wells. Development methods for larger-diameter monitoring wells could include combining or modifying the methods presented herein or the use of some methods applied mainly to water-supply wells, including: (1) double-pipe air-lift pumping and backwashing, with equipment decontamination between wells; (2) overpumping and backwashing (rawhiding), including pump and discharge pipe decontamination between wells; and (3) hydraulic jetting using water foreign to the formation. However, these methods can alter water chemistry in the well and the adjacent formation, and could generate significant volumes of contaminated water, which could create containment and logistical disposal problems and costs. These methods are well documented in U.S. EPA (1975), U.S. Department of Interior (1977), and Driscoll (1986).

The eight monitoring well development methods and guidelines for their use are presented in the following sections. The methods include:

- Airlift surging or pumping with compressed air
- Surface centrifugal pump and mechanical surging
- Submersible pump and mechanical surging
- Mechanical surging with a surge block
- Valved and air-vented surge plunger

- Bailer and mechanical surging
- Manual development
- High-pressure hydraulic jetting using formation water

These methods can be used individually or in any combination appropriate to complete development of a monitoring well, depending upon site conditions. The descriptions of these methods are summarized from the related published literature, the experience of the author, and modification of the procedures in EA Engineering, Science, and Technology, Inc. (1985).

AirLift Surging and Pumping with Compressed Air

Generally, development by surging and pumping with compressed air will work for most monitoring wells no matter what the depth to water below ground surface, assuming sufficient air pressure and volume, and a sufficient height of water column in the well. However, conventional single-line airlift methods that use the well casing as the eductor pipe to bring the discharging air and surging water flow to the surface (i.e., compressed air is blown directly into the well casing) are not appropriate for use in monitoring wells. The disadvantages of using the single-line air-lift method include:

- Introduction of oil into the well if the compressed air is not completely filtered.
- A change in the water chemistry near the well screen due to aeration of the ground water.
- Introduction (entrapment) of air into the filter pack and formation and the slots of the well screen that may reduce the hydraulic conductivity of the formation and hinder hydraulic communication between the well, filter pack, and formation.
- Surface discharge of the water and air is difficult to control, a serious issue to consider if contamination is present.

These concerns can be addressed by using a dual-line airlift system that consists of a smaller-diameter flexible airline (e.g., 0.5-in. polyethylene tubing) within threaded lengths of larger-diameter PVC pipe that together will fit within 2–4-in. diameter wells. As long as the end of the inner (flexible) tubing is kept at least a few feet above the bottom of the 1-in. diameter PVC pipe, the water–air mixture will discharge up the annular space between the tubing and PVC pipe to the ground surface without directly entering the well pipe and screen. A 100–150 cubic feet per minute (cfm) air compressor is generally sufficient for development of 2–4-in. diameter monitoring wells. To develop a well by this method, there must be at least 20%, and preferably 40%, submergence of the air discharge line. For example, if a monitoring well has been installed to a depth of 100 ft below grade, there must be at least a 20-ft (preferably 40-ft) column of water in the well, assuming that the water level will be drawn down during the development activity. Drawdown of the water level will create a lower percentage of submergence than would be calculated using the static water level. This modification of an older air-lift development method offers the advantage of being capable of pumping silt- and sand-laden water from the well that would bind or clog a submersible pump. However, ground-water flow is mostly in one direction through the well screen, except during surging periods when the air-lift pumping is stopped momentarily or fluctuates quickly so drawdown of the water level in the well varies somewhat, creating flow reversals. The effects of this development method can be enhanced by using it in combination with another method (e.g., mechanical surging).

It is imperative that the compressed air discharge line of the air compressor includes a functioning oil–air separator filter, although the effectiveness of such filters is questionable (ASTM, 2004a). The gross effectiveness of such a filter should be checked before and after each well is developed. Such a check can be performed by placing a clean white cloth over the air discharge, opening the discharge valve fully, and then checking the cloth for oil staining. If staining is observed, the problem must be corrected before well development is attempted.

Under no circumstances should the high-pressure hose supplied with an air compressor (especially one that has been rented) be placed within a well. Such hoses are often sheathed with synthetic rubber and have probably laid on the ground at many previous job sites, and thus absorbed a variety of contaminants. New, fresh lengths of flexible polyethylene tubing and small-diameter PVC pipe provide a reasonable alternative for use as the airline and eductor pipe for each well developed. The tubing and pipe are relatively inexpensive. Compressed air discharging at about 100 pounds per square inch (psi) can be dangerous and hoses must be handled carefully. All connections must be securely attached. The tubing and pipe should be stored in large plastic bags or other protective material prior to use, to avoid introducing contamination. Additionally, the tubing and pipe used in the well must be handled with new, clean gloves for each well, and must not be allowed to touch the ground. The method and extent of decontamination needed for any reused down-hole equipment can vary depending upon the contaminant present at a site (see Chapter 20).

Development should begin at the bottom of the well to remove accumulated fines (silt, clay, fine sand), working up to the top of the screen (or open bedrock borehole), and then down to the bottom in increments of 2–5 ft, as many times as necessary. When beginning this well development method, the air pressure should be increased slowly. Development consists of alternate surging and airlift pumping at each interval until the discharged water appears to be clear during surging. The airlifted water discharge should be directed to a containment vessel.

Howard et al. (1988) provide a modification of this method for development of deep wells that uses a dual-line air-lift system along with a double-ball check valve at the bottom to mitigate the potential for discharge of air directly into the well. Their system includes threaded 1-in. diameter schedule 40 PVC pipe and 0.5-in. diameter flexible polyethylene tubing for each well, a double-ball check valve for the base of the PVC pipe, an air compressor capable of generating at least 100 cfm at 100 psi, and an air coalescer unit to filter the compressed air. They reported that one technician could set up and operate the system for wells 100 ft or less deep and that two technicians were needed to set up and operate the system for wells deeper than 100 ft, with an average set-up time of 1 h for a 150-ft well. It was further reported that the overall time necessary for complete development of a well (removing five well volumes or about 275 gallons total per well) averaged 3–4 h.

Nuckols (1990) presented another modification of this method for the development of small-diameter (down to 2-in. diameter) wells using a 2-line (dual-wall pipe) airlift system along with development tooling (a suction tool and a double wiper tool) attached to the end of the dual-wall pipe for a two- or three-step procedure. The system was used successfully on 2.5-in. diameter wells over 600 ft deep. Also, the system is easily adapted for up to 5–8-in. diameter wells. The first step of this development procedure is using the suction tool to clean out debris in the well. The second step is using the double wiper development tool to remove fines from the filter pack and formation. The third step is to return to the first step as needed. Inlet ports located between the two wipers allow for continuous agitation of the filter pack and recovery of water and developed fines. Because the recovered water is from a specific interval, the progress of the development process can be monitored closely and real-time decisions made regarding intervals requiring additional development. The recovered water can be easily and safely discharged to a containment vessel. The key disadvantage of the system is the relatively high initial cost and that the fact that it is not man-portable. The system requires a hoist truck, air compressor, a discharge holding tank, and a two-person crew.

Mechanical Surging and Pumping with a Surface Centrifugal Pump

Monitoring well development using mechanical surging and a surface centrifugal pump can be performed only if the depth to water is within the practical limit of suction lift (i.e., less than about 20 ft below ground surface), and can be effective for low-yield wells. Because operation of a surface centrifugal pump depends upon suction, all fittings on the intake side of the pump must be airtight.

A good, inexpensive choice for the suction line in the well is 1/2- to 3/4-in. diameter flexible polyethylene pipe. A new, unused length should be used for each well. The pipe should be stored in large plastic bags until ready for use, and should be handled with new clean gloves for each well. Additionally, the pipe must not be allowed to become contaminated by touching the ground. The end of the pipe that will be placed into the well should be cut off at an angle to minimize the potential of becoming quickly plugged in potentially accumulated silt at the bottom of the well. Additionally, this end of the pipe should be fitted with one or more large steel washers that are large enough to fit over the polyethylene pipe, but small enough to fit into the well. If the washers are reused, they must be decontaminated before use in each well (see Manual Development section). The washers should be held in place a few inches from the end of the polyethylene pipe by two standard hose clamps tightened by a screwdriver. The washers act as a plunger when repeatedly raised and lowered in 1–2 ft increments within the screened interval or in an open bedrock borehole. Care must be taken during this process to ensure that the well screen is not damaged by the washers. Surging (plunging) will force ground water back and forth through the well screen or bedrock fractures and joints. However, if the well screen is not completely saturated (i.e., submerged below the water table), the surging (plunging) activity must be done carefully to avoid surging the unsaturated portion of the screen, where at least some of the fines-laden water could then flow out into the filter pack and formation instead of being removed from the well. Simultaneous operation of the surface centrifugal pump will remove the turbid water and drilling fluids drawn into the well. The process should begin by pumping from the bottom of the well to remove potentially accumulated fines (e.g., silt, clay, fine sand). The development tool should then be worked up to the top of the saturated portion of the screen (or open bedrock borehole) and back down, repeatedly surging and pumping at intervals of 2–5 ft. This should be repeated as many times as necessary, until the discharged water appears to be clear during surging, field-measured quality (e.g., pH, Eh, conductivity) of the discharged water stabilizes, and the turbidity is reduced to at least 10 NTU. The total volume of water discharged from the well should be at least equal to the estimated volume of fluid lost to the formation during drilling and well installation. For monitoring wells installed in very low-yield formations (with high percentages of silt and clay), the well may quickly pump dry. Therefore, there can be considerable time between each surging and pumping cycle of the development process. The appropriateness of surging wells completed in predominantly fine-grained materials, as investigated by Paul et al. (1988), should be considered before selecting this method.

Mechanical Surging and Pumping with a Submersible Pump

Monitoring well development using mechanical surging in combination with a submersible pump can be performed in a wide variety of depth-to-water conditions, and is not limited by suction lift. The major limiting factors include: (1) well diameter, (2) impeller construction material, and (3) type and concentration of contaminants present in the ground water. Small-diameter (less than 4-in.) wells can be developed using specialty pumps. However, the presence of silt, clay, and fine sand, and high levels of some organic compounds in the well water can quickly clog or damage pumps with plastic impellers, and damage the bladder of bladder pumps. Small-diameter moyno-type, progressing cavity pumps appear to be more durable under these harsh conditions. Submersible pumps that are a nominal 4-in. diameter or larger are available with more durable, stainless steel impellers. Schalla (1986) reports that this method of well development, including the use of the pump itself as a surge block, proved to be the most successful technique for developing 4-in. diameter wells for an investigation conducted in northeastern Alabama.

This method of development is similar to the method using a surface centrifugal pump. The submersible pump must be decontaminated prior to use in each monitoring well. A good, inexpensive choice for the discharge line from the pump is 1/2- to 3/4-in. diameter flexible, polyethylene pipe. A new, unused length should be used for each well. The flexible pipe should be stored in large plastic bags or other protective means until ready for use, and should be handled with new, clean gloves for each well. Additionally, the pipe must not be allowed to become contaminated by touching the ground. A pump of slightly smaller diameter than the inside diameter of the well can act like a surge block when raised and lowered within the interval of the well to be developed. Once the pump has been placed in the monitoring well, it is repeatedly raised and lowered 1–2 ft to impart a plunging action to the water in the screened interval (or saturated open bedrock borehole). Such surging will force ground water back and forth through the well screen (or bedrock fractures and joints). Simultaneous pumping of the submersible pump will remove the turbid water, fines, and drilling/well installation fluids drawn into the well. Well development should begin at the top of the saturated portion of the screened interval (or open bedrock borehole) to prevent sand locking of the pump within the well or borehole. Repeated surging and pumping at intervals of 2–5 ft should be performed from the top of the screen to the bottom of the well and back, as many times as necessary, until the discharged water appears to be clear during surging. Field-measured quality (e.g., pH, Eh, conductivity) of the discharged water should be stable, the turbidity should be reduced to at least 10 NTU, and the total volume of water discharged from the well should be at least equal to the estimated volume of fluid lost to the formation during drilling and well installation.

Mechanical Surging with a Surge Block

Monitoring well development by mechanical surging with a surge block (swab) is described in detail in ASTM D 5521 (ASTM, 2004a). The following summarizes key issues. Although this method can be performed by a drilling contractor, the manual method is described here because, as stated previously, monitoring well development is often completed after all of the wells have been installed and the drilling contractor has left the site. The surge block (swab) may be constructed of a flexible material (i.e., EPDM or Teflon) sandwiched between two pieces of solid material (i.e., stainless steel), or be a solid material only, with or without vents or valves. For deep wells, some mechanical assistance may be necessary, such as a pulley or block and tackle set up with a tripod over the well. The method includes raising and lowering a surge block sized to be within $1/8$ to $1/4$ in. of the inside diameter of a well, and heavy enough to free-fall through the water column to create a surge during its downward stroke. The impact of this manual method is dependent on the length and force of the surging strokes that are limited by the range of motion and strength of the operator or the equipment used for mechanical assistance. This method is very effective in causing water to flow in both directions through the screen and filter pack and drawing fines into the well, thus improving the hydraulic communication between the well and the filter pack, and stabilizing the filter pack and formation. However, because this method involves only surging, with no water or sediment removal, it needs to be combined with a water and sediment removal method (e.g., pumping or bailing) to be effective. Prior to initiating use of a surge block, it is recommended that the well first be bailed or pumped to confirm that it will yield water. If the well will not yield water, surging should not be performed because if a great enough negative pressure is formed during an upward stroke of the surge block, the well screen could collapse. If the well screen is relatively short (e.g., less than 5 ft), surging can be performed within the casing above the screen interval. However, if there are zones of high hydraulic conductivity material within the screened interval, development would preferentially occur there and not result in consistent development of the entire screened interval. If the screen interval is longer (10 ft or more) or there are zones of high hydraulic conductivity material within the screened interval, surging should be performed within the screen, beginning at the top and working downward to mitigate the potential to sand-lock the surge block and damage the screen.

Valved and Air-Vented Surge Plunger

Schalla and Landick (1986) describe the use of valved and air-vented surge plungers for the development of 2-in. diameter monitoring wells. Their study indicated the following important factors in development of the small-diameter surge plunger: length of the cylinder, sufficient weight of the plunger to overcome buoyancy and resistance, the number of water ports, and the number and size of air-vent ports. Schalla and Landick (1986) report the advantages of this device (compared with air-lift pumping) to include: (1) auxillary equipment (e.g., an air compressor) and tools are not required; (2) air is not introduced into the formation as long as surging is performed above the screened interval; (3) set-up, shutdown, and decontamination can be performed in minutes by one person; (4) discharge of hazardous fluids can be piped directly and safely into drums using a "T" bypass; and (5) large volumes of water can be removed in a short period of time. For details of this method using an inexpensive and portable tool, the reader is referred to Schalla and Landick (1986). The equipment for this development method is no longer commercially available, but it could be constructed from the information provided in the referenced article.

Mechanical Surging with a Bailer

Monitoring well development by mechanical surging with a bailer can be particularly useful for developing wells completed in low-yield formations. Bailers used for this method should be dedicated, decontaminated, and bottom-fill type with dedicated, decontaminated line with which to lower the bailer into the well. This method is very labor-intensive and, depending upon the volume of the bailer used and the depth of the well, it may be appropriate to rig a tripod and pulley to aid in lifting the full bailer from the well. Each time the bailer is introduced into and removed from a well, it will impart a limited amount of surging action, which is not as effective as other surging methods, so achieving the objectives of development may require much more time than other methods. Bailing should be performed throughout the screened interval or saturated portion of an open bedrock borehole. Paul et al. (1988) found that for wells completed in fine-grained glacial till, development by bailing only resulted in wells with relatively less turbidity than development by surging and bailing.

Manual Development

The manual development method for a monitoring well can be used when the water level is too deep for a surface centrifugal (suction) pump, where there is insufficient submergence so air-lift pumping cannot be used, or where a submersible pump of appropriate diameter and construction (metal impellers) is not available. This method can be effective in low-yield wells, but is slow and requires considerable physical effort. This method consists of using polyethylene pipe (e.g., $3/4$- to 1-in. diameter) with a foot valve placed on the end of the pipe (i.e., an inertial-lift pump) that will be placed in the well and a washer attached just above the foot-valve for surging (attached as noted previously in the Mechanical Surging and Pumping With a Surface Centrifugal Pump section). The foot-valve keeps water from flowing back out of the pipe once it has entered. By quickly and repeatedly raising and lowering the pipe in increments of about 1–3 ft, the water column is surged and water is forced into and up the pipe until it finally discharges from the other end of the pipe at ground surface. The inclusion of a down-hole pitcher-type pump can reduce the physical effort required to only the surging operation. As with the previously described well development methods, new flexible pipe must be used in each well. The pipe must be carefully handled so it is not contaminated. Any reused attachments (washers, foot valves, etc.) must be decontaminated before use in each well. Finally, development should begin at the bottom of the well to remove potentially accumulated fines (silt, clay, fine sand), working up to the top of the saturated portion of the screen (or open bedrock borehole) and back down as often as necessary to obtain discharged water that is visibly clear.

High-Pressure Water Jetting

Monitoring well development by high-pressure water (or high-velocity hydraulic) jetting is described in ASTM D 5521 (ASTM, 2004a). Because of the size of the required equipment, this development method is typically performed by drilling contractors in 4-in. or larger-diameter wells that have continuous-slot screens and good yields; the conventional equipment is not man-portable. Additionally, it is important to remember that, as stated previously, many monitoring wells are constructed in predominantly fine-grained formation material, in which screens with very small slot sizes (often less than 0.010 in.) are used. A 10-slot slotted casing has approximately 4% open area, while a continuous-slot screen with the same slot size would have approximately 10% open area (see Chapter 10 or Hanson, 2001). This development method would not be very effective on a 10-slot slotted casing because nearly all of the jetting energy would be directed against blank pipe, and the pipe would be damaged very quickly by the high-pressure water.

Dougherty and Paczkowski (1988) provide a modification of this method for development of a wide variety of well sizes down to 2-in. diameter. This modified method was developed to minimize the amount of contaminated fluid produced by using water from the well for jetting. The well should be pumped to remove lost drilling fluid and particulate matter prior to initiating jetting to minimize plugging of the jetting device and the pumping of debris in the well into the screen, filter pack, and formation. This method addresses development of shallow wells (depth to water less than 20 ft below ground surface, i.e., within the limit of suction lift) using a surface centrifugal pump, and deeper wells (depth to water greater than 20 ft below ground surface) using a submersible pump. In both cases, the pump intake is placed within the well with a valve system at ground surface that allows for switching between discharge of the water to an in-well jetting tool or to a container at ground surface for the controlled discharge of water and fines. Dougherty and Paczkowski (1988) recommend that the initial jetting activity be done in short intervals to mitigate pumping loosened fines back into the well screen and filter pack. A pressure gage installed on the discharge side of the pump aids in the monitoring and assessment of the effectiveness of the jetting activity. As development of an interval is completed, the down-well equipment assembly is lowered a few feet to the next interval. Although this method uses no foreign water or fluids, there are two key disadvantages:

- The potential to pump (jet) loosened fines back into the well screen and filter sand.
- If there is stratified dissolved-phase contamination or non-aqueous phase liquid within the screened interval, such contamination would be spread throughout that interval by this development method. Under such conditions, use of this development method would not be recommended.

For details of this method, the reader is referred to Dougherty and Paczkowski (1988). Reichart (1996) provides a reconfiguration of this method of development that combines simultaneous jetting with water and surging using the water in the well. The application presented was for a 6-in. diameter extraction well to be completed in bedrock (sandstone and metamorphic gneiss) with zones of preferential weathering and planned yields approaching 200 gallons per minute. As reported, the well development process was bounded by four constraints: the drilling rig was still over the hole, the addition of foreign fluids was prohibited, generated waste had to be minimized, and specific capacities had to be maximized. Water jetting with recirculated water in the well using a submersible pump was selected. The following disadvantages were noted by Reichart (1996): the method exhibited a tendency to return fines to the formation, and there was a need to test design criteria and parameters for each specific application, which was time-consuming. For details of this method, the reader is referred to Reichart (1996).

Decontamination

It is essential that every effort be made to avoid outside contamination and the cross-contamination of monitoring wells. This can be done by ensuring that all equipment to be introduced into a well is clean. The level of effort for decontamination is a site- and project-specific issue to be resolved individually for each project. The resolution of the decontamination issue and the rationale for selection of site-specific development protocols must be made prior to installation of the monitoring wells. At a minimum, it is recommended that reusable down-hole equipment for developing monitoring wells should be steam-cleaned or washed with methanol and rinsed with clean water prior to use at each well. The reader is referred to Chapter 20 for a detailed presentation of equipment decontamination.

Surveying

Surveying is a necessary part of all ground-water monitoring programs. The locations of monitoring wells must be surveyed so they may be accurately plotted onto maps that will be used to develop and interpret hydrogeologic data. Similarly, the well elevations must be surveyed to help ensure accurate water-level measurements, to allow valid comparisons from well to well across a site, and to allow construction of ground-water contour maps. In many cases, surveying will be required by the local, state, or Federal agency directing the monitoring program, or by the party for whom the program has been implemented. Usually the degree of accuracy and the reference datum, as well as whether the surveying must be performed under the supervision of a licensed surveyor, are set forth in the regulatory requirements.

In ground-water monitoring programs without specific surveying requirements, there is still a need for a minimum amount of surveying. Well locations must be plotted onto a site plan to graphically represent their relative location on the site. For a small program on a small site, a sketch of the site showing the locations relative to site landmarks, as determined with a tape measure and a compass, may be adequate. Likewise, for a small program on a very large site, the approximate locations could be plotted onto a United States Geological Survey (U.S.G.S.) topographic map (usually a $7\frac{1}{2}$-min quadrangle). As the requirement for accuracy in the ground-water monitoring program increases, the survey accuracy should also increase. Because small programs often mushroom into large programs, it may be useful to accurately survey well locations as part of a small program and minimize the problems that could occur as the project grows. Classifications and standards for vertical and horizontal controls have been developed by the Federal Geodetic Control Committee and consist of First, Second, and Third Order Surveys with further divisions into classes (National Oceanic and Atmospheric Administration, 1974). First Order, Class I, is the most accurate and Third Order, Class II, is the least accurate. The details of survey accuracy are presented in Table 12.1; methods for surveying with varying levels of accuracy are described by Moffit and Bouchard (1975).

Monitoring well locations are normally surveyed as accurately as possible. On small to moderately sized sites (up to 100 acres) the ideal horizontal accuracy would be plus or minus one linear foot. On large sites (greater than 100 acres) the ideal horizontal accuracy would be plus or minus two linear feet. The monitoring well locations should be surveyed by reference to a standardized survey grid (i.e., Universal Transverse Mercator or state planar coordinate system).

Monitoring well elevations must be surveyed to a greater degree of accuracy. Without accurate elevations for monitoring wells, the water-level data and interpretation (discussed in Chapter 13) are subject to error. The accuracy of the elevation survey is usually to the nearest 0.01 ft. The elevations should be surveyed using a common datum. The most commonly used datum is the National Geodetic Vertical Datum (NGVD), established by the National Geodetic Survey (NGS), which is part of the National Oceanic and Atmospheric Administration (NOAA). The NGS has benchmarks (permanent landmarks of known position and elevation) throughout the United States from which the elevations of monitoring wells can be surveyed. The locations of some of the benchmarks are shown on $7\frac{1}{2}$-min quadrangle U.S.G.S. topographic maps. The location of the nearest benchmark can be obtained by contacting either the local or national U.S.G.S. office. It is not critical that the NGVD datum be used. However, it is useful if the water-level data obtained from one program are to be compared with data from other programs. Many industrial plants use their own datum and have established their own benchmarks; this is satisfactory and very often the relative difference between the plant datum and NGVD is known, allowing the plant elevations to be converted to NGVD elevations. If it is not practical to reference the elevations to a known datum (due to budget constraints, time, distance, etc.), an assumed datum can be used. For an assumed datum, an arbitrary point is selected (usually one of the monitoring wells) and assigned an elevation. The elevations of the other monitoring wells are then surveyed with reference to this one point. This allows the water levels of the individual monitoring wells to be compared with each other (they are surveyed to a common datum) but they cannot be compared with data from other programs (due to the use of a different datum).

TABLE 12.1
Standards for the Classification of Geodetic Control and Principal Recommended Uses

Classification	First Order	Second Order Class I	Second Order Class II	Third Order Class I	Third Order Class II
Relative accuracy between directly connected adjacent points (at least)	One part in 100,000	One part in 50,000	One part in 20,000	One part in 10,000	One part in 5000
Recommended uses	Primary national network; metropolitan area surveys; scientific studies	Area control which strengthens the national network; subsidiary metropolitan control; vertical control	Area control, which contributes to, but is supplemental to the national network	General control surveys references to the national network; local control surveys	
Relative accuracy between directly connected points or benchmarks (standard error)	Class I (0.5 mm $\sqrt{K}$) Class II (0.7 mm $\sqrt{K}$)	1.0 mm $\sqrt{K}$	1.3 mm $\sqrt{K}$	2.0 mm $\sqrt{K}$	
Recommended uses	Basic framework of the national network and metropolitan area control; regional crustal movement studies; extensive engineering projects; support for subsidiary surveys	Secondary framework of the national network and metropolitan area control; local crustal movement studies; large engineering projects; tidal boundary reference; support for lower-order surveys	Densification within the national network; rapid subsidence studies; local engineering projects; topographic mapping	Small-scale topographic mapping; establishing gradients in mountainous areas; small engineering projects; may or may not be adjusted to the national network	

Note: K is the distance in kilometers between points.
Source: Classification, Standards of Accuracy, and General Specifications of Geodetic Control Surveys, U.S. Department of Commerce, National Oceanic, and Atmospheric Administration, National Ocean Survey, Rockville, MA, February 1974.

Elevations of the protective casing (with the cap off or hinged back), the well casing, and the ground surface should be surveyed for each monitoring well. Water-level data may be measured from either the top of the protective casing or the top of the well casing, depending on the well construction details or the training of the field personnel. The possibility of measuring from a point without a known elevation is eliminated if the elevations are measured for both reference points. The ground surface elevation is usually surveyed to the nearest 0.1 ft. If the top of either the protective casing or well casing is not level (not the same elevation all around the top), a clearly visible mark should be made (generally with an indelible marker) to indicate at what point on the casing the elevation was measured, so that water levels will be measured consistently from the same point. This will help eliminate the possibility of water-level measurement errors.

The use of Global Positioning System (GPS) technology can be used to supplement the surveying discussed earlier. GPS consists of 24 satellites in orbit about the Earth and a network of ground stations used for control and monitoring. A GPS unit can provide horizontal and vertical location information by triangulation of signals from two or more satellites. In general, GPS units (commercially available to the public) may be capable of determining a horizontal location within a few feet. The vertical location ability of hand-held GPS units is no more accurate than the horizontal, and often only half as accurate. Actual precision and accuracy are generally a function of the cost of the hand-held GPS unit, as well as the current state of the technology. GPS, like many emerging technologies, is constantly evolving; as it continues to evolve, the precision and accuracy will only improve.

GPS can be useful on large study areas to provide horizontal location information prior to, during, or immediately after the installation of monitoring wells. Having approximate horizontal location information before wells are surveyed as discussed earlier may prove useful under various situations (i.e., evaluating field-collected data to determine subsequent drilling locations). GPS can also be useful, either before or after monitoring well installation, in study areas with few or no reference points (i.e., heavily wooded areas or large open fields). GPS can be particularly useful to locate wells, after installation, for purposes of sampling (monitoring wells with flush-mounted protective covers can be very difficult to find in large fields, grassy areas, or beneath snow cover).

Well Identification

Identifying a well by placing its number on the protective casing will ensure that the location of water-level measurements and ground-water samples will be recorded correctly. The easiest way to permanently identify each well is to put the well number on the protective well casing in a visible location. Identification can be placed either on the inside or the outside of the protective casing; both have advantages and disadvantages. Placing the well identification on the outside provides the following advantages:

- Identification can be noted while approaching the well.
- As long as the identification is not on a removable cap, the identification will not be mixed with others, particularly at nested or clustered wells.

Placing the well identification on the outside has the following disadvantages:

- Identification may be removed or altered by vandalism.
- Identification may become illegible due to fading (if painted) or weathering (heavy oxidation if punched or etched).

- If the identification is on the cap itself (not permanently attached), there could be a mix-up with other caps, particularly at nested or clustered wells.

The advantages of placing the identification on the inside of the protective casing are as follows:

- Identification cannot be vandalized (provided the well is locked).
- Identification should remain legible.
- If the identification is not on a removable cap, the identification will not be mixed with others, particularly at nested or clustered wells.

The disadvantages of placing the identification on the inside of the protective casing are as follows:

- Identification cannot be noted while approaching the well.
- If the identification is placed on a removable cap it could get mixed with others, particularly at nested or clustered wells.

It is advisable, therefore, that the well identification be placed on both the inside and outside of the protective casing.

Well identifications are usually marked with either paint or a metal punching or etching tool. Paint can be easily seen, but can also fade easily or be chipped off. A common technique that helps to keep the painted identification visible is to paint black numbers or letters on a white background. Punched or etched identification numbers are sometimes difficult to read, and may become illegible due to oxidation. A combination of the two types of identification is recommended, particularly if the monitoring program is to last more than 1 yr.

Confusion can be easily avoided when the identification is marked on a removable cap by simply placing the identification number on the protective casing as well. A common identification problem arises when there are multiple wells within one protective casing. If the protective cap is hinged and therefore secure along one point, the inside of the cap can be used as a guide, as shown in Figure 12.1. This will only work where the cap is secured at one point. If there is a removable cap, some other system must be used. One solution is for each well casing to have a cap that is removable but still secured to the casing by string or wire. This way the caps can be removed, but cannot get mixed up with each other.

A sign may be used to identify a monitoring well number as well as provide other pertinent information, such as:

- Any known hazards
- Permit identification
- Owner's information
- Nonpotable water usage
- Other information that may be required by regulation

Signs are another means of identifying (labeling) monitoring wells. Signs may also help identify a monitoring well location from a distance (i.e., located in a wooded area or in a field of tall grass), but can also attract vandals to the monitoring well location.

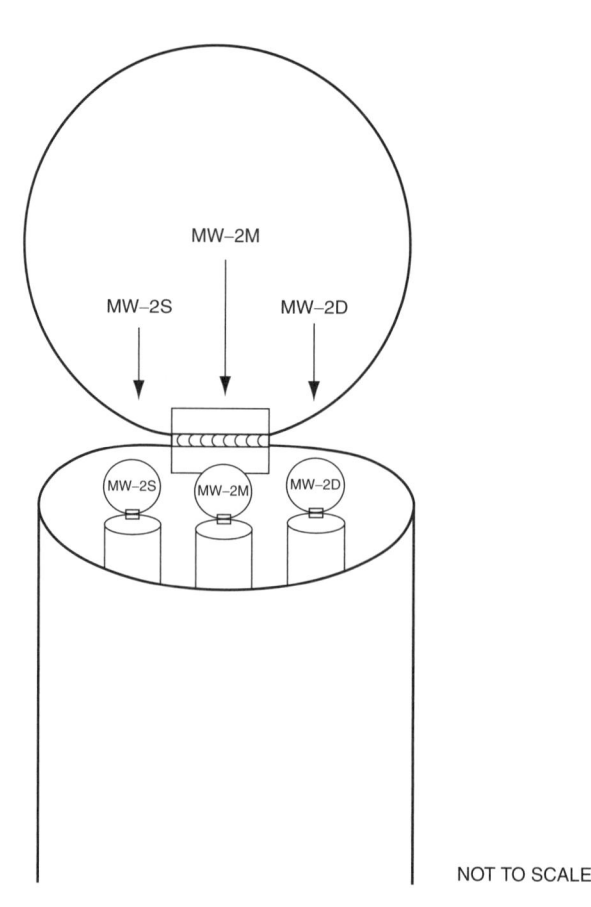

FIGURE 12.1
Monitoring well identification labeling.

Reporting Well Construction Details

Because the results of a ground water-monitoring program can be affected by the details of the monitoring well construction, the construction information should be reported in detail. Many states have regulations requiring the submittal of boring logs and monitoring well construction details. A telephone call to the overseeing state environmental agency should be able to identify all reporting requirements.

Boring logs and monitoring well construction details are often presented in full detail as an appendix to a report. There is usually a significant amount of information to be reported, and it would distract from the report if placed within the text. The well construction details that should be reported include:

- Borehole diameter
- Total depth of borehole (from ground surface)
- Diameter, schedule, and material type of casing
- Diameter, schedule, and material type of screen
- Length of screen

- Screen slot size
- Length of casing (to allow determination of top and bottom screen elevations)
- Length of filter pack
- Description of filter pack (generally referring to mineral content and grain-size distribution)
- Length of bentonite seals
- Description of bentonite seals (pellets vs. chips vs. granules, manufacturer, any special additives)
- Length of grout seal
- Description of grout seal (type of grout [bentonite vs. cement], bentonite–water ratio, cement–water ratio, any special additives)
- Location and description of casing and screen centralizers, if used
- Length and description of surface seal
- Total length and buried length of protective casing
- Diameter and material type of protective casing
- Protective casing elevation
- Well casing elevation
- Ground surface elevation
- Any construction difficulties, and their depth

The most common way of presenting the details is schematically, generally a figure depicting the monitoring well placed into a borehole with the appropriate backfill materials at the correct depths and other specifics presented in the figure. Two very common ways of presenting these details are shown in Figure 12.2 and Figure 12.3. The graphic in Figure 12.2 is not to scale (vertically) and therefore the reader must take some time to determine the exact length of the screen, filter pack, bentonite seal, etc. Additionally, the reader must go to a separate figure that presents the boring log information, and compare it to the well construction details. This is a time-consuming process and very often confusing the reader. However, it does present the necessary information to the reader and is a "form" graphic that can be completed very quickly and inexpensively. The graphic in Figure 12.3 is to scale (vertically), therefore the reader can very quickly see the relative lengths of screen, casing, filter pack, bentonite seal, and other well construction components. The specifics in the lengths and diameters of the well construction are also presented in the graphic. In addition, the reader is simultaneously presented with the boring log so the borehole lithology may be visually compared with the monitoring well construction. This allows the reader to note if the well is screened above the water table, throughout an entire aquifer, or a portion of an aquifer, and whether the bentonite seals have been properly placed.

If the number of monitoring wells is relatively small, the well construction details can also be presented in tabular form. Table 12.2 is typical and includes columns for many of the details presented earlier. This form of presentation is usually not as complete as the graphic presentation, but can be useful in the text of the report, while referring the reader to the appendix for more details. This tabular presentation can be used to focus on specific well construction details that the author of a site-specific report may want to highlight.

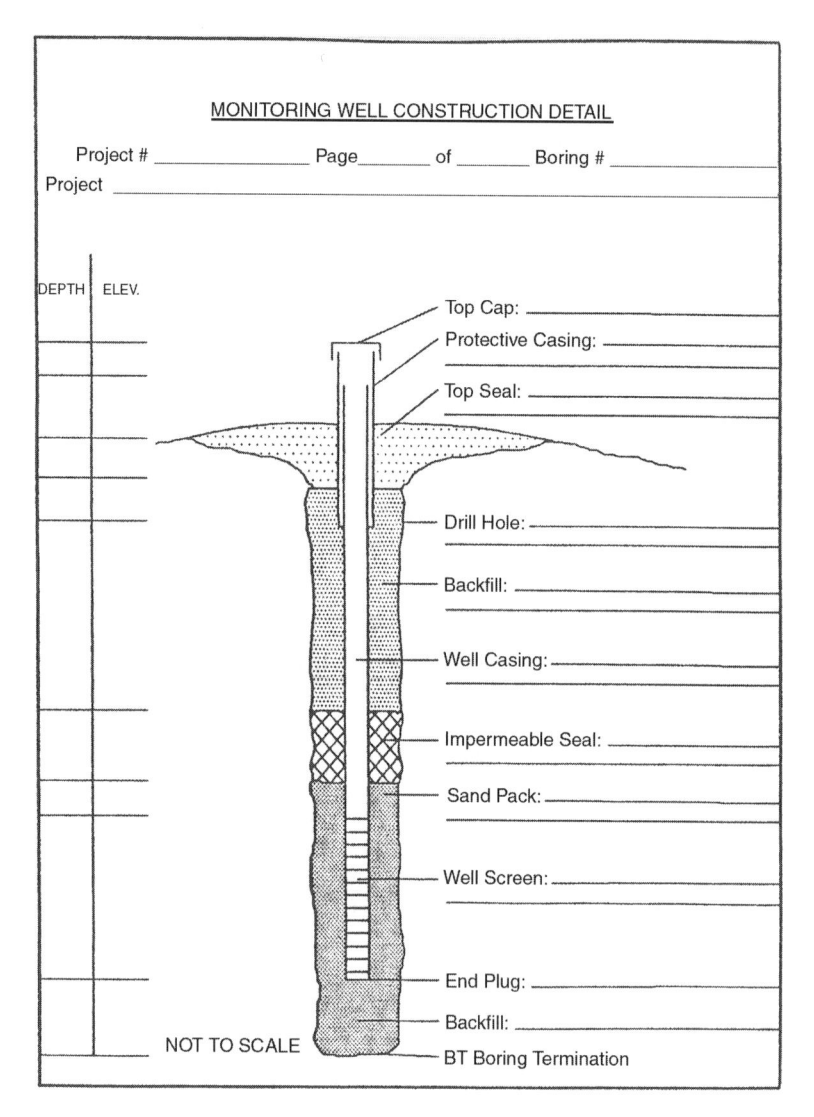

FIGURE 12.2
Monitoring well construction detail (not to scale).

Monitoring Well Maintenance and Rehabilitation

Many ground-water monitoring programs are designed and implemented to collect only a few rounds of samples or water-level data. After the program has served its useful purpose, the monitoring wells are no longer used and often forgotten. For such short-duration monitoring programs, there may be little need for either well maintenance or rehabilitation, although some wells may require rehabilitation shortly after their installation because of improper construction or incomplete development. Wells that are no longer in use should be properly decommissioned (see following section).

Ground-water monitoring programs that are implemented for more than 1 yr should include provisions for the development and implementation of a written well maintenance and rehabilitation program. Generally, the purpose of the well maintenance and rehabilitation program is to ensure that the monitoring wells provide reliable and accurate

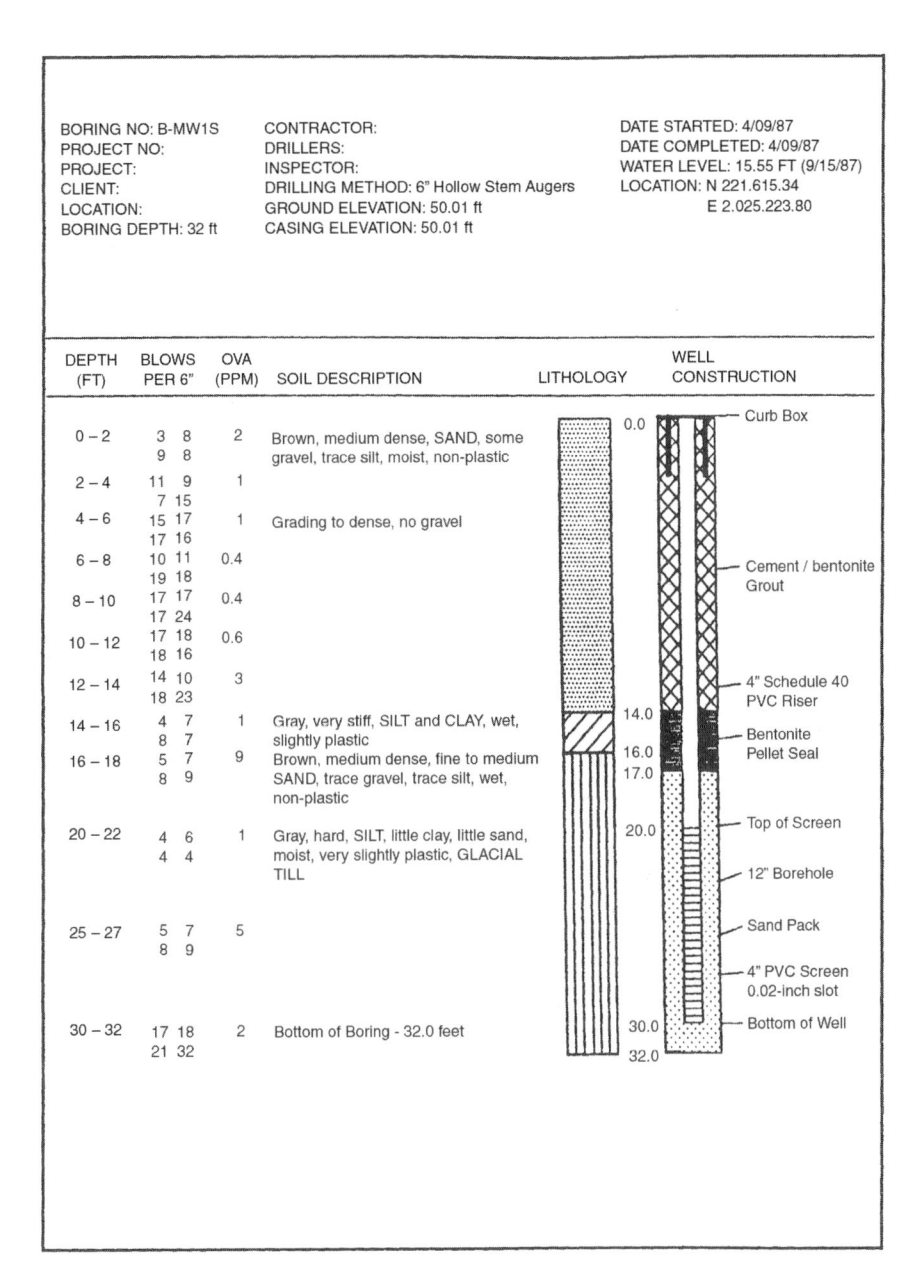

BORING NO: B-MW1S	CONTRACTOR:	DATE STARTED: 4/09/87
PROJECT NO:	DRILLERS:	DATE COMPLETED: 4/09/87
PROJECT:	INSPECTOR:	WATER LEVEL: 15.55 FT (9/15/87)
CLIENT:	DRILLING METHOD: 6" Hollow Stem Augers	LOCATION: N 221.615.34
LOCATION:	GROUND ELEVATION: 50.01 ft	E 2.025.223.80
BORING DEPTH: 32 ft	CASING ELEVATION: 50.01 ft	

DEPTH (FT)	BLOWS PER 6"	OVA (PPM)	SOIL DESCRIPTION	LITHOLOGY	WELL CONSTRUCTION
0 – 2	3 8 / 9 8	2	Brown, medium dense, SAND, some gravel, trace silt, moist, non-plastic	0.0	Curb Box
2 – 4	11 9 / 7 15	1			
4 – 6	15 17 / 17 16	1	Grading to dense, no gravel		
6 – 8	10 11 / 19 18	0.4			Cement / bentonite Grout
8 – 10	17 17 / 17 24	0.4			
10 – 12	17 18 / 18 16	0.6			
12 – 14	14 10 / 18 23	3			4" Schedule 40 PVC Riser
14 – 16	4 7 / 8 7	1	Gray, very stiff, SILT and CLAY, wet, slightly plastic	14.0	Bentonite Pellet Seal
16 – 18	5 7 / 8 9	9	Brown, medium dense, fine to medium SAND, trace gravel, trace silt, wet, non-plastic	16.0 / 17.0	
20 – 22	4 6 / 4 4	1	Gray, hard, SILT, little clay, little sand, moist, very slightly plastic, GLACIAL TILL	20.0	Top of Screen / 12" Borehole
25 – 27	5 7 / 8 9	5			Sand Pack / 4" PVC Screen 0.02-inch slot
30 – 32	17 18 / 21 32	2	Bottom of Boring - 32.0 feet	30.0 / 32.0	Bottom of Well

FIGURE 12.3
Monitoring well construction detail (to scale).

water-level and water-quality data. Monitoring well maintenance and rehabilitation programs should be written and implemented by, or with input from, personnel with specific knowledge and experience with these activities. The program should include an assessment of potential problems, procedures to monitor and evaluate the potential problems, and a process to decide how to proceed to address problems once they have been identified (ASTM, 2004b). The program should incorporate all available site-specific information that may affect the purposes of the monitoring wells, including physical information (aquifer materials and hydraulic properties) and chemical information (ground-water quality data). An ongoing well maintenance and rehabilitation program is usually less costly than replacing the monitoring wells periodically.

TABLE 12.2

Typical Table of Well Construction Details

Well No.	Top of Casing Elevation (NGVD)	Length of Casing (ft)	Top of Screen Elevation (NGVD)	Length of Screen (ft)	Bottom of Screen Elevation (NGVD)	Water Level Elevation 9/11/01 (NGVD)	Water Level Elevation 10/15/01 (NGVD)
MW-1	344.03	14.5	329.53	10.0	319.53	327.08	326.83
MW-2	341.72	15.0	326.72	10.0	316.72	322.64	322.97
MW-3	341.17	15.0	326.17	10.0	316.17	323.19	323.42

Well maintenance, which is described in detail in ASTM Standard D 5978 (Standard Guide for Maintenance and Rehabilitation of Ground-Water Monitoring Wells [ASTM, 2004b]), consists of periodically checking the well conditions, well performance, and ground-water sample quality. Well maintenance requirements will vary from site to site, and in some cases, from well to well. Typical maintenance items that would be checked as part of an ongoing well maintenance program might include the following:

1. Surface observations:
    (a) Visibility (keeping the well easy to find)
    (b) Well location access (maintaining a path to the well location)
    (c) Lock removal (keeping it from rusting or other corrosion)
    (d) Protective cap removal (maintaining the hinge or well cap threads with antioxidizing agents)
    (e) Well identification (confirm legibility of identification)
    (f) Concrete surface seal (checking for loss of integrity of the seal such as cracks in concrete)
    (g) Evidence of frost heave (uplift of concrete surface seal)
    (h) Down-hole borehole observation (using a mirror or narrow-beamed flashlight to confirm the well casing has not been compromised)
2. Subsurface observations:
    (a) Initial static water level (within expected range?)
    (b) Total depth of well (measured from well casing, compared to previous measurements, determine if there is any, and how much, siltation occurring)
    (c) Water-level drawdown (usually recorded during ground-water sampling)
    (d) Ground-water discharge rate (usually recorded during ground-water sampling)
    (e) Pump performance (in monitoring wells with dedicated sampling pumps).
3. Ground-water sample quality:
    (a) Significant changes of field-monitored water-quality parameters (pH, Eh, temperature, conductivity, dissolved oxygen) from historical data
    (b) Significant changes of analytical ground-water quality parameters (site-specific) from historical data
    (c) Significant changes in turbidity
    (d) Presence of slime or microbial growth in ground-water discharge (indication of biofouling)

Almost all well maintenance checks are made during a ground-water sampling event. The checks that are based on water-quality data cannot be made until the laboratory results are completed, but the others can be made at the well location either visually or by comparing new data with previous data. The written well maintenance and rehabilitation program should define the ground-water sampling protocol and specify that the protocol should not be altered unless directed by the proper personnel. Changes in sampling protocol can easily cause significant changes in most of the subsurface observations and ground-water sample quality items presented earlier. If a problem with a monitoring well is identified (as defined by the program), the program should indicate the decision-making process to the appropriate personnel. In some cases well rehabilitation may be required.

Well rehabilitation should be considered if well performance has been reduced significantly or if the ground-water sample quality has significantly changed. A rehabilitation method should return well performance close to its original level, or bring the ground-water quality to within its previously existing range. The most common problem in monitoring wells is siltation of the filter pack or well screen. This occurs when fine-grained materials (clays and silts) within the screened formation migrate into the filter pack or well screen during sampling events and significantly reduce the well efficiency. Rehabilitation for this type of problem most often consists of redevelopment of the well. Several methods of well development are presented earlier in this chapter; one or more of these methods would be used to redevelop the monitoring well. Other problems that might possibly affect the monitoring well performance include:

- Loss of a subsurface seal
- Collapsed or broken well casing or screen
- Incrustation of the well screen
- Corrosion of the well screen

A broken well screen might be diagnosed by conditions similar to siltation, except that coarser material from the filter pack and formation (providing that the screened formation consists of coarser materials) would accumulate at the bottom of the well and be entrained in samples. Incrustation of the well screen can be diagnosed by a significant drop in the well yield, with the well maintaining a sediment-free condition. The loss of a subsurface seal can be diagnosed by changes in water quality (generally, significant increases in pH [from 8.5 to 11] and specific conductance). However, wells that have lost their subsurface seals cannot be rehabilitated; they must be replaced.

The rehabilitation of a broken well screen usually consists of placing an intact, but smaller diameter screen (also called a liner) inside the damaged screen. If the monitoring well is shallow, the smaller-diameter screen may be threaded to the same diameter casing that then comprises a new monitoring well placed inside the old monitoring well. If the well is deep or of a small diameter, a smaller-diameter screen may be placed inside the damaged screen and sealed to the existing casing. A rubber or lead packer may be placed just above the smaller-diameter screen and wedged into the existing casing to form a seal between the screen and the casing.

Well incrustation can significantly reduce the yield of the well, and might also impact the chemical quality of samples collected from the well. There are three general types of incrustation:

- Precipitation of carbonates (or sulfates) of calcium and magnesium
- Precipitation of iron and manganese compounds (i.e., iron oxyhydroxides)
- Build-up of slime produce by iron bacteria or other slime-forming organisms

Well incrustation can be easily remedied by first determining the type of incrustation and then removing it with a combination of chemical treatment, physical removal, and redevelopment. The effect of chemical treatment on water-quality results from a monitoring well may be a problem and should be considered before this type of rehabilitation is attempted. Detailed discussions of the rehabilitation of incrusted wells can be found in Campbell and Lehr (1973), Gass et al. (1982), Helweg et al. (1982), Driscoll (1986) and ASTM (2004b).

In some cases in which rehabilitation is required, particularly where a monitoring well has a small diameter and is relatively shallow, it may be more cost efficient to simply replace the monitoring well. However, when wells are relatively deep and constructed of more expensive materials, even expensive rehabilitation may be warranted. A downhole television camera can be used to provide a videotape in wells with inside diameters as small as 2 in. The pictures can reveal the nature and significance of a problem, the specific location of the problem, and whether or not the problem can be rehabilitated. However, the down-hole television cannot show if, or where, there is a problem outside of the well, such as a loss of a subsurface seal. Some borehole geophysical methods (i.e., a cement bond log or a neutron log) can be used for this purpose (see Chapter 4).

Monitoring Well and Borehole Decommissioning

Proper monitoring well (and borehole) decommissioning (also called "abandonment") is one of the most important post-construction elements of a ground-water monitoring program. The process of decommissioning, and the methods commonly used are described in detail in ASTM Standard D 5299 (Guide for Decommissioning of Ground Water Wells, Vadose Zone Monitoring Devices, Boreholes, and Other Devices for Environmental Activities [ASTM, 2004c]). Proper well and borehole decommissioning can help reduce the threat of cross-contamination, alteration of sampling results, and potential liabilities. Some monitoring wells may simply become outdated (Bergren et al., 1988) and need to be decommissioned. Frequently, little thought is given to the subject of decommissioning when a well is first constructed. Just as frequently, no one follows up with a decommissioning plan when the well has reached the end of its useful life. Perhaps most critical of all may be the need for immediate and proper decommissioning of soil borings, test holes, and improperly constructed or located monitoring wells. It may even be justified to temporarily decommission a monitoring well that is no longer in an active program or that may have been damaged by on-site activities.

Objectives

The two principal objectives of a decommissioning program are to restore the borehole to its original condition and to prevent movement of fluids (cross-contamination) between formations or sampling zones. A monitoring well or boring should never become a conduit for contaminating previously uncontaminated aquifers. The well or borehole decommissioning program should be as well planned and well documented as the original well or boring installation program.

Planning for Decommissioning

Before decommissioning a monitoring well or boring, it is important to address six key elements. First, any state, federal, or local regulations that may control decommissioning of the monitoring well or boring must be taken into account. Table 12.3 lists the agency in

TABLE 12.3

Status of State Well Decommissioning Requirements

State	State-wide Well Decommissioning Laws[a]	State Agency with Program Responsibility
Alabama	Yes	AL Dept. of Environmental Management (334)271-7832
Alaska	Yes	AK Dept. of Environmental Conservation (907)-465-2600
Arizona	Yes	AZ Dept. of Water Resources (602)417-2470
Arkansas	Yes	AR Water Well Construction Commission (501)682-1611
California	Yes	CA Dept. of Water Resources (916)327-8861
Colorado	Yes	CO Div. of Water Resources (303)866-3581
Connecticut	Yes	CT Dept. of Consumer Protection (860)566-3290
Delaware	Yes	DE Div. of Water Resources (302)739-3665
District of Columbia	No	DCRA-Water Resources Management Div. (202)645-6601
Florida	Yes	FL Dept. of Environmental Protection
Georgia	Yes	GA Environmental Protection Div. (404)657-6142
Hawaii	Yes	HI Dept. of Land and Natural Resources (808)587-0263
Idaho	Yes	ID Dept. of Water Resources (208)327-7900
Illinois	Yes	IL Dept. of Public Health (217)782-5830
Indiana	Yes	IN Dept. of Natural Resources, Div. of Water (317)232-4160
Iowa	Yes	IA Dept. of Natural Resources (515)281-7814
Kansas	Yes	KS Dept. of Health and Environment (913)296-3565
Kentucky	Yes	KY Div. of Water (502)-564-3410
Louisiana	Yes	LA Dept. of Transportation and Development, Water Resources Section (504)379-1434
Maine	Yes	ME Dept. of Environmental Protection (207)287-2651
Maryland	Yes	MD Dept. of the Environment (410)631-3784
Massachusetts	Yes	MA Div. of Water Resources (617)727-3267
Michigan	Yes	MI Dept. of Environmental Quality (517)334-6974
Minnesota	Yes	MN Dept. of Health (612)215-0811
Mississippi	Yes	MS Dept. of Environmental Quality (601)961-5200
Missouri	Yes	MO Dept. of Natural Resources, Div. of Geology and Land Survey (573)368-2165
Montana	Yes	MT Board of Water Well Contractors (406)444-6643
Nebraska	Yes	NE Dept. of Health (402)471-2541
Nevada	Yes	NV Div. of Water Resources (702)687-3861
New Hampshire	Yes	NH Water Well Board (603)271-3406
New Jersey	Yes	NJ Dept. of Environmental Protection (609)292-2957
New Mexico	Yes	NM State Engineers Office (505)827-6120
New York	Yes	NY Dept. of Environmental Conservation (518)457-0893
North Carolina	Yes	NC Dept. of Environment, Health and Natural Resources (919)715-6160
North Dakota	Yes	ND Dept. of Health (701)328-5210
Ohio	Yes	OH Environmental Protection Agency (614)644-2752
Oklahoma	Yes	OK Water Resources Board (405)530-8800
Oregon	Yes	OR Water Resources Dept. (503)378-8455
Pennsylvania	No	PA Dept. of Environmental Protection (717)787-5828
Rhode Island	Yes	RI Dept. of Environmental Management (401)277-2234
South Carolina	Yes	SC Dept. of Health and Environmental Control (803)734-5310
South Dakota	Yes	SD Dept. of Environment and Natural Resources (605)773-3352
Tennessee	Yes	TN Div. of Water Supply (615)532-0176
Texas	Yes	TX Commission on Environmental Quality (512)239-0530
Utah	Yes	UT Dept. of Natural Resources, Water Rights Div. (801)538-7382 or (801)538-7416
Vermont	Yes	VT Dept. of Environmental Conservation (802)241-3400
Virginia	Yes	VA Dept. of Environmental Conservation (804)698-4219
Washington	Yes	WA Dept. of Ecology (360)407-6648
West Virginia	Yes	WV Div. of Environmental Protection (304)558-2108
Wisconsin	Yes	WI Dept. of Natural Resources (608)261-6421
Wyoming	Yes	WY State Engineer's Office (307)777-7354

[a]There may be requirements at other levels of government regarding well decommissioning. A check should be made with all appropriate agencies.
Source: Water Well Journal, May, 1996.

each state that oversees well or borehole decommissioning, and notes whether or not the state has decommissioning regulations or guidelines.

A second important consideration is the type of well or borehole to be decommissioned and its design and construction. If it is a monitoring well, was it properly constructed, is it a monitoring well in which the construction is unknown or inadequately documented, is it a monitoring well in which the construction is known to be inadequate, or is it a well that has failed or been damaged? Improperly constructed or damaged monitoring wells will need more extensive decommissioning procedures than monitoring wells that have been properly constructed with suitable materials and have an adequately grouted annular space. All test borings and exploratory boreholes should be decommissioned immediately after installation. Even some properly constructed monitoring wells may have to be decommissioned by destructive methods because of their design. If they have dedicated permanent sampling equipment or are multilevel devices, they may not be amenable to normal open borehole decommissioning methods. Some of the advantages of some of these special monitoring well designs may be detriments at the time of decommissioning.

A third consideration in decommissioning is the hydrogeologic environment. What is the risk of the borehole or monitoring well being a conduit for the movement of ground water or contaminants? Were any low hydraulic conductivity formations penetrated by the well or borehole? What is the minimum hydraulic conductivity of the materials penetrated? What is the depth to the zone of saturation? Are the overburden materials cohesive, granular, or interbedded layers of material? If the well or boring reached or penetrated bedrock, what was the nature of the bedrock? Was it porous, fractured, cavernous, or massive?

A fourth consideration in planning for decommissioning is the chemical environment around the borehole or well. Most decommissioning procedures, including those recommended in this book, recommend or require the use of grouting materials, including sodium bentonite and Portland cement. These materials react unfavorably in highly acidic or alkaline environments (Smith, 1976; Williams and Evans, 1987). Unlike a monitoring well, which may have a life of 30 yr, a properly decommissioned well or boring must be designed to remain sealed forever. If the chemical environment is going to adversely affect the decommissioning material, a different material or procedure may have to be used. Not only is information such as a good boring log or well log important, but sample analytical results from the well or boring are important because they offer clues to possible interference with the setting up or curing of decommissioning materials and to the potential future breakdown of these materials.

The fifth planning consideration is disposal of potentially contaminated materials removed from the borehole or monitoring well. These materials may include soil or rock material, pumps or samplers, pipe or tubing, casing, screens, or the entire monitoring well if the well has to be destructively removed and the resulting borehole backfilled with grout. The handling of contaminated material should receive the same care as when the monitoring well or boring was first constructed. This subject is dealt with more fully elsewhere in this book.

The final planning consideration for decommissioning of monitoring wells or borings is determining the type of equipment and the quantity of grouting or sealing materials that will be needed. In some types of decommissioning, the equipment required may be very simple and the amount of grouting material required may be small. Conversely, wells constructed in cavernous limestone formations or other lost circulation zones, or which require full destruction and removal of all well construction materials, may require significant amounts of grouting material. Where full destruction of the well is needed, drilling equipment capable of drilling a borehole up to $1\frac{1}{2}$ times the size of the original borehole will be needed. Monitoring projects involving only the installation of probes or borings should also have adequate equipment and materials to properly close and seal these holes promptly upon completion of their use.

Location and Inspection

An important step before proceeding with decommissioning is to confirm the location of the monitoring well scheduled for decommissioning. Was sufficient information generated at the time of construction of the well to reidentify the site even after decommissioning? If not, sufficient location information should be generated. A GPS location should be taken in addition to any standard surveying work. If the site does not have significant magnetic interference, it may be desirable to place a permanent magnetic underground marker in the well after it is decommissioned so the site can be relocated in the future.

Another important pre-decommissioning step that should be considered is an inspection of the well itself. A down-hole TV camera survey can help to confirm the condition, depth, and the construction details of the well (the camera should be decontaminated after each use). Even a visual inspection using a flashlight, or sunlight reflected from a mirror, may be helpful. If the monitoring well must be removed by a destructive method and the well is greater than 100 ft deep, a hole deviation survey should be conducted (Bergren et al., 1988).

Decommissioning Materials

To be effective, decommissioning materials must fill the space provided and have an effective hydraulic conductivity less than the original materials through which the well or boring was constructed. Two principal grouting materials, neat cement and high solids bentonite, best meet the needs for decommissioning.

Type I Portland cement (ASTM C-150 [ASTM, 2004d]) is the material most frequently used to make a neat cement decommissioning material. When mixed with no more than 5 or 6 gallons of water per 94-pound bag of cement, and placed in an environment with at least 80% relative humidity, Type I cement will continue to hydrate, and may not shrink substantially. The same mix placed under water will actually expand (Kosmatka and Panarese, 1988). However, in most other environments, Portland cement will shrink around 12% by volume upon setting.

To control shrinkage and improve pumpability, it is necessary to use additives. The most common additive is bentonite. Commonly, between 2 and 6% bentonite, by weight, is prehydrated with the mix water. Table 12.4 lists mixing ratios per 94-pound bag of Portland cement with 0–6% bentonite, and the resulting slurry properties. One benefit of the addition of bentonite is improvement of pumpability of the neat cement (Smith, 1976). Pumpability can also be enhanced and the amount of mix water reduced by 12–30% by use of superplasticizers (Kosmatka and Panarese, 1988). A compressive strength of 500 psi for the set neat cement appears to be adequate when used as a grout or decommissioning material (Smith, 1976). Neat cement can also be modified by adding clean sand and, for larger holes, by adding gravel not larger than 0.5 in. in diameter.

In addition to shrinkage problems that may occur if too much water is added to the cement mix, the cement may also segregate with development of channels and pockets of free water (Coleman and Corrigan, 1941). This problem commonly occurs in deep holes with limited diameter. The permeability of cement is also significantly increased with increased water-to-cement ratios (Williams and Evans, 1987).

The second common decommissioning material is sodium bentonite (API-13A [API, 1983]). The principal asset of this material is its ability to rapidly hydrate to 10 or more

TABLE 12.4

Recommended Slurry Properties of Portland Cement with Bentonite

Percent Bentonite	Type of Cement[a]	Maximum Water Requirements gal/sk	Maximum Water Requirements ft³/sk	Slurry Weight lb/gal	Slurry Weight lb/ft³	Slurry Volume (ft³/sk)
0	I or II	5.2	0.70	15.6	117.0	1.18
0	III	6.3	0.84	14.8	110.7	1.32
2	I or II	6.5	0.87	14.7	110.0	1.36
2	III	7.6	1.02	14.1	105.5	1.51
4	I or II	7.8	1.04	14.1	105.0	1.55
4	III	8.9	1.19	13.5	101.0	1.69
6	I or II	9.1	1.22	13.5	101.0	1.73
6	III	10.2	1.36	13.1	98.0	1.88
8	I or II	10.4	1.39	13.1	98.0	1.92
8	III	NA	NA	NA	NA	NA

[a] ASTM Type I, II, and III used instead of API A, B and C.
Source: After Halliburton Co., Cementing Tables, 1975.

times its original volume. Not only may it be used as a high-solids pumpable grout, but it may also be used in dry form. By adding a quart of polymer to 100 gallons of water it is possible to then add 1.5–2 lb of granular bentonite per gallon and gently pump the resulting mixture (American Colloid Company, 2004). A mixture weighing 9.25–9.4 lb/gal can be achieved by this method. The polymer delays the hydration of the bentonite, permitting approximately a 20-min working time (Gaber and Fisher, 1988).

Where the use of a polymer may be of concern, a similar mixture can be obtained by mixing 40 pounds of high-quality non-polymerized powdered or granular bentonite per 100 gallons of water to form a drilling mud. The bentonite is stirred in just prior to placing the mixture (Gaber and Fisher, 1988). Again, 9.25 and 9.4 lb/gal decommissioning materials with a solid content of 20–22% can be obtained using this mixture.

A modification of grouting and decommissioning materials is the development of a proprietary mixture of sodium and calcium bentonite in which 2.1 lb of bentonite is mixed per gallon of water, followed by the addition of a magnesium oxide compound (Bertane, 1986). This mixture has about a 2-h working time, but does not develop as strong a gel as the pure sodium bentonite mixes (Gaber and Fisher, 1988). The long-term performance of this material is not well known. A second proprietary mixture has also been developed by another bentonite manufacturer. This mixture also uses a retardant and is NSF (National Sanitation Foundation) approved.

Some attempts have also been made to use thickened or "heavy" drilling mud. Unfortunately, this mixture does not contain enough solids to form an effective decommissioning material. Heavy drilling mud may also be the most prone of the fluid-based decommissioning materials to the problems of consolidation noted earlier.

Another alternative to pumped decommissioning material is the use of granular, chip, or pellet forms of bentonite. Where there is little water in the borehole or well, dry bentonite may be an excellent choice. It may also be mixed with sand or gravel in some applications, which helps to ensure full and rapid emplacement. The dry bentonites are probably the most practical choice for decommissioning borings and test holes. With careful placement, even the potential problem of bridging can be overcome.

The use of native materials for decommissioning, such as puddled clay or fine sand, should only be considered in situations where the risk of contamination is effectively nonexistent or can be controlled by careful placement of regular decommissioning materials at selected intervals.

In many cases the materials allowed for use in decommissioning are controlled by state well or boring abandonment rules or regulations.

Placing Abandonment Materials

The most effective way to place the decommissioning material is by tremie pipe. Therefore, the basic equipment for most decommissioning procedures should be the same as for tremie grouting the original well (see Chapter 10). A grout mixer or other method of mixing the grout is needed, along with a positive-displacement pump to deliver the mixture with positive pressure to the bottom of the hole. The mixing of a granular bentonite slurry requires special precautions, however. Mixing of this slurry should be done with a blade- or paddle-type mixer with recirculation (Gaber and Fisher, 1988). A moyno pump with its screw-type pumping action is an ideal grouting pump. Other positive-displacement pumps such as the air diaphragm pump and the typical piston-type mud pump may also be used.

The tremie pipe should be at least 1–1.5 in. in diameter to allow for adequate flow of bentonite or neat cement grout. A combination of rubber hose and short lengths of threaded rigid plastic or metal pipe can work well. Enough hose is needed to run from the grout pump to the well or boring with the pipe used down the well or borehole. If the end of the pipe is to be used to check placement of material, it may be best to install a "T" at the end of the pipe to allow grouting material to discharge from the side of the pipe. During placement of the grout, the tremie pipe should always remain submerged in the grouting material.

When decommissioning with a mixture of neat cement and sand, the previously described tremie set-up will work effectively. Placement of cement with sand, however, will require increasing the size of the tremie pipe and special selection of the pump. When the hole is large enough to permit use of concrete for abandonment, it may be feasible to use a conductor pipe. The concrete can then be fed by gravity. However, the end of the conductor pipe, like the tremie pipe, should remain submerged during placement of the concrete.

Another use of the conductor pipe (Mason, 1988) is for the placement of bentonite pellets for decommissioning of water-filled, uncased holes. The conductor pipe is suspended a short distance off the bottom and water is circulated down the pipe and back up the hole. The pellets can then be added gradually to the circulating water. The pellets drop out below the pipe and fill the hole. The conductor pipe must be raised at regular intervals to prevent plugging by the pellets. It is also important to use only clean pellets (without bentonite dust in the material) so that a drilling mud is not formed by the recirculating water. Some replacement of the water may be necessary.

Direct gravity placement of decommissioning material can also be used satisfactorily. All of the cement-based materials may be placed directly in shallow holes that are free of water and in some similar deeper holes that have sufficient open diameter to prevent bridging. Granular, chip, and pellet bentonite may be used in a similar manner.

With special precautions, chip and pellet bentonite may even be used to abandon deep water-filled boreholes or wells that have sufficient open diameter to prevent bridging. All the fines must be removed to prevent the development of a drilling mud which would prevent proper settlement of the bentonite chips. It is also important to limit the rate of filling to prevent bridging. An effective set up (Figure 12.4) uses a chute formed of 1/4-in. mesh screen. The bentonite is poured down the chute at a rate of not more than

FIGURE 12.4
Filling a large-diameter well with chip bentonite. Bentonite is being poured over 1/4-in. mesh to separate the fines.

20 pounds per minute. Equal amounts of coarse sand or fine rounded stone can also be added at the bottom of the chute if a 100% bentonite fill is not required.

The reader is reminded that use of grout materials and decommissioning abandoned wells and borings is very much like cooking — it is more art than science. What is presented earlier is only a limited survey of available options and materials in a dynamic field. The reader is urged to not only maintain active contact with industry manufacturing representatives, consultants, and others, but also to obtain practical experience with each selected material and to test the long-term effectiveness of these materials in the field. For example, considerable debate exists about mixing bentonite into cement to improve pumpability and reduce shrinkage. Cement is a calcium-based material while bentonite is a sodium-based material, in addition to which cement creates high pH bleed water that prevents the proper hydration of bentonite. On the other hand, adding bentonite will reduce the heat of hydration of cement, which can be a real asset when grouting PVC monitoring well casing, but which may be unimportant in well or boring decommissioning.

Do not do something a particular way just because "the book says so." Develop personal knowledge and experience. This is particularly true for the selection and placement of decommissioning materials.

Procedures for Decommissioning

After review of the type of monitoring well or borehole to be decommissioned, a decision can be made on the type of decommissioning to be performed. Boreholes and properly constructed monitoring wells are relatively simple to decommission. This is particularly true when removal of the well casing or screen is not necessary. However, it is important

Monitoring Well Post-Installation Considerations 879

to recheck all steps in the decommissioning planning process before proceeding. One must be sure of what may be required by the state or local regulatory jurisdiction, and be sure that the procedure and decommissioning materials will do the required job.

The following checklist outlines standard decommissioning procedures:

- Check to be sure that the well or borehole is free of debris.
- Remove any dedicated equipment that will interfere with decommissioning.
- Determine the depth of the well or borehole with a weighted sounding line.
- If there is a screen, it may be desirable to fill the length of the screen with fine sand.
- Proceed to place decommissioning material (Figure 12.5).
- In any large-diameter or deep hole, sound the hole at preset intervals to ensure a complete fill.

FIGURE 12.5
Filling a 2-in. monitoring well with granular bentonite.

The surface finish is frequently determined by the site requirements or by state or local rules or regulations. A preferred final treatment is to cut all casing below land surface, cap and fill the excavation with decommissioning material, and then surface spread native material. Many state regulations require the final seal to be done with cement or concrete. Location measurements and survey markers placed just below the ground surface should be made or emplaced if appropriate.

The other principal type of decommissioning is destructive abandonment. In addition to all the requirements of simple decommissioning is the need to destroy the casing, grout, screen, and any in-place sampling devices. How this decommissioning is carried out will vary greatly depending on construction of the well, purpose or use of the well, site conditions, and drilling equipment available.

Where PVC or similar plastic casing has been used and the well is free of pumps and internal obstructions, a pilot bit inserted in the casing may be used as a guide in reaming out the casing and grout (Bergren et al., 1988). Steel-cased monitoring wells will generally have to be over-reamed and removed. A large-diameter, heavy-duty auger may be the most satisfactory drilling equipment for this operation. An air- or mud-rotary drill with a roller bit may also be satisfactory. If all grout has not been removed with the first pass, a larger reaming bit will need to be used to ensure that all of the remaining material is circulated out of the hole.

It is important to remember that if contamination was detected in the original well or boring, appropriate health and safety requirements should be maintained. Well construction and surrounding soil materials may also need to be handled and disposed of specially.

In some cases it may be desirable to fill the well with neat cement grout if that will assist in its destructive removal. Wells with stainless steel casing present particularly difficult decommissioning problems, as the casing cannot be easily drilled out. Using a hollow-stem auger to drill around (i.e., over the top of) the casing and screen is probably the only practical solution in this case. Another important consideration is keeping the hole open after drilling until the hole is filled with decommissioning material. Once the well has been destructively removed, abandonment can proceed as previously outlined.

A final type of decommissioning that may be appropriate in some cases is temporary decommissioning. In this case the screen or production zone should be filled with medium to fine sand. The rest of the well should then be filled with a bentonite material. If the well needs to be put back into service, the bentonite and sand can be flushed out of the well with water and disposed of. If the well is accidentally destroyed, the potential for ground-water contamination will have been forestalled.

Records and Reports

The last step in a proper decommissioning program is the production of appropriate records and reports. The records should identify the method of decommissioning, the materials used, and the quantity and mixed weight of all materials. The records should clearly establish the location of the decommissioned well or boring. It should be noted if a magnetic survey marker was placed at the well site. Finally, if any materials were disposed of offsite, the location and disposal method should be noted. It may also be desirable to save and store samples of mixed decommissioning material.

Finally, it is important to file required reports with the appropriate state or local agency. A similar or more detailed report and maps should be filed with the site owner.

References

American Colloid Company, *Technical Data Sheets on VOLCLAY Grout*, American Colloid Company, Arlington Heights, IL, 2004.

API, API Specifications for Oil-Well Drilling-Fluid Materials: API Specification 13-A; American Petroleum Institute, Washington, DC, 1983.

ASTM, Standard Guide for Development of Ground-Water Monitoring Wells in Granular Aquifers; ASTM Standard D 5521, ASTM International, West Conshohocken, PA, 2004a.

ASTM, Standard Guide for Maintenance and Rehabilitation of Ground-Water Monitoring Wells; ASTM Standard D 5978, ASTM International, West Conshohocken, PA, 2004b.

ASTM, Standard Guide for Decommissioning of Ground Water Wells, Vadose Zone Monitoring Devices, Boreholes, and Other Devices for Environmental Activities; ASTM Standard D 5299, ASTM International, West Conshohocken, PA, 2004c.

ASTM, Standard Specification for Portland Cement; ASTM Standard C 150, ASTM International, West Conshohocken, PA, 2004d.

Bergren, C.L., J.L. Janssen, and J.D. Heffner, Abandonment of ground water monitoring wells at the Savannah River Plant, Aiken, South Carolina, Proceedings of the Focus Conference on Eastern Regional Ground Water Issues, National Water Well Association, Dublin, OH, 1988, pp. 215–220.

Bertane, M., The Use of Grout for the Sealing of Monitoring Wells; American Colloid Company, Arlington Heights, IL, 1986.

Campbell, M.D. and J.H. Lehr, *Water Well Technology*; McGraw-Hill Book Co., New York, NY, 1973.

Coleman, R.J. and G.L. Corrigan, Fineness and Water-to-Cement Ratio in Relation to Volume and Permeability of Cement, *Petroleum Technology*, Technical Publication No. 1266, 1941, pp. 1–11.

Dougherty, P.J. and M.T. Paczkowski, A Technique to Minimize Contaminated Fluid Produced During Well Development; Proceedings of the Second National Outdoor Action Conference on Aquifer Restoration, Ground Water Monitoring and Geophysical Methods, National Water Well Association, Dublin, OH, 1988, pp. 303–317.

Driscoll, F.G., *Groundwater and Wells*, 2nd ed., Johnson Division, St. Paul, MN, 1986.

EA Engineering, Science, and Technology, Inc., Manual of Standard Operating Procedures for Geotechnical Services, GTS-201, 1985.

Gaber, M.S. and B.O. Fisher, Michigan Water Well Grouting Manual; Ground Water Control Section, Michigan Department of Public Health, Lansing, MI, 1988, 83 pp.

Gass, T.E., T.W. Bennett, J. Miller, and R. Miller, Manual of Water Well Maintenance and Rehabilitation Technology; U.S. Environmental Protection Agency, Robert S. Kerr Environmental Research Laboratory, National Environmental Research Center, Ada, OK, 1982.

Halliburton Company, Halliburton Cementing Tables; Little's, Inc., Duncan, OK, 1975, 501 pp.

Hanson, D.T., Development of wells and its importance to negative coliform, *Water Well Journal*, November, 63–65, 2001.

Helweg, O.J., V.H. Scott, and J.C. Scalmanini, Improving Well and Pump Efficiency, American Water Works Association, Denver, CO, 1982.

Howard, W.O., T.V. Danahy, and M. Ianniello, An Air-Lift Development System for Deep Wells; Proceedings of the Focus Conference on Eastern Regional Ground Issues, National Water Well Association. Dublin, OH, 1988, pp. 559–564.

Kosmatka, S.H. and W.C. Panarese, *Design and Control of Concrete Mixtures*; Portland Cement Association, Skokie, IL, 1988, 205 pp.

Mason, C., Personal communication, N. L. Baroid Company, 1988.

Moffit, H. and H. Bouchard, *Surveying*; Harper and Row, New York, NY, 1975.

National Oceanic and Atmospheric Administration, Classification, Standards of Accuracy, and General Specifications of Geodetic Control Surveys, U.S. Department of Commerce, Washington, DC, 1974.

Nuckols, T.E. Development of Small Diameter Wells, Proceedings of the Fourth National Outdoor Action Conference on Aquifer Restoration, Ground Water Monitoring and Geophysical Methods, National Water Well Association, Dublin, OH, 1990, pp. 193–207.

Paul, D.G., C.D. Palmer, and D.S. Cherkauer, The effect of construction, installation, and development on the turbidity of water in monitoring wells in fine-grained glacial till, *Ground-Water Monitoring Review*, 8(1), 73–82, 1988.

Reichart, R.W., A Different Approach to the Development of Monitoring Wells; *Water Well Journal*, May, 1996, pp. 52–56.

Schalla, R., A Comparison of the Effects of Rotary Wash and Air Rotary Drilling Techniques on Pumping Test Results: Pacific Northwest Laboratory, Proceedings of the Sixth National Symposium and Exposition on Aquifer Restoration and Ground Water Monitoring, National Water Well Association, Worthington, OH, 1986.

Schalla, R. and R.W. Landick, A New Valved and Air-Vented Surge Plunger for Developing Small Diameter Monitor Wells; *Ground-Water Monitoring Review*, 6(2), 1986, pp. 77–80.

Smith, D.K., *Cementing*, Society of Petroleum Engineers Monograph Series, Vol. 4, Society of Petroleum Engineers of AIME, Dallas, TX, 1976.

Treadway, W.C., Development Methods, Part II: Some Solutions; *Water Well Journal*, June, 1992, pp. 56–58.

U.S. Department of the Interior, Ground Water Manual, A Water Resources Technical Publication, chap. 17, U.S. Department of the Interior, Washington, DC, 1977.

U.S. EPA, Manual of Water Well Construction Practices; EPA-570/9-75-001, U.S. Environmental Protection Agency, Office of Water Supply, Washington, DC, 1975.

Williams, C. and L.G. Evans, Guide to the Selection of Cement, Bentonite and Other Additives for Use in Monitoring Well Construction, Proceedings of the First National Outdoor Action Conference on Aquifer Restoration, Ground Water Monitoring and Geophysical Methods, National Water Well Association. Dublin, OH, 1987, pp. 325–343.

13

Acquisition and Interpretation of Water-Level Data

Matthew G. Dalton, Brent E. Huntsman, and Ken Bradbury

CONTENTS

Introduction 884
    Importance of Water-Level Data 884
    Water-Level and Hydraulic-Head Relationships 884
    Hydraulic Media and Aquifer Systems 885
Design Features for Water-Level Monitoring Systems 886
    Piezometers or Wells? 886
    Approach to System Design 887
    Number and Placement of Wells 888
    Screen Depth and Length 888
    Construction Features 890
    Water-Level Measurement Precision and Intervals 891
    Reporting of Data 892
Water-Level Data Acquisition 892
    Manual Measurements in Nonflowing Wells 893
        Wetted Chalked Tape Method 894
        Air-Line Submergence Method 894
        Electrical Methods 895
        Pressure Transducer Methods 896
        Float Method 897
        Sonic or Audible Methods 897
            Popper 897
            Acoustic Probe 897
        Ultrasonic Methods 898
        Radar Methods 898
        Laser Methods 898
    Manual Measurements in Flowing Wells 899
        Casing Extension 899
        Manometers and Pressure Gages 899
        Pressure Transducers 899
    Applications and Limitations of Manual Methods 900
    Continuous Measurements of Ground-Water Levels 900
    Methods of Continuous Measurement 900
        Mechanical: Float Recorder Systems 900
        Electromechanical: Iterative Conductance Probes (Dippers) 901
        Data Loggers 901
Analysis, Interpretation, and Presentation of Water-Level Data 902
    Recharge and Discharge Conditions 903

Approach to Interpreting Water-Level Data 904
Transient Effects 907
Contouring of Water-Level Elevation Data 910
References 910

Introduction

Importance of Water-Level Data

The acquisition and interpretation of water-level data are essential parts of any environmental site characterization or ground-water monitoring program. When translated into values of hydraulic head, water-level measurements are used to determine the distribution of hydraulic head in one or more formations. This information is used, in turn, to assess ground-water flow velocities and directions within a three-dimensional framework. When referenced to changes in time, water-level measurements can reveal changes in ground-water flow regimes brought about by natural or human influences. When measured as part of an *in situ* well or aquifer pumping test, water levels provide information needed to evaluate the hydraulic properties of ground-water systems.

Water-Level and Hydraulic-Head Relationships

Hydraulic head is the driving force for ground-water movement and varies both spatially and temporally. A piezometer is a monitoring device specifically designed to measure hydraulic head at a discrete point in a ground-water system. Figure 13.1 shows water-level and hydraulic-head relationships at a simple vertical standpipe piezometer (A). The piezometer consists of a hollow vertical casing with a short screen open at point P. The piezometer measures total hydraulic head at point P. *Total hydraulic head* (h_t) has two components — *elevation head* (h_e) and *pressure head* (h_p).

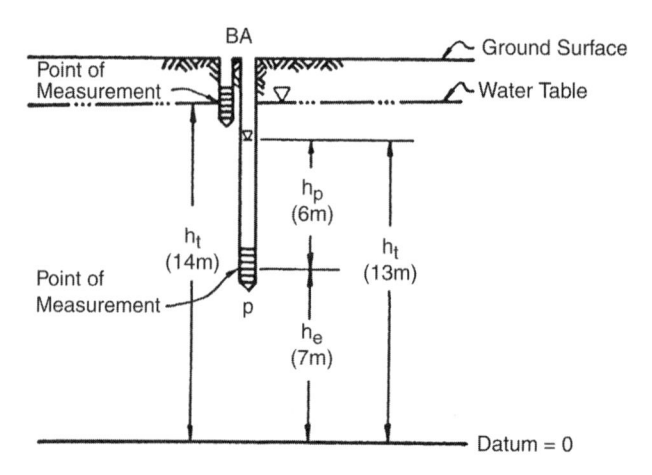

$$h_t = h_e + h_p$$

FIGURE 13.1
Hydraulic-head relationship at a field piezometer. (Adapted from Freeze and Cherry (1979). With permission.)

Elevation head (h_e) refers to the potential energy that ground water possesses by virtue of its elevation above a reference datum. Elevation head is caused by the gravitational attraction between water and earth. In Figure 13.1, the elevation head (h_e) at point P is 7 m.

Pressure head (h_e) refers to the force exerted on water at the measuring point by the height of the static fluid column above it (in this discussion, atmospheric pressure is neglected). In Figure 13.1, the pressure head (h_p) at point P is 6 m. Note that h_p is measured inside the piezometer and corresponds to the distance between point P and the water level in the piezometer.

Total hydraulic head (h_t) is the sum of elevation head (h_e) and pressure head (h_p). The total hydraulic head at point P in Figure 13.1 is $7 + 6 = 13$ m relative to the datum.

The water level in piezometer A is lower than the water level (at the water table) measured in piezometer B. The difference in elevation between the water-table piezometer (B) and the water level in the deeper piezometer (A) corresponds to the hydraulic gradient between the two piezometers. In this case, there is a downward vertical gradient because total hydraulic head decreases from top to bottom.

Hydraulic Media and Aquifer Systems

The "classic" definition of an *aquifer* as "a water-bearing layer of geologic material, which will yield water in a usable quantity to a well or spring" (Heath, 1983) was developed to address water-supply issues, but it is less useful for describing materials in terms of modern ground-water monitoring. Today, ground-water monitoring (including well installation, water-level measurement, and water-quality assessment) occurs in hydrogeologic media ranging from very low hydraulic conductivity shales, clays, and granites to very high hydraulic conductivity sands and gravels. The term aquifer (in ground-water monitoring) is used as a relative term to describe any and all of these materials in various settings.

Aquifers are also generally classified based on where a water level lies with respect to the top of the geologic unit. Figure 13.2 shows an example of layered hydrogeologic media forming both *confined* and *unconfined* aquifers. The confined aquifer is a relatively high hydraulic conductivity unit, bounded on its upper surface by a relatively lower hydraulic conductivity layer. Hydraulic head in the confined aquifer is described by a potentiometric surface, which is an imaginary surface representing the distribution of total hydraulic head (h_t) in the aquifer and which is higher in elevation than the physical top of the aquifer.

The sand layer in the upper part of Figure 13.2 contains an unconfined aquifer, which has the water table as its upper boundary. The water table is a surface corresponding to

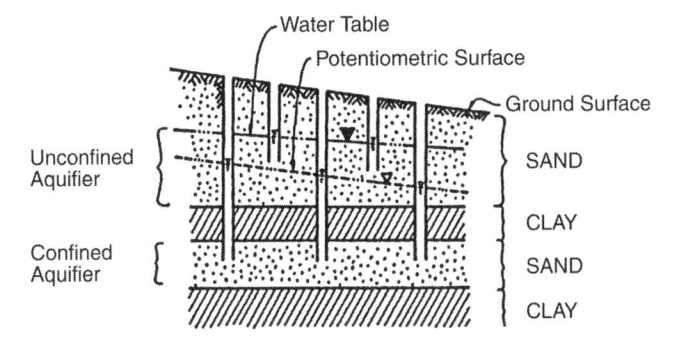

FIGURE 13.2
Unconfined aquifer and its water table; confined aquifer and its potentiometric surface. (Adapted from Freeze and Cherry (1979). With permission.)

the top of the unconfined aquifer where total hydraulic head is zero relative to atmospheric pressure or the hydrostatic pressure is equal to the atmospheric pressure.

Notice that water levels in the piezometers in Figure 13.2 vary with the depth and position of the piezometer. This variation corresponds to the variation of total hydraulic head throughout the saturated system. Hydraulic head often varies greatly in three dimensions over small areas. Thus, the design and placement of water-level monitoring equipment is critical for a proper understanding of the ground-water system.

Design Features for Water-Level Monitoring Systems

An important use of ground-water level (hydraulic head) data from wells or piezometers is assessment of ground-water flow directions and hydraulic gradients. The design of ground-water monitoring systems must usually consider requirements for both water-level monitoring and ground-water sampling. In many cases, both needs can be accommodated with one set of wells and without installing separate systems. However, to collect acceptable water-level data, certain requirements need to be met, which may not always be consistent with the requirements for collecting ground-water samples. For example, additional wells may be required to fully assess the configuration of a water table or potentiometric surface over and above the wells that might be required to collect ground-water samples. Conversely, the design of wells to collect ground-water samples may differ from wells that are used solely to collect ground-water level data.

Water-level monitoring data are generally collected during two phases of a monitoring program. The initial phase is when the site to be monitored is being characterized to provide data to design a monitoring system. The second phase is when water-level data are being collected as part of the actual monitoring program to assess whether changes in ground-water flow directions are occurring and to confirm that wells used to provide ground-water samples are properly located (i.e., hydraulically upgradient and downgradient of a facility that requires monitoring). The latter data also provide a basis to determine the cause of flow-direction changes and to assess whether the monitoring system needs to be reconfigured to account for these changes.

To design a water-level monitoring system, a detailed understanding of the site geology is necessary. The site geology is the physical structure in which ground-water flows and, as such, has a profound influence on water-level data. It is very important that reliable geologic data be collected so that the water-level monitoring system can be properly designed and the water-level data can be accurately interpreted.

Sites at which there is a high degree of geologic variation require more extensive (and costly) water-level monitoring systems than sites that are comparatively more homogeneous in nature. The degree of geologic complexity is often not known or appreciated during the early phases of a site-characterization program, and it may require several stages of drilling, well installation, water-level measurement, and analysis of hydrogeologic data before the required level of understanding is achieved.

Piezometers or Wells?

Ground-water level measurements are typically made in piezometers or wells. Most ground-water monitoring systems associated with assessing ground-water quality are composed of wells rather than piezometers.

Piezometers are specialized monitoring installations; the primary purpose of which is the measurement of hydraulic head. Generally, these installations are relatively small in diameter (less than 1 in. in diameter if a well casing is used), or in some applications, it may not include a well casing and just consist of tubes or electrical wires connected to pressure or electrical transducers. Piezometers are not typically designed to obtain ground-water samples for chemical analysis, although the term piezometer has been applied to pressure measuring devices which have been modified to collect ground-water samples (Maslansky et al., 1987). Piezometers have traditionally had the greatest application in geotechnical engineering for measuring hydraulic heads in dams and embankments.

Wells are normally the primary devices in which water levels are measured as part of a monitoring system. They differ from piezometers in that they are typically designed so ground-water samples can be collected. To accommodate this objective, wells are larger in diameter than piezometers (usually larger than 1.5 in. in diameter), although sampling devices have been developed, which allow ground-water samples to be obtained from small-diameter wells (see Chapter 6 and Chapter 15).

Approach to System Design

Design of a water-level monitoring system should begin with a thorough review of available existing data. This review should be directed toward developing a conceptual model of the site geologic and hydrologic conditions. The conceptual model of the hydrogeologic system is used to determine the locations of an initial array of wells. Tentative decisions regarding drilling depths and the zone or zones to be screened should also be made using existing data. Existing wells may be incorporated into the array if suitable information regarding well construction details is available. Boring and well construction logs, surficial geologic and topographic maps, drainage features, cultural features (e.g., well fields, irrigation, and buried water pipes), and rainfall and recharge patterns (both natural and man-induced) are several of the major factors that need to be assessed as completely as possible.

The available data should be reviewed to identify:

- The depth and characteristics of relatively high hydraulic conductivity geologic materials (aquifers) and low hydraulic conductivity confining beds that may be present beneath a site
- Depth to the water table and the likelihood of encountering perched or intermittently saturated zones above the water table
- Probable ground-water flow directions
- Presence of vertical hydraulic gradients
- Features that might cause ground-water levels to fluctuate, such as well-field pumping, fluctuating river stages, unlined ditches or impoundments, or tides
- Probable frequency of fluctuation
- Existing wells that may be incorporated into the water-level monitoring program

The practical limitations of where wells can be located on a site should not be overlooked during this phase of the system design. Wells can be located almost anywhere on some sites; however, on other sites, buildings, buried utilities, and other site features can impose limitations on siting wells.

Number and Placement of Wells

The number of wells required to assess ground-water flow directions beneath a site is dependent on the size and complexity of the site conditions. Simple and smaller sites require fewer wells than larger or more hydrogeologically complex sites.

Many sites have more than one saturated zone of interest in which ground-water flow directions need to be assessed. High hydraulic conductivity zones may be separated by lower hydraulic conductivity zones. In these cases, several wells screened at different depths may be required at several locations to adequately assess flow directions in, and between, each of the saturated zones of interest.

The minimum number of wells required to estimate a ground-water flow direction within a zone is three (Todd, 1980; Driscoll, 1986). However, the use of just three wells is only appropriate for relatively small sites with very simple geology, where the configuration of the water table or potentiometric surface is essentially planar in nature, as shown in Figure 13.3.

Generally, conditions beneath most sites require more than three wells. Lateral variations in the hydraulic conductivity of subsurface materials, localized recharge patterns, drainage channels, and other factors can cause the potentiometric or water-table surface to be nonplanar.

On large or more geologically complex sites, an initial grid of six to nine wells is usually sufficient to provide a preliminary indication of ground-water flow directions within a target ground-water zone. Such a configuration will generally allow the complexities in the water table or potentiometric surface to be identified. After an initial set of data is collected and analyzed, the need for and placement of additional monitoring installations can be assessed to fill in data gaps or to further refine the assessment of the potentiometric or water-table surface.

Figure 13.4 shows a site at which leakage from a buried pipe has caused a ground-water mound to form. In this situation, a three-well array would not provide sufficient data to detect the presence of the mound and could result in a faulty assessment of the ground-water flow direction beneath the site.

Screen Depth and Length

After well locations are established, well screen depths and lengths should be chosen. Screen depths are generally determined during the drilling operation after a geologic

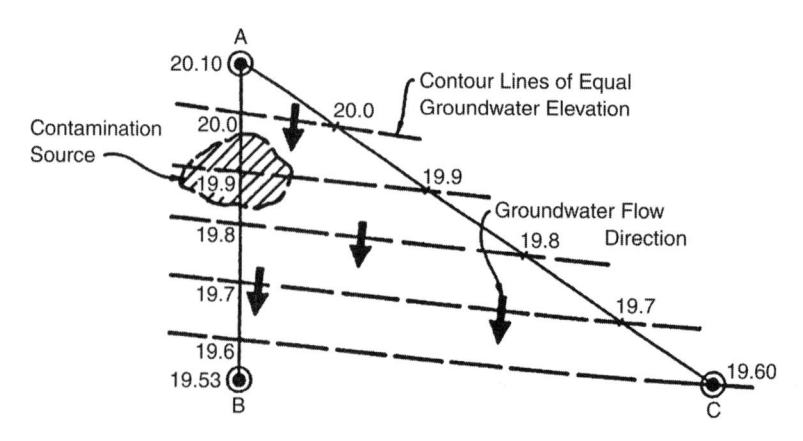

FIGURE 13.3
Assessing ground-water flow directions at a small site with a planar water table surface.

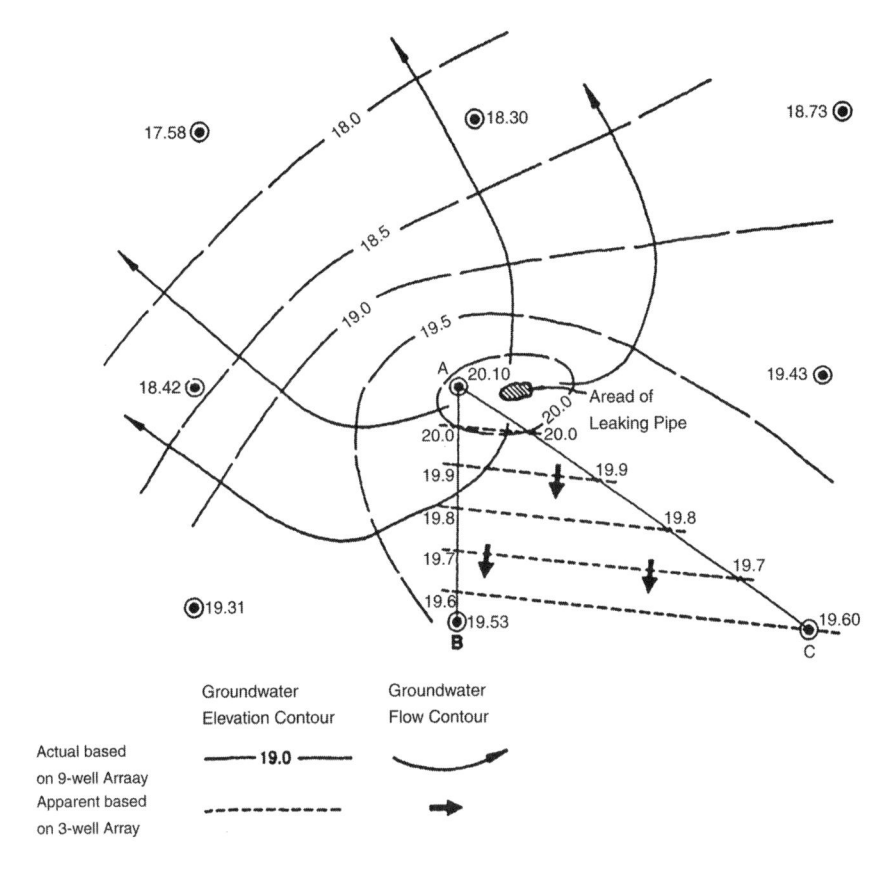

Acquisition and Interpretation of Water-Level Data

FIGURE 13.4
Estimation of ground-water flow directions with a three-well and a nine-well array.

log has been prepared, depending on the amount and quality of the data available prior to drilling.

Wells used to assess flow directions within a zone are usually screened within that zone at similar elevations. Highly layered units may require screens in each depth zone that is isolated by lower hydraulic conductivity layers (Figure 13.5a). Where the units are dipping, it is generally more important to place the screens in the same zone even if the screens are not placed at similar elevations (Figure 13.5b).

Similar well-screen lengths should be used and the screen (and filter pack) should be placed entirely within the zone to be monitored. This will allow field personnel to obtain a water level that is representative of the zone being monitored and will minimize the possibility of allowing contaminants, if present, to migrate between zones screened by the well. If the well screen is open to several zones, then a composite or average water level will be measured, which will not be representative of any single zone, and will add to the difficulty in interpreting the water-level data. Typical commercially available well screens are 5 or 10 ft long, although it is possible to construct wells with longer or shorter screens, to meet specific project objectives.

If multiple saturated zones are present beneath a site, it is generally necessary to install either several wells screened at different depths at a single location or a multilevel monitoring system (Chapter 11). Such installations allow the assessment of both horizontal and vertical hydraulic gradients. If few reliable data are available for a site, it is desirable that the initial hydrologic characterization starts with the uppermost zone of interest. During

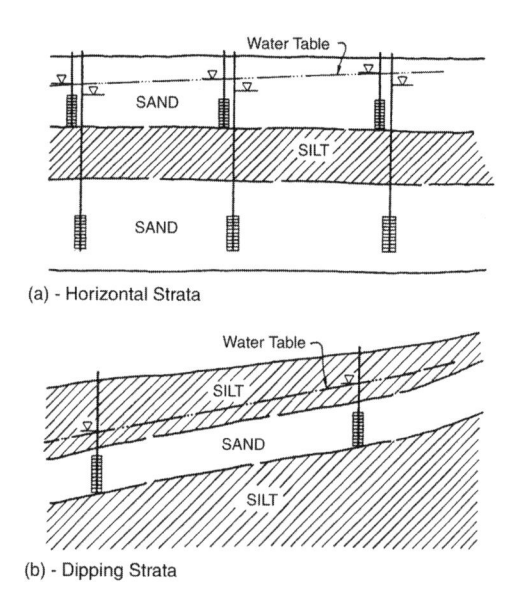

FIGURE 13.5
Well screen placement in horizontal and dipping strata.

this initial work, a limited number of deeper installations can be installed to provide data to assess the need for additional deeper installations. In situations in which contamination is present in a shallow aquifer, extreme care must be exercised with regard to installing deeper wells, to prevent the possible downward movement of contamination into deeper zones.

Construction Features

Water-level monitoring points can be installed using a variety of methods and configurations (Figure 13.6). Typically, the installations are constructed in drilled boreholes,

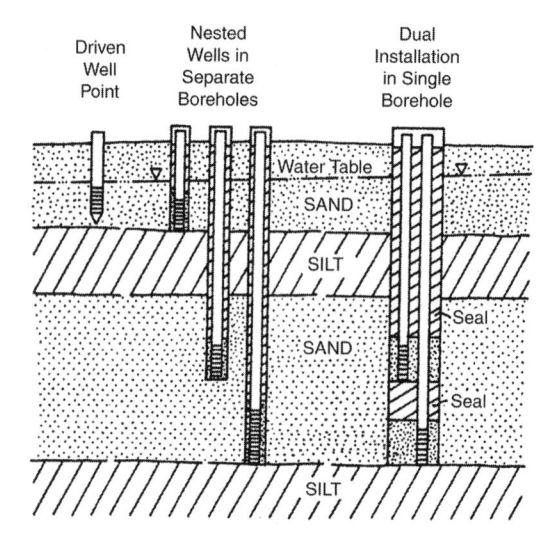

FIGURE 13.6
Typical monitoring well installation configurations.

although driven well points can be used to provide water-level data in shallow, unconfined saturated zones. Drilling and monitoring well installation procedures are discussed in Chapter 5 and Chapter 10, respectively.

At locations where multiple zones are to be monitored, single or multiple installations in the same borehole or multilevel systems can be used. If a single well is installed in a borehole, several boreholes will be necessary to monitor multiple zones.

A single installation in a single borehole is often preferred because it is easier to install a reliable annular seal above the well screen when only one well is completed in a borehole. An annular seal is necessary to ensure that the water-level data are representative of the zone being monitored and to ensure that contaminants do not move between zones within the borehole. In many situations, especially if a hollow-stem auger is being used to install the well, the cost of installing single installations is only marginally higher than multiple installations in a single borehole.

Multiple installations in a single borehole have been installed successfully as long as an adequate borehole or drill casing diameter is used and care is taken in installing the wells. Installing two 2 in. diameter wells per borehole should be feasible within 6 to 8 in. diameter boreholes or drill casings. While multiple installations in the same borehole may be technically feasible, some local well-drilling regulations may preclude or restrict such installations.

Water-Level Measurement Precision and Intervals

Wells should be accurately located horizontally and vertically, although horizontal surveying is not always required, depending on the size of the site and available base maps. The precision of the horizontal locations is generally not as important as the precision of the elevation survey and water-level measurements.

The top of the well casing (or other convenient water-level measuring point) should be surveyed to a common datum (usually National Geodetic Vertical Datum or NGVD) so that water-level measurements can be converted to water-level elevations. The reference point for water-level measurements should be clearly marked at a convenient location on each well casing. This will facilitate reducing measurement error.

The precision of the elevation survey and water-level measurements depends on the slope of the potentiometric or water-table surface and the distance between wells. Greater precision is required at sites where the surface is gradual or the wells are close together. Generally, reference point elevations should be surveyed and water levels measured with a precision ranging between ± 0.1 and ± 0.01 ft.

For example, if water-level fluctuations are occurring over a short period of time, it may be more important to obtain a set of less precise measurements in a short period of time rather than a very precise set of measurements over a longer period of time. In such cases, measurements made to 0.1 ft may be appropriate. In contrast, if the slope of the potentiometric surface or water table surface is very gradual, more precise elevation control and water-level measurements may be required.

Current environmental regulations generally require that water levels be monitored and reported on a quarterly basis. A quarterly monitoring schedule may be appropriate for sites at which water levels fluctuate only in response to seasonal conditions, such as precipitation or irrigation recharge. However, water levels at many sites respond not only to seasonal factors but also to factors of shorter duration or greater frequency. These factors may include fluctuations caused by tides in coastal areas, changes in river stage, and daily well pumping, among others. Separate zones may also respond differently to the cause of the fluctuations.

During site-characterization activities, factors that may cause water levels to fluctuate need to be assessed and their importance evaluated with respect to two issues:

- The time in which a set of water-level measurements needs to be obtained
- How the flow directions may change as the water levels fluctuate

With the advent of computer technology, our ability to analyze complex systems at a reasonable cost has increased dramatically. Microprocessors connected to transducers allow the collection and analysis of water-level data over extended periods of time. To determine a site-specific monitoring interval, continuous monitoring can be economically accomplished in selected wells screened at different depths and at varying distances from the cause of the fluctuation. These data can then be used to determine the time frame and intervals in which to obtain water-level measurements and to determine how the various zones beneath the site respond to the cause of the fluctuation.

The period in which the continuous monitoring should be conducted depends on the frequency and duration of the fluctuations. If possible, monitoring should be conducted at times of representative fluctuation. For example, on sites affected by tides, monitoring over several tidal cycles during relatively high and low tides may be warranted.

Reporting of Data

Interpretation of water-level data requires that information be available about the monitoring installations and the conditions in which the water-level measurements were made. This information includes:

- Monitoring installations
  a. Geologic sequence
  b. Well construction features, especially screen and sand pack length, and geologic strata in which the screen is situated
  c. Depth and elevation of the top and bottom of the screen and sand pack
  d. Measuring point location and elevation
  e. Casing stickup above ground surface
- Water-level data
  a. Date and time of measurement
  b. Method used to obtain the measurement
  c. Other conditions in the area that might be affecting the water-level data, such as tidal or river stage, well pumping, storm events, etc.

Water-Level Data Acquisition

For many purposes in ground-water investigations, the accurate determination of water levels in wells or piezometers is paramount. Without accurate measurements, it is not possible to interpret the data to assess conditions such as ground-water flow directions, ground-water flow velocities, seasonal variations in water levels, aquifer hydraulic conductivity, and other important features.

Acquisition and Interpretation of Water-Level Data

Depending upon the ultimate use of the water-level data, the methods and instruments used to collect and record changes in ground-water levels may vary substantially. Water-level data acquisition techniques are divided into two major categories for discussion purposes: manual measurements or typically nonrecording methods and continuous measurements using instruments that provide a record. Although not exhaustive, the following discussion describes techniques most frequently used by the practicing hydrogeologist. These methods are summarized in Table 13.1.

Manual Measurements in Nonflowing Wells

Accurate manual measurements of water levels in wells and piezometers should be a core skill for any practicing hydrologist, hydrogeologist, ground-water scientist, or technician. Regardless of the method used, repeated measurements of water levels in wells made within a few minutes and within 200 ft of the top of casing should agree within 0.01 or 0.02 ft. As a standard of good practice, Thornhill (1989) suggests that anyone obtaining a water-level measurement in a well should take at least two readings. If they differ by more than 0.02 ft, then continue to measure until the reason for the lack of agreement is determined or until the results are shown to be stable.

TABLE 13.1

Summary of Methods for Manual Measurement of Well-Water Levels in Nonflowing and Flowing Wells

Measurement Method	Measurement Accuracy (ft)	Major Interference or Disadvantage
Nonflowing Wells		
Wetted chalked tape	0.01	Cascading water or casing wall water and chalk in water
Air line	0.01–0.25	Air line or fitting leaks; gage inaccuracies and air source
Electrical probes	0.02–0.1	Cable wear; hydrocarbons on water surface and turbulence
Transducer	0.01–0.1	Temperature changes; electronic drift; blocked capillary
Float	0.02–0.5	Float or cable drag; float size and lag
Popper	0.1	Well noise; well pipes and pumps; well depth
Acoustic probe	0.02	Cascading water; hydrocarbon on water surface
Ultrasonic	0.02–2.4	Temperature changes; well pipes and pumps; casing joints
Radar	0.01–0.02	Temperature; humidity; well pipes and pumps, small wells
Laser	0.01	Nonstraight wells; beam penetration through water
Flowing Wells		
Casing extensions	0.1	Limited range; awkward to implement
Manometer and pressure gage	0.1	Gage inaccuracies; calibration required
Transducers	0.02–0.1	Temperature changes; electronic drift

Wetted Chalked Tape Method

Although less commonly utilized in today's hydrologic assessments, one of the most accurate techniques used to manually measure ground-water levels is the wetted chalked tape method (ASTM D 4750; ASTM, 2004a). The equipment needed to make a measurement using this method consists of a standard steel surveyor's tape, a block of carpenter's chalk, and a slender lead or stainless-steel weight. Steel tapes and hand reels are commercially available in lengths up to 1000 ft. It is recommended, however, that shorter standard lengths (100, 200, 300, and 500 ft) be used because of weight and cost. Steel-tape markings are usually divided only into tenths of feet or inches and fractions of an inch. Interpolation to the nearest 0.01 ft is possible.

The weight is attached to the steel tape end clip with sufficient wire for support, but not enough to be stronger than the tape. This allows the tape to be pulled free if the weight become snagged on something in the well. The bottom 2 or 3 ft of the tape is coated with carpenter's chalk. A water-level measurement is made by lowering the tape slowly into the well, about 1 or 2 ft into the water. It is convenient to lower the tape into the water a sufficient distance to allow the tape to read an even foot mark at the top of the well casing or the reference measuring point at the surface. The water-level measurement is calculated by subtracting the submerged distance, as indicated by the absence of change in chalk color, from the reference point at the top of the well.

The practical limit of measurement precision for this method is ± 0.01 ft (U.S. Geological Survey, 1980). Coefficients of stretch and temperature expansion of the steel tape become a concern when water-level measurements are made in wells that have higher temperatures or at depths greater than 1000 ft (Garber and Koopman, 1968). For most ground-water investigations, corrections for these errors are not necessary.

A disadvantage of using the wetted chalked tape method is that if the approximate depth to water is unknown, too short or too long a length of chalked tape may be lowered into the well, thereby necessitating a number of attempts. In addition, water condensed on the side of the casing, or cascading water, may wet the tape above the actual water level and result in errors in measurement (Everett, 1980). When compared with other manual measurement techniques, this method is more time consuming. Proficiency in obtaining water levels with a wetted chalked tape requires practice. In addition, the introduction of chalk into a well that is used to obtain water-quality samples is discouraged.

Air-Line Submergence Method

The air-line submergence method, although less precise than other manual water-level measurement methods, continues to be a preferred technique in wells that are being pumped. To make an air-line measurement of water level in a well, a straight, small-diameter tube of accurately known length is installed in the well. This tube, usually 0.375 in. or less in diameter, can be made of plastic, copper, or steel. The air line and all connections must be air tight, without bends or kinks, and installed to several feet below the lowest anticipated water level. A pressure gage (preferably calibrated in feet of water), along with a fitting for an air source, is attached to this line. In deep wells or where multiple water-level measurements are needed, a small air compressor is useful. In shallow wells, a hand-operated air pump is typically used.

A water-level measurement is made when air is pumped into the small tube and the pressure is monitored. Air pressure will continue to increase until it expels all water from the line. Air pressure, which is determined when the pressure gage stabilizes, is used to calculate the height of the water in the tube. If the pressure gage is calibrated in pounds per square inch (psi), a conversion is made to feet by multiplying the psi reading by 2.31. The actual water level in the well is determined by subtracting the calculated distance from the air line's length. According to Driscoll (1986), the dependability of measurements made by the air-line method varies with the accuracy of the pressure gage and the care used in determining the initial pressure reading. Depth to water can usually be determined to within 0.25 ft of the true water level. Garber and Koopman (1968) have also shown that the precision of the measurement is mainly dependent upon the accuracy of the pressure gage. They state that even with gages having gradations as small as 0.1 psi, the maximum possible resolution would be 0.23 ft. Digital quartz pressure transducers and specialized data loggers have been tested as replacements for standard pressure gages. Water levels from 0 to 50 ft have been measured with an accuracy of better than 0.01 ft (Paroscientific, Inc., 2002). However, these precision pressure measurement systems are designed for more permanent installations, not for portable applications. Unless the air-line method is used in wells of substantial depth, corrections for thermal expansion, hysteresis, fluid density, and barometric pressure are not necessary.

Electrical Methods

Currently, the most favored technique for manual water-level measurement is the use of an electrical probe. The most widely used instrument of this type is one that operates on the principle that a circuit is completed when two electrodes come in contact with the water surface in the well, which is conductive. Other instruments rely on physical characteristics such as resistance, capacitance, or self-potential to produce a signal. Many of these instruments employ a two-wire conductor that is marked every foot, with minor interval markings of 0.01 ft. Some instruments use vinyl-, epoxy-, or Teflon-clad steel tapes as an insulated electrode and the well casing or grounding wire as the other electrode. Because of weight and the amount of potential cable stretch, most commercial electrical probes are designed for water-level measurements within several hundred feet of the top of casing.

Water-level probes that use self-potential typically have one electrode made of magnesium and the other made of brass or steel. When the probe comes into contact with water, a potential between the two dissimilar metals is measured at the surface on a voltmeter (generally in millivolts).

If a battery is added to the circuit, the two electrodes may be of the same material, usually brass, lead, or ferrous alloy. When the electrodes come into contact with the water surface, the water conducts the current and a meter, light, or buzzer is activated at the ground surface.

The principles of capacitance and inductance have been used by the U.S. Geological Survey to detect water surfaces (Garber and Koopman, 1968). These are basically specialty instruments and few are available commercially for common water-level measurements. However, some units that employ capacitance or inductance are used for detection of water levels and hydrocarbons in wells. These units have the same apparent accuracy and precision as other electrical probes because the sensing elements are suspended in the well via multiwire conductors.

Errors in water-level measurements using electrical probes result from changes in the cable length and diameter as a function of use, depth, and temperature. After repeated use, the markings on the drop line often have a tendency to become loose and slide (if banded) or become illegible from wear (if embossed). Shallow measurements made with well-maintained electrical probes are typically reproducible to within ± 0.02 ft. Because of kinks in the cable and less than vertical suspension in a well, Barcelona et al. (1985) stated that the accuracy of electrical probes is about 0.1 ft. Plazak (1994) showed that even the same make and model of electrical probes may vary ±0.01 to ±0.11 ft in precision and accuracy, depending on the depth of the water-level measurement.

A disadvantage inherent in most electrical probe instruments is that if substantial amounts of oil or other nonconductive materials (i.e., oils) are floating upon the water surface, contact cannot be reliably made. This is a major concern in ground-water investigations involving petroleum hydrocarbon releases. Special sensing probes utilizing an optical or infrared sensor in conjunction with electrical conductivity are commercially available to measure the hydrocarbon−water interface. Because this type of sensing probe is also suspended from multiconductor wire, the same errors as previously discussed for electrical probes apply.

Pressure Transducer Methods

With the advent of reliable silicon-based strain gage pressure sensors and vibrating-wire transducers, a unique type of instrument is being commercially marketed for measuring changes in water levels. These transducers contain a 4−20 mA current transmitter and a strain gage sensor or a vibrating wire in an electromagnetic coil with frequency measurement circuitry. The current transmitter circuitry in both types of transducers prevents measurement sensitivity from being affected by cable length. Because all sensitive electronics are in the transducer and submerged in a constant temperature environment (the well water), errors due to temperature fluctuations are negligible (In Situ, Inc., 1983; Zarriello, 1995). The simultaneous measurement of temperature and water level is becoming a standard feature for most of the transducers used in hydrogeologic studies.

Many transducers used for measuring ground-water levels have a small capillary tube shielded in the support cable leading from one side of a differential pressure sensor. This tube is vented to the atmosphere, which provides automatic compensation of barometric pressure. Care must be taken when working with transducer cables that contain capillaries to avoid kinking, crushing, or allowing condensate to form in the vent tube. Blocked vent tubes may result in erroneous water-level measurements. A signal conditioning unit and a power source are required ancillary equipment to make a water-level measurement.

To avoid the need for electrical cables to transmit signals from the transducer to the data storage unit, some manufacturers have totally sealed the data logger, battery, pressure transducer, and temperature sensor in a small stainless-steel case for total submersion in a well. Communication with the data logger is established via an infrared optical port, either with a cradle component or through extension cables connected to a host computer. These units use an absolute pressure sensor to avoid the need for a vent tube to the surface. However, all water-level readings obtained by this type of monitor will require subsequent corrections for barometric pressure changes (Solinst Ltd., 2001). For a discrete water-level measurement, the transducer is lowered a known distance into the well and allowed to equilibrate to the fluid temperature. The distance of submergence of the transducer is read on the signal conditioning unit and is subtracted from the known cable length referenced at the top of the well.

This technique is easily adaptable to continuous monitoring. It also offers several advantages in ease of accurate measurement in both pumping wells and wells with cascading water. Sources of error in this type of instrument include the electronics (linearity, accuracy, temperature coefficient, etc.), temperature changes, and inappropriate application (i.e., range and material of construction) of a transducer in a given medium (Sheingold, 1980; Zarriello, 1995). Because of the sensitive electronics, rough handling of the transducers in the field or in storage should be avoided.

The accuracy of water-level transducers is dependent upon the type and range (sensitivity) of the device used. Most transducers are rated in terms of a percent of their full-scale capability. For example, a 0 to 5 psi tranducer rated at 0.01% will provide measurements to the nearest 0.01 ft. In contrast, a 0 to 25 psi transducer rated at 0.01% will provide measurements to the nearest 0.05 ft (Barcelona et al., 1985). Standard practices for the static calibration of electronic transducers used for obtaining field pressure measurements have been developed and should be used to document the accuracy of the instrumentation system (ASTM D 5720; ASTM, 2004b). These calibration procedures are typically included in the standard operating procedures prepared for any large-scale hydrogeologic investigation in which electronic pressure transducers are used.

Float Method

As the name implies, a float is attached to a length of steel tape and suspended over a pulley into the well. At the opposite end of the steel tape, a counterweight is attached. The depth to water is read directly from the steel tape at a known reference point at the top of the casing.

To obtain an accurate measurement using this technique, the absolute length of the float assembly must be measured and subtracted from the steel-tape measurement. For greater accuracy, the total amount of float submergence should be calculated and a correction factor applied. This becomes more critical with smaller diameter floats (Leupold and Stevens, Inc., 1978). This method is used principally to obtain continuous water-level measurements. The accuracy and errors in float-operated devices will be discussed in greater detail in the following section.

Sonic or Audible Methods

Virtually every practicing hydrogeologist has (but should not have) dropped a rock down a well, at one time or another, to determine whether water is present and to estimate the depth to water. Stewart (1970) investigated and developed a technique to determine the depth to water by timing the fall of a BB (air rifle shot) or a glass marble and by recording the time of the return sound of impact. This sonic technique will not be discussed here in detail because of the rather large range of error in measurement (± 5 ft), but interested readers are referred to Stewart (1970). Other sonic methods are described subsequently.

Popper

The most simplistic device used to audibly determine the depth to water in a well casing is a popper (also called a plopper). This is a metal cylinder from 1 to 1.5 in. in diameter and generally 2 to 6 in. In length, with a concave bottom. The popper is attached to a steel tape and lowered to within a few inches of the water surface in the well. By repeatedly dropping the popper onto the water surface and noting the tape reading at which a distinctive "pop" is heard, the depth to water is determined (U.S. Bureau of Reclamation, 2001).

Because of noise and the lack of clearance, the use of poppers in pumping wells is limited. The accuracy of water-level measurements made by this technique is highly dependent upon the skill of the measurer and the depth of the well. Determination of the water level to within 0.02 ft is usually the detection limit of this procedure.

Acoustic Probe

A unique adaptation of the popper principle was developed by Schrale and Brandwyk (1979), with the construction of an acoustic probe. This electronic device is attached to a steel tape and lowered into the well until an audible sound is emitted from a battery-powered transducer contained in the probe. The electric circuit is completed when the two electrodes placed in the bottom of the probe come in contact with the water level in the well. As with the previously discussed electrical methods, problems with measurements can occur when hydrocarbons are present or if the well has cascading water. According to the developers of this instrument, a water-level determination is possible to within ±0.02 ft.

Ultrasonic Methods

Instruments that measure the arrival time of a reflected transmitted sonic or ultrasonic wave pulse are becoming more common in the measurement of water levels. These instruments electronically determine the amount of time it takes for a sound wave to travel down the well casing, reflect off the water surface, and return to the surface. Because the electronic circuitry typically uses microprocessors, this signal is transmitted, received, and averaged many times a second. The microprocessor also calculates the depth to water and displays it in various units. Several of the commercially available instruments simply rest on top of the well casing with nothing being lowered into the well. Rapid determination of water depths in deep wells is a distinct advantage of this technique.

The presence of hydrocarbons on the water surface usually has no effect on the measurement. Accuracy can be limited by change of temperature in the path of the sound wave and other reflective surfaces in the well (i.e., pipes, casing burrs, pumps, samplers, crooked casing, etc.). Large variations in humidity will also effect readings. Most commercially available hand-held units can measure the depth to water within 0.1 ft if the well's temperature gradient is uniform. Usually, the greater the depth to water, the less accurate the measurement. One manufacturer reports a ±0.2% accuracy over a range of 25 to 1200 ft. Specialized installations, however, have repeatedly provided water-level measurements accurate to within ±0.02 ft (Alderman, 1986).

Radar Methods

Similar to the ultrasonic measurement instrumentation, radar-based portable units use a pulsed or continuous high-frequency wave to reflect off the water surface in a well. Depth to water is calculated by determining the travel time of the pulse or wave and electronically converting the signal to a depth measurement. Range of measurement to water is typically limited to larger wells and water levels about 100 ft or less from the top of casing. These limitations are the result of a need to maintain a focused beam width.

Accuracy of commercial units is reportedly good, from ±0.01 to ±0.02 ft over the range of measurement. As with other acoustic methods, temperature, humidity, and obstacles in the beam pathway all will have an effect on the quality of the water-level measurement (Ross, 2001).

Laser Methods

Lasers have been used in the food, chemical, and energy industries for over a decade as a method of noncontact level monitoring of liquids and solids in tanks. Advances in laser technology have allowed the manufacturing of battery-powered units potentially capable of obtaining water-level measurements in wells and piezometers. Tests of prototype instrumentation show promise for use in well-monitoring applications, but further development is needed to bring this technology into common use by the ground water professional.

One of the significant advantages of laser technology for obtaining water level measurements is an unparalleled accuracy to depth range. Ross (2001) reported an accuracy of ± 0.01 ft for distances greater than 1000 ft. Because of the very high frequency of the laser pulse, humidity, and temperature variations in a typical well would not significantly effect the signal. However, the use of the laser requires a clear beam pathway. If a well is not plumb or if obstacles in the well prevent a clean line of sight down the well, a measurement cannot be made. Other issues include scattering of the reflected laser beam from the water surface due to turbulence or the beam penetrating through the target water surface without reflection (Ross, 2001).

Manual Measurements in Flowing Wells

Casing Extension

When the pressure of a flowing well is sufficiently low, a simple extension of the well casing allows the water level to stabilize so that a water-level measurement can be made. The direct measurement of the piezometric level by casing extension is practical when the additional height requirement is several feet or less. A water-level measurement using this technique should be accurate to within ± 0.1 ft because flowing well water levels tend to fluctuate.

Manometers and Pressure Gages

If the pressure of the flowing well is sufficiently high, the use of a casing extension is usually not practical. To measure the piezometric level in such circumstances, the well is sealed or "shut-in" and the resulting pressure of the water in the well casing is measured. Two commonly used instruments to monitor the well pressure are manometers and pressure gages.

A mercury manometer, when properly installed and maintained, has a sensitivity of ± 0.005 ft of water, and these devices have been constructed to measure ranges in water levels in excess of 120 ft (Rantz, 1982). When used to monitor shut-in pressure of wells, an accuracy of ± 0.1 ft is typical (U.S. Geological Survey, 1980).

Pressure gages are typically less sensitive to head pressure changes than mercury manometers and, therefore, have only a routine accuracy of ± 0.2 ft under ideal conditions when calibrated to the nearest tenth of a foot of water. According to the U.S. Geological Survey (1980), probable accuracy of measuring the pressure of a shut-in well with pressure gages is about 0.5 ft with these older style units. Many of these less sensitive gages are still in use today. Design advances during the last decade in both mechanical and electronic gages used as replacements for mercury manometers have increased the measurement accuracy to better than ± 0.01 ft of the gage range (Paroscientific, Inc., 2002). However, because well shut-in pressures typically fluctuate, a practical accuracy still remains at about ± 0.1 ft for this technique.

When using either of these instruments to measure well pressure, care should be taken to avoid rapid pressure change caused by opening or closing the valves used in sealing the well. This could create a water-hammer effect and cause subsequent damage to the manometer or pressure gage. In addition, field instruments used to monitor pressure should be checked periodically against master gages and standards.

Pressure Transducers

As previously described, pressure transducers can accurately monitor changes in pressure over a wide range. Transducers have been installed in place of pressure gages to determine the potentiometric level. If the pressure transducer range is carefully matched with the shut-in well pressure, measurements to ±0.02 ft can be obtained. One source of error in these measurements results from changes in temperature in the transducer. Either a transducer unit that has some form of electronic temperature compensation or a unit that is totally submerged in the well should be used. Again, due to fluctuations in well shut-in pressures, the apparent measurement accuracy of this method will be about ±0.1 ft.

Applications and Limitations of Manual Methods

No single method for determining water levels in wells is applicable to all monitoring situations, nor do all monitoring situations require the accuracy and precision of the most sensitive manual measurement technique. The practicing hydrogeologist should become familiar with the various techniques using two or more of these methods to obtain water levels on the same well. By doing so, the strengths and weaknesses of the monitoring methods will quickly become evident.

Table 13.1 is a summary of the manual measurement techniques discussed earlier, with their reported accuracies. Also presented in this summary are several of the principal sources of error or interference relevant to each technique. This table should be used only as a guide because each monitoring application and the skill of the measurer can result in greater or lesser measurement accuracy than stated.

Continuous Measurements of Ground-Water Levels

The collection of long-term water-level data is a necessary component of many hydrogeologic investigations. A commonly employed technique is the use of mechanical float recording systems. These devices typically produce a continuous analog record, usually on a strip chart, which is directly proportional to the water-level change.

Electromechanical instruments that use a conductance probe with a feedback circuit to drive a strip chart or a punched tape can successfully monitor rapid changes in water levels. These are used where float-operated systems fail to follow water-level fluctuations as expected.

With the development of field-operable solid-state data loggers and portable computers, long-term monitoring systems using pressure transducers are favored among those conducting hydrogeologic investigations. As with manual water-level measurements, the type of long-term monitoring system employed is dependent upon the investigator's data needs.

Methods of Continuous Measurement

Mechanical: Float Recorder Systems

Instruments that use a float to operate a chart recorder (a drum or wheel covered with chart paper and containing a time-driven marking pen) have been used to measure water levels since the early 1900s. These devices produce a continuous analog record of water-level change, usually as a graph. Depending upon the gage scale and time-scale gearing, a single chart may record many months of water-level fluctuations. To augment or even replace the analog record of float recorder systems, digital encoders and data loggers have been added to many of these systems. If properly installed and maintained, float recorder systems are very reliable, as is evidenced by their continued use in many municipal well-field monitoring programs. Mechanical systems are also useful when interfering electromagnetic currents or other harsh environmental conditions preclude the use of electronic-based units.

Float-operated devices are subject to several sources of error, which include float lag, line shift, submergence of counterweight, temperature, and humidity. Leupold and Stevens (1978) detail these errors and suggest methods to correct them. The reader should consult this reference for additional details. For purposes of this discussion, it is noted that when smaller floats are used, the magnitude of error is greatest. For example, float lag, or the lag of the indicated water level behind the true water level due to the mechanical work required by the float to move the instrument gears, can be as much as 0.5 ft for a 1.5 in. float if the force to move the instrument is 3 oz. This is contrasted to a 0.07 ft error for a 4 in. float and 0.03 ft error for a 6 in. float on an instrument requiring the same 3 oz of force (Leupold and Stevens, 1978). This error is magnified if the float or float cable is allowed to drag against the well casing. Shuter and Johnson (1961) discuss these problems in measuring water levels in small-diameter wells and offer several devices to improve recorder performance. Because many of the wells constructed in today's ground-water monitoring programs are 2 in. in diameter, caution should be used if a float recording system is installed to obtain continuous water-level measurements.

According to Rantz (1982), if a mechanical float recording system is properly installed and operated, long-term water-level measurements in wells are obtainable to an accuracy of about ± 0.01 ft. This accuracy is based on measurements made in stilling wells used for long-term monitoring of stage height of rivers. Because the piezometers and wells typically utilized in monitoring well networks are smaller in diameter, the accuracy for float recording systems used to measure ground-water fluctuations will usually be greater than ± 0.01 ft.

Electromechanical: Iterative Conductance Probes (Dippers)

Iterative conductance probes, commonly referred to as dipping probes or dippers, are electromechanical devices that use an electronic feedback circuit to measure the water level in a well. A probe is lowered on a wire by a stepping motor until a sensor in the probe makes electrical contact with the water. This generates a signal that causes the motor to reverse and retract the probe slightly. After a set time period, the probe is lowered again until it makes contact with the surface, retracts, etc., thus repeating the iterative cycle. The wire cable is connected to either a drum used for chart recording or a potentiometer whose output signal is proportional to the water level (Grant, 1978).

Dipping probes have several advantages over float recording systems. The well can be of smaller diameter and the system can accommodate some tortuosity in the well casing. Because the sensing probe is electromechanical, greater depths to water can be monitored without the mechanical losses associated with float systems. When water-level fluctuations are cyclic or change moderately rapidly, the dipping probe better reflects the oscillations in the water levels of smaller diameter wells.

Data Loggers

Data loggers consist of microprocessors connected to transducers that are installed in the well. The microprocessors consist of hardware and software that allow the automated collection of water-level data over various time periods. Data can be easily manipulated after transfer to a computer database. The use of this equipment is common, and a variety of equipment systems are commercially available.

Variations of data-logger based systems have been installed to better access and process water-level data. From the transducer at the wellhead, data is transferred to a data logger or signal processor to a central computer via hardwire, line-of-sight radio, satellite radio, or phone lines. At some of these installations, the central computer can query each remote well unit at any desired frequency including a continuous data scan mode (U.S. Bureau of Reclamation, 2001).

Analysis, Interpretation, and Presentation of Water-Level Data

The primary use of ground-water level data is to assess in which direction ground-water is flowing beneath a site. The usual procedure is to plot the location of wells on a base map, convert the depth-to-water measurements to elevations, plot the water-level elevations on the base map, and then construct a ground-water elevation contour map. The direction of ground-water flow is estimated by drawing ground-water flow lines perpendicular to the ground-water elevation contours (Figure 13.4).

The relatively simple approach to estimating ground-water flow directions described earlier is suitable where geologic media are assumed to be isotropic, wells are screened in the same zone, and the flow of ground-water is predominantly horizontal. However, with the increased emphasis on detecting the subsurface positions of contaminant plumes or in predicting possible contaminant migration pathways, it is evident that the assumptions of isotropy and horizontal flow beneath a site are not always valid. Increasingly, flow lines shown on vertical sections are required to complement the planar maps showing horizontal flow directions (Figure 13.7) to illustrate how ground water is flowing either upward or downward beneath a site (Figure 13.8).

Ground water flows in three dimensions and as such can have both horizontal and vertical (either upward or downward) flow components. The magnitude of either the

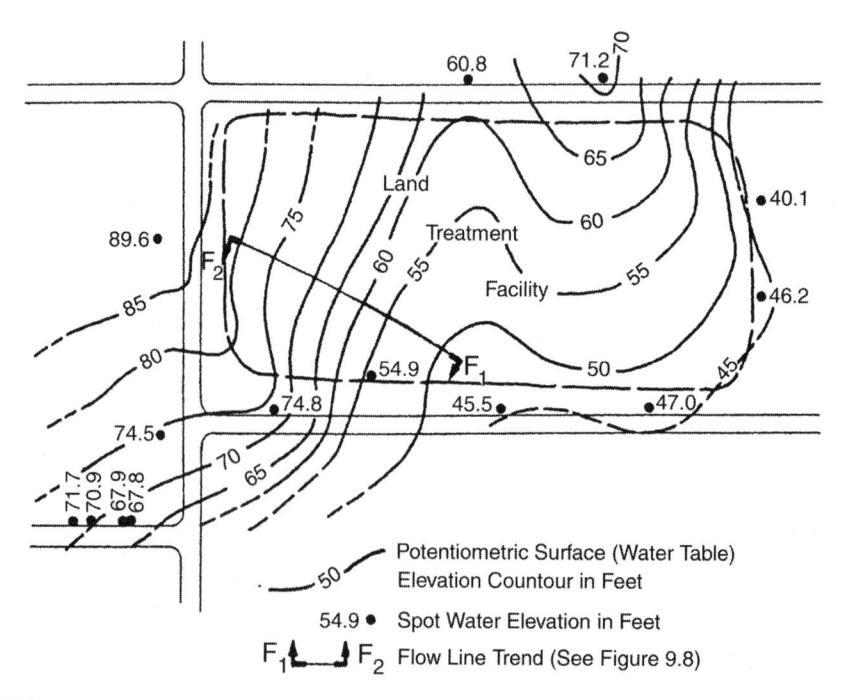

FIGURE 13.7
Potentiometric surface elevation contour map. (Adapted from Rathnayake et al. (1987). With permission.)

Acquisition and Interpretation of Water-Level Data 903

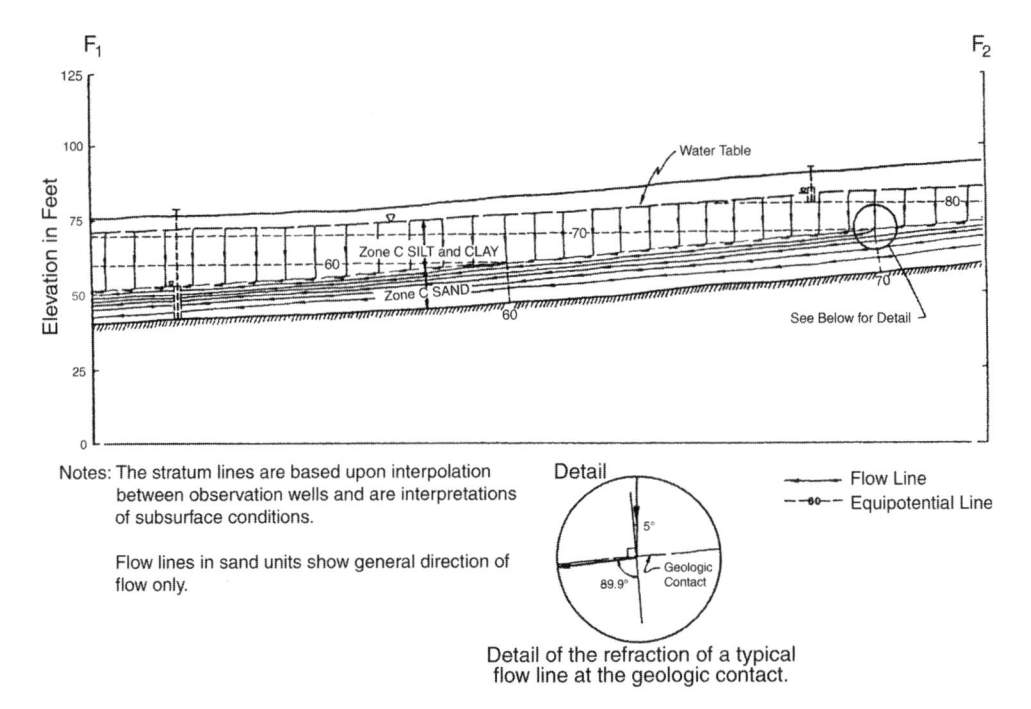

FIGURE 13.8
Cross-section showing vertical flow directions.

horizontal or the vertical flow component and the direction of ground-water flow is dependent on several factors.

Recharge and Discharge Conditions

In recharge areas, ground water flows downward (or away from the water table), while in discharge areas, ground water flows upward (or toward the water table). Ground water migrates nearly horizontally in areas between where recharge or discharge conditions prevail. For example, in Figure 13.9 well cluster A is located in a recharge area, well cluster B is located in an area where flow is predominantly horizontal, and well cluster C is located in a discharge area. Note that in Figure 13.9, wells located adjacent to one another, and at different depths, display different water-level elevations.

Aquifer heterogeneity refers to an aquifer condition in which aquifer properties are dependent on *position* within a geologic formation (Freeze and Cherry, 1979), which is an important consideration when evaluating water-level data. While recharge or discharge may cause vertical gradients to be present within a discrete geologic zone, vertical gradients may be caused by the contrast in hydraulic conductivity between aquifer zones. This is especially evident where a deposit of low hydraulic conductivity material overlies a deposit of relatively higher hydraulic conductivity material, as shown in Figure 13.8.

Aquifer anisotropy refers to an aquifer condition in which aquifer properties vary with direction at a point within a geologic formation (Freeze and Cherry, 1979). For example, many aquifer materials were deposited in more or less horizontal layers, causing the horizontal hydraulic conductivity to be greater than the vertical hydraulic conductivity. This condition tends to create more pronounced vertical gradients (Fetter, 1980) that are not indicative of the actual direction of ground-water flow. In anisotropic zones, flow lines

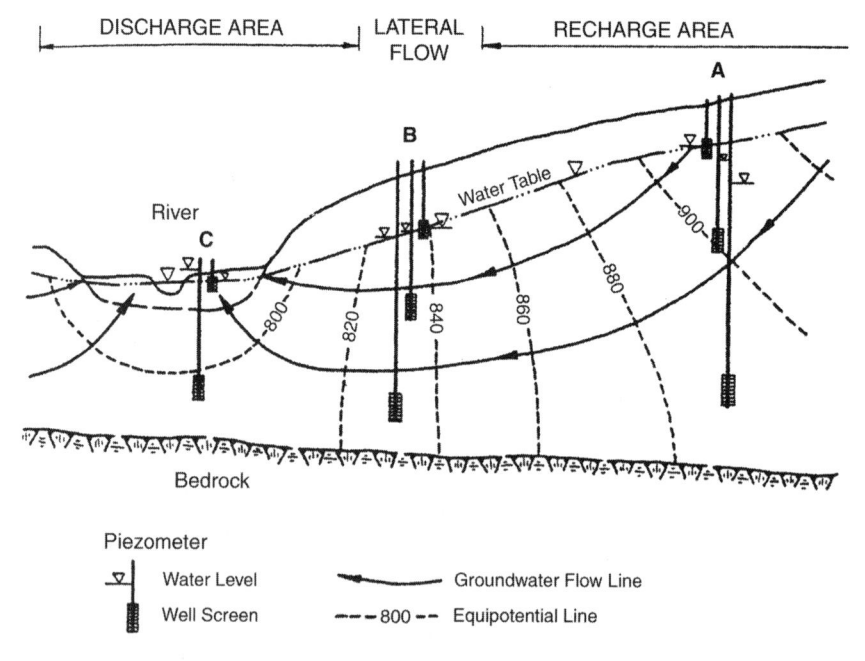

FIGURE 13.9
Ideal flow system showing recharge and discharge relationships. (Adapted from Saines (1981). With permission.)

do not cross potential lines at right angles and flow will be restricted to higher elevations than that in isotropic zones showing the same water-level conditions.

Detailed discussions of each of these factors are beyond the scope of this section. The reader is referred to Fetter (1980) and Freeze and Cherry (1979) for more detailed discussions of the effects of these aquifer conditions on ground-water flow.

The practical significance of the three factors discussed earlier is that ground-water levels can be a function of either well-screen depth or well position along a ground-water flow line or, more commonly, a combination of both. For these reasons, considerable care needs to be taken in evaluating water-level data.

Approach to Interpreting Water-Level Data

The first step in interpreting ground-water-level data is to conduct a thorough assessment of the site geology. The vertical and horizontal extent and relative positions of aquifer zones and the hydrologic properties of each zone should be determined to the extent possible. It is difficult to overemphasize how important it is to have as detailed an understanding of the site geology as possible. Detailed surficial geologic maps and geologic sections should be constructed to provide the framework to interpret ground-water-level data. Man-made features that could influence ground-water levels should also be identified at this stage.

The next step in interpreting ground-water level data is to review monitoring well installation features with respect to screen elevations and the various zones in which the screens are situated. The objective of this review is to identify whether vertical hydraulic gradients are present beneath the site and to determine the probable cause of the gradients.

One method that can be used to assess the distribution of hydraulic head beneath a site is to plot water-level elevations versus screen midpoint elevations. An example of such a plot is shown in Figure 13.10 for wells completed within a layered geologic sequence. Figure 13.10 indicates that a steep downward hydraulic gradient, on the order of 0.85,

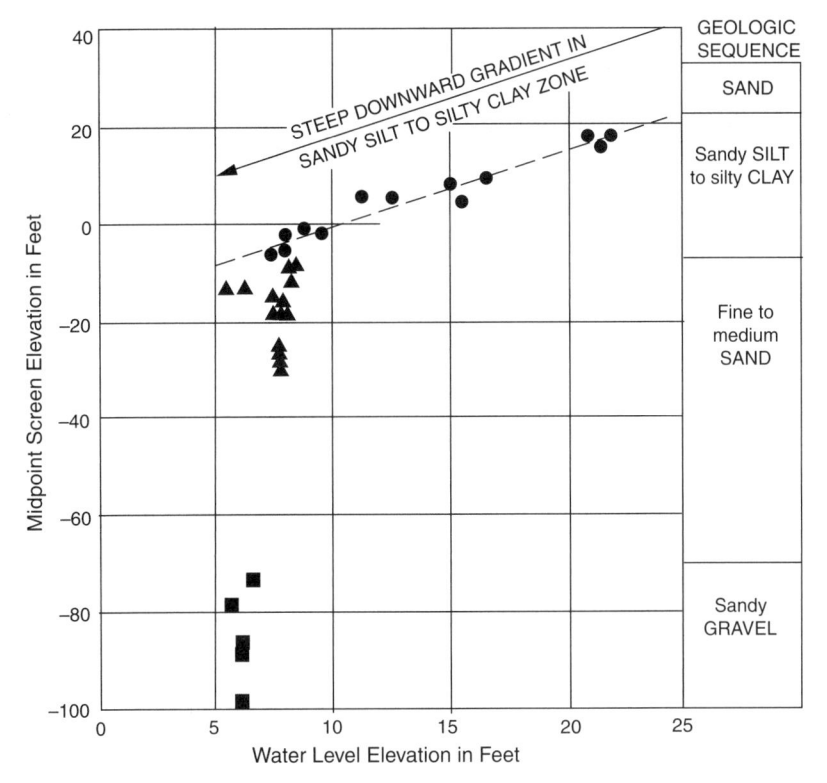

Acquisition and Interpretation of Water-Level Data 905

FIGURE 13.10
Water-level elevation versus midpoint screen elevation for a well screened in a stratified geologic sequence.

exists within the sandy silt to silty clay layer. However, in the lower layers, the vertical component of flow is substantially less both within and between the layers.

Once the presence and magnitude of vertical gradients and the distribution of data with respect to each zone are established, the direction of ground-water flow can be assessed. If the geologic system is relatively simple and if substantial vertical gradients are not present, a planar ground-water elevation contour map can be prepared to show the direction of ground-water flow. However, if multiple zones of differing hydraulic conductivity are present beneath the site, several planar maps may be required to show the horizontal component of flow within each zone (typically the zones of relatively higher hydraulic conductivity). Vertical cross-sections are required to illustrate how ground water flows between each zone.

For the example presented in Figure 13.10, the data indicate that flow is predominantly downward within the upper silt or clay zone. Flow within the lower zone appears to be largely horizontal, although a vertical component of flow is indicated between the sand and the underlying gravel layer.

The examples presented earlier show downward vertical gradients that are indicative of recharge areas. Sites can also be situated within discharge areas where the vertical components of flow are in an upward direction.

The presence of vertical gradients can be anticipated in areas where sites are:

- Underlain by a layered (heterogeneous) geologic sequence, especially where deposits of lower hydraulic conductivity overlie deposits of substantially higher hydraulic conductivity
- Located within recharge or discharge areas

It should be noted that site activities often locally modify site conditions to such an extent that ground water flows in directions contrary to what would be expected for "natural" conditions. Drainage ditches, buried pipelines, and other features can modify flow within near-surface deposits, and facility-induced recharge (e.g., from unlined ponds) can create local downward gradients in regional discharge areas among others.

Figure 13.11 shows the average ground-water elevation contours in a relatively complex hydrogeologic setting. The site lies between two water bodies that are tidally influenced and deep sewer lines are located near the southeast corner of the site. The aquifer of interest lies below a shallow water-table aquifer. A discontinuous aquitard separates the aquifers. The position of the site with respect to the water bodies would suggest that a ground-water divide is present near the site. On the west side of the site, ground water would flow toward the commercial waterway, and on the east side of the divide, ground water would flow toward the river.

Water levels were measured using pressure transducers and data loggers over several days because the site location suggested that tidal fluctuations could affect ground-water levels. Well locations in which transducers were installed are illustrated in Figure 13.11 and some of the transducer data are shown in Figure 13.12. Average water levels and elevations were calculated for each well (see Transient Effects) and were used to construct the ground-water elevation contour map.

Water levels in nested wells screened in the shallow and deeper aquifers indicated the presence of downward vertical gradients (i.e., water-level elevations in the shallower aquifer wells were higher than elevations in wells screened in the deeper aquifer). Analysis of the ground-water contours (for the deeper aquifer) in Figure 13.11 shows that a

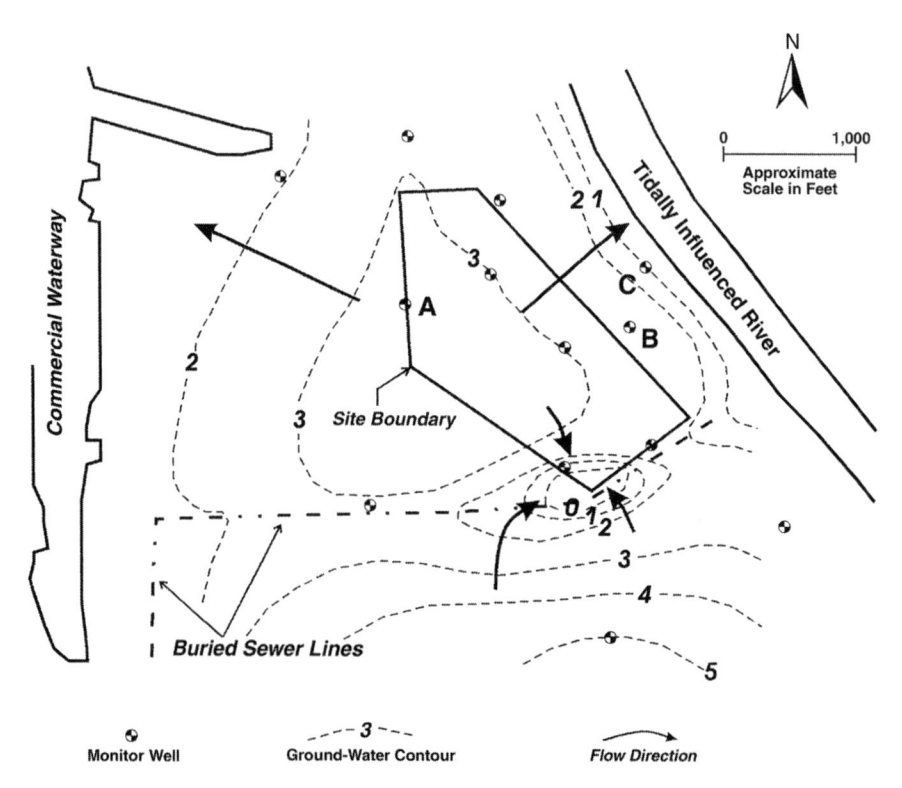

FIGURE 13.11
Average ground-water elevation contours — deeper aquifer.

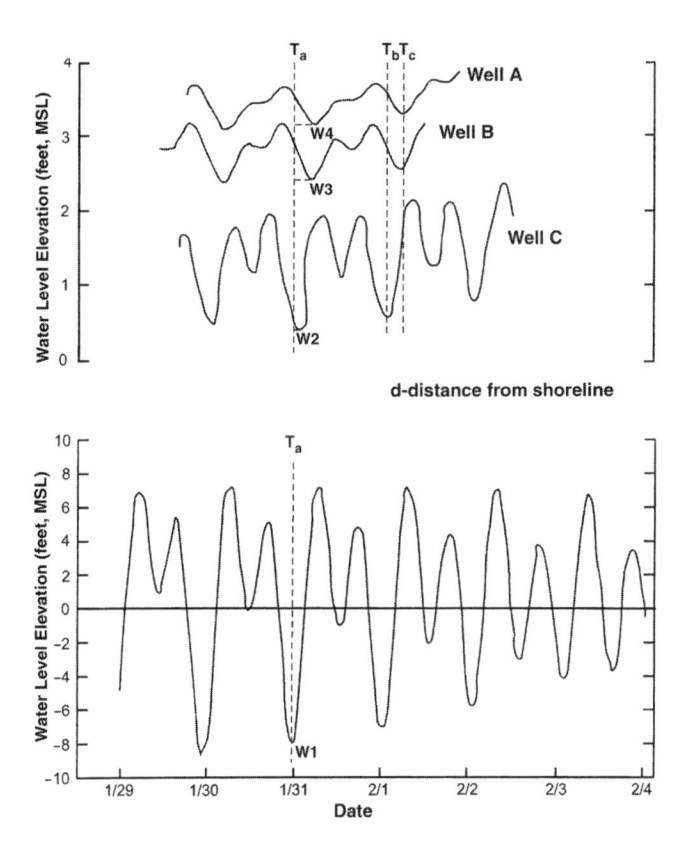

FIGURE 13.12
Influence observed in wells due to tidal fluctuations.

portion of the site (near well A) lies near the center of a ground-water mound generally defined by the 3 ft elevation contour. Evaluation of boring logs indicated that the mound lies in an area where the aquitard appears to be absent. Interpretation of the available data indicates that a partial cause of the ground-water mound was water flowing downward from the shallow aquifer into the deeper aquifer where the aquitard is absent.

As expected, some ground water in the vicinity of the site flows to the east and to the west. However, ground-water contours in the southeastern portion of the site indicated the presence of a low ground-water elevation, where ground water flows in a southerly direction. Two deep buried sewer lines are present near the southeastern site boundary. Review of construction drawings shows that excavation for the sewers penetrated into the deeper aquifer. Interpretation of the water-level elevation data strongly suggests that the sewer lines are acting as drains (i.e., are intercepting ground water). These man-made features appear to have substantially modified the ground-water flow patterns compared to what would be expected under natural conditions.

Transient Effects

Ground-water flow directions and water levels are not static and can change in response to a variety of factors such as seasonal precipitation, irrigation, well pumping, changing river stages, and tidal fluctuations. Fluctuations caused by these factors can modify, or even reverse, horizontal and vertical gradients and thus alter ground-water flow directions. For example, in areas influenced by tides, the net flow of ground water will typically be toward the tidally affected water body. However, during certain portions of the tidal cycle (i.e., during higher tidal levels), there may be a temporary reversal in flow along and some distance inland from the shoreline. Even if significant flow reversals do not occur, hydraulic gradients can change as tidal levels change. Gradients will typically be steeper during lower tides and flatter during periods of higher tides.

Time series water-level data are required to assess how ground-water flow directions change in response to these factors. Figure 13.13 shows data for several wells finished at different depths in an area influenced by changing river stage. The data indicate that river stage affects water levels but that the direction of flow and the horizontal and vertical gradients do not substantially change with river fluctuation. However, the fluctuations do affect the length of time over which each set of ground-water level measurements should be made. In this case, measurements were made in less than 1 h to minimize the effects of the fluctuations on the interpretation of ground-water flow directions.

Figure 13.12 shows hydrographs of water levels in three wells located at varying distances from the shorelines influenced by tides. Well locations are shown in Figure 13.11. Water levels fluctuate in a regular manner but the fluctuations in the wells lag behind the fluctuating tide. In the case illustrated in Figure 13.12, at time T_a, low water levels in the wells occur approximately 2 h (point W2) to 6 h (point W4) after the tidal low (point W1). Several other conclusions can be made using data illustrated in Figure 13.12:

- The mean tidal fluctuation (difference between the mean higher high tide and mean lower low tide) in the area where the data were collected is approximately 11.8 ft. Tidal fluctuations during the measurement period were greater than 15 ft. This means that the water-level measurements are representative of a period of the year when tidal fluctuations are somewhat greater than the mean or average range.

- Water-level elevations in the wells indicate that ground-water flow reversals in the area of interest do not occur. Elevations in well A are always higher than that in well B. Similarly, elevations in well B are always higher than that in well C. This does not mean that flow reversals do not occur nearer to the shoreline, rather it does not occur in the area where the wells are installed.

- Assuming that a sufficient number of wells were instrumented, transducer data can be used to calculate an average water level for each well, and, using the averages, ground-water contour maps can be prepared, which show the average flow direction for the time period in which the data were collected (as shown in Figure 13.11). "Spot" measurements can also be extracted from the hydrographs to construct contour maps representative of tidal highs/lows or ground water highs/lows.

- The time interval in which water-level measurements are taken may affect analysis of flow directions and will affect analysis of hydraulic gradients. For example, if the water level in well A is measured at time T_b, and the water level in well C is measured at time T_c (approximately 5 h later), the water-level elevation will have risen more than 1 ft during the intervening period, which will introduce some error in the analysis.

- Ideally, water levels would be measured in all wells at the same instant (such as at time T_b) to assess flow directions and gradients. As noted earlier, this is a relatively easy matter to resolve if water-level fluctuation data similar to those shown in Figure 13.12 are available for all wells during the same time interval. However, this type of data is seldom available on a routine basis at most

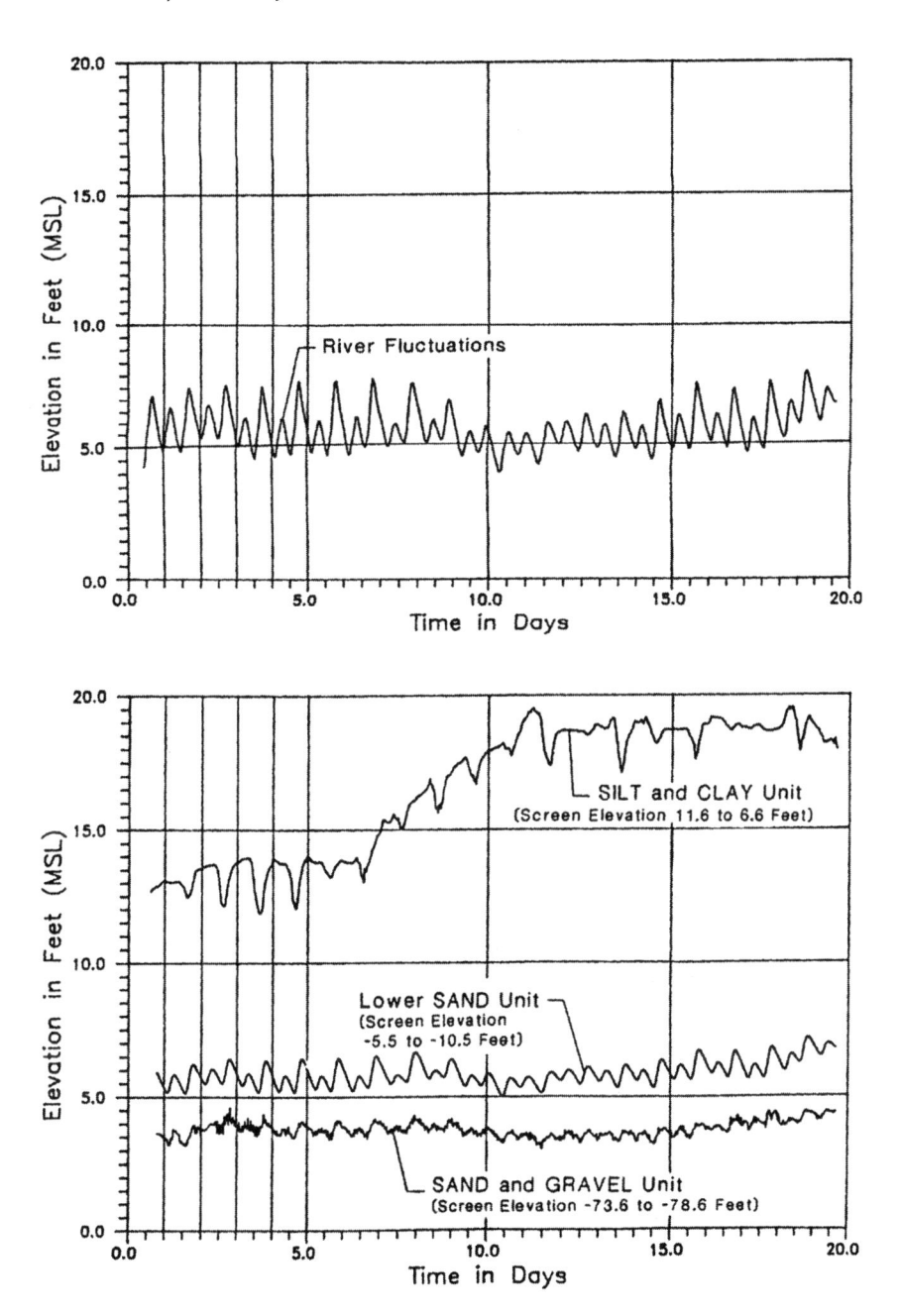

FIGURE 13.13
Influence observed in wells due to river level fluctuations.

sites, especially on a large site with numerous wells. In these cases, if some hydrographic data are available, the data can be used to develop a strategy to minimize the error caused by the regular fluctuations. At the site illustrated in Figure 13.11 and Figure 13.12, these strategies might include:

a. Using several persons to measure water levels in as short a time as possible (i.e., get all of the water-level measurements done before ground-water quality samples are obtained)

b. Selecting a measurement period during a time when the least amount of fluctuation is expected to occur. This might be near tidal high or tidal low periods
c. Initially measuring water levels in wells with the greatest expected fluctuation and moving toward wells where the fluctuations are expected to be less.

Contouring of Water-Level Elevation Data

Typically, ground-water flow directions are assessed after preparing ground-water elevation contour maps. Water-level elevations are plotted on base maps and linear interpolations of data between measuring points are made to construct contours of equal elevation (Figure 13.7). These maps should be prepared using data from measuring points screened in the same zone where the horizontal component of the ground-water flow gradient is greater than the vertical gradient. The greatest amount of interpretation is typically required at the periphery of the data set. A reliable interpretation requires that at least a conceptual analysis of the hydrogeologic system has been conducted. The probable effects of aquifer boundaries, such as valley walls or drainage features, need to be considered.

In areas where substantial vertical gradients are present, the areal ground-water flow maps need to be supplemented with vertical cross-sections that show how ground water flows vertically within and between zones (Figure 13.8). These cross-sections should be oriented parallel to the general direction of ground-water flow and should account for the effects of anisotropy.

Computer contouring and statistical analysis (such as kriging) of water-level elevation data have become more popular (McKown et al., 1987). These tools offer several advantages, especially with large data sets. However, the approach and assumptions that underlie these methods should be thoroughly understood before they are applied and the output from the computer should be critically reviewed. The most desirable approach would be to interpret the water-level data using both manual and computer techniques. If different interpretations result, then the discrepancy between the interpretations should be resolved by further analysis of the geologic and water-level data.

The final evaluation of water-level data should encompass a review of geologic and water-quality data to confirm that a consistent interpretation is being made. For example, at a site where contamination has occurred, wells that are contaminated should be downgradient of the site (based on the water-level data). If this is the case, then a consistent interpretation is indicated. However, if wells that are contaminated are not downgradient of the site, based on water-level data, then further evaluation is required.

References

Alderman, J.W., FM radiotelemetry coupled with sonic transducers for remote monitoring of water levels in deep aquifers, *Ground-water Monitoring Review*, 6(2), 114–116, 1986.

ASTM, Standard Test Method for Determining Subsurface Liquid Levels in a Borehole or Monitoring Well (Observation Well), ASTM Standard D 4750, ASTM International, West Conshohocken, PA, 2004a.

ASTM, Standard Practice for Static Calibration of Electronic Transducer-Based Pressure Measurement Systems for Geotechnical Purposes, ASTM Standard D 5720, ASTM International, West Conshohocken, PA, 2004b.

Barcelona, M.J., J.P. Gibb, J.A. Helfrich, and E.E. Garske, Practical Guide for Groundwater Sampling, EPA-600/2-85-104, U.S. Environmental Protection Agency, Robert S. Kerr Environmental Research Laboratory, Ada, OK, 1985, pp. 78–80.

Driscoll, F.G., *Ground Water and Wells*, Johnson Division, St. Paul, MN, 1089 pp, 1986.
Everett, L.G., *Ground Water Monitoring*, General Electric Company, Schenectady, NY, pp. 196–198, 1980.
Fetter, C.W., *Applied Hydrogeology*, C. E. Merrill Publishing Co., Columbus, OH, 1980.
Freeze, R.A. and J.A. Cherry, *Groundwater*, Prentice Hall, Englewood Cliffs, NJ, 1979.
Garber, M.S. and F.C. Koopman, Methods of measuring water levels in deep wells, *Techniques of Water Resources Investigations*, Book 8, U.S. Geological Survey, Reston, VA, chap. A-1, 1968.
Grant, D.M., *Open Channel Flow Measurement Handbook*, Instrumentation Specialties Company (ISCO, Inc.), Lincoln, NE, pp. 6–7, 1978.
Heath, R.C., Basic Ground-Water Hydrology, U.S. Geological Survey, Water Supply Paper 2220, 1983.
In-Situ, Inc., Owner's Manual: Hydrologic Analysis System, Model SE200, In-Situ, Inc., Laramie, WY, pp. 7–11, 1983.
Leupold and Stevens, Inc., *Stevens Water Resources Data Book*, Leupold & Stevens, Inc., Beaverton, OR, 1978.
Maslansky, S.P., C.A. Kraemer, and J.C. Henningson, An Evaluation of Nested Monitoring Well Systems, Ground-water Monitoring Seminar Series Technical Papers, U.S. Environmental Protection Agency, EPA-CERI-87-7, 1987.
McKown, G.L., G.W. Dawson, and C.J. English, Critical Elements in Site Characterization, Ground-Water Monitoring Seminar Series Technical Papers, U.S. Environmental Protection Agency, EPA-CERI-87-7, 1987.
Paroscientific, Inc., Digital Quartz Pressure Transmitters for Accurate Water Level Measurements, Paroscientific, Inc., pp. 2–7 (Available at http://www.paroscientific.com/waterlevel.htm), 2002.
Plazak, D., Differences between water-level probes, *Ground-water Monitoring and Remediation*, 14(1), 84, 1994.
Rantz, S.E., Measurement and Computation of Streamflow: Volume 1. Measurement of Stage and Discharge, U.S. Geological Survey Water Supply Paper 2175, U.S. Govt. Printing Office, Washington, DC, pp. 63–64, 1982.
Rathnayake, D., C.D. Stanley and D.H. Fujita, Ground water flow and contaminant transport analysis in glacially deposited fine grained soils: a case study, *Proceedings of the FOCUS Conference on Northwestern Ground Water Issues*, National Water Well Association, Dublin, OH, pp. 125–151, 1987.
Ross, J.H., Evaluation of Non-Contact Measurement Instrumentation for Ground Water Wells, Presentation made at the AGWSE Annual Meeting and Conference, Nashville,TN, December, 2001.
Saines, M., Errors in interpretation of ground-water level data, *Ground-water Monitoring Review*, 1(1), 56–61, 1981.
Schrale, G. and J.F. Brandwyk, An acoustic probe for precise determination of deep water levels in boreholes, *Groundwater*, 17(1), 110–111, 1979.
Sheingold, D.H., *Transducer Interfacing Handbook*, Analog Devices, Inc., Norwood, MA, 1980.
Shuter, E. and A.I. Johnson, Evaluation of Equipment for Measurement of Water Levels in Wells of Small Diameter, U.S. Geological Survey Circular 453, 1961.
Solinst, Ltd., Levelogger Series Model 3001 Data Sheet, Solinst Canada Ltd., Georgetown, ON, Canada, 4 p, 2001.
Stewart, D.M., The rock and bong techniques of measuring water levels in wells, *Groundwater*, 8(6), 14–18, 1970.
Thornhill, J.T., Accuracy of depth to water measurements, U.S. EPA Superfund Groundwater Issue, EPA/540/4-89-002, Robert S. Kerr Environmental Research Laboratory, Ada, OK, 3 p, 1989.
Todd, D.K., *Groundwater Hydrology*, 2nd ed., John Wiley and Sons, New York, NY, 1980.
U.S. Bureau of Reclamation, Engineering Geology Field Manual, U.S. Department of Interior, Bureau of Reclamation, U.S. Govt. Printing Office, Denver, CO, Chap. 9, pp. 227–247, 2001.
U.S. Geological Survey, National Handbook of Recommended Methods for Water-Data Acquisition: Chap. 2—Groundwater, U.S. Department of Interior, Geological Survey, Reston, VA, 1980.
Zarriello, P.J., Accuracy, precision, and stability of a vibrating-wire transducer measurement system to measure hydraulic head, *Ground-water Monitoring and Remediation*, 15(2), 157–168, 1995.

14

Methods and Procedures for Defining Aquifer Parameters

John Sevee

CONTENTS

Introduction	914
Bulk Density	915
Mechanical Density Testing	916
Determining Density by Gamma-Ray Attenuation	919
Water Content	919
Laboratory Measurement of Water Content	920
Field Measurement of Water Content	920
Porosity	921
Hydraulic Conductivity and Permeability	923
Laboratory Determination of K	926
Estimation of K from Soil Grain Size	929
Slug Tests	929
Packer Testing	933
Pressure Tests	936
Tracer Tests	937
Vertical Hydraulic Conductivity Using Packer Tests	937
Vertical Hydraulic Conductivity from Seepage Pits	938
Vertical Hydraulic Conductivity from Pumping Tests	939
Direct Measurement of Ground-Water Flow Velocity	940
Specific Storage and Specific Yield	942
Specific Yield	942
Specific Storage	943
Transmissivity	944
Determination of Transmissivity by Pumping Tests	944
Observation Well Positioning	945
Pretest Data Collection	946
Water-level Measurement During Pumping Tests	947
Measuring Well Discharge	947
Analysis of Pumping-Test Data	948
Estimation of Transmissivity	949
Compressibility	950
References	953

Introduction

The storage and movement of ground water through soil and rock obey certain physical laws. These laws are represented mathematically and are used to quantitatively describe the behavior of ground water within a particular hydrogeologic setting. Certain physical parameters, such as hydraulic conductivity, storativity, and aquifer thickness, must be determined in order to solve the mathematical relationships describing ground-water behavior. Determination or measurement of these parameters is a primary purpose of many field investigations. Once defined, the parameters can be utilized with the appropriate mathematical relationships or equations to calculate ground-water flow rate and direction, aquifer yield, or the behavior of chemicals transported in ground water.

Relative to ground-water monitoring, a quantitative description of an aquifer and its properties can be used to accomplish several important objectives. These include: (1) optimizing well placement for detecting a chemical plume; (2) determining the appropriate depth to install a monitoring well to encounter a chemical plume in a layered aquifer system; or (3) deciding the optimum number of wells required to detect a leak from a buried tank for a given probability of detection. For example, to properly locate monitoring wells around a lined landfill for the purpose of detecting liner leakage requires an understanding of the direction of ground-water flow. However, this is not only dependent on the configuration of the water table beneath the landfill, but also, in anisotropic conditions, on the three-dimensional variation in hydraulic conductivity of the geologic materials. Certain aquifer properties can be used to calculate monitoring well recharge rates and zones of pumping influence created during well purging or ground-water sampling. Computer modeling efforts have demonstrated that the zone of pumping influence around a monitoring well varies with the ratio of hydraulic conductivities of the artificial filter pack material to that of the natural geologic material (Cohen and Rabold, 1987). The relative contribution of water to the well from each of these materials and therefore the degree of "representativeness" of the water sample could not be estimated without a quantitative understanding of the aquifer's physical properties. A quantitative description of ground-water and aquifer behavior, therefore, becomes important in establishing a reliable ground-water monitoring program.

This chapter defines the terminology associated with hydrogeologic parameters and presents typical laboratory and field methods for measuring or estimating these parameters. Bulk density, water content, porosity, hydraulic conductivity and permeability, ground-water velocity, specific storage and specific yield, transmissivity, and aquifer compressibility are defined herein. Methods for defining the aquifer parameter known as dispersivity, which is related to the spreading of a contaminant plume moving with ground-water, are not presented. A variety of methods have been presented elsewhere in the literature (Fried and Ungemach, 1971; Sudicky and Cherry, 1979; Bentley and Walter, 1983; Gelhar et al., 1985; Moltz et al., 1986). Dispersivity is both a time- and scale-dependent quantity that describes a process that is presently poorly understood. Determination of dispersivity at typical project scales is extremely costly and requires long periods of time to measure. Because of these constraints, the reader is directed to the literature for descriptions of dispersivity testing methods.

Ideally, all aquifer parameters should be measured in the field (i.e., *in situ*) under the anticipated ground-water conditions. However, some of the parameters can be measured reasonably well in the laboratory on representative samples of unconsolidated geologic material and then applied to the field situation. Representative fractured, jointed, or solution-channeled rock samples are currently impractical to obtain for routine laboratory analysis and the results of testing are difficult to apply to *in situ* conditions. Some of the parameters can be estimated from physical characteristics of the aquifer material. For example, the hydraulic conductivity of sandy materials is often estimated from the results of grain-size distribution analyses of the material. Commonly used laboratory and field methods, which can be used to measure or estimate these parameters, are described herein.

From a practical standpoint, all points within a particular hydrogeologic setting cannot be sampled and tested. Therefore, collected measurements, however many there may be, are typically extrapolated across a study area as a matter of practicality. The more the area that is sampled and tested, the greater the level of confidence in the predictions from any quantitative analyses. Because unlimited testing is not practical, selective testing is used to define average or representative properties for a particular parameter. Sometimes the values used in an analysis are a combination of different types of testing, although this approach must be used with caution. Different test methods may use different size samples or may average values over a greater or lesser volume. Sample volume variation can result in different parameter estimates (Bear, 1972; Parker and Albrecht, 1987). Understanding the geology can help in selecting the sample size or the *in situ* testing methods to be used. Cross-checks between test methods (e.g., comparing slug-test results with aquifer grain-size analyses) can be of value in verifying or evaluating test results or extrapolating parameter estimates over a broader area using a less costly test method (Delhomme, 1974). Geostatistics may be a valuable method for interpolating or extrapolating field or laboratory results and concurrently determining the estimation error (Delhomme, 1978).

The limitations of a particular test method should be incorporated in the selection process of any test method. A parameter test method may be appropriate for one application and not another.

Bulk Density

Hydraulic conductivity of geologic materials is a function of several factors, bulk density being one (Lambe and Whitman, 1969). Hydraulic conductivity typically decreases with increasing bulk density. Bulk density also affects aquifer compressibility and, therefore, aquifer storativity (Lambe and Whitman, 1969). Bulk density, consequently, plays an important role in understanding aquifer properties and behavior.

Total or wet bulk density of a soil or rock is defined as the total weight or mass of soil or rock, including any water, in a unit volume of material. The dry bulk density of a soil or rock is the weight of the dry solids per unit volume of material, and is related to the total or wet bulk density by

$$\gamma_d = \frac{\gamma_t}{1 + w_w} \tag{14.1}$$

where w_w is the gravimetric water content of the soil or rock calculated on a dry-weight basis, expressed as a decimal, γ_d the dry bulk density, weight or mass per unit volume, and γ_t the wet bulk density, weight or mass per unit volume.

Typical units of density are grams per cubic centimeter (g/cm^3) or pounds per cubic foot (lb/ft^3). Methods for determining water content are presented below.

The dry bulk density is related to the specific gravity of the rock or soil solids by

$$\gamma_d = \frac{G_s \gamma_w}{1 + w_w G_s / S} \tag{14.2}$$

where G_s is the specific gravity of the solids (dimensionless), S the degree of saturation expressed as a decimal, γ_w the density of water, weight or mass per unit volume, and w_w the gravimetric water content of the soil or rock calculated on a dry-weight basis, expressed as a decimal.

The degree of saturation represents the fraction of the pore space filled with water. The specific gravities of some typical soil and rock constituents are presented in Table 14.1. A comparison of Equations 14.1 and 14.2 indicates that the dry and wet bulk densities must be less than the particle density ($G_s \times \gamma_w$) of the soil or rock solids. This is due to the incorporation of void space (i.e., pores or fractures) in the bulk material. Density of soil particles or rock matrix (particle density) typically ranges from 160 to 180 pounds per cubic foot, whereas dry bulk densities of soils typically range from 90 to 130 pounds per cubic foot. Unweathered nonporous rock dry bulk densities approach the specific gravity of the constituent minerals in the rock. Weathered or porous rock dry bulk densities may approach that of soils. Typical bulk densities of various geologic materials are given in Table 14.2.

Mechanical Density Testing

Total bulk density of a soil can be measured by obtaining undisturbed cores or block samples. After a sample is obtained, the total bulk density is determined by measuring the volume of the sample and its total weight. The total bulk density is calculated by dividing the sample weight by its volume. Dry bulk density can be obtained by drying the sample after volume measurement but prior to weight measurement. Dry density can

TABLE 14.1

Typical Specific Gravities of Soil and Rock Constituents

Gypsum	2.32
Montmorillonite	2.78
Orthoclase	2.56
Kaolinite	2.6
Illite	2.6–2.86
Chlorite	2.6–3.0
Quartz	2.66
Talc	2.7
Calcite	2.72
Muscovite	2.8–2.9
Dolomite	2.87
Aragonite	2.94
Biotite	3.0–3.1
Augite	3.2–3.4
Hornblende	3.2–3.5
Limonite	3.8
Hematite, hydrous	±4.3
Magnetite	5.2
Hematite	5.17

TABLE 14.2

Natural Bulk Densities of Typical Soils and Rocks

Description	Bulk Density (lb/ft^3) Dry	Wet
Uniform sand, loose	90	118
Uniform sand, dense	109	130
Nonuniform sand, loose	99	124
Nonuniform sand, dense	116	135
Glacial till	132	145
Soft glacial clay	80	110
Stiff glacial clay	100	129
Soft, slightly organic clay	65	98
Soft, very organic clay	45	89
Rock		
Granite	160	170
Dolerite	185	190
Gabbro	185	193
Basalt	175	180
Sandstone	125	162
Shale	125	150
Limestone	135	162
Dolomite	155	162
Quartzite	—	165
Gneiss	180	185
Marble	160	170
Slate	160	170

be calculated by measuring the average water content (see Water Content section) of the sample and using Equation 14.1 once the total bulk density is known.

Core samples of geologic materials are commonly obtained by pushing a thin-walled tube into the soil while installing a borehole (Figure 14.1). ASTM recommends a method (ASTM D 1587; ASTM, 2004a) for obtaining thin-walled tube samples from a borehole. This sampling technique is used principally for soft clays and loose silts. Although the method can be used to sample loose sand under some conditions, significant sample disturbance often occurs and is difficult to prevent. Some disturbance is inevitable with any sampling method and can affect the density of a tube sample (Hvorslev, 1949a). These effects may increase or decrease the measured density depending on the natural density of the soil. Stress relief caused by removal of a soil sample from some depth below the ground surface may also alter the density of the sample. Various versions of the tube sampler are used for soils of differing consistency. A synopsis of sampler types is presented in Winterkorn and Fang (1975).

A block sample is obtained by cutting an undisturbed block of soil from the base or wall of an open excavation. This type of sample is obtained as illustrated in Figure 14.1. The block of soil is typically surrounded by a section of tubing or a square box without covers, and the space between the sample and the container is filled with tamped fine sand or paraffin. A 10–12-in. square box with easily dismantled sides and covers is often used, especially in block sampling of sands. Isolation of the soil block will relieve *in situ* stresses and may cause some expansion of the soil, but block sampling is still the best available method for obtaining large undisturbed samples of very stiff and brittle soils, partially cemented soils, and soils containing coarse gravel and stones. The method can be used in all soils except when cohesion is so poor that a soil block cannot be isolated.

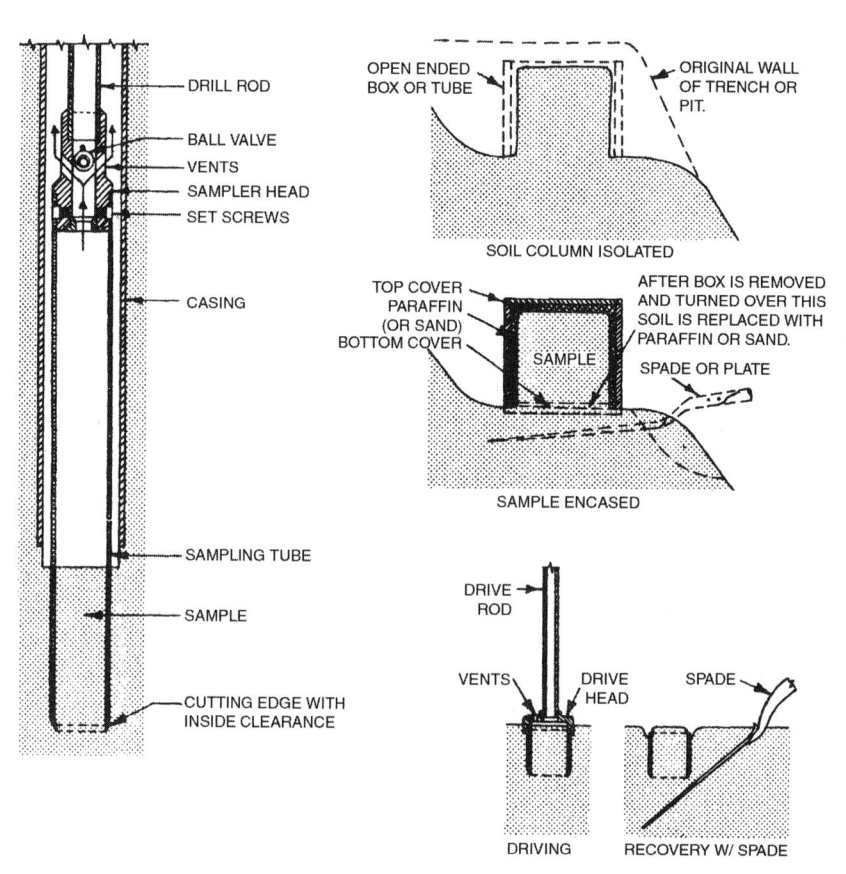

FIGURE 14.1
In Situ soil density sampling methods.

In order to preserve the moisture content of any soil sample, it should be either wrapped in foil or cellophane or covered with paraffin immediately upon removal from the subsurface. The wrapped sample should be kept in a cabinet or room that is maintained at a high humidity to prevent desiccation of the sample, until it is ready for analysis.

ASTM describes other methodologies for obtaining *in situ* bulk-density measurements by mechanical means. These methods involve removal of a volume of soil and measuring the weight and volume of the removed soil. For the drive-cylinder method (ASTM D 2937 [ASTM, 2004b]) a metal cylinder is driven into the ground, as shown in Figure 14.1, and removed with the sample inside. The retrieved sample size and weight are measured and the density is calculated. Drive sampling is primarily suited for soft to stiff cohesive soils, silt, and loose to medium-density fine sand. Compaction of the sample can occur when great force is required to push or drive the sampler into the soil.

The balloon and sand-cone methods involve excavating a hole and placing the excavated soil on a scale for weighing. The volume of the sample is then determined by balancing the scale using water (balloon method) or a clean sand of a known density (sand-cone method). These methods (ASTM D 2167 [ASTM, 2004c] and D 1556 [ASTM, 2004d]) have been used extensively for controlling fill compaction.

ASTM also offers a method for coring rock (ASTM D 2113 [ASTM, 2004e]). However, if the rock is heavily fractured, the measured bulk density of the core may be severely affected by disturbance due to the sampling method. Because coring procedures generally

Methods and Procedures for Defining Aquifer Parameters

require the use of fluid to cool and lubricate the core barrel, erosion of fracture fillings and remolding of the sample can occur if the rock is highly weathered or poorly indurated.

Determining Density by Gamma-Ray Attenuation

Gamma-ray attenuation can be used to determine soil bulk density by placing two boreholes into the ground — one containing a detector and the other containing a gamma-ray source. The distance between the source and the detector tubes is typically of the order of 1–2 ft. The degree of gamma-ray absorption by the soil or rock is a function of the density of material between the source and detector. As the density increases, the degree of gamma-ray absorption also increases. If the boreholes are cased with a metal casing, for example, a correction is required for casing adsorption of the gamma radiation.

Some devices have been developed in which both source and detector are contained in the same common probe and the entire unit is lowered down a single borehole. When using this method, corrections must be made for the degree of gamma-ray absorption by any casing within the borehole. Each individual site requires calibration of the equipment. The gamma-ray method is also affected by the water content of the soil. Therefore, the water content of the soil must be known in order to utilize this method to calculate bulk density.

ASTM D 2922 (ASTM, 2004f) describes the gamma-ray attenuation method used for *in situ* bulk-density determinations. This method involves placing a movable gamma-ray source at depths of up to 1 ft beneath the ground surface. The gamma-ray detector is located within the base of the device, which remains on the ground surface above the source. Therefore, whereas the above-described method measures horizontally between two boreholes, this method measures vertically from the ground surface. The surface may be lowered by excavation and the method repeated.

Water Content

Water content can be used to estimate the soil porosity and density if the degree of saturation of the soil is known. In partially saturated soil, water content is related to the relative permeability and matric suction (Freeze and Cherry, 1979). Because water content reflects the porosity of the soil, it provides a measure of the water held in storage by soil or rock. The change in water content during gravity drainage of an initially saturated sample of soil is a measure of the specific yield.

There are two commonly used definitions for water content: volumetric and gravimetric. The volumetric water content is expressed as the volume of water relative to the total sample volume:

$$w_v = \frac{V_w}{V_t} \tag{14.3}$$

where w_v is the volumetric water content expressed as a decimal and V_w and V_t the volume of water in the sample and the total volume of the sample, respectively.

Water content is typically expressed as a decimal or as a percent (i.e., decimal value multiplied by 100). With this definition, if the material is saturated, the volumetric water content is approximately equal to the total porosity of the soil or rock. Volumetric water content is typically used when examining the behavior of partially saturated soils (Hillel, 1980). Volumetric water content is most easily determined by calculation from the gravimetric water content.

The gravimetric water content (on a dry weight basis), w_w, is determined by dividing the weight of water in a given sample by the dry weight of solids in the sample (i.e., $w_w = W_w/W_s$). The two different water content definitions are related by

$$\frac{w_v}{w_w} = \frac{\gamma_{dry}}{\gamma_w} \tag{14.4}$$

where γ_w is the density of water in units of weight per unit volume and γ_d the bulk dry density of the soil sample in units of weight per unit volume.

Laboratory Measurement of Water Content

The water content of a soil is commonly determined in the laboratory by drying at a temperature of 105–115°C (ASTM D 2216 [ASTM, 2004g]). The method simply involves weighing the moist soil sample, drying the specimen, and then reweighing the sample to determine the weight loss due to drying. For clay and organic soils, drying at temperatures above 115°C may result in the loss of chemically or physically bound water or, in the case of organic soils, weight loss by burning of the organic materials. Certain soils and compounds have water of hydration that can be released at relatively low temperatures, thus exhibiting a water content that is temperature-dependent. For example, gypsum has several different hydrated states, and varying the drying temperature can result in different water contents. The use of a standard temperature range, therefore, provides consistency between measurements.

Field Measurement of Water Content

The water content of soil can also be obtained by various field methods. The most common method utilizes a probe containing a radiation source of fast neutrons (americium or radium) and a detector. The radiation source releases fast neutrons that are decelerated when hydrogen atoms are encountered in the soil or rock. The decelerated neutrons are reflected by the hydrogen to a detector that counts the slowed neutrons. Because water is the primary source of hydrogen, the intensity of the slowed neutron radiation reaching the counter probe is proportional to the water content. ASTM describes the neutron deceleration method for measuring the water content of soil (ASTM D 3017 [ASTM, 2004h]).

Measurable sources of hydrogen can commonly occur in materials such as clay or organic matter, which contain hydrogen within their structures. Furthermore, water may be trapped between clay mineral plates or within unconnected pores in the rock or soil. These factors contribute to errors in measurement of the "mobile" water within the media using the radiation method. Mobile water is that water which can move or drain from the soil or rock under a gravitational pressure gradient.

The neutron deceleration method commonly gives the average water content over a 6-in. diameter sphere around the point of measurement. However, this volume may vary due to the soil density or source strength and the actual volume of measurement is uncertain in most cases. In soils that naturally contain hydrogen, such as organic or clay soils, a lengthy calibration procedure may be required to provide reasonably useful results. Despite the calibration required, two advantages to this method are: (1) a large number of measurements can be made in a short time and (2) the probe can be used in either uncased or cased boreholes. Typically, boreholes are cased with aluminum or steel to provide intimate contact with the surrounding soil when water content determinations with the neutron method are used. This avoids cavities that may fill with water or air between the tube and the soil.

Another common field method consists of using gypsum blocks or tensiometers to infer the water content from calibration curves. These methods are used to monitor moisture changes in the partially saturated zone of soil. In the case of the gypsum block, electrodes running to a power supply are connected to the block that is carefully buried in the soil. Backfilling the block with natural soil is critical to maintain the proper moisture–tension characteristics between the block and the undisturbed soil. The electrical resistance of the block is measured and varies with the water content of the surrounding soil. A careful calibration procedure is required to calibrate block resistance to soil water content. A more detailed description of this and other methods is presented in Gardner (1965).

Porosity

Most rocks and soils are composed of solid mineral particles separated by void spaces. In most soils, the void spaces between particles form a series of interconnected pores. Pores in a rock matrix may not be visible to the naked eye. Primary porosity is that porosity due to voids between the soil or rock grains; root holes, cavities, worm holes, and fractures may cause secondary porosity. Fractures may form as a result of faulting, jointing, foliation, or fissuring.

The volume of the total pore space in a material relative to the overall volume of the rock or soil is termed total porosity, n:

$$n = \frac{V_v}{V_t} \tag{14.5}$$

where V_v is the volume of voids in a sample and V_t the total sample volume.

Total porosity is expressed either as a dimensionless decimal (which must be less than 1) or as a percent (i.e., decimal value times 100). Typical values for total porosity of various geologic materials are presented in Table 14.3.

Ground water moves and is stored within the pores and fractures of soil or rock. Porosity is therefore an important parameter in describing ground water behavior and is quantitatively related to various other ground-water parameters such as hydraulic conductivity, flow velocity, transmissivity, and storativity. Porosity will affect the zone of influence during sampling of a monitoring well by its influence on storativity. Because of its direct influence on flow velocity, the appropriate distance between downgradient monitoring well locations and potential ground-water contaminant sources is related to porosity.

A soil having a broad range of grain sizes generally has a lower porosity than a soil with uniform grain sizes (de Marsily, 1986). This is a result of the finer particles filling in the void spaces between the coarser particles for a soil with a broad range in grain sizes, thus lowering the overall porosity. Clay soils typically have a higher total porosity than sands, silts, and gravels. The porosity of rock is typically much less than that of soils except where the rock is highly weathered or partially dissolved. Karst formations may have very high secondary porosity due to solutioning of the carbonate.

TABLE 14.3

Typical Total Porosities

Material	Total Porosity (%)
Unaltered granite and gneiss	0–2
Quartzites	0–1
Shales, slates, mica-schists	0–10
Chalk	5–40
Sandstones	5–40
Volcanic tuff	30–40
Gravels	25–40
Sands	15–48
Silt	35–50
Clays	40–70
Fractured basalt	5–50
Karst limestone	5–50
Limestone, dolomite	0–20

Water content is related to the porosity by

$$n = \frac{1}{S/G_s w_w + 1} \tag{14.6}$$

where n is the porosity, expressed as a decimal fraction; w_w the the gravimetric water content, expressed as a decimal fraction; S the degree of saturation, if known, typically assumed to be 1 for saturated soils, expressed as a decimal fraction; and G_s the specific gravity of the soil solids or rock.

Dry bulk density is related to porosity by

$$n = 1 - \frac{\gamma_d}{\gamma_w G_s} \tag{14.7}$$

The density of the sample can be determined as discussed earlier. Goodman (1980) describes several methods for laboratory determination of porosity.

A word of caution is warranted relative to porosity. Porosity is used to estimate the saturated flow velocity ($\bar{v}$) of water within pore spaces by the relationship

$$\bar{v} = \frac{Ki}{\bar{n}} \tag{14.8}$$

where K is the hydraulic conductivity, i the hydraulic gradient in the direction of the mean seepage velocity $\bar{v}$, and $\bar{n}$ the effective porosity.

The porosity used in Equation 14.8 is the effective porosity, $\bar{n}$, which always has a lower value than the total porosity. Some water in pore spaces may be held onto soil particles by molecular binding forces (Mitchell, 1976). The soil may contain dead-end pores or unconnected pores which contain water, but through which no water flow is occurring. Therefore, caution should be exercised in selecting effective porosities. Effective porosity is difficult to measure and is typically determined by intuition, experience, or consulting one of many textbooks that have published typical $\bar{n}$ values (e.g., Freeze and Cherry, 1979). Tracer experiments can be used to estimate effective porosities but the procedure is fraught with difficulties. When porosity is very low, laboratory errors may become significant. Field variation in porosity, however, may exceed laboratory errors and suggests that the use of multiple *in situ* test sites may be warranted.

Hydraulic Conductivity and Permeability

The rate of water movement through a soil was first described mathematically by Darcy (1856). By studying the flow of water through sand columns, Darcy developed a relationship between the filtration velocity, the hydraulic gradient, and a coefficient, K, which has come to be known as the hydraulic conductivity. K is a function of both the medium through which the fluid is moving and of the fluid itself. In many engineering texts, K is also known as the coefficient of permeability. As a result, the two terms are used interchangeably in hydrogeologic applications. Hydraulic conductivity is expressed in units of length per unit time such as meters per second (m/s) or feet per day (ft/day).

Another term, intrinsic permeability, k, is used to describe the part of K that depends only on the medium in which a fluid is flowing. Intrinsic permeability has the units of length squared, such as cm^2 or mm^2, or the darcy ($0.987 \times 10^{-12}\,m^2 = 1$ darcy).

For granular porous media, Darcy's law can be written as

$$v_s = Ki \tag{14.9}$$

where v_s is the specific discharge in units of length per time, K the hydraulic conductivity in units of length per time, and i the dimensionless hydraulic gradient in the direction of v_s.

This definition has been modified to describe flow in a fracture (Louis, 1974). Table 14.4 presents typical hydraulic conductivity values for various geologic materials. Generally, the finer the soil particle size, the lower the hydraulic conductivity value. The difference in K ranges between silts or clays and sands is a result of the smaller effective pore sizes in clays and silts than in sands. Soils that contain a broad range of grain sizes, such as a glacial till, typically have lower K values than a uniformly sized soil such as a beach sand. Darcy's law is the cornerstone for evaluating ground-water flow. However, the relationship is valid only as long as the velocity remains within a particular range of values. As the hydraulic gradient is increased, the water velocity increases and friction

TABLE 14.4

Typical Hydraulic Conductivities

Geologic Material	Range of K (m/sec)
Coarse gravels	10^{-1}–10^{-2}
Sands and gravels	10^{-2}–10^{-5}
Fine sands, silts, loess	10^{-5}–10^{-9}
Clay, shale, glacial till	10^{-5}–10^{-13}
Dolomitic limestones	10^{-3}–10^{-5}
Weathered chalk	10^{-3}–10^{-5}
Unweathered chalk	10^{-6}–10^{-9}
Limestone	10^{-3}–10^{-9}
Sandstone	10^{-4}–10^{-10}
Unweathered granite, gneiss, compact basalt	10^{-7}–10^{-13}

loss within the pores or fractures correspondingly increases. This phenomenon is analogous to flow through a pipe. Above a critical velocity, frictional losses are no longer linearly related to i, and Darcy's law must be modified or it becomes invalid. Some authors have suggested that an upper limit to the applicability of Darcy's equation for porous media be established by relating the water velocity to a Reynolds's Number (Bear, 1972). Reynolds's Number for a porous medium can be defined as

$$R_e = \frac{\bar{v} d \rho}{\mu} \tag{14.10}$$

where $\bar{v}$ is the mean velocity of water in pores in units of length per unit time, ρ the fluid unit density in units of mass per unit volume, μ the viscosity of the fluid in units of mass per time-length, and d the mean diameter of the pores as estimated from the effective grain-size diameter in units of length.

In porous media, d is typically selected as the particle size for which 10% of the sample is smaller. In fractured media, d becomes the fracture width. Darcy's law is considered valid up to an R_e of between 1 and 10. Between an R_e of 10 and 100, turbulent flow begins, and beyond 100, turbulence predominates and Darcy's law is invalid.

There also appears to be evidence, although some is conflicting, that for clay or other fine-grained materials, Darcy's law may be invalid for very low gradients (Jacquin, 1965a, 1965b). Desaulniers et al. (1986) performed a field investigation in a thick clayey glacial till that supports the concept of threshold gradients. A relationship for Darcy's law incorporating a threshold gradient is suggested as shown in Figure 14.2. Below a particular threshold gradient, i_0, the hydraulic conductivity may be essentially zero for certain materials. For this case, Darcy's law is revised to

$$v_s = K(i = i_2) \quad \text{for } i_2 > i_1 \tag{14.11a}$$
$$v_s = 0 \quad \text{for } i < i_0 \tag{14.11b}$$

v_s in transition for $i_0 < i < i_1$.

The value of K varies with the type of fluid flowing within the soil or rock and depends on the viscosity and density of the fluid, such that

$$K = \frac{k g \rho}{\mu} = \frac{k g}{v} \tag{14.12}$$

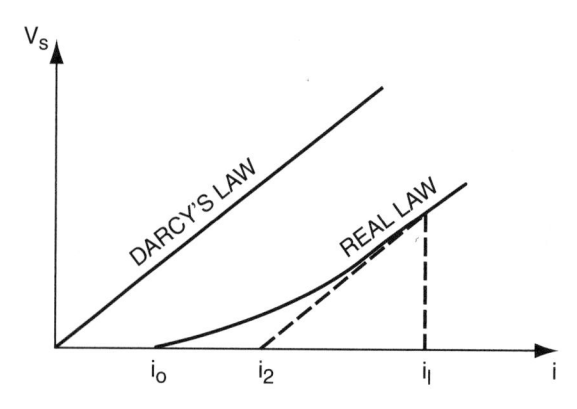

FIGURE 14.2
Darcy's law at low hydraulic gradients.

where ρ is the fluid density in units of mass per volume, μ the viscosity of the fluid in units of mass per time-length, g the acceleration of gravity, e.g., 32 ft/sec^2, v the kinematic viscosity in units of area per time, and k the intrinsic permeability of the media through which the fluid is flowing in units of square length.

Both density and viscosity are a function of temperature. The effects of temperature on the density and viscosity of water is shown in Table 14.5. Of the two parameters, viscosity is the more sensitive to temperature changes.

If the hydraulic conductivity of a medium is known at one temperature, its value can be calculated at different temperatures by using the above relationship. If laboratory determinations of K using water of a known temperature are available, the laboratory

TABLE 14.5

Variation of Properties of Pure Water with Temperature

Temperature (°C)	Density (g/cm³)	Viscosity ($\times 10^{-2}$ dyne sec/cm²)
0	0.99987	1.7921
1	0.99993	1.7313
2	0.99997	1.6728
3	0.99999	1.6191
4	1.00000	1.5674
5	0.99999	1.5188
6	0.99997	1.4728
7	0.99993	1.4284
8	0.99988	1.3860
9	0.99981	1.3462
10	0.99973	1.3077
11	0.99963	1.2713
12	0.99952	1.2363
13	0.99940	1.2028
14	0.99927	1.1709
15	0.99913	1.1404
16	0.99897	1.1111
17	0.99880	1.0828
18	0.99862	1.0559
19	0.99843	1.0299
20	0.99823	1.0050
21	0.99802	0.9810
22	0.99780	0.9579
23	0.99756	0.9358
24	0.99732	0.9142
25	0.99707	0.8937
26	0.99681	0.8737
27	0.99654	0.8545
28	0.99626	0.8360
29	0.99597	0.8180
30	0.99567	0.8007
31	0.99537	0.7840
32	0.99505	0.7679
33	0.99473	0.7523
34	0.99440	0.7371
35	0.99406	0.7225
36	0.99371	0.7085
37	0.99336	0.6947
38	0.99299	0.6814
39	0.99262	0.6685
40	0.99224	0.6560

results can then be used to estimate the field hydraulic conductivity at a different ambient temperature by an adjustment for viscosity, ignoring the slight change in density. The expression

$$K_f = \left(\frac{\mu_l}{\mu_f}\right) K_l \tag{14.13}$$

uses K_l from the laboratory determination; K_f is the field estimation, and μ_f and μ_l are the respective viscosities at the field and laboratory temperatures. *In situ* hydraulic conductivity testing corrections are rarely necessary because ground-water temperatures at depths of up to 200 ft from the ground surface seldom vary more than about 2°C from the mean annual air temperature.

Other factors also affect the magnitude of K; these factors include, but are not limited to, bulk density, grain-size distribution, relative fraction of silt or clay, and, in fractured soil or rock, fracture width and frequency.

Laboratory Determination of *K*

Two laboratory methods are used for measuring hydraulic conductivity — the falling-head and constant-head permeameter test methods. The apparatuses shown in Figure 14.3 illustrate both the falling-head method and the constant-head method. In both tests, water moves through a test specimen under the influence of gravity alone; in both tests, the specimen is placed in a tube or cylinder and is usually remolded in the process of placing it into the cylinder. If an undisturbed sample is placed into a permeameter (either a flexible-wall permeameter, per ASTM D 5084 [ASTM, 2004i], or a rigid-wall permeameter, per ASTM D 5856 [ASTM, 2004j]), a means of assuring no leakage along the boundary between the sample and the cylinder must be devised. This is difficult to accomplish in practice. Rubber membranes, silicon, and wax have been used to seal the sides of specimens.

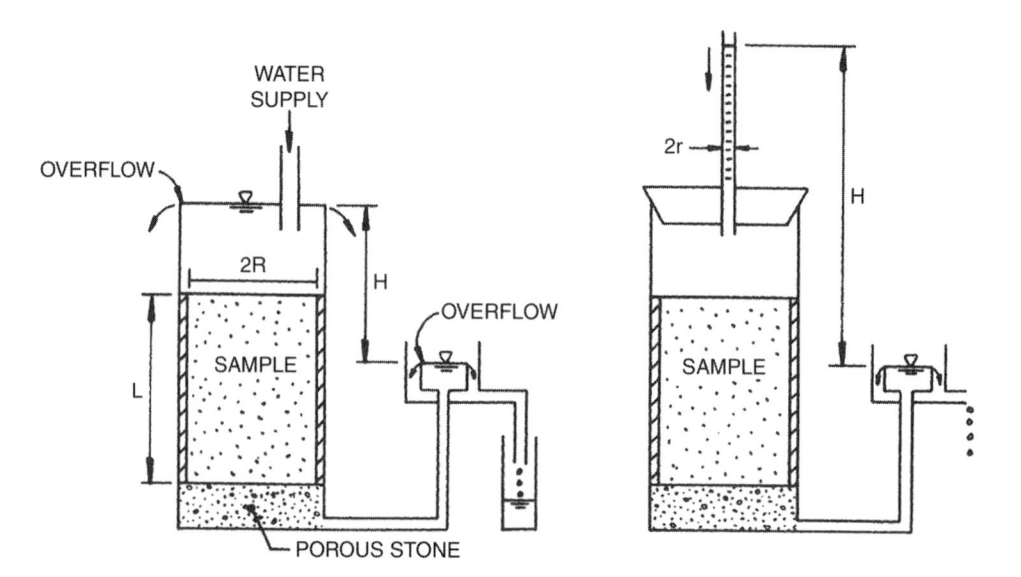

FIGURE 14.3
Constant-head (left) and falling-head (right) permeameters.

In the constant-head test procedure (ASTM D 2434 [ASTM, 2004k]), a known rate of water is allowed to pass through the specimen under a controlled hydraulic head or gradient condition. The hydraulic conductivity can then be computed by using Darcy's law:

$$K = \frac{QL}{AH} \tag{14.14}$$

where H is the total hydraulic head difference between the ends of the soil specimen in units of length, L the length of the soil specimen, A the cross-sectional area of the soil specimen perpendicular to the flow direction, and Q the flow rate in units of volume per unit time.

In the falling-head procedure, the rate of fall of the water level in a tube elevated above the top of the specimen is monitored. The head across the specimen is measured at two different times and inserted into a modified form of Darcy's law:

$$K = \frac{aL}{A(t_2 - t_1)} \ln\left(\frac{h_1}{h_2}\right) \tag{14.15}$$

where a is the cross-sectional area of the water-level-monitoring tube, A the cross-sectional area of the soil sample, L the length of the sample in the direction of flow, and h_1 and h_2 the hydraulic heads, in units of length, across the specimen at times t_1 and t_2, respectively.

The falling-head method is generally applicable for materials with hydraulic conductivities ranging from 10^{-7} to 10^{-3} cm/sec. The constant-head method is generally applicable to materials with hydraulic conductivities ranging from 10^{-3} to 10^{-1} cm/sec.

For materials with very low hydraulic conductivities, a larger hydraulic head difference is required in order to move water through the specimen in a reasonable time for laboratory measurement. Low-permeability material testing is typically carried out in a high-pressure permeameter or a triaxial cell apparatus. Figure 14.4 illustrates a schematic diagram of a triaxial cell permeameter. The sample is placed into the permeameter and a differential fluid pressure (i.e., hydraulic head) is placed across the sample. This apparatus has the advantage of creating a back-pressure condition, which dissolves gas bubbles within the specimen (Bishop and Henkel, 1962). The use of extremely high gradients may cause Darcy's law to become invalid due to turbulence, and also risk hydraulic piping along the sides of the specimen.

The triaxial cell method has the additional advantage of the specimen being placed inside a rubber membrane that is confined by fluid compression to the walls of the specimen. The rubber membrane helps to minimize the leakage along the sides of the specimen — a problem that is inherent to other permeameter methods. Dye can be injected into the test specimen at the end of the test to check for sidewall leakage or piping.

Soil samples are sometimes remolded prior to testing in a triaxial cell using a Harvard miniature mold that has a diameter of about 1.5 in. The specimen is remolded using a tamping rod attached to a spring so that a constant force can be delivered with each stroke of the rod. Soil samples, remolded using this or other compaction methods (ASTM D 698 [ASTM, 2004l] and D 1557 [ASTM, 2004m]), can be trimmed and placed in a triaxial cell. Triaxial or high-pressure permeameters typically are useful on soil or rock with hydraulic conductivities in the range of 10^{-10} to 10^{-4} cm/sec.

Laboratory testing for hydraulic conductivity has several limitations. One major limitation is that a sample with dimensions of the order of a few inches may not be representative of the *in situ* soil. A correction may have to be applied to stony soils because stones within a soil are usually removed prior to remolding and placement of the specimen into a

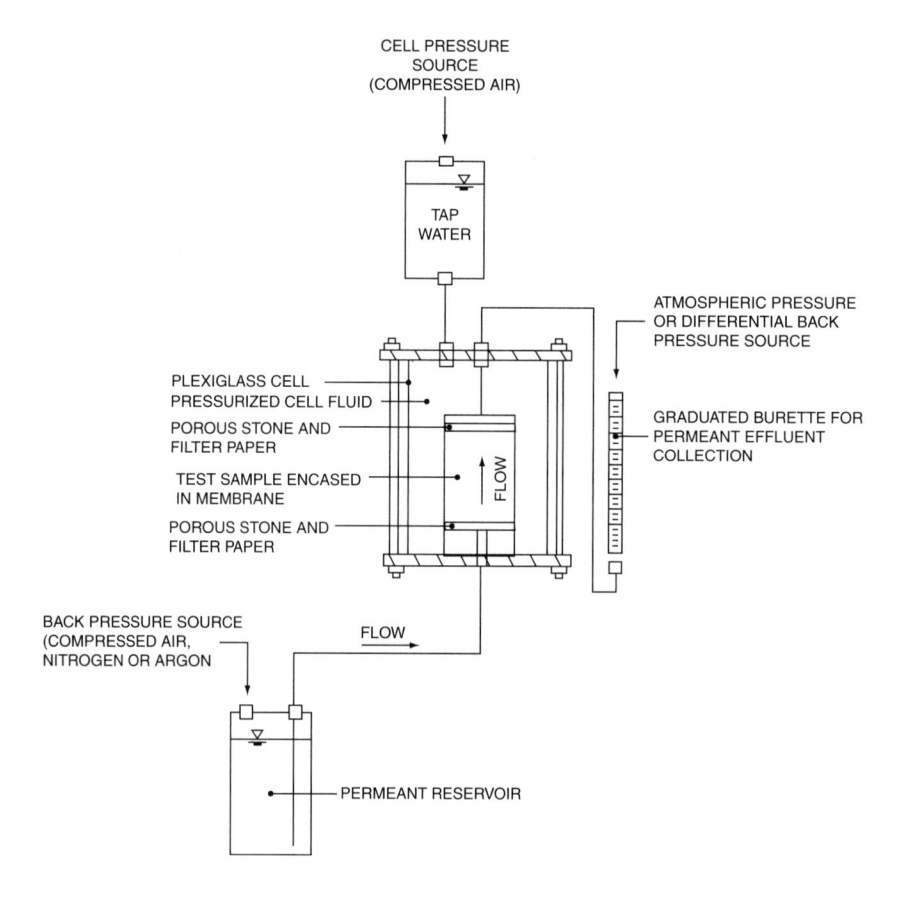

FIGURE 14.4
Schematic of a constant-head triaxial-cell permeameter.

permeameter. In such cases, soil particles should not exceed one third the diameter of the test specimen. Another major limitation is that remolding the soil may remove natural structure, such as root holes, bedding structure, and fissures, which may control the hydraulic conductivity of the soil in its natural setting. Finally, saturation is critical to interpretation of the test results. The degree of saturation has been shown to have a significant effect on the observed hydraulic conductivity (Freeze and Cherry, 1979).

An advantage of laboratory methods for determining hydraulic conductivity, which is difficult to duplicate in the field, is testing of an undisturbed soil sample at various orientations. For example, in a horizontally bedded soil, if flow is forced through a vertically oriented sample, the resulting flow is a measure of K across the bedding. However, if the same sample is trimmed and oriented so that flow during the test is parallel with the bedding planes of a soil, a measurement of the hydraulic conductivity at right angles to the first value can be obtained. If a sample is oriented at an angle other than perpendicular or parallel with the bedding planes, the value must be resolved into the directional components of hydraulic conductivity (Bear, 1972).

When using the falling-head or constant-head methods, it is best to prepare remolded test specimens from slightly moist, undried soils. After the specimen is in place, filling of the permeameter cylinder with water should begin from the bottom of the sample to displace entrapped air. Flushing the sample with carbon dioxide prior to flooding with water will improve the rate of saturation. Oven-drying of natural soils that contain silt, clay, or organic matter prior to specimen preparation and testing may limit the ability to saturate the specimen and modify the structure of the soil to such a degree that an accurate measurement of hydraulic conductivity is unlikely. The preferred saturation method is using back-pressure (Black and Lee, 1973). This can be easily accomplished in a triaxial cell or specially constructed constant-head test cylinders.

Estimation of *K* from Soil Grain Size

Hydraulic conductivity can also be inferred from the grain-size distribution of the soil. Hazen (1911) empirically related the effective particle size to hydraulic conductivity, such that

$$K = C d_{10}^2 \qquad (14.16)$$

where K is the hydraulic conductivity in cm/sec, d_{10} the Particle size (measured in mm) below which 10% (by weight) of the cumulative sample has a smaller size, and C constant which ranges from 1 to 1.2.

This method was developed for estimating the hydraulic conductivity of sand filters. Consequently, its use is generally limited to uniformly graded sands, or sands with a uniformity coefficient of less than 5.0 (Hazen, 1911).

Fair and Hatch (Todd, 1959) proposed another method for estimating hydraulic conductivity that utilizes grain-size data from the entire distribution curve. This method is useful for sandy soils with minor amounts of silt and clay (generally less than 20%). The method assumes that hydraulic conductivity is related to the shape of the grain-size curve and grain characteristics by an empirical mathematical regression:

$$K = \frac{\rho g}{\mu}\left[\frac{n^3}{(1-n)^2}\right]\left[m\left(\frac{\theta}{100}\sum\frac{P}{d_m}\right)^2\right]^{-1} \qquad (14.17)$$

where K is the hydraulic conductivity in units consistent with units of p, g, μ and d_m; μ the viscosity of the fluid in units of mass per time-length; ρ the fluid density in units of mass per unit volume; m the packing factor, 5; θ the sand grain shape factor; 6.0 for spherical grains and 7.7 for angular grains; n the porosity, expressed as a decimal fraction; P the percentage of sand held between adjacent sieves; d_m the geometric mean of rated sizes of adjacent sieves in units of length; and g the acceleration of gravity, e.g., 32 ft/sec^2. This equation is dimensionally correct for any consistent set of units.

The Hazen estimation methods do not account for density effects, which typically cause much less variation in hydraulic conductivity than spatial variation of grain-size distribution. Powers (1981) describes a method for estimating K developed from grain-size analysis and *in situ* density of the sand. From the grain-size analysis, the mean grain size, d_{50}, and the uniformity coefficient, C_u, must be determined (Lambe, 1951). The *in situ* density of the soil can be measured or estimated from standard penetration test results (Gibbs and Holtz, 1975). Using this information, the hydraulic conductivity is estimated from the charts shown in Figure 14.5. This method is useful for sands and gravels.

Slug Tests

In situ slug tests in wells or piezometers are popular for hydraulic conductivity testing in both soil (overburden) and rock. Slug tests involve removing, adding, or displacing a

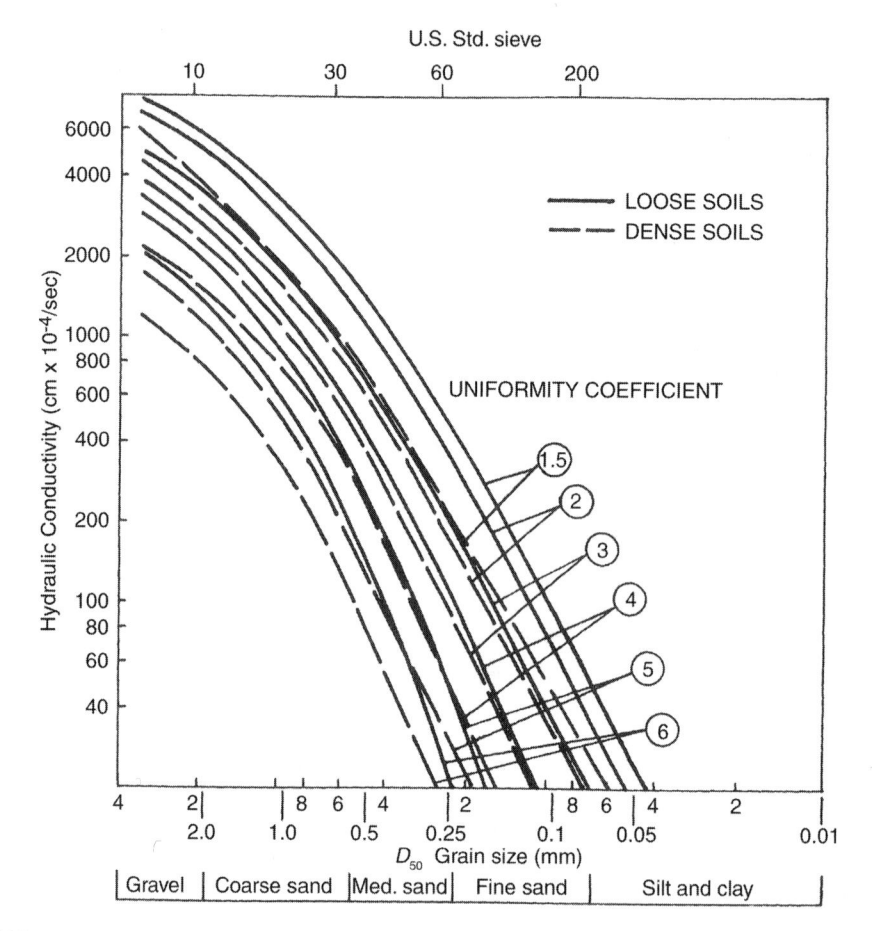

FIGURE 14.5
Estimated hydraulic conductivity of sands and gravels.

quantity of water in a well or piezometer and monitoring the change in water level with time. In vertically oriented wells or piezometers, this method provides a measure of the horizontal hydraulic conductivity.

The slug test (ASTM D 4044 [ASTM, 2004n]) is similar to a falling-head laboratory test in that the rate of water-level decline or increase, as the water level attempts to equilibrate with natural piezometric conditions, is a function of the hydraulic conductivity of the soil and the geometry of the well or screened interval.

In very permeable formations, it may be impossible to measure water-level changes with time because the water-level equilibration is almost immediate. The use of an electronic data logger, with a pressure transducer or strip chart recorder, can facilitate the collection of data in this situation. It is important to note that in some cases the hydraulic conductivity of a formation may be so great (as it is in gravels or limestone solution cavities), that head losses caused by construction of the well or piezometer, rather than the formation's actual hydraulic conductivity, may control the rate of water-level change. Cyclical water-level responses can be observed in highly permeable soils depending on well construction (van der Kamp, 1976). Testing in formations of very low hydraulic conductivity may require long periods of data collection (e.g., days or weeks) or a pressure-test method (see Pressure Tests section).

In situ well or piezometer test methods can be modified so that a constant rate of water is added to or extracted from the casing. This is similar to a constant-head laboratory test and requires knowledge of the flow rate, the head differential from the background piezometric condition, and the well geometry (Hvorslev, 1949b).

Slug-test results are often analyzed using the method of Hvorslev (1949b). This method allows for various well and aquifer geometries, but is based on a quasi-steady-state solution of the flow equations. The lack of conceptual rigor limits the accuracy of the Hvorslev method in some cases (Butler, 1998). Solutions for various well and aquifer geometries are also available (Hvorslev, 1949b; Lambe and Whitman, 1969).

The method developed by Cooper et al. (1967) is based on the analytical solution of transient flow equations. This method was developed for wells that fully penetrate an aquifer, and has been adapted to a type-curve matching procedure. Hydraulic conductivity and storage coefficient can be estimated using method of analysis. The method solves for transmissivity (see Transmissivity section), which, if the aquifer thickness is known, can be used to calculate the hydraulic conductivity. If the well screen partially penetrates the aquifer, the screen length can be substituted for the aquifer thickness and an approximation of the hydraulic conductivity can be obtained. Additional type curves have been developed by Papadopulos et al. (1973) to cover a broader range of storage coefficients and well sizes. This method was updated by Bredehoeft and Papadopulos (1980) for estimating hydraulic conductivity in low-permeability formations. It is described in detail in ASTM D 4104 (ASTM, 2004o).

The Cooper et al. (1967) method utilizes type curves, as shown in Figure 14.6, to match field data of water level above or below static, h, at any time, t, relative to the initial water

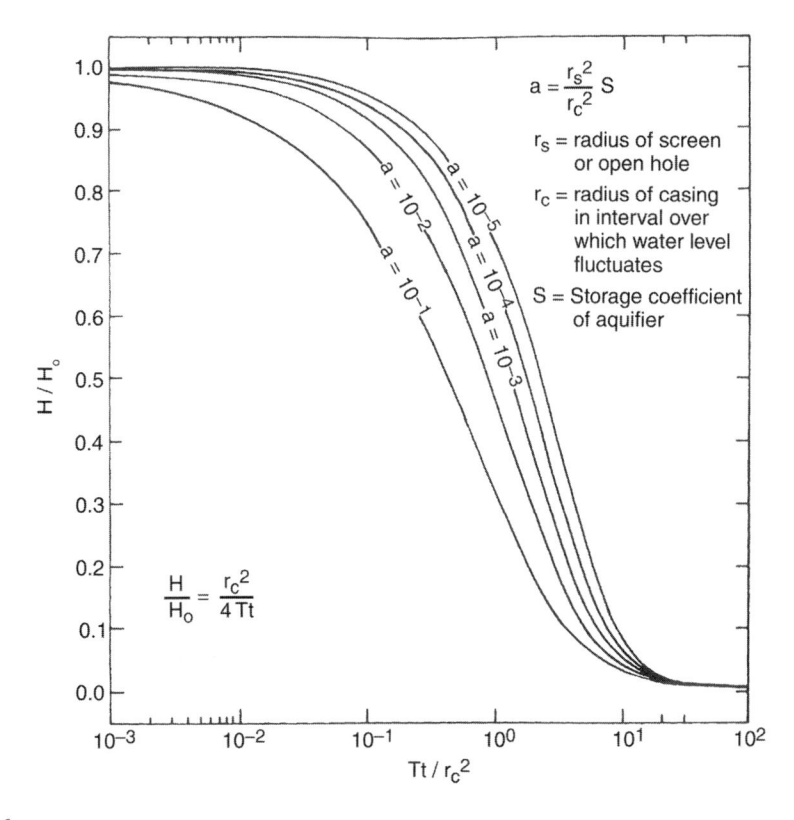

FIGURE 14.6
Slug-test type curves.

level above or below static, h_0, at $t = 0$, to the computed curves presented in dimensionless parameters, $r_w^2 S/r_c^2$ and $2Tt/r_w^2$. The well-screen radius, r_w, and the inside well-casing radius, r_e, are needed to solve for storativity, S, and transmissivity, T. These tests can have a zone of influence of up to several hundred feet (Sageev, 1986), depending on the properties of the soil or rock and the slug volume. Typically, the greater the slug volume and the lower the storativity, the greater the zone of influence. Herzog and Morse (1986) point out that this method may require very long data-collection periods for low-conductivity materials. However, for formation hydraulic conductivities of 10^{-6} cm/sec or greater, testing can typically be completed in 1 day.

Bouwer and Rice (1976) devised a method for analyzing slug-test data from fully or partially penetrating wells completed in an unconfined aquifer. The procedure is based on the Theim equation and assumes negligible drawdown of the water table around the well and no flow above the water table. This test can also be used to estimate hydraulic conductivity of confined aquifers that receive most of their water from an upper confining layer through leakage or compression. A detailed discussion of this method can be found in ASTM D 5912 (ASTM, 2004p).

The method of van der Kamp (1976) allows determination of transmissivity from the measurement of the damped oscillation about the equilibrium water level of a well-aquifer system to a sudden change of water level in a well. Underdamped response of the water level in a well is characterized by oscillatory fluctuation about the static water level with a decrease in the magnitude of fluctuation and recovery to the initial water level. Underdamped response may occur in wells tapping highly transmissive confined aquifers and in deep wells having long water columns. This method of analysis requires that the storage coefficient be known. Assumptions of this method prescribe a fully penetrating well, but it can be adapted to use in partially penetrating wells where the aquifer is stratified and horizontal hydraulic conductivity is much greater than vertical hydraulic conductivity. In such a case, the test would be considered to be representative of the average hydraulic conductivity of the portion of the aquifer adjacent to the open interval of the well. The method assumes laminar flow, and is applicable for a slug test in which the initial water-level displacement is less than 0.1 or 0.2 of the length of the static water column. This analytical procedure, which is described in ASTM D 5785 (ASTM, 2004q), is used in conjunction with ASTM D 4044 (ASTM, 2004n).

The method of Nguyen and Pinder (1984) also allows estimation of both the storage coefficient and the hydraulic conductivity. The procedure is slightly more complicated than the Cooper et al. (1967) method. However, the procedure incorporates constant well discharge or recharge, as well as transient water-level changes. Herzog and Morse (1986) point out that this method has advantages over the Cooper et al. (1967) method. The method was designed for wells screened over only a portion of the aquifer thickness (i.e., partially penetrating wells), which is the more frequent case in practice; the method typically requires less than 1 day of field time to complete. This method can also be used for angled boreholes and, therefore, can be used to evaluate directional effects on the observed hydraulic conductivity. However, Herzog and Morse (1986) caution about careful measurement of water levels, suggesting that accurate recordings of the hydraulic head be taken by using, for instance, pressure transducers with an electronic data logger or strip chart recorder. If discrete water-level measurements are made using a hand-held meter, erratic results can lead to interpretation difficulty.

Kipp (1985) describes a method of type-curve analysis of inertial effects in the response of a well to a slug test that allows determination of transmissivity of confined, non-leaky aquifers. Transmissivity is determined from the measurement of water-level response to a sudden change in water level in a well-aquifer system characterized as being critically damped or in the transition range from being underdamped to overdamped.

Underdamped response is characterized by oscillatory changes in water level; overdamped response is characterized by return of the water level to the initial static water level in an approximately exponential manner. The assumptions of the method prescribe a fully penetrating well, and an aquifer of uniform thickness and of constant homogeneous porosity and constant homogeneous and isotropic hydraulic conductivity. The analytical method is described in detail in ASTM D 5881 (ASTM, 2004r).

Several investigators have analyzed slug-test type curves for fractured media (Gringarten and Ramey, 1974; Karasaki, 1986). These curves often do not have unique solutions unless information on fracture characteristics is known. Alternately, the fractured media can be assumed to be equivalent to a porous medium and the methods described earlier are applicable. Karasaki et al. (1988) examined various fracture conditions local to the test well.

Skin effects involve locally increasing the conductivity near the well by opening fractures or pores (positive skin) or decreasing the conductivity near the well (negative skin) by filling natural fractures with drilling mud or drill cuttings. Negative skin effects can make the apparent measured hydraulic conductivity less than the actual *in situ* hydraulic conductivity of the formation away from the borehole. Negative skin effects can also be created by disturbing a naturally layered soil and forming a more or less uniform soil zone along the wellbore during drilling. Smearing of silts and clays during borehole drilling may create a negative skin. Skin effects can be checked in unfractured formations by performing laboratory tests of the aquifer material, preferably using undisturbed samples. One of the problems with most skin models is the nonuniqueness of the solutions. Knowledge of the skin properties and aquifer storage coefficient can make the solution possible. Faust and Mercer (1984), Ramey and Agarwal (1972), Moench and Hsieh (1985), and Sageev (1986) have examined skin effects. Butler (1998) has developed a method for analyzing slug-test data under a variety of conditions including skin effects.

Either rising-head or falling-head slug tests can be performed. However, the values for the two types of tests in a single well can vary by up to a factor of 100; typically the falling-head result is greater than the rising-head result. The errors in measurement are believed to be associated with well-installation effects, which cause a disturbance of the aquifer material around the borehole. Milligan (1975) suggests that the "best" hydraulic conductivity estimate, based on the two values, is obtained by

$$K = (K_{RH} K_{FH})^{1/2} \qquad (14.18)$$

where K_{RH} is the hydraulic conductivity as determined by rising-head method and K_{FH} the hydraulic conductivity as determined by falling-head method.

Packer Testing

In consolidated rock, hydraulic conductivity is commonly measured using a packer test in an open borehole. This method typically gives a measure of the horizontal hydraulic conductivity. The arrangement for performing packer tests in open boreholes is illustrated in Figure 14.7. This method is used for testing rock in which the walls of the open borehole are stable. Inflatable packers are generally used to isolate the interval of the borehole to be tested. A single packer is used to test a section of borehole between the bottom of the boring and the packer location. Typically, single-packer testing is performed as the hole is advanced. After drilling to a desired depth, the packer is inserted at a selected depth above the bottom of the borehole. The packer is then inflated using water or a gas, and

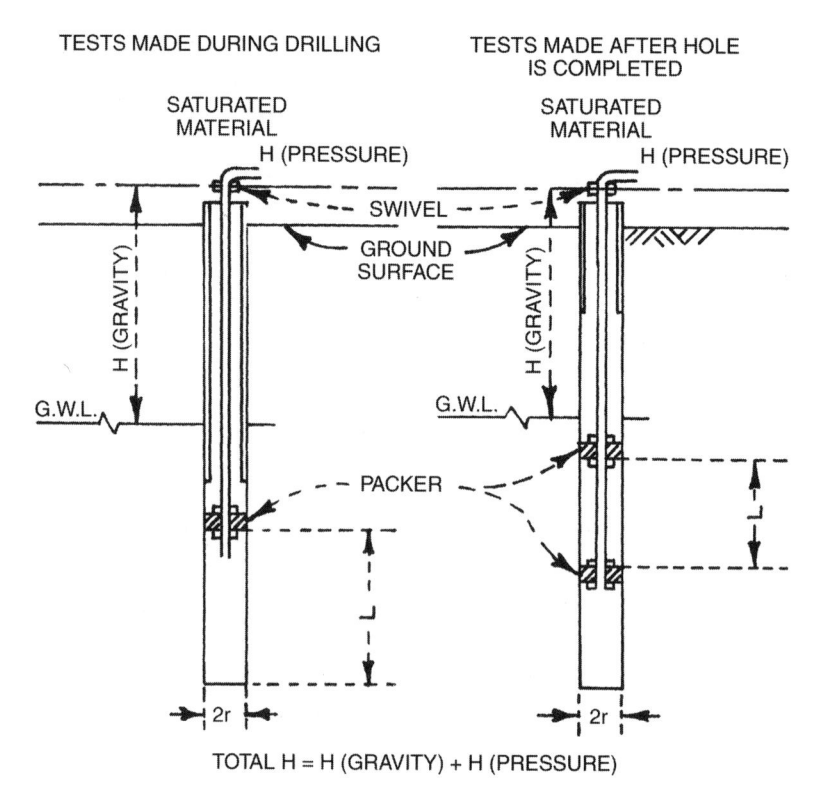

FIGURE 14.7
In Situ packer testing.

water is injected in the borehole for a given length of time to test the "packed-off" portion of the hole. After the test, the packer is removed and the hole is advanced in depth. This procedure can be repeated as many times as desired.

With the use of two packers, any position or discrete interval along the borehole can be selected for testing. The interval may be preselected, or it may be selected based on core descriptions or observed fractures. The two-packer system is usually inserted into the borehole after the entire borehole has been drilled to total depth. The portion of the fill tube between the two packers contains openings through which water can flow once the packers have been inflated. Water is forced under pressure into the fill tube and the test is run for the desired length of time. When testing with two packers, the usual procedure is to begin testing at the bottom of the hole and then proceeds upward. This practice reduces the likelihood that testing may reduce the stability of the hole walls, which could trap the packer test equipment in the borehole. Depending on the flow rate, head losses within the piping system may be critical to the interpretation of the test results. Skin effects due to drilling, particularly when drilling with mud, can also critically influence the results of the packer testing (Faust and Mercer, 1984). Skin effects can also be caused by core removal, which causes stress changes that can close fractures (Neuman, 1987).

Other types of packers, including compression packers and leather cups, are also used. However, these packers are prone to leakage, which may cause an erroneous interpretation of test findings and result in an overestimate of hydraulic conductivity. Inflatable packers generally yield the best results because they can form a tight seal against the borehole wall, even if the hole is rough-walled or out-of-round.

Before performing the test, the borehole should be cleaned of any cuttings or drilling fluids. This involves swabbing or bailing the borehole. The presence of drilling mud or failure to clean the borehole may result in a lower measured hydraulic conductivity than the actual hydraulic conductivity of the rock.

The length of the packer-test section is generally governed by the character of the formation. Typically, a 5–10-ft length is used. However, variations from these standard lengths can be accommodated by modifying the interval between packers. Depending on the hydraulic conductivity of the rock, the test interval may have to be lengthened or shortened in order to obtain a value for the given conditions and the test duration and pressure. The duration of the test should be sufficient to provide measurable flow volumes. The test section should never be shortened to the point at which the ratio of interval length to hole diameter is less than 5, if the standard horizontal flow equations are to be used for analysis. Test results may become invalid when vertical flow becomes an important component.

It is important to measure the piezometric conditions in the packed-off section of the hole prior to beginning the test. If not available prior to the test, the piezometric conditions must be estimated to set a test pressure. Test pressures should not exceed 0.5 psi per foot of depth as measured from the ground surface to the top of the test section. The purpose for not exceeding this pressure is that excessive hydraulic pressure can induce hydraulic fracturing, thus causing an increase in the measured hydraulic conductivity relative to the undisturbed hydraulic conductivity. The test pressure should be measured by a gage located as near the well head as possible. Thus, the pressure at the well head is observed without losses due to meters or pumps or turbulence close to the pump. When multiple tests are being conducted in the same borehole at various depths, residual pressures may occur from previous testing and may have to be accounted for in the testing. Preferably, residual pressures should be allowed to come to equilibrium prior to any subsequent testing.

Typically, packer tests are run for a period from 15 min to 2 h. Generally, the lower the hydraulic conductivity of the formation, the longer the test. During the first 5 min of the test, readings of the flow meter monitoring the amount of injected water should be taken every 30 sec; thereafter, readings should be made at 5-min intervals for the remainder of the test period. During the initial portion of the test, an expansion of the packer device may cause a flow of water that is not indicative of formation hydraulic conductivity. Normally the test is run in two parts at two different pressures. For example, during the first part of the test, the borehole could be pressurized at 15 psi for a period of 15–30 min. After the flow rate has stabilized, the pressure may be increased to 30 psi during the second part of the test.

Packer tests are often performed using pumps supplied on the drilling rig. Depending on the type of pump, water flows of 25–250 gal/min can be achieved. For very low flow rates, a meter calibrated in fractions of a gallon is necessary. However, where high flow rates are expected, a meter calibrated in 5–10 gal/min increments may be satisfactory. Typically, totalizer-type meters are used rather than instantaneous flow meters. However, both meter types have been used in practice. Occasionally, in very low hydraulic conductivity formations, no change in volume will be observed other than that caused during the initial pressurization of the packer system. In these cases, the length of the test can be increased to allow a measurable quantity of water to flow into the rock. An estimate of the upper limit of the hydraulic conductivity can be calculated if, even after extending the test period, no flow is observed.

The data required for computing the hydraulic conductivity include the borehole radius, the pressure at the well head, the depth of the borehole and the packers, the flow rate, and the height of the well head above the ground surface. Figure 14.7 illustrates the information necessary to calculate the hydraulic conductivity. The formulas for calculating hydraulic conductivity are:

$$K = \frac{Q}{2\pi LH} \ln\left(\frac{L}{r}\right), \quad L > 10r \tag{14.19}$$

$$K = \frac{Q}{2\pi LH} \sinh^{-1}\left(\frac{L}{2r}\right), \quad 10r > L > r \tag{14.20}$$

where K is the hydraulic conductivity in units of length per unit time, Q the constant rate of flow into the hole in units of volume per unit time, L the length of the portion of the hole tested, H the total differential head of water in units of length, r the radius of hole tested in units of length, ln the natural logarithm, and $\sinh^{-1}$ the inverse hyperbolic sine.

These formulas are valid for calculating hydraulic conductivity when the thickness of the stratum tested is at least $5L$, and they are considered to be more accurate for tests below the water table than above it. Multiple pressure tests can be utilized to evaluate potential problems, such as leakage, with the packer testing. Tests conducted at three or four different pressures, increasing from zero and decreasing back to zero, are often done.

The constant-head injection test (described in ASTM D 4630 [ASTM, 2004s]) is a form of packer test generally conducted in low-permeability rocks. To accommodate the test, a borehole must be drilled into the rock mass for which hydraulic conductivity, transmissivity, and storativity information are desired. The borehole is cored through the potential zones of interest, and is later subjected to geophysical logging over these intervals. During the test, each interval of interest is packed off at the top and bottom with inflatable packers attached to high-pressure steel tubing. The test involves rapidly applying a constant pressure to the water in the packed-off interval and tubing string, and recording the changes in water flow rate. The water flow rate is measured by one of a series of flow meters of different sensitivities located at the surface. The initial transient water flow rate is dependent on the transmissivity and storativity of the rock surrounding the test interval and on the volume of water contained within the packed-off interval and tubing string. The advantages of this method are: (1) it avoids the effect of wellbore storage; (2) it may be employed over a wide range of rock mass permeabilities; and (3) it is considerably shorter in duration than the conventional pumping tests and slug tests used in more permeable rocks.

Pressure Tests

In formations of very low hydraulic conductivity (i.e., less than 1×10^{-7} cm/sec), pressure tests, sometimes called pulse tests, are more appropriate. Various investigators have examined the use of pressure tests in low-permeability formations (Wang et al., 1977; Forster and Gale, 1981; Neuzil, 1982; Neuman, 1987). The results of these tests are often analyzed using type-curve procedures.

In a pressure test, a packer system is placed into the borehole and the packers inflated. An increment of pressure is applied to the zone between the packers. The decay of pressure is monitored and plotted versus time. The rate of pressure decay is related to the storage coefficient and the hydraulic conductivity of the formation. Pressure response data are typically collected using pressure transducers with electronic data loggers or strip-chart recorders.

The pressure-pulse technique described in ASTM D 4631 (ASTM, 2004t) is carried out in a borehole drilled into the rock mass for which hydraulic conductivity, transmissivity, and storativity data are desired. The borehole is cored through the potential zones of interest, and is later subjected to geophysical logging over these intervals. During the test, each interval of interest is packed off at the top and bottom with inflatable packers attached to high-pressure steel tubing. The test involves applying a pressure pulse to the water in the packed-off interval and tubing string, and recording the resulting pressure transient. A pressure transducer, located in either the packed-off zone or in the tubing at the surface, measures the transient as a function of time. The decay characteristics of the pressure pulse are dependent on the transmissivity and storativity of the rock surrounding the interval being tested and on the volume of water being pulsed. Alternatively, under non-artesian conditions the pulse test may be performed by releasing the pressure on a shut-in well, thereby subjecting the well to a negative pressure pulse. Interpretation of this version of the test is similar to the positive pressure pulse test. This test can generally be conducted in shorter time frames than pumping tests and slug tests used in more permeable formations.

This test is generally only used in low hydraulic conductivity rock formations and must compensate for skin effects and packer adjustment during the application of pressure. An understanding of the presence and orientation of fractures in the borehole is necessary to select an appropriate type curve to analyze test data. The skin effect within the borehole may be critical in evaluating *in situ* hydraulic conductivity test data. The presence of a drilling mud filter cake or the smearing of fine-grained material along the borehole walls, created by the drilling operation, may result in test data that reveal an apparent low hydraulic conductivity. Drill cuttings in an unmudded hole may also create skin effects by washing into joints or fissures or coating the sides of the hole. This may be particularly true with air-rotary drilling in fractured rock. The skin effect may be impossible to detect, as indicated by simulation studies (Faust and Mercer, 1984).

Tracer Tests

Single-well and multi-well tracer tests have been performed to determine horizontal hydraulic conductivity in both unconsolidated deposits and rock. In the single-well test, a tracer is injected into a well, the tracer moves radially away from the well, and then the well is pumped to recover the tracer. The tracer can consist of an easily measured, non-reactive, nondecaying solute. Tritium, visually identifiable dyes, or electrolyte (e.g., chlorides) have been used. The test can involve multi-level injection points to measure formation response at various depths. Alternately, in a multi-well test, sampling wells are placed radially away from a continuously screened injection well and monitored to detect the arrival of tracer. Multi-well tests involve injection of a tracer into one well and extraction by sampling from other wells. Similar to the single-well test, the wells may have multi-level ports for injection and sampling.

These tests require careful planning and experience in interpretation of the results. They can be time-consuming and expensive and, therefore, often are used only on projects in which budgets can accommodate the effort.

Vertical Hydraulic Conductivity Using Packer Tests

Burns (1969) proposed a method for estimating vertical and horizontal hydraulic conductivities in homogeneous granular rock using a procedure similar to a standard packer test. The method involves grouting a well casing into the borehole and then perforating the casing at two depths separated by several feet (see Figure 14.8). A packer is inflated between the two sets of perforations to hydraulically separate them. Greater assurance

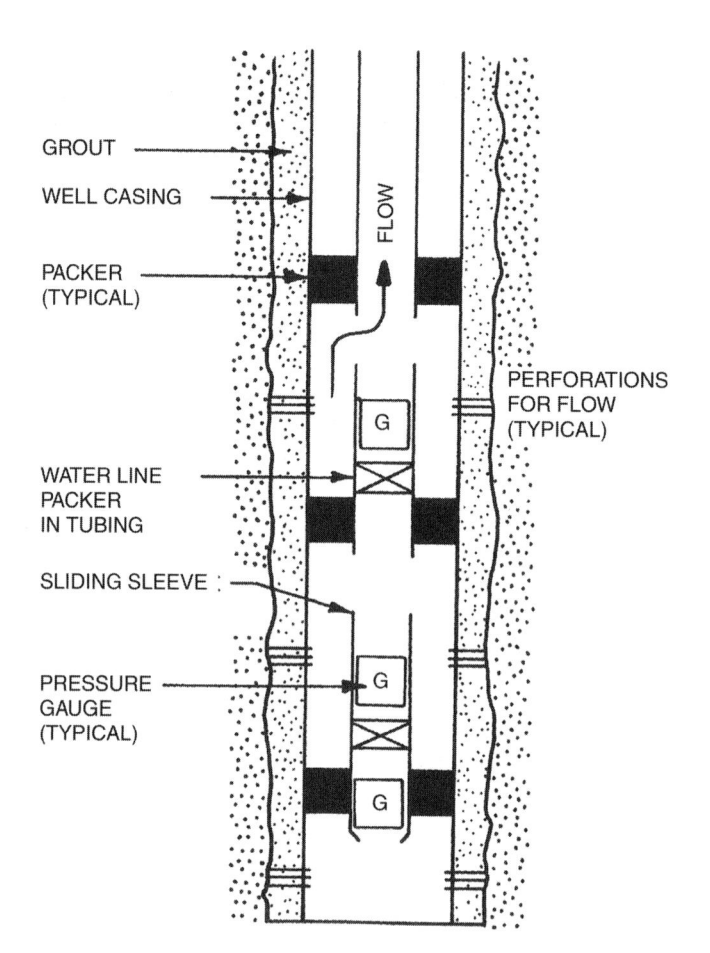

FIGURE 14.8
Vertical permeability well test.

of a seal is obtained through the use of two packers at the end of the interval containing the perforations. Using pressure transducers above and below the packers, the pressures in the borehole are monitored, as a known flow volume is injected through the perforations above the packer. The test is stopped when the lower transducer indicates a pressure response at least 10 times the sensitivity of the gage. The results are calculated using graphical techniques (Burns, 1969).

Packer tests in cased wells rely upon the quality of grouting between the casing and borehole wall. Leaks within the cavity between the perforations will result in an overestimation of the vertical hydraulic conductivity. In low hydraulic conductivity formations, the time required to reach background conditions, which is necessary prior to conducting the test, may be excessive unless a pressure test is performed. Interpretation of the calculated conductivity may be difficult in a fractured medium. Tests similar to the one described above have been developed by Prats (1970) and Hirasaki (1974).

Vertical Hydraulic Conductivity from Seepage Pits

In situations where the water table is very deep, vertical hydraulic conductivity has been measured using seepage pits (see Figure 14.9). One method relies on the development of a

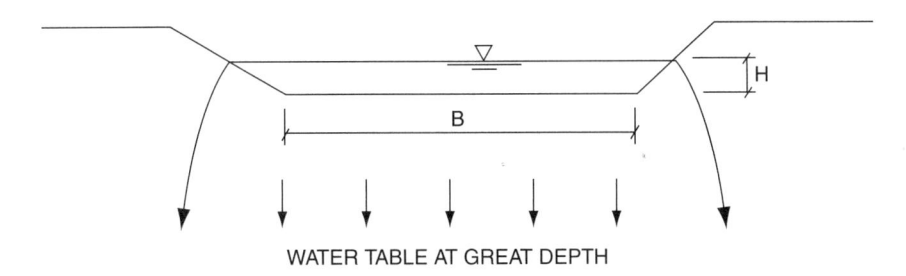

Methods and Procedures for Defining Aquifer Parameters

FIGURE 14.9
Seepage pit for *in situ* measurement of vertical hydraulic conductivity.

steady-state or near-steady-state vertical seepage pattern. The method uses an equation describing the theoretical seepage for a given pit geometry:

$$Q = K_v(B + AH)L \qquad (14.21)$$

where Q is the steady-state flow rate in units of volume per unit time, L the length of trench in units of length, B the pit base width in units of length, H the pond depth in pit in units of length, K_v the vertical hydraulic conductivity in units of length per unit time, and A varies from 2 to 4 depending on the geometry of the pit.

Theoretical development of this equation can be found in Polubarinova-Kochina (1962) and Harr (1962).

During use of the method, precipitation and evaporation must be accounted for because the test may last several days. Also, capillary forces in the unsaturated zone may be important for small basins in which the capillary rise is on the same order as the basin depth. Capillarity at the front of the downward-moving seepage face may also affect the test results during the period near the start of the test.

The test method involves excavating a shallow pit. During excavation, smearing of the soil on the pit walls must be avoided, as this may reduce the apparent hydraulic conductivity value. Excavation by shovel is often required. A major advantage of this method is that the soil or rock is being tested in an essentially undisturbed condition. Therefore, if the soil has secondary permeability due to fissures or rootholes, the test will include their effects on the measured hydraulic conductivity.

Typical pit sizes range from a few square meters up to 100 m^2. The pit is filled with water, but water must be added so as not to erode the pit base and sidewalls, thus creating a suspension of fine soil particles that may settle and form a flow barrier on the base of the pit. The soil below the pond base can be instrumented with piezometers to assure that steady-state seepage has occurred. However, often the rate of seepage is used to determine if steady-state conditions are attained. The test must be monitored continually to maintain the desired water level in the pit. Minor fluctuations in the pit water level may not be important, but large deviations should be avoided. These tests may last several days or up to a week, depending on the soil type.

Vertical Hydraulic Conductivity from Pumping Tests

Vertical hydraulic conductivity of an aquitard can be obtained from pumping test data (see Transmissivity section). Vertical leakage, L, of an aquitard overlying an aquifer can be an important parameter in evaluating the direction of ground-water flow and in estimating the area of influence of a pumping well. This value is related to the vertical hydraulic

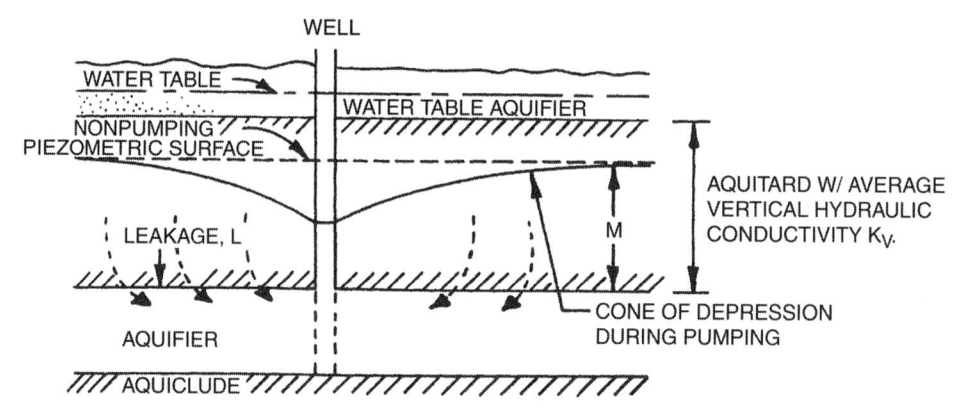

FIGURE 14.10
Definition of vertical leakage.

conductivity and the thickness of the aquitard:

$$L = K_v M \qquad (14.22)$$

where K_v is the average vertical hydraulic conductivity in units of length per time and M the vertical thickness of the aquitard in units of length.

Different methods have evolved for calculating L directly from pumping test data. The first analytical solution to a leaky aquitard problem, which ignored storage in the aquitard and held the piezometric surface constant in the aquitard, was by Hantush (1956). Hantush (1960) developed a method in which storage of the aquitard can be accounted for. Cooley and Case (1973) developed a method for a water-table aquitard. Figure 14.10 illustrates the definition of leakage.

Direct Measurement of Ground-Water Flow Velocity

The need to calculate ground-water flow velocity is one reason that hydraulic conductivity is such an important parameter in hydrogeologic investigations. Ground-water flow velocity is used to determine how rapidly ground water or dissolved constituents in ground water are moving in the ground water system.

Tracer tests have been used to estimate *in situ* flow velocities. The method determines flow velocity, which is related to hydraulic conductivity by the relationship

$$\bar{v} = \frac{KI}{\bar{n}} \qquad (14.23)$$

where $\bar{v}$ is the mean ground water particle velocity in units of length per unit time, I the natural or induced hydraulic gradient expressed as a unitless decimal fraction, $\bar{n}$ the effective porosity of the formation expressed as a unitless decimal fraction, and K the hydraulic conductivity in units of length per time.

This equation can be used for both unconsolidated materials and highly fractured rock. The effective porosity is a measure of only that porosity which contributes to movement of water within the unconsolidated material or rock. It is commonly less than the total porosity due to closed-end pores and double-layer effects.

If the direction of ground-water movement is known, a nonreactive dye, electrolyte, or radionuclide tracer can be injected into the formation at one point, through a well, and its presence monitored at a downgradient point. The amount of fluid that is injected should be controlled to avoid altering the natural hydraulic gradient. Knowing the distance between the two wells and the breakthrough time of the tracer, the ground-water flow velocity can be calculated by dividing the distance by the breakthrough time. Furthermore, by knowing the head difference between the two wells and the effective porosity, the average hydraulic conductivity between the two wells can be estimated using Equation 14.23. The assumptions for a natural gradient tracer slug are that natural gradient is not disturbed, the tracer does not react with the geologic formation, and the conditions between the injection and monitoring wells are uniform so that the flow is directly from one well to the other. The use of dyes or tracers is less useful and more risky in fractured or Karst terrains, where flow directions are controlled by unobservable fractures or solution channels and the actual direction of flow is difficult to estimate.

An electronic velocity meter, which is lowered down a well, can measure both flow direction and rate (Kerfoot, 1982). The meter has a radial array of temperature sensors that extend from the base of the downhole portion of the meter. A prong located at the center of the sensor array is heated by an internal battery source when the test begins. The surrounding sensors are then used to electronically sense temperature changes and calculate the direction and rate at which the heat is being carried by the moving ground water. Testing at various depths in a screened well may indicate a variety of flow directions. These may be due to aquifer heterogeneities or ambient currents within the well. The measured rate of ground-water flow, when the apparatus is used in a well, may also be affected by the converging flow at the well, due to the greater hydraulic conductivity of the well relative to the geologic formation.

The borehole dilution method provides another means of measuring the ground-water flow velocity local to a well. This method involves isolating a section borehole using packers. An electrolyte or radioisotope solution is injected between the packers, but maintained at the same piezometric conditions as the surrounding formation. As ground water flows through the well section, it removes some of the solution causing a dilution with time. The rate of dilution has the relationship to ground-water flow velocity of:

$$\bar{v} = -\frac{V}{n\bar{\alpha}At}\ln\left(\frac{C}{C_0}\right) \qquad (14.24)$$

where $\bar{v}$ is the estimated mean ground-water particle velocity near the well in units of length per unit time, V the volume of isolated well segment, C_0 the initial concentration of solute in units of mass per volume, t the elapsed time from start of test, C the solute concentration at t, in units of mass per unit volume, A the vertical cross-sectional area of isolated well segment, $\bar{n}$ the effective porosity of formation, expressed as a unitless decimal fraction, and $\bar{\alpha}$ the correction factor ranging from 1.5 to 3.

This procedure is limited to use in soil and rock in which the seepage velocity is significant relative to the natural diffusion potential of the solute. $\bar{\alpha}$ varies depending on well design, but is typically about 2.0 for wells without a sand pack.

During the procedure, uniform mixing of the solute must be maintained so that sampling or measurement of the remaining concentration represents an average mix. The density of the mixture should not be so high that density-driven flow from the zone of isolation becomes important. This method is useful in identifying zones of higher seepage velocity where contaminants would move at a greater rate. The method, therefore, can be used to select well screen depths for monitoring well placement. The method is described in detail in Drost et al. (1968).

Specific Storage and Specific Yield

Specific storage and specific yield are measures of the water released or gained by a respective decrease or increase in the piezometric level in an aquifer. They are dependent on the compressibility of the aquifer, the compressibility of water, and the porosity of the formation, whether rock or unconsolidated material. These two terms become important when evaluating transient aquifer behavior, such as determining what portion of an aquifer is contributing to a monitoring well screen during sampling.

Specific Yield

If an initially saturated soil or rock is allowed to drain under gravity alone, the water content will decrease to a certain value which is dependent on the grain size or fracture characteristics of the material. Small pores may not drain because the water retention forces are greater in small pores than in large pores. A clay or peat may drain very little, whereas a sand or gravel may drain almost completely. The amount of water remaining in the soil is a measure of its retention capacity. The amount of water that escapes by gravity drainage is defined as the specific yield or drainable porosity (see Figure 14.11). The specific yield is a measure of the volume of water gained or released from a unit surface area of aquifer due to a respective unit rise or decrease in water-table level. Specific yield is expressed as a decimal or as a percent. Typical values of specific yield are presented in Table 14.6.

The specific yield (also known as drainable porosity) of a soil may be estimated in the laboratory by allowing the soil specimen to free-drain under gravity conditions in a moisture cabinet where the relative humidity is maintained at 100%. The difference between the water contents before and after drainage is an estimate of the drainable porosity. If such a test is carried out, an understanding of the capillary behavior of the soil is necessary. Specific yield or drainable porosity is also a function of time. For example, in a well-sorted sand, one investigation found that 40% of the drainage occurred after the first few hours, but drainage continued for a period of up to 2.5 yr (de Marsily, 1986). Therefore, specific yield can vary in value for a soil depending on the period of interest.

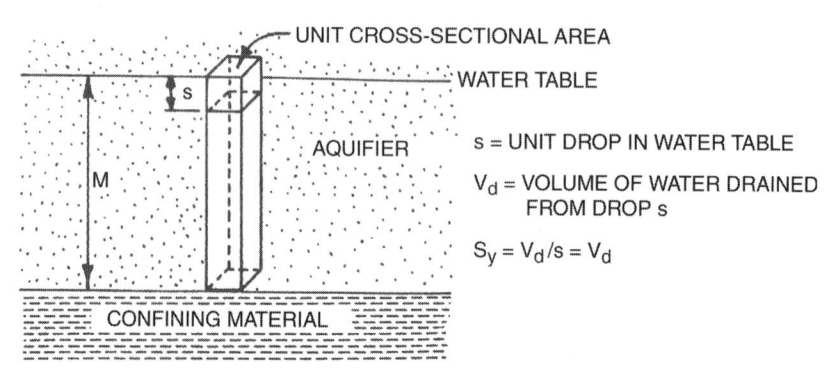

FIGURE 14.11
Definition of specific yield in an unconfined aquifer.

TABLE 14.6

Representative Specific Yields

Rocks	Specific Yield (%)
Clay	1–10
Sand	10–30
Gravel	15–30
Sand and gravel	15–25
Sandstone	5–15
Shale	0.5–5
Limestone	0.5–5

Specific Storage

In confined aquifers, water is released by expansion of the water and compression of the aquifer voids under a decrease in piezometric pressure. The specific storage is a measure of the volume of water released from a unit volume of aquifer per unit decrease in piezometric head. Figure 14.12 illustrates the definition of specific storage. This is a result of: (1) an increase in effective stress between the aquifer particles or blocks when the water level is lowered and (2) an expansion in the volume of water due to the decrease in pressure (see Walton, 1970, or Freeze and Cherry, 1979).

Storativity is equal to the specific storage divided by the saturated aquifer thickness. Specific storage is unitless, whereas storativity has the units of reciprocal length, such as 1/ft or 1/m. The amount of water released by a confined aquifer for a unit decrease in head is usually much less than that for an unconfined aquifer. This is because the amount of water released by aquifer compression and water expansion is small compared with that which drains from a soil under the influence of gravity. Values for the specific storage typically range from 10^{-5} to 10^{-3}, which is a fraction of the specific yield. *In situ* determinations of specific storage and specific yield are obtained from pumping tests (see below). Specific storage can also be estimated from type-curve analysis of slug-test data (see Slug Tests).

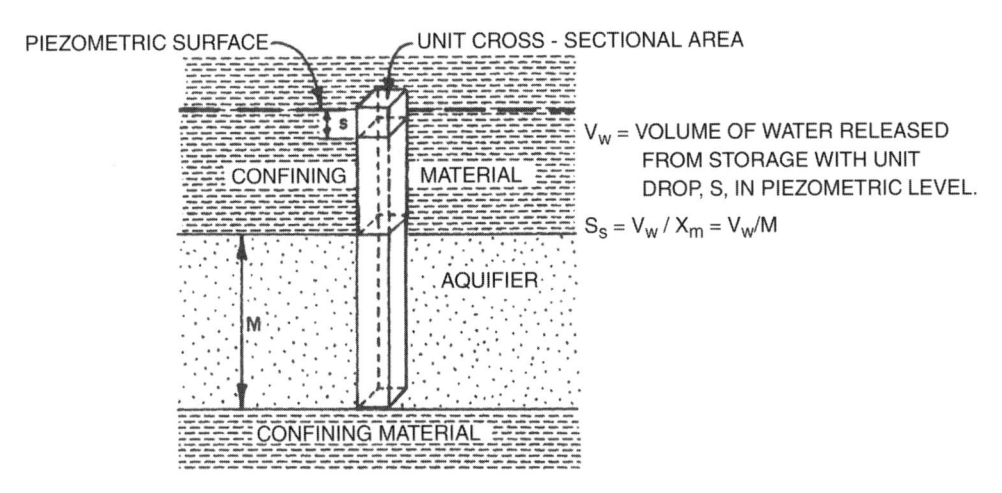

FIGURE 14.12
Definition of specific storage in a confined aquifer.

Transmissivity

A term related to hydraulic conductivity that is commonly used in aquifer evaluations is transmissivity. Transmissivity, like hydraulic conductivity, is useful for calculating ground-water volumetric rates and recharge capacities of wells. This term represents the average water transmission characteristics over the entire aquifer thickness. The term transmissivity was originally introduced by Theis (1935) as the product of hydraulic conductivity and saturated thickness for horizontally confined aquifer flow. Transmissivity, T, is defined as the average of all horizontal hydraulic conductivities at various depths multiplied by the vertical saturated thickness of the aquifer:

$$T = \bar{K}M \qquad (14.25)$$

where T is the transmissivity in units of length squared per unit time, M the saturated thickness in units of length, and $\bar{K}$ the average horizontal hydraulic conductivity.

$\bar{K}$ represents an average horizontal hydraulic conductivity that may vary with horizontal orientation. Relating this definition to Darcy's law, transmissivity becomes the rate of flow under a unit horizontal hydraulic gradient through the entire thickness of an aquifer of unit width perpendicular to the direction of ground-water flow. Accordingly, units for transmissivity are in length squared per unit time, such as ft^2/day or m^2/sec. For convenience, transmissivity is sometimes expressed in units of gallons per day per foot.

Transmissivity can be determined from slug-test data (see Slug Tests section) or aquifer pumping tests, or it can be estimated from laboratory data. In a uniform confined aquifer of constant thickness, transmissivity remains constant provided the piezometric level does not decline below the top of the aquifer. In an unconfined aquifer, transmissivity varies with the saturated thickness, M. As the water table rises or falls, the transmissivity correspondingly increases or decreases. Furthermore, the term loses its value for three-dimensional situations. Because transmissivity carries with it the assumption of horizontal flow, if vertical seepage convergence exists, for example, near a partially penetrating well or in an unconfined aquifer with significant drawdown, the term loses its value in describing flow behavior in such zones.

Determination of Transmissivity by Pumping Tests

A pumping test is a controlled field experiment designed to determine the approximate hydraulic properties of water-bearing geologic material. Pumping tests are the most commonly used methods for determining aquifer transmissivity, hydraulic conductivity, specific storage, and specific yield. Pumping tests are also useful in examining boundary effects from recharge zones (lakes and rivers) and low-conductivity materials (rock walls, clay). But, because a pumping test may influence a relatively large volume of the aquifer some distance from the pumping well (as opposed to a retrieved sample specimen), the resulting aquifer properties that are measured represent an "average" over that volume. The portion of the aquifer influenced by a pumping test can be varied by modifying the pumping rate and the length of the test. In fractured media, pumping tests can be used to evaluate the collective effect of many fractures.

A variety of different testing and analytical methods are available for defining aquifer hydraulic parameters. The procedure for selection of an appropriate method is primarily based on determining which method is most compatible with the hydrogeology of the site at which the test is to be conducted. Secondarily, the method is selected on the basis of the testing conditions specified by the test method, such as the method of stressing or causing water-level changes in the aquifer and the requirements of the test method for observations of water-level response in the aquifer. ASTM D 4043 (ASTM, 2004u) describes in detail the factors that must be considered in selecting a method appropriate for site-specific conditions, and describes the conditions for which each available test method is applicable.

A pumping test is typically conducted using a central pumping well and one or more nearby observation wells (see ASTM D 4050 [ASTM, 2004v]). The decline in the ground-water levels in observation wells is monitored as the central well is pumped at either a constant or variable rate. The location and configuration of the pumping well and observation wells are dependent on aquifer properties. Common types of idealized aquifers include unconfined or water-table aquifers, confined aquifers, leaky aquifers, or multilayered aquifers. The local hydrogeology will determine what portion of the aquifer is screened by the pumping well and how the results should be interpreted.

Boundary conditions, such as an impervious zone or a recharging river, can influence the results of the testing and must be incorporated into the design of the pumping test. If useful estimates of transmissivity and storativity are desired, the pumping well should be located far enough away from boundaries to permit recognition of drawdown trends before boundary conditions influence the drawdown measurements. When more than one boundary is suspected to be present, it is desirable to locate the observation wells so that the effects of encountering the first boundary on well drawdowns are stabilized prior to encountering the second boundary condition. In the case of a recharge boundary, such as a lake or river, if the intent is to induce recharge from the river, then the pumping well should be located as close to the boundary as possible.

To assist in designing a pumping test, preliminary transmissivity (T) and storativity (S) values can be estimated using reasonable ranges of aquifer properties or laboratory test data. Estimates of these values will assist in selecting a pumping rate, observation well locations, and the pumping well location. Because the pumping well's cone of influence expands with time after the beginning of pumping, the distance to observation wells relative to the duration of the test should be considered. The cone of influence depends on the aquifer storativity and transmissivity and the discharge rate of the pumping well. Assuming typical T and S values, estimates can be made before the pumping test to determine the magnitude of drawdown at a particular observation well. The ability to measure drawdown and any outside influences, such as barometric pressure or tidal effects, will determine the desired amount of drawdown at a particular observation well.

Observation Well Positioning

Most pumping test analyses assume that the flow within the aquifer system is predominantly horizontal (Walton, 1970). This condition is not necessarily satisfied in close proximity to a partially penetrating pumping well or in layered aquifers. The degree of vertical ground-water movement also depends on the length of the screen; if the well does not fully penetrate the aquifer, significant head losses due to vertical velocities or turbulence associated with converging flow may exist. In the case of partially penetrating wells, some investigators recommend a minimum distance between the nearest observation well and the pumping well of $1.5M\,(K_H/K_V)^{1/2}$, where M is the thickness of the aquifer in units of length, and K_H and K_V are the horizontal and vertical hydraulic conductivities of aquifer, respectively. Other investigators suggest a minimum distance of at least two times the aquifer thickness between the pumping well and the nearest observation well to avoid the zone of vertically converging flow for a partially penetrating well. For a fully penetrating well, the convergence in the vertical direction is generally not a concern, and closer distances can be utilized. If the above conditions cannot be met, observation wells should be placed in both the upper and lower 15% of the aquifer, and the drawdowns in the observation wells averaged for computational purposes. If the aquifer is stratified, the observation wells can be screened over the same interval as the pumping well, assuming that flow is predominantly horizontal throughout the pumping depth. Analytical solutions for analysis of pumping test data from multi-layered aquifers are available (Neuman and Witherspoon, 1969; Boulton and Streltsova, 1975).

The type of aquifer being tested also influences the distance from the pumping well to the observation wells. In a confined aquifer, the decline in the piezometric surface spreads rapidly because water is coming from storage, which is a function of the aquifer compressibility. However, in an unconfined aquifer, the water moving to the pumping well is principally draining from the pore spaces of the aquifer material at the water-table position and is coming only in part from aquifer compressibility. This results in a slower expansion of the zone influenced by the pumping well. In general, observation wells in a confined aquifer can be spaced further from the pumping well than wells in an unconfined aquifer.

A common well array for a pumping test consists of four observation wells in two sets (two wells per set), with one set placed orthogonally from the other. If anisotropic conditions are suspected, a single row of observation wells cannot be used to estimate the directional dependence of transmissivity. At least three observation wells, none of which are on the same radial, are required to separate out anisotropic behavior (Neuman et al., 1984).

The distance of the observation wells from the pumping well depends on the considerations discussed previously and the anticipated boundary conditions. In an area in which complex boundary conditions are anticipated, additional observation wells may be required for proper interpretation of the test results. The observation wells should be as small in diameter as feasible to minimize the response time. Pneumatic or electronic transducers placed in a sealed borehole are very useful because these instruments respond to pressure changes within the aquifer with virtually no flow equalization time.

Pretest Data Collection

Water levels in the pumping well and observation wells should be measured several days in advance of the pumping test to determine the static water levels within the wells. Because the water levels in the observation wells will be compared with the respective prepumping levels, it is important that the initial water levels and any variability be incorporated into the analysis of the test. If the test is in an area affected by tides, it may be necessary to monitor the daily tidal cycle at each of the observation wells. Each observation well may react differently to the tidal effect. In a shallow water-table aquifer, frozen conditions may influence the results of the pumping test by causing conditions similar to a leaky aquifer or delayed storage effects. Precipitation recharge can cause increasing water levels in observation wells and thus cause an overestimate of transmissivity. It is usually desirable not to conduct a pumping test in a water-table aquifer during a period of heavy rainfall. Evapotranspiration may cause declining water levels in a water-table aquifer for extended-period pumping tests. Water-level measurements made several days prior to and after the pumping test are useful in evaluating evapotranspiration and recharge effects in a shallow water-table aquifer.

If the aquifer is confined, barometric changes may affect the water levels in the wells. An increase in barometric pressure may cause an increase in ground-water levels and, conversely, a decrease in barometric pressure may cause a decrease in water levels. A means for correcting this effect is to monitor the barometric changes at the site of pumping test and the ground-water level changes in observation wells at the pumping test site several days prior to the test. The barometric pressure changes and the associated water-level changes can be correlated by plotting both values on a graph (Walton, 1970). These graphs can be used to correct water-level data to account for barometric pressure effects. Alternately, the ratio of the water-level change to a respective change in barometric pressure can be used directly to correct water-level data. This ratio is called the barometric efficiency of the well. For relatively short tests, i.e., tests of less than 12 h, barometric pressure variations are not of concern unless a significant weather front passes through the area during the test.

In addition to changes in atmospheric conditions, tectonic events or cyclic loading can cause increases or decreases in pore pressures and water levels within an aquifer. The effects of passing railroad trains, automobiles, or earthquakes on water levels in observation wells have been described (Walton, 1970). Stepped-rate pumping tests are useful prior to running the actual pumping test. The step test allows drawdown data to be collected at the pumping well and a pumping rate selected based on drawdown response. A step test is performed by selecting three or more pumping rates that bracket the proposed pumping rate for the test. Each rate is pumped for 1 h or until the water levels in the pumping well have stabilized. The pumping is then stopped and the water levels in the wells allowed to recover for a period of 1 h. Alternately, the pumping rates are increased sequentially after the water level or drawdown has stabilized without the recovery period.

Water-level Measurement During Pumping Tests

During the early part of the pumping test, frequent water-level measurements are required and sufficient manpower or automatic recording devices should be available to obtain the measurements. Defining the time-drawdown curves with accuracy in the early part of the test, when water levels in the wells are changing most rapidly, may require one person to measure water levels at each observation well. Data acquisition systems consisting of pressure transducers in each observation well, attached to a data logger, can accomplish the same goal and provide an increased number of water-level measurements throughout the test (see Chapter 13). Even if an electronic data logger is used, occasional manual measurements are suggested in the event that the transducers or electronics fail.

The length of the test depends on the aquifer setting and boundary conditions. In practice, economic factors and time constraints may also play a role in determining the length of pumping tests. In most cases, the time-versus-drawdown plots for each well should be prepared during the pumping test to assist in determining if the test is successfully achieving its design goals. Graphs constructed on semilog paper with drawdown being plotted on the arithmetic scale and time plotted on the log scale are typically used. Once a straight-line portion of the suggested graph is obtained, estimates of the transmissivity and storativity can be made. This generally occurs within the early part of the test in a confined aquifer. However, in a water-table aquifer, the straight-line condition may require longer pumping periods.

Measuring Well Discharge

For aquifer pumping tests, the discharge rate should be sufficient to cause significant stress to the aquifer without violating test assumptions, meaning that the rate will generally be in the range of tens to hundreds of gallons per minute (perhaps more in high hydraulic conductivity formations). A well discharge measurement method must be selected based on the desired discharge rate (or rates, in the case of step tests), the desired pumping method, the required accuracy and frequency of measurement, the type of pump discharge, and the method of water conveyance to its ultimate discharge point. The most widely used methods for measuring discharge are described in detail in ASTM D 5737 (ASTM, 2004w). These methods include open channel flow methods (weirs [see ASTM D 5242; ASTM, 2004x] and flumes [see ASTM D 5390; ASTM, 2004y]) and closed conduit methods (propeller flow meters, magnetic flow meters, venturi meters, acoustic meters, bucket and stopwatch, orifice bucket, and circular orifice weir [see ASTM D 5716; ASTM, 2004z]).

When setting up the pumping test, the discharge from the pumping well should be located as far as possible from the pumping and observation wells. This will minimize the potential for the discharge water to infiltrate back into the aquifer and affect test results. Preferably, the discharge should be positioned in a stream, river, pond, or some other surface water body.

Analysis of Pumping-Test Data

Analysis of field pumping-test data will generally allow a determination of transmissivity and storativity for the aquifer. Depending on the geologic setting and the design of the pumping test, leakage from a confining layer can also be determined. The reliability of the analysis depends on the design of the pumping test and its principal features such as duration of the test, number of observation wells, and method of analysis. The analysis of the pumping test can utilize steady-state data, near-steady-state data, or transient data.

Steady-state equations can be used when well drawdowns have reached equilibrium and no longer change with time as a result of recharge from a river or lake. In order to apply a steady-state method of analysis, the pumping test must be run long enough so that additional drawdown nearly ceases. Thus, the observed drawdowns in nearby observation wells and in the pumping well itself are in equilibrium with the pumping rate. Analytical and graphical solutions can be applied to the well response data (Walton, 1970; Kruseman and de Ridder, 1991). If the recharge is occurring diffusely from a zone either above or below the formation being pumped through a leaky confining bed, the appropriate steady-state solution must be used (Hantush, 1956). Methods for determining hydraulic properties of confined aquifers under these conditions can be found in ASTM D 6029 (ASTM, 2004aa) and ASTM 6028 (ASTM, 2004bb).

The solution of the ground-water flow equation for transient conditions in a nonleaky confined aquifer was first developed by Theis (1935). Since then, additional solutions have been developed for varying assumptions and boundary conditions. Cooper and Jacob (1946) utilized the series expansion of the well function to develop an alternative method for solving the transient problem. The transmissivity and storage coefficients can also be calculated from the solution of the well function using a type curve or graphical method (Theis, 1935; Jacob, 1940). This method is described in detail in ASTM D 4106 (ASTM, 2004cc); a modified version of this method is described in ASTM D 4105 (ASTM, 2004dd), and a method using data collected during the recovery phase of a pumping test is described in ASTM D 5269 (ASTM, 2004ee).

The solution of the transient flow equation for a uniform, confined aquifer where flow is horizontal and radial toward the well is often used to analyze pumping test data, particularly data from the early part of the test. This solution assumes a homogeneous isotropic and infinite aquifer with a uniform thickness throughout. Furthermore, it assumes that the transmissivity and storage coefficients do not vary with time or distance from the pumping well, and that the aquifer is in equilibrium (except for the pumping). No leakage of water is assumed to occur from either above or below the aquifer. However, the rate of water-level decline decreases as the pumping period extends. The source of water to the well is only that which is released from elastic storage as the aquifer compresses and the water expands due to the decreasing piezometric head. The analysis assumes that the water is released instantaneously as the head declines. The analysis also assumes that the well is of infinitesimal diameter relative to the area influenced by the pumping. Walton (1970) describes this solution as the well function for nonleaking, isotropic, artesian aquifers with fully penetrating wells in constant discharge conditions.

In a water-table aquifer that is underlain by an impervious geologic unit, with a fully penetrating well, water drains from the aquifer as the water table is lowered and enters the well. However, drainage of water from the soil pores may not occur instantaneously with drawdown. Water is assumed to be released instantaneously as the pressure is lowered due to expansion of the water and compression of the aquifer (Boulton, 1963).

It is possible to interpret the results of a pumping test in an unconfined aquifer that is anisotropic in the horizontal plane (Neuman, 1975; Neuman et al., 1984), to determine the transmissivity, storage coefficient, specific yield, and horizontal-to-vertical hydraulic conductivity ratio. In this case, the anisotropic medium is transformed to an equivalent isotropic medium for analysis. The test requires pumping a control well that is open to all or part of an unconfined aquifer at a constant rate for a specified period, and observing the drawdown in at least three observation wells that either partially or fully penetrate the aquifer. This test method may also be used to analyze test data from an injection test with the appropriate change in sign. Two methods of analysis, a type-curve method and a semilogarithmic method, are possible. The analysis is described in detail in ASTM D 5920 (ASTM, 2004ff).

When the pumping test is conducted next to a recharge or impervious boundary, the time-drawdown data may not fit the predicted theoretical curves. In the case of a recharge boundary, the time-drawdown curve will begin to flatten out to a point at which drawdown no longer occurs with time. This is due to recharge to the aquifer for the particular piezometric conditions created by the pumping. If an impervious boundary is encountered by the drawdown, the rate of drawdown will hasten relative to a condition without a boundary. In either case, a more detailed analysis, possibly using image well theory, will be required. The use of image wells is discussed in a number of references, including Walton (1970) and ASTM D 5270 (ASTM, 2004gg).

If the medium is fractured, interpretation of pumping-test data becomes very difficult unless the medium is highly fractured. The degree of drawdown in any particular fracture intercepted by the observation wells is a function of the degree of interconnection with the fractures at the pumping well. The degree of interconnection is a function of the permeability of the individual fractures, their areal extent, and the frequency of fracture intersections. It would not be uncommon in poorly fractured materials, where individual fractures are not well connected, for drawdown in one observation well some distance from the pumping well to exceed that observed at an observation well closer to the pumping well. Methods have been developed for analyzing pumping tests in idealized fractured media cases (Barenblatt et al., 1960; Ramey, 1970; Gringarten and Ramey, 1974; Gringarten et al., 1974; Boulton and Streltsova, 1977; Gringarten, 1982; Karasaki, 1986).

Estimation of Transmissivity

Because transmissivity is the product of hydraulic conductivity of an aquifer and the thickness of that aquifer, individual borehole sample test data can be used to estimate the transmissivity of an aquifer. For instance, if an aquifer is 50 ft thick and samples are taken at 10-ft intervals from a boring through the aquifer, these samples can be tested in the laboratory for hydraulic conductivity. The results can then be multiplied by the sample interval and the products summed for the aquifer. The sum provides an estimate of the entire aquifer transmissivity. Alternately, if the samples represent specific layers within the aquifer, the hydraulic conductivity for each layer can be multiplied by the respective layer thickness and these products summed for all the layers in the aquifer.

Grain-size analysis, with the use of Hazen's approximation, is an economic means of estimating the transmissivity in uniform sand aquifers. Many grain-size tests can be performed quickly, thereby providing a more complete representation of the entire aquifer.

Compressibility

The storage coefficient for a confined aquifer is a function of the elasticity of the water and the aquifer material. The vertical deformation (i.e., subsidence or heave) of an aquifer can be calculated using the formation compressibility and changes in the intergranular pressure due to events such as ground-water extraction or fluid injection into an aquifer. Therefore, as part of a ground-water-monitoring plan, vertical aquifer deformation may be critical to observing the impact of, for example, ground-water pumping on land surface activities. Excessive land surface deformations can lead to building cracking, pipeline breaks, landslides, or unacceptable grade changes.

Documentation of subsidence due to pumping of an aquifer is abundant in the literature (Davis and Rollo, 1969; Poland and Davis, 1969; Riley, 1969; Yerkes and Castle, 1969; Gambolati and Freeze, 1973; Gambolati et al., 1974; Bull, 1975). The amount of subsidence can be calculated if the changes in aquifer water levels are known or can be estimated. The aquifer compressibility relates the change in vertical thickness of the aquifer to a corresponding change in intergranular pressure. A complete review of the theory and methods of calculation can be found in the geotechnical literature (Terzaghi and Peck, 1948; Taylor, 1958; Lambe and Whitman, 1969). An introduction to estimating land subsidence from ground-water level changes is given in Bouwer (1978). Land subsidence can be measured by using settlement platforms, compaction recorders, and electrical strain gages (Lofgren, 1961; U.S. Department of the Interior, 1985). A typical settlement device is shown in Figure 14.13. Such settlement monitoring devices can provide a continuous and long-term record of the rate and magnitude of vertical deformation of an aquifer and any overlying formations.

Vertical deformation can be estimated using undisturbed samples of the formation. The samples are tested for vertical height change under different vertical stress levels in a one-dimensional consolidation test. This test is typically limited to application in silts and clays, because samples of these soils preserve their volume for trimming. However, if representative samples can be obtained, the method is applicable in any geologic material.

The consolidation test has a long history of use in the geotechnical field and a standard method for conducting the test has been specified by ASTM (ASTM D 2435 [ASTM, 2004hh]). This test involves obtaining a representative sample of the soil and placing it in a rigid cylinder as shown in Figure 14.14. A piston is placed on the surface of the confined sample, which is loaded with weights. The amount of deformation for each load is used to calculate the elasticity of the soil. The rate at which deformation occurs is a function of the compressibility of the mineral matrix and the hydraulic conductivity of the

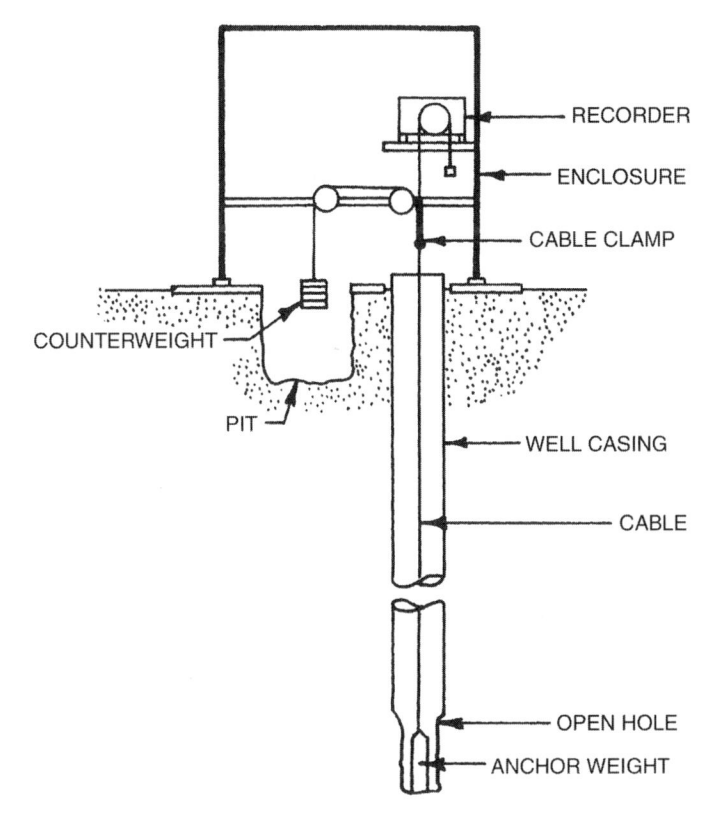

FIGURE 14.13
Ground subsidence monitoring device.

material. The test is typically conducted on a saturated sample and water must be allowed to discharge from the sample in order for deformation to occur. The simplified theory of consolidation assumes that the hydraulic conductivity does not vary with the amount of consolidation, and the water and solid soil elements are incompressible. For limited

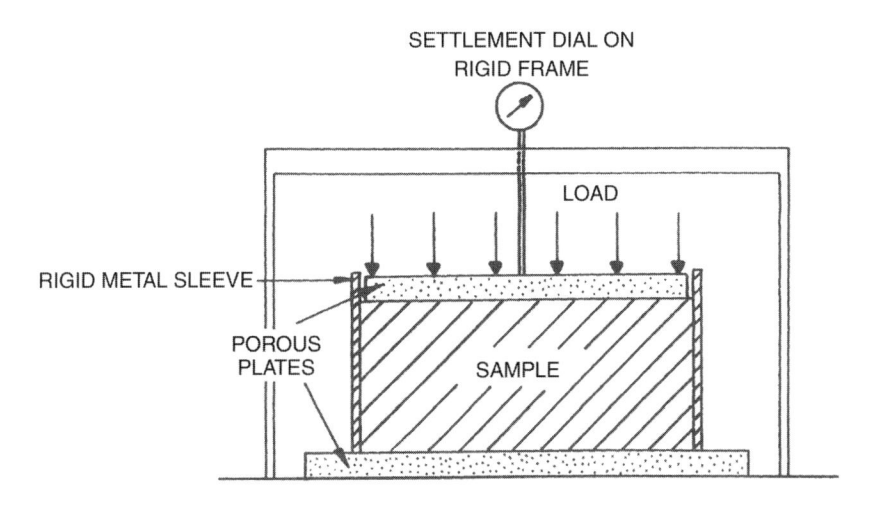

FIGURE 14.14
Schematic of a one-dimensional consolidometer.

TABLE 14.7

Bulk Moduli of Elasticity for Various Soils and Rocks

Material	E (kg/cm^2)
Dense gravel and sand	1,000–10,000
Dense sands	500–2,000
Loose sands	100–200
Dense clays and silts	100–1,000
Medium clays and silts	50–100
Loose clays	10–50
Peat	1–5
Granite	2–6 × 10^{-5}
Microgranite	3–8 × 10^{-5}
Syenite	6–8 × 10^{-5}
Diorite	7–10 × 10^{-5}
Dolerite	8–11 × 10^{-5}
Gabbro	7–11 × 10^{-5}
Basalt	6–10 × 10^{-5}
Sandstone	0.5–8 × 10^{-5}
Shale	1–3.5 × 10^{-5}
Mudstone	2–5 × 10^{-5}
Limestone	1–8 × 10^{-5}
Dolomite	4–8.4 × 10^{-5}
Coal	1–2 × 10^{-5}

deformation, these assumptions are usually reasonable. The modulus of elasticity of the formation is calculated by the following equation from the test results:

$$\frac{dV}{V_o} = \frac{d\sigma}{E} \tag{14.26}$$

where dV and $d\sigma$ are the changes in the volume of the sample for a given change in pressure, V_o the initial volume of the sample, and E the bulk modulus of elasticity in units of force or mass per unit area.

Typical values of E for different materials are given in Table 14.7. This equation assumes that the load due to the consolidation process is entirely transformed to the soil or rock skeleton rather than the water. Furthermore, because the behavior of soil and rock is not totally elastic, the test should be performed over the stress range of interest. A detailed discussion of the test procedure and forms of data presentation is provided by Lambe (1951).

The modulus of elasticity of the formation is also related to the specific storage of the aquifer. Walton (1970) shows that

$$S = \gamma_w n \beta m \left(1 + \frac{\alpha}{n\beta}\right) \tag{14.27}$$

where S is the specific storage, unitless decimal fraction; γ_w the bulk density of water in units of weight or mass per unit volume; n the average porosity of the aquifer, expressed as a unitless decimal fraction; β the reciprocal of the bulk modulus elasticity of the water in units of area per mass or force; α the reciprocal of the bulk modulus of elasticity, E, of the soil or rock comprising the aquifer; and m the aquifer thickness in units of length.

α can be obtained for some materials using a standard laboratory one-dimensional consolidation test as described above. The reciprocal of the bulk modulus for water, β, equals 4.7×10^{-7} cm^2/kg. Thus, from the soil compressibility, the specific storage of the confined aquifer can be estimated.

References

ASTM, Standard Practice for Thin-Walled Tube Sampling of Soils, ASTM Standard D 1587, ASTM International, West Conshohocken, PA, 2004a, 3 pp.
ASTM, Standard Test Method for Density of Soil in Place by the Drive-Cylinder Method, ASTM Standard D 2937, ASTM International, West Conshohocken, PA, 2004b, 4 pp.
ASTM, Standard Test Method for Density and Unit Weight of Soil in Place by the Rubber Balloon Method, ASTM Standard D 2167, ASTM International, West Conshohocken, PA, 2004c, 6 pp.
ASTM, Standard Test Method for Density and Unit Weight of Soil in Place by the Sand Cone Method, ASTM Standard D 1556, ASTM International, West Conshohocken, PA, 2004d, 6 pp.
ASTM, Standard Practice for Diamond Core Drilling for Site Investigation, ASTM Standard D 2113, ASTM International, West Conshohocken, PA, 2004e, 4 pp.
ASTM, Standard Test Method for Density of Soil and Soil Aggregate in Place by Nuclear Methods, ASTM Standard D 2922, ASTM International, West Conshohocken, PA, 2004f, 5 pp.
ASTM, Standard Test Method for Laboratory Determination of Water (Moisture) Content of Soil and Rock, ASTM Standard D 2216, ASTM International, West Conshohocken, PA, 2004g, 4 pp.
ASTM, Standard Test Method for Water Content of Soil and Rock in Place by Nuclear Methods (Shallow Depth), ASTM Standard D 3017, ASTM International, West Conshohocken, PA, 2004h, 5 pp.
ASTM, Standard Test Method for Measurement of Hydraulic Conductivity of Saturated Porous Materials Using a Flexible Wall Permeameter, ASTM Standard D 5084, ASTM International, West Conshohocken, PA, 2004i, 8 pp.
ASTM, Standard Test Method for Measurement of Hydraulic Conductivity of Saturated Porous Material Using a Rigid-Wall, Compaction-Mold Permeameter, ASTM Standard D 5856, ASTM International, West Conshohocken, PA, 2004j, 8 pp.
ASTM, Standard Test Method for Permeability of Granular Soils, ASTM Standard D 2434, ASTM International, West Conshohocken, PA, 2004k, 5 pp.
ASTM, Standard Test Method for Laboratory Compaction Characteristics of Soil Using Standard Effort (12,400 ft-lbs/ft3(600 kN-m/m3)), ASTM Standard D 698, ASTM International, West Conshohocken, PA, 2004l, 8 pp.
ASTM, Standard Test Method for Laboratory Compaction Characteristics of Soil Using Modified Effort (56,000 ft-lbs/ft3(2,700 kN-m/m3)), ASTM Standard D 1557, ASTM International, West Conshohocken, PA, 2004m, 8 pp.
ASTM, Standard Test Method (Field Procedure) for Instantaneous Change in Head (Slug) Tests for Determining Hydraulic Properties of Aquifers, ASTM Standard D 4044, ASTM International, West Conshohocken, PA, 2004n, 3 pp.
ASTM, Standard Test Method (Analytical Procedure) for Determining Transmissivity of Nonleaky Confined Aquifers by Overdamped Well Response to Instantaneous Change in Head (Slug) Tests, ASTM Standard D 4104, ASTM International, West Conshohocken, PA, 2004o, 4 pp.
ASTM, Standard Test Method (Analytical Procedure) for Determining Hydraulic Conductivity of an Unconfined Aquifer by Overdamped Well Response to Instantaneous Change in Head (Slug) Tests, ASTM Standard D 5912, ASTM International, West Conshohocken, PA, 2004p, 4 pp.
ASTM, Standard Test Method (Analytical Procedure) for Determining Transmissivity of Nonleaky Confined Aquifers by Underdamped Well Response to Instantaneous Change in Head (Slug) Tests, ASTM Standard D 5785, ASTM International, West Conshohocken, PA, 2004q, 5 pp.
ASTM, Standard Test Method (Analytical Procedure) for Determining Transmissivity of Nonleaky Confined Aquifers by Critically Damped Well Response to Instantaneous Change in Head (Slug) Tests, ASTM Standard D 5881, ASTM International, West Conshohocken, PA, 2004r, 8 pp.
ASTM, Standard Test Method for Determining Transmissivity and Storage Coefficient of Low-Permeability Rocks by In Situ Measurements Using the Constant-Head Injection Test, ASTM Standard D 4630, ASTM International, West Conshohocken, PA, 2004s, 6 pp.

ASTM, Standard Test Method for Determining Transmissivity and Storage Coefficient of Low-Permeability Rocks by In Situ Measurements Using the Pressure Pulse Technique, ASTM Standard D 4631, ASTM International, West Conshohocken, PA, 2004t, 7 pp.

ASTM, Standard Guide for Selection of Aquifer Test Method in Determining Hydraulic Properties by Well Techniques, ASTM Standard D 4043, ASTM International, West Conshohocken, PA, 2004u, 6 pp.

ASTM, Standard Test Method (Field Procedure) for Withdrawal and Injection Well Tests for Determining Hydraulic Properties of Aquifer Systems, ASTM Standard D 4050, ASTM International, West Conshohocken, PA, 2004v, 4 pp.

ASTM, Standard Guide for Methods for Measuring Well Discharge, ASTM Standard D 5737, ASTM International, West Conshohocken, PA, 2004w, 4 pp.

ASTM, Standard Test Method for Open-Channel Flow Measurement of Water Indirectly at Culverts, ASTM Standard D 5242, ASTM International, West Conshohocken, PA, 2004x, 4 pp.

ASTM, Standard Test Method for Open-Channel Flow Measurement with Palmer-Bowlus Flumes, ASTM Standard D 5390, ASTM International, West Conshohocken, PA, 2004y, 4 pp.

ASTM, Standard Test Method to Measure the Rate of Well Discharge by Circular Orifice Weir, ASTM Standard D 5716, ASTM International, West Conshohocken, PA, 2004z, 4 pp.

ASTM, Standard Test Method (Analytical Procedure) for Determining Hydraulic Properties of a Confined Aquifer and a Leaky Confining Bed With Negligible Storage by the Hantush-Jacob Method, ASTM Standard D 6029, ASTM International, West Conshohocken, PA, 2004aa, 10 pp.

ASTM, Standard Test Method (Analytical Procedure) for Determining Hydraulic Properties of a Confined Aquifer Taking Into Consideration Storage of Water in Leaky Confining Beds by the Modified Hantush Method, ASTM Standard D 6028, ASTM International, West Conshohocken, PA, 2004bb, 9 pp.

ASTM, Standard Test Method (Analytical Procedure) for Determining Transmissivity and Storage Coefficient of Nonleaky Confined Aquifers by the Theis Nonequilibrium Method, ASTM Standard D 4106, ASTM International, West Conshohocken, PA, 2004cc, 5 pp.

ASTM, Standard Test Method (Analytical Procedure) for Determining Transmissivity and Storage Coefficient of Nonleaky Confined Aquifers by the Modified Theis Nonequilibrium Method, ASTM Standard D 4105, ASTM International, West Conshohocken, PA, 2004dd, 5 pp.

ASTM, Standard Test Method (Analytical Procedure) for Determining Transmissivity of Nonleaky Confined Aquifers by the Theis Recovery Method, ASTM Standard D 5269, ASTM International, West Conshohocken, PA, 2004ee, 5 pp.

ASTM, Standard Test Method (Analytical Procedure) for Tests of Anisotropic Unconfined Aquifers by the Neuman Method, ASTM Standard D 5920, ASTM International, West Conshohocken, PA, 2004ff, 8 pp.

ASTM, Standard Test Method (Analytical Procedure) for Determining Transmissivity and Storage Coefficient of Bounded, Nonleaky, Confined Aquifers, ASTM Standard D 5270, ASTM International, West Conshohocken, PA, 2004gg, 7 pp.

ASTM, Standard Test Method for One-Dimensional Consolidation Properties of Soils, ASTM Standard D 2435, ASTM International, West Conshohocken, PA, 2004hh, 10 pp.

Barenblatt, G.I., I.P. Zheltov, and I.N. Kochina, Basic concepts in the theory of seepage of homogeneous liquids in fissured rocks (strata), *Journal of Applied Mathematics and Mechanics*, 24, 1286–1301, 1960.

Bear, J., *Dynamics of Fluids in Porous Media*, American Elsevier, New York, NY, 1972.

Bentley, H.W. and G.R. Walter, Two-Well Recirculating Tracer Test at H-2, Waste Isolation Pilot Plant (WIPP), Southeastern New Mexico, Hydro-Geochem Inc., Tuscon, AZ, 1983.

Bishop, A.W. and D.J. Henkel, *The Measurement of Soil Properties in the Triaxial Test*, St. Martins Press, New York, NY, 1962.

Black, D.K., and K.L. Lee, Saturating laboratory samples by back pressure, *Journal of the Soil Mechanics and Foundation Division*, 99, 1973.

Boulton, N.S., Analysis of Data from Non-Equilibrium Pumping Tests Allowing for Delayed Yield from Storage, *Proceedings*, Institute of Civil Engineers, 26, 1963, 469–482.

Boulton, N.S. and T.D. Streltsova, New equations for determining the formation constants of an aquifer from pumping test data, *Water Resources*, 11(1), 1975.

Boulton, N.S. and T.D. Streltsova, Unsteady flow to a pumped well in a fissured water-bearing formation, *Journal of Hydrology*, 35, 257–269, 1977.

Bouwer, H., *Groundwater Hydrology*, McGraw-Hill, New York, NY, 1978.

Bouwer, H. and R.C. Rice, A slug test for determining hydraulic conductivity of unconfined aquifers with completely or partially penetrating wells, *Water Resources Research*, 12(3), 423–433, 1976.

Bredehoeft, J.D. and I.S. Papadopulos, A method for determining the hydraulic properties of tight formations, *Water Resources Research*, 16(1), 233–238, 1980.

Bull, W.B., Land Subsidence Due to Ground Water Withdrawal in the Los Banos-Kettleman City Area, California. Part 2. Subsidence and Compaction of Deposits, U.S. Geological Survey Professional Paper 437-F, 1975.

Burns, W.A. Jr., New single-well test for determining vertical permeability, Transactions of AIME, 246, 743–752, 1969.

Butler, J. Jr., *The Design, Performance, and Analysis of Slug Tests, Lewis Publishers, Boca Raton, FL, 1998.*

Cohen, R.M. and R.R. Rabold, Numerical Evaluation of Monitoring Well Design, Proceedings of the First National Outdoor Action Conference on Aquifer Restoration, Ground Water Monitoring and Geophysical Methods, National Water Well Association, Dublin, OH, 1987, pp. 267–284.

Cooley, R.L. and C.M. Case, Effect of a water table aquitard on drawdown in an underlying pumped aquifer, *Water Resources Research*, 9(2), 434–447, 1973.

Cooper, H.H., Jr. and C.E. Jacob, A generalized graphical method for evaluating formation constants and summarizing well field history, *Transactions of the American Geophysical Union*, 27, 526–534, 1946.

Cooper, H.H., Jr., J.D. Bredehoeft, and I. S. Papadopulos, Response of a finite diameter well to an instantaneous charge of water, *Water Resources Research*, 3(1), 263–269, 1967.

Darcy, H., Les Fontains Publiques de la Ville de Dijon, Dalmont, Paris, 1856.

Davis, G.H. and J.R. Rollo, Land Subsidence Related to Decline of Artesian Head at Baton Rouge, Lower Mississippi Valley, U.S.A., Proceedings of the Tokyo Symposium on Land Subsidence, IASH-UNESCO, 1969.

Delhomme, J.P. 1974, La Cartographie d'une Grandeur Physique à Partir de Donnees de Differentes Qualites, *Proceedings of the International Association of Hydrogeologists*, 10, No. 1, 1974, pp. 185–194.

Delhomme, J.P., Kriging in the hydrosciences; *Advances in Water Resources*, 1(5), 251–266, 1978.

de Marsily, G., *Quantitative Hydrogeology*, Academic Press, New York, NY, 1986.

Desaulniers, D.E., R.S. Kaufmann, J.A. Cherry, and H.W. Bentley, $^{37}Cl - ^{35}Cl$ variations in a diffusion-controlled ground water system, *Geochimica et Cosmochimica Acta*, 50, 1986.

Drost, W., D. Klotz, A. Koch, H. Moser, F. Neumaier, and W. Rauert, Point dilution methods of investigating ground water flow by means of radioisotopes, *Water Resources Research*, 4, 1968.

Faust, C.R. and J.W. Mercer, Evaluation of slug tests in wells containing a finite-thickness skin, *Water Resources Research*, 20(4), 504–506, 1984.

Forster, C.B. and J.E. Gale, A Field Assessment of the Use of Borehole Pressure Transients to Measure the Permeability of Fractured Rock Masses, Lawrence Berkeley Laboratories, Publication LBL 11829, 1981.

Freeze, R.A. and J.A. Cherry, *Groundwater*, Prentice-Hall, Englewood Cliffs, NJ, 1979.

Fried, J.J. and P. Ungemach, Determination *in-situ* du Coefficient de Dispersion Longitudinale d'un Milieu Poreaux Naturel, Cr. Academic Science Paris, 272, Serie A, 1971.

Gambolati, G. and R.A. Freeze, Mathematical simulation of the subsidence of Venice. 1. Theory, *Water Resources Research*, 9, 721–733, 1973.

Gambolati, G., P. Gatto, and R.A. Freeze, Mathematical simulation of the subsidence of Venice. Part 2. Results, *Water Resources Research*, 10, 563–577, 1974.

Gardner, W.H., Methods of Soils Analysis, Agronomy Monograph No. 9, American Society of gronomy, Madison, WI, 1965, pp. 82–127.

Gelhar, L.W., A. Mantoglou, C. Welty, and K.R. Rehfeldt, *A Review of Field-Scale Physical Solute Transport Processes in Saturated and Unsaturated Porous Media*, EPRI Publication EA-4190, Electric Power Research Institute, Palo Alto, CA, 1985.

Gibbs, H.J. and W.G. Holtz, Research on Determining the Relative Density of Sands by Spoon Penetration Testing, Proceedings of the 4th International Conference on Soil Mechanics and Foundation Engineering, Vol. I, 1975.

Gillham, R.W. and J.A. Cherry, Contaminant Migration in Saturated Unconsolidated Geologic Deposits: Recent Trends in Hydrogeology, Special Paper 189, Geological Society of America, Boulder, CO, 1982.

Goodman, R.E., *Introduction to Rock Mechanics*, John Wiley and Sons, New York, NY, 1980.

Gringarten, A.C., Flow-Test Evaluation of Fractured Reservoirs, Geologic Society of America, Special Paper 189, 1982.

Gringarten, A.C., and H.J. Ramey Jr., Unsteady-state pressure distributions created by a well with a single horizontal fracture, partial penetration, or restricted entry, *Society of Petroleum Engineers Journal*, 14(4), 413–426, 1974.

Gringarten, A.C., H.J. Ramey Jr., and R. Raghavan, Unsteady-state pressure distributions created by a well with a single infinite-conductivity vertical fracture, *Society of Petroleum Engineers Journal*, 14(4), 1974.

Hantush, M.S. Analysis of data from pumping tests in leaky aquifers, *Transactions of the American Geophysical Union*, 37(6), 1956.

Hantush, M.S., Modification of the theory of leaky aquifers, *Journal of Geophysical Research*, 65(11), 3713–3725, 1960.

Hantush, M.S., Hydraulics of Wells, in *Advances in Hydroscience*, Academic Press, Inc., New York, NY, 1964.

Harr, M.E., *Groundwater and Seepage*, McGraw-Hill, New York, NY, 1962.

Hazen, A., Discussion of "Dams on Sand Foundations" by A.C. Koenig, *Transactions of the American Society of Civil Engineers*, 73, 1911.

Herzog, B.L. and W.J. Morse, Hydraulic Conductivity at a Hazardous Waste Disposal Site: Comparison of Laboratory and Field-Determined Values, Waste Management and Research, Vol. 4, Academic Press, London, 1986,

Hillel, D., *Applications of Soil Physics*, Academic Press, New York, NY, 1980.

Hirasaki, G.J., Pulse tests and other early transient pressure analyses for *in situ* estimation of vertical permeability, *Transactions of the American Institute of Mining Engineers*, 257, 75–90, 1974.

Hvorslev, M.J., Subsurface Exploration and Sampling of Soils for Civil Engineering Purposes, U.S. Army Corps of Engineers, Waterways Experiment Station, Vicksburg, MS, 1949a.

Hvorslev, M.J., Time Lag in the Observation of Ground Water Levels and Pressures, U.S. Army Corps of Engineers, Waterways Experiment Station, Vicksburg, MS, 1949b.

Jacob, C.E., On the flow of water in an elastic artesian aquifer, *Transactions of the American Geophysical Union*, 21(2), 574–586, 1940.

Jacob, C.E., and S.W. Lohman, Nonsteady flow to a well of constant drawdown in an extensive aquifer, *Transactions of the American Geophysical Union*, 33(4), 1952.

Jacquin, C., Etude des ecoulements et des equilibres de fluides dans les sables arqileux, *Reveu de l'Institut Francais du Petrole*, 20(4), 1965a.

Jacquin, C., Interactions entre l'argile et les fluides ecoulement á travers les argiles compactes, *Reveu de l'Institut Francais du Petrole*, 20(10), 1965b.

Karasaki, K., Well Test Analysis in Fractured Media, Ph.D. thesis, Lawrence Berkeley Laboratory, University of California, Berkeley, CA, 1986.

Karasaki, K., J.C.S. Long, and P.A. Witherspoon, Analytical models of slug tests, *Water Resources Research*, 24(1), 115–126, 1988.

Kerfoot, W.B., Comparison of 2-D and 3-D Ground Water Flowmeter Probes in Fully-Penetrating Monitoring Wells, Proceedings of the Second National Symposium on Aquifer Restoration and Ground Water Monitoring, National Water Well Association, Dublin, OH, 1982, pp. 264–268.

Kipp, K.L., Jr., Type curve analysis of inertial effects in the response of a well to a slug test, *Water Resources Research*, 21(9), 1397–1408, 1985.

Knutson, G., Tracers for Ground Water Investigations, Proceedings of the International Symposium on Ground Water Problems, Stockholm, Sweden, Pergamon Press, Oxford, England, 1966.

Kruseman, G.P. and N.A. de Ridder, *Analysis and Evaluation of Pumping Test Data*, Publication 47, International Institute for Land Reclamation and Improvement, Wageningen, The Netherlands, 1991, 377 pp.

Lambe, T.W., *Soil Testing for Engineers*, John Wiley & Sons, Inc., New York, NY, 1951.

Lambe, T.W. and R.V. Whitman, *Soil Mechanics*, John Wiley & Sons, Inc., New York, NY, 1969.

Lofgren, B.E., Measurement of Compaction of Aquifer Systems in Areas of Land Subsidence, U.S. Geological Survey Professional Paper 424-B, 1961.

Louis, C., Introduction a l'hydraulique des roches, *Bull. Bur. Rech. Geol. Min., Ser. 2*, Sect. III(4), 1974.

Milligan, V., Field Measurement of Permeability in Soil and Rock, Proceedings of the Conference on *In-Situ* Measurement of Soil Properties, American Society of Civil Engineers, New York, NY, 1975.

Mitchell, J.K., *Fundamentals of Soil Behavior*, John Wiley & Sons, Inc., New York, NY, 1976.

Moench, A.F. and P.A. Hsieh, Slug Testing in Wells with a Finite-Thickness Skin, Proceedings of the 10th Workshop on Geothermal Reservoir Engineering, Stanford, CA, 1985.

Moltz, F.J., O. Guven, J.G. Melville, and J.F. Keely, Performance and Analysis of Aquifer Tests with Implications for Contaminant Transport Modeling, EPA-600/2-86/062, U.S. Environmental Protection Agency, Robert S. Kerr Environmental Research Laboratory, Ada, OK, 1986.

Neuman, S.P., Analysis of pumping test data from anisotropic aquifers considering delayed gravity response, *Water Resources Research*, 11(2), 329–342, 1975.

Neuman, S.P., Theoretical and Practical Considerations of Flow in Fractured Rocks, The 1987 Distinguished Seminar Series on Ground Water Science, National Water Well Association, Dublin, OH, 1987.

Neuman, S.P., and P.A. Witherspoon, Theory of flow in a confined two-layer aquifer system, *Water Resources Research*, 5(4), 803–816, 1969.

Neuman, S.P., G.R. Walter, H.W. Bentley, J.J. Ward, and D.D. Gonzalez, Determination of horizontal aquifer anisotropy with three wells, *Ground Water*, 22(1), 66–72, 1984.

Neuzil, C.E., On conducting the modified "slug" test in tight formations, *Water Resources Research*, 18(2), 439–441, 1982.

Nguyen, V. and G.F. Pinder, Direct Calculation of Aquifer Parameters in Slug Test Analysis, Ground Water Hydraulics, Water Resource Monograph 9, American Geophysical Union, Washington, DC, 1984.

Papadopulos, I.S., J.D. Bredehoeft, and H.H. Cooper, On the analysis of "slug test" data, *Water Resources Research*, 9(4), 1087–1089, 1973.

Parker, J.C. and K.A. Albrecht, Sample volume effects on solute transport predictions, *Water Resources Research*, 23(12), 2293–2301, 1987.

Pickens, J.F. and G.E. Grisak, Scale-dependent dispersion in a stratified aquifer, *Water Resources Research*, 17(4), 1191–1211, 1981.

Poland, J.F. and G.H. Davis, *Land Subsidence Due to the Withdrawal of Fluids: Reviews in Engineering Geology II*, Geological Society of America, Boulder, CO, 1969.

Polubarinova-Kochina, N., *Theory of Ground Water Movement*, R. de Wiest, transl., Princeton University Press, Princeton, NJ, 1962.

Powers, J.P., *Construction Dewatering*, John Wiley and Sons, Inc., New York, NY, 1981.

Prats, M., A method for determining the net vertical permeability near a well from *in-situ* measurements, *Transactions of the American Institute of Mining Engineers*, 249, 637–643, 1970.

Ramey, H.J., Jr., Approximate solutions for unsteady liquid flow in composite reservoirs, *Journal of Canadian Petroleum Technology*, 22(2), 1970.

Ramey, H.J., Jr., and R.G. Agarwal, Annulus unloading rates as influenced by wellbore storage and skin effect, *Transactions of the American Institute of Mining Engineers*, 253, 1972.

Riley, F.S., Analysis of Borehole Extensometer Data From Central California, Proceedings of the Tokyo Symposium on Land Subsidence, IASH-UNESCO, 1969.

Sageev, A., Slug test analysis, *Water Resources Research*, 22(8), 1323–1333, 1986.

Sudicky, E.A. and J.A. Cherry, Field observations of tracer dispersion under natural flow conditions in an unconfined sandy aquifer, *Water Pollution Research Canada*, 14, 1–17, 1979.

Taylor, D.W., *Fundamentals of Soil Mechanics*, John Wiley & Sons, Inc., New York, NY, 1958.

Terzaghi, K. and R.B. Peck, *Soil Mechanics in Engineering Practice*, John Wiley & Sons, Inc., New York, NY, 1948.

Theis, C.V., The relation between the lowering of the piezometric surface and the rate and duration of discharge of a well using ground water storage, *American Geophysical Union Transactions*, 16, 519–524, 1935.

Todd, D.K., *Ground Water Hydrology*, John Wiley & Sons, Inc., New York, NY, 1959.

U.S. Department of the Interior, Earth Manual, U.S. Bureau of Reclamation, U.S. Government Printing Office, Washington, D.C., 1985.

van der Kamp, G., Determining aquifer transmissivity by means of well response tests: the underdamped case, *Water Resources Research*, 12(1), 71–77, 1976.

Walton, W.C., *Ground Water Resource Evaluation*, McGraw-Hill, New York, NY, 1970.

Wang, J.S.Y., T.N. Narasimhan, C.F. Tsang, and P.A. Witherspoon, Transient Flow in Tight Fractures, Proceedings of the International Well Testing Symposium, Publication LBL 7027, Lawrence Berkeley Laboratory, Berkeley, CA, 1977.

Winterkorn, H.F. and H.Y. Fang, *Foundation Engineering Handbook*, Van Nostrand Reinhold, New York, NY, 1975.

Yerkes, R.F. and R.O. Castle, Surface Deformation Associated with Soil and Gas-Field Operations in the United States, Proceedings of the Tokyo Symposium on Land Subsidence, IASH-UNESCO, 1969.

15

Ground-Water Sampling

David M. Nielsen and Gillian L. Nielsen

CONTENTS

Introduction	960
The Science Behind Ground-Water Sampling	961
Objectives of Ground-Water Sampling	961
Regulatory Compliance Monitoring	961
Non-Regulatory Monitoring	962
Collecting "Representative" Samples	962
Ground-Water Sampling and Data Quality	963
Meeting DQOs: A Superfund Project as an Example	964
Factors Affecting the Representative Nature of Ground-Water Samples	965
Factors Related to Formation and Well Hydraulics	965
Understanding Ground-Water Flow	965
Hydraulics and Water Chemistry Between Sampling Events	967
Factors Related to Sampling Point Placement, Design, Installation, and Maintenance	969
Three-Dimensional Placement of the Sampling Point	969
Sampling Point Installation Options	970
Poor Well Design and Construction	973
Inadequate or Improper Well Development	976
Well Maintenance	976
Geochemical Changes in Ground-Water Samples	976
Pressure Changes	977
Temperature Changes	978
Entrainment of Artifactual Particulate Matter During Purging and Sampling	979
Agitation and Aeration of Water Samples During Collection	984
The Ground-Water Sampling and Analysis Plan (SAP): A Road Map to Field Sampling Procedures	987
Objectives of the SAP	987
Preparation of the SAP	987
Selection of Field Protocols to be Incorporated into the SAP	988
Well Headspace Screening	988
Water-Level and Product-Thickness Measurement	989
Field Quality Assurance and Quality Control	991
Purging and Sampling Device Selection and Operation	996
Sampling Point Purging	1035
Field Measurement of Water-Quality Indicator Parameters and Turbidity	1056

    Sample Pretreatment Options 1062
    Field Equipment Cleaning Procedures 1079
    Documenting a Sampling Event 1080
  Conducting a Ground-Water Sampling Event 1086
    Field Preparation for a Ground-Water Sampling Event 1087
     Site Orientation and Sampling Event Preparation 1087
    Conducting the Sampling Event 1087
     Sampling Point Inspection 1087
     Well Headspace Screening 1088
     Water-Level Measurement 1089
     Well Purging and Field Parameter Measurement 1089
     Sample Collection Procedures 1090
      Order of Container Filling 1090
      Sample Collection Protocols 1092
      Protocols for Collecting Field QC Samples 1094
      Ground-Water Sample Pretreatment Procedures 1097
      Physical and Chemical Preservation Procedures for Samples 1100
      Preparation of Sample Containers for Shipment 1100
      Cleanup of the Work Area 1102
      Delivery or Shipment of Samples to the Laboratory 1102
  References 1102

Introduction

Ground-water sampling is a key component of any effective ground-water monitoring program and most environmental site-characterization programs. Developing an effective ground-water sampling program requires an understanding of a variety of factors that can affect the integrity of samples collected from ground-water monitoring wells or other ground-water sampling points and the quality of analytical results or other data generated from those samples. These factors include formation and well hydraulics; sampling point placement, design, installation, and maintenance; purging practices; purging and sampling device selection and operation; and sample collection, pretreatment, and handling procedures. This chapter provides the reader with a detailed discussion of these factors and the processes that occur in a monitoring well between sampling events and during sample collection activities. Incorporating this knowledge should allow readers to make sound technical decisions regarding the choice of sampling procedures to meet the site-specific objectives of a ground-water monitoring or environmental site-characterization program.

  Difficulty in accessing ground water without disturbing ground-water flow patterns, chemistry, microbiology, and the physical and chemical makeup of formation materials has made accurate characterization of *in situ* ground-water conditions a very challenging task. Ground-water monitoring and sampling methodologies have evolved as we have learned more about the significance of subsurface processes (particularly contaminant fate and transport processes), the effects that sampling efforts have on these processes, and the effects that sampling procedures have on sample integrity. Since the first edition of this book was released in 1991, there have been many important developments in the practices, procedures, and equipment options available to ground-water sampling teams. This chapter provides detailed discussions of the science behind ground-water sampling, descriptions of the most important recent developments in ground-water sampling, and explanations of how new sampling protocols can be effectively implemented in the field. It also introduces many ASTM standards that have been developed over the past decade, which provide technical information related to newly developed sampling protocols.

Despite the recent advances in ground-water sampling practices, many regulatory agencies and environmental professionals have become entrenched in older, less-efficient technologies and are reluctant to change. The ultimate objective of this chapter is to allow readers to develop an understanding of why continued use of outdated methods will ultimately call into question the representative nature of samples and the quality of data generated by many sampling programs. This chapter explains in detail why and how updating current sampling protocols will significantly improve the quality of data that can be generated by any ground-water sampling program.

The Science Behind Ground-Water Sampling

Objectives of Ground-Water Sampling

The overall objective of most ground-water sampling programs is to collect samples that are "representative," that is, samples that accurately reflect *in situ* ground-water conditions in the formation of interest at the site under investigation. Ground-water sampling programs are implemented at a variety of locations where they are commonly, although not exclusively designed to characterize or monitor ground-water contamination using traditional monitoring wells. At other sites, single-event or "snapshot" evaluations of ground-water chemistry are increasingly being made using direct-push sampling tools, which permit rapid characterization of ground-water conditions without using traditional monitoring wells. This approach to ground-water characterization is especially viable for properties undergoing real-estate transfers, but it can be used at any site to rapidly collect samples representative of formation conditions.

Regulatory Compliance Monitoring

At many sites, ground-water monitoring programs are required to be implemented under one or more U.S. EPA (or state-equivalent) regulatory programs such as the Resource Conservation and Recovery Act (RCRA), the Comprehensive Environmental Response, Compensation, and Liability Act (CERCLA or "Superfund"), the Clean Water Act (CWA), the Toxic Substances Control Act (TSCA), and the Safe Drinking Water Act (SDWA). In most cases, the objective of such monitoring programs is to provide information on ground-water chemistry, which will help to determine whether a regulated facility is in compliance. These different regulatory programs, discussed in more detail in Chapter 1, regulate a wide variety of sites at which ground-water contamination has a potential to occur, including industrial and municipal solid-waste landfills, chemical and petroleum production facilities, industrial manufacturing facilities, U.S. Department of Defense sites, U.S. Department of Energy sites, uncontrolled hazardous waste sites, and underground storage tank sites. Under structured regulatory programs, either the U.S. EPA or the equivalent state regulatory agency requires that ground-water sampling programs be established as part of a ground-water monitoring program to satisfy a wide range of secondary objectives including determining: (1) whether the operation of a facility has had an effect on ground-water quality (i.e., has resulted in ground-water contamination); (2) the physical and chemical nature of the contamination; (3) the three-dimensional extent of the contamination; (4) the rate and direction of movement of that contamination with respect to other properties or receptors (i.e., water supplies) in the area; (5) what the most appropriate methods of remediating the ground-water contamination may be; (6) the effectiveness of remediation methods implemented at a site; or (7) changes in long-term ground-water quality during or following site remediation or closure.

Non-Regulatory Monitoring

In addition to contaminant characterization and monitoring, ambient monitoring programs are conducted by a variety of government agencies including the U.S. Geological Survey and various state agencies (regulatory and non-regulatory) and tribal governments, as well as regional, county, and municipal government agencies. These programs are not concerned with contaminant fate and transport issues at a particular site, but are focused on large-scale hydrogeologic and geochemical characterization of aquifers to determine their suitability for specific uses or sensitivity to development. In these situations, ground-water sampling programs are implemented to determine the ambient quality of ground water available for public drinking water, agricultural, or industrial uses. For example, the tribal government in the Owens Valley area of California closely monitors water levels and water quality in regional wells to ensure that over-pumping does not occur within the valley. Over-pumping would have a significant and detrimental impact on both the water quality and water supply necessary for local municipal and agricultural use.

Collecting "Representative" Samples

The primary objective of most ground-water sampling programs is to collect samples that are representative of ground water in its *in situ* condition. A representative ground-water sample must accurately reflect the physical and chemical properties of the ground water in that portion of the formation open to the well to be sampled. While this objective is common to most programs, the working definition of "representative" is not. As a consequence, a clearly defined statement of objectives, which includes a site-specific definition of a representative sample, is a critical component of any site-specific sampling and analysis plan (SAP). The term representative may mean different things to different investigators, due mainly to differing project objectives. For example, those interested in characterizing ground-water quality from a water-supply well for the purpose of water-supply evaluation or risk characterization for receptors are most interested in volume-averaged concentrations of chemical constituents in the target water-bearing zone rather than "worst-case" conditions in the heart of a contaminant plume. Samples collected after pumping a significant volume of water from the well may be considered representative for these purposes, whereas samples collected using methods designed to focus on a specific, discrete portion of the formation (i.e., the center of a contaminant plume) may not be representative for that particular application. If contaminants were found in the latter type of sample, the concentrations present would not accurately represent the concentrations to which users of this water-supply well would be exposed; they could be much higher because of the lack of dilution.

To be considered representative for the purpose of contaminant plume characterization, samples should be collected in a manner that maintains the depth-discrete nature of the zone targeted for monitoring, as defined by the site-characterization program. Extensive pumping of contaminant monitoring wells, particularly those in low hydraulic conductivity materials (silts and clays, weakly fractured or solution-channeled rock), can result in volume-averaged samples, which greatly complicate interpretation of data concerning the concentration and spatial distribution of contaminants. If the open portion of the well is long relative to the thickness of the contaminant plume, a negatively biased (underestimated) concentration value would be produced as a result of dilution caused during the sampling process by mixing of water from within the plume with water from uncontaminated portions of the formation. Similarly, in fractured rock, because fracture porosity is typically very small (compared with intergranular porosity), extensive pumping during the sampling process may draw in water from a very large volume of the formation. In an example cited by McCarthy and Shevenell (1998), in a rock with a fracture porosity of 10^{-4}, pumping only 10 l (2.6 gal) of water from a well could conceivably draw in water from a volume of 100 m^3 (3531 ft^3) of rock. Mixing of water from different locations in the formation could greatly dilute the level of contaminants in the sample and the concentrations may not be in equilibrium with the adjoining matrix. Thus, if the well was installed with the intent of monitoring a discrete zone, the sampling process would have the effect of undoing all of the work that went into properly locating the well and the open or screened zone.

Many investigators have acknowledged the difficulty of obtaining samples that are truly representative of subsurface conditions (Grisak et al., 1978; Gibb et al., 1981; Schuller et al., 1981; Claassen, 1982; U.S. EPA, 1982, 1991b; Gillham et al., 1983; Barcelona et al., 1987). There has always been almost universal agreement that to obtain representative samples and prevent sample alteration, subsurface disturbance and sample handling must be kept to a minimum (Puls and Powell, 1992a). However, the procedures that are currently used to access the subsurface to collect ground-water samples make some level of disturbance of subsurface conditions unavoidable. The goal in sample collection should be to use methods that result in the least disturbance or change in the chemical and physical properties of the water collected as the sample and that, therefore, produce the most representative sample possible.

A significant level of effort has been invested over the past few decades in establishing uniform laboratory methods for analyzing samples, while comparatively little effort has been invested in the development of uniform field sampling methodologies (Puls and Powell, 1992a). This deficiency has resulted in the production of large volumes of highly variable and sometimes meaningless data, and a significant waste of time and money. Although ASTM has produced a number of standards related to sampling environmental media, such standards often take many years to be adopted by regulatory agencies and practitioners. Some groups, such as State regulatory agencies and professional trade associations, have considered developing performance-based certification programs for field sampling personnel, but they have recognized what an enormous undertaking developing and administering such programs would be, and none are yet available. Consequently, the discrepancy in standardization of procedures and quality assurance/quality control (QA/QC) measures between the laboratory and the field remains.

Ground-Water Sampling and Data Quality

Because the field and analytical data produced by ground-water sampling programs are often relied upon as the basis for making potentially far-reaching and expensive decisions, it is imperative that these data be of the highest quality possible. Some of the important decisions affected by ground-water sample data analysis from a ground-water monitoring program include:

- Whether a site is in compliance with permit requirements
- Whether a site can be closed and monitoring discontinued
- Whether a site requires active remediation

- Whether a site is a candidate for risk-based corrective action or natural attenuation

To ensure that these decisions are defensible, they must be supported by valid data of suitable quality that are both accurate and reproducible. The chemical analytical data must be representative of formation water quality within an acceptable level of certainty (accuracy), which is defined by data quality objectives (DQOs) (Chapter 2). An important element of obtaining a representative sample is that the sample collection procedures should be reproducible to ensure precision in samples collected for analysis during different sampling events. This aspect of sampling has gained increased emphasis with the requirement to follow Quality Assurance Project Plans (QAPPs) as part of ground-water sampling protocols (Gibs and Imbrigiotta, 1990). The goal of the DQO process and the QAPP is to document protocol and procedures for a given site that will ensure that the data generated are of sufficient quality and quantity to facilitate making sound technical decisions. Furthermore, it is important that the data from well to well and from sampling event to sampling event are reproducible with a minimum amount of bias and with no unexplainable outliers, to permit trend analysis. The key to successfully attaining these goals is to use sampling methodologies that generate data that are both accurate and precise.

Inconsistent and biased ground-water data collection is a serious issue for the interpretation of data for trend analysis and evaluation of remedial action performance (Barcelona et al., 1994). One of the main reasons for this is that the traditional methods used by many investigators for the collection of ground-water samples are not well suited to producing high-quality samples. The accuracy and precision of sample analytical data are only as good as the quality of the samples submitted for analysis, which is strongly controlled by sampling method.

Meeting DQOs: A Superfund Project as an Example

Ground-water sampling project objectives, in the context of a Superfund site investigation, include four main goals: contaminant detection, contaminant assessment, resource evaluation, and corrective action selection and evaluation. Specifically, ground-water samples are first used to determine whether hazardous substances are present in ground water underlying a site under investigation, and at what concentrations. Identifying contamination serves as an important factor in determining locations that require further investigation. If contamination is found to exist, ground-water samples then serve to assess the nature and three-dimensional extent of contamination. This is necessary for ascertaining whether the contamination can be attributed to the site under investigation and for determining the number and types of affected or potentially affected receptors. This information helps to establish the site priority for future remedial response. Whether or not contamination is present, ground-water investigations can establish subsurface characteristics such as aquifer boundaries, ground-water potability, aquifer interconnections, and ground-water flow rate and direction. These parameters are necessary to evaluate the relative risk associated with documented or potential releases (Thornton et al., 1997). If the contamination present at the site poses sufficient risk to receptors to require remediation, data from ground-water samples may then be used to help determine the optimum remedial approach and, after the approach is implemented, to evaluate remedial performance and effectiveness.

Owing to the "screening" nature of the early phase of a Superfund site assessment, data collection focuses on sampling "worst case" concentrations and locations (e.g., searching for highest concentrations within contaminant plumes) rather than determining average or overall site conditions. Because listing of sites on the National Priorities List (NPL) is a process that is subject to legal challenges, the data used to evaluate sites must be scientifically and legally defensible and of known and documented quality. This requires that DQOs be established that the field sample collection, handling, storage, and laboratory analysis procedures be documented as following previously established procedures and that a QC procedure be implemented (Thornton et al., 1997).

Subsequent phases of sampling typically focus on spatial and temporal contaminant plume definition, and finally, on selecting an appropriate remedial strategy and evaluating its effectiveness. This requires samples of a more discrete nature, which represent specific and usually relatively thin target zones within the formation. As some investigators (e.g., Ronen et al., 1987; Cherry, 1992; Puls and McCarthy, 1995) have recognized subsurface heterogeneities often result in distributions of many chemicals of concern (particularly those associated with dense non-aqueous phase liquids [DNAPLs]) that may vary by several orders of magnitude over only a few feet, both vertically and laterally. Ideally, sampling points are constructed so that they are open to and focused on zones that are known through site-characterization activities to be contaminated or that are determined to be the most likely pathways for ground-water and contaminant movement. Examples of preferential flow pathways include sand and gravel lenses within a fine-grained matrix or fractures or solution channels within otherwise competent rock.

A common objective of all phases of Superfund site investigations is the collection of high-quality reproducible data to support decision making at several different stages of the investigation. High-quality data collection implies data of sufficient accuracy, precision, and completeness (i.e., the ratio of valid analytical results to the minimum sample number called for by the sampling program design) to meet program objectives. Accuracy depends on the correct choice of sampling tools and procedures to minimize sample and subsurface disturbance for collection to analysis. Precision depends on the repeatability of sampling and analytical protocols (Puls and Barcelona, 1996).

Factors Affecting the Representative Nature of Ground-Water Samples

A number of factors influence the ability of samplers to collect representative ground-water samples. Table 15.1 provides a summary of the factors that must be evaluated for each site undergoing ground-water sampling to determine how each might affect the representative nature of samples to be collected and the sampling strategy and methods to be used.

Factors Related to Formation and Well Hydraulics

Developing an understanding of site-specific formation hydraulics is critical to the proper design and installation of sampling points used for site characterization and wells used for long-term monitoring and remediation applications, to the ability of investigators to collect the most accurate samples of ground water for chemical analysis, and to the accurate interpretation of geochemical data. Formation hydraulics directly affect a variety of activities, from three-dimensional well placement relative to known or suspected source areas of contamination, to incorporating wells that may not have been designed specifically for the purpose of monitoring contamination (e.g., residential water-supply wells or agricultural irrigation wells) into a monitoring network, to monitoring well intake design (screen slot size selection, screen length, and screen placement).

Understanding Ground-Water Flow

When developing a site-specific ground-water sampling program, it is critical to have an accurate, three-dimensional understanding of the ground-water hydrology of the site under investigation. Investigators must understand how ground water will move across

TABLE 15.1

Factors Influencing the Representative Nature of Ground-Water Samples

---

Factors related to formation and well hydraulics
    Ground-water flow paths and flow through wells
    Hydraulics within a well between sampling events

Factors related to sampling point placement, design, installation, and maintenance
    Placement of the sampling point with respect to the source(s) of contamination being monitored
    Placement of the sampling point intake in the preferential flow pathway (the zone of highest hydraulic conductivity) within the formation(s) of concern
    Installation method used for sampling point construction (i.e., drilling method or direct-push installation method)
    Suitability of sampling point design with regard to material selection, diameter, depth of the sampling point intake, screen length, and screen slot size for the hydrogeologic and geochemical environment being monitored
    Methods used during sampling point construction, including placement of annular seal materials and care in placement of filter pack materials
    Method, timing, and duration of sampling point development
    Long-term maintenance of the sampling point to ensure that the sampling point can continue to provide suitable samples for analysis with regard to representative chemistry that is not impacted by compromises in well integrity (e.g., cracked surface seals)

Factors related to geochemical changes associated with sample collection
    Pressure changes in the sample
    Temperature changes in the sample
    Entrainment of artifactual particulate matter in the sample
    Agitation or aeration of the water column in the well or the sample during sample collection

---

any given site and must identify what factors influence that movement, such as pumping centers, areas of artificial recharge, water-level fluctuations in an adjacent surface–water system, or tides. Samplers must also develop an understanding of how ground water behaves within the sampling point or monitoring well, both between sampling events and during purging and sample collection. Failure to develop this understanding typically results in selection and use of sampling protocols that are not well suited to the hydrogeology or geochemistry of the site. This, in turn, can result in generation of nonrepresentative data.

As illustrated in Figure 15.1, ground water moves through the subsurface in response to differences in hydraulic head, under laminar flow conditions in most hydrogeologic systems. Provided that the difference in hydraulic head is such that horizontal flow dominates in the formation, flow should continue in the same manner through well screens installed in the formation. Robin and Gillham (1987) demonstrated with tracer solutions that formation water moves through the well screen and that this water does not mix with the stagnant water that remains in the casing between sampling rounds. Visual observations of movement of colloidal particles through well screens installed in granular aquifers made by Kearl et al. (1992) demonstrate that horizontal laminar flow in a formation does, in fact, continue through well screens and that water in the screen does not mix with that in the casing. Powell and Puls (1993), using dual tracer tests, also showed that ground water moves through the well screen with little interaction or mixing with the overlying stagnant water in storage in the well casing. Electrical conductivity data presented by Michalski (1989) clearly showed a fresh water zone in the well screen separate from the stagnant water in the casing, further supporting the observations made by Robin and Gillham. Robin and Gillham (1987) also theorized that the continual flow of water through the screen allows chemical reactions, such as desorption and adsorption, between well construction materials (well screen and filter pack) and constituents in ground water to

Ground-Water Sampling

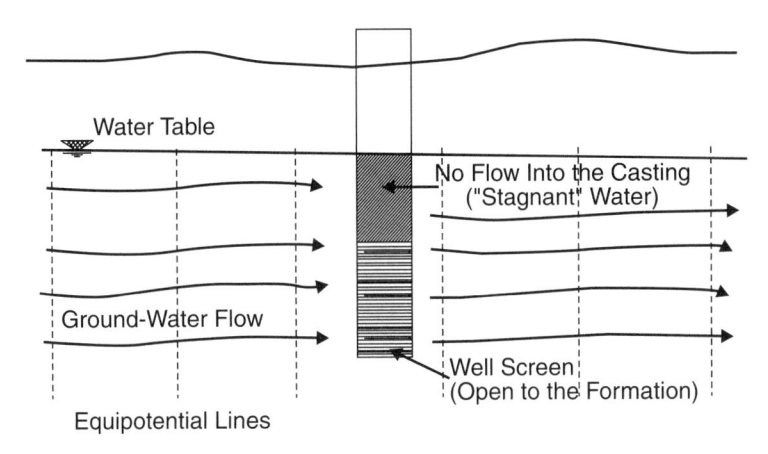

FIGURE 15.1
Movement of ground water in a formation in which horizontal flow dominates. Note that the horizontal flow continues through the screen in a properly designed, constructed, and developed well, but water in the screen does not mix with water in the casing, which is stagnant between sampling events.

approach equilibrium. Work done by Palmer et al. (1987) corroborates this and demonstrates the need to allow time for this equilibration to occur prior to the initial sampling event for a well, which is also suggested by Walker (1983). These studies suggest that for wells in which horizontal flow dominates: (1) water in the screen at any point in time is indeed representative of water in the formation adjacent to the screen; (2) water samples taken directly from the screened interval are representative of ground water in the surrounding formation; and (3) provided that samplers can gain access to the water in the screened interval while minimally disturbing the water column in the well, purging multiple well volumes of water prior to sample collection is unnecessary.

However, where there is a difference in hydraulic head in the formation that results in vertical movement of ground water, and a well is screened across that zone, the well screen effectively acts to short-circuit ground-water flow and directly channels water from the zone with highest hydraulic head to the zone with lowest hydraulic head (Figure 15.2). Several recent studies (i.e., McIlvride and Rector, 1988; Reilly et al., 1989; Church and Granato, 1996; Hutchins and Acree, 2000; Elci et al., 2001, 2003) have documented that, in areas with vertical hydraulic gradients, installation of a monitoring well with a long well screen (i.e., >10 ft long) may set up a localized vertical flow system that renders the well almost useless for sampling because the dilution that occurs in such a well would yield misleading and ambiguous data concerning contaminant concentrations and plume geometry. In some scenarios, installation of a well in this type of setting may result in the spread of contamination to parts of a formation that would not otherwise have become contaminated had the well not been installed. Thus, samplers must be keenly aware of the hydraulic conditions that exist in a well so they can make appropriate decisions on whether a well should be sampled, and, if it is sampled, how to interpret the data generated by each sampling event. More detailed information on this topic is available in Chapter 10 and Chapter 11.

Hydraulics and Water Chemistry Between Sampling Events

It is well recognized that water in storage in a well casing between sampling events (not including water in the well screen) is not representative of formation water quality (Marsh and Lloyd, 1980; Miller, 1982; Gillham et al., 1985; Barcelona and Helfrich, 1986) and thus should not be collected as part of the sample (Barcelona et al., 1985). Water in storage in a well casing between sampling events is physically isolated by the well

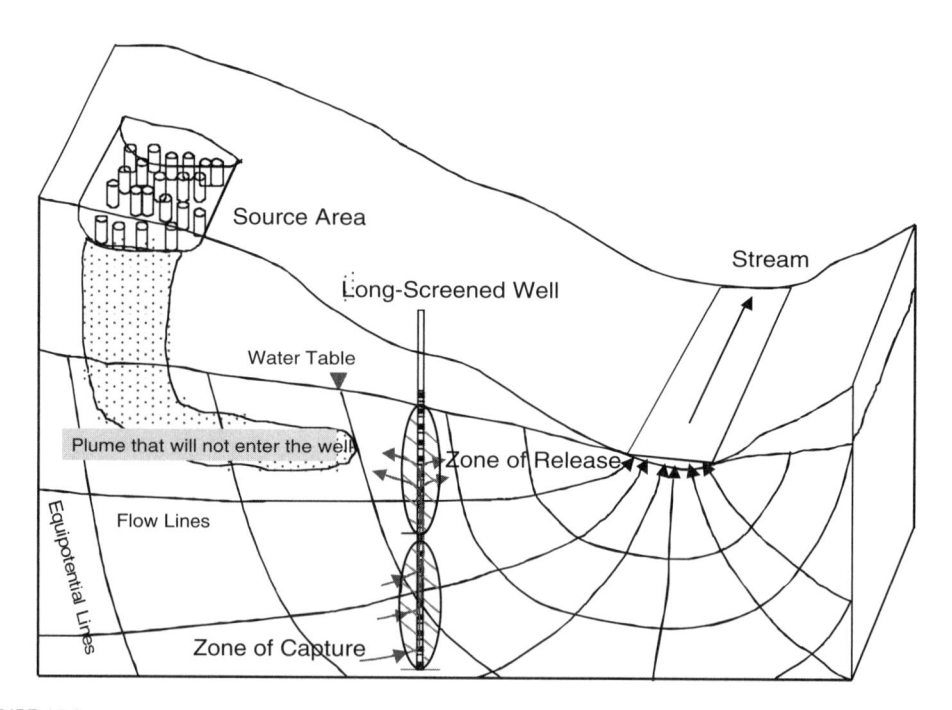

FIGURE 15.2
Movement of ground water in a formation in which there is a vertical difference in hydraulic head. Note that the well screen effectively acts to short-circuit ground-water flow and directly channels water from the zone in the formation with highest hydraulic head to the zone with lowest hydraulic head. (*Source*: McIlvride and Rector, 1988.)

casing from ground water in the formation and cannot interact with "fresh" formation water and is considered to be "stagnant." Changes in water chemistry in the cased portion of the well are caused by a variety of factors as summarized in Table 15.2 and illustrated in Figure 15.3. Purging has historically been implemented as part of routine ground-water sample collection procedures to minimize the bias and error associated with incorporating any portion of this column of stagnant water into samples submitted for analysis.

TABLE 15.2

Factors Affecting the Chemistry of Water in Storage within a Well Casing between Sampling Events

---

The presence of an air–water interface at the top of the water column, which can result the following:
    Creation of a dissolved oxygen (DO) concentration gradient (high to low) with depth
    Increased aerobic microbial activity because of the presence of DO
    Lower pH, due to increased dissolved carbon dioxide (CO_2) at the top of the water column
    Loss of volatile constituents from the water column to the headspace in the well casing

Interactions between the well casing and screen materials and ground-water in storage (leaching from, sorption to, or corrosion of the well construction materials)

Contribution of contaminants from sources above the static water level in the well including:
    Condensation on the inside surface of the well casing
    Water from formations above the zone of interest leaking past joints or cracks in the casing
    Introduction of surface contamination as a result of failure of the surface seal, causing leakage into the well
    Addition of volatile constituents from vadose zone gases (e.g., landfill gases)

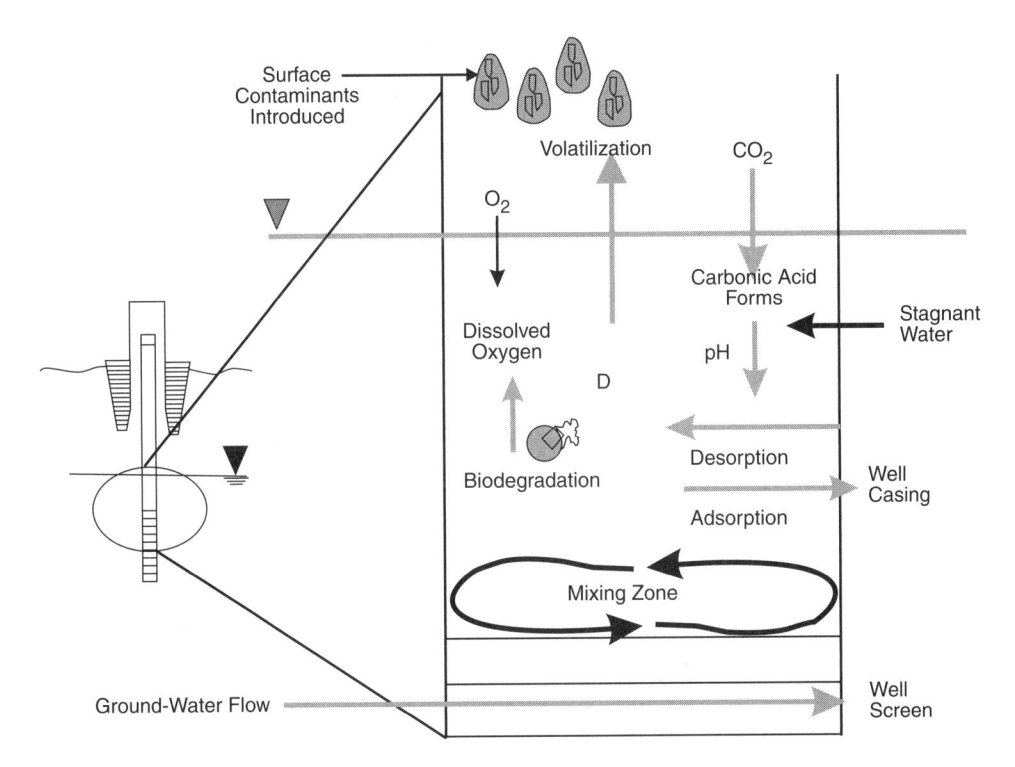

FIGURE 15.3
Changes in water chemistry in the water column within the well casing that occur between sampling events.

However, while water in the well casing may be chemically nonrepresentative of formation water, the research cited in the preceding section has concluded that the water within the screened interval of a short-screened well is representative of formation water quality, provided that the well has been designed, installed, developed, and maintained properly. In such wells, the water in the formation and that within the well screen are able to exchange freely and the well screen is continually "flushed" (Gillham et al., 1985; Robin and Gillham, 1987). That is, water passes continuously through the well screen and surrounding filter pack because they have a higher hydraulic conductivity than the formation, and the well intake is integrated into the ground-water flow field (Figure 15.1). Even in low hydraulic conductivity formations (clays, silty clays, and clayey silts), ground-water flow is often sufficient to maintain a constant (although slow) exchange of water between the formation and the well screen.

Factors Related to Sampling Point Placement, Design, Installation, and Maintenance

Three-Dimensional Placement of the Sampling Point

When collecting ground-water samples from existing wells, it is very important to evaluate the suitability of each individual well with regard to its ability to yield a representative sample from the formation of interest. At the onset of a ground-water sampling program, wells should be evaluated to determine whether they are correctly placed three-dimensionally within the formation to provide the samples required to make decisions at the site. As discussed in Chapter 8, it is critical to understand where the well is placed within the formation and, for contamination investigations, where the well screen is located within the ground-water flow system relative to the known or anticipated source areas of contamination. If the well screen is not in the correct location or if it is too long and not focused on a particular zone of interest, it will be impossible to generate the ground-water data required to satisfy the objectives of the ground-water sampling program, regardless of how carefully the wells are sampled.

If the well is located properly, it is then critical to evaluate the suitability of the design and construction of the well for meeting the objectives of the ground-water sampling program. Design features that must be evaluated include well intake design (i.e., well screen slot size and filter pack grain size), well casing material selection, annular seal construction, surface seal completion, and well security measures. The reader is directed to Chapter 10 for a detailed discussion of how each of these well design and construction features can affect the chemistry of samples collected from the formation under investigation.

Sampling Point Installation Options

Several sampling point options are available for collecting ground-water samples from a zone of interest in subsurface formation materials. The traditional, and still most common approach, is to install 2 or 4 in. diameter ground-water monitoring wells using a conventional drilling method such as hollow-stem auger drilling (refer to Chapter 5 for more detail). As accelerated site-characterization technologies (refer to Chapter 2) are increasingly used to rapidly develop an understanding of ground-water conditions, direct-push technologies are being more widely used to install sampling points. As discussed in Chapter 6, direct-push technologies can facilitate collection of grab samples of ground water from discrete zones to provide a "snapshot" assessment of ground-water chemistry using sealed-screen samplers such as the HydroPunch (Figure 15.4) and exposed-screen samplers such as the Waterloo Profiler. In addition, direct-push technology has been developed to install long-term ground-water monitoring wells, ranging in diameter

FIGURE 15.4
Direct-push ground-water samplers such as the HydroPunch allow collection of samples representative of formation water chemistry without installing a permanent well.

from 0.5 to 2 in. Because the means of accessing formations to collect ground-water samples is no longer limited to traditional ground-water monitoring wells, the term "sampling points" is sometimes used to generically refer to any point of access to the saturated zone. The terms "wells" and "sampling points" will both be used throughout this chapter.

Traditional Drilled Monitoring Wells: An important element of traditional well installation is the drilling method used and the degree of disturbance of the formation that occurs during well installation: As discussed in Chapter 5, a number of drilling methods are available for installation of monitoring wells. Selection of a drilling method is dependent on a variety of site-specific constraints such as total borehole depth; character of geologic materials to be penetrated; depth to ground water; requirements for soil sample collection during drilling; potential impact of introducing drilling fluids (e.g., mud and air) into the formation on soil and ground-water samples; ease of mobilization of the rig to and around the site; and cost. Drilling methods can create sampling artifacts through redistribution of formation materials within the borehole, creation of fine particles by disaggregation and crushing of formation materials (granules, grain coatings, cementing agents, and other solids); introduction of foreign materials (e.g., drilling mud [Fetter, 1983; Brobst and Buszka, 1986] [Figure 15.5], air, and water of a chemistry different from formation water) into the formation; and providing a conduit for atmospheric air to contact ground water, which may result in precipitation of metal oxides and hydroxides.

Direct-Push Sampling Tools and Monitoring Well Installations: Direct-push sampling tools may be advanced into the subsurface with direct-push rigs, manually operated vibratory equipment (e.g., jackhammers), CPT equipment, or conventional drilling rigs. The action of advancing any direct-push sampling tool to the desired sampling depth, whether by pushing, hammering, or vibrating the rod string, displaces formation materials, causing some compaction and some disaggregation of granular materials (and associated breakage of grain coatings and cementing agents, such as iron oxyhydroxides and carbonates). The disturbance caused by installation of direct-push sampling tools is significantly less than that caused by conventional drilling methods. However, some amount of artifactual (i.e., resulting from sampling point installation) particulate matter is created adjacent to the sampling tool. This material enters the sampling tool when it is initially opened,

FIGURE 15.5
Drilling fluid, such as the water-based bentonite fluid used with direct (mud) rotary drilling, can remain in the formation surrounding the well installed in the drilled borehole long after well installation, unless it is properly controlled during drilling and removed during well development. If left in the formation, it can affect ground-water sample chemistry for weeks to months after well installation.

resulting in high initial turbidity levels that represent particulate matter that is not mobile under ambient ground-water flow conditions. Many of these particles are highly surface-reactive and, because of their high surface area per unit mass and volume, have very high sorptive capacities and strong binding capabilities for selected groups of analytes. Some of these materials (the largest fraction) will settle out quickly, but much of it will remain in suspension in the water column and could be collected as part of the sample if the sampling tool is not purged first. Developing the sampling point or using low-flow sampling methods to collect samples (although prolonged pumping may be required) generally results in significantly lower turbidity levels and improved sample quality (U.S. EPA, 1996a).

Most direct-push technologies are also capable of installing small-diameter wells that can be used for either short-term or long-term monitoring of subsurface conditions. Wells installed using direct-push methods have several advantages over traditional drilled monitoring wells, including:

- Minimal disturbance of formation materials during well installation.
- Well intakes are often shorter and more accurately located with respect to installation within a zone of specific interest because direct-push sampling tools can be used to better define and resolve subsurface conditions.
- Well intakes can be installed using "prepacked" or sleeved well screens (Figure 15.6) that allow for superior control over filter pack placement around the well screen and, in some cases, for use of a significantly finer-grained filter

FIGURE 15.6
Prepacked well screens commonly used in direct-push well installations allow for use of fine-grained filter pack materials, which can effectively reduce turbidity in samples, even those collected in wells installed in predominantly fine-grained formation materials (i.e., silty clayey sand).

pack material than is possible for conventional wells. This, in turn, results in samples with noticeably lower turbidity than samples collected from conventional wells installed in the same hydrogeologic setting (McCall et al., 1997). This can be a significant advantage in predominantly fine-grained formations, where conventional well installation technology is typically unsuccessful in retaining formation fines.

Nominal diameters of well casings and screens for direct-push completions range from 0.5 to 2 in. Although 2 in. completions are more typical of conventionally installed monitoring wells, smaller diameter direct-push wells suffer from several constraints, which include difficulty in well development.

Poor Well Design and Construction

The most common problems with well design and construction that contribute sampling artifacts are: inappropriate selection of well casing and screen materials, which results in compromises in water chemistry (through sorption and desorption, corrosion, and degradation); improper well screen length, which can result in collecting samples that may be significantly diluted; incorrect well-screen slot size and filter-pack grain size selection, which can result in sedimentation of the well; improper filter pack installation, which results in excessive filter pack losses to either the well or the formation during development; and improper annular seal material selection and installation, which can result in grout infiltration into the filter pack and well screen. Poor well construction can permanently bias sampling results and impair the usefulness and integrity of wells as sampling points (Fetter, 1983).

If artifactual turbidity is caused by drilling or improper well design and construction, the turbidity in samples will initially be high and may diminish with each sampling event. The identity of drilling or well construction related artifacts can be confirmed by particle analysis (e.g., via scanning electron microscopy) (Backhus et al., 1993).

Improper Selection of Well Construction Materials: Corrosion of steel well casing and screens (Figure 15:7) and chemical degradation of polyvinyl chloride (PVC) casing and screen may result in contributions of both dissolved and particulate matter to samples. Steel and stainless-steel corrosion products, including Fe, Mn, Cu, Pb, Cd, Ni, Cr, and Mo, have been found in samples collected from wells constructed in both neutral and low pH environments (Parker et al., 1990; Parker, 1991; Hewitt, 1992; Oakley and Korte, 1996). PVC degradation products and PVC cementing agents (Figure 15.8) have been found in samples collected from wells exposed to very high concentrations of organic solvents (Sosebee et al., 1982, 1983; Martin and Lee, 1989; Parker et al., 1990; McCaulou et al., 1995).

Well Screen Length: Wells installed for detection monitoring programs may have screens of variable lengths, with length dependent on the degree of heterogeneity in the hydrogeologic setting: In cases where discrete preferential flow pathways are not apparent, long screens may be employed as a screening tool to maximize the likelihood of detecting a potential release from a facility. This approach serves the intended detection monitoring purpose by biasing the sample collection toward the zones that have higher hydraulic conductivity. In other words, the higher hydraulic conductivity zones provide comparatively more sample volume, thus increasing the likelihood of detecting a release. In cases where discrete high hydraulic conductivity zones or preferential flow pathways are readily identified, well screen lengths are often selected to focus on these zones. If contamination is detected in samples above some action level established for a specific site, the focus of the investigation then shifts to spatial and temporal contaminant plume definition and

FIGURE 15.7
Corrosion of steel well construction materials (in this case, galvanized steel) can contribute constituents of the casing or screen to ground-water samples. This possibility must be accounted for in interpretation of analytical data from samples.

assessment monitoring. Wells installed for this purpose and for later corrective action monitoring (if required) are located, and screen positions and lengths are determined by detailed site characterization (e.g., formation profiling), to locate the specific zones of contamination that require monitoring and, possibly, remediation. These wells generally have short screens designed to focus on specific zones within a formation in which contaminants are preferentially transported. Examples of preferential flow pathways include sand and gravel lenses within a fine-grained matrix or fracture zones or solution channels within otherwise competent rock.

Filter-Pack Grain Size and Well-Screen Slot Size: The purpose of a filter pack is to provide a permeable zone between the well screen and the surrounding formation, with the objective of preventing fine-grained materials from entering, and possibly plugging, the well screen: Filter-pack materials need to be appropriately sized and of an inert chemistry to prevent bias of ground-water sample chemistry (Figure 15.9). To that end, a filter pack should consist of a clean, well-rounded silica sand of a selected grain size and gradation suitable to retain the surrounding formation materials. The filter-pack materials should be installed in the annular space between the well screen and the wall of the borehole using a device such as a tremie pipe to ensure controlled placement of the filter-pack material around the well screen.

Closely related to filter-pack design is the selection of an appropriate slot size for the well screen. A common but misguided approach to well screen design is to use a "one slot size fits all" philosophy for all formations. This often results in samples that are highly turbid, because the combination of slot size and filter-pack grain size used is not

FIGURE 15.8
PVC cementing agents (including tetrahydrofuran, methyl ethyl ketone, cyclohexanone, and dimethylformamide) can leach from casing joints into water standing in the well between sampling events, and have been found in samples even after the wells have been purged. If analytical data from samples show constituents of PVC solvent cement, the possibility of false positives attributed to the cementing agents should be considered.

FIGURE 15.9
Filter-pack materials in conventional monitoring wells may contribute analytes to samples if the materials are not inert to formation water chemistry. Materials should consist of at least 95% silica sand to avoid interactions with ground water and possible false positives attributed to filter-pack materials.

appropriate to filter out formation fines. The well-screen slot size should be selected so that the open area is maximized to allow rapid sample recovery, effective well development, and proper conduct of formation hydraulic tests and analysis of hydraulic test data. The slot sizes must be selected so the openings will allow the screen to retain the filter pack (or formation material in the case of naturally developed wells) while still permitting efficient well development. Sedimentation of wells is a common problem in wells in which well-screen or filter-pack design is flawed.

Inadequate or Improper Well Development

A common failure of many ground-water monitoring programs is that most wells are not developed properly. In high hydraulic conductivity formations, development is typically not conducted for a long enough period of time. As discussed in Chapter 12, wells installed in fine-grained formations dominated by clays and silts should not be developed. In these formations, well development may do more harm than good with respect to facilitating collection of representative ground-water samples (Paul et al., 1988; Nielsen, 1995).

Proper well development achieves several objectives including rectifying drilling damage to the borehole caused during well installation, removing fine materials from the formation adjacent to the borehole, stabilizing the filter pack materials installed adjacent to the well screen, retrieving lost drilling fluid from the formation, and optimizing well efficiency and hydraulic communication between the well screen and the adjacent formation. Proper well development is a critical process required to prevent the inclusion of excess turbidity and particulate artifacts in samples during the life of the well. Well development is a component of the well construction process and should not be confused with well purging, which is a component of the ground-water sampling process.

Well Maintenance

A critical error in many ground-water monitoring programs is the failure to adequately maintain the wells during the operating life of the monitoring network. As a standard procedure during ground-water sampling events, samplers should inspect each well to ensure that its structural integrity has not been compromised since the last sampling event. Proper well maintenance can include repair of damaged surface seals, redevelopment of wells in which siltation has occurred and resulted in reduced well recharge rates, or repair of damaged surface protection or security measures. Adhering to a rigorous well maintenance schedule will ensure that the wells will continue to yield representative samples of sufficient volume for the duration of the sampling program.

Geochemical Changes in Ground-Water Samples

Even if a monitoring well is designed correctly, installed in the most appropriate position within the zone of interest, and adequately developed and maintained, ground-water sample chemistry can be affected by the process of sampling the well. During this process, ground water is brought from an *in situ* environment, where pressure and temperature are stable and relatively uniform, to the surface where atmospheric conditions prevail. The resulting changes in pressure and temperature may result in changes to the sample that are manifested in negative biases (underestimation) for some analytes and positive biases (overestimation) for other analytes. Additional important issues include entrainment of artifactual particulate matter in the sample as a result of purging and sampling practices, and agitation and aeration of the water column in the well during sample collection. These issues, which are closely linked to one another, are discussed briefly. The reader is referred to Gibb et al. (1981), Gillham et al. (1983), Barcelona et al. (1985), and Nielsen and Nielsen (2002) for additional background on these issues.

Pressure Changes

Pressure changes to a sample are virtually unavoidable when a sample is brought to the surface from the ground-water environment. In the subsurface, ground water is, by definition, under a pressure greater than atmospheric pressure. When a sample of ground water is brought to the surface, it is collected at atmospheric pressure. The pressure decrease in the sample caused by exposing it to atmospheric conditions is proportional to the height of the water column above the point from which the sample is retrieved. Removing a sample from a depth of a few tens of meters and subjecting it to atmospheric pressure can represent a change in pressure in the sample of several atmospheres (Gillham et al., 1983). Under decreased pressure conditions, the water loses its ability to retain dissolved gases. For example, if a ground-water sample has a high partial pressure of CO_2 (which is true of most ground water), when the pressure on the sample is lowered, it becomes supersaturated with respect to CO_2. The sample will lose CO_2 gas until it reaches equilibrium with atmospheric conditions. This, in turn, may cause a shift in chemical equilibrium, introducing a secondary bias in pH and those parameters for which a change in pH can induce bias (Gillham et al., 1983). Loss of CO_2 causes an increase in sample pH (by between 0.5 and 1.0 pH units), which may then cause various trace metals, including Fe, Mn, Cd, Pb, As, and Zn, to precipitate (Gibb et al., 1981; Unwin, 1982). If volatile organic compounds (VOCs) are present in the sample, pressure decreases can cause their loss from the sample in one or both of two ways. Volatiles may evolve directly from solution, or they may partition into the headspace formed as other dissolved gases (e.g., CO_2) evolve from solution; more volatiles will evolve as the headspace gets larger. The end result is a sample with potentially significant negative biases for a variety of constituents. The magnitude of possible bias is sample-specific and cannot be predicted quantitatively (Gillham et al., 1983). The pressure decrease inherent in bringing a sample to the surface can be exacerbated using sampling devices such as suction-lift pumps (e.g., peristaltic pumps [Figure 15.10]), which may impart substantial negative pressure to the sample during sample collection.

There are also situations during sampling events in which samples may be subjected to increases in pressure. For example, some pumping devices (e.g., electric submersible

FIGURE 15.10
The negative pressure applied to samples collected with peristaltic pumps, particularly at lifts exceeding 15 ft, can significantly alter sample chemistry, especially for ground water with high levels of dissolved gases, as well as VOCs.

centrifugal pumps) pressurize the sample to drive it to the surface, some flow-through cells with a check valve in the discharge line may impart a back-pressure, and some filtration systems use positive pressure to force water through an appropriately sized filter. Positive pressure may cause gases in the air (e.g., O_2) to dissolve in the sample at a higher rate and concentration when compared with simple exposure to atmospheric air at atmospheric pressure, and result in precipitation of some metal species. However, as Gillham et al. (1983) point out, the chemical alterations induced in water as a result of imposing positive pressure are small when compared with the alterations induced by negative pressure.

Temperature Changes

The temperature of shallow ground water is approximately equal to the mean annual air temperature of the locale and is relatively constant year-round, typically varying by less than 2°C (Driscoll, 1986). This stability in temperature results in stability in shallow ground-water chemistry. During the process of collecting samples, however, there may be several points at which significant and potentially damaging temperature increases may occur to a ground-water sample. For example, temperature increases of as much as 10 to 15°C can occur readily when excess sample tubing at the surface (i.e., on a hose reel) or a flow-through cell is exposed to direct sunlight or high ambient temperatures. To avoid such temperature increases, lengths of above-ground discharge tubing should be minimized and flow-through cells should be kept in a shaded and cool environment (Figure 15.11). Some sampling devices can also cause temperature increases in a sample. For example, Pohlmann et al. (1994) and Oneacre and Figueras (1996) found that operation of an electric submersible centrifugal pump (Figure 15.12) at low discharge rates caused noticeable (5 to 7°C) increases in sample discharge temperature. Because this type of pump must be cooled by running water over the motor, and this heated water is eventually collected as the sample, this device is poorly suited to collecting samples to be analyzed for temperature-sensitive parameters. The change in water temperature could alter sample chemistry in a number of ways. Heating water reduces the solubility of dissolved

FIGURE 15.11

To avoid temperature increases in samples, the length of pump discharge tubing should be minimized and flow-through cells (if used) should be kept in the shade.

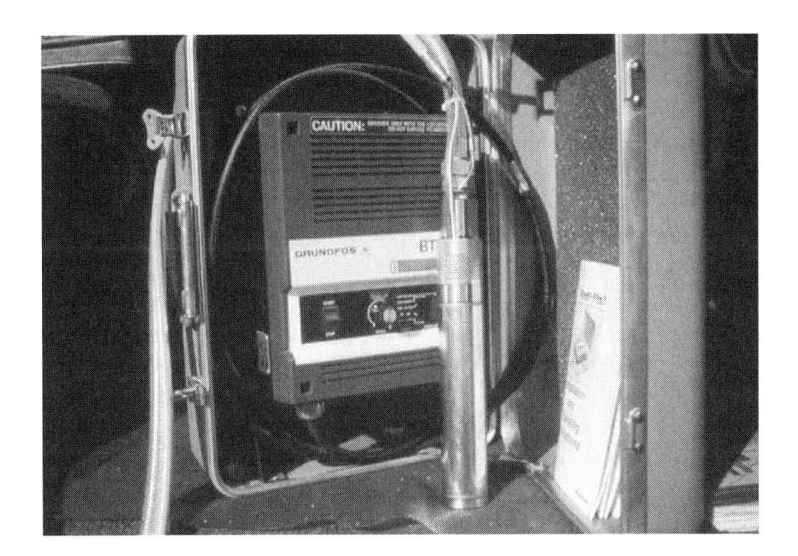

FIGURE 15.12
The motors of most electric submersible centrifugal pumps (such as the one pictured here) are cooled by water in the well that moves up over the motor, into the pump intake, and up to the surface to be collected as the sample. Temperature increases as high as 7°C have been noted in samples collected with these pumps. This large increase in sample temperature can alter sample chemistry for temperature-sensitive parameters.

gases (CO_2 and O_2) in water. The resultant loss of dissolved CO_2 and O_2 can induce a shift in pH and in redox state, which then causes precipitation of carbonates (Ca and Mg) and dissolved metals, most readily Fe. The precipitation of Fe can then cause co-precipitation of other metals such as Ni, Cu, and Cr (Stumm and Morgan, 1996). Heating will also reduce the solubility of VOCs in water, resulting in losses of these constituents of the sample through volatilization (Parker, 1994).

Entrainment of Artifactual Particulate Matter During Purging and Sampling

The significance of disturbance of the water column in the well during purging and sampling caused by the use of sampling devices that agitate and mix the water column (e.g., bailers or inertial-lift pumps) has been the subject of a great deal of research. It is important to understand that a "bulk" ground-water sample should contain not only water obtained from the saturated portion of the formation screened by the well, but also any fine particles, or colloids, that occur under natural conditions and are mobile in the formation under ambient flow conditions. In coarse-grained formations (i.e., coarser than fine sand), fractured bedrock or karst formations, this particulate matter can include colloids that can serve as prime sites for some contaminants to sorb and move with ground-water flow (Dragun, 1988; Mason et al., 1992) and microspheres containing indigenous bacteria (Harvey et al., 1989). This particulate fraction is referred to as the "mobile contaminant load" of a formation and is a legitimate part of a ground-water sample used for assessing contaminant fate and transport and human health risks caused by ingestion or exposure to the water. When a water column is agitated during purging and sampling, however, the samples collected contain not only naturally mobile particulate matter but also larger "artifactual" particles that are not mobile in ground-water systems.

Most investigations are focused strictly on defining the chemistry of the truly dissolved fraction or on determining the presence, concentrations, or fate and transport of specific chemical constituents that are known to be mobile only in the aqueous phase (e.g., highly soluble constituents with very low adsorption coefficients, including low molecular weight organic compounds such as MTBE or some VOCs). However, a number of researchers (O'Melia, 1980, 1990; Eicholz et al., 1982; Robertson et al., 1984; McDowell-Boyer et al., 1986; Cerda, 1987; Gschwend and Reynolds, 1987; Enfield and Bengtsson, 1988; McCarthy and Zachara, 1989; Puls and Barcelona, 1989; Puls, 1990; Ryan and Gschwend, 1990, 1992; Gschwend et al., 1990; Puls and Powell, 1992a, b; Puls et al., 1992; Backhus et al., 1993; Grolimund et al., 1996) have demonstrated that mobile colloid phases (suspended particles in the range of 0.005 to approximately 10 μm) may form in certain subsurface environments and be mobile at ambient ground-water velocities. Many colloidal particles (e.g., secondary clay minerals; hydrous iron, manganese, and aluminum oxides; and dissolved and particulate organic matter), because of their high surface area per unit mass and volume, have enormous sorptive capacities and strong binding capabilities for selected classes of organic and inorganic contaminants (Enfield et al., 1989; Lyman et al., 1992). The colloids present in some aquifers may significantly affect the transport of hydrophobic or strongly sorptive contaminants, including polynuclear or polycyclic aromatic hydrocarbons (PAHs), polychlorinated biphenyls (PCBs), pesticides (e.g., DDT), dioxins and furans, radionuclides (Pu, Am, U, Cs, and Ru), heavy metals and metalloids, and many major ions. Therefore, it is especially important to understand the role of colloids in contaminant transport at sites where low-solubility, highly surface-reactive chemicals are of concern (McCarthy and Zachara, 1989).

Presence and Sources of Artifactual Particulate Matter: For sampling programs in which the primary objective is to determine the "total mobile contaminant load," it is important to ensure that the mobile colloidal fraction is included in the sample: However, while it is important to include naturally mobile colloidal material in samples collected at sites where colloidal transport may be an important contaminant transport mechanism, it is equally important to exclude from all samples particulate or other matter that is clearly artifactual in nature (i.e., due, in some way, to implementing the sampling process). Disturbance caused by practices that significantly agitate and aerate the water column in the well can markedly increase the content of artifactual particulate matter in samples and irreparably change sample chemistry through processes such as oxidation. Sampling artifacts such as these present serious obstacles to proper interpretation of ground-water sampling results.

The most commonly recognized artifact of poor sampling technique is excessive suspended sediment, seen in samples as turbidity. High concentrations of suspended particles in samples collected for chemical analysis are problematic because contaminants potentially present on these particles (e.g., adsorbed to the particle surface) may or may not be mobile in the ground-water system (McCarthy and Zachara, 1989). Because most investigators are interested in determining what is present in the mobile fraction, this compromises the integrity of samples. As Puls et al. (1991) point out, there is a strong inverse correlation between turbidity and representativeness of samples. With many commonly used sampling practices, it is not possible to differentiate between naturally occurring suspended particles and those brought into suspension or created artificially. Therefore, in situations in which it is necessary to assess the total mobile contaminant load, it is important to use sampling practices that do not disturb the water column in the well, to allow collection of representative samples.

Inclusion of artifactual particulate matter in samples can radically alter analytical results, causing spurious increases in concentrations of several classes of analytes. This is particularly true for the major constituents of the aquifer mineral matrix such as Fe, Al, Ca, Mg, Mn, and Si (Powell and Puls, 1997). Maintenance of *in situ* ground-water chemistry conditions during sampling and sample processing at the surface is particularly important in determining mobile colloid concentrations and characteristics. While no effects from artifactual turbidity have been observed for VOCs (Paul and Puls, 1997), large concentration differences between filtered and unfiltered samples have been observed for metals (Puls and Barcelona, 1989; Puls et al., 1992; Pohlmann et al., 1994), PAHs (Backhus et al., 1993), and radionuclides (Buddemeier and Hunt, 1988; Penrose et al., 1990). Failure to address the issue of artifactual particulates can lead to inaccurate estimates of colloid and colloid-associated contaminant concentrations, colloid characteristics, and the distribution of contaminants between the colloidal and dissolved phase (Backhus et al., 1993).

The sources of turbidity caused by implementing the sampling process are many and varied. For example, using a downhole colloidal borescope (similar to a camera), Kearl et al. (1992) observed that a great deal of turbidity measured in the wells they studied was due simply to the initial placement of the sampling device. In this study, colloidal density, quantitatively measured with a turbidimeter and qualitatively measured by visual inspection, was observed to increase significantly with installation of sampling devices and decline exponentially while purging at a low flow rate. The initial disturbance to the well and the high turbidity levels caused by insertion of the sampling device required overnight equilibrium to attain naturally occurring turbidity values and stable flow in the screened interval.

There is substantial evidence to indicate that the activities inherent in bailing (repeated insertion of the bailer in the water column and withdrawal of the bailer to purge the well and collect samples) and pumping at high rates (i.e., in excess of the natural rate of ground-water flow through the well screen) liberate significant quantities of artifactual particles. Using these purging and sampling methods, particulates that would otherwise have remained immobile in the formation under natural ground-water flow conditions are often mobilized into the well. Carefully collected field evidence from a variety of sites in a variety of hydrogeologic settings demonstrates that the use of bailers (Figure 15.13) and high-speed, high-flow-rate pumps (Figure 15.14) disturbs aquifer and filter-pack materials

FIGURE 15.13
Bailers significantly disturb the water column in wells and the formation materials adjacent to the well screen, which results in significant increases in turbidity of samples, much of which is due to the entrainment of artifactual particulate matter. As a result, collection of representative samples for many parameters is very difficult if not impossible using bailers.

FIGURE 15.14
High-speed, high-flow-rate pumps significantly increase the flow velocity in pore spaces in the formation adjacent to the well and create shear that releases substantial amounts of sediment, which is entrained in water collected as samples. (Photo courtesy of Severn-Trent/QED.)

and can disrupt fragile colloidal aggregates, producing samples in which artifactual particulate materials have been entrained (Puls et al., 1992; Backhus et al., 1993; McCarthy and Degueldre, 1993; Powell and Puls, 1993; McCarthy and Wobber, 1996; McCarthy and Shevenell, 1998). Other investigators (Barcelona et al., 1985; Puls and Barcelona, 1989; Puls and Powell, 1992a) have also noted that bailing wells or pumping at high rates during purging or sampling can mobilize particulate matter and significantly increase sample turbidity.

Pohlmann et al. (1994) noted that pumping-induced stresses can cause grain flow within the sand pack of the well and exceed the cohesive forces of aquifer material cementation. Backhus et al. (1993) also cite bailing- and pumping-induced shearing as a source of artifactual turbidity in a series of wells they studied. Shearing, which releases normally immobile solids from the aquifer matrix, is caused by pumping at rates that induce ground-water flow velocities greater than the natural flow velocities; higher pumping rates result in greater shear in the aquifer. Ryan (1998) calculated that in the set of wells he studied, pumping rates greater than 100 ml/min could induce shear capable of mobilizing particles outside of the zone of the aquifer disturbed by well construction. Low pumping rates appeared to provide the best means to minimize collection of artifactual colloids. This was confirmed in tests conducted by Backhus et al. (1993), to determine the effect of higher pumping rates on turbidity levels in several wells at several different sites. In all cases, turbidity increased sharply in response to pumping rate increases after a turbidity plateau had been established at a lower pumping rate. The increase in turbidity reflected greater inclusion of artifactual solids due to increased shear. Their recommendation was that wells not be bailed and that pumping rates used during purging and sampling be minimized (approximately 200 ml/min for the wells they sampled). Rather than specify a particular withdrawal rate to be applied to all wells, Ryan et al. (1995) suggested that a minimal drawdown guideline would be the best way to assure that pumping-induced flow rates near the well were not stressing the aquifer and causing mobilization of artificial particulate matter. On the basis of deleterious impacts of bailing and high-flow-rate pumping on ground-water samples, a number of other field studies (Puls and Barcelona, 1989, 1996; Puls et al., 1991, 1992; Puls and Powell, 1992a) also recommend an approach to sampling that includes:

- Using a low flow rate during both purging and sampling
- Placement of the pump intake within the well screen
- Minimal disturbance of the stagnant water column above the screen
- Monitoring water-quality indicator parameters during purging
- Minimization of atmospheric contact with samples
- Collection of unfiltered samples for metals analysis to estimate total contaminant load in the ground-water system

This sampling approach, referred to as low-flow purging and sampling, is described in detail in later sections of this chapter.

Artifactual colloidal material may also be generated during sampling by exposing ground water containing readily oxidizable materials, such as Fe^{2+}, Mn^{3+}, As^{3+}, Al, and Cd, to atmospheric conditions (Figure 15.15). This increases the redox state of the sample and causes the precipitation of colloid-sized metal oxides and hydroxides (Gillham et al., 1983; Puls et al., 1990; Ryan and Gschwend, 1990; Backhus et al., 1993; McCarthy and Degueldre, 1993). This may occur within the well, as bailing or pumping at rates sufficient to draw water levels down to within the well screen (effectively dewatering the well) exposes anoxic or suboxic ground water to atmospheric air. It may occur at the surface as bailed or pumped water is discharged and exposed to atmospheric conditions. This may also occur in the formation at some distance from a pumped well as dewatering exposes sediments and ground water to atmospheric air and causes mixing of chemically distinct ground waters from different zones. This can impact a much larger segment of the aquifer than the interval of interest (Puls et al., 1991).

The oxidation of iron from a reduced state is particularly important from a sample stability standpoint. As iron is oxidized, precipitating iron oxide and hydroxide, a pH shift (increase) often occurs, which may further impact sample integrity. Iron hydroxide is known to remove other constituents from solution, including Cu, Zn, Co, Cd, As, and Pb (Stoltzenberg and Nichols, 1985, 1986). The kinetics of oxidation-induced precipitation and subsequent sorption processes are such that they can occur within seconds or minutes (Reynolds, 1985; Grundl and Delwiche, 1992; Puls et al., 1992). The end result is that a number of previously dissolved species are removed from solution and may be removed from the sample if it is filtered, resulting in lower concentrations of these analytes than are actually present in ground water.

Inclusion of particulate matter that may have formed or collected in the well between sampling events may also be responsible for increased turbidity in samples. Preventing this material from being included in the sample during a sampling event can be accomplished by minimizing disturbance to the water column during sampling and by positioning the sampling device well above the bottom of the well. Backhus et al. (1993) described a situation where flocculent orange particles visible in water samples collected from a contaminated well were originally settled in the bottom 12 in. of the well and suspended in

FIGURE 15.15
Exposure of ground-water samples to atmospheric air for even a short period of time can result in oxidation of Fe^{2+} and co-precipitation of other metals and some organic species, which can significantly alter the chemical makeup of samples. Note the iron oxyhydroxide precipitate in the bottom of this VOC vial.

the top 2 in. of the water column. The material was probably iron oxide colloids precipitated by oxygenation of the anoxic ground water in the well. In two different wells at the same site, coal tar was found in the bottom of the wells. This was cited as having contributed greatly to observed initial high turbidities and PAH levels in samples collected from these wells by bailers and high-flow-rate pumping. Thus, they recommended cautious insertion of sampling gear into wells and avoidance of contact with the well bottom, to exclude this source of artifactual material.

Agitation and Aeration of Water Samples During Collection

There are two key times during a ground-water sampling event when agitation and aeration of samples can occur: within wells or direct-push sampling tools as they are purged and sampled using sampling devices that agitate and mix the water column (e.g., bailers or inertial-lift pumps), and at the ground surface, as samples are filtered or sample bottles are filled. Agitation of the sample can upset the balance of dissolved gases, pH, dissolved metals, and semivolatiles and can result in the stripping of VOCs from the sample. Agitation also increases turbidity, which can result in positive bias for the parameters associated with artifactual particulate material discussed previously and interfere with some analytical methods (Puls et al., 1992). Aeration of the sample can cause all of the chemical changes to the sample related to oxidation, as discussed in the previous section.

The most significant sources of bias and error due to sample agitation and aeration are those that occur during the sample retrieval process. For example, using bailers (Figure 15.16) or inertial-lift pumps (Figure 15.17) to purge and sample creates a

FIGURE 15.16
The agitation and aeration of the water column caused by bailers is due to the fact that most bailers fit tightly into the well and they must be alternately inserted into and removed from the water column, which creates a surging effect in the well. This results in the collection of samples with high levels of turbidity.

substantial amount of agitation and mixing of the water column, which results in elevated DO levels in samples (and associated chemical changes) and significant increases in turbidity (predominantly artifactual in nature). The agitation and aeration caused by repeated insertion and removal of a bailer often makes it impossible to measure representative values for DO and turbidity (Pennino, 1987; Pohlman et al., 1994), for VOCs (Muska et al., 1986; Imbrigiotta et al., 1988; Gibs et al., 1994), for trace metals (Puls et al., 1992; Heidlauf and Bartlett, 1993), or for PAHs (Backhus et al., 1993). Other sampling devices, provided they are installed carefully and operated properly, do not cause these problems. However, pumps that are pumped at rates high enough to dewater the sampling point can cause significant agitation and aeration of formation water.

Additional but perhaps less significant sources of agitation and aeration include sample bottle filling practices and sample filtration. A high potential for agitation and aeration occurs during sample bottle filling (particularly for VOC vials) using bailers and inertial-lift pumps, as it is often difficult to control the discharge from these devices (Figure 15.18). Bottom-emptying bailers offer an improvement over top-emptying bailers in this regard. For most pumps, whether continuous or cyclic discharge, it is relatively simple to control the discharge rate from the tubing to minimize agitation and aeration concerns. Bottles should be filled at relatively slow rates — about 250 ml/min or less for VOCs and about 500 ml/min or less for most other parameters (Figure 15.19). These rates provide a good compromise between the potential for agitation caused by higher flow rates and the greater exposure to atmospheric air (resulting in aeration) caused by lower flow rates.

FIGURE 15.17
Inertial-lift pumps also significantly agitate and aerate the water column in wells due to the nature of their operation. Pushing the tubing into the water column and pulling it up repeatedly results in discharge of highly turbid water.

FIGURE 15.18
Discharge from inertial-lift pumps (and top-emptying bailers) is difficult to control, which can result in agitation and aeration of samples during bottle filling.

FIGURE 15.19
Bottle filling should be done at a slow, controlled rate, but not so slow that contact of the sample with atmospheric air is prolonged.

The Ground-Water Sampling and Analysis Plan (SAP): A Road Map to Field Sampling Procedures

Objectives of the SAP

As discussed earlier, there is a great deal of complex science behind ground-water sampling. Much of this science drives the selection of sampling methods and equipment for each site. To ensure that sampling teams are aware of requirements for sample collection, a site-specific SAP must be written by an experienced practitioner and followed by the sampling team. The objectives of a typical SAP are summarized in Table 15.3. Implementation of a comprehensive and well thought out SAP should ensure that ground-water sample collection procedures are consistent from one sampling event to the next, thus reducing the potential for sampling team related error and bias. This, in turn, should ensure that the data generated, both in the field and as a result of laboratory analysis of samples, are comparable and without peaks and valleys referred to as "data bounce."

Preparation of the SAP

The SAP should be written by an experienced practitioner who has relevant field experience and who can identify potential sources of error and bias in each component of the ground-water sample collection process. In addition, working experience in a laboratory or working closely with laboratory personnel is an advantage.

Prior to submitting the SAP to regulatory agency personnel for their approval, it is recommended that field sampling team personnel have the opportunity to review and provide honest input on procedures contained within the SAP. Communication between all levels of personnel involved in the sampling program will ensure that the final SAP

TABLE 15.3

Objectives of a Sampling and Analysis Plan

Provide a written statement of objectives of the sampling program
Provide a schedule for sample collection (number of wells, when and how frequently they should be sampled)
Provide detailed procedures for all aspects of the ground-water sample collection process to be implemented by the sampling team to ensure sampling accuracy and precision including:
  – Well integrity inspection
  – Wellhead screening for volatile or combustible vapors
  – Water-level and product-thickness measurement
  – Field QA/QC measures
  – Purging and sampling device selection and operation
  – Well purging procedures and management of purge water
  – Parameters required for ground-water sample analysis
  – Field equipment calibration procedures
  – Field parameter measurement procedures
  – Field equipment decontamination procedures
  – Sample collection procedures, including sample container selection, sample pretreatment requirements, order of sample collection, and sample container filling
  – Sample handling and preparation for shipment
  – Documentation of sampling event activities
Provide written documentation of field procedures for outside evaluation
Provide a vehicle for project management and budgeting

---

is not only technically correct for the site-specific program, but that field protocols are practical and realistic and based on good science. All SAP writers are encouraged to remember this saying: "nothing is impossible for the person who does not have to do it." A procedure may sound good on paper, but if the sampling team finds the procedure to be too cumbersome or too confusing, shortcuts will be made in the field, which can introduce substantial sampling error, imprecision, and bias. Project personnel need to remember that for most ground-water SAPs, once the document has received regulatory approval for implementation, procedural changes cannot be made randomly in the field. Deviation from the documented protocol will call into question all results from field measurements and subsequent data generated by laboratory analysis of collected samples.

Selection of Field Protocols to be Incorporated into the SAP

As illustrated in Table 15.3, the SAP details specific standard operating procedures for a number of work tasks to be implemented during the sampling event. While preparing the SAP, a great deal of thought needs to go into the selection of protocols for each of the work tasks most appropriate for any individual site. This includes methods and equipment for wellhead screening, water-level and product-thickness measurement, field quality assurance (QA) and quality control (QC), sampling point purging and sample collection, field parameter measurement, and field equipment decontamination.

Well Headspace Screening

The first task implemented by the sampling team at each sampling point is opening the well and screening the headspace above the water column in the well for the presence of volatile or combustible gases and vapors. The SAP must detail which instrumentation is required for screening the well headspace (e.g., photoionization detectors [PIDs] [Figure 15.20], flame ionization detectors [FIDs], or combustible gas indicators) and how the data are to be used (e.g., health and safety personal protective equipment [PPE] selection). It must also indicate how the equipment is to be calibrated, operated, and maintained throughout the sampling event. Equipment for headspace screening is selected based on the types of

FIGURE 15.20
Well headspace screening must be done with the appropriate instrument. This PID is used to detect volatile organic compounds (VOCs) associated with petroleum hydrocarbons and organic solvents.

contaminants known to be present at the site. Table 15.4 lists the types of equipment that may be used for well headspace screening and their applications. Chapter 2 describes the use of PIDs and FIDs in more detail for headspace screening, and Chapter 19 describes how to use data from PIDs and FIDs to select the most appropriate PPE.

Factors that must be considered in the selection of headspace screening equipment include: type of data generated (qualitative versus quantitative), ability to detect the parameters of concern, ease of calibration; sources of interference, and ease of use. Readers are directed to Maslansky and Maslansky (1993, 1997) for detailed discussions on the applications and limitations of each of the types of instrumentation presented in Table 15.4 and for guidance on how to incorporate and interpret data from wellhead screening in a health and safety plan.

Water-Level and Product-Thickness Measurement

Following well headspace screening, the next task is to take water-level or product-thickness measurements (Figure 15.21). In this portion of the SAP, specific procedures

TABLE 15.4

Examples of Instrumentation Appropriate for Well Headspace Screening

Instrumentation	Class of Contaminant of Concern	Example Application
PID	Volatile organic vapors and gases and some inorganics	Dry cleaning facility for PCE and TCE
FID	Volatile to semi-volatile organic vapors	Underground storage tanks with jet fuel
Combustible gas indicator (may be combined with an oxygen meter)	Flammable vapors	Landfills where methane may accumulate in wells
Toxic gas sensors	Vapors containing known toxics	Industrial facility where H_2S is generated

FIGURE 15.21
Water-level measurements should be taken in all wells in a sampling program prior to purging and sampling of the wells (to collect data for piezometric surface mapping) and again just before purging (to determine the well volume for well-volume purging or the initial static water level for low-flow purging and sampling). This water-level gauge will be used to track drawdown in a well during low-flow purging and sampling.

for where, when, and how to take water-level and, if applicable, separate-phase product-measurements (LNAPL or DNAPL) must be described in detail. These measurements must be taken in all sampling points prior to any purging and sampling activities, to ensure that the data are collected under as close to the same environmental and atmospheric conditions as possible. This is of particular concern when taking water-level measurements, which are potentially affected by a number of environmental variables including changes in ambient air pressure (especially problematic for shallow, unconfined formations), tides, changes in levels of nearby rivers, precipitation events, and operation of nearby pumping wells.

A number of different methods are available for water-level measurement, including ploppers (or poppers), chalked tapes, electronic water-level gauges, pressure transducers, bubblers, sonic devices, floats, and acoustic probes. The method appropriate for any given site depends on the measurement accuracy required, the depth to static water level in the wells included in the sampling program, well diameter, whether single measurements or continuous data are required, and the possible sources of interference present in the well. ASTM Standard D 4750 (ASTM, 2004a) discusses some of the practices available for measurement of water levels in monitoring wells including ploppers, carpenter's chalk on a steel tape, and a variety of electronic water-level gauges. Chapter 13 discusses these and other water-level measurement methods in detail.

Water-level data are used for a variety of applications including determining the volume of water in the sampling point, determining the direction of ground-water flow and the hydraulic gradient (horizontal and vertical), and calculating the rate of ground-water flow. Readers are directed to Chapter 13 for a detailed discussion on water-level data interpretation and presentation. ASTM Standard D 6000 (ASTM, 2004b) also provides a discussion on methods for presentation of water-level information obtained from ground-water investigations.

At sites with petroleum hydrocarbon contamination, some monitoring wells and direct-push tools may contain a separate-phase layer of product (LNAPL), which must be measured in addition to the water level. The thickness of this layer is referred to as the "apparent product thickness", which reflects the amount of hydrocarbon that has accumulated in the well — not the actual thickness of product in the formation.

To measure the thickness of LNAPL, one of three methods can be used: (1) hydrocarbon-sensitive and water-sensitive pastes applied to a measuring tape; (2) an electronic oil–water interface probe; and (3) a clear, acrylic hydrocarbon bailer (API, 1989). The first method requires the application of a hydrocarbon-sensitive paste to one side of a tape and a water-sensitive paste to the other side. The tape is lowered into the well and as the LNAPL and water contact the pastes, color changes occur. The depths of each color change are measured, with the difference between the changes indicating LNAPL thickness. The depth to ground water, which must be adjusted because of the presence of LNAPL, is indicated by the color change on the side of the tape coated with the water-sensitive paste. Compensation factors and the significance of the LNAPL in the well are discussed in API (1989) and Testa and Winegardner (1991).

Electronic oil–water interface probes are capable of determining the depth to ground water using a conductive probe similar to that used in an electronic water-level gauge. In addition, they are equipped with one of three hydrocarbon-detecting sensors: (1) infrared; (2) optical; or (3) a float. The meter emits different audible tones or different indications via lights (i.e., continuous versus flashing) when the probe contacts the LNAPL and then the LNAPL–water interface. This method is quicker than using pastes but is much more expensive. It is also less accurate than the measuring tape with pastes when measuring thin layers of hydrocarbon. Quoted accuracies and detection capabilities of most oil–water interface probes are usually between 0.12 and 0.06 in.

Clear hydrocarbon bailers are useful for collecting a grab sample of the LNAPL layer, which can be visually examined and measured within the bailer. To measure LNAPL thickness, the bailer is lowered to the point at which the first fluid in the well is encountered, then lowered an amount slightly less than the length of the bailer. To ensure a measurement of the full LNAPL thickness, the bailer must be long enough to ensure that its top will be above the air–LNAPL interface, when the bottom check valve is below the LNAPL–water interface. Of the three methods, this is the least accurate method for LNAPL measurement because the thickness of product in the bailer is never an exact indication of what is in the well. It is always less than the product thickness in the well by a factor that depends on the diameter of the bailer (compared with the diameter of the well) and the design of the check valve.

Field Quality Assurance and Quality Control

A ground-water sampling event generates information and measurements used for making important site-specific decisions such as determining whether a facility design is effective at preventing impacts to the environment, determining whether ground-water contamination is present at a site, determining whether contamination poses an exposure risk to off-site receptors, determining whether natural attenuation is an appropriate remedial alternative for a site, or determining the most effective remedial design for a site. Given the significance of these decisions and their potential associated costs, it is critical that all analytical and field measurement data are not only technically sound but also scientifically and legally defensible. An important component of the SAP is a detailed field QA/QC program which, when implemented, provides sampling teams with the confidence that the results of the sampling event will be technically correct and defensible.

A QA program may be defined as those operations and procedures that are undertaken in the field to provide measurement data of a stated quality with a stated probability of being correct (Taylor, 1985). The QA program documents administrative and field procedures that are designed to monitor management of the project as well as field sample collection and measurement activities. Table 15.5 summarizes the key administrative and field elements of a field QA program for ground-water sampling.

When followed, a field QA/QC program will ensure that all data generated through field measurement and analysis are accurate and precise with a minimum amount of error and bias. The QA/QC program ensures that the data produced are defensible if the project is subject to litigation, because it provides clear documentation of field procedures and outlines the system of checks and balances that were used to verify sampling and field measurement accuracy and precision. ASTM Standard D 7069 provides a standard guide for field QA in a ground-water sampling event (ASTM, 2004c).

Samples referred to as field QC samples should be collected during every sampling event (Table 15.6). The purpose of field QC samples varies with the type of sample collected, but these samples provide a formal means of verifying the precision and accuracy of various components of field sample collection procedures. Field QC samples are commonly collected at one or more designated sampling points after all laboratory samples have been collected at that (those) point(s). This ensures that sufficient sample volume is available for collection of samples for laboratory analysis.

In many cases, regulatory programs dictate which QC samples are required to be included in the sampling protocol. Unfortunately, in many instances, only minimum attention is paid to incorporating these samples into a sampling program. When an SAP is prepared for any site, it is important to incorporate not only those QC elements required to satisfy regulatory requirements, but also those system checks that will facilitate data validation. When determining the level of field QA/QC independently of regulatory guidelines, the political sensitivity of the site being monitored needs to be evaluated. A higher level of QA/QC is generally warranted if the site is operating under a consent order, if there is a strong public opinion about detrimental site impacts, or if the site is about to be sold. QA/QC levels should also be higher than normal at sites where the concentration of contaminants in samples is very close to action levels or analytical method detection limits. Under these circumstances, because critical decisions are made based on comparison of sampling results to action levels, it is important to have the utmost

TABLE 15.5

Key Elements of a Field QA Program

Administrative elements
   Project description and definition of project objectives
   Project fiscal information (travel, support services, expendable supplies, equipment needs)
   Schedule of tasks and products (field activities, analysis, data review, reporting)
   Project organization and responsibility
   Selection of appropriately trained and experienced personnel for field and management roles

Field elements
   Implementation of technically sound SOPs
   Documentation of protocols for operation, calibration, and maintenance of all field instrumentation
   Collection of field QC samples (which, when, how, and how many)
   Adherence to required sample pretreatment methods and holding times
   Use of chain-of-custody procedures
   Record keeping procedures that incorporate good laboratory practices (GLPs)
   Methods for checking accuracy of field parameter measurements
   Description of corrective actions to be implemented if an error is detected at any point in the field

TABLE 15.6

Common Field QC Samples

Type of QC Sample	Recommended Frequency of Collection
Trip blank	One per shipment[a] per day per laboratory
Temperature blank	One per shipping container
Field blank	One per waste management unit
Equipment blank	One for every ten sampling points
Blind duplicate sample	One for every ten sampling points
Spiked sample	One for every ten samples submitted for analysis
Field split sample	Variable, but at least one upgradient location and two downgradient locations should be sampled

[a]The term "shipment" must be defined for each project. For example, if ten coolers are being sent to a laboratory, under one definition, all ten coolers might be defined as a single shipment. In another regulatory program, each cooler could be considered to be an individual shipment.

confidence in the accuracy and precision of the samples and the field data collected. In long-term monitoring programs or in programs where there are sufficient numbers of sampling points to warrant multiple sampling teams, field QA/QC efforts should be elevated to ensure consistency between sampling teams. Without good field QA/QC, it can be difficult to directly compare results generated by two different sampling teams at a single site, even if they are implementing the same sampling protocol.

Table 15.6 presents a list of the various QC samples that may be included in a field QC program, with the typical frequency of collection for each type of sample. If a ground-water sampling program is subject to scrutiny by outside groups, such as attorneys or regulatory agency personnel, the ratio of QC samples to sampling locations will decrease (i.e., instead of a 1:10 ratio, some QC samples may be required at a 1:5 ratio). There is a tendency to try to increase the ratio of sampling locations to QC samples in an effort to save costs. As an example, the authors recently reviewed an SAP that incorporated equipment blanks to verify the effectiveness of field equipment decontamination procedures. The plan indicated that an equipment blank would be collected after every 30 sampling locations (rather than every ten locations) to save money. When laboratory results were reported, a contaminant was detected not only in each of the 30 samples collected but also in the equipment blank. As a consequence, all 30 of the sampling points had to be resampled (rather than ten) to determine whether the contaminant concentrations were real or whether the data indicated that there was a problem with field equipment cleaning procedures.

In addition to selecting which and how many QC samples will be incorporated into the sampling program, it is important to select which parameters should be analyzed on each of the QC samples so that the information derived from their analysis will be meaningful. Ideally, each QC sample will be analyzed for the same parameters as the ground-water samples. This will provide the most meaningful information and will permit identification of any potential problems for any of the analytes of interest. This, however, is also the most expensive strategy and, consequently, the least likely to be implemented. More likely, regulatory guidance will be referenced to put together a short "required" list of analytes. In many cases, regulatory guidance requires that field QC samples be analyzed for volatile constituents because that group of analytes is most sensitive to handling. However, not all sites are concerned with volatiles, so important resources may be expended on potentially inconsequential analyses.

Purposes of Field Quality Control Samples: It is important that sampling team members understand the purpose of each type of QC sample to understand the importance of collecting the samples correctly: The purpose of each of the QC samples presented in Table 15.6 will be described herein. Details on how to collect each QC sample follow in a later section of this chapter.

A trip blank is a blank designed to determine whether anything associated with the preparation of the sample containers by the laboratory, shipment of the empty containers to the field, traveling in the field with the sampling team during the sampling event, and return shipment to the laboratory could have any impact on sample integrity. This blank is prepared by the laboratory and requires no preparation or handling by the field sampling team.

The purpose of a temperature blank is to provide a formal mechanism for the laboratory to confirm actual sample temperatures upon arrival at the laboratory. When samples arrive at the laboratory, temperatures should be checked in sample containers of differing volume and placement within the shipper. This can be done using: (1) a certified thermometer inserted into one or more sample containers; (2) a calibrated infrared gun to determine the surface temperature of individual containers; or (3) a specially prepared 40 ml sentry vial (Figure 15.22) that contains a permanently affixed certified thermometer, usually provided by the laboratory. The latter method, called a temperature blank, is preferred by some laboratories because it does not require that a foreign object (a thermometer) be inserted into the sample container (which could result in sample contamination). The infrared method is preferred by other laboratories because it is fast, does not require that a thermometer be inserted into the sample container, and permits evaluation of a number of sample containers of differing volumes throughout the sample shipper. Regardless of which method is used by the laboratory, when sample temperatures are checked, the laboratory will look for an arrival temperature of $4 \pm 2°C$. If the arrival temperature is outside of this range, the sampling team will be contacted to discuss an appropriate course of action: resample the well or analyze the received sample, with a disclaimer stating that the analyzed sample was not appropriately preserved with respect to temperature.

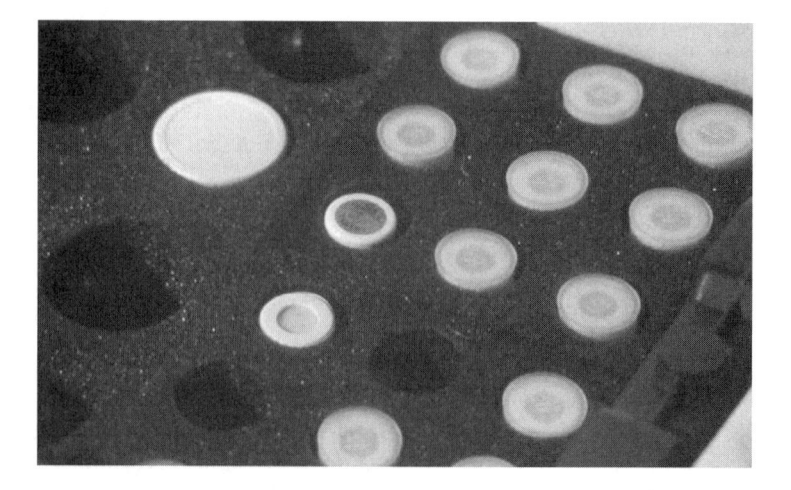

FIGURE 15.22
Temperature blanks (in this case, the vial with the dark dot on the top) are used to provide the laboratory with a convenient means of checking sample temperatures upon arrival of a shipment of samples at the lab.

At some field sites, there is a concern that ambient air contamination levels may be sufficiently high to influence concentrations of contaminants detected in ground-water samples. In some instances, trace quantities of VOCs may be gained by a ground-water sample through contact with atmospheric air, causing a positive bias in their determination (Gillham et al., 1983). A good example of a site where this is a common problem is the corner service station. At these sites, ground-water monitoring wells are commonly installed in the vicinity of pump islands. When samplers purge and sample these wells, the station typically remains open for business, meaning that as wells on one side of the pump island are sampled, station customers may dispense fuel into their vehicle on the other side of the island. Under these circumstances, ambient air is contaminated by fumes from the dispenser and exhaust fumes from adjacent vehicles, which contain the same volatile constituents for which ground-water samples will be analyzed. To quantify the extent to which ambient conditions may influence sample chemistry, a field blank is collected.

A fourth important field QC sample is the equipment blank. Whenever a piece of equipment is used in more than one sampling location, it should be cleaned between locations (Figure 15.23). It will be necessary to collect an equipment blank to verify the effectiveness of field equipment cleaning procedures (see Chapter 20 for more detailed discussions on field equipment cleaning procedures). Two forms of equipment blank can be collected. The first, sometimes referred to as a rinseate blank, is designed to determine whether, following equipment cleaning, soluble contaminants remain on the surfaces of the equipment. In some cases, contaminants may not readily solubilize in control water, but may remain on the surface of cleaned equipment as a residue. To determine whether a residue remains on a surface following cleaning, a wipe or swipe sample may be collected.

FIGURE 15.23
An equipment blank is collected by passing a rinse of final control water over the surface of a piece of equipment after it has been used and then cleaned.

To test the precision of sampling teams, a blind duplicate sample is collected as part of the field QC program. The objective of the blind duplicate is to collect two samples of ground water that are as close to chemically identical as possible. There are two procedures for collecting blind duplicates — one procedure for nonvolatile parameters and a second procedure for parameters requiring the use of 40 ml VOC vials. Detailed procedures on how to collect each type of duplicate sample are provided later in this chapter.

In many regulatory agency guidance documents, a field QC sample called a spiked sample is presented as a means to determine whether there is any matrix activity (most often in the form of microbial activity) that might alter sample chemistry from the time a sample is collected to the time it is extracted or analyzed by the laboratory. Samplers must remember that once filled, sample containers continue to be living, breathing environments that may undergo chemical and physical changes, especially where no chemical preservatives are used. Spiked samples can be an effective means of determining whether field chemical and physical preservation methods are appropriate. Field spiked samples should not be confused with spiked samples used by the laboratory as part of its internal QC program.

The final QC sample that is commonly incorporated into a ground-water sampling program is a field split sample. Field split samples are collected for the purpose of verifying the performance of one laboratory against a laboratory of known performance levels. Typically, field split samples are collected when a regulatory agency wishes to evaluate the performance of a new or unknown laboratory against the regulatory agency's approved laboratory to ensure the accuracy of sample analysis.

Purging and Sampling Device Selection and Operation

Many devices are available for purging and sampling ground water. The SAP must include a description of the devices selected for use on a task-specific basis. It should also include information on how to operate the selected devices and how to maintain the equipment to ensure proper operation in the field. While not always required by regulatory agencies when reviewing SAPs, it is highly recommended that the SAP includes, as an appendix, the operations manuals for each device specified for purging and sampling. This provides sampling team members with critical equipment-specific information such as operating procedures, calibration procedures, troubleshooting tips, spare parts lists, and contact information for equipment repair.

Selection Criteria: One of the most important yet least considered factors that has an effect on the physical and chemical integrity of ground-water samples is the device used for purging and sample collection: Despite being one of the most critical elements of the sampling program, sampling device selection often receives little attention when a sampling plan is prepared. A properly selected sampling device will provide samplers with the most representative sample possible at a reasonable cost; an inappropriate sampling device could have significant and long-lasting effects on sampling results, which may not be immediately apparent.

A number of criteria must be evaluated on a site-specific basis when selecting devices for purging and sample collection. Based on these criteria (summarized in Table 15.7), each device has a unique set of advantages and limitations that define its suitability for use. Sampling device selection criteria are discussed in detail in Nielsen and Yeates (1985), Parker (1994), and ASTM Standard D 6634 (ASTM, 2004d) and summarized subsequently.

Accuracy and Precision of the Device: As stated earlier, a key objective of ground-water sampling programs is to collect representative samples of water in the formation to permit characterization or monitoring of formation chemistry: To do this, it is essential

TABLE 15.7

Selection Criteria for Ground-Water Purging and Sampling Devices

---
Accuracy and precision of samples provided by the device
Materials of construction of the device and accessory equipment
Outside diameter of the device
Lift capability of the device
Flow-rate control and range
Ease of operation and in-field servicing
Portable versus dedicated application
Ease of field decontamination
Reliability and durability
Purchase price and operating costs

---

that sampling devices have the physical capability of moving ground water from *in situ* conditions to ground level in a manner that ensures minimal impact on the physical and chemical properties of the sample (i.e., accuracy). This is essential if analytical data derived from the analysis of samples are to be meaningful. In addition, to ensure that data from different sampling events can be compared, as is essential for long-term monitoring programs, sample collection must be performed in a manner that can be repeated each sampling event, regardless of whether the sampling team members change (i.e., precision). To determine whether this can be accomplished by any individual device, it is important to remember that *in situ* ground water is at a stable temperature and at a pressure that is higher than atmospheric conditions at ground surface.

Materials of Construction: The choice of materials used in the construction of purging and sampling devices should be based on knowledge of the geochemical environment and how the materials may interact with the water collected as a sample via physical, chemical, or biological processes: Materials used in the manufacture of purging and sampling devices and associated tubing, hoses, pipes, and support lines (e.g., rope, cable, or chain) may be a source of bias or error. Materials should not sorb analytes from samples, desorb previously sorbed analytes into samples, leach matrix components of the material that could affect analyte concentrations or cause artifacts, or be physically or chemically degraded by water chemistry.

Materials commonly used in the manufacture of sampling devices include rigid PVC (Type I PVC), stainless steel, polytetrafluorethylene (PTFE),[1] polyethylene (PE), polypropylene (PP), flexible PVC (Type II PVC), fluoroelastomers,[2] polyvinylidene fluoride (PVDF), Buna-N, ethylene–propylene diene monomer (EPDM), and silicone rubber (Figure 15.24). A number of studies indicate the relative sorption and desorption rates of these materials, their potential for alteration of sample chemistry, and their desirability for use in sampling devices (Barcelona et al., 1983; Barcelona et al., 1985; Reynolds and Gillham, 1985; Barcelona and Helfrich, 1986; Holm et al., 1988; Gillham and O'Hannesin, 1990; Parker, 1991).

Other materials-related sources of bias or error include surface traces of organic extrusion aids or mold release compounds used in the extrusion or molding of polymeric materials. In addition, some formulations of polymeric materials may contain fillers or

---
[1] PTFE is also commonly known by the trade name Teflon®, which includes other fluoropolymer formulations. Teflon is a registered trademark of E. I. DuPont De Nemours and Company.
[2] Fluoroelastomers (FPM and FKM) are commonly known by the trade name Viton®, a registered trademark of DuPont.

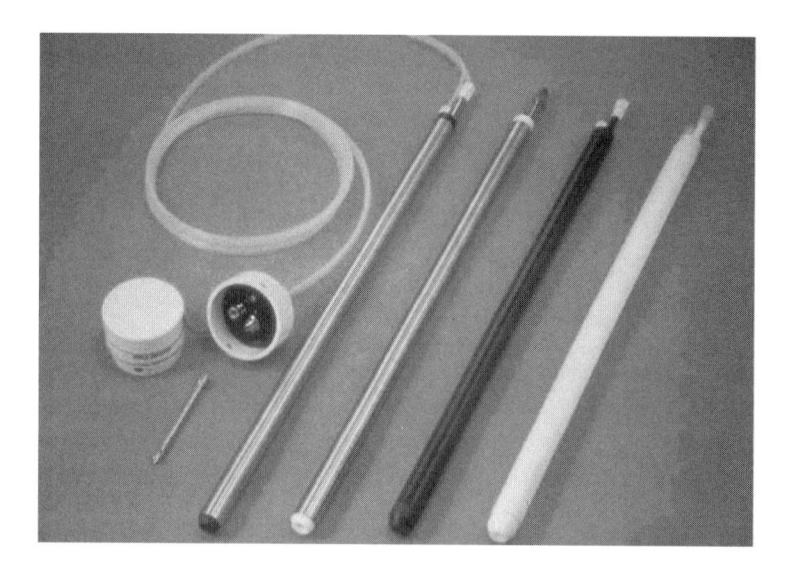

FIGURE 15.24
Some sampling devices can be made of a variety of materials; for others, the materials are fixed. In this case, the dedicated bladder pumps pictured all have PTFE bladders and (left to right) are made of stainless steel and PVC, stainless steel and PTFE, all PVC, and all PTFE. Materials for any particular dedicated application of this type of pump should be selected based on the geochemical environment to which it will be exposed. (Photo courtesy of Severn-Trent/QED.)

processing additives that can leach from the material and alter sample quality. Traces of cutting oils, solvents, or surface coatings may be present on metallic materials. These should be removed by thorough cleaning and, once removed, should not affect sample chemistry. Metallic materials are subject to corrosion, the residues of which could potentially affect sample quality. Electropolishing or other surface passivation processes can improve corrosion resistance. In all cases, the types of materials that may contact the sample should be selected on the basis of the potential for chemical or physical interactions to occur during both short contact times, as in sample collection, and long contact times, typically associated with dedicated equipment submerged in ground water for the duration of the sampling program. The potential interactions that may occur include the following:

- Corrosion of steel parts in water with a low pH, high O_2 content, high CO_2 content, high total dissolved solids content, high H_2S content, or high Cl^- content, or where iron bacteria are present (Figure 15.25) (generally only a concern in long-term contact situations)
- Degradation of plastic materials in the presence of high concentrations of some chlorinated solvents
- Adsorption and later desorption of metals or organic compounds onto or from solid surfaces (including rigid and flexible parts)
- Leaching of compounds from plastic materials

The reader is referred to NWWA/PPI (1981), Barcelona et al. (1983, 1985), Reynolds and Gillham (1985), Barcelona and Helfrich (1986), Holm et al. (1988), Gillham and O'Hannesin (1990), Parker (1991), and Chapter 10 of this book for additional detailed information on materials compatibility with solutes, relative adsorption and desorption rates, potential

FIGURE 15.25
Where steel materials are selected (even stainless steel), the geochemical environment should be evaluated not only for corrosion potential but also for the presence of iron bacteria. These bacteria secrete a material that causes significant corrosion on steel materials from which some dedicated pumps are made. (Photo courtesy of Severn-Trent/QED.)

for alteration of sample chemistry, and material rankings in terms of desirability for use in different sampling applications.

To further minimize the potential for sample alteration caused by sample contact with materials, the surfaces of all materials should be thoroughly cleaned prior to use or installation in a sampling point using an appropriate protocol to avoid possible cross-contamination concerns (refer to Chapter 20 for additional information on equipment decontamination). Several different protocols may be applied at any given site, depending on the oversight agency, the sampling objectives, and other factors. Some of the more common protocols include those outlined in ASTM Standard D 5088 (ASTM, 2004e) or ASTM Standard D 5608 (ASTM, 2004f), U.S. EPA regional office protocols (e.g., Region 4 EISOPQAM [U.S. EPA, 1997]), and state and local regulatory agency protocols.

The materials used in sample discharge tubing (where sampling pumps are used) provide the greatest potential source of bias or error with respect to possible sample alteration, because contact times with the sampled water and surface contact areas are highest with these materials. The most significant interactions are the potential for sorption of metals and/or organic compounds to the tubing, and possible desorption at a later time, and leaching of materials out of the tubing into the sampled water. Ho (1983), Devlin (1987), and Barker and Dickhout (1988) demonstrated that the silicone rubber tubing (Figure 15.26) recommended by manufacturers for use in the pump head of most peristaltic pumps (because of its extreme flexibility) was a significant source of bias for many organic compounds, including VOCs, caused by sorption and desorption processes. They recommended keeping the tubing length in the pump head to a minimum and using more inert tubing materials for the remainder of the tubing run to and from the well. Barcelona et al. (1983, 1985) found that flexible (Type II) PVC tubing could leach plasticizers (primarily phthalate esters) into sampled water and sorb contaminants from and later desorb contaminants into samples. Holm et al. (1988) demonstrated that gas diffusion through flexible polymeric tubing could introduce measurable concentrations of oxygen into initially anoxic water and that the amount of gas transferred is proportional to the tubing length and inversely proportional to the pumping rate. Any of these interactions

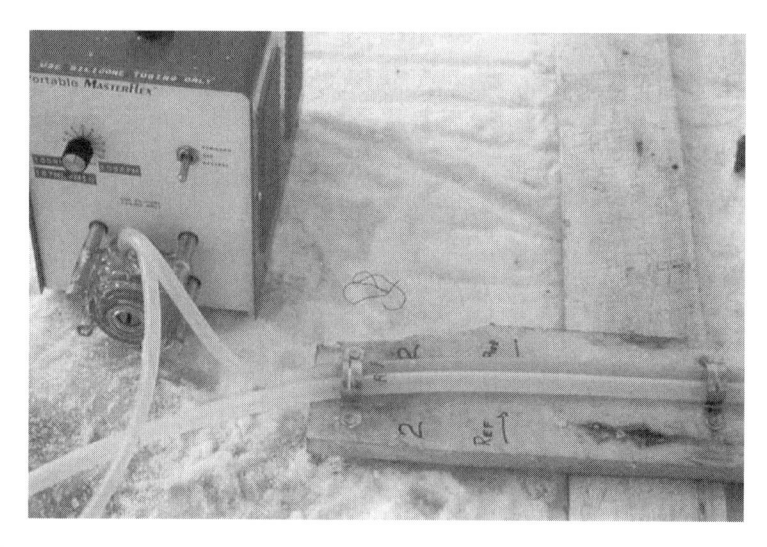

FIGURE 15.26
The silicone rubber tubing typically specified for use with peristaltic pumps can negatively bias samples collected for organic compound analysis.

could introduce significant bias into sampling results, so the sample discharge tubing material should be very carefully chosen to minimize or eliminate this source of bias. PTFE, PE, and PP tubing appear to offer superior performance over other materials.

Outside Diameter of the Device: Historically, the majority of ground-water sampling programs have made use of 2 or 4 in: diameter monitoring wells to facilitate ground-water sample collection. For that reason, the majority of commercially available devices used for purging and sampling wells are designed to fit inside a 2 in. diameter sampling point. However, with increasing use of direct-push sampling tools and smaller diameter monitoring wells to characterize ground-water quality, the need has arisen for smaller diameter devices that will fit inside sampling points with diameters as small as 0.5 in. Not only it is important that these devices be small enough, but they must also deliver samples with the same degree of accuracy and precision afforded by larger diameter devices.

With either traditional monitoring wells or direct-push sampling points, it is also important to consider that casing may not be plumb or may have constrictions limiting the inside diameter (i.e., at rod or casing joints). With direct-push tools, subsurface obstructions, such as boulders, may cause direct-push rod to deflect, resulting in rod that is no longer plumb. In traditional wells, casing may shift over the course of its lifespan to the point where casing joints may fail or where casing is no longer plumb. This is of particular concern at facilities such as landfill sites, where, over time, and under warm ambient subsurface temperatures, well casing can bend or break as wastes settle.

Lift Capability: The term "lift capability" of a device is often misunderstood to mean the maximum depth at which a pump can be installed within a well and still operate: More correctly, the lift capability of a device is linked to the ability of a device to move water from the depth of the static water level within a sampling point (which may be significantly shallower than the intake of the pump) to ground surface. This concept is illustrated in Figure 15.27. The greater the depth to water, the more pumping head the device must overcome to deliver water to the surface. Lift is not related to the depth of the sampling device. The lift capability of the device is a critical selection criterion when deciding whether or not the device is suitable for individual applications, especially when

Ground-Water Sampling

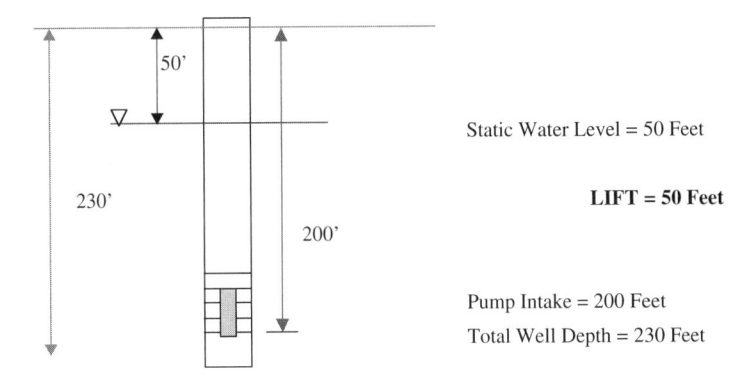

FIGURE 15.27
Lift capability is related to the ability of a pumping device to lift water from the static water level, not the depth at which the pump is set in the well.

working with lifts greater than 200 ft. The selection of available purging and sampling devices is more limited with increased depth to water.

Flow Rate Control and Range: Consideration should be given to appropriate water removal rates when selecting purging and sampling devices: Ideally, the same device will be used for both purging and sampling, therefore, a device should be capable of operating at flow rates suitable for both applications. For sample collection, it is critical to evaluate whether a device can operate at a flow rate suitable for the parameters analyzed in the samples collected. For example, samples collected for analysis of some sensitive parameters (i.e., VOCs and trace metals) should be taken at low flow rates (generally less than 250 ml/min). Sampling rates should be high enough to fill sample containers efficiently and with minimal exposure to atmospheric conditions, but low enough to minimize sample alteration via agitation or aeration. Additionally, the use of low-flow purging and sampling techniques may require adjusting the pumping rate to account for the hydraulic performance of the sampling point. This requires that the flow rate be highly controllable between rates of less than 100 ml/min and more than 500 ml/min (Figure 15.28). Throttling down the device using a valve in the discharge line reduces the flow rate, but creates a pressure drop across the valve, and does not necessarily reduce the speed of the device in the well. Therefore, this method of flow rate control is not recommended. A better method of reducing flow rate is to divert a portion of the discharge stream via the use of a "T" or "Y" fitting.

Ease of Operation and In-Field Servicing: Ease of operation and servicing are important but frequently overlooked practical considerations in the selection of purging and sampling devices: A common source of poor precision in sampling results is a sampling device operating problem (Barcelona et al., 1984). This could be due to any one of several factors including: (1) the device and accessory equipment are too complicated to operate efficiently under field conditions; (2) the operator is not familiar enough with the device to operate it properly; or (3) the operating manual supplied with the device does not clearly outline the procedures for proper use. Thus, it is not only important to select a device that is simple to operate but also to provide proper training for the operators of the device. Because mechanical devices are subject to malfunction or failure, it is desirable to select a device that can be serviced in the field to prevent delays during sampling. An alternative is to have a replacement device available. Some of the devices described subsequently may be too complex for field repairs or may be factory sealed, requiring servicing by the manufacturer or a qualified service facility.

FIGURE 15.28
Flow rate control is important for devices used for both purging and sampling. Control should be exerted directly on the pump driving mechanism, but not through the use of in-line valves.

Operational characteristics, such as solids handling capability, ability to run dry, cooling requirements, and intermittent discharge, must be considered in the application of some purging and sampling devices. Some devices may experience increased wear or damage as solids pass through the device, causing reduced output or failure. Solids may also clog check valves or passages, which can reduce discharge rate or, in the case of grab samplers, cause the sample to leak out.

In wells with little water or low yield, it is important to know whether a device must be fully submerged to operate and what the ramifications of operating the device without water may be. Some devices, such as bailers, are not dependent on submergence to operate, while other devices, such as many (but not all) pumps, do require that they be fully submerged in the water column to operate. A pump running dry can occur when the water level in the sampling point is drawn down below the pump intake. In some pump designs, typically those with rotating or reciprocating mechanisms, this can cause permanent damage to or failure of the device, which is often associated with overheating of the device.

Another operational feature of devices to consider is how water is discharged from the device. In some devices, such as bladder pumps, ground water is brought to the surface in an intermittent or cyclic discharge. Intermittent discharge creates some special considerations for sampling teams to ensure the accuracy of sample collection, particularly for volatile constituents and for accurate measurement of indicator parameters with in-line monitoring devices and flow-through cells. For example, sampling teams must learn how to optimize the discharge from devices with intermittent flow to ensure that sufficient sample volume is available during the pump discharge cycle to fill a 40 ml VOC sample vial without headspace. When taking measurements for indicator parameters in a flow-through cell, care must be taken to ensure that measurements are made during pump discharge cycles while water is running over the sensors. This is especially critical when measuring parameters such as DO or oxidation–reduction potential (ORP). When filtering, care should be taken to prevent air from entering the filter during pump refill cycles.

Portable versus Dedicated Applications: As defined in ASTM Standard D 6634 (ASTM, 2004d), a "dedicated" device is one that is permanently installed in a sampling point and does not come into contact with ambient or atmospheric conditions during operation: The only time dedicated equipment should come into contact with atmospheric conditions is during periodic routine maintenance. A bladder pump (Figure 15.29) is an excellent example of a device ideally suited to a dedicated installation. A "designated" device is a device that is assigned for use at a single sampling point, but which, by virtue of its design, must come into contact with ambient conditions to operate. Grab sampling devices such as bailers can therefore be a designated device but not a dedicated device. The third category of device is a "portable" device. Portable purging and sampling devices are used in multiple sampling points at one or more sites. The device is in contact with ambient conditions between sampling events or between wells, and is typically stored at a location remote from the field site. The electric submersible gear-drive pump (Figure 15.30) is a popular pump that is used portably.

The decision regarding whether to use portable or dedicated purging and sampling equipment is made on a site-by-site basis with a number of considerations in mind, including the number of sampling events anticipated at the site, the level of QA/QC required in the field, ease of equipment decontamination (contaminant concentrations, equipment design features, and ability to thoroughly clean the equipment), and accessibility of sampling points. At sites where a ground-water sampling program requires that six or more sampling events be implemented, it is generally more cost effective to install a dedicated system of sampling pumps. While, in general, it is more expensive initially to purchase and install dedicated equipment for each sampling point, cost savings will be

FIGURE 15.29
Bladder pumps such as the one installed in the well pictured here, are ideally suited to dedicated applications, meaning that the pump is not removed from the well after initial installation except for routine maintenance. (Photo courtesy of Severn-Trent/QED.)

FIGURE 15.30
Portable devices are moved from well to well and cleaned in between uses, so the ideal portable device (and accessory equipment) is lightweight and easy to clean. This electric submersible gear-drive pump is a popular portable purging and sampling device.

realized through significantly lower operating costs associated with equipment setup, decontamination, and collection of fewer QC samples (i.e., equipment blanks). In addition, with dedicated equipment, there is a low potential for the sampling team to be exposed to ground-water contaminants, equipment is not exposed to atmospheric conditions, cross-contamination of sampling points is virtually eliminated, and there is no agitation of the water column associated with equipment deployment into the sampling point. The net result is that more accurate and precise samples can be collected from a dedicated sampling system.

There will be sites, however, where either due to the limited number of sampling events or the temporary nature of sampling points, portable equipment will be the preferred option. In these situations, a common device is used to purge and sample each sampling point. As a consequence, when selecting equipment for portable use, it is critical to focus on the ease with which the device and all accessory equipment (i.e., tubing or tubing bundles, hose reels, battery packs, generators, compressed air source, controlling devices, decontamination equipment and supplies, purge water containers, etc.) can be moved between sampling points — especially in areas of rough terrain. While some devices can be hand-carried to remote sites, some manufacturers have mounted their equipment on backpack frames, small wheeled carts, or specialized vehicles in an effort to improve portability (Figure 15.31). Other equipment is too bulky and heavy to be transported in the field without being vehicle mounted. Another important consideration is the ease with which the portable equipment can be cleaned between sampling points. Manufacturers often design equipment especially for portable applications so it can be easily disassembled in the field for thorough cleaning.

There are several disadvantages to using portable equipment. The equipment will be exposed to surface contaminants and contaminants from other sampling points which,

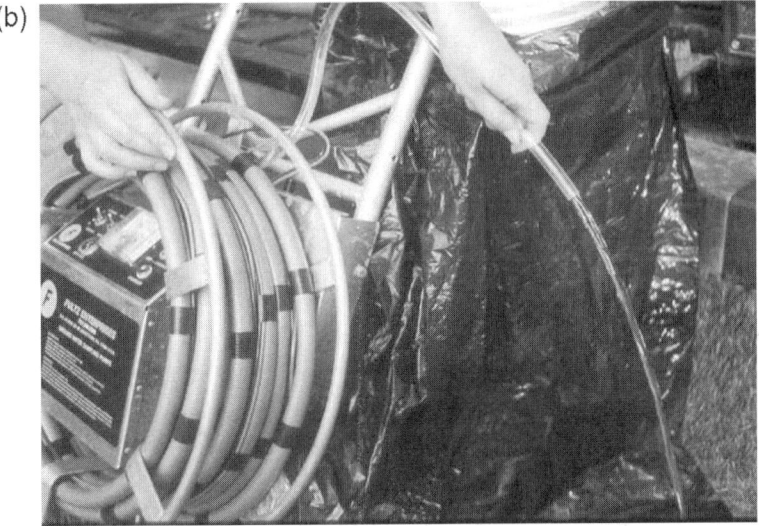

FIGURE 15.31
Portable devices are often mounted on wheeled carts (a), backpack frames (b) or other convenient means of transport.

if not removed during equipment cleaning, can introduce error into samples being collected. In addition, sampling team members are at greater risk of exposure to contaminants in ground water due to their handling of the sampling device and accessory equipment. Of significant concern in many projects is the increase in variability (imprecision) of operation of portable equipment between sampling team members even though they all may be following the same sampling protocol. Issues such as inconsistency in the depth of placement of the pump intake and care with which the device is lowered through the water column can be significant sources of error. As an example, when samplers elect to use portable pumping equipment for low-flow purging and sampling, they must be aware that regardless of how carefully the device is lowered into the sampling point, there will be some disturbance of the water column. This may result in resuspension of sediments that have settled to the bottom of the well screen or sump, which will be measured as higher sample turbidity. To help compensate for this source of error, time must be built into the sampling protocol to permit some equilibration time immediately following pump placement. Some sampling teams may be extremely careful in situating the pump and waiting a prescribed timeframe, while others may ignore these considerations entirely.

The issue of increased turbidity due to pump installation in a well was documented in work done by Kearl et al. (1992) and Puls and Powell (1992a) using a colloidal borescope. Kearl et al. observed a significant increase in particle size and number of particles when a portable pump was installed in a monitoring well prior to purging and sampling. This same effect was observed as increased turbidity measurements in a study conducted by Puls et al. (1992). In both cases, the effects of sampling device insertion decreased with time. In a field study conducted in Kansas City, Missouri, where low-flow purging methods were compared to traditional purging methods, Kearl et al. (1994) determined that the effect of increased particle size and increased number of particles resulting from the insertion of a borescope into a monitoring well, subsided after 30 min. After that time, laminar flow conditions dominated and particle size and number decreased.

Designated equipment is often used in an attempt to have the benefits of both dedicated and portable equipment. In theory, because the device is used in only one sampling point, the potential for cross-contamination of sampling points and samples is minimized. However, introduction of contaminants as a result of improper storage between sampling events and exposure to atmospheric conditions during operation are potential sources of error. In some cases, equipment is stored inside its assigned sampling point (e.g., a designated bailer may be hung inside a well [Figure 15.32]), while in other cases, the equipment is stored remotely (e.g., wrapped in plastic and stored in a field office). As a consequence, field equipment blanks should always be collected to verify cleanliness of designated equipment.

Portable equipment must be cleaned between uses in each sampling point or discarded after use to avoid cross-contamination of sampling points and samples. Thus, the

FIGURE 15.32
A bailer can be designated for use in a well, but not dedicated, because it must be removed from the well and exposed to atmospheric conditions during use. This designated bailer is hung from the well cap between sampling events.

components of portable sampling equipment must be able to withstand the necessary cleaning processes. Some devices, by virtue of their design, are very easy to disassemble for cleaning (Figure 15.33), while others are more difficult (Figure 15.34) and some are impossible to disassemble for cleaning. It may be more practical to clean these devices by circulating cleaning solutions and rinses through the device and any associated tubing, hose, or pipe in accordance with ASTM Standard D 5088 (ASTM, 2004e) or to replace the associated tubing, hose, or pipe. Field decontamination operations can be difficult due to the need for sufficient decontamination supplies, exposure of the equipment to potential contaminants, and handling and disposal of the decontamination waste water and supplies. Where field decontamination is not practical or possible, it may be more practical to use dedicated devices or take a number of portable sampling devices into the field and decontaminate them later at a more appropriate location. Following any cleaning procedure, equipment blanks should be collected to assess the effectiveness of the cleaning procedure. The reader is directed to Chapter 20 of this text for a more detailed discussion of field equipment cleaning procedures.

Reliability and Durability of the Device: A sampling team must be able to rely on the chosen device to perform under field conditions not only for the duration of the sampling event, but also for the lifetime of a monitoring program: A number of practical issues must be considered related to reliability and durability of a device, including battery life, mechanical reliability (moving parts must continue to function for extended periods), ability to withstand chemically aggressive environments, ability to withstand rough handling in the field, ability to maintain water-tight connections, strength of tubing and cables, and sensitivity of the device to outdoor conditions (temperature, light, dust, precipitation, high humidity). Unfortunately, some of these features can only be evaluated during actual field use. Overly optimistic statements in sales literature may not hold true for all operational conditions in the field.

Purchase Price and Operating Costs: Both the initial capital cost and the operating cost (consumable supplies and maintenance costs) of the sampling device and accessory equipment are important considerations: However, cost considerations should not result in the selection of a device that compromises DQOs or sample accuracy and precision.

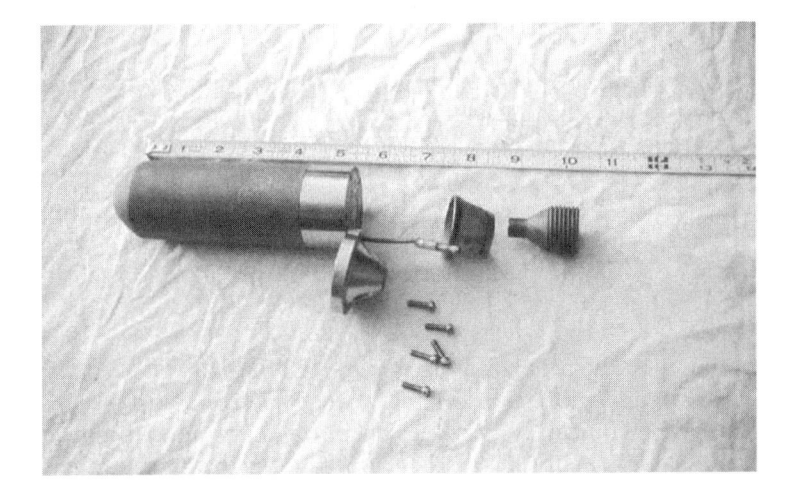

FIGURE 15.33
This electric submersible gear-drive pump is easily disassembled for cleaning, requiring removal of a threaded fitting to take off the inlet screen and five screws to access the internal gears, a process easily accomplished in a few minutes.

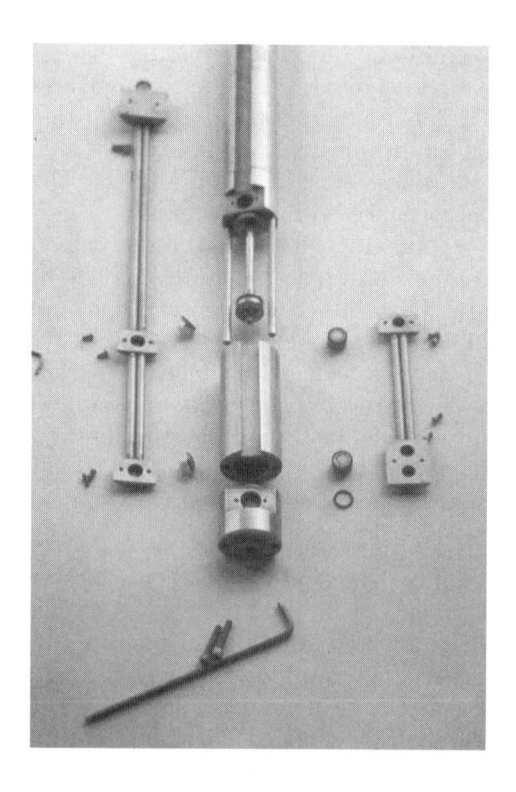

FIGURE 15.34
This double-acting piston pump is more difficult to disassemble for cleaning, as it has multiple valves, o-rings and discharge tubes, a pump cylinder and piston, and an inlet screen, most of which are held in place by multiple hex screws. Make sure you have spare parts, as the pump will not function without all of the pieces in place. (Photo courtesy of Bennett Sample Pumps Inc.)

Proper selection and use of purging and sampling devices will more than pay for the capital and operational costs by providing accurate and precise samples, resulting in cost savings from fewer false positive (or false negative) analytical results, fewer resampling events, and fewer problems in meeting regulatory or scientific goals and objectives.

Purging and Sampling Equipment Options: A wide variety of purging and sampling equipment is available for use in ground-water sampling programs: Available devices can be classified into six general categories: grab sampling mechanisms (including bailers, syringe samplers, and thief samplers), suction-lift mechanisms (including surface centrifugal pumps and peristaltic pumps), electric submersible centrifugal pumps, positive displacement mechanisms (including gas-displacement pumps, bladder pumps, piston pumps, progressing cavity pumps, and electric submersible gear-drive pumps), inertial-lift pumps and passive sampling devices (including passive diffusion bag samplers [PDBSs]). Although frequently used in the ground-water industry for well development, the air-lift pumping method is generally considered unsuitable for purging and sampling because the extensive mixing of drive gas and water is likely to strip dissolved gases from the ground water and significantly alter the concentration of other dissolved constituents (Gillham et al., 1983). This method is not discussed in this chapter for this reason.

Each of the purging and sampling devices described subsequently has specific operational characteristics that partly determine the suitability of each device for specific applications. These operational characteristics are listed in Table 15.8, which summarizes information derived from current manufacturers' specifications for the various devices.

TABLE 15.8
Operational Characteristics of Common Purging and Sampling Devices

Device	Operational Mechanism	Minimum Outside Diameter (in.)	Maximum Lift (ft)	Materials of Construction[a]	Dedicated or Portable	Power Source
Bailer	GS	0.375	Unlimited	Any	P	Manual or mechanical
Peristaltic pump	SL	0.375	25	Variable; tubing must be silicone or flexible PVC	P (pump); tubing may be designated or dedicated	Electric
Electric submersible centrifugal pump	CP	1.75	230	Stainless steel and PTFE	P or D	Electric
Double-acting piston pump	PD	0.625	1000	Stainless steel and PTFE	P or D	Pneumatic
Progressing cavity pump	PD	1.75	180	Stainless steel and EPDM or Viton	P or D	Electric
Electric submersible gear pump	PD	1.75	250	Stainless steel and PTFE	P or D	Electric
Bladder pump	PD	0.375	1000	Stainless steel, PTFE, PVC, and PVDF	P or D	Pneumatic or mechanical
Gas-displacement pump	PD	0.375	300	Stainless steel and PTFE	P or D	Pneumatic
Inertial-lift pump	IL	0.375	260	Acetal thermoplastic stainless steel HDPE and LDPE, PTFE, and PVC	P or D	Manual, mechanical, or electric

[a]For all pumps (except peristaltic), tubing may be composed of any flexible or rigid material.
GS = Grab sampler.
CP = Centrifugal pump.
SL = Suction lift pump.
PD = Positive displacement pump.
IL = Inertial lift pump.

Grab Samplers: Bailers, syringe samplers, and thief samplers (e.g., messenger samplers) are all examples of grab sampling devices used in traditional monitoring wells and direct-push sampling points. These devices are lowered into the sampling point on a cable, rope, string, chain, or tubing to the desired sampling depth and then retrieved for purge water discharge, sample transfer, or direct transport of the device to the laboratory for sample transfer and analysis. Grab sampling devices are generally not limited to a maximum sampling depth, although use in very deep sampling points may be impractical. Because bailers can be manufactured in very small diameters (less than 0.5 in.), they are usually not limited in use to a particular diameter of sampling point; other types of grab samplers require inside diameters of at least 2 in. The rate at which water can be removed with a grab sampler will depend on the volumetric capacity of the device and the time required for lowering, filling, and retrieval and whether or not a bottom-emptying device is used to decant the sample.

Some grab samplers are prone to malfunction or damage by sediment in the well. Operational difficulty may be experienced in sandy or silty water due to check valve or seal leakage. When used portably, the ability to clean or decontaminate a grab sampler between wells will vary depending upon design. Bailers are generally easier to clean than other types of grab sampling devices and are widely available as disposable devices.

Bailers: The most commonly used grab samplers are bailers, in single check valve and dual check valve designs. Bailers are typically constructed of stainless steel, various plastics (e.g., PVC and PE), and fluorocarbon materials (e.g., PTFE) (Figure 15.35) and come in a variety of designs for portable, designated, and disposable applications. Bailers cannot be dedicated, but may be designated for use at a single sampling point.

The single check valve bailer is lowered into the sampling point. The act of lowering the bailer through the water column opens the check valve and water fills the bailer. Upon retrieval, the weight of the water inside the bailer closes the check valve as the bailer exits the water column. The water in the bailer is retained from the greatest depth to which the bailer was lowered. There is some potential for the contents of the bailer to mix with the surrounding water column during deployment or retrieval, depending on the design of the bailer top. A dual check valve bailer is intended to prevent mixing of the sample with the water column upon retrieval. Water passes through the bailer as it is lowered. Upon retrieval, both check valves seat, ideally retaining a depth-discrete aliquot of water inside the bailer.

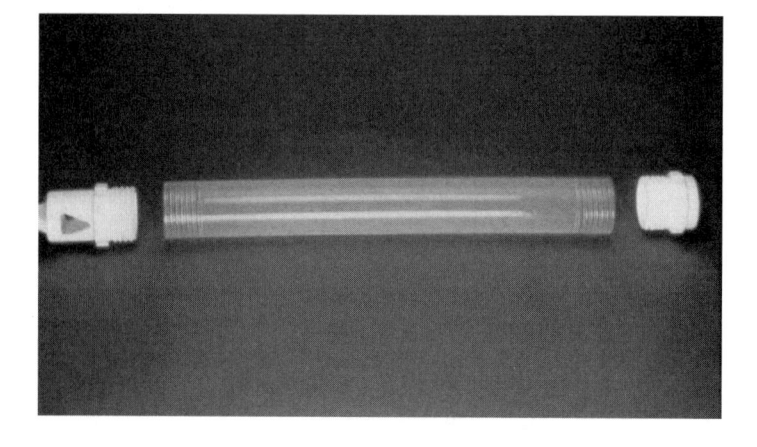

FIGURE 15.35
A bailer is a very simple device, consisting of a tubular body, a lifting bail at the top (left) and a check valve at the bottom (right). Bailers can be made of any material appropriate for contact with samples. This example is an all-PTFE bailer.

Ground-Water Sampling

In the case of both single and dual check valve bailers, the sample water is decanted into a sample container following retrieval of the bailer. A bottom-discharge device with flow control may be used to provide improved control over the discharge of water from the bailer into the sample container. Figure 15.36 illustrates an example of this type of device. A bottom-discharge device may not work with a dual check valve bailer unless the bailer design allows for release of the upper check valve during sample decanting.

Bailers are commonly used for both purging and sampling in small diameter, shallow wells, primarily because of their convenience, ease of use, and low cost. While they have proven to be useful for collecting samples of LNAPLs from the top of the water column of water table wells, they cannot be reliably used to purge wells or to collect representative samples for most parameters, particularly those sensitive to agitation and oxidation.

Based on a large body of scientific evidence gathered in a number of field studies (e.g., Pohlmann and Hess, 1988; Yeskis et al., 1988; Pohlmann et al., 1991, 1994; Puls et al., 1991, 1992; Puls and Powell, 1992a; Backhus et al., 1993; Heidlauf and Bartlett, 1993), it is exceedingly difficult to collect accurate or precise samples with bailers for a wide range of analytical parameters including VOCs, trace metals, colloid-associated analytes, and dissolved gases.

Muska et al. (1986), Imbrigotta et al. (1988), Yeskis et al. (1988), Tai et al. (1991), and Gibs et al. (1994) all concluded that bailers provided poor precision in analytical data for VOCs compared with several different types of pumps, and attributed the results to variability in operator technique. Heidlauf and Bartlett (1993) found that when wells were purged and sampled with bailers, ground-water samples consistently had high turbidity values (in excess of 100 NTUs) and that metal analyte concentrations of unfiltered samples were

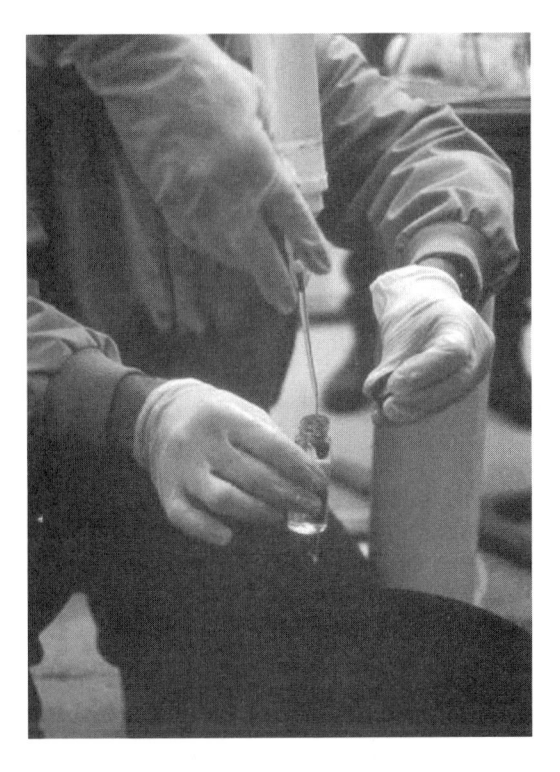

FIGURE 15.36
A bottom-discharge device, such as the one pictured here, makes control of sample discharge much easier than decanting from the top of the bailer.

significantly higher than those of filtered samples. Puls et al. (1992) found that bailed samples had turbidities exceeding 200 NTUs, while samples collected by pumping at low flow rates had turbidities consistently less than 25 NTUs, and that As and Cr levels in bailed samples were five and two times higher, respectively, than in samples collected with a pump. Tests by Backhus et al. (1993) showed that bailing continued to produce samples with high turbidity even after 60 well volumes of purging. Further, they found that bailed and unfiltered samples contained up to 750 times greater concentrations of high molecular weight PAHs than pumped samples from the same wells at a coal tar contaminated site. These PAHs were determined to be sorbed primarily to the immobile aquifer solids and not part of the mobile contaminant load. Puls et al. (1990) found that Cr levels in bailed and unfiltered samples were two to three times higher than levels in samples collected by low-flow rate pumping. Use of data obtained from bailed samples at all of these sites could have led to substantial overestimation of mobile contaminant concentrations in ground water.

Pohlmann et al. (1994) found that particle size distributions of bailed samples were highly skewed toward particles greater than 5 μm in diameter (over 90% in some wells), while pumped samples contained a more uniform distribution of particle sizes. They also determined that bailers produced higher concentrations of particles that could be mobile in ground water (e.g., those in the range of 0.03 to 5.0 μm) than pumping, suggesting that concentrations of colloid-associated contaminants could be positively biased when bailing disturbs the sampling zone and elevates artifactual turbidity. Total particle concentrations in bailed samples were significantly higher (up to 20 times higher) than concentrations in pumped samples. Turbidity values obtained with bailers were as much as two orders of magnitude greater than pumped samples from the same wells. Backhus et al. (1993) also found concentrations the turbidity of bailed samples was several orders of magnitude higher than of pumped samples from the same wells and that there was an enormous difference in the size of particles in bailed samples versus pumped samples. Bailed samples contained particles in the range of 1 to 100 μm in diameter, whereas pumped samples mainly contained particles less than 5 μm in diameter. They determined that particles as large as 100 μm could not have been mobile at ambient ground-water flow rates and concluded that bailing collects particles and particle-associated contaminants that are not representative of *in situ* mobile contaminant loads.

To summarize, the primary drawbacks of using bailers to purge and sample wells include:

- The repeated insertion of the bailer into and withdrawal of the bailer from the water column in the well, even if it is done carefully, aerates the water column and creates a surging effect, which mixes and severely agitates the water column. The surging action that results from using bailers to purge a well creates two-directional flow within the well screen, resulting in continual development or over-development of the well (Puls and Powell, 1992a) and grain flow within the filter pack (Pohlmann et al., 1994). As previously noted, bailing-induced agitation also mobilizes previously immobile aquifer matrix materials, creating substantial artifactual turbidity in water brought into the well. This may occur in wells installed in any type of formation, but is a particular problem in formations that are predominantly fine-grained (i.e., silts and clays). The surging effect also commonly results in resuspension of sediments that have accumulated and settled in the bottom of the well screen or sump, which also results in increased turbidity measurements. Ultimately, decreased life of the well can be expected to occur as a result of repeated bailing.
- Mixing the stagnant water column in the well casing with the dynamic water column in the well screen results in aeration of the entire water column and a

composite averaging effect for all water in the well, which defeats the purpose of purging and makes it impossible to collect a sample representative of formation water chemistry. Puls and Powell (1992a) and Pohlmann et al. (1994) determined that measuring representative values of DO and turbidity during purging with bailers was not possible and that equilibrium conditions for those parameters often could not be reached. DO concentrations in bailed samples are consistently higher than in pumped samples (in the study by Pohlmann et al. [1994], 10 to 20 times higher). Enrichment of DO during the sampling process not only produces samples that are unrepresentative with respect to oxygen content, but also results in oxidation and subsequent precipitation of reduced species, such as Fe^{2+}, and removal of other metal species through adsorption or co-precipitation (Stoltzenberg and Nichols, 1986). Alteration of dissolved constituents in this way during sampling may lead to erroneous conclusions about their concentration or speciation.

- Bailers are highly subject to uncontrolled operator variability, which leads to inconsistency in how wells are purged and samples are collected. This results in poor precision and accuracy (Imbrigotta et al., 1988; Yeskis et al., 1988; Tai et al., 1991; Muska et al., 1986). Operator variability can occur not only between different operators but also with the same operator (Yeskis et al., 1988).

Thief Samplers: Another type of grab sampler, called a thief sampler, employs a mechanical, electrical, or pneumatic trigger to actuate plugs or valves at either end of an open tube to open or close the chamber after lowering it to the desired sampling depth, thus sampling from a discrete interval within the well. Figure 15.37 is an example of this type of sampler called a Kemmerer sampler. This type of device is impractical to use for purging, but can be used to sample from any depth.

Syringe Samplers: The first syringe devices used for ground-water sampling were essentially commercially available medical syringes adapted to sample in a 2 in. diameter well (Gillham, 1982). Several manufacturers have since produced syringe devices specifically for ground-water sampling. The syringe sampler illustrated in Figure 15.38 is divided into two chambers by a moveable piston or float. The upper chamber is attached to a flexible air line that extends to the ground surface. The lower chamber is the sample chamber. The device is lowered into the sampling point and activated by applying suction to the upper chamber, thereby drawing the piston or float upward and allowing water to enter the lower chamber. In situations where the pressure exerted on the lower chamber by submergence is great enough to cause the piston or float to move upward prior to achieving the desired sampling depth, the upper chamber can be pressurized during placement in the well to prevent piston movement. The device is then activated by slowly releasing the pressure from the upper chamber, allowing water to fill the lower chamber under hydrostatic pressure. Syringe devices can be used to sample from any depth and for any parameter, but they are impractical to use for purging.

Other Grab-Sampling Devices: Several other grab-sampling devices have been developed in recent years to sample ground water, including the HydraSleeve™, the Discrete Interval Sampler, the Kabis sampler, the Snap Sampler, and the Pneumo-Bailer. The HydraSleeve is discussed later in this chapter; only very limited information is available on the other devices and their applications and limitations. The interested reader is directed to Parker and Clark (2002), ITRC (2005) and Parsons (2005) for what information is available.

Suction-Lift Devices: Surface centrifugal pumps and peristaltic pumps are two common types of suction-lift pumps: These pumps, typically situated at or above ground level during purging and sampling, draw water to the surface by applying

FIGURE 15.37
This Kemmerer sampler, typically used for surface-water sampling, has been adapted for use in a 2 in. diameter well. It is used in wells to sample from a discrete depth (e.g., from within the well screen), but is impractical to use for purging.

suction to an intake line through the use of impellers or rotors typically driven by an electric motor. In theory, suction-lift pumps are limited to lifting water approximately 34 ft, depending on altitude and barometric pressure. In practice, a lift of 15 to 25 ft is the typical upper limit. The diameter of sampling points to which these devices are applicable is limited only by the size of the intake tubing used; 0.25 or 0.375 in. tubing can be used to sample sampling points as small as 0.5 in. in diameter. Sediment has only a minor effect on suction-lift pumps, although large solids may plug the pump intake line.

The pressure decrease caused by suction-lift devices can result in a number of deleterious effects on samples. The most notable effects include loss of dissolved gases and VOCs and accompanying pH shifts that can affect dissolved metals concentrations. Several researchers have noted losses ranging from as low as 4% to as high as 70% for a variety of VOCs (Ho, 1983; Barcelona et al., 1984; Devlin, 1987; Barker and Dickhout, 1988; Imbrigiotta et al., 1988) and losses from 7 to 17% for several trace metals including B, Ba, Sr, Hg, Mb, and Se (Houghton and Berger, 1984). Generally, losses of VOCs will be greater for compounds with high Henry's law constants. Suction–lift pumps are therefore best suited to collecting samples for parameters that are not sensitive to pressure decreases, such as the very stable major ions, and SVOCs with very low Henry's law constants, such as PCBs and pesticides.

Surface Centrifugal Pumps: Surface centrifugal pumps use impellers typically constructed of metal (brass or mild steel), plastic, or synthetic rubber, usually set within a cast iron or steel pump chamber. Before the pump can operate, it must be primed with

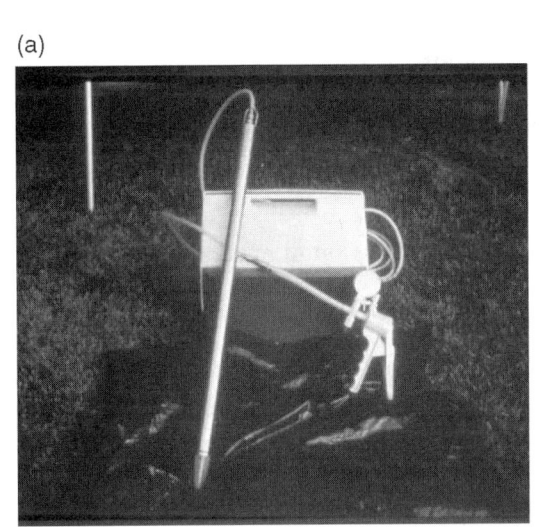

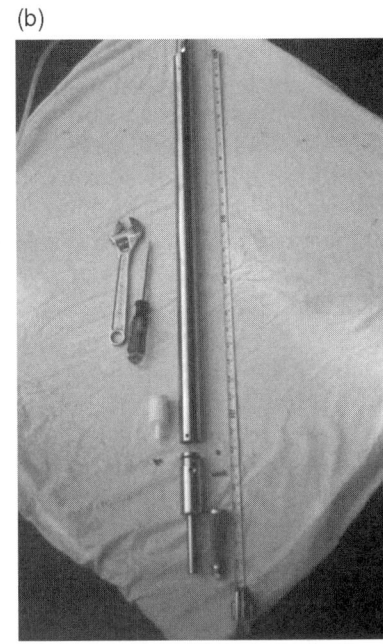

Ground-Water Sampling

FIGURE 15.38
A syringe sampler collects discrete samples by exerting a slight negative pressure on a piston to draw the sample through a bottom check valve into the device. (a) Shows the device ready to install in a well (with a small hand pump to apply suction to the piston). (b) Shows the device disassembled for cleaning between uses. Syringe devices are impractical to use for purging.

water at the surface so it can create the suction necessary to lift water from a well. After priming, the pump can continue to operate without addition of more water. Figure 15.39 shows a representative design for this type of pump. These pumps can pump at rates of 2 to 40 gal/min, with 5 to 15 gal/min being more typical. Because

FIGURE 15.39
Surface centrifugal pumps can be used to purge wells, but they are not appropriate for collecting samples because of their materials of construction and because of the strong negative pressure required to lift the water to the surface. (Photo courtesy of Todd Giddings.)

surface centrifugal pumps operate by suction lift and because they are prone to cavitation, they are not appropriate for collection of samples to be analyzed for pressure-sensitive parameters, such as dissolved gases, VOCs, or trace metals. Because the pumped water contacts the pump mechanism, which is not constructed of inert materials, artifacts from sample contact with these materials should be considered when evaluating these pumps for sampling. In addition, these pumps can mix air from small leaks in the suction circuit into a sample, which can cause sample aeration. These pumps are typically difficult to adequately decontaminate between uses because they are difficult to disassemble. They may be useful for purging, but because of all of their limitations, they are not recommended for sampling.

Peristaltic Pumps: The peristaltic pump is the most common type of suction-lift device used for ground-water sampling in shallow sampling points. A peristaltic pump (Figure 15.40) consists of a rotor with rollers that squeeze flexible tubing as they revolve within a stator housing. This action generates a reduced pressure at one end of the tubing and an increased pressure at the other end. Peristaltic pumps operate at rates of less than 0.01 gal/min to more than 12 gal/min. Peristaltic pumps do not usually cause cavitation but, as in all suction-lift pumps, the application of a reduced pressure on the sample can bias the sample for pressure-sensitive parameters.

Several types of elastomeric material can be used for the pump tubing, although flexible PVC and silicone rubber are most common. The flexible tubing required for use in a peristaltic pump mechanism may cause sample bias. The plasticizers in flexible (Type II) PVC can contaminate samples with phthalate esters (Ho, 1983; Barcelona et al., 1985; Pearsall and Eckhardt, 1987; Barker and Dickout, 1988). The use of silicone rubber tubing, which contains no plasticizers, can obviate this problem. However, the potential for sample bias due to sorption or desorption exists with both materials (Barcelona et al., 1985). These pumps can be used with an intermediate transfer vessel, so the sample contacts only the intake tubing and vessel (which can be made of any appropriate material), avoiding contact with the pump tubing as illustrated in Figure 15.41. Alternatively, using a very

FIGURE 15.40
The pump head of a peristaltic pump contains a set of rollers and a rotor, driven by a pump motor, that squeeze flexible (usually silicone) tubing stretched over the rollers, to create negative pressure on one side of the pump head and positive pressure on the other. The negative pressure can bias samples collected for analysis of pressure-sensitive parameters including dissolved gases, VOCs, and trace metals.

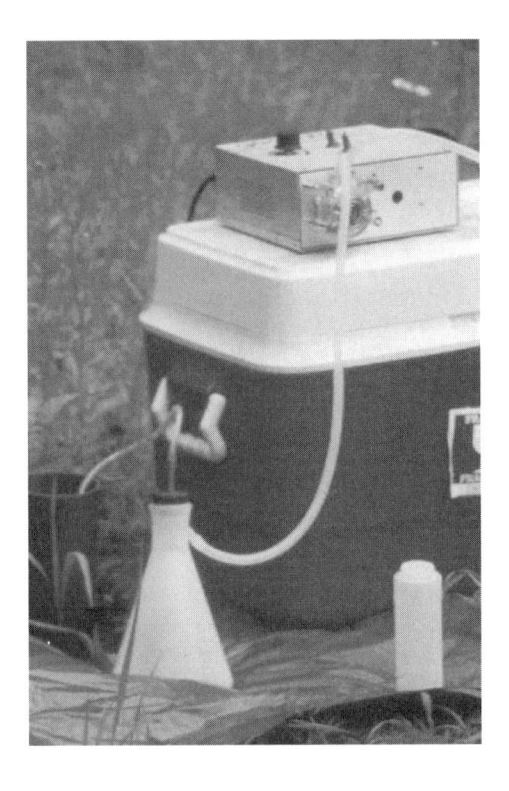

FIGURE 15.41
To avoid contact between the flexible pump tubing and the sample, a sample transfer vessel can be installed between the well and the peristaltic pump. As suction is exerted on the transfer vessel, water is drawn into the vessel through the tubing extended down the well. This configuration results in a slight loss of pumping efficiency, decreasing the lift capability of the pump.

short run of silicone rubber tubing at the pump head only can minimize this problem (Ho, 1983; Barker and Dickhout, 1988). Manufacturers' recommendations for tubing materials should be followed in conjunction with chemical compatibility charts.

Electric Submersible Centrifugal Pumps: An electric submersible centrifugal pump (Figure 15.42) consists of impellers housed within diffuser chambers that are attached to a sealed electric motor, which drives the impellers through a shaft and seal arrangement. Water enters the pump by pressure of submergence, is pressurized by centrifugal force generated by the impellers, and discharged to the surface through tubing, hose, or pipe. An electric submersible centrifugal pump is suspended in a well by its discharge line or a support line. Electric power is supplied to the motor through a braided or flat multiple-conductor insulated cable. Figure 15.43 depicts an electric submersible centrifugal pump. These pumps are available in both fixed-speed and variable-speed configurations.

Electric submersible centrifugal pumps are driven by electric motors. Most designs require that water continually passes over the motor to cool it, while some designs can cool sufficiently by free convection in applications up to 86°F (30°C), provided that the pump motor is installed above the well intake zone. For designs that require flow for cooling, manufacturers of these pumps typically specify a minimum flow rate and velocity over the motor to prevent overheating. If the pump is located within the screened zone of the well, or if the well casing diameter is too large to provide sufficient flow over the motor, the use of a shroud may be required to achieve the necessary flow rate and velocity.

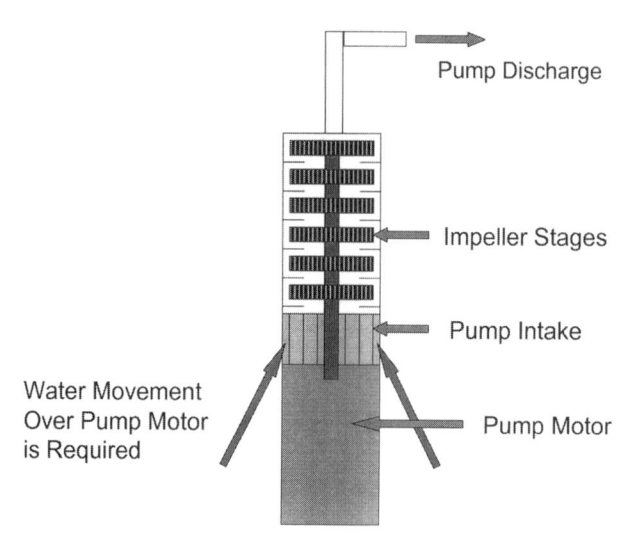

FIGURE 15.42
Schematic of an electric submersible centrifugal pump.

Flow rate and lift capability for pumps of this design are wide ranging. Fixed-speed pumps not specifically designed for ground-water sampling do not have controllable flow rates, and the flow rate is dependent on the horsepower rating of the pump, the number of impellers (stages) in the pump section, and the depth to static water

FIGURE 15.43
An electric submersible centrifugal pump mounted on a rig designed to make purging and sampling, and cleaning of the pump and discharge tubing, convenient for the sampling team.

level (i.e., lift). For variable-speed pumps designed for ground-water sampling, the discharge rate can be altered by regulating the frequency of the electrical power supply and controlling the motor speed to reduce or increase the flow rate. The rate is generally controllable between 0.03 and 8 gal/min. These pumps are capable of pumping at their highest rate at low lifts (10 ft); the rate decreases with increased lift, to a limit of about 230 ft.

With all pumps of this design, heat generated by the motor could increase sample temperature, which could result in loss of dissolved gases and VOCs from the sample and subsequent precipitation of trace metals. The temperature rise in samples is especially acute (between 5 and 7°C [Pohlmann et al., 1994; Oneacre and Figueras, 1996]) at low flow rates. Because the pump's impellers operate at very high speeds (between 2,500 and 3,500 r/min for fixed-speed designs; between 7,500 and 23,000 r/min for variable-speed designs), the accuracy of samples may also be affected by extreme agitation, which may include disaggregation of colloidal or other particulate material that may be entrained in the water. Electric submersible centrifugal pumps are considered acceptable for sampling major ions, radioactive constituents, and some dissolved metals (Pohlman et al., 1994), but they are generally not well-suited for sampling dissolved gases, VOCs, trace metals, or other parameters sensitive to temperature or agitation.

Electric submersible centrifugal pumps are only available in diameters that will fit into sampling points 2 in. or larger in diameter. These pumps can be damaged when used in silty or sandy water, requiring repair or replacement of pump components or the motor. If overheating occurs, there are three possible consequences. First, where the motor has internal water or oil in it for improved cooling, some of this liquid could be released into the sampling point, which could potentially contaminate the sampling point or samples collected from the sampling point. Because of this, motors that contain oil should not be used if the oil could interfere with the analytes of interest. Further, water used in motors should be of known chemistry and should be replaced between uses of the pump (i.e., between wells and between sampling events). Second, when this type of motor eventually cools, it can draw in water from the sampling point, which could cause future cross-contamination problems. Proper decontamination of the pump should include changing internal cooling fluid if the pump is to be used in nondedicated applications. As an alternative, dry sealed motors can be used to avoid these potential problems. Third, extensive or long-term overheating problems may result in motor failure, usually requiring replacement of the motor. Electric submersible centrifugal pumps should not be allowed to operate dry, or damage may occur to the pump seals or motor. Some pump designs may be difficult to disassemble in the field for cleaning or repair. For these pumps, if used portably, cleaning is usually performed by flushing the pump and discharge line and washing the exterior surfaces in accordance with ASTM Standard D 5088 (ASTM, 2004e).

Positive-Displacement Pumps: A number of different devices fall into the category of positive-displacement pumps: gas-displacement pumps, bladder pumps, piston pumps, progressing cavity pumps, and electric submersible gear-drive pumps: Until very recently, all of these devices were limited to use in 2 in. or larger diameter wells. However, some manufacturers of gas-displacement pumps and bladder pumps have developed smaller diameter pump models to fit into smaller direct-push wells and sampling tools.

Gas-Displacement Pumps: A gas-displacement pump (Figure 15.44) forces a discrete column of water to the surface via pressure-induced lift without the extensive mixing of drive gas and water produced by air-lift devices. The principle of operation of a gas-displacement pump is shown schematically in Figure 15.45. Hydrostatic pressure opens

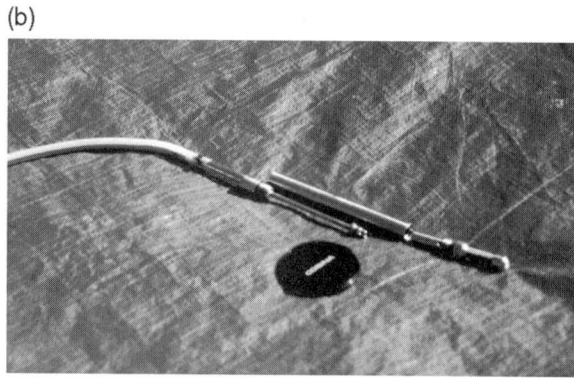

FIGURE 15.44
Gas-displacement pumps range from large, high-volume pumps (a) designed for high-flow-rate pumping to small-diameter, small-volume pumps (b) designed for small-diameter direct-push wells and multilevel monitoring systems. These pumps are easy to operate and easy to disassemble for cleaning.

the inlet check valve and fills the pump chamber (fill cycle). The inlet check valve closes by gravity after the chamber is filled. Pressurized gas is then applied to the chamber, displacing the water up the discharge line (discharge cycle). After releasing the pressure, the cycle can be repeated. A check valve in the discharge line maintains the water in the line above the pump between discharge cycles. A pneumatic logic unit, or controller, is used to control the application and release of the drive-gas pressure. The lift capability of a gas-displacement pump is directly related to the pressure of the drive gas used

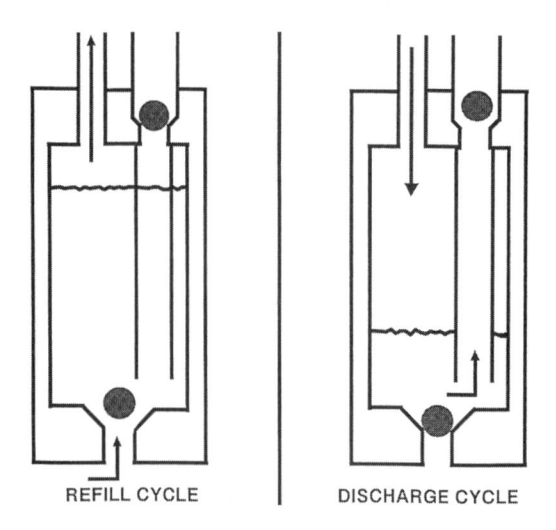

FIGURE 15.45
Schematic representation of a simple gas-displacement pump.

(1 psi of gas pressure = 2.31 ft of lift) and is usually limited by the burst strength of the gas-supply tubing. PE and PTFE are the most commonly used materials for tubing.

The maximum flow rate of a gas-displacement pump is based on the pump chamber volume, the pressure and volume of the drive-gas source, the cycling rate of the pump, the depth of the pump, and the submergence of the pump inlet. The flow rate can be controlled either by adjusting the pressure of the drive gas or the time allowed for the refill or discharge cycles to occur. Typical lifts for gas-displacement pumps rarely exceed 250 ft using single-stage compressors; greater lifts can be achieved using two-stage compressors or compressed-gas cylinders. Gas-displacement pumps are available for sampling points as small as 0.5 in. in diameter.

Gas-displacement pumps are generally not damaged by sediment, although sediment may reduce the maximum flow rate or temporarily clog the check valves and interrupt flow from the pump. These pumps are not damaged by pumping dry, so they are ideally suited for pumping in low-yield sampling points. They are typically easy to disassemble for cleaning, service, or repair.

There is a limited interface between the drive gas and the water in a gas-displacement pump. There is, however, a potential for loss of dissolved gases and VOCs across this interface (Barcelona et al., 1983; Gillham et al., 1983). This potential greatly increases if the pump is allowed to discharge completely, which would cause drive gas to be blown up the discharge line. Contamination of the sample may also result from impurities in the drive gas. It is highly recommended that an inert drive gas, such as nitrogen or helium, be used.

Bladder Pumps: Pneumatic bladder pumps, also known as gas-operated squeeze pumps, Middleburg-type pumps or diaphragm pumps, consist of a flexible membrane (bladder) enclosed by a rigid housing, with check valves on either side of the bladder (Figure 15.46). A schematic is shown in Figure 15.47. In the traditional pneumatic bladder pump, water enters the bladder under hydrostatic pressure through an inlet check valve at the pump bottom. The inlet check valve closes by gravity after the bladder is filled. Compressed gas is applied to the annular space between the outside of the bladder and pump housing, which squeezes the bladder. This action forces the water out of the bladder and up the discharge line to the surface; a check valve in the discharge line prevents discharged water from re-entering the bladder. After releasing the gas pressure, this cycle can be repeated. In some bladder pump designs, the water and air chambers are reversed, with water entering the annular space between the pump housing and bladder; the bladder is then inflated to displace the water. A pneumatic logic controller controls the application and release of drive-gas pressure to the pump. The lift capability of bladder pumps is directly related to the pressure of the drive-gas source and is controlled to some degree by the burst strength of the tubing used and outer materials of pump construction (stainless steel has the greatest depth capability). Bladder pump flow rates are controlled by adjusting the drive-gas pressure or the discharge and refill cycle timing. Flow can be readily controlled, and discharge rates of between 50 ml/min and 4 l/min can be obtained in many applications. When dealing with high-volume sampling points, maximum flow rates from bladder pumps may be too low for high-flow-rate purging. In these situations, secondary purging pumps or packers can be used in conjunction with bladder sampling pumps to reduce purge time requirements or sampling protocols can be modified to include low-flow purging and sampling procedures. Bladder pumps designed to be used in dedicated applications may be difficult to clean if they are used portably. Some manufacturers have developed a portable version of their bladder pump (Figure 15.48) to permit easy disassembly and replacement of the bladders, which is difficult or impossible with the sealed, dedicated

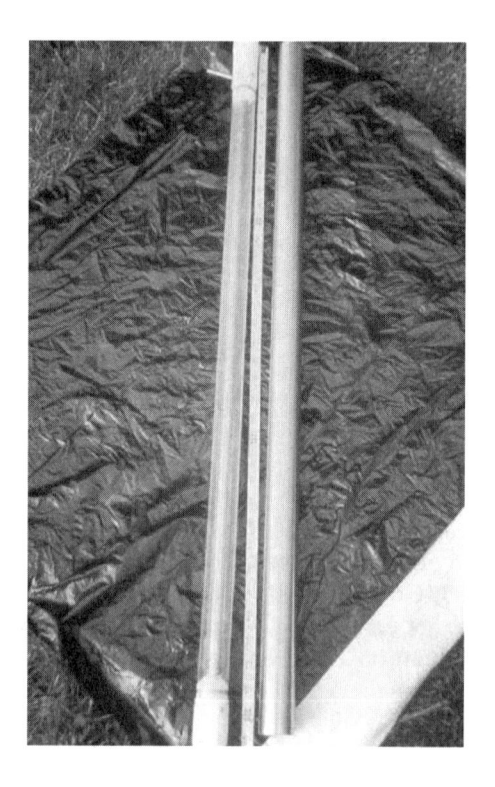

FIGURE 15.46
A dedicated pneumatic bladder pump disassembled to show the pump body (stainless steel) and the bladder (PTFE), with an inlet check valve at the bottom and a discharge check valve at the top (both PTFE). Dedicated pumps of this design are not meant to be disassembled.

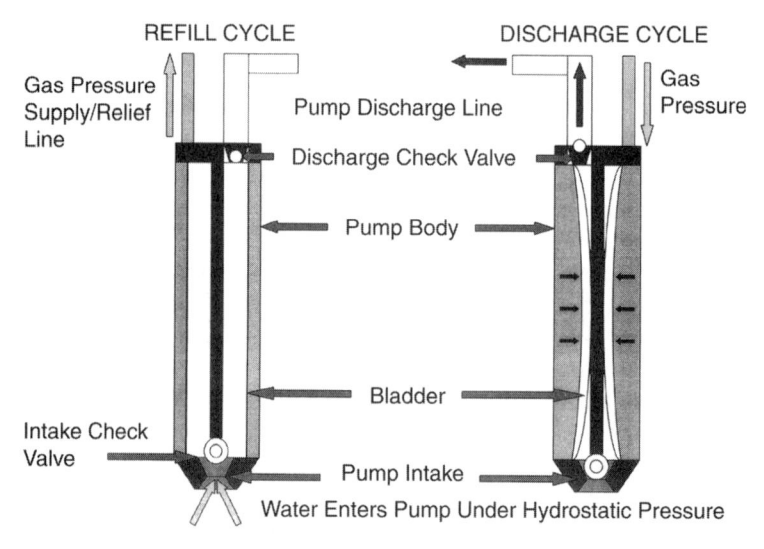

FIGURE 15.47
Schematic of a pneumatic bladder pump.

FIGURE 15.48
A portable pneumatic bladder pump designed to make it easy to disassemble for cleaning.

pump design. Pneumatic bladder pumps are also available for use in sampling points as small as 0.5 in. in diameter. These use the same principle of operation as the larger diameter pumps.

A small-diameter stainless steel mechanical bladder pump has been developed specifically for portable use in small-diameter direct-push wells and sampling tools (Figure 15.49). To operate this device, the pump, attached to an outer tubing of 0.43 in. diameter PE and an inner tubing of 0.25 in. diameter PE or Teflon, is lowered into the sampling point and is held in place by a clamp to prevent the pump from moving within the water column during operation. At the surface, a mechanical actuator employs either a circular stroke or a vertical stroke to retract the inner tubing on the intake stroke. As the inner tube is raised, the upper check ball seats and prevents water from flowing back into the Teflon bladder. The lower check ball opens and water is drawn from the well through the intake screen as the corrugated bladder expands. On the sampling stroke, the outer tubing is held stable and the inner tubing is lowered. As the inner tube is lowered, the upper check ball opens and water is pushed up the inner tube to the surface. The bladder compresses as the inner tube is lowered and water is forced up and out of the bladder. As the bladder is compressed, the lower check ball is closed, preventing water from flowing out of the intake valve.

This mechanical bladder pump does not require the accessory equipment required by traditional bladder pumps, such as generators, compressors, pneumatic controllers, or compressed gases. It was evaluated by the U.S. EPA ETV Program (U.S. EPA, 2003), where it was determined that the pump could be used to collect VOC samples with low turbidity and that it was suitable for low-flow purging and sampling applications. The potential disadvantage of this device is that it is somewhat difficult to thoroughly decontaminate between sampling points, particularly the fine mesh of the inlet screen assembly.

All bladder pumps are susceptible to bladder damage or check valve malfunction caused by sediment. The use of inlet screens, often required by manufacturers as part of

FIGURE 15.49
A mechanical bladder pump designed to operate in small-diameter direct-push rods or wells without reliance on a source of drive gas. (a) Shows the ball check valve on the intake of the pump. (b) Shows the pump being installed in a well. (c) Shows the configuration of the wellhead assembly. (d) Shows the manual cycling mechanism for the pump. (Photos courtesy of Geoprobe Systems.)

their pump warranty program, can minimize or eliminate these problems. Bladder pumps can be run dry without damage.

Bladder pumps provide representative samples under a wide range of field conditions. There is no contact between the drive gas and the water in a pneumatic bladder pump,

Ground-Water Sampling

eliminating the potential for stripping of dissolved gases and VOCs and the potential for sample contamination by the drive gas. Pressure gradients applied to the sample can be controlled by reducing the drive-gas pressure applied to the bladder, thus minimizing disturbance to sample chemistry. Bladder pumps are recommended for sampling all parameters under a wide variety of field conditions (Barcelona et al., 1983; Unwin and Maltby, 1988; Pohlmann et al., 1991, 1994; Tai et al., 1991; Kearl et al., 1992; Puls et al., 1992; Parker, 1994).

Piston Pumps: Piston pumps are a type of pneumatic positive displacement pump that use a drive-gas piston connected via a rod to a pump piston, which reciprocates within a stainless steel pump chamber. Movement of the pump piston draws water into inlet valves on one side of the piston, while simultaneously displacing water through discharge valves on the other side of the piston. Water displaced by the pump is driven to the surface through a discharge tube. The piston is cycled through the use of a pneumatic or mechanical actuator. Figure 15.50 provides a schematic of a piston pump illustrating the flow path of water through the pump.

Double-acting piston pumps (Figure 15.51) are available in several different diameters to fit 2 in. diameter and smaller sampling points. Currently, models are available in the

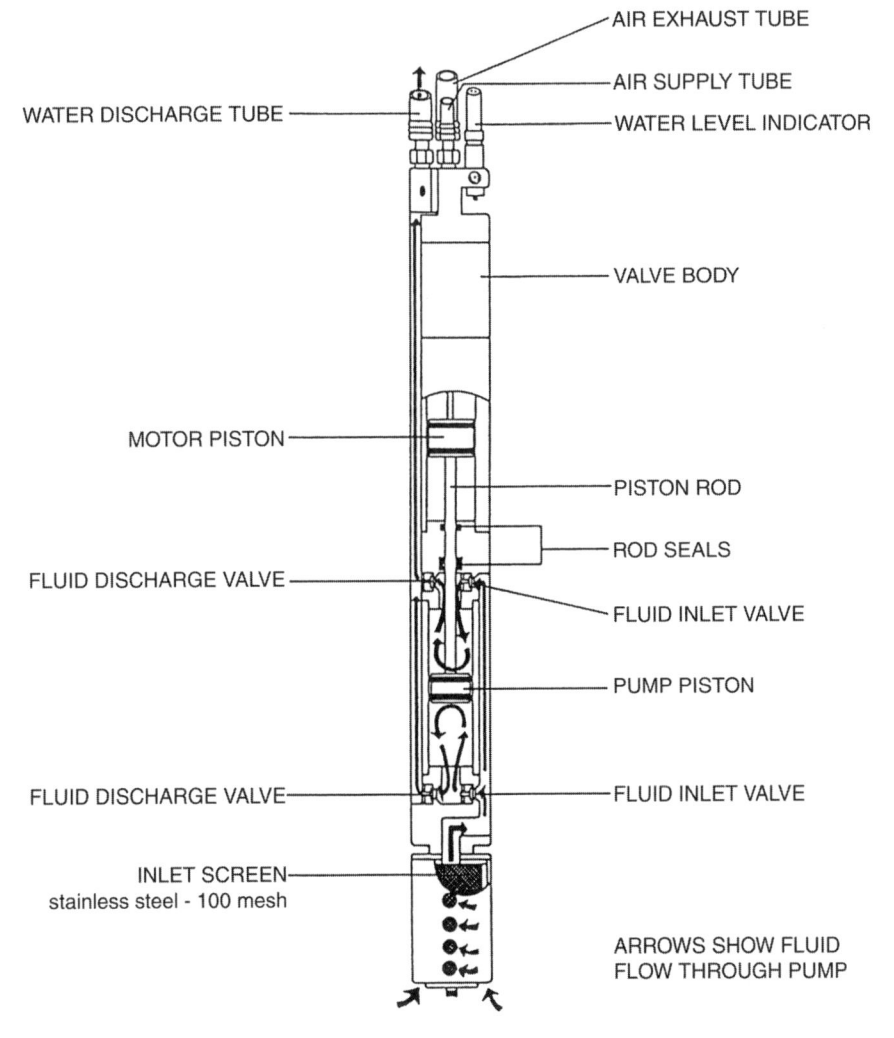

FIGURE 15.50
Schematic of a double-acting piston pump.

FIGURE 15.51
This double-acting piston pump is capable of pumping from lifts up to 1000 ft, at a rate of about a liter per minute. (Photo courtesy of Bennett Sample Pumps, Inc.)

following outside diameters: 0.625, 0.75, 0.875, 1, 1.4, and 1.8 in. The 1.4 and 1.8 in. diameter models are most widely used. The flow rate of a piston pump depends on the inside diameter of the pump cylinder and the stroke length and rate. The ability to control the minimum flow rate for sampling is dependent on the degree to which the stroke rate can be regulated by the controller unit at ground surface.

Owing to the complexity of the internal mechanisms of this pump, it is best suited to dedicated installations; however, it has been used successfully as a portable device. In a portable mode, accessory equipment can become cumbersome (i.e., large bottles of compressed gas or an oil-free compressor; large tubing bundles for deep applications) and will require a suitably sized field vehicle to transport equipment between sampling locations (Figure 15.52). In addition, the pump and tubing must be thoroughly cleaned between wells, which can be difficult because of the complex valving in the pump and the length of the tubing. Pump disassembly is time consuming, so most samplers simply flush the pump and tubing with cleaning solutions. If this is done, equipment blanks must be collected frequently as part of the QC program to confirm no cross-contamination of sampling points.

Piston pumps can provide representative samples for most parameters (Barcelona et al., 1983; Yeskis et al., 1988; Knobel and Mann, 1993), as they are constructed of inert materials, and can deliver samples at a controlled flow rate. Flow rates vary from less than 100 ml/min to up to 5 gal/min. These pumps can be used effectively in high-volume wells for low-flow purging. Dedicated installations of this pump indicate that the device is reliable in long-term monitoring programs. This pump has the greatest lift (over 1000 ft) of any small-diameter pump.

Ground-Water Sampling

FIGURE 15.52
Because the lift capability of the double-acting piston pump allows pumping from depths greater than 1000 ft, portable use often requires that a support vehicle carry the pump and support equipment. The tubing bundle (which consists of three separate lengths of tubing and, optionally, a cord for a pressure transducer) is cumbersome and large, as is the drive-gas source (in this case, a K-size cylinder of nitrogen). (Photo courtesy of Bennett Sample Pumps, Inc.)

There are some concerns about what impact this device may have on sample chemistry due to the slight negative pressure produced during refill of the pump, although this effect is reduced as the pump cycling rate is decreased. Likewise, reducing the pump cycling rate also reduces the pressure applied to the sample, minimizing the potential for sample alteration. If a flow restrictor or valve is used to reduce the discharge rate, then the resultant pressure changes could alter sample chemistry (Barcelona et al., 1983; Gillham et al., 1983). Piston seals and inlet or discharge valves are subject to failure in highly turbid water; inlet screens can reduce or eliminate this damage. These pumps may also be damaged by running dry.

Progressing Cavity or Helical Rotor Pumps: Progressing cavity pumps, also referred to as helical rotor or moyno-type pumps, utilize a rotor driven by an electric motor against a stator assembly to displace water through a discharge line to ground surface. A schematic is shown in Figure 15.53. Rotation of the helical rotor causes the water-filled cavity between the rotor and stator to progress upward, thereby pushing water in a continuous flow upward through the discharge line. In some progressing cavity pumps, the discharge rate can be varied by adjusting the speed of the pump motor between 50 and 500 r/min. The progressing cavity pump (Figure 15.54) is typically suspended in a well by its discharge line or by a suspension cable. A two-conductor electric cable supplies power from a 12 V DC power supply and control box to the pump motor.

Progressing cavity pumps are commonly constructed of stainless steel with PTFE or PE materials used as seals. The rotors are generally constructed of stainless steel, while the stator material may consist of EPDM or Viton.

Progressing cavity pumps require a sampling point diameter of at least 2 in. and will operate at moderate to low flow rates up to a maximum lift of approximately 180 ft. The relatively low discharge rates attainable with most progressing cavity pumps make them most useful in applications where purging does not require removal of large

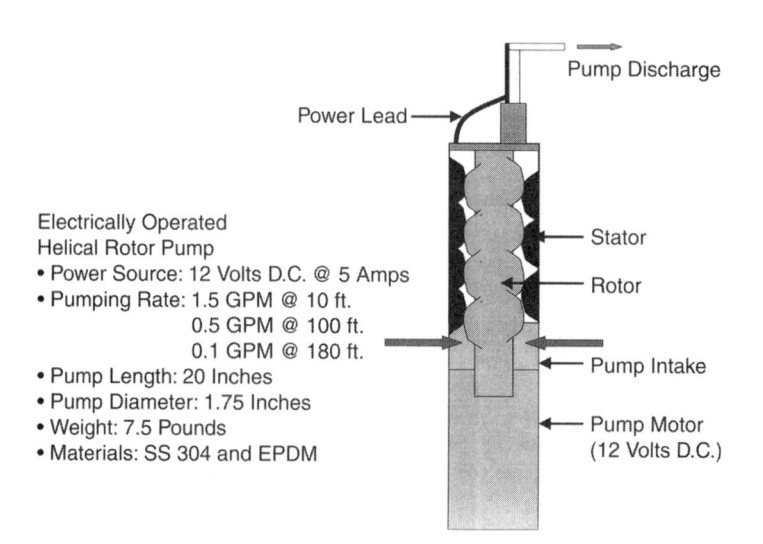

FIGURE 15.53
Schematic of a progressing cavity pump.

volumes of water from monitoring wells. With variable flow rate progressing cavity pumps, once purging is complete, the discharge rate may be reduced before samples are collected. Owing to the sealed nature of the pump and the rotor/stator construction, these devices are well suited to dedicated installations, although they are commonly

FIGURE 15.54
A progressing cavity pump is a relatively portable pump that operates on 12 V DC power supplied by a deep-cycle marine battery.

used portably because they are relatively easy to transport between sampling locations. It is highly recommended when the pump is run using 12 V batteries that a charging and backup battery system be available because the pump quickly drains batteries during use.

When not used frequently, these pumps are subject to locking where the rotor becomes stuck to the stator and is difficult or impossible to turn. This condition is difficult to overcome in the field. Additionally, the rotor and stator may be damaged when pumping silty or sandy water.

The operating principle of progressing cavity pumps makes them suitable for collection of samples for VOCs (Imbrigiotta et al., 1988). There is some evidence that these pumps may not be suitable for sampling trace metals and other inorganic analytes at higher flow rates due to increased turbidity (Barcelona et al., 1983); to control turbidity, a variable speed pump controller can be used to reduce flow rate. The pressure applied to a sample is directly related to the motor speed and can be controlled in designs using variable-speed motor controls. Overheating of the motor may raise the temperature of the sample (Parker, 1994).

Electric Submersible Gear-Drive Pumps: Another type of positive displacement pump is the electric submersible gear-drive pump, shown schematically in Figure 15.55. In this type of pump, an electric motor drives a PTFE gear, which meshes with a second PTFE gear (Figure 15.56). As these gears rotate, their advancing teeth draw water into the pump through the pump intake port and push it through the gears in a continuous flow up the discharge line. The discharge rate can be varied using the pump controls to adjust the speed of the pump motor. As with many other submersible pumps, the gear-drive pump is usually suspended in a well by its discharge line. Electric power is supplied to the 36 V DC motor through a cable from the power source and control box at ground surface. The manufacturer provides either PE or Teflon-lined PE tubing with the pump, although other tubing can be made available if required.

This pump is commonly used portably due to its ease of transport and ease of decontamination between sampling locations (Figure 15.57). It can be installed as a dedicated pump, and a stand-alone control box and special dedicated installation well caps are available from the manufacturer.

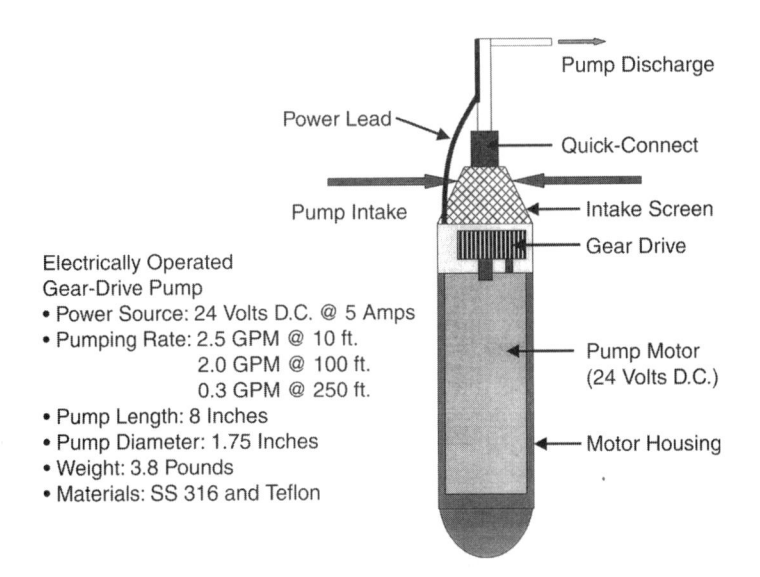

FIGURE 15.55
Schematic of an electric submersible gear-drive pump.

FIGURE 15.56
The PTFE drive gear of a gear-drive pump (center) meshes with a second PTFE gear to displace water (which enters the pump under hydrostatic pressure) through the pump chamber into the discharge tubing to the surface.

Electric submersible gear-drive pumps require a sampling point diameter of at least 2 in. Maximum discharge rates for gear-drive pumps range from more than 3 gal/min at lifts of less than 20 ft to 0.25 gal/min at lifts of 250 ft. Discharge rates are easily controlled using the flow control, which adjusts the power supplied to run the pump motor; pump discharge can be adjusted to less than 50 ml/min.

If electric submersible gear-drive pumps are used extensively for pumping water high in suspended solids, the PTFE gears may clog or wear, thereby reducing the discharge rate.

FIGURE 15.57
Electric submersible gear-drive pumps are highly portable, requiring a 36 V DC gel cell power source, a 110 V AC source connected to a 36 V DC converter, or a 12 V DC marine battery connected to a 110 V AC inverter and a 36 V DC converter (shown here).

Ground-Water Sampling

Disassembly of the pump and replacement of the gears is a procedure easily accomplished in the field in a few minutes. Electric submersible gear-drive pumps are generally very easy to decontaminate when used in a portable mode.

Electric submersible gear-drive pumps provide good sampling accuracy and precision for dissolved gases, VOCs, trace metals, and other inorganics and mobile colloids (Imbrigiotta et al., 1988; Backhus et al., 1993). Cavitation may occur if the pump is run at high speed, which could affect dissolved gases or VOCs. The potential for cavitation can be reduced or eliminated by controlling motor speed. The pressure applied by a gear-drive pump to a sample is directly related to the motor speed and can be controlled using the variable-speed motor controls. Electric submersible gear-drive pumps are constructed of materials acceptable for sampling sensitive ground-water parameters; the pump body is constructed entirely of stainless steel materials, while the gears are constructed of PTFE.

Inertial-Lift Pumps: Inertial-lift pumps (Figure 15.58), also known as tubing-check-valve pumps, consist of a discharge tubing (either flexible tubing or rigid pipe) with a ball-check foot valve attached to the lower end of the tubing. In operation, the tubing is lowered into a water column and cycled through reciprocating motion, either through manual action or through the use of a reciprocating mechanical arm mechanism driven by an electric motor or internal combustion engine, to discharge water. As the tubing is moved upward, water that has entered the tubing under hydrostatic pressure is lifted upward, held in the tubing by the seated foot valve. When the upward motion of the tubing is stopped, the inertia of the water column inside the tubing keeps it moving upward. As the tubing is pushed downward, the foot valve opens, allowing the tubing to refill, and the cycle is repeated to pump water from the sampling point.

FIGURE 15.58
An inertial-lift pump simply consists of a length of rigid or flexible tubing with a foot valve on the bottom. (Photo courtesy of Solinst Canada Ltd.)

Inertial-lift pumps can be constructed of any flexible or rigid tubing material that has sufficient strength to tolerate the pump cycling. Typically, these materials include rigid and flexible PVC, PE, PP, and PTFE. Tubing diameters of 0.25 or 0.375 in. can be used to collect samples from sampling points as small as 0.5 in. in diameter. Inertial-lift pumps are commonly used in small-diameter direct-push sampling tools and wells.

The flow rate of an inertial-lift pump is directly related to the cycling rate. Flexing of the tubing in the sampling point can cause the flow rate to drop. To achieve discharge rates suitable for sample collection, it is necessary to insert a short length of small-diameter flexible tubing into the discharge line to divert a portion of the discharge stream into sample containers. To control pump discharge, the flexible tubing must be held stationary while the discharged water is directed into sample containers.

By nature of their simple design, inertial-lift pumps are not susceptible to damage by suspended solids or dry pumping, although check valve clogging will reduce the flow rate during operation. Some wear or damage may occur on the outer surface of the foot valve or tubing as it comes in contact with the well casing or screen or open borehole during cycling. These pumps are easily disassembled in the field for repairs if needed, although the mechanical cycling mechanisms may be difficult or impossible to repair in the field.

If inertial-lift pumps are cycled rapidly prior to or during sample collection, some loss of VOCs or dissolved gases could occur in the discharge stream. Inertial-lift pumps do not cause pressure changes in the sample. However, the cycling action of an inertial-lift pump in a sampling point can significantly increase sample turbidity and agitate and aerate the water column within the sampling point. This can result in alteration of concentrations of a wide variety of analytes, including dissolved gases, VOCs, and trace metals, and interference with analytical determinations in the laboratory because of the high suspended sediment content of samples.

Passive Diffusion Bag Samplers: A unique device for ground-water sample collection for selected VOCs without purging wells has been developed by the U.S. Geological Survey — the passive diffusion bag sampler (PDBS). A typical PDBS (Figure 15.59) consists of a 3 to 4 mil thick, 18 to 20 in. long, 1.25 in. diameter low-density polyethylene (LDPE) lay-flat tube sealed at both ends and filled with deionized water. The sampler is deployed in a well by attaching a weight to the bottom and a suspension cord to the top (Figure 15.60). It is positioned within the well screen so it is in contact with formation water and so VOCs that are present in formation water can diffuse through the PE bag into the deionized water contained within. The amount of time that the PDBS should be left in the well before retrieval depends on the time required for equilibration to occur and the time required for the well to recover from the disturbance caused by sampler deployment. Laboratory and field data suggest that 2 weeks of equilibration time is adequate for most applications, although in low hydraulic conductivity formations, longer equilibration times may be required (Vroblesky and Campbell, 2001; Vroblesky, 2001a).

Retrieval of a sample from a PDBS consists of pulling the sampler out of the well, puncturing the bag with a sharp rigid PE tube, and transferring the water from the bag into 40 ml sampling vials for later analysis. The concentrations of VOCs in the sample represent an integration of the chemical changes that occurred in the well over the most recent portion (approximately the last 48 to 166 h, depending on the water temperature and the individual compound) of the equilibration period (Vroblesky, 2001a). After samples are collected, the PDBS is discarded and another PDBS is installed in the well and left there until the next sampling event.

The effectiveness of PDBSs is dependent on the assumptions that there is horizontal flow through the well screen and that the water in the screen is representative of the water in the formation adjacent to the screen. PDBSs require sufficient horizontal flow to achieve

Ground-Water Sampling 1033

FIGURE 15.59
A PDBS prepared for deployment in a well. (Photo courtesy of Eon Products.)

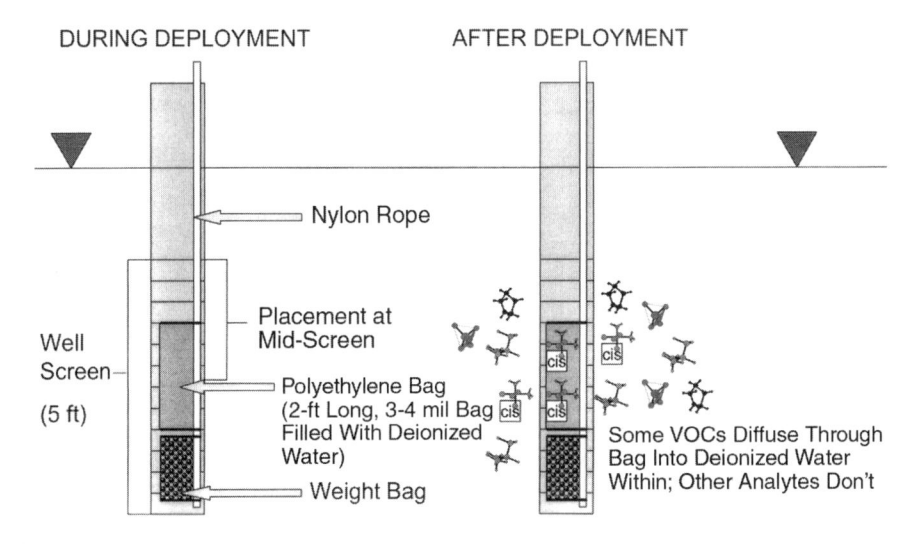

FIGURE 15.60
The PDBS is deployed in the well on a suspension cord, with a weight attached to the bottom and hung within the well screen. This sampler relies on diffusion of VOCs through the thin LDPE bag into the deionized water contained within. After a minimum 2-week equilibration period in the well, the bag is retrieved and the water within is decanted into VOC vials.

chemical and hydraulic equilibrium with the formation. Some studies suggest that with sufficient flow through a properly constructed well, ground water in the well screen may be replaced in as little as 24 h (Robin and Gillham, 1987). For water in the PDBS to be representative of the VOCs in formation water, the rate of solute contribution from the aquifer to the well must equal or exceed the rate of in-well contaminant loss by processes such as volatilization or convection, which may not occur if ground-water velocities are very low or the well has a low yield resulting from low hydraulic conductivity formations or low hydraulic gradients (ITRC, 2004, 2005).

Several studies have determined that PDBSs are effective for determination of concentrations of low-solubility and low vapor pressure VOCs such as benzene, carbon tetrachloride, 1,2-dichlorothane, tetrachloroethylene, and vinyl chloride (Vroblesky, 2001a, b). They are not, however, effective for sampling to high solubility, high vapor pressure organic compounds (such as MTBE, methanol, or acetone), semi-volatile organic compounds (SVOCs), trace metals, or other inorganics that will not diffuse across the PE membrane (Vroblesky and Hyde, 1997; ITRC, 2004, 2005). Therefore, their applications are fairly limited.

These single-use bags are relatively inexpensive and are easy to deploy and retrieve from the well. One person can easily install the PDBS into a well and recover it during the next sampling event without any accessory equipment. Because the devices operate on the principle of diffusion, they are not affected by sediment or high turbidity within the water column. In situations where the objective of sampling is to monitor VOCs in more than one zone within a well, multiple samplers can be suspended in sequence to permit definition of VOC stratification in the water column. It is important to remember, however, that PDBSs are meant to be deployed in wells in which there is no vertical flow within the well. If there is vertical flow, the concentrations represented in the PDBS will be from water flowing vertically past the device and not from a discrete zone within the formation (ITRC, 2004).

PDBSs have a significant number of limitations and, as a result, are not widely accepted by regulatory agencies for use in ground-water sampling programs. The primary limitation is that they are suitable for only a relatively short list of VOCs (Table 15.9). They cannot be used for trace metals or other inorganics, SVOCs, pesticides, PCBs, or highly soluble VOCs. Therefore, they are only applicable to sites at which a limited number of low-solubility, low vapor pressure VOCs are of concern. If any other parameter must also be sampled at the site, a second sampling device must be used. PDBSs can introduce

TABLE 15.9

Compounds Tested Under Laboratory Conditions for Use with PDBSs

Compounds showing good correlation (average differences in concentration of 11% or less versus control)			
Benzene	2-Chlorovinyl ether	cis-1,2-Dichloroethene	1,1,1-Trichloroethane
Bromodichloromethane	Dibromochloromethane	trans-1,2-Dichloroethene	1,1,2-Trichloroethane
Bromoform	Dibromomethane	1,2-Dichloropropane	Trichloroethane
Chlorobenzene	1,2-Dichlorobenzene	cis-Dichloropropene	Trichlorofluoromethane
Carbon tetrachloride	1,3-Dichlorobenzene	1,2-Dibromoethane	1,2,3-Trichloropropane
Chloroethane	1,4-Dichlorobenzene	trans-1,3-Dichloropropene	1,1,2,2-Tetrachloroethane
Chloroform	Dichlorodifluoromethane	Ethyl benzene	Tetrachloroethene
Chloromethane	1,2-Dichloroethane	Naphthalene	Vinyl chloride
	1,1-Dichloroethene	Toluene	Total xylenes

Compounds showing poor correlation (average differences in concentration of greater than 20% versus control)		
Acetone	Methyl-*tert*-butyl ether	Styrene

Source: Vroblesky and Campbell, 2001. With permission.

both negative and positive bias into samples. One source of negative bias is associated with the development of a biofilm on the surface of the bag during long-term deployment, which could reduce diffusion of some compounds through the bag. PDBSs can contribute phthalates to samples (Vroblesky, 2001a), and there is some concern that in highly contaminated ground water, the bags may degrade and contribute degradation products to samples.

PDBSs are available in two forms. One manufacturer sells PDBSs prefilled with deionized water of a known chemistry, while another sells the bags empty (to save costs associated with shipping) and requires the end user to provide the deionized water and seal the end of the device prior to deployment. This introduces a potential source of error associated with the variability in the chemistry of the deionized water used to fill the devices. This, in addition to variability in placement of the device within the water column between sampling events, can introduce imprecision when using PDBSs.

Sampling Point Purging

Objectives of Sampling Point Purging: Most traditional approaches to ground-water sampling are based upon the assumption that all water that resides in sampling points between sampling events is stagnant (i:e., does not interact with formation water) and does not represent the chemistry of water in the formation. Thus, the historical means of meeting the objectives of sampling programs has been to remove all of the water from the well and to induce fresh formation water to enter the well so it can be collected as a sample. This process is referred to as well purging. Several purging strategies, described in detail in the following sections, are in common use.

The SAP must specify which purging strategy or strategies will be used at a site for wells with high yield, as well as for wells with low yield. The SOPs for purging must provide step-by-step procedures that include specifying: water-level measurement requirements and methods; well-depth measurement requirements and methods; the equation used to calculate a well volume (if required); the device used for purging and guidance on placement of the device within the water column, operation of the device, and cleaning of the device if used portably; when and how to measure any required field parameters during purging; and how stabilization of field parameter measurements is defined, if required as part of the purging protocol.

It is apparent from research conducted over the last 20 years that the way in which a well is purged has one of the most significant impacts on sample quality. For example, a study conducted by Barcelona and Helfrich (1986) concluded that variations in water chemistry attributed to well purging were generally greater than errors associated with either sampling mechanism, tubing, or apparent well casing materials effects. This has, therefore, been a well known limitation of ground-water sampling practices for nearly two decades, yet little has been done to improve field methods that have persisted over the years. Traditional purging strategies have encouraged the use of portable devices, particularly grab sampling devices such as bailers, and high-speed submersible pumps.

The difficulty in collecting representative samples using traditional purging methods has been in accessing the water within the well screen (which, as discussed earlier, is representative of water in the formation screened by the well) without disturbing or mixing the water column in the well. Rather than focusing on methods that could be used to access the water in the screen directly, most purging strategies have focused on methods for removing large volumes of water from the well. Although it is important to purge some water from most wells before collecting a sample, purging too much water or purging at too high rate can cause mixing of water from zones of different quality and, potentially, contamination of noncontaminated zones by previously localized or stratified pollutants (Wilson and Rouse, 1983).

Purging strategies, as described in ASTM Standard D 6452 (ASTM, 2004g), are divided into two general categories: those appropriate for high-yield sampling points (points that do not go dry during purging and sampling) and those appropriate for low-yield sampling points (points that may go dry during purging and sampling). Low yield in a well can occur as a result of the low hydraulic conductivity of the formation screened by the sampling point, poor well construction, poor well maintenance, or insufficient sampling point volume. It is imperative that the most appropriate purging strategy is implemented on not just a site-specific basis but, in some instances, on a well-by-well basis.

Purging Strategies for High-Yield Wells:

Traditional Strategies: There are three "traditional" strategies for purging high-yield sampling points: (1) removal of a fixed number of well volumes; (2) purging to stabilization of a predetermined list of field parameters; and (3) in the case of large-diameter wells or wells with large volumes of water, use of inflatable packers to physically reduce the volume of water that must be purged:

Fixed Well-Volume Purging: The most commonly applied purging strategy is to remove a fixed volume of water from the well using a bailer or a pump. Commonly, regulatory guidance mandates that three to five or four to six well volumes of water be removed prior to sample collection. A well volume may have different definitions, but it usually refers to either the total volume of water in the casing and the screen or that volume plus the volume of water contained in the filter pack.

Early research on ground-water sampling (Gibb et al., 1981) seemed to indicate that the removal of three to five well volumes of water from a well was necessary to ensure that a representative sample of formation quality water could be collected. In hindsight, the primary reason for the need to remove this much water during the early studies was the effect that the insertion and use of the purging or sampling device had with respect to mixing of the water column in the well and the release of significant amounts of suspended sediment (turbidity). It often took the removal of three to five (or more) well volumes to negate the effects induced by the method used (Powell and Puls, 1993; Barcelona et al., 1994). As observed by Barcelona et al. (1994), there is no single number of well volumes that should be removed during purging that is best applied to all sites or that is suited to all hydrogeologic conditions. While existing rules requiring removal of a fixed volume of water are administratively convenient and easy to adhere to in theory, their field implementation often results in collection of samples of less than optimum quality that are not easily reproduced and that may not accurately represent formation water chemistry. This is because, with this approach, there is minimal concern about the effect that purging and sampling practices have on three-dimensional well hydraulics and aquifer chemistry, even though the stated objective is to collect a "representative" sample. As discussed earlier, this objective is exceedingly difficult to meet using traditional methods and, as a result, DQOs can rarely be satisfied. Imprecision in well purging and sample collection methods are common issues that call into question the validity of field and lab data generated by sampling events. Without confidence in data quality, decisions based on these data may not be scientifically or legally defensible.

Bailers or portable pumps are the devices used most often for traditional purging and sampling programs. Lowering these devices through the water column during purging causes significant mixing (Gillham et al., 1985; Robin and Gillham, 1987; Keely and Boateng, 1987a, b; Martin-Hayden et al., 1991) and release of considerable sediment (Kearl et al., 1992; Puls et al., 1992). The effects of mixing and the time for re-equilibration are directly related to purging rate, location of the pump intake, and extent of

Ground-Water Sampling

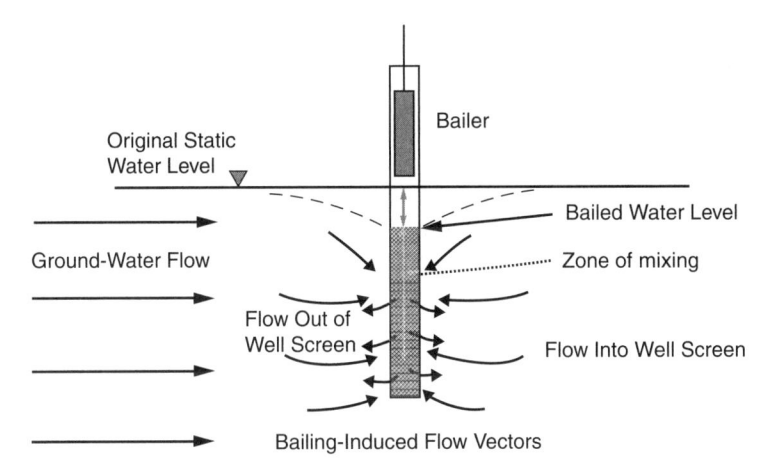

FIGURE 15.61
Conditions that occur in a well during bailing.

disturbance of the stagnant water column in the well (Puls and Powell, 1992a). The way in which these devices have typically been used (bailing rapidly or pumping at a rate higher than the well will yield without appreciable drawdown) cause turbulent flow in the well and the surrounding formation materials and promote mixing of the water column (Figure 15.61 and Figure 15.62). The work of Puls et al. (1992) and Powell and Puls (1993) has clearly demonstrated that bailing and high-flow-rate pumping (in excess of the natural flow rate through the screen) results in significant deleterious effects to the water sample and to the well. These problems include:

- Mixing of water from the zone of the formation targeted by the well screen with water from zones above and below the screen (or zones well beyond the borehole, in the case of fractured or solution-channeled rock [McCarthy and Shevenell, 1998]), resulting in dilution of samples or inclusion of water with constituents that are not the focus of the sampling program, and making data interpretation difficult (Robin and Gillham, 1987; Robbins and Martin-Hayden, 1991).

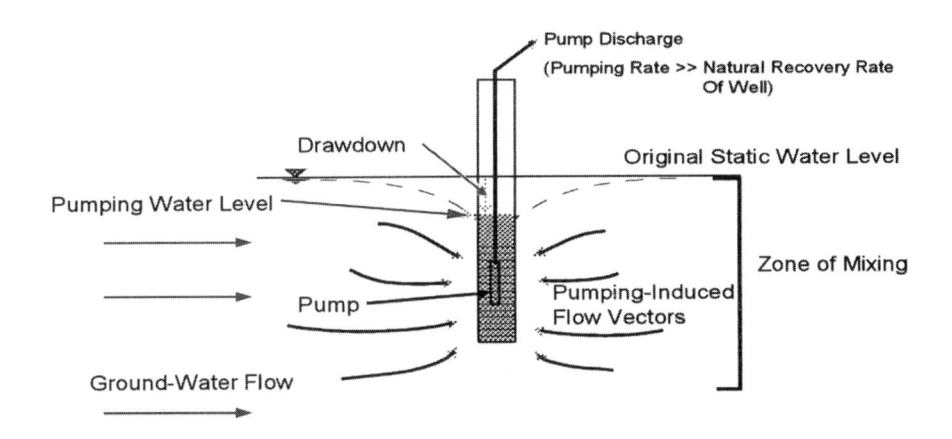

FIGURE 15.62
Conditions that occur in a well during high-flow-rate pumping.

- Mobilization of fine-grained solids from the material surrounding the well, including the formation and the filter pack, and solids settled in the bottom of the well between sampling rounds, resulting in increased sample turbidity and gross overestimation of certain analytes of interest, particularly metals (Kearl et al., 1992; Puls and Paul, 1995). Many of these solids are artifactual (introduced to the subsurface during well construction) or immobile under natural flow conditions and mobilized from the aquifer material by shearing caused by bailing or pumping at high rates (Puls et al., 1992; Backhus et al., 1993).
- Possible dewatering of part of the screen, causing agitation, aeration, and oxidation of formation water and possibly causing precipitation of solids due to shifts in chemical equilibria, resulting in nonrepresentative samples.
- Potential for damage to the monitoring well filter pack (Barcelona et al., 1985, 1994; Pohlman et al., 1994) and disruption of the filter pack and aquifer matrix around the well that exposes fresh, reactive mineral surfaces and increases the possibility of sorption-desorption reactions (Palmer et al., 1987; Powell and Puls, 1993) along with increasing turbidity.
- Generating large amounts of purge water, especially from large diameter or deep wells, that may require management as hazardous waste, resulting in unnecessary expense and the potential for transfer of contaminants from one site to another.

The following discussion, taken from Puls and Paul (1995), illustrates a significant problem with sample accuracy resulting from the use of traditional well-volume purging methods.

> The volume of water purged, the rate at which it is withdrawn, and the location of the sampling device intake all contribute to how well the sample represents the water in the formation. The evacuation of 3 to 5 well volumes in a 200-foot deep well with a 20-foot screen in which the static water level is 20 feet below ground surface removes a large volume of water and averages a large volume of the aquifer in the water sample. Because many contaminant plumes can be narrow or thin, mixing with water from clean portions of the aquifer can provide misleading data concerning contaminant presence and concentration. If the sampling objective is a large, volume-averaged concentration of the water-bearing zone, then a consistently large sample volume, removed in a consistent manner, will generally provide reproducible values. However, if the sampling objective is to provide accurate spatial and temporal plume delineation, then alternative methods of sample collection are necessary.

Following removal of a fixed purge volume, samples are typically collected using either a bailer or a pump operated at a slow discharge rate. The device may or may not be the same device used for purging. Commonly, samples must be filtered to remove the high levels of turbidity produced by purging.

Using a fixed well-volume purging strategy, depending on what type of device is used, where it is placed in the well screen, and how it is used, there is no guarantee that all stagnant water will be removed from the well or that samples uncontaminated by the stagnant water will be obtained. Purging may be insufficient in some cases and much more than necessary in others. One common scenario observed by the authors at dozens of sites across the USA involves purging the well of three to five well volumes at a high rate with a submersible pump set within the well screen, then sampling the water at the top of the water column with a bailer. Considering the hydraulics of the well during pumping, this common practice is not likely to produce samples representative of the formation screened by the well. Most of the water pumped from the well during purging will come from the formation and, unless the water level in the well is drawn down to the intake of the pump (which can only be confirmed by measuring the water level in the well during pumping, which is rarely done), some stagnant water will remain in the well — this is very likely the water sampled by the bailer. Thus, while samplers using this method may adhere to the "letter of the law," they are not collecting samples that are representative of formation water. Only in the case in which the pump is set at the top of the water column at the start of purging and follows the water level down during purging to the point at which the water level stabilizes (and at least one well volume of water is removed from the well), could the pump truly remove all stagnant water from the well to prepare it for subsequent sampling. However, the same concerns regarding pumping at high rates and bailing noted earlier apply here.

In situations in which the static water level in the well is within the screen, removal of multiple well volumes of water is clearly unnecessary, because, as noted earlier, the water within the screen is representative of formation water. Because the top of the water column is in contact with atmospheric conditions (due to the existence of a headspace within the well), some water from the top of the water column should be purged from the well, as it may be chemically different from formation water. This can be confirmed by measurement of indicator parameters either downhole or in a flow-through cell during purging (discussed subsequently). However, application of a strict fixed-volume purging strategy in this type of situation is clearly excessive.

It is clear that the fixed well-volume purging approach has a number of shortcomings, as summarized in Table 15.10. Because of these limitations, the authors recommend that in most cases, use of this purging strategy should be discontinued, and other more appropriate methods (as discussed subsequently) should be used.

Purging to Stabilization of Indicator Parameters: A second approach to purging high-yield sampling points, which is less commonly applied in ground-water sampling programs, is to continuously monitor selected field water-quality indicator parameters during removal of water from the well. The indicator parameters to be measured and the frequency of measurements should be specified in the SAP. The most commonly measured parameters include (but are not limited to) pH, conductivity (or specific conductance), temperature, DO, and ORP. Parameters measured for any given sampling program should be selected based on knowledge of site-specific water chemistry,

TABLE 15.10

Limitations of the Fixed Well-Volume Purging Strategy

No consideration of well-specific hydraulics or site-specific hydrogeology or geochemistry

No specifications on the rate of removal of water during purging, commonly resulting in hydraulically over-purging the well

No standardized definition of "well volume" can lead to errors in calculation of well volume

No measurement of water chemistry to determine, on a chemical basis, when fresh formation water has entered the well

Commonly results in removal of more water than is necessary, leading to decreased efficiency in sampling (increased time) and increased costs (labor and management and disposal of purge water)

Encourages removal of water at rates exceeding natural flow rates for the formation and well screen, which often results in high turbidity samples

Allows use of grab sampling devices and inertial-lift devices, which can severely agitate the water column and release significant turbidity

---

Source: Nielsen and Nielsen, 2002. With permission.

analytes of interest (and their relationship to indicator parameters), and any specific regulatory requirements. Frequency of measurement should be based on purging rate and method used for measurement (i.e., hand-held single- or multiparameter instruments versus sensors in a flow-through cell). The acceptable variation of parameter values to define stabilization and the minimum number of consecutive stable readings within the prescribed variation for each indicator parameter should be defined in the SAP. When the selected parameters stabilize, regardless of the volume of water removed, it is presumed that all stagnant water has been removed from the well and that fresh formation water is available for sampling. As in the fixed-volume strategy described earlier, this strategy permits use of grab samplers and portable pumps, in addition to dedicated pumping systems. The limitations of this purging strategy are summarized in Table 15.11.

Ideally, this purging method is implemented using a dedicated or portable pumping device in conjunction with a flow-through cell, although portable pumps are also commonly used. A flow-through cell is of particular importance if any of the parameters monitored to determine water chemistry stabilization are sensitive to contact with ambient air, such as DO or ORP. Using a pumping device, rather than a grab sampling device, will ensure less disturbance of the water column and reduced turbidity, meaning that stabilization will be reached more quickly.

Use of a Packer During Purging: The third traditional method of purging high-yield wells is a strategy in which the volume to be purged from the well is reduced using a device referred to as a packer. A packer is an expandable device placed within the well casing or competent rock just above the top of the well screen or open zone that, when deployed, physically isolates the water above the packer from the water below the packer. This negates the need to remove the water from above the packer and the need to purge multiple volumes of water from the well prior to sampling. When in use, a pump or pump intake is suspended below the packer (within the screen), to purge and sample the water within the screen without having to remove the overlying stagnant water in the casing (Figure 15.63). Packers are ineffective in isolating sampling zones when installed in the well screen or open zone. Water levels for use in piezometric surface mapping must be taken prior to placement of the pump or packer assembly to avoid bias. Water levels may also be taken in the water column above the packer during pumping to check for leakage of water from within the well casing past the packer; a measurable water-level drop during pumping indicates leakage.

Packers can be used in conjunction with any pumping device (Figure 15.64) in any well in which the static water level is above the top of the well screen, but are most efficient in

TABLE 15.11

Limitations of Purging to Stabilization of Field Indicator Parameters

---

Purging rates are not controlled by, nor are they specific to, formation or well hydraulic conditions
Stabilization of selected field parameters does not necessarily reflect stabilization of contaminant chemistry (Gibs and Imbrigiotta, 1990)
Temperature and pH are not always reliable indicator parameters
Many potential sources of error and bias are associated with field parameter measurement (i.e., measuring DO or ORP in an open container)
Inconsistency in defining stabilization criteria
Method permits the use of grab samplers and inertial-lift devices for purging, which can make it very difficult to achieve stabilization for parameters sensitive to aeration and agitation of the water column in the well

---

Source: Nielsen and Nielsen, 2002. With permission.

Ground-Water Sampling

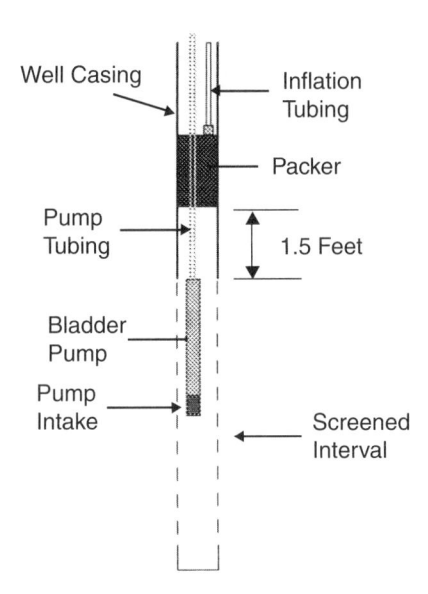

FIGURE 15.63
Placement of a packer to physically isolate the water column in the casing from the water column in the screen.

large-diameter or deep wells in which the water level is significantly above the top of the well screen. Because the packer physically isolates the water within the well intake from the stagnant water in the casing, it is effective in minimizing purge volume and reducing the costs associated with management of purge water. Bailers cannot be used to sample

FIGURE 15.64
A packer used in conjunction with a progressing cavity pump, prior to installation.

when packers are used, and packers are not practical for use in low-yield wells, in which the open interval will dewater during purging.

Low-Flow Purging and Sampling: Researchers have evaluated the shortcomings of traditional purging methods and determined that each method imparts some type of error and bias, which can influence the representative nature of samples. An improved method of purging wells evolved out of research conducted to improve sample accuracy and precision and to encourage the collection of more representative samples. This method of purging is known as low-flow purging and sampling (also referred to as micro-purging, low-stress purging, low-impact purging, or minimal-drawdown purging).

Unlike traditional purging methods, low-flow purging and sampling does not require the removal of large volumes of water from the well. The amount of water purged will vary with well diameter, but is typically less than one half of a well volume (Barcelona et al., 1994) and, in the authors' experience, often less than one third of a well volume. Some investigators have successfully sampled wells after purging as little as one to two times the volume of the sampling system, which includes the pump and discharge tubing (Shanklin et al., 1995). The actual volume purged prior to sampling depends on the time required for water-level stabilization and indicator parameter stabilization to occur and the pumping rate used over that time period, which is specific to each well.

Low-flow purging differs from traditional methods of purging in that its use is based on the observations of many researchers (cited earlier in this chapter) that water moving through the formation also moves through the well screen. Thus, the water in the screen is representative of the formation water surrounding the screen. This assumes that the well has been properly designed, constructed, and developed as described in ASTM Standards D 5092 and D 5521 (ASTM, 2004h, 2004i) and Chapters 10 and 12 of this book. In wells in which the flow through the screen or intake zone is limited by hydraulic conductivity contrasts (e.g., borehole smearing, residual filter cake, filter pack grain size, or well screen open area), the head difference induced by low-flow pumping provides an exchange of water between the formation and the well. Samples collected during low-flow purging and sampling represent the water from the entire screened zone, and the chemistry of the sample represents a weighted average of the concentrations of solutes in the water in the screened interval (Martin-Hayden and Robbins, 1997; Puls and Paul, 1997). The effects of heterogeneities in geologic material screened by the well may change the contributions of various zones to the average, but do not change the overall effect of concentration averaging.

Low-flow purging involves removing water directly from the screened interval without physically or hydraulically disturbing the stagnant water column above the screen. This is done using a dedicated pump (or by very carefully installing a portable pump) with the pump intake set at or near the middle of the screen and pumping the well at a low enough flow rate to maintain a stabilized water level in the well as determined through water-level measurement during pumping. Typically, flow rates on the order of 0.1 to 0.5 l/min are used; however, this is dependent on site-specific and well-specific factors (Puls and Barcelona, 1996). Some very coarse-textured formations have been successfully purged and sampled in this manner at flow rates up to 1 l/min. Pumping water levels in the well and water-quality indicator parameters (such as pH, temperature, specific conductance, DO, and redox potential) are monitored during pumping, with stabilization indicating that purging is completed and sampling can begin (Figure 15.65 and Figure 15.66). Purging at a rate that minimizes drawdown will generally reduce purge volumes and time required to reach stabilization of indicator parameters.

"Low-flow" refers to the velocity that is imparted during pumping to the formation pore water adjacent to the well screen. This velocity must be minimized to avoid turbulent flow through the well screen and to preclude the entrainment of artifactual particulate matter in

Ground-Water Sampling

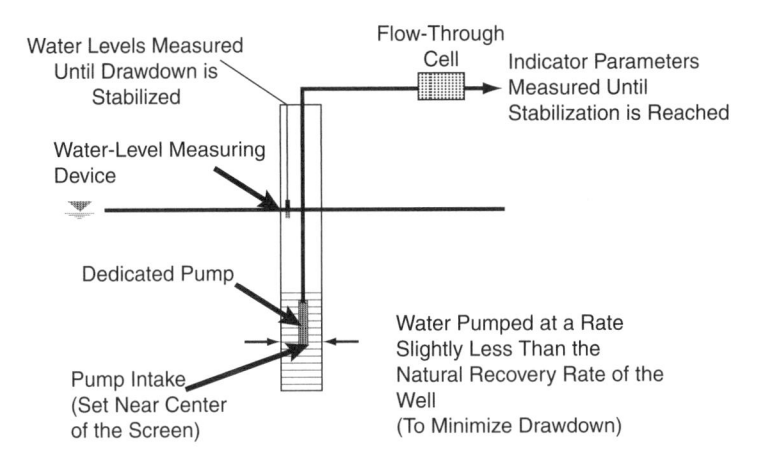

FIGURE 15.65
Illustration of the key components of low-flow purging and sampling.

the water to be collected as a sample. Low-flow does not necessarily refer to the flow rate of water discharged by a pump at the surface, which can be affected by valves, restrictions in the tubing, or flow regulators. "Low stress" or "low impact" refers to the impact of pumping the well on formation hydraulics. Water-level drawdown provides a measurable indication of the stress or impact on a given formation imparted by a pumping device operated at a given flow rate. The objective of low-flow purging is to pump in a manner that minimizes stress or disturbance to the ground-water system to the extent practical. A stabilized water level in a well (not necessarily achieving a particular

FIGURE 15.66
The equipment setup for low-flow purging and sampling includes a pump (in this case, a dedicated bladder pump); a pump controller with drive gas (left); a flow-through cell (center) with sensors for measuring pH, temperature, specific conductance, DO and ORP, and a water-level gauge (right) for measuring drawdown during pumping.

drawdown value) indicates that water subsequently pumped from the well is derived directly from the formation. Because the flow rate used for purging is, in many cases, the same as or only slightly higher than the flow rate used for sampling and because purging and sampling are conducted as one continuous operation in the field, the process is referred to as low-flow purging and sampling.

The most critical aspects of low-flow purging and sampling are summarized in Table 15.12. Low-flow purging requires the use of a pump. Grab samplers, such as bailers, and inertial-lift devices disturb the water column in the well and the formation and cannot reliably provide a representative sample. While dedicated pumps are preferred because they minimize disturbance to the well, portable pumps can be used with some precautions. Portable pumps must be installed carefully and lowered slowly into the screened zone to minimize disturbance of the water column. Even if done with the utmost care, the installation of a portable pump may result in some mixing of the water column above the well screen with that within the screened interval and the release of substantial suspended material (Kearl et al., 1992, 1994; Puls et al., 1992). This usually requires pumping for a longer period of time to achieve stabilization of water-quality indicator parameters and turbidity. Ideally, the pump should remain in place until any turbidity resulting from pump installation has settled and until horizontal flow through the well screen has been reestablished. Carefully lowering the pump into the well, then completing preparation of other equipment and materials to be used in the sampling event, often allows sufficient time for reduction of initial turbidity to acceptable levels. If, after the pump is started, initial turbidity readings are high (>100 NTU), it may be necessary to stop the pump and allow turbidity to settle. The time required for turbidity to settle is well-specific and should be determined on a well-by-well basis.

In low-flow purging, the pump intake should be set either near the middle of the well screen (Figure 15.65), adjacent to the zone within the screen that has the highest hydraulic conductivity, or adjacent to the zone within the screen that has the highest level of contamination (if either of these are known). The well is pumped at a low flow rate, equal to or less than the natural recovery rate of the formation. The pumping flow rate used for any given well is dependent on the hydraulic performance of the well and the site-specific hydrogeology. A volume measuring device (e.g., a graduated cylinder or a container of known volume) and a time piece capable of measuring in seconds is necessary to calculate the flow rate from the discharge tube from the pump.

TABLE 15.12

Summary of the Critical Aspects of Low-Flow Purging and Sampling

---

Requires the use of dedicated or portable pumps capable of operating at low speeds and low discharge rates (less than 1 l/min for most applications)
  –Dedicated pumps are preferred. Portable pumps are appropriate for use, but must be used with some precautions
Bailers, inertial-lift pumps, and high-flow-rate pumps cannot be used for low-flow purging and sampling
Requires knowledge of well construction details, particularly information related to well screen placement and length
Pump intake should be set near the middle of the well screen or adjacent to the zone of highest hydraulic conductivity screened by the well (if it can be identified)
Pumping rate during purging must be equal to or less than the natural recovery rate of the well (minimizing drawdown in the well and allowing drawdown to stabilize prior to sampling)
Requires monitoring of water levels (to determine when the pumping water level has stabilized)
Requires monitoring of field water-quality parameters (to determine when formation water is being sampled)

Source: Nielsen and Nielsen, 2002. With permission.

During low-flow purging and sampling, the water level in the well is continuously monitored using any water-level measurement equipment that does not disturb the water column in the well and that has an accuracy required by the sampling program (generally ±0.01 ft). Devices such as downhole pressure transducers or bubblers (for continuous measurements) or electronic water-level gauges (for periodic measurements) can be used effectively. Water-level measurements should be taken every 1 to 2 min to the point at which the water level in the well has stabilized or at which drawdown ceases. The purging rate is adjusted so drawdown in the well is minimized and stabilized (does not increase over time). This, in effect, hydraulically isolates the column of stagnant water in the well casing from the water in the well screen and negates the need for its removal (Barcelona et al., 1985; Gillham et al., 1985; Robin and Gillham, 1987; Maltby and Unwin, 1992; Powell and Puls, 1993; Nielsen, 1996). After drawdown stabilizes, all water drawn into the pump must come from the formation and water-level measurement can be discontinued. In situations where the sampling rate is less than the purging rate, the well should be recharged during sampling (i.e., drawdown will be decreasing with time), further ensuring that the water sampled is from the formation.

Several researchers have proposed limits on the amount of drawdown that should be allowed before water-level stabilization occurs. In all cases, the proposed limits are arbitrary and no scientific rationale is provided for their adoption. For example, Puls and Barcelona (1996) proposed a limit of less than 0.1 m (0.33 ft or about 4 in.) drawdown for all wells, conceding that this goal may be difficult to achieve under some conditions due to geologic heterogeneities within the screened interval and may require adjustment based on site-specific conditions and personal experience. It should be emphasized that it is far more important to achieve a stabilized water level in a well during purging than to achieve a particular drawdown value, as each well is in a hydrogeologically unique position and thus will respond differently to pumping. In practical terms, to avoid the possibility of drawing stagnant water from the well casing into the pump intake, drawdown should not exceed the distance between the top of the well screen and the pump intake.

In addition to continuous water-level monitoring, low-flow purging and sampling requires continuous or periodic measurement of selected water-quality indicator parameters. The well is considered purged and ready for sampling when the chosen chemical and physical indicator parameters (commonly pH, temperature, specific conductance, and dissolved oxygen, sometimes also including redox potential [ORP] and turbidity) have stabilized, confirming that formation water is being pumped. Continuous monitoring in a closed flow-through cell of known volume generally provides the most consistent and reliable results, especially for DO and ORP, and is the preferred method of measuring indicator parameters. However, individual hand-held instruments designed to measure the most common water-quality indicator parameters (temperature, pH, and conductivity or specific conductance) or turbidity may also be used. DO and ORP measurements made after the purged water is exposed to atmospheric conditions, however, will not accurately reflect *in situ* conditions. All instruments used to measure indicator parameters and turbidity should be properly calibrated and maintained in accordance with manufacturers' instructions at the wellhead at the start of each day of sampling, and calibration should be checked periodically throughout the sampling event. Additional information on field parameter measurement follows later in this chapter.

Low-flow purging and sampling offers a number of benefits over traditional methods, including (ASTM, 2004j):

- Improved sample quality and reduced (or eliminated) need for sample filtration, through minimized disturbance of the well and the formation, which result in greatly reduced artifactual sample turbidity and minimization of false positives for analytes associated with particulate matter

- Improved sample data accuracy and precision and greatly reduced sample variability as a result of reduced stress on the formation, reduced mixing of the water column in the well and dilution of analytes, and reduced potential for sample agitation, aeration, and degassing or volatilization

- Samples represent a smaller section or volume of the formation, representing a significant improvement in the ability to detect and resolve contaminant distributions, which may vary greatly over small distances in three-dimensional space (Puls and McCarthy, 1995)

- Overall, improved sample reproducibility, especially when using dedicated pumps (Karklins, 1996)

- Improved ability to directly quantify the total mobile contaminant load (including mobile colloid-sized particulate matter) without the need for sample filtration (Puls and Barcelona, 1989; Puls et al., 1991, 1992; Puls and Powell, 1992a)

- Increased well life through reduced pumping stress on the well and formation, resulting in greatly reduced movement of fine sediment into the filter pack and well screen

- Greatly reduced purge-water volume (often 90 to 95%), resulting in significant savings of cost related to purge water handling and disposal or treatment, and reduced exposure of field personnel to potentially contaminated purge water

- Reduced purging and sampling time (much reduced at sites using dedicated pumps), resulting in savings of labor cost, depending on the time required for water-quality indicator parameters to stabilize

In addition, this purging technique can be applied to any well that can be pumped at a constant, low flow rate without continuous drawdown of the water level and it can be used in wells where the water table is above or within the well screen. It can also be effectively applied to bedrock well completions.

Although the application of low-flow purging and sampling will improve sampling results and produce significant technical and cost benefits at most sites, not all sites, and not all individual wells within a site, are well suited to this approach. It cannot be applied properly without consideration of site-specific hydrogeology and well-specific hydraulic performance. On a practical basis, low-flow purging and sampling is generally not suitable for use in very low-yield wells (e.g., those that will not yield sufficient water without continued drawdown with pumping over time). As discussed previously, low-flow purging cannot be performed using bailers, or inertial-lift devices, which severely agitate the water column in the well, resulting in significant mixing of the water column and release of considerable sediment, which shows up as increased turbidity in samples.

Low-flow purging and sampling is appropriate for collection of ground-water samples for all categories of aqueous phase contaminants and naturally occurring analytes. This includes VOCs and SVOCs, metals and other inorganics, pesticides, PCBs, other organic compounds, radionuclides, and microbiological constituents. It is particularly well-suited for use where it is desirable to sample aqueous-phase constituents that may sorb or partition to particulate matter. This method is not applicable to the collection of either LNAPLs or DNAPLs.

Four peer-reviewed field studies have been conducted in which analysis of samples collected by low-flow methods and carefully employed conventional methods (well-volume purging) were statistically compared (Powell and Puls, 1993; Kearl et al., 1994;

Shanklin et al., 1995; Serlin and Kaplan, 1996). In each case, the authors reported no significant differences for a variety of analytes, including several VOCs, radionuclides, metals, trace metals, and other inorganics. This indicates that at the very least, the methods produce equivalent results.

In other peer-reviewed field studies, ground-water samples collected using low-flow purging and sampling have demonstrated dramatically lower concentrations of a variety of analytes associated with turbidity. Backhus et al. (1993) and Groher et al. (1990) reported levels of PAHs two to three orders of magnitude lower in low-flow samples than in bailed samples from the same wells. Bangsund et al. (1994) reported levels of some metals (Al and Fe) one to three orders of magnitude lower and dioxins or furans one to three orders of magnitude lower in samples collected using low-flow sampling techniques than in samples collected using traditional purging and sampling (well-volume purging and sampling with a bailer). Several researchers (Puls and Barcelona, 1989; Puls et al., 1992; Hurley and Whitehouse, 1995; Puls and Paul, 1995; McCarthy and Shevenell, 1998) have observed a direct relationship between flow rate (or bailing), turbidity, and metals content in unfiltered samples and a strong inverse correlation between turbidity and sample representativeness. In all instances, wells that had previously produced samples containing significant turbidity (>500 NTUs) using traditional purging and sampling protocols (bailing or high-flow-rate pumping) and produced samples with very low turbidity (typically <10 NTUs) using low-flow purging and sampling. In some cases, suspended solids levels two to three orders of magnitude lower were noted in low-flow samples (Puls et al., 1992; Backhus et al., 1993). This consistently results in samples with much lower concentrations of metals that more accurately reflect true ground-water conditions.

Some ground-water samples collected using low-flow methods may contain higher concentrations of some dissolved species than samples collected using traditional purging and sampling methods, primarily because sampling is more targeted to a specific zone, and samples do not exhibit strong effects of mixing and dilution. The natural tendency for higher horizontal hydraulic conductivity than vertical hydraulic conductivity in most geologic materials tends to reduce the amount of interzonal mixing if a lower pumping rate is used (Wilson and Rouse, 1983). Graham and Goudlin (1996) documented levels of several VOCs (including benzene, toluene, ethylbenzene, and xylenes [BTEX]; 1,1-dichloroethane; 1,2-dichloropropane; and 1,1,2-trichloroethane) that were 2 to 14 times higher in samples collected using low-flow methods than in samples collected using well-volume purging and sampling with bailers. This was apparently due to the lack of mixing between the water in the well screen (which represents formation-quality water) and the water in the well casing (which resulted in dilution when sampling with a bailer). Low-flow sampling in this instance provided more representative samples because it allowed collection of water from only the screened interval, which represented the higher levels of contamination that were the original targets of well screen placement and sampling.

The potential changes in concentrations of target analytes that could result from applying low-flow purging and sampling to wells that were previously sampled via conventional methods may cause some difficulties in data comparison and interpretation of temporal trends. With recognition of the possible changes in contaminant concentration noted earlier, these difficulties are easily overcome. However, anticipation of these difficulties has been offered by some samplers as a reason to continue sampling using traditional methods. Continuing the use of inappropriate sampling methods for the sake of maintaining consistency is not a valid argument; if questionable or bad data were collected before, collection of good data becomes even more important. The cost of making incorrect decisions based on poorly collected samples is much greater than the cost of sampling correctly. It must be recognized that all decisions based on sample data of questionable value will themselves be questionable. The extra time and cost of carefully collecting samples that provide estimates of contaminant concentrations must be weighed against the cost of making a bad decision based on inaccurate estimates. All efforts in sampling ground water should be directed toward collecting the highest quality samples possible, to ensure that decisions based on sampling results are scientifically and legally defensible.

Traditional Purging and Sampling Strategies for Low-Yield Wells: At some facilities, such as landfill sites, it is necessary to install sampling points in fine-grained formations, such as silts and clays, which have inherently low hydraulic conductivity. To date, monitoring well construction and development technologies and standards have not been developed to overcome the problems of sampling from low-yield formations (refer to Chapter 10 for additional information). The approach to purging and sampling wells installed in these formations must be different than approaches used for high-yield wells, as these wells typically run dry prior to removal of a specified number of well volumes. In sampling points with very low yield, even low-flow purging and sampling may not be appropriate because excessive water-level drawdown may result in increased soil particle or colloidal transport into the well (Sevee et al., 2000). As a consequence, in relatively low hydraulic conductivity materials, a sample collected directly from the well screen, without excessively pumping and dewatering the sampling point, will provide the most representative sample of formation-quality water possible.

Regulatory requirements for purging and sampling low-yield monitoring wells are highly variable, but most involve removal of all water from the well during purging, then sampling upon recovery of the well, because these wells will not yield multiple well volumes in a reasonable time. To assist in determining which purging and sampling strategy is most appropriate to use in low-yield wells, some guidance defines "low yield" and "high yield." For example, in current ground-water sampling guidance for San Diego County, California, definitions are provided for both of these terms as follows:

- *Slow Recharging Well [Low-Yield Well]*
  "A well is considered to be slow recharging if recovery to 80% of its static condition takes longer than two hours."
- *Fast Recharging Well [High-Yield Well]*
  "A well is considered to be fast recharging if recovery to 80% or more of its static condition occurs within two hours."

Two traditional purging and sampling strategies have been used in low-yield wells: (1) purging to dryness and sampling during well recovery and (2) purging to the top of the well screen and sampling the water in the well screen. Each of these strategies imparts bias on samples collected following purging; however, they continue to be the most widely used purging strategies implemented in the field for low-yield wells.

Purging to Dryness: The most commonly prescribed strategy for purging and sampling low-yield wells is to purge the well to dryness and then sample during or following well recovery: Examples of sampling strategies following well dewatering are summarized in Table 15.13. The most significant problem encountered in low-yield wells is that dewatering of the well screen (Figure 15.67) results in potentially significant chemical alteration of the water eventually collected as a sample, including the following:

- The time required for sufficient recovery of the well may be excessive, affecting sample chemistry through prolonged exposure of the water in the formation surrounding the well to atmospheric conditions. As Puls et al. (1991) point out, the oxidation reactions that occur under these conditions can dramatically alter

TABLE 15.13
Traditional Strategies for Collecting Samples Following Well Dewatering

Sample after a defined period of time (e.g., 2, 4, 24 h)
Sample after a defined recovery of the water level to a percentage of the initial static water level (e.g., 80% of static)
Sample after a defined volume of water has re-entered the well (e.g. enough water has recharged to permit filling all sample containers required)

Source: Nielsen and Nielsen, 2002. With permission.

sample results for a variety of inorganic analytes. Perhaps the most significant effect is the loss of VOCs from samples collected after recovery of the water level, in some cases, exceeding 70% in only a few hours (McAlary and Barker, 1987; Herzog et al., 1988).

- Purging the well dry causes a significant increase in the hydraulic gradient in formation materials surrounding the well (Figure 15.68 and Figure 15.69), increasing the flow velocity toward the well and resulting in turbulent flow in the formation and filter pack immediately adjacent to the well and mobilization of formation fines, and creating increased turbidity in water entering the well (Giddings, 1983). When this turbid sample is acidified (following standard chemical preservation techniques for metals), metallic ions that were previously adsorbed onto or contained within the suspended clays and silts are released, causing elevated concentrations of metals in the sample. This renders the sample useless for detection of some metals and may also bias other general chemistry parameters.

- Purging the well dry may cause cascading of water as the well recovers (Figure 15.69), resulting in a change in dissolved gases and redox state, and ultimately affecting the concentration of the analytes of interest through the oxidation of dissolved metals and loss of VOCs.

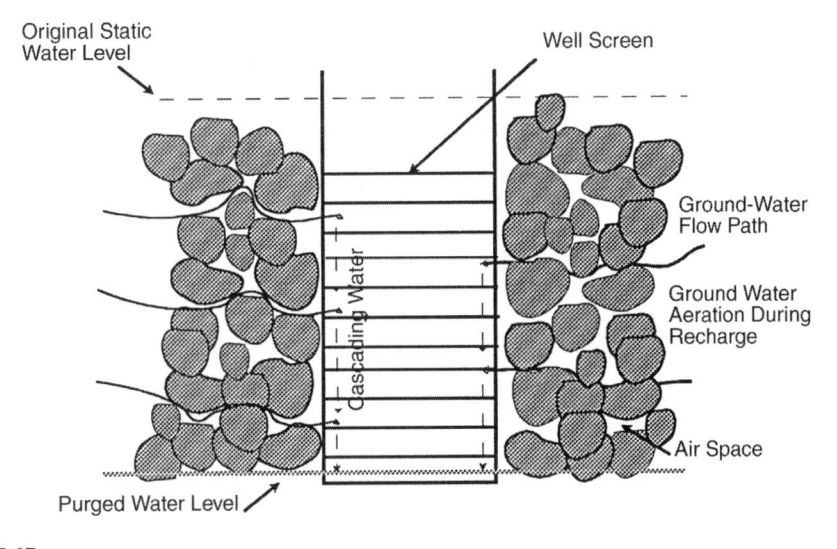

FIGURE 15.67
Well-screen dewatering can cause significant changes in water chemistry for a number of parameters including those associated with artifactual turbidity (i.e., hydrophobic organics, and trace metals) and those affected by exposure to atmospheric air (i.e., VOCs, dissolved gases, and trace metals).

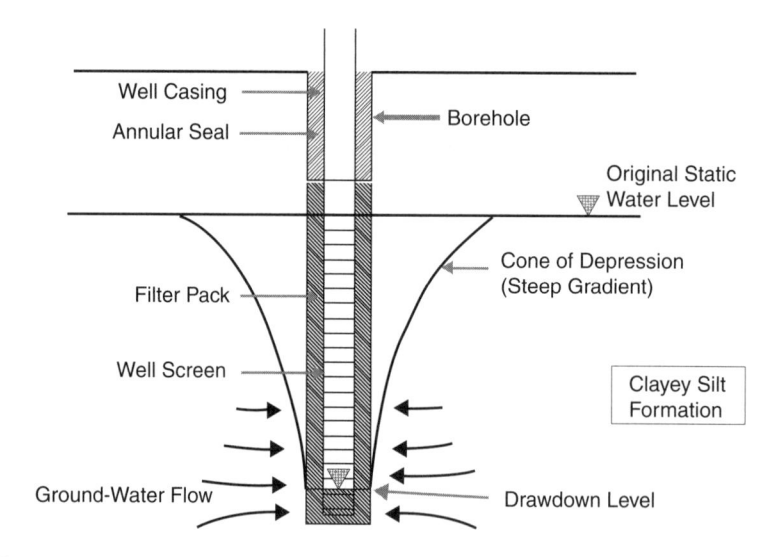

FIGURE 15.68
Purging a low-yield well (in this case, installed in an unconsolidated formation material) often dewaters the well and creates a very steep gradient toward the well. This causes turbulent flow in the formation (at the pore-space level), which results in the mobilization of sediment from the formation that shows up as turbidity in samples.

- Draining water from the filter pack surrounding the screen can result in air being trapped in the pore spaces, with lingering effects on dissolved gas levels and redox state.
- In some cases, the well may not recover sufficiently to produce the sample volume required within a reasonable time period.

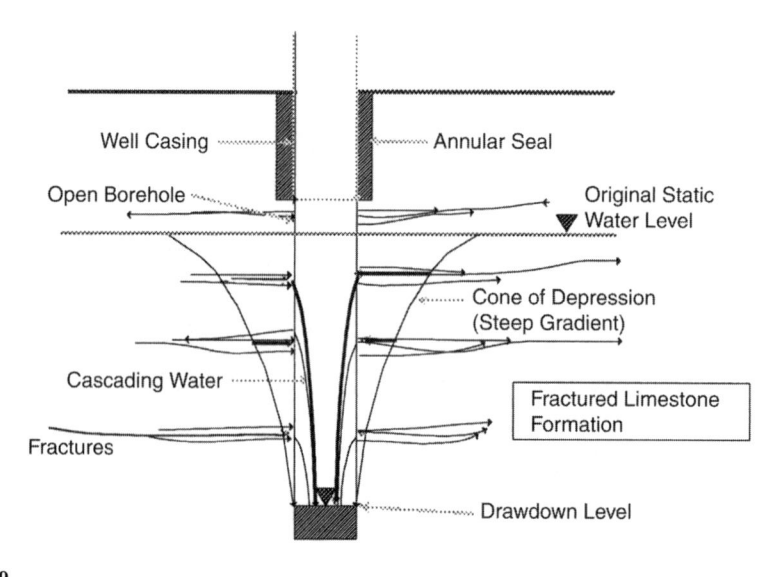

FIGURE 15.69
Purging a low-yield well (in this case, installed in a fractured bedrock) often dewaters the well and causes a very steep gradient toward the well. Discrete flow through fractures can carry sediment into the well and can also cause cascading of water into the well (in this case, an open bedrock borehole), which results in significant aeration and agitation as the water streams down the open borehole.

Thus, while rules requiring complete removal of water from the well during purging are administratively convenient and relatively easy to adhere to, it is clear that this approach results in the collection of samples that are not representative of formation water chemistry. Some state regulatory agencies have recognized this and suggest avoiding pumping wells to dryness, because of the many deleterious effects that this approach has on sensitive analytical parameters (Thurnblad, 1995).

Another problem in sampling low-yield wells is that the approach to determining the volume of water to be removed from a low-yield well prior to sampling is not consistent between regulatory jurisdictions. In some locations, samplers are still required to remove three to five well volumes prior to sample collection. This represents a misapplication of a purging strategy intended for high-yield wells and can result in a tremendous increase in time and expense, with no benefit with respect to ensuring collection of representative samples. It can often require several days to sample a single well using this approach.

Purging Water from the Casing Only: To avoid the problems associated with dewatering the well screen, an alternative approach to purging low-yield wells that is less commonly applied is to remove only that water in storage in the well casing (i.e., only the water above the well screen), then begin sampling after this stagnant water has been removed. This requires the sampler to know the depth to and length of the well screen, to use a purging device that permits accurate and precise placement of the intake relative to the top of the well screen, to measure water levels in the well during purging (to assess drawdown), and to ensure that the water column in the casing and screen is not agitated or mixed during purging and sampling. This precludes the use of bailers or inertial-lift pumps. Using this purging method, the water in storage in the casing is typically removed by pumping from the top of the water column and following the water level down to the top of the screen. Once water in storage is removed, samples can be collected directly from the screened portion of the well, usually using the same device used for purging. The water in the screen comprises the most representative sample that can be collected from a low-yield well.

Other Methods for Purging and Sampling Low-Yield Wells: There are several better alternatives to using these traditional approaches to purge and sample low-yield wells: Two alternative methods that should be considered for use in these wells include (1) minimum-purge sampling (also known as passive sampling) and (2) use of a device to "core" the water column in the well screen without disturbing the water column.

Minimum-Purge Sampling: Minimum-purge sampling or passive sampling, as described in Puls and Barcelona (1996), Powell and Puls (1993), and Nielsen and Nielsen (2002), appears to offer one of the best alternatives for collecting samples from low-yield wells with minimal alteration of formation water chemistry (Figure 15.70). Minimum-purge sampling generally requires the use of a dedicated pump with the pump intake installed near or below the middle of the well screen. It involves the removal of the smallest possible volume of water prior to sample collection. This volume is generally limited to the volume of water in the sampling system, including the volume of water in the dedicated pump chamber and the volume of water in the discharge tubing submerged below the static water level in the well. Immediately following removal of this small volume, samples are collected, with the expectation that the water pumped immediately after evacuation of the sampling system (i.e., the water from the screened zone) represents formation water quality. Flow rates used for minimum-purge sampling are about the same as or slightly lower than those used for low-flow purging and sampling — generally 100 ml/min or less. Because very low hydraulic conductivity formations do not yield sufficient water to satisfy the demands of the pump even at these low flow rates, drawdown cannot be avoided. Thus, to determine the volume of

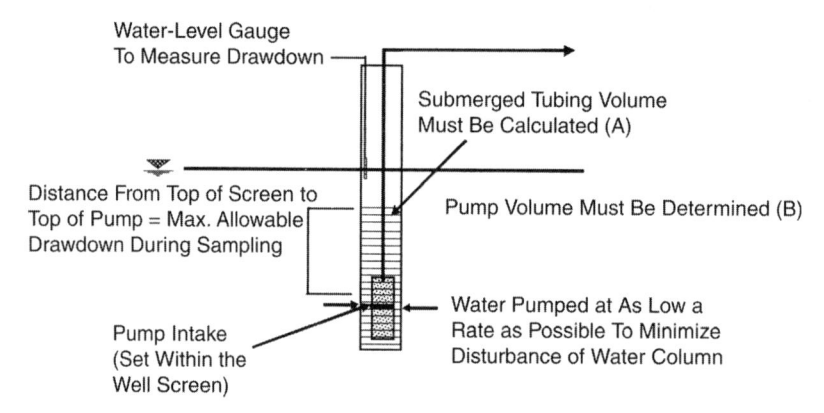

FIGURE 15.70
Illustration of the concepts behind minimum-purge or passive sampling.

water available for sampling, it is necessary to calculate the volume of water within the well screen above the pump intake. Only this volume should be collected as the sample, and sampling must be discontinued once drawdown has reached a prescribed level in the well relative to the pump intake.

The minimum-purge sampling method consists of collecting only the water in the screened zone, thus avoiding the pitfalls of complete evacuation of the well and providing a better opportunity to collect samples that reflect *in situ* water quality. Although the low flow rate of water through the well screen provides only a limited exchange of water with the formation, avoiding the sources of error and bias associated with dewatering the well screen and disturbing the water column is of greater importance.

Minimum-purge sampling can be applied to any well in which there is sufficient water to ensure submergence of the pump intake throughout purging and sample collection, although it is most often applied to wells installed in very low hydraulic conductivity formations. Minimum-purge sampling is appropriate for collection of samples for all categories of naturally occurring analytes and aqueous-phase contaminants including VOCs, SVOCs, metals and other inorganics, pesticides, PCBs, other organic compounds, radionuclides, and microbiological constituents. However, it cannot be used to collect samples of either LNAPLs or DNAPLs.

Minimum-purge sampling differs from low-flow purging and sampling in that while low-speed portable pumps can be used for low-flow purging, they are impractical in most situations where minimum-purge sampling could be applied. Because disturbance of the water column can result in significant mixing of the stagnant water column in the casing with the water column in the well screen, as well as the release of significant turbidity, if a portable pump is used, the pump must be set in the well long before sampling is planned. Generally, the pump must be placed far enough ahead of the time of sampling so that the effect of pump installation has completely dissipated prior to sample collection. The time required will vary from well to well, but may be in excess of 48 h (Kearl et al., 1992; Puls and Barcelona, 1996). For this reason, dedicated pumps are the most practical devices used for nearly all minimum-purge sampling applications. Under no circumstances can grab sampling devices, such as bailers or inertial-lift pumps, be used for minimum-purge sampling.

In minimum-purge sampling, placement of the pump intake depends more on the sample volume required to satisfy the objectives of the sampling program than on other factors. The pump intake should be set within the well screen, but not too close to the bottom of the screen (to avoid drawing in sediment that may have settled in the bottom of the well) or too close to the top of the screen (to avoid incorporating stagnant water from the casing in the sample). Two to three feet above the bottom of the screen is generally sufficient. Because the volume that can be collected as a sample is, for practical purposes, limited to the volume available within the well screen (minus the volume displaced by the pump), it is first necessary to calculate the volume available for pumping and compare that to the volume required for samples—this will help determine the optimum pump setting. To provide a safety factor against including stagnant water from the casing in the sample, drawdown during sampling should be limited to about 2 ft less than the distance between the top of the screen and the pump intake as illustrated in the following example:

Example
In a 2 in. nominal diameter well with a 10 ft long well screen (in which the static water level is above the top of the well screen), if the pump intake (at the bottom of the pump) is set at 2 ft above the bottom of the screen, and the pump is 1 ft long and 1.75 in. O.D., the allowable drawdown is 6 ft, and the effective volume of water available for removal is that within 5 ft of the screen (accounting for the volume displaced by the pump). For a 2 in. well, that volume is 0.16 gal/ft × 5 ft or 0.80 gal (3 l).

In this example, as long as volume requirements for the sampling program are equal to or less than 0.8 gal (3 l), the pump setting will be satisfactory. If additional sample volume appears necessary, the pump should not be set lower in the screen, to avoid the possibility of drawing in sediment that may have settled in the bottom of the screen. Rather, the sampling team should ask the laboratory if they can get by with less volume for each parameter (which is nearly always the case) to ensure that the volume available is sufficient to satisfy sampling and analytical requirements.

With minimum-purge sampling, there is no requirement to monitor indicator parameters to determine chemical stabilization of the well during purging due to the low volume of water collected. Therefore, equipment such as flow-through cells and field parameter instrumentation is not necessary to implement minimum-purge sampling, although some field parameter measurements may still be required by regulatory agencies for other purposes.

"Coring" the Water Column in the Well Screen: A device known as the HydraSleeve (Figure 15.71), a low-cost, disposable grab-sampling device, allows samplers to capture a "core" of water from any discrete interval within the screened portion of the sampling point without purging the well prior to sample collection. During use of this device, there is no change in water level and minimal disturbance to the water column in the well. The device consists of a flexible, lay-flat, PE sleeve at the top of which is a flexible top-loading check valve. Although the standard size of the device is 1.75 in. in diameter and 30 in. in length (holding a volume of about 1 l when full), it can be made in a variety of different lengths and diameters to meet project-specific and sampling-point-specific requirements. To ensure that the device is placed at the bottom of the well screen (so it is in position to core the water in the screen at a later date), a reusable stainless steel weight is attached to the sealed bottom of the flexible sleeve. An optional weight can be attached to the top of the device to compress the sampler in the bottom of the well, if desired. A collar at the top of the device allows attachment of a suspension cable or cord to permit retrieval of the device for sample collection (Figure 15.72). In cases where vertical stratification of contaminants in the well screen is of concern, it is possible to stack several HydraSleeves on the suspension cable to permit vertical profiling of the water column.

FIGURE 15.71
The HydraSleeve is a lightweight disposable sampler, ideally suited for sampling low-yield wells. This photo shows a HydraSleeve after retrieval from a newly installed well that has not yet been developed (thus the high sediment content). In practice, these samplers are capable of collecting sediment-free samples from low-yield wells, by coring the water column in the screen.

During deployment, the HydraSleeve is carefully lowered into the sampling point and positioned at or near the bottom of the well. While it is lowered into the sampling point, hydrostatic pressure keeps the sleeve collapsed and the check valve closed, preventing water from entering the sampler. The slim cross-section of the collapsed sleeve minimizes disturbance of the water column during placement, helping to reduce the time required for the sampling point to return to chemical and hydraulic equilibrium. During equilibration, the sleeve remains collapsed with the check valve closed due to the water pressure difference between the outside and inside of the device. Once the device is situated at the desired sampling interval, the suspension cable is attached to the top of the well (e.g., to the underside of the well cap) to allow samplers to retrieve it after the sampling point has equilibrated (generally between 48 h and 1 week).

To collect a sample after the sampling point has equilibrated, the HydraSleeve is removed from the screened interval by pulling up on the suspension cord at a rate of 1 ft/sec or faster. The upward motion can be accomplished using one long continuous pull, several short rapid strokes, or any combination that moves the device the required distance. When this upward motion is initiated, the check valve at the top of the device opens and the device fills with water from within the screened interval. The total upward distance the device must be pulled to complete the filling process is about two times the length of the sampler. As the device fills with water, there is no change in water level in the well and there is no agitation of the water column or the sample. After the device is full, the check valve closes, thus preventing additional water from overlying portions of the water column from entering the device during retrieval. At ground surface, the sample is

Ground-Water Sampling   1055

ONE

Placing Sampler

HydraSleeve is lowered into place and postioned in the well screen. Water pressure keeps bag collapsed and check valve closed, preventing water from entering sampler. Well is allowed to return to equilibrium.

TWO

Sample Collection

HydraSleeve fills when the check valve is moved upward faster than 1 fps. It can fill by either continous upward movement or cycling the sampler up and down. When moving uipwards, the check valve opens and fluid flows into the bag. There is no change in water level and minimum sample agitation during collection.

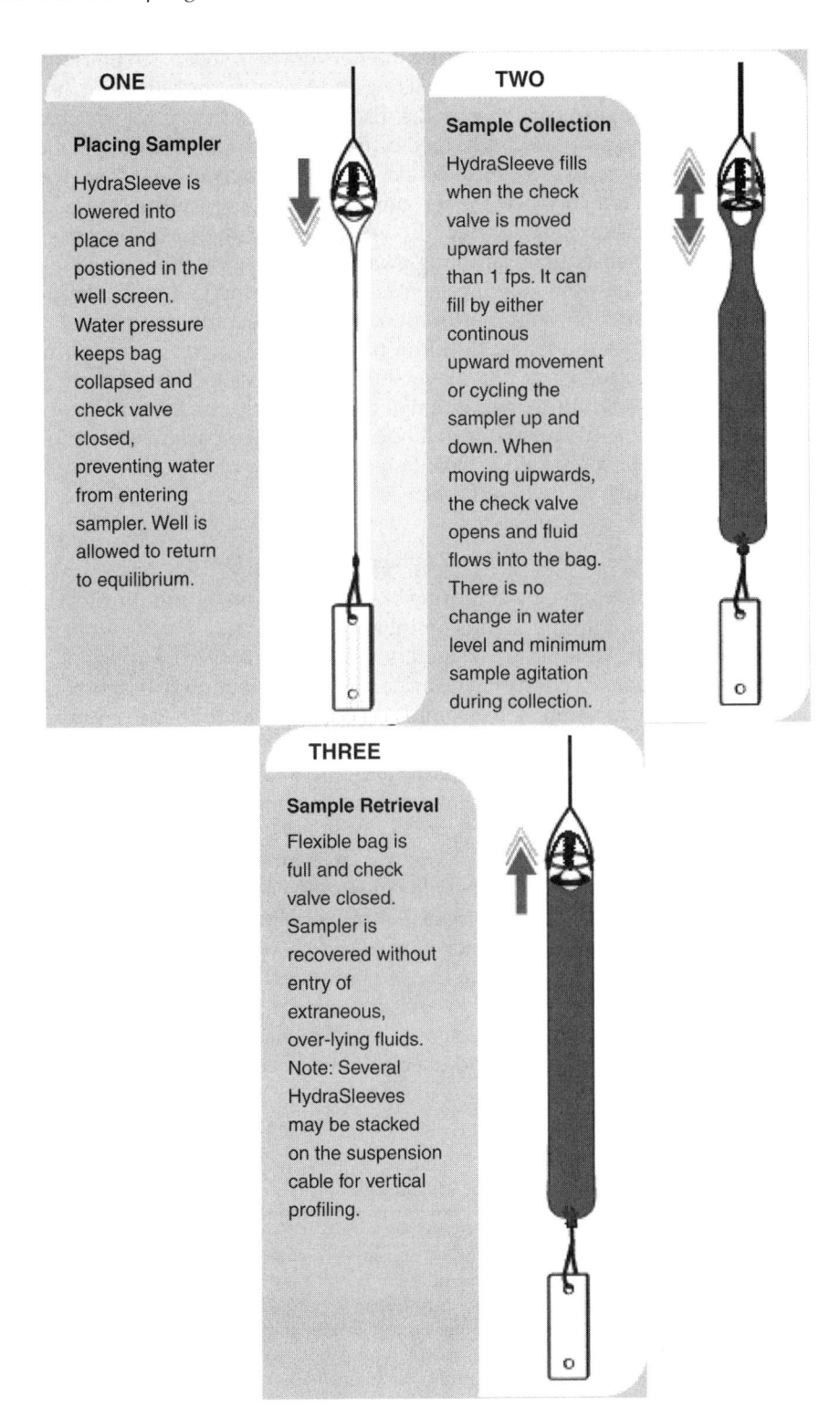

THREE

Sample Retrieval

Flexible bag is full and check valve closed. Sampler is recovered without entry of extraneous, over-lying fluids. Note: Several HydraSleeves may be stacked on the suspension cable for vertical profiling.

FIGURE 15.72
Illustration of the process of collecting samples using a HydraSleeve. (Diagram courtesy of GeoInsight, Inc.)

removed from the HydraSleeve by puncturing the sleeve with a sharp, small-diameter discharge tube at the base of the check valve. By raising and lowering the bottom of the sampler or by pinching the sample sleeve just below the discharge tube, the sample discharge flow rate can be controlled. The sample is discharged directly into any parameter-specific sample container. At the conclusion of sample collection, a new HydraSleeve can be lowered into the sampling point and left in place for the next sampling event. This saves time and improves sampling precision and accuracy by ensuring that the formation has reached equilibrium prior to the subsequent sampling event.

The HydraSleeve can be used to collect representative samples for all chemical parameters due to its construction with inert materials, lack of aeration and agitation of the water column and samples (and resulting low turbidity of samples), and its single use (i.e., disposable) design. Studies to date have shown that because samples are collected at *in situ* pressure with no aeration or agitation of the water column, there are no problems related to loss of volatile constituents or oxidation of sensitive parameters such as trace metals (Parker and Clark, 2002; ITRC, 2005; Parsons, 2005) that are common with other grab-sampling devices such as bailers (Parker and Clark, 2002).

Field Measurement of Water-Quality Indicator Parameters and Turbidity

As discussed earlier in this chapter, monitoring of selected chemical and physical indicator parameters in ground water in the field is an integral component of some strategies used to purge high-yield wells. Chemical parameters most commonly measured include pH, specific conductance (or conductivity) and, more recently, DO and ORP (redox potential, also measured as Eh). In addition, some sampling programs include the measurement of temperature, and some include turbidity, a physical parameter that is an indicator of the disturbance caused to the water column by the purging and sampling method used.

The SAP must detail exactly which indicator parameters are to be measured in the field, and where, when (in the context of well purging and sample collection), and how (related to the type of equipment used) those parameters are to be measured. For applications involving measurement to stabilization, the SAP must also provide an accurate definition of "stabilization" for each parameter, that is specific to the instrumentation used. It is the experience of the authors that this is a common deficiency of many SAPs, which results in a great deal of error in field parameter measurement. Table 15.14 summarizes the common errors in field parameter measurement that lead to inaccuracy and imprecision of data generated in the field.

Some indicator parameter measurements can be made in an open container (Figure 15.73) (e.g., pH, temperature, and specific conductance) if a device other than a

TABLE 15.14

Common Errors in Field Parameter Measurement

No instrument-specific definition of stabilization
No calibration or incorrect calibration of instrumentation under field conditions
Use of expired or incorrect calibration standards
Poor equipment cleaning and maintenance practices
No training of field personnel on the proper use of specified instrumentation
Failure to understand operating ranges, accuracy, resolution, and operational features of individual parameter measurement probes or test methods
Failure to recognize errors in field parameter measurements
Failure to record units of measure and " + " or " − " values for parameters such as ORP
Measurement of DO and ORP in open containers
Taking too long to measure parameters that are temperature sensitive
Errors in collection and handling of subsamples analyzed for turbidity

FIGURE 15.73
Only a few indicator parameters (specific conductance, pH, and temperature) can be reliably measured in an open container at the surface, but only if it is done quickly. Other parameters (DO and ORP) are strongly affected by exposure to atmospheric conditions, and should not be measured in open containers.

pump is used for purging (e.g., a bailer). Other parameters, such as DO and ORP, should never be measured in an open container where there is an air–water interface. In these circumstances, readings will be affected by exposure to atmospheric air and may, in fact, never be accurate and never truly stabilize. DO and ORP should always be measured either downhole (with an *in situ* probe) (Figure 15.74 and Figure 15.75) or in a flow-through cell at the surface (Figure 15.76). The latter option requires that a pump be used for purging and sampling. Several researchers (Puls et al., 1991, 1992; Barcelona et al., 1994; Puls and McCarthy, 1995) have shown temperature and pH (and, in some instances, specific conductance) to be the least sensitive indicators of equilibrated conditions, while ORP and DO are more sensitive indicators.

During purging, the purpose of monitoring field parameters is to determine when formation-quality water is available for sample collection. This is interpreted to occur when the selected indicator parameters have stabilized. The term stabilization must be clearly understood by all samplers to ensure precision between sampling teams and to avoid overpurging or underpurging the sampling point. In the context of low-flow purging and sampling, indicator parameters are considered stable when three consecutive readings made several minutes apart fall within the ranges presented in Table 15.15 (Nielsen and Nielsen, 2002). In the context of traditional purging to stabilization of indicator parameters, stabilization is often defined as the point at which measured values are within $\pm 10\%$ for all parameters for three consecutive readings taken 3 min apart (Nielsen and Nielsen, 2002).

While the criteria in Table 15.15 are reasonable criteria for many hydrogeochemical situations, it should be recognized that firm criteria for indicator parameter stabilization may not be appropriate for some situations because of a variety of factors, including variability

FIGURE 15.74
A multiparameter sonde designed for taking *in situ* measurements of indicator parameters in a 2 in. diameter well. This is the best method for collecting representative data for parameters strongly affected by exposure to atmospheric conditions. This sonde is capable of measuring *in situ* pH, specific conductance, temperature, DO, ORP, and salinity. (Photo courtesy of YSI, Inc.)

FIGURE 15.75
Using a multiparameter sonde designed for collecting *in situ* measurements of indicator parameters allows chemical profiling in the well. The differences in chemistry in the water column are often significant enough to allow accurate location of the top of the well screen or location of contaminated zones within the screen.

FIGURE 15.76
A flow-through cell can also be used to collect representative data on indicator parameters that are affected by exposure to atmospheric conditions. In this case, the same type of equipment shown in Figure 15.74 and Figure 15.75 (with the addition of turbidity as a measurement option) has been adapted for use in a small-volume (250 ml) cell, with the inlet at the bottom and the outlet at the top. This configuration is preferred to achieve flow over the measurement sensors and accurate indicator parameter measurement.

in aquifer properties, monitoring well hydraulics (Robbins and Martin-Hayden, 1991; Barcelona et al., 1994; Robin and Gillham, 1987), and natural spatial and temporal variability in ground-water chemistry and contaminant distribution (Keely and Boateng, 1987a; Huntzinger and Stullken, 1988; Clark and Baxter, 1989). Therefore, the criteria in Table 15.15 should be compared to well-specific measurements to determine whether the site-specific criteria need to be adjusted. Additionally, these criteria should be evaluated to select those that are most important and relevant to meeting the sampling objectives for the specific site. Not all criteria need to be met for all sites. Shanklin et al. (1995) and Puls and McCarthy (1995) point out that stabilization criteria that are too stringent may unnecessarily lead to overpurging of the sampling point and the generation of large amounts of contaminated purge water, without providing the benefit of ensuring that the samples are any more representative. This commonly occurs when stabilization criteria are set at levels that exceed the measurement capability or accuracy of the instrumentation. It is important for sampling team members to understand that just because an instrument displays a measurement to two decimal places (resolution), it does not mean

TABLE 15.15

Example Criteria for Defining Stabilization of Water-Quality Indicator Parameters

Temperature	$\pm 0.2°C$
pH	± 0.2 pH units[a]
Conductivity	$\pm 3\%$ of the reading
DO	$\pm 10\%$ of the reading or ± 0.2 mg/l, whichever is greater[a]
Eh or ORP	± 20 mV[a]

[a] Related to the measurement accuracy of commonly available field instruments.

that the displayed number is accurate to two decimal places. ORP is a good example of a parameter for which this problem is commonly seen. For ORP, samplers need to remember that the recommended stabilization criterion is ± 20 mV.

If stabilization is not achieved, sampling should be done at the discretion of the field sampling team (or in consultation with the project manager). Factors that commonly affect field parameter stabilization and that should be checked prior to sampling without achieving stabilization include improper instrument calibration, failure to allow instrument sensors to warm up and stabilize prior to use, poor condition of instrument sensors, air leaks in the pump tubing or flow-through cell connectors, and too low a flow rate through the flow-through cell. If samples are collected without reaching water-quality indicator parameter stabilization, the conditions under which the samples are collected should be documented.

For in-line flow-through cells, the frequency of the measurements for indicator parameters should be documented in the SAP. The first measurement should generally be made after the volume of the pump and tubing has passed through the flow-through cell. Subsequent measurements should be made based on the volume of the cell and the time required to completely evacuate one volume of the cell at the flow rate used for purging, to ensure that independent measurements are made. For example, a 500 ml flow-through cell in a system pumped at a rate of 250 ml/min will be evacuated in 2 min, so measurements should be made at least 2 min apart. It is important, therefore, that the sampling team establish the following volumes and rates in the field prior to the sampling event:

- Volume of the pump and discharge tubing
- Optimum pump discharge rate
- Volume of the flow-through cell corrected for the displacement volume of the field parameter measurement instrumentation installed inside the flow-through cell

It is also important to know the manufacturers' recommendations for the amount of time required to allow individual sensors (e.g., DO) used to measure field parameters to stabilize to ensure that representative data are collected.

For wells in which dedicated pumps are used, indicator parameters should stabilize shortly after the volume of the pump and tubing has been removed. In a sampling program in which bladder pumps were used for low-flow purging and sampling, Shanklin et al. (1995) found that they achieved indicator parameter stabilization and that they could collect representative samples after purging twice the calculated volume of the pump and discharge line. For wells in which portable pumps are used, the effects of pump installation on the water column in the well usually result in the need to remove significantly more water before chemical indicator parameters (and, as noted subsequently, turbidity) reach stabilization. Owing to the agitation and aeration of the water column caused by a bailer, which results in increased levels of DO and increased amounts of artifactual particulate matter suspended in the water, the volume of water required to be purged and the time required to reach stabilization will be greater than for a scenario in which a pump is used. For some parameters (particularly DO), stabilization may never be reached in bailed samples because of the disturbance caused to the water column in the well (Pohlmann et al., 1994).

Although not a chemical parameter, and not indicative of when formation-quality water is being pumped, turbidity may also be a useful parameter to measure (Figure 15.77). Turbidity is a physical parameter that provides a measure of the suspended particulate matter in the water removed from the well. Turbidity is indicative of pumping stress on

FIGURE 15.77
Field turbidity measurements can provide a useful indication of the degree of disturbance of the water column in the well caused by purging and sampling. In this case, turbidity is being measured using a stand-alone turbidimeter. Some multi-parameter sondes used downhole or in flow-through cells (e.g., the one depicted in Figure 15.76) are also capable of measuring turbidity.

the formation or disturbance to the water column in the well during installation of portable pumps. Sources of turbidity in sampling points can include:

- Naturally occurring colloid-sized or larger solids that may be in transit through the formation
- Artifactual solids from well drilling and installation (e.g., drilling fluids, filter pack, and grout) that have not been effectively removed by well development
- Naturally occurring fine-grained formation materials that are mobilized by agitation of the water column (i.e., by bailing, by installation of a portable pump, or by over-pumping the well)
- Microbial growth that often occurs in monitoring wells in the presence of certain types of contaminants (i.e., petroleum hydrocarbons)
- Precipitation caused by different redox conditions in the sampling point than in the formation

Turbidity levels elevated above the natural formation condition can result in biased analytical results for many parameters, including trace metals, hydrophobic organic compounds, and other strongly sorbed chemical constituents. Naturally occurring turbidity in some ground water can exceed 10 NTU (Puls and Barcelona, 1996) and may be unavoidable. Turbidity in a properly designed, constructed, and developed well is most often a result of significant disturbance of the water column or excessive stress placed on the formation by over-pumping.

The primary reason for minimizing turbidity during purging and sampling is that turbidity can affect the ability to accurately determine the aqueous concentration of some analytes. Analysis of organic analytes can be hampered by the physical presence of suspended solids, and accurate analysis of aqueous inorganics can be affected by stripping of cations, particularly metal species, from the surfaces of suspended clay particles by the sample preservation process (i.e., acidification).

To avoid artifacts in sample analysis, turbidity should be as low as possible when samples are collected. Turbidity measurements should be taken at the same time that chemical indicator parameter measurements are made or, at a minimum, once when pumping is initiated and again just prior to sample collection, after chemical indicator parameters have stabilized. In the context of low-flow purging and sampling, the stabilization criterion for turbidity is $\pm 10\%$ of the prior reading or ± 1.0 NTU, whichever is greater (ASTM, 2004j). If turbidity values are persistently high, the pumping rate should be lowered until turbidity decreases. If high turbidity persists even after lowering the pumping rate, the pump may have to be stopped for a period of time until turbidity settles and the purging process restarted. If this fails to solve the problem, well maintenance or redevelopment may be necessary. In the context of low-flow purging and sampling, difficulties with high turbidity should be identified during pilot tests prior to implementing low-flow purging or during the initial low-flow sampling event, and contingencies should be established to minimize the problem of elevated turbidity.

Sample Pretreatment Options

Another group of parameter-specific field protocols that must be evaluated and included in the SAP are methods for sample pretreatment, including sample filtration and physical and chemical preservation. Sample pretreatment must be performed at the wellhead at the time of sample collection to ensure that physical and chemical changes do not occur in the samples during the time that the sample is collected and after the sample container has been filled and capped. ASTM has published Standard Guides that address both types of sample pretreatment. ASTM Standard D 6564 (ASTM, 2004k) provides a detailed guide for field filtration of ground-water samples, and ASTM Standard D 6517 (ASTM, 2004l) discusses physical and chemical preservation methods for ground-water samples. Each type of sample pretreatment is discussed subsequently.

Filtration: Ground-water sample filtration is a sample pretreatment process implemented in the field for some constituents, when it is necessary to determine whether a constituent is truly "dissolved" in ground water: Filtration involves passing a raw or bulk ground-water sample directly through a filter medium of a prescribed filter pore size either under negative pressure (vacuum) or under positive pressure. Particulates finer than the filter pore size pass through the filter along with the water to form the filtrate, which is submitted to the laboratory for analysis. Particulate matter larger than the filter pore size is retained by the filter medium. In the case of most ground-water monitoring programs, this material is rarely analyzed, although it may be possible to analyze the retained fraction for trace metals or for some strongly hydrophobic analytes such as PCBs or PAHs. Figure 15.78 illustrates a common vacuum filtration setup, and Figure 15.79 illustrates one form of positive pressure filtration.

The most common method for distinguishing between the dissolved and particulate fractions of a sample has historically been filtration with a 0.45 μm filter (see, e.g., U.S. EPA, 1991a). The water that passes through a filter of this pore size has, by default, become the operational definition of the dissolved fraction, even though this pore size

Ground-Water Sampling

FIGURE 15.78
A vacuum filtration system used for ground-water sampling. This practice is not encouraged.

FIGURE 15.79
A positive-pressure filtration system is a better option to use for ground-water sampling. Note the removal of sediment achieved by the cartridge filter.

does not accurately separate dissolved from colloidal matter (Kennedy et al., 1974; Wagemann and Brunskill, 1975; Gibb et al., 1981; Laxen and Chandler, 1982). Some colloidal matter is small enough to pass through this pore size, but this matter cannot be considered dissolved. For this reason, Puls and Barcelona (1989) reported that the use of a 0.45 μm filter was not useful, appropriate, or reproducible in providing information on metals solubility in ground-water systems and that this filter size was not appropriate for determining truly dissolved constituents in ground water.

The boundary between the dissolved phase and the colloidal state is transitional. There is no expressed lower bound for particulate matter and no clear cutoff point to allow selection of the optimum filter pore size to meet the objective of excluding colloidal particles from the sample. The best available evidence indicates that the dissolved phase includes matter that is less than 0.01 μm in diameter (Smith and Hem, 1972; Hem, 1985), suggesting that a filter pore size of 0.01 μm is most appropriate. However, filters with such small pore sizes are subject to rapid plugging, especially if used in highly turbid water, and are not practical to use in the field. Kennedy et al. (1974) and Puls et al. (1991) provide a strong case for the use of a filter pore size of 0.1 μm for field filtration to allow better estimates of dissolved metal concentrations in samples. Puls et al. (1992) and Puls and Barcelona (1996) also recommend the use of 0.1 μm (or smaller) filters for determination of dissolved inorganic constituents in ground water. Such filters are considerably more effective than filters with larger pore sizes (e.g., 0.45 or 1.0 μm) in terms of removing fine particulate matter. These filters are widely available and practical for use in the field for most situations, although in some highly turbid water, filter plugging may make the filtration process difficult and protracted. All factors considered, 0.1 μm field filtration, although it is a compromise, appears to offer the best opportunity for collecting samples that best represent the dissolved fraction.

Yao et al. (1971) indicate that colloids larger than several microns in diameter are probably not mobile in aquifers under natural ground-water flow conditions due to gravitational settling. Puls et al. (1991) also suggest that colloidal materials up to 2 μm are mobile in ground water systems. With the upper bound for colloidal matter described by many investigators as being between 1.0 and 10 μm, it seems reasonable to suggest that a filter pore size of 10 μm would include all potentially mobile colloidal material and exclude the larger, clearly nonmobile artifactual fraction. However, it should be noted that using this filter pore size, artifactual colloidal material that is finer than 10 μm in diameter will be included in the sample. Although this filter pore size is a compromise, it will lead to conservative estimates of total mobile contaminant load while excluding at least a portion of the particulate matter that is artifactual in nature. The collection and analysis of both filtered and unfiltered samples is sometimes suggested as a means of discriminating between natural and artifactual colloidal material or between dissolved and colloidal contaminant concentrations.

Historically, filtration of ground-water samples has served several important functions in ground-water sampling programs. Filtration helps minimize the problem of data bounce, which commonly results from variable levels of suspended particulate matter in samples between sampling events and individual samples, making trend analysis and statistical evaluation of data more reliable. In addition, by reducing suspended particle levels, filtration makes it easier for laboratories to accurately quantify metals concentrations in samples. Perhaps most importantly, filtration of samples makes it possible to determine actual concentrations of dissolved metals in ground water that have not been artificially elevated as a result of sample preservation (acidification), which can leach metals from the surfaces of artifactual or colloidal particles (Nielsen, 1996). The assumption that the separation of suspended particulates from water samples to be analyzed eliminates only matrix-associated (artifactual) constituents may often be incorrect (EPRI, 1985a; Feld et al., 1987), as at least some potentially mobile natural colloidal material will be retained on most commonly used filter pore sizes.

Filtration is often performed as a post-sampling "fix" to exclude from samples any particulate matter that may be an artifact of poor well design or construction, inappropriate sampling methods (use of bailers, inertial-lift pumps, or high-speed, high-flow-rate pumps), or poor sampling techniques (agitating the water column in the well). Filtration may be considered particularly important where turbid conditions caused by high particulate loading might lead to significant positive bias through inclusion of large quantities of matrix metals in the samples (Pohlmann et al., 1994). Alternatively, as discussed earlier, the presence of artifactual particulate matter in samples may also negatively bias analytical results through removal of metal ions from solution during sample shipment and storage as a result of interactions with particle surfaces. However, filtration is not always a valid means of alleviating problems associated with artifactual turbidity, as it often cannot be accomplished without affecting the integrity of the sample in one way or another.

During the planning phase of a ground-water sampling program, each parameter to be analyzed in ground-water samples should be evaluated to determine its suitability for field filtration and the most suitable filtration medium. As a general rule of thumb, parameters that are sensitive to the following effects of filtration should not be filtered in the field:

- Pressure changes that would result in degassing or loss of volatile constituents
- Temperature changes
- Aeration and agitation that may occur during filtration processes

Table 15.16 presents a summary of parameters for which filtration may be used and of parameters for which filtration should not be used in the field.

Samples to be analyzed for alkalinity must be field filtered if significant particulate calcium carbonate is suspected in samples, as this material is likely to impact alkalinity titration results (Puls and Barcelona, 1996). Care should be taken in this instance, however, as filtration may alter the CO_2 content of the sample and, therefore, affect the results.

Filtration is not always appropriate for ground-water sampling programs. If the intent of filtration is to determine truly dissolved constituent concentrations (e.g., for

TABLE 15.16

Analytical Parameter Filtration Recommendations

Examples of parameters that may be field filtered
    Alkalinity
    Trace metals
    Major cations and anions

Examples of parameters that should not be filtered
    VOCs
    TOC
    TOX
    Dissolved gases (e.g., DO and CO_2)
    "Total" analyses (e.g., total arsenic)
    Low molecular weight, highly soluble, and nonreactive constituents
    Parameters for which "bulk matrix" determinations are required

Source: U.S. EPA, 1991a.

geochemical modeling purposes), the inclusion of colloidal matter less than 0.45 μm in the filtrate will result in overestimated values (Wagemann and Brunskill, 1975; Bergseth, 1983; Kim et al., 1984). This result is often obtained with Fe and Al, where "dissolved" values are obtained which are thermodynamically impossible at the sample pH (Puls et al., 1991). Conversely, if the purpose of sampling is to estimate total mobile contaminant load, including both dissolved and naturally occurring colloid-associated constituents, significant underestimates may result from filtered samples, due to the removal of colloidal matter that is larger than 0.45 μm (Puls et al., 1991). A number of researchers have demonstrated that some metal analytes are associated with colloids that are greater than 0.45 μm in size (Gschwend and Reynolds, 1987; Enfield and Bengtsson, 1988; Ryan and Gschwend, 1990) and that these constituents would be removed by 0.45 μm filtration. Kim et al. (1984) found the majority of the concentrations of rare earth elements to be associated with colloidal species that passed through a 0.45 μm filter. Wagemann and Brunskill (1975) found more than twofold differences in total Fe and Al values between 0.05 and 0.45 μm filters of the same type. Some Al compounds, observed by Hem and Roberson (1967) to pass through a 0.45 μm filter, were retained on a 0.10 μm filter. Kennedy et al. (1974) found errors of an order of magnitude or more in the determination of dissolved concentrations of Al, Fe, Mn, and Ti using 0.45 μm filtration as an operational definition for "dissolved." Sources of error were attributed to passage of fine-grained clay particles through the filter.

Evidence from several field studies (Puls et al., 1992; Puls and Powell, 1992a; Backhus et al., 1993; McCarthy and Shevenell, 1998) indicates that field filtration does not effectively remedy the problems associated with artifactual turbidity in samples. These and other studies indicate that filtration may cause concentrations of some analytes to decrease significantly, due to removal of colloidal particles that may be mobile under natural flow conditions. Puls and Powell (1992a) noted that 0.45 μm filtered samples collected with a bailer had consistently lower As concentrations than samples obtained using low-flow-rate pumping. They suggested that the difference may have been due to filter clogging from excessive fines reducing the effective pore size of the filters or adsorption onto freshly exposed surfaces of materials brought into suspension by bailing. Puls et al. (1992) found that high-flow-rate pumping resulted in large differences in metals concentrations between filtered and unfiltered samples, with neither value being representative of values obtained using low-flow-rate sampling. Ambiguous sampling results found by McCarthy and Shevenell (1998) were attributed to analytical values for metals obtained using low-flow sampling that fell between filtered and unfiltered values from samples collected using bailing or high-flow-rate pumping. Discrepancies in analytical values for some metals (Al and Fe) exceeded an order of magnitude in this study. They determined that filtration of turbid samples may have occluded pores in filters, leading to removal of colloidal particles that may be representative of the load of mobile contaminants in ground water. Puls and Barcelona (1989) also point to the removal of potentially mobile species as an effect of filtration, indicating that filtration of ground-water samples for metals analysis will not provide accurate information concerning the mobility of metal contaminants.

If the objective of a ground-water sampling program is to determine the exposure risk of individuals who consume ground-water from private water supply wells, filtration of those samples would not produce meaningful results. To make this type of exposure risk determination, it is important to submit samples for analysis that are representative of water as it is consumed, and, because most people do not have 0.45 μm filters at their taps, unfiltered samples should be collected. In addition, it is important to remember that MCL and MCLG values set for drinking-water standards are based on unfiltered samples (see Chapter 1 for more information on drinking-water standards).

The very act of filtration can introduce significant sources of error and bias into the results obtained from analysis of sample filtrate (Braids et al., 1987). Some of these changes in sample chemistry result from pressure changes in the sample, as well as sample contact with filtration equipment and filter media. It is critical to evaluate the suitability of filtration on a parameter-specific basis and to carefully select filtration methods, equipment, and filtration media when developing site-specific filtration protocols to minimize sample bias caused by filtration. The following sources of negative and positive sample bias need to be considered:

- Potential for negative bias to occur due to adsorption of constituents from the sample (U.S. EPA, 1991a; Horowitz et al., 1996). For example, Puls and Powell (1992a) found that in-line polycarbonate filters adsorbed Cr onto the surface of the filter medium, resulting in an underestimation of Cr concentrations in the ground-water samples being filtered.
- Potential for positive bias to occur due to desorption or leaching of constituents into the sample (Jay, 1985; Puls and Barcelona, 1989; Puls and Powell, 1992a; Horowitz et al., 1996). In the Puls and Powell (1992a) study, K was observed to leach from nylon filters that were not adequately preconditioned prior to use.
- Removal of particulates smaller than the original filter pore size due to filter loading or clogging as filtered particles accumulate on the filter surface (Danielsson, 1982; Laxen and Chandler, 1982) or variable particle size retention characteristics (Sheldon, 1965; Sheldon and Sutcliffe, 1969).
- Removal of particulate matter with freshly exposed reactive surfaces, through particle detachment or disaggregation, that may have sorbed hydrophobic, weakly soluble, or strongly reactive contaminants from the dissolved phase (Puls and Powell, 1992a). This material itself may have been immobile prior to initiation of sampling and mobilized by inappropriate sampling procedures.
- Removal of solids (metal oxides and hydroxides) that may have precipitated during sample collection (particularly where purging or sampling methods that may have agitated or aerated the water column are used) and any adsorbed species that may associate with the precipitates. Such precipitation reactions can occur within seconds or minutes (Reynolds, 1985; Grundl and Delwiche, 1992; Puls et al., 1992), and the resultant solid phase possesses extremely high reactivity (high capacity and rapid kinetics) for many metal species (Puls and Powell, 1992a). Most metal adsorption rates are extremely rapid (Sawhney, 1966; Posselt et al., 1968; Ferguson and Anderson, 1973; Anderson et al., 1975; Forbes et al., 1976; Sparks et al., 1980; Benjamin and Leckie, 1981; Puls, 1986; Barrow et al., 1989). Additionally, increased reaction rates are generally observed with increased sample agitation.
- Exposure of anoxic or suboxic ground water (in which elevated levels of Fe^{2+} are typically present) to atmospheric conditions during filtration can also lead to oxidation of samples, resulting in formation of colloidal precipitates and causing removal of previously dissolved species (NCASI, 1982; EPRI, 1987; Puls and Eychaner, 1990; Puls and Powell, 1992a; Puls and Barcelona, 1996). The precipitation of ferric hydroxide can result in the loss of dissolved metals due to rapid adsorption or co-precipitation potentially affecting As, Cd, Cu, Pb, Ni, and Zn (Kinniburgh et al., 1976; Gillham et al., 1983; Stoltzenburg and Nichols, 1985; Kent and Payne, 1988).

- During sample filtration, care should be taken to minimize sample handling to the extent possible to minimize the potential for aeration. If sample transfer vessels are used, they should be filled slowly and filtration should be done carefully to minimize sample turbulence and agitation. Stoltzenburg and Nichols (1986) demonstrated that the use of sample transfer vessels during filtration imparted significant positive bias for DO and significant negative bias for dissolved metal concentrations. For this reason, the use of transfer vessels is discouraged. In-line filtration is preferred because of the very low potential it poses for sample chemical alteration.

After a decision is made to field filter ground-water samples to meet DQOs for an investigation, decisions must be made regarding selection of the most appropriate field filtration method. The ground-water sample filtration process consists of several phases: (1) selection of a filtration method; (2) selection of filter media (materials of construction, surface area, and pore size); (3) filter preconditioning; and (4) implementation of field filtration procedures. Information on each part of the process must be presented in detail in the SAP to provide step-by-step guidance for sampling teams to implement in the field.

A wide variety of methods are available for field filtration of ground-water samples. In general, filtration equipment can be divided into positive-pressure filtration and vacuum (negative pressure) filtration methods, each with several different filtration medium configurations. As discussed previously, ground-water samples undergo pressure changes as they are brought from the saturated zone (where ground water is under pressure greater than atmospheric pressure) to the surface (where it is under atmospheric pressure), potentially resulting in changes in sample chemistry. The pressure change that occurs when the sample is brought to the surface may cause changes in sample chemistry, which include loss of dissolved gases and precipitation of dissolved constituents such as metals. When handling samples during filtration operations, additional turbulence and mixing of the sample with atmospheric air can cause aeration and oxidation of Fe^{2+} to Fe^{3+}. Fe^{3+} rapidly precipitates as amorphous iron hydroxide and can adsorb other dissolved trace metals (Stolzenburg and Nichols, 1986). Vacuum filtration methods further exacerbate pressure changes and changes due to sample oxidation. For this reason, positive-pressure filtration methods are preferred (Puls and Barcelona, 1989, 1996; U.S. EPA, 1991a).

Table 15.17 presents equipment options available for positive pressure and vacuum filtration of ground-water samples.

TABLE 15.17

Examples of Equipment Options for Positive-Pressure and Vacuum Field Filtration of Ground-Water Samples

---

Positive-pressure filtration equipment
  In-line capsules
    –Attached directly to a pumping device discharge hose
    –Attached to a pressurized transfer vessel
    –Attached to a pressurized bailer
  Free-standing disk filter holders
  Syringe filters
  Zero headspace extraction vessels
Vacuum filtration equipment
  Glass funnel support assembly

When selecting a filtration method, the following criteria should be evaluated on a site-by-site basis:

- Possible effect on sample integrity, considering the potential for the following to occur:
    a. Sample aeration, which may result in sample chemical alteration
    b. Sample agitation, which may result in sample chemical alteration
    c. Change in partial pressure of sample constituents resulting from application of negative pressure to the sample during filtration
    d. Sorptive losses of components from the sample onto the filter medium or components of the filtration equipment (e.g., flasks, filter holders, etc.)
    e. Leaching of components from the filter medium or components of the filtration equipment into the sample
- Volume of sample to be filtered
- Chemical compatibility of the filter medium with ground-water sample chemistry
- Anticipated amount of suspended solids and the attendant effects of particulate loading (reduction in effective filter pore size)
- Time required to filter samples. Short filtration times are recommended to minimize the time available for chemical changes to occur in the sample
- Ease of use
- Availability of an appropriate medium in the desired filter pore size
- Filter surface area
- Use of disposable versus nondisposable equipment
- Ease of cleaning equipment if not disposable
- Potential for sample bias associated with ambient air contact during sample filtration
- Cost, evaluating the costs associated with equipment purchase price, expendable supplies and their disposal, time required for filtration, time required for decontamination of nondisposable equipment, and QC measures.

The filtration method used for any given sampling program should be documented in the site-specific SAP and should be consistent throughout the life of the sampling program to permit comparison of data generated. If an improved method of filtration is determined to be appropriate for a sampling program, the SAP should be revised in lieu of continuing use of the existing filtration method. In this event, the effect on comparability of data needs to be examined and quantified to allow proper data analysis and interpretation. Statistical methods may need to be used to determine the significance of any changes in data resulting from a change in filtration method.

Filtration equipment and filter media are available in a wide variety of materials of construction. Materials of construction should be evaluated in conjunction with parameters of interest being filtered with particular regard to minimizing sources of sample bias, such as adsorption of metals from samples (negative bias) or desorption or leaching of constituents into samples (positive bias). Materials of construction of both the filter holder or support and the filter medium itself need to be carefully selected based on compatibility with the analytes of interest (Puls and Barcelona, 1989). Filter holders that are made of steel are subject to corrosion and may introduce artifactual metals into samples. Glass surfaces may adsorb metals from samples.

Table 15.18 presents a summary of the most commonly used filtration media available for field filtration of water samples. The potential for sample bias for these filter media materials is variable, therefore, filter manufacturers should be consulted to determine recommended applications for specific filtration media and for guidelines on the most effective preconditioning procedures.

Large-diameter filter media (>47 mm) are recommended for ground-water sample filtration (Puls and Barcelona, 1989). Because of the larger surface area of the filter, problems of filter clogging and filter pore size reduction are minimized. High-capacity in-line filters have relatively large filter media surface areas, which may exceed 750 cm^2. This can improve the efficiency of field sample filtration.

Filter media require proper preconditioning prior to sample filtration (Jay, 1985; U.S. EPA, 1995; Puls and Barcelona, 1996; ASTM, 2004k). The purposes of filter preconditioning are: (1) to minimize positive sample bias associated with residues that may exist on the filter surface or constituents that may leach from the filter, and (2) to create a uniform wetting front across the entire surface of the filter to prevent channel flow through the filter and increase the efficiency of the filter surface area. Preconditioning the filter medium may not completely prevent sorptive losses from the sample as it passes through the filter medium.

In most cases, filter preconditioning should be done at the wellhead immediately prior to use (Puls and Barcelona, 1989). In some cases, filter preconditioning must be done in a laboratory prior to use (e.g., GF/F filters must be baked prior to use). Some manufacturers "preclean" filters prior to sale. These filters are typically marked "precleaned" on filter packaging and provide directions for any additional field preconditioning required prior to filter use.

The procedure used to precondition the filter medium is determined by the following: (1) the design of the filter (i.e., filter capsules or disks); (2) the material of construction of the filter medium; (3) the configuration of the filtration equipment; and (4) the parameters of concern for sample analysis. Filtration medium manufacturers' instructions should be followed prior to implementing any filter preconditioning protocols in the field to ensure that proper methods are employed and to minimize potential bias of filtered samples. These instructions will specify filter-specific volumes of water or medium-specific aqueous solutions to be used for optimum filter preconditioning.

The volume of water used in filter preconditioning is dependent on the surface area of the filter and the medium's ability to absorb liquid. Many filter media become fragile when saturated and are highly subject to damage during handling. Therefore, saturated filter media should be handled carefully and are best preconditioned immediately prior to use in the field.

Disk filters (also known as plate filters) should be preconditioned as follows: (1) hold the edge of the filter with filter forceps constructed of materials that are appropriate for the analytes of interest; (2) saturate the entire filter disk with manufacturer-recommended, medium-specific water (e.g., distilled water, deionized water, or sample water) while holding the filter over a containment vessel (not the sample bottle or filter holder) to catch all run-off; (3) then place the saturated filter on the appropriate filter stand or holder in preparation for sample filtration; (4) complete assembly of the filtration apparatus; (5) pass the recommended volume of water through the filter to complete preconditioning; (6) discard preconditioning water; and (7) begin sample filtration using a clean filtration containment vessel or flask. When preconditioning disk filters, care should be taken not to perforate the filter. The filter medium should not be handled with anything other than filter forceps. Otherwise, there may be a reduction in the porosity and

TABLE 15.18
Examples of Common Filter Media Used in Ground-Water Sampling

	Filter Medium							
	Acrylic Copolymer	Glass Fiber	Mixed Cellulose Esters	Nylon	Polycarbonate	Polyethersulfone	Polypropylene
Analytes							
Major ions	X	—	—	—	X	X	X
Minor ions	—	—	—	—	X	X	X
Trace metals	X	—	—	X	X	X	X
Nutrients	X	—	X	—	X	—	—
Organic compounds	—	X	—	—	—	X	—
Filter effective area (cm^2)							
17	X	X	X	X	X	X	—
20	X	X	X	X	X	X	—
64	—	X	—	—	—	X	—
158	X	X	X	—	—	X	—
250	—	—	—	—	—	X	—
600	—	—	—	X	—	X	—
700	X	—	—	—	—	—	—
770	—	—	—	X	—	X	X
Pore size (μm)							
0.1	—	—	X	—	X	X	—
0.2	—	—	X	—	X	X	—
0.45	X	—	X	X	X	X	—
1.0	X	X	X	X	—	X	X
5.0	X	—	X	X	—	X	X
Filter type							
Flat disk	X	X	X	X	X	X	—
Capsule	X	—	—	X	X	X	X
Syringe	X	X	—	X	—	X	—
Funnel	X	—	X	X	X	X	—

Source: ASTM, 2004k.

permeability of the filter medium. In addition, care should be taken to avoid exposure of the filter medium to airborne particulates to minimize introduction of contaminants onto the filter surface.

Preconditioning of capsule filters requires that liquid be passed through the filter prior to sample filtration and collection. A volume of manufacturer-recommended, medium-specific water (e.g., distilled water, deionized water, or sample water) should be passed through the filter while holding the capsule upright, prior to sample collection. In general, large-capacity capsule filters require that 1000 ml of water be passed through the filter prior to sample collection, while small-capacity filters require approximately 500 ml of water to be passed through the filter.

Physical and Chemical Preservation of Ground-Water Samples: The second form of pretreatment of ground-water samples is physical and chemical preservation: As described in ASTM Standard D 6517 (ASTM, 2004l), ground-water samples are subject to unavoidable chemical, physical, and biological changes relative to *in situ* conditions when samples are brought to ground surface during sample collection. These changes result from exposure to ambient conditions, such as pressure, temperature, ultraviolet radiation, atmospheric oxygen, and atmospheric contaminants, in addition to any changes that may be imparted by the sampling device as discussed earlier in this chapter.

The fundamental objective of physical and chemical preservation of samples is to minimize further changes in sample chemistry associated with sample collection and handling from the moment the sample is placed in the sample container to the time it is removed from the container for extraction or analysis in the laboratory. Sample preservation methods are determined on a parameter-specific basis and must be specified in the SAP prior to sample collection. Requirements for sample container type, storage and shipping temperature, and chemical preservatives are specified in the analytical method used for each individual parameter selected for analysis. Sampling team members are encouraged to speak with a laboratory representative prior to the sampling event to ensure that the correct types and numbers of sample containers (along with a few spares) and necessary chemical preservatives are shipped to the field site in sufficient time for the scheduled sampling event. Sampling team members must also learn from the laboratory what the parameter-specific holding times are (the amount of time that can transpire from the moment the sample container is filled to the time the sample is extracted or analyzed by the lab) for each parameter to ensure that samples are received by the laboratory in a timely fashion. This is particularly critical when sampling is conducted late in the work week for parameters that have a short holding time (e.g., 48 h).

Physical Preservation: Physical preservation methods for ground-water samples include: (1) use of appropriate sample containers for each parameter being analyzed; (2) use of appropriate packing and packaging of samples to prevent damage during transport to the laboratory; and (3) temperature control of samples during delivery to the laboratory: Sample containers are specified on a parameter-specific basis by the chosen analytical method for the sampling program (ASTM, U.S. EPA SW846 [U.S. EPA, 1996b], AWWA Standard Methods [APHA, AWWA, and WPCF, 1985]), as well as in Federal (40 CFR Part 136), state, and local regulatory guidelines on ground-water sample collection and preservation. Containers are specified with a number of design criteria in mind, to protect the integrity of the analytes of interest, including shape, volume, gas tightness, materials of construction, use of cap liners, and cap seal or thread design (Figure 15.80). Table 15.19 presents a summary of some of the more common ground-water sample containers used. These required containers are subject to change as methods are revised.

Sampling team members must also be aware that how they package sample containers for either hand delivery to the laboratory or commercial shipping is a critical aspect of

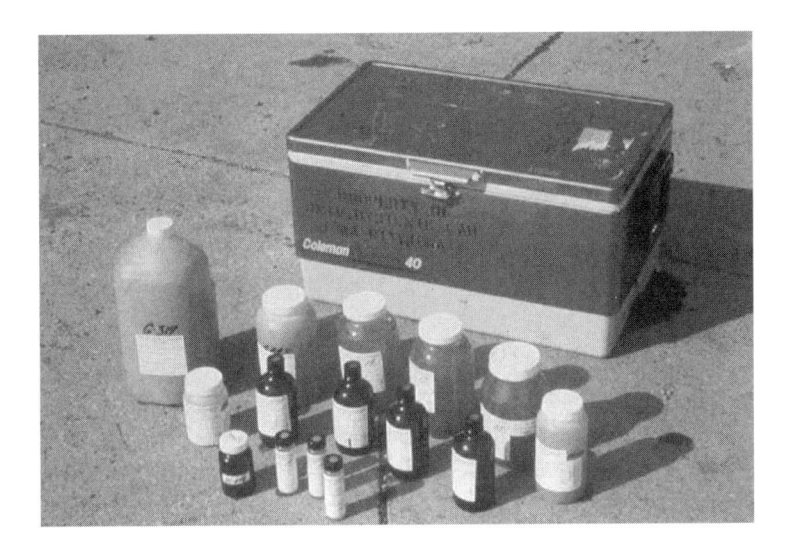

FIGURE 15.80
Sample containers vary in design based on the analytes to be measured in the sample. This example depicts containers used for a full suite of parameters included in the RCRA detection monitoring program as collected from one well.

physical preservation procedures. Field personnel should package and ship samples in compliance with applicable shipping regulations as discussed in ASTM Standard D 6911 (ASTM, 2004m). Shipping regulations such as the U.S. Department of Transportation Title 49 Code of Federal Regulations Part 172 and the International Air Transport Association (IATA) regulations should be consulted by sampling team members prior to a sampling event where ground-water samples may be sufficiently contaminated to require classification as dangerous goods for shipping purposes or where concentrated chemical preservatives require shipment. These regulations will provide definitive instructions on the correct packaging of regulated samples for shipment to the laboratory. Sample containers should be shipped in a manner that will ensure that the samples are received intact by the laboratory at the appropriate temperature as soon as possible after sample collection, to permit sufficient time for the laboratory to perform the requested analyses within the prescribed holding time for each analyte. Care must be taken by sampling team members to ensure that sample containers are packed sufficiently tight within the outer shipping container and to prevent movement during shipment that could result in container breakage. It is also a good practice to avoid packing glass containers against glass containers whenever possible (plastic against glass is a better configuration). Special shock-absorbing sleeves and containers with a plastic coating have been designed to help reduce the incidence of container breakage during shipment and handling. Use of bubble wrap around containers can also minimize container breakage. Commercial carriers often recommend that absorbent pads be placed in the bottom of sample shipping containers and on the top of sample containers after the shipper is filled, to absorb shock during transit.

Another important consideration during handling of samples in the field, following collection and during transport to the laboratory, is temperature control. Many parameters require that samples be stored at 4°C in the field between sample collection locations, during sample shipment (or delivery if by hand), and upon arrival at the laboratory. (4°C is the temperature at which water is at its maximum density and is most chemically stable.) To accomplish this, sample temperatures should be lowered immediately after

TABLE 15.19

Examples of Frequently Used Containers for Ground-Water Samples

	Parameter of Interest	Container	Volume (ml)
Inorganic tests	Chloride	P	125
	Cyanide (total and amenable)	P	1000
	Nitrate	P	125
	Sulfate	P	250
	Sulfide	P	500
Metals	Cr^{6+}	P	500
	Mercury	P	500
	Metals except Cr^{6+} and Hg	P	1000
Organic tests	Acrolein and acrylonitrile	G, PTFE ls	40
	Benzidines	G am, PTFE lc	1000
	Chlorinated hydrocarbons	G am, PTFE lc	1000
	Dioxins and furans	G am, PTFE lc	1000
	Haloethers	G am, PTFE lc	1000
	Nitroaromatics and cyclic ketones	G am, PTFE lc	1000
	Nitrosamines	G am, PTFE lc	1000
	Oil and grease	G am, wm	1000
	Total organic carbon	G am PTFE ls	40
	Organochlorine pesticides	G am, PTFE lc	1000
	Organophosphorus pesticides	G am, PTFE lc	1000
	PCBs	G am, PTFE lc	1000
	Phenols	G am, PTFE lc	1000
	Phthalate esters	G am, PTFE lc	1000
	Polynuclear aromatic hydrocarbons	G am, PTFE lc	1000
	Purgeable aromatic hydrocarbons	G, PTFE ls	40
	Purgeable halocarbons	G, PTFE ls	40
	TOX	G am, PTFE lc	250
Radiological tests	Alpha, beta, and radium	P	1000

Notes: P, high density PE; G, glass; G am, amber glass; wm, wide mouth; PTFE, polytetrafluorethylene (Teflon®); lc, lined cap; ls, lined septum; TOX, total organic halides.
Source: U.S. EPA, 1996b.

sample containers have been filled, labeled, and had any required security seals affixed to them.

Cooling can be accomplished using on-site refrigeration systems if they are available in the field or, more commonly, using wet (natural) ice. Wet ice is the preferred method to cool samples to 4°C. It is inexpensive, readily available, and will not get samples so cold that they will freeze. Wet ice will, however, require replenishment throughout the day to maintain sample temperatures, especially when sampling in warm ambient temperatures. Wet ice should be double bagged to prevent leakage into the shipping container as the ice melts.

Reusable chemical ice packs (also called blue ice packs) (Figure 15.81) are neither suitable for lowering sample temperatures to 4°C from *in situ* temperatures nor suitable for maintaining sample temperatures at 4°C during field handling and shipment. Thus, they are not recommended for use during ground-water sampling (Kent and Payne, 1988; ASTM, 2004l). There is also some concern about what chemicals may be released into the shipping container should a chemical ice pack be punctured or leak during sample shipment.

Dry ice is sometimes specified for use for sample cooling. Unfortunately, when dry ice is used, samples often become too cold and smaller volume containers commonly freeze,

Ground-Water Sampling 1075

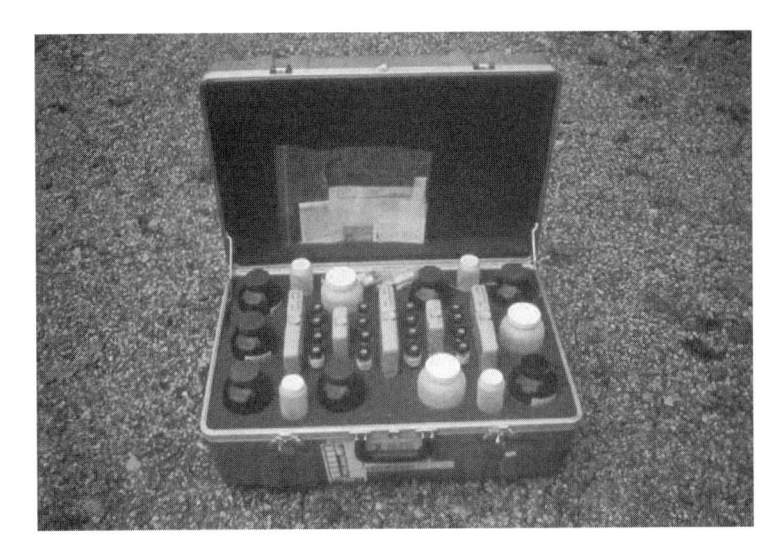

FIGURE 15.81
Reusable chemical ice packs, such as those used in this sample shuttle, are not recommended for ground-water sampling because they are incapable of achieving the desired sample temperature (4°C) in most cases and they may contribute contaminants to samples if they rupture or leak in transit to the laboratory.

resulting in container breakage and sample loss. Dry ice is also relatively expensive and difficult to obtain in the field. It requires special handling procedures in the field and is a regulated substance under shipping regulations.

To verify appropriate temperature control of samples, it is recommended that samplers include a QC sample referred to as a temperature blank (ASTM, 2004l) in the same shipping container as the ground-water samples. Upon receipt at the laboratory, a laboratory representative will check the temperature blank to determine whether samples were approximately preserved with respect to temperature. If the temperature blank is not at the required temperature ($4 \pm 2°C$), the laboratory representative will contact the sampling team to notify them of the sample arrival temperature and to determine an appropriate course of action.

Chemical Preservation: Chemical preservation is an important field procedure that samplers must implement to ensure that chemical change in samples is minimized during sample handling and shipment: Chemical preservation involves the addition of one or more chemicals (reagent grade or better) to the ground-water sample during sample collection. Chemicals can be used to adjust sample pH to keep constituents in solution or to inhibit microbial degradation of samples. Chemical preservatives are specified by each analytical method for each parameter and the preservatives are typically provided by the laboratory. Table 15.20 provides examples of common chemical preservatives used for ground-water samples.

Ground-water samples can be chemically preserved in one of several ways: (1) titration of pH-adjusting compounds (e.g., nitric acid) while monitoring pH change with a pH meter or narrow-range litmus paper; (2) addition of a fixed volume of liquid preservative (e.g., sulfuric acid contained in glass vials or ampoules) to the sample container; (3) addition of a fixed amount of pelletized preservative (e.g., sodium hydroxide) to the sample container; or (4) placement of preservatives in empty sample containers prior to shipment of the containers to the field (i.e., prepreserved sample containers). Titration methods for sample preservation, while theoretically a valid approach, are not always practical under field conditions where samplers are required to handle large volumes of

TABLE 15.20

Examples of Commonly Used Ground-Water Sample Chemical Preservatives and Holding Times

	Parameter of Interest	Preservative	Holding Time
Inorganic tests	Chloride	Cool to 4°C	ASAP
	Cyanide (total and amenable)	Cool to 4°C[a]; if oxidizing agents present, add 5 ml 0.1 N NaAsO^2/l or 0.06 g ascorbic acid/l; pH > 12 with 50% NaOH	14 days
	Nitrate	Cool to 4°C; boric acid for method 9210	ASAP
	Sulfate	Cool to 4°C	ASAP
	Sulfide	Cool to 4°C; pH > 9 with NaOH; Zn acetate; no headspace	7 days
Metals	Chromium^{6+}	Cool to 4°C	1 days
	Mercury	pH < 2 HNO$_3$	28 days
	Metals Except Cr^{6+} and Hg	pH < 2 HNO$_3$	6 months
Organic tests	Acrolien and acrylonitrile	Cool to 4°C; pH 4–5[a] Na$_2$S$_2$O$_3$	14 days
	Benzidines	Cool to 4°C; 0.008% Na$_2$S$_2$O$_3$[a]	7e/40ae
	Chlorinated hydrocarbons	Cool to 4°C; 0.008% Na$_2$S$_2$O$_3$[a]	7e/40ae
	Dioxins and furans	Cool to 4°C; 0.008% Na$_2$S$_2$O$_3$[a]	30e/45ae
	Haloethers	Cool to 4°C; 0.008% Na$_2$S$_2$O$_3$[a]	7e/40ae
	Nitroaromatics and cyclic ketones	Cool to 4°C; 0.008% Na$_2$S$_2$O$_3$[a] dark	7e/40ae
	Nitrosamines	Cool to 4°C; 0.008% Na$_2$S$_2$O$_3$[a] dark	7e/40ae
	Oil and grease	Cool to 4°C; pH<2 HCl	ASAP
	Organic carbon, total (TOC)	Cool to 4°C; pH<2 H$_2$SO$_4$ or HCl; dark	ASAP
	Organochlorine pesticides	Cool to 4°C	7e/40ae
	Organophosphorous pesticides	Cool to 4°C; 0.008% Na$_2$S$_2$O$_3$[a], pH 5–8	7e/40ae
	PCBs	Cool to 4°C; 0.008% Na$_2$S$_2$O$_3$[a]	7e/40ae
	Phenols	Cool to 4°C; 0.008% Na$_2$S$_2$O$_3$[a]	7e/40ae
	Phthalate esters	Cool to 4°C; 0.008% Na$_2$S$_2$O$_3$[a]	7e/40ae
	Polynuclear aromatic hydrocarbons	Cool to 4°C; 0.008% Na$_2$S$_2$O$_3$[a]	7e/40ae
	Purgeable aromatic hydrocarbons	Cool to 4°C; Na$_2$S$_2$O$_3$ pH < 2 HCl or H$_2$SO$_4$ or NAHSO$_4$	14 days
	Purgeable aromatic halocarbons	Cool to 4°C; Na$_2$S$_2$O$_3$ pH < 2 HCl or H$_2$SO$_4$ or NAHSO$_4$	14 days
	TOX	Cool to 4°C; pH < 2 H$_2$SO$_4$; dark; no headspace	28 days
Radiological tests	Alpha, beta, and radium	pH < 2 HNO$_3$	6 months

[a]Only add a reducing agent if the sample contains free or combined chlorine. A field test kit needs to be used for this determination.

7e: sample extraction must be completed within 7 days of sample collection;
40ae: analysis must be completed within 40 days after sample extraction.

Note: ASAP, analysis should be performed as soon as possible. Samplers should discuss with the laboratory how ASAP is to be interpreted on a project-specific basis. In many cases, if the parameter can be analyzed with accuracy and precision in the field that is preferred. Otherwise, many laboratories use a 24 h holding time.
Source: U.S. EPA, 1996a.

concentrated preservatives and work with glass titration apparatus under less than ideal conditions (wind, rain, dust, etc.). A modified version of this method is to use calibrated dropper bottles of preservative rather than glass burettes for titration. This ensures that the correct preservative is titrated into the sample while monitoring pH changes, but in a safer fashion in the field.

Ground-Water Sampling

Vials or ampoules of preservatives are commonly used for sample preservation in the field (Figure 15.82). The laboratory provides the vials or ampoules containing a fixed volume of the required preservative for each sample container requiring chemical preservation. The sampling team should be provided with directions on which preservative must be added to which container on a parameter-specific basis, as well as guidance on whether the preservative should be added to the container before or after filling. One common error made by sampling teams is to assume that the amount of preservative provided in vials or ampoules will always be sufficient to reach the required end pH for the analyte. This is not a safe assumption, especially in situations where ground-water pH may be abnormally high or low based on contaminant chemistry or the natural pH of formation-quality water. For this reason, it is essential that both an initial and final pH measurement be taken to check for pH anomalies and to ensure that the required end pH for the sample has been reached (ASTM, 2004l). To take these measurements, a small aliquot of sample should be decanted into another container (e.g., a clean empty VOC vial without preservative that will not be used for sample collection or a clean small-volume beaker) and the pH measured using either a calibrated pH probe or narrow-range litmus paper. If the sample pH is not at the required endpoint, additional preservative must be added until it is reached. For this reason, sampling teams must ask the laboratory to provide additional preservative, preferably in a vial that can be resealed if only a few extra drops of preservative are required. Sampling teams must resist the temptation to double the preservative required by the method simply for the sake of convenience. This can result in the samples becoming a corrosive liquid for shipping purposes as described in ASTM Standard D 6911 (ASTM, 2004m) and can detrimentally affect the chemistry of the sample.

FIGURE 15.82
Chemical preservation using vials.

It is generally accepted that the sample dilution attributed to the addition of chemical preservatives should be limited to a maximum of 0.5% (ASTM, 2004l). The pH of samples should be checked upon arrival at the laboratory to ensure that appropriate sample preservation procedures were implemented in the field. If the pH is not where it is required to be, the laboratory will consider the sample to be inappropriately preserved and the sampling team will be contacted to discuss an appropriate course of action.

The most convenient method of chemically preserving ground-water samples is to use prepreserved containers (Figure 15.83). Prepreserved containers are either purchased by the laboratory already prepared or they can be prepared by the laboratory. These containers hold a fixed volume of the parameter-specific preservative and are shipped to the sampling team with information about which preservatives have been added to which containers. The advantages of this method of sample preservation are: (1) the sampling team does not have to handle preservatives; (2) theoretically, no errors associated with adding an incorrect preservative to a sample can be made by field personnel; and (3) time savings. There are also several limitations to using prepreserved containers. As when using vials and ampoules to add preservative to samples, the volume of preservative is fixed. Thus, difficulties can arise in the field if field verification of end pH determines that the volume of preservative provided in the container is insufficient. In this situation, sampling teams must have available additional preservative (the same as that used to prepare the container), so the required end pH can be achieved prior to shipping the sample to the laboratory. This may be impossible for the laboratory to provide if they have purchased prepreserved containers from a supplier. From a practical perspective, one common complaint of sampling team members using prepreserved containers is that it is easy to lose preservative from the container, either through accidentally

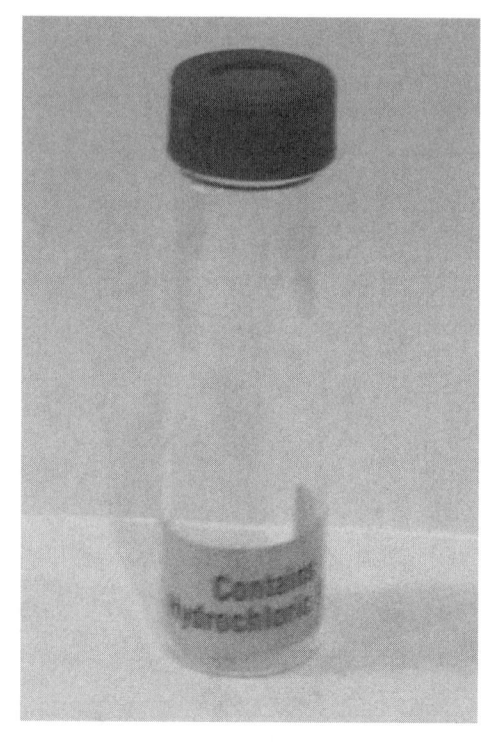

FIGURE 15.83
A prepreserved container used for chemical preservation of samples.

knocking over the container during filling or through overfilling, especially when attempting to collect zero-headspace samples in 40 ml vials.

Another concern over prepreserved containers is related to the fact that concentrated preservatives may react with the empty container during storage prior to sample collection. For example, nitric acid, when in long-term storage in a high-density PE container, will chemically react with the container, resulting in alteration of the container walls (evidenced by orange staining inside the container) and, ultimately, as the authors have observed, failure of the container (the plastic container will crack linearly when squeezed). In glass containers stored in hot ambient conditions, acid preservatives will commonly evaporate to form an acid vapor, which is released to ambient air when the container is opened, leaving little preservative available to lower the pH of the sample and creating a breathing hazard for samplers. In addition, the vapor can deteriorate sample container lid threads — a problem that may not be detected until the container is taken into the field and the cap crumbles into pieces when it is removed for container filling. For these reasons, some sampling protocols do not permit the use of prepreserved containers. In other programs, prepreserved containers are allowed but with storage time restrictions that can vary from days to hours. It is recommended that sampling teams work with the laboratory and regulatory agencies during the planning phases of the ground-water sampling program to determine how long a prepreserved container can remain in storage prior to sample collection.

Field Equipment Cleaning Procedures

The SAP must provide a written protocol appropriate for cleaning field equipment used throughout the ground-water sampling program to prevent cross-contamination of sampling locations and collection of unrepresentative samples. Even in situations where dedicated equipment is being used for purging and sample collection, there will be pieces of equipment taken from sampling point to sampling point, which will require cleaning between locations (Figure 15.84 and Figure 15.85). Examples of equipment that falls into this category are electronic water-level gauges, oil–water interface probes,

FIGURE 15.84
All pieces of field equipment that are used in wells should be thoroughly cleaned before use in the well. Cleaning should also be done as part of periodic maintenance to keep equipment functioning properly.

FIGURE 15.85
A typical setup for cleaning downhole equipment (such as pumps), with a bucket for an initial rinse with control water (foreground), followed by a bucket for washing with a phosphate-free liquid detergent (e.g., Liquinox) (center) and, finally, a bucket for the final rinse with distilled or deionized water (background). For some stringent cleaning programs, additional steps may be required.

flow-through cells, and instrumentation used to measure water-quality indicator parameters (e.g., pH meters and multiparameter sondes). Equipment cleaning is also necessary to ensure that equipment will continue to operate properly in environments with high levels of suspended sediment and aggressive chemical constituents and to ensure that sampling team members are not accidentally exposed to contaminants that may be present on the surfaces of equipment following its use.

To assist in the development of streamlined approaches to equipment cleaning, ASTM has produced standards that address procedures for cleaning field equipment used at non-radioactive sites (ASTM Standard D 5088 [ASTM, 2004e]) and low-level radioactive waste sites (ASTM Standard D 5608 [ASTM, 2004f]). In these standards, field equipment cleaning protocols are described for equipment that contacts samples submitted to the laboratory (such as a pump) and equipment that facilitates sample collection but that does not actually contact the sample (such as a reel used to hold pump tubing). The reader is directed to Chapter 20 of this text for a more detailed discussion of field equipment cleaning protocols.

Documenting a Sampling Event

The procedures and equipment used and the data generated during a ground-water sampling event must be documented in the field at the time of data generation and sample collection. The SAP must detail procedures required for recording field observations and measurements made throughout the sampling event. These records are

Ground-Water Sampling

used to document field conditions that may be important to refer to when interpreting laboratory data and to document exactly what was done in the field during the sampling event, including how the well headspace was screened, how water levels were taken, how sampling points were purged, how samples were collected, what samples were collected at each sampling point, results of all field measurements and field parameter analyses, and how samples were pretreated and prepared for shipment. Specific guidance on what to include in field notes for a ground-water sampling event is provided in ASTM Standard D 6089 (ASTM, 2004n). The suggested content of ground-water sampling event documentation is summarized in Table 15.21.

Three primary mechanisms are available for recording information collected during a sampling event: (1) written records; (2) electronic records; and (3) audio-visual records. When writing the section of the SAP devoted to documentation, project managers are encouraged to seek counsel from in-house or client legal staff for guidance on what are considered to be acceptable practices for field documentation. This is particularly important in cases where a site is undergoing litigation or could be the subject of a corporate sale or merger. It is important to remember that the political climate of a site can change during the course of a ground-water sampling program, so it is often wise to have a higher level of QA/QC associated with field record keeping than might be thought necessary at the time of sample collection.

TABLE 15.21

Information to Document during a Ground-Water Sampling Event

Facility or site name and well identification
Weather conditions
Names and affiliations of sampling team members and others present during the sampling event
Instrumentation calibration results
Well integrity inspection results
Changes in land use or physical conditions at the site since the last sampling event
Results of well headspace screening and details on how screening was performed
Water-level measurement results and product-thickness measurements (if taken), indicating what equipment was used to take the measurements and procedures followed
Well-depth measurement results (if taken)
Description of the well purging method, equipment used, how the equipment was operated, and time required to purge each well
Equation used to calculate a well volume (if relevant) and results of calculations for each well
Total volume of water removed during purging of each well
Description of how purge water was managed
Results of all field parameter measurements and the definition of stabilization used
Description of how it was determined when each well was ready to sample
Description of the sampling device, if different from the device used to purge the well, and a description of how the device was operated during sample collection
Description of the sample containers filled and the order in which containers were filled
Description of the water collected as samples (appearance, odor, and turbidity)
Description of sample pretreatment methods (filtration and chemical or physical preservation) for specific parameters
Description of any problems encountered in the field during the sampling event
Description of the temporary storage method used for samples during the sampling event (including use of sample security seals and tags)
Description of all QC samples collected — types, and how many, and how they were collected
Description of sample preparation for shipment, shipment method, security tag serial numbers, time of sample delivery to the lab (if hand delivered) or to the commercial carrier, and forms accompanying the samples (e.g., chain-of-custody and analysis request form)
Description of any photographs taken during the sampling event
---

Written Records: Written records are the most common form of documentation: Written records generated in the field include site-specific field notebooks (Figure 15.86), sample container labels, sample container security tags and seals, chain-of-custody forms, visitor logs, field equipment calibration and maintenance logs, and commercial sample shipment manifests. In general, all field observations and measurements made during a sampling event should be recorded using black indelible ink on a bound, site-specific field notebook and not on loose pages or forms. Field notebooks, commonly made of water-proof paper, provide a secure and relatively tamper-proof mechanism for retaining all field records. Field notebooks should have preprinted page numbers throughout or should have a printed statement in the front of the book indicating how many pages are contained in the book. This is a critical means of ensuring that pages cannot be removed from or added to the field notes without detection — the major limitation of using a collection of loose-leaf forms.

Many companies have spent a great deal of time developing field forms (Figure 15.87) for sampling teams to complete. The objective of these forms is to ensure precision in information collection during ground-water sampling programs. The result is that many samplers have become dependent on forms to dictate what must be done and which measurements must be taken in the field, so there is sometimes resistance to switch to a field notebook. However, from a legal perspective, field forms are not recommended because they are easy to lose and information is easy to alter and, thus, can be challenged in a legal proceeding. Two alternatives to this problem are: (1) have forms printed on water-proof paper with page or form numbers and bound using saddle-stitching or spiral wire binding (Figure 15.88); or (2) record all original notes in a field notebook and then transfer the information to the form (never the other way around) for easy dissemination of field information back at the office.

Because entries in field notebooks are made in ink, it is important that sampling team members resist the temptation to scratch out or blacken out any errors when entering a correction. Instead, good laboratory practices (GLPs) should be followed. Using GLPs, it is appropriate to use a single stroke through the incorrect information. Next to the line containing the error, the samplers should put their initials, the date of the correction, and an appropriate error code to explain when and why the change was made (Garner et al., 1992). This

FIGURE 15.86
Field notebooks are the preferred method for recording written records in the field.

Ground-Water Sampling

FIGURE 15.87
Field forms are commonly used to record field data, because they prompt the user to collect specific information and they are convenient. However, they are not recommended for recording original data because of the ease with which data can be altered undetectably. Note that, in this case, a pencil is being used to record data — this is strongly discouraged. Field notes should always be recorded in black indelible ink.

makes it much easier to explain entry changes at a later date, as is necessary when field notes are obtained as part of a court case as evidence or through discovery. Table 15.22 provides some examples of common error codes that can be used in written ground-water sampling records.

When recording field notes, samplers must be certain that the information recorded is accurate, factual (not opinionated), and detailed enough so that others can reconstruct and understand what occurred during the sampling event. It is also critical that notes are neat and legible. A good practice for the sampling team to adopt is to submit the field book to the project manager for review immediately at the conclusion of each sampling event so the project manager can review field notes for completeness, clarity, and errors. By conducting this review immediately, the project manager can question the sampling team while memories are still fresh and before they head to another site for field work.

Chain-of-Custody: There are exceptions to the "no loose paper" rule — specifically the chain-of-custody form and shipping manifests: The chain-of-custody form (Figure 15.89) is a loose form typically provided by the laboratory, which is used to document possession of samples in the field during sample collection and to document samples being shipped or delivered to the laboratory. From a legal perspective, the objective of chain-of-custody forms is to provide sufficient evidence of sample integrity to assure legal defensibility of the samples (ASTM, 2004o). Depending upon its design, the chain-of-custody form may also act as a sample analysis request form, or the latter may be a separate form that is sent along with samples to the laboratory. ASTM Standard D 4840 (ASTM, 2004o) provides an example of a chain-of-custody form that has been widely adopted by many laboratories. A common mistake made by sampling teams using chain-of-custody forms is that they wait until the end of the day to complete the form rather than complete it as samples are collected. This defeats one of the purposes of

FIGURE 15.88
If field forms are used, it is preferred that they be printed on water-proof paper and bound into a book like the one pictured here, with pages numbered to prevent the possibility of alteration of information recorded for a sampling event.

having the chain-of-custody form (to document possession of samples during the sampling event) and can result in errors being made when completing the forms at the end of the day. The original chain-of-custody form must accompany the samples to the laboratory and a copy should be retained by the sampling team.

Shipping Manifests: Shipping manifests are used when commercial couriers or carriers are used to transport samples from the field to the laboratory: In most cases, the tracking system used by commercial carriers is considered to be sufficient to document

TABLE 15.22

Examples of Common Error Codes

RE	Recording error
CE	Calculation error
TE	Transcription error
SE	Spelling error
CL	Changed for clarity
DC	Original sample description changed for further clarity
WO	Write over
NI	Not initialed and dated at time of entry
OB	Observation not recorded at time of initial observation

Note: Error codes should be circled when recorded.
Source: Garner et al., 1992. With permission.

PROJ. NO.	PROJECT NAME					NO. OF CON- TAINERS						REMARKS
SAMPLERS: (Signature)												
STA. NO.	DATE	TIME	COMP.	GRAB	STATION LOCATION							

Relinquished by: (Signature)	Date/Time	Received by: (Signature)	Relinquished by: (Signature)	Date/Time	Received by: (Signature)
Relinquished by: (Signature)	Date/Time	Received by: (Signature)	Relinquished by: (Signature)	Date/Time	Received by: (Signature)
Relinquished by: (Signature)	Date/Time	Received for Laboratory by:	Date/Time	Remarks	
Distribution Original Accompanies Shipment Copy to Coordinator Field files					

FIGURE 15.89
A typical chain-of-custody form that should accompany all filled sample containers during the sampling event and while samples are in transit to the laboratory.

possession of a shipment of samples provided samplers retain copies of manifests. It is the responsibility of the sampling team to ensure that the correct type of manifest is used for the samples being shipped and that the manifest is completed correctly. Failure to do this can result in refusal of the carrier to deliver the samples, loss of the samples, or legal action. This is especially critical when samples contain sufficient concentrations of contaminants to require that they be classified as regulated substances for shipping (e.g., a ground-water sample that contains non-aqueous phase gasoline could be considered to be a flammable liquid for shipping purposes).

Electronic Records: Many instruments used in the field for field parameter measurement contain a data-logging component to electronically record all measurements made: While this is certainly convenient for samplers and theoretically should be a more accurate way to record numerical data, there are some concerns from a legal perspective. The greatest concern is that it is possible to change, or lose entirely, an electronic file without keeping a permanent record of the original file. For this reason, in many situations, it is required that original data be recorded in the field notebook and that the data-logging system be used as an electronic backup recording system (not the other way around). For most ground-water sampling applications, this is not problematic due to the timeframe involved between measurements. In situations where pumping tests are being conducted, however, water-level measurements may need to be recorded in a timeframe too fast to be done manually, so exceptions to this rule may be made.

Audio-Visual Records: Audio-visual record keeping is a third mechanism for documenting field activities during ground-water sampling: This may include audio recording of field activities, but more commonly involves taking photographs or video clips of site activities. Written permission must be obtained from authorized facility personnel by sampling teams prior to taking any audio-visual records. This is especially a concern at active facilities where audio-visual records may pose a security risk (e.g., at U.S. Department of Defense or U.S. Department of Energy sites) or breach confidentiality agreements (e.g., at active manufacturing facilities). Options for photographic recording of field activities include 35 mm cameras, digital cameras, cameras with self-developing film, and video cameras (tape and digital formats). In politically sensitive cases, counsel should be consulted during the preparation of the SAP to select the most appropriate type of camera, recording medium (film type or digital disk), camera lens (e.g., fixed, macro or zoom, and wide angle), and method of film development or disk imaging. In addition, some cameras have the capability to date- and time-stamp images, which might be of value in some applications provided the feature is set up correctly.

Conducting a Ground-Water Sampling Event

After the SAP has been prepared, incorporating all of the elements discussed earlier, and has received regulatory and client approval for implementation, the sampling team will travel to the field site to collect ground-water samples using the protocols and procedures contained in the SAP. Each sampling team member should have a personally assigned copy of the SAP on hand in the field to refer to if there is ever a need for clarification of a field procedure. Table 15.23 summarizes the field activities that will be implemented by a sampling team during a typical ground-water sampling event for a long-term ground-water monitoring program. A similar schedule of activities will be followed for

TABLE 15.23

Typical Field Components of a Ground-Water Sampling Event for a Long-Term Monitoring Program

Review of site map for facility orientation
Prepare field notebook for daily entries
Check all sample containers supplied by the laboratory for breakage and to be sure all required containers (and a few extra containers) are available for each well to be sampled
Check all QC sample containers supplied by the laboratory
Calibrate all field equipment (e.g., PID and water-quality instrumentation)
Initial trip to inspect each well for structural integrity, measure headspace levels of volatile or combustible gases or vapors, take water-level or product-thickness measurements, and prepare the wellhead for sample collection by anchoring plastic sheeting on the ground surface (in the order from upgradient to downgradient locations)
Ensure that containment systems are in place to manage any purge water generated (e.g., 55 gal drums at each well) if required
Prepare purging and sampling equipment
Travel to first hydraulically upgradient well to be sampled with all equipment, sample containers, preservatives, filtration equipment, decontamination supplies, field notebook, sample labels, sample security tags and seals, chain-of-custody forms, shipping manifests, PPE, and garbage bags for solid waste
Take a second water-level measurement to provide information necessary for purging method (depth to static water level, height of the water column, and volume of the well)
Purge the well using the prescribed protocol, containing any purge water as necessary
Collect field water-quality data (i.e., indicator parameters)
Collect ground-water samples in the prescribed order of container filling, using required sample collection protocols
Filter and physically and chemically preserve any parameter specified in the SAP
Collect required field QC samples
Complete required labels, seals, and tags, affix to the appropriate sample bottles, and place the sample bottles in shipping containers for storage or transport during the rest of the day's field activities
Complete the chain-of-custody form
Dispose of any solid wastes generated during the sampling event (e.g., disposable gloves, plastic sheeting, disposable filters, disposable tubing, or suspension cord)
Clean any portable equipment prior to transporting it to the next sampling location
At the end of the day, complete the shipping manifest and ship samples to the laboratory

most other ground-water sampling events. These activities are presented in the chronological order in which they would occur in the field. To be effective, sampling team members need to work together and develop a system to ensure efficient use of time and resources. This will come with time and experience as sampling team members work together at each field site. Sampling teams should create a system of checks and balances to ensure that errors are not made as a result of omission, use of improper protocols, unfamiliarity with equipment, or failure to read and understand the SAP. ASTM has published a guide (ASTM Standard D 5903 [ASTM, 2004p]) on planning and preparing for a ground-water sampling event that provides an excellent "to do" list for sampling team members to follow to ensure that the sampling event is well organized and that delays associated with poor preparation are avoided. The guide also provides good check lists of typical sampling equipment and supplies needed in the field.

Field Preparation for a Ground-Water Sampling Event

Site Orientation and Sampling Event Preparation

Upon arrival at a field site, sampling team members should meet with facility personnel to learn of any changes related to facility operations, such as safety procedures, personnel changes, traffic pattern changes, locations of support systems such as fresh water or electricity for the sampling team to use, on-site construction, or other changes that have occurred at the site since the last sampling event. This is also a good time to be updated on field site conditions such as weather or any reports related to structural damage of any of the wells (e.g., if a well has been buried by waste or construction activities and if the protective casing of a well has been damaged by heavy equipment). It is helpful to have a current facility map available during this site orientation meeting to identify areas of concern on the map, especially if new sampling team members or facility personnel are present. This information can be of tremendous value to sampling teams as they plan their day and before they conduct an inspection of all wells scheduled for sampling during the event.

Following site orientation, the sampling team should relocate to either an established field office or to a predetermined area located upgradient and away from high traffic areas to prepare equipment and materials for the day's sampling event. This includes preparation of the field notebook; organization of sample containers, labels, and security tags and seals; organization of QC sample containers; and calibration of field instrumentation that will be used for headspace measurement in sampling points and for water-quality indicator parameter measurement. All instrumentation calibration should be done according to manufacturers' instructions under field conditions. The timing and frequency of calibration should be in accordance with the SAP, which will commonly require, at a minimum, daily calibration and periodic calibration checks of all equipment. After all equipment is checked and calibrated, the sampling team should visit each well to conduct the next phase of the sampling event.

Conducting the Sampling Event

Sampling Point Inspection

Prior to purging and sampling, the sampling team needs to physically inspect each well to ensure that it is structurally sound for sampling. As discussed earlier, reports on possible well damage from facility personnel are very helpful, especially if damage is severe enough to make location of the well difficult. Table 15.24 summarizes features that

TABLE 15.24

Well Inspection Check List

Check identification markings on the well
Check the surface seal to ensure that it is intact with no cracks
Check the above-grade protective casing and the well cap to be sure neither has been damaged (or the cap removed)
Check to be sure that the locking mechanisms are in place and undamaged
Check to be sure that the protective bumper guards are in place and undamaged
Check to be sure that the protective painted surfaces are not weathered or altered or require repainting or etching
Check to be sure that the valve-box covers or vault lids are present and in good condition for flush-to-grade installations
Check to be sure that the vault lid security mechanisms are intact and rust-free for flush-to-grade completions
Check to be sure that the gasket seals in flush-to-grade completions are present and water tight
Check to be sure that there is no standing water inside the flush-to-grade vault; if water is present, note the depth, color, and appearance of any visible contamination of standing water (note especially if the water level is level with the top of the inner well casing cap)
Note if there is any flow of water into the vault from either ground surface or below ground surface (around the vault seal)

---

should be inspected by the sampling team prior to opening the protective casing and the well.

If the well inspection indicates that there is obvious or suspected damage, detailed field notes should be made and photographs taken (if permitted) to describe the damage, including mention of any report made by facility personnel. The SAP should be consulted to determine the most appropriate course of action. In general, if there is any possibility that the observed well damage could result in a detrimental impact on sample chemistry, the location should not be sampled because samples may not be representative of true formation water chemistry. An evaluation should be made as to whether the well can be repaired or whether damage is significant enough to warrant decommissioning of the existing well and replacing it with another. This determination may be made by project managers rather than sampling team personnel, but it should be made in a timely fashion to avoid the possibility of the damaged well acting as a conduit for downward movement of surface contaminants into the formation.

Well Headspace Screening

If the well appears to be structurally sound, instrumentation for well headspace screening should be used to take ambient or background readings and the readings recorded. After background has been established, the well cap should be removed and the probe quickly inserted into the well headspace. After the headspace reading has stabilized or peaked, the probe should be removed from the sampling point and allowed to cycle fresh ambient air through the instrument to purge any vapors that may be present in the instrumentation. The highest reading recorded and the type of instrument response (e.g., a rapid rise and drop or a gradual increase to stabilization) should be documented in the field notes. If readings indicate that volatile or combustible vapors or gases are present, the sampling team must refer to the SAP and the site health and safety plan to determine whether personnel should upgrade their PPE.

In some wells, samplers may hear the sound of air rushing out of the well or a whistling sound when the well cap is removed. This is usually indicative of a well not being properly vented and an air-pressure buildup occurring as a result of a hydraulic pressure increase in the formation since the last sampling event. The sound is the pressurized air being released from the well as it attempts to equilibrate with atmospheric pressure. Equilibration of air pressure should occur within the time taken to screen the wellhead, but the water level in the well may continue to recover for anywhere from several seconds or minutes (in high hydraulic conductivity formations) to several hours or even days (in low hydraulic conductivity formations).

Water-Level Measurement

After the water level in the well has stabilized, water-level measurements should be taken following the protocol documented in the SAP. To take a water-level measurement, samplers need to know the location of the surveyed reference measuring point on each well — this is the point to which all water-level measurements should be made. The reference measuring point should be physically marked on the well casing or outer protective casing but, in some cases, it may not be, so its description should be documented in the SAP. To ensure precision, the sampler should note the units of measure on the gauge tape and test the water-level gauge prior to lowering it into the well. After testing, the sampler should measure the depth to static water level and should take a minimum of three independent water-level readings (for precision) at least 5 to 10 sec apart. This is especially critical in unvented wells in which the water level may still be recovering; multiple measurements with different results indicate that the water level is not stable enough to record representative measurements. Results of water-level measurements should be recorded to an accuracy of ± 0.01 ft in the field notebook. Water levels should be measured in all wells at the site in as short a time interval as possible, before purging and sampling of any of the wells is attempted. These water-level data will be used for determining the direction, gradient, and rate of ground-water flow across the site. On large sites with many wells, a full day or more could be spent implementing this first phase of the ground-water sampling event. Between wells, the water-level gauge must be cleaned following protocols documented in the SAP to prevent potential cross-contamination of sampling locations.

Well Purging and Field Parameter Measurement

After well inspections are complete and the first set of water-level data is collected, the sampling team should return to the first upgradient sampling location to begin well purging and sampling. Prior to purging, a purge-water containment system must be in place. Commonly, this involves placing a 55 gal drum at each well or towing a 500 gal tank on a trailer behind the field truck to contain purge water.

The device selected for purging the well should be lowered into the well (if portable) or, in the case where dedicated pumps are installed, the accessory equipment required for pump operation should be brought to the well and set up on the plastic sheeting placed around it. For portable pumps, an effort should be made to closely match the length of the tubing used for the pump with the depth at which the pump will be set in the well. Excess tubing can affect the temperature of the water sampled and reduce the flow rate. Increases in temperature can affect dissolved gases and trace metals in samples (Parker, 1994; Stumm and Morgan, 1996).

Prior to purging the well, instrumentation to be used for water-quality analysis should be calibrated (Figure 15.90). If a flow-through cell is to be used, it should be assembled with the water-quality instrumentation, typically a multiparameter sonde, installed. The unit should be placed out of direct sunlight to avoid overheating of the cell and sensors. An effort should be made to keep tubing lengths that connect the flow-through cell to the pump discharge as short as possible. As mentioned earlier, excessive lengths of

FIGURE 15.90
Calibration of a multiparameter sonde to be used with a flow-through cell to measure indicator parameters during low-flow purging and sampling.

tubing may result in increases in temperature, which can have a detrimental effect on sample chemistry.

All sampling supplies, such as disposable gloves and paper towels, as well as sample containers, QC sample containers and supplies, preservatives, filtration equipment, labels, security tags and seals, the field notebook, and chain-of-custody forms, should be organized and ready for use at the well following purging. After all equipment is in place, purging should be conducted in accordance with the SAP.

Sample Collection Procedures

After a well has been purged and is deemed ready for sampling, the sampling team must disconnect any flow-through cell equipment from the pump discharge tube and prepare to collect samples for laboratory analysis, and field QC samples. Sampling team members must be consistent in the manner in which they collect and pretreat samples on a parameter-specific basis to ensure both accuracy and precision between sampling events. Table 15.25 provides a check list of sample collection elements that sampling team members should verify at each well during a sampling event. Each of these items must be addressed in the site-specific SAP.

Order of Container Filling

The sampling team should assemble sample containers provided by the laboratory for each parameter or suite of parameters to be analyzed at that particular well and containers required for any field QC samples that will be collected at that location. The sampling team should verify against the SAP that containers provided are correct for the analytes of interest, and they should inspect each container to ensure that it is in good physical condition (clean, not damaged, good fitting caps and seals, etc.) and ensure that there are sufficient numbers of containers to meet the needs of the sampling program. The containers should then be arranged in the correct order for filling, particularly in situations where there may be an insufficient volume of water in the wells to fill all sample containers. U.S. EPA guidance recommends that ground-water samples be collected in a particular order, with

Ground-Water Sampling

TABLE 15.25

Check List of Critical QA/QC Sample Collection Elements

All required water-quality measurements have been made, recorded, and checked for accuracy prior to disconnecting the flow-through cell
Laboratory analyses to be performed on samples from each well are confirmed
The correct sample containers and required sample volumes are checked and confirmed
The order of and methods for sample collection (bottle filling) are clearly documented
Field quality control samples to be collected (which kind, when, where, and how) are documented for each sampling location
The correct types of filtration equipment, including filters of the correct filter pore size, are present at the well head
Filter preconditioning procedures have been followed
The correct type and volumes of chemical preservatives (if required) are present
Procedures have been established to verify arrival temperature and end pH of samples requiring chemical preservation
Sample container labels and security tags and seals (if required) are ready to be completed
The appropriate chain-of-custody forms are available for completion immediately following sampling
Sample shipping containers, compliant with applicable DOT and IATA shipping regulations, are ready for delivering samples to the lab by hand, laboratory courier, or commercial carrier

---

those parameters that are most sensitive to handling being collected first, followed by those less sensitive to handling (U.S. EPA, 1991a). Figure 15.91 illustrates the recommended order for sample container filling.

In addition to the sensitivity of a parameter to handling, it is critical that the relative importance or significance of each parameter be evaluated on a site-by-site basis when

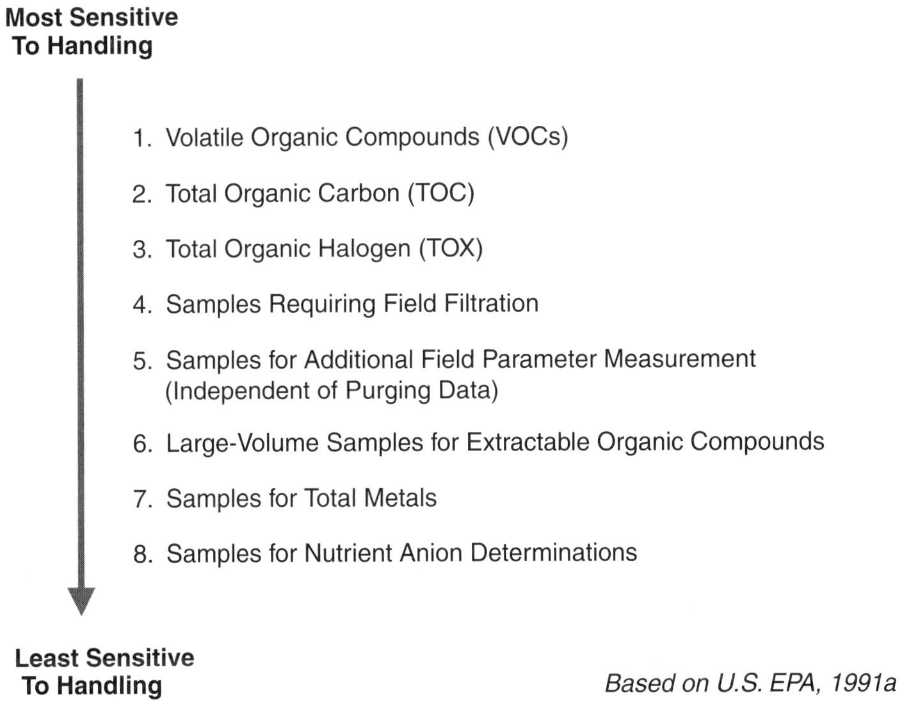

Most Sensitive To Handling

1. Volatile Organic Compounds (VOCs)
2. Total Organic Carbon (TOC)
3. Total Organic Halogen (TOX)
4. Samples Requiring Field Filtration
5. Samples for Additional Field Parameter Measurement (Independent of Purging Data)
6. Large-Volume Samples for Extractable Organic Compounds
7. Samples for Total Metals
8. Samples for Nutrient Anion Determinations

Least Sensitive To Handling

Based on U.S. EPA, 1991a

FIGURE 15.91
U.S. EPA guidelines on the order of sample container filling.

establishing the order of sample collection. For example, at a mining facility, there may be no interest in VOCs at all, and total metals may be the major concern. In this situation, it is wise to collect this parameter first to ensure that a representative sample for the most important constituents can be submitted for analysis rather than leave it toward the end as is suggested by Figure 15.91. This is particularly critical when sampling low-yield wells, which may not have a sufficient volume of water available to fill all sample containers.

Sample Collection Protocols

In addition to establishing the order of sample container filling, sampling teams must follow correct procedures for collecting the ground-water samples. Without exception, ground-water samples should always be collected directly from the discharge tubing from the sampling device and at no time during container filling should anything but sample (e.g., discharge tubing, sampler's gloves, or filtration equipment) be allowed to enter the sample container or contact the mouth of the sample container. Use of funnels and transfer vessels should be avoided during sample collection because, as secondary forms of sample handling, they introduce potential sources of error and bias. Turbulent flow, aeration, and sample cross-contamination can result from the use of funnels and transfer vessels during sample decanting.

Caps should be kept on sample containers until the moment they are ready to be filled, and containers should be resealed immediately upon filling. At no time should the inside of the caps be allowed to come in contact with the ground surface, sampling equipment, or sampler's fingers. This can result in the transfer of contaminants to the inner cap surface and can introduce contaminants into the sample. If a cap is accidentally dropped onto the ground surface, it should be replaced with a new, clean cap.

Once delivered to ground surface, ground-water samples come into contact with atmospheric conditions. Sampling team members should make every attempt to minimize the time during which samples are exposed to atmospheric conditions, as a number of significant changes to the sample, affecting a wide range of analytes, may otherwise occur. Exposure of a sample to atmospheric conditions results in changes in the pressure and temperature of the sample. Additional changes include increases in the levels of DO and other gases and resultant changes in the redox state of ground-water samples. Most ground water is depleted in oxygen content due to chemical and biological reactions that occur during the infiltration process. When a ground-water sample is exposed to atmospheric conditions, the following processes may take place: oxidation of organics; oxidation of sulfide to sulfate; oxidation of ammonium to nitrate; and oxidation of dissolved metals to insoluble precipitates (Stumm and Morgan, 1996). The latter process is very important in terms of sample stability. Multivalent aqueous-phase species, such as Fe, Mn, and As, may be oxidized from a reduced state (Fe^{2+} to Fe^{3+}; Mn^{3+} to Mn^{4+}; and As^{3+} to As^{5+}), causing colloid-sized metal oxides and hydroxides to precipitate (Gillham et al., 1983; Puls et al., 1990; Ryan and Gschwend, 1990; Backhus et al., 1993). Because the kinetics of Mn oxidation are considerably slower than those for Fe, it is possible to collect a sample that is representative for one constituent (Mn) and not for another (Fe), depending on how rapidly the sample is preserved after collection (Gibb et al., 1981).

Because the oxidation of iron is particularly critical to maintaining sample integrity, it is worth discussing further. Under anoxic to suboxic conditions, ground water often contains high concentrations of dissolved iron (Fe^{2+}). Upon exposure to atmospheric conditions, Fe^{2+} oxidizes to Fe^{3+}, which precipitates as iron oxide or iron hydroxide, causes an increase in solution pH, and produces a rust-colored residue of colloid-sized particles in sample bottles. Iron hydroxide is known to adsorb or co-precipitate a number of other metals, including Cu, Zn, Co, Cd, As, Hg, Ag, Pb, V and even some organic species (Gibb et al., 1981; Gillham et al., 1983; Barcelona et al., 1984; Stoltzenburg and Nichols, 1986; Stumm and Morgan, 1996). The kinetics of oxidation-induced precipitation and subsequent sorption processes is such that they can occur within seconds or minutes (Reynolds, 1985; Puls et al., 1992). The end result is that a number of previously dissolved species are removed from solution and may be removed from the sample if it is filtered, resulting in significant negative bias for a number of analytes. Sample preservation methods (e.g., acidification) are meant to prevent such sample alteration, but are only effective if the sample is preserved prior to the occurrence of these reactions. The only commonly analyzed parameters that generally remain unaffected by exposure to atmospheric conditions are major ions.

A special sampling procedure must be used for samples collected for VOC analysis to prevent the loss of volatile constituents from the sample. VOC samples must be collected in specially designed 40 ml vials using a technique that is referred to as zero-headspace sampling, in which sample vials are filled at a relatively slow rate (Figure 15.92). Some U.S. EPA documents (i.e., U.S. EPA, 1986) recommend sampling at a rate of 100 ml/min, which is too slow for many sites, especially those where atmospheric contributions of contaminants are of concern. A sample collection rate of 200 to 250 ml/min is more reasonable for volatile constituents. This rate is fast enough to minimize contact with ambient air, but is not so fast that sample aeration, agitation, or turbulence occur during sample collection. To collect a zero-headspace sample, ground water is collected directly from the pump discharge tubing or the grab sampling device in a controlled manner. To fill the vial, the container is held on an angle and water is allowed to gently flow down the inside wall of the container. As the container fills, it is slowly straightened to vertical. Once vertical, the vial is filled until a positive meniscus forms on top of the water, taking care not to overfill the vial and wash out any chemical preservatives that may be in the vial (Figure 15.93). The cap is then carefully placed on top of the vial without disturbing the meniscus and tightened to the manufacturer-recommended degree of tightness. The samplers should then invert the vial and carefully tap it against the heel of their hand to check for the presence of any bubbles that may have been trapped in the vial during filling. If bubbles are

FIGURE 15.92
Collecting samples for VOC analysis requires use of a slow, controlled discharge rate. A rate of about 200 to 250 ml/min is a good compromise which allows collection of samples without agitation and turbulence, while minimizing the time with which the sample is in contact with atmospheric conditions.

FIGURE 15.93
To fill a VOC vial with zero-headspace, it is necessary to form a positive meniscus on the water surface prior to affixing the cap on the vial. For vials containing a chemical preservative, it is important not to overfill the vial to ensure that the preservative is not washed out.

detected, the sampler must refer to the SAP for procedures on how to handle the vials. If the bubbles are the size of a pinhead or smaller, one trick to implement in the field is to hold the vial vertically, carefully turn the cap one fourth to half turn back (as if taking the cap off) to release the pressure in the vial and then retighten the cap. Using this technique, most small bubbles will be lost and the sample will be zero headspace without having to open the vial and expose the sample to atmospheric conditions again. If, however, the bubbles are larger than pinhead size, it may be necessary to discard the vial and resample. Samplers should refrain from opening a vial with headspace and "topping off" the sample until a zero-headspace sample is collected. This may result in either loss of constituents through volatilization into ambient air or contamination of the sample in some cases where atmospheric levels of volatile constituents are high. In addition, there is a risk that chemical preservatives will be washed out of the vial during this topping-off process, resulting in an improperly preserved sample.

For most parameters other than VOCs, sample collection rates of less than 500 ml/min are appropriate (Puls and Barcelona, 1996).

Protocols for Collecting Field QC Samples

Trip Blanks: A trip blank is prepared and provided by the laboratory as a standard QC sample: The laboratory ships a set of containers prepared for the required list of analytes that are filled with laboratory-prepared water (usually deionized water) of known and documented quality. The containers are labeled by the laboratory as being trip blanks and are shipped to the field along with the empty sample containers. Trip blanks should be documented by the laboratory on the accompanying chain-of-custody forms. After receipt in the field by the sampling team, the trip blanks should be inspected to ensure that all are present, that all containers are in good physical condition, and that there is no headspace in the VOC vials used for trip blanks. Any problems with trip blanks should be documented on the chain-of-custody forms and the laboratory should be notified immediately. The trip blanks should always be kept with the sample containers throughout the sample collection event and should be treated just like samples with regard to temperature control and packaging for shipment to the laboratory. At no time should any of the trip blank containers be opened and exposed to atmospheric conditions. If samples are sent to more than one laboratory, separate trip blanks should be submitted to each laboratory involved in sample analysis.

Temperature Blanks: To confirm that samples are appropriately preserved with respect to temperature ($4 \pm 2°C$), actual sample temperatures are measured upon arrival at the laboratory: Sample temperatures can be checked in one of three ways: (1) using a certified thermometer inserted into one or more sample containers; (2) using a calibrated infrared gun to determine the surface temperature of individual containers; or (3) using a specially prepared 40 ml sentry vial (Figure 15.22) that contains a permanently affixed certified thermometer. Sentry vials, also referred to as temperature blanks, are usually supplied by the laboratory along with trip blanks, and they accompany samples throughout the sampling event. Temperature blanks should be packaged along with the ground-water samples and the rest of the QC samples and shipped to the laboratory. If the temperature measured upon arrival at the laboratory is outside of the allowable range, the sampling team will be contacted to discuss an appropriate course of action.

Field Blanks: To properly collect a field blank, the sampling team must order from the laboratory a set of containers prepared and filled in the same fashion as the trip blank described earlier, but these containers are labeled as field blanks: The deionized water-filled containers are accompanied by an identical but empty set of sample containers. To collect a field blank for parameters other than VOC analysis, the water-filled and empty containers are taken to the point of ground-water sample collection and the volume of water is transferred from the filled containers to the equivalent empty containers. The newly filled containers are labeled and sealed and the original containers are managed as part of the solid waste program for the sampling event. The purpose of this procedure is to expose the entire contents of the sample container to the same atmospheric conditions to which ground-water samples are exposed during sample collection. For VOCs, which require that samples be collected in 40 ml vials, it is strongly recommended that the sampling team make arrangements with the laboratory to have the deionized water sent in 60 ml vials rather than 40 ml vials, to ensure that the volume of water can be transferred with zero headspace. It is impossible to collect a zero-headspace sample when water is directly transferred from one 40 ml vial into another 40 ml vial. It is important that the chemistry of the 60 and 40 ml vials be identical to avoid introduction of a potential source of error.

Equipment Blanks: Two types of equipment blanks may be collected — a rinseate blank or a wipe or swipe blank: To collect a rinseate equipment blank, the field equipment is cleaned following the documented cleaning protocol. At the conclusion of cleaning, an aliquot of the final control water rinse is passed over and through the equipment just cleaned. The rinse water is collected directly into a sample container and is submitted to the laboratory for analysis. To collect a wipe or swipe equipment blank, the sampling team will make arrangements with the laboratory to provide a contaminant-specific wipe kit that contains some form of sterilized gauze or pad that is saturated in a contaminant-specific solvent (e.g., hexane for PCB determination). The sampling team will clean the equipment following the prescribed cleaning protocol. After cleaning, the saturated absorbent material is removed from its container (typically a threaded test tube, can, or vial) using forceps or tweezers and is wiped across the equipment covering a defined surface area (typically 100 cm^2), which may be delineated using a template. The absorbent material is then returned to its container and submitted to the laboratory for analysis. The laboratory analyzes the extract fluid and provides a count per surface area to indicate if a residue is present on the surface of the equipment following cleaning. These types of equipment blanks are appropriate to use on equipment that has a large surface area or has a surface area that is subject to trapping contaminants such as the exterior of pump tubing bundles.

Blind Duplicate Samples: There are two procedures for collecting blind duplicates, one for nonvolatile parameters and the other for volatile parameters requiring the use of 40 ml vials: To collect a blind duplicate sample for nonvolatile parameters, two identical sample containers are alternately filled until both are full. The number of times the samplers go back and forth between the two containers is largely a function of the type of sampling device used. If a pump is used, it is easy to go back and forth with the discharge tubing frequently until both containers are filled. If a grab-sampling device such as a bailer is used, samplers must first determine whether a bottom-emptying device is to be used to decant the sample from the device. If so, then samplers must alternate between containers frequently to avoid filling one container with the bottom portion of the water column in the bailer and the other with water from the top of the water column in the bailer. Alternating between containers will ensure that both containers receive an equal mix of upper and lower portions of the bailer water column. If a top-emptying procedure is used, the water column within the device is mixed during decanting, so the sampler will decant the sample into the two containers by going back and forth between them until they both are filled.

In cases where VOCs are sampled using 40 ml vials, a second strategy for sample collection must be implemented. As already discussed, the key to the successful collection of VOC samples is to ensure that the sample is collected with zero headspace and minimal exposure of the sample to ambient conditions. A typical sample for VOCs will consist of from two to four 40 ml vials from a single sampling point. To collect a duplicate sample for VOCs, the sampling team must assemble all of the sample containers for both the primary set and the duplicate set of samples. They must then alternately fill one vial from the primary set of containers, then the other vial from the duplicate set of containers. This process continues until all of the vials are filled in succession. All samples should be collected using the zero-headspace sampling technique described earlier.

Duplicate samples are referred to as "blind" samples because the sampling team should not indicate on the sample container label or chain-of-custody forms that one sample is a duplicate of another. This is done to prevent possibly biasing the laboratory's handling and analysis of the duplicate sample. It is recommended that a code of some sort be used by samplers to indicate which sample is a duplicate of another. The code selected should be consistent between sampling events and must be documented in field notes. Samplers should resist the temptation to change the time of sample collection for the duplicate sample to ensure that the laboratory will not determine which sample is a duplicate of another. This may be interpreted in a legal context as sample tampering or falsification of data, which can lead to serious consequences for the sampling team if the data are obtained through discovery as evidence in a court case.

When duplicate samples are analyzed, results reported should be within acceptable ranges for the analytical method used. If results do not meet this requirement, the challenge is then to determine why the duplicates were not close. Was there an error in how samples were collected that introduced sampling imprecision? Was there a problem in how the laboratory analyzed the samples? This determination can be difficult, but it is critical to make to assign responsibility and to implement corrective action for future sampling events.

Field Spiked Samples: To prepare a spiked sample, the sampling team collects a second set of duplicate samples as described earlier: To one of the duplicate samples, a spiking solution of one or more known compounds of known concentration is added. Typically, this spiking solution is a commercially prepared and certified matrix of compounds that are close relatives of compounds (e.g., different isomers) analyzed in the ground-water samples by the laboratory. This permits detection of the spiking compounds using the same analytical method used for the samples and avoids masking concentrations of related constituents in the samples. In theory, if there is no microbial activity and the sample has been appropriately preserved, the concentrations of the spiking compounds should be detected during the analysis of the spiked sample and there should be little difference between spiked concentrations and detected concentrations. If there is a large difference between the two values, it is interpreted that microbial activity has occurred and that sample preservation methods need to be improved or altered.

Field Split Samples: Field split samples are collected for the purpose of verifying the performance of one laboratory against a laboratory of known performance: Typically, field split samples are collected when a regulatory agency wishes to evaluate the performance of a new or unknown laboratory against the regulatory agency's approved or internal laboratory to ensure the accuracy of sample analysis. To collect field split samples, a team of samplers from the visiting (i.e., regulatory agency) sampling team will collect duplicate samples with the facility sampling team during a routine sampling event. The facility sampling team collects and handles its samples according to its SAP and sends samples to its laboratory, and the visiting sampling team submits their samples to their "known" laboratory.

Difficulties can arise in cases where sampling points have very low yield or insufficient water volume to permit collection of two full sets of samples. Preliminary planning for the sampling event must include a strategy on how to handle this situation. In most cases, the facility sampling team will have priority over the visiting sampling team to obtain a full set of samples in low-yield or low-volume wells and the visiting sampling team will not collect samples from those locations. This ensures that samples are submitted to the laboratory for analysis so data are reported to the regulatory agency as required for compliance. The alternate approach is to have the facility sampling team to collect all of their samples first and then allow the visiting sampling team to collect their samples from whatever water remains in the wells. Unfortunately, this approach introduces an uncontrollable source of variability, as this water may be of different chemistry and thus may make a meaningful comparison of laboratory results difficult or impossible.

If everything is as it should be, data generated by the two labs should be similar (at least within the performance standards of the analytical method used). If results are not close, then the challenge is to determine why there is a difference. Is it a case where the new laboratory's performance is substandard? Were there differences in laboratory methods used for sample preparation, extraction, or analysis? Where there differences in handling and field pretreatment of samples? It is important to make this determination to implement corrective action prior to the next sampling event.

Ground-Water Sample Pretreatment Procedures

Filtration: As discussed earlier, positive-pressure filtration methods are preferred for ground-water sample filtration (U:S. EPA, 1991a). There are two general categories of positive-pressure filtration equipment: (1) in-line filtration equipment used with pumping devices; and (2) remote pressurized filtration equipment that is not in line with a pumping device.

In-line filtration (Figure 15.94) is recommended for ground-water sample filtration because it provides better consistency through less sample handling and minimized sample exposure to the atmosphere (Stolzenburg and Nichols, 1986; Puls and Barcelona, 1989, 1996). In-line filters eliminate the effects of turbulent discharge and can reveal the

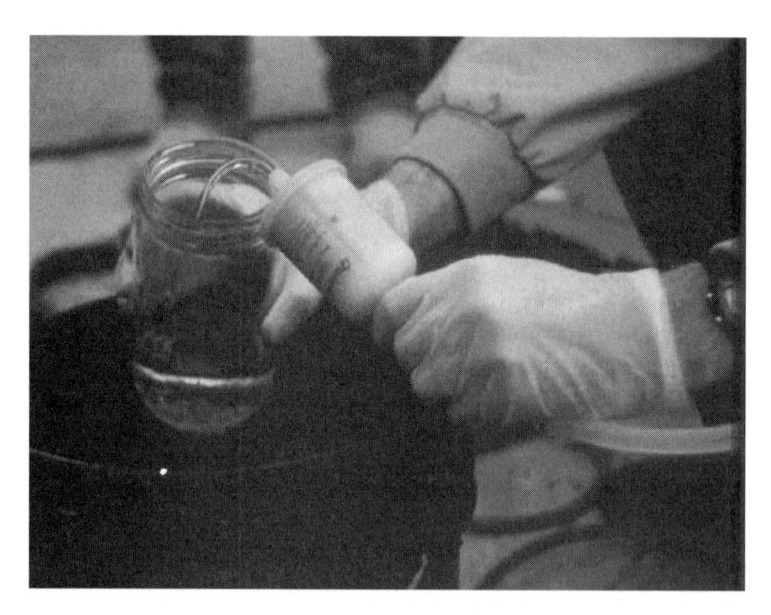

FIGURE 15.94
Positive-pressure filtration using an in-line cartridge filter is the preferred method for sample filtration.

amount of aeration caused solely by the sampling mechanism (Stolzenburg and Nichols, 1986). These filters are available in both disposable (barrel filters) and nondisposable (in-line filter holder using flat membrane filters) formats and in various filter pore sizes (Table 15.18).

Ground-water samples filtered using in-line filtration systems should be collected according to the following sequence:

1a. If using disk filters, assemble the disk filter holder and filtration equipment, so it is leak tight and the filter (handled with forceps) is centered on the holding device. Connect the filtration equipment to the discharge tubing of the pumping device.
1b. If using capsule filters, attach the filter directly to the discharge tubing of the pumping device.
2. Precondition the filter as previously described.
3. Initiate and gradually increase the flow of water through the filter to reach the appropriate rate and pressure, not to exceed the maximum recommended by the filtration equipment manufacturer (e.g., <65 psi) pressure for many capsule filters).
4. Collect the filtered ground-water sample directly into a prepared sample bottle.
5. Preserve the filtered ground-water sample as required on a parameter-specific basis.
6. Release the pressure from the filtration equipment and disconnect it from the sampling device discharge tubing.
7. Discard any disposable materials (e.g., filter media) in accordance with the site-specific waste-management provisions of the SAP.
8. Clean any equipment used for filtration of the next sample following decontamination procedures outlined in the SAP.

Ground-water samples can also be filtered using positive-pressure equipment that is not operated in line with a pumping device. To operate this equipment, the following procedures should be implemented:

1. Assemble the filter holder and support equipment, checking to make sure the system is leak tight.
2. Precondition the filter medium as previously described.
3. Remove a sample of ground water from the monitoring well.
4. Carefully decant the sample into the filtration vessel (if not using the sampling device itself as the vessel) to minimize aeration, agitation, and turbulence and to prevent introduction of airborne contaminants into the sample during transfer.
5. Pressurize the filtration vessel using oil-free inert gas (e.g., nitrogen) or some type of oil-free air pump (e.g., hand-operated positive-pressure pump). Pressure should not exceed manufacturers' guidelines for the equipment in use.
6. Collect the sample directly into a prepared sample container.
7. Preserve the filtered ground-water sample as required on a parameter-specific basis.
8. Release the pressure from the filtration equipment.
9. Discard any disposable materials (e.g., filter media) in accordance with the site-specific waste-management provisions of the SAP.
10. Clean any equipment used for filtration of the next sample, following decontamination procedures outlined in the SAP.

Negative-pressure filtration systems that require applying a vacuum or suction to draw samples through a filter medium are available for ground-water samples. However, because of the detrimental effects on sample chemistry caused by applying negative pressure (discussed earlier in this chapter), vacuum filtration of ground-water samples is not recommended (U.S. EPA, 1991a). If, for some reason, negative-pressure equipment must be used, the following procedures should be implemented:

1. Assemble the filter holder and support equipment making sure it is leak tight.
2. Precondition the filter medium using the methods described previously.
3. Remove a sample of ground water from the monitoring well.
4. Decant the sample into the filtration vessel, taking care not to agitate the sample, increase turbulence, or introduce airborne contaminants into the sample during transfer.
5. Apply a negative pressure to the filtration vessel using a vacuum pump. The negative pressure applied should not exceed manufacturers guidelines for the equipment in use.
6. Collect the filtrate into a flask or other transfer vessel.
7. Release the negative pressure at the vacuum pump connected to the filtration equipment.
8. Transfer the filtrate into a prepared sample container, taking care not to agitate the sample, increase turbulence, or introduce airborne contaminants into the sample.

9. Preserve the filtered ground-water sample as required on a parameter-specific basis.
10. Discard any disposable materials in accordance with the site-specific waste-management provisions of the SAP.
11. Clean any equipment used for filtration of the next sample following decontamination procedures outlined in the SAP.

Physical and Chemical Preservation Procedures for Samples

After samples are collected and, if required, filtered, most must be chemically preserved to protect the physical and chemical integrity of the sample. As mentioned earlier in this chapter, sample chemistry can change between the time of sample container filling and sample extraction or analysis in the laboratory. Samplers must be sure to add the required parameter-specific chemical preservative as documented in the SAP in accordance with the prescribed method (i.e., titration, addition of preservatives using ampoules or vials, or using prepreserved containers). To ensure adequate mixing of the preservative within the sample container, samplers should gently invert (not vigorously shake) the sealed sample container to mix the sample and preservative.

Following mixing the preservative, the end pH of the sample must be verified in the field to ensure that sufficient preservative has been added to those containers requiring pH adjustment. Details on how this should be performed are provided earlier in this chapter.

Preparation of Sample Containers for Shipment

Following field verification of sample pH, the sampling team must prepare all sample containers for shipment or hand delivery to the laboratory. After a full set of samples has been collected from a well, all sample containers must be recorded in the field notebook and on the chain-of-custody form. The exterior surface of the sample containers must be wiped clean to ensure that adhesive labels (that are completed with pertinent sample-related information) will adhere to the container. This can be problematic when condensation forms on the exterior of sample containers. When this occurs, samplers should dry the container surface and then quickly place the label on the dried surface. After the label is in place, it can be covered with a single layer of clear packing tape around the entire container to form a water-proof seal over the label and the tape will help secure the label to the container. Packing tape should never be placed over the top of a container lid, especially two-part septum caps used for VOC vials. This can result in contamination of the sample during sample analysis.

In cases in which there is a need to verify that samples have not been tampered with following collection, it may be necessary to affix a security tag or seal around the lower edge of the container cap. Security seals must be affixed properly. For most sample containers, the seal may be draped over the container cap and attached to the neck of the container. However, for VOC vials and other containers that have a septum in the cap, the seal must be affixed around the cap and must not cover the septum (Figure 15.95). An improperly affixed seal (Figure 15.96) may compromise sample integrity through inclusion of the adhesives on the back of the seal during analysis. Security seals and tags are used as physical deterrents to sample tampering. When containers arrive at the laboratory, seals will be inspected in the sample reception department. If they are broken or missing, that is interpreted by laboratory staff to mean that samples may have been tampered with. At that point, a decision will need to be made by the project manager regarding how (or if) to proceed with sample analysis.

After samples are labeled and sealed, the containers will generally be placed on ice in some form of shipping container, such as a cooler. As indicated earlier in Table 15.20,

Ground-Water Sampling 1101

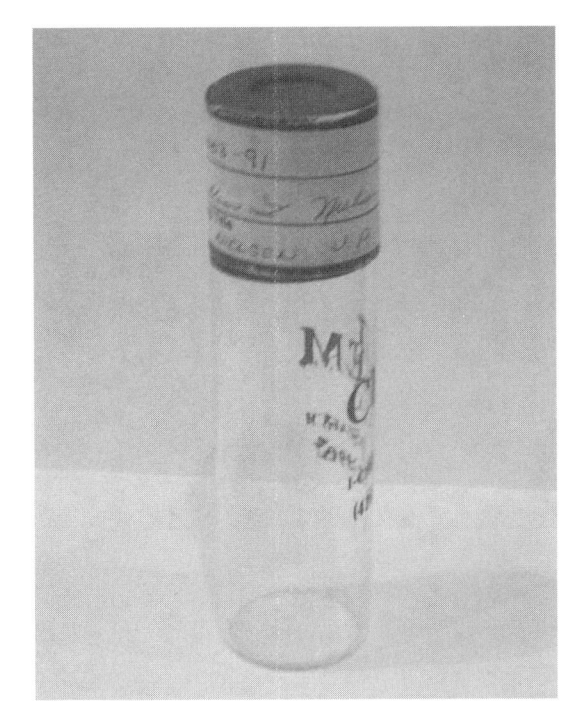

FIGURE 15.95
Sample seals on VOC vials must be affixed around the lid as shown here.

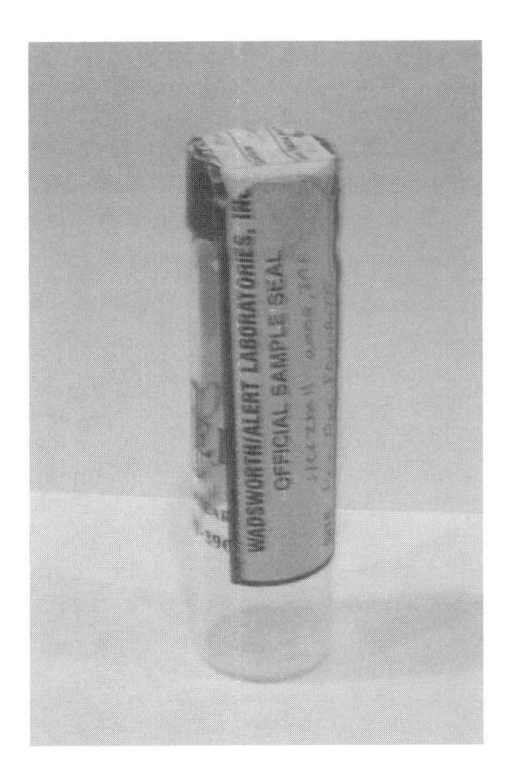

FIGURE 15.96
If the sample seal on a VOC vial covers the septum, the constituents of the adhesive from the seal may be detected in the analysis as the needle from the analytical instrument penetrates the septum to collect the sample.

FIGURE 15.97
All sampling teams should relax after a day of hard work and have a cold beverage to quench their thirst once the samples have been safely delivered to the commercial carrier or lab. Creative application of sampling equipment in this example has uncovered yet another viable use for the equipment. Any equipment used in this manner should be thoroughly cleaned prior to the next day's work.

many parameters are required to be stored and shipped at 4°C. Temperature blanks should be placed in each sample shipper to document arrival temperatures of samples.

Cleanup of the Work Area

Following sample preparation for shipment, the equipment and work area around the sampling point must be cleaned and the well secured for the next sampling event. This typically involves cleaning all water-level measurement devices, purging and sampling devices, and support equipment before it is moved to the next sampling location. In addition, disposable items (e.g., disposable gloves, disposable filters, plastic sheeting, paper towels, etc.) should be collected and managed in accordance with the SAP. If required by the SAP, all purge water should be containerized and records made of the volume of purge water generated. The well should then be locked or otherwise secured.

Delivery or Shipment of Samples to the Laboratory

At the end of the day, samples are either hand delivered (preferred to avoid loss or damage of samples) or shipped to the laboratory for analysis. Samplers must be prepared to have to drive samples to the local station of a commercial carrier, which may be an hour or more from the field site, so time must be well managed in the field. ASTM Standard D 6911 (ASTM, 2004m) provides guidance on appropriate packaging and shipping of both regulated and unregulated ground-water samples.

After samplers have delivered the samples either to the laboratory or to the commercial carrier, they should hit the showers and grab a well-deserved beverage (Figure 15.97) after a day of hard work.

References

APHA, AWWA, and WPCF, *Standard Methods for the Examination of Water and Wastewater*, 17th ed, American Public Health Association, American Water Works Association and Water Pollution Control Federation, Washington, DC, 1989.

Anderson, M.A., J.F. Ferguson, and J. Gavis, Arsenate adsorption on amorphous aluminum hydroxide, *Journal of Colloid and Interface Science*, 54(3), 391–399, 1975.

API, A Guide to the Assessment and Remediation of Underground Petroleum Releases, API Publication 1628, American Petroleum Institute, Washington, DC, 1989, 82 pp.

ASTM, Standard Test Method for Determining Subsurface Liquid Levels in a Borehole or Monitoring Well, ASTM Standard D 4750, ASTM International, West Conshohocken, PA, 2004a, 5 pp.

ASTM, Standard Guide for Presentation of Water-Level Information from Ground-Water Sites, ASTM Standard D 6000, ASTM International, West Conshohocken, PA, 2004b, 16 pp.

ASTM, Standard Guide for Field Quality Assurance in a Ground-Water Sampling Event, ASTM Standard D 7069, ASTM International, West Conshohocken, PA, 2004c, 4 pp.

ASTM, Standard Guide for the Selection of Purging and Sampling Devices for Ground-Water Monitoring Wells, ASTM Standard D 6634, ASTM International, West Conshohocken, PA, 2004d, 14 pp.

ASTM, Standard Practice for Decontamination of Field Equipment Used at Nonradioactive Waste Sites, ASTM Standard D 5088, ASTM International, West Conshohocken, PA, 2004e, 8 pp.

ASTM, Standard Practice for Decontamination of Field Equipment Used at Low-Level Radioactive Waste Sites, ASTM Standard D 5608, ASTM International, West Conshohocken, PA, 2004f, 8 pp.

ASTM, Standard Guide to Purging Methods for Wells Used for Ground-Water Quality Investigations, ASTM Standard D 6452, ASTM International, West Conshohocken, PA, 2004g, 6 pp.

ASTM, Standard Practice for Design and Installation of Ground-Water Monitoring Wells in Aquifers, ASTM Standard D 5092, ASTM International, West Conshohocken, PA, 2004h, 20 pp.

ASTM, Standard Guide to For Development of Ground-Water Monitoring Wells in Granular Aquifers, ASTM Standard D 5521, ASTM International, West Conshohocken, PA, 2004i, 15 pp.

ASTM, Standard Practice for Low-Flow Purging and Sampling for Wells and Devices Used for Ground-Water Quality Investigations, ASTM Standard D 6771, ASTM International, West Conshohocken, PA, 2004j, 7 pp.

ASTM, Standard Guide for Field Filtration of Ground-Water Samples, ASTM Standard D 6564, ASTM International, West Conshohocken, PA, 2004k, 5 pp.

ASTM, Standard Guide for Field Preservation of Ground-Water Samples, ASTM Standard D 6517, ASTM International, West Conshohocken, PA, 2004l, 6 pp.

ASTM, Standard Guide for Packaging and Shipping Environmental Samples for Laboratory Analysis, ASTM Standard D 6911, ASTM International, West Conshohocken, PA, 2004m, 6 pp.

ASTM, Standard Guide for Documenting a Ground-Water Sampling Event, ASTM Standard D 6089, ASTM International, West Conshohocken, PA, 2004n, 3 pp.

ASTM, Standard Guide for Sample Chain-of-Custody Procedures, ASTM Standard D 4840, ASTM International, West Conshohocken, PA, 2004o, 8 pp.

ASTM, Standard Guide for Planning and Preparing for a Ground-Water Sampling Event, ASTM Standard D 5903, ASTM International, West Conshohocken, PA, 2004p, 4 pp.

Backhus, D.A., J.N. Ryan, D.M. Groher, J.K. MacFarlane, and P.M. Gschwend, Sampling colloids and colloid-associated contaminants in ground water, *Ground Water*, 31(3), 466–479, 1993.

Bangsund, W.J., C.G. Peng, and W.R. Mattsfield, Investigation of contaminant migration by low flow rate sampling techniques, Proceedings of the Eighth Annual Outdoor Action Conference, National Ground Water Association, Dublin, OH, 1994, pp. 311–326.

Barcelona, M.J. and J.A. Helfrich, Well construction and purging effects on ground-water samples, *Environmental Science and Technology*, 20(11), 1179–1184, 1986.

Barcelona, M.J., J.A. Helfrich, E.E. Garske, and J.P. Gibb, A laboratory evaluation of ground-water sampling mechanisms, *Ground-Water Monitoring Review*, 4(2), 32–41, 1984.

Barcelona, M.J., J.P. Gibb, and R.A. Miller, A Guide to the Selection of Materials for Monitoring Well Construction and Ground-Water Sampling, Contract Report 327, Illinois State Water Survey, Champaign, IL, 1983, 78 pp.

Barcelona, M.J., J.P. Gibb, J.A. Helfrich, and E.E. Garske, Practical Guide to Ground-Water Sampling, Contract Report 374, Illinois State Geological Survey, Champaign, IL, 1985, 94 pp.

Barcelona, M.J., H.A. Wehrmann, and M.D. Varljen, Reproducible well-purging procedures and VOC stabilization criteria for ground-water sampling, *Ground Water*, 32(1), 12–22, 1994.

Barcelona, M.J., J.F. Keely, W.A. Pettyjohn, and H.A. Wehrmann, *Handbook: Ground Water*, EPA/625/687/016, U.S. Environmental Protection Agency, Center for Environmental Research Information, Cincinnati, OH, 1987, 212 pp.

Barker, J.F. and R. Dickout, An evaluation of some systems for sampling gas-charged ground water for volatile organic analysis, *Ground-Water Monitoring Review*, 8(4), 112–120, 1988.

Barrow, N.J., J. Gerth, and G.W. Brunner, Reaction kinetics of the adsorption and desorption of nickel, zinc, and cadmium by geothite II: modeling the extent and rate of reaction, *Journal of Soil Science*, 40, 437–450, 1989.

Benjamin, M.M. and J.O. Leckie, Multiple-site adsorption of Cd, Cu, Zn, and Pb on amorphous iron oxyhydroxide, *Journal of Colloid and Interface Science*, 79(1), 209–221, 1981.

Bergseth, H., Effect of filter type and filtrate treatment on the measured content of Al and other ions in ground water, *Acta Agriculturae Scandinavica*, 33, 353–359, 1983.

Braids, O.C., R.M. Burger, and J.J. Trela, Should ground-water samples from monitoring wells be filtered before laboratory analysis?, *Ground-Water Monitoring Review*, 7(2), 58–67, 1987.

Brobst, R.B. and P.M. Buszka, The effect of three drilling fluids on ground-water sample chemistry, *Ground-Water Monitoring Review*, 6(1), 62–70, 1986.

Buddemeier, R.W. and J.R. Hunt, Transport of colloidal contaminants in ground water: radionuclide migration at the Nevada test site, *Applied Geochemistry*, 3, 535–548, 1988.

Cerda, C.M., Mobilization of kaolinite fines in porous media, *Colloids and Surfaces*, 27, 219–241, 1987.

Cherry, J.A., Ground water monitoring: some current deficiencies and alternative approaches, in *Hazardous Waste Site Investigations: Toward Better Decisions*, R.B. Gammage and B.A. Berven, Eds., Lewis Publishers, Chelsea, MI, 1992, chap. 13, pp. 119–133.

Church, P.E. and G.E. Granato, Bias in ground-water data caused by well-bore flow in long-screen wells, *Ground Water*, 34(2), 262–273, 1996.

Claassen, H.C., Guidelines and Techniques for Obtaining Water Samples That Accurately Represent the Water Chemistry of an Aquifer, Open-File Report 82-1024, U.S. Geological Survey, Lakewood, CO, 1982, 49 pp.

Clark, L. and K.M. Baxter, Ground-water sampling techniques for organic micropollutants: UK experience, *Quarterly Journal of Engineering Geology*, 22, 159–168, 1989.

Danielsson L.G., On the use of filters for distinguishing between dissolved and particulate fractions in natural waters, *Water Resources*, 16, 179–182, 1982.

Devlin, J.F., Recommendations concerning materials and pumping systems used in the sampling of ground water contaminated with volatile organics, *Water Pollution Research Journal of Canada*, 22(1), 65–72, 1987.

Dragun, J., *The Soil Chemistry of Hazardous Materials*, Hazardous Materials Control Research Institute, Silver Spring, MD, 1988.

Driscoll, F.G., *Ground Water and Wells*, 2nd ed., Johnson Division, St. Paul, MN, 1986, 1089 pp.

Eichholz, G.G., B.G. Wahlig, G.F. Powell, and T.F. Craft, Subsurface migration of radioactive waste materials by particulate transport, *Nuclear Technology*, 58, 511–519, 1982.

Elci, A., G.P. Flach, and F. Molz, Detrimental effects of natural vertical head gradients on chemical and water-level measurements in observation wells: identification and control, *Journal of Contaminant Hydrology*, 28, 70–81, 2003.

Elci, A., F. Molz, and W.R. Waldrop, Implications of observed and simulated ambient flow in monitoring wells, *Ground Water*, 39(6), 853–862, 2001.

EPRI, Field Measurement Methods for Hydrogeologic Investigations: A Critical Review of the Literature, EPRI Report EA-4301, Research Project 2485-7, Electric Power Research Institute, Palo Alto, CA, 1985a, 260 pp.

EPRI, Preliminary Results on Chemical Changes in Ground Water Samples Due to Sampling Devices, EPRI Report EA-4118, Research Project 2485-7, Electric Power Research Institute, Palo Alto, CA, 1985b, 54 pp.

EPRI, Sampling Guidelines for Ground Water Quality, EPRI Report EA-4952, Electric Power Research Institute, Palo Alto, CA, 1987, 47 pp.

Enfield, C.G. and G. Bengtsson, Macromolecular transport of hydrophobic contaminants in aqueous environments, *Ground Water*, 26(1), 64–70, 1988.

Enfield, C.G., G. Bengtsson, and R. Lindquist, Influence of macromolecules on chemical transport, *Environmental Science and Technology*, 23(10), 1278–1286, 1989.

Feld, J., J.P. Connelly, and D.E. Lindorff, Ground water sampling — addressing the turbulent inconsistencies, Proceedings of the First National Outdoor Action Conference on Aquifer

Restoration, Ground Water Monitoring and Geophysical Methods, National Water Well Association, Worthington, OH, 1987, pp. 237–251.

Ferguson, J.F. and M.A. Anderson, Chemical forms of arsenic in water supplies and their removal, in *Chemistry of Water Supply, Treatment, and Distribution*, A.J. Rubin, Ed., Ann Arbor Science Publishers, Ann Arbor, MI, 1973, chap. 7.

Fetter, C.W., Potential sources of contamination in ground-water monitoring, *Ground-Water Monitoring Review*, 3(2), 60–64, 1983.

Forbes, E.A., A.M. Posner, and J.P. Quirk, The specific adsorption of divalent Cd, Co, Cu, Pb, and Zn on geothite, *Journal of Soil Science*, 27, 154–166, 1976.

Garner, W.Y., M.S. Barge, and J.P. Ussary, Eds., *Good Laboratory Practice Standard: Applications for Field and Laboratory Studies*, ACS Professional Reference Book, American Chemical Society, Washington, DC, 1992.

Gibb, J.P., R.M. Schuller, and R.A. Griffin, Procedures for the Collection of Representative Water Quality Data from Monitoring Wells, Cooperative Ground Water Report 7, Illinois State Water and Geological Surveys, Champaign, IL, 1981, 61 pp.

Gibs, J. and T.E. Imbrigiotta, Well-purging criteria for sampling purgeable organic compounds, *Ground Water*, 28(1), 68–78, 1990.

Gibs, J., T.E. Imbrigiotta, J.H. Ficken, J.F. Pankow, and M.E. Rosen, Effect of sample isolation and handling on the recovery of purgeable organic compounds, *Ground-Water Monitoring and Remediation*, 9(2), 142–152, 1994.

Giddings, T., Bore-volume purging to improve monitoring well performance: an often-mandated myth, Proceedings of the Third National Symposium on Aquifer Restoration and Ground Water Monitoring. National Water Well Association, Dublin, OH, 1983, pp. 253–256.

Gillham, R.W., M.J.L. Robin, J.F. Barker, and J.A. Cherry, Ground-Water Monitoring and Sample Bias, American Petroleum Institute Publication, Environmental Affairs Department, Washington, DC, 1983, 206 pp.

Gillham, R.W., M.J.L. Robin, J.F. Barker, and J.A. Cherry, Field Evaluation of Well Purging Procedures, Publication 4405, American Petroleum Institute, Environmental Affairs Department, Washington, DC, 1985, 109 pp.

Gillham, R.W. and S.F. O'Hannesin, Sorption of aromatic hydrocarbons by materials used in construction of ground-water sampling wells, Proceedings, ASTM Symposium on Standards Development for Ground Water and Vadose Zone Monitoring Investigations, Special Technical Publication 1053, D.M. Nielsen and A.J. Johnson, Eds., American Society for Testing and Materials, Philadelphia, PA, 1990, pp. 108–124.

Gillham, R.W., Syringe devices for ground-water sampling, *Ground-Water Monitoring Review*, 2(2), 36–39, 1982.

Graham, B.S. and T.C. Goudlin, Comparison of purge-and-bail sampling to low-stress sampling at an NPL site, Proceedings of the Tenth Annual Outdoor Action Conference, National Ground Water Association, Columbus, OH, 1996, pp. 605–619.

Grisak, G.E., R.E. Jackson, and J.F. Pickens, Monitoring ground water quality: the technical difficulties, in *Establishment of Water Quality Monitoring Programs*, L.G. Everett and K.D. Schmidt, Eds., American Water Resources Association, 1978.

Groher, D., P.M. Gschwend, D. Backhus, and J. MacFarlane, Colloids and sampling ground water to determine subsurface mobile loads, Proceedings of the Environmental Research Conference on Ground-Water Quality and Waste Disposal, I.P. Murarka and S. Cordle, Eds., EPRI EN-6749, Electric Power Research Institute, Palo Alto, CA, 1990, pp. 19.1–19.10.

Grolimund, D., M. Borkovec, K. Barmettler, and H. Sticher, Colloid-facilitated transport of strongly sorbing contaminants in natural porous media: a laboratory column study, *Environmental Science and Technology*, 30(10), 3118–3123, 1996.

Grundl, T. and J. Delwiche, Kinetics of ferric oxyhydroxide precipitation, *Geochimica Cosmochimica Acta*, 56, 1992.

Gschwend, P.M. and M.D. Reynolds, Monodisperse ferrous phosphate colloids in an anoxic ground water plume, *Journal of Contaminant Hydrology*, 1, 309–327, 1987.

Gschwend, P.M., D.A. Backhus, J.K. MacFarlane, and A.L. Page, Mobilization of colloids in ground water due to infiltration of water at a coal ash disposal site, *Journal of Contaminant Hydrology*, 6, 307–320, 1990.

Harvey, R.W., R.L. George, and D.R. LeBlanc, Transport of microspheres and indigenous bacteria through a sandy aquifer: results of natural and forced-gradient tracer experiments, *Environmental Science and Technology*, 23(1), 51–56, 1989.

Heidlauf, D.T. and T.R. Bartlett, Effects of monitoring well purge and sampling techniques on the concentration of metal analytes in unfiltered ground-water samples, Proceedings of the Seventh Outdoor Action Conference and Exposition, National Ground Water Association, Dublin, OH, 1993, pp. 437–450.

Hem, J.D., Study and Interpretation of the Chemical Characteristics of Natural Water, Water Supply Paper 2254, U.S. Geological Survey, Reston, VA, 1985, 263 pp.

Hem, J.D. and C.E. Roberson, Form and Stability of Aluminum Hydroxide Complexes in Dilute Solution, Water Supply Paper 1827-A, U.S. Geological Survey, Reston, VA, 1967.

Herzog, B.L., S.F.J. Chou, J.R. Valkenburg, and R.A. Griffin, Changes in volatile organic chemical concentrations after purging slowly recovering wells, *Ground-Water Monitoring Review*, 8(4), 93–99, 1988.

Hewitt, A.D., Potential of common well casing materials to influence aqueous metal concentrations, *Ground-Water Monitoring and Remediation*, 12(2), 131–136, 1992.

Ho, J.S.Y., Effect of sampling variables on recovery of volatile organics in water, *Journal of the American Water Works Association*, 75(11), 583–386, 1983.

Holm, T.R., G.K. George, and M.J. Barcelona, Oxygen transfer through flexible tubing and its effects on ground-water sampling results, *Ground-Water Monitoring Review*, 8(3), 83–89, 1988.

Horowitz, A.J., K.R. Lum, C. Lemieux, J.R. Garbarino, G.E. Hall, and C.R. Demas, Problems associated with using filtration to define trace element concentrations in natural water samples, *Environmental Science and Technology*, 30(3), 954–963, 1996.

Houghton, R.L. and M.E. Berger, Effects of well-casing composition and sampling method on apparent quality of ground water, Proceedings of the Fourth National Symposium on Aquifer Restoration and Ground Water Monitoring, National Water Well Association, Worthington, OH, 1984, 203–213.

Huntzinger, T.L. and L.E. Stullken, An Experiment in Representative Ground-Water Sampling for Water-Quality Analysis, Water Resources Investigations Report 88-4178, U.S. Geological Survey, Reston, VA, 1988.

Hurley, D.F. and J.M. Whitehouse, ACL monitoring using a low-flow sampling technique: a case study, Proceedings of the Ninth Annual Outdoor Action Conference, National Ground Water Association, Columbus, OH, 1995, pp. 603–616.

Hutchins, S.R. and S.D. Acree, Ground-water sampling bias observed in shallow conventional wells, *Ground-Water Monitoring and Remediation*, 20(1), 86–93, 2000.

Imbrigiotta, T.E., J. Gibs, T.V. Fusillo, G.R. Kish, and J.J. Hochreiter, Field evaluation of seven sampling devices for purgeable organic compounds in ground water, in *Ground-Water Contamination: Field Methods*, A.G. Collins and A.I. Johnson, Eds., Special Technical Publication 963, American Society for Testing and Materials, Philadelphia, PA, 1988, pp. 258–273.

ITRC, Technical and Regulatory Guidance for Using Polyethylene Diffusion Bag Samplers to Monitor VOCs in Ground Water, ITRC Publication, Interstate Technology and Regulatory Council, Washington, DC, 2004.

ITRC, Technical Overview of Passive Sampler Technologies, ITRC Publication, Interstate Technology and Regulatory Council, Washington, DC, 2005.

Jay, P.C., Anion contamination of environmental samples introduced by filter media, *Analytical Chemistry*, 57, 780–782, 1985.

Karklins, S., Ground Water Sampling Desk Reference, Publication PUBL-DG-037 96, Wisconsin Department of Natural Resources, Bureau of Drinking Water and Ground Water, Madison, WI, 1996, 165 pp.

Kearl, P.M., N.E. Korte, M. Stites, and J. Baker, Field Comparison of micropurging vs. traditional ground-water sampling, *Ground-Water Monitoring and Remediation*, 14(4), 183–190, 1994.

Kearl, P.M., N.E. Korte, and T.A. Cronk, Suggested modifications to ground-water sampling procedures based on observations from the colloidal borescope, *Ground-Water Monitoring Review*, 12(2), 155–161, 1992.

Keely, J.F. and K. Boateng, Monitoring well installation, purging, and sampling techniques — Part 1: Conceptualization, *Ground Water*, 25(3), 300–313, 1987a.

Keely, J.F. and K. Boateng, Monitoring well installation, purging, and sampling techniques — Part 2: Case histories, *Ground Water*, 25(4), 427–439, 1987b.

Kennedy, V.C., G.W. Zellweger, and B.F. Jones, Filter pore size effects on the analysis of Al, Fe, Mn, and Ti in water, *Water Resources Research*, 10(4), 785–790, 1974.

Kent, R.T. and K.E. Payne, Sampling ground-water monitoring wells: special quality assurance and quality control considerations, in *Principles of Environmental Sampling*, American Chemical Society, Washington, DC, 1988, pp. 231–246.

Kim, J.I., G. Bachau, F. Baumgartner, H.C. Moon, and D. Lux, Colloid generation and the actinide migration in Gorbelen ground waters, in *Scientific Basis for Nuclear Waste Management*, Vol. 7, G.L. McVay, Ed., Elsevier Publishers, New York, NY, 1984, pp. 31–40.

Kinniburgh, D.G., M.L. Jackson, and J.K. Syers, Adsorption of alkaline earth, transition and heavy metal cations by hydrous gels of iron and aluminum, *Soil Science Society of America Journal*, 40, 796–799, 1976.

Knobel, L.L. and L.J. Mann, Sampling for purgeable organic compounds using positive-displacement piston and centrifugal pumps: a comparative study, *Ground-Water Monitoring and Remediation*, 13(2), 142–148, 1993.

Laxen, D.P.H. and I.M. Chandler, Comparison of filtration techniques for size distribution in fresh waters, *Analytical Chemistry*, 54(8), 1350, 1982.

Lyman, W.J., P.J. Reidy, and B. Levy, Contaminants sorbed onto colloidal particles in water in either the unsaturated or saturated zone, in *Mobility and Degradation of Organic Contaminants in Subsurface Environments*, C.K. Smoley, Inc., 1992, chap. 9, pp. 239–258.

Maltby, V. and J.P. Unwin, A field investigation of ground-water monitoring well purging techniques, in *Current Practices in Ground Water and Vadose Zone Investigations*, Special Technical Publication 1118, D.M. Nielsen and M.N. Sara, Eds., American Society for Testing and Materials, Philadelphia, PA, 1992, pp. 281–299.

Marsh, J.M. and J.W. Lloyd, Details of hydrochemical variations in flowing wells, *Ground Water*, 18(4), 366–373, 1980.

Martin, W.H. and C.C. Lee, Persistent pH and tetrahydrofuran anomalies attributable to well construction, Proceedings of the Third National Outdoor Action Conference on Aquifer Restoration, Ground Water Monitoring and Geophysical Methods, National Water Well Association, Dublin, OH, 1989, pp. 201–213.

Martin-Hayden, J.M. and G.A. Robbins, Plume distortion and apparent attenuation due to concentration averaging in monitoring wells, *Ground Water*, 35(2), 339–346, 1997.

Martin-Hayden, J.M., G.A. Robbins, and R.D. Bristol, Mass balance evaluation of monitoring well purging — Part II: Field tests at a gasoline contamination site, *Journal of Contaminant Hydrology*, 8(1), 225–241, 1991.

Maslansky, C.J. and S.P. Maslansky, *Air Monitoring Instrumentation: A Manual for Emergency, Investigatory and Remedial Responders*, Van Nostrand Reinhold, New York, NY, 1993, 304 pp.

Maslansky, S.P. and C.J. Maslansky, *Health and Safety at Hazardous Waste Sites: An Investigator's and Remediator's Guide to HAZWOPER*, Van Nostrand Reinhold, New York, NY, 1997, 612 pp.

Mason, S.A., J. Barkach, and J. Dragun, Discussion of literature review and model (COMET) for colloidal/metals transport in porous media, *Ground Water*, 30(1), 104–106, 1992.

McAlary, T.A. and J.F. Barker, Volatilization losses of organics during ground-water sampling from low permeability materials, *Ground-Water Monitoring Review*, 7(4), 63–68, 1987.

McCall, W., S. Stover, C. Enos, and G. Fuhrmann, Field comparison of direct-push prepacked screen wells to paired HSA 2″ PVC wells, Proceedings Vol. 2, HazWasteWorld Superfund XVIII Conference, E.J. Krause Co., Washington DC, 1997, pp. 647–655.

McCarthy, J. and L. Shevenell, Obtaining representative ground-water samples in a fractured and karstic formation, *Ground Water*, 36(2), 251–260, 1998.

McCarthy, J.F. and C. Degueldre, Sampling and characterization of colloids and particles in ground water for studying their role in contaminant transport, *Environmental Particles*, J. Buffle and H.P. van Leeuwen, Eds., Lewis Publishers, Chelsea, MI, 1993, pp. 247–315.

McCarthy, J.F. and F.J. Wobber, Transport of Contaminants in the Subsurface: The Role of Organic and Inorganic Colloidal Particles, International Series of Interactive Seminars, Summary Report, ORNL/M-349, Oak Ridge National Laboratory, Oak Ridge, TN, 1996.

McCarthy, J.F. and J.M. Zachara, Subsurface transport of contaminants: binding to mobile and immobile phases in ground water aquifers, *Environmental Science and Technology*, 23(5), 496–504, 1989.

McCaulou, D.R., D.G. Jewett, and S.G. Huling, Non-aqueous Phase Liquids Compatibility with Materials Used in Well Construction, Sampling and Remediation, EPA/540/S-95/503, Ground Water Issue, U.S. Environmental Protection Agency, Office of Research and Development, Washington, DC, 1995, 14 pp.

McDowell-Boyer, L.M., J.R. Hunt, and N. Sitar, Particle transport through porous media, *Water Resources Research*, 22(13), 1901–1921, 1986.

McIlvride, W.A. and B.M. Rector, Comparison of short and long-screened wells in alluvial sediments, Proceedings of the Second Outdoor Action Conference, National Ground-Water Association, Dublin, OH, 1988, pp. 375–390.

Michalski, M., Application of temperature and electrical conductivity logging in ground-water monitoring, *Ground-Water Monitoring Review*, 9(3), 112–118, 1989.

Miller, G.D., Uptake and release of lead, chromium, and trace-level volatile organics exposed to synthetic well casings, Proceedings of the Second National Symposium on Aquifer Restoration and Ground-Water Monitoring, National Water Well Association, Dublin, OH, 1982, pp. 236–245.

Muska, C.F., W.P. Colven, V.D. Jones, J.T. Scogin, B.B. Looney, and V. Price, Jr., Field Evaluation of ground-water sampling devices for volatile organic compounds, Proceedings of the Sixth National Symposium and Exposition on Aquifer Restoration and Ground Water Monitoring, National Water Well Association, Worthington, OH, 1986, pp. 235–246.

NCASI, A Guide to Ground Water Sampling, Technical Bulletin 362 National Council of the Paper Industry for Air and Stream Improvement, 1982, 53 pp.

Nielsen, G.L., Ground-Water Sampling: Field Practices for Regulatory Compliance, Workshop Notebook, Tenth Annual Outdoor Action Conference, National Ground Water Association, Columbus, OH, 1996, pp. 89–97.

Nielsen, D.M., The Relationship of Monitoring Well Design, Construction and Development to Turbidity in Wells, and Related Implications for Ground-Water Sampling, Ground Water Sampling — A Workshop Summary, EPA/600/R-94/205, U.S. Environmental Protection Agency, Office of Research and Development, Washington DC, 1995, 16 pp.

Nielsen, D.M. and G.L. Yeates, A comparison of sampling mechanisms available for small-diameter ground-water monitoring wells, *Ground-Water Monitoring Review*, 5(2), 83–99, 1985.

Nielsen, D.M. and G.L. Nielsen, *Technical Guidance on Low-Flow Purging and Sampling and Minimum-Purge Sampling*, NEFS-TG001-02, Nielsen Environmental Field School, Galena, OH, 2002, 59 pp.

NWWA/PPI, *Manual on the Selection and Installation of Thermoplastic Water Well Casing*, National Water Well Association and Plastic Pipe Institute, Worthington, OH, 1981, 64 pp.

O'Melia, C.R., Aquasols: the behavior of small particles in aquatic systems, *Environmental Science and Technology*, 14(9), 1052–1060, 1980.

O'Melia, C.R., Kinetics of colloid chemical processes in aquatic systems, in *Aquatic Chemical Kinetics*, W. Stumm, Ed., Wiley & Sons, New York, NY, 1990, pp. 447–474.

Oakley, D. and N.E. Korte, Nickel and chromium in ground-water samples as influenced by well construction and sampling methods, *Ground-Water Monitoring and Remediation*, 16(1), 93–99, 1996.

Oneacre, J. and D. Figueras, Ground-water variability at sanitary landfills: causes and solutions, Proceedings, ASCE Conference on Uncertainty in the Geologic Environment, Madison, WI, July 31–August 3, 1996, pp. 965–987.

Palmer, C.D., J.F. Keely, and W. Fish, Potential for solute retardation on monitoring well sand packs and its effect on purging requirements for ground-water sampling, *Ground-Water Monitoring Review*, 7(2), 40–47, 1987.

Parker, L., Suggested guidelines for the use of PTFE, PVC and stainless steel in samplers and well casings, in *Current Practices in Ground Water and Vadose Zone Investigations*, D.M. Nielsen and M.N. Sara, Eds., Special Technical Publication 1118, American Society for Testing and Materials, Philadelphia, PA 1991, pp. 217–229.

Parker, L., The effects of ground-water sampling devices on water quality: a literature review, *Ground-Water Monitoring and Remediation*, 14(2), 130–141, 1994.

Parker, L., A.D. Hewitt, and T.F. Jenkins, Influence of casing materials on trace-level chemicals in well water, *Ground-Water Monitoring and Remediation*, 10(2), 146–156, 1990.

Parker, L.V. and C.H. Clark, Study of Five Discrete Interval Type Ground-Water Sampling Devices, ERDC/CRREL TR-02-12, U.S. Army Corps of Engineers, Cold Regions Research and Engineering Laboratory, Hanover, NH, 2002, 49 pp.

Parsons, Results Report for the Demonstration of No-Purge Ground-Water Sampling Devices at Former McClellan Air Force Base, California; Contract F44650-99-D-0005, Delivery order DKO1, U.S. Army Corps of Engineers (Omaha District), Air Force Center for Environmental Excellence, and Air Force Real Property Agency, 2005.

Paul, C.J. and R.W. Puls, Impact of turbidity on TCE and degradation products in ground water, *Ground-Water Monitoring and Remediation*, 17(1), 128–133, 1997.

Paul, D.G., C.D. Palmer, and D.S. Cherkauer, The effect of construction, installation and development on the turbidity of water in monitoring wells in fine-grained glacial till, *Ground-Water Monitoring Review*, 8(1), 73–82, 1988.

Pearsall, K.A. and D.A. Eckhardt, Effects of selected sampling equipment and procedures on the concentrations of trichloroethylene and related compounds in ground-water samples, *Ground-Water Monitoring Review*, 7(2), 64–73, 1987.

Pennino, J., Dissolved Oxygen Changes in Well Water During Purging and Sampling, Unpublished Report, Ohio Environmental Protection Agency, Dayton, OH, 1987.

Penrose, W.R., W.L. Polzer, E.H. Essington, D.M. Nelson, and K.A. Orlandin, Mobility of plutonium and americium through a shallow aquifer in a semi-arid region, *Environmental Science and Technology*, 24, 228–234, 1990.

Pohlmann, K.F., G.A. Icopini, R.D. McArthur, and C.G. Rosal, Evaluation of Sampling and Field-Filtration Methods for Analysis of Trace Metals in Ground Water, EPA/600/R-94/119, U.S. Environmental Protection Agency, Office of Research and Development, Las Vegas, NV, 1994, 79 pp.

Pohlmann, K.F. and J.W. Hess, Generalized ground-water sampling device matrix, *Ground-Water Monitoring Review*, 8(4), 82–83, 1988.

Pohlmann, K.F., R.P. Blegen, and J.W. Hess, Field Comparison of Ground-Water Sampling Devices for Hazardous Waste Sites: An Evaluation Using Volatile Organic Compounds, EPA/600/4-90/028, U.S. Environmental Protection Agency, Office of Research and Development, Las Vegas, NV, 1991, 102 pp.

Posselt, H.S., F.J. Anderson, and W.J. Weber, Jr., Cation sorption on colloidal hydrous manganese dioxide, *Environmental Science and Technology*, 2, 1087, 1968.

Powell, R.M. and R.W. Puls, Hitting the bull's-eye in ground-water sampling, *Pollution Engineering*, 51–54, 1997.

Powell, R.M. and R.W. Puls, Passive sampling of ground-water monitoring wells without purging: multilevel well chemistry and tracer disappearance, *Journal of Contaminant Hydrology*, 12, 51–77, 1993.

Puls, R.W., Adsorption of Heavy Metals on Soil Clays, Ph.D. dissertation, University of Arizona, Tucson, AZ, 1986, 144 pp.

Puls, R.W., Colloidal considerations in ground-water sampling and contaminant transport predictions, *Nuclear Safety*, 31, 58–65, 1990.

Puls, R.W. and R.M. Powell, Acquisition of representative ground-water quality samples for metals, *Ground-Water Monitoring Review*, 12(3), 167–176, 1992a.

Puls, R.W. and R.M. Powell, Transport of inorganic colloids through natural aquifer material: implications for contaminant transport, *Environmental Science and Technology*, 26(3), 614–621, 1992b.

Puls, R.W. and J.H. Eychaner, Sampling Ground Water for Inorganics — Pumping Rate, Filtration and Oxidation Effects, Proceedings of the Fourth National Outdoor Action Conference on Aquifer Restoration, Ground Water Monitoring and Geophysical Methods, National Water Well Association, Dublin, OH, 1990, pp. 313–327.

Puls, R.W., J.H. Eychaner, and R.M. Powell, Facilitated Transport of Inorganic Contaminants in Ground Water: Part I. Sampling Considerations, EPA/600/M-90/023, U.S. Environmental Protection Agency, Robert S. Kerr Laboratory, Ada, Oklahoma, 1990, 12 pp.

Puls, R.W. and C.J. Paul, Low-flow purging and sampling of ground-water monitoring wells with dedicated systems, *Ground-Water Monitoring and Remediation*, 15(1), 116–123, 1995.

Puls, R.W. and C.J. Paul, Multi-layer sampling in conventional monitoring wells for improved estimation of vertical contaminant distributions and mass, *Journal of Contaminant Hydrology*, 25, 85–111, 1997.

Puls, R.W. and J.F. McCarthy, Well Purging and Sampling (Workshop Group Summary), *Ground Water Sampling — A Workshop Summary*, EPA/600/R-94/205, U.S. Environmental Protection Agency, Office of Research and Development, Washington, DC, 1995, pp. 82–87.

Puls, R.W. and M.J. Barcelona, Ground-Water Sampling for Metals Analysis, EPA/540/4-89/001, Superfund Ground Water Issue, U.S. Environmental Protection Agency, Office of Solid Waste and Emergency Response, Washington, DC, 1989, 6 pp.

Puls, R.W. and M.J. Barcelona, Low-Flow (Minimal-Drawdown) Ground-Water Sampling Procedures, EPA/540/5-95/504, Ground Water Issue, U.S. Environmental Protection Agency, Office of Solid Waste and Emergency Response, Washington, DC, 1996, 12 pp.

Puls, R.W., D.A. Clark, B. Bledsoe, R.M. Powell, and C.J. Paul, Metals in Ground Water: Sampling Artifacts and Reproducibility, *Hazardous Waste and Hazardous Materials*, 9(9), 149–162, 1992.

Puls, R.W., R.M. Powell, D.A. Clark, and C.J. Paul, Facilitated Transport of Inorganic Contaminants in Ground Water: Part II, Colloidal Transport, EPA/600/M-91/040, U.S. Environmental Protection Agency, Robert S. Kerr Environmental Research Laboratory, Ada, OK, 1991.

Reilly, T.E., O.L. Franke, and G.D. Bennett, Bias in ground-water samples caused by well-bore flow, *Journal of Hydrologic Engineering*, 115, 270–276, 1989.

Reynolds, G.W. and R.W. Gillham, Absorption of Halogenated Organic Compounds by Polymer Materials Commonly Used in Ground Water Monitoring, Proceedings of the Second Canadian/American Conference on Hydrogeology, National Water Well Association, Dublin, OH, 1985, pp. 125–132.

Reynolds, M.D., Colloids in Ground Water, Masters thesis, Department of Civil Engineering, Massachusetts Institute of Technology, Cambridge, MA, 1985.

Robbins, G.A. and J.M. Martin-Hayden, Mass balance evaluation of monitoring well purging, Part 1. Theoretical models and implications for representative sampling, *Journal of Contaminant Hydrology*, 8, 203–224, 1991.

Robertson, W.D., J.F. Barker, Y. LeBeau, and S. Marcoux, Contamination of an unconfined sand aquifer by waste pulp liquor: a case study, *Ground Water*, 22(4), 192–197, 1984.

Robin, M.J.L. and R.W. Gillham, Field evaluation of well purging procedures, *Ground-Water Monitoring Review*, 7(4), 85–93, 1987.

Ronen, D., M. Magaritz, H. Gvirtzman, and W. Garner, Microscale chemical heterogeneity in ground water, *Journal of Hydrology*, 92(1/2), 173–178, 1987.

Ryan J.N. and P.M. Gschwend, Effect of iron diagenesis on clay colloid transport in an unconfined sand aquifer, *Geochimica et Cosmochimica Acta*, 56, 1507–1521, 1992.

Ryan, J.N. and P.M. Gschwend, Colloidal mobilization in two atlantic coastal plain aquifers: field studies, *Water Resources Research*, 26, 307–322, 1990.

Ryan, J.N., Ground Water Colloids in Two Atlantic Coastal Plain Aquifers: Colloid Formation and Stability, Masters thesis, Department of Civil Engineering, Massachusetts Institute of Technology, Cambridge, MA, 1988.

Ryan, J.N., S. Mangion, and R. Willey, Turbidity and Colloid Transport (Working Group Summary), EPA/600/R-94/205, Ground Water Sampling — A Workshop Summary, U.S. EPA Office of Research and Development, Washington, DC, 1995, pp. 88–92.

Sawhney, B.L., Kinetics of cesium sorption by clay minerals, *Soil Science Society of America Proceedings*, 30, 565, 1966.

Schuller, R.M., J.P. Gibb, and R.A. Griffin, Recommended sampling procedures for monitoring wells, *Ground-Water Monitoring Review*, 1(2), 42–46, 1981.

Serlin, C.L. and L.M. Kaplan, Field Comparison of Micropurge and Traditional Ground-Water Sampling for Volatile Organic Compounds, Proceedings of the Conference on Petroleum

Hydrocarbons and Organic Chemicals in Ground Water, National Ground Water Association, Westerville, OH, 1996, pp. 177–190.
Sevee, J.E., C.A. White and D.J. Maher, An analysis of low-flow ground-water sampling methodology, *Ground-Water Monitoring and Remediation*, 20(2), 87–93, 2000.
Shanklin, D.E., W.C. Sidle, and M.E. Ferguson, Micropurge low-flow sampling of uranium-contaminated ground water at the Fernald environmental management project, *Ground-Water Monitoring and Remediation*, 15(3), 168–176, 1995.
Sheldon, R.W., Size separation of marine seston by membrane and glass-fiber filters, *Limnology and Oceanography*, 17, 494–498, 1965.
Sheldon, R.W. and W.H. Sutcliffe, Jr., Retention of marine particles by screens and filters, *Limnology and Oceanography*, 14(3), 441–444, 1969.
Smith, R.W. and J.D. Hem, Effects of Aging on Aluminum Hydroxide Complexes in Dilute Aqueous Solutions, Water-Supply Paper 1827-D, U.S. Geological Survey, Reston, VA, 1972, 51 pp.
Sosebee, J.B., P.C. Geiszler, D.L. Winegardner, and C.R. Fisher, Contamination of Ground-Water Samples with PVC Adhesives and PVC Primer from Monitoring Wells, Technical Paper, Environmental Science and Engineering, Englewood, CO, 1983, p. 24.
Sosebee, J.B., P.C. Geiszler, D.L. Winegardner, and C.R. Fisher, Contamination of Ground-Water Samples with PVC Adhesives and PVC Primer from Monitoring Wells, Proceedings of the ASTM Symposium on Hazardous and Solid Waste, American Society for Testing and Materials, West Conshohocken, PA, 1982, 23 pp.
Sparks, D.L., L.W. Zelany, and D.C. Martens, Kinetics of potassium exchange in a paleodult from the coastal plain of virginia, *Soil Science Society of America Journal*, 44(1), 37–40, 1980.
Stolzenburg, T.R. and D.G. Nichols, Preliminary Results on Chemical Changes in Ground-Water Samples Due to Sampling Devices, EPRI EA-4118, Project 2485–7, Electric Power Research Institute, Palo Alto, CA, 1985, 53 pp.
Stolzenburg, T.R. and D.G. Nichols, Effects of Filtration Method and Sampling on Inorganic Chemistry of Sampled Well Water, Proceedings of the Sixth National Symposium and Exposition on Aquifer Restoration and Ground Water Monitoring, National Water Well Association, Dublin, OH, 1986, pp. 216–234.
Stumm, W. and J.J. Morgan, *Aquatic Chemistry: Chemical Equilibria and Rates in Natural Waters*, John Wiley and Sons, Inc., New York, NY, 1996.
Tai, D.Y., K.S. Turner, and L.A. Garcia, The use of a standpipe to evaluate ground-water samplers, *Ground-Water Monitoring Review*, 11(1), 125–132, 1991.
Taylor, J.K., What is Quality Assurance? *Quality Assurance for Environmental Measurements*, Special Technical Publication 867, J.K. Taylor and T.W. Stanley, Eds., American Society of Testing and Materials, Philadelphia, PA, 1985.
Testa, S.M. and D.L. Winegardner, *Restoration of Petroleum-Contaminated Aquifers*, Lewis Publishers, Inc., Chelsea, MI, 1991, 269 pp.
Thornton, D., S. Ita, and K. Larsen, Broader Use of Innovative Ground Water Access Technologies, Proceedings of the HazWasteWorld Superfund XVIII Conference, Vol. 2, E.J. Krause Co., Washington, DC, 1997, pp. 639–646.
Thurnblad, T., Minnesota Pollution Control Agency Ground Water Sampling Guidance: Development of Sampling Plans, Protocols and Reports, Minnesota Pollution Control Agency, St. Paul, MN, 1995.
U.S. EPA, Handbook for Sampling and Sample Preservation of Water and Wastewater, U.S. Environmental Protection Agency, Environmental Monitoring and Support Laboratory, Cincinnati, OH, 1982, pp. 386–397.
U.S. EPA, RCRA Technical Enforcement Guidance Document, OSWER-9950, U.S. Environmental Protection Agency, Office of Waste Programs Enforcement, Office of Solid Waste and Emergency Response, Washington, DC, 1986, p. 208.
U.S. EPA, Compendium of ERT Ground Water Sampling Procedures, EPA/540/P-91/007, U.S. Environmental Protection Agency, Washington, DC, 1991a, 63 pp.
U.S. EPA, Handbook, Ground Water, Volume II: Methodology, EPA/625/6-90/016b, U.S. Environmental Protection Agency, Office of Research and Development, Center for Environmental Research Information, Washington, DC, 1991b, 141 pp.

U.S. EPA, Ground Water Sampling, A Workshop Summary, EPA/600/R-94/205, U.S. Environmental Protection Agency, Office of Research and Development, Washington, DC, 1995, 98 pp.

U.S. EPA, Low Stress (Low Flow) Purging and Sampling Procedures for the Collection of Ground-Water Samples from Monitoring Wells, SOP#GW0001, U.S. Environmental Protection Agency, Region 1, Boston, MA, 1996a, 13 pp.

U.S. EPA, Test Methods for Evaluating Solid Waste, Physical/Chemical Methods, EPA SW846, U.S. Environmental Protection Agency, Office of Solid Waste and Emergency Response, Washington, DC, 1996b.

U.S. EPA, Environmental Investigations Standard Operating Procedures and Quality Assurance Manual, U.S. Environmental Protection Agency Region 4, Athens, GA, 1997.

U.S. EPA, Environmental Technology Verification Report: Geoprobe Inc., Mechanical Bladder Pump, Model MB470, EPA/600R-03/086, U.S. Environmental Protection Agency, Office of Research and Development, Washington, DC, 2003.

Unwin, J.P., A Guide to Ground-Water Sampling, Technical Bulletin 362, National Council of the Paper Industry for Air and Stream Improvement, New York, NY, 1982.

Unwin, J.P. and V. Maltby, Investigations of Techniques for Purging Ground-Water Monitoring Wells and Sampling Ground Water for Volatile Organic Compounds, Special Technical Publication 963, American Society for Testing and Materials, Philadelphia, PA, 1988, pp. 240–252.

Vroblesky, D.A., User's Guide for Polyethylene-Based Passive Diffusion Bag Samplers to Obtain Volatile Organic Compound Concentrations in Wells: Part 1, Development, Recovery, Data Interpretation and Quality Control and Assurance, Water Resources Investigations Report, 01-4060, U.S. Geological Survey, Reston, VA, 2001a.

Vroblesky, D.A., User's Guide for Polyethylene-Based Passive Diffusion Bag Samplers to Obtain Volatile Organic Compound Concentrations in Wells: Part 2, Field Tests, Water Resources Investigations Report, 01-4061, U.S. Geological Survey, Reston, VA, 2001b.

Vroblesky, D.A. and T.R. Campbell, Equilibration times, stability and compound selectivity of diffusion samplers for collection of ground water VOC concentrations, *Advanced Environmental Restoration*, 15(1), 1–12, 2001.

Vroblesky, D.A. and W.T. Hyde, Diffusion samplers as an inexpensive approach to monitoring VOCs in ground water, *Ground-Water Monitoring and Remediation*, 17(3), 177–184, 1997.

Wagemann, R. and G.J. Brunskill, The effect of filter pore size on analytical concentrations of some trace elements in filtrates of natural water, *International Journal of Environmental Analytical Chemistry*, 4, 75–84, 1975.

Walker, S.E., Background Ground-Water Quality Monitoring: Well Installation Trauma, Proceedings of the Third National Symposium on Aquifer Restoration and Ground Water Monitoring, National Water Well Association, Worthington, OH, 1983, pp. 235–246.

Wilson, L.C. and J.V. Rouse, Variations in water quality during initial pumping of monitoring wells, *Ground-Water Monitoring Review*, 3(1), 103–109, 1983.

Yao, K., M.T. Habibian, and C.R. O'Melia, Water and wastewater filtration: concepts and applications, *Environmental Science and Technology*, 5(11), 1105–1112, 1971.

Yeskis, D., K. Chiu, S. Meyers, J. Weiss, and T. Bloom, A Field Study of Various Sampling Devices and Their Effects on Volatile Organic Contaminants, Proceedings of the Second National Conference on Aquifer Restoration, Ground Water Monitoring and Geophysical Methods, National Water Well Association, Dublin, OH, 1988, pp. 471–479.

16

Ground-Water Sample Analysis

Rock J. Vitale and Olin C. Braids

CONTENTS

Introduction	1114
Selection of Analytical Parameters	1114
Ground-Water Investigations Governed by a Regulatory Agency	1115
Analytical Requirements Under RCRA	1115
Analytical Requirements Under a Site-Specific Administrative Consent Order (ACO)	1116
Analytes that are Site-Related	1116
Selection of an Analytical Method	1116
Specific Requirements for an Analytical Method	1117
Description of Analytical Methods	1117
Screening or Diagnostic Tests	1117
Specific Organic Compound Analysis	1118
Volatile Organic Compounds	1118
Semivolatile Organic Compounds	1122
Pesticides, Herbicides, and PCBs	1123
Specific Constituent Inorganic Analysis	1124
Atomic Absorption Spectrophotometry	1124
Atomic Emission Spectroscopy	1126
Other Analyses	1126
Quality Assurance/Quality Control	1127
Selection of an Analytical Laboratory	1127
Preparation of a Quality Assurance Project Plan	1128
Laboratory QA/QC	1129
Chain-of-Custody	1129
Sample Storage and Holding Time Requirements	1129
Sample Preparation	1129
Laboratory QC Samples	1130
Method Blanks	1131
Duplicates	1131
Spiked Samples	1131
Matrix Spikes	1131
Surrogate Spikes	1131
Instrument Calibration	1132
Sample Analysis	1132
Laboratory Validation and Reporting	1132
Documentation and Recordkeeping	1133
Independent Laboratory QA Review	1133

Summary 1134
References 1134

Introduction

With each passing year, advancements in technologies and the resultant analytical capabilities of laboratories have been realized for the handling, preparation, and analysis of water samples. During the early 1970s, while techniques and instrumentation were available for the analysis of common ions and trace metals, analytical techniques and instrumentation for determining specific organic species were extremely limited, both in sensitivity and scope. At that time, general methods (e.g., total organic carbon [TOC], chemical oxygen demand [COD], biochemical oxygen demand [BOD], etc.) were extensively used to approximate the gross amount of carbon in a water sample. By today's standards, these methods, although still used for certain legitimate general water-quality purposes, only provide a general non-compound-specific indication of the presence of organic materials in water samples.

A limited determination of specific organic compounds in water was possible in the early 1970s through the use of gas chromatographs. Earlier organic analytical protocols involved extracting the organic substances from the water using solvents, which would be concentrated and then injected into a gas chromatograph. For volatile organic compounds (VOCs), headspace analysis was the usual approach. Some analysts measured headspace at ambient temperature, while others placed the water sample in a temperature-controlled bath and measured headspace at an elevated temperature. This approach resulted in varying sensitivities with the VOCs because of their differences in aqueous solubility and volatility. Analyte identification depended on matching chromatographic retention times with known standards.

It was not until the research of Bellar and Lichtenberg (1974) resulted in a method for VOCs that released these compounds from water by purging the sample with air, followed by capturing the released compounds on an exchange resin, that the method achieved uniformity. This important development enabled the analysis of VOCs to be done rapidly and with significantly improved sensitivity. Surveys of public water supplies that were made following this analytical development resulted in the detection of VOCs (i.e., chloroform) in many public water supplies in the U.S. (Federal Register, 1985).

The discovery of VOC contaminants in public water supplies and in ground water that had been contaminated by chemicals associated with industrial processes, wastes, and other anthropogenic sources has resulted in continuing developmental challenges to qualitatively and quantitatively detect lower and lower amounts of pollutants, currently at the parts-per-trillion (ppt) and even the parts-per-quadrillion (ppq) level.

Selection of Analytical Parameters

The selection of analytical parameters for a ground-water investigation is primarily driven by the purpose and objectives of the investigation, which is often affected by the site's regulatory status, existing site conditions, knowledge of past site practices, and a number of other considerations. During the past decade, transfers of commercial properties have included due diligence investigations of existing environmental conditions as a condition of sale. Several states have made these investigations mandatory.

Ground-water investigations can be done to determine the natural quality of ground water for academic interest or to evaluate its potential as a potable water supply. Alternatively, ground-water investigations can be done to determine whether chemical contaminants are present and, if so, to what extent. Regardless of the category or reason, the list of analytes may not be appreciably different because anthropogenic sources of contaminants are so widespread that ground water completely unaffected by industrial, agricultural, or municipal practices is extraordinarily rare.

A detailed discussion of the common types of investigations and the typical lists of analytical parameters that are analyzed is presented below. In most cases, the various required parameters consist of a mixture of organic and inorganic constituents in addition to measures of esthetic water-quality parameters such as color, turbidity, and odor.

Ground-Water Investigations Governed by a Regulatory Agency

There are a number of Federal regulations that have established lists of parameters for analysis of ground-water samples including: the Resource Conservation and Recovery Act (RCRA), the Safe Drinking Water Act (SDWA), the Clean Water Act (CWA), and the Comprehensive Environmental Response, Compensation, and Liability Act (CERCLA) and its Superfund Amendments and Reauthorization Act (SARA) (see Chapter 1). Individual state regulatory agencies may also have variations on these lists, separate lists, and analytical method or sensitivity requirements.

Analytical Requirements Under RCRA

RCRA was enacted to regulate activities related to the transport, storage, and disposal of hazardous wastes. As part of the overall regulation, ground water is specifically addressed. Under RCRA, typically hazardous-waste disposal and storage facilities are required to have ground-water monitoring wells. Water-quality parameters required under RCRA are divided into several categories with different requirements for replication and frequency of analysis. The parameters that indicate if ground water is an acceptable drinking water source are included in the U.S. EPA Primary Drinking Water Standards, which were established under the SDWA of 1974 (see Chapter 1). Parameters establishing ground-water quality include analytes such as chloride, iron, manganese, phenols, sodium, and sulfate. Parameters designated as general indicators of ground-water contamination include pH, specific conductance, total organic carbon, and total organic halogen.

Under certain conditions, analytical requirements under RCRA may include the analysis of a very extensive list of organic and inorganic parameters included in RCRA Appendix IX Constituents.

Many states have adopted the National Primary and Secondary Drinking Water Regulations or have modified them in part to become more stringent and applied them to ground-water investigations within the state. Although ground water may not meet drinking-water standards in all places, the objective of applying drinking-water standards is to provide a goal to which ground water should be treated in the event that it has become contaminated.

Under a Remedial Investigation/Feasibility Study (RI/FS) at a Superfund site, the standard analytical suite is presently referred to as the Toxic Compound List (TCL). Although the TCL includes many parameters, additional parameters could be added if there is information that indicates the possible presence of specific compounds at the site (i.e., waste products known to be present at the site). As part of the investigation, records of the potentially responsible parties (PRPs) and those of waste handlers are reviewed to determine the composition of materials that could be present at the site. This information should be used in making decisions on which parameters should be included in (or deleted from) the analytical scheme for the site.

Analytical Requirements Under a Site-Specific Administrative Consent Order (ACO)

The preceding sections have dealt with specific requirements for selecting water-quality parameters under several regulatory programs. These requirements have been developed to provide a broad-based analytical strategy in order to detect and measure chemical species, particularly contaminants that might be present at a site. In some instances, a regulatory agency will require the facility or responsible party to enter into an Administrative Consent Order (ACO). A list of compounds and constituents for analysis under an ACO is developed on a site-specific basis. Because of this, the benefits obtained from historical sampling and analytical events can be significant.

Analytes that are Site-Related

As indicated by the size of the RCRA Appendix IX list, the range of chemicals associated with major manufacturing categories is very broad. The Priority Pollutant list was developed from the chemicals most frequently detected in industrial wastewater effluents. However, those waste streams represent only a fraction of the total number of chemicals that are stored, handled, or discharged by industry. A comprehensive guide to industrial waste chemicals is beyond the scope of this chapter.

Selection of an Analytical Method

Just as important as the selection of the analytical parameters is the selection of the analytical method. The selection of the analytical method is in turn determined by the purpose and objectives of the investigation. For example, if the purpose of the investigation were to determine the presence of a specific organic contaminant within certain concentration bounds, the submission of samples for total organic carbon (a nonselective analysis) would not accomplish the objective. Neither would specifying an analytical method that could not obtain the required detection sensitivity.

After establishing the purpose and objectives of a project, an investigator must select the appropriate analytical methods for the parameters of interest. Quite often, the investigator may not be aware of the differences between methods. In such a case, it is important for the investigator to involve personnel with appropriate chemistry and analytical methods expertise during the planning phase of the investigation.

There may be several analytical methods that are capable of meeting project objectives to choose from for the same parameter. Each method should, ideally, give a similar result. However, due to the variables within each method, the results between various methods can vary dramatically, particularly if the method is operationally defined. For this reason, some methods may be preferred or even mandated, depending on whether the analytical results are to be prepared for, or in conjunction with, a regulatory agency.

Specific Requirements for an Analytical Method

Before ground-water samples are submitted to a laboratory, the specific requirements of the analysis, as dictated by the purpose of the investigation, must be communicated to the laboratory so the investigator does not have to assume that the laboratory understands the requirements of the investigation. Passively allowing the laboratory to conduct an analysis by its standard procedures could lead to production of analytical data that are inappropriate (or useless) for the investigation. Perhaps the most important specific requirement for ground-water investigations is the detection limit that will be reported for the requested analysis.

If ground-water samples are to be taken to show that contamination is not present, the concentration at which contaminants can be detected by current environmental technology must be specified. To say that an analyte is not present is correct only to the quantitative extent that the analysis is capable of detecting the analyte of interest. This minimum detectable level is commonly referred to as a "detection limit." In laypersons' terms, a detection limit is the quantitative point at which the analyte will be detected 99% of the time. Detection limits for aqueous samples are typically reported on a weight-by-volume basis (i.e., $\mu g/l$ or mg/l), or on a statistical basis (i.e., ppb or ppm).

An expensive ground-water sampling and analysis investigation may result in useless information if the detection limits are not low enough to accomplish the objective and satisfy the purpose of the study. An example of this is the analytical detection limit required for a risk determination. Quite often, the primary objective of a ground-water investigation is to assure that human health and the environment are not at risk based upon exposure to analytes of interest that may be present in the ground water. Accordingly, the detection limits that will be needed to accomplish these objectives are levels less than the specific human health-based criteria and environmental-based criteria for the analytes of interest. Obviously the data are of limited usefulness if the resultant analytical detection limits are higher than the most relevant health-based criteria required.

Other specific information that should be discussed with laboratory personnel prior to the sampling and analysis include sample bottle types and volume requirements, field and laboratory quality control (QC) samples, chain-of-custody, hard copy and electronic reporting (documentation) formats, and sample turnaround time.

Description of Analytical Methods

After determining the purpose and objectives of the investigation and defining the specific analytical data requirements, the analytical method can be selected. Some of the most popular references for analytical methods are Standard Methods for the Examination of Water and Wastewater (APHA, AWWA and WPCF, 1989), Methods for Chemical Analysis of Water and Wastes (U.S. EPA, 1979), and Test Methods for Evaluating Solid Waste (SW846) (U.S. EPA, 1986). The latter reference is also available on CD-ROM or from the U.S. EPA Web site (http://www.epa.gov/epaoswer/hazwaste/test/txsw846.htm). The following sections will discuss some of the more general methods available for ground-water investigations, some of the more commonly analyzed organic and inorganic parameters, and the potential benefits and problems associated with the various methods.

Screening or Diagnostic Tests

Screening or diagnostic tests are procedures that provide an initial indication of the quality of water with an economy of time and expense. Although they can seldom be used alone because they are screening methods, they can provide valuable information when sampling a large number of samples in a relatively short period of time. These screening or diagnostic test analytical procedures have traditionally been conducted in the laboratory, although over the last few years they have been more routinely conducted in the field.

Specific Organic Compound Analysis

Organic analyses are typically divided into three fractions: the volatile (VOA) fraction, base–neutral–acid (BNA) extractables (also referred to as the semi-volatile fraction), and the pesticide or polychlorinated biphenyl (PCB) fraction. To facilitate discussion of organic parameter analysis, these will be discussed by fraction. Many of the aspects discussed below are common to all organic analyses and should provide a basis for selecting an appropriate analytical method. Where applicable, the appropriate U.S. EPA method reference will be provided.

Volatile Organic Compounds

The organic fraction analyzed most frequently in ground-water investigations is the volatile fraction. This is particularly true over the last decade, with the detection of methyl-*tert*-butyl-ether (MTBE) and other gasoline-related oxygenates in ground water.

Although many of the VOCs are fairly soluble, the primary fate of VOCs in surface-water systems is loss to the atmosphere. However, many VOCs can be fairly persistent in ground water. Depending on the purpose and requirements of the investigation, different analytical methods can be applied to detect the presence of VOCs.

Because VOCs often present health and safety concerns, it is prudent to use field analytical instruments such as screening devices when sampling for these compounds. This provides a warning to the sampler as well as a preliminary indication of the presence of contamination. An example of such an instrument is the organic vapor analyzer (OVA). An OVA provides an approximation of airborne volatile organics, but is not capable of identifying specific VOCs or their individual concentrations without certain modifications to the instrument since the OVA is calibrated to a specific compound such as isobutylene. The OVA is not ordinarily used as a primary analytical method, but is more appropriately used as a screening tool: (1) to monitor volatile vapor releases when a well head is opened; (2) to assure that vapors are not present in the samplers' ambient breathing zone; and (3) to provide an estimate of relative contaminant concentrations. While this measurement may provide an indication of the presence of volatile contaminants in ground water, it can be deceiving because the measurement is of airborne levels in the well casing and not of the water itself. Other useful screening techniques using the OVA are routinely performed, including, but not limited to, headspace analysis of split-spoon soil samples during borehole drilling and monitoring well installation.

Another field analytical method for VOCs is headspace analysis by portable gas chromatography. Figure 16.1 is a schematic showing the major components of a gas chromatograph (GC). The graphical representation of the compounds as they elute from the GC column are referred to as "peaks" on a gas chromatogram, as represented in Figure 16.2. Peaks are produced during GC analysis by compounds that are present in a sample. Within the limitations and configuration of the GC, if compounds were not present, a flat baseline (without peaks) would be observed on the chromatogram. Chromatographic peaks elute in the order of their boiling points or melting points, with lighter molecular weight

Ground-Water Sample Analysis

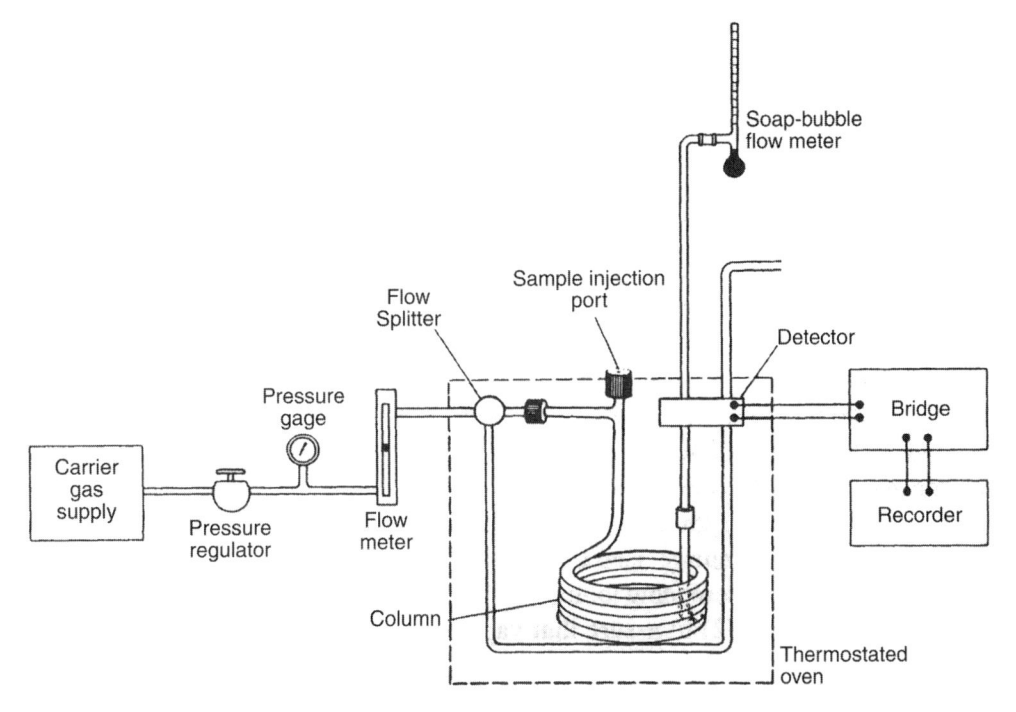

FIGURE 16.1
Schematic diagram of a gas chromatograph. (*Source*: Skoog, D.A., 1985, Principles of Instrumental Analysis, 3rd ed. With permission.)

compounds eluting earlier and heavier compounds eluting later on the chromatogram. Compounds can be selectively detected through the use of different types of detectors located at the end of the GC column. This aids in the identification of specific analytes and is particularly important when analyzing complex samples that contain a variety of organic compounds.

Electron-capture detectors (ECDs) and Hall detectors are sensitive to chlorinated compounds (e.g., trichloroethene) but are not very sensitive to straight-chain (normal) or branched alkanes (petroleum hydrocarbons) such as pentane or hexane. Photoionization detectors (PIDs) are selective to unsaturated hydrocarbon compounds such as mononuclear aromatic compounds (e.g., benzene, toluene, and xylene isomers). Flame ionization detectors (FIDs) are used as non-compound-specific detectors for assessing the presence of a variety of organic compounds.

A field-portable GC, which is commonly used in the early stages of environmental site characterization projects, is calibrated with a mixture of standards for those compounds to be analyzed. Once the GC is calibrated, retention time and response information are established for each compound of interest. A retention time is the specific point in time that a compound (peak) elutes on the gas chromatogram.

The most frequently used GC analytical technique used in the laboratory for VOCs is purge-and-trap (e.g., U.S. EPA Methods 601 and 602; see SW846) (Federal Register, 1984). While the technology of purge and trap concentrators and GCs have evolved considerably over the last decade (and continue to evolve), the basic analytical principle remains the same; analytes are effectively transferred via air sparging from the water sample to the sample headspace above the sample. Figure 16.3 presents a schematic of the purge-and-trap system. Identification of target analytes is based on a single peak that matches the retention time of the compound of interest from previously analyzed

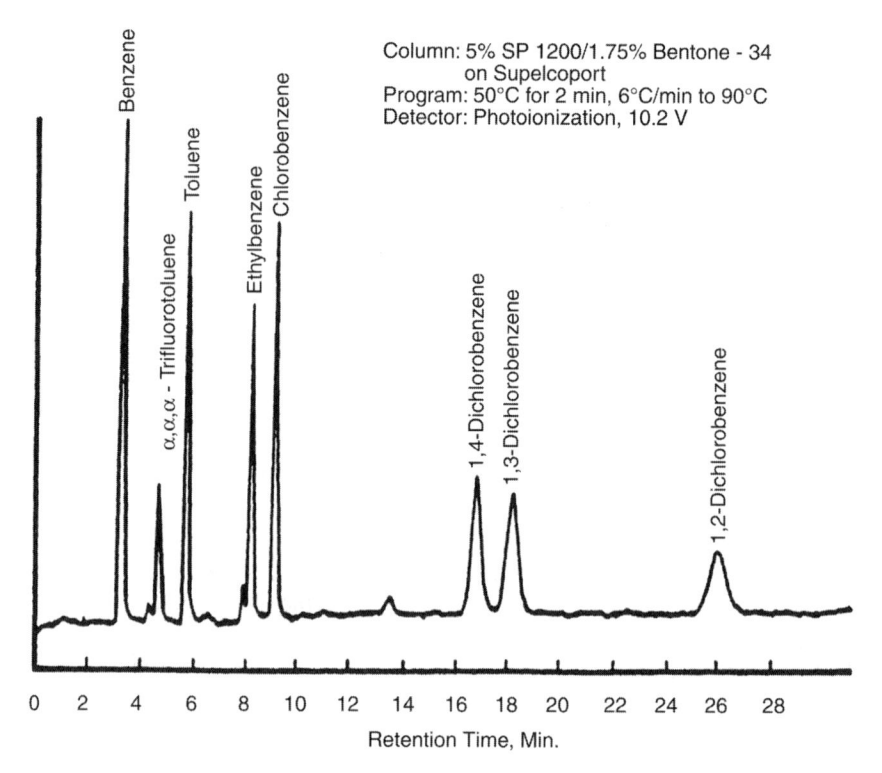

FIGURE 16.2
Gas chromatogram of purgeable aromatics. (*Source*: *Federal Register*, CFR 40 Part 136.)

calibration standards. The method detection limit for most VOCs by purge-and-trap GC is between 0.1 and 1.0 µg/l. The detection limit for some highly water-soluble compounds (e.g., ketones) may be significantly higher. Because of the volatility of this class of organic compounds, samples for this analysis are collected in 40-ml vials with no headspace

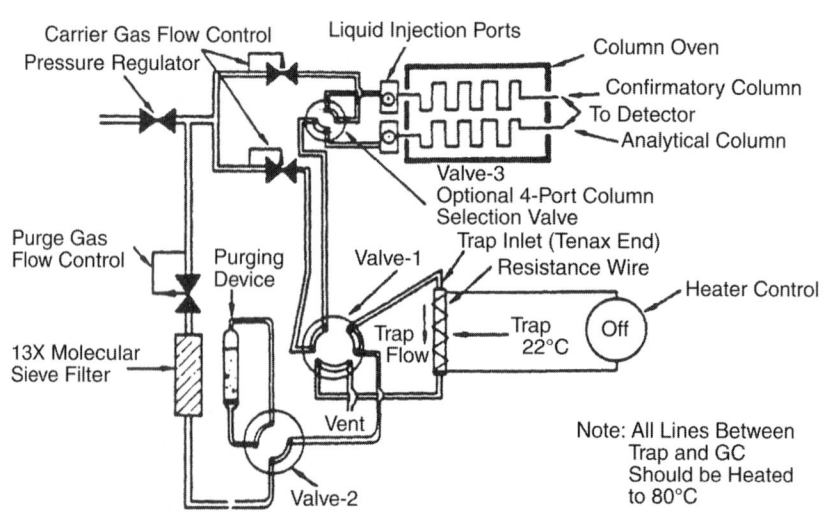

FIGURE 16.3
Purge and trap system. (*Source*: *Federal Register*, CFR 40 Part 136.)

Ground-Water Sample Analysis

(no bubbles). The bubbles will act to liberate the analytes of concern in the same way the analytical method liberates (e.g., sparges) the compounds from the water sample. Holding times are of particular importance due to analyte losses over time. These losses can be attributed to vapor losses through the vial septa, but they have also been shown to be the result of biological degradation.

While GC methods can be used successfully for the analysis of previously characterized ground-water samples, the analytical technique for VOCs that generally provides the most reliable data is a purge-and-trap concentrator, interfaced with a GC, interfaced with a mass spectrometer (MS) (e.g., U.S. EPA Method 624; see SW846) (Federal Register, 1984). This is referred to as a GC–MS (Figure 16.4). Like GCs, GC–MS technology has evolved (and continues to evolve) considerably, but the basic analytical principle remains the same. After organic compounds are separated by the GC column, they are sent through the MS. If the MS detects the presence of a primary mass ion of a targeted analyte, a response will be recorded and processed by the accompanying data system. Typically the MS will listen for the primary mass ion within a certain retention time window, based upon those established during an earlier calibration. With the exception of many isomeric compounds, the identification of target compounds is established confidently because each target compound has a unique mass spectral fingerprint. Because many isomeric compounds have identical mass spectra, isomer specificity enhancement is achieved through GC retention times. Isomeric compounds are compounds that can have several possible orientations

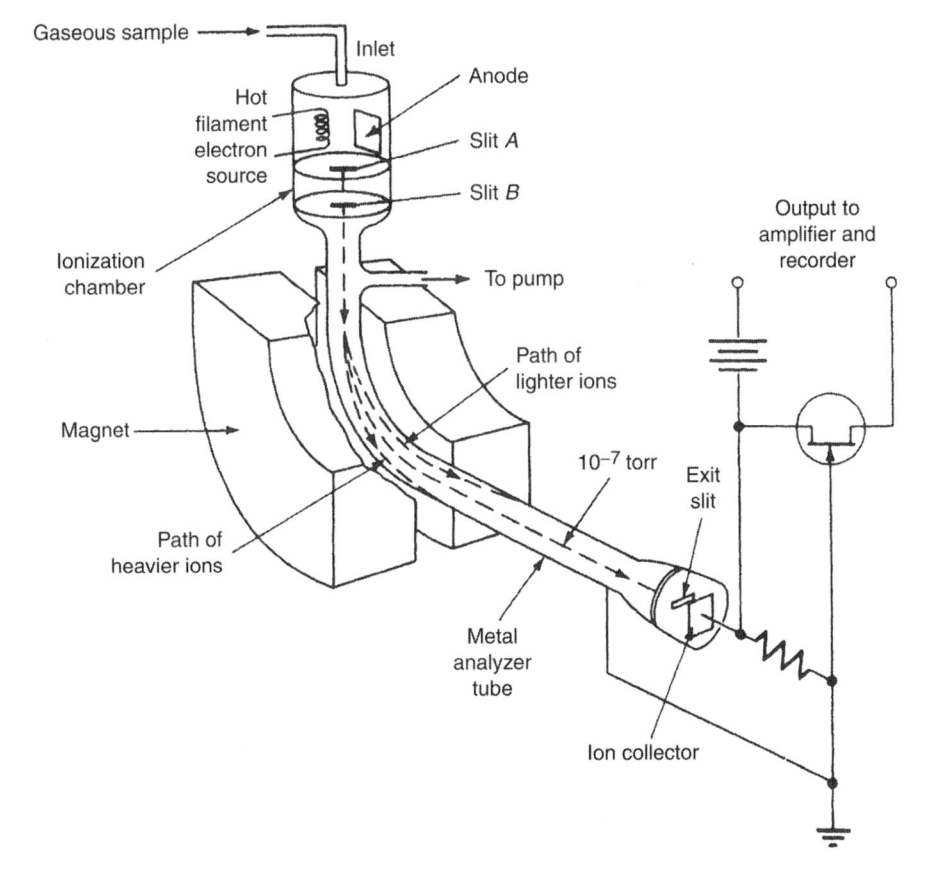

FIGURE 16.4
Schematic diagram of a mass spectrometer. (*Source*: Skoog, D.A., 1985. Principles of Instrumental Analysis, 3rd ed. With permission.)

for the same organic compound. For example, 1,2-xylenes and 1,4-xylenes are both xylene (or dimethyl benzene) isomers. Typically, the detection limits for VOCs by GC–MS are between 1.0 and 5.0 μg/l. Like GC detection limits, the detection limit for some highly water-soluble compounds (e.g., ketones) may be significantly higher.

Quite often, large peaks may be present on the chromatogram (detected by the FID), but the MS has not identified any of the organic compounds as being target analytes. Through the use of the accompanying data system, the mass spectra representing these peaks can be compared with a mass-spectral library in order to attempt to ascertain the identity of these nontarget compounds. Compounds detected during these library searches are referred to as tentatively identified compounds (TICs). These identifications should be considered qualitative to semiquantitative at best, although in some instances reasonably good qualitative mass spectral identifications are possible. It is also important to note that investigators can request these TIC mass spectral library searches to be performed years after the analysis is complete as the relevant data are captured on the accompanying data system during the analysis.

From a cost standpoint, ground-water samples collected for the purpose of quantitative volatile organic analysis should first be characterized by GC–MS techniques. This assures both positive identification and quantification. Analyses for subsequent sampling rounds can then be conducted by less expensive GC techniques. Volatile organic analysis by GC–MS is typically more costly than by GC, although exceptions to this generalization can be found when the analytical laboratory marketplace is extremely competitive.

Semivolatile Organic Compounds

Because the vapor pressures of semivolatile compounds (also referred to as extractable compounds) are lower than those observed for volatile compounds, semivolatile compounds must be removed from ground water samples via solvent extraction. Semivolatile compounds generally have lower solubilities than VOCs, ranging up to tenths of a μg/l. One important variable that governs how semivolatile organic compounds will partition into the solvent is the pH of the sample. The pH of ground-water samples is thus varied during the extraction process to ensure that the target compounds will be extracted. Hence, these compounds are also classified according to the pH at which they were extracted, being either base-, neutral-, or acid-extractable (BNA) compounds.

Ground-water samples for semivolatile organic analyses are typically prepared by taking 1 l of ground water and adjusting the pH at various points during the extraction process. The initial extraction solvent is usually methylene chloride and involves either manual (e.g., separatory funnel) or more extensive automated (e.g., continuous liquid–liquid) extraction techniques. Once the extraction is complete, the extracts are combined and concentrated (evaporated) with a gentle flow of nitrogen or one of several other currently automated solvent concentration techniques. Depending on the type of instrumental analysis being performed, the extract may be exchanged into alternate solvents such as hexane (for pesticides and PCBs by GC), acetonitrile (for polynuclear aromatic hydrocarbons [PAHs] by high-performance liquid chromatography [HPLC] or LC–MS), or toluene (for chlorinated dioxins and furans).

Generally speaking, the most qualitatively reliable analytical method for semivolatile organic compounds is GC–MS for the same reasons previously provided for VOCs. GC–MS semivolatile organic analysis is conducted by injecting microliter amounts of the concentrated methylene chloride extract onto the capillary column. The MS analysis then proceeds in the same manner as for the VOCs, including library search procedures for non-TCL compounds. The typical quantitation limit for most semivolatile compounds by low-resolution GC–MS is 10 μg/l.

There are some specialized GC–MS methods for semivolatiles that can result in substantially lower quantitation limits than 10 µg/l. Methods such as isotope dilution, single-ion monitoring (SIM), and extraordinary concentration techniques can result in parts-per-trillion (ppt) quantitation limits in ground-water samples. Similar quantitation limits can also be achieved by use of a high-resolution MS.

For certain types of semivolatile organics (e.g., PAHs and explosives), the necessity to attain lower detection limits than can be normally attained by GC–MS mandates the use of HPLC and, more recently, liquid chromatography interfaced with a mass spectrometer (LC–MS). HPLC can typically achieve detection limits between 0.1 and 0.3 µg/l. Like a GC, once the HPLC is calibrated, retention time and response information are established for each compound of interest. Identification of target analytes is based on a single peak that matches the retention time of the compound of interest from previously analyzed calibration standards. For PAHs, typically an ultraviolet detector is used with confirmation for positive results by either a fluorescence or diode array. Even with alternate detector and secondary dissimilar column confirmation techniques, as with analysis by GC, false-positive results are possible without a qualitatively definitive mass spectral confirmation.

Pesticides, Herbicides, and PCBs

Chlorinated pesticides, herbicides, and PCBs are classes of semivolatile organic compounds that can be particularly toxic at low concentrations. Because trace concentrations of these chemicals can bio-accumulate and elicit acute and chronic toxic responses in living organisms, the quantitation limits for these compounds must frequently be in the subparts-per-billion range. Most of the routinely analyzed pesticides, herbicides, and PCBs contain at least one chlorine atom (most contain many chlorine atoms), so most are effectively analyzed with the requisite sensitivity by GC with an ECD detector. As stated earlier, an ECD is particularly sensitive to chlorinated compounds. Because these compounds are semivolatile in nature, they must also be extracted and concentrated from ground-water samples by solvent extraction. Although methylene chloride is the solvent of choice for the analysis of semivolatiles, it is inappropriate for GC analysis by ECD, because the ECD is sensitive to chlorine. For the analysis of these compounds, the solvent of choice for GC–ECD analysis is hexane.

Typically, a 1-l ground-water sample is extracted with hexane (or extracted with methylene chloride and then exchanged into hexane). Microliter volumes of hexane are then injected onto the GC column. Quantitation limits for these compounds by GC–ECD are on the order of 10–200 ppt in ground-water samples.

As with the analysis of volatile organics by GC, the primary pitfall of pesticide or herbicide analysis by GC is that identification solely by GC cannot be considered qualitatively confident because the identification of most pesticides and herbicides appears as a single peak on a gas chromatogram. Although these compounds can be further confirmed by analysis on a second dissimilar GC column (i.e., a GC column with a different type of packing material), false-positive results can still be a problem with pesticide or herbicide analysis by GC–ECD.

Some pesticides, such as toxaphene, technical chlordane, and lindane, and all PCBs are mixtures of a base compound with a range of chlorination (congener compounds). The GC analysis of these compounds will result in a unique multi-peak pattern. The positive results for multi-peak pesticides and PCBs can be considered more qualitatively confident (than single-peak pesticides) because the unique multi-peak fingerprint is difficult to randomly generate chromatographically unless the analyte is truly present. Although one column will suffice for multi-peak identifications, the laboratory should be required to confirm the identification on two dissimilar columns. Interpretation of results also has to account for peak shifts or disappearances with the effects of environmental exposure.

Pesticides, herbicides, and PCBs can also be analyzed by GC–MS. These compounds can also be tentatively identified by using the same data system library search procedures as those used for chromatographic peaks from the semivolatile fraction. However, unless extraordinary methods of MS analysis (i.e., SIM, high-resolution MS) are performed, the quantitation limits by GC–MS may not be lower than the health-based criteria for most of these compounds.

Specific Constituent Inorganic Analysis

Inorganic parameters that are typically analyzed in ground-water samples for determination of the presence or absence of contaminants include trace metals and cyanide. The U.S. EPA has established specific holding times, container types, preservatives, and storage requirements for these and other inorganic parameters in ground water (U.S. EPA, 1986).

Trace metals and cyanide can exist in either a nonsoluble solid precipitate (which is immobile in the ground-water system, unless it is in colloidal form and the formation has large enough pore spaces to allow colloidal movement) or in a soluble form. Samples of ground water that are collected by methods that agitate the water column in the well (e.g., bailers or inertial-lift pumps) often contain large quantities of clay, silt, and other solids, and make it difficult to differentiate the forms of metals. In an effort to determine in what form the various inorganic parameters exist, ground-water samples are generally field-filtered through a 0.45-μm filter, though this practice does not always produce the desired result (see Chapter 15). It is important to decide early in the project whether or not samples should be filtered and how the resulting analytical data will subsequently be interpreted. If samples are not filtered, the preservation method for trace metals (addition of nitric acid to a pH of less than 2) may liberate metals sorbed to the surface of formation solids, and produce an erroneous analytical result. Addition of NaOH in a separate bottle to a pH greater than 12, to preserve the sample for cyanide analysis, does not produce the same effect, but may precipitate trace metals out of the sample.

The two methods that are commonly used for metal analysis are atomic absorption spectrophotometry and atomic emission spectroscopy. Recently, mass spectrometers have been put in tandem with atomic emission (e.g., inductively coupled plasma [ICP] or MS). These instrumental methods of analysis require that the sample be prepared so that any metals that are present will be in an ionic form. Metals are converted to an ionic form by sample digestion. Digestion of a ground-water sample is performed by gentle heating with addition of nitric or hydrochloric acid.

Atomic Absorption Spectrophotometry

Atomic absorption (AA) spectrophotometry is based upon a measured difference in electronic signal between instrumental optics induced by the sample, which is present as a gas between these optics (Figure 16.5). There are three types of AA methods: flame AA, graphite furnace AA, and cold vapor AA.

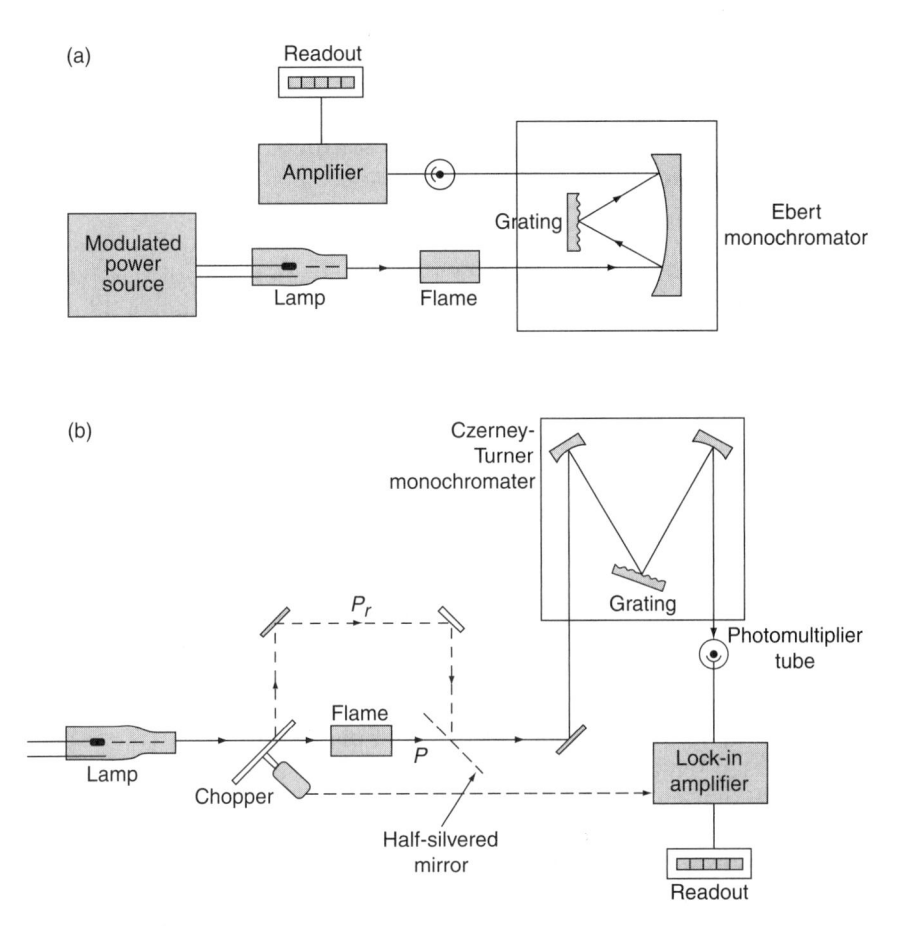

FIGURE 16.5
Typical flame spectrophotometers: single-beam design and double-beam design. (*Source*: Skoog, D.A., 1985. Principles of Instrumental Analysis, 3rd ed. With permission.)

For flame AA analysis, the digested sample is aspirated through a very thin tube, drawn by suction, and introduced into an air–acetylene or nitrous oxide flame. A beam of light from a lamp with a cathode of the metal being analyzed is focused through the flame. Depending on the concentration of the analyte in the digest, the optics will measure an electronic difference in light absorption. Flame AA is used for a wide variety of metals. Some more toxic metals such as arsenic, thallium, lead, and antimony are not sensitive to flame AA methods at concentrations of environmental concern. These elements must be analyzed by graphite furnace AA. Detection limits of 5 pg/l (picogram/l) or less can be achieved for these elements by graphite furnace AA. Typical flame AA detection limits range from 500 to 1000 µg/l for these elements.

The graphite furnace technique involves placing a staged hollow graphite tube in the path of a beam of light set at the wavelength of the analyte of interest. Microliter amounts of the acid digest are placed (or sprayed) into the entrance port of the graphite tube. The temperature of the graphite tube is increased slowly via electrode circuitry. Initially, the liquid is dried within the tube. The temperature is then increased to pyrolysis temperature that will break up various complexes in which the analyte of interest may be tied up. Finally, the temperature is ramped up to the point at which the analyte will be converted to a gaseous form (atomization). When the gas of the analyte of interest passes through the beam of light set at a wavelength unique to that analyte, an electronic difference (absorbance) is measured, which is proportional to the concentration of the analyte.

The cold vapor AA method is used exclusively for the determination of mercury, and can achieve detection limits of 0.1–0.2 µg/l. The theory of cold vapor is similar to that of graphite furnace AA, with one exception. Whereas graphite furnace AA generates the gaseous form of the analyte by a temperature increase, the cold vapor technique generates mercury gas by a chemical reaction. The generation of gaseous elemental mercury is done by the rapid addition of a liquid reagent (stannous chloride) after a complex digestion procedure is carried out in a bottle with a shaved glass stopper (i.e., a BOD bottle). The sample (still in the BOD bottle) is then purged with argon, and any gaseous mercury that is liberated passes quickly across the optics; again, the measured electronic difference is proportional to the concentration of mercury present in the sample.

Atomic Emission Spectroscopy

Just as metals can be measured by an absorbance difference in a gaseous form (some better than others), metals can also be measured by a corresponding emission. The determination of metals by emission spectroscopy can result in much lower detection limits for most metals compared to AA. The two most critical factors in detecting trace-level analytes are the temperature and stability of the flame. To this end, technology has incorporated both these factors by use of ICP and direct-current plasma (DCP) emission spectroscopy.

A plasma is a high-temperature electronic flux that exists at a temperature an order of magnitude higher than conventional flames. Plasmas are extremely stable. The pitfall with ICP (and DCP) is that high concentrations of solids (TDS), salts (sodium, calcium), and common elements (iron, aluminum) can result in severe matrix interferences.

A benefit of the use of ICP, in addition to the lower detection limits, is simultaneous multi-element analysis capability. ICP systems can analyze 15–20 metals in a water sample in a 2-min period. The operation of ICP is similar to that of flame AA. A peristaltic pump draws an acid-digested sample into a chamber, which sprays the sample into the plasma. The optics measure the difference in emissions as intensity at the wavelengths of interest and record the difference in concentration units.

Other Analyses

The preceding discussion focused on the parameters most often required in ground-water investigations in which the emphasis is determining if ground-water contamination is present. There are, of course, many other analytes that may be of interest, depending upon the purpose of the investigation. These include various nitrogen compounds (i.e., ammonia nitrogen, total Kjeldahl nitrogen [TKN], nitrate–nitrogen, and nitrite–nitrogen), sulfur compounds (i.e., sulfate, sulfite, and sulfides), phosphorus (i.e., phosphates), and other soap compounds (i.e., surfactants), cyanide, and total phenols. These organic and inorganic compounds, routinely referred to as "wet chemistries," are analyzed by a variety of colorimetric, potentiometric, titrimetric, gravimetric, and other instrumental techniques (i.e., ion chromatography).

Finally, biological analysis may be of interest in some ground-water investigations, particularly when the ground water is to be used for drinking water. One biological indicator that is commonly used is the analysis of total coliform bacteria. Some coliform bacteria are associated with ground-water contamination by septic systems (i.e., fecal coliform bacteria). Most of the biological parameters are examined by allowing the ground water to incubate with an enriched broth (or administered to agar plates) which the specific microbes of interest can use as a food source. After a designated period of time and temperatures, the culture tubes and agar plates are examined for any activity (i.e., turbidity, the presence of bacteria colonies, etc.).

The procedures for many of these other types of analysis can be found in Standard Methods for the Analysis of Water and Wastewater (APHA, AWWA, and WPCF, 1989).

Quality Assurance/Quality Control

Obviously, selecting the appropriate parameters and methods for analytes of interest are critical steps to properly assessing ground-water quality. However, just as critical is the care taken during sample analysis, the submission of check samples to "test" the sampling process, and a review of the appropriateness of the data after they have been generated. This process is collectively referred to as quality assurance/quality control (QA/QC).

Very often the ability to assess the quality of analytical data is in direct proportion to the degree of confidence that is required in the analytical measurements. For example, taking a pH reading of your garden soil just to satisfy your curiosity does not require stringent QA/QC. However, if you wanted to grow expensive plants that require pH 6–6.5 to survive, more elaborate QA/QC requirements, such as multiple samples, may be required because the measurement will be used to answer a very specific question (i.e., pH between 6 and 6.5). The degree of QA/QC that is implemented should therefore be proportional to the specific requirements with regard to the amount of confidence in the analytical measurements.

The amount of QA/QC that is implemented (bottleware, field notes, duplicate samples, blanks, etc.) is also proportional to the available funds for the investigation. It is intuitive that the costs involved with ground-water investigations will increase as the need for confidence in the data and, hence, the amount of QA/QC in the data increases.

There are various reasons to consider high-confidence analytical data important. Obviously, if expensive decisions are going to be based on analytical data, a high degree of confidence would be desirable. Other reasons to require a high degree of confidence include instances when human health or ecological risks are being assessed or when data are being used for litigation purposes.

Selection of an Analytical Laboratory

A critical quality assurance element is the selection of a qualified analytical laboratory. The selection of a laboratory should be conducted by environmental professionals who are familiar with the types of analysis that are going to be performed, specific methodologies to be used, and various other requirements that are dictated by the purpose of the investigation. The most basic requirement (which should not be assumed) is that the laboratory has the capability (instrumentation, experience, etc.) to perform the specific analyses required for the investigation.

The laboratory should have a good reputation for quality (not to be confused with good service or low prices) and must be willing to cooperate with the investigator, who should set specific requirements for sample analysis. Special requirements such as special holding times, sequence of sample analyses, and frequency of laboratory blanks, duplicates, spikes, calibrations, and data package deliverables are just a few of the requirements that need to be addressed by individuals who are knowledgeable in these areas.

One essential criterion for some projects is that the laboratory be certified under the Contract Laboratory Program (CLP) of the U.S. EPA. A common misconception is that these laboratories are "EPA-certified" laboratories — they are not. This is a contract program that requires an on-site laboratory audit and the successful analysis of a variety of samples submitted by the EPA with known analytical results. This certification is necessary for a laboratory to provide analytical services for projects conducted under the Superfund program.

Another indication of the laboratory's ability to provide quality analytical data is the various state or health department certifications that the laboratory holds. Frequently, laboratories are used outside of the state in which the project is being conducted. However, many states have their own certification programs in order to provide a basis for licensing laboratories. The larger laboratory chains hold licenses in numerous states so they are not geographically confined in terms of accepting samples. Certifications do not exclusively make a good laboratory, but they do indicate that the laboratory is capable (if required) of generating quality work. In order to gain insight into how good (or bad) a laboratory is, an on-site audit should be performed by a competent chemist.

A properly conducted audit will identify those laboratories that can generate high-quality data and those that are not worth further consideration. Additional aspects that should be considered when selecting a laboratory are service considerations, including the availability and flexibility of scheduling sample bottle delivery, sample arrival and analyses, hard copy and electronic reporting formats, and effective and timely communications. For certain ground-water analyses, it is also important to consider the location of the laboratory with regard to the mode of sample shipment, and what impacts, if any, the shipping mode will have on the required sample holding times.

Although the cost of analysis is a consideration, investigators should not select a laboratory solely on a cost basis. Beware of bargain basement prices — more often than not, you get what you pay for. An additional financial consideration that investigators must keep in mind is conducting a due diligence inquiry of the candidate laboratories prior to selection. Many commercial laboratories have been bought or sold or have closed in the middle of a project with little to no notice to their clients, leaving them in a difficult situation.

Preparation of a Quality Assurance Project Plan

For large and small investigations, environmental professionals should consider preparing a Quality Assurance Project Plan (QAPP). In many instances, when the investigation is being conducted under the purview of a state or Federal agency, the preparation of a QAPP is required before work can begin. The purpose of the QAPP is to:

- State data quality objectives as they apply to the investigation.
- State who will perform each task in the investigation (project responsibilities), including the designation of the analytical laboratory.
- Specify what protocols will be used for ground-water sampling.
- Demonstrate sample custody.
- Specify requirements for QC samples.
- Specify analytical methods for each analyte.
- Note holding times.
- Specify sample container and preservative requirements.
- Specify data package requirements.
- Provide data validation and reduction protocols.

- Specify frequency of audits.
- Schedule reports to management.

The ultimate purpose of the QAPP is to describe all the whats, wheres, whens, hows, and whys of the investigation. This is necessary so that all steps in the investigation are understood and nothing is left to the laboratory's interpretation (or imagination). Considering the demonstrated costs associated with disproving bad sample data, a QAPP is a worthwhile effort whether or not a regulatory agency is involved. Nonregulatory QAPPs do not have to be a voluminous, formal document, but rather a concise document that clearly states the needs, requirements, objectives, and logistics of the investigation.

Laboratory QA/QC

The laboratory QA/QC process begins when the laboratory ships sample containers and preservatives to the investigator. The sample containers are usually shipped in insulated shipping containers that will accommodate wet ice for cooling on the return trip. The containers provided by the laboratory will be appropriate for the analytes specified by the investigator. Container labels and chain-of-custody sheets will accompany the sample containers. The topics that follow cover the various aspects of laboratory QA/QC. Figure 16.6 presents the laboratory QA/QC process, beginning with laboratory receipt of samples. The remainder of this chapter briefly discusses each of these elements.

Chain-of-Custody

At the moment the sample bottles are released from the laboratory, a chain-of-custody routinely begins. In other cases, chain-of-custody routinely begins when samples are placed in laboratory bottleware and labeled. When samples are received by the analytical laboratory, chain-of-custody continues with the laboratory Sample Custodian acknowledging receipt of samples. For certain litigation project needs, an internal laboratory chain-of-custody can be requested in which the internal transfer of samples is documented. Chain-of-custody should be considered a fundamental requirement for all investigations. For shipping samples in coolers by a third-party courier, chain-of-custody cannot be defended without the use of custody seals on the shipping container or cooler.

Sample Storage and Holding Time Requirements

One of the most important aspects of laboratory QA/QC is assuring sample integrity and strict adherence to holding times. A holding time is the time that has elapsed from the moment the ground-water sample is collected to the moment of sample preparation or analysis. Holding time requirements are generally specified by the regulatory agency and are published in 40 CFR Part 136. Results from samples analyzed beyond established regulatory holding times should be viewed as questionable and potentially cannot be legally defended, depending on the intended use of the data.

Sample Preparation

Depending on the analysis that has been requested, sample preparation may be required before analysis can proceed. The most common types of sample preparation include extractions, digestions, and distillations. Some holding times are specified from the time of sample collection to the time of sample preparation, and others are specified from the time of sample collection to the time of sample analysis.

```
Laboratory Receives Samples
            │
Samples Logged in, Temperature and pH checked
            │
Paper Work (Chain-of-Custody, Seals) Checked
            │
Laboratory Chain-of-Custody Begins
            │
Samples stored in Refrigerators
            │
Samples Checked out-Sample Preparation begins
            │
Extracts, and/or digests assigned to analysts
            │
Sample Analysis
            │
QC Samples (blanks, duplicates, spikes)
            │
Evaluation and Reporting
            │
Group Leader Validation
            │
Data Package Preparation
            │
Laboratory Manager Signs Data Package
            │
Data Package Submission
            │
Independent Data Validation
            │
USE
```

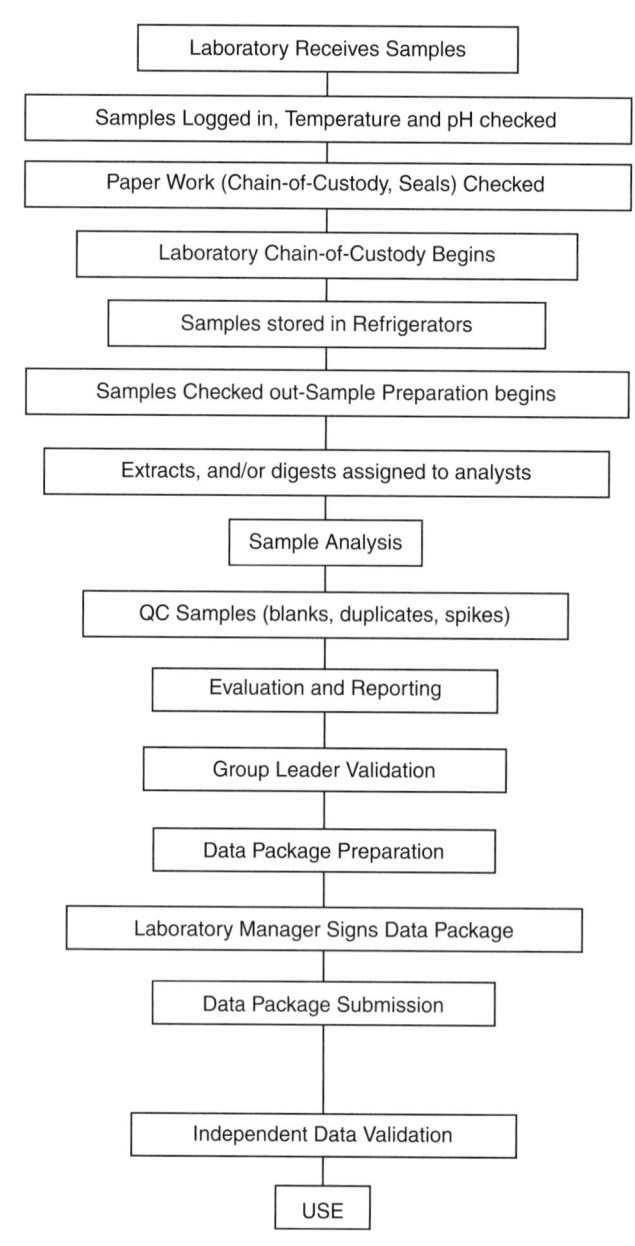

FIGURE 16.6
The laboratory QA/QC process.

Laboratory QC Samples

The analysis of QC samples is another important aspect of laboratory QA/QC and, more specifically, providing a mechanism by which data quality can be assessed. Analysis of certain types of QC samples may require the collection of additional sample volume, which is an important consideration that field samplers must be aware of. The intent of these QC samples is that they must be concurrently prepared and analyzed with the investigative samples with which they are associated. These QC samples, including, but not limited to various types of laboratory blanks, duplicates, and spikes, are described below.

Method Blanks

A method blank is a portion of deionized water that is carried through the entire analytical scheme, including all sample preparation. It is important that the volume used for method blanks be the same as the volume used for the samples. Method blanks must be analyzed (purged) every 12 h for VOC analysis. Method blanks for volatile organic analysis are typically prepared by the laboratory the same day they are analyzed; therefore, they do not monitor contaminants introduced during sample storage as holding blanks do. Method blanks are also referred to as preparation blanks.

Duplicates

Laboratory duplicates are two separate aliquots of the same sample, which have been independently prepared and analyzed for the same parameters to determine the precision of the analytical system. The analytical laboratory should perform a duplicate analysis on a minimum of one sample in 20. Duplicate results are typically compared as relative percent difference (RPD):

$$\text{RPD} = \frac{\text{Sample A} - \text{Sample B}}{\text{Average sample A} + \text{B}} \times 100$$

Spiked Samples

A spike is a sample in which the compound being analyzed is actually added (or spiked) into the sample to determine the accuracy of the analytical system. The question is, can you get back as much as you put in? The results of a spike are expressed in terms of the percent recovery with regard to the amount added. There are two types of spikes that are used by an analytical laboratory: matrix spikes (for both organic and inorganic analyses) and surrogate spikes (organic analysis exclusively).

$$\%\text{Recovery} = \frac{\text{Spike result} - \text{unspiked sample result}}{\text{Concentration added}} \times 100$$

Matrix Spikes

To determine laboratory accuracy and precision, a sample is analyzed unspiked (to determine a baseline). A second portion (or aliquot) of sample is then typically spiked with the target compounds or analytes of interest. The results (percent recoveries) of these spikes are a direct measure of analytical accuracy. A third portion of sample (matrix spike duplicate) is spiked in the same manner (for organic compounds). The comparison of respective recoveries between the two spikes is a measure of analytical precision. Depending on anticipated levels and types of parameters, the laboratory can be instructed to add additional matrix spike parameters at appropriate concentrations.

Surrogate Spikes

Surrogate spikes are added to every sample for organic analysis. A surrogate compound is a special compound, sometimes synthetically prepared and other times a compound rarely observed to be present naturally in environmental samples. Surrogate compounds are added to every sample analyzed for organic compounds to test the analytical procedure. As stated earlier, surrogate compounds are typically not found naturally. However, they are similar in structure to several routinely examined target analytes. There are three typically volatile surrogates, six semivolatiles, and two pesticide surrogates, which are added at predesignated concentrations.

Surrogate compound percent recoveries are calculated concurrently with the target compound of interest. Because the sample characteristics will affect the percent recovery, these percent recoveries indicate the accuracy of the analytical method. More importantly, surrogate recoveries are direct measures of how well the method has worked on each individual sample. If poor recoveries are obtained, the sample should be reextracted and/or reanalyzed a second and possibly third time, using less sample. However, using less sample will result in higher quantitation limits.

The "acceptable" recoveries for surrogate compounds vary depending on the individual compound and the type of compound. The U.S. EPA has generated acceptable ranges for both matrix spike and surrogate recoveries to be used as defaults. These ranges were generated from statistical manipulation involving numerous studies based on recoveries reported from many laboratories. The "acceptable" recoveries for some compounds can be as low as 11%. For some purposes, these recoveries are not acceptable, and the laboratory must be instructed to follow specific criteria for reextraction and reanalysis. One problem, which should be noted with multiple reextractions, is the required sample volume and holding time constraints. For this reason, it is always prudent for samplers to collect and submit large volumes of sample.

Instrument Calibration

Aspects of instrument calibration include: how the instruments used for analysis will be calibrated, how often the calibration will be checked, what actions will be taken if poor calibration checks are obtained, and who prepared the calibration standards. If instruments are calibrated properly using incorrect concentrations, incorrect data will be generated. The source of primary initial calibrations, calibration check standards, and the resident spiking solutions must be different (usually obtained by different vendors) so that the preparation and accuracy of each standard and spiking solution can be a check on the other.

The laboratory must perform an initial multi-point calibration to determine instrument sensitivity and linearity. The linearity of initial multi-point calibrations is typically judged on the basis of the best linear fit on either a relative standard deviation or correlation coefficient basis. Calibration checks are performed to assess instrument stability relative to the previously performed initial multi-point calibration. Calibration checks are typically judged on the basis of percent difference, percent drift, or percent recovery relative to the response of the initial multi-point calibration.

Sample Analysis

Once the instrument has been calibrated and blanks have been analyzed verifying that the instrument is free of contaminants, the analysis of samples can proceed. An important aspect of sample analysis is the analytical sequence. A highly contaminated sample may severely contaminate the instrument. Carryover or memory effects can generate false-positive results for subsequently analyzed samples, particularly during volatile organic analysis.

Laboratory Validation and Reporting

Once the analytical results have been generated, they must be validated. This is typically done by laboratory section heads or the laboratory manager. The QAPP (if prepared) should specify how calculations will be verified and document the procedures that will be used to assess how the results of laboratory QA/QC samples impact the data. The laboratory manager should sign the analytical data before results are released and, quite often, the submission of a complete (inclusive of all "raw data") data package may be required. The submission of a laboratory data package is a decision that the investigator must make. The importance of the investigation and the implications of the results should be considered when making this decision.

Documentation and Recordkeeping

To ensure the integrity of the QA/QC process, full documentation should be provided to account for all laboratory activities. The following are some of the areas requiring laboratory documentation:

- Sample log-in procedures, including assignment of laboratory control number, taking pH and temperature, and verifying field paper work.
- Internal laboratory chain-of-custody — every transfer must be entered.
- Verification of refrigerator storage, including daily temperature verification log.
- Time chronicle to verify holding times.
- Extraction and digestion instrument logs to verify analyses and analytical sequence.
- Lab narrative to describe any problems encountered during the analyses.
- Summary forms allowing brief examination of pertinent QA/QC information.
- Raw data — every item of data (i.e., standards, blanks, spikes, duplicates, and samples) relating to analysis; this includes *all* instrument printouts and copies of analysts' notebooks.
- All information pertaining to the case should be stored in a laboratory file, and all analytical data should be stored on magnetic tape or computer disk for future reference for a period of at least 5 yr.

Independent Laboratory QA Review

Whenever possible, analytical results and laboratory documentation should be independently validated by a qualified, experienced quality-assurance chemist. The QA process does not end when the laboratory delivers the data to the investigator. Laboratory chemists and managers may not give special attention to your samples to make sure that *your* results are valid. Unless an independent data review is performed to make certain the results are correct, it must be assumed that the laboratory made no mistakes. If errors did occur, this may be a costly assumption.

In addition to assessing the validity of the analytical data, an independent data review can also provide the interpretation of analytical bias. Bias is the tendency for the results to be skewed higher or lower than the actual number. For example, if spike recoveries are consistently 130% for benzene and the laboratory reports benzene in a monitoring well at 20 µg/l, intuitively it can be stated that the actual concentration of benzene in the well may be slightly less than the 20 µg/l reported.

Once the data review has been performed, a report should be prepared that qualifies certain areas of the data before the results can be utilized. Another item to consider is data presentation (data reduction). Considering the number of samples and compounds that may be analyzed during an investigation, the number of analytical results can be quite extensive and cumbersome. Typically, a preferred alternative is to reduce the data by use of computer spreadsheets to just those compounds with positive results. Specific codes that provide an indication of data reliability should be placed next to results as appropriate during the independent data validation. Once the validity has been assessed and the positive results have been tabulated, it is highly recommended that individuals familiar with the hydrogeology of the site examine the data tables. This trouble-shooting process can identify anomalies and data gaps.

Summary

The primary thrust of most ground-water investigations is determining ground-water quality. Considering the costs associated with well installation, laboratory analysis, and the implications of discovering human health and ecological threats, it is apparent that all necessary steps should be taken to maximize the results of this effort to assure the quality of the data collected. The quality assurance program starts with the design of the investigation includes the parameters to be analyzed, the analytical methods to be used, and the level of laboratory QA/QC to be followed.

References

APHA, AWWA, and WPCF, *Standard Methods for the Examination of Water and Wastewater*, 17th ed., American Public Health Association, Washington, DC, 1989.

Bellar, T.A. and J.J. Lichtenberg, Determining volatile organics at the microgram-per-liter levels by gas chromatography, *Journal of the American Water Works Association*, December, 739–744, 1974.

Federal Register, Friday, October 26, 1984, Guidelines for Establishing Test Procedures under the Clean Water Act, Part VIII, Environmental Protection Agency, 40 CFR Part 136.

Federal Register, Wednesday, November 13, 1985. National Primary Drinking Water Regulations, Volatile Synthetic Organic Chemicals, Vol. 50, No. 219, Part II, Environmental Protection Agency, 40 CFR Parts 141–142.

Skoog, D.A., *Principles of Instrumental Analysis*, 3rd ed., Saunders College Publishing, 1985.

U.S. EPA, Methods for Chemical Analysis of Water and Wastes, EPA-600/4-79-020, U.S. Environmental Protection Agency, Environmental Monitoring and Support Laboratory, Cincinnati, OH, 1979.

U.S. EPA, Test Methods for Evaluating Solid Wastes, SW-846, U.S. Environmental Protection Agency, Office of Solid Waste and Emergency Response, Washington, DC, 1986.

17

Organization and Analysis of Ground-Water Quality Data

Martin N. Sara and Robert Gibbons

CONTENTS

Introduction	1136
Baseline Water Quality	1139
Selection of Indicator Parameters	1140
Detection Monitoring Indicator Parameters	1141
Complete Detection Parameter List for Sanitary Landfills	1142
Analytical Laboratories	1142
Steps in a Lab Evaluation	1143
SOPs and QAPPs	1143
Custody and Chain-of-Laboratory Security	1145
Facility and Equipment	1145
Data Accuracy and Availability	1146
Data Inquiries	1146
QA Reports to Management and Corrective Action	1146
MDLs, PQLs, IDLs, and EMLRLs	1147
Sample Dilution	1149
Low-Level Organic Chemical Results	1150
Background Water-Quality Evaluation	1150
Monitoring Site Water Quality	1151
Reporting	1151
Significant Digits	1152
Outliers	1152
Units of Measure	1152
Comparisons of Water Quality	1152
Inspection and Comparison	1153
Contour Maps	1155
Time-Series Formats	1157
Histograms	1159
Trilinear Diagrams	1165
Statistical Treatment of Water-Quality Data	1167
Data Independence	1171
Data Normality	1171
Evaluation of Ground-Water Contamination	1172
Types of Statistical Tests	1174
Tests of Central Tendency (Location)	1177
Tests of Trend	1177

Recommended Statistical Methods 1179
Statistical Prediction Intervals 1180
   Single Location and Constituent 1180
   Multiple Locations 1180
   Verification Resampling 1181
   Multiple Constituents 1181
   The Problem of Nondetects 1182
   Nonparametric Prediction Limits 1182
   Intra-Well Comparisons 1183
     Illustration 1183
   Some Methods to be Avoided 1185
     Analysis of Variance — ANOVA 1185
     Cochran's Approximation to the Behrens Fisher t-test 1187
   Summary 1188
Reporting Water-Quality Data to Agencies 1188
References 1189

Introduction

Water-quality analyses and interpretative data summaries are important to Phase II site characterization efforts, but these data are even of greater importance under detection and assessment activities associated with facility compliance. Detection monitoring efforts are performed to verify attainment of performance objectives, and assessment monitoring is made of efforts to identify facility noncompliance in terms of nature, location, and extent of contamination. One should not lose sight of the fact that geologic conditions and observed hydraulic heads are typically more important field data than water-quality data to sort out the contamination flow paths and an ultimate remediation solution for a particular site.

   Hydrogeologists and others who make use of water-quality analyses must incorporate individual values or large numbers of analyses (data sets) into their interpretations. On the basis of these interpretations, final decisions are made regarding detection and assessment monitoring programs. In the last 15 yr, few aspects of hydrogeology have expanded more rapidly than interpretation of water-quality data at and around industrial plants and waste management facilities. The expansion of water-quality programs was based on two factors (McNichols and Davis, 1988):

- Improvements in analytical methods have greatly increased our ability to accurately and precisely analyze a vast number of trace elements and organic compounds in water. Automation of analytical processes now allows statistically significant studies of constituents that formerly were beyond the analytical detection capabilities of all but the most sophisticated instrumentation.

- The expansion of water chemistry technology has occurred in response to public and professional concern about health, particularly as related to analyses of radionuclides and trace-level organic hydrocarbon compounds.

As a result, many comprehensive programs for monitoring water quality at waste management facilities have resulted in analyses of thousands of individual parameters. Interpretation of such massive quantities of data must include attempts to determine correlations among the parameters and demonstration of correlations that exist between water-quality parameters and the hydrogeology of the site. Comparison of water quality in upgradient (background) and downgradient wells may also be necessary as part of detection monitoring programs. In the Superfund program, data are being collected by U.S. EPA regional offices, states, other Federal agencies, potentially responsible parties (PRPs), and contractors. The data are used to support the following functions:

- Waste site characterization
- Risk assessment
- Evaluation of cleanup alternatives
- Monitoring of remedial actions
- Monitoring post-cleanup conditions

In general terms, reports of water quality should contain an organized evaluation of the data, including graphics as necessary, to illustrate important environmental relationships. The recommended procedure for assessment of water-quality baseline and detection monitoring is illustrated in Figure 17.1.

The interpretative techniques and correlation procedures described herein do not require extensive application of chemical principles. The procedures range from simple comparisons and inspection of analytical data to very extensive statistical analyses. Typically the first step in evaluating ground-water quality is to review existing hydrogeologic information and try to define subsurface stratigraphy and ground-water flow. Most regulations require comparisons of data between upgradient to downgradient conditions. This is usually only useful in homogeneous aquifers that have very rapid flow (e.g., hundreds of feet per year). As will be fully explained in the following sections, more than one upgradient well is necessary to account for natural subsurface spatial variability present on most sites. When facilities are located over low-hydraulic-conductivity soils and rock that are heterogeneous in composition, additional spatial variability considerations must be addressed in the evaluation of water quality. Upgradient to downgradient comparisons for natural constituents may not be possible for those sites where vertically downward gradients predominate. These situations require sufficient background sampling points to establish the ambient spatial and seasonal variability. Landfills along hillsides often have recharge and discharge conditions that create different chemical evolution pathways and natural differences in upgradient to downgradient ground-water quality (Freeze and Cherry, 1979). In some cases, wells can be located "side-gradient" (along the downgradient directions of ground-water flow) at these sites if enough land is available to eliminate concerns about landfill impacts. The Federal regulations recognize that if a site is located on a ridge, for example, where there are no upgradient sites for wells available, then wells can be compared to themselves. This comparison is called a trend analysis or intra-well comparison.

Natural ground-water quality is known to vary both spatially between wells and temporally at a single well. Anthropogenic (or man-made) effects also contribute to the variability observed in water-quality data. To evaluate the potential releases from a facility to ground water, the sources of natural variability, and the additional interrelationships of human activities to ground-water quality must be fully understood. Sources of variability and error in ground-water data are listed in Figure 17.2.

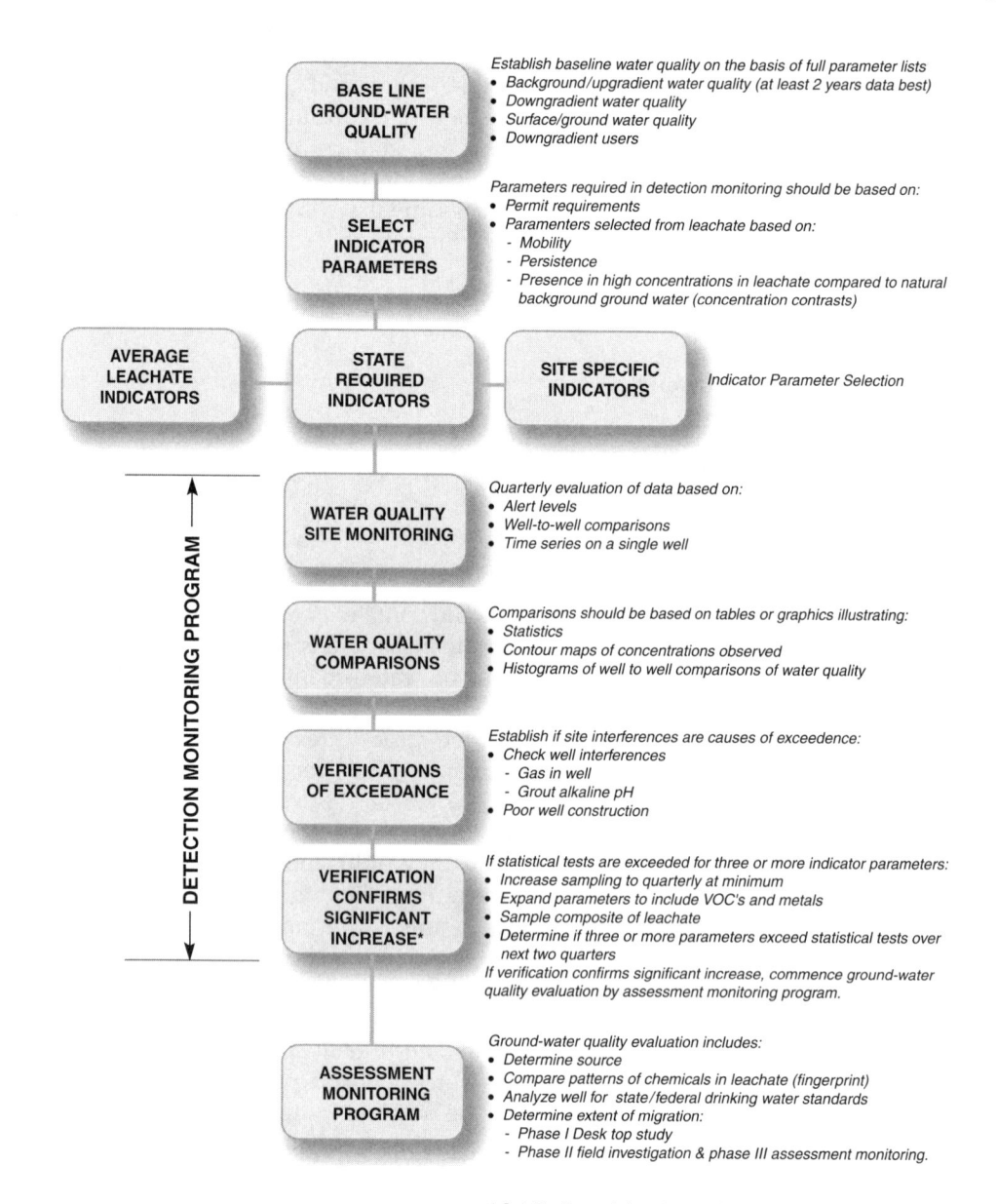

FIGURE 17.1
General water-quality assessment procedure.

Natural spatial variability of ground-water quality is often due to variations in lithology within both aquifers and confining units (Sen, 1982). Soil and rock heterogeneity may cause the chemical composition of ground water to vary even over short distances. As described in previous chapters of this book, spatial variability water-quality data may be additionally affected by variations in well installation and development methods, as well as, the sampling techniques used in the program (Doctor et al., 1985).

Temporal or seasonal effects are usually associated with annual cycles in precipitation recharge events to shallow, unconfined aquifers; these effects are especially pronounced where surface water and aquifer interactions are significant (Harris et al., 1987). Also,

SPATIAL	TEMPORAL	WELL CONSTRUCTION	SAMPLING
GEOLOGIC PROPERTIES	TRENDS	DRILLING PROCESS	COLLECTION
• Lithologic composition sorting and grain size • Structure of lithologic units • Bedding planes • Fractures (joints and faults) • Soil development • Properties of vadose zone HYDRAULIC CONDITIONS • Location of recharge/ discharge zones • Proximity of water • Presence of aquitards • Pumping OTHER • Other chemical sources • Non-point source inputs	SEASONAL • Recharge • Irrigation • Fertilization • Pesticide/herbicide application • Frozen ground PERIODIC • Short-term precipitation • Pumping • River flooding	• Drilling fluids • Type of borehole • Interaquifer transport of materials WELL DESIGN • Casing and screen material • Diameter • Screen length, depth, slot size • Filter pack material • Annular seal WELL DEVELOPMENT	• Purging methods • Purging rate/duration • Sampling apparatus • Cross-contamination between wells • Field versus laboratory measurements • Sample preparation filtering/container/ preservatives/storage time • Operator error • Incomplete well development ANALYTIC ERROR • Analytic methods • Operator experience • Instrument calibration • Interference from other constituents • Holding times • Clerical/transcription errors

FIGURE 17.2
Sources of variability in ground-water data. (*Source*: From Doctor et al., 1985. With permission.)

seasonal pumping for irrigation and high summer recharge from nonpoint pollution sources may be causes for seasonal fluctuations in background water quality (Doctor et al., 1985). A literature review on seasonality in ground-water data is presented by Montgomery et al. (1987).

The relative importance of these sources of variability is clearly site-specific. Doctor et al. (1985) observed that natural temporal and spatial variability was greater in magnitude than sampling and analytical error, unless gross sample contamination or mishandling of the samples occurs. Goals and procedures used in developing a monitoring program (i.e., baseline or detection) and descriptions of tasks are illustrated in Figure 17.1.

Baseline Water Quality

Characterizing the existing ambient or baseline quality of ground water is an important task for a number of reasons. First, existing drinking water quality standards normally define the baseline ground-water conditions, against which risks to human health and the environment are evaluated. Second, existing ground-water quality in part determines current uses and affects potential future uses of the water. In addition, determining ground-water uses is an important initial step in identifying potential exposure pathways downgradient from the site.

In evaluating the background water quality for an area, the investigator must consider possible background concentrations of the selected indicator chemicals and the background concentrations of other potential constituents of leachate. Existing chemical parameters associated with indicator chemicals (i.e., chloride or iron) or other Resource Conservation and Recovery Act (RCRA) hazardous constituents may be due to natural geologic conditions in the area; prior releases from the old, unlined landfills; or prior or current releases from other upgradient sources. Evaluation of water-quality parameters in ground water is necessary to establish an existing baseline of ground-water quality to which the incremental effects of a potential release can be added.

Measuring ambient concentrations of every RCRA-listed hazardous constituent is not feasible during most baseline studies. To adequately assess background ground-water quality, the investigation should attempt to identify other potential sources in the area (e.g., the Comprehensive Environmental Response, Compensation and Liability Act [CERCLA] sites, RCRA facilities, municipal landfills, agricultural areas or NPDES discharges to surface water) and to identify which constituents are most likely to originate from each source. Some of the background chemicals may also be site-specific indicator parameters, particularly if the facility has experienced a prior release. When determining which chemicals to include on a list of background parameters, the investigator should include all indicator chemicals described as baseline water-quality parameters in the next section.

Where sufficient data from historical monitoring are unavailable, the investigator may install a ground-water-monitoring system or expand an existing system in order to adequately assess the background quality of ground water. The design of a monitoring program should be based on guidance in previous chapters and, at a minimum, background water quality should be based upon at least two separate sampling rounds of existing or newly installed monitoring wells.

For facilities that have experienced a prior release, the investigator should also establish the results of any sampling, monitoring, or hydrogeological investigations conducted in connection with the release (if available) and should provide references to any reports prepared in connection with that release.

Selection of Indicator Parameters

The United Nations Statistical Office defines "environmental statistics" as "multidisciplinary in nature, encompassing the natural sciences, sociology, demography and economics. In particular, environmental statistics: (a) cover natural phenomena and human activities that affect the environment and in turn affect human living conditions; (b) refer to the media of the natural environment, i.e., air, water, land or soil and to the man-made environment which includes housing, working conditions and other aspects of human settlements."

Environmental indicators are environmental statistics or aggregations of environmental statistics used in some specific decision-making context to demonstrate environmentally significant trends or relationships. An environmental indicator can be a representative indicator that is selected by some procedure, such as expert opinion or multivariate statistical methods, to reflect the behavior of a larger number of variables, or it can be a composite indicator that aggregates a number of variables into a single quantity (i.e., an index).

The concept of the "indicator parameter" forms the basis for water-quality sampling programs. Because an investigator cannot include all chemical parameters that may be present in a natural or contaminated ground-water system, a selection process must be used to bring the spectrum of chemical parameters down to a workable number. These indicator parameters are selected to provide a representative value that can be used to establish performance of a facility (detection) or quantify rate and extent of contamination (assessment).

Each chemical analysis, with its columns of parameter concentrations reported to two or three significant figures, has an authoritative appearance which can be misleading. Indicator parameters in general terms must represent the movement of ground water or change in water quality in a clear-cut and understandable descriptive presentation.

Detection Monitoring Indicator Parameters

Detection monitoring programs require that individual chemical parameters be selected to represent the natural quality of the water, as well as the chemical parameters that may be changed or adversely affected through facility operation. These parameters, called "indicators," are selected with consideration of a number of criteria:

- Required by permit, state, or federal regulation or regulatory guidance.
- Are mobile (i.e., likely to reach ground water first and be relatively unretarded with respect to ground-water flow), stable, and persistent.
- Do not exhibit significant natural variability in ground water at the site.
- Are correlated with constituents of the wastes that are known to have been disposed at the site are easy to detect and are not subject to significant interferences due to sampling and analysis.
- Are not redundant (i.e., one parameter may sufficiently represent a wider class of potential contaminants).
- Do not create difficulties during interpretation of analyses (e.g., false-positives or false-negatives, caused by common constituents from the laboratory and field).

Selection of indicator parameters should consider natural levels of constituents in the detection process. Because chemical indicators include naturally occurring chemicals, Table 17.1 provides an example indicator parameter list with ranges of values occurring in natural aquifers, as well as the persistent and mobile parameters typically present in leachates from sanitary landfills.

These indicators represent a restricted selection of parameters measurable in an aquifer and limit the ability of an investigator to assess baseline water quality. However, they are the most likely parameters to undergo change when ground water is affected by a chemical release from a solid-waste management facility.

TABLE 17.1

Example Indicator Parameters for Sanitary Landfills

Indicators of Leachate	Ranges in Natural Aquifers
TOC (filtered)	1–10 ppm
pH	6.5–8.5 units
Specific conductance	100–1000 mm/cu.
Manganese (Mn)	0–0.1 ppm
Iron (Fe)	0.01–10 ppm
Ammonium (NH$_4$ as N)	0–2 ppm
Chloride (Cl)	2–200 ppm
Sodium (Na)	1–100 ppm
Volatile organics[a]	<40 ppb

[a] via U.S. EPA Method 624.

TABLE 17.2

A Complete Water Quality Parameter List

Ammonia (as N)	The volatile organic compounds
Bicarbonate (HCO$_3$)	(VOCs) established in Method 624
Calcium	Total organic carbon (TOC)
Chloride	pH
Fluorides (F$^-$)	Arsenic (As)
Iron (Fe)	Barium (Ba)
Magnesium (Mg)	Cadmium (Cd)
Manganese (Mn^{2+})	Chromium (Cr^{3+})
Nitrate (as N)	Cyanide (Cn)
Potassium (K)	Lead (Pb)
Sodium (Na$^+$)	Mercury (Hg)
Sulfate (SO$_4$)	Selenium (Se)
Silicon (H$_2$SiO$_4$)	Silver (Ag)
Chemical oxygen demand (COD)	Nitrogen, dissolved (N$_2$)
Total dissolved solids (TDS)	Oxygen, dissolved (O$_2$)

Complete Detection Parameter List for Sanitary Landfills

Although individual definitions vary, a "complete" analysis of ground water includes those natural constituents that occur commonly in concentrations of 1.0 ppm or more in ground water. Depending on the hydrogeologic setting, a complete analysis is shown in Table 17.2. In general, the investigator should examine closely the water-quality results if these indicators are above the natural ranges of ground water given above. The concentration of total volatile organics (40 ppb) was established from tolerance intervals on numerous upgradient wells at 17 facilities (Hurd, 1986) and includes cross-contamination interferences from the collection and analysis process.

Analytical Laboratories

The importance of laboratory selection for evaluation of water-quality samples cannot be overstressed. Significant legal and technical decisions, many of which will determine the success of the environmental monitoring program, depend on the quality of the lab's work. The choice of a laboratory may ultimately make the difference between a successful project and one that falls into a pattern of persistent failure, frustration, later recrimination, and resampling.

The general requirement of a laboratory program is to determine the types and concentrations of both inorganic and organic indicator parameters present in samples submitted for analysis. Depending on the project requirements, specific laboratory testing methodologies have been approved within the project scope or are specifically required. For example, under Subtitle C of RCRA, analytical methods contained in Test Methods for Evaluating Solid Waste, Physical Chemical Methods (SW-846) (U.S. EPA, 1988a) are specified.

Under the Federal CERCLA or Superfund Amendments and Reauthorization Act (SARA) program, the Contract Laboratory Program (CLP) was established by the EPA in 1980. The CLP program provides standard analytical services and is designed to obtain consistent and accurate results of demonstrated quality through use of extensive quality assurance/quality control (QA/QC) procedures.

The selection of an analytical laboratory service depends primarily on the client needs and the intended end use of the analytical data. While laboratories performing analytical services must use standard methods and employ method-specified quality control procedures, the choice of laboratory may be based on other factors, as described in the following sections.

Laboratory analyses are critical in determining project direction. Therefore, the reliability of the analytical data is essential. The use of QA/QC must be an integral part of laboratory operations and an important element in each phase of the technical review of data and reports.

Steps in a Lab Evaluation

The first step in the laboratory selection process is for the client or for the consultant to organize a detailed document defining the analytical and quality control (QC) requirements of the program determined by the project scope of work. A typical laboratory would be assigned the responsibility to:

- Evaluate the scope of the project
- Confirm its capacity to comply to the program
- Resolve identified discrepancies in the scope of work requirements
- Propose viable analytical alternatives consistent with the data quality objectives (DQOs) of the program
- Confirm project commitment to within the specified turn-around times

Assessment monitoring programs often require that a Quality Assurance Project Plan (QAPP) be approved by the responsible regional EPA office, state regulatory or other regulatory agency. The QAPP documentation describes:

- The full scope of the project field and laboratory activities
- The analytical methods to be used with their QC requirements
- Project reporting and documentation standards

An experienced laboratory will normally perform a complete and independent assessment of the QAPP and document the laboratory's complete understanding of project responsibilities.

Very large or complex projects may require data collection activity over a broad spectrum of soil and water analyses that may require multiple laboratories. These very large projects can be handled in several ways: (1) contract with additional laboratories as needed to encompass the full scope of the project or (2) contract with a primary or lead laboratory, which then has the direct responsibility to obtain subcontracting laboratory services. This is not a job for amateurs; as additional laboratories are added to the project, complexities mount rapidly that require significant experienced project management efforts.

SOPs and QAPPs

The majority of analytical laboratories have standard procedures for how the laboratory conducts its analytical quality and reporting programs just as consulting firms have standard operating procedures (SOPs) for field-testing procedures. Sample and data pathways

TABLE 17.3

Laboratory Quality Assurance Program Plan (QAPP) Guidelines

Title page
Table of contents
Laboratory and quality assurance organization
Facilities and equipment
Personnel training and qualifications
Laboratory safety and security
Sample handling and chain-of-custody
Analytical procedures
Holding times and preservatives
Equipment calibration and maintenance
Detection limits
Quality control objectives for accuracy, precision, and completeness
Analysis of quality control samples and documentation
Data reduction and evaluation
Internal laboratory audits and approvals from other agencies
Quality assurance reports to management
Document control

---

should form part of the documents provided for review from the laboratory. Simple listing of analytical procedures tells only part of the necessary documentation; sample preparation and instrumentation procedures should refer to approved methods (as designated in the QAPP or work plan). Procedures for sample handling and storage, sample tracking, bottle and glassware decontamination, document control, and other important project elements are described in the nonanalytical SOPs.

As with any quality assurance program documents, the laboratory SOPs should employ formal document control procedures so that revision numbers and dates are presented on each page. All SOPs should include the staff position performing the task, the specific analytical and quality procedures involved, and the individual responsible for resolving difficulties before taking corrective action when out-of-control events occur. Formal approval by the designated QA manager and laboratory manager should appear on the SOP permanent training documentation and include each staff member's review and understanding of the SOPs. All copies of earlier revisions of SOPs should also be retained within the laboratory documentation system.

The QAPP is the document that brings together the laboratory QA/QC plans and SOPs and specific project requirements. The QAPP should include, at a minimum, the information presented in Table 17.3. Laboratory quality systems must pay particular attention to data quality assessment and corrective action procedures. The document, through reference to the laboratory SOPs and QA/QC program, specifically addresses the laboratory's mechanisms for a program of QC samples analyzed at the appropriate or predetermined frequencies. The QC sampling requirements within the quality assurance program are usually client-, method- or contract-dependent. The QA plan should specify the mechanisms by which the laboratory identifies these requirements.

Control and reporting of analytical results are important elements of an environmental laboratory's responsibilities. Laboratory data-quality assessment procedures should include:

- General description of all data review levels
- Responsibilities at each level
- Examples of the documentation accompanying the assessment

- Analytical data-quality criteria used by the reviewers
- Final accountability or "sign-off" on the data report

The control and reporting section should also address the use of data qualifiers (tags) and whether or not it is the laboratory's policy to adjust results based on discovery data or observed blank sample contamination. Because very low levels of organic parameters can cause significant data evaluation problems, the policy and procedure used for adjustment of data by the laboratory must be well known by project staff responsible for data interpretation. As general guidance, data tags are generally preferred over data adjustment.

Custody and Chain-of-Laboratory Security

Environmental laboratories should be restricted to authorized personnel only. Security should extend to sample and data storage areas even for the smallest laboratories. The work plan applicable to the project should contain specific chain-of-custody requirements. The basic components for maintaining sample chain of custody are:

- Samples must be delivered into the possession of an authorized laboratory staff member by the sample handling or transporting organization (such as FedEx or specific sampling teams).
- Samples must be within the authorized staff member's line-of-sight.
- Samples must be locked in a secured storage area with restricted access.

Samples should be kept in locked storage with restricted access when not being processed (refrigerated, as required). The chain-of-custody form is used to document the transfer of these sample fractions (such as splits, extracts, or digestates) as part of the permanent sample-processing record.

Facility and Equipment

A quality assurance program typically contains documentation on equipment maintenance and calibration. An analytical laboratory must maintain such documentation as part of its QA/QC program. Standards used in the analytical process must also be traceable to a certified source such as the U.S. EPA, the National Institute of Standards and Technology (NIST), or commercial sources.

A very important part of the success of an environmental sampling program for state or Federal regulatory programs is the turnaround time of the sample. The turnaround time is defined as the time from field sample collection to receiving QA/QC confirmed analytical results usable for evaluating the performance of the facility. Turnaround times provided by laboratories are typically based on the current sample load and capacity, average turnaround times for data delivery, and history in meeting sample-holding times. Holding time is the maximum allowable time between sample collection and analytical testing. Each chemical parameter has a specific holding time attached to the sample, i.e., 24 h, 2 weeks, or 30 days. For most environmental-monitoring projects, data for analytical samples not meeting the required holding times will cause the results to be rejected or, at best, qualified. Exceeding holding times has caused many environmental programs to get into very serious trouble with both permit requirements and stipulated penalties for the project deliverables.

Analytical laboratories are often plagued by persistent low levels of organic parameters such as methylene chloride or acetone. These parameters are common laboratory chemicals used in various organic extraction processes. These organics often show up in analytical results as low background levels. Some laboratories commonly subtract these values from results; other laboratories report the values and let the investigator explain the results to regulatory agencies; others tag the data as background for the lab. Whatever method used by the laboratory, the investigator should expect to see such low levels of common laboratory organic chemicals in analytical results. The laboratory should report in QA/QC plans how they deal with such data.

The laboratory may purchase reagent-grade water or produce its own using a water purification system. A logbook should also be maintained to document checks for water purity, whatever the source. The product water should also be the source for QC method blanks (i.e., samples) in order to verify the absence of organic and inorganic constituents.

Data Accuracy and Availability

Reliability of laboratory-generated environmental data depends on a series of program procedures that include proficiency test samples, mechanisms for handling data inquiries, QA reporting to management, organized ways of handling corrective action, long-term data storage, and access. Initially, analytical results must be reviewed in relationship to the other analytes reported for the project. The purpose of this type of review is to attempt to identify trends, anomalies or interferences that can mislead investigators, or bias the overall use of the data. The technical review process begins with an initial review of the testing program and the overall project requirements. Once samples are analyzed according to project plans and analytical results generated, the laboratory should conduct an initial math check, a QC review, and a laboratory supervisor's technical release of the data. Reviewers consider the relative accuracy and precision of each analyte when interpreting the analytical data. Several alternative methods are available for entering results into a database. Procedures such as double-key entry and internal computer error-checking routines are employed to compare both data entries and generate an exceptions report. Data must be reviewed by qualified staff before changing any analytical or field-generated results. These procedures, along with those described below, are used to establish the reliability of the results before moving to evaluation of the actual project data sets.

Data Inquiries

The mechanisms in place for handling data inquiries are often vital to the success of a project. No matter what the length or the extent of the program, data inquiries will happen on a recurring basis. In general, procedures used in the laboratory should describe how the data are requested from storage, the individuals responsible for resolving the inquiry, and the standard response time.

Expect to see questionable data coming from even the best analytical laboratories. The laboratory should have an SOP in place for responding to client inquiries, both technical and administrative (invoicing, sample shipping logistics, requests for additional copies, etc.).

QA Reports to Management and Corrective Action

When an out-of-control incident is observed on water-quality samples from an environmental-monitoring program, it is essential that the event be documented and a form of corrective action be taken. Out-of-control events may be:

- Isolated to individual QC sample recoveries or calibration criteria failures
- Systematic — having widespread effect on the analytical data generation system

When the sample has triggered an out-of-control action, it may, for example, require reextraction or may require qualification with a notice to the data end user that identifies the criteria that were not met and the effect on data acceptability. When sufficient sample volume is not available to reprocess a sample, resampling may be required for an extreme out-of-control event.

Laboratory records should be archived so that individual reports or project files can be easily retrieved. As with any QA program, access to data must be restricted to specified individuals. If data are also stored on magnetic tape or on computer disks, the tapes or disks should be similarly protected with back-up copies stored at a second location. As part of the QA program, the resumes and qualifications of key technical staff must be maintained along with training records for the staff.

MDLs, PQLs, IDLs, and EMLRLs

Site assessment projects generate a great deal of analytical data that may be reported by the laboratory in numerous ways. These reported values often reference some form of detection limits including: method detection limits (MDLs), instrument detection limits (IDLs), practical quantitation limits (PQLs), or reporting limits (RLs). Each of these limits evolves around a detection limit. These detection limits are only a way of statistically expressing how low a particular measuring system can measure. There are a number of ways to evaluate the limit of detection (LOD) of a particular measuring device. For example, one could take an object for which the weight is known accurately, such as a 10-pound weight. The 10-pound object is weighed a series of times using a typical spring-loaded scale. The results of this process will vary depending on the temperature in the room, how the object is placed on the scale, how accurately the results are read, who reads the results, and the quality of the scale (Jarke, 1989). This is called "variability" of the measuring device.

If, for example, your results were 10.2, 10.4, 10.7, 9.1, 9.8, 9.3, 10.0, then the average value is 10.07 pounds and the standard deviation is 0.4461. In such exercises it is a good practice to carry more figures than are really significant until you make your final calculation, and then report only those figures that are significant.

The U.S. EPA's definition of MDL (40 CFR Part 136 Appendix B) describes the detection limit for this scale as 1.1 lb — any value less than 1.1 lb cannot be determined to be different from zero. Even if the scale shows a value, the significance of this value remains questionable. To obtain a lower MDL result than the 1.1 lb, one must go to a scale with a much lower detection limit to get to an accurate or reliable value.

The example of the simple weight scale is similar in many respects to any measuring device, as every measuring device has a detection limit and every device's detection limit is different depending on who, what, how, when, and where it is used. Because all of these components can vary, detection limits are not constants, especially for analytical instruments.

Every instrumental measuring device used in an analytical laboratory has an inherent minimum LOD, as described above. This LOD is usually referred to as the specific IDL. For simple devices, the IDL is based on the smallest unit of measurement that the device is capable of reporting. For example, if a ruler has markings of a sixteenth of an inch, the IDL (if based on one half of the smallest unit of measure) would be one thirty-second of an inch. While the overall concepts of IDL and MDL are quite similar, IDLs for instruments are generally far below the experimentally determined MDLs. The analytical instrument can be optimized for a specific parameter, with fewer and more easily controlled sources of variability within the IDL procedure. MCL determinations include many more sources of variability and therefore have higher experimentally determined MDLs.

In 1980, U.S. EPA began to administer the RCRA. One of the requirements of this law was that landfills begin to monitor ground water. The agency established MDL in 40 CFR Part 136, Appendix B, to ensure that analytical laboratories were conducting the testing at an acceptable level. This regulation requires that each analytical laboratory must establish MDLs on a routine basis for every analyte, for every analyst, and for every instrument. The goal of the regulation was to demonstrate that the analytical laboratories could obtain results as good as or better than those published with many of the U.S. EPA methods. The U.S. EPA MDL studies are always performed in highly purified water, with only a single known analyte added. The resultant MDLs, therefore, reflect the best performance a laboratory is capable of under the best conditions. Site assessment projects produce environmental samples that do not contain a single known analyte in highly purified water. Rather, samples are delivered to the laboratory containing many types of organic and inorganic parameters, sometimes residing in a significantly concentrated liquid. This produces a matrix effect that can significantly raise MDLs many times over U.S. EPA-reported values. Additional sources of variability presented by real samples can include sampling, site location variability, and interferences that can be caused by compounds in the sample other than the target compound. As one can imagine, the effective MDL for these field samples can be many times larger than those used in establishing laboratory performance.

Although 40 CFR Part 186, Appendix B requirements to establish MDLs are clearly explained, there is little standardization in how the regulation is applied at analytical laboratories. A full spectrum of applications of MCLs is observed applied in analytical laboratory work (Jarke, 1989):

- Laboratories perform MDL studies that meet or exceed the published values but use the published values in their reports.
- Laboratories do not perform MDL studies and assume that if they are using a U.S. EPA-approved method, then the published MDLs can be used in reporting without performing the MDL study.
- Laboratories perform MDL studies and use these as the RLs in their reports.
- Laboratories either do or do not perform MDL studies, but use RLs that are significantly different from the U.S. EPA-published MDLs, such as PQLs or RLs.

Site assessment water-quality evaluations should be based on using analytical laboratories that have performed MDL studies to verify that they can perform a method and provide QA/QC data on how well they are performing that method.

The definition of MDL includes the phrase, "the minimum concentration of a substance that can be measured and reported with 99% confidence that the analyte concentration is greater than zero." Using this definition, if an analytical laboratory reports all results above experimentally determined MDLs, 1% of reported values are false positives. False positives are statistically valid reported values. They appear to be real values, but in reality are not; therefore, many laboratories that perform environmental programs have recognized the need to set meaningful RLs. The CLP, organized by U.S. EPA to control site remediation analytical programs, has also recognized the false MDL rates for analytical data. The methods published for the CLP program use the concept of the PQL.

PQL is considered by the U.S. EPA as the concentration that can be reliably determined within specified limits during routine laboratory operation and is defined as either 5–10 times the MDL or 5–10 times the standard deviation used in calculating the MDL. This definition of an RL still raises technical questions but can be determined experimentally using statistical procedures proposed by Gibbons et al. (1988).

Additional terms have been proposed to address the ability of analytical laboratories to evaluate low levels of chemical parameters. The Environmental Committee of the American Chemical Society (ACS) published a report in 1983 addressing the issue of RLs and detection limits. Figure 17.3 graphically shows this idea. The committee used LOD instead of MDL. A new value, limit of quantitation (LOQ), was defined as 10 times the standard deviation used in the MDL calculation. This value is equal to approximately three times the MDL defined in 40 CFR 136, and is equal to the PQL. The ACS Committee reasoned that data above the LOQ could be reported quantitatively. The region between the LOD and LOQ contained results of 108's uncertain quantitation.

In summary, MDLs should be used in establishing the capability of a laboratory to perform a particular test method in accordance with regulations applicable to the project. The RLs should be established by first determining the intended uses of the data. Reporting any value above the MDL means that some analytical values will still be false positives because they fall in the region of less certain quantification. Each of these detection limit definitions can be summarized using the weight scale example (Jarke, 1989):

- The IDL is the same as the pound scale markings.
- The MDL is determined to be 1.4 lb based on one person (observer) using a single scale.
- The PQL would represent statistically what multiple scales being used by multiple people (observers) could achieve.
- The RL would be a constant value that is above the statistical variation of all people using all similar type scales.

Each type of limit is based on the population observing the operation, from the smallest IDL, where no one is observing, to the single observer (MDL), and finally to the whole population of observers (PQL and RL).

Sample Dilution

In environmental site assessment projects it is often necessary to dilute samples to either eliminate instrument or analyte interferences or to bring down large concentrations to within instrument scale. This reduces the occurrences of "blown columns" during gas chromatographic analysis. Diluting a sample fundamentally affects the MDL first. That is, if the MDL times the dilution factor is still equal to or less than the RL, then the RL remains

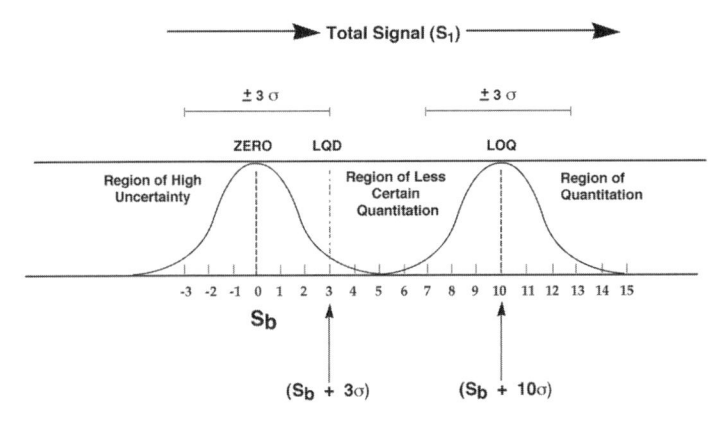

FIGURE 17.3
Relationship of limit of detection (LOD) and limit of quantification (LOQ). (*Source*: Modified from Keith et al., 1983.)

unchanged. If, however, the effect of diluting the sample results in an MDL above the RL, then a new RL must be established. This may seem to be in conflict with the previous discussion. However, if a laboratory is using MDLs as their RLs, then as the sample is diluted, both the MDL and RLs change because they are equal. If a laboratory is using the concept of an RL that is larger than the MDL, then the dilution factor should only affect the MDL until it reaches the value of the RL and then any further dilution should affect the two simultaneously. The client should only be aware of dilution when it affects the RL.

Low-Level Organic Chemical Results

Evaluation of low levels of organic chemicals in ground water presents one of the more common problems in environmental-monitoring programs. The difficulties associated with interpreting low-level analytical results for organic chemicals can be divided into three broad categories:

1. Deficiencies in sampling and analytical methods
2. Background levels for compounds that are commonly present in homes, industrial facilities, transportation facilities, and analytical laboratories
3. Varying significance as well as incomplete data on the significance of organic compounds to public health and the environment

All sampling and analytical methods commonly used for environmental monitoring are subject to variability and error. Replicate samples taken in the field from a single well or samples split in the laboratory will not produce identical analytical results due to:

- Imperfect sampling procedures
- Inability to maintain perfectly constant conditions around a sample point
- Absence of perfect homogeneity in the sample material

Replicate analysis on the same sample by the same method and even by the same analyst will not necessarily produce identical analytical results. At concentrations near the analytical LODs (typically 1.0–10 mg/l for gas chromatography–mass spectrometry [GC–MS] and lower for gas chromatography [GC]) it may be practically impossible to produce two samples that are identical. For ground-water samples, conditions in a well will vary slightly between consecutive sampling events or even during a single sampling process. When a sample is split after sampling, the two splits may not be exposed to the atmosphere in exactly the same way, for exactly the same lengths of time. Furthermore, the slightest amount of suspended solids or turbidity will most likely result in two samples that are not identical. Soil samples can show an extreme lack of sample homogeneity even from samples taken a foot away from a particular coordinate.

The key to evaluation of sampling and analytical data, therefore, is to be cognizant of the types and extent of variability inherent in sampling and analytical methods and to take into account all available QA/QC data when interpreting results.

Background Water-Quality Evaluation

Background refers to chemical parameters introduced into a sample from natural and human-related sources other than those that are the subject of the monitoring program. The problem of background changes in water quality is similar to that of analytical method variability in that it seldom is practical to eliminate it completely. There are many opportunities for a

TABLE 17.4

Examples of Laboratory and Cross-Contamination Compounds

Compound	Typical Sources
Chloroform	Chlorination of drinking water
Phthalates	Plasticizers used in numerous household and industrial products including pipes, shower curtains, car seats, many bottles and containers, etc.
Methylene chloride	Common in paint strippers, household solvents, septic system cleaners, and spray propellants; used extensively in laboratory procedures
Other solvents	Household cleaners, paints and trichloroethylene, paint strippers, septic system tetrachloroethylene, cleaners and to a limited extent toluene, in laboratory procedures dichloroethane
Trichlorofluoromethane	Common refrigerant (freon) found in freezers, refrigerators, and air conditioners

water sample to be exposed to detectable levels of both organic and inorganics at the low detection limits currently available for chemical analysis. As with the problem of method variability, the solution to background sample contamination is to first define to the practical extent the natural variability of the system, then combine these data with documentation of background levels to make reliable interpretations of analytical results.

To give a few examples, some of the most common compounds found as background levels in environmental samples are volatile organics and phthalates. Sources of these compounds include homes, transportation facilities, and analytical laboratories. Some specific examples are included in Table 17.4.

Many laboratories will not even report some of these compounds (e.g., methylene chloride) below certain levels (usually 15–30 ppb) because of assumed laboratory background levels.

Monitoring Site Water Quality

Ground-water data collected during site characterization and detection monitoring is typically restructured or simplified and must be presented in a manner that facilitates verification and interpretation. All analytical data (physical and chemical) are reported through transmittal sheets of laboratory analysis. The data are then compiled into tables and graphic formats that facilitate understanding and correlation of the information. At the very beginning of assessment activities, the investigator should establish common data requirements and standard reporting formats.

A list of all data should be provided for each sampling event and updated as new data become available. The data should include the following: well identification number or alphanumeric designation, date of analysis, name of laboratory, units of measurement, LODs, and chemical concentrations. The data are then categorized and organized into the established format to allow quick reference to specific values. Compilation and evaluation of laboratory data into summary reports must be performed without transcription errors. This task is made more achievable by use of standard formatting procedures.

Reporting

Laboratory results for a given analyte generally are presented as a quantified value or as ND (not detected). All chemical data should be presented according to this protocol. Results are reported either as a quantified concentration or as less than (<) the MDL or threshold value (thus, ND results are shown as < on the summary report). To the extent feasible, all laboratory results should be reported in a manner similar to that described above.

Significant Digits

The number of significant digits reported by the laboratory reflects the precision of the analytical method used. Rounding of values is generally inappropriate because it decreases the number of significant digits and alters the apparent precision of the measurements. Therefore, the investigator retains the number of significant digits in the transcription, evaluation, and compilation of data into secondary reports. Variation in the number of significant digits reported for a given analyte may be unavoidable if there is an order of magnitude change in the concentration of a chemical species from one round of sampling to the next or if the precision of the analytical methodology differs from one round to the next.

Outliers

Unusually high, low, or otherwise unexpected values (i.e., outliers) can be attributed to a number of conditions, including:

- Sampling errors or field contamination
- Analytical errors or laboratory contamination
- Recording or transcription errors
- Faulty sample preparation or preservation or shelf-life exceedance
- Extreme, but accurately detected, environmental conditions (e.g., spills, migration from facility)

Gross outliers may be identified by informal visual scanning of the data. This exercise is facilitated by printouts of high and low values. Formal statistical tests are also available for identification of outliers. When feasible, outliers are corrected (e.g., in the case of transcription errors) and documentation and validation of the reasons for outliers are performed (e.g., review of field blank, trip blank, QA duplicate-sample results, and laboratory QA/QC data). Results of the field and laboratory QA/QC, as well as field and laboratory logs of procedures and environmental conditions, are invaluable in assessing the validity of reported but suspect concentrations. Outliers that can reasonably be shown not to reflect true or accurate environmental conditions are eliminated from statistical analyses, but are permanently flagged and continue to be reported within summaries of data.

Units of Measure

Units of measure must be recorded for each parameter in the laboratory reports. Special care must be taken not to confuse "$\mu g/l$" measurements with "mg/l" measurements when compiling, transcribing, or reporting the data.

Comparisons of Water Quality

The type of interpretation most commonly required of hydrogeologists is preparation of a report summarizing the water quality in an aquifer, a drainage basin, or some other unit that is under study. The author of such a report is confronted with large amounts of data from a few sources and this information must be interpretable. The finished report must convey water-quality information in ways in which it will be understandable by staff of the regulatory agency and technical management staff of the client.

As an aid in interpreting chemical analyses, several approaches will be discussed that can serve to identify chemical relationships and to predict chemical changes in space and in time. Different types of visual aids, which are often useful in reports, will be described. The basic methods used during interpretation are: inspection and simple mathematical or statistical treatment to identify relationships among chemical analyses, procedures for extrapolation of data in space and time, and preparation of graphs, maps, and diagrams to illustrate the relationships.

Inspection and Comparison

A simple inspection of a group of chemical analyses generally will allow distinction of obviously interrelated parameter subgroups. For example, it is easy to group waters that have dissolved solids or chloride concentrations falling within certain ranges. The consideration of dissolved solids, however, should include consideration of the kinds of ions present as well.

Simple visual review of tabulated water-quality data is probably the most frequently used technique by regulatory agencies, to decide if a particular facility is contributing to ground-water contamination. Such analyses commonly exclude consideration of geologic and hydrogeologic conditions at the site. However, placement of water-quality data on maps and cross-sections provides a powerful tool for integration of all chemical and hydrogeologic conditions. These data can be arrayed on maps and cross-sections in a number of ways to enhance interpretation of flow paths and ground-water movement.

Figure 17.4 shows a typical tabular array of water-quality data. Because such a format requires significant efforts to assimilate, it is recommended that alternative formats be employed to display data whenever appropriate for detailed understanding of water-quality information. Water-quality display formats in increasing complexity can be divided into the following categories:

- Tabular presentation
- Contour maps
- Time series displays
- Histograms
- Box plots
- Stiff diagrams
- Scholler diagrams
- Trilinear diagrams
- Correlation coefficients
- Probability plots

Each format can have useful application for understanding variations in water quality and categorization of ground water. Tabular presentations are a necessary evil, the associated tedium of which can be eased by use of summaries and averages. Particular care

Well Number:	\multicolumn{7}{c}{X-106}
ID:	1BN1001	1YN1001	1YN1002	1YN1003	1YN1004	\multicolumn{2}{c}{1YN1005/6}
MEA Sample ID:		N1001	1WN1002	9996.11	9961.03		
Collection Date:	1993	1996	1997	1998	1999	2000	2001
Receipt Date:	26-Mar-93	04/17/96	1997	04/21/98	4/27/1999	4/22/2000	04/27/01
Extraction Date:			03/14/97				
Analysis Date:			03/26/97				05/03/01
Time:		1405	00:27:36				
Matrix:	WATER	WATER	WATER	WATER	WATER	WATER	WATER
Units:	µg/L	µg/L	µg/L	µg/L	µg/L	µg/L	µg/L
Aluminum	209 U*	120 U*	79.7	48.5 U*	67.6 U*	43.2	710
Antimony	30.0 UJ	2.8	1 U	2.8 U*	1.0 U	2.1 U	1.6 U
Arsenic	3.2	3.0 U	3.3	3.2 U	2.6 U*	2.6	3.1
Barium	49.6	59.1	53.3	63.9	27.2 J	83.0	66.4
Beryllium	1.0 U	1.6 U*	1 UJ	0.10 U	0.15 U*	0.20 U	0.40 U
Cadmium	5.0 UJ	1.0	1 U	0.30 UJ	0.59 U*	0.20 U	0.30 U
Calcium	2,570	2,470	178,000	113,000	859	3,890	365,000
Chromium	6.0 U	28.2	17.9	22.7	15.0	72.1	22.9 J
Cobalt	9.0 U	1.8	1 U	0.50 U	0.40 U	1.4	1.1 U*
Copper	4.0 U	4.0	1 U	1.9	4.6 U*	13.3	6.9 J
Iron	112 U*	307 J	596	434	63 J	206	1,250
Lead	4.8 U*	2.7 U*	1 U	1.5	1.0 UJ	1.3 U	2.5 U*
Magnesium	28,500	587	90,800	53,100	33,800	105,000	193,000
Manganese	52.7	608	138	379	12.5	164	142 J
Mercury	0.20 U	0.20 U	0.2 U	0.10 U	0.05 U	0.10 U	0.10 U
Nickel	13.0 U	15.0	13.9	26.2	21.2	53.7	26.4 J
Potassium	36,400	41,100 J	24,400 J	16,300 J	15,800 J	56,500 J	59,200 J
Selenium	3.0 UJ	4.2	2.2	3.2 UJ	2.5 U*	10.4 J	9.9 J
Silver	5.0 UJ	1.0 U	1 U	5.8 U	0.40 U	0.60 U	0.70 U
Sodium	78,000	105,000	97,200 J	60,700	49,000	141,000	120,000
Thallium	3.0 UJ	3.8	3 U	4.1 U	2.1 U	3.2 U	3.5 UJ
Tin	26.0 U						
Vanadium	4.3	13.6	15.1	12.7	53.1	65.6	68.9
Zinc	5.0 U	11.6 U*	11.8 U*	0.50 UJ	0.40 UJ	0.80 UJ	1.0 UJ
Cyanide	5.0 U	10.0 U	10 UJ	10.0 U	4.7 U	6.7 U*	7.2 U*
Chromium (VI)	10 U	100 U	10 U	50 U	10 U	1,000 U	1,000 U
Sulfide	130,000 J	10,600	1,000 U	2,000	200 U	200 U	1,400

Key:
J = The concentration is approximate due to limitations identified during the quality assurance review
U = Indicates the compound was analyzed but not detected. The associated value is the sample quantitation limit
U* = The compound should be considered "not detected" since it was detected in a blank at a similar concentration level
UJ = Indicates the compound was analyzed but not detected. The associated value is an estimated sample quantitation limit based on a bias identified during the quality assurance review
R = The results were considered unusable during the quality assurance review
Blank = The compound was not analyzed for

(1) Values shown are the highest detected between the investigative sample and its unpreserved, duplicate, reanalysis, or dilution sample.

FIGURE 17.4
Typical water-quality tabular data set for inorganic parameters.

should be taken in proofreading sets of compiled or merged data, as massive arrays of data almost always contain errors of transcription. Computer-based spreadsheets can decrease time for data reduction; however, any transcription of data must be carefully checked and rechecked for accuracy.

Organization and Analysis of Ground-Water Quality Data 1155

Contour Maps

Presentation of water-quality base maps has been traditionally handled through contouring of data. The technique of mapping of ground-water quality by drawing lines (isocontours) of equal concentration (isograms) of dissolved solids or of single ions has been used in the scientific literature since the early 1930s (Hem, 1970). The applicability of constructing isogram maps depends on several factors, such as:

- Homogeneity of water composition with depth
- Parameter concentration increment between measuring points

Restriction of sampling point density (i.e., insufficient data points) in either a vertical or horizontal direction will limit the usefulness of this technique. However, if the detection or assessment monitoring system at a typical facility is designed using procedures discussed in Chapter 8, it should provide sufficient data points for construction of isocontour maps. Contour maps can contain either closed isopleths, as shown in Figure 17.5, or open gradient lines, as shown in Figure 17.6. Both these contour maps show isocontours of chlorides. Because chlorides are typically not affected by precipitation or by other reactions that would lower concentrations (decreasing only by dispersion and dilution), this parameter serves as one of the best inorganic parameters to use in contour formats.

Additional parameters such as conductivity, temperature, chemical oxygen demand (COD), or any dissolved parameter with sufficient data density can also be displayed on contour maps. On occasion, lumped organic parameters, such as total volatile organic compounds (VOCs), can also be contoured. Figure 17.7 shows such a presentation.

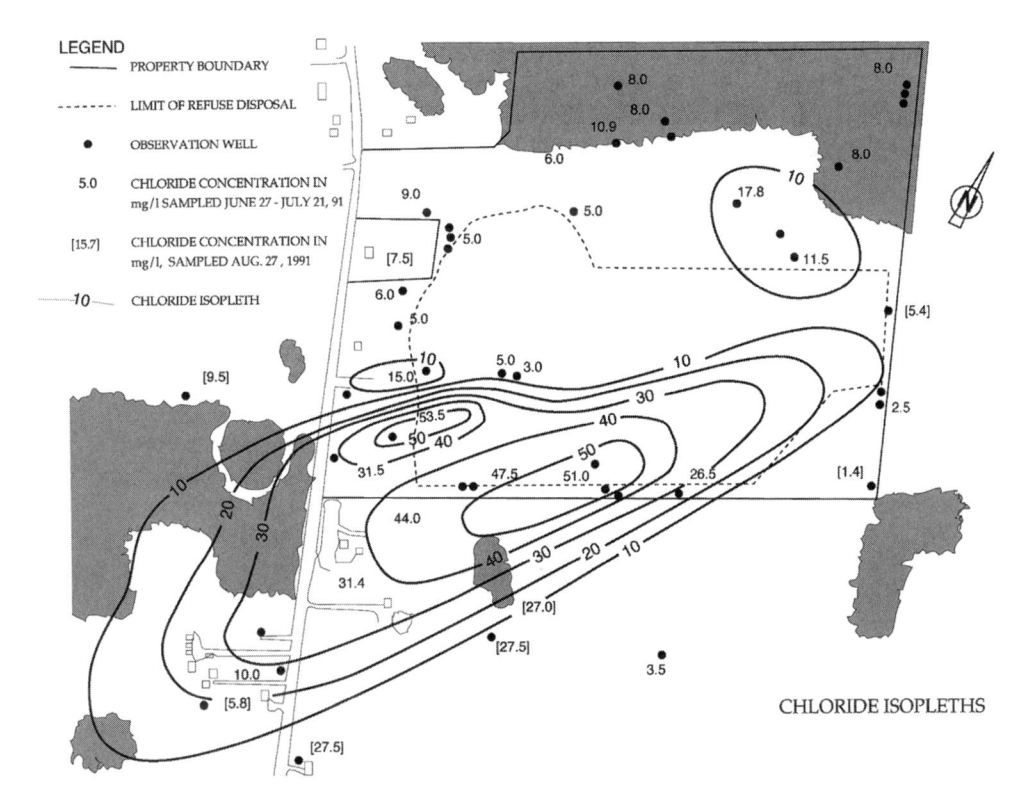

FIGURE 17.5
Closed chloride isopleths (isocontours) at a waste disposal site.

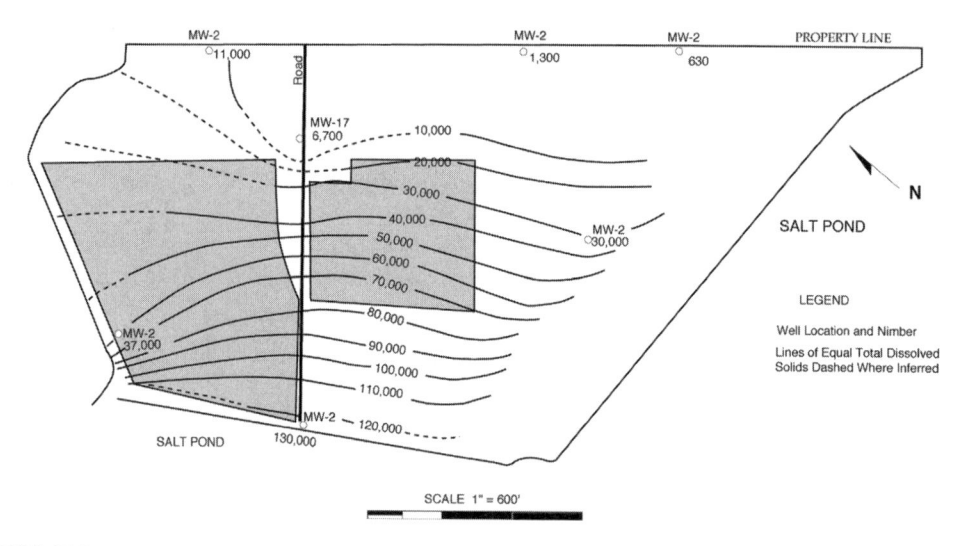

FIGURE 17.6
Open total dissolved solids isopleths (isocontours) at a waste disposal site.

Organic parameters in ground water are difficult to contour effectively due to the typically wide ranges observed in water-quality tests. However, water-quality data from highly concentrated sources, such as product spills or very large-volume, low-concentration organic sources (such as an unlined codisposal facility), may be amenable to such presentations. Questionable data should always be represented by dashed lines on the illustration.

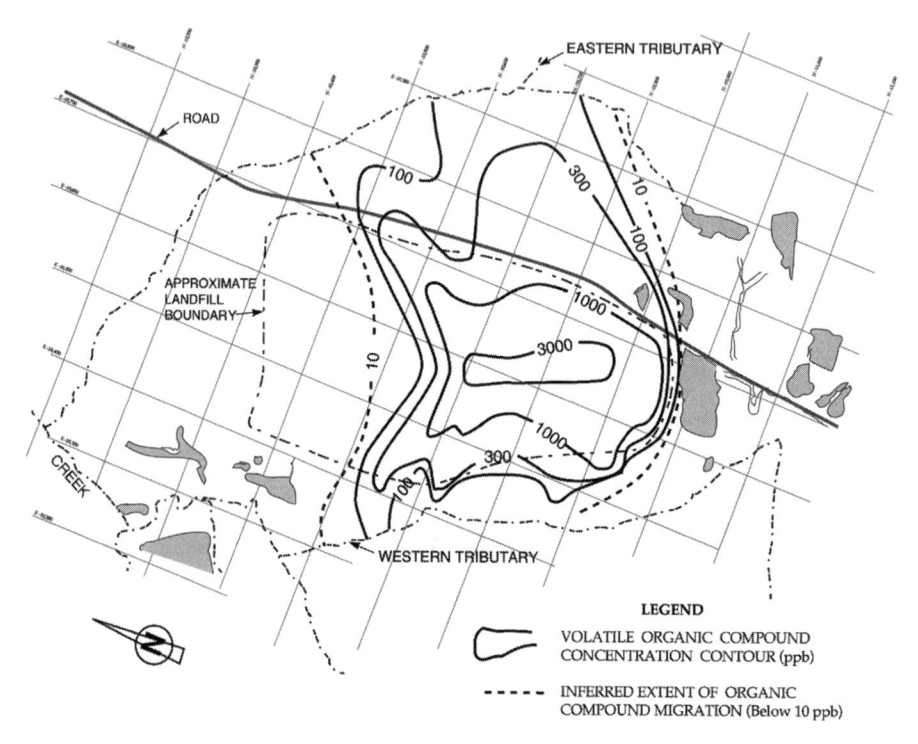

FIGURE 17.7
Isocontour map showing total VOC concentrations.

Time-Series Formats

In water-quality evaluations, there is always a continuing interest in observing parameter concentration change over time. To record such data, the standard approach is to make a series of observations at fixed intervals of time — this describes the time-series format. Such time-series formats have the objective of obtaining an understanding of past events by determining the structure of the data or predicting the future by extrapolating from past data. Those responsible for managing data collection systems can appreciate the difficulties of collection of environmental data at regular time intervals.

The variable (the data point) may be directly related to a defined time interval, such as the high and low temperature for the day. Environmental data may also be continuously changing, as would apply to measurements of hydraulic head in a piezometer. These observations are actually samples of instantaneous values, but are expressed as averages over the measured time interval. Readings taken once per day of a rapidly changing variable establish only a single point on a curve that can vary significantly until the next measurement. Fortunately, water-quality variables obtained from ground water do not vary significantly on a short-term basis due to the typically slow movement of ground water in granular aquifers. Fractured or Karst bedrock may, however, show much faster reaction times both in hydraulic head level changes and variations in water quality. Most detection ground-water monitoring programs sample on a quarterly basis. While a case can be made for somewhat shorter (or longer) sampling periods at some sites, based on ground-water flow rates, these four-times-per-year sampling programs represent a standard period for time-series analysis.

The first step in evaluating a time series is to determine if any structure exists in the data. Structure can be defined as the data behavior at a particular point in time being at least partially predicted by its value at other times. These structure elements in the data can be evaluated by:

- Defining a trend in the data (i.e., do the data increase or decrease with time), using straight lines, higher-order polynomials, or exponential curves.
- Testing for isolated events or unexpected departures from the normal behavior of the data set. This has specific applications for detection-monitoring programs where departures from long-term trends can force environmental programs into assessment actions.

Water quality at a single collection point such as a well or spring should be expected to change with time. Even with the generally slow movement of ground water, long-term detection or assessment monitoring programs can show gradual changes in water quality. These changes can be best illustrated by time-series presentations. Time-series diagrams can be used to compare individual parameters with time (i.e., compare water quality in a well against itself) or can illustrate changes in multiple parameters with time or changes with time for a common parameter in multiple wells. Figure 17.8 shows a comparison of total dissolved solids in a number of wells. All six wells are compared to each other at any displayed point in time. Figure 17.8 also shows changes in TDS with time for each well. Time-series presentations can be ineffective if too large an amount of data is presented on one plot. Figure 17.9 shows a time-series plot for chloride in eight wells; although only a single parameter is displayed, the variable Y-scales used in the presentation make interpretation of trends difficult. Time-series presentations are most effective when single parameters are compared, as shown in Figure 17.10. This illustration includes water-level elevations with chloride concentrations. Whether or not the water-level elevation is related to the chloride

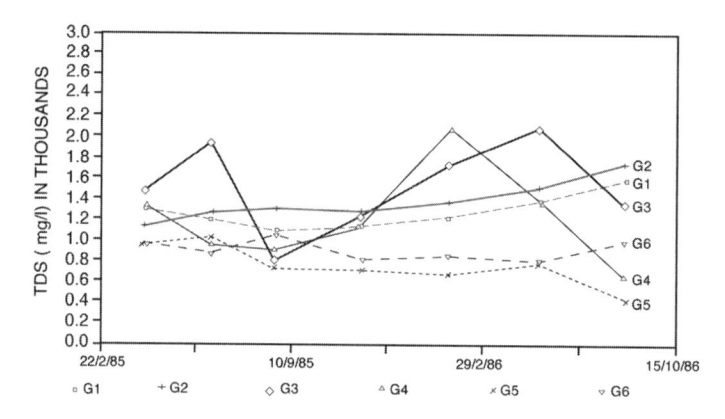

FIGURE 17.8
Time series comparisons for total dissolved solids for six wells.

concentration is a separate question; however, the data are displayed in an easily understood format.

A similar data set is presented in box-and-whisker plots (or "box plots") in Figure 17.11. Box plots are useful statistical tools for evaluating changes in water quality. Complicated site evaluations may require a series of box plots. For example, all wells screened in a hydrostratigraphic unit may be combined on a single plot or data from a number of well "nests" may be shown on one plot to illustrate vertical trends in water quality. The box plot can be considered as an economical graphical method of presenting the constituent summary statistics. The boxes are constructed using the median (middle value of the data) and the interquartile range (IQR; the range of the middle 50% of the data). These plots separate the results of each well and can clearly show the difference in the data distributions. These plots are generated by ranking the data and may be constructed in a number of different ways (McGill et al., 1978). Some box plots constructed by various software programs use the median and the F-spread. The F-spread, or fourth spread, is a function of the data distribution and measures the variability in the water-quality results, similar to the standard deviation. Hoaglin et al. (1983) provide a full discussion of these order statistics. The median and IQR are analogous to the more commonly used mean and standard deviation of a set of data. The mean and median are measures of central tendency or location, whereas the standard deviation and IQR are measures of variability.

Typically the first step in evaluating ground-water quality for box plot presentations is to review existing hydrogeologic information and to try to define subsurface stratigraphy and ground-water flow. The next logical step is to graph the chemical data as concentration versus time-series plots.

Figure 17.12 shows a chart where the mean values (solid circle), ±1 standard deviation error bars (vertical line), are plotted for each well next to each box plot. The plots show that the mean for the data is consistently greater than the median. The two standard deviations for the data are larger than the IQR. High values otherwise described as "outliers" inflate the estimate of the mean and standard deviation in these statistical plots. The median and IQR are based on ranks and are not particularly sensitive to outlying values. Similar to Figure 17.11, the high variability in the impacted data is revealed by the wide error bars.

The box plots are considered more powerful in illustrating impacted water quality than simple error-bar plots because they contain more information about the actual data distribution. The error-bar plots, however, can be applied to parametric statistics evaluations.

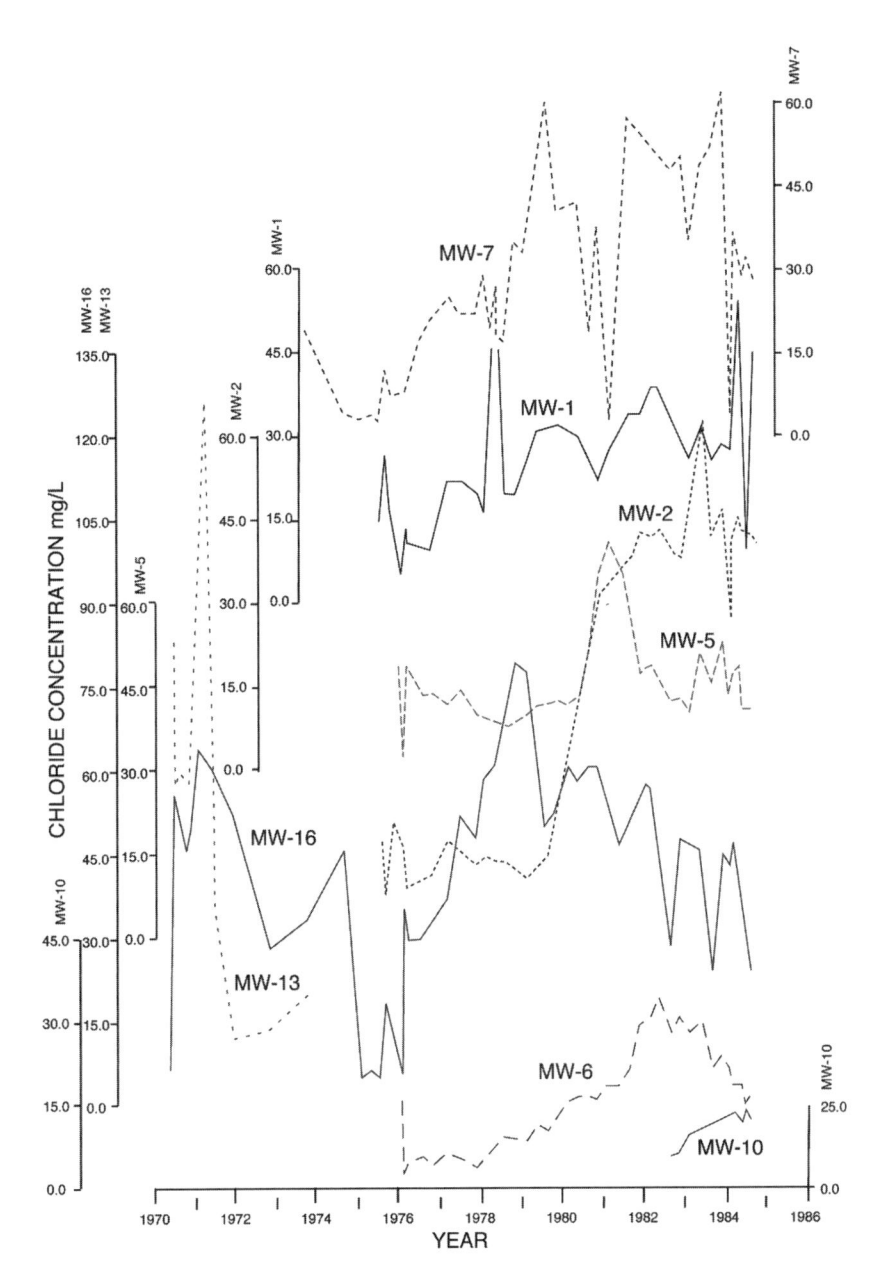

FIGURE 17.9
Time series comparisons for eight wells with sliding scales.

Histograms

The histogram is a two-dimensional graph in which one axis represents the data and the other is the number of samples that have that value. The X- or Y-axis of the plot is frequency expressed in terms of the percentage of total samples, rather than as an absolute count. The process of creating a histogram is primarily a counting process. A number of classes or groupings are defined in terms of subranges of the numeric value. These may be set to cover the complete range of the project data or a restrictive range derived from the mean and standard deviation or from knowledge of data ranges from previous

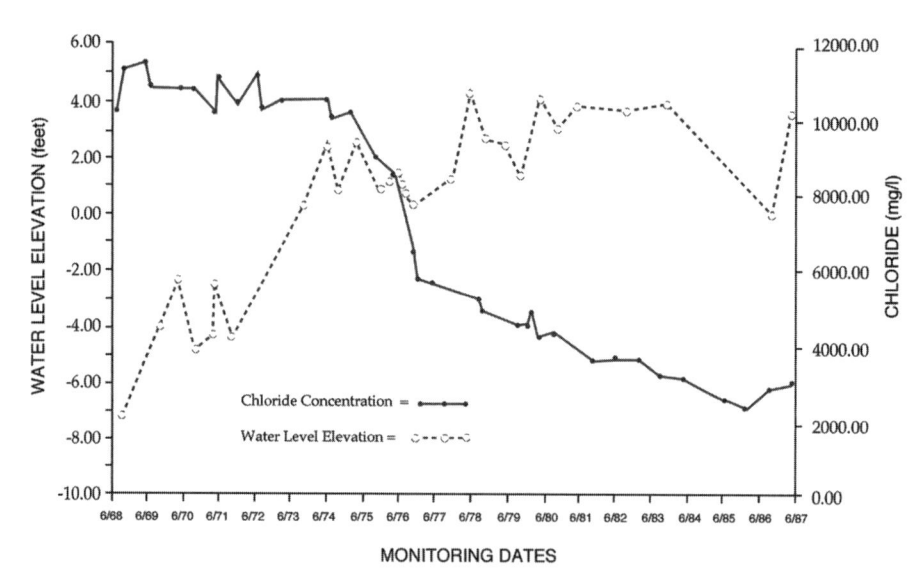

FIGURE 17.10
Time series comparisons for chloride and water levels. (*Source*: Hydro-Search, Inc.)

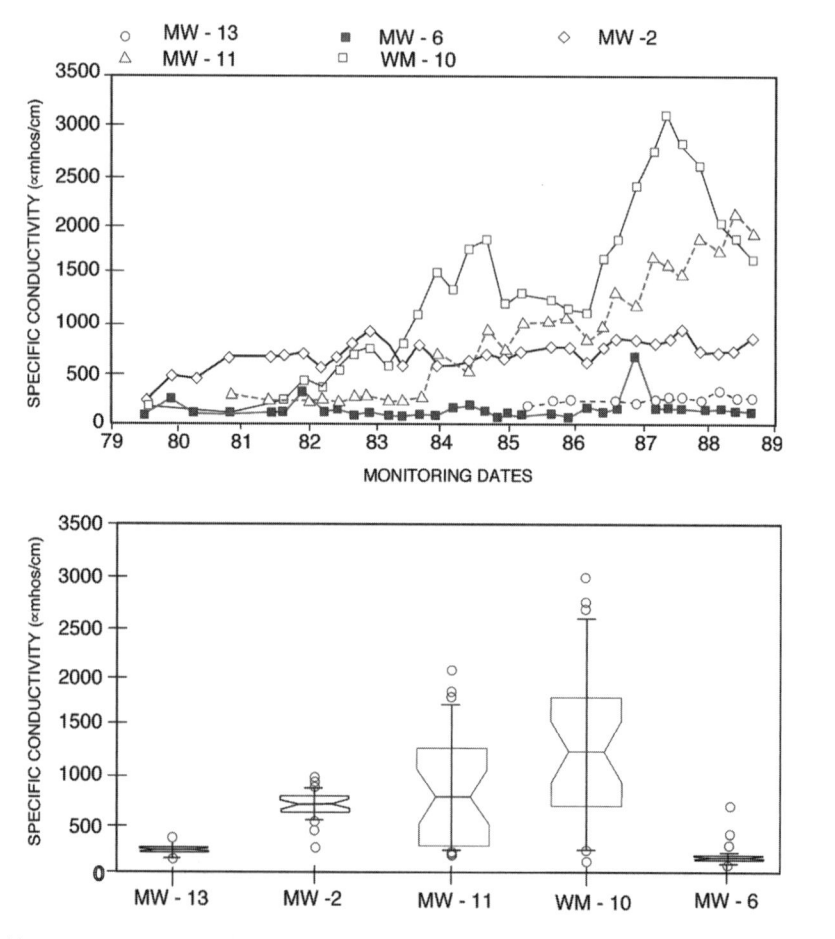

FIGURE 17.11
Time series comparisons with box plots.

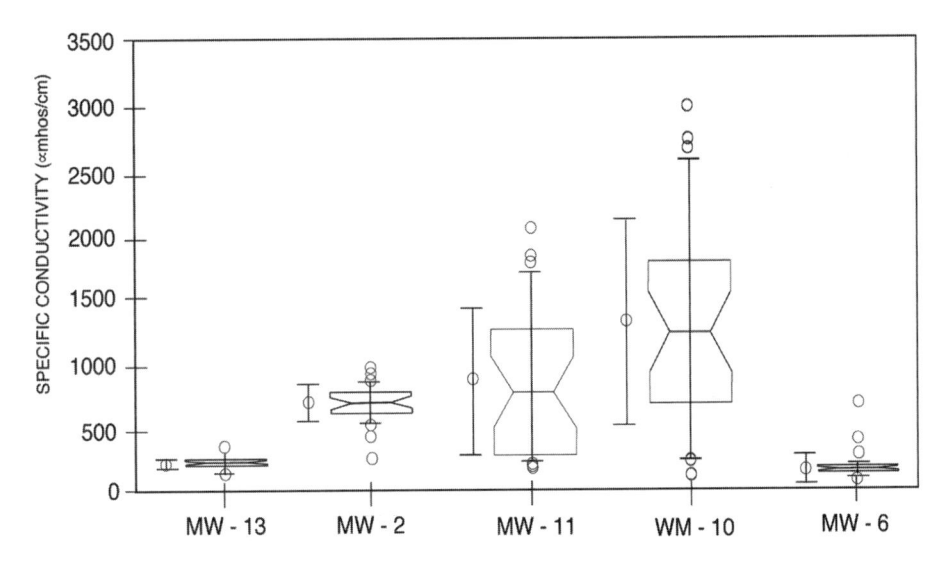

FIGURE 17.12
Error bar plot compared to box and whisker plots.

project data evaluations. With many computer-based spreadsheet programs offering automated histogram production, these project data can be quickly plotted in a histographic format to evaluate the appearance of the figure.

Even with automated histogram production, the basic usefulness of the display can be enhanced by changing the parameters that influence the appearance of the histogram (Green, 1985):

- The range, which includes the minimum and maximum values
- The number of classes used in the counting
- The size of a class, such as the range of numeric values treated as a unit in the value counting
- Transformations of numeric values including scaling, logarithmic, and exponential

As a general guide, at least one histogram should be produced that covers the complete range of data values to evaluate samples outside the main distribution of data sets. It is recommended that all extreme values be investigated as errors or true anomalies. A common problem with histograms that have broad ranges is that the resultant figure will have poor resolution. The majority of the results in these displays are combined into one or two classes, obscuring the details of the distribution. Exclusion of the outliers results in better resolution of the main data sets.

Plotting of chemical data as a series of comparative histograms (or bar graphs) has been a traditional methodology for representation of variability in water quality. Most of the traditional methods are designed to represent the total concentration of solutes and the proportions assigned to each ionic species (for one analysis or group of analyses). The units in which concentrations are expressed in these traditional diagrams are milliequivalents per liter (meq). Hem (1970) provides descriptions of bar graphs, radiating vector plots, circular diagrams, and stiff diagrams — these methods will not be discussed here. Water-quality data collected during detection or assessment monitoring programs traditionally have not been portrayed in a format of "whole" analysis, that is, with anions and cations given in units of milliequivalents per liter. Rather, results of water-quality analyses are presented in milligrams or micrograms per liter and presented in formats including only a few parameters. These data, especially volatile organics and hazardous metals, have been displayed as histogram fingerprints illustrating variations in water quality. Figure 17.13 shows a series of histograms of hazardous metals obtained from analyses of water in individual wells. Similar histograms have also been used to track plumes of VOCs and to compare relative proportions of organic species in water from individual wells.

Tabular summaries of constituents are another form of comparative histograms. Figure 17.14 shows a summary table used to compare organic parameters observed in leachate with organic parameters observed in off-site wells. Many of the constituents in the fingerprint of the landfill leachate are different from those in the off-site monitoring well and thus tend to indicate a nonrelationship. Care must be taken to use indicator parameters that will not change with time and therefore provide a misinterpretation of the water-quality fingerprint.

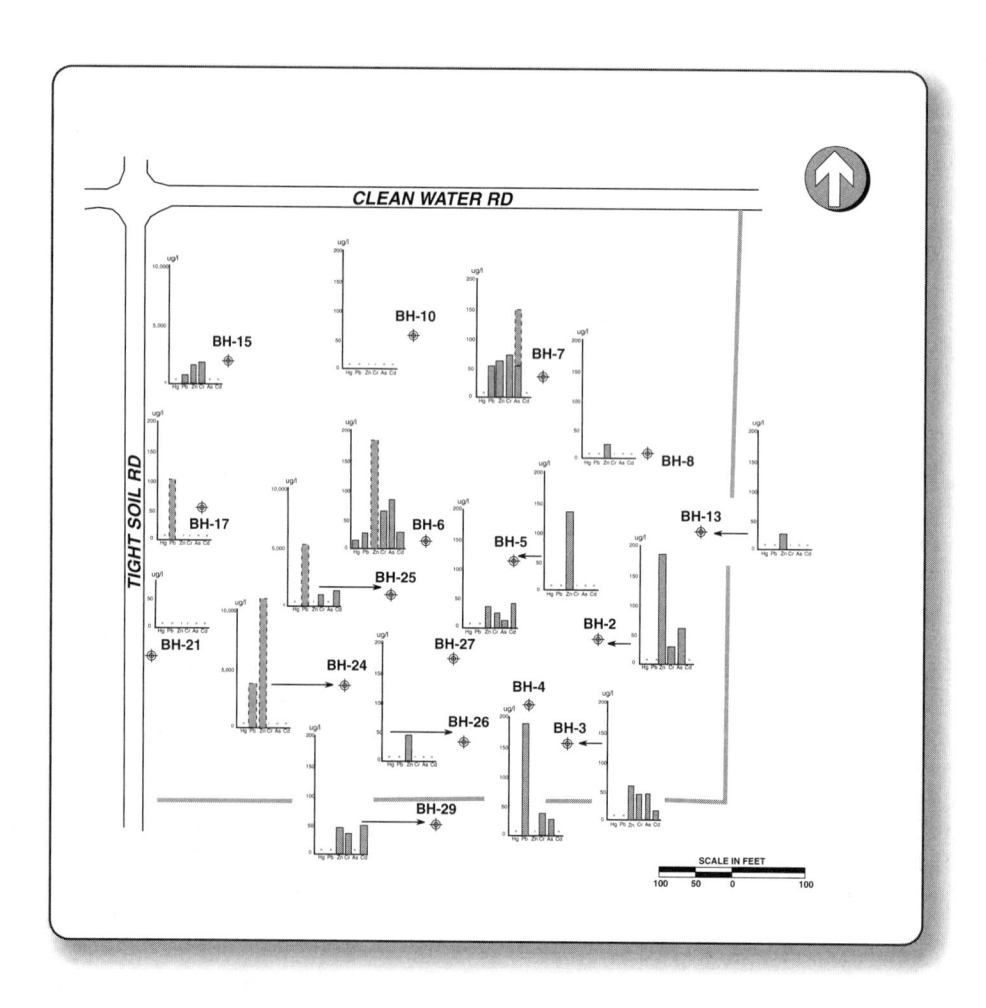

FIGURE 17.13
Histograms of metals data for individual boreholes.

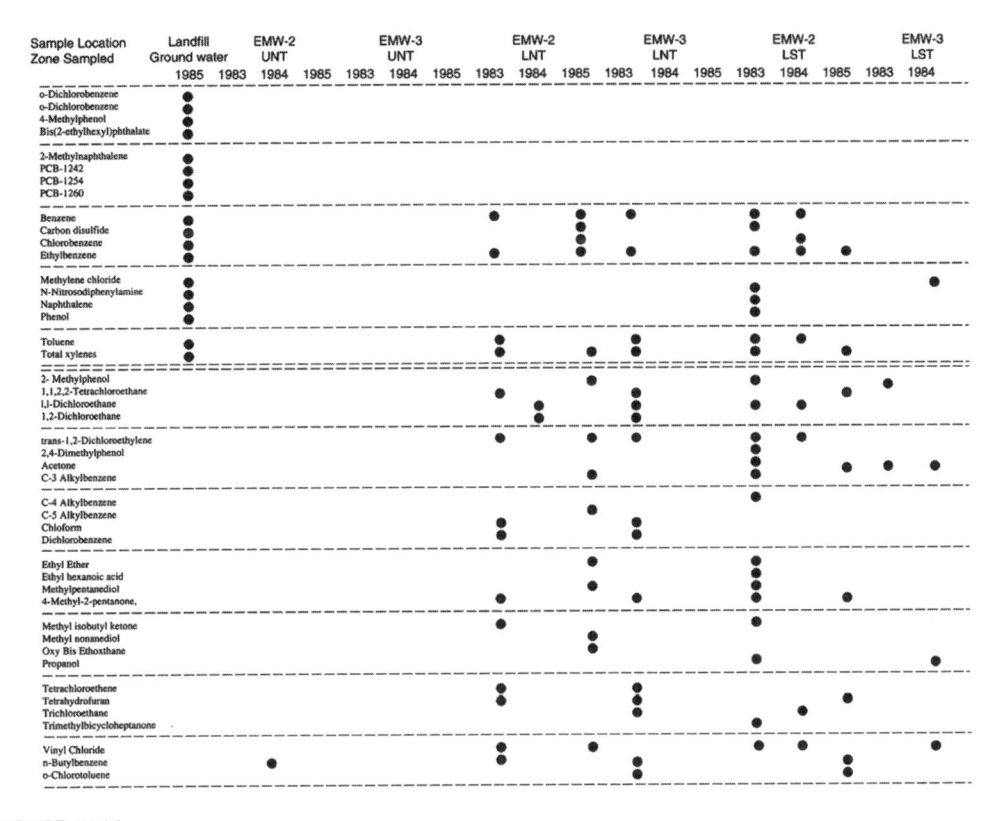

Organization and Analysis of Ground-Water Quality Data 1163

FIGURE 17.14
Histogram of leachate data.

Additional graphical displays of histograms are shown in Figure 17.15. The data shown in Figure 17.15 illustrate over a thousand observations of specific conductivity for two wells. These histograms can be compared to a lognormal distribution (Figure 17.15c) and normal distribution (Figure 17.15d). The histogram construction format for large numbers of observations can be used to investigate the probability distribution of the data. In general terms, the histogram plots values where the higher the bar, the greater the probability that additional measurements will fall in this range. Therefore, the more the sample values are incorporated into the histogram, the closer the graph is to the "true" population distribution. Many statistical tests used in evaluation of water-quality data require knowledge of whether the data come from a normally distributed population. The plotted data distribution illustrated on the histogram can be compared to a normally distributed data set. This provides a qualitative evaluation of the assumption that a normally distributed population is truly represented in the displayed environmental data.

The example project data sets (Figure 17.16) shows that neither of the wells have normally distributed data; both sets of data are skewed to the right. Because the data are not symmetric about the mean, the distribution is considered to be positively skewed. The lognormal distribution is also skewed right as shown. Natural log-scale transformations of positively skewed environmental data can make the data appear more normally distributed. Although histograms represent a good visual tool for evaluation of the probability of the environmental data, Benjamin and Cornell (1970) point out that normal probability plots give a better representation of the data and are easy to construct.

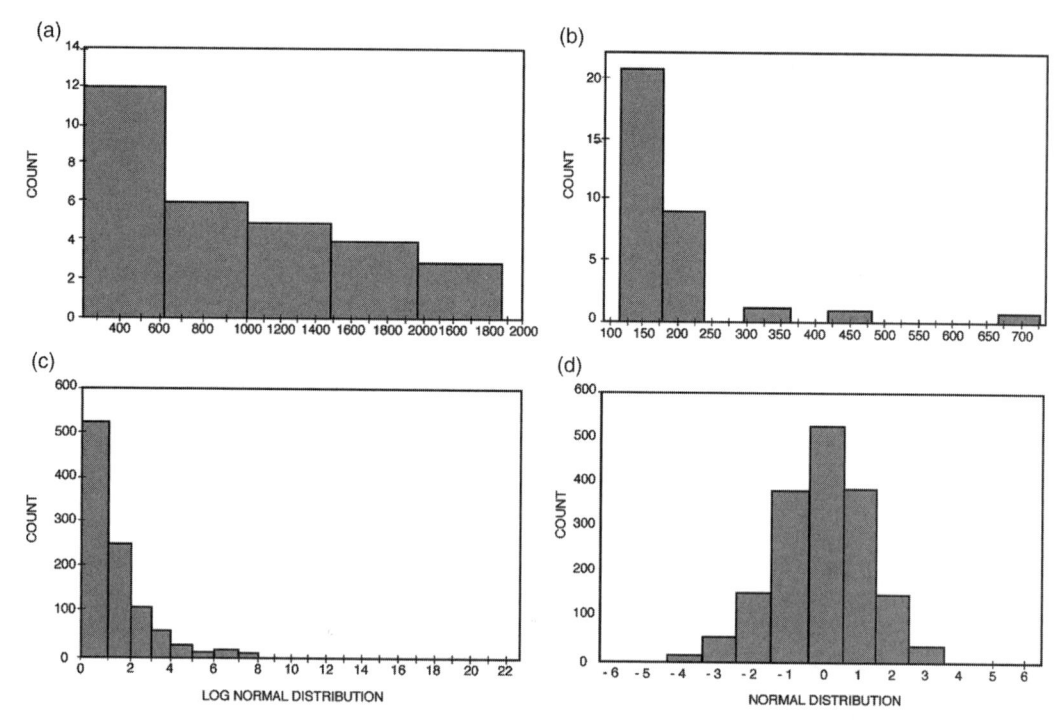

FIGURE 17.15
Histograms used to check data normality.

Normal probability plots provide an excellent technique to compare environmental data to the normal distribution. Figure 17.16 shows a normal probability plot for the same data as in Figure 17.15a. These are constructed by first ordering the raw data from smallest to largest. Let $x[1] < x[2] < \cdots < X[n]$ denote the ordered data. The $X[i]$ are called the order statistics of the data. The $X[i]$ are then plotted on normal probability paper versus the corresponding plotting position $(1/N+1) \times 100$. If the data are from a normal distribution, the plotted points should lie approximately on a straight line (Fisher and Potter, 1989).

Figure 17.16 (top) illustrates that the data do not plot as a straight line; hence, the assumption of normality is in question. Transforming the data to log scale and then replotting as shown in Figure 17.16 (bottom) does not provide for a straighter line; therefore, we cannot conclude that the lognormal distribution is more appropriate for demonstrating normality. In summary, environmental data from a waste disposal facility can be visually presented in a number of ways that assist inspection of the data sets:

- Time versus concentration plots
- Box and whisker plots
- One standard deviation error-bar charts
- Histograms
- Normal probability plots

The first two graphical tools can clearly illustrate qualitatively the relative water quality between wells (known as inter-well comparisons). The error-bar charts may be valuable when working with parametric statistics. Histograms may be used to view the probability distribution of the data. When evaluating the assumption of normality, normal probability

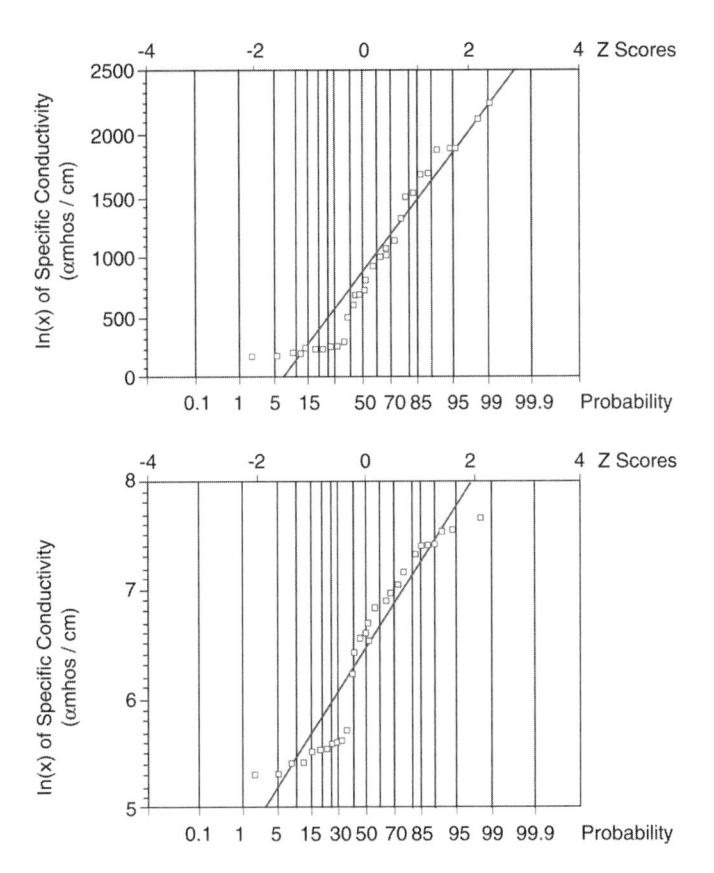

FIGURE 17.16
Probability plots for raw and log-transformed data.

plots are commonly prepared to observe deviations from normality. The example data illustrate several important points:

- Data outliers tend to inflate the mean and standard deviation of the data.
- The median and IQR are good estimates of the central tendency and variation of data sets, particularly when outliers are present.
- Large data variability is usually associated with high medians (i.e., impacted wells). Natural temporal variability is much lower than the variability observed when contamination is present.
- When histograms and normal probability plots show that ground-water data may not be normally distributed, the median and IQR may be better estimates of the central tendency and variability of the data.

Trilinear Diagrams

If one considers only the major, dissolved ionic constituents, in milliequivalents per liter, and lumps potassium and sodium together and fluoride and nitrate with chloride, the composition of most natural water can be illustrated in terms of three cationic and three anionic species. If the values are expressed as percentages of the total milliequivalents per liter of cations and anions, the composition of the water can be represented conveniently by a trilinear plotting technique.

The simplest trilinear plots utilize two triangles, one for anions and one for cations. Each vertex represents 100% of a particular ion or group of ions. The composition of cations is indicated by a point plotted in the cation triangle and the composition of anions by a point plotted in the anion triangle. The coordinates at each point add up to 100%. Most trilinear diagrams are in the form of two triangles bracketing a diamond-shaped plotting field, as first described by Piper (1944).

The trilinear diagram constitutes a useful tool for interpretation of water analysis. Most of the graphical procedures described here are of value in pointing out features of analyses and arrays of data that require closer study. The graphs themselves do not constitute an adequate means of making such studies, however, unless they can demonstrate that certain relationships exist among individual samples. The trilinear diagrams sometimes can be used for this purpose.

Figure 17.17 is a trilinear diagram derived from analyses of water from San Francisco Bay and the Newark Aquifer. In any illustration of water-quality data, a diagram

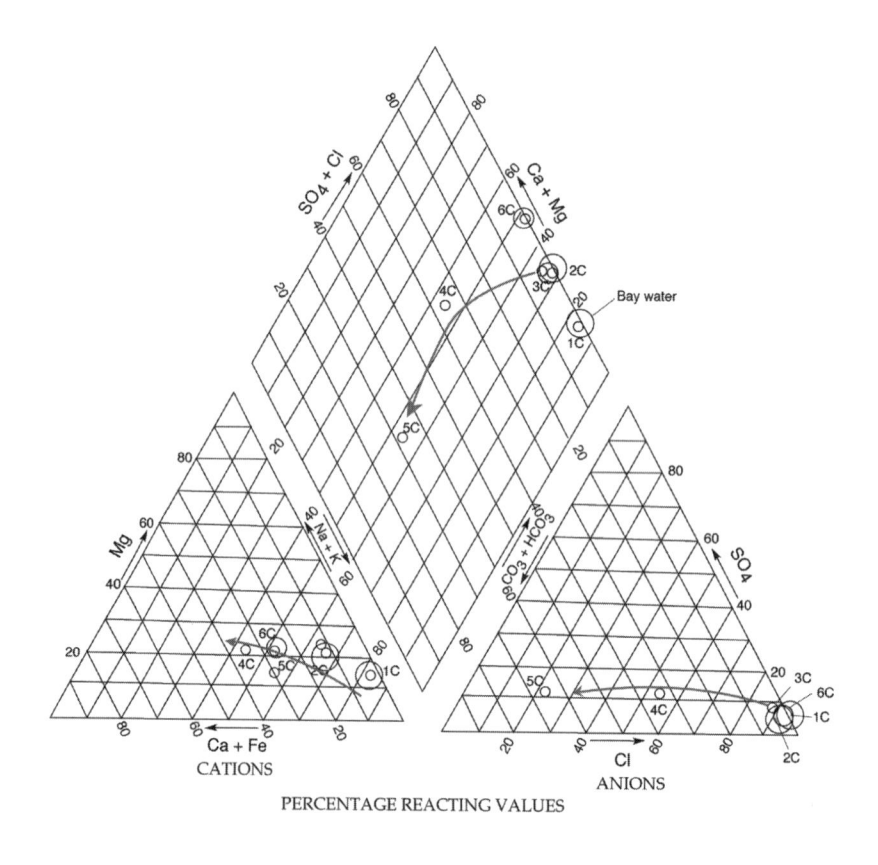

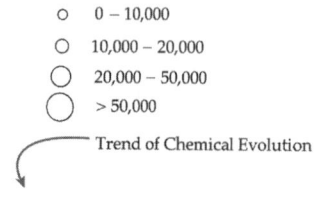

FIGURE 17.17
Trilinear data sets used for comparisons of water quality. (*Source*: Hydro-Search, Inc.)

should aid interpretation by providing a visual clarification of trends or a comparison of differences in water quality. Trilinear diagrams have become so popular that computer programs have been written to automatically calculate and display the data.

Statistical Treatment of Water-Quality Data

Various procedures, such as averaging, determining frequency distributions, and making simple or multiple correlations, are widely used in interpretation of water-quality analyses. Sophisticated applications of statistical methods, particularly procedures that utilize digital computers, are being applied more and more frequently. Some potential applications of these statistical techniques are covered in the following sections. It is essential that proper consideration be given to chemical principles during application of statistical tests of data sets. Most data sets are evaluated using simple averaging and standard deviations, or just represent the water-quality data changes over time as shown in Figure 17.18; however, more complex statistical treatments of analytical data are necessary when state and Federal performance standards are required for waste disposal or remedial cleanup sites.

Federal regulations have been established for statistical determination of compliance for RCRA facilities. Both existing and new hazardous waste facilities are covered by Subtitle C of the RCRA and regulated by 40 CFR Parts 264 and 265. When first issued, Part 264 Subpart F required that Cochran's approximation to the Behrens Fisher Student's t-test (CABF) or an alternative statistical procedure approved by U.S. EPA be used to determine whether there is a statistically significant exceedance. This Part 264 Subpart F regulation and, in particular, the CABF procedure, generated significant technical criticism over use of these statistical procedures for use with ground-water quality data and U.S. EPA proposed a new regulation in response to these concerns (U.S. EPA, August 24, 1987). The proposed regulation was revised based on comments EPA received and was then made final (U.S. EPA, 1988b).

The final regulation (October 11, 1988) describes five performance standards that a statistical procedure must meet. The Federal regulations do recommend four types of statistical procedures to evaluate performance of RCRA facilities for releases to ground water. In addition, U.S. EPA has issued (October 1991) amendments to Subtitle D of RCRA to include criteria for municipal solid waste landfills (MSWLFs). The number of samples collected to establish ground-water quality data must be consistent with the appropriate statistical procedures (discussed below). The sampling procedures are defined in the applicable sections for detection monitoring (§258.54[b]), assessment monitoring (§258.55[b]), and corrective action (§258.56[b]):

- Owner or operator must specify in the operating record one of the following statistical methods to be used in evaluating ground-water monitoring data for each hazardous constituent. The statistical test chosen shall be conducted separately for each hazardous constituent in each well.
    - A parametric analysis of variance followed by multiple comparisons procedures to identify statistically significant evidence of contamination.
    - An analysis of variance based on ranks followed by multiple comparisons procedures to identify statistically significant evidence of contamination.
    - A tolerance or prediction interval procedure in which an interval for each constituent is established from the distribution of the background data and the level of each constituent in each compliance well is compared to the upper tolerance or prediction limit.

Well Number:				MW-16S							MW-16D				
Collection Date:	1993	1996	1997	1998	1999	2000	2001	1993	1996	1997	1998	1999	2000	2001
Units:	µg/L	µg/L	µg/L	µg/L	µg/L	µg/L	µg/L	µg/L	µg/L	µg/L	µg/L	µg/L	µg/L	µg/L
Chloromethane	12,000U	1,000U	250J	2,500U	2,500U	500U	100U	10 U	1 U	0.1 J	1 U	1 UJ	1 U	1 U*
Bromomethane	12,000U	1,000U	2,500UJ	2,500U	2,500U	500U	100U	10 U	1 U	1 U	1 U	1 U	1 U	1 U
Vinyl chloride	2,200U	1,000U	650U	2,500U	2,500U	120J	250	2 U	1 U	0.2 J	1 U	1 U	1 U	1 U
Chloroethane	6,200U	1,000U	2,500U	2,500U	2,500U	500U	100U	19 J	4	6 J	3 J	8	7	6 J
Methylene chloride	9,200U*	2,000U	2,500U*	5,000U*	2,500U	500U	380	7 U*	2.0 U*	1 U*	2 U*	2 U*	2 U*	1 J
Acetone	1,500UR	3,100R	21,000R	22,000U*	11,000J	2,900	2,500J	7 J	R	7 U*	R	R	5 U*	5 U*
Carbon disulfide	6,200U	1,000U	2,500U	2,500U	2,500U	500U	11 J	5 U	1 U	1 U	1 U	1 U	1 U	1 U
1,1-Dichloroethene	1,200U	1,000U	2,500U	2,500U	2,500U	500U	100U	1 U	1 U	1 U	1 U	1 U	1 U	1 U
1,1-Dichloroethane	2,500U	1,000U	320J	2,500U	2,500U	110J	260	2 U	1 U	1 U	1 U	1 U	1 U	1 U
cis-1,2-Dichloroethene	860J	1,000U	620J	2,500	2,500U	500U	110	0.7 J	1 U	1 U	1 U	1 U	1 U	1 U
trans-1,2-Dichloroethene	6,200U	1,000U	2,500U	2,500U	2,500U	500U	17 J	5 U	0.4 J	0.5 J	1 U	1 U	1 U	1 U
Chloroform	1,200U	1,000U	2,500U	2,500U	2,500U	500U	100U*	1 U	1 U	1 U	1 U	1 U	1 U	0.4 J
1,2-Dichloroethane	750U	1,000U	2,500U	2,500U	2,500U	500U	100U	0.6 U	1 U	0.3 J	1 U	1 U	1 U	1 U
2-Butanone	350 UR	3,700J	19,000J	19,000U*	11,000J	3,000J	2,100	2 J	R	2 J	R	R	R	10 U
Bromochloromethane	1,000U	1,000U	2,500U	2,500U	2,500U	2,500U	2,500U		1 U	1 U	1 U	1 U	1 U	1 UJ
1,1,1-Trichloroethane	6,200U	1,000U	2,500U	2,500U	2,500U	500U	100U	5 U	1 U	1 U	1 U	1 U	1 U	1 U
Carbon tetrachloride	1,200U	1,000U	2,500U	2,500U	2,500U	500U	100U	2 U	1 U	1 U	1 U	1 U	1 U	1 UJ
Bromodichloromethane	1,200U	1,000U	2,500U	2,500U	2,500U	500U	100U	1 U	1 U	1 U	1 U	1 U	1 U	1 U
1,2-Dichloropropane	2,500U	1,000U	2,500U	2,500U	2,500U	500U	38 J	2 U	1 U	1 U	1 U	1 U	1 U	1 U
cis-1,3-Dichloropropene	1,200U	1,000U	2,500U	2,500U	2,500U	500U	100U	1 U	1 U	1 U	1 U	1 U	1 U	1 U
Trichloroethene	3,800U	1,000U	2,500U	2,500U	2,500U	500U	100U	0.8 J	1 U	1 U	1 U	1 U	1 U	1 U
Chlorodibromomethane	2,500U	1,000U	2,500U	2,500U	2,500U	500U	100U	2 U	1 U	1 U	1 U	1 U	1 U	1 U
1,1,2-Trichloroethane	620U	1,000U	2,500U	2,500U	2,500U	500U	100U	0.5 U	1 U	1 U	1 U	1 U	1 U	1 U
Benzene	2,500U	1,000U	920J	2,500U	2,500U	83 J	140	3	0.8 J	1	1	1	1	1
trans-1,3-Dichloropropene	1,200U	1,000U	2,500U	2,500U	2,500U	500U	100U	1 U	1 U	1 U	1 U	1 U	1 U	1 U
Bromoform	2,500U	1,000U	2,500U	2,500U	2,500U	500U	100 UJ	2 U	1 U	1 U	1 U	1 U	1 U	1 U
4-Methyl-2-pentanone	13,000	3,000J	14,000	13,000	8,300J	3,400	3,700	5 U	5 U	5 U	5 U	5 U	5 U	5 UJ
2-Hexanone	720 UR	5,000U	R	R	12,000U	230J	430	50 UR	R	R	R	5 U	5 U	5 UJ
Tetrachloroethene	2,500U	1,000U	2,500U	2,500U	2,500U	500U	100U	2 U	1 U	1 U	1 U	1 U	1 U	1 U
1,1,2,2-Tetrachloroethane	620U	1,000U	2,500U	2,500U	2,500U	500U	100U	0.5 U	1 U	1 U	1 U	1 U	1 U	1 UJ
1,2-Dibromoethane	1,200U	1,000U	2,500U	2,500U	2,500U	500U	100U	1 U	1 U	1 U	1 U	1 U	1 U	1 U
Toluene	52,000J	14,000	30,000	44,000	34,000	13,000	12,000	17	1 U	1 U*	1 U	1 U	1 U*	0.1 J
Chlorobenzene	6,200U	1,000U	2,500U	2,500U	2,500U	500U	100U	5 U	1 U	1 U	1 U	1 U	1 U	1 U
Ethyl benzene	2,700J	1,700	2,700	2,500	2,900	1,600	1,800	3 J	1 U	1 UJ	1 U	1 U	1 U	1 U
Styrene	1,200U	1,000U	2,500U	2,500U	2,500U	500U	100U	1 U	1 U	1 U	1 U	1 U	1 U	1 U
Xylenes (total)	14,000	8,700	12,000	13,000	16,000	9,900	11,000	17	1 U	5 U	1 U	1 U	1 U	1 U
1,3-Dichlorobenzene	6,200U	1,000U	2,500U	2,500U	2,500U	500U	100U	5 U	1 U	1 U	1 U	1 U	1 U	1 U
1,4-Dichlorobenzene	6,200U	1,000U	2,500U	2,500U	2,500U	500U	100U	5 U	1 U	1 U	1 U	1 U	1 U	1 U
1,2-Dichlorobenzene	12,000U	1,000U	2,500U	2,500U	2,500U	500U	100U	10 U	1 U	1 U	1 U	1 U	1 U	1 U
1,2-Dibromo-3-chloropropane	5,000UR	R	2,500U	2,500U	2,500U	500U	100 UJ	4 UR	R	1 U	1 U	1 U	1 U	1 U
1,2,4-Trichlorobenzene	2,500U	2,500U	2,500U	2,500U	2,500U	500U	100U							

Key:
J = The concentration is approximate due to limitations identified during the quality assurance review
U = Indicates the compound was analyzed but not detected. The associated value is the sample quantitation limit
U* = The compound should be considered "not detected" since it was detected in a blank at a similar concentration level
UJ = Indicates the compound was analyzed but not detected. The associated value is an estimated sample quantitation limit based on a bias identified during the quality assurance review
R = The results were considered unusable during the quality assurance review
Blank = The compound was not analyzed for

(1) Values shown are the highest detected between the investigative sample and its unpreserved, duplicate, reanalysis, or dilution sample.

FIGURE 17.18
Simple time-series data set for organic parameters.

- A control chart approach that gives control limits for each constituent.
- Another statistical test method that meets the performance standards discussed immediately below.
* Any statistical method chosen shall comply with the following performance standards, as appropriate:
  - The statistical method shall be appropriate for the distribution of chemical parameters or hazardous constituents.
  - If an individual well comparison procedure is used to compare an individual compliance well constituent concentration with background constituent concentrations or a ground-water protection standard, the test shall be done at a Type I error level no less than 0.01 for each testing period. If a multiple comparisons procedure is used, the Type I experiment-wise error rate for each testing period shall be no less than 0.05; however, the Type I error of no less than 0.01 for individual well comparisons must be maintained.
  - If a control chart approach is used to evaluate ground-water monitoring data, the specific type of control chart and its associated parameter values shall be protective of human health and the environment.
  - If a tolerance interval or a prediction interval is used to evaluate ground-water monitoring data, the levels of confidence and, for tolerance intervals, the percentage of the population that the interval must contain shall be protective of human health and the environment.
  - The statistical method shall account for data below the limit of detection with one or more statistical procedures that are protective of human health and the environment. Any practical quantitation limit (PQL) that is used in the statistical method shall be the lowest concentration level that can be reliably achieved within specified limits of precision and accuracy during routine laboratory operating conditions that are available to the facility.
  - If necessary, the statistical method shall include procedures to control or correct for seasonal and spatial variability as well as temporal correlation in the data.
* The owner or operator must determine whether or not there is a statistically significant increase over background values for each parameter or constituent required in the particular ground-water monitoring program that applies to the MSWLF unit.
  - In determining whether a statistically significant increase has occurred, the owner or operator must compare the ground-water quality of each parameter or constituent at each monitoring well to the background value of that constituent.
  - Within a reasonable period of time after completing sampling and analysis, the owner or operator must determine whether there has been a statistically significant increase over background at each monitoring well.

The statistical test requirements are the same as the RCRA Subtitle C final regulation, as the solid waste rules recommend the same four types of procedures.

The performance standards in these Federal rules allow flexibility in designing statistical procedures to site-specific considerations. Selection of an appropriate statistical test must be made based on the quality of the data available, the hydrogeology of the site, and the theoretical properties of the test. As expressed in previous sections, ground-water quality data can be expected to vary temporally and spatially due to natural effects and the results are also affected by sampling and analytical errors. Due to natural variability observed in ground water, the determination of a significant change in water quality is linked to statistical probability theory.

In order to define if there has been a significant change in water quality, comparison must be made between supposedly "clean" background data and possibly impacted data. Both these ground-water classes are subject to temporal and spatial variability as well as sampling and analytical error. Hence, the problem becomes one of evaluation of variable water quality in time and space with potential statistical inferences. A statistical hypothesis is used to compare water quality:

Null hypothesis, H_0: No contamination exists, therefore the facility is in compliance.

Alternative hypothesis, H_1: Contamination exists, therefore the facility is in violation.

A statistical test is made on the null hypothesis and a conclusion is reached that either the facility is or is not in violation. The null hypothesis starts out with the assumption that there is no real difference between the quality of upgradient and downgradient ground water. The assumption is that they are all from the same population. Thus, the difference between the means of the two samples would be just one possible difference from the theoretical distribution where the mean difference is zero.

The assumption is called the *null hypothesis* because it attempts to nullify the difference between the two sample means by suggesting or forming a hypothesis that it is of no statistical difference. If the statistical difference between the two sample means turns out to be too big to be explained by the kind of variation that would often occur by chance between random samples, then one must reject it (the null hypothesis), as it will not explain our observations. The typical alternative hypothesis would be that the two water-quality population means are not equal. In this context, a violation implies that water quality is significantly different from background. Figure 17.19 illustrates the two types of errors associated with hypothesis testing.

Significant technical discussions surround whether a site has observed a false-positive indicating contamination. A Type I error (false-positive) occurs when a site (or well) is actually in compliance but the statistical test is triggered that decides it is in violation. The probability of a Type I error (or) is defined as the controllable significance level of the test. Usually, this is set at 0.05, giving a 1/20 chance that a false-positive conclusion of contamination will occur.

	IN COMPLIANCE	IN VIOLATION
TRUE SITUATION — IN COMPLIANCE	Good Decision $1 - \alpha$	False Positive Decision α Type I error
TRUE SITUATION — IN VIOLATION	False Negative Decision β Type II error	Good Decision $1 - \beta$ Power of Test

FIGURE 17.19
Statistical error in hypothesis testing.

A Type II error (false-negative) occurs when contamination exists but is not detected. The probability of a false-negative conclusion is more difficult to control, is often difficult to calculate, and is dependent on many factors that may include sample size, the overall magnitude of change in parameter concentration, and choice of statistic tested in the decision process.

Statistical hypothesis testing can be divided into two general categories: (1) parametric, or those which rely on the estimation of parameters of a probability distribution (usually the mean and standard deviation of the normal distribution) and (2) nonparametric, or those which do not fit a normal distribution. Nonparametric methods usually rely on test statistics developed from the ordered ranks of the data. The simplest nonparametric evaluation is the median or middle value of a data set. Both parametric and nonparametric statistical tests are reviewed in the context of ground-water monitoring events in later sections of this chapter.

In general terms, the type of statistical test to use for a facility regulated under Federal laws should be consistent with U.S. EPA (1988b) 40 CFR Part 264, Statistical Methods for Evaluating Ground-Water Monitoring from Hazardous Waste Facilities; final rule. (*Federal Register*, 53, 196, 39720–39731). Both RCRA solid waste (Subtitle D) and hazardous waste (Subtitle C) sites are keyed into this code.

Data Independence

Independence of data collected in environmental programs must be evaluated by determining if the data show serial correlation. Serial correlation of ground-water sampling data is most likely to occur from very slow ground-water flow. Even with reasonably permeable aquifers (say with hydraulic conductivity $>10^{-3}$ cm/sec) low gradients can slow ground-water velocity to less than 20 ft per year. When ground-water quality measurements are collected too frequently to be independent of each other, one can observe serial correlation in the data. Independence can often be achieved by increasing the time between observations. Several tests have been reported in the literature to evaluate the presence of serial correlation in ground-water quality data. Montgomery et al. (1987) chose the Lag 1 autocorrelation function (ACF). Goodman and Potter (1987) also used this method as well as the nonparametric auto run (AR) test. The application of the ACF test to ground-water quality data is described in detail by Harris et al. (1987). The AR test was applied to hydrologic data by Sen (1979). Most advanced statistics texts and computer packages include these tests.

An example of data that show serial correlation is provided in Figure 17.8. These results indicate that serial correlation may exist in ground-water quality data even though the sampling was at intervals of 3 months.

The reality of most sampling programs dictates the sampling period required by regulatory standards or permit requirements. It is probably sufficient that one is aware of the potential difficulties associated with serial correlation of the data so that independence of the observations can be checked to help in the selection of the statistical test used in the evaluation of the data.

Data Normality

The normal distribution is perhaps the single most important and widely used probability model in applied statistics. This is because many real systems fluctuate normally about a central mean, that is, measurement error of a random variable is symmetric about a true mean and has a greater probability of being small (close to the mean) than large in the tail of the distribution.

The U.S. EPA's RCRA statistical regulations (40 CFR 264 Subpart F) do not require tests for normality or other distributional assumptions unless:

1. A data transformation is made
2. Nonparametric statistical tests are applied

Data transformations are commonly used to normalize skewed data for parametric tests. Many environmental systems are modeled using the lognormal distribution because: (1) it has a lower bound of zero and (2) it is positively skewed, allowing high values to be included (Benjamin and Cornell, 1970).

The hypothesis of normality can be evaluated through any number of statistical goodness-of-fit tests. These tests are used to mathematically compare the shape of the normal distribution to the data set. Care should be taken to only apply these tests to independent, stationary data sets.

In the ground-water quality literature, Montgomery et al. (1987) tested the normality of ground-water quality data using graphical methods, the chi-square test, and the skewness test. Harris et al. (1987) recommend the skewness test for general use with ground-water quality data.

Evaluation of Ground-Water Contamination

A main objective of a ground-water detection monitoring program is to determine if the facility is affecting ground water. Owner and operators are required in Federal rules to place detection monitoring wells in both upgradient (background) and downgradient locations around the facility and to monitor those wells at regular intervals, typically quarterly or twice per year, for a series of indicator parameters. Subtitle D defines in [§258.53] (Ground-Water Sampling and Analysis Requirements) that:

> The ground-water monitoring program must include consistent sampling and analysis procedures that are designed to ensure monitoring results that provide an accurate representation of ground-water quality at the background and downgradient wells. The owner or operator must notify the State Director that the sampling and analysis program documentation has been placed in the operating record and the program must include procedures and techniques specified in §258.53(a)(1) through (5):
>
> - The ground-water monitoring program must include sampling and analytical methods that are appropriate for ground-water sampling and that accurately measure hazardous constituents and other monitoring parameters in ground-water samples. Ground-water samples shall not be field filtered prior to laboratory analysis.
> - Ground-water elevations must be measured in each well immediately prior to purging each time ground-water is sampled. Ground-water elevations must be measured within a period of time short enough to avoid temporal variations in ground-water flow that could preclude accurate determination of ground-water flow rate and direction.
> - The owner or operator must establish background ground-water quality in a hydraulically upgradient or background well(s) for each of the monitoring parameters or constituents required in the particular ground-water monitoring program that applies to the MSWLF unit.
> - The number of samples collected to establish ground-water quality data must be consistent with the appropriate statistical procedures (discussed below). The sampling procedures are defined in the applicable sections for detection monitoring (§258.54[b]), assessment monitoring (§258.55[b]) and corrective action (§258.56[b]).

The logic of this sampling strategy is that upgradient water quality represents the background conditions for that particular region and downgradient water quality represents background water quality plus any influence produced by the facility. Section 258.40 states, "The relevant point of compliance specified by the Director of an approved State shall be no more than 150 meters from the waste management unit boundary and shall be located on land owned by the owner of the MSWLF unit." This sets the stage for defining a boundary zone in which to locate the monitoring wells.

The detection monitoring program as described in subtitle D for solid waste (§258.54) then defines the following:

- Detection monitoring is required at MSWLF units at all ground-water monitoring wells specified in §258.51(a)(1) and (2) (background and downgradient). At a minimum, the constituents from Appendix I must be included in the program.
- The Director of an approved State may delete any of the Appendix I constituents if it can be shown that the removed constituents are not reasonably expected to be in or derived from the waste contained in the unit.
- A Director of an approved State may establish an alternative list of inorganic indicator parameters for an MSWLF unit, in lieu of some or all of the heavy metals in Appendix I.

Note: Deletion of Appendix I constituents or establishing alternative inorganic parameters is not possible unless the state is approved by the U.S. EPA.

- The monitoring frequency shall be at least semi-annual during the active life of the facility (including closure) and the post-closure period. A minimum of four independent samples from each well (background and downgradient) must be collected and analyzed during the first semi-annual sampling event. At least one sample from each well (background and downgradient) must be collected and analyzed during subsequent semi-annual events. The Director of an approved State may specify an alternative frequency during the active life (including closure) and the post-closure care period. The alternative frequency shall be no less than annual.

Note: An alternative monitoring frequency is not an option unless the state is approved by the U.S. EPA.

- If the owner or operator determines that there is a statistically significant increase over background for one or more of the constituents in Appendix I or an approved alternative list, the owner or operator must:
  - Within 14 days of the finding, place a notice in the operating record indicating which constituents have shown statistically significant changes from background levels and notify the State Director that the notice was placed in the operating record.
  - Establish an assessment monitoring program within 90 days unless the owner or operator can demonstrate that a source other than an MSWLF unit caused the contamination or that a statistically significant increase resulted from an error in sampling, analysis, statistical evaluation or natural variation in ground-water quality. A report documenting the alternate source or error must be certified by a qualified ground-water scientist or the Director of an approved State and placed in the operating record.

As described in Chapter 8, the issue of the regulatory concept of the simple upgradient and downgradient model is rarely observed in real-world monitoring programs; however, the use of background wells as representative of upgradient water can be used in statistical comparisons. In many cases, particularly when adequate background data are available prior to the installation of the facility, intrawell comparisons may be the most successful technique to use (i.e., each well compared to its own history). The major advantage of this approach is that it eliminates the spatial component of variability from the comparison. One is left with evaluating the local effects on the well installation, such as construction, maintenance, and nearby interferences to water quality (such as wells located near roads that are salted during winter, local spills, etc.). The statistical methodology is illustrated in Figure 17.20, which graphically portrays the variable bases for statistical comparisons between wells and for intrawell statistical comparisons.

Detection monitoring programs at waste-disposal facilities require not only that a release to the environment has occurred, but also that the release observed is directly due to discharges from the facility. Water-quality standards are commonly used as a basis for judging if a release to the environment has occurred. Yet, even a water-quality standard exceedance must be compared to background water quality in order to conclude that the facility is responsible for the ground-water impact. In reviewing facility data, one can expect a number of water-quality standards (especially nonorganic parameters) to be exceeded in natural ground water. Thus, comparison of downgradient water quality to known background water quality is an important part of any detection monitoring program. It is imperative that detection monitoring programs not rely on only one background well as the basis for comparisons to downgradient wells. Natural spatial variability within the geologic environment can significantly affect the statistical comparisons necessary for detection monitoring programs. Evaluation and knowledge of ground-water flow and geology sufficient to design detection monitoring systems, together with time-series graphs of ground-water quality, may clearly show a release. Statistical tests applicable to water-quality evaluations currently recommended by U.S. EPA with other methods proposed in the water-quality literature can be used as evaluation tools for the following example facilities:

- Existing MSWLFs, new facilities, and existing facilities with historically clean water quality
- RCRA Subtitle C hazardous waste disposal sites, industrial waste disposal sites, land disposal sites for waste water
- Superfund-type evaluations for assessment and aquifer remediation projects for Federal and State cleanup programs

Types of Statistical Tests

The four general categories of statistical methods used for facility compliance comparisons with ground-water quality RCRA regulations are:

- Tests of central tendency (location)
- Tests of trend
- Prediction, tolerance, and confidence intervals
- Control charts

Statistical tests of central tendency are used to compare the mean or median of two or more sets of data and establish if they are significantly different. Tests of trend evaluate significant increase or decrease in water quality over time. Prediction and tolerance

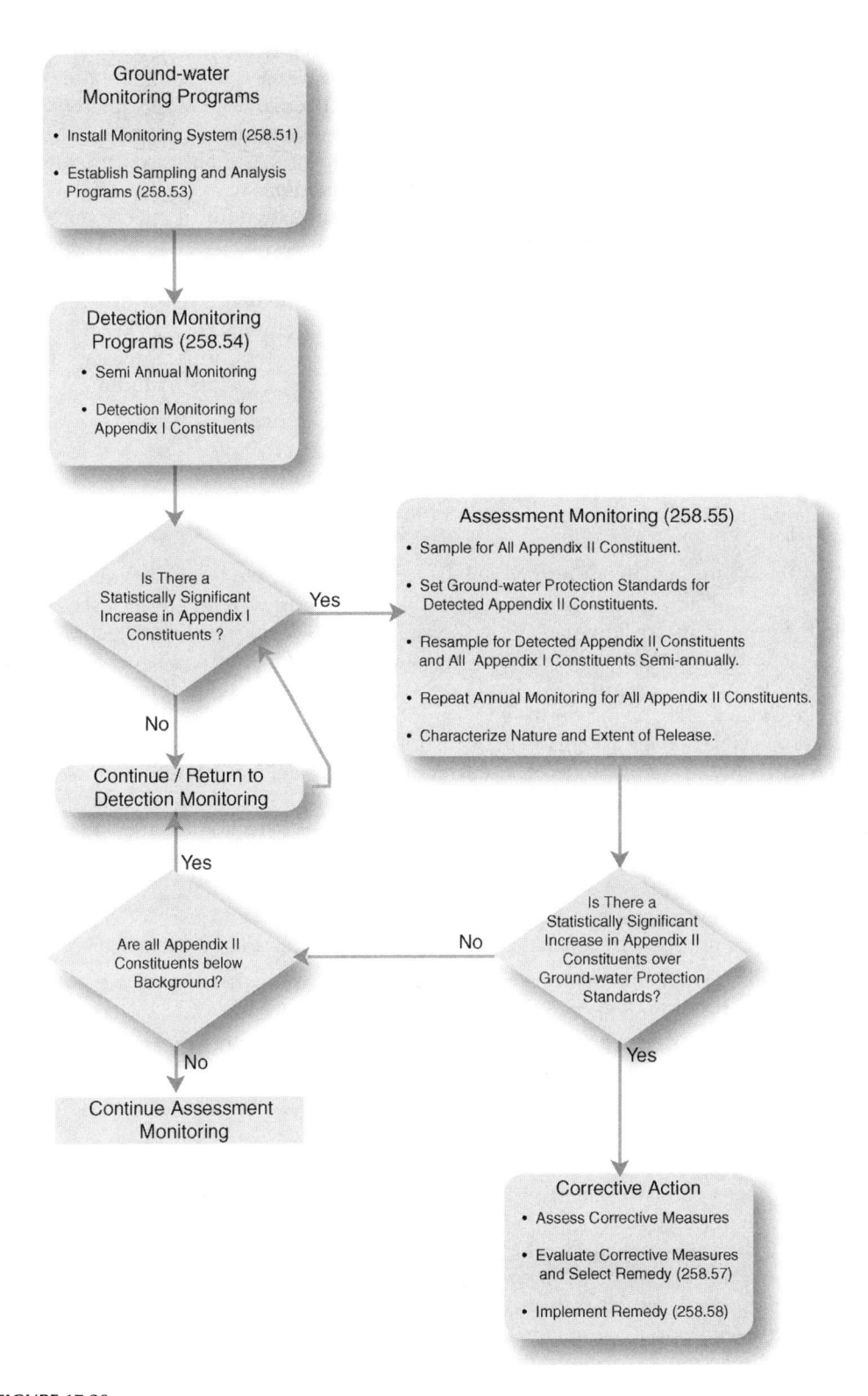

Organization and Analysis of Ground-Water Quality Data

FIGURE 17.20
The variable bases for statistical comparisons between wells and for intra-well statistical comparisons (*Source*: 40 CFR 258.50–258.58).

intervals are statistical methods that set limits for acceptable background water quality based on historic data sets. These interval tests can also be used to define the number of background measurements required to fully establish background water quality. Confidence intervals also set limits for average background water quality. Control charts are widely used graphical methods for industrial engineering quality control and are similar to the prediction, tolerance, and confidence intervals.

To evaluate water-quality data, a series of questions must be formulated to select the appropriate statistical tests to evaluate the data. These statistical tests are designed to evaluate whether or not a significant difference exists between the historical mean and median of background water quality and the mean and median of each downgradient well. The ability of these tests to detect ground-water contamination quickly (i.e., when applied quarterly with detection within one or two quarters after a release occurs) depends on the choice of the statistical test and other data set factors including:

- Length of the unaffected water-quality record
- Variability in the data sets
- Magnitude of the increase in concentration due to the release

Regulatory requirements for immediate detection of releases to the environment can be extremely difficult to demonstrate even with the monitoring well screen located directly within the ground-water flow path from the facility. Also of interest in a detection monitoring program is to establish if specific water-quality standards have been exceeded. If a sample value exceeds a regulatory standard such as maximum contaminant level (MCL) under the Clean Water Act (accounting for sampling and analytic variability), then a determination of a violation may be made by State or Federal regulators. In this situation, a violation means only that a mandated concentration level has been exceeded, not that certain actions must be taken. The problem of regulatory violation of a standard is acute when State or Federal standards are at or approach the level of detection of the contaminant, as is the case with some VOCs. Possible decision approaches may include:

- A regulatory mandated "hard" limit where no data should exceed the water-quality standard with consideration given to sampling and laboratory error.
- The more flexible historic mean concentration at a well where the water-quality standard should not exceed this regulatory limit.
- The moving window approach where the last-year's mean concentration should not exceed the limit
- The statistical limits where 95% of the population must be below the standard.

These provide a number of alternative decision paths for water-quality evaluation or for making decisions on analytical sample values close to a water-quality standard.

Comparison of downgradient water quality to the standard is conceptually straightforward. The indicator parameter is plotted on a simple time-series graph and compared to the concentration called out in the water-quality standard. Background concentrations should be plotted to evaluate if the background water-quality levels exceed the particular standard. If background parameter levels do exceed the relevant standard, the downgradient well parameter concentrations must be evaluated statistically against the background rather than comparing the well data to the water-quality standard. In those cases where background does not exceed the standard, downgradient concentration can be compared to the standard. Parameter trends (especially for inorganic indicator variables) can serve as a useful management tool for implementing corrective actions before primary drinking water standards are exceeded. Once water-quality standards are clearly exceeded and verified through resampling, a release to the environment is confirmed. In the second and third approaches, confidence limits on the mean (where the standard must be below the lower confidence level) can provide the evaluation tool. The last approach can effectively use tolerance interval tests for evaluation of exceedances.

After evaluation of the issues or questions to be answered statistically, the next step is to choose a specific test that answers the question. The test must not only have an appropriate experimental design (i.e., answer the right question), but also the implicit assumptions of the test must not be grossly violated. As previously discussed, ground-water quality data may grossly violate the assumption of normality, even after appropriate data transformation.

A detection monitoring data evaluation must be based on the variable regulatory issues that can change from State to State and from site to site. Some facilities may have specific permit requirements for statistical tests or specified parameter lists that can be significantly different from site to site within a single state. In the following sections, potential releases to ground water can be evaluated using three general types of statistical methods:

1. Tests of central tendency (location)
2. Tests of trend
3. Prediction, tolerance, and confidence intervals

In each section, emphasis is placed on the situations where the type of test is appropriate. The types of water-quality questions these tests can answer are discussed.

Tests of Central Tendency (Location)

The statistical mean and median of water-quality data sets are the most common estimates of central tendency. Tests that compare the mean or median of two or more sets of data are tests of central tendency or tests of location.

The U.S. EPA had previously required that Cochran's approximation to the Student's *t*-test be applied between pooled background water-quality data and each downgradient compliance well. Significant criticism of this procedure (see McBean and Rovers, 1984; Silver, 1986; U.S. EPA, 1988b; and Miller and Kohout, undated) resulted in the Agency change to a parametric one-way analysis of variance (ANOVA) or the nonparametric analog called the Kruskal–Wallis test (U.S. EPA, 1988b). Unfortunately, the ANOVA test also suffers from a high false-positive error rate when many multiple comparisons must be done for sites with, for example, more than five or six compliance wells (see Fisher and Potter, 1989).

Tests of Trend

Tests of trend are commonly used in detection monitoring programs to evaluate whether water-quality parameter values are increasing or decreasing with time. Trend analysis is also useful for evaluating changes in background water quality. Trends in data could be observed as a gradual increase (usually modeled as a linear function) or a step function or even cyclical on a seasonal basis.

Trend evaluations have traditionally been performed by inspection of graphed time concentration plots. Time-series plots can also be used in conjunction with box plots to evaluate trends and seasonal fluctuations. A number of statistical methods can be applied to data sets to evaluate for trends and seasonality. Example procedures such as the Mann–Kendall test for trend evaluates the relative magnitudes of the concentration data with time (Goodman, 1987). The length of time recommended to obtain adequate long-term trends is 2 yr of data (Doctor et al., 1986); for seasonal trends, a much longer period data set may be necessary. Goodman (1987), using a modified Mann–Kendall test, found that at least 10 yr of quarterly data were required for obtaining adequate power to detect seasonal trends. Although few facilities have such a long period of data, the long (post) closure requirements of 10–30 yr in State and Federal regulations will make such evaluations for seasonal trends possible.

Statistical trend tests alone cannot be used to determine compliance with ground-water quality regulations. These tests can only answer the question, "Does a positive or negative trend exist?" The presence of a minor trend should not be construed to mean there has been a release from the facility. Therefore, if a test of trend is used to support the hypothesis of a release, the results must be linked to exceedance of water-quality standards and to likelihood of the release based on review of potential cross-contamination and interferences.

Tests of trend have been commonly used in evaluating the expected effectiveness of remedial action. However, tests of trend should not be used to predict when a target concentration will be reached since aquifer restoration is usually not a linear but rather an asymptotic process.

A common use of trend tests is to evaluate if background water quality is significantly (gradually) changing over time. Hence, the background water quality represents a moving window that will be compared to downgradient water quality. In this case, the background trend should be removed prior to further analysis (Harris et al., 1987). An apparent trend at a downgradient well cannot be confirmed as evidence of contamination, unless it can be shown that the same trend does not exist in background or upgradient wells.

The nonparametric analogs to the linear regression F-test are Kendall's tau statistic and Spearman's (rho) rank correlation coefficient. Usually Kendall's tau is chosen for water-quality data because the test statistic approaches normality at smaller sample sizes than Spearman's rho (Montgomery et al., 1987).

Linear regression is considered a powerful technique of trend, but analysts tend to delete outlying values without physical justification to get a good fit to the data. Also, some users will wrongly try to make predictions of when concentration will return to normal or when a standard will be exceeded. Reviewers should make sure that deletion of data is physically justified. Also, any predictions made with the regression line should be interpreted as no more than a best guess.

Fisher and Potter (1989) reviewed statistical tests for applicability for use in detecting facility ground-water contamination events. They found that tests of central tendency, both parametric and nonparametric, have severe limitations. At least for the cases reviewed, natural spatial variability did not permit ANOVA results to discern between natural variations in the mean and those due to potential contamination. They also observed that ground-water quality data often violated the parametric assumptions of normality for both raw and log-transformed data sets. Even nonparametric tests of central tendency (such as Kruskal–Wallis) are not recommended for detecting contamination but rather should be used for evaluating spatial variability (Fisher and Potter, 1989). Statistical tests based on trend can be used in conjunction with other data evaluation techniques to support the conclusion of observed contamination. Prediction interval tests were recommended by Fisher and Potter (1989) as the most theoretically sound approach to setting background levels, and in the author's opinion interval statistical tests represent the most applicable methods for evaluating detection monitoring programs. As such, the remaining discussion of statistics will concentrate on interval tests.

Recommended Statistical Methods

Ground-water detection monitoring typically involves a series of monitoring wells hydraulically upgradient and downgradient of the facility to compare concentrations of chemical constituents between the upgradient and downgradient locations, assuming that any difference in ground-water quality is caused by leachate released from the facility. However, this assumption is often false because widespread spatial variability in ground-water chemistry exists. In the worst case (often the most typical circumstance), regulations (U.S. EPA, 1988b, 1991) require only one upgradient well and a minimum of three monitoring wells located downgradient from the facility. When a single upgradient well is used to characterize natural variability in background spatial variability, attempts to locate contamination are confounded (i.e., differences between the upgradient and downgradient wells could be due to natural differences between any two locations regardless of their relation to the waste disposal facility). Even with two upgradient wells, characterization of natural background variability may not be possible. That is, two upgradient wells may not display the same amount of variability observed in downgradient wells, which often number between 10 and 100.

Additionally, regulations require each downgradient monitoring well and constituent to be separately tested because releases from a waste disposal facility into ground-water are "plume-shaped," which may influence only a single downgradient well. Pooling data over downgradient wells might mask a release that only affected a single well. In addition, chemical constituents travel at different rates in ground water; the leading edge of the plume may contain only a small number of highly mobile chemical constituents. In many parts of the country, ground water flows quite slowly, in some cases less than a foot per year. Hydrogeologically independent observations from a given monitoring well may be available only quarterly, semiannually, or annually. Pooling data may be impractical because it may result in mixing contaminated and uncontaminated measurements, masking an early-stage release. Therefore, each new datum must be evaluated individually. The two most critical problems are that (a) numerous statistical evaluations must be performed on each monitoring event (typically 100–1000) and (b) environmental data are often censored (i.e., the analyte may or may not be detected when it is present at a level below the capability of the analytical instrument). These two problems complicate analysis of ground-water-monitoring data.

As will be shown in the following sections, solution of these problems leads to the construction of the so-called "prediction limits" adapted to the case of simultaneous statistical inference and sequential testing. Simultaneous statistical inference refers to the construction of limits or bounds that apply simultaneously to all comparisons made on a given monitoring event. In this context, the number of comparisons, which is denoted as k, is the set of all downgradient monitoring wells for all constituents for which statistical evaluation must be performed. Using these methods we can therefore control the overall sitewide false-positive rate (i.e., concluding that there has been an impact when there has not been) at a nominal level (e.g., 5%). As the number of comparisons on any given monitoring event becomes large, however, the associated false-negative rate (i.e., the failure to detect contamination when it is present) also becomes large. To minimize the false-negative rate, we use a sequential testing strategy in which an initial exceedance is then verified by one or more independent verification resamples. In this way, a smaller prediction limit can be used repeatedly, achieving the same site-wide false-positive rate but greatly minimizing the false-negative rate. These ideas are more fully developed in the following sections.

Statistical Prediction Intervals

Single Location and Constituent

If the problem were to set a $(1 - \alpha)100\%$ limit on the next single measurement for one location and one normally distributed constituent, a β-expectation tolerance limit (i.e., a prediction limit — Guttman, 1970; Hahn, 1970) could be computed from n independent background measurements as

$$\bar{x} + t_{(n-1,1-\alpha)} s \sqrt{1 + \frac{1}{n}} \tag{17.1}$$

where concern is that the concentration is elevated above background, x and s are the background sample mean and standard deviation, respectively, and t is the $100(1 - \alpha)$ percentile of Student's t-distribution on $n - 1$ degrees of freedom. If upgradient versus downgradient comparisons are to be performed, then a minimum of two background locations (e.g., wells) should be repeatedly sampled at a time interval sufficient to ensure independence (e.g., quarterly or semiannually). The background time period must include at least 1 yr to ensure that the same seasonal variation present in downgradient locations is rejected in the upgradient background. The reader should note that with multiple upgradient locations, s^2, the traditional estimator of σ^2 is biased (i.e., it is too small) because measurements are nested within upgradient monitoring locations. Alternative estimators for σ^2 based on variance components models have been proposed and should be used where appropriate (Gibbons, 1987a, b, 1994).

Multiple Locations

In practice, multiple comparisons are performed, one for each downgradient monitoring location and constituent. Using the Bonferroni inequality (Miller, 1966), a conservative prediction bound (i.e., the probability of at least one false rejection is at most α) for all kq comparisons (i.e., k locations each tested for q constituents) is

$$\bar{x} + t_{[n-1,(1-\alpha)/kq]} s \sqrt{1 + \frac{1}{n}} \tag{17.2}$$

In the present context, the comparisons are dependent because (1) constituents may be correlated and (2) all downgradient locations are compared to a common background. In this case, the Bonferroni adjustment may be unnecessarily conservative. Some improvement may be gained by adapting the approach of Dunnett and Sobel (1955) originally developed to compare multiple treatment groups to a common control group. The resulting correlation between the comparison of locations i and j to a common background is

$$\frac{\bar{x}_B - \bar{x}_M}{\sqrt{(S_B^2/N_B) + (S_M^2/N_M)}} = \frac{7.62 - 7.40}{0.20\sqrt{(1/4) + 1}} = \frac{0.22}{0.22} = 1.0 \tag{17.3}$$

where n_0 is the number of background measurements, n_j the number of measurements in monitoring location j, and n_i the number of measurements in monitoring location i. In the measurement of ground water the correlation is constant with value $\rho = 1/(n+1)$, since the number of background measurements $n_0 = n$ and $n_i = n_j = 1$ for all i and j (i.e., we are comparing a single new value in each monitoring location to n background measurements). Dunnett (1955) has shown how required values from the multivariate t-distribution can be reduced to evaluation of the equally correlated multivariate normal distribution for which the required probabilities are easily obtained. These critical points have been tabulated by a number of authors (e.g., Gupta and Panchpakesan, 1979; Gibbons, 1994). As shown below, increasing statistical power can be achieved by generalizing the single-stage Dunnett procedure described here to the case of multi-stage sampling using verification resampling. Alternative stage-wise comparison procedures have also been considered (Hochberg and Tamhane, 1987) in the context of multiple comparisons to a common control.

Verification Resampling

As the number of future comparisons increases, the prediction limit increases and false-negative rates can become unacceptably large. Gibbons (1987a, b) and Davis and McNichols (1987) noted this problem and suggested sequential testing of new ground-water monitoring measurements such that the presence of an initial exceedance in a downgradient location requires obtaining one or more independent resamples for that constituent. Failure is indicated only if both initial sample and verification resamples exceed the prediction limit. In this way, fewer samples are required and both false-positive and false-negative rates are controlled at minimum levels. Davis and McNichols (1987) derived simultaneous normal prediction limits for the next r of m measurements at each of k monitoring locations, where in the previous example, $r = 1$ and $m = 2$. Their result is a further generalization to Dunnett's test. The derivation is complicated and is reviewed in Gibbons (1994) and Gibbons and Coleman (2001). Complete tables are provided in Gibbons (1994).

Multiple Constituents

Little is known about the correlation between monitoring constituents, except that the interrelationship is highly variable and that there are too few background measurements to precisely characterize the correlation matrix or to use the matrix to construct accurate multivariate prediction limits. For this reason, the Bonferroni inequality has been used to derive conservative prediction bands. This practice will produce prediction limits larger than required when positive association is present (which, in the authors' experience, appears to be common). For example, in the previous illustration, if we were to monitor 10 constituents, $\alpha = 0.05 = 10 = 0.005$ and the limit

$$\bar{x} + 3.36s$$

would be applied to each location and constituent with an overall site-wide confidence level of 95%. Alternatively, there has been some work on multivariate prediction bounds (Guttman, 1970; Bock, 1975), which might apply to those cases where background sample sizes were sufficiently large to obtain a reasonable estimate of the inter-constituent covariance matrix. Unfortunately, the presence of nondetects (i.e., left-censored distributions) violates the joint normality assumption of the multivariate procedure.

Finally, it should be noted that in all cases, the smallest number of constituents that are indicative of a potential release from the facility should be used. Using fewer constituents will decrease the total number of comparisons and provide more conservative (i.e., smaller) prediction bounds. Also note that some constituents exhibit greater spatial variability than others (e.g., geochemical parameters such as chloride versus metals such as barium) and may be less useful at some facilities with heterogeneous geologic formations.

The Problem of Nondetects

In practice, environmental measurements consist of a mixture of detected and nondetected constituents ranging in detection frequency from 0 to 100%. When the detection frequency is high (e.g., >85%) several studies (Gilliom and Helsel, 1986; Hass and Scheff, 1990; Gibbons, 1994) have shown that most estimates of mean and variance of a left-censored normal or lognormal distribution yield reasonable results. This is not true when detection frequencies are between 50 and 85% (Gibbons, 1994). In this case, available methods include maximum likelihood estimators (MLEs) (Cohen, 1959, 1961), restricted maximum likelihood estimators (Persson and Rootzen, 1977), an estimator based on the Delta distribution which is a lognormal distribution with probability mass at zero (Aitchison, 1955), best linear unbiased estimators (Gupta, 1952; Sarhan and Greenberg, 1962), alternative linear estimators (Gupta, 1952), regression-type estimators (Hashimoto and Trussell, 1983; Gilliom and Helsel, 1986), and substitution of expected values of normal-order statistics (Gleit, 1985). In addition, U.S. EPA has often advocated simple substitution of one-half the MDL. Methods that adequately recover the mean and variance of the underlying distribution from the censored data often inadequately recover the tail probabilities used in computing prediction limits (Gibbons, 1994). In a simulation study (Gibbons, 1994) the MLE was the best overall estimator but the estimator based on the Delta distribution was the best at preserving confidence levels for prediction limits in the presence of censoring.

Nonparametric Prediction Limits

When detection frequency is less than 50%, none of the methods discussed in the previous section work well and an alternative strategy must be employed. In practice, an excellent alternative is to compute a nonparametric prediction limit, which is the maximum of n background measurements. The nonparametric limit is attractive because it makes no distributional assumptions and is defined even if only one of the n background measurements is quantifiable. In some cases, however, the number of background measurements is insufficient to provide a reasonable overall confidence level, therefore the nonparametric prediction limit may not always be an available alternative. Confidence levels for the nonparametric limits are a function of n, kq, and the number of verification resamples similar to the parametric case. For example, let $X_{(max;\ n)}$ represent the maximum value obtained out of a sample of size n and $Y_{(min;\ m)}$ represent the minimum value out of a sample of size m. In the present context, $X_{(max;\ n)}$ is the maximum background concentration and $Y_{(min;\ m)}$ is the minimum of the initial sample and verification resamples for a constituent in a downgradient monitoring location. The objective is to compare $Y_{(min;\ m)}$ with $X_{(max;\ n)}$. The confidence level for the simultaneous upper prediction limit defined as $X_{(max;\ n)}$ is

$$\Pr(Y_{1(min;\ m)} \cdot X_{(max;\ n)};\ Y_{2(min;\ m)} \cdot X_{(max;\ n)},...,Y_{k(min;\ m)} \cdot X_{(max;\ n)}) = 1 - \alpha \qquad (17.4)$$

To achieve a desired confidence level (say $1 - \alpha = 0.95$ for a fixed number of background measurements), m must be adjusted; the more the resamples, the greater the confidence. This probability can be evaluated using a variant of the multivariate hypergeometric distribution (Hall et al., 1975; Chou and Owen, 1986) function as

$$1 - \alpha = \frac{n}{km+n} \sum_{j_1=1}^{m} \sum_{j_2=1}^{m} \cdots \sum_{j_k=1}^{m} \frac{\binom{m}{j_1}\binom{m}{j_2}\cdots\binom{m}{j_k}}{\binom{km+n-1}{\sum_{i=1}^{k} ji + n - 1}} \qquad (17.5)$$

Based on this result, approximate confidence levels have been derived (Gibbons, 1990) for nonparametric prediction limits defined as the maximum of n background samples in which it is required to pass 1 of m samples (i.e., the initial sample or at least one verification resample) at each of k monitoring locations. To incorporate multiple constituents, the confidence level is adjusted to

$$1 - \frac{\alpha}{q} \qquad (17.6)$$

Exact confidence levels for the previous case and approximate confidence levels for the case in which it is required to pass the first or all of m resamples are now also available (Gibbons, 1991). Exact confidence levels for this latter case were recently derived (Willits, 1993; Davis and McNichols, 1994a, b) and extensive tables have been prepared (Gibbons, 1994). The case in which the prediction limit is the second largest measurement has also been considered (Gibbons, 1994; Davis and McNichols, 1994b, 1988).

Intra-Well Comparisons

Upgradient versus downgradient comparisons are often inappropriate (e.g., spatial variability may be present) and some form of intra-well comparisons (i.e., each location compared to its own history) must be performed. Note that intra-well comparisons are only appropriate when (1) predisposal data are available or (2) it can be demonstrated that the facility has not affected that well in the past. In this case, there are two good statistical methods available; combined Shewhart-CUSUM control charts (Lucas, 1982) and intra-well prediction limits (Davis, 1994; Gibbons, 1994). The advantage of the combined Shewhart-CUSUM control chart is that the method is sensitive to both immediate and gradual releases, whereas prediction limits are only sensitive to absolute increases over background. In the intra-well setting, comparisons are independent since each well is compared to its own history. Gibbons (1994, Table 8.3) provides appropriate factors for computing intra-well prediction limits for up to $kq = 500$ future comparisons under a variety of resampling strategies. These factors apply to normally distributed constituents or constituents that can be suitably transformed to approximate normality. In the nonparametric case, selecting a single future sample and setting the confidence level to $(1 - \alpha)/kq$ is also possible; however, overall confidence levels may be poor due to small numbers of background measurements typically available in individual monitoring wells (i.e., generally 8 or fewer). If seasonality is present, adjustments may be required. However, the number of available measurements within a given season is typically one per year, therefore most facilities will have insufficient data to estimate the seasonal effect if present.

Illustration

Consider the data in Table 17.5 for total organic carbon (TOC) measurements from a single well over 2 yr of quarterly monitoring.

TABLE 17.5
Eight Quarterly TOC Measurements

Year	Quarter	TOC (mg/l)
1992	1	10.0
1992	2	11.5
1992	3	11.0
1992	4	10.6
1993	1	10.9
1993	2	12.0
1993	3	11.3
1993	4	10.7

Inspection of the data reveals no obvious trends, and these data have a mean $\bar{x} = 11.0$ and standard deviation $s = 0.61$. The upper 95% point of Student's t-distribution on seven degrees of freedom is $t_{[7.1-0.05]} = 1.895$, therefore the upper 95% confidence normal prediction limit in Equation (1) is given by

$$11.0 + 1.895(0.61)\sqrt{1 + \frac{1}{8}} = 12.22 \text{ mg/l}$$

which is larger than any of the observed values. This limit provides 95% confidence of including the next single observation from a normal distribution for which eight previous measurements have been obtained with observed mean of 11.0 mg/l and standard deviation of 0.61 mg/l. Assuming that spatial variability does not exist (in many cases a demonstrably false assumption), and that values from this single well are representative of values from each of 10 downgradient wells in the absence of contamination, then the corresponding Bonferroni-adjusted 95% confidence normal prediction limit in Equation (2) for the next 10 new downgradient measurements is

$$11.0 + 3.50(0.61)\sqrt{1 + \frac{1}{8}} = 13.26 \text{ mg/l}$$

In contrast, if the dependence introduced by comparing all 10 downgradient wells to the same background were incorporated as in Equation (3), the result of

$$11.0 + 3.31(0.61)\sqrt{1 + \frac{1}{8}} = 13.14 \text{ mg/l}$$

is obtained (see Table 1.4 in Gibbons, 1994). Note that the limit is slightly lower because the multiplier incorporates the dependence introduced by repeated comparison to a common background (i.e., the number of independent comparisons is less than 10 given that they are correlated). Although the Bonferroni-based limit is too conservative, there is little difference in the limits. Extending this result to include the effects of a verification resample as in Equation (4) further decreases the limit to

$$11.0 + 2.03(0.61)\sqrt{1 + \frac{1}{8}} = 12.31 \text{ mg/l}$$

If each of 10 constituents in each of the 10 downgradient wells had been monitored, $\alpha = 0.05 = 10 = 0.005$ and the limit would become

$$11.0 + 3.36(0.61)\sqrt{1 + \frac{1}{8}} = 13.17 \text{ mg/l}$$

(see Table 1.5 in Gibbons, 1994). Note that the verification resample allows application of essentially the same limit derived for 10 wells and 1 constituent (13.14 mg/l) to a problem of 10 wells and 10 constituents (13.17 mg/l). Now, consider the nonparametric alternative of taking the maximum of the initial eight background measurements and applying it to the next future monitoring measurements. In this example, the nonparametric prediction limit is 12.00 mg/l. For a single future measurement, confidence is 88% without a resample and 98% with a resample (see Table 2.5 and Table 2.6 in Gibbons, 1994).

For a single measurement in each of 10 monitoring wells, confidence is 44% without a resample and 84% with a resample (see Table 2.5 and Table 2.6 in Gibbons, 1994). With 10 constituents and 10 monitoring wells, an overall 95% confidence level would be obtained with $n = 60$ background samples for one verification resample (see Table 2.6 in Gibbons, 1994) or $n = 20$ samples for passage of one of two verification resamples (see Table 2.7 in Gibbons, 1994). Note that if either the initial sample or both of two resamples must be passed then $n = 90$ background measurements must be obtained (see Table 2.13 in Gibbons, 1994). Other illustrations and further statistical details are available (Davis, 1994; Gibbons, 1994; Davis and McNichols, 1994b).

Some Methods to be Avoided

Analysis of Variance — ANOVA

In both U.S. EPA Subtitle C and D regulations and associated guidance (U.S. EPA, 1988b, 1989, 1991, 1992), ANOVA is suggested as the statistical method of choice. U.S. EPA's specific recommendation is a one-way fixed-effect model where the upgradient wells are pooled as one level and each downgradient well represents an additional level in the design. A minimum of four samples is obtained from each well within a semiannual period. In the presence of a significant F-statistic, post hoc comparisons (i.e., Fisher's LSD method) between each downgradient well and the pooled upgradient background are performed. Either parametric or nonparametric ANOVA models (i.e., Kruskal–Wallis test) are acceptable. Unfortunately, application of either parametric or nonparametric ANOVA procedures to detection monitoring is inadvisable for the following reasons:

1. Univariate ANOVA procedures do not adjust for multiple comparisons due to multiple constituents. This can be devastating to the site-wide false-positive rate. As such, a site with 10 indicator constituents will have as much as a 40% probability of failing for at least one constituent on every monitoring event by chance alone.

2. ANOVA is more sensitive to spatial variability than to contamination. Spatial variability produces systematic differences between locations that are large relative to within-location variation (i.e., small consistent differences due to spatial variation achieve statistical significance). In contrast, contamination increases variability within the impacted locations, therefore a much larger between-location difference is required to achieve statistical significance. In fact, application of ANOVA methods to predisposal ground-water-monitoring data often results in statistically significant differences between upgradient

and downgradient wells, even when no waste is present (Gibbons, 1994), as illustrated in the example below.

3. Nonparametric ANOVA is often presented as if it would protect the user from all of the weakness of its parametric counterpart; however, the only assumption relaxed is that of normality. The nonparametric ANOVA still assumes independence, homogeneity of variance, and that each measurement is identically distributed. Violation of any of these assumptions can corrupt the power of detection, or the false-positive rate.

4. ANOVA requires pooling of downgradient data. Specifically, U.S. EPA suggests that four samples per semiannual monitoring event be collected (i.e., eight samples per year). However, ANOVA cannot rapidly detect a release since only a subset of the required four semiannual samples will initially be affected by a site impact. This heterogeneity will decrease the mean concentration and increase the variance for the affected location, limiting the ability of the statistical test to detect actual contamination. To illustrate, consider the data in Table 17.6 obtained from a facility in which disposal of waste has not yet taken place (Gibbons, 1994).

Applying both parametric and nonparametric ANOVAs to these predisposal data yielded an effect that approached significance for COD ($p < .072$ parametric and

TABLE 17.6

Raw Data for All Detection Monitoring Wells and Constituents (mg/l) (This Facility has no Waste in It)

	Well Event	TOC	TKN	COD	ALK
MW01	1	5.2000	0.8000	44.0000	58.0000
MW01	2	6.8500	0.9000	13.0000	49.0000
MW01	3	4.1500	0.5000	13.0000	40.0000
MW01	4	15.1500	0.5000	40.0000	42.0000
MW02	1	1.6000	1.6000	11.0000	59.0000
MW02	2	6.2500	0.3000	10.0000	82.0000
MW02	3	1.4500	0.7000	10.0000	54.0000
MW02	4	1.0000	0.2000	13.0000	51.0000
MW03	1	1.0000	1.8000	28.0000	39.0000
MW03	2	1.9500	0.4000	10.0000	70.0000
MW03	3	1.5000	0.3000	11.0000	42.0000
MW03	4	4.8000	0.5000	26.0000	42.0000
MW04	1	4.1500	1.5000	41.0000	54.0000
MW04	2	1.0000	0.3000	10.0000	40.0000
MW04	3	1.9500	0.3000	24.0000	32.0000
MW04	4	1.2500	0.4000	45.0000	28.0000
MW05	1	2.1500	0.6000	39.0000	51.0000
MW05	2	1.0000	0.4000	26.0000	55.0000
MW05	3	19.6000	0.3000	31.0000	60.0000
MW05	4	1.0000	0.2000	48.0000	52.0000
MW06	1	1.4000	0.8000	22.0000	118.0000
MW06	2	1.0000	0.2000	23.0000	66.0000
MW06	3	1.5000	0.5000	25.0000	59.0000
MW06	4	20.5500	0.4000	28.0000	63.0000
P14	1	2.0500	0.2000	10.0000	79.0000
P14	2	1.0500	0.3000	10.0000	96.0000
P14	3	5.1000	0.5000	10.0000	89.0000

Organization and Analysis of Ground-Water Quality Data

$p < .066$ nonparametric) and a significant difference for alkalinity (ALK) ($p < .002$ parametric and $p < .009$ nonparametric). Individually compared (using Fisher's LSD), significantly increased COD levels were found for well MW05 ($p < .026$) and significantly increased ALK was found for wells MW06 ($p < .026$) and P14 ($p < .003$) relative to upgradient wells. These results represent false positives due to spatial variability since no waste has been deposited at this site (i.e., a "greenfield" site).

Most remarkable is the absence of significant results for TOC, notwithstanding the fact that some values are as much as 20 times higher than others. These extreme values increase the within-well variance estimate, rendering the ANOVA powerless to detect differences regardless of magnitude. Elevated TOC data are inconsistent with chance expectations (based on analysis using prediction limits) and should be investigated. In this case, elevated TOC data are likely caused by contamination from insects getting into the wells since this greenfield facility is located in the middle of the Mojave desert.

Cochran's Approximation to the Behrens Fisher t-test

For years the U.S. EPA RCRA regulation (U.S. EPA, 1982) was based on application of the Cochran's approximation to the Behrens Fisher (CABF) *t*-test. The test was incorrectly implemented by requiring that four quarterly upgradient samples from a single well and single samples from a minimum of three downgradient wells each be divided into four aliquots and treated as if there were $4n$ independent measurements. The result was that most hazardous waste disposal facilities regulated under RCRA were declared "leaking." As an illustration, consider the data in Table 17.7. Note that the aliquots are almost perfectly correlated and add virtually no independent information, yet they are assumed by the statistic to be completely independent. The CABF *t*-test is computed as

$$t = \frac{\bar{x}_B - \bar{x}_M}{\sqrt{(S_B^2/N_B) + (S_M^2/N_M)}} = \frac{7.62 - 7.40}{\sqrt{(0.032/16) + (0.004/4)}} = \frac{0.22}{0.05} = 4.82 \qquad (17.7)$$

The associated probability of this test statistic is 1 in 10,000, indicating that the chance that the new monitoring measurement came from the same population as the background

TABLE 17.7

Illustration of pH Data Used in Computing the CABF *t*-Test

	Replicate				
Date	1	2	3	4	Average
Background					
November 1981	7.77	7.76	7.78	7.78	7.77
February 1982	7.74	7.80	7.82	7.85	7.80
May 1982	7.40	7.40	7.40	7.40	7.40
August 1982	7.50	7.50	7.50	7.50	7.50
$\bar{x}_B$		7.62			7.62
SD_B		0.18			0.20
N_B		16			4
Monitoring					
September 1983	7.39	7.40	7.38	7.42	7.40
$\bar{x}_B$		7.40			7.40
SD_B		0.02			
N_B		4			1

measurements is remote. Note that, in fact, the mean concentration of the four aliquots for the new monitoring measurement is identical to one of the four mean values for background, suggesting intuitively that probability is closer to one in four rather than one in 10,000. Averaging the aliquots yields the statistic

$$t = \frac{\bar{x}_B - \bar{x}_M}{S_B\sqrt{(1/N_B)+1}} = \frac{7.62 - 7.40}{0.20\sqrt{(1/4)+1}} = \frac{0.22}{0.22} = 1.0 \qquad (17.8)$$

which has an associated probability of one in two. Had the sample size been increased to $N_B = 20$, the probability would have decreased to one in three. U.S. EPA eliminated this method from the regulation (U.S. EPA, 1988b).

Summary

Protection of our natural resources is critical. However, statistical tools used to make environmental impact decisions are limited and often confusing. The problem is not only interesting regarding development of public policy, but it also contains features of statistical interest such as multiple comparisons, sequential testing, and censored distributions. Highlighting the weaknesses of currently mandated regulations may lead to further critical examination of public policy in the field of ground-water monitoring as well as heightened interest in statistical analysis.

Reporting Water-Quality Data to Agencies

State and Federal regulations require some form of reporting to confirm that the monitoring system is working as required by the codes. Some regulations require the reporting of tabular sets of data on forms or through a formatted electronic media. In general terms, all data should be fully reviewed before transmittal to regulatory agencies. A simple set of guidelines can ease potential errors and embarrassment when submitting water-quality data on your facility:

- Read the permit or waste discharge requirements and follow them.
- Format data as required in a manner that communicates the data most effectively (so everyone reaches the same conclusions).
- If the state requires reporting of exceedances, format the response in a neutral manner:
  - Talk about the specific exceedance issues.
  - Relate progress made on defining causes of the exceedance(s).
  - Propose schedules for establishing the cause of the exceedance or schedules for the remedial actions required.
  - Provide a summary statement on the level of concern
- Maintain consistency and continuity between quarterly reports:
  - Indicator parameter exceedance changes from quarter to quarter.
  - New personnel should review past data.
  - Always cross-check reports from quarter to quarter.

- Explain what will be done with the data.
- Maintain technical standards and textural reporting consistency between sites; always maintain a consistent standard format for reporting water-quality data.

References

Aitchison, L., On the distribution of a positive, random variable having a discrete probability mass at the origin, *Journal of the American Statistical Association*, 50, 901–908, 1955.

Benjamin, J.R. and C.A. Cornell, *Probability Statistics and Decision for Civil Engineers*, McGraw-Hill Publishing Co., New York, NY, 1970.

Bock, R.D., *Multivariate Statistical Methods in Behavioral Research*, McGraw-Hill Publishing Co., New York, NY, 1975.

Chou, Y.M. and D.B. Owen, One-sided distribution-free simultaneous prediction limits for future samples, *Journal of Quality Technology*, 18, 96–98, 1986.

Cohen, A.C., Simplified estimators for the normal distribution when samples are singly censored or truncated, *Technometrics*, 1, 217–237, 1959.

Cohen, A.C., Tables for maximum likelihood estimates: singly truncated and singly censored samples, *Technometrics*, 3, 535–541, 1961.

Davis, C.B., Environmental regulatory statistics, in *Handbook of Statistics: Environmental Statistics*, Patil, G.P. and Rao, C.R. Eds., Vol. 12, Elsevier Press, New York, NY, 1994, Chap. 25.

Davis, C.B. and R.J. McNichols, One-Sided Intervals for at Least p of m Observations From a Normal Population on Each of r Future Occasions, *Technometrics*, 29, 359–370, 1988.

Davis, C.B. and R.J. McNichols, Ground-water monitoring statistics update: I: progress since 1988, *Ground-Water Monitoring and Remediation*, 14(4), 148–158, 1994a.

Davis, C.B. and R.J. McNichols, Ground-water monitoring statistics update: II: non-parametric prediction limits, *Ground-Water Monitoring and Remediation*, 14(4), 159–169, 1994b.

Doctor, P.G., R.O. Gilbert, R.A. Saar, and G. Duffield, Draft Statistical Comparisons of Ground-Water Monitoring Data: Ground-Water Plans and Statistical Procedures to Detect Leaking at Hazardous Waste Facilities, Document PNL-5754, Battelle Pacific Northwest Laboratories, Richland, WA, 1986.

Doctor, P.G., R.O. Gilbert, R.A. Saar, and G. Duffield, An Analysis of Sources of Variation in Ground-Water Monitoring Data of Hazardous Waste Sites, Milestone 1, Revised Draft EPA Contract No. 68-01-6871, Battelle Pacific Northwest Laboratories. Richland, WA, 1985.

Dunnett, C.W., A multiple comparisons procedure for comparing several treatments with a control, *Journal of the American Statistical Association*, 50, 1096–1121, 1955.

Dunnett, C.W. and M. Sobel, Approximations to the probability integral and certain percentage points of a multivariate analogue of student's t-distribution, *Biometrika*, 42, 258–260, 1955.

Fisher, D.A. and K. Potter, Methods for Determining Compliance with Ground Water Quality Regulations at Waste Disposal Facilities, Wisconsin Department of Natural Resources, Madison, WI, 1989, 120 pp.

Freeze, R.A. and J.A. Cherry, *Groundwater*, Prentice-Hall, Inc., Englewood Cliffs, NJ, 1979, 604 pp.

Gibbons, R.D., Statistical prediction intervals for the evaluation of ground-water quality, *Ground Water*, 25(4), 455–465, 1987a.

Gibbons, R.D., Statistical models for the analysis of volatile organic compounds in waste disposal facilities, *Ground Water*, 25(5), 572–580, 1987b.

Gibbons, R.D., A general statistical procedure for ground-water detection monitoring at waste disposal facilities, *Ground Water*, 28(2), 235–243, 1990.

Gibbons, R.D., Some additional nonparametric prediction limits for ground-water monitoring at waste disposal facilities, *Ground Water*, 29(5), 729–736, 1991.

Gibbons, R.D., *Statistical Methods for Ground-Water Monitoring*, John Wiley and Sons, New York, NY, 1994.

Gibbons R.D. and D.E. Coleman, *Statistical Methods for Detection and Quantification of Environmental Contamination*, John Wiley and Sons, New York, NY, 2001.

Gibbons, R.D., F.H. Jarke, and K.P. Stoub, Method Detection Limits, Proceedings of the Fifth Annual U.S. EPA Waste Testing and Quality Assurance Symposium, 1988, pp. 292–319.

Gilliom, R.J. and D.R. Helsel, Estimation of distributional parameters for censored trace-level water-quality data: 1. estimation techniques, *Water Resources Research*, 22, 135–146, 1986.

Gleit, A., Estimation for small normal data sets with detection limits, *Environmental Science and Technology*, 19, 1201–1206, 1985.

Goodman, I., Graphical and Statistical Methods to Assess the Effect of Landfills on Ground-Water Quality, M.S. Thesis, University of Wisconsin Madison, 1987.

Goodman, I. and K. Potter, Graphical and Statistical Methods to Assess the Effects of Landfills on Ground Water Quality, Report to Wisconsin Department of Natural Resources, Bureau of Solid and Hazardous Waste, 1987.

Green, W.R., *Computer-Aided Data Analysis: A Practical Guide*, John Wiley and Sons, New York, NY, 1985, 268 pp.

Gupta, A.K., Estimation of the mean and standard deviation of a normal population from a censored sample, *Biometrika*, 39, 260–273, 1952.

Gupta, S.S. and P. Panchpakesan, *Multiple Decision Procedures*, John Wiley and Sons, Inc., New York, NY, 1979.

Guttman, I., *Statistical Tolerance Regions: Classical and Bayesian*, Hafner Publishing, Darien, CT, 1970.

Haas, C.N. and P.A. Scheff, Estimation of averages in truncated samples, *Environmental Science and Technology*, 24, 912–919, 1990.

Hahn, G.J., Statistical intervals for a normal population, part 1: examples & applications, *Journal of Quality Technology*, 2(3) (July), pp. 15–125, 1970.

Hall, I.J., R.R. Prarie, and C.K. Motlagh, Non-parametric prediction intervals, *Journal of Quality Technology*, 7, 109–114, 1975.

Harris, J., J.C. Loftis, and R.H. Montgomery, Statistical methods for characterizing ground-water quality, *Ground Water*, 25(2), 185–193, 1987.

Hashimoto, L.K. and R.R. Trussell, Evaluating Water Quality Data Near the Detection Limit, Proceedings of the American Water Works Association Advanced Technology Conference, American Water Works Association, Denver, CO, 1983.

Hem, J.D., Study and Interpretation of the Chemical Characteristics of Natural Water, U.S. Geological Survey Water-Supply Paper 1473, 1970, 363 pp.

Hoaglin, D.C., F. Mosteller, and J.W. Tukey, *Understanding Robust and Exploratory Data Analysis*, John Wiley & Sons, New York, NY, 1983.

Hochberg, Y. and A.C. Tamhane, *Multiple Comparison Procedures*, John Wiley and Sons, Inc., New York, NY, 1987.

Hurd, M., Personal communication, U.S. Environmental Protection Agency, 1986.

Jarke, F., Is it possible to understand MDLs, PQLs, IDLs, and EMLRLs? *Lab Notes*, 2, 1989.

Lucas, J.M., Combined shewhart-CUSUM quality control schemes, *Journal of Quality Technology*, 14, 51–59, 1982.

McBean, E. and F.A. Rovers, Alternatives for handling detection limit data in impact assessments, *Ground-Water Monitoring Review*, 4(2), 42–44, 1984.

McGill, R., J.W. Tukey, and W.A. Larsen, Variations of box plots, *The American Statistician*, 32(1), 12–16, 1978.

McNichols, R.J. and C.B. Davis, Statistical issues and problems in ground-water detection monitoring at hazardous waste facilities, *Ground-Water Monitoring Review*, 8(4), 135–150, 1988.

Miller, R.G., *Simultaneous Statistical Inference*, McGraw-Hill Publishing Co., New York, NY, 1966.

Miller, M.D. and F.C. Kohout, RCRA Ground-Water Monitoring Statistical Comparisons: A Better Version of Student's *t*-Test, Mobil Research and Development Corporation, Paulsboro, NJ, undated.

Montgomery, R.H., J.C. Loftis, and J. Harris, Statistical characteristics of ground-water quality variables, *Ground Water*, 25(2), 176–184, 1987.

Persson, T. and H. Rootzen, Simple and highly efficient estimators for a type I censored normal sample, *Biometrika*, 64, 123–128, 1977.

Piper, A.M., A Graphic procedure in the geochemical interpretation of water analysis; transactions of the american geophysical union, 25, 914–923, 1944.

Sarhan, A.E. and B.G. Greenberg, Eds., *Contributions to Order Statistics*, John Wiley and Sons, Inc., New York, NY, 1962.

Sen, Z., Application of the autorun test to hydrologic data, *Journal of Hydrology*, 42, 1–7, 1979.

Sen, Z., Discussion of statistical considerations and sampling techniques for ground-water quality monitoring by J.D. Nelson and R.C. Ward, *Ground Water*, 20(4), 494–495, 1982.

Silver, C.A., Statistical Approaches to Ground-Water Monitoring, Open File Report #7, University of Alabama, Environmental Institute for Waste Management Studies, 1986.

U.S. EPA, Hazardous Waste Management System: Permitting Requirements for Land Disposal Facilities; Federal Register, 47(143), July 26, 1982, 32274–32373.

U.S. EPA, Test Methods for Evaluating Solid Waste, Physical Chemical Methods, SW-846, U.S. Environmental Protection Agency, Washington, DC, 1988a.

U.S. EPA, Statistical Methods for Evaluating Ground Water Monitoring Data From Hazardous Waste Facilities: Final Rule, 40 CFR Part 264, Federal Register, Vol. 53, No. 196, October 11, 1988b, pp. 39720–39731.

U.S. EPA, Interim Final Guidance on Statistical Analysis of Ground-Water Monitoring Data at RCRA Facilities, April, 1989.

U.S. EPA, Solid Waste Disposal Facility Criteria: Final Rule, Federal Register, Vol. 56, No. 196, October 9, 1991, pp. 50978–51119.

U.S. EPA, Addendum to Interim Final Guidance on Statistical Analysis of Ground-Water Monitoring Data at RCRA Facilities, July, 1992.

Willits, N., Personal Communication, University of California-Davis, 1993.

18

Diagnosis of Ground-Water Monitoring Problems

Charles T. Kufs

CONTENTS

Introduction 1193
Sample Space Problems 1194
System Implementation Problems 1196
Program Implementation Problems 1198
Geologic Uniformity Problems 1204
Hydrologic Uniformity Problems 1204
Geochemical Interaction Problems 1208
Diagnosis of Monitoring Problems 1212
References 1216

Introduction

There are many factors to consider in planning and implementing a ground-water monitoring system; therefore, it is not surprising when systems do not function as they were intended to. Sometimes, parts of a system will not function as intended and will have to be replaced if the problem cannot be diagnosed and corrected. In the worst case, a system will appear to be functioning correctly but will actually be producing data that are incorrect, misleading, or uninterpretable.

Monitoring ground water requires using professional judgment, not just following a set of standardized procedures. There are many excellent textbooks available on the subjects of ground-water flow and quality, yet there is no universally accepted procedure for designing monitoring systems, nor could there be. The variability of geologic deposits, the intricacies of contaminant geochemistry, and the under-appreciated micro and macro scales of ground-water hydrology make developing anything beyond a generic approach virtually impossible. This is no consolation either to hydrogeologists who have professional reputations to maintain or to clients who pay for the failures as well as the successes. But it does illustrate why it is essential to be able to prevent, or at least recognize and correct, problems with monitoring systems.

Although the problems that can arise in monitoring ground-water flow and quality are as diverse as the site conditions under which monitoring takes place, it is useful to understand how to address typical classes of monitoring dilemmas. The goal of this chapter is to summarize approaches for the prevention, recognition, and correction of 27 types of ground-water monitoring problems.

Sample Space Problems

Sample space problems refer to situations in which the wells in a system are not in appropriate locations for monitoring an appropriate volume of the aquifer. Typical sample space problems include:

- Wells not positioned for identifying flow directions
- Wells not positioned for evaluating the extent of contamination
- Screen settings not correctly selected
- Screen lengths not correctly selected
- System not designed to accomplish study objectives

Sample space problems can often be attributed to the placement of the wells in a system. Well location is usually the design element that is given the most attention in ground-water monitoring. Nevertheless, several conditions can lead to inadequate arrangement of wells. For example, it is common to place wells along a site boundary downgradient of a contaminant source. However, if there are too few wells placed at appropriate distances upgradient and across the hydraulic gradient, the system will lack dimensionality and will not produce unambiguous flow directions. This problem can occur in a vertical as well as a horizontal plane (Saines, 1981) when well screens are not set to span an appropriate hydrostratigraphic zone (i.e., geologic unit or fracture set). Restricted site access is a common contributor to this type of problem (Figure 18.1).

Ideally, aquifers should be assessed as three-dimensional systems. Pragmatically, technical complexities and cost constraints usually limit initial monitoring efforts to two dimensions, most commonly in the horizontal plane. One-dimensional monitoring systems (i.e., systems in which the majority of the wells are placed in a line) should be avoided.

Individual wells in a monitoring system can also present sample space problems, usually because the size of the aquifer space being monitored is too large for reliable

FIGURE 18.1
Sometimes the ideal well locations you selected at the office do not look so good in the field. Always conduct a site visit before mobilizing heavy equipment. Keep in mind the goals of the monitoring program when you make changes. Remember to clear utilities.

measurement of hydrologic or geochemical parameters. This type of problem can result when the length of the well intake (i.e., well screen or section of open hole) is increased to obtain a predetermined yield. Although this is an appropriate and desirable procedure to follow for water supply wells, it can produce misleading data when applied to the design of monitoring wells if discrete hydrogeochemical zones are mixed during sampling. Issues related to this sample space problem are discussed by Giddings (1986) and Shosky (1986).

A similar situation occurs when excessive volumes of water are pumped prior to sampling, although this procedure may be justifiable for basin-wide studies or for time-series sampling (Keely, 1982). The commonly used practice of purging until the temperature, pH, or specific conductance of ground water stabilize, for instance, will usually yield analytical results that represent the water quality averaged over a relatively large sample space. Depending on the scale of the study area and the objective of the sampling, this approach could result in contaminant concentrations that are negatively biased and are not truly representative of an appropriate finite space in the aquifer.

Consider calculating the size of the aquifer space you might be sampling after you purge the well. For simplicity, you can assume that ground water flows radially to the well from a cylinder or inverted cone. The volume of water in the well will be equal to $\pi r_w^2 h$, where r_w is the radius of the well and h is the height of the water column. Use that volume to estimate the radius of the aquifer (r_a) drained to fill the well from a cylinder ($\sigma \pi r_a^2 h$) or a cone ($1/3 \sigma \pi r_a^2 h$) where σ is an estimate of the aquifer's effective porosity. For example, purge five well volumes from a 4 in. well in an aquifer with an effective porosity of 0.25 and the radius of the area sampled would be between 3 and 10 ft. Purge 50 well volumes (easy to do in some situations if you are purging until field parameters stabilize), and the radius of the area sampled would be between 15 and 50 ft. The difference can be especially consequential when interpreting data from the well as coming from a finite point rather than dispersed over a larger area. It is like treating a composite sample as a grab sample.

A better purging procedure for ground-water contamination studies is to begin purging at a low discharge rate with the pump at the top of the water column and continuously lower the pump as drawdown increases until an appropriate volume of water is purged (Keely and Boateng, 1987). Other issues related to purging before sampling are discussed by Robin and Gillham (1987), Herzog et al. (1988), Gibs and Imbrigiotta (1990), and in Chapter 15 of this book.

Sample space problems also can occur when monitoring programs are not designed appropriately, such as when the objectives of monitoring evolve over time (Herzog et al., 1986). Possibly, the most common type of sample space problem results from the inappropriate use of existing wells. Water-supply wells, in particular, are often used inappropriately because they are frequently the points at which contamination is first identified. Because they are usually screened over large intervals to maximize yields, water-supply wells often tap many different hydrostratigraphic zones. Sampling these wells using traditional methods (e.g., pumps and bailers) can result in a mixture of several different water qualities rather than a true representation of a contaminant plume. Alternative approaches that may be useful in these cases involve the use of passive-membrane samplers (Ronen et al., 1987) and the combination of flowmeter logs and sampling (Sukop, 2000). In general, water-supply wells are appropriate for large-scale aquifer testing and for evaluating risks associated with drinking contaminated ground water but rarely appropriate for evaluating ground-water flow directions or contaminant distribution.

Existing monitoring wells also may be used inappropriately if they were originally designed for a different purpose. For example, monitoring wells installed as part of a

Resource Conservation and Recovery Act (RCRA) detection monitoring program may not be useful for delineating the vertical or horizontal extent of contamination if their locations and screen lengths are not appropriate. Typically, RCRA wells are installed in the uppermost aquifer at the boundary of the waste management unit so that they can detect any leakage of contaminants (Smart and Cook, 1988). However, once contaminants migrate far beyond the locations of the RCRA wells, the wells are of little use in determining the extent of contamination. Nevertheless, there is a tendency to use existing wells, especially monitoring wells, in the mistaken belief that larger data sets produce more certain results even if the quality of the data from some wells may be suspect.

Table 18.1 summarizes methods for the prevention, recognition, and correction of typical sample–space problems.

System Implementation Problems

System implementation problems refer to situations in which the well or other elements of the system do not perform as designed. Typical system implementation problems include:

- Well does not produce sufficient water
- Well silts up after installation
- Sand pack becomes clogged
- Well seals leak
- Well materials degrade
- Well is poorly constructed

System implementation problems are probably the most easily recognizable type of problem. Sometimes, the problems cannot be avoided but can be controlled so that the effects are acceptable. For example, well siltation is difficult to avoid in designing a monitoring well, given that there is rarely any site-specific data available on the aquifer's grain size distribution to calculate an appropriate screen slot size and sand pack gradation. Instead, screen slot sizes are typically selected using general information from regional water-supply studies. Sumps (i.e., blank lengths of well casing generally 1–5 ft long) can be added below the screen to reduce the effects of siltation, and the well can be redeveloped periodically.

Low yields from monitoring wells do not necessarily mean that the well is not usable. Low yields often result when trying to monitor low-permeability zones, or the top few feet of a zone with a fluctuating water table. Sometimes, low yields can be attributed to clogged sand packs or screens, thus requiring extensive redevelopment or, in the extreme, abandonment. This is often true where high concentrations of solvents dissolve aquifer or contaminant solids that subsequently precipitate in sand packs because of changes in flow velocity or solvent concentrations. In most cases, however, these problems can be managed to provide acceptable water for testing.

Although well yield may not necessarily be important for a monitoring well, well efficiency (i.e., the ratio of the actual to the theoretical specific capacity) is an important factor in evaluating the aquifer being pumped. Monitoring wells normally do not have to be highly efficient in providing valid static water levels as long as the system is in equilibrium. However, under dynamic (i.e., pumping) conditions, well efficiency is very important because low-efficiency wells will show excessive drawdowns that will not be

TABLE 18.1
Prevention, Recognition and Correction of Sample–Space Problems

Problem	Prevention	Recognition	Correction
1. Wells not positioned for identifying ground-water flow directions.	Use basic hydrogeologic assumptions to estimate the flow directions. Use ground penetrating radar (GPR), if possible, to evaluate the validity of the assumptions.	Water elevations do not produce a unique contour pattern; too few wells screened in the same zone; wells installed nearly along a line.	Install additional wells or find existing wells screened in the same water-bearing zone.
2. Wells not positioned for evaluating the extent of contamination.	Estimate the distance the contaminant plume may have migrated from the site based on site history, hydrogeology, and contaminant geochemistry. Use aerial images or electromagnetic conductivity (EM) and soil-gas surveys to check estimation.	Contaminant concentrations do not produce a unique contour pattern; the contaminant plume does not appear to be related to the suspected source, or the contaminant pattern suggests undocumented sources.	Install additional wells or find existing wells screened in the same aquifer. In some cases, soil-gas or EM surveys can be used to augment monitor well networks.
3. Screen settings not correctly selected.	Use background geologic and geochemical information and geophysical surveys to project contaminant flow. Compare information to onsite soil samples collected from boreholes.	Water elevations appear to be anomalous, apparent flow directions seem illogical or overly complex; contaminant concentrations are much lower than expected.	Install additional wells or find existing wells screened in the same aquifer. In some cases, packers can be used to test specific zones.
4. Screen length not correctly selected.	Use background information and geophysical surveys to project correct screen length to meet study objectives. Confirm length using soil samples collected from boreholes.	Water elevations appear to be anomalous; contaminant concentrations are lower than expected.	Use packers to isolate zones in open-hole wells. Install additional wells.
5. System not designed to accomplish study objectives.	Identify ultimate use of data and methods of data analysis to estimate minimum sample size.	Ground-water flow or contaminant migration appears to be ambiguous or nonsensical.	Resamples wells and/or install additional wells. Augment direct data with indirect data (e.g., geophysics and soil-gas). Delete anomalous data collected from suspect wells.

representative of the general trend in the aquifer's piezometric surface (i.e., the drawdown in the well will be much larger than the drawdown in the aquifer near the well). This concept is critical if the water elevations are contoured to evaluate flow patterns, especially if the contouring does not reflect the judgment of a competent hydrogeologist.

More serious system implementation problems involve faulty well construction or degradation of well materials by corrosive contaminants or natural ground water. The methods used to install a monitoring well (e.g., hollow-stem augering) can result in both ground-water flow and ground-water quality anomalies (Dunbar et al., 1985; Morin et al., 1988; Paul et al., 1988; Parker et al., 1990). Recovery from well installation trauma can take years in low-permeability formations (Walker, 1983).

Careful and complete evaluation of existing wells is essential before using these wells to collect ground-water information (Ricci, 1985; Knox and Jacobson, 1986). For example, in high enough concentrations, some organic solvents can cause polyvinyl chloride (PVC) well screens to soften and swell shut or to become brittle and crack. Solvents can also degrade bentonite annular seals, thus allowing water to flow freely along the outside of the well casing. Encrustation of steel well screens by bacteria or inorganic salts can reduce well yields and skew sample analyses. These problems are more serious because they are often difficult to recognize, may worsen the spread of contaminants, and invariably result in abandonment of the well. This is particularly true of wells installed into deeper confined aquifers, especially when the uppermost aquifer has not been cased off properly.

Recognizing that a well is defective is not always straightforward. Broken screens or joints may be indicated if aquifer or sand pack particles in the well are larger than the screen's slot size. Encrusted screens or clogged sand packs are often indicated by low well efficiencies, excessive drawdowns during pumping, anomalously low or high water levels, or excessive time required for the well to reach equilibrium relative to the transmissivity of the aquifer. Leaking well seals are the hardest well construction problem to recognize, especially if ground water at the well intake and the seal are at similar piezometric levels. If a well appears to be defective, it should not be used for monitoring and should probably be decommissioned (Perazzo et al., 1984).

Table 18.2 summarizes methods for the prevention, recognition, and correction of typical system implementation problems.

Program Implementation Problems

Program implementation problems refer to situations in which field data collection or laboratory analysis procedures fail to produce high-quality data. Although these problems occur with some frequency, they are often downplayed by investigators because they represent simple and avoidable errors. Typical program implementation problems include:

- Well construction is not adequately documented
- Field data collection procedures are inadequate
- Sample collection procedures are inappropriate
- Sample analysis procedures are inadequate or are undocumented

Most of the data generated during site investigations under Superfund and RCRA are either laboratory chemistry data (e.g., analysis documentation and results) or field geoenvironmental data (e.g., geologic descriptions, well construction information, or water levels). Analytical chemistry data are generated by chemists following EPA-mandated

TABLE 18.2
Prevention, Recognition, and Correction of System Implementation Problems

Problem	Prevention	Recognition	Correction
1. Well does not produce sufficient amounts of water	If consistent with monitoring objectives, screen well in coarse granular or highly fractured medium	Well is dry or recharges too slowly to sample effectively	Redevelop well; deepen bedrock well if consistent with study objectives; redesign new well; reevaluate appropriateness of location
2. Well silts up after installation	Select screen opening size and sand pack gradation to be compatible with geologic materials to be screened; add a sump below the well screen	Water is murky or bottom of well feels "mushy" when sounded	Redevelop well periodically
3. Sand pack becomes clogged	Specify a well-sorted (poorly graded) coarse-grained, washed quartz sand or gravel consistent with the aquifer material	Well recharges much more slowly than expected	Redevelop well periodically; redesign new well
4. Well seals leak	Design seals to be compatible with projected use of well and site hydrogeology and geochemistry; monitor installation of seals closely by repeatedly measuring the depth to the seal	Water elevation and quality on either side of the seal are more similar than expected	Abandon leaking wells to prevent interaquifer leakage, and replace well
5. Well materials are degraded by contaminants or fail structurally	Specify stainless steel for areas of high organic contamination and PVC or Teflon in areas of extreme pH; specify appropriate material strength based on expected loads; screen or overdrill highly fractured bedrock wells	Obstructions are found in the well; aquifer materials that are larger than screen slots enter the well; well yields decrease over time; phthalates or inorganics increase over time	Abandon and replace well
6. Well is poorly constructed	Hire a reliable driller; have an experienced hydrogeologist monitor well installation	Evidence of poor workmanship at surface; well is not plumb and aligned; water levels and quality appear anomalous	Abandon and replace well

protocols using calibrated instruments. Quality is ensured using a variety of quality assurance (QA) samples, lab audits, and an extensive data validation process for every sample. Data are managed from sample collection through data validation using sophisticated laboratory information management systems (LIMs). Geoenvironmental data, on the other hand, can be generated by any of a number of different types of specialists in earth sciences and engineering following one of perhaps half a dozen different procedures. Normally, no special quality checks are applied except on some relatively large programs. Data are handwritten in notebooks (Figure 18.2) in the field and then manually translated into graphical logs. Information is extracted from the logs to form the basis for ground-water models and other methods of hydrogeologic assessment (Kufs et al., 1992).

Of these two types of data, problems with laboratory documentation seem to be cited more commonly than problems with geoenvironmental data. The greater frequency of reported laboratory problems is probably attributable to the fact that sample analysis protocols are more widely accepted and better documented than field protocols. As a consequence, deviations from standard practices are easier to identify. Calculation errors also are sometimes easy to find, especially results that appear to be wrong by a factor of 10. Another aspect of sample analysis problems involves improper specification of analytical procedures. Improper procedure specification occurs most commonly in situations in which inexperienced field personnel specify the analytical procedures to laboratory personnel who do not review the appropriateness of the procedures relative to the objectives of the monitoring program. Very often, a standard protocol is specified without adequate attention given to analytes that may be unnecessary or missing. Certainly, laboratories make mistakes in analyzing samples, and data should be examined carefully to identify possible errors. However, field data collection also should be examined closely.

Descriptions of two types of earth materials, unconsolidated materials and rocks, can be critical in interpreting data from a monitoring system. Rock description is usually straightforward. Although there are hundreds of different types of rocks, there are fewer than two dozen common types. More important than rock type in environmental studies are properties that can influence contaminant migration, such as fractures and other discontinuities, weathering, and porosity. These and other properties are identified in a number

FIGURE 18.2
Geologic descriptions are customarily collected in blank-page notebooks. The quality and completeness of these data will depend largely on the training and experience of the logger. Sampling at large intervals can also degrade the quality of the data. Interpretation of ground-water monitoring results relies on these data.

Diagnosis of Ground-Water Monitoring Problems 1201

of procedures published by the American Society for Testing and Materials (ASTM). Description of unconsolidated materials is more complicated because there are so many different systems in use. A silt with sand is also an ML soil (Unified Soil Classification System; ASTM, 2004) to a geotechnical engineer; a sandy loam (USDA Soil Classification System; U.S. Department of Agriculture, 1975) to a soil scientist; a clayey silt with some sand (Burmeister System; Burmeister, 1951) to a geologist; and a sandy, silty clay (Sheppard System) to an ecologist. Several documents have been developed by EPA, most notably Boulding (1991), Cameron (1991), Breckenridge et al. (1991), and Burden and Sims (1999), that can serve as guides. These documents incorporate elements of the most commonly used classification systems including those developed by ASTM (2004) and the U.S. Department of Agriculture (1975). Other classification systems, such as Burmeister (1951), are also sometimes used for environmental site characterization.

Depending on the technical background and experience of the field staff, important information can be left unrecorded because its potential importance is not understood. Sometimes important data are recorded but then forgotten because they are buried in handwritten strings of text in a mud-covered notebook. Parameters such as sorting, plasticity, and strength are often recorded but are rarely interpreted because it is too time consuming to extract the data from log books and enter the data for computer analysis. Sometimes the absence of a condition (e.g., fracturing, staining) can be important, yet absence information is rarely noted. Loggers may also neglect to record data they do not believe will be important to the study they are conducting. Years later, the information can become critical but is unavailable and has to be recollected. As a result of these problems, considerable amount of money can be spent in developing sample descriptions that neither fulfill the needs of all relevant disciplines nor are useful for any purpose other than that originally intended.

Another program implementation problem that occurs with some frequency is that too few geologic samples are collected or the samples are not representative of the aquifer to be monitored (Figure 18.3). For example, buried sand bodies can sometimes require a large

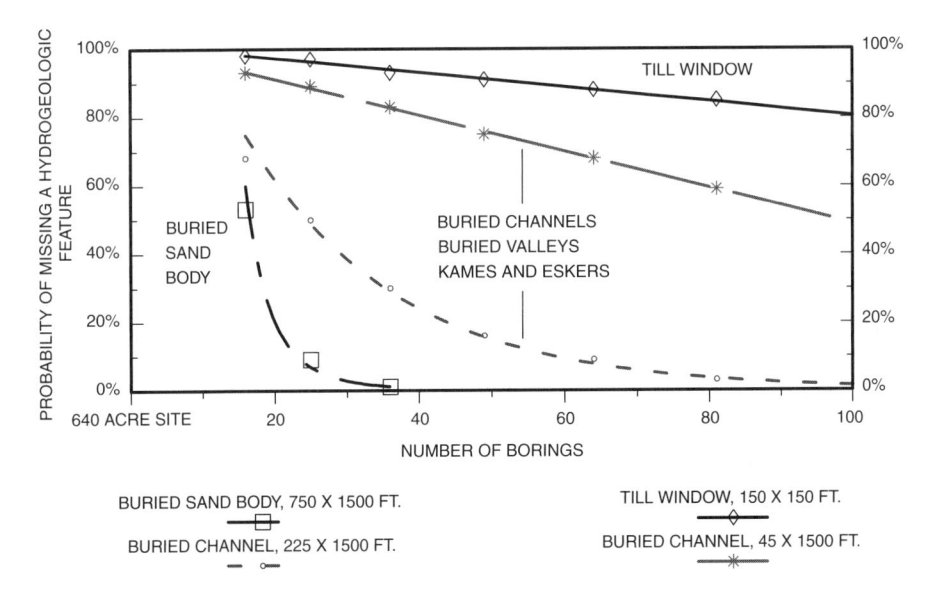

FIGURE 18.3
The probability of finding many types of subsurface geologic features using borings alone is fairly low. The use of supportive studies such as surface geologic mapping, aerial photo interpretation, and surface and borehole geophysics can greatly improve the chances of successfully mapping the subsurface. Without a complete and accurate map of the subsurface, site models can be misleading.

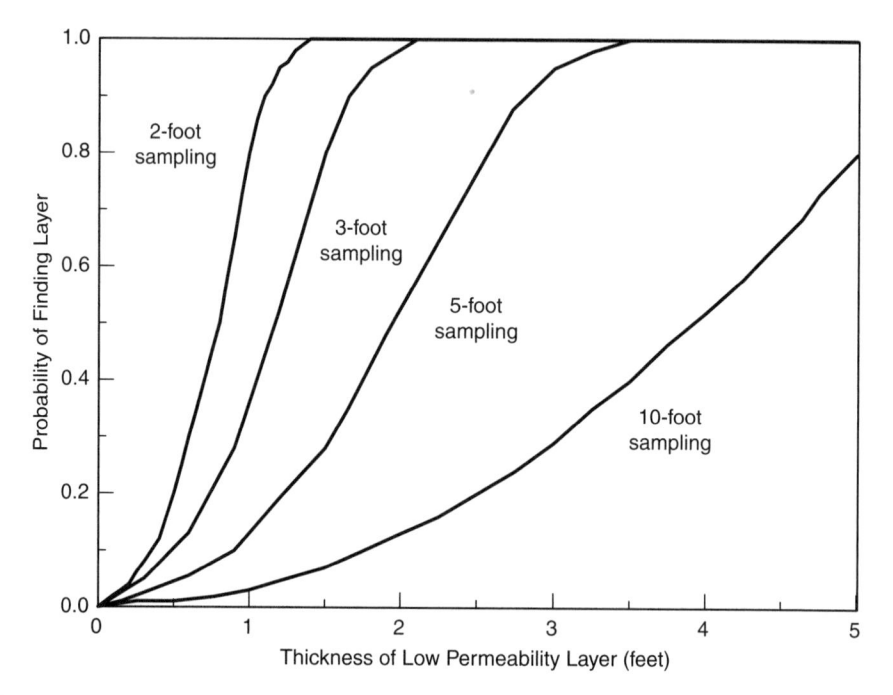

FIGURE 18.4
Sampling at intervals can save money, but the practice creates a risk of missing important geologic units. Continuous sampling is advisable in complex geologic environments and in areas where confining layers may retard the spread of contaminants.

number of borings to identify and delineate even with the support of geophysical studies. Within a boring, samples may be collected at intervals that are too large for the geologic environment being investigated (Figure 18.4). The probability of finding a 2-ft clay layer in a borehole, for instance, is less than 50% if samples are collected at 5-ft intervals. Moreover, the manner in which the sample is collected will dictate whether the sample is representative of the aquifer. Logging from cuttings is rarely an appropriate procedure, for example. Reducing project costs by sacrificing data quality and quantity is ultimately counterproductive.

Other types of field data, such as well construction details, are also frequently cited as being poorly documented. In some cases, geophysical or video logs can be used to determine some well construction specifications. Too often, however, wells must be excluded from a monitoring system because of uncertainties regarding their construction.

Documentation of sample collection procedures is critical. There has been extensive discussion and research on the effects of sample collection procedures on analytical results (Nacht, 1983; Nielsen and Yeates, 1985; Puls and Barcelona, 1989; Maskarinec, et al., 1990; Reynolds et al., 1990). Nevertheless, detailed standardized protocols for ground-water sample collection are not yet generally accepted because of the variety of site conditions and geochemical interactions that can affect a sample. Consequently, ensuring consistency in sample collection procedures for the duration of a long-term monitoring program (e.g., under RCRA) can present formidable problems. Also, it is often true that sample collection is often the responsibility of relatively inexperienced professionals, which can increase procedural inconsistencies. These qualitative factors can make a thorough evaluation of data trends and uncertainties a complex task. Thus, preventing program implementation problems is a much more effective strategy than later diagnosis and correction.

Table 18.3 summarizes methods for the prevention, recognition, and correction of typical program implementation problems.

TABLE 18.3
Prevention, Recognition, and Correction of Program Implementation Problems

Problem	Prevention	Recognition	Correction
1. Well construction not adequately documented	Require geologic descriptions to follow an established procedure; require as-built diagrams of each well installed; have an experienced hydrogeologist monitor installation	Construction details are missing, confusing, or are not consistent with measurements taken on the well	Use downhole TV and geophysical logs to approximate well construction details
2. Field data collection procedures are inadequate	Use trained field staff and detailed protocols; adapt the protocols to the geologic conditions and contaminants expected	Data are missing or are ambiguous	If necessary, resample the well using improved protocols or more experienced personnel
3. Sample collection procedures are inadequate	Use trained field staff and detailed protocols; adapt the protocols to the geologic conditions and contaminants expected	Water quality data are confusing; usually volatile chemicals are at lower concentrations than expected, and other chemicals are present when they were not projected, especially in blanks	If necessary, resample the well using improved protocols or more experienced personnel
4. Sample analysis procedures are inadequate or are undocumented	Work closely with a reputable laboratory to design an appropriate analytical program	Documentation is poor; duplicate samples yield varied results; laboratory blanks are severely contaminated; spike recoveries are poor	If necessary, resample the well and have analyses done by a reputable laboratory

Geologic Uniformity Problems

Geologic uniformity problems refer to situations in which interpretations of ground-water flow are inaccurate because the water-bearing zone being monitored is irregularly shaped or texturally heterogeneous, or ground-water flow in the zone follows irregular, discrete pathways such as fractures or solution channels.

The shape and textural uniformity of a water-bearing zone can greatly affect the effectiveness of a ground-water monitoring system. Unfortunately, it is typically the case that little is known about hydrogeologic conditions at a site until a monitoring system is designed and installed. In the absence of site-specific information, background information on the regional geology of an area, aerial photographs, and geophysical surveys are used. Too often, inappropriate assumptions are made concerning the infinite extent and homogeneous and isotropic nature of water-bearing zones. In practice, these idealized aquifers seldom exist outside of introductory geology textbooks.

Some aquifer shapes are difficult to map in the subsurface (at least at an acceptable cost); consequently, those aquifers are often not monitored effectively. Fault-offset units, shoestring (e.g., fluvial) deposits, and discontinuous lenticular deposits are examples of such shapes. Even relatively continuous, thick, flat-lying deposits can cause problems if they are heterogeneous and anisotropic (Fetter, 1981). Glacial and floodplain deposits are particularly difficult to monitor because of the uncertainty associated with identifying hydrostratigraphically equivalent zones in which to screen wells in a monitoring system.

Selecting appropriate screened intervals (or open-hole intervals in bedrock) is an especially significant problem in zones having secondary permeability (i.e., flow through geologic discontinuities such as fractures or solution channels) or dual permeability (i.e., intergranular or primary permeability in addition to secondary permeability). Faulted, fractured, and jointed rock systems are less difficult to monitor if the fracture orientations are relatively consistent and can be observed directly in outcrops and cores, and as fracture traces in aerial photographs. Furthermore, fractures and other discontinuities are often interconnected, so they can be treated as a single hydrostratigraphic unit exhibiting the characteristics of primary porosity (Figure 18.5). Nativ et al. (1999) describe lessons learned in designing and implementing a monitoring network in a fractured chalk.

Fractured clay systems are more difficult to study because the openings are generally not systematically oriented and do not exhibit surface traces (Figure 18.6). However, this is less important if the fractures are connected so that the unit behaves as a porous medium. True karst systems are the most difficult type of geologic unit to monitor because the orientation, density, and size of the openings can vary greatly. Furthermore, drilling and installing monitoring wells in carbonate terrains is more technically challenging than in almost any other type of geologic material.

Table 18.4 summarizes the prevention, recognition, and correction of typical geologic uniformity problems.

Hydrologic Uniformity Problems

Hydrologic uniformity problems refer to situations in which natural or artificial changes in ground-water elevations or flow directions are not taken into account, thus

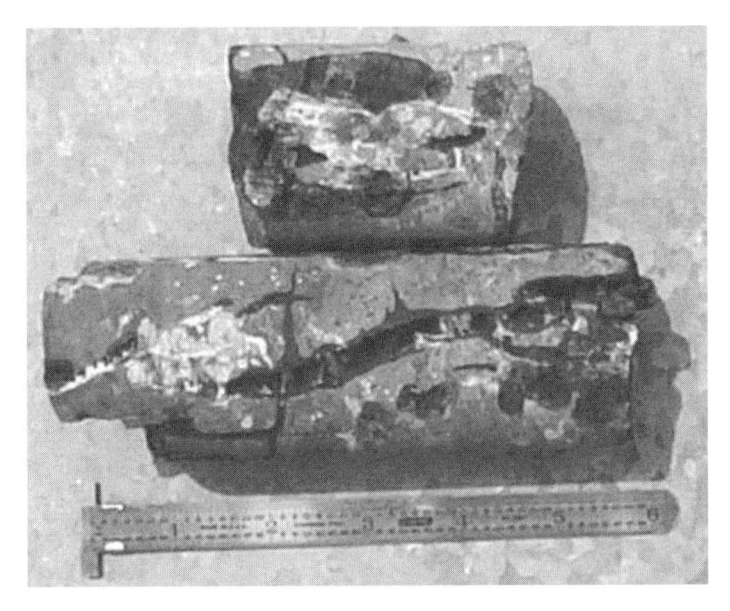

Diagnosis of Ground-Water Monitoring Problems 1205

FIGURE 18.5
Discontinuities that cannot be reliably mapped at the surface are particularly difficult to monitor. Some openings may not be effectively connected and may not be important to monitor. The scale of the discontinuities relative to the scale of the monitoring system is also important.

FIGURE 18.6
Clay units thought to be impermeable can behave like a porous medium if they are cracked or fractured.

TABLE 18.4
Prevention, Recognition, and Correction of Geologic Uniformity Problems

Problem	Prevention	Recognition	Correction
1. Presence of irregularly shaped aquifers	Use background information, geophysical surveys, and soil borings to evaluate aquifer geometry; install monitor wells in phases to optimize effectiveness	Well is installed in a geologic unit likely to be irregularly shaped based on paleoenvironmental genesis; boring logs indicate the presence of irregularities; pumping tests indicate the presence of hydrologic boundaries	Conduct more sophisticated aquifer tests and geophysical surveys, and install additional wells as necessary
2. Contaminant migration follows complex fracture patterns	Evaluate fracture patterns using background information, aerial photographs, measurements of outcrops and cores, downhole flow meters, packer tests, and borehole geophysical devices; install wells in phases to optimize effectiveness	Discontinuities are observed in outcrops or cores; fracture traces are prominent in aerial photographs; transmissivity is much higher than expected and is anisotropic	Conduct additional aquifer tests, especially packer tests, and install more wells as needed
3. Contaminant migration follows complex solution cavities (in karst terrains)	Use background geologic and geomorphological information and geophysical surveys to evaluate aquifer; install wells in phases to optimize effectiveness	Site is situated on carbonates or other soluble bedrock; well recharges much more rapidly than expected from the nature of the aquifer materials; water levels or quality is anomalous	Conduct additional aquifer tests, especially packer tests and tracer tests, and install more wells as needed

reducing the effectiveness of the monitoring system. Typical hydrologic uniformity problems include:

- The water table fluctuates above or below the screened portion of the well
- Periodic changes occur in ground-water flow
- Cyclic variations influence water flow or quality
- Surface water flow changes disrupt ground-water flow
- Pumping wells periodically disrupt ground-water flow patterns

Significant problems with hydrologic uniformity are not as common as other types of monitoring problems possibly because hydrologic processes are less random than geologic processes. However, when they do occur they have major impacts on monitoring system effectiveness. For example, if a well were installed to detect separate-phase hydrocarbons floating on top of the water table, and ground-water levels dropped below the screen, the well would be dry and useless. Conversely, if the water table rose above the top of the screen, the well would not provide information on the presence of the floating product layer. Although these situations should be easy to recognize, this is not always the case.

Recognizing hydrologic uniformity problems can be extremely difficult in three situations. First, long-duration changes in ground-water flow, such as those resulting from excessive pumping, droughts, or land-use modifications, may be difficult to recognize because of their subtlety (Figure 18.7). Second, historic changes resulting from such events as floods or site modifications can be difficult to assess without adequate records. Third, intermittent events such as on-demand pumping of nearby wells, periodic releases from reservoirs, lawn watering, or similar activities can be difficult to identify and plan for because of their variable effects. This problem can be exacerbated if the

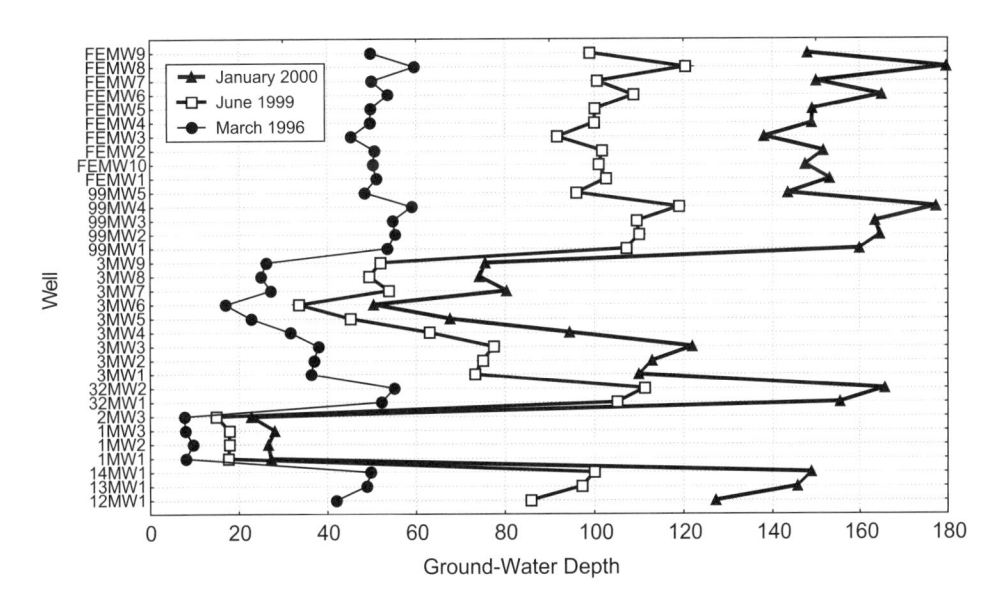

FIGURE 18.7
Hydrographs display changes in water levels in a well or in a surface-water body over time. This type of display usually requires a considerable amount of data overtime from a single well before meaningful conclusions can be made. An alternative approach that is sometimes useful is to plot all the wells in a monitoring system. This graph shows that water levels have declined over 100 feet over 4 years in many of the wells but only 10 feet in others. Such large and uneven change could have a substantial impact on ground-water and contaminant movement.

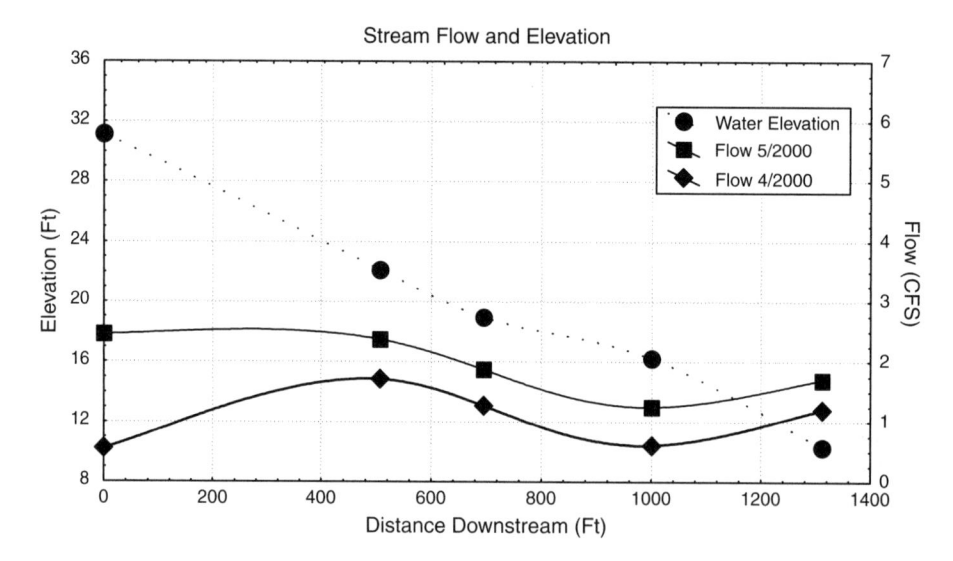

FIGURE 18.8
Hydrographs of surface-water bodies are often useful for assessing interactions between ground water and surface water. This graph shows the stream recharges ground water along one stretch (500 to 1000 ft downstream) and receives ground water discharges along the other stretches. The amount of ground-water discharge to the stream also changes over time. Ground-water monitoring systems need to be designed to be able to assess these changes, especially if the system will be modeled.

water-level changes are subtle and the water-level monitoring sensors are subject to error (Rosenberry, 1990). More complex situations also can occur (Piper, 1991).

Some problems with hydrologic uniformity should be expected when designing any ground-water monitoring system. For example, cyclic changes in ground-water flow and quality can occur from such causes as daily tidal effects and seasonal precipitation or surface water flow changes (e.g., Pettyjohn, 1982). However, these changes are usually easy to recognize and compensate for. Two exceptions to this rule are situations in which surface-water bodies discharge to ground water part of the year and are recharged the other part (Figure 18.8), and situations in which changes in water levels can saturate previously unsaturated geologic units (e.g., buried stream meanders) that impose different flow characteristics.

Table 18.5 summarizes methods for the prevention, recognition, and correction of typical hydrologic uniformity problems.

Geochemical Interaction Problems

Geochemical interaction problems refer to situations in which the sources or characteristics of contaminants confound the interpretation of ground-water data. Typical geochemical interaction problems include:

- Undocumented contaminant sources are not accounted for
- Non-aqueous phase liquids (NAPLs) do not follow expected ground-water flow patterns
- Contaminant transformations complicate data interpretation
- Contaminants react with aquifer materials, which produces unexpected results

TABLE 18.5
Prevention, Recognition, and Correction of Hydrologic Uniformity Problems

Problem	Prevention	Recognition	Correction
1. Water level fluctuates too far above or below the screened portion of the well	Estimate water level fluctuations from historical precipitation records and regional water levels in lakes and existing wells	Well is dry or water level is above the installed elevation of the screen	Schedule sampling to correspond with the appropriate water level or replace the well
2. Periodic changes occur in ground-water flow	Use basic hydrogeologic assumptions and background information to project possible flow changes	Water level elevations collected over time do not produce consistent flow patterns	Measure water levels over a suitable period to evaluate flow changes; install additional wells or find existing wells screened in the same aquifer
3. Cyclic variations influence water flow or quality	Identify possible sources of cyclic fluctuations and measure their effects or obtain estimates from the literature	Water flow or quality fluctuates gradually over time	Measure changes over several cycles to estimate the extent of the effect
4. Surface-water flow changes disrupt ground-water flow patterns	Use historical stream flow records to evaluate fluctuations; measure both surface water and ground-water elevations	Surface water and ground-water elevations change in unison or after a short lag	Record surface water and ground-water levels over a suitable time period to assess flow interactions
5. Pumping wells periodically disrupt flow patterns	Identify the presence and schedule of any pumping wells and estimate their zones of influence; position monitor wells to assess the effects of intermittent pumping	Ground-water flow patterns shift periodically; often the shifts are subtle and are at irregular intervals	Collect water level data at regular intervals over time and attempt to model the site; install more wells as needed

Geochemical interaction problems are often overlooked because their recognition requires knowledge of chemistry as well as hydrogeology. This is especially true in the cases of NAPL flow, contaminant transformations, and contaminant/aquifer interactions.

The migration of NAPLs has been discussed and researched extensively (Jorgensen et al., 1982; Villaume, 1985; Schwille, 1988; Huling and Weaver, 1991; U.S. EPA, 1991; Newell and Ross, 1992; Conrad et al., 1992). However, the subject is still poorly understood by many environmental professionals. A common misconception with NAPL flow is that ground-water samples will have concentrations of a contaminant at the contaminant's solubility limit if NAPL flow exists. In fact, an NAPL may form a relatively thin layer that is diluted to well below the saturation limit during sampling. Furthermore, if the well has not been designed so that the intake section is directly opposite the NAPL layer (i.e., screen slots must begin above the water table for LNAPLs, or be at the top of the low-permeability boundary for DNAPLs), the layer may be missed. Even so, capillary action may retard NAPL movement to a well, thus producing misleading results (Abdul et al., 1989).

Contaminant transformations also present problems in the interpretation of contaminant migration data. For example, trichloroethene (TCE) in ground water will degrade over time to form dichloroethene (DCE) and vinyl chloride. Thus, the true farthest extent of contaminant migration at a site may not be determined if ground-water samples are analyzed only for TCE and not DCE and vinyl chloride. Lipson and Siegel (2000) show how ternary diagrams can be used to diagnose dispersion, sorption, biodegradation, and volatilization.

Inorganic species also change between soluble and insoluble forms with changes in ground-water pH and Eh. Thus, a filtered ground-water sample will display quite a different profile of inorganics over time if the pH is not relatively stable, which is why grout-induced increases in pH should be evaluated carefully before sampling (Dunbar et al., 1985). The transport of organic contaminants on colloids and other particulate fractions also can complicate data interpretation (McCarthy, 1989).

In some cases, contaminants may also interact with aquifer materials and produce unexpected results. For example, organic solvents can degrade clays, thus increasing the bulk permeability of the material. Carbonate aquifers can influence the form of inorganic species to more soluble carbonate salts, or raise the pH of contaminant solutions to form chemical precipitates. Sometimes the form of the interaction involves partitioning and dilution of a contaminant mass because of small-scale differences in flow rates.

Perhaps the most common and significant type of geochemical interaction problem is being unaware of an undocumented source of contaminants (Fetter, 1983; Popkin, 1985). Some undocumented sources of contamination may be regional in extent, such as low pH or inorganic contamination in mining areas, nitrate or pesticide contamination in farming areas, or solvent contamination in industrial areas. Some undocumented sources may be decades old, or the result of illegal activities and, thus, not likely to be traceable without considerable effort.

Undocumented sources can be particularly troublesome if they are located upgradient of a site being monitored such that upgradient contaminant concentrations increase gradually over time. This would be especially important for RCRA sites given that some of the recommended statistical tests do not account for this possibility. Considering the variety and low visibility of many possible contaminant sources such as industrial pipelines and underground tanks, sewer lines, and commercial and private septic tanks, it is easy to understand how offsite contaminant sources can be overlooked in developing a monitoring program.

Table 18.6 summarizes methods for the prevention, recognition, and correction of typical geochemical interaction problems.

Diagnosis of Ground-Water Monitoring Problems 1211

TABLE 18.6
Prevention, Recognition, and Correction of Geochemical Interaction Problems

Problem	Prevention	Recognition	Correction
1. Undocumented contaminant sources are not accounted for	Identify presence of potential contaminant sources from available records and area reconnaissance, and position wells appropriately	Contamination appears in unexpected areas (e.g., upgradient)	Develop chemical signatures for each well to correlate contaminant geochemistry. Install new wells as needed
2. NAPLs do not follow expected flow patterns	Low-density NAPLs: Use soil borings, soil-gas surveys, and geophysical techniques to map the water table as an approximation of contaminant migration patterns. High-density NAPLs: Use GPR, seismic, and resistivity surveys and soil borings to evaluate stratigraphy relative to possible migration patterns	Low-density NAPLs: Can usually be observed in water collected at the top of the water table. High-density NAPLs: Are sometimes difficult to obtain by sampling devices; may have to be inferred from high concentrations in solution	Conduct additional surveys, such as packer tests and tracer studies, and install new wells as needed
3. Contaminant transformations complicate data interpretation	Identify possible degradation products based on primary contaminants	Degradation products (not originally in waste source) are detected, usually at low concentrations	Develop chemical profiles to assess transformations. Install new wells as needed
4. Interactions between the aquifer and the contaminants complicate data interpretation	Identify contaminants of concern and potential breakdown products. conduct laboratory tests (if appropriate) on aquifer or well materials	Contaminants are identified that should not have originated at the site	Use mathematical models or other advanced techniques to evaluate data interrelationships, install additional wells as needed

Diagnosis of Monitoring Problems

Problems in ground-water monitoring usually become apparent for one of two reasons: wells do not yield adequate and appropriate water for sampling or the hydrogeochemical flow interpreted from the data is ambiguous or nonsensical. The first case usually receives more attention because the symptoms are more obvious, and the diagnosis and correction of their causes are more straightforward. However, problems identified during data interpretation will invariably have a much greater impact on the direction and effectiveness of future monitoring activities at a site. Consequently, these situations should receive considerable attention prior to proceeding to subsequent phases of a monitoring program.

If a well does not yield adequate or appropriate water for sampling, the first step would probably be to redevelop the well. If this action does not resolve the problem, further actions would depend on the cause of the problem. There are a number of possible causes to consider. A new well should have adequate information on the site hydrogeology from sampling during well installation, therefore, it should be possible to confirm or eliminate geologic uniformity problems. Such data may not be available for existing wells, thus requiring the use of borehole geophysics or additional borings to resolve the data gaps. Continued water-level monitoring in the area should confirm or eliminate hydrologic uniformity problems.

If both geologic and hydrologic uniformity problems can be eliminated, it should also be possible to evaluate potential sample space problems with the location of the well intake. If sample space problems can be eliminated, it is probable that the problem involves well construction.

New wells should not have degraded materials but well installation trauma should be considered (Walker, 1983). A clogged sand pack, on the other hand, is possible even in a new well if the pack was not designed specifically for the aquifer material. If

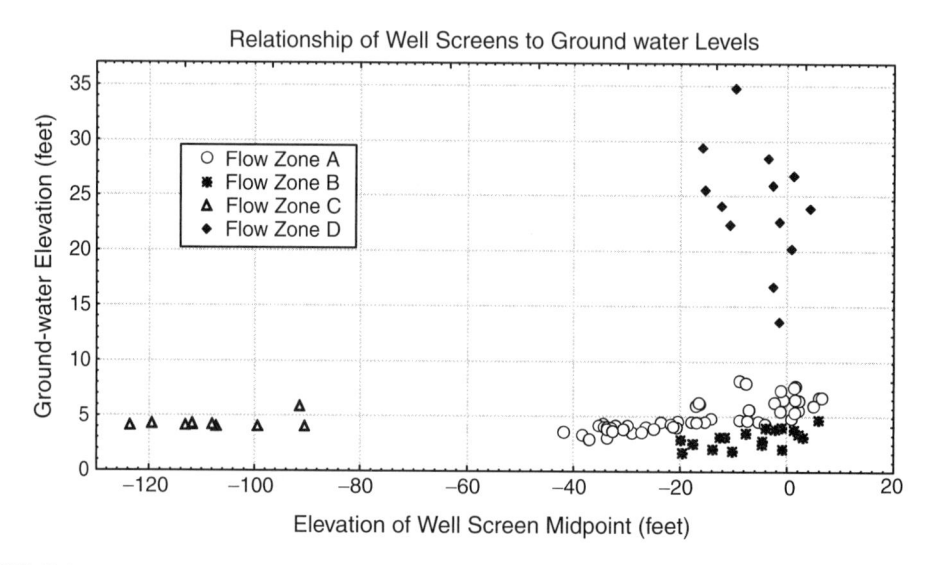

FIGURE 18.9
Plotting ground-water elevations versus the elevation of the water intake can reveal the presence of different flow zones and well construction problems. In this example, four flow zones were detected where geologic analysis suggested only two.

Relationship of Ground-water and Well Screen Elevations

Unconfined Aquifer: Ground-water Elevation = 0.2 + 0.9[Screen Elevation]
Confined Aquifer: Ground-water Elevation = 23.4 + 0.2[Screen Elevation]

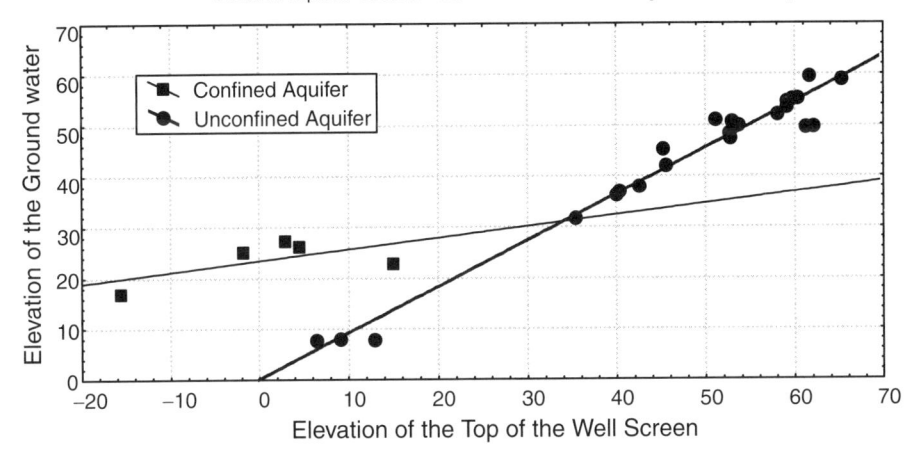

FIGURE 18.10
Ground-water intake elevation plots can reveal where two flow zones may intersect. This example suggests that the 30-ft piezometric contour can be used as the boundary between where ground water migrates upward (graph area to the left of the intersection of the two lines) and where ground water moves downward (graph area right of the intersection). Understanding this type of hydrologic relationship can help explain apparent geochemical anomalies.

redevelopment does not resolve the problem, the well may have to be decommissioned. Borehole videos and geophysical logs may provide information on degraded well materials and leaky well seals.

If hydrologic or geochemical data from the monitoring system cannot be interpreted, the first step is to verify that the wells included in the analysis are all located within the same flow system. Normally, this step should be completed before interpretation of data from a monitoring system is attempted.

Verifying that wells are part of the same flow system usually begins with an evaluation of the site's geology from the descriptions of the soil and rock samples retrieved from boreholes. This evaluation should take into account conditions that would enhance either primary (intergranular) permeability or secondary (discontinuity) permeability. Surface geologic mapping and information from existing geologic reports should be used to support the evaluation. Borehole and surface geophysics can also be helpful.

A second approach for verifying that wells are part of the same flow system involves plotting the ground-water elevation versus the elevation of (or depth to) the top (or midpoint or bottom) of each well's intake section (screen or open hole). Frequently, wells that are located within the same flow zone will plot along a line or curve (Figure 18.9). Sometimes the plots can be used to assess interactions between flow zones (Figure 18.10). This type of plot can also be used to estimate the vertical gradient of ground-water flow in the area (Figure 18.11) (Kufs, 1992). Hydrographs may also be revealing for distinguishing flow zones, but this type of graph usually requires more data over time.

A third approach for verifying that wells are part of the same flow system involves plotting conventional ground-water chemistry parameters. There are several types of plots that can be used for this purpose including Stiff, Piper, and Schoeller diagrams (Hem, 1985) or even simple pie, star, or bar charts. Care should be taken in interpreting these

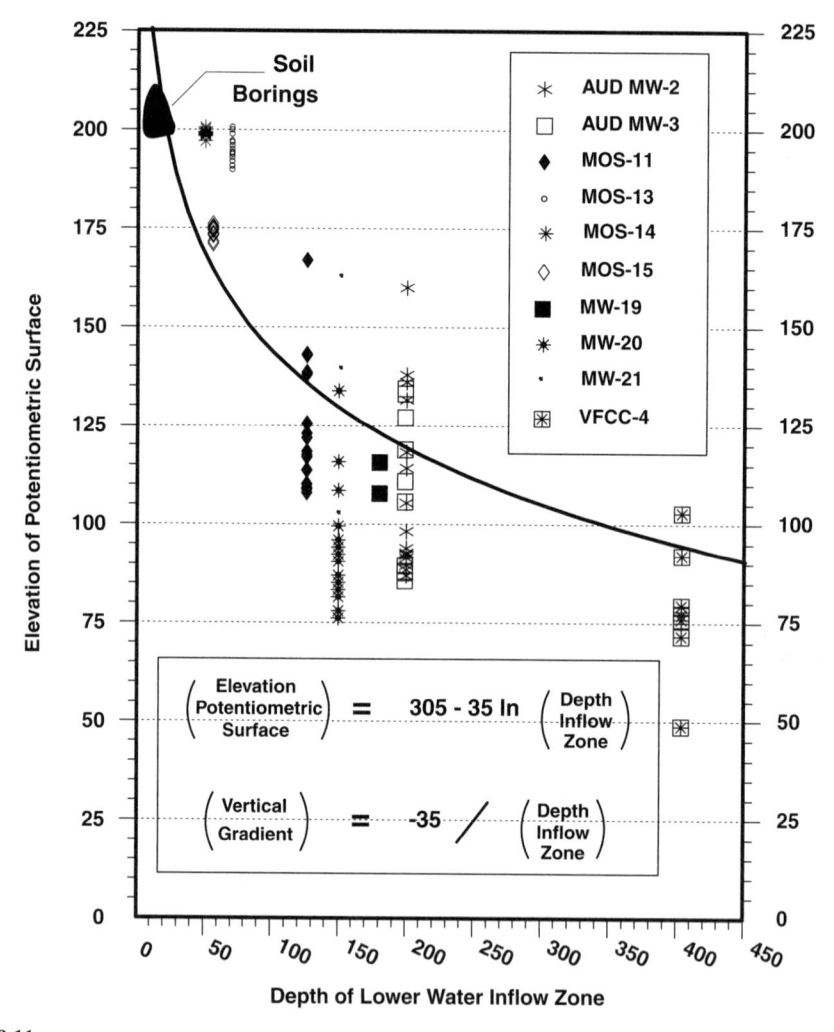

FIGURE 18.11
Plotting ground-water elevations versus the elevation of the water intake can sometimes be used to estimate average vertical gradients. Many ground-water monitoring systems fail to consider the importance of vertical ground water and contaminant movement. This can lead to problems with interpreting hydrogeologic and geochemical data.

plots so that small-scale variations in aquifer geochemistry are not misinterpreted as different flow zones (Figure 18.12).

If it appears certain that all the wells are in the same flow zone, the next step is to recheck the data being interpreted.

- *Coordinates and elevations:* Location coordinates can be verified by preparing maps of the well and sample locations. Gross errors (tens of feet) are usually easy to detect and small errors (less than a foot or so) usually do not matter much. Elevations, on the other hand, usually have to be accurate within an inch or less to evaluate ground-water flow. Surveyed elevations must be checked by the surveyor, as only relatively large errors can currently be detected

Diagnosis of Ground-Water Monitoring Problems

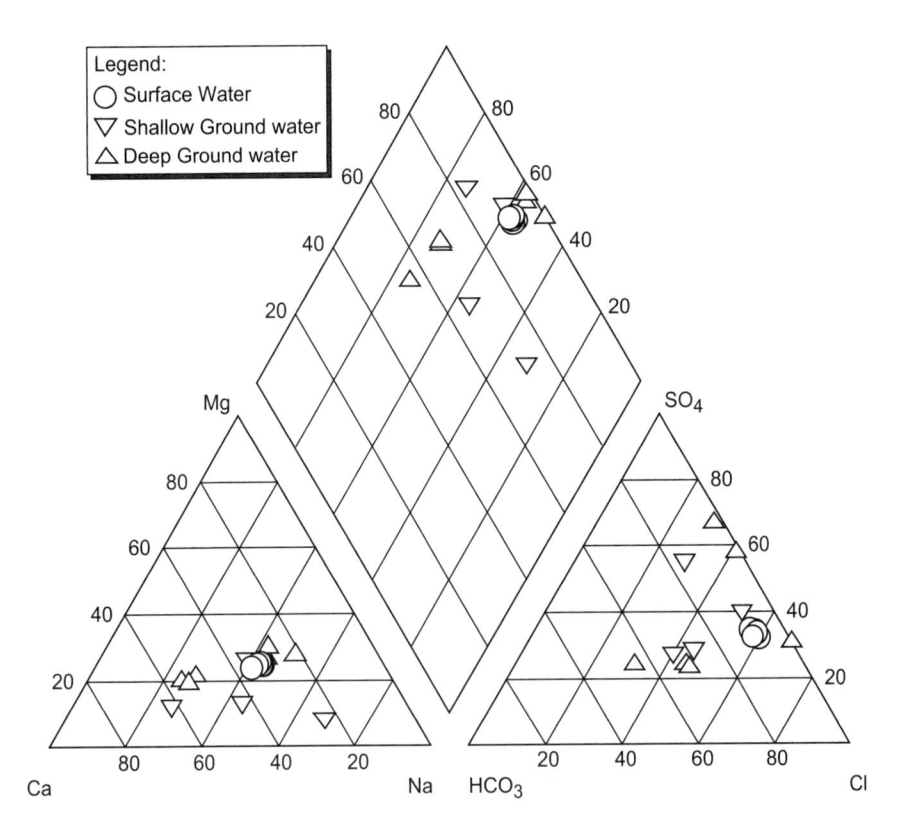

FIGURE 18.12
Geochemical diagrams may be useful for detecting different flow zones. In this example, the diagram suggests that ground waters from the two zones identified in the geologic analysis are mixing.

TABLE 18.7

Diagnosis of Common Problems in Ground-Water Monitoring System Design

Symptom	Possible Cause	Reference for Prevention, Recognition, and Correction
Insufficient water for sampling	Low-yield hydrologic unit	Table 18.1, nos. 1 and 3 Table 18.2, no. 1 Table 18.4, nos. 1, 2, and 3
	Inappropriate well construction	Table 18.1, no. 4 Table 18.2, nos. 2, 3, 4, 5, and 6
	Unexpected conditions	Table 18.5, nos. 1, 2, 3, 4, and 5
Ground-water flow appears anomalous or ambiguous	Poor data collection	Table 18.1, no. 1 Table 18.3, no. 2
	Inappropriate well construction	Table 18.1, nos. 3, 4, and 5 Table 18.2, nos. 2, 3, 4, 5, and 6
	Unexpected conditions	Table 18.4, nos. 1, 2, and 3 Table 18.5, nos. 2, 3, and 4
Contaminant migration appears anomalous or ambiguous	Inappropriate sample space	Table 18.1, nos. 2, 3, and 4 Table 18.3, no. 1
	Poor data collection	Table 18.1, no. 5 Table 18.3, nos. 2, 3, and 4
	Inappropriate well construction	Table 18.2, nos. 2, 4, 5, and 6
	Unexpected hydrologic conditions	Table 18.4, nos. 1, 2, and 3 Table 18.5, nos. 1, 2, 3, and 4
	Unexpected geochemical conditions	Table 18.6, nos. 1, 2, 3, and 4

by GPS devices. Replicate field measurements can reveal inconsistencies in well depth, stickup, and depth to water, especially if different types of devices are used (e.g., using a measuring tape periodically to verify transducer readings).

- *Geology:* Geologic descriptions can be verified by having an independent, experienced geologist reexamining rock and soil samples, provided they were archieved. If no samples are available, logs can be examined for inconsistencies but this approach is less effective. In more extreme cases, borehole geophysics can be used or new borings can be installed.

- *Geochemistry:* Analytical chemistry results are usually verified using mandated data validation procedures. Inorganic results can also be checked by calculating anion–cation balances. The effects of sample collection procedures can usually be assessed by examining the results for blank samples, provided appropriate numbers and types were collected.

If all the data appear to be correct, specific types of monitoring problems should be reviewed for relevance. Table 18.7 summarizes the possible causes of common problems in ground-water monitoring systems.

References

Abdul, A.S., S.F. Kia, and T.L. Gibson, Limitations of monitoring wells for the detection and quantification of petroleum products in soils and aquifers, *Ground-Water Monitoring Review*, 9(2), 90–99, 1989.

ASTM, Standard Practice for Classification of Soils for Engineering Purposes (Unified Soil Classification System), ASTM Standard D 2487, ASTM International, West Conshohocken, PA, 2004, 12 pp.

Boulding, J.R., Description and Sampling of Contaminated Soils, EPA/625/12-91/002, U.S. Environmental Protection Agency, Center for Environmental Research Information, Environmental Monitoring Systems Laboratory, Cincinnati, OH, 1991.

Breckenridge, R.P., J.R. Williams, and J.F. Keck, Characterizing Soils for Hazardous Waste Site Assessments, EPA/600/8-91/008, Superfund Ground-Water Issue Paper, U.S. Environmental Protection Agency, Environmental Monitoring Systems Laboratory, Las Vegas, NV, 1991.

Burden, D.S. and J.L. Sims, Fundamentals of Soil Science as Applicable to Management of Hazardous Wastes, EPA/540/S-98/500, Superfund Ground Water Issue, U.S. Environmental Protection Agency, Office of Research and Development and Office of Solid Waste and Emergency Response, Washington, DC, 1999.

Burmeister, D.M., Identification and Classification of Soils — An Appraisal and Statement of Principles, ASTM STP 113, American Society for Testing and Materials, Philadelphia, PA, 1951.

Cameron, R.E., Guide to Site and Soil Description for Hazardous Waste Site Characterization, EPA/600/4-91/029, U.S. EPA Environmental Monitoring Systems Laboratory, Las Vegas, NV, 1991.

Conrad, S.H., J.L. Wilson, W.R. Mason, and W.J. Peplinski, Visualization of residual organic liquid trapped in aquifers, *Water Resources Research*, 28(2), 467–478, 1992.

Dunbar, D., H. Tuchfeld, R. Siegel, and R. Sterbentz, Ground water quality anomalies encountered during well construction, sampling, and analysis in the environs of a hazardous waste management facility, *Ground-Water Monitoring Review*, 5(3), 70–74, 1985.

Fetter, C.W., Determination of the direction of ground-water flow, *Ground-Water Monitoring Review*, 1(3), 28–31, 1981.

Fetter, C.W., Potential sources of contamination in ground-water monitoring, *Ground-Water Monitoring Review*, 3(2), 60–64, 1983.

Gibs, J. and T.E. Imbrigiotta, Well purging criteria for sampling purgeable organic compounds, *Ground Water*, 28(1), 68–78, 1990.

Giddings, T., Screen length selection for use in detection monitor well networks, Proceedings of the Sixth National Symposium on Aquifer Restoration and Ground Water Monitoring, National Water Well Association, Worthington, OH, 1986, pp. 316–319.

Hem, J.D., *Study and Interpretation of the Chemical Characteristics of Natural Water*, 3rd ed., U.S. Geological Survey Water-Supply Paper 2254, 1985, 264 pp.

Herzog, B.L., B.R. Hensel, E. Mehnert, J.R. Miller, and T.M. Johnson, Evolution and adequacy of ground-water monitoring networks at hazardous waste disposal facilities in Illinois, Proceedings of the Sixth National Symposium on Aquifer Restoration and Ground Water Monitoring, National Water Well Association, Worthington, OH, 1986, pp. 98–118.

Herzog, B.L., S.F.J. Chou, J.R. Valkenburg, and R.A. Griffin, Changes in volatile organic chemical concentrations after purging slowly recovering wells, *Ground-Water Monitoring Review*, 8(4), 93–99, 1988.

Huling, S.G. and J.W. Weaver, Dense Non-Aqueous Phase Liquids, EPA/540/4-91-002, Ground Water Issue, U.S. Environmental Protection Agency, Office of Solid Waste and Emergency Response and Office of Research and Development, Washington, DC, 1991, 21 pp.

Jorgensen, D.G., T. Gogel, and D.C. Signor, Determination of flow in aquifers containing variable-density water, *Ground-Water Monitoring Review*, 2(2), 40–45, 1982.

Keely, J.F. and K. Boateng, Monitoring well installation, purging, and sampling techniques — Part 1: conceptualizations, *Ground Water*, 25(3), 300–313, 1987.

Keely, J.F., Chemical time-series sampling, *Ground-Water Monitoring Review*, 2(4), 29–37, 1982.

Knox, J.N. and P.R. Jacobson, Quality assurance testing of monitoring well integrity, Proceedings of the 7th National Conference on Management of Uncontrolled Hazardous Waste Sites, Hazardous Materials Control Research Institute, Silver Springs, MD, 1986, pp. 233–236.

Kufs, C.T., C.F. Moran, and D.J. Messinger, The future of geoenvironmental logging, *The Professional Geologist*, 29(11), 47, 1992.

Kufs, C.T., Estimating vertical hydraulic gradients without well clusters, *Journal of Environmental Hydrology*, 1(1), 19–22, 1992.

Lipson, D. and D.I. Siegel, Using ternary diagrams to characterize transport and attenuation of BTX, *Ground Water*, 38(1), 106–113, 2000.

Maskarinec, M.P., L.H. Johnson, S.K. Holladay, R.L. Moody, C.K. Bayne, and R.A. Jenkins, Stability of volatile organic compounds in environmental water samples during transport and storage, *Environmental Science and Technology*, 24(11), 1665–1670, 1990.

McCarthy, J.F., The mobility of colloidal particles in the subsurface, *Environmental Science and Technology*, 23, 496–504, 1989.

Morin, R.H., D.R. LeBlanc, and W.E. Teasdale, A statistical evaluation of formation disturbance produced by well casing installation methods, *Ground Water*, 26(2), 207–217, 1988.

Nacht, S.J., Monitoring sampling protocol considerations, *Ground-Water Monitoring Review*, 3(3), 23–29, 1983.

Nativ, R., E.M. Adar, and A. Becker, Designing a monitoring network for contaminated ground water in fractured chalk, *Ground Water*, 37(1), 38–47, 1999.

Newell, C.J. and R.R. Ross, Estimating Potential for Occurrence of DNAPL at Superfund Sites, U.S. Environmental Protection Agency Quick Reference Fact Sheet, Publication 9355.407FS, 1992.

Nielsen, D.M. and G.L. Yeates, A comparison of sampling mechanisms available for small-diameter ground water monitoring wells, *Ground-Water Monitoring Review*, 5(2), 83–99, 1985.

Parker, L.V., A.D. Hewitt, and T.F. Jenkins, Influence of casing materials on trace-level chemicals in well water, *Ground-Water Monitoring Review*, 10(2), 146–156, 1990.

Paul, D.G., C.D. Palmer, and D.S. Cherkauer, The effect of construction, installation, and development on the turbidity of water in monitoring wells in fine-grained glacial till, *Ground-Water Monitoring Review*, 8(1), 73–82, 1988.

Perazzo, J.A., R.C. Dorrier, and J.P. Mack, Long-term confidence in ground water monitoring systems, *Ground-Water Monitoring Review*, 4(4), 119–123, 1984.

Pettyjohn, W.A., Cause and effect of cyclic changes in ground-water quality, *Ground-Water Monitoring Review*, 2(1), 43–49, 1982.

Piper, L.M., Analysis of unexpected results of water level study to determine aquifer interconnection, Proceedings of the Fifth National Outdoor Action Conference on Aquifer Restoration, Ground Water Monitoring, and Geophysical Methods, National Ground-Water Association, Dublin, OH, 1991, pp. 205–219.

Popkin, B.P., Selected waste site puzzles and solutions, *Ground-Water Monitoring Review*, 5(1), 34–37, 1985.

Puls, R.W. and M.J. Barcelona, Ground Water Sampling for Metals Analysis, EPA/540/4-89-001, Superfund Ground Water Issue, U.S. Environmental Protection Agency, Office of Research and Development, Washington, DC, 1989.

Reynolds, G.W., J.T. Hoff, and R.W. Gillham, Sampling bias caused by materials used to monitor halocarbons in ground water, *Environmental Science and Technology*, 24(1), 135–142, 1990.

Ricci, E.D., The evaluation of an existing ground-water monitoring program, Proceedings of the 6th National Conference on the Management of Uncontrolled Hazardous Waste Sites, Hazardous Materials Control Research Institute, Silver Springs, MD, 1985, pp. 84–87.

Robin, M.J.L. and R.W. Gillham, Field evaluation of well purging procedures, *Ground-Water Monitoring Review*, 7(4), 85–93, 1987.

Ronen, D., M. Magaritz, and I. Levy, An *in-situ* multilevel sampler for preventive monitoring and study of hydrochemical profiles in aquifers, *Ground-Water Monitoring Review*, 7(4), 69–74, 1987.

Rosenberry, D.O., Effect of sensor error on interpretation of long-term water-level data, *Ground Water*, 28(6), 927–936, 1990.

Saines, M., Errors in interpretation of ground-water level data, *Ground-Water Monitoring Review*, 1(1), 56–64, 1981.

Schwille, F., *Dense Chlorinated Solvents in Porous and Fractured Media*, Translated by J.F. Pankow, Lewis Publishers, Chelsea, MI, 1988.

Shosky, D.J., A rationale for screen length selection and placement, Proceedings of the Sixth National Symposium and Exposition on Aquifer Restoration and Ground Water Monitoring, National Water Well Association, Worthington, OH, 1986, pp. 320–327.

Smart, G.R. and D.K. Cook, RCRA and CERCLA monitoring well location and sampling requirements, *Hazardous Materials Control*, 1(3), 26–33, 1988.

Sukop, M.C., Estimation of vertical concentration profiles from existing wells, *Ground Water*, 38(6), 836–841, 2000.

U.S. Department of Agriculture, *Soil Taxonomy: A Basic System of Soil Classification for Making and Interpreting Soil Surveys*, USDA Agricultural Handbook No. 436, 1975.

U.S. EPA, Dense Non-Aqueous Phase Liquids A Workshop Summary, EPA/600/R-92-030, U.S. Environmental Protection Agency, Office of Solid Waste and Emergency Response, Washington, DC, 1991.

Villaume, J.F., Investigations at sites contaminated with dense, non-aqueous phase liquids (NAPLs), *Ground-Water Monitoring Review*, 5(2), 60–74, 1985.

Walker, S.E., Background water quality monitoring: well installation trauma, Proceedings of the Third National Symposium on Aquifer Restoration and Ground Water Monitoring, National Water Well Association, Worthington, OH, 1983, pp. 235–246.

19

Health and Safety Considerations in Ground-Water Monitoring Investigations

Steven P. Maslansky and Carol J. Maslansky

CONTENTS

Introduction	1220
Health and Safety Planning	1221
Hazard Identification and Classification	1225
Electrical Hazards	1225
Physical Hazards	1226
Noise	1226
Temperature Stress	1227
Radiation Hazards	1228
Chemical Hazards	1228
Biological Hazards	1229
Toxic Hazards	1230
Exposure Limits	1232
Confined Space Hazards	1232
Risk vs. Hazard	1233
Sources of Information	1235
Respiratory Protection	1236
Air-Purifying Respirators	1236
Atmosphere-Supplying Respirators	1238
Respiratory-Protection Program	1239
Air Monitoring	1240
Protective Clothing	1244
Site Operations	1246
Drilling Techniques	1248
Decontamination	1250
Medical Monitoring	1252
Applicability of HAZWOPER to Ground-Water Investigations	1254
Training Requirements	1254
General Safety and Liability Considerations	1256
Summary	1260
References	1261

Introduction

Environmental investigations on sites with bulging drums and fuming lagoons have become dim memories. Likewise, encountering unexpected light non-aqueous phase liquids (LNAPLs), dense non-aqueous phase liquids (DNAPLs) or drilling into an UST is a rare event these days. Nevertheless today's environmental investigator can still encounter major health and safety hazards.

Many required tasks performed during environmental site characterization studies can be hazardous. Safety must be emphasized, and all personnel must know how to protect themselves, their co-workers, and the equipment they operate. Publications and videos describing safe operating procedures around heavy equipment utilized in environmental site characterization and ground-water monitoring operations are available through the various drilling rig manufacturers and through many trade organizations. Organizations such as the National Ground Water Association (www.ngwa.org), the National Drilling Association (www.4u.com), ADSC: The International Association of Foundation Drilling (www.adsc-iafd.com), Association of Engineering Firms Practicing in the Geosciences (www.asfe.org), International Association of Drilling Contractors (www.iadc.org), Australian Drilling Industry Training Committee (www.aditc.com.au), and the Australian Drilling Industry Association (www.adia.com.au) provide excellent guidance on safety considerations when working around heavy equipment. Safety procedures around heavy equipment thus will not be discussed in detail in this chapter.

One of the most important aspects to consider when conducting environmental investigations is hazard recognition. Look up "hazard" in any dictionary and it will be defined as danger or something that causes danger or difficulty. Hazard can also be defined, however, as a lack of predictability, or as uncertainty. These definitions more aptly define many conditions found in the field at sites where environmental investigations are conducted. The key to safety in the field is an ability to recognize situations that may produce hazardous conditions, and to plan ahead to avoid or mitigate these conditions.

A hazard is a danger that threatens harm to investigators, other workers onsite, and nearby people, property, and the environment. A recognized hazard is one that can be established on the basis of industry recognition (the environmental investigation "industry" as opposed to petrochemical manufacturing), employer recognition, or "commonsense" recognition. Evidence of such recognition may consist of written or oral statements made by the employer or management or supervisory personnel, or instances where employees have clearly called the hazard to the employer's attention. Regulatory and compliance organizations have established standards by which employers can be cited for failing to identify and reduce or eliminate hazards. A review by the authors determined that the most common Occupational Safety and Health Administration (OSHA) violations committed by firms engaged in environmental site characterization and ground-water monitoring (consultants and contractors), for the period 1990–2000, were the following:

General industry standards

- 29 CFR 1910.22–23 — Walking-Working Surfaces: poor housekeeping and guarding holes
- 29 CFR 1910.95 — Occupational Noise Exposure: failure to monitor and attenuate
- 29 CFR 1910.120 — Hazardous Waste Operations and Emergency Response: failure to have or comply with health and safety program

- 29 CFR 1910.132 — Personal Protective Equipment (PPE): failure to assess hazards and select proper PPE, and train employees
- 29 CFR 1910.133 — Eye and Face Protection: failure to use protectors
- 29 CFR 1910.134 — Respiratory Protection: failure to have minimal acceptable program
- 29 CFR 1910.135 — Head Protection: failure to protect head from falling objects
- 29 CFR 1910.138 — Hand Protection: failure to select and use appropriate protection
- 29 CFR 1910.141 — Sanitation: failure to have designated toilet facilities or immediately available transportation to designated facilities
- 29 CFR 1910.146 — Permit-required Confined Spaces: failure to identify spaces and implement program
- 29 CFR 1910.147 — Control of Hazardous Energy: failure to have lockout or tagout program
- 29 CFR 1910.151 — Medical Services and First Aid: failure to provide first-aid kit and eye wash station
- 29 CFR 1910.157 — Portable Fire Extinguishers: failure to provide, inspect, maintain, and test; and train employees in use
- 29 CFR 1910.212 — Machine Guarding: failure to protect employees from rotating parts
- 29 CFR 1910.1030 — Blood-borne Pathogens: failure to have first-aid providers under program
- 29 CFR 1910.1200 — Hazard Communications: failure to document training, improper labeling of containers, failure to have Material Safety Data Sheets (MSDSs) available

Construction industry standards

- 29 CFR 1926.56 — Illumination: failure to have minimum work-site lighting
- 29 CFR 1926.106 — Working Over or Near Water: failure to have personal flotation devices, ring buoys, or lifesaving skiffs
- 29 CFR 1926.152 — Flammable Liquids: failure to use proper containers
- 29 CFR 1926.350 — Gas Welding and Cutting: failure to use valve protection caps and proper storage procedures
- 29 CFR 1926.501 — Fall Protection: failure to protect employees when working 6 ft or more above a lower level
- 29 CFR 1926.651 — Excavations: failure to inspect and protect employees
- 29 CFR 1926.955 — Overhead Lines: failure to avoid contact with energized lines

Health and Safety Planning

The first step in making sure that the investigator, co-workers, employer, other people working onsite, and the public are protected from site hazards and activities is good planning. During the last two decades safety planning documents have been called different things: Health and Safety Plan (HASP), Safety and Health Emergency Response Plan (SHERP), Site-Specific Safety and Health Plan (SSHP), Contingency Plan, or simply Safety Plan. Whatever it is called, the plan should do more than just satisfy a regulatory or contractual requirement. A good plan serves two capacities:

1. As a proactive accident-prevention plan to delineate site hazards, risks to site workers, hazard monitoring, hazard mitigation, and safe operating procedures
2. As a reactive contingency plan to identify procedures to be implemented should something go wrong

A good accident-prevention plan goes a long way toward making the contingency plan unnecessary.

During environmental investigations, every effort must be made to anticipate the unexpected. Contingency planning can be thought of as a method to mitigate the impact of Murphy's Law. A careful site history review or previous monitoring and site characterization activities may minimize the possibility of discovering unknown hazardous materials or unstable conditions.

The very nature of hazardous waste work, however, complicates and exacerbates anticipated field conditions and hazards normally associated with clean water work. The possibility of hitting buried utilities, encountering hazardous substances, initiating an equipment fire, or sustaining personal injury must always be considered. At hazardous substance sites, both the possibility of and severity of emergency situations are enhanced. Those directly responsible or peripherally involved with site safety must ensure that personnel on the job are adequately trained, medically qualified, physically fit, and possess a good mental attitude toward safety.

Logistics must be carefully considered. Workers must have adequate protective clothing (amount, appropriate type, proper sizes) and equipment to do the job safely and efficiently. Temperature stress and fatigue must be anticipated when employing protective equipment. Appropriate real-time air-monitoring devices must be available, and they must be calibrated and working properly. Back-up instrumentation should be available if needed. Because medical emergencies do occur, the site safety officer or some other designated individual should be trained in standard first aid and CPR.

A good site-specific HASP must address a myriad of topics. If properly conceived and written, the plan will address all aspects of the work plan as well as anticipated emergencies. Typical areas that should be addressed in the plan include the following:

1. Safety staff organization, responsibilities of key personnel, and their alternates. This includes identification of the site safety officer and other individuals responsible for implementation and continued enforcement of the on-site health and safety program.
2. Safety and health hazard assessment for site operations. This includes a listing of the known or anticipated site hazards (i.e., chemical, biological, physical), known or anticipated chemical contaminant concentration ranges, and applicable occupational exposure limits. In addition, site operations should be individually evaluated and safety procedures outlined to mitigate specific hazards.
3. PPE requirements. This includes the types of protective clothing and respiratory protection required due to site-specific hazards as well as the development of site-specific action levels to dictate upgrading or downgrading levels of protection.
4. Methods to assess personal and environmental exposure. This includes radiological and meteorological monitoring, real-time or direct-reading air

monitoring, measurement of representative worker exposures using personnel air monitoring, perimeter air monitoring, and time-weighted-average (TWA) air monitoring. Procedures should be outlined regarding sample collection, instrument use, maintenance, calibration, frequency of sampling, and quality assurance/quality control.

5. Standard operating safety procedures, work practices, and engineering controls. This section should contain site rules and prohibitions (i.e., buddy system, site hygiene practices, and eating, drinking, and smoking restrictions), handling procedures for hazardous materials and samples, and sample container handling procedures and precautions. Also included in this section are protocols for activities involving confined spaces, excavations, and welding and cutting. Provisions for night illumination and sanitation facilities are also covered.

6. Site control measures. These measures must include an on-site communication plan, site entry and egress procedures (check-in and check-out), delineation of access points, and site security precautions. A site map should be included that indicates work zones.

7. Personal hygiene and decontamination procedures. Facilities adequate to provide personal hygiene and sanitation for onsite workers must be provided (Figure 19.1). Decontamination protocols for personnel, vehicles, and

FIGURE 19.1
Field sanitation is an OSHA requirement.

equipment should be outlined. The location of decontamination stations should be delineated on the site map.

8. Emergency equipment and medical emergency procedures. The type of emergency equipment should be listed. This equipment should include emergency eyewashes and showers, first-aid supplies including oxygen, fire extinguishers, emergency-use respirators, and spill control equipment and sorbents. First-aid or medical-emergency equipment should be approved by a physician. This portion of the plan may also include provisions for physiological monitoring procedures (heat or cold stress monitoring) and protocols for altering work and rest schedules based on temperature, levels of protection, and field activity.

9. Emergency response plan and contingency procedures. Preplanning and agency contacts should be listed in the event of chemical overexposure, personal injury, fire or explosion, spills or releases, or detection of radioactivity. Instructions for such scenarios should be prepared for posting with a list of all local, municipal, state, and Federal emergency contacts. Criteria and procedures for on-site evacuation and initiation of a community alert should be specified. Preplanning with local agencies will ameliorate local fears regarding the potential onsite hazards for emergency responders, as well as potential health hazards from contaminated equipment and personnel. Also required are decontamination procedures for injured personnel and a route map to the nearest medical facility. The identity, roles, chain of command, and methods of communication between all key personnel must be indicated.

10. Logs, reports, and record-keeping. Examples of all forms, such as air-monitoring data logs, training logs, safety inspection logs, weekly and monthly reports, accident reports, incident reports, employee or visitor register, and medical certification reports should be included. All exposure and medical monitoring records must be maintained according to OSHA 29 CFR 1910.20. Individual responsible and record-keeping methods should be outlined.

The site safety plan should be issued to, reviewed by, and signed by all on-site workers. A statement attached to the plan that the undersigned "... has read, been briefed on, and will comply with all provisions of the plan ..." will aid in minimizing potential problems and misunderstandings; the liability of individual's responsible for health and safety is also lessened. The plan should contain signatures of those who prepared, reviewed, and approved the plan, as well as the time period for which the plan is valid. In a number of incidents investigated by the authors, workers have suffered temperature stress as a direct result of wearing specified PPE that was inappropriate for the time of the year.

A "tailgate" or "toolbox" safety meeting should be held daily to discuss the day's activities; problems encountered the previous day and any changes made to the formal site safety plan should be discussed. It is advisable to keep written documentation of the topics discussed and the workers present at each tailgate briefing. Constant review of the "lines of authority" is necessary to avoid potential liability questions.

The scale of a project should not be used as a guide for the level of effort budgeted to write a good plan. A small site may have as many or more hazards as a large one. The level of detail used to describe the chemical hazards should be appropriate to the concentrations known or anticipated to be present. It is absurd for an HASP to include page after page of pure product MSDSs or discussions of toxicity and flammability of the chemicals that could be encountered at a site when only a few parts per billion of these chemicals are known to be present in the ground water or soil. The same plan may not even mention the presence of the deer ticks found on the poison ivy bushes located on the steep embankment by the railroad tracks and power lines next to where drilling will occur.

Many organizations divide their HASP responsibilities among two or three separate documents including a generic plan and a much shorter site-specific plan that only references the generic plan. Some organizations have prepared task-specific plans that address the overall hazards of a particular activity, such as installing monitoring wells at gasoline stations, marine drilling, or methane recovery well installation at landfills. The task-specific documents are then augmented or annexed to a site-specific plan. All projects, even the simplest field reconnaissance or monitoring well measurement, require a site safety plan even if it is only a one- or two-page checklist.

It is important that some degree of flexibility exists within a plan, particularly in the areas of upgrades and downgrades of PPE, and the use of hoods, tape, and gloves. Flexibility should also be granted for decontamination procedures of personnel and equipment relative to the extent of contamination found or expected.

A model HASP for a ground-water monitoring investigation can be found in Maslansky and Maslansky (1997). Procedures for evaluating HASP and health and safety field reviews can be found in U.S. EPA (1989).

Hazard Identification and Classification

Electrical Hazards

Electrical hazards include electrical wires, buried cables, and generators, all of which pose a danger of shock or electrocution if contacted or severed during site operations. Urban or suburban locations often require drilling adjacent to power lines or buried cables. Electrical shock is an overlooked hazard when installing or testing electrical pumps. Capacitors that retain a charge are a common source of electric shock. Lightning is also an electrical hazard, especially when working around metal equipment. The presence of water on the site compounds these hazards.

Electrocution is a major cause of job-related mortality for drillers and helpers. Contact with overhead wires or drilling into buried cables constitute the majority of cases; line arcing has been implicated in a few cases. The U.S. Bureau of Labor Statistics reported 29 water-well drilling fatalities for the period 1993–1996. Fifteen fatalities were due to electrocution; in 11 of these cases, the drilling rig made contact with overhead power lines (Matetic and Ingram, 2001).

To minimize electrical hazards, low-voltage equipment with ground-fault interrupters and watertight corrosion-resistant connecting cables should always be used. Most regulatory and trade organizations recommend a minimum of 10–20 ft be maintained between drilling equipment and overhead wires. This distance may be increased based upon local utility requirements or state and local regulations; distance requirements are often based on line voltage. Most drilling rig manufacturers now recommend a distance that is much greater than those specified by local utilities or regulatory agencies.

Local utilities should be contacted for information regarding buried cables. In many states, an underground utility protective service, such as One Call or Dig Safe, will notify local utilities about proposed digging activities. However, these services only arrange location of utilities on public property. When working on private property, it is the responsibility of the site owner to map out specific utility locations, usually working from engineering as-built diagrams. The correct location of utilities must be determined prior to drilling or excavating. Personnel representing the owner of the utility should mark locations and advise contractor personnel on specific safety precautions. The latter information is usually not volunteered and it may be necessary to specifically request safety recommendations. Remote sensing or controlled hand-dug pit excavations may have to be used at sites where information on utility locations is limited. It should be remembered that more than one underground line has been found "where it did not belong."

Physical Hazards

Physical hazards are also referred to as general safety hazards. The site itself can be a hazard with unstable slopes, uneven terrain, holes and ditches, steep grades, and slippery, mud-covered surfaces. Sharp debris, such as broken glass and jagged metal, may litter the site. The very act of drilling may increase the slip or trip hazard by creating wet working surfaces. Because drilling and excavating operations do not afford clearance on all sides, avenues of egress may be limited. Extra care must be taken to avoid being struck by or caught between heavy equipment.

Wearing protective equipment further increases the risk of physical and mechanical harm by decreasing hearing, vision, and agility. Constant vigilance is required to avoid injury produced by drilling tools, support equipment, and vehicles.

Good housekeeping around the site under investigation prevents accidents. Likewise, good maintenance of equipment and proper use of hand tools can minimize the potential for personal injury and equipment loss. Personnel should ensure that all emergency shutoffs on heavy equipment are working properly. Care must be exercised around wire-line and rope hoists, and hoisting hardware, and catheads, as well as moving augers and rotary drilling tools. Too many operators, helpers, and inspectors have lost digits, limbs, and lives because they were pulled into moving machinery or struck by objects. The National Drilling Federation's Drilling Safety Guide (NDF, 1986) provides an excellent checklist of safety precautions for working around drilling equipment. Although the NDF no longer exists, the guide may be found in libraries. Contact with the organizations previously mentioned should also be useful.

Noise

Noise is a hazard that is usually overlooked. Noise can produce potential hazards because it interferes with normal communication between workers. It may also startle or distract. Noise can also produce physical damage to the ear that may cause pain and temporary or permanent hearing loss. The effect of noise on hearing depends on the amount and characteristics of the noise as well as the duration of exposure (NSC, 1988).

There are three general classes of noise, all of which are found around environmental site characterization and ground-water monitoring site operations: continuous noise, intermittent noise, and impact-type noise. Continuous noise is noise heard when the drilling rig or excavation equipment is running; intermittent noise can be heard over continuous noise, such as when a compressor or pumping equipment is in use; and impact-type noise is produced by hammers or driving tools.

Sounds vary in intensity and are measured in decibels. As the decibel (dB) level rises, the sound increases more rapidly than is perceived. A sound of 90 dB is twice as loud as an 80-dB sound. Low-intensity sounds such as quiet conversation measure about 40–50 dB and are quite pleasant; city noise is 60–65 dB; heavy equipment, 85–90 dB; a jackhammer produces 100–120 dB. The 120–130-dB sound from a heavy metal rock group can produce discomfort and temporary hearing loss, while a single exposure to a rifle blast can cause pain and permanent hearing loss.

Noise is a pervasive and insidious cause of hearing loss. In most instances, there is no pain. Prolonged exposure to loud noise (85–90 dB) from heavy equipment can produce hearing loss characterized by an inability to hear sounds, as well as difficulty in understanding or distinguishing various sounds. While loudness depends primarily on sound pressure, it is also affected by frequency, which is perceived as pitch. Most sounds contain a mixture of frequencies; sounds that are composed primarily of high-frequency noise are generally more annoying than low-frequency sounds. High-frequency noise also has a greater potential for causing hearing loss.

Hearing loss can be reduced by using hearing protectors, which act as barriers to reduce sound entering the ear. Protectors include disposable or reusable plugs and earmuffs. Manufacturers supply Noise Reduction Ratings (NRR), based on a system that indicates how much noise reduction is attained with each type of protector. Skeptics should note that with hearing protectors, it is easier to hear co-workers over background noise.

OSHA has established guidelines to prevent occupational hearing loss. An exposure of 90 dBA is permissible for 8 h; 95 dBA for 4 h. Whenever employee noise exposures equal or exceed an average of 85 dBA per 8-h day, employers must implement a hearing conservation program. This program is described in OSHA regulation 29 CFR Part 1910.95.

Temperature Stress

Temperature stress includes heat stress as well as cold injury. Heat stress is caused by overheating of the body and loss of fluids through sweating. Left unrecognized, heat stress may progress to heat stroke, which is a life-threatening condition. Heat-related problems usually occur in individuals who are unaccustomed to heavy workloads and heat or who are in poor physical condition. Obesity, alcohol or drug use, age, and the presence of other complicating factors, such as acute and chronic diseases, also affect the way an individual responds to hot working conditions.

Heat stress is a major hazard for workers wearing chemical protective clothing in hot environments. Protective clothing limits the dissipation of body heat and prevents evaporation of moisture. Heat rash or "prickly heat" may result from continuous exposure to hot, humid air conditions found inside protective clothing. Reduced work tolerance and increased risk of heat stress are directly related to the ambient temperature and the amount and type of chemical protective clothing worn. The potential to increase heat stress should be assessed when selecting protective clothing. The frequency of rest periods should be based on anticipated workload, ambient temperatures, worker physical fitness, and protective clothing selected (NIOSH, 1985).

Cold injuries such as frostbite and hypothermia, as well as impaired ability to perform work, are hazards at low temperatures and when there is a significant wind-chill factor. Cold injuries are increased under damp or wet conditions.

Hypothermia may occur in workers wearing protective clothing, especially after episodes of heavy work. Protective clothing offers no insulation and does not retain body heat. Symptoms of hypothermia include shivering, followed by numbness, drowsiness, and progressive loss of coordination. Breathing cold air can rapidly lead to hypothermia and can also cause lung damage. Tanks containing breathable air should not be stored outside during cold weather for this reason.

Frostbite occurs when there is local cooling of the body. Commonly affected are the ears, nose, hands, and feet. Frostnip is the incipient stage of frostbite, and is characterized by numbness. The affected area, which remains soft to the touch, will initially redden and then become waxy-white. Frostbite occurs when the skin freezes and ice crystals form. The skin becomes hard to the touch and turns mottled white or gray.

Radiation Hazards

Radiation hazards can include radioactive materials from industry, laboratories, and hospital wastes. Three types of harmful radiation can be emitted: alpha particles, beta particles, and gamma waves. Alpha radiation has a very limited capacity for penetration and is usually stopped by clothing and the outer layers of skin. Beta radiation has a greater potential for penetration than alpha particles, and can cause burns to the skin and damage to tissue below the skin. While alpha and beta particles pose only a mild to moderate threat outside the body, they can produce significant damage if materials emitting these particles are inhaled or ingested. Use of protective clothing and respirators, in concert with good personal hygiene and decontamination procedures, offers good protection against alpha and beta particles.

Gamma radiation easily passes through clothing and human tissue, and can cause serious damage to the body. Protective clothing affords no protection against gamma radiation. Use of protective clothing and respirators, however, will prevent radiation-emitting materials from entering the body. In the presence of gamma radiation greater than 2 mrem/h (milliroentgen-equivalent-man per hour) above background, all site activities should cease (NIOSH, 1985). The U.S. EPA has established a "backoff" limit of 1 mrem/h (U.S. EPA, 1992).

Chemical Hazards

Chemical hazards are commonly encountered during environmental site characterization and ground-water monitoring investigations. These chemical hazards may include toxic, flammable, explosive, reactive, or corrosive materials. A material may have more than one hazard such as being flammable and toxic. It must be remembered that not all chemical hazards are found in abandoned drums or buried in the ground. Many chemical hazards are brought onto the site. These include: gasoline, diesel, and kerosene fuels; hypochlorites or concentrated bleaching solutions used to kill pathogens in water wells; muriatic acid used in well maintenance work; oxidizers or reducing agents for ground-water treatment; solutions for decontamination of equipment; explosives used in downhole fracturing and some remote-sensing techniques; and compressed gases, such as acetylene, used in cutting and welding.

Flammable liquids and explosive gases or vapors can be encountered at landfills that have received large quantities of organic materials, at hydrocarbon refining or storage areas, and at chemical disposal sites. Of particular concern are methane, a simple asphyxiant gas which is also explosive, and toxic and explosive hydrogen sulfide gas. Care must be taken especially when drilling through the unsaturated zone. Explosive gases may be trapped in areas with natural or artificial deposits of low permeability.

Site history is a valuable aid in determining the types of chemical hazards that may be encountered. It is important to know and understand physical and chemical properties of the anticipated chemical hazards expected, including specific gravity, solubility, boiling point, vapor pressure and density, flash point, and flammable or explosive limits.

Flash point is the minimum temperature that produces sufficient flammable vapor to ignite a substance, given a source of ignition. Field sources of ignition include any open flame, exhaust systems, cigarettes, electrical equipment, a frictional spark (from metal tools or quartz and metal), and static electricity. The lower the flash point, and the closer it is to ambient temperatures, the greater the hazard. *Flammable* or *explosive limits* are the percent concentrations in air that will combust given a source of ignition. Gases with narrow limits, such as methane (5–15% in air), are lesser hazards than those with wide limits, such as acetylene (3–80% in air). Many more fires and explosions could have occurred during landfill and gas station investigations if the common flammable gases and vapors found on-site had wider explosive ranges. Every chemical capable of burning has a flammable range even though it may not be classified as being flammable. In dealing with flammable vapors and gases, one should remember that a lean mixture does not have to get richer, but one that is too rich must eventually lean out, putting the resulting concentration into the explosive range. In most cases, liquids must be raised to their flash point temperature to generate the minimum flammable concentration (the lower explosive limit, or LEL). However, rapid volatilization or aerosolization of a flammable liquid may result in combustion even if the liquid is below its flash-point temperature. This may occur when flammable or combustible liquids are subject to rapid compression and decompression, or if subjected to air sparging or sudden agitation.

Specific gravity is the weight of a liquid substance relative to water (1 g/cm^3). Materials with a specific gravity greater than water will sink in water; those with a specific gravity less than water will float on water. Specific gravity alone is not sufficient for estimating where a chemical will be found in an aqueous environment.

Water solubility is important in determining how much of the material will mix with water. It is critical when drilling in an aqueous environment to know the solubility of the contaminants that could be present. Although a chemical substance may have a low solubility, high concentrations may be recorded from field samples due to sorption of the material onto fine-grained particles or suspension of the material in water. The presence of other chemical species may enhance water solubility.

Vapor pressure indicates the ability of a liquid or solid to evaporate into air; substances with high vapor pressures are highly volatile and evaporate quickly. Vapor pressure is temperature-dependent — as the temperature increases, the vapor pressure increases. Water, for example, has a vapor pressure of 18 mmHg at 20°C. When heated to its boiling point (100°C), water has a vapor pressure of 760 mmHg and quickly evaporates. Most mobile liquids (liquids that readily pour at room temperature) have vapor pressures greater than 1 mmHg. Substances with vapor pressures between 1 and 100 mmHg will release significant vapors into the air; materials with vapor pressures greater than 100 mmHg are highly volatile. *Boiling point* is the temperature at which the vapor pressure of a liquid is equal to atmospheric pressure. Chemicals with low boiling points evaporate very quickly.

Vapor density is important in estimating where vapors will occur. Vapors with densities greater than air will collect close to the ground and in low areas, such as trenches. Heavier-than-air gases can be brought out of a drill hole. A pocket of gas under pressure may be suddenly released and forced out of the hole. Drilling fluids can displace gases, and the removal of drilling tools can create a vacuum that literally sucks gases out of the hole. When working with nested monitoring wells and two or more manhole openings, many field personnel have observed that wind across the annulus of one hole can produce a venturi effect and force gas out of an adjacent hole by eduction. Lighter-than-air gases can act as heavier-than-air gases if they are cooler than ambient temperatures; and heavier-than-air gases can act as lighter-than-air gases if released from a confining or hotter ambient environment, such as is found at most landfills.

Biological Hazards

Biological agents are living organisms or their products that can cause illness or death to the individual exposed. Biological hazards include hospital, medical office, and laboratory material that may contain infectious wastes. This material may contain microorganisms that cause hepatitis, acquired immune deficiency syndrome (AIDS), influenza, and other viral and bacterial diseases. Special care should be exercised around landfills in which biological waste from hospitals and medical laboratories may be deposited. Old (pre-RCRA, before 1980) municipal landfills are especially suspect, because many accepted sewage sludge. Fungal spores are often found in landfills. One variety of spore produces histoplasmosis, a respiratory disease which is usually self-limiting, but may in some cases produce severe symptoms and even death.

Many disease-producing microorganisms require a host, or carrier, which transmits the organisms into humans. These carriers include insects or rodents, both common inhabitants of waste-disposal facilities. Diseases that are transmitted by carriers include: plague, which is transmitted by rodent-borne fleas; Hantavirus spread by the droppings of mice; Rocky Mountain Spotted Fever (typhus fever) which is carried by ticks; Lyme Disease, which is transmitted by the deer tick; and a variety of encephalopathies (inflammation of the brain) transmitted by mosquitoes.

Plants that elicit allergic skin reactions in sensitive individuals, such as poison ivy, oak, and sumac, are biological hazards. Even when not transmitting disease, or producing an allergic response, insects and other invertebrates which produce painful stings or bites should be considered hazardous — bees, wasps, fire ants, and biting flies fall into this category. Biting flies can produce dangerous field situations because they are so distracting.

Many wild animals are attracted to field sites, especially sites that are located in unpopulated areas. Bears, wolves, and wild dogs will investigate equipment left for the night, and may still be present when workers arrive the next morning. Vibrations and water discharge associated with drilling and excavation activities can disturb snakes; snakes are often found in buckets, mud tubs, and other containers left empty overnight.

Common sense can mitigate most biological hazards. Protective clothing should always be worn at landfill locations where waste is exposed. This clothing should include gloves, safety shoes, goggles, coveralls, and a dust mask that covers the nose and mouth when blowing dust is noticed. Insect and invertebrate hazards can be lessened by using repellents; care should be taken to avoid air, soil, or water sample contamination with the repellent. Nests can be avoided or removed. If a wasp or hornet nest is encountered, a carbon dioxide fire extinguisher can be used to incapacitate the residents until the nest is removed. Taping pant legs and sleeves shut lessens ant and tick bite hazards. Plants that provoke allergic reactions should be identified and removed or avoided.

Toxic Hazards

Exposure to toxic materials may occur through inhalation, skin and eye contact, ingestion, or injection. Inhalation is the most common route of exposure. The water solubility of a vapor or gas is important in determining how much of the inhaled material actually reaches the lungs. Highly soluble gases such as ammonia dissolve readily in the upper respiratory tract, while less soluble gases such as phosgene and nitrogen dioxide readily reach the lung. Chemicals rapidly enter the bloodstream after being inhaled into the lungs.

Many chemicals can be absorbed through the skin. The skin and its film of lipid and sweat often act as an effective barrier; absorption is faster through skin that has been damaged by lacerations or abrasions, inflammation, or sunburn. Organic solvents such as acetone and toluene remove lipids from the skin; solvents enhance the permeability of the skin and facilitate skin absorption of other materials. Chemicals can also be absorbed through the eye and enter the blood stream; the eye can also be easily damaged by chemicals.

Ingesting or swallowing a chemical is an unlikely route of exposure. Chemicals can be ingested, however, if they are left on hands or clothing, or if consumption of food or drink is allowed at the work-site. On dusty sites, ingestion of contaminants adsorbed onto particulates is possible if dust filters are not utilized.

Injection exposure is the least common route of exposure. If an open wound is exposed to a chemical, however, direct contact with the blood is possible. A broken or sharp chemical container or ruptured high-pressure line could lead to exposure by injection.

Factors that influence toxicity of chemicals include the amount and duration of exposure, the route of exposure, and the susceptibility of the individual. Individual susceptibility is determined by many factors including age, sex, diet, inherited traits, overall physical health, and use of alcohol, tobacco products, medications, and drugs.

Acute toxicity of a chemical refers to its ability to produce adverse health effects as a result of a one-time exposure of short duration; such exposure often produces an emergency situation. *Chronic toxicity* is the ability of a chemical to produce systemic damage as a result of repeated exposure. Some chronic effects have a very long latency interval and develop gradually, making it difficult to establish a cause-and-effect relationship.

Exposure to multiple chemicals may result in health effects different from the effect of each alone. *Synergism* is a process in which two or more chemicals produce a toxic effect greater than the sum of their individual effects. For example, chlorinated solvents and alcohol can each cause liver damage; exposure to chlorinated solvents while drinking large amounts of alcohol can result in excessive liver damage. *Potentiation* occurs when the toxic effect of one chemical is increased by exposure to another chemical, although the second chemical by itself does not cause the effect. For example, acetone by itself does not damage the liver, but it increases the damage produced by chlorinated solvents.

Systemic toxins are substances that produce damage to specific organs. Many chemicals produce multiple organ effects. Chlorinated solvents, for instance, affect the central and peripheral nervous system, liver, kidney, and heart. *Chemical asphyxiants* are substances that interfere with the transport or use of oxygen by tissues. Carbon monoxide prevents the uptake of oxygen by red blood cells; cyanides poison cellular enzyme systems and prevent tissues from utilizing oxygen.

Irritants produce pain, swelling, and inflammation of exposed tissues. Skin, eyes, lungs, and membranes can all be affected by irritants. Severe eye irritants such as acids and alkalis can cause corneal damage and may impair vision. Pulmonary irritants can produce excessive fluid build-up in the lungs, or pulmonary edema. Severe pulmonary edema, which inhibits oxygen exchange, can be life threatening. Severe pulmonary irritants include chlorine, fluorine, paraquat, sulfur dioxide, and sulfuric acid.

Sensitizers are substances that produce allergic reactions in sensitive individuals. Skin reactions include rashes or blister formation. Itching frequently accompanies allergic skin reactions. Eye irritation is manifested by watery, itching, and reddened eyes. Respiratory sensitizers produce asthma-like reactions or hay-fever symptoms. Common chemical sensitizers are organic amine compounds, epoxy resins, and isocyanates.

Carcinogens are chemicals that cause cancer. *Mutagens* are chemicals that cause genetic change by damaging genes or chromosomes. This type of change is called a mutation. Mutations affect the way cells function and reproduce. Some kinds of mutations can result in cancer; many mutagens are also carcinogens. *Teratogens* are chemicals that cause birth defects by damaging the fetus while it develops in the mother's womb. Some chemicals produce lethal damage and cause the fetus to be aborted. Commonly encountered products such as alcohol, aspirin, and vitamin A can act as teratogens when taken in large quantities.

All unknown chemicals should be considered hazardous until proven otherwise. Care should be taken to prevent contamination and unwarranted exposure. Because even innocuous materials can cause hypersensitivity reactions when in the powdered form, all unknown powders and dusts should be considered hazardous and appropriate protective measures taken to protect skin, lungs, and eyes. Vapors and many powdered substances can react with perspiration to produce localized skin irritation, and in some cases, severe chemical burns.

Exposure Limits

Exposure limits are used to control employee inhalation exposure to specific chemical substances in the workplace. Several organizations recommend exposure levels, including the American Conference of Governmental and Industrial Hygienists (ACGIH), the American Industrial Hygiene Association (AIHA), and the National Institute for Occupational Safety and Health (NIOSH), as well as industrial groups. Employers are required by law (29 CFR 1910.1000) to comply with the exposure limits defined by OSHA. It is important to note that 29 CFR 1910.120 (HAZWOPER Rule) requires that if PPE is to be used, it shall be selected to provide protection below OSHA-permissible exposure limits and published exposure levels. If there is no OSHA value or other published values, the employer can use other published studies as a guide.

Permissible exposure limits (PELs) were established by OSHA in 1971. PELs are defined for approximately 600 substances, and can be changed only by amending the original law by act of Congress. A PEL, as defined in 29 CFR 1910.1000, is the average concentration of a substance to which a typical worker may be exposed to during an 8-h day of a 40-h workweek.

A *short-term exposure limit* (STEL) is defined as the concentration of a substance that a worker may be exposed to for a short interval without experiencing irritation, long-term effects, or acute effects that could interfere with self-rescue. Most STEL exposures do not exceed 15 min and should not be repeated more than four times during a workday.

The ACGIH recommends *threshold limit values* or TLVs. The most popular TLV is the 8-h TWA value, which is the average concentration that nearly all workers can be exposed to during a typical 8-h day and a 40-h workweek, day after day, without suffering any adverse health effects. TLVs are reviewed, updated, and published by ACGIH annually (ACGIH, 2004).

The AIHA recommends *workplace environmental exposure limits* or WEELs. The typical WEEL is an 8-h TWA value. WEELs are reviewed, updated, and published by AIHA annually (AIHA, 2004).

NIOSH is responsible for assisting OSHA in developing new exposure standards. NIOSH publishes criteria documents that discuss health hazards associated with specific substances or work practices and recommends new, usually lower, exposure levels. These are called *recommended exposure limits* or RELs.

Concentration levels above which a substance is considered to be *immediately dangerous to life or health* (IDLH) are also defined by NIOSH. The IDLH level represents a concentration at which exposure can produce severe health effects. The NIOSH Pocket Guide to Chemical Hazards (NIOSH, 1997) contains IDLH and REL values.

Although the odor threshold concentration can be useful in determining if a material is present, great care must be exercised in utilizing the sense of smell in exposure monitoring. Some substances have no perceived odor or are perceived at concentrations exceeding their IDLH. Still other compounds, such as hydrogen sulfide, can be smelled at low concentrations but produce olfactory fatigue with time or at higher concentrations.

Confined Space Hazards

After a 17-yr effort, OSHA published its Permit-Required Confined Space Rule (29 CFR 1910.146) on January 14, 1993. The rule makes a distinction between a confined space and a permit-required confined space for which the rule applies. Those distinctions are as follows.

A confined space:

1. Is large enough for a person to bodily enter it and perform work. The space is big enough to accommodate the person's entire body.
2. Has limited means of entry or exit. It does not have a normal portal, so that a worker must squeeze, bend, step up, or ascend or descend into the space by means other than by stairs.
3. Is not designed for continuous employee occupancy. The space has not been designed with lighting and ventilation.

At environmental site characterization and ground-water monitoring sites, confined spaces include pits, manholes, cisterns, pipe galleries, tanks, and valve boxes. Although a person must fit into the space, the extension of one's head or limb into the space constitutes entry. All three of the above conditions must be satisfied for a space to be considered a confined space, however, any one of the below would constitute a permit-required confined space.

A permit-required confined space:

1. Contains or has the potential to contain a hazardous atmosphere. This includes oxygen deficiency or enrichment, flammable gases and vapors, combustible dusts, toxic materials above their PELs, or any material in excess of its IDLH.
2. Contains a liquid or flowable solid material with the potential to engulf a worker and cause death by its aspiration into the respiratory system or by exerting a constricting or crushing force on the body.
3. Has an internal configuration that would allow a worker to become trapped or asphyxiated by inwardly sloping or converging walls, or it has a floor that slopes or tapers into a smaller cross-section.
4. Contains any other recognized serious safety or health hazards. This would include things such as structural, mechanical, biological, thermal, and radiological hazards.

Permit-required confined-space entry requires a training program for participants, a permit system, air monitoring, ventilation, retrieval systems, and rescue considerations.

Risk vs. Hazard

Most sites will pose multiple hazards to workers. The *degree of hazard* refers to the inherent characteristics of a substance that defines it as being hazardous, i.e., flammable, toxic, reactive, radioactive, carcinogenic, corrosive, and so on. The degree of hazard is a function of the specific hazardous properties of the material and, most importantly, its concentration. For instance, both dioxin (2,3,7,8-TCDD) and saccharin are laboratory animal carcinogens, but dioxin produces its effects at concentrations that are a million times less than those required by saccharin.

The potential harm that may be exerted upon an exposed worker can be considered the *degree of risk*. It is a function not only of the hazards present but also the likelihood that a worker will encounter these hazards. Risk can be minimized by decreasing contact with the hazard, i.e., using PPE to protect workers from chemical and biological hazards. Risk can also be reduced by practicing contamination avoidance and good housekeeping, and eliminating working conditions that may produce physical hazards (Figure 19.2). It is always better to eliminate or minimize hazards through good work practices (administrative controls) and through engineering controls.

For instance, installation of monitoring wells adjacent to leaking underground gasoline storage tanks can be hazardous. Other than the hazards associated with drilling itself, gasoline presents additional flammability and toxic hazards. The hazard is dependent on the concentration of gasoline vapors in air, potential sources of ignition present, and the duration of worker exposure. A concentration of approximately 10,000–80,000 ppm gasoline vapor in air is flammable in the presence of a source of ignition and sufficient oxygen to support combustion. Exposure of unprotected workers to concentrations of several thousand ppm will produce symptoms of toxicity in a short period of time. The hazard can be identified by the use of real-time air-monitoring equipment. The risk can be minimized by purging or venting the borehole, identifying and eliminating potential sources of ignition and, if all else fails, using appropriate PPE.

FIGURE 19.2
A cluttered site is an unsafe site.

Sources of Information

Many reference texts and on-line computer systems are available to assist preparers of HASPs, site health and safety officers and others in assessing potential site hazards and to determine proper control of associated risks. It must be noted that although most reference texts have target audiences (i.e., chemists, toxicologists, firefighters, emergency response personnel, etc.), these texts contain useful information that can be utilized by other disciplines as well.

All texts and computer databases contain errors; some errors are typographical, other errors are derived from incorrect "original" information. For this reason, no less than two sources of information should be consulted, and the latest editions should be utilized, especially when researching industrial hygiene standards or toxicity information. The employment of multiple sources often results in discrepancies between texts regarding chemical and physical parameters and acute toxicity information. In these cases, it is prudent to accept the most conservative number.

There are numerous on-line computer databases that offer information regarding toxicity, chemical and physical properties, and regulatory information on hazardous chemicals. Governmental databases and MSDSs are typically free and discussed below.

The National Library of Medicine (www.nlm.nih.gov) allows free access to health information databases such as MEDLINE, PubMed, and MEDLINEpLUS. MEDLINE databases can be used to search medical journals for information on specific chemicals. Sponsored by the National Library of Medicine, TOXNET (Toxicological Data Network) is a group of databases on toxicology and hazardous chemicals and includes a variety of databases that can be accessed at www.toxnet.nlm.nih.gov. TOXNET databases include the following:

- HSDB (Hazardous Substances Data Bank): broad descriptions of human and animal toxicity, safety and handling procedures, physical and chemical properties, environmental fate, synonyms, and U.S. regulatory information
- IRIS (Integrated Risk Information System): data from the U.S. EPA in support of human health risk assessment, focusing on hazard identification and dose response assessment
- CCRIS (Chemical Carcinogenesis Research Information System): data provided by the National Cancer Institute on carcinogenicity, mutagenicity, and tumor production
- TRI (Toxic Release Inventory): Reporting years 1995–1999
- TOXLINE: Extensive collection of references to literature on biochemical, pharmacological, physiological, and toxicological effects of drugs and chemicals
- EMIC (Environmental Mutagen Information Center): current and older literature on agents tested for genotoxic activity
- DART and ETIC (Developmental and Reproductive Toxicology and Environmental Teratology Information Center): current and older literature on developmental and reproductive toxicology
- ChemIDplus: numerous chemical synonyms, structures, regulatory list information, and links to other databases containing information about chemicals
- HSDB Structures: two-dimensional structural information on chemicals

The U.S. EPA also maintains extensive databases on chemical information. These databases can be freely accessed through U.S. EPA's Envirofacts Warehouse at www.epa.gov/enviro/html. Envirofacts contains chemical data from several different U.S. EPA programs. Of particular use is the EMCI (Envirofacts Master Chemical Integrator) system. Through EMCI one can learn details such as discharge limits and reported releases for specific chemicals.

Respiratory Protection

Respiratory hazards may be particulate or gaseous in nature. Inhalation is a major route of exposure for toxic chemicals, biological hazards, and alpha and beta radiation. Respirators provide protection from hazardous contaminants that may be inhaled. Determining the type of respiratory protection is of primary importance when selecting PPE.

Many workers who have been given respirators to wear (in many cases without fit-testing or training) find them uncomfortable and cumbersome. Respirators interfere with smoking and tobacco or gum chewing. If contaminants have been identified or are suspected and a respirator is deemed necessary, it should be worn. Not wearing a respirator under conditions in which one is recommended is short sighted and stupid. Exposures to high concentrations of toxic substances, even for a short period of time, can cause serious injury or death. There may be no warning signs or symptoms. Exposure to low concentrations can cause damage to lungs or other internal organs. Exposure to some contaminants may impair vision, affect balance, and produce symptoms of intoxication that could endanger the affected worker as well as co-workers.

Respirators, or respiratory-protection devices, are of two basic types: air-purifying and air-supplying. Respirators consist of a face-piece and either an air-purifying device or a source of breathable air.

Air-Purifying Respirators

Air-purifying respirators, or APRs, selectively remove contaminants from the air by filtration, absorption, adsorption, or chemical reaction. The air-purifying device is typically a particulate filter, or a cartridge or canister containing sorbents for specific gaseous contaminants, or a combination of filter and cartridge and canister. Cartridges are usually attached to the face-piece directly; canisters are attached to the chin of the face-piece or are attached by a breathing hose. APRs usually operate in the negative pressure mode; there are also power-assisted APRs, which maintain a positive face-piece pressure during normal breathing conditions.

APRs remove contaminants by passing air through a mechanical filter for particulates, or a cartridge or canister for gases and vapors. These devices are specific for certain types of contaminants, therefore the identity of the hazardous material must be known to select the appropriate cartridge or canister (Figure 19.3). The efficiency of the respirator against the contaminants must also be known. Each mask and cartridge or canister is designed for protection against certain contaminant concentrations. This information is usually available from the manufacturer. Just because a cartridge says it is for use against organic vapors does not mean that it is good for *all* organic vapors. Only NIOSH- or MSHA-approved equipment should be used. NIOSH periodically publishes a list of all approved respirators and respirator components (NIOSH, 2004). The latest list, as well as other NIOSH publications, is available online at www.cdc.gov/niosh.

FIGURE 19.3
PPE must be carefully selected after considering the types of contaminants present and the engineering controls.

Specific information should also be available regarding the known or suspected contaminants, including the following:

1. The physical, chemical, and toxicologic properties of each contaminant
2. Warning properties, including odor, taste, and eye- or respiratory-irritation potential
3. The OSHA PEL, NIOSH REL, ACGIH TLV, or other applicable exposure limits
4. The IDLH concentration or the LEL for flammable materials

Cartridges or canisters maybe used against gases or vapors only if the contaminant in question has "adequate warning properties." Warning properties are considered adequate when odor, taste, or irritant effects are noted and persist at concentrations below the OSHA PEL, NIOSH REL, or ACGIH TLVs. These warning properties are essential to the safe use of APRs, because they alert the user to sorbent exhaustion that allows contaminant breakthrough, poor face-piece fit, or other respirator malfunction. Individuals vary in their ability to detect warning properties. It is prudent to verify, through fit-testing, that respirator users can indeed detect the warning properties of the contaminant in question.

APRs should not be used against identified contaminants that have poor warning properties unless either the respirator is equipped with an approved end-of-service-life indicator (ESLI) or the service life of the sorbent is known.

APRs may not be used in oxygen-deficient atmospheres or in confined spaces. Oxygen deficiency, as defined by OSHA, is a concentration of oxygen in the air that is less than 19.5% — normal oxygen concentration in ambient air is 20.9%. APRs are not permitted in any atmosphere with IDLH concentrations, when the LEL is in excess of 10%, or when the concentration of the contaminant is unknown or when the concentration exceeds the maximum use concentration stipulated by the manufacturer. NIOSH has published a decision logic for selecting suitable classes of respirators for specific contaminants (NIOSH, 1987).

Finally, the use of APRs is prohibited when conditions prevent a good face-piece fit. These conditions include beards, large mustaches, long sideburns, scars, and eyeglass temple bars. Wearing of contact lenses has been debated. Some agencies allow their use while others do not. Because maintaining a leak-free seal is important to the health and safety of the user, all personnel who wear respirators are required by OSHA to pass a fit-test designed to verify the integrity of the seal.

Cartridges and canisters containing chemical sorbents should not be removed from protective packaging until needed. Once opened, they should be used immediately. Efficiency and service life decreases because sorbents begin to absorb humidity and air contaminants even when they are not in use. Cartridges should be changed regularly to prevent sorbent exhaustion and contaminant breakthrough.

The rule of D's is recommended when changing cartridges. That is, cartridges should be changed *daily*, or more often, if any of the following conditions exist: (1) warning properties are *detected* by the user, (2) it becomes *difficult* to breath, or (3) the cartridges are *dirty* or appear *damaged*. Used cartridges should be promptly *discarded*.

Atmosphere-Supplying Respirators

Atmosphere-supplying respirators, or ASRs, supply breathable grade air, not oxygen, to the face-piece via a supply line from a stationary source or from a source carried by the wearer. When air is supplied from a stationary source through a long air line or hose it is called a *supplied-air respirator* (SAR); when the air source is portable and carried by the wearer it is called a *self-contained breathing apparatus* (SCBA).

Atmosphere-supplying respirators, operated in the positive-pressure or pressure-demand mode, are recommended for entry into oxygen-deficient IDLH and LEL atmospheres, for known contaminants for which no suitable cartridge exists, and for entry into atmosphere containing unknown concentrations of contaminants.

SCBAs allow workers unhindered access to nearly all areas of the work-site; in confined spaces, however, worker mobility may be impaired. SCBAs are frequently utilized during initial site surveys, during site characterization, for emergency rescue, or for specific site activities that require mobility, such as sampling or working around heavy equipment.

Operating times for SCBAs are typically between 30 and 60 min. These times vary depending on the size and pressure of the air cylinder, the type of work performed, and the fitness of the wearer. A warning alarm sounds when 20–25% of the air supply remains.

Some of the disadvantages of using SCBAs include the short operating time and the bulk and weight of the air cylinder. The air tank plus backpack, harness, and regulator may weigh up to 35 lb. Although the use of composite materials (carbon, Kevlar, and fiberglass) have reduced the weight of the tanks by over half, SCBAs are still designed primarily for short-time use.

An alternative to the heavy SCBA is the supplied-air or air-line respirator. Air lines allow longer work intervals than SCBAs, are not heavy, and are less bulky. Workers are tethered to a compressor or air cylinders by a long hose, which can decrease mobility. Workers must be careful not to entangle themselves or equipment on the air line, and frequently must retrace their steps when leaving the area. The air line is vulnerable to punctures, kinking, chemical degradation, and damage from equipment, vehicles, and site debris. All potential on-site hazards to the air line should be removed prior to beginning work. As an alternative, air lines can be placed off the ground or encased in protective sleeves. The length of the line should not exceed 300 ft. Experienced workers find that a shorter line is easier to manage.

Air sources for SARs may be stationary compressed-air cylinders or a compressor that purifies and delivers breathable grade air to the face-piece. The grade of air supplied from a compressor or a cylinder should be grade D or better. Lesser grades (A–C) of air may contain unacceptably high concentrations of hydrocarbons, carbon dioxide, carbon monoxide, and nitrogen oxides.

Users of SARs should also be equipped with an escape pack or egress device. The pack supplies the wearer with 5–15 min of breathable air and is designed to allow for emergency exit from the site should the air-line system fail. If exit from the site requires more time than that allowed by the escape pack, SCBA should be utilized.

A combination of SAR and SCBA combines the features of both. This type of respirator allows entry into and exit from the site using the SCBA, as well as an extended work interval within the contaminated area while attached to the air line. This type of respirator is particularly useful when workers must travel well into the site before reaching he work area, and remain there for a prolonged period of time to perform tasks that do not require high mobility. The combination system differs from the escape pack in that it is designed for entry and egress, and provides up to 60 min of air.

Respiratory-Protection Program

Any employer who provides respiratory-protection equipment to his workers is required by law (29 CFR 1910.134) to establish a respiratory-protection program. A minimally acceptable program must include the following:

1. Written procedures describing the selection and use of respirators.
2. Training and instruction in the limitations of respirators, and their proper care and use. All such training should be documented.
3. Selection of respirators based on hazards to which workers are exposed.
4. Regular cleaning, disinfection, and inspection of respirators after each use or at regular intervals when not in use.
5. Surveillance of work-site conditions and extent of worker exposure to ensure proper selection of respirators.
6. Regular inspection and evaluation of the program in order to assess its continued effectiveness.
7. Evaluation by a physician, or other licensed health-care professional, of the medical fitness of each employee to use respiratory protection.

No individual should be assigned tasks involving respirator use until their medical fitness to use respiratory protection has been evaluated by a physician. A written statement by a physician should stipulate the health and physical conditions considered pertinent to each worker's use or nonuse of respiratory-protection equipment.

All aspects of the respiratory-protection program should be documented in writing and retained for future reference. A record of each employee's medical exam, which should include a pulmonary function test and the physician's fitness statement, should be similarly saved. All fit-test records should be documented and saved. Fit-testing and medical examinations should be repeated at least annually.

Air Monitoring

An essential component of all health and safety programs is air monitoring. Air monitoring is an important aspect of the environmental site characterization process that must be performed prior to site entry and then on a regular basis after other activities have been initiated (Figure 19.4). Site characterization is required by the OSHA rule covering hazardous materials site workers. This information is used to assess the hazards and associated risks to site workers as well as to off-site receptors. Identification and quantification of air contaminants is required in order to select appropriate PPE and define areas where protective equipment is required. Air monitoring may also be helpful in determining the effectiveness of mitigative activities. It is absolutely necessary, from the standpoints of both fire safety and worker exposure, to monitor the air during drilling operations at sites that are known or suspected to have hazardous substances.

Various monitoring devices can be employed around drilling operations including fixed or portable survey instruments and dosimeters. The devices include instruments for measuring oxygen deficiency, combustible or explosive atmospheres, toxic substances, and radiation. It is very important that workers operating monitoring equipment be thoroughly trained in the use, limitations, and operating characteristics of each piece of equipment.

Instruments selected for use in the field must be capable of generating reliable and useful information. They should be capable of selectively detecting the contaminants of interest, and sensitive at a useful concentration range. Instruments should have a good battery life; they should also be portable, weather-resistant, and easy to operate, calibrate, and maintain in the field. It is recommended that intrinsically safe instruments be used

FIGURE 19.4
Air monitoring is an essential component of an HASP.

when available. Costs of instruments vary greatly, depending on the function of the instrument, its sophistication, desired accessories, and calibration equipment.

There are three different methods used during air monitoring. The method used for any given application is dependent upon the type of equipment, the number and training of personnel, and the degree of hazard known or anticipated onsite. Intermittent monitoring involves readings taken when targets of opportunity present themselves or when there is a change in field conditions. Semi-continuous monitoring is utilized when readings are required on a regular basis, i.e., each time drilling tools are removed from the borehole. Continuous monitoring constantly assesses site conditions, i.e., during drilling or excavation activities.

Direct-reading instruments provide information at the time of sampling. Many such instruments can detect contaminants at concentrations as low as 0.1 ppm. However, quantitative data are difficult to obtain when multiple compounds are present. High background concentrations caused by heavy equipment exhaust can mask the benefits of such a low reading instrument.

Combustible gas indicators (CGIs) measure the risk of fire and explosion from flammable vapors. Readings may be in ppm, percent LEL, or percent combustible gas by volume. It is important to know which type of meter is being used and to understand what the meter readings mean. A CGI may also be used as a toxic meter for flammable materials if the substance is known and the response efficiency of the meter is adequate around a predetermined exposure limit (OSHA PEL, ACGIH TLVs, NIOSH REL, or IDLH). These instruments are internally calibrated for normal oxygen atmospheres, although most will work properly when oxygen levels are somewhat reduced (i.e. <20% but >10%). Enriched oxygen (>21%) concentrations may give false high readings. The manufacturer should be contacted for specific information regarding the minimum and maximum concentrations of oxygen necessary for accurate readings. Acid gases and organic lead, sulfur, and silicon compounds can damage the sensor element.

CGIs measure the total amount of combustible vapor present and cannot differentiate between multiple compounds. Although the instrument is calibrated for a specific flammable gas (the calibrant gas), its relative response to other gases will be different. Temperature also affects response efficiency. Manufacturers supply information on the limitations and relative response efficiency of their instruments to frequently encountered flammable gases.

Oxygen deficiency meters are used to assess the air for oxygen content to determine if respiratory protection is necessary. Oxygen content less than 19.5% requires the use of atmosphere-supplying respirators. Oxygen meters are also used to indicate increased oxygen conditions; such conditions may be due to the presence of chemical oxidizers. Oxygen-enriched (>25%) atmospheres increase the risk of combustion. These meters may also be used indirectly to detect the presence of other contaminants. A decrease in oxygen content that is not due to consumption (i.e., combustion or chemical reaction) is generally due to displacement by another substance that may be hazardous.

The oxygen sensor relies on atmospheric oxygen pressure; pressure decreases as elevation increases. To obtain an accurate reading, instrument calibration using clean ambient air should be performed at the altitude at which the instrument is used. Temperature can also affect the response. The normal operating range is usually between 32 and 120°F. High concentrations of carbon dioxide and other acid gases shorten the life span of the sensor. Exhaling into the meter to test its function is therefore not recommended. Some manufacturers offer combination meters that detect combustible gases and oxygen deficiency, and common gases such as carbon monoxide and hydrogen sulfide. The electrochemical sensors for oxygen, carbon monoxide, and hydrogen sulfide have limited life spans of typically 1 yr.

Radiation meters detect the presence of ionizing radiation. Three types of radiation are of major concern: alpha particles, beta particles, and gamma rays. Radiation survey instruments are designed to detect one or more types of radiation. Ion detector tubes are used for measuring high levels of gamma radiation. Proportional detector tubes detect only alpha radiation. Geiger–Mueller tubes are sensitive, and used to detect low levels of beta or gamma radiation. Scintillation detectors are sensitive to low levels of alpha and gamma radiation.

Radiation instruments typically measure exposure rates in milliroentgens/hour (mR/h), or roentgens/hour. Normal background radiation exposure rates are 0.01 to 0.05 mR/h. Monitoring for ionizing radiation is required by OSHA at uncontrolled hazardous waste sites, such as during invasive activities at landfills.

Detector tubes consist of a glass tube filled with an indicating chemical matrix that changes color in the presence of a specific contaminant or type of contaminant. The length of color change is proportional to the concentration present. The tube is connected to a bellows or piston pump, and air is drawn through the tube by the pump. A long probe or hose can be placed between the tube and pump for sampling remote locations. Tubes may be specific for one contaminant or for a class of contaminants. Some manufacturers produce a "polytube" which is designed to detect the presence of an air contaminant. Additional tubes are used following a decision matrix to aid in the identification of the contaminant. Multi-tube racks are also available allowing the measurement or detection of several chemical families at the same time.

Detector tubes are inexpensive and easy to use, and can be useful as a screening tool to determine the presence of organic and inorganic contaminants. Accuracy of detector tubes varies between tubes and manufacturers; some manufacturers report error factors of up to 50% for some tubes. All tubes have expiration dates; shelf life can be reduced by temperature fluctuations. Refrigeration of tubes (at 40°F) will minimize this problem. Temperature, humidity, and interfering substances present at time of use can affect accuracy, and manufacturers' instructions usually list limitations for each tube. The presence of interfering substances, i.e., other compounds that also produce a color reaction, often makes interpretation of results difficult. Tubes that produce poor color changes are difficult to use in the field; some tube protocols require multiple pumps strokes and take an inordinate amount of time to achieve a single reading. It is recommended that tube protocols and efficiency be evaluated prior to use under field conditions.

Personal monitors for specific hazards may be worn by individual workers. Monitors are available for combustible gases and oxygen deficiency as well as toxic gases such as carbon monoxide, hydrogen sulfide, cyanides, phosgene, and so on. High-risk workers, i.e., those closest to the potential source of the hazard, are likely candidates for personal monitors. Personal monitors should have an audible alarm, be lightweight and easy to carry, and have good battery life. Dosimeters are also available for organic vapor screening.

Survey instruments are used to detect the presence and total concentration of organic gases or vapors in air. Contaminants are detected at the same time, and there is no identification of individual compounds. Two types of survey instruments are commonly used — the photoionization detector and the flame ionization detector.

Photoionization detectors (PIDs) are capable of monitoring many organic and some inorganic vapors and gases. A PID uses ultraviolet light to ionize gas or vapor molecules. Ions are collected and produce a current; the measured current is proportional to the number of ionized molecules present. The energy required for ionization, measured in electron volts (eV), is called the ionization potential (IP).

A variety of ultraviolet lamps are available, depending on the manufacturer. More commonly used lamps are 9.5, 10.0, 10.2, 10.6, and 11.7 eV. In order to detect a specific chemical, the energy generated by the ultraviolet lamp of the PID must be equal to or greater than the IP of the chemical. Hexane, with an IP of 10.2 eV, is detected with a 10.2 eV lamp; however, the response efficiency is very poor. A better response is obtained with a higher energy lamp, such as 10.6 or 11.7 eV. Hexane would not be detected by a lamp with an energy less than 10.2 eV. PIDs cannot detect light hydrocarbons such as methane. When present at low ppm concentrations, chemical that are not ionized do not interfere with readings from ionizable substances. However, high concentrations of nonionizable vapors can mask the presence of ionizable contaminants. Relatively high concentrations of methane (>5000 ppm) PID responses to ionizable gases and vapors can be severely decreased or even absent (Maslansky and Maslansky, 1993). For this reason, the use of PIDs at landfills is not recommended unless augmented with other detection devices.

Dust in the air can collect on the ultraviolet lamp, interfere with light transmission, and reduce instrument readings. Most lamps are easily accessible and should be cleaned periodically. Some instruments are equipped with particulate filters; these are useful but must be cleaned or changed frequently. Contaminants may adsorb onto particulates trapped in the filter and interfere with subsequent readings.

High humidity can condense on the lamp and decrease the amount of light reaching the air sample. Humidity also reduces the ionization of chemicals and thereby decreases instrument readings. If water is drawn into the ionization chamber, the lamp will short out and must be replaced. PIDs are normally factory calibrated to one chemical; calibration kits are usually supplied by the manufacturer. The instrument's response to other chemicals varies, depending on the molecular configuration, concentration, and IP. The intensity of the lamp declines slowly with age, however the ionization energy remains unchanged. The same effect is seen in light bulbs; the intensity of a bulb may decrease with age, but the wattage remains the same. Decreases in ultraviolet light intensity will be perceived during calibration; instrument settings should be adjusted to compensate.

Flame ionization detectors (FIDs) use a hydrogen-fed flame to ionize organic vapors and gases. When the vapors burn, positively charged carbon-containing ions are produced and collected. A current is generated that is proportional to the ions collected. Unlike PIDs, FIDs are capable of detecting virtually all compounds that contain carbon—hydrogen or carbon—carbon bonds. FIDs respond differently to different compounds, however there is less variability in sensitivity between different substances when compared to a PID. FIDs do not detect inorganic compounds.

Most FIDs can be operated in the survey mode or, with appropriate attachments, in the gas chromatographic (GC) mode. The GC mode is capable of separating components of an air sample using a GC column packed with an inert solid. With proper standards, each constituent can be identified and quantified. The identity of the chemicals of interest must be determined in order to prepare standards. Training and experience are required to successfully operate the GC option.

FIDs are less susceptible than PIDs to high humidity; however, very high humidity conditions will reduce the relative response. A supply of ultra-pure hydrogen is required. FIDs can detect methane and methane is frequently the factory calibrant. At landfills or other field situations where light hydrocarbon gases are found, the FID is not useful for detecting toxic air contaminants. In these cases, the relative responses of both an FID and PID should be assessed in order to determine the source of the readings. The FID will detect methane, while the PID does not. On the other hand, a PID will detect inorganic contaminants while the FID will not. Additional monitoring options for landfills can be found in Maslansky and Maslansky (1993).

Portable, programmable GCs are available. These instruments are not designed for survey work, but they can be valuable for identifying and quantifying on-site contamination. Programmable GCs have also become popular for soil gas and headspace analyses. Relatively light-weight 35–50 lb, but expensive ($75,000–125,000) portable GC–MS units for quantifying and speciating VOCs made their debut in the late 1990s.

Other instruments that have garnered favor include real-time particulate and aerosol monitors and infrared sensors. Particulate monitors are particularly useful on sites with dust-borne heavy metals and other toxics, and where total dust exposure is used for level of protection determination. Infrared sensors are the easiest way to monitor for carbon dioxide at landfill investigations.

Individuals should be trained in the use, maintenance, and calibration of the instruments they operate. Operators should have their own copy of the instrument manual, which should be thoroughly read and understood before attempting to use the instrument. Experience is the best teacher — it is advisable to allow the operator to take the instrument home and experiment with it. Obtaining readings in the garage or kitchen often gives a different perspective to their meaning. That is, not everything that gives a reading is hazardous. Finally, it must be remembered that: (1) no instrument has been designed to detect all possible contaminants and (2) a zero reading indicates a lack of instrument response, not zero contamination. In the words of Carl Sagan, "... the absence of evidence is not evidence of absence ..." (Sagan, 1977).

Protective Clothing

Site workers must be protected against potential hazards. PPE is utilized to decrease exposure to biological and chemical hazards and to shield against physical hazards. PPE should be considered the last line of defense. Administrative controls (work practices) and engineering controls should be attempted first. The employment of PPE does nothing to reduce or eliminate the hazards that might be present. Proper selection and use of PPE should protect the respiratory system, eyes, skin, face, hands, feet, body, and hearing.

The nature of the hazard, based on physical, chemical, or biological properties, and the expected concentrations of contaminants known or anticipated to be present determine the combination of protective clothing and equipment that will be used. The U.S. EPA Levels of Protection system, which has been incorporated into OSHA rules, is used by most organizations when dealing with hazardous materials.

Level A is worn when the highest level of respiratory, skin, and eye protection is required. A Level A ensemble consists of a pressure-demand atmosphere-supplying respirator, fully encapsulated chemical-resistant suit, inner and outer chemical-resistant gloves, chemical-resistant safety boots (steel toe, shank, and metatarsal protection), and hard hat. Optional equipment might include a cooling system in hot weather, flash-over protection, abrasive-resistant gloves, disposable oversuit and boot covers, communication equipment, and safety line. Drillers use Level A protection only in situations where remote techniques cannot be employed and the site is contaminated with highly toxic or corrosive materials (e.g. nerve agents or hydrofluoric acid).

Level B protection is utilized in areas in which full respiratory protection is warranted, but a lower level of skin and eye protection is adequate. Level B consists of a pressure-demand atmosphere-supplying respirator, splash suit (one- or two-piece) or disposable chemical-resistant coveralls (with or without hood), inner and outer chemical-resistant gloves, chemical-resistant safety boots, and hard hat with face shield. Optional items include glove and boot covers, and inner chemical- and flash-resistant fabric coveralls. Many monitoring and recovery wells have been installed in Level B protection, particularly at abandoned hazardous waste sites or at spill sites where the concentration, lack of warning properties, or breakthrough characteristics of the contaminants precluded the use of APRs. Level B protection is required by OSHA during characterization activities for sites with unknown hazards.

Level C permits the use of APRs. Level B body, foot, and hand protection is normally maintained. Many organizations will permit only the use of approved full-face respirators equipped with a chin- or harness-mounted canister. However, many drillers prefer to wear half-mask cartridge-equipped respirators. If allowed by the client, the decision of which type to use becomes a trade-off of decreased protection versus increased comfort.

Level D protection consists of a standard work uniform of coveralls, gloves, safety shoes or boots, hard hat, and goggles or safety glasses. Some organizations require personnel who are outfitted in Level D or C protection to carry emergency escape masks which supply 5–15 min of air, or place these masks around the work site and next to vehicles.

Protective clothing is selected to guard against vapors, splash, flash, and physical contact with chemicals. Clothing should prevent or minimize penetration, permeation, and degradation by the contaminants encountered. Penetration is breakthrough of a chemical through seams, zippers, buttonholes, and the like. Permeation occurs when a chemical soaks into the fabric without altering its physical properties. Degradation occurs when a chemical changes the physical properties of the fabric. One reference, *Guidelines for the Selection of Chemical Protective Clothing* (Schwope, 1987), provides a matrix of clothing material recommendations for approximately 300 chemicals, based upon permeation and degradation data. Another popular book is Forsberg and Mandorf's Quick Selection Guide to Chemical Protective Clothing, 3rd Edition published by John Wiley & Sons (1997). Protective clothing manufacturers also supply chemical compatibility data for their products. Companies such as Dupont (www.DuPontProtectiveApprl.com) and Kappler (www.kappler.com) have CDs and interactive Web sites for help in selection of protective clothing. Chemical compatibility is not a major concern if site contaminants are found at low (ppm) levels.

It is advisable to wear disposable clothing over reusable clothing for additional protection and to minimize decontamination procedures. Disposable suits are highly recommended for drilling work. One-piece saran-coated or polyethylene-coated Tyvek overalls offer good splash and vapor protection for low-level contamination. Duct tape may be used to seal openings around disposable boots and gloves. Taping increases heat load, however, and can contribute to heat stress. Uncoated Tyvek overalls offer no splash or vapor resistance and should be reserved for training or for sites at which dry particles are the only problem.

Disposable gloves and boot covers may be worn over expensive or leather counterparts that cannot be easily decontaminated. Disposable boots can increase the slip and trip hazard unless they have aggressive soles. Thin boot covers with flat soles are useless in the field and should not be considered. Disposables should be large enough to fit but not so large as to interfere with a worker's ability to perform assigned tasks. Disposable items come in different sizes; the proper sizes should be supplied to all site personnel.

No one PPE ensemble, no matter what combination is used, will be capable of protecting against all hazards. To be effective, PPE must be used in concert with other protective methods as well as good field practices. PPE is not a suit of armor; one should not feel invincible when wearing PPE. Indeed, the use of PPE creates significant worker hazards, such as heat stress, loss of mobility, difficulty in communicating, and impaired vision. It is important, therefore, to select the appropriate level of PPE without overprotecting the worker, because the greater the level of PPE, the greater the associated risk (Figure 19.5). Health and safety are not mutually inclusive.

FIGURE 19.5
Health and safety are not mutually inclusive.

Site Operations

When working at known or suspected hazardous substance sites, work zones should be established to protect site workers and to minimize the potential risk of injury to workers as well as authorized and unauthorized visitors and untrained support staff. Minimizing the risk of injury also limits potential liability. Site control areas or work zones are established which allow site procedures to be safely conducted while reducing the potential for contacting any contamination present and minimizing the possibility of removing contamination through personnel or equipment leaving the site.

Procedures for minimizing exposure to or transfer of potentially hazardous materials are numerous, and include the following:

1. Elimination of unnecessary personnel in the general area and reducing the amount of workers and equipment onsite to that consistent with effective and safe operations.

2. Establishment of site security and physical barriers to exclude unnecessary personnel and vehicles from the general area.
3. Establishment of work zones and control points to regulate access to the site.
4. Implementation of an appropriate contamination avoidance program in order to reduce personnel and equipment exposure to surface and subsurface contamination, and to minimize airborne dispersion of contamination.

Various terms have been used to describe work zones at hazardous materials sites (NIOSH, 1985; HAZWOPER, 1989). The *Exclusion Zone* (Zone A, Zone 1, Hot Zone, Red Zone) is the area of contamination, and is marked off by the Hot Line. This line is a real or imaginary barrier determined by instrumentation, visual observation, fragmentation distance, or most commonly, by "geopolitical-institutional-topographic boundaries"; these boundaries are usually a fence, road, stream, or some other preexisting natural or man-made boundary.

As a minimum around drilling equipment, the exclusion zone should be defined as the height of the boom, derrick, or mast forming the radius of the zone. For excavation equipment, the minimum radius of the zone is defined as the maximum extension of the boom plus 20–30 ft.

The *Contamination Reduction Zone* (Zone B, Zone 2, Warm Zone, Yellow Zone) acts as a buffer between contaminated areas and clean areas. It is initially established in a noncontaminated or clean area. It is in this zone that personnel are decontaminated at a Personnel Decontamination Station (PDS). Equipment is decontaminated at a separate Equipment Decontamination Station (EDS). The dimensions of the zone are site-specific, not only in terms of logistics, but also in terms of specific site hazards. At a minimum, the zone should be large enough to comfortably contain both the PDS and the EDS.

The *Support Zone* (Zone C, Zone 3, Cold Zone, Cool Zone, Green Zone, Staging Area) is the outermost portion of the site, and is considered clean or noncontaminated. This zone contains support equipment, trailers, and parking areas. The location of the zone may ideally be established upwind of the prevailing winds, but is usually practically based on site access and location of available utilities and resources.

Level D protection, as a minimum, should be employed around drilling or other heavy equipment. The need for higher levels of protection in each of the work zones is a function of known or suspected hazards and their associated risks. The job function of site workers, the potential length of exposure, weather conditions, and the types of exploration or excavation equipment employed must also be examined before determining the level of protection. For example, on a site with a known contaminant, a drilling activity may be conducted in Level C with drillers protected against splash and vapors. The site geologist, several feet away from the rig, may also be in Level C respiratory protection; however, he may require less skin protection against splash and vapor contamination.

Although many personnel, equipment, and site restrictions may be employed while conducting environmental site characterization or ground-water monitoring investigations at a known or suspected hazardous substance site, at a minimum, personnel must adhere to the following:

1. Workers should wear properly selected and fitted protective clothing and respirators at all times when required. Personnel must be given suitable training in the use, limitations, maintenance, cleaning, and storage of protective clothing and equipment.

2. Personnel should not eat, drink, chew gum or tobacco, smoke, take medicines, or perform any other practice onsite that might increase hand to mouth transfer of potentially toxic materials from gloves, unwashed hands, or equipment.
3. Personnel should not have excessive facial hair in the form of heavy mustaches, long sideburns, or beards, which can prevent the proper fit of respirators.
4. Personnel should avoid unnecessary contact with hazardous materials by staying clear of puddles, vapors, mud, discolored surfaces, and containers or site debris.

Many symptoms of temperature stress and toxic overexposure are not readily apparent. Site workers should observe each other for any signs of physical or mental abnormalities such as confusion, complexion change, lack of coordination, or changes in speech patterns. Workers should immediately inform each other if they experience headache, dizziness, blurred vision, cramps, nausea, respiratory distress, irritation to eyes, nose, or mouth, or any other signs of distress. It should be noted that many exposures can produce symptoms that may be delayed hours or days after contact.

Drilling Techniques

During the 1970s, it was common practice for test boring contractors and monitoring well installers to preflash drill holes while engaged in landfill operations. This was done in particular before any welding or cutting took place around the holes, and was typically accomplished by throwing a lit traffic flare in the hole. The preflashing resulted in "controlled" methane fires and explosions, and led to some very tall tales about launching rods and augers. Without a doubt, preflashing and the unintended sparking of a hole were very unsafe procedures; many drillers can attest to burns and lost drilling rigs. As Chapter 5 explains, the selection of an appropriate drilling method for a site is a complex issue, with the choice based on the site's geology, potential impact on sample integrity, required hole size and depth, equipment availability and cost, among other factors. However, certain drilling methods are inherently safer than others when doing work that would encounter flammable gases and vapors, such as during landfill or petrochemical site investigation and remediation projects.

Typically, the critical zone for an explosion or a flash to occur extends 1 ft into the hole and 1 ft above the hole, or at the top of the drill casing or hollow-stem auger. At depths greater than 1 ft, the gas or vapor concentrations tend to become too rich, and the oxygen concentration is too low for combustion. Conditions 1 ft or so above the hole or equipment opening, assuming normal outside ventilation, usually produce rapid dispersion of lighter-than-air gases and either dispersion or a cascading of heavier-than-air gases. The authors (Maslansky and Maslansky, 1996) have measured methane concentration of dozens of wells in excess of 600,000 ppm (60% by volume) immediately inside the bore or casing annulus. These concentrations were found to be reduced an order of magnitude within 1 ft in all directions above the hole, and to low ppm levels within 3 ft above the hole. This testing was done under various wind speeds conditions, including no wind.

A dangerous condition can exist if heavier-than-air gases or vapors settle into low or confined areas. For example, gasoline vapors can collect under vehicles at service station tank pulls and then explode upon vehicle start-up. Similarly, invasive work at landfills can release carbon dioxide that may migrate into low areas producing asphyxiating conditions.

Most drilling equipment represents a potential source of ignition from frictional sparking and power trains. The venting of aerosolized hydraulic fluids during sonic drilling operations can increase the fire potential should a flash-over occur. Although the use of an air rotary rig utilizing a compressor may reduce some sources of ignition, it can blow gases and vapors out of the hole, aerosolize liquids and release them from the hole, and produce static charges. The application of drilling foams can lessen these potential problems. Auger rigs allow for the placement of dry ice around the auger base, so as the dry ice is covered with spoil, the off-gassing, heavier-than-air, carbon dioxide fills the voids near the top of the hole and produces a cascading effect down the hole. The use of hollow-stem augers facilitates monitoring as well as the introduction of inert gases. Mud-rotary drilling techniques are used to suppress vapors and ignition sources, but the fluids present can bring nonaqueous-phase liquids to the surface. Cable tool drills can generate frictional sparking hazards and can also displace or cause upward migration of gases and vapors, as the tools are placed into and removed the hole. This problem is of particular concern when a drive casing is not utilized.

Other methods of hazard control include inerting and purging. *Inerting* is the introduction of a nonflammable gas (argon, helium, nitrogen, or carbon dioxide) into a space to displace oxygen and thus make combustion less likely. *Purging* is the introduction of a nonflammable gas or fresh air into a space to remove flammable gases or vapors. Inerting is typically a down-hole process, whereas purging takes place around the hole or underneath vehicles.

It may be useful, particularly on jobs of short duration (where the rig is moved every day), to take advantage of the prevailing wind direction. The rig position depends on a number of factors, including: the drilling technique, where the driller stands or sits, whether a helper is present while tools are being advanced, whether the hole is "producing," air temperature compared to ground temperature, and whether or not a heavier- or lighter-than-air gas or vapor is present. In dealing with heavier-than-air vapors (i.e. gasoline and most liquid hydrocarbons), the rig should be positioned so vapors are blown away from it and do not accumulate underneath the rig. Typically, this would mean the cab end of the drilling rig would face into the wind. When employing long-trailered rigs and rigs nested with a support tender, perpendicular positioning can be utilized.

Portable ventilation devices, commonly referred to as blowers, have been utilized at many drilling sites. In particular, they have been used successfully at landfill investigations and during the installation of gas extraction or recovery wells. Blowers can be used with or without duct work and typically move between 1000 and 5000 cfm. Because the production of static is of great concern, many units have built-in grounding and bonding systems or come equipped with a grounding lug. Venturi-style or pneumatic systems with a low noise muffler and compressor are the most commonly employed. Blowing across the hole to push gases or vapors away from the rig is considered positive ventilation. As with the wind blowing across open nested monitoring wells or manholes, such an action can cause more material to be educted out of the hole. Pulling gases and vapors into ductwork and discharging away (downwind) from the rig is considered negative ventilation, and while more material is moved, control is better. Air monitoring should be conducted at the discharge location as well as around the hole.

It may be necessary on jobs with high concentrations of heavier-than-air flammable vapors to utilize vertical exhausts and spark arrestors. Although smoking should be prohibited while working, common sense dictates that the no-smoking zone be extended to at least a "shadow of the mast" distance away. If welding and cutting tasks must be performed at or by the hole, they should be initiated only after air monitoring has been done and, if necessary, after inerting or purging has taken place. Remember that PVC pipe glue is flammable and can produce explosive concentrations in a well.

Extra care must be exercised when working in topographic lows and at the toes of embankments and landfills (Figure 19.6), and in employing wind screens and barriers.

FIGURE 19.6
This driller is literally caught between a rock and a hard place.

Heavier-than-air gases and vapors may collect at such locations. Of course, working inside structures also poses numerous hazards. The importance of good housekeeping cannot be overemphasized. The presence of gasoline cans near the hole and excessive oil and grease on rig components will only compound the problem, should a hole flash.

Decontamination

Although contamination avoidance is the best procedure at a hazardous materials site, workers' protective clothing is likely to come into contact with contaminated vapors, particulates, drilling mud, and ground water. Walking or driving on site may contaminate vehicles, equipment, and clothing. During installation and testing of recovery wells, drillers may come into contact with relatively high concentrations of chemical compounds. Drilling tools, as well as development and sampling equipment, may become unavoidably contaminated. These items must be properly decontaminated before being removed from the site or, in the case of sampling equipment, be thoroughly cleaned before the next use. The decontamination procedure may vary greatly depending on the size, condition, and status of the site, the nature of the hazardous materials, and the nature of site activities. In general, the more harmful the contaminant, the more extensive or thorough the decontamination procedure should be.

Decontamination is a process during which a hazard is reduced to some predetermined safe level (usually normal background concentrations) by removal, neutralization, absorption, chemical degradation, dilution, covering, or weathering. Decontamination, or "decon," can be broken down into three general categories: environmental decon, safety decon, and health decon.

Environmental decon is performed to protect some aspect of the environment, such as soil, air, or a water supply, from low concentrations of a pollutant that may have long-term environmental consequences, and to minimize cross-contamination. This includes contaminants present in the parts-per-million (ppm) or parts-per-billion (ppb) range of contamination.

Safety decon is conducted when a substance is not overly toxic or hazardous, but produces safety problems, such as slippery walking or riding surfaces. Many mild alkalis and diesel fuel fall into this category.

Health decon is performed when the contaminant, by virtue of its toxic, reactive, flammable, or corrosive properties, presents a hazard to site workers or equipment.

When determining proper procedures for decon, both the degree of hazard and the degree of risk to personnel or equipment should be considered. Many sites involve multiple contaminants, and information regarding specific chemicals may not be available for uncharacterized sites. In these cases it is necessary to select decon procedures that will cover a broad range of contaminants. The decon protocol initially assumes that personnel or equipment working in the hot or exclusion zone are contaminated until instrumentation reading or visual observation indicate differently. The decon area in the warm zone should be large enough to handle personnel and equipment. Decon procedures can later be downgraded as more information is gathered on the type of contamination and its volume and concentration. The time constraints imposed by adhering to decontamination protocols must always be considered when preparing site work plans.

Contamination avoidance will help to minimize later decon procedures. Equipment, personnel, and the decon area should be kept upwind of the contaminated area if possible. Covering monitoring instruments, tools, and equipment with plastic sheets and tarpaulins can also minimize the need for subsequent cleaning. Care must be taken when placing plastic near hot or moving parts. Disposable clothing, gloves, and boots may be worn over reusables to reduce decon and extend the wear life of more expensive equipment. Porous items such as wooden truck beds or pallets, cloth hoses, hemp ropes, and wooden handles cannot, in many cases, be properly cleaned — these items should be considered expendable and discarded.

Site workers should normally go through a PDS, which consists of several cleaning and rinsing stations as well as clothing and equipment removal stations. Contaminated clothing and equipment should not be taken off site where others may be exposed to hazardous substances.

In general, wet contamination should be kept wet, and dry contamination should be left dry. Some dry compounds may form solid oxides or other reaction products when wetted; these products may complicate decon procedures or may be more difficult to remove.

Decon operations should start with the simplest methods. For vapors or wet contamination, a general spraying will remove the bulk of contamination, followed by scrubbing of difficult areas if necessary. Dry material should be brushed or scraped off — the contaminated area can then be sprayed and scrubbed if needed. These procedures avoid unnecessary contact with contaminated material by decon personnel.

Organic solvents used for the decontamination of drilling tools and sampling equipment include acetone, *n*-hexane, and various alcohols. These materials are often used in large quantities and typically present a greater flammability and toxicity hazard than on-site contaminants.

Trisodium phosphate (TSP) has been used as a general decontamination agent. TSP, however, has a high pH in solution. As a solid, or in concentrated solution, TSP is a skin and eye irritant; in areas where phosphates have been banned, decon solutions normally considered nonhazardous but containing TSP may have to be stored and disposed of elsewhere. A nonionic, anionic, cationic surfactant solution, otherwise known as liquid detergent, is the best overall decon agent available. Low sudsing liquid laundry detergents are readily available, easily stored and transported in concentrated form; they are nontoxic and go into solution easily. Detergent solutions work well against most contaminants encountered. In some cases, it may be necessary to utilize special neutralization solutions or solutions containing solvents to effect a thorough decontamination. Whatever decon agent is used, its possible reactivity and suitability for the hazardous materials involved must be evaluated. Not all decontaminants are compatible with each other. It is important that decon personnel understand the potential hazards of the contaminants, as well as those associated with cleaning equipment and decon solutions.

Common sense must be used when determining procedures for decontaminating sampling equipment. Procedures appropriate when working with ppb or ppm concentrations of aqueous-phase contaminants will be considerably different than when working with sludges containing high concentrations of toxic materials. Low boiling point, high vapor pressure contaminants found in low concentrations (ppb or ppm), such as chlorinated solvents in ground water, rarely contaminate equipment except by sorption. These substances can be easily removed by a low-sudsing detergent solution followed by a water rinse.

Wet decon solutions and water rinses create unwanted runoff that spreads contamination. It is generally a good practice to limit the amount of water utilized. The use of large wash tubs or children's wading pools for decon of personnel can aid in the collection of decon water. Likewise, using small pneumatic garden sprayers as a water source will help minimize the volume of water employed. Remember that it is the pressure rather than the volume that exerts the cleansing action.

The decontamination of vehicles and large pieces of equipment, such as pumps and augers, may be performed on a wash pad constructed so that cleaning solutions and wash water can be recycled or collected for proper disposal. A raised graveled area lined with polyethylene is a lower cost alternative.

It is important that all equipment surfaces, including undercarriage, wheels or tracks, chassis, and cab are thoroughly cleaned. Air filters on equipment operating in the hot zone should be considered highly contaminated and treated as such; contaminated filters should be removed and replaced before equipment leaves the site.

Steam cleaning or high-pressure spraying utilizing low volumes of water is the decontamination method of choice for equipment and vehicles. Lower pressure units (90–120 psi) use low-sudsing detergent or special neutralizing solutions at a rate generally from 3 to 5 gpm. High-pressure spray units (up to 5000 psi) normally do not require decon solutions. Similarly, hot water (120–180°F) is usually not necessary when using high-pressure sprayers. Field experience has shown that spray units operating at 500–800 psi with a flow rate of 3–5 gpm are satisfactory; units offering pressures of 1000 psi and flow rates of 1–2 gpm are considered ideal. Stream cleaning or high-pressure sprayers can also remove necessary lubricants, so fluid reservoirs on cleaned equipment should be checked regularly. Wire ropes, bearings, and other vital components must be inspected and relubricated as necessary after decontamination.

It is imperative that decon activities be practiced before they are executed. A quick walk-through of decon procedures at the personnel decon station should be conducted before workers go into the hot zone. For workers utilizing atmosphere-supplying respirators, this can help ensure that sufficient time is allotted to proceed through the decon line. This is especially important on sites where workers will perform self-decon.

Medical Monitoring

The OSHA rule covering hazardous materials workers (29 CFR 1910.120) stipulates that a comprehensive medical surveillance program must be provided to all employees who have been or expected to be exposed to hazardous substances for 30 or more days during a 12-month interval. In addition, workers who wear respirators for any part of

30 days during a 1-yr period must also be included in a medical monitoring program. Medical monitoring programs are used to establish baseline data that can verify the adequacy of protective methods and determine if exposures have adversely affected the health and well-being of the worker.

The medical tests appropriate to the monitoring program should be determined by a physician, based upon the information provided by the employer regarding potential and actual exposure, respirator and protective clothing use, and job descriptions. Where job duties and exposures are substantially different, several different monitoring protocols may be appropriate. It should be noted that the tests are supposed to be based upon anticipated exposures and not previous exposures. To assist the physician in formulating a medical monitoring program, the OSHA rule recommends that the employer provide the physician with a copy of the OSHA rule and the NIOSH/OSHA/USCG/EPA Occupational Safety and Health Guidance Manual for Hazardous Waste Site Activities (NIOSH, 1985). The NIOSH/OSHA guidance manual can be obtained from the NIOSH publication office or ordered through the GPO. The OSHA rule for hazardous materials can be obtained from the OSHA consumer affairs office.

All aspects of the medical monitoring program should be conducted at reasonable times and locations so workers are not discouraged from participating. The employer pays for the program — if given during working hours, the employee must receive normal pay for that time, and if given outside normal working hours, the employee must be paid regular wages for the time involved taking and waiting for the examination.

Employees have full access to all aspects of their medical records, including results of medical tests, exposure records, medical opinions, and recommended restrictions. The examining physician must submit a written report to the employer on each employee certifying the presence or absence of medical conditions which may pose a health risk when working at hazardous materials sites. The physician must also document the fitness of the employee to use different types of respiratory protection and personal protective clothing. Any recommended limitations regarding use of protective equipment or on-site activities should also be documented. All diagnoses and medical conditions that are unrelated to employment must remain as confidential information between the physician and the employee.

An initial baseline medical exam is required, which includes a full medical history. The exam should be performed prior to any exposures, so changes in baseline parameters can be more easily associated with hazardous substance exposure. The medical history should include questions concerning family history of specific organ disease (heart, liver, kidney) and cancer, sexual history (onset of menses for women, number of children, fertility problems), dietary habits (food preferences, drinking habits), tobacco use (smoking and chewing) and over-the-counter and prescription drug usage (birth control pills for women, analgesics, antipyretics, antiseptics, cathartics, stimulants, and vitamins).

It is recommended that a lifestyle section be included to document activities that may augment exposures to hazardous substances, such as making model airplanes (exposure to solvents, epoxy resins, and catalysts), mechanical engine repair (petroleum products, combustion products, asbestos, solvents), painting or wood-working (solvents, thinners, lacquers, wood dust), gardening (pesticides), photography (solvents), glass making (solvents, lead fumes), painting (solvents, thinners, pigments), and sculpting (dusts, wood preservatives, thinners, lacquers).

The physical exam should focus on the pulmonary, cardiovascular, and musculoskeletal systems. Conditions that may predispose an individual to heat stress, including obesity and lack of physical fitness, should be noted. To assess a worker's capacity to perform while wearing PPE, a pulmonary function test and electrocardiogram should be performed. A stress test may be performed at the discretion of the examining physician.

Conditions that may affect performance while wearing protective equipment, such as facial scars, poor eyesight, or orthopedic problems, should be noted.

Medical screening tests frequently employed by examining physicians include blood tests to evaluate liver and kidney function, urinalysis, and complete blood count with differential and platelet evaluation. Specialized tests may be appropriate for workers working with known hazards, i.e., blood tests for lead or other heavy metal contamination, cholinesterase activity (when working with organophosphates), or PCBs.

Periodic exams are required at least yearly; more frequent supplemental exams may be appropriate depending on the type of exposure involved or if symptoms of exposure are noted. An exit exam is also required at the termination of employment. All medical records are to be made available to the worker and maintained by the employer for the period of employment plus 30 yr. Specific medical monitoring requirements can be found in the OSHA Hazardous Waste Operations and Emergency Response standard (29 CFR 1910.120).

Applicability of HAZWOPER to Ground-Water Investigations

The OSHA Hazardous Waste Operations and Emergency Response Rule (HAZWOPER) has many times in the past been invoked when it has not been applicable. This has been done out of ignorance or because of perceived liability concerns. Many a client has required 40-h HAZWOPER trained personnel to sample wells containing just a few parts per billion of some contaminant. The rule covers:

1. Initial investigations and clean-up operations at government-identified uncontrolled hazardous waste sites, as well as those sites that would be eligible for listing upon discovery of hazardous wastes.
2. Corrective actions involving clean-up operations at RCRA sites.
3. Emergency responses for releases or substantial threats of releases of hazardous materials or hazardous wastes.

In order for the rule to apply, the task undertaken by workers must be directly related to the above operations, and that task is anticipated to expose the workers to contaminants found on site. As stated in 29 CFR 1910.120(a), if the employer "can demonstrate that the operation does not involve employee exposure or the reasonable possibility for employee exposure to safety or health hazards," the rule does not apply. This clause has been become known in the industry as the "employer's out." Applying the rule needlessly can greatly increase personnel costs through unnecessary training, medical, and administrative requirements. Many an investigator working at a characterized site with a few parts per million of something in the soil or a few parts per billion of something in the ground water has sat through a 40-h class when all that was necessary was a pledge by the workers that they would not eat the soil or drink the water on site.

Training Requirements

Should the HAZWOPER rule be applicable or invoked contractually, a number of training requirements must be followed. In the past, safety training was often perfunctory or nonexistent for those involved in sample collection or subsurface investigations. Since the promulgation of the OSHA Hazardous Waste Operations and Emergency Response standard, however, employers now have a legal obligation to ensure that workers are trained to work safely at sites containing hazardous materials and wastes.

The OSHA standard requires training for all workers involved in hazardous materials operations at CERCLA (Superfund), RCRA, emergency response sites, and sites designated for cleanup by state or local governments. A minimum of 40 h of off-site training is required for workers at uncontrolled hazardous materials sites; 24 h is required for workers performing routine operations at RCRA facilities, or sites where hazards have been identified and respiratory protection is not required. Workers on site occasionally for a specific limited task, such as ground-water monitoring, surveying, or for geophysical investigations, would fall under the 24-h training provision unless the "employer's out" was invoked. RCRA sites are considered to have more stable working conditions and better identified hazards. An RCRA site with uncharacterized hazards is considered an uncontrolled site; workers involved in the investigation of such a site require 40 h of training.

During the basic 24 or 40 h, workers must be trained to recognize hazards and be provided with the skills necessary to minimize those hazards. Workers must be made familiar with the use and limitations of safety equipment and PPE that may be required on site.

Workers requiring 40 h of basic training also must receive a minimum of 3 days of field training or on-the-job training under the direct supervision of a trained, experienced supervisor. The 24-h option requires 1 day of field experience. All personnel must receive a minimum of 8 h of retraining on an annual basis. Employers need to document the initial and refresher training, as well as supervised field experience.

Supervisors and managers must receive an additional 8 h of training on managing the employers health and safety programs. A supervisor can be defined as any individual who has responsibility for, or who makes decisions regarding, the health and safety of personnel in the field.

Individuals working at hazardous waste sites have different roles and levels of responsibility. Site hazards and conditions can vary greatly between sites or portions of the same site. Training must therefore be organized to meet the specific needs and levels of comprehension of the individual, the organization, the work assignment, and the specific requirements of a particular site. In order to maximize the needs of the individual and minimize budget and time constraints, a tiered training approach is usually the most effective and efficient.

The first level of a recommended training program is *overview training*. This is equivalent to the initial training specified by OSHA. Overview training serves as an introduction to the nature and types of hazards that may be encountered in the field. Although basic in nature, overview training should be designed for a specific audience, and its needs and level of comprehension, in order to be successful. Unfortunately, most HAZWOPER training programs are designed for a generic audience; such training frequently does not meet the needs of ground-water investigators. As a result, the supervised field experience and refresher training has become more important.

Certain supervisory, technical support, or field personnel may require intensive training in specific areas of responsibility, such as incident management, respiratory protection, radioactive materials handling, or sample and decontamination procedures. Most organizations do not have the capabilities to develop such *discipline-intensive training* programs. Intensive training programs are sponsored by NIOSH, U.S. EPA, and private training organizations.

The final type of training, and probably the most important, is *site-specific training*. During this phase the individual receives detailed training in the actual conditions and hazards that may be encountered while performing specific tasks. During this training workers should also receive on-site instructions regarding decontamination procedures, emergency escape routes, communications, and the location of emergency equipment.

A record of training for each individual should be maintained to confirm that all workers have received adequate and appropriate training for the tasks assigned, and that each worker's training is up-to-date and in compliance with applicable regulations.

Those working at nonhazardous waste sites still need safety training. All too often, those with experience do not relate the potential for safety and health hazards to the "new hire"; likewise some of the "old hands" have been doing things the wrong way all along. The reader is referred back to page 1 for some of the organizations that can provide general safety training support. A good reference for general field safety is the American Geological Institute's Planning for Field Safety (AG1, 1992).

General Safety and Liability Considerations

There are many risks associated with environmental site characterization and ground-water monitoring at potentially contaminated sites. In addition to the presence of hazardous substances, drillers, equipment operators, field engineers, scientists, and technicians may be exposed to a myriad of other hazards and associated risks. However, informal surveys conducted by the authors during the past two decades have shown fewer injuries and ailments on hazardous waste sites (other than temperature stress), and more injuries per capita on clean water jobs. Most of the injuries and ailments were due to biological hazards and physical hazards of slip, trip, and fall (Figure 19.7). Vehicular accidents are always high on the list.

The operation of heavy equipment, with moving parts, electrical systems, and high-pressure lines, is dangerous regardless of the nature of the site. Site topography, layout, and the presence of surface and subsurface debris can increase the likelihood of an accident. It is not unusual that the largest single overhead expense is insurance. Workmen's compensation, heavy equipment coverage, general liability, medical, life, unemployment compensation, disability — the list goes on and on.

When an accident occurs, no one is immune to a liability suit. A driller's helper fatality that occurred when the rig overturned spawned suits against the driller, the drilling company, the consulting firm, and the contracting state agency. All parties were considered potentially responsible despite the specification that required that "...the driller will ensure that the rig is in a stable position prior to its operation." Even if the outcome is favorable to the defendants, they still must pay the costs, in terms of time and money, to defend themselves.

It is difficult sometimes to distinguish between safety and technical specifications when the contract states "... the driller will supply all equipment in a safe working condition ..." Is the inspecting hydrogeologist responsible for ensuring that the driller follows the site safety plan? Is this person also the health and safety officer? The inspector is at least responsible to point out major violations of the plan, as well as any condition or activity that jeopardizes the health and safety of those involved. The inspector's supervisor should also be notified, and detailed documentation of each violation and specific actions taken should be noted in his or her daily log or field journal. The inspector should have the authority to shut down a job because of unsafe conditions. Was the inspector only an observer? Have titles been changed because of liability concerns? Not only have our monitoring techniques become more complex, so has our response to health and safety concerns.

During the hazardous waste site investigation era of the mid-1970s to the late-1980s, it was common practice for the site consultant (typically an environmental consulting firm) to write an HASP which was authored by the corporate or regional Health and Safety Officer and approved by corporate management. The corporate Health and Safety

FIGURE 19.7
Fall protection is an important consideration.

Officer in most cases had a certification (CIH, CSP, CSS, CHCM, etc.) or was a scientist or engineer with additional health and safety training and experience. Safety in the field was the responsibility of a field supervisor or inspector (geologist, engineer) who acted as the Site Safety Officer. Their training was minimal. Their qualifications improved with EPA-required training of the early 1980s and the OSHA-required training in 1987, which consisted of a 40-h HAZWOPER class and the 8-h supervisor's training. Back then, the consultant working for the site owner or government agency controlling the site hired the subcontractors (e.g. drilling, backhoe, surveying, laboratory). Payment would go through the consultant and a markup taken. The consultant would prepare the site safety plan and it was considered a "good" plan if it addressed all activities and tasks performed by all parties. In theory, a plan organized as a single document and addressing all tasks and operations performed on the site would promote the "4Cs": completeness, clarity, coordination, and competency in performance. A plan may or may not have been approved by the client. For plans that were not approved, work would start when "no more comments" to the plan were received. A lack of critical comments was considered analogous to tacit approval of the plan.

As liability issues continued to arise and many sites entered in a remedial phase, site safety issues also changed. HASPs were often prepared as bid documents rather than written after the job was awarded. The bid document was then modified and made organization-specific by the selected bidder. The plan may have been written by a third party and enforced by the same or a different party. It has been common practice, since the early 1990s, for the consultant to prepare an HASP that covers only their employees. The plan is typically then taken by other site contractors or subcontractors and incorporated into their own HASP with specific requirements for their own employees based on their job and site requirements. Health and safety specialty firms may be brought in to assist subcontractors. In general, each functional area (e.g., drilling, sampling, geophysics, treatment) has its own plan and site safety enforcement. It is not unusual to see several site safety plans and site safety officers for the same site. Although this has resulted in more expertise (at least theoretically) being on site, it has been bad for overall health and safety management of the site.

Primary and ultimate responsibility for the health and safety of workers rests with the employer. This is the prime tenet of all occupational safety and health laws and regulations. Section 5(a) of the OSHA Act of 1970 states that: Each employer

1. shall furnish to each of his employees employment and a place of employment which are free from recognized hazards that are causing or are likely to cause death or serious physical harm to his employees; and
2. shall comply with occupational safety and health standards promulgated under this Act.

The employee also has an obligation to provide for a safe workplace. Section 5(b) of the OSHA Act of 1970 states that: Each employee

shall comply with occupational safety and health standards and all rules, regulations, and orders issued pursuant to this act which are applicable to his own actions and conduct.

OSHA further defines employee (OSHA Field Inspection Reference Manual, Chapter III, C 1.b. (1), September, 1994):

Whether or not exposed persons are employees of an employer depends on several factors, the most important of which is who controls the manner in which the employees perform their assigned work. The question of who pays these employees may not be the determining factor.

It has been recognized that at sites with complex hazards, such as high concentrations of multiple chemicals, the employer and his or her employees may not have the knowledge, training, and the experience to monitor the workplace for all hazards, particularly chemical. Many of the crafts utilized, such as water well or test boring contractors, may not have hazardous waste health and safety skills in-house and expect to rely on others. For example, a well driller should be familiar with the safety hazards associated with the installation of a monitoring well. He or she may not be familiar with the chemical hazards associated with the site, how to monitor for those hazards, and how to minimize hazards and risks through engineering controls or the employment of PPE. In this case, a health and safety officer from another organization may be assigned to conduct air monitoring and report workplace or down-hole concentrations, or perhaps to actually have contractual authority to dictate work functions including stopping work. Likewise, the hydrogeologist and safety officer may be clueless when comes to the safe operation of a drilling rig.

Improper use of job titles, misunderstanding of positions and authority, and implied responsibilities have resulted in more than one unexpected lawsuit. Three titles have come into common use during environmental and ground-water monitoring investigations: supervisor, inspector, and observer. Typically their function and authority is established contractually, but "standard of practice" issues have also been in effect. At ground-water investigations, the supervisor is typically an employee who oversees employees from his or her own company. He or she might be a project manager, senior geologist, supervising engineer, team leader, etc. As with construction sites, sometimes the supervisor is hired specifically for the project. The supervisor maybe considered the OSHA "competent person" for a particular task or for the entire site.

An inspector performs quality assurance/quality control functions. This person is usually not from the same organization that is doing the work but represent the designer, consultant, client or owner, or a regulatory agency. For example, the driller is the supervisor, overseeing the helper, while the hydrogeologist is the inspector ensuring that plans and specifications are being complied with. Contractually the inspector may or may not have safety as an inspection item and may or may not be empowered to stop work due to safety concerns. There may be several different inspectors on the job due to size, complexity, duration, expertise available, and need.

An observer is usually on site to report deficiencies or problems to the owner or another third party such as a regulatory agency. This person may have duties similar to that of an inspector, but is not empowered to stop the job, and typically does not consult with those observed unless requested by the owner or client. The observer may act as an independent auditor but usually reports general observations and does not conduct formal inspections. Many consultants and governmental organizations call their field people observers rather than inspectors in the hope of minimizing liability. It is not the title but rather the function of the person that is important. If polled, most consultants would say that observers play an "oversight" role, a term many have discovered in court to be defined by dictionaries and opposition lawyers as "supervision."

OSHA also defines site positions. A "competent person" is defined by OSHA in 29 CFR 1926.32(f) as one:

> who is capable of identifying existing and predictable hazards in the surroundings or working conditions which are unsanitary, hazardous, or dangerous to employees, and who has authorization to take prompt corrective action measures to eliminate them.

A test pit excavation 4 ft or deeper in which an employee works would require the presence of a "competent person." A site safety officer, field technician, geologist or engineer, or perhaps the backhoe operator may be considered to be the "competent person," but this person may not have "authorization to take prompt corrective action measures" to eliminate hazards.

Another OSHA term utilized in the environmental field is that of "qualified" person. 29 CFR 1926.32(m) defines a "qualified person" as:

> one who, by possession of a recognized degree, certificate, or professional standing, or by extensive knowledge, training, or experience, has successfully demonstrated his ability to solve or resolve problems relating to the subject matter, the work or the project.

This individual might have more technical expertise than a "competent person." He or she might be the world's greatest expert on lacustrine deposits, but would not necessarily have the skills in hazard recognition and the authority to correct the recognized hazards.

Obviously, even an observer who does not have the authority to stop work should make an effort to not allow an imminent hazard to continue. One tactic that is being more widely used is inserting wording into an agreement that states: "work can only be undertaken when an observer (from such and such organization) is present." Although an observer may not have the authority to stop work upon witnessing an unsafe act, leaving the site may force work to cease. Any person on a work site has an obligation to report what they consider to be an imminent hazard. They may report the condition to their supervisor, to their client, to the contractor's supervisor responsible for taking corrective actions, or they may feel strong enough to report to OSHA or the appropriate state agency, or to a local government authority. Many an environmental investigation at a gas station has been stopped because someone smelled gasoline and called the fire department.

Negligence is viewed from the standpoint of personal and organizational responsibilities and authority. This includes regulations, statutes, codes, ordinances, and standards that govern the activities of employees for whom an individual is responsible. If no regulations have been violated, then the question must be asked if all the terms outlined in the contracts or agreements have been complied with. Have industry consensus standards been followed? Finally, did the responsible individual exercise prudence as compared to someone of similar training and experience. Responsibility should be carefully defined in both the contract and the site-specific safety plan. These documents should spell out who is responsible or not responsible for certain actions.

Prequalification of subcontractors with good safety histories and records of adhering to both technical and safety specifications minimizes the potential of a job shutdown. Although not always feasible during the prequalification process, the investigator may find it worthwhile to visit typical job sites and observe the applicant firms' crews at work. Warning flags that suggest a firm may be a safety risk include: workers not wearing hard hats and safety shoes while working on or around the rig; not using goggles and gloves while welding, cutting, or grinding; operating equipment with worn cables, pins, sheaves, and air and hydraulic lines; not cleaning up the work site; using improper tools or tools in need of repair; and allowing unauthorized access to the job site. Some practices should immediately disqualify a firm. These include operating equipment with excessively worn parts or with any guard removed, or using improperly assembled, poorly maintained, or unsafe equipment. A list of firms by Standard Industrial Code that have had OSHA complaints, visits, or had a reportable accident, is available on OSHA's Web site (www.osha.gov).

Familiarity with all appropriate Federal, state, and industry standards is also necessary. Failure to enforce standards increases the risk of injury and liability. Applicable standards should be reviewed periodically. Federal standards include the OSHA General Industry Standard (29 CFR 1910), OSHA Construction Industry Standard (29 CFR 1926), and some state OSHA standards that augment or supersede Federal OSHA regulations. Many companies and governmental organizations, such as the U.S. EPA and the U.S. Army Corps of Engineers, have published their own safety and health requirements; contractors to these agencies must comply with these requirements.

There are also industry guidelines that are published on a regular basis. Failure to adhere to these voluntary standards may not only increase the risk of accident or injury, but may increase a firm's liability should a mishap occur.

Summary

The health and safety requirements that must be met before performing environmental site characterization or ground-water monitoring tasks may seem unwieldy and costly.

They have been imposed, however, to safeguard the health and safety of on-site workers. Field personnel face additional hazards because of the nature of the mechanical and electrical equipment used in environmental investigations. As professionals, environmental investigators should understand why they are wearing a particular item of protective clothing, the limitations of the monitoring equipment being used, why decontamination procedures are being implemented, and how to plan for uncertain contingencies.

References

ACGIH, Threshold Limit Values and Biological Exposure Indices for 2004, American Conference of Governmental Industrial Hygienists, Cincinnati, OH, 2004.
AGI, *Planning for Field Safety*, American Geological Institute, Alexandria, VA, 1992.
AIHA, *Emergency Response Planning Guides and Workplace Environmental Exposure Level Guides*, American Industrial Hygiene Association, Fairfax, VA, 2004.
Forsberg, S. and B. Mandorf, *Quick Selection Guide to Chemical Protective Clothing*, 3rd ed., John Wiley & Sons, New York, NY, 1997.
HAZWOPER, Hazardous Waste Operations and Emergency Response, Final Rule, 29 CFR 1910.120, 54 FR 9294, March 6, 1989.
Maslansky, C.J. and S.P. Maslansky, *Air Monitoring Instrumentation-A Manual for Emergency, Investigatory, and Remedial Responders*, John Wiley and Sons, New York, NY, 1993.
Maslansky, S.P. and C.J. Maslansky, Drilling in Flammable Environments, *Workshop Proceedings of the Tenth Outdoor Action Conference and Exposition*, National Ground Water Association, Dublin, OH, 1996, pp. 47–56.
Maslansky, S.P. and C.J. Maslansky, *Health and Safety at Hazardous Waste Sites-An Investigators and Remediators Guide to HAZWOPER*, John Wiley and Sons, New York, NY, 1997.
Matetic, R.J. and D.K. Ingram, Preventing high insurance premiums and on-the-job injuries, *Water Well Journal*, 55(13), 10–13, 2001.
NDF, *Drilling Safety Guide*, National Drilling Federation, Columbia, SC, 1986.
NIOSH, NIOSH/OSHA/USCG/EPA Occupational Safety and Health Guidance for Hazardous Waste Site Activities, DHHS (NIOSH) Publication No. 85-115, U.S. Government Printing Office, Washington, DC, 1985.
NIOSH, NIOSH Respirator Decision Logic, DHHS (NIOSH) Publication No. 87-108, U.S. Government Printing Office, Washington, DC, 1987.
NIOSH, NIOSH Pocket Guide to Chemical Hazards, DHHS (NIOSH) Publication No. 97-140, U.S. Government Printing Office, Washington, DC, 1997.
NIOSH, NIOSH Certified Equipment List, DHHS (NIOSH) Publication No. 2001-139, U.S. Government Printing Office, Washington, DC, 2004.
NSC, *Fundamentals of Industrial Hygiene*, 3rd ed., National Safety Council, Chicago, IL, 1988.
Sagan, C., *The Dragons of Eden: Speculations on the Evolution of Human Intelligence*, Random House, Inc., New York, NY, 1977.
Schwope, A.D., *Guidelines for the Selection of Chemical Protective Clothing*, 3rd ed., American Conference of Governmental Industrial Hygienists, Cincinnati, Ohio, 1987.
U.S. EPA Health and Safety Audit Guidelines, SARA Title I Section 126, EPA/540/G-89/010, 1989.
U.S. EPA Standard Operating Safety Guides, Publication 9285.1-03, PB92-963414, 1992.

20

Decontamination of Field Equipment Used in Environmental Site Characterization and Ground-Water Monitoring Projects

Gillian L. Nielsen

CONTENTS

Introduction	1263
Objectives of Equipment Decontamination	1264
Current Status of Equipment Decontamination Protocols	1264
Preparing an Effective Decontamination Protocol for Field Equipment	1265
What Equipment Requires Field Decontamination?	1265
Using Disposable Equipment to Avoid Equipment Decontamination Issues	1267
When and Where Should Equipment Be Decontaminated?	1268
Remote Equipment Cleaning	1268
Field Equipment Cleaning	1269
Selecting an Appropriate Decontamination Protocol	1272
Factors to Evaluate on a Task-Specific Basis	1272
Available Decontamination Procedures	1274
Methods for Larger Support Equipment	1274
Methods for Sample Collection or Analysis Equipment	1276
Inherent Problems with Decontamination Techniques	1278
Quality Assurance/Quality Control Components of Decontamination Protocols	1279
Summary	1280
References	1280

Introduction

As discussed throughout this text, many different types of samples and field data are collected during the course of environmental site characterization and ground-water monitoring projects. In many cases, the primary objectives of these projects are to determine the presence or absence of subsurface contamination, assess the three-dimensional extent of contamination in a variety of media, and determine the environmental and health risk associated with that contamination. To meet these objectives, it is critical that samples obtained for field or laboratory analysis be representative, accurate, and precise, and not be influenced by bias or error associated with sample collection. The economic and technical consequences associated with making decisions based on field and laboratory analyses of samples that are not representative can be substantial.

One major source of bias or error that has the potential to influence the quality and representative nature of samples collected for chemical and physical analyses, is the presence of contamination on field equipment. If the equipment used to collect samples or generate field analytical data is not appropriately cleaned to remove potential contaminants, the data collected with that equipment could be erroneous. Failure to adequately clean equipment used to collect environmental samples between sampling points, such as split-spoon samplers or ground-water sampling pumps, could result in the cross-contamination of individual samples or sampling locations. This would, in turn, make any information obtained from these samples unrepresentative of actual *in situ* physical and chemical properties of the material being sampled. Data derived from the analysis of these samples would not accurately reflect actual site conditions and would, therefore, be virtually meaningless to interpret and use for the purpose of making important decisions for a site under investigation.

This chapter will focus on the objectives and methods available for decontamination of field equipment. Because Chapter 19 discusses personnel decontamination practices in detail, this subject will not be addressed here.

Objectives of Equipment Decontamination

An effective equipment decontamination protocol must be designed to meet the following objectives:

- Prevent introduction of contaminants from one site to another site.
- Prevent contamination of areas on a site designated as being "clean" work or equipment storage areas.
- Prevent cross-contamination of individual sampling locations at a single site.
- Prevent cross-contamination of individual samples from a single sampling location as a result of using common or portable sampling equipment at more than one sampling location or to collect more than one sample at a single location.
- Ensure proper operation of equipment.
- Prevent accidental exposure of workers to contaminants that may be distributed on equipment through unprotected handling of equipment.

If these objectives are successfully met, samples should not be impacted by either negative or positive bias associated with poor equipment cleaning practices, provided the chosen cleaning protocols are effective and implemented correctly. Consequently, field or lab analysis of samples should accurately reflect *in situ* chemistry.

Current Status of Equipment Decontamination Protocols

Environmental scientists have many sources of protocols for decontamination of field equipment. Protocols are available from several Federal as well as most state regulatory and non-regulatory agencies, equipment manufacturers, corporate standard operating procedures, and manufacturers of cleaning solutions and equipment. Unfortunately, there is a lack of continuity between these protocols, making it difficult to establish a single standard protocol to follow (Mickam et al., 1989; Parker, 1995). Extensive surveys of Federal and state agencies across the U.S. have been conducted to evaluate the status of current field equipment decontamination procedures. The need for some form of standardization of decontamination methods became readily apparent during these surveys. Of the Federal agencies interviewed (including the Department of the Army, Office of the Chief of Engineers, the Nuclear Regulatory Commission, and the National Science Foundation), none had any specific guidance addressing field equipment decontamination protocol (Parker, 1995). This survey also found that the U.S. Environmental Protection Agency (U.S. EPA, 1996) had no document used nationally to furnish guidance on field equipment decontamination protocols. While this survey found there were numerous decontamination methods in the published literature, there was a significant disparity between the protocols and there was no systematic study on the relative effectiveness of the various procedures (Parker, 1995).

In response to the lack of a standard equipment decontamination protocol, ASTM International developed two standards on field equipment decontamination in the early 1990s. The primary objective of these standards is to provide a basis for standardized protocols for effective equipment decontamination that could be used at a wide variety of facilities for a wide variety of equipment. The ASTM standards on equipment decontamination are:

- ASTM Standard D 5088—Standard Guide for Decontamination of Field Equipment Used at Non-Radioactive Waste Sites (ASTM, 2004a).
- ASTM Standard D 5608—Standard Guide for Decontamination of Field Equipment Used at Low-Level Radioactive Waste Sites (ASTM, 2004b).

These standards are now widely referred to by regulatory agencies as standard methods for equipment decontamination for use in a wide variety of environmental projects. Their use requires an understanding of which procedures are most appropriate for specific applications at any given site.

Preparing an Effective Decontamination Protocol for Field Equipment

The field equipment decontamination protocol is an important component of any site-specific sampling and analysis plan, regardless of the simplicity or intricacy of the environmental investigation. The decontamination protocol must provide easily understood and implemented procedures for all aspects of equipment cleaning, including: (1) what equipment should be cleaned; (2) whether disposable equipment can be used in lieu of cleaning between uses; (3) when and where equipment cleaning should take place; (4) what cleaning protocols (equipment, solutions, other materials, techniques) should be used on a parameter- and equipment-specific basis; and (5) what should be done with any waste materials generated by equipment-cleaning activities. The decontamination protocol must be written in sufficient detail to ensure that the selected protocol will be effectively and consistently implemented. As part of a field quality assurance/quality control (QA/QC) program, it is essential to include the collection of equipment blanks to verify the effectiveness of the decontamination protocol.

What Equipment Requires Field Decontamination?

A typical environmental investigation involves several different phases of activity in the field. In each phase of investigation, a wide variety of equipment may be used, and most of this equipment requires decontamination at one point or another in the investigation. During the development of a field equipment decontamination program, it is important to develop an itemized list of equipment that will require field decontamination. Equipment should be evaluated with regard to its role in the investigation from the perspective of actual or potential contact with a sample analyzed either in a lab or in the field. Three general categories can be created for the evaluation of how field equipment is used: (1) equipment that directly contacts a sample being collected for physical or chemical analysis; (2) equipment that facilitates sample collection but does not contact the sample directly; and (3) equipment that is used for measurement or analysis of some type of parameter. Table 20.1 provides examples of equipment that would require field decontamination for each of these three equipment-use categories.

TABLE 20.1

Examples of Field Equipment That May Require Decontamination

Examples of equipment that contacts samples collected for physical or chemical analysis	
Soil sampling	Split-spoon samplers
	Thin-wall (Shelby) tube samplers
	Direct-push soil samplers
	Hand auger barrels or bits
	Continuous tube samplers
	Sample inspection tools (e.g., knives, metal spatulas)
Monitoring well installation	Well screen
	Well casing
	Well screen centralizers
	Field sieves for determining grain-size distribution
Ground-water sample collection	Well purging and sampling devices
	Pump tubing
	Sample filtration apparatus
Equipment that facilitates sample collection but does not contact the sample	
Soil sampling	Drilling rig and drill rod
	Hand auger rods and handles
	Direct-push rig and rod
Monitoring well installation	Drilling rig and associated tools
	Auger flights
	Well development equipment
Ground-water sample collection	Reels for pump tubing
	Support vehicle
	Suspension cable
	Rope, cord or line attached to grab sampling devices
	Flow-through cell and associated discharge tubing
Equipment used for field parameter measurement or analysis	
Soil sample collection	Tape measure
	X-ray fluorescence devices
	Field-portable analytical balance
Monitoring well installation	Tape measure
	Borehole TV camera
	Borehole geophysical equipment
	Pump discharge flow gauges
Ground-water sample collection	Flow-through cells
	Multi-parameter sondes
	Single-parameter meters
	Beakers or open containers
	Water-level gauges
	Oil–water interface probes

Whether or not equipment contacts a sample directly is one criterion used to determine the most appropriate method for equipment decontamination. This is discussed in greater detail later in this chapter.

Using Disposable Equipment to Avoid Equipment Decontamination Issues

In some programs where the level of contamination is high, it is often desirable to identify ways to minimize or eliminate the need for field equipment decontamination to prevent cross-contamination of samples being collected or measurements being taken. There are two primary options available to meet this objective. One option is to use disposable equipment that is brought to the site in a sealed package or container as shipped by the manufacturer, and then is used to collect one sample only, after which the equipment is discarded. Two common examples of disposable equipment used for ground-water sample collection are disposable bailers and disposable filtration media (see Chapter 15 for more information on ground-water sampling equipment and sample pretreatment methods). Examples of disposable equipment used in soil sample collection are polyvinyl chloride (PVC) or Teflon™ core barrel liners and sample retainer baskets used in devices such as split-spoon samplers. Disposable equipment can offer a number of advantages as summarized in Table 20.2.

One of the major limitations of attempting to incorporate disposable equipment in ground-water investigations is that disposable equipment options are somewhat limited. In general, support equipment such as drilling rigs, sampling vehicles, sample collection devices involving pumping mechanisms, and field parameter measurement instrumentation are not disposable and must therefore be cleaned. Other limitations of disposable equipment are presented in Table 20.3.

The second alternative to field equipment decontamination is to use equipment that is "dedicated" or "designated" for use at a single location at a single site. Dedicated equipment is equipment that is permanently installed within a single monitoring or sampling location and is never exposed to atmospheric conditions during operation or use of that equipment (ASTM, 2004c). Using dedicated equipment can virtually eliminate the potential for cross-contamination of sampling locations and samples associated with contact with the sampling device itself. Examples of dedicated equipment used in ground-water investigations include: dedicated bladder pumps installed in a ground-water monitoring well, or a bubbler system permanently installed in a monitoring well for long-term water-level measurement.

Designated equipment is defined by ASTM (ASTM, 2004c) as equipment that is restricted to use at a single location. Designated equipment is differentiated from

TABLE 20.2

Advantages of Using Disposable Field Equipment

Saves time associated with field equipment cleaning
Reduces the number of field quality control samples required to verify the effectiveness of field equipment cleaning
Minimizes the potential for cross-contamination of samples and sampling locations
Reduces the volume of liquid waste generated by field decontamination activities
Equipment is generally simple to operate
Precleaned equipment options may be available
Some equipment is available in a variety of different materials (e.g., polytetrafluoroethylene [PTFE], PVC, high-density polyethylene), making it possible to select equipment with chemical compatibility in mind
Convenience

TABLE 20.3

Limitations of Disposable Field Equipment

Increased volume of solid waste generated in the field, which may require handling and disposal as a hazardous waste, and may increase overall costs of the ground-water investigation

May become very expensive when a large number of samples must be collected

Potential for residual contamination as a result of manufacture of the equipment (i.e., extrusion agents or mold-release compounds for plastic equipment) if not precleaned prior to use

Some cleaning protocols may require the collection of a rinseate blank on a per-lot or per-manufacturer basis to quantify the presence of any surface residues (if any) on equipment prior to use

Limited selection of types of disposable equipment available

Very rarely can all samples or field data measurements be collected using disposable equipment entirely; therefore, there will still be a need to implement some level of field equipment decontamination for every project

There is a temptation to try to clean and reuse disposable equipment to save money

---

dedicated equipment in that designated equipment typically comes in contact with atmospheric conditions during use or storage between sampling events. The advantage of using designated equipment is that it removes the potential for introduction of contaminants from a remote site to the site under investigation and helps to ensure precision in field measurements and sample collection. In some long-term environmental investigations, it is common to assign specific pieces of equipment to be used at that single site exclusively. This can include equipment such as photoionization detectors for sample screening, pH meters for sample analysis, or water-level gauges. On a sample-collection location basis, it is common to designate lengths of tubing to individual wells when a portable pump is used for purging and sample collection at a number of wells. This eliminates the need to try to clean lengths of pump tubing between wells, which can be very difficult to do successfully. Bailers are also commonly designated, although it is a common error to refer to these devices as dedicated equipment. Under the ASTM definition, bailers cannot be dedicated because they must come into contact with atmospheric conditions during use.

When equipment is designated for use at a specific location, it is critical that control is maintained over the equipment when it is in storage to ensure that it does not become contaminated as a result of contact with atmospheric or surface contaminants, or use at a location other than the one for which it is intended. At sites where atmospheric contributions of contaminants are of concern, it may be necessary to clean at least the exterior surfaces of designated equipment prior to use and prior to putting it into the storage container after use. It is necessary to address these issues on a site-specific basis within the field QA/QC program.

When and Where Should Equipment Be Decontaminated?

Remote Equipment Cleaning

Equipment decontamination can be performed in a remote location such as a laboratory. When equipment is cleaned in a remote location, it is precleaned prior to shipment to or use in the field. In theory, this approach ensures the highest level of equipment cleaning possible because the equipment is cleaned in a controlled indoor environment with ideal facilities for both chemical and physical cleaning procedures (McLaughlin and Levin, 1995). During cleaning at a remote location, a piece of equipment can also be inspected and repaired as necessary prior to shipment to the field. This should ensure optimal operation and performance of any piece of field equipment. From the field perspective, remote cleaning is perhaps the most convenient option for field equipment cleaning. Precleaning equipment can save time in the field for investigators because they do not need to create a formal decontamination area at a site, haul and store cleaning and rinsing solutions into the field, or deal with the generation of investigation-derived waste (IDW) that may need to be containerized and managed as a hazardous waste.

There are several drawbacks to this approach to cleaning. Typically, during environmental investigations, multiple samples are collected daily, therefore, if equipment is cleaned prior to use in the field exclusively, more equipment will need to be cleaned and shipped to a site during the investigation because the field team will not have cleaning solutions in the field. Without having cleaning solutions and rinse water available in the field, it would be impossible to clean equipment that has been accidentally contaminated. An example of accidental contamination would be if a precleaned recovery auger was accidentally dropped onto the ground in an area that is not going to be sampled or if a sampling device was exposed to high concentrations of atmospheric contamination prior to use. While the field team may not generate IDW as a result of equipment cleaning, the remote cleaning location would, and it may not be as well equipped to manage this waste. Equipment that has been used in the field would be returned to the remote location for cleaning after a single use. This would increase the potential of worker exposure to contaminants because the equipment will not have been cleaned prior to transport to and storage at the remote location. It also represents a real potential for contamination of a previously uncontaminated work area. Cost evaluations of remote vs. in-field cleaning need to be conducted to determine which option is least expensive.

Field Equipment Cleaning

More commonly, field equipment is cleaned in the field, either at a designated decontamination area located at some point on site that is determined to be free of contamination (normally a central location), or at the point of equipment use. The decision regarding which approach to take is largely dependent on the nature of the equipment to be cleaned and the characteristics of the contaminants of concern. For example, if heavy equipment such as a drilling rig is to be cleaned, it may be necessary to construct a centrally located decontamination pad with access to power and water supplies to support a portable power washer unit. Many of these decontamination pads are designed to facilitate complete containment and collection of any IDW generated, so it may be subjected to on-site or off-site treatment or disposal (see Figure 20.1).

In cases where sample-collection equipment is being cleaned, a smaller-scale decontamination area is created at each sampling location to facilitate cleaning of all equipment immediately prior to use or movement to the next sampling location. As illustrated in Figure 20.2, these decontamination areas typically consist of a series of buckets or pails placed on heavy-gage plastic sheeting. All cleaning supplies such as detergent solutions and disposable supplies are also placed on this plastic sheeting. One common error observed in the field when using this type of set up is that sampling team members sometimes mistakenly run equipment through the decontamination line backwards (i.e., they begin equipment cleaning at the point of the final control water rinse and end in the bucket containing the detergent solutions and most-contaminated control water). Such an error can result in use of improperly decontaminated equipment and cross-contamination of samples. One solution to this problem is to use color-coded pails where, in the example illustrated in Figure 20.3, equipment cleaning begins in the red pail containing a detergent solution, progressing to the yellow pail with a rinse solution, and finally the green pail where the final equipment rinse water is contained. If this is not

FIGURE 20.1
This decontamination area was designed to contain all waste water generated from cleaning heavy equipment used during site remediation. A french drain system was built beneath the gravel pad in which a sump pump was used to transfer all waste water into the 500-gallon poly tank seen to the right of the decontamination pad.

an option, an alternative is to create a directional arrow using duct tape on the plastic sheeting to direct personnel through the decontamination line correctly.

Advantages of field-cleaning equipment rather than cleaning it in a remote location include: (1) ability to clean equipment that may have been accidentally contaminated in the field; (2) ability to adjust cleaning protocols if in-field QC samples indicate that the equipment cleaning protocol being used proves to be ineffective; (3) sampling team

FIGURE 20.2
A series of buckets or pails is commonly used in the field for containing various cleaning solutions when equipment is cleaned at the point of use rather than remotely. Often the buckets are similarly colored, which can lead to confusion and error, resulting in sampling team members going through the decontamination line backwards.

FIGURE 20.3
A series of color-coded buckets is a good solution to prevent errors in equipment decontamination. The red bucket in the foreground indicates that that is the pail with the detergent solution and is the place to begin equipment cleaning. The middle pail is yellow, indicating that it contains control water rinse liquids; the green pail at the opposite end of the decontamination line indicates that that is the location of the final control water rinse. In this example, chemical desorbing agents were not required.

members are assured that equipment is correctly cleaned immediately prior to use; (4) it is commonly less expensive to clean equipment in the field because it is not necessary to have as many pieces of equipment as is necessary when all equipment is cleaned remotely; and (5) sampling team members are not required to move amounts of increasingly contaminated equipment in field vehicles to a remote location for cleaning, thereby reducing the potential for accidental personnel exposure to contaminants, contamination of the field vehicle and support materials and supplies that may be stored in the field support vehicle, and cross-contamination of samples that may be transported in the field vehicle.

Field decontamination of equipment does, however, require that sampling team members spend time cleaning equipment under field conditions, which are often not ideal. Problems associated with poor weather (e.g., wind, precipitation, extreme temperatures, high relative humidity) and less than ideal support facilities (e.g., unsuitable water supplies, lack of electricity) can result in less than optimal effectiveness of equipment cleaning. In some cases, for example, when collecting continuous split-spoon soil samples, it may be necessary to have extra personnel in the field to clean equipment, to prevent down time in the field, which can cause major cost overruns on some projects.

Sampling team members will typically need to make arrangements for managing and testing waste water generated by equipment-cleaning activities.

Selecting an Appropriate Decontamination Protocol

Factors to Evaluate on a Task-Specific Basis

A number of different protocols exist for equipment decontamination. It is the responsibility of the project manager to determine which method or methods are most appropriate for all pieces of equipment that are to be used for sample collection, field measurements, or sampling point construction (e.g., ground-water monitoring wells). To make those decisions, a number of factors must be evaluated. These factors are summarized in Table 20.4.

Of utmost importance in developing an effective decontamination program is establishing the purpose of the environmental investigation. This is directly linked to the level of QA/QC demanded by the investigation. For example, during the installation of a ground-water monitoring system to act as a leak-detection system around a newly installed underground storage tank system, the required level of QA/QC might be low. No subsurface contamination would be expected at the site unless the tanks were being installed as replacements for old tanks. Therefore, decontamination of equipment used at the site may not be an issue. In contrast, however, installation of ground-water monitoring wells at hazardous waste (Comprehensive Environmental Response, Compensation and Liability Act [CERCLA], Resource Conservation and Recovery Act [RCRA]) sites would require a higher level of QA/QC throughout every aspect of the investigation to ensure the collection of representative data. The level of QA/QC can be further intensified when an investigation is conducted at a site under litigation. Under these circumstances, not only is sample integrity of concern, but also all data must prove to be legally defensible in terms of validity and reproducibility.

TABLE 20.4

Criteria for Selection of Field Equipment Decontamination Protocol

---

Existence of Federal, state, and/or regional regulatory guidelines that must be followed
Purpose of the investigation (e.g., initial site assessment, long-term monitoring, site remediation design, litigation-driven monitoring or sampling)
Media to be sampled (soil, soil gas, ground water, surface water, waste)
Does the equipment requiring cleaning actually contact the sample
Does the equipment requiring cleaning facilitate sample collection but does not contact the sample itself
Nature and anticipated concentrations of expected contaminants (chemical species, carrier chemicals such as petroleum hydrocarbons or solvents, chemical properties of contaminants)
How will the contaminants be physically distributed (i.e., in air, fill, soil, or ground water)
Physical features of the equipment to be cleaned and its associated support equipment including:
    Materials of construction, including inner parts, seals, and external components
    Ease of disassembly and reassembly for cleaning
    Ability to withstand the rigors of cleaning
    Size
    Dedicated versus portable devices
Management of decontamination wastes generated
Site support equipment requirements (e.g., power, water, site security)
Site accessibility (political and physical, seasonal variability)
Cost
Health and safety concerns when using chemical desorbing agents such as acids or solvents

After the purpose of the investigation has been established, the suspected site contaminants must be identified and a representative list of parameters that will be analyzed must be established. This process requires an evaluation of a number of contaminant-specific physical and chemical properties to provide information that will be incorporated into the decontamination protocol design. These properties include:

- Contaminant physical distribution (i.e., in air, fill, soil, ground water, or surface water)
- Chemical matrix (i.e., interfering constituents, or carrier chemicals such as petroleum hydrocarbons or solvents)
- Chemical species (i.e., volatile organics, non-volatile organics, heavy metals, inorganic nonmetals, or others)
- Physical properties (density, volatility, flammability, corrosivity, viscosity, reactivity, natural decomposition, or transformation rates)

These are critical factors that direct the selection of decontamination procedures. Decontamination activities must be selected based upon chemical suitability and compatibility with the constituents to be removed during decontamination, and with the concentrations of constituents anticipated. For example, decontamination protocols for an investigation being conducted at a fuel distribution terminal would have to incorporate decontamination procedures such as solvent rinses or special degreasing detergents effective in removing oily substances from all equipment used in the project. However, for an investigation conducted at a metal sludge surface impoundment, metals would be of prime concern and would require different decontamination procedures such as dilute acid rinses. Method selection can become more complicated if more than one contaminant group is of concern at a site. For example, if petroleum hydrocarbons and solvents are of concern in an investigation, use of solvents such as acetone or hexane to degrease oily equipment may interfere with efforts to characterize solvent contamination at the site.

Relative concentrations of contaminants influence many components of an environmental investigation, from establishing health and safety guidelines, to sample collection techniques, to decontamination of personnel and equipment. Anticipated concentration levels (i.e., percent-range concentrations versus parts-per-million or parts-per-billion levels) must be considered along with project objectives and QA/QC controls to identify any physical limitations to decontamination.

Matteoli and Noonan (1987) determined through controlled field testing that the time required for effective decontamination was directly related to the construction materials of the equipment being decontaminated. They found that more than 3 hours of rinsing with clean water was required to lower trichloroethylene (TCE) levels below detection limits for a submersible pump equipped with a rubber discharge hose, compared to 90 min for the same pump equipped with a Teflon® hose. They also determined that individual parameters had unique responses to decontamination. For example, they found that decontamination times for Freon 113 were longer than those for TCE, while decontamination times for 1,1-dichloroethylene, selenium, and chromium were shorter.

Identifying physical limitations or logistical problems associated with decontamination for a project is essential if the decontamination protocol is to be workable. Many very elaborate decontamination procedures developed and approved for theoretical use may not actually be implemented on a project, due to factors such as time and budget constraints, incompatibility with equipment to be decontaminated, inconvenience, and inability to manage the wastes generated. To avoid this problem, a realistic protocol must be developed that recognizes and accounts for myriad project-specific logistical constraints listed in Table 20.4. It is neither logistically reasonable nor safe, for example, to specify that a drilling rig and all necessary support tools and equipment undergo decontamination by solvent or acid washing on a project. Conversely, it is also not reasonable to expect a drilling rig to be "dedicated" to a site unless the project is conducted under the most drastic conditions, such as a high-level radioactive waste site, where it may not be physically possible to decontaminate the rig sufficiently to permit demobilization to another site. Some element of compromise must be incorporated into the decontamination protocol to make the program workable and allow it to meet the objectives of the investigation.

As a case in point, Keely and Boateng (1987) found that procedures originally developed for field decontamination of an electric submersible pump were not workable due to the time and patience required on the part of the sampling team to completely disassemble the sampling pump, scrub each individual component, and then reassemble the pump for use in the next monitoring well. The original decontamination procedures were modified to permit circulation of decontamination solutions through the pump and tubing, thereby avoiding the need to disassemble the multi-component submersible pump. The compromise, however, as indicated by the authors, was the potential for carryover of the solvent (acetone) used in the decontamination procedure. The argument used to justify the change in protocol was that the amount of carryover was limited to a few milliliters of water wetting the surface of the pump, and that the device would be immersed in many gallons of water in the casing of the next monitoring well, resulting in potential residual concentrations of acetone in the sub-parts-per-trillion level. This type of compromise may jeopardize the ability to meet the objectives of QA/QC standards of a highly sensitive analytical program. However, it does illustrate the need to anticipate actual field conditions as opposed to ideal field conditions. Under actual field sampling conditions, performance factors such as weather, operator skill, and pressure to complete assigned tasks within a limited time frame can have a significant impact on the quality of actual decontamination techniques.

The most widely applied solution to this problem is to use dedicated equipment, thereby avoiding all but initial equipment decontamination prior to installation. The second alternative is to reevaluate and modify the program's original decontamination QA/QC specifications.

It is possible to avoid the need to decontaminate well construction materials, such as casing and screen, prior to installation of a well. Many manufacturers can supply precleaned casing and screen that is delivered to a job site in sealed containers. While the initial costs of these precleaned materials may be somewhat high, the higher cost associated with on-site cleaning of these materials is eliminated, as long as the sealed containers are not opened until use.

It is evident that a variety of factors must be considered when developing a decontamination program. It is also readily apparent that those factors are not only site-specific, but are work-task-specific as well. For this reason, it is often necessary to prepare a document that details decontamination procedures on a task-by-task basis. Some states, such as Florida, require that these work plans go one step further and address decontamination procedures for equipment before it is brought to a site for use, as well as on-site field procedures. In a case study of decontamination procedures implemented at a Superfund site in Indiana, Fetter and Griffin (1988) discussed no fewer than eight task-specific components that were incorporated into the field investigation, from well installation to ground-water sample collection.

Available Decontamination Procedures

Methods for Larger Support Equipment

Most larger equipment, such as a drilling rig or a support truck, is cleaned between sampling locations using a high-pressure, hot-water power wash system as described in ASTM

FIGURE 20.4
Power washers such as the one illustrated here are commonly used to decontaminate larger support equipment like drilling rigs and field vehicles. Often confused with "steam cleaning," these power washers are effective at removing particulate contaminants as well as chemical contaminants from surfaces. This cleaning method is not appropriate for smaller, more delicate instrumentation such as pH meters or for personnel.

Standard D5608 (ASTM, 2004a). This system frequently employs an initial soapy wash and scrubbing with brushes to remove larger soil particles and contaminants from surface areas. This is followed by a high-pressure, clean-water (potable water) wash to remove soap and contaminants (see Figure 20.4). Equipment is usually allowed to air dry before being placed onto elevated racks or plastic sheeting for storage before use at the next location.

This type of power wash system should not be confused with or replaced by the type of power wash system found at the local car wash. Most car washes recirculate water and use additives such as a glycerin to provide a shine to cleaned vehicles. Both these practices make it impossible to have good quality control on decontamination efforts in an environmental investigation.

Hot-water power wash systems are often incorrectly referred to as "steam cleaning," probably because the fine, mist-like hot-water spray that is generated by the high-pressure spray nozzle resembles steam. Steam cleaning technically refers to the application of high-pressure steam to remove contaminants and solid particles from larger pieces of field equipment such as drilling rigs or direct-push rods. The primary advantage of steam cleaning is that the volume of decontamination wastewater is minimized. There are several disadvantages associated with steam cleaning, however, which include: (1) the lack of adequate pressure to effectively dislodge large particles or sticky substances such as clayey soils or residues such as oils; (2) logistical difficulties in generating a sufficient source of steam under field conditions; and (3) temperatures that are so high as to cause degradation of flexible materials such as hydraulic lines or seals on equipment and failure of the equipment.

Most state and Federal regulatory agencies will specify the use of "steam cleaning" or high-pressure hot-water power wash methods for drilling rigs, associated tools, and support vehicles. There are exceptions, however, Mickam et al. (1989) indicate that in one state's decontamination program, drilling tools, including augers and split-spoon samplers, must undergo the following decontamination procedures before steam-cleaning:

- Immersion and scrubbing in a mixture of detergent and water
- Rinse with clean water
- Rinse with isopropyl alcohol, methanol, or acetone
- Multiple rinses with distilled water

This may be a manageable procedure for small equipment such as split-spoon samplers and wrenches. However, a significant exposure hazard can be associated with working with the large volumes of concentrated solvents necessary for decontaminating larger equipment such as augers, drill rod, and bits. In addition, large quantities of potentially hazardous waste would be generated by this cleaning procedure; these materials would be costly to manage and dispose of properly.

Methods for Sample Collection or Analysis Equipment

The variability in decontamination procedures employed for cleaning small equipment, such as sampling devices, is considerable. Some agencies have developed analyte-specific protocols, while others have developed procedures on the basis of equipment type and materials of construction of the equipment to be cleaned. Usually, one or a combination of several of five solutions are typically specified in field decontamination procedures: tap (potable) water, dilute acid, solvent, distilled or deionized water, and laboratory-grade phosphate-free detergent. The number and sequence of use of these solutions is usually the largest source of variation between decontamination protocols.

To generalize, for sampling devices used to sample for inorganics, such as metals, the most commonly used decontamination procedure is as follows:

- Initial wash with water (tap, distilled, or deionized) and laboratory-grade detergent
- Rinse with control water (i.e., water of a known and acceptable chemistry for a particular application)
- Rinse with dilute acid solution (10% nitric acid or hydrochloric acid), followed by
- Final rinses with distilled water or deionized water

When sampling devices are used to sample for organics, the decontamination procedure typically includes:

- Initial wash with water (tap, distilled, or deionized) and laboratory-grade detergent
- Rinse with control water
- Solvent rinse (pesticide-grade acetone, hexane, isopropyl alcohol, methanol alone, or in some combination) then
- Final rinses with distilled water or deionized water rinses

It is important to understand the purposes of each individual step in these protocols. Aqueous cleaning is used to initially remove gross contamination and particles from equipment. The control water used to make up the detergent solutions acts as a solvent for water-soluble contaminants and as a dispersal medium for insoluble substances that can be carried in suspension (Parker and Ranney, 1997a, 2000). Detergents are used to improve the "wetting" ability of the cleaning solution and to aid cleaning by separating the contaminant from the solid surface and keeping contaminants in suspension to prevent redeposition onto the equipment. The detergent should be selected with consideration of:

- Method of cleaning and detergent use (manual vs. machine, water temperature)
- Will the detergent be able to remove contaminants of concern from equipment at the anticipated contaminant concentrations
- Does the detergent pose any health hazards associated with its use (e.g., high alkalinity, corrosivity, flammability, reactivity)
- Will the detergent be able to physically remove particulates from equipment (e.g., mild alkaline cleaners containing a blend of surfactants and sequestering agents are effective at removing a broad range of organic and inorganic soils)
- Will the detergent leave a residue that could potentially interfere with sample analyses (McLaughlin and Levin, 1995)

In some cases, it may be necessary to use a chemical desorbing agent to thoroughly remove a contaminant that a detergent is unable to fully remove from a piece of equipment. Acid rinses are typically used to desorb metal ions from non-metal surfaces such as polymers and glass. Commonly used acids include dilute (5–10%) solutions of nitric acid and hydrochloric acid. Organic solvent rinses are used to remove residual organic contaminants by dissolving them. The general rule of thumb when selecting an appropriate chemical desorbing agent is "like dissolves like." That is, polar solvents dissolve polar contaminants and non-polar solvents dissolve non-polar or less polar solvents (Parker and Ranney, 1997b, 2000). Because water is a very polar solvent, non-polar or less polar solvents are typically used to remove residual non-polar organic contaminants such as oils and tars. Commonly used solvents include pesticide-grade isopropanol, methanol, acetone, and hexane.

Considering the number of subsurface investigations associated with underground petroleum product storage systems, pipelines, and terminals, it is unfortunate that decontamination procedures addressing petroleum hydrocarbons are commonly overlooked in decontamination protocols. Solvents such as acetone and hexane are commonly used to degrease equipment. However, this practice can cause interference in analyses for dissolved-phase hydrocarbons and product-specific additives. Other problems include generating hazardous wastes, the potential for the solvents to degrade flexible (i.e., rubber or plastic) parts on some field equipment, and the potential to damage sensitive hydrocarbon sensors on oil/water interface probes. Specially formulated laboratory-grade detergents, such as "Detergent 8," are more effective than most solvents at removing petroleum hydrocarbon residues from the surfaces of field equipment (ASTM, 2004a). Additionally, use of detergents eliminates personnel exposure hazards associated with solvents and greatly reduces the cost associated with wastewater disposal.

For practical purposes, decontamination programs can often be simplified to using an initial rinse with water (tap, distilled, or deionized), followed by a wash and scrubbing with a phosphate-free laboratory-grade detergent (e.g., Liquinox, or Detergent 8), followed by two to three rinses with distilled water. This method has been used in ground-water monitoring programs for inorganics, organics, pesticides, and petroleum hydrocarbons, and for concentrations of these compounds ranging from percent to low parts-per-billion levels. Equipment blanks should be used to determine the effectiveness of this and other decontamination practices (refer to Chapter 15 for additional detail on how to collect equipment blanks). Table 20.5 presents a summary of decontamination procedures that have been implemented during a variety of field investigations.

TABLE 20.5

Currently Available Decontamination Protocol Options

Physical decontamination
- Air blasting
- Wet blasting
- Dry ice blasting
- High-pressure Freon cleaning
- Ultrasonic cleaning
- Vacuum cleaning
- Physical removal or scrubbing

Chemical decontamination
- Water wash
- Tap water followed by deionized and distilled water rinses
- Water, laboratory-grade detergent wash, distilled and deionized water rinses
- Pesticide-grade solvent rinses in combination with distilled and deionized water rinses
- High-pressure steam cleaning
- High-pressure hot-water power wash
- Hydrolazer
- Acid rinses in combination with distilled or deionized water rinses

Inherent Problems with Decontamination Techniques

A review of the various Federal and state guidelines demonstrates that materials that are themselves hazardous by definition (i.e., acids and solvents) are commonly incorporated into many decontamination procedures. Other methods, such as steam cleaning or hot-water power washes, require the use of potable water supplies and a variety of support equipment. It is, therefore, apparent that the method of decontamination selected for a given project must be evaluated with respect to its potential to impact sample integrity by contributing potential contaminants to the equipment being cleaned. Table 20.6 presents a summary of some of the more common potential sources of contamination associated with decontamination procedures.

With any decontamination program, solid and/or liquid wastes will be generated. The exact nature and volume of the waste generated is dependent on the equipment decontaminated, the method of decontamination, QA/QC performance standards, the amount of decontamination required, and the contaminants of concern. Depending upon the decontamination methods used and the type of investigation being conducted, it is possible that

TABLE 20.6

Potential Sources of Contamination Associated with Decontamination Procedures

- Use of "contaminated" potable water supplies (i.e., contaminated due to the presence of bacteria, organic compounds, metals, or other objectionable substances)
- Use of contaminated supplies of commercially prepared distilled water (i.e., contaminated due to the presence of plasticizers and organic compounds)
- Contamination of samples with residues of fluids used for decontamination such as dilute acids or solvents
- Oil spray from unfiltered exhaust from generators used as power sources
- Volatile organic contamination associated with equipment exhaust systems (steam cleaners, generators, support trucks, etc.)
- Use of ethylene glycol (antifreeze) in decontamination equipment (e.g., hot-water power wash spray nozzles and hoses) to prevent freezing in extremely cold weather
- Hydraulic fluids, oils, gasoline, or diesel fuel used to operate generators and other support equipment

decontamination wastes may be classified as hazardous by virtue of the decontamination fluids used (i.e., nitric acid, acetone, methanol, etc.) and contaminants encountered during the investigation. Generally, when working at hazardous waste sites, management of hazardous decontamination wastes is not a major issue, although it can result in substantial budget increases when specially designed decontamination facilities must be constructed and wastes must be containerized and sampled and the samples analyzed to provide for appropriate disposal. However, at non-hazardous sites where the monitoring program is designed to monitor mainly non-hazardous constituents (i.e., inorganic parameters such as chloride or nitrate), the potentially hazardous wastes generated by decontamination procedures can pose a major problem. Compliance with federal hazardous waste regulations becomes a major issue that must be resolved. This is typically very time-consuming and costly and, consequently, can have the undesired effect of encouraging improper management of hazardous materials and wastes. Because of these potentially significant problems, an attempt is usually made to avoid the use of solvents and acids in decontamination programs.

Quality Assurance/Quality Control Components of Decontamination Protocols

As with all other components of an environmental investigation, it is necessary to monitor the effectiveness of decontamination protocols. This is done to verify that the contaminants of concern are removed from all equipment being decontaminated so that any data generated from samples collected for chemical analysis during the investigation can be considered valid and uncompromised.

The QA/QC segment of the decontamination protocol must address several key issues including:

- Location and construction of a decontamination area.
- Movement of clean and contaminated equipment in and out of the decontamination area.
- Preliminary cleaning of all equipment to be used in a project at an off-site location prior to being permitted access to the site.
- Segregation of clean and contaminated equipment.
- Controls to ensure that cleaned equipment does not become contaminated prior to use (i.e., placing cleaned auger flights on elevated racks or plastic sheeting).
- Controls to ensure that the equipment used for decontamination will not in itself act as a source of contamination (i.e., installing exhaust collectors on generators).
- Chemical verification of suitability of the potable water supply.
- Use of rinse (equipment) blanks and "wipe" samples to verify the effectiveness of decontamination procedures, some of which may be analyzed in the field to provide real-time indications of the effectiveness of the method being implemented.

Should any of the approved decontamination procedures be modified during the course of the field investigation, it is critical to thoroughly document all changes, and provide justification for any changes. Under these circumstances, the use of rinse blanks or wipe samples can be even more critical.

Summary

It is not possible to write a single decontamination program that will be applicable to every field situation. Each environmental investigation will be governed by site-specific physical and chemical variables that will direct the process of selection of the most effective decontamination method on a task-specific and project-specific basis. Once these variables have been defined, however, it is possible to develop a workable decontamination program incorporating procedures that will ensure that all data generated by the investigation are representative of site conditions and that the results of the study are not compromised.

References

ASTM, Standard Practice for Decontamination of Field Equipment Used at Non-Radioactive Waste Sites, ASTM Standard D 5088, ASTM International, West Conshohocken, PA, 2004a, 8 pp.

ASTM, Standard Practice for Decontamination of Field Equipment Used at Low-Level Radioactive Waste Sites, ASTM Standard D 5608, ASTM International, West Conshohocken, PA, 2004b, 8 pp.

ASTM, Standard Guide for the Selection of Purging and Sampling Devices for Ground-Water Monitoring Wells, ASTM Standard D 6634, ASTM International, West Conshohocken, PA, 2004c, 14 pp.

Fetter, C.W. and R.A. Griffin, Field Verification of Noncontaminating Methodology for Installation of Monitoring Wells and Collection of Ground Water Samples, Proceedings of the Second National Outdoor Action Conference on Aquifer Restoration, Ground Water Monitoring and Geophysical Methods, Vol. 1, National Water Well Association, Dublin, OH, 1988, pp. 437–444.

Keely, J.F. and K. Boateng, Monitoring Well Installation, Purging and Sampling Techniques Part 1: Conceptualizations, Proceedings of the FOCUS Conference on Northwestern Ground Water Issues, National Water Well Association, Dublin, OH, 1987, pp. 443–472.

Matteoli, R.J. and J.M. Noonan, Decontamination of Rubber Hose and "Teflon" Tubing for Ground Water Sampling, Proceedings of the First National Outdoor Action Conference on Aquifer Restoration, Ground Water Monitoring and Geophysical Methods, National Water Well Association, Dublin, OH, 1987, pp. 159–183.

McLaughlin, M. and P. Levin, Reduce the risk of cross-contamination, *International Ground Water Technology*, 13–15, March, 1995.

Mickam, J.T., R. Bellandi, and E.C. Tifft, Jr., Equipment decontamination procedures for ground water and vadose zone monitoring programs: status and prospects, *Ground-Water Monitoring Review*, 9(2), 100–121, 1989.

Parker, L.V., A Literature Review on Decontaminating Ground-Water Sampling Devices: Organic Pollutants, CRREL Report 95-14, U.S. Army Corps of Engineers, Cold Regions Research and Engineering Lab, Hanover, NH, 1995, 22 pp.

Parker, L.V. and T.A. Ranney, Decontaminating Ground-Water Sampling Devices, Special Report 97-25, U.S. Army Corps of Engineers, Cold Regions Research and Engineering Lab, Hanover, NH, 1997a, 28 pp.

Parker, L.V. and T.A. Ranney, Decontaminating Materials Used in Ground-Water Sampling Devices, Special Report 97-24, U.S. Army Corps of Engineers, Cold Regions Research and Engineering Lab, Hanover, NH, 1997b, 35 pp.

Parker, L.V. and T.A. Ranney, Decontaminating materials used in ground-water sampling devices: organic compounds, *Ground-Water Monitoring and Remediation*, 20(1), 56–68, 2000.

U.S. EPA, Test Methods for Evaluating Solid Waste: Physical/Chemical Methods, SW 846, U.S. Environmental Protection Agency, Office of Solid Waste and Emergency Response, Washington, DC, 1996.

Index

AA. *see* Atomic absorption (AA) spectrophotometry
Above-ground
  storage tanks, 125f
Above-ground
  well protection, 454f
Accelerated Site Characterization (ASC), 68
  approach
    conventional, 84
  approaches
    comprehensive final report, 175
    conventional drilling methods, 143
    DFA, 84
    DMP, 170
    field work project, 176
    project management, 179
    soil gas surveys, 154
    systematic project planning, 79
    technologies and methods, 138
    traditional, 84
  petroleum, 69
  process
    analytical data, 137
    primary focus, 172
  program
    activities, 69
    advantages, 70
    field-based analytical methods, 149
    flow chart, 70f
    iterative approach, 69
    sampling tools, 71f
AccuSensor field test kit, 187
Acetylene
  definition, 622
ACFEE. *see* Air Force Center for environmental Excellence (AFCEE)
ACGIH. *see* American Conference of Governmental and Industrial Hygienists (ACGIH)
Acid-extractable compounds (BNA), 1122
ACO. *see* Administrative Consent Order (ACO)
Acquired immune deficiency syndrome (AIDS), 1229

Acquired immunodeficiency syndrome (AIDS). *see* Acquired immune deficiency syndrome (AIDS)
ACR. *see* Autocorrelation function (ACF)
Active remediation technologies
  interactions, 629t, 630t
Acute toxicity 1231
Adaptive sampling and analysis plans (ASAPs), 161
  CSM, 161
  decision rule, 165
  program
    GIS, 162
    U.S. DOE, 163, 164
Adjacent monitoring programs
  separation, 566–571
    gradient control, 566, 567
Administrative Consent Order (ACO), 1116
Advancement grouting, 461
Aerial photo, 105f
  black and white, 104f
  color, 101f
  historical, 106f
  tributary drainage ways, 106f
Aerial photography
  information use, 109
Aerial photos
  black and white, 95f
  fracture trace analysis, 109
  fracture traces, 111
  historical sequence, 103f
  source availability, 108, 109
  stereo pair, 107f
Aerobic respiration, 618
AIHA. *see* American Industrial Hygiene Association (AIHA)
Air compressors
  high-pressure hose, 856
Air drilling, 314
Air Force Center for Environmental Excellence (AFCEE), 576
Airline submergence method, 894, 895

1281

Air monitoring, 1240–1244
  HASP, 1239f
Air-purifying devices, 1236
Air-purifying respirators, 1236–1238
Air-purifying respirators (APRs), 1236
Air rotary
  down-hole hammer, 153f
Air rotary drilling, 152f
Air-vented surge plunger, 859
Aliquots, 1188
Alluvial aquifer site
  geologic cross-section, 424f
Ambient vertical groundwater flow, 814
American Conference of Governmental and Industrial Hygienists (ACGIH), 1232
American Geological Institute Planning for Field Safety, 1256
American Industrial Hygiene Association (AIHA), 1232
American Society for Testing and Materials, 1201
Analysis of variance (ANOVA), 1177
  nonparametric, 1186
  U.S. EPA, 1185
Analytes
  analytical methods, 1116–1119
    description, 1117
    requirements, 1119
  site related, 1116
Analytical Protocol, 571
Annular seal
  construction, 970
  design
    installation, 754–777
  dry injection systems, 772f
  materials, 645, 755, 756
    installation methods, 769–777
    neat cement, 768
  monitoring well, 755
  volumes, 738t
ANOVA. see Analysis of variance (ANOVA)
Anoxic ground water, 1067
APR. see Air-purifying respirators (APRs)
Aquifers, 1152
  anisotropy, 903
  calculating space, 1195
  definition, 885
  heterogeneity, 903
  hydraulic conductivity, 892
  parameters, 914
    methods and procedure definitions, 913–953
  piezometer nest
    unconfined, 537f
  pumping tests, 947

shapes, 1204
  unconfined, 886f
Artifactual colloidal material, 983
Artifactual particles, 979
ASAP. see Adaptive sampling and analysis plans (ASAPs)
ASC. see Accelerated Site Characterization (ASC)
Asphalt patches
  contamination, 121f
ASR. see Atmosphere supplying respirators (ASRs)
ASTM Standard Guides, 185, 210, 1062, 1265
  CPT, 420t
  purging strategies, 1036
  soil gas sampling, 381
  soil sample collection, 654t
Atmosphere supplying respirators (ASRs), 1238, 1239
Atmospheric air
  bottle filling, 987
Atomic absorption (AA) spectrophotometry, 1124–1126
Atomic emission spectroscopy, 1126
Auger drill cutter heads, 302
Auger drilling, 300–305
  bucket auger, 301
  continuous-flight solid-stem augers, 301, 302
  hollow-stem augers, 302–305
  rigs, 318
Australian Drilling Industry Association, 1220
Autocorrelation function (ACF), 1171

Background samples
  functions, 136
Background water quality
  trend tests, 1178
Back-tracking
  dissolved well concentration, 503, 504
Back washing method
  filter-pack materials, 740f
Bacteria
  alkaline pH, 621
Bailers, 62f, 412, 1006f, 1010f, 1036
  grab samplers, 1010
  mechanical surging, 859, 860
  sample composition, 63f
  wells, 1012, 1013, 1037f
Baseline approach, 501
Baseline water quality, 1139, 1140
  parameters indication selection, 1140–1142
Base station magnetometer, 275
Bat Enviroprobe, 402
Bedding characteristics, 338

Index

Bedrock
  borehole, 666f
  completion, 659–661
  fractures, 112
  well completion types, 660f
Behrens Fisher t-test
  Cochran's approximation, 1187, 1188
Bentonite, 756–763
  boreholes, 769
  chips, 758f
    cold climates, 780f
    wells, 878f
  drilling, 673f
  grout, 763f, 773
    neat cement, 775
    weighing, 760f
  hydrous aluminum, 756
  material
    chemical considerations, 761
  monitoring wells, 756
  pellets, 757, 762, 877
    cold climates, 780f
    neat cement grout, 771f
    placement, 769f
    placement rate, 770
    tremie pipe, 770f
  properties, 876t
  seals, 775f
    DNAPL, 762
Biochemical oxygen demand (BOD), 1114
Biodegradable drilling fluid, 835
Biological agents, 1229
Biological hazards, 1229, 1230
Black box technology, 251
Bladder pumps, 414–416, 1003f, 1021, 1022
  bladder damage, 1023
  low-flow groundwater purging, 416
  mechanical design, 1024f
  representative samples, 1024
  small-diameter, 415f
Blank-page notebooks
  geologic descriptions, 1200f
Blind duplicate samples
  collection procedures, 1096
BOD. *see* Biochemical oxygen demand (BOD)
Bonferroni inequality, 1180, 1181
Borehole
  bedrock drilling, 833
  bentonite, 769
  casings, 738t
  centering casing, 709–716
  characteristics
    primary factors, 675
  cored, 654f
  dilution method, 941

geophysical measurements
  types, 257f
geophysical surveys, 653
granular bentonite, 772f
installation, 833
log, 280f, 335f
  form, 334
logging, 333–338
  definition, 333
  drilling information, 336–338
  log completion information, 334
  log heading information, 334
  sample information, 334, 335
  soil and rock descriptions, 335, 336
metal histograms, 1162
silt and clay deposits, 833
site-specific factors, 755
Bottom-discharge device, 1011
Bouguer anomaly, 273
Bow-type centralizers, 711f
Box plots
  error bar plots, 1161f
  time series comparisons, 1160f
Bridge-slot well screen, 748f
Brown fields, 28, 29
Brownfields Initiative, 28
  categories, 29
Brownfields National Partnership, 29
Bucket auger, 300f
Bulk density, 915–919
Bumper guards, 787f
Bundle piezometers, 655f
Bundle wells, 825f, 826f
  installation, 824–829
Burial trench
  metal detector survey, 274f
Burn pit activities, 105

Cable-tool drilling, 154f, 672f, 676
California Department of Water Resources, 826
Caliper logs, 283
Capacitively couple resistivity, 268
Capillary pressure
  formula, 216
Capsule filters
  preconditioning, 1072
Carbon dioxide, 622
Carbon steels
  production, 693
Carcinogens, 1231
Cased hole, 457
Casehardened steel lock, 785f
Casing
  centering guides
    types, 710f

Casing (*Contd.*)
   chemical resistance, 682, 683
   extension, 899
   hammer, 671f
   joints
      effective seals, 701
   materials
      cleaning requirements, 713, 714
      drilling cost, 709
      preinstallation, 713
      unit costs, 709
   placement, 736f
   slot widths
      commercial availability, 723t
   string
      installation, 709
   tensile strength, 679
   types, 683–697
Cation exchange capacity (CEC), 761
CDP. *see* Common depth point (CDP)
CEC. *see* Cation exchange capacity (CEC)
Cement, 764f
   grout
      mixture, 766
   slurry
      hydration process, 766
   surface-seals, 778
Centralizer, 829f
CERCLA. *see* Comprehensive Environmental Response, Compensation, and Liability Act (CERCLA)
CGI. *see* Combustible gas indicators (CGIs)
CH. *see* Cyclohexanone (CH)
Chain-of-custody, 1129
   form, 331, 332f
      example, 1085f
Chain-of-laboratory security
   custody, 1145
Chemical data
   plotting, 1161
Chemical degradation
   screen material, 715
Chemical hazards, 1228, 1229
Chemical ice packs
   reusable, 1075f
Chemical industries
   lasers, 898
Chemical oxygen demand (COD), 1114
Chemical preservation
   methods, 1078, 1079
   vials, 1077f
Chemical preservation procedures, 1100
Chemical sample preservation, 1072
Chemicals of concern (COC), 43
Chemical storage areas

TCE, 125f
Chloride
   time series comparisons, 1160f
Chloride isopleths, 1155f
Chlorinated hydrocarbons
   biodegradation, 624
Chlorinated pesticides, 1123
Chlorine
   halogens, 624
Circulation fluids, 305
Clay units
   impermeable, 1206f
Cleaning solution
   wetting ability, 1277
Clean water act, 10, 11
Clean Water Act (CWA), 961, 1115
   National Contingency Plan, 11
   water pollution, 10
Closed-piston samplers, 371
CLP. *see* Contract Laboratory Program (CLP)
Cluttered sites
   hazard, 1234f
CMT. *see* Continuous multichannel tubing (CMT)
Coal tar
   image, 438f
COC. *see* Chemicals of concern (COC)
Cochran's approximation
   Behrens Fisher t-test, 1187, 1188
COD. *see* Chemical oxygen demand (COD)
Coefficient of variation (CV) measurements, 610
Color infrared photography, 102f
Combined sampling tool
   Simulprobe, 403
Combustible gas indicators (CGIs), 1241
Common depth point (CDP), 270
Comparison measurements, 255f
Compliance Monitoring Program, 6
Comprehensive Environmental Response, Compensation, and Liability Act (CERCLA), 347, 808, 1115, 1140
Comprehensive State Groundwater Protection Programs (CSGWPPs)
   groundwater protection goals, 32
Compressed air
   airlift surging, 855–857
   pumping, 855–857
Computer literature searches
   representative sampling databases, 99f
Computer literature search system
   advantages, 99
   conflicts, 100
Conceptual site model (CSM), 129f
   ASAPs, 161

data collection program, 135
development, 77, 132
  environmental site characterization, 22–135
  process and revision, 134f
dynamic work plan development, 73
example, 131
groundwater systems, 135
initial components, 134
Concrete patches
  contamination, 121f
Concrete surface seal, 796f
  drainage channels, 792
Concrete surface seals, 781f
Cone penetration testing (CPT), 76f, 135, 349
  approaches, 504, 505
    electrochemical sensor, 507
    fluorescence techniques, 505
    GeoVIS, 505, 506
    hydrosparge, 505
    LIF/GeoVIS, 505, 506
    LIF/Raman, 506, 507
    MIP, 505
    PIX, 508
    Raman spectroscopy, 506
    waterloo profiler, 507, 508
  ASTM standards, 420t
  cabling, 438
  data acquisition, 418, 419
  data interpretation, 419, 420
  deployed probes and sensors, 432, 433
  direct-push apparatus techniques, 485
  DP soil sampling methods, 416
    dual-tube soil-sampling methods, 377, 378
  equipment, 971
  grouting, 462
  instrumentation
    development, 423
  interwell DNAPL
    co-solvent injection, 490, 491
    electrochemical sensor, 489, 490
    fluorescence techniques, 487, 488
    GeoVIS, 488
    hydrosparge, 486, 487
    LIF/GeoVIS, 488, 489
    LIF/Raman, 489
    permeable membrane, 486
    Raman Spectroscopy, 489
    waterloo profiler, 490
  investigations, 420
  platforms, 416, 417
  probe, 418f
    environmental sites, 437
    sensor measurements, 418
  soil moisture, 424
  systems, 417, 418
  probes
    U.S. Department of Defense, 432
    Raman spectroscopy, 430, 506
    resistivity module, 421f
  rigs, 140f, 354
    DP technology, 357–360
  sampling, 507
  sensor approaches, 509
  subsurface conditions, 419
  systems, 416–437
    components, 416
  tools, 349
  truck, 357, 358
  vehicles, 417f
  vs. Wireline CPT system, 377
  wireline soil sampler, 378
ConeSipper, 408, 815
  groundwater sampling methods
    dual-tube methods, 407
Confined aquifer
  piezometer nest, 537f
  specific storage, 943f
Confined space hazards, 1232, 1233
Congener compounds, 1123
Consolidation test, 950
Consortium for Site Characterization Technology (CSCT), 178
Constant head injection tests, 936
Constant head permeameters, 926f
Constant head triaxial-cell permeameters, 928f
Container filling
  order, 1090
Contaminants
  concentration plots
    sampling locations, 603f, 604f
  distribution, 813
  pathways, 754f
  plumes
    surface geophysical methods, 285
Contamination, 44f
  areas
    petroleum product storage, 53
  avoidance, 1251
Contamination Reduction Zone, 1247
Contingency plans, 627–631
Contingency wells, 586
Continuous EM profile measurements, 263f
Continuous-flight solid-stem auger, 301f
Continuous measurement
  methods, 900
Continuous multichannel tubing (CMT), 841

Continuous multichannel tubing
(CMT) (Contd.)
   multilevel monitoring system, 843
   multilevel wells, 843
   Solinst system, 841–843, 842f
   tubing
      construction, 841
Continuous sampling, 379, 1202
Continuous slot screen
   slotted pipe
      comparison, 747t
Continuous-slot wire-wound well screen
   intake area, 745t
   PVC, 746f
   V-shaped wire, 746f
Continuous tube sampler, 148f, 149f, 323f
Continuous vapor profiling, 426, 427
Contouring methods, 173
Contract Laboratory Program (CLP), 1128, 1142
   U.S. EPA, 1148
      protocols, 45f
Conventional drilling rigs, 360f
   DP sampling, 356
Coopers formation, 558
Core barrel
   continuous samples, 146f
   retrieval, 146
Core boxes, 332
Cored boreholes, 654f
Core drilling
   samples, 327
Core log
   objective, 329
Core sample, 252
Core technical team
   personnel components, 83
Coring systems
   types, 327
Corrective Action Program, 6
Corrosion
   downhole camera photo, 698f
CPT. see Cone penetration testing (CPT)
Cradle-to-grave concept, 3
Critical exposure pathways, 126f
Critical test statistic values, 614t
Cross-contamination compounds
   laboratory examples, 1151
CSCT. see Consortium for Site Characterization
   Technology (CSCT)
CSGWPPs. see Comprehensive State
   Groundwater Protection Programs
   (CSGWPPs)
CSM. see Conceptual site model (CSM)
Custody sample chain
   components, 1145

CV. see Coefficient of variation (CV)
   measurements
CWA. see Clean Water Act (CWA)
Cyanide, 1124
Cyclical water level responses, 930
Cyclohexanone (CH), 685

Darcy's equation, 228
Darcy's law, 923, 927
   continuity, 220
   formula, 220
   hydraulic conductivity, 228, 927
   hydraulic gradients, 924f
   revision, 924
Data
   graphical presentations, 173
Data Base, 100
Data bounce, 987
Data collection program
   background samples, 136
   objectives, 136
Data loggers, 901, 902
   components, 901
Data Management Criteria, 571
Data Management Plan (DMP), 78, 170
   aspect equality, 172
   data verification and validation, 171
   development, 171
   dynamic work plan, 166
   ESC approach, 170
   Triad approach, 170
Data quality levels, 410t
Data quality objectives (DQOs), 49, 409, 964
   process
      decision making process, 88a
      definition, 87
      procedural steps, 87
   programs, 1143
   regulatory agency levels, 177
   sampling devices, 412
   Superfund project, 964, 965
   universal, 86
Data visual scanning, 1152
Daughter product data
   evaluation, 617
DCA. see Dichloroethane (DCA)
DCE. see Dichloroethene (DCE)
DCP. see Direct-current plasma (DCP)
Decommissioning materials
   hydraulic conductivity, 875
Decommissioning procedures
   granular bentonite, 879
Decommissioning program
   objectives, 871
Decontamination, 1250–1252

Index   1287

areas, 1270f
disposable equipment, 1267, 1268
equipment, 1268–1272
field equipment, 1266t
field equipment requirements, 1265–1267
procedures, 1273, 1274–1278
  larger support equipment, 1274–1276
  potential sources, 1278t
  sample collection, 1276–1280
programs, 1277
protocol, 1272–1278
  field equipment, 1265–1279
  options, 1278t
  QA, 1279, 1280
  QC, 1279, 1280
techniques
  inherent problems, 1278, 1279
Deep lysimeters, 234
Deep wells
  method development, 856
Degradation studies
  PVC, 687t
Delineating subsurface contamination
  tools, 425–432
Denaturing gradient gel electrophoresis,
  626, 627
Dense nonaqueous phase liquids (DNAPLs),
  474, 564, 965, 990, 1046
  approach and cost consumption, 493t–499t
  baseline methods, 475–483
  bentonite seals, 762
  characterization approach
    applicable approach costs, 510f, 511f
    costs, 510f
  characterization methods and approaches
    performance and cost comparisons,
      473–512
  contaminant data, 584f
  contaminant source zones, 508, 509
  contamination, 485
  cost comparisons, 500t
  generic cost estimates, 500t
  geochemical data, 584f
  geophysical techniques, 485
  LIF, 488
  PITT method, 501
  PIX, 490
  plume delineation, 475
  plumes
    dual-probe system, 488
  site-characterization methods, 476t–481t
  site-characterization techniques, 475–491
  soil gas surveys, 483
  source zone methods, 484
  VOC, 482

water -partition
  saturation relationship, 483
Density control, 563–566
Depth-discrete samples
  collection, 822
Detection monitoring data
  evaluation, 1177
Detection Monitoring Program, 5
Detection monitoring wells
  raw data, 1186t
Detector tubes, 1242
DFA. *see* Dynamic Field Activities (DFA)
Dialog
  computer database source, 100
Diamond core drilling, 329
Dichloroethane (DCA), 243
Dichloroethene (DCE), 243, 1210
Diffusion multilevel system, 820, 821
Diffusion multilevel system (DMLS), 820
  sample collection, 821
Diffusion samplers
  influential performance factors, 820
  multiple, 820
Diffusive transport, 221
Digital cone systems, 419
Dinitrotoluene (DNT), 688
Dipping probes, 901
Direct-current plasma (DCP), 1126
Direct current resistivity measurements, 267
Directional hydraulic heads, 539
Direct mud rotary
  drilling, 150f, 670f
  methods, 144
Direct-push (DP)
  continuous *vs.* discrete sampling, 379
  dual-tube soil sampling
    split-barrel samplers, 374
  dual-tube soil-sampling methods, 373–378
    CPT, 377, 378
    dual-tube precore system, 376, 377
    solid-barrel samplers, 374–376
    split-barrel samplers, 374
    wireline CPT, 377, 378
  dual-tube system
    components, 366f
  electrical conductivity log, 422f
  electrical conductivity logging system, 422f
  e-logs, 423f
  groundwater samplers, 970f
  groundwater sampling tools, 401f
    applications and limitations, 407–411
    consequences, 409
    depth-discrete samples, 394
    drive tips, 410
  installation methods, 830

Direct-push (DP) (*Contd.*)
  machines, 351, 668f
    development, 445
    percussion hammer, 421
    setup and installation, 351
    track-mounted, 362f
    typical percussion-type, 361f
  methods, 816
    cone penetration testing (CPT), 416
    HSAs, 381
    split-spoon samplers, 381
  monitoring wells
    performance comparing, 450
  open *vs.* sealed samplers, 378
  probe holes
    sealing methods, 458–462
  probes, 424
    specialized, 416–438
  rigs, 155f
    capabilities, 141, 142f
    platforms, 143f
  rods, 356
    gas-drive pumps, 414
    sample recovery, 411–416
  samplers, 813
  sampling
    conventional drilling rigs, 356
    equipment, 824
    rigs, 834
  sampling tools, 352, 356, 815
    installation mode, 411
  single-rod methods, 369–373
    nonsealed soil samplers, 369–371
    sealed soil samplers, 371–373
  single-rod *vs.* dual-tube method, 378, 379
  single-rod *vs.* sealed samplers, 378
  soil gas sampling point
    configurations, 385f
  soil gas sampling tools, 381–393
  soil sampler integrity tests
    results, 380t
  soil sampling methods, 368–381
    applications and limitations, 378, 379
    cone penetration testing (CPT), 416
  specialized tools, 403f
  systems, 139f
  tools, 356, 437–439
    gas-drive pumps, 414
    sample recovery, 411–416
  unit
    pickup truck-mounted, 361f
  well construction
    open-hole, 444f
  well construction methods
    cased-hole methods, 457
  well installation
    advantages and limitations, 441, 442
    applications and limitations, 454–457
    case-method, 445
    driven well points, 442
    drive rods, 447f
    filter packed wells, 447, 448
    mandrel-pushed screen and casing, 443–445
    methods, 455, 850
    naturally developed wells, 445–447
    open hole procedure, 442, 443
    prepacked wells, 448–450
  wells, 456f
    chromium results, 455f
    conventional well comparison, 450–454
    turbidity measurements, 456f
Direct-push (DP) technologies, 65, 652f, 972
  advantages, 352
  advantages and limitations, 350–354
  annular seal and grouting requirements, 440, 441
  application use, 354, 355
  conventional drilling rigs, 356, 357
  cost-effective profiling capability, 352
  CPT rigs, 357–360
  definition, 348–350, 355, 356
  depth constraints, 353, 354
  depth-discrete sampling, 352, 353
  environmental site characterization
    groundwater monitoring, 345–433
  geological constraints, 353
  improved site access, 351
  investigation-derived wastes, 351
  limitations, 353, 354
  lower cost, 352
  manual and mechanical methods, 356
  monitoring well installation, 438, 439
  objectives, 355
  percussion machines, 360–362
  rapid sampling, 351, 352
  requirements, 438–457
  rods and tools, 356
  rod systems, 363, 364
  sampling methods, 367, 368
    presampling considerations, 368
  soil compaction and smearing, 354
  specific site application, 354
  subsurface disturbance, 351
  unfavorable conditions, 353
  vibrator heads, 362, 363
Direct sampling ion trap mass spectrometers (DSITMS), 427
Disc auger, 300f

Index

Discharging areas
  gradient comparisons, 544f
Discharging sand layers, 540f, 541f
Discharging sand lenses, 540f, 541f
Discharging sand units
  conceptual model, 543f
Discolored soil
  uncovering process, 115f
Discontinuities
  mapping, 1205f
Discrete-interval sampling, 379
Disk filters, 1070
Disposable field equipment, 1267t
  limitations, 1268t
Dissertation Abstracts
  geology and groundwater information, 98
Dissolved oxygen
  electron acceptor, 618
  nitrate, 619
Dissolved phase
  colloidal state, 1064
DMLS. *see* Diffusion multilevel system (DMLS)
DMP. *see* Data Management Plan (DMP)
DNA
  PCR, 626
DNAPL. *see* Dense nonaqueous phase liquids (DNAPLs)
DNT. *see* Dinitrotoluene (DNT)
Double-acting piston pump, 1008f, 1026f
  schematic diagram, 1025f
Downgradient measurement equations, 1184
Downhole colloidal borescope, 981
Downhole equipment, 706–708
  cleaning, 1079f
Downhole geophysical logs
  general characteristics, 278t
Downhole geophysical measurements, 276–292
  geophysical methods
    applications, 283–291
    buried wastes and utilities, 290, 291
    contaminant plumes, 287–290
    hydrogeologic condition assessment, 283–297
  nonnuclear logs, 279–283
    borehole conditions, 283
    fluid conductivity, 283
    fluid flow, 282, 283
    induction log, 279–281
    mechanical caliper log, 283
    resistance log, 282
    resistivity log, 281, 282
    spontaneous-potential log, 282
    temperature log, 282
  nuclear logs, 277–279
    gamma-gamma log, 277
    natural gamma log, 277
    neutron-neutron log, 277–279
Downhole logging techniques, 276
Downhole TV camera survey, 875
DP. *see* Direct-push (DP)
DQO. *see* Data quality objectives (DQOs)
Drainage basins, 1152
Drilling
  installation considerations, 833
  well installation, 891
Drilling contract
  technical specifications, 341
Drilling contracts, 338–342
  agreement, 339
  articles, 340
  clauses, 340
  conditions, 339, 340
  construction drawings, 340, 341
  definition, 338
  documents, 342
  special conditions, 341, 342
  specifications, 341
Drilling fluid, 971f
Drilling information
  types, 336
Drilling method
  aquifer characteristics, 316
  lithology, 316
Drilling methods, 299
  borehole damage, 676
  circulation fluids, 305–313
  cost, 317, 318
  direct-push displacement boring, 300
  drilling depth, 316, 317
  lubricants, 674
  no circulation fluids, 299–305
  probing, 299, 300
  selection, 313–318
    access and noise, 315
    fluids and cuttings disposal, 315, 316
    health and safety, 314, 315
    lithology and aquifer characteristics, 316, 317
Drilling projects
  construction projects, 340
Drilling techniques
  1970's, 1248
Drilling well installation methods, 850
Drill-string lubricants
  synthetic, 675
Drill-through casing hammer, 311f
  dual rotary drilling, 311
Drinking water regulations, 1115

Drive-cylinder method, 918
Drive rods
  DP well installation, 447f
  prepacked screens, 450f
Driving casing
  methods, 446f
Drolleries, 1250f
Dry bulk density, 916
Dry injection systems
  annular seals, 772f
DSITMS. see Direct sampling ion trap mass spectrometers (DSITMS)
Dual-line airlift system, 855
Dual-rod system, 367f
Dual rotary drilling, 311, 312
Dual-tube DP soil sampling
  operational steps, 376f
Dual-tube groundwater profiler system
  basic steps, 406f
Dual-tube groundwater profiling tool, 405
Dual-tube groundwater sampling device, 404f
Dual-tube reverse-circulation air rotary rig, 151f
Dual-tube reverse circulation drilling, 308, 309, 309f
Dual-tube reverse-circulation rotary drilling, 671f
Dual-tube sampling, 370f
  system
    The Enviro-Core, 405
Dual-tube soil sampling
  activities
    solid barrel, 373
    split barrel, 373
  system
    benefits, 379
Dual-tube system
  types, 404
Dual-tube systems, 375
Dumping area, 120f
Duping, 1245
Dynamic Field Activities (DFA), 68
  approach
    comprehensive final report, 175
    conventional drilling methods, 143
    DMP, 170
    field work project, 176
    major steps, 77f
    project management, 179
    soil gas surveys, 154
    systematic project planning, 79
    technologies and methods, 138
  approach vs. traditional approach, 84
  hazardous waste site assessment, 75
  personnel requirements, 82
  process
    analytical data, 137
    primary focus, 172
    systematic planning, 76
  program
    field-based analytical methods, 149
Dynamic work plan
  DMP, 166
  PSP, 166
  QAPP, 166
Dynamic work strategies, 160

EA Engineering, Science, and Technology, Inc., 855
ECD. see Electron capture detector (ECD)
EKG. see Electrocardiogram (EKG)
Electrical conductivity (EC), 232
Electrical hazards, 1225, 1226
  minimization, 1225
Electrical log
  depth measurements, 426f
Electrical logs
  surface elevations, 425f
Electrical methods, 895, 896
Electrical resistivity, 161f
Electric oil-water interface probe, 991
Electric submersible centrifugal pumps
  motors, 979f
Electric submersible gear-drive pumps, 1004f, 1007f, 1029, 1030f
  schematic diagram, 1029f
  VOCs, 1031
Electrocardiogram (EKG), 1253
Electrochemical sensors
  hydrocarbon organic vapors, 489
Electrocution, 1225
Electromagnetic (EM)
  conductivity, 290
  instrumentation
    types, 265
  measurements, 255, 269
  method, 258
  profiling data, 265
  resistivity, 288
  resistivity measurements, 268, 269
  spectrum portions
    geophysical measurement, 259f
  surface geophysical methods
    resistivity methods, 263–269
Electromagnetic conductivity, 159f
Electron acceptor data
  dissolved oxygen, 618, 619
  evaluation, 617, 618
  nitrate, 619
  sulfate and sulfide, 619, 620

Index

Electron capture detector (ECD), 238, 1119
Electronic submersible centrifugal pumps, 1017
  diameter availability, 1019
  electric motors, 1017, 1018f
  schematic diagram, 1018f
Electronic velocity meter, 941
Electropolishing, 998
EM. *see* Electromagnetic (EM)
EMLRLs, 1147–1149
Endangered Species Act, 30
End-of-service-life indicator (ESLI), 1238
Energy industries
  lasers, 898
Enviro-Core
  screen, 407f
Envirol Quick Test, 187
Environmental contamination
  transportation accidents, 38f
Environmental data
  evaluation quality, 48
Environmental decontamination, 1250
Environmental drilling
  borehole logging, 297–343
  rock coring, 297–343
  soil sampling, 297–343
  well installation, 297–343
Environmental investigations
  drilling, 315
  primary purpose, 379, 380
  sampling, 315
  selecting drilling methods, 313
Environmental Protection Agency (EPA), 1265. *see also* U.S. Environmental Protection Agency (U.S. EPA)
  GC, 197
  groundwater classification guidelines, 30
  Hazard Ranking System, 9
  HSP, 170
  immunoassay-based field test methods, 189
  mass spectrometry (MS) technology, 197
  natural attenuation, 575
  RCRA, 30
Environmental remediation programs
  primary failure reasons, 41
Environmental response liability act
  compensation, 8–10
Environmental site, 35–198
  accelerated site characterization, 69–72
  accurate data collection, 54
  approaches, 57–79
    phase I, 57–65
    phase II and beyond, 65, 66
  background, 36–39
  characterization programs, 53f
  data type requirements, 55f, 56f
  data types, 54
  environmental contamination, 57
  phased approach objectives, 57
  phase I, 57
  surface geophysical methods, 152
  conducting site reconnaissance, 114–122
  conventional approach, 57
  CSM development, 122–135
  data analysis
    evaluation, 173–177
  data collection, 40
    program design, 135–137
  data requirements, 54–57
  data sets, 50f
  data validation
    considerations, 172
  dynamic field activities, 76–79
  dynamic work plans, 155–173
    adaptive sampling and plans, 160–167
  environmental project foundation, 40
  expedited site characterization, 72, 73
  field analytical program development, 180–198
    field analysis, 181, 182
  field-based analytical technologies, 177, 178
  field methods
    DP technologies, 139–142
    drilling methods, 144–148
    evaluation and selection, 137–177
    sample collection methods, 138, 139
    sonic drilling, 142–144
  Freedom Information Act, 113
  heterogeneity problem, 41–49
  historical problems, 41–51
  importance, 39–41
  improved approaches, 66, 79
    background, 66–68
    description, 68–69
  information collection methods, 97–114
    aerial photos, 100–109
    computer searches, 99, 100
    fracture trace, 109–111
    literature searches, 97–99
    site owner files, 112, 113
    state and local files, 111, 112
  investigation
    selecting equipment factors, 137
  investigation methods, 151–155
    soil gas surveys, 151–153
    surface geophysics, 151–153
  modern approach, 58f
  new-generation approaches, 66
  objectives, 51–54

Environmental site (*Contd.*)
    programs
        data use, 41f
        decision definition, 85, 86
        elements, 79
        establishing DQOs, 86–89
        personnel selection, 82–84
        project goals and objectives, 84, 85
        resource conservation, 51
        reviewing existing site information, 89–114
            sources, 96, 97
            types, 89–96
        sample analysis methods, 148–151
            field-based methods, 148–151
        sample and data problems, 49–51
        sample types, 318, 319
        site owner and operator information, 114
        site physical condition understanding, 51
        supporting work plans, 167
            data management plan, 171–173
            health and safety plan, 171
            quality assurance project plan, 169–171
            sampling and analysis plan, 167–169
        systematic project planning, 79–82
        steps, 80
    traditional approach, 58f
    triad approach, 73–76
Environmental technology verification (ETV), 178
Environment Committee of the American Chemical Society, 1149
Environment portable equipment, 1005f
    cleaning, 1006, 1007
    disadvantages, 1004, 1005
    durability, 1007
    reliability, 1007
Environment site
    characterization
        program, 44f
            sample contamination, 60f
            small-volume samples, 45f
EPA. *see* Environmental Protection Agency (EPA)
Equipment blank, 995
Equipment blanks
    types, 1095
Equipment cleaning
    methods, 1080
Equipment decontamination
    objectives, 1264
Equipment decontamination protocols
    current status, 1264, 1265
Error codes
    examples, 1084t
ESC. *see* Expedited Site Characterization (ESC)

ESLI. *see* End-of-service-life indicator (ESLI)
EVT. *see* Environmental technology verification (ETV)
Exit strategy
    development focus, 634
Expedited Site Characterization (ESC), 68
    approach
        traditional, 84
    approaches
        comprehensive final report, 175
        conventional, 84
        conventional drilling methods, 143
        field work project, 176
        project management, 179
        soil gas surveys, 154
        systematic project planning, 79
        technologies and methods, 138
    personnel requirements, 82
    process
        analytical data, 137
        focus areas, 72
        origin, 72
        primary focus, 172
    program
        field-based analytical methods, 149
Expendable points
    depth, 449f
Exploratory boreholes, 874
Exposed-screen samplers, 396
Exposed-screen tools, 394
Exposure limits, 1232

Facility
    conceptual view, 568f
    cross-section, 569f
    flow-net construction, 570f
    plain view, 568f
Facility monitoring density considerations, 565f
Falling-head permeameters, 926f
Fast gas chromatographs (GCs), 427
Federal groundwater quality standards, 29, 30
    groundwater classification, 30
Federal Remediation Technologies Roundtable (FRTR), 179
FFD. *see* Fuel fluorescence detector (FFD)
Fickian transport, 221
Field analytical methods
    reagent test kits, 185
Field analytical program
    development criteria, 180t
    QC samples, 181t
Field analytical program development
    calibration standards, 182

Index

confirmatory samples, 184
control samples, 184
duplicate samples, 183
instrument blanks, 183
matrix spike
   surrogate spikes, 183, 184
method blanks, 182, 183
technology selection, 184–186
technology summary, 186–193
   field-portable X-ray, 191–193
   head space screening, 193
   reagent test kits, 186–191
Field analytical quality control (QC) program
  duplicate samples, 182
Field analytical technologies
  XRF, 190
Field-based analytical equipment, 71f
Field-based analytical methods, 148, 155f
Field-based analytical technologies
  real-time data
    field sample analysis, 179, 180
    strategy development, 180
  regulatory agencies, 178, 179
Field-checking fracture traces, 112
Field equipment
  buckets, 1270f
  cleaning, 1078f
  color-coded buckets, 1271f
  pails, 1270f
Field equipment cleaning, 1269–1272
Field equipment cleaning procedures, 1079, 1080
Field equipment decontamination
  alternatives, 1267
Field equipment decontamination protocol
  criteria, 1272t
Field forms, 1083f, 1084f
Field hydraulic conductivity test results
  anisotropic conditions, 546
Field investigation
  underground utilities, 94f
Field notebooks, 1082f
Field operation, 434, 435
Field parameter measurement
  common errors, 1056t
Field-portable gas chromatograph (GC), 157f
Field-portable radioisotopes
  examples, 191t
Field-portable x-ray fluorescence (XRF) devices, 158f
Field quality assurance (QA), 988
Field quality assurance and quality control, 991–996
Field quality assurance (QA) program
  elements, 992t

Field quality control (QC), 988
Field quality control (QC) samples, 993t
Field screening tools
  contaminants, 60f
Field sieve analysis, 719f
  formation materials, 720f
  grain-size distribution, 722f
  graphed results, 721f
  sample amount retained, 720f
  sample formation, 719f
  sample size effectiveness, 722f
  uniformity coefficient sample, 723f
Field spike samples
  preparation, 1096, 1097
Field split samples
  difficulties, 1097
Field test kits, 156f
Field test method
  control samples, 183
Filter pack
  installation effects, 725
Filter-pack characteristics, 728t
Filter-pack design, 643
Filter-pack distribution
  objectives, 726
Filter-packed well
  installation, 723
Filter-packed wells, 723–741
  well screen design, 725–741
Filter-pack grain size, 725
Filter-pack material, 726f
  fine-grained formation, 724
  geologic situations, 724
Filter-pack materials, 644f, 731f, 975f
  back washing method, 740f
  commercial, 730
  installation, 748f
  processing facility, 733
  quartz sand, 733f
  reverse-circulation method, 739f
Filter-pack placement
  progression, 737f
  tremie-pipe method, 736f
Filter-pack sand
  placement, 734f
Filter preconditioning, 1070
Filtration, 1065
  field studies, 1066
Filtration equipment, 1069
Filtration methods
  criteria, 1069
Filtration parameters
  recommendations, 1065t
Fixed well-volume purging, 1036
Fixed well-volume purging strategy, 1039t

Flame ionization detector (FID), 238, 1243
    compound examples, 194f
    disadvantages, 195
    PIDs, 1243
Flame spectrophotometers, 1125f
Flammable liquids, 1228
Flash point, 1228
Flexible linear underground technology. see
        FLUTe (flexible linear underground
        technology)
Flexible Liner Underground Technologies, Ltd.
    FLUTe system, 475, 844
Flexible pump tubing, 1017f
Float instruments, 900
Float method, 897
Float recorder systems, 900, 901
Flow nets
    confined, 535f
    unconfined, 535f
Flow rate control, 1002f
Flow rates
    minimum-purge sampling, 1051
Flow-through cell
    atmospheric conditions, 1058f
    schematic diagram, 596f
Flow-through cells
    SAP, 1060
Flow velocity
    vadose zone, 229
Flush-joint threaded casing, 701
Flush-mount well protection, 454f
Flush-to-grade completion, 789, 791f, 795f
    popularity, 784
Flush-to-grade installation, 793f
Flush-to-grade monitoring well completion,
        787f
Flush-to-grade well vault, 788, 789f
FLUTe (flexible linear underground
        technology)
    approach, 511
        baseline cost, 512
    device, 491
    membrane, 491–501
        cost analysis, 491–501
    system
        Flexible Liner Underground Technologies,
            Ltd., 844
Flu-to-grade completion, 792f
Flux
    vadose zone, 229
Food industries
    lasers, 898
Foundation excavation, 130f
Fractured rock
    groundwater flow, 560f

local flow pathways, 561
pathway flow distribution, 562
porosity data, 212f
Fractured rock environments, 560
Fracture trace analysis, 109
Freedom Information Act requests, 113
Free-drainage samplers, 236
Frostbite, 1227
FRTR. see Federal Remediation Technologies
        Roundtable (FRTR)
Fuel fluorescence detector, 430–432
Fuel fluorescence detector (FFD), 350
    commercial, 431
    systems, 431
Full protection, 1257f

Gage tensiometer
    schematic diagram, 223f
Gamma-gamma log
    lithology, 277
Gamma radiation, 1228
Gamma-ray attenuation, 919
    determining density, 919
    disadvantages, 225
Garber-Wellington Formation, 569
Gas chromatography (GC), 1118, 1243
    analytical technique, 1119
    detectors
        method examples, 196
        EPA, 197
    mass spectrometry, 196
    organic compounds, 195
    principles, 195
    purgeable aromatics, 1120f
    schematic diagram, 1119t
Gas-displacement pumps, 1020f
    schematic representation, 1020f
    sediments, 1021
Gas-drive pumps, 414, 415f
GC. see Gas chromatography (GC)
Gear drive pump, 1030f
Gear pumps, 774
Geochemical data, 1213
Geochemical diagrams, 1215f
Geochemical interaction problems,
        1208–1212
    prevention, 1211t
Geodetic control
    classification standards, 863t
Geoelectric cross-section
    2D resistivity imaging, 267
Geologic maps
    regional, 90f
Geologic materials
    core samples, 917

Index 1295

Geologic uniformity problems, 1204
   prevention, 1206
Geomorphic features, 118
Geophysical measurements
   application, 256
   application modes, 256
   EM spectrum portions, 259f
Geophysical methods
   application areas, 256
   selection, 292
   volumes comparison, 253f
Geophysical surveys, 254
Geophysics, 504
Geoprobe Groundwater Profiler, 815
Global Positioning System (GPS), 864
GLP. *see* Good laboratory practices (GLPs)
Gonzales Amendment, 12
Good laboratory practices (GLPs), 1082
GPS. *see* Global Positioning System (GPS)
Grab samplers
   thief samplers, 1013
Gradient-controlled groundwater contours, 566f
Gradient controlled sites
   guidance, 534, 535
   procedures, 534–543
Grain-size distribution
   curve, 210t
   elements, 210
Granular bentonite, 757
   boreholes, 772f
Granular filter packs
   designation, 727
Graphite furnace technique, 1125
Gravel
   estimated hydraulic conductivity, 930f
Gravimetric water content, 920
Gravity pouring, 459
Gravity techniques, 291
Ground subsidence monitoring device, 951f
Groundwater, 921
   carbon dioxide, 622
   complete analysis definition, 1142
   concentration
      seasonal variability, 612
      statistical analysis, 607
   contamination, 10, 117f, 752, 1172–1174
   development, 33
   exploitation, 33
   federal legislators, 2
   long-term monitoring, 585f
   natural resource, 1
   organic chemicals, 1150
   sampling locations, 585f
   state legislators, 2
   temperature, 978
   terrorism, 33
Groundwater contour map, 570f
Groundwater contour plan, 569
Groundwater data, 1151
   sources, 1139f
Groundwater detection, 1179
Groundwater discharge, 557f
Groundwater elevation
   contours, 906f
Groundwater elevation plots, 1213f
Groundwater elevations
   plotting, 1212f, 1214f
Groundwater flow
   assessing, 888f
   confined aquifer, 535
   structural control, 556
   understanding, 965–967
   velocity relationship, 941
Groundwater flow conditions
   complex, 569–571
Groundwater flow directions, 654, 907
   estimation, 889f
Groundwater flow equation
   solution, 948
Groundwater flow pathways, 650
Groundwater flow velocity, 940–942
   hydraulic conductivity, 940
   specific storage, 942, 943
   specific yields, 942, 943
Groundwater geochemical data
   evaluation, 616–627
Groundwater investigations, 411, 892, 1127
   ACO requirements, 1116
   alternatives, 1115
   general safety, 1256–1260
   HAZWOPER, 1254–1260
   liability considerations, 1256–1260
   objectives, 1134
   RCRA requirements, 1115, 1116
   regulatory agencies, 1115
   training requirements, 1254–1256
Groundwater level data
   water flow, 902
Groundwater levels
   continuous measurements, 900
Groundwater monitoring
   components, 520
   conductivity, 621
   federal regulatory mandates, 2, 3
   federal regulatory programs, 3
   multilevel, 807–845
   multilevel options, 819
   oxidation-reduction potential, 620, 621
   parameter evaluation, 620, 621

Groundwater monitoring (*Contd.*)
  pH, 621
  placement, 816–819
  primary failure reasons, 41
  problems, 1212
  problems diagnosis, 1193–1216
  RCRA citations, 7t
  RCRA requirements, 7t
  regulatory mandates, 1–33
  relativity, 914
  resource protection, 1, 2
  temperature, 621
  treatment, storage, and disposal facilities (TSDFs), 3
  water quality, 30
Groundwater monitoring investigations
  drilling techniques, 1248–1250
  health and safety considerations, 1219–1261
  information sources, 1235, 1236
  site operations, 1246–1250
Groundwater monitoring programs, 862
  data objectives, 640
  failure, 976
Groundwater monitoring projects
  field equipment decontamination, 1263–1280
Groundwater monitoring system
  data analysis requirements, 519–566
    down gradient wells, 528
    flat gradients, 531–534
    gradients, 528–531
    steep gradients, 531–534
    target monitoring zones, 526–528
  design, 517–571
  design criteria, 571
  design summary, 525f
  geologic structural control, 554–556
  geologic structures, 555
  hydrogeologic framework elements, 526
  location selection, 518
  monitoring system design, 527f
  regulatory concepts, 519–525
  system attributes, 518
  target monitoring zones, 526
    background monitoring wells, 529
    selection process, 528
Groundwater monitoring system design
  diagnosis, 1215t
Groundwater monitoring systems
  cross-contamination, 836, 837
  geologic controls, 543–545
  installation requirements, 4
  multilevel, 830–845
  multilevel comparisons, 831t, 832t
  multilevel systems
    advantages, 830–833
    disadvantages, 833
    WHP, 22
Groundwater monitoring well design
  basic requirements, 641
  data needs, 651t
Groundwater monitoring well installation
  data needs, 651t
Groundwater monitoring wells, 61f, 779f
  design and installation, 639–797, 645
  groundwater-monitoring system requirements, 524, 525
  purposes, 648
  site characterization, 650–657
  surface completion types, 777
  United States, 816
Groundwater movement
  geologic movements, 543
  geologic structures, 554
  types, 559
Groundwater movement pathways
  fracture, 111
Groundwater Protection Strategy, 29
Groundwater pump-and-treat system, 631
Groundwater purging
  selection criteria, 997t
Groundwater quality, 1155
  analytical laboratories, 1142–1151
  anomalies, 1198
  background water quality evaluation, 1150, 1151
  comparisons, 1152, 1153
  contour maps, 1155–1157
  data accuracy, 1146
  data inquiries, 1146
  data organization and analysis, 1135–1189
  detection monitoring, 1141, 1142
  facility and equipment, 1145, 1146
  histograms, 1159–1165
  inspection, 1153–1155
  lab evaluation steps, 1143
  low-level organic chemical results, 1150
  measurement units, 1152
  monitoring, 1151, 1152
  outliers, 1152
  reporting, 1151, 1152
  sample dilution, 1149, 1150
  significant digits, 1152
  time-series formats, 1157–1159
Groundwater Quality Assessment Program, 7
Groundwater quality monitoring wells
  general construction requirements, 439
  nominal diameters, 705
  regulatory requirements, 439

Index 1297

Groundwater quality standards
   discussion, 30–32
Groundwater recharge zones, 128f
Groundwater researchers, 824
Groundwater sample analysis, 1113, 1134
   diagnostic tests, 1117, 1118
   parameters, 1114, 1115
Groundwater sample collection, 64
Groundwater sample filtration, 1062
Groundwater samplers
   active methods, 821
Groundwater samples, 964, 1068t
   container examples, 1074t
   field filtration, 1068
   filtration, 1064
   functions, 1064
   influential factors, 966f
Groundwater sampling, 959–1102
   analysis requirements, 1172
   applications, 1085
   chemical preservation, 1075, 1076t
   collection, 962, 963
   collection procedures, 1090–1102
   collection protocols, 1092–1094
   data quality, 963, 964
   document information, 1081t
   documenting field activities, 1085
   examples, 1071t
   exposure, 984f
   field components, 1086t
   geochemical changes, 976–987
   groundwater monitoring, 960
   high-flow-rate pumping, 983
   influential factors, 965–987
   in-line filtration systems, 1098
   laboratory shipment, 1102
   methods, 1092
   non regulatory compliance monitoring, 962
   objectives, 961
   pressure changes, 977, 978
   pretreatment procedures, 1097–1100
   purging and sampling, 979–984
   regulatory compliance monitoring, 961, 962
   science, 961
   site-specific decisions, 991
   temperature changes, 978–999
   vacuum filtration, 1099
   well point driven, 442f
Groundwater sampling and analysis plan (SAP), 987
Groundwater sampling device, 62
Groundwater sampling event
   conduction, 1086–1102
   site orientation, 1087

Groundwater sampling investigation
   cost, 1117
Groundwater sampling methods, 393–416
   combined sampling tools, 403, 404
   dual-tube methods, 404–407
   exposed screen sampling tools, 395–398
   sealed screen samplers, 398–403
   single-rod methods, 394, 395
Groundwater sampling procedures, 595
Groundwater sampling program
   objectives, 1066
   planning phase, 1065
   QC sample, 996
   site-specific, 965
Groundwater sampling programs
   objectives, 961
   site characterization data, 42f
Groundwater sampling techniques, 595–598
   bailer sampling, 596, 597
   considerations, 597, 598
   diffusion samplers, 597
   peristaltic pump sampling, 595
   sampling frequency, 598, 599
   submersible pump sampling, 595, 596
Groundwater sampling tools, 140f
Groundwater supplemental data
   evaluation, 616–627
Groundwater system
   shallow discharging, 538f
Groundwater systems
   confined, 522f, 523f
   unconfined, 522f, 523f
   unconfined and confined, 522f, 523f
Groundwater temperature, 621
Grout mixture, 775f
Guidelines for the Selection of Chemical Protective Clothing, 1245
Gypsum blocks, 230

Halogenated chemicals
   stability, 242
Halogen specific detector (XSD), 486
Hanby Field Test Kit, 186
Handbook Suggested Practices Design and Installation of Ground Water Monitoring Wells, 827
Hand-slotted casing
   well screens, 743f
HASP. *see* Health and Safety Plan (HASP)
Hazard control
   methods, 1249
Hazard identification
   classification, 1225–1235
Hazardous materials
   minimizing exposure, 1246

Hazardous waste disposal facilities
    groundwater-monitoring system
        requirements, 523, 524
Hazardous waste lagoon, 122f
Hazardous waste management facilities, 519
Hazardous waste site investigation, 1256
Hazardous waste sites
    controlled and uncontrolled, 51
    uncontrolled, 38f
Hazardous Waste Trust Fund, 8
Hazards
    degree, 1233
    versus risk, 1233–1235
    risk degree, 1233
Hazen's approximation, 950
Head space screening
    FID, 193
    field sample analysis, 192
    PID, 193
Health
    safety, 1246f
Health and Safety Plan (HASP), 169, 1222
    air monitoring, 1239f
    plan objectives, 170
    U.S. EPA, 170
Health and safety planning, 1221–1225
Health and safety programs
    air monitoring, 1239
Health and safety specialty firms, 1258
Heavier-than-air gases, 1248
Heavy drilling mud, 876
Helical rotor pumps, 1027
Herbicides, 1124
Heterogeneous soils, 46f
High-flow rate pumps
    high speed, 982f
High-performance liquid chromatography
        (HPLC), 1122
High-pressure grouting equipment, 459f
High-pressure grout pumps, 441, 453f
High-pressure jetting, 860, 861
High-solids bentonite grout, 760f
High-solids bentonite slurries, 758
Histograms
    data normality, 1164f
Hollow-stem auger (HSA), 147f, 303f, 348, 351,
        736f, 737f, 834
    continuous tube samplers, 321
    drilling, 58f, 670f
    successful method use, 304
    wells, 456f
        turbidity measurements, 456f
Homogeneous granular rock
    hydraulic conductivity estimation, 937
Hot-water power wash systems, 1275

HPLC. *see* High-performance liquid
        chromatography (HPLC)
Hybrid wells, 715, 716
HydraSleeve, 1054f
    sample collection, 1055f
Hydrated bentonite grout, 776f
Hydration heat methods, 768
Hydration process
    cement slurry, 766
Hydraulic conductivity, 923t
    calculation, 230
    consolidated rock, 933
    establishment criteria, 538
    formulas, 936
    geologic materials, 915
    grain-size distribution, 929
    laboratory methods, 926, 928
    laboratory testing, 927
    mathematical regression, 929
    Nguyen and Pinder method, 932
    permeability, 923–940
    secondary, 559–563
    transmissivity, 944
    vadose zone, 229
Hydraulic conductivity testing, 649f, 656f
    alternatives, 656
    methods, 656
    rock, 929
    soil, 929
Hydraulic conductivity zones, 888
Hydraulic head
    definition, 818
Hydraulic head data, 886
Hydraulic head distribution, 904
Hydraulic head relationship
    piezometer, 884f
Hydraulic heads
    groundwater movement, 968f
    measurement, 842, 843
    measurement technology, 819
Hydraulic media
    aquifer system, 885, 886
Hydraulic-powered percussion hammers, 360f
Hydraulics
    site-specific formation, 965
    water chemistry, 967–969
Hydrocarbons
    biodegradation, 288
Hydrogen concentration
    measuring methods, 623
Hydrogeologic conditions
    assessment, 256
Hydrogeologic cross-section
    facility, 567f
Hydrogeologic environments, 874

Index 1299

Hydrogeologic site conditions
   objectives, 540
Hydrogeologists, 1152, 1193
Hydrographs, 1207f
   surface-water bodies, 1208
Hydrologic data, 1213
Hydrologic uniformity
   problems, 1207, 1209t
Hydrologic uniformity problems, 1204–1208
Hydrosparge field sampling, 505
Hydrostatic law
   formula, 214
Hydrostatic pressure
   external, 682
Hydrous aluminum
   bentonite, 756
Hypergeometric distribution function, 1183
Hypothesis testing
   statistical error, 1170f
Hysteresis
   ink bottle effect, 219
   suction-moisture content relationship, 218f

ICP. *see* Inductively coupled plasma (ICP)
IDLH. *see* Immediately dangerous to life or health (IDLH)
IDW. *see* Investigation-derived waste (IDW)
Imaging methods
   picture surface, 258
Immediately dangerous to life or health (IDLH), 1147–1149, 1232
Immunoassay-based field test methods
   U.S. EPA, 189
Immunoassay methods, 188
   types, 188
Immunoassay test kits, 156f
Induction logs, 281f
Inductively coupled plasma (ICP), 1124
Industrial-solid waste landfills, 52f
Inertial-lift pumps, 413, 986f, 1031, 1032, 1031f
   discharge, 986f
Inertial-lift (tubing check-valve) pumps, 412, 413
Infiltration rate
   cumulative infiltration relation, 226
Infiltrometers, 226
Information collecting
   methods, 97
Information-gathering stage, 96f
Inner barrel protrusion
   relationship, 327f
Innocuous materials, 120
In situ
   Latin definition, 319
Instrumentation calibration, 1087

Instrument calibration
   aspects, 1132
International Association of Drilling Contractors, 1220
Interwell dense nonaqueous phase liquids (DNAPLs)
   cone penetrometer testing methods, 485, 486
   CPT methods
      co-solvent injection, 490, 491
      electrochemical sensor, 489, 490
      fluorescence techniques, 487, 488
      GeoVIS, 488
      hydrosparge, 486, 487
      LIF/GeoVIS, 488, 489
      LIF/Raman, 489
      permeable membrane, 486
      Raman Spectroscopy, 489
      waterloo profiler, 490
   geophysical surveys, 485
   partitioning interwell tracer tests, 483, 484
   radon flux rates, 484
   VOC concentrations, 484, 485
Intra-well statistical comparisons, 1175f
Investigation-derived waste (IDW), 61f, 65
Ion trap mass spectrometers (ITMS), 486
Iridescent fluid sheen, 119f
Iron bacteria, 698
Iron oxidation, 1092
Isopleth maps, 602f
Isopleths
   dissolved solid, 1156f
Isotopes
   stable, 627
Iterative conductance probes, 901
   definition, 901
ITMS. *see* Ion trap mass spectrometers (ITMS)

Joining casing
   coupling procedures, 697–705
Joining polyvinyl chloride (PVC) casing, 698–703
Joining steel casing, 703–705
Joint steel well casings
   options, 703

Karst terrain
   flow pathways, 562
   guidance, 563
   sampling water quality, 563
Kemmerer sampler, 1014f

Laboratories, 158f
   health department certifications, 1128
Laboratory data-quality assessment
   procedures, 1144, 1145

Laboratory documentation, 1133
Laboratory hydraulic conductivity
    test results
        anisotropic conditions, 546
Laboratory information management systems
    (LIMs), 1200
Laboratory microcosm study, 624
Laser beam generator, 141f
Laser induced fluorescence (LIF),
    183, 487
    DNAPL, 488
    techniques
        Raman techniques, 488
    tool
        schematic, 428f, 429f
Laser induced fluorescence (LIF) spectroscopy,
    427–430
Laser methods, 898
Lasers
    chemical industries, 898
    energy industries, 898
    food industries, 898
Layered deposits
    computer model flow, 549f
    conceptual flow, 548f
    conceptual model, 548f
LDPE. *see* Low-density polyethylene (LDPE)
Leachate data
    histograms, 1163f
Leachate movement
    geologic effect, 555
Leachate plume
    resistivity map, 264f
    time sequence, 547f
Leachate seeps, 118f, 119f
Leachate water quality plume, 545f
Lift capability, 1001f
    definition, 1000
Light nonaqueous-phase phase liquids
    (LNAPLs), 753, 990, 1046, 1220
Limit of detection (LOD), 1147
Limit of quantification (LOQ), 1149
LIMs. *see* Laboratory information management
    systems (LIMs)
Lineaments, 110
Linear regression
    trend tests, 1178
LNAPLs. *see* Light nonaqueous-phase phase
    liquids (LNAPLs)
Local flow cells
    conceptual model, 533f
    map view, 534f
Local land-use maps, 95f
LOD. *see* Limit of detection (LOD)
Log normal distribution, 1172
Logs and interpretation, 435–437

Log-transformed data
    probability plots, 1165, 1165f
Long-screen wells, 664
Long-term monitoring frequency, 599
Long-term monitoring plan
    data interpretation, 599, 600
Long-term monitoring programs
    analytical protocol, 586
    exit strategy, 631
Long well screens, 644f, 665f
LOQ. *see* Limit of quantification (LOQ)
Losing stream target monitoring zones
    cross-sectional view, 531f
    plain view, 531f
Louvered well screen, 747f
Low-density polyethylene (LDPE), 1032
    PDBS, 1032
Low-flow purging, 1042, 1043f
    equipment, 1043f
    pump intake, 1044
    water level monitoring, 1045
    water levels, 1045
Low-flow sampling, 1042, 1043f
    equipment, 1043f
    water level monitoring, 1045
    water levels, 1045
Low-hydraulic conductivity
    non-discharging sand lenses, 539f
Low-hydraulic conductivity environments,
    551–554
Low-hydraulic conductivity materials, 554f
    channel deposits, 553f
    drain effects, 552f
Lysimeters cups
    types, 235

Magnetic gradient, 275f
Magnetic techniques, 291
Magnetometer, 274–276
Mandrel-driven screens
    driven well point, 457
Man-made excavations
    regional and local geology, 129f
Mann-Kendall test, 607, 616
    example calculation, 608
    null probabilities, 609t, 610t
    station homogeneity, 611
Mann-Whitney U-test, 613
Manometers, 899
Manufacturing process site, 122f
Mass spectrometry (MS)
    schematic diagram, 1121f
    technology
        EPA, 197
MASW. *see* Multichannel analysis of surface
    waves (MASW)

Index 1301

Material Safety Data Sheets (MSDS), 96
Matrix heterogeneity
　field studies, 43
Maximum contaminant level (MCL), 1148
Maximum contaminant level goals (MCLGs), 31
MCL. *see* Maximum contaminant level (MCL)
MCLGs. *see* Maximum contaminant level goals (MCLGs)
MDL. *see* Method detection limits (MDLs)
Mechanical auger, 300f
Mechanical density testing, 916–919
Mechanical jack, 358f
Mechanical surging
　bailer, 859
　monitoring well development, 857
　　submersible pumps, 858
　submersible pumps
　　monitoring well development, 858
Medical monitoring, 1252–1254
Medical screening tests, 1254
Megahertz (MHz), 260
MEK. *see* Methyl ethyl ketone (MEK)
Membrane interface probe (MIP), 350, 433, 434, 486
　electrical log, 437f
　logging operations, 436f
　schematic representation, 433f, 434f
　system
　　components, 433, 435f
　system components, 433, 434
Membrane interface probe (MIP)-flame ionization detector (FID) log, 436, 436f
Metabolic byproduct data
　ethane, 620
　ethene, 620
　evaluation, 620
　Fe(II), 620
　methane, 620
Metabolic byproducts
　electron acceptors, 594
Metal detection, 273, 274
Metal detectors, 290
Metallic materials, 692–697
Methane concentrations, 390f
Method blank
　instrument bank, 182
Method detection limits (MDLs), 1147–1149
Methyl ethyl ketone (MEK), 685
Methyl tertiary butyl ether (MTBE), 36, 37, 809, 1118
　construction details, 812f
Microbial characterization techniques
　practical considerations, 627
Microbial communities
　diversity, 626

Microcosm studies, 624
Microgravity, 273
　measurements, 273
Microgravity profile, 273f
Mill-slotted well points, 395, 395f
Minimum purge sampling, 1053
Minimum purge sampling method, 1052
　pump placement, 1052
MIP. *see* Membrane interface probe (MIP)
Modular well seal
　foam bridge, 452f
Moisture content
　hydraulic conductivity relationship, 221f
Monitoring plan
　analytical protocol, 586–595
　design elements, 579, 580
　location and placement, 580–582
Monitoring problems
　diagnosis, 1212–1216
Monitoring programs
　bailers, 596
　goals, 1194f
Monitoring well casing
　installation, 678f
　logistical factors, 678
　screen materials, 677, 678
　site-specific, 678
Monitoring well completion, 781f
　multiple vertically separated zones, 661–667
　types, 657–659
Monitoring well construction
　detail, 869
　details, 868f
　volume calculation information, 739t
Monitoring well design and installation
　technology limitations, 647
Monitoring well design components, 647f
Monitoring well design flaws
　types, 641, 642
Monitoring well design practices, 646
Monitoring well development, 850–866, 856
　considerations, 850–855
　decontamination, 861
　goals, 850, 851
　manual development, 860
　manually, 860
　mechanical surging, 857
　screen and filter pack, 851
　submersible pumps
　　mechanical surging, 858
　surveying, 861–864
　well casing, 851
　well identification, 864–866

Monitoring well installations
  conceptual site development, 650
  drilling method, 317
Monitoring well intakes, 716–741
Monitoring well maintenance, 868–872
Monitoring well post-installation
  considerations, 849–881
Monitoring well rehabilitation, 868–872
Monitoring wells
  abandonment material placement, 877, 878
  annular seal, 755, 758f
  bentonite, 756
  borehole, 667
  borehole decommissioning, 872
    objectives, 872
  boring, 872
  bow-type centralizers, 711f
  casing, 683
  chemical monitoring, 276
  cluster, 830f
  contaminant data, 583f
  cross-contamination, 861
  decommissioning, 872, 874
  decommissioning plans, 872–875
  decommissioning procedures, 878–870
  detection purposes, 752
  factors, 707f
  geochemical data, 583f
  ground water, 648
  identification, 795f
  installation purposes, 709
  location, 587f
  location and inspection, 875
  multilevel purpose, 836
  multilevel sampling, 819
  multiple diffusion sampler installation,
    819, 820
  physical location, 794
  protective well casing, 796f
  purposes and objectives, 648–650
  records and reports, 880
  stainless steel casings, 704
  steel casing, 704f
  steel screen, 704f
  surface seal, 778f
  water level data, 649f
  water quality analysis, 853
  water-supply wells
    types, 729
  well screen types, 742–748, 744f
Monitoring well screens
  installation, 678f
  materials, 682
  open areas, 751
Monitoring well technology, 716

MS. see Mass spectrometry (MS)
MSDS. see Material Safety Data Sheets (MSDS)
MSWLF. see Municipal solid waste landfill
  (MSWLF)
MTBE. see Methyl tertiary butyl ether (MTBE)
Multichannel analysis of surface waves
  (MASW), 272
Multilevel monitoring devices, 815, 816
Multilevel monitoring systems, 142f, 664–667
  borehole and drilling impacts, 667–677
Multiparameter sonde, 1058f, 1090f
  chemical profiling, 1058f
Multiple aquifers, 549–551
Multiple-screen wells, 662
  single-casing, 663f
Multiple single-screen wells
  single borehole, 663
Multi-well tracer tests, 937
Municipal solid waste landfill (MSWLF), 52f,
  521, 569, 1167
Murphy's Law, 1222

NAPL. see Nonaqueous-phase liquid (NAPL)
National Drilling Association, 1220
National Geodetic Survey (NGS), 862
National Geodetic Vertical Datum (NGVD),
  559, 891
National Ground Water Association, 1220
National Institute for Occupational Safety and
  Health (NIOSH), 1232
National Institute of Standards and
  Technology (NIST), 1145
National Library of Medicine, 1235
National Oceanic and Atmospheric
  Administration (NOAA), 862
National Pipe Thread (NPT), 700
National Pollutant Discharge Elimination
  System (NPDES), 11
National primary drinking water standards,
  14t–21t
National Priorities List (NPL), 9
National Sanitation Foundation (NSF),
  690, 877
  chemical parameters, 691t
National secondary drinking water standards,
  22t
National Technical Information Service (NTIS),
  98
Natural attenuation
  compliance monitoring, 578, 579
  conceptual diagram, 581
  contaminant data, 600–616
  contingency monitoring, 578, 579
  designing monitoring programs, 573–634
  exit strategies, 631–634

Index 1303

groundwater analytes, 589f, 590f
  long-term evaluation, 588
groundwater analytical parameters, 588
hydrologic systems, 574
institutional factors, 580
interactions, 629t, 630t
long-term monitoring, 578, 579
long-term monitoring data, 600
long-term monitoring program, 631
  criteria, 632
  technical basis, 632
long-term monitoring program duration, 631
monitoring, 576
monitoring data, 600–616
monitoring duration, 631–634
monitoring purpose, 576, 577
monitoring reasons, 600
monitoring types, 577–579
performance, 588
performance monitoring, 578
performance objectives, 631
regulatory factors, 580
statistical methods, 615
supplemental groundwater analytes, 591f–593f
technical factors, 580
three-dimensional approach, 580
U.S. EPA, 575
Natural drainage ways, 107f
Natural gamma logs, 279f
Naturally developed well
  well screen design, 718–723
Natural well development
  limitations, 445
Near-surface Dawson clay stones, 553
Neat cement, 763–769, 766
  annular seal material, 768
  bentonite grout, 775
  composition, 763, 764
  high-shear paddle mixer, 767f
  mixture, 767f
Neat cement grout, 777
  bentonite pellets, 771f
Negative-pressure equipment
  procedures, 1099, 1100
Negative-pressure filtration systems, 1099
Nephelometric Turbidity Units (NTU), 856
Nested wells, 662–664, 823, 824, 823f, 827f
  conceptual design, 824
  installation, 829, 830
  tag line, 825
Neutron access tubes, 230
Neutron-neuron log
  water table moisture content, 277
NGS. *see* National Geodetic Survey (NGS)

NGVD. *see* National Geodetic Vertical Datum (NGVD)
NIOSH. *see* National Institute for Occupational Safety and Health (NIOSH)
NIST. *see* National Institute of Standards and Technology (NIST)
Nitrate
  iron and sulfate reduction, 619
NOAA. *see* National Oceanic and Atmospheric Administration (NOAA)
Noise, 1226, 1227
Noise reduction ratings (NPR), 1227
No loose paper rule, 1083
Nonaqueous-phase liquid (NAPL), 43, 394, 588, 808, 1208
  saturation
    expression, 484
  source area sampling, 588–594
  subsurface, 588
Non-discharging sand layers, 540f, 541f
Non-discharging sand lenses, 540f, 541f
Nonhalogenated chemicals, 242
NPDES. *see* National Pollutant Discharge Elimination System (NPDES)
NPL. *see* National Priorities List (NPL)
NPR. *see* Noise reduction ratings (NPR)
NPT. *see* National Pipe Thread (NPT)
NSF. *see* National Sanitation Foundation (NSF)
NTIS. *see* National Technical Information Service (NTIS)
NTU. *see* Nephelometric Turbidity Units (NTU)
Nuclear logging tools, 424, 425
Nuclear moisture logging equipment, 225
Nuclear Regulatory Commission, 2
Null distributions
  606
Null hypothesis, 605
  statistical test, 1170

Observational method, 66
  applications, 67
  information limitations, 68
  key elements, 66, 67
Occupational Safety and Health Administration (OSHA), 169, 1220, 1227, 1238, 1253, 1259
  field sanitation, 1223f
  hazardous materials, 1252
  Hazardous Waste Operations and Emergency Response Rule (HAZWOPER), 1254, 1258
  NIOSH, 1253
  RCRA, 1255
  site positions, 1259

Occupational Safety and Health
    Administration (OSHA) (Contd.)
  U.S. EPA, 1253
  USCG, 1253
Occupational Safety and Health
    Administration (OSHA) Act of 1970,
    1258
Occupational Safety and Health
    Administration (OSHA) Construction
    Industry Standard, 1260
Office of Solid Waste and Emergency Response
    (OSWER), 574
Office of Surface Mining (OSM), 2
One-dimensional consolidometer
  schematic, 951f
On Line, 100
Open containers
  atmospheric conditions, 1057f
Open dumps, 4
Open hole, 457
Open-tube sampler, 377f
Organic carbon biodegradation, 622
Organic chemicals
  ground water, 1150
Organic polymer drilling fluids, 674f
Organic sampling devices, 1276
Organic vapor profile, 289
O-ring elastomer materials
  chemical media, 702t
O-ring seal, 703
ORP. *see* Oxidation-reduction potential (ORP)
OSHA. *see* Occupational Safety and Health
    Administration (OSHA)
OSM. *see* Office of Surface Mining (OSM)
Osmotic potential
  definition sketch, 220f
OSWER. *see* Office of Solid Waste and
    Emergency Response (OSWER)
Overview training, 1255
Oxidation-reduction potential (ORP), 407, 594,
    1002
  ground water, 620
  measurement, 620
  measurements
    thermodynamic equilibrium, 623
Oxygen concentrations, 390f
Oxygen deficiency meters, 1241

Packers
  placement, 1041f
  progressing cavity pump, 1041f
  purging, 1040
  test section length, 935
  types, 934
Packer testing, 933–936, 934f

Packer tests, 938
  hydraulic conductivity, 935
  hydraulic conductivity data, 935
  vertical hydraulic conductivity, 937, 938
PAH. *see* Polycyclic aromatic hydrocarbons
    (PAHs)
Panchromatic photos
  black and white, 101f
Parametric statistics, 607
Particle roundness
  powers scale, 732f
Particle segregation, 730
Partitioning interwell tracer test (PITT), 483,
    501–503
  preliminary steps, 502
Passive diffusion bag (PDB), 821
Passive diffusion bag samplers (PDBS), 821,
    1032, 1033f
  availability, 1035
  compounds tested, 1034f
  limitations, 1034
  single-use bags, 1034
  VOC, 821
  well screens, 1033f
Passive diffusion samplers, 819
Passive sampling, 1052f
PCE. *see* Perchloroethylene (PCE)
PDB. *see* Passive diffusion bag (PDB)
PDBS. *see* Passive diffusion bag samplers
    (PDBS)
PEL. *see* Permissible exposure limits (PELs)
Perched ground water, 556–559
Perched structural control
  groundwater flow, 557f
Perchloroethylene (PCE), 430
Percussion drilling, 309–311, 310
Performance monitoring wells (PMWs), 578
  contingency monitoring wells, 582
Peristaltic pump
  pump head, 1016f
Peristaltic pumps, 414f
  negative pressure, 977f
Permeable reactive barriers (PRBs), 813
Permissible exposure limits (PELs),
    1232, 1233
Personal protective equipment (PPE),
    988, 1088
  employment, 1244
  engineering controls, 1237f
Pesticides, 1124
PetroFlag test kit, 187
Petroleum distribution terminal, 126f
Petroleum hydrocarbon-based drill string
    lubricants, 674f
Petroleum-stained soil, 117f

Phase II
  site characterization, 1136
PH data
  illustration, 1187t
Phospholipid fatty acids, 625, 626
Photoionization detector (PID), 59, 157f, 238, 988, 1119, 1242
  compound examples, 193f
Physical hazards, 1226
  prevention, 1226
Physical preservation methods, 1072
PID. *see* Photoionization detector (PID)
Piezometers, 886, 887
  aquifers
    unconfined, 537f
  bundle, 655f
  groundwater level measurements, 886
Pipe centering guides
  types, 710f
Piston pumps, 1026
  double-acting, 1027f
Piston samplers, 325f
  components, 323
Pitcher rotary soil core sampler, 324f
PITT. *see* Partitioning interwell tracer test (PITT)
PIX. *see* Precision injection/extraction (PIX)
Plastic casings
  PVC, 880
Plume
  graphical evaluating methods, 600–604
  hydraulic capture, 810f, 811f
  statistical evaluating methods, 604–607
    data group test differences, 612–615
    results, 615, 616
    trend tests, 607–612
Plumes
  dissolved characterization, 818
Plume stability plots, 605f
PMW. *see* Performance monitoring wells (PMWs)
Pneumatic bladder pump, 1022f
  portable, 1023f
  schematic diagram, 1022f
Pneumatic hammers
  varieties, 358f
Polycyclic aromatic hydrocarbons (PAHs), 980
Polyvinyl chloride (PVC), 389
  aromatic hydrocarbons, 687
  casing
    non-steel, 677
    water-supply wells, 700f
  cementing agents, 975f
  centralizer, 828
  chemical degradation, 973

chemical interference effects, 691
chemical resistance, 685
continuous-slot wire-wound well screen, 746f
degradation studies, 687t
materials
  pipe casing, 690t
  well casing, 690t
material strength data, 680t
monitoring well casing, 684f
monitoring well casings, 878
monitoring well screens, 684f
plastic casings, 880
PTFE, 689
PTFE casing materials, 689
screen, 402f
stainless steel, 714
well casing, 702f, 703, 707f
well casing materials
  physical properties, 685t
well-casing materials, 688
well casings
  dimensions, 686t
  hydraulic collapse pressure, 686t
  unit weight, 686t
well screen, 702f
well screens, 835f
Porosity, 921
  clay soils, 921
  definition, 921
  dry bulk density, 922
  typical total, 922
Porosity logs, 281f
Porous cup size
  effect, 235
Portable grouting units, 774f
Portable pumps, 1036
Portable ventilation devices, 1249
Portland cement
  properties, 876t
  water mixture, 764
Positive-displacement pumps, 1019, 1020
Positive-pressure equipment
  groundwater samples, 1099
Positive-pressure filtration methods, 1097, 1098f
Positive-pressure filtration system, 1063f
Potassium
  laboratory determination, 926–929
Potentially responsible parties (PRPs), 1137
Potentiometric surface elevation contour map, 902f
Powdered bentonite, 759
Power washers, 1275f
PPE. *see* Personal protective equipment (PPE)

PQL. *see* Practical quantitation limit (PQL)
Practical quantitation limit (PQL), 1147–1149, 1169
PRB. *see* Permeable reactive barriers (PRBs)
Precision injection/extraction (PIX), 475
Prepacked well screen
   installation, 668
Prepacked well screens
   components, 750f
Pressure gages, 899
   head pressure changes, 899
Pressure tests, 936, 937
Pressure transducer methods, 896, 897
Pressure transducers, 899, 900, 937
   pressure gauges, 899, 900
Pressure-vacuum lysimeters
   schematic diagram, 233
Probe configurations, 382–391
   cased systems, 386, 387
   expendable drive point samplers, 382–384
   multiple depth sampling, 385, 386
   retractable drive point samplers, 384, 385
   retraction distance, 385
   soil gas implants, 387
Probe rods
   manual driver, 357f
Profile types
   information needs associated, 133t
Program implementation problems, 1198–1204
   prevention, 1203t
Progressing cavity pumps, 1027, 1028f
   packer, 1041f
   schematic diagram, 1028f
Project planning
   technical and project management skills, 80
Property ownership maps, 95f
Protected-screen groundwater sampler, 400f
Protected-screen groundwater sampling tools
   single-tube, 400f
Protected-screen tools, 394
Protected-screen well point
   installation, 443
Protective clothing, 1244–1246
PRP. *see* Potentially responsible parties (PRPs)
Pump discharge tubing
   temperature increases, 978f
Pumping
   compressed air development, 855
Pumping devices, 1008, 1009
Pumping tests, 936
   aquifers, 947
   aquifer systems, 945
   data analysis, 948, 949
   definition, 944
   design, 945

   vertical hydraulic conductivity, 939, 940
   water level measurement, 947
   well array, 946
Purge system, 1120f
Purge volume, 708, 709
Purging
   limitations, 1040t
   low-yield monitoring wells
      requirements, 1048
   low-yield wells, 1050f
   well flow, 1050f
Purging collection
   device selection, 996
Purging devices, 1008, 1009
   characteristics, 1009t
   selection and operation, 996–1035
Purging methods
   portable pumping devices, 1040
Purging strategies, 1036
Pushing
   samplers, 326
PVC. *see* Polyvinyl chloride (PVC)

QA. *see* Quality assurance (QA)
QAPP. *see* Quality Assurance Project Plan (QAPP)
QC. *see* Quality control (QC)
Quality assurance (QA), 1127–1134
   controls, 1273
   independent laboratory review, 1133, 1134
   laboratory process, 1129, 1130f
   measures, 963
   programs, 181, 992, 1265
      confirmatory samples, 183
   reports, 1146, 1147
   sample collection elements, 1091t
   segment, 1279
Quality Assurance Project Plan (QAPP), 964, 1128, 1143–1145
   classifications, 169
   guidelines, 1144t
   project managers, 168
   QA, 168
   QC, 168
Quality control (QC), 81, 1127–1134
   controls, 1273
   field analytical
      duplicate samples, 182
   laboratory process, 1129, 1130f
   laboratory program requirements, 1143
   measures, 963
   program, 181, 992, 1026, 1265
      confirmatory samples, 183
   sample collection elements, 1091t
   sample containers, 1087, 1090

Index

samples
  collection, 1095–1097
  duplicates, 1131
  equipment blank, 995
  instrument calibration, 1132
  laboratory, 1130–1132
  laboratory validation, 1132, 1133
  matrix spikes, 1131, 1132
  method blanks, 1131
  sample analysis, 1132
  sampling programs, 993
  spiked samples, 1131
  surrogate spikes, 1132
segment, 1279
Quarries
  unconsolidated materials, 130
Quartz filter-pack sand, 734f
Quartz sand
  filter-pack materials, 733f
Quartz sand over clay
  radar profile, 262

Radar
  ground-penetrating, 159f
  surface geophysical methods, 262
Radar methods, 898
Radiation hazards, 1228
Radiation instruments, 1242
Radiation meters, 1242
Radon flux approach
  PITT approach, 503
Radon flux rates, 503
Raman data profiles, 507
Raman spectroscopy, 430
Rapid Site Assessment approach, 78
RCRA. *see* Resource Conservation and Recovery Act (RCRA)
RDNA. *see* Ribosomal RNA (rDNA)
RDP. *see* Ribosomal Database Project (RDP)
Reaction grouting, 461
Real-time measurement technologies, 75f
Recharging areas
  gradient comparisons, 544f
Recharging conditions
  conceptual, 544f
Recovery rate, 709
Reentry grouting, 459, 460
Refraction method, 270
Relative percent difference (RPD), 1131
Remote equipment cleaning, 1268, 1269
Remote sensing
  airborne geophysical methods, 258
    imaging methods, 258–260
    non imaging methods, 260, 261
  geophysical methods, 249–292

Representative elemental volume (REV), 211, 211f
Representativeness concept, 49
Residual vinyl chloride monomer (RVCM), 690
Resistivity geoelectric section, 266
Resistivity imaging, 267, 268, 268f
  2D geoelectric cross-section, 267
Resistivity techniques, 267, 291
Resource Conservation and Recovery Act (RCRA), 3–7, 347, 517, 808, 1115, 1169, 1196, 1199
  groundwater monitoring system types, 5
  Groundwater Monitoring Technical Enforcement Guidance Document (TEGD), 520f
  hazardous constituents, 1139
  hazardous waste, 4
  regulations, 1174
  sites, 1255
  subtitle 1, 7, 8
  subtitle C, 4–7
  subtitle D, 4
  water cleanup levels, 12
Respiratory protection, 1236–1240
Respiratory-protection program, 1239, 1240
  requirements, 1239
Responsible parties (RPs), 815
Retractable drive point soil gas sampler
  cross section, 387f
Retraction grouting, 460, 461
REV. *see* Representative elemental volume (REV)
Reverse circulation method, 308
Reverse-circulation method
  filter-pack materials, 739f
Reverse circulation rotary drilling, 308
Reynold's Number
  definition, 924
Ribbon NAPL Sampler FLUTe (flexible linear underground technology) method, 508
Ribbon nonaqueous-phase liquid (NAPL) Sampler FLUTe, 508
Ribosomal Database Project (RDP), 626
Ribosomal RNA (rDNA)
  PCR amplification process, 626
RNA
  PCR, 626
Rock constituents, 916
Rock core boxes
  labeling, 333f
Rock core log form, 330f
Rock coring
  objectives, 326
Rock outcrops, 127f

Rocks
   elasticity, 952t
   natural bulk densities, 917t
   saturation
      drainage, 942
Rotary drilling, 306–308
Rotary drilling rigs, 307f, 318
Rotary drills
   rotation speeds, 307
Rotary methods
   cobbles and boulders, 316
Rotasonic drilling method, 363f
RPD. *see* Relative percent difference (RPD)
RPs. *see* Responsible parties (RPs)
RVCM. *see* Residual vinyl chloride monomer (RVCM)

Safe Drinking Water Act (SDWA), 11–26, 1115
   drinking water quality standards, 11, 12
   public drinking water contaminants, 11
   Secondary Drinking Water Regulations, 31
   sole-source aquifer program, 12
   standard terms, 31
   underground injection control program, 26
   well head protection program, 12–26
Safety and Health Emergency Response Plan (SHERP), 1222
Salton Sea
   satellite imagery, 102f
Sample collection equipment, 1269
Sample collection procedures, 1202
Sample containers, 1073f
   shipment, 1100–1102
Sample cooling
   dry ice, 1074
Sample discharge tubing
   materials, 999
Sample filtration, 1070
Sample integrity
   chemical analysis, 379–381
Sample preparation, 1129, 1130
Sample space problems, 1194–1196
   prevention, 1197f
Sample storage
   holding time requirements, 1129
Sampling
   chemical preservation, 1075
   low-yield monitoring wells
      requirements, 1048
Sampling and analysis plan (SAP), 1035, 1056, 1088
   cleaning field equipment, 1078
   device description, 996
   DQOs, 166
   elements, 167
   field protocols, 988
   objectives, 987, 988t
   potential cross-contamination, 1089
   preparation, 987
   site-specific, 1069
Sampling collection
   device selection, 996
Sampling devices, 998f
   characteristics, 1009t
   DQOs, 412
   selection and operation, 996–1035
   selection criteria, 997t
Sampling equipment
   application, 1102f
Sampling error, 47
Sampling event
   conduction, 1087
   documentation, 1080–1086
Sampling frequency
   chemical characteristics, 597
Sampling methods, 299
   inappropriate use, 1047
Sampling point inspection, 1087, 1088
Sampling point installation
   options, 970–973
Sampling point placement
   factors, 969–976
   three-dimensional, 969, 970
Sampling point purging, 1035–1056
Sampling procedures
   assessment monitoring, 1167
   detection monitoring, 1167
Sampling Protocol, 571
Sampling rod, 388f
Sand
   porosity data, 212f
Sands
   estimated hydraulic conductivity, 930f
Sanitary landfills
   detection parameters, 1142
   parameter indications, 1141t
SAP. *see* Sampling and analysis plan (SAP); Site-specific sampling and analysis plan (SAP)
SAR. *see* Supplied-air respirator (SAR)
SARA. *see* Superfund Amendments and Reauthorization Act (SARA)
SASW. *see* Spectral analysis of surface waves (SASW)
SBCA. *see* Self-contained breathing apparatus (SBCA)
SCAPS. *see* Site Characterization and Analysis Penetrometer system (SCAPS)
Screen
   physical strength, 678–682

Index

Screen formation collapse, 451f
Screen materials, 714f
   chemical degradation, 715
   chemical resistance, 682, 683
   cleaning requirements, 713, 714
   composition, 696t
   environment corrosion, 692
   installation, 681f, 748f
   tensile strength, 679
   types, 683–697
Screens
   joint types, 699f
Screen slot sizes
   characteristics, 728t
SCS. *see* U.S. Soil Conservation Service (SCS)
SDWA. *see* Safe Drinking Water Act (SDWA)
Sealed double-ring infiltrometers
   schematic diagram, 228
Sealed-screen samplers
   hydrostatic pressure, 398
Sealed-screen sampling tool
   BAT Enviroprobe, 401
Seasonal Kendall test, 612
   limitations, 612
Sedimentary bedrock flatland, 110
Sedimentary deposits
   grain-size distribution, 209
   installation, 833–836
Sediment sump
   installation, 712
Sediment sumps, 712, 713
   installation, 712f
Sediment trap
   installation, 712
Seepage pits
   vertical hydraulic conductivity, 938, 939
Seismic methods, 269–273
   reflection, 270–272
   refraction, 270
   surface wave analysis, 272, 273
Seismic reflection data, 272f
Seismic reflection method, 270
Seismic refraction, 160f
Seismic refraction survey, 271f
Seismic techniques, 291
Seismic waves, 269
Self-contained breathing apparatus (SBCA), 1238
Semipermeable membrane, 219
Semivolatile organic compounds, 1122, 1123
Sensitizes, 1231
Separate-phase petroleum hydrocarbons, 425
Shallow monitoring wells, 708
Shearing, 982
Shelby-tube samplers, 372f

SHERP. *see* Safety and Health Emergency Response Plan (SHERP)
Short screen well, 658f
Short-term exposure limit (STEL), 1232
Shutter-type well screen, 747f
Side-looking airborne radar (SLAR), 260
Silicone rubber tubing, 1000f
SIM. *see* Single-ion monitoring (SIM)
Simulprobe
   combined sampling tool, 403
Single aquifer
   variable hydraulic conductivity, 546–549
Single-aquifer flow systems, 529
Single homogeneous aquifer, 545, 546
Single-interval wells, 821, 822
Single-ion monitoring (SIM), 1123
Single-rod exposed-screen sampling tools, 407
Single-rod sampling system, 364f
Single-rod soil-sampling system
   benefits, 378, 379
Single-screen well, 658f, 661f, 665f
   multiple-casing, 659f
Single-screen wells, 659
Single-tube sampling
   reentry grouting, 460
Single-tube soil sampler
   DP methods, 365f
   open barrel, 371f
Single-tube soil sampling, 369f
Single-tube soil sampling tool
   components, 373f
Single-well tracer tests, 937
Single-well tracer-test theory, 822
SITE. *see* Superfund Innovative Technology Evaluation (SITE)
Site Characterization and Analysis Penetrometer system (SCAPS), 486
Site characterization monitoring, 577, 578
Site characterization program, 50
Site environment conditions
   information types and formats, 89
Site investigations
   direct sampling methods, 251
Site plan maps, 93f
Site reconnaissance, 114
   field investigation, 115f
      obstacles, 115f
   hazardous materials, 123f
   objectives, 127f
   solid-waste landfill areas, 124f
   utility company scheduling, 115
Site reconnaissance activities
   contaminants, 121f
Site-specific decision strategy, 86
Site-specific filtration protocols, 1067

Site-Specific Safety and Health Plan (SSHP), 1222, 78
Site-specific sampling and analysis plan (SAP), 962
Site-specific water chemistry, 1039, 1040
Slam bars, 348
SLAR. *see* Side-looking airborne radar (SLAR)
Sleeved well screen
    components, 749
Sleeved well screens, 748–751, 749
    components, 751
Slug test
    specific storage, 943
Slug tests, 929–933, 936
    curve type analysis, 933
    curve types, 931
    falling-head, 933
    rising-head, 933
Small-diameter bailers, 412f
Small-diameter nylon tremie tubes, 463f
SMCRA
    environmental protection performance standards, 28
    purposes, 27
Soil
    dry bulk density, 915
    saturation
        drainage, 942
Soil boring
    well-construction logs, 92f
Soil classification samples, 59f
Soil classification system, 337f
Soil constituents
    specific gravity, 916t
Soil description, 652f
Soil electrical resistivity
    conductivity probes, 420–423
Soil excavation
    site characterization data, 43f
Soil gas
    PCE
        hydrocarbon concentration, 240
    properties, 213
Soil gas contaminant investigation technology
    representative application, 239f
    schematic diagram, 237
Soil gas implant installation, 391f
Soil gas measurements
    rainfall, 244
Soil gas probes, 163
Soil gas samples, 164
Soil gas sampling
    cased system, 389f
    DP equipment, 367

lost point probing tool, 384f
QA plan, 393
QC plan, 393
quality assurance and quality control procedures, 392, 393
Soil gas sampling implants, 390f
Soil gas sampling point and rod, 384f
Soil gas sampling system
    cross-section, 386f
Soil gas sampling tool
    expendable-point configuration, 383
Soil gas sampling train, 391, 392, 392
Soil gas surveys, 162f, 501
    VOCs, 153, 483
Soil grain size
    potassium estimation, 929
Soil lithologic data, 506
Soil matrix
    probe insertion, 381, 382
Soil moisture flux
    calculation, 230
Soil moisture probes, 423, 424
Soil porosity
    water content, 919
Soil probe, 299f
Soil properties
    tools, 420–425
Soils
    bulk densities, 917t
    density sampling methods, 918f
    elasticity, 952t
    geological materials, 47
    water content, 920
        field methods, 920, 921
Soil sample, 252
Soil sampling, 318–326
    advancing and retrieving, 359
    bulk samples, 319
    composite samples, 319, 320
    cuttings samples, 333
    DP equipment, 368
        methods, 368
    representatives samples, 319
    rock core samples, 332, 333
    rock coring, 326–331
        core losses, 329
    rock coring logs, 329–331
    sampling methods, 324–326
        driving, 326
        pushing, 326
        pushing rotations, 326
    soil and rock samples, 331–333
    soil sampler types, 320–324
        continuous tube samplers, 321–323
        rotary samplers, 323

Index

solid-barrel samplers, 320
split-barrel samplers, 320, 321
thin-wall tube samplers, 321
types, 318, 319
undisturbed samples, 319
Soil sampling method
split spoon samples, 59f
Soil sampling tool, 139
Soil surveys, 91f
Soil water
properties, 213
Soil-water system
nondestructive techniques, 226
Sole-Source Aquifer Program, 12
Solid-barrel samplers
split-barrel samplers, 320
Solid phase
skeletal structure, 209
Solid-stem augers
drilling rigs, 304
Solinst CMT system, 841–843, 842f
Solinst Waterloo multilevel groundwater monitoring system, 840f
Solinst Waterloo system, 830
multi-screened wells, 839
Solinst waterloo system, 839–841
Solute plume, 633
Solute plume behavior, 601f
Sonic drilling, 144f, 145f, 312, 313, 672f
problems, 145
Sonic drilling rigs, 147f, 312f
Sonic methods, 897
acoustic probe, 897, 898
popper, 897
SOP. *see* Standard operating procedures (SOPs)
SP. *see* Spontaneous-potential (SP)
Spatial heterogeneity, 48
Spatial sampling requirements, 252f
Specific constituent inorganic analysis, 1124–1127
Specific gravity, 1229
Specific organic compound analysis, 1118–11124
Spectral analysis of surface waves (SASW), 272
Split-barrel drive sampler (SPT), 320f
Split-barrel samplers, 374f
Split-spoon sample
sub sample, 46f
Split-spoon sampler, 718f
components, 372f
Split-spoon samplers, 1276
Spontaneous-potential (SP), 282
SPT. *see* Split-barrel drive sampler (SPT); Standard penetration test (SPT)

SSHP. *see* Site-Specific Safety and Health Plan (SSHP)
Stabilization
water quality indicator parameters, 1058t
Stainless corrosive environments, 693
Stainless steel
material strength data, 679t
PVC, 714
Stainless steel casings
monitoring wells, 704
Stainless steel protective casings, 782f
Stainless steel well casing, 697f
burst pressure, 695t
composition, 696
dimensions, 695t
hydraulic collapse, 695t
threaded joints, 705f
unit weight, 695t
Stainless steel well screen
threaded joints, 705f
Standard Guide for Acquisition of File Aerial Photography, 109
Standard Guide for Development of Groundwater Monitoring Wells in Granular Aquifers, 851
Standard Guide for Selecting Surface Geophysical Methods, 292
Standard Methods for the Analysis of Water and Wastewater, 1127
Standard operating procedures (SOPs), 1143–1145
Standard penetration test (SPT), 320
Standard Practice for Design and Installation of Groundwater Monitoring Wells, 641
State groundwater protection programs, 32, 33
Static fluid
pressure fluid, 214f
Station measurements, 255f
Statistical hypothesis testing, 1171
Steam cleaning, 1275
Steel
corrosion, 692
Steel casing
production, 713
welding, 704f
Steel corrosion
potential determination, 692
Steel materials
geochemical environment, 999f
Steel well casing
corrosion, 973
environment casing, 692
Steel well construction materials
corrosion, 974f

Steep gradient facilities
   example, 533f
STEL. *see* Short-term exposure limit (STEL)
Submersible pump
   mechanical pumping, 857, 858
   mechanical surging, 857, 858
Submersible pumps
   monitoring well development
      mechanical surging, 858
Suboxic ground water, 1067
Subsurface condition evaluation
   background, 251, 252
   continuation measurements, 254
   downhole geophysics, 256–258
   geophysical method use, 253, 254
   sample density, 252, 253
   site investigation methods, 255, 256
   station measurements, 254
Subsurface conditions
   investigations, 963
Subsurface conditions evaluation
   mapping buried wastes, 256
   mapping contaminant plumes, 256
Subsurface geological materials
   displacement, 354
Subsurface geologic features, 1201f
Suction-lift pumps, 413, 414
Suction-lift sampling technology, 595
Sulfate concentrations
   ramifications, 619
Superfund, 1198
   abandoned waste sites, 8
   cleanup, 8
Superfund Amendments and Reauthorization
   Act (SARA), 1115, 1143
Superfund Innovative Technology Evaluation
   (SITE), 380
Superfund site investigation, 965
Superfund sites, 52f
Supplemental daughter product data, 622
Supplemental geochemical data, 622–624
Supplemental monitoring parameters,
   594, 595
Supplied-air respirator (SAR), 1238
   SCBA, 1239
Support Zone, 1247
Surface Centrifugal Pump
   pumping, 860
   surging, 860
Surface centrifugal pump
   mechanical pumping, 857
   mechanical surging, 857
Surface centrifugal pumps, 1014, 1015, 1015f
Surface geophysical measurements
   modes, 257f

Surface geophysical methods, 261–276, 655f
   buried wastes and utilities, 286f
   electromagnetic, 265–267
   EM
      resistivity methods, 263–269
   evaluation, 284t
   ground-penetrating radar, 261–263
   resistivity, 267–269
Surface geophysical surveys, 653
Surface geophysics
   IDW, 151
Surface mining control
   reclamation act, 26–28
Surface sampling locations, 587f
Surface seals, 646f, 777–779
   above-grade completion, 779–784
   damage, 779f–780f
   flush-to-grade completion, 784–792
Surface-seals
   cement, 778
Surface-water bodies
   discharges, 123f
   plumes, 582–586
Surface-water systems
   industrial discharges, 38f
Surge block
   mechanical surging, 858, 859
Surging
   compressed air development, 855
Survey instruments, 1242
SVOCs, 192
Synergism, 1231
Syringe sampler, 1015f
Systematic project planning
   data collection, 81
   project implementation relationship, 81f
System implementation problems, 1196–1198,
   1199t

Tailgate safety meeting, 1223
Target monitoring zones
   cross-section, 530f
   potential areas, 530f
   procedures, 528
Task-specific basis
   factor evaluation, 1272–1274
TCA. *see* Trichloroethane (TCA)
TDS. *see* Total dissolved solids (TDS)
TEGD, 524
   upgradient well definition, 529
Temperature blanks, 994f
Temperature stress, 1227, 1228, 1248
Tensiometers, 230
   disadvantages, 224
   negative pressures, 224

Index 1313

positive pressures, 224
pressure, 222
Tentatively identified compounds (TICs), 1122
Test borings, 874
Tetrachloroethane, 238
Tetrahydrofuran (THF), 685
Thermocouple psychrometers
   disadvantages, 225
Thermoplastic materials, 683–692
   definition, 683
Thermoplastic well casing, 683
THF. *see* Tetrahydrofuran (THF)
Thief samplers
   grab samplers, 1013
Thin-walled samplers, 372f
Thin-wall tube sampler, 322f
Threaded joints, 704f
   stainless steel well casing, 705f
   stainless steel well screen, 705f
Three-dimensional plume delineation, 813, 814
Three-dimensional seismic surveying
   technology, 504
Three-layer flow model, 551f
Threshold joints
   types, 700f
Threshold limit values (TLV), 1232
TICs. *see* Tentatively identified compounds (TICs)
Tidal fluctuation, 906f
Time-series data
   analysis methods, 174
   organic data, 1168f
Time-weighted-average, 1223
TKN. *see* Total Kjeldahl nitrogen (TKN)
TLV. *see* Threshold limit values (TLV)
TNT. *see* Trinitrotoluene (TNT)
TOC. *see* Total organic carbon (TOC)
Toolbox safety meeting, 1223
Topographic maps, 90f
   current and historical, 91f
Total dissolved solids (TDS), 761
Total Kjeldahl nitrogen (TKN), 1126
Total mobile contaminant load, 980
Total organic carbon (TOC), 1114, 1183, 1184t
Toxic hazards, 1230–1232
Toxic materials, 1230
Toxicological Data Network (TOXNET), 1235
Toxic overexposure, 1248
Toxic substance control act, 10
Toxic Substances Control Act (TSCA), 961
   chemical industry, 10
TOXNET. *see* Toxicological Data Network (TOXNET)
Trace metals, 1124
Tracer tests, 937

single well, 937
Transducers
   microprocessors, 892
Transformation products
   contaminants, 594
Transformers, 124f
Transmissivity, 944–950
   estimation, 949, 950
   hydraulic conductivity, 944, 949
   pretest data collection, 946
   pumping test, 944, 945
   slug test data, 944
   van der Kamp method, 932
Trap system, 1120f
Travel time
   predicted expression, 229
Tremie grouting, 461, 462
Tremie pipe
   bentonite pellets, 770f
Tremie-pipe method, 735
   filter-pack placement, 736f
Tremie-pipe placement
   grout, 776f
Triad approach
   comprehensive final report, 175
   decision list, 84
   DMP, 170
   elements, 74f
   field work project, 176
   focus areas, 73
   major components, 74t
   project management, 179
   soil gas surveys, 154
   systematic project planning, 79
   technical staff, 82
   traditional approach
      conventional approach, 84
   U.S. EPA program, 75
Triad approaches
   conventional drilling methods, 143
Triad process
   analytical data, 137
   primary focus, 172
Triad program
   field-based analytical
      methods, 149
Triaxial cell method, 927
Trichloroethane (TCA), 238, 617
Trichloroethene, 238, 1210
Trichloroethylene, 187, 430
Trilinear data
   water quality, 1166f
Trilinear diagrams, 1167
Trilinear groundwater quality
   trilinear diagrams, 1165–1167

Trinitrotoluene (TNT), 688
Trisodium phosphate (TSP), 1251
TSCA. *see* Toxic Substances Control
    Act (TSCA)
TSDF
  regulations requirements, 5
TSP. *see* Trisodium phosphate (TSP)
Turbidity, 981
  field measurements, 1061f
  natural formation, 1061
  pumping stress, 1060, 1061
  pump installation, 1006
2D resistivity imaging, 267, 268, 268f
  2D geoelectric cross-section, 267
Two-layer flow model, 550f

UIC. *see* Underground injection control (UIC)
    program
Ultrasonic methods, 898
Ultraviolet (UV) lamp
  shake tests, 501
UMS. *see* Unit model scenarios (UMS)
Unconfined aquifer
  specific yield definition, 942f
  specific yield representation, 943t
Underground injection control (UIC) program
  SDWA, 26
Underground storage tanks (USTs), 37
  systems, 37f
Unified Soil Classification System (USCS), 336
Uniform sand aquifers
  transmissivity, 950
United Nations Statistical Office, 1140
United States
  groundwater monitoring wells, 816
Unit model scenarios (UMS), 491
  predetermined parameters, 492t
University Microfilms International, 98
University of Waterloo Center for Ground
    Water Research, 490
Unprotected-screen profiling tools, 397f
U.S. Army Corps of Engineers, 822, 826
  facility profile types, 132
  SAPs, 168
U.S. Department of Agriculture
  soil survey maps, 258
U.S. Department of Defense
  CPT probes, 432
U.S. Department of Energy (U.S. DOE),
    38, 377
  ASAP, 163, 164
  probes, 432
U.S. Department of the Interior
  SMCRA, 26
U.S. Department of Transportation, 1073

U.S. DOE. *see* U.S. Department of Energy
    (U.S. DOE)
U.S. Environmental Protection Agency
    (U.S. EPA), 2, 1090, 1132, 1137, 1145,
    1148, 1167
  alternative monitoring frequency, 1174
  analytical data, 177
  ANOVA, 1185
  borehole grouting, 458
  chemical information, 1236
  CLP, 1148
    protocols, 45f
  Cochran's approximation, 1177
  colorimetric test methods, 186t
  contingency measurements, 628
  contingency plan, 627
  Dexsil Corporation, 186
  down gradient wells, 519
  DQOs, 86
    approach, 165
  environmental monitoring types, 577
  federal regulations, 524
  guidelines, 1091f
  hazardous waste cleanup information, 475
  Hazardous Waste Facilities, 1171
  hazardous waste sites, 37
  injection well classifications, 27t
  Levels of Protection system, 1244
  MDL, 1182
  minimum goals, 577
  policy directives, 177
  RCRA, 1172
    regulations, 1187
  Regional Administration, 6
  Sole-Source Aquifer designations,
    23t–25t
  systematic planning, 77
  Triad approach, 73
  Tribal Nations
    partnerships, 32
  VOCs, 475
  well grouting, 458
U.S. EPA. *see* U.S. Environmental Protection
    Agency (U.S. EPA)
U.S. Geological Survey (USGS), 828, 962
  drills, 828
U.S. Postal Service
  ground water
    terrorism, 33
U.S. Soil Conservation Service (SCS), 97
USCS. *see* Unified Soil Classification System
    (USCS)
USGS. *see* U.S. Geological Survey (USGS)
UST. *see* Underground storage
    tanks (USTs)

Index

Utility corridors
  underground, 94f
UV. *see* Ultraviolet (UV) lamp

Vacuum filtration methods, 1068
Vacuum filtration system, 1063f
Vacuum gage tensiometers, 223
Vadose
  definition, 208
Vadose Zone
  capillarity, 215, 216
  characteristics, 208–222
  definitions and terminology, 208, 209
  energy potential, 218–220
  flow, 220
    relative permeability, 221, 222
    vapors, 221
    water, 220, 221
  fluid continuum, 215
  fluid properties, 212
  gas/vapor phase, 213
  hydrostatics, 214, 215
  hysteresis, 217, 218
  immiscible fluid, 213
  immiscible fluids, 213, 214
  moisture, 216
  moisture and energy, 214–218
  monitoring and sampling, 207–244
  monitoring methods, 222–226
    electrical resistance blocks, 224, 225
    gamma-ray attenuation, 225
    nuclear moisture logging, 225, 226
    storage properties, 222
    tensiometers, 222–224
    thermocouple psychrometers, 225
  monitoring transmission properties, 226–232
    borehole permeameters, 231
    Darcy's Law, 229
    Green-Ampt wetting front model, 229, 230
    internal drainage method, 230, 231
    theoretical perspective, 229
    tracer movement measurement, 231, 232
    water flux characteristics, 229–232
  monitoring water quality, 232–236
    electrical properties measurements, 232
    method types, 232
  multiple-phase components, 209–220
  physical properties, 222
  pore space, 216
  saturation example, 217f
  soil gas monitoring technology, 236–244
    applications, 239, 240
    case study, 239, 240
    compounds, 242, 243
    data interpretation, 243, 244
    geologic barriers, 240–242
    halo carbon *versus* petroleum hydrocarbons, 240
    methodology background, 237
    problems, 240–244
    quality assurance and control procedures, 238, 239
    sampling and analytical procedures, 238, 239
    summary, 244
  soil mass illustration, 382f
  soil sampling and water sampling
    pan lysimeters, 236
    pore water extraction, 232, 233
    suction lysimeters, 233–236
  solid phase, 209
    fractured rock, 211, 212
    sedimentary deposits, 209–211
  suction, 216, 217
  suction-moisture content relationship, 218
  suction profile, 217f
  water
    hydrostatic pressure, 219
  zone water, 212, 213
Validation monitoring, 578
Valved surge plungers, 859
Vapor pressure, 1229
VC. *see* Vinyl chloride (VC)
Vertical gradients
  groundwater flow directions, 910
  site areas, 905, 906
Vertical hydraulic conductivity, 939f
  pumping test data, 939
Vertical hydraulic gradients
  definition, 814
Vertical hydraulic has, 814, 815
Vertical hydraulic heads
  measurement, 815
Vertical leakage
  definition, 940f
Vertical permeability well test, 838
VFAs
  ion chromatography, 625
Video imaging systems, 437, 438
Vinyl chloride (VC), 617
  concentrations, 243t
VOC. *see* Volatile organic compound (VOC)
Volatile fatty acids, 625
Volatile organic compound (VOC), 192, 236, 350, 1114, 1118–1122, 1176
  agitation and aeration, 984
  contours, 818f
  diffusion samplers, 597

Volatile organic compound (VOC) (Contd.)
  geologic barriers, 240
  groundwater sample, 995
  groundwater sampling, 984
  isocontour map, 1156f
  PDB, 821
  sample collection, 1029, 1093, 1093f
  sample collection rates, 1094
  vapor migration, 240
  vials, 1094, 1094f
    packing tape, 1100
    sample seals, 1101f
    security seals, 1100
    septum, 1101f
Volatile organic compounds (VOC), 59
Volumetric wetness, 231

Wash boring, 306
  borehole method, 306
Waste disposal facilities
  presentation, 1164
Water
  bulk density, 952
  temperature properties, 925t
  values, 216
Water-based drilling fluid, 673f
Water-bearing properties
  geologic formulations, 525
  hydrologic factors, 525
Water chemistry
  water column, 969f
Water content, 920, 921
  field measurement, 920, 921
  laboratory measurement, 920
Water flow
  vertical directions, 903f
Water flow systems
  recharge and discharge relationship, 904
Water FLUTe (flexible linear underground technology), 843–845, 844f
Water level
  hydraulic-head relationships, 884, 885
  pumping well observation, 946
Water level and product-thickness measurement, 989–991
Water level data
  acquisition, 892–902
  acquisition and interpretation, 884–910
  analysis, interpretation, and presentation, 902, 903
  importance, 884
  interpretation approach, 904
  long-term collection, 900
  recharge and discharge conditions, 903–919
  time-series, 908

  transient effects, 907–910
Water level elevation, 905f
  computer contouring, 910
  statistical analysis, 910
Water level elevation data
  contouring, 910
Water level measurement, 62f, 1089
  data reporting, 892
  precision and intervals, 891, 892
Water level measurements, 990, 1089
Water level monitoring
  low-flow purging, 1045
Water level monitoring systems
  construction features, 890, 891
  design approaches, 887, 888
  design features, 886–992
  design guidelines, 886
  geologic sites, 886
  screen depth and length, 888–890
Water levels
  barometric changes, 946
  time series comparisons, 1160f
Water level transducers, 897
Waterloo profiler, 398f, 399
Water measurements, 276
Water pathways, 754f
Water quality, 28
  category display format, 1153–1155
  detection monitoring programs, 1141
  evaluations, 1157
  field measurement, 1056–1062
  parameter list, 1142t
  parameters, 594
  sample collection, 1157
  sample pretreatment options, 1062–1079
  tabular data, 1154f
Water quality assessment procedure, 1138f
Water quality data, 1153
  agencies, 1188, 1189
  central tendency tests, 1177
  data independence, 1171
  data normality, 1171, 1172
  intra-well comparisons, 1183–1185
    illustrations, 1183–1185
  methods, 1185
  multiple constituents, 1181, 1182
  multiple locations, 1180, 1181
  nondetects problem, 1182
  nonparametric prediction limits, 1182, 1183
  recommended statistical methods, 1179, 1180
  single location, 1180
  statistical prediction intervals, 1180–1188
  statistical test types, 1174–1177

statistical treatment, 1167–1188
trend tests, 1177–1179
variance analysis, 1185–1187
verification resembling, 1181
Water quality monitoring well
cross-section, 440f
Water quality samples
out-of-control incident, 1146
Water samples
agitation and aeration, 984–987
Water solubility, 1229
Water supply wells, 851
monitoring wells
types, 729
Water surfaces
survey detection, 895
Water-table aquifer, 949
Watertight joints, 703
Weathering degree, 338t
Well casing, 714f
physical strength, 678–682
slotted, 745f
Well casing diameters
factors affecting, 705–709
Well casing installation, 681f
Well casing materials, 713
comparative strengths, 680t
weight ratios, 681t
Well casings, 782f, 788
chemistry factors, 968
inside diameters, 706t
joint types, 699f
outside diameters, 706t
PVC
methods, 698
wall thickness, 706f
watertight locking cap, 784f
Well clusters, 661, 662, 662f, 823f
Well completion types
application recommendations, 657t
Well construction
detail reports, 866–868
details, 870t
problems, 973
reporting details, 866, 867
tremie grouting method, 461
Well construction details
documentation, 797
Well construction materials, 642f
Well construction reports, 797t
Well decommissioning
requirements, 873t
Well depth
casing strength, 708
Well design process, 651f

Well designs, 653f
prepacked well screens, 668f
problems, 973
Well development, 708
cost-benefit ratio, 852
gradiation, 717f
inadequate, 976
naturally, 717, 718
Well development activity
components, 851
Well development methods
guidelines, 854, 855
Well development monitoring
influential factors, 852
methods, 854
water column surging, 853
Well discharge
measurement, 947, 948
Well Head Protection Areas (WHPAs), 12
methods, 22, 23
Well headspace screening, 988, 989, 989f, 1088, 1089
instrumentation examples, 989t
Well hydraulic formation
related factors, 965–969
Well identification
disadvantages, 864, 865
labeling information, 865, 866f
markings, 865
surveying and alignment testing, 792–797
Well incrustation, 871
Well installation method, 708
Well intake technology, 716
Well maintenance, 870t
Well monitoring development
methods, 856
Well nest
short well screens, 664f
Well positioning
observation, 945, 946
Well purging, 1089, 1090
instrumentation, 1089
Well rehabilitation, 871
Wells, 886, 887
bailers, 981f, 1012, 1013, 1037f
barometric efficiency, 947
bentonite chips, 878f
cluster disadvantages, 829
design and construction, 973–976
dewatering, 1049t
flood plain, 783f
groundwater flow, 888
groundwater level measurements, 886
high-flow-rate pumping, 1037f
influence observations, 909f

Wells, (*Contd.*)
  inspection check lists, 1088t
  maintenance, 976
  manual measurement, 899, 900
  manual methods
    applications and limitation, 900
  measurement techniques, 896
  monitoring, 1195
  multilevel development, 837
  multilevel transect placement, 817
  multilevel transects, 816, 817f
  non flowing manual measurements, 893, 894
  number and placement, 888
  protection structures, 786f
  protective casings, 786f
  purging, 1049
  recovery rates, 709
  security levels, 785f
  sliding scales, 1159f
  time series comparisons, 1158f
  water depth devices, 897
  yield monitoring, 1196
Well screen, 669f
  installation, 444f
  prepacked, 448f
Well screens, 643f, 741–754, 1049f
  commercially manufactured, 743
  construction, 742
  coring, 1053
  design, 741
  different depths, 822, 823
  filter pack, 732
  flushed, 969
  hand-slotted casing, 743f
  hydraulic efficiency, 742, 743
  hydraulic performance, 751f
  length, 742, 751–754, 752
  length effect, 809f
  low-yield wells, 1051
  nonstandard materials, 643f
  placement, 736f, 890f
  prepacked, 748f, 972f
  purposes, 741
  PVC, 835f
  selection, 751
  slot
    size, 725
  slot size, 727f
  slot sizes, 721f
Well-type lysimeters, 234
Well vaults, 793f, 794f
  concrete seal, 790f
Well-volume purging, 63f
Well-volume purging methods, 1038
Well-water levels
  non flowing manual measurement, 893t
West bay casing system, 837
West bay instrumentation, 838
West bay MP system, 830, 837–839, 838f, 839f
West bay system
  geologic environments, 838
Wetted chalked tap method, 894
Wetted chalk tape method
  disadvantage, 894
Whisker plots
  error bar plots, 1161f
WHPAs. *see* Well Head Protection Areas (WHPAs)
Wilcoxon signed rank test, 613
  calculations examples, 613t
  monitoring wells
    paired data, 614t
Wireline CPT soil sampler, 378
Wireline CPT system
  convention CPT system comparison, 377
Wire-line rock core barrel, 328f
Wire-wound screens, 397f
Workable federal groundwater program
  mechanisms, 32
Work area cleanup, 1102

X-ray fluorescence (XRF), 65, 190, 355
  components, 190
  technology, 190
XRF. *see* X-ray fluorescence (XRF)